CAMPBELL BIOLOGY
CONCEPTS & CONNECTIONS

EIGHTH EDITION

JANE B. REECE *Berkeley, California*
MARTHA R. TAYLOR *Ithaca, New York*
ERIC J. SIMON *New England College*
JEAN L. DICKEY *Clemson University*
KELLY HOGAN *University of North Carolina, Chapel Hill*

PEARSON

Boston Columbus Indianapolis New York San Francisco Upper Saddle River
Amsterdam Cape Town Dubai London Madrid Milan Munich Paris Montréal Toronto
Delhi Mexico City São Paulo Sydney Hong Kong Seoul Singagore Taipei Tokyo

Editor-in-Chief:
Beth Wilbur

Executive Director of Development:
Deborah Gale

Acquisitions Editor:
Alison Rodal

Executive Editorial Manager:
Ginnie Simione Jutson

Editorial Project Manager:
Debbie Hardin

Development Editors:
Debbie Hardin, Susan Teahan

Editorial Assistant:
Libby Reiser

Senior Supplements Project Editor:
Susan Berge

Supplements Production Project Manager:
Jane Brundage

Manager, Text Permissions:
Tim Nicholls

Project Manager, Text Permissions:
Alison Bruckner

Text Permissions Specialist:
James Toftness, Creative Compliance, LLC

Director of Production:
Erin Gregg

Managing Editor:
Michael Early

Production Project Manager:
Lori Newman

Production Management and Composition:
S4Carlisle Publishing Services

Design Manager:
Marilyn Perry

Cover and Interior Designer:
Hespenheide Design

Illustrations:
Precision Graphics

Development Artists:
Kelly Murphy; Andrew Recher, Precision Graphics

Senior Photo Editor:
Donna Kalal

Photo Researcher:
Kristin Piljay

Photo Permissions Management:
Bill Smith Group

Director of Editorial Content MasteringBiology®:
Tania Mlawer

Development Editor, MasteringBiology®:
Juliana Tringali

Senior Mastering® Media Producer:
Katie Foley

Associate Mastering® Media Producer:
Taylor Merck

Editorial Media Producer:
Daniel Ross

Senior Manager Web Development:
Steve Wright

Web Development Lead:
Dario Wong

Vice President of Marketing:
Christy Lesko

Executive Marketing Manager:
Lauren Harp

Senior Marketing Manager:
Amee Mosley

Manufacturing Buyer:
Jeffrey Sargent

Cover Printer:
Lehigh-Phoenix

Text Printer:
Courier/Kendallville

PEARSON

Cover Photo Credit:
Andy Rouse/Nature Picture Library

Credits and acknowledgments for materials borrowed from other sources and reproduced, with permission, in this textbook appear on the appropriate page within the text or on p. A-26.

Library of Congress Cataloging-in-Publication Data

Reece, Jane B.
Campbell biology: concepts and connections / Jane B. Reece [and four others].—Eighth edition.
pages cm
Previous edition: Campbell biology: concepts & connections, 2012.
ISBN 978-0-321-88532-6
1. Biology. I. Title. II. Title: Biology.
QH308.2.B56448 2013
570—dc23
2013024409

ISBN 10: 0-321-88532-5; ISBN 13: 978-0-321-88532-6 (Student Edition)
ISBN 10: 0-321-94668-5; ISBN 13: 978-0-321-94668-3 (Books a la Carte Edition)
2 3 4 5 6 7 8 9 10—CRK—18 17 16 15 14

www.pearsonhighered.com

About the Authors

Jane B. Reece has worked in biology publishing since 1978, when she joined the editorial staff of Benjamin Cummings. Her education includes an A.B. in biology from Harvard University, an M.S. in microbiology from Rutgers University, and a Ph.D. in bacteriology from the University of California, Berkeley. At UC Berkeley, and later as a postdoctoral fellow in genetics at Stanford University, her research focused on genetic recombination in bacteria. Dr. Reece taught biology at Middlesex County College (New Jersey) and Queensborough Community College (New York). During her 12 years as an editor at Benjamin Cummings, she played a major role in a number of successful textbooks. She is coauthor of *Campbell Biology*, Tenth Edition, *Campbell Biology in Focus*, *Campbell Essential Biology*, and *Campbell Essential Biology with Physiology*, Fourth Edition.

Martha R. Taylor has been teaching biology for more than 35 years. She earned her B.A. in biology from Gettysburg College and her M.S. and Ph.D. in science education from Cornell University. At Cornell, she has served as assistant director of the Office of Instructional Support and has taught introductory biology for both majors and nonmajors. Most recently, she was a lecturer in the Learning Strategies Center, teaching supplemental biology courses. Her experience working with students in classrooms, in laboratories, and with tutorials has increased her commitment to helping students create their own knowledge of and appreciation for biology. She has been the author of the *Student Study Guide* for all ten editions of *Campbell Biology*.

Eric J. Simon is a professor in the Department of Biology and Health Science at New England College (Henniker, New Hampshire). He teaches introductory biology to science majors and nonscience majors, as well as upper-level courses in tropical marine biology and careers in science. Dr. Simon received a B.A. in biology and computer science and an M.A. in biology from Wesleyan University, and a Ph.D. in biochemistry from Harvard University. His research focuses on innovative ways to use technology to improve teaching and learning in the science classroom, particularly for nonscience majors. Dr. Simon is the lead author of the introductory nonmajors biology textbooks *Campbell Essential Biology,* Fifth Edition, and *Campbell Essential Biology with Physiology,* Fourth Edition, and the author of the introductory biology textbook *Biology: The Core.*

Jean L. Dickey is Professor Emerita of Biological Sciences at Clemson University (Clemson, South Carolina). After receiving her B.S. in biology from Kent State University, she went on to earn a Ph.D. in ecology and evolution from Purdue University. In 1984, Dr. Dickey joined the faculty at Clemson, where she devoted her career to teaching biology to nonscience majors in a variety of courses. In addition to creating content-based instructional materials, she developed many activities to engage lecture and laboratory students in discussion, critical thinking, and writing, and implemented an investigative laboratory curriculum in general biology. Dr. Dickey is author of *Laboratory Investigations for Biology*, Second Edition, and coauthor of *Campbell Essential Biology*, Fifth Edition, and *Campbell Essential Biology with Physiology*, Fourth Edition.

Kelly Hogan is a faculty member in the Department of Biology at the University of North Carolina at Chapel Hill, teaching introductory biology and introductory genetics to science majors. Dr. Hogan teaches hundreds of students at a time, using active-learning methods that incorporate technology such as cell phones as clickers, online homework, and peer evaluation tools. Dr. Hogan received her B.S. in biology at the College of New Jersey and her Ph.D. in pathology at the University of North Carolina, Chapel Hill. Her research interests relate to how large classes can be more inclusive through evidence-based teaching methods and technology. She provides faculty development to other instructors through peer-coaching, workshops, and mentoring. Dr. Hogan is the author of *Stem Cells and Cloning,* Second Edition, and is lead moderator of the *Instructor Exchange*, a site within MasteringBiology® for instructors to exchange classroom materials and ideas.

Neil A. Campbell (1946–2004) combined the inquiring nature of a research scientist with the soul of a caring teacher. Over his 30 years of teaching introductory biology to both science majors and nonscience majors, many thousands of students had the opportunity to learn from him and be stimulated by his enthusiasm for the study of life. While he is greatly missed by his many friends in the biology community, his coauthors remain inspired by his visionary dedication to education and are committed to searching for ever better ways to engage students in the wonders of biology.

To the Student: How to use this book and MasteringBiology®

Make important connections between biological concepts and your life

NEW! Each chapter opens with a **high-interest question** to spark your interest in the topic. Questions are revisited later in the chapter, in either a Scientific Thinking or Evolution Connection module.

CHAPTER 12

DNA Technology and Genomics

? *Are genetically modified organisms safe?*

Papaya fruit, shown in the photograph below, are sweet and loaded with vitamin C. They are borne on a rapidly growing treelike plant (*Carica papaya*) that grows only in tropical climates. In Hawaii, papaya is both a dietary staple and a valuable export crop.

Although thriving today, Hawaii's papaya industry seemed doomed just a few decades ago. A deadly pathogen called the papaya ringspot virus (PRV) had spread throughout the islands and appeared poised to completely eradicate the papaya plant population. But scientists from the University of Hawaii were able to rescue the industry by creating new, genetically engineered PRV-resistant strains of papaya. Today, the papaya industry is once again vibrant—and the vast majority of Hawaii's papayas are genetically modified organisms (GMOs).

However, not everyone is happy about the circumstances surrounding the recovery of the Hawaiian papaya industry. Although genetically modified papayas are approved for consumption in the United States (as are many other GMO fruits and vegetables), some critics have raised safety concerns—for the people who eat them and for the environment. On three occasions over a three-year

span, thousands of pa
down under the cove
GMO crops. Althoug
should we in fact be
question continues t

In addition to GM
in many other ways:
dustrial products, D
ence, new technolo
and DNA can even
chapter, we'll discu
specific techniques
legal, and ethical is

230

The New York Times

May 12, 2013

Seeking Clues to Heart Disease in DNA of an Unlucky Family

By GINA KOLATA

Early heart disease ran in Rick Del Sontro's family, and every time he went for a run, he was scared his heart would betray him. So he did all he could to improve his odds. He kept himself lean, stayed away from red meat, spurned cigarettes and exercised intensely, even completing an Ironman Triathlon.

SUPER
SALMON?
abc NEWS
.com

MasteringBiology®

◁ ABC News Videos and Current Events articles from The *New York Times* connect what you learn in biology class to fascinating stories in the news.

BIG IDEAS

Gene Cloning (12.1–12.5)

A variety of laboratory techniques can be used to copy and combine DNA molecules.

Genetically Modified Organisms (12.6–12.10)

Transgenic cells, plants, and animals are used in agriculture and medicine.

DNA Profiling (12.11–12.16)

Genetic markers can be used to definitively match a DNA sample to an individual.

Genomics (12.17–12.21)

The study of complete DNA sets helps us learn about evolutionary history.

231

d of Hawaii were hacked as a protest against h criminal behavior, y of GMO crops? This te and disagreement. nologies affect our lives roduce medical and in- the field of forensic sci- for biological research, orical questions. In this ns. We'll also consider the d, and some of the social, new technologies.

◁ **Big Ideas** help you connect the overarching concepts that are explored in the chapter.

CONNECTION

▽ **Connection** modules in every chapter relate biology to your life and the world outside the classroom.

16.5 Biofilms are complex associations of microbes

CONNECTION

In many natural environments, prokaryotes attach to surfaces in highly organized colonies called **biofilms**. A biofilm may consist of one or several species of prokaryotes, and it may include protists and fungi as well. Biofilms can form on almost any support, including rocks, soil, organic material (including living tissue), metal, and plastic. You have a biofilm on your teeth—dental plaque is a biofilm that can cause tooth decay. Biofilms can even form without a solid foundation, for example, on the surface of stagnant water.

Biofilm formation begins when prokaryotes secrete signaling molecules that attract nearby cells into a cluster. Once the cluster becomes sufficiently large, the cells produce a gooey coating that glues them to the support and to each other, making the biofilm extremely difficult to dislodge. For example, if you don't scrub your shower, you could find a biofilm growing around the drain—running water alone is not strong enough to wash it away. As the biofilm gets larger and more complex, it becomes a "city" of microbes. Communicating by chemical signals, members of the community coordinate the division of labor, defense against invaders, and other activities. Channels in the biofilm allow nutrients to reach cells in the interior and allow wastes to leave, and a variety of environments develop within it.

Biofilms are common among bacteria that cause disease in humans. For instance, ear infections and urinary tract infections are often the result of biofilm-forming bacteria. Cystic fibrosis patients are vulnerable to pneumonia caused by bac-... their lungs. Biofilms of harmful bacteria can also form on implanted medical devices such as catheters, replacement joints, or pacemakers. The complexity of biofilms makes these infections especially difficult to defeat. Antibiotics may not be able to penetrate beyond the outer layer of cells, leaving much of the community intact. For example, some biofilm bacteria produce an enzyme that breaks down penicillin faster than it can diffuse inward.

Biofilms that form in the environment can be difficult to eradicate, too. A variety of industries spend billions of dollars every year trying to get rid of biofilms that clog and corrode pipes, gum up filters and drains, and coat the hulls of ships (**Figure 16.5**). Biofilms in water distribution pipes may survive chlorination, the most common method of ensuring that drinking water does not contain any harmful microorganisms. For example, biofilms of *Vibrio cholera*, the bacterium that causes cholera, found in water pipes were capable of withstanding levels of chlorine 10 to 20 times higher than the concentrations routinely used to chlorinate drinking water.

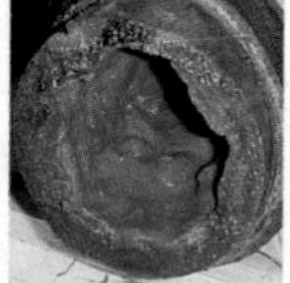

▲ **Figure 16.5** A biofilm fouling the insides of a pipe

? **Why are biofilms difficult to eradicate?**

... biofilm stick to each other; the outer layer of cells may prevent antimicrobial substances from penetrating into the interior of the biofilm.

EVOLUTION CONNECTION

◁ **Evolution Connection** modules present concrete examples of the evidence for evolution within each chapter, providing you with a coherent theme for the study of life.

10.19 Emerging viruses threaten human health

EVOLUTION CONNECTION

Emerging viruses are ones that seem to burst on to the scene, becoming apparent to the medical community quite suddenly. There are many familiar examples, such as the 2009 H1N1 influenza virus (discussed in the chapter introduction). Another example is **HIV** (human immunodeficiency virus), the virus that causes **AIDS** (acquired immunodeficiency syndrome). HIV appeared in New York and California in the early 1980s, seemingly out of nowhere. Yet another example is the deadly Ebola virus, recognized initially in 1976 in central Africa; it is one of several emerging viruses that cause hemorrhagic fever, an often fatal syndrome characterized by fever, vomiting, massive bleeding, and circulatory system collapse. A number of other dangerous newly recognized viruses cause encephalitis, inflammation of the brain. One example is the West Nile virus, which appeared in North America in 1999 and has since spread to all 48 contiguous U.S. states. West Nile virus is spread primarily by mosquitoes, which carry the virus in blood sucked from one victim and can transfer it to another victim. West Nile virus cases surged in 2012, especially in Texas. Severe acute respiratory syndrome (SARS) first appeared in China in 2002. Within eight months, about 8,000 ... and 10% died. Researchers quickly ...

Why are viral diseases such a constant threat?

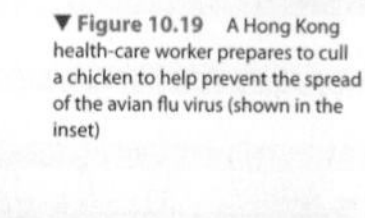

Colorized TEM 180,000×

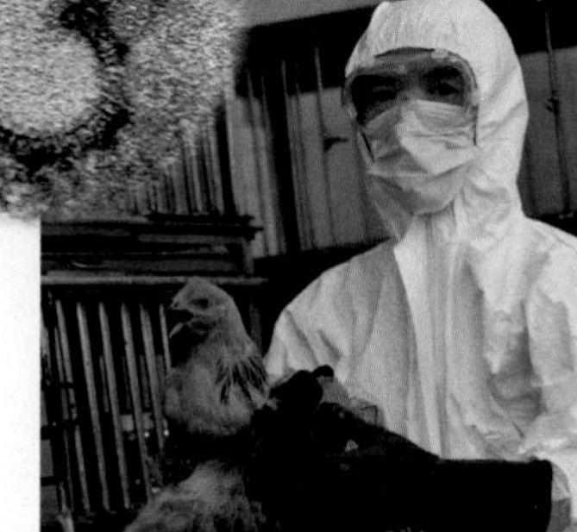

▼ **Figure 10.19** A Hong Kong health-care worker prepares to cull a chicken to help prevent the spread of the avian flu virus (shown in the inset)

To the Student: How to use this book and MasteringBiology®

Stay focused on the key concepts

Central concepts summarize the key topic of each module, helping you stay focused as you study.

Checkpoint questions at the end of each module help you stay on track.

NEW and revised art provides clear visuals to help you understand key topics. Selected figures include numbered steps that are keyed to explanations in the text.

4.9 The Golgi apparatus modifies, sorts, and ships cell products

After leaving the ER, many transport vesicles travel to the **Golgi apparatus**. Using a light microscope and a staining technique he developed, Italian scientist Camillo Golgi discovered this membranous organelle in 1898. The electron microscope confirmed his discovery more than 50 years later, revealing a stack of flattened sacs, looking much like a pile of pita bread. A cell may contain many, even hundreds, of these stacks. The number of Golgi stacks correlates with how active the cell is in secreting proteins—a multistep process that, as you have just seen, is initiated in the rough ER.

The Golgi apparatus serves as a molecular warehouse and processing station for products manufactured by the ER. You can follow these activities in **Figure 4.9**. Note that the flattened Golgi sacs are not connected, as are ER sacs. ❶ One side of a Golgi stack serves as a receiving dock for transport vesicles produced by the ER. ❷ A vesicle fuses with a Golgi sac, adding its membrane and contents to the "receiving" side. ❸ Products of the ER are modified as a Golgi sac progresses through the stack. ❹ The "shipping" side of the Golgi functions as a depot, dispatching its products in vesicles that bud off and travel to other sites.

How might ER products be processed during their transit through the Golgi? Various Golgi enzymes modify the carbohydrate portions of the glycoproteins made in the ER, removing some sugars and substituting others. Molecular identification tags, such as phosphate groups, may be added that help the Golgi sort molecules into different batches for different destinations.

Finished secretory products, packaged in transport vesicles, move to the plasma membrane for export from the cell. Alternatively, finished products may become part of the plasma membrane itself or part of another organelle, such as a lysosome, which we discuss next.

? **What is the relationship of the Golgi apparatus to the ER in a protein-secreting cell?**

● The Golgi receives transport vesicles budded from the ER that contain proteins synthesized by bound ribosomes. The Golgi finishes processing the proteins and dispatches transport vesicles to the plasma membrane, where the proteins are secreted.

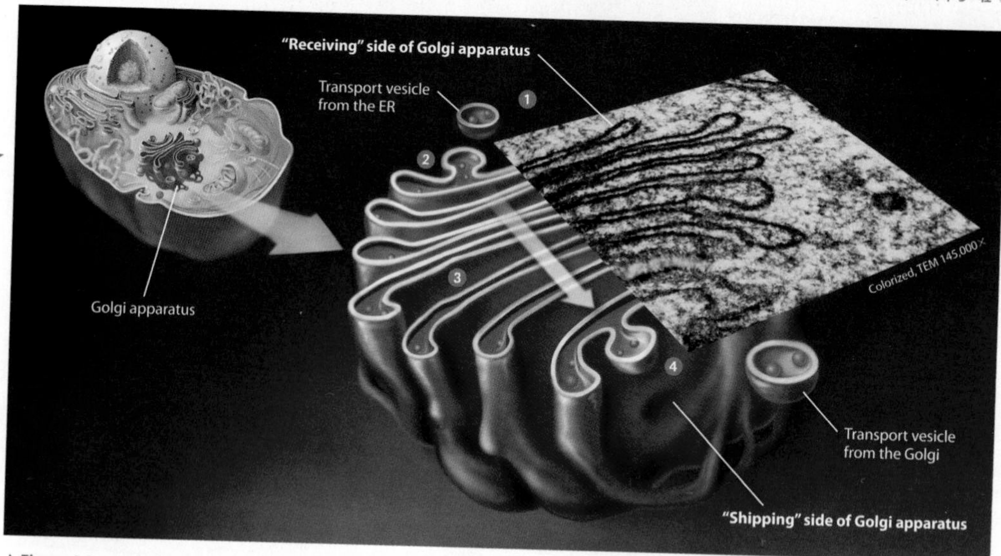
"Receiving" side of Golgi apparatus
Transport vesicle from the ER
1
2
3
4
Golgi apparatus
Colorized, TEM 145,000×
Transport vesicle from the Golgi
"Shipping" side of Golgi apparatus

▲ **Figure 4.9** The Golgi apparatus receiving, processing, and shipping products

The Endomembrane System **61**

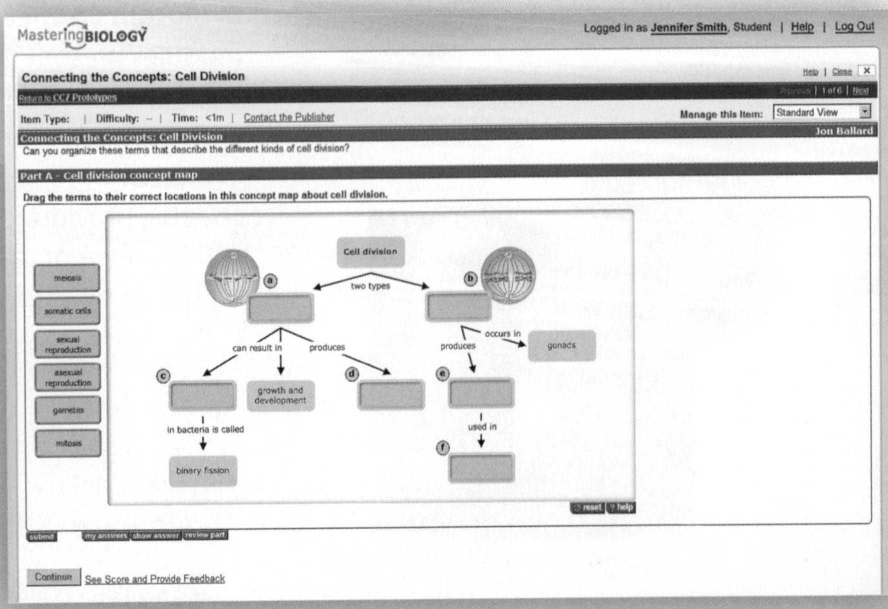
MasteringBIOLOGY
Logged in as Jennifer Smith, Student | Help | Log Out
Connecting the Concepts: Cell Division
Connecting the Concepts: Cell Division
Can you organize these terms that describe the different kinds of cell division?
Part A - Cell division concept map
Drag the terms to their correct locations in this concept map about cell division.
meiosis
somatic cells
sexual reproduction
asexual reproduction
gametes
mitosis
Cell division
two types
can result in
produces
produces
occurs in
gonads
growth and development
in bacteria is called
binary fission
used in
Continue
See Score and Provide Feedback

MasteringBiology®

◁ **Connecting the Concepts** activities link one biological concept to another.

Learn how to to think like a scientist

▷ **New Scientific Thinking** modules explore how scientists use the processes of science for discovery. Each module concludes with a question that challenges you to think like a scientist.

SCIENTIFIC THINKING

▷ **New Scientific Thinking** topics include:

- **Module 2.15** — Scientists study the effects of rising atmospheric CO_2 on coral reef ecosystems
- **Module 8.10** — Tailoring treatment to each patient may improve cancer therapy
- **Module 25.3** — Coordinated waves of movement in huddles help penguins thermoregulate
- **Module 26.3** — A widely used weed killer demasculinizes male frogs
- **Module 29.2** — The model for magnetic sensory reception is incomplete

12.9 Genetically modified organisms raise health concerns

SCIENTIFIC THINKING

Are genetically modified organisms safe?

As soon as scientists realized the power of DNA technology, they began to worry about potential dangers. Early concerns focused on the possibility that recombinant DNA technology might create new pathogens. To guard against rogue microbes, scientists developed a set of guidelines including strict laboratory safety and containment procedures, the genetic crippling of transgenic organisms to ensure that they cannot survive outside the laboratory, and a prohibition on certain dangerous experiments. Today, most public concern centers on GMOs used for food.

Human Safety Genetically modified organisms are used in crop production because they are more nutritious or because they are cheaper to produce. But do these advantages come at a cost to the health of people consuming GMOs? When investigating complex questions like this one, scientists often use multiple experimental methods. A 2012 animal study involved 104 pigs that were divided into two groups: The first was fed a diet containing 39% GMO corn and the other a closely related non-GMO corn. The health of the pigs was measured over the short term (31 days), the medium term (110 days), and the normal generational life span. The researchers reported no significant differences between the two groups and no traces of foreign DNA in the slaughtered pigs.

Although pigs are a good model organism for human digestion, critics argue that human data are required to draw conclusions about the safety of dietary GMOs for people. The results of one human study, conducted jointly by Chinese and ... were published in 2012. Sixty-eight Chi... (ages 6–8) were fed Golden Rice, spinach ...eta-carotene), or a capsule containing ...Over 21 days, blood samples were drawn ...h vitamin A the body produced from ... data show that the beta-carotene in ...d the capsules was converted to vitamin ...milar efficiency, while the beta-carotene ...ificantly less vitamin A (**Figure 12.9**). ...chers to conclude that GMO rice can ...preventing vitamin A deficiency. ...findings, this study caused an uproar. ...lled the study an unethical "scandal," ...scientists had used Chinese schoolchil...jects. The project leaders countered ...n and consent had been obtained in ...nited States. The controversy highlights ...n conducting research on human ...ies are of limited value, but human ...al. To date, no study has documented ...from GMO foods, and there is gen... scientists that the GMO foods on the ...er, it is not yet possible to measure the ...y) of GMOs on human health.

Advocates of a cautious approach ... that transgenic plants might pass their new genes to related species in nearby wild areas, disturbing the composition of the natural ecosystem. Critics of GMO crops can point to several studies that do indeed show unintended gene transfer from engineered crops to nearby wild relatives. But GMO advocates counter that no lasting or detrimental effects from such transfers have been demonstrated, and that some GMOs (such as bacteria engineered to break down oil spills) can actively help the environment.

Labeling Although the majority of several staple crops grown in the United States—including corn and soybeans—are genetically modified, products made from GMOs are not required to be labeled in any way. Chances are you ate a food containing GMOs today, but the lack of labeling means you probably can't say for certain. Labeling of foods containing more than trace amounts of GMOs is required in Europe, Japan, Australia, China, Russia, and other countries. Labeling advocates point out that the information would allow consumers to decide for themselves whether they wish to be exposed to GMO foods. Some biotechnology advocates, however, respond that similar demands were not made when "transgenic" crop plants produced by traditional breeding techniques were put on the market. For example, triticale (a crop used primarily in animal feed but also in some human foods) was created decades ago by combining the genomes of wheat and rye—two plants that do not interbreed in nature. Triticale is now sold worldwide without any special labeling.

Scientists and the public need to weigh the possible benefits versus risks on a case-by-case basis. The best scenario would be to proceed with caution, basing our decisions on sound scientific information rather than on either irrational fear or blind optimism.

? **Why might crop plants engineered to be resistant to weed killer pose a danger to the environment?**

The genes for herbicide resistance could transfer to closely related weeds, which could themselves then become resistant.

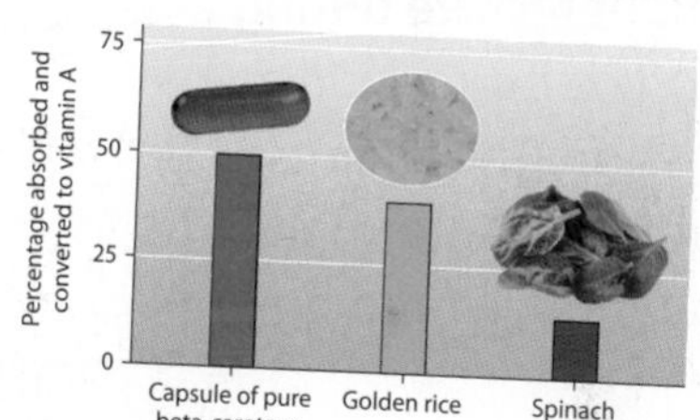

Percentage absorbed and converted to vitamin A
75
50
25
0
Capsule of pure beta-carotene
Golden rice
Spinach

▲ **Figure 12.9** Vitamin A production after consumption of different sources of beta-carotene

Data from G. Tang et al., Beta-carotene in Golden Rice is as good as beta-carotene in oil at providing vitamin A to children, *American Journal of Clinical Nutrition* 96(3): 658–64 (2012).

...ology and Genomics

Scientific Thinking | Signed in as Libby Reiser, Instructor | Help | Close

What Roles Do Diet and the Microbial ... › Scientific Thinking: What Roles Do Diet and the Microbial Community in the In...

Item Type: Coaching Activities | **Difficulty:** -- | **Time:** -- | Contact the Publisher | **Manage this Item:** Standard View

Scientific Thinking: What Roles Do Diet and the Microbial Community in the Intestines Play in Obesity?

Fast foods, cookies and ice cream, sodas and energy drinks--Americans eat a lot of processed foods high in fats and simple sugars. Not surprisingly, this type of diet can lead to weight gain and is one of the main culprits in the obesity epidemic in this country. But, is there more to this story?

The foods you eat serve as food for the community of microorganisms that inhabit your digestive tract. Those microbes have their own food "preferences," metabolizing different types of food molecules and releasing their byproducts, which your body then absorbs.

Scientists have hypothesized that a high-fat, high sugar diet actually alters the composition of the microbial community that inhabits the beginning of the large intestine, which contributes to obesity. Because of the difficulties of carrying out experiments on humans, scientists have used mice as an animal model in which to test this hypothesis.

Part A - Designing a controlled experiment

In one experiment, scientists raised mice in germ-free conditions (who therefore lacked intestinal microbes). The mice were fed a low-fat diet rich in the complex plant polysaccharides often called fiber.

When the mice were 12 weeks old, the scientists transplanted the microbial community from the intestine of a single "donor" mouse into all of the germ-free mice. Then they divided the mice randomly into two groups and fed each group a different diet.

- Group 1 (the control group) continued to eat a low-fat, high-fiber diet.
- Group 2 (the experimental group) ate a high-fat, high-sugar diet.

"Donor" mouse
Germ-free mice
Group 1: Control group
Low-fat, high-fiber diet
High-fat, high-sugar diet
Group 2: Experimental group

MasteringBiology®

◁ **NEW! Scientific Thinking activities** teach you how to practice important scientific skills like understanding variables and making predictions. Specific wrong-answer feedback coaches you to the correct response.

To the Student: How to use this book and MasteringBiology®

Maximize your learning and success

- **New Visualizing the Concept** modules walk you through challenging concepts and complex processes.
- The brief narrative works together with the artwork to help you visualize and understand the topic.

Hints embedded within the module emulate the guidance that you might receive during instructor office hours or in a tutoring session. These hints provide additional information to deepen your understanding of the topic.

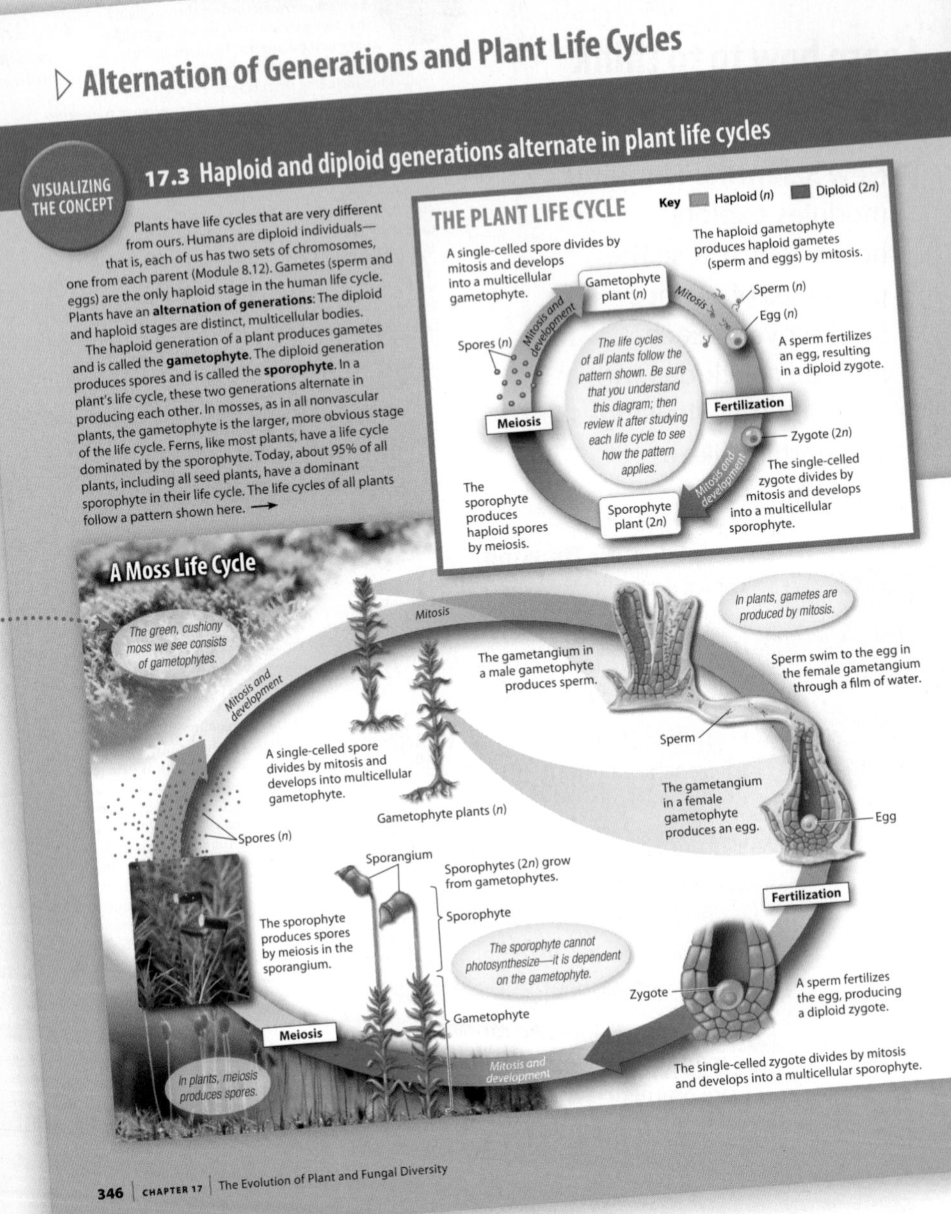

▷ Alternation of Generations and Plant Life Cycles

VISUALIZING THE CONCEPT

17.3 Haploid and diploid generations alternate in plant life cycles

Plants have life cycles that are very different from ours. Humans are diploid individuals—that is, each of us has two sets of chromosomes, one from each parent (Module 8.12). Gametes (sperm and eggs) are the only haploid stage in the human life cycle. Plants have an **alternation of generations**: The diploid and haploid stages are distinct, multicellular bodies.

The haploid generation of a plant produces gametes and is called the **gametophyte**. The diploid generation produces spores and is called the **sporophyte**. In a plant's life cycle, these two generations alternate in producing each other. In mosses, as in all nonvascular plants, the gametophyte is the larger, more obvious stage of the life cycle. Ferns, like most plants, have a life cycle dominated by the sporophyte. Today, about 95% of all plants, including all seed plants, have a dominant sporophyte in their life cycle. The life cycles of all plants follow a pattern shown here. ⟶

THE PLANT LIFE CYCLE
Key Haploid (n) Diploid (2n)
A single-celled spore divides by mitosis and develops into a multicellular gametophyte.
The haploid gametophyte produces haploid gametes (sperm and eggs) by mitosis.
Gametophyte plant (n)
Mitosis
Mitosis and development
Sperm (n)
Egg (n)
Spores (n)
A sperm fertilizes an egg, resulting in a diploid zygote.
The life cycles of all plants follow the pattern shown. Be sure that you understand this diagram; then review it after studying each life cycle to see how the pattern applies.
Fertilization
Meiosis
Zygote (2n)
The single-celled zygote divides by mitosis and develops into a multicellular sporophyte.
The sporophyte produces haploid spores by meiosis.
Sporophyte plant (2n)
A Moss Life Cycle
The green, cushiony moss we see consists of gametophytes.
Mitosis
In plants, gametes are produced by mitosis.
The gametangium in a male gametophyte produces sperm.
Sperm swim to the egg in the female gametangium through a film of water.
Mitosis and development
A single-celled spore divides by mitosis and develops into multicellular gametophyte.
Sperm
The gametangium in a female gametophyte produces an egg.
Egg
Gametophyte plants (n)
Spores (n)
Sporangium
Sporophytes (2n) grow from gametophytes.
Fertilization
Sporophyte
The sporophyte produces spores by meiosis in the sporangium.
The sporophyte cannot photosynthesize—it is dependent on the gametophyte.
Zygote
A sperm fertilizes the egg, producing a diploid zygote.
Gametophyte
Meiosis
In plants, meiosis produces spores.
Mitosis and development
The single-celled zygote divides by mitosis and develops into a multicellular sporophyte.
346 CHAPTER 17 The Evolution of Plant and Fungal Diversity

MasteringBiology®

- **NEW! Visualizing the Concept Activities** include interactive videos that were created and narrated by the authors of the text.

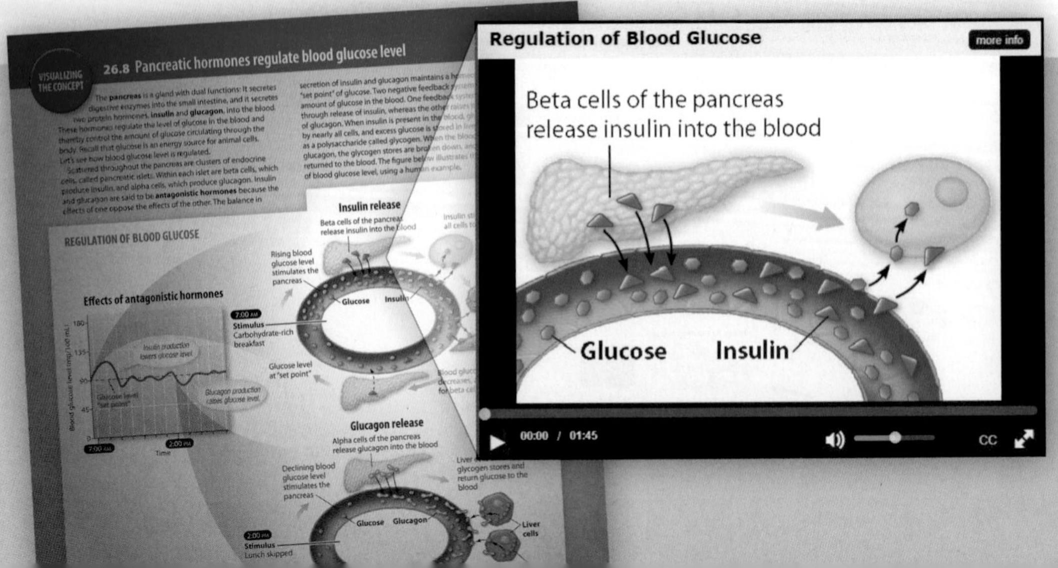

VISUALIZING THE CONCEPT
26.8 Pancreatic hormones regulate blood glucose level
REGULATION OF BLOOD GLUCOSE
Effects of antagonistic hormones
Insulin release
Glucagon release
Regulation of Blood Glucose
more info
Beta cells of the pancreas release insulin into the blood
Glucose
Insulin
00:00 / 01:45
CC

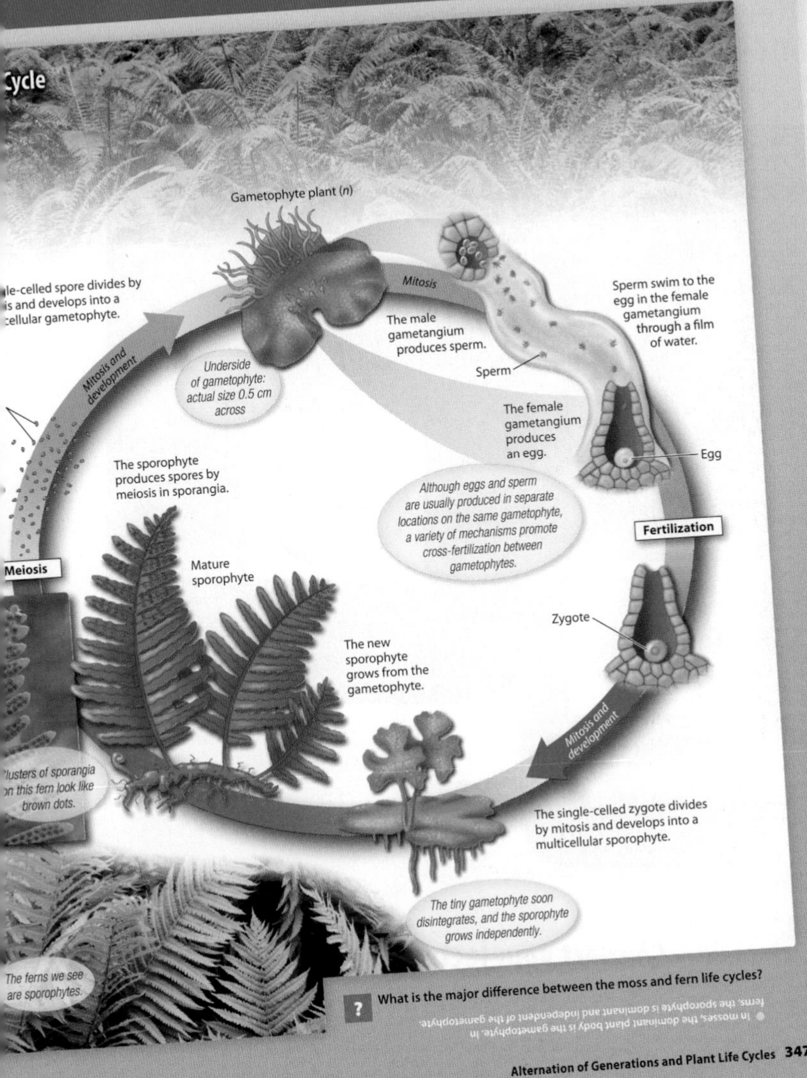

Gametophyte plant (n)
Mitosis
The male gametangium produces sperm.
Sperm swim to the egg in the female gametangium through a film of water.
Sperm
Underside of gametophyte: actual size 0.5 cm across
Mitosis and development
The female gametangium produces an egg.
Egg
The sporophyte produces spores by meiosis in sporangia.
Although eggs and sperm are usually produced in separate locations on the same gametophyte, a variety of mechanisms promote cross-fertilization between gametophytes.
Fertilization
Meiosis
Mature sporophyte
Zygote
The new sporophyte grows from the gametophyte.
Mitosis and development
The single-celled zygote divides by mitosis and develops into a multicellular sporophyte.
The tiny gametophyte soon disintegrates, and the sporophyte grows independently.
The ferns we see are sporophytes.
What is the major difference between the moss and fern life cycles?
Alternation of Generations and Plant Life Cycles 347

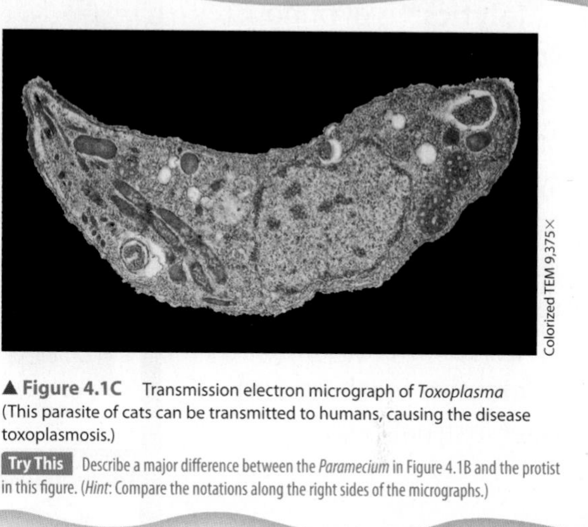

Colorized TEM 9,375×
▲ Figure 4.1C Transmission electron micrograph of Toxoplasma (This parasite of cats can be transmitted to humans, causing the disease toxoplasmosis.)
Try This Describe a major difference between the Paramecium in Figure 4.1B and the protist in this figure. (Hint: Compare the notations along the right sides of the micrographs.)

New! Try This activities help you actively engage with the figures and develop positive study habits.

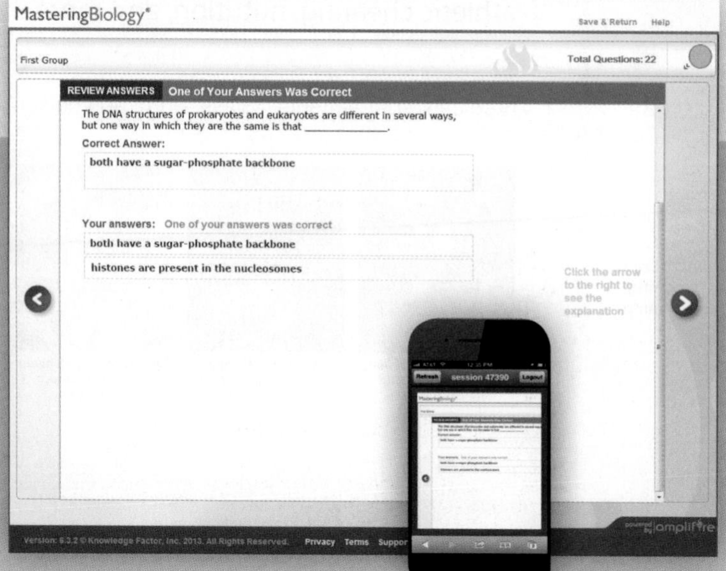

MasteringBiology®
Save & Return Help
First Group
Total Questions: 22
REVIEW ANSWERS One of Your Answers Was Correct
The DNA structures of prokaryotes and eukaryotes are different in several ways, but one way in which they are the same is that ___________.
Correct Answer:
both have a sugar-phosphate backbone
Your answers: One of your answers was correct
both have a sugar-phosphate backbone
histones are present in the nucleosomes
Click the arrow to the right to see the explanation

MasteringBiology®

New Dynamic Study Modules enable you to study effectively on your own and more quickly learn the information. These modules can be accessed on smartphones, tablets, and computers.

Resources save you hours of time preparing for class

▷ **NEW! Learning Catalytics™** is a "bring your own device" student engagement, assessment, and classroom intelligence system. This technology has grown out of twenty years of cutting-edge research, innovation, and implementation of interactive teaching and peer instruction.

learning catalytics
Courses Questions Classrooms Licenses Tour Help
Student view
my courses > test fall 2013 demo course > test module
current session: 47390 | 1 student
Reactions Stop session Edit Show session ID window Ask a new question on the fly
Jump to
Stop delivery Deliver again Assign groups Show all results
Where is the nucleus of the cell?
Round 1
90 responses, 88% correct
Answer
session 47390
Refresh
Logout
region question
Where is the nucleus of the cell?
Submit response
I don't understand

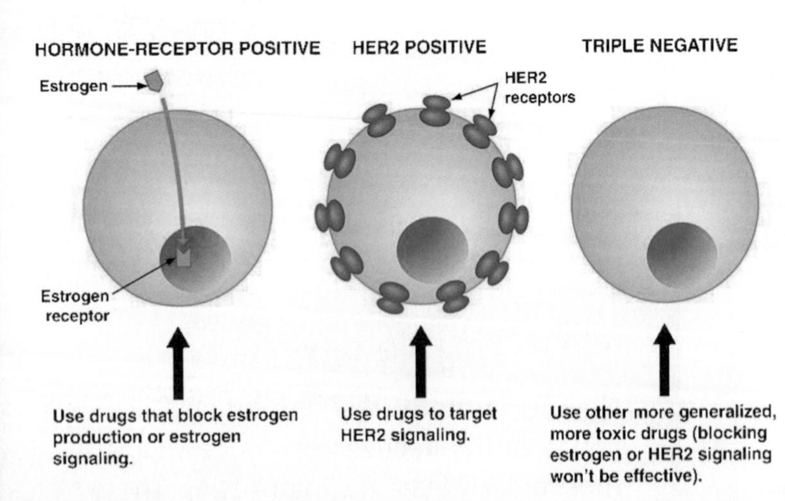
Three classes of breast cancer tumors lead to more personalized therapy
HORMONE-RECEPTOR POSITIVE
HER2 POSITIVE
TRIPLE NEGATIVE
Estrogen
HER2 receptors
Estrogen receptor
Use drugs that block estrogen production or estrogen signaling.
Use drugs to target HER2 signaling.
Use other more generalized, more toxic drugs (blocking estrogen or HER2 signaling won't be effective).

Connect your lectures to current topics

◁ **Campbell Current Topics PowerPoint** slides help you prepare a high-impact lecture developed around current issues. Topics include cancer, global climate change, athletic cheating, nutrition, and more.

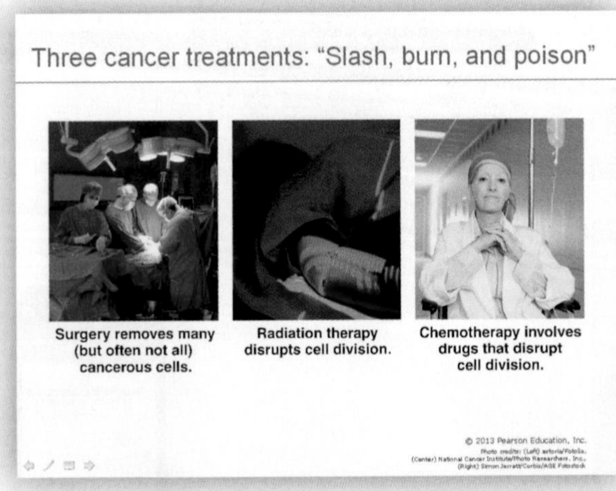
Three cancer treatments: "Slash, burn, and poison"
Surgery removes many (but often not all) cancerous cells.
Radiation therapy disrupts cell division.
Chemotherapy involves drugs that disrupt cell division.
© 2013 Pearson Education, Inc.

MasteringBiology®

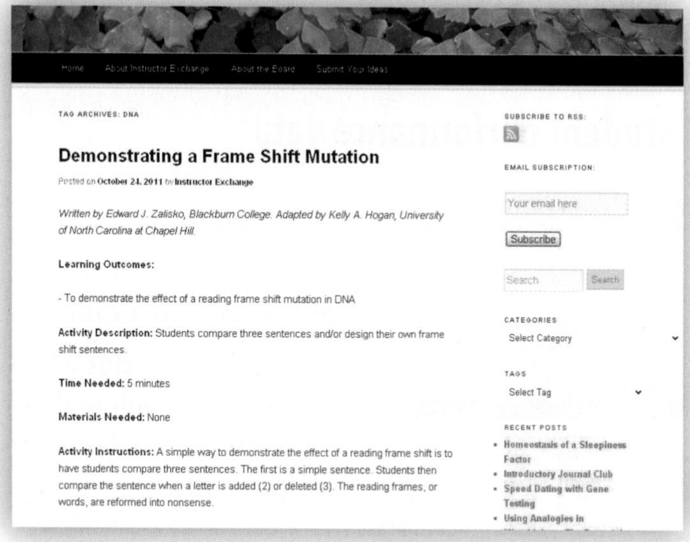
Home About Instructor Exchange About the Board Submit Your Ideas
TAG ARCHIVES: DNA
Demonstrating a Frame Shift Mutation
Posted on October 24, 2011 by Instructor Exchange
Written by Edward J. Zalisko, Blackburn College. Adapted by Kelly A. Hogan, University of North Carolina at Chapel Hill.
Learning Outcomes:
- To demonstrate the effect of a reading frame shift mutation in DNA
Activity Description: Students compare three sentences and/or design their own frame shift sentences.
Time Needed: 5 minutes
Materials Needed: None
Activity Instructions: A simple way to demonstrate the effect of a reading frame shift is to have students compare three sentences. The first is a simple sentence. Students then compare the sentence when a letter is added (2) or deleted (3). The reading frames, or words, are reformed into nonsense.
SUBSCRIBE TO RSS:
EMAIL SUBSCRIPTION:
Your email here
Subscribe
Search
Search
CATEGORIES
Select Category
TAGS
Select Tag
RECENT POSTS
Homeostasis of a Sleepiness Factor
Introductory Journal Club
Speed Dating with Gene Testing
Using Analogies in

◁ **Instructor Exchange,** moderated by co-author Kelly Hogan, offers a library of active learning strategies contributed by instructors from across the country.

▽ **BioFlix activities** offer students 3-D animations to help them visualize and learn challenging topics.

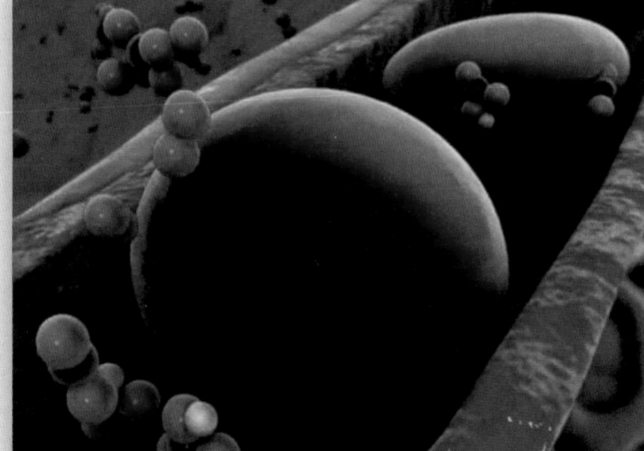

Assign tutorials to help students prepare for class

▽ **Video Tutor Sessions and MP3 Tutor Sessions,** hosted by co-author Eric Simon, provide on-the-go tutorials focused on key concepts and vocabulary.

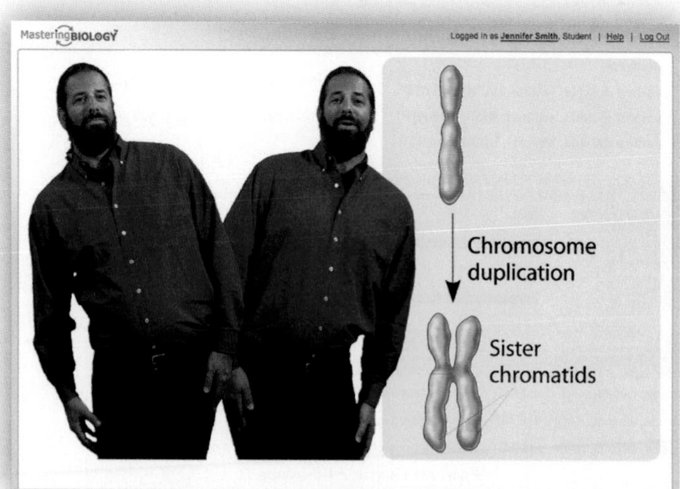
MasteringBIOLOGY
Logged in as Jennifer Smith, Student | Help | Log Out
Chromosome duplication
Sister chromatids

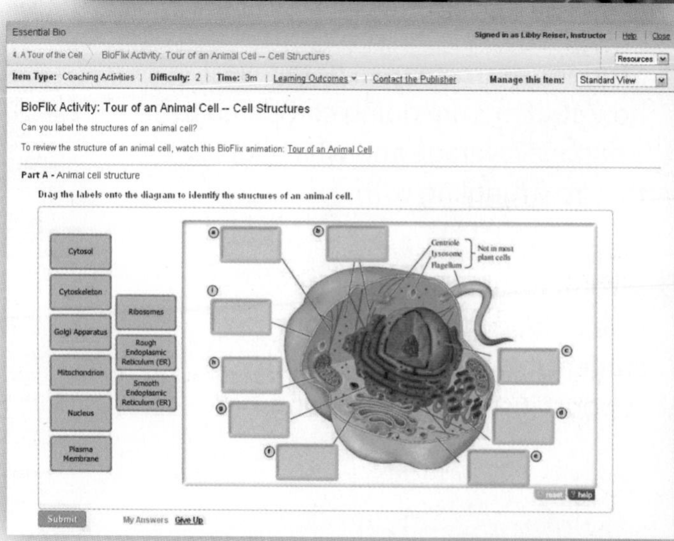
Essential Bio
Signed in as Libby Reiser, Instructor | Help | Close
4. A Tour of the Cell
BioFlix Activity: Tour of an Animal Cell -- Cell Structures
Resources
Item Type: Coaching Activities | Difficulty: 2 | Time: 3m | Learning Outcomes | Contact the Publisher
Manage this Item: Standard View
BioFlix Activity: Tour of an Animal Cell -- Cell Structures
Can you label the structures of an animal cell?
To review the structure of an animal cell, watch this BioFlix animation: Tour of an Animal Cell
Part A - Animal cell structure
Drag the labels onto the diagram to identify the structures of an animal cell.
Cytosol
Cytoskeleton
Golgi Apparatus
Mitochondrion
Nucleus
Plasma Membrane
Ribosomes
Rough Endoplasmic Reticulum (ER)
Smooth Endoplasmic Reticulum (ER)
Centriole
Lysosome
Flagellum
Not in most plant cells
reset
? help
Submit
My Answers Give Up

MasteringBiology® is an online assessment and tutorial system designed to help you teach more efficiently. It offers a variety of interactive activities to engage students and help them to succeed in the course.

Access students' results with easy-to-interpret student performance data

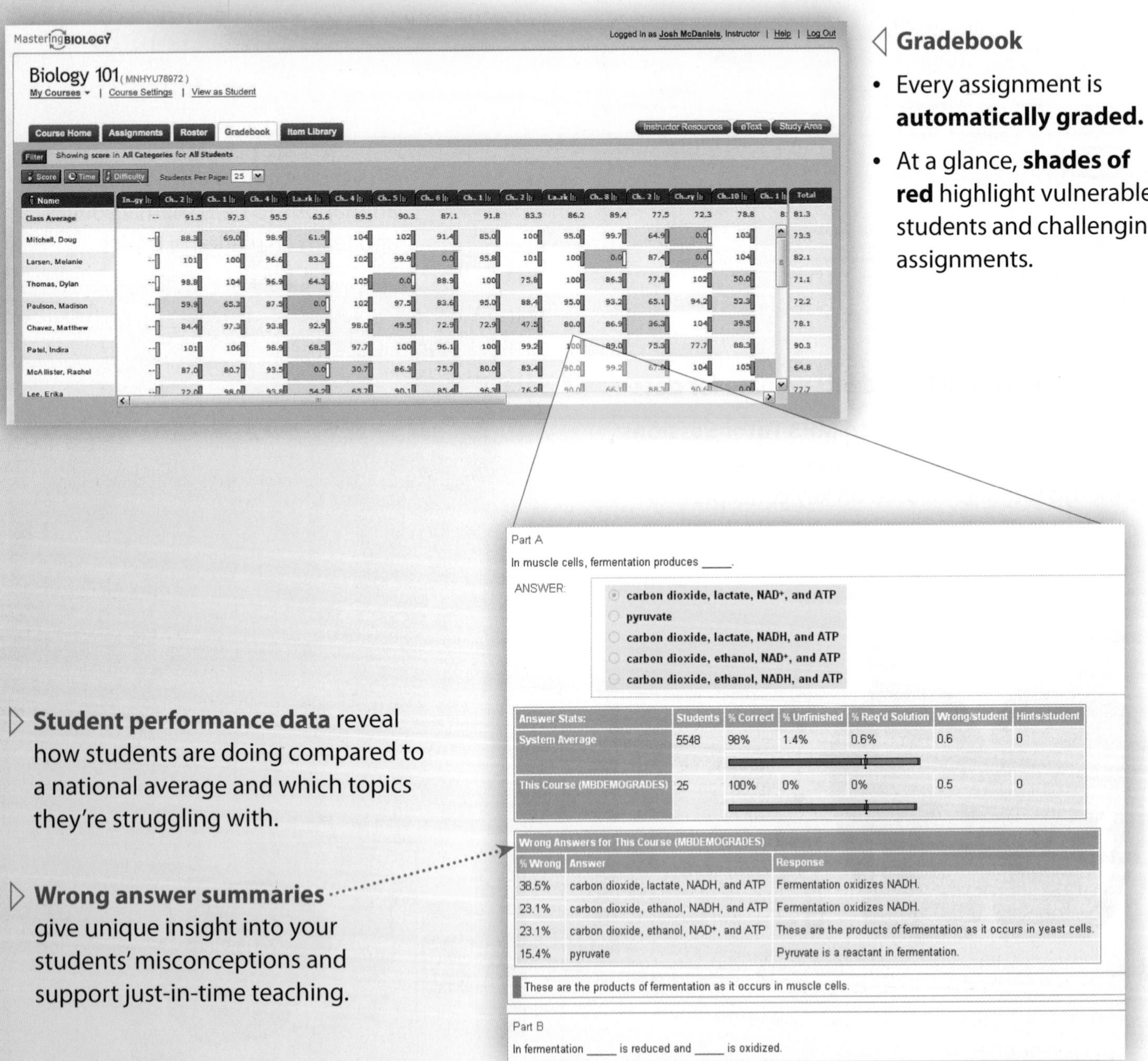
MasteringBIOLOGY
Logged in as Josh McDaniels, Instructor | Help | Log Out
Biology 101 (MNHYU78972)
My Courses | Course Settings | View as Student
Course Home
Assignments
Roster
Gradebook
Item Library
Instructor Resources
eText
Study Area
Filter Showing score in All Categories for All Students
Score
Time
Difficulty
Students Per Page: 25
Name
Total
Class Average
Mitchell, Doug
Larsen, Melanie
Thomas, Dylan
Paulson, Madison
Chavez, Matthew
Patel, Indira
McAllister, Rachel
Lee, Erika
Part A
In muscle cells, fermentation produces _____.
ANSWER:
carbon dioxide, lactate, NAD+, and ATP
pyruvate
carbon dioxide, lactate, NADH, and ATP
carbon dioxide, ethanol, NAD+, and ATP
carbon dioxide, ethanol, NADH, and ATP
Answer Stats: | Students | % Correct | % Unfinished | % Req'd Solution | Wrong/student | Hints/student
System Average | 5548 | 98% | 1.4% | 0.6% | 0.6 | 0
This Course (MBDEMOGRADES) | 25 | 100% | 0% | 0% | 0.5 | 0
Wrong Answers for This Course (MBDEMOGRADES)
% Wrong | Answer | Response
38.5% | carbon dioxide, lactate, NADH, and ATP | Fermentation oxidizes NADH.
23.1% | carbon dioxide, ethanol, NADH, and ATP | Fermentation oxidizes NADH.
23.1% | carbon dioxide, ethanol, NAD+, and ATP | These are the products of fermentation as it occurs in yeast cells.
15.4% | pyruvate | Pyruvate is a reactant in fermentation.
These are the products of fermentation as it occurs in muscle cells.
Part B
In fermentation _____ is reduced and _____ is oxidized.

◁ Gradebook

- Every assignment is **automatically graded.**
- At a glance, **shades of red** highlight vulnerable students and challenging assignments.

▷ **Student performance data** reveal how students are doing compared to a national average and which topics they're struggling with.

▷ **Wrong answer summaries** give unique insight into your students' misconceptions and support just-in-time teaching.

Gain insight into student progress at a glance

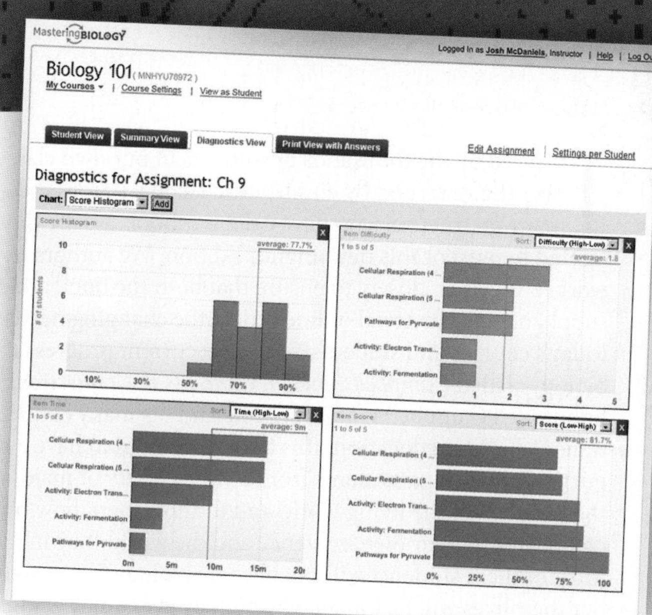
MasteringBIOLOGY
Logged in as Josh McDaniels, Instructor | Help | Log Out
Biology 101 (MNHYU78972)
My Courses | Course Settings | View as Student
Student View
Summary View
Diagnostics View
Print View with Answers
Edit Assignment | Settings per Student
Diagnostics for Assignment: Ch 9
Chart: Score Histogram
Add
average: 77.7%
Cellular Respiration (4 ...
Cellular Respiration (5 ...
Pathways for Pyruvate
Activity: Electron Trans...
Activity: Fermentation
Time (High-Low)
Score (Low-High)
Difficulty (High-Low)
average: 81.7%

▷ Get daily diagnostics.

Gradebook Diagnostics provide unique insight into class performance. With a single click, see a summary of how your students are struggling or progressing.

MasteringBiology® is easy for you and your students to use

MasteringBIOLOGY
Make Learning Part of the Grade™
Home
Product Info
Tours and Training
Proven Results
Books Available
Support
Register
New Students
New Instructors
Log In
Forgot your password?
MasteringBiology
Engages and motivates students to learn
Provides students personalized coaching and feedback
Quickly monitors student learning
The Mastering platform is the most effective and widely used online science tutorial, homework, and assessment system for the sciences.
Play Movie
Prepare for class
Make sure your students have the Course ID to register for class
Register for online training session
See what's new in this release
Over ONE MILLION STUDENTS Across the Sciences
View all Disciplines
Proven Results
Gerry Rayner, Monash University
Shannon Datwyler, California State University, Sacramento
Scott Bowling, Auburn University
"MasteringBiology has quickly become an integral part of my class."
Read More
PEARSON

◁ **The Mastering platform** is the most effective and widely used online tutorial, homework, and assessment system for the sciences.

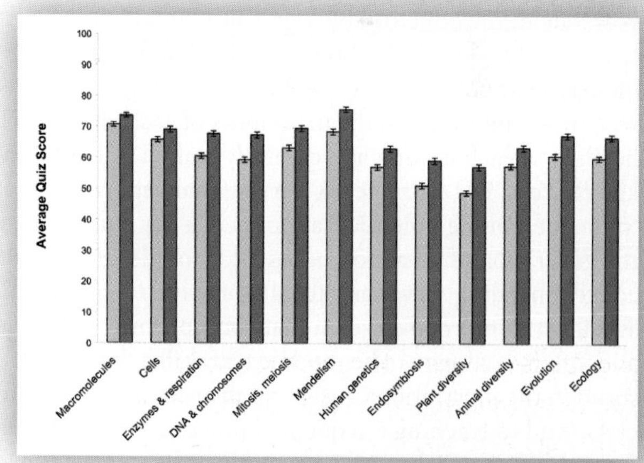
Average Quiz Score
0
10
20
30
40
50
60
70
80
90
100
Macromolecules
Cells
Enzymes & respiration
DNA & chromosomes
Mitosis, meiosis
Mendelism
Human genetics
Endosymbiosis
Plant diversity
Animal diversity
Evolution
Ecology

△ Efficacy studies

Go to the **"Proven Results"** tab at www.masteringbiology.com to see efficacy studies.

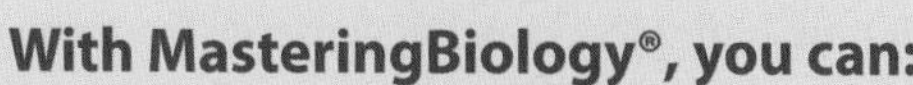

With MasteringBiology®, you can:

- **Assign** publisher-created pre-built assignments to get started quickly.
- **Easily edit** any of our questions or answers to match the precise language you use.
- **Import your own questions** and begin compiling meaningful data on student performance.
- **Easily export grades** to Microsoft® Excel or other course-management systems.

Preface

Inspired by the thousands of students in our own classes over the years and by enthusiastic feedback from the many instructors who have used our book, we are delighted to present this new, Eighth Edition. We authors have worked together closely to ensure that both the book and the supplementary material online reflect the changing needs of today's courses and students, as well as current progress in biology. Titled *Campbell Biology: Concepts & Connections* to honor Neil Campbell's founding role and his many contributions to biology education, this book continues to have a dual purpose: to engage students from a wide variety of majors in the wonders of the living world and to show them how biology relates to their own existence and the world they inhabit. Most of these students will not become biologists themselves, but their lives will be touched by biology every day. Understanding the concepts of biology and their connections to our lives is more important than ever. Whether we're concerned with our own health or the health of our planet, a familiarity with biology is essential. This basic knowledge and an appreciation for how science works have become elements of good citizenship in an era when informed evaluations of health issues, environmental problems, and applications of new technology are critical.

Concepts and Connections

Concepts Biology is a vast subject that gets bigger every year, but an introductory biology course is still only one or two semesters long. This book was the first introductory biology textbook to use concept modules to help students recognize and focus on the main ideas of each chapter. The heading of each module is a carefully crafted statement of a key concept. For example, "A nerve signal begins as a change in the membrane potential" announces a key concept about the generation of an action potential (Module 28.4). Such a concept heading serves as a focal point, and the module's text and illustrations converge on that concept with explanation and, often, analogies. The module text walks the student through the illustrations, just as an instructor might do in class. And in teaching a sequential process, such as the one diagrammed in Figure 28.4, we number the steps in the text to correspond to numbered steps in the figure. The synergy between a module's narrative and graphic components transforms the concept heading into an idea with meaning to the student. The checkpoint question at the end of each module encourages students to test their understanding as they proceed through a chapter. Finally, in the Chapter Review, all the key concept statements are listed and briefly summarized under the overarching section titles, explicitly reminding students of what they've learned.

Connections Students are more motivated to study biology when they can connect it to their own lives and interests—for example, when they are able to relate science to health issues, economic problems, environmental quality, ethical controversies, and social responsibility. In this edition, blue Connection icons mark the numerous application modules that go beyond the core biological concepts. For example, the new Connection Module 26.12 describes the potential role oxytocin plays in human–dog bonding. In addition, our Evolution Connection modules, identified by green icons, connect the content of each chapter to the grand unifying theme of evolution, without which the study of life has no coherence. Explicit connections are also made between the chapter introduction and either the Evolution Connection module or the new Scientific Thinking module in each chapter; new high-interest questions introduce each chapter, drawing students into the topic and encouraging a curiosity to explore the question further when it appears again later in the chapter.

New to This Edition

New Scientific Thinking Modules In this edition we placed greater emphasis on the process of scientific inquiry through the addition to each chapter of a new type of module called Scientific Thinking, which is called out with a purple icon. These modules cover recent scientific research as well as underscore the spirit of inquiry in historical discoveries. All Scientific Thinking modules strive to demonstrate to students what scientists do. Each of these modules identifies key attributes of scientific inquiry, from the forming and testing of hypotheses to the analysis of data to the evaluation and communication of scientific results among scientists and with society as a whole. For example, the new Module 2.15 describes how scientists use both controlled experiments and observational field studies to document the effects of rising atmospheric CO_2 on coral reef ecosystems. Module 13.3 describes the scientific search for the common ancestor of whales, using different lines of inquiry from early fossil clues, molecular comparisons, and a series of transitional fossils that link whales to cloven-hoofed mammals, animals that live on land. And to prepare students for the renewed focus in the book on how biological concepts emerge from the process of science, we have significantly revised the introduction in Chapter 1, Biology: Exploring Life. These changes will better equip students to think like scientists and emphasize the connections between discovery and the concepts explored throughout the course.

New Visualizing the Concept Modules Also new to this edition are modules that raise our hallmark art–text integration to a new level. These Visualizing the Concept modules take challenging concepts or processes and walk students through them in a highly visual manner, using engaging, attractive art; clear and concise labels; and instructor "hints" called out in light blue bubbles. These short hints emulate the one-on-one coaching an instructor might provide to a students during office hours and help students make key connections within the figure. Examples of this new feature include Module 9.8, which demonstrates to students the process of reading and analyzing a family pedigree; Module 17.3, which introduces the concept of plant life cycles through a combination of photographs and detailed life cycle art displayed across an impressive two-page layout; and Module 26.8, which walks students through the concept of homeostatic controls in blood glucose levels.

New "Try This" Tips One theme of the revision for the Eighth Edition is to help all students learn positive study habits they can take with them throughout their college careers and, in particular, to encourage them to be active in their reading and studying. To foster good study habits, several figures in each chapter feature a new "Try This" study tip. These action-oriented statements or questions direct students to study a figure more closely and explain, interpret, or extend what the figure presents. For example, in Figure 3.13B, students are asked to "Point out the bonds and functional groups that make the R groups of these three amino acids either hydrophobic or hydrophilic." Figure 6.10B is a new figure illustrating the molecular rotary motor ATP synthase, and the accompanying Try This tip asks students to "Identify the power source that runs this motor. Explain where this 'power' comes from." Figure 36.7, on the effect of predation on the life history traits of guppies, offers the following Try This tip: "Use the figure to explain how the hypothesis was tested."

Improvements to End-of-Chapter Section The Testing Your Knowledge questions are now arranged to reflect Bloom's Taxonomy of cognitive domains. Questions and activities are grouped into Level 1: Knowledge/Comprehension, Level 2: Application/Analysis, and Level 3: Synthesis/Evaluation. In addition, a new Scientific Thinking question has been added to each chapter that connects to and extends the topic of the Scientific Thinking module. Throughout the Chapter Review, new questions have been added that will help students better engage with the chapter topic and practice higher-level problem solving.

New Design and Improved Art The fresh new design used throughout the chapters and the extensive reconceptualization of many figures make the book even more appealing and accessible to visual learners. The cellular art in Chapter 4, A Tour of the Cell, for example, has been completely reimagined for more depth perspective and richer color. The new big-picture diagrams of the animal and plant cells are vibrant and better demonstrate the spatial relationships among the cellular structures with an almost three-dimensional style. The illustrations of cellular organelles elsewhere in Chapter 4 include electron micrographs overlaid on diagrams to emphasize the connection between the realistic micrograph depiction and the artwork. Figure 4.9, for example, shows a micrograph of an actual Golgi apparatus paired with an illustration; an accompanying orientation diagram—a hallmark of *Concepts and Connections*—continues to act as a roadmap that reminds students of how an organelle fits within the overall cell structure. Finally, throughout the book we have introduced new molecular art; for example, see Figure 10.11B for a new representation of a molecule of tRNA binding to an enzyme molecule.

The Latest Science Biology is a dynamic field of study, and we take pride in our book's currency and scientific accuracy. For this edition, as in previous editions, we have integrated the results of the latest scientific research throughout the book. We have done this carefully and thoughtfully, recognizing that research advances can lead to new ways of looking at biological topics; such changes in perspective can necessitate organizational changes in our textbook to better reflect the current state of a field. You will find a unit-by-unit account of new content and organizational improvements in the "New Content" section on pp. xvii–xviii following this Preface.

New MasteringBiology® A specially developed version of MasteringBiology, the most widely used online tutorial and assessment program for biology, continues to accompany *Campbell Biology: Concepts & Connections*. In addition to 170 author-created activities that help students learn vocabulary, extend the book's emphasis on visual learning, demonstrate the connections among key concepts (helping students grasp the big ideas), and coach students on how to interpret data, the Eighth Edition features two additional new activity types. New Scientific Thinking activities encourage students to practice the basic science skills explored in the in-text Scientific Thinking feature, allowing students to try out thinking like a scientist and allowing instructors to assess this understanding; new Visualizing the Concept activities take students on an animated and narrated tour of select Visualizing the Concept modules from the text, offering students the chance to review key concepts in a digital learning modality. MasteringBiology® for *Campbell Biology: Concepts & Connections*, Eighth Edition, will help students to see strong connections through their print textbook, and the additional practice available online allows instructors to capture powerful data on student performance, thereby making the most of class time.

This Book's Flexibility

Although a biology textbook's table of contents is by design linear, biology itself is more like a web of related concepts without a single starting point or prescribed path. Courses can navigate this network by starting with molecules, with ecology, or somewhere in between, and courses can omit topics. *Campbell Biology: Concepts & Connections* is uniquely suited to offer flexibility and thus serve a variety of courses. The seven units of the book are largely self-contained, and in a number of the units, chapters can be assigned in a different order without much loss of coherence. The use of numbered modules makes it easy to skip topics or reorder the presentation of material.

■ ■ ■

For many students, introductory biology is the only science course that they will take during their college years. Long after today's students have forgotten most of the specific content of their biology course, they will be left with general impressions and attitudes about science and scientists. We hope that this new edition of *Campbell Biology: Concepts & Connections* helps make those impressions positive and supports instructors' goals for sharing the fun of biology. In our continuing efforts to improve the book and its supporting materials, we benefit tremendously from instructor and student feedback, not only in formal reviews but also via informal communication. Please let us know how we are doing and how we can improve the next edition of the book.

Jane Reece, janereece@cal.berkeley.edu

Martha Taylor (Chapter 1 and Unit I), mrt2@cornell.edu

Eric Simon (Units II and VI and Chapters 21 and 27), esimon@nec.edu

Jean Dickey (Units III, IV, and VII and Chapters 22, 23, and 30), dickeyj@clemson.edu

Kelly Hogan (Chapters 20, 24–26, 28, and 29), leek@email.unc.edu

New Content

Below are some important highlights of new content and organizational improvements in *Campbell Biology: Concepts & Connections*, Eighth Edition.

Chapter 1, Biology: Exploring Life The snowy owl, our cover organism for the Eighth Edition, is featured in the chapter introduction. The discussion of the evolutionary adaptations of these owls to life on the arctic tundra links to a new Scientific Thinking module on testing the hypothesis that camouflage coloration protects some animals from predation. An expanded module on evolution as the core theme of biology now includes a phylogenetic tree of elephants to enhance the discussion of the unity and diversity of life.

Unit I, The Life of the Cell Throughout the Eighth Edition, the themes introduced in new chapter introductions are expanded and further explored in either Scientific Thinking or Evolution Connection modules. For instance, in this unit, Chapter 5, The Working Cell, begins with the question "How can water flow through a membrane?" and an essay that describes the role these water channels play in kidney function; the essay is illustrated with a computer model of aquaporins spanning a membrane. Module 5.7, a Scientific Thinking module, then details the serendipitous discovery of aquaporins and presents data from a study that helped identify their function. Chapter 7, Photosynthesis: Using Light to Make Food, begins with the question "Will global climate change make you itch?" and uses the example of proliferation of poison ivy to introduce this chapter on photosynthesis. Then, Module 7.13, another Scientific Thinking module, explores various ways that scientists test the effects of rising atmospheric CO_2 levels on plant growth and presents results from a study on poison ivy growth. The Scientific Thinking question at the end of the chapter continues this theme, with data from a study on pollen production by ragweed under varying CO_2 concentrations, beginning with the question "Will global climate change make you sneeze as well as itch?" This unit also has three of the new Visualizing the Concept modules: Module 3.14: A protein's functional shape results from four levels of structure; Module 5.1: Membranes are fluid mosaics of lipids and proteins with many functions; and Module 7.9: The light reactions take place within the thylakoid membranes. These modules use both new and highly revised art to guide students through these challenging topics in a visual, highly intuitive manner. Chapter 6, How Cells Harvest Chemical Energy, now includes a new figure and expanded explanation of the amazing molecular motor, ATP synthase. The art program in Chapter 4, A Tour of the Cell, has been completely reimagined and revised. The beautiful new diagrams of animal and plant cells and their component parts are designed to help students appreciate the complexities of cell structure and explore the relationship between structure and function.

Unit II, Cellular Reproduction and Genetics The purpose of this unit is to help students understand the relationship between DNA, chromosomes, and organisms and to help them see that genetics is not purely hypothetical but connects in many important and interesting ways to their lives, human society, and other life on Earth. In preparing this edition, we worked to clarify difficult concepts, enhancing text and illustrations and providing timely new applications of genetic principles. The content is reinforced with updated discussions of relevant topics, such as personalized cancer therapy, the H1N1 and H5N1 influenza viruses, umbilical cord blood banking, and the science and controversy surrounding genetically modified foods. This edition includes discussion of many recent advances in the field. Some new topics concern our basic understanding of genetics and the cell cycle, such as how sister chromatids are physically attached during meiosis, how chemical modifications such as methylation and acetylation affect inheritance, and the roles of activators and enhancers in controlling gene expression. Other topics include recent advances in our understanding of genetics, such as the analysis of recent human evolution of high-altitude-dwelling Sherpas, expanded roles for microRNAs in the control of genetic information, and our improved understanding of the cellular basis of health problems in cloned animals. In some cases, sections within chapters have been reorganized to present a more logical flow of materials. Examples of new organization include the discussion of human karyotypes and the diagnosis of chromosomal abnormalities (Modules 8.18–8.20) and the processes of reproductive and therapeutic cloning (Modules 11.12–11.14). Material throughout the unit has been updated to reflect recent data, such as the latest cancer statistics and results from whole-genome sequencing.

Unit III, Concepts of Evolution This unit presents the basic principles of evolution and natural selection, the overwhelming evidence that supports these theories, and their relevance to all of biology—and to the lives of students. A new chapter introduction in Chapter 13, How Populations Evolve, highlights the role that evolution plays in thwarting human attempts to eradicate disease. The chapter has been reorganized so that the opening module on Darwin's development of the theory of evolution is followed immediately by evidence for evolution, including a Scientific Thinking module on fossils of transitional forms. Another new module (13.4) assembles evidence from homologies, including an example of "pseudogenes." New material in this unit also supports our goal of directly addressing student misconceptions about evolution. For example, a new chapter introduction and Scientific Thinking module in Chapter 14, The Origin of Species, tackle the question "Can we observe speciation occurring?" and a new chapter introduction in Chapter 15, Tracing Evolutionary History, poses the question (answered in Module 15.12) "How do brand-new structures arise by evolution?"

Unit IV, The Evolution of Biological Diversity The diversity unit surveys all life on Earth in less than a hundred pages! Consequently, descriptions and illustrations of the unifying characteristics of each major group of organisms, along with a small sample of its diversity, make up the bulk of the content. Two recurring elements are interwoven with these descriptions: evolutionary history and examples of relevance to our everyday lives and society at large. For the Eighth Edition, we have improved and updated those two elements. For example, Chapter 16, Microbial Life: Prokaryotes and Protists, opens with a new introduction on human microbiota and the question "Are antibiotics making us fat?" The related Scientific Thinking module (16.11) updates the story of Marshall's discovery of the role of *Helicobacter pylori* in ulcers with a new hypothesis about a possible connection between *H. pylori* and obesity. A new chapter introduction and Scientific Thinking module in Chapter 17, The Evolution of Plant and Fungal Diversity, highlight the interdependence of plants and fungi. The alternation of generations and the life cycle in mosses and ferns are presented in an attractive two-page Visualizing the Concept module (17.3), while details of the pine life cycle have been replaced with a new Module 17.5 that emphasizes pollen and seeds as key adaptations for terrestrial life. The animal diversity chapters (18, The Evolution of Invertebrate Diversity; and 19, The Evolution of Vertebrate Diversity) also have new opening essays. A Visualizing the Concept module (18.3) beautifully illustrates features of the animal body plan. A new Module 18.16 calls attention to the value of invertebrate diversity. Chapter 19 includes a Visualizing the Concept module (19.9) on primate diversity and also updates the story of hominin evolution, including the recently described *Australopithecus sediba*.

Unit V, Animals: Form and Function This unit combines a comparative approach with an exploration of human anatomy and physiology. Many chapters begin with an overview of a general problem that animals face and a comparative discussion of how different animals address this problem, all framed within an evolutionary context. For example, the introduction to Chapter 20, Unifying Concepts of Animal Structure and Function, begins with the question "Does evolution lead to the perfect animal form?" Module 20.1 is a new Evolution Connection that discusses the long, looped laryngeal nerve in vertebrates (using the giraffe as an example) to illustrate that a structure in an ancestral organism can become adapted to function in a descendant organism without being "perfected," thereby combating common student misconceptions about evolution. The main portion of every chapter is devoted to detailed presentations of human body systems, frequently illuminated by discussion of the health consequences of disorders in those systems. For example, Chapter 28, Nervous Systems, includes new material describing a genetic risk for developing Alzheimer's disease, the long-term consequences of traumatic brain injury, and how some antidepressants may not be as effective at combating depression as once thought. In many areas, content has been updated to reflect newer issues in biology. The chapter introduction and new Scientific Thinking module in Chapter 26, Hormones and the Endocrine System, discuss the consequences of endocrine disruptors in the environment. The Scientific Thinking module in Chapter 23 describes large clinical trials investigating the hypothesis that heart attacks are caused by the body's inflammatory response. Chapter 27, Reproduction and Embryonic Development, has a new chapter introduction on viral STDs, improved figures presenting embryonic development, as well as a Visualizing the Concept module on human pregnancy. Improvements to this unit also include a significant revision to the presentation of nutrition in Modules 21.14 to 21.21 and a reorganization of text and art in Modules 25.6 and 25.7 to guide students through the anatomy and physiology of the kidneys.

Unit VI, Plants: Form and Function To help students gain an appreciation of the importance of plants, this unit presents the anatomy and physiology of angiosperms with frequent connections to the importance of plants to society. New Connections in this edition include an increased discussion of the importance of agriculture to human civilization (including presentation of genomic data investigating this question) in Chapter 31, issues surrounding organic farming (including presentation of data on the nutritional value of organic versus conventionally grown produce) in Chapter 32, an expanded discussion of phytoestrogens, as well as a new discussion on the production of seedless vegetables in Chapter 33. Throughout the unit, the text has been revised with the goal of making the material more engaging and accessible to students. For example, the difficult topic of transpiration is now presented in an entirely new, visual style within a Visualizing the Concept module (Module 32.3), and streamlined and simplified discussions were written for such topics as the auxin hormones and phytochromes. All of these changes are meant to make the point that human society is inexorably connected to the health of plants.

Unit VII, Ecology In this unit, students learn the fundamental principles of ecology and how these principles apply to environmental problems. Along with a new introduction in each chapter, the Eighth Edition features many new photos and two Visualizing the Concept modules (35.7 and 37.9)—one focuses on whether animal movement is a response to stimuli or requires spatial learning and the other explores the interconnection of food chains and food webs. Scientific Thinking modules sample the variety of approaches to studying ecology, including the classic field study that led to the concept of keystone species (37.11); the "natural experiment" of returning gray wolves to the Yellowstone ecosystem (38.11); and the combination of historical records, long-term experimentation, and modern technology to investigate the snowshoe hare–lynx population cycle (36.6). The pioneering work of Rachel Carson (34.2) and Jane Goodall (35.22) is also described in Scientific Thinking modules. Modules that present data on human population (36.3, 36.9–36.11), declining biodiversity (38.1), and global climate change (38.3, 38.4) have all been updated.

Acknowledgments

This Eighth Edition of *Campbell Biology: Concepts & Connections* is a result of the combined efforts of many talented and hardworking people, and the authors wish to extend heartfelt thanks to all those who contributed to this and previous editions. Our work on this edition was shaped by input from the biologists acknowledged in the reviewer list on pages xx–xxii, who shared with us their experiences teaching introductory biology and provided specific suggestions for improving the book. Feedback from the authors of this edition's supplements and the unsolicited comments and suggestions we received from many biologists and biology students were also extremely helpful. In addition, this book has benefited in countless ways from the stimulating contacts we have had with the coauthors of *Campbell Biology*, Tenth Edition.

We wish to offer special thanks to the students and faculty at our teaching institutions. Marty Taylor thanks her students at Cornell University for their valuable feedback on the book. Eric Simon thanks his colleagues and friends at New England College, especially within the collegium of Natural Sciences and Mathematics, for their continued support and assistance. Jean Dickey thanks her colleagues at Clemson University for their expertise and support. And Kelly Hogan thanks her students for their enthusiasm and thanks her colleagues at the University of North Carolina, Chapel Hill, for their continued support.

We thank Paul Corey, president, Science, Business, and Technology, Pearson Higher Education. In addition, the superb publishing team for this edition was headed up by acquisitions editor Alison Rodal, with the invaluable support of editor-in-chief Beth Wilbur. We cannot thank them enough for their unstinting efforts on behalf of the book and for their commitment to excellence in biology education. We are fortunate to have had once again the contributions of executive director of development Deborah Gale and executive editorial manager Ginnie Simione Jutson. We are similarly grateful to the members of the editorial development team—Debbie Hardin, who also served as the day-to-day editorial project manager, and Susan Teahan—for their steadfast commitment to quality. We thank them for their thoroughness, hard work, and good humor; the book is far better than it would have been without their efforts. Thanks also to senior supplements project editor Susan Berge for her oversight of the supplements program and to editorial assistants Rachel Brickner, Katherine Harrison-Adcock, and Libby Reiser for the efficient and enthusiastic support they provided.

This book and all the other components of the teaching package are both attractive and pedagogically effective in large part because of the hard work and creativity of the production professionals on our team. We wish to thank managing editor Mike Early and production project manager Lori Newman. We also acknowledge copyeditor Joanna Dinsmore, proofreader Pete Shanks, and indexer Lynn Armstrong. We again thank senior photo editor Donna Kalal and photo researcher Kristin Piljay for their contributions, as well as project manager for text permissions Alison Bruckner. S4Carlisle Publishing Services was responsible for composition, headed by senior project editor Emily Bush, with help from paging specialist Donna Healy; and Precision Graphics, headed by project manager Amanda Bickel, was responsible for rendering new and revised illustrations. We also thank manufacturing buyer Jeffrey Sargent.

We thank Gary Hespenheide for creating a beautiful and functional interior design and a stunning cover, and we are again indebted to design manager Marilyn Perry for her oversight and design leadership. The new Visualizing the Concept modules benefited from her vision, as well as from the early input of art editor Elisheva Marcus and the continuing contributions of artist Andrew Recher of Precision Graphics. Art editor Kelly Murphy envisioned the beautiful new cell art throughout the book.

The value of *Campbell Biology: Concepts & Connections* as a learning tool is greatly enhanced by the hard work and creativity of the authors of the supplements that accompany this book: Ed Zalisko (*Instructor's Guide* and *PowerPoint® Lecture Presentations*); Jean DeSaix, Tanya Smutka, Kristen Miller, and Justin Shaffer (*Test Bank*); Dana Kurpius (*Active Reading Guide*); Robert Iwan and Amaya Garcia (*Reading Quizzes* and media correlations); and Shannon Datwyler (*Clicker Questions* and *Quiz Shows*). In addition to senior supplements project editor Susan Berge, the editorial and production staff for the supplements program included supplements production project manager Jane Brundage, *PowerPoint® Lecture Presentations* editor Joanna Dinsmore, and project manager Sylvia Rebert of Progressive Publishing Alternatives. And the superlative MasteringBiology® program for this book would not exist without Lauren Fogel, Stacy Treco, Tania Mlawer, Katie Foley, Sarah Jensen, Juliana Tringali, Daniel Ross, Dario Wong, Taylor Merck, Caroline Power, and David Kokorowski and his team. And a special thanks to Sarah Young-Dualan for her thoughtful work on the Visualizing the Concepts interactive videos.

For their important roles in marketing the book, we are very grateful to senior marketing manager Amee Mosley, executive marketing manager Lauren Harp, and vice president of marketing Christy Lesko. We also appreciate the work of the executive marketing manager for MasteringBiology®, Scott Dustan. The members of the Pearson Science sales team have continued to help us connect with biology instructors and their teaching needs, and we thank them.

Finally, we are deeply grateful to our families and friends for their support, encouragement, and patience throughout this project. Our special thanks to Paul, Dan, Maria, Armelle, and Sean (J.B.R.); Josie, Jason, Marnie, Alice, Jack, David, Paul, Ava, and Daniel (M.R.T.); Amanda, Reed, Forest, and dear friends Jamey, Nick, Jim, and Bethany (E.J.S.); Jessie and Katherine (J.L.D.); and Tracey, Vivian, Carolyn, Brian, Jake, and Lexi (K.H.)

Jane Reece, Martha Taylor, Eric Simon, Jean Dickey, and Kelly Hogan

Reviewers

Visualizing the Concept Review Panel, Eighth Edition

Erica Kipp, *Pace University*
David Loring, *Johnson County Community College*
Sheryl Love, *Temple University*
Sukanya Subramanian, *Collin County Community College*
Jennifer J. Yeh, *San Francisco, California*

Reviewers of the Eighth Edition

Steven Armstrong, *Tarrant County College*
Michael Battaglia, *Greenville Technical College*
Lisa Bonneau, *Metropolitan Community College*
Stephen T. Brown, *Los Angeles Mission College*
Nancy Buschhaus, *University of Tennessee at Martin*
Glenn Cohen, *Troy University*
Nora Espinoza, *Clemson University*
Karen E. Francl, *Radford University*
Jennifer Greenwood, *University of Tennessee at Martin*
Joel Hagen, *Radford University*
Chris Haynes, *Shelton State Community College*
Duane A. Hinton, *Washburn University*
Amy Hollingsworth, *The University of Akron*
Erica Kipp, *Pace University*
Cindy Klevickis, *James Madison University*
Dubear Kroening, *University of Wisconsin, Fox Valley*
Dana Kurpius, *Elgin Community College*
Dale Lambert, *Tarrant County College*
David Loring, *Johnson County Community College*
Mark Meade, *Jacksonville State University*
John Mersfelder, *Sinclair Community College*
Andrew Miller, *Thomas University*
Zia Nisani, *Antelope Valley College*
Camellia M. Okpodu, *Norfolk State University*
James Rayburn, *Jacksonville State University*
Ashley Rhodes, *Kansas State University*
Lori B. Robinson, *Georgia College & State University*
Ursula Roese, *University of New England*
Doreen J. Schroeder, *University of St. Thomas*
Justin Shaffer, *North Carolina A&T State University*
Marilyn Shopper, *Johnson County Community College*
Ayesha Siddiqui, *Schoolcraft College*
Ashley Spring, *Brevard Community College*
Thaxton Springfield, *St. Petersburg College*
Linda Brooke Stabler, *University of Central Oklahoma*
Patrick Stokley, *East Central Community College*
Lori Tolley-Jordan, *Jacksonville State University*
Jimmy Triplett, *Jacksonville State University*
Lisa Weasel, *Portland State University*
Martin Zahn, *Thomas Nelson Community College*

Reviewers of Previous Editions

Michael Abbott, *Westminster College*
Tanveer Abidi, *Kean University*
Daryl Adams, *Mankato State University*
Dawn Adrian Adams, *Baylor University*
Olushola Adeyeye, *Duquesne University*
Shylaja Akkaraju, *Bronx Community College*
Felix Akojie, *Paducah Community College*
Dan Alex, *Chabot College*
John Aliff, *Georgia Perimeter College*
Sylvester Allred, *Northern Arizona University*
Jane Aloi-Horlings, *Saddleback College*
Loren Ammerman, *University of Texas at Arlington*
Dennis Anderson, *Oklahoma City Community College*
Marjay Anderson, *Howard University*
Bert Atsma, *Union County College*
Yael Avissar, *Rhode Island College*
Gail Baker, *LaGuardia Community College*
Caroline Ballard, *Rock Valley College*
Andrei Barkovskii, *Georgia College and State University*
Mark Barnby, *Ohlone College*
Chris Barnhart, *University of San Diego*
Stephen Barnhart, *Santa Rosa Junior College*
William Barstow, *University of Georgia*
Kirk A. Bartholomew, *Central Connecticut State University*
Michael Battaglia, *Greenville Technical College*
Gail Baughman, *Mira Costa College*
Jane Beiswenger, *University of Wyoming*
Tania Beliz, *College of San Mateo*
Lisa Bellows, *North Central Texas College*
Ernest Benfield, *Virginia Polytechnic Institute*
Rudi Berkelhamer, *University of California, Irvine*
Harry Bernheim, *Tufts University*
Richard Bliss, *Yuba College*
Lawrence Blumer, *Morehouse College*
Dennis Bogyo, *Valdosta State University*
Lisa K. Bonneau, *Metropolitan Community College, Blue River*
Mehdi Borhan, *Johnson County Community College*
Kathleen Bossy, *Bryant College*
William Bowen, *University of Arkansas at Little Rock*
Robert Boyd, *Auburn University*
Bradford Boyer, *State University of New York, Suffolk County Community College*
Paul Boyer, *University of Wisconsin*
William Bradshaw, *Brigham Young University*
Agnello Braganza, *Chabot College*
James Bray, *Blackburn College*
Peggy Brickman, *University of Georgia*
Chris Brinegar, *San Jose State University*
Chad Brommer, *Emory University*
Charles Brown, *Santa Rosa Junior College*
Carole Browne, *Wake Forest University*
Becky Brown-Watson, *Santa Rosa Junior College*
Delia Brownson, *University of Texas at Austin and Austin Community College*
Michael Bucher, *College of San Mateo*
Virginia Buckner, *Johnson County Community College*
Joseph C. Bundy, Jr., *University of North Carolina at Greensboro*
Ray Burton, *Germanna Community College*
Warren Buss, *University of Northern Colorado*
Linda Butler, *University of Texas at Austin*
Jerry Button, *Portland Community College*
Carolee Caffrey, *University of California, Los Angeles*
George Cain, *University of Iowa*
Beth Campbell, *Itawamba Community College*
John Campbell, *Northern Oklahoma College*
John Capeheart, *University of Houston, Downtown*
James Cappuccino, *Rockland Community College*
M. Carabelli, *Broward Community College*
Jocelyn Cash, *Central Piedmont Community College*
Cathryn Cates, *Tyler Junior College*
Russell Centanni, *Boise State University*
David Chambers, *Northeastern University*
Ruth Chesnut, *Eastern Illinois University*
Vic Chow, *San Francisco City College*
Van Christman, *Ricks College*
Craig Clifford, *Northeastern State University, Tahlequah*
Richard Cobb, *South Maine Community College*
Mary Colavito, *Santa Monica College*
Jennifer Cooper, *Itawamba Community College*
Bob Cowling, *Ouachita Technical College*
Don Cox, *Miami University*
Robert Creek, *Western Kentucky University*
Hillary Cressey, *George Mason University*
Norma Criley, *Illinois Wesleyan University*
Jessica Crowe, *South Georgia College*
Mitch Cruzan, *Portland State University*
Judy Daniels, *Monroe Community College*
Michael Davis, *Central Connecticut State University*
Pat Davis, *East Central Community College*
Lewis Deaton, *University of Louisiana*
Lawrence DeFilippi, *Lurleen B. Wallace College*
James Dekloe, *Solano Community College*
Veronique Delesalle, *Gettysburg College*
Loren Denney, *Southwest Missouri State University*
Jean DeSaix, *University of North Carolina at Chapel Hill*
Mary Dettman, *Seminole Community College of Florida*
Kathleen Diamond, *College of San Mateo*
Alfred Diboll, *Macon College*
Jean Dickey, *Clemson University*
Stephen Dina, *St. Louis University*
Robert P. Donaldson, *George Washington University*
Gary Donnermeyer, *Iowa Central Community College*
Charles Duggins, *University of South Carolina*
Susan Dunford, *University of Cincinnati*
Lee Edwards, *Greenville Technical College*
Betty Eidemiller, *Lamar University*
Jamin Eisenbach, *Eastern Michigan University*
Norman Ellstrand, *University of California, Riverside*
Thomas Emmel, *University of Florida*
Cindy Erwin, *City College of San Francisco*
Gerald Esch, *Wake Forest University*
David Essar, *Winona State University*

Cory Etchberger, *Longview Community College*
Nancy Eyster-Smith, *Bentley College*
William Ezell, *University of North Carolina at Pembroke*
Laurie Faber, *Grand Rapids Community College*
Terence Farrell, *Stetson University*
Shannon Kuchel Fehlberg, *Colorado Christian University*
Jerry Feldman, *University of California, Santa Cruz*
Eugene Fenster, *Longview Community College*
Dino Fiabane, *Community College of Philadelphia*
Kathleen Fisher, *San Diego State University*
Edward Fliss, *St. Louis Community College, Florissant Valley*
Linda Flora, *Montgomery County Community College*
Dennis Forsythe, *The Citadel Military College of South Carolina*
Robert Frankis, *College of Charleston*
James French, *Rutgers University*
Bernard Frye, *University of Texas at Arlington*
Anne Galbraith, *University of Wisconsin*
Robert Galbraith, *Crafton Hills College*
Rosa Gambier, *State University of New York, Suffolk County Community College*
George Garcia, *University of Texas at Austin*
Linda Gardner, *San Diego Mesa College*
Sandi Gardner, *Triton College*
Gail Gasparich, *Towson University*
Janet Gaston, *Troy University*
Shelley Gaudia, *Lane Community College*
Douglas Gayou, *University of Missouri at Columbia*
Robert Gendron, *Indiana University of Pennsylvania*
Bagie George, *Georgia Gwinnett College*
Rebecca German, *University of Cincinnati*
Grant Gerrish, *University of Hawaii*
Julie Gibbs, *College of DuPage*
Frank Gilliam, *Marshall University*
Patricia Glas, *The Citadel Military College of South Carolina*
David Glenn-Lewin, *Wichita State University*
Robert Grammer, *Belmont University*
Laura Grayson-Roselli, *Burlington County College*
Peggy Green, *Broward Community College*
Miriam L. Greenberg, *Wayne State University*
Sylvia Greer, *City University of New York*
Eileen Gregory, *Rollins College*
Dana Griffin, *University of Florida*
Richard Groover, *J. Sargeant Reynolds Community College*
Peggy Guthrie, *University of Central Oklahoma*
Maggie Haag, *University of Alberta*
Richard Haas, *California State University, Fresno*
Martin Hahn, *William Paterson College*
Leah Haimo, *University of California, Riverside*
James Hampton, *Salt Lake Community College*
Blanche Haning, *North Carolina State University*
Richard Hanke, *Rose State College*
Laszlo Hanzely, *Northern Illinois University*
David Harbster, *Paradise Valley Community College*
Sig Harden, *Troy University Montgomery*
Reba Harrell, *Hinds Community College*
Jim Harris, *Utah Valley Community College*
Mary Harris, *Louisiana State University*
Chris Haynes, *Shelton State Community College*
Janet Haynes, *Long Island University*
Jean Helgeson, *Collin County Community College*
Ira Herskowitz, *University of California, San Francisco*
Paul Hertz, *Barnard College*
Margaret Hicks, *David Lipscomb University*
Jean Higgins-Fonda, *Prince George's Community College*
Phyllis Hirsch, *East Los Angeles College*
William Hixon, *St. Ambrose University*
Carl Hoagstrom, *Ohio Northern University*
Kim Hodgson, *Longwood College*
Jon Hoekstra, *Gainesville State College*
Kelly Hogan, *University of North Carolina at Chapel Hill*
John Holt, *Michigan State University*
Laura Hoopes, *Occidental College*
Lauren Howard, *Norwich University*
Robert Howe, *Suffolk University*
Michael Hudecki, *State University of New York, Buffalo*
George Hudock, *Indiana University*
Kris Hueftle, *Pensacola Junior College*
Barbara Hunnicutt, *Seminole Community College*
Brenda Hunzinger, *Lake Land College*
Catherine Hurlbut, *Florida Community College*
Charles Ide, *Tulane University*
Mark Ikeda, *San Bernardino Valley College*
Georgia Ineichen, *Hinds Community College*
Robert Iwan, *Inver Hills Community College*
Mark E. Jackson, *Central Connecticut State University*
Charles Jacobs, *Henry Ford Community College*
Fred James, *Presbyterian College*
Ursula Jander, *Washburn University*
Alan Jaworski, *University of Georgia*
R. Jensen, *Saint Mary's College*
Robert Johnson, *Pierce College, Lakewood Campus*
Roishene Johnson, *Bossier Parish Community College*
Russell Johnson, *Ricks College*
John C. Jones, *Calhoun Community College*
Florence Juillerat, *Indiana University at Indianapolis*
Tracy Kahn, *University of California, Riverside*
Hinrich Kaiser, *Victor Valley College*
Klaus Kalthoff, *University of Texas at Austin*
Tom Kantz, *California State University, Sacramento*
Jennifer Katcher, *Pima Community College*
Judy Kaufman, *Monroe Community College*
Marlene Kayne, *The College of New Jersey*
Mahlon Kelly, *University of Virginia*
Kenneth Kerrick, *University of Pittsburgh at Johnstown*
Joyce Kille-Marino, *College of Charleston*
Joanne Kilpatrick, *Auburn University, Montgomery*
Stephen Kilpatrick, *University of Pittsburgh at Johnstown*
Lee Kirkpatrick, *Glendale Community College*
Peter Kish, *Southwestern Oklahoma State University*
Cindy Klevickis, *James Madison University*
Robert Koch, *California State University, Fullerton*
Eliot Krause, *Seton Hall University*
Dubear Kroening, *University of Wisconsin, Fox Valley*
Kevin Krown, *San Diego State University*
Margaret Maile Lam, *Kapiolani Community College*
MaryLynne LaMantia, *Golden West College*
Mary Rose Lamb, *University of Puget Sound*
Dale Lambert, *Tarrant County College, Northeast*
Thomas Lammers, *University of Wisconsin, Oshkosh*
Carmine Lanciani, *University of Florida*
Vic Landrum, *Washburn University*
Deborah Langsam, *University of North Carolina at Charlotte*
Geneen Lannom, *University of Central Oklahoma*
Brenda Latham, *Merced College*
Liz Lawrence, *Miles Community College*
Steven Lebsack, *Linn-Benton Community College*
Karen Lee, *University of Pittsburgh at Johnstown*
Tom Lehman, *Morgan Community College*
William Lemon, *Southwestern Oregon Community College*
Laurie M. Len, *El Camino College*
Peggy Lepley, *Cincinnati State University*
Richard Liebaert, *Linn-Benton Community College*
Kevin Lien, *Portland Community College*
Harvey Liftin, *Broward Community College*
Ivo Lindauer, *University of Northern Colorado*
William Lindsay, *Monterey Peninsula College*
Kirsten Lindstrom, *Santa Rosa Junior College*
Melanie Loo, *California State University, Sacramento*
David Loring, *Johnson County Community College*
Eric Lovely, *Arkansas Tech University*
Paul Lurquin, *Washington State University*
James Mack, *Monmouth University*
David Magrane, *Morehead State University*
Joan Maloof, *Salisbury State University*
Joseph Marshall, *West Virginia University*
Presley Martin, *Drexel University*
William McComas, *University of Iowa*
Steven McCullagh, *Kennesaw State College*
Mitchell McGinnis, *North Seattle Community College*
James McGivern, *Gannon University*
Colleen McNamara, *Albuquerque TVI Community College*
Caroline McNutt, *Schoolcraft College*
Scott Meissner, *Cornell University*
Joseph Mendelson, *Utah State University*
Timothy Metz, *Campbell University*
Iain Miller, *University of Cincinnati*
Robert Miller, *University of Dubuque*
V. Christine Minor, *Clemson University*
Brad Mogen, *University of Wisconsin, River Falls*
James Moné, *Millersville University*
Jamie Moon, *University of North Florida*
Juan Morata, *Miami Dade College*
Richard Mortensen, *Albion College*
Henry Mulcahy, *Suffolk University*
Christopher Murphy, *James Madison University*
Kathryn Nette, *Cuyamaca College*
James Newcomb, *New England College*
Zia Nisani, *Antelope Valley College*
James Nivison, *Mid Michigan Community College*
Peter Nordloh, *Southeastern Community College*
Stephen Novak, *Boise State University*
Bette Nybakken, *Hartnell College*
Michael O'Donnell, *Trinity College*
Steven Oliver, *Worcester State College*
Karen Olmstead, *University of South Dakota*

Steven O'Neal, *Southwestern Oklahoma State University*
Lowell Orr, *Kent State University*
William Outlaw, *Florida State University*
Phillip Pack, *Woodbury University*
Kevin Padian, *University of California, Berkeley*
Kay Pauling, *Foothill College*
Mark Paulissen, *Northeastern State University, Tahlequah*
Debra Pearce, *Northern Kentucky University*
David Pearson, *Bucknell University*
Patricia Pearson, *Western Kentucky University*
Kathleen Pelkki, *Saginaw Valley State University*
Andrew Penniman, *Georgia Perimeter College*
John Peters, *College of Charleston*
Gary Peterson, *South Dakota State University*
Margaret Peterson, *Concordia Lutheran College*
Russell L. Peterson, *Indiana University of Pennsylvania*
Paula Piehl, *Potomac State College*
Ben Pierce, *Baylor University*
Jack Plaggemeyer, *Little Big Horn College*
Barbara Pleasants, *Iowa State University*
Kathryn Podwall, *Nassau Community College*
Judith Pottmeyer, *Columbia Basin College*
Donald Potts, *University of California, Santa Cruz*
Nirmala Prabhu, *Edison Community College*
Elena Pravosudova, *University of Nevada, Reno*
James Pru, *Belleville Area College*
Rongsun Pu, *Kean University*
Charles Pumpuni, *Northern Virginia Community College*
Kimberly Puvalowski, *Old Bridge High School*
Rebecca Pyles, *East Tennessee State University*
Shanmugavel Rajendran, *Baltimore City Community College*
Bob Ratterman, *Jamestown Community College*
Jill Raymond, *Rock Valley College*
Michael Read, *Germanna Community College*
Brian Reeder, *Morehead State University*
Bruce Reid, *Kean College*
David Reid, *Blackburn College*
Stephen Reinbold, *Longview Community College*
Erin Rempala, *San Diego Mesa College*
Michael Renfroe, *James Madison University*
Tim Revell, *Mt. San Antonio College*
Douglas Reynolds, *Central Washington University*
Fred Rhoades, *Western Washington University*
John Rinehart, *Eastern Oregon University*
Laura Ritt, *Burlington County College*
Lynn Rivers, *Henry Ford Community College*
Bruce Robart, *University of Pittsburgh at Johnstown*
Jennifer Roberts, *Lewis University*
Laurel Roberts, *University of Pittsburgh*
Luis A. Rodriguez, *San Antonio Colleges*
Duane Rohlfing, *University of South Carolina*
Jeanette Rollinger, *College of the Sequoias*
Steven Roof, *Fairmont State College*
Jim Rosowski, *University of Nebraska*
Stephen Rothstein, *University of California, Santa Barbara*
Donald Roush, *University of North Alabama*
Lynette Rushton, *South Puget Sound Community College*
Connie Rye, *East Mississippi Community College*
Linda Sabatino, *State University of New York, Suffolk County Community College*
Douglas Schamel, *University of Alaska, Fairbanks*
Douglas Schelhaas, *University of Mary*
Beverly Schieltz, *Wright State University*
Fred Schindler, *Indian Hills Community College*
Robert Schoch, *Boston University*
Brian Scholtens, *College of Charleston*
John Richard Schrock, *Emporia State University*
Julie Schroer, *Bismarck State College*
Fayla Schwartz, *Everett Community College*
Judy Shea, *Kutztown University of Pennsylvania*
Daniela Shebitz, *Kean University*
Thomas Shellberg, *Henry Ford Community College*
Cara Shillington, *Eastern Michigan University*
Lisa Shimeld, *Crafton Hills College*
Brian Shmaefsky, *Kingwood College*
Mark Shotwell, *Slippery Rock University*
Jane Shoup, *Purdue University*
Michele Shuster, *New Mexico State University*
Linda Simpson, *University of North Carolina at Charlotte*
Gary Smith, *Tarrant County Junior College*
Marc Smith, *Sinclair Community College*
Michael Smith, *Western Kentucky University*
Phil Snider, *University of Houston*
Sam C. Sochet, *Thomas Edison Career and Technical Education High School*
Gary Sojka, *Bucknell University*
Ralph Sorensen, *Gettysburg College*
Ruth Sporer, *Rutgers University*
Linda Brooke Stabler, *University of Central Oklahoma*
David Stanton, *Saginaw Valley State University*
Amanda Starnes, *Emory University*
John Stolz, *Duquesne University*
Ross Strayer, *Washtenaw Community College*
Donald Streuble, *Idaho State University*
Megan Stringer, *Jones County Junior College*
Mark Sugalski, *New England College*
Gerald Summers, *University of Missouri*
Marshall Sundberg, *Louisiana State University*
Christopher Tabit, *University of West Georgia*
David Tauck, *Santa Clara University*
Hilda Taylor, *Acadia University*
Franklin Te, *Miami Dade College*
Gene Thomas, *Solano Community College*
Kenneth Thomas, *Northern Essex Community College*
Kathy Thompson, *Louisiana State University*
Laura Thurlow, *Jackson Community College*
Anne Tokazewski, *Burlington County College*
John Tolli, *Southwestern College*
Bruce Tomlinson, *State University of New York, Fredonia*
Nancy Tress, *University of Pittsburgh at Titusville*
Donald Trisel, *Fairmont State College*
Kimberly Turk, *Mitchell Community College*
Virginia Turner, *Harper College*
Mike Tveten, *Pima College*
Michael Twaddle, *University of Toledo*
Rani Vajravelu, *University of Central Florida*
Leslie VanderMolen, *Humboldt State University*
Cinnamon VanPutte, *Southwestern Illinois College*
Sarah VanVickle-Chavez, *Washington University*
John Vaughan, *Georgetown College*
Martin Vaughan, *Indiana University*
Mark Venable, *Appalachian State University*
Ann Vernon, *St. Charles County Community College*
Rukmani Viswanath, *Laredo Community College*
Frederick W. Vogt, *Elgin Community College*
Mary Beth Voltura, *State University of New York, Cortland*
Jerry Waldvogel, *Clemson University*
Robert Wallace, *Ripon College*
Dennis Walsh, *MassBay Community College*
Patricia Walsh, *University of Delaware*
Lisa Weasel, *Portland State University*
James Wee, *Loyola University*
Harrington Wells, *University of Tulsa*
Jennifer Wiatrowski, *Pasco-Hernando Community College*
Larry Williams, *University of Houston*
Ray S. Williams, *Appalachian State University*
Lura Williamson, *University of New Orleans*
Sandra Winicur, *Indiana University, South Bend*
Robert R. Wise, *University of Wisconsin Oshkosh*
Mary E. Wisgirda, *San Jacinto College*
Mary Jo Witz, *Monroe Community College*
Neil Woffinden, *University of Pittsburgh at Johnstown*
Michael Womack, *Macon State University*
Patrick Woolley, *East Central College*
Maury Wrightson, *Germanna Community College*
Tumen Wuliji, *University of Nevada, Reno*
Mark Wygoda, *McNeese State University*
Tony Yates, *Seminole State College*
William Yurkiewicz, *Millersville University of Pennsylvania*
Gregory Zagursky, *Radford University*
Martin Zahn, *Thomas Nelson Community College*
Edward J. Zalisko, *Blackburn College*
David Zeigler, *University of North Carolina at Pembroke*
Uko Zylstra, *Calvin College*

Detailed Contents

4 A Tour of the Cell 50

5 The Working Cell 72

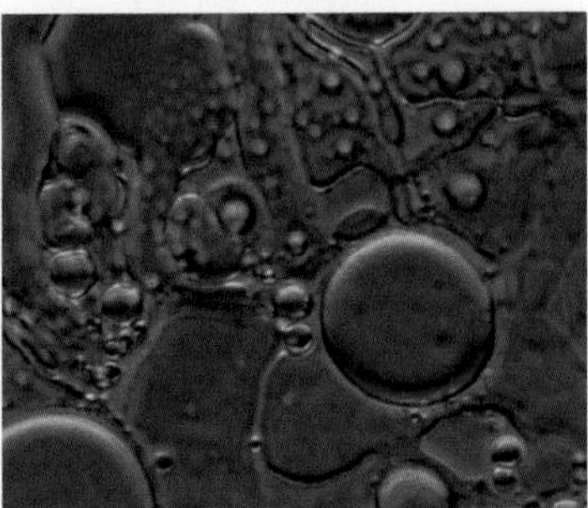

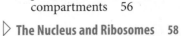

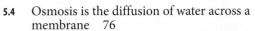

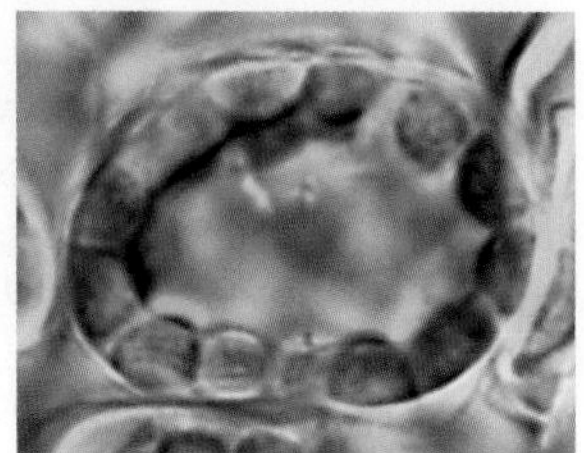

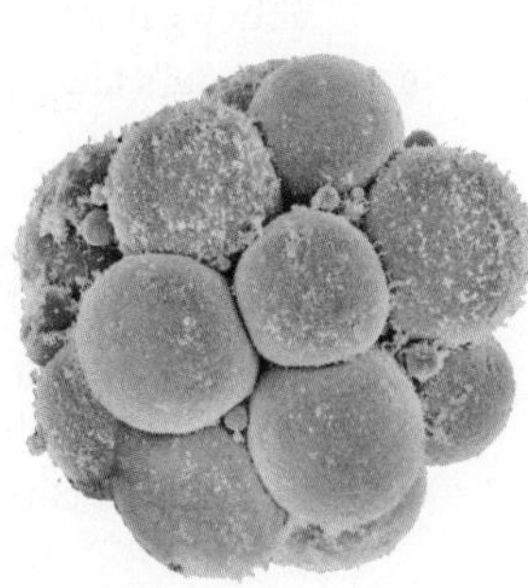

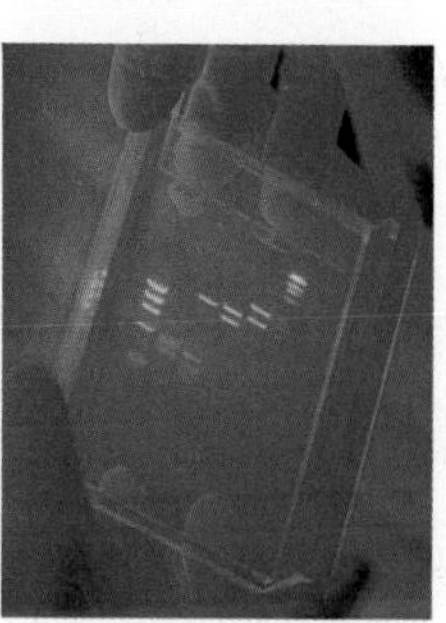

UNIT V

Animals: Form and Function 411

20 Unifying Concepts of Animal Structure and Function 412

21 Nutrition and Digestion 428

22 Gas Exchange 452

23 Circulation 466

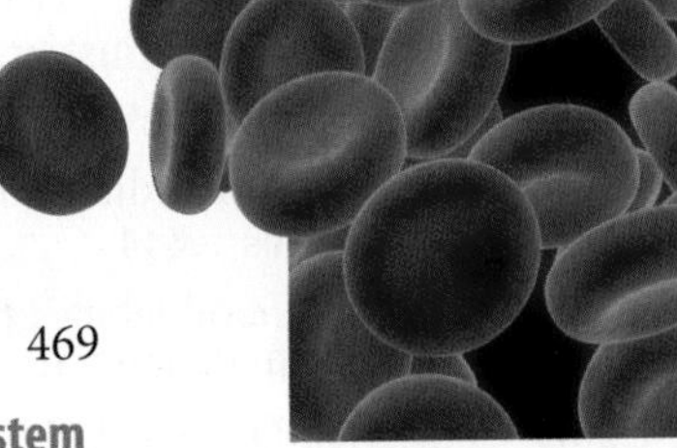

24 The Immune System 484

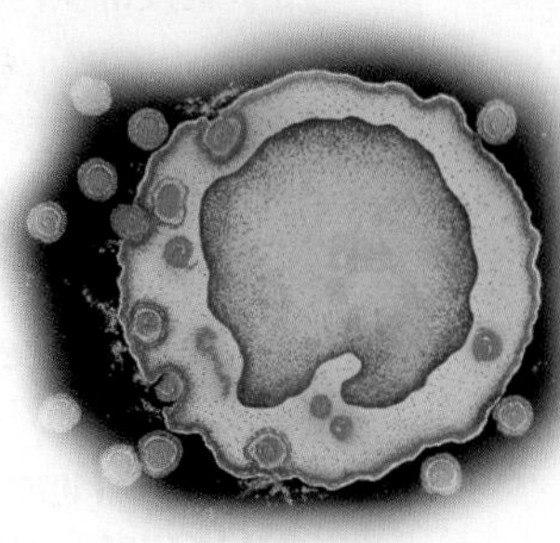

25 Control of Body Temperature and Water Balance 504

29 The Senses 586

30 How Animals Move 602

UNIT VI

Plants: Form and Function 619

31 Plant Structure, Growth, and Reproduction 620

32 Plant Nutrition and Transport 642

33 Control Systems in Plants 660

UNIT VII

Ecology 677

34 The Biosphere: *An Introduction to Earth's Diverse Environments* 678

35 Behavioral Adaptations to the Environment 698

36 Population Ecology 722

37 Communities and Ecosystems 738

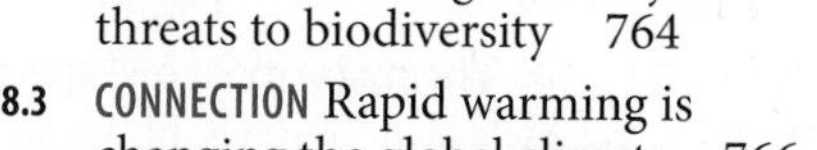

38 Conservation Biology 760

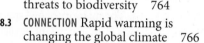

CHAPTER 1

Biology: *Exploring Life*

Snowy owls (*Bubo scandiacus*), such as the one on the cover of this textbook and pictured below, are strikingly beautiful owls with bright orange eyes and wingspans as wide as five feet. These swift and silent predators exhibit remarkable adaptations for life in their frozen, barren habitat. The layers of fine feathers on their face, body, legs, and even their feet provide insulation in subzero weather. They breed on the Arctic tundra, nesting on open ground. The female broods the eggs and young, while the male provides a steady supply of food. His keen vision and acute hearing help him locate small mammals such as voles and lemmings, which he then snatches in mid-flight with his sharp talons.

? ***Why do so many animals match their surroundings?***

The majority of owl species are nocturnal. But during the endless days of arctic summers, snowy owls hunt in daylight. Projecting upper eyelids help shield their eyes from bright sun. As with all owls, the overlapping fields of vision of their forward-facing eyes provide superior depth perception. These large eyes cannot move, so an owl must turn its whole head to follow a moving object. This is not a problem for an owl, as you can see in the photo below, because adaptations of its neck

vertebrae enable it to rotate its head a full 270 degrees. Imagine being able to look over your left shoulder by turning your head to the right!

You may think of owls in general in shades of brown, nesting in tree cavities and blending in with their surroundings. And with snowy owls, you may think of Harry Potter's white-feathered companion. In real life, these owls also blend in with their wintry habitat. Later in this chapter, you will read about an experiment that tests the hypothesis that camouflage coloration protects animals from predators.

The amazing adaptations of snowy owls are the result of evolution, the process that has transformed life from its earliest beginnings to the astounding array of organisms living today. In this chapter, we begin our exploration of biology—the scientific study of life.

BIG IDEAS

Themes in the Study of Biology
(1.1–1.4)

Common themes help to organize the study of life.

Evolution, the Core Theme of Biology
(1.5–1.7)

Evolution accounts for the unity and diversity of life and the evolutionary adaptations of organisms to their environment.

The Process of Science
(1.8–1.9)

In studying nature, scientists make observations, form hypotheses, and test predictions.

Biology and Everyday Life
(1.10–1.11)

Learning about biology helps us understand many issues involving science, technology, and society.

▷ Themes in the Study of Biology

1.1 All forms of life share common properties

Defining **biology** as the scientific study of life raises the obvious question: What is *life*? Even a small child realizes that a bug or a flower is alive, whereas a rock or a car is not. But the phenomenon we call life defies a simple, one-sentence definition. We recognize life mainly by what living things do. **Figure 1.1** highlights seven of the properties and processes that we associate with life.

1. *Order.* This sunflower illustrates the ordered structure that typifies life. Living cells make up this complex organization.
2. *Reproduction.* Organisms reproduce their own kind. Here a baby African elephant walks beneath its mother.
3. *Growth and development.* Inherited information in the form of DNA controls the pattern of growth and development of all organisms, including this hatching crocodile.
4. *Energy processing.* This caterpillar will use the chemical energy stored in the plant it is eating to power its own activities and chemical reactions.
5. *Regulation.* Many types of mechanisms regulate an organism's internal environment, keeping it within limits that sustain life. Pictured here is a lizard "sunbathing"—which helps raise its body temperature on cool mornings.
6. *Response to the environment.* All organisms respond to environmental stimuli. This Venus flytrap closed its trap rapidly in response to the stimulus of a damselfly landing on it.
7. *Evolutionary adaptation.* A snowy owl's sharp talons facilitate prey capture and its feathered feet keep it warm in its cold habitat. Such adaptations evolve over many generations as individuals with traits best suited to their environment have greater reproductive success and pass their traits to offspring.

Figure 1.1 reminds us that the living world is wondrously varied. How do biologists make sense of this diversity and complexity, and how can you? Indeed, biology is a subject of enormous scope that gets bigger all the time. One of the ways to help you organize this information is to connect what you learn to a set of themes that you will encounter throughout your study of life. The next few modules introduce several important themes: novel properties emerging at each level of biological organization, the correlation of structure and function, and the exchange of matter and energy as organisms interact with the environment. We then focus on the core theme of biology—evolution, the theme that makes sense of both the unity and diversity of life.

Let's begin our journey with a tour through the levels of the biological hierarchy.

? **How would you define life?**

● Life can be defined by a set of common properties such as those described in this module.

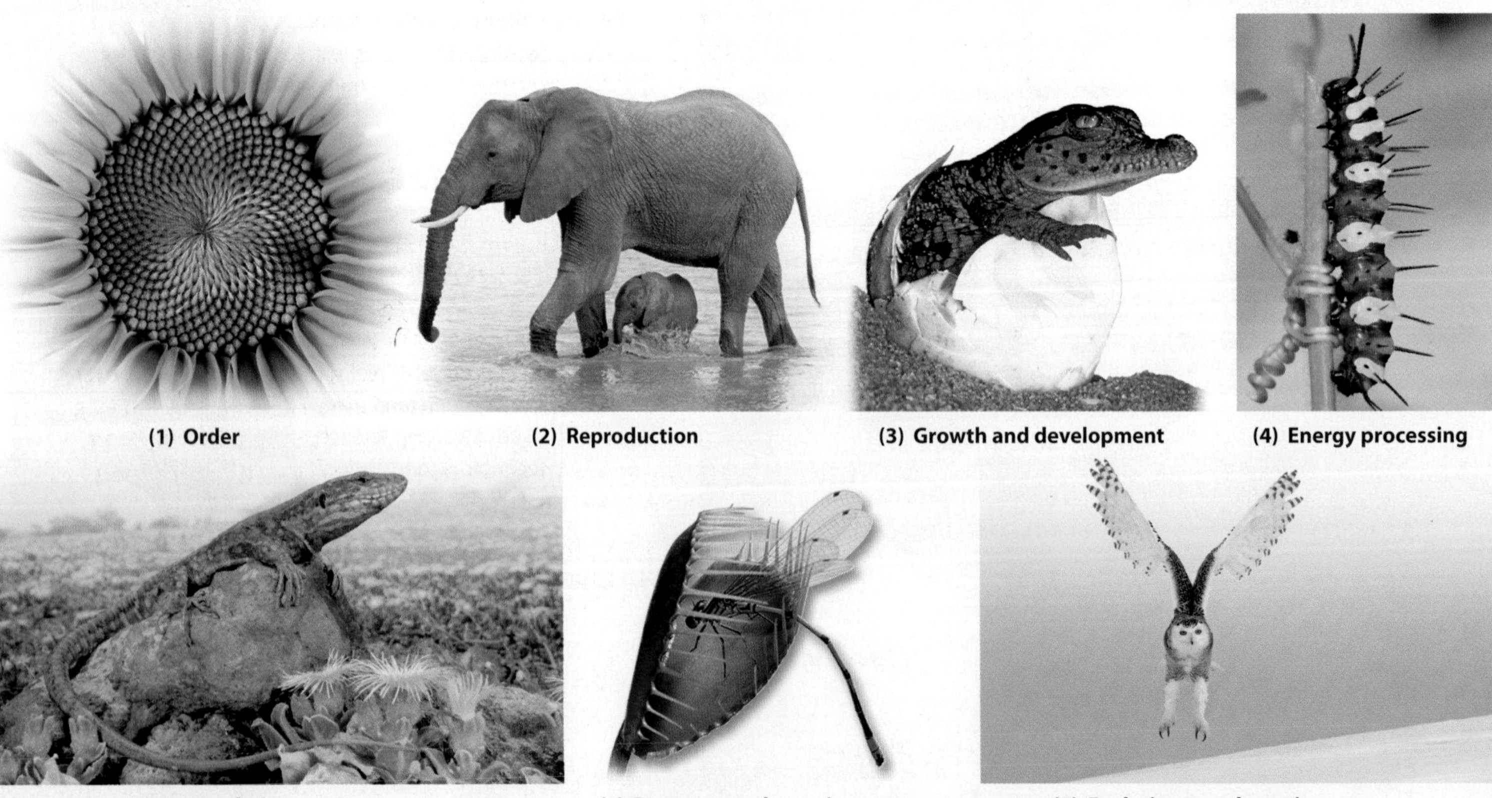
(1) Order
(2) Reproduction
(3) Growth and development
(4) Energy processing
(5) Regulation
(6) Response to the environment
(7) Evolutionary adaptation

▲ **Figure 1.1** Some important properties of life

1.2 In life's hierarchy of organization, new properties emerge at each level

As **Figure 1.2** illustrates, the study of life extends from the global scale of the biosphere to the microscopic level of molecules. At the upper left we take a distant view of the **biosphere**, all of the environments on Earth that support life. These include most regions of land, bodies of water, and the lower atmosphere. A closer look at one of these environments brings us to the level of an **ecosystem**, which consists of all the organisms living in a particular area, as well as the physical components with which the organisms interact, such as air, soil, water, and sunlight.

The entire array of organisms in an ecosystem is called a **community**. In this community, we find alligators and snakes, herons and egrets, myriad insects, trees and other plants, fungi, and enormous numbers of microorganisms. Each unique form of life is called a species.

A **population** includes all the individuals of a particular species living in an area. Next in the hierarchy is the **organism**, an individual living thing, such as an alligator.

Within a complex organism, life's hierarchy continues to unfold. An **organ system**, such as the circulatory system or nervous system, consists of several organs that cooperate in a specific function. For instance, the organs of the nervous system are the brain, the spinal cord, and the nerves. An alligator's nervous system controls all its actions.

An **organ** is made up of several different **tissues**, each in turn made up of a group of similar cells that perform a specific function. A **cell** is the fundamental unit of life. In the nerve cell shown here, you can see several organelles, such as the nucleus. An **organelle** is a membrane-enclosed structure that performs a specific function within a cell.

Finally, we reach the level of molecules in the hierarchy. A **molecule** is a cluster of small chemical units called atoms held together by chemical bonds. Our example in Figure 1.2 is a computer graphic of a section of DNA (deoxyribonucleic acid)—the molecule of inheritance.

Now let's work our way in the opposite direction in Figure 1.2, moving up life's hierarchy from molecules to the biosphere. At each higher level, there are novel properties that arise, properties that were not present at the preceding level. For example, life emerges at the level of the cell—a test tube full of organelles is not alive. Such **emergent properties** represent an important theme of biology. The familiar saying that "the whole is greater than the sum of its parts" captures this idea. The emergent properties of each level result from the specific arrangement and interactions of its parts.

? Which of these levels of biological organization includes all others in the list: cell, molecule, organ, tissue?

● Organ

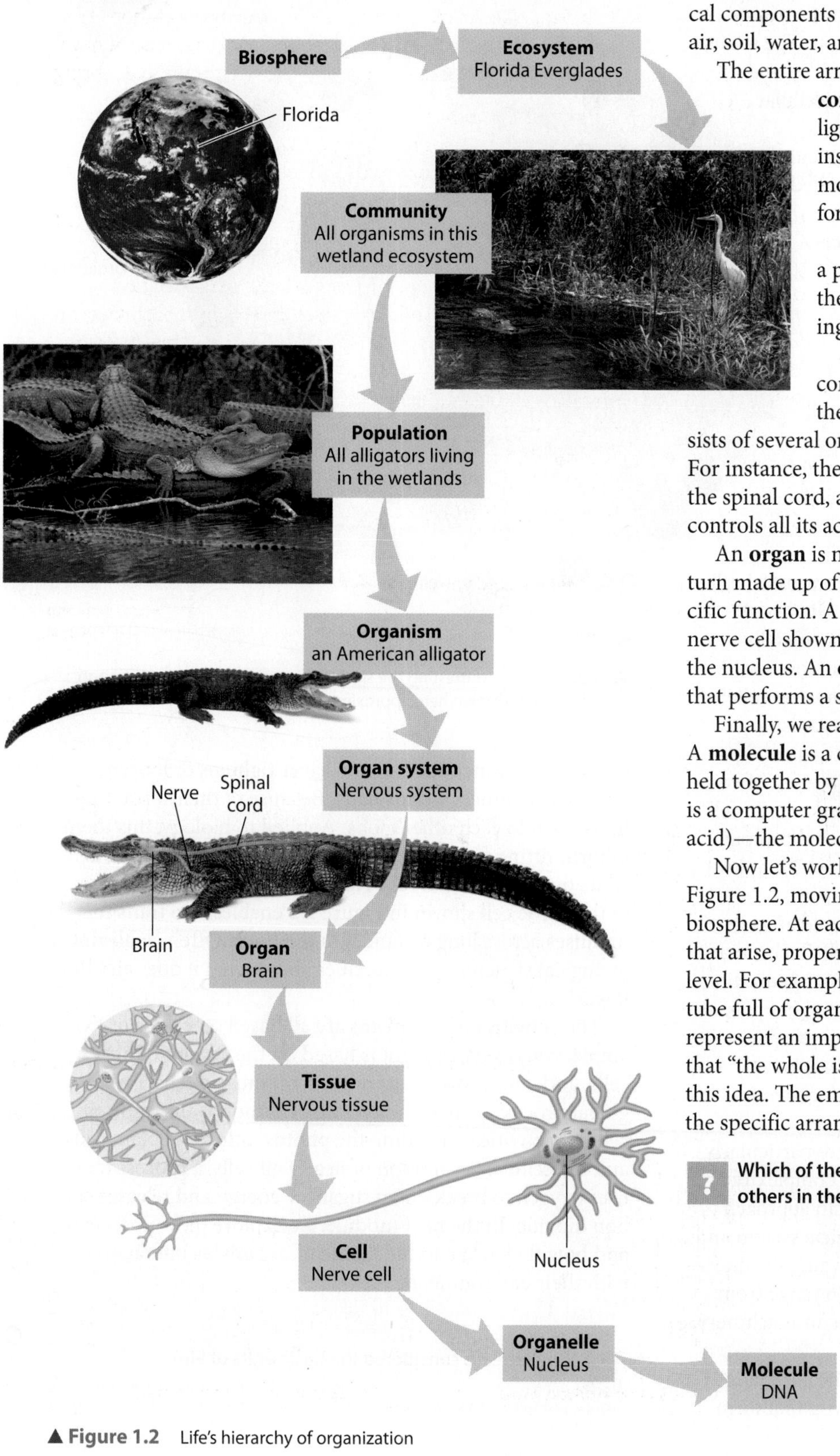

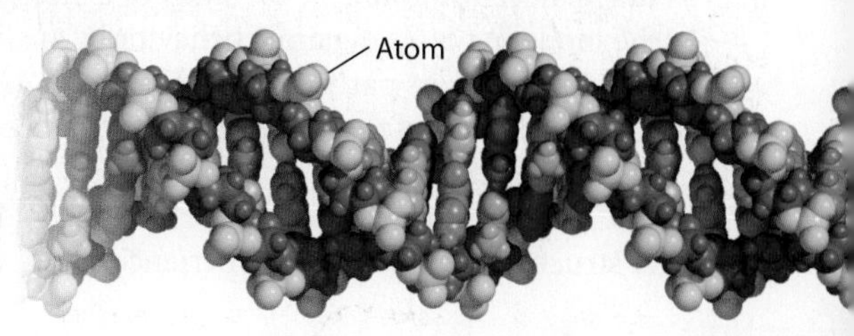

Biosphere
Florida
Ecosystem
Florida Everglades
Community
All organisms in this wetland ecosystem
Population
All alligators living in the wetlands
Organism
an American alligator
Organ system
Nervous system
Nerve
Spinal cord
Brain
Organ
Brain
Tissue
Nervous tissue
Cell
Nerve cell
Nucleus
Organelle
Nucleus
Molecule
DNA
Atom

▲ **Figure 1.2** Life's hierarchy of organization

1.3 Cells are the structural and functional units of life

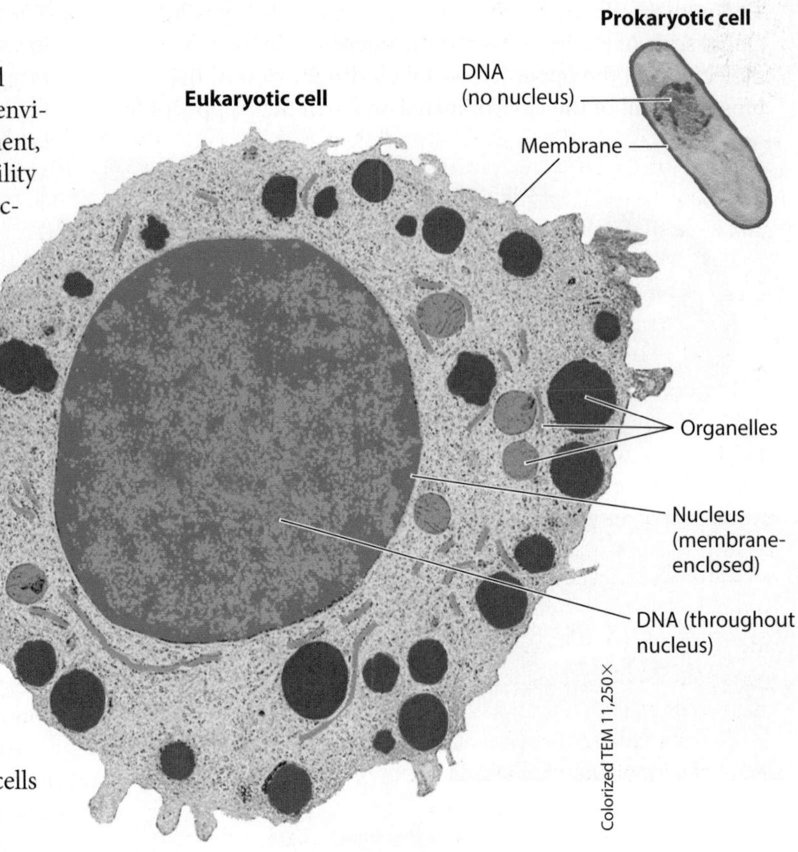

Prokaryotic cell
Eukaryotic cell
DNA
(no nucleus)
Membrane
Organelles
Nucleus
(membrane-
enclosed)
DNA (throughout
nucleus)
Colorized TEM 11,250×

▲ **Figure 1.3** Contrasting the size and complexity of prokaryotic and eukaryotic cells (shown here approximately 11,250 times their real size)

The cell has a special place in the hierarchy of biological organization. It is the level at which the properties of life emerge—the lowest level of structure that can perform all activities required for life. A cell can regulate its internal environment, take in and use energy, respond to its environment, and build and maintain its complex organization. The ability of cells to give rise to new cells is the basis for all reproduction and also for the growth and repair of multicellular organisms.

All organisms are composed of cells. They occur singly as a great variety of unicellular (single-celled) organisms, such as amoebas and most bacteria. And cells are the subunits that make up multicellular organisms, such as owls and trees. Your body consists of trillions of cells of many different kinds.

All cells share certain characteristics. For example, every cell is enclosed by a membrane that regulates the passage of materials between the cell and its surroundings. And every cell uses DNA as its genetic information. However, we can distinguish between two main forms of cells. **Prokaryotic cells** were the first to evolve and were Earth's sole inhabitants for more than 1.5 billion years. Fossil evidence indicates that **eukaryotic cells** evolved from prokaryotic ancestral cells about 1.8 billion years ago.

Figure 1.3 shows these two types of cells as artificially colored photographs taken with an electron microscope. A prokaryotic cell is much simpler and usually much smaller than a eukaryotic cell. The cells of the microorganisms we call bacteria are prokaryotic. Plants, animals, fungi, and protists (mostly unicellular organisms) are all composed of eukaryotic cells. As you can see in Figure 1.3, a eukaryotic cell is subdivided by membranes into various functional compartments, or organelles. These include a nucleus, which houses the cell's DNA.

The properties of life emerge from the ordered arrangement and interactions of the structures of a cell. Such a combination of components forms a more complex organization that we can call a *system*. Systems and their emergent properties are not unique to life. Consider a box of bicycle parts. When all of the individual parts are properly assembled, the result is a mechanical system you can use for exercise or transportation.

The emergent properties of life, however, are particularly challenging to study because of the unrivaled complexity of biological systems. Biologists today often use an approach called **systems biology**—the study of a biological system and the modeling of its dynamic behavior by analyzing the interactions among its parts. Biological systems can range from the functioning of the biosphere to the molecular machinery of an organelle.

Cells illustrate another theme of biology: the correlation of structure and function. Experience shows you that form generally fits function. A screwdriver tightens or loosens screws, a hammer pounds nails. Because of their form, these tools can't do each other's jobs. Applied to biology, this theme of form fitting function is a guide to the structure of life at all its organizational levels. For example, the long extension of the nerve cell shown in Figure 1.2 enables it to transmit impulses across long distances in the body. Often, analyzing a biological structure gives us clues about what it does and how it works.

The activities of organisms are all based on cells. For example, your every thought is based on the actions of nerve cells, and your movements depend on muscle cells. Even a global process such as the cycling of carbon is the result of cellular activities, including the photosynthesis of plant cells and the cellular respiration of nearly all cells, a process that uses oxygen to break down sugar for energy and releases carbon dioxide. In the next module, we explore these processes and how they relate to the theme of organisms interacting with their environments.

? **Why are cells considered the basic units of life?**

● They are the lowest level in the hierarchy of biological organization at which the properties of life emerge.

1.4 Organisms interact with their environment, exchanging matter and energy

An organism interacts with its environment, and that environment includes other organisms as well as physical factors. **Figure 1.4** is a simplified diagram of such interactions taking place in a forest in Canada. Plants are the producers that provide the food for a typical ecosystem. A tree, for example, absorbs water (H_2O) and minerals from the soil through its roots, and its leaves take in carbon dioxide (CO_2) from the air. In photosynthesis, a tree's leaves use energy from sunlight to convert CO_2 and H_2O to sugar and oxygen (O_2). The leaves release O_2 to the air, and the roots help form soil by breaking up rocks. Thus, both organism and environment are affected by the interactions between them.

The consumers of a ecosystem eat plants and other animals. The moose in Figure 1.4 eats the grasses and tender shoots and leaves of trees in a forest ecosystem in Canada. To release the energy in food, animals (as well as plants and most other organisms) take in O_2 from the air and release CO_2. An animal's wastes return other chemicals to the environment.

Another vital part of the ecosystem includes the small animals, fungi, and bacteria in the soil that decompose wastes and the remains of dead organisms. These decomposers act as recyclers, changing complex matter into simpler chemicals that plants can absorb and use.

The dynamics of ecosystems include two major processes—the recycling of chemicals and the flow of energy. These processes are illustrated in Figure 1.4. The most basic chemicals necessary for life—carbon dioxide, oxygen, water, and various minerals—cycle within an ecosystem from the air and soil to plants, to animals and decomposers, and back to the air and soil (shown with blue arrows in the figure).

By contrast, an ecosystem gains and loses energy constantly. Energy flows into the ecosystem when plants and other photosynthesizers absorb light energy from the sun (yellow arrow) and convert it to the chemical energy of sugars and other complex molecules. Chemical energy (orange arrow) is then passed through a series of consumers and, eventually, to decomposers, powering each organism in turn. In the process of these energy conversions between and within organisms, some energy is converted to heat, which is then lost from the system (red arrow). In contrast to chemicals, which recycle within an ecosystem, energy flows through an ecosystem, entering as light and exiting as heat.

In this first section, we have touched on several themes of biology, from emergent properties in the biological hierarchy of organization, to cells as the structural and functional units of life, to the exchange of matter and energy as organisms interact with their environment. In the next section, we begin our exploration of evolution, the core theme of biology.

? Explain how the photosynthesis of plants functions in both the cycling of chemicals and the flow of energy in an ecosystem.

Photosynthesis uses light energy to convert carbon dioxide and water to energy-rich food, making it the pathway by which both chemicals and energy become available to most organisms.

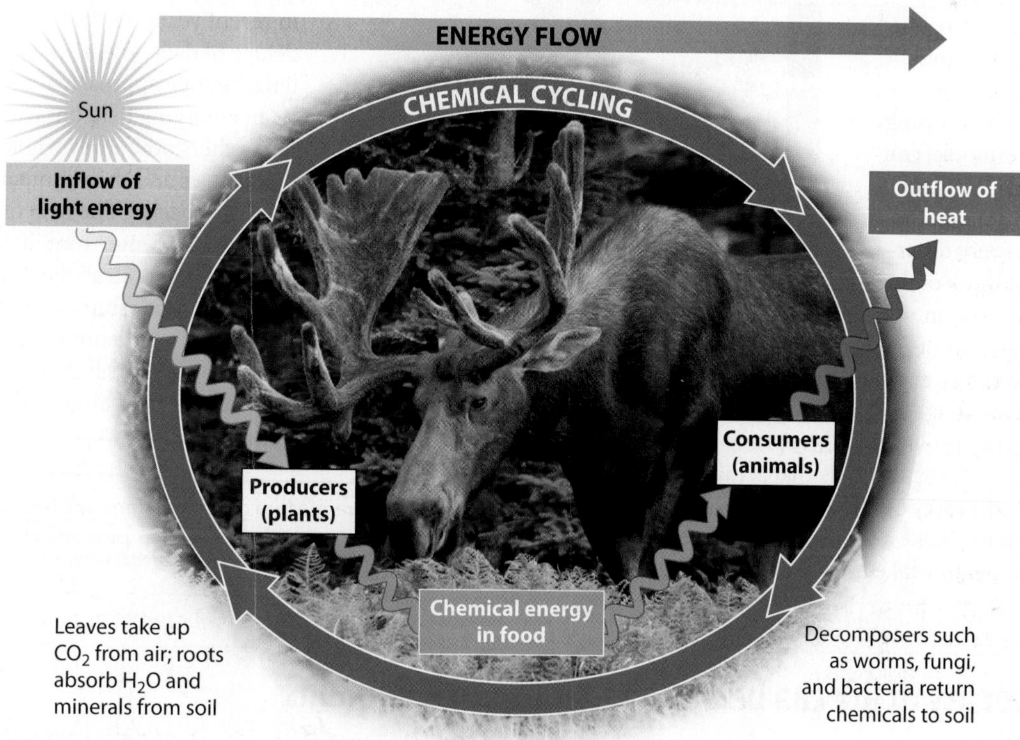

ENERGY FLOW
Sun
CHEMICAL CYCLING
Inflow of light energy
Outflow of heat
Consumers (animals)
Producers (plants)
Chemical energy in food
Leaves take up CO_2 from air; roots absorb H_2O and minerals from soil
Decomposers such as worms, fungi, and bacteria return chemicals to soil

▲ **Figure 1.4** The cycling of chemicals and flow of energy in an ecosystem

▷ Evolution, the Core Theme of Biology

1.5 The unity of life is based on DNA and a common genetic code

All cells have DNA, and the continuity of life depends on this universal genetic material. DNA is the chemical substance of **genes**, the units of inheritance that transmit information from parents to offspring. Genes, which are grouped into very long DNA molecules called chromosomes, also control all the activities of a cell.

How does the molecular structure of DNA account for its ability to encode and transmit information? Each DNA molecule is made up of two long chains, called strands, coiled together into a double helix. The strands are made up of four kinds of chemical building blocks. **Figure 1.5** (left side) illustrates these four building blocks, called nucleotides, with different colors and letter abbreviations of their names. The right side of the figure shows a short section of a DNA double helix.

Each time a cell divides, its DNA is first replicated, or copied—the double helix unzips and new complementary strands assemble along the separated strands. Thus, each new cell inherits a complete set of DNA, identical to that of the parent cell. You began as a single cell stocked with DNA inherited from your two parents. The replication of that DNA during each round of cell division transmitted copies of the DNA to what eventually became the trillions of cells of your body.

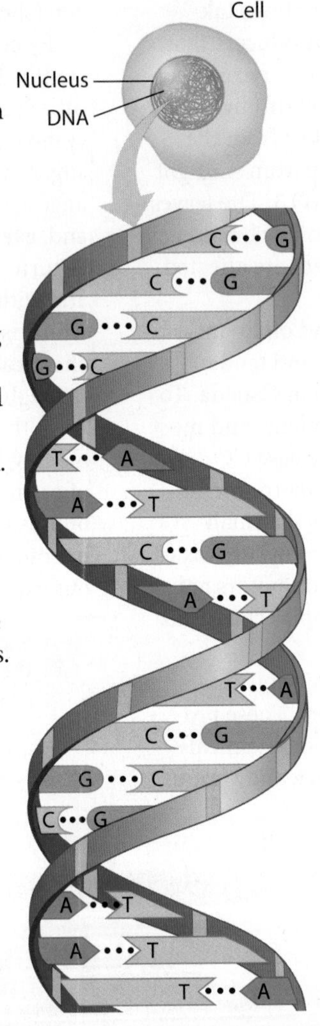

Cell
Nucleus
DNA
A
T
C
G

▲ **Figure 1.5** The four building blocks of DNA (left); part of a DNA double helix (right)

The way DNA encodes a cell's information is analogous to the way we arrange letters of the alphabet into precise sequences with specific meanings. The word *rat*, for example, conjures up an image of a rodent; *tar* and *art*, which contain the same letters, mean very different things. We can think of the four building blocks as the alphabet of inheritance. Specific sequential arrangements of these four chemical letters encode precise information in genes, which are typically hundreds or thousands of "letters" long.

The DNA of genes provides the blueprints for making proteins, and proteins serve as the tools that actually build and maintain the cell and carry out its activities. A bacterial gene may direct the cell to "Make a yellow pigment." A particular human gene may mean "Make the hormone insulin." All forms of life use essentially the same genetic code to translate the information stored in DNA into proteins. This makes it possible to engineer cells to produce proteins normally found only in some other organism. Thus, bacteria can be used to produce insulin for the treatment of diabetes by inserting a gene for human insulin into bacterial cells.

The diversity of life arises from differences in DNA sequences—in other words, from variations on the common theme of storing genetic information in DNA. Bacteria and humans are different because they have different genes. But both sets of instructions are written in the same language.

The entire "library" of genetic instructions that an organism inherits is called its **genome**. A typical human cell has two similar sets of chromosomes, and each set contains about 3 billion nucleotide pairs. In recent years, scientists have determined the entire sequence of nucleotides in the human genome, as well as the genomes of thousands of other species. More species continue to be added to the list of species whose genomes have been sequenced as the rate at which sequencing can be done has accelerated rapidly in recent years. To deal with the resulting deluge of data, scientists are applying a systems biology approach at the molecular level. In an emerging field known as genomics, researchers now study whole sets of genes in a species and then compare genes across multiple species. The benefits from such an approach range from identifying genes that may be implicated in human cancers to revealing the evolutionary relationships among diverse organisms based on similarities in their genomes. Genomics affirms the unity of life based on the universal genetic material—DNA.

In the next module, we see how biologists attempt to organize the diversity of life.

? **What are the two main functions of DNA?**

● DNA is the genetic material that is passed from parents to offspring, and it codes for proteins that control the activity of cells.

1.6 The diversity of life can be arranged into three domains

We can think of biology's enormous scope as having two dimensions. The "vertical" dimension, which we examined in Module 1.2, is the size scale that stretches from molecules to the biosphere. But biology also has a "horizontal" dimension, spanning across the great diversity of organisms existing now and over the long history of life on Earth.

Diversity is a hallmark of life. Biologists have so far identified and named about 1.8 million species. Estimates of the total number of species range from 10 million to more than 100 million.

There seems to be a human tendency to group things, such as owls or butterflies, although we recognize that each group includes many different species. And then we cluster groups into broader categories, such as birds and insects. Taxonomy, the branch of biology that names and classifies species, arranges species into a hierarchy of broader and broader groups: genus, family, order, class, phylum, and kingdom.

Historically, biologists divided all of life into five kingdoms. But new methods for assessing evolutionary relationships, such as comparisons of DNA sequences, have led to an ongoing reevaluation of the number and boundaries of kingdoms. Although the debate on such divisions continues, there is consensus among biologists that life can be organized into three higher levels called **domains**. Figure 1.6 shows representatives of domains Bacteria, Archaea, and Eukarya.

Domains **Bacteria** and **Archaea** both consist of prokaryotes, organisms with prokaryotic cells. Bacteria are the most diverse and widespread prokaryotes. Many of the prokaryotes known as archaea live in Earth's extreme environments, such as salty lakes and boiling hot springs. Each rod-shaped or round structure in the photos of the prokaryotes in Figure 1.6 is a single cell. These photos were made with an electron microscope, and the number along the side indicates the magnification of the image.

All the eukaryotes, organisms with eukaryotic cells, are grouped in domain **Eukarya**. Protists are a diverse collection of mostly single-celled organisms. Pictured in Figure 1.6 is an assortment of protists in a drop of pond water. Biologists are currently assessing how to group the protists to reflect their evolutionary relationships.

The three remaining groups within Eukarya are distinguished partly by their modes of nutrition. Kingdom Plantae consists of plants, which produce their own food by photosynthesis. The plant pictured in Figure 1.6 is a tropical bromeliad, a plant native to the Americas.

Kingdom Fungi, represented by the mushrooms in Figure 1.6, is a diverse group whose members mostly decompose the remains of dead organisms and organic wastes and absorb the nutrients into their cells.

Animals obtain food by eating other organisms. The sloth in Figure 1.6 resides in South American rain forests. There are actually members of two other groups in the sloth photo. The sloth is clinging to a tree (kingdom Plantae), and the greenish tinge in its hair is a luxuriant growth of photosynthetic prokaryotes (domain Bacteria). This photograph exemplifies a theme reflected in our book's title: connections between living things. The sloth depends on trees for food and shelter; the tree uses nutrients from the decomposition of the sloth's feces; the prokaryotes gain access to the sunlight necessary for photosynthesis by living on the sloth; and the sloth is camouflaged from predators by its green coat.

Domain Bacteria

Colorized SEM 7,500×

Bacteria

Domain Archaea

Colorized SEM 10,000×

Archaea

Domain Eukarya

LM 340×

Protists (multiple kingdoms)

Kingdom Plantae

Kingdom Fungi

Kingdom Animalia

▲ **Figure 1.6** The three domains of life

The diversity of life and its interconnectedness are evident almost everywhere. Earlier we looked at life's unity in its shared properties and common genetic code. In the next module, we explore how evolution explains both the unity and the diversity of life.

? To which of the three domains of life do we belong?

Eukarya

1.7 Evolution explains the unity and diversity of life

Evolution can be defined as the process of change that has transformed life on Earth from its earliest beginnings to the diversity of organisms living today. The fossil record documents the fact that life has been evolving on Earth for billions of years, and patterns of ancestry can be traced through this record. For example, the mammoth being excavated in **Figure 1.7A** is clearly related to present-day elephants. We can explain the shared traits of mammoths and elephants with the premise that they descended from a common ancestor in the distant past. Their differences reflect the evolutionary changes that occurred within their separate lineages during the history of their existence on Earth. Thus, evolution accounts for life's dual nature of kinship and diversity.

▲ **Figure 1.7A** Excavation of 26,000-year-old fossilized mammoth bones from a site in South Dakota

This evolutionary view of life came into sharp focus in November 1859, when Charles Darwin **(Figure 1.7B)** published one of the most important and influential books ever written. Entitled *On the Origin of Species by Means of Natural Selection*, Darwin's book was an immediate bestseller and soon made his name synonymous with the concept of evolution.

▲ **Figure 1.7B** Charles Darwin in 1859

As a young man, Darwin made key observations that greatly influenced his thinking. During a five-year, around-the-world voyage, he collected and documented plants and animals in widely varying locations—from the isolated Galápagos Islands off the coast of Ecuador to the heights of the Andes mountains to the jungles of Brazil. He was particularly struck by the adaptations of these varied organisms that fit them to their diverse habitats. After returning to England, Darwin spent more than two decades continuing his observations, performing experiments, corresponding with other scientists, and refining his thinking before he finally published his work.

The first of two main points that Darwin presented in *The Origin of Species* was that species living today arose from a successor of ancestors that differed from them. Darwin called this process "descent with modification." It was an insightful phrase, because it captured both the unity of life (descent from a common ancestor) and the diversity of life (modifications that evolved as species diverged from their ancestors). **Figure 1.7C** illustrates this unity and diversity among birds. These three birds all have a common "bird" body plan of wings, beak, feet, and feathers, but these features are highly specialized for each bird's unique lifestyle.

Darwin's second point was to propose a mechanism for evolution, which he called **natural selection**. Darwin started with two observations, from which he drew two inferences.

OBSERVATION #1: *Individual variation.* Individuals in a population vary in their traits, many of which are inherited from parents to offspring.

OBSERVATION #2: *Overproduction of offspring.* All species can produce far more offspring than the environment can support. Competition for resources is thus inevitable, and many of these offspring fail to survive and reproduce.

INFERENCE #1: *Unequal reproductive success.* Individuals with heritable traits best suited to the local environment are more likely to survive and reproduce than are less well-suited individuals.

INFERENCE #2: *Accumulation of favorable traits over time.* As a result of this unequal reproductive success over many generations, a higher and higher proportion of individuals in the population will have the advantageous traits.

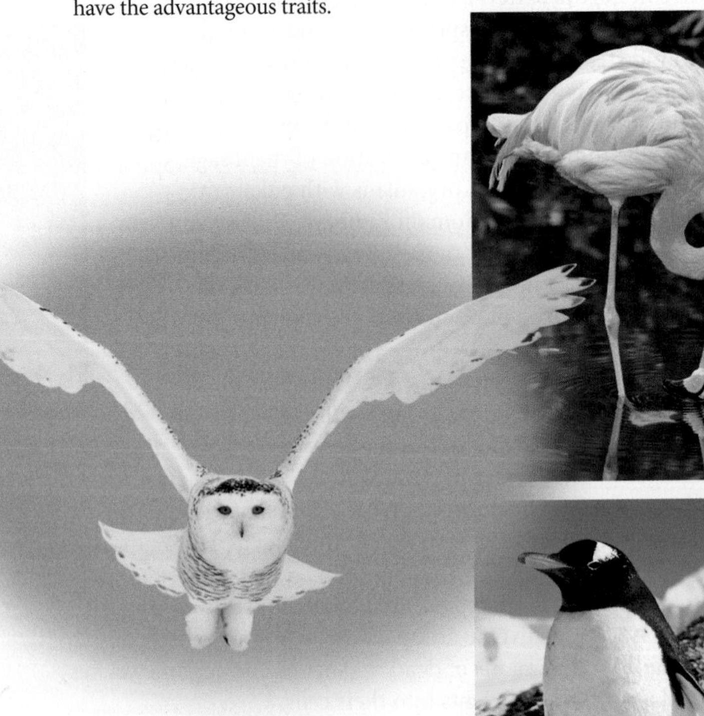

▲ **Figure 1.7C** Unity and diversity among birds

Try This For each bird, describe some adaptations that fit it to its environment and way of life.

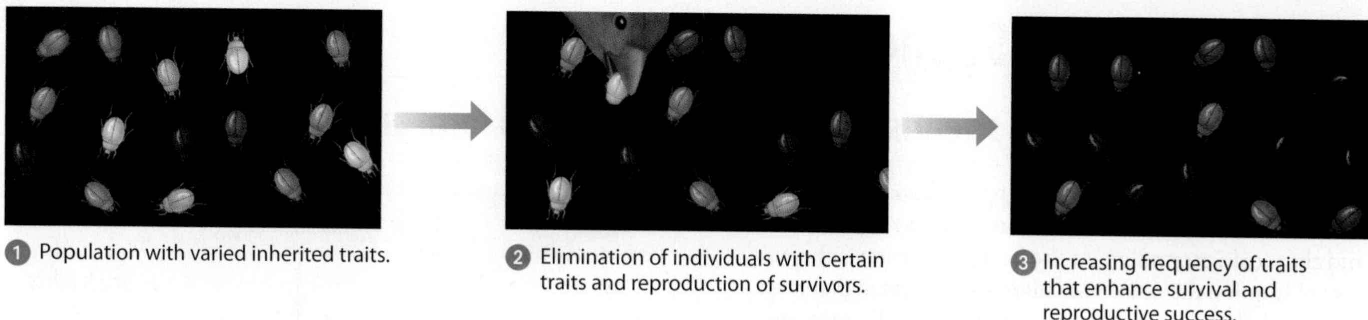
1 Population with varied inherited traits.
2 Elimination of individuals with certain traits and reproduction of survivors.
3 Increasing frequency of traits that enhance survival and reproductive success.

▲ **Figure 1.7D** An example of natural selection in action

Try This Describe what might happen if some of these beetles colonized a sand dune habitat.

Figure 1.7D uses a simple example to show how natural selection works. ❶ An imaginary beetle population has colonized an area where the soil has been blackened by a recent brush fire. Initially, the population varies extensively in the inherited coloration of individuals, from very light gray to charcoal. ❷ A bird eats the beetles it sees most easily, the light-colored ones. This selective predation reduces the number of light-colored beetles and favors the survival and reproductive success of the darker beetles, which pass on the genes for dark coloration to their offspring. ❸ After several generations, the population is quite different from the original one. As a result of natural selection, the frequency of the darker-colored beetles in the population has increased.

Darwin realized that numerous small changes in populations as a result of natural selection could eventually lead to major alterations of species. He proposed that new species could evolve as a result of the gradual accumulation of changes over long periods of time. This could occur, for example, if one population fragmented into subpopulations isolated in different environments. In these separate arenas of natural selection, one species could gradually divide into multiple species as isolated populations adapted over many generations to different sets of environmental factors.

The fossil record provides evidence of such diversification of species from ancestral species. **Figure 1.7E** traces an evolutionary tree of elephants and some of their relatives. (Biologists' diagrams of evolutionary relationships generally take the form of branching trees, usually turned sideways and read from left to right.) You can see that the three living species of elephants are very similar because they shared a recent common ancestor (dating to about 3 million years ago, which is relatively recent in an evolutionary timeframe). Notice that all the other close relatives of elephants are extinct—their branches do not extend to the present. (The mammoth being excavated in Figure 1.7A belonged to the genus *Mammuthus*, whose members became extinct less than 10,000 years ago.) If we were to trace this family tree back to about 60 million years ago, however, you would find a common ancestor that connects elephants to their closest living relatives—the manatees and hyraxes. The fossil record, along with other evidence such as comparisons of DNA, allows scientists to trace the evolutionary history of life back through time.

All of life is connected, and the basis for this kinship is evolution—the core theme that makes sense of everything we

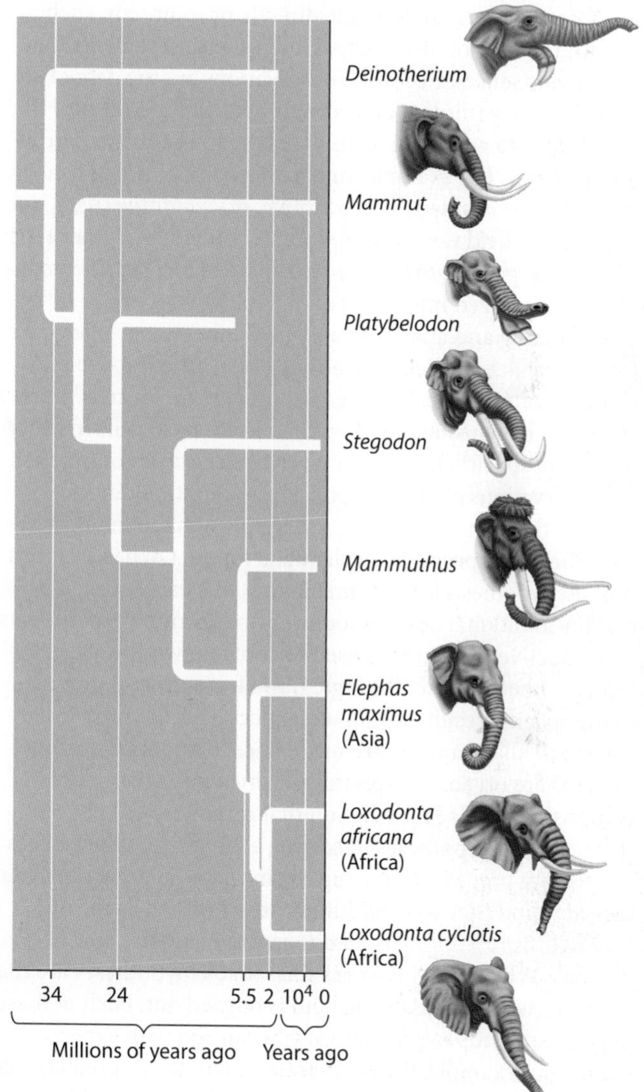
Deinotherium
Mammut
Platybelodon
Stegodon
Mammuthus
Elephas maximus (Asia)
Loxodonta africana (Africa)
Loxodonta cyclotis (Africa)
34
24
5.5
2
10⁴
0
Millions of years ago
Years ago

▲ **Figure 1.7E** An evolutionary tree of elephants

Try This Use this tree to determine when mastadons (in the genus *Mammut*) last shared a common ancestor with African elephants.

know and learn about life. In the next module, we introduce scientific inquiry, the process we use to study the natural world.

? **Explain the cause and effect of unequal reproductive success.**

● Those individuals with heritable traits best suited to the local environment produce the greatest number of offspring. Over many generations, the frequency of those adaptive traits increases in the population.

▷ The Process of Science

1.8 In studying nature, scientists make observations and form and test hypotheses

Science is a way of knowing—an approach to understanding the natural world. It stems from our curiosity about ourselves and the world around us. At the heart of science is the process of inquiry, the search for information and explanations of natural phenomena. Scientific inquiry usually involves making observations, forming hypotheses, and testing them.

Observations may be made directly or indirectly, such as with the help of microscopes and other instruments that extend our senses. Recorded observations are the data of science. You may think of data as numbers, but a great deal of scientific data are in the form of detailed, carefully recorded observations. For example, much of our knowledge of snowy owl behavior is based on descriptive, or *qualitative*, data, documented in field notes, photographs, and videos. Other types of data are *quantitative*, such as numerical measurements that may be organized into tables and graphs.

Collecting and analyzing a large number of specific observations can lead to generalizations based on inductive reasoning. For example, "All organisms are made of cells" is an inductive conclusion based on the discovery of cells in every biological specimen observed over two centuries of time.

Observations often prompt us to ask questions and seek answers through the forming and testing of hypotheses. A **hypothesis** is a proposed explanation for a set of observations. A good hypothesis leads to predictions that can be tested by making additional observations or by performing experiments.

Deductive reasoning is used to come up with ways to test hypotheses. Here, the logic flows from general premises to the specific results we should expect if the premises are correct. *If* all organisms are made of cells (premise 1), *and* humans are organisms (premise 2), *then* humans should be composed of cells (a prediction that can be tested).

We all use hypotheses in solving everyday problems. Let's say you are preparing for a big storm that is approaching your area and find that your flashlight isn't working. That your flashlight isn't working is an observation, and the question is obvious: Why doesn't it work? Reasonable hypotheses are that the batteries are dead or the bulb is burned out. Each of these hypotheses leads to predictions you can test with experiments. For example, the dead-battery hypothesis predicts that replacing the batteries with new ones will fix the problem. **Figure 1.8** diagrams the results of testing these hypotheses.

An important point about scientific inquiry is that we can never *prove* that a hypothesis is true. As shown in Figure 1.8, the burned-out bulb hypothesis is the more likely explanation in our hypothetical scenario. But perhaps the old bulb was simply loose and the new bulb was inserted correctly. We could test this hypothesis by trying another experiment—carefully reinstalling the original bulb. If the flashlight still doesn't work, the burned-out bulb hypothesis is supported by another line of evidence. Testing a hypothesis in various ways provides additional support for a hypothesis and increases our confidence in it.

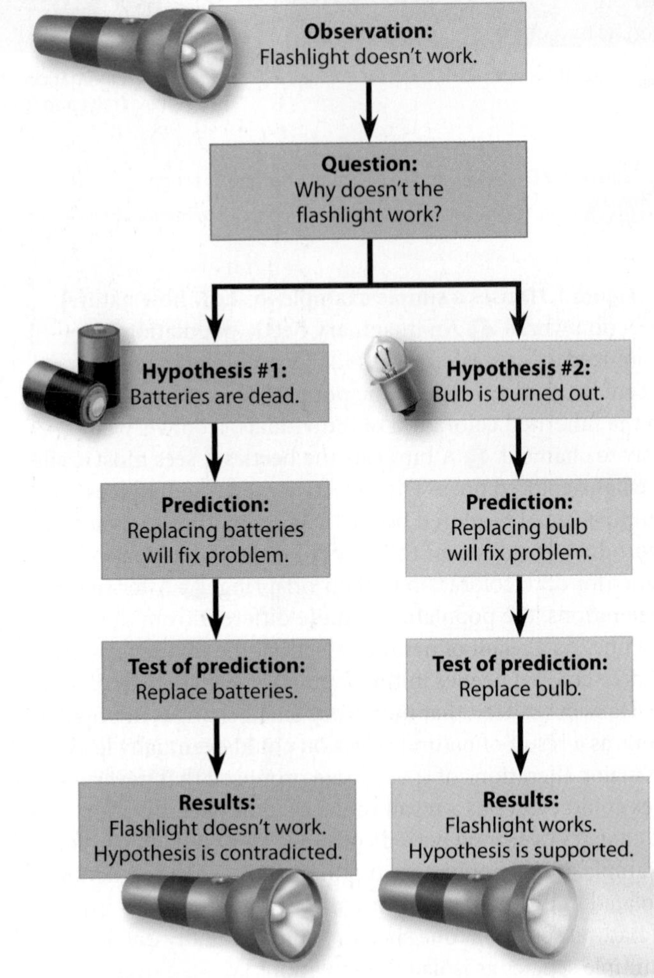
Observation:
Flashlight doesn't work.
Question:
Why doesn't the flashlight work?
Hypothesis #1:
Batteries are dead.
Hypothesis #2:
Bulb is burned out.
Prediction:
Replacing batteries will fix problem.
Prediction:
Replacing bulb will fix problem.
Test of prediction:
Replace batteries.
Test of prediction:
Replace bulb.
Results:
Flashlight doesn't work.
Hypothesis is contradicted.
Results:
Flashlight works.
Hypothesis is supported.

▲ **Figure 1.8** An everyday example of forming and testing hypotheses

A scientific **theory** is much broader in scope than a hypothesis and is supported by a large and usually growing body of evidence. For example, the theory of evolution explains a great diversity of observations and is supported by multiple lines of evidence. In addition, the theory of evolution has not been contradicted by any scientific data.

Another important aspect of science is that it is necessarily repetitive: In testing a hypothesis, researchers may make observations that call for rejecting the hypothesis or at least revising and further testing it. This process allows biologists to circle closer and closer to their best estimation of how nature works. As in all quests, science includes elements of challenge, adventure, and luck, along with careful planning, reasoning, creativity, cooperation, competition, and persistence.

Science is a social activity, with most scientists working in teams, which often include graduate and undergraduate students. Scientists share information through peer-reviewed publications, seminars, meetings, and personal communication. Scientists build on what has been learned from earlier research and often check each other's claims by attempting to confirm observations or repeat experiments.

To help you better understand what scientists do, we include a Scientific Thinking module in each chapter. These discussions will encompass several broad activities of science: the forming and testing of hypotheses using various research methods; the analysis and evaluation of data; the use of tools and technologies that have built and continue to expand scientific knowledge; and the communication of the results of scientific studies and the evaluation of their implications for society as a whole.

What is the main criterion for a scientific hypothesis?

It must generate predictions that can be tested.

1.9 Hypotheses can be tested using controlled field studies

SCIENTIFIC THINKING

You have undoubtedly observed that many animals match their environment: white snowy owls in their arctic habitat, toads the color of dead leaves, flounders that blend in with the sandy sea floor. From these observations, you might hypothesize that such color patterns have evolved as adaptations that protect animals from predation. Can scientists test this camouflage hypothesis? Let's consider an experiment with two populations of mice that belong to the same species (*Peromyscus polionotus*) but live in different environments.

The beach mouse lives along the Florida seashore, a habitat of white sand dunes with sparse clumps of beach grass. The inland mouse lives on darker soil farther inland. As you can see in **Figure 1.9**, there is a striking match between mouse coloration and habitat. In 2010, biologist Hopi Hoekstra of Harvard University and a group of her students headed to Florida to test the camouflage hypothesis. They reasoned that *if* camouflage coloration protects mice from predators, *then* mice with coloration that did not match their habitat would be preyed on more heavily than the native mice that were well-matched to their environment.

The researchers built 250 plastic models of mice and painted them to resemble either beach or inland mice. Equal numbers of models were placed randomly in both habitats. The models resembling the native mice in each habitat were the control group, and the mice with the non-native coloration were the experimental group. Signs of predation were recorded for three days. Judging by the bite marks and surrounding tracks, the researchers determined the predators were likely foxes, coyotes, owls, herons, and hawks.

Why do so many animals match their surroundings?

As you can see by the results presented in **Table 1.9**, the noncamouflaged models had a much higher percentage of predation attacks in both the beach and inland habitats. The data thus fit the key prediction of the camouflage hypothesis.

TABLE 1.9 | RESULTS FROM CAMOUFLAGE EXPERIMENT

	Number of Attacks		
Habitat	On Camouflaged Models	On Non-camouflaged Models	% Attacks on Non-camouflaged Models
Beach (light habitat)	2	5	71%
Inland (dark habitat)	5	16	76%

Data from S. N. Vignieri et al., The selective advantage of crypsis in mice, *Evolution* 64: 2153–2158 (2010).

This study is an example of a **controlled experiment**, one that is designed to compare an experimental group (the noncamouflaged mice models) with a control group (the camouflaged models that matched the mice native in each area). Ideally, in a controlled experiment the two groups differ only in the one factor the experiment is designed to test—in this case, coat color and its effect on the success of predators. The experimental design left coloration as the only factor that could account for the higher predation rate on the noncamouflaged mice in both the beach and the inland habitats. This study is also an example of a field study, one not done in a laboratory but out in nature. Researchers tested their hypothesis using the natural habitat of the mice and their predators.

These two populations of mice belong to the same species, yet they have very different coloration. How does natural selection explain these differences?

Camouflaged mice are more likely to survive and reproduce, passing their protective coloration to their offspring.

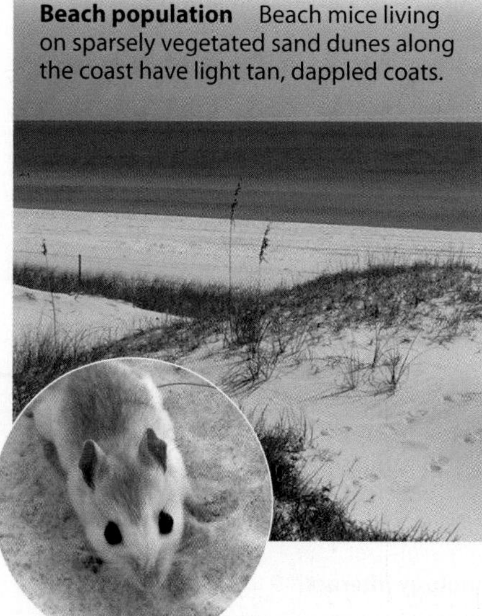

Beach population Beach mice living on sparsely vegetated sand dunes along the coast have light tan, dappled coats.

Inland population Members of the same species living about 30 km inland are darker in color.

▲ **Figure 1.9** Beach mouse and inland mouse with their native habitat

Biology and Everyday Life

1.10 Evolution is connected to our everyday lives

EVOLUTION CONNECTION

To emphasize evolution as the core theme of biology, we include an Evolution Connection module in each chapter in this text. But how is evolution connected to your everyday life?

You just learned that natural selection is the primary mechanism of evolution, in which the environment "selects" for adaptive traits when organisms with such traits are better able to survive and reproduce. Through the selective breeding of plants and animals, humans are also an agent of evolution. As a result of **artificial selection**, our crops, livestock, and pets bear little resemblance to their wild ancestors. Humans have been modifying species for millennia, and recent advances in biotechnology have increased our capabilities. Plant biologists using genomics can identify beneficial genes in relatives of our crop plants, enabling the breeding or genetic engineering of enhanced crops. Genes from totally unrelated species have also been inserted into plants. For example, genes for such traits as drought or flood tolerance, improved growth, and increased nutrition have been engineered into rice plants (**Figure 1.10**).

But humans also affect evolution unintentionally. The impact of habitat loss and global climate change can be seen in the loss of species. Indeed, scientists estimate that the current rate of extinction is 100 to 1,000 times the typical rate seen in the fossil record. Our actions are also driving evolutionary changes in species. For example, our widespread use of antibiotics and pesticides has led to the evolution of antibiotic resistance in bacteria and pesticide resistance in insects.

▲ **Figure 1.10** Researcher working with transgenic rice

How can evolutionary theory help address such worldwide problems? Understanding evolution can help us develop strategies for conservation efforts and prompt us to be more judicious in our use of antibiotics and pesticides. It can also help us create flu vaccines and HIV drugs by tracking the rapid evolution of these viruses. Identifying shared genes and studying their actions in closely related organisms may produce new knowledge about cancer or other diseases and lead to new medical treatments. New sources of drugs may be found by tracing the evolutionary history of medicinal plants and identifying beneficial compounds in their relatives. Our understanding of evolution can yield many beneficial results.

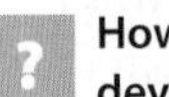

How might an understanding of evolution contribute to the development of new drugs?

● As one example, we can test the actions of potential drugs in organisms that share our genes and similar cellular processes.

1.11 Biology, technology, and society are connected in important ways

CONNECTION

Many of the current issues facing society are related to biology, and they often involve our expanding technology. What are the differences between science and technology? The goal of science is to understand natural phenomena. In contrast, the goal of **technology** is to apply scientific knowledge for some specific purpose. Scientists usually speak of "discoveries," whereas engineers more often speak of "inventions." These two fields, however, are interdependent. Scientists use new technology in their research, and scientific discoveries often lead to the development of new technologies.

The potent combination of science and technology can have dramatic effects on society. For example, the discovery of the structure of DNA by Watson and Crick 60 years ago and subsequent advances in DNA science led to the technologies of DNA manipulation that today are transforming applied fields such as medicine, agriculture, and forensics.

Technology has improved our standard of living in many ways, but not without consequences. Technology has helped Earth's population to grow tenfold in the past three centuries and more than double to 7 billion in just the past 40 years. Global climate change, toxic wastes, deforestation, and nuclear accidents are just some of the repercussions of more and more people wielding more and more technology. Science can help identify problems and provide insight into how to slow down or prevent further damage. But solutions to these problems have as much to do with politics, economics, and cultural values as with science and technology. Every citizen has a responsibility to develop a reasonable amount of scientific literacy to be able to participate in the debates regarding science and technology. The crucial science-technology-society relationship is a theme we will return to throughout this text.

We hope this book will help you develop an appreciation for biology and help you apply your new knowledge to evaluating issues ranging from your personal health to the well-being of the whole world.

How do science and technology interact?

● New scientific discoveries may lead to new technologies; new technologies may increase the ability of scientists to discover new knowledge.

CHAPTER 1 REVIEW

For practice quizzes, BioFlix animations, MP3 tutorials, video tutors, and more study tools designed for this textbook, go to MasteringBiology®

Reviewing the Concepts

Themes in the Study of Biology (1.1–1.4)

1.1 All forms of life share common properties. Biology is the scientific study of life. Properties of life include order, reproduction, growth and development, energy processing, regulation, response to the environment, and evolutionary adaptation.

1.2 In life's hierarchy of organization, new properties emerge at each level. Biological organization unfolds as follows: biosphere > ecosystem > community > population > organism > organ system > organ > tissue > cell > organelle > molecule. Emergent properties result from the interactions among component parts.

1.3 Cells are the structural and functional units of life. Eukaryotic cells contain membrane-enclosed organelles, including a nucleus. Prokaryotic cells lack such organelles. Structure is related to function at all levels of organization. Systems biology models the complex behavior of biological systems.

1.4 Organisms interact with their environment, exchanging matter and energy. Ecosystems are characterized by the cycling of chemicals from the atmosphere and soil through producers, consumers, decomposers, and back to the environment. Energy flows one way through an ecosystem—entering as sunlight, converted to chemical energy by producers, passed on to consumers, and exiting as heat.

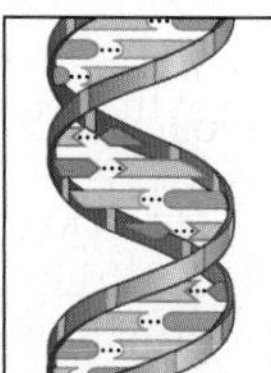
ENERGY FLOW
CHEMICAL CYCLING
Light
Heat
Producers
Consumer
Chemical energy

Evolution, the Core Theme of Biology (1.5–1.7)

1.5 The unity of life is based on DNA and a common genetic code. DNA is responsible for heredity and for programming the activities of a cell. A species' genes are coded in the sequences of the four building blocks making up DNA's double helix. Genomics is the analysis and comparison of genomes.

1.6 The diversity of life can be arranged into three domains. Taxonomists name species and classify them into a system of broader groups. Domains Bacteria and Archaea consist of prokaryotes. The eukaryotic domain, Eukarya, includes various protists and the kingdoms Fungi, Plantae, and Animalia.

1.7 Evolution explains the unity and diversity of life. Darwin synthesized the theory of evolution by natural selection.

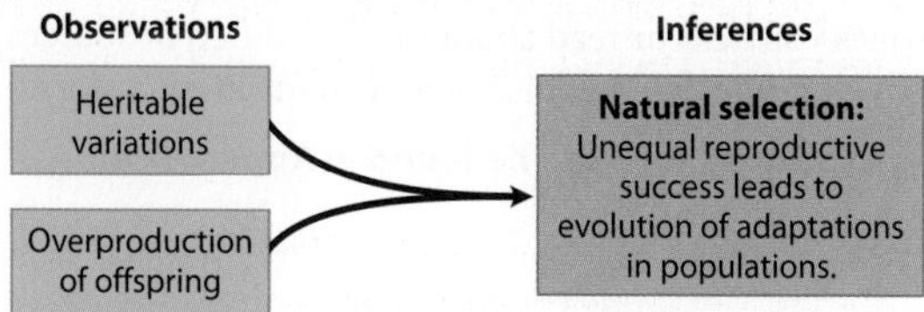
Observations
Inferences
Heritable variations
Overproduction of offspring
Natural selection: Unequal reproductive success leads to evolution of adaptations in populations.

The Process of Science (1.8–1.9)

1.8 In studying nature, scientists make observations and form and test hypotheses. Scientists use inductive reasoning to draw general conclusions from many observations. They form hypotheses and use deductive reasoning to make predictions, which can be tested with experiments or additional observations. Data may be qualitative or quantitative. A scientific theory is broad in scope and is supported by a large body of evidence.

1.9 Hypotheses can be tested using controlled field studies. Researchers found that mice models that did not match their habitat had higher predation rates than camouflaged models. In a controlled experiment, the use of control and experimental groups can demonstrate the effect of a single variable.

Biology and Everyday Life (1.10–1.11)

1.10 Evolution is connected to our everyday lives. Evolutionary theory is useful in medicine, agriculture, forensics, and conservation. Human-caused environmental changes are powerful selective forces that affect the evolution of many species.

1.11 Biology, technology, and society are connected in important ways. Technological advances stem from scientific research, and research benefits from new technologies.

Connecting the Concepts

1. Complete the following map organizing some of biology's major concepts.

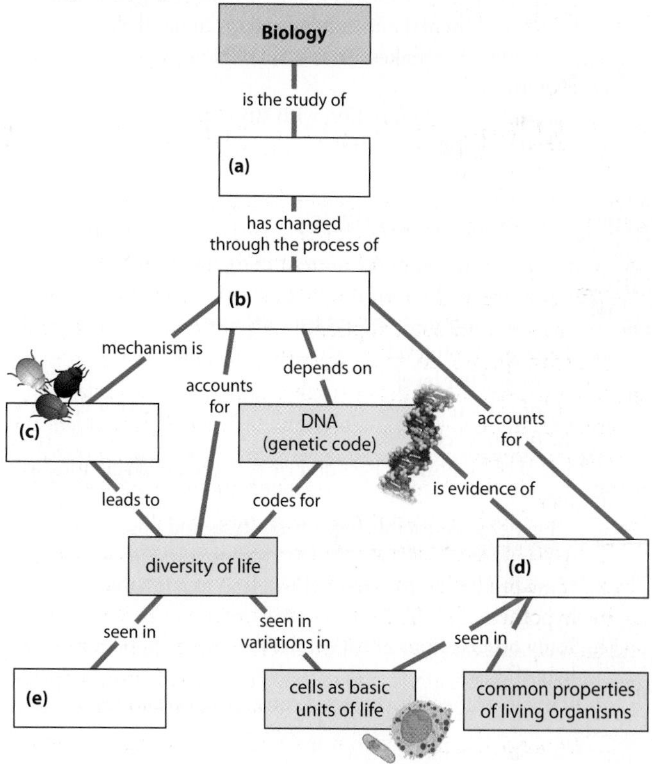
Biology
is the study of
(a)
has changed through the process of
(b)
mechanism is
accounts for
depends on
accounts for
(c)
DNA (genetic code)
leads to
codes for
is evidence of
diversity of life
(d)
seen in
seen in variations in
seen in
(e)
cells as basic units of life
common properties of living organisms

Testing Your Knowledge

Level 1: Knowledge/Comprehension

2. All the organisms on your campus make up
   a. an ecosystem.
   b. a community.
   c. a population.
   d. the domain Eukarya.
3. Single-celled amoebas and bacteria are grouped into different domains because
   a. amoebas eat bacteria.
   b. bacteria are not made of cells.
   c. bacterial cells lack a membrane-enclosed nucleus.
   d. amoebas are motile; bacteria are not.
4. Which of the following statements best distinguishes hypotheses from theories in science?
   a. Theories are hypotheses that have been proved.
   b. Hypotheses usually are narrow in scope; theories have broad explanatory power.
   c. Hypotheses are tentative guesses; theories are correct answers to questions about nature.
   d. Hypotheses and theories are different terms for essentially the same thing in science.
5. Which of the following best demonstrates the unity among all living organisms?
   a. descent with modification
   b. DNA and a common genetic code
   c. emergent properties
   d. natural selection
6. A controlled experiment is one that
   a. proceeds slowly enough that a scientist can make careful records of the results.
   b. keeps all variables constant.
   c. is repeated many times to make sure the results are accurate.
   d. tests experimental and control groups in parallel.
7. The core idea that makes sense of all of biology is
   a. evolution.
   b. the correlation of function with structure.
   c. systems biology.
   d. the process of science.

Level 2: Application/Analysis

8. A biologist studying interactions among the protists in an ecosystem could *not* be working at which level in life's hierarchy? (*Choose carefully and explain your answer.*)
   a. the population level
   b. the molecular level
   c. the organism level
   d. the organ level
9. Which of the following best describes the logic of scientific inquiry?
   a. If I generate a testable hypothesis, tests and observations will support it.
   b. If my prediction is correct, it will lead to a testable hypothesis.
   c. If my observations are accurate, they will support my hypothesis.
   d. If my hypothesis is correct, I can expect certain test results.
10. In an ecosystem, how is the movement of energy similar to that of chemicals, and how is it different?
11. Explain the role of heritable variations in Darwin's theory of natural selection.
12. Describe the process of scientific inquiry and explain why it is not a rigid method.
13. Contrast technology with science. Give an example of each to illustrate the difference.
14. Biology can be described as having both a vertical scale and a horizontal scale. Explain what that means.

Level 3: Synthesis/Evaluation

15. Explain what is meant by this statement: Natural selection is an editing mechanism rather than a creative process.
16. The graph below shows the results of an experiment in which mice learned to run through a maze.

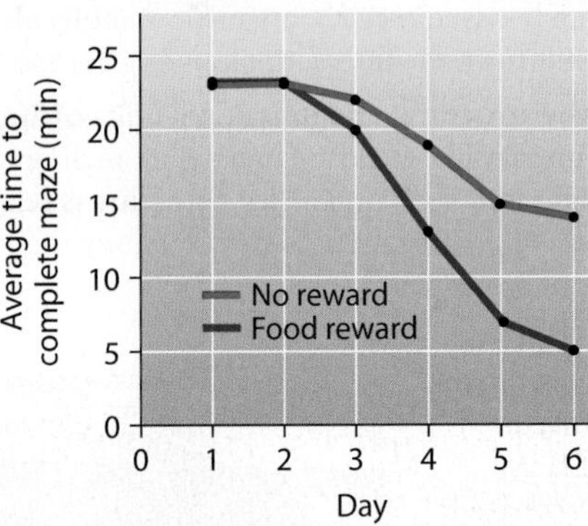

Average time to complete maze (min)
25
20
15
10
5
0
No reward
Food reward
0 1 2 3 4 5 6
Day

   a. State the hypothesis and prediction that you think this experiment tested.
   b. Which was the control group and which the experimental? Why was a control group needed?
   c. List some variables that must have been controlled so as not to affect the results.
   d. Do the data support the hypothesis? Explain.
17. **SCIENTIFIC THINKING** Suppose that in an experiment similar to the camouflage experiment described in Module 1.9, a researcher observed and recorded more total predator attacks on dark-model mice in the inland habitat than on dark models in the beach habitat. From comparing these two pieces of data, the researcher concluded that the camouflage hypothesis is false. Do you think this conclusion is justified? Why or why not?
18. The fruits of wild species of tomato are tiny compared to the giant beefsteak tomatoes available today. This difference in fruit size is almost entirely due to the larger number of cells in the domesticated fruits. Plant biologists have recently discovered genes that are responsible for controlling cell division in tomatoes. Why would such a discovery be important to producers of other kinds of fruits and vegetables? To the study of human development and disease? To our basic understanding of biology?
19. The news media and popular magazines frequently report stories that are connected to biology. In the next 24 hours, record the ones you hear or read about in three different sources and briefly describe the biological connections in each story.

Answers to all questions can be found in Appendix 4.

UNIT I

The Life of the Cell

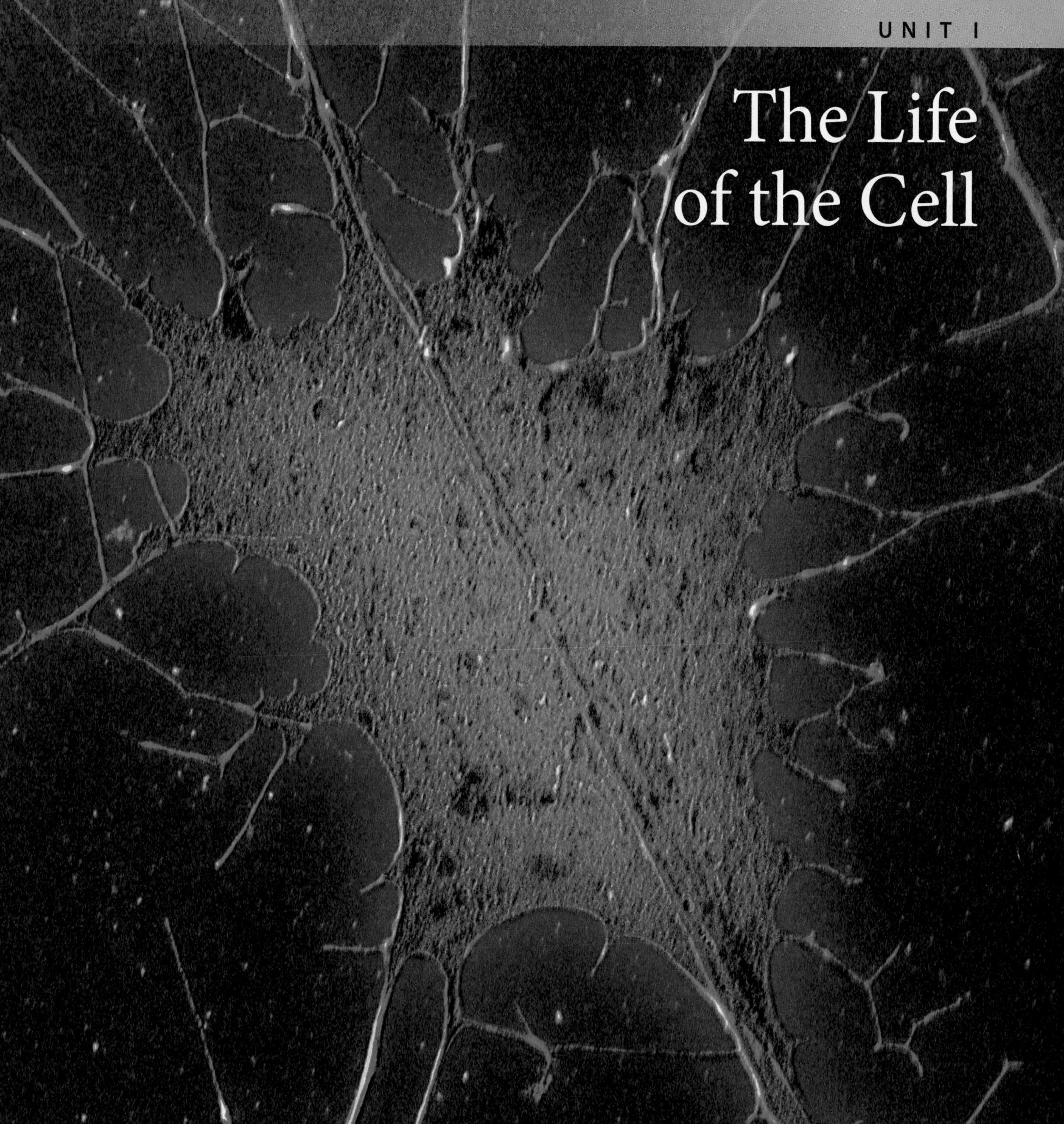

CHAPTER 2

The Chemical Basis of Life

Coral reefs are among the most diverse ecosystems on Earth. They are formed from the gradual buildup of the calcium carbonate skeletons of small coral animals. As you can see in the photo below, these structurally diverse habitats provide havens for a huge diversity of fish and other marine organisms. But in recent years, something in the air is threatening coral reefs. How might a chemical compound in the air harm such a vibrant ecosystem? The answer is chemistry. When carbon dioxide dissolves in water, it reacts with water to form an acid, which then makes the water more acidic. Later in the chapter we will see how scientists are exploring the effects of this ocean acidification on coral reefs.

? ***Will rising atmospheric CO_2 harm coral reefs?***

Why do we begin our study of biology with a chapter on chemistry? Well, chemistry is the basis of life—it explains how elements combine into the compounds that make up your body and the bodies of all other living organisms and how chemical reactions underlie the functions of all cells.

Life and its chemistry are tied to water. Life began in water and evolved there for 3 billion years before spreading onto land. And all life, even land-dwelling life, is still dependent on water. Your

cells are about 75% water, and that is where the chemical reactions of your body take place. What properties of the simple water molecule make it so indispensable to life on Earth? You'll find out in this chapter.

This chapter will also make connections to one of the main themes in biology—the organization of life into a hierarchy of structural levels, with new properties emerging at each successive level (see Chapter 1). You will see that emergent properties are apparent even at the lowest levels of biological organization—the ordering of atoms into molecules and the interactions of those molecules. Thus we begin our study of biology with some basic concepts of chemistry that will apply throughout our study of life.

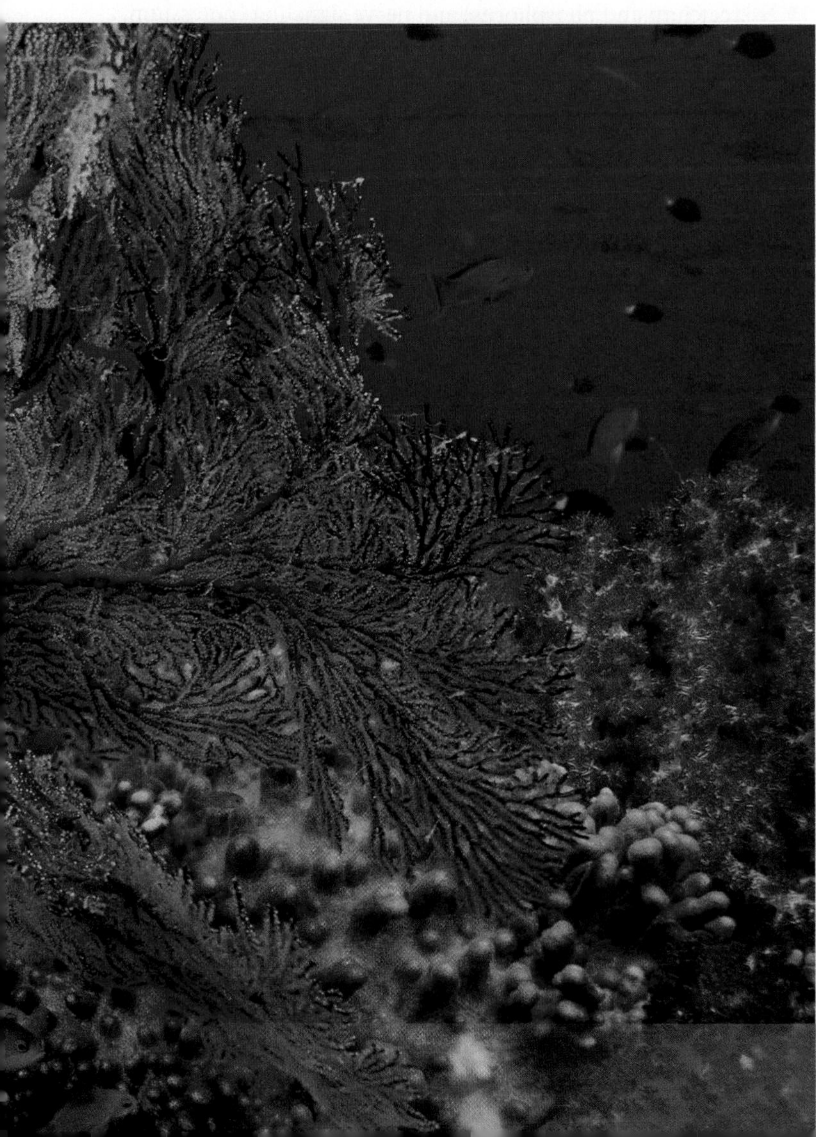

BIG IDEAS

Elements, Atoms, and Compounds
(2.1–2.4)

Living organisms are made of atoms of certain elements, mostly combined into compounds.

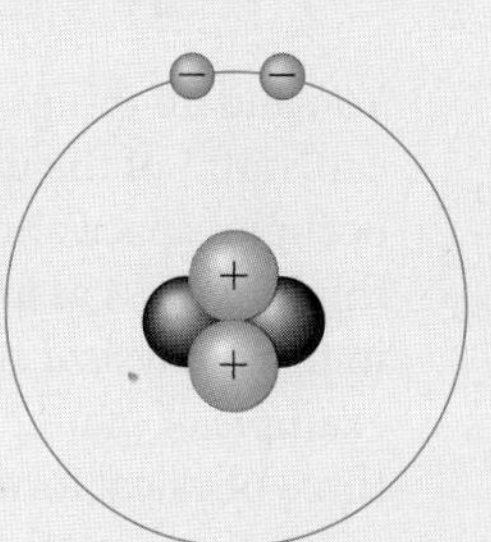
+
+

Chemical Bonds
(2.5–2.9)

The structure of an atom determines what types of bonds it can form with other atoms.

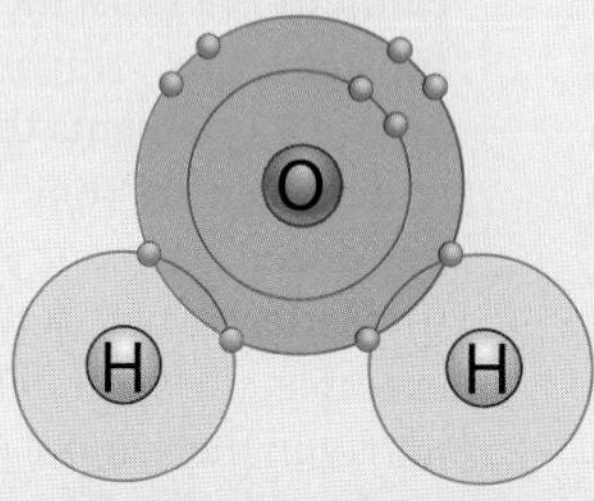
O
H
H

Water's Life-Supporting Properties
(2.10–2.16)

The unique properties of water derive from the polarity and hydrogen bonding of water molecules.

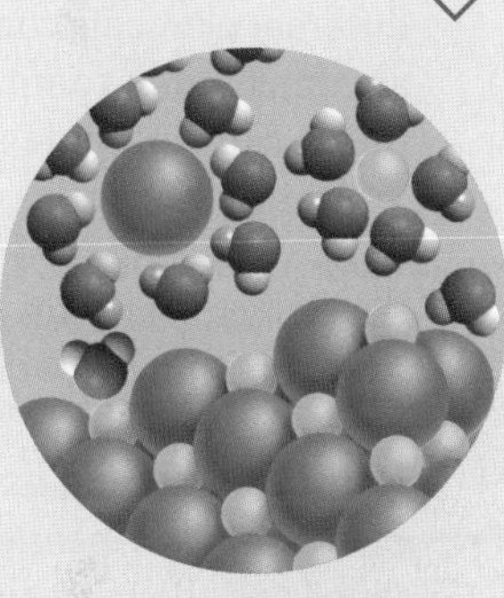

▷ Elements, Atoms, and Compounds

2.1 Organisms are composed of elements, in combinations called compounds

You and all things around you are made of matter—the physical "stuff" of the universe. **Matter** is defined as anything that occupies space and has mass. (In everyday language, we can think of mass as an object's weight.) Matter is found on Earth in three physical states: solid, liquid, and gas. Water is a rare example of matter that exists in the natural environment in all three physical forms: as ice, liquid water, and water vapor.

Types of matter as diverse as water, rocks, air, and biology students are all composed of chemical elements. An **element** is a substance that cannot be broken down to other substances by ordinary chemical means. Today, chemists recognize 92 elements that occur in nature; gold, copper, carbon, and oxygen are some examples. Chemists have also made a few dozen synthetic elements. Each element has a symbol, the first letter or two of its English, Latin, or German name. For example, the symbol for sodium, Na, is from the Latin word *natrium*; the symbol O comes from the English word *oxygen*.

A **compound** is a substance consisting of two or more different elements combined in a fixed ratio. Compounds are much more common than pure elements. In fact, few elements exist in a pure state in nature.

Many compounds consist of only two elements; for instance, table salt (sodium chloride, NaCl) has equal parts of the elements sodium (Na) and chlorine (Cl). Pure sodium is a metal and pure chlorine is a poisonous gas. Chemically combined, however, they form an edible compound **(Figure 2.1)**. Hydrogen (H) and oxygen (O) are elements that typically exist as gases. Chemically combined in a ratio of 2:1, however, they form the most abundant compound on the surface of Earth—water (H_2O). These are simple examples of organized matter having emergent properties: A compound has characteristics different from those of its elements.

Most of the compounds in living organisms contain at least three or four elements. Sugar, for example, is formed of carbon (C), hydrogen, and oxygen. Proteins are compounds containing carbon, hydrogen, oxygen, nitrogen (N), and a small amount of sulfur (S). Different arrangements of the atoms of these elements give rise to the unique properties of each compound.

How many of the 92 natural elements are essential for life? The requirements are similar among organisms, but there is some variation. For example, humans need 25 elements, but plants need only 17. Four elements—oxygen, carbon, hydrogen, and nitrogen—make up about 96% of all living matter. These elements are the main ingredients of biological molecules. As you can see in **Table 2.1**, which lists the 25 elements found in humans, calcium (Ca), phosphorus (P), potassium (K), sulfur, sodium, chlorine, and magnesium (Mg) account for most of the remaining 4% of your body. These elements are involved in such important functions as bone formation (calcium and phosphorus) and nerve signaling (potassium, sodium, calcium, and chlorine).

The **trace elements** listed at the bottom of the table are essential for humans, but only in minute quantities. Some trace elements, such as iron (Fe), are needed by all forms of life. Iron makes up only about 0.004% of your body weight but is vital for energy processing and for transporting oxygen in your blood. Other trace elements are required only by certain species. For example, iodine is an essential element only for vertebrates—animal with backbones, which, of course, includes you. We explore the importance of trace elements to your health next.

TABLE 2.1 | ELEMENTS IN THE HUMAN BODY

Element	Symbol	Percentage of Body Weight (Including Water)	
Oxygen	O	65.0	
Carbon	C	18.5	96.3%
Hydrogen	H	9.5	
Nitrogen	N	3.3	
Calcium	Ca	1.5	
Phosphorus	P	1.0	
Potassium	K	0.4	
Sulfur	S	0.3	3.7%
Sodium	Na	0.2	
Chlorine	Cl	0.2	
Magnesium	Mg	0.1	

Trace elements, less than 0.01% of human body weight: Boron (B), chromium (Cr), cobalt (Co), copper (Cu), fluorine (F), iodine (I), iron (Fe), manganese (Mn), molybdenum (Mo), selenium (Se), silicon (Si), tin (Sn), vanadium (V), zinc (Zn)

? Explain how table salt illustrates the theme of emergent properties.

● The elements that make up the edible crystals of table salt, sodium and chlorine, are in pure form a metal and a poisonous gas.

+
Sodium (Na)
Chlorine (Cl)
Sodium chloride (NaCl)

▲ **Figure 2.1** The emergent properties of the edible compound sodium chloride

2.2 Trace elements are common additives to food and water

CONNECTION

Trace elements are required in very small quantities, but, in some cases, even those small requirements are difficult to fulfill.

Iodine is an essential ingredient of a hormone produced by the thyroid gland, which is located in your neck. You need to ingest only a tiny speck of iodine each day, about 0.15 milligram (mg). An iodine deficiency in the diet causes the thyroid gland to grow to abnormal size, a condition called goiter **(Figure 2.2A)**. The most serious effects of iodine deficiency take place during fetal development and childhood, leading to miscarriages, poor growth, and mental impairment. A global strategy to eliminate iodine deficiency involves universal iodization of all salt used for human and animal consumption. Unfortunately, about 30% of global households still do not have access to iodized salt, and an estimated 2 billion people are still at risk of iodine deficiency. Seafood, kelp, dairy products, and dark, leafy greens are good natural sources. Thus, deficiencies are often found in inland regions, especially in areas where the soil is lacking in iodine. Although most common in developing nations, iodine deficiencies may also result from excessive consumption of highly processed foods (which often use non-iodized salt) and low-salt diets intended to lower the risk of cardiovascular disease.

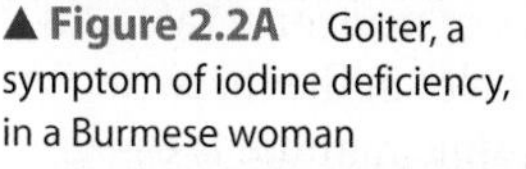

▲ **Figure 2.2A** Goiter, a symptom of iodine deficiency, in a Burmese woman

Iodine is just one example of a trace element added to food or water to improve health. For more than 60 years, the American Dental Association has supported fluoridation of community drinking water as a public health measure. Fluoride is a form of fluorine (F), an element in Earth's crust that is found in small amounts in all water sources. In many communities, fluoride is added during the municipal water treatment process to raise levels to a concentration that can reduce tooth decay. If you mostly drink bottled water, however, your fluoride intake may be reduced, although some bottled water now contains added fluoride. Fluoride is also frequently added to dental products, such as toothpaste and mouthwash **(Figure 2.2B)**.

Chemicals are added to food to help preserve it, make it more nutritious, or simply make it look better. Read the nutrition facts label from the side of the cereal box in **Figure 2.2C** to see a familiar example of how foods are fortified with mineral elements. Iron, for example, is commonly added to foods. (You can actually see that iron has been added to a fortified cereal by crushing the cereal and then stirring a magnet through it.) Also note that the nutrition facts label lists numerous vitamins that are added to improve the nutritional value of the cereal. For instance, the cereal in this example supplies 10% of the recommended daily value for vitamin A. Vitamins consist of more than one element and are examples of compounds.

In the next module, we explore the chemical properties of elements and how the structure of an atom—the smallest unit of an element—determines those properties.

◀ **Figure 2.2B** Mouthwash and toothpaste with added fluoride

? **In addition to iron, what other trace elements are found in the cereal in Figure 2.2C? Does one serving provide the total daily amount needed of these elements?**

● Zinc and copper: one serving provides 100% of the zinc but only 4% of the copper needed in a day.

100% WHOLE GRAIN
Cereal
Crunchy wheat flakes
100% Nutritious
Full of Vitamins and Minerals
NET WT 1 LB (16 OZ) (453G)

Nutrition Facts

Serving Size ¾ cup (30g)
Servings Per Container about 17

Amount Per Serving	Whole Grain Cereal	with ½ cup skim milk
Calories	100	140
Calories from Fat	5	10
		% Daily Value**
Total Fat 0.5g*	**1%**	**1%**
Saturated Fat 0g	**0%**	**0%**
Trans Fat 0g		
Polyunsaturated Fat 0g		
Monounsaturated Fat 0g		
Cholesterol 0mg	**0%**	**1%**
Sodium 135mg	**6%**	**9%**
Potassium 125mg	**4%**	**10%**
Total Carbohydrate 23g	**8%**	**10%**
Dietary Fiber 3g	**10%**	**10%**
Sugars 5g		
Other Carbohydrate 15g		
Protein 2g		
Vitamin A	10%	15%
Vitamin C	100%	100%
Calcium	100%	110%
Iron	100%	100%
Vitamin D	10%	25%
Vitamin E	100%	100%
Thiamin	100%	100%
Riboflavin	100%	110%
Niacin	100%	100%
Vitamin B_6	100%	100%
Folic Acid	100%	100%
Vitamin B_{12}	100%	110%
Pantothenic Acid	100%	100%
Phosphorus	8%	20%
Magnesium	6%	10%
Zinc	100%	100%
Copper	4%	4%

*Amount in cereal. A serving of cereal plus skim milk provides 1g total fat, less than 5mg cholesterol, 260mg sodium, 290mg potassium, 29g total carbohydrate (11g sugars) and 7g protein.

**Percent Daily Values are based on a 2,000 calorie diet. Your daily values may be higher or lower depending on your calorie needs:

	Calories	2,000	2,500
Total Fat	Less than	65g	80g
Sat Fat	Less than	20g	25g
Cholesterol	Less than	300mg	300mg
Sodium	Less than	2,400mg	2,400mg
Potassium		3,500mg	3,500mg
Total Carbohydrate		300g	375g
Dietary Fiber		25g	30g

▲ **Figure 2.2C** Nutrition facts from a fortified cereal

2.3 Atoms consist of protons, neutrons, and electrons

Each element has its own type of atom, which is different from the atoms of other elements. An **atom**, named from a Greek word meaning "indivisible," is the smallest unit of matter that still retains the properties of an element. Atoms are so small that it would take about a million of them to stretch across the period printed at the end of this sentence.

Subatomic Particles Physicists have split the atom into more than a hundred types of subatomic particles. However, only three kinds of particles are relevant here. A **proton** is a subatomic particle with a single positive electrical charge (+). An **electron** is a subatomic particle with a single negative charge (−). A **neutron**, as its name implies, is electrically neutral (has no charge).

Figure 2.3 shows two very simple models of an atom of the element helium (He), the "lighter-than-air" gas that makes balloons rise. Notice that two protons (⊕) and two neutrons (●) are tightly packed in the atom's central core, or **nucleus**. Two electrons (⊖) form a sort of cloud of negative charge around the nucleus. The attraction between the negatively charged electrons and the positively charged protons holds the electrons near the nucleus. The left-hand model shows the two electrons on a circle around the nucleus. The right-hand model, slightly more realistic, shows a spherical cloud of negative charge created by the two rapidly moving electrons. Neither model is drawn to scale. In real atoms, the electrons are very much smaller than the protons and neutrons, and the electron cloud is very much bigger compared to the nucleus. Imagine that this atom was the size of a baseball stadium: The nucleus would be the size of a pea in center field, and the electrons would be like two tiny gnats buzzing around the stadium.

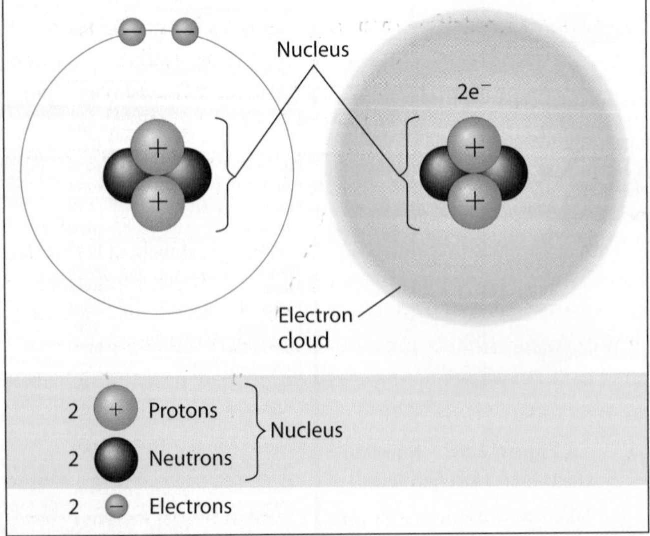

Nucleus
$2e^-$
Electron cloud
2 Protons
2 Neutrons
Nucleus
2 Electrons

▲ **Figure 2.3** Two models of a helium atom. (Note that these models are not to scale; they greatly overestimate the size of the nucleus in relation to the electron cloud.)

Atomic Number and Mass Number So what makes the atoms of different elements different? All atoms of a particular element have the same unique number of protons. This number is the element's **atomic number**. Thus, an atom of helium, with 2 protons, has an atomic number of 2. Unless otherwise indicated, an atom has an equal number of protons and electrons, and thus its net electrical charge is 0 (zero).

What other numbers are associated with an atom? An atom's **mass number** is the sum of the number of protons and neutrons in its nucleus. For helium, the mass number is 4. The mass of a proton and the mass of a neutron are almost identical and are expressed in a unit of measurement called the dalton. Protons and neutrons each have masses close to 1 dalton. An electron has only about 1/2,000 the mass of a proton, so it contributes very little to an atom's mass. Thus, an atom's **atomic mass** (or weight) is approximately equal to its mass number—the sum of its protons and neutrons—in daltons.

Isotopes All atoms of an element have the same atomic number, but some atoms of that element may differ in mass number. The different **isotopes** of an element have the same number of protons and behave identically in chemical reactions, but they have different numbers of neutrons. **Table 2.3** shows the numbers of subatomic particles in the three isotopes of carbon. Note that carbon's atomic number is 6—all of its atoms have 6 protons. Carbon-12 (named for its mass number), with 6 neutrons, accounts for about 99% of naturally occurring carbon. Most of the remaining 1% consists of carbon-13, with a mass number of 13 and thus 7 neutrons. A third isotope, carbon-14, with 8 neutrons, occurs in minute quantities. Of course, all three isotopes have 6 protons—otherwise, they would not be carbon.

Both carbon-12 and carbon-13 are stable isotopes, meaning that their nuclei remain intact more or less forever. The isotope carbon-14, on the other hand, is unstable, or radioactive. A **radioactive isotope** is one in which the nucleus decays spontaneously, giving off particles and energy. Radiation from decaying isotopes can damage cellular molecules and thus can pose serious risks to living organisms. But radioactive isotopes can be helpful, as in their use in dating fossils (see Module 15.5). They are also used in biological research and medicine, as we see next.

TABLE 2.3 | ISOTOPES OF CARBON

	Carbon-12		Carbon-13		Carbon-14	
Protons	6	Mass number 12	6	Mass number 13	6	Mass number 14
Neutrons	6		7		8	
Electrons	6		6		6	

? **A nitrogen atom has 7 protons, and its most common isotope has 7 neutrons. A radioactive isotope of nitrogen has 9 neutrons. What is the atomic number and mass number of this radioactive nitrogen?**

● Atomic number = 7; mass number = 16

2.4 Radioactive isotopes can help or harm us

CONNECTION

Living cells cannot readily distinguish between isotopes of the same element. Consequently, organisms take up and use compounds containing radioactive isotopes in the usual way. Because radioactivity is easily detected and measured by instruments, radioactive isotopes are useful as tracers—biological spies, in effect—for monitoring the fate of atoms in living organisms.

Basic Research Biologists often use radioactive tracers to follow molecules as they undergo chemical changes in an organism. For example, researchers have used carbon dioxide (CO_2) containing the radioactive isotope carbon-14 to study photosynthesis. Using sunlight to power the conversion, plants take in CO_2 from the air and use it to make sugar molecules. Radioactively labeled CO_2 has enabled researchers to trace the sequence of molecules made by plants in the chemical route from CO_2 to sugar.

Medical Diagnosis and Treatment Radioactive isotopes may also be used to tag chemicals that accumulate in specific areas of the body, such as phosphorus in bones. After injection of such a tracer, a special camera produces an image of where the radiation collects. In most diagnostic uses, the patient receives only a tiny amount of an isotope.

Sometimes radioactive isotopes are used for treatment. As you learned in Module 2.2, the body uses iodine to make a thyroid hormone. Because radioactive iodine accumulates in the thyroid, it can be used to kill cancer cells there.

Substances that the body metabolizes, such as glucose or oxygen, may also be labeled with a radioactive isotope. **Figure 2.4A** shows a patient being examined by a PET (positron-emission tomography) scanner, which can produce three-dimensional images of areas of the body with high metabolic activity. PET is useful for diagnosing certain heart disorders and cancers and for basic research on the brain.

The early detection of Alzheimer's disease may be a new use for such techniques. This devastating illness gradually destroys a person's memory and ability to think. As the disease progresses, the brain becomes riddled with deposits (plaques) of a protein called beta-amyloid. Researchers have identified a protein molecule called PIB that binds to beta-amyloid. PIB contains a radioactive isotope that can be detected on a PET scan. **Figure 2.4B** shows PET images of the brains of a healthy person (left) and a person with Alzheimer's (right) injected with PIB. Notice that the brain of the Alzheimer's patient has high levels of PIB (red and yellow areas), whereas the unaffected person's brain has lower levels (blue). New therapies are focused on limiting the production of beta-amyloid or clearing it from the brain. A diagnostic test using PIB would allow researchers to monitor the effectiveness of new drugs in people living with the disease.

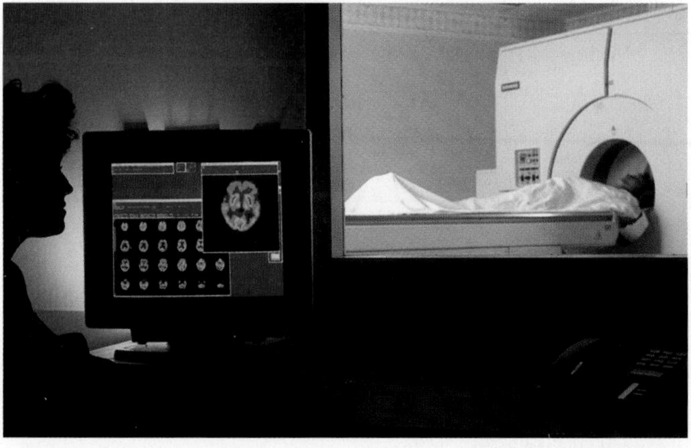

▲ **Figure 2.4A** Technician monitoring the output of a PET scanner

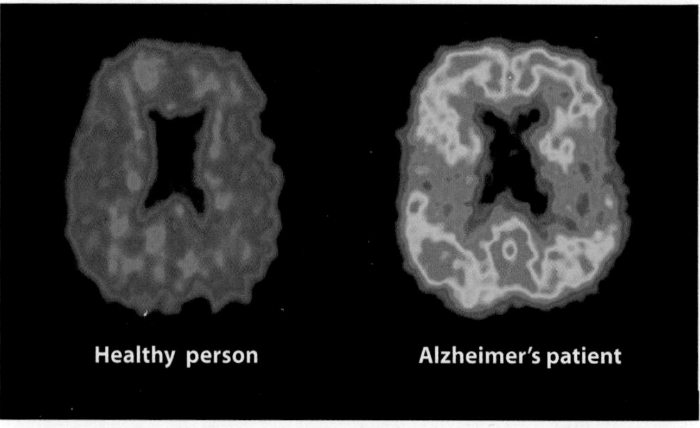
Healthy person
Alzheimer's patient

▲ **Figure 2.4B** PET images of brains of a healthy person (left) and a person with Alzheimer's disease (right). Red and yellow colors indicate high levels of PIB bound to beta-amyloid plaques.

Dangers Although radioactive isotopes have many beneficial uses, uncontrolled exposure to them can harm living organisms by damaging molecules, especially DNA. The particles and energy thrown off by radioactive atoms can break chemical bonds and also cause abnormal bonds to form. The explosion of a nuclear reactor in Chernobyl, Ukraine, in 1986 released large amounts of radioactive isotopes into the environment, which drifted over large areas of Russia, Belarus, and Europe. A few dozen people died from acute radiation poisoning, and more than 100,000 people were evacuated from the immediate area. Increased rates of thyroid cancer in children exposed to the radiation have been reported. Likewise, scientists will carefully monitor the long-term health consequences of the 2011 post-tsunami Fukushima nuclear disaster in Japan.

Natural sources of radiation can also pose a threat. Radon, a radioactive gas, may be a cause of lung cancer. Radon can contaminate buildings in regions where underlying rocks naturally contain uranium, a radioactive element. Homeowners can buy a radon detector or hire a company to test their home to ensure that radon levels are safe. If levels are found to be unsafe, technology exists to remove radon from homes.

? **Why are radioactive isotopes useful as tracers in research on the chemistry of life?**

● Organisms incorporate radioactive isotopes of an element into their molecules, and researchers can use special scanning devices to detect the presence of these isotopes in biological pathways or locations in the body.

Chemical Bonds

2.5 The distribution of electrons determines an atom's chemical properties

To understand how atoms interact with each other, we need to explore atomic structure further. Of the three subatomic particles—protons, neutrons, and electrons—only electrons are directly involved in the chemical activity of an atom.

If you glance back to the model of the helium atom in Figure 2.3, you see that its 2 electrons are shown together on a circle around the nucleus. But where should the electrons be shown in an atom with more than 2 electrons—say, in carbon, whose atomic number is 6? As you see in **Figure 2.5A**, 2 electrons are still shown on an inner circle, but the next 4 are placed on a larger outside circle. It turns out that electrons can be located in different **electron shells**, each with a characteristic distance from the nucleus. Depending on an element's atomic number, an atom may have one, two, or more electron shells.

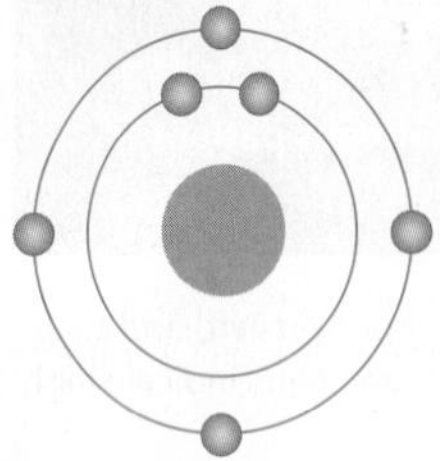

▲ **Figure 2.5A** An electron distribution model of carbon

Figure 2.5B is an abbreviated version of the periodic table of the elements (see Appendix 2 for the complete table). The figure shows the distribution of electrons for the first 18 elements, arranged in rows according to the number of electron shells (one, two, or three). Within each shell, electrons travel in different *orbitals*, which are discrete volumes of space in which electrons are most likely to be found. Each orbital can hold a maximum of 2 electrons. The first electron shell has only one orbital and can hold only 2 electrons. Thus, hydrogen and helium are the only elements in the first row. For the second and third rows, the outer shell has four orbitals and can hold up to 8 electrons (four pairs).

It is the number of electrons present in the outermost shell, called the valence shell, that determines the chemical properties of an atom. Atoms whose outer shells are not full tend to interact with other atoms in ways that enable them to complete or fill their valence shells.

Look at the electron shells of the atoms of the four elements that are the main components of biological molecules (highlighted in green in Figure 2.5B). Because their outer shells are incomplete, all these atoms react readily with other atoms. The hydrogen atom has only 1 electron in its single electron shell, which can accommodate 2 electrons. Atoms of carbon, nitrogen, and oxygen also have unpaired electrons and incomplete shells. In contrast, the helium atom has a first-level shell that is full with 2 electrons. Neon and argon also have full outer shells. As a result, these elements are chemically inert (unreactive).

How do chemical interactions between atoms enable them to fill their outer electron shells? When two atoms with incomplete outer shells react, each atom will share, donate, or receive electrons, so that both partners end up with completed outer shells. These interactions usually result in atoms staying close together, held by attractions known as **chemical bonds**. In the next two modules, we look at two important types of chemical bonds.

? **How many electrons and electron shells does a sodium atom have? How many electrons are in its valence shell?**

● 11 electrons; 3 electron shells; 1 electron in the outer shell

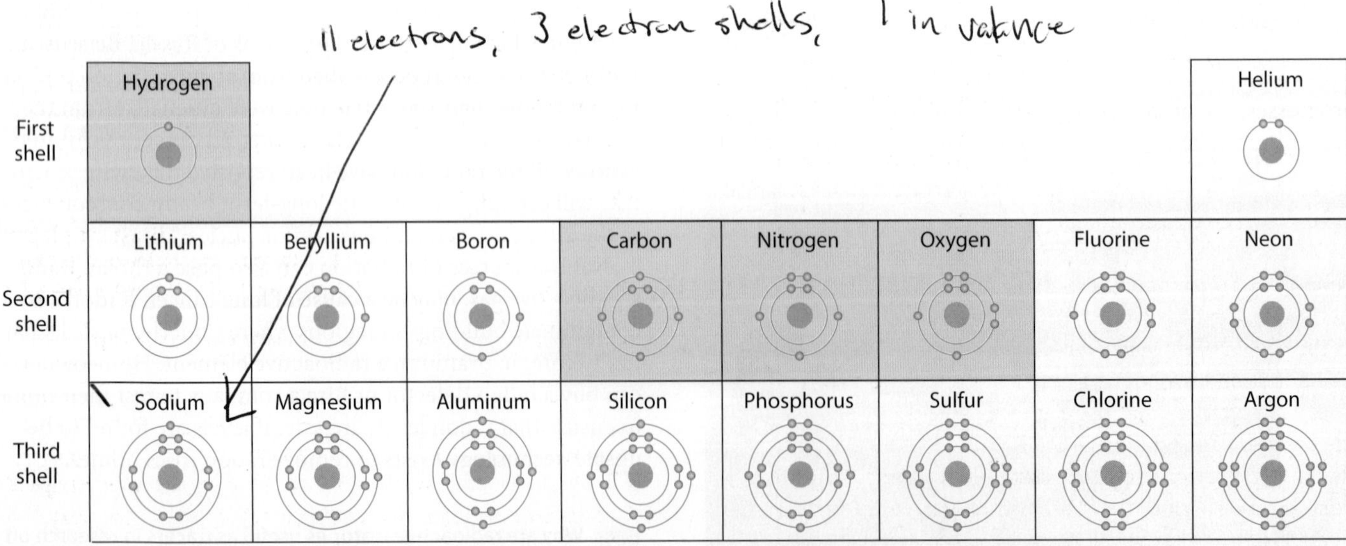

First shell
Second shell
Third shell
Hydrogen
Helium
Lithium
Beryllium
Boron
Carbon
Nitrogen
Oxygen
Fluorine
Neon
Sodium
Magnesium
Aluminum
Silicon
Phosphorus
Sulfur
Chlorine
Argon

▲ **Figure 2.5B** The electron distribution diagrams of the first 18 elements in the periodic table

Try This As you read from left to right across each row, describe how the number of electrons changes. Note that the electrons don't pair up until all orbitals have at least one electron.

2.6 Covalent bonds join atoms into molecules through electron sharing

In a **covalent bond**, two atoms, each with an unpaired electron in its outer shell, actually *share* a pair of electrons. Sharing one or more pairs of electrons enables atoms to complete their outer shells. Atoms held together by covalent bonds form a **molecule**. For example, a covalent bond connects two hydrogen atoms in a molecule of the gas H_2, and a covalent bond connects each of two hydrogen atoms to an oxygen atom in a molecule of water (H_2O).

How many covalent bonds can an atom form? It depends on the number of additional electrons needed to fill its valence shell. This number is called the valence, or bonding capacity, of an atom. Look back at the electron distribution diagrams in Figure 2.5B and see if you can determine how many covalent bonds hydrogen, oxygen, nitrogen, and carbon can form.

Figure 2.6 shows how molecules can be represented in several different ways. Let's see what we can learn from this figure. As you will notice in the electron distribution diagram, the hydrogen atoms in H_2 are held together by a pair of shared electrons. But two atoms can share more than just one pair of electrons. In an oxygen molecule (O_2), for example, the two oxygen atoms share two pairs of electrons, forming a double bond. A double bond is indicated in a structural formula by a pair of lines.

H_2 and O_2 are molecules composed of only one element. Methane (CH_4) and water (H_2O) are compounds. Methane is a major component of natural gas. As shown in Figure 2.6, it takes four hydrogen atoms to satisfy carbon's valence of 4.

Atoms in a molecule are in a constant tug-of-war for the shared electrons of their covalent bonds. An atom's attraction for shared electrons is called its **electronegativity**. The more electronegative an atom, the more strongly it pulls shared electrons toward its nucleus. In molecules of only one element, such as H_2 and O_2, the two identical atoms exert an equal pull on the electrons. The bonds in such molecules are said to be **nonpolar covalent bonds** because the electrons are shared equally between the atoms. Compounds such as methane also have nonpolar bonds, because the atoms of carbon and hydrogen are not substantially different in electronegativity.

Water, on the other hand, is composed of atoms with quite different electronegativities. Oxygen is one of the most electronegative of the elements. As indicated by the arrows in the blowup of a water molecule in Figure 2.6, oxygen attracts the shared electrons in H_2O much more strongly than does hydrogen, so that the electrons spend more time near the oxygen atom than near the hydrogen atoms. This unequal sharing of electrons produces a **polar covalent bond**. In a polar covalent bond, the pulling of shared, negatively charged electrons closer to the more electronegative atom makes that atom partially negative and the other atom partially positive. Thus, in H_2O, the oxygen atom actually has a slight negative charge and each hydrogen atom a slight positive charge.

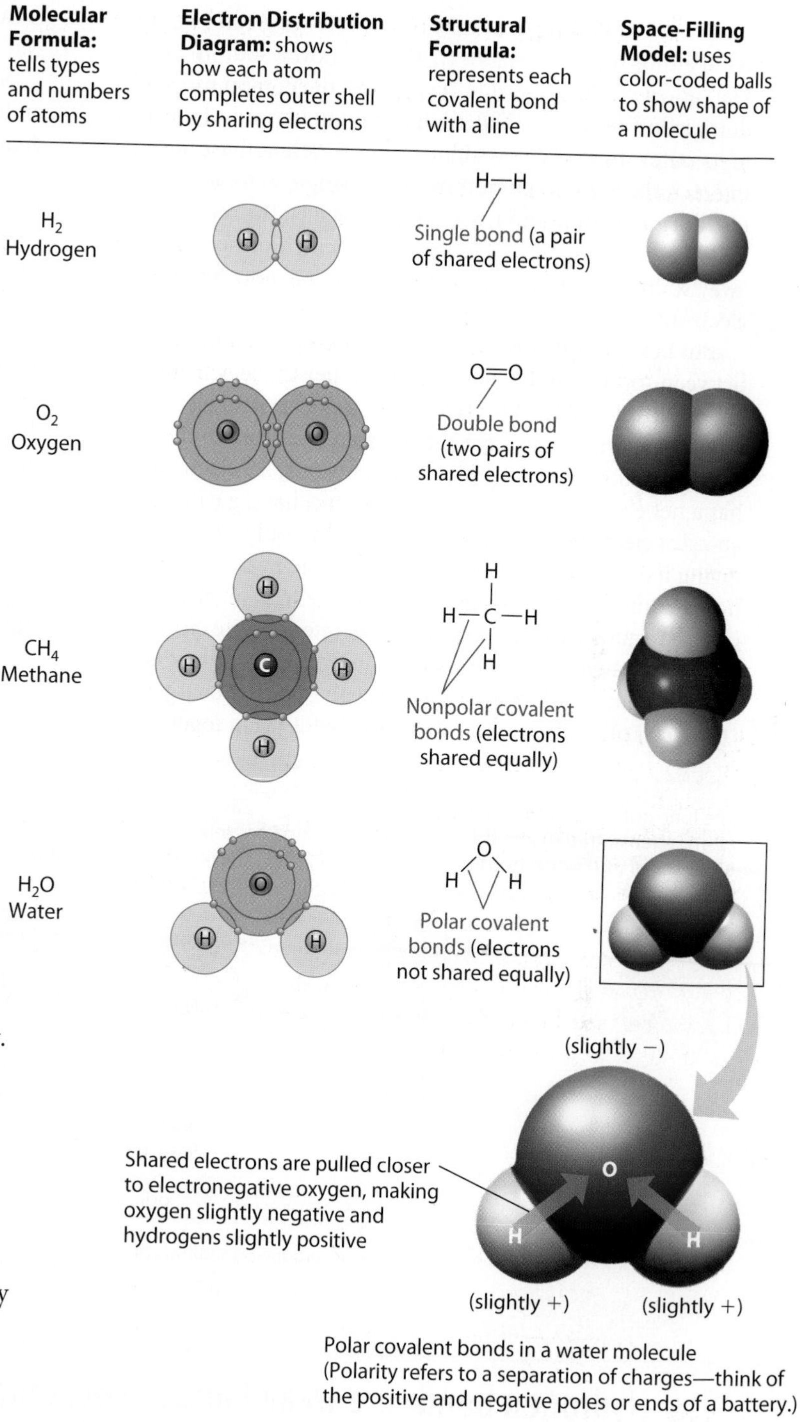

Molecular Formula: tells types and numbers of atoms	**Electron Distribution Diagram:** shows how each atom completes outer shell by sharing electrons	**Structural Formula:** represents each covalent bond with a line	**Space-Filling Model:** uses color-coded balls to show shape of a molecule
H_2 Hydrogen	H H	H—H Single bond (a pair of shared electrons)	
O_2 Oxygen	O O	O=O Double bond (two pairs of shared electrons)	
CH_4 Methane	H H C H H	H–C(H)(H)–H Nonpolar covalent bonds (electrons shared equally)	
H_2O Water	O H H	H–O–H Polar covalent bonds (electrons not shared equally)	

(slightly −)

Shared electrons are pulled closer to electronegative oxygen, making oxygen slightly negative and hydrogens slightly positive

O H H

(slightly +) (slightly +)

Polar covalent bonds in a water molecule (Polarity refers to a separation of charges—think of the positive and negative poles or ends of a battery.)

▲ **Figure 2.6** Alternative ways to represent four common molecules

In some cases, two atoms are so unequal in their attraction for electrons that the more electronegative atom strips an electron completely away from its partner, as we see next.

? **What is chemically nonsensical about this structure?**

H—C=C—H

● Each C has only three bonds instead of the four required by its valence.

2.7 Ionic bonds are attractions between ions of opposite charge

Table salt is an example of how the transfer of electrons can bond atoms together. **Figure 2.7A** shows how a sodium atom (Na) and a chlorine atom (Cl) can form the compound sodium chloride (NaCl). Notice that sodium has only 1 electron in its outer shell, whereas chlorine has 7. When these atoms interact, the sodium atom transfers its single outer electron to chlorine. Sodium now has only two shells, the second shell having a full set of 8 electrons. When chlorine strips away sodium's electron, its own outer shell is now full with 8 electrons.

But how does this electron transfer result in an ionic bond between sodium and chlorine? Remember that electrons are negatively charged particles. The transfer of an electron moves one unit of negative charge from one atom to the other. Sodium, with 11 protons but now only 10 electrons, has a net electrical charge of 1+. Chlorine, having gained an extra electron, now has 18 electrons but only 17 protons, giving it a net electrical charge of 1−. In each case, an atom has become an **ion**—an atom or molecule with an electrical charge resulting from a gain or loss of one or more electrons. (Note that the names of negatively charged ions often end in *-ide*, such as *chloride*.) Two ions with opposite charges attract each other. When the attraction holds them together, it is called an **ionic bond**. The resulting compound, in this case NaCl, is electrically neutral.

Sodium chloride is a familiar type of **salt**, a synonym for an ionic compound. Salts often exist as crystals in nature. **Figure 2.7B** shows the ions Na^+ and Cl^- in a crystal of sodium chloride. An NaCl crystal can be of any size (there is no fixed number of ions), but sodium and chloride ions are always present in a 1:1 ratio. The ratio of ions can differ in the various kinds of salts.

The environment affects the strength of ionic bonds. In a dry salt crystal, the bonds are so strong that it takes a hammer and chisel to break enough of them to crack the crystal. If the same salt crystal is placed in water, however, the ionic bonds break when the ions interact with water molecules and the salt dissolves, as we'll discuss in Module 2.13. Most drugs are manufactured as salts because they are quite stable when dry but can dissolve easily in water.

? **Explain what holds together the ions in a crystal of table salt (NaCl).**

● Opposite charges attract. The positively charged sodium ions (Na^+) and the negatively charged chloride ions (Cl^-) are held together by ionic bonds, attractions between oppositely charged ions.

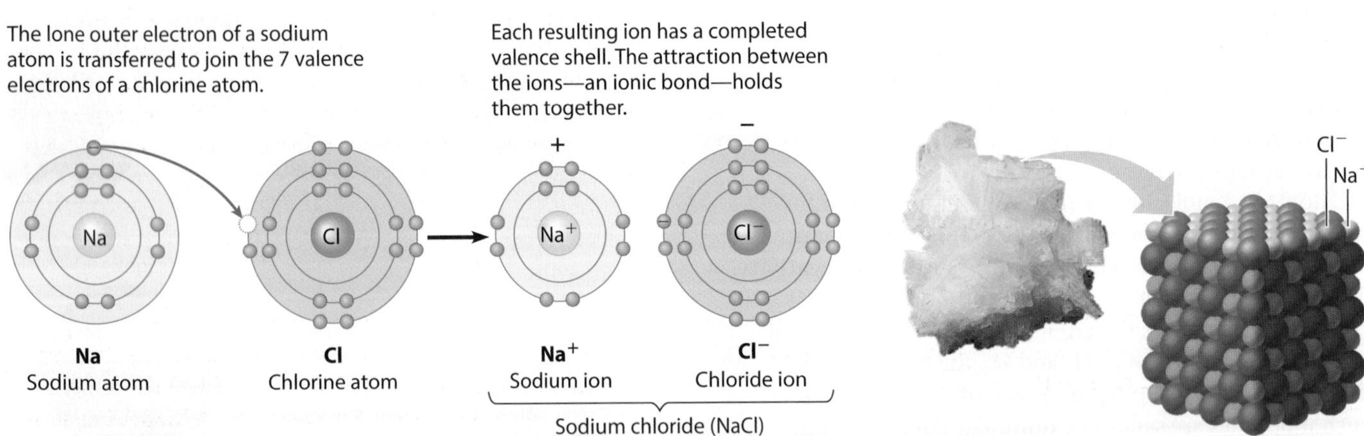
The lone outer electron of a sodium atom is transferred to join the 7 valence electrons of a chlorine atom.
Each resulting ion has a completed valence shell. The attraction between the ions—an ionic bond—holds them together.
Na
Cl
Na+
Cl−
+
−
Na
Sodium atom
Cl
Chlorine atom
Na+
Sodium ion
Cl−
Chloride ion
Sodium chloride (NaCl)
Cl−
Na+

▲ **Figure 2.7A** Formation of an ionic bond, producing sodium chloride

▲ **Figure 2.7B** A crystal of sodium chloride

2.8 Hydrogen bonds are weak bonds important in the chemistry of life

In living organisms, most of the strong chemical bonds are covalent, linking atoms to form a cell's molecules. But crucial to the functioning of a cell are weaker bonds within and between molecules, such as the ionic bonds we just discussed. One of the most important types of weak bonds is the **hydrogen bond**, which is best illustrated with water molecules.

As you saw in Figure 2.6, the hydrogen atoms of a water molecule are attached to oxygen by polar covalent bonds. Because of these polar bonds and the wide V shape of the molecule, water is a **polar molecule**—that is, it has an unequal distribution of charges. It is slightly negative at the oxygen end of the molecule (the point of the V) and slightly positive at each of the two hydrogen ends. This partial positive charge allows each hydrogen to be attracted to—in a sense, to "flirt" with—a nearby atom (often an oxygen or nitrogen) that has a partial negative charge.

Figure 2.8, on the next page, illustrates how these weak bonds form between water molecules. They are called

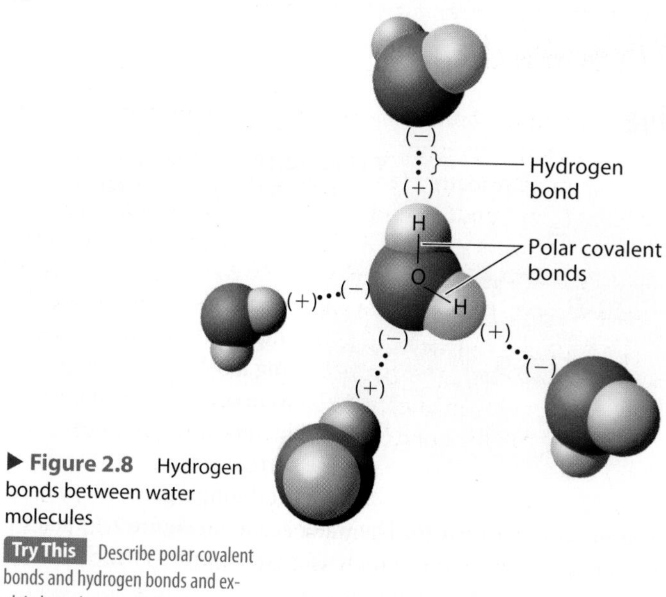

(−)
(+)
Hydrogen bond
H
O
H
Polar covalent bonds
(+)
(−)
(−)
(+)
(+)
(−)

▶ **Figure 2.8** Hydrogen bonds between water molecules

Try This Describe polar covalent bonds and hydrogen bonds and explain how they are related.

hydrogen bonds because the positively charged atom in this type of attraction is always a hydrogen atom. As Figure 2.8 shows, each hydrogen atom of a water molecule can form a hydrogen bond (depicted by dotted lines) with a nearby partially negative oxygen atom of another water molecule. And the negative (oxygen) pole of a water molecule can form hydrogen bonds to two hydrogen atoms. Thus, each water molecule can hydrogen-bond to as many as four partners.

You will learn later how hydrogen bonds help to create a protein's shape (and thus its function) and hold the two strands of a DNA molecule together (see Chapter 3). Later in this chapter, we explore how water's polarity and hydrogen bonds give it unique, life-supporting properties. But first we discuss how the making and breaking of bonds change the composition of matter.

? **What enables neighboring water molecules to hydrogen-bond to one another?**

● The molecules are polar, with each positive end (hydrogen end) of one molecule attracted to the negative end (oxygen end) of another molecule.

2.9 Chemical reactions make and break chemical bonds

Your cells are constantly rearranging molecules in **chemical reactions**—breaking existing chemical bonds and forming new ones. A simple example of a chemical reaction is the reaction between hydrogen gas and oxygen gas that forms water (this is an explosive reaction, which, fortunately, does not occur in your cells):

$$2\,H_2 + O_2 \rightarrow 2\,H_2O$$

In this case, two molecules of hydrogen ($2\,H_2$) react with one molecule of oxygen (O_2) to produce two molecules of water ($2\,H_2O$). The arrow in the equation indicates the conversion of the starting materials, called the **reactants**, to the **product**, the material resulting from the chemical reaction. Notice that the same *numbers* of hydrogen and oxygen atoms appear on the left and right sides of the arrow, although they are grouped differently. Chemical reactions do not create or destroy matter; they only rearrange it in various ways. As shown in **Figure 2.9**, the covalent bonds (represented here as white "sticks" between atoms) holding hydrogen atoms together in H_2 and holding oxygen atoms together in O_2 are broken, and new bonds are formed to yield the H_2O product molecules.

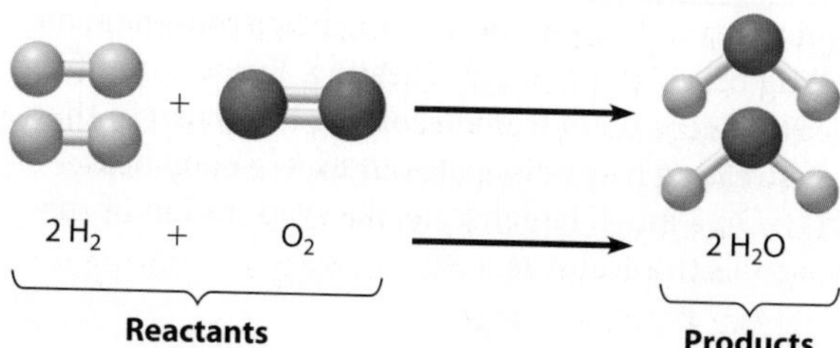

+
$2\,H_2$
+
O_2
$2\,H_2O$
Reactants
Products

▲ **Figure 2.9** Breaking and making of bonds in a chemical reaction

Organisms cannot make water from H_2 and O_2, but they do carry out a great number of chemical reactions that rearrange matter in significant ways. Let's examine a chemical reaction that is essential to life on Earth: photosynthesis. The raw materials of photosynthesis are carbon dioxide (CO_2), which is taken from the air, and water (H_2O), which plants absorb from the soil. Within green plant cells, sunlight powers the conversion of these reactants to the sugar product glucose ($C_6H_{12}O_6$) and oxygen (O_2), a by-product that the plant releases into the air. The following chemical shorthand summarizes the process:

$$6\,CO_2 + 6\,H_2O \rightarrow C_6H_{12}O_6 + 6\,O_2$$

Although photosynthesis is actually a sequence of many chemical reactions, we see that we end up with the same number and kinds of atoms we started with. Matter has simply been rearranged, with an input of energy provided by sunlight.

Your body routinely carries out thousands of chemical reactions. These reactions take place in the watery environment of your cells. We look at the life-supporting properties of water next.

? **Fill in the blanks with the correct numbers in the following chemical process:**

$$C_6H_{12}O_6 + \underline{6}\,O_2 \rightarrow \underline{6}\,CO_2 + \underline{6}\,H_2O$$

What process do you think this reaction represents? (*Hint:* Think about how your cells use these reactants to produce energy.)

● $C_6H_{12}O_6 + 6\,O_2 \rightarrow 6\,CO_2 + 6\,H_2O$; the breakdown of sugar in the presence of oxygen to carbon dioxide and water, with the release of energy that the cell can use

Water's Life-Supporting Properties

2.10 Hydrogen bonds make liquid water cohesive

We can trace water's life-supporting properties to the structure and interactions of its molecules—their polarity and resulting hydrogen bonding between molecules (review Figure 2.8).

Hydrogen bonds between molecules of liquid water last for only a few trillionths of a second, yet at any instant, many molecules are hydrogen-bonded to others. This tendency of molecules of the same kind to stick together, called **cohesion**, is much stronger for water than for most other liquids. The cohesion of water is important in the living world. Trees, for example, depend on cohesion to help transport water and nutrients from their roots to their leaves. The evaporation of water from a leaf exerts a pulling force on water within the veins of the leaf. Because of cohesion, the force is relayed all the way down to the roots. **Adhesion**, the clinging of one substance to another, also plays a role. The adhesion of water to the cell walls of a plant's thin veins helps counter the downward pull of gravity.

Related to cohesion is **surface tension**, a measure of how difficult it is to stretch or break the surface of a liquid. Hydrogen bonds give water unusually high surface tension, making it behave as though it were coated with an invisible film. You can observe the surface tension of water by slightly overfilling a glass; the water will stand above the rim. The water strider in **Figure 2.10** takes advantage of the high surface tension of water to "stride" across ponds without breaking the surface.

▼ **Figure 2.10** Surface tension allowing a water strider to walk on water

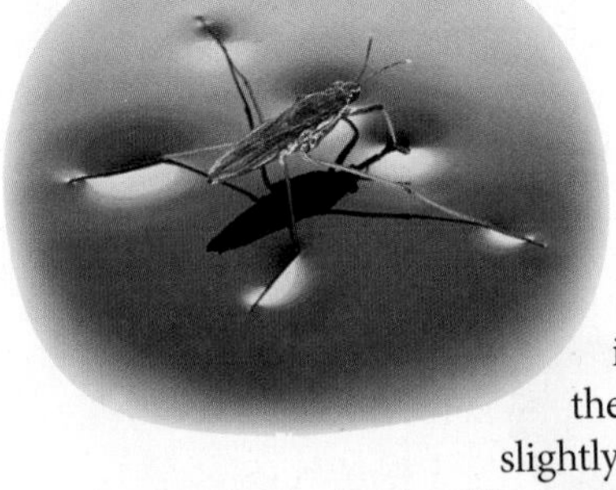

? After a hard workout, you may notice "beads" of sweat on your face. Can you explain what holds the sweat in droplet form?

● The cohesion of water molecules and its high surface tension hold water in droplets. The adhesion of water to your skin helps hold the beads in place.

2.11 Water's hydrogen bonds moderate temperature

Thermal energy is the energy associated with the random movement of atoms and molecules. Thermal energy in transfer from a warmer to a cooler body of matter is defined as **heat**. **Temperature** measures the intensity of heat—that is, the *average* speed of molecules in a body of matter. If you have ever burned your finger on a metal pot while waiting for the water in it to boil, you know that water heats up much more slowly than metal. In fact, because of hydrogen bonding, water has a stronger resistance to temperature change than most other substances.

▲ **Figure 2.11** Sweating as a mechanism of evaporative cooling

Heat must be absorbed to break hydrogen bonds, and heat is released when hydrogen bonds form. To raise the temperature of water, hydrogen bonds between water molecules must be broken before the molecules can move faster. Thus, water absorbs a large amount of heat (much of it used to disrupt hydrogen bonds) while warming up only a few degrees. Conversely, when water cools, water molecules slow down and more hydrogen bonds form, releasing a considerable amount of heat.

Earth's giant water supply moderates temperatures, helping to keep temperatures within limits that permit life. Oceans, lakes, and rivers store a huge amount of heat from the sun during warm periods. Heat given off from gradually cooling water warms the air. That's why coastal areas generally have milder climates than inland regions. Water's resistance to temperature change also stabilizes ocean temperatures, creating a favorable environment for marine life. Because water accounts for approximately 66% of your body weight, it also helps moderate your temperature.

When a substance evaporates (changes physical state from a liquid to a gas), the surface of the liquid that remains behind cools down. This **evaporative cooling** occurs because the molecules with the greatest energy (the "hottest" ones) leave. It's as if the 10 fastest runners on the track team left school, lowering the average speed of the remaining team. Evaporative cooling helps prevent some land-dwelling organisms from overheating. Evaporation from a plant's leaves keeps them from becoming too warm in the sun, just as sweating helps dissipate our excess body heat (**Figure 2.11**). On a much larger scale, the evaporation of surface waters cools tropical seas.

? Explain the popular adage "It's not the heat, it's the humidity."

● High humidity hampers cooling by slowing the evaporation of sweat.

2.12 Ice floats because it is less dense than liquid water

Water exists on Earth in three forms: gas (water vapor), liquid, and solid. Unlike most substances, water is less dense as a solid than as a liquid. As you might guess, this unusual property is due to hydrogen bonds.

As water freezes, each molecule forms stable hydrogen bonds with its neighbors, holding them at "arm's length" and creating a three-dimensional crystal. In **Figure 2.12**, compare the spaciously arranged molecules in the ice crystal with the more tightly packed molecules in the liquid water. The ice crystal has fewer molecules than an equal volume of liquid water. Therefore, ice is less dense and floats on top of liquid water.

If ice sank, then eventually ponds, lakes, and even oceans would freeze solid. Instead, when a body of water cools, the floating ice insulates the water below from colder air above. This "blanket" of ice prevents the water from freezing and allows fish and other aquatic forms of life to survive under the frozen surface.

In the Arctic, this frozen surface serves as the winter hunting ground for polar bears (Figure 2.12). The shrinking of this ice cover as a result of global climate change may doom these bears.

? **Explain how freezing water can crack boulders.**

● Water in the crevices of a boulder expands as it freezes because the water molecules become spaced farther apart in forming ice crystals, which can crack the rock.

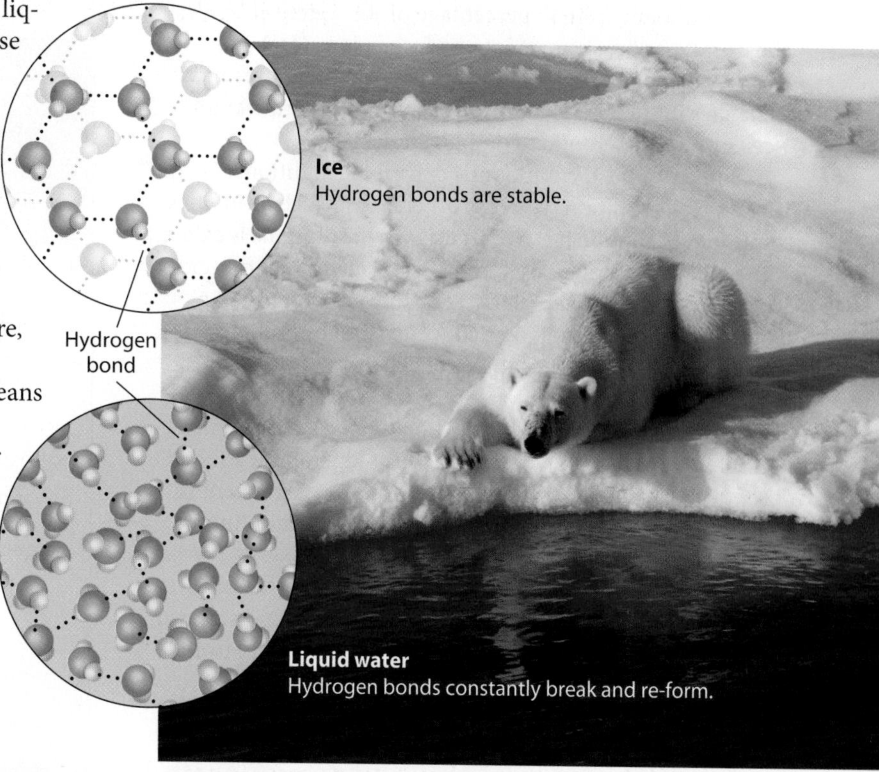

Ice
Hydrogen bonds are stable.
Hydrogen bond
Liquid water
Hydrogen bonds constantly break and re-form.

▲ **Figure 2.12** Hydrogen bonds between water molecules in ice and water

2.13 Water is the solvent of life

If you add a teaspoon of table salt to a glass of water, the salt will eventually dissolve, forming a solution. A **solution** is a liquid consisting of a uniform mixture of two or more substances. The dissolving agent (in our example, water) is the **solvent**, and a substance that is dissolved (in this case, salt) is a **solute**. An **aqueous solution** (from the Latin *aqua*, water) is one in which water is the solvent.

Water's versatility as a solvent results from the polarity of its molecules. **Figure 2.13** shows how a teaspoon of salt dissolves in water. At the surface of each grain, or crystal, the sodium and chloride ions are exposed to water. These ions and the water molecules are attracted to each other due to their opposite charges. The oxygen ends of the water molecules have a partial negative charge and cling to the positive sodium ions (Na^+). The hydrogen ends of the water molecules, with their partial positive charge, are attracted to the negative chloride ions (Cl^-). Working inward from the surface of each salt crystal, water molecules eventually surround and separate all the ions. Water dissolves other ionic compounds as well. Seawater, for instance, contains a great variety of dissolved ions, as do your cells.

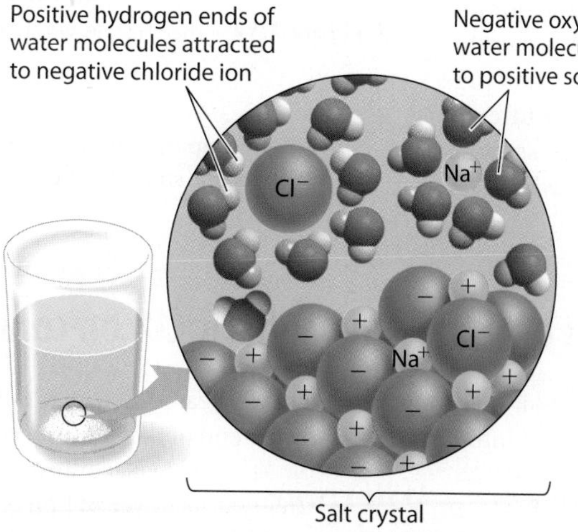

Positive hydrogen ends of water molecules attracted to negative chloride ion

Negative oxygen ends of water molecules attracted to positive sodium ion

Salt crystal

▲ **Figure 2.13** A crystal of salt (NaCl) dissolving in water

A compound doesn't need to be ionic to dissolve in water. A spoonful of sugar will also dissolve in a glass of water. Polar molecules such as sugar dissolve as water molecules surround them and form hydrogen bonds with their polar regions. Even large molecules, such as proteins, can dissolve if they have ionic or polar regions on their surface. As the solvent inside all cells, in blood, and in plant sap, water dissolves an enormous variety of solutes necessary for life.

? **Why are blood and most other biological fluids classified as aqueous solutions?**

● The solvent in these fluids is water.

2.14 The chemistry of life is sensitive to acidic and basic conditions

In liquid water, a small percentage of the water molecules dissociate or break apart into hydrogen ions (H^+) and hydroxide ions (OH^-). These ions are very reactive, and changes in their concentrations can drastically affect a cell's proteins and other complex molecules.

Some chemical compounds contribute additional H^+ to an aqueous solution, whereas others remove H^+ from it. A substance that donates hydrogen ions to solutions is called an **acid**. An example of a strong acid is hydrochloric acid (HCl), the acid in the gastric juice in your stomach. An acidic solution has a higher concentration of H^+ than OH^-.

A **base** is a substance that reduces the hydrogen ion concentration of a solution. Some bases, such as sodium hydroxide (NaOH), do this by donating OH^-; the OH^- combines with H^+ to form H_2O, thus reducing the H^+ concentration. Sodium hydroxide is a common ingredient in oven cleaners. Other bases accept H^+ ions from solution, resulting in a higher OH^- concentration.

We use the **pH scale** to describe how acidic or basic a solution is (pH stands for potential of hydrogen). As shown in **Figure 2.14**, the scale ranges from 0 (most acidic) to 14 (most basic). Each pH unit represents a 10-fold change in the concentration of H^+ in a solution. For example, lemon juice at pH 2 has 10 times more H^+ than an equal amount of a cola at pH 3 and 100 times more H^+ than tomato juice at pH 4.

Pure water and aqueous solutions that are neither acidic nor basic are said to be neutral; they have a pH of 7, and the concentrations of H^+ and OH^- are equal. The pH inside most cells is close to 7.

The pH of human blood plasma (the fluid portion of the blood) is very close to 7.4. A person cannot survive for more than a few minutes if the blood pH drops to 7.0 or rises to 7.8. How can your body maintain a relatively constant pH in your cells and blood? Biological fluids contain **buffers**, substances that minimize changes in pH. They do so by accepting H^+ when it is in excess and donating H^+ when it is depleted.

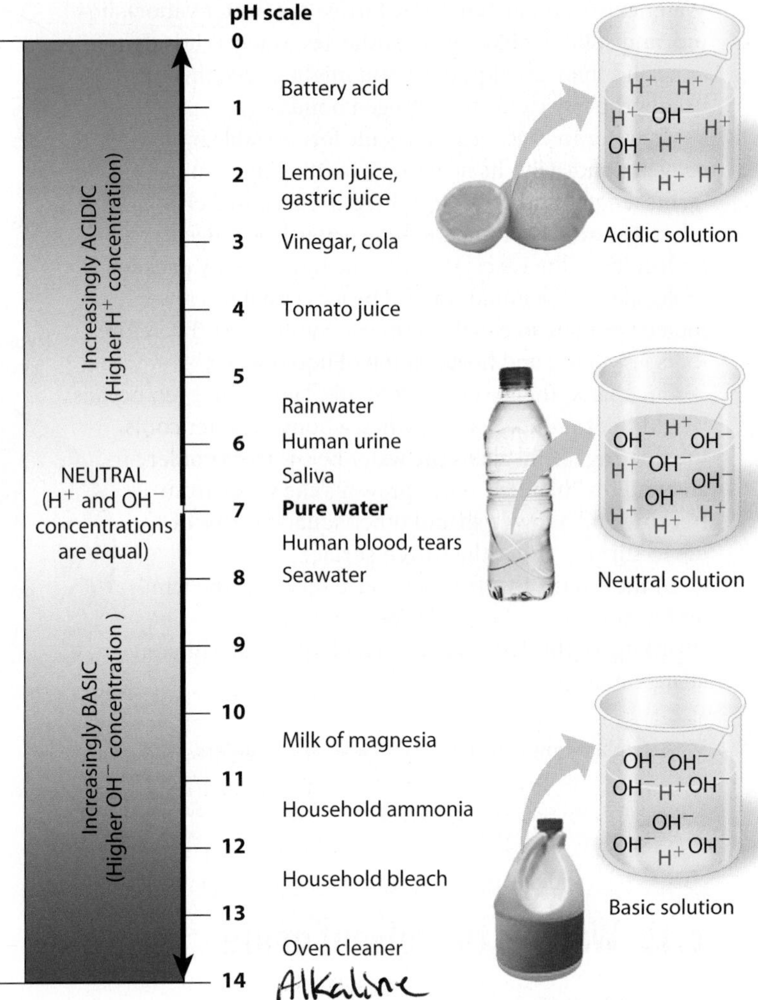
pH scale
0
Battery acid
1
2
Lemon juice, gastric juice
3
Vinegar, cola
4
Tomato juice
5
Rainwater
6
Human urine
Saliva
7
Pure water
Human blood, tears
8
Seawater
9
10
Milk of magnesia
11
Household ammonia
12
Household bleach
13
Oven cleaner
14
Increasingly ACIDIC (Higher H+ concentration)
NEUTRAL (H+ and OH− concentrations are equal)
Increasingly BASIC (Higher OH− concentration)
Acidic solution
Neutral solution
Basic solution

▲ **Figure 2.14** The pH scale, which reflects the relative concentrations of H^+ and OH^-

? **Compared to a basic solution at pH 9, the same volume of an acidic solution at pH 4 has ______ times more H^+.**

100,000

2.15 Scientists study the effects of rising atmospheric CO_2 on coral reef ecosystems

SCIENTIFIC THINKING

Will rising atmospheric CO_2 harm coral reefs?

Carbon dioxide is the main product of fossil fuel combustion, and its steadily increasing release into the atmosphere is linked to global climate change (see Module 38.4). About 25% of this CO_2 is absorbed by the oceans—and this naturally occurring remedy to excess CO_2 would seem to be a good thing. However, as CO_2 levels on the planet continue to rise, the increasing absorption of CO_2 is expected to change ocean chemistry and harm marine life and ecosystems.

In **ocean acidification**, CO_2 dissolving in seawater lowers the pH of the ocean. Recent studies estimate that the pH of the oceans is 0.1 pH unit lower now than at any time in the past 420,000 years and may drop from the current level of 8.1 to 7.8 by the end of this century. How will this affect marine organisms?

Several studies investigating the impact of a lower pH on coral reef ecosystems have looked at the process called calcification, in which coral animals combine calcium and carbonate ions to form their calcium carbonate skeletons. As seawater acidifies, the extra hydrogen ions (H^+) combine with carbonate ions (CO_3^{2-}) to form bicarbonate ions (HCO_3^-). This reaction reduces the carbonate ion concentration available to corals and other shell-building animals. Scientists predict that ocean acidification will cause the carbonate ion concentration to decrease by 40% by the year 2100.

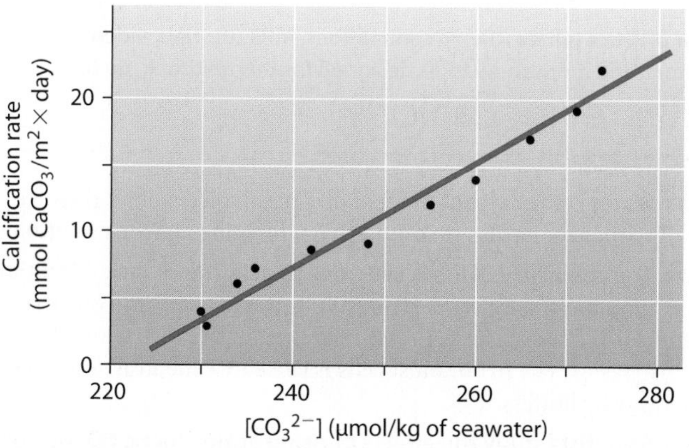

Calcification rate
(mmol CaCO$_3$/m^2 × day)
20
10
0
220
240
260
280
[CO$_3^{2-}$] (μmol/kg of seawater)

Source: Adaptation of figure 5 from "Effect of Calcium Carbonate Saturation State on the Calcification Rate of an Experimental Coral Reef" by C. Langdon, et al., from *Global Biogeochemical Cycles*, June 2000, Volume 14(2). Copyright © 2000 by American Geophysical Union. Reprinted with permission of Wiley Inc.

▲ **Figure 2.15A** The effect of carbonate ion concentration on calcification rate in an artificial coral reef system. (The independent variable shown on the *x* axis is the concentration of carbonate ions, which the researchers manipulated. The dependent variable—the calcification rate, shown on the *y* axis—is what was measured in the experiment and was predicted to "depend on" or respond to the experimental treatment.)

Rising CO_2 bubbles lower the pH of the water

▲ **Figure 2.15B** A "champagne" reef with bubbles of CO_2 rising from a volcanic seep

In a controlled experiment, scientists looked at the effect of decreasing carbonate ion concentration on the rate of calcium deposition by reef organisms. The Biosphere 2 aquarium in Arizona contains a large coral reef system that behaves like a natural reef. Researchers measured how the calcification rate changed with differing amounts of dissolved carbonate ions. **Figure 2.15A** presents the results of one set of experiments, in which pH, temperature, and concentration of calcium ions were held constant while the carbonate ion concentration of the seawater was varied. As you can see from the graph, the lower the concentration of carbonate ions, the lower the rate of calcification, and the slower the growth of coral animals.

Controlled studies such as this one have provided evidence that ocean acidification and the resulting reduction in carbonate ion concentration will negatively affect coral reefs. But scientists have also looked to natural habitats to study how ocean acidification affects coral reef ecosystems. A 2011 study looked at three volcanic seeps in Papua New Guinea. As you can see in **Figure 2.15B**, bubbles of CO_2 released from underwater volcanoes around such "champagne reefs" lower the pH of the water. Researchers surveyed three study sites in which the pH naturally varied from 8.1 to 7.8. They found reductions in coral diversity and the recruitment of juvenile coral as the pH of the sites declined, both of which undermine the resiliency of a reef community. Researchers also found a shift to less structurally complex and slower growing corals. The structural complexity of coral reef ecosystems makes them havens for a great diversity of organisms.

Scientists often synthesize their conclusions using multiple lines of evidence. The results from both controlled experimental studies and observational field studies of sites where pH naturally varies have dire implications for the health of coral reefs and the diversity of organisms they support.

? **What is the relationship between fossil fuel consumption and coral reefs?**

● Some of the increased CO_2 released by burning fossil fuels dissolves in and lowers the pH of the oceans. A lower pH reduces levels of carbonate ions, which then lowers the rate of calcification by coral animals. A lower pH also changes the composition and resiliency of coral reefs.

2.16 The search for extraterrestrial life centers on the search for water

EVOLUTION CONNECTION

When astrobiologists search for signs of extraterrestrial life on distant planets, they look for evidence of water. Why? As we've seen in this chapter, the emergent properties of water support life on Earth in many ways. Is it possible that some form of life has evolved on other planets that have water in their environment? Scientists with the National Aeronautics and Space Administration (NASA) are looking into this possibility.

Like Earth, Mars has an ice cap at both poles, and scientists have found signs that water may exist elsewhere on the planet. In 2008, the robotic spacecraft *Phoenix* landed on Mars and sent back images showing that ice is present just under Mars's surface. Then, in 2011, high-resolution images sent to Earth from the Mars Reconnaissance Orbiter showed evidence for liquid water beneath the surface. Distinctive streaks form along steep slopes during the Mars spring and summer, which then vanish during the winter. Careful study of these images over time has led scientists to conclude that these streaks are most likely seasonal streams of flowing water resulting when subsurface ice melts during the warm season. This exciting finding has reinvigorated the search for signs of life, past or present, on Mars and other planets. If any life-forms or fossils are found, their study will shed light on the process of evolution from an entirely new perspective.

? **Why is the presence of water important in the search for extraterrestrial life?**

● Water plays important roles in life as we know it, from moderating temperatures on the planet to functioning as the solvent of life.

CHAPTER 2 REVIEW

For practice quizzes, BioFlix animations, MP3 tutorials, video tutors, and more study tools designed for this textbook, go to MasteringBiology®

Reviewing the Concepts

Elements, Atoms, and Compounds (2.1–2.4)

2.1 Organisms are composed of elements, in combinations called compounds. Oxygen, carbon, hydrogen, and nitrogen make up about 96% of living matter.

2.2 Trace elements are common additives to food and water.

2.3 Atoms consist of protons, neutrons, and electrons.

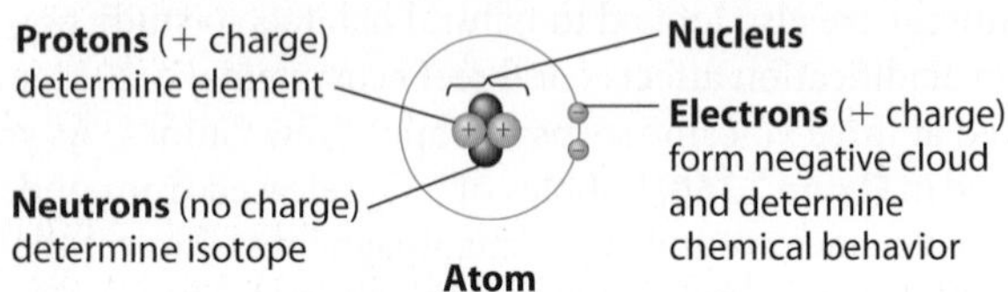

Protons (+ charge) determine element
Nucleus
Electrons (+ charge) form negative cloud and determine chemical behavior
Neutrons (no charge) determine isotope
Atom

2.4 Radioactive isotopes can help or harm us. Radioactive isotopes are valuable in basic research and medicine.

Chemical Bonds (2.5–2.9)

2.5 The distribution of electrons determines an atom's chemical properties. An atom whose outer electron shell is not full tends to interact with other atoms and share, gain, or lose electrons, resulting in attractions called chemical bonds.

2.6 Covalent bonds join atoms into molecules through electron sharing. In a nonpolar covalent bond, electrons are shared equally. In polar covalent bonds, such as those found in water, electrons are pulled closer to the more electronegative atom.

2.7 Ionic bonds are attractions between ions of opposite charge. Electron gain and loss create charged atoms, called ions.

2.8 Hydrogen bonds are weak bonds important in the chemistry of life. The slightly positively charged H atoms in one polar molecule may be attracted to the partial negative charge of an oxygen or nitrogen atom in a neighboring molecule.

2.9 Chemical reactions make and break chemical bonds. The composition of matter is changed as bonds are broken and formed to convert reactants to products.

Water's Life-Supporting Properties (2.10–2.16)

2.10 Hydrogen bonds make liquid water cohesive. Cohesion creates surface tension and allows water to move from plant roots to leaves.

2.11 Water's hydrogen bonds moderate temperature. Heat is absorbed when hydrogen bonds break and released when hydrogen bonds form. This helps keep temperatures relatively steady. As the most energetic water molecules evaporate, the surface of a substance cools.

2.12 Ice floats because it is less dense than liquid water. Floating ice protects lakes and oceans from freezing solid, which in turn protects aquatic life.

Liquid water: Hydrogen bonds constantly break and re-form

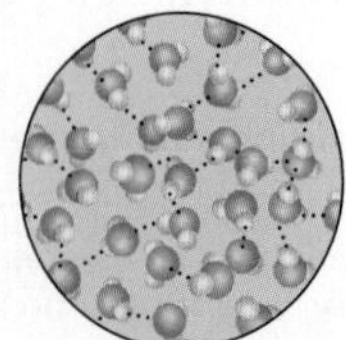

Ice: Stable hydrogen bonds hold molecules apart

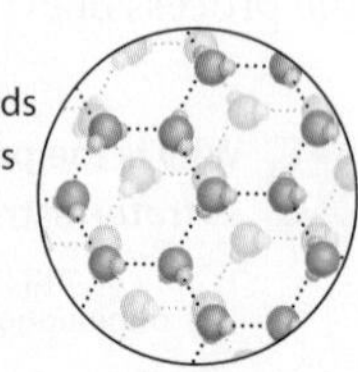

2.13 Water is the solvent of life. Polar or charged solutes dissolve when water molecules surround them, forming aqueous solutions.

2.14 The chemistry of life is sensitive to acidic and basic conditions. A compound that releases H^+ in solution is an acid, and one that accepts H^+ is a base. The pH scale ranges from 0 (most acidic) to 14 (most basic). The pH of most cells is close to 7 (neutral) and is kept that way by buffers.

2.15 Scientists study the effects of rising atmospheric CO_2 on coral reef ecosystems. The acidification of the ocean threatens coral reefs and other marine organisms.

2.16 The search for extraterrestrial life centers on the search for water. The emergent properties of water support life on Earth and may contribute to the potential for life to have evolved on other planets.

Connecting the Concepts

1. Fill in the blanks in this concept map to help you tie together the key concepts concerning elements, atoms, and molecules.

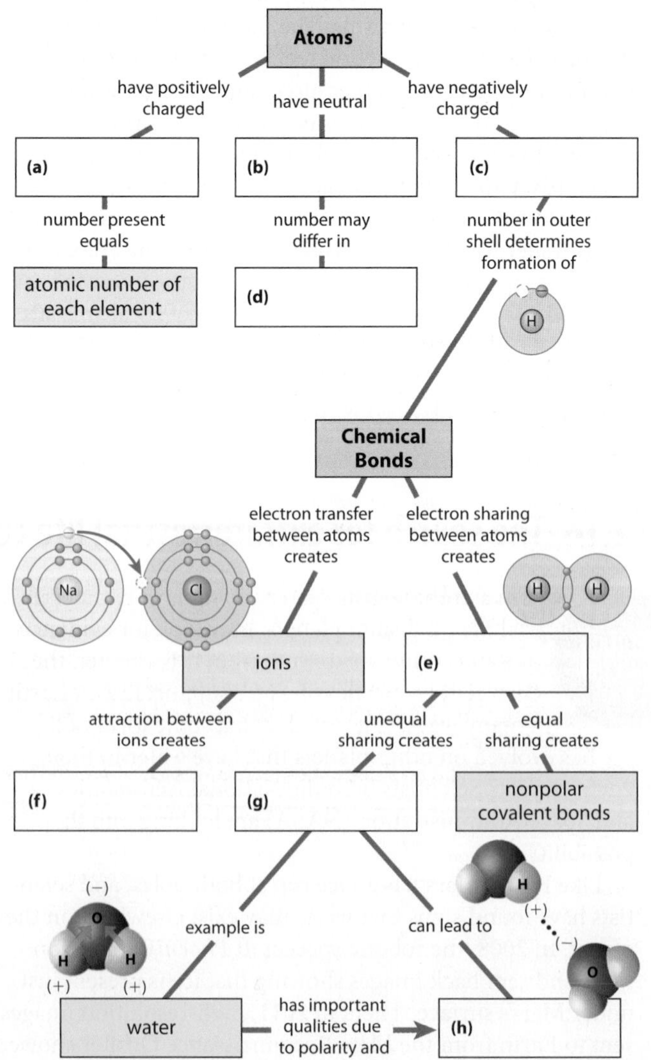

Atoms
have positively charged
have neutral
have negatively charged
(a)
(b)
(c)
number present equals
number may differ in
number in outer shell determines formation of
atomic number of each element
(d)
H
Chemical Bonds
electron transfer between atoms creates
electron sharing between atoms creates
Na
Cl
H
H
ions
(e)
attraction between ions creates
unequal sharing creates
equal sharing creates
(f)
(g)
nonpolar covalent bonds
(−)
O
H
H
(+)
(+)
example is
can lead to
H
(+)
(−)
O
water
has important qualities due to polarity and
(h)

2. Create a concept map to organize your understanding of the life-supporting properties of water. A sample map is in the answer section, but the value of this exercise is in the thinking and integrating you must do to create your own map.

Testing Your Knowledge

Level 1: Knowledge/Comprehension

3. Changing the _______________ would change it into an atom of a different element.
   a. number of electrons surrounding the nucleus of an atom
   b. number of protons in the nucleus of an atom
   c. electrical charge of an atom
   d. number of neutrons in the nucleus of an atom
4. A solution at pH 6 contains _______________ H^+ than the same amount of a solution at pH 8.
   a. 20 times more
   b. 100 times more
   c. 2 times less
   d. 100 times less
5. Most of the unique properties of water result from the fact that water molecules
   a. are the most abundant molecules on Earth's surface.
   b. are held together by covalent bonds.
   c. are constantly in motion.
   d. are polar and form hydrogen bonds.
6. A can of cola consists mostly of sugar dissolved in water, with some carbon dioxide gas that makes it fizzy and makes the pH less than 7. In chemical terms, you could say that cola is an aqueous solution where water is the _______________, sugar is a _______________, and carbon dioxide makes the solution _______________.
   a. solvent . . . solute . . . basic
   b. solute . . . solvent . . . basic
   c. solvent . . . solute . . . acidic
   d. solute . . . solvent . . . acidic

Level 2: Application/Analysis

7. The atomic number of sulfur (S) is 16. Sulfur combines with hydrogen by covalent bonding to form a compound, hydrogen sulfide. Based on the number of valence electrons in a sulfur atom, predict the molecular formula of the compound. (*Explain your answer.*)
   a. HS
   b. H_2S
   c. H_4S_2
   d. H_4S
8. In what way does the need for iodine or iron in your diet differ from your need for calcium or phosphorus?
9. Use carbon-12, the most common isotope of carbon, to define these terms: atomic number, mass number, valence. Which of these numbers is most related to the chemical behavior of an atom? Explain.
10. In terms of electron sharing between atoms, compare nonpolar covalent bonds, polar covalent bonds, and ions.
11. The diagram below shows the arrangement of electrons around the nucleus of a fluorine and a potassium atom. What kind of bond do you think would form between these two atoms?

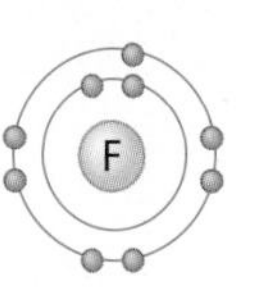
F
Fluorine atom

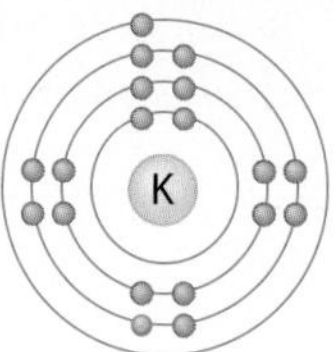
K
Potassium atom

Level 3: Synthesis/Evaluation

12. Look back at the abbreviated periodic table of the elements in Figure 2.5B. If two elements are in the same row, what do they have in common? If two elements are in the same column, what do they have in common? Would you predict that elements in the same row or the same column will have similar chemical properties? Explain.
13. **SCIENTIFIC THINKING** A recent experimental study looked at the combined effects of ocean acidification (see Module 2.15) and increased ocean temperatures, both aspects of global climate change, on the growth of polyps, juvenile coral animals. Researchers reported the average polyp biomass (in μg/polyp) after 42 days of growth under four treatments: a control with pH and temperature maintained close to normal reef conditions, a pH lowered by 0.2 units, a temperature raised by 1°C, and a combined lower pH and higher temperature. The results showed that polyp biomass was reduced somewhat in both the low-pH and high-temperature treatments, but the combined treatment resulted in a reduction in growth by almost a third—a statistically significant result. Experiments often look at the effects of changing one variable at a time, while keeping all other variables constant. Explain why this experiment considered two variables—both a higher temperature and a lower pH—at the same time.
14. In agricultural areas, farmers pay close attention to the weather forecast. Right before a predicted overnight freeze, farmers spray water on crops to protect the plants. Use the properties of water to explain how this method works. Be sure to mention why hydrogen bonds are responsible for this phenomenon.
15. This chapter explains how the emergent properties of water contribute to the suitability of the environment for life. Until fairly recently, scientists assumed that other physical requirements for life included a moderate range of temperature, pH, and atmospheric pressure. That view has changed with the discovery of organisms known as extremophiles, which have been found flourishing in hot, acidic sulfur springs and around hydrothermal vents deep in the ocean. What does the existence of life in such environments say about the possibility of life on other planets?

Answers to all questions can be found in Appendix 4.

CHAPTER 3

The Molecules of Cells

Is a big glass of milk a way to a healthy diet—or an upset stomach? Quite often, the answer is the latter. Most of the world's adult populations cannot easily digest milk-based foods. Such people suffer from lactose intolerance, the inability to properly break down lactose, the main sugar found in milk. Almost all infants are able to drink breast milk or other dairy products, benefiting from the proteins, fats, and sugars in this nutritious food. But as they grow older, many people find that drinking milk comes with a heavy dose of digestive discomfort.

What does evolution have to do with drinking milk?

The young man in the photograph below can enjoy drinking milk because his body continues to produce lactase—the enzyme that speeds the digestion of lactose into smaller sugars that his digestive system can absorb. In most human populations, the production of this enzyme begins to decline after the age of 2. In the United States, as many as 80% of African Americans and Native Americans and 90% of Asian Americans are lactase-deficient once they reach their teenage years. Americans of northern European descent make up one of the few groups in which lactase production continues

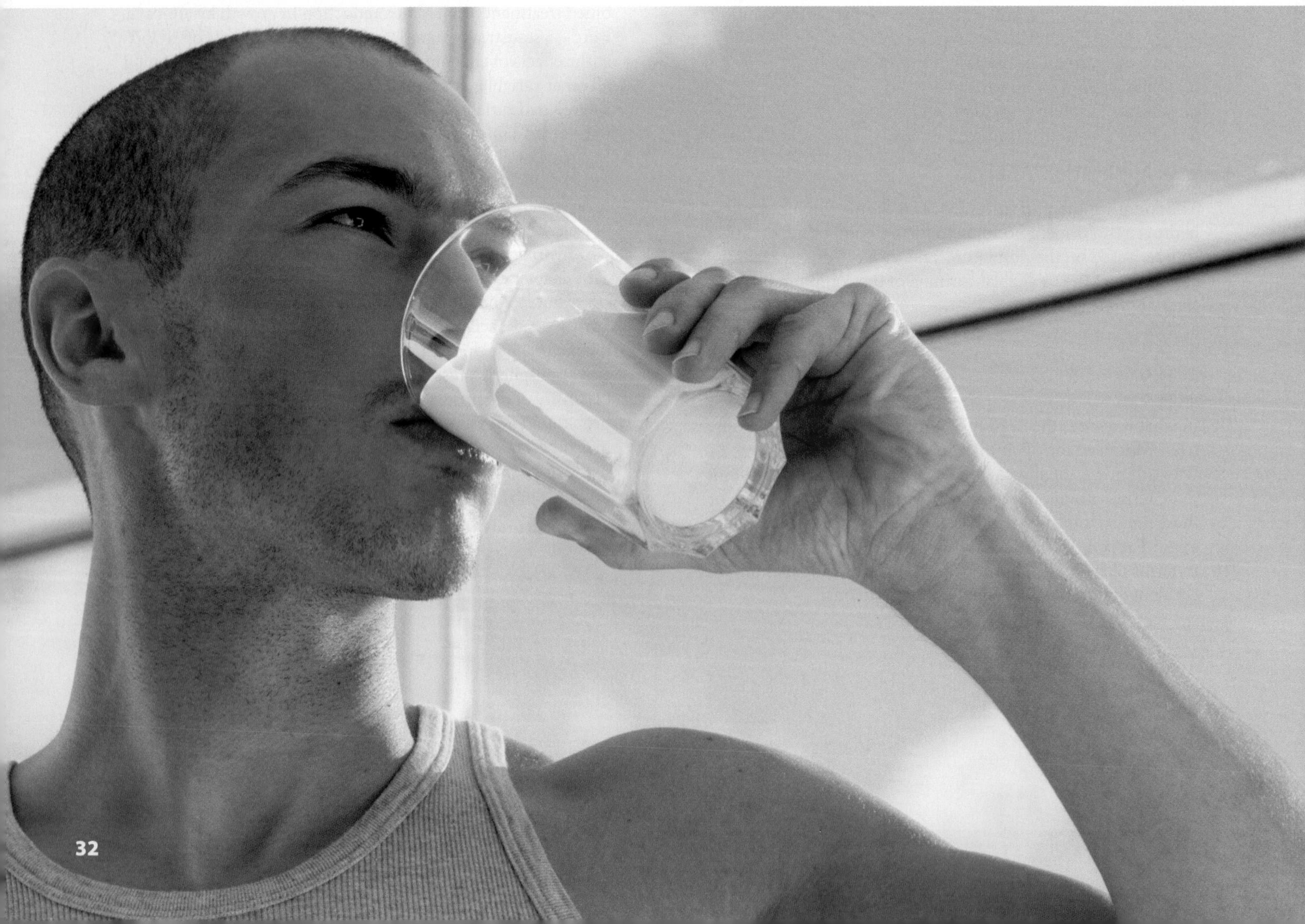

into adulthood. Why are some people lactose tolerant while others are not? As you'll find out later in this chapter, the answer has to do with evolution and the inheritance of a genetic mutation that occurred in the ancestors of certain groups.

In people who easily digest milk, lactose (a sugar) is broken down by lactase (a protein), which is coded for by a gene made of DNA (a nucleic acid). Such molecular interactions, repeated in countless variations, drive all biological processes. In this chapter, we explore the structure and function of sugars, proteins, fats, and nucleic acids—the biological molecules that are essential to life. We begin with a look at carbon, the versatile atom at the center of life's molecules.

BIG IDEAS

Introduction to Organic Compounds

(3.1–3.3)

Carbon-containing compounds are the chemical building blocks of life.

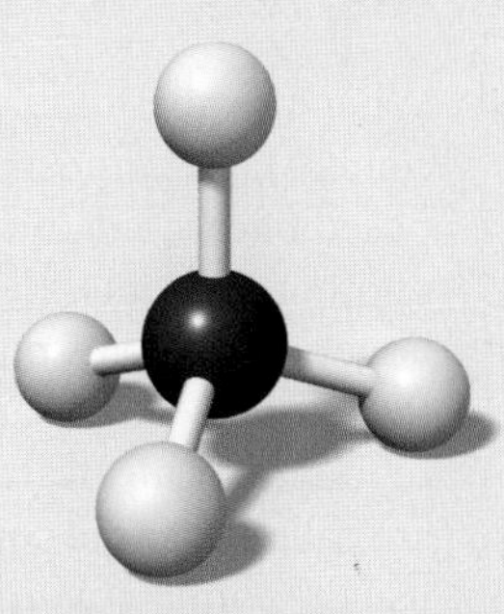

Carbohydrates

(3.4–3.7)

Carbohydrates serve as a cell's fuel and building material.

Lipids

(3.8–3.11)

Lipids are hydrophobic molecules with diverse functions.

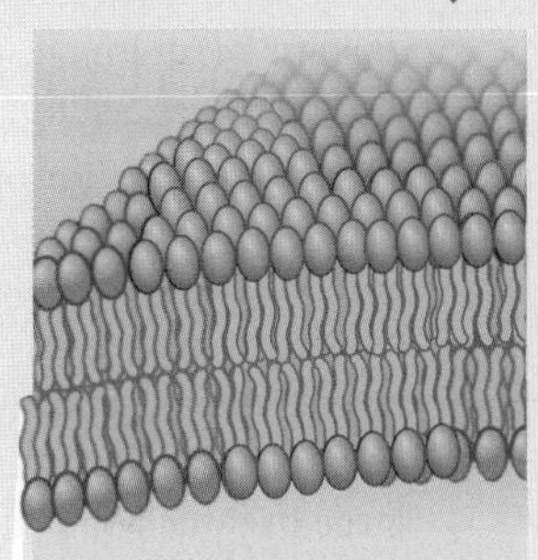

Proteins

(3.12–3.14)

Proteins are essential to the structures and functions of life.

Nucleic Acids

(3.15–3.17)

Nucleic acids store, transmit, and help express hereditary information.

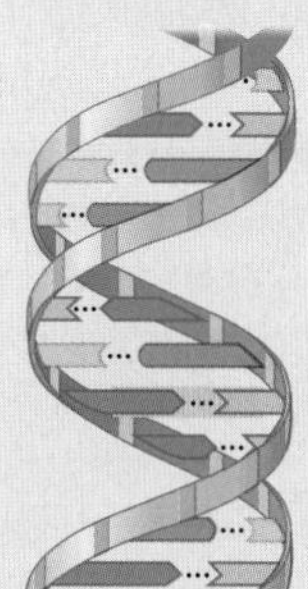

▷ Introduction to Organic Compounds

3.1 Life's molecular diversity is based on the properties of carbon

When it comes to making molecules, carbon usually takes center stage. Almost all the molecules a cell makes are composed of carbon atoms bonded to one another and to atoms of other elements. Carbon is unparalleled in its ability to form large and complex molecules, which build the structures and carry out the functions required for life. Carbon-based molecules are called **organic compounds**, and they usually contain hydrogen atoms in addition to carbon.

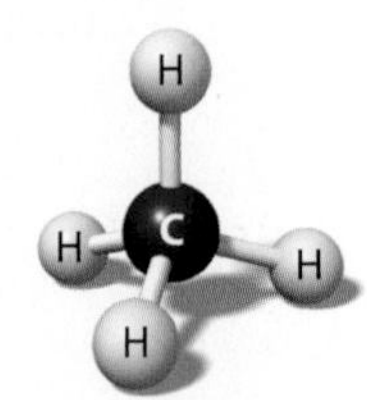

H
H
C
H
H

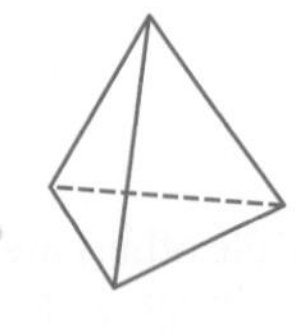

The four single bonds of carbon point to the corners of a tetrahedron.

▲ **Figure 3.1A** A model of methane (CH_4) and the tetrahedral shape of a molecule in which a carbon atom forms four single bonds to other atoms

Why are carbon atoms the lead players in the chemistry of life? Remember that the number of electrons in the outermost shell determines an atom's chemical properties. A carbon atom has 4 electrons in a valence shell that holds 8. Carbon completes its outer shell by sharing electrons with other atoms in four covalent bonds (see Module 2.6).

Figure 3.1A presents a ball-and-stick model of methane (CH_4), one of the simplest organic molecules. It shows that carbon's four bonds (the white "sticks") angle out toward the corners of an imaginary tetrahedron (an object with four triangular sides, as illustrated in red to the right of the model). This shape occurs wherever a carbon atom participates in four single bonds. In molecules with more than one carbon, each carbon atom is a connecting point from which a molecule can branch in up to four directions. In addition, different shapes occur when carbon atoms form double bonds. Thanks to the geometry of carbon's single and double bonds, large organic molecules can have very elaborate shapes. And as you will see repeatedly, a molecule's shape usually determines its function.

Carbon chains form the backbone of most organic molecules. **Figure 3.1B** illustrates four ways in which such "carbon skeletons" (shaded in gray in the figure) can vary. They may differ in length and can be straight, branched, or arranged in rings. Carbon skeletons may also include double bonds, which can vary in number and location.

Notice that the two compounds on the bottom left of Figure 3.1B, butane and isobutane, have the same molecular formula, C_4H_{10}. They differ, however, in the arrangement of their carbon skeleton. The two molecules on the top right also have the same numbers of atoms (C_4H_8), but they have different three-dimensional shapes because of the location of the double bond. Compounds with the same formula but different structural arrangements are called **isomers**. The different shapes of isomers result in unique properties and add greatly to the diversity of organic molecules.

Isomers can also result from the different spatial arrangements that can occur when four different partners are bonded to a carbon atom. This type of isomer is important in the pharmaceutical industry, because the two isomers of a drug may not be equally effective or may have different (and sometimes harmful) effects.

Methane (Figure 3.1A) and the compounds illustrated in Figure 3.1B are composed of only carbon and hydrogen—they are called **hydrocarbons**. The majority of naturally occurring hydrocarbons are found in crude oil and natural gas and provide most of the world's energy. Hydrocarbons are rare in living organisms, but hydrocarbon chains are found in regions of some molecules. For instance, fats contain hydrocarbon chains that provide fuel to your body.

In the next module, we see how attaching atoms other than just hydrogen to carbon skeletons produces a huge diversity of biological molecules.

? **One isomer of methamphetamine is the addictive illegal drug known as "crank." The other is a medicine for sinus congestion. How can you explain the differing effects of the two isomers?**

● Isomers have different structures, or shapes, and the shape of a molecule usually helps determine the way it functions in the body.

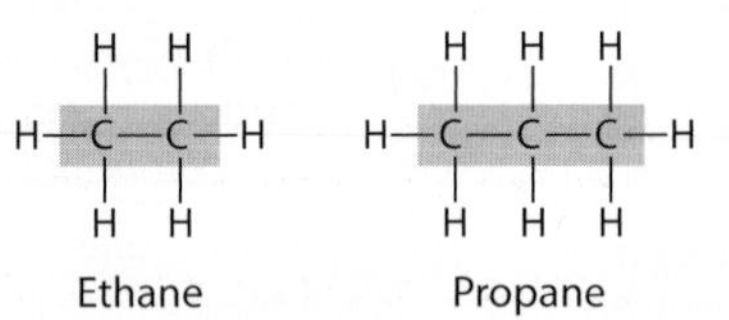

Ethane
Propane

Length: Carbon skeletons vary in length.

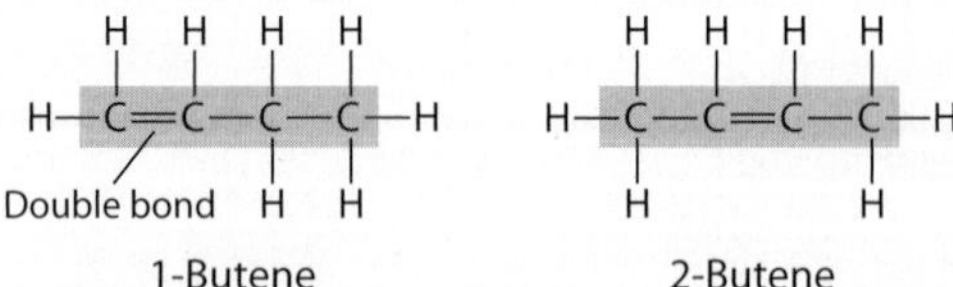

Double bond
1-Butene
2-Butene

Double bonds: Carbon skeletons may have double bonds, which can vary in location.

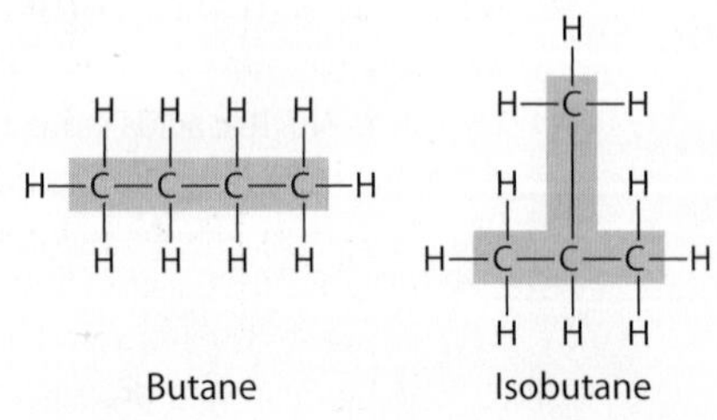

Butane
Isobutane

Branching: Carbon skeletons may be unbranched or branched.

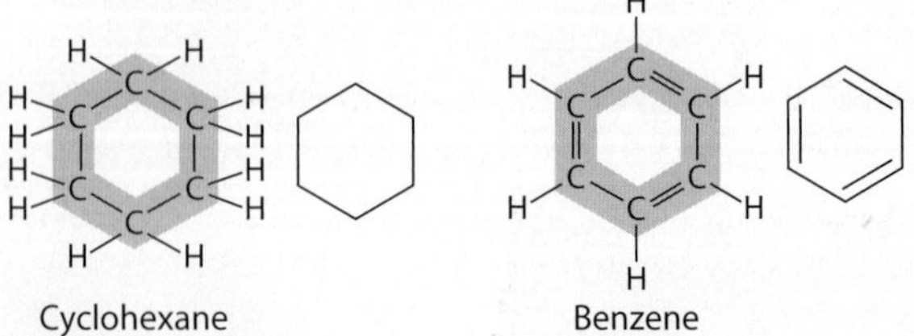

Cyclohexane
Benzene

Rings: Carbon skeletons may be arranged in rings. (In the abbreviated ring structures, each corner represents a carbon and its attached hydrogens.)

▲ **Figure 3.1B** Four ways in which carbon skeletons can vary

3.2 A few chemical groups are key to the functioning of biological molecules

The unique properties of an organic compound depend not only on the size and shape of its carbon skeleton but also on the groups of atoms that are attached to that skeleton.

Figure 3.2 shows what a difference chemical groups can make. The hormones testosterone and estradiol (a type of estrogen) differ only in the groups of atoms highlighted here with colored boxes. These subtle differences affect the functioning of these molecules, helping to produce male and female features in lions, humans, and other vertebrates.

Table 3.2 illustrates six important chemical groups. The first five are called **functional groups**. They affect a molecule's function by participating in chemical reactions. These groups are polar, which tends to make compounds containing them **hydrophilic** (water-loving) and therefore soluble in water—a necessary condition for their roles in water-based life. The sixth group, a methyl group, is nonpolar and not reactive, but it affects molecular shape and thus function.

A **hydroxyl group** consists of a hydrogen atom bonded to an oxygen atom, which in turn is bonded to the carbon skeleton. Ethanol, shown in the table, and other organic compounds containing hydroxyl groups are called alcohols.

In a **carbonyl group**, a carbon atom is linked by a double bond to an oxygen atom. If the carbonyl group is at the end of a carbon skeleton, the compound is called an aldehyde; if it is within the chain, the compound is called a ketone. Simple sugars contain a carbonyl group and several hydroxyl groups.

A **carboxyl group** consists of a carbon double-bonded to an oxygen atom and also bonded to a hydroxyl group. The carboxyl group acts as an acid by contributing an H^+ to a solution (see Module 2.14) and thus becoming ionized. Compounds with carboxyl groups are called carboxylic acids.

An **amino group** has a nitrogen bonded to two hydrogens and the carbon skeleton. It acts as a base by picking up an H^+ from a solution. Organic compounds with an amino group are called amines. The building blocks of proteins—amino acids—contain an amino and a carboxyl group.

Testosterone
OH
CH_3
CH_3
O
Estradiol
OH
CH_3
HO

▲ **Figure 3.2** Differences in the chemical groups of sex hormones

TABLE 3.2 IMPORTANT CHEMICAL GROUPS OF ORGANIC COMPOUNDS

Chemical Group	Examples
Hydroxyl group —OH	Alcohol
Carbonyl group >C=O	Aldehyde; Ketone
Carboxyl group —COOH	Carboxylic acid; Ionized
Amino group —NH_2	Amine; Ionized
Phosphate group —OPO_3^{2-}	Adenosine —O—P—O—P—O—P—O⁻ — Organic phosphate
Methyl group —CH_3	Methylated compound

A **phosphate group** consists of a phosphorus atom bonded to four oxygen atoms. It is usually ionized and attached to the carbon skeleton by one of its oxygen atoms. Compounds with phosphate groups are called organic phosphates and are often involved in energy transfers, as is the energy-rich compound ATP, shown in the table.

A **methyl group** consists of a carbon bonded to three hydrogen atoms. The methylated compound in the table—a component of DNA—affects the expression of genes.

You will meet these chemical groups again as you learn about the four major classes of organic molecules. But first, let's see how your cells make large molecules out of smaller ones.

? **Identify the chemical groups that do *not* contain carbon.**

● The hydroxyl, amino, and phosphate groups

3.3 Cells make large molecules from a limited set of small molecules

Given the rich complexity of life on Earth, we might expect there to be an enormous diversity of types of molecules. Remarkably, however, the important molecules of all living things—from bacteria to elephants—fall into just four main classes: carbohydrates, lipids, proteins, and nucleic acids. On a molecular scale, molecules of three of these classes—carbohydrates, proteins, and nucleic acids—can be gigantic; in fact, biologists call them **macromolecules**. For example, a protein may consist of thousands of atoms. How does a cell make such a huge molecule?

Cells make most of their macromolecules by joining smaller molecules into chains called **polymers** (from the Greek *polys*, many, and *meros*, part). A polymer is a long molecule consisting of many identical or similar building blocks strung together, much as a train consists of a chain of cars. The building blocks of polymers are called **monomers**.

Making Polymers Cells link monomers together to form polymers by a **dehydration reaction**, a reaction that removes a molecule of water as two molecules become bonded together. Each monomer contributes part of the water molecule that is released during the reaction. As you can see on the left side of **Figure 3.3**, one monomer (the one at the right end of the short polymer in this example) loses a hydroxyl group and the other monomer loses a hydrogen atom to form H_2O. As this occurs, a new covalent bond forms, linking the two monomers. Dehydration reactions are the same regardless of the specific monomers and the type of polymer the cell is producing.

Breaking Polymers Cells not only make macromolecules but also have to break them down. For example, most of the organic molecules in your food are in the form of polymers that are much too large to enter your cells. You must digest these polymers to make their monomers available to your cells. This digestion process is called **hydrolysis**. Essentially the reverse of a dehydration reaction, hydrolysis means to break (*lyse*) with water (*hydro-*). As the right side of Figure 3.3 shows, the bond between monomers is broken by the addition of a water molecule, with the hydroxyl group from the water attaching to one monomer and a hydrogen attaching to the adjacent monomer.

The lactose-intolerant individuals you learned about in the chapter introduction are unable to hydrolyze such a bond in the sugar lactose because they lack the enzyme lactase. Both dehydration reactions and hydrolysis require the help of enzymes to make and break bonds. **Enzymes** are specialized macromolecules that speed up chemical reactions in cells.

The Diversity of Polymers The diversity of macromolecules in the living world is vast. Surprisingly, a cell makes all its thousands of different macromolecules from a small list of ingredients—about 40 to 50 common components and a few others that are rare. Proteins, for example, are built from only 20 kinds of amino acids. Your DNA is built from just four kinds of monomers called nucleotides. The key to the great diversity of polymers is arrangement—variation in the sequence in which monomers are strung together.

The variety in polymers accounts for the uniqueness of each organism. The monomers themselves, however, are essentially universal. Your proteins and those of a tree or an ant are assembled from the same 20 amino acids. Life has a simple yet elegant molecular logic: Small molecules common to all organisms are ordered into large molecules, which vary from species to species and even from individual to individual in the same species.

In the remainder of the chapter, we explore each of the four classes of large biological molecules. Like water and simple organic molecules, large biological molecules have unique emergent properties arising from the orderly arrangement of their atoms. As you will see, for these molecules of life, structure and function are inseparable.

? **Suppose you eat some cheese. What reactions must occur for the protein of the cheese to be broken down into its amino acid monomers and then for these monomers to be converted to proteins in your body?**

● In digestion, the proteins are broken down into amino acids by hydrolysis. New proteins are formed in your body cells from these monomers in dehydration reactions.

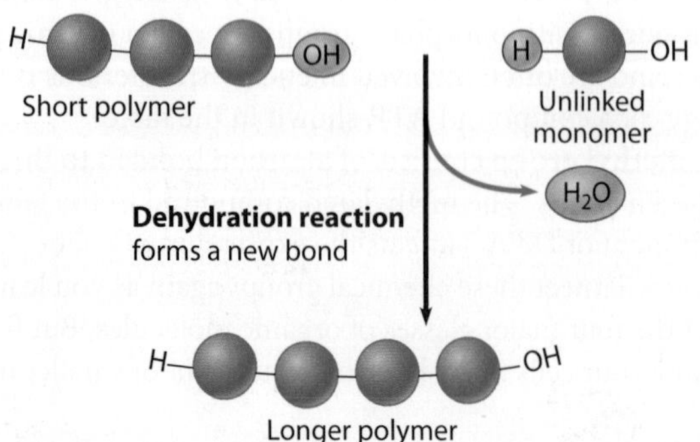

H
OH
Short polymer
H
OH
Unlinked monomer
H₂O
Dehydration reaction
forms a new bond
H
OH
Longer polymer

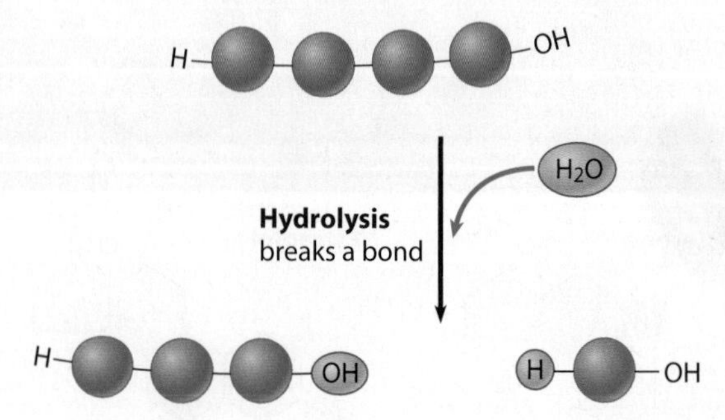

H
OH
H₂O
Hydrolysis
breaks a bond
H
OH
H
OH

▲ **Figure 3.3** Dehydration reaction building a polymer (left); Hydrolysis breaking down a polymer (right)

▷ Carbohydrates

3.4 Monosaccharides are the simplest carbohydrates

Let's start our survey of large biological molecules with **carbohydrates**, the class of molecules that range from small sugar molecules, such as those dissolved in soft drinks, to large polysaccharides, such as the starch molecules we consume in pasta and potatoes.

Simple sugars, or **monosaccharides** (from the Greek *monos*, single, and *sacchar*, sugar), are the monomers of carbohydrates. The honey shown in **Figure 3.4A** consists mainly of monosaccharides called glucose and fructose. These and other single-unit sugars can be hooked together by dehydration reactions to form more complex sugars and polysaccharides.

Monosaccharides generally have molecular formulas that are some multiple of CH_2O. For example, the formula for **glucose**, a common monosaccharide of central importance in the chemistry of life, is $C_6H_{12}O_6$. **Figure 3.4B** illustrates the molecular structure of glucose, with its carbons numbered 1 to 6. This structure also shows the two trademarks of a sugar: a number of hydroxyl groups (—OH) and a carbonyl group (>C=O, highlighted in blue). The hydroxyl groups make a sugar an alcohol, and the carbonyl group, depending on its location, makes it either an aldehyde sugar (glucose) or a ketone sugar (fructose).

If you count the numbers of different atoms in the fructose molecule in Figure 3.4B, you will find that its molecular formula is $C_6H_{12}O_6$, identical to that of glucose. Thus, glucose and fructose are isomers; they differ only in the arrangement of their atoms (in this case, the positions of the carbonyl groups). Because the shape of molecules is so important, seemingly minor differences like this give isomers different properties, such as how they react with other molecules. These differences also make fructose taste considerably sweeter than glucose.

▲ **Figure 3.4A** Bees with honey, a mixture of two monosaccharides

The carbon skeletons of both glucose and fructose are six carbon atoms long. Other monosaccharides may have three to seven carbons. Five-carbon sugars, called pentoses, and six-carbon sugars, called hexoses, are among the most common. (Note that most names for sugars end in *-ose*. Also, as you saw with the enzyme lactase, which digests the sugar lactose, the names for most enzymes end in *-ase*.)

It is convenient to draw sugars as if their carbon skeletons were linear, but in aqueous solutions, most five- and six-carbon sugars form rings, as shown for glucose in **Figure 3.4C**. To form the glucose ring, carbon 1 bonds to the oxygen attached to carbon 5. As shown in the middle representation, the ring diagram of glucose and other sugars may be abbreviated by not showing the carbon atoms at the corners of the ring. Also, the bonds in the ring are often drawn with varied thickness, indicating that the ring is a relatively flat structure with attached atoms extending above and below it. The simplified ring symbol on the right is often used in this book to represent glucose.

Monosaccharides, particularly glucose, are the main fuel molecules for cellular work. Because cells release energy from glucose when they break it down, an aqueous solution of glucose (often called dextrose) may be injected into the bloodstream of sick or injured patients; the glucose provides an immediate energy source to tissues in need of repair. Cells also use the carbon skeletons of monosaccharides as raw material for making other kinds of organic molecules, such as amino acids and fatty acids. Sugars not used in these ways may be incorporated into disaccharides and polysaccharides, as we see next.

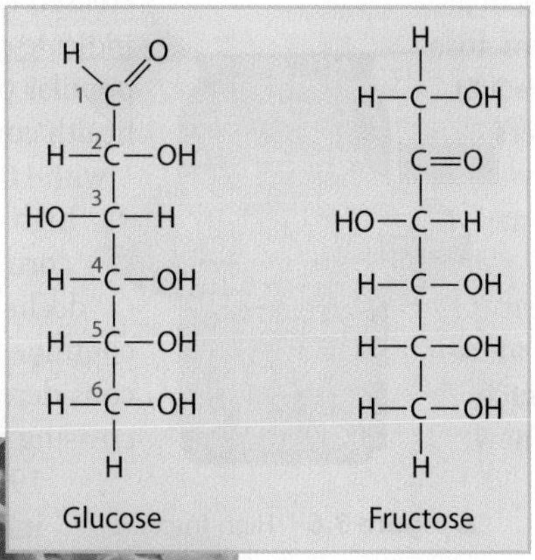

? **Write the formula for a monosaccharide that has three carbons.**

● $C_3H_6O_3$

Glucose   Fructose

◀ **Figure 3.4B** Structures of glucose and fructose

Try This Identify the two functional groups that are characteristic of a monosaccharide.

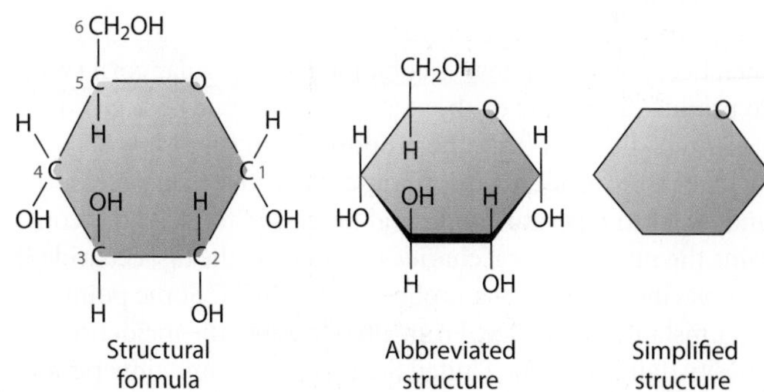
6 CH2OH
5 C
O
4 C
1 C
3 C
2 C
H
OH
HO
Structural formula
Abbreviated structure
Simplified structure

▲ **Figure 3.4C** Three representations of the ring form of glucose

3.5 Two monosaccharides are linked to form a disaccharide

Cells construct a **disaccharide** from two monosaccharide monomers by a dehydration reaction. **Figure 3.5** shows how maltose, also called malt sugar, is formed from two glucose monomers. One monomer gives up a hydroxyl group and the other gives up a hydrogen atom. As H_2O is released, an oxygen atom is left, linking the two monomers. Malt sugar, which is common in germinating seeds, is used in making beer, malt whiskey, and malted milk candy.

Sucrose is the most common disaccharide. It is made of a glucose monomer linked to a fructose monomer. Transported in plant sap, sucrose provides a source of energy and raw materials to all the parts of the plant. We extract it from the stems of sugarcane or the roots of sugar beets to use as table sugar.

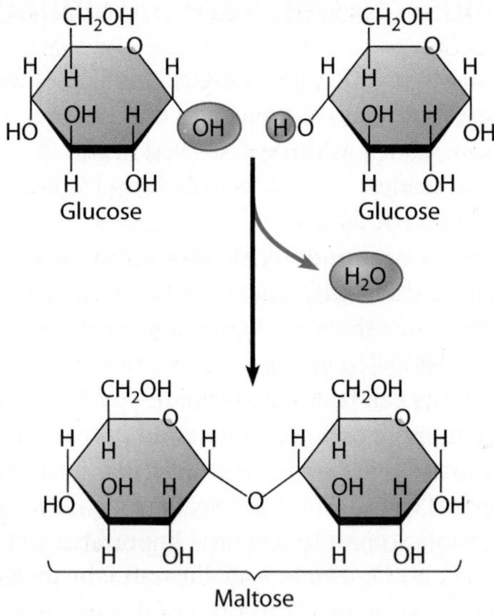

Glucose
Glucose
H_2O
Maltose

▲ **Figure 3.5** Disaccharide formation by a dehydration reaction

? Lactose, as you read in the chapter introduction, is the disaccharide sugar in milk. It is formed from glucose and galactose. The formula for both these monosaccharides is $C_6H_{12}O_6$. What is the formula for lactose?

● $C_{12}H_{22}O_{11}$

3.6 What is high-fructose corn syrup, and is it to blame for obesity?

CONNECTION

If you want to sweeten your coffee or tea, you probably reach for sugar—the disaccharide sucrose. But if you drink sodas, you're probably consuming the monosaccharides of sucrose in the form of high-fructose corn syrup. In fact, if you look at the label of almost any processed food, you will see high-fructose corn syrup listed as one of the ingredients (**Figure 3.6**).

What is high-fructose corn syrup (HFCS)? Let's start with the corn syrup part. The main carbohydrate in corn is starch, a polysaccharide built from glucose monomers. Industrial processing hydrolyzes starch into these monomers, producing corn syrup. Glucose, however, does not taste as sweet to us as sucrose. Fructose, on the other hand, tastes much sweeter than both glucose and sucrose. When a new process was developed in the 1970s that used an enzyme to rearrange the atoms of glucose into the sweeter isomer, fructose (see Figure 3.4B), the high-fructose corn syrup industry was born. (High-fructose corn syrup is a bit of a misnomer, because the fructose is combined with regular corn syrup to produce a mixture of about 55% fructose and 45% glucose, not much different from the proportions in sucrose.)

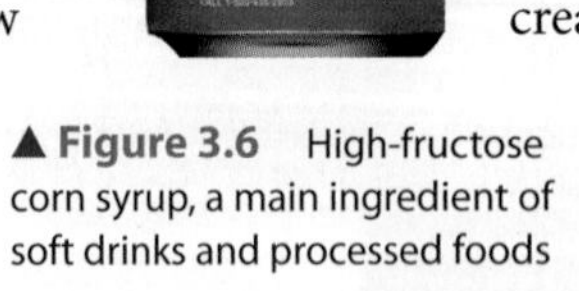

CARBONATED WATER, HIGH FRUCTOSE
CORN SYRUP, CARAMEL COLOR,
HOSPHORIC ACID, NATURAL
CAFFEINE

▲ **Figure 3.6** High-fructose corn syrup, a main ingredient of soft drinks and processed foods

The resulting clear, goopy liquid is cheaper than sucrose and easier to mix into drinks and processed food. And it contains the same monosaccharides as sucrose, the disaccharide it is replacing. So is there a problem with HFCS? Some point to circumstantial evidence. From 1980 to 2000, the incidence of obesity doubled in the United States. In that same time period, the consumption of HFCS more than tripled, whereas the consumption of refined cane and beet sugar decreased 21%.

Is high-fructose corn syrup to blame for the current "obesity epidemic" with its attendant increases in type 2 diabetes, high blood pressure, and other chronic diseases associated with increased weight? Does this *correlation* between increased HFCS consumption and increased obesity indicate *causation*? In spite of alarming claims in the popular press, most scientific studies have not shown health consequences associated with replacing sucrose with HFCS. Data also show that some countries with high obesity rates consume little high-fructose corn syrup. And although HFCS consumption has declined somewhat in recent years, obesity rates continue to rise, with almost 36% of U.S. adults now considered obese. Alternative hypotheses for our increasing obesity abound, including the fact that, from 1980 to 2000, the U.S. per capita daily caloric intake increased 23%.

It does not appear that high-fructose corn syrup is the "smoking gun" responsible for the obesity crisis. There is solid evidence, however, that overconsumption of sugar and/or HFCS along with dietary fat and decreased physical activity contribute to weight gain. In addition, high sugar consumption tends to replace eating more varied and nutritious foods. For good health, you require proteins, fats, vitamins, and minerals, as well as complex carbohydrates, the topic of the next module.

? How is high-fructose corn syrup made from corn?

● Corn starch is hydrolyzed to glucose; then enzymes convert glucose to fructose. This fructose is combined with corn syrup to produce HFCS.

3.7 Polysaccharides are long chains of sugar units

Polysaccharides are macromolecules, polymers of hundreds to thousands of monosaccharides linked together by dehydration reactions. Polysaccharides may function as storage molecules or as structural compounds. **Figure 3.7** illustrates three common types: starch, glycogen, and cellulose.

Starch, a storage polysaccharide in plants, consists of long chains of glucose monomers. Starch molecules coil into a helical shape and may be unbranched (as shown in the figure) or branched. Starch granules serve as carbohydrate "banks" from which plant cells can withdraw glucose for energy or building materials. Humans and most other animals have enzymes that can hydrolyze plant starch to glucose. Potatoes and grains, such as wheat, corn, and rice, are the major sources of starch in the human diet.

Animals store glucose in a polysaccharide called **glycogen**. Glycogen is more highly branched than starch, as shown in the figure. Most of your glycogen is stored as granules in your liver and muscle cells, which hydrolyze the glycogen to release glucose when it is needed.

Cellulose, the most abundant organic compound on Earth, is a major component of the tough walls that enclose plant cells. Cellulose is also a polymer of glucose, but its monomers are linked together in a different orientation. (Carefully compare the oxygen "bridges" highlighted in yellow in the figure between glucose monomers in starch, glycogen, and cellulose.) Arranged parallel to each other, cellulose molecules are joined by hydrogen bonds, forming cable-like microfibrils. Layers of microfibrils combine with other polymers, producing strong support for trees and the structures we build with lumber.

Animals do not have enzymes that can hydrolyze the glucose linkages in cellulose. Therefore, cellulose is not a nutrient for humans, although it does contribute to digestive health. The cellulose that passes unchanged through your digestive tract is referred to as "insoluble fiber." Fresh fruits, vegetables, and whole grains are rich in fiber.

Some microorganisms do have enzymes that can hydrolyze cellulose. Cows and termites house such microorganisms in their digestive tracts and are thus able to derive energy from cellulose. Decomposing fungi also digest cellulose, helping to recycle its chemical elements within ecosystems.

Chitin is a structural polysaccharide used by insects and crustaceans to build their exoskeleton, the hard case enclosing the animal. Chitin is also found in the cell walls of fungi.

Almost all carbohydrates are hydrophilic owing to the many hydroxyl groups attached to their sugar monomers (see Figure 3.4B). Thus, cotton bath towels, which are mostly cellulose, are quite water absorbent due to the water-loving nature of cellulose. As you'll see next, not all biological molecules "love water."

? **Compare and contrast starch and cellulose, two plant polysaccharides.**

Both are polymers of glucose, but the bonds between glucose monomers have different shapes. Starch functions mainly for sugar storage. Cellulose is a structural polysaccharide that is the main material of plant cell walls.

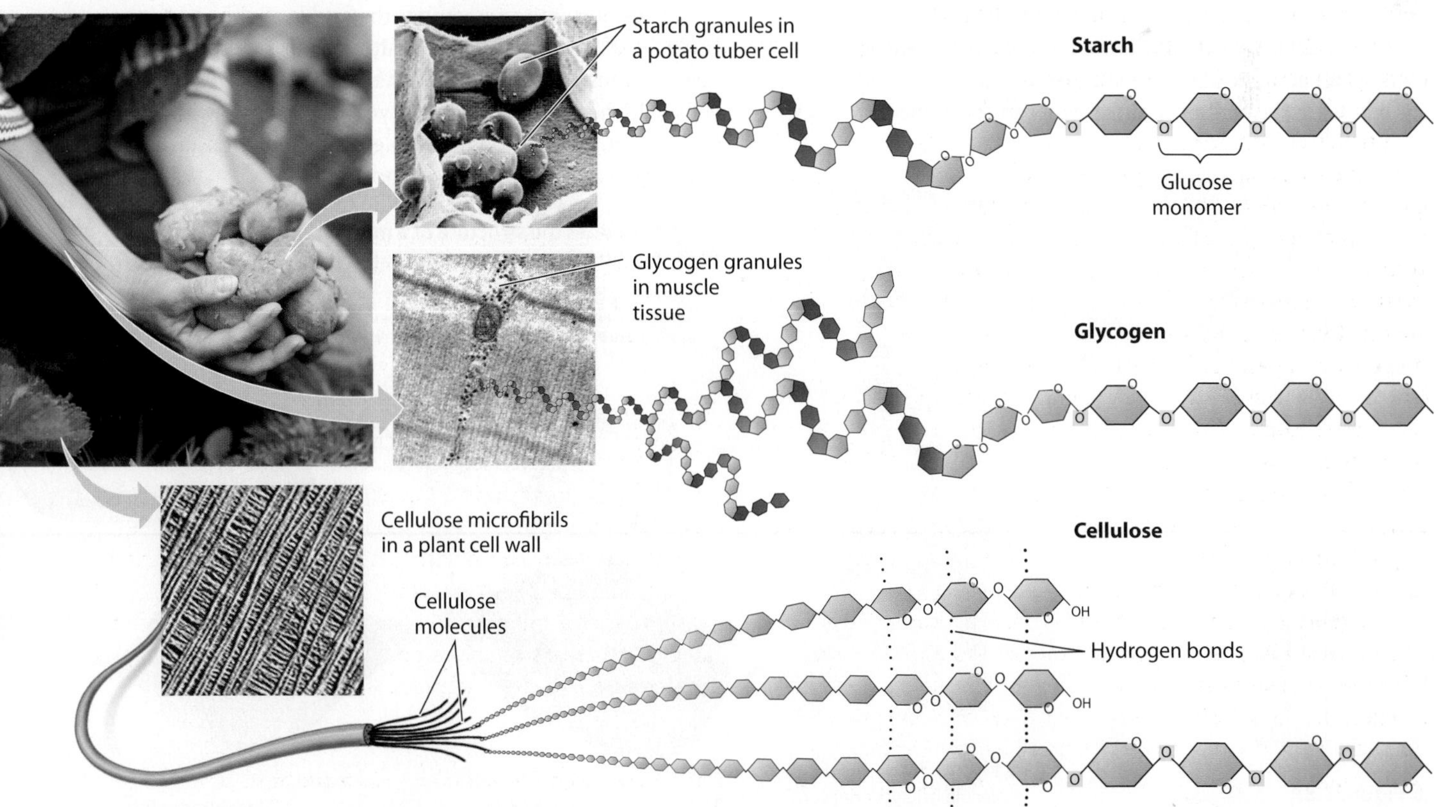
Starch granules in a potato tuber cell
Starch
Glucose monomer
Glycogen granules in muscle tissue
Glycogen
Cellulose microfibrils in a plant cell wall
Cellulose
Cellulose molecules
Hydrogen bonds
OH
OH

▲ **Figure 3.7** Polysaccharides of plants and animals

▷ Lipids

3.8 Fats are lipids that are mostly energy-storage molecules

Lipids are a diverse group of molecules that are classified together because they share one trait: They do not mix well with water. In contrast to carbohydrates and most other biological molecules, lipids are **hydrophobic** (water-fearing). You can see this chemical behavior in an unshaken bottle of salad dressing. The oil (a type of lipid) separates from the vinegar (which is mostly water).

Lipids also differ from carbohydrates, proteins, and nucleic acids in that they are neither huge macromolecules nor polymers built from similar monomers. In this and the next few modules, we consider the structures and functions of three important types of lipids: fats, phospholipids, and steroids.

A **fat** is a large lipid made from two kinds of smaller molecules: glycerol and fatty acids. Shown at the top in **Figure 3.8A**, glycerol consists of three carbons, each bearing a hydroxyl group (—OH). A fatty acid consists of a carboxyl group (the functional group that gives these molecules the name fatty *acid*, —COOH) and a hydrocarbon chain, usually 16 or 18 carbon atoms in length. The nonpolar C—H bonds in the hydrocarbon chains are the reason fats are hydrophobic.

Figure 3.8A shows how one fatty acid molecule can link to a glycerol molecule by a dehydration reaction. Linking three fatty acids to glycerol produces a fat, as illustrated in **Figure 3.8B**. A synonym for fat is *triglyceride*, a term you may see on food labels or on medical tests for fat in the blood.

A fatty acid whose hydrocarbon chain contains one or more double bonds is called an **unsaturated fatty acid**. Each carbon atom connected by a double bond has one fewer hydrogen atom attached to it. These double bonds usually cause kinks (or bends) in the carbon chain, as you can see in the third fatty acid in Figure 3.8B. A fatty acid that has no double bonds in its hydrocarbon chain has the maximum number of hydrogen atoms attached to each carbon atom (its carbons are "saturated" with hydrogen) and is called a **saturated fatty acid**.

Most animal fats are saturated: Their hydrocarbon chains—the "tails" of their fatty acids—lack double bonds and thus pack closely together, making them solid at room temperature (**Figure 3.8C**). In contrast, the fats of plants and fishes generally contain unsaturated fatty acids. The kinks in their unsaturated fatty acid tails prevent them from packing tightly together. Thus, unsaturated fats are usually liquid at room temperature and are referred to as oils. When you see "hydrogenated vegetable oils" on a margarine label, it means that unsaturated fats have been converted to saturated fats by adding hydrogen. Unfortunately, the process of hydrogenation also creates **trans fats**, a form of fat that recent research associates with health risks. We will discuss some of that research in Module 3.9.

The main function of fats is long-term energy storage. A gram of fat stores more than twice as much energy as a gram of polysaccharide. For immobile plants, the bulky energy storage form of starch is not a problem. (Vegetable oils are generally obtained from seeds, where more compact energy storage is a benefit.) Mobile animals, such as humans, can get around much more easily carrying their energy stores in the form of fat. Of course, the downside of this energy-packed storage form is that it takes more effort for a person to "burn off" excess fat.

It is important to remember that a reasonable amount of body fat is both normal and healthy. You stock these long-term fuel reserves in specialized reservoirs called adipose cells, which swell and shrink as you deposit and withdraw fat from them. In addition to storing energy, fatty tissue cushions vital organs and insulates the body.

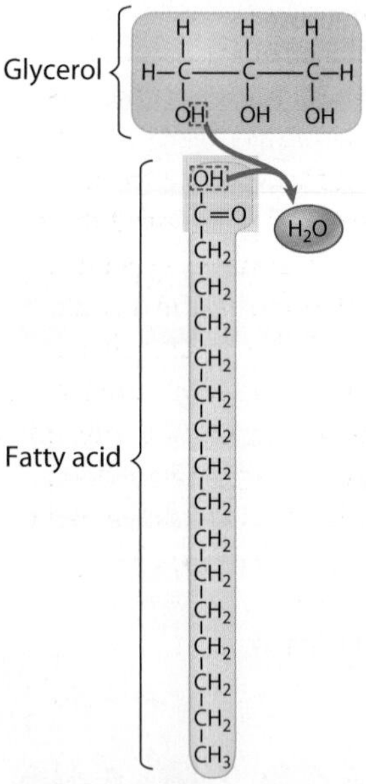

Glycerol

Fatty acid

H_2O

▲ **Figure 3.8A** A dehydration reaction that will link a fatty acid to glycerol

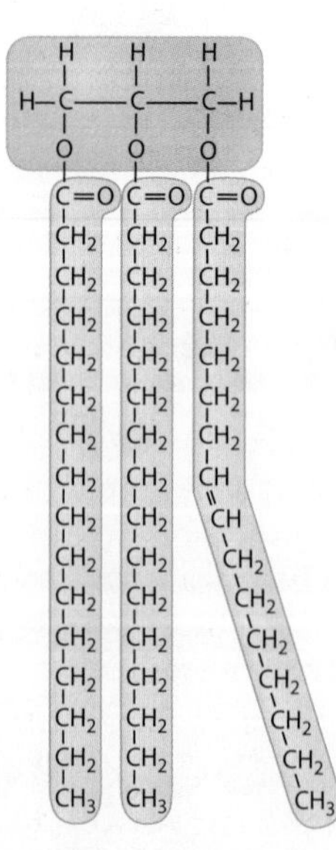

▲ **Figure 3.8B** A fat molecule (triglyceride) consisting of three fatty acids linked to glycerol

? How does the structure of a monounsaturated fat differ from a polyunsaturated fat?

● A monounsaturated fat has a fatty acid with a single double bond in its carbon chain. A polyunsaturated fat contains a fatty acid with more than one double bond.

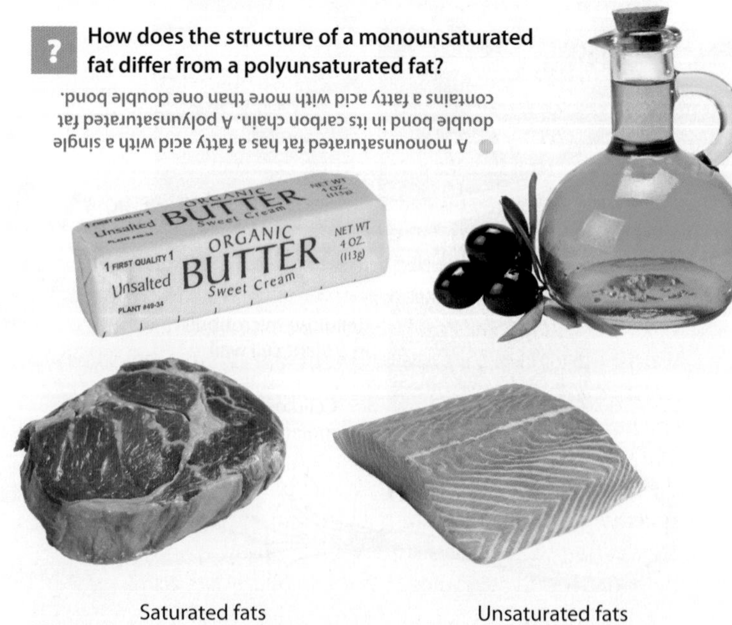
1 FIRST QUALITY 1
Unsalted
ORGANIC
BUTTER
Sweet Cream
NET WT 4 OZ. (113g)
PLANT #40-34
Saturated fats
Unsaturated fats

▲ **Figure 3.8C** Types of fats

3.9 Scientific studies document the health risks of trans fats

SCIENTIFIC THINKING

In the previous module, you learned about the difference between vegetable oils and animal fats and their unsaturated versus saturated fatty acids. In the 1890s, a process was invented that added hydrogen atoms to the double-bonded carbon atoms of unsaturated fats, producing partially hydrogenated vegetable oils. These new fats had several desirable traits: They didn't spoil as quickly as oils and could withstand repeated reheating for frying. In addition, in the 1950s and 1960s, scientific studies began to associate saturated fats with an increased risk of heart disease, leading to a public health campaign to reduce consumption of animal fats (such as butter) and replace them with unsaturated oils and the supposedly healthier partially hydrogenated vegetable oils (such as margarine).

Jump ahead to the 1990s, and partially hydrogenated oils were found in myriad foods—cookies, crackers, snacks, baked goods, and fried foods. But new research began to show that the trans fats produced in the process of hydrogenation were an even greater health risk than were saturated fats. One study estimated that eliminating trans fats from the food supply could prevent up to one in five heart attacks! In 2006, the U.S. Food and Drug Administration required the listing of trans fat on food labels. Because foods sold in restaurants and schools do not come with labels, many cities and states since have passed laws to eliminate trans fats in these foods. And an increasing number of countries have banned trans fats.

The scientific studies establishing the risks of trans fats were of two types: experimental and observational. In experimental controlled feeding trials, the diets of participants contained different proportions of saturated, unsaturated, and partially hydrogenated fats. The hypothesis of these studies was that trans fats adversely affect cardiovascular health; the prediction was that the more trans fats in the diet, the greater the risk. But how does one measure risk? Should the study proceed until participants start having heart attacks? For both ethical and practical reasons, controlled feeding trials are usually fairly short in duration, involve only limited dietary changes, generally use healthy individuals, and measure intermediary risk factors, such as changes in cholesterol levels, rather than actual disease outcomes.

Many scientific studies on dietary health effects are observational. The advantages of such studies are that they can extend over a longer time period, use a more representative population, and measure disease outcomes as well as risk factors. Observational studies may be retrospective (looking backward): Present health status is documented, and participants report their prior eating habits. Two difficulties with retrospective studies are that people may not accurately remember and report their dietary histories, and anyone who has already died, say, of a heart attack, is not included in the study. Prospective studies, on the other hand, look forward. Researchers conducting such studies enlist a study group, quantify participants' health attributes, and then collect data on the group over many years. Diet, lifestyle habits, risk factors, and disease outcomes can all be recorded and then analyzed.

TABLE 3.9 RISK OF HEART DISEASE ASSOCIATED WITH INCREASES IN SPECIFIC TYPES OF FAT CONSUMED

Variable	Relative Risk*
Saturated fat (each increase of 5% of energy	1.17
Monounsaturated fat (each increase of 5% of energy)	0.81
Polyunsaturated fat (each increase of 5% of energy)	0.62
Trans fat (each increase of 2% of energy	1.93

* A relative risk (RR) of 1 means that there is no difference in risk of coronary heart disease when compared to an equivalent intake of carbohydrate; RR of less than 1 means there is a decreased risk; RR greater than 1, there is a greater risk.

Data from F. B. Hu et al. Dietary fat intake and the risk of coronary heart disease in women, *New England Journal of Medicine* 337: 1491–9 (1997).

A landmark example of a prospective study is the Nurses' Health Study, begun in 1976 with more than 120,000 female nurses. In a portion of the study that looked at dietary fat intake, 80,082 women were followed from 1980 to 1994. The researchers estimated the relative risk of coronary heart disease associated with the intake of different types of fats. As you can see in **Table 3.9**, for each 5% increase in energy consumed in the form of saturated fat, as compared to an equivalent energy intake from carbohydrates, the relative risk rises to 1.17—or a 17% increase in the risk of heart disease. For each 2% increase in the amount of energy consumed in the form of trans fat, however, there is a 93% increase in risk. Trans fats are indeed a greater health risk than saturated fats.

Based on an accumulation of scientific evidence from many studies, U.S. governmental agencies have revised their policies—from promoting partially hydrogenated vegetable oils as a healthful alternative to saturated fats in the middle of the 20th century to today regulating and increasingly banning the trans fats that are produced when such oils are hydrogenated. Such changes in policy reflect changes in our understanding based on current research. Scientific knowledge both expands and is revised as new questions are asked, new studies are done, and new evidence accumulates.

In the next module, we consider two other important types of lipids: phospholipids and steroids.

Data from D. Mozaffarian et al., Trans fatty acids and cardiovascular disease, *The New England Journal of Medicine* 354: 1601–13 (2006).

? How does the relative risk of coronary heart disease change with a 5% increase in polyunsaturated fat in the diet?

The relative risk drops to 0.62. In other words, there is a 38% decrease in risk.

3.10 Phospholipids and steroids are important lipids with a variety of functions

Cells could not exist without **phospholipids**, the major component of cell membranes. Phospholipids are structurally similar to fats, except that they contain only two fatty acids attached to glycerol instead of three. As shown in **Figure 3.10A**, a negatively charged phosphate group (shown as a yellow circle in the figure and linked to another small molecule) is attached to glycerol's third carbon.

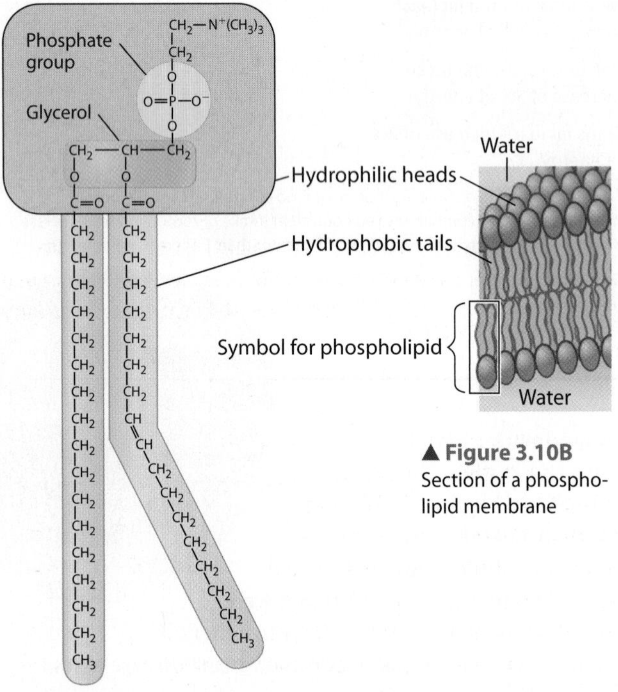

Phosphate group
Glycerol
Hydrophilic heads
Hydrophobic tails
Water
Symbol for phospholipid
Water

▲ **Figure 3.10B** Section of a phospholipid membrane

▲ **Figure 3.10A** Chemical structure of a phospholipid molecule

Try This Explain why the gray region of this phospholipid is hydrophilic and why the yellow tails are hydrophobic.

The structure of phospholipids provides a classic example of how form fits function. The two ends of a phospholipid have different relationships with water, resulting in the aggregation of multiple phospholipid molecules into a membrane (**Figure 3.10B**). The hydrophobic tails of the fatty acids cluster together in the center, excluded from water, and the hydrophilic phosphate heads face the watery environment on either side of the membrane. Each gray-headed, yellow-tailed structure in the membrane shown here represents a phospholipid; this visual representation is used throughout this book. (We will explore the structure and function of biological membranes in more detail in Chapter 5.)

Steroids are lipids in which the carbon skeleton contains four fused rings, as shown in the structural formula of cholesterol in **Figure 3.10C**. (The diagram omits the carbons making up the rings and most of the chain and also their attached hydrogen atoms.) **Cholesterol** is a common component in animal cell membranes and is also the precursor for making other steroids, including sex hormones. Different steroids vary in the chemical groups attached to the rings, as you saw in Figure 3.2. Too much cholesterol in the blood may contribute to atherosclerosis.

? Compare the structure of a phospholipid with that of a fat (triglyceride).

A phospholipid has two fatty acids and a phosphate group attached to glycerol. Three fatty acids are attached to the glycerol of a fat molecule.

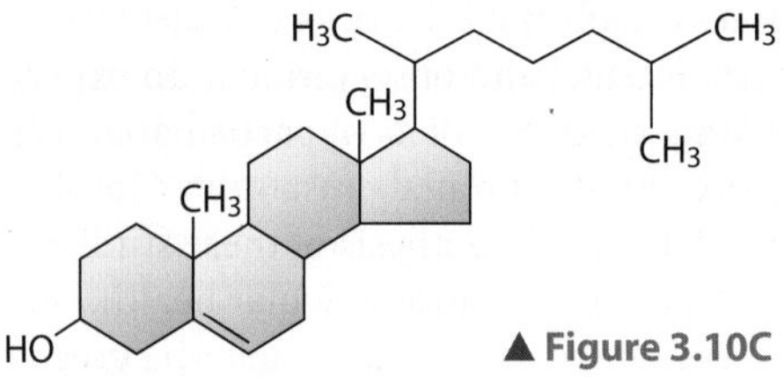

▲ **Figure 3.10C** Cholesterol, a steroid

3.11 Anabolic steroids pose health risks

CONNECTION

Anabolic steroids are synthetic variants of the male hormone testosterone. Testosterone causes a general buildup of muscle and bone mass in males during puberty and maintains masculine traits throughout life. Because anabolic steroids structurally resemble testosterone, they also mimic some of its effects. (The word *anabolic* comes from *anabolism*, the building of substances by the body.)

Anabolic steroids are used to treat general anemia and diseases that destroy body muscle. Some athletes use these drugs to build up their muscles quickly and enhance their performance. But at what cost? Steroid abuse may cause violent mood swings ("roid rage"), depression, liver damage or cancer, and high cholesterol levels and blood pressure. Use of these drugs often makes the body reduce its output of natural male sex hormones, which can cause shrunken testicles, reduced sex drive, infertility, and breast enlargement in men. Use in women has been linked to menstrual cycle disruption and development of masculine characteristics. An effect in teens is that bones may stop growing.

Despite the risks, some athletes continue to abuse synthetic steroids, and unscrupulous chemists, trainers, and coaches try to find ways to avoid their detection. Meanwhile, the U.S. Congress, professional sports authorities, and high school and college athletic programs ban the use of anabolic steroids, implement drug testing, and penalize violators in an effort to keep the competition fair and protect the health of athletes.

? Explain why fats and steroids, which are structurally very different, are both classed as lipids?

Fats and steroids are hydrophobic molecules, the key characteristic of lipids.

▷ Proteins

3.12 Proteins have a wide range of functions and structures

Nearly every dynamic function in your body depends on proteins. A **protein** is a polymer of small building blocks called amino acids. Of all of life's molecules, proteins are structurally and functionally the most elaborate and varied.

You have tens of thousands of different proteins in your body. What do they all do? Probably their most important role is as enzymes, the chemical catalysts that speed and regulate virtually all chemical reactions in your cells. Lactase, which you read about in the chapter introduction, is just one example of an enzyme.

Other types of proteins include transport proteins that are embedded in cell membranes and move sugar molecules and other nutrients into your cells. Moving through your blood stream are defensive proteins, such as the antibodies of the immune system, and signal proteins, such as many of the hormones and other chemical messengers that help coordinate your body's activities. Receptor proteins built into cell membranes receive and transmit such signals into your cells.

Muscle cells are packed with contractile proteins, and structural proteins are found in the fibers that make up your tendons and ligaments. Indeed, the structural protein collagen, which forms the long, strong fibers of connective tissues, accounts for 40% of the protein in your body.

Some proteins are storage proteins, which supply amino acids to developing embryos. The proteins found in eggs and seeds are examples.

The functions of all of these different types of proteins depend on their individual shapes. **Figure 3.12A** shows a ribbon model of lysozyme, an enzyme found in your sweat, tears, and saliva. Lysozyme consists of one long polymer of amino acids, represented by the purple ribbon. Lysozyme's general shape is called globular. This overall shape is more apparent in **Figure 3.12B**, a space-filling model of lysozyme. In that model, the colors represent the different atoms of carbon, oxygen, nitrogen, and hydrogen. The barely visible yellow balls represent sulfur atoms that form the stabilizing bonds shown as yellow lines in the ribbon model. Most enzymes and many other proteins are globular. Structural proteins, such as those making up hair, tendons, and ligaments, are typically long and thin and are called fibrous proteins. **Figure 3.12C** shows a spider's web, made up of fibrous silk proteins. The structural arrangement within these proteins makes each silk fiber stronger than a steel strand of the same weight.

Descriptions such as *globular* and *fibrous* refer to a protein's general shape. Each protein also has a much more specific shape. The coils and twists of lysozyme's ribbon in Figure 3.12A may appear haphazard, but they represent the molecule's specific, three-dimensional shape. Nearly all proteins must recognize and bind to some other molecule to function. Lysozyme can destroy bacterial cells, but first it must bind to molecules on the bacterial cell surface. Lysozyme's specific shape enables it to recognize and attach to its molecular target, which fits into the groove you see on the right in the figures.

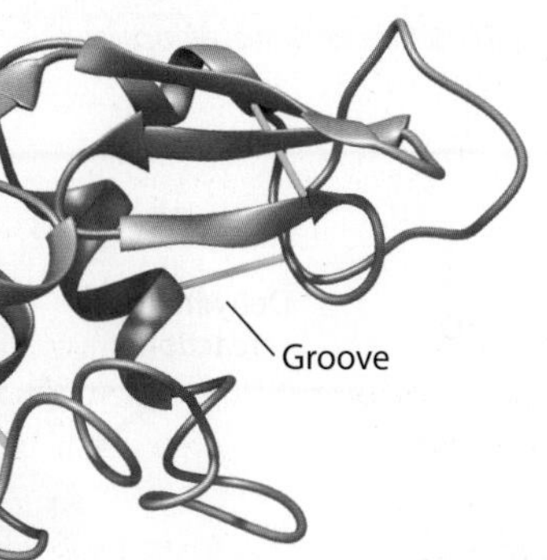
Groove

▲ **Figure 3.12A** Ribbon model of the protein lysozyme

Groove

▲ **Figure 3.12B** Space-filling model of the protein lysozyme

The dependence of protein function on a protein's shape becomes clear when a protein is altered. In a process called **denaturation**, a protein unravels, losing its specific shape and, as a result, its function. Excessive heat can denature many proteins. For example, visualize what happens when you fry an egg. Heat quickly denatures the clear proteins surrounding the yolk, making them solid, white, and opaque.

Given the proper cellular environment, a newly synthesized amino acid chain spontaneously folds into its functional shape. What happens if a protein doesn't fold correctly? Many diseases, such as Alzheimer's and Parkinson's, involve an accumulation of misfolded proteins. Prions are infectious misshapen proteins that are associated with serious degenerative brain diseases such as mad cow disease (see Module 10.21). Such diseases reinforce the theme that structure fits function: A protein's unique three-dimensional shape determines its proper functioning. In the next two modules, we'll learn how a protein's structure takes shape.

▲ **Figure 3.12C** Fibrous silk proteins of a spider's web

? **Why does a denatured protein no longer function normally?**

● The function of each protein is a consequence of its specific shape, which is lost when a protein denatures.

3.13 Proteins are made from amino acids linked by peptide bonds

Now let's see what the monomers of proteins look like. **Amino acids** all have an amino group and a carboxyl group (which makes it an acid, hence the name amino *acid*). As you can see in the general structure shown in **Figure 3.13A**, both of these functional groups are covalently bonded to a central carbon atom. Also bonded to this carbon is a hydrogen atom. The fourth bond of the central carbon is to a variable chemical group symbolized by the letter R. In the simplest amino acid (glycine), the R group is just a hydrogen atom. In all others, the R group consists of one or more carbon atoms with various functional groups attached.

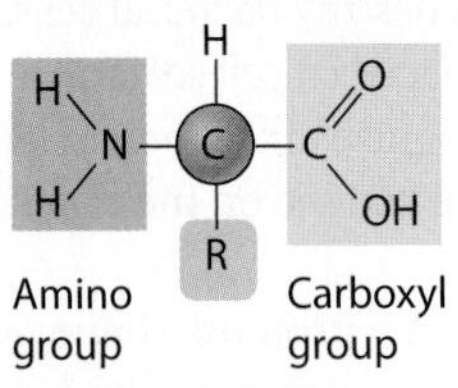

Amino group
Carboxyl group

▲ **Figure 3.13A** General structure of an amino acid

All 20 amino acids are included in Appendix 3, grouped according to whether their R groups are hydrophobic or hydrophilic. **Figure 3.13B** shows representatives of these two main types. Hydrophobic amino acids have nonpolar R groups—note the nonpolar C—H bonds in the R group of leucine (abbreviated Leu) shown in the figure. The R groups of hydrophilic amino acids may be polar or charged. R groups that contain acidic or basic groups are charged at the pH of a cell. Indeed, the amino and carboxyl groups attached to the central carbon of all amino acids are also usually ionized at cellular pH (see ionized forms in Table 3.2).

Now that we have examined amino acids, let's see how they are linked to form polymers. Can you guess? Cells join amino acids together in a dehydration reaction that links the carboxyl group of one amino acid to the amino group of the next amino acid as a water molecule is removed (**Figure 3.13C**). The resulting covalent linkage is called a **peptide bond**. The product of the reaction shown in the figure is called a *di*peptide, because it was made from *two* amino acids. Additional amino acids can be added by the same process to form a chain of amino acids, a **polypeptide**.

How is it possible to make thousands of different kinds of proteins from just 20 amino acids? The answer has to do with sequence. You know that thousands of English words can be made by varying the sequence of letters and word length. Although the protein "alphabet" is slightly smaller (just 20 "letters," rather than 26), the "words" are much longer. Most polypeptides are at least 100 amino acids in length; some are 1,000 or more. Each polypeptide has a unique sequence of amino acids.

But a long polypeptide chain of specific sequence is not the same as a protein, any more than a long strand of yarn is the same as a sweater that can be knit from that yarn. What are the stitches that coil and fold a polypeptide chain into a unique three-dimensional shape? This is where the R groups of the constituent amino acids play their role in influencing protein structure. Hydrophobic amino acids may cluster together in the center of a globular protein, while hydrophilic amino acids face the outside, helping proteins dissolve in the aqueous solution of a cell. Hydrogen bonds and ionic bonds between hydrophilic R groups help determine a protein's shape, as do covalent bonds called disulfide bridges between sulfur atoms in some R groups. (Look back at the yellow lines in Figure 3.12A.) The unique sequence of the various types of amino acids in a polypeptide determines how a protein takes shape. Let's visualize this process in the next module.

? By what process do you digest the proteins you eat into their individual amino acids?

● By hydrolysis, adding a molecule of water back to break each peptide bond

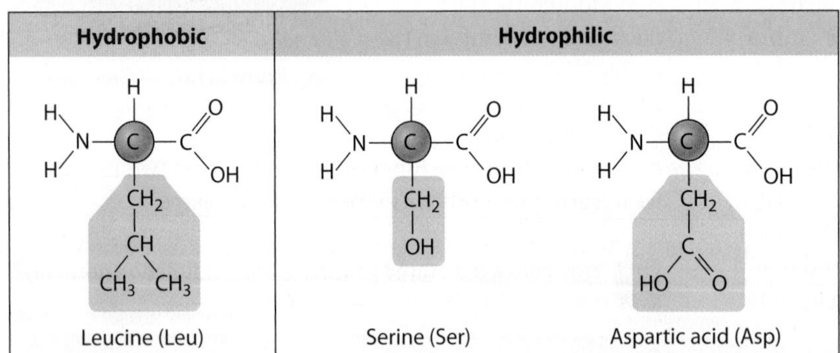

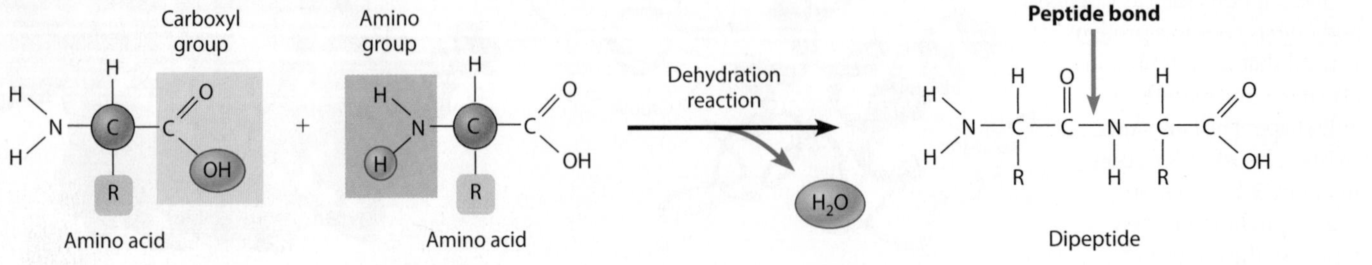

Hydrophobic
Hydrophilic
Leucine (Leu)
Serine (Ser)
Aspartic acid (Asp)

▲ **Figure 3.13B** Examples of amino acids with hydrophobic and hydrophilic R groups

Try This Point out the bonds and functional groups that make the R groups of these three amino acids either hydrophobic or hydrophilic.

Carboxyl group
Amino group
Dehydration reaction
H_2O
Peptide bond
Amino acid
Amino acid
Dipeptide

▲ **Figure 3.13C** Peptide bond formation

VISUALIZING THE CONCEPT

3.14 A protein's functional shape results from four levels of structure

The **primary structure** of a protein is the precise sequence of amino acids in the polypeptide chain. Segments of the chain then coil or fold into local patterns called **secondary structure**. The overall three-dimensional shape of a protein is called **tertiary structure**. Proteins with more than one polypeptide chain have **quaternary structure.**

To help you visualize how these structural levels are superimposed on each other to form a functional protein, let's look at transthyretin, an important transport protein found in your blood. Its specific shape enables it to transport vitamin A and one of the thyroid hormones throughout your body.

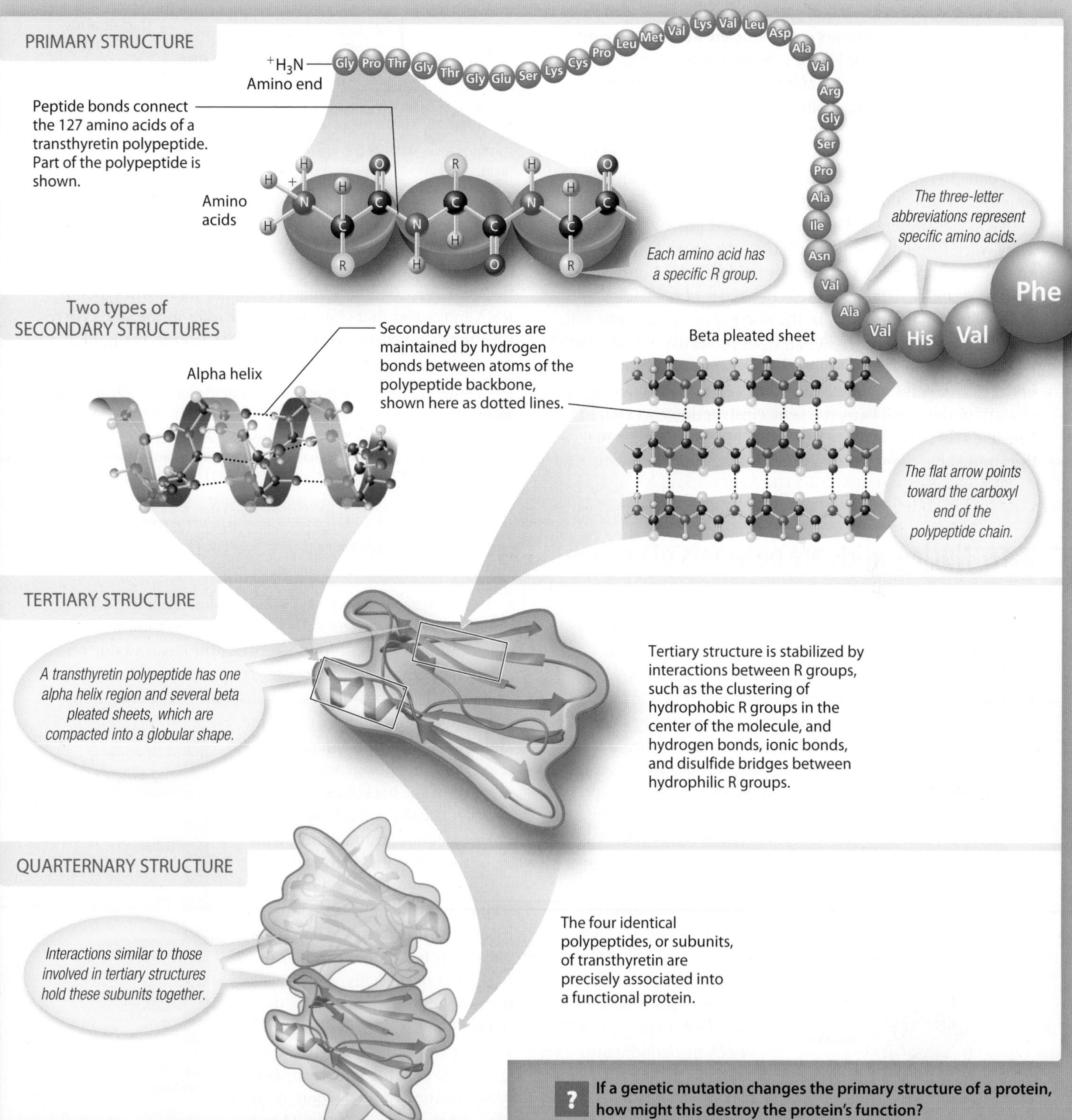

PRIMARY STRUCTURE
$^{+}H_3N$
Amino end
Gly Pro Thr Gly Thr Gly Glu Ser Lys Cys Pro Leu Met Val Lys Val Leu Asp Ala Val Arg Gly Ser Pro Ala Ile Asn Val Ala Val His Val Phe
Peptide bonds connect the 127 amino acids of a transthyretin polypeptide. Part of the polypeptide is shown.
Amino acids
Each amino acid has a specific R group.
The three-letter abbreviations represent specific amino acids.
Two types of SECONDARY STRUCTURES
Alpha helix
Secondary structures are maintained by hydrogen bonds between atoms of the polypeptide backbone, shown here as dotted lines.
Beta pleated sheet
The flat arrow points toward the carboxyl end of the polypeptide chain.
TERTIARY STRUCTURE
A transthyretin polypeptide has one alpha helix region and several beta pleated sheets, which are compacted into a globular shape.
Tertiary structure is stabilized by interactions between R groups, such as the clustering of hydrophobic R groups in the center of the molecule, and hydrogen bonds, ionic bonds, and disulfide bridges between hydrophilic R groups.
QUARTERNARY STRUCTURE
Interactions similar to those involved in tertiary structures hold these subunits together.
The four identical polypeptides, or subunits, of transthyretin are precisely associated into a functional protein.

? If a genetic mutation changes the primary structure of a protein, how might this destroy the protein's function?

Primary structure determines the secondary and tertiary structure due to the chemical nature of the polypeptide backbone and R groups of the amino acids positioned along the chain. Even a slight change may affect a protein's shape and thus its function.

▷ Nucleic Acids

3.15 DNA and RNA are the two types of nucleic acids

As we just saw, the primary structure of a polypeptide determines the shape of a protein. But what determines the primary structure? The amino acid sequence of a polypeptide is programmed by a discrete unit of inheritance known as a **gene**. Genes consist of **DNA (deoxyribonucleic acid)**, one of the two types of polymers called **nucleic acids**. The name *nucleic* comes from their location in the nuclei of eukaryotic cells. The genetic material that humans and other organisms inherit from their parents consists of DNA. Unique among molecules, DNA provides directions for its own replication. Thus, as a cell divides, its genetic instructions are passed to each daughter cell. These instructions program all of a cell's activities by directing the synthesis of proteins.

The genes present in DNA do not build proteins directly. They work through an intermediary—the second type of nucleic acid, known as **ribonucleic acid (RNA)**. Figure 3.15 illustrates the main roles of these two types of nucleic acids in the production of proteins. In the nucleus of a eukaryotic cell, a gene directs the synthesis of an RNA molecule. We say that DNA is transcribed into RNA. The RNA molecule moves out of the nucleus and interacts with the protein-building machinery of the cell. There, the gene's instructions, written in "nucleic acid language," are translated into "protein language," the amino acid sequence of a polypeptide. (In prokaryotic cells, which lack nuclei, both transcription and translation take place within the cytoplasm of the cell.)

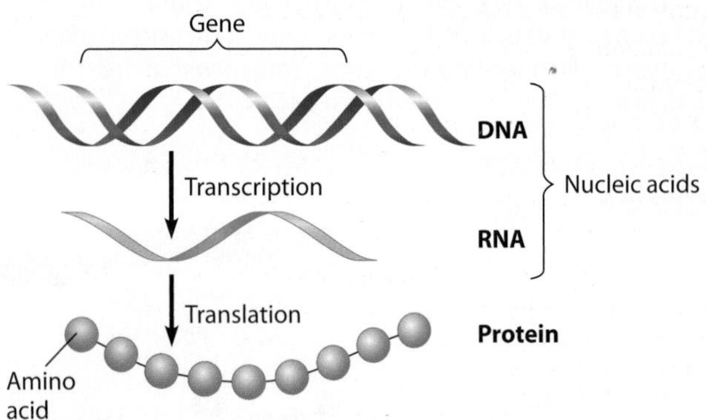
Gene
DNA
Transcription
Nucleic acids
RNA
Translation
Protein
Amino acid

▲ **Figure 3.15** The flow of genetic information in the building of a protein

Recent research has found previously unknown types of RNA molecules that play many other roles in the cell. (We return to the functions of DNA and RNA later in the book.)

? **How are the two types of nucleic acids functionally related?**

● The hereditary material of DNA contains the instructions for the primary structure of polypeptides. RNA is the intermediary that conveys those instructions to the protein-making machinery that assembles amino acids in the designated order.

3.16 Nucleic acids are polymers of nucleotides

The monomers that make up nucleic acids are **nucleotides**. As indicated in Figure 3.16A, each nucleotide contains three parts. At the center of a nucleotide is a five-carbon sugar (blue); the sugar in DNA is deoxyribose, whereas RNA has a slightly different sugar called ribose. Linked to one side of the sugar in both types of nucleotides is a negatively charged phosphate group (yellow). Linked to the sugar's other side is a nitrogenous base (green), a molecular structure containing nitrogen and carbon. (The nitrogen atoms tend to take up H^+ in aqueous solutions, which explains why it is called a nitrogenous *base.*) Each DNA nucleotide has one of four different nitrogenous bases: adenine (A), thymine (T), cytosine (C), and guanine (G). Thus, all genetic information is written in a four-letter alphabet. RNA nucleotides also contain the bases A, C, and G; but the base uracil (U) is found instead of thymine.

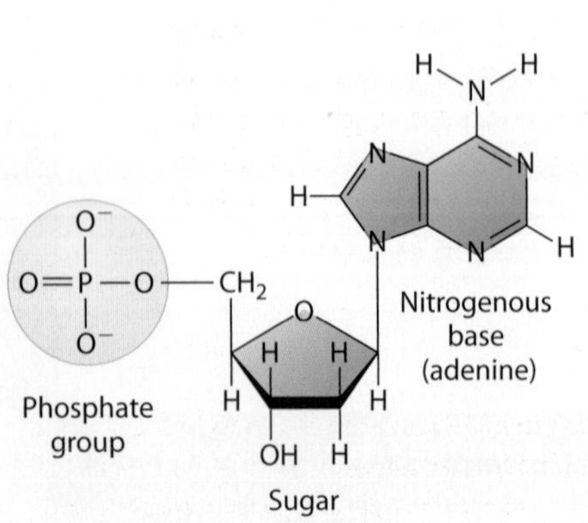
Phosphate group
Nitrogenous base (adenine)
Sugar

▲ **Figure 3.16A** A nucleotide

Like polysaccharides and polypeptides, a nucleic acid polymer—a polynucleotide—is built from its monomers by dehydration reactions. In this process, the sugar of one nucleotide bonds to the phosphate group of the next monomer. The result is a repeating sugar-phosphate backbone in the polymer, as represented by the blue and yellow ribbon in Figure 3.16B. (Note that the nitrogenous bases are not part of the backbone.)

RNA usually consists of a single polynucleotide strand, but DNA is a **double helix**, in which two polynucleotides wrap around each other (Figure 3.16C). The nitrogenous bases protrude from the two sugar-phosphate backbones and pair in the center of the helix. As shown by their diagrammatic shapes in the figure, A always pairs with T, and C always pairs with G. The two DNA chains are held together by hydrogen bonds (indicated by the dotted lines) between their paired bases. These bonds are individually weak, but

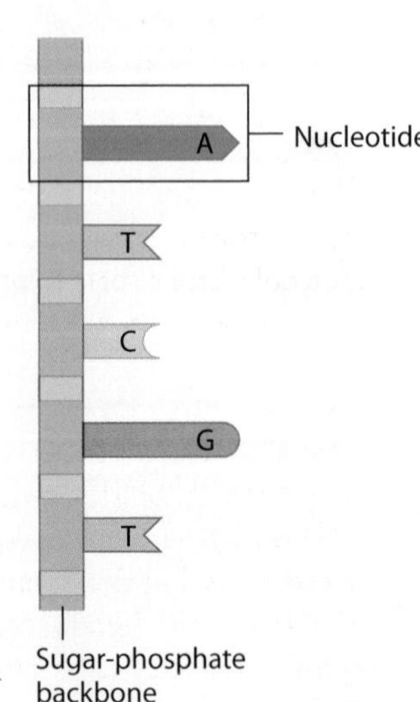
A
Nucleotide
T
C
G
T
Sugar-phosphate backbone

▲ **Figure 3.16B** Part of a polynucleotide

collectively they zip the two strands together into a very stable double helix. Most DNA molecules have thousands or even millions of base pairs.

Because of the base-pairing rules, the two strands of the double helix are said to be *complementary*, each a predictable counterpart of the other. Thus, if a stretch of nucleotides on one strand has the base sequence –AGCACT–, then the same stretch on the other strand must be –TCGTGA–. Complementary base pairing is the key to how a cell makes two identical copies of each of its DNA molecules every time it divides. Thus, the structure of DNA accounts for its function of transmitting genetic information whenever a cell reproduces. The same base-pairing rules (with the exception that U nucleotides of RNA pair with A nucleotides of DNA) also

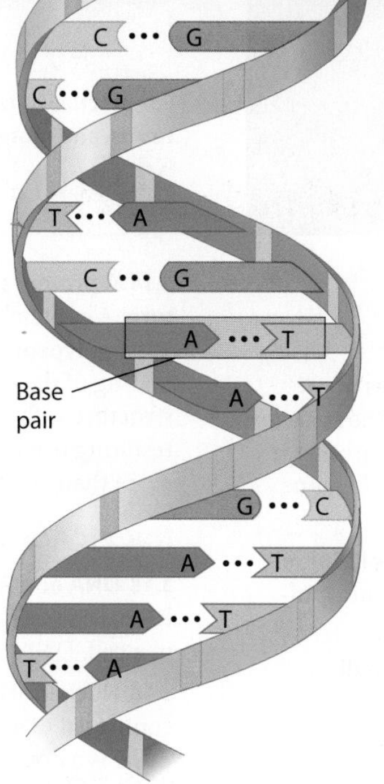
C ··· G
C ··· G
T ··· A
C ··· G
A ··· T
Base pair
A ··· T
G ··· C
A ··· T
A ··· T
T ··· A

▲ **Figure 3.16C** DNA double helix

account for the precise transcription of information from DNA to RNA. (The details of gene transcription and translation are covered in detail in Chapter 10.)

An organism's genes determine the proteins and thus the structures and functions of its body. Let's return to the subject of the chapter introduction—lactose intolerance—to conclude our study of biological molecules. (In the next chapter, we move up in the biological hierarchy to the level of the cell.)

? What roles do complementary base pairing play in the functioning of nucleic acids?

● Complementary base pairing makes possible the precise replication of DNA, ensuring that genetic information is faithfully transmitted every time a cell divides. It also ensures that RNA molecules carry accurate instructions for the synthesis of proteins.

3.17 Lactose tolerance is a recent event in human evolution

EVOLUTION CONNECTION

As you'll recall from the chapter introduction, the majority of people stop producing the enzyme lactase in early childhood and thus do not easily digest the milk sugar lactose. Researchers were curious about the genetic and evolutionary basis for the regional distribution of lactose tolerance and intolerance.

What does evolution have to do with drinking milk?

In 2002, a group of scientists completed a study of the genes of 196 lactose-intolerant adults of African, Asian, and European descent. They determined that lactose intolerance is actually the human norm. It is "lactose tolerance" that represents a relatively recent mutation in the human genome.

The ability to make lactase into adulthood is concentrated in people of northern European descent, and the researchers speculated that lactose tolerance became widespread among this group because it offered a survival advantage. In northern Europe's relatively cold climate, only one harvest a year is possible. Therefore, animals were a main source of food for early humans in that region. Cattle were first domesticated in northern Europe about 9,000 years ago (**Figure 3.17**). With milk and other dairy products at hand year-round, natural selection would have favored anyone with a mutation that kept the lactase gene switched on into adulthood.

Researchers wondered whether the lactose tolerance mutation found in Europeans might be present in other cultures that kept dairy herds. Indeed, a 2006 study compared the genetic makeup and lactose tolerance of 43 ethnic groups in East Africa. The researchers identified three mutations, all different from each other and from the European mutation, that keep the lactase gene permanently turned on. The

▲ **Figure 3.17** A prehistoric European cave painting of cattle

mutations appear to have occurred beginning around 7,000 years ago, around the time that archaeological evidence shows the domestication of cattle in these African regions.

Mutations that conferred a selective advantage, such as surviving cold winters or withstanding drought by drinking milk, spread rapidly in these early pastoral peoples. Their evolutionary and cultural history is thus recorded in their genes and in their continuing ability to digest milk.

Data from S. A. Tishkoff et al., Convergent adaptation of human lactase persistence in Africa and Europe, *Nature Genetics* 39: 31–40 (2006).

? Explain how lactose tolerance involves three of the four major classes of biological macromolecules.

● Lactose, milk sugar, is a carbohydrate that is hydrolyzed by the enzyme lactase, a protein. The ability to make this enzyme and the regulation of when it is made is coded for in DNA, a nucleic acid.

CHAPTER 3 REVIEW

For practice quizzes, BioFlix animations, MP3 tutorials, video tutors, and more study tools designed for this textbook, go to MasteringBiology®

Reviewing the Concepts

Introduction to Organic Compounds (3.1–3.3)

3.1 Life's molecular diversity is based on the properties of carbon. Carbon's ability to bond with four other atoms is the basis for building large and diverse organic compounds. Hydrocarbons are composed of only carbon and hydrogen. Isomers have the same molecular formula but different structures.

3.2 A few chemical groups are key to the functioning of biological molecules. Hydrophilic functional groups give organic molecules specific chemical properties.

3.3 Cells make large molecules from a limited set of small molecules.

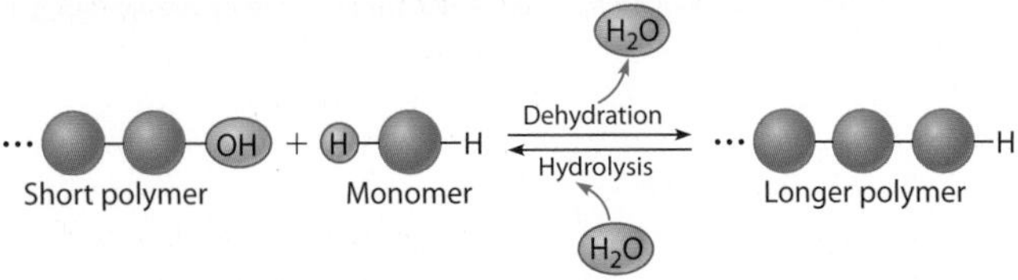

H_2O
Dehydration
Hydrolysis
H_2O
OH
H
H
H
Short polymer
Monomer
Longer polymer

Carbohydrates (3.4–3.7)

3.4 Monosaccharides are the simplest carbohydrates. A monosaccharide has a formula that is a multiple of CH_2O and contains hydroxyl groups and a carbonyl group.

3.5 Two monosaccharides are linked to form a disaccharide.

3.6 What is high-fructose corn syrup, and is it to blame for obesity? HFCS, a mixture of glucose and fructose derived from corn, is commonly added to drinks and processed foods.

3.7 Polysaccharides are long chains of sugar units. Starch and glycogen are storage polysaccharides; cellulose is structural, found in plant cell walls. Chitin is a component of insect exoskeletons and fungal cell walls.

Lipids (3.8–3.11)

3.8 Fats are lipids that are mostly energy-storage molecules. Lipids are diverse, hydrophobic compounds composed largely of carbon and hydrogen. Fats (triglycerides) consist of glycerol linked to three fatty acids. Saturated fatty acids are found in animal fats; unsaturated fatty acids are typical of plant oils.

3.9 Scientific studies document the health risks of trans fats.

3.10 Phospholipids and steroids are important lipids with a variety of functions. Phospholipids are components of cell membranes. Steroids include cholesterol and some hormones.

3.11 Anabolic steroids pose health risks.

Proteins (3.12–3.14)

3.12 Proteins have a wide range of functions and structures. Proteins are involved in almost all of a cell's activities; as enzymes, they regulate chemical reactions.

3.13 Proteins are made from amino acids linked by peptide bonds. Protein diversity is based on different sequences of amino acids, monomers that contain an amino group, a carboxyl group, an H atom, and an R group, all attached to a central carbon. The R groups distinguish 20 amino acids, each with specific properties.

3.14 A protein's functional shape results from four levels of structure. A protein's primary structure is the sequence of amino acids in its polypeptide chain. Its secondary structure is the coiling or folding of the chain, stabilized by hydrogen bonds. The tertiary structure is the overall three-dimensional shape of a polypeptide, resulting from interactions among R groups. Proteins made of more than one polypeptide have quaternary structure.

Nucleic Acids (3.15–3.17)

3.15 DNA and RNA are the two types of nucleic acids. DNA and RNA serve as the blueprints for proteins and thus control the life of a cell. DNA is the molecule of inheritance.

3.16 Nucleic acids are polymers of nucleotides. Nucleotides are composed of a sugar, a phosphate group, and a nitrogenous base. DNA is a double helix; RNA is a single polynucleotide chain.

3.17 Lactose tolerance is a recent event in human evolution. Mutations in DNA have led to lactose tolerance in several human groups whose ancestors raised dairy cattle.

Connecting the Concepts

1. Complete the table to help review the structures and functions of the four classes of organic molecules.

Classes of Molecules and Their Components	Functions	Examples
Carbohydrates	Energy for cell, raw material	a. ________
	b. ________	Starch, glycogen
Monosaccharide	Plant cell support	c. ________
Lipids (don't form polymers)	Energy storage	d. ________
Glycerol; Fatty acid	e. ________	Phospholipids
Components of a fat molecule	Hormones	f. ________
Proteins	j. ________	Lactase
g. ________ h. ________	k. ________	Hair, tendons
	l. ________	Muscle proteins
	Transport	m. ________
	Communication	Signal proteins
	n. ________	Antibodies
i. ________	Storage	Proteins in seeds
Amino acid	Receive signals	Receptor protein
Nucleic Acids p. ________ o. ________	Heredity	r. ________
Nucleotide q. ________	s. ________	DNA and RNA

Testing Your Knowledge

Level 1: Knowledge/Comprehension

2. A glucose molecule is to starch as (*Explain your answer.*)
   a. a steroid is to a lipid.
   b. a protein is to an amino acid.
   c. a nucleic acid is to a polypeptide.
   d. a nucleotide is to a nucleic acid.
3. What makes a fatty acid an acid?
   a. It does not dissolve in water.
   b. It is capable of bonding with other molecules to form a fat.
   c. It has a carboxyl group that can donate an H^+ to a solution.
   d. It contains only two oxygen atoms.
4. Cows can derive nutrients from cellulose because
   a. they produce enzymes that recognize the shape of the glucose-glucose bonds and hydrolyze them.
   b. they re-chew their cud to break down cellulose fibers.
   c. their digestive tract contains prokaryotes that can hydrolyze the bonds of cellulose.
   d. they convert cellulose to starch and can digest starch.
5. Of the following functional groups, which is/are polar, tending to make organic compounds hydrophilic?
   a. carbonyl
   b. amino
   c. hydroxyl
   d. all of the above
6. Unsaturated fats
   a. have double bonds in their fatty acid chains.
   b. have fewer fatty acid molecules per fat molecule.
   c. are associated with greater health risks than are saturated fats.
   d. are more common in animals than in plants.

Level 2: Application/Analysis

7. A shortage of phosphorus in the soil would make it especially difficult for a plant to manufacture
   a. DNA.
   b. proteins.
   c. cellulose.
   d. sucrose.
8. Which of the following substances is a major component of the cell membrane of a fungus?
   a. cellulose
   b. chitin
   c. cholesterol
   d. phospholipids
9. Which structural level of a protein would be *least* affected by a disruption in hydrogen bonding?
   a. primary structure
   b. secondary structure
   c. tertiary structure
   d. quaternary structure
10. Circle and name the functional groups in this organic molecule. What type of compound is this? For which class of macromolecules is it a monomer?
11. Most proteins are soluble in the aqueous environment of a cell. Knowing that, where in the overall three-dimensional shape of a protein would you expect to find amino acids with hydrophobic R groups?
12. Sucrose is broken down in your intestine to the monosaccharides glucose and fructose, which are then absorbed into your blood. What is the name of this type of reaction? Using this diagram of sucrose, show how this would occur.

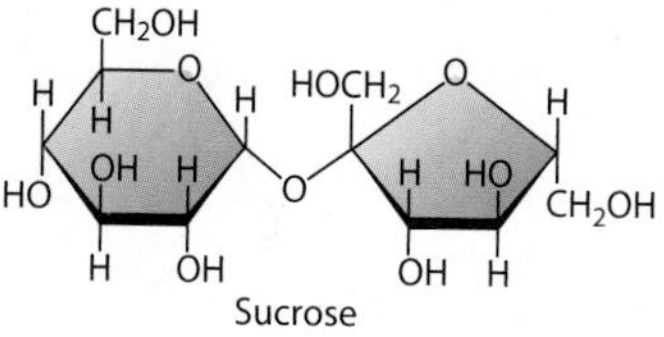

CH_2OH
$HOCH_2$
CH_2OH
Sucrose

Level 3: Synthesis/Evaluation

13. The diversity of life is staggering. Yet the molecular logic of life is simple and elegant: Small molecules common to all organisms are ordered into unique macromolecules. Explain why carbon is central to this diversity of organic molecules. How do carbon skeletons, chemical groups, monomers, and polymers relate to this molecular logic of life?
14. How can a cell make many different kinds of proteins out of only 20 amino acids? Of the myriad possibilities, how does the cell "know" which proteins to make?
15. Given that the function of egg yolk is to nourish and support the developing chick, explain why egg yolks are so high in fat, protein, and cholesterol.
16. Enzymes usually function best at an optimal pH and temperature. The following graph shows the effectiveness of two enzymes at various temperatures.

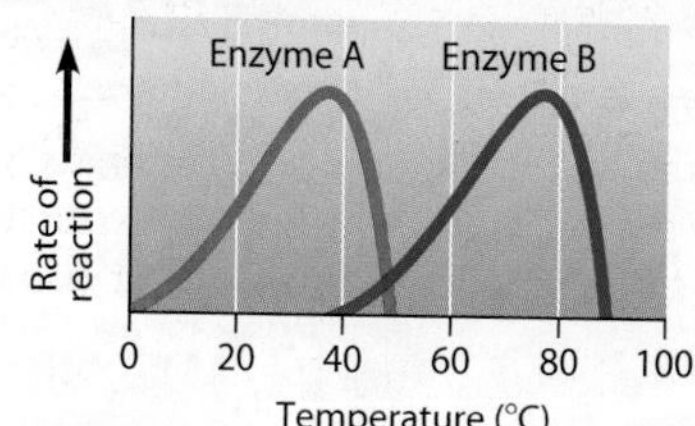

Enzyme A
Enzyme B
Rate of reaction
0
20
40
60
80
100
Temperature (°C)

   a. At which temperature does enzyme A perform best? Enzyme B?
   b. One of these enzymes is found in humans and the other in thermophilic (heat-loving) bacteria. Which enzyme would you predict comes from which organism?
   c. From what you know about enzyme structure, explain why the rate of the reaction catalyzed by enzyme A slows down at temperatures above 40°C (140°F).

17. **SCIENTIFIC THINKING** Another aspect of the Nurses' Health Study introduced in Module 3.9 looked at the percentage of change in the risk of coronary heart disease associated with substituting one dietary component for another. These results estimated that replacement of 5% of energy from saturated fat in the diet with unsaturated fats would reduce the risk by 42%, and that the replacement of 2% of energy from trans fat with unsaturated fats would reduce the risk by 53%. Explain what these numbers mean.

Answers to all questions can be found in Appendix 4.

CHAPTER 4

A Tour of the Cell

? *How has our knowledge of cells grown?*

You can probably identify the blue blobs in this beautiful micrograph as the nuclei of the cells it depicts. But did you know that the brightly colored strands you also see form a cell's skeleton? These structures are part of a system of protein fibers called the cytoskeleton. Much like the way your skeleton provides support and also enables you to move, the cytoskeleton provides structural support to a cell and allows some cells to crawl and others to swim. But even stationary cells have movement: Many of their internal parts bustle about, often traveling on cytoskeletal "roads." Later in the chapter you will learn more about the cytoskeleton and how our knowledge of its structures and functions has grown. As you will see, our understanding of nature often goes hand in hand with the invention and refinement of instruments that extend our senses. This certainly applies to how cells were first discovered.

In 1665, Robert Hooke used a crude microscope to examine a piece of cork. Hooke compared the structures he saw to "little rooms"—*cellulae* in Latin—and the term *cell* stuck. His contemporary, Antoni van Leeuwenhoek, working with more refined lenses, examined numerous subjects, from

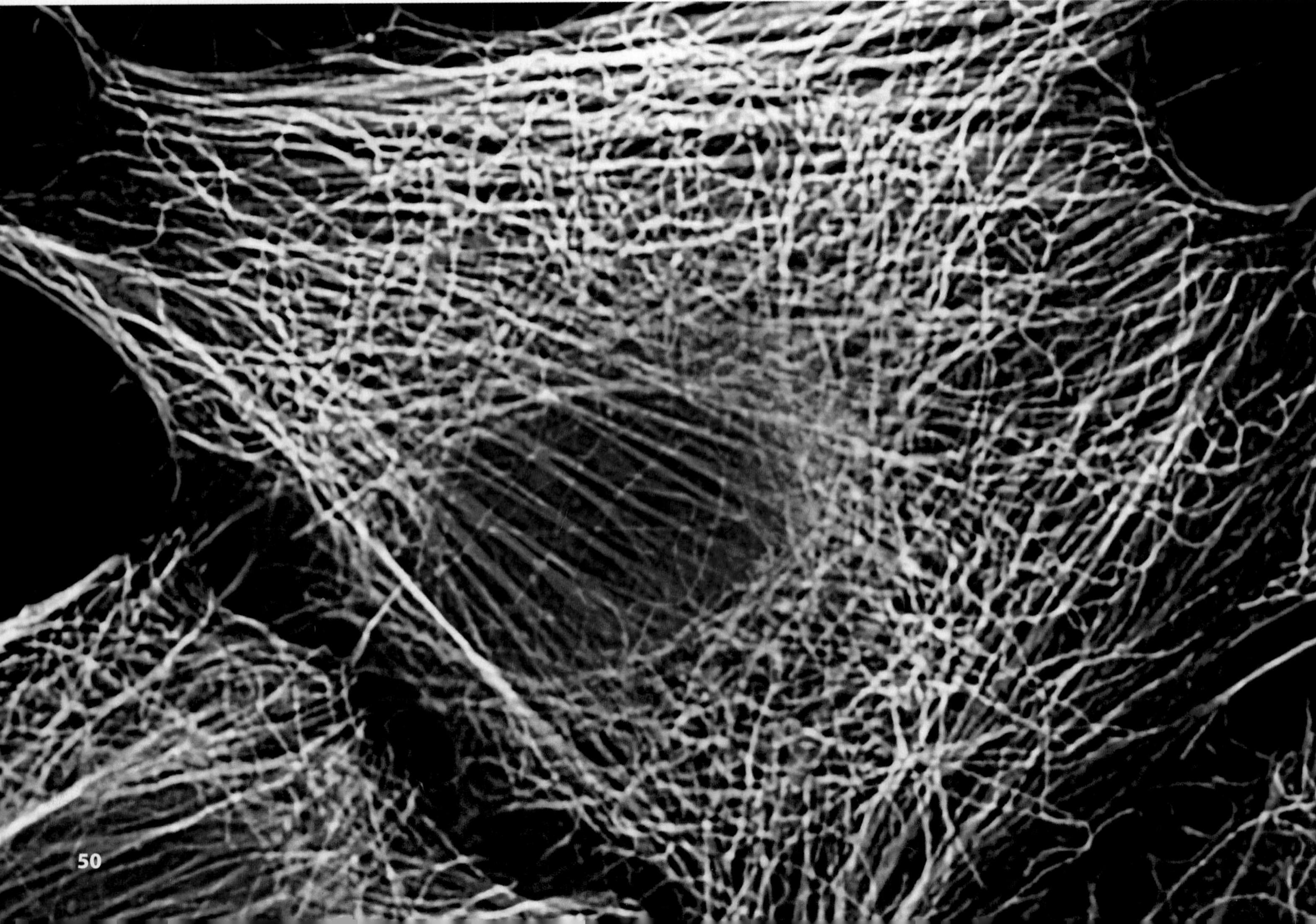

blood and sperm to pond water. He produced drawings and enthusiastic descriptions of his discoveries, such as the tiny "animalcules, very prettily a-moving" he found in the scrapings from his teeth.

Since the days of Hooke and Leeuwenhoek, improved microscopes and techniques have vastly expanded our view of the cell. For example, fluorescently colored stains reveal the cytoskeleton in the cells pictured below. In this chapter, you will see many micrographs using such techniques, and they will often be paired with drawings that help emphasize specific details.

Neither drawings nor micrographs allow you to see the dynamic nature of living cells. For that, you need to look through a microscope or view videos. As you study the images in this chapter, keep in mind that the parts of a cell are moving and interacting. Indeed, the phenomenon we call life emerges from the interactions of the many components of a cell.

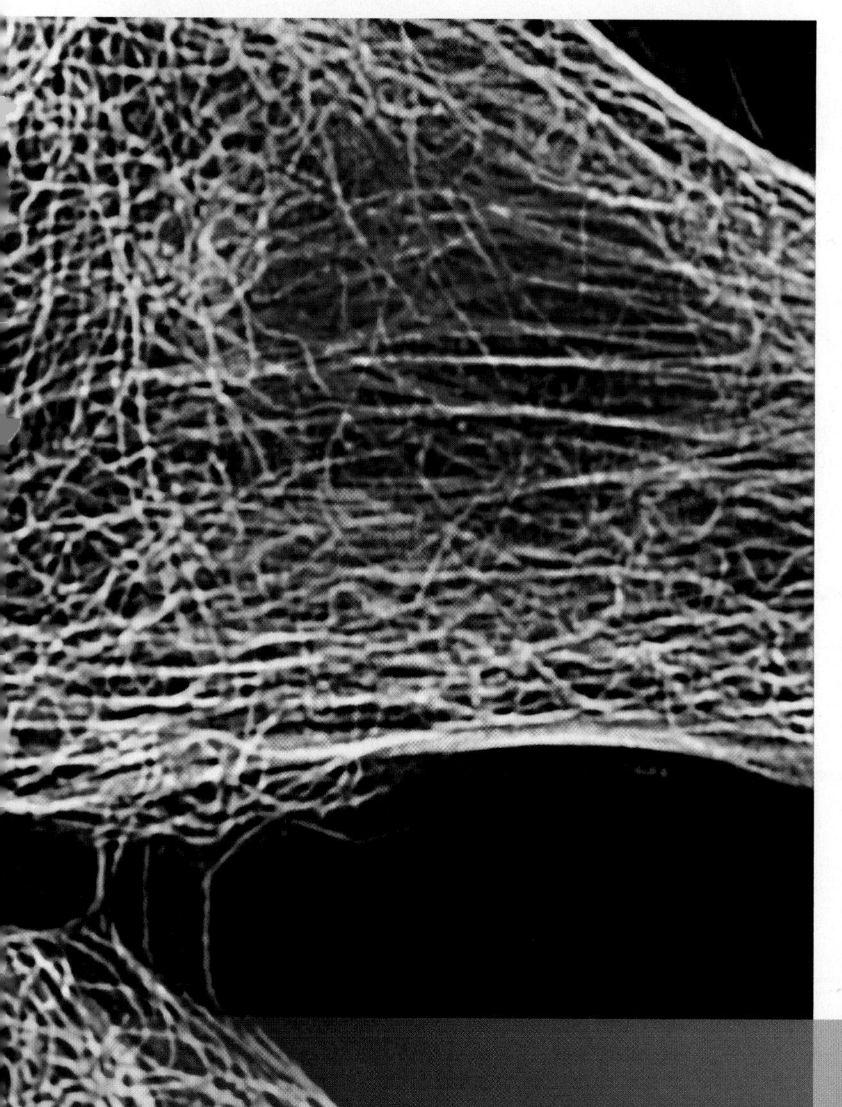

BIG IDEAS

Introduction to the Cell
(4.1–4.4)

Microscopes reveal the structures of cells—the fundamental units of life.

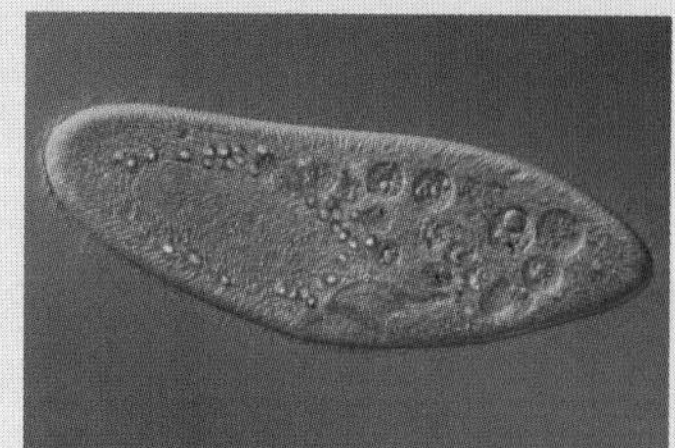

The Nucleus and Ribosomes
(4.5–4.6)

A cell's genetic instructions are housed in the nucleus and carried out by ribosomes.

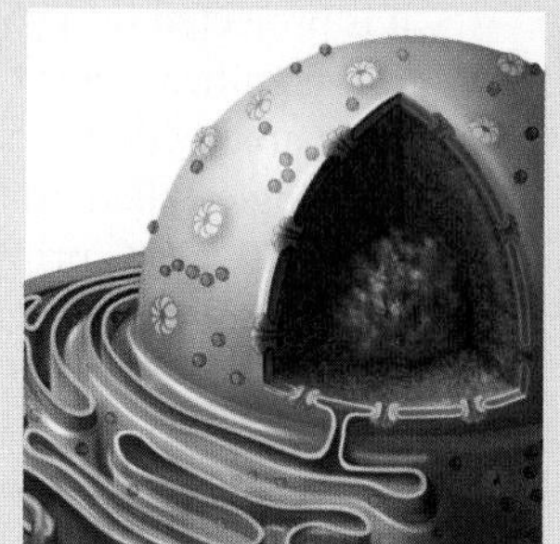

The Endomembrane System
(4.7–4.12)

The endomembrane system participates in the manufacture, distribution, and breakdown of materials.

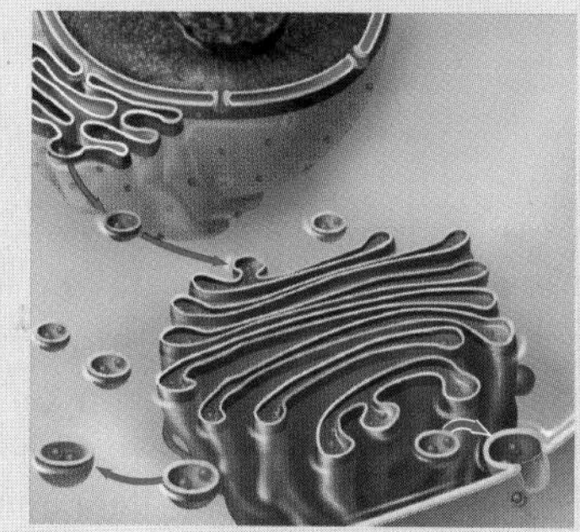

Energy-Converting Organelles
(4.13–4.15)

Mitochondria in all eukaryotic cells and chloroplasts in plant cells function in energy processing.

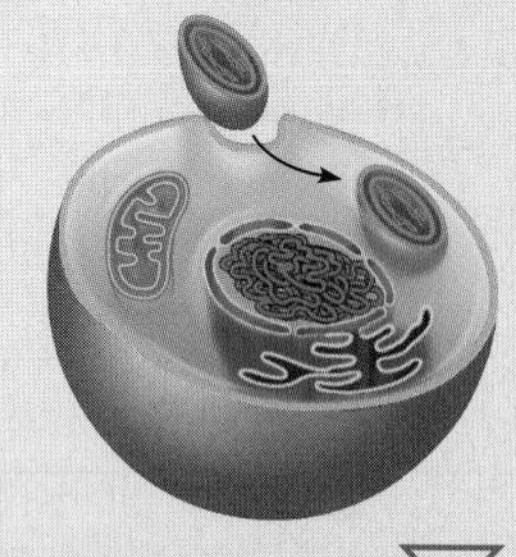

The Cytoskeleton and Cell Surfaces
(4.16–4.22)

The cytoskeleton and extracellular components provide support, motility, and functional connections.

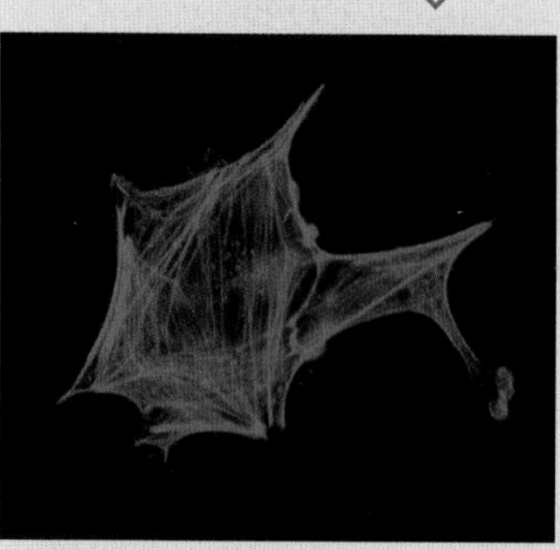

Introduction to the Cell

4.1 Microscopes reveal the world of the cell

Before microscopes were first used in the 1600s, no one knew that living organisms were composed of the tiny units we call cells. The first microscopes were light microscopes, like the ones you may use in a biology laboratory. In a **light microscope (LM)**, visible light is passed through a specimen, such as a microorganism or a thin slice of animal or plant tissue, and then through glass lenses. The lenses bend the light in such a way that the image of the specimen is magnified as it is projected into your eye or a camera.

Magnification is the increase in an object's image size compared with its actual size. **Figure 4.1A** shows a micrograph of a single-celled organism called *Paramecium*. The notation "LM 230×" printed along the right edge tells you that this photograph was taken through a light microscope and that the image is 230 times the actual size of the organism. This *Paramecium* is about 0.33 millimeter in length. **Table 4.1** shows the most common units of length that biologists use.

An important factor in microscopy is resolution, a measure of the clarity of an image. Resolution is the ability to distinguish two nearby objects as separate. For example, what you see as a single star in the sky may be resolved as twin stars with a telescope. Each optical instrument—be it an eye, a telescope, or a microscope—has a limit to its resolution. The human eye can distinguish points as close together as 0.1 millimeter (mm), about the size of a very fine grain of sand. A typical light microscope cannot resolve detail finer than about 0.2 micrometer (μm), about the size of the smallest bacterium. No matter how many times the image of such a small cell is magnified, the light microscope cannot resolve the details of its structure. Indeed, light microscopes can effectively magnify objects only about 1,000 times.

From the time that Hooke discovered cells in 1665 until the middle of the 1900s, biologists had only light microscopes for viewing cells. With these microscopes and various staining techniques to increase contrast between parts of cells, these early biologists discovered microorganisms, animal and plant cells, and even some structures within cells. By the mid-1800s, this accumulation of evidence led to the **cell theory**, which states that all living things are composed of cells and that all cells come from other cells.

Our knowledge of cell structure took a giant leap forward as biologists began using the electron microscope in the 1950s. Instead of using light, an **electron microscope (EM)** focuses a beam of electrons through a specimen or onto its surface. Electron microscopes can distinguish biological structures as small as about 2 nanometers (nm), a 100-fold improvement over the light microscope. This high resolution has enabled biologists to explore cell ultrastructure, the complex internal anatomy of a cell. **Figures 4.1B** and **4.1C** show images produced by two kinds of electron microscopes.

TABLE 4.1 | METRIC MEASUREMENT EQUIVALENTS

1 meter (m) = 100 cm = 1,000 mm = 39.4 inches
1 centimeter (cm) = 10^{-2} (1/100) m = 0.4 inch
1 millimeter (mm) = 10^{-3} (1/1,000) m = 10^{-1} (1/10) cm
1 micrometer (μm) = 10^{-6} m = 10^{-3} mm
1 nanometer (nm) = 10^{-9} m = 10^{-3} μm

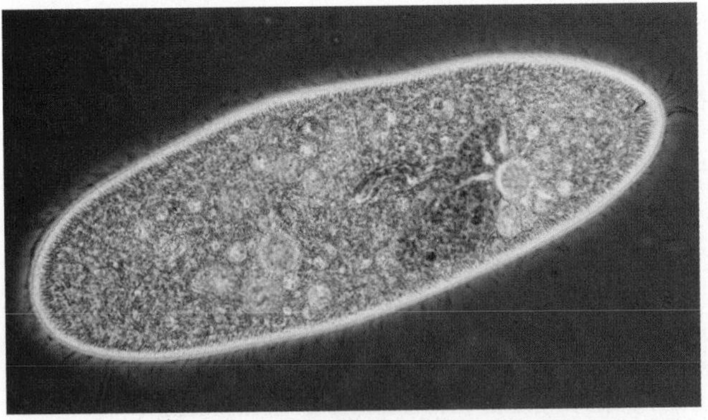
LM 230×

▲ **Figure 4.1A** Light micrograph of a unicellular organism, *Paramecium*

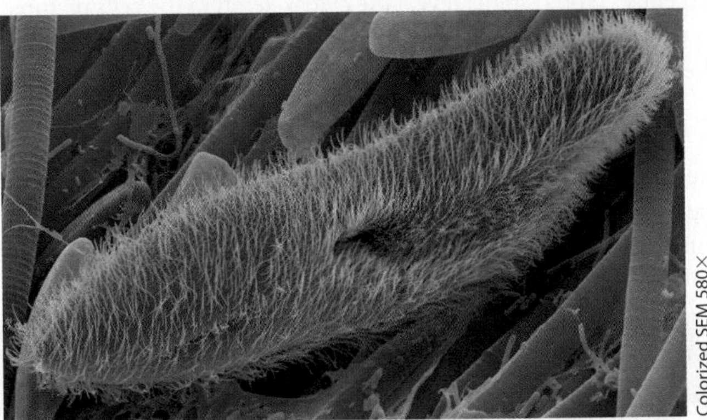
Colorized SEM 580×

▲ **Figure 4.1B** Scanning electron micrograph of *Paramecium*

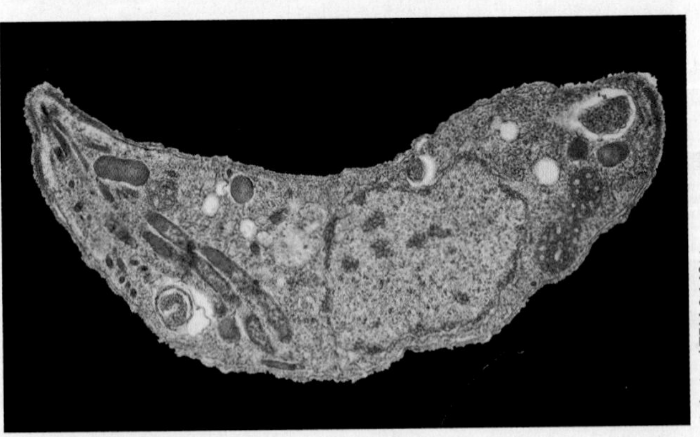
Colorized TEM 9,140×

▲ **Figure 4.1C** Transmission electron micrograph of *Toxoplasma* (This parasite of cats can be transmitted to humans, causing the disease toxoplasmosis.)

Try This Describe a major difference between the *Paramecium* in Figure 4.1B and the protist in this figure. (*Hint*: Compare the notations along the right sides of the micrographs.)

Biologists use the **scanning electron microscope (SEM)** to study the detailed architecture of cell surfaces. The SEM uses an electron beam to scan the surface of a cell or other sample, which is usually coated with a thin film of gold. The beam excites electrons on the surface, and these electrons are then detected by a device that translates their pattern into an image projected onto a video screen. The scanning electron micrograph in Figure 4.1B highlights the numerous cilia on *Paramecium*, projections it uses for movement. Notice the indentation, called the oral groove, through which food enters the cell. As you can see, the SEM produces images that look three-dimensional.

The **transmission electron microscope (TEM)** is used to study the details of internal cell structure. The TEM aims an electron beam through a very thin section of a specimen, just as a light microscope aims a beam of light through a specimen. The section is stained with atoms of heavy metals, which attach to certain cellular structures more than others. Electrons are scattered by these more dense parts, and the image is created by the pattern of transmitted electrons. Instead of using glass lenses, both the SEM and TEM use electromagnets as lenses to bend the paths of the electrons, magnifying and focusing the image onto a monitor. The transmission electron micrograph in Figure 4.1C shows internal details of a protist called *Toxoplasma*. SEMs and TEMs are initially black and white but are often artificially colorized, as they are in these figures, to highlight or clarify structural features.

Electron microscopes have truly revolutionized the study of cells and their structures. Nonetheless, they have not replaced the light microscope: Electron microscopes cannot be used to study living specimens because the methods used to prepare the specimen kill the cells. For a biologist studying a living process, such as the movement of *Paramecium*, a light microscope equipped with a video camera is more suitable than either an SEM or a TEM.

There are different types of light microscopy, and major technical advances in the past several decades have greatly expanded our ability to visualize cells. **Figure 4.1D** shows *Paramecium* as seen using differential interference contrast microscopy. This optical technique amplifies differences in density so that the structures in living cells appear almost three-dimensional. Other techniques use fluorescent stains that selectively bind to various cellular molecules (see the chapter introduction).

You will see many beautiful and illuminating examples of microscopy in this textbook. But even with the magnification shown beside each micrograph, it is often hard to imagine just how small cells are. **Figure 4.1E** shows the size range of cells compared with objects both larger and smaller and the optical instrument that allows us to view them. Notice that the scale along the left side of the figure is logarithmic to accommodate the range of sizes shown. Starting at the top with 10 meters (m), each reference measurement marks a 10-fold decrease in length. Most cells are between 1 and 100 μm in diameter (yellow region of the figure) and are therefore visible only with a microscope. Certain bacteria are as small as 0.2 μm and can barely be seen with a light microscope, whereas bird eggs are large enough to be seen with the unaided eye. A single nerve cell running from the base of your spinal cord to your big toe may be 1 m in length, although it is so thin you would still need a microscope to see it. In the next module, we explore why cells are so small.

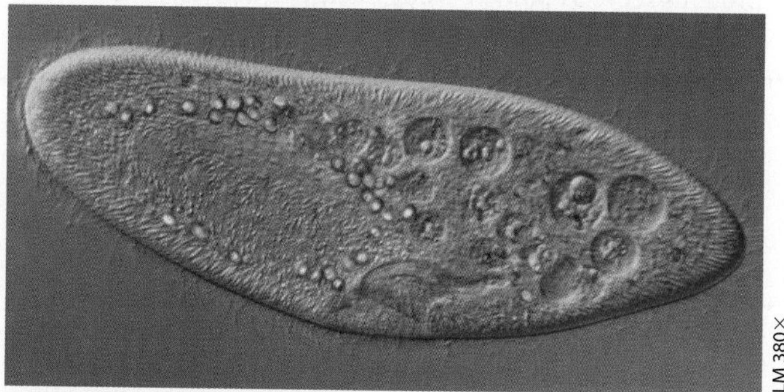
LM 380×

▲ **Figure 4.1D** Differential interference contrast micrograph of *Paramecium*

? Which type of microscope would you use to study (a) the changes in shape of a living human white blood cell; (b) the finest details of surface texture of a human hair; (c) the detailed structure of an organelle in a liver cell?

(a) Light microscope; (b) scanning electron microscope; (c) transmission electron microscope

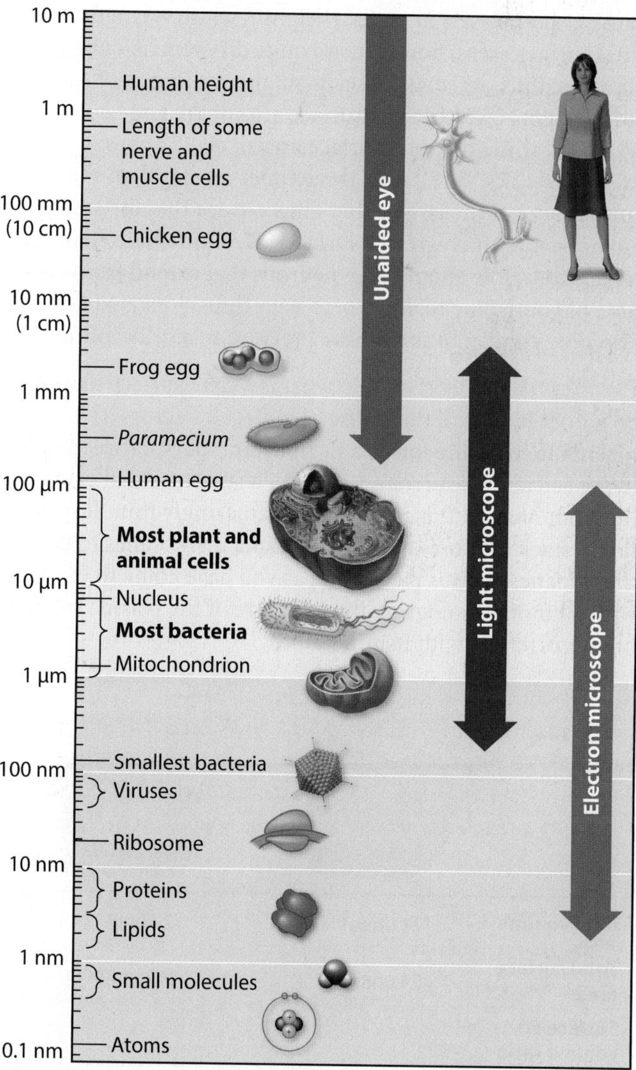
10 m
1 m
100 mm (10 cm)
10 mm (1 cm)
1 mm
100 μm
10 μm
1 μm
100 nm
10 nm
1 nm
0.1 nm
Human height
Length of some nerve and muscle cells
Chicken egg
Frog egg
Paramecium
Human egg
Most plant and animal cells
Nucleus
Most bacteria
Mitochondrion
Smallest bacteria
Viruses
Ribosome
Proteins
Lipids
Small molecules
Atoms
Unaided eye
Light microscope
Electron microscope

▲ **Figure 4.1E** The size range of cells and related objects

4.2 The small size of cells relates to the need to exchange materials across the plasma membrane

As you saw in Figure 4.1E, most cells are microscopic. Are there advantages to being so small? The logistics of carrying out a cell's functions appear to set both lower and upper limits on cell size. At minimum, a cell must be large enough to house enough DNA, protein molecules, and structures to survive and reproduce. But why aren't most cells as large as chicken eggs? The maximum size of a cell is influenced by geometry—the need to have a surface area large enough to service the volume of a cell. Active cells have a huge amount of traffic across their outer surface. A chicken egg cell isn't very active, but once a chick embryo starts to develop, the egg is divided into many microscopic cells, each bounded by a membrane that allows the essential flow of oxygen, nutrients, and wastes across its surface.

Surface-to-Volume Ratio Large cells have more surface area than small cells, but they have a much smaller surface area relative to their volume than small cells. **Figure 4.2A** illustrates this by comparing one large cube to 27 small ones. Using arbitrary units of measurement, the total volume is the same in both cases: 27 units^3 (height $\times$ width $\times$ length). The total surface areas, however, are quite different. A cube has six sides; thus, its surface area is six times the area of each side (height $\times$ width). The surface area of the large cube is 54 units^2, while the total surface area of all 27 cubes is 162 units^2 ($27 \times 6 \times 1 \times 1$), three times greater than the surface area of the large cube. Thus, we see that the smaller cubes have a much greater surface-to-volume ratio than the large cube. How about those neurons that extend from the base of your spine to your toes? Very thin, elongated shapes also provide a large surface area relative to a cell's volume.

The Plasma Membrane So what is a cell's surface like? And how does it control the traffic of molecules across it? The **plasma membrane** forms a flexible boundary between the living cell and its surroundings. For a structure that separates life from nonlife, this membrane is amazingly thin. It would take a stack of more than 8,000 plasma membranes to equal the thickness of this page. And, as you have come to expect with all things biological, the structure of the plasma membrane correlates with its function.

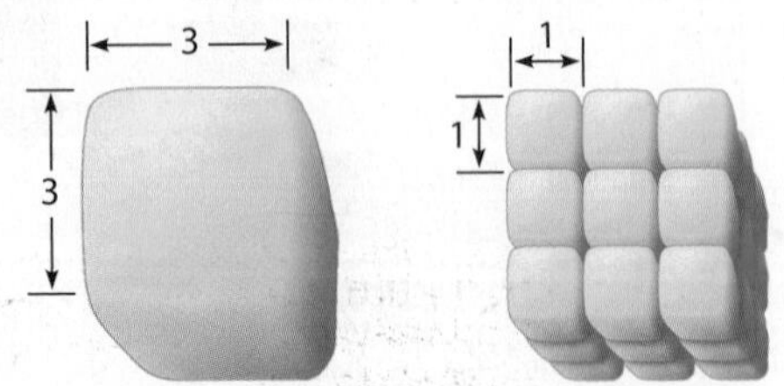

3
3
1
1

Total volume	27 units^3	27 units^3
Total surface area	54 units^2	162 units^2
Surface-to-volume ratio	2	6

▲ **Figure 4.2A** Effect of cell size on surface area

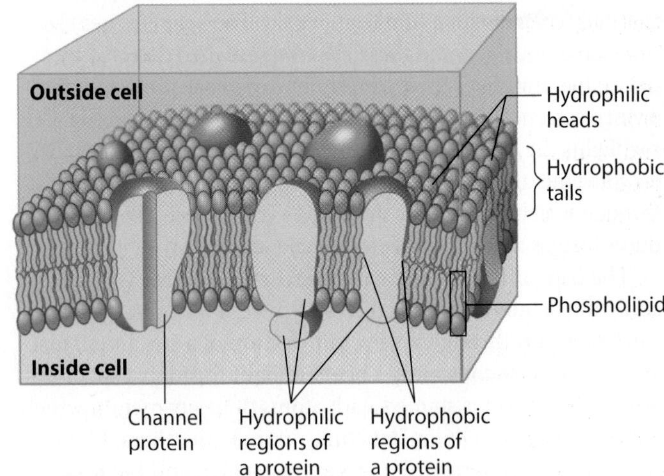

Outside cell
Inside cell
Hydrophilic heads
Hydrophobic tails
Phospholipid
Channel protein
Hydrophilic regions of a protein
Hydrophobic regions of a protein

▲ **Figure 4.2B** A plasma membrane: a phospholipid bilayer with associated proteins

The structure of phospholipid molecules is well suited to their role as a major constituent of biological membranes. Each phospholipid is composed of two distinct regions—a head with a negatively charged phosphate group and two nonpolar fatty acid tails (see Module 3.10). Phospholipids group together to form a two-layer sheet called a phospholipid bilayer. As you can see in **Figure 4.2B**, the phospholipids' hydrophilic (water-loving) heads face outward, exposed to the aqueous solutions on both sides of a membrane. Their hydrophobic (water-fearing) tails point inward, mingling together and shielded from water. Embedded in this lipid bilayer are diverse proteins, floating like icebergs in a phospholipid sea. The regions of the proteins within the center of the membrane are hydrophobic; the exterior sections exposed to water are hydrophilic.

Now let's see how the properties of the phospholipid bilayer and the proteins suspended in it relate to the plasma membrane's job as a traffic cop, regulating the flow of material into and out of the cell. Nonpolar molecules, such as O_2 and CO_2, can easily move across the membrane's hydrophobic interior. Some of the membrane's proteins form channels (tunnels) that shield ions and polar molecules as they pass through the hydrophobic center of the membrane. Still other proteins serve as pumps, using energy to actively transport molecules into or out of the cell.

We will return to the structure and function of biological membranes later (see Chapter 5). In the next module, we consider other features common to all cells and take a closer look at the prokaryotic cells of domains Bacteria and Archaea.

? To convince yourself that a small cell has a greater surface area relative to volume than a large cell, compare the surface-to-volume ratios of the large cube and one of the small cubes in Figure 4.2A.

● Large cube: 54/27 = 2; small cube: 6/1 = 6 (surface area is $1 \times 1 \times 6$ sides = 6 units^2; volume is $1 \times 1 \times 1$ unit^3)

4.3 Prokaryotic cells are structurally simpler than eukaryotic cells

Cells are of two distinct types: prokaryotic and eukaryotic. The microorganisms placed in domains Bacteria and Archaea consist of **prokaryotic cells**. These organisms are known as prokaryotes. All other forms of life (protists, fungi, plants, and animals) are placed in domain Eukarya and are composed of **eukaryotic cells**. They are referred to as eukaryotes.

Eukaryotic cells are distinguished by having a membrane-enclosed nucleus, which houses most of their DNA, and many membrane-enclosed organelles that perform specific functions. Prokaryotic cells are smaller and simpler in structure.

Both types of cells, however, share certain basic features. In addition to being bounded by a plasma membrane, the interior of all cells is filled with a thick, jellylike fluid called **cytosol**, in which cellular components are suspended. All cells have one or more **chromosomes**, which carry genes made of DNA. They also contain **ribosomes**, tiny structures that make proteins according to instructions from the genes. The inside of both types of cells is called the **cytoplasm**. However, in eukaryotic cells, this term refers only to the region between the nucleus and the plasma membrane.

Figure 4.3 explores the structure of a generalized prokaryotic cell. Notice that the DNA is coiled into a region called the **nucleoid** ("nucleus-like"), but no membrane surrounds the DNA. The ribosomes of prokaryotes (shown here in brown) are smaller and differ somewhat from those of eukaryotes. These molecular differences are the basis for the action of some antibiotics, such as tetracycline and streptomycin, which target prokaryotic ribosomes. Thus, protein synthesis can be blocked for the bacterium that's invaded you, but not for you, the eukaryote who is taking the drug.

Outside the plasma membrane (shown here in gray) of most prokaryotes is a fairly rigid, chemically complex cell wall (orange). The wall protects the cell and helps maintain its shape. Some antibiotics, such as penicillin, prevent the formation of these protective walls. Again, because your cells don't have such walls, these antibiotics can kill invading bacteria without harming your cells. Certain prokaryotes have a sticky outer coat called a capsule (yellow) around the cell wall, helping to glue the cells to surfaces or to other cells in a colony. In addition to capsules, some prokaryotes have surface projections. Short projections help attach prokaryotes to each other or their substrate. Longer projections called **flagella** (singular, *flagellum*) propel a prokaryotic cell through its liquid environment.

It takes an electron microscope to see the internal details of any cell, and this is especially true of prokaryotic cells. Notice that the TEM of the bacterium in Figure 4.3 has a magnification of 4,700×. Most prokaryotic cells are about one-tenth the size of a typical eukaryotic cell (see Figure 1.3). (Prokaryotes will be described in more detail later; see Chapter 16.) Eukaryotic cells are the main focus of this chapter, so we turn to these next.

? List three features that are common to prokaryotic and eukaryotic cells. List three features that differ.

● Both types of cells have plasma membranes, chromosomes containing DNA, and ribosomes. Prokaryotic cells are smaller, do not have a nucleus that houses their DNA or other membrane-enclosed organelles, and have smaller, somewhat different ribosomes.

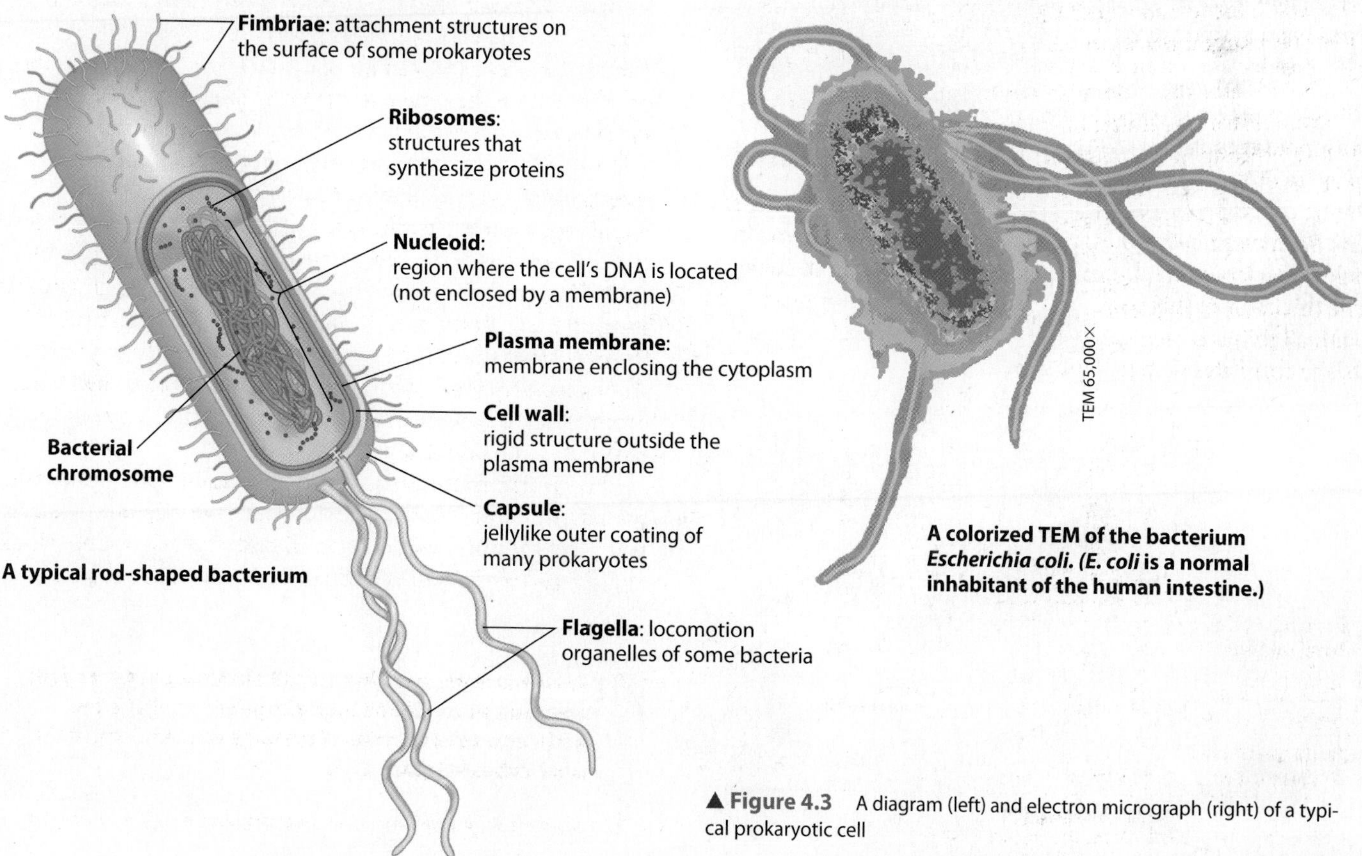

Fimbriae: attachment structures on the surface of some prokaryotes
Ribosomes: structures that synthesize proteins
Nucleoid: region where the cell's DNA is located (not enclosed by a membrane)
Plasma membrane: membrane enclosing the cytoplasm
Cell wall: rigid structure outside the plasma membrane
Capsule: jellylike outer coating of many prokaryotes
Flagella: locomotion organelles of some bacteria
Bacterial chromosome
A typical rod-shaped bacterium
TEM 65,000×
A colorized TEM of the bacterium *Escherichia coli*. (*E. coli* is a normal inhabitant of the human intestine.)

▲ **Figure 4.3** A diagram (left) and electron micrograph (right) of a typical prokaryotic cell

4.4 Eukaryotic cells are partitioned into functional compartments

All eukaryotic cells—whether from animals, plants, protists, or fungi—are fundamentally similar to one another and profoundly different from prokaryotic cells. Let's look at an animal cell and a plant cell as representatives of the eukaryotes.

Figure 4.4A is a diagram of a generalized animal cell, and **Figure 4.4B** shows a generalized plant cell. We color-code the various organelles and other structures in the diagrams for easier identification, and you will see miniature versions of these cells to orient you during our in-depth tour in the rest of the chapter. But no cells would look exactly like these. For one thing, cells have multiple copies of all of these structures (except for the nucleus). Your cells have hundreds of mitochondria and millions of ribosomes. A plant cell may have 30 chloroplasts packed inside. Cells also have different shapes and relative proportions of cell parts, depending on their specialized functions.

The nucleus is the most obvious difference between a prokaryotic and eukaryotic cell. A eukaryotic cell also contains various other **organelles** ("little organs"), which perform specific functions in the cell. Just as the cell itself is wrapped in a membrane made of phospholipids and proteins that perform various functions, each organelle is bounded by a membrane with a lipid and protein composition that suits its function.

The organelles and other structures of eukaryotic cells can be organized into four basic functional groups: (1) The nucleus and ribosomes carry out the genetic control of the cell. (2) Organelles involved in the manufacture, distribution, and breakdown of molecules include the endoplasmic reticulum, Golgi apparatus, lysosomes, vacuoles, and peroxisomes. (3) Mitochondria in all cells and chloroplasts in plant cells function in energy processing. (4) Structural support, movement, and communication between cells are the functions of the cytoskeleton, plasma membrane, and plant cell wall. The cellular components identified in these two figures will be examined in detail in the modules that follow.

In essence, the internal membranes of a eukaryotic cell partition it into compartments. Many of the chemical

▼ **Figure 4.4A** A generalized animal cell

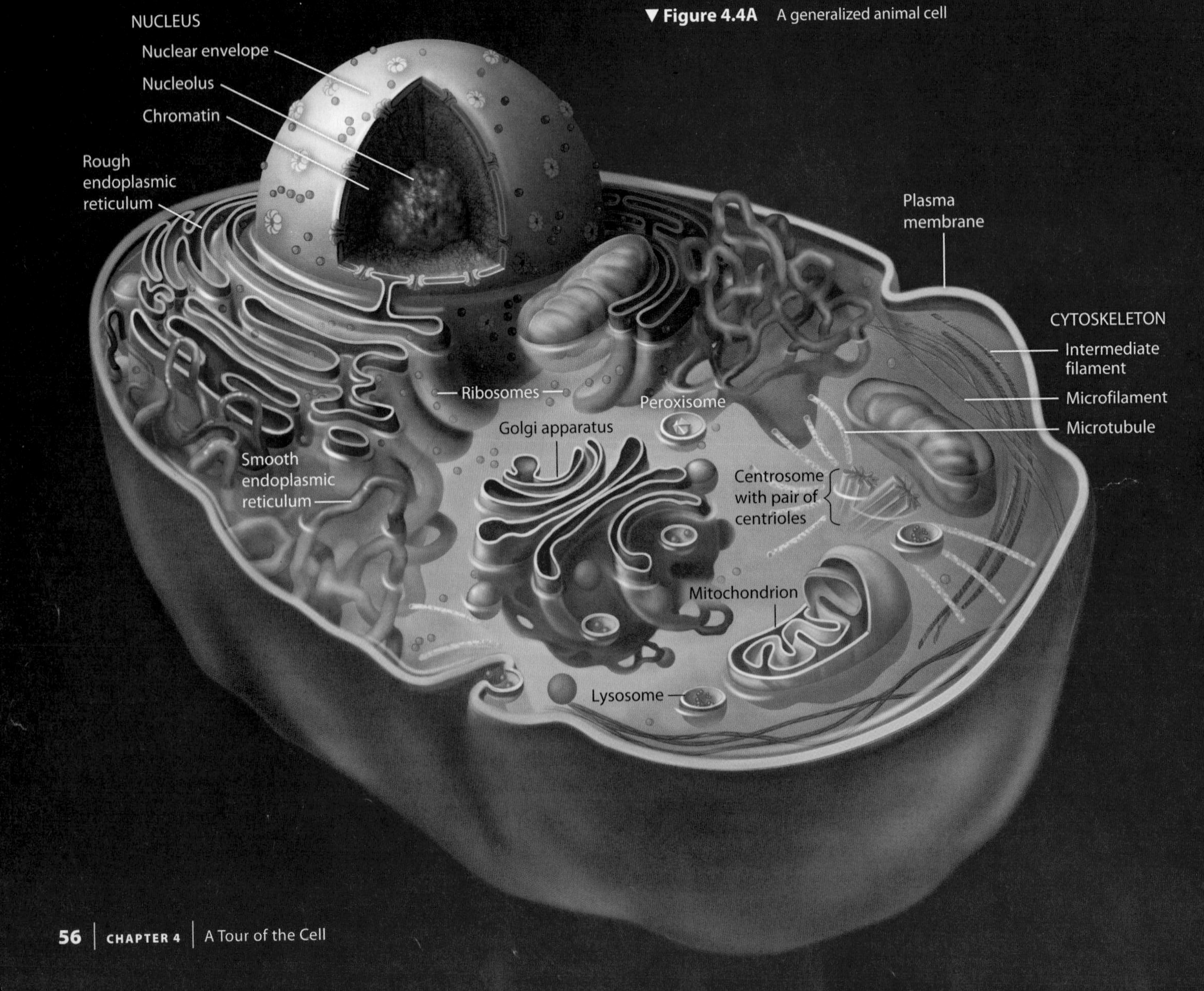
NUCLEUS
Nuclear envelope
Nucleolus
Chromatin
Rough endoplasmic reticulum
Plasma membrane
CYTOSKELETON
Intermediate filament
Microfilament
Microtubule
Ribosomes
Peroxisome
Golgi apparatus
Smooth endoplasmic reticulum
Centrosome with pair of centrioles
Mitochondrion
Lysosome

activities of cells—collectively, **cellular metabolism**—occur within organelles. In fact, many enzymes essential for metabolic processes are built into the membranes of organelles. The fluid-filled spaces within organelles are important as sites where specific chemical conditions are maintained. These conditions vary among organelles and favor the metabolic processes occurring in each. For example, while a part of the endoplasmic reticulum is engaged in making hormones, neighboring peroxisomes may be detoxifying harmful compounds and making hydrogen peroxide (H_2O_2) as a poisonous by-product of their activities. But because the H_2O_2 is confined within the peroxisomes, where it is converted to H_2O by resident enzymes, the rest of the cell is protected.

Except for lysosomes and centrosomes, the organelles and other structures of animal cells are found in plant cells. Also, although some animal cells have flagella or cilia (not shown in Figure 4.4A), among plants, only the sperm cells of a few species have flagella.

A plant cell (Figure 4.4B) also has some structures that an animal cell lacks. For example, a plant cell has a rigid, rather thick cell wall. Cell walls protect cells and help maintain their shape. Chemically different from prokaryotic cell walls, plant cell walls contain the polysaccharide cellulose. Plasmodesmata (singular, plasmodesma) are cytoplasmic channels through cell walls that connect adjacent cells. An important organelle found in plant cells is the chloroplast, where photosynthesis occurs. Unique to plant cells is a large central vacuole, a compartment that stores water and a variety of chemicals.

Eukaryotic cells contain nonmembranous structures as well. The cytoskeleton, which you were introduced to in the chapter introduction, is composed of different types of protein fibers that extend throughout the cell. These networks provide for support and movement. And ribosomes occur throughout the cytosol, as they do in prokaryotic cells. In addition, eukaryotic cells have many ribosomes attached to membranes.

After you preview these cell diagrams, let's move to the first stop on our detailed tour of the eukaryotic cell—the nucleus.

? Identify the structures in the plant cell that are not present in the animal cell.

● Chloroplasts, central vacuole, cell wall, and plasmodesmata

▼ Figure 4.4B A generalized plant cell

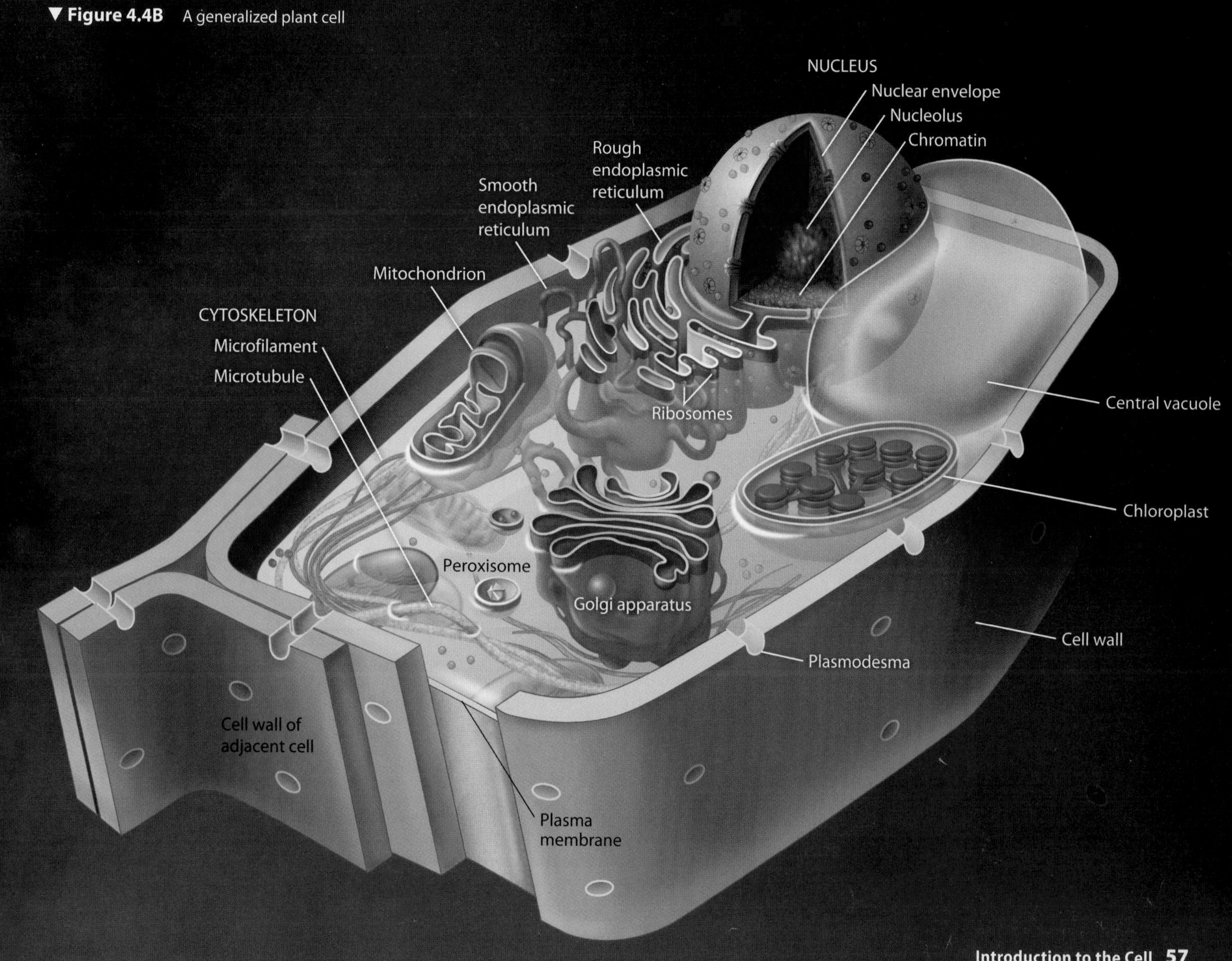
NUCLEUS
Nuclear envelope
Nucleolus
Chromatin
Rough endoplasmic reticulum
Smooth endoplasmic reticulum
Mitochondrion
CYTOSKELETON
Microfilament
Microtubule
Ribosomes
Central vacuole
Chloroplast
Peroxisome
Golgi apparatus
Cell wall
Plasmodesma
Cell wall of adjacent cell
Plasma membrane

The Nucleus and Ribosomes

4.5 The nucleus contains the cell's genetic instructions

You just saw a preview of the many intricate structures that can be found in a eukaryotic cell. A cell must build and maintain these structures and also process energy to support its work of transport, movement, and communication. But who is in charge of this bustling factory? Who stores the master plans, gives the orders, changes course in response to environmental input, and, when called on, makes another factory just like itself? The cell's nucleus functions as this command center.

The **nucleus** contains the cell's genetic instructions encoded in DNA. These master plans control the cell's activities by directing protein synthesis. The DNA is associated with many proteins and organized into structures called chromosomes. The proteins help coil these long DNA molecules. Indeed, the DNA of the 46 chromosomes in one of your cells laid end to end would stretch to a length of more than 2 m, but it must coil up to fit into a nucleus only 5 μm in diameter. When a cell is not dividing, this complex of proteins and DNA, called **chromatin**, appears as a diffuse mass within the nucleus, as shown in the TEM (right half) and diagram (left half) of a nucleus in **Figure 4.5**.

As a cell prepares to divide, the DNA is copied so that each daughter cell can later receive an identical set of genetic instructions. Just prior to cell division, the thin chromatin fibers coil up further, becoming thick enough to be visible with a light microscope as the familiar separate structures you would probably recognize as chromosomes.

Enclosing the nucleus is a double membrane called the **nuclear envelope**. Each of the two membranes is a separate phospholipid bilayer with associated proteins. Similar in function to the plasma membrane, the nuclear envelope controls the flow of materials into and out of the nucleus. As you can see in the diagram of a nucleus in Figure 4.5, the nuclear envelope is perforated with protein-lined pores. These pores regulate the entry and exit of large molecules and also connect with the cell's network of membranes called the endoplasmic reticulum.

The **nucleolus**, a prominent structure in the nucleus, is the site where a special type of RNA called ribosomal RNA(rRNA) is synthesized according to instructions in the DNA. Proteins brought in from the cytoplasm are assembled with this rRNA to form the subunits of ribosomes. These subunits then exit to the cytoplasm, where they will join to form functional ribosomes.

Another type of RNA, messenger RNA (mRNA), directs protein synthesis. Essentially, mRNA is a transcription of protein-synthesizing instructions written in a gene's DNA (see Figure 10.7). The mRNA moves into the cytoplasm, where ribosomes translate it into the amino acid sequences of proteins. Let's look at ribosomes next.

? **Describe the processes that occur in the nucleus.**

DNA is copied and passed on to daughter cells in cell division; rRNA is made and ribosomal subunits assembled; protein-making instructions in DNA are transcribed into mRNA.

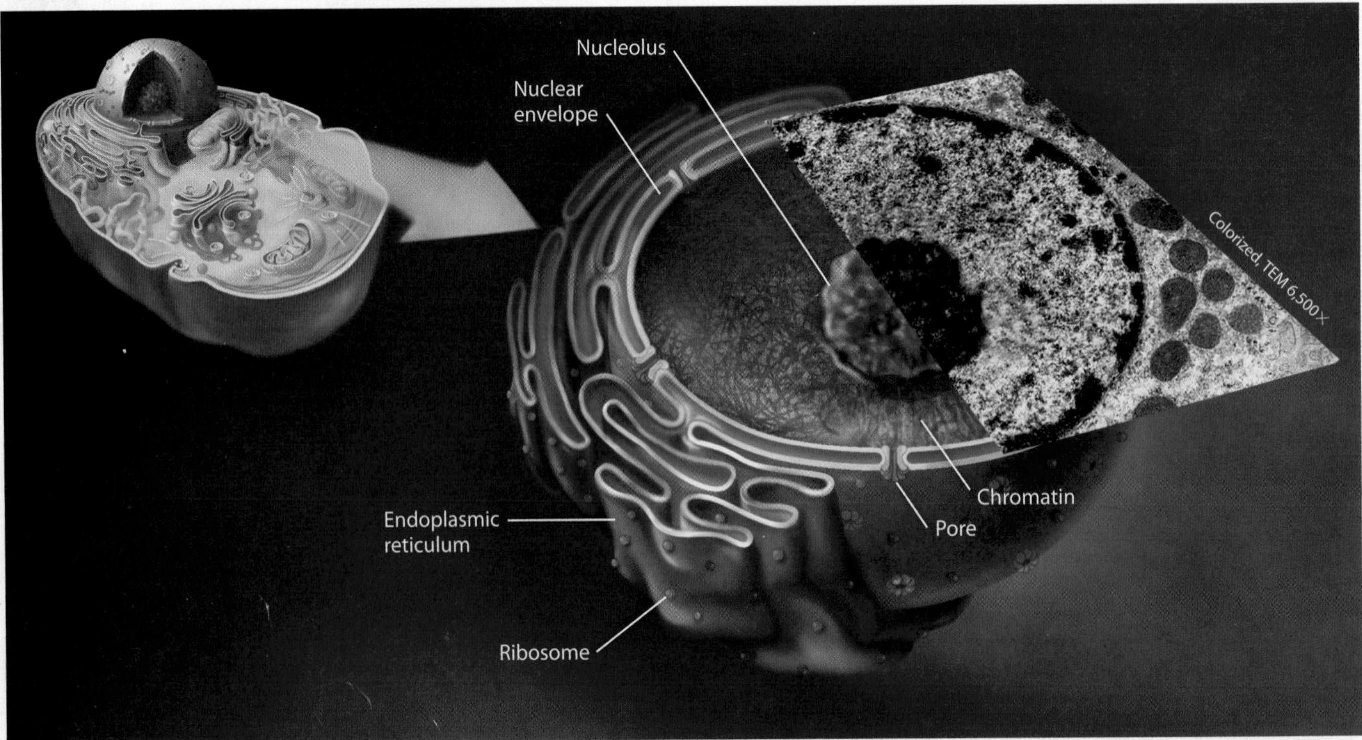

Nucleolus
Nuclear envelope
Colorized, TEM 6,500×
Endoplasmic reticulum
Chromatin
Pore
Ribosome

▲ **Figure 4.5** A diagram with a superimposed TEM of the nucleus

4.6 Ribosomes make proteins for use in the cell and for export

If the nucleus is the cell's command center, then ribosomes are the machines that carry out those commands. Ribosomes are the cellular components that use instructions from the nucleus, written in mRNA, to build proteins. Cells that make a lot of proteins have a large number of ribosomes. For example, a cell in your pancreas that produces digestive enzymes may contain a few million ribosomes. What other structure would you expect to be prominent in cells that are active in protein synthesis? Remember that the nucleolus in the nucleus is the site where the subunits of ribosomes are assembled.

As shown in **Figure 4.6**, ribosomes are found in two locations in the cell. Free ribosomes are suspended in the cytosol, while bound ribosomes are attached to the outside of the endoplasmic reticulum or nuclear envelope. Free and bound ribosomes are structurally identical, and they can move about and function in either location.

Most of the proteins made on free ribosomes function within the cytosol; examples are enzymes that catalyze the first steps of sugar breakdown for cellular respiration. In Module 4.8, you will see how bound ribosomes make proteins that will be inserted into membranes, packaged in certain organelles, or exported from the cell.

At the bottom right in Figure 4.6, you see how ribosomes interact with messenger RNA (carrying the instructions from a gene) to build a protein. The nucleotide sequence of an mRNA molecule is translated into the amino acid sequence of a polypeptide. (Protein synthesis is explored in more detail in Chapter 10.) Next let's look at more of the manufacturing equipment of the cell.

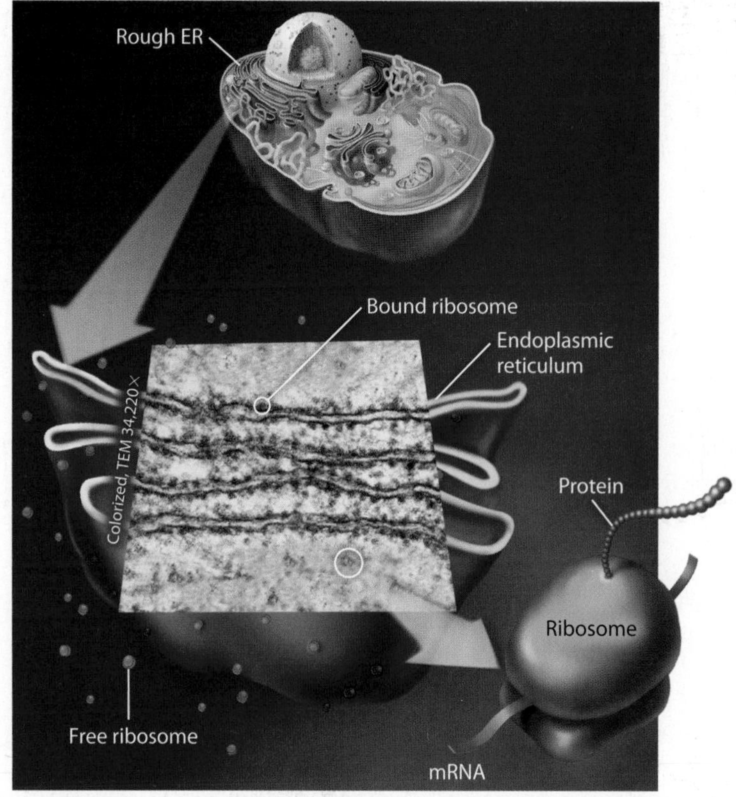

Rough ER
Bound ribosome
Endoplasmic reticulum
Colorized, TEM 34,220×
Protein
Ribosome
Free ribosome
mRNA

▲ **Figure 4.6** The locations and structure of ribosomes

? What role do ribosomes play in carrying out the genetic instructions of a cell?

● Ribosomes synthesize proteins according to the instructions of messenger RNA, which was transcribed from DNA in the nucleus.

▷ The Endomembrane System

4.7 Many organelles are connected in the endomembrane system

Ribosomes may be a cell's protein-making machines, but running a factory as complex as a cell requires infrastructure and many different departments that perform separate but related functions. Internal membranes, a distinguishing feature of eukaryotic cells, are involved in most of a cell's functions. Many of the membranes of the eukaryotic cell are part of an **endomembrane system**. Some of these membranes are physically connected and others are linked when tiny **vesicles** (sacs made of membrane) transfer membrane segments between them.

The endomembrane system includes the nuclear envelope, endoplasmic reticulum, Golgi apparatus, lysosomes, vacuoles, and the plasma membrane. (The plasma membrane is not exactly an *endo* membrane in physical location, but it is related to the other membranes by the transfer of vesicles). Many of these organelles interact in the synthesis, distribution, storage, and export of molecules.

The largest component of the endomembrane system is the **endoplasmic reticulum (ER)**, an extensive network of flattened sacs and tubules. (The term *endoplasmic* means "within the cytoplasm," and *reticulum* is Latin for "little net.") The ER is a prime example of the direct and indirect interrelatedness of parts of the endomembrane system. As shown in Figure 4.5 on the facing page, membranes of the ER are continuous with the nuclear envelope. And when vesicles bud from the ER, they travel to many other components of the endomembrane system.

The tubules and sacs of the ER enclose a space that is separate from the cytosol. Dividing the cell into functional compartments, each of which may require different conditions, is an important aspect of the endomembrane system.

? Which structure includes all others in the list: rough ER, smooth ER, endomembrane system, nuclear envelope?

● Endomembrane system

4.8 The endoplasmic reticulum is a biosynthetic workshop

One of the major manufacturing sites in a cell is the endoplasmic reticulum. The diagram in **Figure 4.8A** shows a cutaway view of the interconnecting membranes of the smooth and rough ER, which can be distinguished in the superimposed electron micrograph. **Smooth endoplasmic reticulum** is called *smooth* because its outer surface lacks attached ribosomes. **Rough endoplasmic reticulum** has bound ribosomes that stud the outer surface of the membrane; thus, it appears *rough* in the electron micrograph.

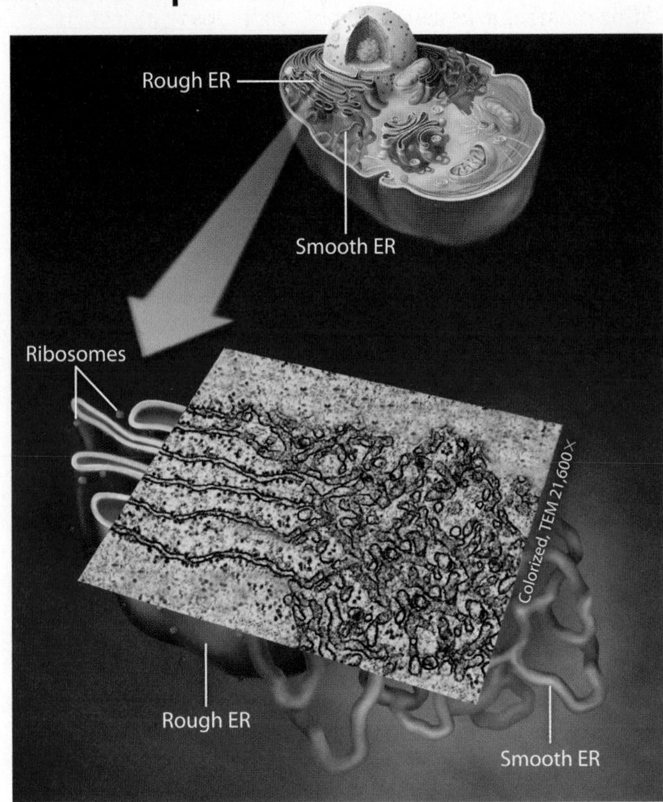

Rough ER
Smooth ER
Ribosomes
Colorized, TEM 21,600×
Rough ER
Smooth ER

▲ **Figure 4.8A** A diagram and TEM of smooth and rough endoplasmic reticulum

Smooth ER The smooth ER of various cell types functions in a variety of metabolic processes. Enzymes of the smooth ER are important in the synthesis of lipids, including oils, phospholipids, and steroids. In vertebrates, for example, cells of the ovaries and testes synthesize the steroid sex hormones. These cells are rich in smooth ER, a structural feature that fits their function by providing ample machinery for steroid synthesis.

Our liver cells also have large amounts of smooth ER, with enzymes that help process drugs, alcohol, and other potentially harmful substances. The sedative phenobarbital and other barbiturates are examples of drugs detoxified by these enzymes. As liver cells are exposed to such chemicals, the amount of smooth ER and its detoxifying enzymes increases, thereby increasing the rate of detoxification and thus the body's tolerance to the drugs. The result is a need for higher doses of a drug to achieve a particular effect, such as sedation. Also, because detoxifying enzymes often cannot distinguish among related chemicals, the growth of smooth ER in response to one drug can increase the need for higher doses of other drugs. Barbiturate abuse, for example, can decrease the effectiveness of certain antibiotics and other useful drugs.

Smooth ER has yet another function, the storage of calcium ions. In muscle cells, for example, a specialized smooth ER membrane pumps calcium ions into the interior of the ER. When a nerve signal stimulates a muscle cell, calcium ions rush from the smooth ER into the cytosol and trigger contraction of the cell.

Rough ER Many types of cells secrete proteins produced by ribosomes attached to rough ER. An example of a secretory protein is insulin, a hormone produced and secreted by certain cells of the pancreas and transported in the bloodstream. Type 1 diabetes results when these cells are destroyed and a lack of insulin disrupts glucose metabolism in the body.

Figure 4.8B follows the synthesis, modification, and packaging of a secretory protein. ❶ As the polypeptide is synthesized by a bound ribosome following the instructions of an mRNA, it is threaded into the cavity of the rough ER. As it enters, the new protein folds into its three-dimensional shape. ❷ Short chains of sugars are often linked to the polypeptide, making the molecule a **glycoprotein** (*glyco* means "sugar"). ❸ When the molecule is ready for export from the ER, it is packaged in a **transport vesicle**, a vesicle that moves from one part of the cell to another. ❹ This vesicle buds off from the ER membrane.

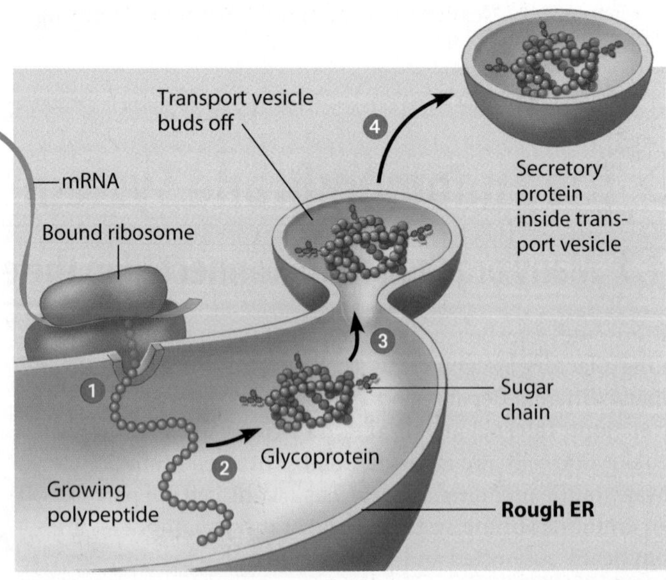

Transport vesicle buds off
4
mRNA
Secretory protein inside transport vesicle
Bound ribosome
3
1
Sugar chain
2
Glycoprotein
Growing polypeptide
Rough ER

▲ **Figure 4.8B** Synthesis and packaging of a secretory protein by the rough ER

Try This Explain where the protein-making instructions carried by the mRNA came from.

The vesicle now carries the protein to the Golgi apparatus for further processing. From there, a transport vesicle containing the finished molecule makes its way to the plasma membrane and releases its contents from the cell.

In addition to making secretory proteins, rough ER is a membrane-making machine for the cell. It grows in place by adding membrane proteins and phospholipids to its own membrane. As polypeptides destined to be membrane proteins grow from bound ribosomes, they are inserted into the ER membrane. Phospholipids are made by enzymes of the rough ER and also inserted into the membrane. Thus, the ER membrane grows, and portions of it are transferred to other components of the endomembrane system in the form of transport vesicles.

Now let's follow a transport vesicle carrying products of the rough ER to the Golgi apparatus.

? Explain why we say that the endoplasmic reticulum is a biosynthetic workshop.

● The ER produces a huge variety of molecules, including phospholipids for cell membranes, steroid hormones, and proteins (synthesized by bound ribosomes) for membranes, other organelles, and secretion by the cell.

4.9 The Golgi apparatus modifies, sorts, and ships cell products

After leaving the ER, many transport vesicles travel to the **Golgi apparatus**. Using a light microscope and a staining technique he developed, Italian scientist Camillo Golgi discovered this membranous organelle in 1898. The electron microscope confirmed his discovery more than 50 years later, revealing a stack of flattened sacs, looking much like a pile of pita bread. A cell may contain many, even hundreds, of these stacks. The number of Golgi stacks correlates with how active the cell is in secreting proteins—a multistep process that, as you have just seen, is initiated in the rough ER.

The Golgi apparatus serves as a molecular warehouse and processing station for products manufactured by the ER. You can follow these activities in **Figure 4.9**. Note that the flattened Golgi sacs are not connected, as are ER sacs. ❶ One side of a Golgi stack serves as a receiving dock for transport vesicles produced by the ER. ❷ A vesicle fuses with a Golgi sac, adding its membrane and contents to the "receiving" side. ❸ Products of the ER are modified as a Golgi sac progresses through the stack. ❹ The "shipping" side of the Golgi functions as a depot, dispatching its products in vesicles that bud off and travel to other sites.

How might ER products be processed during their transit through the Golgi? Various Golgi enzymes modify the carbohydrate portions of the glycoproteins made in the ER, removing some sugars and substituting others. Molecular identification tags, such as phosphate groups, may be added that help the Golgi sort molecules into different batches for different destinations.

Finished secretory products, packaged in transport vesicles, move to the plasma membrane for export from the cell. Alternatively, finished products may become part of the plasma membrane itself or part of another organelle, such as a lysosome, which we discuss next.

? What is the relationship of the Golgi apparatus to the ER in a protein-secreting cell?

● The Golgi receives transport vesicles budded from the ER that contain proteins synthesized by bound ribosomes. The Golgi finishes processing the proteins and dispatches transport vesicles to the plasma membrane, where the proteins are secreted.

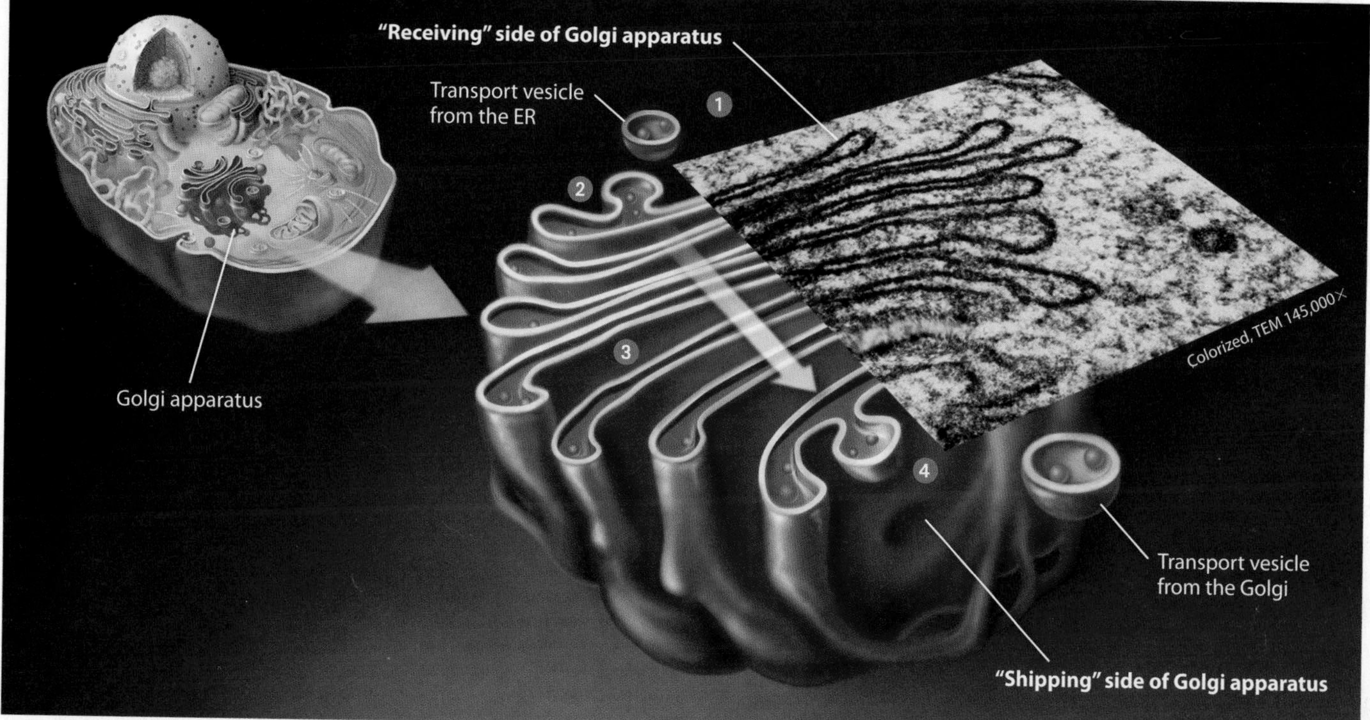

"Receiving" side of Golgi apparatus
Transport vesicle from the ER
1
2
3
4
Golgi apparatus
Colorized, TEM 145,000×
Transport vesicle from the Golgi
"Shipping" side of Golgi apparatus

▲ **Figure 4.9** The Golgi apparatus receiving, processing, and shipping products

4.10 Lysosomes are digestive compartments within a cell

A **lysosome** is a membrane-enclosed sac of digestive enzymes. The name *lysosome* is derived from two Greek words meaning "breakdown body." The enzymes and membranes of lysosomes are made by rough ER and processed in the Golgi apparatus. Illustrating a main theme of eukaryotic cell structure—compartmentalization—a lysosome provides an acidic environment for its enzymes, while safely isolating them from the rest of the cell.

Lysosomes have several types of digestive functions. Many protists engulf food particles into membranous sacs called food vacuoles. As Figure 4.10A shows, lysosomes fuse with food vacuoles and digest the food. The nutrients are then released into the cytosol. Our white blood cells engulf bacteria and then destroy them using lysosomes. Lysosomes also serve as recycling centers. Cells enclose damaged organelles or small amounts of cytosol in membrane sacs. A lysosome fuses with such a vesicle (Figure 4.10B) and dismantles its contents, making organic molecules available for reuse. With the help of lysosomes, a cell continually renews itself.

The cells of people with inherited lysosomal storage diseases lack one or more lysosomal enzymes. The lysosomes become engorged with undigested material, eventually interfering with cellular function. In Tay-Sachs disease, for example, a lipid-digesting enzyme is missing, and brain cells become impaired by an accumulation of lipids. Fortunately, lysosomal storage diseases are rare, as they are often fatal in early childhood.

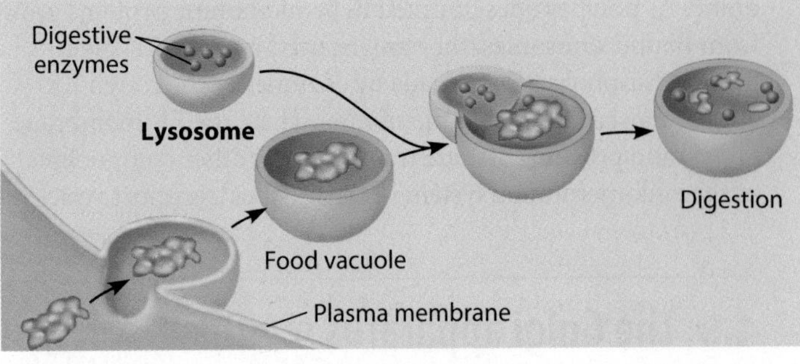
Digestive
enzymes
Lysosome
Digestion
Food vacuole
Plasma membrane

▲ **Figure 4.10A** Lysosome fusing with a food vacuole and digesting food, after which nutrients are released to the cytosol

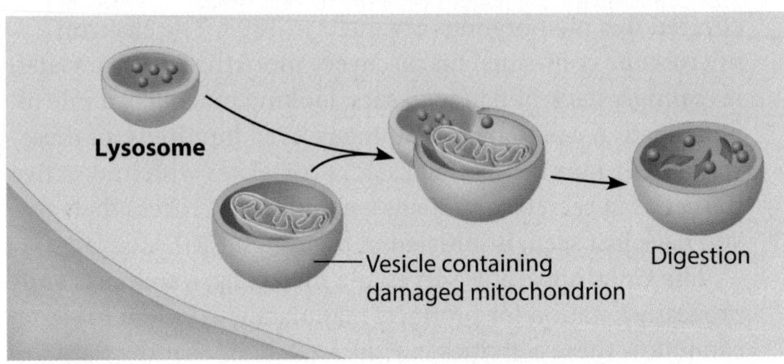
Lysosome
Vesicle containing
damaged mitochondrion
Digestion

▲ **Figure 4.10B** Lysosome fusing with a vesicle containing a damaged organelle and then digesting and recycling its contents

? **How is a lysosome like a recycling center?**

● It breaks down damaged organelles and recycles their molecules.

4.11 Vacuoles function in the general maintenance of the cell

Vacuoles are large vesicles that have a variety of functions. In Figure 4.10A, you saw how a food vacuole forms as a cell ingests food. Figure 4.11A shows two contractile vacuoles in the protist *Paramecium*, looking somewhat like wheel hubs with radiating spokes. The "spokes" collect water from the cell, and the hub expels it to the outside. Freshwater protists constantly take in water from their environment. Without a way to get rid of the excess water, the cell would swell and burst.

In plants, some vacuoles have a digestive function similar to that of lysosomes in animal cells. Vacuoles in flower petals contain pigments that attract pollinating insects. Vacuoles may also contain poisons or unpalatable compounds that protect the plant against herbivores; examples include nicotine, caffeine, and various chemicals we use as pharmaceutical drugs.

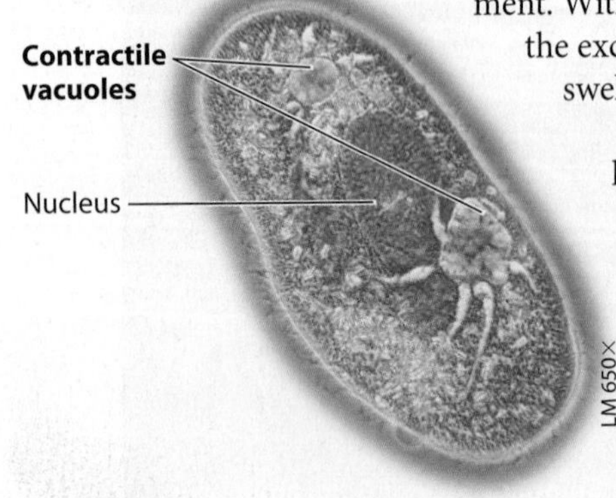
Contractile
vacuoles
Nucleus
LM 650×

▲ **Figure 4.11A** Contractile vacuoles in *Paramecium*, a unicellular eukaryote

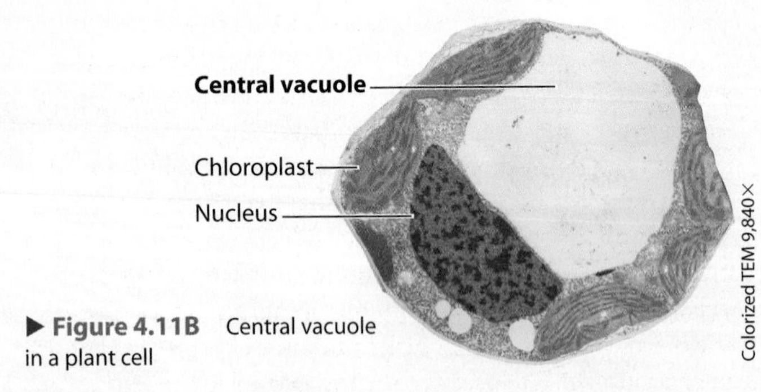
Central vacuole
Chloroplast
Nucleus
Colorized TEM 9,840×

▶ **Figure 4.11B** Central vacuole in a plant cell

Figure 4.11B shows a plant cell's large **central vacuole**, which helps the cell grow in size by absorbing water and enlarging. It also stockpiles vital chemicals and may act as a trash can, safely storing toxic waste products.

? **Is a food vacuole part of the endomembrane system? Explain.**

● Yes; it forms by pinching in from the plasma membrane, which is part of the endomembrane system.

4.12 A review of the structures involved in manufacturing and breakdown

Figure 4.12 summarizes the relationships within the endomembrane system. You can see the direct *structural* connections between the nuclear envelope, rough ER, and smooth ER. The red arrows show the *functional* connections, as membranes and proteins produced by the ER travel in transport vesicles to the Golgi and on to other destinations. Some vesicles develop into lysosomes or vacuoles. Others travel to and fuse with the plasma membrane, secreting their contents and adding their membrane to the plasma membrane.

Peroxisomes (see Figures 4.4A and B) are metabolic compartments that do not originate from the endomembrane system. In fact, how they are related to other organelles is still unknown. Some peroxisomes break down fatty acids to be used as cellular fuel. In your liver, peroxisomes detoxify harmful compounds. In these processes, enzymes transfer hydrogen from the compounds to oxygen, producing hydrogen peroxide (H_2O_2). Other enzymes in the peroxisome convert this toxic by-product to water—another example of the importance of a cell's compartmental structure.

A cell requires a continuous supply of energy to perform the work of life. Next we consider two organelles that act as cellular power stations—mitochondria and chloroplasts.

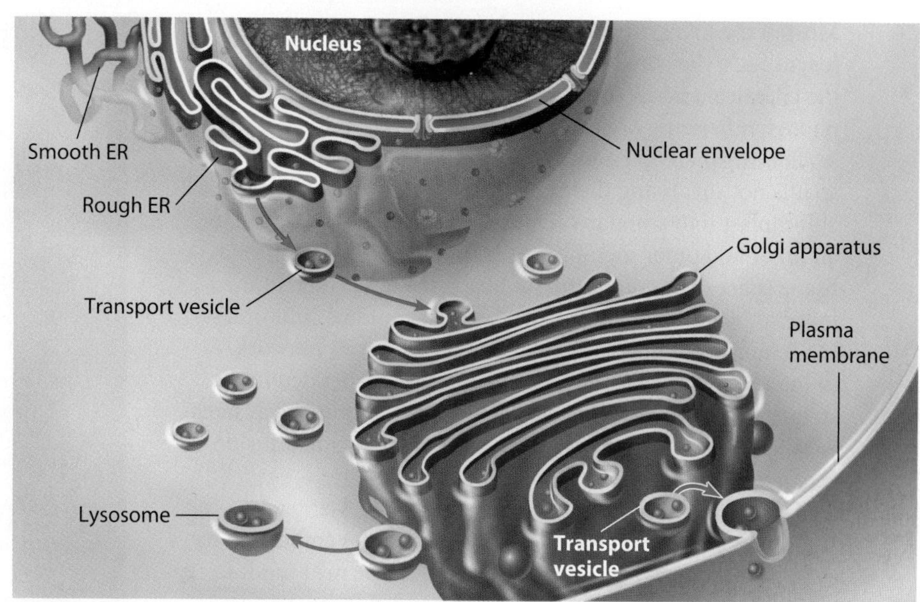
Nucleus
Smooth ER
Nuclear envelope
Rough ER
Golgi apparatus
Transport vesicle
Plasma membrane
Lysosome
Transport vesicle

▲ **Figure 4.12** Review of the endomembrane system

Try This Explain the role of transport vesicles in the endomembrane system.

? **How do transport vesicles help tie together the endomembrane system?**

● Transport vesicles move membranes and the substances they enclose between components of the endomembrane system.

▷ Energy-Converting Organelles

4.13 Mitochondria harvest chemical energy from food

Mitochondria (singular, *mitochondrion*) are organelles that carry out cellular respiration in nearly all eukaryotic cells, converting the chemical energy of foods such as sugars to the chemical energy of the molecule called ATP (adenosine triphosphate). ATP is the main energy source for cellular work.

A mitochondrion is enclosed by two membranes, each a phospholipid bilayer with a unique collection of embedded proteins (**Figure 4.13**). The mitochondrion has two internal compartments. The first is the intermembrane space, the narrow region between the inner and outer membranes. The inner membrane encloses the second compartment, the **mitochondrial matrix**, which contains mitochondrial DNA and ribosomes, as well as enzymes that catalyze some of the reactions of cellular respiration. The inner membrane is highly folded and contains many embedded protein molecules that function in ATP synthesis. The folds, called **cristae**, increase the membrane's surface area, enhancing the mitochondrion's ability to produce ATP.

? **What is cellular respiration?**

● A process that converts the chemical energy of sugars and other food molecules to the chemical energy of ATP

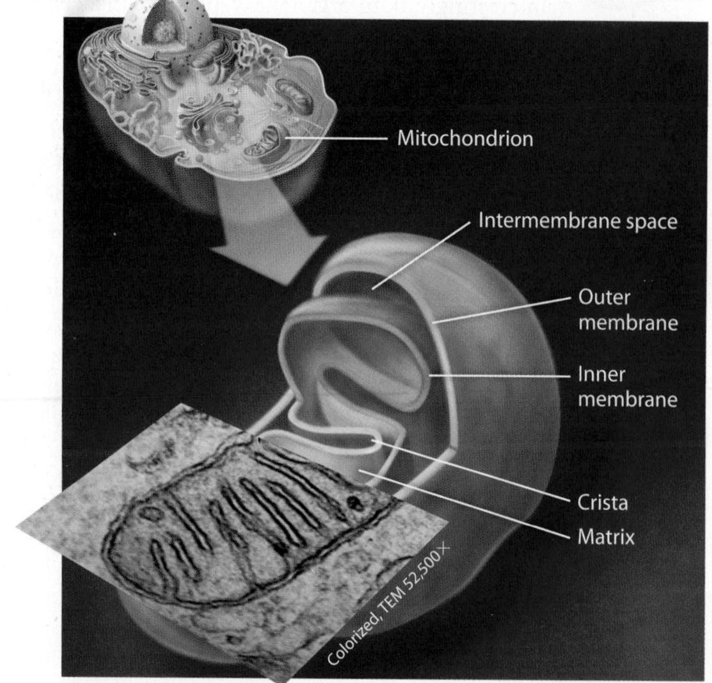
Mitochondrion
Intermembrane space
Outer membrane
Inner membrane
Crista
Matrix
Colorized, TEM 52,500×

▲ **Figure 4.13** The mitochondrion, site of cellular respiration

4.14 Chloroplasts convert solar energy to chemical energy

Most of the living world runs on the energy provided by photosynthesis, the conversion of light energy from the sun to the chemical energy of sugar molecules. **Chloroplasts** are the photosynthesizing organelles of plants and algae.

This organelle carries out complex, multistep processes, so it is not surprising that internal membranes partition the chloroplast into compartments (**Figure 4.14**). It is enclosed by an inner and outer membrane separated by a thin intermembrane space. The compartment inside the inner membrane holds a thick fluid called **stroma**, which contains chloroplast DNA and ribosomes as well as many enzymes. A network of interconnected sacs called **thylakoids** is suspended in the stroma. The sacs are often stacked like poker chips; each stack is called a **granum** (plural, *grana*). The compartment inside the thylakoids is called the thylakoid space.

The thylakoids are the chloroplast's solar power packs—the sites where the green chlorophyll molecules embedded in thylakoid membranes trap solar energy. In the next module, we explore the origin of mitochondria and chloroplasts.

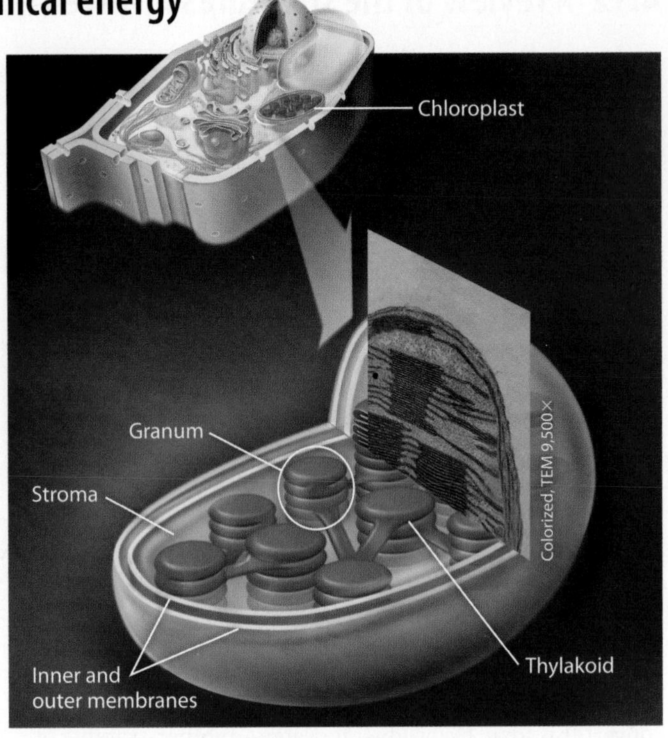

Chloroplast
Granum
Stroma
Colorized, TEM 9,500×
Inner and outer membranes
Thylakoid

▲ **Figure 4.14** The chloroplast, site of photosynthesis

? Which membrane in a chloroplast appears to be the most extensive? Why might this be so?

● The thylakoid membranes contain chlorophyll for photosynthesis.

4.15 Mitochondria and chloroplasts evolved by endosymbiosis

EVOLUTION CONNECTION

Mitochondria and chloroplasts contain a single circular DNA molecule, similar in structure to a prokaryotic chromosome, and ribosomes more similar to prokaryotic ribosomes than to eukaryotic ones. Interestingly, both organelles reproduce in a cell by a process resembling that of certain prokaryotes.

The **endosymbiont theory** states that mitochondria and chloroplasts were formerly small prokaryotes that began living within larger cells. These prokaryotes may have gained entry to the larger cell as undigested prey or parasites (**Figure 4.15**).

We can hypothesize how the symbiosis could have been beneficial. In a world that was becoming increasingly aerobic from the oxygen-generating photosynthesis of prokaryotes, a host would have benefited from an endosymbiont that was able to use oxygen to release large amounts of energy from organic molecules. Over the course of evolution, the host cell and its endosymbiont merged into a single organism—a eukaryotic cell with a mitochondrion. If one of these cells acquired a photosynthetic prokaryote, the prokaryote could provide the host cell with nourishment. An increasingly interdependent host and endosymbiont, over many generations, could become a eukaryotic cell containing chloroplasts.

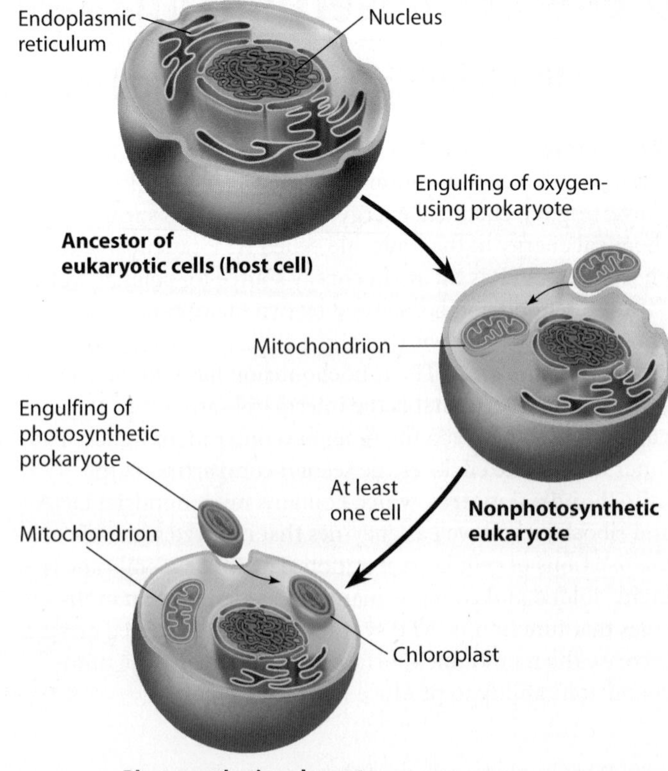

Endoplasmic reticulum
Nucleus
Ancestor of eukaryotic cells (host cell)
Engulfing of oxygen-using prokaryote
Mitochondrion
Engulfing of photosynthetic prokaryote
At least one cell
Nonphotosynthetic eukaryote
Mitochondrion
Chloroplast
Photosynthetic eukaryote

▲ **Figure 4.15** Endosymbiotic origin of mitochondria and chloroplasts

? All eukaryotes have mitochondria, but not all have chloroplasts. What is the evolutionary explanation?

● The first endosymbiosis would have given rise to eukaryotic cells containing mitochondria. A second endosymbiotic event gave rise to cells containing chloroplasts as well as mitochondria.

The Cytoskeleton and Cell Surfaces

4.16 The cell's internal skeleton helps organize its structure and activities

As you saw in the chapter introduction, networks of protein fibers extend throughout a cell. Collectively called the **cytoskeleton**, these fibers play a major role in organizing the structures and activities of the cell. And like a skeleton, they provide for structural support as well as movement, including both the internal movement of cell parts and the swimming or crawling motility of some cells.

Three main kinds of fibers make up the cytoskeleton: microtubules, the thickest fiber; microfilaments, the thinnest; and intermediate filaments, in between in thickness. Figure 4.16 shows three micrographs of cells of the same type, each stained with a different fluorescent dye that selectively highlights one of these types of fibers.

Microtubules are straight, hollow tubes composed of globular proteins called tubulins. As indicated in the bottom left of Figure 4.16, microtubules elongate by the addition of tubulin proteins. They are readily disassembled, and their subunits can be reused elsewhere in the cell. In animal cells, microtubules grow out from a region near the nucleus called the **centrosome**, which contains a pair of centrioles, each composed of a ring of microtubules (see Figure 4.4A). Plant cells lack centrosomes with centrioles and organize microtubules by other means.

Microtubules shape and support the cell and also act as tracks along which organelles equipped with motor proteins move. For example, a lysosome might use its motor protein "feet" to "walk" along a microtubule to reach a food vacuole. Microtubules also guide the movement of chromosomes when cells divide, and they are the main components of cilia and flagella. We will return to the structure of these locomotive appendages in Module 4.18.

Intermediate filaments are found in the cells of most animals. They are made of various fibrous proteins that supercoil into cables. Intermediate filaments reinforce cell shape and anchor some organelles. For example, the nucleus typically sits in a cage made of intermediate filaments. While microtubules may be disassembled and reassembled elsewhere, intermediate filaments are often more permanent fixtures in the cell. The outer layer of your skin consists of dead skin cells packed full of intermediate filaments.

Microfilaments, also called actin filaments, are solid rods composed mainly of globular proteins called actin, arranged in a twisted double chain (bottom right of Figure 4.16). Microfilaments form a three-dimensional network just inside the plasma membrane that helps support the cell's shape. This is especially important for animal cells, which lack cell walls.

Microfilaments are also involved in cell movements. Actin filaments and thicker filaments made of a type of motor protein called myosin interact to cause contraction of muscle cells (see Figure 30.9B). Localized contractions brought about by actin and myosin are involved in the amoeboid (crawling) movement of the protist *Amoeba* and some of your white blood cells.

In the next module, we survey some of the techniques that led to the discovery of the cytoskeleton.

? **Which component of the cytoskeleton is most important in (a) holding the nucleus in place within the cell; (b) guiding transport vesicles from the Golgi to the plasma membrane; (c) contracting muscle cells?**

(a) Intermediate filaments; (b) microtubules; (c) microfilaments

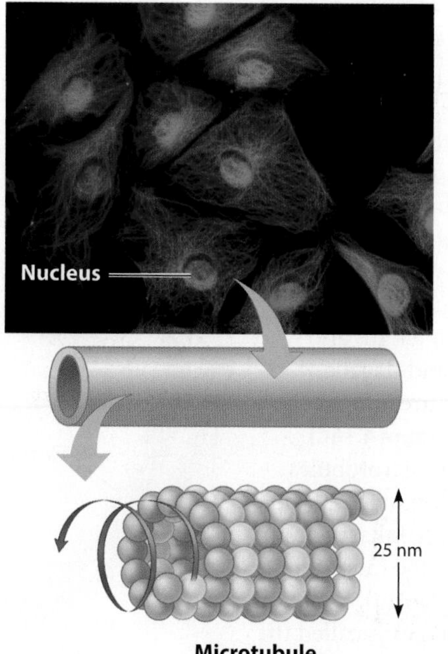
Nucleus

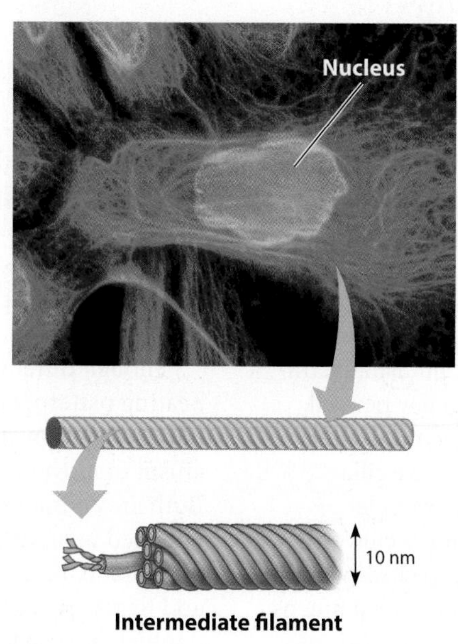
Nucleus

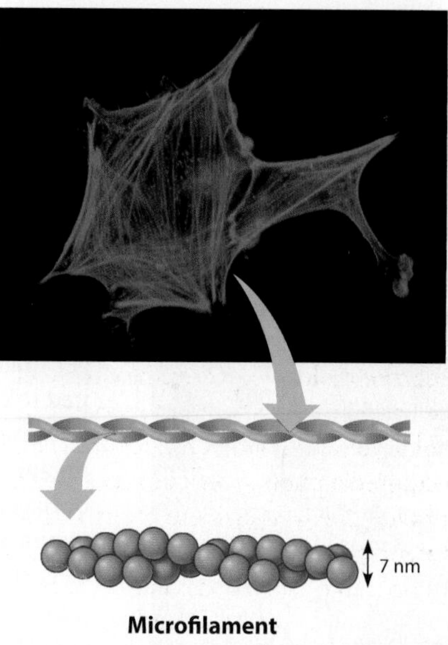

25 nm

Microtubule

10 nm

Intermediate filament

7 nm

Microfilament

▲ **Figure 4.16** Three types of fibers of the cytoskeleton: microtubules labeled with green fluorescent molecules (left), intermediate filaments labeled yellow-green (center), and microfilaments labeled red (right)

4.17 Scientists discovered the cytoskeleton using the tools of biochemistry and microscopy

SCIENTIFIC THINKING

How has our knowledge of cells grown?

As you learned in Module 4.1, improvements in microscopes and staining techniques led to the discovery of organelles. But biologists originally thought that these structures floated freely in the cell. Let's trace the progressive sequence of new techniques that led to the discovery of microfilaments, the component of the cytoskeleton built from actin.

In the 1940s, biochemists first isolated and identified the proteins actin and myosin from muscle cells. In 1954, scientists, using newly developed techniques of microscopy, established how filaments of actin and myosin interact in muscle contraction. In the next decade, researchers developed a technique to stain and identify actin filaments with the electron microscope. Imagine their surprise when they found actin not just in the muscle cells they were studying but also in other cells present in their samples. Further study identified actin filaments in all types of cells.

Today we take for granted our ability to "see" the cytoskeleton (as you saw in the chapter introduction). But intact networks of microfilaments were not visualized in cells until 1974. Using a technique called immunofluorescence microscopy, scientists developed antibody proteins that would bind to actin and attached fluorescent molecules to them. When fluorescent molecules absorb light, they "glow" because they emit light of a specific wavelength or color.

When injected into cells, the fluorescent antibodies revealed a remarkable and beautiful web of microfilaments. **Figure 4.17** shows how different fluorescent tags can attach to various components of the cytoskeleton.

Researchers then tagged actin proteins themselves with fluorescent molecules and injected them into living cells. Instead of just marking where proteins are found, this technique, known as molecular cytochemistry, enabled scientists to visualize the dynamic behavior of cytoskeletal proteins in living cells.

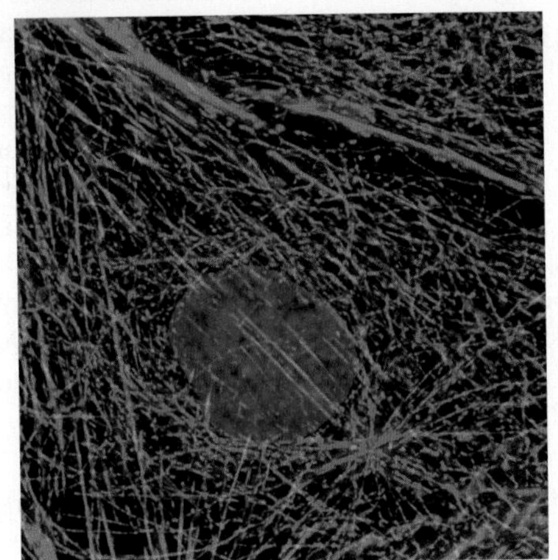

▲ **Figure 4.17** A fluorescence micrograph of the cytoskeleton (microtubules are green, microfilaments are red)

In the early 1980s, biologists first paired a video camera with a microscope. Suddenly they could "watch" what was happening in cells over time and follow the changing architecture of the cytoskeleton.

As scientists develop new techniques, our understanding of the structure and function of the cytoskeleton will continue to grow.

? **What is the difference between immunofluorescence microscopy and molecular cytochemistry?**

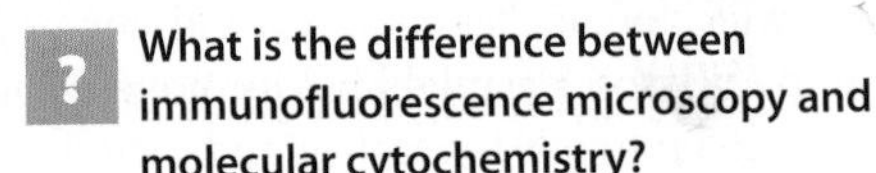
In the former, fluorescently labeled antibodies show the locations of specific molecules. In the latter, the molecules themselves are labeled, and their behavior within a living cell can be tracked.

4.18 Cilia and flagella move when microtubules bend

The role of the cytoskeleton in movement is clearly seen in the motile appendages that protrude from certain cells. The short, numerous appendages that propel protists such as *Paramecium* (see Figure 4.1B) are called **cilia** (singular, *cilium*). Other protists may move using flagella, which are longer than cilia and usually limited to one or a few per cell.

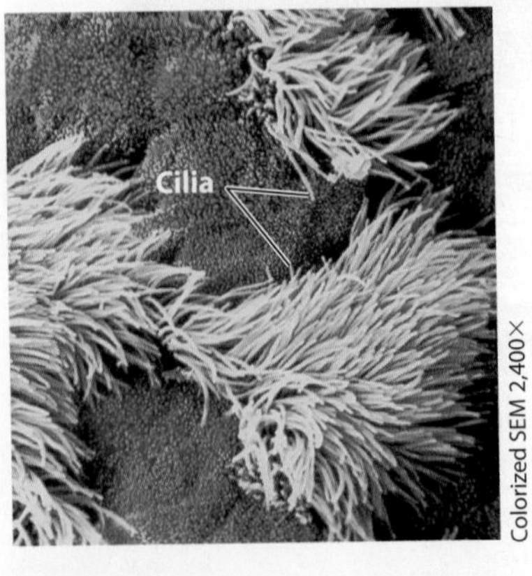

Some cells of multicellular organisms also have cilia or flagella. For example, **Figure 4.18A** shows cilia on cells lining the trachea (windpipe). These cilia sweep mucus containing trapped debris out of your lungs. (This cleaning function is impaired by cigarette smoke, which paralyzes the cilia.) Most animals and some plants have flagellated sperm. A flagellum, shown in **Figure 4.18B**, propels the cell by an undulating whiplike motion. In contrast, cilia work more like the coordinated oars of a rowing team.

▲ **Figure 4.18A** Cilia on cells lining the respiratory tract

Though different in length and beating pattern, cilia and flagella have a common structure and mechanism of movement **(Figure 4.18C)**. Both are composed of microtubules wrapped in an extension of the plasma membrane. In nearly all eukaryotic cilia and flagella, a ring of nine microtubule doublets surrounds a central pair of microtubules. This arrangement is called the "9 + 2" pattern. The microtubule assembly is anchored in the cell by a

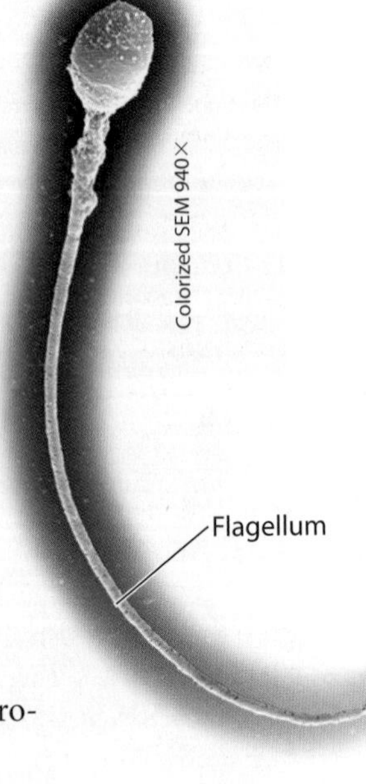
Colorized SEM 940×
Flagellum

▲ **Figure 4.18B** Undulating flagellum on a human sperm cell

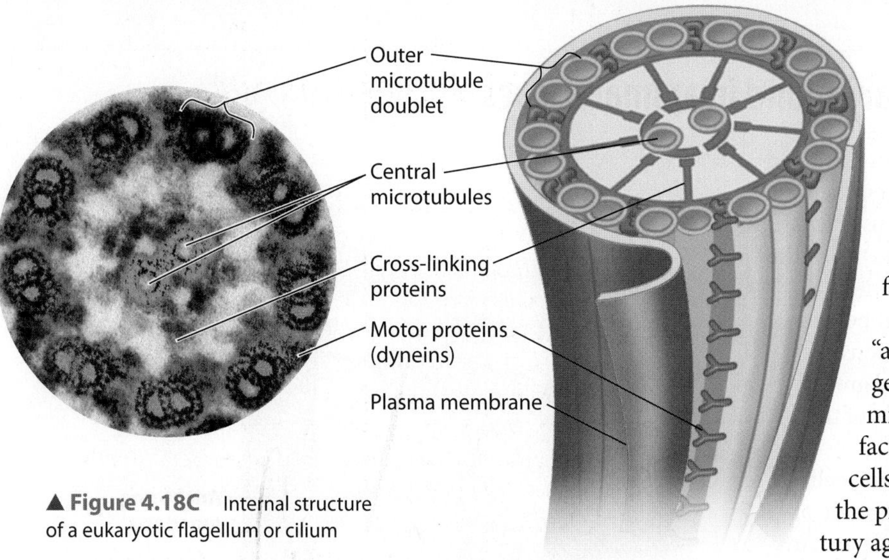
Colorized TEM 290,000×
Outer microtubule doublet
Central microtubules
Cross-linking proteins
Motor proteins (dyneins)
Plasma membrane

▲ **Figure 4.18C** Internal structure of a eukaryotic flagellum or cilium

basal body (not shown in the figure), which is structurally very similar to a centriole. In fact, in humans and many other animals, the basal body of the fertilizing sperm's flagellum enters the egg and becomes a centriole.

How does the microtubule assembly shown in Figure 4.18C produce the movement of cilia and flagella? Large motor proteins called dyneins (red in the figure) are attached along each outer microtubule doublet. A dynein protein has two "feet" that "walk" along an adjacent doublet. The walking movement is coordinated so that it happens on one side at a time. The microtubules are held together by flexible cross-linking proteins (purple in the diagram). If the doublets were not held in place, they would slide past each other. Instead, the "walking" of the dynein feet causes the microtubules—and consequently the cilium or flagellum—to bend.

A cilium may also serve as a signal-receiving "antenna" for the cell. Cilia with this function are generally nonmotile (they lack the central pair of microtubules), and there is only one per cell. In fact, in vertebrate animals, it appears that almost all cells have what is called a *primary cilium*. Although the primary cilium was discovered more than a century ago, its importance to embryonic development, sensory reception, and cell function is only now being recognized. Defective primary cilia have been linked to polycystic kidney disease and other human disorders.

? **Primary ciliary dyskinesia (PCD), also known as immotile cilia syndrome, is a fairly rare disease in which cilia and flagella are lacking motor proteins. PCD is characterized by recurrent respiratory tract infections and immotile sperm. How would you explain these seemingly unrelated symptoms?**

● Without motor proteins, microtubules cannot bend. Thus cilia cannot cleanse the respiratory tract, and sperm cannot swim.

4.19 The extracellular matrix of animal cells functions in support and regulation

The plasma membrane is usually regarded as the boundary of the cell, but most cells synthesize and secrete materials that are external to the plasma membrane. These extracellular structures are essential to many cell functions.

Animal cells produce an **extracellular matrix (ECM)** (**Figure 4.19**). This elaborate layer helps hold cells together in tissues and protects and supports the plasma membrane. The main components of the ECM are glycoproteins, proteins bonded with carbohydrates. The most abundant glycoprotein is collagen, which forms strong fibers outside the cell. In fact, collagen accounts for about 40% of the protein in your body. The collagen fibers are embedded in a network woven from other types of glycoproteins. Large complexes form when hundreds of small glycoproteins connect to a central long polysaccharide molecule (shown as green in the figure). The ECM may attach to the cell through other glycoproteins that then bind to membrane proteins called **integrins**. Integrins span the membrane, attaching on the other side to proteins connected to microfilaments of the cytoskeleton.

As their name implies, integrins have the function of integration: They transmit signals between the ECM and the cytoskeleton. Thus, the cytoskeleton can influence the organization of the ECM and vice versa. For example, research shows that the ECM can regulate a cell's behavior, directing the path along which embryonic cells move and even influencing the activity of genes through the signals it relays. Genetic changes in cancer cells may result in a change in the composition of the ECM they produce, causing such cells to lose their connections and spread to other tissues.

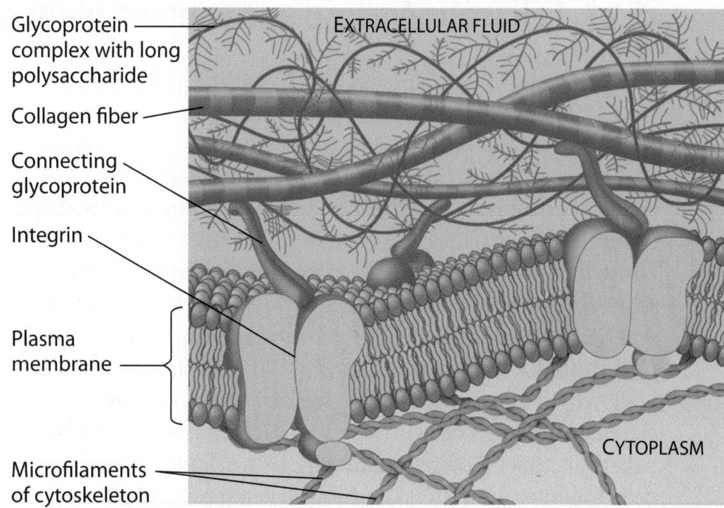
Glycoprotein complex with long polysaccharide
EXTRACELLULAR FLUID
Collagen fiber
Connecting glycoprotein
Integrin
Plasma membrane
CYTOPLASM
Microfilaments of cytoskeleton

▲ **Figure 4.19** The extracellular matrix (ECM) of an animal cell

? **Referring to Figure 4.19, describe the structures that provide support to the plasma membrane.**

● The membrane is attached through membrane proteins (integrins) to microfilaments in the cytoskeleton and collagen fibers of the ECM.

4.20 Three types of cell junctions are found in animal tissues

Neighboring cells in animal tissues often adhere, interact, and communicate through specialized junctions between them. **Figure 4.20** uses cells lining the digestive tract to illustrate three types of cell junctions. (The projections at the top of the cells increase the surface area for absorption of nutrients.)

At tight junctions, the plasma membranes of neighboring cells are knit tightly together by proteins. Tight junctions prevent leakage of fluid across a layer of cells. The dotted green arrows show how tight junctions prevent the contents of the digestive tract from leaking into surrounding tissues.

Anchoring junctions function like rivets, fastening cells together into strong sheets. Intermediate filaments made of sturdy proteins anchor these junctions in the cytoplasm. Anchoring junctions are common in tissues subject to stretching or mechanical stress, such as skin and muscle.

Gap junctions, also called communicating junctions, are channels that allow small molecules to flow through protein-lined pores between cells. The flow of ions through gap junctions in the cells of heart muscle coordinates their contraction. Gap junctions are common in embryos, where communication between cells is essential for development.

A muscle tear injury would probably involve the rupture of which type of cell junction?

Anchoring junction

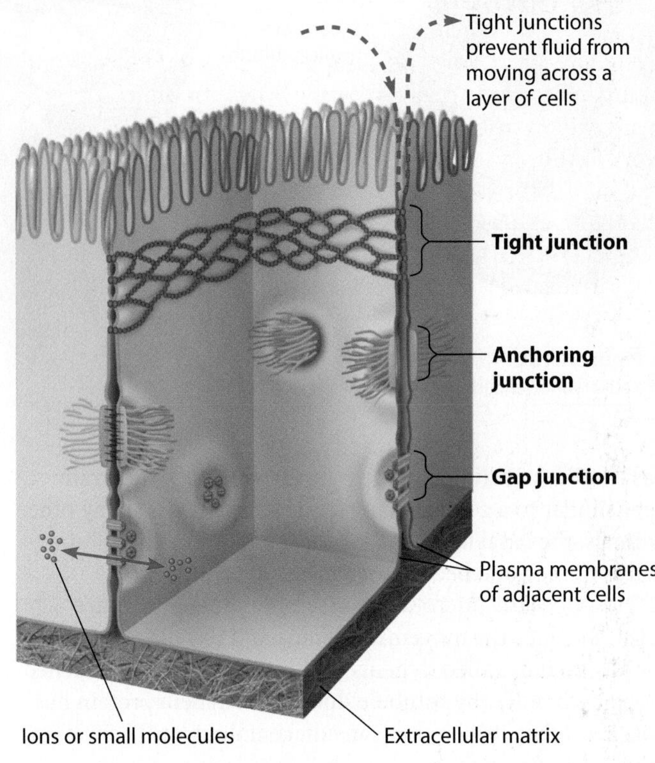
Tight junctions prevent fluid from moving across a layer of cells
Tight junction
Anchoring junction
Gap junction
Plasma membranes of adjacent cells
Ions or small molecules
Extracellular matrix

▲ **Figure 4.20** Three types of cell junctions in animal tissues

4.21 Cell walls enclose and support plant cells

The **cell wall** is one of the features that distinguishes plant cells from animal cells. This rigid extracellular structure not only protects the cells but also provides the skeletal support that keeps plants upright on land. Plant cell walls consist of fibers of cellulose (see Figure 3.7) embedded in a matrix of other polysaccharides and proteins. This fibers-in-a-matrix construction resembles that of steel-reinforced concrete, which is also noted for its strength.

Figure 4.21 shows the layered structure of plant cell walls. Cells initially lay down a relatively thin and flexible primary wall, which allows the growing cell to continue to enlarge. Between adjacent cells is a layer of sticky polysaccharides called pectins (shown here in dark brown), which glue the cells together. (Pectin is used to thicken jams and jellies.) When a cell stops growing, it strengthens its wall. Some cells add a secondary wall deposited in laminated layers next to the plasma membrane. Wood consists mainly of secondary walls, which are strengthened with rigid molecules called lignin.

Despite their thickness, plant cell walls do not totally isolate the cells from each other. Figure 4.21 shows the numerous channels that connect adjacent plant cells, called **plasmodesmata** (singular, *plasmodesma*). Cytosol passing through the plasmodesmata allows water and other small molecules to freely move from cell to cell. Through plasmodesmata, the cells of a plant tissue share water, nourishment, and chemical messages.

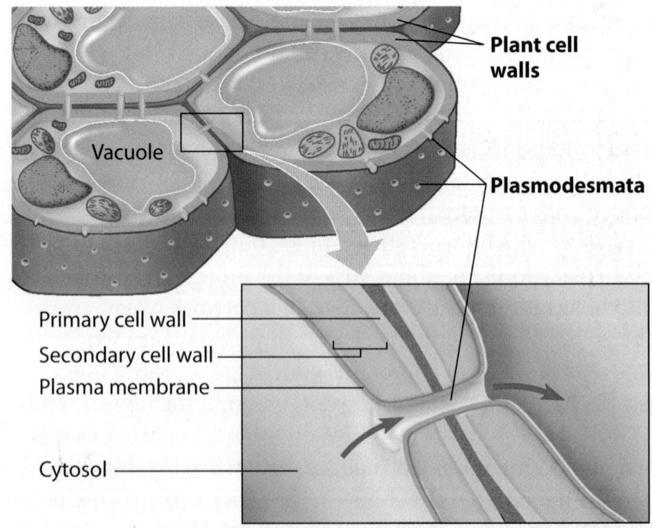
Plant cell walls
Vacuole
Plasmodesmata
Primary cell wall
Secondary cell wall
Plasma membrane
Cytosol

▲ **Figure 4.21** Plant cell walls and plasmodesmata

Which animal cell junction is analogous to a plasmodesma?

A gap junction

4.22 Review: Eukaryotic cell structures can be grouped on the basis of four main functions

Congratulations: You have completed the grand tour of the cell. In the process, you have been introduced to many important cell structures. To provide a framework for this information and reinforce the theme that structure is correlated with function, we have grouped the eukaryotic cell structures into four categories by general function, as reviewed in **Table 4.22**.

The first category is genetic control. Here we include the nucleus that houses a cell's genetic instructions and the ribosomes that produce the proteins coded for in those instructions. The second category includes organelles of the endomembrane system that are involved in the manufacture, distribution, and breakdown of materials. The third category includes the two energy-processing organelles, mitochondria and chloroplasts. And the fourth category—structural support, movement, and intercellular communication—includes the cytoskeleton, extracellular structures, and connections between cells.

Within most of these categories, a structural similarity underlies the general function of each component. Manufacturing depends heavily on a network of structurally and functionally connected membranes. All the organelles involved in the breakdown or recycling of materials are membranous sacs, inside of which enzymatic digestion can safely occur. In the energy-processing category, expanses of metabolically active membranes and intermembrane compartments within the organelles enable chloroplasts and mitochondria to perform the complex energy conversions that power the cell. Even in the diverse fourth category, there is a common structural theme in the various protein fibers of most of these cellular systems.

We can summarize further by noting that the overall structure of a cell is closely related to its specific function. Thus, cells that produce proteins for export contain a large quantity of ribosomes and rough ER, while muscle cells are packed with microfilaments, myosin motor proteins, and mitochondria. And, finally, let us emphasize that these cellular structures form an integrated team—with the property of life emerging at the level of the cell from the coordinated functions of the team members.

? **How do mitochondria, smooth ER, and the cytoskeleton all contribute to the contraction of a muscle cell?**

Mitochondria supply energy in the form of ATP. The smooth ER helps regulate contraction by the uptake and release of calcium ions. Microfilaments function in the actual contractile apparatus.

TABLE 4.22 | EUKARYOTIC CELL STRUCTURES AND FUNCTIONS

Structure	Function
1. Genetic Control	
Nucleus 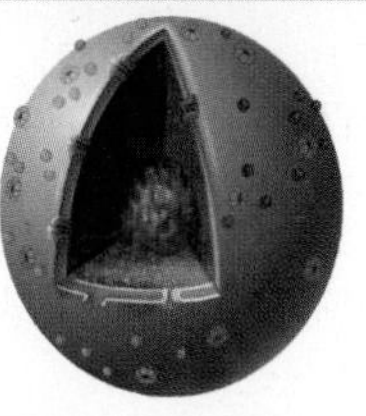	DNA replication, RNA synthesis; assembly of ribosomal subunits (in nucleolus)
Ribosomes	Polypeptide (protein) synthesis
2. Manufacturing, Distribution, and Breakdown	
Rough ER	Synthesis of membrane lipids and proteins, secretory proteins, and hydrolytic enzymes; formation of transport vesicles
Smooth ER 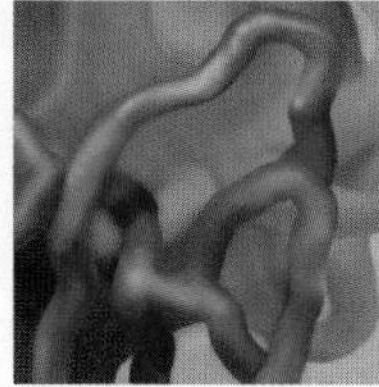	Lipid synthesis; detoxification in liver cells; calcium ion storage
Golgi apparatus	Modification and sorting of macromolecules; formation of lysosomes and transport vesicles
Lysosomes (in animal cells and some protists)	Digestion of ingested food or bacteria and recycling of a cell's damaged organelles and macromolecules
Vacuoles	Digestion (food vacuole); storage of chemicals and cell enlargement (central vacuole); water balance (contractile vacuole)
Peroxisomes (not part of endomembrane system)	Diverse metabolic processes, with breakdown of toxic hydrogen peroxide by-product
3. Energy Processing	
Mitochondria 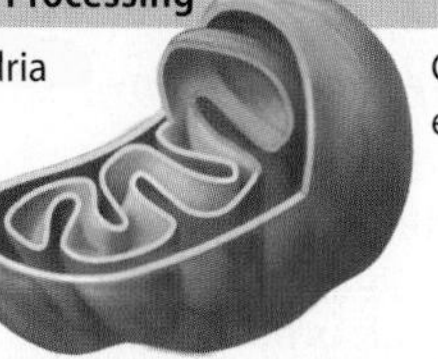	Conversion of chemical energy in food to chemical energy of ATP
Chloroplasts (in plants and algae)	Conversion of light energy to chemical energy of sugars
4. Structural Support, Movement, and Communication Between Cells	
Cytoskeleton (microfilaments, intermediate filaments, and microtubules)	Maintenance of cell shape; anchorage for organelles; movement of organelles within cells; cell movement (crawling, muscle contraction, bending of cilia and flagella)
Plasma membrane	Regulate traffic in and out of cell
Extracellular matrix (in animals)	Support; regulation of cellular activities
Cell junctions	Communication between cells; binding of cells in tissues
Cell walls (in plants) 	Support and protection; binding of cells in tissues

CHAPTER 4 REVIEW

For practice quizzes, BioFlix animations, MP3 tutorials, video tutors, and more study tools designed for this textbook, go to MasteringBiology®

Reviewing the Concepts

Introduction to the Cell (4.1–4.4)

4.1 Microscopes reveal the world of the cell. The light microscope can display living cells. The greater magnification and resolution of the scanning and transmission electron microscopes reveal the ultrastructure of cells.

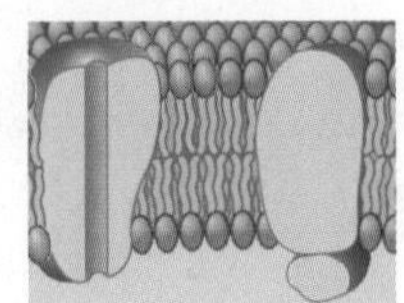

4.2 The small size of cells relates to the need to exchange materials across the plasma membrane. The microscopic size of most cells provides a large surface-to-volume ratio. The plasma membrane is a phospholipid bilayer with embedded proteins.

4.3 Prokaryotic cells are structurally simpler than eukaryotic cells. All cells have a plasma membrane, DNA, ribosomes, and cytosol. Prokaryotic cells lack organelles.

4.4 Eukaryotic cells are partitioned into functional compartments. Membrane-enclosed organelles compartmentalize a cell's activities.

The Nucleus and Ribosomes (4.5–4.6)

4.5 The nucleus contains the cell's genetic instructions. The nucleus houses the cell's DNA, which directs protein synthesis via messenger RNA. Subunits of ribosomes are assembled in the nucleolus.

4.6 Ribosomes make proteins for use in the cell and for export. Composed of ribosomal RNA and proteins, ribosomes synthesize proteins according to directions from DNA.

The Endomembrane System (4.7–4.12)

4.7 Many organelles are connected in the endomembrane system.

4.8 The endoplasmic reticulum is a biosynthetic workshop. The ER is a membranous network of tubes and sacs. Smooth ER synthesizes lipids and processes toxins. Rough ER produces membranes, and ribosomes on its surface make membrane and secretory proteins.

4.9 The Golgi apparatus modifies, sorts, and ships cell products. The Golgi apparatus consists of stacks of sacs in which products of the ER are processed and then sent to other organelles or to the cell surface.

4.10 Lysosomes are digestive compartments within a cell. Lysosomes house enzymes that break down ingested substances and damaged organelles for recycling.

4.11 Vacuoles function in the general maintenance of the cell. Some protists have contractile vacuoles. Plant cells contain a large central vacuole that stores molecules and wastes and facilitates growth.

4.12 A review of the structures involved in manufacturing and breakdown. The organelles of the endomembrane system are interconnected structurally and functionally.

Energy-Converting Organelles (4.13–4.15)

4.13 Mitochondria harvest chemical energy from food.

4.14 Chloroplasts convert solar energy to chemical energy.

4.15 Mitochondria and chloroplasts evolved by endosymbiosis. These organelles originated from prokaryotic cells that became residents in a host cell.

The Cytoskeleton and Cell Surfaces (4.16–4.22)

4.16 The cell's internal skeleton helps organize its structure and activities. The cytoskeleton includes microfilaments, intermediate filaments, and microtubules. Their functions include muscle contraction, anchorage and movement of organelles, and maintenance of cell shape.

4.17 Scientists discovered the cytoskeleton using the tools of biochemistry and microscopy.

4.18 Cilia and flagella move when microtubules bend. Eukaryotic cilia and flagella are locomotor appendages made of microtubules in a "9 + 2" arrangement.

4.19 The extracellular matrix of animal cells functions in support and regulation. The ECM consists mainly of glycoproteins, which bind tissue cells together, support the membrane, and communicate with the cytoskeleton.

4.20 Three types of cell junctions are found in animal tissues. Tight junctions bind cells to form leakproof sheets. Anchoring junctions rivet cells into strong tissues. Gap junctions allow substances to flow from cell to cell.

4.21 Cell walls enclose and support plant cells. Plant cell walls are made largely of cellulose. Plasmodesmata are connecting channels between cells.

4.22 Review: Eukaryotic cell structures can be grouped on the basis of four main functions. These functions are (1) genetic control; (2) manufacturing, distribution, and breakdown; (3) energy processing; and (4) structural support, movement, and communication between cells.

Connecting the Concepts

1. Label the structures in this diagram of an animal cell. Review the functions of each of these organelles.

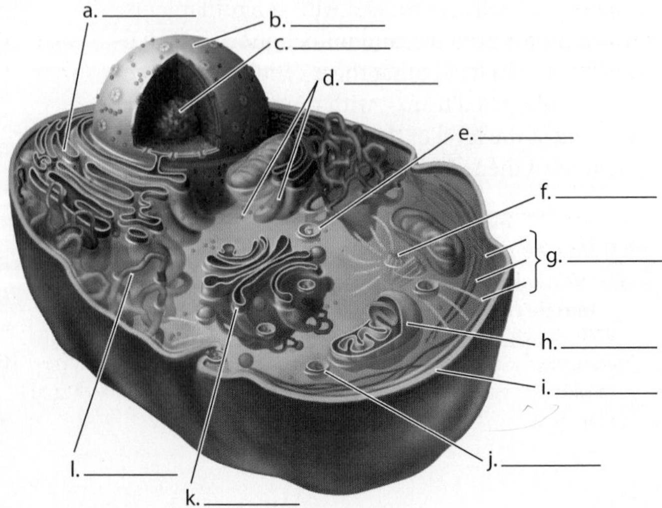
a. ________
b. ________
c. ________
d. ________
e. ________
f. ________
g. ________
h. ________
i. ________
j. ________
k. ________
l. ________

Testing Your Knowledge

Level 1: Knowledge/Comprehension

2. The ultrastructure of a chloroplast is best studied using a
   a. light microscope.
   b. scanning electron microscope.
   c. transmission electron microscope.
   d. light microscope and fluorescent dyes.
3. The cells of an ant and an elephant are, on average, the same small size; an elephant just has more of them. What is the main advantage of small cell size? (*Explain your reasoning.*)
   a. A small cell has a larger plasma membrane surface area than does a large cell.
   b. Small cells can better take up sufficient nutrients and oxygen to service their cell volume.
   c. It takes less energy to make an organism out of small cells.
   d. Small cells require less oxygen than do large cells.
4. Which of the following clues would tell you whether a cell is prokaryotic or eukaryotic?
   a. the presence or absence of a rigid cell wall
   b. whether or not the cell is partitioned by internal membranes
   c. the presence or absence of ribosomes
   d. Both a and b are important clues.
5. Which of the following is one of the major components of the plasma membrane of a plant cell?
   a. phospholipids
   b. cellulose fibers
   c. collagen fibers
   d. pectins
6. What four cellular components are shared by prokaryotic and eukaryotic cells?
7. Describe two different ways in which cilia can function in organisms.

Level 2: Application/Analysis

Choose from the following cells for questions 8–11:
   a. pancreatic cell that secretes digestive enzymes
   b. ovarian cell that produces estrogen (a steroid hormone)
   c. muscle cell in the thigh of a long-distance runner
   d. white blood cell that engulfs bacteria

8. In which cell would you find the most lysosomes?
9. In which cell would you find the most smooth ER?
10. In which cell would you find the most rough ER?
11. In which cell would you find the most mitochondria?
12. In what ways do the internal membranes of a eukaryotic cell contribute to the functioning of the cell?
13. Is this statement true or false? "Animal cells have mitochondria; plant cells have chloroplasts." Explain your answer, and describe the functions of these organelles.
14. Describe the structure of the plasma membrane of an animal cell. What would be found directly inside and outside the membrane?
15. Imagine a spherical cell with a radius of 10 µm. What is the cell's surface area in μm^2? Its volume, in μm^3? (*Note*: For a sphere of radius r, surface area $= 4\pi r^2$ and volume $= \pi r^3$. Remember that the value of π is 3.14.) What is the ratio of surface area to volume for this cell? Now do the same calculations for a second cell, this one with a radius of 20 µm. Compare the surface-to-volume ratios of the two cells. How is this comparison significant to the functioning of cells?
16. Describe the pathway of the protein hormone insulin from its gene to its export from a cell of your pancreas.

Level 3: Synthesis/Evaluation

17. How might the phrase "ingested but not digested" be used in a description of the endosymbiotic theory?
18. Cilia are found on cells in almost every organ of the human body, and the malfunction of cilia is involved in several human disorders. During embryological development, for example, cilia generate a leftward flow of fluid that initiates the left-right organization of the body organs. Some individuals with primary ciliary dyskinesia (see Module 4.18 checkpoint question) exhibit *situs inversus*, in which internal organs such as the heart are on the wrong side of the body. Explain why this reversed arrangement may be a symptom of PCD.
19. **SCIENTIFIC THINKING** Microtubules often produce movement through their interaction with motor proteins. But in some cases, microtubules move cell components when the length of the microtubule changes. Through a series of experiments, researchers determined that microtubules grow and shorten as tubulin proteins are added or removed from their ends. Other experiments showed that microtubules make up the spindle apparatus that "pulls" chromosomes toward opposite ends (poles) of a dividing cell. The figures below describe a clever experiment done in 1987 to determine whether a spindle microtubule shortens (depolymerizes) at the end holding a chromosome or at the pole end of a dividing cell.

    Experimenters labeled the microtubules of a dividing cell from a pig kidney with a yellow fluorescent dye. As shown on the left half of the diagram below, they then marked a region halfway along the microtubules by using a laser to eliminate the fluorescence from that region. They did not mark the other side of the spindle (right side of the figure).

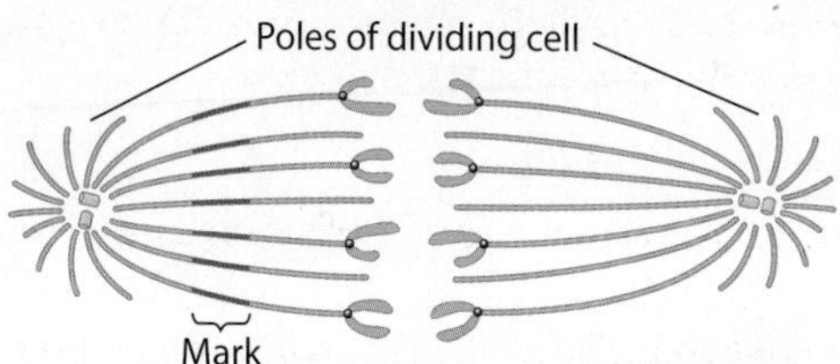
Poles of dividing cell
Mark

The figure below illustrates the results they observed as the chromosomes moved toward the opposite poles of the cell.

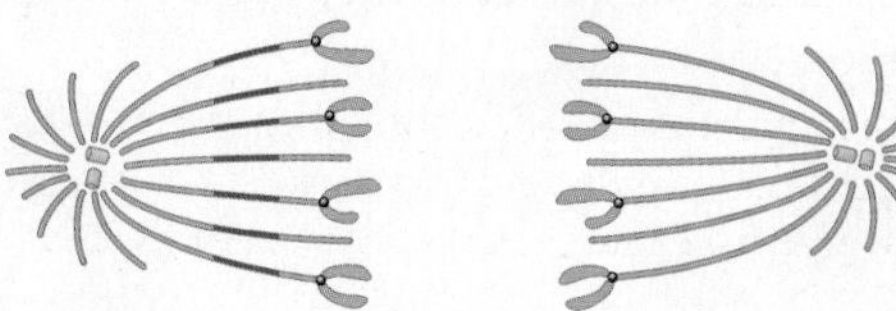

Describe these results. What would you conclude about where the microtubules depolymerize from comparing the length of the microtubules on either side of the mark? How could the experimenters determine whether this is the mechanism of chromosome movement in all cells?

Source: G. J. Gorbsky et al. Chromosomes move poleward in anaphase along stationary microtubules that coordinately disassemble from their kinetochore ends, *Journal of Cell Biology* 104:9–18 (1987).

Answers to all questions can be found in Appendix 4.

CHAPTER 5

The Working Cell

> ? *How can water flow through a membrane?*

The illustration below is beautiful and intriguing—but what does it represent? This computer model shows a small section of a membrane in a human cell; notice the water molecules (depicted with red and gray balls) streaming single file across the membrane. Notice also the phospholipids that make up the lipid bilayer of this membrane: The yellow balls represent the phosphate heads and the green squiggles are the fatty acid tails of the phospholipids. The blue ribbons embedded in the membrane represent regions of a membrane protein called aquaporin that function as water channels. Just one molecule of this protein enables billions of water molecules to flow through the membrane every second—many more than could wander through the lipid bilayer on their own.

Aquaporins are common in cells involved in water balance. For example, your kidneys must filter and reabsorb many liters of water a day, and aquaporins are vital to their proper functioning. There are rare cases of people with defective aquaporins whose kidneys can't reabsorb water and must drink 20 liters of water every day to prevent dehydration. On the other hand, if kidney cells have

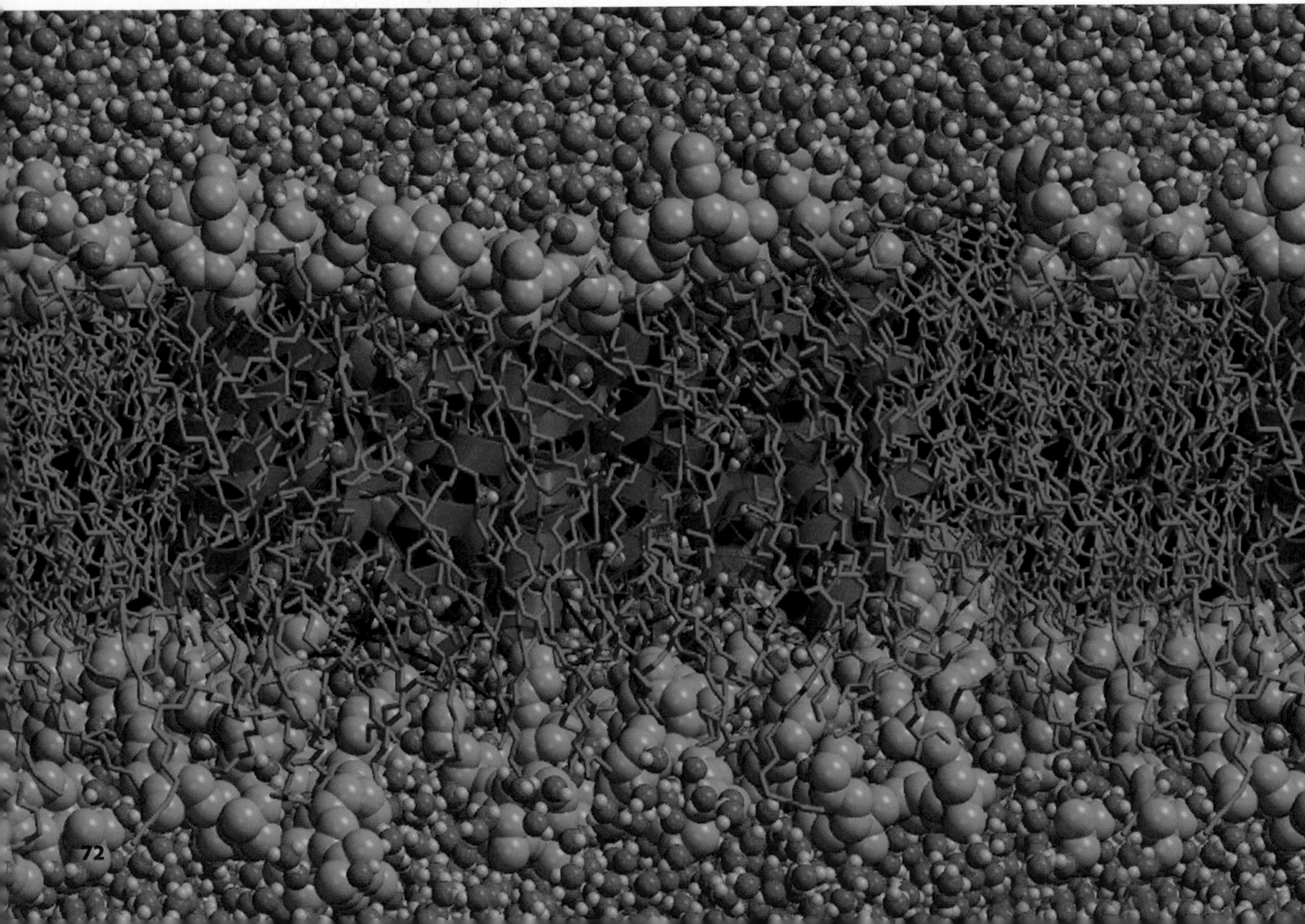

too many aquaporins, excess water is reabsorbed and body tissues may swell. A common complication of pregnancy is fluid retention, and it is likely caused by increased synthesis of aquaporin proteins. Later in the chapter you will learn about the serendipitous discovery of these water channels.

But aquaporins are only one example of how the plasma membrane and its proteins enable cells to survive and function. We begin this chapter by examining membranes. A cell expends energy to build membranes, and many of a membrane's functions require energy. A cell's energy conversions involve enzymes, which control all of its chemical reactions. Indeed, everything that is depicted in this computer model of water molecules zipping through a membrane relates to how working cells use membranes, energy, and enzymes—which are the topics of this chapter.

BIG IDEAS

Membrane Structure and Function

(5.1–5.9)

A cell membrane's structure enables its many functions, such as regulating traffic across the membrane.

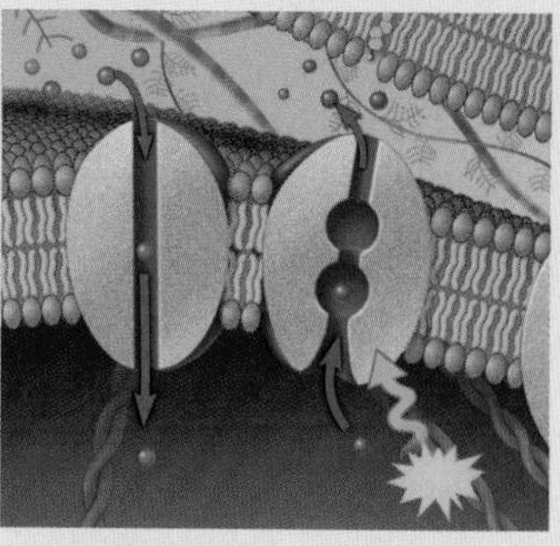

Energy and the Cell

(5.10–5.12)

A cell's metabolic reactions transform energy, using ATP to drive cellular work.

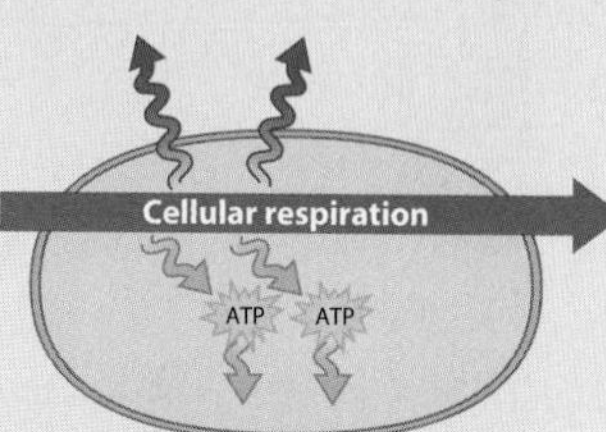
Cellular respiration
ATP
ATP

How Enzymes Function

(5.13–5.16)

Enzymes speed up a cell's chemical reactions and provide precise control of metabolism.

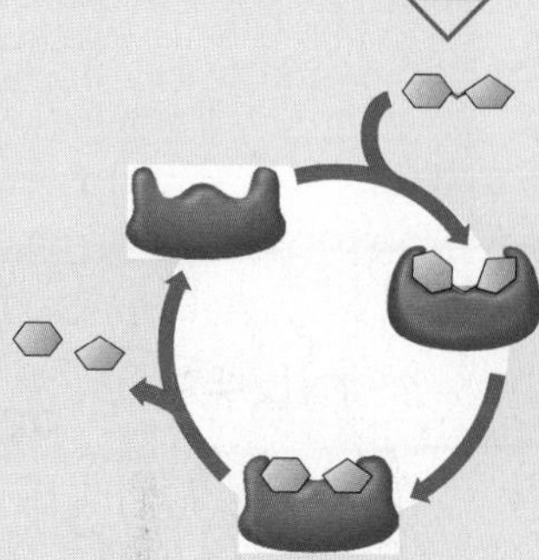

▷ Membrane Structure and Function

VISUALIZING THE CONCEPTS

5.1 Membranes are fluid mosaics of lipids and proteins with many functions

Biologists use the **fluid mosaic model** to describe a membrane's structure—diverse protein molecules suspended in a fluid phospholipid bilayer. This module illustrates the structure and function of a plasma membrane, the boundary that encloses a living cell. Like all cellular membranes, the plasma membrane exhibits **selective permeability**; that is, it allows some substances to cross more easily than others. But the plasma membrane does more than just regulate the exchange of materials. This figure will help you visualize all the activity taking place in and across the membranes of two adjacent cells.

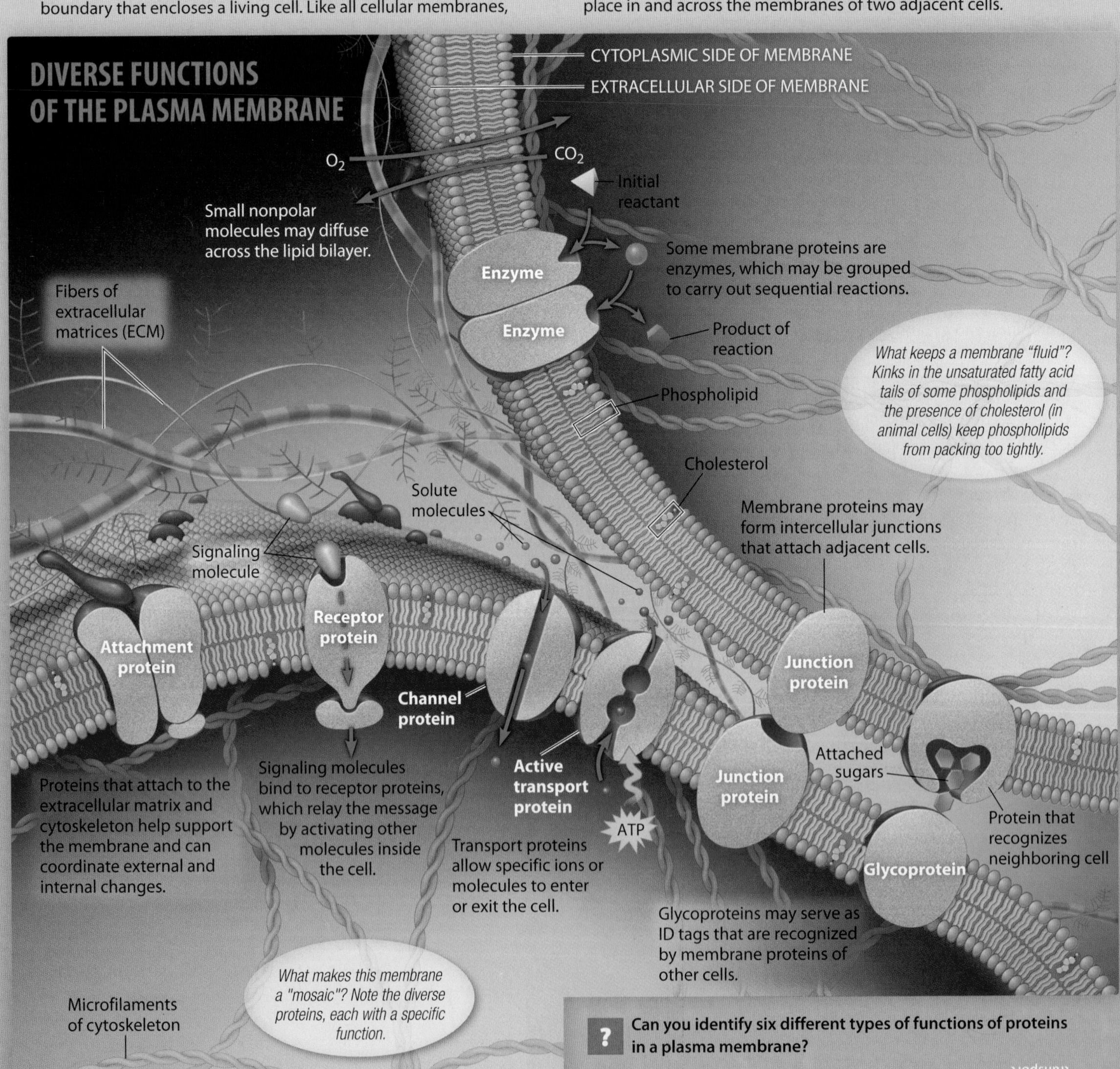

DIVERSE FUNCTIONS OF THE PLASMA MEMBRANE
CYTOPLASMIC SIDE OF MEMBRANE
EXTRACELLULAR SIDE OF MEMBRANE
O_2
CO_2
Small nonpolar molecules may diffuse across the lipid bilayer.
Initial reactant
Enzyme
Enzyme
Some membrane proteins are enzymes, which may be grouped to carry out sequential reactions.
Product of reaction
Fibers of extracellular matrices (ECM)
Phospholipid
What keeps a membrane "fluid"? Kinks in the unsaturated fatty acid tails of some phospholipids and the presence of cholesterol (in animal cells) keep phospholipids from packing too tightly.
Cholesterol
Solute molecules
Membrane proteins may form intercellular junctions that attach adjacent cells.
Signaling molecule
Receptor protein
Attachment protein
Channel protein
Junction protein
Junction protein
Attached sugars
Active transport protein
ATP
Protein that recognizes neighboring cell
Glycoprotein
Proteins that attach to the extracellular matrix and cytoskeleton help support the membrane and can coordinate external and internal changes.
Signaling molecules bind to receptor proteins, which relay the message by activating other molecules inside the cell.
Transport proteins allow specific ions or molecules to enter or exit the cell.
Glycoproteins may serve as ID tags that are recognized by membrane proteins of other cells.
Microfilaments of cytoskeleton
What makes this membrane a "mosaic"? Note the diverse proteins, each with a specific function.

? **Can you identify six different types of functions of proteins in a plasma membrane?**

● Attachment to the cytoskeleton and ECM, signal reception and relay, enzymatic activity, cell-cell recognition, intercellular joining, and transport

5.2 The spontaneous formation of membranes was a critical step in the origin of life

EVOLUTION CONNECTION

Phospholipids, the key ingredients of biological membranes, were probably among the first organic molecules that formed from chemical reactions on early Earth (see Module 15.2). These lipids could spontaneously self-assemble into simple membranes, as can be demonstrated in a test tube. When a mixture of phospholipids and water is shaken, the phospholipids organize into bilayers surrounding water-filled bubbles (**Figure 5.2**). This assembly requires neither genes nor other information beyond the properties of the phospholipids themselves.

The formation of membrane-enclosed collections of molecules would have been a critical step in the evolution of the first cells. A membrane can enclose a solution that is different in composition from its surroundings. A plasma membrane that allows cells to regulate their chemical exchanges with the environment is a basic requirement for life. Indeed, all cells are enclosed by a plasma membrane that is similar in structure and function—illustrating the evolutionary unity of life.

Water-filled
bubble made of
phospholipids

▲ **Figure 5.2** Artificial membrane-bounded sacs

? In the origin of a cell, why would the formation of a simple lipid bilayer membrane not be sufficient? What else would have to be part of such a membrane?

● The membrane would need embedded proteins that could regulate the movement of substances into and out of the cell.

5.3 Passive transport is diffusion across a membrane with no energy investment

Molecules have a type of energy called thermal energy, due to their constant motion. One result of this motion is **diffusion**, the tendency for particles of any substance to spread out into the available space. How might diffusion affect the movement of substances into or out of a cell?

The figures to the right will help you visualize diffusion across a membrane. **Figure 5.3A** shows a solution of green dye separated from pure water by an artificial membrane. Assume that this membrane has microscopic pores through which dye molecules can move. Thus, we say the membrane is permeable to the dye. Although each molecule moves randomly, there will be a *net* movement from the side of the membrane where dye molecules are more concentrated to the side where they are less concentrated. Put another way, the dye diffuses down its **concentration gradient**. Eventually, the solutions on both sides will have equal concentrations of dye. At this dynamic equilibrium, molecules still move back and forth, but there is no *net* change in concentration on either side of the membrane.

Figure 5.3B illustrates the important point that two or more substances diffuse independently of each other; that is, each diffuses down its own concentration gradient.

Because a cell does not have to do work when molecules diffuse across its membrane, such movement across a membrane is called **passive transport**. Much of the traffic across cell membranes occurs by diffusion. For example, diffusion down concentration gradients is the sole means by which oxygen (O_2), essential for metabolism, enters your cells and carbon dioxide (CO_2), a metabolic waste, passes out of them.

Both O_2 and CO_2 are small, nonpolar molecules that diffuse easily across the phospholipid bilayer of a membrane. But can ions and polar molecules also diffuse across the hydrophobic interior of a membrane? They can if they are moving down their concentration gradients and if they have transport proteins to help them cross.

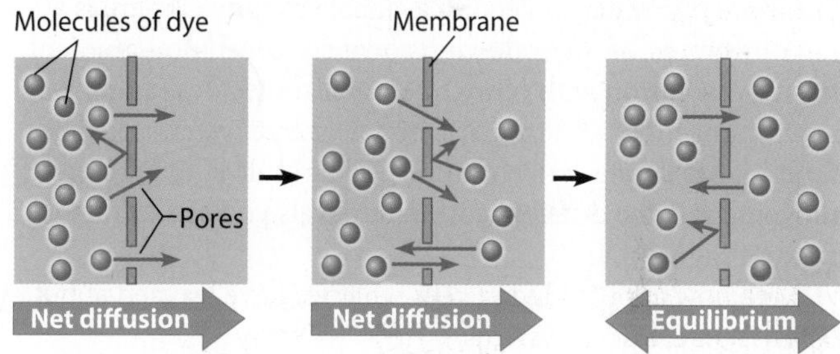
Molecules of dye
Membrane
Pores
Net diffusion
Net diffusion
Equilibrium

▲ **Figure 5.3A** Diffusion of one type of molecule across a membrane

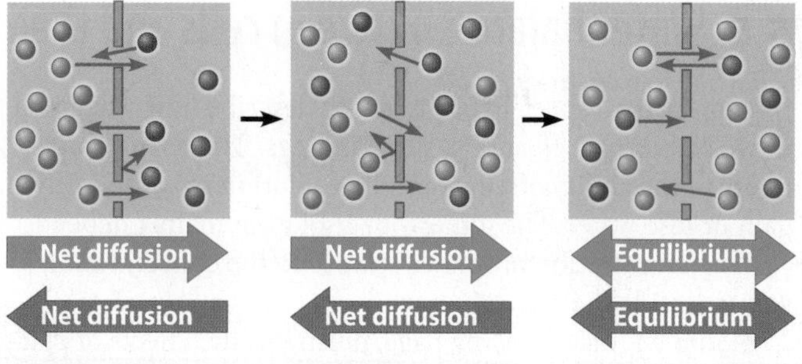
Net diffusion
Net diffusion
Net diffusion
Net diffusion
Equilibrium
Equilibrium

▲ **Figure 5.3B** Diffusion of two types of molecules across a membrane

Try This Explain why these two types of molecules are moving in opposite directions.

? Why is diffusion across a membrane called passive transport?

● The cell does not expend energy to transport substances that are diffusing down their concentration gradients.

5.4 Osmosis is the diffusion of water across a membrane

One of the most important substances that crosses membranes by passive transport is water. In the next module, we consider the critical balance of water between a cell and its environment. But first let's explore a physical model of the diffusion of water across a selectively permeable membrane, a process called **osmosis**. Remember that a selectively permeable membrane allows some substances to cross more easily than others.

The top of **Figure 5.4** shows what happens if a membrane permeable to water but not to a solute (such as glucose) separates two solutions that have different concentrations of solute. (A solute is a substance that dissolves in a liquid solvent. The resulting mixture is a solution.) The solution on the right side initially has a higher concentration of solute than that on the left. As you can see, water crosses the membrane until the solute concentrations are more nearly equal on both sides.

In the close-up view at the bottom of Figure 5.4, you can see what happens at the molecular level. Polar water molecules cluster around hydrophilic (water-loving) solute molecules. The effect is that on the right side, there are fewer water molecules that are *free* to cross the membrane. The less concentrated solution on the left has fewer solute molecules but more *free* water molecules available to move. There is a net movement of water down its own concentration gradient, from the solution with more free water molecules (and lower solute concentration) to that with fewer free water molecules (and higher solute concentration). The result of this water movement is the difference in water levels you see at the top right of Figure 5.4.

Let's now apply to living cells what we have learned about osmosis in artificial systems.

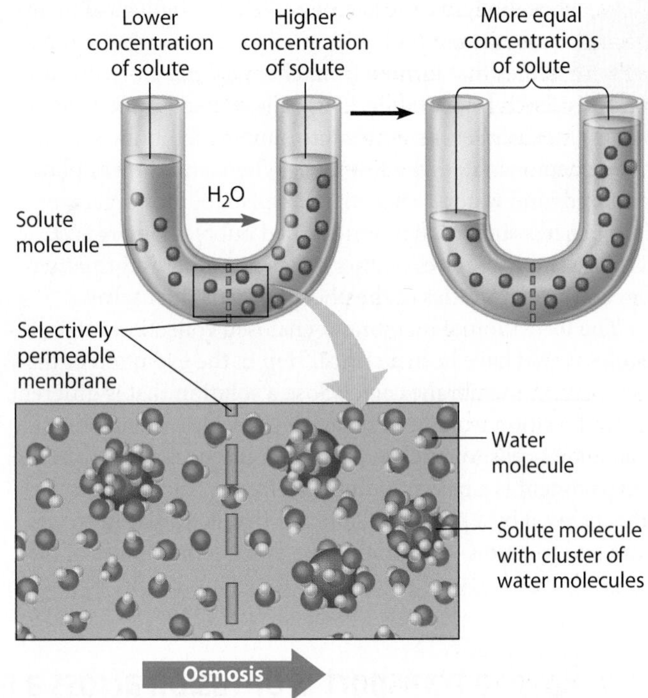

Lower concentration of solute
Higher concentration of solute
More equal concentrations of solute
H_2O
Solute molecule
Selectively permeable membrane
Water molecule
Solute molecule with cluster of water molecules
Osmosis

▲ **Figure 5.4** Osmosis, the diffusion of water across a membrane

Try This Identify the solution that has more free water molecules. Predict which way water will move.

? **Indicate the direction of net water movement between two solutions—a 0.5% sucrose solution and a 2% sucrose solution—separated by a membrane not permeable to sucrose.**

● From the 0.5% sucrose solution (lower solute concentration) to the 2% sucrose solution (higher solute concentration)

5.5 Water balance between cells and their surroundings is crucial to organisms

Biologists use a special vocabulary to describe how water will move between a cell and its surroundings. The term **tonicity** refers to the ability of a surrounding solution to cause a cell to gain or lose water. The tonicity of a solution mainly depends on its concentration of solutes relative to the concentration of solutes inside the cell.

Figure 5.5, on the facing page, illustrates the effects of placing animal and plant cells in solutions of various tonicities.

When an animal cell, such as the red blood cell shown in the top center of the figure, is immersed in a solution that is **isotonic** to the cell (*iso*, same, and *tonos*, tension), the cell's volume remains constant. The solute concentration of a cell and its isotonic environment are essentially equal, and the cell gains water at the same rate that it loses it. In your body, red blood cells are transported in the isotonic plasma of the blood. Intravenous (IV) fluids administered in hospitals must also be isotonic to blood cells. The body cells of most animals are bathed in an extracellular fluid that is isotonic to the cells. And seawater is isotonic to the cells of many marine animals, such as sea stars and crabs.

What happens when an animal cell is placed in a **hypotonic** solution (*hypo*, below), a solution with a solute concentration lower than that of the cell? As shown in the upper left of the figure, the cell gains water, swells, and may burst (lyse) like an overfilled balloon. The upper right shows the opposite case—an animal cell placed in a **hypertonic** solution (*hyper*, above), a solution with a higher solute concentration. In which direction will water move? The cell shrivels and can die from water loss.

For an animal to survive in a hypotonic or hypertonic environment, it must have a way to prevent excessive uptake or loss of water and regulate the solute concentration of its body fluids. The control of water balance is called **osmoregulation**. For example, a freshwater fish, which lives in a hypotonic environment, takes up water by osmosis across the cells of its gills. Its kidneys work constantly to remove excess water from the body. (We will discuss osmoregulation further in Module 25.4.)

Water balance issues are somewhat different for the cells of plants, prokaryotes, and fungi because of their cell walls. As shown in the bottom of Figure 5.5, in a hypotonic environment a plant cell is turgid (very firm), which is the healthy state for most plant cells. Although the plant cell swells as water enters by osmosis, the cell wall exerts a back pressure, called turgor pressure, which prevents the cell from taking in too much water and bursting. Plants that are not woody, such as most houseplants, depend on their turgid cells for mechanical support. In contrast, when a plant cell is surrounded by an isotonic solution, there is no net movement of water into the cell, and the cell is flaccid (limp).

In a hypertonic environment (bottom right), a plant cell is no better off than an animal cell. As a plant cell loses water, it shrivels, and its plasma membrane pulls away from the cell wall. This process, called plasmolysis, causes the plant to wilt and can be lethal to the cell and the plant. The walled cells of bacteria and fungi also plasmolyze in hypertonic environments. Thus, meats and other foods can be preserved with concentrated salt solutions because the cells of food-spoiling bacteria or fungi become plasmolyzed and eventually die.

In the next module, we explore how water and other polar solutes move across cell membranes.

Animal cell

Plant cell

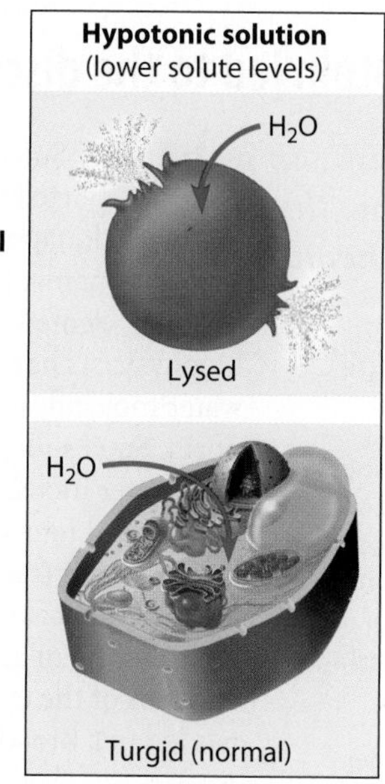

Hypotonic solution
(lower solute levels)
H_2O
Lysed
H_2O
Turgid (normal)

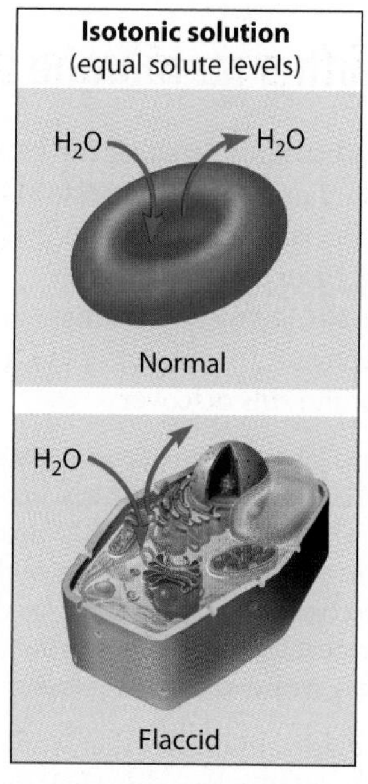

Isotonic solution
(equal solute levels)
H_2O
H_2O
Normal
H_2O
Flaccid

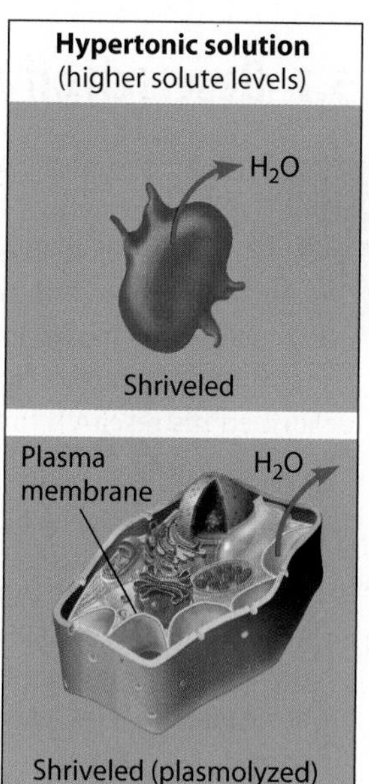

Hypertonic solution
(higher solute levels)
H_2O
Shriveled
Plasma membrane
H_2O
Shriveled (plasmolyzed)

▲ **Figure 5.5** How animal and plant cells react to changes in tonicity

? **Explain the function of the contractile vacuoles in a freshwater *Paramecium* (shown in Figure 4.11A) in terms of what you have just learned about water balance in cells.**

● The pond water in which *Paramecium* lives is hypotonic to the cell. The contractile vacuoles expel the water that constantly enters the cell by osmosis.

5.6 Transport proteins can facilitate diffusion across membranes

Recall that nonpolar molecules, such as O_2 and CO_2, can dissolve in the lipid bilayer of a membrane and diffuse through it with ease. But how do polar or charged substances make it past the hydrophobic center of a membrane? Hydrophilic molecules and ions require the help of specific transport proteins to move across a membrane. This assisted transport, called **facilitated diffusion**, is a type of passive transport because it does not require energy. As in all passive transport, the driving force is the concentration gradient.

Figure 5.6 shows a common type of transport protein, which provides a channel that specific molecules or ions use as a passageway through a membrane. Another type of transport protein binds its passenger, changes shape, and releases the transported molecule on the other side. In both cases, the transport protein helps a specific substance diffuse across the membrane down its concentration gradient and, thus, requires no input of energy.

Substances that use facilitated diffusion for crossing cell membranes include a number of sugars, amino acids, ions—and even water. The water molecule is very small, but because it is polar (see Module 2.6), its diffusion through a membrane's hydrophobic interior is relatively slow. For many cells, this slow diffusion of water is adequate. Cells such as plant cells, red blood cells, and the cells lining your kidney tubules, however, have greater water-permeability needs. As you saw in the chapter introduction, the very rapid diffusion of water into and out of such cells is made possible by a protein channel called an **aquaporin**. In the next module, we explore the discovery of these transport proteins.

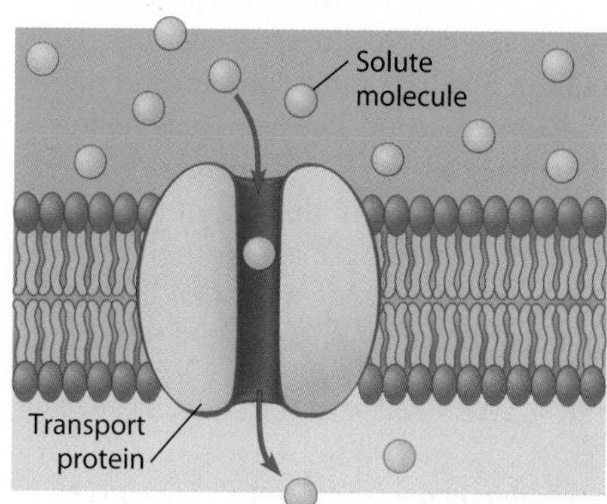

Solute molecule
Transport protein

◀ **Figure 5.6** Transport protein providing a channel for the diffusion of a specific solute across a membrane

? **How do transport proteins contribute to a membrane's selective permeability?**

● Because they are specific for the solutes they transport, the numbers and kinds of transport proteins affect a membrane's permeability to various solutes.

5.7 Research on another membrane protein led to the discovery of aquaporins

SCIENTIFIC THINKING

How can water flow through a membrane?

Sometimes major advances in science occur when a scientist is studying something else but makes the wise decision to explore an unexpected finding. Peter Agre received the 2003 Nobel Prize in Chemistry for this sort of discovery of aquaporins. In an interview, Dr. Agre described his research that led to this discovery:

> When I joined the faculty at the Johns Hopkins School of Medicine, I began to study the Rh blood antigens. Rh is of medical importance . . . when Rh-negative mothers have Rh-positive babies. Membrane-spanning proteins are really messy to work with. But we worked out a method to isolate the Rh protein. Our sample seemed to consist of two proteins, but we were sure that the smaller one was just a breakdown product of the larger one. We were completely wrong.

Dr. Agre's research team made antibodies that would specifically bind to and label this smaller protein. They found two interesting results: The antibody did not bind to any part of the Rh protein, indicating that the smaller protein wasn't part of the Rh protein. And the antibody did bind in huge quantities to red blood cells, showing that this new protein is one of the most abundant proteins in red cell membranes. Agre and his team also determined that the protein was identical to and even more abundant in certain kidney cells. But they didn't know what this protein did.

A colleague suggested that the protein might be the elusive water channel that physiologists had predicted would explain the rapid transport of water in some cells. To test this hypothesis, the researchers injected messenger RNA for the protein into frog eggs, whose plasma membranes are known to be quite water impermeable. Biochemical tests showed that within 72 hours, the frog egg cells had translated the mRNA into the new protein. They transferred a group of RNA-injected frog eggs and a control group of eggs injected with only a buffer solution to a hypotonic solution and monitored the eggs with videomicroscopy. The osmotic swelling of RNA-injected and control cells is plotted in **Figure 5.7**. The experimental egg cells exploded in three minutes; the control eggs showed minimal swelling, even for time periods exceeding an hour. The researchers concluded that the newly discovered protein enabled the rapid movement of water into the cells.

Since the results of that experiment were reported in 1992, much research has been done on aquaporins, determining their structure and dynamic functioning. The chapter introduction presented a model of aquaporin structure. Molecular biophysicists have produced computer simulations that show water molecules flipping their way single file through an aquaporin. Such simulations have revealed how aquaporins allow only water molecules to pass through them. Aquaporins have been found in bacteria, plants, and animals, and evolutionary biologists are tracing the relationships of these various aquaporins. Medical researchers study the function and occasional malfunction of aquaporins in the human kidney, lungs, brain, and lens of the eye. The serendipitous discovery of aquaporins has led to a broad range of scientific research.

? **Why are aquaporins important in kidney cells?**

Kidney cells must reabsorb a large amount of water when producing urine.

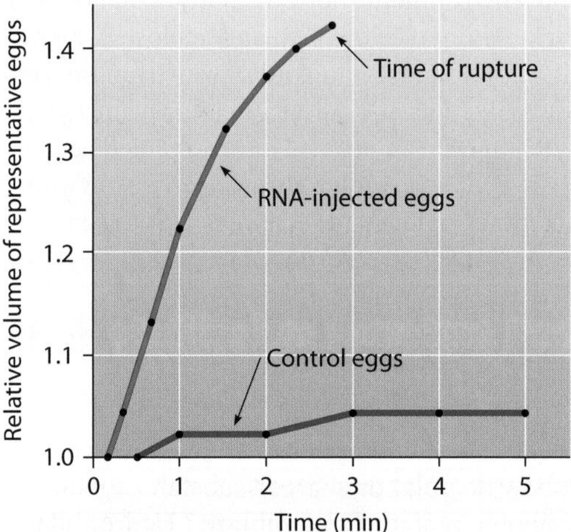

Relative volume of representative eggs
1.4
1.3
1.2
1.1
1.0
Time of rupture
RNA-injected eggs
Control eggs
0
1
2
3
4
5
Time (min)

Source: Adaptation of Figure 2A from "Appearance of Water Channels in Xenopus Oocytes Expressing Red Cell CHIP28 Protein" by Gregory Preston et al., from *Science*, April 1992, Volume 256(5055). Copyright © 1992 by AAAS. Reprinted with permission.

▲ **Figure 5.7** Osmotic swelling of representative aquaporin RNA-injected and control-injected oocytes following transfer to a hypotonic medium

5.8 Cells expend energy in the active transport of a solute

In **active transport**, a cell must expend energy to move a solute *against* its concentration gradient—that is, across a membrane toward the side where the solute is more concentrated. The energy molecule ATP (described in more detail in Module 5.12) supplies the energy for most active transport.

Active transport allows a cell to maintain internal concentrations of small molecules and ions that are different from concentrations in its surroundings. For example, the inside of an animal cell has a higher concentration of potassium ions (K^+) and a lower concentration of sodium ions (Na^+) than the solution outside the cell. The generation of nerve signals depends on these concentration differences, which a transport protein called the sodium-potassium pump maintains by actively moving Na^+ out of the cell and K^+ into the cell.

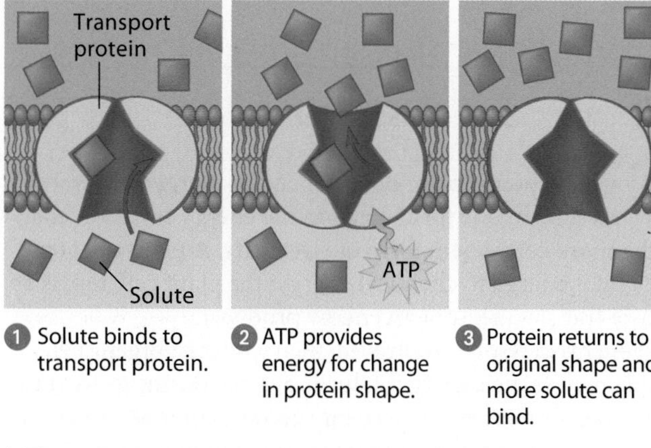
Transport protein
Solute
ATP
1 Solute binds to transport protein.
2 ATP provides energy for change in protein shape.
3 Protein returns to original shape and more solute can bind.

▲ **Figure 5.8** Active transport of a solute across a membrane

Figure 5.8 shows a simple model of an active transport system that pumps a solute out of the cell against its concentration gradient. ❶ The process begins when solute molecules on the cytoplasmic side of the plasma membrane attach to specific binding sites on the transport protein. ❷ With energy provided by ATP, the transport protein changes shape in such a way that the solute is released on the other side of the membrane. ❸ The transport protein returns to its original shape, ready for its next passengers.

? Cells actively transport Ca^{2+} out of the cell. Is calcium more concentrated inside or outside of the cell? Explain.

● Outside: Active transport moves calcium against its concentration gradient.

5.9 Exocytosis and endocytosis transport large molecules across membranes

So far, we've focused on how water and small solutes enter and leave cells. The story is different for large molecules.

A cell uses the process of **exocytosis** (from the Greek *exo*, outside, and *kytos*, cell) to export bulky materials such as proteins or polysaccharides. A transport vesicle filled with macromolecules buds from the Golgi apparatus and moves to the plasma membrane (see Figure 4.12). Once there, the vesicle fuses with the plasma membrane, and the vesicle's contents spill out of the cell when the vesicle membrane becomes part of the plasma membrane. For example, the cells in your pancreas that manufacture the hormone insulin secrete it into the extracellular fluid by exocytosis, where it is picked up by the bloodstream.

Endocytosis (*endo*, inside) is a transport process through which a cell takes in large molecules. Figure 5.9 shows two kinds of endocytosis. The top diagram illustrates **phagocytosis**, or "cellular eating." A cell engulfs a particle by wrapping extensions called pseudopodia around it and packaging it within a membrane-enclosed sac called a vacuole. The vacuole then fuses with a lysosome, whose hydrolytic enzymes digest the contents of the vacuole (see Figure 4.10A). Protists such as amoeba take in food particles this way, and some of your white blood cells engulf invading bacteria via phagocytosis.

The bottom diagram illustrates **receptor-mediated endocytosis**, which enables a cell to acquire specific solutes. Receptor proteins for specific molecules are embedded in regions of the membrane that are lined by a layer of coat proteins. The plasma membrane indents to form a coated pit, whose receptor proteins pick up particular molecules from the extracellular fluid. The coated pit pinches closed to form a vesicle, which then releases the molecules into the cytoplasm.

Your cells use receptor-mediated endocytosis to take in cholesterol from the blood for synthesis of membranes and as a precursor for other steroids. Cholesterol circulates in the blood in particles called low-density lipoproteins (LDLs). LDLs bind to receptor proteins and then enter cells by endocytosis. In humans with the inherited disease familial hypercholesterolemia, LDL receptor proteins are defective or missing. Cholesterol accumulates to high levels in the blood, leading to atherosclerosis, the buildup of fatty deposits in the walls of blood vessels (see Module 9.11).

Phagocytosis

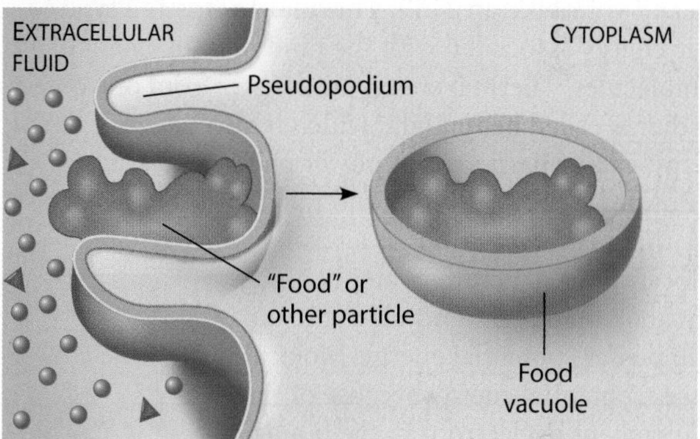
EXTRACELLULAR FLUID
CYTOPLASM
Pseudopodium
"Food" or other particle
Food vacuole

Receptor-mediated endocytosis

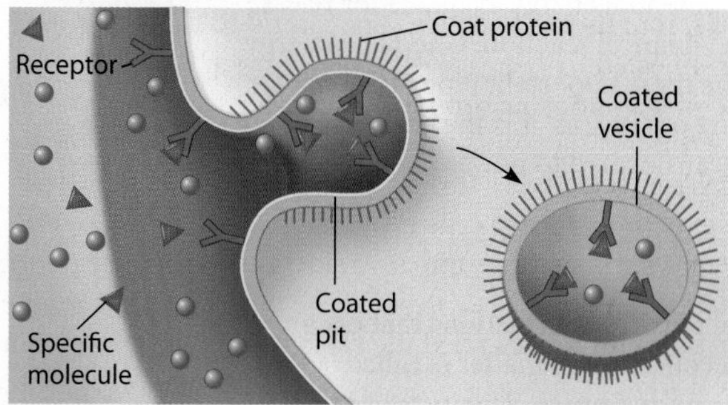
Receptor
Coat protein
Coated vesicle
Coated pit
Specific molecule

▲ **Figure 5.9** Two kinds of endocytosis

? As a cell grows, its plasma membrane expands. Does this involve endocytosis or exocytosis? Explain.

● Exocytosis: When a transport vesicle fuses with the plasma membrane, its contents are released and the vesicle membrane adds to the plasma membrane.

Energy and the Cell

5.10 Cells transform energy as they perform work

The title of this chapter is "The Working Cell." But just what type of work does a cell do? You just learned that a cell can actively transport substances across membranes. The cell also builds those membranes and the proteins embedded in them. A cell is a miniature chemical factory in which thousands of reactions occur within a microscopic space. Some of these reactions release energy; others require energy. But before you can understand how the cell works, you must have a basic knowledge of energy.

Forms of Energy **Energy** is the capacity to cause change or to perform work. There are two basic forms of energy: kinetic energy and potential energy. **Kinetic energy** is the energy of motion. Moving objects can perform work by transferring motion to other matter. For example, the movement of your legs can push bicycle pedals, turning the wheels and moving you and your bike up a hill. **Thermal energy** is a type of kinetic energy associated with the random movement of atoms or molecules. Thermal energy in transfer from one object to another is called **heat**. Light, which is also a type of kinetic energy, can be harnessed to power photosynthesis.

Potential energy, the second main form of energy, is energy that matter possesses as a result of its location or structure. Water behind a dam and you on your bicycle at the top of a hill possess potential energy. Molecules possess potential energy because of the arrangement of electrons in the bonds between their atoms. **Chemical energy** is the potential energy available for release in a chemical reaction. Chemical energy is the most important type of energy for living organisms; it is the energy that can be transformed to power the work of the cell.

Energy Transformations The study of energy transformations that occur in a collection of matter is called **thermodynamics**. Scientists use the word *system* for the matter under study and refer to the rest of the universe—everything outside the system—as the *surroundings*. A system can be an electric power plant, a single cell, or the entire planet. An organism is an open system; that is, it exchanges both energy and matter with its surroundings.

The **first law of thermodynamics**, also known as the law of energy conservation, states that the energy in the universe is constant. Energy can be transferred and transformed, but it cannot be created or destroyed. A power plant does not create energy; it merely converts it from one form (such as the energy stored in coal) to the more convenient form of electricity. A plant cell converts light energy to chemical energy; the plant cell, too, is an energy transformer, not an energy producer.

If energy cannot be destroyed, then why can't organisms simply recycle their energy? It turns out that during every transfer or transformation, some energy becomes unavailable to do work—it is converted to thermal energy (random molecular motion) and released as heat. Scientists use a quantity called **entropy** as a measure of disorder, or randomness. The more randomly arranged a collection of matter is, the greater its entropy. According to the **second law of thermodynamics**, energy conversions increase the entropy (disorder) of the universe.

Figure 5.10 compares a car and a cell to show how energy can be transformed and how entropy increases as a result. Automobile engines and cells use the same basic process to make the chemical energy of their fuel available for work. The engine mixes oxygen with gasoline in an explosive chemical reaction that pushes the pistons, which eventually move the wheels. The waste products emitted from the exhaust pipe are carbon dioxide and water, energy-poor, simple molecules. Only about

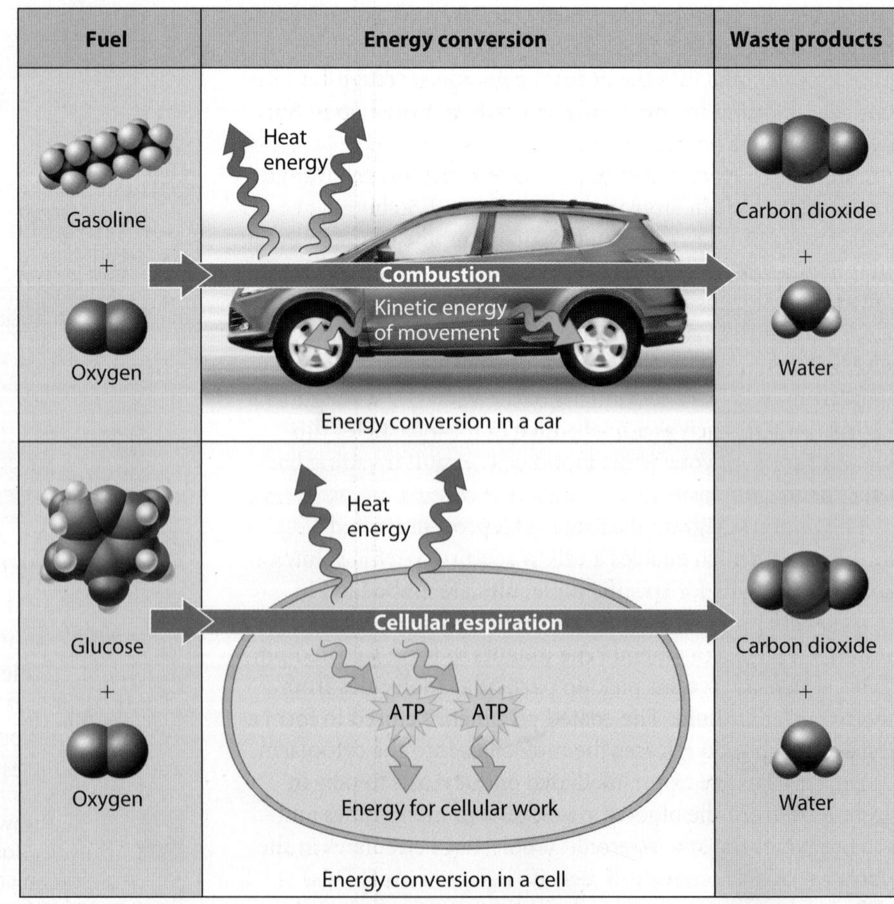
Fuel
Energy conversion
Waste products
Heat energy
Gasoline
+
Oxygen
Combustion
Kinetic energy of movement
Carbon dioxide
+
Water
Energy conversion in a car
Heat energy
Glucose
+
Oxygen
Cellular respiration
ATP
ATP
Energy for cellular work
Carbon dioxide
+
Water
Energy conversion in a cell

▲ **Figure 5.10** Energy transformations in a car and a cell

25% of the energy stored in gasoline is converted to the kinetic energy of the car's movement; the rest is lost as heat.

Cells also use oxygen in reactions that release energy from fuel molecules. In the process called **cellular respiration**, the chemical energy stored in organic molecules is used to produce ATP, which the cell can use to perform work. Just like for the car, the waste products are carbon dioxide and water. Cells are more efficient than cars, however, converting about 34% of the chemical energy in their fuel to energy for cellular work. The other 66% generates heat, which explains why vigorous exercise makes you so warm.

According to the second law of thermodynamics, energy transformations result in the universe becoming more disordered. How, then, can we account for biological order? Although the intricate structures of a cell correspond to a decrease in entropy, their production is accomplished at the expense of ordered forms of matter and energy taken in from the surroundings. As shown in Figure 5.10, cells extract the chemical energy of glucose and return disordered heat and lower-energy carbon dioxide and water to the surroundings. In a thermodynamic sense, a cell is an island of low entropy in an increasingly random universe.

? **How does the second law of thermodynamics explain the diffusion of a solute across a membrane?**

● Diffusion across a membrane results in equal concentrations of solute, which is a more disordered arrangement (higher entropy) than a high concentration on one side and a low concentration on the other.

5.11 Chemical reactions either release or store energy

Chemical reactions are of two types: exergonic or endergonic. An **exergonic reaction** releases energy (*exergonic* means "energy outward"). As **Figure 5.11A** shows, an exergonic reaction begins with reactants whose covalent bonds contain more potential energy than those in the products. The reaction releases to the surroundings an amount of energy equal to the difference in potential energy between the reactants and the products.

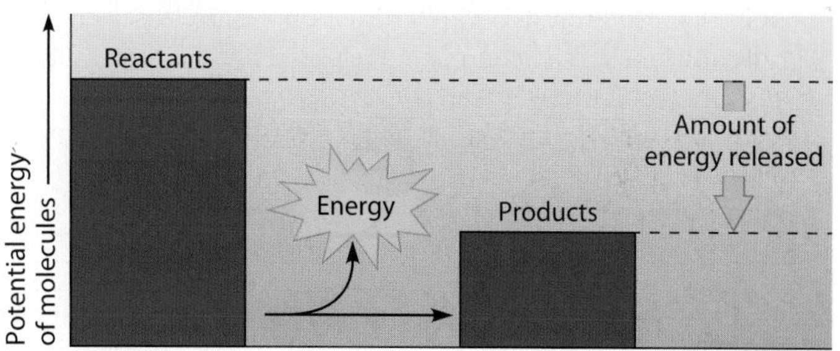
Reactants
Amount of energy released
Energy
Products
Potential energy of molecules
▲ **Figure 5.11A** Exergonic reaction, energy released

Consider what happens when wood burns. One of the major components of wood is cellulose, a large energy-rich carbohydrate composed of many glucose monomers. Burning wood releases the energy of glucose as heat and light. Carbon dioxide and water are the products of the reaction.

As you learned in Module 5.10, cells release energy from fuel molecules in the process called cellular respiration. Burning and cellular respiration are alike in being exergonic. They differ in that burning is essentially a one-step process that releases all of a substance's energy at once. Cellular respiration, on the other hand, involves many steps, each a separate chemical reaction; you can think of it as a "slow burn." Some of the energy released by cellular respiration escapes as heat, but a substantial amount is stored in ATP, the immediate source of energy for a cell.

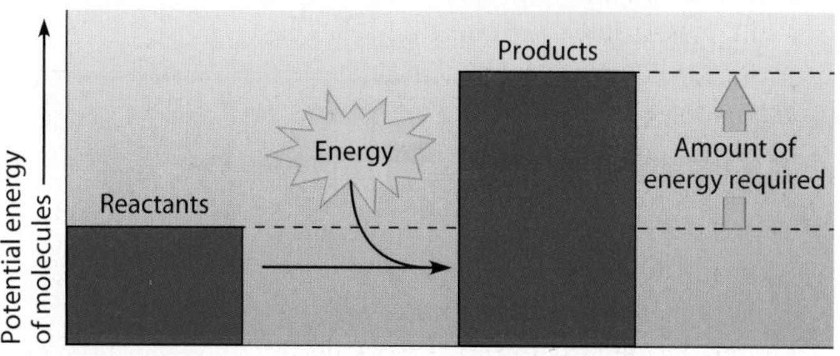
Products
Energy
Amount of energy required
Reactants
Potential energy of molecules
▲ **Figure 5.11B** Endergonic reaction, energy required

Endergonic reactions require a net input of energy and yield products that are rich in potential energy (*endergonic* means "energy inward"). As shown in **Figure 5.11B**, an endergonic reaction starts with reactants that contain relatively little potential energy. Energy is absorbed from the surroundings as the reaction occurs, so the products of an endergonic reaction contain more chemical energy than the reactants did.

Photosynthesis, the process by which plant cells make sugar, is an example of an endergonic process. Photosynthesis starts with energy-poor reactants (carbon dioxide and water molecules) and, using energy absorbed from sunlight, produces energy-rich sugar molecules.

Living cells carry out thousands of exergonic and endergonic reactions. The total of an organism's chemical reactions is called **metabolism**. We can picture a cell's metabolism as a road map of thousands of chemical reactions arranged as intersecting highways or metabolic pathways. A **metabolic pathway** is a series of chemical reactions that either builds a complex molecule or breaks down a complex molecule into simpler compounds. The "slow burn" of cellular respiration is an example of a metabolic pathway in which a sequence of reactions slowly releases the potential energy stored in sugar.

All of an organism's activities require energy, which is obtained from sugar and other molecules by the exergonic reactions of cellular respiration. Cells then use that energy in endergonic reactions to make molecules and do the work of the cell. **Energy coupling**—the use of energy released from exergonic reactions to drive endergonic reactions—is crucial in all cells. ATP molecules are the key to energy coupling. In the next module, we explore the structure and function of ATP.

? **Cellular respiration is an exergonic process. Remembering that energy must be conserved, what do you think becomes of the energy extracted from food during this process?**

● Some of it is stored in ATP molecules; the rest is released as heat.

5.12 ATP drives cellular work by coupling exergonic and endergonic reactions

ATP powers nearly all forms of cellular work. The abbreviation ATP stands for adenosine triphosphate, and as **Figure 5.12A** shows, ATP consists of an organic molecule called adenosine and a triphosphate tail of three phosphate groups (each symbolized by (P)). All three phosphate groups are negatively charged (see Table 3.2). These like charges are crowded together, and their mutual repulsion makes the triphosphate tail of ATP the chemical equivalent of a compressed spring.

As a result, the bonds connecting the phosphate groups are unstable and can readily be broken by hydrolysis, the addition of water. Notice in Figure 5.12A that when the bond to the third group breaks, a phosphate group leaves ATP—which becomes ADP (adenosine diphosphate)—and energy is released.

Thus, the hydrolysis of ATP is exergonic—it releases energy. How does a cell couple this reaction to an endergonic (energy-requiring) reaction? It usually does so by transferring a phosphate group from ATP to another molecule. This phosphate transfer is called **phosphorylation**, and most cellular work depends on ATP energizing molecules by phosphorylating them.

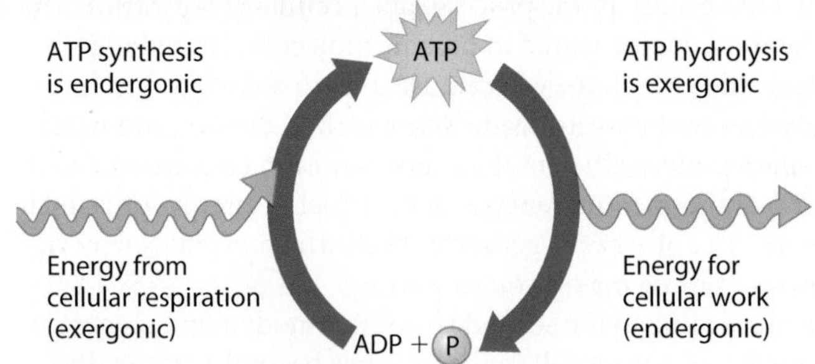
ATP synthesis is endergonic
ATP
ATP hydrolysis is exergonic
Energy from cellular respiration (exergonic)
ADP + (P)
Energy for cellular work (endergonic)

▲ **Figure 5.12C** The ATP cycle

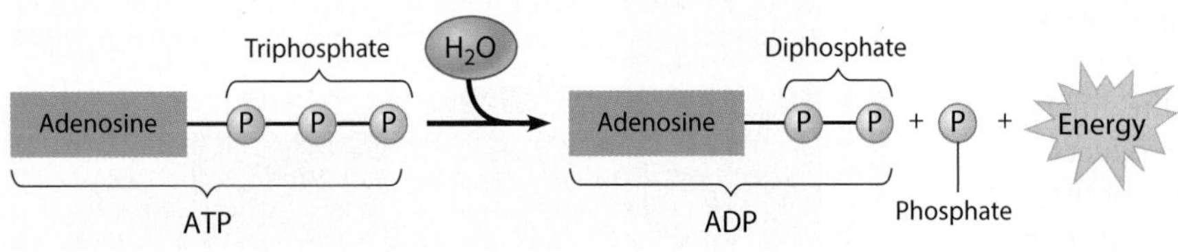
Triphosphate
H_2O
Diphosphate
Adenosine
(P)–(P)–(P)
Adenosine
(P)–(P)
+ (P) +
Energy
ATP
ADP
Phosphate

▲ **Figure 5.12A** The hydrolysis of ATP yielding ADP, a phosphate group, and energy.

What types of work does a cell do? As **Figure 5.12B** shows, the chemical, transport, and mechanical work of a cell are all driven by ATP. In chemical work, the phosphorylation of reactants provides energy to drive the endergonic synthesis of products. In transport work, ATP drives the active transport of solutes across a membrane against their concentration gradients by phosphorylating transport proteins. And in an example of mechanical work, the transfer of phosphate groups to special motor proteins in muscle cells causes the proteins to change shape and pull on other protein filaments, in turn causing the cells to contract.

ATP is a renewable resource. A cell uses and regenerates ATP continuously. **Figure 5.12C** shows the ATP cycle. Each side of this cycle illustrates energy coupling. Energy released in exergonic reactions, such as the breakdown of glucose during cellular respiration, is used to generate ATP from ADP. In this endergonic process, a phosphate group is bonded to ADP, forming ATP. The hydrolysis of ATP releases energy that drives endergonic reactions. The ATP cycle runs at an astonishing pace. In fact, a working muscle cell may consume and regenerate 10 million ATP molecules each second.

Chemical work

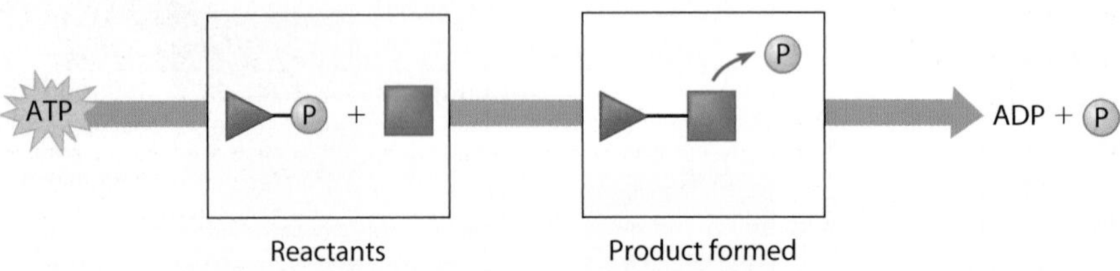
ATP
(P) +
(P)
ADP + (P)
Reactants
Product formed

Transport work

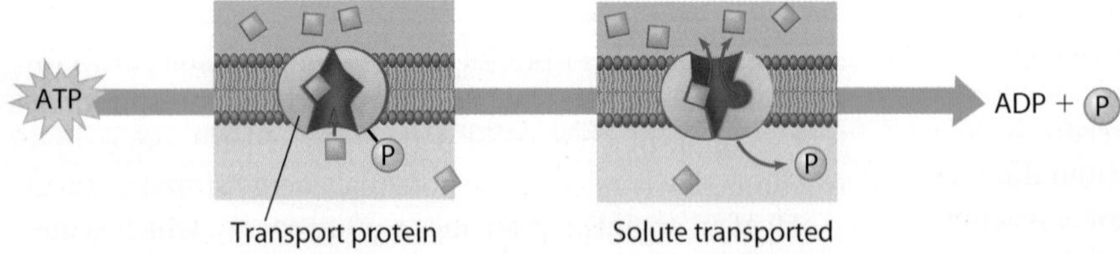
ATP
(P)
(P)
ADP + (P)
Transport protein
Solute transported

Mechanical work

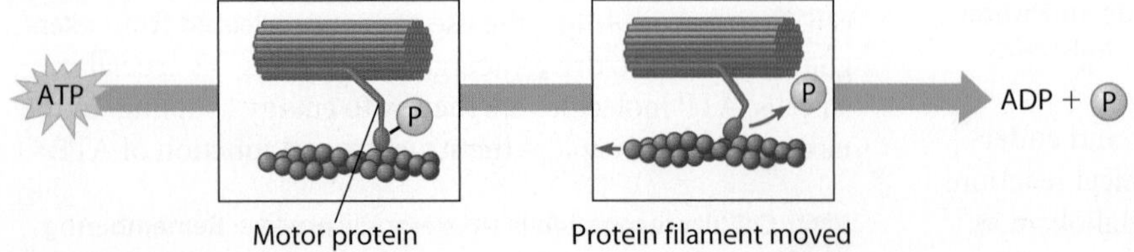
ATP
(P)
(P)
ADP + (P)
Motor protein
Protein filament moved

▲ **Figure 5.12B** How ATP powers cellular work

Try This Examine this figure and identify the way in which all three types of cellular work are similar.

? **Explain how ATP transfers energy from exergonic to endergonic processes in the cell.**

● Exergonic processes phosphorylate ADP to form ATP. ATP transfers energy to endergonic processes by phosphorylating other molecules.

How Enzymes Function

5.13 Enzymes speed up the cell's chemical reactions by lowering energy barriers

Your room gets messier; water flows downhill; sugar crystals dissolve in your coffee. Ordered structures tend toward disorder, and high-energy systems tend to change toward a more stable state of low energy. Proteins, DNA, carbohydrates, lipids—most of the complex molecules of your cells are rich in potential energy. Why don't these high-energy, ordered molecules spontaneously break down into less ordered, lower-energy molecules? They remain intact for the same reason that wood doesn't normally burst into flames or the gas in an automobile's gas tank doesn't spontaneously explode.

There is an energy barrier that must be overcome before a chemical reaction can begin. Energy must be absorbed to contort or weaken bonds in reactant molecules so that they can break and new bonds can form. We call this the **activation energy** (because it activates the reactants). We can think of activation energy as the amount of energy needed for reactant molecules to move "uphill" to a higher-energy, unstable state so that the "downhill" part of a reaction can begin.

The activation energy barrier protects the highly ordered molecules of your cells from spontaneously breaking down. But now we have a dilemma. Life depends on countless chemical reactions that constantly change a cell's molecular makeup. Most of the essential reactions of metabolism must occur quickly and precisely for a cell to survive. How can the specific reactions that a cell requires get over that energy barrier?

One way to speed reactions is to add heat. Heat speeds up molecules and agitates atoms so that bonds break more easily and reactions can proceed. Certainly, adding a match to kindling will start a fire, and the firing of a spark plug ignites gasoline in an engine. But heating a cell would speed up all chemical reactions, not just the necessary ones, and too much heat would kill the cell.

The answer to this dilemma lies in **enzymes**—molecules that function as biological catalysts, increasing the rate of a reaction without being consumed by the reaction. Almost all enzymes are proteins. (Some RNA molecules can also function as enzymes.) An enzyme speeds up a reaction by lowering the activation energy needed for a reaction to begin. **Figure 5.13** compares a reaction without an enzyme (left) and with an enzyme (right). Notice how much easier it is for the reactant to get over the activation energy barrier when an enzyme is involved. In the next module, we explore how the structure of an enzyme enables it to lower the activation energy, allowing a reaction to proceed.

? The graph below illustrates the course of a reaction with and without an enzyme. Which curve represents the enzyme-catalyzed reaction? What energy changes are represented by the lines labeled a, b, and c?

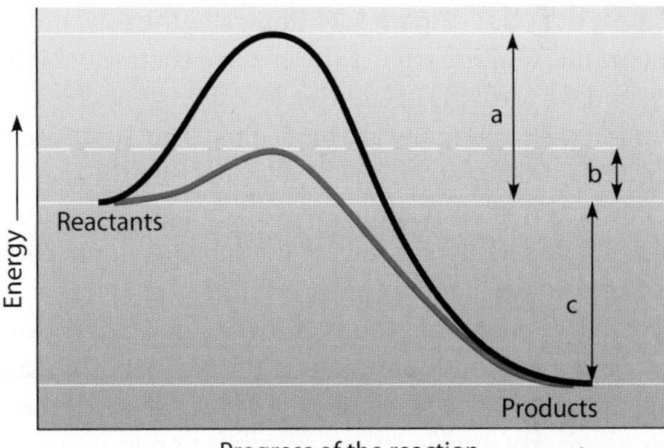

a
b
Reactants
c
Products
Energy
Progress of the reaction

The lower (red) curve. Line a is the activation energy without enzyme; b is the activation energy with enzyme; c is the change in energy between reactants and products, which is the same for both the catalyzed and uncatalyzed reactions.

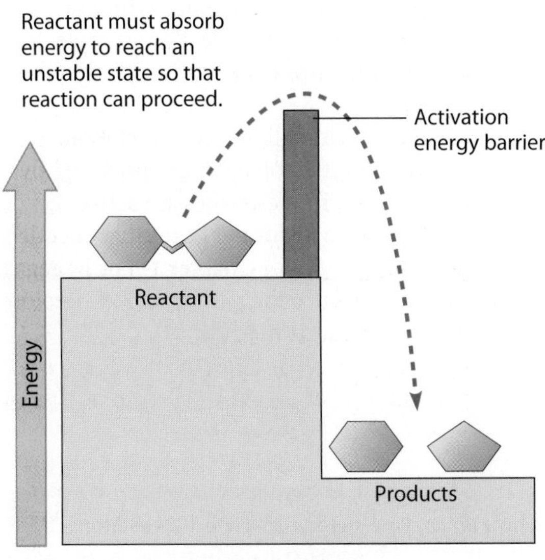

Reactant must absorb energy to reach an unstable state so that reaction can proceed.
Activation energy barrier
Reactant
Products
Energy
Without enzyme

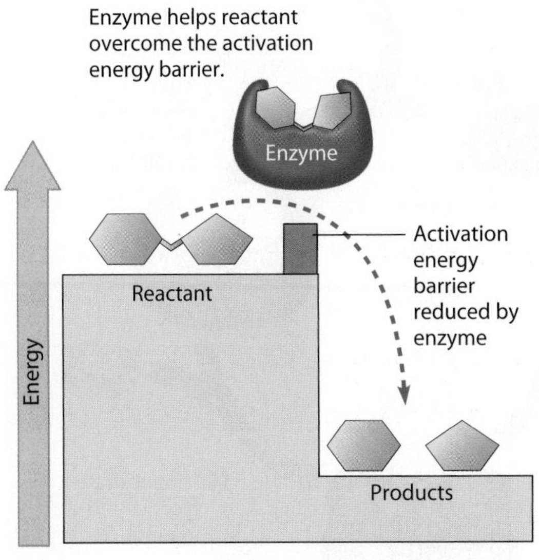

Enzyme helps reactant overcome the activation energy barrier.
Enzyme
Activation energy barrier reduced by enzyme
Reactant
Products
Energy
With enzyme

▲ **Figure 5.13** The effect of an enzyme in lowering the activation energy

5.14 A specific enzyme catalyzes each cellular reaction

You just learned that an enzyme catalyzes a reaction by lowering the activation energy barrier. How does it do that? With the aid of an enzyme, the bonds in a reactant are contorted into the higher-energy, unstable state from which the reaction can proceed. Without an enzyme, the activation energy barrier might never be breached. For example, a solution of sucrose (table sugar) can sit for years at room temperature with no appreciable hydrolysis into its components glucose and fructose. But if we add a small amount of the enzyme sucrase, all the sucrose will be hydrolyzed within seconds.

An enzyme is very selective in the reaction it catalyzes. As a protein, an enzyme has a unique three-dimensional shape, and that shape determines the enzyme's specificity. The specific reactant that an enzyme acts on is called the enzyme's **substrate**. A substrate fits into a region of the enzyme called the **active site**—typically a pocket or groove on the surface of the enzyme. Enzymes are specific because only specific substrate molecules fit into their active sites.

The Catalytic Cycle **Figure 5.14** illustrates the catalytic cycle of an enzyme. Our example is the enzyme sucrase, which catalyzes the hydrolysis of sucrose. (Most enzymes have names that end in *-ase*, and many are named for their substrate.) ❶ The enzyme starts with an empty active site. ❷ Sucrose enters the active site, attaching by weak bonds. The active site changes shape slightly, embracing the substrate more snugly, like a firm handshake. This **induced fit** may contort substrate bonds or place chemical groups of the amino acids making up the active site in position to catalyze the reaction. (In reactions involving two or more reactants, the active site holds the substrates in the proper orientation for a reaction to occur.)

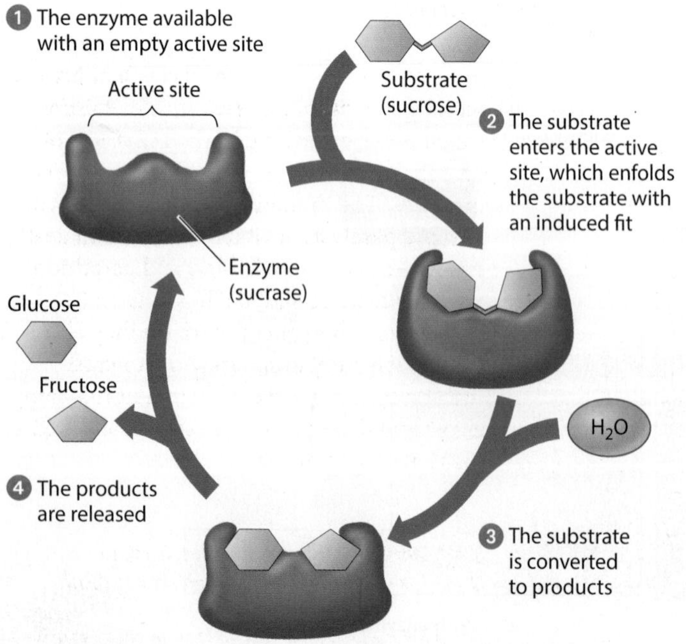

❶ The enzyme available with an empty active site
Active site
Enzyme (sucrase)
Substrate (sucrose)
❷ The substrate enters the active site, which enfolds the substrate with an induced fit
H_2O
❸ The substrate is converted to products
❹ The products are released
Glucose
Fructose

▲ **Figure 5.14** The catalytic cycle of an enzyme

❸ The strained bond of sucrose reacts with water, and the substrate is converted (hydrolyzed) to the products glucose and fructose. ❹ The enzyme releases the products and emerges unchanged from the reaction. Its active site is now available for another substrate molecule, and another round of the cycle can begin. A single enzyme molecule may act on thousands or even millions of substrate molecules per second.

Optimal Conditions for Enzymes As with all proteins, an enzyme's shape is central to its function, and this three-dimensional shape is affected by the environment. For every enzyme, there are optimal conditions under which it is most effective. Temperature, for instance, affects molecular motion, and an enzyme's optimal temperature produces the highest rate of contact between reactant molecules and the enzyme's active site. Higher temperatures denature the enzyme, altering its specific shape and destroying its function. Most human enzymes work best at 35–40°C (95–104°F), close to our normal body temperature of 37°C. Prokaryotes that live in hot springs, however, contain enzymes with optimal temperatures of 70°C (158°F) or higher. Scientists make use of the enzymes of these bacteria in a technique that rapidly replicates DNA sequences from small samples (see Module 12.12).

The optimal pH for most enzymes is near neutrality, in the range of 6–8. There are exceptions, of course. Pepsin, a digestive enzyme in your stomach, works best at pH 2. Such an environment would denature most enzymes, but the structure of pepsin is most stable and active in this acidic environment.

Cofactors Many enzymes require nonprotein helpers called **cofactors**, which bind to the active site and function in catalysis. The cofactors of some enzymes are inorganic, such as the ions of zinc, iron, and copper. If the cofactor is an organic molecule, it is called a **coenzyme**. Most vitamins are important in nutrition because they function as coenzymes or raw materials from which coenzymes are made. For example, folic acid is a coenzyme for a number of enzymes involved in the synthesis of nucleic acids.

Chemical chaos would result if all of a cell's metabolic pathways were operating simultaneously. A cell must tightly control when and where its various enzymes are active. It does this either by switching on or off the genes that encode specific enzymes (as you will learn in Chapter 11) or by regulating the activity of enzymes once they are made. We explore this second mechanism in the next module.

? **Explain how an enzyme speeds up a specific reaction.**

● An enzyme lowers the activation energy needed for a reaction when its specific substrate enters its active site. With an induced fit, the enzyme strains bonds that need to break or positions substrates in an orientation that aids the conversion of reactants to products.

5.15 Enzyme inhibition can regulate enzyme activity in a cell

A chemical that interferes with an enzyme's activity is called an inhibitor. Scientists have learned a great deal about enzyme function by studying the effects of such chemicals. Some inhibitors resemble the enzyme's normal substrate and compete for entry into the active site. As shown in the lower left of **Figure 5.15A**, such a **competitive inhibitor** reduces an enzyme's productivity by blocking substrate molecules from entering the active site. Competitive inhibition can be overcome by increasing the concentration of the substrate, making it more likely that a substrate molecule rather than an inhibitor will be nearby when an active site becomes vacant.

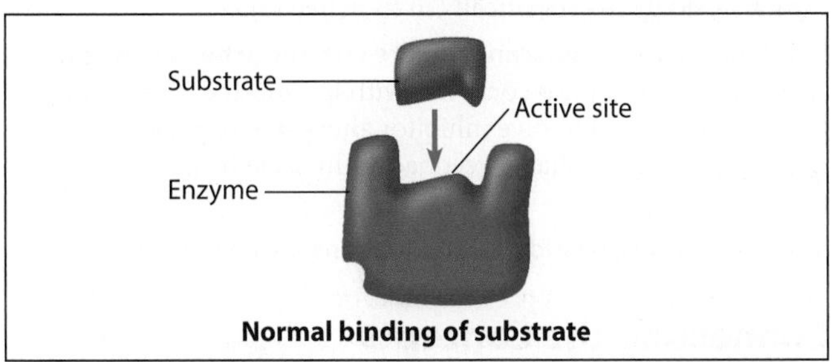

Substrate
Active site
Enzyme
Normal binding of substrate

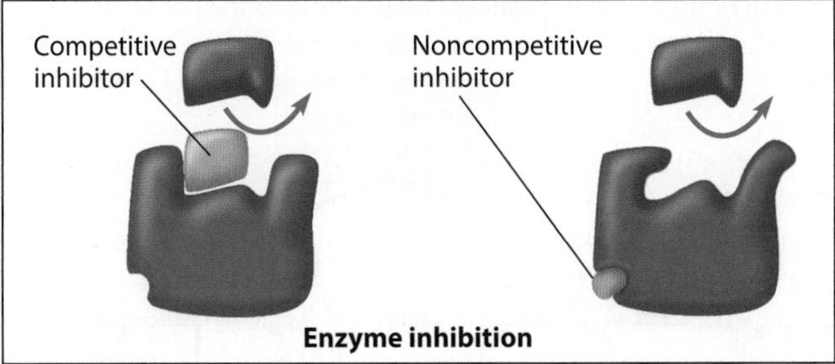

Competitive inhibitor
Noncompetitive inhibitor
Enzyme inhibition

▲ **Figure 5.15A** How inhibitors interfere with substrate binding

In contrast, a **noncompetitive inhibitor** does not enter the active site. Instead, it binds to a site elsewhere on the enzyme, and its binding changes the enzyme's shape so that the active site no longer fits the substrate (lower right of Figure 5.15A).

Although enzyme inhibition sounds harmful, cells use inhibitors as important regulators of cellular metabolism. Many of a cell's chemical reactions are organized into metabolic pathways in which a molecule is altered in a series of steps, each catalyzed by a specific enzyme, to form a final product. If a cell is producing more of that product than it needs, the product may act as an inhibitor of one of the enzymes early in the pathway. **Figure 5.15B** illustrates this sort of inhibition, called **feedback inhibition**. Because only weak interactions bind inhibitor and enzyme, this inhibition is reversible. When the product is used up by the cell, the enzyme is no longer inhibited and the pathway functions again.

In the next module, we explore some uses that people make of enzyme inhibitors.

? **Explain an advantage of feedback inhibition to a cell.**

● It prevents the cell from wasting valuable resources by synthesizing more of a particular product than is needed.

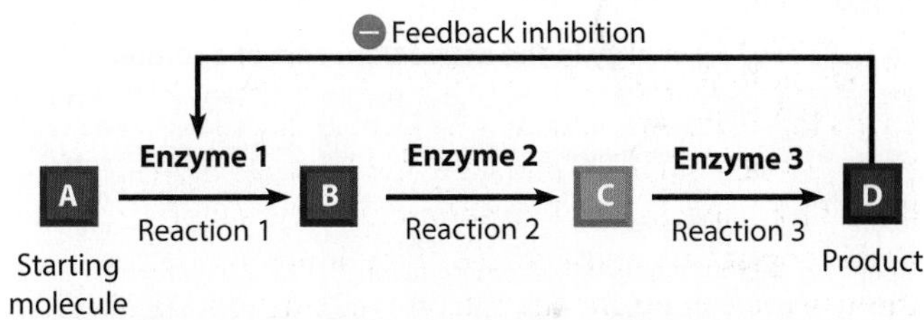

Feedback inhibition
Enzyme 1
Enzyme 2
Enzyme 3
A
B
C
D
Reaction 1
Reaction 2
Reaction 3
Starting molecule
Product

▲ **Figure 5.15B** Feedback inhibition of a metabolic pathway in which product D acts as an inhibitor of enzyme 1

5.16 Many drugs, pesticides, and poisons are enzyme inhibitors

CONNECTION

Many beneficial drugs act as enzyme inhibitors. Ibuprofen (**Figure 5.16**) is a common drug that inhibits an enzyme involved in the production of prostaglandins—messenger molecules that increase the sensation of pain and inflammation. Other drugs that function as enzyme inhibitors include some blood pressure medicines and antidepressants. Many antibiotics work by inhibiting enzymes of disease-causing bacteria. Penicillin, for example, blocks the active site of an enzyme that many bacteria use in making cell walls. Protease inhibitors are HIV drugs that target a key viral enzyme. And many cancer drugs are inhibitors of enzymes that promote cell division.

Humans have developed enzyme inhibitors as pesticides, and occasionally as deadly poisons for use in warfare. Poisons often attach to an enzyme by covalent bonds, making the inhibition irreversible. Poisons called nerve gases bind in the active site of an enzyme vital to the transmission of nerve impulses. The inhibition of this enzyme leads to rapid paralysis of vital functions and death. Pesticides such as malathion and parathion are toxic to insects (and dangerous to the people who apply them) because they also irreversibly inhibit this enzyme. Interestingly, some drugs reversibly inhibit this same enzyme and are used in anesthesia and treatment of certain diseases.

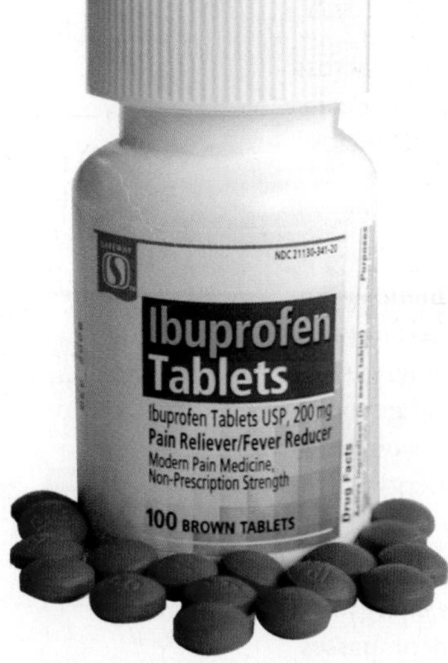

Ibuprofen Tablets
Ibuprofen Tablets USP, 200 mg
Pain Reliever/Fever Reducer
Modern Pain Medicine,
Non-Prescription Strength
100 BROWN TABLETS

▲ **Figure 5.16** Ibuprofen, an enzyme inhibitor

? **What determines whether enzyme inhibition is reversible or irreversible?**

● If the inhibitor binds to the enzyme with covalent bonds, the inhibition is usually irreversible. When weak chemical interactions bind inhibitor and enzyme, the inhibition is reversible.

CHAPTER 5 REVIEW

For practice quizzes, BioFlix animations, MP3 tutorials, video tutors, and more study tools designed for this textbook, go to MasteringBiology®

Reviewing the Concepts

Membrane Structure and Function (5.1–5.9)

5.1 Membranes are fluid mosaics of lipids and proteins with many functions. The proteins embedded in a membrane's phospholipid bilayer perform various functions.

5.2 The spontaneous formation of membranes was a critical step in the origin of life.

5.3 Passive transport is diffusion across a membrane with no energy investment. Solutes diffuse across membranes down their concentration gradients.

5.4 Osmosis is the diffusion of water across a membrane.

5.5 Water balance between cells and their surroundings is crucial to organisms. Cells shrink in a hypertonic solution and swell in a hypotonic solution. In isotonic solutions, animal cells are normal, but plant cells are flaccid.

5.6 Transport proteins can facilitate diffusion across membranes.

5.7 Research on another membrane protein led to the discovery of aquaporins.

5.8 Cells expend energy in the active transport of a solute.

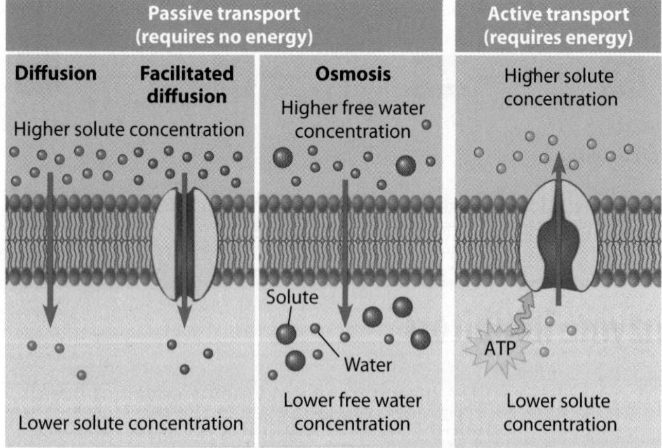
Passive transport (requires no energy)
Active transport (requires energy)
Diffusion
Facilitated diffusion
Osmosis
Higher solute concentration
Higher free water concentration
Higher solute concentration
Solute
Water
ATP
Lower solute concentration
Lower free water concentration
Lower solute concentration

5.9 Exocytosis and endocytosis transport large molecules across membranes. A vesicle may fuse with the membrane and expel its contents (exocytosis), or the membrane may fold inward, enclosing material from the outside (endocytosis).

Energy and the Cell (5.10–5.12)

5.10 Cells transform energy as they perform work. Kinetic energy is the energy of motion. Potential energy is energy stored in the location or structure of matter and includes chemical energy. According to the laws of thermodynamics, energy can change form but cannot be created or destroyed, and energy transfers or transformations increase disorder, or entropy, with some energy being lost as heat.

5.11 Chemical reactions either release or store energy. Exergonic reactions release energy. Endergonic reactions require energy and yield products rich in potential energy. Metabolism encompasses all of a cell's chemical reactions.

5.12 ATP drives cellular work by coupling exergonic and endergonic reactions. The transfer of a phosphate group from ATP is involved in chemical, transport, and mechanical work.

How Enzymes Function (5.13–5.16)

5.13 Enzymes speed up the cell's chemical reactions by lowering energy barriers. Enzymes are protein catalysts that decrease the activation energy needed to begin a reaction.

5.14 A specific enzyme catalyzes each cellular reaction. An enzyme's substrate fits specifically in its active site.

5.15 Enzyme inhibition can regulate enzyme activity in a cell. A competitive inhibitor competes with the substrate for the active site. A noncompetitive inhibitor alters an enzyme's function by changing its shape. Feedback inhibition helps regulate metabolism.

5.16 Many drugs, pesticides, and poisons are enzyme inhibitors.

Connecting the Concepts

1. Fill in the following concept map to review the processes by which molecules move across membranes.

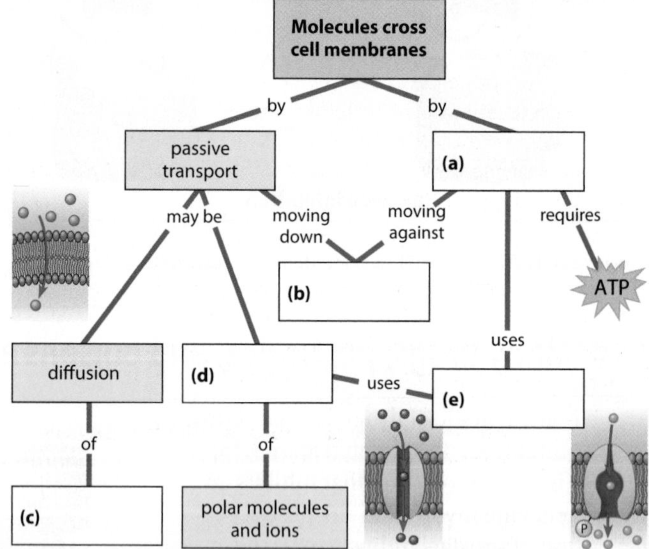
Molecules cross cell membranes
by
by
passive transport
(a)
may be
moving down
moving against
requires
(b)
ATP
uses
diffusion
(d)
uses
(e)
of
of
(c)
polar molecules and ions

2. Label the parts of the following diagram illustrating the catalytic cycle of an enzyme.

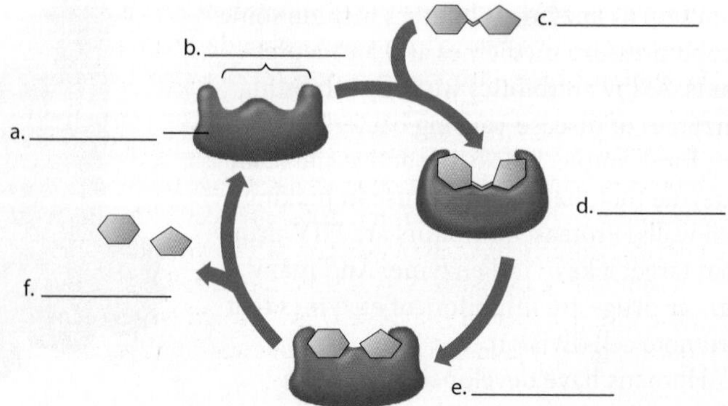
a. ________
b. ________
c. ________
d. ________
e. ________
f. ________

Testing Your Knowledge

Level 1: Knowledge/Comprehension

3. Which best describes the structure of a cell membrane?
   a. proteins between two bilayers of phospholipids
   b. proteins embedded in a bilayer of phospholipids
   c. a bilayer of protein coating a layer of phospholipids
   d. cholesterol embedded in a bilayer of phospholipids
4. A plant cell placed in distilled water will ______________; an animal cell placed in distilled water will ______________.
   a. burst . . . burst
   b. become flaccid . . . shrivel
   c. become turgid . . . be normal in shape
   d. become turgid . . . burst
5. The sodium concentration in a cell is 10 times less than the concentration in the surrounding fluid. How can the cell move sodium out of the cell? (*Explain your answer.*)
   a. passive transport
   b. receptor-mediated endocytosis
   c. active transport
   d. facilitated diffusion
6. The synthesis of ATP from ADP and Ⓟ
   a. stores energy in a form that can drive cellular work.
   b. involves the hydrolysis of a phosphate bond.
   c. transfers a phosphate, priming a protein to do work.
   d. is an exergonic process.
7. Facilitated diffusion across a membrane requires ______________ and moves a solute ______________ its concentration gradient.
   a. transport proteins . . . up (against)
   b. transport proteins . . . down
   c. energy and transport proteins . . . up
   d. energy and transport proteins . . . down
8. What are the main types of cellular work? How does ATP provide the energy for this work?

Level 2: Application/Analysis

9. Why is the barrier of the activation energy beneficial for cells? Explain how enzymes lower activation energy.
10. Relate the laws of thermodynamics to living organisms.
11. How do the components and structure of cell membranes relate to the functions of membranes?
12. Sometimes inhibitors can be harmful to a cell; often they are beneficial. Explain.

Level 3: Synthesis/Evaluation

13. Cells lining kidney tubules function in the reabsorption of water from urine. In response to chemical signals, they reversibly insert additional aquaporins into their plasma membranes. In which of these situations would your tubule cells have the most aquaporins: after a long run on a hot day, right after a large meal, or after drinking a large bottle of water? Explain.
14. **SCIENTIFIC THINKING** Mercury is known to inhibit the permeability of water channels. To help establish that the protein isolated by Agre's group was a water channel (see Module 5.7), the researchers incubated groups of RNA-injected oocytes (which then made aquaporin proteins) in four different solutions: plain buffer, low concentration and high concentration of a mercury chloride ($HgCl_2$) solution, and a low concentration of a mercury solution followed by an agent (ME) known to reverse the effects of mercury. The water permeability of the cells was determined by the rate of their osmotic swelling. Interpret the results of this experiment, which are presented in the graph below.

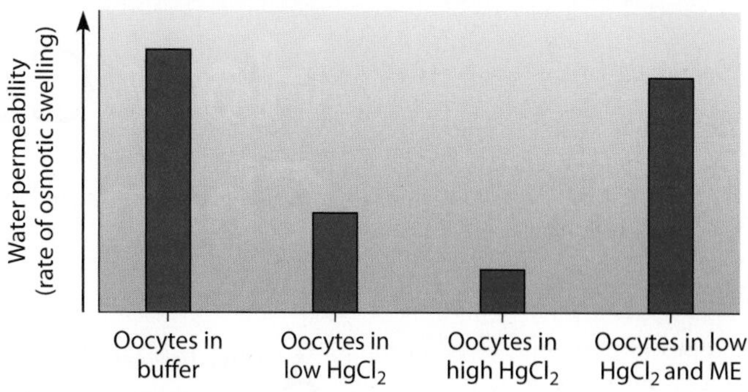

Water permeability
(rate of osmotic swelling)
Oocytes in buffer
Oocytes in low HgCl2
Oocytes in high HgCl2
Oocytes in low HgCl2 and ME

Data from G. M. Preston et al., Appearance of water channels in *Xenopus* oocytes expressing red cell CHIP28 protein, *Science* 256: 3385–7 (1992).

Control oocytes not injected with aquaporin RNA were also incubated with buffer and the two concentrations of mercury. Predict what the results of these treatments would be.

15. A biologist performed two series of experiments on lactase, the enzyme that hydrolyzes lactose to glucose and galactose. First, she made up 10% lactose solutions containing different concentrations of enzyme and measured the rate at which galactose was produced (grams of galactose per minute). Results of these experiments are shown in Table A below. In the second series of experiments (Table B), she prepared 2% enzyme solutions containing different concentrations of lactose and again measured the rate of galactose production.

Table A Rate and Enzyme Concentration

Lactose concentration	10%	10%	10%	10%	10%
Enzyme concentration	0%	1%	2%	4%	8%
Reaction rate	0	25	50	100	200

Table B Rate and Substrate Concentration

Lactose concentration	0%	5%	10%	20%	30%
Enzyme concentration	2%	2%	2%	2%	2%
Reaction rate	0	25	50	65	65

   a. Graph and explain the relationship between the reaction rate and the enzyme concentration.
   b. Graph and explain the relationship between the reaction rate and the substrate concentration. How and why did the results of the two experiments differ?
16. Organophosphates (organic compounds containing phosphate groups) are commonly used as insecticides to improve crop yield. Organophosphates typically interfere with nerve signal transmission by inhibiting the enzymes that degrade transmitter molecules. They affect humans and other vertebrates as well as insects. Thus, the use of organophosphate pesticides poses some health risks. On the other hand, these molecules break down rapidly upon exposure to air and sunlight. As a consumer, what level of risk are you willing to accept in exchange for an abundant and affordable food supply?

Answers to all questions can be found in Appendix 4.

CHAPTER 6

How Cells Harvest Chemical Energy

A baby's first cry! This welcome sound shows that the baby is breathing and taking in oxygen. But why is oxygen necessary for life? Oxygen is a reactant in cellular respiration—the process that breaks down sugar and other food molecules and generates ATP, the energy currency of cells.

Cellular respiration occurred in this baby's cells before she was born, but the oxygen and sugar her cells required were delivered from her mother's blood. Now this baby takes in her own oxygen—although she still can't obtain her own food. The process of cellular respiration produces heat as well as ATP, which helps maintain a warm body temperature. But if this baby is exposed to the cold, she can't keep herself warm. If you get cold, you put on more clothes, move to a warmer place, or shiver—generating heat as your contracting muscles increase their production of ATP and heat. This baby can't do any of those things yet. Instead, along her back she has a layer of a special kind of "baby fat," called brown fat, that helps keep her warm. The cells of brown fat have a "short circuit" in their cellular respiration—they consume oxygen and burn fuel, but generate only heat, not ATP.

? ***Can brown fat keep a newborn warm* and *help keep an adult thin?***

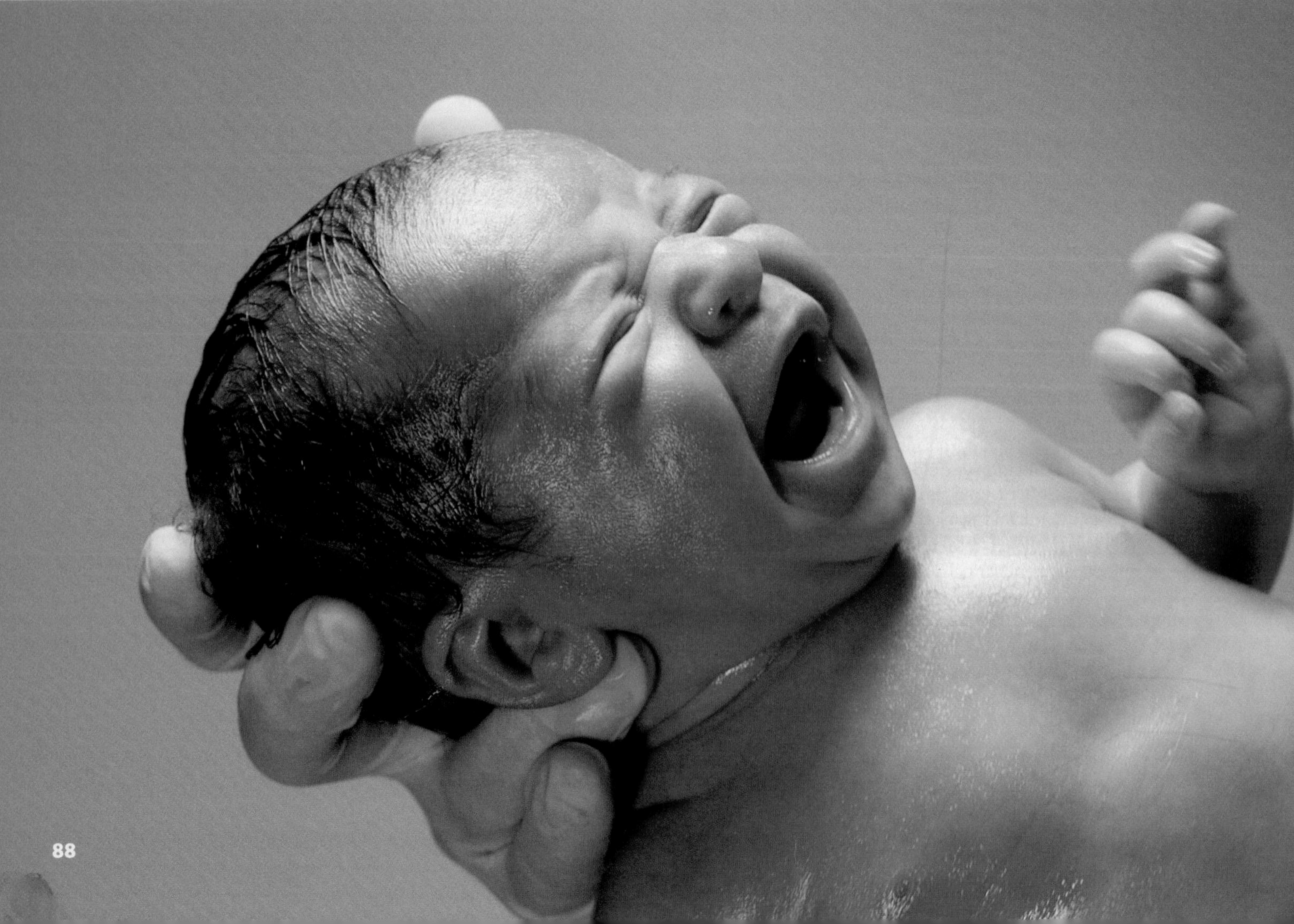

Scientists have long known that brown fat is important for heat production in small mammals, hibernating bears, and newborn infants. Studies have also shown brown fat to be involved in weight regulation in mice. As you will learn later in the chapter, brown fat deposits have only recently been discovered in adult humans. Scientists are now exploring whether this heat-generating, calorie-burning tissue may be tapped in the fight against obesity.

We begin this chapter with an overview of cellular respiration and then focus on its stages: glycolysis, the citric acid cycle, and oxidative phosphorylation. We also consider fermentation, an extension of glycolysis that has deep evolutionary roots. We complete the chapter with a comparison of the metabolic pathways that break down and build up the organic molecules of your body.

BIG IDEAS

Cellular Respiration: Aerobic Harvesting of Energy

(6.1–6.5)

Cellular respiration oxidizes fuel molecules and generates ATP for cellular work.

Stages of Cellular Respiration

(6.6–6.12)

The main stages of cellular respiration are glycolysis, the citric acid cycle, and oxidative phosphorylation.

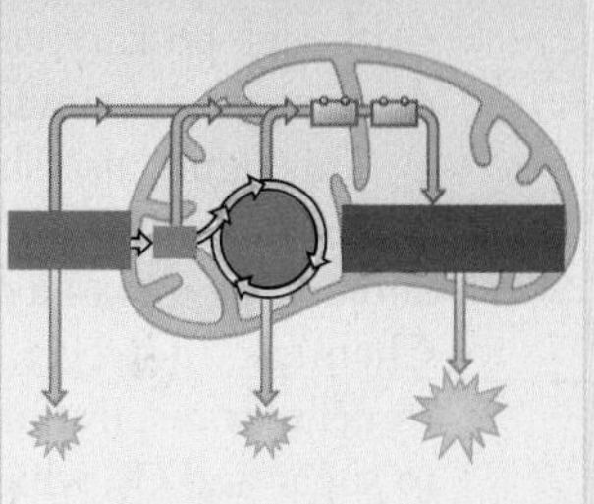

Fermentation: Anaerobic Harvesting of Energy

(6.13–6.14)

Fermentation regenerates NAD^+, allowing glycolysis and ATP production to continue without oxygen.

Connections Between Metabolic Pathways

(6.15–6.16)

The breakdown pathways of cellular respiration intersect with biosynthetic pathways.

▷ Cellular Respiration: Aerobic Harvesting of Energy

6.1 Photosynthesis and cellular respiration provide energy for life

Life requires energy. In almost all ecosystems, that energy ultimately comes from the sun. (Photosynthesis, the process by which the sun's energy is captured, will be explored later, in Chapter 7.) **Figure 6.1** illustrates how photosynthesis and cellular respiration together provide energy for living organisms. In photosynthesis, which takes place in a plant cell's chloroplasts, the energy of sunlight is used to rearrange the atoms of carbon dioxide (CO_2) and water (H_2O) to produce sugar and oxygen (O_2). In **cellular respiration**, O_2 is consumed as sugar is broken down to CO_2 and H_2O; the cell captures the energy released in ATP. Cellular respiration takes place in the mitochondria of almost all eukaryotic cells—in the cells of plants, animals, fungi, and protists. (Although prokaryotes don't have mitochondria, some do break down sugar in a similar type of oxygen-using respiration.)

This figure also shows that in these energy conversions, some energy is lost as heat. Life on Earth is solar powered, and energy makes a one-way trip through an ecosystem. Chemicals, however, are recycled. The CO_2 and H_2O released by cellular respiration are converted through photosynthesis to sugar and O_2, which are then used in respiration.

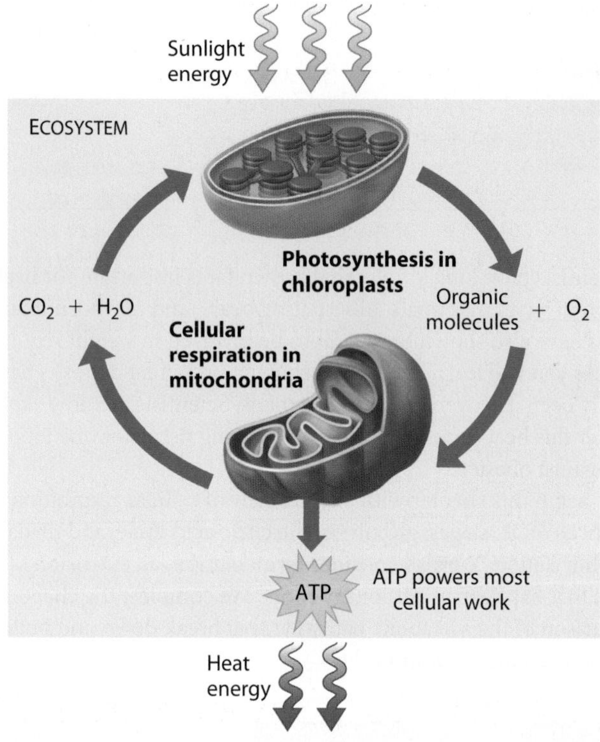

Sunlight energy
ECOSYSTEM
Photosynthesis in chloroplasts
CO_2 + H_2O
Organic molecules + O_2
Cellular respiration in mitochondria
ATP
ATP powers most cellular work
Heat energy

▲ **Figure 6.1** The connection between photosynthesis and cellular respiration

? **What is misleading about the following statement? "Plant cells perform photosynthesis, and animal cells perform cellular respiration."**

● The statement implies that cellular respiration does not occur in plant cells. In fact, almost all eukaryotic cells use cellular respiration to obtain energy for their cellular work.

6.2 Breathing supplies O_2 for use in cellular respiration and removes CO_2

We often use the word *respiration* as a synonym for "breathing," the meaning of its Latin root. In that case, respiration refers to an exchange of gases: An organism obtains O_2 from its environment and releases CO_2 as a waste product. Biologists also define respiration as the aerobic (oxygen-requiring) harvesting of energy from food molecules by cells. This process is called cellular respiration to distinguish it from breathing.

Breathing and cellular respiration are closely related. As the runner in **Figure 6.2** breathes in air, her lungs take up O_2 and pass it to her bloodstream. The bloodstream carries the O_2 to her muscle cells. Mitochondria in the muscle cells use the O_2 in cellular respiration to harvest energy from glucose and other organic molecules and generate ATP. Muscle cells use ATP to fuel contractions. The runner's bloodstream and lungs also perform the vital function of disposing of the CO_2 waste, which is produced in cellular respiration. You can see the roles of O_2 and CO_2 in the equation for cellular respiration at the bottom of the figure.

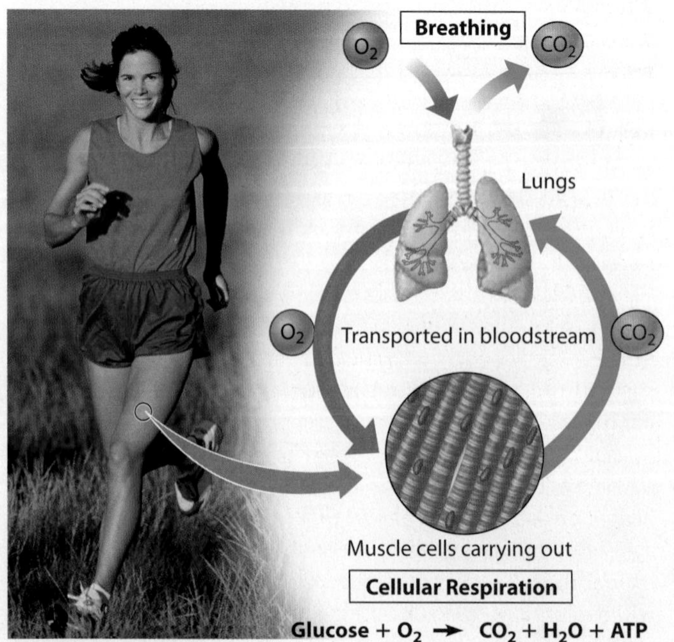

Breathing
O_2
CO_2
Lungs
O_2
Transported in bloodstream
CO_2
Muscle cells carrying out
Cellular Respiration
Glucose + O_2 → CO_2 + H_2O + ATP

▲ **Figure 6.2** The connection between breathing and cellular respiration

? **How is your breathing related to your cellular respiration?**

● In breathing, CO_2 and O_2 are exchanged between your lungs and the air. In cellular respiration, cells use the O_2 obtained through breathing to break down fuel, releasing CO_2 as a waste product.

6.3 Cellular respiration banks energy in ATP molecules

You breathe air and eat food to supply your cells with the reactants for cellular respiration—the process that generates ATP for cellular work.

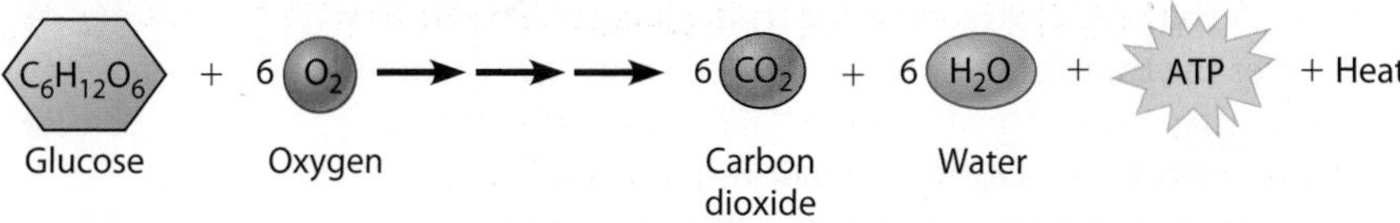

$C_6H_{12}O_6$ + 6 O_2 → → → 6 CO_2 + 6 H_2O + ATP + Heat
Glucose
Oxygen
Carbon dioxide
Water

▲ **Figure 6.3** Summary equation for cellular respiration

The chemical equation in **Figure 6.3** summarizes cellular respiration. The simple sugar glucose ($C_6H_{12}O_6$) is the fuel that cells use most often, although other organic molecules can also be "burned" in cellular respiration. The equation tells us that the atoms of the reactant molecules $C_6H_{12}O_6$ and O_2 are rearranged to form the products CO_2 and H_2O. In this exergonic (energy-releasing) process, the chemical energy of the bonds in glucose is released and stored (or "banked") in the chemical bonds of ATP (see Module 5.12). The series of arrows in Figure 6.3 indicates that cellular respiration consists of many steps, not just a single reaction.

Cellular respiration can produce up to 32 ATP molecules for each glucose molecule, a capture of about 34% of the energy originally stored in glucose. The rest of the energy is released as heat (see Module 5.10). This may seem inefficient, but it compares very well with the efficiency of most energy-conversion systems. For instance, the average automobile engine is able to convert only about 25% of the energy in gasoline to the kinetic energy of movement. And, as you learned in the chapter introduction, heat released in cellular respiration helps maintain your warm body temperature.

How great are the energy needs of a cell? If ATP could not be regenerated through cellular respiration, you would use up nearly your body weight in ATP each day. Let's consider the energy requirements for various human activities next.

? Why are sweating and other body-cooling mechanisms necessary during vigorous exercise?

● The demand for ATP is supported by an increased rate of cellular respiration, but about 66% of the energy released from food produces heat instead of ATP.

6.4 The human body uses energy from ATP for all its activities

CONNECTION

Your body requires a continuous supply of energy just to stay alive—to keep your heart pumping and to keep you breathing. Your brain especially requires a huge amount of energy; its cells burn about 120 grams (g)—a quarter of a pound!—of glucose a day, accounting for about 15% of total oxygen consumption. Maintaining brain cells and other life-sustaining activities uses as much as 75% of the energy a person takes in as food during a typical day.

Above and beyond the energy you need for body maintenance, cellular respiration provides energy for voluntary activities. **Figure 6.4** shows the amount of energy it takes to perform some of these activities. The energy units are **kilocalories (kcal)**, a measure of the quantity of heat required to raise the temperature of 1 kilogram (kg) of water by 1°C. (The "Calories" listed on food packages are actually kilocalories, usually signified by a capital C.) The values shown do not include the energy the body consumes for its basic life-sustaining activities. Even sleeping or lying quietly requires energy for metabolism.

The U.S. National Academy of Sciences estimates that the average adult needs to take in food that provides about 2,200 kcal of energy per day, although the number varies based on age, sex, and activity level. A balance of energy intake and expenditure is required to maintain a healthy weight. (We will explore nutritional needs further in Chapter 21.) Now we begin the study of how cells liberate the energy stored in fuel molecules to produce the ATP used to power the work of your cells and thus the activities of your body.

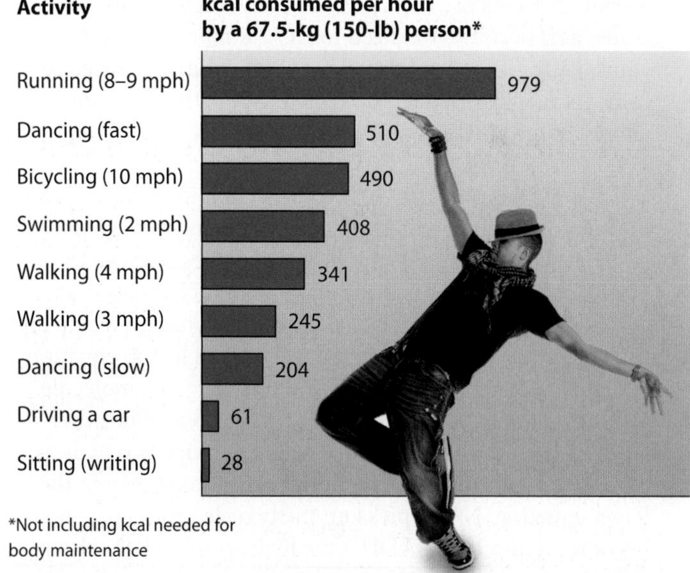

Activity
kcal consumed per hour by a 67.5-kg (150-lb) person*
Running (8–9 mph) 979
Dancing (fast) 510
Bicycling (10 mph) 490
Swimming (2 mph) 408
Walking (4 mph) 341
Walking (3 mph) 245
Dancing (slow) 204
Driving a car 61
Sitting (writing) 28
*Not including kcal needed for body maintenance

▲ **Figure 6.4** Energy consumed by various activities

? Walking at 3 mph, how far would you have to travel to "burn off" the equivalent of an extra slice of pizza, which has about 475 kcal? How long would that take?

● You would have to walk about 6 miles, which would take you about 2 hours. (Now you understand why the most effective exercise for losing weight is pushing away from the table!)

6.5 Cells capture energy from electrons "falling" from organic fuels to oxygen

How do your cells extract energy from glucose? The answer involves the transfer of electrons during chemical reactions.

Redox Reactions During cellular respiration, electrons are transferred from glucose to oxygen, releasing energy. Oxygen attracts electrons very strongly, and an electron loses potential energy when it is transferred to oxygen. If you burn a cube of sugar, this electron "fall" happens very rapidly, releasing energy in the form of heat and light. Cellular respiration is a more controlled descent of electrons—more like rolling down an energy hill, with energy released in small amounts that can be stored in the chemical bonds of ATP.

The movement of electrons from one molecule to another is an oxidation-reduction reaction, or **redox reaction** for short. In a redox reaction, the loss of electrons from one substance is called **oxidation**, and the addition of electrons to another substance is called **reduction**. A molecule is said to become oxidized when it loses one or more electrons and reduced when it gains one or more electrons. Because an electron transfer requires both a donor and an acceptor, oxidation and reduction always go together.

In the cellular respiration equation in **Figure 6.5A** below, you cannot see any electron transfers. What you do see are changes in the location of hydrogen atoms. These hydrogen movements represent electron transfers because each hydrogen atom consists of an electron (e^-) and a proton (hydrogen ion, or H^+). Glucose ($C_6H_{12}O_6$) loses hydrogen atoms (electrons) as it becomes oxidized to CO_2; simultaneously, O_2 gains hydrogen atoms (electrons) as it becomes reduced to H_2O. As they pass from glucose to oxygen, the electrons lose energy, some of which cells capture.

NADH and Electron Transport Chains An important player in the process of oxidizing glucose is a coenzyme called **NAD^+**, which accepts electrons and becomes reduced to NADH. NAD^+ (nicotinamide adenine dinucleotide) is an organic molecule that cells make from the vitamin niacin and use to shuttle electrons in redox reactions. The top equation in **Figure 6.5B** depicts the oxidation of an organic molecule. We show only its three carbons (●) and a few of its other atoms. An enzyme called dehydrogenase strips two hydrogen atoms from this molecule. Simultaneously, as shown in the lower equation, NAD^+ picks up the two electrons (⊖) and becomes reduced to NADH. One hydrogen ion (H^+) also becomes part of the NADH, and the other is released. (NADH is represented throughout this chapter as a light brown box carrying two blue electrons.)

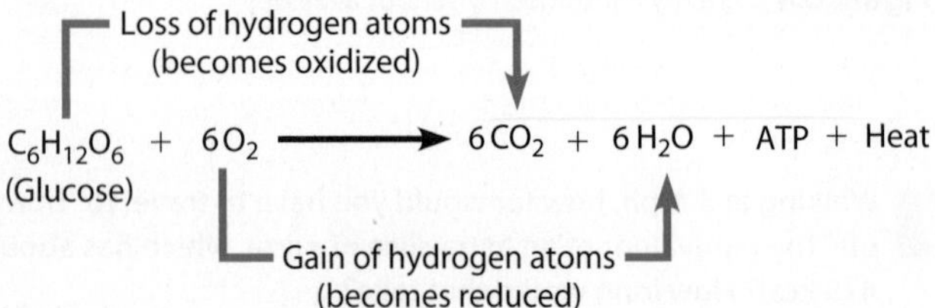

Loss of hydrogen atoms
(becomes oxidized)
$C_6H_{12}O_6 + 6O_2 \longrightarrow 6CO_2 + 6H_2O + ATP + Heat$
(Glucose)
Gain of hydrogen atoms
(becomes reduced)

▲ **Figure 6.5A** Rearrangement of hydrogen atoms (with their electrons) in the redox reactions of cellular respiration

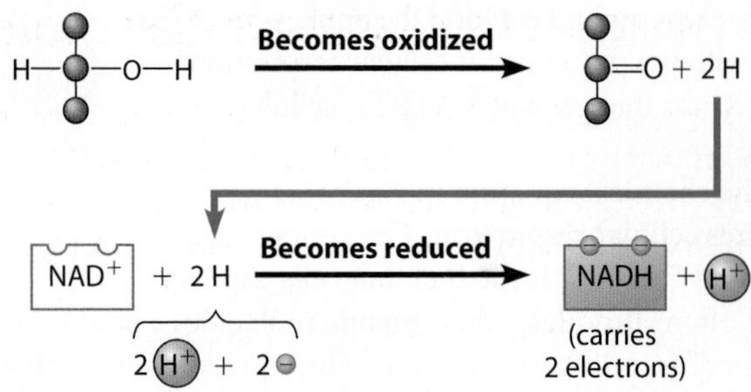

H—C—O—H
Becomes oxidized
C=O + 2 H
NAD+
+ 2 H
Becomes reduced
NADH
+ H+
2 H+ + 2 e
(carries
2 electrons)

▲ **Figure 6.5B** A pair of redox reactions occurring simultaneously

Try This Circle the two atoms that will be removed from the molecule on the left as it becomes oxidized to the molecule on the right. Explain why we say that a hydrogen atom consists of a hydrogen ion (H^+) and an electron.

Using the energy hill analogy for electrons rolling from glucose to oxygen, the transfer of electrons from an organic molecule to NAD^+ is just the beginning. **Figure 6.5C** shows NADH delivering these electrons to a string of electron carrier molecules, shown here as purple ovals, that lead down the hill. At the bottom of the hill is oxygen ($\frac{1}{2}\,O_2$), which accepts two electrons, picks up two H^+, and becomes reduced to water.

These carrier molecules form an **electron transport chain**. In a cell, they are built into the inner membrane of a mitochondrion. Through a series of redox reactions, electrons are passed from carrier to carrier, releasing energy in amounts small enough to be used by the cell to make ATP.

With an understanding of this basic mechanism of electron transfer and energy release, we can now explore cellular respiration in more detail.

? **What chemical characteristic of the element oxygen accounts for its function in cellular respiration?**

● Oxygen is extremely electronegative (see Module 2.6), making it very powerful in pulling electrons down the electron transport chain.

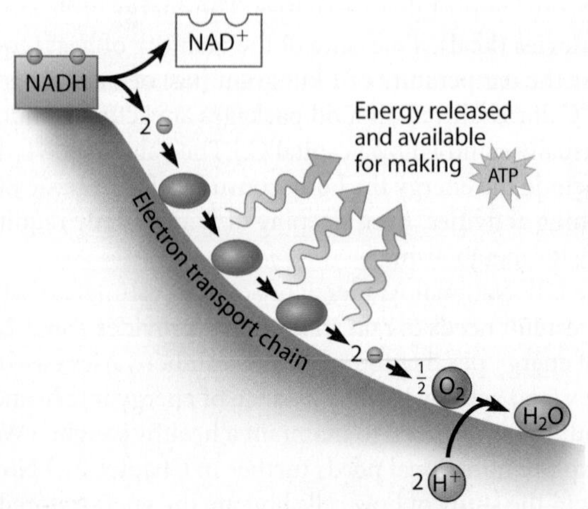

NAD+
NADH
2 e
Energy released
and available
for making
ATP
Electron transport chain
2 e
1/2 O2
H2O
2 H+

▲ **Figure 6.5C** Electrons releasing energy for ATP synthesis as they roll down an energy hill from NADH through an electron transport chain to O_2

▷ Stages of Cellular Respiration

6.6 Overview: Cellular respiration occurs in three main stages

Cellular respiration consists of a sequence of many chemical reactions that we can divide into three main stages. **Figure 6.6** gives an overview of these stages and shows where they occur in a eukaryotic cell. (In prokaryotic cells that use aerobic respiration, these steps occur in the cytosol, and the electron transport chain is built into the plasma membrane.)

Stage 1: Glycolysis (shown with a teal background throughout this chapter) occurs in the cytosol of the cell. Glycolysis begins cellular respiration by breaking glucose into two molecules of a three-carbon compound called pyruvate.

Stage 2: Pyruvate oxidation and the **citric acid cycle** (shown in shades of orange) take place within the mitochondria. Pyruvate is oxidized to a two-carbon compound. The citric acid cycle then completes the breakdown of glucose to carbon dioxide. Thus, the CO_2 that you exhale is formed in the mitochondria of your cells during this second stage of respiration.

As suggested by the smaller ATP symbols in the diagram, the cell makes a small amount of ATP during glycolysis and the citric acid cycle. The main function of these first two stages, however, is to supply the third stage of respiration with electrons (shown with gold arrows).

Stage 3: Oxidative phosphorylation (purple background) involves electron transport and a process known as chemiosmosis. NADH and a related electron carrier, $FADH_2$ (flavin adenine dinucleotide), shuttle electrons to an electron transport chain embedded in the inner mitochondrial membrane. Most of the ATP produced by cellular respiration is generated by oxidative phosphorylation, which uses the energy released by the downhill fall of electrons from NADH and $FADH_2$ to oxygen to phosphorylate ADP. (Recall from Module 5.12 that cells generate ATP by adding a phosphate group to ADP.)

What couples the electron transport chain to ATP synthesis? As the electron transport chain passes electrons down the energy hill, it also pumps hydrogen ions (H^+) across the inner mitochondrial membrane into the narrow intermembrane space. The result is a concentration gradient of H^+ across the membrane. In **chemiosmosis**, the potential energy of this concentration gradient is used to make ATP. The details of this process are explored in Module 6.10.

The small amount of ATP produced in glycolysis and the citric acid cycle is made by substrate-level phosphorylation, a process we discuss in the next module. In the next several modules, we look more closely at the stages of cellular respiration and the mechanisms of ATP synthesis.

? Of the three main stages of cellular respiration, which is the only one that uses oxygen?

● Oxidative phosphorylation, in which the electron transport chain ultimately transfers electrons to oxygen

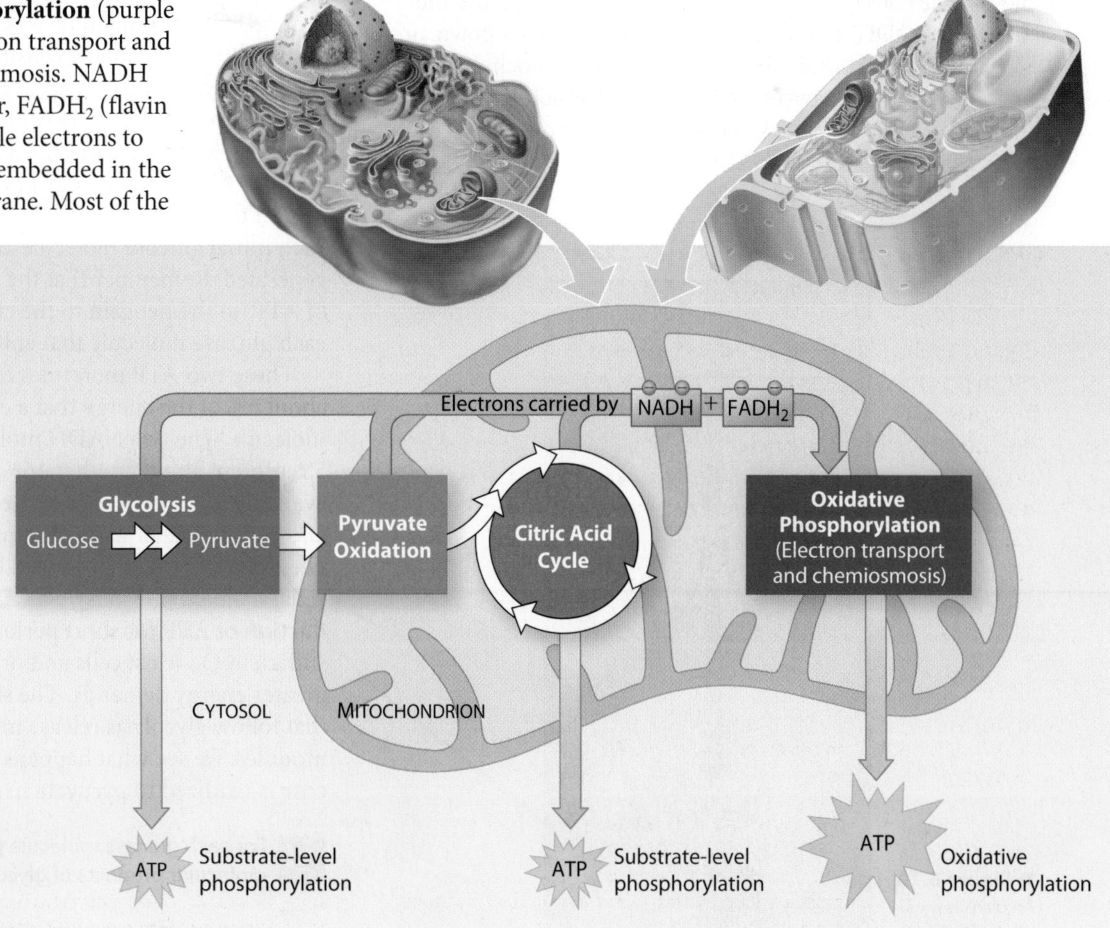

Electrons carried by NADH + $FADH_2$
Glycolysis
Glucose
Pyruvate
Pyruvate Oxidation
Citric Acid Cycle
Oxidative Phosphorylation
(Electron transport and chemiosmosis)
CYTOSOL
MITOCHONDRION
ATP
Substrate-level phosphorylation
ATP
Substrate-level phosphorylation
ATP
Oxidative phosphorylation

▶ **Figure 6.6** An overview of cellular respiration

6.7 Glycolysis harvests chemical energy by oxidizing glucose to pyruvate

Now that you have been introduced to the major players and processes, it's time to focus on the individual stages of cellular respiration. The term for the first stage, *glycolysis*, means "splitting of sugar" (*glyco*, sweet, and *lysis*, split), and that's exactly what happens during this phase.

Figure 6.7A below gives an overview of glycolysis, which begins with a single molecule of glucose and concludes with two molecules of pyruvate. (Pyruvate is the ionized form of pyruvic acid.) Each ● represents a carbon atom in the molecules; glucose has six carbons, and these same six carbons end up in the two molecules of pyruvate (three carbons in each). The straight arrow shown running from glucose to pyruvate actually represents nine chemical steps, each catalyzed by its own enzyme. As these reactions occur, two molecules of NAD^+ are reduced to two molecules of NADH, and a net gain of two molecules of ATP is produced.

Figure 6.7B illustrates how ATP is formed in glycolysis by the process called **substrate-level phosphorylation**. In this process, an enzyme transfers a phosphate group (Ⓟ) from a substrate molecule directly to ADP, forming ATP. You will come across substrate-level phosphorylation again in our discussion of the citric acid cycle, in which a small amount of ATP is generated by this process.

The oxidation of glucose to pyruvate during glycolysis releases energy, which is stored in ATP and in NADH. The cell can use the energy in ATP immediately, but for it to use the energy in NADH, electrons from NADH must pass down an electron transport chain located in the inner mitochondrial membrane. And the pyruvate molecules still hold most of the energy of glucose; these molecules will be oxidized in the citric acid cycle.

Let's take a closer look at glycolysis. **Figure 6.7C**, on the next page, names and shows simplified structures for all the organic compounds that form in the nine chemical reactions of glycolysis. Commentary on the left highlights the main features of these reactions.

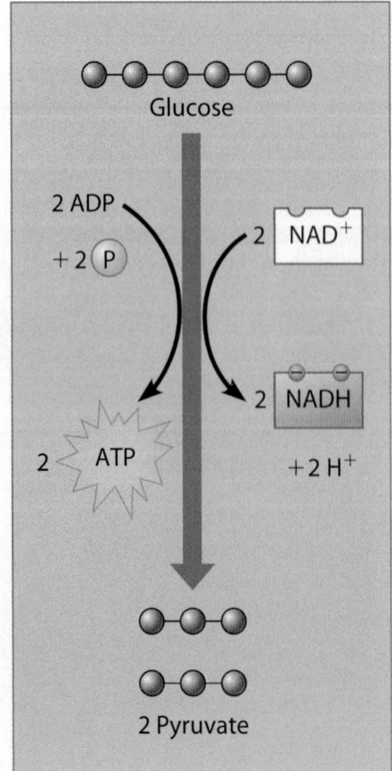
Glucose
2 ADP
+2 P
2 NAD+
2 NADH
2 ATP
+2 H+
2 Pyruvate

▶ **Figure 6.7A** An overview of glycolysis

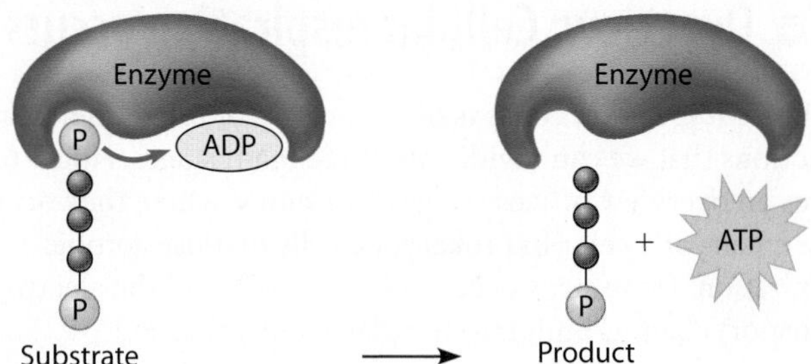
Enzyme
P
ADP
P
Substrate
Enzyme
ATP
P
Product

▲ **Figure 6.7B** Substrate-level phosphorylation: transfer of a phosphate group from a substrate to ADP, producing ATP

Compounds that form between the initial reactant, glucose, and the final product, pyruvate, are known as **intermediates**. Glycolysis is an example of a metabolic pathway in which each chemical step feeds into the next one. For instance, the intermediate glucose 6-phosphate is the product of step 1 and the reactant for step 2. Similarly, fructose 6-phosphate is the product of step 2 and the reactant for step 3. Also essential are the specific enzymes that catalyze each chemical step; however, the figure does not include the enzymes.

As indicated in Figure 6.7C, the steps of glycolysis can be grouped into two main phases. Steps ①–④, the energy investment phase, actually *consume* energy. In this phase, two molecules of ATP are used to energize a glucose molecule, which is then split into two small sugars. The figure follows each of these three-carbon sugars through the second phase.

Steps ⑤–⑨, the energy payoff phase, *yield* energy for the cell. In this phase, two NADH molecules are produced for each initial glucose molecule, and four ATP molecules are generated. Remember that the first phase used two molecules of ATP, so the net gain to the cell is two ATP molecules for each glucose molecule that enters glycolysis.

These two ATP molecules from glycolysis account for only about 6% of the energy that a cell can harvest from a glucose molecule. The two NADH molecules generated during step 5 represent about another 16%, but their stored energy is not available for use without oxygen. Some organisms—yeasts and certain bacteria, for instance—can satisfy their energy needs with the ATP produced by glycolysis alone. And some cells, such as your muscle cells, may use this anaerobic production of ATP for short periods when they do not have sufficient O_2. Most cells and organisms, however, have far greater energy demands. The stages of cellular respiration that follow glycolysis release much more energy. In the next modules, we see what happens in most organisms after glucose is oxidized to pyruvate in glycolysis.

? For each glucose molecule processed, what are the net molecular products of glycolysis?

● Two molecules of pyruvate, two molecules of ATP, and two molecules of NADH

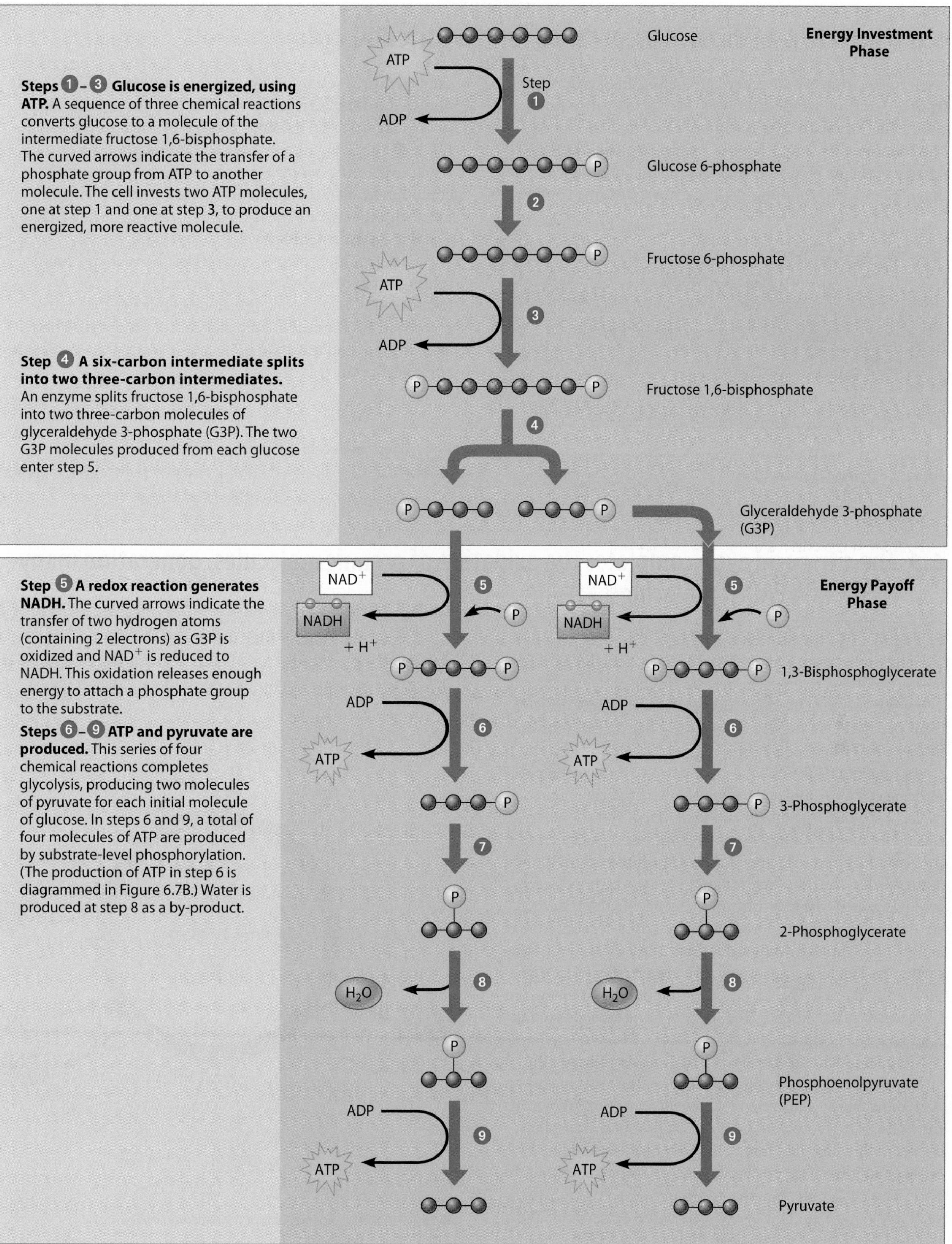

Energy Investment Phase
Steps 1–3 Glucose is energized, using ATP. A sequence of three chemical reactions converts glucose to a molecule of the intermediate fructose 1,6-bisphosphate. The curved arrows indicate the transfer of a phosphate group from ATP to another molecule. The cell invests two ATP molecules, one at step 1 and one at step 3, to produce an energized, more reactive molecule.
Step 4 A six-carbon intermediate splits into two three-carbon intermediates. An enzyme splits fructose 1,6-bisphosphate into two three-carbon molecules of glyceraldehyde 3-phosphate (G3P). The two G3P molecules produced from each glucose enter step 5.
Glucose
ATP
ADP
Step 1
Glucose 6-phosphate
2
Fructose 6-phosphate
3
Fructose 1,6-bisphosphate
4
Glyceraldehyde 3-phosphate (G3P)
P
Energy Payoff Phase
Step 5 A redox reaction generates NADH. The curved arrows indicate the transfer of two hydrogen atoms (containing 2 electrons) as G3P is oxidized and NAD+ is reduced to NADH. This oxidation releases enough energy to attach a phosphate group to the substrate.
Steps 6–9 ATP and pyruvate are produced. This series of four chemical reactions completes glycolysis, producing two molecules of pyruvate for each initial molecule of glucose. In steps 6 and 9, a total of four molecules of ATP are produced by substrate-level phosphorylation. (The production of ATP in step 6 is diagrammed in Figure 6.7B.) Water is produced at step 8 as a by-product.
NAD+
NADH
+ H+
5
1,3-Bisphosphoglycerate
6
3-Phosphoglycerate
7
2-Phosphoglycerate
8
H2O
Phosphoenolpyruvate (PEP)
9
Pyruvate

▲ **Figure 6.7C** Details of glycolysis

6.8 Pyruvate is oxidized in preparation for the citric acid cycle

As pyruvate forms at the end of glycolysis, it is transported from the cytosol, where glycolysis takes place, into a mitochondrion, where the citric acid cycle and oxidative phosphorylation will occur. Pyruvate itself does not enter the citric acid cycle. As shown in **Figure 6.8**, it first undergoes some major chemical "grooming." A large, multi-enzyme complex catalyzes three reactions: ❶ A carboxyl group ($—COO^-$) is removed from pyruvate and given off as a molecule of CO_2 (this is the first step in which CO_2 is released during respiration); ❷ the two-carbon compound remaining is oxidized while a molecule of NAD^+ is reduced to NADH; and ❸ a compound called coenzyme A, derived from a B vitamin, joins with the two-carbon group to form a molecule called acetyl coenzyme A, abbreviated **acetyl CoA**.

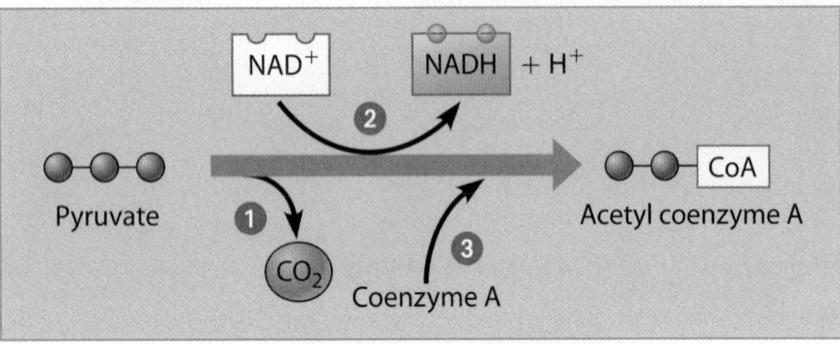
NAD+
NADH + H+
2
Pyruvate
1
CO2
3
Coenzyme A
CoA
Acetyl coenzyme A

▲ **Figure 6.8** The link between glycolysis and the citric acid cycle: the oxidation of pyruvate to acetyl CoA

These grooming steps—a chemical "haircut and conditioning" of pyruvate—set up the second major stage of cellular respiration. For each molecule of glucose that enters glycolysis, two molecules of pyruvate are produced. These are oxidized, and then two molecules of acetyl CoA enter the citric acid cycle.

? Which molecule in Figure 6.8 has been reduced?

NAD^+ has been reduced to NADH.

6.9 The citric acid cycle completes the oxidation of organic molecules, generating many NADH and $FADH_2$ molecules

The citric acid cycle is often called the Krebs cycle in honor of Hans Krebs, the German-British researcher who worked out much of this pathway in the 1930s. The cycle functions as a metabolic furnace that oxidizes the acetyl CoA derived from pyruvate. We present an overview figure first, followed by a more detailed look at this cycle.

As shown in **Figure 6.9A**, only the two-carbon acetyl part of the acetyl CoA molecule actually enters the citric acid cycle; coenzyme A splits off and is recycled. Not shown in this figure are the multiple steps that follow, each catalyzed by a specific enzyme located in the mitochondrial matrix or embedded in the inner membrane. The two-carbon acetyl group is joined to a four-carbon molecule. As the resulting six-carbon molecule is processed through a series of redox reactions, two carbon atoms are removed as CO_2, and the four-carbon molecule is regenerated; this regeneration accounts for the word *cycle*. The six-carbon compound first formed in the cycle is citrate, the ionized form of citric acid; hence the name *citric acid cycle*.

Compared with glycolysis, the citric acid cycle pays big energy dividends to the cell. Each turn of the cycle makes one ATP molecule by substrate-level phosphorylation (shown at the bottom of Figure 6.9A). But it also produces four other energy-rich molecules: three NADH molecules and one molecule of another electron carrier, $FADH_2$. Remember that the citric acid cycle processes two molecules of acetyl CoA for each initial glucose. Thus, two turns of the cycle occur, and the overall yield per molecule of glucose is 2 ATP, 6 NADH, and 2 $FADH_2$.

So how many energy-rich molecules have been produced by processing one molecule of glucose through glycolysis and the citric acid cycle? Up to this point, the cell has gained a

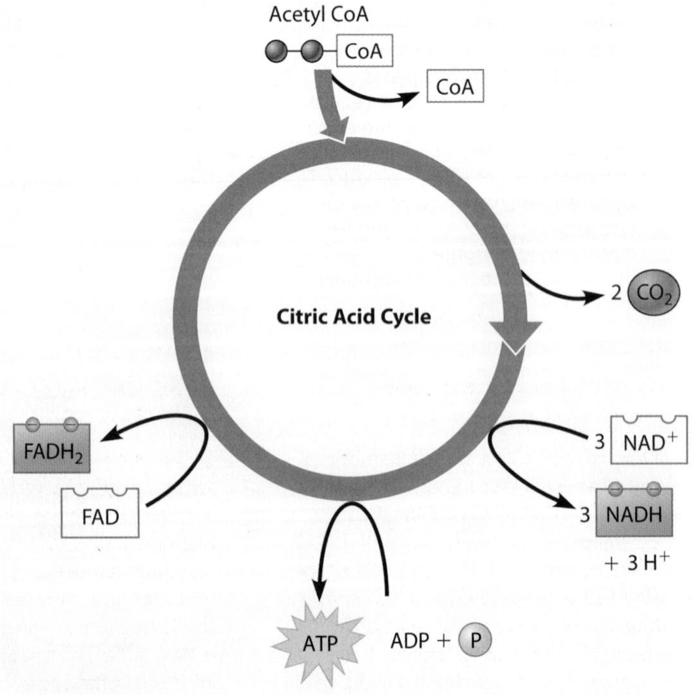
Acetyl CoA
CoA
CoA
2 CO2
Citric Acid Cycle
FADH2
FAD
3 NAD+
3 NADH
+ 3 H+
ATP
ADP + P

▲ **Figure 6.9A** An overview of the citric acid cycle

Try This Remember that 2 acetyl CoA are produced from each glucose. Use this figure to determine the per-glucose return from the citric acid cycle.

total of 4 ATP (all from substrate-level phosphorylation), 10 NADH, and 2 $FADH_2$. For the cell to be able to harvest the energy banked in NADH and $FADH_2$, these molecules must shuttle their high-energy electrons to an electron transport chain. There the energy from the *oxidation* of organic molecules is used to *phosphorylate* ADP to ATP—hence the name *oxidative phosphorylation*. Before we look at how oxidative phosphorylation works, you may want to examine the inner workings of the citric acid cycle in **Figure 6.9B**, below.

? **What is the total number of NADH and $FADH_2$ molecules generated during the complete breakdown of one glucose molecule to six molecules of CO_2? (Hint: Combine the outputs discussed in Modules 6.7–6.9.)**

10 NADH: 2 from glycolysis; 2 from the oxidation of pyruvate; 6 from the citric acid cycle; and 2 $FADH_2$ from the citric acid cycle. (Did you remember to double the output after to the sugar-splitting step of glycolysis?)

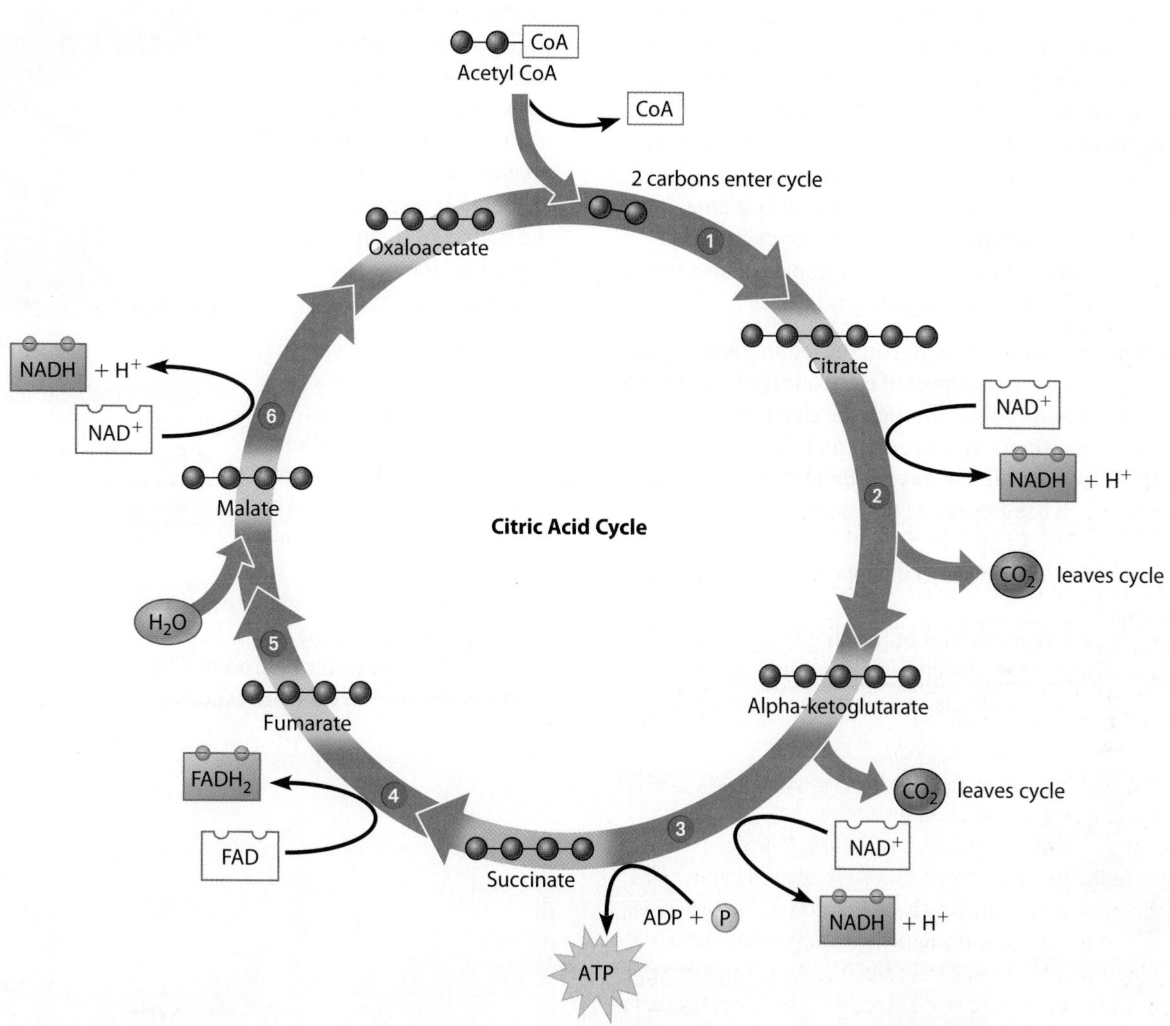

CoA
Acetyl CoA
CoA
2 carbons enter cycle
1
Oxaloacetate
Citrate
NAD+
NADH + H+
2
CO2 leaves cycle
Alpha-ketoglutarate
CO2 leaves cycle
NAD+
NADH + H+
3
ADP + P
ATP
Succinate
4
FADH2
FAD
Fumarate
5
H2O
Malate
6
NAD+
NADH + H+
Citric Acid Cycle

Step 1
Acetyl CoA stokes the furnace.

A turn of the citric acid cycle begins (top center) as enzymes strip the CoA portion from acetyl CoA and combine the remaining two-carbon group with the four-carbon molecule oxaloacetate (top left) already present in the mitochondrion. The product of this reaction is the six-carbon molecule citrate. All the acid compounds in this cycle exist in the cell in their ionized form, hence the suffix *-ate*.

Steps 2–3
NADH, ATP, and CO_2 are generated during redox reactions.

Successive redox reactions harvest energy by stripping hydrogen atoms from citrate and then alpha-ketoglutarate and producing energy-laden NADH molecules. In two places, an intermediate compound loses a CO_2 molecule. Energy is harvested by substrate-level phosphorylation of ADP to produce ATP. A four-carbon compound called succinate emerges at the end of step 3.

Steps 4–6
Further redox reactions generate $FADH_2$ and more NADH.

Succinate is oxidized as the electron carrier FAD is reduced to $FADH_2$. Fumarate is converted to malate, which is then oxidized as one last NAD^+ is reduced to NADH. One turn of the citric acid cycle is completed with the regeneration of oxaloacetate, which is then ready to start the next cycle by accepting an acetyl group from acetyl CoA.

▲ **Figure 6.9B** A closer look at the citric acid cycle. (Remember that the cycle runs two times for each glucose molecule oxidized.)

6.10 Most ATP production occurs by oxidative phosphorylation

Your main objective in this chapter is to learn how cells harvest the energy of glucose to make ATP. But so far, you've seen the production of only 4 ATP per glucose molecule. Now it's time for the big energy payoff. The final stage of cellular respiration is oxidative phosphorylation, which uses the electron transport chain and chemiosmosis—a process introduced in Module 6.6. Oxidative phosphorylation clearly illustrates the concept of structure fitting function: The arrangement of electron carriers built into a membrane makes it possible to create an H^+ concentration gradient across the membrane and then use the energy of that gradient to drive ATP synthesis.

Figure 6.10A shows how an electron transport chain is arranged in the inner membrane of the mitochondrion. The folds (cristae) of this membrane enlarge its surface area, providing space for thousands of copies of the chain. Also embedded in the membrane are multiple copies of an enzyme complex called **ATP synthase**, which synthesizes ATP.

Electron Transport Chain Starting on the left in Figure 6.10A, the gold arrow traces the transport of electrons from the shuttle molecules NADH and $FADH_2$ through the electron transport chain to oxygen, the final electron acceptor. It is in this end stage of cellular respiration that oxygen finally steps in to play its critical role. Each oxygen atom ($\frac{1}{2}O_2$) accepts 2 electrons from the chain and picks up 2 H^+ from the surrounding solution, forming H_2O. You can see that happening in the center right of the figure.

Most of the carrier molecules of the chain reside in four main protein complexes (labeled I to IV in the diagram), while two mobile carriers transport electrons between the complexes. All of the carriers bind and release electrons in redox reactions, passing electrons down the "energy hill." Three of the protein complexes use the energy released from these electron transfers to actively transport H^+ across the membrane, from where H^+ is less concentrated to where it is more concentrated. The green vertical arrows show H^+ being transported from the matrix of the mitochondrion into the narrow intermembrane space.

Chemiosmosis Recall that chemiosmosis is a process that uses the energy stored in a hydrogen ion gradient across a membrane to drive ATP synthesis. **Figure 6.10B** shows the role of ATP synthase in chemiosmosis. The H^+ concentration gradient across the membrane stores potential energy, much the way a dam stores energy by holding back the elevated water behind it. The energy stored by a dam can be harnessed to do work (such as generating electricity)

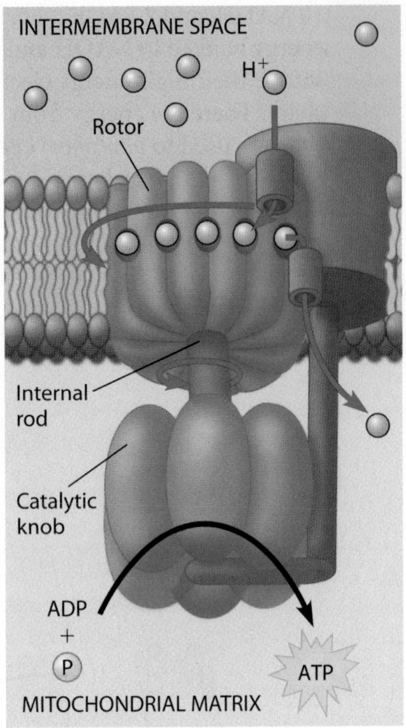

▲ **Figure 6.10B** ATP synthase—a molecular rotary motor

Try This Identify the power source that runs this motor. Explain where this "power" comes from.

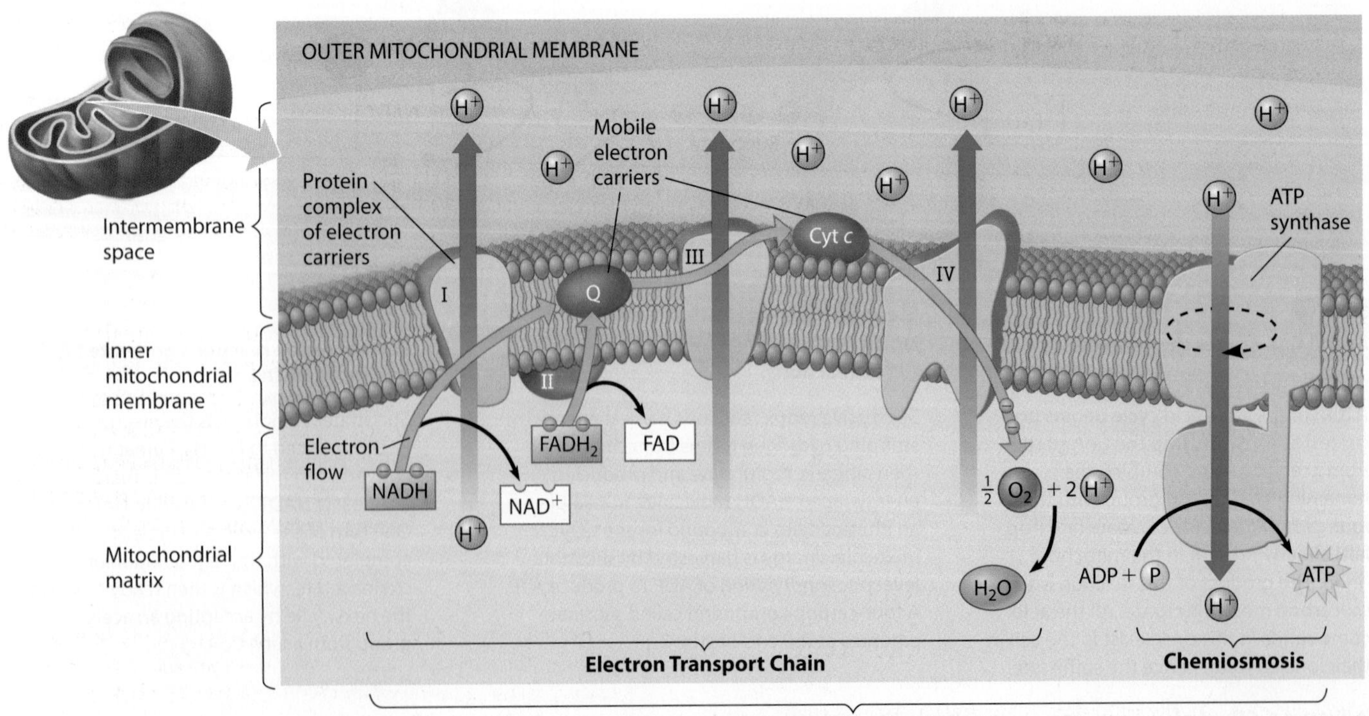

OUTER MITOCHONDRIAL MEMBRANE
Intermembrane space
Inner mitochondrial membrane
Mitochondrial matrix
Protein complex of electron carriers
Mobile electron carriers
ATP synthase
I
II
III
IV
Q
Cyt c
Electron flow
NADH
NAD+
FADH2
FAD
½ O2 + 2 H+
H2O
ADP + P
ATP
H+
Electron Transport Chain
Chemiosmosis
Oxidative Phosphorylation

▲ **Figure 6.10A** Oxidative phosphorylation: electron transport and chemiosmosis in a mitochondrion

when the water is allowed to rush downhill, turning giant wheels called turbines. The ATP synthases built into the inner mitochondrial membrane act like miniature turbines, with the rush of H^+ ions down their concentration gradient turning the wheels. Indeed, ATP synthase is considered the smallest molecular rotary motor known in nature.

As Figure 6.10B shows, hydrogen ions move one by one into binding sites within this protein complex (the rotor), causing it to spin. After once around, they are spit out into the mitochondrial matrix. The spinning rotor turns an internal rod, which activates sites in the catalytic knob that phosphorylate ADP to ATP.

We will make a final tally of ATP production in Module 6.12. But first, let's consider how cellular respiration can sometimes be used primarily to generate heat.

? What effect would an absence of oxygen (O_2) have on the process illustrated in Figure 6.10A?

● Without oxygen to "pull" electrons down the electron transport chain, the energy stored in NADH and $FADH_2$ could not be harnessed for ATP synthesis.

6.11 Scientists have discovered heat-producing, calorie-burning brown fat in adults

SCIENTIFIC THINKING

Ordinary body fat, called white fat, has little metabolic activity. Each cell is filled with a single large droplet of fat. Brown fat, on the other hand, actively burns energy. You learned in the chapter introduction that brown fat helps keep infants warm. Brown fat is named for its color, which comes from the brownish mitochondria that pack its cells. These mitochondria are unique in that they can burn fuel and produce heat without making ATP. How can they do that? Look back at Figure 6.10A and imagine ion channels spanning the inner mitochondrial membrane that allow H^+ to flow freely across the membrane. Such channels would dissipate the H^+ gradient that the electron transport chain had produced. Without that gradient, ATP synthase could not make ATP, and all the energy from the burning of fuel molecules would be released as heat. The mitochondria of brown fat cells have just such channels.

Can brown fat keep a newborn warm and help keep an adult thin?

Until recently, brown fat in humans was thought to disappear after infancy. The presence of unidentified tissue in the PET scans of cancer patients, however, caused researchers to question that conclusion. Could the tissues in the scans be brown fat? To test that hypothesis, researchers analyzed 3,640 PET-CT scans that had been performed on 1,972 patients for various diagnostic reasons. PET is a technique that identifies areas with high uptake of radioactively labeled glucose, and CT scans can detect adipose (fat) tissue. The combined PET-CT scans revealed small areas in the neck and chest of some patients that fit the criteria for brown fat—adipose tissue that was metabolically active (burning glucose). The researchers correlated the presence or absence of brown fat with patients' sex, age, weight, and other parameters, including the outdoor temperature. The results showed that 7.5% of the women and 3% of the men examined had deposits of brown fat. The tissues were found to be more prevalent both in patients who were thinner and when the scans had been taken in cold weather.

As is typical in science, the results from one study led to new questions and new research. Is brown fat activated by cold temperatures and, thus, could a much higher percentage of adults have brown fat than shown in scans of patients who were presumably *not* cold? Is the prevalence of this fat-burning tissue in thinner individuals related to why some people are thin and others are obese?

A second study involving 24 men looked at the presence and activity of brown fat during cold exposure. Ten participants were classified as lean (based on a BMI [body mass index] of less than 25) and 14 were identified as overweight or obese. Combined PET-CT scans were taken of all research participants following a two-hour exposure at 16°C (60.8°F). The scans of all but one participant (the one with the highest BMI) revealed activated brown fat tissues. As shown in **Figure 6.11**, the measured brown fat activity of the lean group was significantly higher than that of the overweight/obese group.

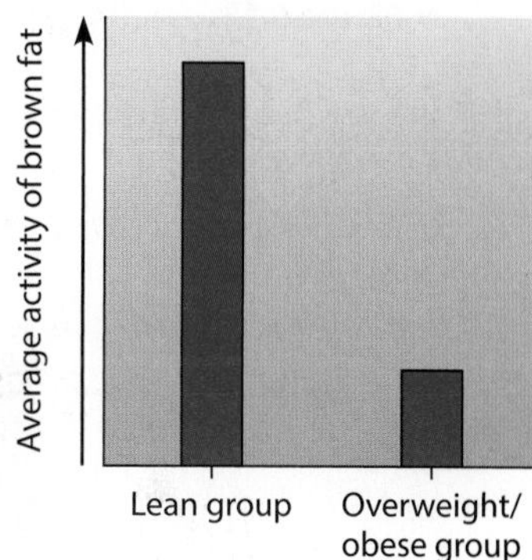

Average activity of brown fat
Lean group
Overweight/
obese group

▲ **Figure 6.11** Activity level of brown fat of lean and overweight/obese participants after cold exposure

Data from W. D. van Marken Lichtenbelt et al., Cold-activated brown adipose tissue in healthy men, *New England Journal of Medicine* 360: 1500–8 (2009).

These results indicate that brown fat may be present in most people, and, when activated by cold, the brown fat of lean individuals is more active (burns more calories). What questions does this study raise? Does the more active brown fat of thin individuals help keep them thin? Are there other ways to turn on brown fat besides exposure to cold? Could brown fat be a target for obesity-fighting drugs?

Research continues on brown fat and on the signals (temperature, hormonal, and nervous) that activate it. But as the popular press reports this potential link between calorie-burning brown fat and weight loss, be prepared to see new diets, supplements, exercise routines, and perhaps even cold spa treatments that promise to rev up your brown fat furnace. As always when evaluating such information, look for the science behind those claims.

? The initial study discussed identified brown fat in less than 10% of the patients whose scans were analyzed. The second study identified brown fat in 96% of participants. What accounts for this difference?

● Brown fat is activated in response to cold temperature, and the second study involved cold treatment.

Data from A. M. Cypess et al., Identification and importance of brown adipose tissue in adult humans, *New England Journal of Medicine* 360: 1509–17 (2009).

6.12 Review: Each molecule of glucose yields many molecules of ATP

Let's review what you have learned about cellular respiration by following the oxidation of one molecule of glucose. Starting on the left in **Figure 6.12**, glycolysis, which occurs in the cytosol, oxidizes glucose to two molecules of pyruvate, produces 2 NADH, and produces a net of 2 ATP by substrate-level phosphorylation. Within the mitochondrion, the oxidation of 2 pyruvate yields 2 NADH and 2 acetyl CoA. The 2 acetyl CoA feed into the citric acid cycle, which yields 6 NADH and 2 $FADH_2$, as well as 2 ATP by substrate-level phosphorylation. NADH and $FADH_2$ deliver electrons to the electron transport chain, where they are finally passed to O_2, forming H_2O. The electron transport chain pumps H^+ into the intermembrane space. The resulting H^+ gradient is tapped by ATP synthase to produce about 28 molecules of ATP by oxidative phosphorylation (according to current experimental data). Thus, the total yield of ATP molecules per glucose is about 32.

The number of ATP molecules cannot be stated exactly for several reasons. The NADH produced in glycolysis passes its electrons across the mitochondrial membrane to either NAD^+ or FAD. Because $FADH_2$ adds its electrons farther along the electron transport chain (see Figure 6.10A), it contributes less to the H^+ gradient and thus generates less ATP. In addition, some of the energy of the H^+ gradient may be used for work other than ATP production, such as the active transport of pyruvate into the mitochondrion.

Because most of the ATP generated by cellular respiration results from oxidative phosphorylation, the ATP yield depends on an adequate supply of oxygen to the cell. Without oxygen to function as the final electron acceptor, electron transport and ATP production stop. But as we see next, some cells can oxidize organic fuel and generate ATP *without* oxygen.

? Explain where O_2 is used and CO_2 is produced in cellular respiration.

● O_2 accepts electrons at the end of the electron transport chain. CO_2 is released during the oxidation of intermediate compounds in pyruvate oxidation and the citric acid cycle.

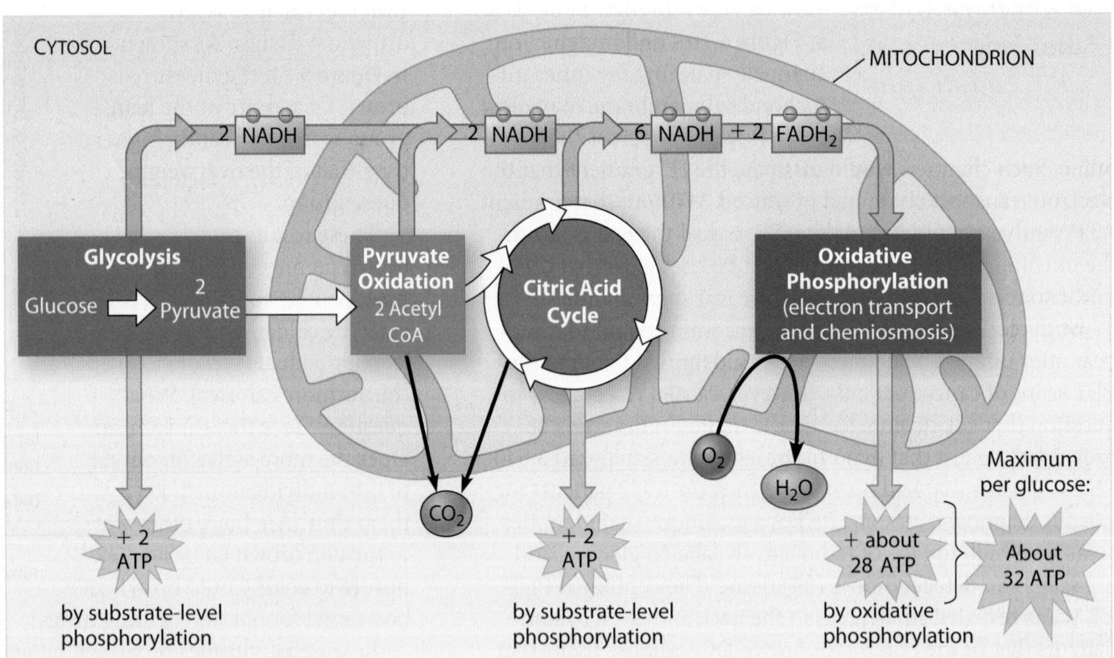

CYTOSOL
MITOCHONDRION
2 NADH
2 NADH
6 NADH + 2 $FADH_2$
Glycolysis
Glucose
2 Pyruvate
Pyruvate Oxidation
2 Acetyl CoA
Citric Acid Cycle
Oxidative Phosphorylation (electron transport and chemiosmosis)
O_2
H_2O
CO_2
+ 2 ATP
by substrate-level phosphorylation
+ 2 ATP
by substrate-level phosphorylation
+ about 28 ATP
by oxidative phosphorylation
Maximum per glucose:
About 32 ATP

▶ **Figure 6.12** An estimated tally of the ATP produced per molecule of glucose by substrate-level and oxidative phosphorylation in cellular respiration

▷ Fermentation: Anaerobic Harvesting of Energy

6.13 Fermentation enables cells to produce ATP without oxygen

Fermentation is a way of harvesting chemical energy that does not require oxygen. The metabolic pathway that generates ATP during fermentation is glycolysis, the same pathway that functions in the first stage of cellular respiration. Remember that glycolysis uses no oxygen; it simply generates a net gain of 2 ATP while oxidizing glucose to two molecules of pyruvate and reducing NAD^+ to NADH. The yield of 2 ATP is certainly a lot less than the possible 32 ATP per glucose generated during aerobic respiration, but it is enough to keep your muscles contracting for a short period of time when oxygen is scarce. And many microorganisms supply all their energy needs with the 2 ATP per glucose yield of glycolysis.

There is more to fermentation, however, than just glycolysis. To oxidize glucose in glycolysis, NAD^+ must be present as an electron acceptor. This is no problem under aerobic conditions, because the cell regenerates its pool of NAD^+ when NADH passes its electrons into the mitochondrion, to be transported to the electron transport chain. Fermentation provides an anaerobic path for recycling NADH back to NAD^+.

Lactic Acid Fermentation One common type of fermentation is called **lactic acid fermentation**. Your muscle cells and certain bacteria can regenerate NAD^+ by this process, as illustrated in **Figure 6.13A**. You can see that NADH is oxidized back to NAD^+ as pyruvate is reduced to lactate (the ionized form of lactic acid). Muscle cells can switch to lactic acid fermentation when the need for ATP outpaces the delivery of O_2 via the bloodstream. The lactate that builds up in muscle cells during strenuous exercise was previously thought to cause muscle fatigue and pain, but research now indicates that increased levels of other ions may be to blame. In any case, the lactate is gradually carried away by the blood to the liver, where it is converted back to pyruvate and oxidized in the mitochondria of liver cells.

The dairy industry uses lactic acid fermentation by bacteria to make cheese and yogurt. Other types of microbial fermentation turn soybeans into soy sauce and cabbage into sauerkraut.

Alcohol Fermentation For thousands of years, people have used **alcohol fermentation** in brewing, winemaking, and baking. Yeasts are single-celled fungi that normally use aerobic respiration to process their food. But they are also able to survive in anaerobic environments. Yeasts and certain bacteria recycle their NADH back to NAD^+ while converting pyruvate to CO_2 and ethanol (**Figure 6.13B**). The CO_2 provides the bubbles in beer and champagne. Bubbles of CO_2 generated by baker's yeast cause bread dough to rise. Ethanol (ethyl alcohol), the two-carbon end product, is toxic to the organisms that produce it. Yeasts release their alcohol wastes to their surroundings, where it usually diffuses away. When yeasts are confined in a wine vat, they die when the alcohol concentration reaches 14%.

Types of Anaerobes Unlike muscle cells and yeasts, many prokaryotes that live in stagnant ponds and deep in the soil are *obligate anaerobes*, meaning they require anaerobic conditions and are poisoned by oxygen. Yeasts and many other bacteria are facultative anaerobes. A *facultative anaerobe* can make ATP either by fermentation or by oxidative phosphorylation, depending on whether O_2 is available. On the cellular level, our muscle cells behave as facultative anaerobes.

For a facultative anaerobe, pyruvate is a fork in the metabolic road. If oxygen is available, the organism will always use the more productive aerobic respiration. Thus, to make wine and beer, yeasts must be grown anaerobically so that they will ferment sugars and produce ethanol. For this reason, the wine barrels and beer fermentation vats in **Figure 6.13C** are designed to keep air out.

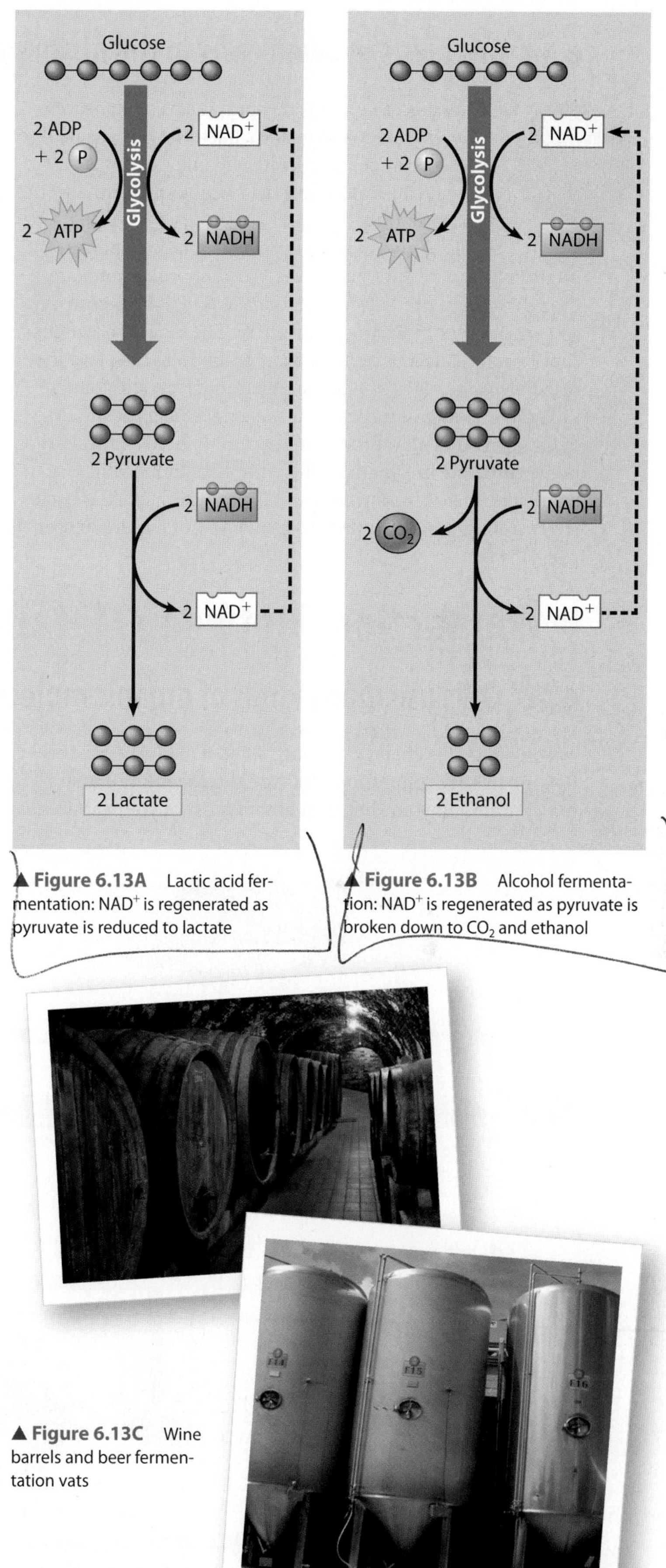

Glucose
2 ADP
+ 2 P
2 ATP
Glycolysis
2 NAD+
2 NADH
2 Pyruvate
2 NADH
2 NAD+
2 Lactate
Glucose
2 ADP
+ 2 P
2 ATP
Glycolysis
2 NAD+
2 NADH
2 Pyruvate
2 CO2
2 NADH
2 NAD+
2 Ethanol

▲ **Figure 6.13A** Lactic acid fermentation: NAD^+ is regenerated as pyruvate is reduced to lactate

▲ **Figure 6.13B** Alcohol fermentation: NAD^+ is regenerated as pyruvate is broken down to CO_2 and ethanol

▲ **Figure 6.13C** Wine barrels and beer fermentation vats

? A glucose-fed yeast cell is moved from an aerobic environment to an anaerobic one. For the cell to continue generating ATP at the same rate, how would its rate of glucose consumption need to change?

● The cell would have to consume glucose at a rate about 16 times the consumption rate in the aerobic environment (2 ATP per glucose molecule is made by fermentation versus 32 ATP by cellular respiration).

6.14 Glycolysis evolved early in the history of life on Earth

EVOLUTION CONNECTION

Glycolysis is the universal energy-harvesting process of life. If you looked inside a bacterial cell, inside one of your body cells, or inside virtually any other living cell, you would find the metabolic machinery of glycolysis.

The role of glycolysis in both fermentation and respiration has an evolutionary basis. Ancient prokaryotes are thought to have used glycolysis to make ATP long before oxygen was present in Earth's atmosphere. The oldest known fossils of bacteria date back more than 3.5 billion years, and they resemble some types of photosynthetic bacteria still found today. The evidence indicates, however, that significant levels of O_2, formed as a by-product of bacterial photosynthesis, did not accumulate in the atmosphere until about 2.7 billion years ago. Thus, early prokaryotes most likely generated ATP exclusively from glycolysis, a process that does not require oxygen.

The fact that glycolysis is the most widespread metabolic pathway found in Earth's organisms today suggests that it evolved very early in the history of life. The location of glycolysis within the cell also implies great antiquity; the pathway does not require any of the membrane-enclosed organelles of the eukaryotic cell, which evolved about a billion years after the prokaryotic cell. Glycolysis is a metabolic heirloom from early cells that continues to function in fermentation and as the first stage in the breakdown of organic molecules by cellular respiration.

? List some of the characteristics of glycolysis that indicate that it is an ancient metabolic pathway.

Glycolysis occurs universally (functioning in both fermentation and respiration), does not require oxygen, and does not occur in a membrane-enclosed organelle.

▷ Connections Between Metabolic Pathways

6.15 Cells use many kinds of organic molecules as fuel for cellular respiration

Throughout this chapter, we have spoken of glucose as the fuel for cellular respiration. But free glucose molecules are not common in your diet. You obtain most of your calories as carbohydrates (such as sucrose and other disaccharide sugars and starch, a polysaccharide), fats, and proteins. You consume all three of these classes of organic molecules when you eat a handful of peanuts, for instance.

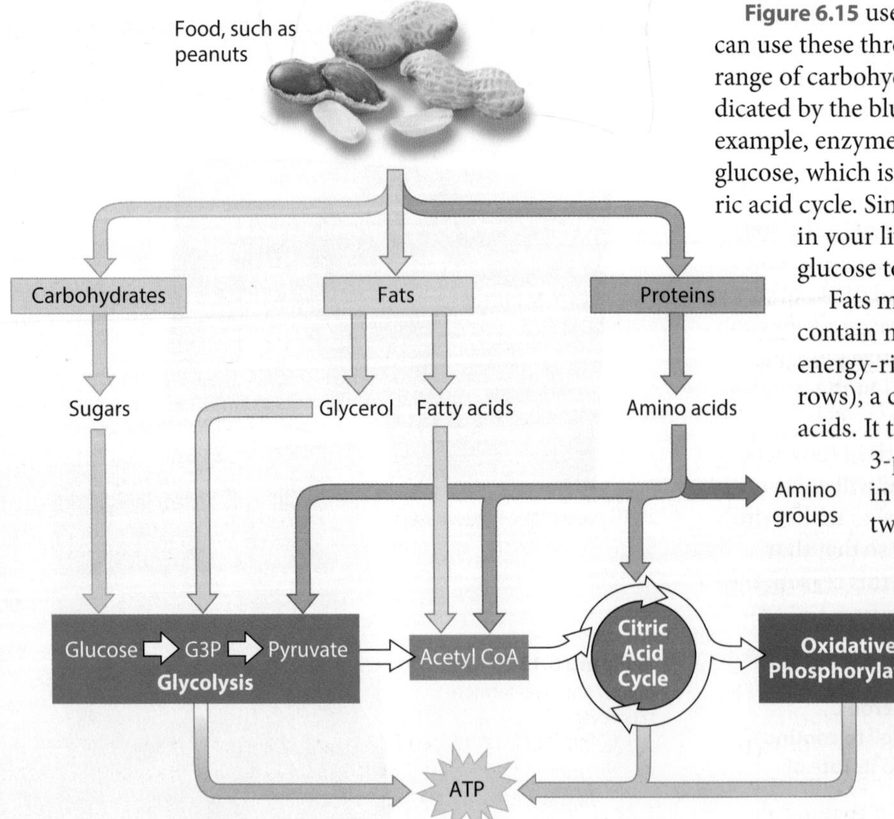
Food, such as peanuts
Carbohydrates
Fats
Proteins
Sugars
Glycerol
Fatty acids
Amino acids
Amino groups
Glucose
G3P
Pyruvate
Glycolysis
Acetyl CoA
Citric Acid Cycle
Oxidative Phosphorylation
ATP

▲ **Figure 6.15** Pathways that break down various food molecules

Figure 6.15 uses color-coded arrows to illustrate how a cell can use these three types of molecules to make ATP. A wide range of carbohydrates can be funneled into glycolysis, as indicated by the blue arrows on the far left of the diagram. For example, enzymes in your digestive tract hydrolyze starch to glucose, which is then broken down by glycolysis and the citric acid cycle. Similarly, glycogen, the polysaccharide stored in your liver and muscle cells, can be hydrolyzed to glucose to serve as fuel between meals.

Fats make excellent cellular fuel because they contain many hydrogen atoms and thus many energy-rich electrons. As the diagram shows (tan arrows), a cell first hydrolyzes fats to glycerol and fatty acids. It then converts the glycerol to glyceraldehyde 3-phosphate (G3P), one of the intermediates in glycolysis. The fatty acids are broken into two-carbon fragments that enter the citric acid cycle as acetyl CoA. A gram of fat yields more than twice as much ATP as a gram of carbohydrate. Because so many calories are stockpiled in each gram of fat, you must expend a large amount of energy to burn fat stored in your body. This helps explain why it is so difficult for a dieter to lose excess fat.

Proteins (purple arrows in Figure 6.15) can also be used for fuel, although your body preferentially burns sugars and fats first. To be oxidized as fuel, proteins must first be digested to their constituent amino acids. Typically, a cell will use most of these amino acids to make its own proteins. Enzymes can convert excess amino acids to intermediates of glycolysis or the citric acid cycle, and their energy is then harvested by cellular respiration. During the conversion, the amino groups are stripped off and later disposed of in urine.

? Animals store most of their energy reserves as fats, not as polysaccharides. What is the advantage of this mode of storage for an animal?

● Most animals are mobile and benefit from a compact and concentrated form of energy storage. Also, because fats are hydrophobic, they can be stored without extra water associated with them (see Module 3.8).

6.16 Organic molecules from food provide raw materials for biosynthesis

Not all food molecules are destined to be oxidized as fuel for making ATP. Food also provides the raw materials your cells use for biosynthesis—the production of organic molecules using energy-requiring metabolic pathways. A cell must be able to make its own molecules to build its structures and perform its functions. Some raw materials, such as amino acids, can be incorporated directly into your macromolecules. However, your cells also need to make molecules that are not present in your food. Indeed, glycolysis and the citric acid cycle function as metabolic interchanges that enable your cells to convert some kinds of molecules to others as you need them.

Figure 6.16 outlines the pathways by which your cells can make three classes of organic molecules using some of the intermediate molecules of glycolysis and the citric acid cycle. By comparing Figures 6.15 and 6.16, you can see clear connections between the energy-harvesting pathways of cellular respiration and the biosynthetic pathways used to construct the organic molecules of the cell.

Basic principles of supply and demand regulate these pathways. If there is an excess of a certain amino acid, for example, the pathway that synthesizes it is switched off. The most common mechanism for this control is feedback inhibition: The end product inhibits an enzyme that catalyzes an early step in the pathway (see Module 5.15). Feedback inhibition also controls cellular respiration. If ATP accumulates in a cell, it inhibits an early enzyme in glycolysis, slowing down respiration and conserving resources. On the other hand, the same enzyme is activated by a buildup of ADP in the cell, signaling the need for more energy.

The cells of all living organisms—including those of the panda shown in Figure 6.16 and the plants they eat—have the ability to harvest energy from the breakdown of organic molecules. When the process is cellular respiration, the atoms of the starting materials end up in carbon dioxide and water. In contrast, the ability to make organic molecules from carbon dioxide and water is not universal. Animal cells lack this ability, but plant cells can actually produce organic molecules from inorganic ones using the energy of sunlight in the process of photosynthesis. (We explore photosynthesis in Chapter 7.)

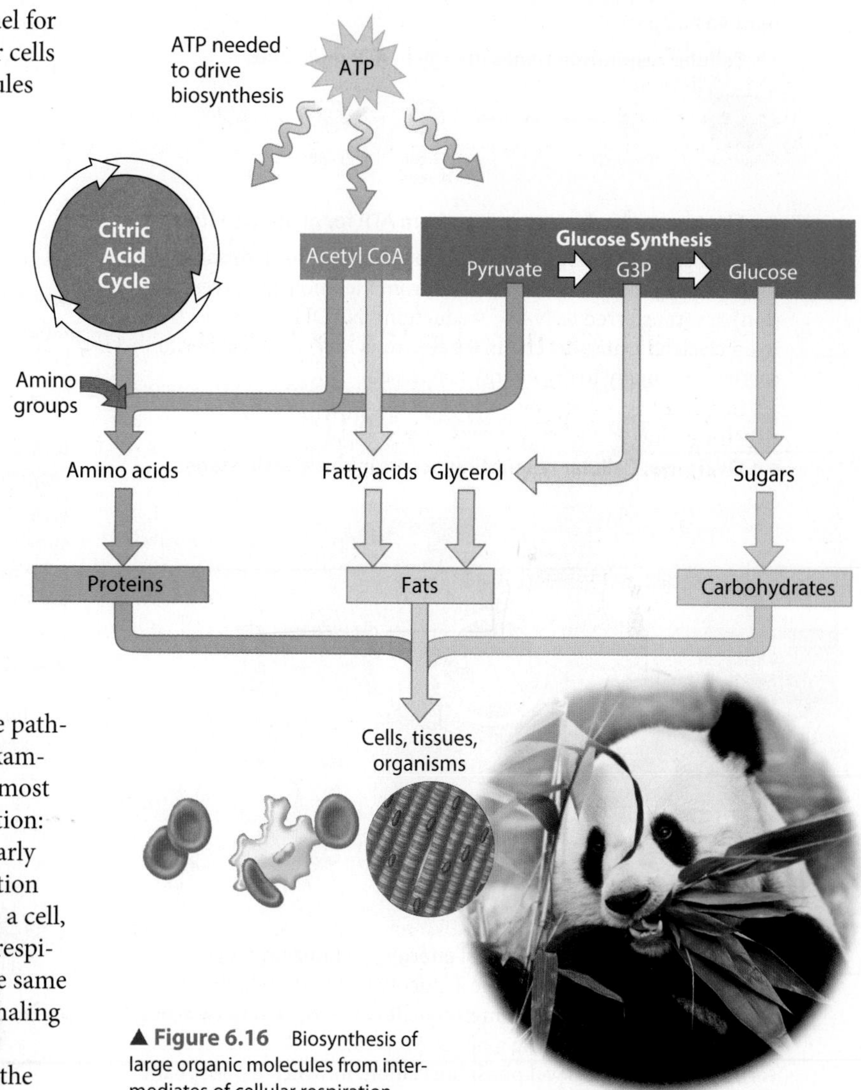

ATP needed to drive biosynthesis
ATP
Citric Acid Cycle
Acetyl CoA
Glucose Synthesis
Pyruvate
G3P
Glucose
Amino groups
Amino acids
Fatty acids
Glycerol
Sugars
Proteins
Fats
Carbohydrates
Cells, tissues, organisms

▲ **Figure 6.16** Biosynthesis of large organic molecules from intermediates of cellular respiration

? Explain how someone can gain weight and store fat even when on a low-fat diet. (Hint: Look for G3P and acetyl CoA in Figures 6.15 and 6.16.)

● If caloric intake is excessive, body cells use metabolic pathways to convert the excess to fat. The glycerol and fatty acids of fats are made from G3P and acetyl CoA, respectively, both produced from the oxidation of carbohydrates.

CHAPTER 6 REVIEW

For practice quizzes, BioFlix animations, MP3 tutorials, video tutors, and more study tools designed for this textbook, go to MasteringBiology®

Reviewing the Concepts

Cellular Respiration: Aerobic Harvesting of Energy (6.1–6.5)

6.1 Photosynthesis and cellular respiration provide energy for life. Photosynthesis uses solar energy to produce glucose and O_2 from CO_2 and H_2O. In cellular respiration, O_2 is consumed during the breakdown of glucose to CO_2 and H_2O, and energy is released.

6.2 Breathing supplies O_2 for use in cellular respiration and removes CO_2.

6.3 Cellular respiration banks energy in ATP molecules.

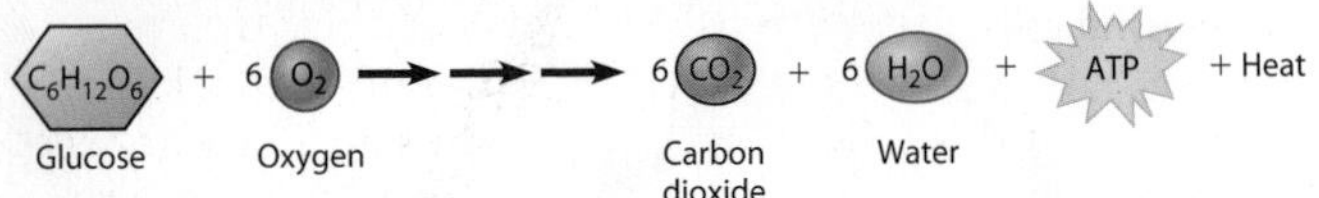

$C_6H_{12}O_6$ + 6 O_2 → 6 CO_2 + 6 H_2O + ATP + Heat
Glucose
Oxygen
Carbon dioxide
Water

6.4 The human body uses energy from ATP for all its activities.

6.5 Cells capture energy from electrons "falling" from organic fuels to oxygen. Electrons removed from fuel molecules (oxidation) are transferred to NAD^+ (reduction). NADH passes electrons to an electron transport chain. As electrons "fall" from carrier to carrier and finally to O_2, energy is released.

Stages of Cellular Respiration (6.6–6.12)

6.6 Overview: Cellular respiration occurs in three main stages.

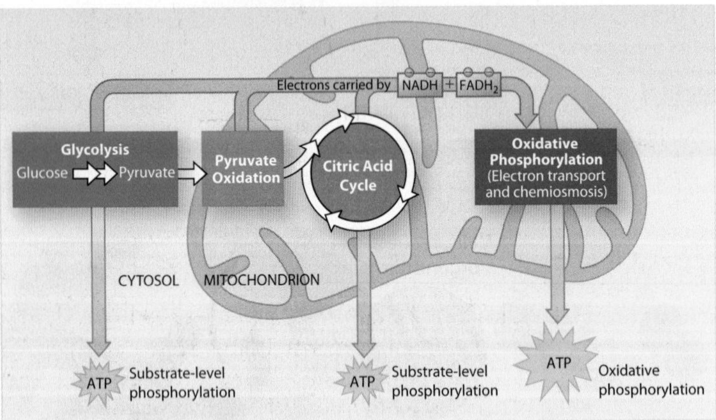

Electrons carried by NADH + $FADH_2$
Glycolysis
Glucose
Pyruvate
Pyruvate Oxidation
Citric Acid Cycle
Oxidative Phosphorylation (Electron transport and chemiosmosis)
CYTOSOL
MITOCHONDRION
ATP
Substrate-level phosphorylation
ATP
Substrate-level phosphorylation
ATP
Oxidative phosphorylation

6.7 Glycolysis harvests chemical energy by oxidizing glucose to pyruvate. ATP is used to prime a glucose molecule, which is split in two. These three-carbon intermediates are oxidized to two molecules of pyruvate, yielding a net of 2 ATP and 2 NADH. ATP is formed by substrate-level phosphorylation, in which a phosphate group is transferred from an organic molecule to ADP.

6.8 Pyruvate is oxidized in preparation for the citric acid cycle. The oxidation of pyruvate yields acetyl CoA, CO_2, and NADH.

6.9 The citric acid cycle completes the oxidation of organic molecules, generating many NADH and $FADH_2$ molecules. For each turn of the cycle, two carbons from acetyl CoA are added and 2 CO_2 are released.

6.10 Most ATP production occurs by oxidative phosphorylation. In mitochondria, electrons from NADH and $FADH_2$ are passed down the electron transport chain to O_2, which picks up H^+ to form water. Energy released by these redox reactions is used to pump H^+ into the intermembrane space. In chemiosmosis, the H^+ gradient drives H^+ back through ATP synthase complexes in the inner membrane, synthesizing ATP.

6.11 Scientists have discovered heat-producing, calorie-burning brown fat in adults.

6.12 Review: Each molecule of glucose yields many molecules of ATP. Substrate-level phosphorylation and oxidative phosphorylation produce up to 32 ATP molecules for every glucose molecule oxidized in cellular respiration.

Fermentation: Anaerobic Harvesting of Energy (6.13–6.14)

6.13 Fermentation enables cells to produce ATP without oxygen. Under anaerobic conditions, muscle cells, yeasts, and certain bacteria produce ATP by glycolysis. NAD^+ is recycled from NADH as pyruvate is reduced to lactate (lactic acid fermentation) or alcohol and CO_2 (alcohol fermentation).

6.14 Glycolysis evolved early in the history of life on Earth. Glycolysis occurs in the cytosol of the cells of nearly all organisms and is thought to have evolved in ancient prokaryotes.

Connections Between Metabolic Pathways (6.15–6.16)

6.15 Cells use many kinds of organic molecules as fuel for cellular respiration.

6.16 Organic molecules from food provide raw materials for biosynthesis. Cells use intermediates from cellular respiration and ATP for biosynthesis of other organic molecules. Metabolic pathways are often regulated by feedback inhibition.

Connecting the Concepts

1. Fill in the blanks in this summary map to help you review the key concepts of cellular respiration.

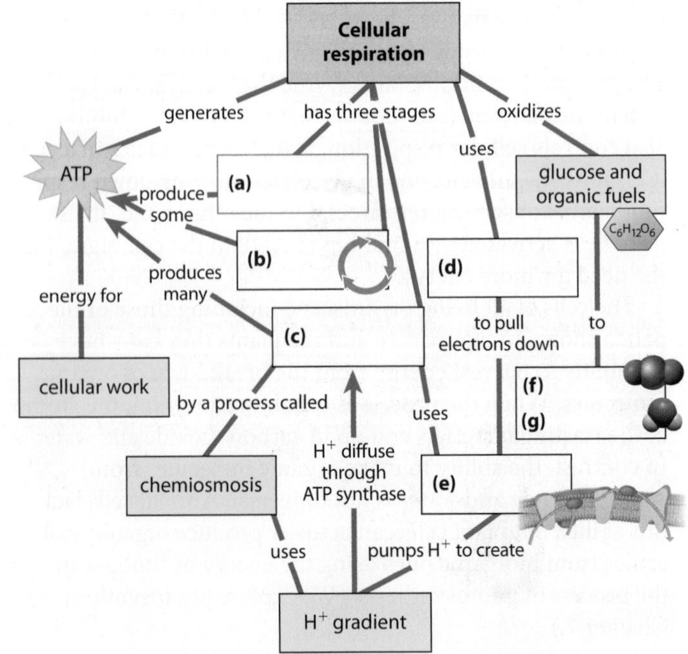

Cellular respiration
generates
has three stages
oxidizes
uses
ATP
produce some
(a)
glucose and organic fuels
$C_6H_{12}O_6$
(b)
produces many
(d)
energy for
(c)
to pull electrons down
to
cellular work
(f)
by a process called
uses
(g)
H^+ diffuse through ATP synthase
chemiosmosis
(e)
uses
pumps H^+ to create
H^+ gradient

Testing Your Knowledge

Level 1: Knowledge/Comprehension

2. A biochemist wanted to study how various substances were used in cellular respiration. In one experiment, she allowed a mouse to breathe air containing O_2 "labeled" by a particular isotope. In the mouse, the labeled oxygen first showed up in
   a. ATP.
   b. NADH.
   c. CO_2.
   d. H_2O.
3. In glycolysis, ____________ is oxidized and ____________ is reduced.
   a. NAD^+ . . . glucose
   b. glucose . . . oxygen
   c. ATP . . . ADP
   d. glucose . . . NAD^+
4. Which of the following is the most immediate source of energy for making most of the ATP in your cells?
   a. the transfer of (P) from intermediate substrates to ADP
   b. the movement of H^+ across a membrane down its concentration gradient
   c. the splitting of glucose into two molecules of pyruvate
   d. electrons moving through the electron transport chain
5. Which of the following is a true distinction between cellular respiration and fermentation?
   a. NADH is oxidized by passing electrons to the electron transport chain in respiration only.
   b. Only respiration oxidizes glucose.
   c. Substrate-level phosphorylation is unique to fermentation; cellular respiration uses oxidative phosphorylation.
   d. Fermentation is the metabolic pathway found in prokaryotes; cellular respiration is unique to eukaryotes.

Level 2: Application/Analysis

6. The poison cyanide binds to an electron carrier within the electron transport chain and blocks the movement of electrons. When this happens, glycolysis and the citric acid cycle soon grind to a halt as well. Why do you think these other two stages of cellular respiration stop? (*Explain your answer.*)
   a. They run out of ATP.
   b. Unused O_2 interferes with cellular respiration.
   c. They run out of NAD^+ and FAD.
   d. Electrons are no longer available.
7. In which of the following is the first molecule becoming reduced to the second molecule?
   a. pyruvate $\rightarrow$ acetyl CoA
   b. pyruvate $\rightarrow$ lactate
   c. glucose $\rightarrow$ pyruvate
   d. $NADH + H^+ \rightarrow NAD^+ + 2\,H$
8. Which of the three stages of cellular respiration is considered the most ancient? Explain your answer.
9. Compare and contrast fermentation as it occurs in your muscle cells and in yeast cells.
10. Explain how your body can convert excess carbohydrates in the diet to fats. Can excess carbohydrates be converted to protein? What else must be supplied?
11. An average adult human requires 2,200 kcal of energy per day. Suppose your diet provides an average of 2,300 kcal per day. How many hours per week would you have to walk to burn off the extra calories? Swim? Run? (See Figure 6.4.)
12. Your body makes NAD^+ and FAD from two B vitamins, niacin and riboflavin. The Recommended Dietary Allowance for niacin is 20 mg and for riboflavin, 1.7 mg. These amounts are thousands of times less than the amount of glucose your body needs each day to fuel its energy needs. Why is the daily requirement for these vitamins so small?

Level 3: Synthesis/Evaluation

13. In the citric acid cycle, an enzyme oxidizes malate to oxaloacetate, with the production of NADH and the release of H^+. You are studying this reaction using a suspension of bean cell mitochondria and a blue dye that loses its color as it takes up H^+. You know that the higher the concentration of malate, the more rapid the decolorization of the dye. You set up reaction mixtures with mitochondria, dye, and three different concentrations of malate (0.1 mg/L, 0.2 mg/L, and 0.3 mg/L). Which of the following graphs represents the results you would expect, and why?

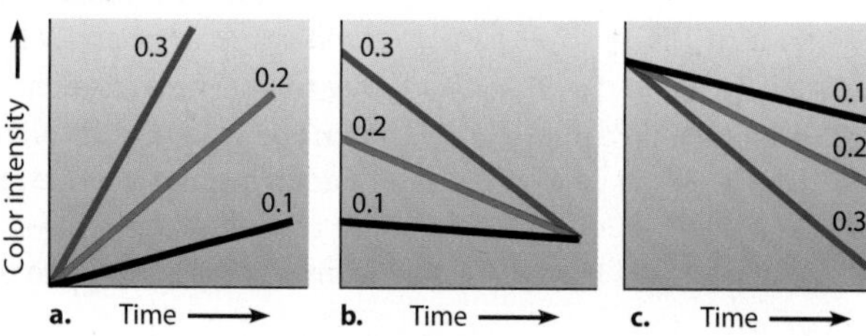
Color intensity
0.3
0.2
0.1
a. Time
0.3
0.2
0.1
b. Time
0.1
0.2
0.3
c. Time

14. ATP synthase enzymes are found in the prokaryotic plasma membrane and in the inner membrane of a mitochondrion. What does this suggest about the evolutionary relationship of this eukaryotic organelle to prokaryotes?
15. **SCIENTIFIC THINKING** Several studies have found a correlation between the activity levels of brown fat tissue in research participants following exposure to cold and their percentage of body fat (see Module 6.11). Devise a graph that would present the results from such a study, labeling the axes and drawing a line to show whether the results show a positive or negative correlation between the variables. Propose two hypotheses that could explain these results.
16. For a short time in the 1930s, some physicians prescribed low doses of a compound called dinitrophenol (DNP) to help patients lose weight. This unsafe method was abandoned after some patients died. DNP uncouples the chemiosmotic machinery by making the inner mitochondrial membrane leaky to H^+. Explain how this drug could cause profuse sweating, weight loss, and possibly death.
17. Explain how the mechanism of brown fat metabolism is similar to the effect that the drug DNP described above has on mitochondria. Pharmaceutical companies may start targeting brown fat for weight loss drugs. How might such drugs help patients lose weight? What dangers might such drugs pose?

Answers to all questions can be found in Appendix 4.

CHAPTER 7

Photosynthesis: *Using Light to Make Food*

? ***Will global climate change make you itch?***

If you are among the 80% of people allergic to poison ivy, the thick patch of three-leaved plants pictured below may make you want to scrub with soap and water and rush to find calamine lotion. A close encounter with this noxious weed often leads to itchy and oozing blisters that can last for weeks. The allergic component of poison ivy sap, urushiol, binds to skin, clothing, and pet fur on contact, where it remains active until washed off. Even dead leaves or vines retain active urushiol for several years.

Poison ivy is found throughout much of North America, often growing along the ground in both woods and open areas. It can also grow as a vine, climbing high up trees with its lateral branches that are sometimes mistaken for tree limbs. The rhymes "hairy vine, no friend of mine" and "raggy rope, don't be a dope" help alert hikers to the danger around them when the characteristic shiny leaves are hidden high in the tree foliage.

Like all plants, poison ivy produces energy for its growth by photosynthesis, the process that converts light energy to the chemical energy of sugar. Photosynthesis removes CO_2 from the

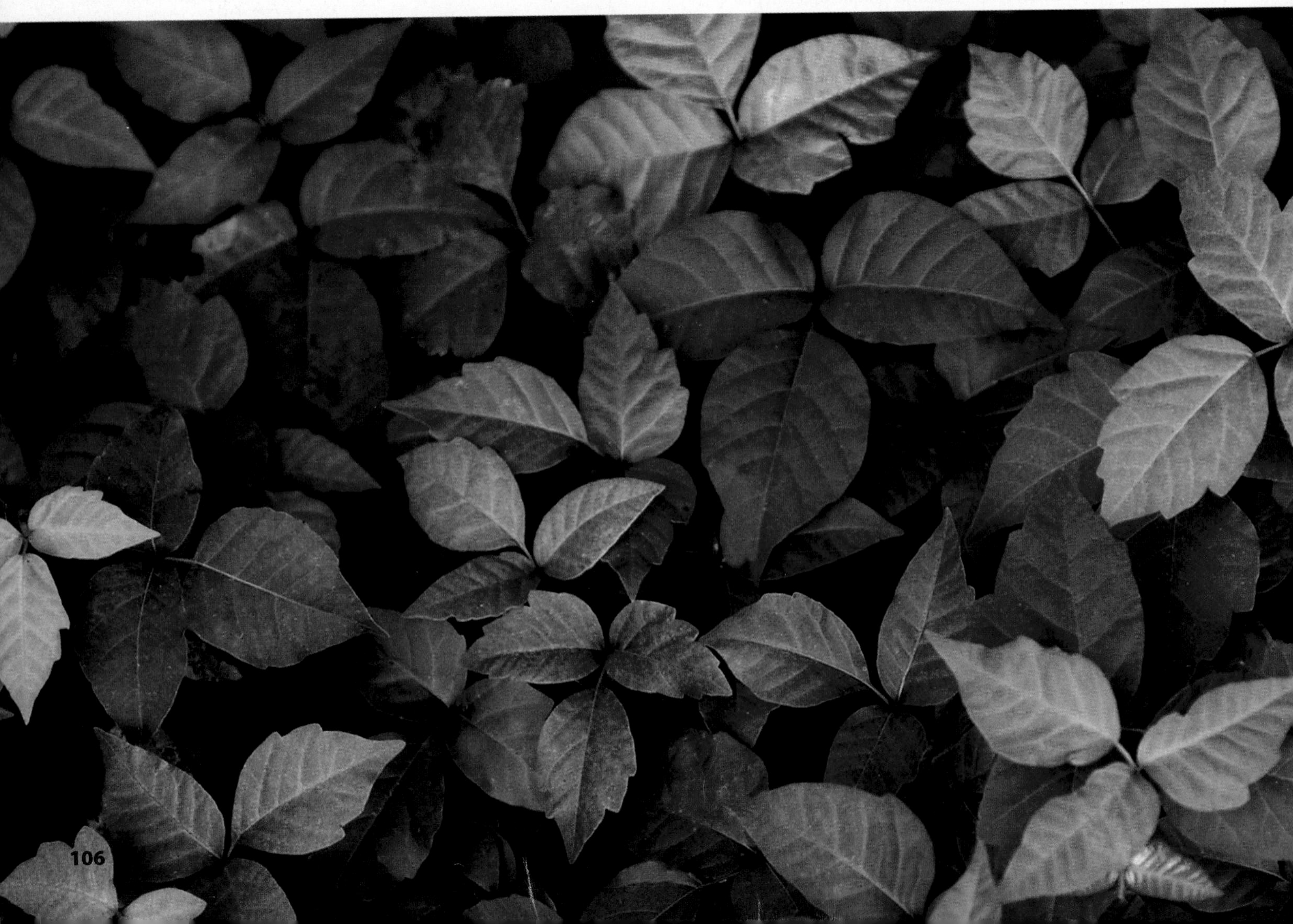

atmosphere and stores it in plant matter. The burning of sugar in the cellular respiration of almost all organisms releases CO_2 back to the environment. Burning fossil fuels and deforestation also release CO_2, and these activities are contributing to the current rise in atmospheric CO_2 and the accompanying global climate change. How might higher CO_2 levels affect plant growth? Unfortunately, many studies indicate that weeds grow faster under such conditions than do our crop plants or trees. Later in the chapter we will discuss one such study concerning the growth of poison ivy.

But first, let's learn how photosynthesis works. We begin with some basic concepts and then look more closely at the two stages of photosynthesis: the light reactions and the Calvin cycle. Finally, we explore ways in which photosynthesis affects our global environment.

BIG IDEAS

An Introduction to Photosynthesis
(7.1–7.5)

Plants and other photoautotrophs use the energy of sunlight to convert CO_2 and H_2O to sugar and O_2.

The Light Reactions: Converting Solar Energy to Chemical Energy
(7.6–7.9)

In the thylakoids of a chloroplast, the light reactions generate ATP and NADPH.

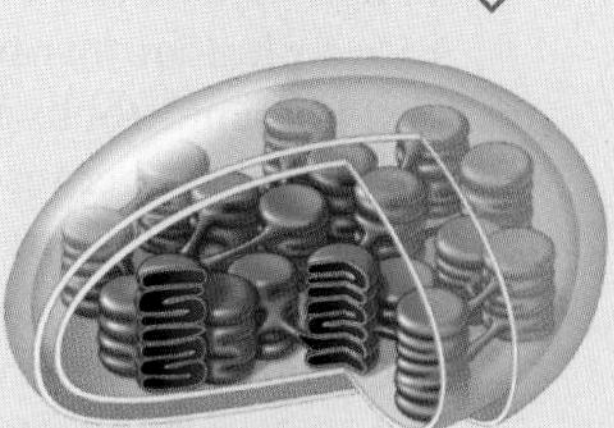

The Calvin Cycle: Reducing CO_2 to Sugar
(7.10–7.11)

The Calvin cycle, which takes place in the stroma of the chloroplast, uses ATP and NADPH to reduce CO_2 to sugar.

The Global Significance of Photosynthesis
(7.12–7.14)

Photosynthesis provides the energy and building material for ecosystems. It also affects global climate and the ozone layer.

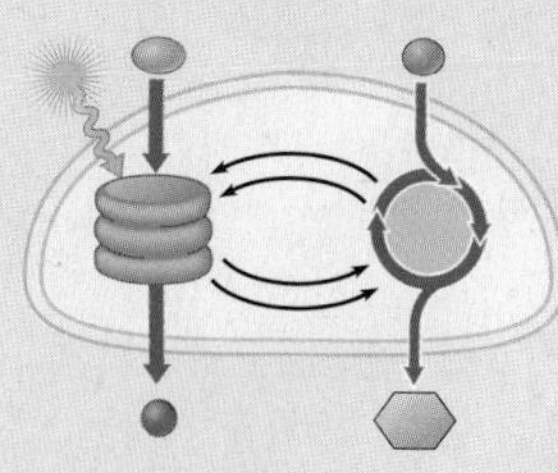

An Introduction to Photosynthesis

7.1 Photosynthesis fuels the biosphere

Life on Earth is solar powered. The chloroplasts in plant cells capture light energy that has traveled 150 million kilometers from the sun. Through the process of **photosynthesis**, plants use solar energy to convert CO_2 and H_2O to sugars and other organic molecules, and they release O_2 as a by-product. Plants are **autotrophs** (meaning "self-feeders" in Greek) in that they make their own food. Autotrophs not only feed themselves, but they are the ultimate source of organic molecules for almost all other organisms. Because they use the energy of light, plants and other photosynthesizers are specifically called **photoautotrophs**.

Photoautotrophs are often referred to as the producers of the biosphere because they produce its food supply. (In Chapter 16, you will learn about chemoautotrophs—prokaryotes that use inorganic chemicals as their energy source and are the producers in deep-sea vent communities.) Producers feed the consumers of the biosphere—the **heterotrophs** that cannot make their own food but must consume plants or animals or decompose organic material (*hetero* means "other"). You and almost all other heterotrophs are completely dependent on photoautotrophs for the raw materials and organic fuel necessary to maintain life and for the oxygen required to burn that fuel in cellular respiration.

Photoautotrophs not only feed us; they also clothe us (think cotton), house us (think wood), and provide energy for warmth, light, transport, and manufacturing. The fossil fuels we use as energy sources represent stores of the sun's energy captured by photoautotrophs in the far distant past.

The photographs shown on this page illustrate some of the diversity among today's photoautotrophs. On land, plants, such as those in the tropical forest in **Figure 7.1A**, are the producers. In aquatic environments, photoautotrophs include unicellular and multicellular algae, as well as photosynthetic prokaryotes. **Figure 7.1B** shows kelp, a large alga that forms extensive underwater "forests" off the coast of California. **Figure 7.1C** is a micrograph of cyanobacteria, which are important producers in freshwater and marine ecosystems.

In this chapter, we focus on photosynthesis in plants, which takes place in chloroplasts. The remarkable ability of these organelles to harness light energy and use it to drive the synthesis of organic compounds emerges from their structural organization: Photosynthetic pigments and enzymes are grouped together in membranes or compartments, facilitating the complex series of chemical reactions in photosynthesis. Photosynthetic bacteria have infolded regions of the plasma membrane containing such clusters of pigments and enzymes. In fact, according to the widely accepted theory of endosymbiosis, chloroplasts originated from a photosynthetic prokaryote that took up residence inside a eukaryotic cell (see Module 4.15).

Let's begin our study of photosynthesis with an overview of the location and structure of plant chloroplasts.

? **What do "self-feeding" photoautotrophs require from the environment to make their own food?**

● Light, carbon dioxide, and water. (Minerals are also required; you'll learn about the needs of plants in Chapter 32.)

▲ **Figure 7.1A** Tropical forest plants

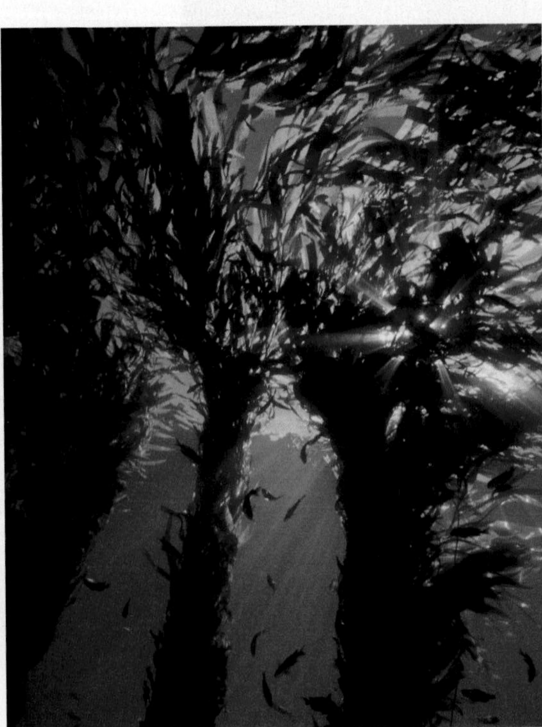

▲ **Figure 7.1B** Kelp, a large, multicellular alga

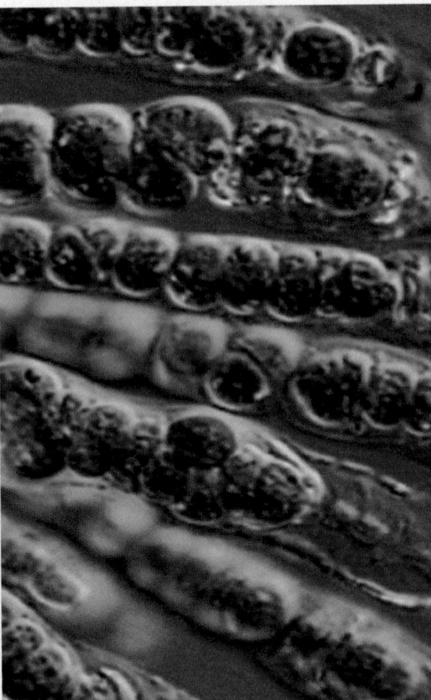

▲ **Figure 7.1C** Cyanobacteria (photosynthetic bacteria) LM 980X

7.2 Photosynthesis occurs in chloroplasts in plant cells

All green parts of a plant have chloroplasts in their cells, but leaves are the major sites of photosynthesis in most plants. Indeed, a section of leaf with a top surface area of 1 mm^2 has about a half million chloroplasts. A leaf's green color comes from **chlorophyll**, a light-absorbing pigment in the chloroplasts that plays a central role in converting solar energy to chemical energy.

Figure 7.2 zooms in on a leaf to show the actual sites of photosynthesis. As you can see in the leaf cross section, chloroplasts are concentrated in the cells of the **mesophyll**, the green tissue in the interior of the leaf. Carbon dioxide enters the leaf, and oxygen exits, by way of tiny pores called **stomata** (singular, *stoma*, meaning "mouth"). Water absorbed by the roots is delivered to the leaves in veins. Leaves also use veins to export sugar to roots and other parts of the plant.

As you will notice in the light micrograph of a single mesophyll cell, each cell has numerous chloroplasts. A typical mesophyll cell has about 30 to 40 chloroplasts. The bottom drawing and the electron micrograph show the structures in a single chloroplast. Membranes in the chloroplast form the framework within which many of the reactions of photosynthesis occur, just as mitochondrial membranes are the site for much of the energy-harvesting machinery in cellular respiration (see Module 6.10). In the chloroplast, an envelope of two membranes encloses an inner compartment, which is filled with a thick fluid called **stroma**. Suspended in the stroma is a system of interconnected membranous sacs, called **thylakoids**, which enclose another internal compartment, called the thylakoid space. (As you will see later, this thylakoid space plays a role analogous to the intermembrane space of a mitochondrion in the generation of ATP.) In many places, thylakoids are concentrated in stacks called grana (singular, *granum*). Built into the thylakoid membranes are the chlorophyll molecules that capture light energy. The thylakoid membranes also house much of the machinery that converts light energy to chemical energy, which is used in the stroma to make sugar.

Later in the chapter, we examine the function of these structures in more detail. But first, let's look more closely at the general process of photosynthesis.

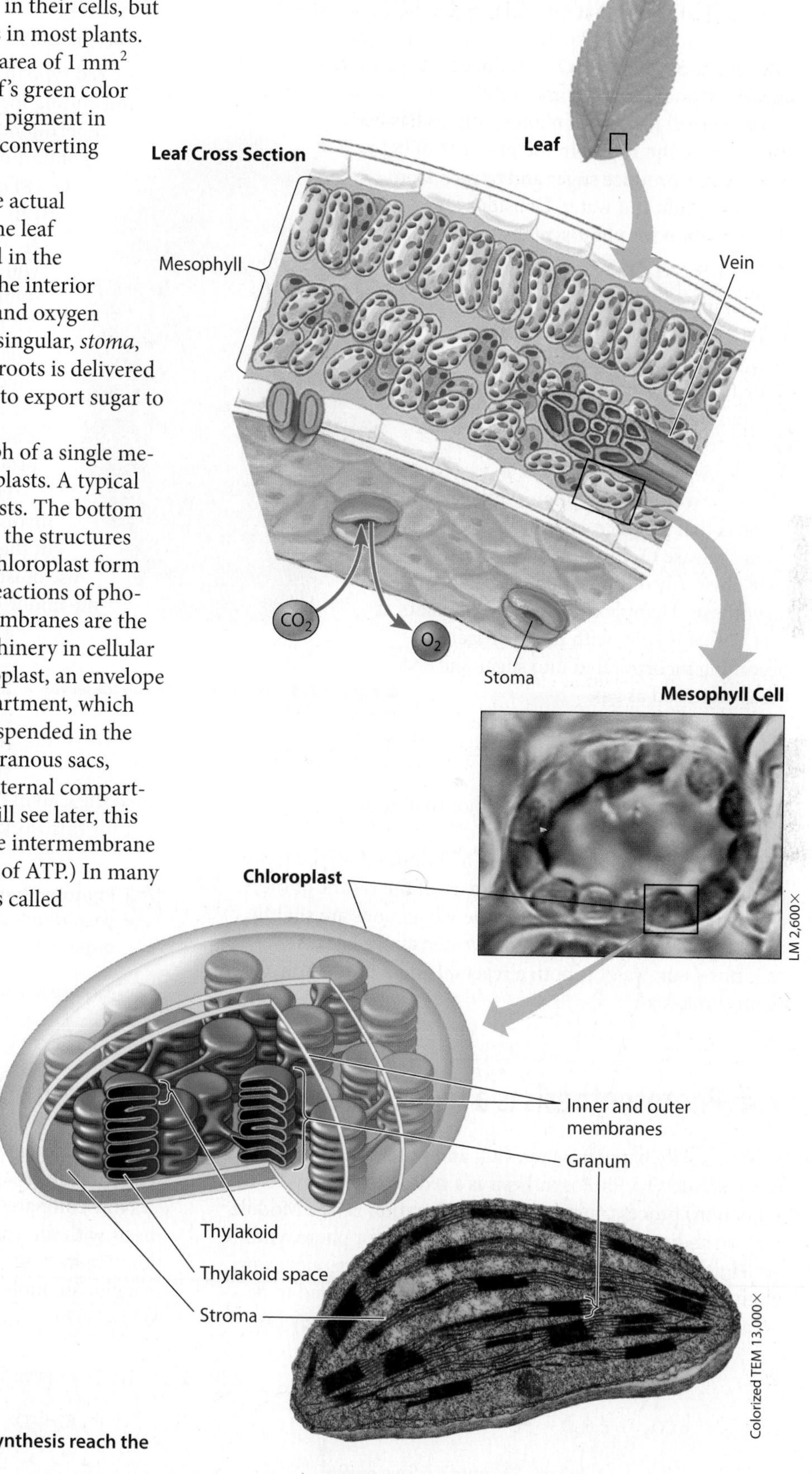

Leaf Cross Section
Leaf
Mesophyll
Vein
CO_2
O_2
Stoma
Mesophyll Cell
Chloroplast
LM 2,600×
Inner and outer membranes
Granum
Thylakoid
Thylakoid space
Stroma
Colorized TEM 13,000×

▲ **Figure 7.2** Zooming in on the location and structure of chloroplasts

? **How do the reactant molecules of photosynthesis reach the chloroplasts in leaves?**

● CO_2 enters leaves through stomata, and H_2O enters the roots and is carried to leaves through veins.

7.3 Scientists traced the process of photosynthesis using isotopes

The leaves of plants that live in lakes and ponds are often covered with bubbles like the ones shown in **Figure 7.3**. The bubbles are oxygen gas (O_2) produced during photosynthesis. But where does this O_2 come from?

The overall process of photosynthesis has been known since the 1800s: In the presence of light, green plants produce sugar and oxygen from carbon dioxide and water. Consider the basic equation for photosynthesis:

$$6\ CO_2 + 6\ H_2O \rightarrow C_6H_{12}O_6 + 6\ O_2$$

Looking at this equation, you can understand why scientists hypothesized that photosynthesis first splits carbon dioxide ($CO_2 \rightarrow C + O_2$), releasing oxygen gas, and then adds water (H_2O) to the carbon to produce sugar. In the 1930s, this idea was challenged by C. B. van Niel, who was working with photosynthesizing bacteria that produce sugar from CO_2 but do not release O_2 in the process. These bacteria obviously did not split CO_2 in their photosynthesis. He hypothesized that in plants, it is H_2O that is split, with the hydrogen becoming incorporated into sugar and the O_2 released as gas.

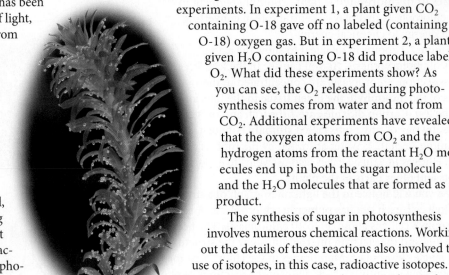

▲ **Figure 7.3** Oxygen bubbles on the leaves of an aquatic plant

In the 1950s, scientists confirmed van Niel's hypothesis by using a heavy isotope of oxygen, O-18, to follow the fate of oxygen atoms during photosynthesis. Isotopes are atoms with differing numbers of neutrons: O-18 has two more neutrons in the nucleus of its atom than the more common isotope O-16. The summary of the results of these experiments follows. (Note that these equations are slightly more detailed than the equation written above because, as it turns out, water is both a reactant and a product in photosynthesis.)

Experiment 1: $6\ CO_2 + 12\ H_2O \rightarrow C_6H_{12}O_6 + 6\ H_2O + 6\ O_2$

Experiment 2: $6\ CO_2 + 12\ H_2O \rightarrow C_6H_{12}O_6 + 6\ H_2O + 6\ O_2$

The red type in the equations above denotes the source and ending location of O-18, the tracer used in these experiments. In experiment 1, a plant given CO_2 containing O-18 gave off no labeled (containing O-18) oxygen gas. But in experiment 2, a plant given H_2O containing O-18 did produce labeled O_2. What did these experiments show? As you can see, the O_2 released during photosynthesis comes from water and not from CO_2. Additional experiments have revealed that the oxygen atoms from CO_2 and the hydrogen atoms from the reactant H_2O molecules end up in both the sugar molecule and the H_2O molecules that are formed as a product.

The synthesis of sugar in photosynthesis involves numerous chemical reactions. Working out the details of these reactions also involved the use of isotopes, in this case, radioactive isotopes. In the mid-1940s, American biochemist Melvin Calvin and his colleagues began using radioactive C-14 to trace the sequence of intermediates formed in the cyclic pathway that produces sugar from CO_2. (See Module 2.3 to review radioactive isotopes.) They worked for 10 years to elucidate this cycle, which is now called the Calvin cycle. Calvin received the Nobel Prize in 1961 for this work.

? Photosynthesis produces billions of tons of carbohydrate a year. Where does most of the mass of this huge amount of organic matter come from?

● Mostly from CO_2 in the air, which provides both the carbon and oxygen in carbohydrate. Water supplies only the hydrogen.

7.4 Photosynthesis is a redox process, as is cellular respiration

What actually happens when CO_2 and water are converted to sugar and O_2? Photosynthesis is a redox (oxidation-reduction) process, just as cellular respiration is (see Module 6.5). As indicated in the summary equation for photosynthesis (**Figure 7.4A**), CO_2 becomes reduced to sugar as electrons, along with hydrogen ions (H^+) from water, are added to it. Meanwhile, water molecules are oxidized; that is, they lose electrons, along with hydrogen ions. Recall that oxidation and reduction always go hand in hand.

Now compare the food-producing equation for photosynthesis with the energy-releasing equation for cellular respiration (**Figure 7.4B**). Cellular respiration harvests energy stored in a glucose molecule by oxidizing the sugar and reducing O_2 to H_2O. This process involves a number of energy-releasing

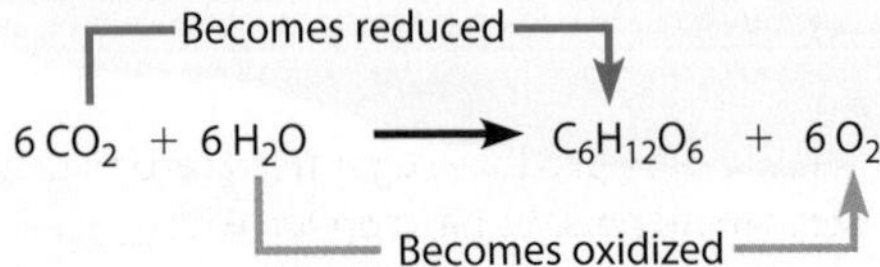
Becomes reduced
6 CO2 + 6 H2O → C6H12O6 + 6 O2
Becomes oxidized

▲ **Figure 7.4A** Photosynthesis (uses light energy)

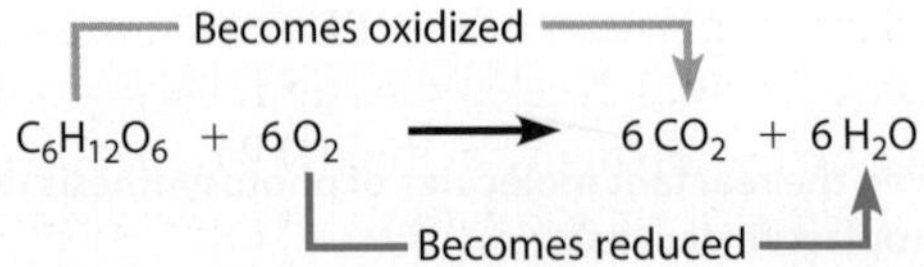
Becomes oxidized
C6H12O6 + 6 O2 → 6 CO2 + 6 H2O
Becomes reduced

▲ **Figure 7.4B** Cellular respiration (releases chemical energy)

redox reactions, with electrons losing potential energy as they are passed down an electron transport chain to O_2. Along the way, the mitochondrion uses some of the energy to synthesize ATP.

In contrast, the food-producing redox reactions of photosynthesis require energy. The potential energy of electrons increases as they move from H_2O to CO_2 during photosynthesis. The light energy captured by chlorophyll molecules in the chloroplast provides this energy boost. Photosynthesis converts light energy to chemical energy and stores it in the chemical bonds of sugar molecules, which can provide energy for later use or raw materials for biosynthesis.

? Which redox process, photosynthesis or cellular respiration, is endergonic? (*Hint*: See Module 5.11.)

● Photosynthesis

7.5 The two stages of photosynthesis are linked by ATP and NADPH

The equation for photosynthesis is a simple summary of a rather complex process. Photosynthesis occurs in two stages, each with multiple steps. Let's begin our study of photosynthesis with an overview of the two stages. **Figure 7.5** shows the inputs and outputs of the light reactions and the Calvin cycle and how these two stages are related.

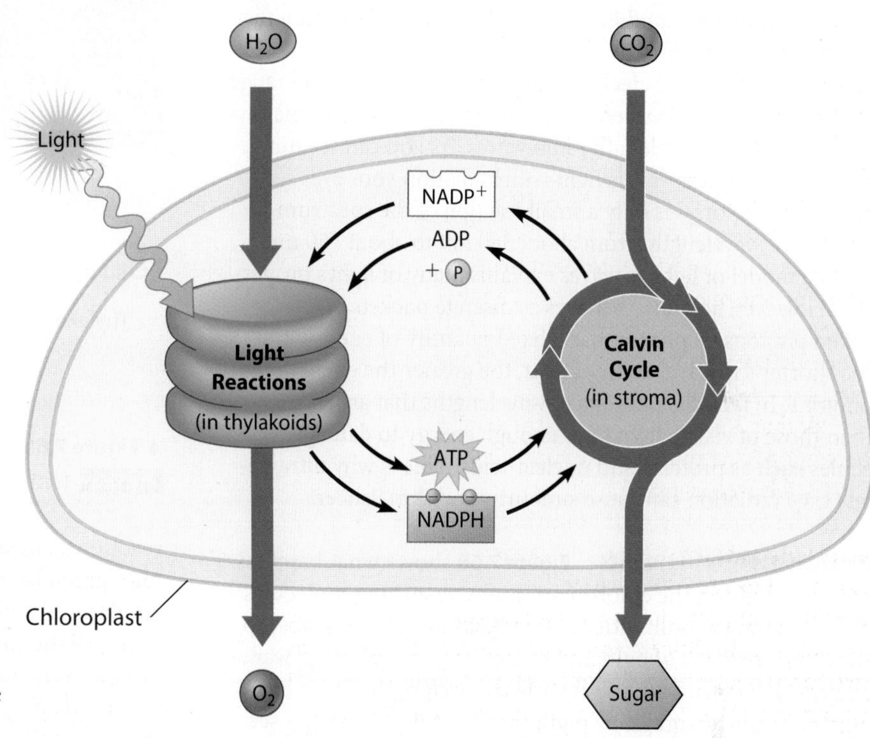
H_2O
CO_2
Light
$NADP^+$
ADP
+ P
Light
Reactions
(in thylakoids)
Calvin
Cycle
(in stroma)
ATP
NADPH
Chloroplast
O_2
Sugar

▲ **Figure 7.5** An overview of the two stages of photosynthesis in a chloroplast

Try This Relate the summary equation for photosynthesis to this overview diagram.

The **light reactions**, which occur in the thylakoids, include the steps that convert light energy to chemical energy and release O_2. Water is split, providing a source of electrons and giving off O_2 as a by-product. Light energy absorbed by chlorophyll molecules built into the thylakoid membranes is used to drive the transfer of electrons and H^+ from water to the electron acceptor $\mathbf{NADP^+}$, reducing it to NADPH. NADPH is first cousin to NADH, which transports electrons in cellular respiration; the two differ only in the extra phosphate group in NADPH. NADPH temporarily stores electrons and hydrogen ions and provides "reducing power" to the Calvin cycle. The light reactions also generate ATP from ADP and a phosphate group.

In summary, the light reactions absorb solar energy and convert it to chemical energy stored in both ATP and NADPH. Notice that these reactions produce no sugar; sugar is not made until the Calvin cycle, which is the second stage of photosynthesis.

The **Calvin cycle** occurs in the stroma of the chloroplast. It is a cyclic series of reactions that assembles sugar molecules using CO_2 and the energy-rich products of the light reactions. The incorporation of carbon from CO_2 into organic compounds, shown in the figure as CO_2 entering the Calvin cycle, is called **carbon fixation**. After carbon fixation, the carbon compounds are reduced to sugars.

As the figure suggests, it is NADPH produced by the light reactions that provides the electrons for reducing carbon compounds in the Calvin cycle. And ATP from the light reactions provides chemical energy that powers several of the steps of the Calvin cycle. The Calvin cycle is sometimes referred to as the dark reactions, or light-independent reactions, because none of the steps requires light directly. However, in most plants, the Calvin cycle occurs during daylight, when the light reactions power the cycle's sugar assembly line by supplying it with NADPH and ATP.

The word *photosynthesis* encapsulates the two stages. *Photo*, from the Greek word for "light," refers to the light reactions; *synthesis*, meaning "putting together," refers to sugar construction by the Calvin cycle. In the next several modules, we look at these two stages in more detail. But first, let's consider some of the properties of light, the energy source that powers photosynthesis.

? For chloroplasts to produce sugar from carbon dioxide in the dark, they would need to be supplied with ____________ and ____________.

● ATP . . . NADPH

▷ The Light Reactions: Converting Solar Energy to Chemical Energy

7.6 Visible radiation absorbed by pigments drives the light reactions

What do we mean when we say that photosynthesis is powered by light energy from the sun?

The Nature of Sunlight Sunlight is a type of energy called electromagnetic energy or radiation. Electromagnetic energy travels in space as rhythmic waves analogous to those made by a pebble dropped in a puddle of water. The distance between the crests of electromagnetic waves is called a **wavelength**. **Figure 7.6A** shows the **electromagnetic spectrum**, the full range of electromagnetic wavelengths from the very short gamma rays to the very long-wavelength radio waves. As you can see in the center of the figure, visible light—the radiation your eyes see as different colors—is only a small fraction of the spectrum. It consists of wavelengths from about 380 nm to about 750 nm.

The model of light as waves explains many of light's properties. However, light also behaves as discrete packets of energy called photons. A **photon** has a fixed quantity of energy, and the shorter the wavelength of light, the greater the energy of its photons. In fact, the photons of wavelengths that are shorter than those of visible light have enough energy to damage molecules such as proteins and nucleic acids. This is why ultraviolet (UV) radiation can cause sunburns and skin cancer.

Photosynthetic Pigments **Figure 7.6B** shows what happens to visible light in the chloroplast. Light-absorbing molecules called pigments, built into the thylakoid membranes, absorb some wavelengths of light and reflect or transmit other wavelengths. We do not see the absorbed wavelengths; their energy has been absorbed by pigment molecules. What we see when we look at a leaf are the green wavelengths that are not absorbed but are transmitted and reflected by the pigments.

Different pigments absorb light of different wavelengths, and chloroplasts contain more than one type of pigment. Chlorophyll *a*, which participates directly in the light reactions, absorbs mainly blue-violet and red light. It looks blue-green because it reflects mainly green light. A very similar molecule, chlorophyll *b*, absorbs mainly blue and orange light and reflects (appears) olive green. Chlorophyll *b* broadens the range of light that a plant can use by conveying absorbed energy to chlorophyll *a*, which then puts the energy to work in the light reactions.

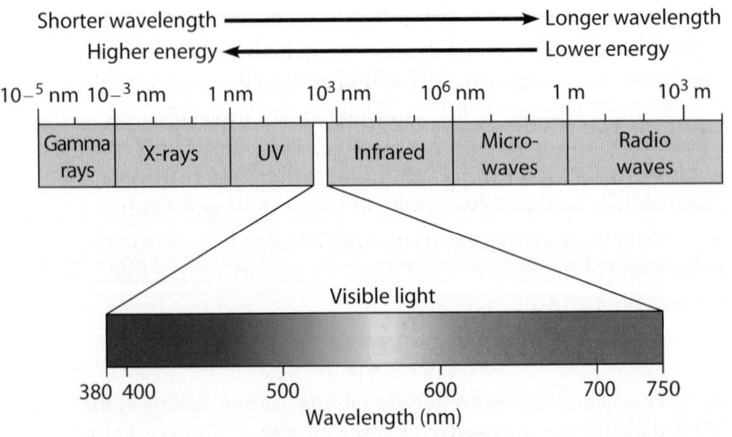

Shorter wavelength
Longer wavelength
Higher energy
Lower energy
10^{-5} nm
10^{-3} nm
1 nm
10^3 nm
10^6 nm
1 m
10^3 m
Gamma rays
X-rays
UV
Infrared
Micro-waves
Radio waves
Visible light
380
400
500
600
700
750
Wavelength (nm)

▲ **Figure 7.6A** The electromagnetic spectrum and the wavelengths of visible light

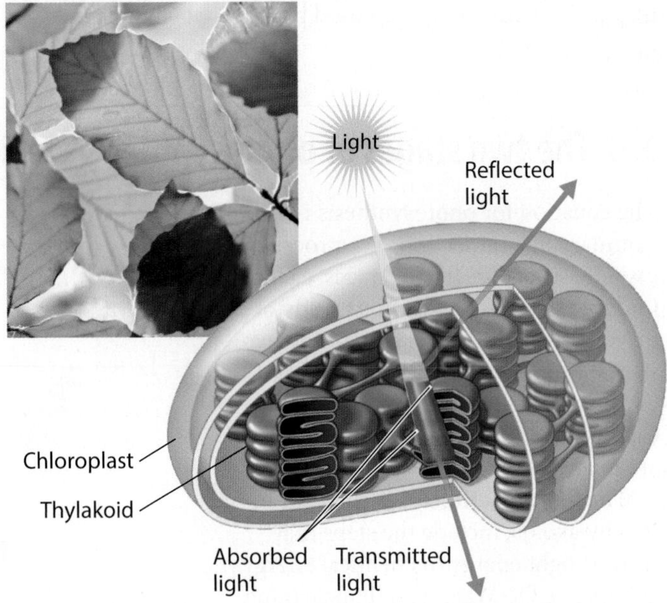

Light
Reflected light
Chloroplast
Thylakoid
Absorbed light
Transmitted light

▲ **Figure 7.6B** The interaction of light with chlorophyll in a chloroplast

Try This Use this diagram to explain why leaves are green.

Chloroplasts also contain pigments called carotenoids, which are various shades of yellow and orange. The spectacular colors of fall foliage in certain parts of the world are due partly to the yellow-orange hues of longer-lasting carotenoids that show through once the green chlorophyll breaks down. Carotenoids may broaden the spectrum of colors that can drive photosynthesis. However, a more important function seems to be photoprotection: Some carotenoids absorb and dissipate excessive light energy that would otherwise damage chlorophyll or interact with oxygen to form reactive oxidative molecules that can damage cell molecules. Similar carotenoids, which we obtain from carrots and other vegetables and fruits, have a photoprotective role in our eyes.

Each type of pigment absorbs certain wavelengths of light because it is able to absorb the specific amounts of energy in those photons. Next we see what happens when a pigment molecule such as chlorophyll absorbs a photon of light.

? **What color of light is least effective at driving photosynthesis? Explain.**

Green, because it is mostly transmitted and reflected—not absorbed—by photosynthetic pigments.

7.7 Photosystems capture solar energy

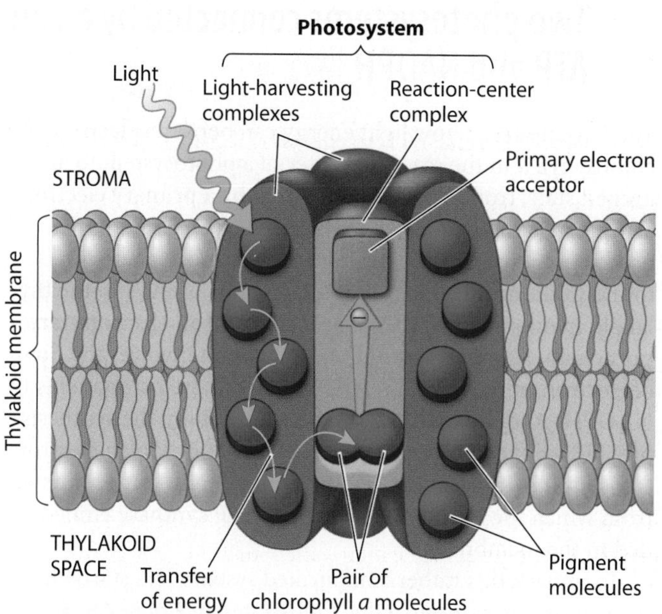

Photosystem
Light
Light-harvesting complexes
Reaction-center complex
Primary electron acceptor
STROMA
Thylakoid membrane
THYLAKOID SPACE
Transfer of energy
Pair of chlorophyll *a* molecules
Pigment molecules

▲ **Figure 7.7B** A light-excited pair of chlorophyll molecules in the reaction center of a photosystem passing an excited electron to a primary electron acceptor

Energy cannot be created or destroyed, but it can be transferred or transformed (see Module 5.10). Let's examine how light energy can be transformed to other types of energy. When a pigment molecule absorbs a photon of light, one of the pigment's electrons jumps to an energy level farther from the nucleus. In this location, the electron has more potential energy, and we say that the electron has been raised from a ground state to an excited state. The excited state, like all high-energy states, is unstable. Generally, when isolated pigment molecules absorb light, their excited electrons drop back down to the ground state in a billionth of a second, releasing their excess energy as heat. This conversion of light energy to heat is what makes a black car so hot on a sunny day (black pigments absorb all wavelengths of light).

Some isolated pigments, including chlorophyll, emit light as well as heat after absorbing photons. We can demonstrate this phenomenon in the laboratory with a chlorophyll solution, as shown on the left in **Figure 7.7A**. When brightly illuminated, the chlorophyll emits photons of light that produce a reddish afterglow called fluorescence. The right side of Figure 7.7A illustrates what happens in fluorescence: An absorbed photon boosts an electron of chlorophyll to an excited state, from which it falls back to the ground state, emitting its energy as heat and light.

But chlorophyll behaves very differently in isolation than it does in an intact chloroplast. In their native habitat of the thylakoid membrane, chlorophyll and other pigments that absorb photons transfer the energy to other pigment molecules and eventually to a special pair of chlorophyll molecules. This pair passes off an excited electron to a neighboring molecule before it has a chance to drop back to the ground state.

In the thylakoid membrane, chlorophyll molecules are organized along with other pigments and proteins into clusters called photosystems (**Figure 7.7B**). A **photosystem** consists of a number of light-harvesting complexes surrounding a reaction-center complex. A light-harvesting complex contains various pigment molecules bound to proteins. Collectively, the light-harvesting complexes function as a light-gathering antenna. The pigments absorb photons and pass the energy from molecule to molecule, somewhat like a human "wave" at a sporting event, until it reaches the reaction center. The reaction-center complex contains a pair of special chlorophyll *a* molecules and a molecule called the primary electron acceptor, which is capable of accepting electrons and becoming reduced. The solar-powered transfer of an electron from the reaction-center chlorophyll *a* pair to the primary electron acceptor is the first step in the transformation of light energy to chemical energy in the light reactions.

Two types of photosystems have been identified, and they cooperate in the light reactions. They are referred to as photosystem I and photosystem II, in order of their discovery, although photosystem II actually functions first in the sequence of steps that make up the light reactions. Each photosystem has a characteristic reaction-center complex, with a special pair of chlorophyll *a* molecules associated with a particular primary electron acceptor. Now let's see how the two photosystems work together in the light reactions to generate ATP and NADPH.

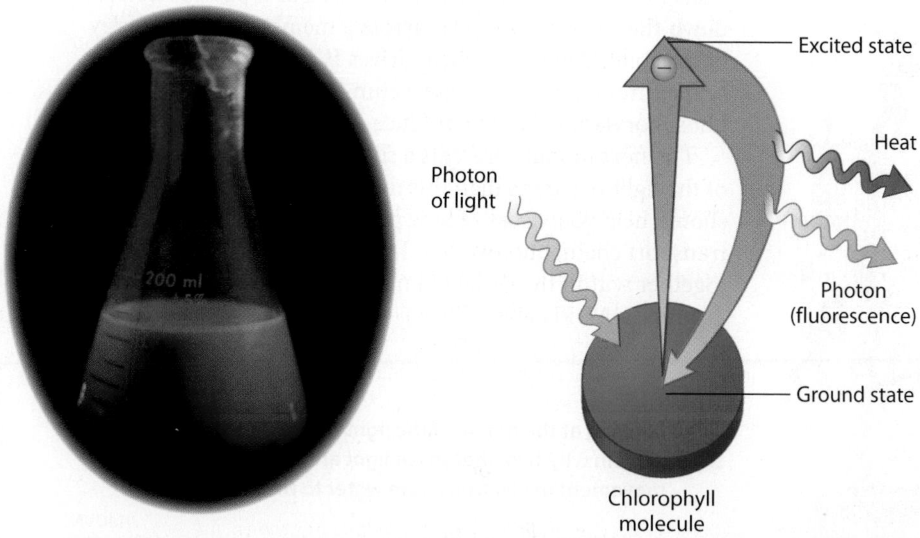

200 ml
Excited state
Heat
Photon of light
Photon (fluorescence)
Ground state
Chlorophyll molecule

▲ **Figure 7.7A** A solution of chlorophyll glowing red when illuminated (left); a diagram of an isolated, light-excited chlorophyll molecule that releases heat and a photon of red light when it falls back to ground state (right)

? Compared with a solution of isolated chlorophyll, why do intact chloroplasts release less heat and fluorescence when illuminated?

● In the chloroplasts, a light-excited electron from the reaction-center chlorophyll molecules is trapped by a primary electron acceptor rather than giving up its energy as heat and light.

7.8 Two photosystems connected by an electron transport chain generate ATP and NADPH

You have just seen how light energy can boost an electron of chlorophyll *a* in the reaction center of a photosystem to an excited state, from which it is captured by a primary electron acceptor. But how do these captured electrons lead to the production of ATP and NADPH? Part of the explanation is found in the arrangement of photosystems II and I in the thylakoid membrane and their connection via an electron transport chain. Another part of the explanation involves the flow of electrons removed from H_2O through these components to NADPH. And the final part of the explanation, the synthesis of ATP, is linked (as it is in cellular respiration) to an electron transport chain pumping H^+ into a membrane compartment, from which the ions flow through an ATP synthase embedded in the membrane.

To unpack this rather complicated system, let's start with the simple mechanical analogy illustrated in **Figure 7.8**. Starting on the left, you see that the large yellow photon mallet provides the energy to boost an electron from photosystem II to a higher energy level, where it is caught by the primary electron acceptor standing on the platform. The electron is loaded onto an electron transport chain "ramp" leading to photosystem I. (Recall that photosystem II precedes photosystem I in the light reactions.) As electrons roll down the ramp, they release energy that is used for the production of ATP. When an electron reaches photosystem I, another photon mallet pumps it up to a higher energy level, where it is caught by a primary electron acceptor on the photosystem I platform. From there, the photoexcited electrons are thrown into a bucket to produce NADPH. This construction analogy shows how the coupling of two photosystems and an electron transport chain can transform the energy of light to the chemical energy of ATP and NADPH.

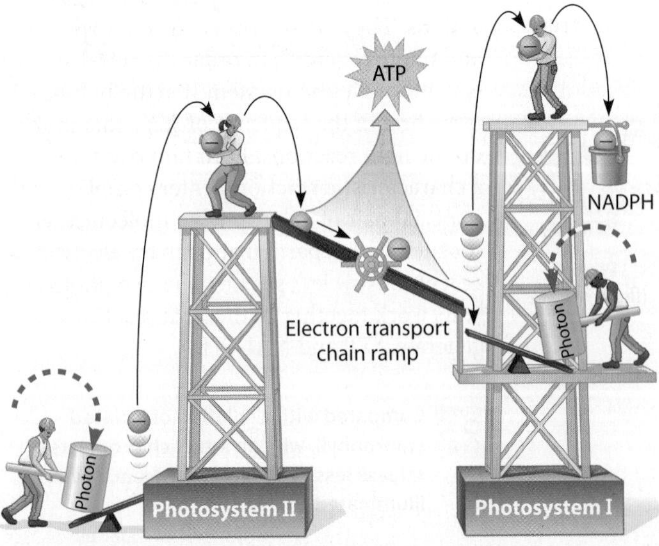
ATP
NADPH
Electron transport chain ramp
Photon
Photon
Photosystem II
Photosystem I

▲ **Figure 7.8** A mechanical analogy of the light reactions

The simple analogy in Figure 7.8 does leave a few important unanswered questions: What is the source of the electrons that are moving through the photosystems to NADPH? Don't the light reactions produce O_2—where does that happen? And how does the flow of electrons down that ramp produce ATP?

The electrons that end up reducing $NADP^+$ to NADPH originally come from water. An enzyme in the thylakoid space splits H_2O into 2 electrons, 2 hydrogen ions (H^+), and 1 oxygen atom ($\frac{1}{2}$ O_2). The H^+ stay in the thylakoid space. The oxygen atom immediately joins with another oxygen to form O_2. As you learned in Module 7.3, water is the source of the O_2 produced in photosynthesis, and these oxygen molecules diffuse out of the thylakoids, the chloroplast, and the plant cell, finally exiting the leaf through its stomata. The all-important electrons from water are passed, one by one, to the reaction center chlorophyll *a* molecules in photosystem II, replacing the photoexcited electron that was just captured by the primary electron acceptor. From photosystem II, the electrons pass through an electron transport chain to the reaction center chlorophyll *a* molecules in photosystem I, again replacing photoexcited electrons that had been captured by its primary electron acceptor. Although the illustration shows these electrons being dropped in a bucket, they actually are passed through a short electron transport chain to $NADP^+$, reducing it to NADPH.

Now that we have accounted for NADPH and O_2, all that is left is ATP. Making ATP in the light reactions involves an electron transport chain and chemiosmosis—the same players and process you met in the synthesis of ATP in cellular respiration. Recall that in chemiosmosis, the potential energy of a concentration gradient of H^+ across a membrane powers ATP synthesis. This gradient is created when an electron transport chain uses the energy released as it passes electrons down the chain to pump H^+ across a membrane. The energy of the concentration gradient drives H^+ back across the membrane through ATP synthase, spinning this rotary motor and phosphorylating ADP to produce ATP (see Figure 6.10B).

The next module presents a slightly more realistic model of the light reactions than this mechanical analogy, which should help you visualize how photosystem II, the electron transport chain, photosystem I, and ATP synthase function together within the thylakoid membranes of a chloroplast to produce NADPH and ATP.

? Looking at the model of the light reactions in Figure 7.8, explain why two photons of light are required in the movement of electrons from water to NADPH.

● One photon excites an electron from photosystem II, which is then passed down an electron transfer chain to photosystem I. A second photon excites an electron from photosystem I, which is then used in the reduction of $NADP^+$ to NADPH.

VISUALIZING THE CONCEPT

7.9 The light reactions take place within the thylakoid membranes

The diagram below shows the relationship between chloroplast structure and function in the light reactions.

Depicted here is a small portion of a thylakoid sac showing how the two photosystems and electron transport chain are embedded in a thylakoid membrane. All of the components shown here are present in numerous copies in each thylakoid. Moving from left to right, you can see how light energy absorbed by the two photosystems drives the flow of electrons from water to NADPH. The electron transport chain helps to produce the concentration gradient of H^+ across the thylakoid membrane, which drives H^+ through ATP synthase, producing ATP. Because the initial energy input is light (*photo-*), this chemiosmotic production of ATP is called **photophosphorylation**.

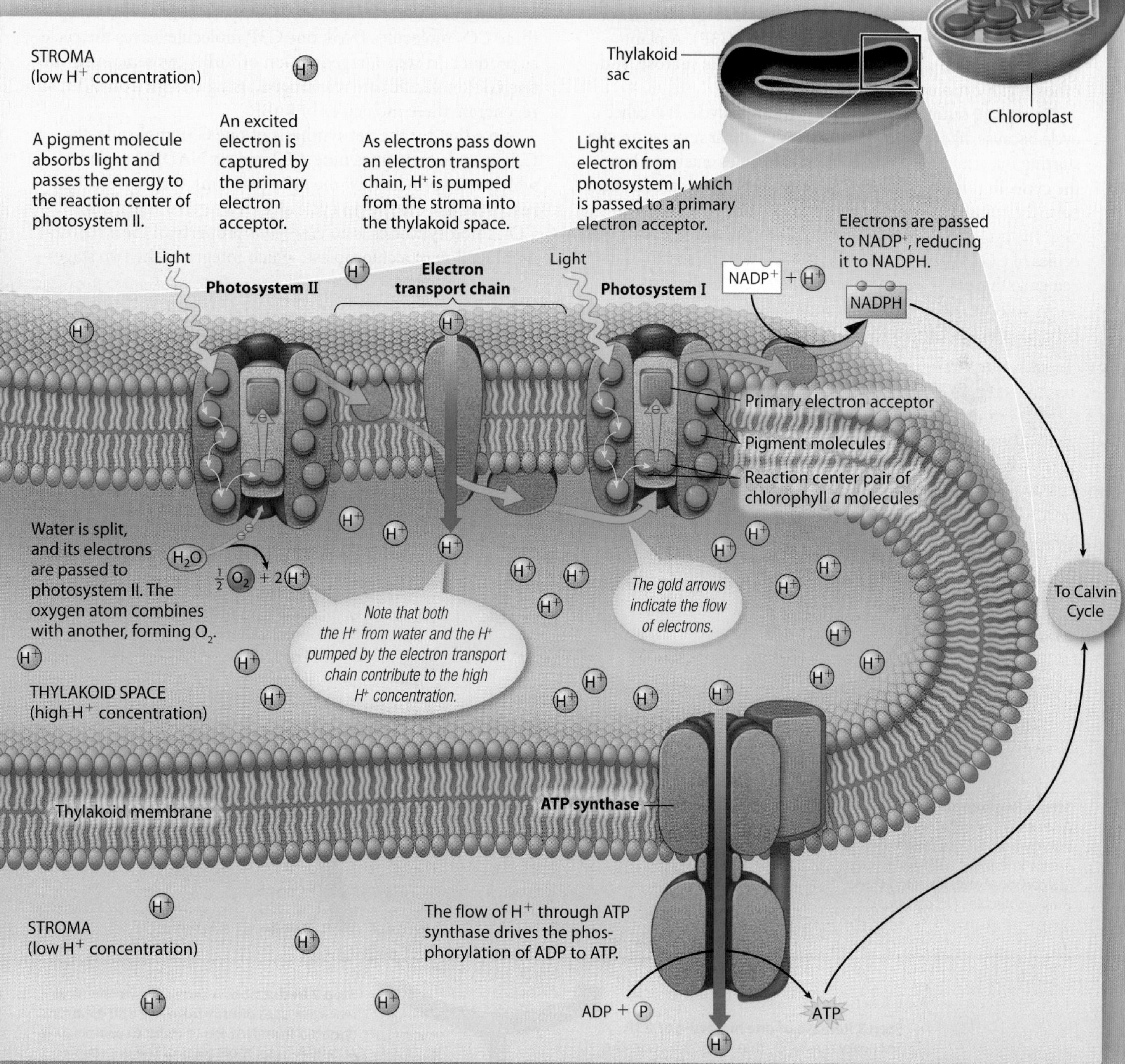

STROMA
(low H^+ concentration)
Thylakoid sac
Chloroplast
A pigment molecule absorbs light and passes the energy to the reaction center of photosystem II.
An excited electron is captured by the primary electron acceptor.
As electrons pass down an electron transport chain, H^+ is pumped from the stroma into the thylakoid space.
Light excites an electron from photosystem I, which is passed to a primary electron acceptor.
Electrons are passed to $NADP^+$, reducing it to NADPH.
Light
Photosystem II
Electron transport chain
Light
Photosystem I
$NADP^+$ + H^+
NADPH
Primary electron acceptor
Pigment molecules
Reaction center pair of chlorophyll *a* molecules
Water is split, and its electrons are passed to photosystem II. The oxygen atom combines with another, forming O_2.
H_2O
$\frac{1}{2}O_2 + 2H^+$
Note that both the H^+ from water and the H^+ pumped by the electron transport chain contribute to the high H^+ concentration.
The gold arrows indicate the flow of electrons.
To Calvin Cycle
THYLAKOID SPACE
(high H^+ concentration)
Thylakoid membrane
ATP synthase
STROMA
(low H^+ concentration)
The flow of H^+ through ATP synthase drives the phosphorylation of ADP to ATP.
ADP + P
ATP

? **What is the advantage of the light reactions producing NADPH and ATP on the stroma side of the thylakoid membrane?**

● The Calvin cycle, which uses the NADPH and ATP, occurs in the stroma.

▷ The Calvin Cycle: Reducing CO_2 to Sugar

7.10 ATP and NADPH power sugar synthesis in the Calvin cycle

The Calvin cycle functions like a sugar factory within a chloroplast. The inputs to this all-important food-making process are CO_2 (from the air) and ATP and NADPH (both generated by the light reactions). ATP is used as an energy source and NADPH provides high-energy electrons for reducing CO_2 to sugar. The output of the Calvin cycle is an energy-rich, three-carbon sugar, glyceraldehyde 3-phosphate (G3P). A plant cell uses G3P to make glucose, the disaccharide sucrose, and other organic molecules as needed.

Figure 7.10 outlines the steps of the Calvin cycle. It is called a cycle because, like the citric acid cycle in cellular respiration, the starting material is regenerated after molecules enter and leave the cycle. In this case, the starting material is a five-carbon sugar named ribulose bisphosphate (RuBP). To make a molecule of G3P, the cycle must turn three times, incorporating three molecules of CO_2. We show the cycle starting with three CO_2 molecules so that we end up with a complete G3P molecule.

As you can see in step 1, carbon fixation, the enzyme rubisco attaches CO_2 to RuBP. (Recall that carbon fixation refers to the initial incorporation of CO_2 into organic compounds.) This unstable six-carbon molecule splits into two three-carbon molecules. In step 2, reduction, ATP and NADPH are used to reduce the three-carbon molecule to G3P.

For this to be a cycle, RuBP must be regenerated. In step 3, release of one molecule of G3P, you can see that for every three CO_2 molecules fixed, one G3P molecule leaves the cycle as product. In step 4, regeneration of RuBP, the remaining five G3P molecules are rearranged, using energy from ATP, to regenerate three molecules of RuBP.

Note that for the net synthesis of one G3P molecule, the Calvin cycle consumes nine ATP and six NADPH molecules, which were provided by the light reactions. Neither the light reactions nor the Calvin cycle alone can make sugar from CO_2. Photosynthesis is an emergent property of the structural organization of a chloroplast, which integrates the two stages of photosynthesis.

? To synthesize one glucose molecule, the Calvin cycle uses __________ CO_2, __________ ATP, and __________ NADPH. Explain why this high number of ATP and NADPH molecules is consistent with the value of glucose as an energy source.

● 6 . . . 18 . . . 12. Glucose is a highly reduced molecule, storing lots of potential energy in its electrons. The more energy a molecule stores, the more energy and reducing power required to produce that molecule.

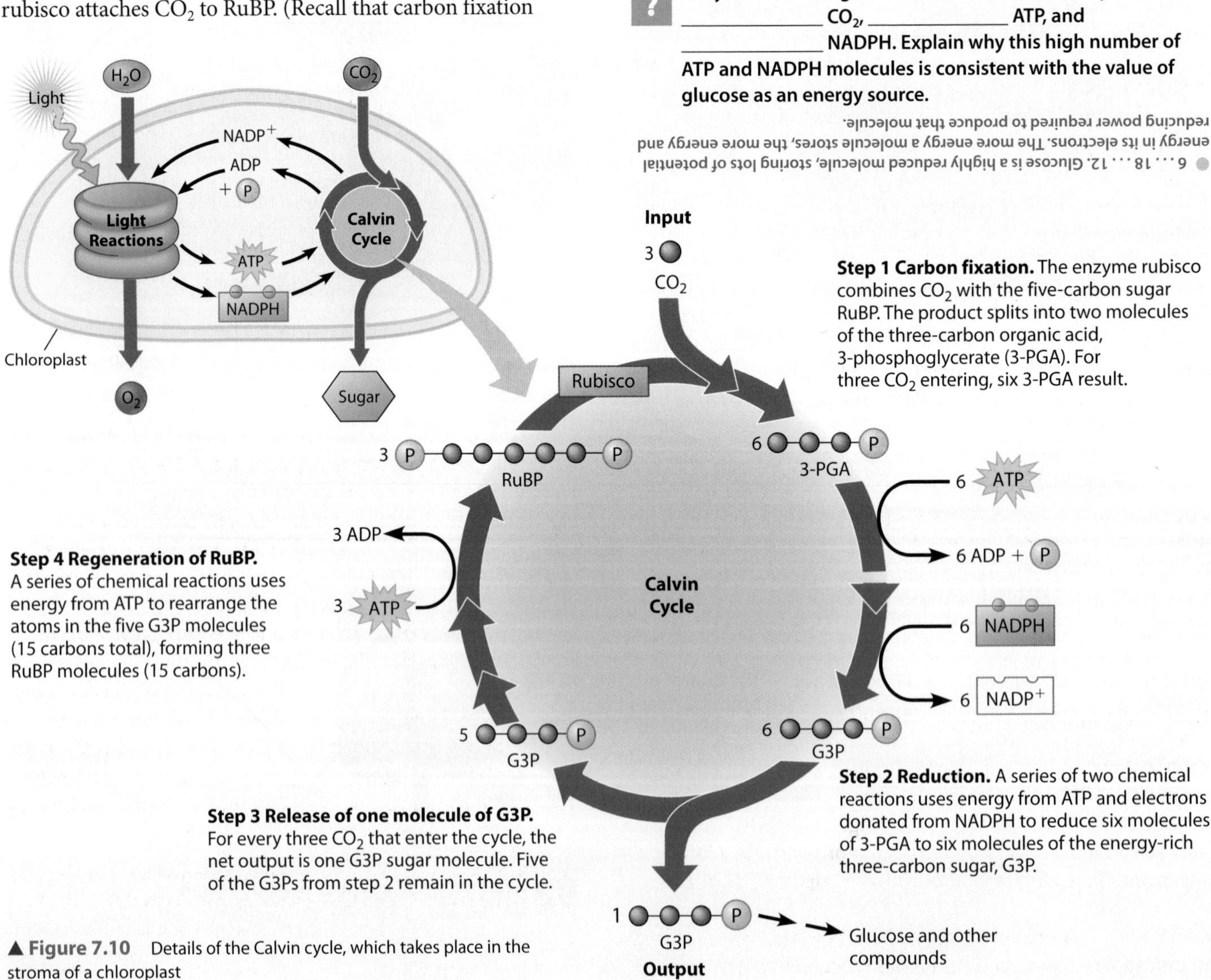

Light
H_2O
CO_2
NADP+
ADP
+ P
Light Reactions
Calvin Cycle
ATP
NADPH
Chloroplast
O_2
Sugar
Input
3
CO_2
Step 1 Carbon fixation. The enzyme rubisco combines CO_2 with the five-carbon sugar RuBP. The product splits into two molecules of the three-carbon organic acid, 3-phosphoglycerate (3-PGA). For three CO_2 entering, six 3-PGA result.
Rubisco
3 P P
RuBP
6 P
3-PGA
6 ATP
6 ADP + P
6 NADPH
6 NADP+
Calvin Cycle
3 ADP
3 ATP
Step 4 Regeneration of RuBP. A series of chemical reactions uses energy from ATP to rearrange the atoms in the five G3P molecules (15 carbons total), forming three RuBP molecules (15 carbons).
5 P
G3P
6 P
G3P
Step 2 Reduction. A series of two chemical reactions uses energy from ATP and electrons donated from NADPH to reduce six molecules of 3-PGA to six molecules of the energy-rich three-carbon sugar, G3P.
Step 3 Release of one molecule of G3P. For every three CO_2 that enter the cycle, the net output is one G3P sugar molecule. Five of the G3Ps from step 2 remain in the cycle.
1 P
G3P
Output
Glucose and other compounds

▲ **Figure 7.10** Details of the Calvin cycle, which takes place in the stroma of a chloroplast

7.11 Other methods of carbon fixation have evolved in hot, dry climates

EVOLUTION CONNECTION

As you learned in the previous module, the first step of the Calvin cycle is carbon fixation. Most plants use CO_2 directly from the air, and carbon fixation occurs when the enzyme rubisco adds CO_2 to RuBP (see step 1 of Figure 7.10). Such plants are called **C_3 plants** because the first product of carbon fixation is the three-carbon compound 3-PGA. C_3 plants are widely distributed; they include such important agricultural crops as soybeans, oats, wheat, and rice. One problem that farmers face in growing C_3 plants is that hot, dry weather can decrease crop yield. In response to such conditions, plants close their stomata, the pores in their leaves. This adaptation reduces water loss and helps prevent dehydration, but it also prevents CO_2 from entering the leaf and O_2 from exiting. As a result, CO_2 levels get very low in the leaf and photosynthesis slows. And the O_2 released from the light reactions begins to accumulate, creating another problem.

As O_2 builds up in a leaf, rubisco adds O_2 instead of CO_2 to RuBP. A two-carbon product of this reaction is then broken down in the cell. This process is called **photorespiration** because it occurs in the light and, like respiration, it consumes O_2 and releases CO_2. But unlike cellular respiration, it uses ATP instead of producing it; and unlike photosynthesis, it yields no sugar. Photorespiration can, however, drain away as much as 50% of the carbon fixed by the Calvin cycle.

According to one hypothesis, photorespiration is an evolutionary relic from when the atmosphere had less O_2 than it does today. In the ancient atmosphere that prevailed when rubisco first evolved, the inability of the enzyme's active site to exclude O_2 would have made little difference. It is only after O_2 became so concentrated in the atmosphere that the "sloppiness" of rubisco presented a problem. New evidence also indicates that photorespiration may play a protective role when the products of the light reactions build up in a cell (as occurs when the Calvin cycle slows due to a lack of CO_2).

C_4 Plants In some plant species found in hot, dry climates, alternate modes of carbon fixation have evolved that minimize photorespiration and optimize the Calvin cycle. **C_4 plants** are so named because they first fix CO_2 into a four-carbon compound. When the weather is hot and dry, a C_4 plant keeps its stomata mostly closed, thus conserving water. It continues making sugars by photosynthesis using the pathway and the two types of cells shown on the left side of **Figure 7.11**. An enzyme in the mesophyll cells has a high affinity for CO_2 and can fix carbon even when the CO_2 concentration in the leaf is low. The resulting four-carbon compound then acts as a carbon shuttle; it moves into bundle-sheath cells, which are packed around the veins of the leaf, and releases CO_2. Thus, the CO_2 concentration in these cells remains high enough for the Calvin cycle to make sugars and avoid photorespiration. Corn and sugarcane are examples of agriculturally important C_4 plants.

CAM Plants A second photosynthetic adaptation has evolved in pineapples, many cacti, and other succulent (water-storing) plants, such as aloe and jade plants. Called **CAM plants**, these species are adapted to very dry climates. A CAM plant (right side of Figure 7.11) conserves water by opening its stomata and admitting CO_2 only at night. CO_2 is fixed into a four-carbon compound, which banks CO_2 at night and releases it during the day. Thus, the Calvin cycle can operate, even with the leaf's stomata closed during the day.

In C_4 plants, carbon fixation and the Calvin cycle occur in different types of cells. In CAM plants, these processes occur in the same cells, but at different times of the day. Keep in mind that CAM, C_4, and C_3 plants all eventually use the Calvin cycle to make sugar from CO_2. The C_4 and CAM pathways are two evolutionary adaptations that minimize photorespiration and maximize photosynthesis in hot, dry climates.

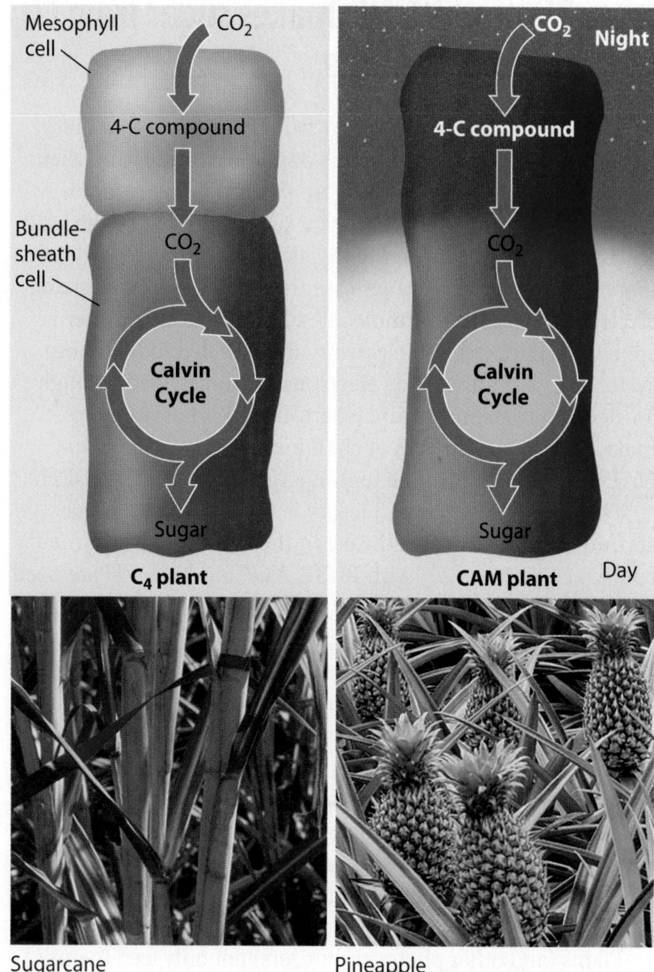

Mesophyll cell
CO₂
4-C compound
Bundle-sheath cell
CO₂
Calvin Cycle
Sugar
C₄ plant
CO₂
Night
4-C compound
CO₂
Calvin Cycle
Sugar
CAM plant
Day
Sugarcane
Pineapple

▲ **Figure 7.11** Adaptations for photosynthesis in hot, dry climates

Try This Use these diagrams to explain the differences between C_4 and CAM photosynthesis.

? **Why would you expect photorespiration on a hot, dry day to occur less in C_4 and CAM plants than in C_3 plants?**

● Because of their initial fixing of carbon, both C_4 and CAM plants can supply rubisco with CO_2. When a C_3 plant closes its stomata, CO_2 levels drop and O_2 rises, making it more likely that rubisco will add O_2 to RuBP.

▷ The Global Significance of Photosynthesis

7.12 Photosynthesis makes sugar from CO_2 and H_2O, providing food and O_2 for almost all living organisms

Now that we have made our way from photons to food, let's step back and review the process of photosynthesis and then discuss its importance. Starting on the left of the overview diagram shown in **Figure 7.12**, you see a summary of the light reactions, which occur in the thylakoid membranes. Two photosystems in the membranes capture solar energy, energizing electrons in chlorophyll molecules. Simultaneously, water is split, O_2 is released, and electrons are funneled to the photosystems. The photoexcited electrons are transferred through an electron transport chain, where energy is harvested to make ATP by the process of chemiosmosis, and finally to $NADP^+$, reducing it to the high-energy compound NADPH.

The chloroplast's sugar factory is the Calvin cycle, the second stage of photosynthesis. In the stroma, the enzyme rubisco combines CO_2 with RuBP. ATP and NADPH are used to reduce 3-PGA to G3P. Sugar molecules made from G3P serve as a plant's own food supply.

About 50% of the carbohydrate made by photosynthesis is consumed as fuel for cellular respiration in the mitochondria of plant cells. Sugars also serve as starting material for making other organic molecules, such as a plant's proteins and lipids. Many glucose molecules are linked together to make cellulose, the main component of cell walls. Cellulose is the most abundant organic molecule in a plant—and probably on the surface of the planet. Most plants make much more food each day than they need. They store the excess in roots, tubers, seeds, and fruits.

Plants (and other photosynthesizers) not only feed themselves but also are the ultimate source of food for virtually all other organisms. Humans and other animals make none of their own food and are totally dependent on the organic matter made by photosynthesizers. Even the energy we acquire when we eat meat was originally captured by photosynthesis. The energy in a steak, for instance, came from sunlight that was originally converted to a chemical form in the grasses eaten by cattle.

The collective productivity of the tiny chloroplasts is truly amazing: Photosynthesis makes an estimated 150 billion metric tons of carbohydrate per year (about 165 billion tons). That's equivalent in mass to a stack of about 100 trillion copies of this textbook—10 stacks of books reaching from Earth to the sun!

The products of photosynthesis provide us with more than just food. For most of human history, burning plant material has been a major source of heat, light, and cooking fuel. The use of fossil fuels is a relatively recent development, and these sources of energy come from the remains of ancient organisms that removed CO_2 from the atmosphere by photosynthesis over the course of hundreds of millions of years. The burning of these ancient carbon stores is increasing the atmospheric level of CO_2, which has risen more than 40% since 1850, the start of the Industrial Revolution, changing the global climate and affecting current-day photosynthesizers. In the next module, we explore how scientists study the effects of rising CO_2 levels on plants.

? Explain this statement: No process is more important to the welfare of life on Earth than photosynthesis.

● Photosynthesis is the ultimate source of the food for almost all organisms and the oxygen they need for cellular respiration.

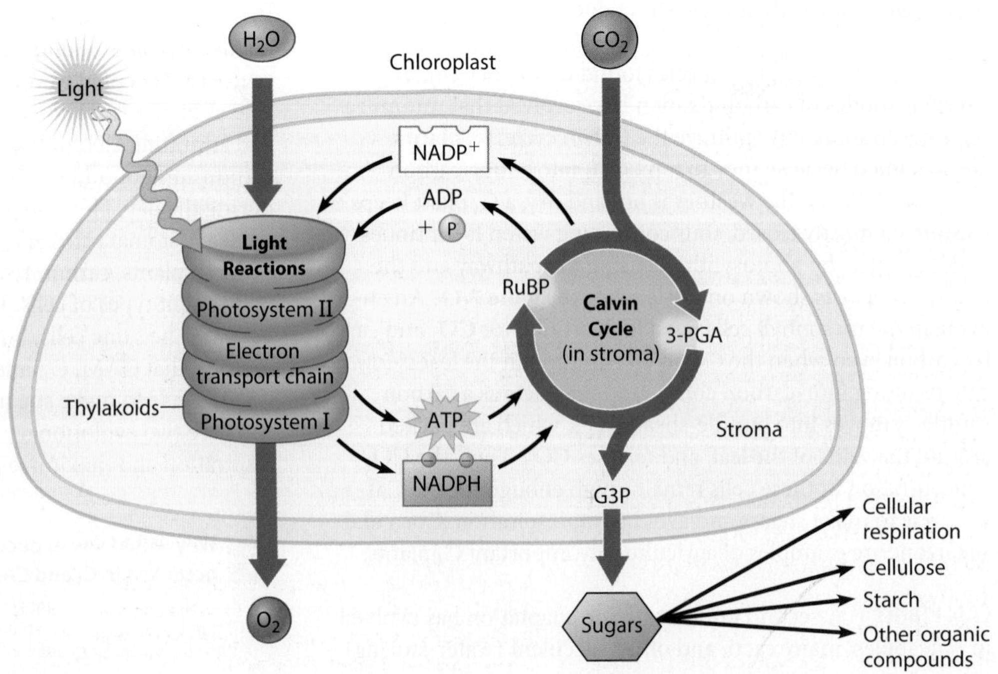

H_2O
CO_2
Chloroplast
Light
$NADP^+$
ADP
+ P
Light Reactions
Photosystem II
Electron transport chain
Thylakoids
Photosystem I
ATP
NADPH
RuBP
Calvin Cycle (in stroma)
3-PGA
Stroma
G3P
O_2
Sugars
Cellular respiration
Cellulose
Starch
Other organic compounds

▶ **Figure 7.12** A summary of photosynthesis

7.13 Rising atmospheric levels of carbon dioxide and global climate change will affect plants in various ways

SCIENTIFIC THINKING

How is the increase in atmospheric carbon dioxide affecting Earth's climate? First, let's consider the role of CO_2 as a so-called greenhouse gas. As you probably know, greenhouses are used to grow plants when the weather outside is too cold. Solar radiation can pass through their transparent walls, and much of the heat that accumulates inside is trapped.

An analogous process, called the **greenhouse effect**, operates on a global scale. Solar radiation passes through the atmosphere and warms Earth's surface. Heat radiating from the warmed planet is absorbed by greenhouse gases, such as CO_2, water vapor, and methane, which then reflect some of the heat back to Earth. Without this natural heating effect, the average air temperature would be a frigid −18°C (−0.4°F), and most life as we know it could not exist. But this insulating blanket of greenhouse gases is starting to warm Earth *too* much.

Increasing concentrations of greenhouse gases have been linked to **global climate change**, of which one major aspect is global warming. The predicted consequences of global climate change include melting of polar ice, rising sea levels, extreme weather patterns, droughts, increased extinction rates, and the spread of tropical diseases. Indeed, many of these effects are already being documented.

How may global climate change affect plants? You might predict that, as a raw material for photosynthesis, increasing CO_2 levels would increase plant productivity. Indeed, research has documented such an increase, although results often indicate that the growth rates of weeds, such as the poison ivy described in the chapter introduction, increase more than those of crop plants and trees.

How do scientists study the effects of increasing CO_2 on plants? As is so often the case, scientists use different types of experiments to test their hypotheses. Many experiments are done in small growth chambers in which variables can be carefully controlled. But the availability of facilities and resources often limits such studies in scope and length. Some creative researchers have made use of study areas that naturally vary in CO_2 levels, such as comparing plant diversity and growth in experimental plots set in urban, suburban, and country locations.

Other scientists are turning to long-term field studies that include large-scale manipulations of CO_2 levels. In the Free-Air CO_2 Enrichment (FACE) experiment set up in Duke University's experimental forest, scientists monitored the effects of elevated CO_2 levels on an intact forest ecosystem over a period of 15 years. Six study sites were established, each 30 m in diameter and ringed by 16 towers (**Figure 7.13A**). In three of the plots, the towers released air containing CO_2 concentrations about $1\frac{1}{2}$ times present-day levels. Monitoring instruments on a tall tower in the center of each plot adjusted the distribution of CO_2 to maintain a stable concentration. All other factors, such as temperature, precipitation, and wind patterns, varied normally for both experimental plots and adjacent control plots.

▲ **Figure 7.13A** Large-scale experiment in the Duke University Experimental Forest on the effects of elevated CO_2 concentration. (Rings of towers emit CO_2-enriched air in three of the plots.)

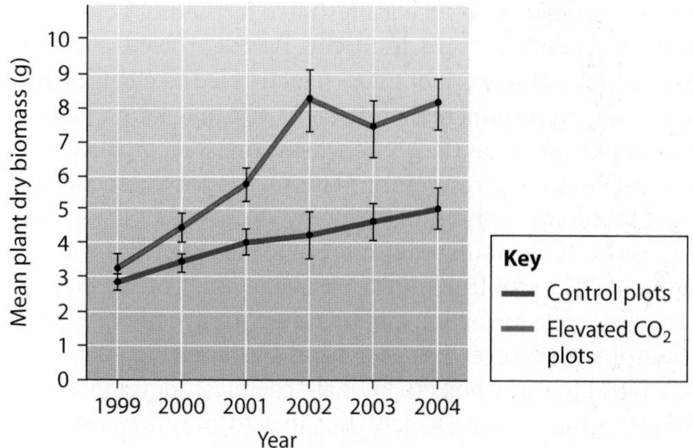
Mean plant dry biomass (g)
10
9
8
7
6
5
4
3
2
1
0
1999 2000 2001 2002 2003 2004
Year
Key
Control plots
Elevated CO2 plots

Source: Adaptation of Figure 1A from "Biomass and Toxicity Responses of Poison Ivy (*Toxicodendron Radicans*) to Elevated Atmospheric CO_2," by Jacqueline E. Mohan, et al., from *PNAS*, June 2006, Volume 103(24). Copyright © 2006 by National Academy of Sciences. Reprinted with permission.

▲ **Figure 7.13B** The mean poison ivy biomass in control plots and elevated CO_2 plots (with error bars showing the variation around the mean)

Figure 7.13B shows some results from a study that compared the growth of poison ivy in experimental and control plots. The poison ivy in the elevated CO_2 plots showed an average annual growth increase of 149% compared to control plots. This increase is much greater than the increase for woody plants that similar studies have documented. Indeed, over a 12-year monitoring period, the trees in the FACE experimental plots produced only about 15% more wood per year than those in the control plots.

Will global climate change make you itch?

There was one other significant finding of the poison ivy study. A chemical analysis showed that the high-CO_2 plants produced a more potent form of poison ivy's allergenic compound. Thus, poison ivy is predicted to become both more abundant and more toxic ("itchy") as CO_2 levels rise.

? **Describe three research methods that scientists use to test the hypothesis that increasing CO_2 levels will affect the growth of plants.**

● Laboratory growth chambers, field studies in areas where CO_2 levels vary naturally, and large-scale field studies in which CO_2 levels are manipulated

7.14 Scientific research and international treaties have helped slow the depletion of Earth's ozone layer

The importance of science is illustrated by the story of how synthetic chemicals were destroying Earth's protective ozone layer and how the work of many scientists led to changes in worldwide environmental policies. As you now know, photosynthesis produces the O_2 on which almost all organisms depend for cellular respiration. This O_2 has another benefit: High in the atmosphere, high-energy solar radiation converts it to ozone (O_3). Acting as sunscreen for the planet, the ozone layer shields Earth from ultraviolet radiation. The balance between ozone formation and its natural destruction in the atmosphere, however, has been upset by human actions.

Chlorofluorocarbons (CFCs) are chemicals developed in the 1930s that became widely used in aerosol sprays, refrigerators, and Styrofoam production. In 1970, a scientist wondered whether CFCs were accumulating in the environment and sent a homemade detector on a boat trip to Antarctica. He found CFCs in the air all along the journey. When he reported his findings at a scientific meeting in 1972, two chemists, Sherwood Rowland and Mario Molina, further wondered what happened to CFCs once they entered the atmosphere. A search of the literature found that these chemicals were not broken down in the lower atmosphere.

But the intense solar radiation in the upper atmosphere could break down CFCs, releasing chlorine atoms. Molina learned that chlorine reacts with ozone, reducing it to O_2. Other reactions liberate the chlorine, allowing it to destroy more ozone. In 1974, Molina and Rowland published their work predicting that the release of CFCs would damage the ozone layer.

As other researchers tested the CFC-ozone depletion hypothesis, the evidence accumulated, and more people became concerned about ozone depletion. Others worried about the economic impact of banning CFCs, and many chemical manufacturers mounted a campaign to cast doubt on the Molina-Rowland hypothesis in any way they could.

Then, in 1985, scientists from the British Antarctic Survey published their observations that the ozone level above Antarctica had decreased drastically. A reanalysis of data collected by the National Aeronautics and Space Administration (NASA) over that period confirmed a gigantic hole in the ozone layer. This hole has appeared every spring over Antarctica since the late 1970s and continues today. **Figure 7.14A** shows an image produced from atmospheric data from 2012. Dark blue colors show where there is the least ozone. The ozone depletion turned out to be much greater than had been predicted by Molina and Rowland. How could that be explained?

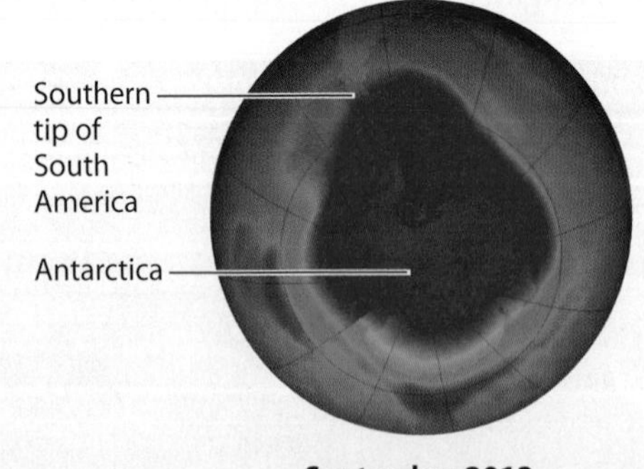
Southern tip of South America
Antarctica
September 2012

▲ **Figure 7.14A** The ozone hole in the Southern Hemisphere, fall 2012

Susan Solomon (**Figure 7.14B**), with the National Oceanic and Atmospheric Administration (NOAA), developed a hypothesis that the unique ice clouds that develop during the cold winter in Antarctica could be involved in speeding up the reactions that destroy ozone. The then 30-year-old scientist led two research expeditions to the Antarctic. The data that she and her team collected indicated that chemical reactions occurring on the icy particles when the spring sun hits them made CFCs destroy ozone at an astonishing rate. Once the clouds warm and disperse, these reactions slow.

▲ **Figure 7.14B** Susan Solomon at her cold research site

Further laboratory tests and field measurements taken from the ground, balloons, and airplanes supported Solomon's hypothesis. In response to these scientific findings, the first treaty to address Earth's environment was signed in 1987. In the original Montreal Protocol, more than two dozen nations agreed to phase out CFCs. As research continued to establish the extent and danger of ozone depletion, these agreements were strengthened in 1990, and now nearly 200 nations participate in the protocol. In 1995, Molina and Rowland shared a Nobel Prize for their work in determining how CFCs were damaging the atmosphere.

Global emissions of CFCs are near zero now, but because these compounds are so stable, recovery of the ozone layer is not expected until around 2060. Meanwhile, unblocked UV radiation is predicted to increase skin cancer and cataracts, as well as damage crops and phytoplankton in the oceans.

In addition to being ozone destroyers, CFCs are also potent greenhouse gases. The phaseout of CFCs has avoided what would have been the equivalent of adding 10 gigatons of CO_2 to the atmosphere. (For comparison, the Kyoto Protocol of 1997 set a 2012 target of a reduction of 2 gigatons of CO_2 emissions.)

Whether an environmental problem involves CFCs or CO_2, the scientific research is often complicated and the solutions complex. The connections between science, technology, and society, so clearly exemplified by the work of the scientists studying the ozone layer, are a major theme of this book.

? **Where does the ozone layer come from, and why is it so important to life on Earth?**

● High in the atmosphere, radiation from the sun converts O_2 to ozone. The ozone layer absorbs potentially damaging UV radiation before it can reach Earth's surface.

CHAPTER 7 REVIEW

For practice quizzes, BioFlix animations, MP3 tutorials, video tutors, and more study tools designed for this textbook, go to MasteringBiology®

Reviewing the Concepts

An Introduction to Photosynthesis (7.1–7.5)

7.1 Photosynthesis fuels the biosphere. Plants, algae, and some bacteria are photoautotrophs, the producers of food consumed by virtually all heterotrophic organisms.

7.2 Photosynthesis occurs in chloroplasts in plant cells. Chloroplasts are surrounded by a double membrane and contain stacks of thylakoids and a thick fluid called stroma.

7.3 Scientists traced the process of photosynthesis using isotopes. Experiments using both heavy and radioactive isotopes helped determine the details of the process of photosynthesis.

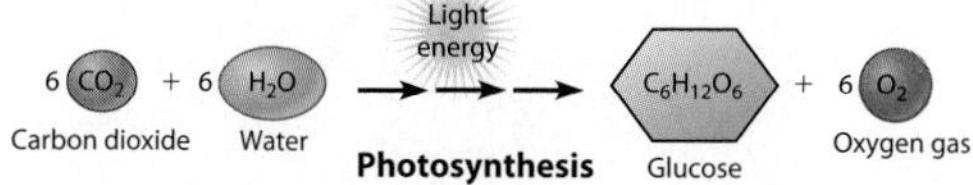

Light energy
6 CO_2 + 6 H_2O → $C_6H_{12}O_6$ + 6 O_2
Carbon dioxide
Water
Photosynthesis
Glucose
Oxygen gas

7.4 Photosynthesis is a redox process, as is cellular respiration. In photosynthesis, H_2O is oxidized and CO_2 is reduced.

7.5 The two stages of photosynthesis are linked by ATP and NADPH. The light reactions occur in the thylakoids, producing ATP and NADPH for the Calvin cycle, which takes place in the stroma.

The Light Reactions: Converting Solar Energy to Chemical Energy (7.6–7.9)

7.6 Visible radiation absorbed by pigments drives the light reactions. Certain wavelengths of visible light are absorbed by chlorophyll and other pigments. Carotenoids also function in photoprotection from excessive light.

7.7 Photosystems capture solar energy. Thylakoid membranes contain photosystems, each consisting of light-harvesting complexes and a reaction-center complex. A primary electron acceptor receives photoexcited electrons from chlorophyll.

7.8 Two photosystems connected by an electron transport chain generate ATP and NADPH. Electrons shuttle from photosystem II to photosystem I, providing energy to make ATP, and then reduce $NADP^+$ to NADPH. Photosystem II regains electrons as water is split and O_2 released.

7.9 The light reactions take place within the thylakoid membranes. In photophosphorylation, the electron transport chain pumps H^+ into the thylakoid space. The concentration gradient drives H^+ back through ATP synthase, powering the synthesis of ATP.

The Calvin Cycle: Reducing CO_2 to Sugar (7.10–7.11)

7.10 ATP and NADPH power sugar synthesis in the Calvin cycle. The steps of the Calvin cycle include carbon fixation, reduction, release of G3P, and regeneration of RuBP. Using carbon from CO_2, electrons from NADPH, and energy from ATP, the cycle constructs G3P, which is used to build glucose and other organic molecules.

7.11 Other methods of carbon fixation have evolved in hot, dry climates. In C_3 plants, a drop in CO_2 and rise in O_2 when stomata close divert the Calvin cycle to photorespiration. C_4 plants and CAM plants first fix CO_2 into four-carbon compounds that provide CO_2 to the Calvin cycle even when stomata close on hot, dry days.

The Global Significance of Photosynthesis (7.12–7.14)

7.12 Photosynthesis makes sugar from CO_2 and H_2O, providing food and O_2 for almost all living organisms.

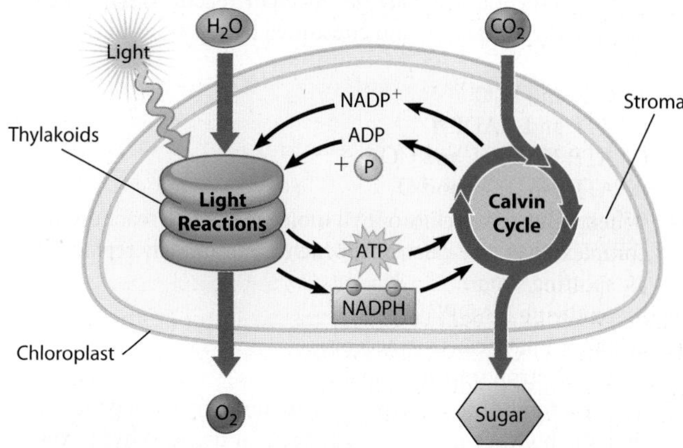

Light
H_2O
CO_2
Stroma
Thylakoids
$NADP^+$
ADP
+ P
Light Reactions
Calvin Cycle
ATP
NADPH
Chloroplast
O_2
Sugar

7.13 Rising atmospheric levels of carbon dioxide and global climate change will affect plants in various ways. Scientists study the effects of rising CO_2 levels using laboratory growth chambers and field studies. Long-term field projects enable scientists to assess the effects of CO_2 levels on natural ecosystems.

7.14 Scientific research and international treaties have helped slow the depletion of Earth's ozone layer. Solar radiation converts O_2 high in the atmosphere to ozone (O_3), which shields organisms from damaging UV radiation. Industrial chemicals called CFCs caused dangerous thinning of the ozone layer, but international restrictions on CFC use are allowing its recovery.

Connecting the Concepts

1. Complete this summary map of photosynthesis.

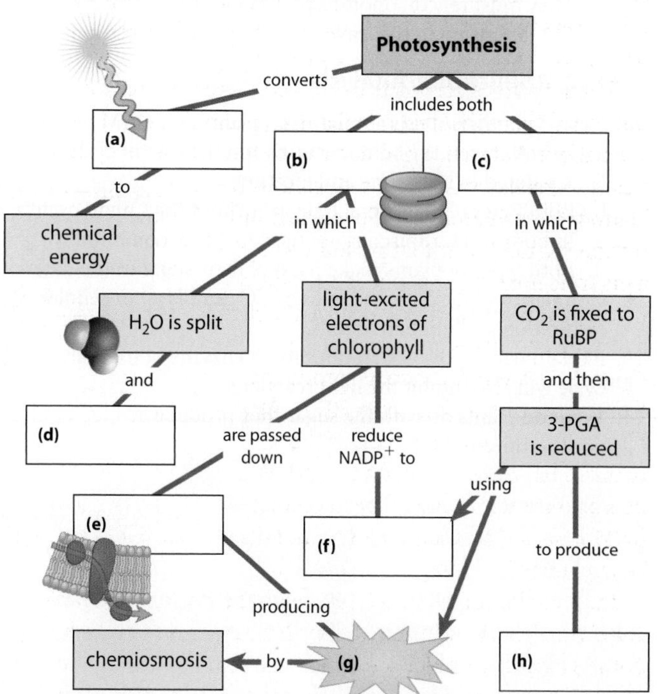

Photosynthesis
converts
includes both
(a)
(b)
(c)
to
chemical energy
in which
in which
H_2O is split
light-excited electrons of chlorophyll
CO_2 is fixed to RuBP
and
and then
(d)
are passed down
reduce $NADP^+$ to
3-PGA is reduced
using
(e)
(f)
to produce
producing
chemiosmosis
by
(g)
(h)

Testing Your Knowledge

Level 1: Knowledge/Comprehension

2. In photosynthesis, ______________ is oxidized and ______________ is reduced.
   a. water . . . oxygen
   b. carbon dioxide . . . water
   c. water . . . carbon dioxide
   d. glucose . . . carbon dioxide
3. Which of the following are produced by reactions that take place in the thylakoids and consumed by reactions in the stroma?
   a. CO_2 and H_2O
   b. ATP and NADPH
   c. ATP, NADPH, and CO_2
   d. ATP, NADPH, and O_2
4. When light strikes chlorophyll molecules in the reaction-center complex, they lose electrons, which are ultimately replaced by
   a. splitting water.
   b. oxidizing NADPH.
   c. the primary electron acceptor.
   d. the electron transport chain.
5. The reactions of the Calvin cycle are not directly dependent on light, but they usually do not occur at night. Why? (*Explain your answer.*)
   a. It is often too cold at night for these reactions to take place.
   b. Carbon dioxide concentrations decrease at night.
   c. The Calvin cycle depends on products of the light reactions.
   d. Plants usually close their stomata at night.
6. Which of the following does *not* occur during the Calvin cycle?
   a. carbon fixation
   b. oxidation of NADPH
   c. consumption of ATP
   d. release of oxygen
7. Why is it difficult for C_3 plants to carry out photosynthesis in very hot, dry environments such as deserts?
   a. The light is too intense and destroys the pigment molecules.
   b. The closing of stomata keeps CO_2 from entering and O_2 from leaving the plant.
   c. They must rely on photorespiration to make ATP.
   d. CO_2 builds up in the leaves, blocking carbon fixation.

Level 2: Application/Analysis

8. How is photosynthesis similar in C_4 plants and CAM plants?
   a. In both cases, the light reactions and the Calvin cycle are separated in both time and location.
   b. Both types of plants make sugar without the Calvin cycle.
   c. In both cases, rubisco is not used to fix carbon initially.
   d. Both types of plants make most of their sugar in the dark.
9. Compare and describe the roles of CO_2 and H_2O in cellular respiration and photosynthesis.
10. Explain why a poison that inhibits an enzyme of the Calvin cycle will also inhibit the light reactions.
11. What do plants do with the sugar they produce in photosynthesis?

Level 3: Synthesis/Evaluation

12. The following diagram compares the chemiosmotic synthesis of ATP in mitochondria and chloroplasts. Identify the components that are shared by both organelles and indicate which side of the membrane has the higher H^+ concentration. Then label on the right the locations within the chloroplast.

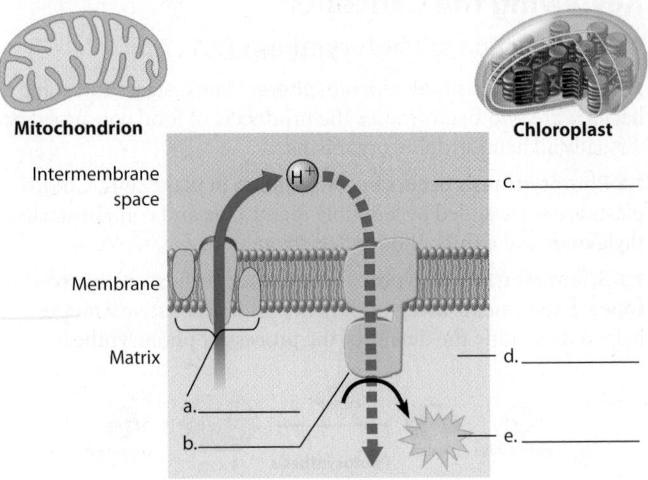

Mitochondrion
Chloroplast
Intermembrane space
Membrane
Matrix
H+
a.
b.
c.
d.
e.

13. Continue your comparison of electron transport and chemiosmosis in mitochondria and chloroplasts. In each case,
    a. where do the electrons come from?
    b. how do the electrons get their high potential energy?
    c. what picks up the electrons at the end of the chain?
    d. how is the energy given up by the electrons used?
14. **SCIENTIFIC THINKING** Will global climate change make you sneeze as well as itch? Scientists studying the effects of rising CO_2 levels have looked at ragweed, whose pollen is the primary allergen for fall hay fever. They grew ragweed in three levels of CO_2: a pre-industrial concentration of 280 ppm, a year 2000 level of 370 ppm, and a projected level of 600 ppm. They found that pollen production increased by 131% and 320% in the plants exposed to the recent and projected CO_2 levels, respectively. What was the hypothesis of this experiment? Do the results support the hypothesis? Given what you know about global climate change, what other variables would you like to test, and what other measurements would you like to take?
15. Most experts agree that global climate change is already occurring and that global warming will increase rapidly in this century. Recent international negotiations, however, including a 2012 meeting in Doha, Qatar, have yet to reach a global consensus on how to reduce greenhouse gas emissions. Some countries have resisted taking action because a very few scientists and policymakers think that the warming trend may be just a random fluctuation and/or not related to human activities or that cutting CO_2 emissions would sacrifice economic growth. Do you think we need more evidence before taking action? Or is it better to act now to reduce CO_2 emissions? What are the possible costs and benefits of each of these two strategies?

Answers to all questions can be found in Appendix 4.

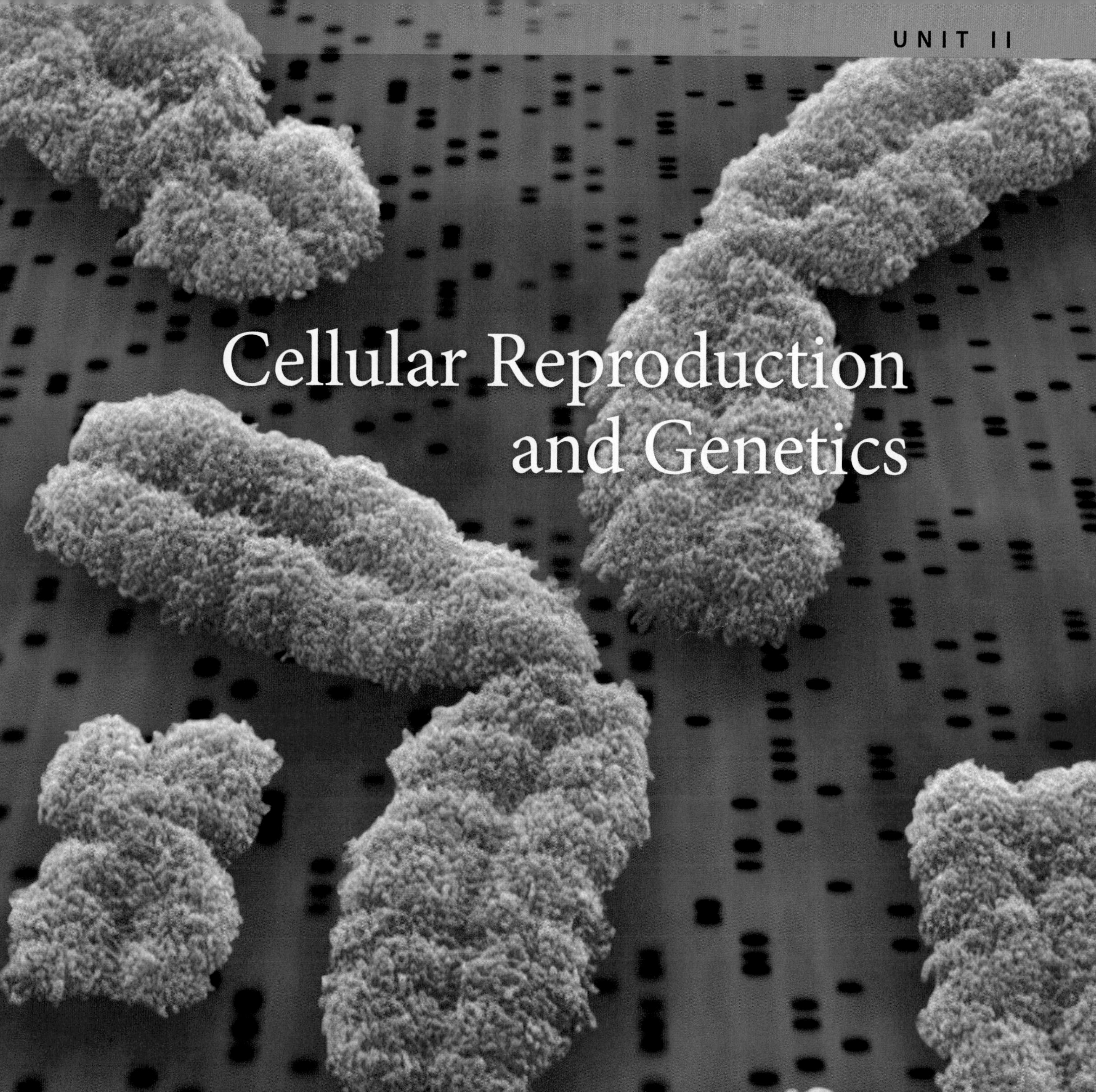

Cellular Reproduction and Genetics

CHAPTER 8

The Cellular Basis of Reproduction and Inheritance

The photo below shows a cancer cell dividing. Cancer cells start as normal body cells, but genetic mutations cause them to lose the ability to regulate the tempo of their own division. Like a car careening downhill without brakes, unconstrained body cells will likely wreak havoc. If left untreated, cancer cells will continue to divide and spread, invading other tissues, disrupting organ function, and eventually killing the host.

How can cancer be stopped? Most cancer treatments seek to disrupt one or more steps in cell division. Some anticancer drugs target dividing DNA; others disrupt the cellular structures that assist in cell division. Recent advances in cancer therapy have sought to match particular patients with specific therapies. For example, about two-thirds of human breast tumors contain cells that bear receptors for the sex hormone estrogen, which enhances breast cell division, thereby accelerating the growth of the tumor. Patients with such tumors may be treated with tamoxifen, a drug that specifically blocks estrogen receptors. Women who lack estrogen receptors on their tumor cells (about one-third of breast cancer patients)

? *Can cancer therapy be personalized?*

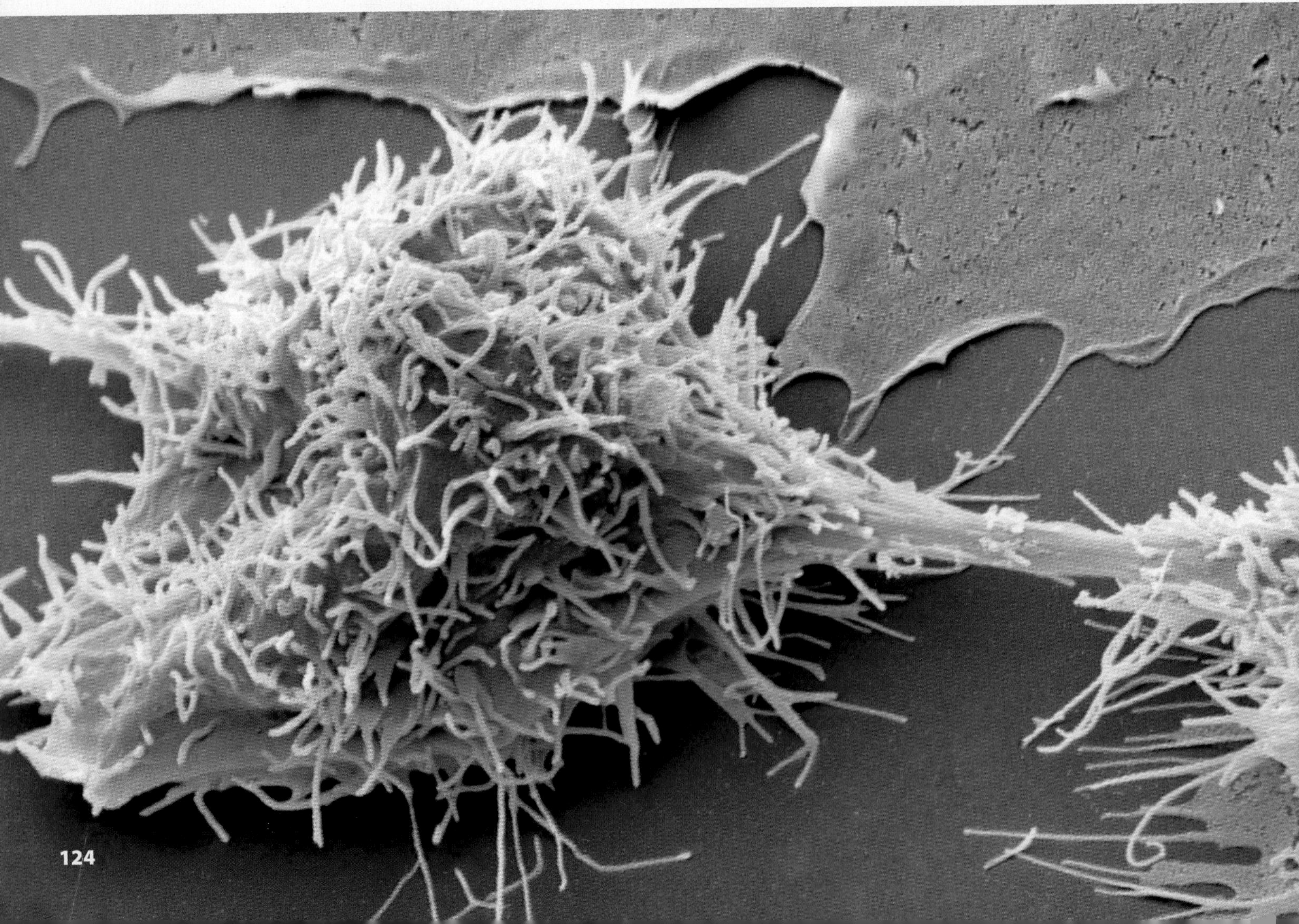

will not respond to this therapy. As more is learned about the underlying biology of cancer cells, cancer therapy will become even more personalized, with the most reliable drugs chosen for each patient.

Although uncontrolled cell division in cancer cells is harmful, normal cell division is necessary in all forms of life. Some organisms, such as single-celled prokaryotes, reproduce themselves via cell division, creating two genetically identical offspring. In the bodies of all multicellular organisms, cell division allows for growth, replacement of damaged cells, and development of an embryo into an adult. In sexually reproducing organisms, eggs and sperm are produced by a particular type of cell division. In this chapter, we discuss the two main types of cell division—mitosis and meiosis—and explore how they function within organisms.

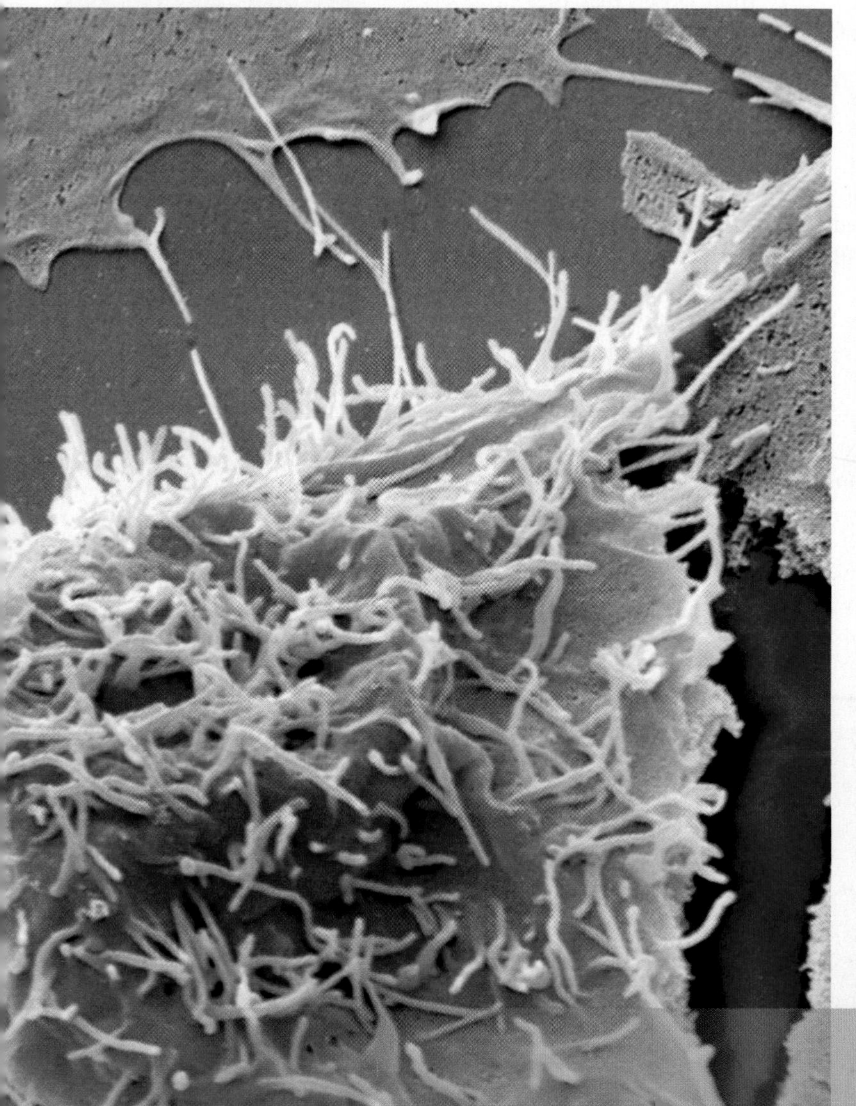

BIG IDEAS

Cell Division and Reproduction (8.1–8.2)

Cell division underlies many of life's important processes.

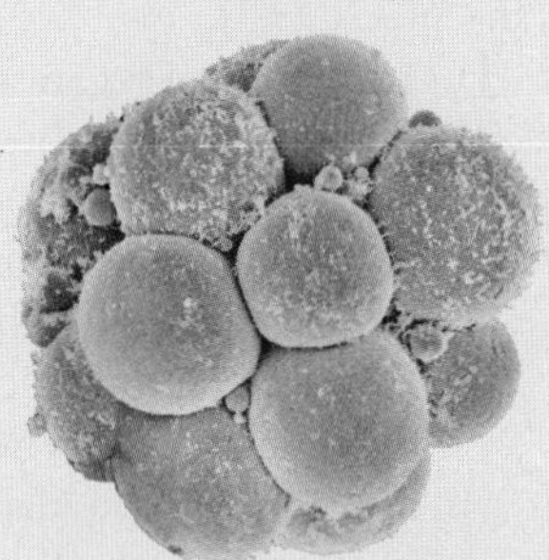

The Eukaryotic Cell Cycle and Mitosis (8.3–8.10)

Cells produce genetic duplicates through an ordered, tightly controlled series of steps.

Meiosis and Crossing Over (8.11–8.17)

The process of meiosis produces genetically varied haploid gametes from diploid cells.

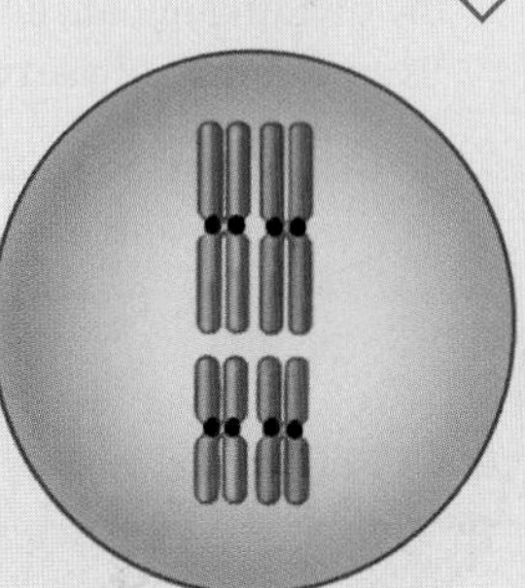

Alterations of Chromosome Number and Structure (8.18–8.23)

Errors in cell division can produce organisms with abnormal numbers of chromosomes.

▷ Cell Division and Reproduction

8.1 Cell division plays many important roles in the lives of organisms

The ability of organisms to reproduce their own kind is the characteristic that best distinguishes living things from nonliving matter (see Module 1.1 to review the characteristics of life). Only amoebas produce more amoebas, only people make more people, and only maple trees produce more maple trees. However, reproduction actually occurs much more often at the cellular level. When a cell undergoes reproduction, or **cell division**, the two "daughter" cells that result are genetically identical to each other and to the original "parent" cell. (Biologists traditionally use the word *daughter* in this context; it does not imply gender.) Before the parent cell splits into two, it duplicates its **chromosomes**, the structures that contain most of the cell's DNA. Then, during the division process, one set of chromosomes is distributed to each daughter cell. As a rule, the daughter cells receive identical sets of chromosomes from the lone, original parent cell. Each offspring cell will thus be genetically identical to the other and to the original parent cell.

Sometimes, cell division results in the reproduction of a whole organism. Many single-celled organisms, such as prokaryotes or the eukaryotic yeast cell in **Figure 8.1A**, reproduce by dividing in half, and the offspring are genetic replicas. This is an example of **asexual reproduction**, the creation of genetically identical offspring by a single parent, without the participation of sperm and egg. Many multicellular organisms can reproduce asexually as well. For example, some sea star species have the ability to grow new individuals from fragmented pieces (**Figure 8.1B**). And if you've ever grown a houseplant from a clipping, you've observed asexual reproduction in plants (**Figure 8.1C**). In asexual reproduction, there is one simple principle of inheritance: The lone parent and each of its offspring have identical genes.

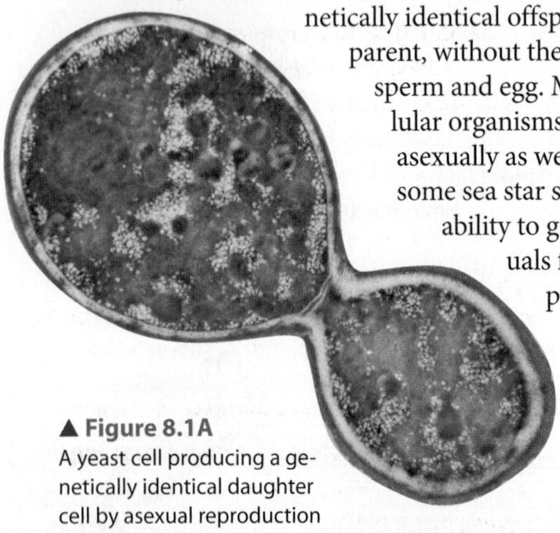

Colorized TEM 5,000×

▲ **Figure 8.1A** A yeast cell producing a genetically identical daughter cell by asexual reproduction

▲ **Figure 8.1B** A sea star reproducing asexually via fragmentation and regeneration of the body from the fragmented arm

Sexual reproduction is different; it requires fertilization of an egg by a sperm. The production of gametes—egg and sperm—involves a particular type of cell division that occurs only in reproductive organs (testes and ovaries in humans). A gamete has only half as many chromosomes as the parent cell that gave rise to it (see Module 8.13), and these chromosomes contain unique combinations of genes. Therefore, offspring produced by sexual reproduction generally resemble their parents more closely than they resemble unrelated individuals of the same species, but they are not identical to their parents or (with the exception of identical twins) to each other (**Figure 8.1D**). Each offspring inherits a unique combination of genes from its two parents, and this one-and-only set of genes programs a unique combination of traits. As a result, sexual reproduction can produce great variation among offspring.

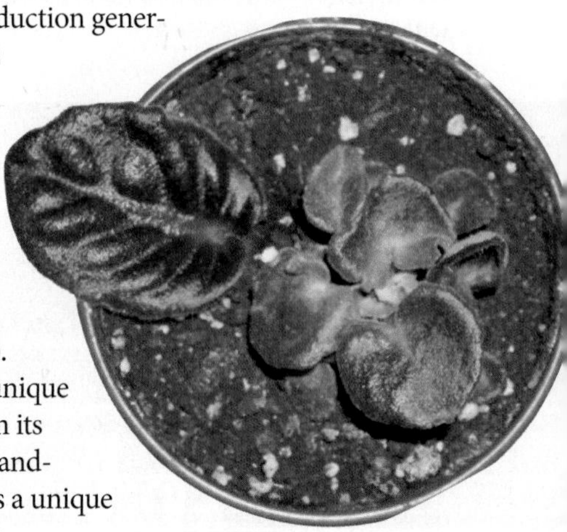

▲ **Figure 8.1C** An African violet reproducing asexually from a cutting (the large leaf on the left)

▼ **Figure 8.1D** Sexual reproduction produces offspring with unique combinations of genes

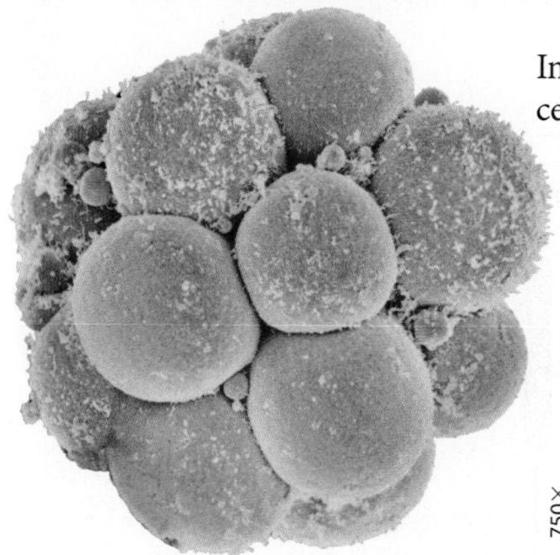
750×

▲ **Figure 8.1E** Dividing cells in an early human embryo

In multicellular organisms, cell division plays other important roles, in addition to the production of gametes. Cell division enables sexually reproducing organisms to develop from a single cell—the fertilized egg, or zygote (**Figure 8.1E**)—into an adult organism. All of the trillions of cells in your body arose via repeated cell divisions that began in your mother's body with a single fertilized egg cell. After an organism is fully grown, cell division continues to function in renewal and repair, replacing cells that die from normal wear and tear or from accidents. Within your body, millions of cells must divide every second to replace damaged or lost cells (**Figure 8.1F**). For example, dividing cells within your epidermis continuously replace dead cells that slough off the surface of your skin.

The type of cell division responsible for the growth and maintenance of multicellular organisms and for asexual reproduction involves a process called mitosis. The production of egg and sperm cells involves a different type of cell division called meiosis. In the remainder of this chapter, you will learn the details of both mitosis and meiosis. To start, we'll look briefly at prokaryotic cell division.

LM 850×

▲ **Figure 8.1F** A human kidney cell dividing

? **What function does cell division play in an amoeba (a single-celled protist)? What functions does it play in your body?**

● Reproduction; development, growth, and repair

8.2 Prokaryotes reproduce by binary fission

Prokaryotes (single-celled bacteria and archaea) reproduce by a type of cell division called **binary fission**, a term that means "dividing in half." In typical prokaryotes, the majority of genes are carried on a single circular DNA molecule that, with associated proteins, constitutes the organism's chromosome. Although prokaryotic chromosomes are much smaller than those of eukaryotes, duplicating them in an orderly fashion and distributing the copies equally to two daughter cells are still formidable tasks. Consider, for example, that when stretched out, the chromosome of the bacterium *Escherichia coli* (*E. coli*) is about 500 times longer than the cell itself. It is no small achievement to accurately replicate this molecule when it is coiled and packed inside the cell.

Figure 8.2A illustrates binary fission in a prokaryote. ❶ As the chromosome is duplicating, the copies move toward the opposite ends of the cell. ❷ Meanwhile, the cell elongates. ❸ When chromosome duplication is complete and the cell has reached about twice its initial size, the plasma membrane grows inward and more cell wall is made, which eventually divides the parent cell into two daughter cells (**Figure 8.2B**).

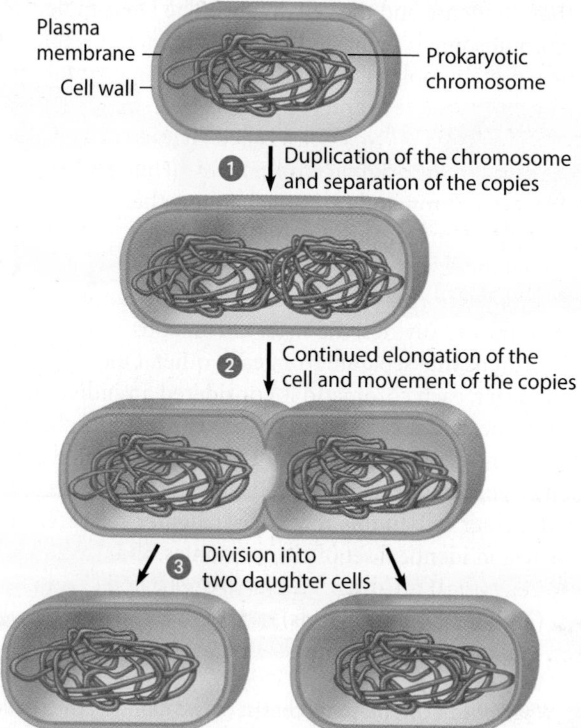
Plasma membrane
Cell wall
Prokaryotic chromosome
1 Duplication of the chromosome and separation of the copies
2 Continued elongation of the cell and movement of the copies
3 Division into two daughter cells

▲ **Figure 8.2A** Binary fission of a prokaryotic cell

? **Why is binary fission classified as asexual reproduction?**

● Because the genetically identical offspring inherit their DNA from a single parent

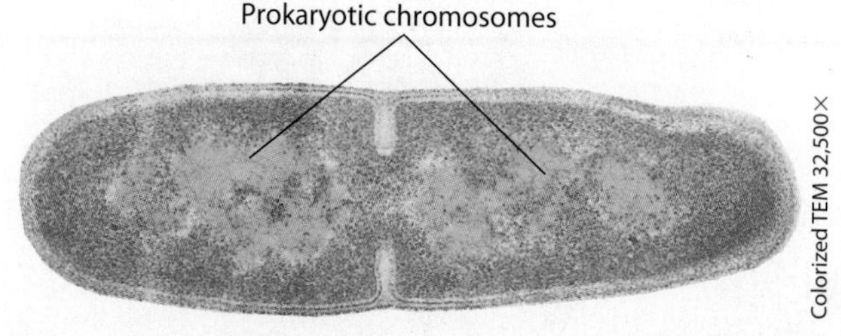
Prokaryotic chromosomes
Colorized TEM 32,500×

▲ **Figure 8.2B** An electron micrograph of a bacterium in a late stage of dividing

▷ The Eukaryotic Cell Cycle and Mitosis

8.3 The large, complex chromosomes of eukaryotes duplicate with each cell division

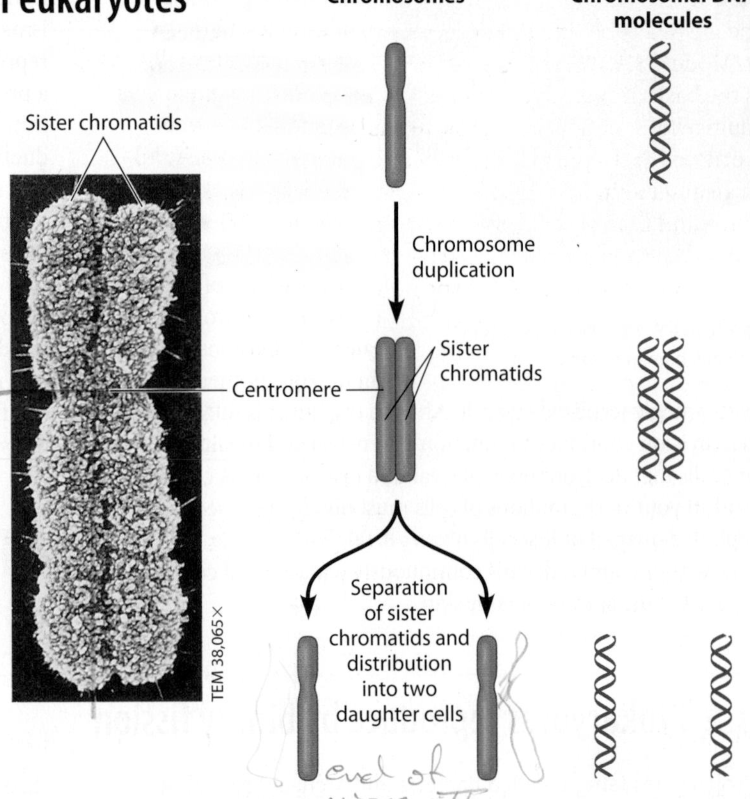
Sister chromatids
Centromere
TEM 38,065×
Chromosomes
Chromosomal DNA molecules
Chromosome duplication
Sister chromatids
Separation of sister chromatids and distribution into two daughter cells

▲ **Figure 8.3B** Chromosome duplication and distribution

Eukaryotic cells, in general, are more complex and much larger than prokaryotic cells. In addition, eukaryotic cells usually have many more genes, the units of information that specify an organism's inherited traits. Human cells, for example, carry just under 21,000 genes, versus about 3,000 for a typical bacterium. Almost all the genes in the cells of humans, and in all other eukaryotes, are found in the cell nucleus, grouped into multiple chromosomes. (The exceptions include genes on the small DNA molecules within mitochondria and, in plants, within chloroplasts.) Each eukaryotic species has a characteristic number of chromosomes in each cell nucleus. For example, human body cells have 46 chromosomes, while the body cells of a dog have 78 and those of a hedgehog have 90.

Each eukaryotic chromosome consists of one long DNA molecule—bearing hundreds or thousands of genes—and a number of protein molecules, which are attached to the DNA. The proteins help maintain the chromosome's structure and control the activity of its genes. Together, the entire complex—consisting of roughly equal amounts of DNA and protein—is called **chromatin**.

Most of the time, chromatin exists as a diffuse mass of long, thin fibers that, if stretched out, would be far too long to fit in a cell's nucleus. In fact, the total length of DNA in just one of your cells exceeds your height! Chromatin in this state is too thin to be seen using a light microscope.

As a cell prepares to divide, its chromatin coils up, forming tight, distinct chromosomes that are visible under a light microscope. Why is it necessary for a cell's chromosomes to be compacted in this way? Imagine an analogy from your own life: Your belongings are spread throughout your home, but as you prepare to move, you gather them up and pack them into small containers to make them more easily sorted and transported. Similarly, before a cell can undergo division, it must compact all its DNA into manageable packages. **Figure 8.3A** is a micrograph of a plant cell that is about to divide; each thick purple thread is actually an individual chromosome consisting of a single DNA molecule tightly wrapped around proteins.

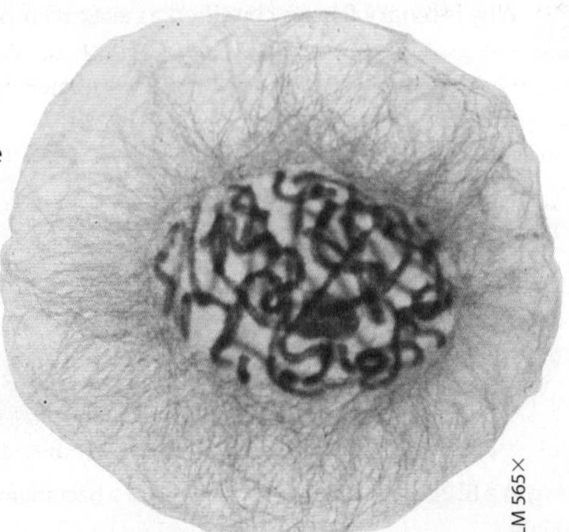
LM 565×

▶ **Figure 8.3A** A plant cell (from an African blood lily (*Scadoxus multiflorus*) just before cell division

The chromosomes of a eukaryotic cell are duplicated before they condense and the cell divides. The DNA molecule of each chromosome is replicated (as you'll learn in Chapter 10), and new protein molecules attach as needed to maintain the chromosome's structure and regulate its genes. Each chromosome now consists of two copies called **sister chromatids**, joined copies of the original chromosome (**Figure 8.3B**). The two sister chromatids are attached together along their lengths by proteins and are cinched especially tightly at a region called the **centromere** (visible as a narrow "waist" near the center of each chromosome shown in the figure).

When the cell divides, the sister chromatids of a duplicated chromosome separate from each other. Once separated from its sister, each chromatid is considered an individual chromosome, and it is identical to the cell's original chromosome. During cell division, one of the newly separated chromosomes goes to one daughter cell, and the other goes to the other daughter cell. In this way, each daughter cell receives a complete and identical set of chromosomes. In humans, for example, a typical dividing cell has 46 duplicated chromosomes (and thus 92 chromatids), and each of the two daughter cells that results from it has 46 single chromosomes.

? **When does a chromosome consist of two identical chromatids?**

● When the cell is preparing to divide and has duplicated its chromosomes but before the duplicates actually separate

8.4 The cell cycle includes growing and division phases

How do chromosome duplication and cell division fit into the life of a cell—and the life of an organism? As discussed in Module 8.1, all life depends on cell division: Cell division is the basis of reproduction for every organism; it enables a multicellular organism to grow to adult size; and it replaces worn-out or damaged cells, keeping the total number of cells in an adult animal relatively constant. In your body, for example, millions of cells must divide every second to maintain the total number of about 10 trillion cells. Some cells divide once a day, others less often; and highly specialized cells, such as mature muscle cells, do not divide at all. The fact that some mature cells never divide explains why some kinds of damage—such as the death of cardiac muscle during a heart attack or the death of brain cells during a stroke—can never be reversed.

The process of cell division is a key component of the **cell cycle**, an ordered sequence of events that extends from the instant a cell is first formed from a dividing parent cell until its own division into two cells. The cell cycle consists of two main stages: a growing stage (called interphase), during which the cell approximately doubles everything in its cytoplasm and precisely replicates its chromosomal DNA, and the actual cell division (called the mitotic phase).

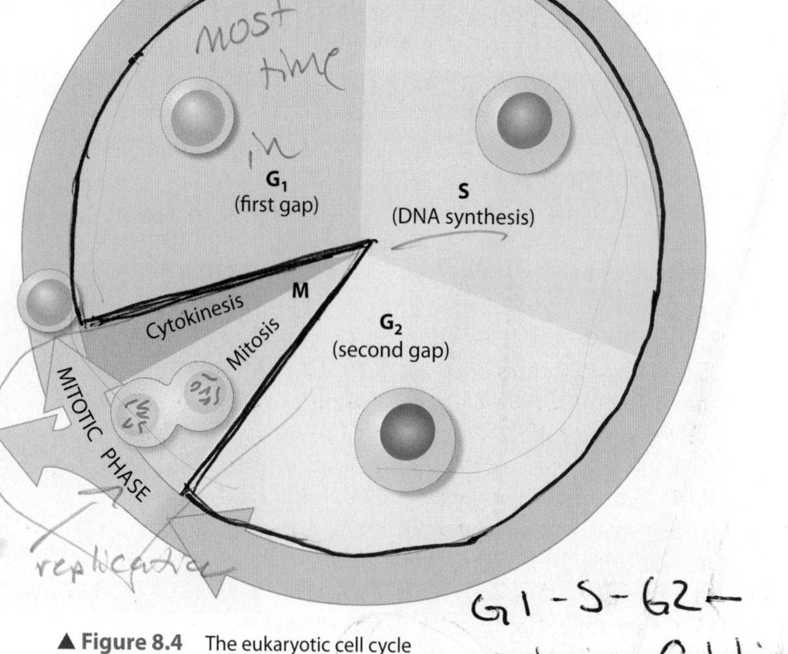

INTERPHASE
G_1
(first gap)
S
(DNA synthesis)
M
Cytokinesis
Mitosis
G_2
(second gap)
MITOTIC PHASE

▲ **Figure 8.4** The eukaryotic cell cycle

As **Figure 8.4** indicates, most of the cell cycle is spent in **interphase**. This is an interval when a cell's metabolic activity is very high and the cell performs its normal functions. For example, a cell in your small intestine might release digestive enzymes and absorb nutrients. Your intestinal cell also grows in size during interphase, making more cytoplasm, increasing its supply of digestive proteins, and creating more cytoplasmic organelles (such as mitochondria and ribosomes). In addition, the cell duplicates its chromosomes during this period. Typically, interphase lasts for at least 90% of the total time required for the cell cycle.

Interphase (illustrated in the beige portion of the figure) can be divided into three subphases: the G_1 phase ("first gap"), the S phase ("synthesis" of DNA—also known as DNA replication), and the G_2 phase ("second gap"). During all three subphases, the cell grows. The chromosomes are duplicated during the S phase: At the beginning of the S phase, each chromosome is single. At the end of this subphase, after DNA replication, the chromosomes are doubled, each consisting of two sister chromatids joined along their lengths. During the G_2 phase, the cell grows more as it completes preparations for cell division.

The **mitotic phase** (**M phase**; illustrated in the blue portion of the figure), the interval of the cell cycle when the cell physically divides, accounts for only about 10% of the total time required for the cell cycle. The mitotic phase is divided into two overlapping stages, called mitosis and cytokinesis. In **mitosis**, the nucleus and its contents—most important, the duplicated chromosomes—divide and are evenly distributed, forming two daughter nuclei. During **cytokinesis**, which usually begins before mitosis ends, the cytoplasm is divided in two. The combination of mitosis and cytokinesis produces two genetically identical daughter cells, each with a single nucleus, surrounding cytoplasm stocked with organelles, and a plasma membrane. Each newly produced daughter cell may then proceed through G_1 and repeat the cycle.

Mitosis is unique to eukaryotes and is the evolutionary solution to the problem of allocating an identical copy of the whole set of chromosomes to two daughter cells. Mitosis is a remarkably accurate mechanism. Experiments with yeast, for example, indicate that an error in chromosome distribution occurs only once in about 100,000 cell divisions.

The extreme accuracy of mitosis is essential to the development of your own body. You began as a single cell. Mitotic cell division ensures that all your body cells receive copies of the 46 chromosomes that were found in this original cell. Thus, every one of the trillions of cells in your body today can trace its ancestry back through mitotic divisions to that first cell produced when your father's sperm and mother's egg fused about nine months before your birth.

During the mitotic phase, a living cell viewed through a light microscope undergoes dramatic changes in the appearance of the chromosomes and other structures. In the next module, we'll use these visible changes as a guide to the stages of mitosis.

? **A researcher treats cells with a chemical that prevents DNA synthesis from starting. This treatment would trap the cells in which part of the cell cycle?**

● G_1

8.5 Cell division is a continuum of dynamic changes

Figure 8.5 illustrates the cell cycle for an animal cell using micrographs from a newt (with chromosomes shown in blue and the mitotic spindle stained green) and drawings (simplified to include just four chromosomes). Interphase is illustrated here, but the emphasis is on the dramatic changes that occur during cell division, the mitotic phase. Mitosis is a continuum, but biologists can distinguish five main stages: **prophase**, **prometaphase**, **metaphase**, **anaphase**, and **telophase**.

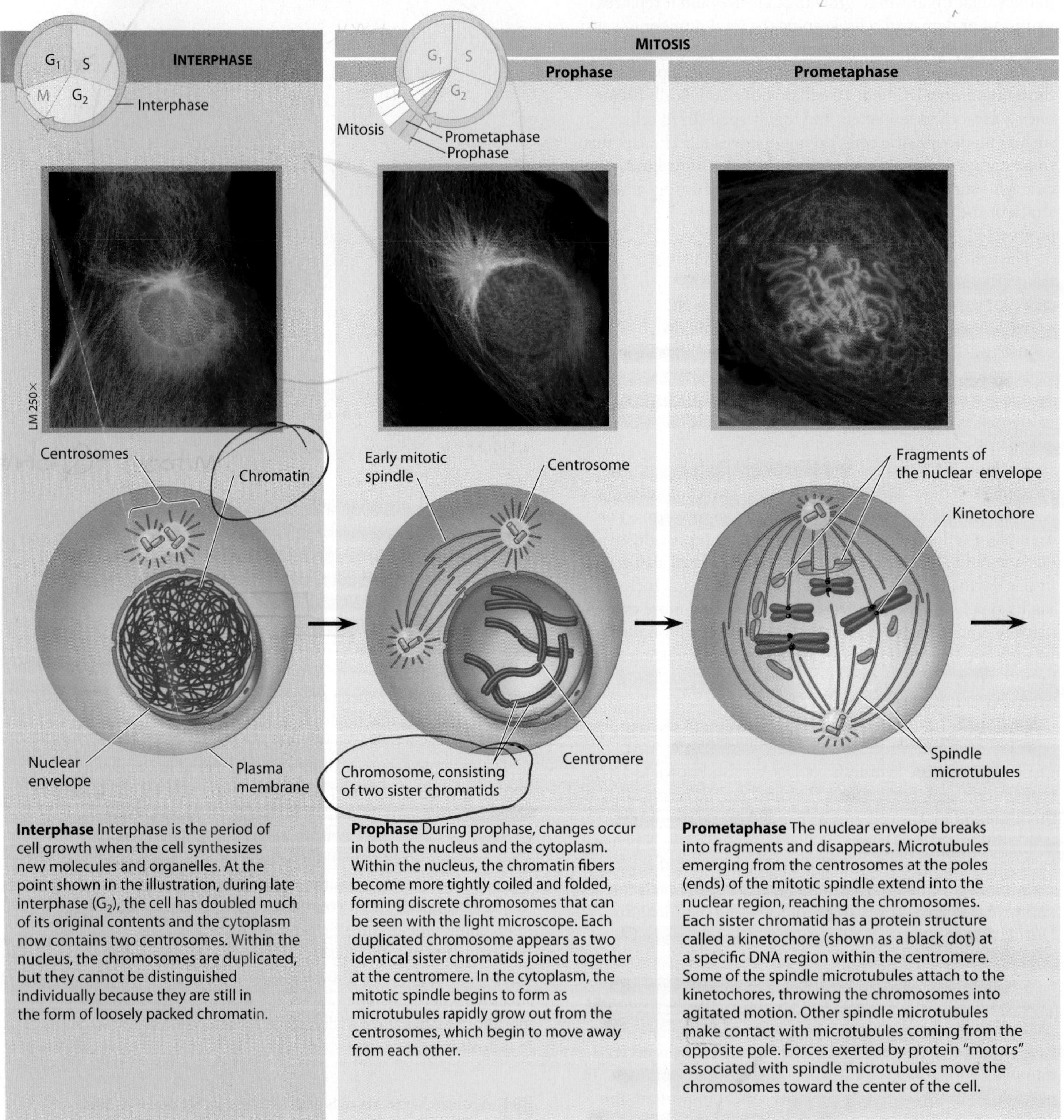

INTERPHASE
G1
S
G2
M
Interphase
MITOSIS
Prophase
Prometaphase
Mitosis
Prometaphase
Prophase
LM 250×
Centrosomes
Chromatin
Nuclear envelope
Plasma membrane
Early mitotic spindle
Centrosome
Centromere
Chromosome, consisting of two sister chromatids
Fragments of the nuclear envelope
Kinetochore
Spindle microtubules

Interphase Interphase is the period of cell growth when the cell synthesizes new molecules and organelles. At the point shown in the illustration, during late interphase (G_2), the cell has doubled much of its original contents and the cytoplasm now contains two centrosomes. Within the nucleus, the chromosomes are duplicated, but they cannot be distinguished individually because they are still in the form of loosely packed chromatin.

Prophase During prophase, changes occur in both the nucleus and the cytoplasm. Within the nucleus, the chromatin fibers become more tightly coiled and folded, forming discrete chromosomes that can be seen with the light microscope. Each duplicated chromosome appears as two identical sister chromatids joined together at the centromere. In the cytoplasm, the mitotic spindle begins to form as microtubules rapidly grow out from the centrosomes, which begin to move away from each other.

Prometaphase The nuclear envelope breaks into fragments and disappears. Microtubules emerging from the centrosomes at the poles (ends) of the mitotic spindle extend into the nuclear region, reaching the chromosomes. Each sister chromatid has a protein structure called a kinetochore (shown as a black dot) at a specific DNA region within the centromere. Some of the spindle microtubules attach to the kinetochores, throwing the chromosomes into agitated motion. Other spindle microtubules make contact with microtubules coming from the opposite pole. Forces exerted by protein "motors" associated with spindle microtubules move the chromosomes toward the center of the cell.

▲ **Figure 8.5** The stages of cell division by mitosis

Try This Using simple drawings, illustrate the stages of mitosis for a cell that has six chromosomes.

The chromosomes are the stars of the mitotic drama. Their movements depend on the **mitotic spindle**, a football-shaped structure of microtubules and associated proteins that guides the separation of the two sets of daughter chromosomes. The spindle microtubules emerge from two **centrosomes**, microtubule-organizing regions in the cytoplasm of eukaryotic cells.

? **You view an animal cell through a microscope and observe dense, duplicated chromosomes scattered throughout the cell. Which state of mitosis are you witnessing?**

● Prophase (because the chromosomes are condensed but not yet aligned)

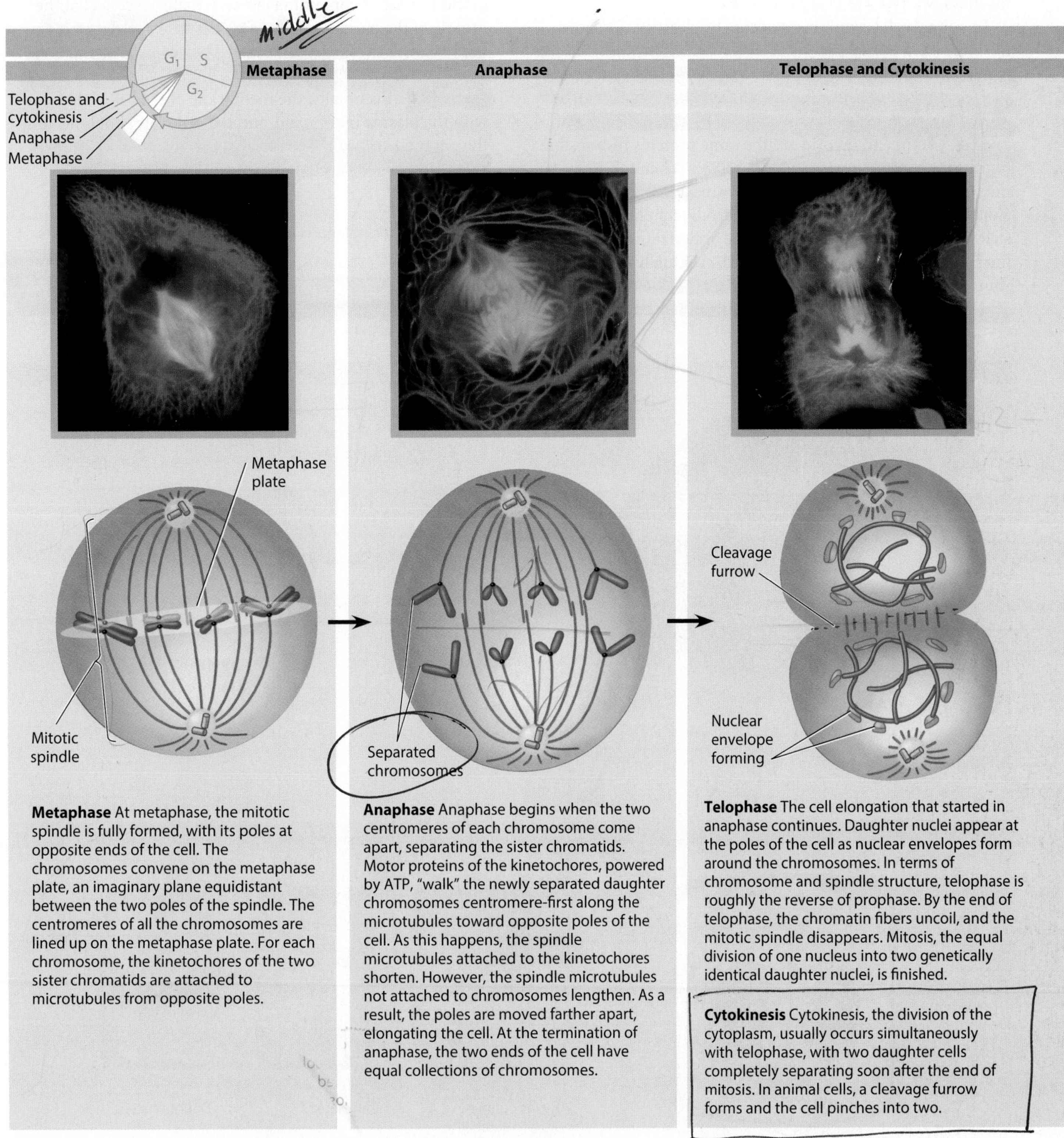

G1
S
G2
Telophase and cytokinesis
Anaphase
Metaphase
Metaphase
Anaphase
Telophase and Cytokinesis
Metaphase plate
Mitotic spindle
Separated chromosomes
Cleavage furrow
Nuclear envelope forming

Metaphase At metaphase, the mitotic spindle is fully formed, with its poles at opposite ends of the cell. The chromosomes convene on the metaphase plate, an imaginary plane equidistant between the two poles of the spindle. The centromeres of all the chromosomes are lined up on the metaphase plate. For each chromosome, the kinetochores of the two sister chromatids are attached to microtubules from opposite poles.

Anaphase Anaphase begins when the two centromeres of each chromosome come apart, separating the sister chromatids. Motor proteins of the kinetochores, powered by ATP, "walk" the newly separated daughter chromosomes centromere-first along the microtubules toward opposite poles of the cell. As this happens, the spindle microtubules attached to the kinetochores shorten. However, the spindle microtubules not attached to chromosomes lengthen. As a result, the poles are moved farther apart, elongating the cell. At the termination of anaphase, the two ends of the cell have equal collections of chromosomes.

Telophase The cell elongation that started in anaphase continues. Daughter nuclei appear at the poles of the cell as nuclear envelopes form around the chromosomes. In terms of chromosome and spindle structure, telophase is roughly the reverse of prophase. By the end of telophase, the chromatin fibers uncoil, and the mitotic spindle disappears. Mitosis, the equal division of one nucleus into two genetically identical daughter nuclei, is finished.

Cytokinesis Cytokinesis, the division of the cytoplasm, usually occurs simultaneously with telophase, with two daughter cells completely separating soon after the end of mitosis. In animal cells, a cleavage furrow forms and the cell pinches into two.

8.6 Cytokinesis differs for plant and animal cells

As discussed in the previous module, cytokinesis typically overlaps with telophase. Given the differences between animal and plant cells—particularly the stiff cell wall in plant cells—it isn't surprising that cytokinesis proceeds differently for these two types of eukaryotic cells.

In animal cells, cytokinesis occurs by cleavage. As shown in **Figure 8.6A**, the first sign of cleavage in animal cells is the appearance of a **cleavage furrow**, a shallow groove in the cell surface. At the site of the furrow, the cytoplasm has a ring of microfilaments made of actin, associated with molecules of myosin. (Actin and myosin are the same proteins responsible for muscle contraction; see Module 30.8.) When the actin microfilaments interact with the myosin, the ring contracts. Contraction of the myosin ring is much like pulling a drawstring on a hooded sweatshirt: As the drawstring is pulled, the ring of the hood contracts inward, eventually pinching shut. Similarly, the cleavage furrow deepens and eventually pinches the parent cell in two, resulting in two completely separate daughter cells, each with its own nucleus and share of cytoplasm.

Cytokinesis is markedly different in plant cells, which possess stiff cell walls (**Figure 8.6B**). During telophase, membranous vesicles containing cell wall material collect at the middle of the parent cell. The vesicles fuse, forming a membranous disk called the **cell plate**. The cell plate grows outward, accumulating more cell wall materials as more vesicles fuse with it. Eventually, the membrane of the cell plate fuses with the plasma membrane, and the cell plate's contents join the parental cell wall. The result is two daughter cells, each bounded by its own plasma membrane and cell wall.

? **Contrast cytokinesis in animals with cytokinesis in plants.**

In animals, cytokinesis involves a cleavage furrow in which contracting microfilaments pinch the cell in two. In plants, it involves formation of a cell plate, a fusion of vesicles that forms new plasma membranes and new cell walls between the cells.

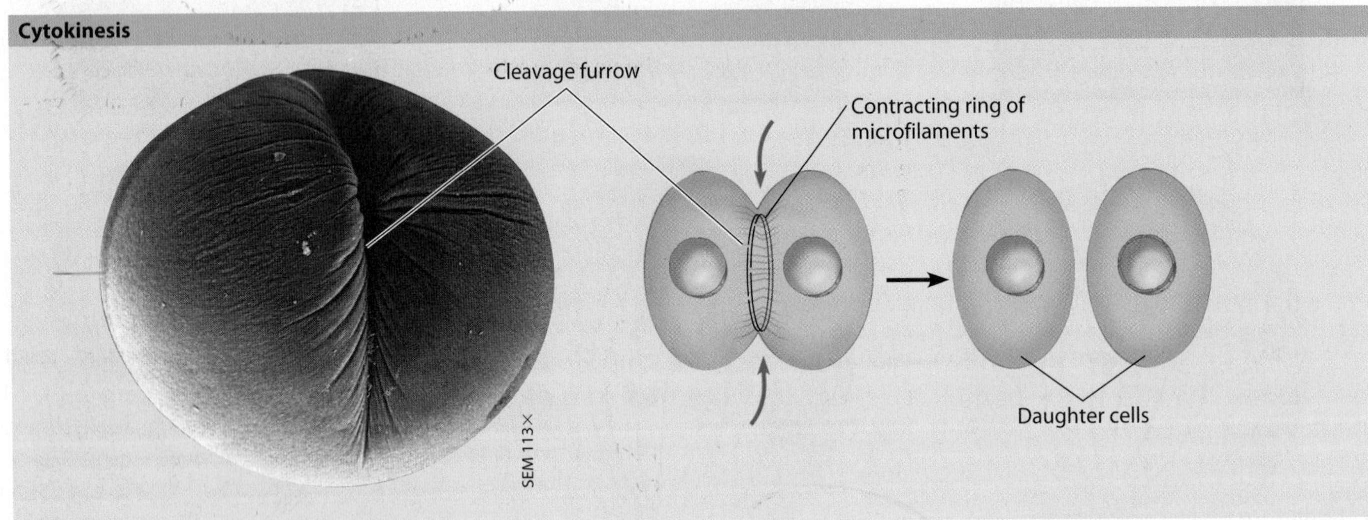
Cytokinesis
Cleavage furrow
Contracting ring of microfilaments
SEM 113×
Daughter cells

▲ **Figure 8.6A** Cleavage of an animal cell

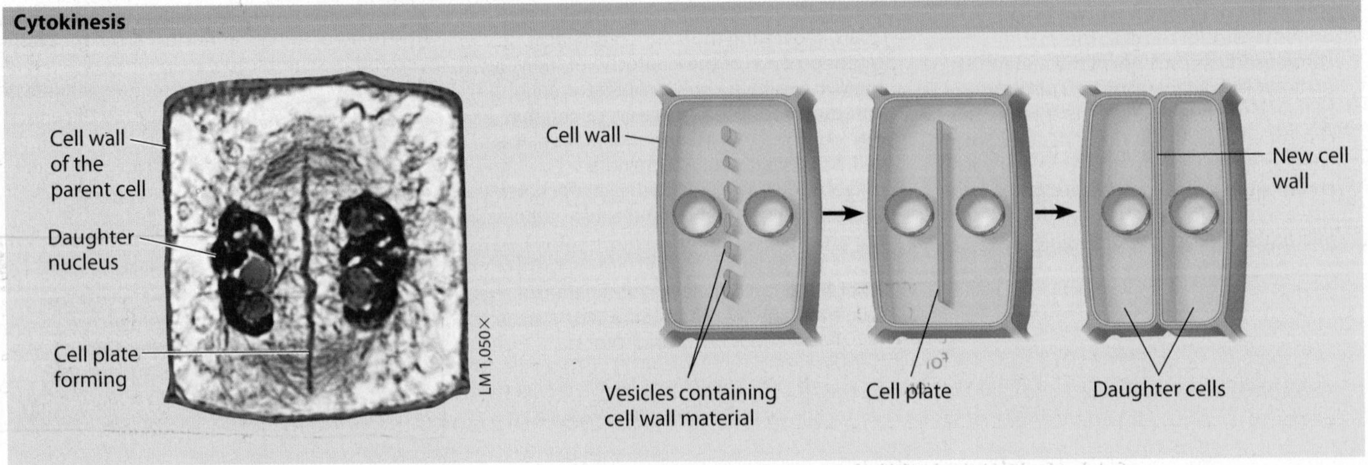
Cytokinesis
Cell wall of the parent cell
Daughter nucleus
Cell plate forming
LM 1,050×
Cell wall
Vesicles containing cell wall material
Cell plate
New cell wall
Daughter cells

▲ **Figure 8.6B** Cell plate formation in a plant cell

8.7 Anchorage, cell density, and chemical growth factors affect cell division

For a plant or an animal to grow, develop normally, and maintain its tissues once fully grown, it must be able to control the timing of cell division in different parts of its body. For example, in your body, skin cells and the cells lining your digestive tract divide frequently, replacing cells that are constantly being abraded and sloughed off. In contrast, cells in your liver usually do not divide unless the liver is damaged.

By growing animal cells in culture, researchers have been able to identify many factors, both chemical and physical, that influence cell division. For example, most animal cells exhibit **anchorage dependence**; they must be in contact with a solid surface—such as the inside of a culture dish or the extracellular matrix of a tissue—to divide. One of the characteristics that distinguishes cancerous cells from normal body cells is their failure to exhibit anchorage dependence; they grow whether or not they are in contact with a suitable surface.

Another physical factor that can regulate growth rate is **density-dependent inhibition**, a phenomenon in which crowded cells stop dividing (**Figure 8.7A**). Animal cells growing on the surface of a dish multiply to form a single layer and usually stop dividing when they touch one another. If some cells are removed, those bordering the open space begin dividing again and continue until the vacancy is filled. What actually causes the inhibition of growth? Studies of cultured cells suggest that physical contact of cell-surface proteins between adjacent cells is responsible for inhibiting cell division. As with anchorage dependence, density-dependent inhibition fails in tumors; cancer cells continue to divide even at high densities, piling up on one another (bottom of Figure 8.7A).

Chemical factors can also influence the rate of cell growth. For example, when grown in the laboratory, cells fail to divide if an essential nutrient is left out of the culture medium. And most types of mammalian cells divide in culture only if certain specific growth factors are included. A **growth factor** is a protein secreted by certain body cells that stimulates other cells to divide (**Figure 8.7B**). Researchers have discovered at least 50 different growth factors that can trigger cell division. Different cell types respond specifically to certain growth factors or a combination of growth factors. For example, injury to the skin causes blood platelets to release a protein called platelet-derived growth factor. This protein promotes the rapid growth of connective tissue cells that help seal the wound. Another well-studied example is a protein called vascular endothelial growth factor (VEGF), which stimulates the growth of new blood vessels during fetal development and after injury. Interestingly, VEGF overproduction is a hallmark of many dangerous cancers; several anticancer drug therapies work by inhibiting the action of VEGF.

How do growth factors work? We will explore this question in the next module.

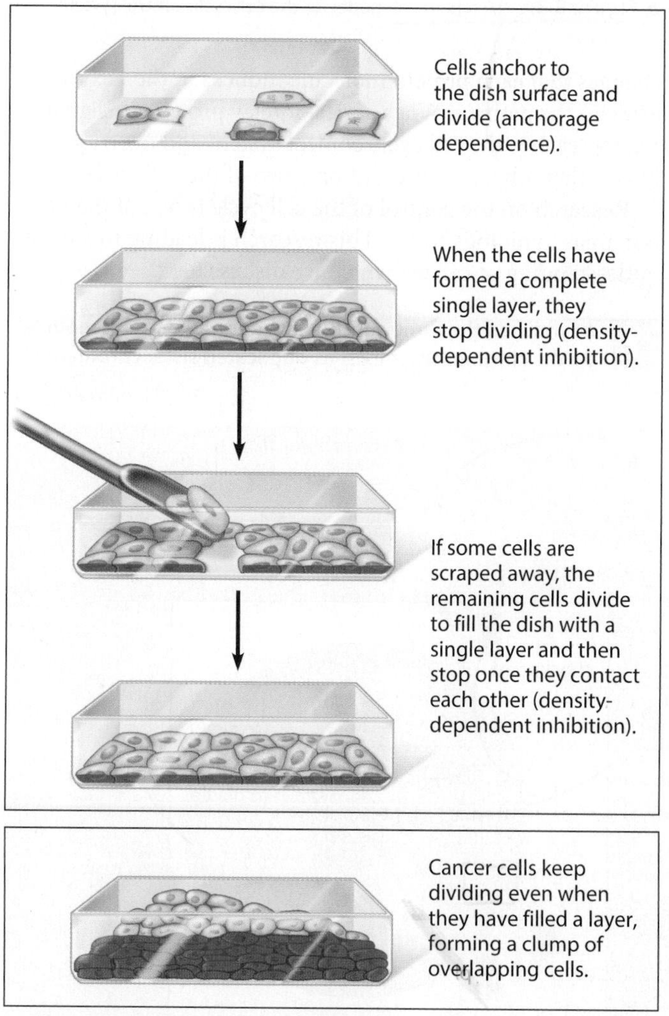

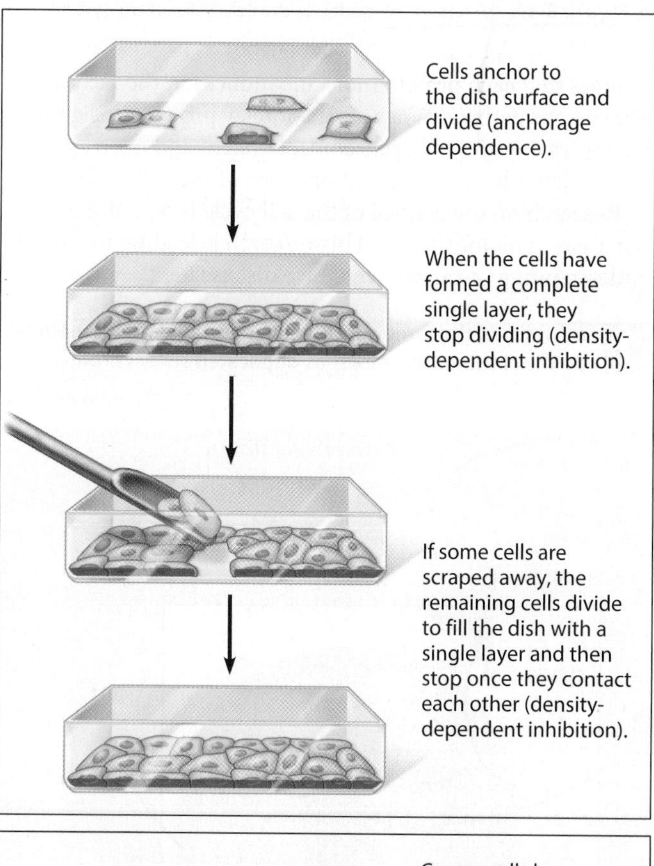

Cells anchor to the dish surface and divide (anchorage dependence).
When the cells have formed a complete single layer, they stop dividing (density-dependent inhibition).
If some cells are scraped away, the remaining cells divide to fill the dish with a single layer and then stop once they contact each other (density-dependent inhibition).
Cancer cells keep dividing even when they have filled a layer, forming a clump of overlapping cells.

▲ **Figure 8.7A** An experiment demonstrating density-dependent inhibition, using animal cells grown in culture

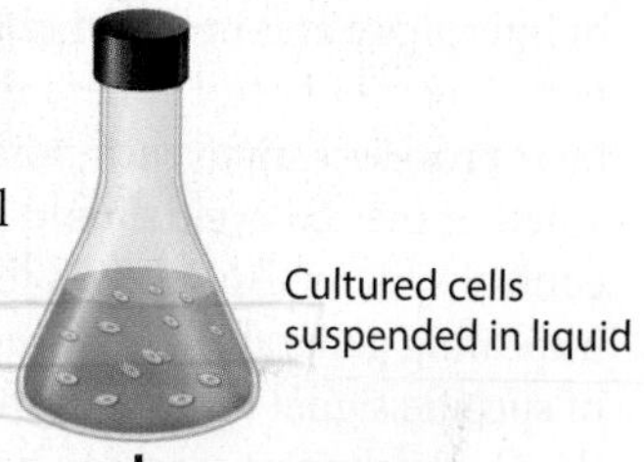

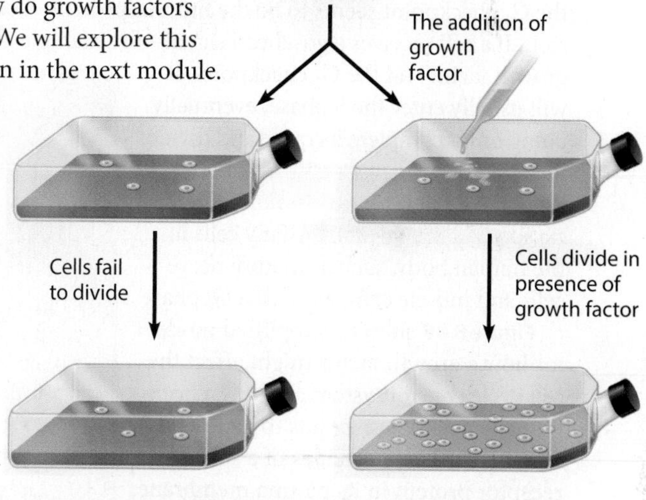

Cultured cells suspended in liquid
The addition of growth factor
Cells fail to divide
Cells divide in presence of growth factor

▲ **Figure 8.7B** An experiment demonstrating the effect of growth factors on the division of cultured animal cells

? Compared to a control culture, the cells in an experimental culture are fewer but much larger in size when they cover the dish surface and stop growing. What is a reasonable hypothesis for this difference?

● The experimental culture is deficient in one or more growth factors.

8.8 Growth factors signal the cell cycle control system

In a living animal, most cells are anchored in a fixed position and bathed in a solution of nutrients supplied by the blood, yet they usually do not divide unless they are signaled by other cells to do so. Growth factors are the main signals, and their role in promoting cell division leads us back to our earlier discussion of the cell cycle.

The sequential events of the cell cycle, represented by the circle in **Figure 8.8A**, are directed by a distinct cell cycle control system, represented by the gray circle in the center of the art. The thin gray bar extending from the circle represents the current position in the cell cycle. The **cell cycle control system** is a cyclically operating set of molecules in the cell that both triggers and coordinates key events in the cell cycle. The cell cycle is *not* like a row of falling dominoes, with each event causing the next one in line. Within the M phase, for example, metaphase does not automatically lead to anaphase. Instead, proteins of the cell cycle control system must trigger the separation of sister chromatids that marks the start of anaphase.

A checkpoint in the cell cycle is a critical control point where stop and go-ahead signals (represented by stop/go lights in the figure) can regulate the cycle. The default action in most animal cells is to halt the cell cycle at these checkpoints unless overridden by specific go-ahead signals.

The red and white gates in Figure 8.8A represent major checkpoints in the cell cycle: during the G_1 and G_2 subphases of interphase and in the M phase. Intracellular signals detected by the control system tell the system whether key cellular processes up to each point have been completed and whether the cell cycle should proceed past that point. The control system also receives messages from outside the cell, indicating general environmental conditions and the presence of specific signal molecules from other cells. For many cells, the G_1 checkpoint seems to be the most important in cell division. If a cell receives a go-ahead signal—for example, from a growth factor—at the G_1 checkpoint, it will usually enter the S phase, eventually going on to complete its cycle and divide. If such a signal never arrives, the cell will switch to a permanently nondividing state called the G_0 phase. Many cells in the human body, such as mature nerve cells and muscle cells, are in the G_0 phase.

Figure 8.8B shows a simplified model for how a growth factor might affect the cell cycle control system at the G_1 checkpoint. A cell that responds to a growth factor (▼) has molecules of a specific receptor protein in its plasma membrane. Binding of the growth factor to the receptor (●) triggers a signal transduction pathway in the cell. A signal transduction pathway is a series of protein molecules that conveys a message (see Modules 5.1 and 11.10). In this case, that message leads to cell division. The "signals" are changes that each protein molecule induces in the next molecule in the pathway. Via a series of relay proteins, a signal finally reaches the cell cycle control system and overrides the brakes that otherwise prevent progress of the cell cycle.

Research on the control of the cell cycle is one of the hottest areas in biology today. This research is leading to a better understanding of cancer, which we discuss next.

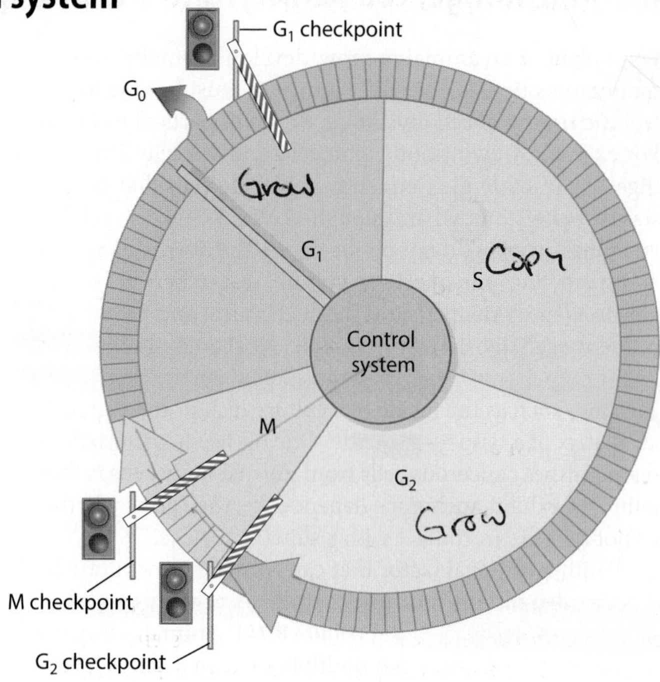
G_1 checkpoint
G_0
G_1
S
Control system
M
G_2
M checkpoint
G_2 checkpoint

▲ **Figure 8.8A** A schematic model for the cell cycle control system

? **At which of the three checkpoints described in this module do the chromosomes exist as duplicated sister chromatids?**

● G_2 and M checkpoints

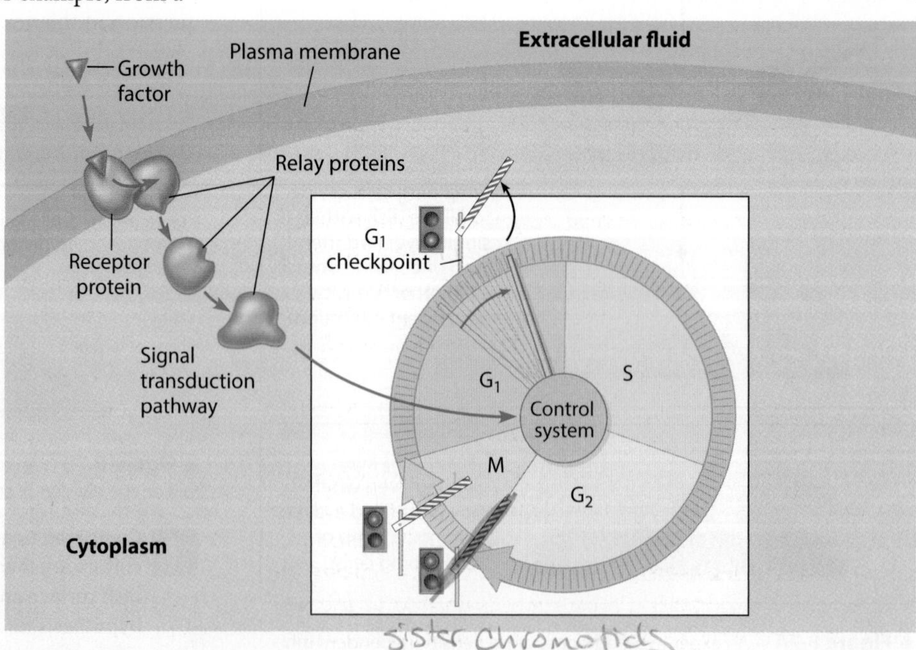
Extracellular fluid
Plasma membrane
Growth factor
Relay proteins
G_1 checkpoint
Receptor protein
Signal transduction pathway
G_1
S
Control system
M
G_2
Cytoplasm

▲ **Figure 8.8B** How a growth factor signals the cell cycle control system

8.9 Growing out of control, cancer cells produce malignant tumors

CONNECTION

Cancer, which claims the lives of one out of every five people in the United States, is a disease of the cell cycle. Cancer cells do not heed the normal signals that regulate the cell cycle; they divide excessively and invade other tissues of the body. If unchecked, cancer cells may continue to grow and spread until they kill the organism.

The abnormal behavior of cancer cells begins when a single cell undergoes transformation, a process that converts a normal cell to a cancer cell. Transformation occurs following a mutation in one or more genes that encode for proteins in the cell cycle control system. Because a transformed cell grows abnormally, the immune system usually recognizes it as such and destroys it. However, if the cell evades destruction, it may multiply to form a **tumor**, a mass of abnormally growing cells within otherwise normal tissue. If the abnormal cells remain at their original site, the lump is called a **benign tumor**. Benign tumors can cause problems if they grow in and disrupt certain organs, such as the brain, but often they can be completely removed by surgery or even (in cases in which they pose no imminent threat) left alone.

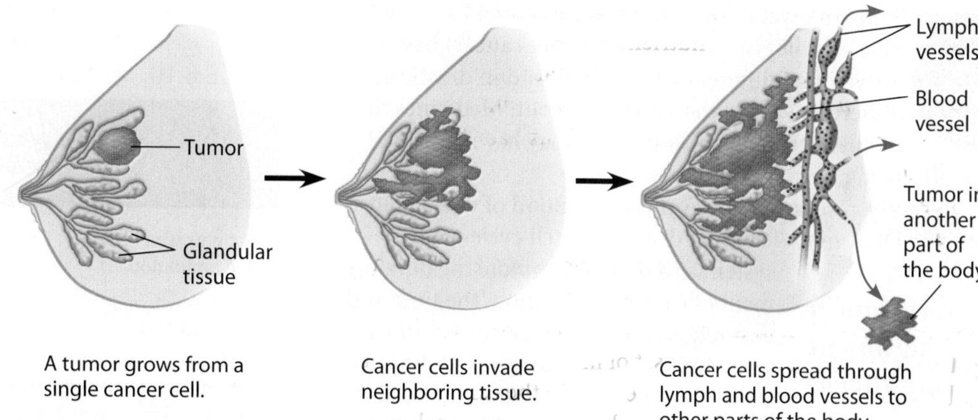

Tumor
Glandular tissue
Lymph vessels
Blood vessel
Tumor in another part of the body
A tumor grows from a single cancer cell.
Cancer cells invade neighboring tissue.
Cancer cells spread through lymph and blood vessels to other parts of the body.

▲ **Figure 8.9** Growth and metastasis of a malignant tumor of the breast

In contrast, a **malignant tumor** can spread into neighboring tissues and invade other parts of the body, displacing normal tissue and interrupting organ function as it grows (**Figure 8.9**). An individual with a malignant tumor is said to have **cancer**. Cancer cells may separate from the original tumor or secrete signal molecules that cause blood vessels to grow toward the tumor. A few tumor cells may then enter the blood and lymph vessels and move to other parts of the body, where they may proliferate and form new tumors. The spread of cancer cells beyond their original site is called **metastasis**.

Cancers are named according to the organ or tissue in which they originate. Liver cancer, for example, starts in liver tissue and may or may not spread from there. Carcinomas are cancers that originate in the external or internal coverings of the body, such as the skin or the lining of the intestine. Leukemia is a broad term covering a number of diseases that originate in immature white blood cells within the blood or bone marrow.

From studying cancer cells in culture, researchers have learned that cancer cells do not heed the normal signals that regulate the cell cycle. For example, many cancer cells have defective cell cycle control systems that proceed past checkpoints even in the absence of growth factors. Other cancer cells synthesize growth factors themselves, causing the cells to divide continuously. If cancer cells do stop dividing, they seem to do so at random points in the cell cycle, rather than at the normal cell cycle checkpoints. Moreover, in the laboratory, cancer cells are "immortal"; they can go on dividing indefinitely, as long as they have a supply of nutrients (whereas normal mammalian cells divide only about 20 to 50 times before they stop). A striking example of the immortality of cancer cells is a line that has been continuously multiplying in culture since 1951. Cells of this line are called HeLa cells, named for the original donor, Henrietta Lacks, who died of cervical cancer more than 60 years ago.

Luckily, many tumors can be successfully treated. A tumor that appears to be localized may be removed surgically. Alternatively, it can be treated with concentrated beams of high-energy radiation, which usually damages DNA in cancer cells more than it does in normal cells, perhaps because cancer cells have lost the ability to repair such damage. However, radiation also damages normal body cells, producing harmful side effects. For example, radiation damage to cells of the ovaries or testes can lead to sterility.

Chemotherapy is used to treat widespread or metastatic tumors. During periodic chemotherapy treatments, drugs are administered that disrupt specific steps in the cell cycle. For instance, the drug paclitaxel (trade name Taxol) freezes the mitotic spindle after it forms, which stops actively dividing cells from proceeding past metaphase. Vinblastin, a chemotherapeutic drug first obtained from the periwinkle plant, prevents the mitotic spindle from forming.

The side effects of chemotherapy are due to the drugs' effects on normal cells that rapidly divide. Nausea results from chemotherapy's effects on intestinal cells; hair loss comes from effects on hair follicle cells; and susceptibility to infection results from effects on immune cell production. (We will return to the topic of cancer—specifically, how mutations in genes that control cell division can lead to cancer—in Chapter 11.)

In the next module, you'll learn how understanding such mutations may aid in cancer treatment.

What is metastasis?

● Metastasis is the spread of cancer cells from their original site of formation to other sites in the body.

8.10 Tailoring treatment to each patient may improve cancer therapy

SCIENTIFIC THINKING

Although several forms of cancer treatment are available, oncologists (doctors who treat cancer) have observed that different patients respond in drastically different ways to the same treatment. Medicines that are ineffective for most patients may have profound effects in a few, and vice versa.

Recently, researchers have amassed a flood of data about specific DNA mutations that disable the cell cycle control system and thereby promote the development of cancer. In addition, the time and cost required to sequence genes from a given individual have decreased dramatically. Thus, with increased understanding of DNA mutations and better access to gene sequencing, it is increasingly possible for physicians to personalize cancer treatment. The goal in moving toward personalized treatment is to identify and administer drugs that have the best chance of success in combating a given tumor's specific genetic profile.

Can cancer therapy be personalized?

A study that examined personalized cancer treatment was conducted at Memorial Sloan-Kettering Cancer Center in New York in 2012. Researchers sequenced the complete genome of tumor cells taken from a patient with metastasized bladder cancer. This patient, unlike most individuals with this type of cancer, responded well to a drug called everolimus. The researchers hoped that a genetic profile could provide insight into this unexpected result. Their next step was to identify 140 distinct mutations within the protein-coding genes of this patient's tumor cells. One of these mutations—in a gene called *TSC1*—affects a pathway that is targeted by the drug. The researchers hypothesized that mutations in this gene might be predictive of the drug's effect in cancer patients; those with the mutation would fare well when taking everolimus, whereas those who lack the mutation would not.

Because their hypothesis was based on a single case—and researchers are careful never to draw broad conclusions from just one case—the Sloan-Kettering researchers expanded their study by analyzing the *TSC1* gene in tumors from 13 additional bladder cancer patients. Their results **(Table 8.10)** suggest that the drug may indeed be most effective in patients who harbor mutations in the *TSC1* gene: three out of four patients with the mutation responded to the drug, whereas only one out of nine patients without the mutation responded.

Critics caution that the *TSC1* mutation is sufficiently rare that enough data to draw reliable conclusions may never be obtained. The nature of the evidence—scant and difficult to generalize—is typical of personalized cancer therapy trials. But small advances are continuously being made that may lead to improved treatment for a variety of cancer patients.

TABLE 8.10 NUMBERS OF CANCER PATIENTS RESPONDING TO THE DRUG EVEROLIMUS ($N = 13$)

	Patients Responding to Everolimus	Patients *Not* Responding to Everolimus
Patients who had *TSC1* mutation	3	1
Patients who *did not* have *TSC1* mutation	1	8

Source: Data from G. Iyer et al., Genome sequencing identifies a basis for everolimus sensitivity, *Science* 338: 221 (2012).

? Think critically about the data presented in Table 8.10: Which data in Table 8.10 do not support the hypothesis that everolimus works *only* in patients who harbor a *TSC1* mutation?

One patient without the mutation responded to the drug, and one patient with the mutation did not respond to the drug.

▷ Meiosis and Crossing Over

8.11 Chromosomes are matched in homologous pairs

In humans, a typical body cell, called a **somatic cell**, has 46 chromosomes. When chromosomes from a cell in metaphase (when the chromosomes are most condensed) are viewed with a microscope, they can be arranged into matching pairs; **Figure 8.11** illustrates one pair of metaphase chromosomes consisting of two joined sister chromatids. A human cell at metaphase contains 23 sets of duplicated chromosomes. Other species have different numbers of chromosomes, but these, too, usually occur in matched pairs. Moreover, when treated with special dyes, the chromosomes of a pair display matching staining patterns (represented by colored stripes in Figure 8.11).

Almost every chromosome has a twin that resembles it in length, centromere position, and staining (coloration) pattern. The two chromosomes of such a matching pair are **homologous chromosomes** (or homologs) because each chromosome carries genes controlling the same inherited characteristics. For example, if a gene for freckles is located at a particular place, or **locus** (plural, *loci*), on one chromosome—within the

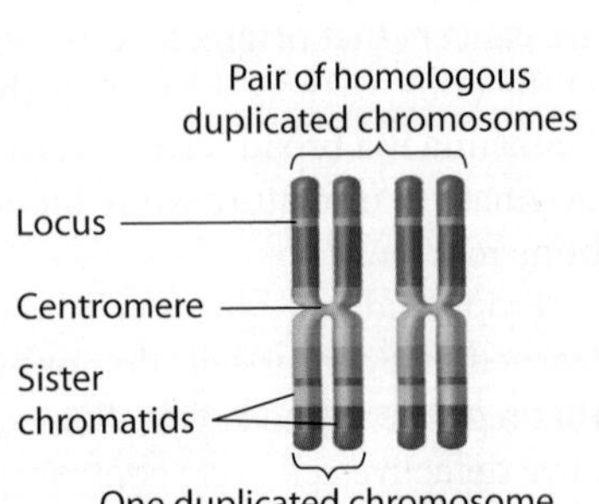

Pair of homologous duplicated chromosomes
Locus
Centromere
Sister chromatids
One duplicated chromosome

▲ Figure 8.11 A pair of homologous chromosomes

Try This Cover the figure, then draw a pair of homologous chromosomes and label sister chromatids, the centromere, and one chromosome.

narrow orange band in our drawing—then the homologous chromosome has that same gene at that same locus. However, the two chromosomes of a homologous pair may have different versions of the same gene.

The two distinct chromosomes referred to as X and Y are an important exception to the general pattern of homologous chromosomes in human somatic cells. Human females have a homologous pair of X chromosomes (XX), but males have one X and one Y chromosome (XY). Only small parts of the X and Y are homologous. Most of the genes carried on the X chromosome do not appear on the tiny Y, and the Y chromosome has genes not present on the X. Because they determine an individual's sex, the X and Y chromosomes are called **sex chromosomes**. Chromosomes other than sex chromosomes are called **autosomes**.

? Are all of *your* chromosomes fully homologous?

If you are female, yes; if you are male, no.

8.12 Gametes have a single set of chromosomes

The development of a fertilized egg into a new adult organism is one phase of a multicellular organism's **life cycle**, the sequence of stages leading from the adults of one generation to the adults of the next (**Figure 8.12A**). Having two sets of chromosomes, one inherited from each parent, is a key factor in the life cycle of all species that reproduce sexually.

Most animals and plants are said to be **diploid** organisms because all somatic cells contain pairs of homologous chromosomes. The total number of chromosomes is called the diploid number (abbreviated $2n$). For humans, the diploid number is 46; that is, $2n = 46$. The exceptions are the egg and sperm cells, collectively known as **gametes**. Each gamete has a single set of chromosomes: 22 autosomes plus a sex chromosome, either X or Y. A cell with a single chromosome set is called a **haploid** cell; it has only one member of each homologous pair. For humans, the haploid number (abbreviated n) is 23; that is, $n = 23$.

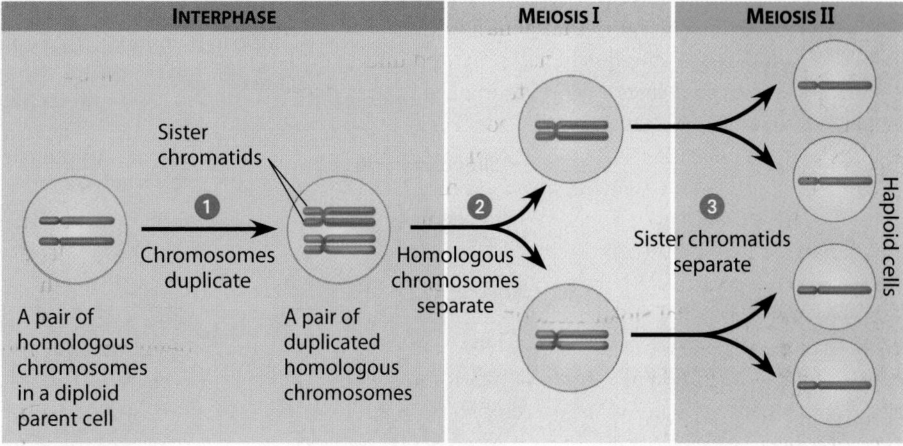

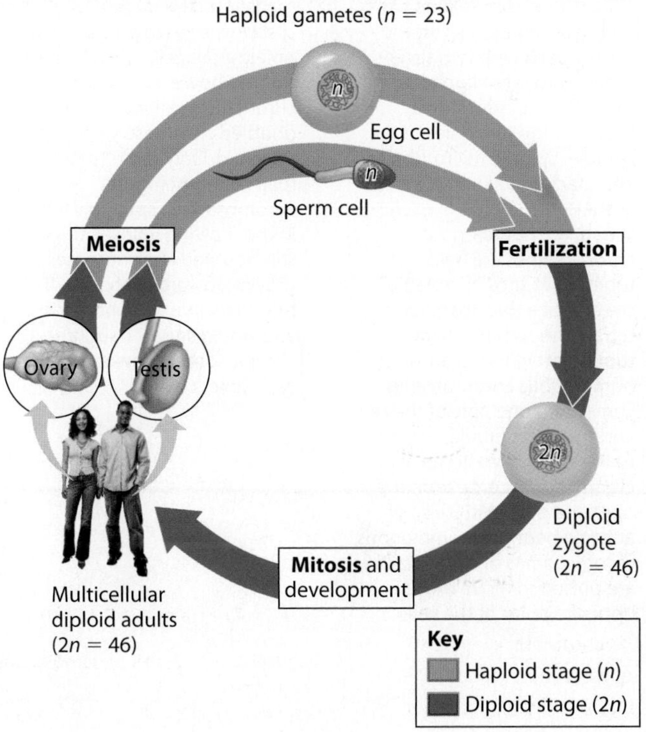

INTERPHASE
MEIOSIS I
MEIOSIS II
Sister chromatids
1
Chromosomes duplicate
2
Homologous chromosomes separate
3
Sister chromatids separate
A pair of homologous chromosomes in a diploid parent cell
A pair of duplicated homologous chromosomes
Haploid cells

▲ **Figure 8.12B** How meiosis halves chromosome number through two sequential divisions

Haploid gametes ($n = 23$)

Egg cell

Sperm cell

Meiosis

Fertilization

Ovary

Testis

Diploid zygote ($2n = 46$)

Mitosis and development

Multicellular diploid adults ($2n = 46$)

Key: Haploid stage (n); Diploid stage ($2n$)

▲ **Figure 8.12A** The human life cycle

The human life cycle begins when a haploid sperm cell from the father fuses with a haploid egg cell from the mother in the process of **fertilization**. The resulting fertilized egg, called a **zygote**, has one set of homologous chromosomes from each parent, and so is diploid. As a human develops into an adult, mitosis of the zygote and its descendants generates all the somatic cells.

The only cells of the human body not produced by mitosis are the gametes. Gametes are made by a different form of cell division called meiosis, which occurs only in reproductive organs. Whereas mitosis produces daughter cells with the same number of chromosomes as the parent cell, meiosis reduces the chromosome number by half. **Figure 8.12B** tracks one pair of homologous chromosomes through the two divisions of meiosis. **1** Each of the chromosomes is duplicated during interphase (before meiosis). **2** The first division, meiosis I, segregates the two chromosomes of the homologous pair, packaging them in separate (haploid) daughter cells. But each chromosome is still doubled. **3** Meiosis II separates the sister chromatids. Each of the four daughter cells is haploid and contains only a single chromosome from the homologous pair.

Next, we'll take a closer look at the process of meiosis.

? How many autosomes are found in a human sperm cell? How many and which sex chromosomes?

22 autosomes plus either an X or Y sex chromosome

8.13 Meiosis reduces the chromosome number from diploid to haploid

Meiosis is a type of cell division that produces haploid gametes in diploid organisms. Two haploid gametes may then combine via fertilization to restore the diploid state in the zygote. Fertilization and meiosis alternate in sexual life cycles, which serves to maintain a constant number of chromosomes in each species from one generation to the next.

Many of the stages of meiosis closely resemble corresponding stages in mitosis. Meiosis, like mitosis, is preceded by the duplication of chromosomes. However, this single duplication is followed by not one but two consecutive cell divisions, called meiosis I and meiosis II. Because one duplication of the chromosomes is followed by two divisions, the result is four daughter cells, each with half as many chromosomes as the parent cell. The illustrations in **Figure 8.13** show the two meiotic divisions for an animal cell with a diploid number of 6. The members of a pair of homologous chromosomes in Figure 8.13 (and later figures) are colored red and blue to help distinguish them. (Imagine that the red chromosomes were inherited from the mother and the blue chromosomes from the father.) One of the most important events in meiosis

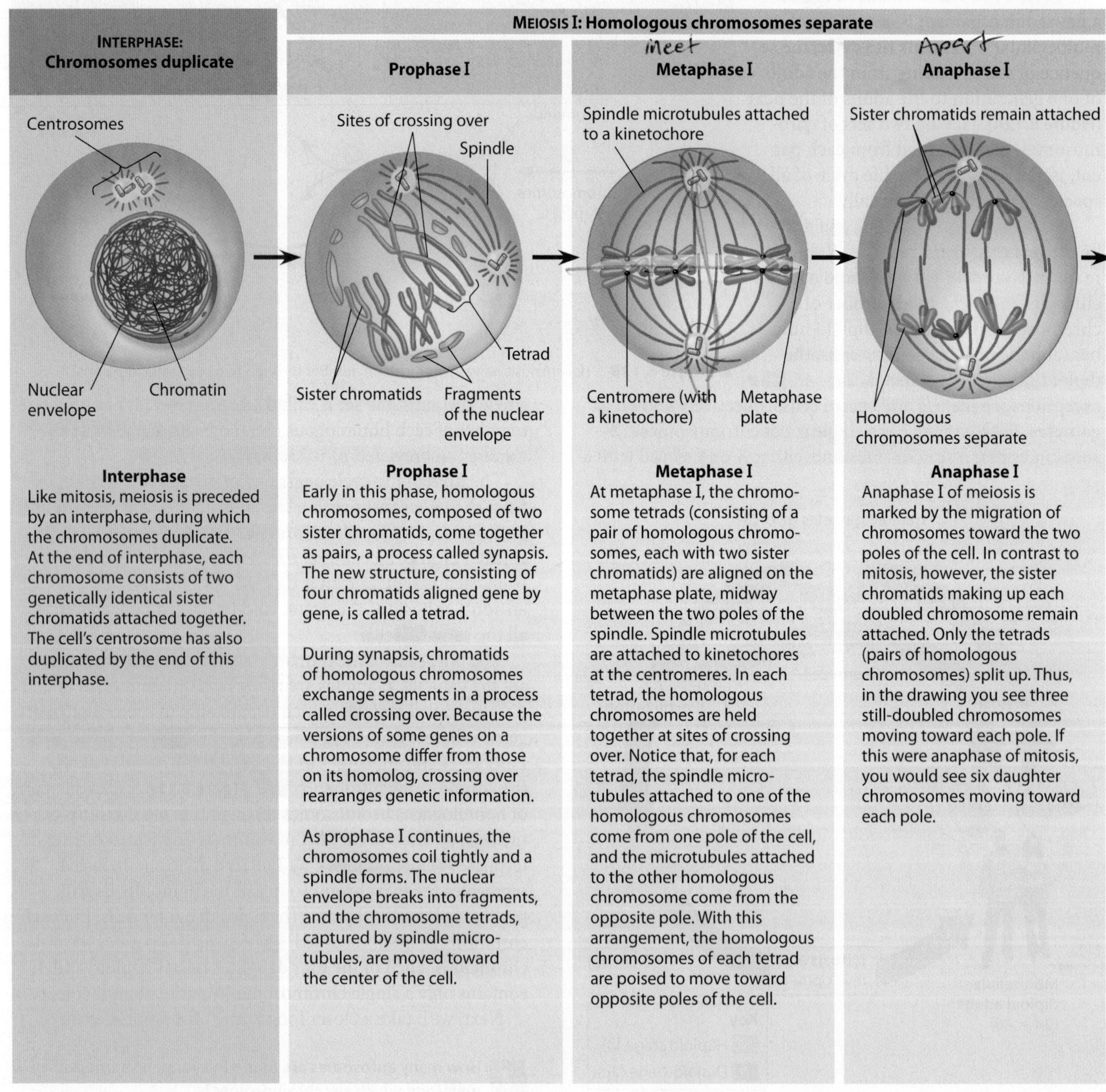

INTERPHASE: Chromosomes duplicate
MEIOSIS I: Homologous chromosomes separate
Prophase I
Metaphase I
Anaphase I
Centrosomes
Nuclear envelope
Chromatin
Sites of crossing over
Spindle
Tetrad
Sister chromatids
Fragments of the nuclear envelope
Spindle microtubules attached to a kinetochore
Centromere (with a kinetochore)
Metaphase plate
Sister chromatids remain attached
Homologous chromosomes separate
Interphase
Like mitosis, meiosis is preceded by an interphase, during which the chromosomes duplicate. At the end of interphase, each chromosome consists of two genetically identical sister chromatids attached together. The cell's centrosome has also duplicated by the end of this interphase.
Prophase I
Early in this phase, homologous chromosomes, composed of two sister chromatids, come together as pairs, a process called synapsis. The new structure, consisting of four chromatids aligned gene by gene, is called a tetrad.
During synapsis, chromatids of homologous chromosomes exchange segments in a process called crossing over. Because the versions of some genes on a chromosome differ from those on its homolog, crossing over rearranges genetic information.
As prophase I continues, the chromosomes coil tightly and a spindle forms. The nuclear envelope breaks into fragments, and the chromosome tetrads, captured by spindle microtubules, are moved toward the center of the cell.
Metaphase I
At metaphase I, the chromosome tetrads (consisting of a pair of homologous chromosomes, each with two sister chromatids) are aligned on the metaphase plate, midway between the two poles of the spindle. Spindle microtubules are attached to kinetochores at the centromeres. In each tetrad, the homologous chromosomes are held together at sites of crossing over. Notice that, for each tetrad, the spindle microtubules attached to one of the homologous chromosomes come from one pole of the cell, and the microtubules attached to the other homologous chromosome come from the opposite pole. With this arrangement, the homologous chromosomes of each tetrad are poised to move toward opposite poles of the cell.
Anaphase I
Anaphase I of meiosis is marked by the migration of chromosomes toward the two poles of the cell. In contrast to mitosis, however, the sister chromatids making up each doubled chromosome remain attached. Only the tetrads (pairs of homologous chromosomes) split up. Thus, in the drawing you see three still-doubled chromosomes moving toward each pole. If this were anaphase of mitosis, you would see six daughter chromosomes moving toward each pole.

▲ **Figure 8.13** The stages of meiosis

occurs during prophase I. At this stage, four chromatids (two sets of sister chromatids) are aligned and physically touching each other. When in this configuration, nonsister chromatids may trade segments. As you will learn in Module 8.17, this exchange of chromosome segments—called crossing over—shuffles genes, making an important contribution to the genetic variability that results from sexual reproduction.

? A cell has the haploid number of chromosomes, but each chromosome has two chromatids. The chromosomes are arranged singly at the center of the spindle. What is the meiotic stage?

● Metaphase II (because the chromosomes line up two by two in metaphase I)

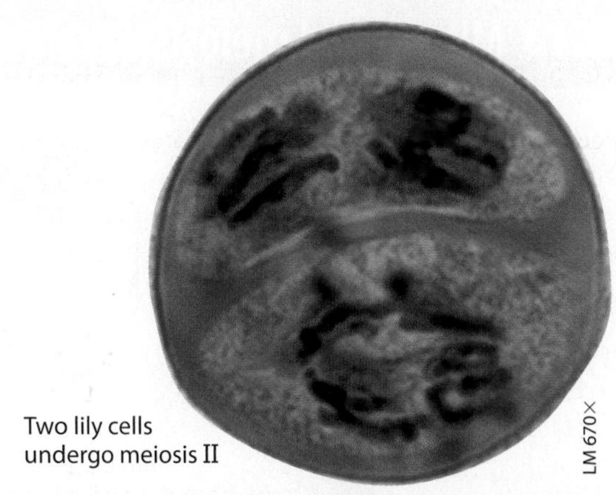
LM 670×

Two lily cells undergo meiosis II

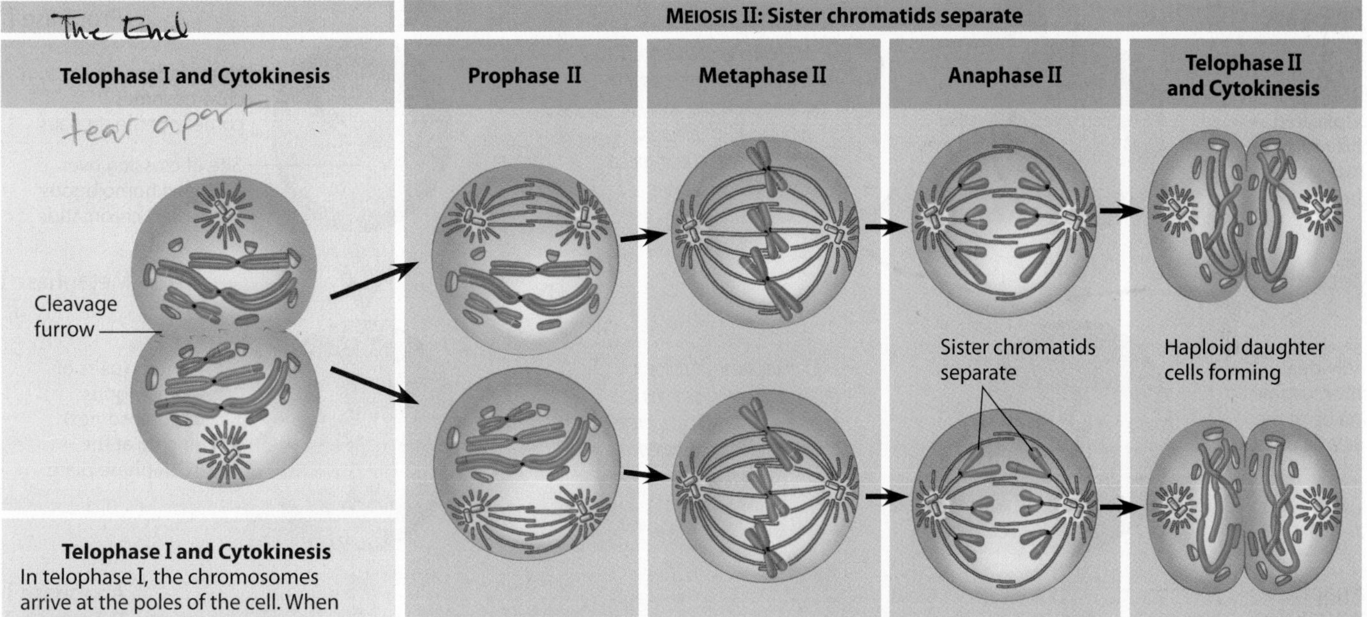
Meiosis II: Sister chromatids separate
Telophase I and Cytokinesis
Prophase II
Metaphase II
Anaphase II
Telophase II and Cytokinesis
Cleavage furrow
Sister chromatids separate
Haploid daughter cells forming

Telophase I and Cytokinesis

In telophase I, the chromosomes arrive at the poles of the cell. When the chromosomes finish their journey, each pole of the cell has a haploid chromosome set, although each chromosome is still in duplicate form (with two sister chromatids) at this point. Usually, cytokinesis (division of the cytoplasm) occurs along with telophase I, and two haploid daughter cells are formed.

Following telophase I in some organisms, there is an interphase between telophase I and meiosis II. In other species, meiosis I immediately leads to meiosis II. In either case, no chromosome duplication occurs between telophase I and the onset of meiosis II.

Meiosis II

Meiosis II is essentially the same as mitosis. The important difference is that meiosis II starts with a haploid cell.

During prophase II, a spindle forms and moves the chromosomes toward the middle of the cell. During metaphase II, the chromosomes are aligned on the metaphase plate as they are in mitosis, with the kinetochores of the sister chromatids of each chromosome pointing toward opposite poles. In anaphase II, the centromeres of sister chromatids separate, and the sister chromatids of each pair, now individual chromosomes, move toward opposite poles of the cell. In telophase II, nuclei form at the cell poles, and cytokinesis occurs at the same time. There are now four daughter cells, each with the haploid number of (single) chromosomes.

VISUALIZING THE CONCEPT

8.14 Mitosis and meiosis have important similarities and differences

Carefully review this module, which compares mitosis and meiosis starting from a diploid parent cell with four chromosomes. Homologous chromosomes match in size. Color distinguishes the two chromosomes of each homologous pair.

MITOSIS | **MEIOSIS I**

Parent cell
(before chromosome duplication)
$2n = 4$

Chromosome duplication
(Occurs once, during S phase of preceding interphase)

Prophase

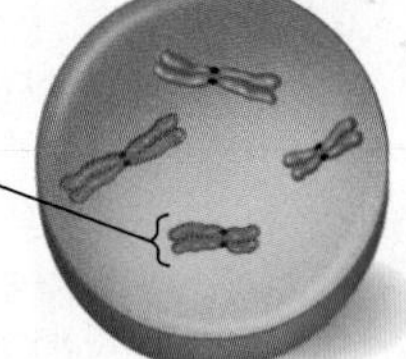

Duplicated chromosome (two sister chromatids)

In prophase of mitosis, each duplicated chromosome remains separate, while in prophase I of meiosis, chromosomes are associated with their homologs.

Prophase I

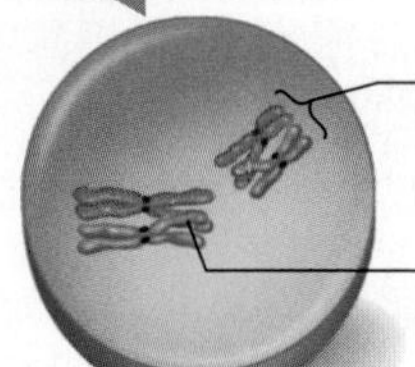

Homologous chromosomes come together in pairs

Site of crossing over between homologous (nonsister) chromatids

Metaphase

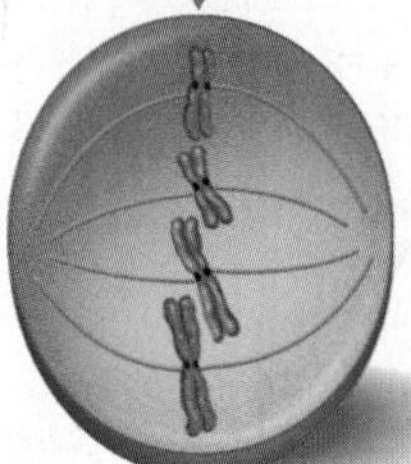

Individual chromosomes line up at the metaphase plate

In metaphase of mitosis, duplicated chromosomes line up singly, while in metaphase I of meiosis, duplicated homologous chromosomes line up in pairs.

Metaphase I

Tetrads (pairs of homologous chromosomes) line up at the metaphase plate

Anaphase Telophase

In anaphase of mitosis, sister chromatids separate, while in anaphase I of meiosis, pairs of homologous chromosomes separate.

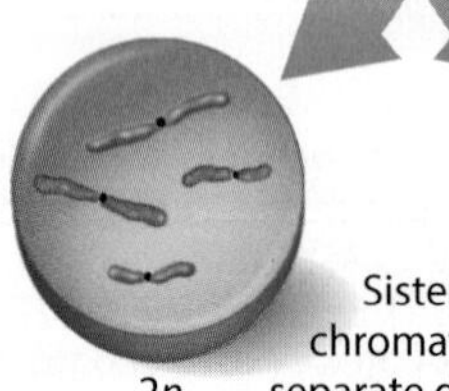

$2n$ — Sister chromatids separate during anaphase — $2n$

Anaphase I Telophase I

Homologous chromosomes separate during anaphase I; sister chromatids remain attached

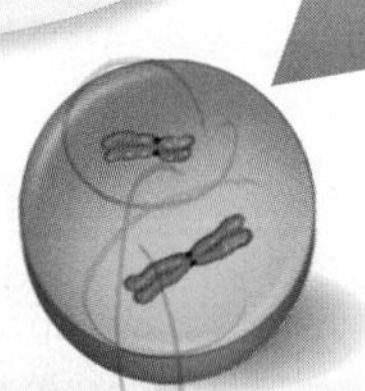

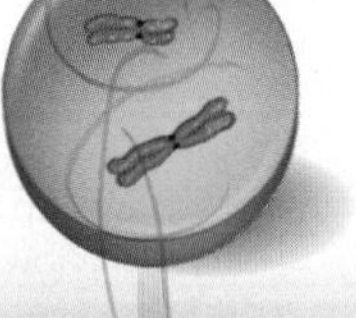

$n = 2$

Mitosis involves one division of the nucleus and cytoplasm, while meiosis involves two divisions.

MEIOSIS II

Sister chromatids separate

Sister chromatids separate during anaphase II

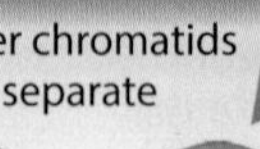

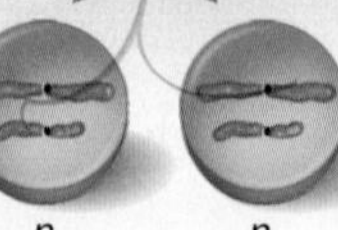

n n n n

Result:	Two genetically identical diploid cells
Used for:	Growth, tissue repair, asexual reproduction

Result:	Four genetically unique haploid cells
Used for:	Sexual reproduction

? Which stage of meiosis shown here most closely resembles mitosis?

● The movement of chromosomes during meiosis II very closely matches mitosis (except with half as many chromosomes).

8.15 Independent orientation of chromosomes in meiosis and random fertilization lead to varied offspring

Offspring that result from sexual reproduction are highly varied (see Module 8.1); they are genetically different from their parents and from one another. How does this genetic variation result from meiosis?

Figure 8.15 illustrates one way: The arrangement of homologous chromosome pairs at metaphase I affects the resulting gametes. Once again, our example is from a diploid organism with four chromosomes (two homologous pairs, with one set larger than the other to help make them distinct), and red represents chromosomes inherited from the mother, whereas blue represents chromosomes inherited from the father.

Recall that joined homologous chromosomes form tetrads, a set of four chromatids. At metaphase, the orientation of these tetrads—whether the maternal or paternal chromosome is closer to a given pole—is as random as the flip of a coin. Thus, there is a 50% chance that a particular daughter cell will get the maternal chromosome of a certain homologous pair and a 50% chance that it will receive the paternal chromosome. In this example, there are two possible ways that the two tetrads can align during metaphase I. In possibility A, the tetrads are oriented with both red chromosomes on the same side of the metaphase plate. Therefore, the gametes produced in possibility A can each have either two red *or* two blue chromosomes (bottom row, combinations 1 and 2).

In possibility B, the tetrads are oriented differently (blue/red and red/blue). This arrangement produces gametes that each have one red and one blue chromosome. Furthermore, half the gametes have a big blue chromosome and a small red one (combination 3), and half have a big red chromosome and a small blue one (combination 4).

So we see that for this example, four chromosome combinations are possible in the gametes. In fact, the organism will produce gametes of all four types in equal quantities. For a species with more than two pairs of chromosomes, such as humans, *all* the chromosome pairs orient independently at metaphase I. (Chromosomes X and Y behave as a homologous pair in meiosis.)

For any species, the total number of combinations of chromosomes that meiosis can produce in gametes is 2^n, where n is the haploid number. For the organism in this figure, $n = 2$, so the number of chromosome combinations is 2^2, or 4. For a human ($n = 23$), there are 2^{23}, or about 8 million possible chromosome combinations! This means that each gamete you produce contains one of roughly 8 million possible combinations of chromosomes.

How many possibilities are there when a gamete from one individual unites with a gamete from another individual in fertilization? In humans, the random fusion of a single sperm with a single egg during fertilization will produce a zygote with any of about 64 trillion (8 million × 8 million) combinations of chromosomes! Although the random nature of fertilization adds a huge amount of potential variability to the offspring of sexual reproduction, there is in fact even more variety created during meiosis, as we see in the next two modules.

? A particular species of worm has a diploid number of 10. How many chromosomal combinations are possible for gametes formed by meiosis?

● 32; $2n = 10$, so $n = 5$ and $2^n = 32$

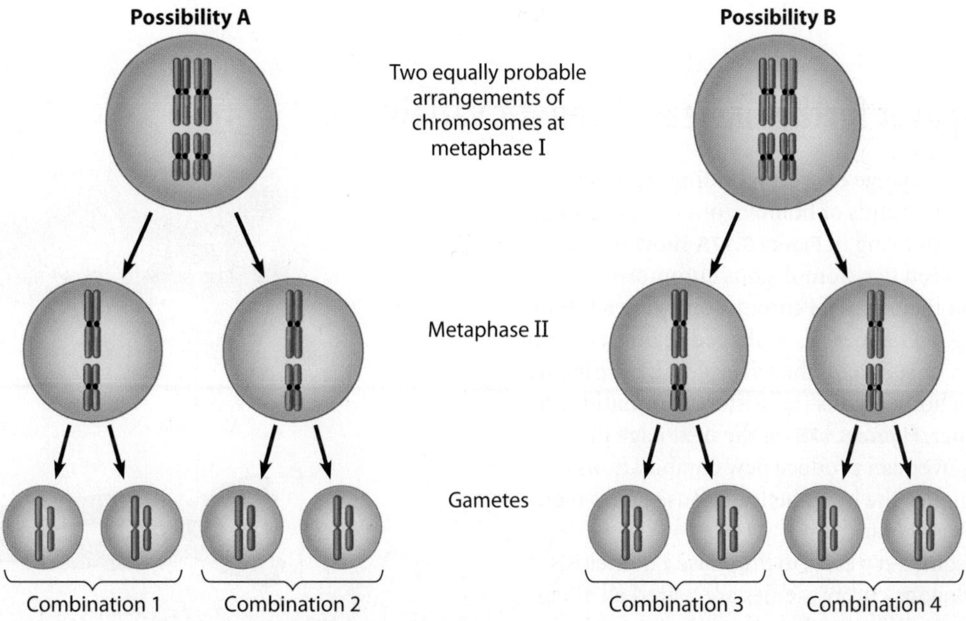

Possibility A
Possibility B
Two equally probable arrangements of chromosomes at metaphase I
Metaphase II
Gametes
Combination 1
Combination 2
Combination 3
Combination 4

▲ **Figure 8.15** Results of the independent orientation of chromosomes at metaphase I

8.16 Homologous chromosomes may carry different versions of genes

So far, we have focused on genetic variability in gametes and zygotes at the whole-chromosome level. We have yet to discuss the actual genetic information—the genes—contained in the chromosomes. The question we need to answer now is: What is the significance of the independent orientation of metaphase chromosomes at the level of genes?

Let's take a simple example, the single tetrad in **Figure 8.16**. The letters on the homologous chromosomes represent genes. Recall that homologous chromosomes have genes for the same characteristic at corresponding loci. Our example involves hypothetical genes controlling the appearance of mice. *C* and *c* indicate different versions of a gene for one characteristic, coat color; *E* and *e* are different versions of a gene for another characteristic, eye color. (As you'll learn in later chapters, different versions of a gene contain slightly different nucleotide sequences in the chromosomal DNA.)

Let's say that *C* represents the gene for a brown coat and that *c* represents the gene for a white coat. In the chromosome diagram, notice that *C* is at the same locus on the red chromosome as *c* is on the blue one. Likewise, gene *E* (for black eyes) is at the same locus as *e* (pink eyes).

The fact that homologous chromosomes can bear two different kinds of genetic information for the same characteristic (for instance, coat color) is what really makes gametes—and therefore offspring—different from one another. In our example, a gamete carrying a red chromosome would have genes specifying brown coat color and black eye color, whereas a gamete with the homologous blue chromosome would have genes for white coat and pink eyes. Thus, we see that a tetrad can yield two genetically different kinds of gametes. In the next module, we go a step further and see how this same tetrad can actually yield *four* different kinds of gametes.

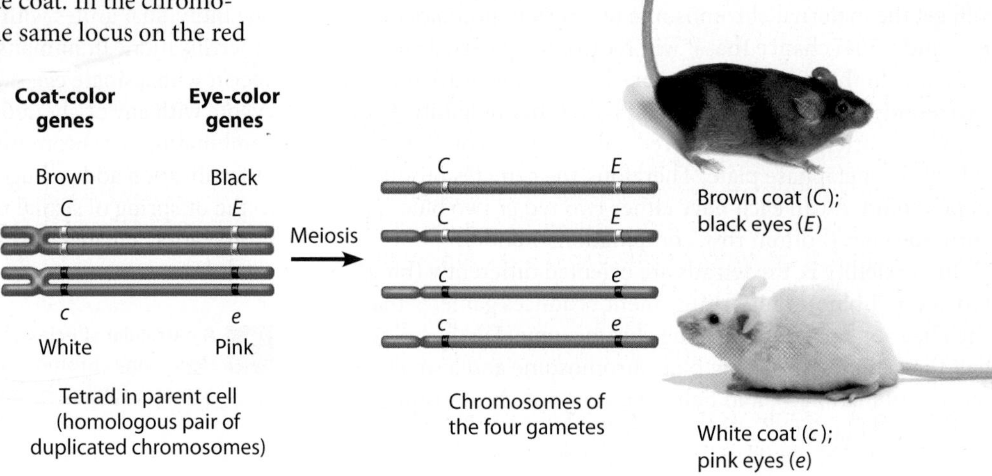
Coat-color genes
Eye-color genes
Brown
Black
C
E
c
e
White
Pink
Tetrad in parent cell (homologous pair of duplicated chromosomes)
Meiosis
C
E
C
E
c
e
c
e
Chromosomes of the four gametes
Brown coat (C); black eyes (E)
White coat (c); pink eyes (e)

▲ **Figure 8.16** Differing genetic information (coat color and eye color) on homologous chromosomes

? **In the tetrad of Figure 8.16, use labels to distinguish the pair of homologous chromosomes from sister chromatids.**

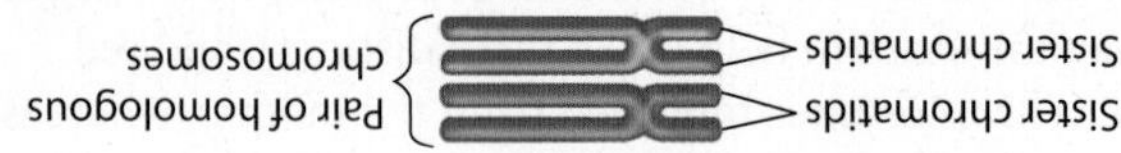
Sister chromatids
Sister chromatids
Pair of homologous chromosomes

8.17 Crossing over further increases genetic variability

Crossing over is an exchange of corresponding segments between nonsister chromatids of homologous chromosomes. The micrograph and drawing in **Figure 8.17A** show the results of crossing over between two homologous chromosomes during prophase I of meiosis. The chromosomes are a tetrad. The sites of crossing over appear as X-shaped regions; each of these sites is called a **chiasma**. A chiasma (plural, *chiasmata*) is a place where two homologous (nonsister) chromatids are attached to each other. **Figure 8.17B** on the next page illustrates how crossing over can produce new combinations of genes, using as examples the hypothetical mouse genes mentioned in the previous module.

Crossing over begins very early in prophase I of meiosis. At that time, homologous chromosomes are paired all along their lengths, with a precise gene-by-gene alignment. Notice

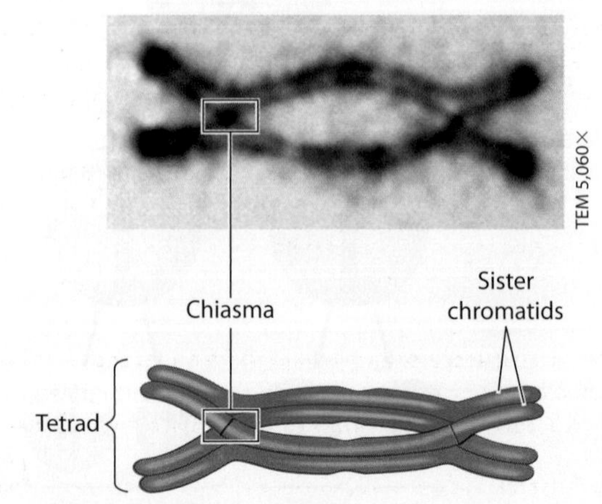
TEM 5,060×
Chiasma
Sister chromatids
Tetrad

▲ **Figure 8.17A** Chiasmata, the sites of crossing over

at the top of the figure a tetrad with coat-color (*C, c*) and eye-color (*E, e*) genes labeled. ❶ The DNA molecules of two nonsister chromatids—one maternal (red) and one paternal (blue)—break at the same place. ❷ Immediately, the two broken chromatids join together in a new way (red to blue and blue to red). In effect, the two homologous segments trade places, or cross over, producing hybrid chromosomes (red/blue and blue/red) with new combinations of maternal and paternal genes. ❸ When the homologous chromosomes separate in anaphase I, each contains a new segment originating from its homolog. ❹ Finally, during anaphase II, the sister chromatids separate, each going to a different gamete.

In this example, if there were no crossing over, meiosis could produce only two genetic types of gametes. These would be the ones ending up with the "parental" types of chromosomes (either all blue or all red), carrying either genes *C* and *E* or genes *c* and *e*. These are the same two kinds of gametes we saw in Figure 8.16. With crossing over, two other types of gametes can result, ones that are part blue and part red. One of these carries genes *C* and *e* and the other carries genes *c* and *E*. Chromosomes with these combinations of genes would not exist if not for crossing over. They are called "recombinant" because they result from **genetic recombination**, the production of gene combinations different from those carried by the original parental chromosomes.

In meiosis in humans, an average of one to three crossover events occur per chromosome pair. Thus, if you were to examine a chromosome from one of your gametes, you would most likely find that it is not exactly like any one of your own chromosomes. Rather, it is probably a patchwork of segments derived from a pair of homologous chromosomes, in essence cut and pasted together to form a hybrid chromosome with a unique combination of genes.

We have now examined three sources of genetic variability in sexually reproducing organisms: independent orientation of chromosomes at metaphase I, random fertilization, and crossing over during prophase I of meiosis. The different versions of genes that homologous chromosomes may have at each locus originally arise from mutations (changes in the sequence of DNA), so mutations are ultimately responsible for genetic diversity in living organisms. Once these differences arise, reshuffling of the different versions during sexual reproduction increases genetic variation. (When we discuss natural selection and evolution in Unit III, we will see that this genetic variety in offspring is the raw material for natural selection.)

Our discussion of meiosis to this point has focused on the process as it normally occurs. In the next section, we consider some of the consequences of errors in the process.

? **Describe how crossing over and the random alignment of homologous chromosomes on the metaphase I plate account for the genetic variation among gametes formed by meiosis.**

● Crossing over creates recombinant chromosomes having a combination of genes that were originally on different, though homologous, chromosomes. Homologous chromosome pairs are oriented randomly at metaphase of meiosis I.

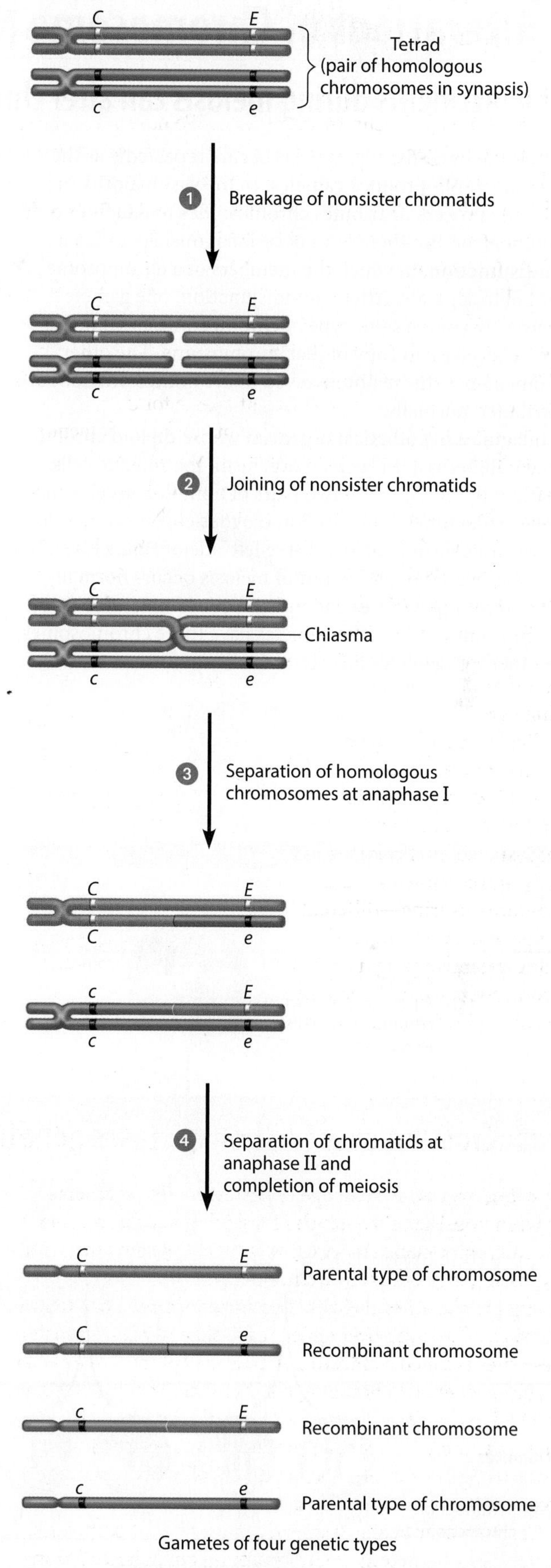

C
E
c
e
Tetrad
(pair of homologous chromosomes in synapsis)
❶ Breakage of nonsister chromatids
❷ Joining of nonsister chromatids
Chiasma
❸ Separation of homologous chromosomes at anaphase I
❹ Separation of chromatids at anaphase II and completion of meiosis
Parental type of chromosome
Recombinant chromosome
Recombinant chromosome
Parental type of chromosome
Gametes of four genetic types

▲ **Figure 8.17B** How crossing over leads to genetic recombination

▷ Alterations of Chromosome Number and Structure

8.18 Accidents during meiosis can alter chromosome number

Within the human body, meiosis occurs repeatedly as the testes or ovaries produce gametes. In the vast majority of cases, the process distributes chromosomes to daughter cells without error. But there is an occasional mishap, called a **nondisjunction**, in which the members of a chromosome pair fail to separate. After a nondisjunction, one gamete receives two of the same type of chromosome and another gamete receives no copy of that chromosome. The other chromosomes (those not involved in the nondisjunction) are distributed normally.

Imagine a hypothetical organism whose diploid chromosome number is 4. In such an organism, the somatic cells are diploid ($2n = 4$), with two pairs of homologous chromosomes. Sometimes, a pair of homologous chromosomes does not separate during meiosis I (see left side of **Figure 8.18**). In this case, even though the rest of meiosis occurs normally, all the resulting gametes end up with abnormal numbers of chromosomes. Two of the gametes have three chromosomes; the other two gametes have only one chromosome each.

Alternatively, sometimes meiosis I procedes normally, but one pair of sister chromatids fails to separate during meiosis II (see right side of Figure 8.18). In this case, two of the resulting gametes are abnormal—one with an extra chromosome and one that is missing a chromosome; the other two gametes are normal.

If an abnormal gamete produced by nondisjunction unites with a normal gamete during fertilization, the result is a zygote with an abnormal number of chromosomes. Mitosis will then transmit the mistake to all embryonic cells. If the organism survived, it would most likely display a syndrome of disorders caused by the abnormal number of genes. Biologists can detect such syndromes by taking an inventory of the chromosomes in a person's cells, as we'll see next.

? Explain how nondisjunction could result in a diploid gamete.

● A diploid gamete would result if the nondisjunction affected all the chromosomes during one of the meiotic divisions.

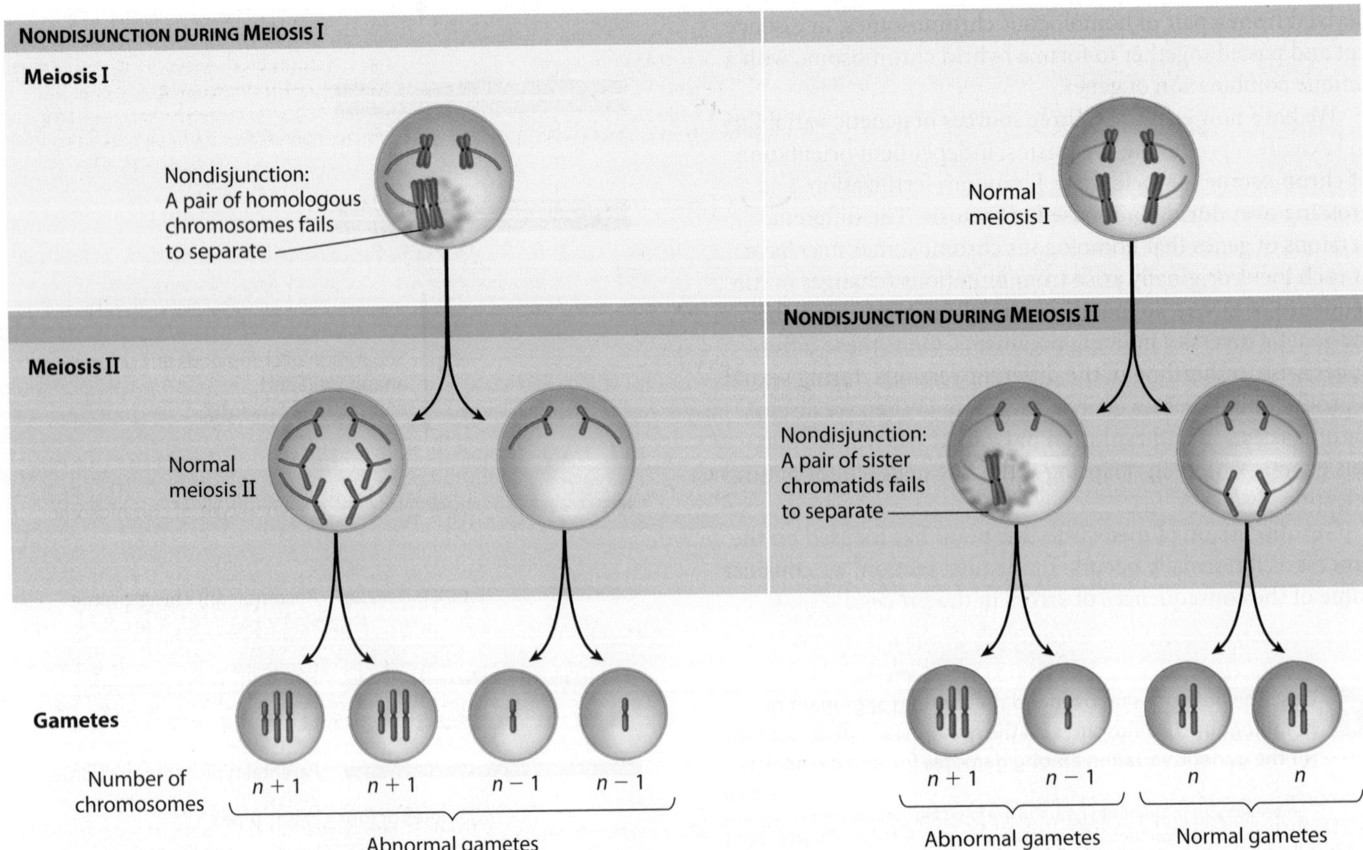

NONDISJUNCTION DURING MEIOSIS I
Meiosis I
Nondisjunction: A pair of homologous chromosomes fails to separate
Meiosis II
Normal meiosis II
Gametes
Number of chromosomes
n + 1
n + 1
n − 1
n − 1
Abnormal gametes
Normal meiosis I
NONDISJUNCTION DURING MEIOSIS II
Nondisjunction: A pair of sister chromatids fails to separate
n + 1
n − 1
n
n
Abnormal gametes
Normal gametes

▲ **Figure 8.18** Nondisjunction in meiosis I and meiosis II

8.19 A karyotype is a photographic inventory of an individual's chromosomes

Chromosomal abnormalities can be readily detected in a **karyotype**, an ordered display of magnified images of an individual's chromosomes arranged in pairs. A karyotype shows the chromosomes condensed and doubled, as they appear in metaphase of mitosis.

To prepare a karyotype, scientists often use lymphocytes, a type of white blood cell. A blood sample is treated with a chemical that stimulates mitosis. After growing in culture for several days, the cells are treated with another chemical to arrest mitosis at metaphase, when the chromosomes, each consisting of two joined sister chromatids, are most highly condensed. **Figure 8.19** outlines the steps of one method for the preparation of a karyotype from a blood sample.

The photograph on the right shows the karyotype of a normal human male. Images of the 46 chromosomes from a single diploid cell are arranged in 23 homologous pairs: autosomes numbered from 1 to 22 (starting with the largest) and one pair of sex chromosomes (X and Y in this case). The chromosomes have been stained to reveal band patterns, which are helpful in differentiating the chromosomes and in detecting structural abnormalities. Among the alterations that can be detected by karyotyping is trisomy 21, the basis of Down syndrome, which we discuss next.

? How would the karyotype of a human female differ from the male karyotype in Figure 8.19?

● Instead of an XY combination for the sex chromosomes, there would be a homologous pair of X chromosomes (XX).

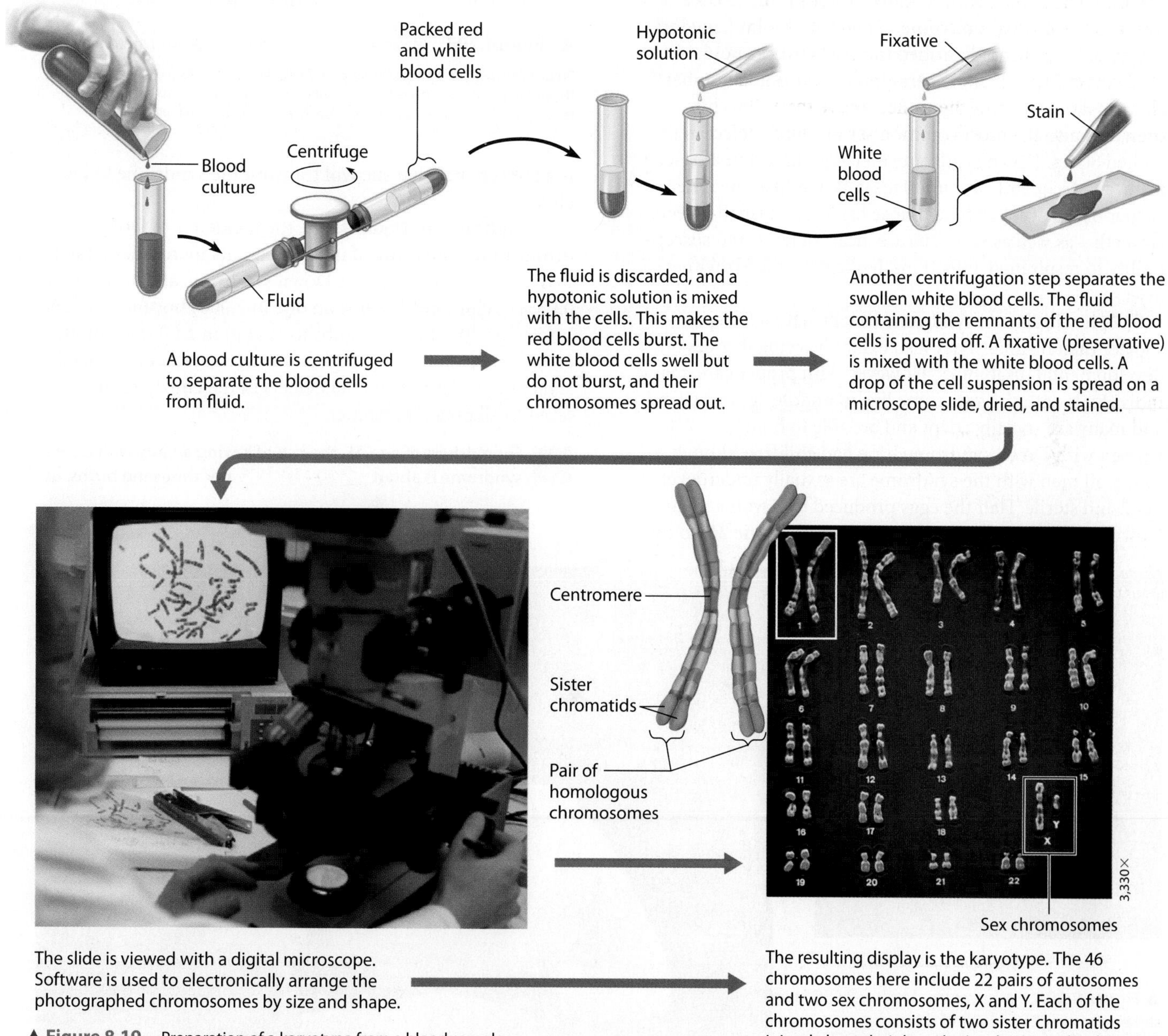

Packed red and white blood cells
Hypotonic solution
Fixative
Stain
Blood culture
Centrifuge
White blood cells
Fluid
A blood culture is centrifuged to separate the blood cells from fluid.
The fluid is discarded, and a hypotonic solution is mixed with the cells. This makes the red blood cells burst. The white blood cells swell but do not burst, and their chromosomes spread out.
Another centrifugation step separates the swollen white blood cells. The fluid containing the remnants of the red blood cells is poured off. A fixative (preservative) is mixed with the white blood cells. A drop of the cell suspension is spread on a microscope slide, dried, and stained.
Centromere
Sister chromatids
Pair of homologous chromosomes
1
2
3
4
5
6
7
8
9
10
11
12
13
14
15
16
17
18
Y
X
19
20
21
22
3,330×
Sex chromosomes
The slide is viewed with a digital microscope. Software is used to electronically arrange the photographed chromosomes by size and shape.
The resulting display is the karyotype. The 46 chromosomes here include 22 pairs of autosomes and two sex chromosomes, X and Y. Each of the chromosomes consists of two sister chromatids joined along their lengths (as shown in the diagram).

▲ **Figure 8.19** Preparation of a karyotype from a blood sample

8.20 An extra copy of chromosome 21 causes Down syndrome

CONNECTION

The karyotype in Figure 8.19 shows the normal human complement of 23 pairs of chromosomes. Compare this figure with the karyotype shown in **Figure 8.20A**; besides having two X chromosomes (because it's from a female), notice that there are three number 21 chromosomes, making 47 chromosomes in total. This condition is called **trisomy 21**.

In most cases, an abnormal number of chromosomes is so harmful to development that an affected embryo is spontaneously aborted (miscarried) long before birth. But some aberrations in chromosome number, including trisomy 21, appear to upset the genetic balance less drastically, and individuals carrying such chromosomal abnormalities can survive into adulthood. Individuals with chromosomal abnormalities have a characteristic set of symptoms, collectively called a syndrome. A person with trisomy 21, for instance, has a condition called **Down syndrome**, named after John Langdon Down, a doctor who described the syndrome in 1866.

Trisomy 21 is the most common chromosome number abnormality. Affecting about one out of every 700 children, it is also the most common serious birth defect in the United States. Down syndrome includes characteristic facial features—frequently a round face, a skin fold at the inner corner of the eye, a flattened nose bridge, and small, irregular teeth—as well as short stature, heart defects, and susceptibility to respiratory infections, leukemia, and Alzheimer's disease.

People with Down syndrome usually have a life span shorter than normal. They also exhibit varying degrees of developmental disabilities. However, with proper care, many individuals with the syndrome live to middle age or beyond, and many are socially adept and are able to hold jobs. A few women with Down syndrome have had children, though nearly all men with the syndrome are sexually underdeveloped and sterile. Half the eggs produced by a woman with Down syndrome will have the extra chromosome 21, so there is a 50% chance that she will transmit the syndrome to her child.

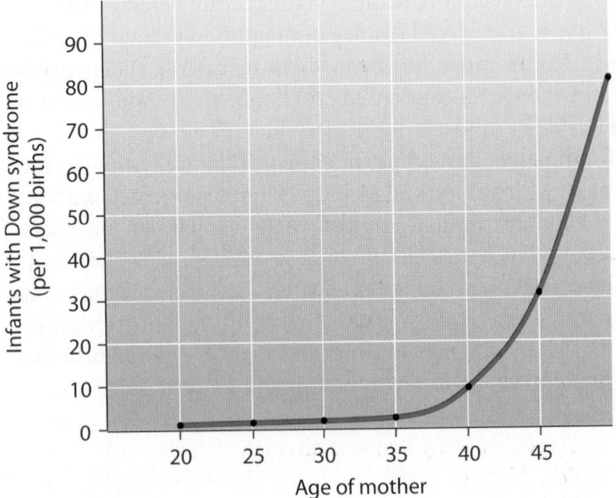

Infants with Down syndrome (per 1,000 births)
90
80
70
60
50
40
30
20
10
0
20
25
30
35
40
45
Age of mother

▲ **Figure 8.20B** Maternal age and incidence of Down syndrome

Source: Adapted from C. A. Huether et al., Maternal age specific risk rate estimates for Down syndrome among live births in whites and other races from Ohio and Metropolitan Atlanta, 1970–1989, *Journal of Medical Genetics* 35: 482–90 (1998).

As indicated in **Figure 8.20B**, the incidence of Down syndrome in the offspring of normal parents increases markedly with the age of the mother. Down syndrome affects less than 0.05% of children (fewer than one in 2,000) born to women under age 30. The risk climbs to 1% (10 in 1,000) for mothers at age 40 and is even higher for older mothers. Prenatal screening for chromosomal defects in the embryo is now offered to all pregnant women.

? **For mothers of age 47, the risk of having a baby with Down syndrome is about ____________ per thousand births, or ____________ %.**

40 . . . 4

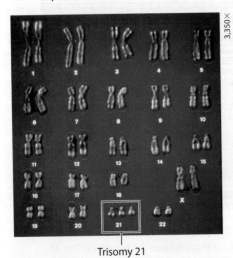

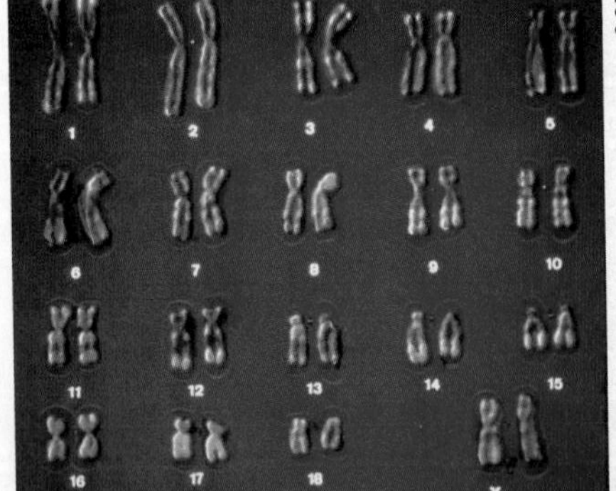

1
2
3
4
5
6
7
8
9
10
11
12
13
14
15
16
17
18
X
19
20
21
22
Trisomy 21

3,350×

▲ **Figure 8.20A** A karyotype showing trisomy 21 and an individual with Down syndrome

8.21 Abnormal numbers of sex chromosomes do not usually affect survival

CONNECTION

Nondisjunction can result in abnormal numbers of sex chromosomes, X and Y. Unusual numbers of sex chromosomes seem to upset the genetic balance less than unusual numbers of autosomes. This may be because the Y chromosome is very small and carries relatively few genes. Furthermore, mammalian cells usually operate with only one functioning X chromosome because other copies of the chromosome become inactivated in each cell (as you'll learn in Module 11.2).

Table 8.21 lists the most common human sex chromosome abnormalities. An extra X chromosome in a male, making him XXY, occurs approximately once in every 1,000 live male births. Men with this disorder, called Klinefelter syndrome, have male sex organs, but the testes are abnormally small, the individual is sterile, and he often has female body characteristics. Affected individuals may have subnormal intelligence. Klinefelter syndrome is also found in individuals with more than three sex chromosomes, such as XXYY, XXXY, or XXXXY. These abnormal numbers of sex chromosomes result from multiple nondisjunctions; such men are more likely to have developmental disabilities than XY or XXY individuals.

Human males with an extra Y chromosome do not have any well-defined syndrome, although they tend to be taller than average. Females with an extra X chromosome cannot be distinguished from XX females except by karyotype.

Females who lack an X chromosome are designated XO; the O indicates the absence of a second sex chromosome. These women have Turner syndrome. They have a characteristic appearance, including short stature and often a web of skin extending between the neck and the shoulders. Women with Turner syndrome are sterile because their sex organs do not fully mature at adolescence. If left untreated, girls with Turner syndrome have poor development of breasts and other secondary sexual characteristics. Artificial administration of estrogen can alleviate these symptoms. Women with Turner syndrome have normal intelligence. The XO condition is the sole known case where having only 45 chromosomes is not fatal in humans.

? What is the total number of autosomes you would expect to find in the karyotype of a female with Turner syndrome?

● 44 (plus one sex chromosome)

TABLE 8.21 | ABNORMALITIES OF SEX CHROMOSOME NUMBER IN HUMANS

Sex Chromosomes	Syndrome	Origin of Nondisjunction	Frequency in Population
XXY	Klinefelter syndrome (male)	Meiosis in egg or sperm formation	1/1,000 live male births
XYY	None (normal male)	Meiosis in sperm formation	1/1,000 live male births
XXX	None (normal female)	Meiosis in egg or sperm formation	1/1,000 live female births
XO	Turner syndrome (female)	Meiosis in egg or sperm formation	1/2,500 live female births

8.22 New species can arise from errors in cell division

EVOLUTION CONNECTION

Errors in meiosis or mitosis do not always lead to problems. In fact, biologists hypothesize that such errors have been instrumental in the evolution of many species. Such new species are polyploid, meaning that they have more than two sets of homologous chromosomes in each somatic cell. At least half of all species of flowering plants are polyploid, including such crops as wheat, potatoes, and cotton.

Let's consider one scenario by which a diploid ($2n$) plant species might generate a tetraploid ($4n$) plant. Imagine that, like many plants, our diploid plant produces both sperm and egg cells and can self-fertilize. If meiosis fails to occur in the plant's reproductive organs and gametes are instead produced by mitosis, the gametes will be diploid. The union of a diploid ($2n$) sperm with a diploid ($2n$) egg during self-fertilization will produce a tetraploid ($4n$) zygote, which may develop into a mature tetraploid plant that can itself reproduce by self-fertilization. The tetraploid plants will constitute a new species, one that has evolved in just one generation. Although polyploid animal species are less common than polyploid plants, they are known to occur among the fishes and amphibians (**Figure 8.22**). Moreover, researchers in Chile have identified the first candidate for polyploidy among the mammals—the Viscacha rat (*Tympanoctomys barrerae*), a rodent whose cells seem to be tetraploid. Tetraploid organisms are sometimes strikingly different from their recent diploid ancestors. Scientists don't yet understand exactly how polyploidy brings about such differences.

▲ **Figure 8.22** The gray tree frog (*Hyla versicolor*), a tetraploid organism

? What is a polyploid organism?

● An organism with more than two sets of homologous chromosomes in its body cells

8.23 Alterations of chromosome structure can cause birth defects and cancer

CONNECTION

Errors in meiosis or damaging agents such as radiation can cause a chromosome to break, which can lead to four types of changes in chromosome structure (**Figure 8.23A**). A **deletion** occurs when a chromosomal fragment (along with its genes) becomes detached. The "deleted" fragment may disappear from the cell, or it may become attached as an extra segment to its sister chromatid or a homologous chromosome, producing a **duplication**. A chromosomal fragment may also reattach to the original chromosome but in the reverse orientation, producing an **inversion**. A fourth possible result of chromosomal breakage is for the fragment to join a nonhomologous chromosome, a rearrangement called a **translocation**. As shown in the figure, a translocation may be reciprocal; that is, two nonhomologous chromosomes may exchange segments.

Inversions are less likely than deletions or duplications to produce harmful effects, because in inversions all genes are still present in their normal number. Many deletions in human chromosomes, however, cause serious physical and mental problems. One example is a specific deletion in chromosome 5 that causes *cri du chat* ("cat-cry") syndrome. A child born with this syndrome has severe developmental disabilities, a small head with unusual facial features, and a cry that sounds like the mewing of a distressed cat. Such individuals usually die in infancy or early childhood.

Like inversions, translocations may or may not be harmful. Some people with Down syndrome have only part of a third chromosome 21; as the result of a translocation, this partial chromosome is attached to another (nonhomologous) chromosome. Other chromosomal translocations have been implicated in certain cancers, including chronic myelogenous leukemia (CML). CML develops after a reciprocal translocation during mitosis of cells that will become white blood cells. In these cells, the exchange of a large portion of chromosome 22 with a small fragment from a tip of chromosome 9 produces a much shortened, easily recognized chromosome 22 (**Figure 8.23B**). Such an exchange causes cancer by activating a gene that leads to uncontrolled cell cycle progression.

Because the chromosomal changes in cancer are usually confined to somatic cells, cancer is not usually inherited. (We'll return to cancer in Chapter 11.) We continue our study of genetic principles (in Chapter 9), looking first at the historical development of the science of genetics and then at the rules governing the way traits are passed from parents to offspring.

? **How is reciprocal translocation different from crossing over?**

Reciprocal translocation swaps chromosome segments between nonhomologous chromosomes. Crossing over exchanges corresponding segments between homologous chromosomes.

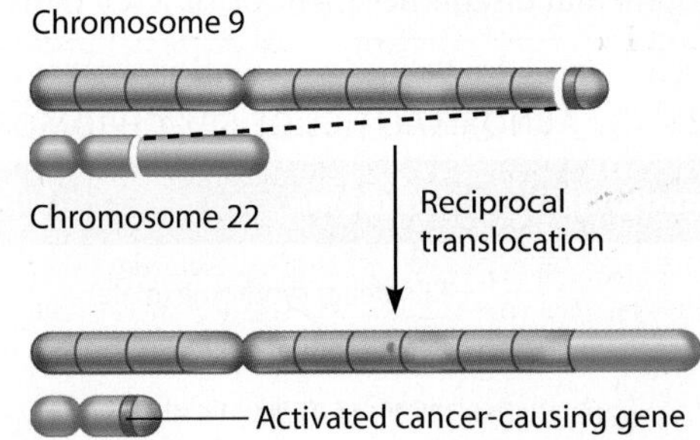
Chromosome 9
Chromosome 22
Reciprocal translocation
Activated cancer-causing gene

▲ **Figure 8.23B** The translocation associated with chronic myelogenous leukemia

Deletion

A segment of a chromosome is removed

Inversion

A segment of a chromosome is removed and then reinserted opposite to its original orientation

Duplication

A segment of a chromosome is copied and inserted into the homologous chromosome

Homologous chromosomes

Reciprocal translocation

Segments of two nonhomologous chromosomes swap locations with each other

Nonhomologous chromosomes

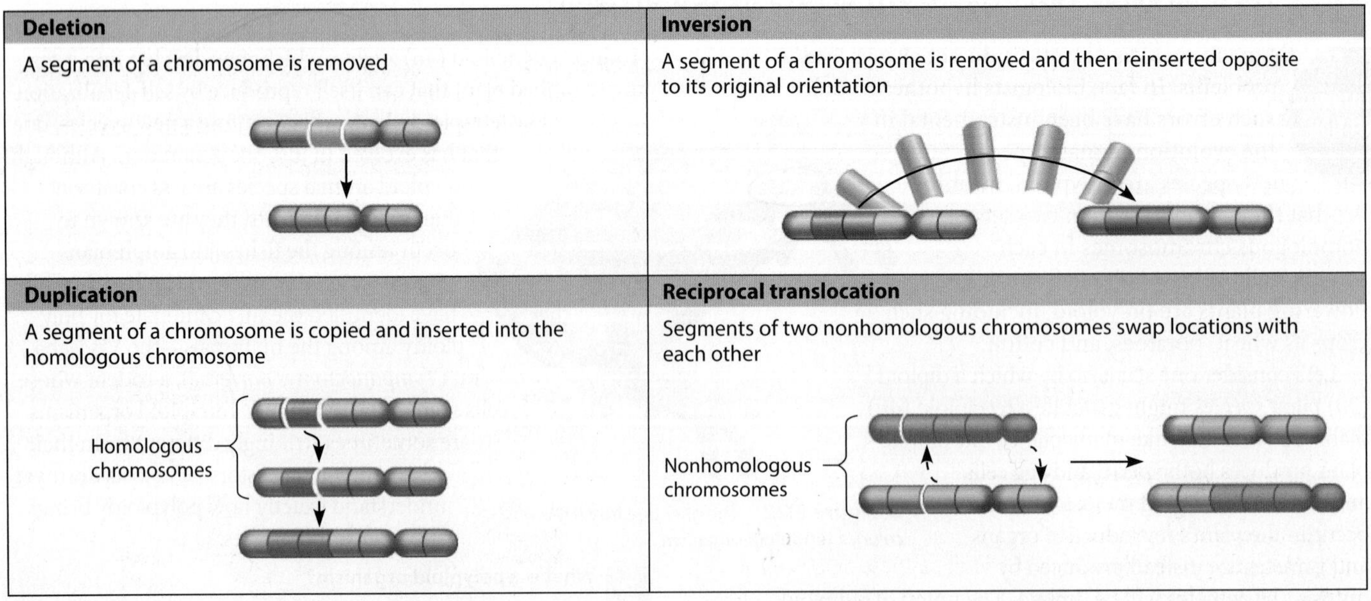

▲ **Figure 8.23A** Alterations of chromosome structure

For practice quizzes, BioFlix animations, MP3 tutorials, video tutors, and more study tools designed for this textbook, go to MasteringBiology®

Reviewing the Concepts

Cell Division and Reproduction (8.1–8.2)

8.1 Cell division plays many important roles in the lives of organisms. Cell division is at the heart of the reproduction of cells and organisms because cells originate only from preexisting cells. Some organisms reproduce through asexual reproduction, and in such instances their offspring are all genetic copies of the parent and identical to each other. Other organisms reproduce through sexual reproduction, creating a variety of offspring.

8.2 Prokaryotes reproduce by binary fission. Prokaryotic cells reproduce asexually by cell division. As the cell replicates its single chromosome, the copies move apart; the growing membrane then divides the cell.

The Eukaryotic Cell Cycle and Mitosis (8.3–8.10)

8.3 The large, complex chromosomes of eukaryotes duplicate with each cell division. A eukaryotic cell has many more genes than a prokaryotic cell, and they are grouped into multiple chromosomes in the nucleus. Each chromosome contains one long DNA molecule. Individual chromosomes are visible under a light microscope only when the cell is in the process of dividing; otherwise, chromosomes are thin, loosely packed chromatin fibers too small to be seen. Before a cell starts dividing, the chromosomes duplicate, producing sister chromatids (containing identical DNA) that are joined together along their lengths. Cell division involves the separation of sister chromatids and results in two daughter cells, each containing a complete and identical set of chromosomes.

8.4 The cell cycle includes growing and division phases.

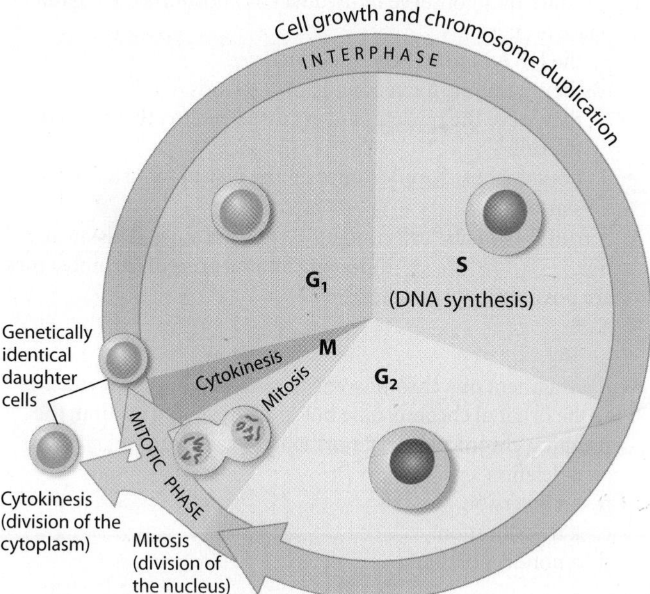
Cell growth and chromosome duplication
INTERPHASE
G_1
S
(DNA synthesis)
M
G_2
Cytokinesis
Mitosis
MITOTIC PHASE
Genetically identical daughter cells
Cytokinesis (division of the cytoplasm)
Mitosis (division of the nucleus)

8.5 Cell division is a continuum of dynamic changes. Mitosis distributes duplicated chromosomes into two daughter nuclei. After the chromosomes are coiled up, a mitotic spindle made of microtubules moves the chromosomes to the middle of the cell. The sister chromatids then separate and move to opposite poles of the cell, at which point two new nuclei form.

8.6 Cytokinesis differs for plant and animal cells. Cytokinesis, in which the cell divides in two, overlaps the end of mitosis. In animals, cytokinesis occurs when a cell constricts, forming a cleavage furrow. In plants, a membranous cell plate forms and then splits the cell in two.

8.7 Anchorage, cell density, and chemical growth factors affect cell division. In laboratory cultures, most normal cells divide only when attached to a surface. The cultured cells continue dividing until they touch one another. Most animal cells divide only when stimulated by growth factors, and some do not divide at all. Growth factors stimulate other cells to divide.

8.8 Growth factors signal the cell cycle control system. A set of proteins within the cell controls the cell cycle. Signals affecting critical checkpoints in the cell cycle determine whether a cell will go through the complete cycle and divide. The binding of growth factors to specific receptors on the plasma membrane is usually necessary for cell division.

8.9 Growing out of control, cancer cells produce malignant tumors. Cancer cells divide excessively to form masses called tumors. Malignant tumors can invade other tissues. Radiation and chemotherapy are effective as cancer treatments because they interfere with cell division.

8.10 Tailoring treatment to each patient may improve cancer therapy. Determination of specific mutations within the cells of an individual tumor may aid in treatment. Although reliable data are scant today, such personalized medicine may prove to be an important treatment regimen in the future.

Meiosis and Crossing Over (8.11–8.17)

8.11 Chromosomes are matched in homologous pairs. The somatic (body) cells of each species contain a specific number of chromosomes; for example, human cells have 46, consisting of 23 pairs of homologous chromosomes. The chromosomes of a homologous pair of autosomes carry genes for the same characteristics at the same place, or locus.

8.12 Gametes have a single set of chromosomes. Cells with two sets of homologous chromosomes are diploid. Gametes—eggs and sperm—are haploid cells with a single set of chromosomes. Sexual life cycles involve the alternation of haploid and diploid stages.

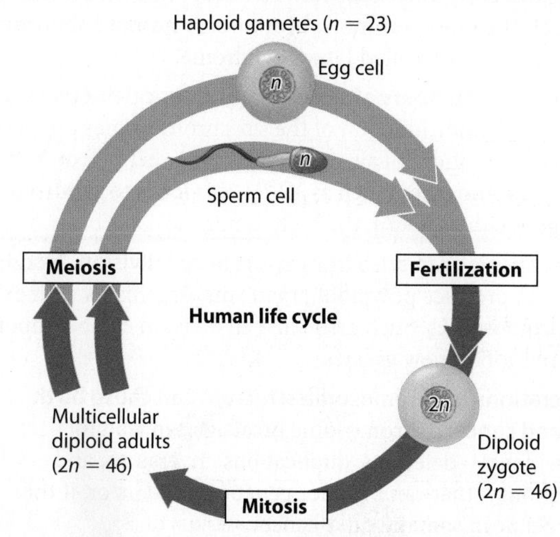
Haploid gametes ($n = 23$)
Egg cell
n
Sperm cell
n
Meiosis
Fertilization
Human life cycle
$2n$
Multicellular diploid adults ($2n = 46$)
Diploid zygote ($2n = 46$)
Mitosis

8.13 Meiosis reduces the chromosome number from diploid to haploid. Meiosis, like mitosis, is preceded by chromosome duplication, but in meiosis, the cell divides twice to form four daughter cells. The first division, meiosis I, starts with the pairing of homologous chromosomes. In crossing over, homologous chromosomes exchange corresponding segments. Meiosis I separates the members of each homologous pair and produces two daughter cells, each with one set of chromosomes. Meiosis II is essentially the same as mitosis: In each of the cells, the sister chromatids of each chromosome separate. The result is a total of four haploid cells.

8.14 Mitosis and meiosis have important similarities and differences. Both mitosis and meiosis begin with diploid parent cells that have chromosomes duplicated during the previous interphase. Mitosis produces two genetically identical diploid somatic daughter cells, whereas meiosis produces four genetically unique haploid gametes.

8.15 Independent orientation of chromosomes in meiosis and random fertilization lead to varied offspring. Each chromosome of a homologous pair differs at many points from the other member of the pair. Random arrangements of chromosome pairs at metaphase I of meiosis lead to many different combinations of chromosomes in eggs and sperm. Random fertilization of eggs by sperm greatly increases this variation.

8.16 Homologous chromosomes may carry different versions of genes. The differences between homologous chromosomes come from the fact that they can bear different versions of genes at corresponding loci.

8.17 Crossing over further increases genetic variability. Genetic recombination, which results from crossing over during prophase I of meiosis, increases variation still further.

Alterations of Chromosome Number and Structure (8.18–8.23)

8.18 Accidents during meiosis can alter chromosome number. An abnormal chromosome count is the result of nondisjunction, which can result from the failure of a pair of homologous chromosomes to separate during meiosis I or from the failure of sister chromatids to separate during meiosis II.

8.19 A karyotype is a photographic inventory of an individual's chromosomes. To prepare a karyotype, white blood cells are isolated, stimulated to grow, arrested at metaphase, and photographed under a microscope. The chromosomes are arranged into ordered pairs so that any chromosomal abnormalities can be detected.

8.20 An extra copy of chromosome 21 causes Down syndrome. Trisomy 21, the most common chromosome number abnormality, results in a condition called Down syndrome.

8.21 Abnormal numbers of sex chromosomes do not usually affect survival. Nondisjunction of the sex chromosomes during meiosis can result in individuals with a missing or extra X or Y chromosome. In some cases (such as XXY), this leads to syndromes; in other cases (such as XXX), the body is normal.

8.22 New species can arise from errors in cell division. Nondisjunction can produce polyploid organisms, organisms with extra sets of chromosomes. Such errors in cell division can be important in the evolution of new species.

8.23 Alterations of chromosome structure can cause birth defects and cancer. Chromosome breakage can lead to rearrangements—deletions, duplications, inversions, and translocations—that can produce genetic disorders or, if the changes occur in somatic cells, cancer.

Connecting the Concepts

1. Complete the following table to compare mitosis and meiosis.

	Mitosis	Meiosis
Number of chromosomal duplications	1	1
Number of cell divisions	1	2
Number of daughter cells produced	2	4
Number of chromosomes in the daughter cells	diploid 2n	haploid n
How the chromosomes line up during metaphase	singly	tetrads I singly II
Genetic relationship of the daughter cells to the parent cell	indentical	unique
Functions performed in the human body	growth development repair	production of gamets

Testing Your Knowledge

Level 1: Knowledge/Comprehension

2. If an intestinal cell in a grasshopper contains 24 chromosomes, then a grasshopper sperm cell contains ____12____ chromosomes.
   a. 6
   b. 12
   c. 24
   d. 48

3. Which of the following is *not* a function of mitosis in humans?
   a. repair of wounds
   b. growth
   c. production of gametes from diploid cells
   d. replacement of lost or damaged cells

4. It is difficult to observe individual chromosomes during interphase because
   a. the DNA has not been replicated yet.
   b. they are in the form of long, thin strands.
   c. they leave the nucleus and are dispersed to other parts of the cell.
   d. homologous chromosomes do not pair up until division starts.

5. A fruit fly somatic cell contains 8 chromosomes. This means that ______________ different combinations of chromosomes are possible in its gametes.
   a. 8
   b. 16
   c. 32
   d. 64

6. If a fragment of a chromosome breaks off and then reattaches to the original chromosome but in the reverse direction, the resulting chromosomal abnormality is called
   a. a deletion.
   b. an inversion.
   c. a translocation.
   d. a nondisjunction.

Level 2: Application/Analysis

7. Which of the following phases of mitosis is essentially the opposite of prophase in terms of changes within the nucleus?
   a. telophase
   b. metaphase
   c. interphase
   d. anaphase

8. A biochemist measured the amount of DNA in cells growing in the laboratory and found that the quantity of DNA in a cell doubled
   a. between prophase and anaphase of mitosis.
   b. between the G_1 and G_2 phases of the cell cycle.
   c. during the M phase of the cell cycle.
   d. between prophase I and prophase II of meiosis.
9. A micrograph of a dividing cell from a mouse showed 19 chromosomes, each consisting of two sister chromatids. During which of the following stages of cell division could such a picture have been taken? (*Explain your answer.*)
   a. prophase of mitosis
   b. telophase II of meiosis
   c. prophase I of meiosis
   d. prophase II of meiosis
10. Cytochalasin B is a chemical that disrupts microfilament formation. This chemical would interfere with
    a. DNA replication.
    b. formation of the mitotic spindle.
    c. cleavage.
    d. formation of the cell plate.
11. Why are individuals with an extra chromosome 21, which causes Down syndrome, more numerous than individuals with an extra chromosome 3 or chromosome 16?
    a. There are probably more genes on chromosome 21 than on the others.
    b. Chromosome 21 is a sex chromosome and chromosomes 3 and 16 are not.
    c. Down syndrome is not more common, just more serious.
    d. Extra copies of the other chromosomes are probably fatal.
12. In the light micrograph below of dividing cells near the tip of an onion root, identify a cell in interphase, prophase, metaphase, anaphase, and telophase. Describe the major events occurring at each stage.

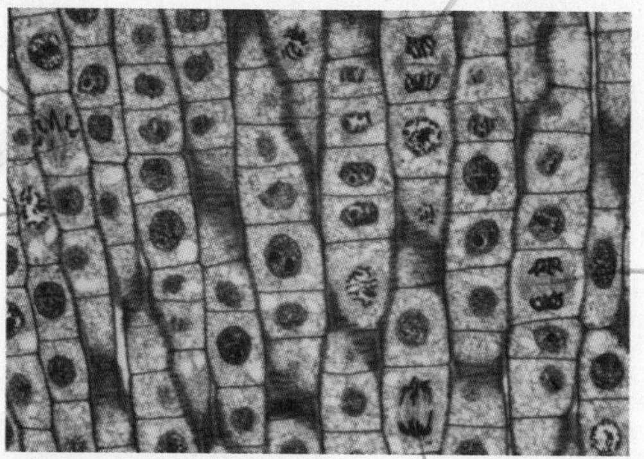

Level 3: Synthesis/Evaluation

13. An organism called a plasmodial slime mold is one large cytoplasmic mass with many nuclei. Explain how such a "megacell" could form.
14. Briefly describe how three different processes that occur during a sexual life cycle increase the genetic diversity of offspring.
15. Discuss the factors that control the division of eukaryotic cells grown in the laboratory. Cancer cells are easier to grow in the lab than other cells. Why do you suppose this is?
16. Compare cytokinesis in plant and animal cells. In what ways are the two processes similar? In what ways are they different?
17. Sketch a cell with three pairs of chromosomes undergoing meiosis, and show how nondisjunction can result in the production of gametes with extra or missing chromosomes.
18. Suppose you read in the newspaper that a genetic engineering laboratory has developed a procedure for fusing two gametes from the same person (two eggs or two sperm) to form a zygote. The article mentions that an early step in the procedure prevents crossing over from occurring during the formation of the gametes in the donor's body. The researchers are in the process of determining the genetic makeup of one of their new zygotes. Which of the following predictions do you think they would make? Justify your choice, and explain why you rejected each of the other choices.
    a. The zygote would have 46 chromosomes, all of which came from the gamete donor (its one parent), so the zygote would be genetically identical to the gamete donor.
    b. The zygote *could* be genetically identical to the gamete donor, but it is much more likely that it would have an unpredictable mixture of chromosomes from the gamete donor's parents.
    c. The zygote would not be genetically identical to the gamete donor, but it would be genetically identical to one of the donor's parents.
    d. The zygote would not be genetically identical to the gamete donor, but it would be genetically identical to one of the donor's grandparents.
19. Bacteria are able to divide on a faster schedule than eukaryotic cells. Some bacteria can divide every 20 minutes, while the minimum time required by eukaryotic cells in a rapidly developing embryo is about once per hour, and most cells divide much less often than that. State several testable hypotheses explaining why bacteria can divide at a faster rate than eukaryotic cells.
20. Red blood cells, which carry oxygen to body tissues, live for only about 120 days. Replacement cells are produced by cell division in bone marrow. How many cell divisions must occur each second in your bone marrow just to replace red blood cells? Here is some information to use in calculating your answer: There are about 5 million red blood cells per cubic millimeter (mm^3) of blood. An average adult has about 5 L (5,000 cm^3) of blood. (*Hint*: What is the total number of red blood cells in the body? What fraction of them must be replaced each day if all are replaced in 120 days?)
21. A mule is the offspring of a horse and a donkey. A donkey sperm contains 31 chromosomes and a horse egg cell contains 32 chromosomes, so the zygote contains a total of 63 chromosomes. The zygote develops normally. The combined set of chromosomes is not a problem in mitosis, and the mule combines some of the best characteristics of horses and donkeys. However, a mule is sterile; meiosis cannot occur normally in its testes (or ovaries). Explain why mitosis is normal in cells containing both horse and donkey chromosomes but the mixed set of chromosomes interferes with meiosis.
22. **SCIENTIFIC THINKING** The personalized cancer therapy study described in Module 8.10 began with a single bladder cancer patient. This patient, unlike many others, responded well to treatment with the drug everolimus. Many scientific studies begin with a detailed examination of an unusual case like this; the goal is to understand the underlying cause and use this information to understand (and help) more individuals. How could this approach be extended from the data presented in Table 8.10? Do you see any unusual results that could prompt another round of study? How might such a study be designed?

Answers to all questions can be found in Appendix 4.

CHAPTER 9

Patterns of Inheritance

The people of Tibet occupy a high-altitude region of the Himalayan mountains and are renowned for their mountaineering skills. For centuries, westerners have marveled at their uncanny ability to live and work at altitudes above 13,000 feet, where the amount of oxygen that reaches the blood is 40% less than at sea level. Without extensive conditioning, most people from other regions can hardly walk under such conditions, let alone carry heavy loads.

What makes the Tibetan people so able to tolerate their harsh surroundings? The answer lies, at least in part, in their genes: Over the last several thousand years, the Tibetan population has accumulated several dozen genetic mutations that affect their circulatory and respiratory systems. Tibetans commonly have versions of these genes that are rare among their low-dwelling Chinese neighbors. A newborn with such mutations is three times more likely to survive a high-altitude infancy, an advantage that clearly increases fitness in this environment and so is favored by natural selection.

? *Are humans evolving?*

In this chapter, we'll examine the rules that govern how inherited traits are passed from parents to offspring. We'll look at several different patterns of inheritance and investigate how we can predict the ratios of offspring with particular traits. Most important, we'll uncover a basic biological concept: how the behavior of chromosomes during gamete formation and fertilization (discussed in Chapter 8) accounts for the patterns of inheritance we observe. Along the way, we'll consider many examples of how genetic principles can help us understand the biology of humans, plants, and many other familiar creatures.

BIG IDEAS

Mendel's Laws

(9.1–9.10)

A few simple and long-established rules explain many aspects of heredity.

Variations on Mendel's Laws

(9.11–9.15)

Some inheritance patterns are more complex than the ones described by Mendel.

The Chromosomal Basis of Inheritance

(9.16–9.19)

Hereditary rules can be understood by following the behavior of chromosomes.

Sex Chromosomes and Sex-Linked Genes

(9.20–9.23)

Genes found on sex chromosomes display unique patterns of inheritance.

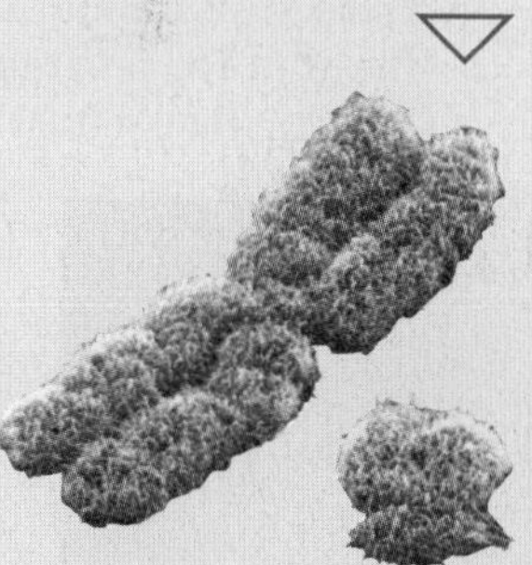

Mendel's Laws

9.1 The study of genetics has ancient roots

Attempts to explain inheritance date back at least to the ancient Greek physician Hippocrates (**Figure 9.1**). He suggested that particles called "pangenes" travel from each part of an organism's body to the eggs or sperm and then are passed to the next generation; moreover, Hippocrates argued, changes that occur in the body during an organism's life are passed on in this way.

Hippocrates's idea is incorrect in several respects. The reproductive cells are not composed of particles from somatic (body) cells, and changes in somatic cells do not influence eggs and sperm. For instance, no matter how many years you endure orthodontic braces, cells in your mouth do not transmit genetic information to your gametes, and there is no higher likelihood that your offspring will have straight teeth just because you wore braces. This may seem like common sense today, but the idea that traits acquired during an individual's lifetime are passed on to offspring prevailed well into the 19th century.

▲ **Figure 9.1** Hippocrates (approximately 460–370 BCE)

By observing inheritance patterns in ornamental plants, biologists of the early 19th century established that offspring inherit traits from both parents. The favored explanation of inheritance then became the "blending" hypothesis, the idea that the hereditary materials contributed by the male and female parents mix in forming the offspring similar to the way that blue and yellow paints blend to make green. For example, according to this hypothesis, after the genetic information for the colors of black and chocolate brown Labrador retrievers is blended, the colors should be as inseparable as paint pigments. But this is not what happens: Instead, the offspring of a purebred black Lab and a purebred brown Lab will all be black, but some of the dogs in the next generation will be brown (you'll learn why in Module 9.5). The blending hypothesis was finally rejected because it does not explain how traits that disappear in one generation can reappear in later ones.

? Imagine you have two different houseplants of the same species, one of which blooms with white flowers and the other with red flowers. Design a simple experiment to test the blending hypothesis.

● Cross the two plants and observe the resulting flower color in the offspring. The blending hypothesis predicts the appearance of pink flowers.

9.2 The science of genetics began in an abbey garden

Heredity is the transmission of traits from one generation to the next. The field of **genetics**, the scientific study of heredity, began in the 1860s, when an Augustinian monk named Gregor Mendel (**Figure 9.2A**) deduced the fundamental principles of genetics by breeding garden peas. Mendel lived and worked in an abbey in Brunn, Austria (now Brno, in the Czech Republic). His research was strongly influenced by his study of physics, mathematics, and chemistry at the University of Vienna; the research was both experimentally and mathematically rigorous, and these qualities were largely responsible for his success.

▲ **Figure 9.2A** Gregor Mendel

In a classic paper published in 1866, Mendel correctly argued that parents pass discrete "heritable factors" on to their offspring. (It is interesting to note that Mendel's landmark publication appeared just seven years after Darwin's 1859 publication of *The Origin of Species*, making the 1860s a banner decade in the development of modern biology.) Mendel stressed that the heritable factors, today called genes, retain their individuality generation after generation. That is, genes are like playing cards: A deck may be shuffled, but the cards always retain their original identities, and no card is ever blended with another. Similarly, genes may be sorted, but each gene permanently retains its identity.

Mendel probably chose to study garden peas because they had short generation times, produced large numbers of offspring from each mating, and came in many readily distinguishable varieties. For example, one variety has purple flowers, and another variety has white flowers. A heritable feature that varies among individuals, such as flower color, is called a **character**. Each variant for a character, such as purple or white flowers, is called a **trait**. Perhaps the most important advantage of pea plants as an experimental model was that Mendel could strictly control matings. As **Figure 9.2B** shows, the petals of the pea flower almost completely enclose the reproductive organs: the stamens and carpel.

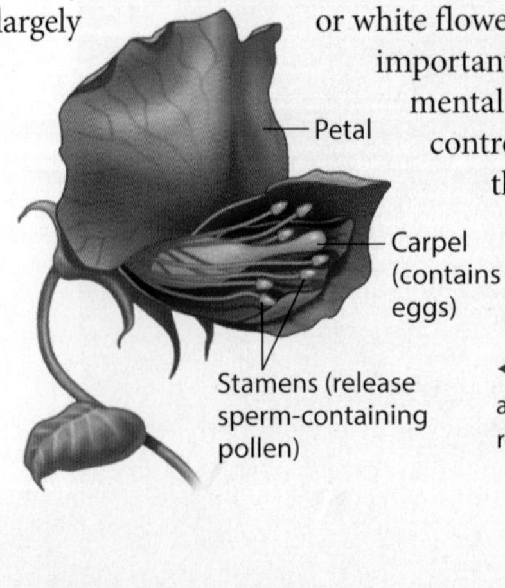

Petal
Carpel (contains eggs)
Stamens (release sperm-containing pollen)

◀ **Figure 9.2B** The anatomy of a garden pea flower (with one petal removed to improve visibility)

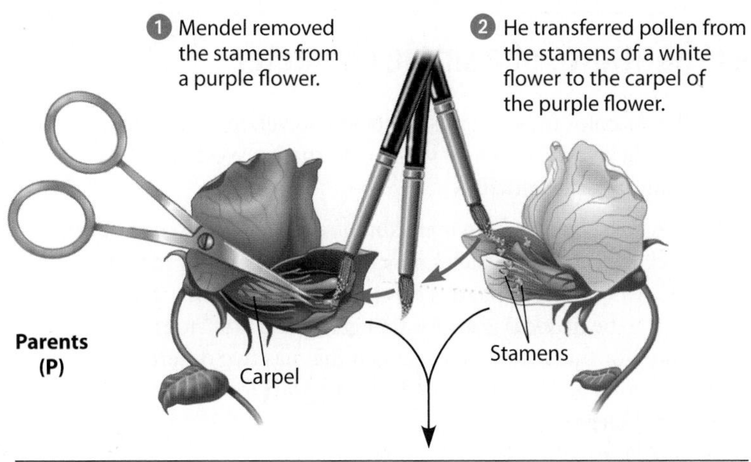

1 Mendel removed the stamens from a purple flower.
2 He transferred pollen from the stamens of a white flower to the carpel of the purple flower.
Parents (P)
Carpel
Stamens

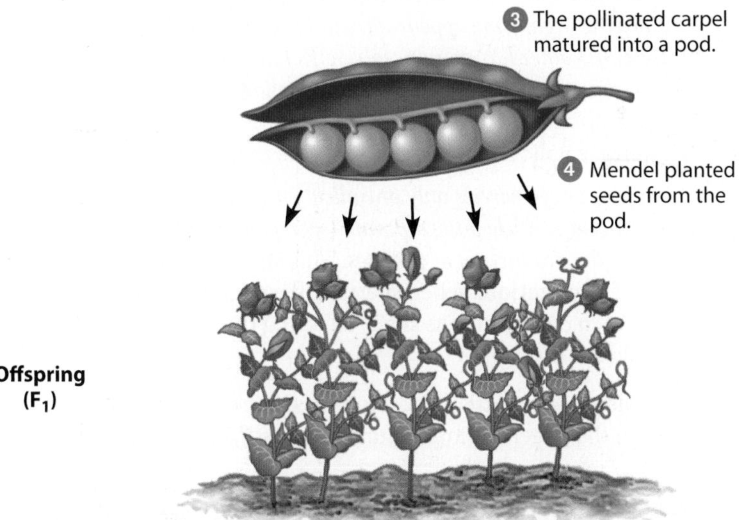

3 The pollinated carpel matured into a pod.
4 Mendel planted seeds from the pod.
Offspring (F_1)

▲ **Figure 9.2C** Mendel's technique for cross-fertilization of pea plants

Consequently, pea plants usually are able to self-fertilize in nature. That is, sperm-carrying pollen grains released from the stamens land on the egg-containing carpel of the same flower. Mendel could ensure self-fertilization by covering a flower with a small bag so that no pollen from another plant could reach the carpel. When he wanted cross-fertilization (fertilization of one plant by pollen from a different plant), he used the method shown in **Figure 9.2C**. 1 He prevented self-fertilization by cutting off the immature stamens of a plant before they produced pollen. 2 To cross-fertilize the stamenless flower, he dusted its carpel with pollen from another plant. After pollination, 3 the carpel developed into a pod, containing seeds (peas) that 4 he later planted. The seeds grew into offspring plants. Through these methods, Mendel could always be sure of the parentage of new plants.

Mendel's success was due to his experimental approach and choice of organism and to his selection of characters to study. He chose to observe seven characters, each of which occurred as two distinct traits (**Figure 9.2D**). Mendel worked with his plants until he was sure he had **true-breeding** varieties—that is, varieties for which self-fertilization produced offspring all identical to the parent. For instance, he identified a purple-flowered variety that, when self-fertilized, produced offspring plants that all had purple flowers.

Mendel was then ready to ask what would happen when he crossed his different true-breeding varieties with each other. For example, what offspring would result if plants with purple flowers and plants with white flowers were cross-fertilized? The offspring of two different varieties are called **hybrids**, and the cross-fertilization itself is referred to as a hybridization, or simply a genetic **cross**. The true-breeding parents are called the **P generation** (P for parental), and their hybrid offspring are called the **F_1 generation** (F for *filial*, from the Latin word for "son"). When F_1 plants self-fertilize or fertilize each other, their offspring are the **F_2 generation**.

Mendel's quantitative analysis of the F_2 plants from thousands of genetic crosses allowed him to deduce the fundamental principles of heredity. We turn to Mendel's results next.

? Describe three generations of your own family using the terminology of a genetic cross (P, F_1, F_2).

The P generation is your grandparents, the F_1 your parents, and the F_2 is you (and any siblings).

Character	Traits	
	Dominant	**Recessive**
Flower color	Purple	White
Flower position	Axial	Terminal
Seed color	Yellow	Green
Seed shape	Round	Wrinkled
Pod shape	Inflated	Constricted
Pod color	Green	Yellow
Stem length	Tall	Dwarf

▲ **Figure 9.2D** The seven pea characters studied by Mendel

9.3 Mendel's law of segregation describes the inheritance of a single character

Mendel performed many experiments in which he tracked the inheritance of characters that occur in two forms, such as flower color. The results led him to formulate several hypotheses about inheritance. Let's look at some of his experiments and follow the reasoning that led to his hypotheses.

Figure 9.3A starts with a cross between a true-breeding pea plant with purple flowers and a true-breeding pea plant with white flowers. This is an example of a **monohybrid cross** because it follows just one character—flower color. Mendel observed that F_1 plants all had purple flowers. Was the white-flowered plant's genetic contribution to the hybrids lost? By mating the F_1 plants with each other, Mendel found the answer to be no. Out of 929 F_2 plants, 705 (about $\frac{3}{4}$) had purple flowers and 224 (about $\frac{1}{4}$) had white flowers. That is, there are about three plants with purple flowers for every one with white flowers, or a 3:1 ratio of purple to white. Mendel reasoned that the heritable factor for white flowers did not disappear in the F_1 plants but was masked when the purple-flower factor was present. He also deduced that the F_1 plants must have carried two factors for the flower-color character, one for purple and one for white. From these results and others, Mendel developed four hypotheses, described here using modern terminology, such as "gene" instead of "heritable factor."

1. *There are alternative versions of genes that account for variations in inherited characters.* For example, the gene for flower color in pea plants exists in two versions: one for purple and the other for white. Alternative versions of a gene are called **alleles**.
2. *For each character, an organism inherits two alleles, one from each parent.* These alleles may be identical or they may differ. An organism that has two identical alleles for a gene is said to be **homozygous** for that gene (and is a "homozygote" for that trait). An organism that has two different alleles for a gene is said to be **heterozygous** for that gene (and is a "heterozygote").
3. *If the two alleles of an inherited pair differ, then one determines the organism's appearance and is called the* **dominant allele**; *the other has no noticeable effect on the organism's appearance and is called the* **recessive allele**. Geneticists use uppercase italic letters to represent dominant alleles and lowercase italic letters to represent recessive alleles.
4. *A sperm or egg carries only one allele for each inherited character because allele pairs separate (segregate) from each other during the production of gametes.* This statement is called the **law of segregation**. When sperm and egg unite at fertilization, each contributes its allele, restoring the paired condition in the offspring.

Do Mendel's hypotheses account for the 3:1 ratio he observed in the F_2 generation? **Figure 9.3B** illustrates Mendel's law

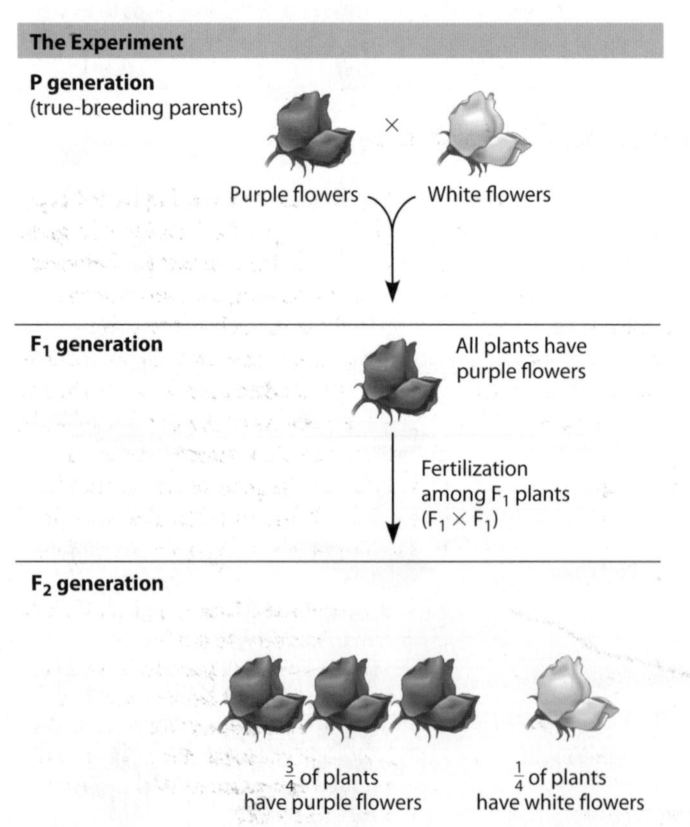

The Experiment
P generation
(true-breeding parents)
×
Purple flowers
White flowers
F1 generation
All plants have
purple flowers
Fertilization
among F1 plants
(F1 × F1)
F2 generation
3/4 of plants
have purple flowers
1/4 of plants
have white flowers

▲ **Figure 9.3A** A monohybrid cross that tracks one character (flower color)

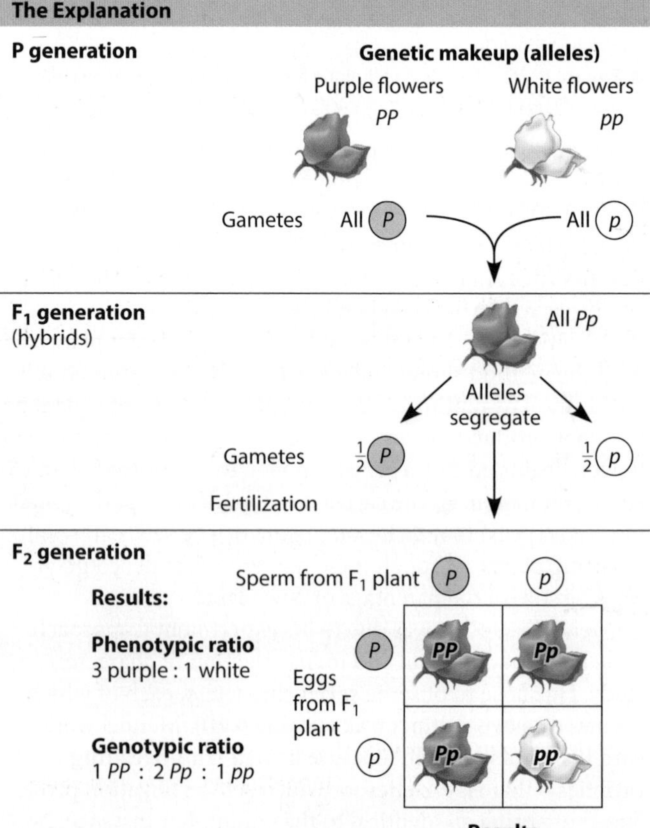

The Explanation
P generation
Genetic makeup (alleles)
Purple flowers
PP
White flowers
pp
Gametes
All P
All p
F1 generation
(hybrids)
All Pp
Alleles
segregate
Gametes
1/2 P
1/2 p
Fertilization
F2 generation
Sperm from F1 plant
P
p
Results:
Phenotypic ratio
3 purple : 1 white
Eggs
from F1
plant
P
p
PP
Pp
Pp
pp
Genotypic ratio
1 PP : 2 Pp : 1 pp
Results

▲ **Figure 9.3B** An explanation of the crosses in Figure 9.3A

of segregation, which explains the inheritance pattern shown in Figure 9.3A. Mendel's hypotheses predict that when alleles segregate during gamete formation in the F_1 plants, half the gametes will receive a purple-flower allele (*P*) and the other half a white-flower allele (*p*). During pollination among the F_1 plants, the gametes unite randomly. An egg with a purple-flower allele has an equal chance of being fertilized by a sperm with a purple-flower allele or one with a white-flower allele (that is, a *P* egg may fuse with a *P* sperm or a *p* sperm). Because the same is true for an egg with a white-flower allele (a *p* egg with a *P* sperm or *p* sperm), there are a total of four equally likely combinations of sperm and egg in the F_2 generation.

The diagram at the bottom of Figure 9.3B, called a **Punnett square**, repeats the cross shown in Figure 9.3A in a way that highlights the four possible combinations of gametes and the resulting four possible offspring in the F_2 generation. Each square represents an equally probable product of fertilization. For example, the box in the upper right corner of the Punnett square shows the genetic combination resulting from a *p* sperm fertilizing a *P* egg.

According to the Punnett square, what will be the physical appearance of these F_2 offspring? One-fourth of the plants have two alleles specifying purple flowers (*PP*); clearly, these plants will have purple flowers. One-half (two-fourths) of the F_2 offspring have inherited one allele for purple flowers and one allele for white flowers (*Pp*); like the F_1 plants, these plants will also have purple flowers, the dominant trait. (Note that *Pp* and *pP* are equivalent and usually written as *Pp*.) Finally, one-fourth of the F_2 plants have inherited two alleles specifying white flowers (*pp*) and will express this recessive trait. Thus, Mendel's model accounts for the 3:1 ratio that he observed in the F_2 generation.

Because an organism's appearance does not always reveal its genetic composition, geneticists distinguish between an organism's physical traits, called its **phenotype** (such as purple or white flowers), and its genetic makeup, its **genotype** (in this example, *PP*, *Pp*, or *pp*). So now we see that Figure 9.3A shows just phenotypes, whereas Figure 9.3B shows both phenotypes and genotypes in our sample crosses. For the F_2 plants, the ratio of plants with purple flowers to those with white flowers (3:1) is called the phenotypic ratio. The genotypic ratio, as shown by the Punnett square, is 1*PP* : 2*Pp* : 1*pp*.

Mendel found that each of the seven characters he studied exhibited the same inheritance pattern: One parental trait disappeared in the F_1 generation, only to reappear in $\frac{1}{4}$ of the F_2 offspring. The mechanism underlying this inheritance pattern is stated by Mendel's law of segregation: *Pairs of alleles segregate (separate) during gamete formation; the fusion of gametes at fertilization creates allele pairs once again.* Research since Mendel's time has established that the law of segregation applies to all sexually reproducing organisms, including humans.

Later in this chapter, we'll return to Mendel and his experiments with pea plants (in Module 9.5). But first, we'll investigate how cell division (the topic of Chapter 8) fits with what we've learned about genetics so far.

? How can two plants with different genotypes for a particular inherited character be identical in phenotype?

● One could be homozygous for the dominant allele and the other heterozygous.

9.4 Homologous chromosomes bear the alleles for each character

Figure 9.4 shows a pair of homologous chromosomes. The chromosomes in a homologous pair—chromosomes that carry alleles of the same genes—are also called homologs. Recall that every diploid cell, whether from a pea plant or a person, has pairs of homologous chromosomes. One member of each pair comes from the organism's female parent and the other member of each pair comes from the male parent.

Each labeled band on the chromosomes in Figure 9.4 represents a gene **locus** (plural, *loci*), a specific location of a gene along the chromosome. You can see the connection between Mendel's law of segregation and homologous chromosomes: Alleles (alternative versions) of a gene reside at the same locus on homologous chromosomes. However, the two chromosomes may bear either identical alleles at a locus (as in the *P*/*P* and *a*/*a* loci) or different alleles (as in the *B*/*b* locus)—the organisms may be homozygous or heterozygous for the gene at any particular locus. We will return to the chromosomal basis of Mendel's law later in the chapter.

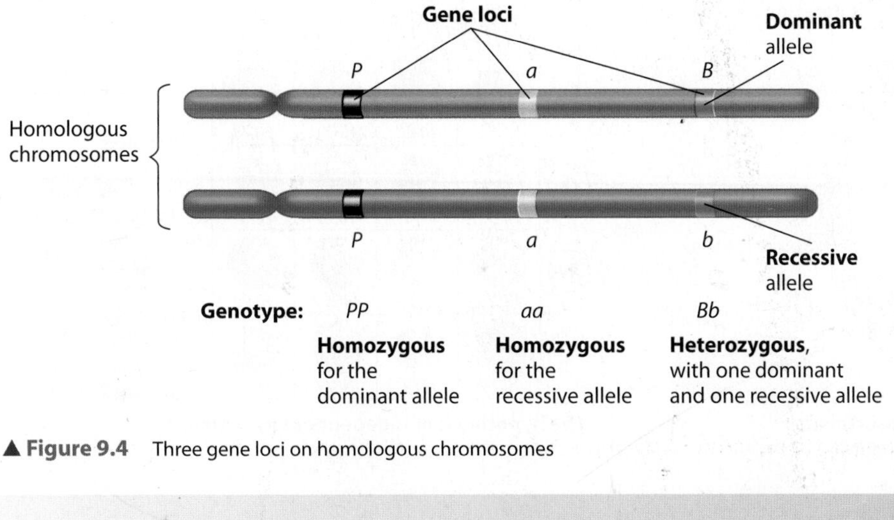

Gene loci
Dominant allele
P a B
Homologous chromosomes
P a b
Recessive allele
Genotype: PP aa Bb
Homozygous for the dominant allele
Homozygous for the recessive allele
Heterozygous, with one dominant and one recessive allele

▲ **Figure 9.4** Three gene loci on homologous chromosomes

? An individual is heterozygous, *Bb*, for a gene. According to the law of segregation, each gamete formed by this individual will have *either* the *B* allele *or* the *b* allele. Which step in the process of meiosis is the physical basis for this segregation of alleles? (*Hint*: See Figure 8.12.)

● The *B* and *b* alleles are located at the same gene locus on homologous chromosomes, which separate during meiosis I and are packaged in separate gametes during meiosis II.

9.5 The law of independent assortment is revealed by tracking two characters at once

Recall from Module 9.3 that Mendel deduced his law of segregation by following through the F_1 and F_2 generations one character from the P generation. From such monohybrid crosses, Mendel knew that the allele for round seed shape (designated *R*) was dominant to the allele for wrinkled seed shape (*r*) and that the allele for yellow seed color (*Y*) was dominant to the allele for green seed color (*y*). Mendel wondered: What would happen if he crossed plants that differ in both seed shape and seed color?

To find out, Mendel set up a **dihybrid cross**, a mating of parental varieties differing in two characters. Mendel crossed homozygous plants having round yellow seeds (genotype *RRYY*) with plants having wrinkled green seeds (*rryy*). Mendel knew that an *RRYY* plant would produce only gametes with *RY* alleles; an *rryy* plant would produce only gametes with *ry* alleles. Therefore, Mendel knew there was only one possible outcome for the F_1 generation: The union of *RY* and *ry* gametes would yield hybrids heterozygous for both characters (*RrYy*)—that is, dihybrids. All of these *RrYy* offspring would have round yellow seeds, the double dominant phenotype.

The F_2 generation is trickier to predict. To find out if genes for seed color and shape would be transmitted as a package, Mendel crossed the *RrYy* F_1 plants with each other. He hypothesized two outcomes from this experiment: Either the dihybrid cross would exhibit *dependent* assortment, with the alleles for seed color and seed shape inherited together as they came from the P generation, or it would exhibit *independent* assortment, with the genes inherited independently.

As shown on the left side of **Figure 9.5A**, the hypothesis of dependent assortment leads to the prediction that each F_2 plant would inherit one of two possible sperm (*RY* or *ry*) and one of two possible eggs (*RY* or *ry*), for a total of four combinations. The Punnett square shows that there could be only two F_2 phenotypes—round yellow or wrinkled green—in a 3:1 ratio. However, when Mendel actually performed this cross, he did not obtain these results, thus refuting the hypothesis of dependent assortment.

The alternative hypothesis—that the genes would exhibit independent assortment—is shown on the right side of Figure 9.5A. This leads to the prediction that the F_1 plants would produce four different gametes: *RY*, *rY*, *Ry*, and *ry*. Each F_2 plant would inherit one of four possible sperm and one of four possible eggs, for a total of 16 possible combinations. Fertilization among these gametes would lead to four different seed phenotypes—round yellow, round green, wrinkled yellow, or wrinkled green—in a 9:3:3:1 ratio. In fact, Mendel observed such a ratio in the F_2 plants, indicating that each pair of alleles segregates independently of the other.

From the 9:3:3:1 ratio, we can see that there are 12 plants with round seeds to 4 with wrinkled seeds, and 12 yellow-seeded plants to 4 green-seeded ones. These 12:4 ratios each reduce to 3:1, which is the F_2 ratio for a monohybrid cross. In other words, an independent monohybrid cross is occurring for each of the two characters. Mendel tried his seven pea characters in various dihybrid combinations and always obtained data close to the predicted 9:3:3:1 ratio. These results supported the hypothesis that each pair of alleles segregates independently of other pairs of alleles during gamete formation. Put another way, the inheritance of one character has no effect on the inheritance of another. This is referred to as Mendel's **law of independent assortment**.

Figure 9.5B shows how this law applies to the inheritance of two characters controlled by separate genes in Labrador retrievers: black versus chocolate coat color and normal vision versus progressive retinal atrophy (PRA), an eye disorder that leads to blindness. Black Labs have at least one copy of an allele, *B*, that gives their hairs densely packed granules of a dark pigment. The *B* allele is dominant to the *b* allele, which leads to a less tightly packed distribution of pigment. As a result, the coats of dogs with genotype *bb* are chocolate in color. The

▼ Figure 9.5A Two hypotheses for segregation in a dihybrid cross

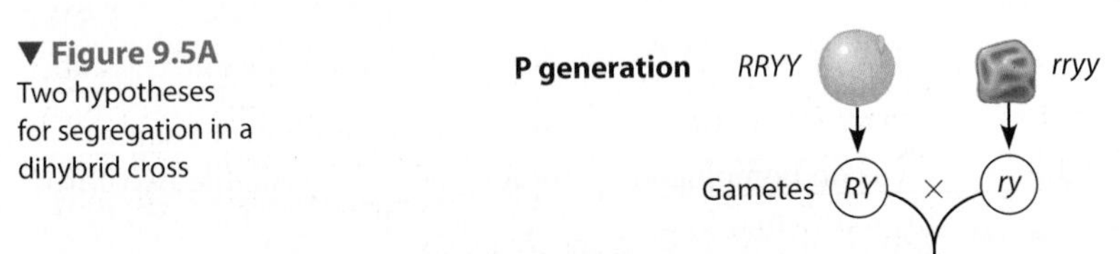
P generation
RRYY
rryy
Gametes
RY
×
ry
F₁ generation
RrYy

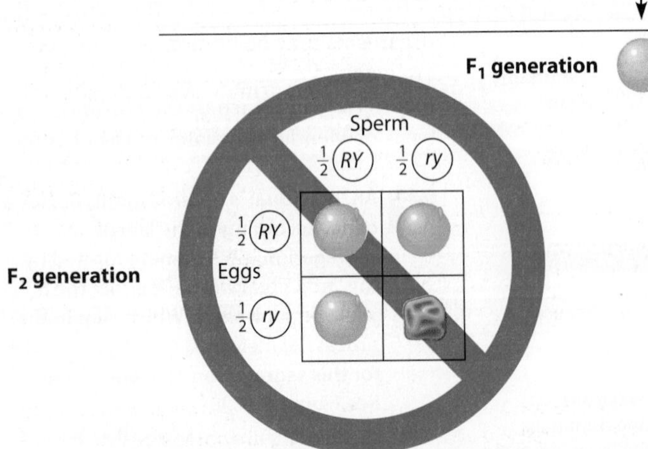
F₂ generation
Sperm
½ RY
½ ry
½ RY
Eggs
½ ry

The hypothesis of dependent assortment
Not actually seen; hypothesis refuted

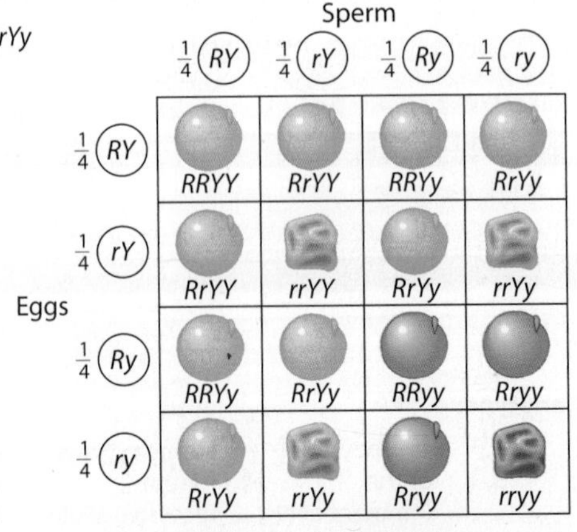
Sperm
¼ RY
¼ rY
¼ Ry
¼ ry
Eggs
¼ RY
¼ rY
¼ Ry
¼ ry
RRYY
RrYY
RRYy
RrYy
RrYY
rrYY
RrYy
rrYy
RRYy
RrYy
RRyy
Rryy
RrYy
rrYy
Rryy
rryy
Results:
9/16 Yellow round
3/16 Green round
3/16 Yellow wrinkled
1/16 Green wrinkled

The hypothesis of independent assortment
Actual results; hypothesis supported

allele that causes PRA, *n*, is recessive to allele *N*, which is necessary for normal vision. Thus, only dogs of genotype *nn* become blind from PRA. In the top of this figure, blanks in the genotypes are used where a particular phenotype may result from multiple genotypes. For example, a black Lab may have either genotype *BB* or *Bb*, which we abbreviate as *B_*.)

The lower part of Figure 9.5B shows what happens when we mate two heterozygous Labs, both of genotype *BbNn*. The F_2 phenotypic ratio will be nine black dogs with normal eyes to three black with PRA to three chocolate with normal eyes to one chocolate with PRA. These 9:3:3:1 results are analogous to the results in Figure 9.5A and demonstrate that the alleles for the *B* and *N* genes are inherited independently.

? Predict the phenotypes of offspring obtained by mating a black Lab homozygous for both coat color and normal eyes with a chocolate Lab that is blind from PRA.

All offspring would be black with normal eyes (*BBNN* × *bbnn* → *BbNn*).

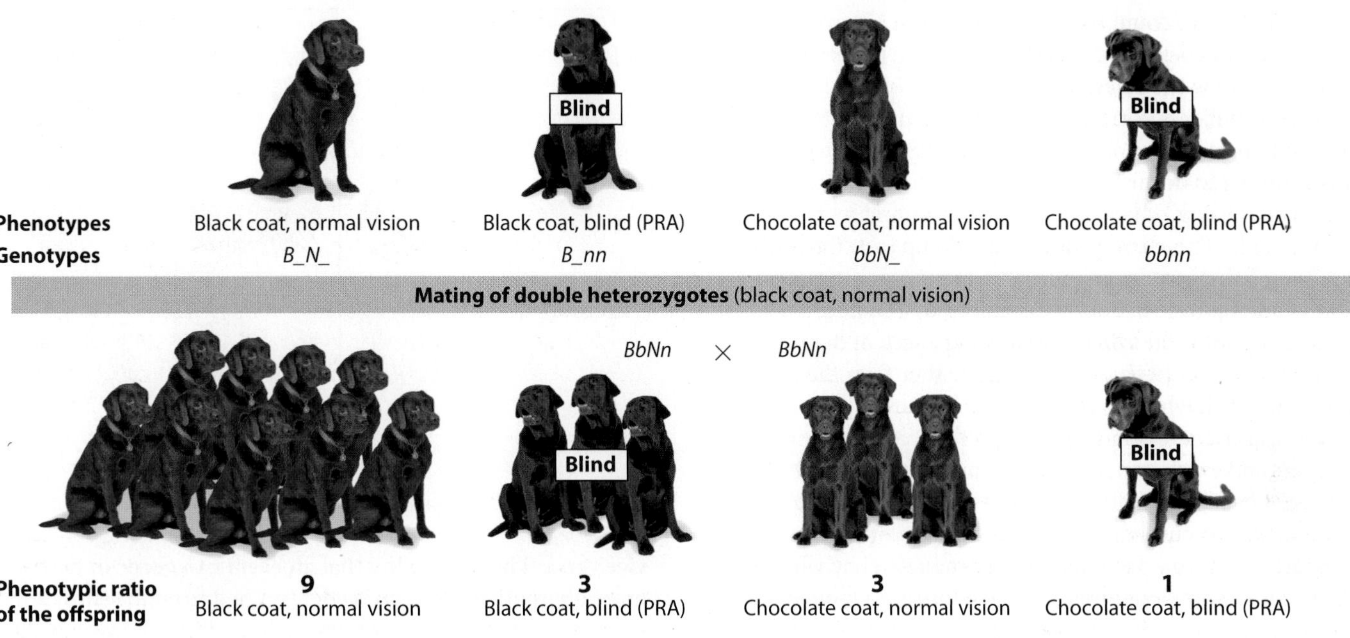
Blind
Blind
Phenotypes
Genotypes
Black coat, normal vision
B_N_
Black coat, blind (PRA)
B_nn
Chocolate coat, normal vision
bbN_
Chocolate coat, blind (PRA)
bbnn
Mating of double heterozygotes (black coat, normal vision)
BbNn × BbNn
Blind
Blind
Phenotypic ratio of the offspring
9
Black coat, normal vision
3
Black coat, blind (PRA)
3
Chocolate coat, normal vision
1
Chocolate coat, blind (PRA)

▲ **Figure 9.5B** Independent assortment of two genes in Labrador retrievers

Try This Rewrite the cross shown in this figure using a Punnett square, like the one used in the previous figure. You should get the same results!

9.6 Geneticists can use a testcross to determine unknown genotypes

Suppose you have a chocolate Lab. Referring to Figure 9.5B, you can tell that its genotype must be *bb*. But what if you had a black Lab? It could have one of two possible genotypes—*BB* or *Bb*—and there is no way to tell simply by looking at the dog. To determine your dog's genotype, you could perform a **testcross**, a mating between an individual of unknown genotype (your black Lab) and a homozygous recessive (*bb*) individual—in this case, a chocolate Lab.

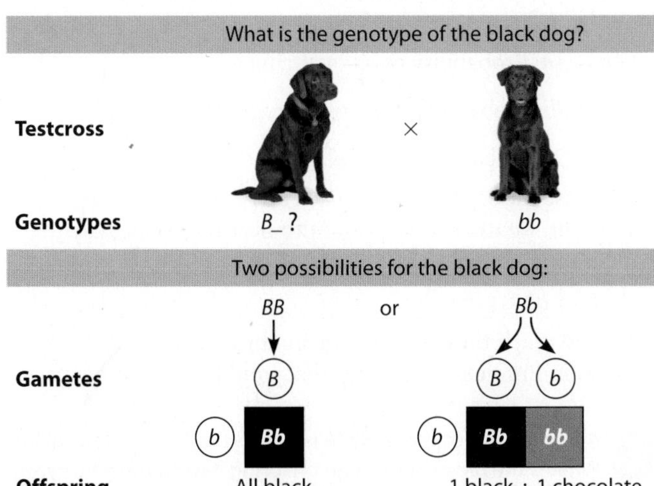
What is the genotype of the black dog?
Testcross
×
Genotypes
B_?
bb
Two possibilities for the black dog:
BB
or
Bb
Gametes
B
B
b
b
Bb
b
Bb
bb
Offspring
All black
1 black : 1 chocolate

▲ **Figure 9.6** Using a testcross to determine genotype

Figure 9.6 shows the offspring that could result from such a mating. If, as shown on the bottom left, the black-coated parent's genotype is *BB*, all the offspring would be black because a cross between genotypes *BB* and *bb* can produce only *Bb* offspring. On the other hand, if the black parent is *Bb*, as shown on the bottom right, we would expect both black (*Bb*) and chocolate (*bb*) offspring. Thus, the appearance of the offspring reveals the original black dog's genotype.

To understand the results of any genetic cross, you need to understand the rules of probability, our next topic.

? You use a testcross to determine the genotype of a Lab with normal eyes. Half of the offspring are normal and half develop PRA. What is the genotype of the normal parent?

Heterozygous (*Nn*)

9.7 Mendel's laws reflect the rules of probability

Mendel's strong background in mathematics served him well in his studies of inheritance. He understood, for instance, that the segregation of allele pairs during gamete formation and the re-forming of pairs at fertilization obey the rules of probability—the same rules that apply to the tossing of coins, the rolling of dice, and the drawing of cards. Mendel also appreciated the statistical nature of inheritance. He knew that he needed to obtain large samples—to count many offspring from his crosses—before he could begin to interpret inheritance patterns.

Let's see how the rules of probability apply to inheritance. The probability scale ranges from 0 to 1. An event that is certain to occur has a probability of 1, whereas an event that is certain *not* to occur has a probability of 0. For example, a tossed coin has a $\frac{1}{2}$ chance of landing heads and a $\frac{1}{2}$ chance of landing tails. These two possibilities add up to 1; the probabilities of all possible outcomes for an event to occur must always add up to 1. In another example, in a standard deck of 52 playing cards, the chance of drawing a jack of diamonds is $\frac{1}{52}$ and the chance of drawing any card other than the jack of diamonds is $\frac{51}{52}$, which together add up to 1.

An important lesson we can learn from coin tossing is that for each and every toss of the coin, the probability of heads is $\frac{1}{2}$. Even if heads has landed five times in a row, the probability of the next toss coming up heads is still $\frac{1}{2}$. In other words, the outcome of any particular toss is unaffected by what has happened on previous attempts. Each toss is an independent event. If two coins are tossed simultaneously, the outcome for each coin is an independent event, unaffected by the other coin. What is the chance that both coins will land heads up when tossed together? The probability of such a dual event is the product of the separate probabilities of the independent events; for the coins, $\frac{1}{2} \times \frac{1}{2} = \frac{1}{4}$. This statistical principle is called the **rule of multiplication**, and it holds true for independent events in genetics as well as coin tosses.

Figure 9.7 offers a visual analogy of a cross between F_1 Labrador retrievers that have the *Bb* genotype for coat color. The genetic cross is portrayed by the tossing of two coins that stand in for the two gametes (a dime for the egg and a penny for the sperm); the heads side of each coin stands for the dominant *B* allele and the tails side of each coin the recessive *b* allele. What is the probability that a particular F_2 dog will have the *bb* genotype? To produce a *bb* offspring, both egg and sperm must carry the *b* allele. The probability that an egg will have the *b* allele is $\frac{1}{2}$, and the probability that a sperm will have the *b* allele is also $\frac{1}{2}$. By the rule of multiplication, the probability that the two *b* alleles will come together at fertilization is $\frac{1}{2} \times \frac{1}{2} = \frac{1}{4}$. This is exactly the answer given by the Punnett square in Figure 9.7. If we know the genotypes of the parents, we can predict the probability for any genotype among the offspring.

Now consider the probability that an F_2 Lab will be heterozygous for the coat-color gene. As Figure 9.7 shows, there are two ways in which F_1 gametes can combine to produce a heterozygous offspring. The dominant (*B*) allele can come from the egg and the recessive (*b*) allele from the sperm, or

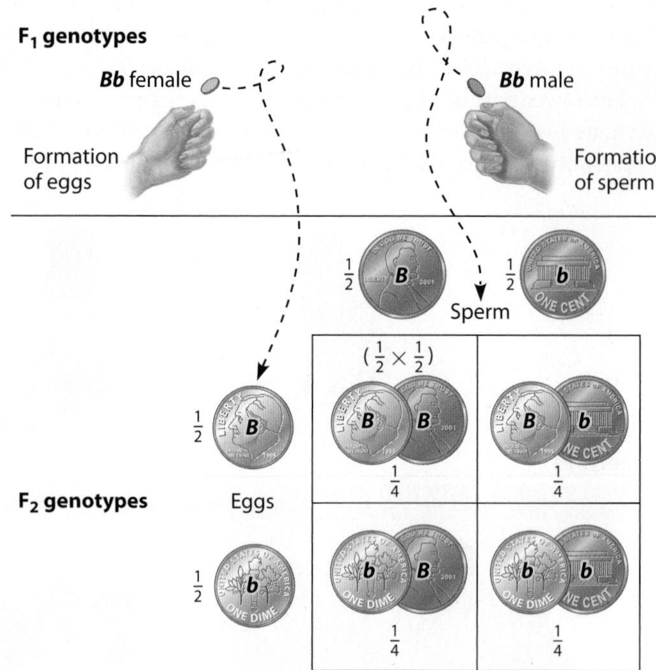

F1 genotypes
Bb female
Bb male
Formation of eggs
Formation of sperm
1/2
B
1/2
b
Sperm
(1/2 × 1/2)
1/2
B
B B
1/4
B b
1/4
F2 genotypes
Eggs
1/2
b
b B
1/4
b b
1/4

▲ **Figure 9.7** Segregation and fertilization as chance events

vice versa. The probability that an event can occur in two or more alternative ways can be determined from the sum of the separate probabilities of the alternatives; this is known as the **rule of addition**. Using this rule, we can calculate the probability of an F_2 heterozygote as $\frac{1}{4} + \frac{1}{4} = \frac{1}{2}$.

By applying the rules of probability to segregation and independent assortment, we can solve some rather complex genetics problems. For instance, we can predict the results of trihybrid crosses, in which three different characters are involved. Consider a cross between two organisms that both have the genotype *AaBbCc*. What is the probability that an offspring from this cross will be a recessive homozygote for all three genes (*aabbcc*)? Because each allele pair assorts independently, we can treat this trihybrid cross as three separate monohybrid crosses:

Aa × *Aa*: Probability of *aa* offspring $= \frac{1}{4}$

Bb × *Bb*: Probability of *bb* offspring $= \frac{1}{4}$

Cc × *Cc*: Probability of *cc* offspring $= \frac{1}{4}$

Because the segregation of each allele pair is an independent event, we use the rule of multiplication to calculate the probability that the offspring will be *aabbcc*:

$\frac{1}{4}aa \times \frac{1}{4}\,bb \times \frac{1}{4}cc = \frac{1}{64}$

We could reach the same conclusion by constructing a 64-section Punnett square, but that would take a lot of space!

? A plant of genotype *AABbCC* is crossed with an *AaBbCc* plant. What is the probability of an offspring having the genotype *AABBCC*?

● $\frac{1}{16}$ (that is, $\frac{1}{2} \times \frac{1}{4} \times \frac{1}{2}$).

VISUALIZING THE CONCEPT

9.8 Genetic traits in humans can be tracked through family pedigrees

Although Mendel developed his laws of inheritance while working with peas, these principles apply to the inheritance of many human traits just as well. How are human traits studied? We obviously cannot perform testcrosses on people, so geneticists must analyze the results of matings that have already occurred. First, a geneticist collects information about a family's history for a trait. This information is assembled into a family tree, called a **pedigree**, that describes the traits of parents and children across generations. Here, you can see a pedigree that traces the incidence of straight hairline versus a "widow's peak" (pointed) hairline through three generations of a hypothetical family. Notice that Mendel's laws and simple logic enable us to deduce the genotypes for nearly every person in the pedigree.

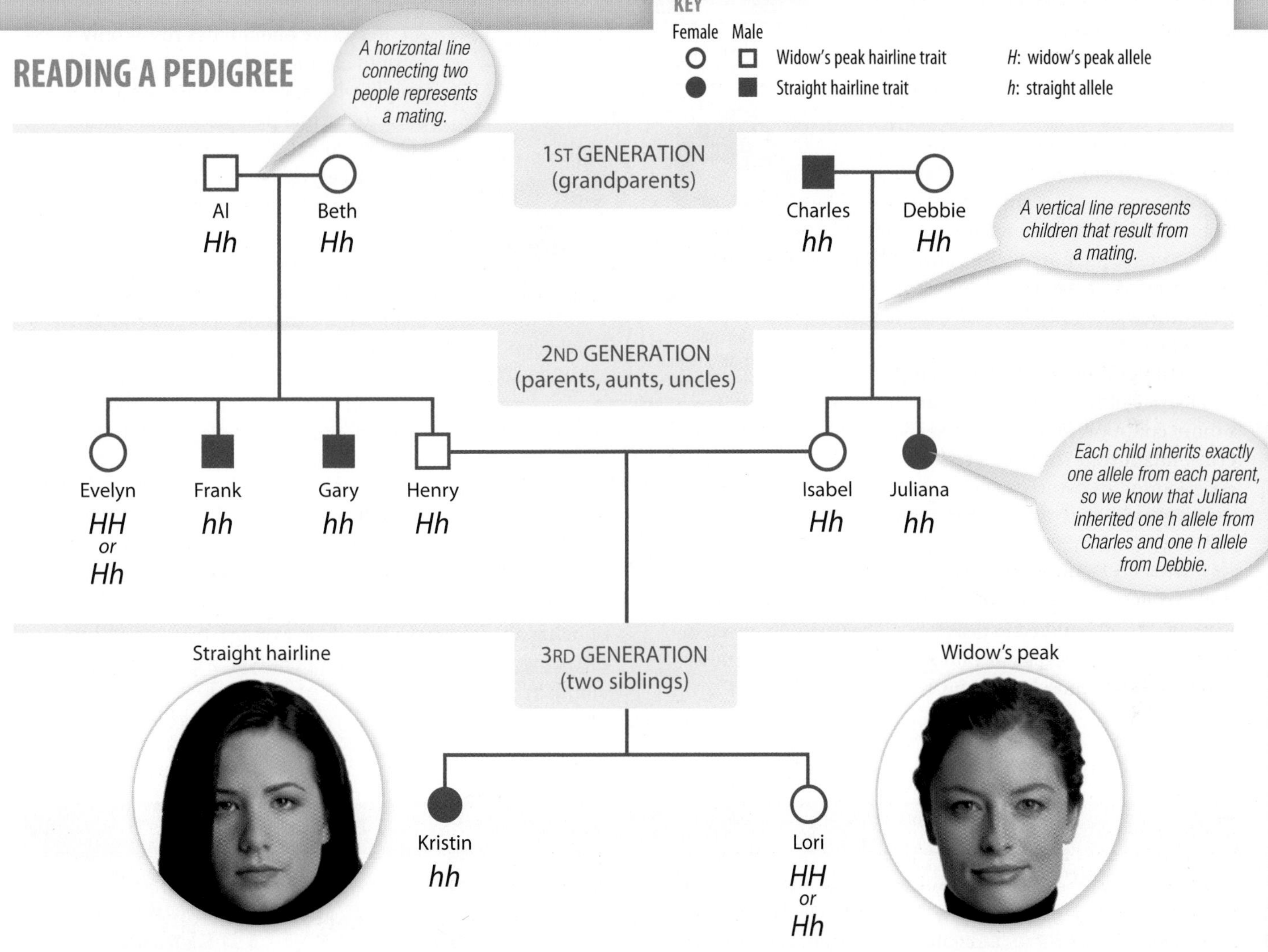

KEY
Female Male
Widow's peak hairline trait
Straight hairline trait
H: widow's peak allele
h: straight allele
READING A PEDIGREE
A horizontal line connecting two people represents a mating.
1ST GENERATION (grandparents)
Al Hh
Beth Hh
Charles hh
Debbie Hh
A vertical line represents children that result from a mating.
2ND GENERATION (parents, aunts, uncles)
Evelyn HH or Hh
Frank hh
Gary hh
Henry Hh
Isabel Hh
Juliana hh
Each child inherits exactly one allele from each parent, so we know that Juliana inherited one h allele from Charles and one h allele from Debbie.
3RD GENERATION (two siblings)
Straight hairline
Widow's peak
Kristin hh
Lori HH or Hh

- Kristin has a straight hairline but neither of her parents (Henry and Isabel) do. This is only possible if the trait is recessive. We therefore know that Kristin and every other individual with a straight hairline must be homozygous recessive *hh*.
- Henry and Isabel must each have a copy of the *h* allele, because they each passed one on to daughter Kristin. And because they both have widow's peaks, they must each be heterozygous (*Hh*).
- Grandparents Al and Beth must both be *Hh* because they both had widow's peaks, but two of their sons (Frank and Gary) had straight hairlines and must therefore be *hh*.
- We cannot deduce the genotype of every member of the pedigree. Lori must have at least one *H* allele (since she has a widow's peak), but she could be either *HH* or *Hh*. We cannot distinguish between these two possibilities using the available data.

? If Lori had a child, which phenotype would allow her to deduce her own genotype for certain?

● If her child had a straight hairline (*hh*), then Lori would know that she herself must be *Hh*.

9.9 Many inherited traits in humans are controlled by a single gene

CONNECTION

In the previous module, you studied an example of a human character (widow's peak) controlled by simple dominant-recessive inheritance of one gene. **Figure 9.9A** shows two more examples. (The genetic underpinnings of many other human characters, such as eye and hair color, are more complex and as yet poorly understood.) A trait being dominant does not mean that it is "normal" or more common than a recessive trait; **wild-type traits**—those seen most often in nature—are not necessarily specified by dominant alleles. Rather, dominance means that a heterozygote (*Aa*) displays the dominant phenotype. By contrast, the phenotype of a recessive allele is seen only in a homozygote (*aa*). Recessive traits may in fact be more common in the population than dominant ones. For example, the absence of freckles (a dominant trait) is more common than their presence. The term mutant trait refers to a trait that is less common in nature.

The genetic disorders listed in **Table 9.9** are known to be inherited as dominant or recessive traits controlled by a single gene. These human disorders show simple inheritance patterns like the traits Mendel studied in pea plants. The genes discussed in this module are all located on autosomes, chromosomes other than the sex chromosomes X and Y (see Module 8.11).

Dominant Traits
Recessive Traits
Freckles
No freckles
Normal pigmentation
Albinism
Key
Wild-type (more common) trait
Mutant (less common) trait

▲ **Figure 9.9A** Examples of single-gene inherited traits in humans

Recessive Disorders Thousands of human genetic disorders—ranging in severity from relatively mild, such as albinism, to invariably fatal, such as cystic fibrosis—are inherited as recessive traits. Remember that the dominant phenotype results from either the homozygous genotype *AA* or the heterozygous genotype *Aa*. Recessive phenotypes result only from the homozygous genotype *aa*. Most people who have recessive disorders are born to normal parents who are both heterozygotes—that is, those parents who are **carriers** of the recessive allele for the disorder but are phenotypically normal.

Using Mendel's laws, we can predict the fraction of affected offspring likely to result from a mating between two carriers (**Figure 9.9B**). Suppose two people who are heterozygous carriers for albinism (*Aa*) had a child. What is the probability that this child would display albinism? Each child of two carriers has a $\frac{1}{4}$ chance of inheriting two recessive alleles. To put it another way, we can say that about one-fourth of the children from such a mating are predicted to display albinism. We can also say that a child with normal pigmentation has a $\frac{2}{3}$ chance of being an *Aa* carrier; that is, on average, two out of three offspring with the pigmented phenotype will be carriers for albinism.

TABLE 9.9 | SOME AUTOSOMAL DISORDERS IN HUMANS

Disorder	Major Symptoms	Incidence	Comments
Recessive Disorders			
Albinism	Lack of pigment in the skin, hair, and eyes	1/22,000	Prone to skin cancer
Cystic fibrosis	Excess mucus in the lungs, digestive tract, liver; increased susceptibility to infections; death in early childhood unless treated	1/2,500 Caucasians	See Module 9.9
Phenylketonuria (PKU)	Accumulation of phenylalanine in blood; lack of normal skin pigment; developmental disabilities	1/10,000 in United States and Europe	See Module 9.10
Sickle-cell disease	Sickled red blood cells; damage to many tissues	1/400 African Americans	See Module 9.13
Tay-Sachs disease	Lipid accumulation in brain cells; mental deficiency; blindness; death in childhood	1/3,600 Jews from central Europe	See Module 4.10
Dominant Disorders			
Achondroplasia	Dwarfism	1/25,000	See Module 9.9
Huntington's disease	Developmental disabilities and uncontrollable movements; strikes in middle age	1/25,000	See Module 9.9
Hypercholesterolemia	Excess cholesterol in the blood; heart disease	1/500 are heterozygous	See Module 9.11

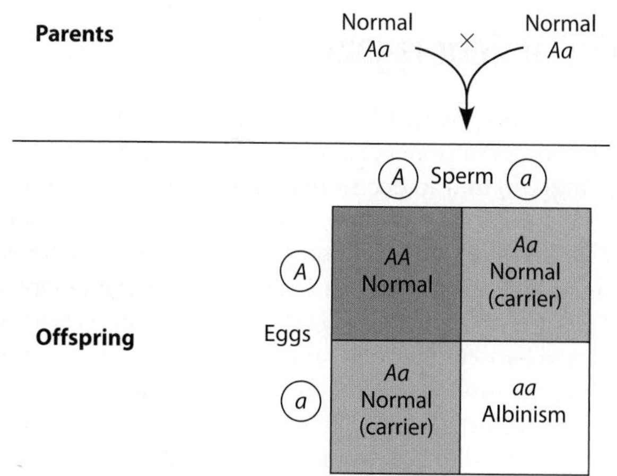
Parents
Normal
Aa
×
Normal
Aa
Sperm
A
a
Offspring
Eggs
A
a
AA
Normal
Aa
Normal
(carrier)
Aa
Normal
(carrier)
aa
Albinism

▲ **Figure 9.9B** Offspring produced by parents who are both carriers for albinism, a recessive disorder

The most common lethal genetic disease in the United States is cystic fibrosis (CF). Affecting about 30,000 Americans, the recessive CF allele is carried by about one in 31 Americans. A person with two copies of this allele has cystic fibrosis, which is characterized by an excessive secretion of very thick mucus from the lungs and other organs. This mucus can interfere with breathing, digestion, and liver function and makes the person vulnerable to recurrent bacterial infections. Although there is no cure for CF, strict adherence to a daily health regimen—including gentle pounding on the chest and back to clear the airway, inhaled antibiotics, and a special diet—can have a profound impact on the health of the affected person. CF was once invariably fatal in childhood, but tremendous advances in treatment have raised the median survival age of Americans with CF to 37.

Cystic fibrosis is most common in Caucasians. In fact, most genetic disorders are not evenly distributed across all ethnic groups. Such uneven distribution is the result of prolonged geographic isolation of certain populations. Isolation (as with settlers of a new island, for example) can lead to matings between close blood relatives. People with recent common ancestors are more likely to carry the same recessive alleles than are unrelated people. Therefore, matings between close relatives may cause the frequency of a rare allele (and the disease it causes) to increase within that community. Geneticists have observed increased incidence of harmful recessive traits among many types of inbred animals. For example, the detrimental effects of inbreeding are seen in some endangered species that recovered from small populations (see Module 13.11). With the increased mobility in most human populations today, it is relatively unlikely that two people who carry a rare, harmful allele will meet and mate.

▲ **Figure 9.9C** Dr. Michael C. Ain, a specialist in the repair of bone defects caused by achondroplasia and related disorders

Dominant Disorders Although most harmful alleles are recessive, a number of human disorders are caused by dominant alleles. Some are harmless conditions, such as extra fingers and toes (called polydactyly) or webbed fingers and toes.

A more serious dominant disorder is achondroplasia, a form of dwarfism in which the head and torso of the body develop normally but the arms and legs are short **(Figure 9.9C)**. The homozygous dominant genotype (*AA*) causes death of the embryo; therefore, only heterozygotes (*Aa*) have this disorder. (This also means that a person with achondroplasia has a 50% chance of passing the condition on to any children.) Therefore, all those who do not have achondroplasia, more than 99.99% of the population, are homozygous for the recessive allele (*aa*). This example makes it clear that a dominant allele is not necessarily more common in a population than a corresponding recessive allele.

Dominant alleles that cause lethal diseases are much less common than recessive alleles that cause lethal diseases. One reason is that the dominant lethal allele cannot be carried by heterozygotes without affecting them. Many lethal dominant alleles result from mutations in a sperm or egg that subsequently kill the embryo. And if the afflicted individual is born but does not survive long enough to reproduce, he or she will not pass on the lethal allele to future generations. This is in contrast to lethal recessive mutations, which are perpetuated from generation to generation by healthy heterozygous carriers.

A lethal dominant allele can escape elimination, however, if it does not cause death until a relatively advanced age. One such example is the allele that causes **Huntington's disease**, a degenerative disorder of the nervous system that usually does not appear until middle age. Once the deterioration of the nervous system begins, it is irreversible and inevitably fatal. Because the allele for Huntington's disease is dominant, any child born to a parent with the allele has a 50% chance of inheriting the allele and the disorder. This example makes it clear that a dominant allele is not necessarily "better" than a corresponding recessive allele.

Until relatively recently, the onset of symptoms was the only way to know if a person had inherited the Huntington's allele. This is no longer the case. A genetic test is now available that can detect the presence of the Huntington's allele in an individual's genome. This is one of several genetic tests currently available. We'll explore the topic of personal genetic screening in the next module.

? Peter is a 30-year-old man whose father died of Huntington's disease. Neither Peter's mother nor a much older sister shows any signs of Huntington's. What is the probability that Peter has inherited Huntington's disease?

● Since his father had the disease, there is a $\frac{1}{2}$ chance that Peter received the gene. (The genotype of his sister is irrelevant.)

9.10 New technologies can provide insight into one's genetic legacy

CONNECTION

Some prospective parents are aware that they have an increased risk of having a baby with a genetic disorder. For example, many pregnant women over age 35 know that they have a heightened risk of bearing children with Down syndrome (see Module 8.20), and some couples are aware that certain genetic diseases run in their families. These prospective parents may want to learn more about their own and their baby's genetic makeup. Modern technologies offer ways to obtain such information.

Genetic Testing Because most children with recessive disorders are born to healthy parents, the genetic risk for many diseases is determined by whether the prospective parents are carriers of the recessive allele. For an increasing number of genetic disorders, including Tay-Sachs disease, sickle-cell disease, and one form of cystic fibrosis, tests are available that can distinguish between individuals who have no disease-causing alleles and those who are heterozygous carriers. Other parents may know that a dominant but late-appearing disease, such as Huntington's disease, runs in their family. Such people may benefit from genetic tests for dominant alleles. Information from genetic testing (also called genetic screening) can inform decisions about family planning.

Fetal Testing Several technologies are available for detecting genetic conditions in a fetus. Genetic testing before birth requires the collection of fetal cells. In **amniocentesis**, performed between weeks 14 and 20 of pregnancy, a physician carefully inserts a needle through the abdomen and into the mother's uterus while watching an ultrasound imager to guide the needle away from the fetus (**Figure 9.10A**, left). The physician extracts about 10 milliliters (2 teaspoons) of the amniotic fluid that bathes the developing fetus. Tests for genetic disorders can be performed on fetal cells that have been isolated from the fluid. These cells are usually cultured in the laboratory for several weeks. By then, enough dividing cells can be harvested to allow karyotyping (see Module 8.19) that will detect chromosomal abnormalities such as Down syndrome. Biochemical tests can also be performed on the cultured cells, revealing conditions such as Tay-Sachs disease.

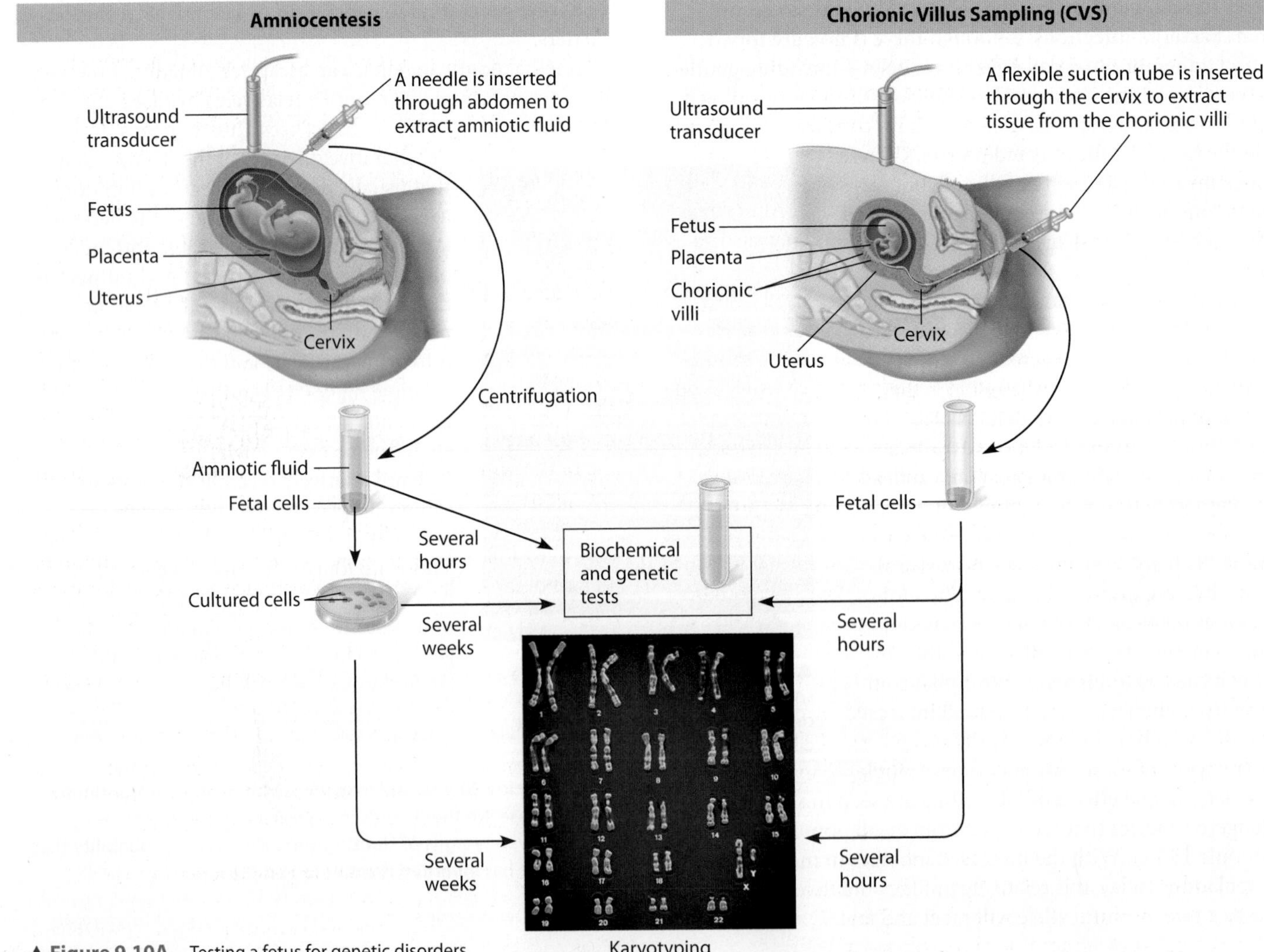
Amniocentesis
Ultrasound transducer
A needle is inserted through abdomen to extract amniotic fluid
Fetus
Placenta
Uterus
Cervix
Centrifugation
Amniotic fluid
Fetal cells
Several hours
Cultured cells
Several weeks
Biochemical and genetic tests
Several weeks
Chorionic Villus Sampling (CVS)
Ultrasound transducer
A flexible suction tube is inserted through the cervix to extract tissue from the chorionic villi
Fetus
Placenta
Chorionic villi
Uterus
Cervix
Fetal cells
Several hours
Several hours
Karyotyping

▲ **Figure 9.10A** Testing a fetus for genetic disorders

In another procedure, **chorionic villus sampling (CVS)**, a physician extracts a tiny sample of chorionic villus tissue from the placenta, the organ that carries nourishment and wastes between the fetus and the mother. The tissue can be obtained using a narrow, flexible tube inserted through the mother's vagina and cervix into the uterus (Figure 9.10A, right). Results of karyotyping and some biochemical tests can be available within 24 hours. The speed of CVS is an advantage over amniocentesis. Another advantage is that CVS can be performed as early as the 8th week of pregnancy.

Unfortunately, both amniocentesis and CVS pose some risk of complications, such as maternal bleeding, miscarriage, or premature birth. Complication rates for amniocentesis and CVS are about 1% and 2%, respectively. Because of the risks, these procedures are usually reserved for situations in which the possibility of a genetic disease is significantly higher than average. Newer genetic screening procedures involve isolating tiny amounts of fetal cells or DNA released into the mother's bloodstream. Although few reliable tests are yet available using this method, this promising and complication-free technology may soon replace more invasive procedures.

Blood tests on the mother at 15 to 20 weeks of pregnancy can help identify fetuses at risk for certain birth defects—and thus candidates for further testing that may require more invasive procedures (such as amniocentesis). The most widely used blood test measures the mother's blood level of alpha-fetoprotein (AFP), a protein produced by the fetus. High levels of AFP may indicate a neural tube defect in the fetus. (The neural tube is an embryonic structure that develops into the brain and spinal cord.) Low levels of AFP may indicate Down syndrome. For a more complete risk profile, a woman's doctor may order a "triple screen test," which measures AFP as well as two other hormones produced by the placenta. Abnormal levels of these substances in the maternal blood may also point to a risk of Down syndrome.

Fetal Imaging Other techniques enable a physician to examine a fetus directly for anatomical deformities. The most common procedure is **ultrasound imaging**, which uses sound waves to produce a picture of the fetus. **Figure 9.10B** shows an ultrasound scanner, which emits high-frequency sounds, beyond the range of hearing. When the sound waves bounce off the fetus, the echoes produce an image on the monitor. The inset image in Figure 9.10B shows a fetus at about 20 weeks. Traditional ultrasound imaging is noninvasive—no foreign objects are inserted into the mother's body—and has no known risk. Transvaginal ultrasound imaging—during which a probe is placed in the woman's vagina—can be used to provide clear images in early pregnancy. In another imaging method, fetoscopy, a needle-thin tube containing a fiber-optic viewing scope is inserted into the uterus. Fetoscopy can provide highly detailed images of the fetus but, unlike ultrasound, carries risk of complications.

Newborn Screening Some genetic disorders can be detected at birth by simple tests that are now routinely performed in most hospitals in the United States. One common screening program is for phenylketonuria (PKU), a recessively inherited disorder that occurs in about one out of every 10,000 births in the United States. Children with this disease cannot properly break down the naturally occurring amino acid phenylalanine, and an accumulation of phenylalanine may lead to developmental disabilities. However, if the deficiency is detected in the newborn, a special diet low in phenylalanine can usually prevent symptoms. Unfortunately, few other genetic disorders are currently treatable.

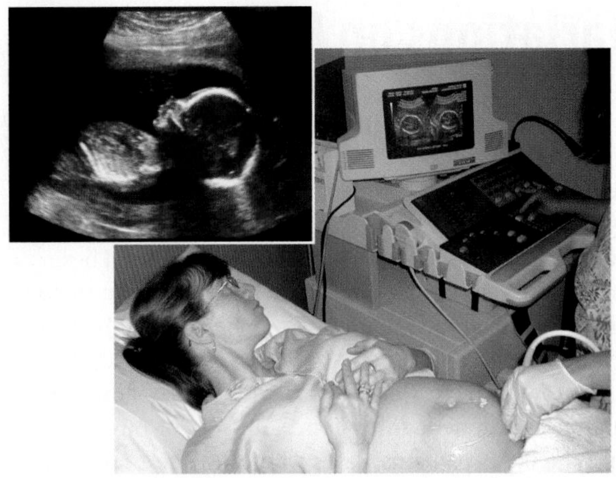

▲ **Figure 9.10B** Traditional ultrasound scanning of a fetus

Ethical Considerations As new technologies such as fetal imaging and testing become more widespread, geneticists are working to make sure that they do not cause more problems than they solve. Consider the tests for identifying carriers of recessive diseases. Such information may enable people with family histories of genetic disorders to make informed decisions about having children. But these new methods for genetic screening pose problems, too. If confidentiality is breached, will carriers be stigmatized? For example, will they be denied health or life insurance, even though they themselves are healthy? Geneticists stress that patients seeking genetic testing should receive counseling both before and after to clarify their family history, to explain the test, and to help them cope with the results. But with a wealth of genetic information increasingly available, a full discussion of the meaning of the results might be time-consuming and costly, raising the question of who should pay for such counseling.

Couples at risk for conceiving children with genetic disorders may now learn a great deal about their unborn children. In particular, CVS gives parents a chance to become informed very early in pregnancy. What is to be done with such information? If fetal tests reveal a serious disorder, the parents must choose between terminating the pregnancy and preparing themselves for a baby with severe problems. Identifying a genetic disease early can give families time to prepare—emotionally, medically, and financially. The dilemmas posed by human genetics reinforce one of this book's central themes: the immense social implications of biology.

? **What is the primary benefit of genetic screening by CVS? What is the primary risk?**

● CVS allows genetic screening to be performed very early in pregnancy and provides quick results, but it carries a risk of miscarriage.

▷ Variations on Mendel's Laws

9.11 Incomplete dominance results in intermediate phenotypes

Mendel's two laws explain inheritance in terms of discrete factors—genes—that are passed along from generation to generation according to simple rules of probability. These laws are valid for all sexually reproducing organisms, including garden peas, Labradors, and human beings. But just as the basic rules of musical harmony cannot account for all the rich sounds of a symphony, Mendel's laws stop short of explaining some patterns of genetic inheritance. In fact, for most sexually reproducing organisms, cases where Mendel's laws can strictly account for the patterns of inheritance are relatively rare. More often, the observed inheritance patterns are more complex, as we will see in this and the next four modules.

The F_1 offspring of Mendel's pea crosses always looked like one of the two parental varieties. In this situation—called **complete dominance**—the dominant allele has the same phenotypic effect whether present in one or two copies. But for some characters, the appearance of F_1 hybrids falls between the phenotypes of the two parental varieties, an effect called **incomplete dominance**. For instance, when red snapdragons are crossed with white snapdragons, all the F_1 hybrids have pink flowers (**Figure 9.11A**). This third phenotype results from flowers of the heterozygote having less red pigment than the red homozygotes.

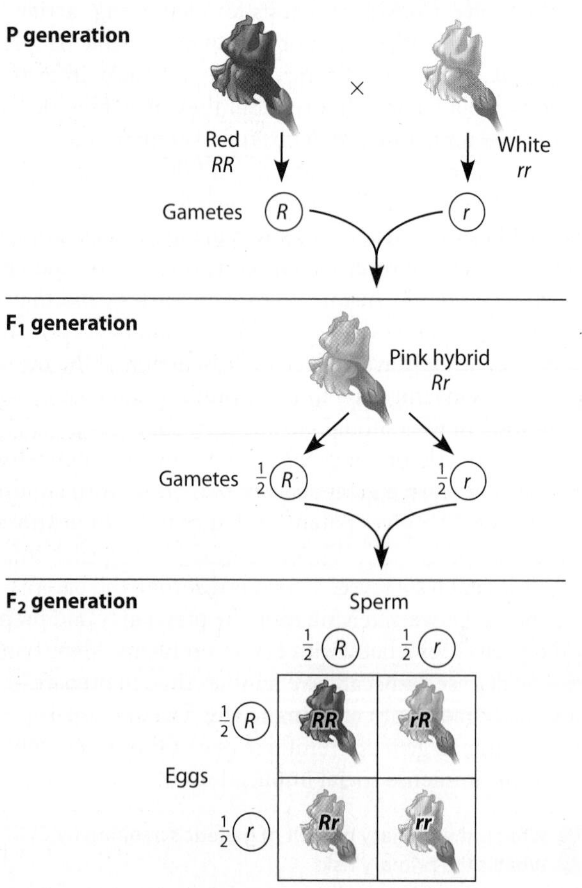
P generation
Red
RR
White
rr
Gametes
R
r
F1 generation
Pink hybrid
Rr
Gametes 1/2 R
1/2 r
F2 generation
Sperm
1/2 R
1/2 r
Eggs
1/2 R
1/2 r
RR
rR
Rr
rr

▲ **Figure 9.11A** Incomplete dominance in snapdragon flower color

As the Punnett square at the bottom of Figure 9.11A shows, the F_2 offspring appear in a phenotypic ratio of one red to two pink to one white, as the red and white alleles segregate during gamete formation in the pink F_1 hybrids. In incomplete dominance, the phenotypes of heterozygotes differ from the two homozygous varieties, and the genotypic ratio and the phenotypic ratio are both 1:2:1 in the F_2 generation.

We also see examples of incomplete dominance in humans. One case involves a recessive allele (*h*) that can cause hypercholesterolemia, dangerously high levels of cholesterol in the blood. Normal individuals are homozygous dominant (*HH*). Heterozygotes (*Hh*; about one in 500 people) have blood cholesterol levels about twice normal. They are unusually prone to atherosclerosis, the blockage of arteries by cholesterol buildup in artery walls, and they may have heart attacks from blocked heart arteries by their mid-30s. This form of the disease can often be controlled through changes in diet and by taking statins, a class of medications that can significantly lower blood cholesterol. Hypercholesterolemia is even more serious in homozygous recessive individuals (*hh*; about one in a million people). Homozygotes have about five times the normal amount of blood cholesterol and may have heart attacks as early as age 2. Homozygous hypercholesterolemia is harder to treat; options include high doses of statin drugs, organ surgeries or transplants, or physically filtering lipids from the blood.

Figure 9.11B illustrates the molecular basis for hypercholesterolemia. The dominant allele (*H*), which normal individuals carry in duplicate (*HH*), specifies a cell-surface receptor protein called an LDL receptor. Low-density lipoprotein (LDL, sometimes called "bad cholesterol") is transported in the blood. In certain cells, the LDL receptors mop up excess LDL particles from the blood and promote their breakdown. This process helps prevent the accumulation of cholesterol in arteries. Heterozygotes (*Hh*) have only half the normal

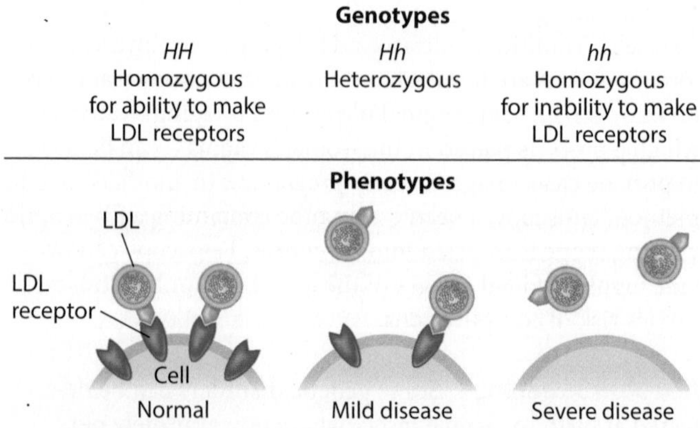
Genotypes
HH
Homozygous
for ability to make
LDL receptors
Hh
Heterozygous
hh
Homozygous
for inability to make
LDL receptors
Phenotypes
LDL
LDL
receptor
Cell
Normal
Mild disease
Severe disease

▲ **Figure 9.11B** Incomplete dominance in human hypercholesterolemia number of LDL receptors, and homozygous recessives (*hh*) have none. A lack of receptors prevents the cells from removing much of the excess cholesterol from the blood. The resulting buildup of LDL in the blood can be lethal.

? **Why doesn't the cross shown in Figure 9.11A support the blending hypothesis (see Module 9.1)?**

● Although two of the F_2 offspring show a "blended" phenotype (pink flowers), the other two do not, and the white and red alleles are not lost to future generations.

9.12 Many genes have more than two alleles in the population

So far, we have discussed inheritance patterns involving only two alleles per gene (*H* versus *h*, for example). But most genes can be found in populations in more than two versions, known as multiple alleles. Although each individual carries, at most, two different alleles for a particular gene, in cases of multiple alleles, more than two possible alleles exist in the population.

For instance, the **ABO blood group** phenotype in humans involves three alleles of a single gene. Various combinations of three alleles—called I^A, I^B, and *i*—produce four phenotypes: A person's blood type may be A, B, AB, or O **(Figure 9.12)**. These letters refer to two carbohydrates, designated A and B, that may be found on the surface of red blood cells. A person's red blood cells may be coated with carbohydrate A (in which case they are said to have type A blood), carbohydrate B (type B), both carbohydrates (type AB), or neither carbohydrate (type O). (In case you are wondering, the "positive" and "negative" notations on blood types—referred to as the Rh blood group system—are due to inheritance of a separate, unrelated gene.)

Matching compatible blood types is critical for safe blood transfusions. If a donor's blood cells have a carbohydrate (A or B) that is foreign to the recipient, then the recipient's immune system produces proteins called antibodies (see Module 24.9) that bind specifically to the foreign carbohydrates and cause the donor blood cells to clump together, potentially killing the recipient. The clumping reaction is also the basis of a blood-typing test performed in the laboratory. In Figure 9.12, notice that AB individuals can receive blood from anyone without fear of clumping, making them "universal recipients," while donated type O blood never causes clumping, making those with type O blood "universal donors."

The four blood groups result from various combinations of the three different alleles: I^A (for an enzyme referred to as *I*, which adds carbohydrate A to red blood cells), I^B (which adds carbohydrate B), and *i* (which adds neither A nor B carbohydrate). Each person inherits one of these alleles from each parent. Because there are three alleles, there are six possible genotypes, as illustrated in the figure. Both the I^A and I^B alleles are dominant to the *i* allele. Thus, I^AI^A and I^Ai people have type A blood, and I^BI^B and I^Bi people have type B. Recessive homozygotes, *ii*, have type O blood, with neither carbohydrate. The I^A and I^B alleles are **codominant**: Both alleles are expressed in heterozygous individuals (I^AI^B), who have type AB blood. Be careful to distinguish codominance (the expression of both alleles) from incomplete dominance (the expression of one intermediate trait).

? **Maria has type O blood, and her sister has type AB blood. The girls know that both of their maternal grandparents are type A. What are the genotypes of the girls' parents?**

● Their mother is I^Ai; their father is I^Bi.

Blood Group (Phenotype)	Genotypes	Carbohydrates Present on Red Blood Cells	Antibodies Present in Blood	Reaction When Blood from Groups Below Is Mixed with Antibodies from Groups at Left: O	A	B	AB
A	I^AI^A or I^Ai	Carbohydrate A	Anti-B				
B	I^BI^B or I^Bi	Carbohydrate B	Anti-A				
AB	I^AI^B	Carbohydrate A and Carbohydrate B	None				
O	*ii*	Neither	Anti-A Anti-B				

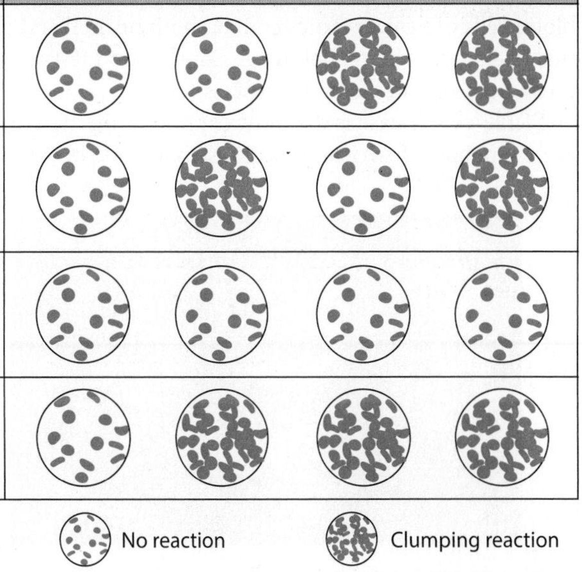

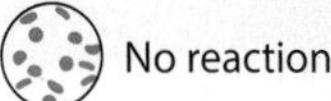
No reaction

Clumping reaction

▲ **Figure 9.12** Multiple alleles for the ABO blood groups

9.13 A single gene may affect many phenotypic characters

All of our genetic examples to this point have been cases in which each gene specifies only one hereditary character. In many cases, however, one gene influences multiple characters, a property called **pleiotropy**.

An example of pleiotropy in humans is **sickle-cell disease** (sometimes called sickle-cell anemia). The direct effect of the sickle-cell allele is to make red blood cells produce abnormal hemoglobin proteins. These molecules tend to link together and crystallize, especially when the oxygen content of the blood is lower than usual because of high altitude, overexertion, or respiratory ailments. As the hemoglobin crystallizes, the normally disk-shaped red blood cells deform to a sickle shape with jagged edges (**Figure 9.13A**). Sickled cells are destroyed rapidly by the body, and their destruction may seriously lower the individual's red cell count, causing anemia and general weakening of the body. Also, because of their angular shape, sickled cells do not flow smoothly in the blood and tend to accumulate and clog tiny blood vessels. Blood flow to body parts is reduced, resulting in periodic fever, weakness, severe pain, and damage to various organs, including the heart, brain, and kidneys. The overall result is a disorder characterized by the cascade of symptoms shown in **Figure 9.13B**. Blood transfusions and drug treatment may relieve some of the symptoms, but there is no cure; sickle-cell disease kills about 100,000 people each year.

In most cases, only people who are homozygous for the sickle-cell allele have sickle-cell disease. Heterozygotes, who have one sickle-cell allele and one normal allele, are usually healthy—hence, the disease is considered recessive. However, in rare cases, heterozygotes may experience some effects of the disease when oxygen in the blood is severely reduced, such as at very high altitudes. Thus, at the organismal level, a heterozygote displays incomplete dominance for the sickle-cell trait, with a phenotype between the homozygous dominant and homozygous recessive phenotypes. At the molecular level, however, the two alleles are actually codominant; the blood cells of heterozygotes contain both normal and abnormal (sickle-cell) hemoglobins. A simple blood test can distinguish homozygotes from heterozygotes.

Sickle-cell disease is the most common inherited disorder among people of African descent, striking one in 400 African Americans. About one in ten African Americans is a heterozygous carrier. Among Americans of other ancestry, the sickle-cell allele is extremely rare.

One in ten is an unusually high frequency of carriers for an allele with such harmful effects in homozygotes. We might expect that the frequency of the sickle-cell allele in the population would be much lower because many homozygotes die before passing their genes to the next generation. The high frequency appears to be a vestige of the ancestral roots of African Americans. Sickle-cell disease is most common in tropical Africa, where the deadly disease malaria is also prevalent. The parasite that causes malaria spends part of its life cycle inside red blood cells. When it enters those of a person with the sickle-cell allele, it triggers sickling. The body destroys most of the sickled cells, killing the parasite with them. Consequently, sickle-cell carriers have increased resistance to malaria, and in many parts of Africa, they live longer and have more offspring than noncarriers who are exposed to malaria. In this way, malaria has kept the frequency of the sickle-cell allele relatively high in much of the African continent. To put it in evolutionary terms, as long as the environment harbors malaria, individuals with one sickle-cell allele will have a selective advantage.

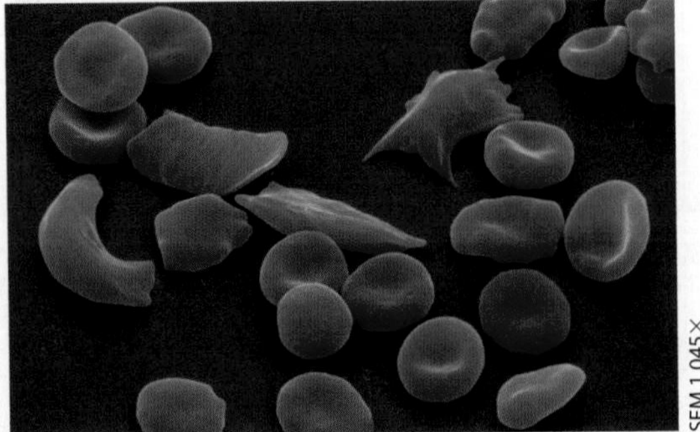
SEM 1,045×

▲ **Figure 9.13A** In this micrograph, you can see several jagged sickled cells in the midst of normal red blood cells.

? **Why is the sickle-cell trait considered codominant at the molecular level?**

Codominance means that both traits are expressed; a carrier for the sickle-cell allele produces both normal and abnormal hemoglobin.

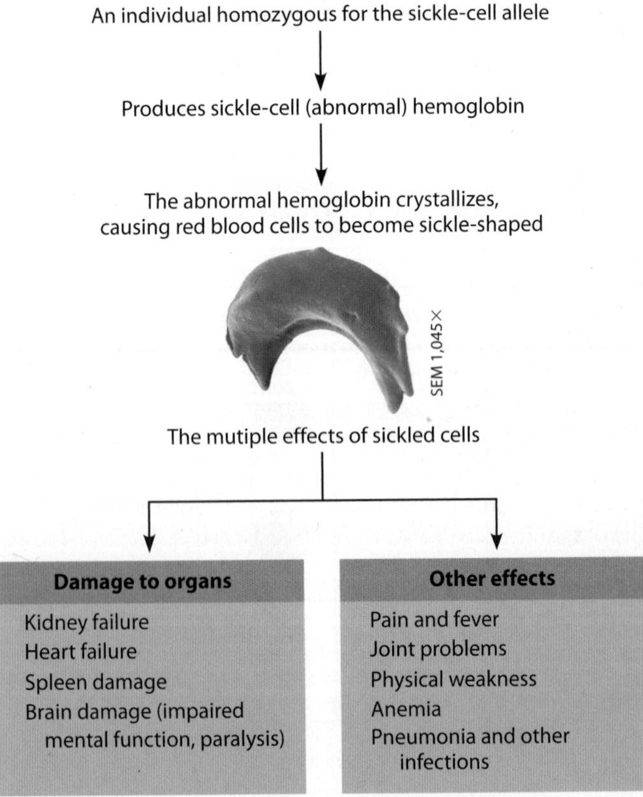
An individual homozygous for the sickle-cell allele
Produces sickle-cell (abnormal) hemoglobin
The abnormal hemoglobin crystallizes, causing red blood cells to become sickle-shaped
SEM 1,045×
The mutiple effects of sickled cells
Damage to organs
Kidney failure
Heart failure
Spleen damage
Brain damage (impaired mental function, paralysis)
Other effects
Pain and fever
Joint problems
Physical weakness
Anemia
Pneumonia and other infections

▲ **Figure 9.13B** Sickle-cell disease, an example of pleiotropy

9.14 A single character may be influenced by many genes

Mendel studied genetic characters that could be classified on an either-or basis, such as purple or white flower color. However, many characters, such as human skin color and height, vary in a population along a continuum. Many such features result from **polygenic inheritance**, the additive effects of two or more genes on a single phenotypic character. (This is the opposite of pleiotropy, in which one gene affects several characters.)

Let's consider a hypothetical example. Assume that the continuous variation in human skin color is controlled by three genes that are inherited separately, like Mendel's pea genes. (Actually, this character is probably affected by a great many genes, but for our purposes we'll simplify.) The "dark-skin" allele for each gene (*A*, *B*, or *C*) contributes one "unit" of darkness to the phenotype and is incompletely dominant to the other allele (*a*, *b*, or *c*). An *AABBCC* person would be very dark, whereas an *aabbcc* individual would be very light. An *AaBbCc* person would have skin of an intermediate shade. Because the alleles have an additive effect, the genotype *AaBbCc* would produce the same skin color as any other genotype with just three dark-skin alleles, such as *AABbcc*, because both of these individuals have three "units" of darkness.

The Punnett square in **Figure 9.14** shows all possible genotypes from a mating of two triple heterozygotes (*AaBbCc*). The row of squares below the Punnett square shows the seven skin pigmentation phenotypes that would theoretically result from this mating. The seven bars in the graph at the bottom of the figure depict the relative numbers of each of the phenotypes in the F_2 generation. This hypothetical example shows how inheritance of three genes could lead to a wide variety of pigmentation phenotypes. As we will see in the next module, in actual human populations, skin color has even more variations than shown in the figure.

Up to this point in the chapter, we have presented four types of inheritance patterns that are extensions of Mendel's laws of inheritance: incomplete dominance, codominance, pleiotropy, and polygenic inheritance. It is important to realize that these patterns are extensions of Mendel's model, rather than exceptions to it. From Mendel's pea garden experiments came data supporting the idea that genes are transmitted according to the same rules of chance that govern the tossing of coins. This basic idea of genes as discrete units of inheritance holds true for all inheritance patterns, even the patterns that are more complex than the ones originally considered by Mendel. In the next module, we consider another important source of deviation from Mendel's standard model: the effect of the environment.

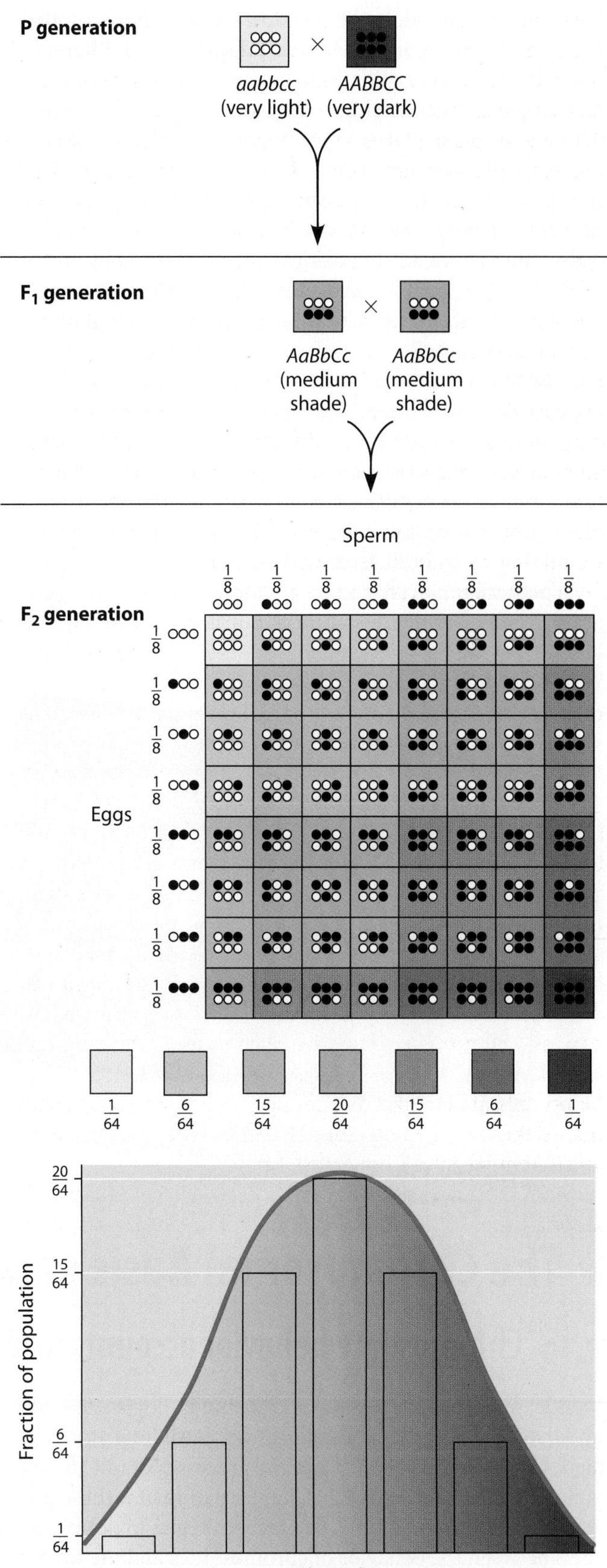

P generation
aabbcc (very light)
AABBCC (very dark)
F1 generation
AaBbCc (medium shade)
AaBbCc (medium shade)
Sperm
F2 generation
Eggs
1/8
1/64
6/64
15/64
20/64
15/64
6/64
1/64
Fraction of population
Skin color

▲ **Figure 9.14** A model for polygenic inheritance of skin color

? **An *AaBbcc* individual would be indistinguishable in phenotype from which of the following individuals: *AAbbcc*, *aaBBcc*, *AabbCc*, *Aabbcc*, or *aaBbCc*?**

All except *Aabbcc*

9.15 The environment affects many characters

In the previous module, we saw how a set of three hypothetical human skin-color genes could produce seven different phenotypes for skin color. But, of course, if we examine a real human population for skin color, we would see more shades than just seven. The true range might be similar to the entire spectrum of color under the bell-shaped curve in Figure 9.14. In fact, no matter how carefully we characterize the genes for skin color, a purely genetic description will always be incomplete. This is because skin color is also influenced by environmental factors, such as exposure to the sun (**Figure 9.15A**).

Many characters result from a combination of heredity and environment. For humans, nutrition influences height; exercise alters build; sun-tanning darkens the skin; experience improves performance on intelligence tests; and social and cultural forces greatly affect appearance. As geneticists learn more and more about our genes, it is becoming clear that many human characters—such as risk of heart disease and cancer and susceptibility to alcoholism and schizophrenia—are influenced by both genes and environment.

Whether human characters are more influenced by genes or by the environment—nature or nurture—is a very hotly contested debate. For some characters, such as the ABO blood group, a given genotype mandates a very specific phenotype, and the environment plays no role whatsoever. In contrast, how many red blood cells are circulating in your body is significantly influenced by environmental factors such as the altitude of your environment and your overall health.

It is important to realize that the individual features of any organism arise from a combination of genetic and environmental factors. Simply spending time with identical twins will convince anyone that environment, and not just genes, affects a person's traits (**Figure 9.15B**). Next, we turn to a discussion of the cellular basis of heredity: the behavior of chromosomes.

? If most characters result from a combination of environment and heredity, why was Mendel able to ignore environmental influences in his pea plants?

● The characters he chose for study were all entirely genetically determined and all his test subjects were raised in a similar environment.

▲ **Figure 9.15A** The effect of genes and sun exposure on the skin of one of this book's authors and his family

▲ **Figure 9.15B** Varying phenotypes due to environmental factors in genetically identical twins

▷ The Chromosomal Basis of Inheritance

9.16 Chromosome behavior accounts for Mendel's laws

Mendel published his results in 1866, but biologists did not understand the significance of his work until long after he died. Cell biologists worked out the processes of mitosis and meiosis by the late 1800s (see Chapter 8 to review these processes). Then, around 1902, researchers began to notice parallels between the behavior of chromosomes and the behavior of Mendel's "heritable factors" (what we now call genes). By combining this new understanding of mitosis and meiosis with an increasing understanding of genes, one of biology's most important concepts was formulated: The **chromosome theory of inheritance** holds that genes occupy specific loci (positions) on chromosomes, and it is the chromosomes that undergo segregation and independent assortment during meiosis. Thus, it is the behavior of chromosomes during meiosis and fertilization that accounts for inheritance patterns.

We can see the chromosomal basis of Mendel's laws by following the fates of two genes during meiosis and fertilization in pea plants. In **Figure 9.16**, the genes for seed shape (alleles

R and *r*) and seed color (*Y* and *y*) are shown as black bars on different chromosomes. Notice that the Punnett square is repeated from Figure 9.5A; we will now follow the chromosomes to see how they account for the results of the dihybrid cross shown in the Punnett square. We start with the F_1 generation, in which all plants have the *RrYy* genotype. To simplify the diagram, we show only two of the seven pairs of pea chromosomes and three of the stages of meiosis.

To see the chromosomal basis of the law of segregation (which states that pairs of alleles separate from each other during gamete formation via meiosis; see Module 9.3), let's follow just the homologous pair of long chromosomes, the ones carrying *R* and *r*, taking either the left or the right branch from the F_1 cell. Whichever arrangement the chromosomes assume at metaphase I, the two alleles segregate as the homologous chromosomes separate in anaphase I. And at the end of meiosis II, a single long chromosome ends up in each of the gametes. Fertilization then randomly recombines the two alleles, resulting in F_2 offspring that are $\frac{1}{4}$ *RR*, $\frac{1}{2}$ *Rr*, and $\frac{1}{4}$ *rr*. The ratio of round to wrinkled phenotypes is thus 3:1 (12 round to 4 wrinkled), the ratio Mendel observed, as shown in the Punnett square in the figure.

To see the chromosomal basis of the law of independent assortment (which states that each pair of alleles sorts independently of other pairs of alleles during gamete formation; see Module 9.5), follow both the long and short (nonhomologous) chromosomes through the figure. Two alternative arrangements of tetrads can occur at metaphase I. The nonhomologous chromosomes (and their genes) assort independently, leading to four gamete genotypes. Random fertilization leads to the 9:3:3:1 phenotypic ratio in the F_2 generation.

? Which of Mendel's laws have their physical basis in the following phases of meiosis: (a) the orientation of homologous chromosome pairs in metaphase I; (b) the separation of homologs in anaphase I?

(a) The law of independent assortment; (b) the law of segregation

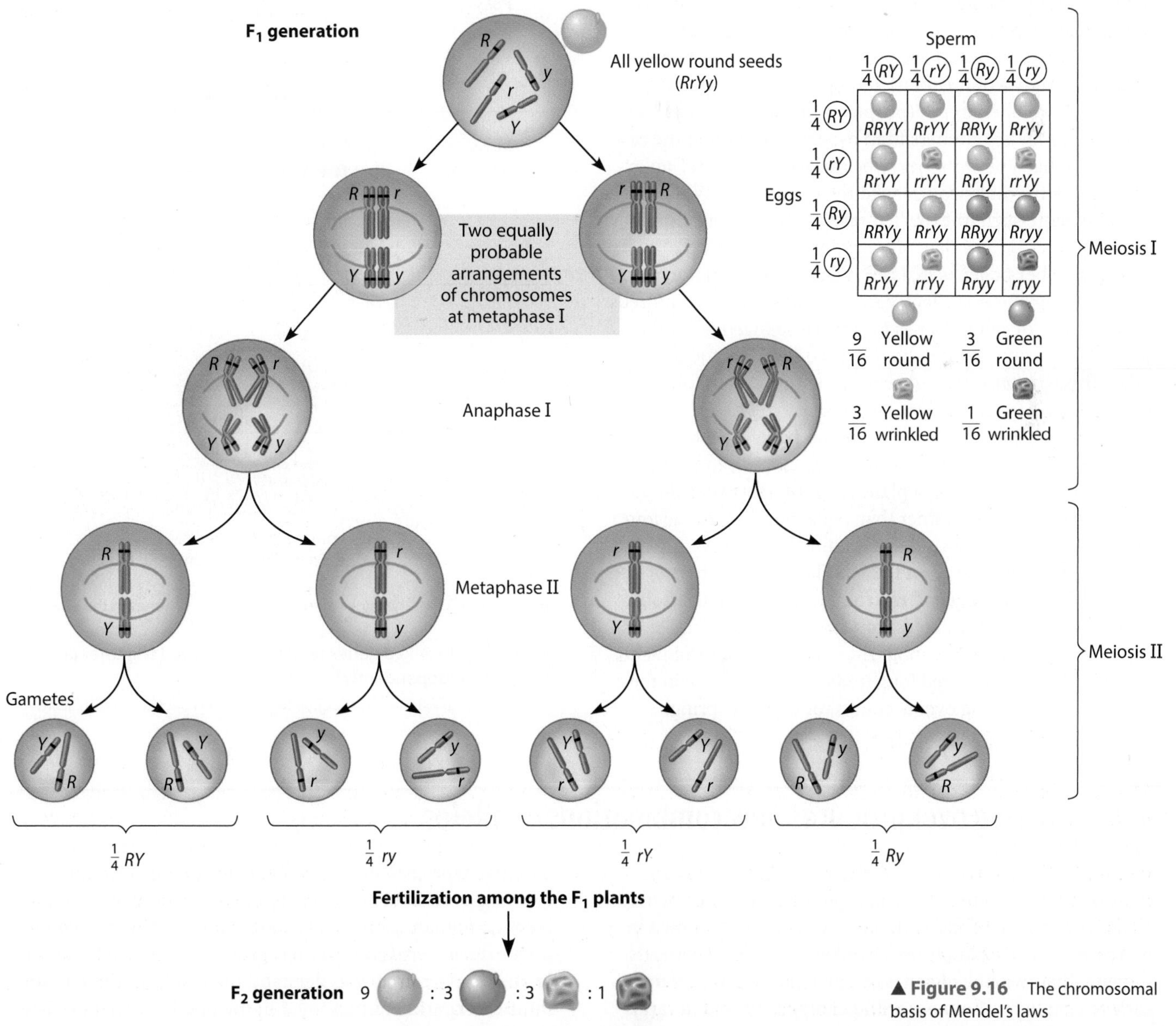

F_1 generation
All yellow round seeds (*RrYy*)
Sperm
Eggs
RRYY RrYY RRYy RrYy
RrYY rrYY RrYy rrYy
RRYy RrYy RRyy Rryy
RrYy rrYy Rryy rryy
9/16 Yellow round
3/16 Green round
3/16 Yellow wrinkled
1/16 Green wrinkled
Meiosis I
Two equally probable arrangements of chromosomes at metaphase I
Anaphase I
Metaphase II
Meiosis II
Gametes
1/4 RY
1/4 ry
1/4 rY
1/4 Ry
Fertilization among the F_1 plants
F_2 generation 9 : 3 : 3 : 1

▲ **Figure 9.16** The chromosomal basis of Mendel's laws

9.17 Genes on the same chromosome tend to be inherited together

SCIENTIFIC THINKING

In 1908, British biologists William Bateson and Reginald Punnett (originator of the Punnett square) observed an inheritance pattern that seemed inconsistent with Mendelian laws. Bateson and Punnett were working with two characters in sweet peas: flower color and pollen shape. They crossed doubly heterozygous plants (*PpLl*) that exhibited the dominant traits: purple flowers (expression of the *P* allele) and long pollen grains (expression of the *L* allele). The corresponding recessive traits are red flowers (in *pp* plants) and round pollen (in *ll* plants).

The top part of **Figure 9.17** illustrates Bateson and Punnett's experiment. When they looked at just one of the two characters (that is, either cross *Pp* × *Pp* or cross *Ll* × *Ll*), they recorded a phenotypic ratio of approximately 3:1 for the offspring, in agreement with Mendel's law of segregation. However, when the biologists combined their data for the two characters, they did not see the 9:3:3:1 ratio predicted by Mendel's law of independent assortment. Instead, as shown in the table, they found a disproportionately large number of plants with just two of the predicted phenotypes: purple long (almost 75% of the total) and red round (about 14%). The other two phenotypes (purple round and red long) were found in far fewer numbers than expected. It is often the case in science that a new discovery begins with a "failure," an experiment with results contrary to those expected. When scientists explore such unexpected results, the investigation may lead to deeper insight than was originally anticipated.

The number of genes in a cell is far greater than the number of chromosomes; in fact, each chromosome has hundreds or thousands of genes. Genes located close together on the same chromosome tend to be inherited together and are called **linked genes**. Linked genes generally do not follow Mendel's law of independent assortment.

Sweet-pea genes for flower color and pollen shape are located on the same chromosome. Thus, meiosis in the heterozygous (*PpLl*) sweet-pea plant yields mostly two genotypes of gametes (*PL* and *pl*) rather than equal numbers of the four types of gametes that would result if the flower-color and pollen-shape genes were not linked. The large numbers of plants with purple long and red round traits in the Bateson-Punnett experiment resulted from fertilization among the *PL* and *pl* gametes. But what about the smaller numbers of plants with purple round and red long traits? As you will see in the next module, crossing over accounts for these offspring.

The Experiment

Purple flower

PpLl × *PpLl*

Long pollen

Phenotypes	Observed offspring	Prediction (9:3:3:1)
Purple long	284	215
Purple round	21	71
Red long	21	71
Red round	55	24

The Explanation: Linked Genes

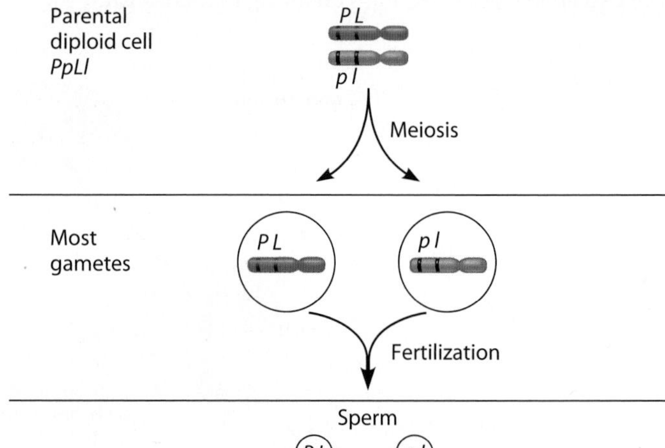
Parental diploid cell *PpLl*
P L
p l
Meiosis
Most gametes
P L
p l
Fertilization

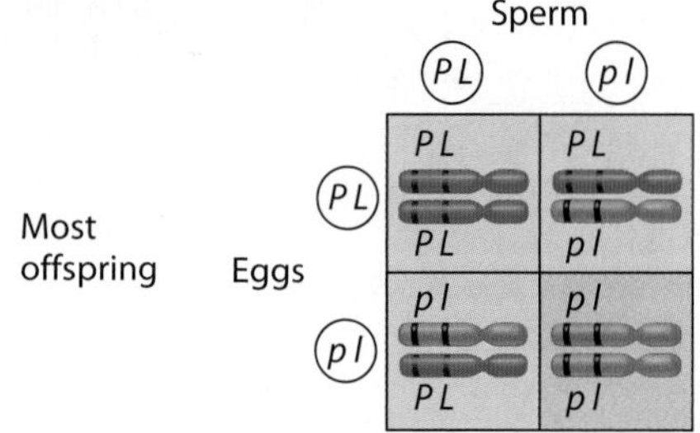
Sperm
P L
p l
Most offspring
Eggs
P L
p l
P L
P L
P L
p l
p l
P L
p l
p l

3 purple long : 1 red round
Not accounted for: purple round and red long

▲ **Figure 9.17** The experiment revealing linked genes in the sweet pea

? Why do linked genes tend to be inherited together and not sort independently?

● Because they are located close together on the same chromosome

9.18 Crossing over produces new combinations of alleles

During meiosis, crossing over between homologous chromosomes produces new combinations of alleles in gametes (as we saw in Module 8.17). Using the experiment shown in Figure 9.17 as an example, **Figure 9.18A** reviews this process, showing that two linked genes can give rise to four different gamete genotypes. Gametes with genotypes *PL* and *pl* carry parental-type chromosomes that have not been altered by crossing over. In contrast, gametes with genotypes *Pl* and *pL* are recombinant gametes. The exchange of chromosome segments during crossing over has produced new combinations of alleles. We can now understand the results of the Bateson-Punnett experiment presented in the previous module: The

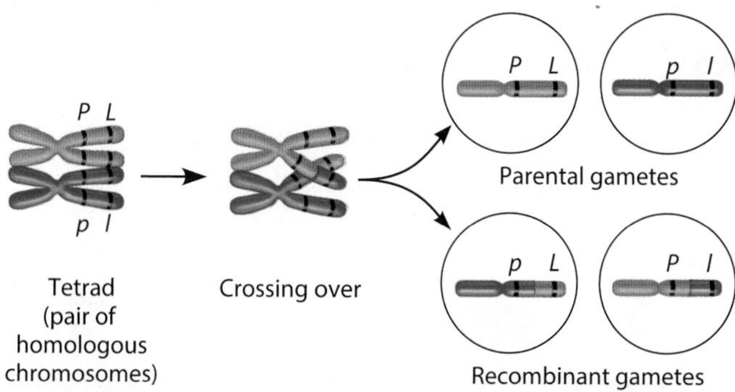
P L
p l
Tetrad (pair of homologous chromosomes)
Crossing over
P L
p l
Parental gametes
p L
P l
Recombinant gametes

▲ **Figure 9.18A** Review: the production of recombinant gametes

small fraction of offspring with recombinant phenotypes (purple round and red long) must have resulted from fertilization involving recombinant gametes.

The discovery of how crossing over creates gamete diversity confirmed the relationship between chromosome behavior and heredity. Some of the most important early studies of crossing over were performed in the laboratory of American embryologist Thomas Hunt Morgan in the early 1900s. Morgan and his colleagues used the fruit fly *Drosophila melanogaster* in many of their experiments (**Figure 9.18B**). *Drosophila* is a good research animal for genetic studies because it can be bred easily and inexpensively, producing each new generation in two weeks.

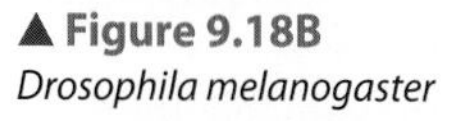

▲ **Figure 9.18B** *Drosophila melanogaster*

Figure 9.18C shows one of Morgan's experiments. This cross involves a wild-type fruit fly—recall from Module 9.8 that "wild-type" refers to the traits most common in nature, in this case, gray body and long wings—and a fly with a black body and vestigial wings. (Used here, the term *vestigial* describes the undeveloped, shrunken appearance of the wings and should not be confused with the evolutionary use of the word *vestigial.*) Morgan knew the genotypes of these flies from previous studies. Here we use the following gene symbols: G = gray body (dominant), g = black body (recessive), L = long wings (dominant), l = vestigial wings (recessive).

In mating a heterozygous gray fly with long wings (genotype *GgLl*) with a black fly with vestigial wings (genotype *ggll*), Morgan performed a testcross (see Module 9.6). If the genes were not linked, then independent assortment would produce offspring in a phenotypic ratio of 1:1:1:1 ($\frac{1}{4}$ gray body, long wings; $\frac{1}{4}$ black body, vestigial wings; $\frac{1}{4}$ gray body, vestigial wings; and $\frac{1}{4}$ black body, long wings). But because these genes are linked, Morgan obtained the results shown in the top part of Figure 9.18C: Most of the offspring had parental phenotypes, but 17% of the offspring flies were recombinants. The percentage of recombinant offspring among the total is called the **recombination frequency**.

The lower part of Figure 9.18C explains Morgan's results in terms of crossing over. A crossover between chromatids of homologous chromosomes in parent *GgLl* broke linkages between the *G* and *L* alleles and between the *g* and *l* alleles, forming the recombinant chromosomes *Gl* and *gL*. Later steps in meiosis distributed the recombinant chromosomes to gametes, and random fertilization produced the four kinds of offspring Morgan observed.

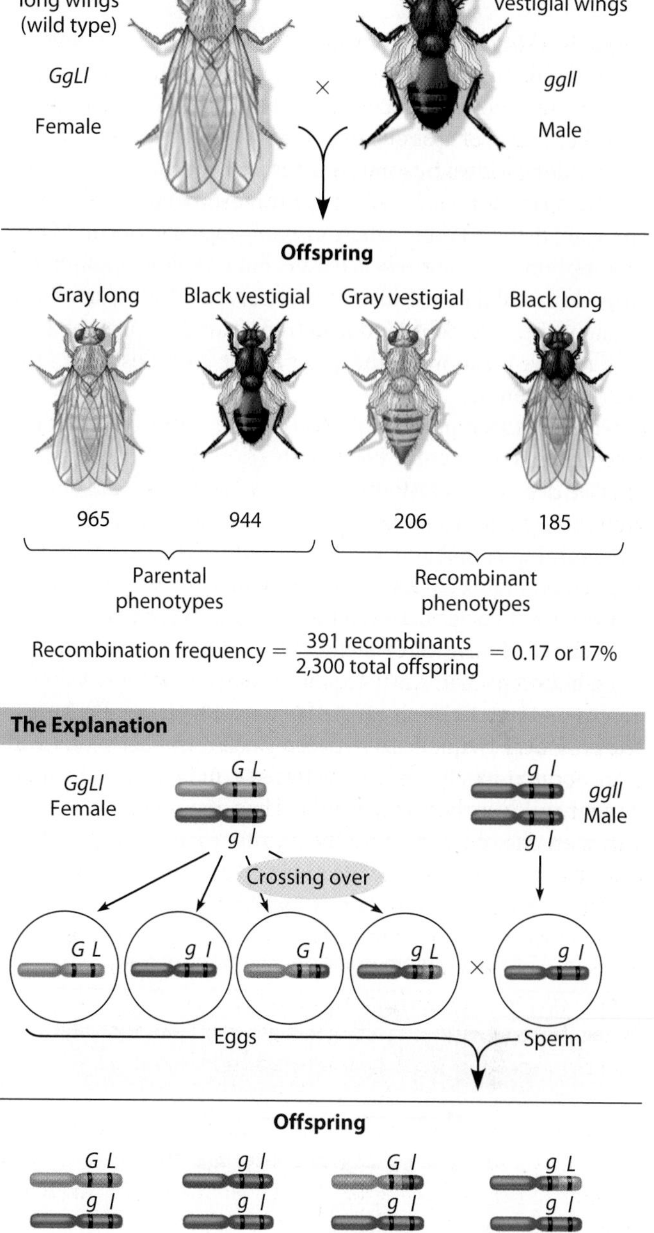
The Experiment
Gray body, long wings (wild type)
GgLl
Female
Black body, vestigial wings
ggll
Male
Offspring
Gray long
Black vestigial
Gray vestigial
Black long
965
944
206
185
Parental phenotypes
Recombinant phenotypes
Recombination frequency = 391 recombinants / 2,300 total offspring = 0.17 or 17%
The Explanation
GgLl
Female
G L
g l
ggll
Male
g l
g l
Crossing over
G L
g l
G l
g L
g l
Eggs
Sperm
Offspring
G L
g l
g l
g l
G l
g l
g L
g l
Parental
Recombinant

▲ **Figure 9.18C** A fruit fly experiment demonstrating the role of crossing over in inheritance

From T. H. Morgan and C. J. Lynch, The linkage of two factors in *Drosophila* that are not sex-linked, *Biological Bulletin* 23: 174–82 (1912).

? Return to the data in Figure 9.17. What is the recombination frequency between the flower-color and pollen-length genes?

● 11% (42/381)

9.19 Geneticists use crossover data to map genes

While working with *Drosophila*, Alfred H. Sturtevant, one of Morgan's students, developed a way to use crossover data to map gene loci. This technique is based on the assumption that the chance of crossing over is approximately equal at all points along a chromosome. Sturtevant hypothesized that the farther apart two genes are on a chromosome, the more points there are between them where crossing over can occur. (This assumption is not entirely accurate, but it is good enough to provide useful data.) With this principle in mind, Sturtevant began using recombination data from fruit fly crosses to assign relative positions of the genes on the chromosomes—that is, to map genes.

Figure 9.19A represents a part of the chromosome that carries the linked genes for black body (*g*) and vestigial wings (*l*) that we described in Module 9.18. This same chromosome also carries a gene that has a recessive allele (we'll call it *c*) determining cinnabar eye color, a brighter red than the wild-type color. Figure 9.19A shows the actual crossover (recombination) frequencies between these alleles, taken two at a time: 17% between the *g* and *l* alleles, 9% between *g* and *c*, and 9.5% between *c* and *l*. Sturtevant reasoned that these values represent the relative distances between the genes. Because the crossover frequencies between *g* and *c* and between *l* and *c* are approximately half that between *g* and *l*, gene *c* must lie roughly midway between *g* and *l*. Thus, the sequence of these genes on one of the fruit fly chromosomes must be *g-c-l*. Such a diagram of relative gene locations is called a **linkage map**.

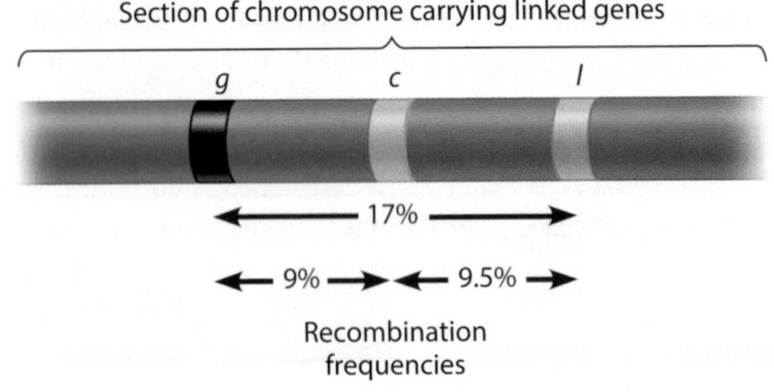

Section of chromosome carrying linked genes
g
c
l
17%
9%
9.5%
Recombination frequencies

▲ **Figure 9.19A** Mapping genes from crossover data

Mutant (less common) phenotypes

Short aristae | Black body (*g*) | Cinnabar eyes (*c*) | Vestigial wings (*l*) | Brown eyes

Long aristae (appendages on head) | Gray body (*G*) | Red eyes (*C*) | Normal wings (*L*) | Red eyes

Wild-type (more common) phenotypes

▲ **Figure 9.19B** A partial linkage map of a fruit fly chromosome

Although based on some approximations, Sturtevant's method of mapping genes helped establish the relative positions of many other fruit fly genes. Eventually, enough data were accumulated to reveal that *Drosophila* has four groups of genes, corresponding to its four pairs of homologous chromosomes. **Figure 9.19B** is a genetic map showing just five of the gene loci on part of one chromosome: the loci labeled *g*, *c*, and *l* and two other traits, aristae and brown eyes.

The linkage-mapping method has proved valuable in establishing the relative positions of many genes in many organisms. The real beauty of the technique is that a wealth of information about genes can be learned simply by breeding and observing the organisms; no fancy equipment is required.

? **You design *Drosophila* crosses to provide recombination data for a gene not included in Figure 9.19A. The gene has recombination frequencies of 3% with the vestigial-wing (*l*) locus and 7% with the cinnabar-eye (*c*) locus. Where is it located on the chromosome?**

The gene is located between the vestigial and cinnabar loci, a bit closer to the vestigial-wing locus (because the vestigial-wing locus has a lower recombination frequency).

▷ Sex Chromosomes and Sex-Linked Genes

9.20 Chromosomes determine sex in many species

Many animals, including fruit flies and all mammals, have a pair of **sex chromosomes**, designated X and Y, that determine an individual's sex (**Figure 9.20A**). Among humans, individuals with one X chromosome and one Y chromosome are males; XX individuals are females. In addition, human males and females both have 44 autosomes (nonsex chromosomes). After meiosis, each gamete contains one sex chromosome and a haploid set of autosomes (22 in humans). All eggs contain a single X chromosome. Of the sperm cells, half contain an X chromosome and half contain a Y chromosome. An offspring's sex depends on whether the sperm cell that fertilizes the egg bears an X chromosome or a Y chromosome (**Figure 9.20B**).

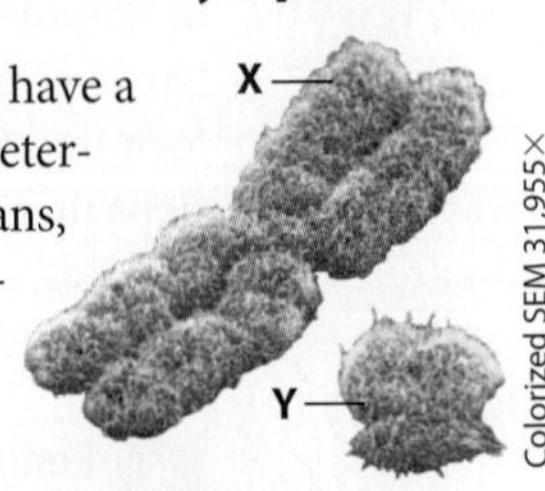

X
Y
Colorized SEM 31,955×

▲ **Figure 9.20A** The human sex chromosomes

The genetic basis of sex determination in humans is not yet completely understood, but one gene on the Y chromosome plays a crucial role.

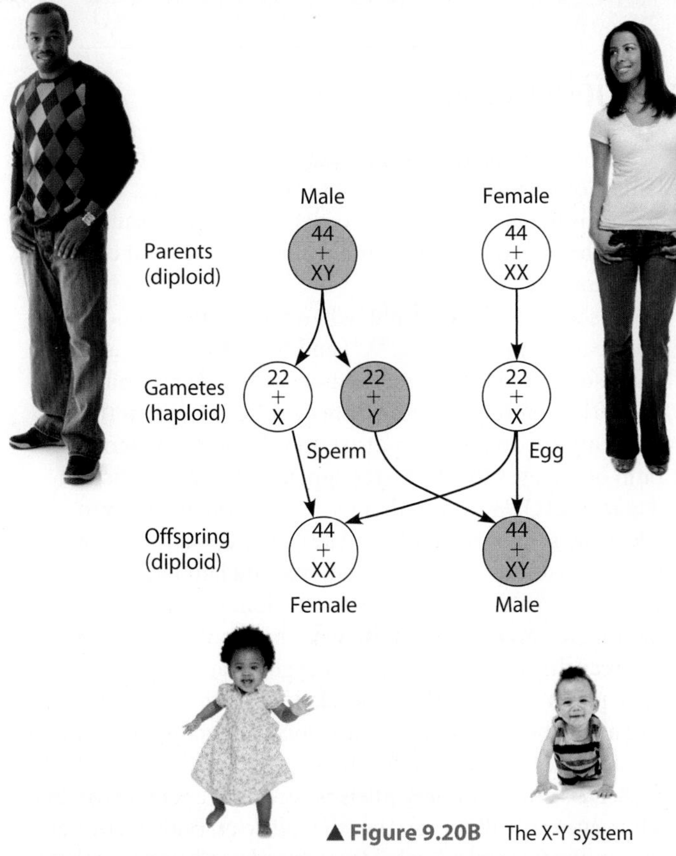
Male
Female
Parents
(diploid)
44
+
XY
44
+
XX
Gametes
(haploid)
22
+
X
22
+
Y
22
+
X
Sperm
Egg
Offspring
(diploid)
44
+
XX
44
+
XY
Female
Male

▲ **Figure 9.20B** The X-Y system

This gene is called *SRY* (for sex-determining region of Y) and triggers testis development. In the absence of *SRY*, ovaries develop rather than testes. *SRY* codes for proteins that regulate other genes on the Y chromosome. These genes in turn produce proteins necessary for normal testis development.

The X-Y system is only one of several sex-determining systems (**Table 9.20** summarizes three other systems). For example, grasshoppers, roaches, and some other insects have an X-O system, in which O stands for the absence of a sex chromosome. Females have two X chromosomes (XX); males have only one sex chromosome (XO). Males produce two classes of sperm: Half bear an X and half lack a sex chromosome. In this case, as in humans, sperm cells determine the sex of the offspring at fertilization.

In contrast to the X-Y and X-O systems, eggs determine sex in certain fishes, butterflies, and birds. The sex chromosomes in these animals are designated Z and W. Males have the genotype ZZ; females are ZW. In this system, sex is determined by whether the egg carries a Z or a W.

Some organisms lack sex chromosomes altogether. In most ants and bees, sex is determined by chromosome number rather than by sex chromosomes. Females develop from fertilized eggs and thus are diploid. Males develop from unfertilized eggs—they are fatherless—and are haploid.

Most animals have two separate sexes; that is, individuals are either male or female. Many plant species have sperm-bearing and egg-bearing flowers found on different individuals. Some plant species, such as date palms, have the X-Y system of sex determination; others, such as the wild strawberry, have the Z-W system. However, most plant species and some animal species have individuals that produce both sperm and eggs. In such species, all individuals have the same complement of chromosomes.

In Module 9.15, we discussed the role that environment plays in determining many characters. Among some animals, environment can even determine sex. For some species of reptiles, the temperature at which eggs are incubated during a specific period of embryonic development determines whether that embryo will develop into a male or female. For example, if green sea turtle hatchlings incubate above 30°C (86°F), nearly all the resulting turtles will be males. (Some worry that global climate change might therefore affect the makeup of turtle populations.) Such temperature-dependent sex determination is an extreme example of the environment affecting the phenotype of an individual.

TABLE 9.20 THREE SYSTEMS OF SEX DETERMINATION

System	Example Organism	Genetic Makeup: Males	Genetic Makeup: Females
X-O		22 + X	22 + XX
Z-W		76 + ZZ	76 + ZW
Chromosome number		16	32

? King Henry VIII of England was quick to blame his wives for bearing him only daughters. Explain how, from a genetic point of view, his thinking was wrong.

● The male sperm bears either an X or Y, thereby determining the sex of the offspring; his wives' eggs always carried an X.

9.21 Sex-linked genes exhibit a unique pattern of inheritance

Besides bearing genes that determine sex, the sex chromosomes also contain genes for characters unrelated to femaleness or maleness. A gene located on either sex chromosome is called a **sex-linked gene**. Because the human X chromosome contains many more genes than the Y, the vast majority of sex-linked genes are on the X chromosome. Be careful not to confuse the term *sex-linked gene*, which refers to a gene on a sex chromosome, with the term *linked genes*, which refers to genes on the same chromosome that tend to be inherited together.

The figures in this module illustrate inheritance patterns for white eye color in the fruit fly, an X-linked recessive trait. Wild-type fruit flies have red eyes; white eyes are very rare (**Figure 9.21A**). We use the uppercase letter *R* for the dominant, wild-type, red-eye allele and *r* for the recessive white-eye allele. Because these alleles are carried on the X chromosome, we show them as superscripts to the letter X. Thus, red-eyed male fruit flies have the genotype X^RY; white-eyed males are X^rY. The Y chromosome does not have a gene locus for eye color; therefore, the male's phenotype results entirely from his single X-linked gene. In the female, X^RX^R and X^RX^r flies have red eyes, and X^rX^r flies have white eyes.

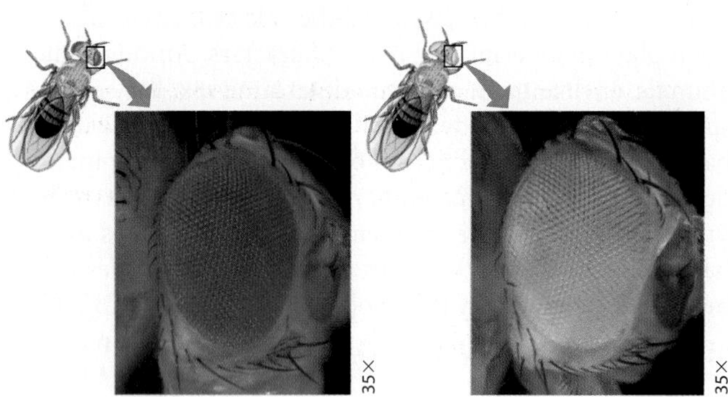
35×
35×

▲ **Figure 9.21A** Fruit fly eye color determined by sex-linked gene

A white-eyed male (X^rY) will transmit his X^r to all of his female offspring but to none of his male offspring. This is because his female offspring, in order to be female, must inherit his X chromosome, but his male offspring must inherit his Y chromosome.

As shown in **Figure 9.21B**, when the female parent is a dominant homozygote (X^RX^R) and the male parent is X^rY, all the offspring have red eyes, but the female offspring are all carriers of the allele for white eyes (X^RX^r). When those offspring are bred to each other, the classic 3:1 phenotypic ratio of red eyes to white eyes appears among the offspring (**Figure 9.21C**). However, there is a twist: The white-eyed trait shows up only in males. All the females have red eyes, whereas half the males have red eyes and half have white eyes. All females inherit at least one dominant allele (from their male parent); half of them are homozygous dominant, whereas the other half are heterozygous carriers, like their female parent. Among the males, half of them inherit the recessive allele their mother was carrying, producing the white-eye phenotype.

Because the white-eye allele is recessive, a female will have white eyes only if she receives that allele on both X chromosomes. For example, if a heterozygous female mates with a white-eyed male, there is a 50% chance that each offspring will have white eyes (resulting from genotype X^rX^r or X^rY), regardless of sex (**Figure 9.21D**). Female offspring with red eyes are heterozygotes, whereas red-eyed male offspring completely lack the recessive allele.

? **A white-eyed female *Drosophila* is mated with a red-eyed (wild-type) male. What result do you predict for the numerous offspring?**

● All female offspring will be red-eyed but heterozygous (X^RX^r); all male offspring will be white-eyed (X^rY).

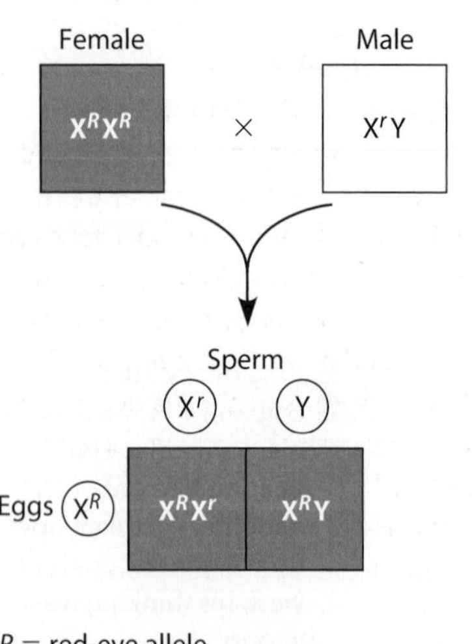
Female
Male
X^RX^R
×
X^rY
Sperm
X^r
Y
Eggs
X^R
X^RX^r
X^RY
R = red-eye allele
r = white-eye allele

▲ **Figure 9.21B** A homozygous, red-eyed female crossed with a white-eyed male

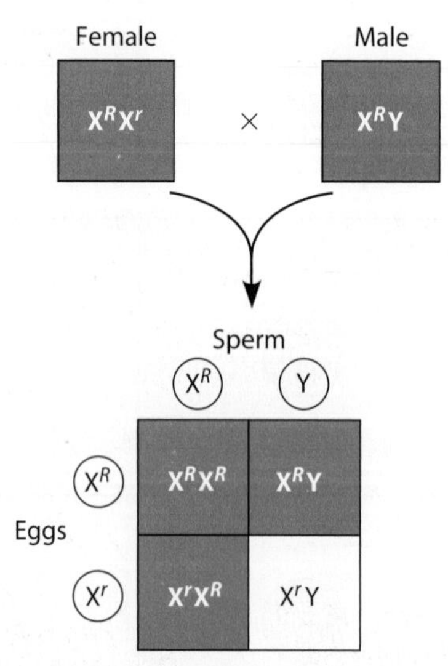
Female
Male
X^RX^r
×
X^RY
Sperm
X^R
Y
Eggs
X^R
X^r
X^RX^R
X^RY
X^rX^R
X^rY

▲ **Figure 9.21C** A heterozygous female crossed with a red-eyed male

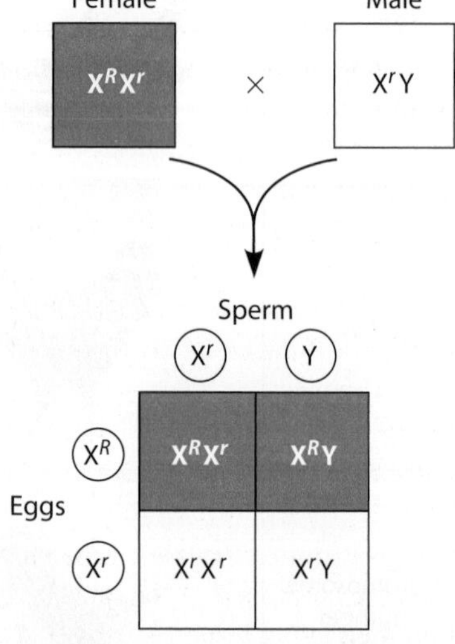
Female
Male
X^RX^r
×
X^rY
Sperm
X^r
Y
Eggs
X^R
X^r
X^RX^r
X^RY
X^rX^r
X^rY

▲ **Figure 9.21D** A heterozygous female crossed with a white-eyed male

9.22 Human sex-linked disorders affect mostly males

CONNECTION

A number of human conditions result from sex-linked recessive alleles located on the X chromosome. If a man inherits only one X-linked recessive allele—from his mother—the allele will be expressed. In contrast, a woman has to inherit two such alleles—one from each parent—to exhibit the trait. Thus, recessive X-linked traits are expressed much more frequently in men than in women.

Hemophilia is an X-linked recessive trait with a well-documented history. Hemophiliacs bleed excessively when injured because they lack one or more of the proteins required for blood clotting. A high incidence of hemophilia plagued the royal families of Europe. Queen Victoria (1819–1901) of England was a carrier of the hemophilia allele. She passed it on to one of her sons and two of her daughters. Through marriage, her daughters then introduced the disease into the families of Prussia, Russia, and Spain. The pedigree in **Figure 9.22** traces the disease through one branch of the royal family. As you can see, Alexandra, like her mother and grandmother, was a carrier, and Alexis had the disease.

Another human X-linked recessive disorder is Duchenne muscular dystrophy, a condition characterized by a progressive weakening of the muscles and loss of coordination. The first symptoms appear in early childhood, when the child begins to have difficulty standing up. Eventually, muscle tissue becomes severely wasted, the individual becomes wheelchair-bound, and normal breathing becomes difficult. Affected individuals rarely live past their early 20s. Researchers have traced the disorder to a recessive mutation in a gene on the X chromosome that codes for a muscle protein.

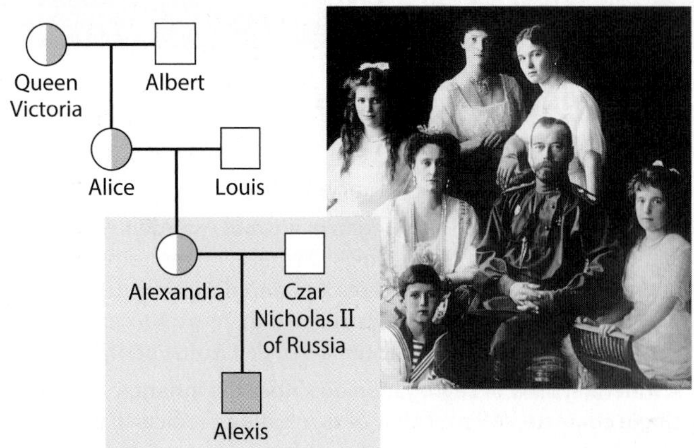
Queen Victoria
Albert
Alice
Louis
Alexandra
Czar Nicholas II of Russia
Alexis

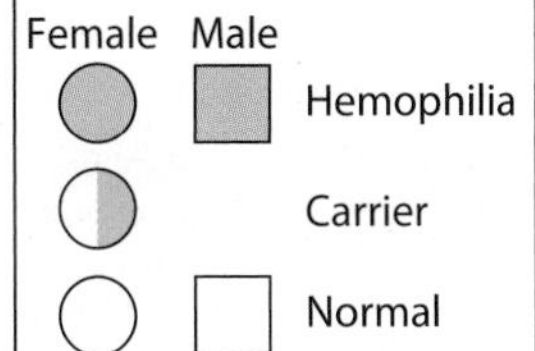
Female
Male
Hemophilia
Carrier
Normal

▲ **Figure 9.22** Hemophilia in the royal family of Russia

Try This Alexis must have had an X^h chromosome (because he had hemophilia). Use your finger to trace back this mutant chromosome through three generations of his ancestors.

? **Neither Tom nor Sue has hemophilia, but their first son does. If the couple has a second child, what is the probability that he or she will also have the disease?**

● $\frac{1}{4}$ ($\frac{1}{2}$ chance of a male child × $\frac{1}{2}$ chance that he will inherit the mutant X)

9.23 The Y chromosome provides clues about human male evolution

EVOLUTION CONNECTION

As you learned in the chapter-opening story about Tibetans, our genes often bear evidence of recent human evolution. The Y chromosome can be particularly useful for tracing our evolutionary past because, barring mutations, the human Y chromosome passes essentially intact from father to son. By analyzing Y DNA, researchers can learn about the ancestry of human males.

In 2003, geneticists discovered that about 8% of males currently living in central Asia have Y chromosomes of striking genetic similarity. Further analysis traced their common genetic heritage to a single man living about 1,000 years ago. In combination with historical records, the data led to the speculation that the Mongolian ruler Genghis Khan **(Figure 9.23)** may be responsible for the spread of the telltale chromosome to nearly 16 million men living today. A similar study of Irish men suggested that nearly 10% of them were descendants of Niall of the Nine Hostages, a warlord who lived during the 5th century.

▲ **Figure 9.23** Genghis Khan

Another study of Y DNA seemed to confirm the claim by the Lemba people of southern Africa that they are descended from ancient Jews. Sequences of Y DNA distinctive of a particular Jewish priestly caste are found at high frequencies among the Lemba.

Are humans evolving?

The discovery of the sex chromosomes and their pattern of inheritance was one of many breakthroughs in understanding how genes are passed from one generation to the next. During the first half of the 20th century, geneticists rediscovered Mendel's work, reinterpreted his laws in light of chromosomal behavior during meiosis, and firmly established the chromosome theory of inheritance. This work set the stage for discoveries in molecular genetics, an area we explore in the next three chapters.

? **Why is the Y chromosome particularly useful in tracing recent human heritage?**

● Because it is passed directly from father to son, forming an unbroken chain of male lineage

CHAPTER 9 REVIEW

For practice quizzes, BioFlix animations, MP3 tutorials, video tutors and more study tools designed for this textbook, go to MasteringBiology®

Reviewing the Concepts

Mendel's Laws (9.1–9.10)

9.1 The study of genetics has ancient roots.

9.2 The science of genetics began in an abbey garden. The science of genetics began with Gregor Mendel's quantitative experiments. Mendel crossed pea plants and traced traits from generation to generation. He hypothesized that there are alternative versions of genes (alleles), the units that determine heritable traits.

9.3 Mendel's law of segregation describes the inheritance of a single character. Mendel's law of segregation predicts that each set of alleles will separate as gametes are formed.

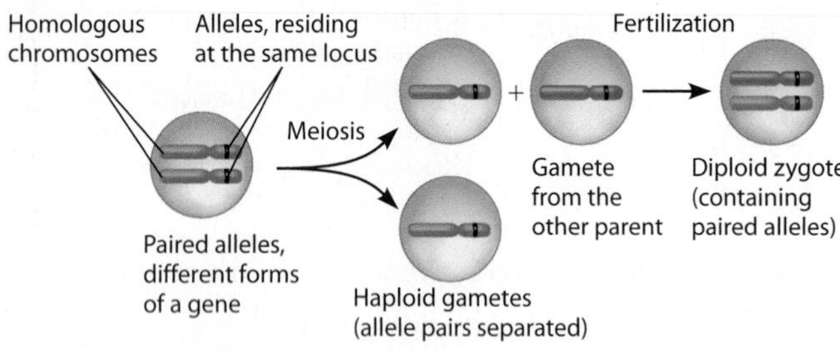

Homologous chromosomes
Alleles, residing at the same locus
Fertilization
Meiosis
Gamete from the other parent
Diploid zygote (containing paired alleles)
Paired alleles, different forms of a gene
Haploid gametes (allele pairs separated)

9.4 Homologous chromosomes bear the alleles for each character. When the two alleles of a gene in a diploid individual are different, the dominant allele determines the inherited trait, whereas the recessive allele has no effect.

9.5 The law of independent assortment is revealed by tracking two characters at once. Mendel's law of independent assortment states that the alleles of a pair segregate independently of other allele pairs during gamete formation.

9.6 Geneticists can use a testcross to determine unknown genotypes. The offspring of a testcross, a mating between an individual of unknown genotype and a homozygous recessive individual, can reveal the unknown's genotype.

9.7 Mendel's laws reflect the rules of probability. The rule of multiplication calculates the probability of two independent events both occurring. The rule of addition calculates the probability of an event that can occur in alternative ways.

9.8 Genetic traits in humans can be tracked through family pedigrees. The inheritance of many human traits follows Mendel's laws. Family pedigrees can help determine individual genotypes.

9.9 Many inherited disorders in humans are controlled by a single gene.

9.10 New technologies can provide insight into one's genetic legacy. Carrier screening, fetal testing, fetal imaging, and newborn screening can provide information for reproductive decisions but may create ethical dilemmas.

Variations on Mendel's Laws (9.11–9.15)

9.11 Incomplete dominance results in intermediate phenotypes. Mendel's laws are valid for all sexually reproducing species, but genotype often does not dictate phenotype in the simple way Mendel's laws describe.

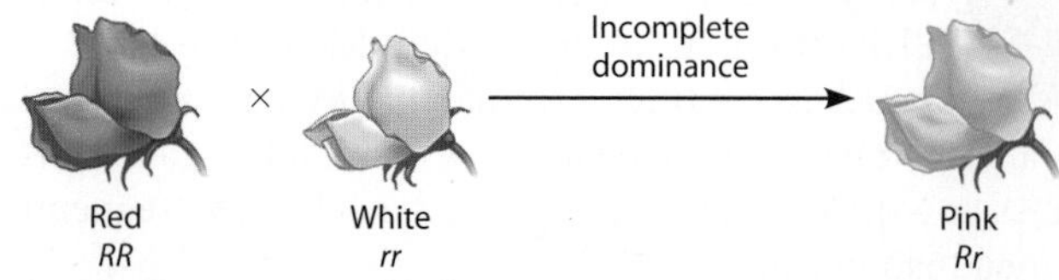

Incomplete dominance
×
Red
RR
White
rr
Pink
Rr

9.12 Many genes have more than two alleles in the population. For example, the ABO blood group phenotype in humans is controlled by three alleles that produce a total of four phenotypes.

9.13 A single gene may affect many phenotypic characters.

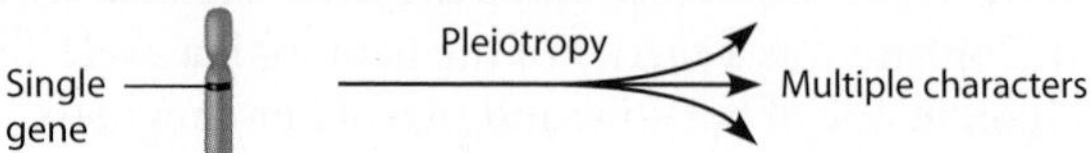

Single gene
Pleiotropy
Multiple characters

9.14 A single character may be influenced by many genes.

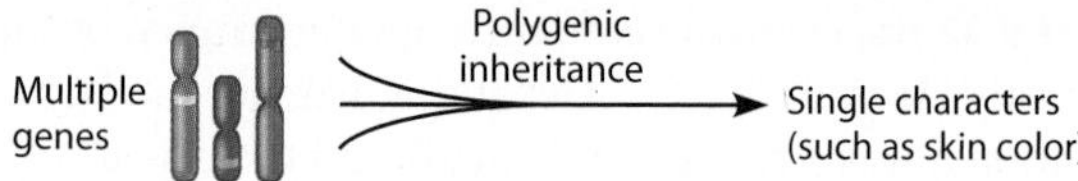

Multiple genes
Polygenic inheritance
Single characters (such as skin color)

9.15 The environment affects many characters. Many traits are affected, in varying degrees, by both genetic and environmental factors.

The Chromosomal Basis of Inheritance (9.16–9.19)

9.16 Chromosome behavior accounts for Mendel's laws. Genes are located on chromosomes, whose behavior during meiosis and fertilization accounts for inheritance patterns.

9.17 Genes on the same chromosome tend to be inherited together. Such genes are said to be linked; they display non-Mendelian inheritance patterns.

9.18 Crossing over produces new combinations of alleles. Crossing over can separate linked alleles, producing gametes with recombinant chromosomes.

9.19 Geneticists use crossover data to map genes. Recombination frequencies can be used to map the relative positions of genes on chromosomes.

Sex Chromosomes and Sex-Linked Genes (9.20–9.23)

9.20 Chromosomes determine sex in many species. In mammals, a male has XY sex chromosomes, and a female has XX. The Y chromosome has genes for the development of testes, whereas an absence of the Y allows ovaries to develop. Other systems of sex determination exist in other animals and plants.

9.21 Sex-linked genes exhibit a unique pattern of inheritance. All genes on the sex chromosomes are said to be sex-linked. However, the X chromosome carries many genes unrelated to sex.

9.22 Human sex-linked disorders affect mostly males. Most sex-linked (X-linked) human disorders are due to recessive alleles and are seen mostly in males. A male receiving a single X-linked recessive allele from his mother will have the disorder; a female must receive the allele from both parents to be affected.

9.23 The Y chromosome provides clues about human male evolution. Because they are passed on intact from father to son, Y chromosomes can provide data about recent human evolutionary history.

Connecting the Concepts

1. Complete this concept map to help you review some key concepts of genetics.

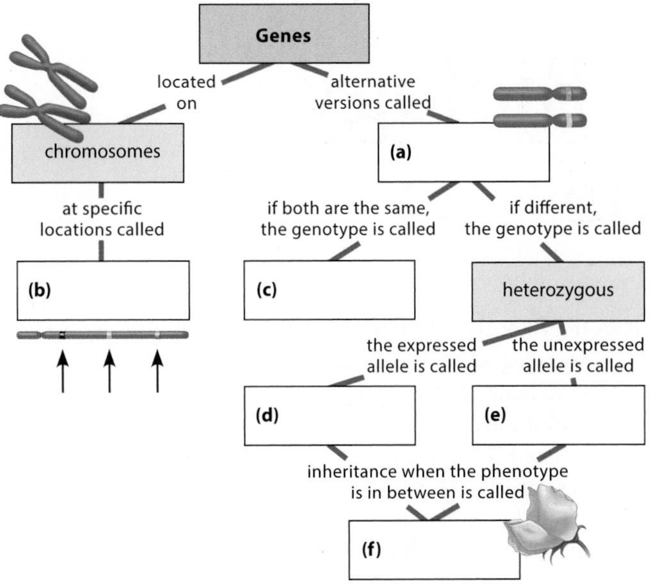

Genes
located on
alternative versions called
chromosomes
(a)
at specific locations called
if both are the same, the genotype is called
if different, the genotype is called
(b)
(c)
heterozygous
the expressed allele is called
the unexpressed allele is called
(d)
(e)
inheritance when the phenotype is in between is called
(f)

Testing Your Knowledge

Level 1: Knowledge/Comprehension

2. Whether an allele is dominant or recessive depends on
   a. how common the allele is, relative to other alleles.
   b. whether it is inherited from the mother or the father.
   c. whether it or another allele determines the phenotype when both are present.
   d. whether or not it is linked to other genes.
3. Edward was found to be heterozygous (*Ss*) for sickle-cell trait. The alleles represented by the letters *S* and *s* are
   a. linked.
   b. on homologous chromosomes.
   c. both present in each of Edward's sperm cells.
   d. on the same chromosome but far apart.

Level 2: Application/Analysis

4. Two fruit flies with eyes of the usual red color are crossed, and their offspring are as follows: 77 red-eyed males, 71 ruby-eyed males, 152 red-eyed females. The allele for ruby eyes is
   a. autosomal (carried on an autosome) and dominant.
   b. autosomal and recessive.
   c. sex-linked and dominant.
   d. sex-linked and recessive.
5. A man with type B blood and a woman who has type A blood could have children of which of the following phenotypes?
   a. A or B only
   b. AB only
   c. AB or O
   d. A, B, AB, or O
6. Tim and Jan both have freckles (see Module 9.9), but their son Mike does not. Show with a Punnett square how this is possible. If Tim and Jan have two more children, what is the probability that both will have freckles?
7. Both Tim and Jan (problem 6) have a widow's peak (see Module 9.8), but Mike has a straight hairline. What are their genotypes? What is the probability that Tim and Jan's next child will have freckles and a straight hairline?
8. In rabbits, black hair depends on a dominant allele, *B*, and brown hair on a recessive allele, *b*. Short hair is due to a dominant allele, *S*, and long hair to a recessive allele, *s*. If a true-breeding black short-haired male is mated with a brown long-haired female, describe their offspring. What will be the genotypes of the offspring? If two of these F_1 rabbits are mated, what phenotypes would you expect among their offspring? In what proportions?
9. A fruit fly with a gray body and red eyes (genotype *BbPp*) is mated with a fly having a black body and purple eyes (genotype *bbpp*). What ratio of offspring would you expect if the body-color and eye-color genes are on different chromosomes (unlinked)? When this mating is actually carried out, most of the offspring look like the parents, but 3% have a gray body and purple eyes, and 3% have a black body and red eyes. Are these genes linked or unlinked? What is the recombination frequency?
10. A series of matings shows that the recombination frequency between the black-body gene (problem 9) and the gene for dumpy (shortened) wings is 36%. The recombination frequency between purple eyes and dumpy wings is 41%. What is the sequence of these three genes on the chromosome?
11. A couple are both phenotypically normal, but their son suffers from hemophilia, a sex-linked recessive disorder. What fraction of their children are likely to suffer from hemophilia? What fraction are likely to be carriers?

Level 3: Synthesis/Evaluation

12. Why do more men than women have colorblindness?
13. In fruit flies, the genes for wing shape and body stripes are linked. In a fly whose genotype is *WwSs*, *W* is linked to *S*, and *w* is linked to *s*. Show how this fly can produce gametes containing four different combinations of alleles. Which are parental-type gametes? Which are recombinant gametes? How are the recombinants produced?
14. Adult height in humans is at least partially hereditary; tall parents tend to have tall children. But humans come in a range of sizes, not just tall and short. Which extension of Mendel's model accounts for the hereditary variation in human height?
15. Heather was surprised to discover she suffered from red-green colorblindness. She told her biology professor, who said, "Your father is colorblind, too, right?" How did her professor know this? Why did her professor not say the same thing to the colorblind males in the class?
16. In 1981, a stray black cat with unusual rounded, curled-back ears was adopted by a family in Lakewood, California. Suppose you owned the first curl cat and wanted to breed it to develop a true-breeding variety. Describe tests that would determine whether the curl gene is dominant or recessive and whether it is autosomal or sex-linked. Explain why you think your tests would be conclusive. Describe a test to determine that a cat is true-breeding.
17. **SCIENTIFIC THINKING** The breakthrough that led Bateson and Punnett to recognize the existence of linked genes (Module 9.17) was the appearance of unexpected results after they crossed double heterozygous pea plants (*PpLl*) with each other. Imagine that you have a group of Labrador retrievers that are all heterozygous for both coat color and blindness (*BbNn*). If you used this group of dogs to produce 160 puppies, how many puppies of each phenotype do you expect to get if the genes are not linked? How would the results differ if the genes are in fact linked?

Answers to all questions can be found in Appendix 4.

CHAPTER 10

Molecular Biology of the Gene

Why are viral diseases such a constant threat?

The electron micrograph below shows the 2009 H1N1 influenza virus, an infectious microbe that first appeared in a cluster of flu cases diagnosed in and near Mexico City. Once the threat had been identified, the city was virtually shut down in an effort to contain the outbreak. Despite these efforts, the new viral strain spread so quickly that within months the World Health Organization (WHO) declared H1N1 a pandemic (global epidemic). A vaccine against the 2009 H1N1 strain was rushed into production and became available in the fall of that year. By the time the WHO declared the pandemic over in August 2010, the disease had reached 207 countries, infecting more than 600,000 and killing an estimated 20,000 people.

The 2009 H1N1 influenza virus is just one example in a long, bleak history of human viral diseases. The flu virus, like all viruses, consists of a relatively simple structure of protein and nucleic acid (RNA in this case). Viruses share some of the characteristics of living organisms, but are generally not considered alive because they are not cellular and cannot reproduce on their own. So how

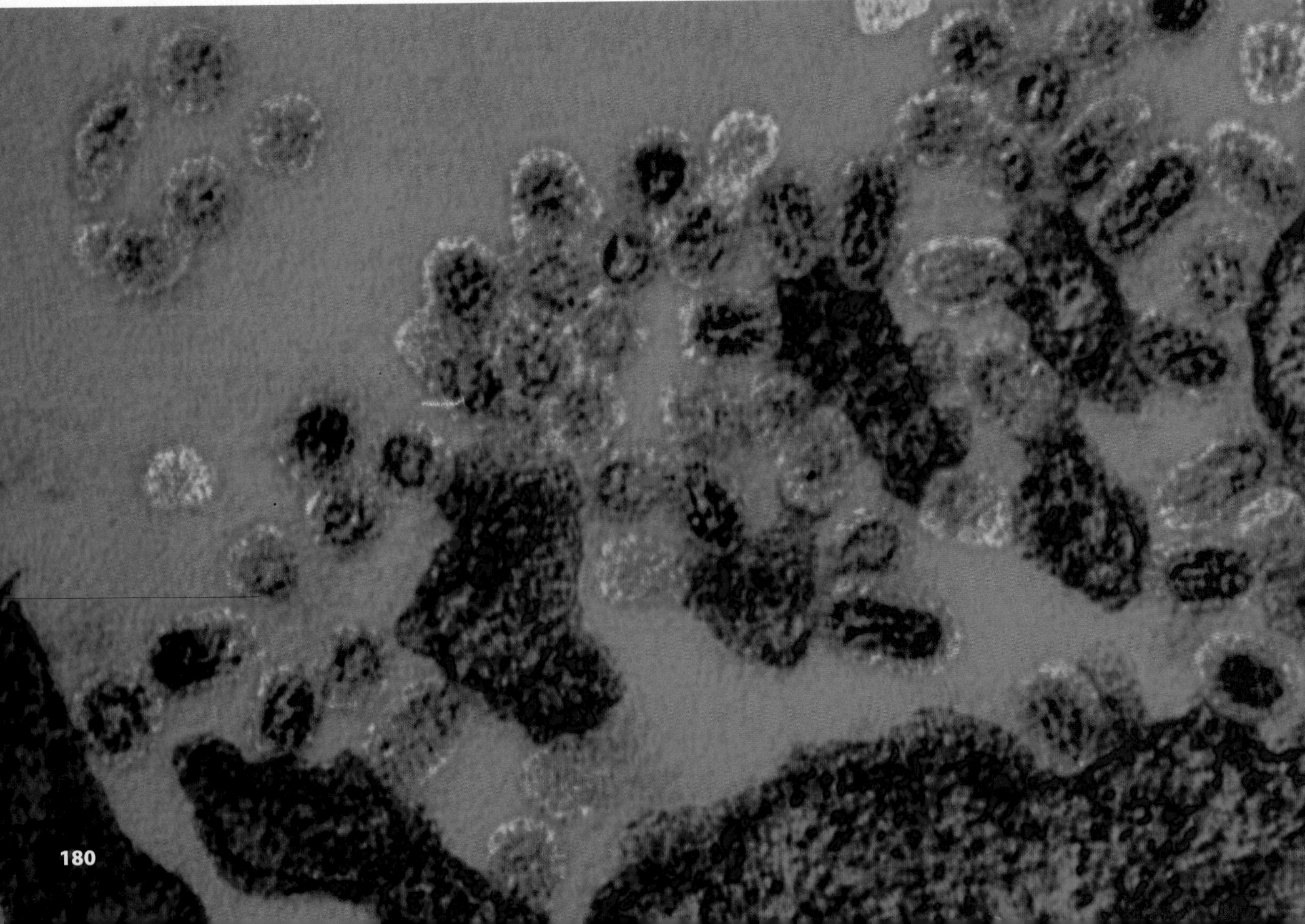

can such a simple pathogen e
The key to the elusiveness of
rapidly. The 2009 H1N1 strai
shuffling of multiple flu virus
and pigs.

Combating any virus requi
biology, the study of DNA and
fact, we owe our first glimpses
controls hereditary traits, to th
plore the structure of DNA, ho
by directing RNA and protein
genetics of viruses and bacteri

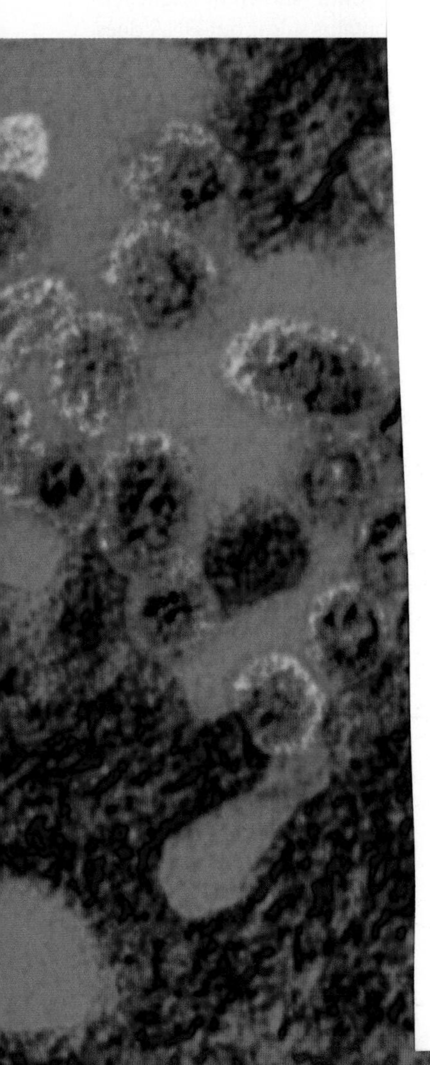

1. ____is the diffusion of water in a selectively permeable membrane.
2. ____is the movement of solutes from an area of greater concentration to an area of lesser concentration.
3. ____is the most important key ingredient of biological membranes
4. __is the process when a cell does not have to do work when molecules diffuse across membranes.
5. ____a process that requires energy to move molecules across membranes.
6. ____is a solution in which the volume of a cell remains constant.
7. ____is a solution in the cell in which the solute concentration is lower than that of the cell.
8. ____Is a solution in which the solute concentration is higher than that of the cell.
9. ____is a region along which the density of a chemical substance increases or decreases.
10. __is a diffusion that does not require energy but requires a protein to carry molecules to the plasma membrane.

BIG IDEAS

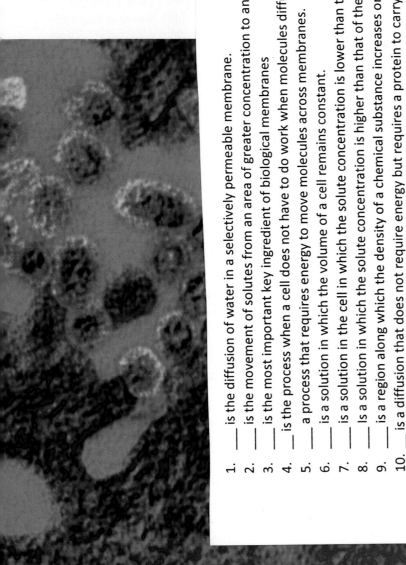

The Structure of the Genetic Material
(10.1–10.3)

A series of experiments established DNA as the molecule of heredity.

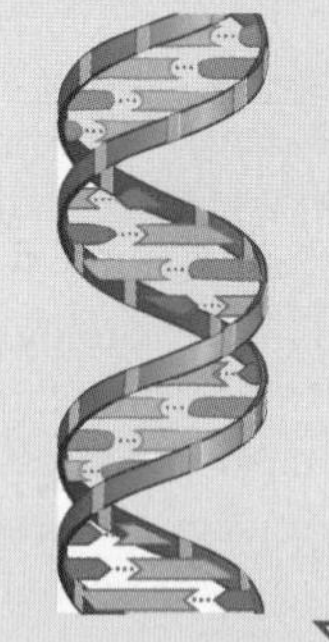

DNA Replication
(10.4–10.5)

Each DNA strand can serve as a template for another.

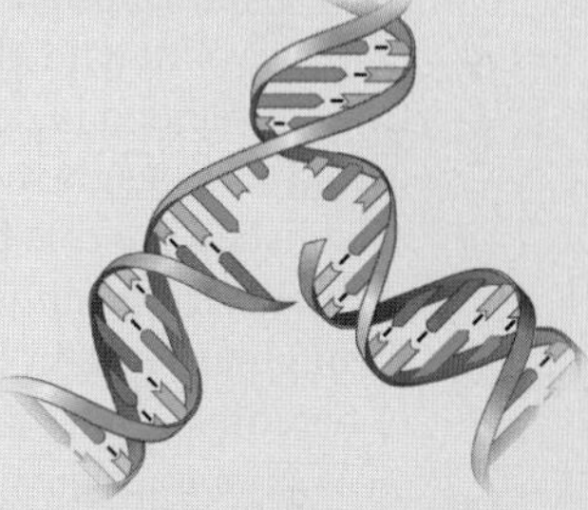

The Flow of Genetic Information from DNA to RNA to Protein
(10.6–10.16)

Genotype controls phenotype through the production of proteins.

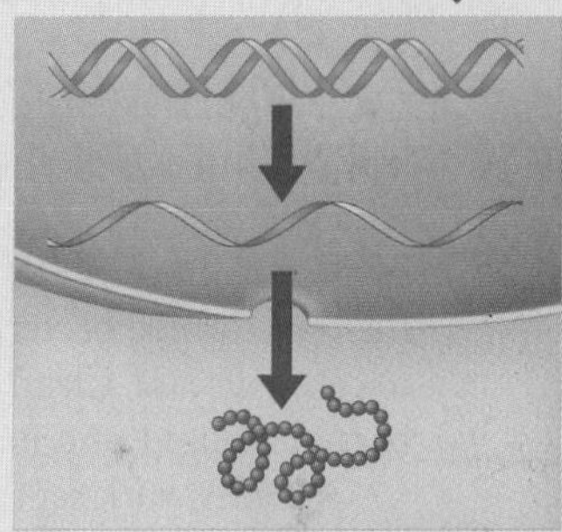

The Genetics of Viruses and Bacteria
(10.17–10.23)

Viruses and bacteria are useful model systems for the study of molecular biology.

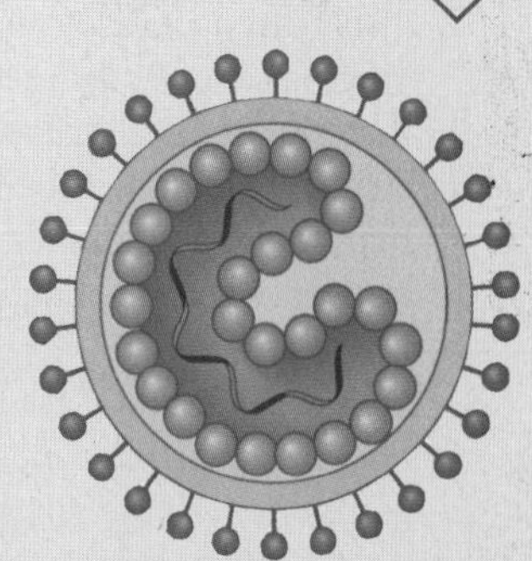

▷ The Structure of the Genetic Material

10.1 Experiments showed that DNA is the genetic material

SCIENTIFIC THINKING

Today, scientists routinely manipulate DNA in the laboratory and use it to change the heritable characteristics of cells. Early in the 20th century, however, the molecular basis for inheritance was a mystery. Biologists did know that genes were located on chromosomes. The two chemical components of chromosomes—DNA and protein—were therefore the leading candidates to be the genetic material. Until the 1940s, the case for proteins seemed stronger because proteins appeared to be more structurally complex: Proteins were known to be made from 20 different amino acid building blocks, whereas DNA was known to be made from just four kinds of nucleotides. It seemed logical that the more complex molecule would serve as the hereditary material. Biologists finally established the role of DNA in heredity through experiments with bacteria and the viruses that infect them. This breakthrough ushered in the field of **molecular biology**, the study of heredity at the molecular level.

We can trace the discovery of the genetic role of DNA to 1928. British medical officer Frederick Griffith was studying two strains (varieties) of a bacterium: a harmless strain and a pathogenic (disease-causing) strain that causes pneumonia. Griffith was surprised to find that when he killed the pathogenic bacteria and then mixed the bacterial remains with living harmless bacteria, some living bacterial cells became pathogenic. Furthermore, all of the descendants of the transformed bacteria inherited the newly acquired ability to cause disease. Clearly, some chemical component of the dead bacteria caused a heritable change in live bacteria.

Griffith's work set the stage for a race to discover the identity of the chemical basis of heredity. In 1952, American biologists Alfred Hershey and Martha Chase performed a very convincing set of experiments that showed DNA to be the genetic material of T2, a virus that infects the bacterium *Escherichia coli* (*E. coli*), a microbe normally found in the intestines of mammals (including humans). Viruses that exclusively infect bacteria are called **bacteriophages** ("bacteria-eaters"), or **phages** for short.

Hershey and Chase knew that T2 could reprogram its host cell to produce new phages, but they did not know what component of the virus conferred this capability. At the time, it was known that the structure of phage T2 consists solely of two types of molecules (**Figure 10.1A**): DNA (blue in the figure) and protein (gold). The researchers took advantage of this fact to devise an elegantly simple experiment that determined which of these molecules the phage transferred to *E. coli* during infection.

The Hershey and Chase experiment illustrates a point that arises repeatedly in the history of science: the importance of carefully designing an experiment and choosing the correct model organism to study. Like Mendel's garden peas (Module 9.2) and Morgan's fruit flies (Module 9.18), bacteriophage T2 had just the right properties (in this case, a simple structure consisting of just two contrasting elements) to allow for the design of a simple but conclusive experiment.

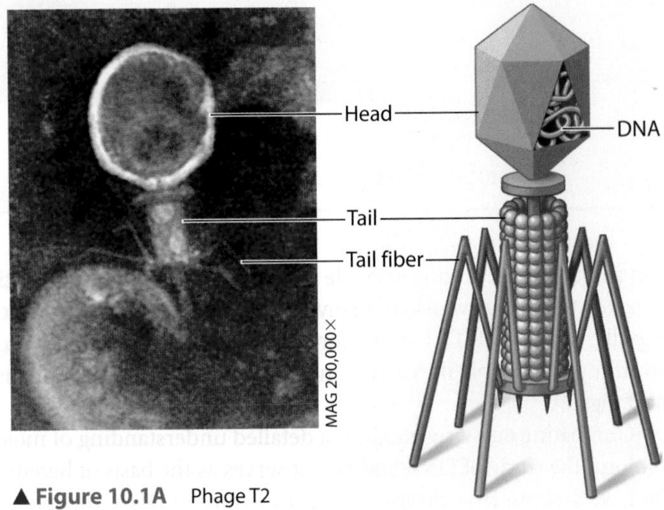
Head
DNA
Tail
Tail fiber
MAG 200,000×

▲ **Figure 10.1A** Phage T2

To begin, Hershey and Chase grew T2 with *E. coli* in a solution containing radioactive sulfur (shown in bright yellow in **Figure 10.1B**). Protein contains sulfur but DNA does not, so as new phages were made, the radioactive sulfur atoms were incorporated only into the proteins of the bacteriophage. The researchers grew a second batch of phages in a solution containing radioactive phosphorus (green). Because nearly all the phage's phosphorus is in DNA, this labeled only the phage DNA.

Armed with the two batches of labeled T2, Hershey and Chase were ready to perform the experiment outlined in Figure 10.1B. They allowed the two batches of T2 to infect separate samples of nonradioactive bacteria. Shortly after the onset of infection, they agitated the cultures in an ordinary kitchen blender to shake loose any parts of the phages that remained outside the bacterial cells. Then, they collected the mixtures in tubes and spun the tubes in a centrifuge. The cells were deposited as a solid pellet at the bottom of the centrifuge tubes, but phages and parts of phages—because they were lighter—remained suspended in the liquid. The researchers then measured the radioactivity in the pellet and in the liquid.

Hershey and Chase found that when the bacteria had been infected with T2 phages containing labeled protein, as in batch 1, the radioactivity ended up mainly in the solution within the centrifuge tube, which contained phages but not bacteria. This result suggested that the phage protein did not enter the cells. But when the bacteria had been infected with phages whose DNA was tagged, as in batch 2, most of the radioactivity was in the pellet of bacterial cells at the bottom of the centrifuge tube. Furthermore, when these bacteria were returned to a liquid growth medium, they soon lysed, or broke open, releasing new phages that contained some radioactive phosphorus in their DNA in their proteins.

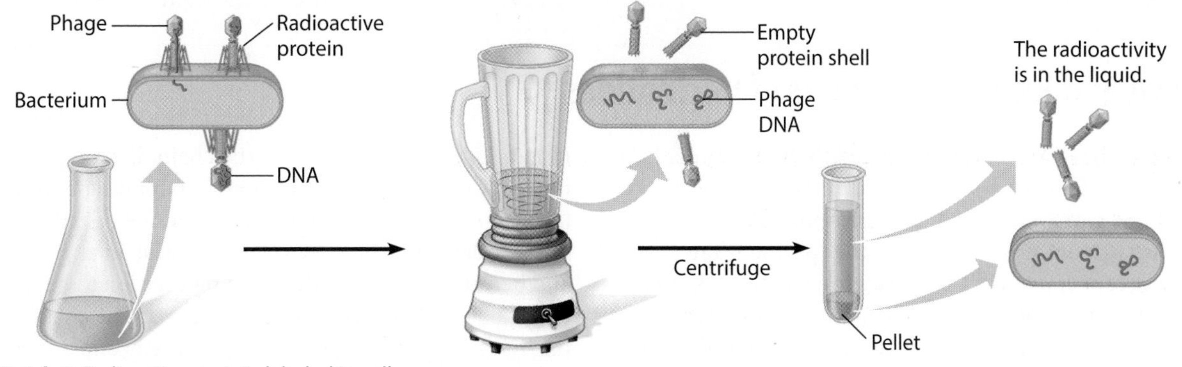
Phage
Radioactive protein
Bacterium
DNA
Empty protein shell
Phage DNA
Centrifuge
Pellet
The radioactivity is in the liquid.
Batch 1: Radioactive protein labeled in yellow

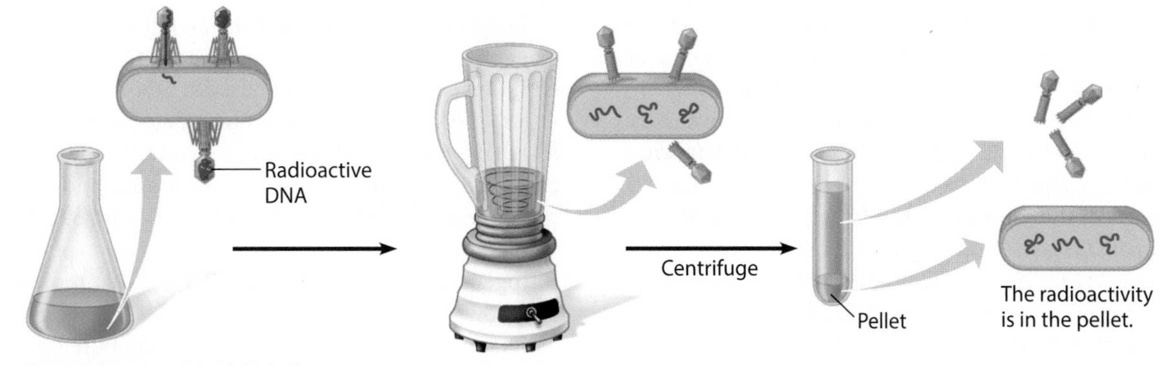
Radioactive DNA
Centrifuge
Pellet
The radioactivity is in the pellet.
Batch 2: Radioactive DNA labeled in green
Hershey and Chase mixed radioactively labeled phages with bacteria. The phages infected the bacterial cells.
They agitated the cultures in a blender to separate the phages outside of the bacteria from the cells and their contents.
They centrifuged the mixture so that the bacteria formed a pellet at the bottom of the test tube.
Finally, they measured the radioactivity in the pellet and in the liquid.

▲ **Figure 10.1B** The Hershey-Chase experiment

Figure 10.1C outlines our current understanding of the replication cycle of phage T2. After the virus ❶ attaches to the host bacterial cell, it ❷ injects its DNA into the host. Notice that virtually all of the viral protein (yellow) is left outside (which is why the radioactive protein did not show up in the host cells during the experiment shown at the top of Figure 10.1B). Once injected, the viral DNA causes the bacterial cells to ❸ produce new phage proteins and DNA molecules—indeed, complete new phages—which soon ❹ cause the cell to lyse, releasing the newly produced phages, which may then attach to other host bacterial cells. As Hershey and Chase discovered, it is the viral DNA that contains the instructions for making phages.

Once DNA was shown to be the molecule of heredity, understanding its structure became the most important quest in biology. In the next two modules, we'll review the structure of DNA and discuss how it was discovered.

? **What convinced Hershey and Chase that DNA, rather than protein, is the genetic material of phage T2?**

● Radioactively labeled phage DNA, but not labeled protein, entered the host cell during infection and directed the synthesis of new viruses.

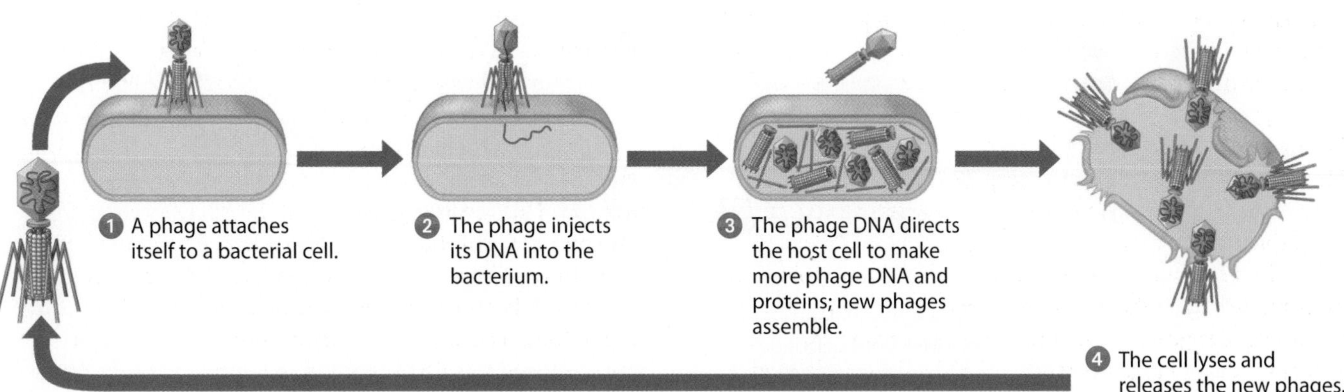
❶ A phage attaches itself to a bacterial cell.
❷ The phage injects its DNA into the bacterium.
❸ The phage DNA directs the host cell to make more phage DNA and proteins; new phages assemble.
❹ The cell lyses and releases the new phages.

▲ **Figure 10.1C** A phage replication cycle

10.2 DNA and RNA are polymers of nucleotides

By the time Hershey and Chase performed their experiments, much was already known about DNA. Scientists had identified all its atoms and knew how they were bonded to one another. What was not understood was the specific three-dimensional arrangement of atoms that gave DNA its unique properties—the capacity to store genetic information, copy it, and pass it from generation to generation. However, only one year after Hershey and Chase published their results, scientists figured out the structure of DNA and the basic strategy of how it works. We will examine that momentous discovery in Module 10.3, but first, let's look at the underlying chemical structure of DNA and its chemical cousin, RNA.

DNA and RNA are nucleic acids, consisting of long chains (polymers) of chemical units (monomers) called **nucleotides** (see Module 3.16). **Figure 10.2A** shows four representations of various parts of the same molecule. At left is a view of a DNA double helix. One of the strands is opened up (center) to show two different views of an individual DNA **polynucleotide**, a nucleotide polymer (chain). The view on the far right zooms in to a single nucleotide from the chain. Each type of DNA nucleotide has a different nitrogen-containing base: adenine (A), cytosine (C), thymine (T), or guanine (G). Because nucleotides can occur in a polynucleotide in any sequence and because polynucleotides can be very long, the number of possible polynucleotides is enormous. The chain shown in this figure has the sequence ACTGG, only one of many possible arrangements of the four types of nucleotides that make up DNA.

Looking more closely at our polynucleotide, we see in the center of Figure 10.2A that each nucleotide consists of three components: a nitrogenous base (in DNA: A, C, T, or G), a sugar (shown in blue), and a phosphate group (shown in yellow). The nucleotides are joined to one another by covalent bonds between the sugar of one nucleotide and the phosphate of the next, which forms a **sugar-phosphate backbone** with a repeating pattern of sugar-phosphate-sugar-phosphate. The nitrogenous bases are arranged like ribs that project from the backbone.

Examining a single nucleotide in even more detail (on the right in Figure 10.2A), you can see the chemical structure of its three components. The phosphate group has a phosphorus atom (P) at its center and is the source of the word *acid* in *nucleic acid*. The sugar has five carbon atoms, shown in red here for emphasis—four in its ring and one extending above the ring. The ring also includes an oxygen atom. The sugar is called deoxyribose because, compared with the sugar

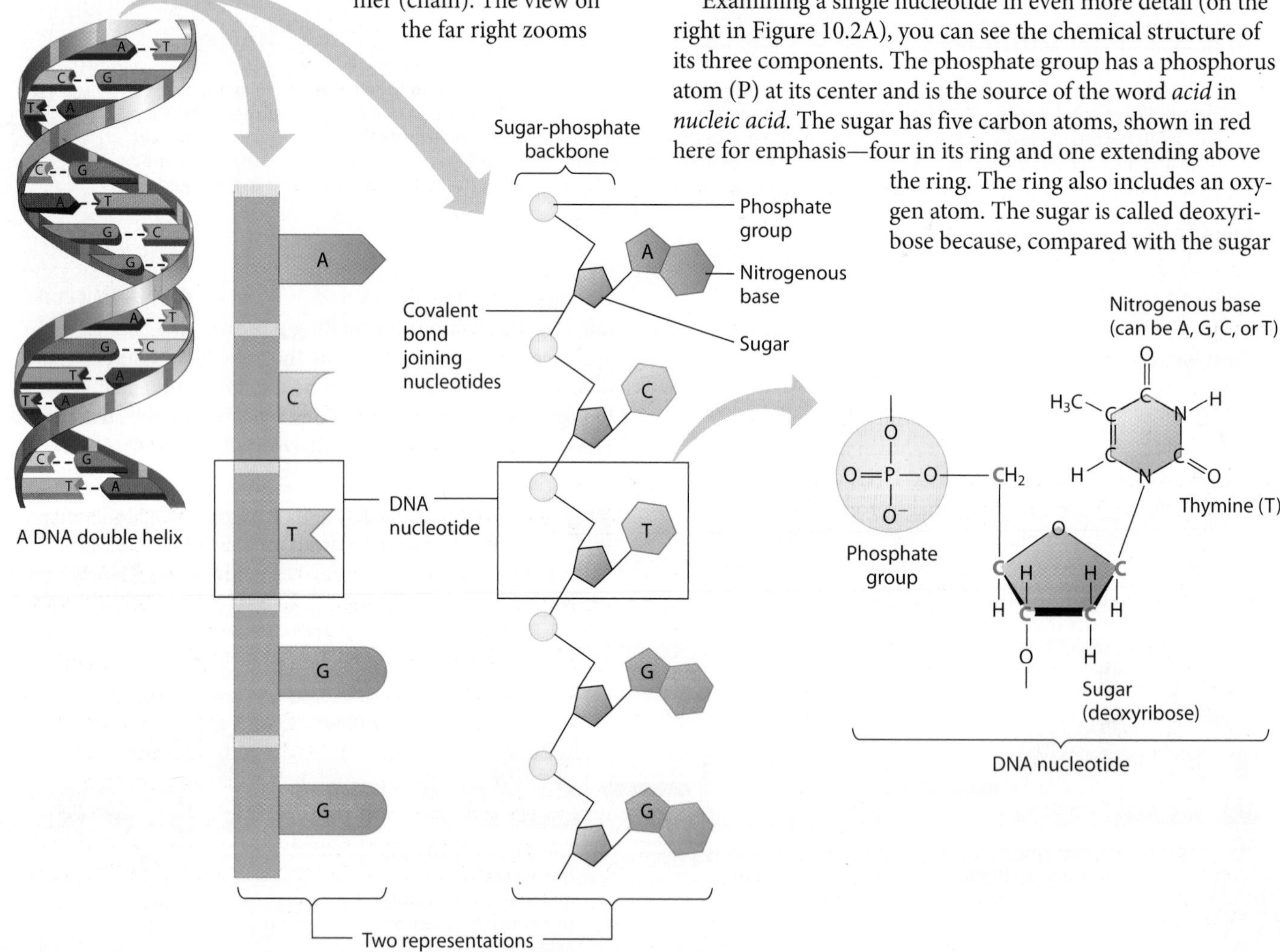

A DNA double helix
Sugar-phosphate backbone
Phosphate group
Nitrogenous base
Sugar
Covalent bond joining nucleotides
DNA nucleotide
Two representations of a DNA polynucleotide
Nitrogenous base (can be A, G, C, or T)
Thymine (T)
Phosphate group
Sugar (deoxyribose)
DNA nucleotide

▲ **Figure 10.2A** Breaking down the structure of DNA

Try This Use your finger to trace each part of the nucleotide—sugar, phosphate, and base—in all four parts of this figure.

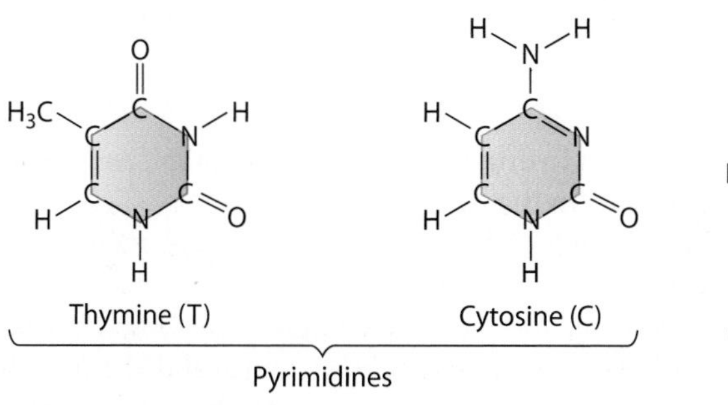
Thymine (T)
Cytosine (C)
Pyrimidines

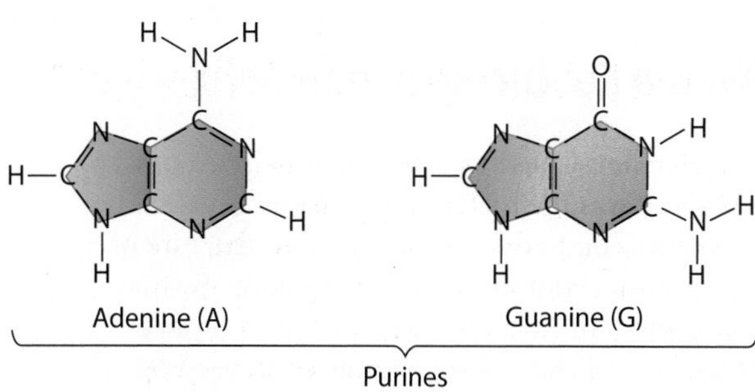
Adenine (A)
Guanine (G)
Purines

▲ **Figure 10.2B** The nitrogenous bases of DNA

ribose, it is missing an oxygen atom. Notice that the C atom in the lower right corner of the ring is bonded to an H atom instead of to an —OH group, as it is in ribose; see Figure 10.2C. Hence, DNA is "deoxy"—which means "without an oxygen"—compared to RNA.

The full name for **DNA** is **deoxyribonucleic acid**, with *nucleic* referring to DNA's location in the nuclei of eukaryotic cells. Each nitrogenous base (thymine, in our example at the right in Figure 10.2A) has a single or double ring consisting of nitrogen and carbon atoms with various functional groups attached **(Figure 10.2B)**. Recall that a functional group is a chemical group that affects a molecule's function by participating in specific chemical reactions (see Module 3.2). In the case of DNA, the main role of the functional groups is to determine which other kind of bases each base can form hydrogen bonds with. For example, the NH_2 group hanging off cytosine is capable of forming a hydrogen bond to the C═O group hanging off guanine, but not with the NH_2 group protruding from adenine. The chemical groups of the bases are therefore responsible for DNA's most important property, which you will learn more about in the next module. In contrast to the acidic phosphate group, nitrogenous bases are basic—hence their name.

The four nucleotides found in DNA differ only in the structure of their nitrogenous bases. At this point, the structural details are not as important as the fact that the bases are of two types. **Thymine (T)** and **cytosine (C)** are single-ring structures called pyrimidines. **Adenine (A)** and **guanine (G)** are larger, double-ring structures called purines. The one-letter abbreviations can be used either for the bases alone or for the nucleotides containing them.

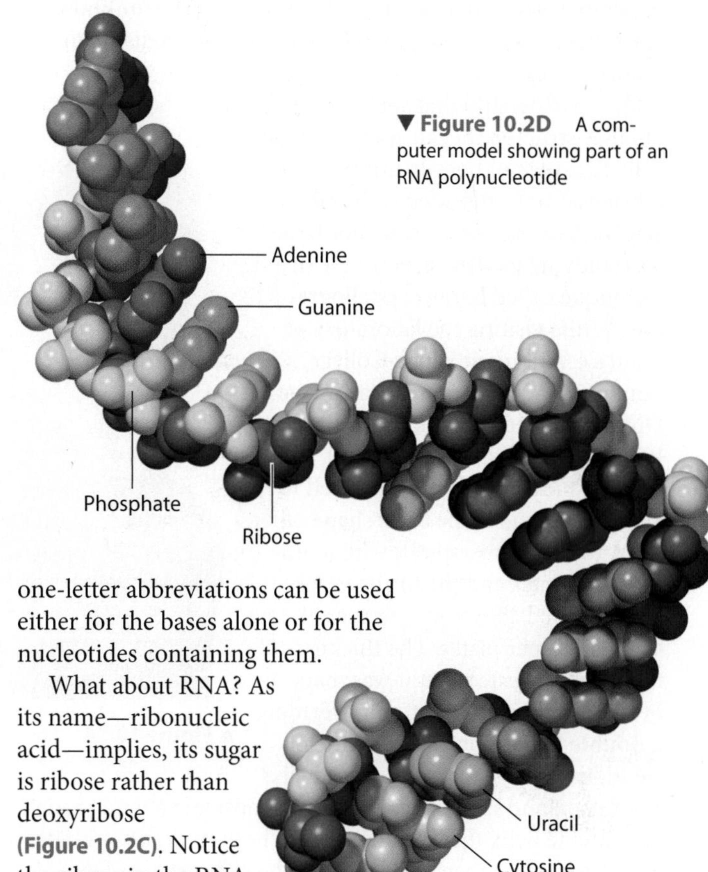
Adenine
Guanine
Phosphate
Ribose
Uracil
Cytosine

▼ **Figure 10.2D** A computer model showing part of an RNA polynucleotide

What about RNA? As its name—ribonucleic acid—implies, its sugar is ribose rather than deoxyribose **(Figure 10.2C)**. Notice the ribose in the RNA nucleotide; unlike deoxyribose, the sugar ring has an —OH group attached to the C atom at its lower-right corner. Another difference between RNA and DNA is that instead of thymine, RNA has a nitrogenous base called **uracil (U)**. (You can see the structure of uracil in Figure 10.2C; it is very similar to thymine.) Except for the presence of ribose and uracil, an RNA polynucleotide chain is identical to a DNA polynucleotide chain. **Figure 10.2D** is a computer graphic of a piece of RNA polynucleotide about 20 nucleotides long. In this three-dimensional view, each sphere represents an atom; notice that the color scheme is the same as in the other figures in this module. The yellow phosphate groups and blue ribose sugars make it easy to spot the sugar-phosphate backbone. In the next module, we'll see how two DNA polynucleotides join together in a molecule of DNA.

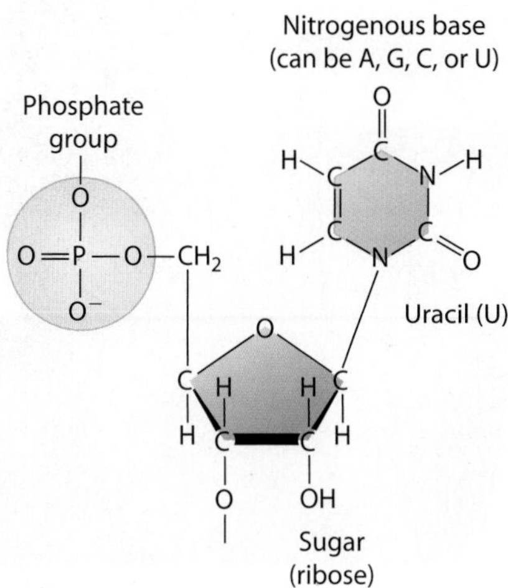
Nitrogenous base
(can be A, G, C, or U)
Phosphate
group
Uracil (U)
Sugar
(ribose)

▲ **Figure 10.2C** An RNA nucleotide

? **Compare and contrast DNA and RNA polynucleotides.**

● Both are polymers of nucleotides consisting of a sugar, a nitrogenous base, and a phosphate. In RNA, the sugar is ribose; in DNA, it is deoxyribose. Both RNA and DNA have the bases A, G, and C, but DNA has a T and RNA has a U.

10.3 DNA is a double-stranded helix

After the 1952 Hershey-Chase experiment convinced most biologists that DNA was the material that stored genetic information, a race was on to determine how the structure of this molecule could account for its role in heredity. By that time, the arrangement of covalent bonds in a nucleic acid polymer was well established, and researchers focused on discovering the three-dimensional shape of DNA. First to the finish line were two scientists who were relatively unknown at the time—American James D. Watson and Englishman Francis Crick.

The partnership that solved the puzzle of DNA structure began soon after Watson, a 23-year-old newly minted Ph.D., journeyed to Cambridge University in England, where the more senior Crick was studying protein structure with a technique called X-ray crystallography. While visiting the laboratory of Maurice Wilkins at King's College in London, Watson saw an X-ray image of DNA produced by Wilkins's colleague, Rosalind Franklin (**Figure 10.3A**). A careful study of the image enabled Watson to deduce the basic shape of DNA to be a helix (spiral) with a uniform diameter and the nitrogenous bases located above one another like a stack of dinner plates. The thickness of the helix suggested that it was made up of two polynucleotide strands, forming a **double helix**. But how were the nucleotides arranged in the double helix?

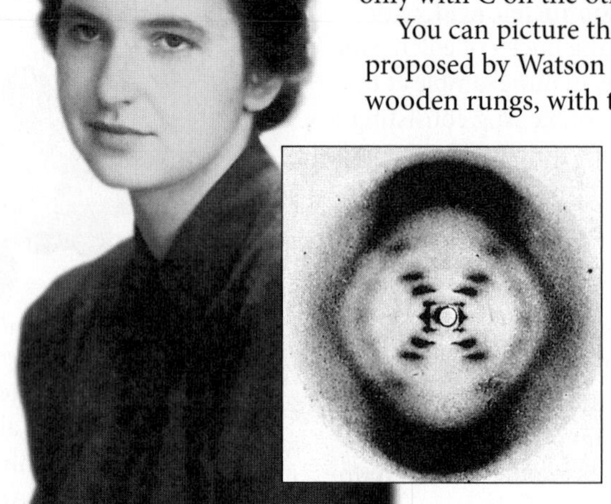

▲ **Figure 10.3A** Rosalind Franklin and her X-ray image of DNA

Watson and Crick began trying to construct a wire model of a double helix that would conform both to Franklin's data and to what was then known about the chemistry of DNA (**Figure 10.3B**). They knew that Franklin had concluded that the sugar-phosphate backbones must be on the outside of the double helix, forcing the nitrogenous bases to swivel to the interior of the molecule. But how were the bases arranged in the interior of the double helix?

At first, Watson and Crick imagined that the bases paired like with like—for example, A with A and C with C. But that kind of pairing did not fit the X-ray data, which suggested that the DNA molecule has a uniform diameter. An A-A pair, with two double-ring bases, would be almost twice as wide as a C-C pair, made of two single-ring bases. It soon became apparent that a double-ringed base (purine) on one strand must always be paired with a single-ringed base (pyrimidine) on the opposite strand to produce a molecule of uniform thickness. After considerable trial and error, Watson and Crick realized that the chemical structures of the bases dictated the pairings even more specifically. As discussed in the previous module, each base has protruding functional groups that can best form hydrogen bonds with just one appropriate partner (to review the hydrogen bond, see Module 2.8). Adenine can best form hydrogen bonds with thymine, and guanine with cytosine. In the biologist's shorthand, A pairs with T, and G pairs with C. A is also said to be "complementary" to T and G to C.

Watson and Crick's pairing scheme both fit what was known about the physical attributes and chemical bonding of DNA and explained some data obtained several years earlier by American biochemist Erwin Chargaff. Chargaff had discovered that the amount of adenine in the DNA of any one species was equal to the amount of thymine and that the amount of guanine was equal to that of cytosine. Chargaff's rules, as they are called, are explained by the fact that A on one of DNA's polynucleotide chains always pairs with T on the other polynucleotide chain, and G on one chain pairs only with C on the other chain.

You can picture the model of the DNA double helix proposed by Watson and Crick as a rope ladder with wooden rungs, with the ladder twisting into a spiral (**Figure 10.3C**). The side ropes represent the sugar-phosphate backbones, and the rungs represent pairs of nitrogenous bases joined by hydrogen bonds.

Figure 10.3D shows three representations of the double helix. The shapes of the base symbols in the ribbonlike diagram on the left indicate the bases' complementarity; notice that the shape of any kind of base matches only one other kind of base. In the center of the diagram is an atomic-level version showing four base pairs, with the helix untwisted and the hydrogen bonds specified by dotted

▲ **Figure 10.3B** Watson and Crick in 1953 with their model of the DNA double helix

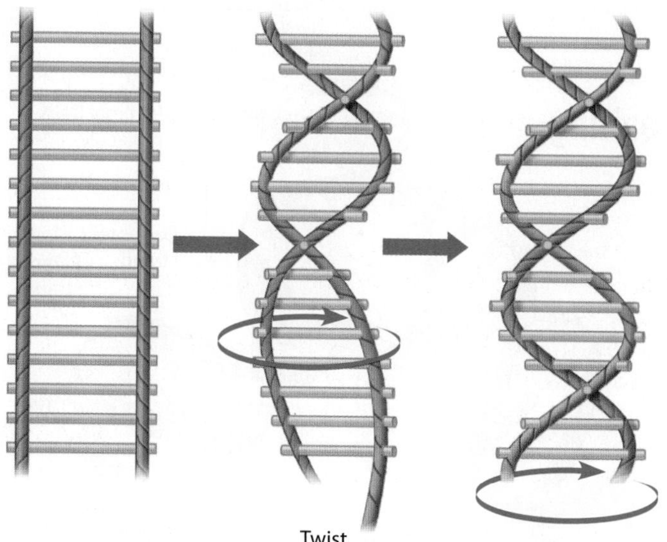
Twist

▲ **Figure 10.3C** A rope ladder analogy for the double helix

lines. Notice that a C-G base pair has functional groups that form three hydrogen bonds, whereas an A-T base pair has functional groups that form two hydrogen bonds. This difference means that C-G base pairs are somewhat stronger than A-T base pairs. You can see that the two sugar-phosphate backbones of the double helix are oriented in opposite directions. (Notice that the sugars on the two strands are upside down with respect to each other.) On the right is a computer graphic showing most of the atoms of part of a double helix. The atoms that compose the deoxyribose sugars are shown as blue, phosphate groups as yellow, and nitrogenous bases as shades of green and orange.

Although the Watson-Crick base-pairing rules dictate the side-by-side combinations of nitrogenous bases that form the rungs of the double helix, they place no restrictions on the sequence of nucleotides along the length of a DNA strand. In fact, the sequence of bases can vary in countless ways, and each gene has a unique order of nucleotides, or base sequence.

In April 1953, Watson and Crick rocked the scientific world with a succinct paper explaining their molecular model for DNA in the British scientific journal *Nature*. In 1962, Watson, Crick, and Wilkins received the Nobel Prize for their work. (Sadly, Rosalind Franklin died in 1958 at the age of 38 and was thus ineligible for the prize.) Few milestones in the history of biology have had as broad an impact as the discovery of the double helix, with its A-T and C-G base pairing.

The Watson-Crick model gave new meaning to the words *genes* and *chromosomes*—and to the chromosome theory of inheritance (see Module 9.16). With a complete picture of DNA, we can see that the genetic information in a chromosome must be encoded in the nucleotide sequence of the molecule. One powerful aspect of the Watson-Crick model is that the structure of DNA suggests a molecular explanation for genetic inheritance, as we will see in the next module.

? Along one strand of a double helix is the nucleotide sequence GGCATAGGT. What is the complementary sequence for the other DNA strand?

● CCGTATCCA

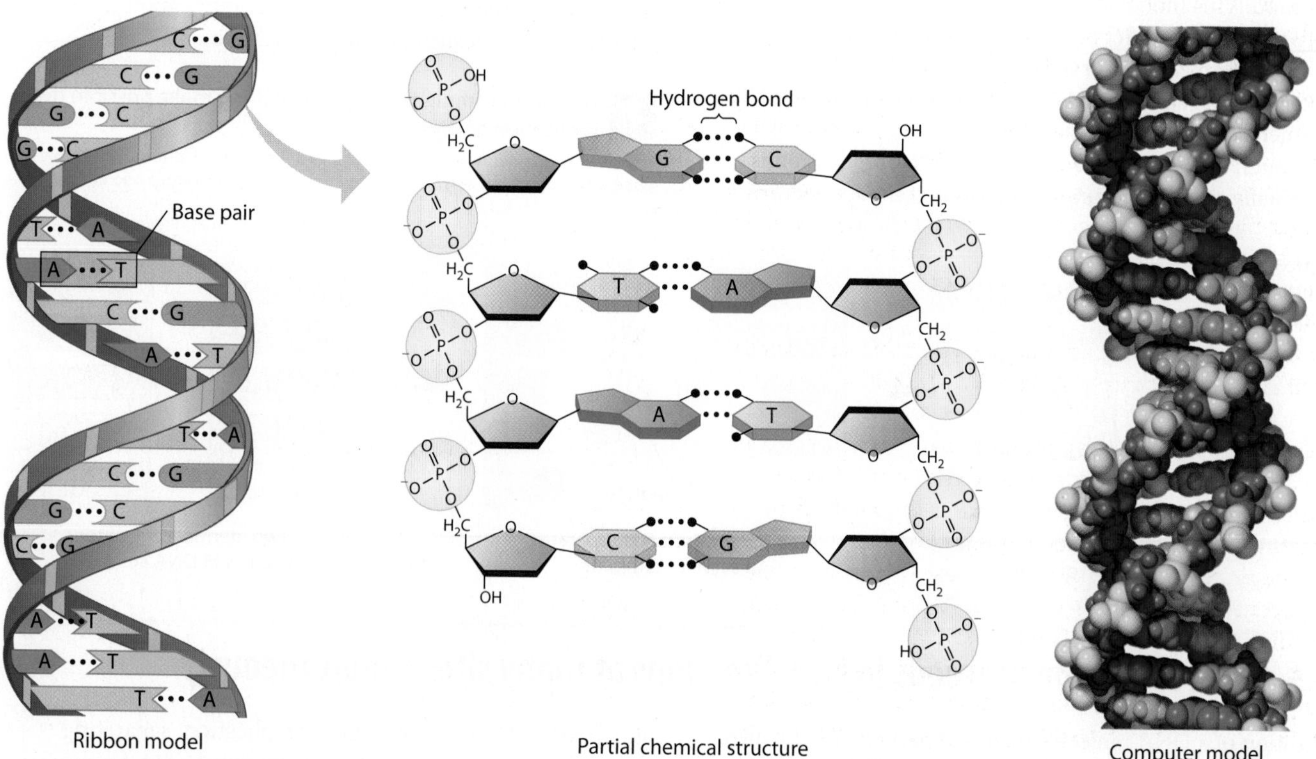
C G
C G
G C
G C
Base pair
T A
A T
C G
A T
T A
C G
G C
C G
A T
A T
T A
Ribbon model
Hydrogen bond
OH
G
C
T
A
A
T
C
G
OH
Partial chemical structure
Computer model

▲ **Figure 10.3D** Three representations of DNA

▷ DNA Replication

10.4 DNA replication depends on specific base pairing

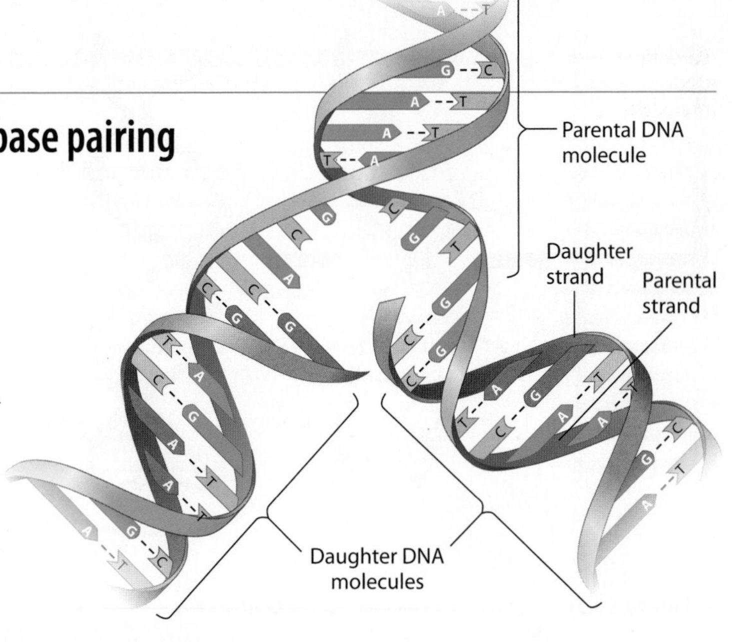

Parental DNA molecule
Daughter strand
Parental strand
Daughter DNA molecules

▲ **Figure 10.4B** The untwisting and replication of DNA

One of biology's overarching themes—the relationship between structure and function—is evident in the double helix. The idea that there is specific pairing of bases in DNA was the flash of inspiration that led Watson and Crick to the correct structure of the DNA molecule. At the same time, they saw the functional significance of the base-pairing rules.

The logic behind the Watson-Crick proposal for how DNA is copied—by specific pairing of complementary bases—is quite simple. You can see this by covering one of the strands in the parental DNA molecule in **Figure 10.4A**. You can determine the sequence of bases in the covered strand by applying the base-pairing rules to the unmasked strand: A pairs with T (and T with A), and G pairs with C (and C with G).

Watson and Crick predicted that a cell applies the same rules when copying its genes during each turn of the cell cycle (Module 8.3). As shown in Figure 10.4A, the two strands of parental DNA (blue) separate. Each strand becomes a template for the assembly of a complementary strand from a supply of free nucleotides (gray) available within the nucleus. The nucleotides line up one at a time along the template strand in accordance with the base-pairing rules. Enzymes link the nucleotides to form the new DNA strands. The completed new molecules, identical to the parental molecule, are known as daughter DNA (although no gender should be inferred).

Watson and Crick's model predicts that when a double helix replicates, each of the two daughter molecules will have one old strand from the parental molecule and one newly created strand. This model for DNA replication is known as the **semiconservative model** because half of the parental molecule is maintained (conserved) in each daughter molecule. The semiconservative model of replication was confirmed by experiments performed in the 1950s.

Although the general mechanism of DNA replication is conceptually simple, the actual process is complex, requiring the coordination of more than a dozen enzymes and other proteins. Some of the complexity arises from the need for the helical DNA molecule to untwist as it replicates and for the two new strands to be made roughly simultaneously (**Figure 10.4B**). Another challenge is the speed of the process. *E. coli*, with about 4.6 million DNA base pairs, can copy its entire genome in less than an hour. Human cells, with more than 6 billion base pairs in 46 chromosomes, require only a few hours. Despite this speed, the process is amazingly accurate; typically, only about one DNA nucleotide per several billion is incorrectly paired. In the next module, we take a closer look at the mechanisms of DNA replication that allow it to proceed with such speed and accuracy.

? How does complementary base pairing make possible the replication of DNA?

● When the two strands of the double helix separate, free nucleotides can base-pair along each strand, leading to the synthesis of new complementary strands.

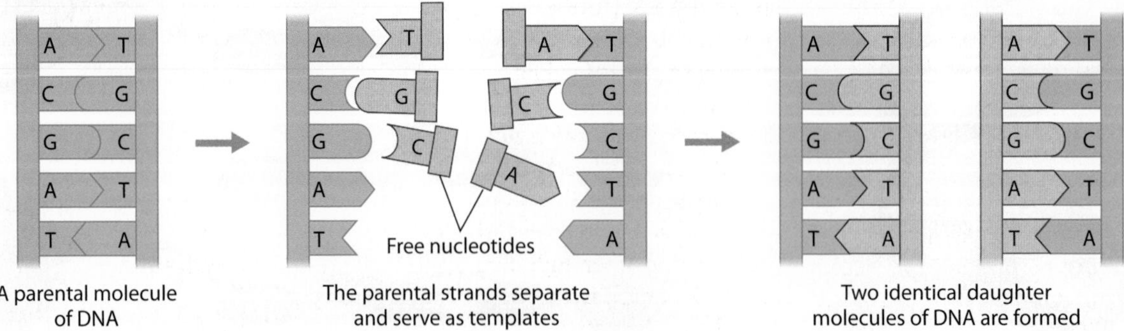

A parental molecule of DNA
The parental strands separate and serve as templates
Free nucleotides
Two identical daughter molecules of DNA are formed

▶ **Figure 10.4A** A template model for DNA replication

10.5 DNA replication proceeds in two directions at many sites simultaneously

Replication of a DNA molecule begins at particular sites called origins of replication, short stretches of DNA having a specific sequence of nucleotides. Proteins that initiate DNA replication attach to the DNA at the origin of replication, separating the two strands of the double helix (**Figure 10.5A**). Replication then proceeds in both directions, creating replication

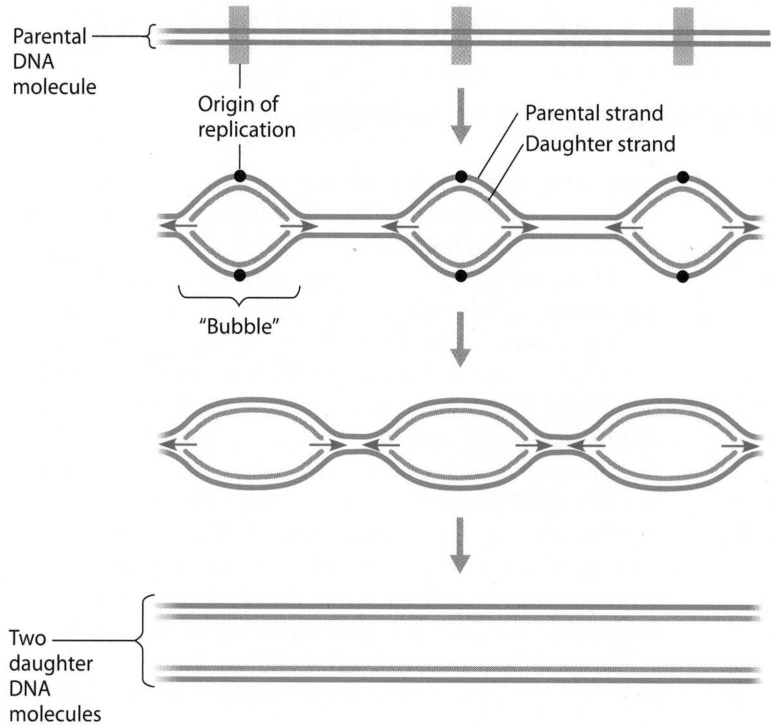
Parental DNA molecule
Origin of replication
Parental strand
Daughter strand
"Bubble"
Two daughter DNA molecules

▲ **Figure 10.5A** Multiple replication bubbles in DNA

"bubbles." The parental DNA strands (blue) open up as daughter strands (gray) elongate on both sides of each bubble. The DNA molecule of a eukaryotic chromosome has many origins where replication can start simultaneously. Thus, hundreds or thousands of bubbles can be present at once, shortening the total time needed for replication. Eventually, all the bubbles fuse, yielding two completed, double-stranded daughter DNA molecules (see the bottom of Figure 10.5A).

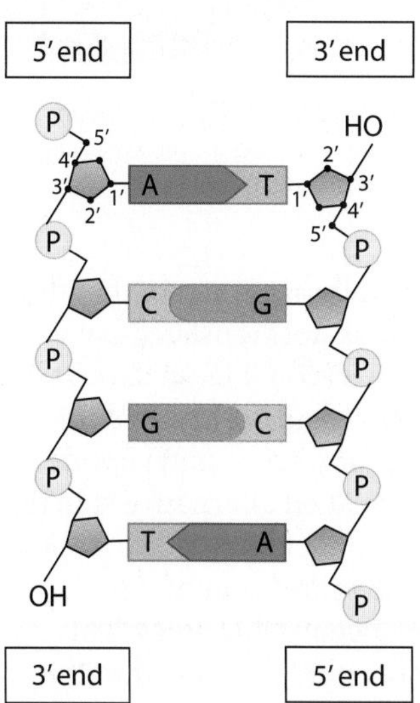
5′ end
3′ end
P
5′
4′
3′
2′
1′
A
T
HO
P
C
G
G
C
T
A
OH
3′ end
5′ end

▲ **Figure 10.5B** The opposite orientations of DNA strands

Try This On each polynucleotide strand, identify one complete nucleotide by drawing a circle around it. Notice that in one strand, the phosphate is on top, and on the other strand, it's on the bottom.

Figure 10.5B shows the molecular building blocks of a tiny segment of DNA. Notice that the sugar-phosphate backbones run in opposite directions. As a result, each strand has a 3′ ("three-prime") end and a 5′ ("five-prime") end. The primed numbers refer to the carbon atoms of the nucleotide sugars. At one end of each DNA strand, the sugar's 3′ carbon atom is attached to an —OH group; at the other end, the sugar's 5′ carbon is attached to a phosphate group.

The opposite orientation of the strands is important in DNA replication. The enzymes that link DNA nucleotides to a growing daughter strand, called **DNA polymerases**, add nucleotides only to the 3′ end of the strand, never to the 5′ end. Thus, a daughter DNA strand can only grow in the 5′ → 3′ direction. You see the consequences of this enzyme specificity in **Figure 10.5C**, where the forked structure represents one side of a replication bubble. One of the daughter strands (shown in gray) can be synthesized in one continuous piece by a DNA polymerase working toward the forking point of the parental DNA. However, to make the other daughter strand, polymerase molecules must work outward from the forking point. The only way this can be accomplished is if the new strand is synthesized in short pieces as the fork opens up. These pieces are called Okazaki fragments, after the Japanese husband and wife team of molecular biologists who discovered them. Another enzyme, called **DNA ligase**, then links, or ligates, the pieces together into a single DNA strand.

In addition to their roles in adding nucleotides to a DNA chain, DNA polymerases carry out a proofreading step that quickly removes nucleotides that have base-paired incorrectly during replication. DNA polymerases and DNA ligase are also involved in repairing DNA damaged by harmful radiation, such as ultraviolet light and X-rays, or toxic chemicals in the environment, such as those found in tobacco smoke.

DNA replication ensures that all the somatic cells in a multicellular organism carry the same genetic information. It is also the means by which genetic instructions are copied for the next generation of the organism. In the next module, we begin to pursue the connection between DNA instructions and an organism's phenotypic traits.

? **What is the function of DNA polymerase in DNA replication?**

● As free nucleotides base-pair to a parental DNA strand, the enzyme covalently bonds them to the 3′ end of a growing daughter strand.

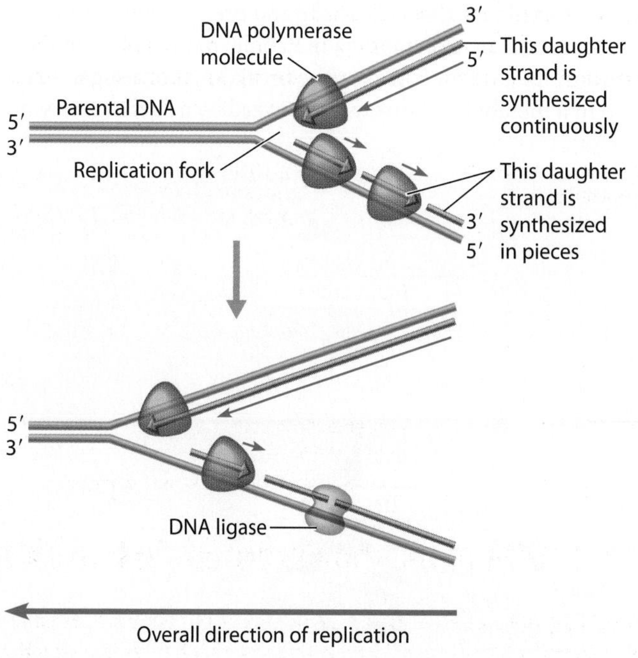
DNA polymerase molecule
3′
5′
This daughter strand is synthesized continuously
Parental DNA
5′
3′
Replication fork
This daughter strand is synthesized in pieces
3′
5′
5′
3′
DNA ligase
Overall direction of replication

▲ **Figure 10.5C** How daughter DNA strands are synthesized

▷ The Flow of Genetic Information from DNA to RNA to Protein

10.6 Genes control phenotypic traits through the expression of proteins

We can now define genotype and phenotype in terms of the structure and function of DNA. An organism's genotype, its genetic makeup, is the heritable information contained in the sequence of nucleotide bases in DNA. The phenotype is the organism's physical traits. So what is the molecular connection between genotype and phenotype?

The answer is that the DNA inherited by an organism specifies traits by dictating the synthesis of proteins. In other words, proteins are the links between genotype and phenotype. However, a gene does not build a protein directly. Rather, a gene dispatches instructions in the form of RNA, which in turn programs protein synthesis. This fundamental concept in biology is summarized in **Figure 10.6A**. The molecular "chain of command" is from DNA in the nucleus of the cell to RNA to protein synthesis in the cytoplasm. The two main stages are **transcription**, the synthesis of RNA under the direction of DNA, and **translation**, the synthesis of protein under the direction of RNA.

The relationship between genes and proteins was first proposed in 1909, when English physician Archibald Garrod suggested that genes dictate phenotypes through enzymes, proteins that catalyze chemical reactions. Garrod hypothesized that an inherited disease reflects a person's inability to make a particular enzyme. He gave as one example the hereditary condition called alkaptonuria, in which the urine is dark because it contains a chemical called alkapton. Garrod reasoned that most people have an enzyme that breaks down alkapton, whereas people with alkaptonuria inherited an inability to make the enzyme. Garrod's hypothesis was ahead of its time, but research conducted decades later supported his hypothesis. In the intervening years, biochemists accumulated evidence that cells make and break down biologically important molecules via metabolic pathways, as in the synthesis of an amino acid or the breakdown of a sugar. Each step in a metabolic pathway is catalyzed by a specific enzyme (as we described in Unit I; see Module 5.15, for example). Therefore, individuals lacking one of the enzymes for a pathway are unable to complete it.

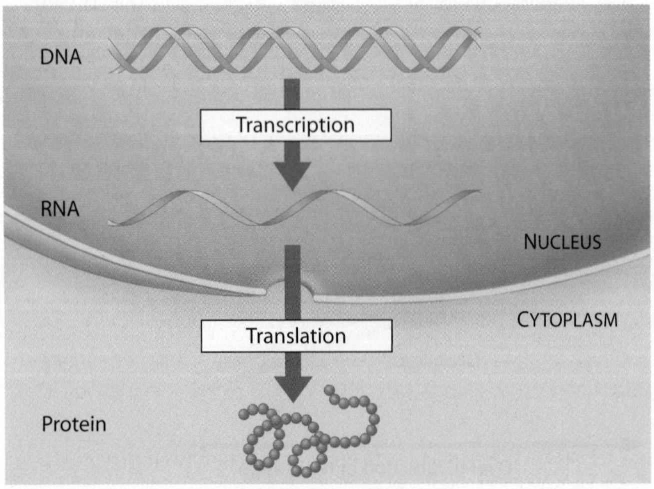

DNA
Transcription
RNA
NUCLEUS
CYTOPLASM
Translation
Protein

▲ **Figure 10.6A** The flow of genetic information in a eukaryotic cell

The major breakthrough in demonstrating the relationship between genes and enzymes came in the 1940s from the work of American geneticists George Beadle and Edward Tatum with the bread mold *Neurospora crassa* (**Figure 10.6B**). Beadle and Tatum studied strains of the mold that were unable to grow on a simple growth medium. Each of these so-called nutritional mutants turned out to lack an enzyme in a metabolic pathway that synthesized some molecule the mold needed, such as an amino acid. Beadle and Tatum also showed that each mutant was defective in a single gene. Accordingly, they hypothesized that the function of an individual gene is to dictate the production of a specific enzyme.

The "one gene–one enzyme hypothesis" has since been modified. First, it was extended beyond enzymes to include *all* types of proteins. For example, keratin (the structural protein of hair) and the hormone insulin are two examples of proteins that are not enzymes. In addition, many proteins are made from two or more polypeptide chains, with each polypeptide specified by its own gene. For example, hemoglobin, the oxygen-transporting protein in your red blood cells, is built from two kinds of polypeptides, encoded by two different genes. Thus, Beadle and Tatum's hypothesis is now stated as follows: The function of a gene is to dictate the production of a polypeptide.

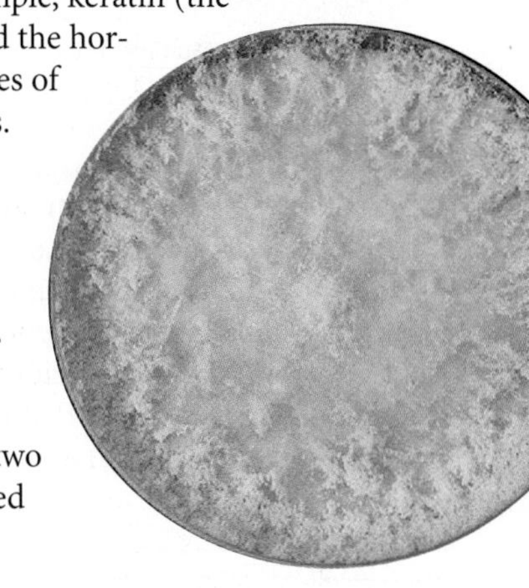

▲ **Figure 10.6B** The bread mold *Neurospora crassa* growing in a culture dish

Even this description is not entirely accurate, in that the RNA transcribed from some genes is not translated but nonetheless has important functions (you'll learn about two such kinds of RNA in Modules 10.11 and 10.12). In addition, many eukaryotic genes code for a set of polypeptides (rather than just one) by a process called alternative splicing (discussed in Module 11.4). The more biologists learn about the ways that genes act in cells, the more complicated the picture—and the very concept of a gene—becomes. This topic continues to be an active area of research within the biology community. But for now, we'll focus on genes that do code for polypeptides. The nature of that code is our next topic.

? **What are the functions of transcription and translation?**

● Transcription is the transfer of information from DNA to RNA. Translation is the use of the information in RNA to make a polypeptide.

10.7 Genetic information written in codons is translated into amino acid sequences

Genes provide the instructions for making specific proteins. But a gene does not build a protein itself. As you have learned, the bridge between DNA and protein synthesis is the nucleic acid RNA: DNA is transcribed into RNA, which is then translated into protein. Put another way, information within the cell flows as DNA ⟶ RNA ⟶ protein. This can be stated as: "DNA makes RNA makes protein."

Transcription and translation are linguistic terms, and it is useful to think of nucleic acids and proteins as having languages. To understand how genetic information passes from genotype to phenotype, we need to see how the chemical language of DNA is translated into the different chemical language of proteins.

What exactly is the language of nucleic acids? Both DNA and RNA are polymers made of nucleotide monomers strung together in specific sequences that convey information, much as specific sequences of letters convey information in written language. In DNA, there are four types of nucleotides, which differ in their nitrogenous bases (A, T, C, and G). The same is true for RNA, although it has the base U instead of T.

Figure 10.7 focuses on a small region of one gene (gene 3, shown in light blue) on a DNA molecule. DNA's language is written as a linear sequence of nucleotide bases on a polynucleotide, a sequence such as the one you see on the enlarged DNA segment in the figure. Specific sequences of bases, each with a beginning and an end, make up the genes on a DNA strand. A typical gene consists of hundreds or thousands of nucleotides in a specific sequence.

The pink strand underneath the enlarged DNA segment represents the results of transcription: an RNA molecule. The process is called transcription because the nucleic acid language of DNA has been rewritten (transcribed) as a sequence of bases on RNA. Notice that the language is still that of nucleic acids, although the nucleotide bases on the RNA molecule are complementary to those on the DNA strand. As we will see in Module 10.9, this is because the RNA was synthesized using the DNA as a template.

The purple chain at the bottom of Figure 10.7 represents the results of translation, the conversion of the nucleic acid language to the polypeptide language. Like nucleic acids, polypeptides are polymers, but the monomers that compose them are the 20 different kinds of amino acids. Again, the language is written in a linear sequence, and the sequence of nucleotides of the RNA molecule dictates the sequence of amino acids of the polypeptide. The RNA acts as a messenger carrying genetic information from DNA.

During translation, there is a change in language from the nucleotide sequence of the RNA to the amino acid sequence of the polypeptide. How is this translation achieved? Recall that there are only four different kinds of nucleotides in DNA (A, G, C, T) and in RNA (A, G, C, U). In translation, these four nucleotides must somehow specify all 20 amino acids. Consider if each single nucleotide base were to specify one amino acid. In this case, only four of the 20 amino acids could be accounted for, one for each type of base. What if the language consisted of two-letter code words? If we read the bases of a gene two at a time—AG, for example, could specify one amino acid, whereas AT could designate a different amino acid—then only 16 arrangements would be possible (4^2), which is still not enough to specify all 20 amino acids. However, if the base code in DNA consists of a triplet, with each arrangement of three consecutive bases specifying an amino acid—AGT specifies one amino acid, for example, while AGA specifies a different one—then there can be 64 (that is, 4^3) possible code words, more than enough to specify the 20 amino acids. Thus, triplets of bases are the smallest "words" of uniform length that can specify all the amino acids (see the brackets below the strand of RNA in Figure 10.7). Indeed, the 64 triplets allow for more than one to represent an amino acid. For example, the base triplets AAT and AAC could both code for the same amino acid.

Experiments have verified that the flow of information from gene to protein is based on a **triplet code**: The genetic instructions for the amino acid sequence of a polypeptide chain are written in DNA and RNA as a series of nonoverlapping three-base "words" called **codons**. Notice in the figure that three-base codons in the DNA are transcribed into complementary three-base codons in the RNA, and then the RNA codons are translated into amino acids that form a polypeptide. We turn to the codons themselves in the next module.

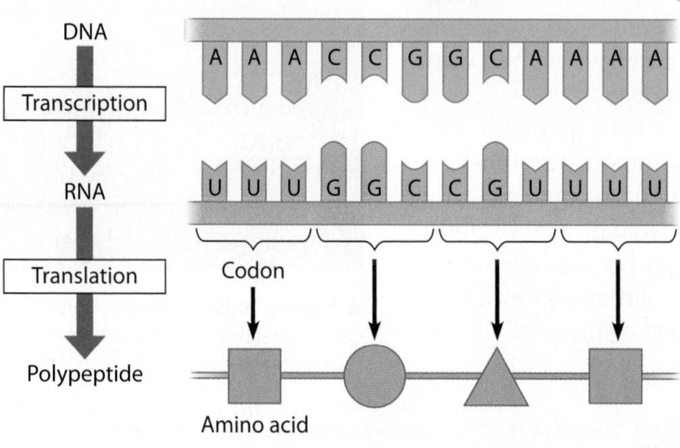
DNA molecule
Gene 1
Gene 2
Gene 3
DNA
A A A C C G G C A A A A
Transcription
RNA
U U U G G C C G U U U U
Translation
Codon
Polypeptide
Amino acid

▲ **Figure 10.7** Transcription and translation of codons

? **What is the minimum number of nucleotides necessary to code for 100 amino acids?**

300

10.8 The genetic code dictates how codons are translated into amino acids

During the 1960s, molecular biologists used a series of elegant experiments to crack the **genetic code**, the amino acid translations of each of the nucleotide triplets. The first codon was deciphered in 1961 by American biochemist Marshall Nirenberg. He synthesized an artificial RNA molecule by linking together identical RNA nucleotides having uracil as their only base. No matter where this message started or stopped, it could contain only one type of triplet codon: UUU. Nirenberg added this "poly-U" to a test-tube mixture containing ribosomes and the other ingredients required for polypeptide synthesis. His artificial system translated the poly-U into a polypeptide containing a single kind of amino acid, phenylalanine (Phe). Thus, Nirenberg learned that the RNA codon UUU specifies the amino acid phenylalanine. By variations on this method, the amino acids specified by all the codons were soon determined.

As shown in **Figure 10.8A**, 61 of the 64 triplets code for amino acids. The triplet AUG (green box in the figure) has a dual function: It codes for the amino acid methionine (Met) and also can provide a signal for the start of a polypeptide chain. Three codons (UAA, UGA, and UAG, shown in red) do not designate amino acids but serve as stop codons that mark the end of translation.

Notice in Figure 10.8A that there is redundancy in the code but no ambiguity. For example, although codons UUU and UUC both specify phenylalanine (redundancy), neither of them ever represents any other amino acid (no ambiguity). The codons in the figure are the triplets found in RNA. They have a straightforward, complementary relationship to the codons in DNA, with UUU in the RNA matching AAA in the DNA, for example. The nucleotides making up the codons occur in a linear order along the DNA and RNA, with no gaps separating the codons.

As an exercise in translating the genetic code, consider the 12-nucleotide segment of DNA in **Figure 10.8B**. Let's read this as a series of triplets. Using the base-pairing rules (with U in RNA instead of T), we see that the RNA codon corresponding to the first transcribed DNA triplet, TAC, is AUG. As you can see in Figure 10.8A, AUG specifies, "Place Met as the first amino acid in the polypeptide." The second DNA triplet, TTC, dictates RNA codon AAG, which designates lysine (Lys) as the second amino acid. We continue until we reach a stop codon (UAG in this example).

The genetic code is nearly universal, shared by organisms from the simplest bacteria to the most complex plants and animals. Such universality is key to modern DNA technologies because it allows scientists to mix and match genes from various species (**Figure 10.8C**; see Chapter 12). A language shared by all living things must have evolved early enough in the history of life to be present in the common ancestors of all modern organisms. A shared genetic vocabulary is a reminder of the evolutionary kinship that connects all life on Earth.

Second base of RNA codon

First base of RNA codon	U	C	A	G	Third base of RNA codon
U	UUU Phe	UCU Ser	UAU Tyr	UGU Cys	U
	UUC Phe	UCC Ser	UAC Tyr	UGC Cys	C
	UUA Leu	UCA Ser	UAA Stop	UGA Stop	A
	UUG Leu	UCG Ser	UAG Stop	UGG Trp	G
C	CUU Leu	CCU Pro	CAU His	CGU Arg	U
	CUC Leu	CCC Pro	CAC His	CGC Arg	C
	CUA Leu	CCA Pro	CAA Gln	CGA Arg	A
	CUG Leu	CCG Pro	CAG Gln	CGG Arg	G
A	AUU Ile	ACU Thr	AAU Asn	AGU Ser	U
	AUC Ile	ACC Thr	AAC Asn	AGC Ser	C
	AUA Ile	ACA Thr	AAA Lys	AGA Arg	A
	AUG Met or start	ACG Thr	AAG Lys	AGG Arg	G
G	GUU Val	GCU Ala	GAU Asp	GGU Gly	U
	GUC Val	GCC Ala	GAC Asp	GGC Gly	C
	GUA Val	GCA Ala	GAA Glu	GGA Gly	A
	GUG Val	GCG Ala	GAG Glu	GGG Gly	G

▲ **Figure 10.8A** The genetic code used to translate RNA codons to amino acids

Try This Identify the DNA triplet that produces an RNA start codon, and identify the three DNA triplets that produce an RNA stop codon.

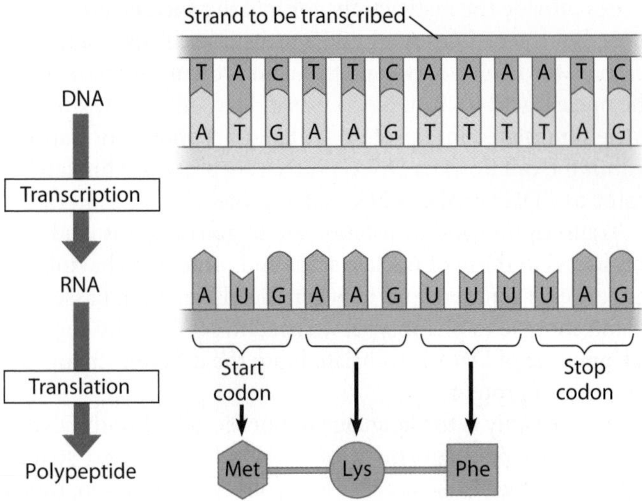

Strand to be transcribed
DNA
T A C T T C A A A A T C
A T G A A G T T T T A G
Transcription
RNA
A U G A A G U U U U A G
Translation
Start codon
Stop codon
Polypeptide
Met
Lys
Phe

▲ **Figure 10.8B** Deciphering the genetic information in DNA

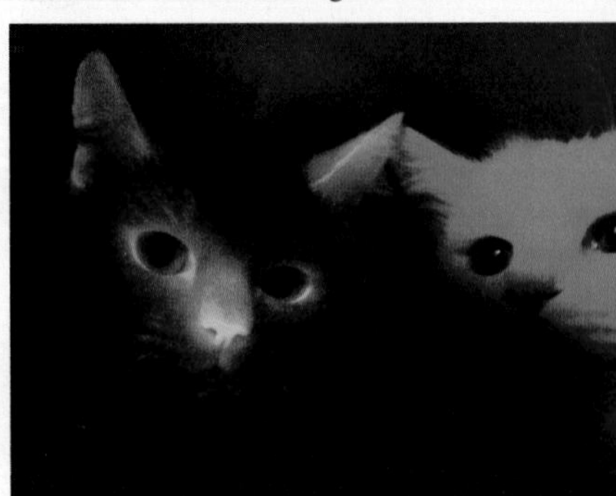

▲ **Figure 10.8C** The cat on the left was engineered to express a protein that fluoresces red when exposed to UV light

? Translate the RNA sequence CCAUUUACG into the corresponding amino acid sequence.

● Pro-Phe-Thr

VISUALIZING THE CONCEPT

10.9 Transcription produces genetic messages in the form of RNA

In eukaryotic cells, transcription, the transfer of genetic information from DNA to RNA, occurs in the nucleus. Here we focus on transcription in prokaryotic cells, which is a simpler process than in eukaryotes.

After separation of the two DNA strands, one strand serves as a template for a new RNA molecule; the other DNA strand is unused. The transcription enzyme **RNA polymerase** moves along the gene, forming a new RNA strand by following the base-pairing rules—but remember that in RNA, uracil replaces thymine. A specific nucleotide sequence called a **promoter** acts as a binding site for RNA polymerase and determines where transcription starts and on which strand. RNA polymerase adds RNA nucleotides until it reaches a sequence of DNA bases called the **terminator**, which signals the end of the gene.

TRANSCRIPTION OF A GENE

Initiation

Initiation involves the attachment of RNA polymerase to the promoter and the start of RNA synthesis.

RNA polymerase

DNA of gene

Promoter

The promoter is the binding site for RNA polymerase. This marks the start of the gene.

Terminator DNA

The terminator is a sequence of bases that marks the end of a gene.

Once attached, RNA polymerase starts to synthesize RNA.

Direction of transcription

Unused strand of DNA

Newly formed RNA

Template strand of DNA

Elongation

During elongation, the newly formed RNA strand grows. As synthesis continues, the growing RNA molecule peels away from its DNA template, allowing the two separated DNA strands to come back together in the region already transcribed.

Newly made RNA

Direction of transcription

Free nucleotides form hydrogen bonds with the nucleotide bases of the template DNA.

Free RNA nucleotide

DNA strands reunite

T C A T C C A A T G

G A U C C A A U

A T A G G T T A

DNA strands separate

Notice that the RNA nucleotides follow the base-pairing rules, with U (rather than T) pairing with A.

RNA polymerase adds new RNA nucelotides.

Termination

When RNA polymerase reaches the terminator DNA (which signals the end of the gene), the polymerase molecule detaches from the newly made RNA strand and the gene.

Terminator DNA

RNA polymerase detaches

Completed RNA

? **How does RNA polymerase recognize the start and end of the gene?**

● Special DNA sequences mark the start (promoter) and end (terminator) of a gene.

10.10 Eukaryotic RNA is processed before leaving the nucleus as mRNA

The kind of RNA that encodes amino acid sequences is called **messenger RNA (mRNA)** because it conveys genetic messages from DNA to the translation machinery of the cell. Messenger RNA is transcribed from DNA, and the information in the mRNA is then translated into polypeptides. In prokaryotic cells, which lack nuclei, transcription and translation occur in the same place: the cytoplasm. In eukaryotic cells, however, mRNA molecules must exit the nucleus via the nuclear pores and enter the cytoplasm, where the machinery for polypeptide synthesis is located.

Before leaving the nucleus as mRNA, eukaryotic transcripts are modified, or processed, in several ways (**Figure 10.10**). One kind of RNA processing is the addition of extra nucleotides to the ends of the RNA transcript. These additions include a small cap (a modified form of a G nucleotide) at the 5′ end and a long tail (a chain of 50 to 250 A nucleotides) at the 3′ end. The cap and tail (yellow in the figure) facilitate the export of the mRNA from the nucleus, protect the mRNA from degradation by cellular enzymes, and help ribosomes bind to the mRNA. The cap and tail themselves are not translated into protein.

Another type of RNA processing is made necessary in eukaryotes by noncoding stretches of nucleotides that interrupt the nucleotides that actually code for amino acids. It is as if nonsense words were randomly interspersed in a story. Most genes of plants and animals include such internal noncoding regions, which are called **introns** ("intervening sequences"). The coding regions—the parts of a gene that are expressed—are called **exons**. As Figure 10.10 shows, both exons (shown in a darker color) and introns (in a lighter color) are transcribed from DNA into RNA. However, before the RNA leaves the nucleus, the introns are removed, and the exons are joined to produce an mRNA molecule with a continuous coding sequence. (The short noncoding regions just inside the cap and tail are considered parts of the first and last exons.) This cutting-and-pasting process is called **RNA splicing**. In most cases, RNA splicing is catalyzed by a complex of proteins and small RNA molecules. RNA splicing also provides a means to produce multiple polypeptides from a single gene (as we will see in the next chapter; see Module 11.4). In fact, RNA splicing is believed to play a significant role in humans in allowing our approximately 21,000 genes to produce many thousands more polypeptides. This is accomplished by varying the exons that are included in the final mRNA.

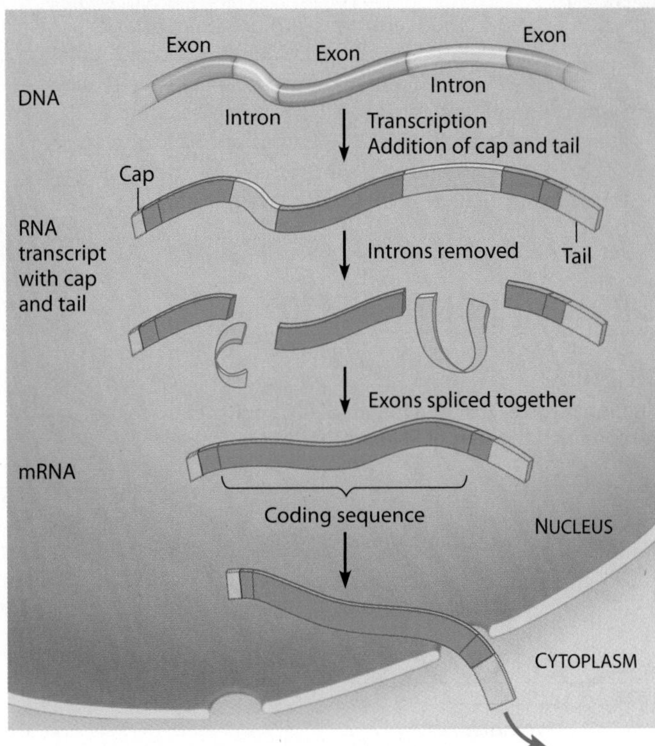

DNA
Exon
Exon
Exon
Intron
Intron
Transcription
Addition of cap and tail
Cap
RNA transcript with cap and tail
Introns removed
Tail
Exons spliced together
mRNA
Coding sequence
NUCLEUS
CYTOPLASM

▲ **Figure 10.10** The production of eukaryotic mRNA

As we have discussed, translation is a conversion between different languages—from the nucleic acid language to the protein language—and it involves more elaborate machinery than transcription. The first important ingredient required for translation is the processed mRNA. Once it is present, the machinery used to translate mRNA requires enzymes and sources of chemical energy, such as ATP. In addition, translation requires two heavy-duty components: ribosomes and a kind of RNA called transfer RNA, the subject of the next module.

? **Explain why most eukaryotic genes are longer than the mRNA that leaves the nucleus.**

● These genes have introns, noncoding sequences of nucleotides that are spliced out of the initial RNA transcript to produce mRNA.

10.11 Transfer RNA molecules serve as interpreters during translation

Translation of any language into another language requires an interpreter, someone or something that can recognize the words of one language and convert them to another. Translation of a genetic message carried in mRNA into the amino acid language of proteins also requires an interpreter. To convert the three-letter "words" of nucleic acids (codons) to the amino acid "words" of proteins, a cell uses a molecular interpreter, a special type of RNA called **transfer RNA (tRNA)**.

A cell that is producing proteins has in its cytoplasm a supply of amino acids, either obtained from food or manufactured by the cell. But amino acids themselves cannot recognize the codons in the mRNA. It is up to the cell's molecular interpreters, tRNA molecules, to match amino acids to the appropriate codons to form the new polypeptide. To perform this task, tRNA molecules must carry out two functions: (1) picking up the appropriate amino acids and (2) recognizing

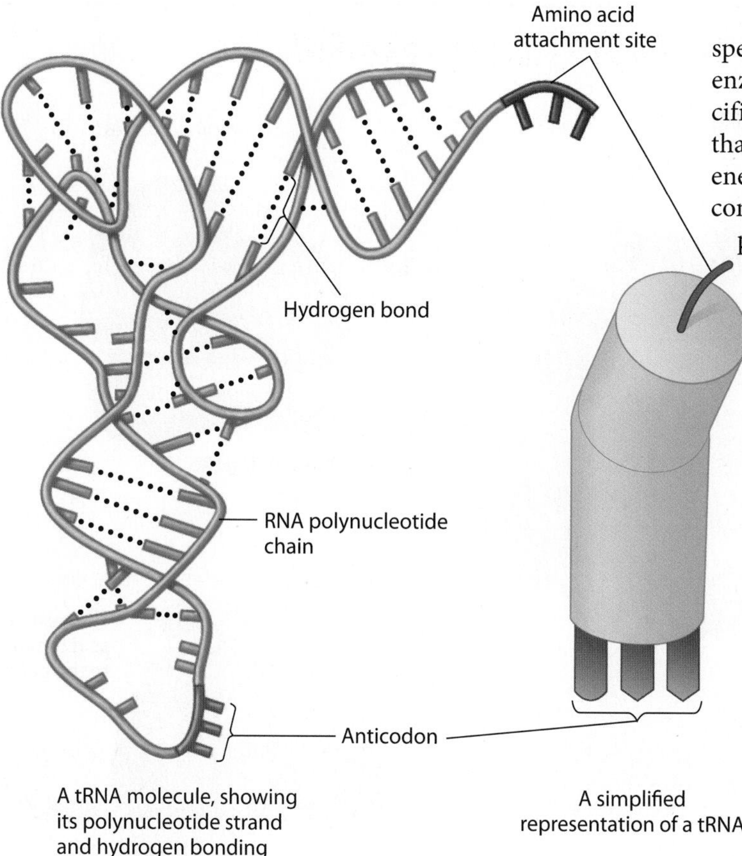
Amino acid attachment site
Hydrogen bond
RNA polynucleotide chain
Anticodon
A tRNA molecule, showing its polynucleotide strand and hydrogen bonding
A simplified representation of a tRNA

▲ **Figure 10.11A** The structure of tRNA

the appropriate codons in the mRNA. The unique structure of tRNA molecules enables them to perform both tasks.

Figure 10.11A shows two representations of a tRNA molecule. The structure on the left shows the backbone and bases, with hydrogen bonds between bases shown as dotted lines. The structure on the right is a simplified schematic that emphasizes the most important parts of the structure. Notice from the structure on the left that a tRNA molecule is made of a single strand of RNA—one polynucleotide chain—consisting of about 80 nucleotides. By twisting and folding upon itself, tRNA forms several double-stranded regions in which short stretches of RNA base-pair with other stretches via hydrogen bonds. A single-stranded loop at one end of the folded molecule contains a special triplet of bases called an **anticodon**. The anticodon triplet is complementary to a codon triplet on mRNA. During translation, the anticodon on the tRNA recognizes a particular codon on the mRNA by using base-pairing rules. At the other end of the tRNA molecule is a site where one specific kind of amino acid attaches.

In the modules that follow, we represent tRNA with the simplified shape shown on the right in Figure 10.11A. This shape emphasizes the two parts of the molecule—the anticodon and the amino acid attachment site—that give tRNA its ability to match a particular nucleic acid word (a codon in mRNA) with its corresponding protein word (an amino acid). Although all tRNA molecules are similar, there is a slightly different variety of tRNA for each amino acid.

Each amino acid is joined to the correct tRNA by a specific enzyme. There is a family of 20 versions of these enzymes, one enzyme for each amino acid. Each enzyme specifically binds one type of amino acid to all tRNA molecules that code for that amino acid, using a molecule of ATP as energy to drive the reaction. The resulting amino acid–tRNA complex can then furnish its amino acid to a growing polypeptide chain, a process that we describe in Module 10.12.

The computer graphic in **Figure 10.11B** shows a tRNA molecule (green) and an ATP molecule (yellow) bound to the enzyme molecule (blue). (To help you see the two distinct molecules, the tRNA molecule is shown with a stick representation, while the enzyme is shown as space-filling spheres.) In this figure, you can see the proportional sizes of these three molecules. The amino acid that would attach to the tRNA is not shown; it would be less than half the size of the ATP.

Once an amino acid is attached to its appropriate tRNA, it can be incorporated into a growing polypeptide chain. This is accomplished within ribosomes, the cellular structures directly responsible for the synthesis of protein. We examine ribosomes in the next module.

? **What is an anticodon, and what is its function?**

● It is the base triplet of a tRNA molecule that couples the tRNA to a complementary codon in the mRNA. This is a key step in translating mRNA to polypeptide.

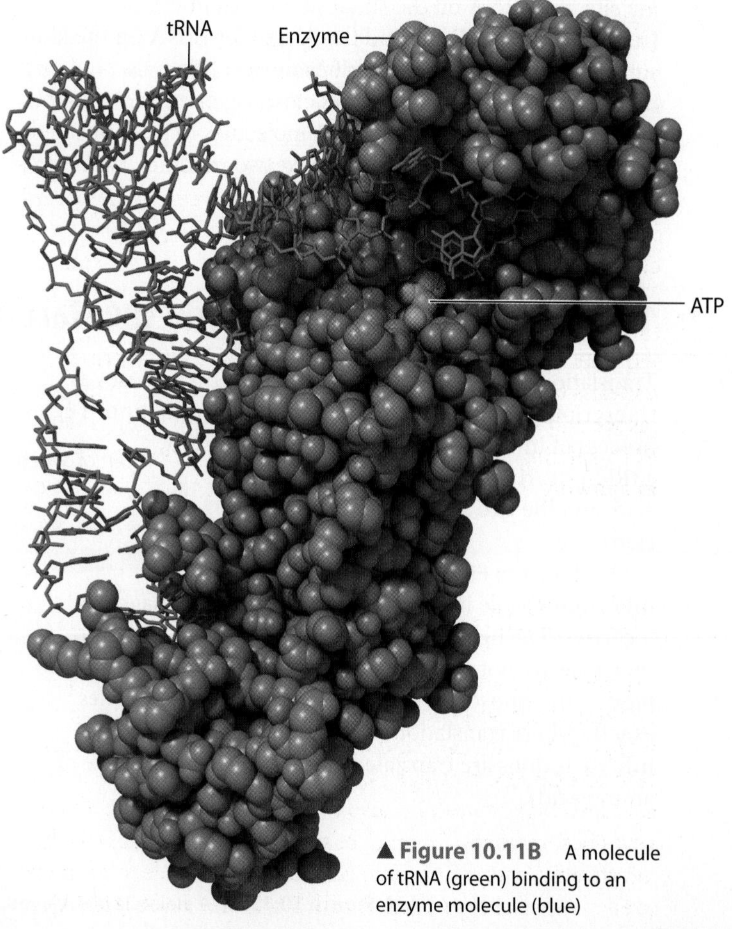
tRNA
Enzyme
ATP

▲ **Figure 10.11B** A molecule of tRNA (green) binding to an enzyme molecule (blue)

10.12 Ribosomes build polypeptides

We have now looked at many of the components a cell needs to carry out translation: instructions in the form of mRNA molecules, tRNA to interpret the instructions, a supply of amino acids and enzymes (for attaching amino acids to tRNA), and ATP for energy. The final components are the **ribosomes**, structures in the cytoplasm that coordinate the functioning of mRNA and tRNA and catalyze the synthesis of polypeptides **(Figure 10.12)**. A ribosome consists of two subunits—a large subunit and a small subunit—each made up of proteins and a kind of RNA called **ribosomal RNA (rRNA)**.

The ribosomes of bacteria and eukaryotes are very similar in function, but those of eukaryotes are slightly larger and different in composition. The differences are medically significant. Certain antibiotic drugs can inactivate bacterial ribosomes while leaving eukaryotic ribosomes unaffected. These drugs, such as tetracycline and streptomycin, are used to combat bacterial infections.

The simplified drawings on the right side of Figure 10.12 indicate how tRNA anticodons and mRNA codons fit together on ribosomes. A fully assembled ribosome has a binding site for mRNA on the small subunit and binding sites (referred to as the P site and the A site) for tRNA on the large subunit. The subunits of the ribosome act like a vise, holding the tRNA and mRNA molecules close together, allowing the amino acids carried by the tRNA molecules to be connected into a polypeptide chain. In the next two modules, we examine the steps of translation in detail.

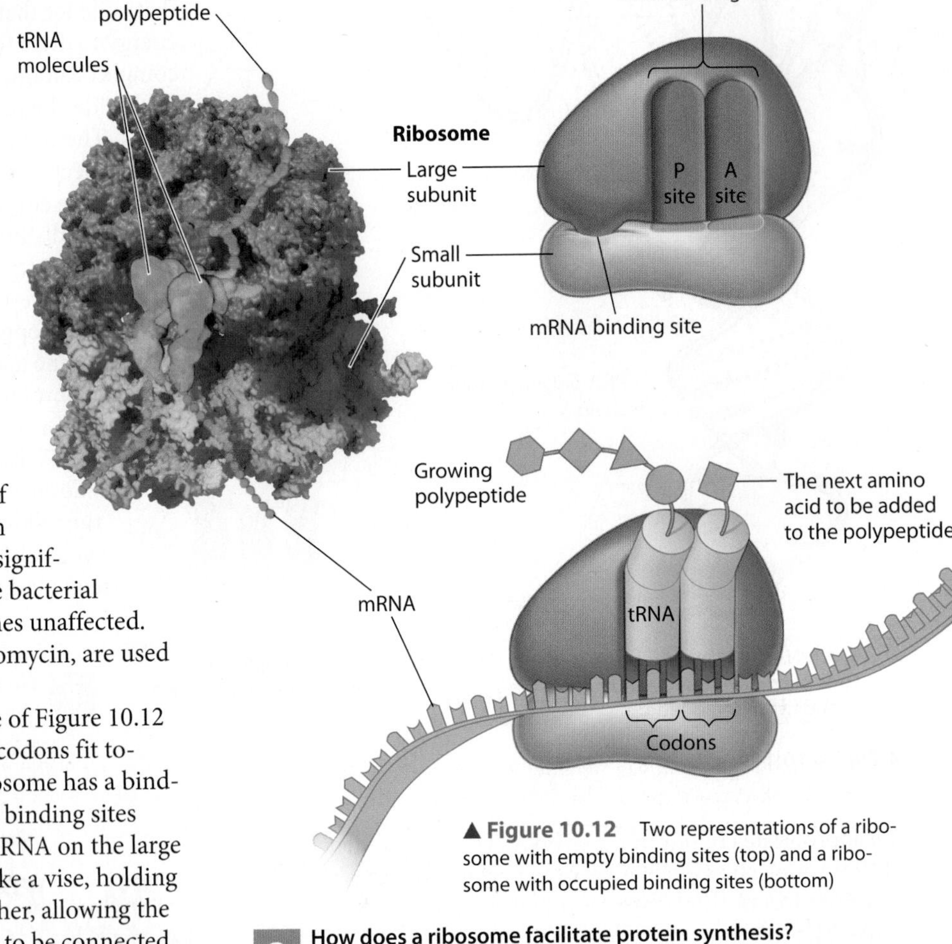

Growing polypeptide
tRNA molecules
tRNA binding sites
Ribosome
Large subunit
Small subunit
P site
A site
mRNA binding site
Growing polypeptide
The next amino acid to be added to the polypeptide
mRNA
tRNA
Codons

▲ **Figure 10.12** Two representations of a ribosome with empty binding sites (top) and a ribosome with occupied binding sites (bottom)

? How does a ribosome facilitate protein synthesis?

A ribosome holds mRNA and tRNAs together and connects amino acids from the tRNAs to the growing polypeptide chain.

10.13 An initiation codon marks the start of an mRNA message

Translation can be divided into the same three phases as transcription: initiation, elongation, and termination. The process of initiation brings together the mRNA, a tRNA bearing the first amino acid, and the two subunits of a ribosome.

As shown in **Figure 10.13A**, an mRNA molecule is longer than the genetic message it carries. The light pink nucleotides at either end of the molecule are not part of the message but help the mRNA to bind to the ribosome. The initiation process establishes exactly where translation will begin, ensuring that the mRNA codons are translated into the correct sequence of amino acids.

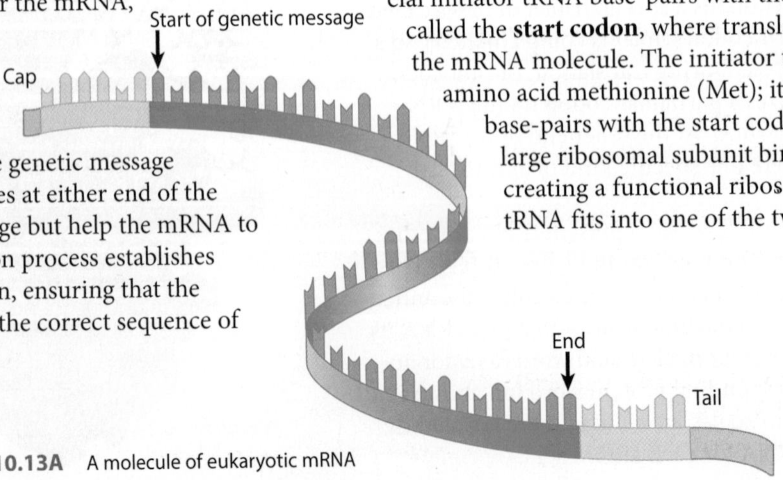

Start of genetic message
Cap
End
Tail

▲ **Figure 10.13A** A molecule of eukaryotic mRNA

Initiation occurs in two steps **(Figure 10.13B)**. ❶ An mRNA molecule binds to a small ribosomal subunit. A special initiator tRNA base-pairs with the specific codon, called the **start codon**, where translation is to begin on the mRNA molecule. The initiator tRNA carries the amino acid methionine (Met); its anticodon, UAC, base-pairs with the start codon, AUG. ❷ Next, a large ribosomal subunit binds to the small one, creating a functional ribosome. The initiator tRNA fits into one of the two tRNA binding sites on the ribosome. This site, called the **P site**, will hold the growing polypeptide. The other tRNA binding site, called the **A site**, is shown vacant and ready for the next amino-acid-bearing tRNA.

What would happen if a genetic mutation in a gene changed a start codon to some other codon?

The messenger RNA transcribed from the mutated gene would be nonfunctional because ribosomes could not initiate translation correctly.

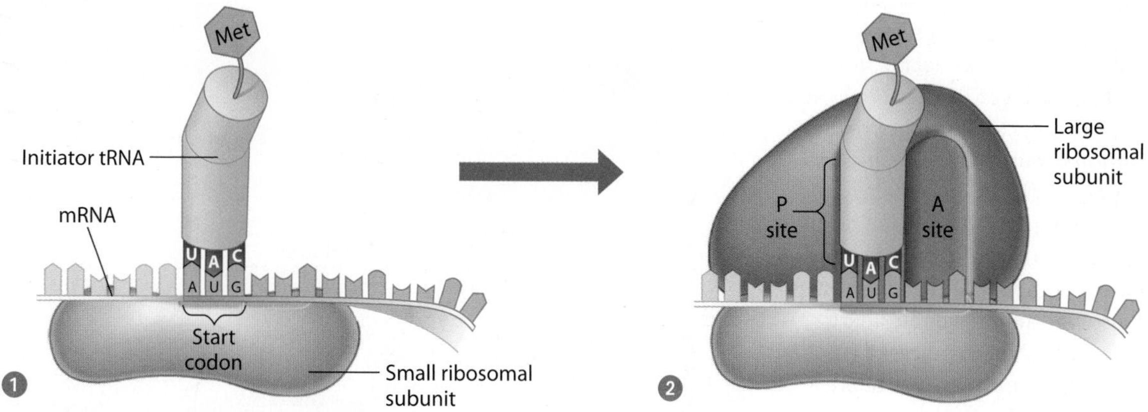

Met
Initiator tRNA
mRNA
U A C
A U G
Start codon
Small ribosomal subunit
1
Met
Large ribosomal subunit
P site
A site
U A C
A U G
2

▲ **Figure 10.13B** The initiation of translation

10.14 Elongation adds amino acids to the polypeptide chain until a stop codon terminates translation

Once initiation is complete, amino acids are added one by one to the previous amino acid. Each addition occurs in a three-step elongation process **(Figure 10.14)**. ❶ The anticodon of an incoming tRNA molecule, carrying its amino acid, pairs with the mRNA codon in the A site of the ribosome. ❷ The polypeptide separates from the tRNA in the P site and attaches by a new peptide bond to the amino acid carried by the tRNA in the A site. The ribosome catalyzes formation of the peptide bond, adding one more amino acid to the growing polypeptide chain. ❸ Then, the P site tRNA (which is now lacking an amino acid) leaves the ribosome, and the ribosome translocates (moves) the remaining tRNA (which holds the growing polypeptide) from the A site to the P site. The codon and anticodon remain hydrogen-bonded, and the mRNA and tRNA move as a unit. This movement brings into the A site the next mRNA codon to be translated, and the process can start again with step 1.

Elongation continues until a **stop codon** reaches the ribosome's A site. As discussed earlier, stop codons—UAA, UAG, and UGA—do not code for amino acids but instead act as signals to stop translation. This is the termination stage of translation. The completed polypeptide is freed from the last tRNA, and the ribosome splits back into its separate subunits.

What would happen if a codon in the middle of an mRNA mutated from UUA to UAA?

Translation would stop prematurely, because a stop codon was introduced.

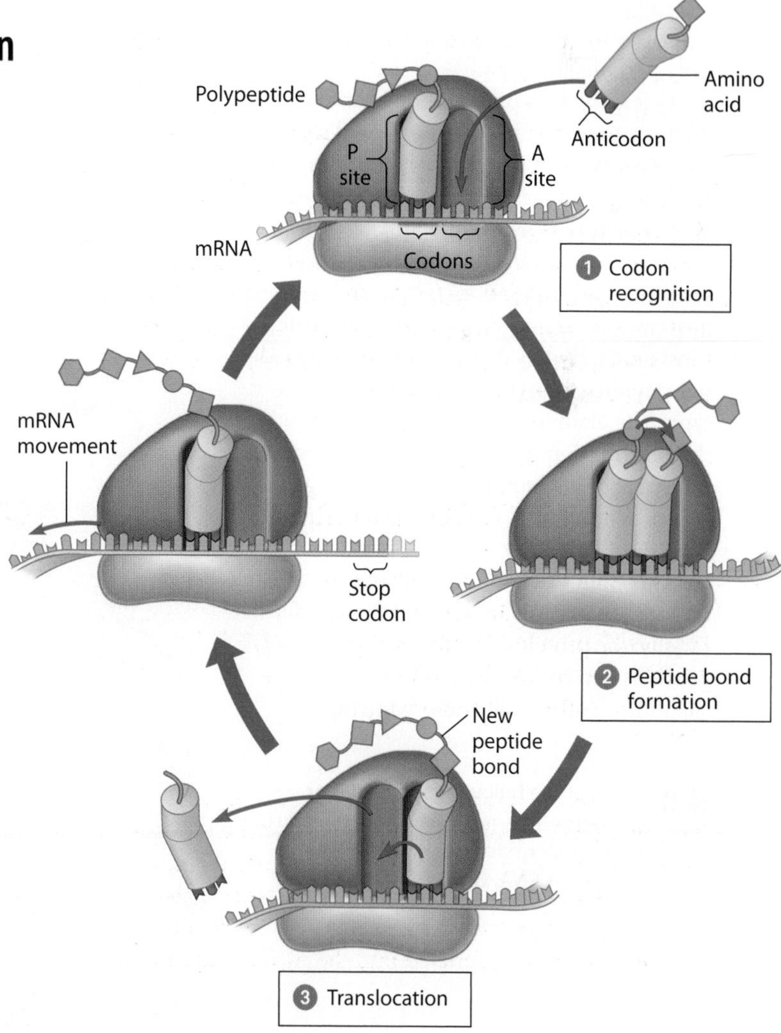

Polypeptide
Amino acid
Anticodon
P site
A site
mRNA
Codons
1 Codon recognition
mRNA movement
Stop codon
2 Peptide bond formation
New peptide bond
3 Translocation

▲ **Figure 10.14** Polypeptide elongation; the small green arrows indicate movement

10.15 Review: The flow of genetic information in the cell is DNA → RNA → protein

Figure 10.15 summarizes the flow of genetic information from DNA to RNA to protein. ❶ In transcription (DNA → RNA), the mRNA is synthesized on a DNA template. In eukaryotic cells, transcription occurs in the nucleus, and the messenger RNA is processed before it travels to the cytoplasm. In prokaryotes, transcription occurs in the cytoplasm.

❷–❺ Translation (RNA → protein) can be divided into four steps, all of which occur in the cytoplasm in eukaryotic cells. When the polypeptide is complete at the end of step 5, the two ribosomal subunits come apart, and the tRNA and mRNA are released (not shown in this figure). Translation is rapid; a single ribosome can make an average-sized polypeptide in less than a minute. Typically, an mRNA molecule is translated simultaneously by a number of ribosomes. Once the start codon emerges from the first ribosome, a second ribosome can attach to it; thus, several ribosomes may trail along on the same mRNA molecule.

As it is made, a polypeptide coils and folds, assuming a three-dimensional shape, its tertiary structure. Several polypeptides may come together, forming a protein with quaternary structure (see Module 3.14).

What is the overall significance of transcription and translation? These are the main processes whereby genes control the structures and activities of cells—or, more broadly, the way the genotype produces the phenotype. The chain of command originates with the information in a gene, a specific linear sequence of nucleotides in DNA. The gene serves as a template, dictating transcription of a complementary sequence of nucleotides in mRNA. In turn, mRNA dictates the linear sequence of amino acids in a polypeptide. Finally, the proteins that form from the polypeptides determine the appearance and the capabilities of the cell and organism.

? Which of the following molecules or structures does not participate directly in translation: ribosomes, transfer RNA, messenger RNA, DNA?

● DNA

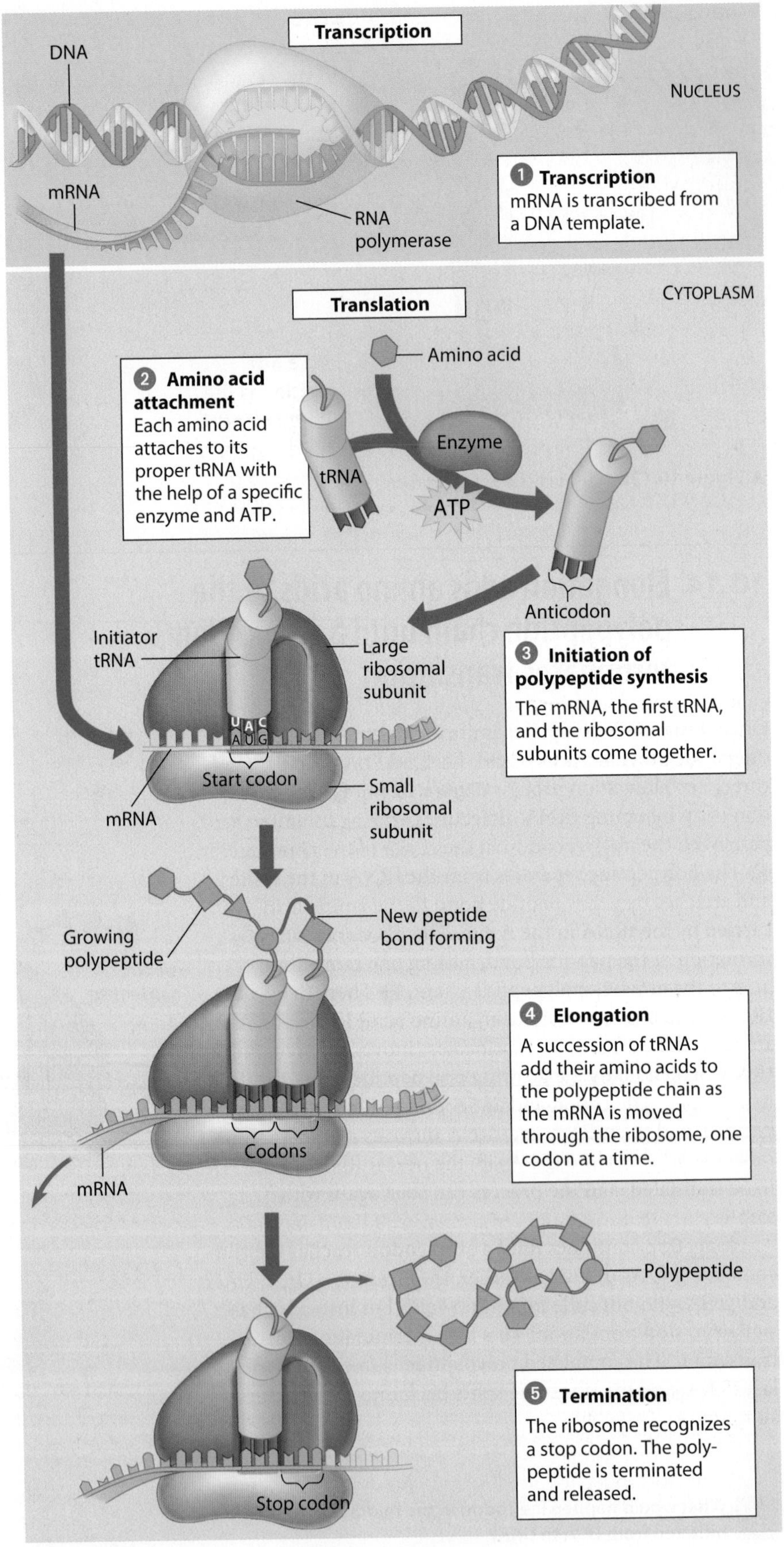

Transcription
DNA
NUCLEUS
mRNA
RNA polymerase
❶ Transcription
mRNA is transcribed from a DNA template.
Translation
CYTOPLASM
Amino acid
❷ Amino acid attachment
Each amino acid attaches to its proper tRNA with the help of a specific enzyme and ATP.
tRNA
Enzyme
ATP
Anticodon
Initiator tRNA
Large ribosomal subunit
❸ Initiation of polypeptide synthesis
The mRNA, the first tRNA, and the ribosomal subunits come together.
U A C
A U G
Start codon
Small ribosomal subunit
mRNA
Growing polypeptide
New peptide bond forming
❹ Elongation
A succession of tRNAs add their amino acids to the polypeptide chain as the mRNA is moved through the ribosome, one codon at a time.
Codons
mRNA
Polypeptide
❺ Termination
The ribosome recognizes a stop codon. The polypeptide is terminated and released.
Stop codon

▲ **Figure 10.15** A summary of transcription and translation

10.16 Mutations can affect genes

Many inherited traits can be understood in molecular terms. For instance, sickle-cell disease can be traced to a change in a single amino acid in one of the polypeptides in the hemoglobin protein (see Module 9.13). This difference is caused by a single nucleotide difference in the DNA coding for that polypeptide **(Figure 10.16A)**. In the double helix, one nucleotide *pair* is changed.

Any change in the nucleotide sequence of a cell's DNA is called a **mutation**. Mutations can involve large regions of a chromosome or just a single nucleotide pair, as in sickle-cell disease.

Mutations within a gene can be divided into two general categories: nucleotide substitutions and nucleotide insertions or deletions **(Figure 10.16B)**. A nucleotide substitution is the replacement of one nucleotide and its base-pairing partner with another pair of nucleotides. For example, in the second row in Figure 10.16B, A replaces G in the fourth codon of the mRNA. What effect can a substitution have? Because the genetic code is redundant, some substitution mutations have no effect at all. For example, if a mutation causes an mRNA codon to change from GAA to GAG, no change in the protein product would result because GAA and GAG both code for the same amino acid (Glu; see Figure 10.8A). Such a change is called a **silent mutation**.

In contrast, a **missense mutation** changes one amino acid to another one. For example, if a mutation causes an mRNA codon to change from GGC to AGC, as in the second row of Figure 10.16B, the resulting protein will have a serine (Ser) instead of a glycine (Gly) at this position. Some missense mutations have little or no effect on the resulting protein, but others, as in the case of sickle-cell disease, prevent the protein from performing its normal function.

Some substitutions, called **nonsense mutations**, change an amino acid codon into a stop codon. For example, if an AGA (Arg) codon is changed to a UGA (stop) codon, the result will be a prematurely terminated protein, which probably will not function properly.

Mutations involving the insertion or deletion of one or more nucleotides in a gene, called **frameshift mutations**, often have disastrous effects. Because mRNA is read as a series of nucleotide triplets (codons) during translation, adding or subtracting nucleotides may alter the reading frame (triplet grouping) of the genetic message. All the nucleotides

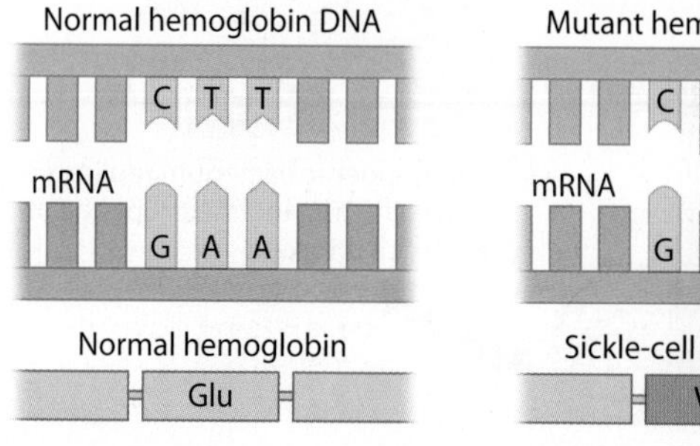
Normal hemoglobin DNA
C T T
mRNA
G A A
Normal hemoglobin
Glu
Mutant hemoglobin DNA
C A T
mRNA
G U A
Sickle-cell hemoglobin
Val

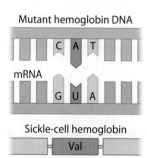

▲ **Figure 10.16A** The molecular basis of sickle-cell disease

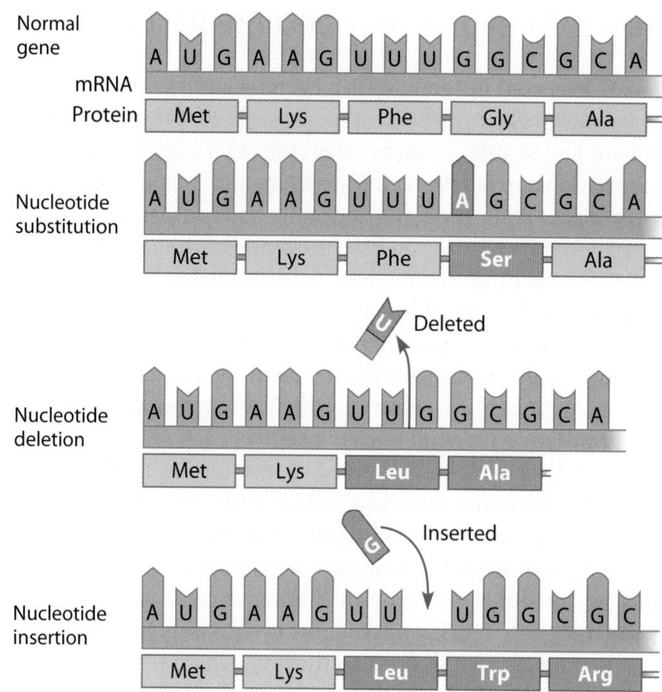
Normal gene
mRNA
Protein
A U G A A G U U U G G C G C A
Met Lys Phe Gly Ala
Nucleotide substitution
A U G A A G U U U A G C G C A
Met Lys Phe Ser Ala
Deleted
U
Nucleotide deletion
A U G A A G U U G G C G C A
Met Lys Leu Ala
Inserted
G
Nucleotide insertion
A U G A A G U U U G G C G C
Met Lys Leu Trp Arg

▲ **Figure 10.16B** Types of mutations and their effects

after the insertion or deletion will be regrouped into different codons (Figure 10.16B, bottom two rows). Consider this example in the English language: The red cat ate the big rat. Deleting the second letter produces an entirely nonsensical message: Ter edc ata tet heb igr at. Frameshift mutations will most likely produce a nonfunctional polypeptide.

The production of mutations, called **mutagenesis**, can occur in a number of ways. Spontaneous mutations result from errors during DNA replication or recombination. Other mutations are caused by physical or chemical agents called **mutagens**. High-energy radiation, such as X-rays or ultraviolet light, is a physical mutagen. One class of chemical mutagens consists of chemicals that are similar to normal DNA bases but pair incorrectly or are otherwise disruptive when incorporated into DNA. For example, the anti-AIDS drug AZT works because its structure is similar enough to thymine that viral polymerases incorporate it into newly synthesized DNA but different enough that the drug blocks further replication.

Occasionally, a mutation leads to a protein that enhances the success of the mutant organism and its descendants. Much more often, mutations are harmful to an organism. Mutations are, however, an important source of the rich diversity of genes in the living world, a diversity that makes evolution by natural selection possible. And within the laboratory, mutations are essential tools for geneticists, creating the different alleles needed for genetic research.

? How could a single nucleotide substitution result in a shortened protein product?

● A substitution that changes an amino acid codon into a stop codon would produce a prematurely terminated polypeptide.

▷ The Genetics of Viruses and Bacteria

10.17 Viral DNA may become part of the host chromosome

As we discussed in Module 10.1, bacteria and viruses served as models in experiments that uncovered the molecular details of heredity. Now let's take a closer look at viruses, focusing on the relationship between viral structure and the processes of nucleic acid replication, transcription, and translation.

In a sense, a **virus** is an infectious particle consisting of little more than "genes in a box": a bit of nucleic acid wrapped in a protein coat called a **capsid** and, in some cases, a membrane envelope. Viruses are parasites that can replicate (reproduce) only inside cells. In fact, the host cell provides most of the components necessary for replicating, transcribing, and translating the viral nucleic acid.

In Figure 10.1C, we described the replication cycle of phage T2. This sort of cycle is called a **lytic cycle** because it results in the lysis (breaking open) of the host cell and the release of the viruses that were produced within the cell. Some phages can also replicate by an alternative route called the lysogenic cycle. During a **lysogenic cycle**, viral DNA replication occurs without destroying the host cell.

Figure 10.17 illustrates the two kinds of cycles for a phage called lambda that infects *E. coli*. Both cycles begin when the phage DNA ❶ enters the bacterium and ❷ forms a loop. The DNA then embarks on one of the two pathways. In the lytic cycle (left), ❸ lambda's DNA immediately turns the cell into a virus-producing factory, and ❹ the cell soon lyses and releases its viral products, which may then infect another cell.

In the lysogenic cycle, ❺ viral DNA is inserted by genetic recombination into the bacterial chromosome. Once inserted, the phage DNA is referred to as a **prophage**, and most of its genes are inactive. ❻ Every time the *E. coli* cell prepares to divide, it replicates the phage DNA along with its own chromosome and passes the copies on to daughter cells. A single infected bacterium can thereby quickly give rise to a large population of bacterial cells that all carry a prophage. The lysogenic cycle enables viruses to spread without killing the host cells on which they depend. The prophages may remain in the bacterial cells indefinitely. Occasionally, however, an environmental signal—typically, one that indicates an unfavorable turn in the environment, such as an increase in radiation, drought, or certain toxic chemicals—triggers a switchover from the lysogenic cycle to the lytic cycle. This causes the viral DNA to be excised from the bacterial chromosome, eventually leading to death of the host cell.

Sometimes, the few prophage genes active in a lysogenic bacterium can cause medical problems. For example, the bacteria that cause diphtheria, botulism, and scarlet fever would be harmless to people if it were not for the prophage genes they carry. Certain of these genes direct the bacteria to produce the toxins responsible for making people ill. In the next module, we will explore viruses that infect animals and plants.

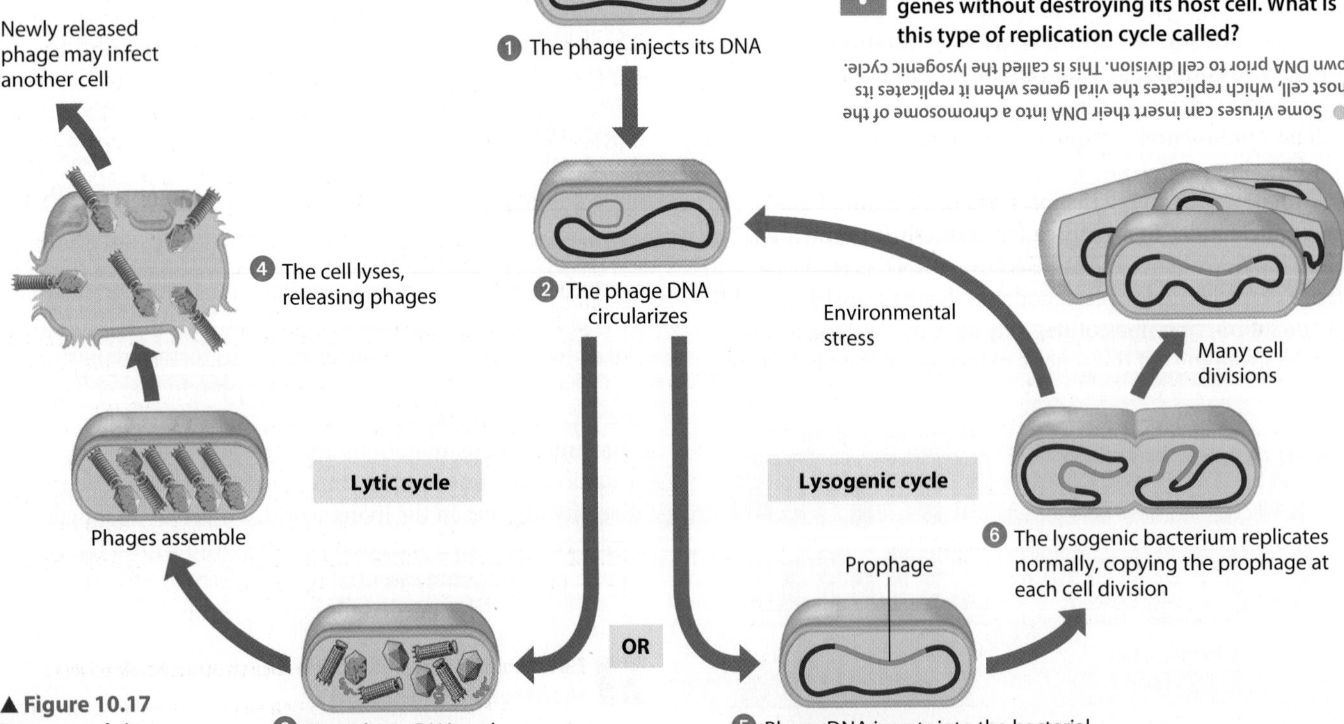
Phage
Attaches to cell
Phage DNA
Bacterial chromosome
❶ The phage injects its DNA
Newly released phage may infect another cell
❹ The cell lyses, releasing phages
❷ The phage DNA circularizes
Environmental stress
Many cell divisions
Lytic cycle
Lysogenic cycle
Phages assemble
❻ The lysogenic bacterium replicates normally, copying the prophage at each cell division
Prophage
OR
❸ New phage DNA and proteins are synthesized
❺ Phage DNA inserts into the bacterial chromosome by recombination

▲ **Figure 10.17** Two types of phage replication cycles

? **Describe one way a virus can perpetuate its genes without destroying its host cell. What is this type of replication cycle called?**

Some viruses can insert their DNA into a chromosome of the host cell, which replicates the viral genes when it replicates its own DNA prior to cell division. This is called the lysogenic cycle.

10.18 Many viruses cause disease in animals and plants

CONNECTION

Viruses can cause disease in both animals and plants. A typical animal virus has a membranous outer envelope and projecting spikes of glycoprotein (protein molecules with attached sugars). The envelope helps the virus enter and leave the host cell. Many animal viruses have RNA rather than DNA as their genetic material. Examples of RNA viruses include those that cause the common cold, measles, mumps, polio, and AIDS. Examples of diseases caused by DNA viruses include hepatitis, chicken pox, and herpes infections.

Figure 10.18 shows the replication cycle of a typical enveloped RNA virus: the mumps virus. Once a common childhood disease characterized by fever and painful swelling of the salivary glands, mumps has become quite rare in industrialized nations thanks to widespread vaccination. When the mumps virus contacts a susceptible cell, the glycoprotein spikes attach to receptor proteins on the cell's plasma membrane. The viral envelope fuses with the cell's membrane, allowing the protein-coated RNA to ❶ enter the cytoplasm. ❷ Enzymes (not shown) then remove the protein coat. ❸ An enzyme that entered the cell as part of the virus uses the virus's RNA genome as a template for making complementary strands of RNA (shown in pink). The new strands have two functions: ❹ They serve as mRNA for the synthesis of new viral proteins, and they serve as templates for synthesizing new viral genome RNA. ❺ The new coat proteins assemble around the new viral RNA. ❻ Finally, the viruses leave the cell by cloaking themselves in the host cell's plasma membrane. Thus, the virus obtains its envelope from the host cell, leaving the cell without necessarily lysing it.

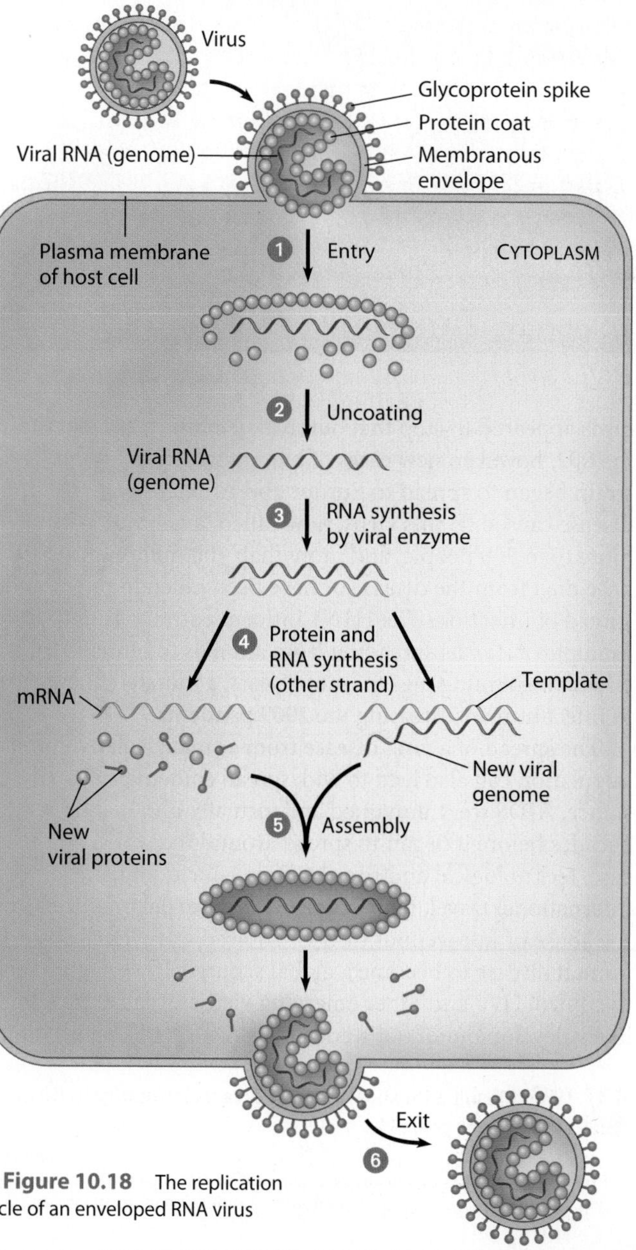

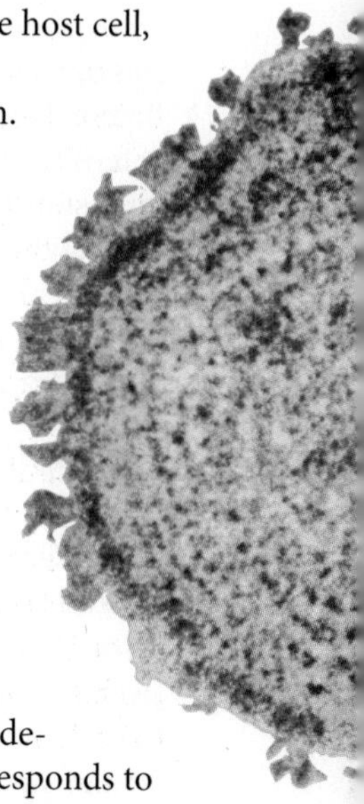

Not all animal viruses replicate in the cytoplasm. For example, herpesviruses—which cause chicken pox, shingles, cold sores, and genital herpes—are enveloped DNA viruses that replicate in the host cell's nucleus; they acquire their envelopes from the cell's nuclear membranes. While inside the nuclei of certain nerve cells, herpesvirus DNA may remain permanently dormant, without destroying these cells. From time to time, physical stress, such as a cold or sunburn, or emotional stress may stimulate the herpesvirus DNA to begin production of the virus, which then infects cells at the body's surface and causes unpleasant symptoms.

The amount of damage a virus causes our body depends partly on how quickly our immune system responds to fight the infection and partly on the ability of the infected tissue to repair itself. We usually recover completely from colds because our respiratory tract tissue can efficiently replace damaged cells by mitosis. In contrast, the poliovirus attacks nerve cells, which are not usually replaceable. The damage to such cells, unfortunately, is permanent. In such cases, we try to prevent the disease with vaccines (see Module 24.3).

Plants, like animals, are susceptible to viral infections. Viruses that infect plants can stunt plant growth and diminish crop yields. Most known plant viruses are RNA viruses. To infect a plant, a virus must first get past the plant's outer protective layer of cells (the epidermis). Once a virus enters a plant cell and begins replicating, it can spread throughout the entire plant through plasmodesmata, the cytoplasmic connections that penetrate the walls between adjacent plant cells (see Figure 4.21). Plant viruses may spread to other plants by insects, herbivores, humans, or farming tools. As with animal viruses, there are no cures for most viral diseases of plants. Agricultural scientists focus instead on preventing infections and on breeding resistant varieties of crop plants.

Virus
Glycoprotein spike
Protein coat
Viral RNA (genome)
Membranous envelope
Plasma membrane of host cell
CYTOPLASM
1 Entry
2 Uncoating
Viral RNA (genome)
3 RNA synthesis by viral enzyme
4 Protein and RNA synthesis (other strand)
mRNA
Template
New viral genome
New viral proteins
5 Assembly
Exit
6

▲ **Figure 10.18** The replication cycle of an enveloped RNA virus

? **Explain how some viruses replicate without having DNA.**

● The genetic material of these viruses is RNA, which is replicated inside the host cell by special enzymes encoded by the virus. The viral genome (or its complement) serves as mRNA for the synthesis of viral proteins.

10.19 Emerging viruses threaten human health

EVOLUTION CONNECTION

Emerging viruses are ones that seem to burst on to the scene, becoming apparent to the medical community quite suddenly. There are many familiar examples, such as the 2009 H1N1 influenza virus (discussed in the chapter introduction). Another example is **HIV** (human immunodeficiency virus), the virus that causes **AIDS** (acquired immunodeficiency syndrome). HIV appeared in New York and California in the early 1980s, seemingly out of nowhere. Yet another example is the deadly Ebola virus, recognized initially in 1976 in central Africa; it is one of several emerging viruses that cause hemorrhagic fever, an often fatal syndrome characterized by fever, vomiting, massive bleeding, and circulatory system collapse. A number of other dangerous newly recognized viruses cause encephalitis, inflammation of the brain. One example is the West Nile virus, which appeared in North America in 1999 and has since spread to all 48 contiguous U.S. states. West Nile virus is spread primarily by mosquitoes, which carry the virus in blood sucked from one victim and can transfer it to another victim. West Nile virus cases surged in 2012, especially in Texas. Severe acute respiratory syndrome (SARS) first appeared in China in 2002. Within eight months, about 8,000 people were infected, and 10% died. Researchers quickly identified the infectious agent as a previously unknown, single-stranded RNA coronavirus, so named for its crown-like "corona" of spikes.

Why are viral diseases such a constant threat?

How do such viruses emerge suddenly, giving rise to new diseases? And why are viral diseases so hard to eradicate? Three processes contribute to the emergence of viral diseases: mutation, contact among species, and spread from isolated populations.

The mutation of existing viruses is a major source of new viral diseases. RNA viruses tend to have unusually high rates of mutation because errors in replicating their RNA genomes are not subject to the kind of proofreading and repair mechanisms that help reduce errors in DNA replication. Some mutations change existing viruses into new strains (genetic varieties) that can cause disease in individuals who have developed immunity to ancestral strains. That is why we need yearly flu vaccines: Mutations create new influenza virus strains to which previously vaccinated people have no immunity.

New viral diseases often arise from the spread of existing viruses from one host species to another. Scientists estimate that about three-quarters of new human diseases have originated in other animals. For example, in 1997, at least 18 people in Hong Kong were infected with a strain of flu virus called H5N1, which was previously seen only in birds. A mass culling of all of Hong Kong's 1.5 million domestic birds appeared to stop that outbreak (**Figure 10.19**). Beginning in 2002, however, new cases of human infection by this bird strain began to spread to Europe and Africa. As of 2012, the disease caused by this virus, now called "avian flu," has killed more than 300 people, and more than 100 million birds either have died from the disease or have been killed to prevent the spread of infection. The H1N1 influenza strain is another example of a virus spreading from animals to humans; after circulating among pigs for many years, a mutated form began to infect humans, creating the 2009 pandemic.

Colorized TEM 180,000×

▼ **Figure 10.19** A Hong Kong health-care worker prepares to cull a chicken to help prevent the spread of the avian flu virus (shown in the inset)

The spread of a viral disease from a small, isolated human population can also lead to widespread epidemics. For instance, AIDS went unnamed and virtually unnoticed for decades before it began to spread around the world. In this case, technological and social factors—including affordable international travel, blood transfusions, sexual practices, and the abuse of intravenous drugs—allowed a previously rare human disease to become a global scourge. If we ever manage to control HIV and other emerging viruses, that success will likely develop out of our understanding of molecular biology.

? **Why doesn't a flu shot one year give us immunity to flu in subsequent years?**

Influenza viruses evolve rapidly by frequent mutation; thus, the strains that infect us later will most likely be different from the ones to which we've been vaccinated.

10.20 The AIDS virus makes DNA on an RNA template

HIV, the virus that causes AIDS, is an RNA virus with some special properties. In outward appearance, HIV resembles the flu or mumps virus (**Figure 10.20A**). Its membranous envelope and glycoprotein spikes enable HIV to enter and leave a host cell much the way the mumps virus does (see Figure 10.18). Notice, however, that HIV contains two identical copies of its RNA instead of one. HIV also has a different mode of replication. It is a **retrovirus**, an RNA virus that reproduces by means of a DNA molecule. Retroviruses are so named because they reverse the usual DNA → RNA flow of genetic information. These viruses carry molecules of an enzyme called **reverse transcriptase**, which catalyzes reverse transcription: the synthesis of DNA on an RNA template.

Figure 10.20B illustrates what happens after HIV RNA is uncoated in the cytoplasm of a host cell. ❶ Reverse transcriptase (●) uses the RNA as a template to make a DNA strand and then ❷ adds a second, complementary DNA strand. ❸ The resulting double-stranded viral DNA enters the cell's nucleus and inserts itself into the chromosomal DNA, becoming a provirus (analogous to a prophage). The host's RNA polymerase ❹ transcribes the proviral DNA into RNA, which can then be ❺ translated by ribosomes into viral proteins. ❻ New viruses assembled from these components leave the cell and can infect other cells.

HIV infects and kills white blood cells that play important roles in the body's immune system. The loss of such cells causes the body to become susceptible to other infections that it would normally be able to fight off. Such secondary infections cause the syndrome (a collection of symptoms) that can kill an AIDS patient. (We discuss AIDS in more detail when we take up the immune system in Chapter 24.)

? Why is HIV reverse transcriptase a good target for anti-AIDS drug therapy?

● Reverse transcriptase is unique to HIV; we do not normally copy genetic information from RNA to DNA, so disabling reverse transcriptase would not affect a human.

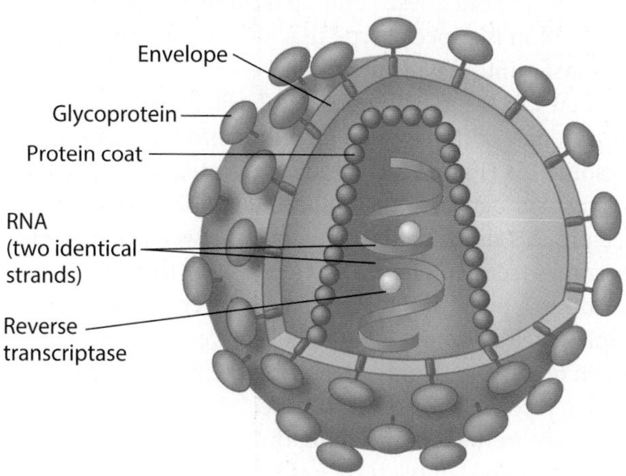

Envelope
Glycoprotein
Protein coat
RNA (two identical strands)
Reverse transcriptase

▲ **Figure 10.20A** A model of HIV structure

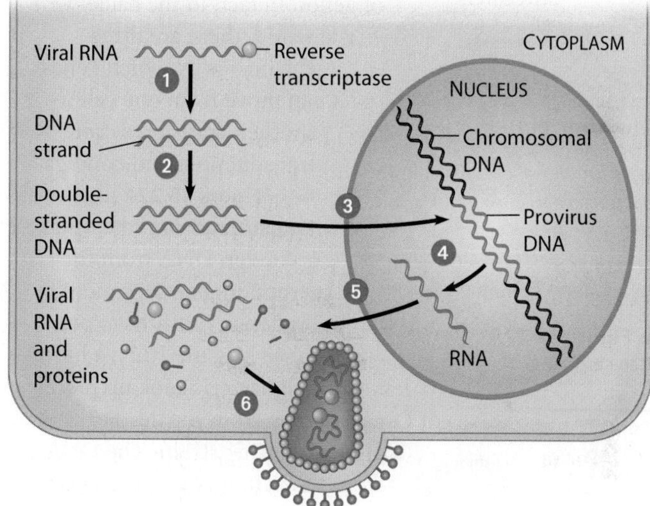

Viral RNA
Reverse transcriptase
CYTOPLASM
NUCLEUS
DNA strand
Chromosomal DNA
Double-stranded DNA
Provirus DNA
Viral RNA and proteins
RNA
1
2
3
4
5
6

▲ **Figure 10.20B** The behavior of HIV nucleic acid in a host cell

10.21 Viroids and prions are formidable pathogens in plants and animals

Viruses may be small and simple, but they are huge in comparison with another class of pathogens: viroids. **Viroids** are small, circular RNA molecules that infect plants. Unlike the nucleic acid of a virus, viroids do not encode proteins but can replicate in host plant cells, apparently using cellular enzymes. These small RNA molecules seem to interfere with regulatory systems that control plant growth. The typical signs of viroid diseases are abnormal development and stunted growth.

An important lesson to learn from viroids is that a single molecule can be an infectious agent. Viroids consist solely of nucleic acid, whose ability to be replicated is well known. Even more surprising are infectious proteins called **prions**, which cause a number of degenerative brain diseases in various animal species, including scrapie in sheep and goats, chronic wasting disease in deer and elk, mad cow disease (bovine spongiform encephalopathy, or BSE), and Creutzfeldt-Jakob disease in humans (which is exceedingly rare).

A prion is thought to be a misfolded form of a protein normally present in brain cells. When a prion enters a cell containing the normal form of protein, the prion somehow converts the normal protein molecules to the misfolded prion versions. The abnormal proteins clump together, which may lead to loss of brain tissue (although *how* this occurs is the subject of much debate and ongoing research). There is no cure for prion diseases, and the only hope for developing effective treatments lies in understanding and preventing the process of infection.

? What makes prions different from all other known infectious agents?

● Prions are proteins and have no nucleic acid.

10.22 Bacteria can transfer DNA in three ways

By studying viral replication, researchers also learn about the mechanisms that regulate DNA replication and gene expression in living cells. Bacteria are equally valuable as microbial models in genetics research. As prokaryotic cells, bacteria allow researchers to investigate molecular genetics in the simplest living organisms.

Most of a bacterium's DNA is found in a single chromosome, a closed loop of DNA with associated proteins. In the diagrams here, we show the chromosome much smaller than it actually is relative to the cell. A bacterial chromosome is hundreds of times longer than its cell; it fits inside the cell because it is tightly folded.

Bacterial cells reproduce by replication of the bacterial chromosome followed by binary fission (see Module 8.2). Because binary fission is an asexual process involving only a single parent, the bacteria in a colony are genetically identical to the parental cell. But this does not mean that bacteria lack ways to produce new combinations of genes. In fact, in the bacterial world, there are three mechanisms by which genes can move from one cell to another: transformation, transduction, and conjugation. **Figure 10.22A** illustrates **transformation**, the uptake of foreign DNA from the surrounding environment. In Frederick Griffith's experiments (see Module 10.1), a harmless strain of bacteria took up pieces of DNA left over from the dead cells of a disease-causing strain. The DNA from the pathogenic bacteria carried a gene that made the cells resistant to an animal's defenses, and when the previously harmless bacteria acquired this gene and replaced its own with the pathogenic version, it caused pneumonia in infected animals.

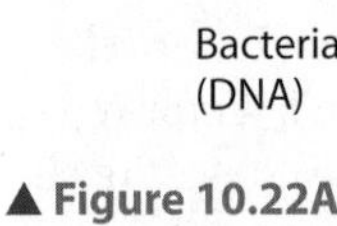
DNA enters cell
A fragment of DNA from another bacterial cell
Bacterial chromosome (DNA)

▲ **Figure 10.22A** Transformation

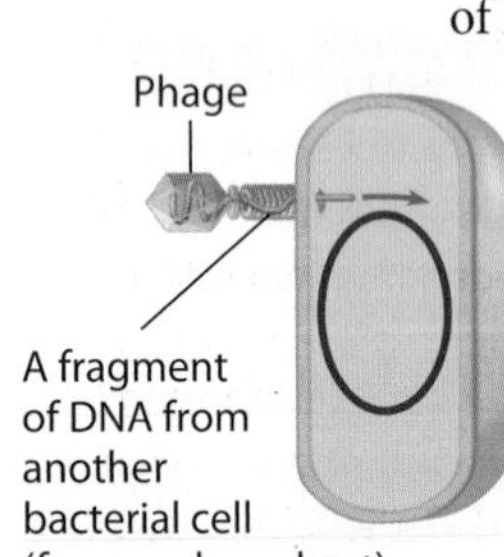
Phage
A fragment of DNA from another bacterial cell (former phage host)

▲ **Figure 10.22B** Transduction

Bacteriophages, the viruses that infect bacteria, provide the second means of bringing together genes of different bacteria. The transfer of bacterial genes by a phage is called **transduction**. During a lytic infection, when new viruses are being assembled in an infected bacterial cell, a fragment of DNA belonging to the host cell may be mistakenly packaged within the phage's coat instead of, or along with, the phage's DNA. When the phage infects a new bacterial cell, the DNA stowaway from the former host cell is injected into the new host (**Figure 10.22B**).

Figure 10.22C is an illustration of what happens at the DNA level when two bacterial cells "mate." This physical union of two bacterial cells—of the same or different species—and the DNA transfer between them is called **conjugation**. The donor cell has hollow appendages called sex pili, one of which is attached to the recipient cell in the figure. After attachment, the pilus retracts, pulling the two cells together, much like a grappling hook. The donor then transfers DNA (light blue in the figure) to the recipient. The donor cell replicates its DNA as it transfers it, so the cell doesn't end up lacking any genes. The DNA replication is a special type that allows one copy to peel off and transfer into the recipient cell.

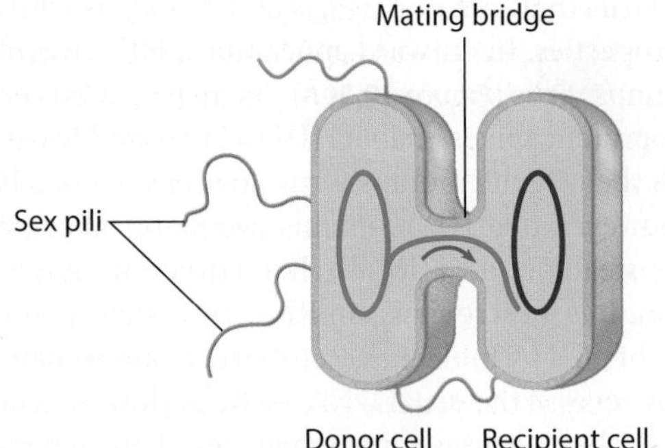
Mating bridge
Sex pili
Donor cell
Recipient cell

▲ **Figure 10.22C** Conjugation

Once new DNA gets into a bacterial cell, by whatever mechanism, part of it may then integrate into the recipient's chromosome. As **Figure 10.22D** indicates, integration occurs by crossing over between the donor and recipient DNA molecules, a process similar to crossing over between eukaryotic chromosomes (see Module 8.17). Here we see that two crossovers result in a piece of the donated DNA replacing part of the recipient cell's original DNA. The leftover pieces of DNA are broken down and degraded, leaving the recipient bacterium with a recombinant chromosome.

As we'll see in the next module, the transfer of genetic material between bacteria has important medical consequences.

? **The three modes of gene transfer between bacteria are ____________, which is transfer via a virus; ____________, which is the uptake of DNA from the surrounding environment; and ____________, which is bacterial "mating."**

transduction . . . transformation . . . conjugation

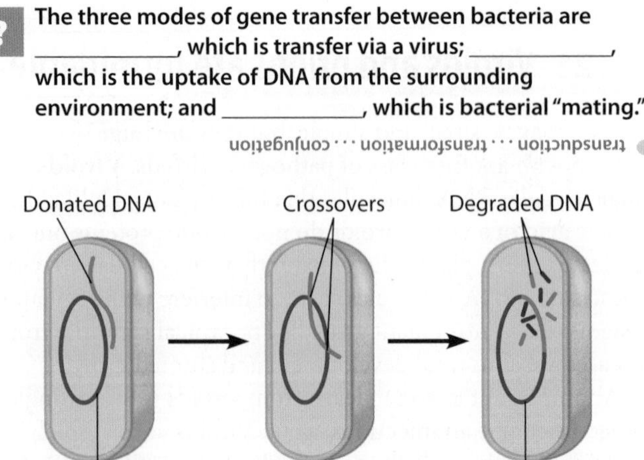
Donated DNA
Crossovers
Degraded DNA
Recipient cell's chromosome
Recombinant chromosome

▲ **Figure 10.22D** The integration of donated DNA into the recipient cell's chromosome

10.23 Bacterial plasmids can serve as carriers for gene transfer

The ability of a donor *E. coli* cell to carry out conjugation is usually due to a specific piece of DNA called the **F factor** (F for *fertility*). The F factor carries about 25 genes for making sex pili and other requirements for conjugation; it also contains an origin of replication, where DNA replication starts.

Let's see how the F factor behaves during conjugation. In **Figure 10.23A**, the F factor (light blue) is integrated into the donor bacterium's chromosome. When this cell conjugates with a recipient cell, the donor chromosome starts replicating at the F factor's origin of replication, indicated in the figure by the blue dot on the DNA. The growing copy of the DNA peels off the chromosome and heads into the recipient cell. Thus, part of the F factor serves as the leading end of the transferred DNA, but right behind it are genes from the donor's original chromosome. The rest of the F factor stays in the donor cell. Once inside the recipient cell, the transferred donor genes can recombine with the corresponding part of the recipient chromosome by crossing over. If crossing over occurs, the recipient cell may be genetically changed, but it usually remains a recipient because the two cells break apart before the rest of the F factor transfers.

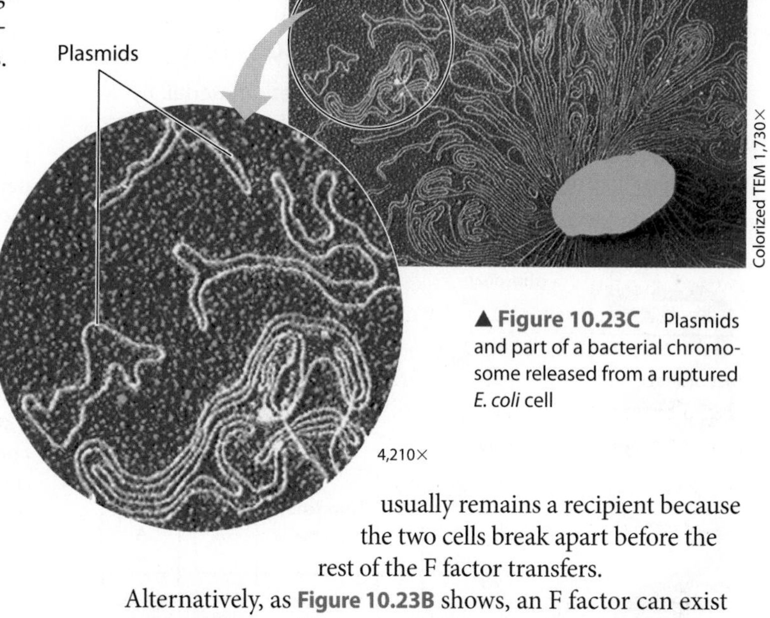

Plasmids
Colorized TEM 1,730×
4,210×

▲ **Figure 10.23C** Plasmids and part of a bacterial chromosome released from a ruptured *E. coli* cell

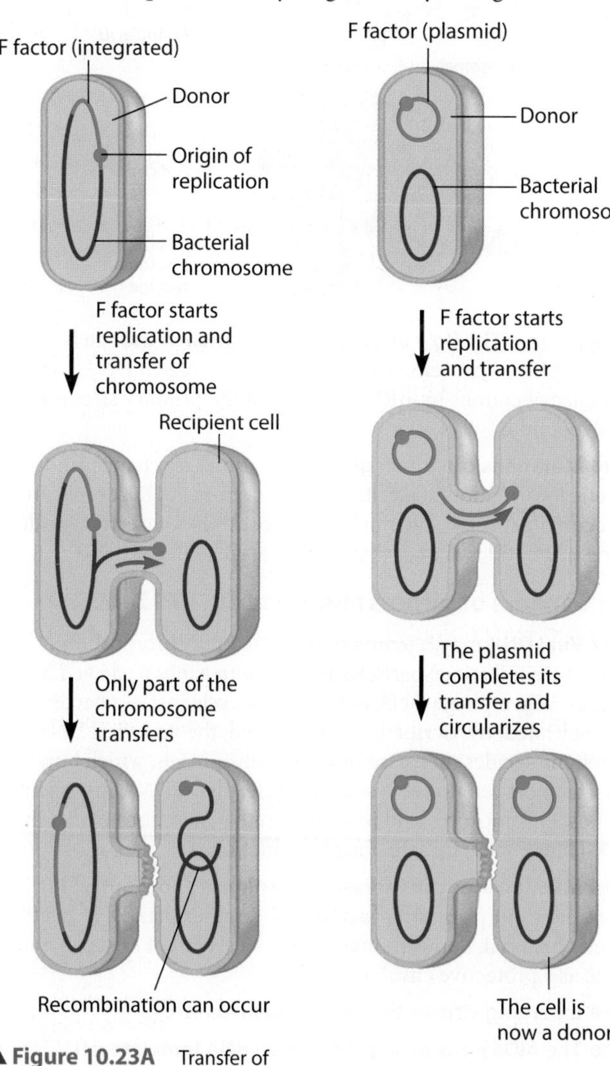

F factor (integrated)
Donor
Origin of replication
Bacterial chromosome
F factor starts replication and transfer of chromosome
Recipient cell
Only part of the chromosome transfers
Recombination can occur
F factor (plasmid)
Donor
Bacterial chromosome
F factor starts replication and transfer
The plasmid completes its transfer and circularizes
The cell is now a donor

▲ **Figure 10.23A** Transfer of chromosomal DNA by an integrated F factor

▲ **Figure 10.23B** Transfer of an F factor plasmid

Alternatively, as **Figure 10.23B** shows, an F factor can exist as a **plasmid**, a small, circular DNA molecule separate from the bacterial chromosome. Every plasmid has an origin of replication, required for its replication within the cell. Some plasmids, including the F factor plasmid, can bring about conjugation and move to another cell. When the donor cell in Figure 10.23B mates with a recipient cell, the F factor replicates and at the same time transfers one whole copy of itself, in linear rather than circular form, to the recipient cell. The transferred plasmid re-forms a circle in the recipient cell, and the cell becomes a donor.

E. coli and other bacteria have many different kinds of plasmids. You can see several from one cell in **Figure 10.23C**, along with part of the bacterial chromosome, which extends in loops from the ruptured cell. Some plasmids carry genes that can affect the survival of the cell. Plasmids of one class, called **R plasmids**, pose serious problems for human medicine. Transferable R plasmids carry genes for enzymes that destroy antibiotics such as penicillin and tetracycline. Bacteria containing R plasmids are resistant (hence the designation R) to antibiotics that would otherwise kill them. The widespread use of antibiotics in medicine and agriculture has tended to kill off bacteria that lack R plasmids, whereas those with R plasmids have multiplied. As a result, an increasing number of bacteria that cause human diseases, such as food poisoning and gonorrhea, are becoming resistant to antibiotics (see Module 13.15).

We'll continue our study of molecular genetics and explore what is known about genes (see Chapter 11) and return to our discussion of plasmids (Chapter 12) in later chapters.

Plasmids are useful tools for genetic engineering. Can you guess why?

● Scientists can take advantage of the ability of plasmids to carry foreign genes, to replicate, and to be inherited by progeny cells.

CHAPTER 10 REVIEW

For practice quizzes, BioFlix animations, MP3 tutorials, video tutors, and more study tools designed for this textbook, go to MasteringBiology®

Reviewing the Concepts

The Structure of the Genetic Material (10.1–10.3)

10.1 Experiments showed that DNA is the genetic material. By carefully choosing their model organism, Hershey and Chase were able to show that certain phages (bacterial viruses) reprogram host cells to produce more phages by injecting their DNA.

10.2 DNA and RNA are polymers of nucleotides.

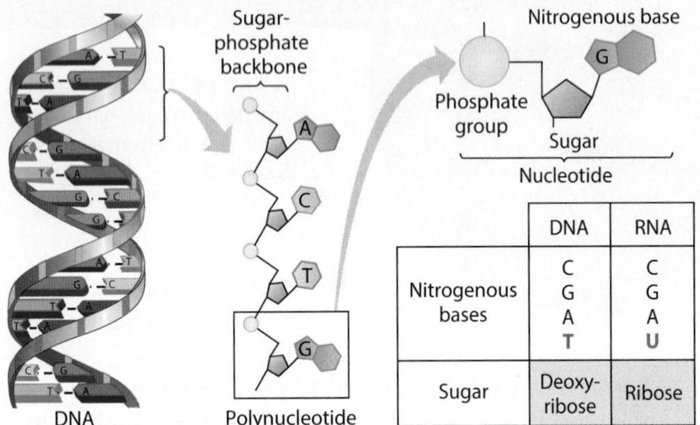
Sugar-phosphate backbone
Nitrogenous base
Phosphate group
Sugar
Nucleotide
DNA
Polynucleotide

	DNA	RNA
Nitrogenous bases	C G A T	C G A U
Sugar	Deoxy-ribose	Ribose

10.3 DNA is a double-stranded helix. Watson and Crick worked out the three-dimensional structure of DNA: two polynucleotide strands wrapped around each other in a double helix. Hydrogen bonds between bases hold the strands together. Each base pairs with a complementary partner: A with T, G with C.

DNA Replication (10.4–10.5)

10.4 DNA replication depends on specific base pairing. DNA replication starts with the separation of DNA strands. Enzymes then use each strand as a template to assemble new nucleotides into a complementary strand.

10.5 DNA replication proceeds in two directions at many sites simultaneously. Using the enzyme DNA polymerase, the cell synthesizes one daughter strand as a continuous piece. The other strand is synthesized as a series of short pieces, which are then connected by the enzyme DNA ligase.

The Flow of Genetic Information from DNA to RNA to Protein (10.6–10.16)

10.6 Genes control phenotypic traits through the expression of proteins. The DNA of a gene—a linear sequence of many nucleotides—is transcribed into RNA, which is translated into a polypeptide.

10.7 Genetic information written in codons is translated into amino acid sequences. Codons are base triplets.

10.8 The genetic code dictates how codons are translated into amino acids. Nearly all organisms use an identical genetic code to convert the mRNA codons transcribed from a gene to the amino acid sequence of a polypeptide.

10.9 Transcription produces genetic messages in the form of RNA. In the nucleus, the DNA helix unzips, and RNA nucleotides line up and hydrogen-bond along one strand of the DNA, following the base-pairing rules.

10.10 Eukaryotic RNA is processed before leaving the nucleus as mRNA. Noncoding segments of RNA (introns) are spliced out, and a cap and tail are added to the ends of the mRNA.

10.11 Transfer RNA molecules serve as interpreters during translation. Translation takes place in the cytoplasm. A ribosome attaches to the mRNA and translates its message into a specific polypeptide, aided by transfer RNAs (tRNAs). Each tRNA is a folded molecule bearing a base triplet called an anticodon on one end and a specific amino acid attachment site at the other end.

10.12 Ribosomes build polypeptides. Made of rRNA and proteins, ribosomes have binding sites for tRNAs and mRNA.

10.13 An initiation codon marks the start of an mRNA message.

10.14 Elongation adds amino acids to the polypeptide chain until a stop codon terminates translation. As the mRNA moves one codon at a time relative to the ribosome, a tRNA with a complementary anticodon pairs with each codon, adding its amino acid to the growing polypeptide chain.

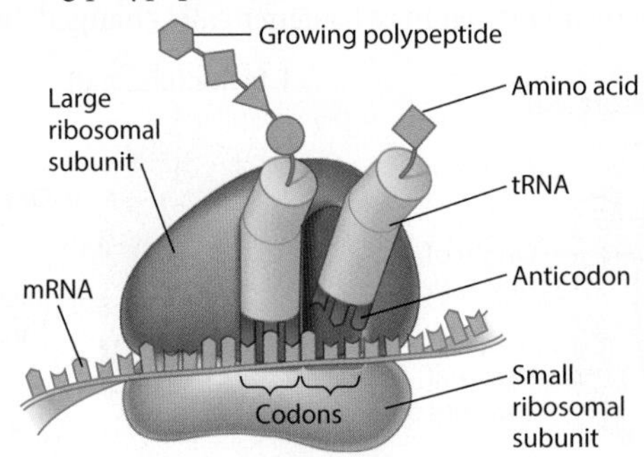
Growing polypeptide
Amino acid
Large ribosomal subunit
tRNA
mRNA
Anticodon
Codons
Small ribosomal subunit

10.15 Review: The flow of genetic information in the cell is DNA → RNA → protein. The sequence of codons in DNA, via the sequence of codons in mRNA, spells out the primary structure of a polypeptide.

10.16 Mutations can affect genes. Mutations are changes in the genetic information of a cell, caused by errors in DNA replication or recombination, or by mutagens. Substituting, inserting, or deleting nucleotides alters a gene, with varying effects.

The Genetics of Viruses and Bacteria (10.17–10.23)

10.17 Viral DNA may become part of the host chromosome. Viruses are infectious particles that contain genes packaged in protein. When phage DNA enters a lytic cycle inside a bacterium, it is replicated, transcribed, and translated; the new viral DNA and protein molecules then assemble into new phages, which burst from the host cell. In the lysogenic cycle, phage DNA inserts into the host chromosome and is passed on to generations of daughter cells. Later, it may initiate phage production.

10.18 Many viruses cause disease in animals and plants. Flu viruses and most plant viruses have RNA, rather than DNA, as their genetic material. Some animal viruses steal a bit of host cell membrane as a protective envelope.

10.19 Emerging viruses threaten human health.

10.20 The AIDS virus makes DNA on an RNA template. HIV is a retrovirus: It uses RNA as a template for making DNA, which then inserts into a host chromosome.

10.21 Viroids and prions are formidable pathogens in plants and animals. Viroids are RNA molecules that can infect plants. Prions are infectious proteins that can cause brain diseases in animals.

10.22 Bacteria can transfer DNA in three ways. Bacteria can transfer genes from cell to cell by transformation, transduction, or conjugation.

10.23 Bacterial plasmids can serve as carriers for gene transfer. Plasmids are small, circular DNA molecules separate from the bacterial chromosome.

Connecting the Concepts

1. Check your understanding of the flow of genetic information through a cell by filling in the blanks.

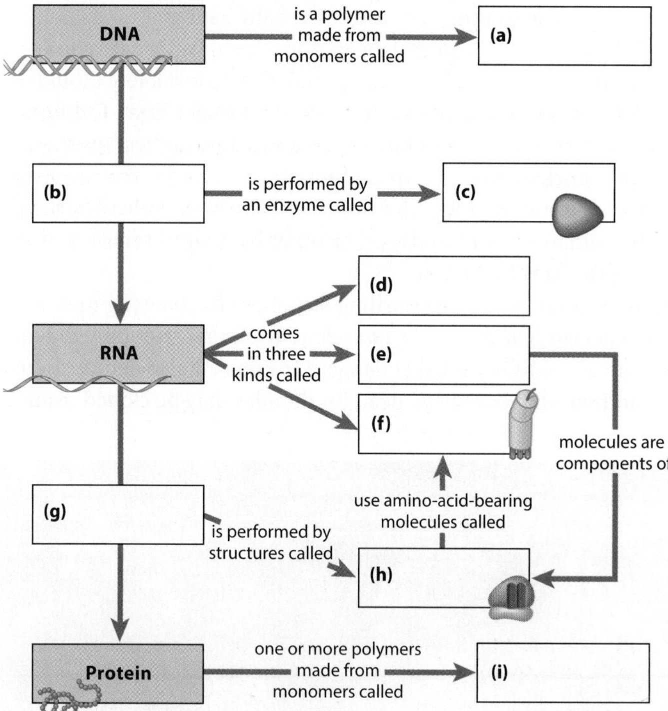

DNA
is a polymer made from monomers called
(a)
(b)
is performed by an enzyme called
(c)
(d)
RNA
comes in three kinds called
(e)
(f)
molecules are components of
use amino-acid-bearing molecules called
(g)
is performed by structures called
(h)
Protein
one or more polymers made from monomers called
(i)

Testing Your Knowledge

Level 1: Knowledge/Comprehension

2. Which of the following correctly ranks the structures in order of size, from largest to smallest?
   a. gene-chromosome-nucleotide-codon
   b. chromosome-gene-codon-nucleotide
   c. nucleotide-chromosome-gene-codon
   d. chromosome-nucleotide-gene-codon
3. Describe the process of DNA replication: the ingredients needed, the steps in the process, and the final product.
4. What is the name of the process that produces RNA from a DNA template? What is the name of the process that produces a polypeptide from an RNA template?

Level 2: Application/Analysis

5. Scientists have discovered how to put together a bacteriophage with the protein coat of phage T2 and the DNA of phage lambda. If this composite phage were allowed to infect a bacterium, the phages produced in the host cell would have ____________________________. (*Explain your answer.*)
   a. the protein of T2 and the DNA of lambda
   b. the protein of lambda and the DNA of T2
   c. the protein and DNA of T2
   d. the protein and DNA of lambda
6. A geneticist found that a particular mutation had no effect on the polypeptide encoded by a gene. This mutation probably involved
   a. deletion of one nucleotide.
   b. alteration of the start codon.
   c. insertion of one nucleotide.
   d. substitution of one nucleotide.
7. Describe the process by which the information in a eukaryotic gene is transcribed and translated into a protein. Correctly use these words in your description: tRNA, amino acid, start codon, transcription, RNA splicing, exons, introns, mRNA, gene, codon, RNA polymerase, ribosome, translation, anticodon, peptide bond, stop codon.

Level 3: Synthesis/Evaluation

8. The nucleotide sequence of a DNA codon is GTA. A messenger RNA molecule with a complementary codon is transcribed from the DNA. In the process of protein synthesis, a transfer RNA pairs with the mRNA codon. What is the nucleotide sequence of the tRNA anticodon?
   a. CAT
   b. CUT
   c. GUA
   d. CAU
9. A cell containing a single chromosome is placed in a medium containing radioactive phosphate so that any new DNA strands formed by DNA replication will be radioactive. The cell replicates its DNA and divides. Then the daughter cells (still in the radioactive medium) replicate their DNA and divide, and a total of four cells are present. Sketch the DNA molecules in all four cells, showing a normal (nonradioactive) DNA strand as a solid line and a radioactive DNA strand as a dashed line.
10. The base sequence of the gene coding for a short polypeptide is CTACGCTAGGCGATTGACT. What would be the base sequence of the mRNA transcribed from this gene? Using the genetic code in Figure 10.8A, give the amino acid sequence of the polypeptide translated from this mRNA. (*Hint*: What is the start codon?)
11. Researchers working on the Human Genome Project have determined the nucleotide sequences of human genes and in many cases identified the proteins encoded by the genes. Knowledge of the nucleotide sequences of genes might be used to develop lifesaving medicines or treatments for genetic defects. In the United States, both government agencies and biotechnology companies have applied for patents on their discoveries of genes. In Britain, the courts have ruled that a naturally occurring gene cannot be patented. Do you think individuals and companies should be able to patent genes and gene products? Before answering, consider the following: What are the purposes of a patent? How might the discoverer of a gene benefit from a patent? How might the public benefit? What might be some positive and negative results of patenting genes?
12. **SCIENTIFIC THINKING** The success of an experiment often depends on choosing an appropriate organism to study. For example, Gregor Mendel was able to deduce the fundamental principles in genetics in part because of his choice of the pea plant. What properties of bacteriophage T2 allowed Hershey and Chase (Module 10.1) to identify DNA as the genetic materials in their famous experiment?

Answers to all questions can be found in Appendix 4.

CHAPTER 11

How Genes Are Controlled

Can life be rebooted?

Even in a world in which cloning has become routine, the case of the baby banteng stands out. The Java banteng (*Bos javanicus*), shown below, is an endangered species of cattle native to Indonesia. Hunting and habitat destruction have reduced their wild population to just a few thousand.

Scientists obtained banteng skin tissue from "The Frozen Zoo," a facility in San Diego, California, where samples from rare or endangered animals are stored for conservation. Researchers inserted nuclei from these frozen skin cells into nucleus-free eggs from domestic beef cattle. The resulting embryos were implanted into surrogate domestic cows. Months later, two bantengs were born via Caesarian section, but one had to be euthanized within days. The other banteng survived, and as of 2013, remains a healthy inhabitant of the San Diego Zoo.

Although cloning mammals is now commonplace, one startling fact about the banteng makes this case unique: *The banteng that was cloned had died more than 20 years earlier*. Not only did this undertaking make the banteng the first successfully cloned endangered animal, it also demonstrated for the first time that well-preserved animals that have been dead for decades may be cloned using

closely related animals as surrogates. Biologists hope that such efforts may increase the populations of threatened species or even recently extinct species.

How can a single cell be "rebooted" to develop into an entire organism? To be successful, the starting cell must contain a complete set of genes capable of directing the production of all the cell types in an organism. The development of different types of cells must therefore depend on turning on and off different genes in different cells—the control of gene expression.

We begin this chapter with examples of how and where cells may alter their patterns of gene expression. Next we look at the methods and applications of plant and animal cloning. Finally, we discuss cancer, a disease that can be caused by changes in gene expression.

BIG IDEAS

Control of Gene Expression (11.1–11.11)

Cells can turn genes on and off through a variety of mechanisms.

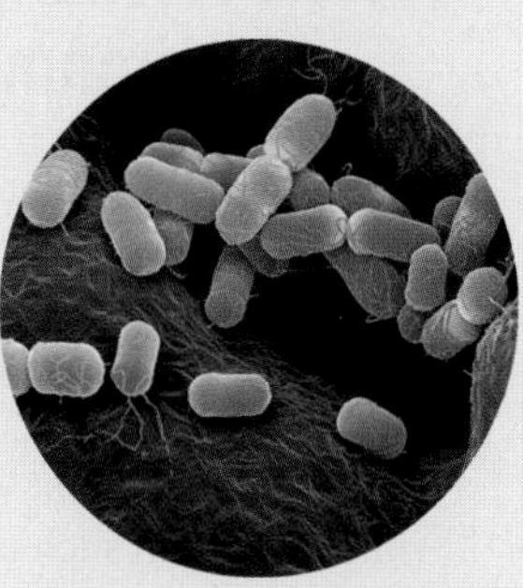

Cloning of Plants and Animals (11.12–11.14)

Cloning demonstrates that many body cells retain their full genetic potential.

The Genetic Basis of Cancer (11.15–11.18)

Changes in genes that control gene expression can lead to out-of-control cell growth.

▷ Control of Gene Expression

11.1 Proteins interacting with DNA turn prokaryotic genes on or off in response to environmental changes

Picture an *Escherichia coli* (*E. coli*) bacterium living in your intestine (**Figure 11.1A**). Its environment changes continuously, depending on your dietary whims. For example, if you eat a sweet roll for breakfast, the bacterium will be bathed in sugars and broken-down fats. Later, if you have a salad for lunch, the *E. coli*'s environment will change drastically. How can a bacterium cope with such a constantly shifting flow of resources?

The answer is that **gene regulation**—the turning on and off of genes—can help organisms respond to environmental changes. What does it mean to turn a gene on or off? Genes determine the nucleotide sequences of specific messenger RNA (mRNA) molecules (as we saw in Chapter 10), and mRNA in turn determines the sequences of amino acids in protein molecules (DNA → RNA → protein). Thus, a gene that is turned on is being transcribed into mRNA, and that message is being translated into specific protein molecules. The overall process by which genetic information flows from genes to proteins—that is, from genotype to phenotype—is called **gene expression**. The control of gene expression makes it possible for cells to produce specific kinds of proteins when and where they are needed.

Let's think back to our example of a bacterium living in your digestive system. It's no coincidence that we used *E. coli* as our example. Our earliest understanding of gene control came from studies of this bacterium by French biologists François Jacob and Jacques Monod. *E. coli* has a remarkable ability to change its metabolic activities in response to changes in its environment. For example, *E. coli* produces enzymes needed to metabolize a specific nutrient only when that nutrient is available. Bacterial cells that can conserve resources and energy have an advantage over cells that are unable to do so. Thus, natural selection has favored bacteria that express only the genes whose products are needed by the cell. Let's look at how the regulation of gene transcription helps *E. coli* efficiently use available resources.

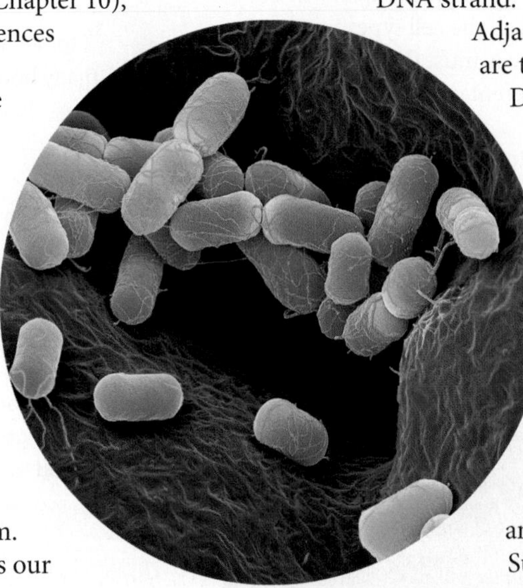

▲ **Figure 11.1A** Cells of *E. coli* bacteria

The *lac* Operon Imagine the bacterium in your intestine soon after you drink a glass of milk. One of the main nutrients in milk is the disaccharide sugar lactose. When lactose is plentiful in the intestine, *E. coli* makes the enzymes necessary to absorb the sugar and use it as an energy source. Conversely, when lactose is not plentiful, *E. coli* does not waste its energy producing these enzymes.

Recall that most enzymes are proteins; their production is an outcome of gene expression. *E. coli* can make lactose-utilization enzymes because it has genes that code for these enzymes. **Figure 11.1B** presents a model (first proposed in 1961 by Jacob and Monod) to explain how an *E. coli* cell can turn genes coding for lactose-utilization enzymes off or on, depending on whether lactose is available.

E. coli uses three enzymes to take up and start metabolizing lactose, and the genes coding for these three enzymes are regulated as a single unit. The DNA at the top of Figure 11.1B represents a small segment of the bacterium's chromosome. Notice that the three genes that code for the lactose-utilization enzymes (light blue) are situated next to each other along the DNA strand.

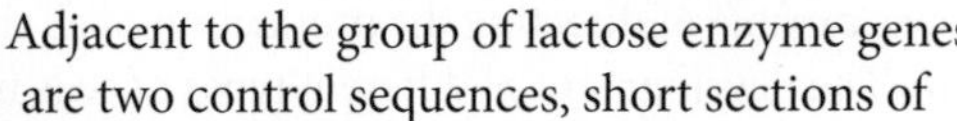

Adjacent to the group of lactose enzyme genes are two control sequences, short sections of DNA that help control the expression of these genes. One control sequence is a **promoter**, a site where the transcription enzyme, RNA polymerase, attaches and initiates transcription—in this case, transcription of all three lactose enzyme genes (as depicted in the bottom panel of Figure 11.1B). Between the promoter and the enzyme genes, a DNA control sequence called an **operator** acts as a switch. The operator determines whether RNA polymerase can attach to the promoter and start transcribing the genes. Such a cluster of genes with related functions, along with the control sequences—in this instance, the entire stretch of DNA required for enzyme production—is called an **operon**; with rare exceptions, operons exist only in prokaryotes. The key advantage to the grouping of related genes into operons is that a single "on-off switch" can control the whole cluster. The operon discussed here is called the *lac* operon, short for lactose operon. When an *E. coli* bacterium encounters lactose, all the enzymes needed for its metabolism are made at once because the operon's genes are all controlled by a single switch, the operator. But what determines whether the operator switch is on or off?

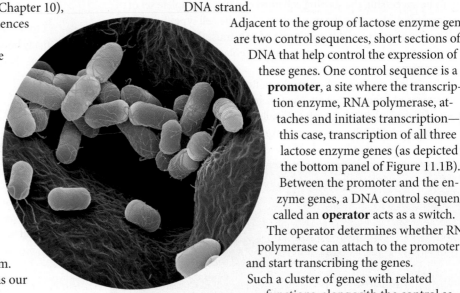

The top panel of Figure 11.1B shows the *lac* operon in "off" mode, its status when there is no lactose in the cell's environment. Transcription is turned off because a protein called a **repressor** ( ) binds to the operator ( ) and physically blocks the attachment of RNA polymerase ( ) to the promoter ( ). On the left side of the figure, you can see where the repressor comes from. A gene called a **regulatory gene** (dark blue), located outside the operon, codes for the repressor. The regulatory gene is expressed continually, so the cell always has a small supply of repressor molecules.

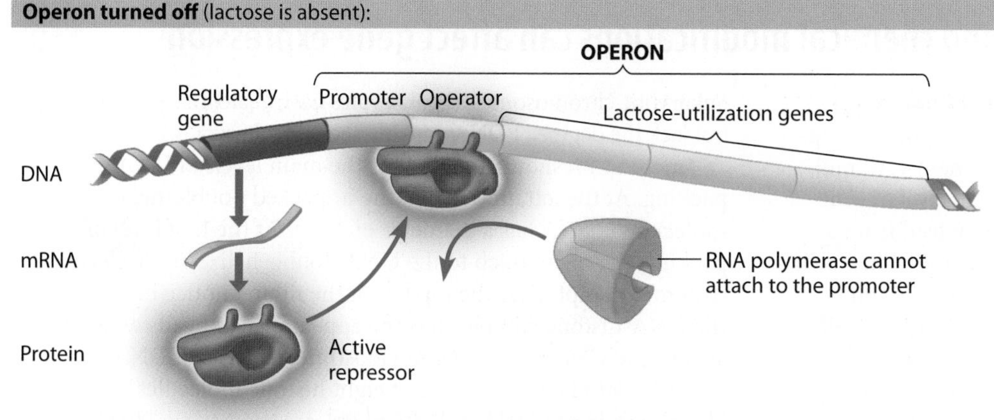

Operon turned off (lactose is absent):
OPERON
Regulatory gene
Promoter
Operator
Lactose-utilization genes
DNA
mRNA
Protein
Active repressor
RNA polymerase cannot attach to the promoter

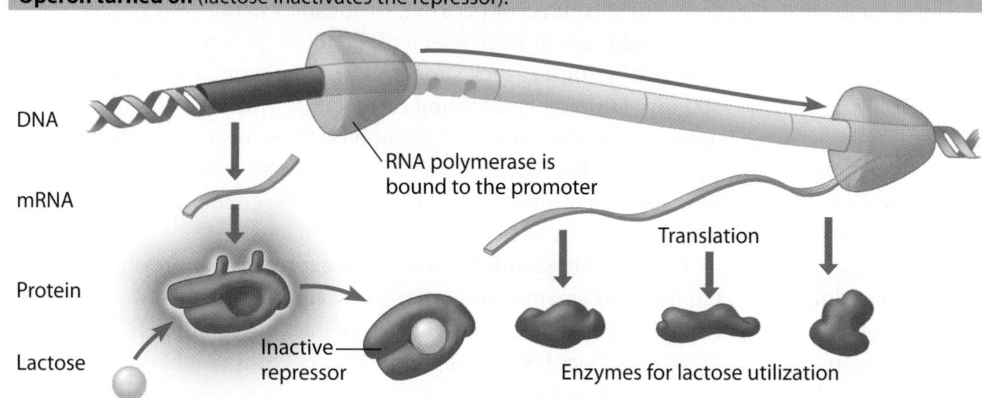

Operon turned on (lactose inactivates the repressor):
DNA
mRNA
Protein
Lactose
RNA polymerase is bound to the promoter
Translation
Inactive repressor
Enzymes for lactose utilization

▲ **Figure 11.1B** The *lac* operon

Try This Compare the structure and function of the *lac* repressor protein when lactose is absent (top) and present (bottom).

How can an operon be turned on if its repressor is always present? As the bottom panel of Figure 11.1B indicates, lactose ( ) interferes with the attachment of the *lac* repressor to the operator by binding to the repressor and changing its shape. With its new shape ( ), the repressor cannot bind to the operator, and the operator switch remains on. RNA polymerase is then able to bind to the promoter (because it is no longer being blocked) and from there transcribes the genes of the operon. The resulting mRNA carries coding sequences for all three enzymes needed for lactose metabolism. The cell can translate the message in this single mRNA into three separate polypeptides because the mRNA has multiple codons signaling the start and stop of translation.

The *lac* operon is so efficient that the addition of lactose to a bacterium's environment results in a thousandfold increase in lactose utilization enzymes in just 15 minutes. The newly produced mRNA and protein molecules will remain intact for only a short time before cellular enzymes break them down. When the synthesis of mRNA and protein stops because lactose is no longer present, the existing molecules are quickly degraded.

Other Kinds of Operons The *lac* operon is an example of an inducible operon (**Figure 11.1C**, left), one that is usually turned off but can be stimulated (induced) by a molecule—in this case, by lactose. Such operons usually operate as part of a pathway that breaks down a nutrient to simpler molecules. By producing the digestive enzymes only when the nutrient is available, the cell avoids wasting resources. A second type of operon, represented here by the *trp* operon, is called a repressible operon, because it is normally turned on but can be inhibited (repressed) when a specific molecule is present in abundance. In our example, the molecule is tryptophan (Trp), an amino acid essential for protein synthesis. *E. coli* can make tryptophan from scratch, using enzymes encoded in the *trp* operon. But *E. coli* will stop making tryptophan and simply absorb it from the surroundings whenever possible. When *E. coli* is bathed in tryptophan in the intestines (as occurs when you eat foods such as milk and poultry), the tryptophan binds to the repressor of the *trp* operon. This activates the *trp* repressor, enabling it to switch off the operon. Thus, this type of operon allows bacteria to stop making certain essential molecules when the molecules are already present in the environment, saving materials and energy for the cells.

Another type of operon control involves **activators**, proteins that turn operons *on* by binding to DNA and stimulating gene transcription. Activator proteins act by making it easier for RNA polymerase to bind to the promoter, rather than by blocking RNA polymerase, as repressors do. Activators help control a wide variety of operons.

Armed with a variety of operons regulated by repressors and activators, *E. coli* and other prokaryotes can thrive in frequently changing environments. Next we examine how more complex eukaryotes regulate their genes.

? **A certain mutation in *E. coli* impairs the ability of the *lac* repressor to bind to the *lac* operator. How would this affect the cell?**

● The cell would wastefully produce the enzymes for lactose metabolism continuously, even when lactose is not present.

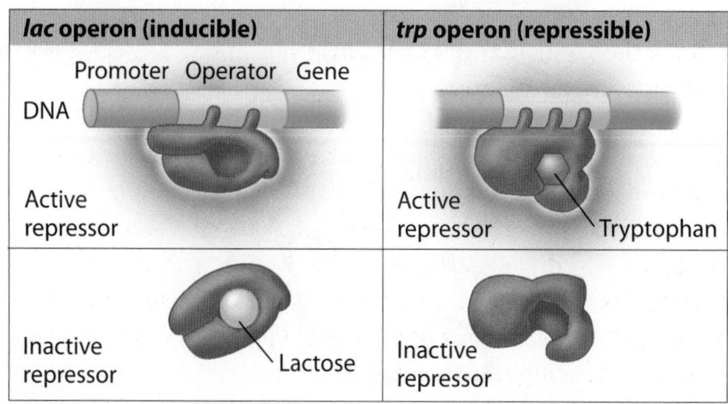

lac operon (inducible)
trp operon (repressible)
Promoter
Operator
Gene
DNA
Active repressor
Active repressor
Tryptophan
Inactive repressor
Lactose
Inactive repressor

▲ **Figure 11.1C** Two types of repressor-controlled operons

11.2 Chromosome structure and chemical modifications can affect gene expression

The cells of all organisms, whether prokaryotes or eukaryotes, must be able to turn genes on and off in response to signals from their external and internal environments. All multicellular eukaryotes also require an additional level of gene control: During the repeated cell divisions that lead from a zygote to an adult in a multicellular organism, individual cells must undergo **differentiation**—that is, they must become specialized in structure and function, with each type of cell fulfilling a distinct role. Your body, for example, contains hundreds of different types of cells. What makes a kidney cell different from, say, a bone cell?

To perform its specialized role, each cell type must maintain a specific program of gene expression in which some genes are expressed and others are not. Almost all the cells in an organism contain an identical genome, yet the subset of genes expressed in each cell type is unique, reflecting its specific function. Each adult human cell expresses only a small fraction of its total genes at any given time. And even one particular cell type can change its pattern of gene expression over time in response to developmental signals or other changes in the environment.

The differences between cell types, therefore, are not due to different genes being present but instead due to selective gene expression. In this module, we begin our exploration of gene regulation in eukaryotes by looking at the structure of chromosomes.

DNA Packing The DNA of each human chromosome is thousands of times longer than the diameter of the nucleus. All of this DNA can fit within the nucleus because of an elaborate, multilevel system of packing, coiling, and folding. A crucial aspect of DNA packing is the association of the DNA with small proteins called **histones**. In fact, histone proteins account for about half the mass of eukaryotic chromosomes. (Prokaryotes have analogous proteins, but lack the degree of DNA packing seen in eukaryotes.)

Figure 11.2A shows a model for the main levels of DNA packing. At the left, notice that the unpacked double-helical molecule of DNA has a diameter of 2 nm. At the first level of packing, histones attach to the DNA double helix. In the electron micrograph near the top left of the figure, notice how the DNA-histone complex has the appearance of beads on a string. Each "bead," called a **nucleosome**, consists of DNA wound around a protein core of eight histone molecules. Short stretches of DNA, called linkers, are the "strings" that join consecutive "beads" of nucleosomes.

At the next level of packing, the beaded string is wrapped into a tight helical fiber with a diameter of 30 nm. This fiber coils further into a thick supercoil with a diameter of about 300 nm. Looping and folding can compact the DNA even more, as you can see in the metaphase chromosome on the right side of the figure. Figure 11.2A gives an overview of how successive levels of coiling and folding enable a huge amount of DNA to fit into a tiny cell nucleus.

DNA packing tends to block gene expression by preventing RNA polymerase and other transcription proteins from contacting the DNA. Higher levels of packing can therefore inactivate genes for the long term. Genes within highly compacted chromatin, as seen in mitotic chromosomes—such as the duplicated chromosome shown on the right side of the figure—and in varying regions of interphase chromosomes are generally not expressed at all.

Chemical Modifications and Epigenetic Inheritance Eukaryotic chromosomes can be chemically modified in ways that help regulate gene expression. For example, the addition of methyl (CH_3) groups to some of the amino acids in histone proteins can cause the chromosomes to become more

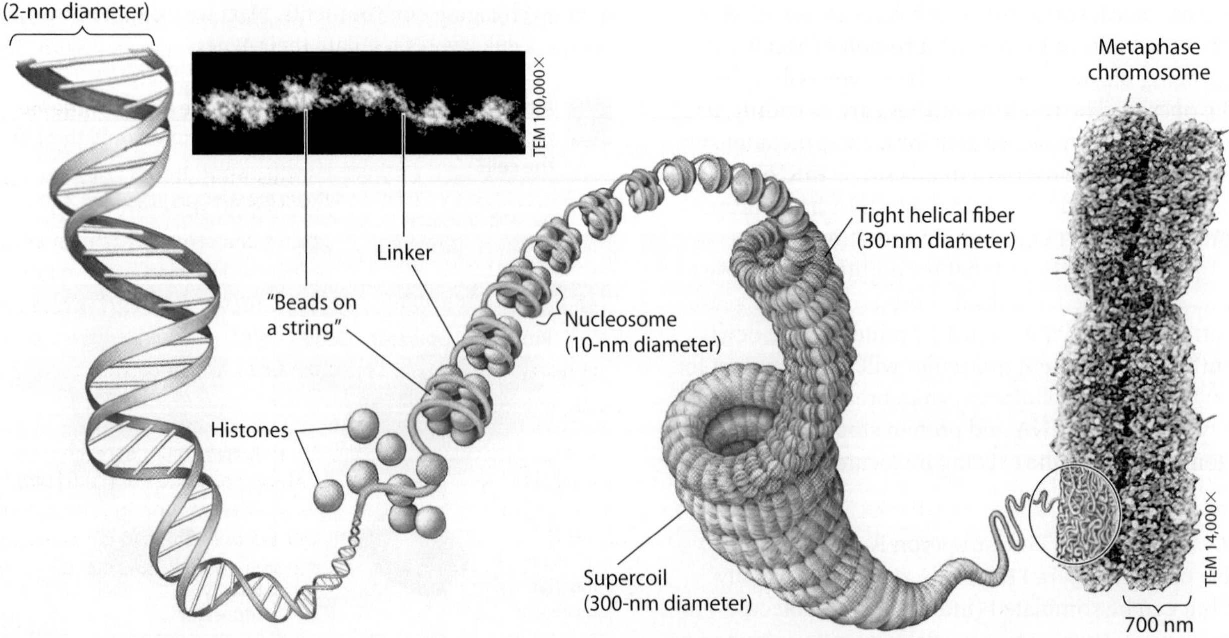
DNA double helix
(2-nm diameter)
TEM 100,000×
Linker
"Beads on
a string"
Nucleosome
(10-nm diameter)
Histones
Tight helical fiber
(30-nm diameter)
Supercoil
(300-nm diameter)
Metaphase
chromosome
TEM 14,000×
700 nm

▲ **Figure 11.2A** DNA packing in a eukaryotic chromosome

compact, leading to reduced transcription. Conversely, adding acetyl groups (—$COCH_3$) opens up the chromatin structure, promoting transcription.

DNA can also be chemically modified. For example, certain enzymes add a methyl group to DNA bases, usually cytosine, without changing the actual sequence of the bases. Individual genes are usually more heavily methylated in cells in which they are not expressed, and removing the extra methyl groups can turn on some of these genes. Thus, DNA methylation appears to play a role in turning genes off. At least in some species, DNA methylation seems to be essential for the long-term inactivation of genes. Such modifications are a normal and necessary mechanism for the regulation of gene expression, and improper methylation can lead to problems for the organism. For example, insufficient DNA methylation can lead to abnormal embryonic development in many species.

Once methylated, genes usually stay that way through successive cell divisions in a given individual. During replication, when a methylated stretch of DNA is duplicated, enzymes methylate the corresponding daughter strands to match. Methylation patterns are therefore passed on, allowing cells that form specialized tissues to keep a chemical record of what occurred during embryonic development. Thus, modifications to DNA and histones can be passed along to future generations of cells—that is, they can be inherited. Inheritance of traits transmitted by mechanisms not directly involving the nucleotide sequence is called **epigenetic inheritance**. Whereas mutations in the DNA are permanent changes, modifications to the chromatin, which do not affect the sequence of DNA itself, can be reversed by processes that are not yet fully understood.

Researchers are amassing more evidence for the importance of epigenetic information in the regulation of gene expression. Epigenetic variations might help explain differences in identical twins. For example, it is often the case that one identical twin acquires a genetically influenced disease, such as schizophrenia, but the other does not, despite their identical genomes. Researchers suspect that epigenetics may be behind such differences. Alterations in normal patterns of DNA methylation are seen in some cancers, where they are associated with inappropriate gene expression. Evidently, enzymes that modify chromatin structure are integral parts of the eukaryotic cell's machinery for regulating transcription.

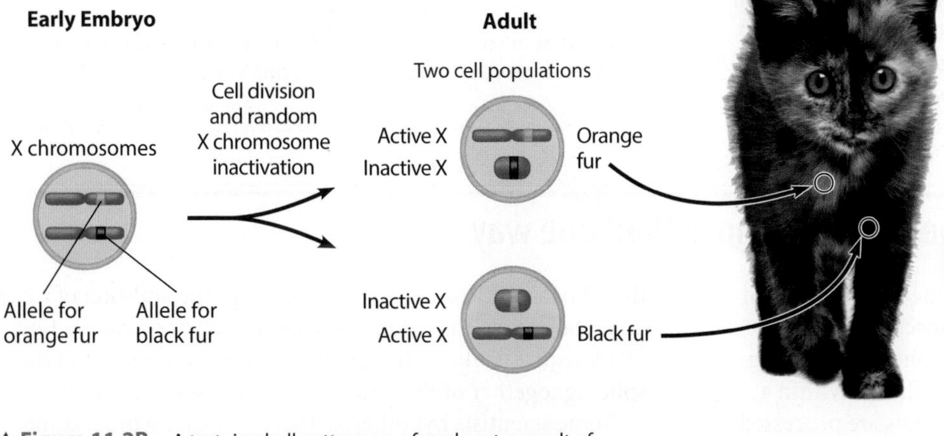

X Inactivation Gene regulation sometimes occurs at the whole chromosome level. For example, female mammals (including humans) inherit two X chromosomes, whereas males inherit only one. So why don't females make twice as much of the proteins encoded by genes on the X chromosome compared to the amounts in males? It turns out that in female mammals, one X chromosome in each somatic (body) cell is chemically modified and highly compacted, rendering it almost entirely inactive. Inactivation of an X chromosome involves modification of the DNA (by, for example, methylation) and the histone proteins that help compact it. A specific gene on the X chromosomes ensures that only one of them will be inactivated. This **X chromosome inactivation** is initiated early in embryonic development, when one of the two X chromosomes in each cell is inactivated at random. As a result, the cells of females and males have the same effective dose (one copy) of these genes. The inactive X in each cell of a female condenses into a compact object called a **Barr body**.

Which X chromosome is inactivated is a matter of chance in each embryonic cell, but once an X chromosome is inactivated, all descendant cells have the same copy turned off—an example of epigenetic inheritance. Consequently, females consist of a mosaic of two types of cells: those with the active X derived from the father and those with the active X derived from the mother. If a female is heterozygous for a gene on the X chromosome (a sex-linked gene; see Module 9.21), about half her cells will express one allele, while the others will express the alternate allele.

A striking example of this mosaic phenomenon is the tortoiseshell cat, which has orange and black patches of fur (**Figure 11.2B**). The relevant fur-color gene is on the X chromosome, and the tortoiseshell phenotype requires the presence of two different alleles, one for orange fur and one for black fur. Normally, only females can have both alleles because only they have two X chromosomes. If a female is heterozygous for the tortoiseshell gene, she will have the tortoiseshell phenotype. Orange patches are formed by populations of cells in which the X chromosome with the orange allele is active; black patches have cells in which the X chromosome with the black allele is active.

In this module, we have seen how the physical structure of chromosomes can affect which genes are expressed in a cell. In the next module, we discuss mechanisms for regulating genes in active, unpacked chromosomes.

Early Embryo
X chromosomes
Allele for orange fur
Allele for black fur
Cell division and random X chromosome inactivation
Adult
Two cell populations
Active X
Inactive X
Orange fur
Inactive X
Active X
Black fur

▲ **Figure 11.2B** A tortoiseshell pattern on a female cat, a result of X chromosome inactivation

? **In your body, a nerve cell has a very different structure and performs very different functions than a skin cell. Because the two cell types have the same genes, how can the cells be so different?**

● Each cell type must be expressing certain genes that are present in, but not expressed in, the other cell type.

11.3 Complex assemblies of proteins control eukaryotic transcription

The process of packing and unpacking of chromosomal DNA provides a coarse adjustment for eukaryotic gene expression by making a region of DNA either more or less available for transcription, the synthesis of RNA. The fine-tuning begins with the initiation of transcription. In both prokaryotes and eukaryotes, the initiation of transcription (whether transcription starts or not) is the most important stage for regulating gene expression.

Like prokaryotes (see Module 11.1), eukaryotes use regulatory proteins—activators and repressors—that bind to specific segments of DNA and either promote or block the binding of RNA polymerase, turning the transcription of genes on or off. However, most eukaryotic genes have individual promoters and other control sequences and are not clustered together as in operons.

The current model for the initiation of eukaryotic transcription features an intricate array of regulatory proteins that interact with DNA and with one another to turn genes on or off. In eukaryotes, activator proteins seem to be more important than repressors. A typical animal or plant cell needs to turn on (transcribe) only a small percentage of its genes, those required for the cell's specialized structure and function. Therefore, in multicellular eukaryotes, the "default" state for most genes seems to be "off." Important exceptions include housekeeping genes, those continually active in virtually all cells for routine activities such as glycolysis, which may be in an "on" state by default.

To function, eukaryotic RNA polymerase requires the assistance of proteins called **transcription factors**. In the model depicted in **Figure 11.3**, the first step in initiating gene transcription is the binding of activator proteins ( ) to DNA control sequences called **enhancers** ( ). In contrast to the operators found within prokaryotic operons, enhancers are usually located far away on the chromosome from the gene they help regulate. Next, a DNA-bending protein brings the bound activators closer to the promoter. Once the DNA is bent, the bound activators interact with other transcription factor proteins ( ), which then bind as a complex at the gene's promoter ( ). This large assembly of proteins facilitates the correct attachment of RNA polymerase to the promoter and the initiation of transcription. Only when the complete complex of proteins has assembled can the polymerase begin to move along the gene, producing an RNA strand. As shown in the figure, multiple enhancers and activators may be involved in turning on a single gene.

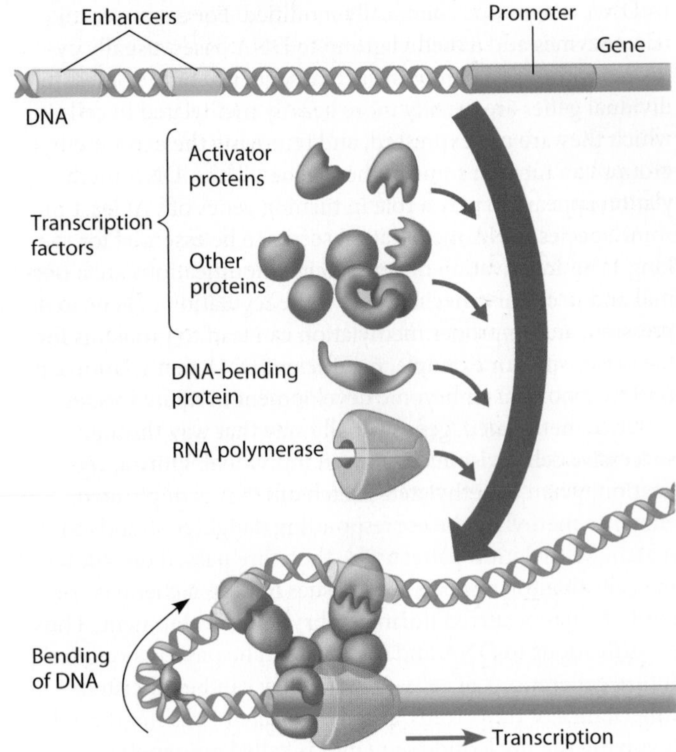

Enhancers
Promoter
Gene
DNA
Transcription factors
Activator proteins
Other proteins
DNA-bending protein
RNA polymerase
Bending of DNA
Transcription

▲ **Figure 11.3** A model for the turning on of a eukaryotic gene

Try This Explain how enhancer sequences can promote transcription of a gene that is located far away on the DNA molecule.

Genes coding for the enzymes of a metabolic pathway are often scattered across different chromosomes. How can a eukaryotic cell turn on or off all functionally related genes at the same time? The key to coordinated gene expression in eukaryotes is often the association of a specific combination of control sequences with every gene of a particular metabolic pathway. Copies of the activators that recognize these control sequences bind to them all at once (because they are all identical), promoting simultaneous transcription of the genes, no matter where they are in the genome. In the next module, we consider another method of gene regulation that is unique to eukaryotes.

? **What must occur before RNA polymerase can bind to a promoter and transcribe a specific eukaryotic gene?**

● Transcription factors must bind to enhancers to facilitate the attachment of RNA polymerase to the promoter.

11.4 Eukaryotic RNA may be spliced in more than one way

Although regulation of transcription is the most important step in gene regulation in most cells, transcription alone does not equal gene expression. Several other points along the path from DNA to protein can be regulated. Within a eukaryotic cell, for example, RNA transcripts are processed into mRNA before moving to the cytoplasm for translation by the ribosomes. RNA processing includes the addition of a cap and a tail, as well as the removal of any introns—noncoding DNA segments that interrupt the genetic message—and the splicing together of the remaining exons (see Module 10.10).

Some scientists hypothesize that the splicing process may help control the flow of mRNA from nucleus to cytoplasm because until splicing is completed, the RNA is attached to the molecules of the splicing machinery and cannot pass through the nuclear pores. Moreover, in some cases, the cell can carry out splicing in more than one way, generating different mRNA molecules from the same RNA transcript. Notice in **Figure 11.4**, for example, that one mRNA molecule ends up with the green exon and the other with the brown exon. With this sort of **alternative RNA splicing**, an organism can produce more than one type of polypeptide from a single gene.

One interesting example of two-way splicing is found in the *Drosophila* fruit fly, where the differences between males and females are largely due to different patterns of RNA splicing. In addition, researchers have found a gene in *Drosophila* that, through the alternate splicing of many exons, produces more than 17,500 proteins, each of which is found in the membrane of a different nerve cell where it acts as an identification marker. In humans, more than 90% of protein-coding genes appear to undergo alternate splicing.

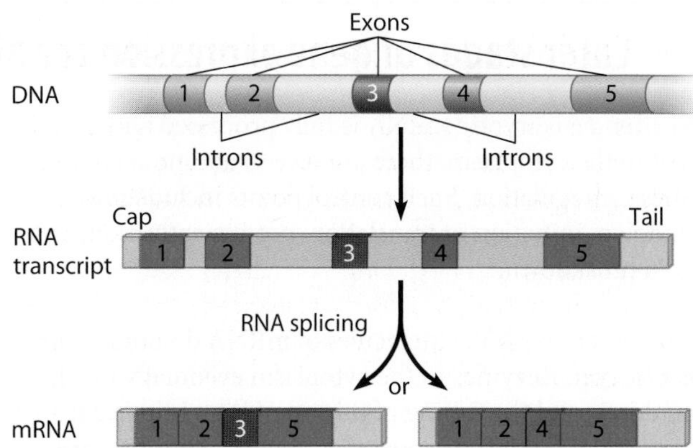
Exons
DNA
1
2
3
4
5
Introns
Introns
Cap
Tail
RNA transcript
RNA splicing
or
mRNA

▲ **Figure 11.4** The production of two different mRNAs from the same gene

? How is it possible that just under 21,000 human genes can produce more than 100,000 polypeptides?

● Through alternate splicing: Each kind of polypeptide is encoded by an mRNA molecule containing a different combination of exons.

11.5 Small RNAs play multiple roles in controlling gene expression

Genome research has revealed that only 1.5% of the human genome—and a similarly small percentage of the genomes of many other multicellular eukaryotes—codes for proteins. Another very small fraction of DNA consists of genes for ribosomal RNA and transfer RNA. Until recently, most of the remaining DNA was thought to be untranscribed and therefore considered to be lacking any genetic information. However, a flood of recent data has contradicted this view. It turns out that a significant amount of the genome is transcribed into functioning but non–protein-coding RNAs, including a variety of small RNAs. Although many questions about the functions of these RNAs remain unanswered, researchers are uncovering more evidence of their biological roles every day.

In 1993, researchers discovered small RNA molecules, called **microRNAs (miRNAs)**, that can bind to complementary sequences on mRNA molecules **(Figure 11.5)**. Each miRNA, typically about 22 nucleotides long, ❶ forms a complex with one or more proteins. The miRNA-protein complex can ❷ bind to any mRNA molecule with 7 to 8 nucleotides of complementary sequence. ❸ Then the complex either degrades the target mRNA or blocks its translation. It has been estimated that miRNAs may regulate the expression of at least one-half of all human genes, a striking figure given that miRNAs were unknown 20 years ago.

Researchers can take advantage of miRNA to artificially control gene expression. For example, injecting miRNA into a cell can turn off expression of a gene with a sequence that matches the miRNA, a procedure called **RNA interference (RNAi)**. The RNAi pathway may have evolved as a natural defense against infection by certain viruses with RNA genomes (see Chapter 10). In 2006, two American researchers were awarded a Nobel Prize for their discovery and categorization of RNA interference.

Biologists are excited about these recent discoveries, which hint at a large, diverse population of RNA molecules in the cell that play crucial roles in regulating gene expression—and have gone largely unnoticed until very recently. Our new understanding may lead to important clinical applications. For example, in 2009, researchers discovered a particular miRNA that is essential for the proper functioning of the pancreas. Without it, insulin-producing beta cells die, which can lead to diabetes.

? If a gene has the sequence AATTCGCG, what would be the sequence of an miRNA that turns off the gene?

● The gene will be transcribed as the mRNA sequence UUAAGCGC; an miRNA of sequence AAUUCGCG would bind to and disable this mRNA.

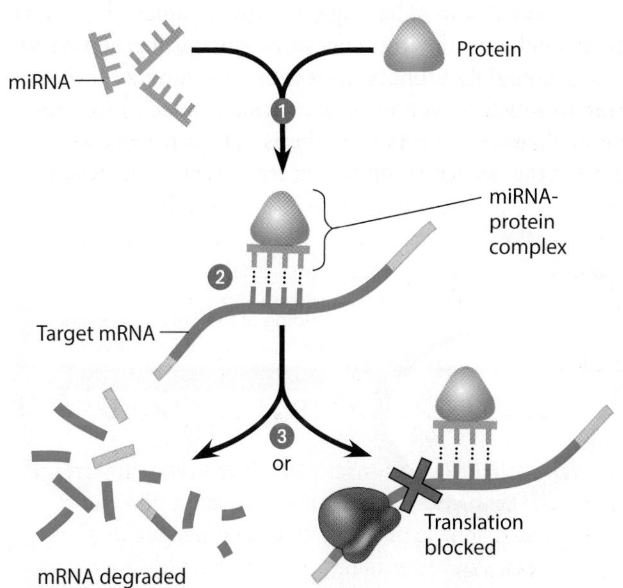
miRNA
Protein
1
miRNA-protein complex
2
Target mRNA
3
or
mRNA degraded
Translation blocked

▲ **Figure 11.5** Mechanisms of RNA interference

11.6 Later stages of gene expression are also subject to regulation

Even after a eukaryotic mRNA is fully processed and transported to the cytoplasm, there are several additional opportunities for regulation. Such control points include mRNA breakdown, initiation of translation, protein activation, and protein breakdown.

Breakdown of mRNA Molecules of mRNA do not remain intact forever. Enzymes in the cytoplasm eventually break them down, and the timing of this event is an important factor regulating the amounts of various proteins that are produced in the cell. Long-lived mRNAs can be translated into many more protein molecules than short-lived ones. Prokaryotic mRNAs have very short lifetimes; they are typically degraded by enzymes within a few minutes after their synthesis. This is one reason bacteria can change their protein production so quickly in response to environmental changes. In contrast, the mRNAs of eukaryotes have lifetimes from hours to weeks.

A striking example of long-lived mRNA is found in vertebrate red blood cells, which manufacture large quantities of the protein hemoglobin. In most species of vertebrates, the mRNAs for hemoglobin are unusually stable. They probably last as long as the red blood cells that contain them—about a month or a bit longer in reptiles, amphibians, and fishes—and are translated again and again. Mammals are an exception. When their red blood cells mature, they lose their ribosomes (along with their other organelles) and thus cease to make new hemoglobin. However, mammalian hemoglobin itself lasts about as long as the red blood cells last: around four months.

Initiation of Translation The process of translating an mRNA into a polypeptide also offers opportunities for regulation. Among the molecules involved in translation are a great many proteins that control the start of polypeptide synthesis. Red blood cells, for instance, have an inhibitory protein that prevents translation of hemoglobin mRNA unless the cell has a supply of heme, the iron-containing chemical group essential for hemoglobin function. (It is the iron atom of the heme group to which oxygen molecules actually attach.) By controlling the start of protein synthesis, cells can avoid wasting energy if the needed components are currently unavailable.

Protein Activation After translation is complete, some polypeptides require alterations before they become functional. Post-translational control mechanisms in eukaryotes often involve the cleavage (cutting) of a polypeptide to yield a smaller final product that is the active protein, able to carry out a specific function in the organism. In **Figure 11.6**, we see the example of the protein hormone insulin. Insulin is synthesized in the cells of the pancreas as one long polypeptide that has no hormonal activity. After translation is completed, the polypeptide folds up, and covalent bonds form between the sulfur (S) atoms of sulfur-containing amino acids (see Figure 3.12A, which shows S—S bonds in another protein). Two H atoms are lost as each S—S bond forms, linking together parts of the polypeptide in a specific way. Finally, a large center portion is cut away, leaving two shorter chains held together by the sulfur linkages. This combination of two shorter polypeptides is the form of insulin that functions as a hormone. By controlling the timing of such protein modifications, the rate of insulin synthesis can be fine-tuned.

Protein Breakdown The final control mechanism operating after translation is the selective breakdown of proteins. Although mammalian hemoglobin may last as long as the red blood cell housing it, the lifetimes of many other proteins are closely regulated. Some of the proteins that trigger metabolic changes in cells are broken down within a few minutes or hours. This regulation allows a cell to adjust the kinds and amounts of its proteins in response to changes in its environment. It also enables the cell to maintain its proteins in prime working order. Indeed, when proteins are damaged, they are usually broken down right away and replaced by new ones that function properly.

Over the last five modules, you have learned about several ways that eukaryotes can control gene expression. The next module summarizes all of these processes.

? Review Figure 11.6. If the enzyme responsible for cleaving inactive insulin is deactivated, what effect will this have on the form and function of insulin?

● The final molecule will have a shape different from that of active insulin and therefore will not be able to function as a hormone.

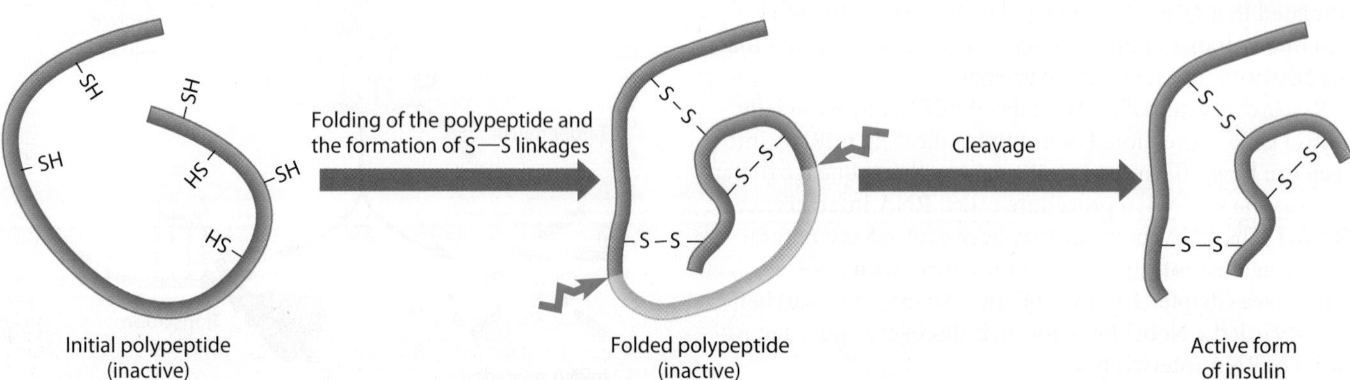

SH
SH
SH
HS
HS
SH
Folding of the polypeptide and the formation of S—S linkages
S–S
S–S
S–S
Cleavage
S–S
S–S
S–S
Initial polypeptide (inactive)
Folded polypeptide (inactive)
Active form of insulin

▲ **Figure 11.6** Protein activation: the role of polypeptide cleavage in producing the active insulin protein

VISUALIZING THE CONCEPT

11.7 Multiple mechanisms regulate gene expression in eukaryotes

This summary of eukaryotic gene expression highlights the multiple control points where the process can be turned on or off, speeded up, or slowed down. Although many control points are shown, only a few of them may be important for any particular protein.

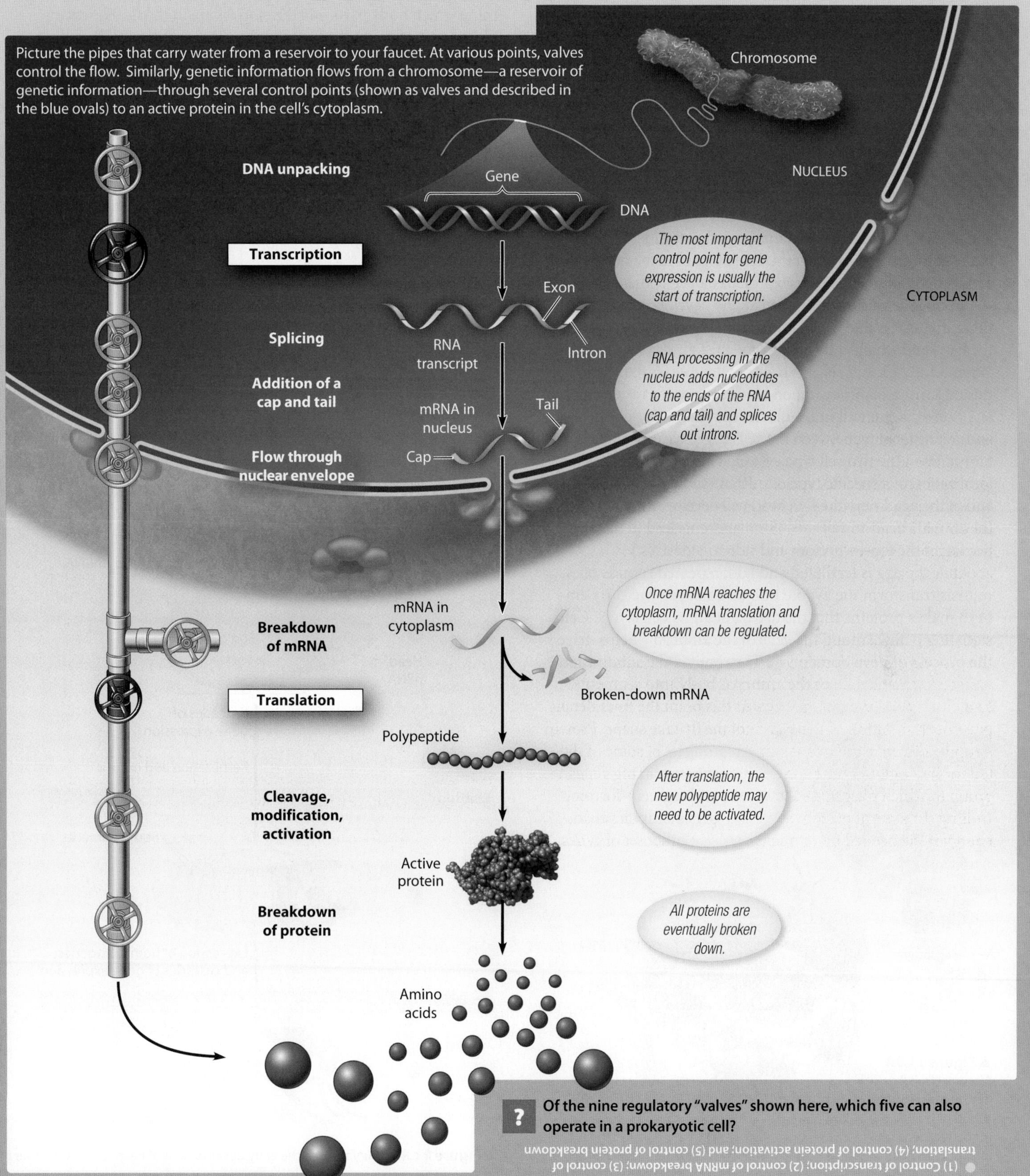

Picture the pipes that carry water from a reservoir to your faucet. At various points, valves control the flow. Similarly, genetic information flows from a chromosome—a reservoir of genetic information—through several control points (shown as valves and described in the blue ovals) to an active protein in the cell's cytoplasm.
Chromosome
NUCLEUS
CYTOPLASM
DNA unpacking
Gene
DNA
Transcription
The most important control point for gene expression is usually the start of transcription.
Exon
Splicing
RNA transcript
Intron
Addition of a cap and tail
mRNA in nucleus
Tail
RNA processing in the nucleus adds nucleotides to the ends of the RNA (cap and tail) and splices out introns.
Flow through nuclear envelope
Cap
mRNA in cytoplasm
Breakdown of mRNA
Once mRNA reaches the cytoplasm, mRNA translation and breakdown can be regulated.
Broken-down mRNA
Translation
Polypeptide
Cleavage, modification, activation
After translation, the new polypeptide may need to be activated.
Active protein
Breakdown of protein
All proteins are eventually broken down.
Amino acids

? Of the nine regulatory "valves" shown here, which five can also operate in a prokaryotic cell?

(1) Control of transcription; (2) control of mRNA breakdown; (3) control of translation; (4) control of protein activation; and (5) control of protein breakdown

11.8 Cell signaling and waves of gene expression direct animal development

In eukaryotes, cellular differentiation results from the selective turning on and off of genes. During the life cycle of a multicellular eukaryote, cellular differentiation by selective gene expression is most vital during the development of an embryo from a zygote. Waves of gene expression, with the protein products of one set of genes activating other sets of genes, are a common mechanism of development.

Some of the first glimpses into the relationship between gene expression and embryonic development came from studies of mutants of the fruit fly *Drosophila melanogaster* (see Module 9.18). **Figure 11.8A** shows the heads of two fruit flies. The one on the right, a mutant, developed in a strikingly abnormal way: It has two legs where its antennae should be! Research on this and other developmental mutants has led to the identification of many of the genes that program development in the normal fly. This genetic approach has revolutionized developmental biology.

Among the earliest events in fruit fly development are those that determine which end of the egg cell will become the head and which end will become the tail. As you can see in **Figure 11.8B**, ❶ these events occur in the ovaries of the mother fly and involve communication between an unfertilized egg cell and cells adjacent to it in its follicle (egg chamber). The back-and-forth signaling between the cells triggers expression of certain genes in the two cell types. ❷ One important result is the localization of a specific type of mRNA (shown in pink) at the end of the egg where the fly's head will develop, thus defining the animal's head-to-tail axis. (Similar events lead to the positioning of the top-to-bottom and side-to-side axes.)

After the egg is fertilized and laid, repeated rounds of mitosis transform the zygote into an embryo. The early embryo makes proteins that diffuse through its cell layers. Cell signaling—now among the cells of the embryo—helps drive the process of development. ❸ The result is the subdivision of the embryo's body into segments. At this point the finer details of the fly take shape. Protein products of some of the axis-specifying genes and segment-forming genes activate yet another set of genes, called homeotic genes. A **homeotic gene** is a master control gene that regulates the "batteries" of other genes that determine the anatomy of parts of the body and in this example determine which body parts will develop where in the fly. For example, one set of homeotic genes in fruit flies instructs cells in the segments of the head and thorax (midbody) to form antennae and legs, respectively. (See Module 27.14 for further discussion of homeotic genes.) ❹ The eventual outcome is an adult fly. Notice that the adult's body segments correspond to those of the embryo in step 3. It was mutation of a homeotic gene that was responsible for the abnormal fly in Figure 11.8A.

How can scientists study the expression of genes within living systems? In the next module, we'll look at how DNA technology can help elucidate gene expression in any cell.

Eye

Antenna

SEM 50×

Extra pair of legs

SEM 50×

▲ **Figure 11.8A** A normal fruit fly (left) compared with a mutant fruit fly (right) with legs coming out of its head

? **What determines which end of a developing fruit fly will become the head?**

● A specific kind of mRNA localizes at the end of the unfertilized egg that will become the head.

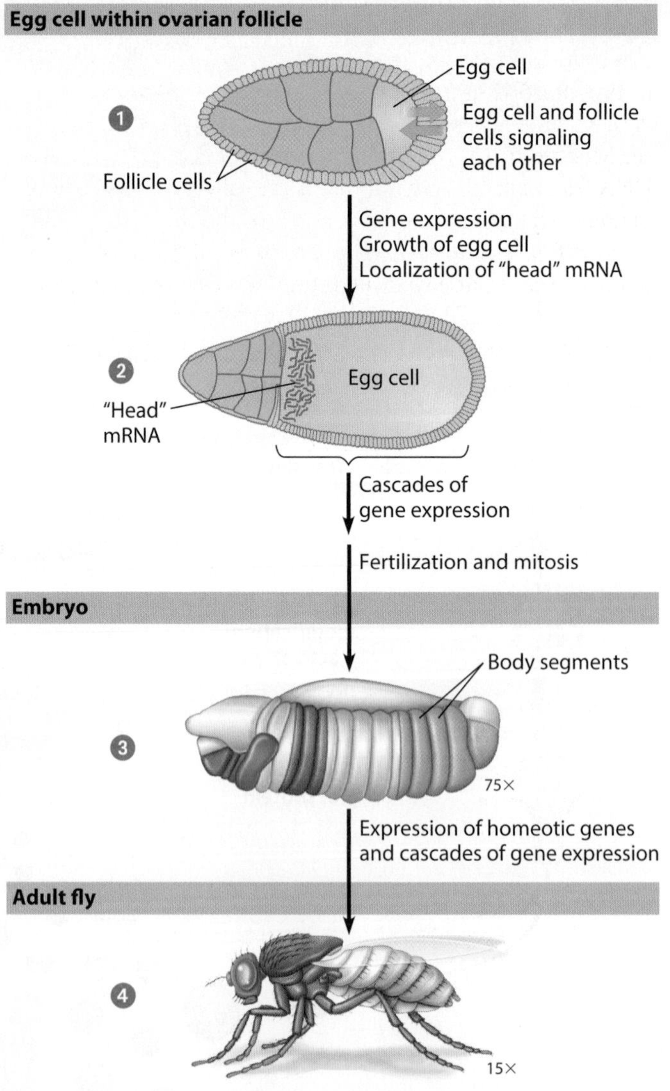

Egg cell within ovarian follicle
1
Egg cell
Egg cell and follicle cells signaling each other
Follicle cells
Gene expression
Growth of egg cell
Localization of "head" mRNA
2
"Head" mRNA
Egg cell
Cascades of gene expression
Fertilization and mitosis
Embryo
Body segments
3
75×
Expression of homeotic genes and cascades of gene expression
Adult fly
4
15×

▲ **Figure 11.8B** Key steps in the early development of head-tail axis in a fruit fly

11.9 Scientists use DNA microarrays to test for the transcription of many genes at once

CONNECTION

Biologists today are working hard to learn how genes act together within a functioning organism. Now that a number of whole genomes have been sequenced (see Module 12.17), it is possible to study the expression of large groups of genes. For example, researchers can investigate which genes are transcribed in different situations, such as in different tissues or at different stages of development. They can also look for groups of genes that are expressed in a coordinated manner, with the aim of identifying networks of gene expression across an entire genome.

Genome-wide expression studies are made possible by DNA microarrays. A **DNA microarray** consists of a glass or plastic surface with tiny amounts of thousands of different kinds of single-stranded DNA fragments attached to microscopic wells in a tightly spaced array, or grid. (A DNA microarray is also called a DNA chip or gene chip by analogy to a computer chip.) Each fixed DNA fragment is obtained from a particular gene; a single microarray thus carries DNA from thousands of genes, perhaps even all the genes of an organism.

Figure 11.9 outlines how microarrays are used. ❶ A researcher collects all of the mRNA transcribed from genes in a particular type of cell at a given moment. ❷ This collection of mRNAs is mixed with reverse transcriptase (a viral enzyme that produces DNA from an RNA template; see Module 10.20) to produce a mixture of single-stranded DNA fragments. These fragments are called cDNAs (complementary DNAs) because each one is complementary to one of the mRNAs. The cDNAs are produced in the presence of nucleotides that have been modified to fluoresce (glow). The fluorescent cDNA collection thus represents all of the genes that are being actively transcribed in that particular cell at that particular time. ❸ A small amount of the fluorescently labeled cDNA mixture is added to each of the wells in the microarray. If a molecule in the cDNA mixture is complementary to a DNA fragment at a particular location on the grid, the cDNA molecule binds to it, becoming fixed there. ❹ After unbound cDNA is rinsed away, the remaining cDNA produces a detectable glow in the microarray. The pattern of glowing spots enables the researcher to determine which genes were being transcribed in the starting cells.

DNA microarrays hold great promise in medical research. Many cancers have a variety of subtypes with different patterns of gene expression that can be identified with DNA microarrays. For example, one study showed that DNA microarray data can classify different types of leukemia into specific subtypes based on the activity of 17 genes. This information can be used to predict which of several available regimens of chemotherapy is likely to be most effective. In addition, comparing patterns of gene expression in breast cancer tumors and noncancerous breast tissue has resulted in more informed and effective treatment protocols. Some oncologists predict that DNA microarrays will usher in a new era in which medical treatment is personalized to each patient (see Module 8.10).

DNA microarrays can also reveal general profiles of gene expression over the lifetime of an organism. For example, researchers performed DNA microarray experiments on more than 90% of the genes of the nematode worm *Caenorhabditis elegans* during every stage of its life cycle. The results showed that expression of nearly 60% of the *C. elegans* genes changed dramatically during development. This study supported the model held by most developmental biologists that embryonic development of multicellular eukaryotes involves a complex and elaborate program of gene expression, rather than simply the expression of a small number of important genes.

? **What can be learned from a DNA microarray?**

● Which genes are active (transcribed) in a particular sample of cells

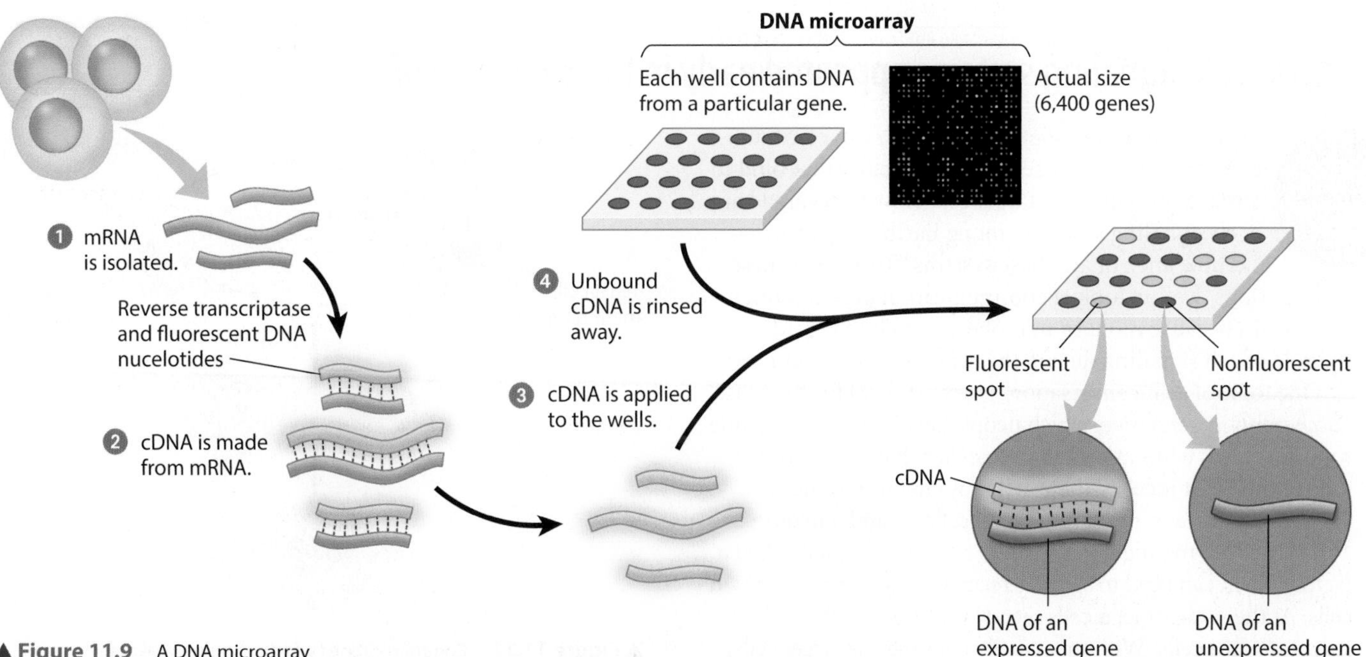
DNA microarray
Each well contains DNA from a particular gene.
Actual size (6,400 genes)
❶ mRNA is isolated.
Reverse transcriptase and fluorescent DNA nucelotides
❷ cDNA is made from mRNA.
❸ cDNA is applied to the wells.
❹ Unbound cDNA is rinsed away.
Fluorescent spot
Nonfluorescent spot
cDNA
DNA of an expressed gene
DNA of an unexpressed gene

▲ **Figure 11.9** A DNA microarray

11.10 Signal transduction pathways convert messages received at the cell surface to responses within the cell

Within a multicellular organism, cells must be able to communicate messages that will coordinate gene expression. Cell-to-cell signaling via proteins or other kinds of molecules carrying messages from signaling cells to receiving (target) cells is a key mechanism in the coordination of cellular activities. In most cases, a signaling molecule acts by binding to a receptor protein in the plasma membrane and initiating a signal transduction pathway in the target cell.

A **signal transduction pathway** is a series of molecular changes that converts a signal on a target cell's surface to a specific response inside the cell (**Figure 11.10**). ❶ The cell sending a message secretes a signaling molecule. ❷ This molecule binds to a specific receptor protein embedded in the target cell's plasma membrane. ❸ The binding activates the first in a series of relay proteins within the target cell. Each relay molecule activates another. ❹ The last relay molecule in the series activates a transcription factor that ❺ triggers transcription of a specific gene. ❻ Translation of the mRNA produces a protein that performs the function originally called for by the signal.

Signal transduction pathways are crucial to many cellular functions. Throughout your study of biology, you'll see the importance of signal transduction pathways again and again. We encountered them when we studied the cell cycle control system (Module 8.8); we'll revisit them when we discuss cancer later in this chapter (see, for example, Module 11.17); and we'll see how they relate to hormone function in animals (Chapter 26) and plants (Chapter 33).

? To turn on a gene, must a signal molecule actually enter a target cell?

No: a signal molecule can bind to a receptor protein in the outer membrane of the target cell and trigger a signal transduction pathway that activates transcription factors.

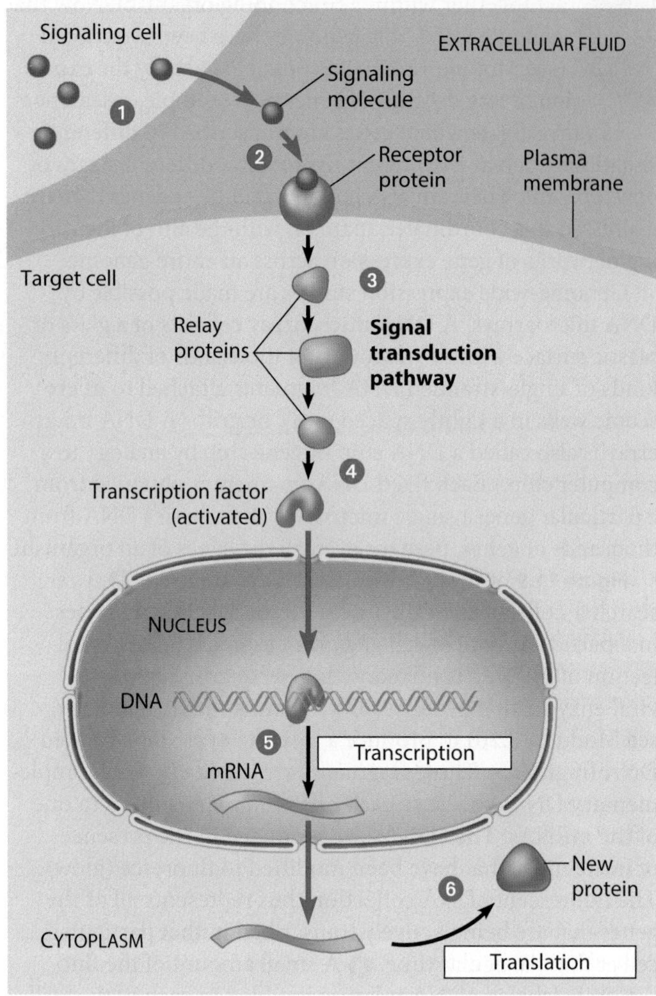

Signaling cell
EXTRACELLULAR FLUID
1
Signaling molecule
2
Receptor protein
Plasma membrane
Target cell
3
Relay proteins
Signal transduction pathway
4
Transcription factor (activated)
NUCLEUS
DNA
5
mRNA
Transcription
6
New protein
CYTOPLASM
Translation

▲ **Figure 11.10** A signal transduction pathway that turns on a gene

11.11 Cell-signaling systems appeared early in the evolution of life

EVOLUTION CONNECTION

As explained in Module 11.10, one cell can communicate with another by secreting molecules that bind to surface proteins on a target cell. How widespread are such signaling systems among Earth's organisms—and how ancient are these systems? To answer these questions, we can look at communication between microorganisms, because modern microbes offer clues regarding the role of cell signaling during the evolution of life on Earth.

One topic of cell "conversation" is sex—at least for the yeast *Saccharomyces cerevisiae,* which people have used for millennia to make bread, wine, and beer. Researchers have learned that cells of this yeast identify their mates by chemical signaling. There are two sexes, or mating types, called **a** and **α** (**Figure 11.11**). Cells of mating type **a** secrete a chemical signal called **a** factor, which can bind to specific receptor proteins on nearby **α** cells. At the same time, **α** cells secrete α factor, which binds to receptors on **a** cells. Without actually entering the target cells,

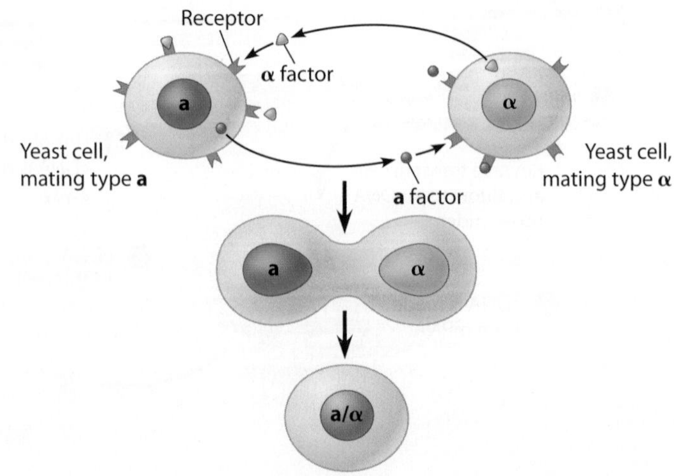

Receptor
α factor
a
α
Yeast cell, mating type **a**
Yeast cell, mating type α
a factor
a
α
a/α

▲ **Figure 11.11** Communication between mating yeast cells

the two mating factors cause the cells to grow toward each other and bring about other cellular changes. The result is the fusion, or mating, of two cells of opposite type. The resulting **a/α** cell contains all the genes of both original cells, a combination of genetic resources that provides advantages to the cell's descendants, which arise by subsequent cell divisions.

Extensive studies across different species have revealed that the molecular details of signal transduction in yeast and mammals are strikingly similar, even though the last common ancestor of these two groups of organisms lived more than a billion years ago. These similarities—and others more recently uncovered between signaling systems in bacteria and plants—suggest that early versions of the cell-signaling mechanisms used today evolved well before the first multicellular creatures appeared on Earth. Scientists think that signaling mechanisms evolved first in ancient prokaryotes and single-celled eukaryotes and then became adapted for new uses in their multicellular descendants.

? In what sense is the joining of yeast mating types "sex"?

● The process results in the creation of a diploid cell that is a genetic blend of two parental haploid cells.

▷ Cloning of Plants and Animals

11.12 Plant cloning shows that differentiated cells may retain all of their genetic potential

One of the most important take-home lessons from this chapter is that most cells express only a small percentage of their genes. If all genes are still present but some are turned off, have the unexpressed genes become permanently disabled? Or do all genes (even the unexpressed ones) retain the potential to be expressed?

One way to approach these questions is to determine if a differentiated cell can be stimulated to generate a whole new organism. In plants, this ability is common. In fact, if you have ever grown a plant from a small cutting, you've seen evidence that a differentiated plant cell can undergo cell division and give rise to all the tissues of an adult plant. On a larger scale, the technique described in **Figure 11.12** can be used to produce hundreds or thousands of genetically identical plants from the cells of a single plant. For example, when cells from a carrot are transferred to a culture medium, a single cell can begin dividing and eventually grow into an adult plant, a genetic replica of the original. Such an organism, produced through asexual reproduction from a single parent, is called a **clone**. In this context, the term clone refers to an individual created by asexual reproduction (that is, reproduction of a single individual that does not involve fusion of sperm and egg). The fact that a mature plant cell can dedifferentiate, or reverse its differentiation, and then give rise to all the different kinds of specialized cells of a new plant shows that differentiation does not necessarily involve irreversible changes in the plant's DNA.

Plant cloning is used extensively in agriculture. Seedless plants (such as seedless grapes and watermelons) cannot reproduce sexually, leaving cloning as the sole means of mass-producing these common foods. Other plants, such as orchids, reproduce poorly in artificial settings, leaving cloning as the only commercially practical means of production. In other cases, cloning has been used to reproduce a plant with desirable traits, such as high fruit yield or resistance to disease.

But is this sort of cloning possible in animals? A good indication that differentiation need not impair an animal cell's genetic potential is the natural process of **regeneration**, the regrowth of lost body parts. When a salamander loses a leg, for example, certain cells in the leg stump dedifferentiate, divide, and then redifferentiate, giving rise to a new leg. Many animals, especially among the invertebrates (sea stars, for example), can regenerate lost parts. In a few relatively simple animals (such as some sponges), isolated differentiated cells can dedifferentiate and then develop into an entire organism (see Module 27.1). Additional evidence for the complete genetic potential of animal cells comes from cloning experiments, our next topic.

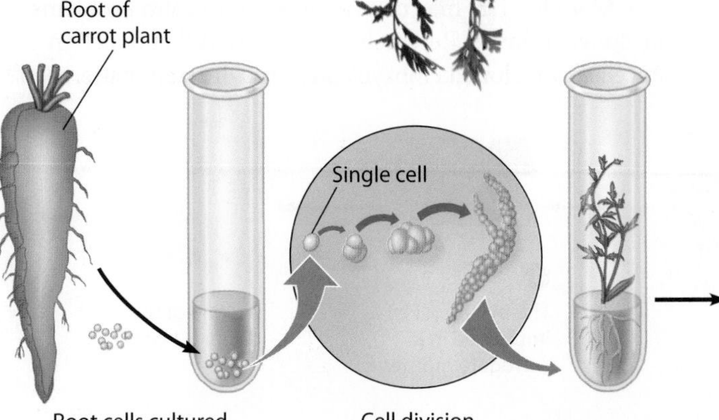
Root of
carrot plant
Single cell
Root cells cultured
in growth medium
Cell division
in culture
Plantlet
Adult plant

▶ Figure 11.12
Growth of a carrot plant from a differentiated root cell

? How does the cloning of plants from differentiated cells support the view that differentiation is based on the control of gene expression rather than on irreversible changes in the genome?

● Cloning shows that all the genes of a fully differentiated plant cell are still present.

11.13 Biologists can clone animals via nuclear transplantation

SCIENTIFIC THINKING

Animal cloning has been achieved through **nuclear transplantation** (Figure 11.13). This method involves 1 replacing the nucleus of an egg cell or a zygote with 2 a nucleus removed from an adult somatic cell. If properly stimulated, the recipient cell may then begin to divide. 3 After a few days, repeated cell divisions form a blastocyst, a hollow ball of about 100 cells. 4 If the animal being cloned is a mammal, the blastocyst is then implanted into the uterus of a surrogate mother. The cloned animal will be genetically identical to the donor of the nucleus—in other words, a clone of the donor. This type of cloning is called **reproductive cloning** because it results in the birth of a new living individual.

Can life be rebooted?

Nuclear transplantation was first performed in the 1950s using cells from frog embryos. In later decades, scientists successfully cloned mammals starting from embryonic cells. However, investigators had success only with very young embryos; they found that the older a donor nucleus, the less chance it could be used to successfully clone an animal. In fact, attempts to clone using nuclei from adult cells failed repeatedly. Something within the nucleus of adult cells was preventing them from being "rebooted" into a cell that would give rise to a new, living animal.

A major breakthrough in cloning came in 1996 when Scottish researcher Ian Wilmut and his colleagues cloned a sheep named Dolly, the first mammal successfully cloned from an adult cell. How did they achieve success? Wilmut and his team hypothesized that the significant changes that occur to chromosomes over the cell cycle require that the phases of the donor nucleus and recipient egg be matched during nuclear transplantation. To achieve synchronization, the researchers grew both nucleus-donor mammary gland cells and nucleus-recipient egg cells in a growth medium that contained only 1/20th the normal nutrients. Faced with starvation, all of the cells switched into the dormant G_0 phase of the cell cycle (see Module 8.8). The researchers then removed the nuclei from the dormant eggs, fused these empty egg cells with nuclei from dormant udder cells, and zapped them with electricity to fuse and reboot them. After several days of growth, the resulting embryos were implanted in the uteruses of surrogate sheep mothers. One of the embryos developed into Dolly.

The pioneering cloning work of Wilmut and colleagues demonstrates how scientific success is often preceded by numerous failures. After decades of failures in previous experiments, the Scottish team produced a total of 277 zygotes, of which only 29 survived to implantation, resulting in just one live birth.

Dolly demonstrated that the differentiation of animal cells is achieved by changes in gene expression, rather than by permanent changes in the genes themselves. This conclusion has numerous practical implications. Since Dolly's landmark birth, researchers have cloned many other mammals, including mice, cats, horses, cows, mules, pigs, rabbits, ferrets, and dogs. Why bother cloning animals? Scientists working in agriculture are cloning farm animals with specific sets of desirable traits in the hope of creating high-yielding herds. The pharmaceutical industry is experimenting with cloning mammals for the production of potentially valuable drugs. For example, researchers have produced pig clones that lack a gene for a protein that can cause immune system rejection in humans. Organs from such pigs may one day be used in human patients who require life-saving transplants.

Conservation biologists hope that reproductive cloning can be used to restock the populations of endangered animals. In addition to the banteng discussed in the chapter introduction, other rare animals have been cloned, including a wild mouflon (a small European sheep), a gaur (an Asian ox), and gray wolves. Despite cloning's potential for increasing the numbers of endangered animals, some conservationists object, arguing that cloning may detract from efforts to preserve natural habitats. Such critics correctly point out that cloning does not increase genetic diversity and is therefore not as beneficial to endangered species as natural reproduction.

Another important consideration when cloning animals is the health of the offspring produced. An increasing body of evidence suggests that cloned animals may be less healthy than those arising from a fertilized egg. In 2003, Dolly was euthanized after suffering complications from a lung disease usually seen only in much older sheep. She was 6 years old, while her breed has a life expectancy of 12 years. Other cloned animals have exhibited arthritis, susceptibility to obesity, pneumonia, liver failure, and premature death. Recent research suggests that the methylation of chromatin (see Module 11.2) may be responsible for health problems in cloned animals. Researchers have found that the DNA in cells from cloned embryos often has different patterns of

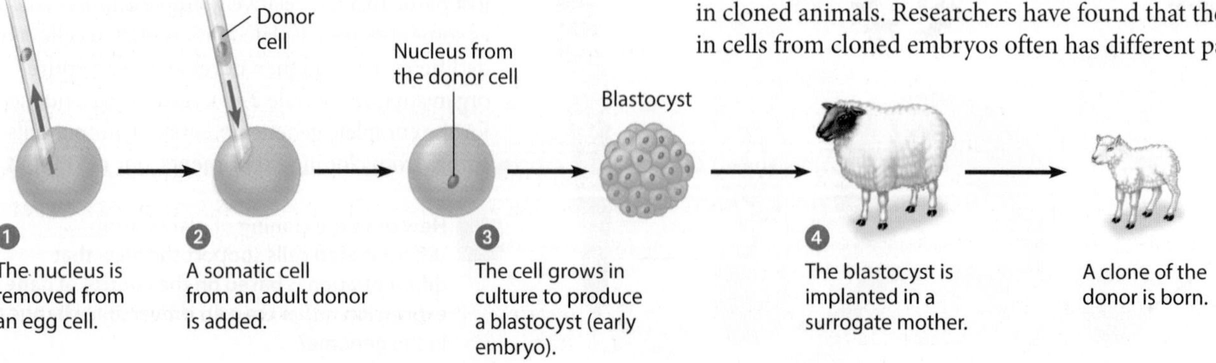
Donor cell
Nucleus from the donor cell
Blastocyst
1
The nucleus is removed from an egg cell.
2
A somatic cell from an adult donor is added.
3
The cell grows in culture to produce a blastocyst (early embryo).
4
The blastocyst is implanted in a surrogate mother.
A clone of the donor is born.

▲ **Figure 11.13** Reproductive cloning via nuclear transplantation methylation than does the DNA in equivalent cells from normal embryos of the same species. Because DNA methylation helps regulate gene expression, misplaced methyl groups may interfere with the pattern of gene expression necessary for normal embryonic development. Researchers are investigating whether chromatin in a donor nucleus can be artificially "rejuvenated" to resemble that of a newly fertilized egg.

The successful cloning of mammals has heightened speculation that humans could be cloned. Critics point out that there are many obstacles—both practical and ethical—to human cloning. Practically, animal cloning is extremely difficult and inefficient. Only a small percentage of cloned embryos (usually less than 10%) develop normally, and those cloned animals that do develop are less healthy than naturally born kin. Indeed, Dolly's creators have since predicted that their cloning technique will never be sufficiently efficient to attempt in humans. Ethically, the discussion about whether people should be cloned, and under what circumstances, is far from settled. Ethical questions also surround the outcomes of therapeutic cloning, our next topic.

? It took three sheep to create Dolly: A blackface sheep donated the egg, a white-faced sheep donated the mammary cells from which the nucleus was taken, and a blackface sheep served as surrogate. What color face should Dolly have?

● White, reflecting the genetic makeup of the nucleus donor

11.14 Therapeutic cloning can produce stem cells with great medical potential

CONNECTION

A blastocyst, made via natural sexual reproduction or produced via nuclear transplantation (see Figure 11.13), can provide **embryonic stem cells (ES cells)**. Within the embryo, ES cells differentiate to give rise to all the specialized cell types of the body. When grown in laboratory culture, embryonic stem cells can divide indefinitely. The right conditions—such as the presence of certain growth factors—can (hypothetically) induce changes in gene expression that cause differentiation of ES cells into a particular cell type (**Figure 11.14**). When the goal is to produce ES cells to use in therapeutic treatments, this process is called **therapeutic cloning**.

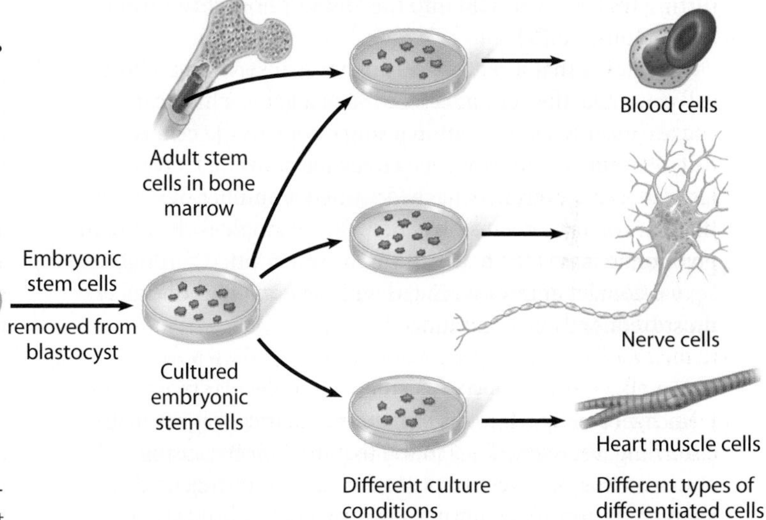

Adult stem cells in bone marrow
Embryonic stem cells
removed from blastocyst
Cultured embryonic stem cells
Blood cells
Nerve cells
Heart muscle cells
Different culture conditions
Different types of differentiated cells

▲ **Figure 11.14** Therapeutic cloning using stem cells

Embryonic stem cells are not the only stem cells available to researchers. Blood collected from the umbilical cord and placenta at birth contains stem cells that are partially differentiated. In 2005, doctors reported that an infusion of umbilical cord blood stem cells from a compatible (but unrelated) donor appeared to cure some babies of Krabbe's disease, a usually fatal inherited disorder of the nervous system. To date, however, most attempts at umbilical cord blood therapy have not been successful. At present, the American Academy of Pediatrics recommends cord blood banking only for babies born into families with a known genetic risk.

The adult body also has stem cells, which serve to replace nonreproducing specialized cells as needed. Because **adult stem cells** are farther along the road to differentiation than ES cells, they can give rise to only a few related types of cells. Adult animals have only tiny numbers of stem cells, but scientists are learning to identify and isolate these cells from various tissues and, in some cases, to grow them in culture. For example, bone marrow contains several types of stem cells, including one that can generate all the different kinds of blood cells. Adult stem cells from donor bone marrow have long been used as a source of immune system cells in patients whose own immune systems have been destroyed by genetic disorders or radiation treatments for cancer. More recently, clinical trials using bone marrow stem cells have shown slight success in promoting regeneration of heart tissue in patients whose hearts have been damaged by heart attacks.

The ultimate aim of therapeutic cloning is to supply cells for the repair of damaged or diseased organs. Some people speculate, for example, that ES cells may one day be used to replace cells damaged by spinal cord injuries or heart attacks. In the future, a donor nucleus from a patient could allow production of embryonic stem cells that are an exact genetic match for that patient and are thus not rejected by his or her immune system.

Opinions vary widely about the morality of therapeutic cloning using embryonic stem cells, which require destruction of the embryo upon harvesting. Because no embryonic tissue is involved in their harvest, adult stem cells are less ethically problematic in therapy than ES cells. On the other hand, many researchers hypothesize that only the more versatile ES cells are likely to lead to groundbreaking advances in human health. The study of stem cells emphasizes the importance of understanding the control of gene expression. In the next section, we'll explore another important implication of gene regulation to human health: cancer.

? In nature, how do embryonic stem cells differ from adult stem cells?

● Embryonic cells give rise to all the different kinds of cells in the body. Adult stem cells generate only a few related types of cells.

▷ The Genetic Basis of Cancer

11.15 Cancer results from mutations in genes that control cell division

Cancer is a set of diseases in which the control mechanisms that normally limit cellular growth have malfunctioned (see Module 8.9). Scientists have learned that such malfunction is often due to changes in gene expression.

The abnormal behavior of cancer cells was observed years before anything was known about the cell cycle, its control, or the role genes play in making cells cancerous. One of the earliest clues to the cancer puzzle was the discovery, in 1911, of a virus that causes cancer in chickens. Recall that viruses are simply molecules of DNA or RNA surrounded by protein and in some cases a membranous envelope. Viruses that cause cancer can become permanent residents in host cells by inserting their nucleic acid into the DNA of host chromosomes (see Module 10.17).

The genes that a cancer-causing virus inserts into a host cell can make the cell cancerous. Such a gene, which can cause cancer when present in a single copy in the cell, is called an **oncogene** (from the Greek *onco*, tumor). Over the last century, researchers have identified a number of viruses that harbor cancer-causing genes. One example is the human papillomavirus (HPV), which can be transmitted through sexual contact and is associated with several types of cancer, most frequently cervical cancer.

Proto-oncogenes In 1976, American molecular biologists J. Michael Bishop, Harold Varmus, and their colleagues made a startling discovery. They found that the cancer-causing chicken virus discovered in 1911 contains an oncogene that is an altered version of a normal chicken gene. Subsequent research has shown that the genomes of many animals, including humans, contain genes that can be converted to oncogenes. A normal gene that has the potential to become an oncogene is called a **proto-oncogene**. (These terms can be confusing, so they bear repeating: a *proto-oncogene* is a normal, healthy gene that, if changed, can become a cancer-causing *oncogene*.) A cell can acquire an oncogene from a virus or from the mutation of one of its own proto-oncogenes.

Searching for their normal role, researchers found that many proto-oncogenes code for proteins that affect the cell cycle. When these proteins are functioning normally, in the right amounts at the right times, they help properly control cell division and cellular differentiation. But changes in these proteins can result in out-of-control growth.

How might a proto-oncogene—a gene that has an essential function in normal cells—become a cancer-causing oncogene? In general, an oncogene arises from a genetic change that leads to an increase either in the amount of the proto-oncogene's protein product or in the activity of each protein molecule. **Figure 11.15A** illustrates three kinds of changes in DNA that can produce oncogenes. Let's assume that the starting proto-oncogene codes for a growth factor, a protein that stimulates cell division. On the left in the figure, a mutation (shown in green) in the proto-oncogene creates an oncogene that codes for a hyperactive protein, one whose stimulating effect is stronger than normal. An error in DNA replication or recombination can generate multiple copies of the gene (as shown in the center of the figure), which are all transcribed and translated; the result is an excess of the normal stimulatory protein. On the right in the figure, the proto-oncogene has been moved from its normal location in the cell's DNA to another location. At its new site, the gene is under the control of a different promoter, one that causes it to be transcribed more often than normal; the normal protein is again made in excess. So in all three cases, normal gene expression is changed, and the cell is stimulated to divide excessively.

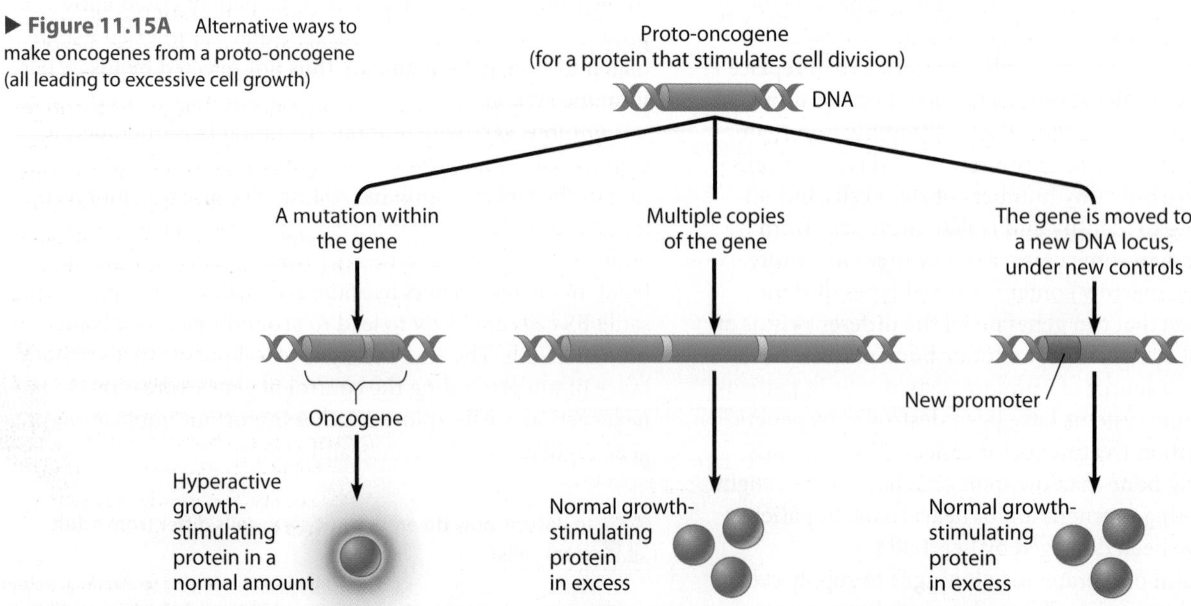
Proto-oncogene
(for a protein that stimulates cell division)
DNA
A mutation within the gene
Multiple copies of the gene
The gene is moved to a new DNA locus, under new controls
Oncogene
New promoter
Hyperactive growth-stimulating protein in a normal amount
Normal growth-stimulating protein in excess
Normal growth-stimulating protein in excess

▶ **Figure 11.15A** Alternative ways to make oncogenes from a proto-oncogene (all leading to excessive cell growth)

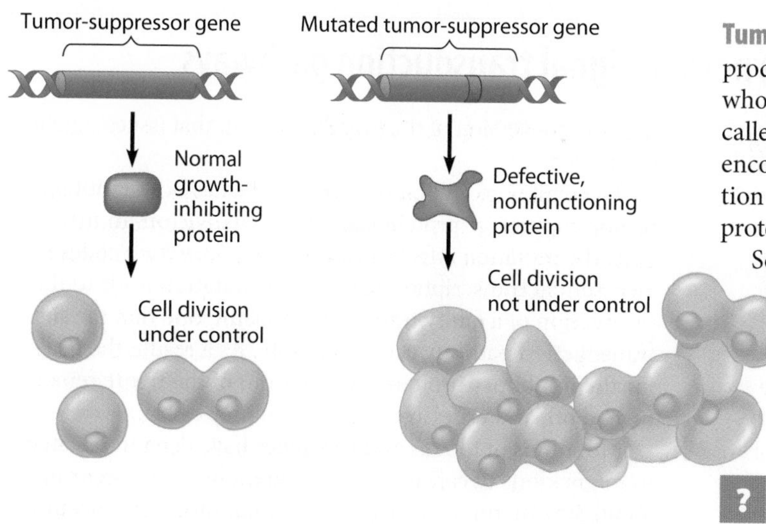

Tumor-suppressor gene
Mutated tumor-suppressor gene
Normal growth-inhibiting protein
Defective, nonfunctioning protein
Cell division under control
Cell division not under control

▲ **Figure 11.15B** The effect of a mutation in a tumor-suppressor gene

Tumor-Suppressor Genes In addition to genes whose products normally *promote* cell division, cells contain genes whose normal products *inhibit* cell division. Such genes are called **tumor-suppressor genes** because the proteins they encode help prevent uncontrolled cell growth. Any mutation that decreases the normal activity of a tumor-suppressor protein may contribute to the onset of cancer (**Figure 11.15B**). Scientists have also discovered a class of tumor-suppressor genes that function in the repair of damaged DNA. When these genes are mutated, other cancer-causing mutations are more likely to accumulate.

How do such DNA mutations lead to the progression of disease? We consider that question next.

? How do proto-oncogenes relate to oncogenes?

A proto-oncogene is a normal gene that, if mutated, can become a cancer-causing oncogene.

11.16 Multiple genetic changes underlie the development of cancer

More than 100,000 Americans will be stricken by cancer of the colon (the main part of the large intestine) this year. One of the best-understood types of human cancer, colon cancer illustrates an important principle about how cancer develops: More than one somatic mutation is needed to produce a full-fledged cancer cell. As in many cancers, the development of malignant (spreading) colon cancer is a gradual process. (See Module 8.9 to review cancer terms.)

Figure 11.16A illustrates the gradual progression from somatic mutation to cancer using colon cancer as an example. ❶ Colon cancer begins when an oncogene arises or is activated through mutation, causing unusually frequent division of apparently normal cells in the colon lining. ❷ Later, additional DNA mutations, such as the inactivation of a tumor-suppressor gene, cause the growth of a small benign tumor (a polyp) in the colon wall. ❸ Still more mutations eventually lead to formation of a malignant tumor, a tumor that has the potential to metastasize (spread). The requirement for several mutations—the actual number is usually around six—explains why cancers can take a long time to develop.

Thus, the development of a malignant tumor is paralleled by a gradual accumulation of mutations that convert proto-oncogenes to oncogenes and knock out tumor-suppressor genes.

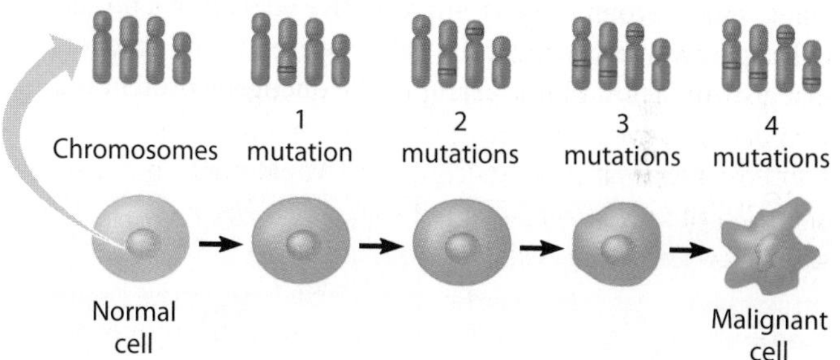

Chromosomes
1 mutation
2 mutations
3 mutations
4 mutations
Normal cell
Malignant cell

▲ **Figure 11.16B** Accumulation of mutations in a cancer cell

Multiple changes must occur at the DNA level for a cell to become fully cancerous. In **Figure 11.16B**, colors distinguish the normal cell (tan) from cells with one or more mutations leading to increased cell division and cancer (red). Once a cancer-promoting mutation occurs (the red band on the chromosome), it is passed to all the descendants of the cell carrying it.

The fact that more than one somatic mutation is generally needed to produce a full-fledged cancer cell may help explain why the incidence of cancer increases with age. If cancer results from an accumulation of mutations that occur throughout life, then the longer we live, the more likely we are to develop cancer. Researchers are steadily cataloguing mutations that cause cancer and placing them in public databases. The hope is that such data will lead to improved treatment strategies and perhaps, someday, a cure.

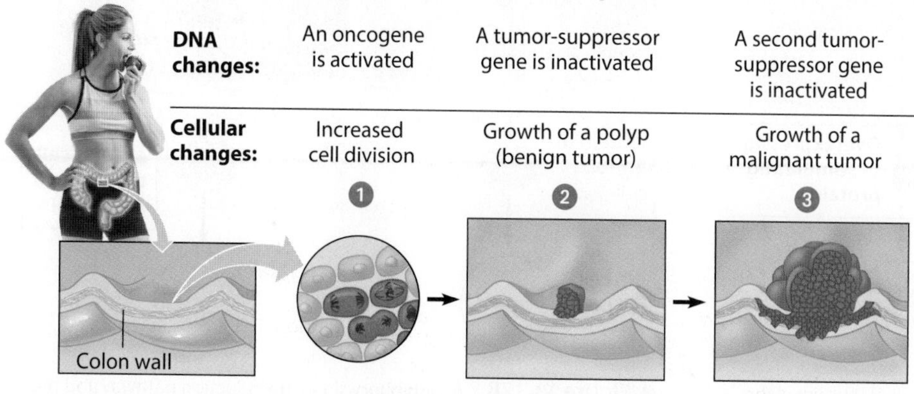

DNA changes:
An oncogene is activated
A tumor-suppressor gene is inactivated
A second tumor-suppressor gene is inactivated
Cellular changes:
Increased cell division
Growth of a polyp (benign tumor)
Growth of a malignant tumor
1
2
3
Colon wall

▲ **Figure 11.16A** Stepwise development of a typical colon cancer

? Epithelial cells, those that line body cavities, are frequently replaced and so divide more often than most other types of body cells. Will epithelial cells become cancerous more or less frequently than other types of body cells?

More frequent cell divisions will result in more frequent mutation and thus a greater chance of cancer.

11.17 Faulty proteins can interfere with normal signal transduction pathways

The figures below (excluding, for the moment, the white boxes) illustrate two types of signal transduction pathways leading to the synthesis of proteins that influence the cell cycle. In **Figure 11.17A**, the pathway leads to the stimulation of cell division. The initial signal is a growth factor (●), and the target cell's ultimate response is the production of a protein that stimulates the cell to divide. By contrast, **Figure 11.17B** shows an inhibitory pathway, in which a growth-*inhibiting* factor (▼) causes the target cell to make a protein that inhibits cell division. In both cases, the newly made proteins function by interacting with components of the cell cycle control system.

Now, let's see what can happen when the target cell undergoes a cancer-causing mutation. The white box in Figure 11.17A shows the protein product of an oncogene resulting from mutation of a proto-oncogene called *ras*. The normal product of *ras* is a relay protein. Ordinarily, a stimulatory pathway like this will not operate unless the growth factor is available. However, an oncogene protein that is a hyperactive version of a protein in the pathway may trigger the pathway even in the absence of a growth factor. In this example, the oncogene protein is a hyperactive version of the *ras* relay protein that issues signals on its own.

The white box in Figure 11.17B indicates how a mutant tumor-suppressor protein can affect cell division. In this case, the mutation affects a gene called *p53*, which codes for an essential transcription factor. This mutation leads to the production of a faulty transcription factor, one that the signal transduction pathway cannot activate. As a result, the gene for the inhibitory protein at the bottom of the figure remains turned off, and excessive cell division may occur.

Mutations of the *ras* and *p53* genes have been implicated in many kinds of cancer. In fact, mutations in *ras* occur in about 30% of human cancers, and mutations in *p53* occur in more than 50%. As we see next, carcinogens are responsible for many mutations that lead to cancer.

? **Contrast the action of an oncogene with that of a cancer-causing mutation in a tumor-suppressor gene.**

● An oncogene encodes an abnormal protein that stimulates cell division via a signal transduction pathway; a mutant tumor-suppressor gene encodes a defective protein unable to function in a pathway that normally inhibits cell division.

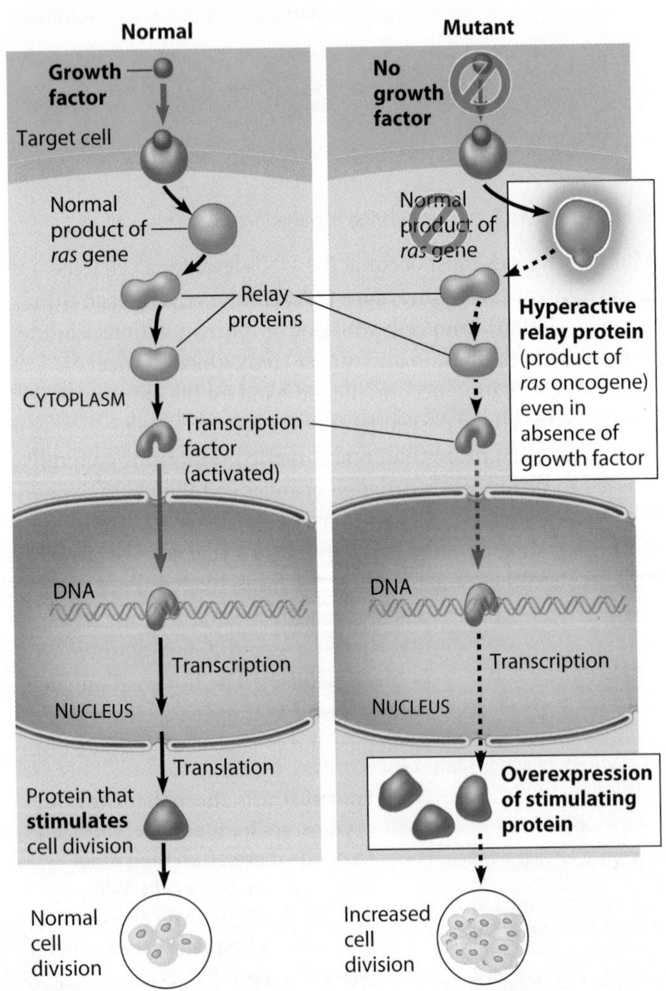

Normal
Mutant
Growth factor
Target cell
No growth factor
Normal product of ras gene
Normal product of ras gene
Relay proteins
Hyperactive relay protein (product of ras oncogene) even in absence of growth factor
CYTOPLASM
Transcription factor (activated)
DNA
DNA
Transcription
Transcription
NUCLEUS
NUCLEUS
Translation
Protein that stimulates cell division
Overexpression of stimulating protein
Normal cell division
Increased cell division

▲ **Figure 11.17A** A stimulatory signal transduction pathway and the effect of an oncogene protein

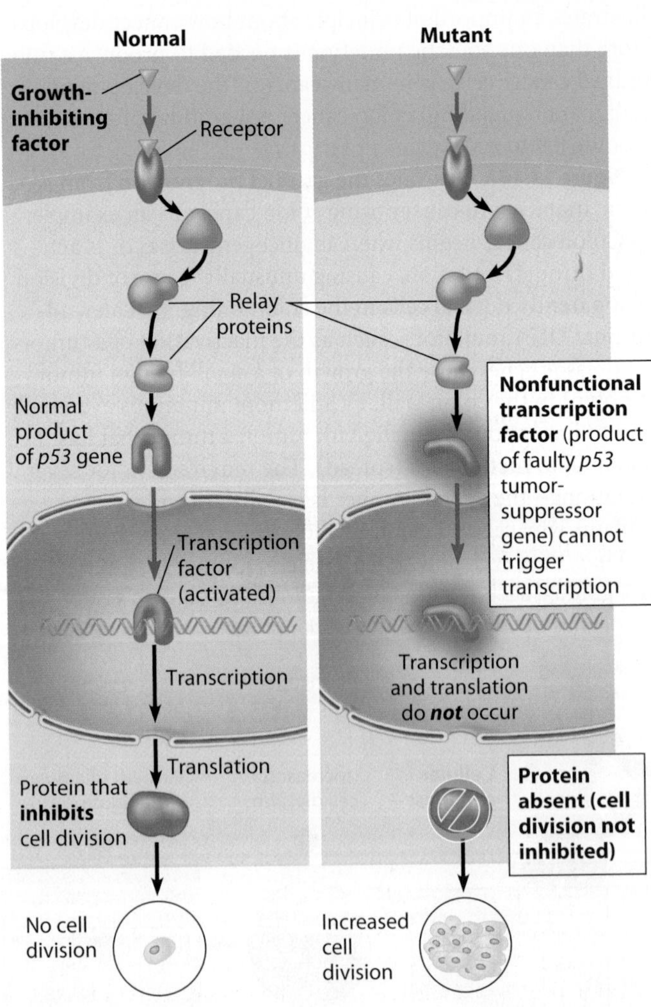

Normal
Mutant
Growth-inhibiting factor
Receptor
Relay proteins
Normal product of p53 gene
Nonfunctional transcription factor (product of faulty p53 tumor-suppressor gene) cannot trigger transcription
Transcription factor (activated)
Transcription
Transcription and translation do not occur
Translation
Protein that inhibits cell division
Protein absent (cell division not inhibited)
No cell division
Increased cell division

▲ **Figure 11.17B** An inhibitory signal transduction pathway and the effect of a faulty tumor-suppressor protein

11.18 Lifestyle choices can reduce the risk of cancer

CONNECTION

Cancer is the second-leading cause of death (after heart disease) in most industrialized nations. Death rates due to certain forms of cancer—including stomach, cervical, and uterine cancers—have decreased in recent years, but the overall cancer death rate is on the rise, currently increasing at about 1% per decade. **Table 11.18** lists the most common cancers in the United States and associated risk factors for each.

The fact that multiple genetic changes are required to produce a cancer cell helps explain the observation that cancers can run in families. An individual inheriting an oncogene or a mutant allele of a tumor-suppressor gene is one step closer to accumulating the necessary mutations for cancer to develop than an individual without any such mutations.

But the majority of cancers are not associated with a mutation that is passed from parent to offspring; they arise from new mutations caused by environmental factors. Agents that alter DNA and make cells cancerous are called **carcinogens**. Most mutagens, substances that cause mutations, are carcinogens. Two of the most potent mutagens are X-rays and ultraviolet radiation in sunlight. X-rays are a significant cause of leukemia and brain cancer. Exposure to UV radiation from the sun is known to cause skin cancer, including a deadly type called melanoma.

The one substance known to cause more cases and types of cancer than any other single agent is tobacco. More people die of lung cancer (nearly 160,000 Americans in 2013) than any other form of cancer. Most tobacco-related cancers come from smoking, but the passive inhalation of secondhand smoke is also a risk. As Table 11.18 indicates, tobacco use, sometimes in combination with alcohol consumption, causes a number of other types of cancer in addition to lung cancer. In nearly all cases, cigarettes are the main culprit, but smokeless tobacco products, such as snuff and chewing tobacco, are linked to cancer of the mouth and throat.

How do carcinogens cause cancer? In many cases, the genetic changes that cause cancer result from decades of exposure to the mutagenic effects of carcinogens. Carcinogens can also produce their effect by promoting cell division. Generally, the higher the rate of cell division, the greater the chance for mutations resulting from errors in DNA replication or recombination. Some carcinogens seem to have both effects. For instance, the hormones linked to breast and uterine cancers promote cell division and may also cause mutations that lead to cancer. In other cases, several different agents, such as viruses and one or more carcinogens, may together produce cancer.

Avoiding carcinogens is not the whole story, because there is growing evidence that some food choices significantly reduce the risk of some cancers. For instance, eating 20–30 grams (g) of plant fiber daily—roughly equal to the amount of fiber in four slices of whole-grain bread, 1 cup of bran flakes, one apple, and $\frac{1}{2}$ cup of carrots combined—and at the same time reducing animal fat intake may help prevent colon cancer. There is also evidence that other substances in fruits and vegetables, including vitamins C and E and certain compounds related to vitamin A, may offer protection against a variety of cancers. Cabbage and its relatives, such as broccoli and cauliflower (see Figure 13.2), are thought to be especially rich in substances that help prevent cancer, although the identities of these substances are not yet established.

The battle against cancer is being waged on many fronts, and there is reason for optimism. It is especially encouraging that we can help reduce our risk of acquiring—and increase our chance of surviving—some of the most common forms of cancer by the choices we make in our daily life. Not smoking, exercising adequately, avoiding overexposure to the sun, and eating a high-fiber, low-fat diet can all help prevent cancer. Furthermore, seven types of cancer can be easily detected: cancers of the skin and oral cavity (via physical exam), breast (via self-exams and mammograms for higher-risk women), prostate (via rectal exam), cervix (via Pap smear), testes (via self-exam), and colon (via colonoscopy). Regular visits to the doctor can help identify tumors early, thereby significantly increasing the possibility of successful treatment.

TABLE 11.18 | CANCER IN THE UNITED STATES

Cancer	Risk Factors	New Cases 2013 (est.)
Prostate	African heritage; possibly dietary fat	239,000
Breast	Estrogen	235,000
Lung	Tobacco smoke	228,000
Colon, rectum	High dietary fat; tobacco smoke; alcohol	143,000
Lymphomas	Viruses (for some types)	79,000
Melanoma of the skin	Ultraviolet light	77,000
Urinary bladder	Tobacco smoke	73,000
Kidney	Tobacco smoke	65,000
Uterus	Estrogen	62,000
Leukemias	X-rays; benzene; virus (for one type)	49,000
Pancreas	Tobacco smoke; obesity	45,000
Oral cavity	Tobacco in various forms; alcohol	41,000
Liver	Alcohol; hepatitis viruses	31,000
Brain and nerve	Trauma; X-rays	23,000
Ovary	Obesity; many ovulation cycles	22,000
Stomach	Table salt; tobacco smoke	22,000
Cervix	Sexually transmitted viruses; tobacco smoke	12,000
Total, including all other types		1,660,000

? **Which of the most common cancers listed in Table 11.18 affect primarily males? Which affect primarily females?**

Males: prostate; females: breast, uterus, cervix

CHAPTER 11 REVIEW

For practice quizzes, BioFlix animations, MP3 tutorials, video tutors, and more study tools designed for this textbook, go to MasteringBiology®

Reviewing the Concepts

Control of Gene Expression (11.1–11.11)

11.1 Proteins interacting with DNA turn prokaryotic genes on or off in response to environmental changes. In prokaryotes, genes for related enzymes are often controlled together in units called operons. Regulatory proteins bind to control sequences in the DNA and turn operons on or off in response to environmental changes.

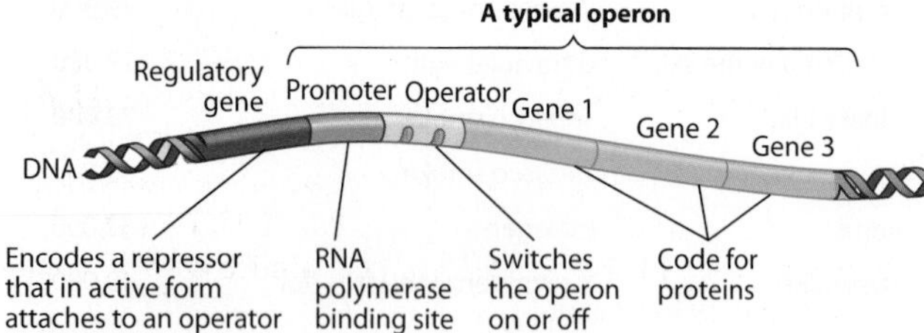
A typical operon
Regulatory gene
Promoter
Operator
Gene 1
Gene 2
Gene 3
DNA
Encodes a repressor that in active form attaches to an operator
RNA polymerase binding site
Switches the operon on or off
Code for proteins

11.2 Chromosome structure and chemical modifications can affect gene expression. In multicellular eukaryotes, different types of cells make different proteins because different combinations of genes are active in each type. A chromosome contains DNA wound around clusters of histone proteins, forming a string of bead-like nucleosomes. DNA packing tends to block gene expression by preventing access of transcription proteins to the DNA. One example of DNA packing is X chromosome inactivation in the cells of female mammals. Chemical modification of DNA bases or histone proteins can result in epigenetic inheritance.

11.3 Complex assemblies of proteins control eukaryotic transcription. A variety of regulatory proteins interact with DNA and with each other to turn the transcription of eukaryotic genes on or off.

11.4 Eukaryotic RNA may be spliced in more than one way. After transcription, alternative RNA splicing may generate two or more types of mRNA from the same transcript.

11.5 Small RNAs play multiple roles in controlling gene expression. MicroRNAs, bound to proteins, can prevent gene expression by forming complexes with mRNA molecules.

11.6 Later stages of gene expression are also subject to regulation. The lifetime of an mRNA molecule helps determine how much protein is made, as do factors involved in translation. A protein may need to be activated in some way, and eventually the cell will break it down.

11.7 Multiple mechanisms regulate gene expression in eukaryotes. Gene expression can be regulated multiple ways within both the nucleus and cytoplasm.

11.8 Cell signaling and waves of gene expression direct animal development. A series of RNAs and proteins produced in the embryo control the development of an animal from a fertilized egg.

11.9 Scientists use DNA microarrays to test for the transcription of many genes at once. Scientists can use a DNA microarray to gather data about which genes are turned on or off in a particular cell. A glass slide containing DNA fragments from thousands of genes can be used to test which of those genes are being expressed in a particular cell type.

11.10 Signal transduction pathways convert messages received at the cell surface to responses within the cell.

11.11 Cell-signaling systems appeared early in the evolution of life. Similarities among organisms suggest that signal transduction pathways evolved early in the history of life on Earth.

Cloning of Plants and Animals (11.12–11.14)

11.12 Plant cloning shows that differentiated cells may retain all of their genetic potential. A clone is an individual created by asexual reproduction and thus genetically identical to a single parent.

11.13 Biologists can clone animals via nuclear transplantation. Inserting DNA from a donor cell into a nucleus-free host egg can result in an early embryo that is a clone of the DNA donor. Implanting a blastocyst into a surrogate mother can lead to the birth of a cloned mammal.

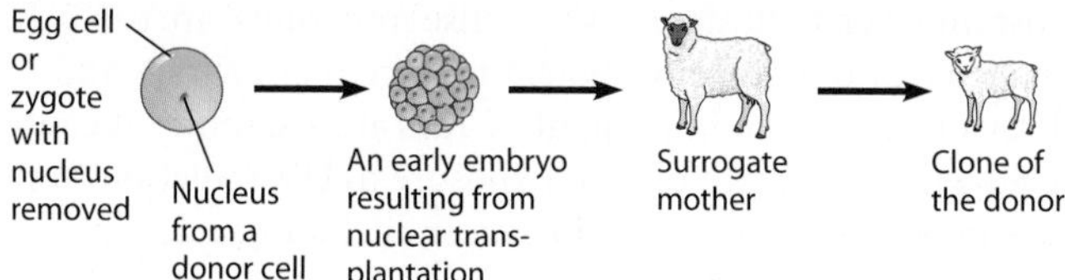
Egg cell or zygote with nucleus removed
Nucleus from a donor cell
An early embryo resulting from nuclear transplantation
Surrogate mother
Clone of the donor

11.14 Therapeutic cloning can produce stem cells with great medical potential. The goal of therapeutic cloning is to produce embryonic stem cells. Such cells may eventually be used for a variety of therapeutic purposes. Like embryonic stem cells, adult stem cells can both perpetuate themselves in culture and give rise to differentiated cells. Unlike embryonic stem cells, adult stem cells normally give rise to only a limited range of cell types.

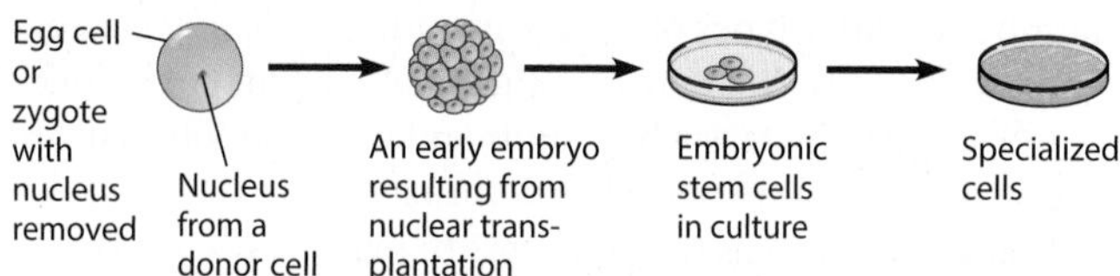
Egg cell or zygote with nucleus removed
Nucleus from a donor cell
An early embryo resulting from nuclear transplantation
Embryonic stem cells in culture
Specialized cells

The Genetic Basis of Cancer (11.15–11.18)

11.15 Cancer results from mutations in genes that control cell division. Cancer cells, which divide uncontrollably, result from mutations in genes whose protein products affect the cell cycle. A mutation can change a proto-oncogene, a normal gene that helps control cell division, into an oncogene, which causes cells to divide excessively. Mutations that inactivate tumor-suppressor genes have similar effects.

11.16 Multiple genetic changes underlie the development of cancer. Cancers result from a series of genetic changes.

11.17 Faulty proteins can interfere with normal signal transduction pathways. Many proto-oncogenes and tumor-suppressor genes code for proteins active in signal transduction pathways regulating cell division.

11.18 Lifestyle choices can reduce the risk of cancer. Reducing exposure to carcinogens, which induce cancer-causing mutations, and making other lifestyle choices can help reduce cancer risk.

Connecting the Concepts

1. Complete the following concept map to test your knowledge of gene regulation.

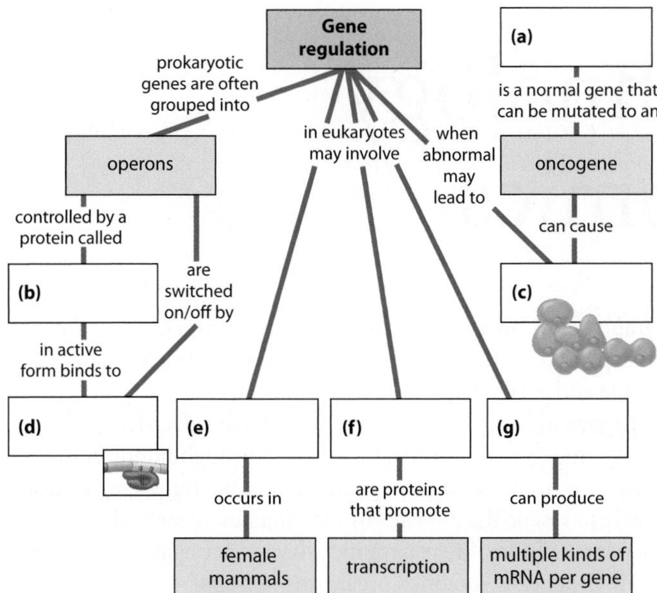

Gene regulation
prokaryotic genes are often grouped into
operons
controlled by a protein called
(b)
in active form binds to
(d)
are switched on/off by
in eukaryotes may involve
when abnormal may lead to
(a)
is a normal gene that can be mutated to an
oncogene
can cause
(c)
(e)
occurs in
female mammals
(f)
are proteins that promote
transcription
(g)
can produce
multiple kinds of mRNA per gene

Testing Your Knowledge

Level 1: Knowledge/Comprehension

2. Which of the following methods of gene regulation do eukaryotes and prokaryotes have in common?
   a. elaborate packing of DNA in chromosomes
   b. activator and repressor proteins, which attach to DNA
   c. the addition of a cap and tail to mRNA after transcription
   d. *lac* and *trp* operons
3. A homeotic gene does which of the following?
   a. It serves as the ultimate control for prokaryotic operons.
   b. It regulates the expression of groups of other genes during development.
   c. It represses the histone proteins in eukaryotic chromosomes.
   d. It helps splice mRNA after transcription.
4. Which of the following is a valid difference between embryonic stem cells and the stem cells found in adult tissues?
   a. In laboratory culture, only adult stem cells are immortal.
   b. In nature, only embryonic stem cells give rise to all the different types of cells in the organism.
   c. Only adult stem cells can differentiate in culture.
   d. Embryonic stem cells are generally more difficult to grow in culture than adult stem cells.

Level 2: Application/Analysis

5. The control of gene expression is more complex in multicellular eukaryotes than in prokaryotes because ______________. (*Explain your answer.*)
   a. eukaryotic cells are much smaller
   b. in a multicellular eukaryote, different cells are specialized for different functions
   c. prokaryotes are restricted to stable environments
   d. eukaryotes have fewer genes, so each gene must do several jobs
6. Your bone cells, muscle cells, and skin cells look different because
   a. each cell contains different kinds of genes.
   b. they are present in different organs.
   c. different genes are active in each kind of cell.
   d. they contain different numbers of genes.
7. All your cells contain proto-oncogenes, which can change into cancer-causing oncogenes. Why do cells possess such potential time bombs?
   a. Viruses infect cells with proto-oncogenes.
   b. Proto-oncogenes are genetic "junk" with no known function.
   c. Proto-oncogenes are unavoidable environmental carcinogens.
   d. Proto-oncogenes normally control cell division.
8. You obtain an egg cell from the ovary of a white mouse and remove the nucleus from it. You then obtain a nucleus from a liver cell from an adult black mouse. You use the methods of nuclear transplantation to insert the nucleus into the empty egg. After some prompting, the new zygote divides into an early embryo, which you then implant into the uterus of a brown mouse. A few weeks later, a litter of mice is born. What color will they be? Why?
9. Mutations can alter the function of the *lac* operon (see Module 11.1). Predict how the following mutations would affect the function of the operon in the presence and absence of lactose:
   a. Mutation of regulatory gene; repressor cannot bind to lactose.
   b. Mutation of operator; repressor will not bind to operator.
   c. Mutation of regulatory gene; repressor will not bind to operator.
   d. Mutation of promoter; RNA polymerase will not attach to promoter.

Level 3: Synthesis/Evaluation

10. A mutation in a single gene may cause a major change in the body of a fruit fly, such as an extra pair of legs or wings. Yet it probably takes the combined action of hundreds or thousands of genes to produce a wing or leg. How can a change in just one gene cause such a big change in the body?
11. A chemical called dioxin is produced as a by-product of some chemical manufacturing processes. This substance was present in Agent Orange, a defoliant sprayed on vegetation during the Vietnam War. There has been a continuing controversy over its effects on soldiers exposed to it during the war. Animal tests have suggested that dioxin can be lethal and can cause birth defects, cancer, organ damage, and immune system suppression. But its effects on humans are unclear, and even animal tests are inconclusive. Researchers have discovered that dioxin enters a cell and binds to a protein that in turn attaches to the cell's DNA. How might this mechanism help explain the variety of dioxin's effects? How might you determine whether a particular individual became ill as a result of exposure to dioxin?
12. **SCIENTIFIC THINKING** Each scientist works as part of a broader community of scientists, building on the work of others. Scientific advances often depend on the application of new technologies and/or on new techniques applied to an existing problem. What improvements to existing cloning methods did Wilmut make that allowed him to successfully clone Dolly the sheep from an adult cell?

Answers to all questions can be found in Appendix 4.

CHAPTER

12 DNA Technology and Genomics

Papaya fruit, shown in the photograph below, are sweet and loaded with vitamin C. They are borne on a rapidly growing treelike plant (*Carica papaya*) that grows only in tropical climates. In Hawaii, papaya is both a dietary staple and a valuable export crop.

Although thriving today, Hawaii's papaya industry seemed doomed just a few decades ago. A deadly pathogen called the papaya ringspot virus (PRV) had spread throughout the islands and appeared poised to completely eradicate the papaya plant population. But scientists from the University of Hawaii were able to rescue the industry by creating new, genetically engineered PRV-resistant strains of papaya. Today, the papaya industry is once again vibrant—and the vast majority of Hawaii's papayas are genetically modified organisms (GMOs).

? *Are genetically modified organisms safe?*

However, not everyone is happy about the circumstances surrounding the recovery of the Hawaiian papaya industry. Although genetically modified papayas are approved for consumption in the United States (as are many other GMO fruits and vegetables), some critics have raised safety concerns—for the people who eat them and for the environment. On three occasions over a three-year

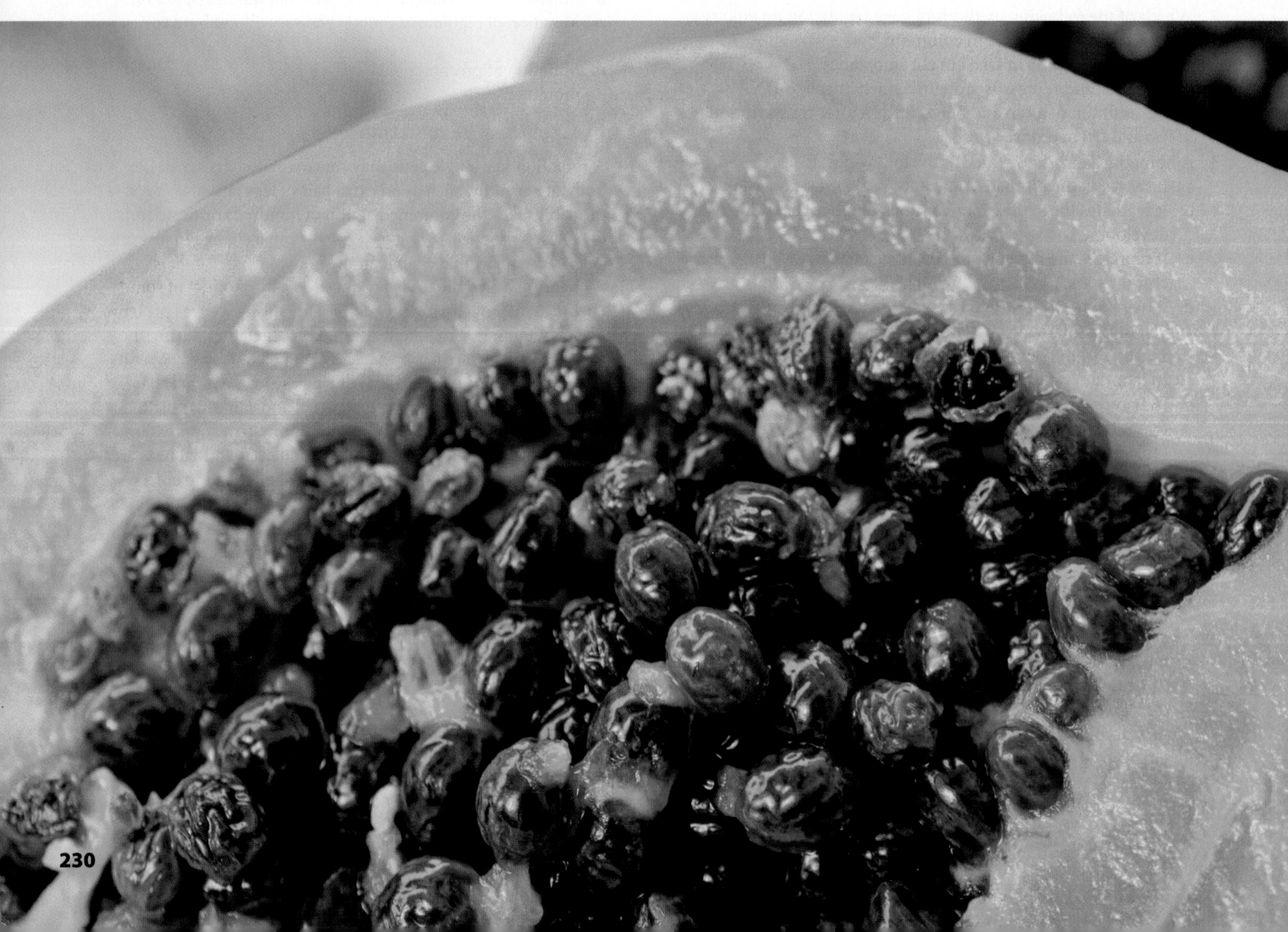

span, thousands of papaya trees on the big island of Hawaii were hacked down under the cover of darkness, presumably as a protest against GMO crops. Although few would condone such criminal behavior, should we in fact be concerned about the safety of GMO crops? This question continues to foster considerable debate and disagreement.

In addition to GMOs in our diet, DNA technologies affect our lives in many other ways: Gene cloning is used to produce medical and industrial products, DNA profiling has changed the field of forensic science, new technologies produce valuable data for biological research, and DNA can even be used to investigate historical questions. In this chapter, we'll discuss each of these applications. We'll also consider the specific techniques used, how they are applied, and some of the social, legal, and ethical issues that are raised by the new technologies.

BIG IDEAS

Gene Cloning
(12.1–12.5)

A variety of laboratory techniques can be used to copy and combine DNA molecules.

Genetically Modified Organisms
(12.6–12.10)

Transgenic cells, plants, and animals are used in agriculture and medicine.

DNA Profiling
(12.11–12.16)

Genetic markers can be used to definitively match a DNA sample to an individual.

Genomics
(12.17–12.21)

The study of complete DNA sets helps us learn about evolutionary history.

▷ Gene Cloning

12.1 Genes can be cloned in recombinant plasmids

Although it may seem like a modern field, **biotechnology**, the manipulation of organisms or their components to make useful products, actually dates back to the dawn of civilization. Consider such ancient practices as the use of yeast to make beer, wine, and bread and the selective breeding of livestock, dogs, and other animals. But when people use the term *biotechnology* today, they are usually referring to **DNA technology**, modern laboratory techniques for studying and manipulating genetic material. Using these methods, scientists can, for instance, modify specific genes and move them between organisms as different as *Escherichia coli* bacteria, papaya, and fish.

In the 1970s, the field of biotechnology exploded with the invention of methods for making recombinant DNA in a test tube. **Recombinant DNA** is formed when scientists combine pieces of DNA from two different sources—often different species—*in vitro* (in a test tube) to form a single DNA molecule. Today, recombinant DNA technology is widely used for **genetic engineering**, the direct manipulation of genes for practical purposes. Scientists have genetically engineered bacteria to mass-produce a variety of useful chemicals, from cancer drugs to pesticides. Scientists have also transferred genes from bacteria into plants and from one animal species into another (**Figure 12.1A**).

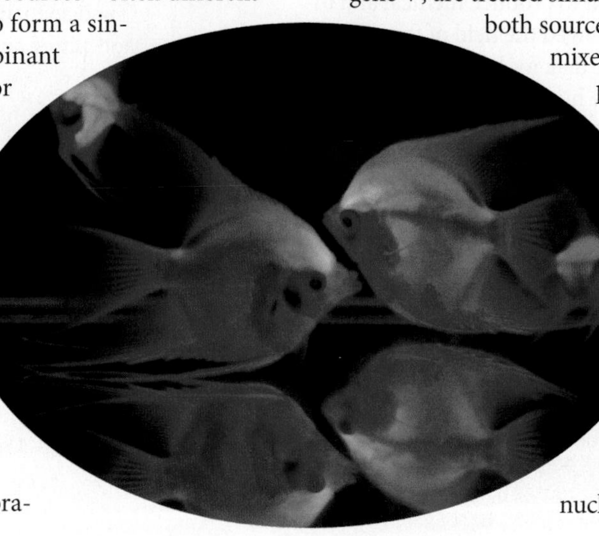

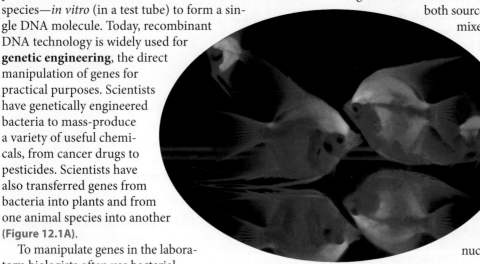

▲ **Figure 12.1A** Glowing angelfish produced by transferring a gene originally obtained from a jelly (cnidarian)

To manipulate genes in the laboratory, biologists often use bacterial **plasmids**, small, circular DNA molecules that replicate (duplicate) separately from the much larger bacterial chromosome (see Module 10.23). Plasmids typically carry only a few genes and are passed from one generation of bacteria to the next. Because plasmids are easily manipulated to carry virtually any genes, they are key tools for **gene cloning**, the production of many identical copies of a gene-carrying piece of DNA. Gene-cloning methods are central to the production of useful products via genetic engineering.

Consider a typical genetic engineering challenge: A molecular biologist at a pharmaceutical company has identified a gene that codes for a valuable product, a hypothetical substance called protein V. The biologist wants to manufacture the protein on a large scale. The biggest challenge in such an effort is of the "needle in a haystack" variety: The gene of interest is one relatively tiny segment embedded in a much longer DNA molecule. **Figure 12.1B** illustrates how the techniques of gene cloning can be used to find the desired gene and copy it.

To begin, the biologist isolates two kinds of DNA: ❶ a bacterial plasmid (usually from the bacterium *E. coli*) that will serve as the **vector**, or gene carrier, and ❷ the DNA from another organism that includes the gene that codes for protein V (gene *V*) along with other, unwanted genes. The DNA containing gene *V* could come from a variety of sources, such as a different bacterium, a plant, a nonhuman animal, or even human tissue cells growing in laboratory culture.

❸ The researcher treats both the plasmid and the gene *V* source DNA with an enzyme that cuts DNA. An enzyme is chosen that cleaves the plasmid in only one place. ❹ The source DNA, which is usually much longer in sequence than the plasmid, may be cut into many fragments, one of which carries gene *V*. The figure shows the processing of just one DNA fragment and one plasmid, but actually, millions of plasmids and DNA fragments, most of which do not contain gene *V*, are treated simultaneously. ❺ The cut DNA from both sources—the plasmid and target gene—are mixed. The single-stranded ends of the plasmid base-pair with the complementary ends of the target DNA fragment (see Module 10.4 if you need a refresher on the DNA base-pairing rules). ❻ The enzyme **DNA ligase** joins the two DNA molecules by way of covalent bonds. This enzyme, which the cell normally uses in DNA replication (see Module 10.5), is a "DNA pasting" enzyme that catalyzes the formation of covalent bonds between adjacent nucleotides, joining the strands. The result is a recombinant DNA plasmid containing gene *V*, as well as many other recombinant DNA plasmids carrying other genes not shown here.

❼ The recombinant plasmid containing the targeted gene is mixed with bacteria. Under the right conditions, a bacterium takes up the plasmid DNA by transformation (see Module 10.22). ❽ The recombinant bacterium then reproduces to form a **clone** of cells, each carrying a copy of gene *V*. (In this context, the term *clone* refers to a group of identical cells descended from a single ancestral cell.) This step is the actual gene cloning. In our example, the biologist will eventually grow a cell clone large enough to produce protein *V* in marketable quantities.

❾ Gene cloning can be used to produce a variety of desirable products. Copies of the gene itself can be the immediate product, to be used in additional genetic engineering projects. For example, a pest-resistance gene present in one plant species might be cloned and transferred into plants of another species. Other times, the protein product of the cloned gene is harvested and used. For example, an enzyme that creates a faded look in blue jeans can be harvested in large quantities from bacteria carrying the cloned gene (in

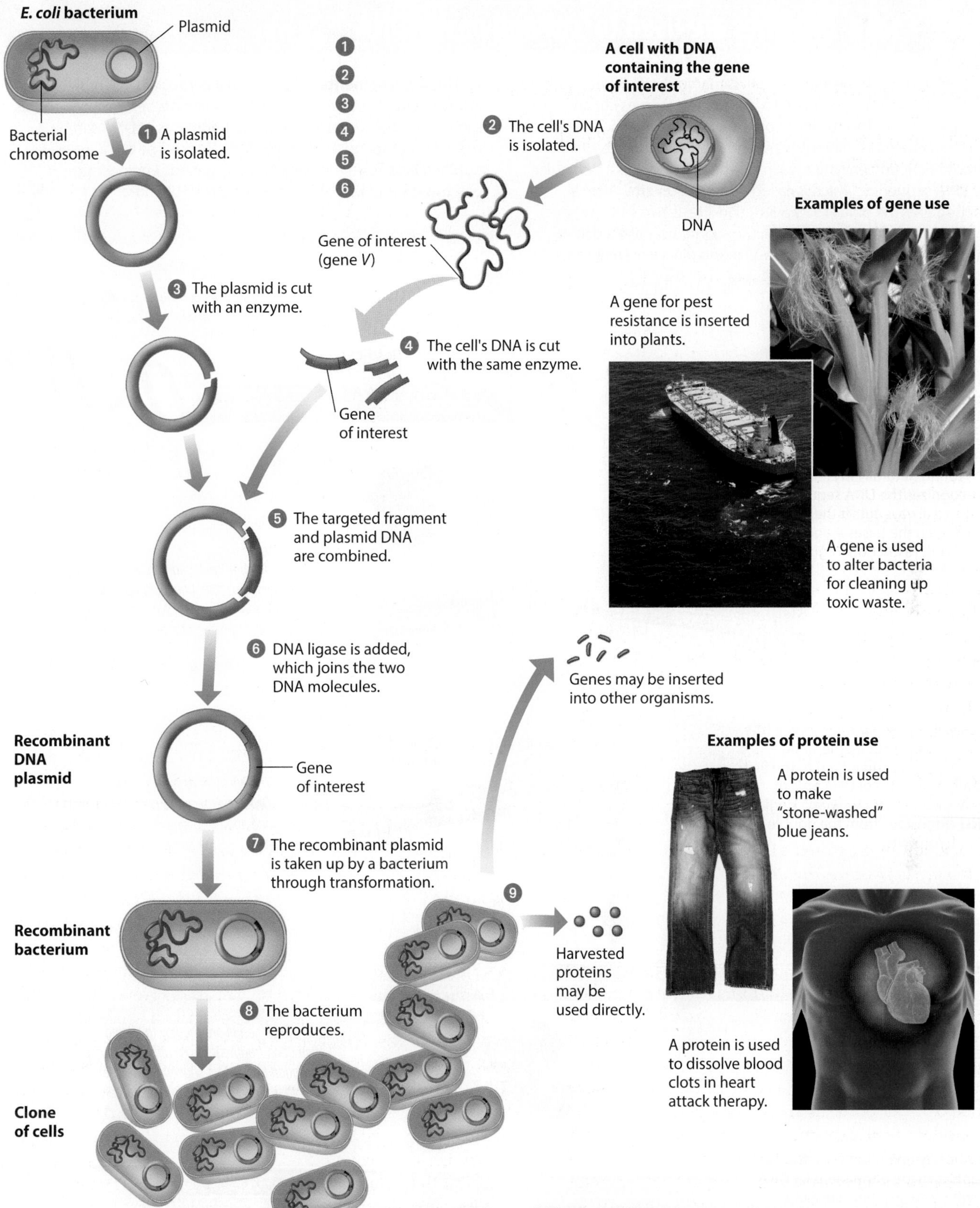

***E. coli* bacterium**

Plasmid

Bacterial chromosome

1 A plasmid is isolated.

A cell with DNA containing the gene of interest

2 The cell's DNA is isolated.

DNA

Gene of interest (gene *V*)

3 The plasmid is cut with an enzyme.

4 The cell's DNA is cut with the same enzyme.

Gene of interest

5 The targeted fragment and plasmid DNA are combined.

6 DNA ligase is added, which joins the two DNA molecules.

Recombinant DNA plasmid

Gene of interest

7 The recombinant plasmid is taken up by a bacterium through transformation.

Recombinant bacterium

8 The bacterium reproduces.

Clone of cells

9

Genes may be inserted into other organisms.

Harvested proteins may be used directly.

Examples of gene use

A gene for pest resistance is inserted into plants.

A gene is used to alter bacteria for cleaning up toxic waste.

Examples of protein use

A protein is used to make "stone-washed" blue jeans.

A protein is used to dissolve blood clots in heart attack therapy.

▲ Figure 12.1B An overview of gene cloning

Try This Examine the shapes of the cut regions of the plasmid and the gene of interest, paying particular attention to the complementary shapes of the pieces that allow them to bond together.

other words, no stones are harmed in the making of stone-washed jeans!).

In the next four modules, we discuss the methods outlined in Figure 12.1B. You may find it useful to turn back to this summary figure as each technique is discussed.

In the example shown in Figure 12.1, what is the vector?

A plasmid isolated from an *E. coli* bacterium

VISUALIZING THE CONCEPT

12.2 Enzymes are used to "cut and paste" DNA

To understand how DNA is manipulated in the laboratory, you need to learn how enzymes cut and paste DNA. The cutting tools are bacterial enzymes called **restriction enzymes**. Biologists have identified hundreds of different restriction enzymes, each of which recognizes a particular short DNA sequence, which is called a **restriction site**. After a restriction enzyme binds to its restriction site, it cuts both strands of the DNA at precise points within the sequence—like a pair of highly specific molecular scissors—yielding pieces of DNA called **restriction fragments**. All copies of a particular DNA molecule always yield the same set of DNA fragments when exposed to the same restriction enzyme. Once cut, fragments of DNA can be pasted together by the enzyme **DNA ligase**. The techniques outlined here form the basis of many genetic engineering procedures that involve combining DNA from different sources.

A restriction site is usually 4–8 nucleotide pairs long.

DNA

Restriction site

GAATTC
CTTAAG

The restriction enzyme shown here, called EcoRI, is found naturally in *E. coli* bacteria. EcoRI recognizes the DNA sequence GAATTC and always cuts it the same way—between the bases A and G—producing restriction fragments.

Restriction enzyme

G / CTTAA   AATTC / G

In bacteria, restriction enzymes play a defensive role, chopping up foreign DNA; the cell's own DNA is protected by the addition of methyl groups.

A piece of DNA from another source (the gene of interest) is cut by the same restriction enzyme and added to the first DNA. Both molecules of DNA are cut unevenly, yielding "sticky ends," single-stranded extensions from the double-stranded fragments.

Sticky end

Gene of interest

AATTC / G   G / CTTAA

Sticky end

The sticky ends from the two different DNA molecules are complementary to one another because they were cut by the same enzyme.

The complementary ends on the two different fragments stick together by base pairing.

GAATTC / CTTAAG   GAATTC / CTTAAG

Sticky ends are the key to joining restriction fragments from different sources: Hydrogen bonds (not shown) form base pairs that hold the two strands together.

DNA ligase

The temporary union between the DNA fragments is made permanent by DNA ligase, which creates new covalent bonds that join the sugar-phosphate backbones of the DNA strands.

Recombinant DNA

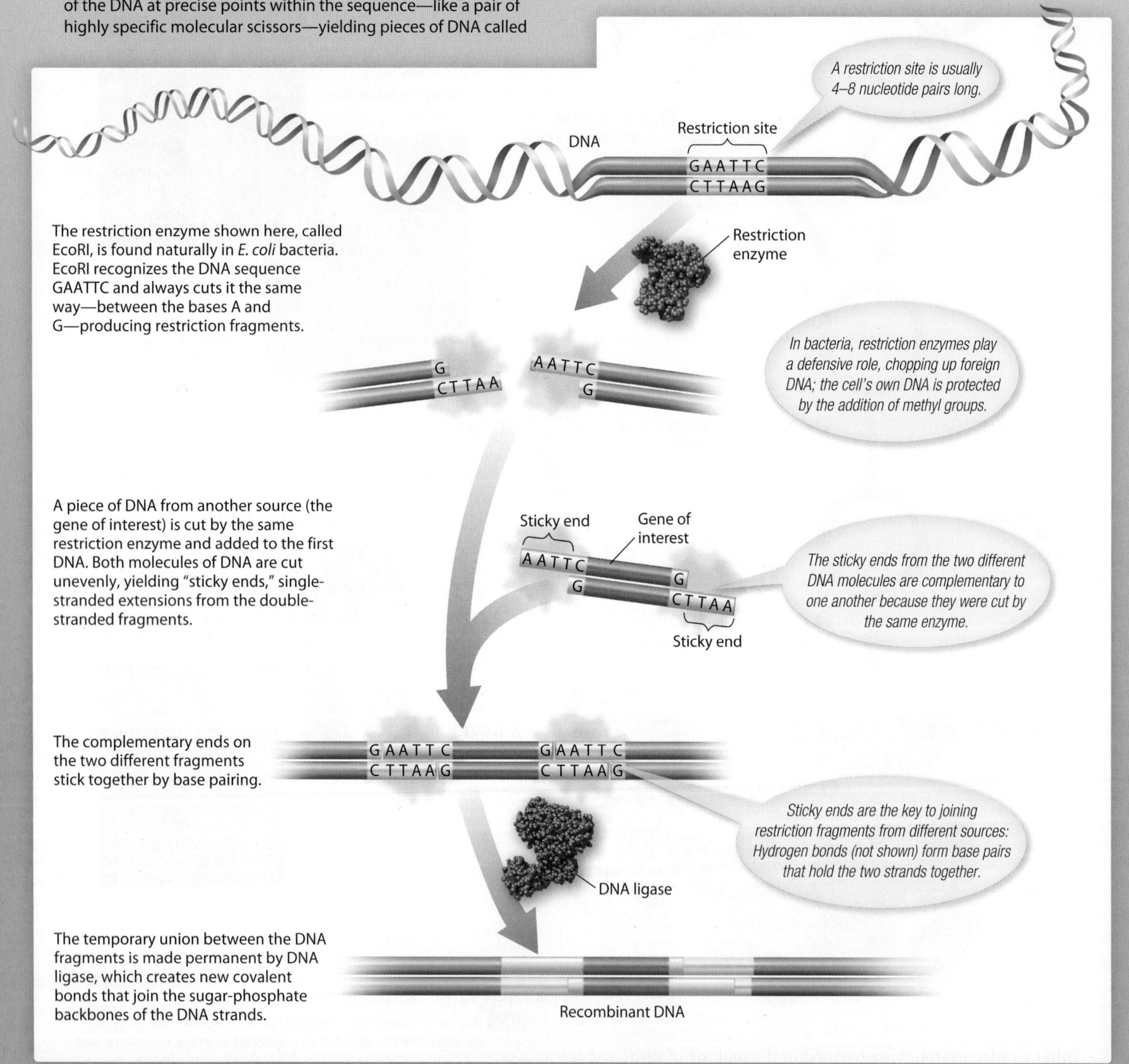

? **What are "sticky ends"?**

Answer: Single-stranded regions whose unpaired bases can hydrogen-bond to the complementary sticky ends of other fragments created by the same restriction enzyme

12.3 Cloned genes can be stored in genomic libraries

Each bacterial clone created using the procedure described in Figure 12.1B consists of identical cells with plasmids carrying one particular fragment of target DNA. A collection of cloned DNA fragments that includes an organism's entire genome is called a **genomic library**. On the left side of **Figure 12.3**, the red, yellow, and green DNA segments represent three of the thousands of different library "books" that are "shelved" in plasmids inside bacterial cells. A typical cloned DNA fragment is big enough to carry one or a few genes, and together, the fragments include the entire genome of the organism from which the DNA was derived.

▼ **Figure 12.3** Genomic libraries

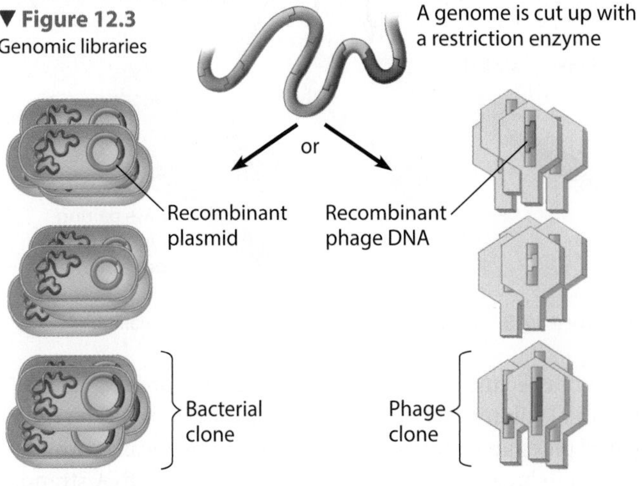

A genome is cut up with a restriction enzyme
or
Recombinant plasmid
Recombinant phage DNA
Bacterial clone
Phage clone
Plasmid library
Phage library

Bacteriophages (also called phages)—viruses that infect bacteria—can also serve as vectors when cloning genes (Figure 12.3, right). When a phage is used as a vector, the DNA fragments are inserted into phage DNA molecules. The recombinant phage DNA can then be introduced into a bacterial cell through the normal infection process. Inside the cell, phage DNA is replicated, producing new phage particles carrying the foreign DNA. In the next module, we look at another source of DNA for cloning: eukaryotic mRNA.

? In what sense does a genomic library have multiple copies of each "book"?

Each "book"—a piece of DNA from the genome that was the source of the library—is present in every recombinant bacterium or phage in a clone.

12.4 Reverse transcriptase can help make genes for cloning

Rather than starting with an entire eukaryotic genome, a researcher can focus on the genes expressed in a particular kind of cell by using its mRNA as the starting material for cloning (**Figure 12.4**). ❶ The chosen cells transcribe their genes within the nucleus and ❷ process the transcripts, removing introns and splicing exons together, producing mRNA. ❸ The researcher isolates the mRNA in a test tube. ❹ Single-stranded DNA transcripts are made from the mRNA using **reverse transcriptase**, a viral enzyme that can synthesize DNA from an RNA template (gold in the figure; see Module 10.20). ❺ Another enzyme is added to break down the mRNA, and ❻ DNA polymerase (the enzyme that replicates DNA; see Module 10.5) is used to synthesize a second DNA strand.

The DNA that results from such a procedure, called **complementary DNA (cDNA)**, represents only the subset of genes that had been transcribed into mRNA in the starting cells. Among other purposes, such a cDNA library is useful for studying the genes responsible for the specialized functions of a particular cell type, such as brain or liver cells.

? Why is a cDNA gene made using reverse transcriptase often shorter than the natural form of the gene?

Because cDNAs are made from spliced mRNAs, which lack introns

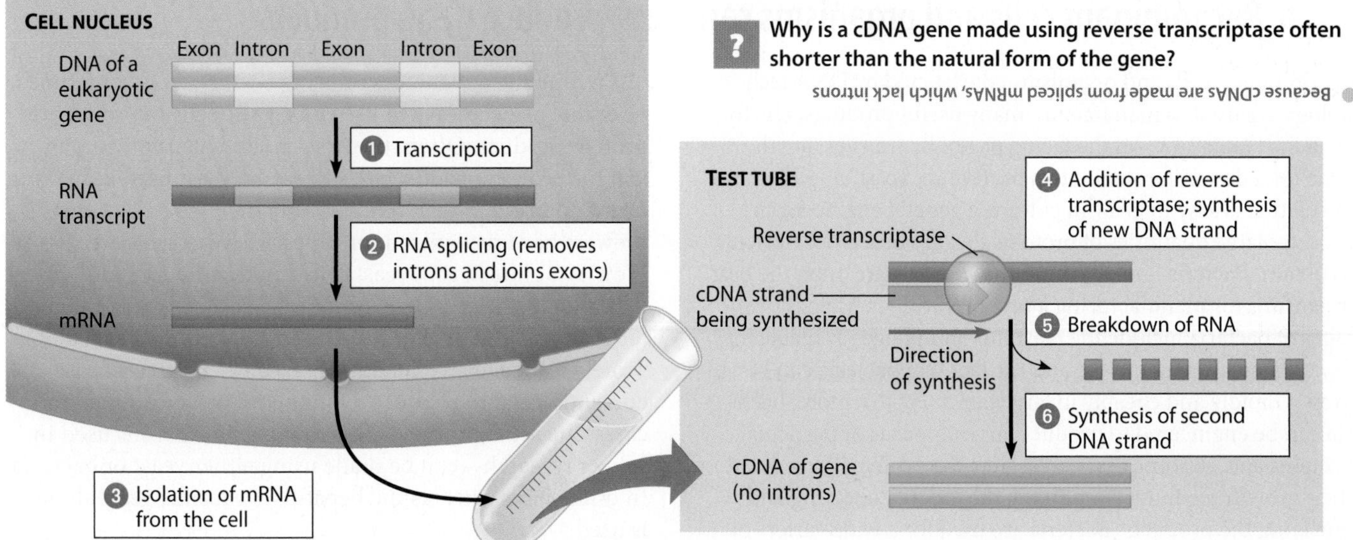

CELL NUCLEUS
DNA of a eukaryotic gene
Exon Intron Exon Intron Exon
1 Transcription
RNA transcript
2 RNA splicing (removes introns and joins exons)
mRNA
3 Isolation of mRNA from the cell
TEST TUBE
Reverse transcriptase
4 Addition of reverse transcriptase; synthesis of new DNA strand
cDNA strand being synthesized
Direction of synthesis
5 Breakdown of RNA
6 Synthesis of second DNA strand
cDNA of gene (no introns)

▲ **Figure 12.4** Making complimentary DNA (cDNA) from eukaryotic mRNA

12.5 Nucleic acid probes identify clones carrying specific genes

Often, the most difficult task in gene cloning is finding the right "books" in a genomic library—that is, identifying only the clones that contain a desired gene from among all those created. For example, a researcher might want to pull out just the clone of bacteria carrying what is depicted as the red gene in Figure 12.3. If bacterial clones containing a specific gene actually translate the gene into protein, they can be identified by testing for the protein product. However, not every desired gene produces detectable proteins. In such cases, researchers can also test directly for the gene itself.

Methods for detecting a gene directly depend on base pairing between the gene and a complementary sequence on another nucleic acid molecule, either DNA or RNA. When at least part of the nucleotide sequence of a gene is known, this information can be used to a researcher's advantage. For example, if we know that a gene contains the sequence TAGGCT, a biochemist can synthesize a short single strand of DNA with the complementary sequence (ATCCGA) and label it with a radioactive isotope or fluorescent tag. This labeled, complementary molecule is called a **nucleic acid probe** because it is used to find a specific gene or other nucleotide sequence within a mass of DNA. (In actual practice, probe molecules are considerably longer than six nucleotides.)

Figure 12.5 shows how a probe works. The DNA sample to be tested is treated with heat or chemicals to separate the DNA strands. When the radioactive DNA probe is added to these strands, it tags the correct molecules—that is, it finds the correct books in the genomic library—by hydrogen-bonding to the complementary sequence in the gene of interest. Such a probe can be simultaneously applied to many bacterial clones to screen all of them at once for a desired gene.

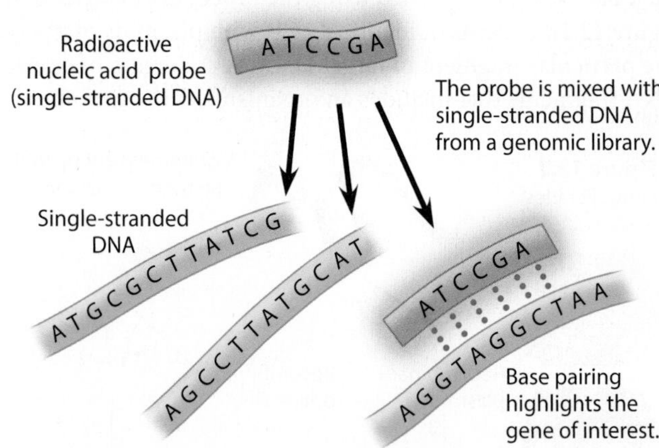

Radioactive nucleic acid probe (single-stranded DNA)
ATCCGA
The probe is mixed with single-stranded DNA from a genomic library.
Single-stranded DNA
ATGCGCTTATCG
AGCCTTATGCAT
ATCCGA
AGGTAGGCTAA
Base pairing highlights the gene of interest.

▲ **Figure 12.5** How a DNA probe tags a gene by base pairing

In one technique, a piece of filter paper is pressed against bacterial colonies (clones) growing on a petri dish. The filter paper picks up cells from each colony. A chemical treatment is used to break open the cells and separate the DNA strands. The DNA strands are then soaked in probe solution. Any bacterial colonies carrying the gene of interest will be tagged on the filter paper, marking them for easy identification. Once the researcher identifies a colony carrying the desired gene, the cells can be grown further, and the gene of interest, or its protein product, can be collected in large amounts.

? How does a probe consisting of radioactive DNA or RNA enable a researcher to find the bacterial clones carrying a particular gene?

The probe molecules bind to and label DNA only from the cells containing the gene of interest, which has a complementary DNA sequence.

▷ Genetically Modified Organisms

12.6 Recombinant cells and organisms can mass-produce gene products

Recombinant cells and organisms constructed by DNA technology are used to manufacture many useful products, chiefly proteins (**Table 12.6**, on the facing page). By transferring the gene for a desired protein into a bacterium, yeast, or other kind of cell that is easy to grow in culture, a genetic engineer can produce large quantities of proteins that are otherwise difficult to obtain. Bacteria—most commonly *E. coli*—are often the best organisms for manufacturing a protein product. Major advantages of bacteria include the plasmids and phages available for use as gene-cloning vectors and the fact that bacteria can be grown rapidly and cheaply in large tanks. Furthermore, bacteria can be engineered to produce large amounts of particular proteins and, in some cases, to secrete the proteins directly into their growth medium, simplifying the task of collecting and purifying the products. Bacteria are used for a wide variety of purposes, from producing valuable human drugs to enzymes used in making cheese and processed fruit juice.

Despite the advantages of using bacteria, it is sometimes desirable or necessary to use eukaryotic cells to produce a protein product. Often, the first-choice eukaryotic organism for protein production is the same yeast used in making bread and beer, *Saccharomyces cerevisiae*. As bakers and brewers have recognized for centuries, yeast cells are easy to grow. And like *E. coli*, yeast cells can take up foreign DNA and integrate it into their genomes. Yeast cells are often better than bacteria at synthesizing and secreting eukaryotic proteins, such as the hepatitis B vaccine. *S. cerevisiae* is currently used to produce a number of proteins. In certain cases, the same product—for example, interferons used in cancer research—can be made using either yeast or bacteria. In other cases, such as the hepatitis B vaccine, yeast alone is used.

The cells of choice for making some gene products come from mammals. Many proteins that mammalian cells

TABLE 12.6 | SOME PROTEIN PRODUCTS OF RECOMBINANT DNA TECHNOLOGY

Product	Made by	Use
Human insulin	*E. coli*	Treatment for diabetes
Human growth hormone (HGH)	*E. coli*	Treatment for growth defects
Epidermal growth factor (EGF)	*E. coli*	Treatment for burns, ulcers
Interleukin-2 (IL-2)	*E. coli*	Possible treatment for cancer
Bovine growth hormone (BGH)	*E. coli*	Improving weight gain in cattle
Cellulase	*E. coli*	Breaking down cellulose for animal feeds
Taxol	*E. coli*	Treatment for ovarian cancer
Interferons (alpha and gamma)	*S. cerevisiae*, *E. coli*	Possible treatment for cancer and viral infections
Hepatitis B vaccine	*S. cerevisiae*	Prevention of viral hepatitis
Erythropoietin (EPO)	Mammalian cells	Treatment for anemia
Factor VIII	Mammalian cells	Treatment for hemophilia
Tissue plasminogen activator (TPA)	Mammalian cells	Treatment for heart attacks and some strokes

normally secrete are glycoproteins, proteins with chains of sugars attached. Because only mammalian cells can attach the sugars correctly, mammalian cells must be used for making these products. For example, recombinant mammalian cells growing in laboratory cultures are currently used to produce human erythropoietin (EPO), a hormone that stimulates the production of red blood cells. EPO can save lives as a treatment for anemia. However, EPO is also abused by some athletes who seek the advantage of artificially high levels of oxygen-carrying red blood cells (called "blood doping"; see Module 23.13). EPO is one of several drugs bicyclist Lance Armstrong and his teammates have admitted to abusing during his historic string of victories in the Tour de France bicycle race. After years of denial, Armstrong admitted his abuse of EPO (and other performance-enhancing substances) in 2012, leading to a lifetime ban from sanctioned athletic competition.

Recently, pharmaceutical researchers have been exploring the mass production of gene products by whole animals or plants rather than cultured cells. Genetic engineers have used recombinant DNA technology to insert genes for desired human proteins into other mammals, where the protein encoded by the recombinant gene may be secreted in the animal's milk. For example, a gene for antithrombin—a human protein that helps prevent inappropriate blood clotting—has been inserted into the genome of a goat (**Figure 12.6A**); isolated from the milk, the protein reduces the risk of life-threatening blood clots during surgery or childbirth. The sheep in **Figure 12.6B** have been genetically modified to produce a human protein called AAT; this protein can be supplied to patients to treat a hereditary form of emphysema. Other

▲ **Figure 12.6A** A goat carrying a gene for a human blood protein that is secreted in the milk

mammals have been modified to serve as models for human diseases or to improve the health of livestock.

However, genetically engineered animals are difficult and costly to produce. Typically, a biotechnology company starts by injecting the desired DNA into a large number of embryos, which are then implanted into surrogate mothers. With luck, one or a few recombinant animals may result, but success rates for such procedures are very low. Once a recombinant organism is successfully produced, it may be cloned. The result can be a genetically identical herd—a grazing pharmaceutical "factory" of "pharm" animals that produce otherwise rare biological substances for medical use.

We continue an exploration of the medical applications of DNA technology in the next module.

? **Why can't all human proteins be synthesized in *E. coli*?**

● Because bacteria cannot correctly produce many eukaryotic (and mammalian) proteins, such as ones that require the attachment of sugar groups

▲ **Figure 12.6B** Sheep that have been genetically modified to produce a useful human protein

12.7 DNA technology has changed the pharmaceutical industry and medicine

CONNECTION

DNA technology, and gene cloning in particular, is widely used to produce medicines and to diagnose diseases.

Therapeutic Hormones Consider the first two products in Table 12.6 on the previous page—human insulin and human growth hormone. Insulin, normally secreted by the pancreas, is a hormone that helps regulate the levels of glucose in the blood. About 2 million Americans with diabetes depend on insulin injections. Before 1982, the main sources of this hormone were slaughtered pigs and cattle. Insulin extracted from these animals is chemically similar, but not identical, to human insulin, and it causes allergic reactions in some people. Genetic engineering has largely solved this problem by developing bacteria that synthesize and secrete the human form of insulin. In 1982, Humulin (**Figure 12.7A**)—human insulin produced by bacteria—became the first recombinant DNA drug approved by the U.S. Food and Drug Administration.

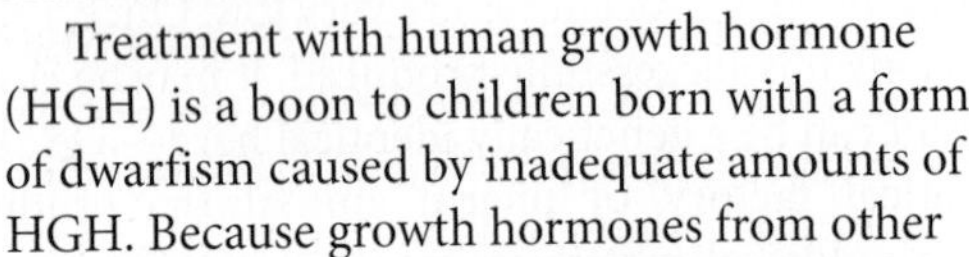

▲ **Figure 12.7A** Human insulin produced by bacteria

Treatment with human growth hormone (HGH) is a boon to children born with a form of dwarfism caused by inadequate amounts of HGH. Because growth hormones from other animals are not effective in humans, children with HGH deficiency historically have had to rely on scarce and expensive supplies from human cadavers. In 1985, however, molecular biologists made an artificial gene for HGH by joining a human DNA fragment to a chemically synthesized piece of DNA; using this gene, they were able to produce HGH in *E. coli*. HGH from recombinant bacteria is now widely used.

Another important pharmaceutical product produced by genetic engineering is tissue plasminogen activator (TPA). If administered soon after a heart attack, this protein helps dissolve blood clots and reduces the risk of subsequent heart attacks.

Diagnosis of Disease DNA technology can also serve as a diagnostic tool. Among the hundreds of genes for human diseases that have been identified are those for sickle-cell disease, hemophilia, cystic fibrosis, and Huntington's disease. Affected individuals with such diseases often can be identified before the onset of symptoms, even before birth. It is also possible to identify symptomless carriers of potentially harmful recessive alleles (see Module 9.9). In addition, DNA technology can pinpoint infections. For example, DNA analysis can help track down and identify elusive viruses such as HIV, the virus that causes AIDS.

Vaccines DNA technology is also helping medical researchers develop vaccines. A **vaccine** is a harmless variant (mutant) or derivative of a pathogen—usually a bacterium or virus—that is used to stimulate the immune system to mount a lasting defense against that pathogen, thereby preventing disease (see Module 24.3). For many viral diseases (such as measles, mumps, and polio), prevention by vaccination is the only medical way to prevent illness among those exposed to the virus.

Genetic engineering can be used in several ways to make vaccines. One approach is to use genetically engineered cells or organisms to produce large amounts of a protein molecule that is found on the pathogen's outside surface. This method has been used to make the vaccine against hepatitis B, a disabling and sometimes fatal liver disease. **Figure 12.7B** shows a tank for growing yeast cells that have been engineered to carry the gene for the hepatitis B virus's surface protein. This protein, which is made by the yeast, will be the main ingredient of the vaccine.

Another way to use DNA technology in vaccine development is to make a harmless artificial mutant of the pathogen by altering one or more of its genes. When a harmless mutant is used as a so-called live vaccine, it multiplies in the body and may trigger a strong immune response. Artificial-mutant vaccines may cause fewer side effects than vaccines that have traditionally been made from natural mutants.

Yet another method for making vaccines uses a virus related to the one that causes smallpox. Smallpox was once a dreaded human disease, but it was eradicated worldwide in the 1970s by widespread vaccination with a harmless variant of the smallpox virus. Using this harmless virus, genetic engineers could replace some smallpox genes with genes that induce immunity to other diseases. In fact, the virus could be engineered to carry genes needed to vaccinate against several diseases simultaneously. In the future, one inoculation may prevent a dozen diseases.

Genetic engineering rapidly transformed the field of medicine and continues to do so today. But this new technology affects our lives in other ways, as we'll see next.

? **Human growth hormone and insulin produced by DNA technology are used in the treatment of __________ and __________, respectively.**

● dwarfism . . . diabetes

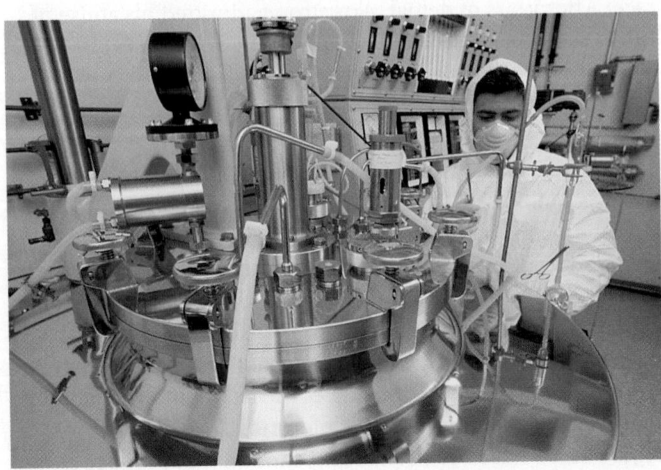

▲ **Figure 12.7B** Equipment used in the production of a vaccine against hepatitis B

12.8 Genetically modified organisms are transforming agriculture

CONNECTION

Since ancient times, people have selectively bred agricultural crops to make them more useful (see Chapter 31 for a discussion of plant domestication). Today, DNA technology is quickly replacing traditional breeding programs as scientists work to improve the productivity of agriculturally important plants and animals. Genetic engineers have produced many different varieties of **genetically modified organisms (GMOs)**, organisms that have acquired one or more genes by artificial means. If the new gene is from another organism, typically of another species, the recombinant organism is called a **transgenic organism**.

A common vector used to introduce new genes into plant cells is a plasmid from the soil bacterium *Agrobacterium tumefaciens* called the Ti plasmid **(Figure 12.8A)**. ❶ With the help of a restriction enzyme and DNA ligase, the gene for the desired trait (indicated in red) is inserted into a modified version of the plasmid. ❷ Then the recombinant plasmid is put into a plant cell, where the DNA carrying the new gene integrates into one of the plant's chromosomes. ❸ Finally, the recombinant cell is cultured and grown into a plant.

With an estimated 1 billion people facing malnutrition, GMO crops may be able to help a great many hungry people by improving food production, shelf life, pest resistance, and the nutritional value of crops. The story of the Hawaiian papaya industry (see the chapter introduction) provides one dramatic example. In India, the insertion of a natural but rare salinity-resistance gene has enabled new varieties of rice to grow in water three times as salty as seawater, allowing food to be grown in drought-stricken or flooded areas. Similar research is under way in Australia to help improve wheat yields in salty soil. Golden Rice, a transgenic variety created in 2000 with a few daffodil genes, produces yellow grains containing beta-carotene, which our body uses to make vitamin A **(Figure 12.8B)**. A newer strain (Golden Rice 2) uses corn genes to boost beta-carotene levels even higher. Golden rice could help prevent vitamin A deficiency, which causes blindness in a quarter million children each year.

In addition to agricultural applications, genetic engineers are now creating plants that make human proteins for medical use. A recently developed transgenic rice strain harbors genes for milk proteins that can be used in rehydration formulas to treat infant diarrhea, a serious problem in developing countries. Other pharmaceutical trials under way involve using modified corn to treat cystic fibrosis, and duckweed to treat hepatitis. Although these trials seem promising, no plant-made drugs intended for use by humans have been approved.

Agricultural researchers are producing transgenic animals by injecting cloned genes directly into the nuclei of fertilized eggs. Some of the cells integrate the foreign DNA into their genomes. The engineered embryos are then surgically implanted into a surrogate mother. If an embryo develops successfully, the resulting animal will contain a gene from a third "parent"—the gene donor—which may even be of another species.

▲ **Figure 12.8B** Golden Rice

The goals in creating a transgenic animal are often the same as the goals of traditional breeding—for instance, to make a sheep with better quality wool or a cow that will mature in less time. In 2006, researchers genetically modified pigs to carry a roundworm gene whose protein converts less healthy fatty acids to omega-3 fatty acids. Meat from the modified pigs contains four to five times as much healthy omega-3 fat as regular pork. Atlantic salmon have been genetically modified by the addition of a more active promoter of a growth hormone gene from Chinook salmon. Such fish can mature in half the time of conventional salmon and grow to twice the size. In late 2012, the FDA found that the modified salmon had no significant effect on humans or the environment, clearing an important hurdle to their sale as food. As of mid–2013, public hearings on whether to allow the salmon to become the first transgenic animal sold as food were ongoing. As we'll discuss next, some people worry that GMOs are not safe for human health or the environment.

? What is the function of the Ti plasmid in the creation of transgenic plants?

● It is used as the vector for introducing foreign genes into a plant cell.

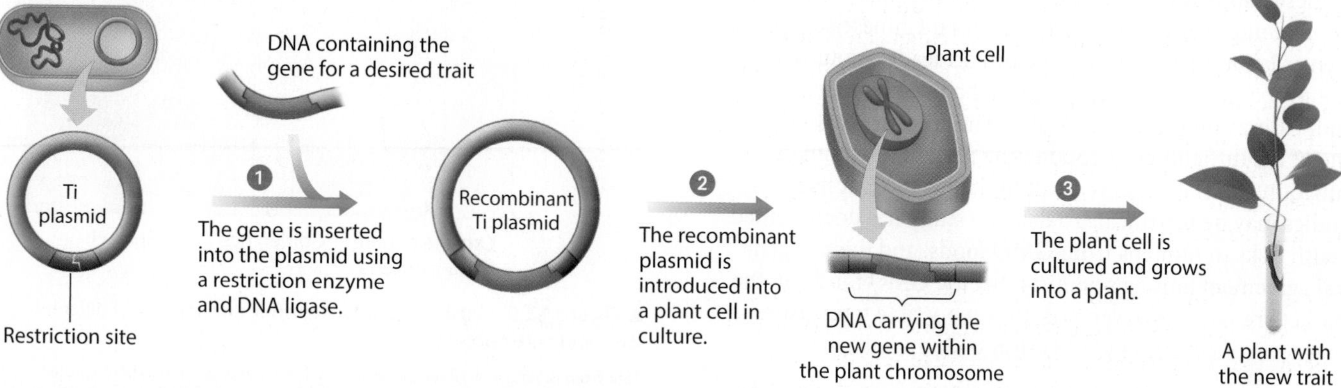

Agrobacterium tumefaciens
DNA containing the gene for a desired trait
Ti plasmid
Restriction site
1
The gene is inserted into the plasmid using a restriction enzyme and DNA ligase.
Recombinant Ti plasmid
2
The recombinant plasmid is introduced into a plant cell in culture.
Plant cell
DNA carrying the new gene within the plant chromosome
3
The plant cell is cultured and grows into a plant.
A plant with the new trait

▲ **Figure 12.8A** Using the Ti plasmid to genetically engineer plants

12.9 Genetically modified organisms raise health concerns

SCIENTIFIC THINKING

Are genetically modified organisms safe?

As soon as scientists realized the power of DNA technology, they began to worry about potential dangers. Early concerns focused on the possibility that recombinant DNA technology might create new pathogens. To guard against rogue microbes, scientists developed a set of guidelines including strict laboratory safety and containment procedures, the genetic crippling of transgenic organisms to ensure that they cannot survive outside the laboratory, and a prohibition on certain dangerous experiments. Today, most public concern centers on GMOs used for food.

Human Safety Genetically modified organisms are used in crop production because they are more nutritious or because they are cheaper to produce. But do these advantages come at a cost to the health of people consuming GMOs? When investigating complex questions like this one, scientists often use multiple experimental methods. A 2012 animal study involved 104 pigs that were divided into two groups: The first was fed a diet containing 39% GMO corn and the other a closely related non-GMO corn. The health of the pigs was measured over the short term (31 days), the medium term (110 days), and the normal generational life span. The researchers reported no significant differences between the two groups and no traces of foreign DNA in the slaughtered pigs.

Although pigs are a good model organism for human digestion, critics argue that human data are required to draw conclusions about the safety of dietary GMOs for people. The results of one human study, conducted jointly by Chinese and American scientists, were published in 2012. Sixty-eight Chinese schoolchildren (ages 6–8) were fed Golden Rice, spinach (a natural source of beta-carotene), or a capsule containing pure beta-carotene. Over 21 days, blood samples were drawn to measure how much vitamin A the body produced from each food source. The data show that the beta-carotene in both Golden Rice and the capsules was converted to vitamin A in the body with similar efficiency, while the beta-carotene in spinach led to significantly less vitamin A (**Figure 12.9**). The results led researchers to conclude that GMO rice can indeed be effective in preventing vitamin A deficiency.

Despite its positive findings, this study caused an uproar. Chinese authorities called the study an unethical "scandal," complaining that U.S. scientists had used Chinese schoolchildren as laboratory subjects. The project leaders countered that proper permission and consent had been obtained in both China and the United States. The controversy highlights one of the difficulties in conducting research on human nutrition: Animal studies are of limited value, but human studies may be unethical. To date, no study has documented health risks in humans from GMO foods, and there is general agreement among scientists that the GMO foods on the market are safe. However, it is not yet possible to measure the long-term effects (if any) of GMOs on human health.

Environmental Safety Advocates of a cautious approach toward GMO crops fear that transgenic plants might pass their new genes to related species in nearby wild areas, disturbing the composition of the natural ecosystem. Critics of GMO crops can point to several studies that do indeed show unintended gene transfer from engineered crops to nearby wild relatives. But GMO advocates counter that no lasting or detrimental effects from such transfers have been demonstrated, and that some GMOs (such as bacteria engineered to break down oil spills) can actively help the environment.

Labeling Although the majority of several staple crops grown in the United States—including corn and soybeans—are genetically modified, products made from GMOs are not required to be labeled in any way. Chances are you ate a food containing GMOs today, but the lack of labeling means you probably can't say for certain. Labeling of foods containing more than trace amounts of GMOs is required in Europe, Japan, Australia, China, Russia, and other countries. Labeling advocates point out that the information would allow consumers to decide for themselves whether they wish to be exposed to GMO foods. Some biotechnology advocates, however, respond that similar demands were not made when "transgenic" crop plants produced by traditional breeding techniques were put on the market. For example, triticale (a crop used primarily in animal feed but also in some human foods) was created decades ago by combining the genomes of wheat and rye—two plants that do not interbreed in nature. Triticale is now sold worldwide without any special labeling.

Scientists and the public need to weigh the possible benefits versus risks on a case-by-case basis. The best scenario would be to proceed with caution, basing our decisions on sound scientific information rather than on either irrational fear or blind optimism.

? Why might crop plants engineered to be resistant to weed killer pose a danger to the environment?

The genes for herbicide resistance could transfer to closely related weeds, which could themselves then become resistant.

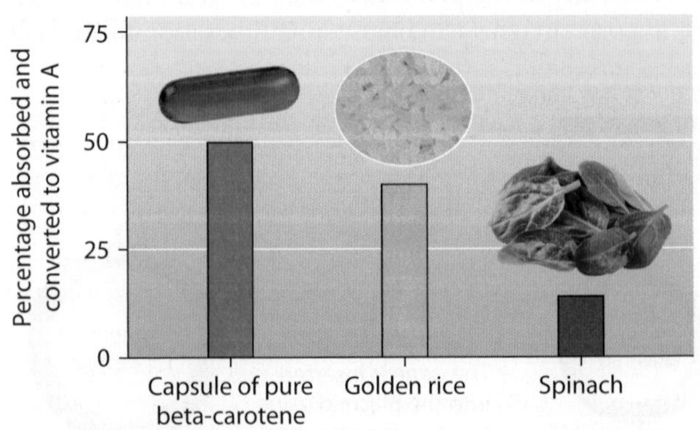

Percentage absorbed and converted to vitamin A
75
50
25
0
Capsule of pure beta-carotene
Golden rice
Spinach

▲ **Figure 12.9** Vitamin A production after consumption of different sources of beta-carotene

Data from G. Tang et al., Beta-carotene in Golden Rice is as good as beta-carotene in oil at providing vitamin A to children, *American Journal of Clinical Nutrition* 96(3): 658–64 (2012).

12.10 Gene therapy may someday help treat a variety of diseases

CONNECTION

So far in this chapter, we have discussed transgenic viruses, bacteria, yeast, plants, and animals. What about transgenic humans? Why would anyone want to insert genes into a living person?

One reason to tamper with the human genome is the potential of **gene therapy**—alteration of a diseased individual's genes for therapeutic purposes. In people afflicted with disorders caused by a single defective gene, it might be possible to replace or supplement the defective gene by inserting a normal allele into cells of the tissue affected by the disorder. Once there, the normal allele might be expressed, potentially offering a permanent cure after just one treatment.

For gene therapy to be permanent, the normal allele would have to be transferred to cells that multiply throughout a person's life. Bone marrow cells, which include the stem cells that give rise to all the cells of the blood and immune system, are prime candidates **(Figure 12.10)**. ❶ A gene from a healthy person is cloned, converted to an RNA version, and then inserted into the RNA genome of a harmless virus. ❷ Bone marrow cells are taken from the patient and infected with the recombinant virus. ❸ The virus inserts a DNA version of its genome, including the normal human gene, into the cells' DNA. ❹ The engineered cells are then injected back into the patient. If the procedure succeeds, the cells will multiply throughout the patient's life and produce a steady supply of the missing protein, curing the patient.

The promise of gene therapy thus far exceeds actual results, but there have been some successes. From 2000 to 2011, gene therapy cured 22 children with severe combined immunodeficiency (SCID), a fatal inherited disease caused by a defective gene that prevents development of the immune system, requiring patients to live within protective "bubbles." Unless treated with a bone marrow transplant, which is effective only 60% of the time, SCID patients quickly die from infections by microbes that most of us fend off. Although the gene therapy treatment cured the patients of SCID, there were some serious side effects: Four of the treated patients developed leukemia, and one died after the inserted gene turned the blood cells cancerous.

A 2009 gene therapy trial involved a disease called Leber's congenital amaurosis (LCA). People with one form of LCA produce abnormal rhodopsin, a pigment that enables the eye to detect light. In such people, photoreceptor cells gradually die, causing progressive blindness. Researchers found that a single injection—containing a virus carrying the normal gene—into one eye of affected children improved vision in that eye, sometimes enough to allow normal functioning.

Gene therapy raises difficult ethical questions. Some critics suggest that tampering with human genes in any way will inevitably lead to eugenics, the deliberate effort to control the genetic makeup of human populations. Other observers see no fundamental difference between the transplantation of genes into somatic cells and the transplantation of organs.

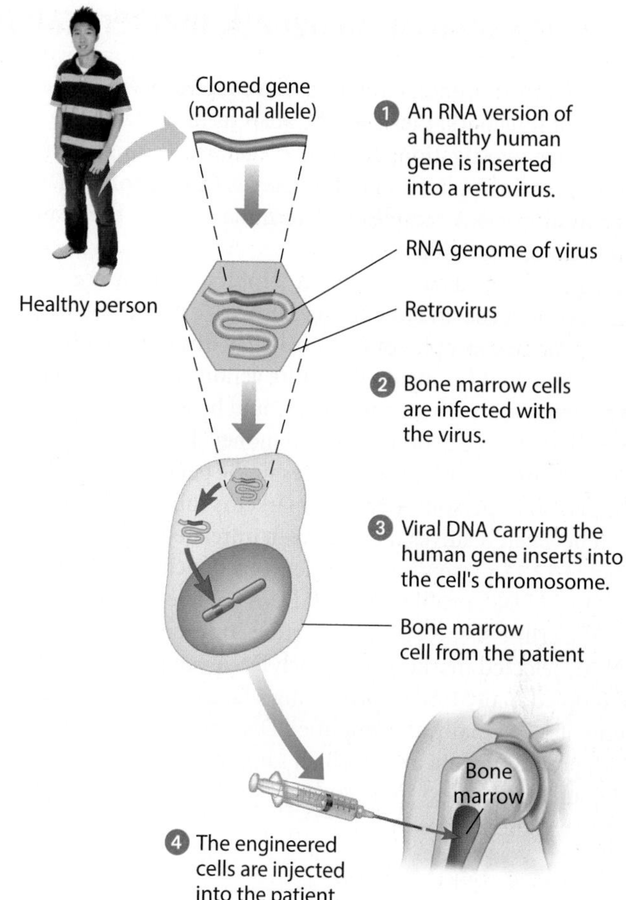

Cloned gene (normal allele)
❶ An RNA version of a healthy human gene is inserted into a retrovirus.
Healthy person
RNA genome of virus
Retrovirus
❷ Bone marrow cells are infected with the virus.
❸ Viral DNA carrying the human gene inserts into the cell's chromosome.
Bone marrow cell from the patient
Bone marrow
❹ The engineered cells are injected into the patient.

▲ **Figure 12.10** One type of gene therapy procedure

The implications of genetically manipulating gamete-forming cells or zygotes (already accomplished in lab animals) are more problematic. This possibility raises the most difficult ethical questions of all: Should we try to eliminate genetic defects in our children and their descendants? Should we interfere with evolution in this way? From a biological perspective, the elimination of unwanted alleles from the gene pool could backfire. Genetic variety is a necessary ingredient for the survival of a species as environmental conditions change with time. Genes that are damaging under some conditions may be advantageous under others (one example is the sickle-cell allele; see Module 9.13). Are we willing to risk making genetic changes that could be detrimental to our species in the future? We may have to face this question soon.

? Why does bone marrow make a good target for gene therapy?

● Bone marrow cells multiply throughout a person's life and contain stem cells that give rise to different kinds of blood cells.

▷ DNA Profiling

12.11 The analysis of genetic markers can produce a DNA profile

Modern DNA technology methods have rapidly transformed the field of **forensics**, the scientific analysis of evidence for crime scene investigations and other legal proceedings. The most important application to forensics is **DNA profiling**, the analysis of DNA samples to determine whether they came from the same individual.

Imagine that you have two DNA samples, perhaps one from a crime scene and one from a suspect. How do you test whether the two samples of DNA originate from the same person? You could compare the entire genomes found in the two samples, but such an approach would be extremely impractical, requiring a lot of time and money. Instead, scientists compare genetic markers, sequences in the genome that vary from person to person. A genetic marker is more likely to be a match between relatives than between unrelated individuals.

Figure 12.11 presents an overview of a typical investigation involving a DNA profile. ❶ First, DNA samples are isolated from the crime scene, suspects, victims, or other evidence. ❷ Next, selected markers from each DNA sample are amplified (copied many times), producing a large sample of DNA fragments. ❸ Finally, the amplified DNA markers are compared, providing data about which samples are from the same individual. In the next four modules, we'll explore the methods behind these steps in detail.

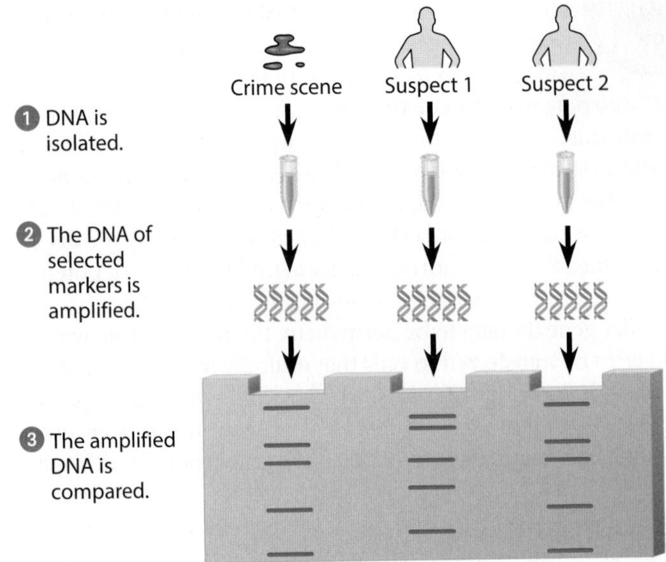

Crime scene
Suspect 1
Suspect 2
1 DNA is isolated.
2 The DNA of selected markers is amplified.
3 The amplified DNA is compared.

▲ **Figure 12.11** An overview of DNA profiling

? **According to the data presented in Figure 12.11, which suspect left DNA at the crime scene?**

● Suspect 2: Notice that the number and location of the DNA markers in Suspect 2's DNA and the crime scene DNA match.

12.12 The PCR method is used to amplify DNA sequences

The **polymerase chain reaction (PCR)** is a technique by which a specific segment of a DNA molecule can be targeted and quickly amplified in the laboratory. Starting with a minute sample, automated PCR can generate billions of copies of a DNA segment in just a few hours, producing enough DNA to allow a DNA profile to be constructed.

In principle, PCR is fairly simple (**Figure 12.12**). A repeated cycle brings about a chain reaction that doubles the population of identical DNA molecules during each round. The key to amplifying one particular segment of DNA and no others is the use of **primers**, short (usually 15–20 nucleotides long), chemically synthesized single-stranded DNA molecules with

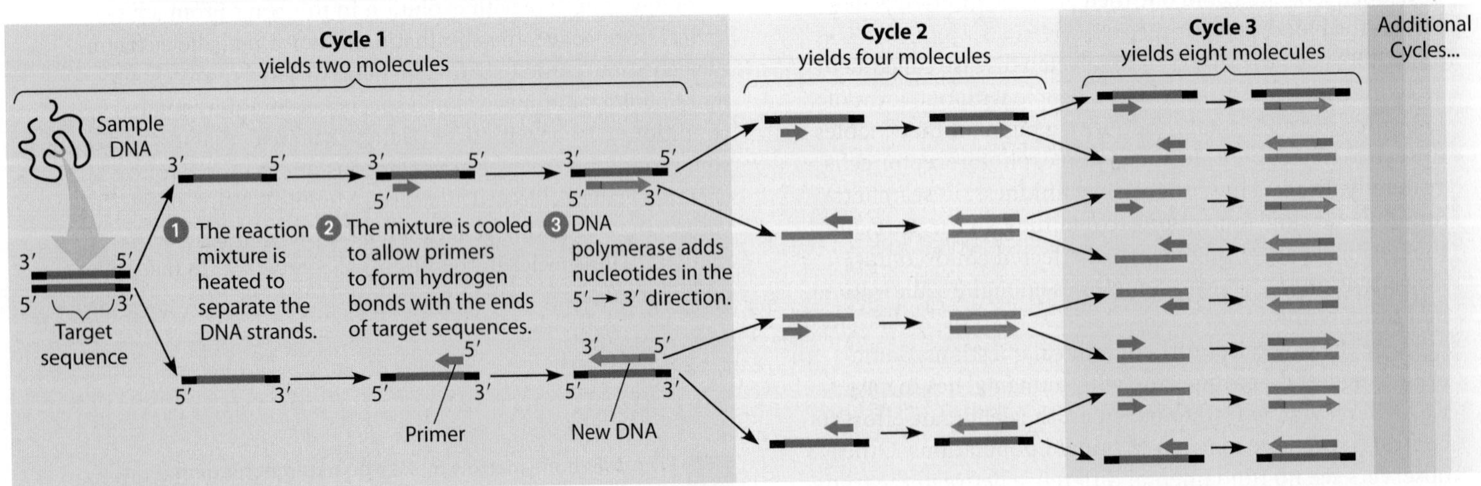

Cycle 1
yields two molecules
Cycle 2
yields four molecules
Cycle 3
yields eight molecules
Additional Cycles...
Sample DNA
Target sequence
1 The reaction mixture is heated to separate the DNA strands.
2 The mixture is cooled to allow primers to form hydrogen bonds with the ends of target sequences.
3 DNA polymerase adds nucleotides in the 5′→ 3′ direction.
Primer
New DNA

▲ **Figure 12.12** DNA amplification by PCR

Try This If each cycle takes 15 minutes, calculate the number of copies of the original DNA molecule that will be present after 6 hours.

sequences that are complementary to sequences at each end of the target sequence. One primer is complementary to one strand at one end of the target sequence; the second primer is complementary to the other strand at the other end of the sequence. The primers thus bind to sequences that flank the target sequence, marking the start and end points for the segment of DNA being amplified.

❶ In the first step of each PCR cycle, the reaction mixture is heated to separate the strands of the DNA double helices. ❷ Next, the strands are cooled. As they cool, primer molecules hydrogen-bond to their target sequences on the DNA. ❸ Then a heat-stable DNA polymerase builds new DNA strands by extending the primers in the $5' \rightarrow 3'$ direction. These three steps are repeated over and over, doubling the amount of DNA after each three-step cycle. A key prerequisite for automating PCR was the discovery of an unusual DNA polymerase, first isolated from a bacterium living in hot springs, that could withstand the heat at the start of each cycle. Without such a heat-stable polymerase, PCR would not be possible because standard DNA polymerases would denature (unfold) during the heating step of each cycle.

Only minute amounts of DNA need to be present in the starting material, and this DNA can even be in a partially degraded state. The key to the high sensitivity is the primers. Because the primers only bind the sequences associated with the target, the DNA polymerase duplicates only the desired segments of DNA. Other DNA will not be bound by primers and thus not copied by the DNA polymerase.

Devised in 1985, PCR has had a major impact on biological research and biotechnology. PCR has been used to amplify DNA from a wide variety of sources: fragments of ancient DNA from a mummified human, a 40,000-year-old frozen woolly mammoth, and a 30-million-year-old plant fossil; DNA from fingerprints or from tiny amounts of blood, tissue, or semen found at crime scenes; DNA from single embryonic cells for rapid prenatal diagnosis of genetic disorders; and DNA of viral genes from cells infected with viruses that are difficult to detect, such as HIV.

? Why does PCR amplify only one specific region of DNA rather than all of it?

● The primers mark the ends, ensuring that only the DNA within the region between is amplified.

12.13 Gel electrophoresis sorts DNA molecules by size

Many DNA technology applications rely on **gel electrophoresis**, a method that separates macromolecules—usually proteins or nucleic acids—on the basis of size, electrical charge, or other physical properties. A gel is a thin rectangle of jellylike material often made from agarose, a carbohydrate polymer extracted from seaweed. Because agarose contains a tangle of cable-like threads, it can act as a molecular sieve.

Figure 12.13 outlines how gel electrophoresis can be used to separate mixtures of DNA fragments obtained from three different sources. A DNA sample from each source is placed in a separate well (hole) at one end of a gel that is suspended in liquid. A negatively charged electrode from a power supply is attached near the end of the gel containing the DNA, and a positive electrode is attached near the far end. Because all nucleic acid molecules carry negative charges on their phosphate groups (PO_4^-; see Module 10.2), the DNA molecules all travel through the gel toward the positive pole. However, longer DNA fragments are held back by the thicket of polymer fibers within the gel, so they move more slowly than the shorter fragments. Over time, shorter molecules move farther through the gel than longer fragments. Gel electrophoresis thus separates DNA fragments by length, with shorter molecules migrating toward the bottom faster.

When the current is turned off, a series of bands is left in each "lane" of the gel. Each band is a collection of DNA fragments of the same length. The bands can be made visible by staining, by exposure onto photographic film (if the DNA is radioactively labeled), or by measuring fluorescence (if the DNA is labeled with a fluorescent dye).

? What causes DNA molecules to move toward the positive pole during electrophoresis? Why do large molecules move more slowly than smaller ones?

● The negatively charged phosphate groups of the DNA are attracted to the positive pole; the gel restricts the movement of longer fragments more.

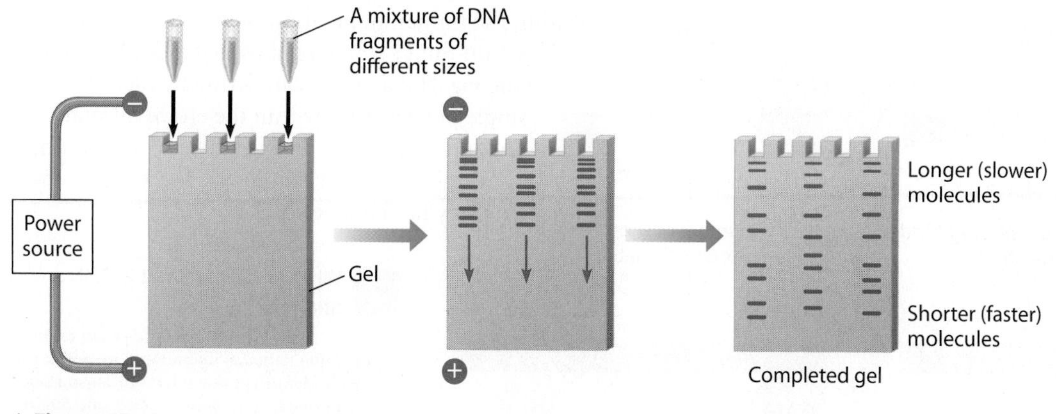
A mixture of DNA fragments of different sizes
Power source
Gel
Longer (slower) molecules
Shorter (faster) molecules
Completed gel

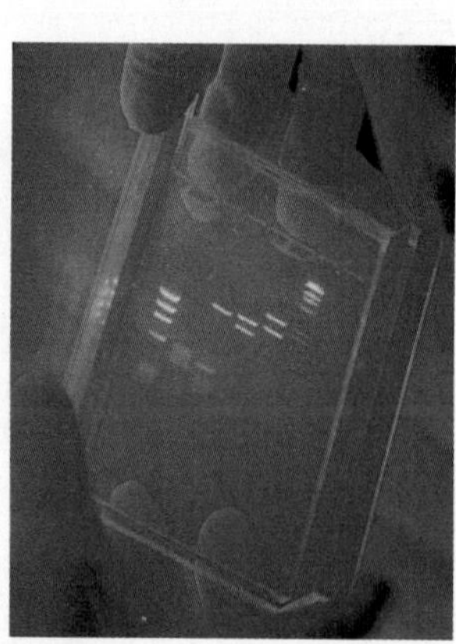

▲ **Figure 12.13** Gel electrophoresis of DNA

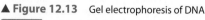

Try This Explain why the shortest DNA molecules end up at the bottom of the gel.

12.14 Short tandem repeat analysis is commonly used for DNA profiling

Now that we've learned about DNA amplification by PCR and gel electrophoresis, let's see how these methods can be combined in DNA profiling. To create a DNA profile, a forensic scientist gathers data about a predefined set of genetic markers. The genetic markers most often used in DNA profiling are inherited variations in the lengths of repetitive DNA segments. **Repetitive DNA** consists of nucleotide sequences that are present in multiple copies in the genome; much of the DNA that lies between genes in humans is of this type. Some regions of repetitive DNA vary considerably from one individual to the next.

The repetitive DNA used in genetic DNA profiles consists of short sequences repeated many times in a row; such a series of repeats is called a **short tandem repeat (STR)**. For example, one person might have the sequence AGAT repeated 12 times in a row at one place in the genome, the sequence GATA repeated 45 times in a row at a second place, and so on. Another person has the same sequences at the same places but with different numbers of repeats. These stretches of repetitive DNA, like any genetic marker, are more likely to be an exact match between relatives than between unrelated individuals.

STR analysis is a method of DNA profiling that compares the lengths of STR sequences at specific sites in the genome. The current standard for DNA profiling in forensic and legal systems compares the number of repeats of specific four-nucleotide DNA sequences at 13 sites scattered throughout the genome. Each of these repeat sites, which typically contain from 3 to 50 four-nucleotide repeats in a row, vary widely from person to person. In fact, some of the STRs used in the standard procedure can be found in up to 80 different variations in the human population. In the United States, the number of repeats at each site is entered into a database called CODIS (Combined DNA Index System) administered by the Federal Bureau of Investigation (FBI). Law enforcement agencies around the world can access CODIS to search for matches to DNA samples they have obtained from crime scenes or suspects.

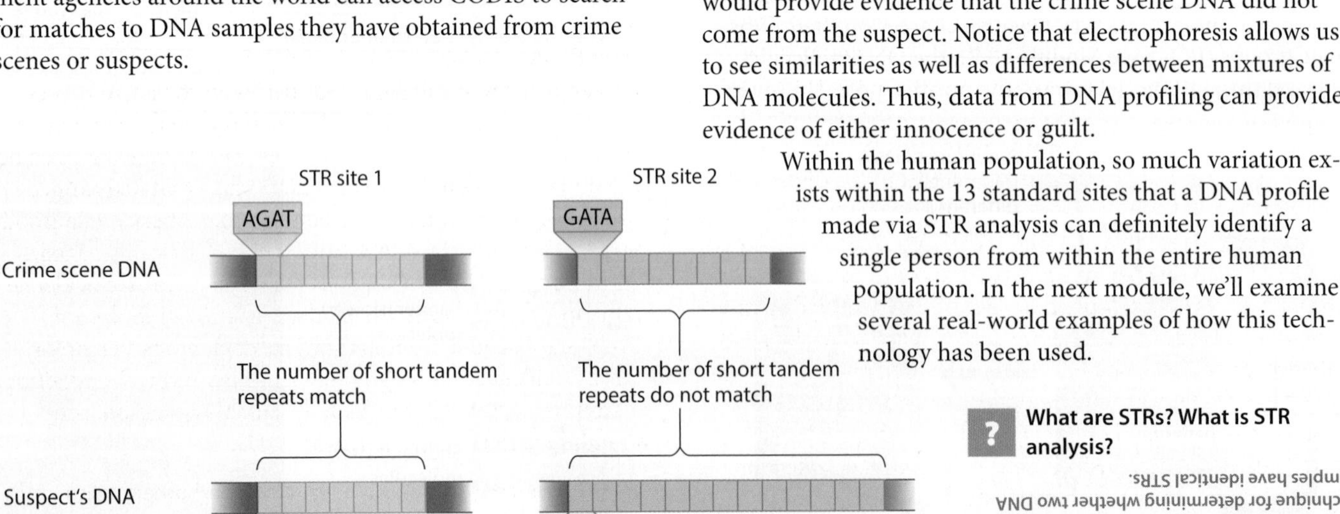
STR site 1
STR site 2
AGAT
GATA
Crime scene DNA
The number of short tandem repeats match
The number of short tandem repeats do not match
Suspect's DNA
AGAT
GATA

▲ **Figure 12.14A** Two representative STR sites from crime scene DNA samples

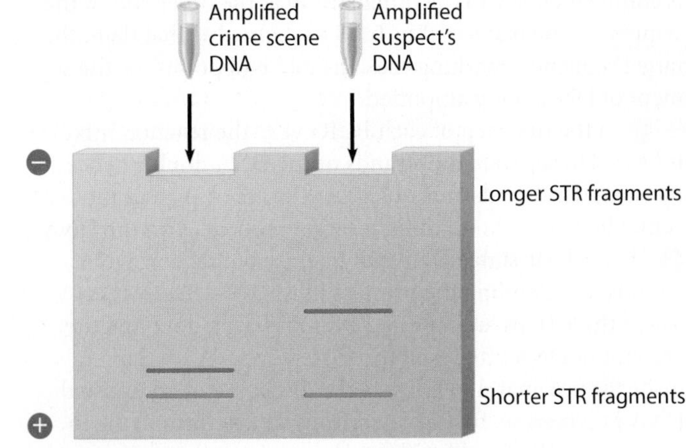
Amplified crime scene DNA
Amplified suspect's DNA
Longer STR fragments
Shorter STR fragments

▲ **Figure 12.14B** DNA profiles generated from the STRs in Figure 12.14A

Consider the two samples of DNA shown in **Figure 12.14A**, where the top DNA was obtained at a crime scene and the bottom DNA from a suspect. The two segments have the same number of repeats at the first site: 7 repeats of the four-nucleotide DNA sequence AGAT (shown in orange). Notice, however, that they differ in the number of repeats at the second site: 8 repeats of GATA (shown in purple) in the crime scene DNA, compared with 13 repeats in the suspect's DNA. To create a DNA profile, a scientist uses PCR to specifically amplify the regions of DNA that include these STR sites. This can be done by using primers matching nucleotide sequences known to flank the STR sites. The resulting DNA molecules are then compared by gel electrophoresis.

Figure 12.14B shows a gel that could have resulted from the STR fragments in Figure 12.14A. The differences in the locations of the bands reflect the different lengths of the DNA fragments. (A gel from an actual DNA profile would typically contain more than just two bands in each lane.) This gel would provide evidence that the crime scene DNA did not come from the suspect. Notice that electrophoresis allows us to see similarities as well as differences between mixtures of DNA molecules. Thus, data from DNA profiling can provide evidence of either innocence or guilt.

Within the human population, so much variation exists within the 13 standard sites that a DNA profile made via STR analysis can definitely identify a single person from within the entire human population. In the next module, we'll examine several real-world examples of how this technology has been used.

? What are STRs? What is STR analysis?

● STRs are regions of the genome that contain varying numbers of sequential repeats of a short nucleotide sequence; STR analysis is a technique for determining whether two DNA samples have identical STRs.

12.15 DNA profiling has provided evidence in many forensic investigations

CONNECTION

When a violent crime is committed, body fluids or small pieces of tissue may be left at the crime scene or on the clothes of the victim or assailant. If rape has occurred, semen may be recovered from the victim. DNA profiling can match such samples to the person they came from with a high degree of certainty because the DNA sequence of every person is unique (except for identical twins). PCR amplification allows tissue samples comprising as few as 20 cells to be tested.

Since its introduction in 1986, DNA profiling has become a standard tool of forensics and has provided crucial evidence in many famous cases. DNA profiling first gained wide public attention during the O. J. Simpson murder trial, when DNA analysis proved that blood in Simpson's car belonged to the victims and that blood at the crime scene belonged to Simpson. (The jury in this case did not find the DNA evidence alone to be sufficient, and Simpson was found not guilty.) During the investigation that led to his impeachment, President Bill Clinton repeatedly denied that he had sexual relations with Monica Lewinsky—until DNA profiling proved that his semen was on her dress.

DNA evidence can prove innocence as well as guilt. Lawyers at the Innocence Project, a nonprofit organization dedicated to overturning wrongful convictions, have used DNA technology and legal work to exonerate more than 300 convicted criminals since 1989, including 17 who were on death row (**Figure 12.15A**). In more than a third of these cases, DNA profiling also identified the true perpetrators.

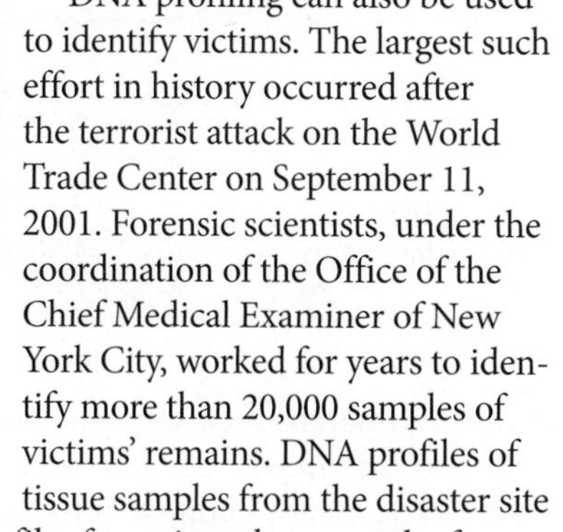

▲ **Figure 12.15A** Earl Washington, a convicted murderer who was freed after 17 years in prison thanks to an STR analysis that proved his innocence

DNA profiling can also be used to identify victims. The largest such effort in history occurred after the terrorist attack on the World Trade Center on September 11, 2001. Forensic scientists, under the coordination of the Office of the Chief Medical Examiner of New York City, worked for years to identify more than 20,000 samples of victims' remains. DNA profiles of tissue samples from the disaster site were matched to DNA profiles from tissue known to be from the victims. If no sample of a victim's DNA was available, blood samples from close relatives were used to confirm identity through near matches. More than half of the identified victims at the World Trade Center site were recognized solely by DNA evidence, providing closure to many grieving families.

The use of DNA profiling extends beyond crimes. For instance, a comparison of the DNA of a child and the purported father can conclusively settle a question of paternity. Sometimes, paternity is of historical interest: DNA profiling proved that Thomas Jefferson or a close male relative of Jefferson's fathered a child with his slave Sally Hemings. Going back much further, one of the strangest cases of DNA profiling is that of Cheddar Man, a 9,000-year-old skeleton found in a cave near Cheddar, England (**Figure 12.15B**). DNA was extracted from his tooth and analyzed. The DNA profile showed that Cheddar Man was a direct ancestor—through approximately 300 generations—of a present-day schoolteacher who lived only a half mile from the cave!

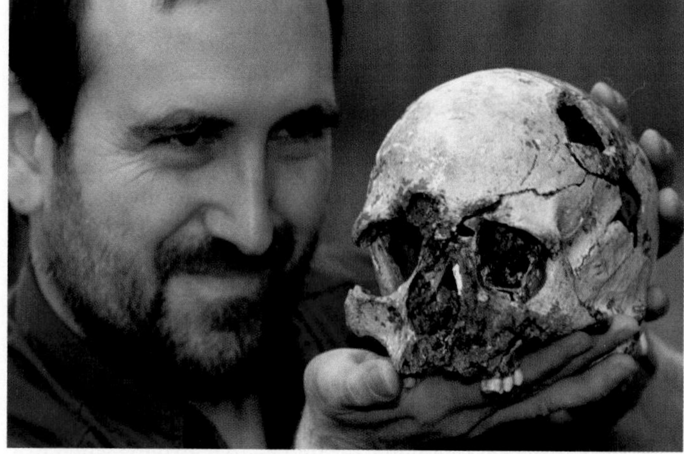

▲ **Figure 12.15B** Cheddar Man and one of his modern-day descendants

Just how reliable is DNA profiling? When the standard CODIS set of 13 STR sites (see Module 12.14) is used, the probability of finding the same DNA profile in randomly selected, unrelated individuals is less than one in 10 billion. Put another way, a standard DNA profile can provide a statistical match of a particular DNA sample to just one living human. For this reason, DNA analyses are now accepted as compelling evidence by legal experts and scientists alike. In fact, DNA analysis on stored forensic samples has provided the evidence needed to solve many "cold cases" in recent years.

DNA analysis has also been used to probe the origin of nonhuman materials. For example, examination of DNA can prove the origin of food, as with a U.S. Fish and Wildlife Service program that tests caviar to determine if the fish eggs originate from the species claimed on the label. In addition, DNA profiling can help protect endangered species by conclusively providing the origin of contraband animal products, allowing for increased vigilance in endangered regions. Animals can also be the subject of research, as with a 2005 study that determined that DNA extracted from a 27,000-year-old Siberian mammoth was 98.6% identical to DNA from modern African elephants.

Although DNA profiling has provided definitive evidence in many investigations, the method is far from foolproof. Problems can arise from insufficient data, human error, or flawed evidence. Although the science behind DNA profiling is irrefutable, the human element remains a possible confounding factor.

? **In what way is DNA profiling valuable for determining innocence as well as guilt?**

● A DNA profile can prove with near certainty that a sample of DNA does not come from a particular individual. DNA profiling therefore can provide evidence in support of guilt or innocence.

12.16 RFLPs can be used to detect differences in DNA sequences

Recall that a genetic marker is a DNA sequence that varies in a population. Like different alleles of a gene, the DNA sequence at a specific place on a chromosome may exhibit small nucleotide differences, or polymorphisms (from the Greek for "many forms"). Geneticists have cataloged many single-base-pair variations in the genome. Such a variation found in at least 1% of the population is called a **single nucleotide polymorphism (SNP**, pronounced "snip"). SNPs occur on average about once in 100 to 300 base pairs in the human genome, both in the coding sequences of genes and in noncoding sequences between genes. Scientists have identified several million SNP sites, and more are discovered each year.

SNPs may alter a restriction site—the sequence recognized by a restriction enzyme. Such alterations change the lengths of the restriction fragments formed by that enzyme when it cuts the DNA. A sequence variation of this type is called a **restriction fragment length polymorphism (RFLP**, pronounced "rif-lip"). Thus, RFLPs can serve as genetic markers for particular loci in the genome. RFLPs have many uses. For example, disease-causing alleles can be diagnosed with reasonable accuracy if a closely linked RFLP marker has been found. Alleles for a number of genetic diseases were first detected by means of RFLPs in this indirect way.

Restriction fragment analysis involves two of the methods you have learned about: DNA fragments produced by restriction enzymes (Module 12.2) are sorted by gel electrophoresis (Module 12.13). The number of restriction fragments and their sizes reflect the specific sequence of nucleotides in the starting DNA.

At the top of **Figure 12.16**, you can see corresponding segments of DNA from two DNA samples prepared from human tissue. Notice that the DNA sequences differ by a single base pair (highlighted in yellow). In this case, the restriction enzyme cuts DNA between two cytosine (C) bases in the sequence CCGG and in its complement, GGCC. Because DNA from the first sample has two recognition sequences for the restriction enzyme, it is cleaved in two places, yielding three restriction fragments (labeled *w*, *x*, and *y*). DNA from the second sample, however, has only one recognition sequence and yields only two restriction fragments (*z* and *y*). Notice that the lengths of restriction fragments, as well as the number of fragments, differ, depending on the exact sequence of bases in the DNA.

To detect the differences between the collections of restriction fragments, we need to separate the restriction fragments in the two mixtures and compare their lengths. This process, called RFLP analysis, is accomplished through gel electrophoresis. As shown in the bottom of the figure, the three kinds of restriction fragments from sample 1 separate into three bands in the gel, whereas those from sample 2 separate into only two bands. Notice that the shortest fragment from sample 1 (*y*) produces a band at the same location as the identical short fragment from the sample 2. So you can see that electrophoresis allows us to see similarities as well as differences between mixtures of restriction fragments—and similarities as well as differences between the base sequences in DNA from two individuals. The restriction fragment analysis in Figure 12.16 clearly shows that the two DNA samples differ in sequence. Although RFLP analysis is rarely used today for identification, this method was vital in some of the earliest discoveries of disease-causing genes. For example, the gene for Huntington's disease was found after researchers used RFLPs to track a genetic marker that was closely associated with the disorder.

▶ **Figure 12.16** RFLP analysis

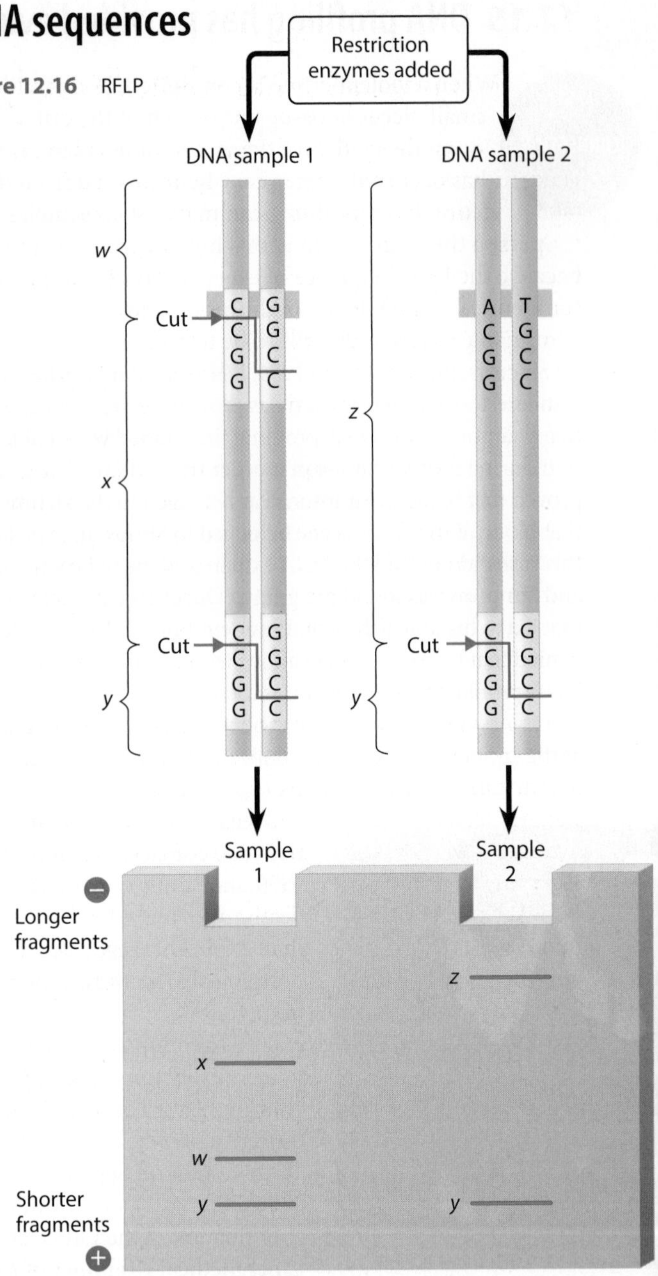

? You use a restriction enzyme to cut a DNA molecule that has three copies of the enzyme's recognition sequence clustered near one end. When you separate the restriction fragments by gel electrophoresis, how do you expect the bands to look?

● There should be three bands near the positive pole at the bottom of the gel (small fragments) and one band near the negative pole at the top of the gel (large fragment).

Genomics

12.17 Genomics is the scientific study of whole genomes

By the 1980s, biologists were using RFLPs to map important genes in humans and some other organisms. But it didn't take long for biologists to think on a larger scale. In 1995, a team of scientists determined the nucleotide sequence of the entire genome of *Haemophilus influenzae*, a bacterium that can cause several human diseases, including pneumonia and meningitis. **Genomics**, the study of complete sets of genes (genomes) and their interactions, was born.

Since 1995, researchers have used the tools and techniques of DNA technology to develop more and more detailed maps of the genomes of a number of species. The first targets of genomics research were bacteria, which have relatively little DNA. The *H. influenza* genome, for example, contains only 1.8 million nucleotides and 1,709 genes. But soon, the attention of genomics researchers turned toward more complex organisms with much larger genomes. As of 2013, the genomes of nearly 7,000 species have been completed, and thousands more are in progress. **Table 12.17** lists some of the completed genomes; for diploids, the size refers to the haploid genome. The majority of genomes under study are from prokaryotes, including *E. coli* and several thousand other bacteria (some of medical importance), and more than 200 Archaea. Over 300 eukaryotic species have been sequenced, including vertebrates, invertebrates, fungi, and plants.

Baker's yeast (*Saccharomyces cerevisiae*) was the first eukaryote to have its full sequence determined, and the roundworm *Caenorhabditis elegans* was the first multicellular organism. Other sequenced animals include the fruit fly (*Drosophila melanogaster*) and the laboratory mouse (*Mus musculus*), both model organisms for genetics research. Plants, such as one type of mustard (*Arabidopsis thaliana*, an important research organism) and rice (*Oryza sativa*, one of the world's most economically important crops), have also been completed. Other recently completed eukaryotic genomes include sorghum (another important commercial crop), the honeybee, dog, wallaby (a marsupial), turkey, and bottlenose dolphin.

In 2005, researchers completed the genome sequence for our closest living relative on the evolutionary tree of life, the chimpanzee (*Pan troglodytes*). Comparisons with human DNA revealed that we share 96% of our genome with our closest animal relative. As you will see in Module 12.21, genomic scientists are currently finding and studying the important differences, shedding scientific light on the age-old question of what makes us human.

Why map so many genomes? Not only are all genomes of interest in their own right, but comparative analysis provides invaluable insights into the evolutionary relationships among organisms. Also, having maps of a variety of genomes helps scientists interpret the human genome. For example, when scientists find a nucleotide sequence in the human genome similar to a yeast gene whose function is known, they have a valuable clue to the function of the human sequence. In fact, the roles of several human disease-causing genes were determined by studying their yeast counterparts. Indeed, many genes of disparate organisms are being found to be astonishingly similar: Some researchers joke that fruit flies can even be thought of as "little people with wings."

? Does a larger genome always correlate with more genes?

No. Compare, for example, the genomes of rice and mice.

TABLE 12.17 | SOME IMPORTANT COMPLETED GENOMES

Organism	Year Completed	Size of Haploid Genome (in Base Pairs)	Approximate Number of Genes
Haemophilus influenzae (bacterium)	1995	1.8 million	1,700
Saccharomyces cerevisiae (yeast)	1996	12 million	6,300
Escherichia coli (bacterium)	1997	4.6 million	4,400
Caenorhabditis elegans (nematode)	1998	100 million	20,100
Drosophila melanogaster (fruit fly)	2000	165 million	14,000
Arabidopsis thaliana (mustard plant)	2000	120 million	27,000
Mus musculus (mouse)	2001	2.6 billion	22,000
Oryza sativa (rice)	2002	430 million	42,000
Homo sapiens (humans)	2003	3.0 billion	21,000
Rattus norvegicus (lab rat)	2004	2.8 billion	25,000
Pan troglodytes (chimpanzee)	2005	3.1 billion	22,000
Macaca mulatta (macaque)	2007	2.9 billion	22,000
Macropus eugenii (wallaby)	2012	2.9 billion	18,000

12.18 The Human Genome Project revealed that most of the human genome does not consist of genes

CONNECTION

The **Human Genome Project (HGP)** was a massive, long-term scientific endeavor with the goals of determining the nucleotide sequence of all DNA in the human genome and identifying the location and sequence of every gene. The HGP began in 1990 and was completed in 2003. More than 99% of the human genome has been determined to 99.999% accuracy. (There remain a few hundred gaps of unknown sequences within the human genome that remain elusive.) The DNA sequences determined by the HGP and related projects have been deposited in a publicly available database called GenBank.

The chromosomes in the human genome—22 autosomes plus the X and Y sex chromosomes—contain approximately 3 billion nucleotide pairs of DNA. To get a sense of this much DNA, imagine that its nucleotide sequence is printed in letters (A, T, C, and G) like the letters in this book. At this size, the sequence would fill a stack of books 18 stories high! The biggest surprise from the HGP is the small number of human genes. The current estimate is just below 21,000 genes—very close to the number found in a microscopic worm. How, then, do we account for human complexity? Part of the answer may lie in alternative RNA splicing (see Module 11.4); scientists think that a typical human gene specifies several different polypeptides.

In humans, as in most complex eukaryotes, only a small amount of our total DNA (about 1.5%) is contained in genes that code for proteins, tRNAs, or rRNAs (**Figure 12.18**). Most multicellular eukaryotes have a huge amount of noncoding DNA; about 98.5% of human DNA is of this type. About one-quarter of our DNA consists of introns and gene control sequences such as promoters, enhancers, and microRNAs (see Chapter 11). The remaining noncoding DNA had been dubbed "junk DNA," a tongue-in-cheek way of saying that scientists don't fully understand its functions, but it is now generally accepted that much of this DNA probably plays some role. This stands in sharp contrast to a typical prokaryotic genome, in which the vast majority of DNA serves an obvious function; the human genome contains roughly 10,000 times more noncoding DNA than the *E. coli* genome.

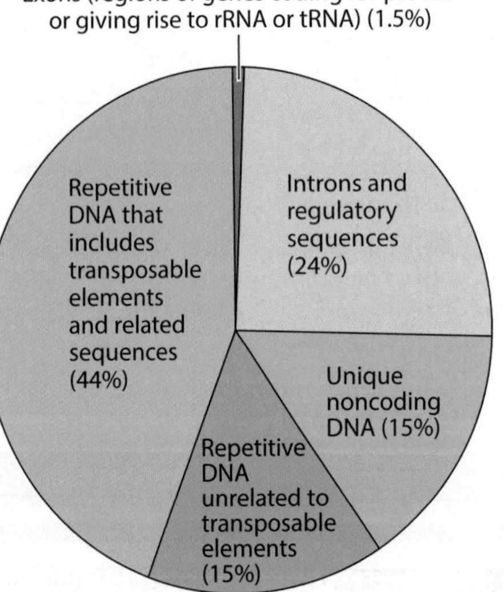

Exons (regions of genes coding for protein or giving rise to rRNA or tRNA) (1.5%)
Repetitive DNA that includes transposable elements and related sequences (44%)
Introns and regulatory sequences (24%)
Unique noncoding DNA (15%)
Repetitive DNA unrelated to transposable elements (15%)

▲ **Figure 12.18** Composition of the human genome

Much of the DNA between genes consists of repetitive DNA, nucleotide sequences present in many copies in the genome. The repeated units of some of this DNA are short, such as the STRs used in DNA profiling (see Module 12.14). Stretches of DNA with thousands of short repetitions are also prominent at the centromeres and ends of chromosomes—called **telomeres**—suggesting that this DNA plays a role in chromosome structure.

In the second main type of repetitive DNA, each repeated unit is hundreds of nucleotides long, and the copies are scattered around the genome. Most of these sequences seem to be associated with **transposable elements** ("jumping genes"), DNA segments that can move or be copied from one location to another in a chromosome and even between chromosomes.

The potential benefits of having a complete map of the human genome are enormous. For instance, hundreds of disease-associated genes have been identified. One example is the gene that is mutated in an inherited type of Parkinson's disease, a debilitating brain disorder that causes tremors of increasing severity. Until recently, Parkinson's disease was not known to have a hereditary component. But data from the Human Genome Project mapped a small number of cases of Parkinson's disease to a specific gene. Interestingly, an altered version of the protein encoded by this gene has also been tied to Alzheimer's disease, suggesting a previously unknown link between these two brain disorders. Moreover, the same gene is also found in rats, where it plays a role in the sense of smell, and in zebra finches, where it is thought to be involved in song learning. Cross-species comparisons such as these may uncover clues about the role played by the normal version of the protein in the human brain. And such knowledge could eventually lead to treatment for the half million Americans with Parkinson's disease.

One interesting question about the Human Genome Project is, Whose genome was sequenced? The first human genome to be sequenced was actually a reference genome compiled from a group of individuals. At nearly the same time, a private biotechnology company sequenced the genome of the company's president. These representative sequences will serve as standards so that comparisons of individual differences and similarities can be made. Starting in 2007, the genomes of a number of other individuals—the first was James Watson, codiscoverer of the structure of DNA—have also been sequenced. These sequences are part of a larger effort to collect information on all of the genetic variations that affect human characteristics. As the amount of sequence data multiplies, the small differences that account for individual variation within our species will come to light.

? **The haploid human genome consists of about __________ base pairs and __________ genes spread over __________ different chromosomes (provide three numbers).**

● 3 billion . . . 21,000 . . . 24 (22 autosomes plus 2 sex chromosomes)

12.19 The whole-genome shotgun method of sequencing a genome can provide a wealth of data quickly

Sequencing an entire genome is a complex task that requires careful work. The Human Genome Project proceeded through three stages that provided progressively more detailed views of the human genome. First, geneticists combined pedigree analyses of large families to map more than 5,000 genetic markers (mostly RFLPs) spaced throughout all of the chromosomes. The resulting low-resolution linkage map (see Module 9.19) provided a framework for mapping other markers and for arranging later, more detailed maps of particular regions. Next, researchers determined the number of base pairs between the markers in the linkage map. These data helped them construct a physical map of the human genome. Finally came the most arduous part of the project: determining the nucleotide sequences of the set of DNA fragments that had been mapped. Advances in automated DNA sequencing were crucial to this endeavor.

This three-stage approach is logical and thorough. However, in 1992, molecular biologist J. Craig Venter proposed an alternative strategy called the **whole-genome shotgun method** and set up the company Celera Genomics to implement it. His idea was essentially to skip the genetic and physical mapping stages and start directly with the sequencing step. In the whole-genome shotgun method, an entire genome is chopped by restriction enzymes into fragments that are cloned and sequenced in just one stage **(Figure 12.19)**. High-performance computers running specialized mapping software can assemble the millions of overlapping short sequences into a single continuous sequence for every chromosome—an entire genome.

Today, the whole-genome shotgun approach is the method of choice for genomic researchers because it is fast and relatively inexpensive. However, recent research has revealed some limitations of this method, such as difficulties with repetitive sequences, suggesting that a hybrid approach that combines whole-genome shotgunning with physical or genetic maps may prove to be the most useful method.

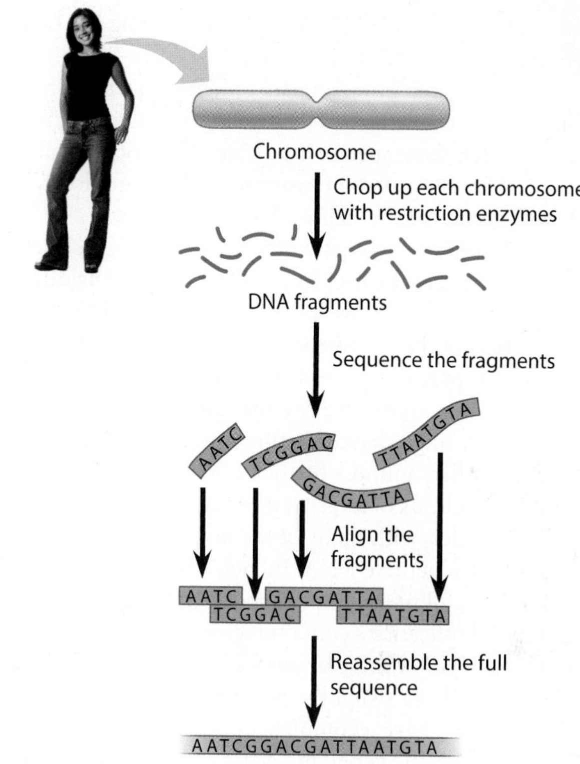

Chromosome
Chop up each chromosome with restriction enzymes
DNA fragments
Sequence the fragments
AATC
TCGGAC
TTAATGTA
GACGATTA
Align the fragments
AATC
GACGATTA
TCGGAC
TTAATGTA
Reassemble the full sequence
AATCGGACGATTAATGTA

▲ **Figure 12.19** The whole-genome shotgun method

? **What are the primary advantages of the whole-genome shotgun method?**

● It is faster and cheaper than the three-stage method of genome sequencing.

12.20 Proteomics is the scientific study of the full set of proteins encoded by a genome

The successes in the field of genomics have encouraged scientists to begin similar systematic studies of the full protein sets (proteomes) encoded by genomes, an approach called **proteomics**. The number of different proteins in humans far exceeds the number of different genes—about 100,000 proteins versus about 21,000 genes. And because proteins, not genes, actually carry out most of the activities of the cell, scientists must study when and where proteins are produced in an organism and how they interact to understand the functioning of cells and organisms. Given the huge number of proteins and the myriad ways that their production can be controlled, assembling and analyzing proteomes pose many experimental challenges. Ongoing advances are beginning to provide the tools to meet those challenges.

Genomics and proteomics are enabling biologists to approach the study of life from an increasingly holistic (whole-system) perspective. Biologists are now compiling catalogs of genes and proteins—that is, listings of all the "parts" that contribute to the operation of cells, tissues, and organisms. As such catalogs become complete, researchers are shifting their attention from the individual parts to how these parts function together in biological systems.

? **If every protein is encoded by a gene, how can humans have many more proteins than genes?**

● The RNA transcribed from one gene may be spliced several different ways to produce different mRNAs that are translated into different proteins (see Module 11.4).

12.21 Genomes hold clues to human evolution

EVOLUTION CONNECTION

Scientists are accumulating genomic sequence data at an astonishing pace. As of 2013, the publicly accessible GenBank included DNA sequences totaling 150 billion base pairs, and this total is doubling every 18 months. Geneticists can now compare genome sequences from many species, allowing hypotheses about evolutionary relationships between those species to be tested. The more similar in sequence the same gene is in two species, the more closely related those species are in their evolutionary history.

The small number of genetic differences between closely related species makes it easier to correlate phenotypic differences between the species with particular genetic differences. The completion of the chimpanzee genome in 2005 has allowed us to compare our genome with that of our primate cousins. Such an analysis revealed that these two genomes differ by 1.2% in single-base substitutions. Researchers were surprised when they found a further 2.7% difference due to insertions or deletions of larger regions in the genome of one or the other species, with many of the insertions being duplications or other repetitive DNA. In fact, a third of the human duplications are not present in the chimpanzee genome, and some of these duplications contain regions associated with human diseases. All of these observations provide clues to the forces that might have swept the two genomes along different paths, but we don't have a complete understanding yet.

What about specific genes and types of genes that differ between humans and chimpanzees? Using evolutionary analyses, biologists have identified a number of genes that have evolved faster in humans. Among them are genes involved in defense against malaria and tuberculosis and a gene regulating brain size. One gene that changed rapidly in the human lineage is *FOXP2*, a gene implicated in speech and vocalization. Differences between the *FOXP2* gene in humans and chimpanzees may play a role in the ability of humans, but not chimpanzees, to communicate by speech.

Neanderthals (*Homo neanderthalensis*) were humans' closest relatives (**Figure 12.21**). First appearing at least 300,000 years ago, Neanderthals lived in Europe and Asia until suddenly going extinct a mere 30,000 years ago. Modern humans (*Homo sapiens*) first appeared in Africa around 200,000 years ago and spread into Europe and Asia around 50,000 years ago (see Module 19.14)—meaning that modern humans and Neanderthals most likely comingled for some time.

▲ **Figure 12.21** Reconstruction of a Neanderthal female, based on a 36,000-year-old skull

A 2009 rough draft of a 60%-complete Neanderthal genome has, for the first time, allowed detailed genomic comparisons between two species in the genus *Homo*. Using 38,000-year-old thigh bone fossils of two *Homo neanderthalensis* females discovered in a Croatian cave, genomic analysis confirmed Neanderthals as a separate species and as our closest relatives (much closer than chimpanzees). Further comparisons, completed in 2010, suggested that Neanderthals and some *H. sapiens* probably did interbreed. Analysis of the sequence of the *FOXP2* gene showed that Neanderthals had the same allele as modern humans, hinting that Neanderthals may have had the ability to speak. Other genetic analyses of less complete Neanderthal genomes revealed one male to have an unusual allele for a pigment gene that would have given him pale skin and red hair. And, interestingly, analysis of the lactase gene suggests that Neanderthals, like the majority of modern humans, were lactose intolerant as adults (see Chapter 3).

Comparisons with Neanderthals and chimpanzees are part of a larger effort to learn more about the human genome. Other research efforts are extending genomic studies to many more species. These studies will advance our understanding of all aspects of biology, including health, ecology, and evolution. In fact, comparisons of the completed genome sequences of bacteria, archaea, and eukaryotes supported the theory that these are the three fundamental domains of life —a topic we discuss further in the next unit.

? How can cross-species comparisons of the nucleotide sequences of a gene provide insight into evolution?

● Similarities in gene sequences correlate with evolutionary relatedness; greater genetic similarities reflect a more recent shared ancestry.

CHAPTER 12 REVIEW

For practice quizzes, BioFlix animations, MP3 tutorials, video tutors, and more study tools designed for this textbook, go to MasteringBiology®

Reviewing the Concepts

Gene Cloning (12.1–12.5)

12.1 Genes can be cloned in recombinant plasmids. Gene cloning is one application of biotechnology, the manipulation of organisms or their components to make useful products. Researchers can manipulate bacterial plasmids so that they contain genes from other organisms. These recombinant DNA plasmids can then be inserted into bacteria. If the recombinant bacteria multiply into a clone, the foreign genes are also duplicated and copies of the gene or its protein product can be harvested.

12.2 Enzymes are used to "cut and paste" DNA. Restriction enzymes cut DNA at specific sequences, forming restriction fragments. DNA ligase "pastes" DNA fragments together.

12.3 Cloned genes can be stored in genomic libraries. Genomic libraries, sets of DNA fragments containing all of an organism's genes, can be constructed and stored for use in DNA technology applications.

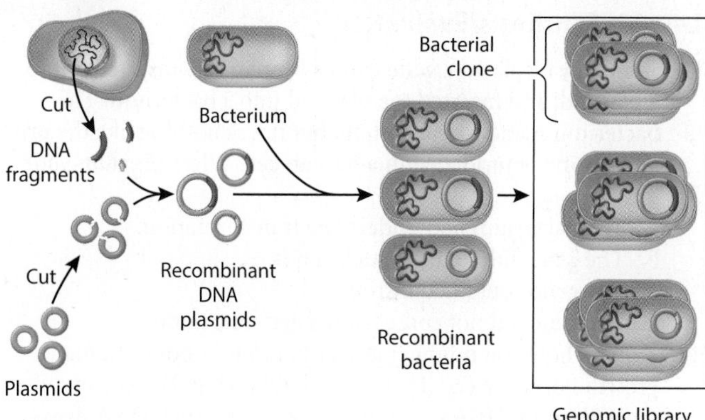
Cut
DNA fragments
Cut
Plasmids
Bacterium
Recombinant DNA plasmids
Recombinant bacteria
Bacterial clone
Genomic library

12.4 Reverse transcriptase can help make genes for cloning. cDNA libraries contain only the genes that are transcribed by a particular type of cell.

12.5 Nucleic acid probes identify clones carrying specific genes. A short, single-stranded molecule of labeled DNA or RNA can tag a desired gene in a library.

Genetically Modified Organisms (12.6–12.10)

12.6 Recombinant cells and organisms can mass-produce gene products. Bacteria, yeast, cell cultures, and whole animals can be used to make products for medical and other uses.

12.7 DNA technology has changed the pharmaceutical industry and medicine. Researchers use gene cloning to produce hormones, diagnose diseases, and produce vaccines.

12.8 Genetically modified organisms are transforming agriculture. A number of important crop plants are genetically modified.

12.9 Genetically modified organisms raise health concerns. Genetic engineering involves potential risks to human health and the environment.

12.10 Gene therapy may someday help treat a variety of diseases.

DNA Profiling (12.11–12.16)

12.11 The analysis of genetic markers can produce a DNA profile. DNA technology—methods for studying and manipulating genetic material—has revolutionized the field of forensics. DNA profiling can determine whether two samples of DNA come from the same individual.

12.12 The PCR method is used to amplify DNA sequences. The polymerase chain reaction (PCR) can be used to amplify a DNA sample. The use of specific primers that flank the desired sequence ensures that only a particular subset of the DNA sample will be copied.

12.13 Gel electrophoresis sorts DNA molecules by size.

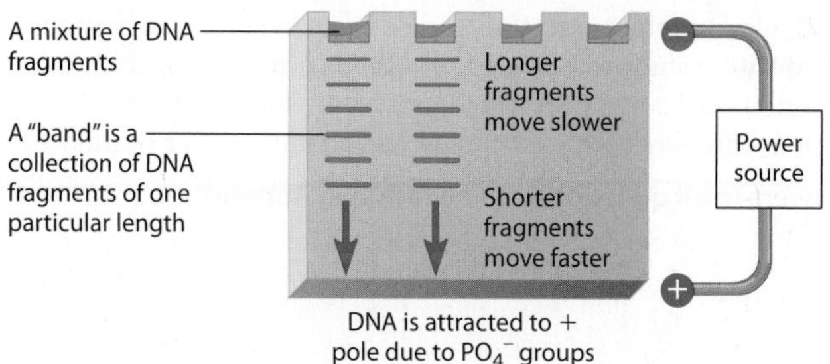
A mixture of DNA fragments
A "band" is a collection of DNA fragments of one particular length
Longer fragments move slower
Shorter fragments move faster
Power source
DNA is attracted to + pole due to PO_4^- groups

12.14 Short tandem repeat analysis is commonly used for DNA profiling. Short tandem repeats (STRs) are stretches of DNA that contain short nucleotide sequences repeated many times in a row. DNA profiling by STR analysis involves amplifying 13 STRs.

12.15 DNA profiling has provided evidence in many forensic investigations. The applications of DNA profiling include helping to solve crimes, establishing paternity, and identify victims.

12.16 RFLPs can be used to detect differences in DNA sequences. Restriction fragment length polymorphisms (RFLPs) reflect differences in the sequences of DNA samples.

Genomics (12.17–12.21)

12.17 Genomics is the scientific study of whole genomes. Genomics researchers have sequenced many prokaryotic and eukaryotic genomes. Besides being of interest in their own right, nonhuman genomes can be compared with the human genome.

12.18 The Human Genome Project revealed that most of the human genome does not consist of genes. Data from the Human Genome Project (HGP) revealed that the human genome contains just under 21,000 genes and a huge amount of noncoding DNA, much of which consists of repetitive nucleotide sequences.

12.19 The whole-genome shotgun method of sequencing a genome can provide a wealth of data quickly. The HGP used genetic and physical mapping of chromosomes followed by DNA sequencing. Modern genomic analysis often uses the faster whole-genome shotgun method.

12.20 Proteomics is the scientific study of the full set of proteins encoded by a genome.

12.21 Genomes hold clues to human evolution.

Connecting the Concepts

1. Imagine you have found a small quantity of DNA. Fill in the following diagram, which outlines a series of DNA technology experiments you could perform to study this DNA.

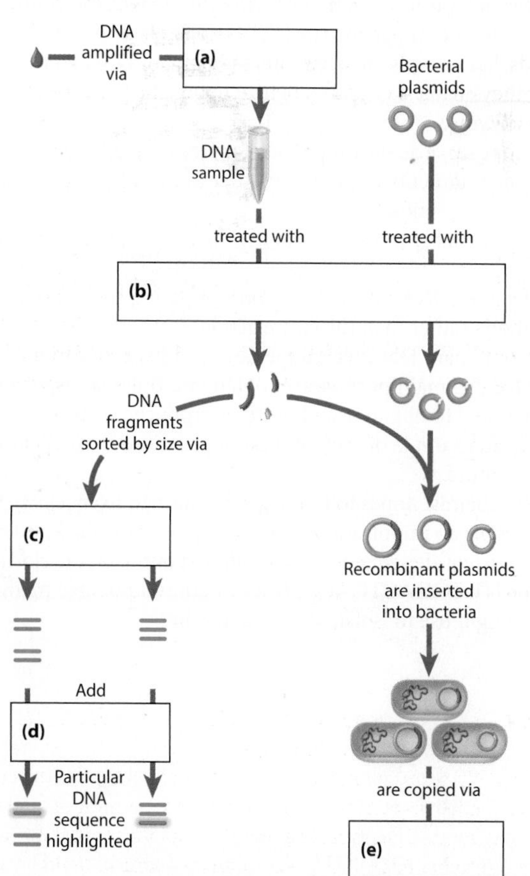
DNA amplified via
(a)
DNA sample
Bacterial plasmids
treated with
treated with
(b)
DNA fragments sorted by size via
(c)
Add
(d)
Particular DNA sequence highlighted
Recombinant plasmids are inserted into bacteria
are copied via
(e)

Testing Your Knowledge

Level 1: Knowledge/Comprehension

2. Which of the following would be considered a transgenic organism?
   a. a bacterium that has received genes via conjugation
   b. a human given a corrected human blood-clotting gene
   c. a fern grown in cell culture from a single fern root cell
   d. a rat with rabbit hemoglobin genes
3. The DNA profiles used as evidence in a murder trial look something like supermarket bar codes. The pattern of bars in a DNA profile shows
   a. the order of bases in a particular gene.
   b. the presence of various-sized fragments of DNA.
   c. the presence of dominant or recessive alleles for particular traits.
   d. the order of genes along particular chromosomes.
4. A paleontologist has recovered a tiny bit of organic material from the 400-year-old preserved skin of an extinct dodo. She would like to compare DNA from the sample with DNA from living birds. Which of the following would be most useful for increasing the amount of DNA available for testing?
   a. restriction fragment analysis
   b. polymerase chain reaction
   c. molecular probe analysis
   d. electrophoresis
5. How many genes are there in a human sperm cell?
   a. 23
   b. 46
   c. about 21,000
   d. about 3 billion

Level 2: Application/Analysis

6. When a typical restriction enzyme cuts a DNA molecule, the cuts are uneven, giving the DNA fragments single-stranded ends. These ends are useful in recombinant DNA work because
   a. they enable a cell to recognize fragments produced by the enzyme.
   b. they serve as starting points for DNA replication.
   c. the fragments will bond to other fragments with complementary ends.
   d. they enable researchers to use the fragments as molecular probes.
7. Why does DNA profiling rely on comparing specific genetic markers rather than the entire genome?
8. Recombinant DNA techniques are used to custom-build bacteria for two main purposes: to obtain multiple copies of certain genes and to obtain useful proteins produced by certain genes. Give an example of each of these applications in medicine and agriculture.
9. A biochemist hopes to find a gene in human liver cells that codes for an important blood-clotting protein. She knows that the nucleotide sequence of a small part of the blood-clotting gene is CTGGACTGACA. Briefly outline a possible method she might use to isolate the desired gene.

Level 3: Synthesis/Evaluation

10. A biologist isolated a gene from a human cell, inserted it into a plasmid, and inserted the plasmid into a bacterium. The bacterium made a new protein, but it was nothing like the protein normally produced in a human cell. Why? (*Explain your answer.*)
    a. The bacterium had undergone transformation.
    b. The gene did not have sticky ends.
    c. The gene contained introns.
    d. The gene did not come from a genomic library.
11. Explain how you might engineer *E. coli* to produce human growth hormone (HGH) using the following: *E. coli* containing a plasmid, DNA carrying the gene for HGH, DNA ligase, a restriction enzyme, equipment for manipulating and growing bacteria, a method for extracting and purifying the hormone, and an appropriate DNA probe. (Assume that the human HGH gene lacks introns.)
12. What is left for genetic researchers to do now that the Human Genome Project has determined nearly complete nucleotide sequences for all of the human chromosomes? Explain.
13. Today, it is fairly easy to make transgenic plants and animals. What are some important safety and ethical issues raised by this use of recombinant DNA technology? What are some of the possible dangers of introducing genetically engineered organisms into the environment? What are some reasons for and against leaving decisions in these areas to scientists? To business owners and executives? What are some reasons for and against more public involvement? How might these decisions affect you? How do you think these decisions should be made?
14. In the not-too-distant future, gene therapy may be an option for the treatment and cure of some inherited disorders. What do you think are the most serious ethical issues that must be dealt with before human gene therapy is used on a large scale? Why do you think these issues are important?
15. The possibility of extensive genetic testing raises questions about how personal genetic information should be used. For example, should employers or potential employers have access to such information? Why or why not? Should the information be available to insurance companies? Why or why not? Is there any reason for the government to keep genetic files? Is there any obligation to warn relatives who might share a defective gene? Might some people avoid being tested for fear of being labeled genetic outcasts? Or might they be compelled to be tested against their wishes? Can you think of other reasons to proceed with caution?
16. **SCIENTIFIC THINKING** Scientists investigate hypotheses using a variety of methods, depending on the circumstances behind the research. Human nutrition studies (such as those studying whether GMO foods have any health effects) are particularly problematic. Can you design a hypothetical human nutrition study to test whether GMO corn is less healthy than traditional corn? Can you identify real-world problems that may interfere with your design and confound your results?

Answers to all questions can be found in Appendix 4.

Concepts of Evolution

CHAPTER 13

How Populations Evolve

What does actor George Clooney have in common with George Washington, Ernest Hemingway, Christopher Columbus, and Mother Teresa? They all survived bouts with malaria, a disease caused by a protozoan parasite that is one of the worst killers in human history. In the 1960s, the World Health Organization (WHO) launched a campaign to eradicate malaria. Their strategy focused on killing the mosquitoes that carry the parasite from person to person. DDT, a widely used pesticide, was deployed in massive spraying operations. But in one location after another, early success was followed by rebounding mosquito populations in which resistance to DDT had evolved. Malaria continued to spread. Today, malaria causes more than a million deaths and 250 million cases of miserable illness each year.

? ***How does evolution hinder attempts to eradicate disease?***

Evolution has also hindered efforts to help malaria victims, such as the child shown in the photo below. At the same time that DDT was being celebrated as a miracle pesticide in the war against malaria, a drug called chloroquine was hailed as the miracle cure. But its effectiveness has diminished over

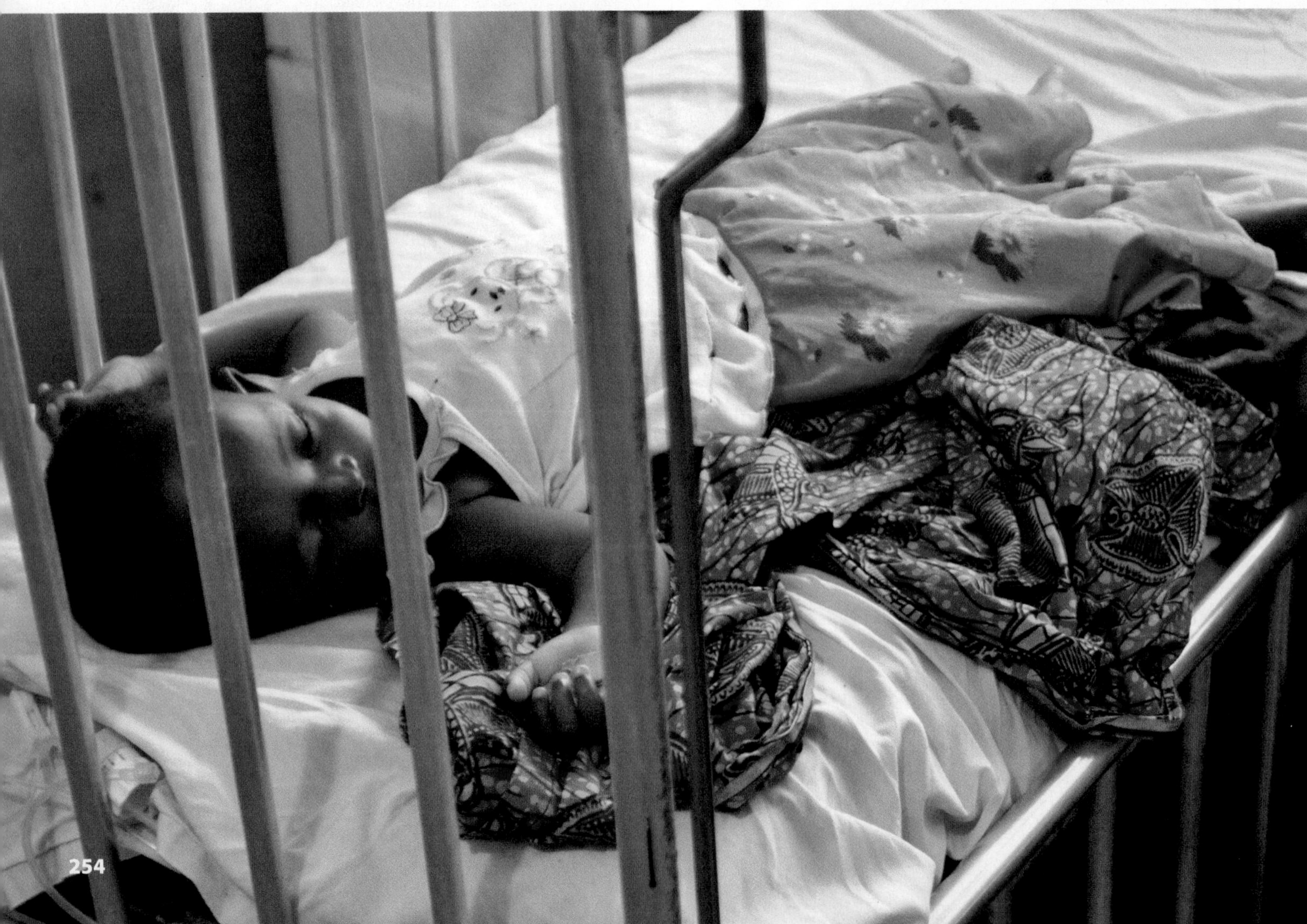

time, as resistance to the drug has evolved in parasite populations. In some regions, chloroquine is powerless against the disease. The most effective antimalarial drug now is artemisinin, a compound extracted from a plant used in traditional Chinese medicine. But the effectiveness of this drug will eventually succumb to the power of evolution, too. Cases of malaria that don't respond to artemisinin have already appeared in Southeast Asia.

An understanding of evolution informs all of biology, from exploring life's molecules to analyzing ecosystems. Applications of evolutionary biology are transforming fields as diverse as medicine, agriculture, and conservation biology. In this chapter, we begin our study of evolution with the enduring legacy of Charles Darwin's explanation for the unity and diversity of life. We also delve into the nitty-gritty of natural selection, the mechanism for evolution that Darwin proposed.

BIG IDEAS

Darwin's Theory of Evolution
(13.1–13.7)

Darwin's theory of evolution explains the adaptations of organisms and the unity and diversity of life.

The Evolution of Populations
(13.8–13.11)

Genetic variation makes evolution possible within a population.

Mechanisms of Microevolution
(13.12–13.18)

Natural selection, genetic drift, and gene flow can alter gene pools; natural selection leads to adaptive evolution.

▷ Darwin's Theory of Evolution

13.1 A sea voyage helped Darwin frame his theory of evolution

If you have heard of the theory of evolution, you have probably heard of Charles Darwin. Although Darwin was born more than 200 years ago, his work had such an extraordinary impact that many biologists mark his birthday—February 12, the same as Abraham Lincoln's—with a celebration of his contributions to science. The publication of Darwin's best-known book, *On the Origin of Species by Means of Natural Selection*, commonly referred to as *The Origin of Species*, launched the era of evolutionary biology.

Darwin's Cultural and Scientific Context Darwin's early career gave no hint of his future fame. As a boy, he was fascinated with nature. When not reading books about nature, he was fishing, hunting, and collecting insects. His father, an eminent physician, could see no future for his son as a naturalist and sent him to medical school. But Darwin, finding medicine boring and surgery before the days of anesthesia horrifying, quit medical school. His father then enrolled him at Cambridge University with the intention that he should become a clergyman. Thus, Darwin's education was typical for a young man of his social class.

The cultural and scientific context of his time also instilled Darwin with a conventional view of Earth and its life. Most scientists accepted the views of the Greek philosopher Aristotle, who generally held that species are fixed, permanent forms that do not evolve. Judeo-Christian culture fortified this idea with a literal interpretation of the biblical book of Genesis, which tells the story of each form of life being individually created in its present-day form. In the 1600s, religious scholars used biblical accounts to estimate the age of Earth at 6,000 years. Thus, the idea that all living species came into being relatively recently and are unchanging in form dominated the intellectual climate of the Western world at the time.

Darwin's radical thinking stemmed from his post-college life, when he returned to his childhood interests rather than following the career path mapped out by his father. At the age of 22, Darwin took a position on HMS *Beagle*, a survey ship preparing for a long expedition to chart poorly known stretches of the South American coast **(Figure 13.1A)**.

Darwin's Sea Voyage As the ship's naturalist (field biologist), Darwin spent most of his time on shore collecting thousands of specimens of fossils and living plants and animals. He also kept detailed journals of his observations. For a naturalist from a small, temperate country, seeing the glorious diversity of unfamiliar life-forms on other continents was a revelation. He carefully noted the characteristics of plants and animals that made them well suited to such diverse environments as the jungles of Brazil, the grasslands of Argentina, the towering peaks of the Andes, and the desolate and frigid lands at the southern tip of South America.

Many of Darwin's observations indicated that geographic proximity is a better predictor of relationships among organism than similarity of environment. For example, the plants

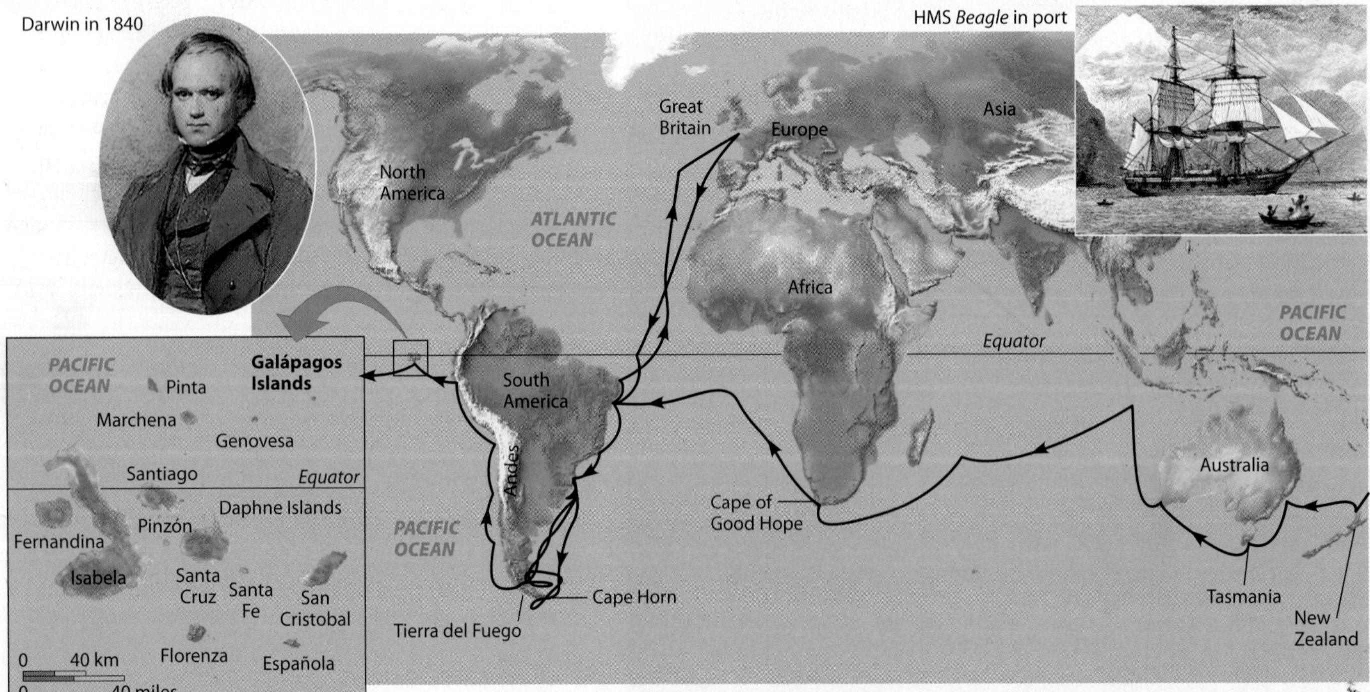

Darwin in 1840
HMS *Beagle* in port
Great Britain
Europe
Asia
North America
ATLANTIC OCEAN
Africa
PACIFIC OCEAN
Equator
South America
Andes
PACIFIC OCEAN
Cape of Good Hope
Australia
Tasmania
New Zealand
Cape Horn
Tierra del Fuego
PACIFIC OCEAN
Galápagos Islands
Pinta
Marchena
Genovesa
Santiago
Equator
Daphne Islands
Pinzón
Fernandina
Isabela
Santa Cruz
Santa Fe
San Cristobal
Florenza
Española
0 40 km
0 40 miles

▲ **Figure 13.1A** The voyage of the *Beagle* (1831–1836), with insets showing a young Charles Darwin and the ship on which he sailed

and animals living in temperate regions of South America more closely resembled species living in tropical regions of that continent than species living in temperate regions of Europe. And the South American fossils Darwin found, though clearly species different from living ones, were distinctly South American in their resemblance to the contemporary plants and animals of that continent. For instance, he collected fossilized armor plates resembling those of living armadillo species. Paleontologists later reconstructed the creature to which the armor belonged—an extinct armadillo the size of a Volkswagen Beetle.

Darwin was particularly intrigued by the geographic distribution of organisms on the Galápagos Islands. The Galápagos are relatively young volcanic islands about 900 kilometers (540 miles) off the Pacific coast of South America. Most of the animals that inhabit these remote islands are found nowhere else in the world, but they resemble South American species. For example, Darwin noticed that Galápagos marine iguanas—with a flattened tail that aids in swimming—are similar to, but distinct from, land-dwelling iguanas on the islands and on the South American mainland (**Figure 13.1B**). Furthermore, each island had its own distinct variety of giant tortoise (**Figure 13.1C**), the strikingly unique inhabitant for which the islands were named (galápago means "tortoise" in Spanish).

▲ Figure 13.1B A marine iguana in the waters around the Galápagos Islands

▲ Figure 13.1C A giant tortoise, one of the unique inhabitants of the Galápagos Islands

While on his voyage, Darwin was strongly influenced by the newly published *Principles of Geology*, by Scottish geologist Charles Lyell. The book presented the case for an ancient Earth sculpted over millions of years by gradual geologic processes that continue today. Having witnessed an earthquake that raised part of the coastline of Chile almost a meter, Darwin realized that natural forces gradually changed Earth's surface and that these forces still operate. Thus, the growth of mountains as a result of earthquakes could account for the presence of marine snail fossils he collected on mountaintops in the Andes.

By the time Darwin returned to Great Britain five years after the *Beagle* first set sail, he had begun to seriously doubt that Earth and all its living organisms had been specially created only a few thousand years earlier. As he reflected on his observations, analyzed his collections, and discussed his work with colleagues, he concluded that the evidence was better explained by the hypothesis that present-day species are the descendants of ancient ancestors that they still resemble in some ways. Over time, differences gradually accumulated by a process that Darwin called "descent with modification," his phrase for evolution. Darwin did not originate the concept of evolution; other scientists had explored the idea that organisms had changed over time. Unlike the others, however, Darwin also proposed a scientific mechanism for how life evolves, a process he called natural selection (see Module 1.7). He hypothesized that as the descendants of a remote ancestor spread into various habitats over millions and millions of years, they accumulated diverse modifications, or **adaptations**, that fit them to specific ways of life in their environment.

Darwin's Writings By the early 1840s, Darwin had composed a long essay describing the major features of his theory of evolution by natural selection. He realized that his ideas would cause an uproar, however, and he delayed publishing his essay. Even as he procrastinated, Darwin continued to compile evidence in support of his hypothesis. In 1858, Alfred Wallace, a British naturalist doing fieldwork in Indonesia, conceived a hypothesis almost identical to Darwin's. Faced with the possibility that Wallace's work would be published first, Darwin finally released his essay to the scientific community.

The following year, Darwin published *The Origin of Species*, a book that supported his hypothesis with immaculate logic and hundreds of pages of evidence drawn from observations and experiments in biology, geology, and paleontology. The hypothesis of evolution set forth in *The Origin of Species* also generated predictions that have been tested and verified by more than 150 years of research. Consequently, scientists regard Darwin's concept of evolution by means of natural selection as a **theory**—a widely accepted explanatory idea that is broader in scope than a hypothesis, generates new hypotheses, and is supported by a large body of evidence.

Next, we examine lines of evidence for Darwin's theory of **evolution**, the idea that living species are descendants of ancestral species that were different from present-day ones. We then return to the second main point Darwin made in *The Origin of Species*, that natural selection is the mechanism for evolutionary change. With our current understanding of how this mechanism works, we extend Darwin's definition of evolution to include "genetic changes in a population from generation to generation."

? **What was Darwin's phrase for evolution? What does it mean?**

● Descent with modification. An ancestral species could diversify into many descendant species by the accumulation of adaptations to various environments.

13.2 The study of fossils provides strong evidence for evolution

Fossils—imprints or remains of organisms that lived in the past—document differences between past and present organisms and the fact that many species have become extinct. The organic substances of a dead organism usually decay rapidly, but the hard parts of an animal that are rich in minerals, such as the bones and teeth of vertebrates and the shells of clams and snails, may remain as fossils. For example, the fossilized skull in **Figure 13.2A** is from one of our early relatives, *Homo erectus*, who lived some 1.5 million years ago in Africa.

Some fossils are not the actual remnants of organisms. The 375-million-year-old fossils shown in **Figure 13.2B** are casts of ammonites, shelled marine animals related to the present-day nautilus (see Figure 18.9E). Casts form when a dead organism captured in sediment decomposes and leaves an empty mold that is later filled by minerals dissolved in water. The minerals harden, making a replica of the organism. Fossils may also be imprints that remain after the organism decays. Footprints, burrows, and fossilized feces (known as coprolites) provide evidence of an ancient organism's behavior.

In rare instances, an entire organism, including its soft parts, is encased in a medium that prevents bacteria and fungi from decomposing the corpse. Examples include insects trapped in amber (fossilized tree resin) and mammoths, bison, and even prehistoric humans frozen in ice or preserved in bogs.

Many fossils are found in fine-grained sedimentary rocks formed from the sand or mud that settles to the bottom of seas, lakes, swamps, and other aquatic habitats. New layers of sediment cover older ones and compress them into layers of rock called **strata** (singular, *stratum*). The fossils in a particular stratum provide a glimpse of some of the organisms that lived in the area at the time the layer formed. Because younger strata are on top of older ones, the relative ages of fossils can be determined by the layer in which they are found. Thus, the sequence in which fossils appear within layers of sedimentary rocks is a historical record of life on Earth.

Paleontologists (scientists who study fossils) sometimes gain access to very old fossils when erosion carves through upper (younger) strata, revealing deeper (older) strata that had been buried. **Figure 13.2C** shows strata of sedimentary rock at the Grand Canyon. The Colorado River has cut through more than 2,000 m (more than a mile) of rock, exposing sedimentary layers that can be read like huge pages from the book of life. Scan the canyon wall from rim to floor, and you look back through hundreds of millions of years. Each layer entombs fossils that represent some of the organisms from that period of Earth's history.

Of course, the **fossil record**—the chronicle of evolution over millions of years of geologic time engraved in the order in which fossils appear in rock strata—is incomplete. Many of Earth's organisms did not live in areas that favor fossilization. Many fossils that did form were in rocks later distorted or destroyed by geologic processes. Furthermore, not all fossils that have been preserved are accessible to paleontologists. Even with its limitations, however, the fossil record is remarkably detailed.

▲ **Figure 13.2A** Skull of *Homo erectus*

▲ **Figure 13.2B** Ammonite casts

▲ **Figure 13.2C** Strata of sedimentary rock at the Grand Canyon

? What types of animals do you think would be most represented in the fossil record? Explain your answer.

● Animals with hard parts, such as shells or bones that readily fossilize, and those that lived in areas where sedimentary rock may form

13.3 Fossils of transitional forms support Darwin's theory of evolution

SCIENTIFIC THINKING

In *The Origin of Species*, Darwin predicted the existence of fossils of transitional forms linking very different groups of organisms. For example, if his hypothesis that whales evolved from land-dwelling mammals was correct, then fossils should show a series of changes in a lineage of mammals adapted to a fully aquatic habitat. Although Darwin lacked evidence with which to test this prediction, thousands of fossil discoveries have since shed light on the evolutionary origins of many groups of plants and animals, including the transition of fish to amphibian (see Module 19.4), the origin of birds from a lineage of dinosaurs (see Module 19.7), and the evolution of mammals from a reptilian ancestor. If Darwin were alive today, he would surely be delighted to know that evidence discovered over the past few decades has made the origin of whales from terrestrial mammals one of the best-documented evolutionary transitions to date.

Whales are cetaceans, a group that also includes dolphins and porpoises. They have forelimbs in the form of flippers but lack hind limbs (**Figure 13.3A**). If cetaceans evolved from four-legged land animals, then transitional forms should have reduced hind limb and pelvic bones. In the 1960s, observations of fossil teeth led paleontologists to hypothesize that whales were the descendants of primitive hoofed, wolflike carnivores. However, few fossil whales were available to study.

Beginning in the late 1970s, paleontologists unearthed an extraordinary series of transitional fossils in Pakistan and Egypt. The 50-million-year-old *Pakicetus* ("whale of Pakistan"; **Figure 13.3B**) was a carnivorous four-legged mammal that, remarkably, had distinctively cetacean ear structures. *Ambulocetus* ("walking whale"), roughly 48 million years old, was a perfect intermediate between modern whales and their land-dwelling ancestors. Its legs were short and sturdy. The wrist and elbow joints of the forelimbs suggested mobility on land, while a powerful tail and large, paddle-like hind feet suggested the ability to swim. Like *Pakicetus*, *Ambulocetus* had a whalelike ear, as did *Rodhocetus*, another mammal that apparently spent time both on land and in the water. The fossil genus *Dorudon*, which lived between 40 and 35 million years ago, had completed the transition to aquatic life. The wrist and elbow joints of its paddle-like forelimbs could not have been used for walking, and its tiny hind limbs were not connected to the vertebral column.

The new fossil discoveries were consistent with the earlier hypothesis, and paleontologists became more firmly convinced that whales did indeed arise from a wolflike carnivore. But molecular biologists were testing an alternative hypothesis using DNA analysis to infer relationships among living animals. They found a close relationship between whales and hippopotamuses, which are members of a group of mostly herbivorous, cloven-hoofed mammals that includes pigs, deer, and camels. Consequently, these researchers hypothesized that whales and hippos were both descendants of a cloven-hoofed ancestor.

Paleontologists were taken aback by the contradictory results, but openness to new evidence is a hallmark of science. They turned their attention to seeking a fossil that would resolve the issue. Cloven-hoofed mammals have a unique ankle bone. If the ancestor of whales was a wolflike carnivore, then the shape of its ankle bone would be similar to most present-day mammals. Two fossils discovered in 2001 provided the answer. Both *Pakicetus* and *Rodhocetus* had the distinctive ankle bone of a cloven-hoofed mammal. Thus, as is often the case in science, scientists are becoming more certain about the evolutionary origin of whales as mounting evidence from different lines of inquiry converge.

? **What anatomical feature did scientists predict in fossils of species transitional between terrestrial and aquatic mammals?**

● Reduced hind limb and pelvic bones

▲ **Figure 13.3A** A killer whale (*Orcinus orca*)

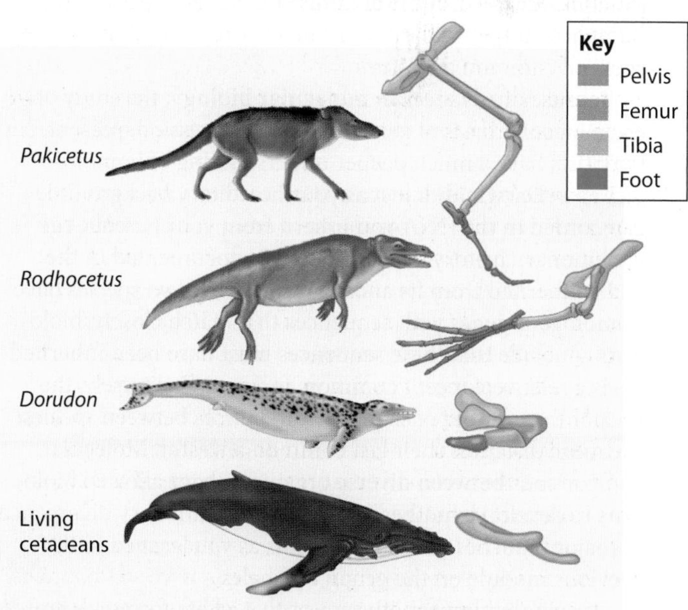
Key
Pelvis
Femur
Tibia
Foot
Pakicetus
Rodhocetus
Dorudon
Living cetaceans

▲ **Figure 13.3B** The transition to life in the sea

Try This List the animals shown, and describe how the structure of each animal's hind limbs reflects their function.

13.4 Homologies provide strong evidence for evolution

A second type of evidence for evolution comes from analyzing similarities among different organisms. Evolution is a process of descent with modification—characteristics present in an ancestral organism are altered over time by natural selection as its descendants face different environmental conditions. In other words, evolution is a remodeling process. As a result, related species can have characteristics that have an underlying similarity yet function differently. Similarity resulting from common ancestry is known as **homology**.

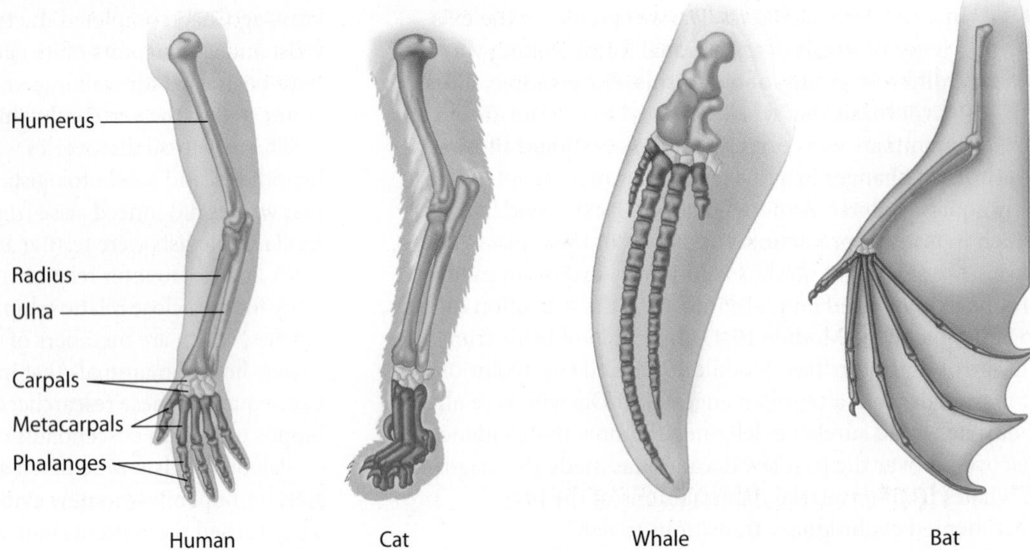

Humerus
Radius
Ulna
Carpals
Metacarpals
Phalanges
Human
Cat
Whale
Bat

▲ **Figure 13.4A** Homologous structures: vertebrate forelimbs

Darwin cited the anatomical similarities among vertebrate forelimbs as evidence of common ancestry. As **Figure 13.4A** shows, the same skeletal elements make up the forelimbs of humans, cats, whales, and bats. The functions of these forelimbs differ. A whale's flipper does not do the same job as a bat's wing, so if these structures had been uniquely engineered, then we would expect that their basic designs would be very different. The logical explanation is that the arms, forelegs, flippers, and wings of these different mammals are variations on an anatomical structure of an ancestral organism that over millions of years has become adapted to different functions. Biologists call such anatomical similarities in different organisms **homologous structures**—features that often have different functions but are structurally similar because of common ancestry.

Because of advances in **molecular biology**, the study of the molecular basis of genes and gene expression, present-day scientists have a much deeper understanding of homologies than Darwin did. Just as your hereditary background is recorded in the DNA you inherit from your parents, the evolutionary history of each species is documented in the DNA inherited from its ancestral species. If two species have homologous genes with sequences that match closely, biologists conclude that these sequences must have been inherited from a relatively recent common ancestor. Conversely, the greater the number of sequence differences between species, the more distant is their last common ancestor. Molecular comparisons between diverse organisms have allowed biologists to develop hypotheses about the evolutionary divergence of major branches on the tree of life, as you learned in the previous module on the origin of whales.

Darwin's boldest hypothesis was that all life-forms are related. Molecular biology provides strong evidence for this claim: All forms of life use the same genetic language of DNA and RNA, and the genetic code—how RNA triplets are translated into amino acids—is essentially universal. Thus, it is likely that all species descended from common ancestors that used this code. Because of these homologies, bacteria engineered with human genes can produce human proteins such as insulin and human growth hormone (see Module 12.6). But molecular homologies go beyond a shared genetic code. For example, organisms as dissimilar as humans and bacteria share homologous genes inherited from a very distant common ancestor.

An understanding of homology can also explain observations that are otherwise puzzling. For example, comparing early stages of development in different animal species reveals similarities not visible in adult organisms. At some point in their development, all vertebrate embryos have a tail posterior to the anus, as well as structures called pharyngeal (throat) pouches. These pouches are homologous structures that ultimately develop to have very different functions, such as gills in fishes and parts of the ears and throat in humans. Note the pharyngeal pouches and tails of the bird embryo (left) and the human embryo (right) in **Figure 13.4B**.

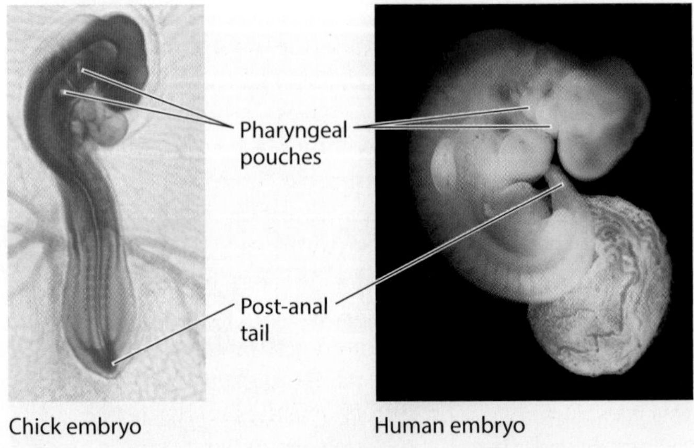

Pharyngeal pouches
Post-anal tail
Chick embryo
Human embryo

▲ **Figure 13.4B** Homologous structures in vertebrate embryos

Some of the most interesting homologies are "leftover" structures that are of marginal or perhaps no importance to the organism. These **vestigial structures** are remnants of features that served important functions in the organism's ancestors. For example, the small pelvis and hind-leg bones of ancient whales are vestiges (traces) of their walking ancestors. The eye remnants that are buried under scales in blind species of cave fishes—a vestige of their sighted ancestors—are another example.

Organisms may also retain genes that have lost their function, even though homologous genes in related species are fully functional. Researchers have identified many of these inactive "pseudogenes" in humans. One such gene encodes an enzyme known as GLO that is used in making vitamin C. Almost all mammals have a metabolic pathway to synthesize this essential vitamin from glucose. Although humans and other primates have functional genes for the first three steps in the pathway, the inactive GLO gene prevents vitamin C from being made—we must get sufficient amounts in our diet to maintain health.

Next we see how homologies help us trace evolutionary descent.

? What is homology? How does the concept of homology relate to molecular biology?

● Homology is similarity in different species due to evolution from a common ancestor. Similarities in DNA sequences or proteins reflect the evolutionary relationship that is the basis of homology.

13.5 Homologies indicate patterns of descent that can be shown on an evolutionary tree

Darwin was the first to view the history of life as a tree, with multiple branchings from a common ancestral trunk to the descendant species at the tips of the twigs. Biologists represent these patterns of descent with an **evolutionary tree** (see Figure 14.1), although today they often turn the trees sideways.

Homologous structures, both anatomical and molecular, can be used to determine the branching sequence of such a tree. Some homologous characters, such as the genetic code, are shared by all species because they date to the deep ancestral past. In contrast, characters that evolved more recently are shared only within smaller groups of organisms. For example, all tetrapods (from the Greek *tetra*, four, and *pod*, foot) possess the same basic limb bone structure illustrated in Figure 13.4A, but their ancestors do not.

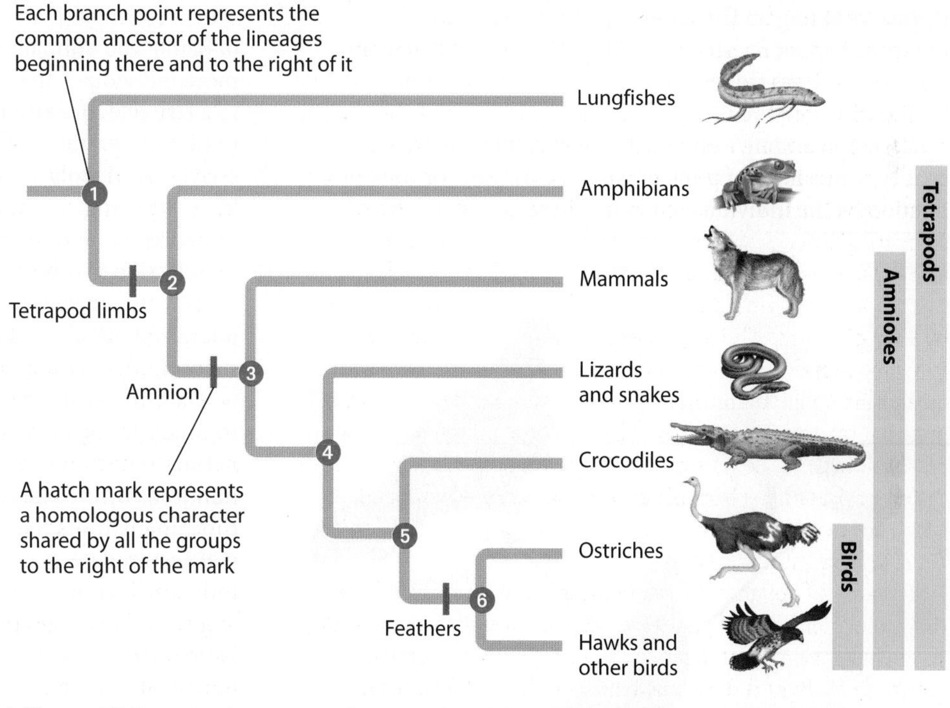

Each branch point represents the common ancestor of the lineages beginning there and to the right of it
1
2
3
4
5
6
Tetrapod limbs
Amnion
Feathers
A hatch mark represents a homologous character shared by all the groups to the right of the mark
Lungfishes
Amphibians
Mammals
Lizards and snakes
Crocodiles
Ostriches
Hawks and other birds
Tetrapods
Amniotes
Birds

▲ **Figure 13.5** An evolutionary tree for tetrapods and their closest living relatives, the lungfishes

Figure 13.5 is an evolutionary tree of tetrapods (amphibians, mammals, and reptiles, including birds) and their closest living relatives, the lungfishes. In this diagram, each branch point represents the common ancestor of all species that descended from it. For example, lungfishes and all tetrapods descended from ancestor ①, whereas crocodiles and birds descended from ancestor ⑤. Three homologies are shown by the purple hatch marks on the tree—tetrapod limbs, the amnion (a protective embryonic membrane), and feathers. Tetrapod limbs were present in ancestor ② and hence are found in all of its descendants. The amnion was present only in ancestor ③ and thus is shared only by mammals and reptiles. Feathers were present only in ancestor ⑥ and hence are found only in birds.

Evolutionary trees are hypotheses reflecting our current understanding of patterns of evolutionary descent. Some trees, such as the one in Figure 13.5, are supported by a strong combination of fossil, anatomical, and molecular data. Others are more speculative because few data are available.

Now that you have learned about Darwin's view of evolution as descent with modification, let's examine the mechanism he proposed for how life evolves—natural selection.

? Refer to the evolutionary tree in Figure 13.5. Are crocodiles more closely related to lizards or birds?

● Look for the most recent common ancestor of these groups. Crocodiles are more closely related to birds because they share a more recent common ancestor with birds (ancestor ⑤) than with lizards (ancestor ④).

13.6 Darwin proposed natural selection as the mechanism of evolution

Darwin's greatest contribution to biology was his explanation of *how* life evolves. Because he thought that species formed gradually over long periods of time, he knew that he would not be able to study the evolution of new species by direct observation. But he did have a way to gain insight into the process of incremental change—the practices used by plant and animal breeders.

All domesticated plants and animals are the products of selective breeding from wild ancestors. For example, the baseball-size tomatoes grown today are very different from their Peruvian ancestors, which were not much larger than blueberries, and dachshunds bear little resemblance to the wolves from which they were bred. Having conceived the notion that **artificial selection**—the selective breeding of domesticated plants and animals to promote the occurrence of desirable traits in the offspring—was the key to understanding evolutionary change, Darwin bred fancy pigeons **(Figure 13.6)** to gain firsthand experience. He also talked to farmers about livestock breeding. He learned that artificial selection has two essential components, variation and heritability.

Variation among individuals, for example, differences in coat type in a litter of puppies, size of corn ears, or milk production by the individual cows in a herd, allows the breeder to select the animals or plants with the most desirable combination of characters as breeding stock for the next generation. Heritability refers to the transmission of a trait from parent to offspring. Despite their lack of knowledge of the underlying genetics, breeders had long understood the importance of heritability in artificial selection.

Unlike most naturalists, who sought consistency of traits in order to classify organisms, Darwin was a careful observer of variations between individuals. He knew that individuals in natural populations have small but measurable differences. But what forces in nature played the role of the breeder by choosing which individuals became the breeding stock for the next generation?

Darwin found inspiration in an essay written by economist Thomas Malthus, who contended that much of human suffering—disease, famine, and war—was the consequence of human populations increasing faster than food supplies and other resources. Darwin applied Malthus's idea to populations of plants and animals. He deduced that the production of more individuals than the limited resources can support leads to a struggle for existence, with only some offspring surviving in each generation. Of the many eggs laid, young born, and seeds spread, only a tiny fraction complete development and leave offspring. The rest are eaten, starved, diseased, unmated, or unable to reproduce for other reasons. The essence of natural selection is this unequal reproduction. Individuals whose traits better enable them to obtain food or escape predators or tolerate physical conditions will survive and reproduce more successfully, passing these adaptive traits to their offspring.

Darwin reasoned that if artificial selection can bring about so much change in a relatively short period of time, then natural selection could modify species considerably over hundreds or thousands of generations. Over vast spans of time, many traits that adapt a population to its environment will accumulate. If the environment changes, however, or if individuals move to a new environment, natural selection will select for adaptations to these new conditions, sometimes producing changes that result in the origin of a completely new species in the process.

It is important to emphasize three key points about evolution by natural selection. First, although natural selection occurs through interactions between individual organisms and the environment, individuals do not evolve. Rather, it is the population—the group of organisms—that evolves over time as adaptive traits become more common in the group and other traits change or disappear.

Second, natural selection can amplify or diminish only heritable traits. Certainly, an organism may become modified through its own interactions with the environment during its lifetime, and those acquired characteristics may help the organism survive. But unless coded for in the genes of an organism's gametes, such acquired characteristics cannot be passed on to offspring. Thus, a championship female bodybuilder will not give birth to a muscle-bound baby.

Fantail

Frillback

Rock pigeon

Old Dutch Capuchine

Trumpeter

▲ **Figure 13.6** Artificial selection: fancy pigeon varieties bred from the rock pigeon

Third, evolution is not goal directed; it does not lead to perfectly adapted organisms. Whereas artificial selection is a deliberate attempt by humans to produce individuals with specific traits, natural selection is the result of environmental factors that vary from place to place and over time. A trait that is favorable in one situation may be useless—or even detrimental—in different circumstances. And as you will see, adaptations are often compromises. Now let's look at some examples of natural selection.

? Compare artificial selection and natural selection.

● In artificial selection, humans choose the desirable traits and breed only organisms with those traits. In natural selection, the environment does the choosing: Individuals with traits best suited to the environment survive and reproduce most successfully, passing those adaptive traits to offspring.

13.7 Scientists can observe natural selection in action

Look at any natural environment, and you will see the products of natural selection—adaptations that suit organisms to their environment. But can we see natural selection in action?

Indeed, biologists have documented evolutionary change in thousands of scientific studies. A classic example comes from work that Peter and Rosemary Grant and their students did with finches in the Galápagos Islands over more than 30 years. As part of their research, they measured changes in beak size in a population of a ground finch species. These birds eat mostly small seeds. In dry years, when all seeds are in short supply, birds must eat more large seeds. Birds with larger, stronger beaks have a feeding advantage and greater reproductive success, and the Grants measured an increase in the average beak depth for the population. During wet years, smaller beaks are more efficient for eating the now abundant small seeds, and the Grants found a decrease in average beak depth.

An unsettling example of natural selection in action is the evolution of pesticide resistance in hundreds of insect species. Pesticides control insects and prevent them from eating crops or transmitting diseases. Whenever a new type of pesticide is used to control pests, the story is similar (**Figure 13.7**): A relatively small amount of poison initially kills most of the insects, but subsequent applications are less and less effective. The few survivors of the first pesticide wave are individuals that are genetically resistant, carrying an allele (alternative form of a gene, colored red in the figure) that somehow enables them to survive the chemical attack. So the poison kills most members of the population, leaving the resistant survivors to reproduce and pass the alleles for pesticide resistance to their offspring. The proportion of pesticide-resistant individuals thus increases in each generation.

The World Health Organization's campaign against malaria described in the chapter introduction is a real-world example of the evolution of pesticide resistance. Some mosquitoes in the populations that were sprayed with DDT carried an allele that codes for an enzyme that detoxifies the pesticide. When the presence of DDT changed the environment, the individuals carrying that allele survived to leave offspring, while nonresistant individuals did not. Thus, the process of natural selection defeated the efforts of WHO to control the spread of malaria by using DDT to kill mosquitoes.

These examples of evolutionary adaptation highlight two important points about natural selection. First, natural selection is more an editing process than a creative mechanism. A pesticide does not create new alleles that allow insects to survive. Rather, the presence of the pesticide leads to natural selection for insects in the population that already have those alleles. Second, natural selection is contingent on time and place: It favors those heritable traits in a varying population that fit the current, local environment. If the environment changes, different traits may be favored.

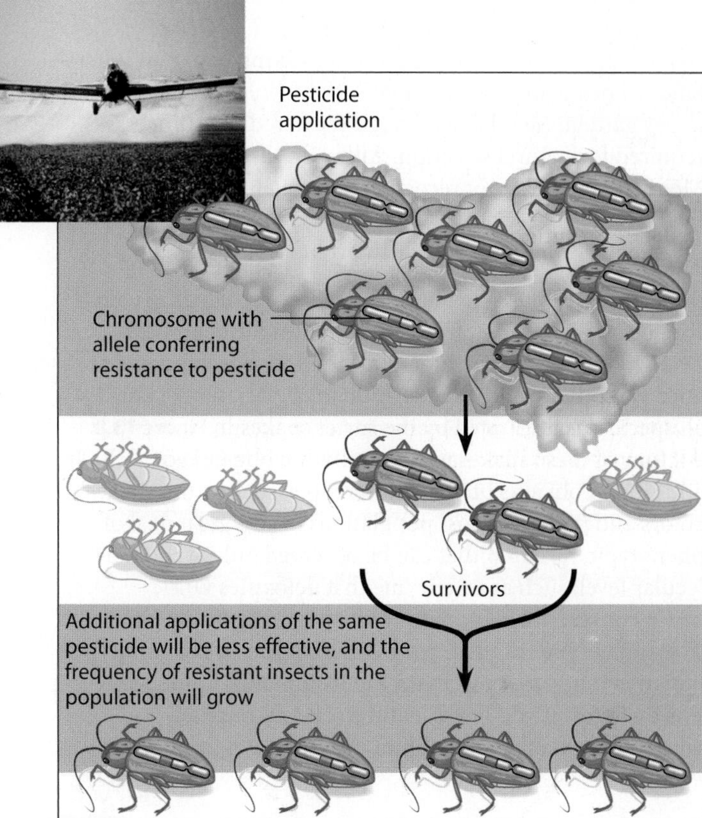

Pesticide application
Chromosome with allele conferring resistance to pesticide
Survivors
Additional applications of the same pesticide will be less effective, and the frequency of resistant insects in the population will grow

▲ **Figure 13.7** Evolution of pesticide resistance in an insect population

In the next few modules, we examine the genetic basis of evolution more closely.

? In what sense is natural selection more an editing process than a creative process?

● Natural selection cannot create beneficial traits on demand but instead "edits" variation in a population by selecting for individuals with those traits that are best suited to the current environment.

▷ The Evolution of Populations

13.8 Mutation and sexual reproduction produce the genetic variation that makes evolution possible

In *The Origin of Species*, Darwin provided evidence that life on Earth has evolved over time, and he proposed that natural selection, in favoring some heritable traits over others, was the primary mechanism for that change. But he could not explain the cause of variation among individuals, nor could he account for how those variations passed from parents to offspring.

Just a few years after the publication of *The Origin of Species*, Gregor Mendel wrote a groundbreaking paper on inheritance in pea plants (see Module 9.2). By breeding peas in his abbey garden, Mendel discovered the hereditary processes required for natural selection. Although the significance of Mendel's work was not recognized during his or Darwin's lifetime, its rediscovery in 1900 set the stage for understanding the genetic differences on which evolution is based.

Genetic Variation You have no trouble recognizing your friends in a crowd. Each person has a unique genome, reflected in individual phenotypic variations such as appearance and other traits. Indeed, individual variation occurs in all species, as illustrated by the garter snakes in **Figure 13.8**. All four of these snakes were captured in one Oregon field. In addition to obvious physical differences, such as the snakes' colors and patterns, most populations have a great deal of phenotypic variation that can be observed only at the molecular level, such as an enzyme that detoxifies DDT.

Of course, not all variation in a population is heritable. The phenotype results from a combination of the genotype, which is inherited, and many environmental influences. For instance, if you have dental work to straighten and whiten your teeth, you will not pass your environmentally produced smile to your offspring. Only the genetic component of variation is relevant to natural selection.

Many of the characters that vary in a population result from the combined effect of several genes. Polygenic inheritance produces characters that vary more or less continuously—in human height, for instance, from very short individuals to very tall ones (see Module 9.14). By contrast, other features, such as Mendel's purple and white pea flowers or human blood types, are determined by a single gene locus, with different alleles producing distinct phenotypes. But where do these alleles come from?

Mutation New alleles originate by mutation, a change in the nucleotide sequence of DNA. Thus, mutation is the ultimate source of the genetic variation that serves as raw material for evolution. In multicellular organisms, however, only mutations in cells that produce gametes can be passed to offspring and affect a population's genetic variability.

A change as small as a single nucleotide in a protein-coding gene can have a significant effect on phenotype, as in sickle-cell disease (see Module 9.13). An organism is a refined product of thousands of generations of past selection, and a random change in its DNA is not likely to improve its genome any more than randomly changing some words on a page is likely to improve a story. In fact, mutation that affects a protein's function will probably be harmful. On rare occasions, however, a mutated allele may actually improve the adaptation of an individual to its environment and enhance its reproductive success. This kind of effect is more likely when the environment is changing in such a way that mutations that were once disadvantageous are favorable under the new conditions. For instance, mutations that endow houseflies with resistance to the pesticide DDT also reduce their growth rate. Before DDT was introduced, such mutations were a handicap to the flies that had them. But once DDT was part of the environment, the mutant alleles were advantageous, and natural selection increased their frequency in fly populations.

Chromosomal mutations that delete, disrupt, or rearrange many gene loci at once are almost certain to be harmful. But duplication of a gene or small pieces of DNA through errors in meiosis can provide an important source of genetic variation. If a repeated segment of DNA can persist over the generations, mutations may accumulate in the duplicate copies without affecting the function of the original gene, eventually leading to new genes with novel functions. This process may have played a major role in evolution. For

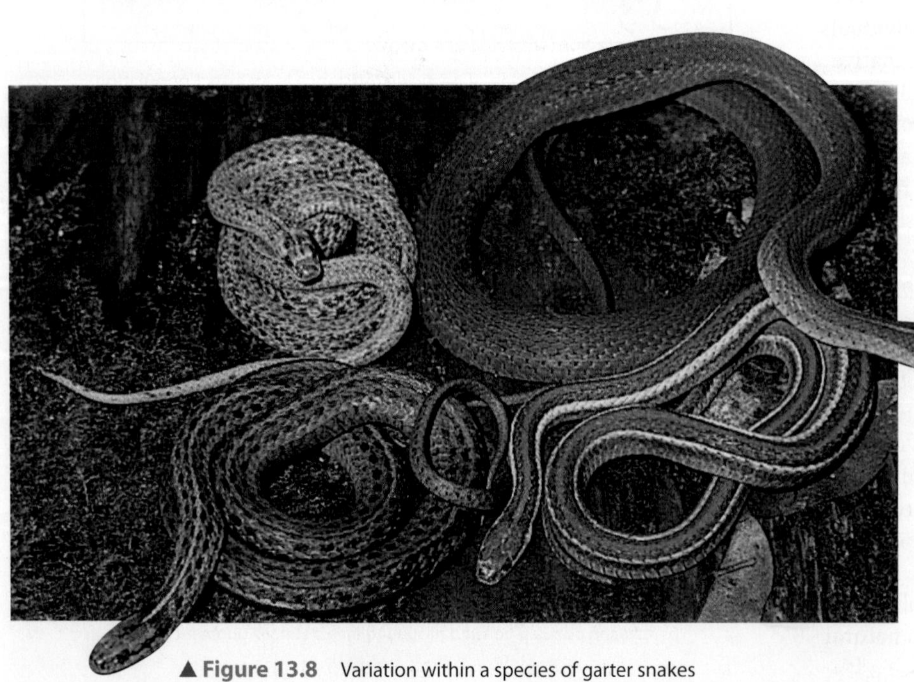

▲ **Figure 13.8** Variation within a species of garter snakes

example, the remote ancestors of mammals carried a single gene for detecting odors that has since been duplicated repeatedly. As a result, mice have about 1,300 different olfactory receptor genes. It is likely that such dramatic increases helped early mammals by enabling them to distinguish among many different smells. And repeated duplications of genes that control development are linked to the origin of vertebrate animals from an invertebrate ancestor (see Module 15.11).

In prokaryotes, mutations can quickly generate genetic variation. Because bacteria multiply so rapidly, a beneficial mutation can increase in frequency in a matter of hours or days. And because bacteria are haploid, with a single allele for each gene, a new allele can have an effect immediately.

Mutation rates in animals and plants average about one in every 100,000 genes per generation. For these organisms, low mutation rates, long time spans between generations, and diploid genomes prevent most mutations from significantly affecting genetic variation from one generation to the next.

Sexual Reproduction In organisms that reproduce sexually, most of the genetic variation in a population results from the unique combination of alleles that each individual inherits. (Of course, the origin of those allele variations is past mutations.)

Fresh assortments of existing alleles arise every generation from three random components of sexual reproduction: crossing over, independent orientation of homologous chromosomes at metaphase I of meiosis, and random fertilization (see Modules 8.15 and 8.17). During meiosis, pairs of homologous chromosomes, one set inherited from each parent, trade some of their genes by crossing over. These homologous chromosomes separate into gametes independently of other chromosome pairs. Thus, gametes from any individual vary extensively in their genetic makeup. Finally, each zygote made by a mating pair has a unique assortment of alleles resulting from the random union of sperm and egg.

Now let's see why genetic variation is such an essential element of evolution.

? **What is the ultimate (original) source of genetic variation? What is the source of most genetic variation in a population that reproduces sexually?**

Mutation; unique combinations of alleles resulting from sexual reproduction

13.9 Evolution occurs within populations

One common misconception about evolution is that individual organisms evolve during their lifetimes. It is true that natural selection acts on individuals: Each individual's combination of traits affects its survival and reproductive success. But the evolutionary impact of natural selection is only apparent in the changes in a population of organisms over time.

A **population** is a group of individuals of the same species that live in the same area and interbreed. We can measure evolution as a change in the prevalence of certain heritable traits in a population over a span of generations. The increasing proportion of resistant insects in areas sprayed with pesticide is one example. Natural selection favored insects with alleles for pesticide resistance; these insects left more offspring than nonresistant individuals, changing the genetic makeup of the population.

Different populations of the same species may be geographically isolated from each other to such an extent that an exchange of genetic material never or only rarely occurs. Such isolation is common in populations confined to different lakes (Figure 13.9) or islands. For example, each population of Galápagos tortoises is restricted to its own island. Not all populations have such sharp boundaries; however, members of a population typically breed with one another and are therefore more closely related to each other than they are to members of a different population.

In studying evolution at the population level, biologists focus on the **gene pool**, which consists of all copies of every type of allele at every locus in all members of the population. For many loci, there are two or more alleles in the gene pool. For example, in a mosquito population, there may be two alleles relating to DDT breakdown, one that codes for an enzyme that breaks down DDT and one for a version of the enzyme that does not. In populations living in fields sprayed with DDT, the allele for the enzyme conferring resistance will increase in frequency and the other allele will decrease in frequency. When the relative frequencies of alleles in a population change like this over a number of generations, evolution is occurring on its smallest scale. Such a change in a gene pool is often called **microevolution**.

In the next module, we'll explore how to test whether evolution is occurring in a population.

? **Why can't an individual evolve?**

Evolution involves changes in the genetic makeup of a population over time. An individual's genetic makeup rarely changes during its lifetime.

▲ **Figure 13.9** Isolated lakes in Denali National Park, Alaska

13.10 The Hardy-Weinberg equation can test whether a population is evolving

To understand how microevolution works, let's first examine a simple population in which evolution is not occurring and thus the gene pool is not changing. Consider an imaginary population of iguanas with individuals that differ in foot webbing (**Figure 13.10A**). Let's assume that foot webbing is controlled by a single gene and that the allele for nonwebbed feet (W) is completely dominant to the allele for webbed feet (w). The term *dominant* (see Module 9.3) may seem to suggest that over many generations, the W allele will somehow come to "dominate," becoming more and more common at the expense of the recessive allele. In fact, this is not what happens. The shuffling of alleles that accompanies sexual reproduction does not alter the genetic makeup of the population. In other words, no matter how many times alleles are segregated into different gametes and united in different combinations by fertilization, the frequency of each allele in the gene pool will remain constant unless other factors are operating. This equilibrium is known as the **Hardy-Weinberg principle**, named for the two scientists who derived it independently in 1908.

No webbing

Webbing

▲ **Figure 13.10A** Imaginary iguanas, with and without foot webbing

To test the Hardy-Weinberg principle, let's look at two generations of our imaginary iguana population. **Figure 13.10B** shows the frequencies of alleles in the gene pool of the original population. We have a total of 500 animals; of these, 320 have the genotype WW (nonwebbed feet), 160 have the heterozygous genotype, Ww (also nonwebbed feet, because the nonwebbed allele W is dominant), and 20 have the genotype ww (webbed feet). The proportions or frequencies of the three genotypes are shown in the middle of Figure 13.10B: 0.64 for WW ($\frac{320}{500}$), 0.32 for Ww ($\frac{160}{500}$), and 0.04 for ww ($\frac{20}{500}$).

From these genotype frequencies, we can calculate the frequency of each allele in the population. Because these are diploid organisms, this population of 500 has a total of 1,000 alleles for foot type. To determine the number of W alleles, we add the number in the WW iguanas, $2 \times 320 = 640$, to the number in the Ww iguanas, 160. The total number of W alleles is thus 800. The frequency of the W allele, which we will call p, is $\frac{800}{1{,}000}$, or 0.8. We can calculate the frequency of the w allele in a similar way; this frequency, called q, is 0.2. The letters p and q are often used to represent allele frequencies. Notice that $p + q = 1$. The combined frequencies of all alleles for a gene in a population must equal 1. If there are only two alleles and you know the frequency of one allele, you can calculate the frequency of the other.

What happens when the iguanas of this parent population form gametes? At the end of meiosis, each gamete has one allele for foot type, either W or w. The frequencies of the two alleles in the gametes will be the same as their frequencies in the gene pool of the parental population, 0.8 for W and 0.2 for w.

Figure 13.10C shows a Punnett square that uses these gamete allele frequencies and the rule of multiplication (see Module 9.7) to calculate the frequencies of the three genotypes in the next generation. The probability of producing a WW individual (by combining two W alleles from the pool of gametes) is $p \times p = p^2$, or $0.8 \times 0.8 = 0.64$. Thus, the frequency of WW iguanas in the next generation would be 0.64. Likewise, the frequency of ww individuals would be $q^2 = 0.04$. For heterozygous individuals, Ww, the genotype can form in two ways, depending on whether the sperm or egg supplies the dominant allele. In other words, the frequency of Ww would be $2pq = 2 \times 0.8 \times 0.2 = 0.32$. Do these frequencies look familiar? Notice that the three genotypes have the same frequencies in the next generation as they did in the parent generation.

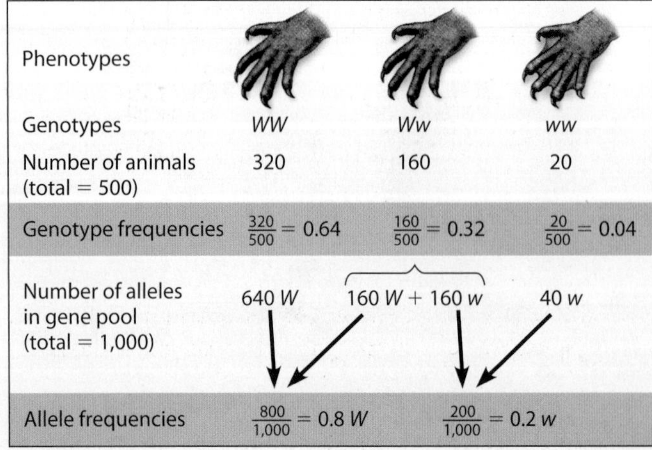

Phenotypes
Genotypes
WW
Ww
ww
Number of animals (total = 500)
320
160
20
Genotype frequencies
320/500 = 0.64
160/500 = 0.32
20/500 = 0.04
Number of alleles in gene pool (total = 1,000)
640 W
160 W + 160 w
40 w
Allele frequencies
800/1,000 = 0.8 W
200/1,000 = 0.2 w

▲ **Figure 13.10B** Gene pool of the original population of imaginary iguanas

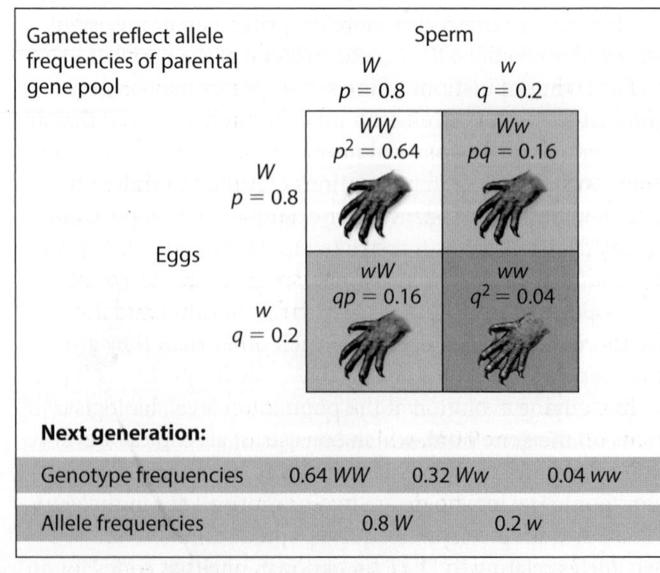

Gametes reflect allele frequencies of parental gene pool
Sperm
W
p = 0.8
w
q = 0.2
Eggs
W
p = 0.8
w
q = 0.2
WW
p² = 0.64
Ww
pq = 0.16
wW
qp = 0.16
ww
q² = 0.04
Next generation:
Genotype frequencies
0.64 WW
0.32 Ww
0.04 ww
Allele frequencies
0.8 W
0.2 w

▲ **Figure 13.10C** Gene pool of next generation of imaginary iguanas

Finally, what about the frequencies of the alleles in this new generation? Because the genotype frequencies are the same as in the parent population, the allele frequencies p and q are also the same. In fact, we could follow the frequencies of alleles and genotypes through many generations, and the results would continue to be the same. Thus, the gene pool of this population is in a state of equilibrium—Hardy-Weinberg equilibrium.

Now let's write a general formula for calculating the frequencies of genotypes in a population from the frequencies of alleles in the gene pool. In our imaginary iguana population, the frequency of the W allele (p) is 0.8, and the frequency of the w allele (q) is 0.2. Again note that $p + q = 1$. Also notice in Figures 13.10B and 13.10C that the frequencies of the three possible genotypes in the populations also add up to 1 (that is, $0.64 + 0.32 + 0.04 = 1$). We can represent these relationships symbolically with the Hardy-Weinberg equation:

p^2	+	$2pq$	+	q^2	$= 1$
Frequency of homozygous dominants		Frequency of heterozygotes		Frequency of homozygous recessives	

If a population is in Hardy-Weinberg equilibrium, allele and genotype frequencies will remain constant generation after generation. The Hardy-Weinberg principle tells us that something other than the reshuffling processes of sexual reproduction is required to change allele frequencies in a population. One way to find out what factors *can* change a gene pool is to identify the conditions that must be met if genetic equilibrium is to be maintained. For a population to be in Hardy-Weinberg equilibrium, it must satisfy five main conditions:

1. Very large population. The smaller the population, the more likely that allele frequencies will fluctuate by chance from one generation to the next.
2. No gene flow between populations. When individuals move into or out of populations, they add or remove alleles, altering the gene pool.
3. No mutations. By changing alleles or deleting or duplicating genes, mutations modify the gene pool.
4. Random mating. If individuals mate preferentially, such as with close relatives (inbreeding), random mixing of gametes does not occur, and genotype frequencies change.
5. No natural selection. The unequal survival and reproductive success of individuals (natural selection) can alter allele frequencies.

Rarely are all five conditions met in real populations; thus, allele and genotype frequencies often do change. The Hardy-Weinberg equation can be used to test whether evolution is occurring in a population. The equation also has medical applications, as we see next.

? Which is *least* likely to alter allele and genotype frequencies in a few generations of a large, sexually reproducing population: gene flow, mutation, or natural selection? Explain.

● Mutation. Because mutations are rare, their effect on allele and genotype frequencies from one generation to the next is likely to be small.

13.11 The Hardy-Weinberg equation is useful in public health science

CONNECTION

Public health scientists use the Hardy-Weinberg equation to estimate how many people carry alleles for certain inherited diseases. Consider the case of phenylketonuria (PKU), an inherited inability to break down the amino acid phenylalanine that results in brain damage if untreated. Newborns are routinely screened for PKU, which occurs in about one out of 10,000 babies born in the United States. The health problems associated with PKU can be prevented by strict adherence to a diet that limits the intake of phenylalanine. Packaged foods with ingredients such as aspartame, a common artificial sweetener that contains phenylalanine, must be labeled clearly (**Figure 13.11**).

PKU is due to a recessive allele, so the frequency of individuals born with PKU corresponds to the q^2 term in the Hardy-Weinberg equation. Given one PKU occurrence per 10,000 births, $q^2 = 0.0001$. Therefore, the frequency of the recessive allele for PKU in the population, q, equals the square root of 0.0001, or 0.01. And the frequency of the dominant allele, p, equals $1 - q$, or 0.99. The frequency of carriers, heterozygous people who do not have PKU but may pass the PKU allele on to offspring, is $2pq$, which equals $2 \times 0.99 \times 0.01$, or 0.0198. Thus, the equation tells us that about 2% (actually 1.98%) of the U.S. population are carriers of the PKU allele. Estimating the frequency of a harmful allele is part of any public health program dealing with genetic diseases.

INGREDIENTS: SORBITOL, MAGNESIUM STEARATE, ARTIFICIAL FLAVOR, ASPARTAME† (SWEETENER), ARTIFICIAL COLOR (YELLOW 5 LAKE, BLUE 1 LAKE), ZINC GLUCONATE.
†PHENYLKETONURICS: CONTAINS PHENYLALANINE
Diet
Soda

▲ **Figure 13.11** A warning to individuals with PKU

? Which term in the Hardy-Weinberg equation—p^2, $2pq$, or q^2—corresponds to the frequency of individuals who have no alleles for the disease PKU?

● The frequency of individuals with no PKU alleles is p^2.

▷ Mechanisms of Microevolution

13.12 Natural selection, genetic drift, and gene flow can cause microevolution

Deviations from the five conditions named in Module 13.10 for Hardy-Weinberg equilibrium can alter allele frequencies in a population (microevolution). Although new genes and new alleles originate by mutation, these random and rare events probably change allele frequencies little within a population of sexually reproducing organisms. Nonrandom mating can affect the frequencies of homozygous and heterozygous genotypes, but by itself usually does not affect allele frequencies. The three main causes of evolutionary change are natural selection, genetic drift, and gene flow.

Natural Selection The condition for Hardy-Weinberg equilibrium that there be no natural selection—that all individuals in a population be equal in ability to reproduce—is probably never met in nature. Populations consist of varied individuals, and some variants leave more offspring than others. In our imaginary iguana population, individuals with webbed feet (genotype *ww*) might survive better and produce more offspring because they are more efficient at swimming and catching food than individuals that lack webbed feet. Genetic equilibrium would be disturbed as the frequency of the *w* allele increased in the gene pool from one generation to the next.

Genetic Drift Flip a coin a thousand times, and a result of 700 heads and 300 tails would make you suspicious about that coin. But flip a coin 10 times, and an outcome of 7 heads and 3 tails would seem within reason. The smaller the sample, the more likely that chance alone will cause a deviation from an idealized result—in this case, an equal number of heads and tails. Let's apply that logic to a population's gene pool. The frequencies of alleles will be more stable from one generation to the next when a population is large. In a process called **genetic drift**, chance events can cause allele frequencies to fluctuate unpredictably from one generation to the next. The smaller the population, the more impact genetic drift is likely to have. In fact, an allele can be lost from a small population by such chance fluctuations. Two situations in which genetic drift can have a significant impact on a population are those that produce the bottleneck effect and the founder effect.

Catastrophes such as hurricanes, floods, or fires may kill large numbers of individuals, leaving a small surviving population that is unlikely to have the same genetic makeup as the original population. Such a drastic reduction in population size is called a **bottleneck effect**. Analogous to shaking just a few marbles through a bottleneck (**Figure 13.12A**), certain alleles (purple marbles) may be present at higher frequency in the surviving population than in the original population, others (green marbles) may be present at lower frequency, and some (orange marbles) may not be present at all. After a population is drastically reduced, genetic drift may continue for many generations until the population is again large enough for fluctuations due to chance to have less of an impact. Even if a population that has passed through a bottleneck ultimately recovers its size, it may have low levels of genetic variation—a legacy of the genetic drift that occurred when the population was small.

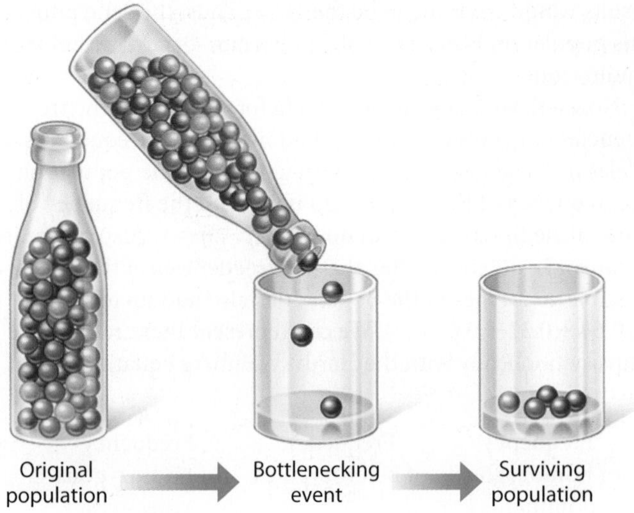

Original population
Bottlenecking event
Surviving population

▲ **Figure 13.12A** The bottleneck effect

One reason it is important to understand the bottleneck effect is that human activities such as overhunting and habitat destruction may create severe bottlenecks for other species. Examples of species affected by bottlenecks include the endangered Florida panther, the African cheetah, and the greater prairie chicken (**Figure 13.12B**). Millions of these birds once lived on the prairies of Illinois. But as their habitat was converted to farmland and other uses during the 19th and 20th centuries, the number of greater prairie chickens plummeted. By 1993, only two Illinois populations remained, with a total of fewer than 50 birds. Less than 50% of the eggs of these birds hatched. Researchers compared the DNA of the 1993 population with DNA extracted from museum specimens dating back to the 1930s. They surveyed six gene loci and found that the modern birds had lost 30% of the alleles that were present

▲ **Figure 13.12B** Greater prairie chicken (*Tympanuchus cupido*)

in the museum specimens. Thus, genetic drift as a result of the bottleneck reduced the genetic variation of the population and may have increased the frequency of harmful alleles, leading to the low egg-hatching rate.

Genetic drift is also likely when a few individuals colonize an island or other new habitat, producing what is called the **founder effect**. The smaller the group, the less likely that the genetic makeup of the colonists will represent the gene pool of the larger population they left.

The founder effect explains the relatively high frequency of certain inherited disorders among some human populations established by small numbers of colonists. For example, in 1814, 15 people founded a colony on Tristan da Cunha, a group of small islands in the middle of the Atlantic Ocean. Apparently, one of the colonists carried a recessive allele for retinitis pigmentosa, a progressive form of blindness. Of the 240 descendants who still lived on the islands in the 1960s, four had retinitis pigmentosa, and at least nine others were known to be heterozygous carriers of the allele. The frequency of this allele is 10 times higher on Tristan da Cunha than in the British population from which the founders came.

Gene Flow Allele frequencies in a population can also change as a result of **gene flow**, by which a population may gain or lose alleles when fertile individuals move into or out of a population or when gametes (such as plant pollen) are transferred between populations. Gene flow tends to reduce differences between populations. For example, humans today move more freely about the world than in the past, and gene flow has become an important agent of evolutionary change in previously isolated human populations.

Let's return to the Illinois greater prairie chickens and see how gene flow improved their fate. To counteract the lack of genetic diversity, researchers added a total of 271 birds from neighboring states to the Illinois populations. This strategy worked. New alleles entered the population, and the egg-hatching rate improved to more than 90%.

? **How might gene flow between populations living in different habitats actually interfere with each population's adaptation to its local environment?**

● The introduction of alleles that may not be beneficial in a particular habitat prevents the population living there from becoming fully adapted to its local conditions.

13.13 Natural selection is the only mechanism that consistently leads to adaptive evolution

Genetic drift, gene flow, and even mutation can cause microevolution. But only by chance could these events result in improving a population's fit to its environment. In natural selection, on the other hand, only the events that produce genetic variation (mutation and sexual reproduction) are random. The process of natural selection, in which better-adapted individuals are more likely to survive and reproduce, is not random. Consequently, only natural selection consistently leads to adaptive evolution—evolution that results in a better fit between organisms and their environment.

▲ **Figure 13.13** Blue-footed booby (*Sula nebouxii*)

The adaptations of organisms include many striking examples. Consider some of the features that make the blue-footed booby **(Figure 13.13)** suited to its home on the Galápagos Islands. The bird's body and bill are streamlined like a torpedo, minimizing friction as it dives from heights up to 24 m (over 75 feet) into the shallow water below. To pull out of this high-speed dive once it hits the water, the booby uses its large tail as a brake. Its large, webbed feet make great flippers, propelling the bird through the water at high speeds—a huge advantage when hunting fish.

Such adaptations are the result of natural selection. By consistently favoring some alleles over others, natural selection improves the match between organisms and their environment. However, the environment may change over time. As a result, what constitutes a "good match" between an organism and its environment is a moving target, making adaptive evolution a continuous, dynamic process.

Let's take a closer look at natural selection. The commonly used phrases "struggle for existence" and "survival of the fittest" are misleading if we take them to mean direct competition between individuals. There *are* animal species in which individuals lock horns or otherwise do combat to determine mating privilege. But reproductive success is generally more subtle and passive. In a varying population of moths, certain individuals may produce more offspring than others because their wing colors hide them from predators better. Plants in a wildflower population may differ in reproductive success because some attract more pollinators, owing to slight variations in flower color, shape, or fragrance. In a given environment, such traits can lead to greater **relative fitness**: the contribution an individual makes to the gene pool of the next generation relative to the contributions of other individuals. The fittest individuals in the context of evolution are those that produce the largest number of viable, fertile offspring and thus pass on the most genes to the next generation.

? **Explain how the phrase "survival of the fittest" differs from the biological definition of relative fitness.**

● Survival alone does not guarantee reproductive success. An organism's relative fitness is determined by its number of fertile offspring and thus its relative contribution to the gene pool of the next generation.

13.14 Natural selection can alter variation in a population in three ways

Evolutionary fitness is related to genes, but it is an organism's phenotype—its physical traits, metabolism, and behavior—that is directly exposed to the environment. Let's see how natural selection can affect the distribution of phenotypes using an imaginary mouse population that has a heritable variation in fur coloration. The bell-shaped curve in the top graph of **Figure 13.14** depicts the frequencies of individuals in an initial population in which fur color varies along a continuum from very light (only a few individuals) through various intermediate shades (many individuals) to very dark (a few individuals). The bottom graphs show three ways in which natural selection can alter the phenotypic variation in the mouse population. The blue downward arrows symbolize the pressure of natural selection working against certain phenotypes.

Stabilizing selection favors intermediate phenotypes. In the mouse population depicted in the graph on the bottom left, stabilizing selection has eliminated the extremely light and dark individuals, and the population has a greater number of intermediate phenotypes, which may be best suited to an environment with medium gray rocks. Stabilizing selection typically reduces variation and maintains the status quo for a particular character. For example, this type of selection keeps the majority of human birth weights in the range of 3–4 kg (6.5–9 pounds). For babies a lot smaller or larger than this, infant mortality may be greater.

Directional selection shifts the overall makeup of the population by acting against individuals at *one* of the phenotypic extremes. For the mouse population in the bottom center graph, the trend is toward darker fur color, as might occur if a fire darkened the landscape so that darker fur would more readily camouflage the animal. Directional selection is most common during periods of environmental change or when members of a species migrate to some new habitat with different environmental conditions. The changes we described in populations of insects exposed to pesticides are an example of directional selection. Another example is the increase in beak depth in a population of Galápagos finches following a drought, when the birds that were better able to eat larger seeds were more likely to survive.

Disruptive selection typically occurs when environmental conditions vary in a way that favors individuals at *both* ends of a phenotypic range over individuals with intermediate phenotypes. For the mice in the graph on the bottom right, individuals with light and dark fur have increased numbers. Perhaps the mice colonized a patchy habitat where a background of light soil was studded with areas of dark rocks. Disruptive selection can lead to two or more contrasting phenotypes in the same population. For example, in a population of African black-bellied seedcracker finches, large-billed birds, which specialize in cracking hard seeds, and small-billed birds, which feed mainly on soft seeds, survive better than birds with intermediate-sized bills, which are fairly inefficient at cracking both types of seeds.

Next we consider a special case of selection, one that leads to phenotypic differences between males and females.

? **What type of selection probably resulted in the color variations evident in the garter snakes in Figure 13.8?**

Disruptive selection

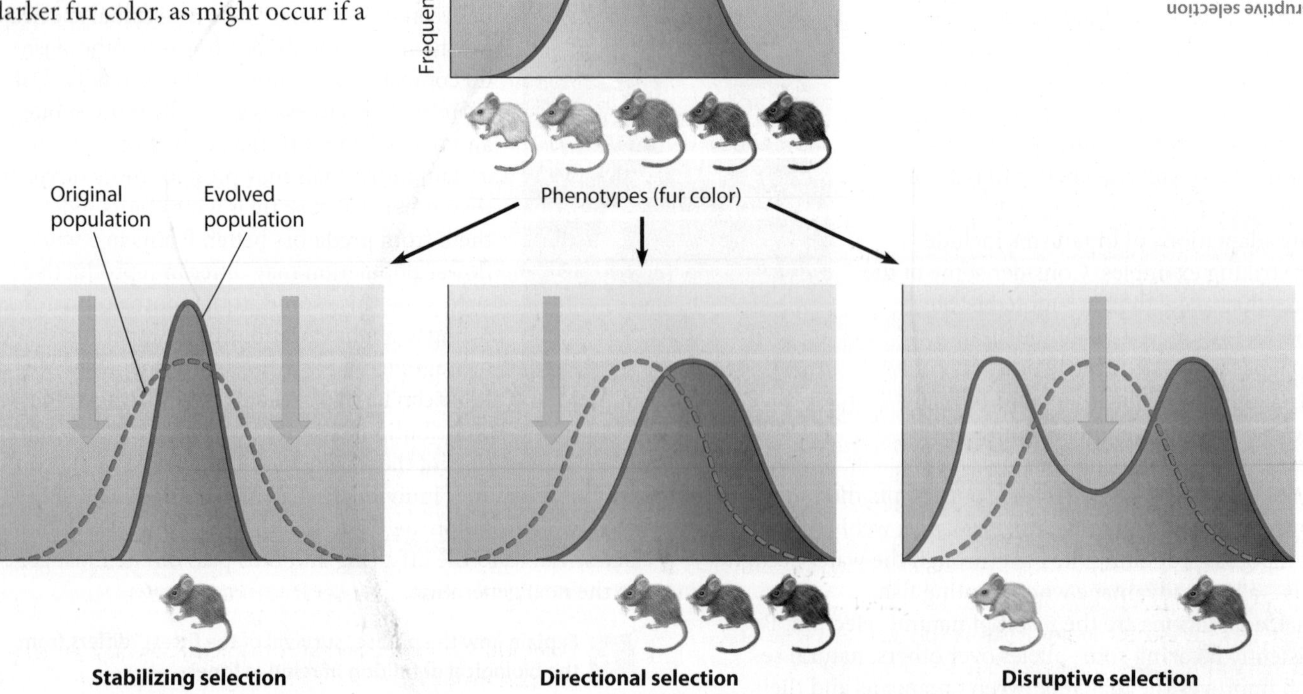

Frequency of individuals
Original population
Phenotypes (fur color)
Original population
Evolved population
Stabilizing selection
Directional selection
Disruptive selection

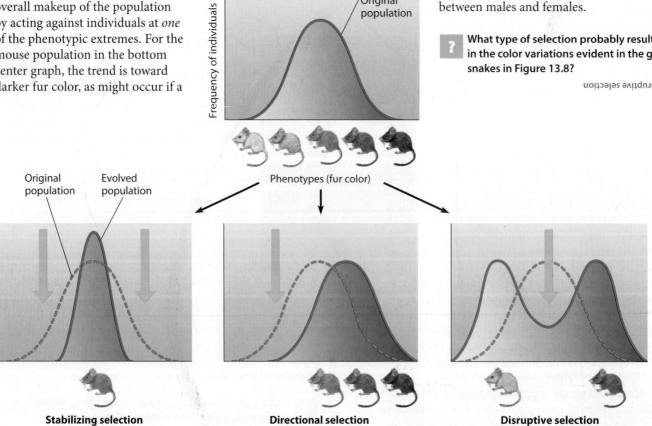

▲ **Figure 13.14** Three possible effects of natural selection on a phenotypic character

Try This Propose hypotheses to explain how natural selection on the original population resulted in each of the evolved populations.

13.15 Sexual selection may lead to phenotypic differences between males and females

Darwin was the first to examine **sexual selection**, a form of natural selection in which individuals with certain traits are more likely than other individuals to obtain mates. The males and females of an animal species obviously have different reproductive organs. But they may also have secondary sexual characteristics, noticeable differences not directly associated with reproduction or survival. This distinction in appearance, called **sexual dimorphism**, is often manifested in a size difference, but it can also include forms of adornment, such as manes on lions or colorful plumage on birds **(Figure 13.15A)**. Males are usually the showier sex, at least among vertebrates.

In some species, individuals compete directly with members of the same sex for mates **(Figure 13.15B)**. This type of sexual selection is called intrasexual selection (within the same sex, most often the males). Contests may involve physical combat, but are more often ritualized displays (see Module 35.19). Intrasexual selection is frequently found in species where the winning individual acquires a harem of mates.

In a more common type of sexual selection, called intersexual selection (between sexes) or mate choice, individuals of one sex (usually females) are choosy in selecting their mates. Males with the largest or most colorful adornments are often the most attractive to females. The extraordinary feathers of a peacock's tail are an example of this sort of "choose me" statement. What intrigued Darwin is that some of these mate-attracting features do not seem to be otherwise adaptive and may in fact pose some risks. For example, showy plumage may make male birds more visible to predators. But if such secondary sexual characteristics help a male gain a mate, then they will be reinforced over the generations for the most Darwinian of reasons—because they enhance reproductive success. Every time a female chooses a mate based on a certain appearance or behavior, she perpetuates the alleles that influenced her to make that choice and allows a male with that particular phenotype to perpetuate his alleles.

What is the advantage to females of being choosy? One hypothesis is that females prefer male traits that are correlated with "good genes." In several bird species, research has shown that traits preferred by females, such as bright beaks or long tails, are related to overall male health. The "good genes" hypothesis was also tested in gray tree frogs. Female frogs prefer to mate with males that give long mating calls **(Figure 13.15C)**. Researchers collected eggs from wild gray tree frogs. Half of each female's eggs were fertilized with sperm from long-calling males, and the others with sperm from short-calling males. The offspring of long-calling male frogs grew bigger, grew faster, and survived better than their half-siblings fathered by short-calling males. The duration of a male's mating call was shown to be indicative of the male's overall genetic quality, supporting the hypothesis that female mate choice can be based on a trait that indicates whether the male has "good genes."

Next we return to the concept of directional selection, focusing on the evolution of drug resistance in microorganisms that cause disease.

▲ **Figure 13.15A** Extreme sexual dimorphism (peacock and peahen)

▲ **Figure 13.15B** A contest for access to mates between two male elks

▲ **Figure 13.15C** A male gray tree frog calling for mates

? Males with the most elaborate ornamentation may garner the most mates. How might choosing such a mate be advantageous to a female?

● An elaborate display may signal good health and therefore good genes, which in turn could be passed along to the female's offspring.

13.16 The evolution of drug-resistant microorganisms is a serious public health concern

EVOLUTION CONNECTION

Antibiotics are drugs that kill infectious microorganisms. Penicillin, the first antibiotic to be developed, has been widely prescribed since the 1940s. A revolution in human health followed its introduction, rendering many previously fatal diseases easily curable. During the 1950s, some medical experts even thought the age of human infectious diseases would soon be over.

How does evolution hinder attempts to eradicate disease?

Why didn't that optimistic forecast come true? It did not take into account the force of evolution. In the same way that pesticides select for resistant insects, antibiotics select for resistant bacteria. A gene that codes for an enzyme that breaks down an antibiotic or a mutation that alters the site where an antibiotic binds can make a bacterium and its offspring resistant to that antibiotic. Again we see both the random and nonrandom aspects of natural selection—the random genetic mutations in bacteria and the nonrandom selective effects as the environment favors the antibiotic-resistant phenotype.

In what ways do we contribute to the problem of antibiotic resistance? Livestock producers add antibiotics to animal feed as a growth promoter and to prevent illness. These practices may select for bacteria that are resistant to standard antibiotics. Doctors may overprescribe antibiotics—for example, to patients with viral infections, which do not respond to antibiotic treatment. And patients may misuse prescribed antibiotics by prematurely stopping the medication because they feel better. This allows mutant bacteria that may be killed more slowly by the drug to survive and multiply. Subsequent mutations in such bacteria may lead to full-blown antibiotic resistance.

Difficulty in treating certain bacterial infections is a serious public health concern. Penicillin is virtually useless today in its original form. New drugs have been developed, but they are rendered ineffective as resistant bacteria evolve. Natural selection for antibiotic resistance is particularly strong in hospitals, where antibiotic use is extensive. A formidable "superbug" known as MRSA (methicillin-resistant *Staphylococcus aureus*) can cause "flesh-eating disease" **(Figure 13.16)** and potentially fatal systemic (whole-body) infections. Incidents of MRSA infections in both hospital and community settings continue to increase.

MRSA is not the only antibiotic-resistant microorganism—medical and pharmaceutical researchers are engaged in a race against the powerful force of evolution on many fronts. In 2013, the Centers for Disease Control reported that drug-resistant microorganisms infect more than two million people in the United States each year, and 23,000 people die from their infections. At least a dozen bacterial infections are no longer treatable with standard antibiotics. The most recent "superbug" to emerge is a strain of the bacteria that causes gonorrhea, a sexually transmitted disease. Public health officials fear that as this strain spreads, gonorrhea will become an incurable disease. And, as you learned in the chapter introduction, the decreasing effectiveness of chloroquine against malaria also resulted from the evolution of drug resistance in populations of the parasite that causes the disease. Experts know that resistance to artemisinin, currently the most effective drug, is only a matter of time. Indeed, artemisinin-resistant malaria has already been detected in Southeast Asia.

? Explain why the following statement is incorrect: "Antibiotics have created resistant bacteria."

The use of antibiotics did not cause bacteria to make new alleles. Rather, antibiotic use has increased the frequency of alleles for resistance that were already naturally present in bacterial populations.

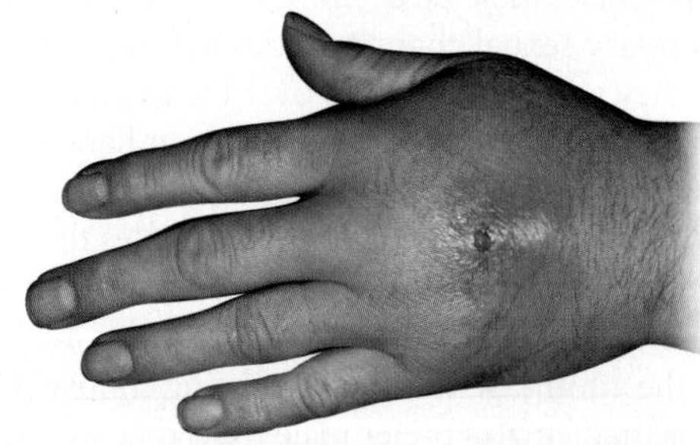

▲ **Figure 13.16** A MRSA skin infection

13.17 Diploidy and balancing selection preserve genetic variation

Natural selection acting on some variants within a population adapts that population to its environment. But what prevents natural selection from eliminating all variation as it selects against unfavorable genotypes? Why aren't less adaptive alleles eliminated as the "best" alleles are passed to the next generation? It turns out that the tendency for natural selection to reduce variation in a population is countered by mechanisms that maintain variation.

Most eukaryotes are diploid. Having two sets of chromosomes helps to prevent populations from becoming genetically uniform. As you know, natural selection acts on the phenotype, and recessive alleles only influence the phenotype of a homozygous recessive individual. In a heterozygote, a recessive allele is, in effect, protected from natural selection. The "hiding" of recessive alleles in heterozygotes can maintain a huge pool of alleles that may not be favored under present conditions but that could be advantageous if the environment changes.

In some cases, genetic variation is preserved rather than reduced by natural selection. **Balancing selection** occurs when natural selection maintains stable frequencies of two or more phenotypic forms in a population.

Heterozygote advantage is a type of balancing selection in which heterozygous individuals have greater reproductive success than either type of homozygote, with the result that two or more alleles for a gene are maintained in the population. An example of heterozygote advantage is the protection from malaria conferred by sickle hemoglobin (see Module 9.13). The frequency of the sickle-cell allele is generally highest in areas where malaria is a major cause of death, such as West Africa. Heterozygotes are protected from the most severe effects of malaria. Individuals who are homozygous for the normal hemoglobin allele are selected against by malaria. Individuals homozygous for the sickle-cell allele are selected against by sickle-cell disease. Thus, sickle hemoglobin is an evolutionary response to a fatal disease that first emerged in the environment of humans around 10,000 years ago. Notice that it is not an ideal solution—even heterozygotes may have health problems—but adaptations are often compromises.

Frequency-dependent selection is a type of balancing selection that maintains two different phenotypic forms in a population. In this case, selection acts against either phenotypic form if it becomes too common in the population. An example of frequency-dependent selection is a scale-eating fish in Lake Tanganyika, Africa, which attacks other fish from behind, darting in to remove a few scales from the side of its prey (**Figure 13.17**). These fish are either "left-mouthed" or "right-mouthed," a heritable character. Because its mouth twists to the left, a left-mouthed fish always attacks its prey's right side—try twisting your lower jaw and lips to the left and imagine which side of a fish you could take a bite from. Similarly, a right-mouthed fish attacks from the left. Prey fish guard more effectively against attack from whichever phenotype is most common. As a result, scale-eating fish with the less common phenotype have a feeding advantage that enhances survival and reproductive success. According to a recent study, frequency-dependent selection keeps each phenotype close to 50%.

Some of the genetic variation in a population probably has little or no impact on reproductive success. But even if only a fraction of the variation in a gene pool affects reproductive success, that is still an enormous resource of raw material for natural selection and the adaptive evolution it brings about.

? **Why would natural selection tend to reduce genetic variation more in populations of haploid organisms than in populations of diploid organisms?**

● All alleles in a haploid organism are phenotypically expressed and are hence screened by natural selection.

"Left-mouthed"
"Right-mouthed"

▲ **Figure 13.17** Left-mouthed and right-mouthed scale-eating fish (*Perissodus microlepis*)

13.18 Natural selection cannot fashion perfect organisms

Though natural selection leads to adaptation, there are several reasons why nature abounds with organisms that seem to be less than ideally "engineered" for their lifestyles.

1. *Selection can act only on existing variations.* Natural selection favors only the fittest variants from the phenotypes that are available, which may not be the ideal traits. New, advantageous alleles do not arise on demand.
2. *Evolution is limited by historical constraints.* Each species has a legacy of descent with modification from ancestral forms. Evolution does not scrap ancestral anatomy and build each new complex structure from scratch; it co-opts existing structures and adapts them to new situations. Thus, as birds and bats evolved from four-legged ancestors, their existing forelimbs took on new functions for flight and each lineage was left with only two limbs for walking.
3. *Adaptations are often compromises.* Each organism must do many different things. A blue-footed booby uses its webbed feet to swim after prey in the ocean, but these same feet make for clumsy travel on land.
4. *Chance, natural selection, and the environment interact.* Chance events often affect the genetic makeup of populations. When a storm blows insects over an ocean to an island, the wind does not necessarily transport the individuals that are best suited to the new environment. In small populations, genetic drift can result in the loss of beneficial alleles. In addition, the environment may change unpredictably from year to year, again limiting the extent to which adaptive evolution results in a close match between organisms and the environment.

With all these constraints, we cannot expect evolution to craft perfect organisms. Natural selection operates on a "better than" basis. Evidence for evolution is seen in the imperfections of the organisms it produces as well as in adaptations.

? **Humans owe much of their physical versatility and athleticism to their flexible limbs and joints. But we are prone to sprains, torn ligaments, and dislocations. Why?**

● Adaptations are compromises: Structural reinforcement has been compromised as agility was selected for.

CHAPTER 13 REVIEW

For practice quizzes, BioFlix animations, MP3 tutorials, video tutors, and more study tools designed for this textbook, go to MasteringBiology®

Reviewing the Concepts

Darwin's Theory of Evolution (13.1–13.7)

13.1 A sea voyage helped Darwin frame his theory of evolution. Darwin's theory differed greatly from the long-held notion of a young Earth inhabited by unchanging species. Darwin called his theory descent with modification, which explains that all of life is connected by common ancestry and that descendants have accumulated adaptations to changing environments over vast spans of time.

13.2 The study of fossils provides strong evidence for evolution. The fossil record reveals the historical sequence in which organisms have evolved.

13.3 Fossils of transitional forms support Darwin's theory of evolution.

13.4 Homologies provide strong evidence for evolution. Structural and molecular homologies reveal evolutionary relationships.

13.5 Homologies indicate patterns of descent that can be shown on an evolutionary tree.

13.6 Darwin proposed natural selection as the mechanism of evolution.

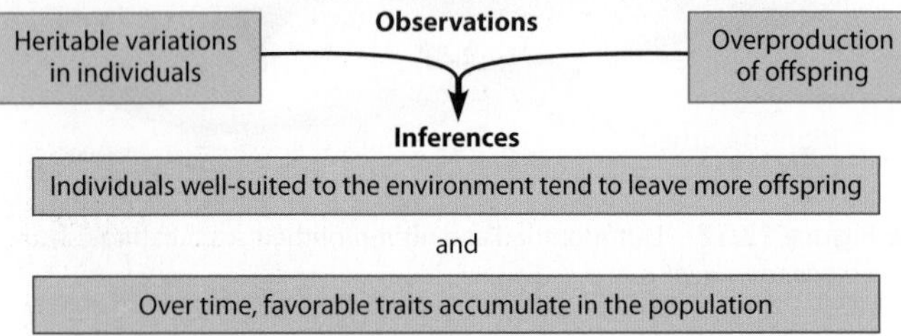

Observations
Heritable variations in individuals
Overproduction of offspring
Inferences
Individuals well-suited to the environment tend to leave more offspring
and
Over time, favorable traits accumulate in the population

13.7 Scientists can observe natural selection in action.

The Evolution of Populations (13.8–13.11)

13.8 Mutation and sexual reproduction produce the genetic variation that makes evolution possible.

13.9 Evolution occurs within populations. Microevolution is a change in the frequencies of alleles in a population's gene pool.

13.10 The Hardy-Weinberg equation can test whether a population is evolving. The Hardy-Weinberg principle states that allele and genotype frequencies will remain constant if a population is large, mating is random, and there is no mutation, gene flow, or natural selection.

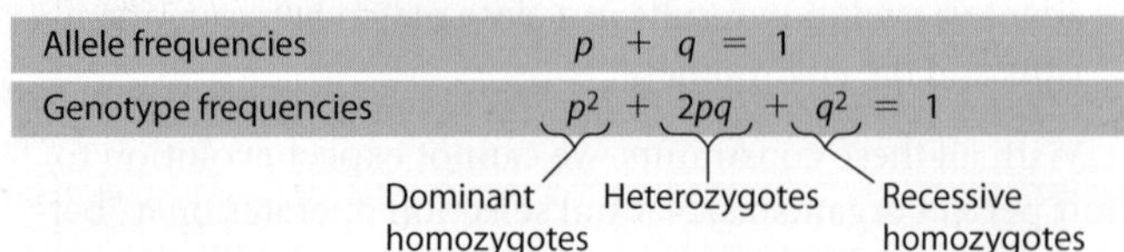

Allele frequencies $p + q = 1$
Genotype frequencies $p^2 + 2pq + q^2 = 1$
Dominant homozygotes
Heterozygotes
Recessive homozygotes

13.11 The Hardy-Weinberg equation is useful in public health science.

Mechanisms of Microevolution (13.12–13.18)

13.12 Natural selection, genetic drift, and gene flow can cause microevolution. The bottleneck effect and founder effect lead to genetic drift.

13.13 Natural selection is the only mechanism that consistently leads to adaptive evolution. Relative fitness is the relative contribution an individual makes to the gene pool of the next generation. As a result of natural selection, favorable traits increase in a population.

13.14 Natural selection can alter variation in a population in three ways.

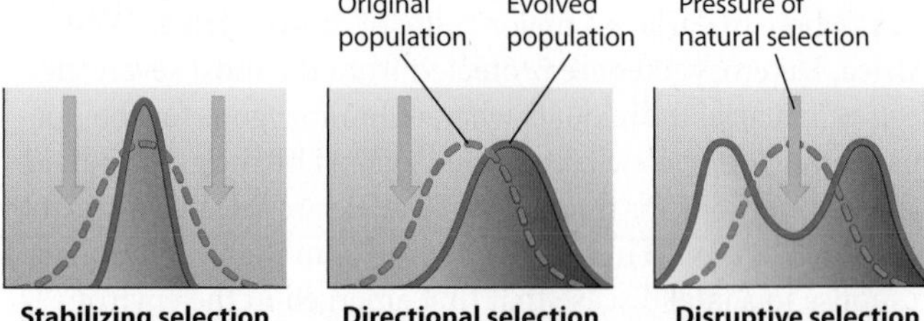

Original population
Evolved population
Pressure of natural selection
Stabilizing selection
Directional selection
Disruptive selection

13.15 Sexual selection may lead to phenotypic differences between males and females. Secondary sex characteristics can give individuals an advantage in mating.

13.16 The evolution of drug-resistant microorganisms is a serious public health concern.

13.17 Diploidy and balancing selection preserve genetic variation. Diploidy preserves variation by "hiding" recessive alleles. Balancing selection may result from heterozygote advantage or frequency-dependent selection.

13.18 Natural selection cannot fashion perfect organisms. Natural selection can act only on available variation; anatomical structures result from modified ancestral forms; adaptations are often compromises; and chance, natural selection, and the environment interact.

Connecting the Concepts

1. Summarize the key points of Darwin's theory of descent with modification, including his proposed mechanism of evolution.
2. Complete this concept map describing potential causes of evolutionary change within populations.

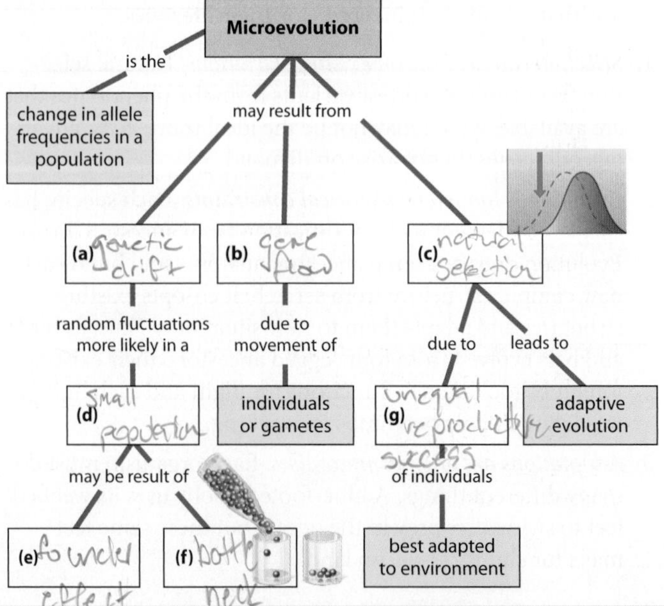

Microevolution
is the
change in allele frequencies in a population
may result from
(a) genetic drift
(b) gene flow
(c) natural selection
random fluctuations more likely in a
due to movement of
due to
leads to
(d) small population
individuals or gametes
(g) unequal reproductive success
of individuals
adaptive evolution
may be result of
(e) founder effect
(f) bottle neck
best adapted to environment

Testing Your Knowledge

Level 1: Knowledge/Comprehension

3. Which of the following did not influence Darwin as he synthesized the theory of evolution by natural selection?
   a. examples of artificial selection that produce large and relatively rapid changes in domesticated species
   b. Lyell's *Principles of Geology*, on gradual geologic changes
   c. comparisons of fossils with living organisms
   d. Mendel's paper describing the laws of inheritance
4. Natural selection is sometimes described as "survival of the fittest." Which of the following best measures an organism's fitness?
   a. how many fertile offspring it produces
   b. how strong it is when pitted against others of its species
   c. its ability to withstand environmental extremes
   d. how much food it is able to make or obtain
5. In an area of erratic rainfall, a biologist found that grass plants with alleles for curled leaves reproduced better in dry years, and plants with alleles for flat leaves reproduced better in wet years. This situation would tend to ______________. (*Explain your answer.*)
   a. cause genetic drift in the grass population.
   b. preserve genetic variation in the grass population.
   c. lead to stabilizing selection in the grass population.
   d. lead to uniformity in the grass population.
6. If an allele is recessive and lethal in homozygotes before they reproduce,
   a. the allele will be removed from the population by natural selection in approximately 1,000 years.
   b. the allele will likely remain in the population at a low frequency because it cannot be selected against in heterozygotes.
   c. the fitness of the homozygous recessive genotype is 0.
   d. both b and c are correct.
7. In a population with two alleles, *B* and *b*, the allele frequency of *b* is 0.4. *B* is dominant to *b*. What is the frequency of individuals with the dominant phenotype if the population is in Hardy-Weinberg equilibrium?
   a. 0.16
   b. 0.36
   c. 0.48
   d. 0.84
8. Within a few weeks of treatment with the drug 3TC, a patient's HIV population consists entirely of 3TC-resistant viruses. How can this result best be explained?
   a. HIV can change its surface proteins and resist vaccines.
   b. The patient must have become reinfected with a resistant virus.
   c. A few drug-resistant viruses were present at the start of treatment, and natural selection increased their frequency.
   d. HIV began making drug-resistant versions of its enzymes in response to the drug.

Level 2: Application/Analysis

9. In the late 18th century, machines that could blast through rock to build roads and railways were invented, exposing deep layers of rocks. How would you expect this development to aid the science of paleontology?
10. Write a paragraph briefly describing the kinds of scientific evidence for evolution.
11. In the early 1800s, French naturalist Jean Baptiste Lamarck suggested that the best explanation for the relationship of fossils to current organisms is that life evolves. He proposed that by using or not using its body parts, an individual may change its traits and then pass those changes on to its offspring. He suggested, for instance, that the ancestors of the giraffe had lengthened their necks by stretching higher and higher into the trees to reach leaves. Evaluate Lamarck's hypotheses from the perspective of present-day scientific knowledge.
12. Sickle-cell disease is caused by a recessive allele. Roughly one out of every 400 African Americans (0.25%) is afflicted with sickle-cell disease. Use the Hardy-Weinberg equation to calculate the percentage of African Americans who are carriers of the sickle-cell allele. (*Hint*: $q^2 = 0.0025$.)
13. It seems logical that natural selection would work toward genetic uniformity; the genotypes that are most fit produce the most offspring, increasing the frequency of adaptive alleles and eliminating less adaptive alleles. Yet there remains a great deal of genetic variation within populations. Describe some of the factors that contribute to this variation.

Level 3: Synthesis/Evaluation

14. **SCIENTIFIC THINKING** Cetaceans are fully aquatic mammals that evolved from terrestrial ancestors. Gather information about the respiratory system of cetaceans and describe how it illustrates the statement made in Module 13.18 that "Evolution is limited by historical constraints."
15. A population of snails is preyed on by birds that break the snails open on rocks, eat the soft bodies, and leave the shells. The snails occur in both striped and unstriped forms. In one area, researchers counted both live snails and broken shells. Their data are summarized below:

	Striped	Unstriped	Total	Percent Striped
Living	264	296	560	47.1
Broken	486	377	863	56.3

Which snail form seems better adapted to this environment? Why? Predict how the frequencies of striped and unstriped snails might change in the future.

16. Advocates of "scientific creationism" and "intelligent design" lobby school districts for such things as a ban on teaching evolution, equal time in science classes to teach alternative versions of the origin and history of life, or disclaimers in textbooks stating that evolution is "just a theory." They argue that it is only fair to let students evaluate both evolution and the idea that all species were created by God as the Bible relates or that, because organisms are so complex and well adapted, they must have been created by an intelligent designer. Do you think that alternative views of evolution should be taught in science courses? Why or why not?

Answers to all questions can be found in Appendix 4.

CHAPTER 14

The Origin of Species

Can we observe speciation occurring?

Compared to many male birds that sport brilliant plumage—the shimmering eyes of the peacock's tail or the fire-engine red feathers of the cardinal, for example—the Vogelkop bowerbird (*Amblyornis inornata*) is a rather dull fellow. However, he does have a unique talent: He's a fabulous decorator. Bowerbirds, which are native to New Guinea and Australia, are named for the structure, called a bower, that the male weaves from twigs and grasses to attract females. The hut-style bower shown in the photo below, built by a Vogelkop bowerbird, is about two meters (6.5 feet) wide and one meter high.

After completing his elaborate construction project, the male bowerbird collects objects such as fruits, seeds, insect parts, rocks, flowers, and leaves and arranges them artfully by color and type. Individual males differ in their preferences for certain colors and arrangements of objects. Females tour the bowers of local males, inspecting each bower carefully while its owner courts her with a song and dance. A female may visit promising candidates multiple times before finally mating with one.

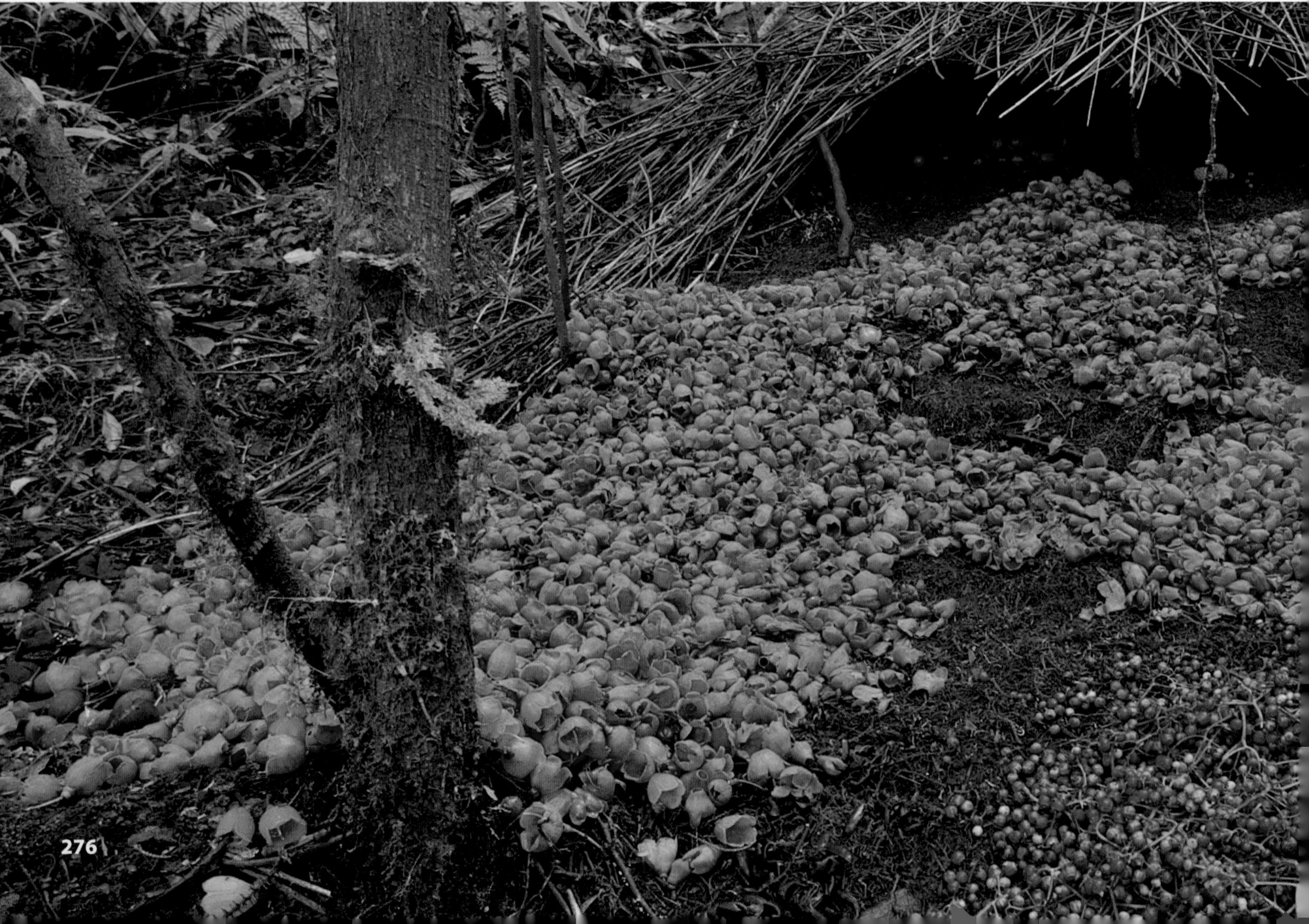

Not all Vogelkop bowerbirds construct displays like the one in the photo. The males of another population build a simpler structure consisting of sticks loosely woven around a central sapling. The objects ornamenting the display are all drab-colored. Researchers hypothesize that this divergence in display preferences has started the two populations on separate evolutionary paths, with each path leading to a new species, or speciation.

In this chapter, we explore how natural selection, which adapts a population to its environment, and other processes may lead to speciation—the origin of new species. Speciation is responsible for the amazing diversity of life on Earth. We begin with the biological definition of a species and describe the mechanisms through which new species may evolve. We also explore some of the evidence for speciation and how scientists study this evolutionary process.

BIG IDEAS

Defining Species
(14.1–14.3)

A species can be defined as a group of populations whose members can produce fertile offspring.

Mechanisms of Speciation
(14.4–14.11)

Speciation can take place with or without geographic isolation, as long as reproductive barriers evolve that keep species separate.

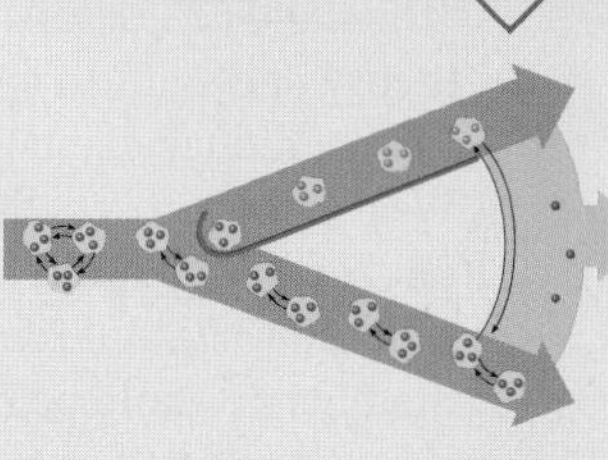

Defining Species

14.1 The origin of species is the source of biological diversity

Darwin was eager to explore landforms newly emerged from the sea when he came to the Galápagos Islands. He noted that these volcanic islands, despite their geologic youth, were teeming with plants and animals found nowhere else in the world. He realized that these species, like the islands, were relatively new. He wrote in his diary: "Both in space and time, we seem to be brought somewhat near to that great fact—that mystery of mysteries—the first appearance of new beings on this Earth."

Even though Darwin titled his seminal work *On the Origin of Species by Means of Natural Selection*, most of his theory of evolution focused on the role of natural selection in the gradual adaptation of a population to its environment. We call this process microevolution—changes in the gene pool of a population from one generation to the next (see Module 13.9). But if microevolution were *all* that happened, then Earth would be inhabited only by a highly adapted version of the first form of life.

The "mystery of mysteries" that fascinated Darwin is **speciation**, the process by which one species splits into two or more species. He envisioned the history of life as a tree, with multiple branchings from a common trunk out to the tips of the youngest twigs (**Figure 14.1**).

Each time speciation occurs, the diversity of life increases. Over the course of 3.5 billion years, an ancestral species first gave rise to two or more different species, which then branched to new lineages, which branched again, until we arrive at the millions of species that live, or once lived, on Earth. This origin of species explains both the diversity and the unity of life. When one species splits into two, the new species share many characteristics because they are descended from a common ancestor.

? **How does microevolution differ from speciation?**

● Microevolution involves evolutionary changes within a population; speciation occurs when a population changes enough that it diverges from its parent species and becomes a new species.

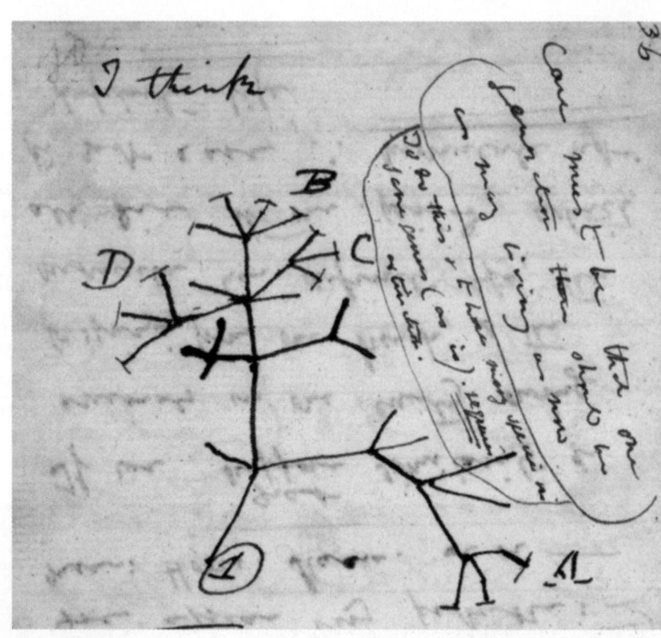

36
I think
B
C
D
1

▲ **Figure 14.1** Sketch made by Darwin as he pondered the origin of species

14.2 There are several ways to define a species

The word *species* is from the Latin for "kind" or "appearance," and indeed, even young children learn to distinguish between kinds of plants and animals—between roses and dandelions or dogs and cats—from differences in their appearance. Although the basic idea of species as distinct life-forms seems intuitive, devising a more formal definition is not so easy.

In many cases, the differences between two species are obvious. In other cases, the differences between two species are not so obvious. Although the two birds in **Figure 14.2A** look much the same, they are different species—the one on the left is an eastern meadowlark (*Sturnella magna*); the bird on the right is a western meadowlark (*Sturnella neglecta*). They are distinct species because their songs and other behaviors are different enough that each type of meadowlark breeds only with individuals of its own species.

▲ **Figure 14.2A** Similarity between two species: the eastern meadowlark (left) and western meadowlark (right)

How similar are members of the same species? Whereas the individuals of many species exhibit fairly limited variation in physical appearance, certain other species—our own, for example—seem extremely varied. The physical diversity within our species (partly illustrated in **Figure 14.2B**, on the facing page) might lead you to guess that there are several human species. Despite these outward appearances, however, humans all belong to the same species, *Homo sapiens*.

▲ **Figure 14.2B** Diversity within one species

The Biological Species Concept How then do biologists define a species? And what keeps one species distinct from others? The primary definition of species used in this book is called the **biological species concept**. It defines a **species** as a group of populations whose members have the potential to interbreed in nature and produce fertile offspring (offspring that themselves can reproduce). A businesswoman in Manhattan may be unlikely to meet a dairy farmer in Mongolia, but if the two should happen to meet and mate, they could have viable babies that develop into fertile adults. Thus, members of a biological species are united by being reproductively compatible, at least potentially.

Members of different species do not usually mate with each other. In effect, **reproductive isolation** prevents genetic exchange (gene flow) and maintains a boundary between species. But there are some pairs of clearly distinct species that do occasionally interbreed. The resulting offspring are called **hybrids**. An example is the grizzly bear (*Ursus arctos*) and the polar bear (*Ursus maritimus*), whose hybrid offspring have been called "grolar bears" (**Figure 14.2C**). The two species have been known to interbreed in zoos, and DNA testing confirmed that a bear shot in 2006 in the Canadian Arctic was a wild polar bear–grizzly offspring. Another grolar bear was shot in 2010, and more sightings have been reported as melting polar sea ice brings the two species into contact more often. Hybridization is much more common in plants than animals and is a major factor in speciation. Clearly, identifying species solely on the basis of reproductive isolation can be more complex than it may seem.

Grizzly bear

Polar bear

Hybrid "grolar" bear

▲ **Figure 14.2C** Hybridization between two species of bears

There are other instances in which applying the biological species concept is problematic. For example, there is no way to determine whether organisms that are now known only through fossils were once able to interbreed. Also, this criterion is useless for organisms such as prokaryotes that reproduce asexually. Because of such limitations, alternative species concepts are useful in certain situations.

Other Definitions of Species For most organisms—sexual, asexual, and fossils alike—classification is based mainly on physical traits such as shape, size, and other features of morphology (form). This **morphological species concept** has been used to identify most of the 1.8 million species that have been named to date. The advantages of this concept are that it can be applied to asexual organisms and fossils and does not require information on possible interbreeding. The disadvantage, however, is that this approach relies on subjective criteria, and researchers may disagree on which features distinguish a species.

Another species definition, the **ecological species concept**, identifies species in terms of their ecological niches, focusing on unique adaptations to particular roles in a biological community (see Module 37.3). For example, two species of fish may be similar in appearance but distinguishable based on what they eat or the depth of water in which they are usually found.

Finally, the **phylogenetic species concept** defines a species as the smallest group of individuals that share a common ancestor and thus form one branch on the tree of life. Biologists trace the phylogenetic history of such a species by comparing its characteristics, such as morphology, DNA sequences, or biochemical pathways, with those of other organisms. These sorts of analyses can distinguish groups that are generally similar yet different enough to be considered separate species. Of course, agreeing on the amount of difference required to establish separate species remains a challenge.

Each species definition is useful, depending on the situation and the questions being asked. The biological species concept, however, helps focus on how these discrete groups of organisms arise and are maintained by reproductive isolation. Because reproductive isolation is an essential factor in the evolution of many species, we look at it more closely next.

? Which species concepts could you apply to both asexual and sexual species? Explain.

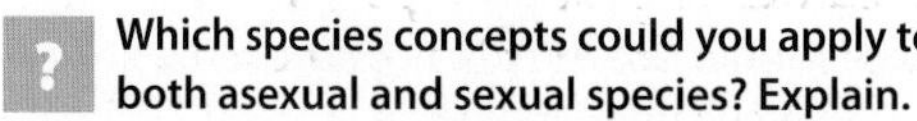
The morphological, ecological, and phylogenetic species concepts could all be used because they do not rely on the criterion of reproductive isolation.

VISUALIZING THE CONCEPT

14.3 Reproductive barriers keep species separate

Clearly, a fly will not mate with a frog or a fern. But what prevents species that are closely related from interbreeding? Reproductive isolation depends on one or more types of reproductive barriers—biological features of the organism that prevent individuals of different species from interbreeding. The various types of reproductive barriers that isolate the gene pools of species can be categorized as either prezygotic or postzygotic, depending on whether they function before or after zygotes (fertilized eggs) form. **Prezygotic barriers** prevent mating or fertilization between species. **Postzygotic barriers** operate after hybrid zygotes have formed.

PREZYGOTIC BARRIERS

The garter snake *Thamnophis atratus* lives mainly in water.

The eastern spotted skunk (*Spilogale putorius*) breeds in late winter.

The blue-footed booby (*Sula nebouxii*) performs an elaborate courtship dance.

Heliconia pogonantha is pollinated by hummingbirds with long, curved bills.

Type of isolation	Habitat	Temporal	Behavioral	Mechanical
	Lack of opportunities to encounter each other	Breeding at different times or seasons	Failure to send or receive appropriate signals	Physical incompatibility of reproductive parts

The garter snake *Thamnophis sirtalis* lives on land.

The western spotted skunk (*Spilogale gracilis*) breeds in the fall.

The masked booby (*Sula dactylatra*) performs a different courtship ritual.

Heliconia latispatha is pollinated by hummingbirds with short, straight bills.

Species are not necessarily separated by obvious physical barriers. These snakes occupy different habitats in the same area.

Temporal isolation also happens in plants that flower during different seasons or open flowers at different times during the day.

In another example, male fireflies signal to females of the same species by blinking their lights in the particular rhythm of their species. Females respond only to that rhythm.

Pollinators pick up pollen from the male parts of one flower and transfer it to the female parts of another flower. Floral characteristics determine the best fit between pollinator and flower.

POSTZYGOTIC BARRIERS

Purple sea urchin (*Strongylocentrotus purpuratus*)

Gametic
Molecular incompatibility of eggs and sperm or pollen and stigma

Red sea urchin (*Strongylocentrotus franciscanus*)

Sea urchins release their gametes into the water. Surface proteins prevent the gametes of different species from binding to each other.

Reduced hybrid viability
Interaction of parental genes impairs the hybrid's development or survival.

Some species of salamander can hybridize, but their offspring do not develop fully or, like this one, are frail and will not survive long enough to reproduce.

Reduced hybrid fertility
Hybrids are vigorous but cannot produce viable offspring.

The hybrid offspring of a horse and a donkey is a mule, which is robust but sterile.

Hybrid breakdown
Hybrids are viable and fertile, but their offspring are feeble or sterile.

The rice hybrids on the left and right are fertile, but plants of the next generation (middle) are sterile.

If chromosomes of the parent species differ in number or structure, meiosis in hybrids may fail to produce normal gametes.

? **Two closely related fish live in the same lake, but one feeds along the shoreline and the other is a bottom feeder in deep water. This is an example of ______________ isolation, which is a ______________ reproductive barrier.**

habitat . . . prezygotic

Mechanisms of Speciation

14.4 In allopatric speciation, geographic isolation leads to speciation

A key event in the origin of a new species is the separation of a population from other populations of the same species. With its gene pool isolated, the splinter population can follow its own evolutionary course. Changes in allele frequencies caused by natural selection, genetic drift, and mutation will not be diluted by alleles entering from other populations (gene flow). The initial block to gene flow may come from a geographic barrier that isolates a population. This mode of speciation is called **allopatric speciation** (from the Greek *allos*, other, and *patra*, fatherland). Populations separated by a geographic barrier are known as allopatric populations.

Geographic Barriers Several geologic processes can isolate populations. A mountain range may emerge and gradually split a population of organisms that can inhabit only lowlands. A large lake may subside until there are several smaller lakes, isolating certain fish populations. On a larger scale, continents themselves can split and move apart (see Module 15.7). Allopatric speciation can also occur when individuals colonize a remote area and become geographically isolated from the parent population.

How large must a geographic barrier be to keep allopatric populations apart? The answer depends on the ability of the organisms to move. Birds, mountain lions, and coyotes can easily cross mountain ranges. The windblown pollen of trees is not hindered by such barriers, and the seeds of many plants may be carried back and forth by animals. In contrast, small rodents may find a canyon or a wide river a formidable barrier. The Grand Canyon and Colorado River (**Figure 14.4A**) separate two species of antelope squirrels. Harris's antelope squirrel (*Ammospermophilus harrisii*) inhabits the south rim. Just a few kilometers away on the north rim, but separated by the deep and wide canyon, lives the closely related white-tailed antelope squirrel (*Ammospermophilus leucurus*).

Evidence of Allopatric Speciation Many studies provide evidence that speciation has occurred in allopatric populations. An interesting example is the 30 species of snapping shrimp in the genus *Alpheus* that live off the Isthmus of Panama, the land bridge that connects South and North America (**Figure 14.4B**). Snapping shrimp are named for the snapping together of their single oversized claw, which creates a high-pressure blast that stuns their prey. Morphological and genetic data group these shrimp into 15 pairs of species, with the members of each pair being each other's closest relative. In each case, one member of the pair lives on the Atlantic side of the isthmus, while the other lives on the Pacific side, strongly suggesting that geographic separation of the ancestral species of these snapping shrimp led to allopatric speciation.

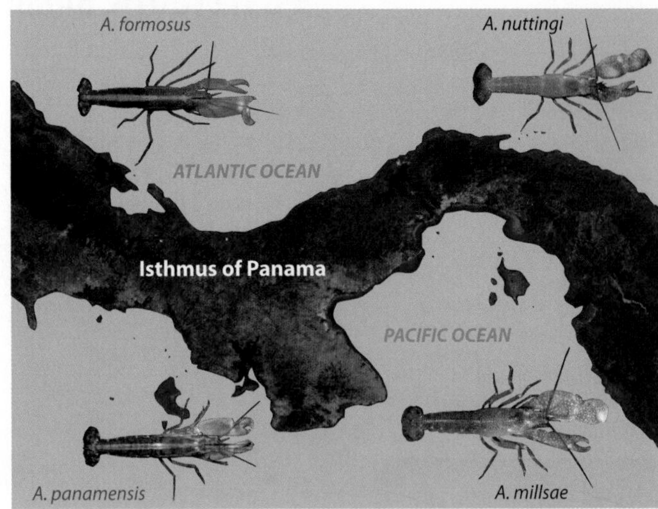
A. formosus
A. nuttingi
ATLANTIC OCEAN
Isthmus of Panama
PACIFIC OCEAN
A. panamensis
A. millsae

▲ **Figure 14.4B** Allopatric speciation in snapping shrimp: 2 of the 15 pairs of shrimp species that are separated by the Isthmus of Panama

? **Geologic evidence indicates that the Isthmus of Panama gradually closed about 3 million years ago. Genetic analyses indicate that the various species of snapping shrimp originated from 9 to 3 million years ago, with the species pairs that live in deepest water diverging first. How would you interpret these data?**

● The deeper species would have been separated into two isolated populations first, which enabled them to diverge into new species first.

South rim
North rim
A. harrisii
A. leucurus

▲ **Figure 14.4A** Allopatric speciation of geographically isolated antelope squirrels

14.5 Reproductive barriers can evolve as populations diverge

Geographic isolation creates opportunities for speciation, but it does not necessarily lead to new species. Speciation occurs only when the gene pool undergoes changes that establish reproductive barriers such as those described in Module 14.3. What might cause such barriers to arise? The environment of an isolated population may include different food sources, different types of pollinators, and different predators. As a result of natural selection acting on preexisting variations (or as a result of genetic drift or mutation), a population's traits may change in ways that also establish reproductive barriers.

Researchers have successfully documented the evolution of reproductive isolation with laboratory experiments. While at Yale University, Diane Dodd tested the hypothesis that reproductive barriers can evolve as a by-product of changes in populations as they adapt to different environments. Dodd raised fruit flies on different food sources. Some populations were fed starch; others were fed maltose. After about 40 generations, populations raised on starch digested starch more efficiently, and those raised on maltose digested maltose more efficiently.

Dodd then combined flies from various populations in mating experiments. **Figure 14.5A** shows some of her results. When flies from "starch populations" were mixed with flies from "maltose populations," the flies mated more frequently with partners raised on the same food source (left grid), even when the partners came from different populations. In one of the control tests (right grid), flies taken from different populations adapted to starch were about as likely to mate with each other as with flies from their own populations. The mating preference shown in the experimental group is an example of a prezygotic barrier. The reproductive barrier was not absolute—some mating between maltose flies and starch flies did occur—but reproductive isolation was under way as these allopatric populations became adapted to different environments.

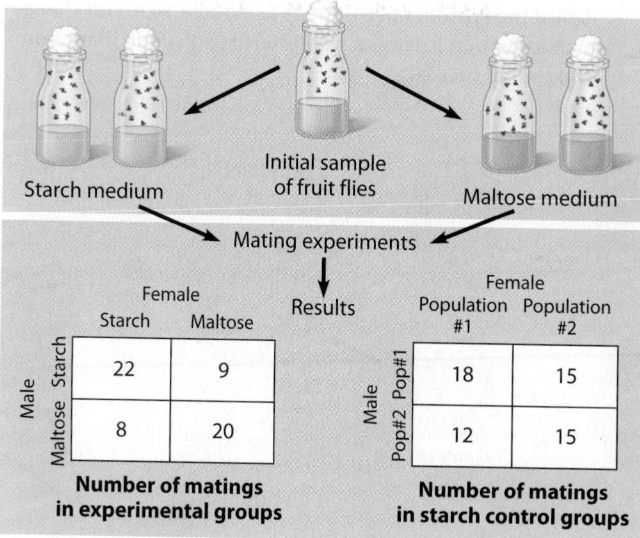

Initial sample of fruit flies
Starch medium
Maltose medium
Mating experiments
Results
Female
Starch
Maltose
Male
Starch
Maltose
22
9
8
20
Number of matings in experimental groups
Female
Population #1
Population #2
Male
Pop#1
Pop#2
18
15
12
15
Number of matings in starch control groups

▲ **Figure 14.5A** Evolution of reproductive barriers in laboratory populations of fruit flies adapted to different food sources

Try This In your own words, explain how the experiment was performed and interpret the results.

In plants, pollinator choice is often a reproductive barrier. Perhaps populations of an ancestral species became separated in environments that had either more hummingbirds than bees or vice versa. Flower color and shape would evolve through natural selection in ways that attracted the most common pollinator, and these changes would help separate the species should they later share the same region. For example, two closely related species of monkey flower are found in the same area of the Sierra Nevada, but they rarely interbreed. Bumblebees prefer the pink-flowered *Mimulus lewisii*, and hummingbirds prefer the red-flowered *Mimulus cardinalis*. Scientists experimentally exchanged the alleles for flower color between these two species. As a result, *M. lewisii* produced light orange flowers **(Figure 14.5B)** that received many more visits from hummingbirds than did the normal pink-flowered *M. lewisii*. *M. cardinalis* plants with the *M. lewisii* allele produced pinker flowers that received many more visits from bumblebees than the normal red-flowered plants. Thus, a change in flower color influenced pollinator preference, which normally provides a reproductive barrier between these two species.

Sometimes reproductive barriers can arise even when populations are not geographically separated, as we see next.

? **Females of the Galápagos finch *Geospiza difficilis* respond to the songs of males from their island but ignore songs of males from other islands. How would you interpret these findings?**

● Behavioral barriers to reproduction have begun to develop in these allopatric (geographically separated) finch populations.

Pollinator choice in typical monkey flowers

Typical *M. lewisii* (pink)

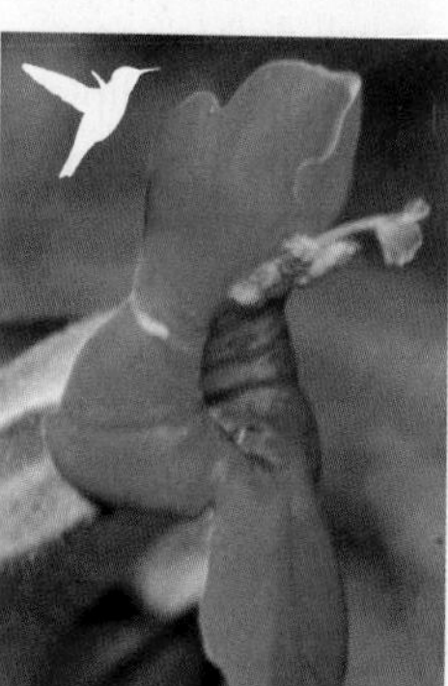

Typical *M. cardinalis* (red)

Pollinator choice after color allele transfer

M. lewisii with red-color allele

M. cardinalis with pink-color allele

▲ **Figure 14.5B** Effect of changing color of monkey flowers on pollinator choice

14.6 Sympatric speciation takes place without geographic isolation

In **sympatric speciation** (from the Greek *syn*, together, and *patra*, fatherland), a new species arises within the same geographic area as its parent species. How can reproductive isolation develop when members of sympatric populations remain in contact with each other? Sympatric speciation may occur when mating and the resulting gene flow between populations are reduced by factors such as polyploidy, habitat differentiation, and sexual selection.

Many plant species have originated from sympatric speciation that occurs when accidents during cell division result in extra sets of chromosomes. New species formed in this way are **polyploid**, meaning that their cells have more than two complete sets of chromosomes. **Figure 14.6A** shows one way in which a tetraploid plant ($4n$, with four sets of chromosomes) can arise from a parent species that is diploid. ❶ A failure of cell division after chromosome duplication could double a cell's chromosomes. ❷ If this $4n$ cell gives rise to a tetraploid branch, flowers produced on this branch would produce diploid gametes. ❸ If self-fertilization occurs, as it commonly does in plants, the resulting tetraploid zygotes would develop into plants that can produce fertile tetraploid offspring by self-pollination or by mating with other tetraploids.

A tetraploid cannot, however, produce fertile offspring by mating with a parent plant. The fusion of a diploid ($2n$) gamete from the tetraploid plant and a haploid (n) gamete from the diploid parent would produce triploid ($3n$) offspring. Triploid individuals are sterile; they cannot produce normal gametes because the odd number of chromosomes cannot form homologous pairs and separate normally during meiosis (see Module 8.13). Thus, the formation of a tetraploid ($4n$) plant is an instantaneous speciation event: A new species, reproductively isolated from its parent species, is produced in just one generation.

Most polyploid species, however, arise from hybridization of two different species. **Figure 14.6B** illustrates one way in which this can happen. ❶ When haploid gametes from two different species combine, the resulting hybrid is normally sterile because its chromosomes cannot pair during meiosis. ❷ However, the hybrid may reproduce asexually, as many plants can do. ❸ Subsequent errors in cell division may produce chromosome duplications that result in a diploid set of chromosomes ($2n = 10$). Now chromosomes *can* pair in meiosis, and haploid gametes will be produced; thus, a fertile polyploid species has formed. The new species has a chromosome number equal to the sum of the diploid chromosome numbers of its parent species. Again, this new species is reproductively isolated, this time from both parent species. Biologists have identified several plant species that originated through these mechanisms within the past 150 years—virtually instantaneously on an evolutionary time scale.

Does polyploid speciation occur in animals? It appears to happen occasionally. For example, the gray tree frog (see Figure 13.15C) is thought to have originated in this way. However, sympatric speciation in animals is more likely to happen through habitat differentiation or sexual selection than by polyploidy. Both habitat differentiation and sexual selection may have been involved in the origin of several hundred species of small fish called cichlids in Lake Victoria in East Africa. Adaptations for exploiting different food sources may have evolved in different subgroups of the original cichlid population. If those sources were in different habitats, mating between the populations would become rare, isolating their gene pools as each population becomes adapted to a different resource. As you will learn in Module 14.9, speciation in these brightly colored fish may also have been driven by the type of sexual selection in which females choose mates based on coloration. Such mate choice can contribute to reproductively isolating populations, keeping the gene pools of newly forming species separate. Of course, both habitat differentiation and sexual selection can also contribute to the formation of reproductive barriers between allopatric populations.

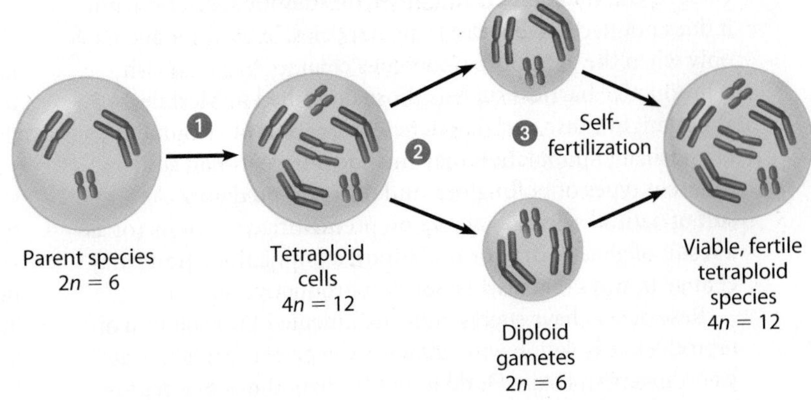

1
2
3
Self-fertilization
Parent species
$2n = 6$
Tetraploid cells
$4n = 12$
Diploid gametes
$2n = 6$
Viable, fertile tetraploid species
$4n = 12$

▲ **Figure 14.6A** Sympatric speciation by polyploidy within a single species

? Revisit the reproductive barriers in Module 14.3, and choose the barrier that isolates a viable, fertile polyploid plant from its parental species.

● Reduced hybrid fertility

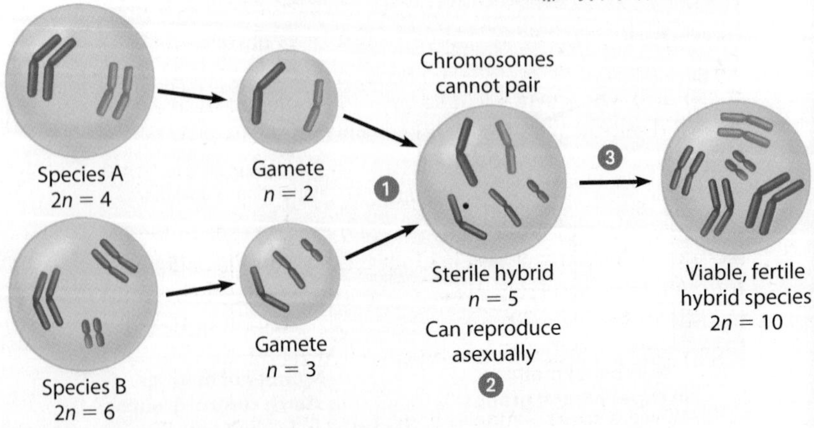

Species A
$2n = 4$
Gamete
$n = 2$
Species B
$2n = 6$
Gamete
$n = 3$
Chromosomes cannot pair
Sterile hybrid
$n = 5$
Can reproduce asexually
Viable, fertile hybrid species
$2n = 10$

▲ **Figure 14.6B** Sympatric speciation producing a hybrid polyploid from two different species

Try This Explain how a new species produced by the process shown in this figure differs from a new species produced by the process shown in Figure 14.6A.

14.7 The origin of most plant species can be traced to polyploid speciation

EVOLUTION CONNECTION

Plant biologists estimate that 80% of living plant species are descendants of ancestors that formed by polyploid speciation. Hybridization between two species accounts for most of these species, perhaps because of the adaptive advantage of the diverse genes a hybrid inherits from different parental species.

Many of the plants we grow for food are polyploids, including oats, potatoes, bananas, peanuts, barley, plums, apples, sugarcane, coffee, and wheat. Cotton, also a polyploid, provides one of the world's most popular clothing fibers.

Wheat, the most widely cultivated plant in the world, occurs as 20 different species of *Triticum*. We know that humans were cultivating wheat at least 10,000 years ago because wheat grains of *Triticum monococcum* ($2n = 14$) have been found in the remains of Middle Eastern farming villages from that time. This species has small seed heads and is not highly productive, but some varieties are still grown in the Middle East.

Our most important wheat species is bread wheat (*Triticum aestivum*), a polyploid with 42 chromosomes. **Figure 14.7** illustrates how this species may have evolved; the uppercase letters represent not genes but *sets of chromosomes* that have been traced through the lineage.

1 The process may have begun with hybridization between two wheats, the cultivated species *T. monococcum* (AA) and one of several wild species that probably grew as weeds at the edges of fields (BB). Chromosome sets A and B of the two species would not have been able to pair at meiosis, making the AB hybrid sterile. 2 However, an error in cell division and self-fertilization would have produced a new species (AABB) with 28 chromosomes. Today, we know this species as emmer wheat (*T. turgidum*), varieties of which are grown widely in Eurasia and western North America. Emmer wheat is used mainly for making macaroni and other noodle products because its proteins hold their shape better than bread-wheat proteins.

The final steps in the evolution of bread wheat are thought to have occurred in early farming villages on the shores of European lakes more than 8,000 years ago. 3 The cultivated emmer wheat, with its 28 chromosomes, hybridized spontaneously with the closely related wild species *T. tauschii* (DD), which has 14 chromosomes. The hybrid (ABD, with 21 chromosomes) was sterile, 4 but a cell division error in this hybrid and self-fertilization doubled the chromosome number to 42. The result was bread wheat, with two each of the three ancestral sets of chromosomes (AABBDD).

Today, plant geneticists generate new polyploids in the laboratory by using chemicals that induce meiotic and mitotic cell division errors. Researchers can produce new hybrids with special qualities, such as a hybrid combining the high yield of wheat with the hardiness of rye.

? **Why are errors in mitosis or meiosis a necessary part of speciation by hybridization between two species?**

● If a hybrid has a single copy of the chromosomes from two species, homologous pairs cannot join and separate during meiosis to produce gametes. Errors in mitosis or meiosis must somehow duplicate chromosomes so that there is a diploid number of each set and normal gametes can form.

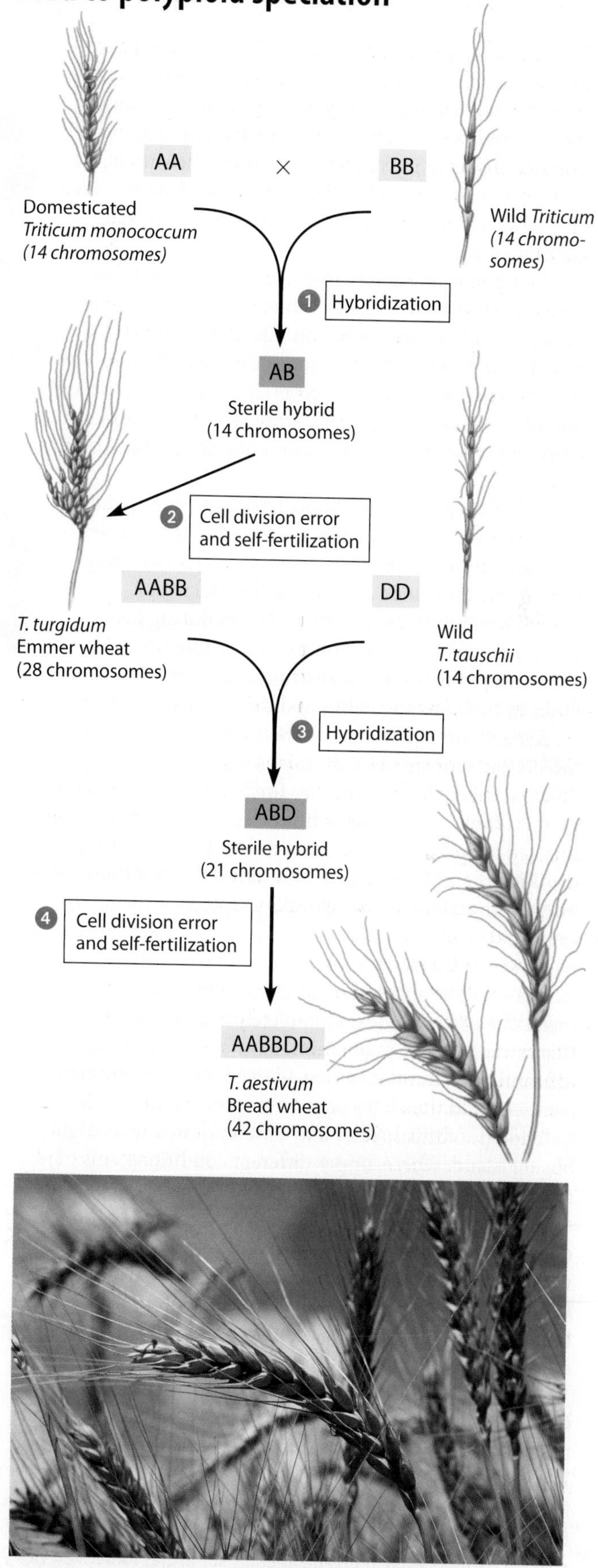

AA
×
BB
Domesticated
Triticum monococcum
(14 chromosomes)
Wild *Triticum*
(14 chromosomes)
1 Hybridization
AB
Sterile hybrid
(14 chromosomes)
2 Cell division error and self-fertilization
AABB
DD
T. turgidum
Emmer wheat
(28 chromosomes)
Wild
T. tauschii
(14 chromosomes)
3 Hybridization
ABD
Sterile hybrid
(21 chromosomes)
4 Cell division error and self-fertilization
AABBDD
T. aestivum
Bread wheat
(42 chromosomes)

▲ **Figure 14.7** The evolution of bread wheat, *Triticum aestivum*

14.8 Isolated islands are often showcases of speciation

Isolated island chains are often inhabited by unique collections of species. Islands that have physically diverse habitats and that are far enough apart to permit populations to evolve in isolation but close enough to allow occasional dispersions to occur are often the sites of multiple speciation events. The evolution of many diverse species from a common ancestor is known as **adaptive radiation**.

The Galápagos Archipelago, located about 900 km (560 miles) west of Ecuador, is one of the world's great showcases of adaptive radiation. Each island was born naked from underwater volcanoes from 5 million to 1 million years ago and was gradually covered by plants, animals, and microorganisms derived from strays that rode the ocean currents and winds from other islands and the South American mainland.

The Galápagos Islands today have numerous plants, snails, reptiles, and birds that are found nowhere else on Earth. For example, they have 14 species of closely related finches, which are often called Darwin's finches because Darwin collected them during his around-the-world voyage on the *Beagle* (see Module 13.1). These birds share many finch-like traits, but they differ in their feeding habits and their beaks, which are specialized for what they eat. Their various foods include insects, large or small seeds, cactus fruits, and even eggs of other species. The woodpecker finch uses cactus spines or twigs as tools to pry insects from trees. The "vampire" finch is noted for pecking wounds on the backs of seabirds and drinking their blood. Figure 14.8 shows some of these birds, with their distinctive beaks adapted for their specific diet.

How might Darwin's finch species have evolved from a small population of ancestral birds that colonized one of the islands? Completely isolated on the island, the founder population may have changed significantly as natural selection adapted it to the new environment, and thus it became a new species. Later, a few individuals of this species may have migrated to a neighboring island, where, under different conditions, this new founder population was changed enough through natural selection to become another new species. Some of these birds may then have recolonized the first island and coexisted there with the original ancestral species if reproductive barriers kept the species distinct. Multiple colonizations and speciations on the many separate islands of the Galápagos probably followed.

Today, each of the Galápagos Islands has several species of finches, with as many as 10 on some islands. The effects of the adaptive radiation of Darwin's finches are evident not just in their many types of beaks but also in their different habitats—some live in trees and others spend most of their time on the ground. Reproductive isolation due to species-specific songs helps keep the species separate. However, occasional interbreeding happens when a male sings the song of a different species. For example, a cactus finch nestling whose father dies may learn a neighbor's song, even if the neighbor is not a cactus finch.

Cactus-seed-eater (cactus finch)

Tool-using insect-eater (woodpecker finch)

Seed-eater (large ground finch)

▲ **Figure 14.8** Examples of differences in beak shape and size in Galápagos finches, each adapted for a specific diet

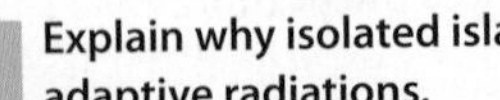

? Explain why isolated island chains provide opportunities for adaptive radiations.

● The chance colonization of an island often presents a species with new resources and an absence of predators. Through natural selection acting on existing variation, the colonizing population becomes adapted to its new habitat and may evolve into a new species. Subsequent colonizations of nearby islands would provide additional opportunities for adaptation and genetic drift, which could lead to further speciations.

14.9 Lake Victoria is a living laboratory for studying speciation

SCIENTIFIC THINKING

Can we observe speciation occurring?

In contrast to microevolutionary change, which may be apparent in a population within a few generations, the process of speciation is generally extremely slow. So you may be surprised to learn that we *can* see speciation occurring. Consider that life has been evolving over hundreds of millions of years and will continue to evolve. The species living today represent a snapshot, a brief instant in this vast span of time. The environment continues to change—sometimes rapidly due to human impact—and natural selection continues to act on affected populations. It is reasonable to assume that some of these populations are changing in ways that could eventually lead to speciation. Studying populations as they diverge gives biologists a window on the process of speciation. Researchers have documented at least two dozen cases in which populations are diverging as they exploit different food resources or breed in different habitats.

The bowerbirds you read about in the chapter introduction provide an example of another means by which populations can diverge—sexual selection (see Module 13.15). Sexual selection is a form of natural selection in which individuals with certain traits are more likely to obtain mates. The authors of the bowerbird study hypothesized that the differences in male displays of the allopatric bowerbird populations resulted from changes in female preferences. Biologists have also identified several other animal populations that are diverging as a result of differences in how males attract females or how females choose mates. Because of its direct effect of reproductive success, sexual selection can interrupt gene flow within a population and may therefore be an important factor in sympatric speciation.

Biologists can also test hypotheses about the process of speciation by studying species that arose recently. Let's look at a series of investigations into the role of sexual selection in the adaptive radiation of cichlids in Lake Victoria (**Figure 14.9A**).

▲ Figure 14.9A Map of East Africa showing Lake Victoria

Cichlids are a family of fishes that live in tropical lakes and rivers. They come in all colors of the rainbow, making them favorites of the aquarium trade. Among evolutionary biologists, they are renowned for the spectacular adaptive radiations that stocked the large lakes of East Africa with more than a thousand species of cichlids in less than 100,000 years. In the largest of these lakes, Lake Victoria, roughly 500 species evolved in about 15,000 years. For comparison, there are approximately 525 species of fish in all the lakes and rivers of Europe combined. How can a single body of water host such diversity? The answer lies partly in the heterogeneity of the environment. Various species have adaptations that suit them to inhabit the lake's rocky shores, muddy bottom, or open water. Specialized feeding adaptations abound. For example, there are algae-scrapers, snail-crushers, leaf-biters, insect-eaters, and fish-hunters. The visual environment, including predominant wavelengths of light and water clarity, is also heterogeneous, a fact that is crucial to speciation via sexual selection.

In Lake Victoria, there are pairs of closely related cichlid species that differ in color but nothing else. Breeding males of *Pundamilia nyererei* have a bright red back and dorsal fin, while *Pundamilia pundamilia* males are metallic blue-gray (**Figure 14.9B**). Researchers hypothesized that sexual selection—divergent female preference for red or blue mates—led to reproductive isolation. Let's examine the evidence for this hypothesis.

Pundamilia nyererei

Pundamilia pundamilia

▲ Figure 14.9B Males of *Pundamilia nyererei* and *Pundamilia pundamilia*

Pundamilia females prefer brightly colored males. Mate-choice experiments performed in the laboratory showed that *P. nyererei* females prefer red males over blue males, and *P. pundamilia* females prefer blue males over red males. Furthermore, the vision of *P. nyererei* females is more sensitive to red light than blue light; *P. pundamilia* females are more sensitive to blue light. Researchers also demonstrated that this color sensitivity is heritable.

As mentioned above, the visual environment varies in Lake Victoria. As light travels through water, suspended particles selectively absorb and scatter the shorter (blue) wavelengths, so light becomes increasingly red with increasing depth. Thus, in deeper waters *P. nyererei* males are pleasingly apparent to females with red-sensitive vision but virtually invisible to *P. pundamilia* females. Accordingly, we would expect the two species to breed in different areas of the lake—and they do. When biologists sampled cichlid populations in Lake Victoria, they found that *P. nyererei* breeds in deep water, while *P. pundamilia* inhabits shallower habitats where the blue males shine brightly. As a consequence of their mating behavior, the two species encounter different environments that may result in further divergence.

In recent years, new environmental factors have had a dramatic impact on cichlids. Hybridization is rampant; a multitude of cichlid species have been genetically homogenized. You'll learn why in the next module.

? **Why was it important for researchers to establish that cichlid color sensitivity is heritable?**

● Sensitivity to red is the phenotypic trait that allows *P. nyererei* females to choose *P. nyererei* males as mates, ensuring reproductive isolation.

14.10 Hybrid zones provide opportunities to study reproductive isolation

What happens when separated populations of closely related species come back into contact with one another? Will reproductive barriers be strong enough to keep the species separate? Or will the two species interbreed and become one? Biologists attempt to answer such questions by studying **hybrid zones**, regions in which members of different species meet and mate, producing at least some hybrid offspring.

Figure 14.10A illustrates the formation of a hybrid zone, starting with the ancestral species. ❶ Three populations are connected by gene flow. ❷ A barrier to gene flow separates one population. ❸ Over time, this population diverges from the other two. ❹ Later, gene flow is reestablished in the hybrid zone. Let's consider possible outcomes for this hybrid zone over time.

Three populations of a species
❶
❷
❸
Newly formed species
Hybrid zone
❹
Gene flow
Hybrid individual
Gene flow
Population
Barrier to gene flow

▲ **Figure 14.10A** Formation of a hybrid zone

Reinforcement

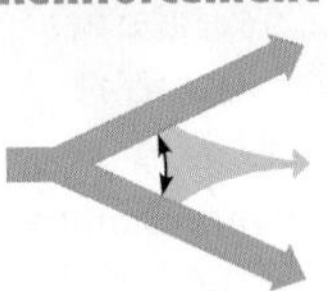

When hybrid offspring are less fit than members of both parent species, we might expect natural selection to strengthen, or *reinforce*, reproductive barriers, thus reducing the formation of unfit hybrids. And we would predict that barriers between species should be stronger where the species overlap (that is, where the species are sympatric).

As an example, consider the closely related collared flycatcher and pied flycatcher illustrated in **Figure 14.10B**.

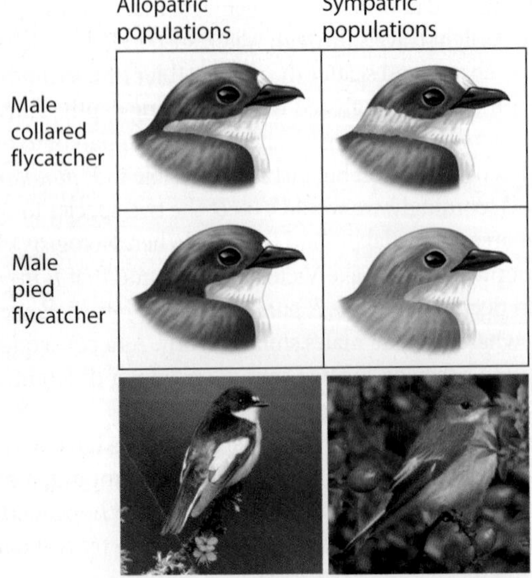
Allopatric populations
Sympatric populations
Male collared flycatcher
Male pied flycatcher
Pied flycatcher from allopatric population
Pied flycatcher from sympatric population

▲ **Figure 14.10B** Reinforcement of reproductive barriers

When populations of these two species do not overlap (that is, when they are allopatric), males closely resemble each other, with similar black and white coloration (see left side of Figure 14.10B). However, when populations of the two species are sympatric, male collared flycatchers are still black but with enlarged patches of white, whereas male pied flycatchers are a dull brown (see right side of Figure 14.10B). The photographs at the bottom of the figure show two pied flycatchers, the one on the left from a population that has no overlap with collared flycatchers and the one on the right from a population in an area where both species coexist. When scientists performed mate choice experiments, they found that female flycatchers frequently made mistakes when presented with males from allopatric populations, which look similar. But females never selected mates from the other species when presented with males from sympatric populations, which look different. Thus, reproductive barriers are reinforced when populations of these two species overlap.

Fusion

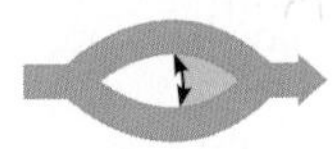

What happens when the reproductive barriers between species are not strong and the species come into contact in a hybrid zone? So much gene flow may occur that the speciation process reverses, causing the two hybridizing species to fuse into one.

Such a situation has been occurring among the cichlid species in Lake Victoria that we discussed in Module 14.9. Since the 1980s, as many as 200 species of cichlids have disappeared from Lake Victoria. Some species were driven to extinction by an introduced predator, the Nile perch. But many species not eaten by Nile perch are also disappearing. Pollution caused by development along the shores of Lake Victoria has turned the water murky. To understand how water clarity affects sexual selection in cichlids, think about how your eyes work in different lighting. It's easy to distinguish colors in bright light, but difficult in a dimly lit room. What happens when *P. nyererei* or *P. pundamilia* females can't tell red males from blue males? The behavioral barrier crumbles. Many viable hybrid offspring are produced by interbreeding, and the once isolated gene pools of the parent species are combining—two species fusing into a single hybrid species (**Figure 14.10C**).

Hybrid: *Pundamilia "turbid water"*

▲ **Figure 14.10C** Fusion: hybrid of *Pundamilia nyererei* and *Pundamilia pundamilia* from an area with turbid water

Recently, pollution in Lake Victoria has been greatly reduced, and cichlid numbers—though not diversity—have rebounded. By mixing the unique alleles of separate species into a single gene pool, hybridization can increase a population's genetic variation, which in turn increases the phenotypic variation on which natural selection can act. If environmental conditions continue to improve, biologists may have an opportunity to study a new radiation of cichlid diversity.

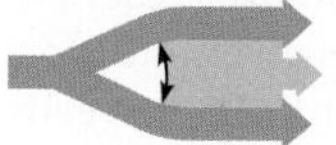

Stability One might predict that either reinforcement of reproductive barriers or fusion of gene pools would occur in a hybrid zone. However, many hybrid zones are fairly stable, and hybrids continue to be produced. Although these hybrids allow for some gene flow between populations, each species maintains its own integrity. The island inhabited by two finch species that occasionally interbreed (see Module 14.8) is an example of a stable hybrid zone.

? **Why might hybrid zones be called "natural laboratories" in which to study speciation?**

By studying the fate of hybrids over time, scientists can directly observe factors that cause (or fail to cause) reproductive isolation.

14.11 Speciation can occur rapidly or slowly

Biologists continue to make field observations and devise experiments to study evolution in progress. However, much of the evidence for evolution comes from the fossil record. What does this record say about the process of speciation?

Many fossil species appear suddenly in a layer of rock and persist essentially unchanged through several layers (strata) until disappearing just as suddenly. Paleontologists coined the term **punctuated equilibria** to describe these long periods of little apparent morphological change (equilibria) interrupted (punctuated) by relatively brief periods of sudden change. **Figure 14.11** (top) illustrates the evolution of two lineages of butterflies in a punctuated pattern. Notice that the butterfly species change little, if at all, once they appear.

Other fossil species appear to have diverged gradually over long periods of time. As shown in Figure 14.11 (bottom), differences gradually accumulate, and new species (represented by the two butterflies at the far right) evolve gradually from the ancestral population.

Even when fossil evidence points to a punctuated pattern, species may not have originated as rapidly as it appears. Suppose that a species survived for 5 million years but that most of the changes in its features occurred during the first 50,000 years of its existence. Time periods this short often cannot be distinguished in fossil strata. And should a new species originate from a small, isolated population—as no doubt many species have—the chances of fossils being found are low.

But what about the total length of time between speciation events—between when a new species forms and when its populations diverge enough to produce another new species? In one survey of 84 groups of plants and animals, this time ranged from 4,000 to 40 million years. Overall, the time between speciation events averaged 6.5 million years. Such long time frames tell us that it has taken vast spans of time for life on Earth to evolve.

As you've seen, speciation may begin with small differences. However, as speciation occurs again and again, these differences accumulate and may eventually lead to new groups that differ greatly from their ancestors, as in the origin of cetaceans from four-legged land animals (see Module 13.3), The cumulative effects of multiple speciations, as well as extinctions, have shaped the dramatic changes documented in the fossil record. (Such macroevolutionary changes are the subject of our next chapter.)

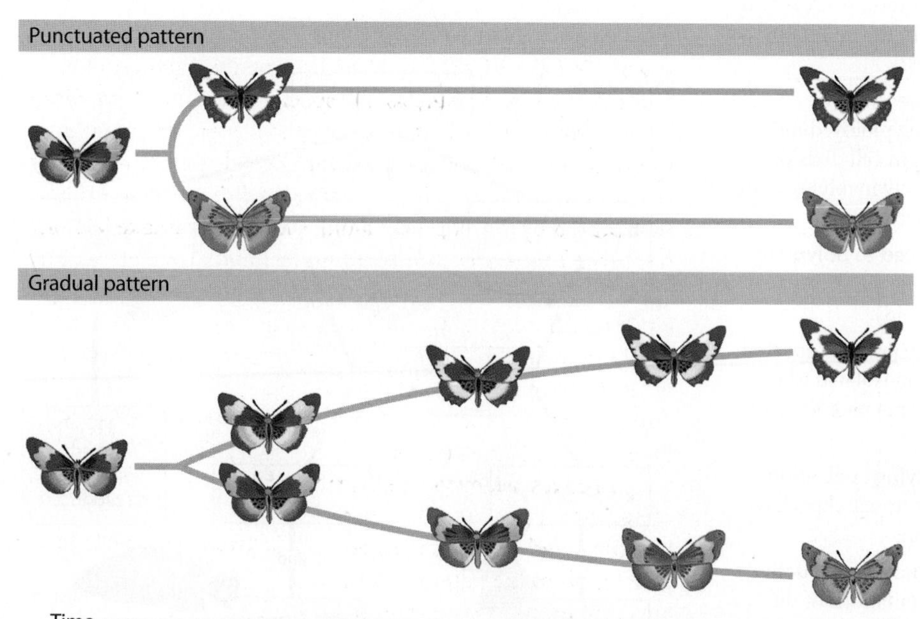
Punctuated pattern
Gradual pattern
Time

▲ **Figure 14.11** Two models for the tempo of speciation

? **How does the punctuated equilibrium model account for the relative rarity of transitional fossils linking newer species to older ones?**

If speciation takes place in a relatively short time or in a small isolated population, the transition of one species to another may be difficult to find in the fossil record.

CHAPTER 14 REVIEW

For practice quizzes, BioFlix animations, MP3 tutorials, video tutors, and more study tools designed for this textbook, go to MasteringBiology®

Reviewing the Concepts

Defining Species (14.1–14.3)

14.1 The origin of species is the source of biological diversity. Speciation, the process by which one species splits into two or more species, accounts for both the unity and diversity of life.

14.2 There are several ways to define a species. The biological species concept holds that a species is a group of populations whose members can interbreed and produce fertile offspring with each other but not with members of other species. This concept emphasizes reproductive isolation. Most organisms are classified based on observable traits—the morphological species concept.

14.3 Reproductive barriers keep species separate. Such barriers isolate a species' gene pool and prevent interbreeding.

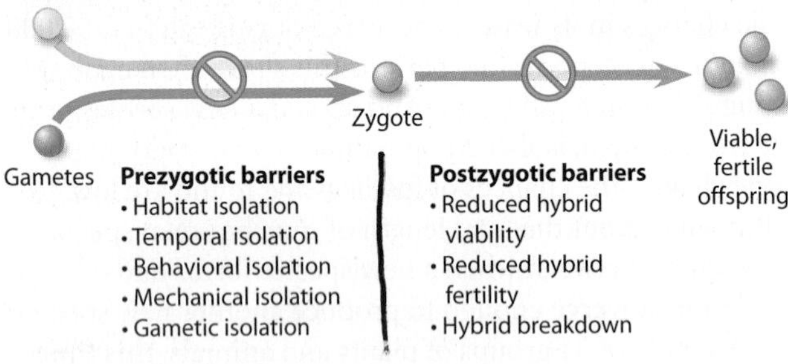
Gametes
Zygote
Viable, fertile offspring
Prezygotic barriers
• Habitat isolation
• Temporal isolation
• Behavioral isolation
• Mechanical isolation
• Gametic isolation
Postzygotic barriers
• Reduced hybrid viability
• Reduced hybrid fertility
• Hybrid breakdown

Mechanisms of Speciation (14.4–14.11)

14.4 In allopatric speciation, geographic isolation leads to speciation. Geographically separated from other populations, a small population may become genetically unique as its gene pool is changed by natural selection, mutation, or genetic drift.

14.5 Reproductive barriers can evolve as populations diverge. Researchers have documented the beginning of reproductive isolation in fruit fly populations adapting to different food sources and have identified a gene for flower color involved in the pollinator choice that helps separate monkey flower species.

14.6 Sympatric speciation takes place without geographic isolation. Many plant species have evolved by polyploidy, duplication of the chromosome number due to errors in cell division. Habitat differentiation and sexual selection, usually involving mate choice, can lead to sympatric (and allopatric) speciation.

14.7 The origin of most plant species can be traced to polyploid speciation. Many plants, including food plants such as bread wheat, are the result of hybridization and polyploidy.

14.8 Isolated islands are often showcases of speciation. Repeated isolation, speciation, and recolonization events on isolated island chains have led to adaptive radiations of species, many of which are found nowhere else in the world.

14.9 Lake Victoria is a living laboratory for studying speciation. Through the rapid adaptive radiation of cichlids, researchers have gained insight into speciation via sexual selection.

14.10 Hybrid zones provide opportunities to study reproductive isolation. Hybrid zones are regions in which populations of different species overlap and produce at least some hybrid offspring. Over time, reinforcement may strengthen barriers to reproduction, or fusion may reverse the speciation process as gene flow between species increases. In stable hybrid zones, a limited number of hybrid offspring continue to be produced.

14.11 Speciation can occur rapidly or slowly. The punctuated equilibria model, which states that species change most as they arise from an ancestral species and then change relatively little for the rest of their existence, draws on the fossil record. Other species appear to have evolved more gradually. The time interval between speciation events varies from a few thousand years to tens of millions of years.

Connecting the Concepts

1. Name the two types of speciation represented by this diagram. For each type, describe how reproductive barriers may develop between the new species.

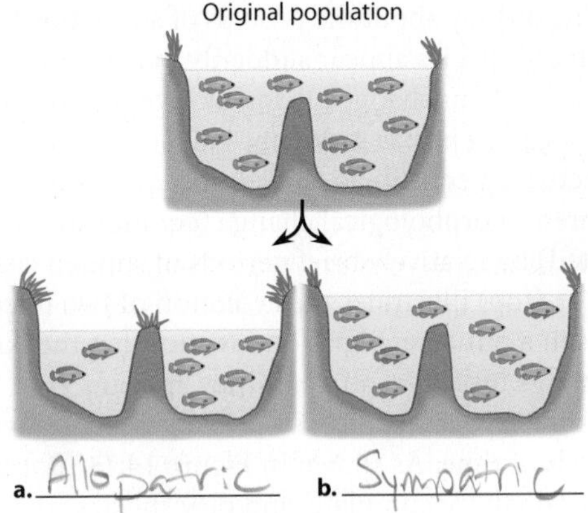
Original population
a. Allopatric
b. Sympatric

2. Fill in the blanks in the following concept map.

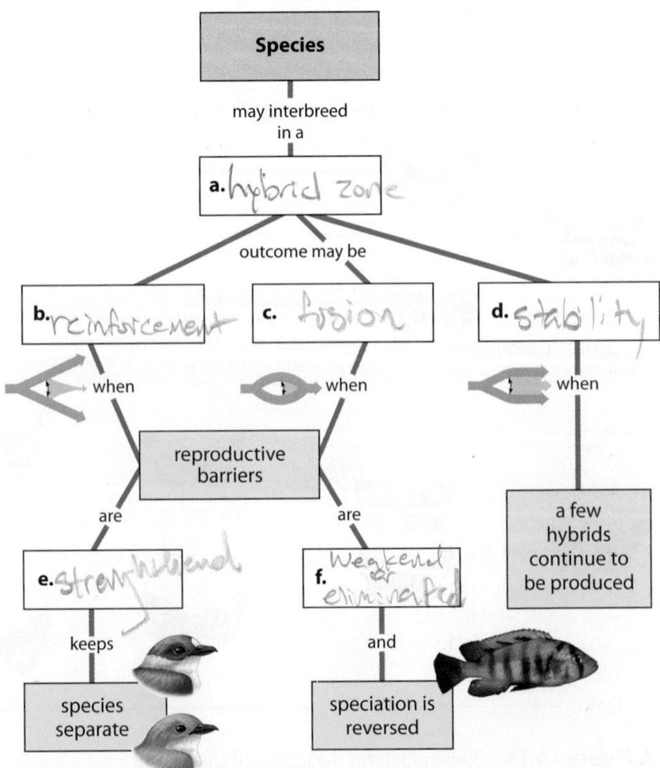
Species
may interbreed in a
a. hybrid zone
outcome may be
b. reinforcement
c. fusion
d. stability
when
when
when
reproductive barriers
are
are
e. strengthened
f. weakened or eliminated
keeps
and
species separate
speciation is reversed
a few hybrids continue to be produced

Testing Your Knowledge

Level 1: Knowledge/Comprehension

3. Which concept of species would be most useful to a field biologist identifying new plant species in a tropical forest?
   a. biological
   b. ecological
   c. morphological
   d. phylogenetic
4. The *largest* unit within which gene flow can readily occur is a
   a. population.
   b. species.
   c. genus.
   d. phylum.
5. Bird guides once listed the myrtle warbler and Audubon's warbler as distinct species that lived side by side in parts of their ranges. However, recent books show them as eastern and western forms of a single species, the yellow-rumped warbler. Most likely, it has been found that these two kinds of warblers
   a. live in similar habitats and eat similar foods.
   b. interbreed often in nature, and the offspring are viable and fertile.
   c. are almost identical in appearance.
   d. have many genes in common.
6. Which of the following is an example of a postzygotic reproductive barrier?
   a. One *Ceanothus* shrub lives on acid soil, another on alkaline soil.
   b. Mallard and pintail ducks mate at different times of year.
   c. Two species of leopard frogs have different mating calls.
   d. Hybrid offspring of two species of jimsonweeds always die before reproducing.
7. Biologists have found more than 500 species of fruit flies on the various Hawaiian Islands, all apparently descended from a single ancestor species. This example illustrates
   a. polyploidy.
   b. temporal isolation.
   c. adaptive radiation.
   d. sympatric speciation.
8. A new plant species C, which formed from hybridization of species A ($2n = 16$) with species B ($2n = 12$), would probably produce gametes with a chromosome number of
   a. 12.
   b. 14.
   c. 16.
   d. 28.
9. A horse ($2n = 64$) and a donkey ($2n = 62$) can mate and produce a mule. How many chromosomes would there be in a mule's body cells?
   a. 31
   b. 62
   c. 63
   d. 126
10. What prevents horses and donkeys from hybridizing to form a new species?
    a. limited hybrid fertility
    b. limited hybrid viability
    c. hybrid breakdown
    d. gametic isolation
11. When hybrids produced in a hybrid zone can breed with each other and with both parent species, and they survive and reproduce as well as members of the parent species, one would predict that
    a. the hybrid zone would be stable.
    b. sympatric speciation would occur.
    c. reinforcement of reproductive barriers would keep the parent species separate.
    d. reproductive barriers would lessen and the two parent species would fuse.
12. Which of the following factors would *not* contribute to allopatric speciation?
    a. A population becomes geographically isolated from the parent population.
    b. The separated population is small, and genetic drift occurs.
    c. The isolated population is exposed to different selection pressures than the parent population.
    d. Gene flow between the two populations continues to occur.

Level 2: Application/Analysis

13. Explain how each of the following makes it difficult to clearly define a species: variation within a species, geographically isolated populations, asexual species, fossil organisms.
14. Explain why allopatric speciation would be less likely on an island close to a mainland than on a more isolated island.
15. What does the term *punctuated equilibria* describe?
16. Can factors that cause sympatric speciation also cause allopatric speciation? Explain.

Level 3: Synthesis/Evaluation

17. Cultivated American cotton plants have a total of 52 chromosomes ($2n = 52$). In each cell, there are 13 pairs of large chromosomes and 13 pairs of smaller chromosomes. Old World cotton plants have 26 chromosomes ($2n = 26$), all large. Wild American cotton plants have 26 chromosomes, all small. Propose a testable hypothesis to explain how cultivated American cotton probably originated.
18. **SCIENTIFIC THINKING** Explain how the murky waters of Lake Victoria may be contributing to the decline in cichlid species. How might these polluted waters affect the formation of new species?
19. The red wolf, *Canis rufus,* which was once widespread in the southeastern and south central United States, was declared extinct in the wild by 1980. Saved by a captive breeding program, the red wolf has been reintroduced in areas of eastern North Carolina. The current wild population estimate is about 100 individuals. It is presently being threatened with extinction due to hybridization with coyotes, *Canis latrans,* which have become more numerous in the area. Red wolves and coyotes differ in terms of morphology, DNA, and behavior, although these differences may disappear if interbreeding continues. Although the red wolf has been designated as an endangered species under the Endangered Species Act, some people think that its endangered status should be withdrawn and resources should not be spent to protect what is not a "pure" species. Do you agree? Why or why not?

Answers to all questions can be found in Appendix 4.

CHAPTER 15

Tracing Evolutionary History

The feathered flight of birds is a perfect marriage of structure and function. The skeleton, muscles, nervous system, internal organs, and especially feathers of birds, including those of the roseate spoonbill below, are marvelously adapted for life on the wing. Let's consider the evolution of feathers. Clearly, these structures are essential to avian aeronautics. In a flight feather, separate filaments called barbs emerge from a central shaft that runs from base to tip. Each barb is linked to the next by tiny hooks that act much like the teeth of a zipper. The result is a tightly connected sheet of barbs that is strong but flexible. In flight, the shapes and arrangements of various feathers produce lift, smooth airflow, and help with steering and balance. Layered like shingles over the bird's body, feathers also provide a waterproof, lightweight covering. How did such a beautifully intricate structure evolve? You'll learn the answer to this question later in this chapter, but here's a clue: Birds were not the first feathered animals on Earth—dinosaurs were.

? ***How do brand-new structures arise by evolution?***

The first feathered dinosaur to be discovered, a 130-million-year-old fossil found in northeastern China, was named *Sinosauropteryx* ("Chinese lizard-wing"). About the size of a turkey, it had short

arms and ran on its hind legs, using its long tail for balance. Its unimpressive plumage consisted of a downy covering of hairlike feathers. Since the discovery of *Sinosauropteryx*, thousands of fossils of feathered dinosaurs have been found and classified into more than 30 different species. Although none was unequivocally capable of flying, many of these species had elaborate feathers that would be the envy of any modern bird.

The evolution of birds is an example of macroevolution, the major changes recorded in the history of life over vast tracts of time. In this chapter, we turn our attention to macroevolution and explore some of the mechanisms responsible for such changes. Finally, we consider how scientists organize the amazing diversity of life according to evolutionary relationships. To approach these wide-ranging topics, we begin with the most basic of questions: How did life first arise on planet Earth?

BIG IDEAS

Early Earth and the Origin of Life
(15.1–15.3)

Scientific experiments can test the four-stage hypothesis of how life originated on early Earth.

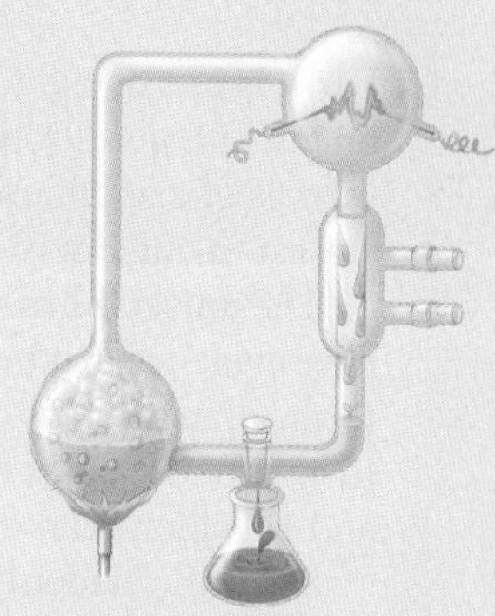

Major Events in the History of Life
(15.4–15.6)

The fossil record and radiometric dating establish a geologic record of key events in life's history.

Mechanisms of Macroevolution
(15.7–15.13)

Continental drift, mass extinctions, adaptive radiations, and changes in developmental genes have all contributed to macroevolution.

Phylogeny and the Tree of Life
(15.14–15.19)

The evolutionary history of a species is reconstructed using fossils, homologies, and molecular systematics.

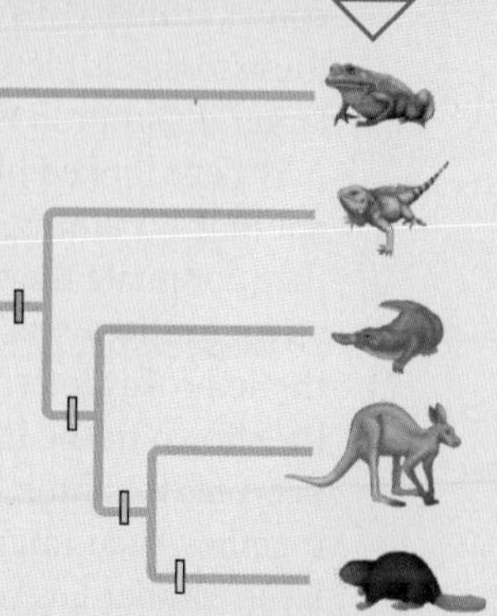

▷ Early Earth and the Origin of Life

15.1 Conditions on early Earth made the origin of life possible

Earth is one of eight planets orbiting the sun, and the sun is one of billions of stars in the Milky Way. The Milky Way, in turn, is one of billions of galaxies in the universe. The star closest to our sun is 40 trillion kilometers away.

The universe has not always been so spread out. Physicists have evidence that before the universe existed in its present form, all matter was concentrated in one mass. The mass seems to have blown apart with a "big bang" sometime between 12 and 14 billion years ago and has been expanding ever since.

Scientific evidence indicates that Earth formed about 4.6 billion years ago from a vast swirling cloud of dust that surrounded the young sun. As gases, dust, and rocks collided and stuck together, larger bodies formed, and the gravity of the larger bodies in turn attracted more matter, eventually forming Earth and other planets.

Conditions on Early Earth Immense heat would have been generated by the impact of meteorites and compaction by gravity, and young planet Earth probably began as a molten mass. The mass then sorted into layers of varying densities, with the least dense material on the surface, solidifying into a thin crust.

As the bombardment of early Earth slowed about 3.9 billion years ago, conditions on the planet were extremely different from those of today. The first atmosphere was probably thick with water vapor, along with various compounds released by volcanic eruptions, including nitrogen and its oxides, carbon dioxide, methane, ammonia, hydrogen, and hydrogen sulfide. As Earth slowly cooled, the water vapor condensed into oceans. Not only was the atmosphere of young Earth very different from the atmosphere we know today, but lightning, volcanic activity, and ultraviolet radiation were much more intense.

When Did Life Begin? The earliest evidence of life on Earth comes from fossils that are about 3.5 billion years old. One of these fossils is pictured in the inset in **Figure 15.1**; the larger illustration is an artist's rendition of what Earth may have looked like at that time. Life is already present in this painting, as shown by the "stepping stones" that dominate the shoreline. These rocks, called **stromatolites**, were built up by ancient photosynthetic prokaryotes. As evident in the fossil stromatolite shown in the inset, the rocks are layered. The prokaryotes that built them bound thin films of sediment together, then migrated to the surface and started the next layer. Similar layered mats are still being formed today by photosynthetic prokaryotes in a few shallow, salty bays, such as Shark Bay, in western Australia.

Photosynthesis is not a simple process, so it is likely that significant time had elapsed before life as complex as the organisms that formed the ancient stromatolites had evolved. The evidence that these prokaryotes lived 3.5 billion years ago is strong support for the hypothesis that life in a simpler form arose much earlier, perhaps as early as 3.9 billion years ago.

How Did Life Arise? From the time of the ancient Greeks until well into the 1800s, it was commonly believed that nonliving matter could spontaneously generate living organisms. Many people believed, for instance, that flies come from rotting meat and fish from ocean mud. Experiments by the French scientist Louis Pasteur in 1862, however, confirmed that all life arises only by the reproduction of preexisting life.

Pasteur ended the argument over spontaneous generation of present-day organisms, but he did not address the question of how life arose in the first place. To attempt to answer that question, for which there is no fossil evidence available, scientists develop hypotheses and test their predictions.

Scientists hypothesize that chemical and physical processes on early Earth could have produced very simple cells through a sequence of four main stages:

1. The abiotic (nonliving) synthesis of small organic molecules, such as amino acids and nitrogenous bases
2. The joining of these small molecules into polymers, such as proteins and nucleic acids (see Module 3.3)
3. The packaging of these molecules into "protocells," droplets with membranes, maintained an internal chemistry different from that of their surroundings
4. The origin of self-replicating molecules that eventually made inheritance possible

In the next two modules, we examine some of the experimental evidence for each of these four stages.

? **Why do 3.5-billion-year-old stromatolites suggest that life originated *before* 3.5 billion years ago?**

● If photosynthetic prokaryotes existed by 3.5 billion years ago, a simpler, nonphotosynthetic cell probably originated well before that time.

▲ **Figure 15.1** A depiction of Earth about 3 billion years ago (inset: photo of a cross section of a fossilized stromatolite)

15.2 Experiments show that the abiotic synthesis of organic molecules is possible

SCIENTIFIC THINKING

Organic molecules are essential to the structures and functions of life, but Earth and its atmosphere are made up of inorganic molecules. How did the first organic molecules arise?

In the 1920s, Russian chemist A. I. Oparin and British scientist J. B. S. Haldane independently proposed that conditions on early Earth could have generated organic molecules. They reasoned that present-day conditions on Earth do not allow the spontaneous synthesis of organic compounds simply because the atmosphere is now rich in oxygen. As a strong oxidizing agent, O_2 tends to disrupt chemical bonds. However, before the early photosynthetic prokaryotes added O_2 to the air, Earth may have had a reducing (electron-adding) atmosphere. The energy for this abiotic synthesis of organic compounds could have come from lightning and intense UV radiation.

In 1953, Stanley Miller, then a graduate student in the laboratory of Nobel laureate Harold Urey, tested the Oparin-Haldane hypothesis. Miller devised the apparatus shown in **Figure 15.2**. A flask of warmed water represented the primeval sea. ❶ The water was heated so that some vaporized and moved into a second, higher flask. ❷ The "atmosphere" in this higher flask consisted of water vapor, hydrogen gas (H_2), methane (CH_4), and ammonia (NH_3)—the gases that scientists at the time thought prevailed in the ancient world. Electrodes discharged sparks into the flask to mimic lightning. ❸ A condenser with circulating cold water cooled the atmosphere, raining water and any dissolved compounds back down into the miniature sea. ❹ As material cycled through the apparatus, Miller periodically collected samples for chemical analysis.

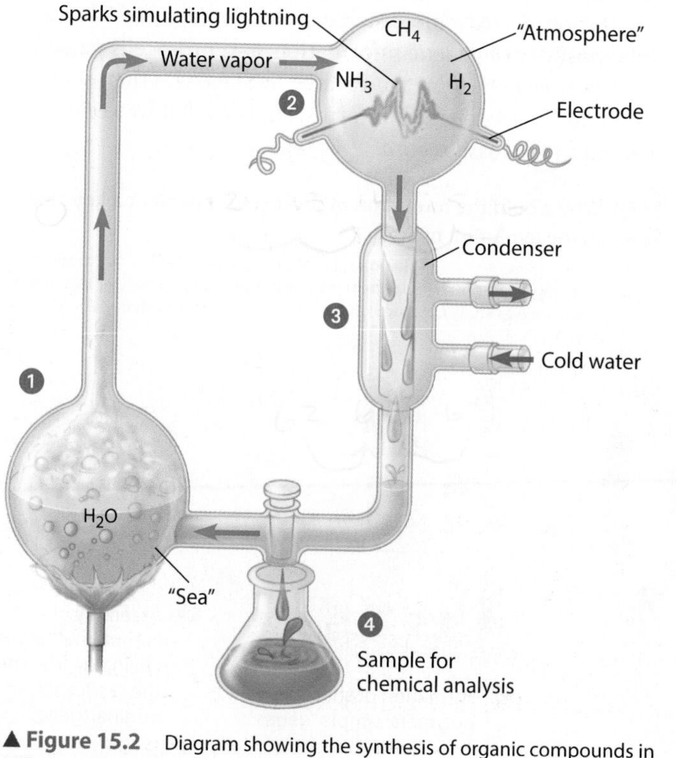

Miller identified a variety of organic molecules that are common in organisms, including hydrocarbons (long chains of carbon and hydrogen) and some of the amino acids that make up proteins. His results—the first evidence that the molecules of life could have arisen spontaneously from inorganic precursors—attracted global attention. Many laboratories have since repeated Miller's classic experiment using various atmospheric mixtures and produced organic compounds.

Recent evidence indicates that the early atmosphere may not have been as strongly reducing as once assumed. However, results from experiments using such atmospheres have also produced organic molecules, corroborating Miller's results. And it is possible that small "pockets" of the early atmosphere—perhaps near volcanic openings—were reducing. In 2008, a former graduate student of Miller's discovered some samples from an experiment that Miller had designed to mimic volcanic conditions. Reanalyzing these samples using modern equipment, he identified additional organic compounds that had been synthesized. Indeed, 22 amino acids had been produced under Miller's simulated volcanic conditions, compared with the 11 produced with the atmosphere in his original 1953 experiment. (Miller had only found 5 amino acids using the analytical methods available to him at the time.)

Scientists continue to generate alternative hypotheses for the origin of organic molecules on Earth. Some researchers are exploring the hypothesis that life may have begun in submerged volcanoes or deep-sea hydrothermal vents, gaps in Earth's crust where hot water and minerals gush into deep oceans. These environments, among the most extreme environments in which life exists today, could have provided the initial chemical resources for life.

Another hypothesis proposes that meteorites were the source of Earth's first organic molecules. Fragments of a 4.5-billion-year-old meteorite that fell to Earth in Australia in 1969 contain more than 80 types of amino acids, some in large amounts. Recent studies have shown that this meteorite also contains other key organic molecules, including lipids, simple sugars, and nitrogenous bases such as uracil. Chemical analyses show that these organic compounds are not contaminants from Earth.

Research will continue on the possible origins of organic molecules on early Earth. We next turn our attention to the subsequent stages that scientists hypothesize gave rise to the earliest cells.

▲ **Figure 15.2** Diagram showing the synthesis of organic compounds in Miller's 1953 experiment

? **Which of the four stages in the hypothetical scenario of the origin of simple cells was Stanley Miller testing with his experiments?**

Stage 1: Conditions on early Earth favored abiotic synthesis of organic molecules important to life, such as amino acids, from simpler ingredients.

15.3 Stages in the origin of the first cells probably included the formation of polymers, protocells, and self-replicating RNA

The abiotic synthesis of small organic molecules would have been a first step in the origin of life. But what is the evidence that the next three stages—synthesis of polymers, formation of protocells, and self-replicating RNA—could have occurred on early Earth?

Abiotic Synthesis of Polymers In a cell, enzymes catalyze the joining of monomers to build polymers. But could this happen without enzymes? Scientists have produced polymers in the laboratory by dripping dilute solutions of amino acids or RNA monomers onto hot sand, clay, or rock. The heat vaporizes the water and concentrates the monomers, some of which then spontaneously bond together in chains. A similar reaction might have happened on early Earth, when waves splashed organic monomers onto fresh lava or other hot rocks and then rinsed polypeptides and other polymers back into the sea.

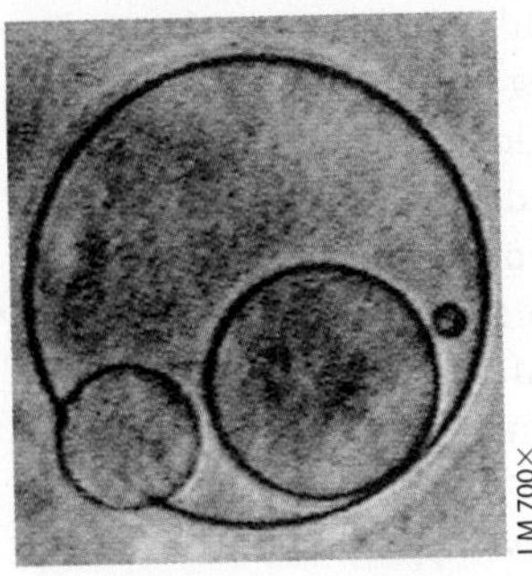
LM 700×

▲ **Figure 15.3A** Microscopic vesicle, with membranes made of lipids, "giving birth" to smaller vesicles

Formation of Protocells A key step in the origin of life would have been the isolation of a collection of organic molecules within a membrane-enclosed compartment. Laboratory experiments demonstrate that small membrane-enclosed sacs or vesicles form when lipids are mixed with water (see Module 5.2). When researchers add to the mixture a type of clay thought to have been common on early Earth, such vesicles form at a faster rate. Organic molecules become concentrated on the surface of this clay and thus more easily interact. As shown by the smaller droplets forming in **Figure 15.3A**, these abiotically created vesicles are able to grow and divide (reproduce). Researchers have shown that these vesicles can absorb clay particles to which RNA and other molecules are attached. In a similar fashion, protocells on early Earth may have been able to form, reproduce, and create and maintain an internal environment different from their surroundings.

Self-Replicating RNA Today's cells store their genetic information as DNA, transcribe the information into RNA, and then translate RNA messages into proteins. This DNA → RNA → protein assembly system is extremely intricate (as we saw in Chapter 10). Most likely, it emerged gradually through a series of refinements of much simpler processes.

What were the first genes like? One hypothesis is that they were short strands of self-replicating RNA. Laboratory experiments have shown that short RNA molecules can assemble spontaneously from nucleotide monomers. Furthermore, when RNA is added to a solution containing a supply of RNA monomers, new RNA molecules complementary to parts of the starting RNA sometimes assemble. So we can imagine a scenario on early Earth like the one in **Figure 15.3B**: ① RNA monomers adhere to clay particles and become concentrated. ② Some monomers spontaneously join, which form the first small "genes." ③ Then an RNA chain complementary to one of these genes assembles. If the new chain, in turn, serves as a template for another round of RNA assembly, the result is a replica of the original gene.

This replication process could have been aided by the RNA molecules themselves, acting as catalysts for their own replication. The discovery that some RNAs, which scientists call **ribozymes**, can carry out enzyme-like functions supports this hypothesis. Thus, the "chicken and egg" paradox of which came first, genes or enzymes, may be solved if the chicken and egg came together in the same RNA molecules. Scientists use the term "RNA world" for the hypothetical period in the evolution of life when RNA served as both rudimentary genes and catalytic molecules.

Once some protocells contained self-replicating RNA molecules, natural selection would have begun to shape their properties. Those that contained genetic information that helped them grow and reproduce more efficiently than others would have increased in number, passing their abilities on to subsequent generations. Mutations, errors in copying RNA "genes," would result in additional variation on which natural selection could work. At some point during millions of years of selection, DNA, a more stable molecule, replaced RNA as the repository of genetic information, and protocells passed a fuzzy border to become true cells. The stage was then set for the evolution of diverse life-forms, changes that we see documented in the fossil record.

? **Why would the formation of protocells represent a key step in the evolution of life?**

● Segregating mixtures of molecules within compartments could concentrate organic molecules and facilitate chemical reactions. Natural selection could act on protocells once self-replicating "genes" evolved.

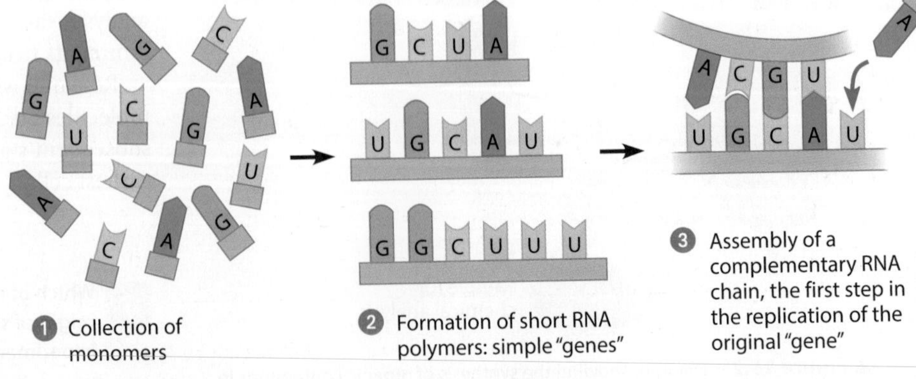
① Collection of monomers
② Formation of short RNA polymers: simple "genes"
③ Assembly of a complementary RNA chain, the first step in the replication of the original "gene"

▲ **Figure 15.3B** A hypothesis for the origin of the first genes

▷ Major Events in the History of Life

15.4 The origins of single-celled and multicellular organisms and the colonization of land were key events in life's history

We now begin our study of **macroevolution**, the broad pattern of changes in life on Earth. **Figure 15.4** shows a timeline from the origin of Earth 4.6 billion years ago to the present. Earth's history can be divided into four eons of geologic time. The Hadean, Archaean, and Proterozoic eons together lasted about 4 billion years. The Phanerozoic eon includes the last half billion years.

Origin of Prokaryotes The earliest evidence of life comes from fossil stromatolites formed by ancient photosynthetic prokaryotes (see Figure 15.1). Prokaryotes (the gold band in Figure 15.4) were Earth's sole inhabitants from at least 3.5 billion years ago to about 2 billion years ago. During this time they transformed the biosphere. As a result of prokaryotic photosynthesis, oxygen saturated the seas and began to appear in the atmosphere 2.7 billion years ago (the green band). By 2.2 billion years ago, atmospheric O_2 began to increase rapidly, causing an "oxygen revolution." Many prokaryotes, unable to live in this aerobic environment, became extinct. Some species survived in anaerobic habitats, where their descendants live today. The evolution of cellular respiration, which uses O_2 in harvesting energy from organic molecules, allowed other prokaryotes to flourish.

Origin of Single-celled Eukaryotes The oldest widely accepted fossils of eukaryotes are about 1.8 billion years old (the orange band). The more complex eukaryotic cell originated when small prokaryotic cells capable of aerobic respiration or photosynthesis took up life inside larger cells (as you learned in Module 4.15). After the first eukaryotes appeared, a great range of unicellular forms evolved, giving rise to the diversity of single-celled eukaryotes that continue to flourish today.

Origin of Multicellular Eukaryotes Another wave of diversification followed: the origin of multicellular forms whose descendants include a variety of algae, plants, fungi, and animals. Molecular comparisons suggest that the common ancestor of multicellular eukaryotes arose about 1.5 billion years ago (the light blue band). The oldest known fossils of multicellular eukaryotes are of relatively small algae that lived 1.2 billion years ago.

Larger and more diverse multicellular organisms do not appear in the fossil record until about 600 million years ago. A great increase in the diversity of animal forms occurred 535–525 million years ago, in a period known as the Cambrian explosion. Animals are shown on the timeline by the bright blue band.

Colonization of Land There is fossil evidence that photosynthetic prokaryotes coated damp terrestrial surfaces well over a billion years ago. However, larger forms of life did not begin to colonize land until about 500 million years ago (the purple band).

Plants colonized land in the company of fungi. Even today, the roots of most plants are associated with fungi that aid in absorption of water and minerals; the fungi receive nutrients in return (see Module 17.12).

The most widespread and diverse land animals are arthropods (particularly insects and spiders) and tetrapods (vertebrates with four appendages). Tetrapods include humans, but we are late arrivals on the scene—the human lineage diverged from other primates around 6 to 7 million years ago, and our own species originated about 195,000 years ago. If the clock of Earth's history were rescaled to represent an hour, humans appeared less than 0.2 second ago! In the next two modules, we see how scientists have determined when these key episodes in Earth's history have occurred in geologic time.

? **For how long did life on Earth consist solely of single-celled organisms?**

● More than 2 billion years: From the first fossils of prokaryotes (3.5 billion years old) until the oldest known fossils of multicellular eukaryotes (1.2 billion years old)

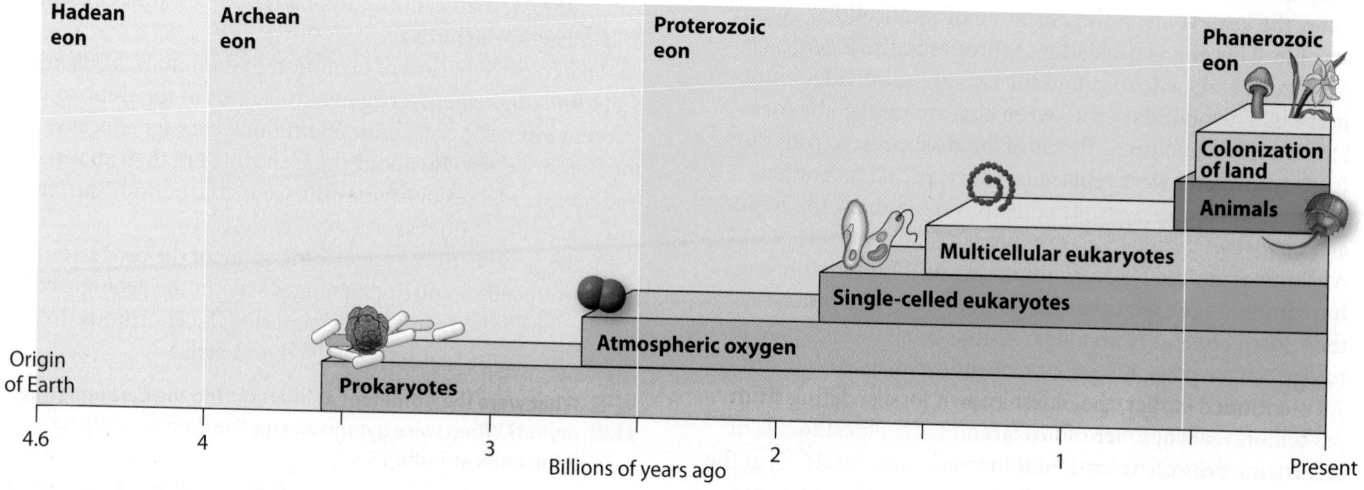

Hadean eon
Archean eon
Proterozoic eon
Phanerozoic eon
Colonization of land
Animals
Multicellular eukaryotes
Single-celled eukaryotes
Atmospheric oxygen
Prokaryotes
Origin of Earth
4.6
4
3
2
1
Present
Billions of years ago

▲ **Figure 15.4** Some key events in the history of life on Earth

15.5 The actual ages of rocks and fossils mark geologic time

Geologists use several techniques to determine the ages of rocks and the fossils they contain. The method most often used, called **radiometric dating**, is based on the decay of radioactive isotopes (unstable forms of an element; see Module 2.3). Fossils contain isotopes of elements that accumulated when the organisms were alive. For example, a living organism contains both the common isotope carbon-12 and the radioactive isotope carbon-14 in the same ratio as is present in the atmosphere. Once an organism dies, it stops accumulating carbon, and the stable carbon-12 in its tissues does not change. Its carbon-14, however, starts to decay to another element. The rate of decay is expressed as a half-life, the time required for 50% of the isotope in a sample to decay. Carbon-14 has a half-life of 5,730 years, so half the carbon-14 in a specimen decays in about 5,730 years, half the remaining carbon-14 decays in the next 5,730 years, and so on (**Figure 15.5**). Knowing both the half-life of a radioactive isotope and the ratio of radioactive to stable isotope in a fossil enables us to determine the age of the fossil.

Carbon-14 is useful for dating relatively young fossils—up to about 75,000 years old. Radioactive isotopes with longer half-lives are used to date older fossils.

There are indirect ways to estimate the age of much older fossils. For example, potassium-40, with a half-life of 1.3 billion years, can be used to date volcanic rocks hundreds of

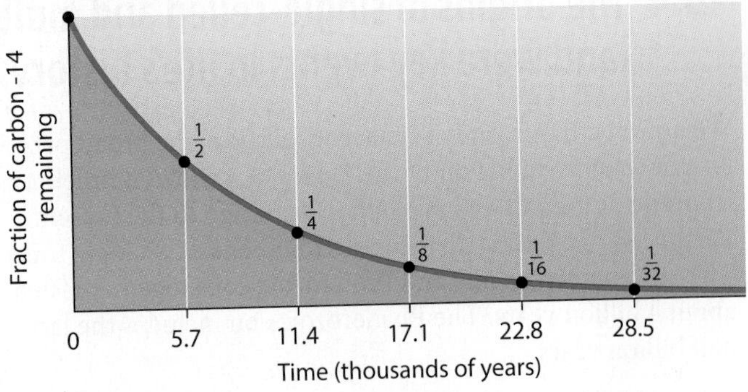

Fraction of carbon-14 remaining
1/2
1/4
1/8
1/16
1/32
0
5.7
11.4
17.1
22.8
28.5
Time (thousands of years)

▲ **Figure 15.5** Radiometric dating using carbon-14

millions of years old. A fossil's age can be inferred from the ages of the rock layers above and below the stratum in which it is found.

By dating rocks and fossils, scientists have established a geologic record of Earth's history.

? Estimate the age of a fossil found in a sedimentary rock layer between two layers of volcanic rock that are determined to be 530 and 520 million years old.

● We can infer that the organism lived between 530 and 520 million years ago.

15.6 The fossil record documents the history of life

The fossil record, the sequence in which fossils appear in rock strata, is an archive of evolutionary history (see Module 13.2). Based on this sequence and the ages of rocks and fossils, geologists have established a **geologic record**, as shown in **Table 15.6**, on the facing page. As you saw in Figure 15.4, Earth's history is divided into four eons, the Hadean, Archaean, Proterozoic, and Phanerozoic. The timeline in Table 15.6 shows the lengths and ages (in millions of years ago) of these eons. Note that the Phanerozoic eon, which is only the last 542 million years, is expanded in the table to show the key events in the evolution of multicellular eukaryotic life. This eon is divided into three eras: the Paleozoic, Mesozoic, and Cenozoic, and the eras are subdivided into periods. The boundaries between eras are marked by mass extinctions, when many forms of life disappeared from the fossil record and were replaced by species that diversified from the survivors. Lesser extinctions often mark the boundaries between periods.

Rocks from the Hadean, Archaean, and Proterozoic eons have undergone extensive change over time, and much of their fossil content is no longer visible. Nonetheless, paleontologists have pieced together ancient events in life's history. As mentioned earlier, the oldest known fossils, dating from 3.5 billion years ago, are of prokaryotes; the oldest fossils of eukaryotic cells are from 1.8 billion years ago. Strata from the Ediacaran period (635–542 million years ago) bear diverse fossils of multicellular algae and soft-bodied animals.

Dating from about 542 million years ago, rocks of the Paleozoic ("ancient animal") era contain fossils of lineages that gave rise to present-day organisms, as well as many lineages that have become extinct. During the early Paleozoic, virtually all life was aquatic, but by about 400 million years ago, plants and animals were well established on land.

The Mesozoic ("middle animal") era is also known as the age of reptiles because of its abundance of reptilian fossils, including those of the dinosaurs. The Mesozoic era also saw the first mammals and flowering plants (angiosperms). By the end of the Mesozoic, dinosaurs had become extinct except for one lineage—the birds.

An explosive period of evolution of mammals, birds, insects, and angiosperms began at the dawn of the Cenozoic ("recent animal") era, about 65 million years ago. Because much more is known about the Cenozoic era than about earlier eras, our table subdivides the Cenozoic periods into finer intervals called epochs.

In the next section, we examine some of the processes that have produced the distinct changes seen in the geologic record. (The chapters in Unit IV describe the enormous diversity of life-forms that have evolved on Earth.)

? What were the dominant animals during the Carboniferous period? When were gymnosperms the dominant plants? (*Hint*: Look at Table 15.6.)

● Amphibians. Gymnosperms were dominant during the Triassic and Jurassic periods (251–145.5 million years ago).

TABLE 15.6 | THE GEOLOGIC RECORD

Relative Duration of Eons	Era	Period	Epoch	Age (millions of years ago)	Important Events in the History of Life
Phanerozoic	Cenozoic	Quaternary	Holocene		Historical time
				0.01	
			Pleistocene		Ice ages; origin of genus *Homo*
				2.6	
		Tertiary	Pliocene		Appearance of bipedal human ancestors
				5.3	
			Miocene		Continued radiation of mammals and angiosperms; earliest direct human ancestors
				23	
			Oligocene		Origins of many primate groups
				33.9	
			Eocene		Angiosperm dominance increases; continued radiation of most present-day mammalian orders
				55.8	
			Paleocene		Major radiation of mammals, birds, and pollinating insects
				65.5	
	Mesozoic	Cretaceous			Flowering plants (angiosperms) appear and diversify; many groups of organisms, including most dinosaurs, become extinct at end of period
				145.5	
		Jurassic			Gymnosperms continue as dominant plants; dinosaurs abundant and diverse
				199.6	
		Triassic			Cone-bearing plants (gymnosperms) dominate landscape; dinosaurs evolve and radiate; origin of mammals
				251	
	Paleozoic	Permian			Radiation of reptiles; origin of most present-day groups of insects; extinction of many marine and terrestrial organisms at end of period
				299	
		Carboniferous			Extensive forests of vascular plants form; first seed plants appear; origin of reptiles; amphibians dominant
				359	
		Devonian			Diversification of bony fishes; first tetrapods and insects appear
				416	
		Silurian			Diversification of early vascular plants
				444	
		Ordovician			Marine algae abundant; colonization of land by diverse fungi, plants, and animals
				488	
		Cambrian			Sudden increase in diversity of many animal phyla (Cambrian explosion)
				542	
Proterozoic		Ediacaran			Diverse algae and soft-bodied invertebrate animals appear
				635	
				1,800	Oldest fossils of eukaryotic cells appear
				2,500	
Archaean				2,700	Concentration of atmospheric oxygen begins to increase
				3,500	Oldest fossils of cells (prokaryotes) appear
				3,850	Oldest known rocks on Earth's surface
				4,000	
Hadean				Approx. 4,600	Origin of Earth

Mechanisms of Macroevolution

15.7 Continental drift has played a major role in macroevolution

The fossil record documents macroevolution, the major events in the history of life on Earth. In this section, we explore some of the factors that helped shape these evolutionary changes, such as plate tectonics, mass extinctions, and adaptive radiations.

Plate Tectonics If photographs of Earth were taken from space every 10,000 years and then spliced together, it would make a remarkable movie. The seemingly "rock solid" continents we live on move over time. Since the origin of multicellular eukaryotes roughly 1.5 billion years ago, there have been three occasions—1.1 billion, 600 million, and 250 million years ago—in which the landmasses of Earth came together to form a supercontinent, and later broke apart. Each time the landmasses split, they yielded a different configuration of continents. Geologists estimate that the continents will come together again and form a new supercontinent roughly 250 million years from now.

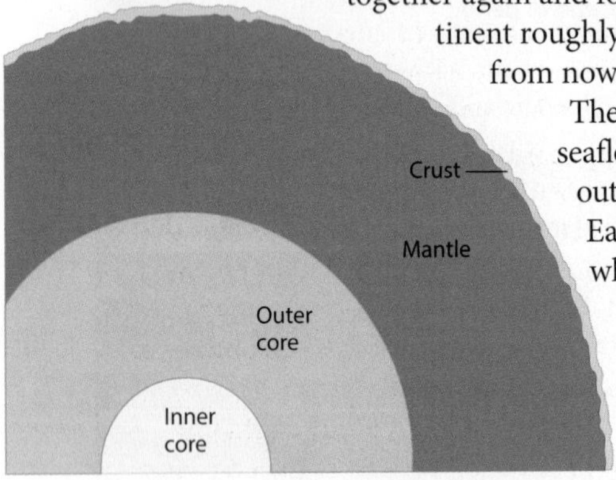

The continents and seafloors form a thin outer layer of planet Earth, called the crust, which covers a mass of hot, viscous material called the mantle. The outer core is liquid and the inner core is solid (**Figure 15.7A**). According to the theory of **plate tectonics**, Earth's crust is divided into giant, irregularly shaped plates (outlined in black in **Figure 15.7B**) that essentially float on the underlying mantle. In a process called continental drift, movements in the mantle cause the plates to move (black arrows in the figure). In some cases, the plates are moving away from each other. North America and Europe, for example, are drifting apart at a rate of about 2 cm per year. In other cases, two plates are sliding past each other, forming regions where earthquakes are common. In still other cases, two plates are colliding. Massive upheavals may occur, forming mountains along the plate boundaries. The red dots in Figure 15.7B indicate zones of violent geologic activity, most of which are associated with plate boundaries.

Crust
Mantle
Outer core
Inner core

▲ **Figure 15.7A** Cross-sectional view of Earth (with the thickness of the crust exaggerated)

Consequences of Continental Drift Throughout Earth's history, continental drift has reshaped the physical features of the planet and altered the habitats in which organisms live. **Figure 15.7C**, on the facing page, shows continental movements that greatly influenced life during the Mesozoic and Cenozoic eras. ❶ About 250 million years ago, near the end of the Paleozoic era, plate movements brought all the previously separated landmasses together into a supercontinent we call **Pangaea**, meaning "all land." When the landmasses fused, ocean basins became deeper, lowering the sea level and draining the shallow coastal seas. Then, as now, most marine species inhabited shallow waters, and much of that habitat was destroyed. The interior of the vast continent was cold and dry. Overall, the formation of Pangaea had a tremendous impact on the physical environment and climate. As the fossil record documents, biological diversity was reshaped. Many species were driven to extinction, and new opportunities arose for organisms that survived the crisis.

During the Mesozoic era, Pangaea started to break apart, causing a geographic isolation of colossal proportions. As the continents drifted apart, each became a separate evolutionary arena—a huge island on which organisms evolved in isolation from their previous neighbors. ❷ At first, Pangaea split into northern and southern landmasses, which we call Laurasia and Gondwana, respectively. ❸ By the end of the Mesozoic era, some 65 million years ago, the modern continents were beginning to take shape. Note that at that time Madagascar became isolated and India was still a large island. Then, around 45 million years ago, the India plate collided with the Eurasian plate, and the slow, steady buckling at the plate boundary formed the Himalayas, the tallest and youngest of Earth's mountain ranges. ❹ The continents continue to drift today, and the Himalayas are still growing (about 1 cm per year).

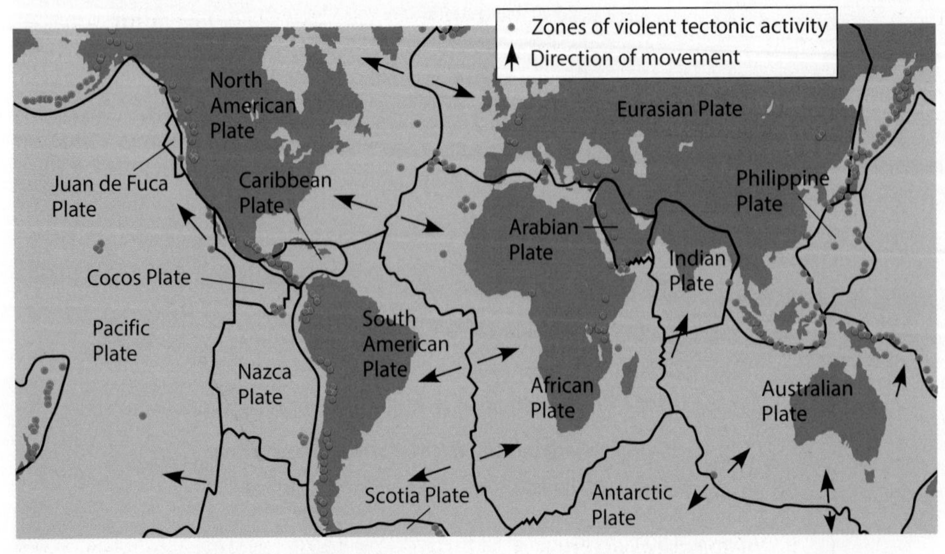
• Zones of violent tectonic activity
Direction of movement
North American Plate
Eurasian Plate
Juan de Fuca Plate
Caribbean Plate
Philippine Plate
Arabian Plate
Indian Plate
Cocos Plate
Pacific Plate
South American Plate
Nazca Plate
African Plate
Australian Plate
Scotia Plate
Antarctic Plate

▲ **Figure 15.7B** Earth's tectonic plates

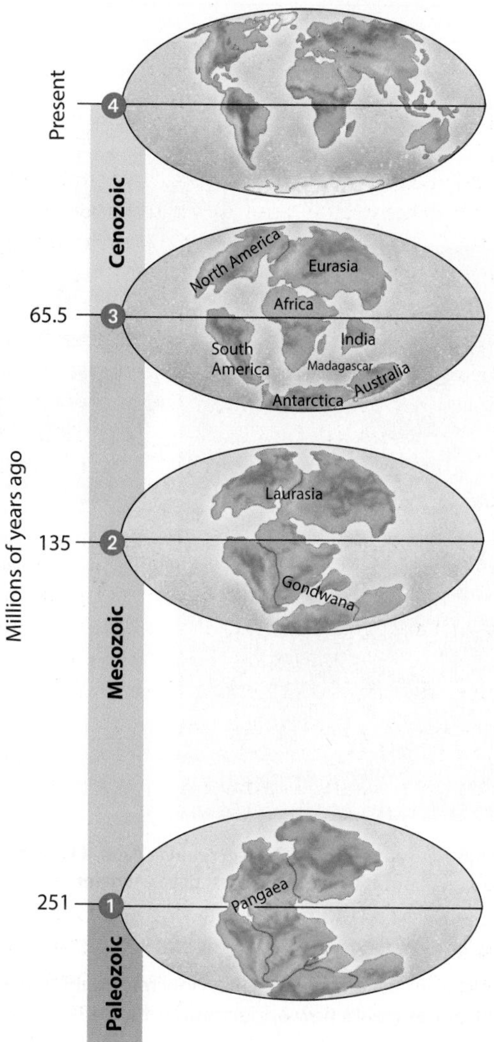
Present
Cenozoic
North America
Eurasia
Africa
65.5
South America
India
Madagascar
Australia
Antarctica
Millions of years ago
Laurasia
135
Gondwana
Mesozoic
251
Pangaea
Paleozoic
1
2
3
4

▲ **Figure 15.7C** Continental drift during the Phanerozoic eon

Try This Use Table 15.6 to identify important events in the history of life that occurred while the continents occupied the positions shown at 1, 2, and 3.

The history of continental mergers and separations explains many patterns of **biogeography**, the study of the past and present distribution of organisms. For example, almost all the animals and plants that live on the island of Madagascar are unique—they diversified from ancestral populations after Madagascar was isolated from Africa and India. As in the Galápagos Islands, adaptive radiations (see Module 14.8) occurred in many groups. The more than 50 species of lemurs that currently inhabit Madagascar, for instance, evolved from a common ancestor over the past 40 million years.

Continental drift solves the mystery of marsupials, mammals whose young complete their embryonic development in a pouch outside the mother's body, such as kangaroos, koalas, and wombats. Australia and its neighboring islands are home to more than 200 species of marsupials, most of which are found nowhere else in the world **(Figure 15.7D)**.

What accounts for the predominance of marsupials in Australia, while the rest of the world is dominated by eutherian (placental) mammals, whose young complete their development in the mother's uterus? Looking at a current map of the world, you might hypothesize that marsupials evolved only on this island continent. But marsupials are not unique to Australia. More than a hundred species live in Central and South America **(Figure 15.7E)**; North America is home to only a few, including the Virginia opossum **(Figure 15.7F)**. The distribution of marsupials only makes sense in the context of continental drift—marsupials must have originated when the continents were joined. Fossil evidence suggests that marsupials originated in what is now Asia and later dispersed to the tip of South America while it was still connected to Antarctica. They made their way to Australia before continental drift separated Antarctica from Australia, setting it "afloat" like a great raft of marsupials. The few early eutherians that lived there became extinct, while on other continents, most marsupials became extinct. Isolated on Australia, marsupials evolved and diversified, filling ecological roles analogous to those filled by eutherians on other continents.

Continental drift solves puzzles about the geographic distribution of extinct organisms as well as living ones. For example, paleontologists have discovered fossils of the same species of Permian freshwater reptiles in West Africa and Brazil, regions now separated by 3,000 km of ocean.

In the next module, we consider some of the perils associated with the movements of Earth's crustal plates.

? **If marsupials originated in Asia and reached Australia via South America, where else should paleontologists find fossil marsupials? (*Hint*: Look at Figure 15.7C.)**

● Antarctica

▲ **Figure 15.7D** Greater bilby (*Macrotis lagotis*), an Australian marsupial

▲ **Figure 15.7E** Mexican mouse opossum (*Marmosa mexicana*)

▲ **Figure 15.7F** Virginia opossum female with young (*Didelphis virginiana*)

15.8 Plate tectonics may imperil human life

CONNECTION

Not only do moving crustal plates cause continents to collide, pile up, and build mountain ranges; they also produce volcanoes and earthquakes. The boundaries of plates are hot spots of such geologic activity. California's frequent earthquakes are a result of movement along the infamous San Andreas Fault, part of the border where the Pacific and North American plates grind together and gradually slide past each other **(Figure 15.8)**—in what we can call a strike-slip fault. Two major earthquakes have occurred in the region in the past century: the San Francisco earthquake of 1906 and the 1989 Loma Prieta earthquake, also near San Francisco.

In such a strike-slip fault, the two plates do not slide smoothly past each other. They often stick in one spot until enough pressure builds along the fault that the landmasses suddenly jerk forward, releasing massive amounts of energy and causing the surrounding area to move or shake. A strike-slip fault runs under Haiti and is responsible for the devastating magnitude 7.0 earthquake of January 2010. In Haiti, the North American plate is moving west past the Caribbean plate (see Figure 15.7B). Undersea earthquakes can cause giant waves, such as the massive 2011 tsunami in Japan, a seismically active area where four tectonic plates meet.

A volcano is a rupture that allows hot, molten rock, ash, and gases to escape from beneath Earth's crust. Volcanoes are often found where tectonic plates are diverging or converging, as opposed to sliding past each other. Volcanoes can cause tremendous devastation, as when Mt. Vesuvius in southern Italy erupted in 79 AD, burying Pompeii in a layer of ash. But sometimes volcanoes imperil more than just local life, as we see in the next module.

Pacific Plate
North American Plate
San Francisco
San Andreas Fault
Los Angeles
California

▲ **Figure 15.8** An aerial view of the San Andreas Fault, a boundary between two crustal plates, about 100 miles northwest of Los Angeles

? **Volcanoes usually destroy life. How might undersea volcanoes create new opportunities for life?**

● By creating new landmasses on which life can evolve, such as the Galápagos and Hawaiian Islands

15.9 During mass extinctions, large numbers of species are lost

Extinction is inevitable in a changing world. Indeed, the fossil record shows that the vast majority of species that have ever lived are now extinct. A species may become extinct because its habitat has been destroyed, because of unfavorable climatic changes, or because of changes in its biological community, such as the evolution of new predators or competitors. Extinctions occur all the time, but extinction rates have not been steady.

Mass Extinctions The fossil record chronicles a number of occasions when global environmental changes were so rapid and disruptive that a majority of species were swept away in a relatively short amount of time. Five mass extinctions have occurred over the past 500 million years. In each of these events, 50% or more of Earth's species became extinct.

Of all the mass extinctions, the ones marking the ends of the Permian and Cretaceous periods have received the most attention. The Permian extinction, which occurred about 251 million years ago and defines the boundary between the Paleozoic and Mesozoic eras, claimed about 96% of marine animal species and took a tremendous toll on terrestrial life as well. This mass extinction occurred in less than 500,000 years, and possibly in just a few thousand years—an instant in the context of geologic time.

At the end of the Cretaceous period, about 65 million years ago, the world again lost an enormous number of species—more than half of all marine species and many lineages of terrestrial plants and animals. At that point, dinosaurs had dominated the land and pterosaurs had ruled the air for some 150 million years. After the Cretaceous mass extinction, the pterosaurs and almost all the dinosaurs were gone, leaving behind only the descendants of one lineage, the birds.

Causes of Mass Extinctions The Permian mass extinction occurred at a time of enormous volcanic eruptions in what is now Siberia. Vast stretches of land were covered with lava hundreds to thousands of meters thick. Besides spewing lava and ash into the atmosphere, the eruptions may have produced enough carbon dioxide to warm the global climate by an estimated 6°C. Reduced temperature differences between the equator and the poles would have slowed the mixing of

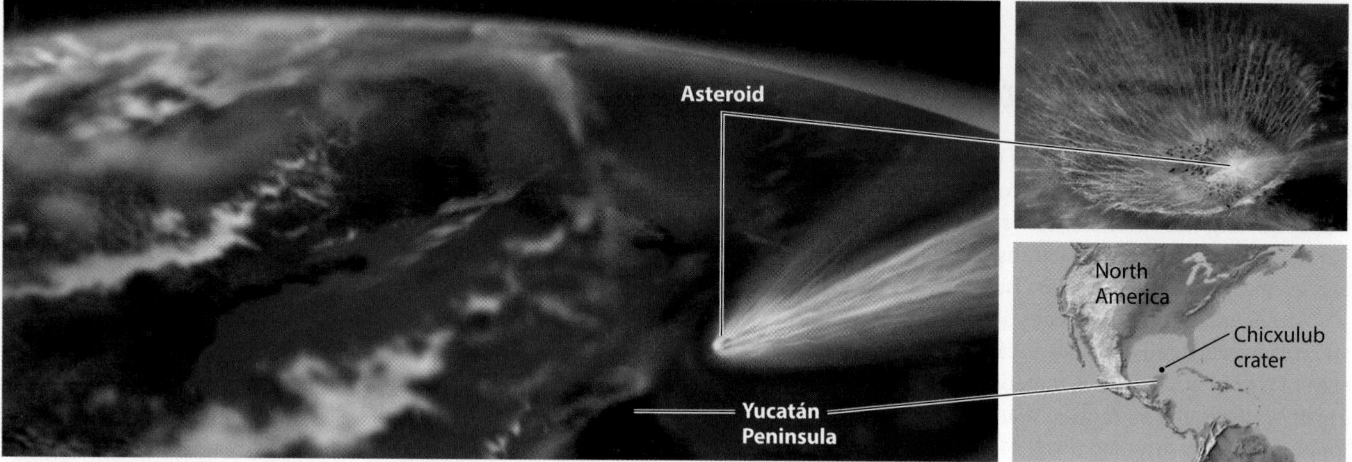
Asteroid
North America
Chicxulub crater
Yucatán Peninsula

▲ **Figure 15.9** The impact hypothesis for the Cretaceous mass extinction

ocean water, leading to a widespread drop in oxygen concentration in the water. This oxygen deficit would have killed many marine organisms and promoted the growth of anaerobic bacteria that emit a poisonous by-product, hydrogen sulfide. As this gas bubbled out of the water, it would have killed land plants and animals and initiated chemical reactions that would destroy the protective ozone layer. Thus, a cascade of factors may have contributed to the Permian extinction.

One clue to a possible cause of the Cretaceous mass extinction is a thin layer of clay enriched in iridium that separates sediments from the Mesozoic and Cenozoic eras. Iridium is an element very rare on Earth but common in meteorites and other extraterrestrial objects that occasionally fall to Earth. The rocks of the Cretaceous boundary layer have many times more iridium than normal Earth levels. Most paleontologists conclude that the iridium layer is the result of fallout from a huge cloud of dust that billowed into the atmosphere when an asteroid or large comet hit Earth. The cloud would have blocked light and severely disturbed the global climate for months.

Is there evidence of such an asteroid? A large crater, the 65-million-year-old Chicxulub impact crater, has been found in the Caribbean Sea near the Yucatán Peninsula of Mexico **(Figure 15.9)**. About 180 km wide (about 112 miles), the crater is the right size to have been caused by an object with a diameter of 10 km (about 6 miles). The horseshoe shape of the crater and the pattern of debris in sedimentary rocks indicate that an asteroid or comet struck at a low angle from the southeast. The artist's interpretation in Figure 15.9 represents the impact and its immediate effect—a cloud of hot vapor and debris that could have killed most of the plants and animals in North America within hours. The collision is estimated to have released more than a billion times the energy of the nuclear bombs dropped in Japan during World War II.

In March 2010, an international team of scientists reviewed two decades' worth of research on the Cretaceous extinction and endorsed the asteroid hypothesis as the triggering event. Nevertheless, research will continue on other contributing causes and the multiple and interrelated effects of this major ecological disaster.

Consequences of Mass Extinctions Whatever their causes, mass extinctions affect biological diversity profoundly. By removing large numbers of species, a mass extinction can decimate a thriving and complex ecological community. Mass extinctions are random events that act on species indiscriminately. They can permanently remove species with highly advantageous features and change the course of evolution forever. Consider what would have happened if our early primate ancestors living 65 million years ago had died out in the Cretaceous mass extinction—or if a few large, predatory dinosaurs had *not* become extinct!

How long does it take for life to recover after a mass extinction? The fossil record shows that it typically takes 5–10 million years for the diversity of life to return to previous levels. In some cases, it has taken much longer: It took about 100 million years for the number of marine families to recover after the Permian mass extinction.

Is a Sixth Mass Extinction Under Way? Human actions that result in habitat destruction and climate change are modifying the global environment to such an extent that many species are currently threatened with extinction (as we'll explore in Chapter 38). In the past 400 years, more than a thousand species are known to have become extinct. Scientists estimate that this rate is 100 to 1,000 times the normal rate seen in the fossil record. Does this represent the beginning of a sixth mass extinction?

This question is difficult to answer, partly because it is hard to document both the total number of species on Earth and the number of extinctions that are occurring. It is clear that losses have not reached the level of the other "big five" extinctions. Monitoring, however, does show that many species are declining at an alarming rate, suggesting that a sixth (human-caused) mass extinction could occur within the next few centuries or millennia. And as seen with prior mass extinctions, it may take millions of years for life on Earth to recover.

But the fossil record also shows a creative side to the destruction. Mass extinctions can pave the way for adaptive radiations in which new groups rise to prominence, as we see next.

? **The Permian and Cretaceous mass extinctions mark the ends of the ______________ and ______________ eras, respectively. (*Hint*: Refer to Table 15.6.)**

● Paleozoic . . . Mesozoic

15.10 Adaptive radiations have increased the diversity of life

Adaptive radiations are periods of evolutionary change in which many new species evolve from a common ancestor, often following the colonization of new, unexploited areas (Module 14.8). Adaptive radiations have also followed each mass extinction, when survivors became adapted to the many vacant ecological roles, or niches, in their communities.

For example, fossil evidence indicates that mammals underwent a dramatic adaptive radiation after the extinction of terrestrial dinosaurs 65 million years ago (**Figure 15.10**). Although mammals originated 180 million years ago, fossils older than 65 million years indicate that they were mostly small and not very diverse. Early mammals may have been eaten or outcompeted by the larger and more diverse dinosaurs. With the disappearance of the dinosaurs (except for the bird lineage), mammals expanded greatly in both diversity and size, filling the ecological roles once occupied by dinosaurs.

The history of life has also been altered by radiations that followed the evolution of new adaptations. Major new adaptations facilitated the colonization of land by plants, insects, and tetrapods. The radiation of land plants, for example, was associated with features such as stems that supported the plant against gravity and a waxy coat that protected leaves from water loss. Finally, note that organisms that arise in an adaptive radiation can serve as a new source of food for still other organisms. In this way, the diversification of land plants stimulated a series of adaptive radiations in insects that ate or pollinated plants—helping to make insects the most diverse group of animals on Earth today.

Now that we've looked at geologic and environmental influences, let's consider how genes can affect macroevolution.

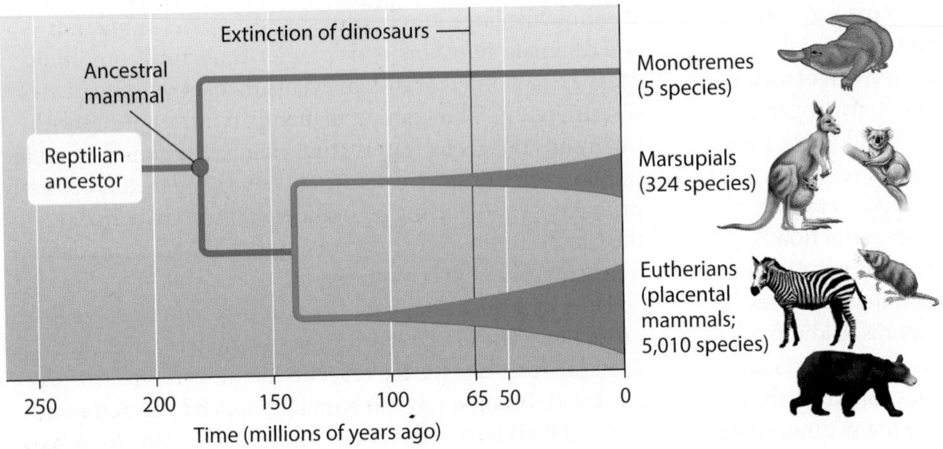

? **In addition to the new resources of plants, what other factors likely promoted the adaptive radiation of insects on land?**

Many unfilled ecological roles on land and the evolution of wings and a supportive, protective, and waterproof exoskeleton

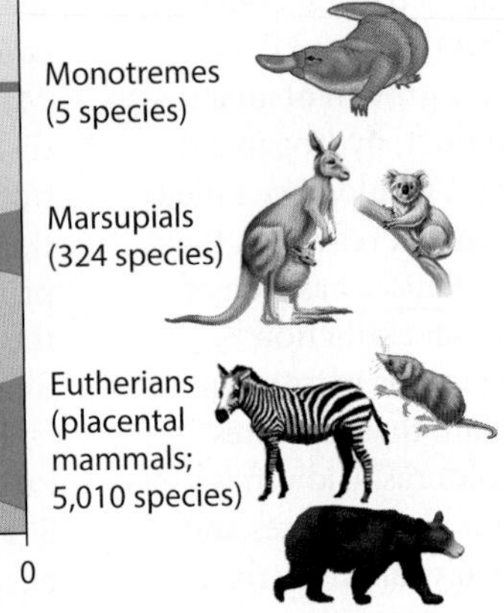

Extinction of dinosaurs
Ancestral mammal
Reptilian ancestor
Monotremes (5 species)
Marsupials (324 species)
Eutherians (placental mammals; 5,010 species)
250
200
150
100
65 50
0
Time (millions of years ago)

▲ **Figure 15.10** Adaptive radiation of mammals (width of line reflects numbers of species).

15.11 Genes that control development play a major role in evolution

The fossil record can tell us *what* the great events in the history of life have been and *when* they occurred. Continental drift, mass extinctions, and adaptive radiation provide a big-picture view of *how* those changes came about. But now we are increasingly able to understand the basic biological mechanisms that underlie the changes seen in the fossil record.

Scientists working at the interface of evolutionary biology and developmental biology—the research field abbreviated **"evo-devo"**—are studying how slight genetic changes can become magnified into major morphological differences between species. Genes that program development control the rate, timing, and spatial pattern of change in an organism's form as it develops from a zygote into an adult. A great many of these genes appear to have been conserved throughout evolutionary history: The same or very similar genes are involved in the development of form across multiple lineages.

Changes in Rate and Timing Many striking evolutionary transformations are the result of a change in the rate or timing of developmental events. **Figure 15.11A** shows a photograph of an axolotl, a salamander that illustrates a phenomenon called **paedomorphosis** (from the Greek *paedos*, of a child, and *morphosis*, formation), the retention in the adult of body structures that were juvenile features in an ancestral species. Most salamander species have aquatic larvae (with gills) that undergo metamorphosis in becoming terrestrial adults (with lungs). The axolotl is a salamander that grows to a sexually mature adult while retaining gills and other larval features.

Gills

▲ **Figure 15.11A** An axolotl, a paedomorphic salamander

Slight changes in the relative growth of different body parts can change an adult form substantially. As the skulls and photo in **Figure 15.11B** on the next page show, humans and chimpanzees are much more alike as fetuses than they are as adults. As development proceeds, accelerated growth in the jaw produces the elongated skull, sloping forehead, and

Chimpanzee infant
Chimpanzee adult

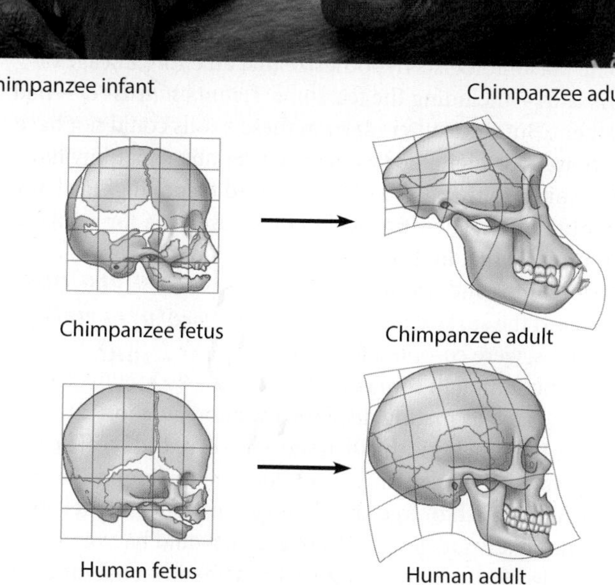
Chimpanzee fetus
Chimpanzee adult
Human fetus
Human adult

▲ **Figure 15.11B** Chimpanzee and human skull shapes compared

massive jaws of an adult chimpanzee. In the human lineage, genetic changes that slowed the growth of the jaw relative to other parts of the skull produced an adult whose head proportions still resembled that of a child (and that of a baby chimpanzee). Our large skull and complex brain are among our most distinctive features. Compared to the slow growth of a chimpanzee brain after birth, our brain continues to grow at the rapid rate of a fetal brain for the first year of life.

Changes in Spatial Pattern Homeotic genes, the master control genes, determine such basic features as where a pair of wings or legs will develop on a fruit fly (see Module 11.8). Changes in homeotic genes or in how or where such genes are expressed can have a profound impact on body form. Consider, for example, the evolution of snakes from a four-limbed lizard-like ancestor. Researchers have found that one pattern of expression of two homeotic genes in tetrapods results in the formation of forelimbs and of vertebrae with ribs, whereas a different pattern of expression of these two genes results in the development of vertebrae with ribs but no limbs, as in snakes (see Figure 30.3C).

New Genes and Changes in Genes New developmental genes that arose as a result of gene duplications may have facilitated the origin of new body forms. For example, a fruit fly (an invertebrate) has a single cluster of several homeotic genes that direct the development of major body parts. A mouse (a vertebrate) has four clusters of very similar genes that occur in the same linear order on chromosomes and direct the development of the same body regions as the fly genes (see Figure 27.14B). Two duplications of these gene clusters appear to have occurred in the evolution of vertebrates from invertebrate animals. Mutations in these duplicated genes may then have led to the origin of novel vertebrate characters, such as a backbone, jaws, and limbs.

Changes in Gene Regulation Researchers are finding that changes in the form of organisms often are caused by mutations that affect the regulation of developmental genes. As we just discussed, such a change in gene expression was shown to correlate with the lack of forelimbs in snakes.

Additional evidence for this type of change in gene regulation is seen in studies of the threespine stickleback fish. In western Canada, these fish live in the ocean and also in lakes that formed when the coastline receded during the past 12,000 years. Ocean populations have bony plates that make up a kind of body armor and a set of pelvic spines that help deter predatory fish. The body armor and pelvic spines are reduced or absent in threespine sticklebacks living in lakes that lack predatory fishes and that are also low in calcium. In the absence of predators, spineless sticklebacks may have a selective advantage because the limited calcium is needed for purposes other than constructing spines. **Figure 15.11C** shows specimens of an ocean and a lake stickleback, which have been stained to highlight their bony plates and spines.

Researchers have identified a key gene that influences the development of these spines. Was the reduction of spines in lake populations due to changes in the gene itself or to changes in how the gene is expressed? It turns out that the gene is identical in the two populations, and it is expressed in the mouth region and other tissues of embryos from both populations. Studies have shown, however, that while the gene is also expressed in the developing pelvic region of ocean sticklebacks, it is not turned on in the pelvic region in lake sticklebacks. This example shows how morphological change can be caused by altering the expression of a developmental gene in some parts of the body but not others.

? **Research shows that many differences in body form are caused by changes in gene regulation and not changes in the nucleotide sequence of the developmental gene itself. Why might this be the case?**

● A change in sequence may affect a gene's function wherever that gene is expressed—with potentially harmful effects. Changes in the regulation of gene expression can be limited to specific areas in a developing embryo.

▲ **Figure 15.11C** Stickleback fish from ocean (top) and lake (bottom), stained to show bony plates and spines. (Arrow indicates the absence of the pelvic spine in the lake fish.)

15.12 Novel traits may arise in several ways

EVOLUTION CONNECTION

Let's see how the Darwinian theory of gradual change can account for the evolution of intricate structures such as eyes or of novel body structures such as wings (that is, new kinds of structures). Most complex structures have evolved in increments from simpler versions having the same basic function—a process of refinement.

Consider the amazing camera-like eyes of vertebrates and squids. Although these complex eyes evolved independently, the origin of both can be traced from a simple ancestral patch of photoreceptor cells through a series of incremental modifications that benefited their owners at each stage. Indeed, there appears to have been a single evolutionary origin of light-sensitive cells, and all animals with eyes—vertebrates and invertebrates alike—share the same master genes that regulate eye development.

Figure 15.12 illustrates the range of complexity in the structure of eyes among molluscs living today. Simple patches of pigmented cells enable limpets to distinguish light from dark, and they cling more tightly to their rock when a shadow falls on them—a behavioral adaptation that reduces the risk of being eaten. Other molluscs have eyecups that have no lenses or other means of focusing images but can indicate light direction. In those molluscs that do have complex eyes, the organs probably evolved in small steps of adaptation.

Although eyes have retained their basic function of vision throughout their evolutionary history, evolutionary novelty can also arise when structures that originally played one role gradually acquire a different one. Structures that evolve in one context but become co-opted for another function are called *exaptations*. However, exaptation does not mean that a structure evolves in anticipation of future use. Natural selection cannot predict the future; it can only improve an existing structure in the context of its current use. Novel features can arise gradually via a series of intermediate stages, each of which has some function in the organism's current situation.

The evolution of feathers is a good example of exaptation. Some paleontologists hypothesize that an entire lineage of dinosaurs—including the fearsome *Tyrannosaurus rex*—had feathers. But the feathers seen in these fossils could not have been used for flight, nor would their reptilian anatomy have been suited to flying. If feathers evolved before flight, what was their function? Their first utility may have been for insulation. It is possible that longer, winglike forelimbs and feathers, which increased the surface area of these forelimbs, were co-opted for flight after functioning in some other capacity, such as mating displays, thermoregulation, or camouflage (all functions that feathers still serve today). The first flights may have been only short glides to the ground or from branch to branch in tree-dwelling species. Once flight itself became an advantage, natural selection would have gradually remodeled feathers and wings to fit their additional function.

How do brand-new structures arise by evolution?

The flippers of penguins are another example of the modification of existing structures for different functions. Penguins cannot fly, but their modified wings are powerful oars that make them strong, fast underwater swimmers.

? Explain why the concept of exaptation does not imply that a structure evolves in anticipation of some future environmental change.

● Although a structure is co-opted for new or additional functions in a new environment, the structure existed because it worked as an adaptation in the old environment.

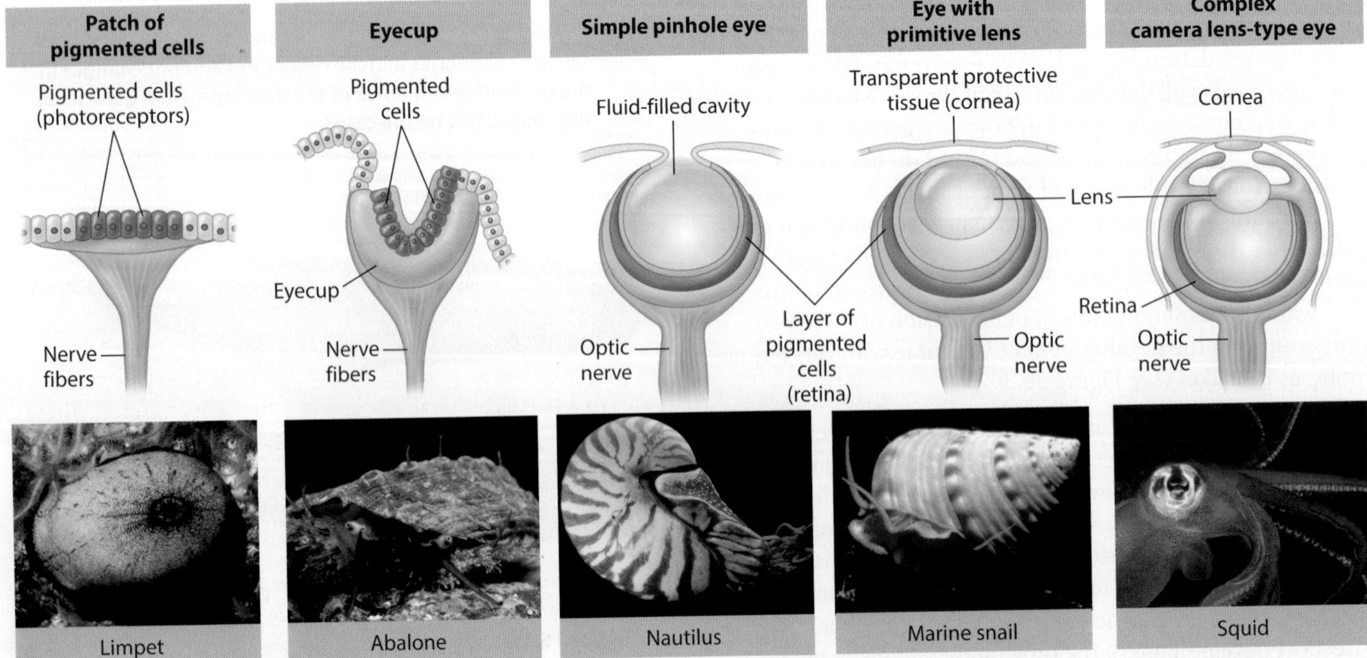

Patch of pigmented cells
Pigmented cells (photoreceptors)
Nerve fibers
Eyecup
Pigmented cells
Eyecup
Nerve fibers
Simple pinhole eye
Fluid-filled cavity
Optic nerve
Layer of pigmented cells (retina)
Eye with primitive lens
Transparent protective tissue (cornea)
Lens
Optic nerve
Complex camera lens-type eye
Cornea
Retina
Optic nerve
Limpet
Abalone
Nautilus
Marine snail
Squid

▲ **Figure 15.12** A range of eye complexity among molluscs

15.13 Evolutionary trends do not mean that evolution is goal directed

The fossil record seems to show trends in the evolution of many species, for example, toward larger or smaller body size. Let's look at apparent trends in the evolution of the modern horse (genus *Equus*), from an ancestor known as *Hyracotherium* that lived some 55 million years ago. *Hyracotherium*, which was about the size of a large dog, had four toes on its front feet and three toes on its hind feet. Its teeth were adapted to browsing on shrubs and trees. In contrast, the present-day horse has only one toe on each foot (the hoof) and teeth modified for grazing on grasses.

Did the horse lineage progress gradually toward larger size, reduced number of toes, and teeth adapted to grazing? **Figure 15.13** shows the fossil record of horses, with the vertical bars representing the period of time each group persisted in the record. If you follow the fossil species highlighted in yellow from the bottom to the top of Figure 15.13, it appears that modern horses evolved linearly from *Hyracotherium* to *Equus* through a series of intermediate forms. However, if we consider *all* fossil horses known today, this apparent trend vanishes. The genus *Equus* actually descended through a series of speciation episodes, not all of which led to large, one-toed grazers. The present-day horse is the only surviving twig of an evolutionary tree with many divergent branches.

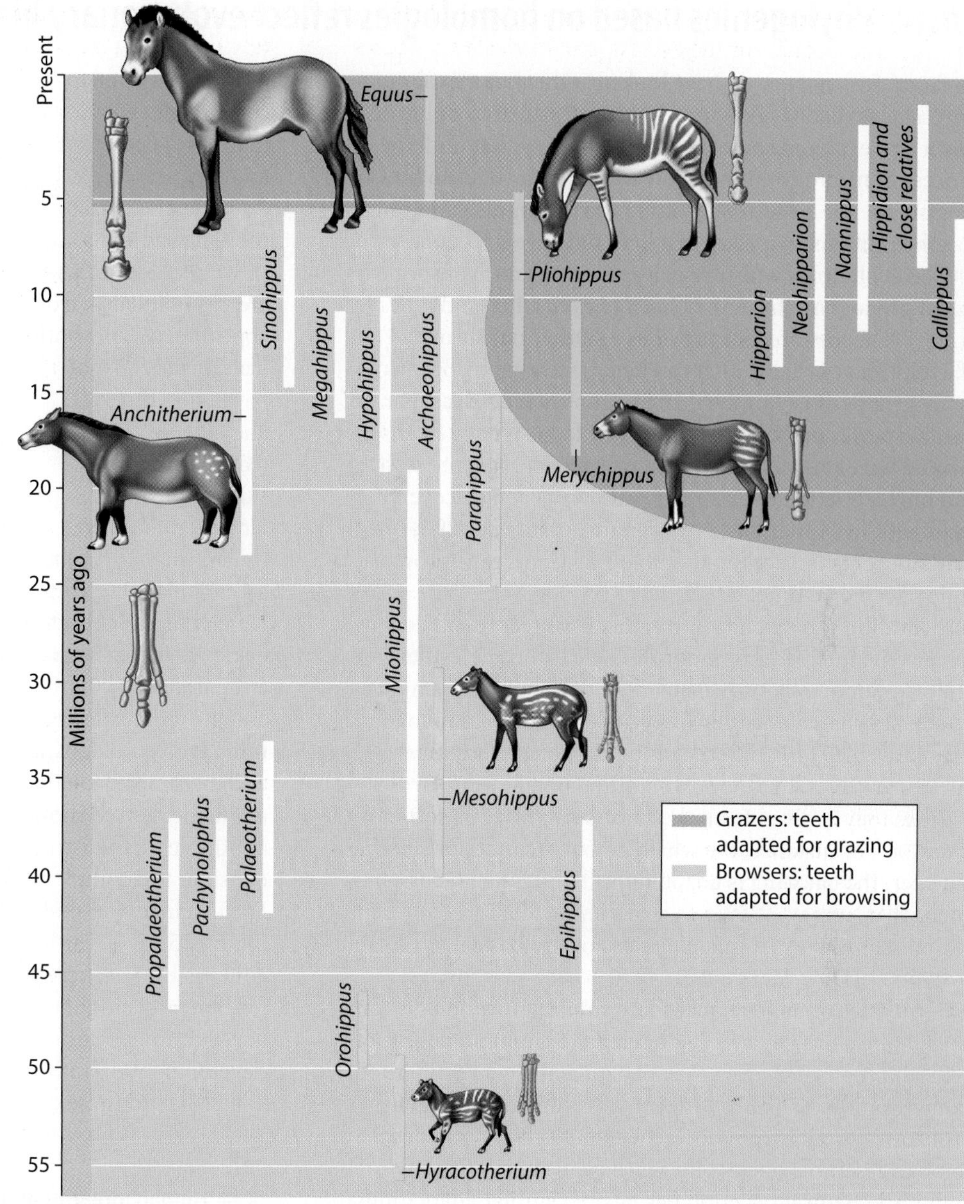

Present
Millions of years ago
5
10
15
20
25
30
35
40
45
50
55
Equus
Pliohippus
Hipparion
Neohipparion
Nannippus
Hippidion and close relatives
Callippus
Sinohippus
Megahippus
Hypohippus
Archaeohippus
Anchitherium
Parahippus
Merychippus
Miohippus
Mesohippus
Propalaeotherium
Pachynolophus
Palaeotherium
Epihippus
Orohippus
Hyracotherium
Grazers: teeth adapted for grazing
Browsers: teeth adapted for browsing

▲ **Figure 15.13** The branched evolution of horses

Branching evolution *can* lead to a real evolutionary trend, however. One model of long-term trends compares species to individuals: Speciation is their birth, extinction their death, and new species that diverge from them are their offspring. According to this model of species selection, unequal survival of species and unequal generation of new species play a role in macroevolution similar to the role of unequal reproduction in microevolution. In other words, the species that generate the greatest number of new species determine the direction of major evolutionary trends.

Evolutionary trends can also result directly from natural selection. For example, when horse ancestors invaded the grasslands that spread during the mid-Cenozoic, there was strong selection for grazers that could escape predators by running faster. This trend would not have occurred without open grasslands.

Whatever its cause, it is important to recognize that an evolutionary trend does not imply that evolution progresses toward a particular goal. Evolution is the result of interactions between organisms and the current environment. If conditions change, an apparent trend may cease or even reverse itself.

In the final section, we explore how biologists arrange life's astounding diversity into an evolutionary tree of life.

? **A trend in the evolution of mammals was toward a larger brain size. Use the species selection model to explain how such a trend could occur.**

● Those species with larger brains persisted longer before extinction and gave rise to more "offspring" species than did species with smaller brains.

▷ Phylogeny and the Tree of Life

15.14 Phylogenies based on homologies reflect evolutionary history

So far in this chapter, we have looked at the major evolutionary changes that have occurred during the history of life on Earth and explored some of the mechanisms that underlie the process of macroevolution. Now we shift our focus to how biologists use the pattern of evolution to distinguish and categorize the millions of species that live, and have lived, on Earth.

The evolutionary history of a species or group of species is called **phylogeny** (from the Greek *phylon*, tribe, and *genesis*, origin). The fossil record provides a substantial chronicle of evolutionary change that can help trace the phylogeny of many groups. It is, however, an incomplete record, as many of Earth's species probably never left any fossils; many fossils that formed were probably destroyed by later geologic processes; and only a fraction of existing fossils have been discovered. Even with its limitations, however, the fossil record is a remarkably detailed account of biological change over the vast scale of geologic time.

In addition to evidence from the fossil record, phylogeny can also be inferred from morphological and molecular homologies among living organisms. Homologies are similarities due to shared ancestry (see Module 13.4). Homologous structures may look different and function differently in different species, but they exhibit fundamental similarities because they evolved from the same structure in a common ancestor. For instance, the whale limb is adapted for steering in water; the bat wing is adapted for flight. Nonetheless, the bones that support these two structures, which were present in their common mammalian ancestor, are basically the same (see Figure 13.4A).

Generally, organisms that share similar morphologies are likely to be closely related. The search for homologies is not without pitfalls, however, for not all likenesses are inherited from a common ancestor. In a process called **convergent evolution**, species from different evolutionary branches may come to resemble one another if they live in similar environments and natural selection has favored similar adaptations. In such cases, body structures and even whole organisms may resemble each other.

Similarity due to convergent evolution is called **analogy**. For example, the two mole-like animals shown in **Figure 15.14** are very similar in external appearance. They both have enlarged front paws, small eyes, and a pad of protective thickened skin on the nose. However, the Australian "mole" (top) is a marsupial; the North American mole (bottom) is a eutherian. Genetic and fossil evidence indicates that the last common ancestor of these two animals lived 140 million years ago. And in fact, that ancestor and most of its descendants were not mole-like. Analogous traits evolved independently in these two mole lineages as they each became adapted to burrowing lifestyles.

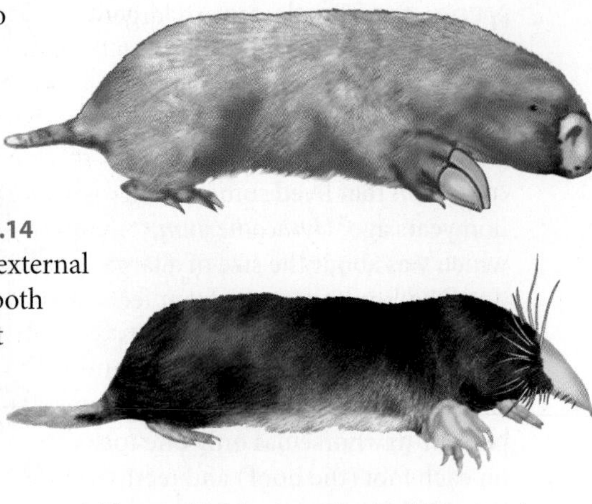

▲ **Figure 15.14** Australian "mole" (top) and North American mole (bottom)

In addition to molecular comparisons and fossil evidence, another clue to distinguishing homology from analogy is to consider the complexity of the structure being compared. For instance, the skulls of a human and a chimpanzee (see Figure 15.11B) consist of many bones fused together, and the composition of these skulls matches almost perfectly, bone for bone. It is highly improbable that such complex structures have separate origins. More likely, the genes involved in the development of both skulls were inherited from a common ancestor, and these complex structures are homologous.

? **Human forearms and a bat's wings are __________. A bat's wings and a bee's wings are __________.**

● homologous . . . analogous

15.15 Systematics connects classification with evolutionary history

Systematics is a discipline of biology that focuses on classifying organisms and determining their evolutionary relationships. In the 18th century, Carolus Linnaeus introduced a system of naming and classifying species, a discipline we call **taxonomy**. Although Linnaeus's system was not based on evolutionary relationship, many of its features, such as the two-part Latin names for species, remain useful in systematics.

Common names such as squirrel and daisy may work well in everyday communication, but they can be ambiguous because there are many species of each of these kinds of organisms. In addition, people in different regions may use the same common name for different species. For example, the flowers called bluebells in Scotland, England, Texas, and the eastern United States are actually four unrelated species. And some common names are downright misleading. Consider these three "fishes": jellyfish (a cnidarian), crayfish (a crustacean), and silverfish (an insect).

To avoid such confusion, biologists assign each species a two-part scientific name, or **binomial**. The first part is the **genus** (plural, *genera*), a group of closely related species. For example, the genus of tree squirrels is *Sciurus*. The second part of the binomial, often called the specific epithet, is used to distinguish species within a genus. The scientific name for the Eastern gray squirrel is *Sciurus carolinensis*; the fox squirrel is *Sciurus niger*. The first part of the scientific name is analogous to a person's surname in that it is shared by close relatives. The specific epithet is analogous to a person's first name—unrelated people often have the same first name. For example, "*carolinensis*" is the second part of the binomial of diverse species: *Poecile carolinensis* is the Carolina chickadee and *Anolis carolinensis* is a type of lizard. Thus, both parts must be used together to name a species. Notice that the first letter of the genus name is capitalized and that the binomial is italicized and Latinized.

In addition to naming species, Linnaeus also grouped them into a hierarchy of categories. Beyond the grouping of species within genera, the Linnaean system extends to progressively broader categories of classification. It places related genera in the same **family**, puts families into **orders**, orders into **classes**, classes into **phyla** (singular, *phylum*), phyla into **kingdoms**, and kingdoms into **domains**.

Figure 15.15A uses the domestic cat (*Felis catus*) to illustrate this progressively more inclusive classification system. The genus *Felis* includes the domestic cat and several closely related species of small wild cats, represented by small yellow boxes in the figure. The genus *Felis* is placed in the cat family, Felidae, along with other genera of cats, such as the genus *Panthera*, which includes the tiger, leopard, jaguar, and African lion. Family Felidae belongs to the order Carnivora, which also includes the family Canidae (for example, the wolf and coyote) and several other families. Order Carnivora is grouped with many other orders in the class Mammalia, the mammals. Class Mammalia is one of the classes belonging to the phylum Chordata in the kingdom Animalia, which is one of several kingdoms in the domain Eukarya. Each taxonomic unit at any level—family Felidae or class Mammalia, for instance—is called a **taxon** (plural, *taxa*).

Grouping organisms into broader categories seems to come naturally to humans—it is a way to structure our world. Classifying species into higher (more inclusive) taxa, however, is ultimately arbitrary. Higher classification levels are generally defined by morphological characters chosen by taxonomists rather than by quantitative measurements that could apply to the same taxon level across all lineages.

Ever since Darwin, systematics has had a goal beyond simple organization: to have classification reflect evolutionary relationships. Biologists traditionally use **phylogenetic trees** to depict hypotheses about the evolutionary history of species. These branching diagrams reflect the hierarchical classification of groups nested within more inclusive groups. **Figure 15.15B** illustrates the connection between classification and phylogeny. This tree shows the classification of some of the taxa in the order Carnivora and the probable evolutionary relationships among these groups. Note that such a phylogenetic tree does not indicate when a particular species evolved but only the pattern of descent from the last common ancestors of the species shown.

? **How much of the classification in Figure 15.15A do we share with the domestic cat?**

● We are classified the same from the domain to the class level: Both cats and humans are mammals. We do not belong to the same order.

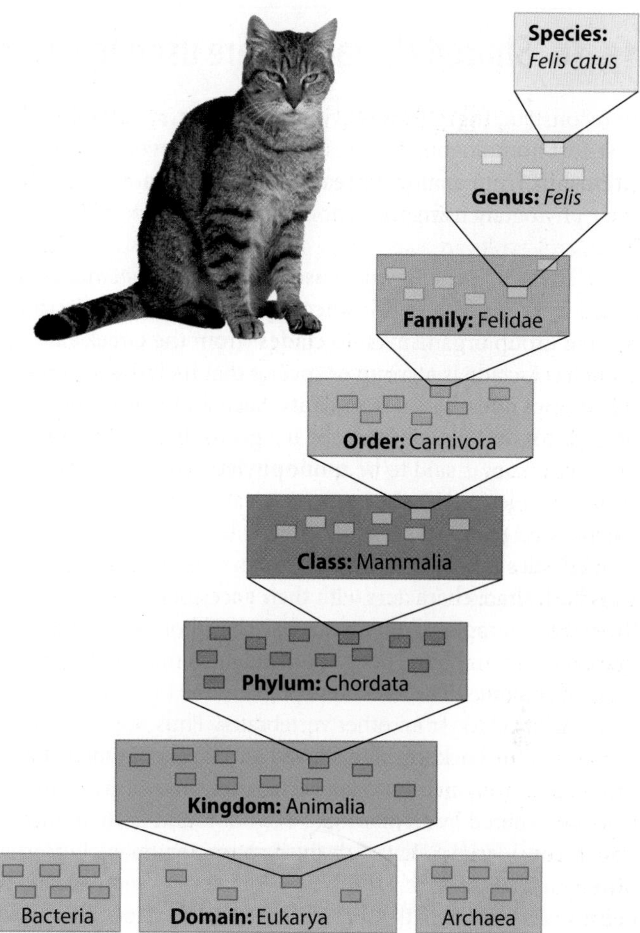

Species: Felis catus
Genus: Felis
Family: Felidae
Order: Carnivora
Class: Mammalia
Phylum: Chordata
Kingdom: Animalia
Bacteria
Domain: Eukarya
Archaea

▲ **Figure 15.15A** Hierarchical classification of the domestic cat

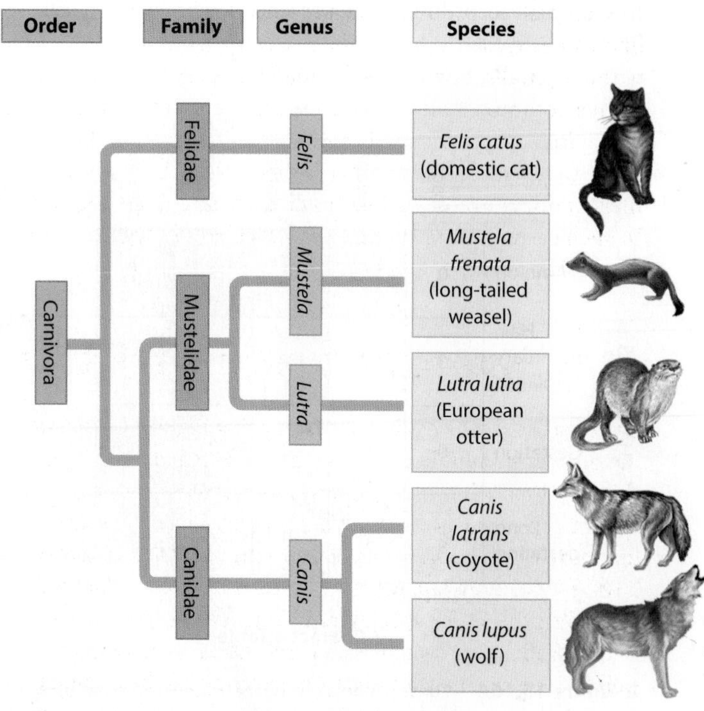

Order
Family
Genus
Species
Carnivora
Felidae
Mustelidae
Canidae
Felis
Mustela
Lutra
Canis
Felis catus (domestic cat)
Mustela frenata (long-tailed weasel)
Lutra lutra (European otter)
Canis latrans (coyote)
Canis lupus (wolf)

▲ **Figure 15.15B** Relating classification to phylogeny

15.16 Shared characters are used to construct phylogenetic trees

In reconstructing a group's evolutionary history, biologists first sort homologous features, which reflect evolutionary relationship, from analogous features, which do not. They then infer phylogeny using these homologous characters.

Cladistics The most widely used method in systematics is called **cladistics**. Common ancestry is the primary criterion used to group organisms into **clades** (from the Greek *clados*, branch). A clade is a group of species that includes an ancestral species and all its descendants. Such an inclusive group of ancestor and descendants, be it a genus, family, or some broader taxon, is said to be **monophyletic** (meaning "single tribe"). Clades reflect the branching pattern of evolution and can be used to construct phylogenetic trees.

Cladistics is based on the Darwinian concept that organisms both share characters with their ancestors and differ from them. For example, all mammals have backbones, but the presence of a backbone does not distinguish mammals from other vertebrates. The backbone predates the branching of the mammalian clade from other vertebrates. Thus, we say that for mammals, the backbone is a **shared ancestral character** that originated in an ancestor of all vertebrates. In contrast, hair, a character shared by all mammals but not found in their ancestors, is considered a **shared derived character**, an evolutionary novelty unique to mammals. Shared derived characters distinguish clades and thus the branch points in the tree of life.

Inferring Phylogenies Using Shared Characters The simplified example in **Figure 15.16A** illustrates that the sequence in which shared derived characters appear can be used to construct a phylogenetic tree. The figure compares five animals according to the presence or absence of a set of characters. An important part of cladistics is a comparison between a so-called ingroup and an outgroup. The **outgroup** (in this example, the frog) is a species from a lineage that is known to have diverged before the lineage that includes the species we are studying, the **ingroup**.

In our example, the frog (representing amphibians, the outgroup) and the other four animals (collectively the ingroup) are all related in that they are tetrapods (vertebrates with limbs). By comparing members of the ingroup with each other and with the outgroup, we can determine which characters are the derived characters—evolutionary innovations—that define the sequence of branch points in the phylogeny of the ingroup.

In the character table in Figure 15.16A, 0 indicates that a particular character is not present in a group; 1 indicates that the character is present. Let's work through this example step by step. All the animals in the ingroup have an amnion, a membrane that encloses the embryo in a fluid-filled sac. The outgroup does not have this character. Now consider the next character—hair and mammary glands. This character is present in all three mammals (the duck-billed platypus, kangaroo, and beaver) but not the iguana or frog. The third character in the table is gestation, the carrying of developing offspring within the uterus of the female parent. Both the outgroup and iguanas do not exhibit gestation. Instead, frogs release their eggs into the water, and iguanas and most other reptiles lay eggs with a shell. One of the mammals, the duck-billed platypus, also lays eggs with a shell; and from this we might infer that the duck-billed platypus represents an early branch point in the mammalian clade. In fact, this hypothesis is strongly supported by structural, fossil, and molecular evidence. The final character is long gestation, in which an offspring completes its embryonic development within the uterus. This is

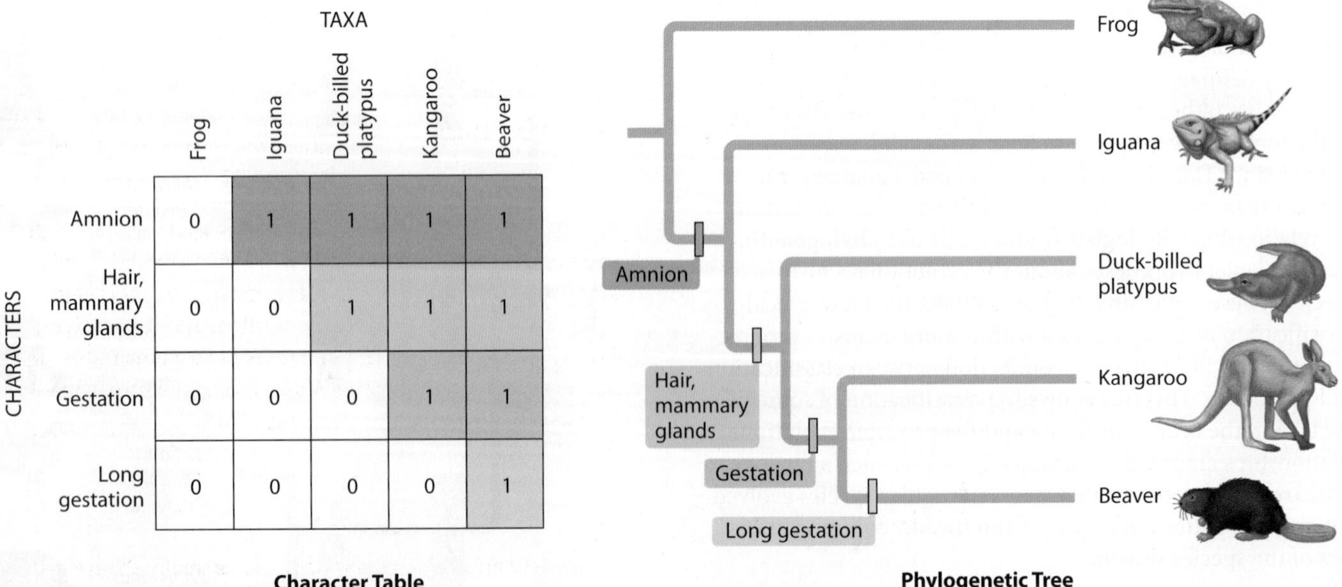

CHARACTERS \ TAXA	Frog	Iguana	Duck-billed platypus	Kangaroo	Beaver
Amnion	0	1	1	1	1
Hair, mammary glands	0	0	1	1	1
Gestation	0	0	0	1	1
Long gestation	0	0	0	0	1

Character Table
Frog
Iguana
Duck-billed platypus
Kangaroo
Beaver
Amnion
Hair, mammary glands
Gestation
Long gestation
Phylogenetic Tree

▲ **Figure 15.16A** Constructing a phylogenetic tree using cladistics

Try This Label the outgroup and the ingroup. Circle the branch point that represents the most recent common ancestor of kangaroos and beavers, and name the derived character that defines this branch point.

the case for a beaver, but a kangaroo has a very short gestation period and completes its embryonic development while nursing in its mother's pouch.

We can now translate the data in our table of characters into a phylogenetic tree. Such a tree is constructed from a series of two-way branch points (see Module 13.5). Each branch point (also called a node) represents the divergence of two groups from a common ancestor and the emergence of a lineage possessing a new set of derived characters. By tracing the distribution of shared derived characters, you can see how we inferred the sequence of branching and the evolutionary relationships of this group of animals.

Parsimony Useful in many areas of science, **parsimony** is the adoption of the simplest explanation for observed phenomena. Systematists use the principle of parsimony to construct phylogenetic trees that require the smallest number of evolutionary changes. For instance, parsimony leads to the hypothesis that a beaver is more closely related to a kangaroo than to a platypus, because in both the beaver and the kangaroo, embryos begin development within the female uterus. It is possible that gestation evolved twice, once in the kangaroo lineage and independently in the beaver lineage, but this explanation is more complicated and therefore less likely. Typical cladistic analyses involve much more complex data sets than that presented in Figure 15.16A (often including comparisons of DNA sequences) and are usually handled by computer programs designed to construct parsimonious trees.

Phylogenetic Trees as Hypotheses Systematists use many kinds of evidence, such as structural and developmental features, molecular data, and behavioral traits, to reconstruct evolutionary histories. However, even the best tree represents only the most likely hypothesis based on available evidence. As new data accumulate, hypotheses may be revised and new trees drawn.

An example of a redrawn tree is shown in **Figure 15.16B**. In traditional vertebrate taxonomy, crocodiles, snakes, lizards,

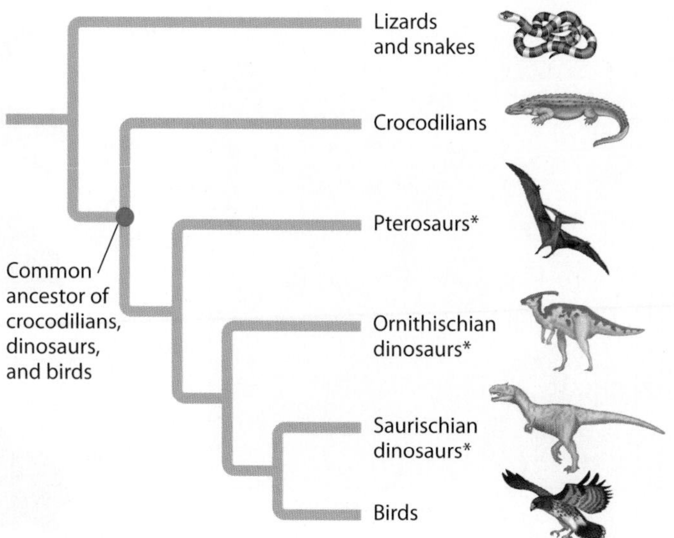
Lizards and snakes
Crocodilians
Pterosaurs*
Ornithischian dinosaurs*
Saurischian dinosaurs*
Birds
Common ancestor of crocodilians, dinosaurs, and birds

▲ **Figure 15.16B** A phylogenetic tree of reptiles (* indicates extinct lineages)

and other reptiles were classified in the class Reptilia, while birds were placed in the separate class Aves. However, such a reptilian clade is not monophyletic—in other words, it does not include an ancestral species and all of its descendants, one group of which includes the birds. Many lines of evidence support the tree shown in Figure 15.16B, showing that birds belong to the clade of reptiles.

Thinking of phylogenetic trees as hypotheses allows us to use them to make and test predictions. For example, if our phylogeny is correct, then features shared by two groups of closely related organisms should be present in their common ancestor. Using this reasoning, consider the novel predictions that can be made about dinosaurs. As seen in the tree in Figure 15.16B, the closest *living* relatives of birds are crocodiles. Birds and crocodiles share numerous features: They have four-chambered hearts, they "sing" to defend territories and attract mates (although a crocodile "song" is more like a bellow), and they build nests. Both birds and crocodiles care for and warm their eggs by brooding. Birds brood by sitting on their eggs, whereas crocodiles cover their eggs with their neck. Reasoning that any feature shared by birds and crocodiles is likely to have been present in their common ancestor (denoted by the red circle in Figure 15.16B) and all of its descendants, biologists hypothesize that dinosaurs had four-chambered hearts, sang, built nests, and exhibited brooding.

Internal organs such as hearts rarely fossilize, and it is, of course, difficult to determine whether dinosaurs sang. However, fossilized dinosaur nests have been found. **Figure 15.16C** shows a fossil of an *Oviraptor* dinosaur thought to have died in a sandstorm while incubating or protecting its eggs. The hypothesis that dinosaurs built nests and exhibited brooding has been further supported by additional fossils that show other species of dinosaurs caring for their eggs.

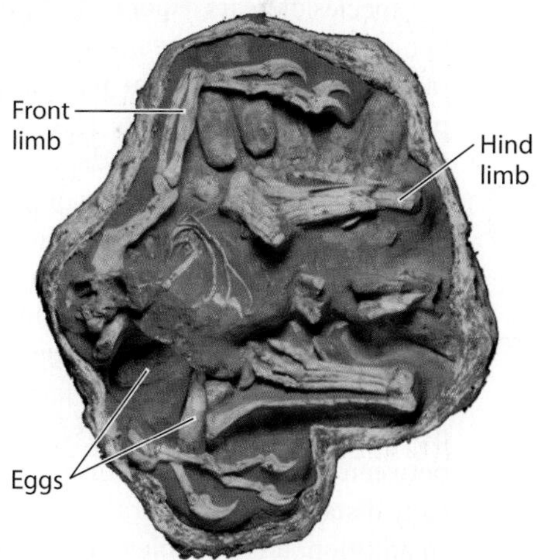
Front limb
Hind limb
Eggs

▲ **Figure 15.16C** Fossil remains of *Oviraptor* and eggs (The orientation of the bones, which surround the eggs, suggests that the dinosaur died while incubating or protecting its eggs.)

The more we know about an organism and its relatives, the more accurately we can portray its phylogeny. In the next module, we consider how molecular biology is providing valuable data for tracing evolutionary history.

? **To distinguish a particular clade of mammals within the larger clade that corresponds to class Mammalia, why is hair not a useful character?**

Hair is a shared ancestral character common to all mammals and thus is not helpful in distinguishing different mammalian subgroups.

15.17 An organism's evolutionary history is documented in its genome

The more recently two species have branched from a common ancestor, the more similar their DNA sequences should be. The longer two species have been on separate evolutionary paths, the more their DNA is expected to have diverged.

Molecular Systematics A method called **molecular systematics**, which uses DNA or other molecules to infer relatedness, is a valuable approach for tracing phylogeny. Scientists have sequenced more than 153 billion bases of DNA from thousands of species. This enormous database has fueled a boom in the study of phylogeny and clarified many evolutionary relationships. Consider the red panda, an endangered, Southeast Asian tree-dwelling mammal that feeds mostly on bamboo. It was initially classified as a close relative of the giant panda, and then as a member of the raccoon family. Recent molecular studies, however, suggest that the red panda represents a separate group, which diverged from the lineage that led to the raccoon and the weasel families. Molecular evidence has also begun to sort out the relationships among the species of bears. Figure 15.17 presents a phylogenetic hypothesis for the lineages that include bears, raccoons, weasels, and the red panda. Notice that the phylogenetic tree in Figure 15.17 includes a timeline, which is based on fossil evidence and molecular data that can estimate when many of these divergences occurred. (Most of the phylogenetic trees we have seen so far indicate only the relative order in which lineages diverged; they do not show the timing of those events.)

Bears, raccoons, and the red panda are closely related mammals that diverged fairly recently. But biologists can also use DNA analyses to assess relationships between groups of organisms that are so phylogenetically distant that structural similarities are absent. In addition, it is possible to reconstruct phylogenies among groups of present-day prokaryotes and other microorganisms for which we have no fossil record at all. Molecular biology has helped to extend systematics to the extremes of evolutionary relationships far above and below the species level, ranging from major branches of the tree of life to its finest twigs.

The observation that different genes evolve at different rates allows scientists to use molecular systematics for constructing phylogenetic trees that encompass both long and short periods of time. The DNA specifying ribosomal RNA (rRNA) changes relatively slowly, so comparisons of DNA sequences in these genes are useful for investigating relationships between taxa that diverged hundreds of millions of years ago. Studies of the genes for rRNA have shown, for example, that fungi are more closely related to animals than to green plants—something that certainly could not have been deduced from morphological comparisons alone. In contrast, the DNA in mitochondria (mtDNA) evolves relatively rapidly and can be used to investigate more recent evolutionary events. For example, researchers have used mtDNA sequences to study the relationships between Native American groups. Their studies support earlier evidence that the Pima of Arizona, the Maya of Mexico, and the Yanomami of Venezuela are closely related, probably descending from the first wave of immigrants to cross the Bering Land Bridge from Asia to the Americas about 15,000 years ago.

Genome Evolution Now that we can compare entire genomes, including our own, some interesting facts have emerged. As you may have heard, the genomes of humans and chimpanzees are strikingly similar. An even more

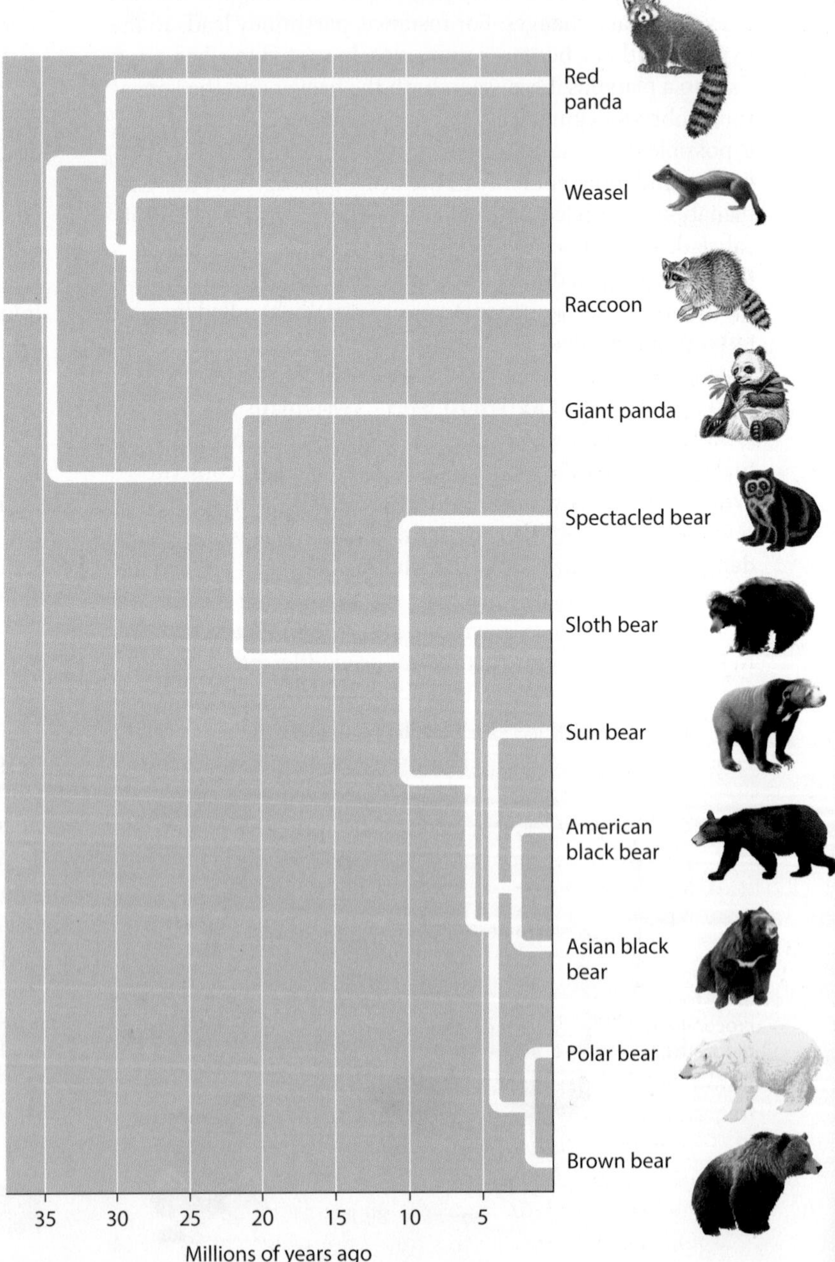

Red panda
Weasel
Raccoon
Giant panda
Spectacled bear
Sloth bear
Sun bear
American black bear
Asian black bear
Polar bear
Brown bear
35
30
25
20
15
10
5
Millions of years ago

▲ **Figure 15.17** A phylogenetic tree based on molecular data remarkable fact is that homologous genes (similar genes that species share because of descent from a common ancestor) are widespread and can extend over huge evolutionary distances. While the genes of humans and mice are certainly not identical, 99% of them are detectably homologous. And 50% of human genes are homologous with those of yeast. This remarkable commonality demonstrates that all living organisms share many biochemical and developmental pathways and provides overwhelming support for Darwin's theory of "descent with modification."

Gene duplication has played a particularly important role in evolution because it increases the number of genes in the genome, providing additional opportunities for further evolutionary changes (see Module 15.11). Molecular techniques now allow scientists to trace the evolutionary history of such duplications—in which lineage they occurred and how the multiple copies of genes have diverged from each other over time.

Another interesting fact evident from genome comparisons is that the number of genes has not increased at the same rate as the complexity of organisms. Humans have only about four times as many genes as yeasts. Yeasts are simple, single-celled eukaryotes; humans have a complex brain and a body that contains more than 200 different types of tissues. Evidence is emerging that many human genes are more versatile than those of yeast, but explaining the mechanisms of such versatility remains an exciting scientific challenge.

? What types of molecules should be compared to help determine whether fungi are more closely related to plants or to animals?

● Because these organisms diverged so long ago, scientists should compare molecules that change or evolve very slowly, such as the DNA that specifies rRNA.

15.18 Molecular clocks help track evolutionary time

The longer two groups have been separated, the greater the divergence of their genes. For example, sharks and tunas have been on separate evolutionary paths for more than 420 million years, whereas dolphins and bats diverged about 60 million years ago. Despite the obvious differences between dolphins and bats, their homologous genes are much more alike than are such genes in sharks and tuna. Indeed, molecular changes have kept better track of time than have changes in morphology. Biologists have found that some genes or other regions of genomes appear to accumulate changes at constant rates. Such observations form the basis for the concept of a **molecular clock**, a method that estimates the time required for a given amount of evolutionary change.

The molecular clock of a gene shown to have a reliable average rate of change can be calibrated in actual time by graphing the number of nucleotide differences against the dates of evolutionary branch points known from the fossil record. The graph line can then be used to estimate the dates of other evolutionary episodes not documented in the fossil record.

Molecular clocks have been used to date a wide variety of events. In one fascinating example published in 2011, researchers studied the divergence of human body lice (**Figure 15.18**) from head lice. Lice are tiny, blood-sucking insects that live in the fur of most mammal species. Early in human evolution, the loss of body hair restricted lice to the head—bare skin deprived the parasites of their refuge. When clothing offered a new habitat, populations diverged into two types, head lice and body lice, each with adaptations specific to its habitat. (Pubic lice have a different evolutionary history and are members of a different genus.) By comparing data from four different DNA sequences in head lice and body lice, the researchers estimated that people began to wear clothing between 83,000 and 170,000 years ago.

Some biologists are skeptical about the accuracy of molecular clocks because the rate of molecular change may vary at different times, in different genes, and in different groups

▲ **Figure 15.18** Human body louse (*Pediculus humanus*)

of organisms. In some cases, problems may be avoided by calibrating molecular clocks with many genes rather than just one or a few genes. One group of researchers used sequence data from 658 genes to construct a molecular clock that covered almost 600 million years of vertebrate evolution. Their estimates of divergence times agreed closely with fossil-based estimates. An abundant fossil record extends back only about 550 million years, and molecular clocks have been used to date evolutionary divergences that occurred a billion or more years ago. But the estimates assume that the clocks have been constant for all that time. Thus, such estimates are highly uncertain.

Evolutionary theory holds that all of life has a common ancestor. Molecular systematics is helping to link all living organisms into a comprehensive tree of life, as we see next.

? What is a molecular clock? What assumption underlies the use of such a clock?

● A molecular clock estimates the actual time of evolutionary events based on the number of DNA changes. It is based on the assumption that some regions of genomes evolve at constant rates.

15.19 Constructing the tree of life is a work in progress

Phylogenetic trees are hypotheses about evolutionary history. Like all hypotheses, they are revised, or in some cases rejected, in accordance with new evidence. As you have learned, molecular systematics and cladistics are remodeling some trees.

Over the years, many schemes have been proposed for classifying all of life. Historically, a two-kingdom system divided all organisms into plants and animals. But it was beset with problems. Where do bacteria fit? Or photosynthetic unicellular organisms that move? And what about the fungi?

By the late 1960s, many biologists recognized five kingdoms: Monera (prokaryotes), Protista (a diverse kingdom consisting mostly of unicellular eukaryotes), Plantae, Fungi, and Animalia. However, molecular studies highlighted fundamental flaws in the five-kingdom system. Biologists have since adopted a **three-domain system**, which recognizes three basic groups: two domains of prokaryotes, Bacteria and Archaea, and one domain of eukaryotes, called Eukarya. Kingdoms Fungi, Plantae, and Animalia are still recognized, but kingdoms Monera and Protista are obsolete because they are not monophyletic.

Molecular and cellular evidence indicates that the two lineages of prokaryotes (bacteria and archaea) diverged very early in the evolutionary history of life. Molecular evidence also suggests that archaea are more closely related to eukaryotes than to bacteria. **Figure 15.19A** is an evolutionary tree based largely on rRNA genes. As you just learned, rRNA genes have evolved so slowly that homologies between distantly related organisms can still be detected. This tree shows that ❶ the first major split in the history of life was the divergence of the bacteria from the other two domains, followed by the divergence of domains Archaea and Eukarya.

Comparisons of complete genomes from the three domains, however, show that, especially during the early history of life, there have been substantial interchanges of genes between organisms in the different domains. These took place through **horizontal gene transfer**, a process in which genes are transferred from one genome to another through mechanisms such as plasmid exchange and viral infection (see Modules 10.22 and 10.23) and even through the fusion of different organisms. Figure 15.19A shows two major episodes of horizontal gene transfer: ❷ gene transfer between a mitochondrial ancestor and the ancestor of eukaryotes and ❸ gene transfer between a chloroplast ancestor and the ancestor of green plants. (Module 4.15 describes the endosymbiont theory for the origin of mitochondria and chloroplasts.)

Some scientists have argued that horizontal gene transfers were so common that the early history of life should be represented as a tangled network of connected branches. Others have suggested that the early history of life is best represented by a ring, not a tree (**Figure 15.19B**). Based on an analysis of hundreds of genes, some researchers have hypothesized that the eukaryote lineage (gold in the figure) arose when an early archaean (blue-green) fused with an early bacterium (purple). In this model, eukaryotes are as closely related to bacteria as they are to archaea—an evolutionary relationship that can best be shown in a ring of life. As new data and new methods for analyzing that data emerge, constructing a comprehensive tree of life will continue to challenge and intrigue scientists.

In the next unit, we examine the enormous diversity of organisms that have populated Earth since life first arose more than 3.5 billion years ago.

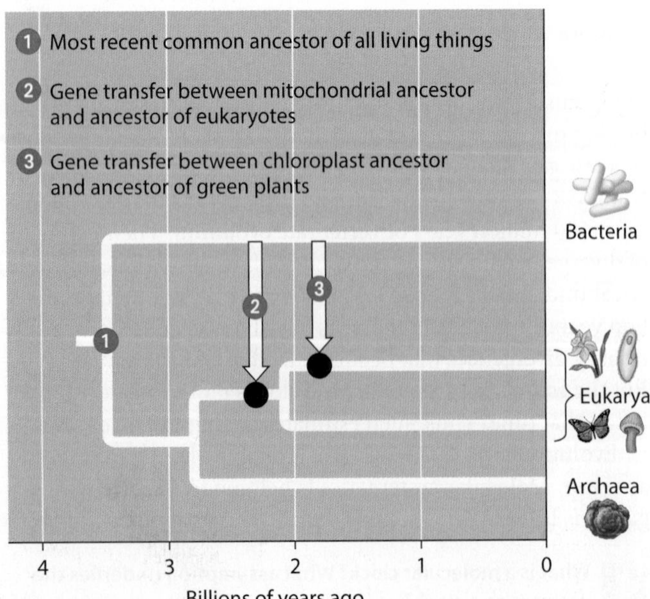
1 Most recent common ancestor of all living things
2 Gene transfer between mitochondrial ancestor and ancestor of eukaryotes
3 Gene transfer between chloroplast ancestor and ancestor of green plants
Bacteria
Eukarya
Archaea
4
3
2
1
0
Billions of years ago

▲ **Figure 15.19A** Two major episodes of horizontal gene transfer in the history of life (dates are uncertain)

? **Why might the evolutionary history of the earliest organisms be best represented by a ring of life rather than a tree?**

There appear to have been multiple horizontal gene transfers among the earliest organisms before the three domains of life eventually emerged from the ring and gave rise to Earth's diversity of life.

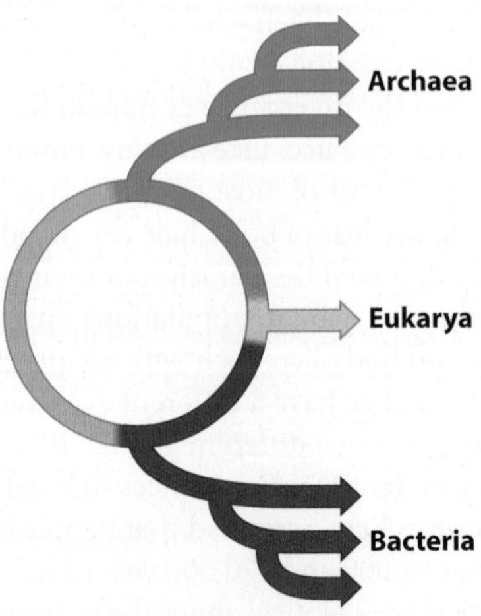
Archaea
Eukarya
Bacteria

▲ **Figure 15.19B** Is the tree of life really a ring of life? In this model, eukaryotes arose when an early archaean fused with an early bacterium.

CHAPTER 15 REVIEW

For practice quizzes, BioFlix animations, MP3 tutorials, video tutors, and more study tools designed for this textbook, go to MasteringBiology®

Reviewing the Concepts

Early Earth and the Origin of Life (15.1–15.3)

15.1 Conditions on early Earth made the origin of life possible. Earth formed some 4.6 billion years ago. Fossil stromatolites formed by prokaryotes date back 3.5 billion years.

15.2 Experiments show that the abiotic synthesis of organic molecules is possible.

15.3 Stages in the origin of the first cells probably included the formation of polymers, protocells, and self-replicating RNA. Natural selection could have acted on protocells that contained self-replicating molecules.

Major Events in the History of Life (15.4–15.6)

15.4 The origins of single-celled and multicellular organisms and the colonization of land were key events in life's history.

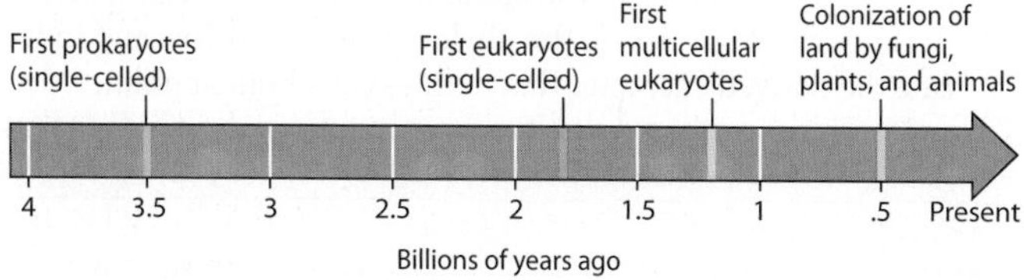

First prokaryotes (single-celled)
First eukaryotes (single-celled)
First multicellular eukaryotes
Colonization of land by fungi, plants, and animals
4
3.5
3
2.5
2
1.5
1
.5
Present
Billions of years ago

15.5 The actual ages of rocks and fossils mark geologic time. Radiometric dating can date rocks and fossils.

15.6 The fossil record documents the history of life. In the geologic record, eras and periods are separated by major transitions in life-forms, often caused by extinctions.

Mechanisms of Macroevolution (15.7–15.13)

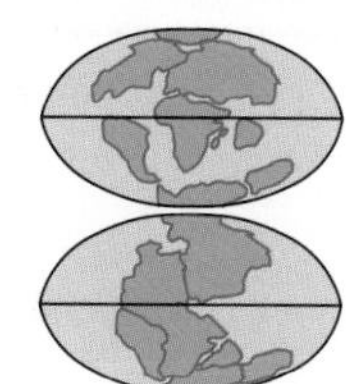

15.7 Continental drift has played a major role in macroevolution. The formation and split-up of Pangaea affected the distribution and diversification of organisms.

15.8 Plate tectonics may imperil human life. Volcanoes and earthquakes often occur at the boundaries of Earth's plates.

15.9 During mass extinctions, large numbers of species are lost. The Permian extinction is linked to the effects of extreme volcanic activity, and the Cretaceaous extinction, which included most dinosaurs, may have been caused by the impact of an asteroid.

15.10 Adaptive radiations have increased the diversity of life. The origin of many new species often follows mass extinctions, colonization of new habitats, and the evolution of new adaptations.

15.11 Genes that control development play a major role in evolution. "Evo-devo" combines evolutionary and developmental biology. New forms can evolve by changes in the number, sequences, or regulation of developmental genes.

15.12 Novel traits may arise in several ways. Complex structures may evolve in stages from simpler versions with the same basic function or from the gradual adaptation of existing structures to new functions.

15.13 Evolutionary trends do not mean that evolution is goal directed. An evolutionary trend may be a result of species selection or natural selection in changing environments.

Phylogeny and the Tree of Life (15.14–15.19)

15.14 Phylogenies based on homologies reflect evolutionary history. Homologous structures and molecular sequences provide evidence of common ancestry.

15.15 Systematics connects classification with evolutionary history. Taxonomists assign each species a binomial—a genus and species name. Genera are grouped into progressively broader categories. A phylogenetic tree is a hypothesis of evolutionary relationships.

15.16 Shared characters are used to construct phylogenetic trees. Cladistics uses shared derived characters to define clades. A parsimonious tree requires the fewest evolutionary changes.

15.17 An organism's evolutionary history is documented in its genome. Molecular systematics uses molecular comparisons to build phylogenetic trees. Homologous genes are found across distantly related species.

15.18 Molecular clocks help track evolutionary time. Regions of DNA that change at a constant rate can provide estimated dates of past events.

15.19 Constructing the tree of life is a work in progress. Evidence of multiple horizontal gene transfers suggests that the early history of life may be best represented by a ring, from which domains Bacteria, Archaea, and Eukarya emerge.

Connecting the Concepts

1. Using the figure below, describe the stages that may have led to the origin of life.

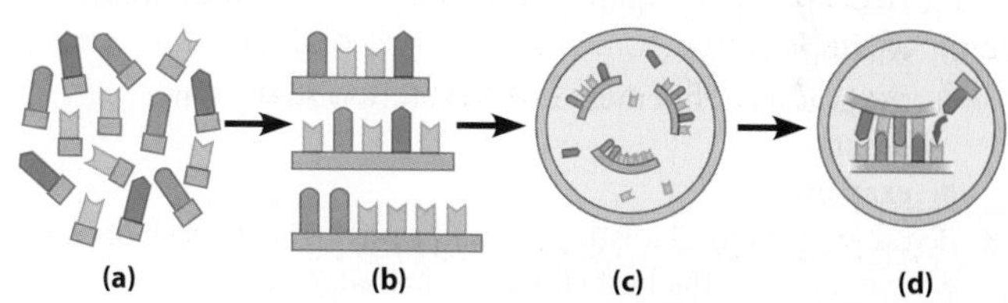

(a)
(b)
(c)
(d)

2. Fill in this concept map about systematics.

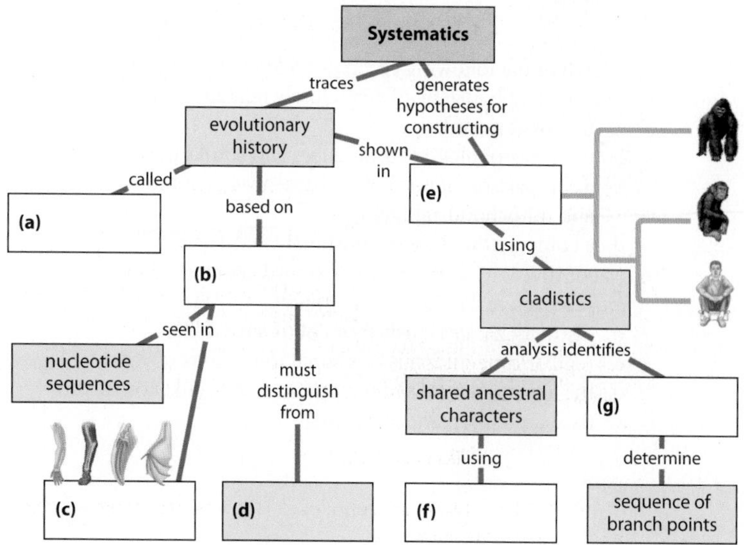

Systematics
traces
generates hypotheses for constructing
evolutionary history
shown in
called
(a)
based on
(e)
using
(b)
cladistics
seen in
nucleotide sequences
must distinguish from
analysis identifies
shared ancestral characters
(g)
using
determine
(c)
(d)
(f)
sequence of branch points

Testing Your Knowledge

Level 1: Knowledge/Comprehension

3. You set your time machine for 3 billion years ago and push the start button. When the dust clears, you look out the window. Which of the following describes what you would probably see?
   a. a cloud of gas and dust in space
   b. green scum in the water
   c. land and water sterile and devoid of life
   d. an endless expanse of red-hot molten rock
4. Ancient photosynthetic prokaryotes were very important in the history of life because they
   a. produced the oxygen in the atmosphere.
   b. are the oldest known archaea.
   c. were the first multicellular organisms.
   d. showed that life could evolve around deep-sea vents.
5. The animals and plants of India are very different from the species in nearby Southeast Asia. Why might this be true?
   a. India was once covered by oceans and Asia was not.
   b. India is in the process of separating from the rest of Asia.
   c. Life in India was wiped out by ancient volcanic eruptions.
   d. India was a separate continent until about 45 million years ago.
6. Adaptive radiations may be promoted by all of the following *except* one. Which one?
   a. mass extinctions that result in vacant ecological niches
   b. colonization of an isolated region with few competitors
   c. a gradual change in climate
   d. a novel adaptation
7. A swim bladder is a gas-filled sac that helps fish maintain buoyancy. Evidence indicates that early fish gulped air into primitive lungs, helping them survive in stagnant waters. The evolution of the swim bladder from lungs of an ancestral fish is an example of
   a. an evolutionary trend.
   b. paedomorphosis.
   c. the gradual refinement of a structure with the same function.
   d. exaptation.
8. If you were using cladistics to build a phylogenetic tree of cats, which would be the best choice for an outgroup?
   a. kangaroo
   b. leopard
   c. domestic cat
   d. iguana
9. Which of the following could provide the best data for determining the phylogeny of very closely related species?
   a. the fossil record
   b. their morphological differences and similarities
   c. a comparison of nucleotide sequences in homologous genes and mitochondrial DNA
   d. a comparison of their ribosomal DNA sequences
10. Major divisions in the geologic record are marked by
    a. radioactive dating.
    b. distinct changes in the types of fossilized life.
    c. regular time intervals measured in millions of years.
    d. the appearance, in order, of prokaryotes, eukaryotes, protists, animals, plants, and fungi.

Level 2: Application/Analysis

11. Distinguish between microevolution and macroevolution.
12. Which are more likely to be closely related: two species with similar appearance but divergent gene sequences or two species with different appearances but nearly identical genes? Explain.
13. How can the Darwinian concept of descent with modification explain the evolution of such complex structures as an eye?
14. Explain why changes in the regulation of developmental genes may have played such a large role in the evolution of new forms.
15. What types of molecular comparisons are used to determine the very early branching of the tree of life? Explain.

Level 3: Synthesis/Evaluation

16. Measurements indicate that a fossilized skull you unearthed has a carbon-14/carbon-12 ratio about one-sixteenth that of the skulls of present-day animals. What is the approximate age of the fossil? (The half-life of carbon-14 is 5,730 years.)
17. A paleontologist compares fossils from three dinosaurs and *Archaeopteryx*, the earliest known bird. The following table shows the distribution of characters for each species, where 1 means that the character is present and 0 means it is not. The outgroup (not shown in the table) had none of the characters. Arrange these species on the phylogenetic tree below and indicate the derived character that defines each branch point.

Trait	*Velociraptor*	*Coelophysis*	*Archaeopteryx*	*Allosaurus*
Hollow bones	1	1	1	1
Three-fingered hand	1	0	1	1
Half-moon-shaped wrist bone	1	0	1	0
Reversed first toe	0	0	1	0

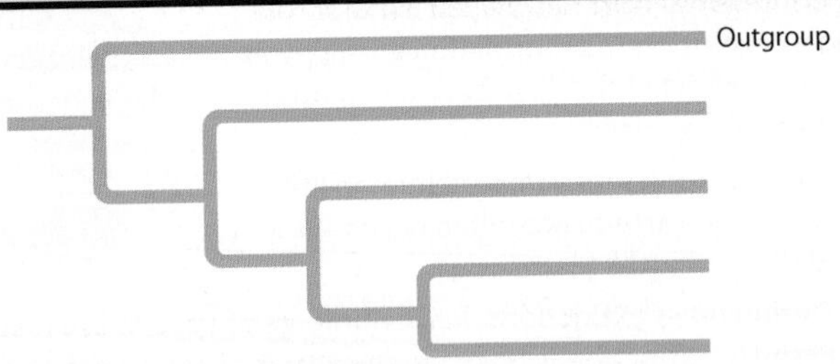
Outgroup

18. **SCIENTIFIC THINKING** When Stanley Miller's experiment was published in 1953, his results made global headlines. The general public thought Miller had answered the question of how life on Earth began by creating life in a test tube. However, scientists understood that Miller's experiment was neither a final answer nor a recipe for life. Rather, it was the first test of a long-standing hypothesis about the origin of life. Using the information in Module 15.2 (and additional research, if you wish) as an example, write an essay describing how the process of science progresses over time toward understanding how nature works. (You will find Module 1.8 helpful.)

Answers to all questions can be found in Appendix 4.

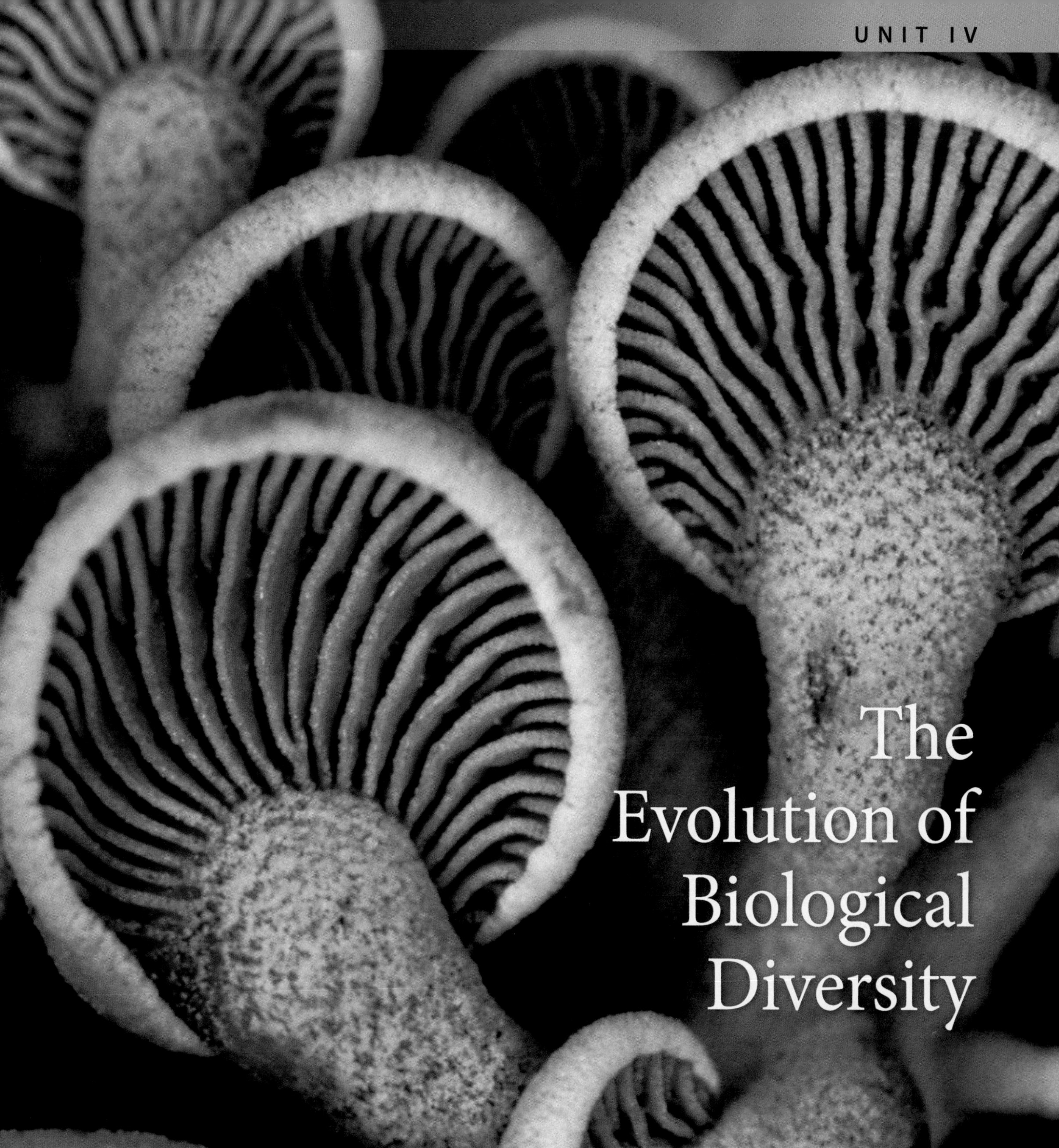

The Evolution of Biological Diversity

CHAPTER 16

Microbial Life: *Prokaryotes and Protists*

Are antibiotics making us fat?

You know that your body contains trillions of individual cells, but did you know that they aren't all "you"? In fact, microorganisms residing in and on your body outnumber your own cells 10 to 1—100 trillion bacteria (including *Helicobacter pylori,* shown below), archaea, and protists call your body home. Your skin, mouth, nasal passages, and digestive and urogenital tracts are prime real estate for these microorganisms. Although each individual is so tiny that it would have to be magnified hundreds of times for you to see it, the weight of your microbial residents totals two to five pounds.

We acquire our microbial communities during the first two years of life, and they remain fairly stable thereafter. However, modern life is taking a toll on that stability. We alter the makeup of these communities by taking antibiotics, purifying our water, sterilizing our food, attempting to germ-proof our surroundings, and scrubbing our skin and teeth. Scientists hypothesize that disrupting our microbial communities may increase our susceptibility to infectious diseases, predispose us to

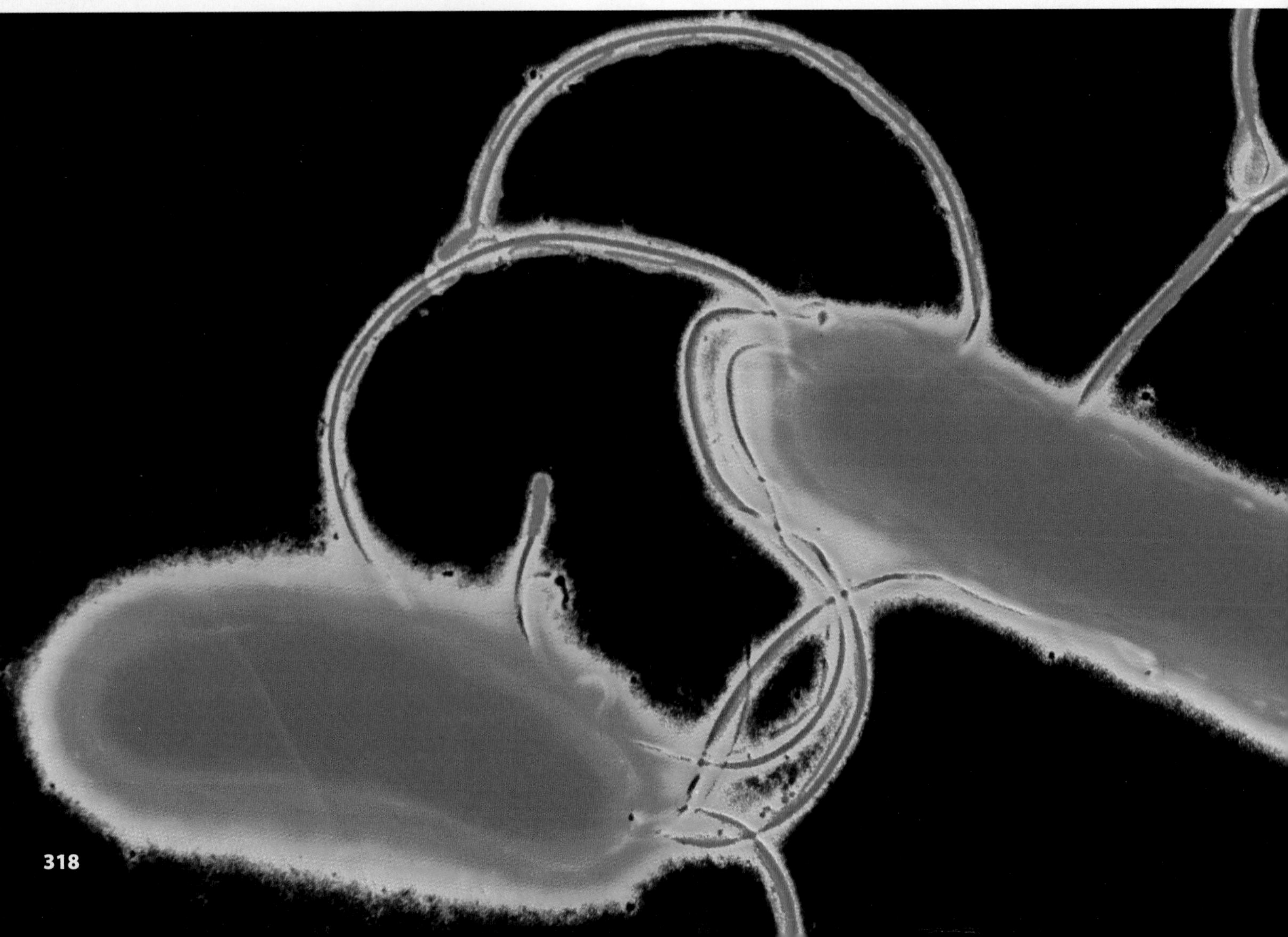

certain cancers, and contribute to conditions such as asthma and other allergies, irritable bowel syndrome, Crohn's disease, and autism. One of the most intriguing hypotheses, as you'll learn later in this chapter, is that obesity results from changes in the species composition of the stomach.

In this chapter, you will learn some of the benefits and drawbacks of human-microbe interactions. You will also sample a bit of the remarkable diversity of prokaryotes and protists.

Our exploration of the magnificent diversity of life begins with this chapter. And so it is fitting that we begin with the prokaryotes, Earth's first life-form, and the protists, the bridge between unicellular eukaryotes and multicellular plants, fungi, and animals.

BIG IDEAS

Prokaryotes
(16.1–16.11)

Prokaryotes, the smallest organisms known, are extraordinarily diverse.

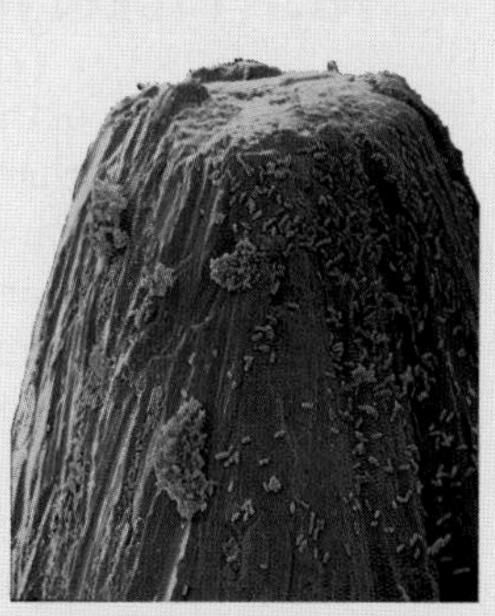

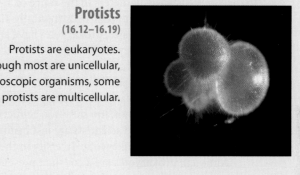

Protists
(16.12–16.19)

Protists are eukaryotes. Though most are unicellular, microscopic organisms, some protists are multicellular.

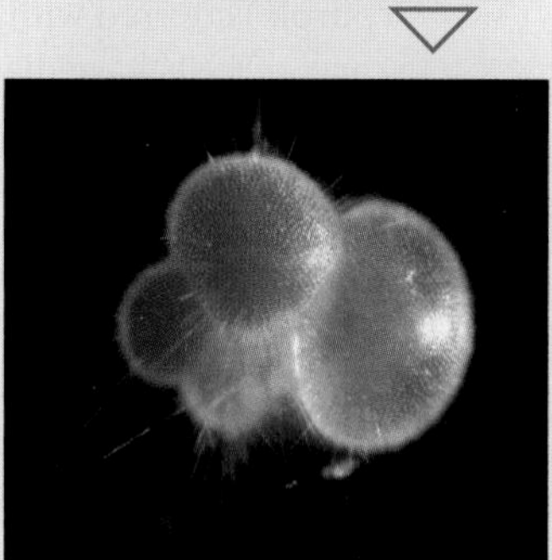

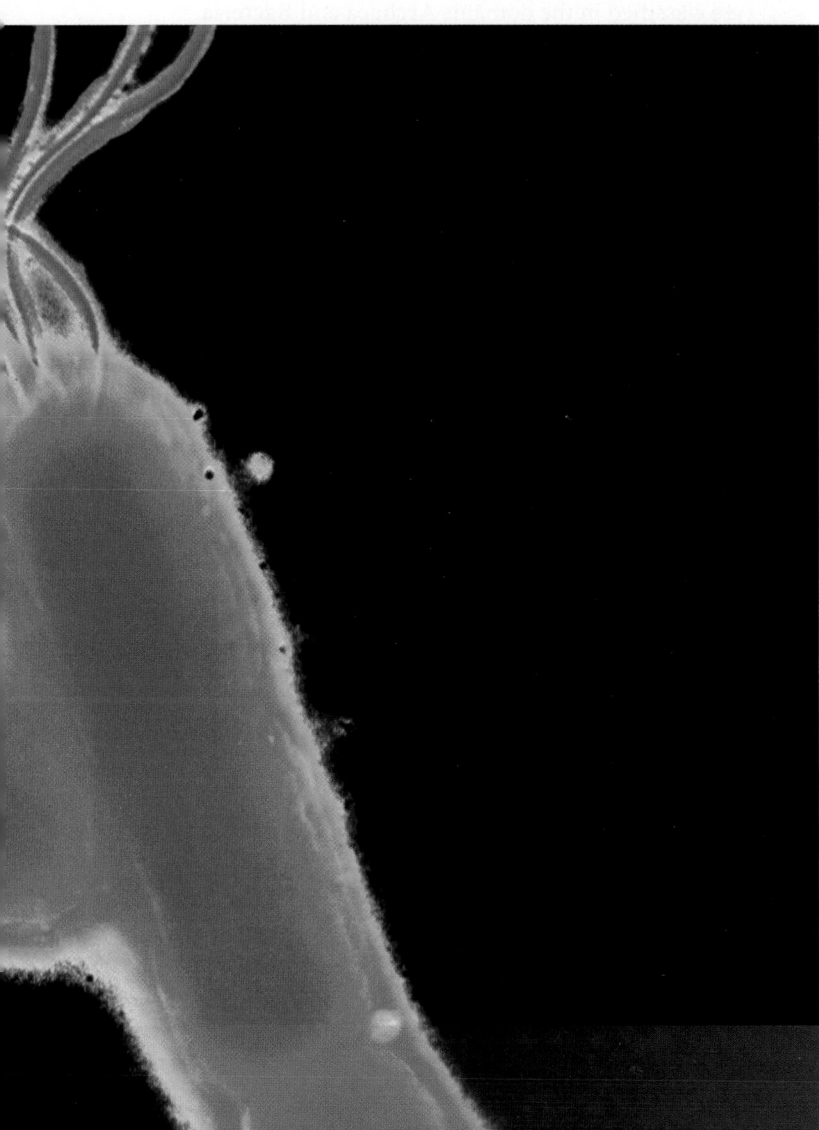

▷ Prokaryotes

16.1 Prokaryotes are diverse and widespread

In the first half of this chapter, you will learn about prokaryotes, organisms that have a cellular organization fundamentally different from that of eukaryotes (see Modules 4.3 and 4.4). Whereas eukaryotic cells have a membrane-enclosed nucleus and numerous other membrane-enclosed organelles, prokaryotic cells lack these structural features. Prokaryotes are also typically much smaller than eukaryotes. You can get an idea of the size of most prokaryotes from **Figure 16.1**, a colorized scanning electron micrograph of the point of a pin (purple) covered with numerous bacteria (orange). Most prokaryotic cells have diameters in the range of 1–5 μm, much smaller than most eukaryotic cells (typically 10–100 μm).

Despite their small size, prokaryotes have an immense impact on our world. They are found wherever there is life, including in and on the bodies of multicellular organisms. The collective biological mass (biomass) of prokaryotes is at least 10 times that of all eukaryotes! Prokaryotes also thrive in habitats too cold, too hot, too salty, too acidic, or too alkaline for any eukaryote. And scientists are just beginning to investigate the extensive prokaryotic diversity in the oceans.

Although prokaryotes are a constant presence in our environment, we hear most about the relatively few species that cause illnesses. We focus on bacterial **pathogens**, disease-causing agents, in Module 16.10. But harmless or beneficial prokaryotes are far more common than harmful prokaryotes. The chapter introduction introduced our **microbiota**, the community of microorganisms that live in and on our bodies. Each of us harbors several hundred different species and genetic strains of prokaryotes, including a few whose positive effects are well studied. For example, some of our intestinal inhabitants supply essential vitamins and enable us to extract nutrition from food molecules that we can't otherwise digest.

Colorized SEM 525×

▶ **Figure 16.1** Bacteria on the point of a pin

Many of the bacteria that live on our skin perform helpful housekeeping functions such as decomposing dead skin cells. Prokaryotes also guard the body against pathogenic intruders.

Prokaryotes are also essential to the health of the environment. They help to decompose dead organisms and other organic waste material, returning vital chemical elements to the environment. They are indispensable components of the chemical cycle that makes nitrogen available to plants and other organisms. If prokaryotes were to disappear, the chemical cycles that sustain life would halt, and all forms of eukaryotic life would also be doomed. In contrast, prokaryotic life would undoubtedly persist in the absence of eukaryotes, as it once did for billions of years.

There are two very different kinds of prokaryotes, which are classified in the domains **Archaea** and **Bacteria** (Module 15.19). In the next several modules, we describe the features that have made prokaryotes so successful, followed by a look at the diversity of each domain.

? **The number of bacterial cells that live in and on our body is greater than the number of eukaryotic cells that make up the body. Why aren't we aware of these trillions of cells?**

● We can't sense our own eukaryotic cells individually, and bacterial cells are much smaller than that. Also, our microbiota are adapted for coexisting with us.

16.2 External features contribute to the success of prokaryotes

Some of the diversity of prokaryotes is evident in their external features, including shape, cell walls, and projections such as flagella. These features are useful for identifying prokaryotes as well as helping the organisms survive in their environments.

Cell Shape Determining cell shape by microscopic examination is an important step in identifying prokaryotes. The micrographs in **Figure 16.2A** show three of the most common prokaryotic cell shapes. Spherical prokaryotic cells are called **cocci** (singular, *coccus*). Cocci that occur in chains, like the ones in the left photo, are called streptococci (from the Greek *streptos*, twisted). The bacterium that causes strep throat in humans is a streptococcus. Other cocci occur in clusters; they are called staphylococci (from the Greek *staphyle*, cluster of grapes).

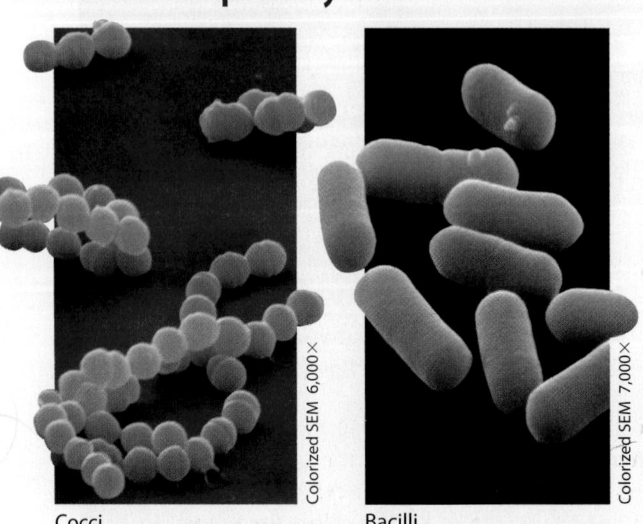

Colorized SEM 6,000×
Cocci
Colorized SEM 7,000×
Bacilli
Colorized SEM 5,000×
Spirochete

▲ **Figure 16.2A** Three common shapes of prokaryotes

Rod-shaped prokaryotes are called **bacilli** *(singular, bacillus)*. Most bacilli occur singly, like the *Escherichia coli* cells in the middle photo in Figure 16.2A. However, the cells of some species occur in pairs or chains of rods. Bacilli may also be threadlike, or filamentous.

A third prokaryotic cell shape is spiral, like a corkscrew. Spiral prokaryotes that are relatively short and rigid are called *spirilla*; those with longer, more flexible cells, like the one shown on the right in Figure 16.2A, which causes Lyme disease, are called *spirochetes*. The bacterium that causes syphilis is also a spirochete. Spirochetes include some giants by prokaryotic standards—cells 0.5 mm long (though very thin).

Cell Wall Nearly all prokaryotes have a cell wall, a feature that enables them to live in a wide range of environments. The cell wall provides physical protection and prevents the cell from bursting in a hypotonic environment (see Module 5.5). The cell walls of bacteria fall into two general types, which scientists can identify with a technique called the **Gram stain** (**Figure 16.2B**). Gram-positive bacteria have simpler walls with a relatively thick layer of a unique material called **peptidoglycan**, a polymer of sugars cross-linked by short polypeptides. The walls of gram-negative bacteria stain differently. They have less peptidoglycan and are more complex, with an outer membrane that contains lipids bonded to carbohydrates. The cell walls of archaea do not contain peptidoglycan, but can also be gram-positive or gram-negative.

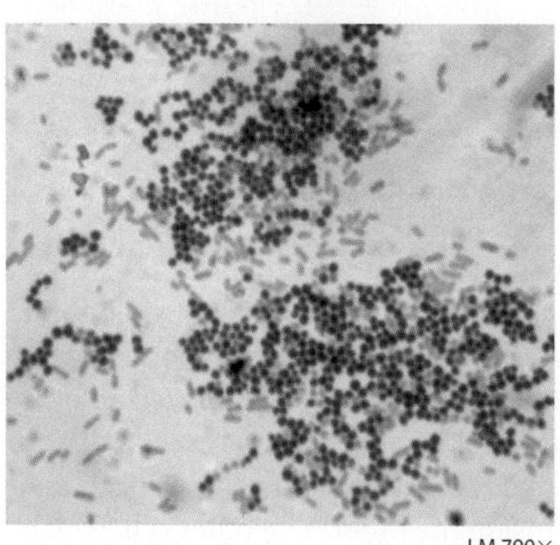
LM 790×

▲ Figure 16.2B Gram-positive (purple) and gram-negative (pink) bacteria

In medicine, Gram stains are often used to detect the presence of bacteria and indicate the type of antibiotic to prescribe. Among disease-causing bacteria, gram-negative species are generally more threatening than gram-positive species because lipid molecules of the outer membrane of gram-negative bacteria are often toxic. The membrane also protects the gram-negative bacteria against the body's defenses and hinders the entry of antibiotic drugs into the bacterium.

The cell wall of many prokaryotes is covered by a capsule, a sticky layer of polysaccharide or protein. The capsule enables prokaryotes to adhere to a surface or to other individuals in a colony. Capsules can also shield pathogenic prokaryotes from attacks by their host's immune system. The capsule surrounding the *Streptococcus* bacterium shown in **Figure 16.2C** enables it to attach to cells that line the human respiratory tract—in this image, a tonsil cell.

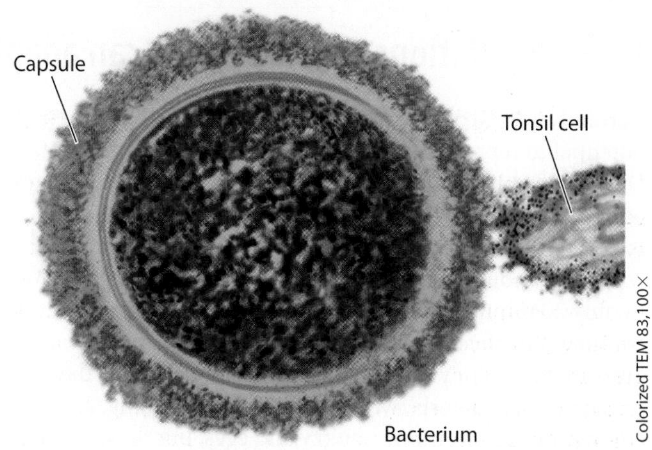
Capsule
Tonsil cell
Bacterium
Colorized TEM 83,100×

▲ Figure 16.2C A capsule attaching a bacterium to a host cell

Projections Some prokaryotes have external structures that extend beyond the cell wall. Many bacteria and archaea are equipped with flagella, adaptations that enable them to move about in response to chemical or physical signals in their environment. For example, prokaryotes can move toward nutrients or other members of their species or away from a toxic substance. Flagella may be scattered over the entire cell surface or concentrated at one or both ends of the cell. Unlike the flagellum of eukaryotic cells (described in Module 4.18), the prokaryotic flagellum is a naked protein structure that lacks microtubules. The flagellated bacterium in **Figure 16.2D** is *E. coli*, as seen in a TEM.

Figure 16.2D also illustrates the hairlike projections called **fimbriae** that enable some prokaryotes to stick to a surface or to one another. Fimbriae allow many pathogenic bacteria to latch onto the host cells they colonize. For example, *Neisseria gonorrhoeae*, which causes the sexually transmitted infection gonorrhea, uses fimbriae to attach to cells in the reproductive tract. During sexual intercourse, *N. gonorrhoeae* bacteria may also attach to sperm cells and travel to a woman's oviducts; an infection in these narrow tubes can impair fertility.

? **How could a microscope help you distinguish the cocci that cause "staph" infections from those that cause "strep" throat?**

● It would show clusters of cells for staphylococcus and chains of cells for streptococcus.

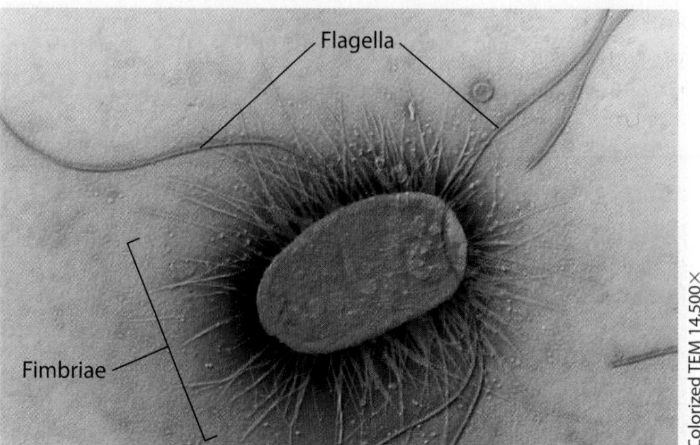
Flagella
Fimbriae
Colorized TEM 14,500×

▲ Figure 16.2D Flagella and fimbriae

16.3 Populations of prokaryotes can adapt rapidly to changes in the environment

Certainly a large part of the success of prokaryotes is their potential to reproduce quickly in a favorable environment. Dividing by binary fission (see Module 8.2), a single prokaryotic cell becomes 2 cells, which then become 4, 8, 16, and so on. While many prokaryotes produce a new generation within 1–3 hours, some species can produce a new generation in only 20 minutes under optimal conditions. If reproduction continued unchecked at this rate, a single prokaryote could give rise to a colony outweighing Earth in only three days!

Salmonella bacteria, which cause food poisoning, are commonly found on raw poultry and eggs, but the bacterial population is often too small to cause symptoms. Refrigeration slows (but does not stop) bacterial reproduction. However, when raw poultry is left in the warm environment of the kitchen, bacteria multiply rapidly and can quickly reach a risky population size. Similarly, bacteria that remain on the counter, cutting board, or kitchen implements may continue to reproduce. So be sure to cook poultry thoroughly (an internal temperature of 165°F is considered safe), and clean anything that has come into contact with raw poultry with soap and hot water or an antimicrobial cleaner.

Each time DNA is replicated prior to binary fission, a few spontaneous mutations occur. As a result, rapid reproduction generates a great deal of genetic variation in a prokaryote population. If the environment changes, an individual that possesses a beneficial allele can quickly take advantage of the new conditions. For example, exposure to antibiotics may select for antibiotic resistance in a bacterial population (see Module 13.16).

The amount of DNA in a prokaryotic cell is on average only about one-thousandth as much as that in a eukaryotic cell. The genome of a typical prokaryote is one long, circular chromosome (**Figure 16.3A**). (In an intact cell, it is packed into a distinct region; see Figure 4.3.) Many prokaryotes also have additional small, circular DNA molecules called plasmids, which replicate independently of the chromosome (see Module 10.23). Some plasmids carry genes that enhance survival under certain conditions. For example, plasmids may provide resistance to antibiotics, direct the metabolism of rarely encountered nutrients, or have other "contingency" functions. The ability of many prokaryotes to transfer plasmids within and even between species provides another rapid means of adaptation to changes in the environment.

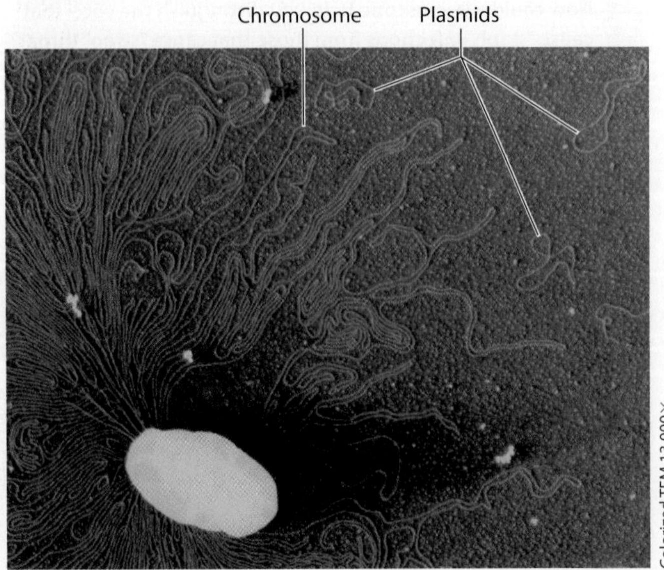

Chromosome
Plasmids
Colorized TEM 13,000×

▲ **Figure 16.3A** DNA released from a ruptured bacterial cell

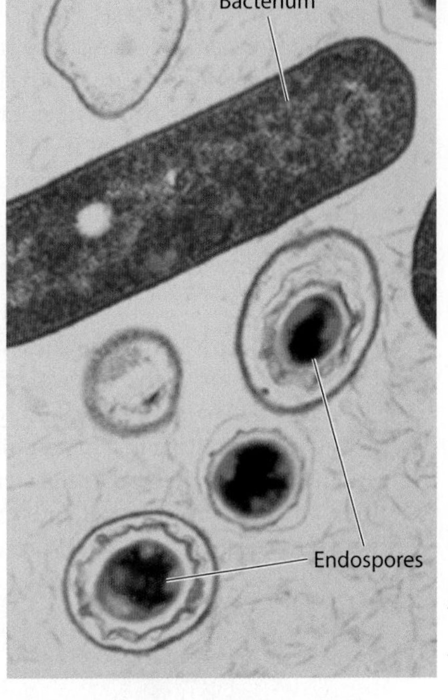

Bacterium
Endospores
Colorized TEM 21,500×

▲ **Figure 16.3B** Endospores of anthrax bacteria

If environmental conditions become too harsh to sustain active metabolism—for example, when food or moisture is depleted—some prokaryotes form specialized resistant cells. **Figure 16.3B** shows an example of such an organism, *Bacillus anthracis*, the bacterium that causes a disease called anthrax in cattle, sheep, and humans. There are actually two cells here, one inside the other. The outer cell, which will later disintegrate, produced the specialized inner cell, called an **endospore**. The endospore, which has a thick, protective coat, dehydrates and becomes dormant. It can survive all sorts of trauma, including extreme heat or cold. When the endospore receives environmental cues that conditions have improved, it absorbs water and resumes growth.

Some endospores can remain dormant for centuries. Not even boiling water kills most of these resistant cells, making it difficult to get rid of spores in a contaminated area. An island off the coast of Scotland that was used for anthrax testing in 1942 was finally declared safe 48 years later, after tons of formaldehyde were applied and huge amounts of topsoil were removed. The food-canning industry kills endospores of dangerous bacteria such as *Clostridium botulinum*, the source of the potentially fatal disease botulism, by heating the food to a temperature of 110–150°C (230–300°F) with high-pressure steam.

Another feature that contributes to the success of prokaryotes is the diversity of ways in which they obtain their nourishment, which we consider in the next module.

? **Why does rapid reproduction produce high genetic variation in populations of prokaryotes?**

● Each time DNA replicates, spontaneous mutations may occur.

16.4 Prokaryotes have unparalleled nutritional diversity

One way to organize the vast diversity of prokaryotes is by their mode of nutrition—how they obtain energy for cellular work and carbon to build organic molecules. Prokaryotes exhibit much more nutritional diversity than eukaryotes. This allows them to inhabit almost every nook and cranny on Earth.

Source of Energy As shown in **Figure 16.4**, two sources of energy can be used by prokaryotes. Like plants, prokaryotic *phototrophs* capture energy from sunlight. Prokaryotic cells do not have chloroplasts, but some prokaryotes have thylakoid membranes where photosynthesis takes place.

Prokaryotes called *chemotrophs* harness the energy stored in chemicals, either organic molecules or inorganic chemicals such as hydrogen sulfide (H_2S), elemental sulfur (S), iron (Fe)-containing compounds, or ammonia (NH_3).

Source of Carbon Organisms that make their own organic compounds from inorganic sources are autotrophic (see Module 7.1). Autotrophs, including plants and some prokaryotes and protists, obtain their carbon atoms from carbon dioxide (CO_2). Most prokaryotes, as well as animals, fungi, and some protists, are heterotrophs, meaning they obtain their carbon atoms from the organic compounds of other organisms.

Mode of Nutrition The terms used to describe how an organism obtains energy and carbon are combined to describe its mode of nutrition (see Figure 16.4).

Photoautotrophs harness sunlight for energy and use CO_2 for carbon. Cyanobacteria, such as the *Oscillatoria* shown in Figure 16.4, are photoautotrophs. As in plants, photosynthesis in cyanobacteria uses chlorophyll *a* and produces O_2 as a by-product.

Photoheterotrophs obtain energy from sunlight but get their carbon atoms from organic sources. This unusual mode of nutrition is found in only a few types of bacteria called purple nonsulfur bacteria. Many of them, including *Rhodopseudomonas,* the example shown in Figure 16.4, are found in aquatic sediments.

Chemoautotrophs harvest energy from inorganic chemicals and use carbon from CO_2 to make organic molecules. Because they don't depend on sunlight, chemoautotrophs can thrive in conditions that seem totally inhospitable to life. Near hydrothermal vents, where scalding water and hot gases surge into the sea more than a mile below the surface, chemoautotrophic bacteria use sulfur compounds as a source of energy. The organic molecules they produce using CO_2 from the seawater support diverse animal communities. The chemoautotrophs shown in Figure 16.4 live between layers of rocks buried 100 m below Earth's surface. Chemoautotrophs are also found in more predictable habitats, such as the soil.

Chemoheterotrophs, which acquire both energy and carbon from organic molecules, are by far the largest and most diverse group of prokaryotes. Almost any organic molecule is food for some species of chemoheterotrophic prokaryote.

? **Which term would describe your mode of nutrition?**

● Chemoheterotrophy

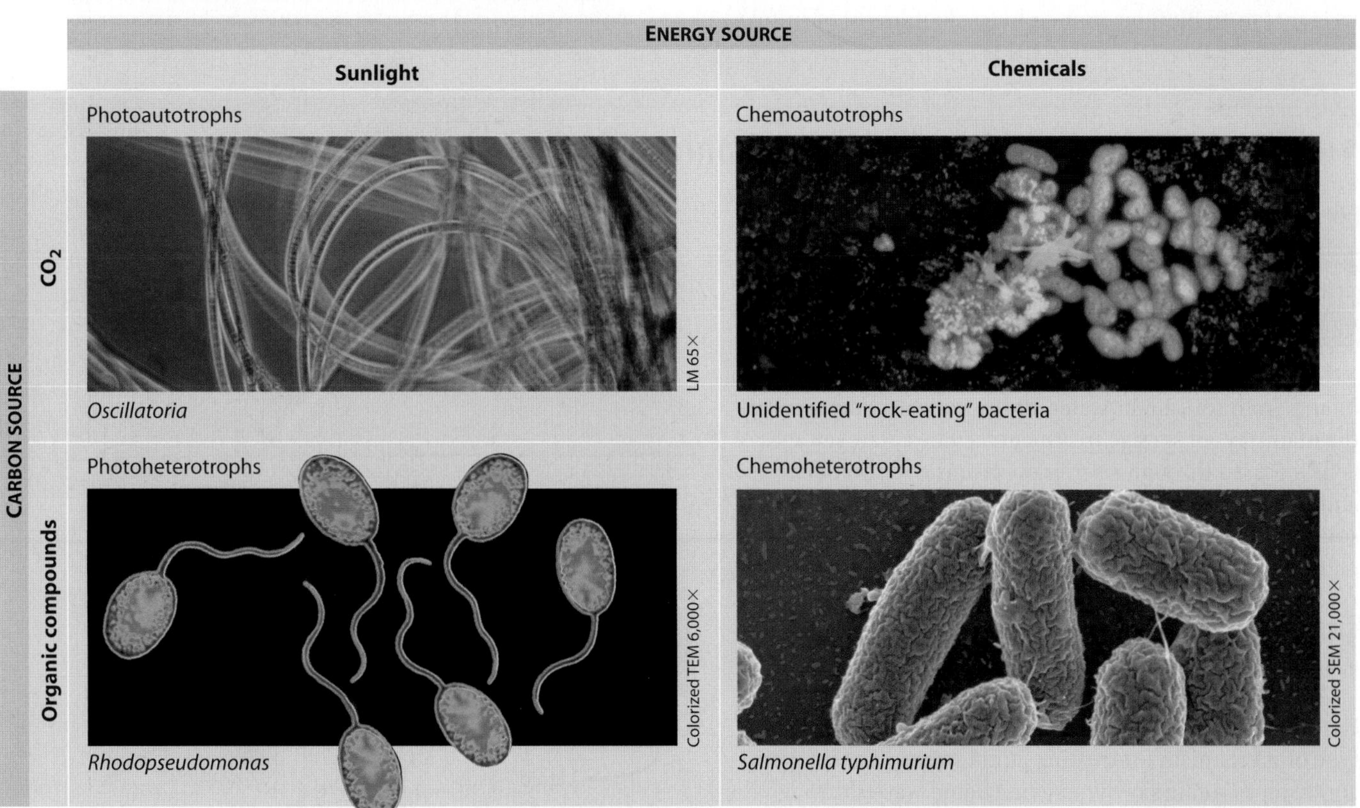
ENERGY SOURCE
Sunlight
Chemicals
CARBON SOURCE
CO_2
Organic compounds
Photoautotrophs
Oscillatoria
LM 65×
Chemoautotrophs
Unidentified "rock-eating" bacteria
Photoheterotrophs
Rhodopseudomonas
Colorized TEM 6,000×
Chemoheterotrophs
Salmonella typhimurium
Colorized SEM 21,000×

▲ **Figure 16.4** Sources of energy and carbon in prokaryotic modes of nutrition

16.5 Biofilms are complex associations of microbes

CONNECTION

In many natural environments, prokaryotes attach to surfaces in highly organized colonies called **biofilms**. A biofilm may consist of one or several species of prokaryotes, and it may include protists and fungi as well. Biofilms can form on almost any support, including rocks, soil, organic material (including living tissue), metal, and plastic. You have a biofilm on your teeth—dental plaque is a biofilm that can cause tooth decay. Biofilms can even form without a solid foundation, for example, on the surface of stagnant water.

Biofilm formation begins when prokaryotes secrete signaling molecules that attract nearby cells into a cluster. Once the cluster becomes sufficiently large, the cells produce a gooey coating that glues them to the support and to each other, making the biofilm extremely difficult to dislodge. For example, if you don't scrub your shower, you could find a biofilm growing around the drain—running water alone is not strong enough to wash it away. As the biofilm gets larger and more complex, it becomes a "city" of microbes. Communicating by chemical signals, members of the community coordinate the division of labor, defense against invaders, and other activities. Channels in the biofilm allow nutrients to reach cells in the interior and allow wastes to leave, and a variety of environments develop within it.

Biofilms are common among bacteria that cause disease in humans. For instance, ear infections and urinary tract infections are often the result of biofilm-forming bacteria. Cystic fibrosis patients are vulnerable to pneumonia caused by bacteria that form biofilms in their lungs. Biofilms of harmful bacteria can also form on implanted medical devices such as catheters, replacement joints, or pacemakers. The complexity of biofilms makes these infections especially difficult to defeat. Antibiotics may not be able to penetrate beyond the outer layer of cells, leaving much of the community intact. For example, some biofilm bacteria produce an enzyme that breaks down penicillin faster than it can diffuse inward.

Biofilms that form in the environment can be difficult to eradicate, too. A variety of industries spend billions of dollars every year trying to get rid of biofilms that clog and corrode pipes, gum up filters and drains, and coat the hulls of ships (**Figure 16.5**). Biofilms in water distribution pipes may survive chlorination, the most common method of ensuring that drinking water does not contain any harmful microorganisms. For example, biofilms of *Vibrio cholera*, the bacterium that causes cholera, found in water pipes were capable of withstanding levels of chlorine 10 to 20 times higher than the concentrations routinely used to chlorinate drinking water.

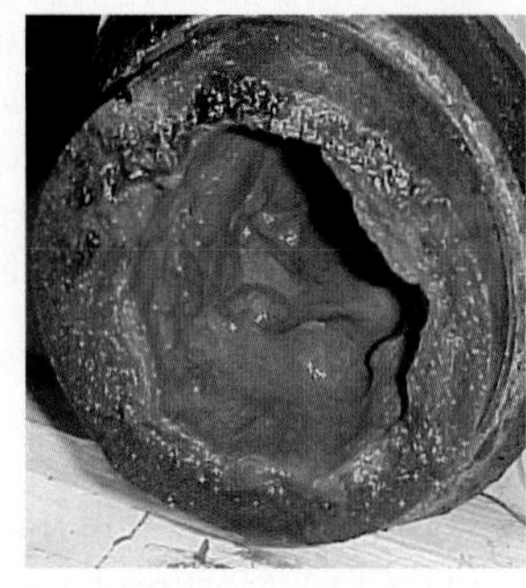

▲ **Figure 16.5** A biofilm fouling the insides of a pipe

? **Why are biofilms difficult to eradicate?**

● The biofilm sticks to the surface it resides on, and the cells that make up the biofilm stick to each other; the outer layer of cells may prevent antimicrobial substances from penetrating into the interior of the biofilm.

16.6 Prokaryotes help clean up the environment

CONNECTION

The characteristics that have made prokaryotes so widespread and successful—their nutritional diversity, adaptability, and capacity for forming biofilms—also make them useful for cleaning up contaminants in the environment.

Bioremediation is the use of organisms to remove pollutants from soil, air, or water. Prokaryotic decomposers are the mainstays of sewage treatment facilities. Raw sewage is first passed through a series of screens and shredders, and solid matter settles out from the liquid waste. This solid matter, called sludge, is then gradually added to a culture of anaerobic prokaryotes, including both bacteria and archaea. The microbes decompose the organic matter in the sludge into material that can be placed in a landfill or used as fertilizer.

Liquid wastes are treated separately from the sludge. In **Figure 16.6A**, you can see a trickling filter system, one type of mechanism for treating liquid wastes. The long horizontal pipes rotate slowly, spraying liquid wastes through the air onto a thick bed of rocks, the filter. Biofilms of aerobic bacteria and fungi growing on the rocks remove much of the

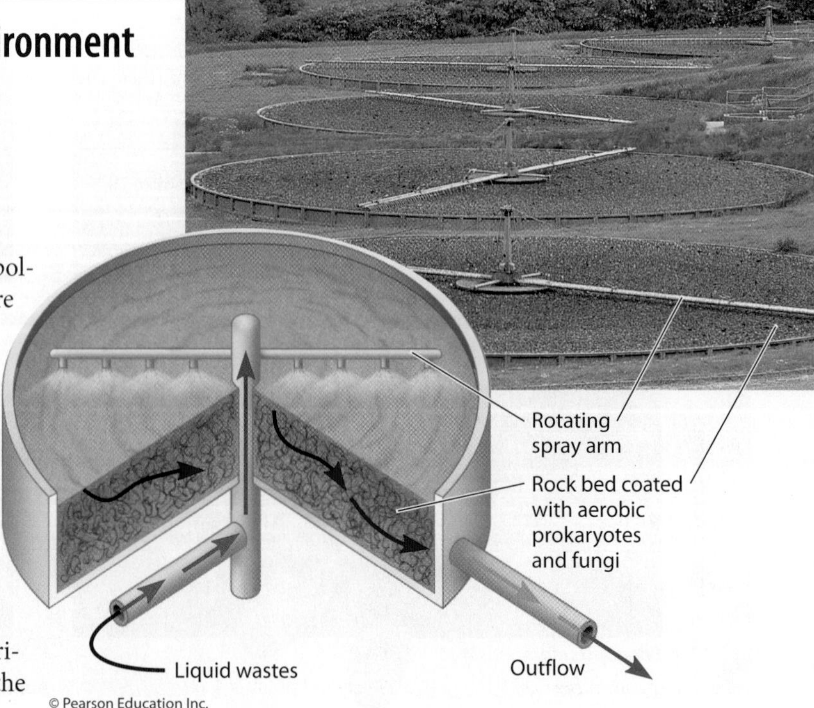
Rotating spray arm
Rock bed coated with aerobic prokaryotes and fungi
Liquid wastes
Outflow

▲ **Figure 16.6A** The trickling filter system at a sewage treatment plant organic material dissolved in the waste. Outflow from the rock bed is sterilized and then released, usually into a river or ocean.

Bioremediation has also become a useful tool for cleaning up toxic chemicals released into the soil and water. Naturally occurring prokaryotes capable of degrading pollutants such as oil, solvents, and pesticides are often present in contaminated soil and water, but environmental workers may use methods of speeding up their activity. In **Figure 16.6B**, an airplane is spraying chemical dispersants on oil from the disastrous 2010 Deepwater Horizon spill in the Gulf of Mexico. Like detergents that help clean greasy dishes, these chemicals break oil into smaller droplets that offer more surface area for microbial attack.

▲ **Figure 16.6B** Spraying chemical dispersants on oil spill in the Gulf of Mexico, 2010

? What is bioremediation?

● The use of organisms to clean up pollution

16.7 Bacteria and archaea are the two main branches of prokaryotic evolution

Researchers recently discovered that many prokaryotes once classified as bacteria are actually more closely related to eukaryotes and belong in a domain of their own (as you learned in Module 15.19). As a result, prokaryotes are now classified in two domains: Bacteria and Archaea (from the Greek *archaios*, ancient). Many bacterial and archaeal genomes have now been sequenced. When compared with each other and with the genomes of eukaryotes, these genome sequences strongly support the three-domain view of life. Some genes of archaea are similar to bacterial genes, others to eukaryotic genes, and still others seem to be unique to archaea.

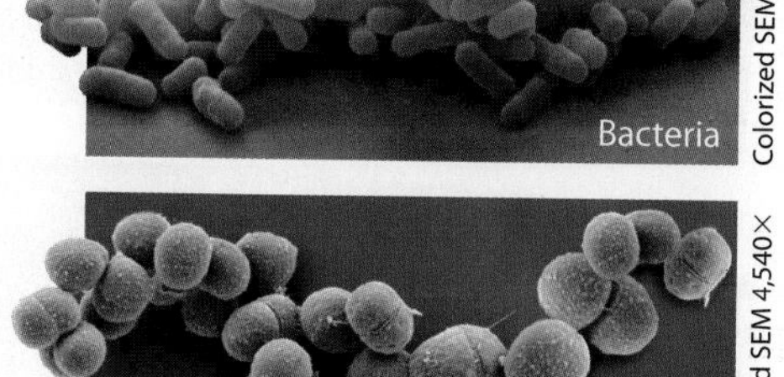
Bacteria
Colorized SEM 4,290×
Archaea
Colorized SEM 4,540×

Table 16.7 summarizes some of the main differences between the three domains. Differences between the ribosomal RNA (rRNA) sequences provided the first clues of a deep division among prokaryotes. Other differences in the cellular machinery for gene expression include differences in RNA polymerases (the enzymes that catalyze the synthesis of RNA) and in the presence of introns within genes. The cell walls and membranes of bacteria and archaea are also distinctive. Bacterial cell walls contain peptidoglycan (see Module 16.2), while archaea do not. Furthermore, the lipids forming the backbone of plasma membranes differ between the two domains. Intriguingly, for most of the features listed in the table, archaea have at least as much in common with eukaryotes as they do with bacteria.

Now that you are familiar with the general characteristics of prokaryotes and the features underlying their spectacular success, let's take a look at prokaryotic diversity. We begin with domain Archaea.

? As different as bacteria and archaea are, both groups are characterized by ______________ cells, which lack nuclei and other membrane-enclosed organelles.

● prokaryotic

TABLE 16.7 | DIFFERENCES BETWEEN THE DOMAINS BACTERIA, ARCHAEA, AND EUKARYA

Characteristic	Bacteria	Archaea	Eukarya
rRNA sequences	Some unique to bacteria	Some unique to archaea; some match eukaryotic sequences	Some unique to eukaryotes; some match archaeal sequences
RNA polymerase	One kind; relatively small and simple	Several kinds; complex	Several kinds; complex
Introns	Rare	In some genes	Present
Peptidoglycan in cell wall	Present	Absent	Absent
Histones associated with DNA	Absent	Present in some species	Present

16.8 Archaea thrive in extreme environments—and in other habitats

Archaea are abundant in many habitats, including places where few other organisms can survive. The archaeal inhabitants of extreme environments have unusual proteins and other molecular adaptations that enable them to metabolize and reproduce effectively. Scientists are only beginning to learn about these adaptations.

A group of archaea called the **extreme halophiles** ("salt lovers") thrive in very salty places, such as the Great Salt Lake in Utah, the Dead Sea, and seawater-evaporating ponds used to produce salt. Many species flourish when the salinity of the water is 15–30% and can tolerate even higher salt concentrations. Because seawater, with a salt concentration of about 3%, is hypertonic enough to shrivel most cells, these archaea have very little competition from other organisms. Extremely salty environments may turn red, purple, or yellow as a result of the dense growth and colorful pigments of halophilic archaea.

Another group of archaea, the **extreme thermophiles** ("heat lovers"), thrive in very hot water; some even live near deep-ocean vents, where temperatures are above 100°C, the boiling point of water at sea level! One such habitat is the Nevada geyser shown in **Figure 16.8A**. Other thermophiles thrive in acid. Many hot, acidic pools in Yellowstone National Park harbor such archaea, which give the pools a vivid greenish color. One of these organisms, *Sulfolobus*, can obtain energy by oxidizing sulfur or a compound of sulfur and iron; the mechanisms involved may be similar to those used billions of years ago by the first cells.

▲ **Figure 16.8A** Orange and yellow colonies of heat-loving archaea growing in a Nevada geyser

A third group of archaea, the **methanogens**, live in anaerobic (oxygen-lacking) environments and give off methane as a waste product. Many thrive in anaerobic mud at the bottom of lakes and swamps. You may have seen methane, also called marsh gas, bubbling up from a swamp. A large amount of methane is generated in solid waste landfills, where methanogens flourish in the anaerobic conditions. Many municipalities collect this methane and use it as a source of energy (**Figure 16.8B**). Great numbers of methanogens also inhabit the digestive tracts of cattle, deer, and other animals that depend heavily on cellulose for their nutrition. Because methane is a greenhouse gas (see Module 38.3), landfills and livestock contribute significantly to global warming.

Accustomed to thinking of archaea as inhabitants of extreme environments, scientists have been surprised to discover their abundance in more moderate conditions, especially in the oceans. Archaea live at all depths, making up a substantial fraction of the prokaryotes in waters more than 150 m beneath the surface and half of the prokaryotes that live below 1,000 m. Archaea are thus one of the most abundant cell types in Earth's largest habitat.

Because bacteria have been the subject of most prokaryotic research for over a century, much more is known about them than about archaea. Now that the evolutionary and ecological importance of archaea has come into focus, we can expect research on this domain to turn up many more surprises about the history of life and the roles of microbes in ecosystems.

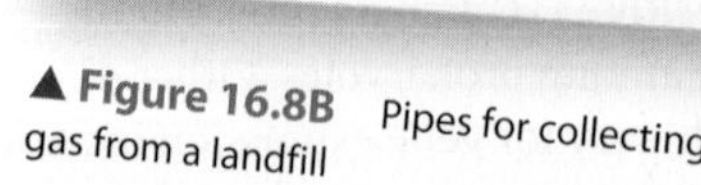

▲ **Figure 16.8B** Pipes for collecting gas from a landfill

? Some archaea are referred to as "extremophiles." Why?

Because they can thrive in extreme environments that are too hot, too salty, or too acidic for other organisms

16.9 Bacteria include a diverse assemblage of prokaryotes

Domain Bacteria is currently divided into five groups based on comparisons of genetic sequences. In this module, we sample some of the diversity in each group.

Proteobacteria are all gram-negative and share a particular rRNA sequence. With regard to other characteristics, however, this large group encompasses enormous diversity. For example, all four modes of nutrition are represented.

Chemoheterotrophic proteobacteria include pathogens such as *Vibrio cholerae*, which causes cholera. *Escherichia coli* (see Figure 16.2A), which is a common resident of the intestines of humans and other mammals and a favorite research organism, is also a member of this group.

Thiomargarita namibiensis (**Figure 16.9A**), an example of a photoautotrophic species of proteobacteria, uses H_2S to generate organic molecules from CO_2. The small greenish globules you see in the photo are sulfur wastes. Other proteobacteria, including *Rhodopseudomonas* (see Figure 16.4), are photoheterotrophs; they cannot convert CO_2 to sugars.

Chemoautotrophic soil bacteria such as *Nitrosomonas* obtain energy by oxidizing inorganic nitrogen compounds. These and related species of proteobacteria are essential to the chemical cycle that makes nitrogen available to plants.

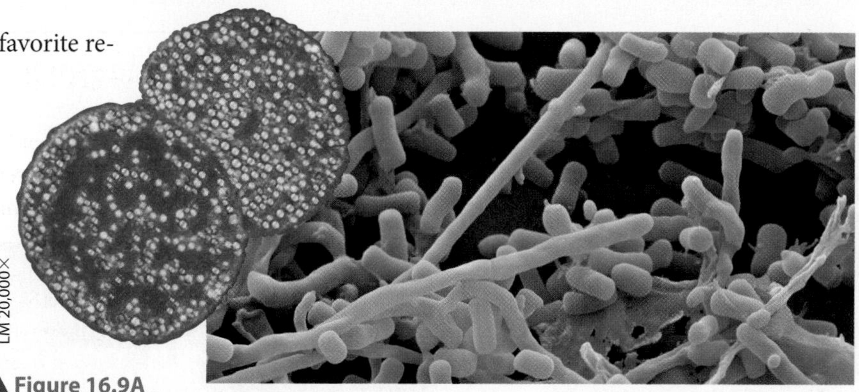
LM 20,000×
SEM 6,650×

▲ **Figure 16.9A** *Thiomargarita namibiensis*

▲ **Figure 16.9B** *Streptomyces*, the source of many antibiotics

Proteobacteria also include *Rhizobium* species that live symbiotically in root nodules of legumes such as soybeans and peas (see Figure 32.13B). **Symbiosis** is a close association between organisms of two or more species, and *endo*symbiosis refers to one species, called the endosymbiont, living *within* another. *Rhizobium* endosymbionts convert atmospheric nitrogen gas to a form usable by their legume host.

A second major group of bacteria, **gram-positive bacteria**, rivals the proteobacteria in diversity. One subgroup, the actinomycetes (from the Greek *mykes*, fungus, for which these bacteria were once mistaken), forms colonies of branched chains of cells. Actinomycetes are very common in the soil, where they decompose organic matter. Soil-dwelling species in the genus *Streptomyces*, shown in **Figure 16.9B**, are cultured by pharmaceutical companies as a source of many antibiotics, including streptomycin. Gram-positive bacteria also include the pathogens *Staphylococcus* and *Streptococcus* as well as many solitary species, such as *Bacillus anthracis* (see Figure 16.3B).

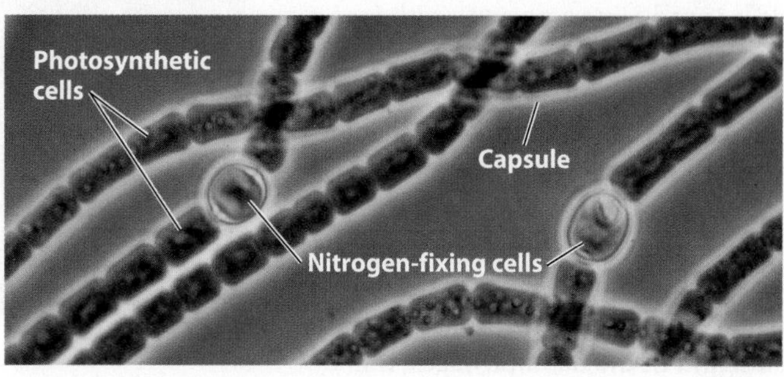
Photosynthetic cells
Capsule
Nitrogen-fixing cells
LM 650×

▲ **Figure 16.9C** *Anabaena*, a filamentous cyanobacterium

The **cyanobacteria** are the only group of prokaryotes with plantlike, oxygen-generating photosynthesis. Ancient cyanobacteria generated the oxygen that changed Earth's atmosphere more than two billion years ago. Today, cyanobacteria provide an enormous amount of food for freshwater and marine ecosystems. Some species, such as the cyanobacterium *Anabaena* in **Figure 16.9C**, have specialized cells that fix nitrogen. Many species of cyanobacteria have symbiotic relationships with organisms such as fungi, mosses, and a variety of marine invertebrates.

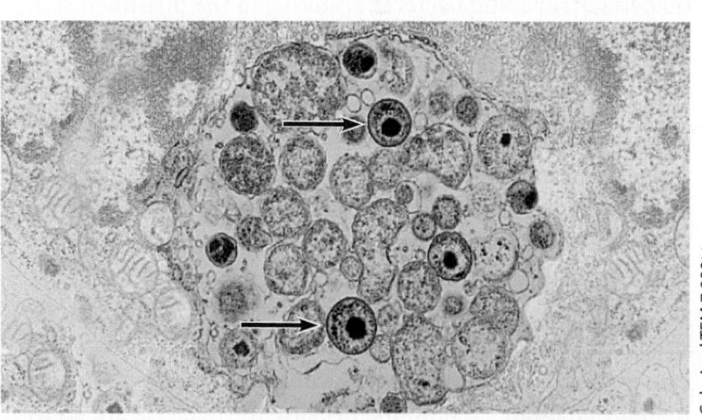
Colorized TEM 7,000×

▲ **Figure 16.9D** *Chlamydia* cells (arrows) inside an animal cell

The **chlamydias**, which live inside eukaryotic host cells, form a fourth bacterial group (**Figure 16.9D**). *Chlamydia trachomatis* is a common cause of blindness in developing countries and also causes nongonococcal urethritis, the most common sexually transmitted disease in the United States.

Spirochetes, the fifth group, are helical bacteria that spiral through their environment by means of rotating, internal filaments. Some spirochetes are notorious pathogens: *Treponema pallidum*, shown in **Figure 16.9E**, causes syphilis, and *Borrelia burgdorferi* (see Figure 16.2A) causes Lyme disease.

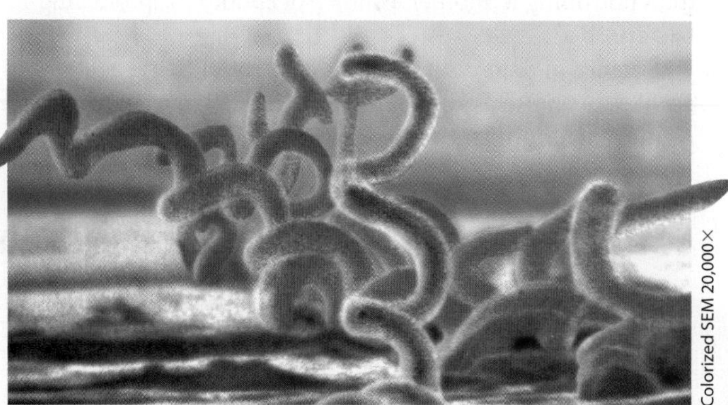
Colorized SEM 20,000×

▲ **Figure 16.9E** *Treponema pallidum*, the spirochete that causes syphilis

? **How are *Thiomargarita namibiensis* similar to the cyanobacteria?**

● They are both photoautotrophic.

16.10 Some bacteria cause disease

CONNECTION

All organisms, people included, are almost constantly exposed to pathogenic bacteria. Most often, our body's defenses prevent pathogens from affecting us. Occasionally, however, a pathogen establishes itself in the body and causes illness. Even some of the bacteria that are normal residents of the human body can make us ill when our immune system is compromised by poor nutrition or by a viral infection.

Most bacteria that cause illness do so by producing a poison—either an exotoxin or an endotoxin. **Exotoxins** are proteins that bacterial cells secrete into their environment. They include some of the most powerful poisons known. For example, *Staphylococcus aureus*, shown in **Figure 16.10A**, produces several exotoxins. Although *S. aureus* is commonly found on the skin and in the nasal passages, if it enters the body through a wound, it can cause serious disease. One of its exotoxins destroys the white blood cells that attack invading bacteria, resulting in the pus-filled skin bumps characteristic of methicillin-resistant *S. aureus* infections (MRSA; see Module 13.16). Food may also be contaminated with *S. aureus* exotoxins, which are so potent that less than a millionth of a gram causes vomiting and diarrhea.

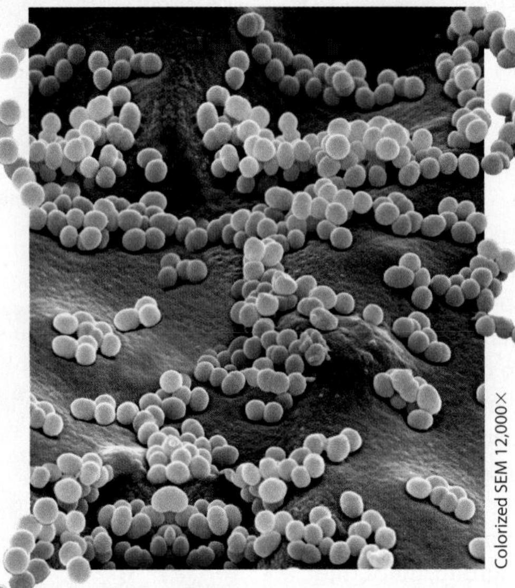
Colorized SEM 12,000×

▲ **Figure 16.10A** *Staphylococcus aureus*, an exotoxin producer

Endotoxins are lipid components of the outer membrane of gram-negative bacteria that are released when the cell dies or is digested by a defensive cell. All endotoxins induce the same general symptoms: fever, aches, and sometimes a dangerous drop in blood pressure (septic shock). Septic shock triggered by an endotoxin of *Neisseria meningitidis*, which causes bacterial meningitis, can kill a healthy person in a matter of days or even hours. Because the bacteria are easily transmitted among people living in close contact, many colleges require students to be vaccinated against this disease. The species of *Salmonella* (shown in Figure 16.4) that causes food poisoning is another example of endotoxin-producing bacteria.

Because of their disease-causing potential, some bacteria have been used as biological weapons. *Bacillus anthracis*, the bacterium that causes anthrax, and the exotoxin of *Clostridium botulinum* are among the biological agents that are considered the highest-priority threats.

Bacillus anthracis forms hardy endospores (see Figure 16.3B) that are commonly found in the soil of agricultural regions. "Weaponizing" anthrax involves manufacturing a preparation of endospores that disperses easily in the air, where they will be inhaled by the target population. Endospores germinate in the lungs, and the bacteria multiply, producing an exotoxin that eventually accumulates to lethal levels in the blood. Although antibiotics can kill the bacteria, they can't eliminate the toxin already in the body. As a result, weaponized anthrax has a very high death rate.

The weapon form of *C. botulinum* is the exotoxin it produces, botulinum, rather than the living microbes. Botulinum is the deadliest poison known. Thirty grams of pure toxin, a bit more than an ounce, could kill every person in the United States. Botulinum blocks transmission of the nerve signals that cause muscle contraction, resulting in paralysis of the muscles required for breathing. This effect is also responsible for a more benign use of botulinum —relaxing facial muscles that cause wrinkles **(Figure 16.10B)**.

? **Contrast exotoxins with endotoxins.**

Exotoxins are proteins secreted by pathogenic bacteria; endotoxins are components of the outer membranes of pathogenic bacteria.

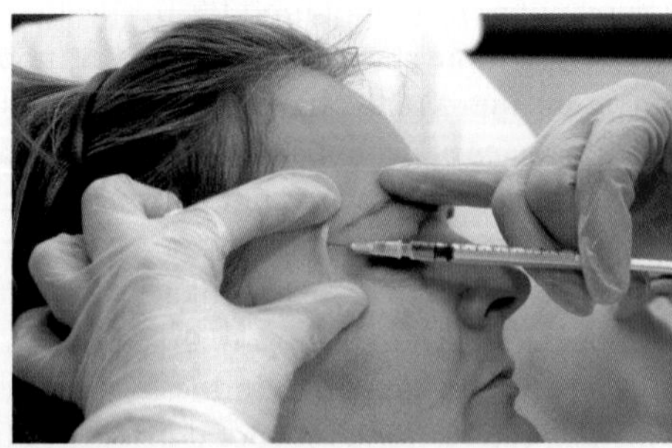

▲ **Figure 16.10B** Injecting Botox, which contains a minute amount of botulinum, to smooth wrinkles

16.11 Stomach microbiota affect health and disease

SCIENTIFIC THINKING

In the chapter introduction, you learned that each of us houses trillions of bacteria that are harmless, or even beneficial. In the previous module, you learned about bacteria that cause disease. How do scientists determine which is which?

To test the hypothesis that a certain bacterium is the cause of a disease, a researcher must satisfy four conditions. This method of hypothesis testing, formulated by microbiologist Robert Koch in the late 19th century, is known as *Koch's postulates*. For a human disease, the researcher must be able to (1) find the candidate bacterium in every case of the disease; (2) isolate the bacterium from a person who has the disease and grow it in pure culture; (3) show that the cultured bacterium causes the disease when transferred to a healthy subject (usually an animal); and (4) isolate the bacterium from the experimentally infected subject. So when Australian microbiologist Barry Marshall hypothesized that chronic gastritis (an inflammation of the stomach lining that can lead to ulcers) was caused by a bacterium called *Helicobacter pylori*, he knew he would need to fulfill these criteria.

Are antibiotics making us fat?

Over the course of several years, Marshall satisfied the first two requirements, but his efforts to infect animals failed to produce results. Although he continued to accumulate evidence supporting his hypothesis, the scientific community was skeptical of Marshall's idea and he had difficulty obtaining funding for his research. Frustrated by watching so many patients suffer life-threatening complications from peptic (stomach) ulcers when his research might yield a simple cure, Marshall decided to take a radical course of action—he would experiment on himself. He concocted a nasty brew of *H. pylori* and swallowed it. Several days later, he became ill from gastritis (step 3 of Koch's postulates). His stomach lining proved to be teeming with *H. pylori* (step 4). Marshall then cleared up his infection with antibiotics. He continued to make progress in his research, and other scientists followed up with further studies. Several years after Marshall's big gulp, antibiotics became a standard treatment for ulcer patients **(Figure 16.11A)**.

Since Marshall's breakthrough work, scientists have learned that our relationship with *H. pylori* is ancient—at least 50,000 years old—and it's complicated. Only a particular genetic strain causes ulcers; other strains are harmless members of our microbiota. In fact, some scientists hypothesize that the *absence* of *H. pylori* can cause problems. Fifty years ago, *H. pylori* was present in most Americans, but its prevalence has been steadily declining. Researchers are investigating a possible connection between this decline and the high rate of obesity. *H. pylori* is thought to affect the stomach's production of a hormone called ghrelin that sends hunger signals to the brain **(Figure 16.11B)**. Ghrelin output should decrease after a meal, ending the urge to eat. Studies have linked the absence of *H. pylori* to continued ghrelin output after eating. In other words, the brain doesn't get the message that you've had dinner, which leads to overeating. Investigations have also suggested a correlation between an absence of *H. pylori* and increased body mass index.

▲ Figure 16.11A Barry Marshall (left) and collaborator Robin Warren were awarded the 2005 Nobel Prize in Medicine for their discovery of *H. pylori* and its role in peptic ulcers

Is a simple, microbe-based cure for obesity just around the corner? Probably not. *H. pylori* is just one member of a diverse microbial community within the complex ecosystem of the human body. While the results obtained so far are intriguing, they are characteristic of the early stages of scientific investigation—preliminary, tentative, and sometimes even contradictory.

? According to a study published in 2012, infants treated with antibiotics before the age of 6 months were more likely to be overweight at the age of 3. Do these results support the hypothesis that an absence of *H. pylori* is a factor in causing obesity?

● This evidence tentatively supports the hypothesis that disturbing the body's microbial community is a factor in causing obesity (other explanations for the results are possible), but the study did not look specifically at *H. pylori*.

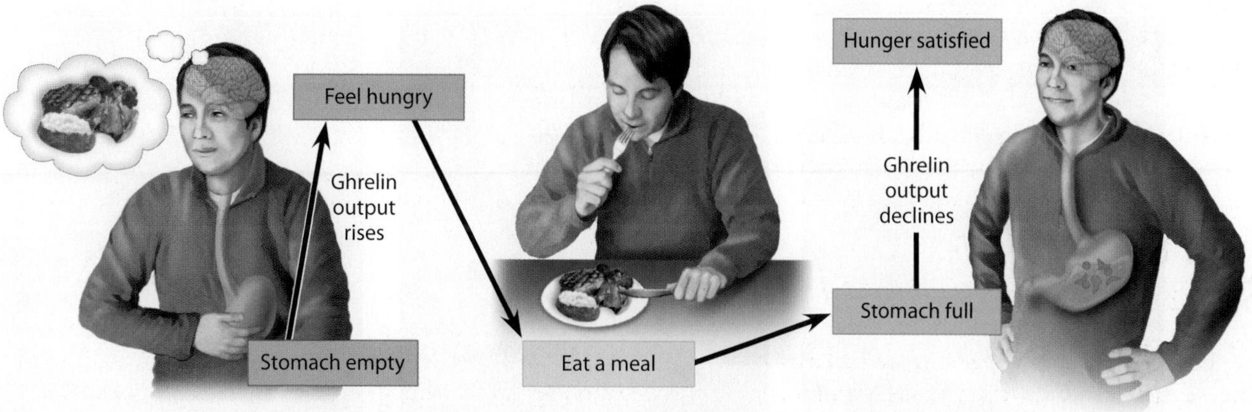
Feel hungry
Ghrelin output rises
Stomach empty
Eat a meal
Stomach full
Ghrelin output declines
Hunger satisfied

▲ Figure 16.11B Effect of ghrelin on hunger

Try This Use the diagram to explain the hypothesis linking obesity to *H. pylori*.

Protists

16.12 Protists are an extremely diverse assortment of eukaryotes

Protists are a diverse collection of mostly unicellular eukaryotes. Biologists used to classify all protists in a kingdom called Protista, but now it is clear that these organisms constitute multiple kingdoms within domain Eukarya. While our knowledge of the evolutionary relationships among these diverse groups remains incomplete, *protist* is still useful as a convenient term to refer to eukaryotes that are not plants, animals, or fungi.

Protists obtain their nutrition in a variety of ways (**Figure 16.12A**). Some protists are autotrophs, producing their food by photosynthesis; these are called **algae** (another useful term that is not taxonomically meaningful). Many algae, including the one shown on the left in Figure 16.12A, are multicellular. Other protists, informally called **protozoans**, are heterotrophs, eating bacteria and other protists. Some heterotrophic protists are fungus-like and obtain organic molecules by absorption, and some are parasitic. **Parasites** derive their nutrition from a living host, which is harmed by the interaction. *Giardia*, shown in the middle of Figure 16.12A, is a human parasite. Still other protists are **mixotrophs**, capable of both photosynthesis and heterotrophy, depending on availability of light and nutrients. An example is *Euglena*, shown on the right in Figure 16.12A.

Protist habitats are also diverse. Most protists are aquatic, and they are found almost anywhere there is moisture, including terrestrial habitats such as damp soil and leaf litter. Others inhabit the bodies of various host organisms. For example, **Figure 16.12B** shows one of the protists that are endosymbionts in the intestinal tract of termites. Termite endosymbionts digest the tough cellulose in the wood eaten by their host. Some of these protists even have endosymbionts of their own—prokaryotes that metabolize the cellulose.

As eukaryotes, protists are more complicated than any prokaryotes. Their cells have a membrane-enclosed nucleus (containing multiple chromosomes) and other organelles characteristic of eukaryotic cells. The flagella and cilia of protistan cells have a 9 + 2 pattern of microtubules, another typical eukaryotic trait (see Module 4.18).

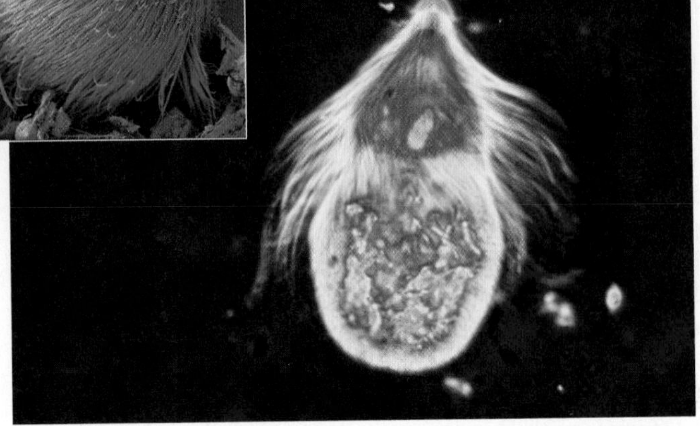
SEM 560×
LM 325×

▼ Figure 16.12B A protist from a termite gut covered by thousands of flagella, viewed with scanning electron microscope (left) and light microscope (below)

Because most protists are unicellular, they are justifiably considered the simplest eukaryotes. However, the cells of many protists are among the most elaborate in the world. This level of cellular complexity is not really surprising, for each unicellular protist is a complete eukaryotic organism analogous to an entire animal or plant.

With their extreme diversity, protists are difficult to categorize. Recent molecular and cellular studies have shaken the foundations of protistan taxonomy as much as they have that of the prokaryotes. Intuitive groupings such as protozoans and algae are phylogenetically meaningless because the nutritional modes used to categorize them are spread across

▶ Figure 16.12A Protist modes of nutrition

Autotrophy

Caulerpa, a green alga

Heterotrophy

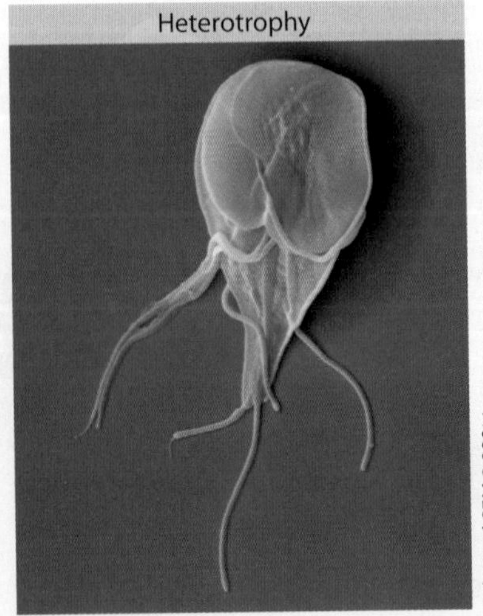
Colorized SEM 2,000×

Giardia, a parasite

Mixotrophy

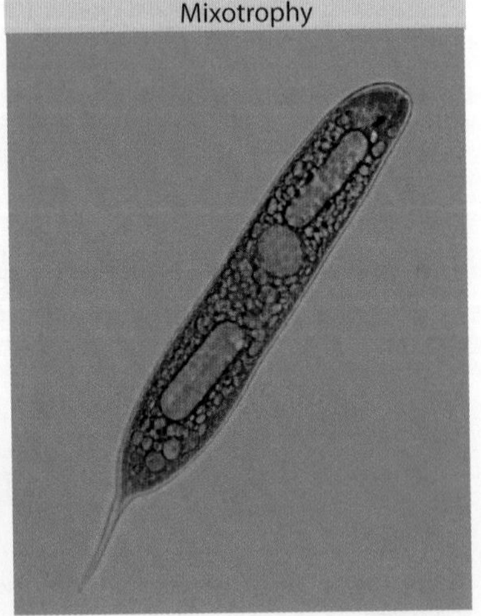
LM 700×

Euglena

many different lineages. It is now clear that there are multiple clades of protists, with some lineages more closely related to plants, fungi, or animals than they are to other protists. We have chosen to organize our brief survey of protist diversity using one current hypothesis of protist phylogeny, which proposes four monophyletic "supergroups." The largest and most diverse supergroup is "SAR," which contains three clades: Stramenopila, Alveolata, and Rhizaria. The other supergroups are Excavata, Unikonta, and Archaeplastida.

While there is general agreement on some of these groupings, others are hotly debated—the classification of protists is very much a work in progress. Before embarking on our tour of protists, however, let's consider how their extraordinary diversity originated.

? **What is a general definition for "protist"?**

A eukaryote that is not an animal, fungus, or plant

16.13 Endosymbiosis of unicellular algae is the key to much of protist diversity

EVOLUTION CONNECTION

As Module 16.12 indicates, protists are bewilderingly diverse. What is the origin of this enormous diversity? To explain, let's first review the theory of endosymbiosis for the origin of mitochondria and chloroplasts in eukaryotes (see Module 4.15). According to this theory, oxygen-using prokaryotes established residence within other, larger cells. These endosymbionts evolved into mitochondria, giving rise to heterotrophic eukaryotes.

As shown in **Figure 16.13**, autotrophic eukaryotes also arose through endosymbiosis of a prokaryote by a eukaryote after ① a heterotrophic eukaryote engulfed an autotrophic cyanobacterium. If the cyanobacterium continued to function within its host cell, its photosynthesis would have provided a steady source of food for the heterotrophic host and thus given it a significant selective advantage. And because the cyanobacterium had its own DNA, it could reproduce to make multiple copies of itself within the host cell. In addition, cyanobacteria could be passed on when the host reproduced. Over time, ② the descendants of the original cyanobacterium evolved into chloroplasts. The chloroplast-bearing lineage of eukaryotes later diversified into ③ the autotrophs green algae and red algae (see Module 16.18).

On subsequent occasions during eukaryotic evolution ④ green algae and red algae themselves became endosymbionts following ingestion by different heterotrophic eukaryotes. The heterotrophic host cells enclosed the algal cells in food vacuoles ⑤ but the algae—or parts of them—survived and became cellular organelles. The presence of the endosymbionts, which also had the ability to replicate themselves, gave their hosts a selective advantage. Figure 16.13 shows how endosymbiosis of green algae could give rise to mixotrophs, such as the *Euglena* in Figure 16.12A. Endosymbiosis of red algae led to nutritional diversity in other groups of protists. Thus, endosymbiosis of unicellular algae appears to explain a large part of protist diversity.

? **How did the endosymbiosis that resulted in chloroplasts differ from the endosymbiosis that led to protist diversity?**

Chloroplasts evolved from endosymbiosis of a photosynthetic prokaryote. Much of protist diversity resulted from endosymbiosis of photosynthetic eukaryotes.

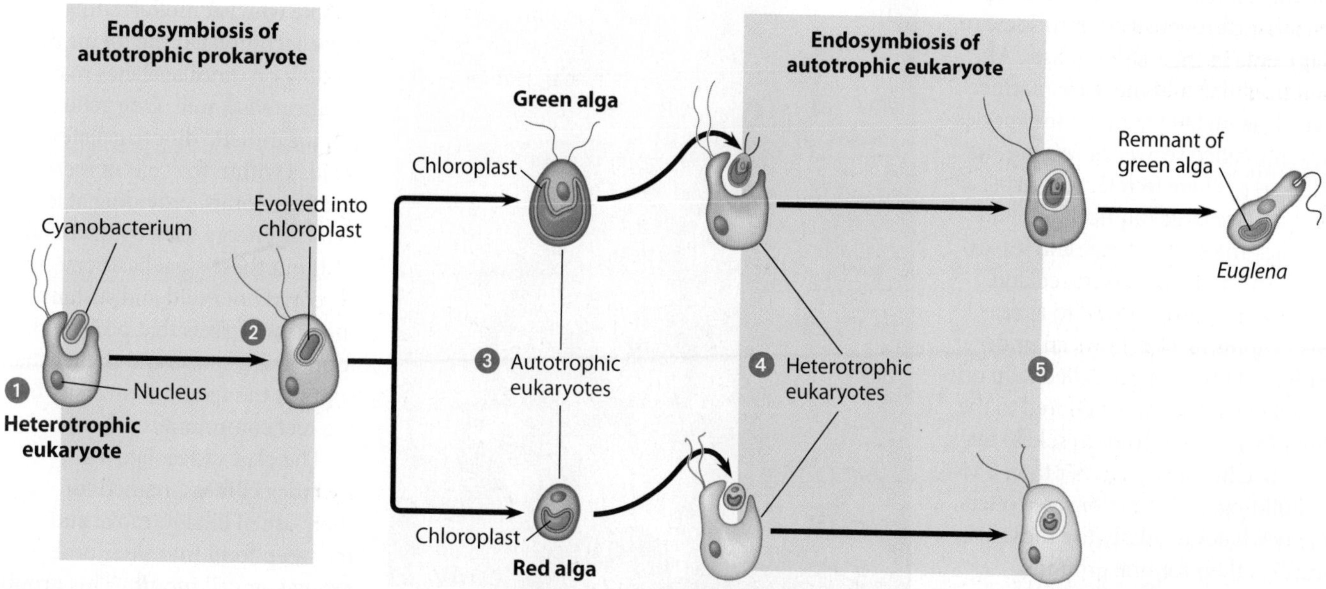

Endosymbiosis of autotrophic prokaryote
Cyanobacterium
Evolved into chloroplast
1
Nucleus
Heterotrophic eukaryote
2
Green alga
Chloroplast
3 Autotrophic eukaryotes
Chloroplast
Red alga
Endosymbiosis of autotrophic eukaryote
4 Heterotrophic eukaryotes
5
Remnant of green alga
Euglena

▲ **Figure 16.13** The theory of the origin of protistan diversity through endosymbiosis (mitochondria not shown)

Try This Trace the cyanobacterium from the left side of the figure to the protists at step 5.

16.14 The "SAR" supergroup represents the range of protist diversity

Our sample of protist diversity begins with **SAR**, recently proposed as a monophyletic supergroup on the basis of genomic studies. "SAR" stands for **Stramenopila**, **Alveolata**, and **Rhizaria**, the three clades that make up this huge, extremely diverse group.

Stramenopiles Diatoms and brown algae are two examples of autotrophic stramenopiles. **Diatoms**, unicellular algae that are one of the most important photosynthetic organisms on Earth, have a unique glassy cell wall containing silica. The cell wall of a diatom consists of two halves that fit together like the bottom and lid of a shoe box (**Figure 16.14A**). Both freshwater and marine environments are rich in diatoms, and the organic molecules these microscopic algae produce are a key source of food in all aquatic environments. Some diatoms store food reserves in the form of lipid droplets as well as carbohydrates. In addition to being a rich source of energy, the lipids make the diatoms buoyant, which keeps them floating near the surface in the sunlight. Massive accumulations of fossilized diatoms make up thick sediments known as diatomaceous earth, which is mined for use as a filtering medium and as a grinding and polishing agent.

Colorized SEM 200×

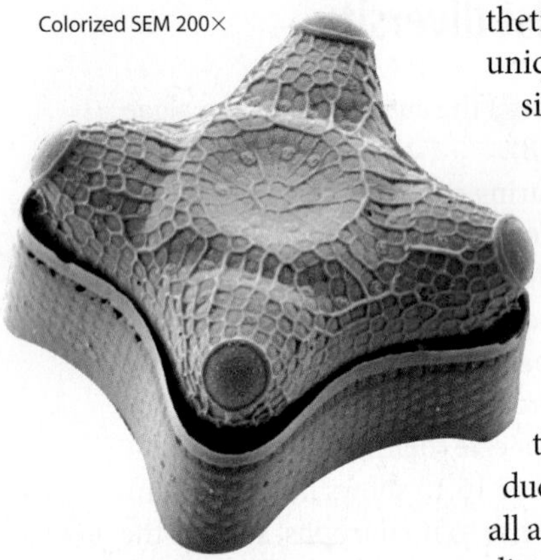

▲ **Figure 16.14A** Diatom, a unicellular alga

Brown algae are large, complex stramenopiles. Brown algae owe their characteristic brownish color to some of the pigments in their chloroplasts. All are multicellular, and most are marine. Brown algae include many of the species commonly called seaweeds. We use the word *seaweeds* here to refer to marine algae that have large multicellular bodies but lack the roots, stems, and leaves found in most plants. (Some red and green algae are also referred to as seaweeds.) **Figure 16.14B** shows an underwater bed of brown algae called **kelp** off the coast of California. Anchored to the seafloor by rootlike structures, kelp may grow to heights of 60 m, taller than a 15-story building. Fish, sea lions, sea otters, and gray whales regularly use these kelp "forests" as their feeding grounds.

▲ **Figure 16.14B** Brown algae: a kelp "forest"

Water molds are heterotrophic unicellular stramenopiles that typically decompose dead plants and animals in freshwater habitats. Because many species resemble fungi (**Figure 16.14C**), water molds were classified as fungi until molecular comparisons revealed their kinship to protists. Parasitic water molds sometimes grow on the skin or gills of fish. Water molds also include plant parasites called downy mildews. "Late blight" of potatoes, a disease caused by a downy mildew, led to a devastating famine in Ireland in the mid-1800s. A closely related pathogen has swept through tomato crops in the eastern United States, depriving fast-food burgers of a standard topping and home gardeners of a favorite summertime treat.

▲ **Figure 16.14C** Water mold (white threads) decomposing a goldfish

Alveolates **Dinoflagellates**, a diverse group that includes unicellular autotrophs, heterotrophs, and mixotrophs, are also very common components of marine and freshwater plankton (communities of microorganisms that live near the water's surface). Blooms—population explosions—of autotrophic dinoflagellates sometimes cause warm coastal waters to turn pinkish orange, a phenomenon known as "red tide" (**Figure 16.14D**, on the facing page). Toxins produced by some red-tide dinoflagellates have killed large numbers of fish. People who eat molluscs that have accumulated the toxins by feeding on dinoflagellates may be affected as well. One genus of photosynthetic dinoflagellates resides within the cells of reef-building corals, providing at least half the energy used by the corals. Without these algal partners, corals could not build and sustain the massive reefs that provide the food, living space, and shelter that support the splendid diversity of the reef community.

The clade Alveolata also includes **ciliates**, named for their use of cilia to move and to sweep food into their oral groove, or cell mouth. This group of unicellular protists includes

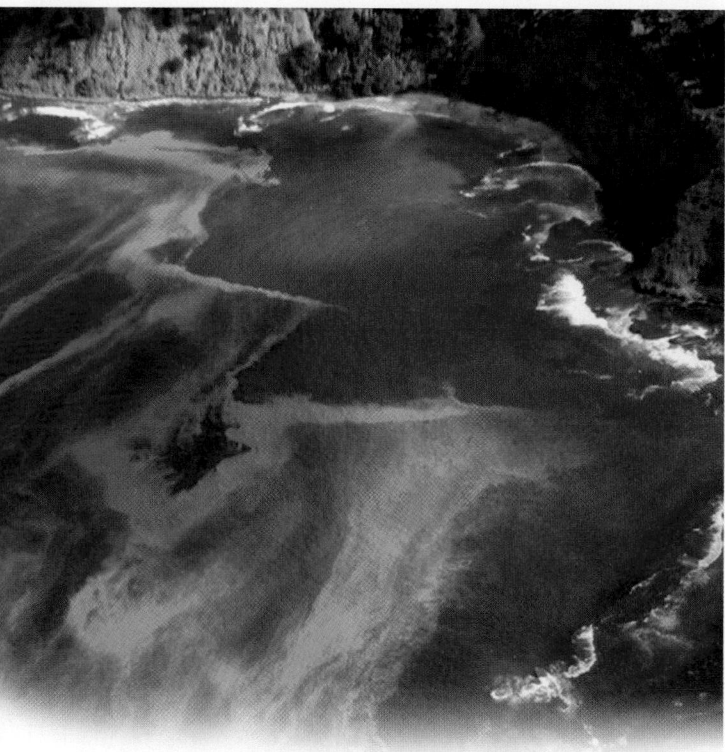

▲ **Figure 16.14D** A red tide caused by *Gymnodinium*, a dinoflagellate

heterotrophs and mixotrophs. You may have seen the common freshwater protist *Paramecium* **(Figure 16.14E)** in a biology lab. Like many ciliates, *Paramecium* swims by beating its cilia in a wavelike motion. Other ciliates "crawl" over a surface using cilia that are arranged in bundles along the length of the cell.

Another subgroup of alveolates is made up of parasites, including some that cause serious diseases in humans. For example, *Plasmodium*, which causes malaria, kills nearly a million people a year. Some stages of *Plasmodium's* complex life cycle take place in certain species of mosquitoes, which transmit the parasite to humans.

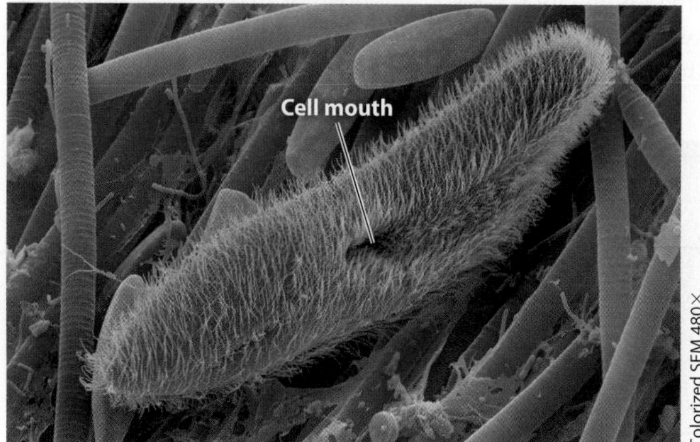
Cell mouth
Colorized SEM 480×

▲ **Figure 16.14E** A freshwater ciliate, *Paramecium,* showing cilia distributed over the cell surface (The photo also includes other unicellular organisms.)

Rhizaria The two largest groups in Rhizaria, foraminiferans and radiolarians, are among the organisms referred to as amoebas. **Amoebas** move and feed by means of **pseudopodia** (singular, *pseudopodium*), which are temporary extensions of the cell. Molecular systematics now indicates that many different taxonomic groups include organisms that share this means of movement and feeding. Most of the amoebas in Rhizaria are distinguished from other amoebas by their threadlike (rather than lobe-shaped) pseudopodia.

Foraminiferans (forams) **(Figure 16.14F)** are found both in the ocean and in fresh water. They have porous shells, called *tests,* composed of organic material hardened by calcium carbonate. The pseudopodia, which function in feeding and locomotion, extend through small pores in the test (see Figure 16.14F, inset). Ninety percent of forams that have been identified are fossils. The fossilized tests, which are a component of sedimentary rock, are excellent markers for correlating the ages of rocks in different parts of the world.

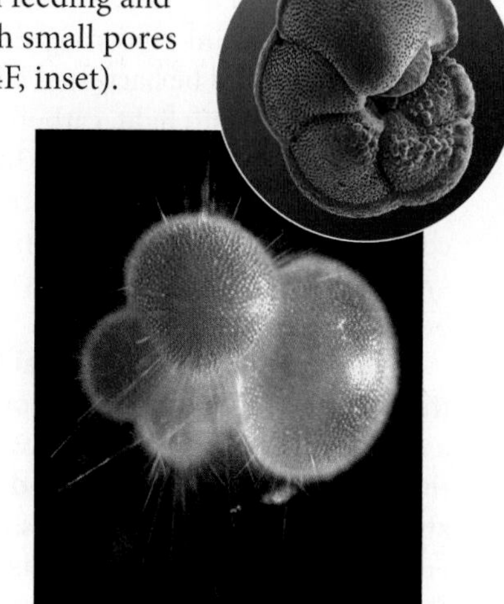

▲ **Figure 16.14F** A foraminiferan (inset SEM shows a foram test)

Like forams, **radiolarians** produce a mineralized support structure, in this case an internal skeleton made of silica **(Figure 16.14G)**. The cell is also surrounded by a test composed of organic material. Most species of radiolarians are marine. When they die, their hard parts, like those of forams, settle to the bottom of the ocean and become part of the sediments. In some areas, radiolarians are so abundant that sediments, known as radiolarian ooze, are hundreds of meters thick.

? **Which groups of Stramenopila, Alveolata, and Rhizaria include autotrophs?**

Stramenopila: diatoms, brown algae; Alveolata: dinoflagellates; Rhizaria: none

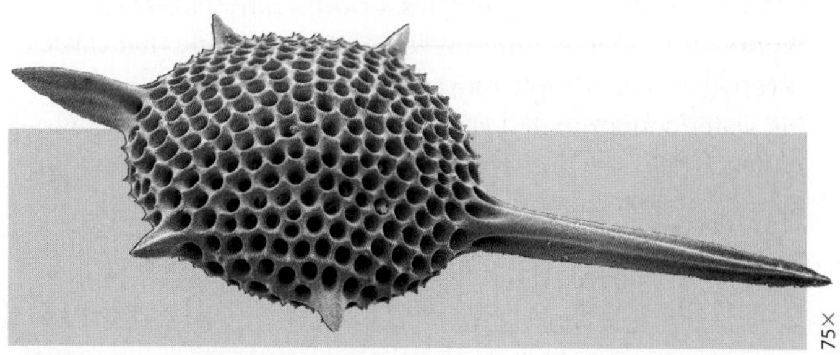
75×

▲ **Figure 16.14G** A radiolarian skeleton

16.15 Can algae provide a renewable source of energy?

CONNECTION

Have you ever wondered what the "fossils" are in fossil fuels? They are organic remains of organisms that lived hundreds of millions of years ago. Diatoms are thought to be the main source of oil, while coal was formed from primitive plants (see Module 17.4). However, rapid consumption is depleting the world's supply of readily accessible fossil fuels.

Entrepreneurs are now eying the lipid droplets in diatoms and other algae as a renewable source of energy. After all, the energy we extract from fossil fuels was originally stored in organisms through the process of photosynthesis. Why wait millions of years? If unicellular algae could be grown on a large scale, the oil could be harvested and processed into biodiesel. When supplied with light, carbon dioxide, and nutrients, unicellular algae reproduce rapidly. In one scenario, algae could be grown indoors in closed "bioreactor" vessels under tightly controlled environmental conditions **(Figure 16.15)**. Outdoor systems using closed bioreactors or open-air ponds are also being developed.

There are numerous technical hurdles to overcome before the industrial-scale production of biofuel from algae becomes a reality. Investigators must identify the most productive of the hundreds of algal species and test whether they are suitable for mass culturing methods. With further research, scientists may be able to improve desirable characteristics such as growth rate or oil yield through genetic engineering. In addition, manufacturers need to develop cost-effective methods of harvesting the algae and extracting and processing the oil.

▲ **Figure 16.15** Green algae in a bioreactor

Nevertheless, there might be an alga-powered vehicle in your future.

? What characteristics of unicellular algae make them attractive candidates for the production of biofuels?

Rapid reproduction; would not occupy farmland needed to grow food crops

16.16 Some excavates have modified mitochondria

Excavata, the second supergroup in our survey of protists, has recently been proposed as a clade on the basis of molecular and morphological similarities. The name refers to an "excavated" feeding groove possessed by some members of the group. Many excavates have modified mitochondria that lack functional electron transport chains and use anaerobic pathways such as glycolysis to extract energy. Heterotrophic excavates include the termite endosymbiont shown in Figure 16.12B. There are also autotrophic species and mixotrophs, such as *Euglena* (see Figure 16.12A).

Some excavates are parasites. *Giardia intestinalis* (see Figure 16.12A) is a common waterborne parasite that causes severe diarrhea. People most often pick up *Giardia* by drinking water contaminated with feces containing the parasite. For example, a swimmer in a lake or river might accidentally ingest water contaminated with feces from infected animals, or a hiker might drink contaminated water from a seemingly pristine stream. (Boiling the water first will kill *Giardia.*)

Another excavate, *Trichomonas vaginalis* **(Figure 16.16A)**, is a common sexually transmitted parasite that causes an estimated 5 million new infections each year. The parasite travels through the reproductive tract by moving its flagella and undulating part of its membrane. In women, the protists feed on white blood cells and bacteria living on the cells lining the vagina. *T. vaginalis* also infects the cells lining the male reproductive tract, but limited availability of food results in very small population sizes. Consequently, males typically have no symptoms of infection, although they can repeatedly infect their female partners. The only treatment available is a drug

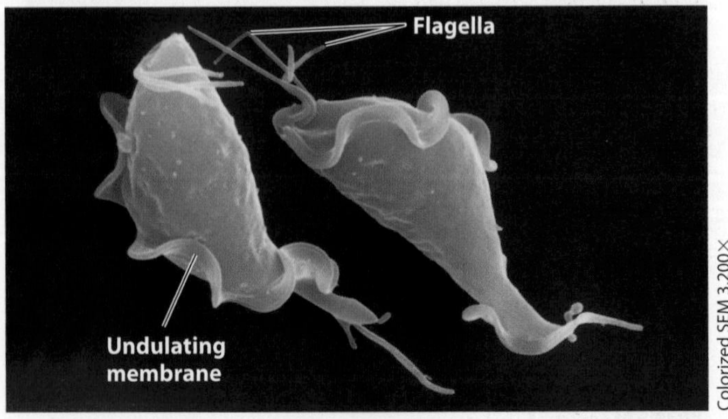
Flagella
Undulating membrane
Colorized SEM 3,200×

▲ **Figure 16.16A** A parasitic excavate: *Trichomonas vaginalis* called metronidazole. Disturbingly, drug resistance seems to be evolving in *T. vaginalis*, especially on college campuses.

Members of the excavate genus *Trypanosoma* are parasites that can be transmitted to humans by insects. For instance, the trypanosome shown in **Figure 16.16B** causes sleeping sickness, a potentially fatal disease spread by the African tsetse fly. The squiggly "worms" in the photo are cells of *Trypanosoma;* the circular cells are human red blood cells.

? How do the nutritional modes of *Euglena* and *Trichomonas* differ?

Euglena is mixotrophic; *Trichomonas* is strictly heterotrophic.

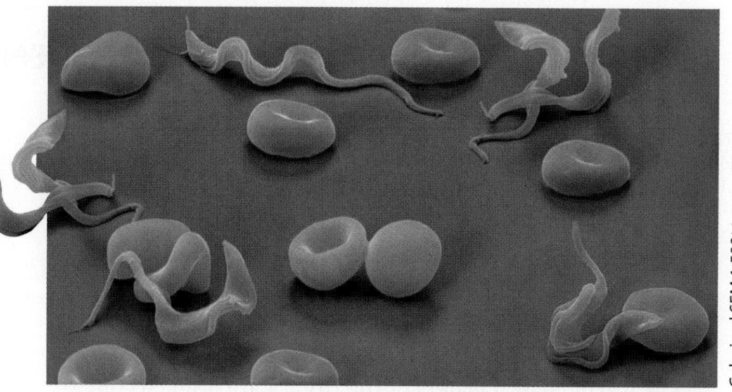
Colorized SEM 1,500×

▲ Figure 16.16B A parasitic excavate: *Trypanosoma* (with blood cells)

16.17 Unikonts include protists that are closely related to fungi and animals

Unikonta is a controversial grouping that joins two well-established clades: **amoebozoans**, which are protists, and a second clade that includes animals and fungi. You'll learn about the amoebozoans in this module, then return to the second clade in the last module of this chapter.

Amoebozoans, including many species of free-living amoebas, some parasitic amoebas, and the slime molds, have lobe-shaped pseudopodia. The amoeba in **Figure 16.17A** is poised to ingest an alga. Its pseudopodia arch around the prey and will enclose it in a food vacuole (see Figure 5.9). Free-living amoebas creep over rocks, sticks, or mud at the bottom of a pond or ocean. A parasitic species of amoeba causes amoebic dysentery, a potentially fatal diarrheal disease.

Colorized TEM 2,000×

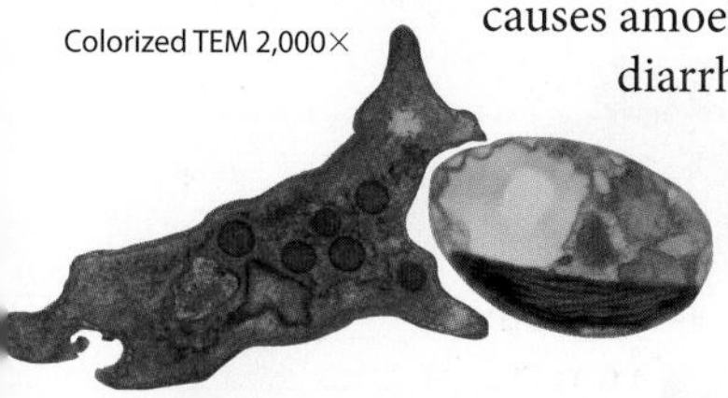

▲ Figure 16.17A An amoeba beginning to ingest an algal cell

The yellow growth creeping over on the dead log in **Figure 16.17B** is an amoebozoan called a **plasmodial slime mold**. These protists are common where there is moist, decaying organic matter and are often brightly pigmented. Although it is large and has many extensions, the organism in Figure 16.17B is not multicellular. Rather, it is a **plasmodium**, a single, multinucleate mass of cytoplasm undivided by plasma membranes. (Don't confuse this word with the alveolate *Plasmodium*, which causes malaria.) Because most of the nuclei go through mitosis at the same time, plasmodial slime molds are used to study molecular details of the cell cycle.

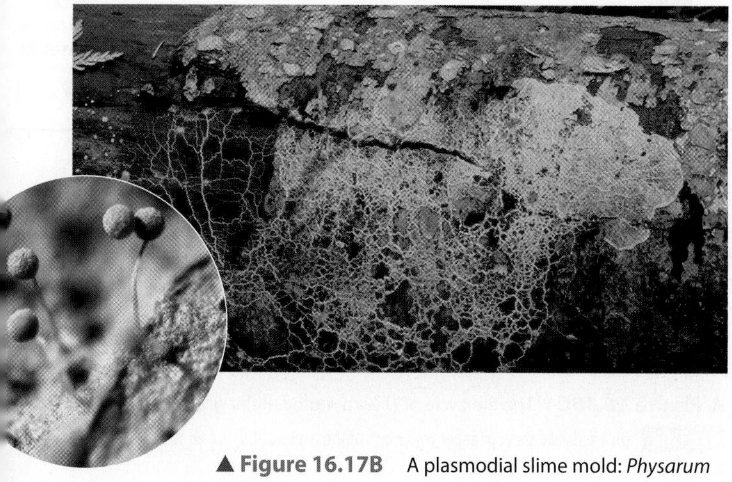

▲ Figure 16.17B A plasmodial slime mold: *Physarum*

The plasmodium extends pseudopodia through soil and rotting logs, engulfing food by phagocytosis as it grows. Within the fine channels of the plasmodium, cytoplasm streams first one way and then the other in pulsing flows that probably help distribute nutrients and oxygen. When food and water are in short supply, the plasmodium stops growing and differentiates into reproductive structures (shown in the inset in Figure 16.17B) that produce spores. When conditions become favorable, the spores release haploid cells that fuse to form a zygote, and the life cycle continues.

Cellular slime molds are also common on rotting logs and decaying organic matter. Most of the time, these organisms exist as solitary amoeboid cells. When food is scarce, the amoeboid cells swarm together, forming a slug-like aggregate that wanders around for a short time. Some of the cells then dry up and form a stalk supporting an asexual reproductive structure in which yet other cells develop into spores. The cellular slime mold *Dictyostelium*, shown in **Figure 16.17C**, is a useful model for researchers studying the genetic mechanisms and chemical changes underlying cellular differentiation.

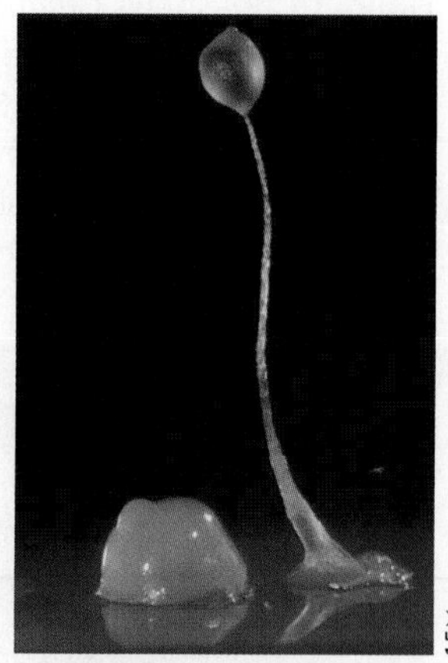
15×

▲ Figure 16.17C An aggregate of amoeboid cells (left) and the reproductive structure of a cellular slime mold, *Dictyostelium*

? Contrast the plasmodium of a plasmodial slime mold with the slug-like stage of a cellular slime mold.

A plasmodium is not multicellular, but is one cytoplasmic mass with many nuclei; the slug-like stage of a cellular slime mold consists of many cells.

16.18 Archaeplastids include red algae, green algae, and land plants

Almost all the members of the supergroup **Archaeplastida** are autotrophic. As you learned in Module 16.13, autotrophic eukaryotes are thought to have arisen by primary endosymbiosis of a cyanobacterium that evolved into chloroplasts. The descendants of this ancient protist evolved into red algae and green algae, which are key photosynthesizers in aquatic food webs. Archaeplastida also includes land plants, which evolved from a group of green algae.

The warm coastal waters of the tropics are home to the majority of species of **red algae**. Their red color comes from an accessory pigment that masks the green of chlorophyll. Although a few species are unicellular, most red algae are multicellular. Multicellular red algae are typically soft-bodied, but some have cell walls encrusted with hard, chalky deposits (**Figure 16.18A**). Encrusted species are common on coral reefs, and their hard parts are important in building and maintaining the reef. Other red algae are commercially important. Carrageenan, a gel that is used to stabilize many products, including ice cream, chocolate milk, and pudding, is derived from species of red algae. Sheets of a red alga, known as nori, are used to wrap sushi. Agar, a polysaccharide used as a substrate for growing bacteria, also comes from red algae.

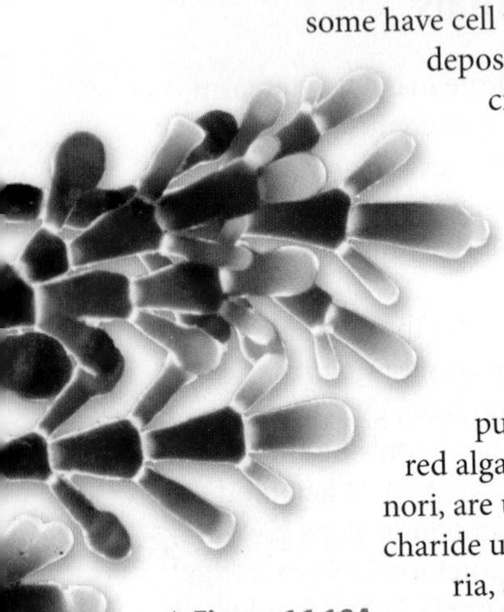

▲ **Figure 16.18A** An encrusted red alga

Green algae, which are named for their grass-green chloroplasts, include unicellular and colonial species as well as multicellular seaweeds. The micrograph on the right in **Figure 16.18B** shows *Chlamydomonas*, a unicellular alga common in freshwater lakes and ponds. It is propelled through the water by two flagella. (Such cells are said to be biflagellated.) *Volvox*, shown on the left, is a colonial green alga. Each *Volvox* colony is a hollow ball composed of hundreds or thousands of biflagellated cells. As the flagella move, the colony tumbles slowly through the water. Some of the large colonies shown here contain small daughter colonies that will eventually be released.

Volvox LM 29× *Chlamydomonas* Colorized SEM 2,600×

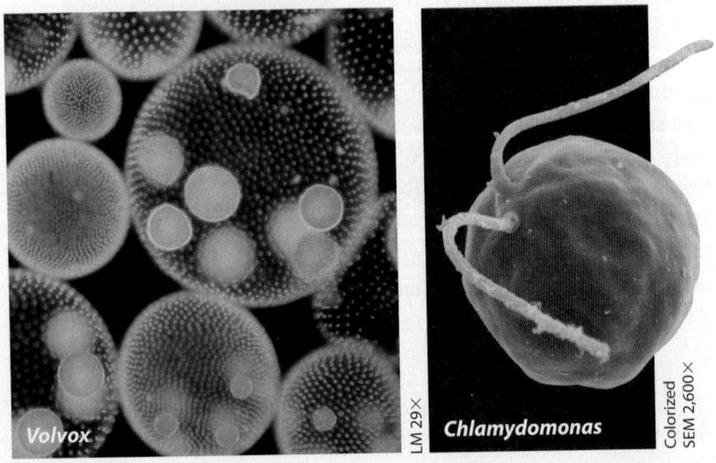

▲ **Figure 16.18B** Green algae, colonial (left) and unicellular (right)

Ulva, or sea lettuce, is a multicellular green alga. Like many multicellular algae and all land plants, *Ulva* has a complex life cycle that includes an **alternation of generations** (**Figure 16.18C**). In this type of life cycle, a multicellular diploid ($2n$) form alternates with a multicellular haploid (n) form. Notice in the figure that multicellular diploid forms are called **sporophytes**, because they produce spores. The sporophyte generation alternates with a haploid generation that features a multicellular haploid form called a **gametophyte**, which produces gametes. In *Ulva*, the gametophyte and sporophyte organisms are identical in appearance; both look like the one in the photograph, although they differ in chromosome number. The haploid gametophyte produces gametes by mitosis, and fusion of the gametes begins the sporophyte generation. In turn, cells in the sporophyte undergo meiosis and produce haploid, flagellated spores. The life cycle is completed when a spore settles to the bottom of the ocean and develops into a gametophyte. (In Chapter 17, you will learn about the alternation of generation life cycles in land plants.)

? **How does chromosome number differ in the gametophyte and sporophyte in the alternation of generations life cycle?**

● The gametophyte is haploid (n); the sporophyte is diploid ($2n$).

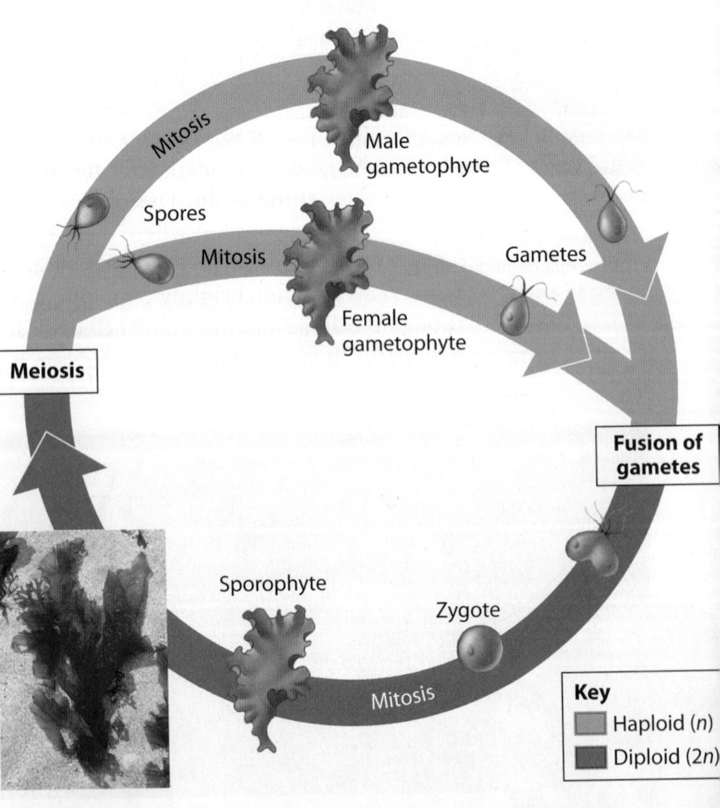

Mitosis
Male gametophyte
Spores
Mitosis
Gametes
Female gametophyte
Meiosis
Fusion of gametes
Sporophyte
Zygote
Mitosis
Key
Haploid (n)
Diploid (2n)

▲ **Figure 16.18C** The life cycle of *Ulva*, a multicellular green alga

Try This On a separate sheet of paper, make lists of the haploid and diploid structures in the life cycle.

16.19 Multicellularity evolved several times in eukaryotes

EVOLUTION CONNECTION

Increased complexity often makes more variations possible. Thus, the origin of the eukaryotic cell led to an evolutionary radiation of new forms of life. As you have seen in this chapter, unicellular protists, which are structurally complex eukaryotic cells, are much more diverse in form than the simpler prokaryotes. The evolution of multicellular bodies broke through another threshold in structural organization.

Multicellular organisms—seaweeds, plants, animals, and most fungi—are fundamentally different from unicellular ones. In a unicellular organism, all of life's activities occur within a single cell. In contrast, a multicellular organism has various specialized cells that perform different functions and are dependent on one another. For example, some cells give the organism its shape, while others make or procure food, transport materials, enable movement, or reproduce.

As you have seen in this chapter, multicellular organisms have evolved in three different ancestral lineages: stramenopiles (brown algae), unikonts (fungi and animals), and archaeplastids (red algae and green algae). **Figure 16.19A** summarizes some current hypotheses for the early phylogeny of land plants and animals, which are all multicellular, and fungi, which are mostly multicellular.

According to one hypothesis, two separate unikont lineages led to fungi and animals. Based on molecular clock calculations (see Module 15.18), scientists estimate that the ancestors of animals and fungi diverged more than 1 billion years ago. A combination of morphological and molecular evidence suggests that a group of unikonts called *choanoflagellates* are the closest living protist relatives of animals. The bottom half of **Figure 16.19B** shows that the cells of choanoflagellates strongly resemble the "collar cells" with which sponges, the group that is closest to the root of the animal tree, obtain food. Similar cells have been found in other animals, but not in fungi or plants. Some species of choanoflagellates live as colonies, federations of independent cells sticking loosely together. Scientists hypothesize that the common ancestor of living animals may have been a stationary colonial choanoflagellate similar to the one shown in Figure 16.19B.

A different group of unikont protists is thought to have given rise to the fungi. Molecular evidence suggests that a group of single-celled protists called *nucleariids*, amoebas that feed on algae and bacteria, are the closest living relatives of fungi (top of Figure 16.19B).

A group of green algae called *charophytes* are the closest living relatives of land plants. Around 500 million years ago, the move onto land began, probably as green algae living along the edges of lakes gave rise to primitive plants. In the next chapter, we trace the long evolutionary movement of plants onto land and their diversification there. (After that, we pick up the thread of animal evolution in Chapter 18.)

? **In what way do multicellular organisms differ fundamentally from unicellular ones?**

In unicellular organisms, all the functions of life are carried out within a single cell. Multicellular organisms have specialized cells that perform different functions.

Ancestral eukaryote
Archaeplastids
Red algae
Green algae
Other green algae
Charophytes
Land plants
Unikonts
Amoebozoans
Nucleariids
Fungi
Choanoflagellates
Animals

Key
All unicellular
Both unicellular and multicellular
All multicellular

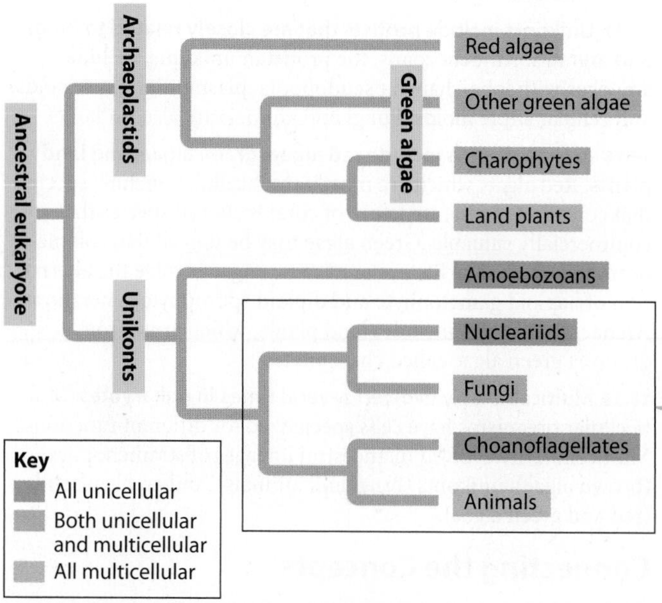

▲ **Figure 16.19A** A hypothesis for the phylogeny of plants, fungi, and animals

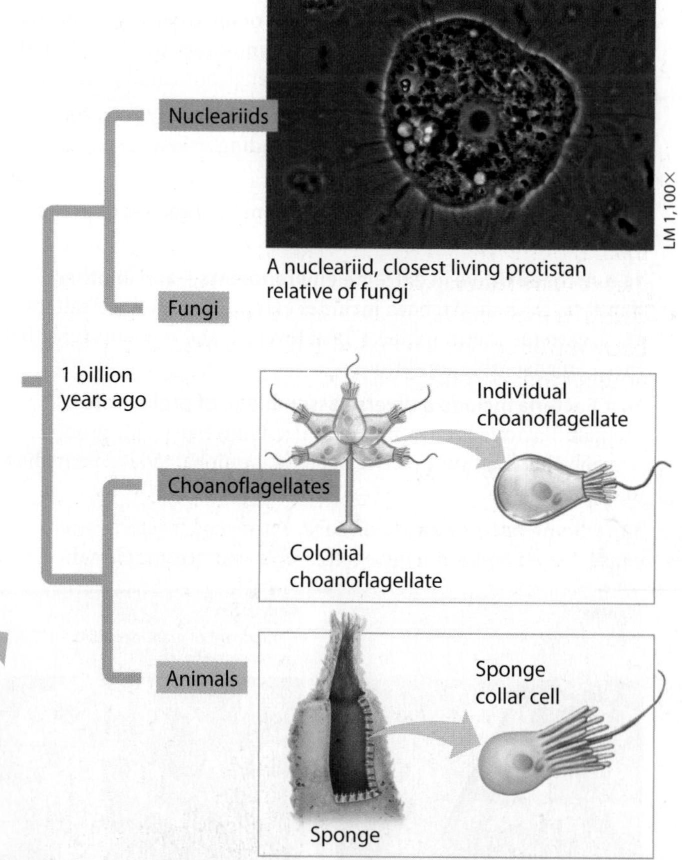
Nucleariids
Fungi
1 billion years ago
LM 1,100×
A nucleariid, closest living protistan relative of fungi
Individual choanoflagellate
Choanoflagellates
Colonial choanoflagellate
Animals
Sponge collar cell
Sponge

▲ **Figure 16.19B** The closest living protist relatives of fungi (top) and animals (bottom)

CHAPTER 16 REVIEW

For practice quizzes, BioFlix animations, MP3 tutorials, video tutors, and more study tools designed for this textbook, go to MasteringBiology®

Reviewing the Concepts

Prokaryotes (16.1–16.11)

16.1 Prokaryotes are diverse and widespread. Prokaryotes are the most numerous organisms. Although small, they have an immense impact on the environment and on our own health.

16.2 External features contribute to the success of prokaryotes. Prokaryotes can be classified by shape and by reaction to a Gram stain. Almost all prokaryotes have a cell wall. Other features may include a sticky capsule, flagella, or fimbriae.

16.3 Populations of prokaryotes can adapt rapidly to changes in the environment. Rapid prokaryote population growth generates a great deal of genetic variation, increasing the likelihood that the population will persist in a changing environment. Some prokaryotes form endospores that remain dormant through harsh conditions.

16.4 Prokaryotes have unparalleled nutritional diversity.

Nutritional mode	Energy source	Carbon source
Photoautotroph	Sunlight	CO_2
Chemoautotroph	Inorganic chemicals	
Photoheterotroph	Sunlight	Organic compounds
Chemoheterotroph	Organic compounds	

16.5 Biofilms are complex associations of microbes. Prokaryotes attach to surfaces and form biofilm communities that are difficult to eradicate, causing both medical and environmental problems.

16.6 Prokaryotes help clean up the environment. Prokaryotes are often used for bioremediation, including in sewage treatment facilities.

16.7 Bacteria and archaea are the two main branches of prokaryotic evolution.

16.8 Archaea thrive in extreme environments—and in other habitats. Domain Archaea includes extreme halophiles ("salt lovers"), extreme thermophiles ("heat lovers"), and methanogens that thrive in anaerobic conditions.

16.9 Bacteria include a diverse assemblage of prokaryotes. Domain Bacteria is currently organized into five major groups: proteobacteria, gram-positive bacteria, cyanobacteria, chlamydias, and spirochetes.

16.10 Some bacteria cause disease. Pathogenic bacteria often cause disease by producing exotoxins or endotoxins. Certain bacteria, such as the species that causes anthrax, and bacterial toxins, such as botulinum, can be used as biological weapons.

Exotoxin
Secreted by cell

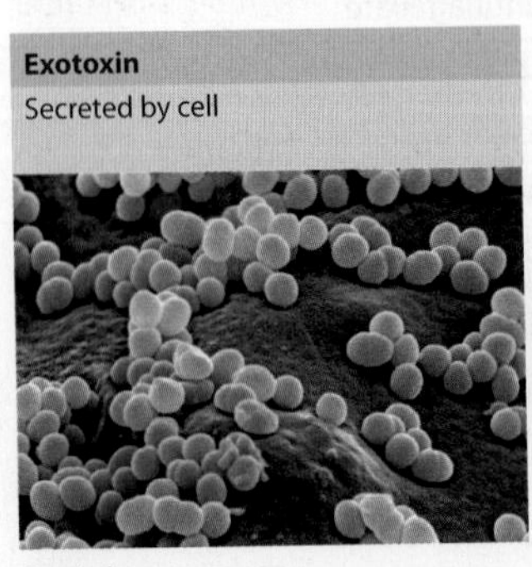
Colorized SEM 12,400×

Staphylococcus aureus

Endotoxin
Component of gram-negative plasma membrane

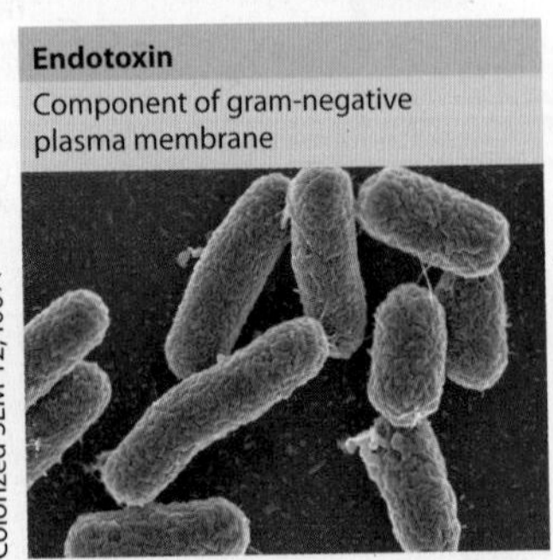
Colorized SEM 7,840×

Salmonella typhimurium

16.11 Stomach microbiota affect health and disease. Barry Marshall used Koch's postulates to show that peptic ulcers are usually caused by a bacterium, *Helicobacter pylori*. Researchers are now beginning to learn that *H. pylori* may also have beneficial roles in the stomach microbiota.

Protists (16.12–16.19)

16.12 Protists are an extremely diverse assortment of eukaryotes. Protists are mostly unicellular eukaryotes that are found in a variety of aquatic or moist habitats. They may be autotrophic, heterotrophic, or mixotrophic. Molecular systematists are exploring protistan phylogeny, but at present it is highly tentative.

16.13 Endosymbiosis of unicellular algae is the key to much of protist diversity. Endosymbiosis of prokaryotic cells resulted in the evolution of eukaryotic cells containing mitochondria. By a similar process, heterotrophic eukaryotic cells engulfed cyanobacteria, which became chloroplasts. Endosymbiosis of red and green algae by eukaryotic cells gave rise to diverse lineages of protists.

16.14 The "SAR" supergroup represents the range of protist diversity. The three clades that make up this supergroup are Stramenopila (including diatoms, brown algae, and water molds), Alveolata (including dinoflagellates, ciliates, and certain parasites), and Rhizaria (including forams and radiolarians).

16.15 Can algae provide a renewable source of energy? Researchers are working on methods of growing diatoms and other algae as a source of biofuels.

16.16 Some excavates have modified mitochondria. Some excavates are anaerobic protists that have modified mitochondria; they include parasitic *Giardia*, *Trichomonas vaginalis*, and *Trypanosomas*. Other excavates include *Euglena*, a mixotroph, and termite endosymbionts.

16.17 Unikonts include protists that are closely related to fungi and animals. Amoebozoans, the protistan unikonts, include amoebas with lobe-shaped pseudopodia, plasmodial slime molds, and cellular slime molds. Fungi and animals are also unikonts.

16.18 Archaeplastids include red algae, green algae, and land plants. Red algae, which are mostly multicellular, include species that contribute to the structure of coral reefs and species that are commercially valuable. Green algae may be unicellular, colonial, or multicellular. The life cycles of many algae involve the alternation of haploid gametophyte and diploid sporophyte generations. Archaeplastida also includes land plants, which arose from a group of green algae called charophytes.

16.19 Multicellularity evolved several times in eukaryotes. Multicellular organisms have cells specialized for different functions. Multicellularity evolved in ancestral lineages of stramenopiles (brown algae), unikonts (fungi and animals), and archaeplastids (red and green algae).

Connecting the Concepts

1. Explain how each of the following characteristics contributes to the success of prokaryotes: cell wall, capsule, flagella, fimbriae, endospores.

2. Fill in the blanks on the phylogenetic tree to show current hypotheses for the origin of multicellular organisms.

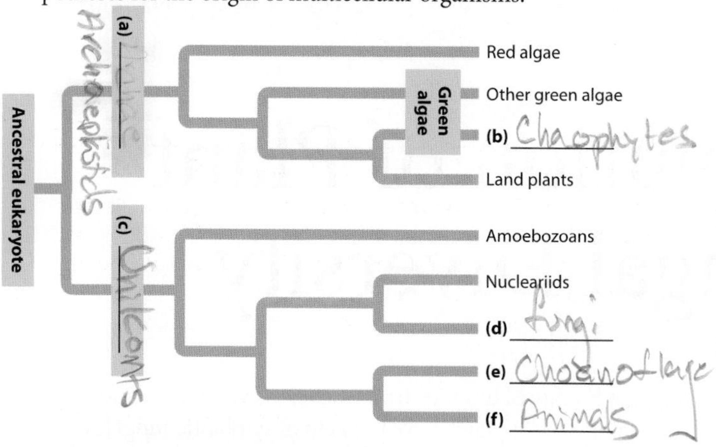
Ancestral eukaryote
(a) ________
Red algae
Green algae
Other green algae
(b) ________
Land plants
(c) ________
Amoebozoans
Nucleariids
(d) ________
(e) ________
(f) ________

Testing Your Knowledge

Level 1: Knowledge/Comprehension

3. In terms of nutrition, autotrophs are to heterotrophs as
   a. kelp are to diatoms.
   b. archaea are to bacteria.
   c. slime molds are to algae.
   d. algae are to slime molds.
4. A new organism has been discovered. Tests have revealed that it is unicellular, is autotrophic, and has a cell wall that contains peptidoglycan. Based on this evidence, it should be classified as a(n)
   a. alga.
   b. archaean.
   c. protist.
   d. bacterium.
5. Which pair of protists has support structures composed of silica?
   a. dinoflagellates and diatoms
   b. diatoms and radiolarians
   c. radiolarians and forams
   d. forams and amoebozoans
6. Which of the following members of the SAR supergroup is incorrectly paired with its clade?
   a. stramenopiles—brown algae
   b. alveolates—parasites such as *Plasmodium*
   c. alveolates—dinoflagellates
   d. Rhizaria—diatoms
7. Which of the following prokaryotes is not pathogenic?
   a. *Chlamydia*
   b. *Rhizobium*
   c. *Streptococcus*
   d. *Salmonella*
8. Explain why prokaryote populations can adapt rapidly to changes in their environment.
9. What characteristic distinguishes true multicellularity from colonies of cells?
10. *Chlamydomonas* is a unicellular green alga. How does it differ from a photosynthetic bacterium, which is also single-celled? How does it differ from a protozoan, such as an amoeba? How does it differ from larger green algae, such as sea lettuce (*Ulva*)?

Level 2: Application/Analysis

11. The bacteria that cause tetanus can be killed only by prolonged heating at temperatures considerably above boiling. This suggests that tetanus bacteria
    a. have cell walls containing peptidoglycan.
    b. secrete endotoxins.
    c. are autotrophic.
    d. produce endospores.
12. Which of the following experiments could test the hypothesis that bacteria cause ulcers in humans? (Assume each experiment includes a control group.) Explain what evidence would be provided by the results of the experiment.
    a. Identify the microbes found in the stomachs of ulcer patients.
    b. Treat a group of ulcer patients with antibiotics.
    c. Place a group of ulcer patients on a strict low-acid diet.
    d. Obtain stomach fluid from ulcer patents and feed it to mice.
13. *Euglena* (an excavate) and *Gymnodinium* (an alveolate) are both capable of photosynthesis. However, researchers have identified differences in their chloroplasts. Use the theory of endosymbiosis to explain the reason for these differences. (*Hint*: Review Figure 16.13.)

Level 3: Synthesis/Evaluation

14. **SCIENTIFIC THINKING** Probiotics, foods and supplements that contain living microorganisms, are thought to cure problems of the digestive tract by restoring the natural balance of its microbial community. Sales of these products total billions of dollars a year. Explore the topic of probiotics and evaluate the scientific evidence for their beneficial effects. A good starting point is the website of the U.S. Food and Drug Administration, which regulates advertising claims of health benefits of dietary supplements (www.fda.gov/Food/DietarySupplements/default.htm. **Source** U.S. Food and Drug Administration website, 2013.).
15. Imagine you are on a team designing a moon base that will be self-contained and self-sustaining. Once supplied with building materials, equipment, and organisms from Earth, the base will be expected to function indefinitely. One of the team members has suggested that everything sent to the base be sterilized so that no bacteria of any kind are present. Do you think this is a good idea? Predict some of the consequences of eliminating all bacteria from an environment.
16. The buildup of CO_2 in the atmosphere resulting from the burning of fossil fuels is regarded as a major contributor to global warming (see Module 38.3). Diatoms and other microscopic algae in the oceans counter this buildup by using large quantities of atmospheric CO_2 in photosynthesis, which requires small quantities of iron. Experts suspect that a shortage of iron may limit algal growth in the oceans. Some scientists have suggested that one way to reduce CO_2 buildup might be to fertilize the oceans with iron. The iron would stimulate algal growth and thus the removal of more CO_2 from the air. A single supertanker of iron dust, spread over a wide enough area, might reduce the atmospheric CO_2 level significantly. Do you think this approach would be worth a try? Why or why not?

Answers to all questions can be found in Appendix 4.

CHAPTER 17

The Evolution of Plant and Fungal Diversity

Do fungi feed the world?

Plants get all the credit for giving us a bounty of luscious fruits, delicious vegetables, and nutritious grains, but plants don't do it alone. Most plants have the help of symbiotic fungi known as mycorrhizae (meaning "fungus root") that thread their way into the roots and grow small, bushy projections within plant cells. These intimate associations allow the plant to tap a vast underground network of fungal filaments into which water and mineral nutrients flow. Plants return the favor by supplying the fungi with sugars and other organic molecules. It's a win-win situation for plant and fungus, and people benefit, too. The growth of almost every plant you eat—oranges, peaches, strawberries, cherries, potatoes, tomatoes, corn, and wheat, to name just a few—was assisted by these unseen fungal partners. Coffee (shown in the photo below) is highly dependent on its hidden fungal allies, and we have mycorrhizae to thank for chocolate and cashews as well. In fact, if all the mycorrhizae of food plants were to disappear from the planet, our most productive crops would be some of the less popular ones—cabbage, broccoli, and beets, for example.

The dependence of food plants on mycorrhizae is not unique—at least 90% of all plants form such relationships. Mutually beneficial symbioses between plants and fungi began 500 million years ago, when plants first occupied land. As you'll learn later in this chapter, researchers hypothesize that these symbioses were crucial to the colonization of land by plants. Accordingly, although the lineages that gave rise to plants and fungi diverged more than a billion years ago, we explore fungal diversity in this chapter along with evolution in the plant kingdom.

We begin the chapter by exploring structural and reproductive adaptations that equip plants for life on land. From a modest beginning, plants diversified into the 290,000 present-day species with adaptations that enable them to live in all kinds of environments.

BIG IDEAS

Plant Evolution and Diversity

(17.1–17.2)

A variety of adaptations enable plants to live on land.

Alternation of Generations and Plant Life Cycles

(17.3–17.11)

Plant life cycles alternate haploid (gametophyte) and diploid (sporophyte) generations.

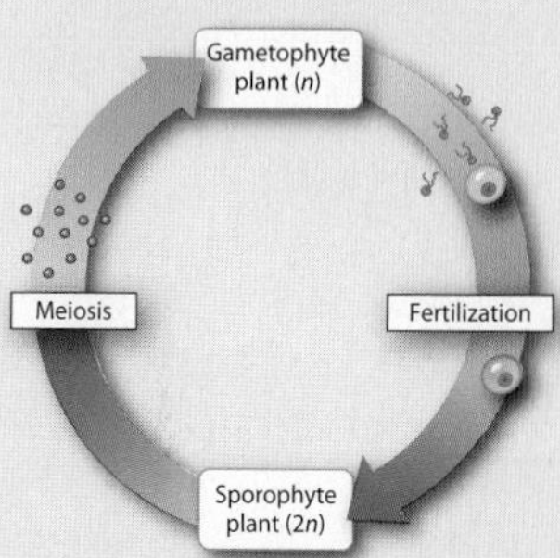
Gametophyte plant (n)
Meiosis
Fertilization
Sporophyte plant (2n)

Diversity of Fungi

(17.12–17.19)

Fungi are a diverse group of organisms that acquire nutrients through absorption. Many fungi have complex life cycles.

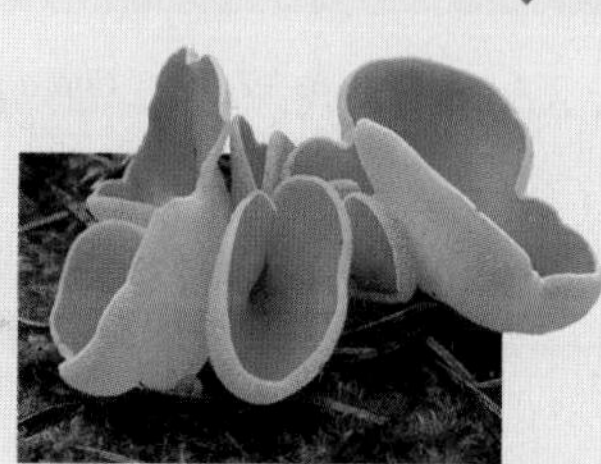

▷ Plant Evolution and Diversity

17.1 Plants have adaptations for life on land

Plants and green algae called charophytes are thought to have evolved from a common ancestor (Module 16.19). Like plants, charophytes are photosynthetic eukaryotes, and many species have complex, multicellular bodies (**Figure 17.1A**). As you will see, many of the adaptations that evolved after plants diverged from algae facilitated survival and reproduction on dry land. Some plant groups, including water weeds such as *Anacharis* that are used in aquariums, returned to aquatic habitats during their evolution. However, most present-day plants live in terrestrial environments.

▲ **Figure 17.1A** *Chara*, an elaborate charophyte

The algal ancestors of plants may have carpeted moist fringes of lakes or coastal salt marshes over 500 million years ago. These shallow-water habitats were subject to occasional drying, and natural selection would have favored algae that could survive periodic droughts. Some species accumulated adaptations that enabled them to live permanently above the water line. The modern-day green alga *Coleochaete* (**Figure 17.1B**), which grows at the edges of lakes as disklike, multicellular colonies, may resemble one of these early plant ancestors.

Adaptations making life on dry land possible had accumulated by about 470 million years ago, the age of the oldest known land plant fossils. The evolutionary novelties of these first land plants opened the new frontier of a terrestrial habitat. Early plant life would have thrived in the new environment. Bright sunlight was virtually limitless on land; the atmosphere had an abundance of carbon dioxide (CO_2); and at first there were relatively few pathogens and plant-eating animals.

On the other hand, life on land poses a number of challenges. Because terrestrial organisms are surrounded by air rather than water, they must be able to maintain moisture inside their cells, support the body in a nonbuoyant medium, and reproduce and disperse offspring without water. As nonmotile organisms, plants must also anchor their bodies in the soil and obtain resources from both soil and air. Thus, the water-to-land transition required fundamental changes in algal structure and life cycle.

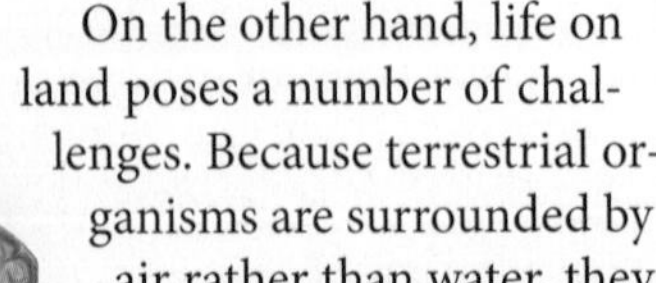

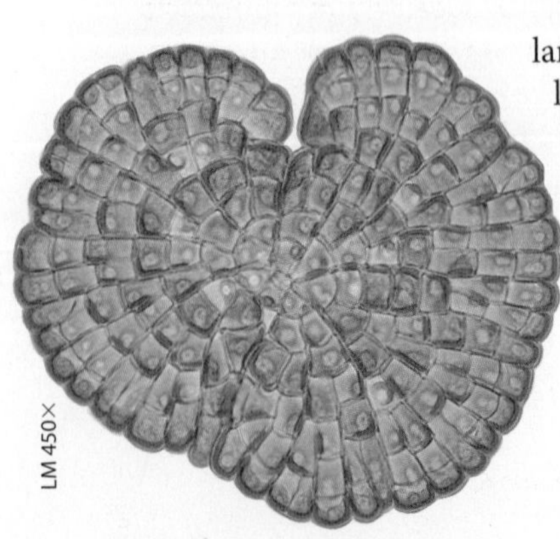
LM 450×

▲ **Figure 17.1B** *Coleochaete*, a simple charophyte

Figure 17.1C compares how multicellular algae like *Chara* differ from plants such as mosses, ferns, and pines. Many algae are anchored by a holdfast, but generally they have no rigid tissues and are supported by the surrounding water. The whole algal body obtains CO_2 and minerals directly from the water. Almost all of the organism receives light and can perform photosynthesis. For reproduction, flagellated sperm swim to fertilize an egg. The offspring are dispersed by water as well.

Maintaining Moisture The aboveground parts of most land plants are covered by a waxy cuticle that prevents water loss. Gas exchange cannot occur directly through the cuticle, but CO_2 and O_2 diffuse across the leaf surfaces through the tiny pores called stomata (see Module 7.2). Two surrounding cells regulate each stoma's opening and closing. Stomata are usually open during sunlight hours, allowing gas exchange, and closed at other times, preventing water loss by evaporation.

Obtaining Resources from Two Locations A typical plant must obtain chemicals from both soil and air, two very different media. Water and mineral nutrients are mainly found in the soil; light and CO_2 are available above ground. Most plants have discrete organs—roots, stems, and leaves—that help meet this resource challenge.

Plant roots provide anchorage and absorb water and mineral nutrients from the soil. Above ground, a plant's stems bear leaves, which obtain CO_2 from the air and light from the sun, enabling them to perform photosynthesis. Growth-producing regions of cell division, called **apical meristems**, are found near the tips of stems and roots. The elongation and branching of a plant's stems and roots maximize exposure to the resources in the soil and air.

A plant must be able to connect its subterranean and aerial parts, conducting water and minerals upward from its roots to its leaves and distributing sugars produced in the leaves throughout its body. Most plants, including ferns, pines, and flowering plants, have **vascular tissue**, a network of thick-walled cells joined into narrow tubes that extend throughout the plant body (traced in red in Figure 17.1C). The photograph of part of an aspen leaf in **Figure 17.1D**, at the bottom of the next page, shows the leaf's network of veins, which are fine branches of the vascular tissue. There are two types of vascular tissue. **Xylem** includes dead cells that form microscopic pipes conveying water and minerals up from the roots. **Phloem**, which consists entirely of living cells, distributes sugars throughout the plant. In contrast to plants with elaborate vascular tissues, mosses lack a complex transport system (although some mosses do have simple vascular tissue). With limited means for distributing water and minerals from soil to the leaves, the height of nonvascular plants is severely restricted.

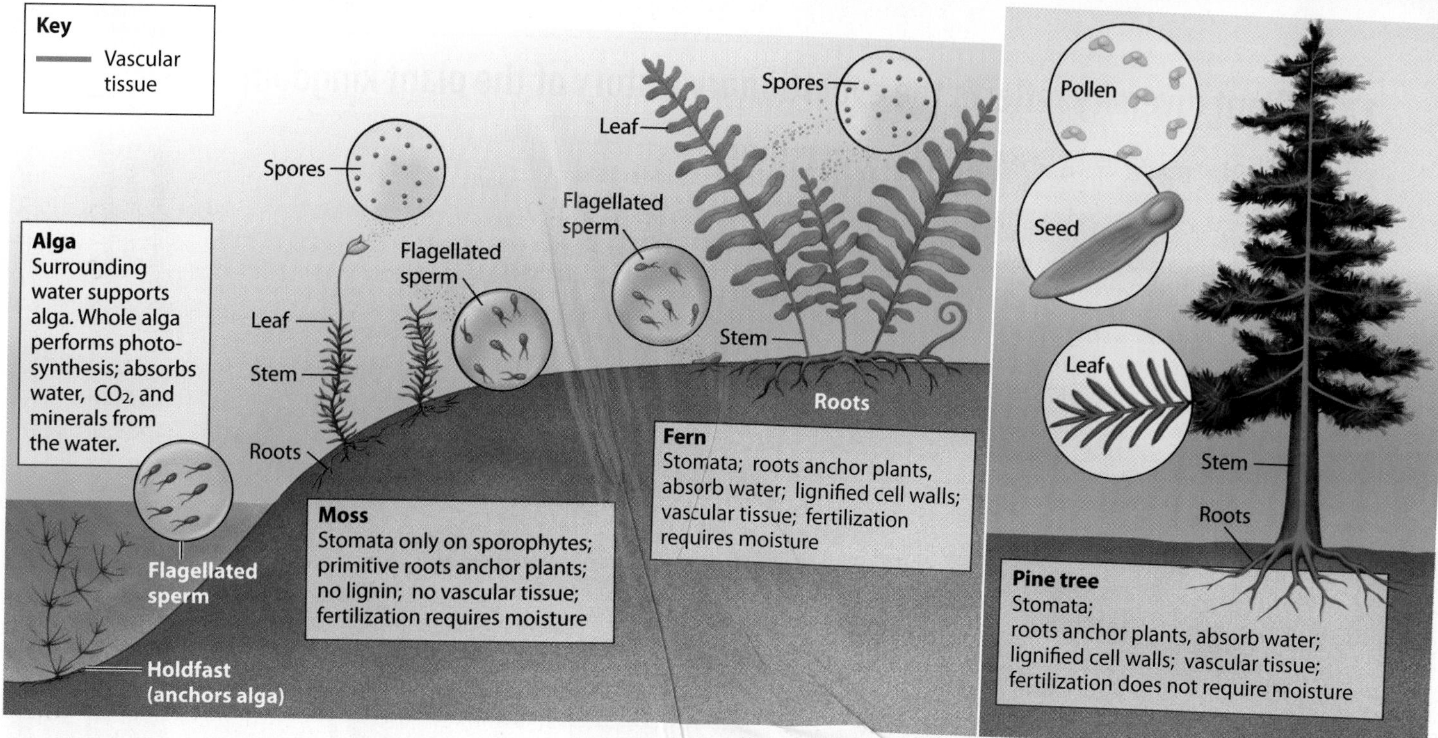
Key
Vascular tissue
Alga
Surrounding water supports alga. Whole alga performs photosynthesis; absorbs water, CO_2, and minerals from the water.
Flagellated sperm
Holdfast (anchors alga)
Spores
Leaf
Stem
Roots
Flagellated sperm
Moss
Stomata only on sporophytes; primitive roots anchor plants; no lignin; no vascular tissue; fertilization requires moisture
Spores
Leaf
Flagellated sperm
Stem
Roots
Fern
Stomata; roots anchor plants, absorb water; lignified cell walls; vascular tissue; fertilization requires moisture
Pollen
Seed
Leaf
Stem
Roots
Pine tree
Stomata;
roots anchor plants, absorb water; lignified cell walls; vascular tissue; fertilization does not require moisture

▲ **Figure 17.1C** Comparing the aquatic adaptations of *Chara*, a multicellular green alga, with the terrestrial adaptations of moss, fern, and pine

Try This Construct a table showing adaptations for maintaining moisture, obtaining resources, support, reproduction, and dispersal in each plant represented (multicellular alga, moss, fern, and pine).

Supporting the Plant Body Because air provides much less support than water, plants must be able to hold themselves up against the pull of gravity. The cell walls of some plant tissues, including xylem, are thickened and reinforced by a chemical called **lignin**. The absence of lignified cell walls in mosses and other plants that lack vascular tissue is another limitation on their height.

Reproduction and Dispersal Reproduction on land presents complex challenges. For *Chara* and other algae, the surrounding water ensures that gametes and offspring stay moist. Plants, however, must keep their gametes and developing embryos from drying out in the air. Like the earliest land plants, mosses and ferns produce gametes in male and female structures called **gametangia** (singular, *gametangium*), which consist of protective jackets of cells surrounding the gamete-producing cells. The egg remains in the female gametangium and is fertilized there by a sperm that swims through a film of water. As a result, mosses and ferns can only reproduce in a moist environment. Pines and flowering plants have **pollen grains**, structures that contain the sperm-producing cells. Pollen grains are carried close to the egg by wind or animals; moisture is not required for bringing sperm and egg together.

In all plants, the fertilized egg (zygote) develops into an embryo while attached to and nourished by the parent plant. This multicellular, dependent embryo is the basis for designating plants as **embryophytes** (*phyte* means "plant"), distinguishing them from algae.

The life cycles of all plants involve the alternation of a haploid generation, which produces eggs and sperm, and a diploid generation, which produces spores within protective structures called **sporangia** (singular, *sporangium*). A **spore** is a cell that can develop into a new organism without fusing with another cell. The earliest land plants relied on tough-walled spores for dispersal, a trait retained by mosses and ferns today. Pines and flowering plants have seeds for launching their offspring. Seeds are elaborate embryo-containing structures that are well protected from the elements and are dispersed by wind or animals. Plants that disperse their offspring as spores are often referred to as seedless plants.

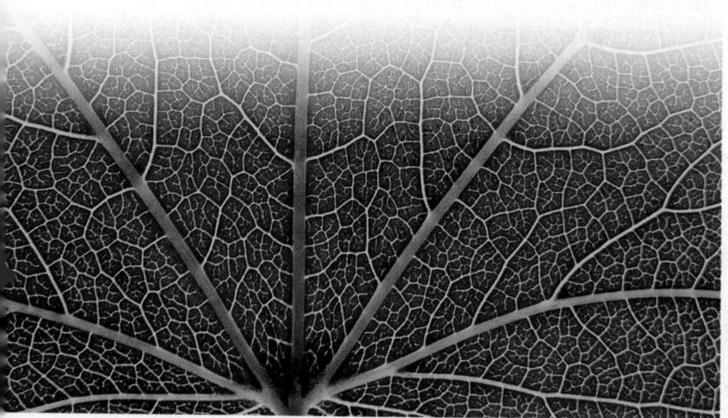

▲ **Figure 17.1D** The network of veins in a leaf

? **What adaptations enable plants to grow tall?**

● Vascular tissues to transport materials from belowground parts to aboveground parts and vice versa; lignified cell walls to provide structural support

17.2 Plant diversity reflects the evolutionary history of the plant kingdom

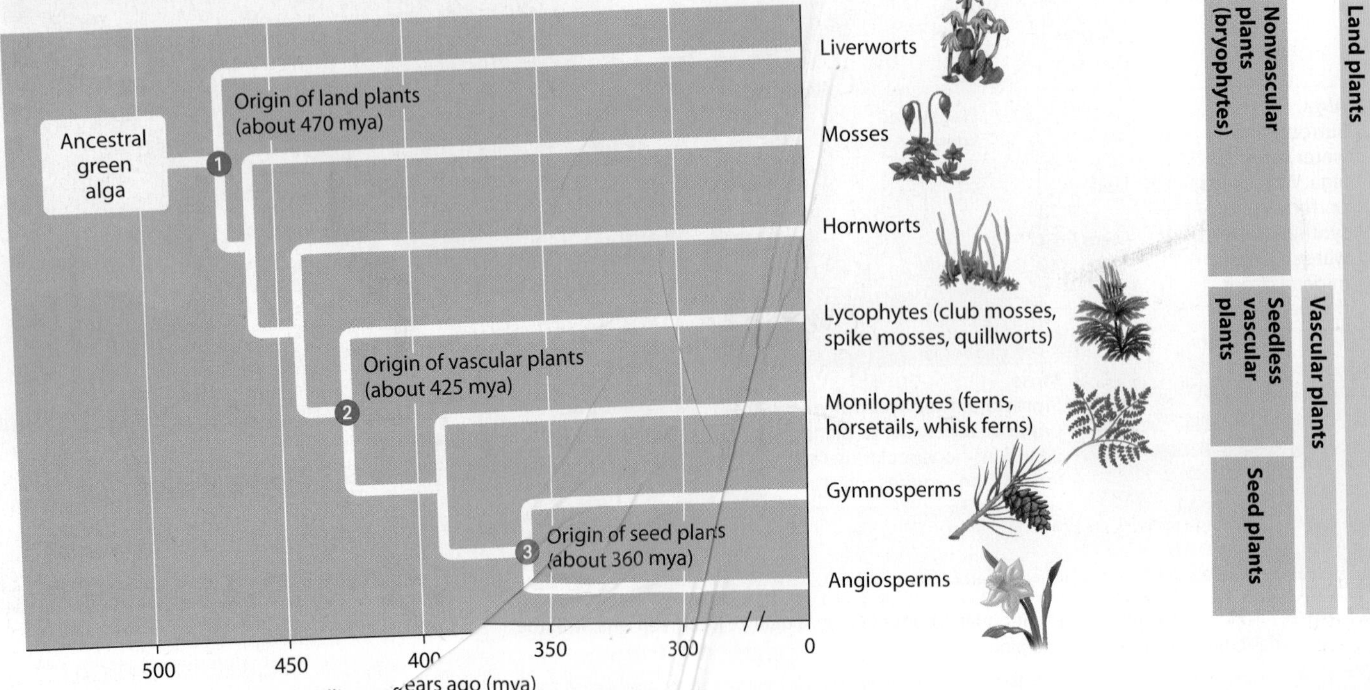

Ancestral green alga
1
Origin of land plants (about 470 mya)
2
Origin of vascular plants (about 425 mya)
3
Origin of seed plants (about 360 mya)
Liverworts
Mosses
Hornworts
Lycophytes (club mosses, spike mosses, quillworts)
Monilophytes (ferns, horsetails, whisk ferns)
Gymnosperms
Angiosperms
Nonvascular plants (bryophytes)
Seedless vascular plants
Seed plants
Vascular plants
Land plants
500
450
400
350
300
0
Millions of years ago (mya)

▲ **Figure 17.2A** Some highlights of plant evolution

Figure 17.2A highlights some of the major events in the history of the plant kingdom and presents a widely held view of the relationships between surviving lineages of plants.

❶ After plants originated from an algal ancestor approximately 470 million years ago, early diversification gave rise to seedless, nonvascular plants, including mosses, liverworts, and hornworts (**Figure 17.2B**). These plants, which are informally called **bryophytes**, resemble other plants in having apical meristems and embryos that are retained on the parent plant, but they lack true roots and leaves. Without lignified cell walls, bryophytes with an upright growth habit lack support.

Moss

Liverwort

Hornwort

▲ **Figure 17.2B** Bryophytes

A mat of moss actually consists of many plants growing in a tight pack. Like people crowded together in a small space, their bodies hold one another upright. The mat is spongy and retains water. Other bryophytes grow flat against the ground. Growth in dense mats facilitates fertilization by flagellated sperm swimming through a film of water left by rain or dew.

❷ The origin of **vascular plants** occurred about 425 million years ago. Their lignin-hardened vascular tissues provide strong support, enabling stems to stand upright and grow tall on land. Two clades of vascular plants are informally called **seedless vascular plants** (**Figure 17.2C**, on the facing page): the lycophytes (such as club mosses) and the widespread monilophytes (ferns and their relatives). A fern has well-developed roots and rigid stems. Ferns are common in temperate forests, but they are most diverse in the tropics. In some tropical species, called tree ferns, upright stems can grow several meters tall. Like bryophytes, however, ferns and club mosses require moist conditions for fertilization, and they disperse their offspring as spores that are carried by air currents.

❸ The first vascular plants with seeds evolved about 360 million years ago. Today, the seed plant lineage accounts for over 90% of the approximately 290,000 species of living plants. Seeds and pollen are key adaptations that improved the ability of plants to diversify in terrestrial habitats. A **seed** consists of an embryo packaged with a food supply within a protective covering. This survival packet facilitates wide dispersal of plant embryos. As you learned in Module 17.1,

Club moss (a lycophyte). Spores are produced in the upright tan-colored structures.

Fern (a monilophyte)

▲ **Figure 17.2C** Seedless vascular plants

pollen brings sperm-producing cells into contact with egg-producing parts without water. And unlike flagellated sperm, which can swim a few centimeters at most, pollen can travel great distances.

Gymnosperms (from the Greek *gymnos*, naked, and *sperma*, seed) were among the earliest seed plants. Seeds of gymnosperms are said to be "naked" because they are not produced in specialized chambers. The largest clade of gymnosperms is the conifers, consisting mainly of cone-bearing trees, such as pine, spruce, and fir. (The term *conifer* means "cone-bearing.") Some examples of gymnosperms that are less common are the ornamental ginkgo tree, the palmlike cycads, and desert shrubs in the genus *Ephedra* (**Figure 17.2D**). Gymnosperms flourished alongside the dinosaurs in the Mesozoic era.

The most recent major episode in plant evolution was the appearance of flowering plants, or **angiosperms** (from the Greek *angion*, container, and *sperma*, seed), at least 140 million years ago. Flowers are complex reproductive structures that develop seeds within protective chambers. The great majority of living plants—some 250,000 species—are angiosperms, which include a wide variety of plants, such as grasses, flowering shrubs, and flowering trees (**Figure 17.2E**).

In summary, four key adaptations for life on land distinguish the main lineages of the plant kingdom. (1) Dependent embryos are present in all plants. (2) Lignified vascular tissues mark a lineage that gave rise to most living plants. (3) Seeds are found in a lineage that includes all living gymnosperms and angiosperms and that dominates the plant kingdom today. (4) Flowers mark the angiosperm lineage. As you will see in the next module, the life cycles of living plants reveal additional details about plant evolution.

▼ **Figure 17.2D** Gymnosperms

Ginkgo

Cycad

Ephedra (Mormon tea)

A conifer

A jacaranda tree

Green foxtail, a grass

▲ **Figure 17.2E** Angiosperms

? Identify which of the following traits is shared by all plants: flowers, seeds, retained embryo, vascular tissue.

An embryo retained on parent plant

▷ Alternation of Generations and Plant Life Cycles

VISUALIZING THE CONCEPT

17.3 Haploid and diploid generations alternate in plant life cycles

Plants have life cycles that are very different from ours. Humans are diploid individuals—that is, each of us has two sets of chromosomes, one from each parent (Module 8.12). Gametes (sperm and eggs) are the only haploid stage in the human life cycle. Plants have an **alternation of generations**: The diploid and haploid stages are distinct, multicellular bodies.

The haploid generation of a plant produces gametes and is called the **gametophyte**. The diploid generation produces spores and is called the **sporophyte**. In a plant's life cycle, these two generations alternate in producing each other. In mosses, as in all nonvascular plants, the gametophyte is the larger, more obvious stage of the life cycle. Ferns, like most plants, have a life cycle dominated by the sporophyte. Today, about 95% of all plants, including all seed plants, have a dominant sporophyte in their life cycle. The life cycles of all plants follow a pattern shown here. ⟶

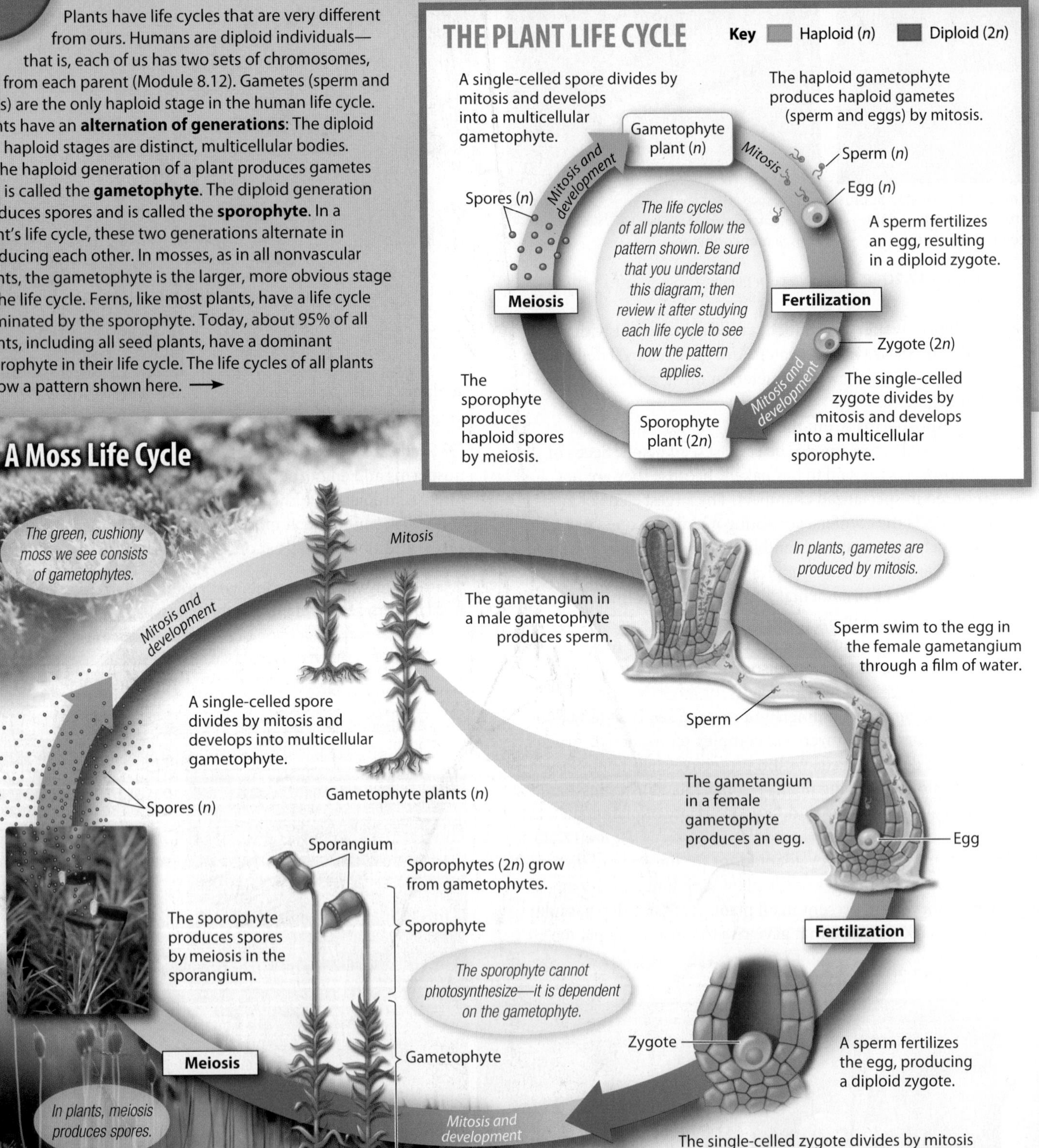

THE PLANT LIFE CYCLE
Key
Haploid (*n*)
Diploid (2*n*)
A single-celled spore divides by mitosis and develops into a multicellular gametophyte.
Gametophyte plant (*n*)
Mitosis and development
Mitosis
The haploid gametophyte produces haploid gametes (sperm and eggs) by mitosis.
Sperm (*n*)
Egg (*n*)
Spores (*n*)
The life cycles of all plants follow the pattern shown. Be sure that you understand this diagram; then review it after studying each life cycle to see how the pattern applies.
A sperm fertilizes an egg, resulting in a diploid zygote.
Meiosis
Fertilization
Zygote (2*n*)
The sporophyte produces haploid spores by meiosis.
Sporophyte plant (2*n*)
Mitosis and development
The single-celled zygote divides by mitosis and develops into a multicellular sporophyte.
A Moss Life Cycle
The green, cushiony moss we see consists of gametophytes.
Mitosis
Mitosis and development
In plants, gametes are produced by mitosis.
The gametangium in a male gametophyte produces sperm.
Sperm swim to the egg in the female gametangium through a film of water.
A single-celled spore divides by mitosis and develops into multicellular gametophyte.
Sperm
Spores (*n*)
Gametophyte plants (*n*)
The gametangium in a female gametophyte produces an egg.
Egg
Sporangium
Sporophytes (2*n*) grow from gametophytes.
Sporophyte
The sporophyte produces spores by meiosis in the sporangium.
Fertilization
The sporophyte cannot photosynthesize—it is dependent on the gametophyte.
Zygote
Gametophyte
A sperm fertilizes the egg, producing a diploid zygote.
Meiosis
In plants, meiosis produces spores.
Mitosis and development
The single-celled zygote divides by mitosis and develops into a multicellular sporophyte.

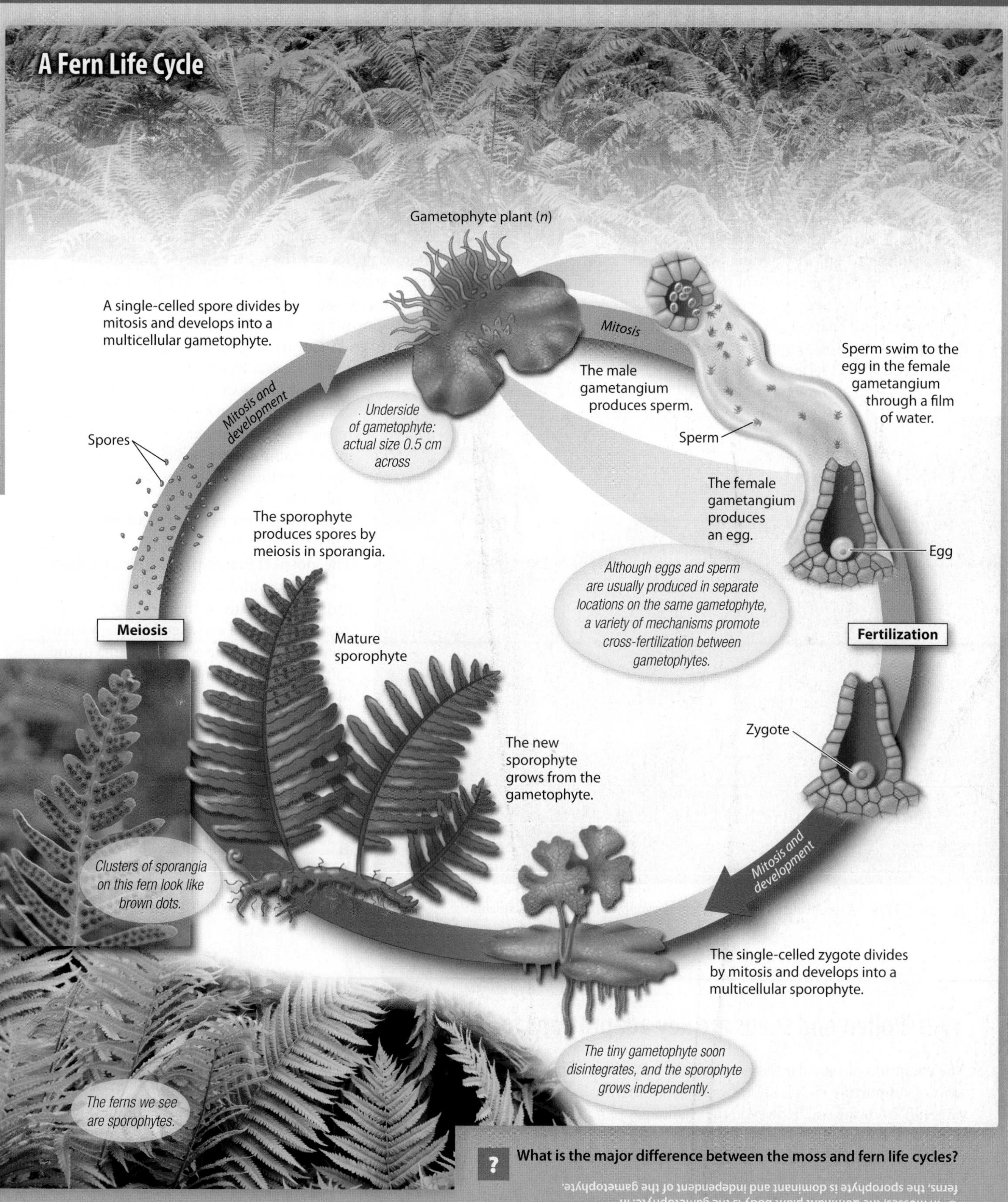

A Fern Life Cycle
Gametophyte plant (n)
A single-celled spore divides by mitosis and develops into a multicellular gametophyte.
Mitosis
The male gametangium produces sperm.
Sperm swim to the egg in the female gametangium through a film of water.
Mitosis and development
Underside of gametophyte: actual size 0.5 cm across
Spores
Sperm
The female gametangium produces an egg.
The sporophyte produces spores by meiosis in sporangia.
Egg
Although eggs and sperm are usually produced in separate locations on the same gametophyte, a variety of mechanisms promote cross-fertilization between gametophytes.
Meiosis
Fertilization
Mature sporophyte
The new sporophyte grows from the gametophyte.
Zygote
Clusters of sporangia on this fern look like brown dots.
Mitosis and development
The single-celled zygote divides by mitosis and develops into a multicellular sporophyte.
The tiny gametophyte soon disintegrates, and the sporophyte grows independently.
The ferns we see are sporophytes.

? **What is the major difference between the moss and fern life cycles?**

In mosses, the dominant plant body is the gametophyte. In ferns, the sporophyte is dominant and independent of the gametophyte.

17.4 Seedless vascular plants dominated vast "coal forests"

During the Carboniferous period (about 360–299 million years ago), the two clades of seedless vascular plants, lycophytes (for example, club mosses) and monilophytes (such as ferns and their relatives), grew in vast forests in the low-lying wetlands of what is now Eurasia and North America. At that time, these continents were close to the equator and had warm, humid climates that supported broad expanses of lush vegetation.

Figure 17.4 shows an artist's reconstruction of one of these forests based on fossil evidence. Tree ferns are visible in the foreground. Most of the large trees are lycophytes, giants that grew as tall as a 12-story building, with diameters of more than 2 m (6 feet). (For a sense of scale, dragonflies such as the one shown in the foreground had wingspans of up to 1 m.) Vertebrates were adapting to terrestrial habitats in parallel with plants; amphibians and early reptiles lived among these trees (see Module 19.4).

Photosynthesis in these immense swamp forests fixed large amounts of carbon from CO_2 into organic molecules, dramatically reducing CO_2 levels in the atmosphere. Because atmospheric CO_2 traps heat, this change caused global cooling. Photosynthesis generated great quantities of organic matter. As the plants died, they fell into stagnant swamps and did not decay completely. Their remains formed thick organic deposits called peat. Later, seawater covered the swamps, marine sediments covered the peat, and pressure and heat gradually converted the peat to coal, black sedimentary rock made up of fossilized plant material. Coal deposits from the Carboniferous period are the most extensive ever formed. (The name Carboniferous comes from the Latin *carbo*, coal, and *fer*, bearing.) Coal, oil, and natural gas are called **fossil fuels** because they were formed from the remains of ancient organisms. (Oil and natural gas were formed from marine organisms.) Since the Industrial Revolution, coal has been a crucial source of energy for human society. However, burning these fossil fuels releases CO_2 and other greenhouse gases into the atmosphere, which are now causing a warming climate (see Modules 38.3 and 38.4).

As temperatures dropped during the late Carboniferous period, glaciers formed. The global climate turned drier, and the vast swamps and forests began to disappear. The climate change provided an opportunity for the early seed plants, which grew along with the seedless plants in the Carboniferous swamps. With their wind-dispersed pollen and protective seeds, seed plants could complete their life cycles on dry land.

▲ **Figure 17.4** A reconstruction of an extinct forest dominated by seedless plants

? **How did the tropical swamp forests contribute to global cooling in the Carboniferous period?**

● Photosynthesis by the abundant plant life reduced atmospheric CO_2, a gas that traps heat.

17.5 Pollen and seeds are key adaptations for life on land

The evolution of vascular tissue solved the terrestrial problems of supporting the plant body and obtaining water and minerals from the soil. However, the challenges of reproduction and dispersing offspring on dry land remained. In contrast to seedless plants, which produce flagellated sperm that need moisture to reach an egg, seed plants—gymnosperms and angiosperms—have pollen grains that carry their sperm-producing cells through the air. In addition, the offspring of seedless plants are sent off into the world as haploid, single-celled spores that must survive independently as gametophytes before producing the next sporophyte generation. Seed plants launch next-generation sporophytes that are ready to grow. Let's look at how these adaptations fit into the life cycle of a gymnosperm.

In seed plants, a specialized structure within the sporophyte houses all reproductive stages, including spores, eggs, sperm, zygotes, and embryos. In gymnosperms such as pines and other conifers, this structure is called a cone. If you look at the longitudinal section of the cones in **Figure 17.5A**, you'll see that the cone resembles a short stem bearing thick leaves. The resemblance is not surprising—cones are modified shoots that serve a reproductive function. Each "leaf," or scale, of the cone contains sporangia that produce spores by meiosis. Unlike seedless plants, however, the spores of seed plants are not released. Rather, spores give rise to gametophytes within the shelter of the sporophyte.

Longitudinal section of ovulate cone
Sporangia
Longitudinal section of pollen cone

▲ Figure 17.5A Male and female pine cones

In the male reproductive structures of seed plants, haploid spores develop into pollen grains, which are male gametophytes enclosed within a tough wall. Many species, including all conifers and many flowering trees, release millions of microscopic pollen grains in great clouds. You may have seen yellowish pollen covering cars or floating on puddles after a spring rain. If a pollen grain lands on a compatible female structure, an event known as **pollination**, it undergoes mitosis to produce a sperm. In seed plants, the sperm is reduced to a nucleus.

Haploid spores in female reproductive structures develop into **ovules**, which contain the egg-producing female gametophytes (**Figure 17.5B**). If pollination has occurred, the pollen grain grows a tiny tube that enters the ovule and releases the sperm nucleus. Only then does fertilization occur—pollination and fertilization are separate events. The resulting diploid zygote undergoes mitoses and becomes a sporophyte embryo. The ovule and its surrounding tissues mature into a seed consisting of an embryo, a food supply to sustain it until it is capable of photosynthesis, and a tough **seed coat**. In many plants, including pines, the seed coat is extended into a winglike structure that catches the breeze and carries the seed far from the parent plant. You'll learn about other adaptations for seed dispersal in Module 17.8.

Next we consider the reproductive adaptations of flowering plants, the most diverse and geographically widespread of all plants. Angiosperms dominate most landscapes today, and it is their flowers that account for their unparalleled success.

? How do pollen and seeds increase the reproductive success of seed plants?

● Pollen transfers sperm to eggs without the need for water. Seeds protect, nourish, and help disperse plant embryos.

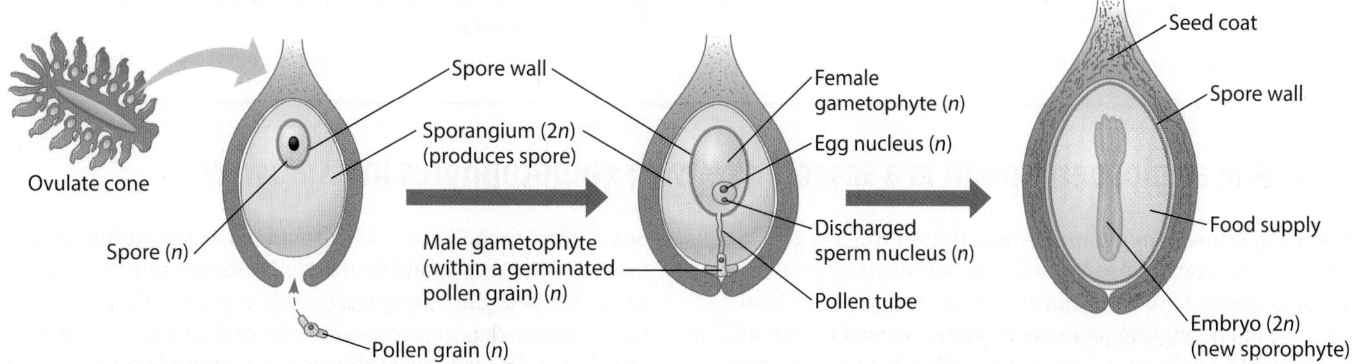
Ovulate cone
Spore wall
Sporangium (2*n*) (produces spore)
Spore (*n*)
Pollen grain (*n*)
Male gametophyte (within a germinated pollen grain) (*n*)
Female gametophyte (*n*)
Egg nucleus (*n*)
Discharged sperm nucleus (*n*)
Pollen tube
Seed coat
Spore wall
Food supply
Embryo (2*n*) (new sporophyte)

▲ Figure 17.5B From ovule to seed in a gymnosperm

17.6 The flower is the centerpiece of angiosperm reproduction

No organisms make a showier display of their sex life than angiosperms (**Figure 17.6A**). From roses to cherry blossoms, flowers are the sites of pollination and fertilization. Like pine cones, flowers house separate male and female sporangia and gametophytes, and the mechanisms of sexual reproduction, including pollination and fertilization, are similar. And like cones, flowers are also short stems bearing modified leaves. However, as you can see in **Figure 17.6B**, the modifications are quite different from the scales of a pine cone. Each floral structure is highly specialized for a different function, and the structures are attached in a circle to a receptacle at the base of the flower. The outer layer of the circle consists of the **sepals**, which are usually green. They enclose the flower before it opens. When the sepals are peeled away, the next layer is the **petals**, which are often conspicuous and attract animal pollinators. As we'll explore further in Module 17.10, showy petals are a major reason for the overwhelming success of angiosperms.

Stigma
Style
Ovary
Carpel
Anther
Filament
Stamen
Petal
Sepal
Ovule
Receptacle

▲ **Figure 17.6B** The parts of a flower

Plucking off a flower's petals reveals the filaments of the **stamens**. The **anther**, a sac at the top of each filament, contains male sporangia and will eventually release pollen. At the center of the flower is the **carpel**, the female reproductive structure. It includes the stigma, the style, and the **ovary**, a unique angiosperm adaptation that encloses the ovules. If you cut open the ovary of a flower, you can see its white, egg-shaped ovules. As in pines, each ovule contains a sporangium that will produce a female gametophyte and eventually become a seed. The ovary matures into a fruit, which aids in seed dispersal, as we'll discuss shortly. In the next module, you will learn how the alternation of generations life cycle proceeds in angiosperms.

Coral bean
Foxglove (a cluster of flowers)
Passionflower

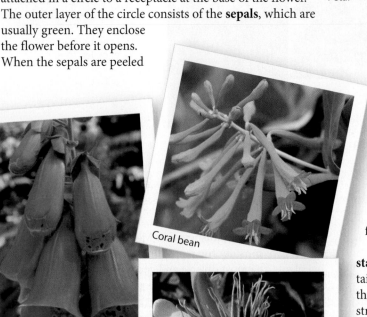

▲ **Figure 17.6A** Some examples of floral diversity

? Where are the male and female sporangia located in flowering plants?

● Anthers contain the male sporangia; ovules contain the female sporangia.

17.7 The angiosperm plant is a sporophyte with gametophytes in its flowers

In angiosperms, as in gymnosperms, the sporophyte generation is dominant and produces the gametophyte generation within its body. **Figure 17.7** illustrates the life cycle of a flowering plant and highlights features that have been especially important in angiosperm evolution. (We will discuss these features, as well as double fertilization in angiosperms, in more detail in Modules 31.9–31.13.) Starting at the "Meiosis" box at the top of Figure 17.7, ❶ meiosis in the anthers of the flower produces haploid spores that undergo mitosis and form the male gametophytes, or pollen grains. ❷ Meiosis in the ovule produces a haploid spore that undergoes mitosis and forms the few cells of the female gametophyte, one of which becomes an egg. ❸ Pollination occurs when a pollen grain, carried by the wind or an animal, lands on the stigma.

As in gymnosperms, a tube grows from the pollen grain to the ovule, and a sperm fertilizes the egg, ④ forming a zygote. Also as in gymnosperms, ⑤ a seed develops from each ovule. Each seed consists of an embryo (a new sporophyte) surrounded by a food supply and a seed coat derived from the tissues surrounding the ovule. While the seeds develop, ⑥ the ovary's wall thickens, forming the fruit that encloses the seeds. When conditions are favorable, ⑦ a seed **germinates**, which means it begins to grow. As the embryo begins to grow, it uses the food supply from the seed until it can begin to photosynthesize. Eventually, it develops into a mature sporophyte plant, completing the life cycle.

The evolution of flowers that attract animals, which carry pollen more reliably than the wind, was a key adaptation of angiosperms. The success of angiosperms was also enhanced by their ability to reproduce rapidly. Fertilization in angiosperms usually occurs about 12 hours after pollination, making it possible for the plant to produce seeds in only a few days or weeks. A typical pine takes years to produce seeds. Rapid seed production is especially advantageous in harsh environments such as deserts, where growing seasons are extremely short.

Another feature contributing to the success of angiosperms is the development of fruits, which protect and help disperse the seeds, as we see in the next module.

? **What is the difference between pollination and fertilization?**

● Pollination is the transfer of pollen by wind or animals from stamens to the tips of carpels. Fertilization is the union of egg and sperm; the sperm are released from the pollen tube after the tube grows and makes contact with an ovule.

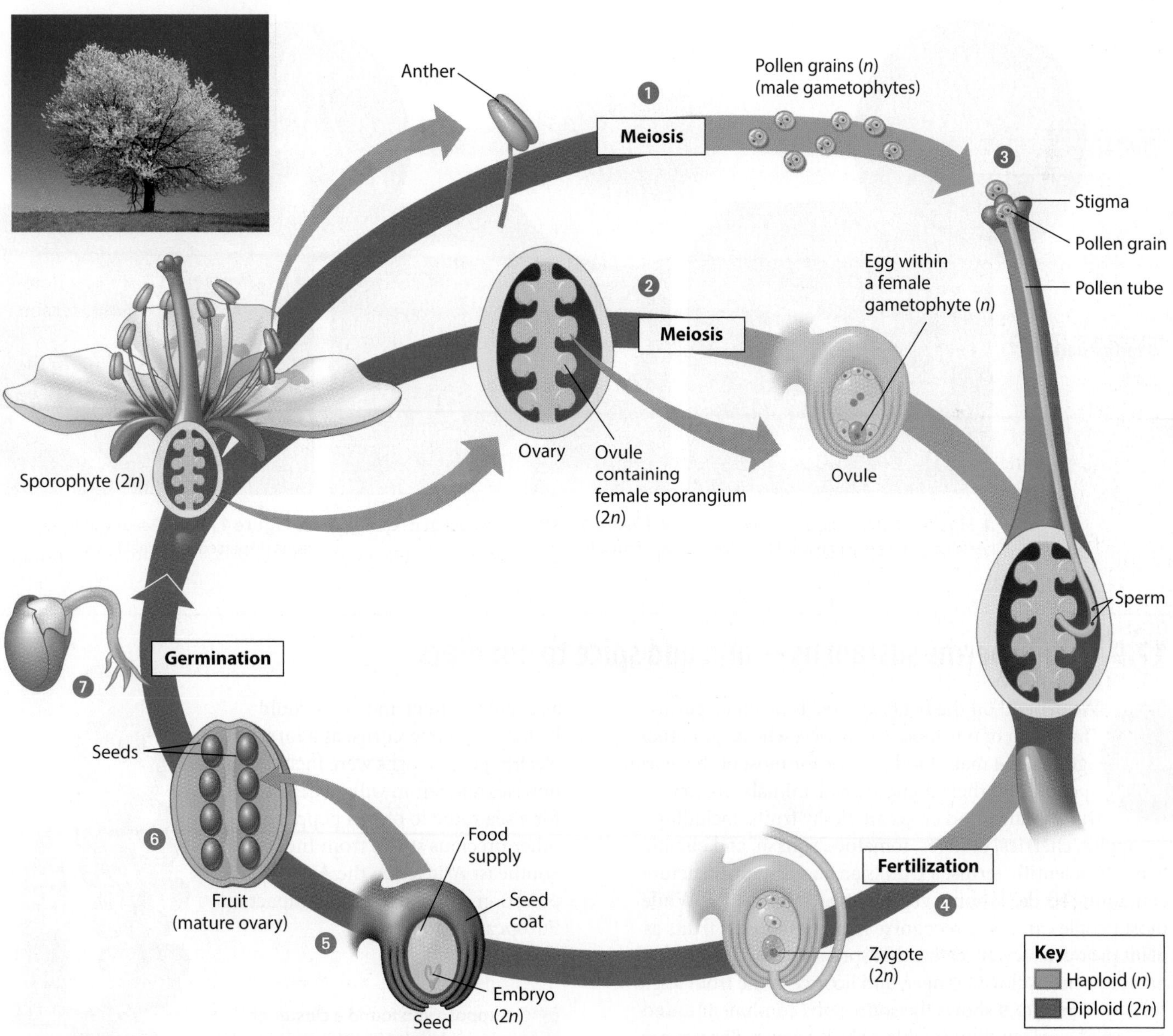

Anther
1
Meiosis
Pollen grains (n)
(male gametophytes)
3
Stigma
Pollen grain
Pollen tube
2
Meiosis
Egg within
a female
gametophyte (n)
Sporophyte ($2n$)
Ovary
Ovule
containing
female sporangium
($2n$)
Ovule
Sperm
Germination
7
Seeds
6
Fruit
(mature ovary)
Food
supply
Seed
coat
5
Embryo
($2n$)
Seed
Fertilization
4
Zygote
($2n$)
Key
Haploid (n)
Diploid ($2n$)

▲ **Figure 17.7** Life cycle of an angiosperm

Try This Each time you encounter a circled number in the text, find the corresponding number on the figure and identify the structures described.

17.8 The structure of a fruit reflects its function in seed dispersal

A **fruit**, the ripened ovary of a flower, is an adaptation that helps disperse seeds. Some angiosperms depend on wind for seed dispersal. For example, the fruit of a dandelion (**Figure 17.8A**) has a parachute-like extension that carries the tiny seed away from the parent plant on wind currents. Hook-like modifications of the outer layer of the fruit or seed coat allow some angiosperms to hitch a ride on animals. The fruits of the cocklebur plant (**Figure 17.8B**), for example, may be carried for miles before they open and release their seeds.

Many angiosperms produce fleshy, edible fruits that are attractive to animals as food. While the seeds are developing, these fruits are green and effectively camouflaged against green foliage. When ripe, the fruit turns a bright color, such as red or yellow, advertising its presence to birds and mammals. When the catbird in **Figure 17.8C** eats a berry, it digests the fleshy part of the fruit, but most of the tough seeds pass unharmed through its digestive tract. The bird may then deposit the seeds, along with a supply of natural fertilizer, some distance from where it ate the fruit.

The dispersal of seeds in fruits is one of the main reasons that angiosperms are so successful. Humans have also made extensive use of fruits and seeds, as we see next.

? **What is a fruit?**

● A ripened ovary of a flower, which contains, protects, and aids in the dispersal of seeds

Fruit
Seed dispersal

▲ **Figure 17.8A** Dandelion seeds launching into the air on a light breeze

▲ **Figure 17.8B** Cocklebur fruits carried by animal fur

▲ **Figure 17.8C** Seeds within edible fruits dispersed in animal feces

17.9 Angiosperms sustain us—and add spice to our diets

CONNECTION

We depend on the fruits and seeds of angiosperms for much of our food. Corn, rice, wheat, and other grains, the main food sources for most of the world's people and their domesticated animals, are dry fruits. Many food crops are fleshy fruits, including apples, cherries, oranges, tomatoes, squash, and cucumbers. (In scientific terms, a fruit is an angiosperm structure containing seeds, so some vegetables are also fruits.) While most people can easily recognize grains and fleshy fruits as plant products, fewer realize that spices such as nutmeg, cinnamon, cumin, cloves, ginger, and licorice come from angiosperms. **Figure 17.9** shows the source of a condiment found on most American dinner tables: black pepper. The pepper fruits are harvested before ripening, then dried and ground into powder or sold whole as "peppercorns." In medieval Europe, peppercorns were so valuable that they were used as currency. Rent and taxes could be paid in peppercorns; as a form of wealth, peppercorns were included in dowries and left in wills. The search for a sea route to obtain pepper and other precious spices from India and Southeast Asia led to the Age of Exploration and had a lasting impact on European history.

? **Suppose you found a cluster of pepper berries like the ones in Figure 17.9. How would you know that they are fruits?**

● Each berry has seeds inside it.

▲ **Figure 17.9** Berries (fruits) on *Piper nigrum*

17.10 Pollination by animals has influenced angiosperm evolution

EVOLUTION CONNECTION

Most of us associate flowers with colorful petals and sweet fragrances, but not all flowers have these accessories. **Figure 17.10A**, for example, shows flowers of a red maple, which have many stamens but no petals (carpels are borne on separate flowers). Compare those flowers to the large, vibrantly colored columbine in **Figure 17.10B**. Such an elaborate flower costs the columbine an enormous amount of energy to produce, but the investment pays off when a pollinator, attracted by the flower's color or scent, carries the plant's pollen to another flower of the same species. Red maple, on the other hand, devotes substantial energy to making massive amounts of pollen for release into the wind, a far less certain method of pollination. Both species have adaptations to achieve pollination, which is necessary for reproductive success, but they allocate their resources differently.

▲ **Figure 17.10A** Flowers of red maple, whose pollen is carried by the wind

Plant scientists estimate that about 90% of angiosperms employ animals to transfer their pollen. Birds, bats, and many different species of insects (notably bees, butterflies, moths, and beetles) serve as pollinators. These animals visit flowers in search of a meal, which the flowers provide in the form of nectar, a high-energy fluid. For pollinators, the colorful petals and alluring odors are signposts that mark food resources.

The cues that flowers offer are keyed to the sense of sight and smell of certain types of pollinators. For example, birds are attracted by bright red and orange flowers but not to particular scents, while most beetles are drawn to fruity odors but are indifferent to color. The petals of bee-pollinated flowers may be marked with guides in contrasting colors that lead to the nectar. In some flowers, the nectar guides are pigments that reflect ultraviolet light, a part of the electromagnetic spectrum that is invisible to us and most other animals, but readily apparent to bees. Flowers that are pollinated by night-flying animals such as bats and moths typically have large, light-colored, highly scented flowers that can easily be found at night. Some flowers even produce an enticing imitation of the smell of rotting flesh, thereby attracting pollinators such as carrion flies and beetles.

Many flowers have additional adaptations that improve pollen transfer, and thus reproductive success, once a pollinator arrives. The location of the nectar, for example, may manipulate the visitor's position in a way that maximizes pollen pickup and deposition. In **Figure 17.10C**, the pollen-bearing stamens of a scotch broom flower arch over the bee as it harvests nectar. Some of the pollen the bee picks up here will rub off onto the stigmas of other flowers it visits. In the columbine, as well as in many other flowers, the nectar can only be reached by pollinators with long tongues, a group that includes butterflies, moths, birds, and some bees.

▲ **Figure 17.10B** Showy columbine flower

Pollination is only effective if the pollen transfer occurs between members of the same species, but relatively few pollinators visit one species of flower exclusively. Biologists hypothesize that pollinators may benefit from sequential visits to a single species. It takes time, through trial and error, for a pollinator to learn to extract nectar from a flower. Insects, for instance, can only remember one extraction technique at a time. Thus, pollinators are most successful at obtaining food if they visit another flower with the same cues immediately after mastering a technique for nectar extraction. Natural selection has also favored floral adaptations that increase pollinator fidelity to a species. For example, some plant species may strengthen pollinator fidelity by spiking their nectar with caffeine. According to a study published in 2013, the caffeine buzz enhances the ability of honeybees to recall the scent of flowers that provided a nectar reward.

▲ **Figure 17.10C** A bee picking up pollen from a scotch broom flower as it feeds on nectar

Although floral characteristics are adaptations that attract pollinators, they are a source of enjoyment to us, as well. People use flowering plants, including many species of trees, for a variety of purposes. As we consider in the next module, however, the most essential role of plants in our lives is as food.

? **What type of pollinator do you think would be attracted to the columbine in Figure 17.10B?**

● Long-tongued birds, because the flower is red and has long floral tubes

17.11 Plant diversity is vital to the future of the world's food supply

CONNECTION

Plants have always been a mainstay of the human diet, but the way we obtain our plant food has changed enormously. Early humans made use of any edible plant species that was available, probably eating different plants as the seasons changed, as do present-day hunter-gatherers. During the development of agriculture, people in different parts of the world domesticated the tastiest, most easily cultivated species, gradually increasing their productivity through generation after generation of artificial selection. In modern agriculture, plant-breeding techniques have further narrowed the pool of food plant diversity by creating a select few genotypes possessing the most desirable characteristics. Most of the world's population is now fed by varieties of rice, wheat, corn, and soybeans that require specific cultivation techniques.

Agriculture has also changed the landscape. Over thousands of years, the expanding human population created farms by clear-cutting or burning forests. More recently, deforestation has accelerated to replace the vast expanses of cropland that have been severely degraded by unsustainable agricultural practices. But converting more land to farms is not the only way to ensure an adequate food supply for the future. Plant diversity offers possibilities for developing new crops and improving existing ones.

Some new crops may come from the hundreds of species of nutritious fruits, nuts, and grains that people gather and use locally. In a recent study of potential food sources in Africa, scientists identified dozens of wild plants that might be suitable for domestication and regional production **(Figure 17.11A)**. Promising candidates include intriguingly named fruits such as chocolate berries, gingerbread plums, and monkey oranges. In addition, some regions already have unique domesticated or semi-domesticated crops with the potential for greater production, especially in marginal farmland. For example, some African grains tolerate heat and drought, and many grow better on infertile soil than grains cultivated elsewhere. One species is so tough that it grows on sand dunes where the annual rainfall is less than 70 mm (about 2.5 inches) per year!

Modifying crops to enable them to thrive in less than ideal conditions—through either traditional breeding methods or biotechnology—is another approach. Genes that enable plants to grow in salty soil, to resist pests, or to tolerate heat and drought would also be useful. But where might such genes be found? All crop plants were originally derived from wild ancestors. Those ancestors and their close relatives are a rich source of genetic diversity that could be used to bolster existing crops.

Both of these approaches to crop diversification are undermined by the ongoing loss of natural plant diversity caused by habitat destruction. Besides clear-cutting to create farmland, forests are being lost to logging, mining, and air pollution. Roads slice vast tracts of lands into fragments that are too small to support a full array of species **(Figure 17.11B)**. As a result, species that could potentially be domesticated are instead being lost. So are many wild relatives of crop species, and with them a priceless pool of genetic diversity. These losses are just one example of the impact of declining biodiversity on the future of our species. (We will return to this problem—and possible solutions—in Chapter 38.)

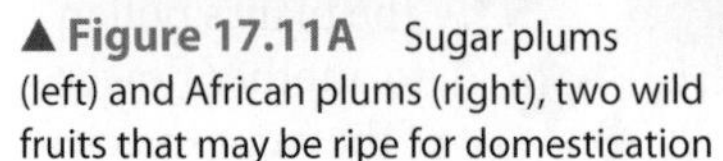

▲ **Figure 17.11A** Sugar plums (left) and African plums (right), two wild fruits that may be ripe for domestication

? Name two ways in which the loss of plant diversity might affect the world's future supply of food.

● Potential crop plants could be lost; potentially useful genes carried by wild ancestors or close relatives of crop plants could be lost.

Rio Candeias
B R A Z I L
Rondônia
Rio Branco
São João
João Filipe
Uru-Eu-Wau-Wau
Indigenous
Area
Pacaás Novos
National Park

▲ **Figure 17.11B** Satellite photo of Amazonian rain forest (The "fishbone" pattern marks a network of roads carved through the forest when farmers and loggers came to the area).

▷ Diversity of Fungi

17.12 Fungi absorb food after digesting it outside their bodies

You have probably seen members of the kingdom **Fungi** at some time, whether they were mushrooms sprouting in a lawn, bracket fungi attached to a tree like small shelves, or fuzzy patches of mold on leftover food. Despite the differences in their visible body forms, each of these fungi is obtaining food from its substrate in the same way. All fungi are heterotrophs that acquire their nutrients by **absorption**. They secrete powerful enzymes that digest macromolecules into monomers and then absorb the small nutrient molecules into their cells. For example, enzymes secreted by fungi growing on a loaf of bread digest the bread's starch into glucose molecules, which the fungal cells absorb. Some fungi produce enzymes that digest cellulose and lignin, the major structural components of plants. Consequently, fungi are essential decomposers in most ecosystems.

The feeding structures of a fungus are a network of threadlike filaments called **hyphae** (singular *hypha*). Hyphae branch repeatedly as they grow, forming a mass known as a **mycelium** (plural, *mycelia*) **(Figure 17.12A)**. The "umbrellas" that you recognize as fungi, such as the ones in **Figure 17.12B**, are reproductive structures made up of tightly packed hyphae. In the type of fungus shown, mushrooms arise as small buds on a mycelium that extends throughout the food source, hidden from view. When a bud has developed sufficiently, the rapid absorption of water (for example, after a rainfall) creates enough hydraulic pressure to pop the mushroom to the surface. Above ground, the mushroom produces tiny reproductive cells called spores at the tips of specialized hyphae, and the spores are then dispersed on air currents.

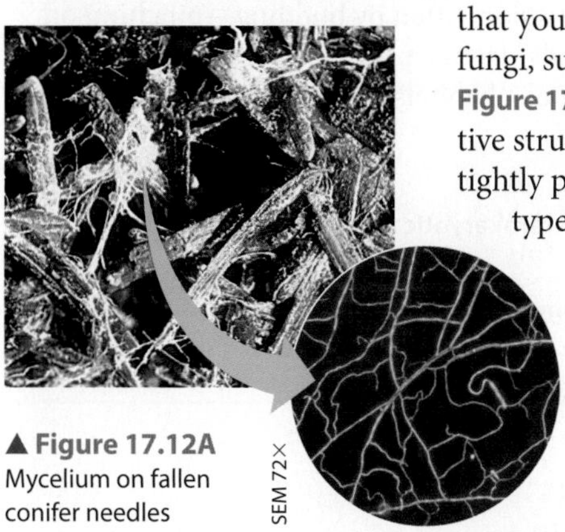

▲ **Figure 17.12A** Mycelium on fallen conifer needles

SEM 72×

Fungal hyphae are surrounded by a cell wall. Unlike plants, which have cellulose cell walls, most fungi have cell walls made of **chitin**, a strong, flexible nitrogen-containing polysaccharide, identical to the chitin found in the external skeletons of insects. In most fungi, the hyphae consist of chains of cells separated by cross-walls that have pores large enough to allow ribosomes, mitochondria, and even nuclei to flow from cell to cell. Some fungi lack cross-walls entirely and have many nuclei within a single mass of cytoplasm.

Fungi cannot run or fly in search of food. But their mycelium makes up for the lack of mobility by being able to grow at a phenomenal rate, branching throughout a food source and extending its hyphae into new territory. If you were to break open the log beneath the mushrooms in Figure 17.12B, for example, you would see strands of hyphae throughout the wood. Because its hyphae grow longer without getting thicker, the fungus develops a huge surface area from which it can secrete digestive enzymes and through which it can absorb food.

Not all fungi are decomposers. Some fungi live symbiotically with other organisms. As you learned in the chapter introduction, the symbiosis between fungi and plant roots, called a **mycorrhiza** (plural, *mycorrhizae*), is of special significance. The hyphae of some mycorrhizal fungi branch into the root cells; other species surround the root but don't penetrate its living cells. Both types of mycorrhizae absorb phosphorus and other essential minerals from the soil and make them available to the plant. Sugars produced by the plant through photosynthesis nourish the fungus, making the relationship mutually beneficial. The mycorrhizal partnership is thought to have played a crucial role in the success of early land plants, a topic we'll examine in Module 17.18. And in Module 17.19, you will learn about fungi that are **parasites**, obtaining their nutrients at the expense of living plants or animals.

? **Contrast how fungi digest and absorb their food with the way humans eat and digest their food.**

● A fungus digests its food externally by secreting enzymes onto the food and then absorbing the small nutrients that result from digestion. In contrast, humans and most other animals eat relatively large pieces of food and digest the food within their bodies.

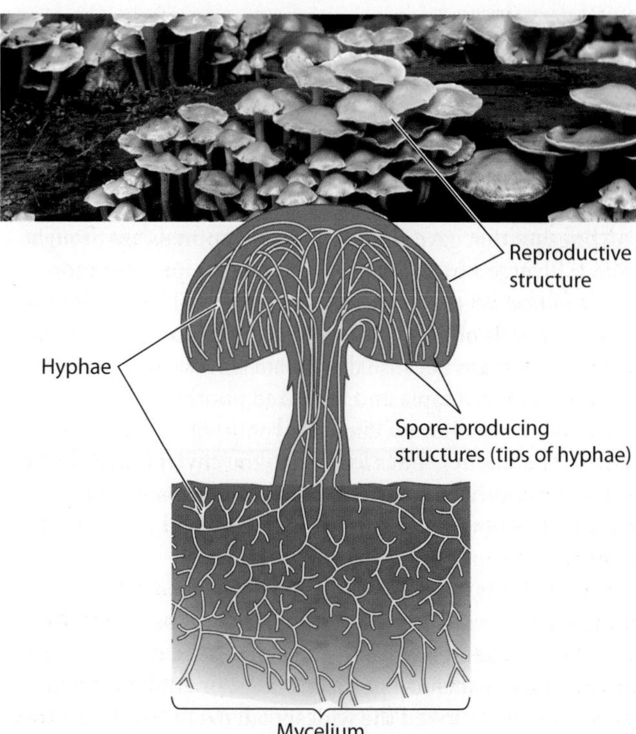

Reproductive structure
Hyphae
Spore-producing structures (tips of hyphae)
Mycelium

▲ **Figure 17.12B** Fungal reproductive and feeding structures

17.13 Fungi produce spores in both asexual and sexual life cycles

Fungal reproduction typically involves the release of vast numbers of haploid spores, which are transported easily over great distances by wind or water. A spore that lands in a moist place where food is available germinates and produces a new haploid fungus by mitosis. As you can see in **Figure 17.13**, however, spores can be produced either sexually or asexually.

In many fungi, sexual reproduction involves mycelia of different mating types. Hyphae from each mycelium release signaling molecules and grow toward each other. ❶ When the hyphae meet, their cytoplasms fuse. But this fusion of cytoplasm is often not followed immediately by the fusion of "parental" nuclei. Thus, many fungi have what is called a **heterokaryotic stage** (from the Greek, meaning "different nuclei"), in which cells contain two genetically distinct haploid nuclei. Hours, days, or even centuries may pass before the parental nuclei fuse, ❷ forming the usually short-lived diploid phase. ❸ Zygotes undergo meiosis, producing haploid spores. As you'll learn in the next module, the specialized structures in which these spores are formed are used to classify fungi.

In asexual reproduction, ❹ spore-producing structures arise from haploid mycelia that have undergone neither a heterokaryotic stage nor meiosis. Many fungi that reproduce sexually can also produce spores asexually. In addition, asexual reproduction is the only known means of spore production in some fungi, informally known as **imperfect fungi**. Many species commonly called molds and yeasts are imperfect fungi. The term **mold** refers to any rapidly growing fungus that reproduces asexually by producing spores, often at the tips of specialized hyphae. These familiar furry carpets often appear on aging fruit and bread and in seldom-cleaned shower stalls. The term **yeast** refers to any single-celled fungus. Yeasts reproduce asexually by cell division, often by budding—pinching off small "buds" from a parent cell. Yeasts inhabit liquid or moist habitats, such as plant sap and animal tissues.

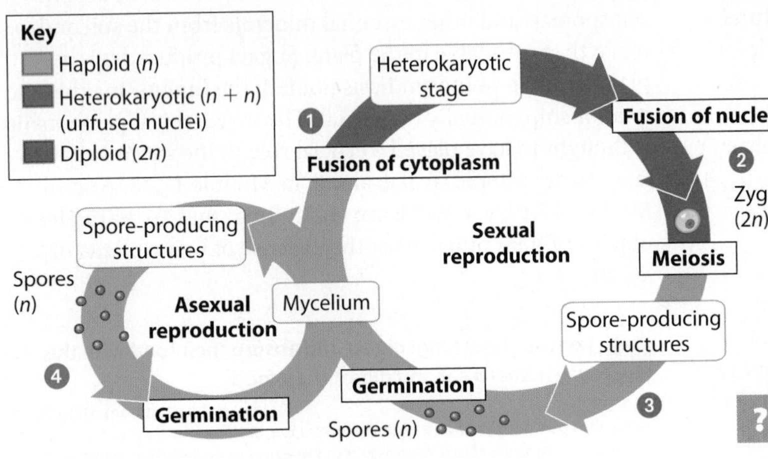

Key
Haploid (n)
Heterokaryotic ($n + n$) (unfused nuclei)
Diploid ($2n$)
Heterokaryotic stage
Fusion of nuclei
1
Fusion of cytoplasm
2
Zygote ($2n$)
Sexual reproduction
Meiosis
Spore-producing structures
Spores (n)
Asexual reproduction
Mycelium
Spore-producing structures
4
Germination
Germination
Spores (n)
3

▲ **Figure 17.13** Generalized life cycle of a fungus

? **What is the heterokaryotic stage of a fungus?**

● The stage in which each cell has two different nuclei (from two different parents), with the nuclei not yet fused

17.14 Fungi are classified into five groups

Biologists who study fungi have described over 100,000 species, but there may be as many as 1.5 million. Molecular analysis has helped scientists understand the phylogeny of fungi. The lineages that gave rise to fungi and animals are thought to have diverged from a flagellated unikont ancestor more than 1 billion years ago (see Module 16.19). The oldest undisputed fossils of fungi, however, are only about 460 million years old, perhaps because the aquatic ancestors of terrestrial fungi were microscopic and fossilized poorly.

Figure 17.14A shows a current hypothesis of fungal phylogeny. The multiple lines leading to the chytrids and the zygomycetes indicate that these groups are probably not monophyletic. For now, though, most biologists still talk in terms of the five groups of fungi shown here.

The **chytrids**, the only fungi with flagellated spores, are thought to represent the earliest lineage of fungi. They are common in lakes, ponds, and soil. Some species are decomposers; others parasitize protists, plants, or animals. Some researchers have linked the widespread decline of amphibian species to a highly infectious fungal disease caused by a species of chytrid. Populations of frogs in mountainous regions of Central America and Australia have suffered massive mortality from this emerging disease.

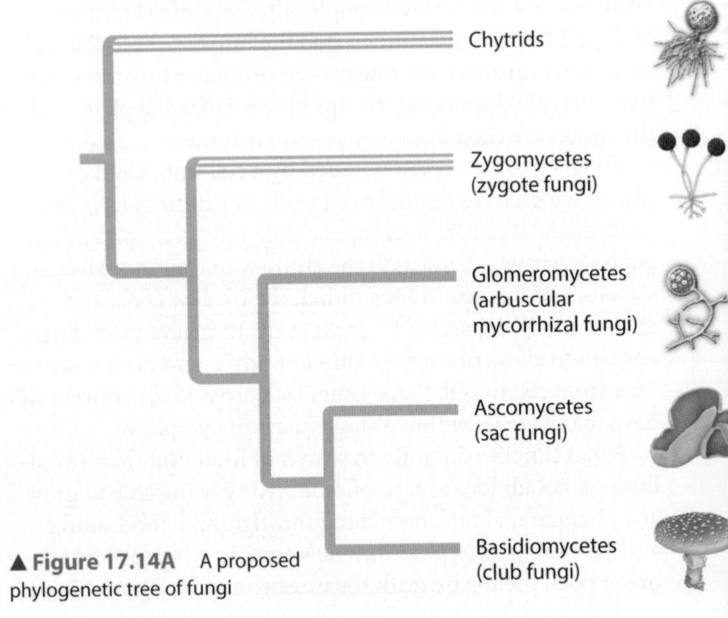

Chytrids
Zygomycetes (zygote fungi)
Glomeromycetes (arbuscular mycorrhizal fungi)
Ascomycetes (sac fungi)
Basidiomycetes (club fungi)

▲ **Figure 17.14A** A proposed phylogenetic tree of fungi

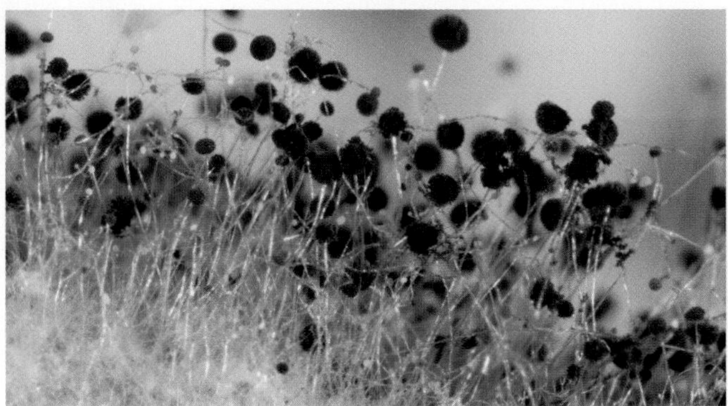

▲ **Figure 17.14B** Zygomycete: *Rhizopus stolonifer*, black bread mold

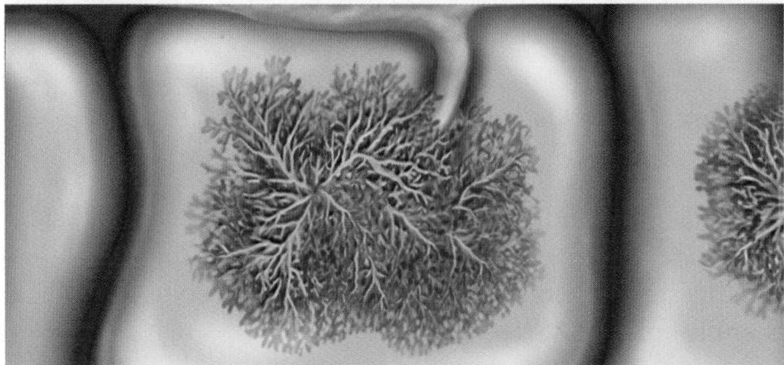

▲ **Figure 17.14C** Glomeromycete: a drawing of an arbuscule in a root cell

Mushrooms
A puffball
Shelf fungi

▲ **Figure 17.14E** Basidiomycetes (club fungi)

The **zygomycetes** are characterized by their protective zygosporangium, where zygotes produce haploid spores by meiosis. This diverse group includes fast-growing molds, such as black bread mold **(Figure 17.14B)** and molds that rot produce such as peaches, strawberries, and sweet potatoes. Some zygote fungi are parasites on animals.

The **glomeromycetes** (from the Latin *glomer*, ball) form a distinct type of mycorrhiza in which hyphae invade plant root cells, where they branch into tiny treelike structures known as arbuscules **(Figure 17.14C)**. About 80% of all plants have such symbiotic partnerships with glomeromycetes, which deliver phosphate and other minerals to plants while receiving organic nutrients in exchange.

The **ascomycetes**, or **sac fungi**, are named for saclike structures called asci (from the Greek *asco*, pouch) that produce spores in sexual reproduction. They live in a variety of marine, freshwater, and terrestrial habitats and range in size from unicellular yeasts to elaborate morels and cup fungi **(Figure 17.14D)**. Ascomycetes include some of the most devastating plant pathogens. Other species of ascomycetes live with green algae or cyanobacteria in symbiotic associations called lichens, which we discuss in Module 17.17.

When you think of fungi, you probably picture mushrooms, puffballs, or shelf fungi **(Figure 17.14E)**. These are examples of **basidiomycetes**, or **club fungi**. They are named for their club-shaped, spore-producing structure, called a basidium (meaning "little pedestal" in Latin; plural, *basidia*). Many basidiomycete species excel at breaking down the lignin found in wood and thus play key roles as decomposers. For example, shelf fungi often break down the wood of weak or damaged trees and continue to decompose the wood after the tree dies. The basidiomycetes also include two groups of particularly destructive plant parasites, the rusts and smuts, which we discuss in Module 17.19.

Edible morels

Cup fungus

▲ **Figure 17.14D** Ascomycetes (sac fungi)

? What is one reason that chytrids are thought to have diverged earliest in fungal evolution?

● Chytrids are the only fungi that have flagellated spores, a characteristic of the ancestor of fungi.

17.15 Fungi have enormous ecological benefits

CONNECTION

Fungi have been major players in terrestrial communities ever since they moved onto land in the company of plants. As symbiotic partners in mycorrhizae, fungi supply essential nutrients to plants and are enormously important in natural ecosystems and agriculture.

Fungi, along with prokaryotes, are essential decomposers in ecosystems, breaking down organic matter and restocking the environment with vital nutrients essential for plant growth. So many fungal spores are in the air that as soon as a leaf falls or an insect dies, it is covered with spores and is soon infiltrated by fungal hyphae (**Figure 17.15**). If fungi and prokaryotes in a forest suddenly stopped decomposing, leaves, logs, feces, and dead animals would pile up on the forest floor, and plants—and the animals that eat plants—would starve because elements taken from the soil would not be replenished through decomposition.

Almost any organic (carbon-containing) substance can be consumed by fungi. During World War II, the moist tropical heat of Southeast Asia and islands in the Pacific Ocean provided ideal conditions for fungal decomposition of wood and natural fibers such as canvas and cotton. Packing crates, military uniforms, and tents quickly disintegrated, causing supply problems for the military forces. Synthetic substances are more resistant to fungal attack, but some fungi have the useful ability to break down toxic pollutants, including the pesticide DDT and certain chemicals that cause cancer. Scientists are also investigating the possibility of using fungi that can digest petroleum products to clean up oil spills and other chemical messes.

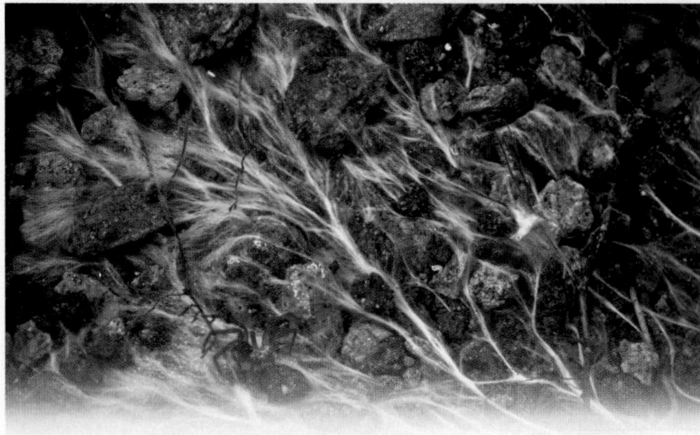

▲ **Figure 17.15** A fungal mycelium

? Name two essential roles that fungi play in terrestrial ecosystems.

● In mycorrhizae, fungi help plants acquire nutrients from the soil. When soil fungi decompose dead animals, fallen leaves, and other organic materials, they release nutrients that fertilize plant growth.

17.16 Fungi have many practical uses

CONNECTION

Fungi have a number of culinary uses. Most of us have eaten mushrooms, although we may not have realized that we were ingesting reproductive structures of subterranean fungi. The distinctive flavors of certain cheeses, including Roquefort and blue cheese (**Figure 17.16A**), come from fungi used to ripen them. Truffles, which are produced by certain mycorrhizal fungi associated with tree roots, are highly prized by gourmets. And humans have used yeasts for thousands of years to produce alcoholic beverages and cause bread to rise.

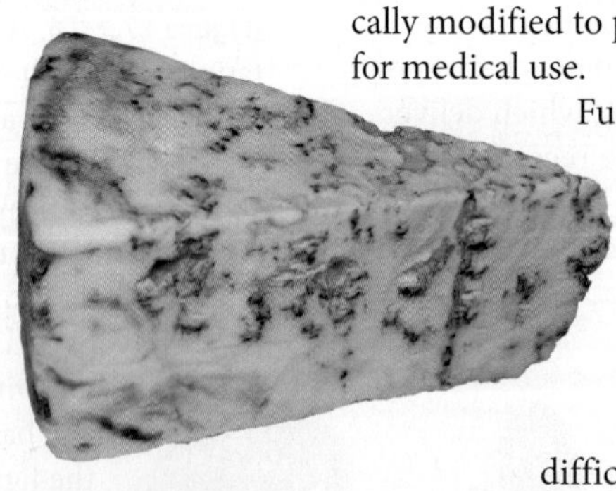

▲ **Figure 17.16A** Blue cheese

Fungi are medically valuable as well. Like the bacteria called actinomycetes (see Module 16.9), some fungi produce antibiotics that we use to treat bacterial diseases. In fact, the first antibiotic discovered was penicillin, which is made by the common mold called *Penicillium*. In **Figure 17.16B**, the clear area between the mold and the bacterial growth is where the antibiotic produced by *Penicillium* has inhibited the growth of the bacteria (*Staphylococcus aureus*).

Fungi also figure prominently in research in molecular biology and in biotechnology. Researchers often use yeasts to study the molecular genetics of eukaryotes because they are easy to culture and manipulate. Yeasts have also been genetically modified to produce human proteins for research and for medical use.

Fungi may play a major role in the future production of biofuels from plants. Ideally, the biofuels would be derived from plants or plant parts that could not be used to feed people or livestock, such as straw, certain grasses, and wood. These plant materials are primarily made up of cellulose and lignin, large molecules that are difficult to decompose into smaller molecules

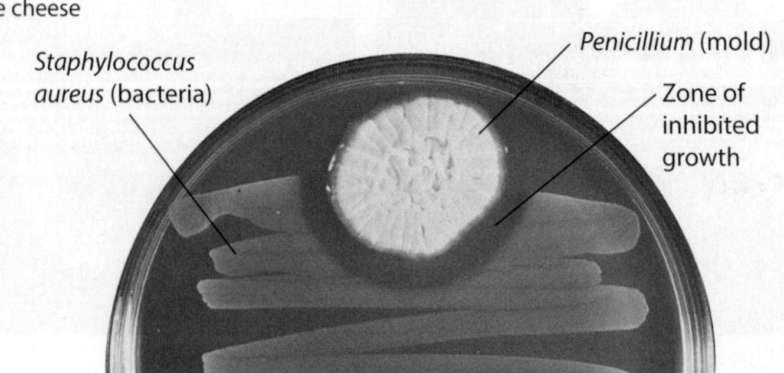
Staphylococcus aureus (bacteria)
Penicillium (mold)
Zone of inhibited growth

▲ **Figure 17.16B** A culture of *Penicillium* and bacteria

that can be processed to make biofuel. Researchers are currently investigating a variety of fungi that produce enzymes capable of digesting the toughest plant parts. Basidiomycetes called white rot fungi (because the enzymatic breakdown of lignin bleaches the brown color out of wood) are promising candidates for providing these enzymes **(Figure 17.16C)**.

? What do you think is the function of the antibiotics that fungi produce in their natural environments?

● The antibiotics probably block the growth of microorganisms, especially prokaryotes that compete with the fungi for nutrients and other resources.

▲ **Figure 17.16C** White rot fungus

17.17 Lichens are symbiotic associations of fungi and photosynthetic organisms

The rock in **Figure 17.17A** is covered with a living crust of **lichens**, symbiotic associations of millions of unicellular green algae or cyanobacteria held in a mass of fungal hyphae. The partners are so closely entwined that they appear to be a single organism. How does this merger come about? When the growing hyphal tips of a lichen-forming fungus come into contact with a suitable partner, the hyphae quickly fork into a network of tendrils that encircle and overgrow the algal cells **(Figure 17.17B)**. The fungus invariably benefits from the symbiosis, receiving food from its photosynthetic partner. In fact, fungi with the ability to form lichens rarely thrive on their own in nature. In many lichens that have been studied, the alga or cyanobacterium also benefits, as the fungal mycelium provides a suitable habitat that helps the alga or cyanobacterium absorb and retain water and minerals. In other lichens, it is not clear whether the relationship benefits the photosynthetic partner or is only advantageous to the fungus. In any case, the two symbionts are so completely intertwined that lichens are named as if they were a single organism.

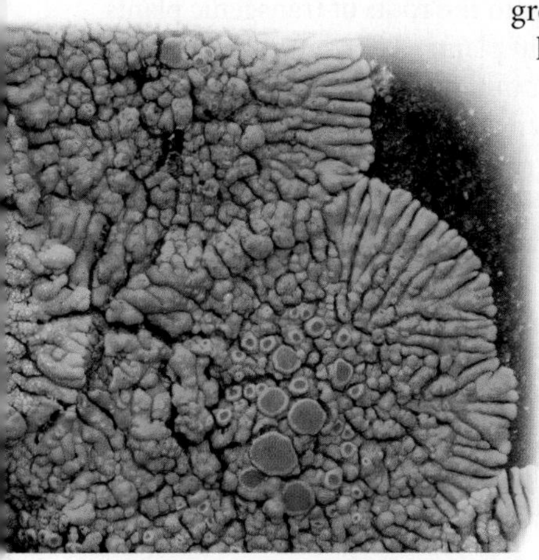

▲ **Figure 17.17A** Two species of lichens commonly found on coastal rocks

Algal cell

Fungal hyphae

Colorized SEM 1,000×

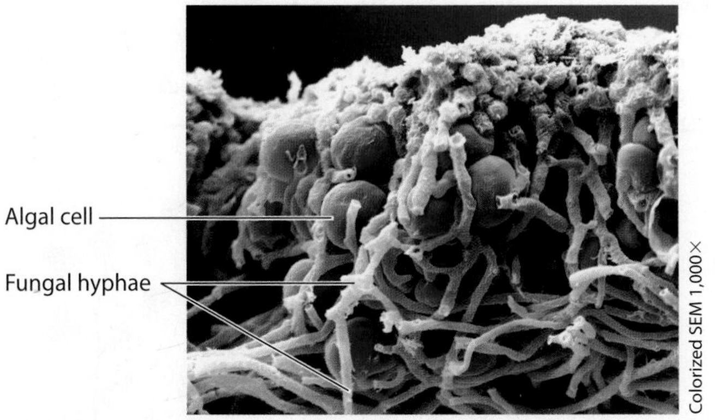

▲ **Figure 17.17B** The close relationship between fungal and algal partners in a lichen

▲ **Figure 17.17C** Reindeer moss, a lichen

Lichens are rugged and able to live where there is little or no soil. As a result, they are important pioneers on new land. Lichens grow into tiny rock crevices, where the acids they secrete help to break down the rock to soil, paving the way for future plant growth. Some lichens can tolerate severe cold, and carpets of them cover the arctic tundra. Caribou eat lichens known as reindeer "moss" **(Figure 17.17C)**, which grow in their winter feeding grounds of Alaska.

Lichens can also withstand severe drought. They are opportunists, growing in spurts when conditions are favorable. When it rains, a lichen quickly absorbs water and photosynthesizes at a rapid rate. In dry air, it dehydrates and photosynthesis may stop, but the lichen remains alive more or less indefinitely. Some lichens are thousands of years old, rivaling the longevity of the oldest plants and fungi.

As tough as lichens are, many do not withstand air pollution. Because they get most of their minerals from the air, in the form of dust or compounds dissolved in raindrops, lichens are very sensitive to airborne pollutants. The death of lichens is often a sign that air quality in an area is deteriorating.

? What benefit do fungi in lichens receive from their partners?

● Access to the sugar molecules that algae or cyanobacteria produce by photosynthesis

17.18 Mycorrhizae may have helped plants colonize land

SCIENTIFIC THINKING

The complementary abilities of fungi and unicellular photosynthesizers make lichens well-adapted for pioneering new land. When plants began to colonize terrestrial habitats, they were also pioneers. Away from water's edge, conditions were harsh. Could fungal partners have facilitated the transition by helping plants obtain water and scarce mineral nutrients from the soil? Scientists have proposed that symbioses with fungi were crucial to the colonization of land by plants. To test this hypothesis, researchers have pursued three lines of evidence, including present-day mycorrhizal relationships, fossils of early land plants, and molecular genetics

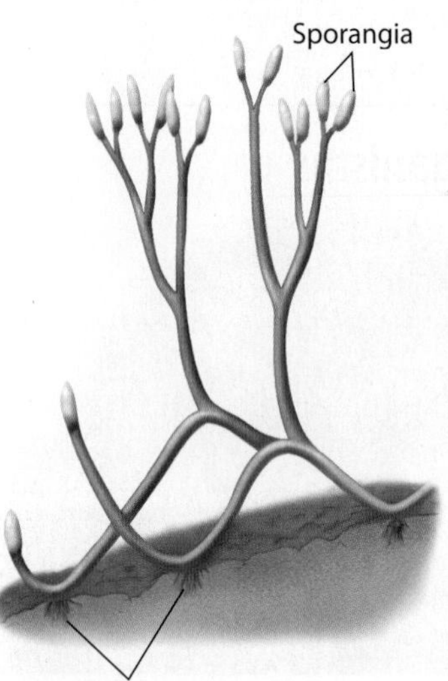
Sporangia
Root-like structures
▲ **Figure 17.18A** *Aglaophyton*, an early land plant

As you learned in Module 17.14, about 80% of all plant species establish symbioses with glomeromycetes, mycorrhizal fungi that form bushy structures called arbuscules in root cells. Arbuscular fungi are also associated with present-day liverworts and hornworts, demonstrating that glomeromycetes can successfully form relationships with haploid plant tissue that lacks true roots—characteristics of the earliest plants (see Module 17.2). The presence of mycorrhizal associations in almost all major lineages of present-day plants suggests an ancient origin for plant-fungus symbioses.

Fossil evidence supporting the hypothesis comes from a rock formation in Scotland known as the Rhynie chert. (Chert refers to a type of fine-grained sedimentary rock.) Both plants and fungi have been preserved in exquisite detail, providing an extraordinary look at a 400-million-year-old ecosystem. **Figure 17.18A** shows a reconstruction of an early land plant called *Aglaophyton* from the Rhynie chert. Although this small plant lacked leaves and roots, microscopic examination of ultrathin sections of *Aglaophyton* fossils revealed fungal hyphae winding among the plant's cells. Arbuscules identical to those found in living mycorrhizal associations (see Figure 17.14C) are visible within some of the cells. Additional fossil evidence from the Rhynie chert and elsewhere indicates an ancient origin for glomeromycete fungi.

Recently, studies using molecular genetics have introduced another line of evidence. The ability to establish a symbiotic relationship requires prospective partners to first identify each other by exchanging signals of mutual recognition and acceptance. Scientists have studied three of the genes, called *sym* (for symbiosis) genes, that encode the plant's side of the molecular "handshake" with mycorrhizal fungi. In a paper published in 2010, researchers reported that *sym* genes are found in all major lineages of land plants. These results imply that the genes were present in the common ancestor of land plants. Furthermore, a phylogeny based on analysis of the three genes proved to be virtually identical to the widely accepted phylogeny shown in Figure 17.2A.

The hypothesized importance of *sym* genes to plants also predicts that their function should have changed little over time. That is, the *sym* genes of liverworts should work similarly to those of flowering plants. Scientists tested this prediction by ❶ using *Agrobacterium* to introduce a functional liverwort *sym* gene into an angiosperm possessing a nonfunctional version of the gene that could not form mycorrhizae (**Figure 17.18B**; see Figure 12.8A). The investigators then ❷ applied fungal spores to the roots of transgenic plants and a control group of mutant plants. After several weeks, ❸ mycorrhizae were present in the transgenic plants—the result of the liverwort *sym* gene functioning normally. The control plants had no mycorrhizae.

By investigating the evolutionary history of mycorrhizae, researchers are gaining insight into how these relationships can be initiated in an agricultural setting to help plants obtain key minerals and tolerate environmental stresses. This knowledge is especially critical in developing sustainable methods of agriculture. It also underscores the fact that neither plants nor fungi can be given full credit for feeding the world—both are needed to produce the bountiful harvests we reap.

Do fungi feed the world?

? **What prediction is tested by the fossil evidence discussed in this module?**

● If plant-fungus symbioses were crucial to plant colonization of land, then these associations should be found in the earliest land plants.

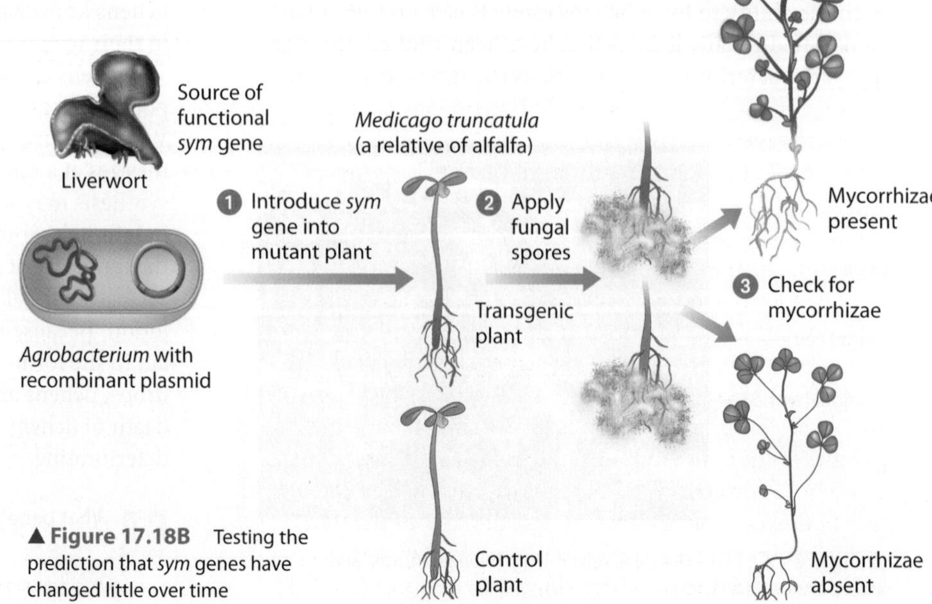
Source of functional *sym* gene
Liverwort
Agrobacterium with recombinant plasmid
Medicago truncatula (a relative of alfalfa)
❶ Introduce *sym* gene into mutant plant
❷ Apply fungal spores
Transgenic plant
Control plant
❸ Check for mycorrhizae
Mycorrhizae present
Mycorrhizae absent
▲ **Figure 17.18B** Testing the prediction that *sym* genes have changed little over time

17.19 Parasitic fungi harm plants and animals

CONNECTION

Despite the many ecological and economic benefits of fungi described in the previous modules, some species are harmful. Of the 100,000 known species of fungi, about 30% make their living as parasites or pathogens, mostly in or on plants. In some cases, fungi have literally changed the landscape. In 1926, a fungus that causes Dutch elm disease was accidentally introduced into the United States on logs sent from Europe to make furniture. (The name refers to the Netherlands, where the disease was first identified; the fungus originated in Asia.) Over the course of several decades, the fungus destroyed around 70% of the elm trees across the eastern United States. English elms (a different species), such as those in **Figure 17.19A**, fared even worse. They were completely annihilated. Recently, scientists studying the DNA of English elms have found evidence that they were all genetically identical, derived by asexual reproduction from a single ancestor brought to England by the Romans 2,000 years ago. As a result, they were all equally susceptible to the ravages of Dutch elm disease.

Fungi are serious agricultural pests. Crop fields typically contain genetically identical individuals of a single species planted close together—ideal conditions for the spread of disease. About 80% of plant diseases are caused by fungi, causing tremendous economic losses each year. Between 10% and 50% of the world's fruit harvest is lost each year to fungal attack. A variety of fungi, including smuts and rusts, are common on grain crops. The ear of corn shown in **Figure 17.19B** is infected with a club fungus called corn smut. The grayish growths, known as galls, are made up of hyphae that invade a developing corn kernel and eventually displace it. When a gall matures, it breaks open and releases thousands of blackish spores. Although most farmers try to eradicate corn smut, some have found a silver lining to the pest. In parts of Central America, the fungus is cooked and eaten as a delicacy known as huitlacoche. When scientists analyzed its chemical composition recently, they found that corn smut is full of healthy nutrients. Gourmets in the United States have recently discovered huitlacoche and are willing to pay exorbitant prices for infected corn.

Some of the fungi that attack food crops are toxic to humans. The seed heads of many kinds of grain, including rye, wheat, and oats, may be infected with fungal growths called ergots, the dark structures on the seed head of rye shown in **Figure 17.19C**. Consumption of flour made from ergot-infested grain can cause gangrene, nervous spasms, burning sensations, hallucinations, temporary insanity, and death. Several toxins have been isolated from ergots. One, called lysergic acid, is the raw material from which the hallucinogenic drug LSD is made.

Animals are much less susceptible to parasitic fungi than are plants. Only about 500 species of fungi are known to be parasitic in humans and other animals. In humans, fungi cause infections ranging from annoyances such as athlete's foot to deadly lung diseases. Fungal diseases of the skin include ringworm, so named because it can appear as circular red areas on the skin. The ringworm fungus can infect virtually any skin surface. Most commonly, it attacks the feet, causing the intense itching and sometimes blistering known as athlete's foot. Systemic fungal infections spread throughout the body, usually from spores that are inhaled. These can be very serious diseases. Coccidioidomycosis is a systemic fungal infection that produces tuberculosis-like symptoms in the lungs. It is so deadly that it is considered a potential biological weapon.

The yeast that causes vaginal infections (*Candida albicans*) is an example of an opportunistic fungal pathogen—a normal member of the microbiota that causes problems only when some change in the body's microbiology, chemistry, or immunology allows the yeast to grow unchecked.

Fungi are the third group of eukaryotes we have surveyed so far. (Protists and plants were the first two groups.) Strong evidence suggests that fungi evolved from unikont protists, a group that also gave rise to the fourth and most diverse group of eukaryotes, the animals, which we study next.

▲ **Figure 17.19A** Stately English elms in Australia, unaffected by Dutch elm disease

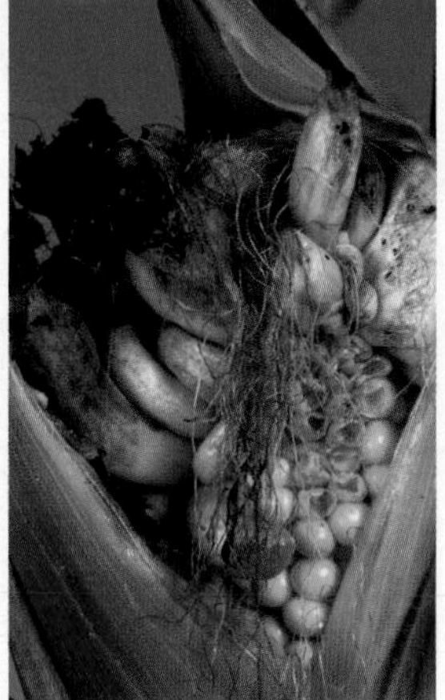

▲ **Figure 17.19B** Corn smut

Ergots

▲ **Figure 17.19C** Ergots on rye

? Name three fungal infections that affect people.

● Ringworm (athlete's foot); coccidioidomycosis; yeast infections

CHAPTER 17 REVIEW

For practice quizzes, BioFlix animations, MP3 tutorials, video tutors, and more study tools designed for this textbook, go to MasteringBiology®

Reviewing the Concepts

Plant Evolution and Diversity (17.1–17.2)

17.1 Plants have adaptations for life on land.

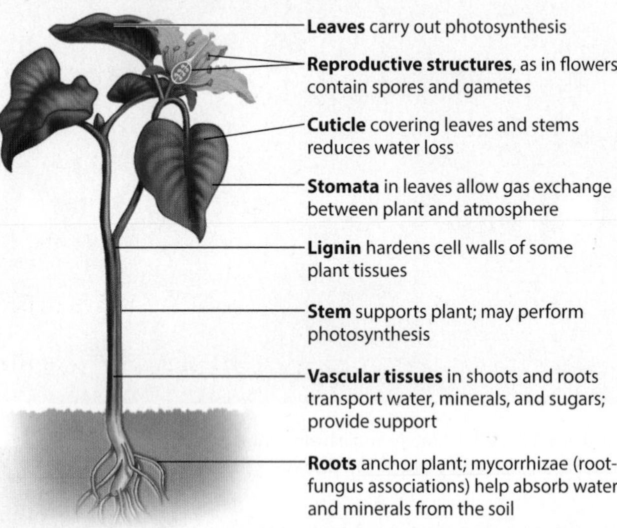

Leaves carry out photosynthesis
Reproductive structures, as in flowers, contain spores and gametes
Cuticle covering leaves and stems reduces water loss
Stomata in leaves allow gas exchange between plant and atmosphere
Lignin hardens cell walls of some plant tissues
Stem supports plant; may perform photosynthesis
Vascular tissues in shoots and roots transport water, minerals, and sugars; provide support
Roots anchor plant; mycorrhizae (root-fungus associations) help absorb water and minerals from the soil

17.2 Plant diversity reflects the evolutionary history of the plant kingdom. Nonvascular plants (bryophytes) include the mosses, hornworts, and liverworts. Vascular plants have supportive conducting tissues. Ferns are seedless vascular plants with flagellated sperm. Seed plants have sperm-transporting pollen grains and protect embryos in seeds. Gymnosperms, such as pines, produce seeds in cones. The seeds of angiosperms develop within protective ovaries.

Alternation of Generations and Plant Life Cycles (17.3–17.11)

17.3 Haploid and diploid generations alternate in plant life cycles. The haploid gametophyte produces eggs and sperm by mitosis. The zygote develops into the diploid sporophyte, in which meiosis produces haploid spores. Spores grow into gametophytes. The life cycle of a moss is dominated by the gametophyte. Ferns, like most plants, have a life cycle dominated by the sporophyte. In both mosses and ferns, sperm swim to the egg.

17.4 Seedless vascular plants dominated vast "coal forests."

17.5 Pollen and seeds are key adaptations for life on land. In seed plants, reproduction does not require moisture; pollen grains carry the male gametophyte to the female gametophyte. The zygote develops into a sporophyte embryo, and the ovule becomes a seed, with stored food and a protective coat.

17.6 The flower is the centerpiece of angiosperm reproduction. A flower usually consists of sepals, petals, stamens (produce pollen), and carpels (produce ovules).

17.7 The angiosperm plant is a sporophyte with gametophytes in its flowers. The sporophyte is independent, with tiny, dependent gametophytes protected in flowers. Ovules become seeds, and ovaries become fruits.

17.8 The structure of a fruit reflects its function in seed dispersal.

17.9 Angiosperms sustain us—and add spice to our diets.

17.10 Pollination by animals has influenced angiosperm evolution. Flowers attract pollinators by color and scent. Visiting pollinators are rewarded with nectar and pollen.

17.11 Plant diversity is vital to the future of the world's food supply. As plant biodiversity is lost through extinction and habitat destruction, potential crop species and valuable genes are lost.

Diversity of Fungi (17.12–17.19)

17.12 Fungi absorb food after digesting it outside their bodies. Fungi are heterotrophic eukaryotes that digest their food externally and absorb the resulting nutrients. A fungus usually consists of a mass of threadlike hyphae, called a mycelium.

17.13 Fungi produce spores in both asexual and sexual life cycles. In some fungi, fusion of haploid hyphae produces a heterokaryotic stage containing nuclei from two parents. After the nuclei fuse, meiosis produces haploid spores.

17.14 Fungi are classified into five groups. Fungi evolved from a protist ancestor. Fungal groups include chytrids, zygomycetes, glomeromycetes, ascomycetes, and basidiomycetes.

17.15 Fungi have enormous ecological benefits. Fungi are essential decomposers and also participate in mycorrhizae.

17.16 Fungi have many practical uses. Some fungi provide food or antibiotics.

17.17 Lichens are symbiotic associations of fungi and photosynthetic organisms. The photosynthesizers are algae or cyanobacteria.

17.18 Mycorrhizae may have helped plants colonize land. The hypothesis that plant-fungus symbioses facilitated the evolution of land plants is tested by investigating three lines of evidence: present-day mycorrhizal relationships, fossils of early land plants, and molecular homologies.

17.19 Parasitic fungi harm plants and animals.

Connecting the Concepts

1. In this abbreviated diagram, identify the four major plant groups and the key terrestrial adaptation associated with each of the three major branch points.

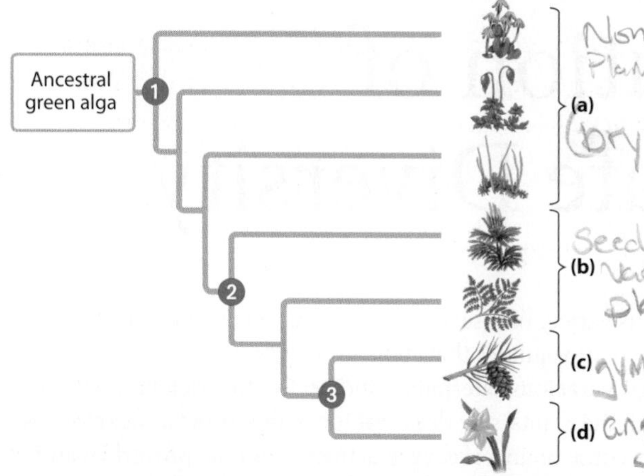
Ancestral green alga
1
2
3
(a)
(b)
(c)
(d)

2. Identify the cloud seen in each photograph. Describe the life cycle events associated with each cloud.

(a) Pine tree, a gymnosperm

(b) Puffball, a club fungus

Testing Your Knowledge

Level 1: Knowledge/Comprehension

3. Angiosperms are different from all other plants because only they have
   a. a vascular system.
   b. flowers.
   c. seeds.
   d. a dominant sporophyte phase.
4. Which of the following produce eggs and sperm? (*Explain your answer.*)
   a. fern sporophytes
   b. moss gametophytes
   c. the anthers of a flower
   d. moss sporangia
5. The eggs of seed plants are fertilized within ovules, and the ovules then develop into
   a. seeds.
   b. spores.
   c. fruit.
   d. sporophytes.
6. The diploid sporophyte stage is dominant in the life cycles of all of the following *except*
   a. a pine tree.
   b. a rose bush.
   c. a fern.
   d. a moss.
7. Which of the following terms includes all the others?
   a. angiosperm
   b. gymnosperm
   c. vascular plant
   d. fern
   e. seed plant
8. Under a microscope, a piece of a mushroom would look most like
   a. jelly.
   b. a tangle of string.
   c. grains of sugar or salt.
   d. foam.
9. Which of the following groups is made up exclusively of fungi that form symbioses with plant roots?
   a. ascomycetes
   b. basidiomycetes
   c. glomeromycetes
   d. zygomycetes

Level 2: Application/Analysis

10. Compare a seed plant with an alga in terms of adaptations for life on land versus life in the water.
11. How do animals help flowering plants reproduce? How do the animals benefit?
12. Why are fungi and plants classified in different kingdoms?

Level 3: Synthesis/Evaluation

13. Truffles (the fungi, not the chocolates) are the reproductive bodies of ascomycetes that form mycorrhizae with certain tree species. They are highly prized by gourmets for the delicious scent they add to food. Because truffles grow underground, they are difficult to find—human noses are not sensitive enough to locate them. Many animals, however, are excellent truffle hunters and eagerly consume the fungi. Why would these fungi produce a scent that attracts fungus-eating animals?
14. In April 1986, an accident at a nuclear power plant in Chernobyl, Ukraine, scattered radioactive fallout for hundreds of miles. In assessing the biological effects of the radiation, researchers found mosses to be especially valuable as organisms for monitoring the damage. As mentioned in Module 10.16, radiation damages organisms by causing mutations. Explain why it is faster to observe the genetic effects of radiation on mosses than on plants from other groups. Imagine that you are conducting tests shortly after a nuclear accident. Using potted moss plants as your experimental organisms, design an experiment to test the hypothesis that the frequency of mutations decreases with the organism's distance from the source of radiation.
15. **SCIENTIFIC THINKING** As you learned in Module 17.18, symbiotic relationships with mycorrhizal fungi are found in almost all present-day plant lineages. Mosses are a major exception—most mosses lack mycorrhizal associations. Assuming that mycorrhizae were a key factor in the colonization of land by plants, propose an explanation for the absence of mycorrhizae in present-day moss lineages.

Answers to all questions can be found in Appendix 4.

CHAPTER 18

The Evolution of Invertebrate Diversity

Why so many insects?

Did you ever notice the tremendous variety of insects? To name just a few that you may have encountered, there are bloodthirsty mosquitoes that spoil a summer evening; bees, wasps, and hornets that deliver painful stings; voracious caterpillars and beetles that defoliate garden plants and thousands of aphids that suck the plants' sap; fleas that torment our pets; cockroaches that scuttle for shelter when the lights come on in a skeevy apartment; and the proverbial ants at every picnic. Not all insects bug us, however. Many are colorful, graceful, and a pleasure to behold—for instance, butterflies, dragonflies, and fireflies.

The vast diversity of insects encompasses a wide variety of shapes and sizes, habitats, diets, mating habits, and other characteristics. Consider, for example, their huge range in body length—the longest known species (a walking stick) measures more than 3,500 times the length of the smallest (a parasitic wasp). For almost any food resource, there is an insect prepared to make use of it. Blowflies, which can smell a dead animal from a mile away, arrive within minutes to deposit eggs that will

hatch into corpse-consuming maggots. Overwatered houseplants may host a swarm of fungus gnats that feed on the algae and fungi growing in the damp soil. And if you plant milkweed, you may reap the reward of watching monarch butterflies sip nectar from its flowers.

With more than a million species—nearly three-quarters of all animal species—insects are exemplars of animal diversity, the subject of this chapter. In our brief tour of the animal kingdom, we will sample just 9 of the roughly 35 animal phyla. Along the way, you will encounter a dazzling variety of forms ranging from corals to cuttlefish to chordates. You will also learn the secret to the spectacular success of insects. But first let's define what an animal is!

BIG IDEAS

Animal Evolution and Diversity
(18.1–18.4)

Animal body plans and molecular data can be used to build a phylogenetic tree.

Invertebrate Diversity
(18.5–18.16)

In this chapter, you will learn about animals without backbones, which make up the overwhelming majority of all animals.

Animal Evolution and Diversity

18.1 What is an animal?

▲ **Figure 18.1A** Ingestion, the animal way of life

Animals are multicellular, heterotrophic eukaryotes that (with a few exceptions) obtain nutrients by ingestion. Now that's a mouthful—and speaking of mouthfuls, **Figure 18.1A** shows a rock python just beginning to ingest a gazelle. **Ingestion** means eating food. This mode of nutrition contrasts animals with fungi, which absorb nutrients after digesting food outside their body. Animals digest food within their body after ingesting other organisms, dead or alive, whole or by the piece.

Animals also have cells with distinctive structures and specializations. Animal cells lack the cell walls that provide strong support in the bodies of plants and fungi. Animal cells are held together by extracellular structural proteins, the most abundant of which is collagen, and by unique types of intercellular junctions (see Module 4.20). In addition, all but the simplest animals have muscle cells for movement and nerve cells for conducting impulses.

Other unique features are seen in animal reproduction and development. Most animals are diploid and reproduce sexually; eggs and sperm are the only haploid cells, as shown in the life cycle of a sea star in **Figure 18.1B**. ❶ Male and female adult animals make haploid gametes by meiosis, and ❷ an egg and a sperm fuse, producing a zygote. ❸ The zygote divides by mitosis, ❹ forming an early embryonic stage called a **blastula**, which is usually a hollow ball of cells. ❺ In the sea star and most other animals, one side of the blastula folds inward, forming a stage called a **gastrula**. ❻ The internal sac formed by gastrulation becomes the digestive tract, lined by a cell layer called the **endoderm**. The embryo also has an **ectoderm**, an outer cell layer that gives rise to the outer covering of the animal and, in some phyla, to the central nervous system. Most animals have a third embryonic layer, known as the **mesoderm**, which forms the muscles and most internal organs.

After the gastrula stage, many animals develop directly into adults. Others, such as the sea star, ❼ develop into one or more larval stages first. A **larva** is an immature individual that looks different from the adult animal. ❽ The larva undergoes a major change of body form, called **metamorphosis**, in becoming an adult capable of reproducing sexually.

This transformation of a zygote into an adult animal is controlled by clusters of homeotic genes (see Modules 11.8 and 15.11). The study of these master control genes has helped scientists investigate the phylogenetic relationships among the highly diverse animal forms we are about to survey.

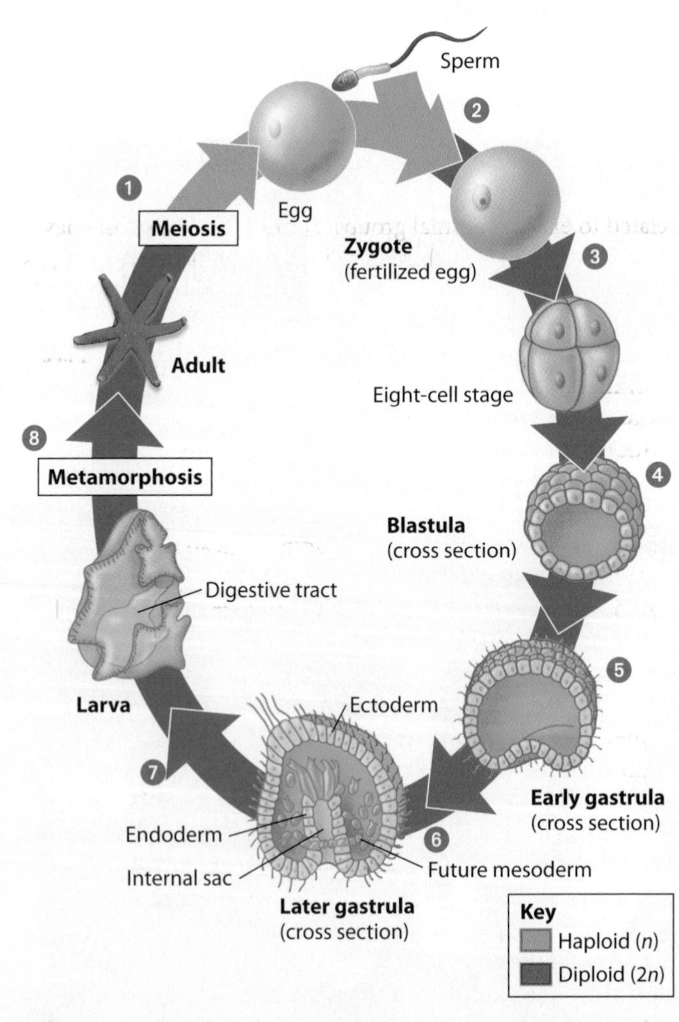

Sperm
2
1
Egg
Meiosis
Zygote
(fertilized egg)
3
Adult
Eight-cell stage
8
Metamorphosis
4
Blastula
(cross section)
Digestive tract
5
Larva
Ectoderm
7
Early gastrula
(cross section)
Endoderm
6
Internal sac
Future mesoderm
Later gastrula
(cross section)
Key
Haploid (n)
Diploid (2n)

▲ **Figure 18.1B** The life cycle of a sea star

Try This On a piece of scratch paper, list and describe the zygote, blastula, gastrula, and larval stages of animal development.

? **List the distinguishing characteristics of animals.**

● Bodies composed of multiple eukaryotic cells; ingestion of food (heterotrophic nutrition); absence of cell walls; unique cell junctions; nerve and muscle cells (generally); sexual reproduction and life cycles with unique embryonic stages; unique developmental (homeotic) genes; gametes alone representing the haploid stage of the life cycle

18.2 Animal diversification began more than half a billion years ago

The lineage that gave rise to animals is thought to have diverged from a flagellated unikont ancestor more than 1 billion years ago (see Module 16.19). This ancestor may have resembled modern choanoflagellates, colonial protists that are the closest living relatives of animals. But despite the molecular data indicating this early origin of animals, the oldest generally accepted animal fossils that have yet been found are 575–550 million years old, from the late Ediacaran period. These fossils were first discovered in the 1940s, in the Ediacara Hills of Australia (hence the name). Similar fossils have since been found in Asia, Africa, and North America. All are impressions of soft-bodied animals that varied in shape and ranged in length from 1 cm to 1 m (**Figure 18.2A**). Although some of the fossils may belong to groups of invertebrates that still exist today, such as sponges and cnidarians, most do not appear to be related to any living organism.

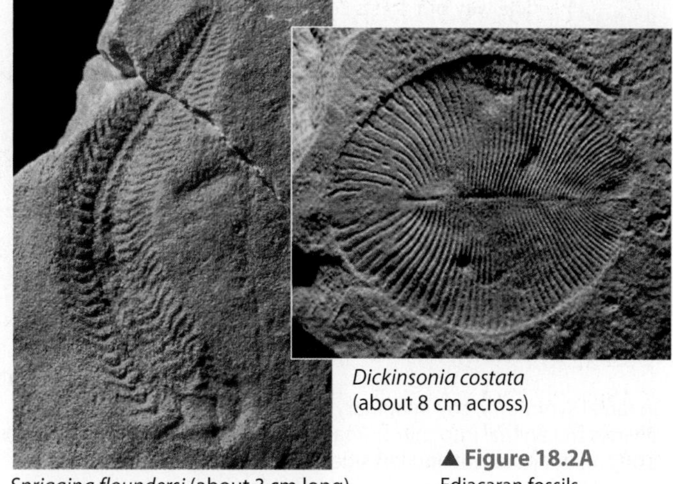

Dickinsonia costata (about 8 cm across)

Spriggina floundersi (about 3 cm long)

▲ **Figure 18.2A** Ediacaran fossils

Animal diversification appears to have accelerated rapidly from 535 to 525 million years ago, during the Cambrian period. Because many animal body plans and new phyla appear in the fossils from such an evolutionarily short time span, biologists call this episode the Cambrian explosion. The most celebrated source of Cambrian fossils is a fine-grained deposit of sedimentary rock in British Columbia. The Burgess Shale, as it is known, provided a cornucopia of perfectly preserved animal fossils. In contrast to the uniformly soft-bodied Ediacaran animals, many Cambrian animals had hard body parts such as shells and spikes. Many of these fossils are clearly related to existing animal groups. For example, scientists have classified more than a third of the species found in the Burgess Shale as arthropods, including the one labeled in **Figure 18.2B**. (Present-day arthropods include crabs, shrimp, and insects.) Another striking fossil represented in this reconstruction is an early member of our own phylum, Chordata. Other fossils are more difficult to place, and some are downright weird, such as the spiky creature called *Hallucigenia* and the formidable predator *Anomalocaris* (dominating the left half of the illustration), which grew to an estimated 2 feet in length. The circular structure on the underside of the animal's head is its mouth.

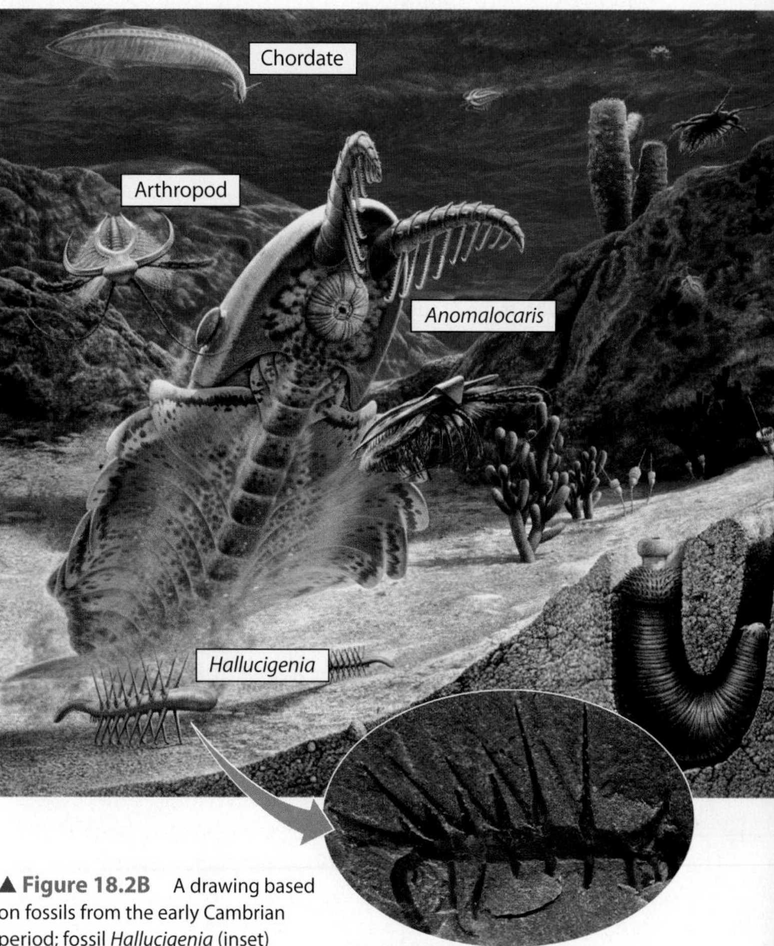
Chordate
Arthropod
Anomalocaris
Hallucigenia

▲ **Figure 18.2B** A drawing based on fossils from the early Cambrian period; fossil *Hallucigenia* (inset)

What ignited the Cambrian explosion? Scientists have proposed several hypotheses, including increasingly complex predator-prey relationships and an increase in atmospheric oxygen. But whatever the cause of the rapid diversification, it is highly probable that the set of homeotic genes—the genetic framework for complex bodies—was already in place. Much of the diversity in body form among the animal phyla is associated with variations in where and when homeotic genes are expressed within developing embryos (see Module 15.11). The role of these master control genes in the evolution of animal diversity will be discussed further in Module 18.13.

Of the 35 or so animal phyla (systematists disagree on the precise number), only one phylum includes **vertebrates**, animals with a backbone. The members of all other animal phyla—roughly 96% of animals—are **invertebrates**, animals that lack a backbone. Now let's look at some of the anatomical features biologists use to classify this vast animal diversity.

? What are two major differences between the fossil animals from the Ediacaran and the Cambrian periods?

● Ediacaran animals were all soft-bodied; many Cambrian animals had hard parts such as shells. Few Ediacaran animals can be classified as members of present-day groups; many Cambrian animals are clearly related to present-day groups.

VISUALIZING THE CONCEPT

18.3 Animals can be characterized by basic features of their "body plan"

One way biologists categorize animals is by certain general features of body structure, which together describe what is called an animal's "body plan." Distinctions between body plans help biologists infer the phylogenetic relationships between animal groups, as you will see in the next module.

One prominent feature of a body plan is symmetry. In **radial symmetry**, the body parts radiate from the center like the spokes of a bicycle wheel. An animal with **bilateral symmetry** has mirror-image right and left sides; a distinct head, or **anterior**, end; a tail, or **posterior**, end; a back, or **dorsal**, surface; and a bottom, or **ventral**, surface.

Body plans also vary in tissue organization. A tissue is an integrated group of cells with a common function, structure, or both. True tissues are collections of specialized cells, usually isolated from other tissues by membrane layers, that perform specific functions. An example is the nervous tissue of your brain and spinal cord. Sponges lack true tissues, but in other animals, the cell layers formed during gastrulation (see Figure 18.1B) give rise to true tissues and organs.

TYPE OF SYMMETRY

In radial symmetry, any imaginary slice through the central axis divides the animal into mirror images. The animal has a top and a bottom, but not right and left sides. In bilateral symmetry, the animal has mirror-image right and left sides.

Radial symmetry

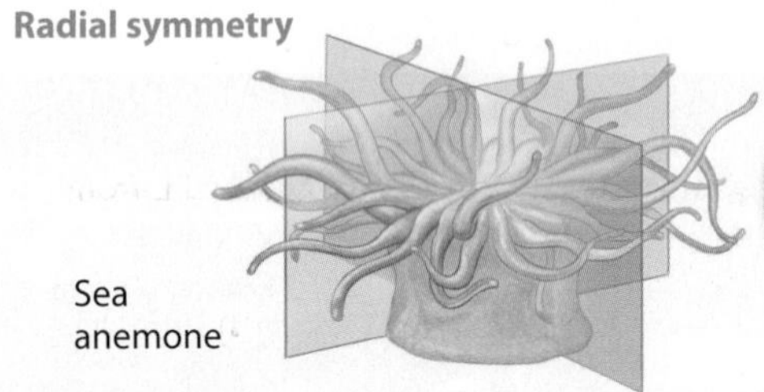
Sea anemone

A radial animal is typically sedentary or passively drifting, meeting its environment equally on all sides.

Bilateral symmetry

Bilateral symmetry, with the brain, sense organs, and mouth usually located in the head, facilitates mobility. As the animal travels headfirst through the environment, its eyes and other sense organs contact the environment first.

Lobster

EMBRYONIC DEVELOPMENT

Animals with three tissue layers are divided into two groups based on details of their later embryonic development. In **protostomes**, the first opening that forms during gastrulation becomes the mouth; in **deuterostomes**, this opening becomes the anus and the mouth forms from a second opening.

In addition, most animals with three tissue layers have a **body cavity**, or *coelom*, a fluid-filled space between the digestive tract and outer body wall in which the internal organs are suspended.

Gastrulation

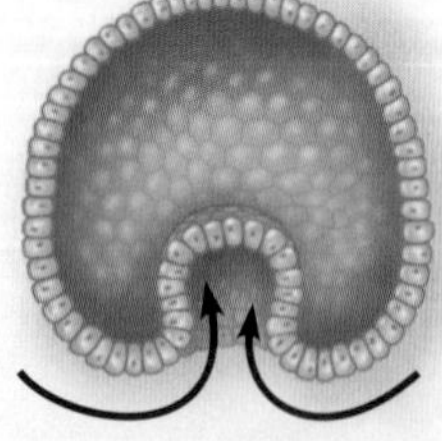
First opening in embryo

Gastrulation forms just two tissue layers in some animals, but most have three layers.

Two tissue layers

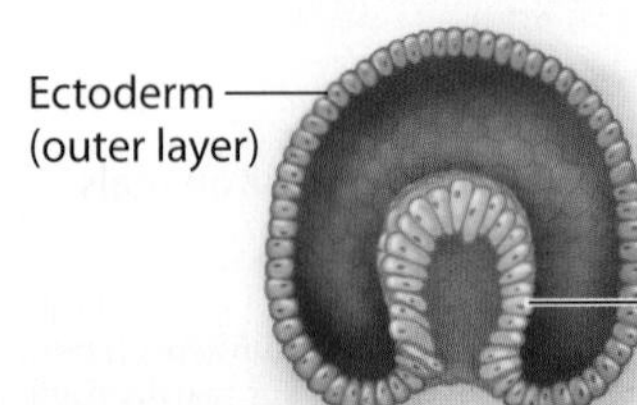
Ectoderm (outer layer)
Endoderm (inner layer)

Three tissue layers

Protostome

From the Greek "protos," first, and "stoma," mouth

Deuterostome

From the Greek "deutero," second, and "stoma," mouth

Future anus
Future mouth
Mesoderm
Future digestive tract
Future mouth
Future anus

Body cavity

The body cavity helps protect the suspended organs from injury.

Body covering (from ectoderm)
Body cavity
Digestive tract (from endoderm)
Tissue layer lining body cavity (from mesoderm)

? **List four features of an animal's body plan.**

Symmetry, number of embryonic tissue layers, fate of first opening formed by gastrulation, and presence of body cavity

18.4 Body plans and molecular comparisons of animals can be used to build phylogenetic trees

Biologists traditionally used evidence from body plan characteristics and the fossil record to make hypotheses about the phylogeny of animal groups. Molecular data, chiefly DNA sequences, have provided new opportunities to test these hypotheses. As a result, scientists have recently revised the animal phylogenetic tree to reflect the new information. **Figure 18.4** shows a current hypothesis of the evolutionary history of nine major animal phyla.

At the far left of Figure 18.4 is the hypothetical ancestral colonial protist (see Module 16.19). The tree has a series of branch points that represent shared derived characters. The first branch point splits the sponges, which lack true tissues, from the clade of **eumetazoans** ("true animals"), the animals with true tissues. The eumetazoans split into two distinct lineages that differ in body symmetry and the number of cell layers formed in gastrulation. Members of phylum Cnidaria are radially symmetric and have two cell layers. The other lineage consists of animals with bilateral symmetry, the **bilaterians.** This clade contains most animal phyla.

Animals with deuterostome embryonic development, which include the echinoderms and chordates, are recognized as one lineage of bilaterians. The other two branch points are the ecdysozoans and lophotrochozoans, most of which have protostome development. The ecdysozoans include the nematodes and arthropods, which have external skeletons that must be shed for the animal to grow. The shedding process, called ecdysis, is the basis for the name ecdysozoan. Lophotrochozoans, which include the flatworms, molluscs, annelids, and many other phyla not represented in this figure, are grouped on the basis of genetic similarities. The cumbersome name of the lineage comes from the feeding apparatus (called a lophophore) of some phyla in the group and from the trochophore larva found in molluscs and annelids (see Figure 18.9B).

Now that you have a framework for organizing animal diversity, let's look at the unique characteristics of each phylum and meet some examples.

? **What shared derived character separates bilaterians from cnidarians?**

● Bilateral symmetry

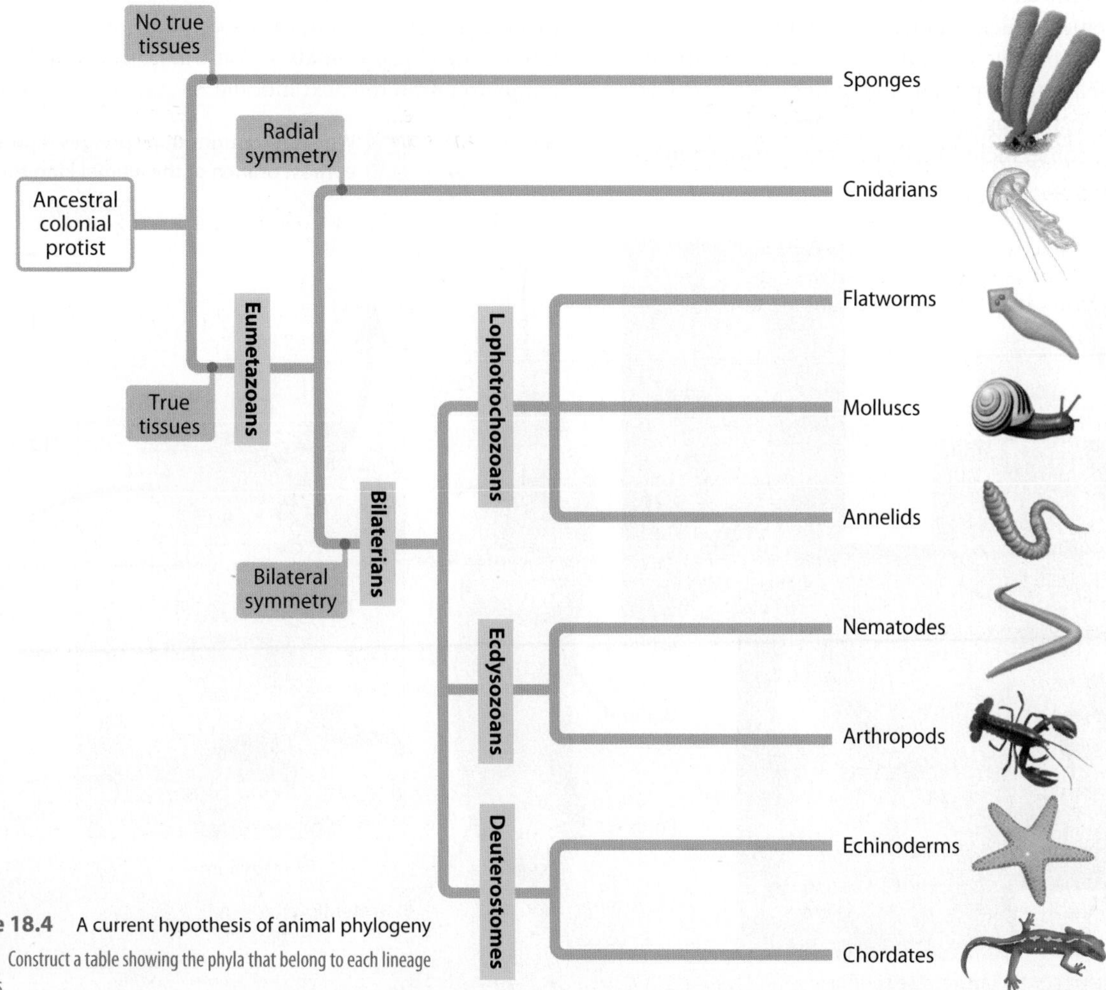
No true tissues
Sponges
Radial symmetry
Cnidarians
Ancestral colonial protist
Flatworms
Eumetazoans
Lophotrochozoans
True tissues
Molluscs
Annelids
Bilaterians
Bilateral symmetry
Nematodes
Ecdysozoans
Arthropods
Deuterostomes
Echinoderms
Chordates

▲ **Figure 18.4** A current hypothesis of animal phylogeny

Try This Construct a table showing the phyla that belong to each lineage of bilaterians.

▷ Invertebrate Diversity

18.5 Sponges have a relatively simple, porous body

Sponges (phylum Porifera) are the simplest of all animals. They have no nerves or muscles, though their individual cells can sense and react to changes in the environment. The majority of species are marine, although some are found in fresh water. Some sponges are radially symmetric, but most lack body symmetry. There is variation in the size and internal structure of sponges. For example, the purple tube sponge shown in **Figure 18.5A** can reach heights of 1.5 m (about 5 feet). The body of *Scypha,* a small sponge only about 1–3 cm tall resembles a simple sac. Other sponges, such as the azure vase sponge, have folded body walls and irregular shapes.

A simple sponge resembles a thick-walled sac perforated with holes. (*Porifera* means "pore-bearer" in Latin.) Water enters through the pores into a central cavity, and then flows out through a larger opening **(Figure 18.5B)**. More complex sponges have branching water canals.

The body of a sponge consists of two layers of cells separated by a gelatinous region. Because the cell layers are loose associations of cells, they are not considered true tissues. The inner cell layer consists of flagellated "collar" cells called **choanocytes** (tan in Figure 18.5B), which help to sweep water through the sponge's body. (Notice the resemblance of choanocytes to the choanoflagellate in Figure 16.19B.) **Amoebocytes** (blue), which wander through the middle body region, produce supportive skeletal fibers (yellow) composed of a flexible protein called spongin and mineralized particles called spicules. Most sponges have both types of skeletal components, but some, including those used as bath sponges, only contain spongin.

Sponges are examples of **suspension feeders**, animals that collect food particles from water passed through some type of food-trapping equipment. Sponges feed by collecting food particles suspended in the water that streams through their porous bodies. To obtain enough food to grow by 100 g (about 3 ounces), a sponge must filter roughly 1,000 kg (about 275 gallons) of seawater. Choanocytes trap food particles in mucus on the membranous collars that surround the base of their flagella and then engulf the food by phagocytosis (see Module 5.9). Amoebocytes pick up food packaged in food vacuoles from choanocytes, digest it, and carry the nutrients to other cells.

Adult sponges are **sessile**, meaning they are anchored in place—they cannot escape from predators. Researchers have found that sponges produce defensive compounds such as toxins and antibiotics that deter pathogens, parasites, and predators. Some of these compounds may prove useful to humans as new drugs.

Biologists hypothesize that sponge lineages arose very early from the multicellular organisms that gave rise to the animal kingdom. The choanocytes of sponges and the cells of living choanoflagellates are similar, supporting the molecular evidence that animals evolved from a colonial protist ancestor. Sponges are the only animal phylum covered in this book that lack true tissues and thus are not members of the clade Eumetazoa ("true animals"). You will learn about the simplest Eumetazoans in the next module.

? **Why is it thought that sponges represent the earliest branch of the animal kingdom?**

● Sponges lack true tissues, and their choanocytes resemble certain flagellated protists.

A purple tube sponge
An azure vase sponge
Scypha
Central cavity
Water flow
Pores
Water flow
Skeletal fiber
Water flow
Choanocyte in contact with an amoebocyte
Pore
Choanocyte
Amoebocyte
Flagellum

▲ **Figure 18.5A** Sponges

▲ **Figure 18.5B** The structure of a simple sponge

18.6 Cnidarians are radial animals with tentacles and stinging cells

Among eumetazoans, one of the oldest groups is phylum Cnidaria, which includes the hydras, sea anemones, corals, and jellies (also called "jellyfish"). **Cnidarians** are characterized by radial symmetry and bodies arising from only two tissue layers. The simple body of most cnidarians has an outer epidermis and an inner cell layer that lines the digestive cavity. A jelly-filled middle region may contain scattered amoeboid cells. Contractile tissues and nerves occur in their simplest forms in cnidarians.

Cnidarians exhibit two kinds of radially symmetric body forms. Hydras, common in freshwater ponds and lakes, and sea anemones have a cylindrical body with tentacles projecting from one end. This body form is a **polyp** (**Figure 18.6A**). The other type of cnidarian body is the **medusa**, exemplified by the marine jelly in **Figure 18.6B**. While polyps are mostly stationary, medusae move freely about in the water. Medusae are shaped like an umbrella with a fringe of tentacles around the lower edge. A few jellies have tentacles 60–70 m long dangling from an umbrella up to 2 m in diameter, but the diameter of most jellies ranges from 2 to 40 cm.

Some cnidarians pass sequentially through both a polyp stage and a medusa stage in their life cycle. Others exist only as medusae; still others, including hydras and sea anemones, exist only as polyps.

Despite their flowerlike appearance, cnidarians are carnivores that use their tentacles to capture small animals and protists and to push the prey into their mouths. In a polyp, the mouth is on the top of the body, at the hub of the radiating tentacles (see Figure 21.3A). In a medusa, the mouth is in the center of the undersurface. The mouth leads into a multifunctional compartment called a **gastrovascular cavity** (from the Greek *gaster*, belly, and Latin *vas*, vessel), where food is digested. The mouth is the only opening in the body, so it is also the exit for undigested food and other wastes. Fluid in the gastrovascular cavity circulates nutrients and oxygen to internal cells and removes metabolic wastes (hence the "vascular" in gastrovascular; see Module 23.1). The pressure of the fluid supports the body and helps to give a cnidarian its shape, much like water can give shape to a balloon. When the animal closes its mouth, the volume of the cavity is fixed. Then contraction of muscle cells in the body wall can shorten or lengthen the body. Some polyps can use these contractions to produce movement.

Phylum Cnidaria (from the Greek *cnide*, nettle, a stinging plant) is named for its unique stinging cells, called **cnidocytes**, that function in defense and in capturing prey. Each cnidocyte contains a fine thread coiled within a capsule (**Figure 18.6C**). When it is discharged, the thread can sting or entangle prey. Some large marine cnidarians use their stinging threads to catch fish. A group of cnidarians called cubozoans have highly toxic cnidocytes. The sea wasp, a cubozoan found off the coast of northern Australia, is the deadliest organism on Earth: One animal may produce enough poison to kill as many as 60 people.

Coral animals are polyp-form cnidarians (see Figure 37.4) that secrete a hard external skeleton. Each generation builds on top of the skeletons of previous generations, constructing the characteristic shapes of "rocks" we call coral. Reef-building corals depend on sugars produced by symbiotic algae to supply them with enough energy to maintain the reef structure in the face of erosion and reef-boring animals.

? **What are three functions of a cnidarian's gastrovascular cavity?**

● Digestion, circulation (transport of oxygen, nutrients, and wastes), and physical support and movement

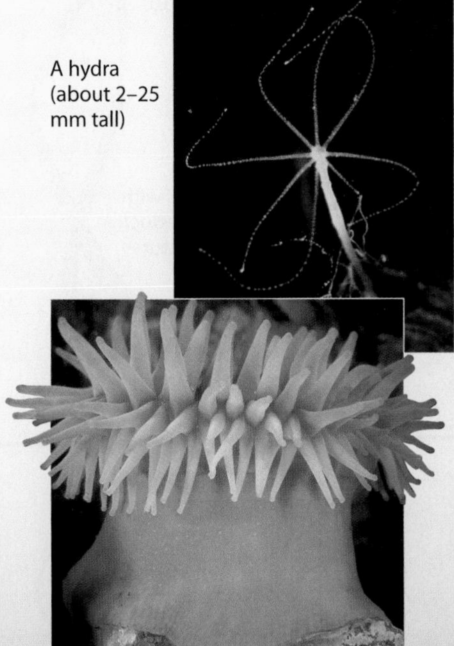
A hydra (about 2–25 mm tall)

A sea anemone (about 6 cm in diameter)

▲ **Figure 18.6A** Polyp body form

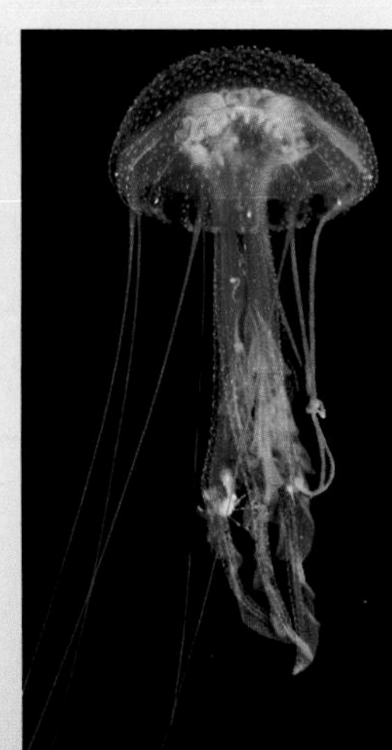
A marine jelly (about 6 cm in diameter)

▲ **Figure 18.6B** Medusa body form

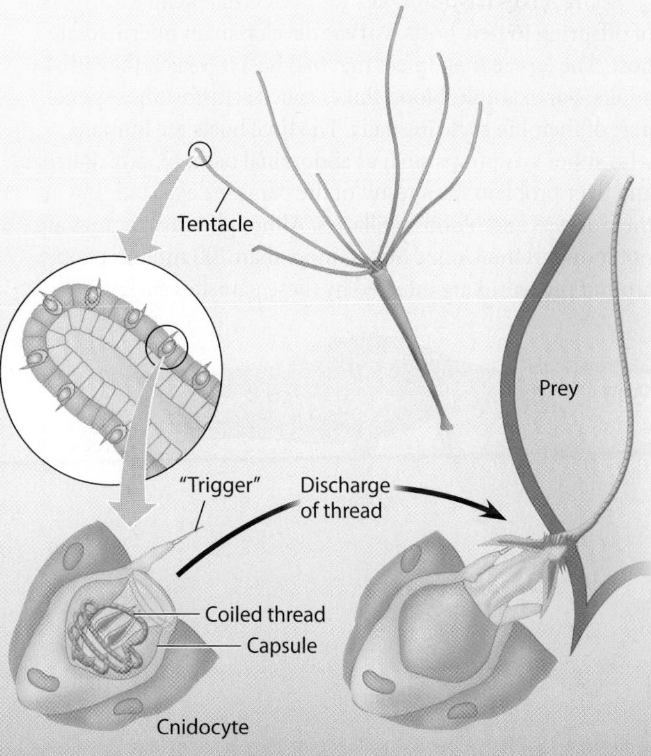
Tentacle
Prey
"Trigger"
Discharge of thread
Coiled thread
Capsule
Cnidocyte

▲ **Figure 18.6C** Cnidocyte action

18.7 Flatworms are the simplest bilateral animals

Flatworms, phylum Platyhelminthes (from the Greek *platys*, flat, and *helmis*, worm), belong to the lophotrochozoan lineage of bilaterians. Along with other bilaterians, flatworms have bilateral symmetry and three tissue layers. However, they lack a body cavity. These thin, often ribbonlike animals range in length from about 1 mm to 20 m and live in marine, freshwater, and damp terrestrial habitats. In addition to free-living forms, there are many parasitic species. Like cnidarians, most flatworms have a gastrovascular cavity with only one opening. Fine branches of the gastrovascular cavity distribute food throughout the animal.

There are three major groups of flatworms. Worms called planarians represent the **free-living flatworms** (**Figure 18.7A**). Planarians live on the undersurfaces of rocks in freshwater ponds and streams. A planarian has a head with a pair of light-sensitive eyecups and a flap at each side that detects chemicals. Dense clusters of nerve cells form a simple brain, and a pair of nerve cords connect with small nerves that branch throughout the body. The location of the brain, sense organs, and mouth in the anterior end is characteristic of bilaterally symmetric animals. This arrangement facilitates mobility. As the animal travels headfirst through the environment, its eyes and other sense organs contact the environment first. When a planarian feeds, it sucks food in through a mouth at the tip of a muscular tube that projects from a surprising location—the mid-ventral surface of the body (as shown in the figure).

A second group of flatworms, the **flukes**, live as parasites in other animals. Many flukes have suckers that attach to their host and a tough protective covering. Reproductive organs occupy nearly the entire interior of these worms.

Many flukes have complex life cycles that facilitate dispersal of offspring to new hosts. Larvae develop in an intermediate host. The larvae then infect the final host in which they live as adults. For example, blood flukes called schistosomes spend part of their life cycle in snails. The final hosts are humans, who suffer symptoms such as abdominal pain, bloody diarrhea, and liver problems as a result of the parasite's eggs lodging in their organs and blood capillaries. Although schistosomes are not found in the United States, more than 200 million people around the world are infected by these parasites each year.

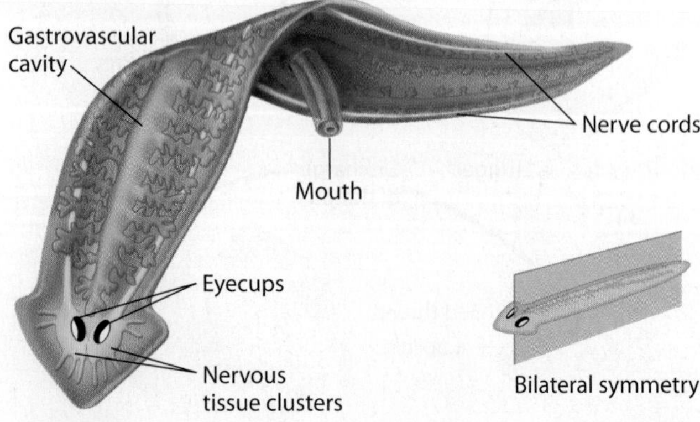

Gastrovascular cavity
Nerve cords
Mouth
Eyecups
Nervous tissue clusters
Bilateral symmetry

▲ **Figure 18.7A** A free-living flatworm, the planarian (most are about 5–10 mm long)

Tapeworms are another parasitic group of flatworms. Adult tapeworms inhabit the digestive tracts of vertebrates, including humans. In contrast with planarians and flukes, most tapeworms have a very long, ribbonlike body with repeated units. As **Figure 18.7B** shows, the anterior end, called the scolex, is armed with hooks and suckers that grasp the host. Notice that there is no mouth. Bathed in the partially digested food in the intestines of their hosts, tapeworms simply absorb nutrients across their body surface and have no digestive tract. Because of this adaptation to their parasitic lifestyle, tapeworms are an exception to our definition of animals in Module 18.1; other animals ingest nutrients. Behind the scolex is a long ribbon of repeated units filled with both male and female reproductive structures. The units at the posterior end, which are full of ripe eggs, break off and pass out of the host's body in feces.

Like parasitic flukes, tapeworms have a complex life cycle, usually involving more than one host. Most species take advantage of the predator-prey relationships of their hosts. A prey species—a sheep or a rabbit, for example—may become infected by eating grass contaminated with tapeworm eggs. Larval tapeworms develop in these hosts, and a predator—a coyote or a dog, for instance—becomes infected when it eats an infected prey animal. The adult tapeworms then develop in the predator's intestines. Humans can be infected with tapeworms by eating undercooked beef, pork, or fish infected with tapeworm larvae. The larvae are microscopic, but the adults of some species can reach lengths of 2 m in the human intestine.

Flatworms are just one of three major animal phyla known as worms. You'll learn about roundworms in the next module and segmented worms in Module 18.10.

? **Flatworms and cnidarians differ in symmetry, with flatworms being __________ and cnidarians being __________, but the animals of both phyla have a __________.**

● bilateral . . . radial . . . gastrovascular cavity

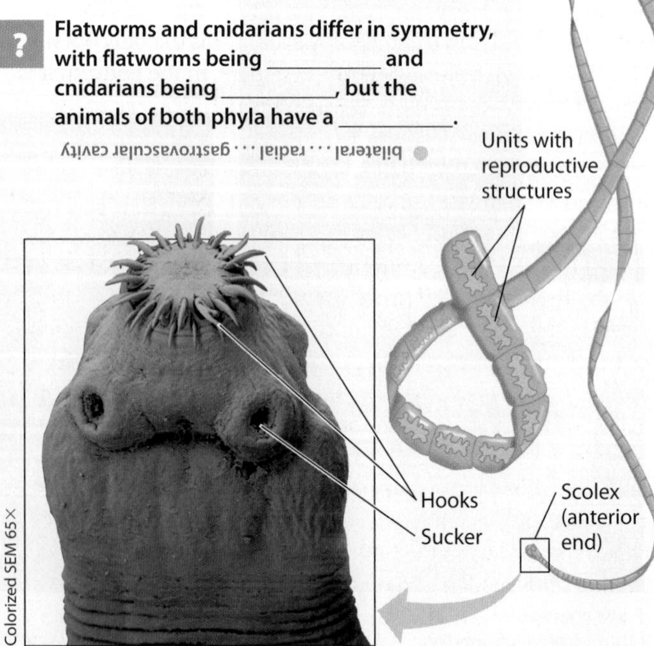

Units with reproductive structures
Hooks
Sucker
Scolex (anterior end)
Colorized SEM 65×

▲ **Figure 18.7B** A tapeworm, a parasitic flatworm

18.8 Nematodes have a body cavity and a complete digestive tract

Nematodes, also called roundworms, make up the phylum Nematoda. As bilaterians, these animals have bilateral symmetry and an embryo with three tissue layers. In contrast with flatworms, roundworms have a fluid-filled body cavity and a digestive tract with two openings.

Nematodes are cylindrical with a blunt head and tapered tail. They range in size from less than 1 mm to more than a meter. Several layers of tough, nonliving material called a **cuticle** cover the body and prevent the nematode from drying out. What looks like a corduroy coat on the nematode in Figure 18.8A is its cuticle. In parasitic species, the cuticle protects the nematode from the host's digestive system. When the worm grows, it periodically sheds its cuticle (molts) and secretes a new, larger one. Thus, nematodes are members of the ecdysozoan lineage of the clade Bilateria.

You can also see the mouth at the tip of the blunt anterior end of the nematode in Figure 18.8A. Nematodes have a **complete digestive tract**, extending as a tube from the mouth to the anus near the tip of the tail. Food travels only one way through the system and is processed as it moves along. In animals with a complete digestive tract, the anterior regions of the tract churn and mix food with enzymes, while the posterior regions absorb nutrients and then dispose of wastes. This division of labor makes the process more efficient and allows each part of the digestive tract to be specialized for its particular function.

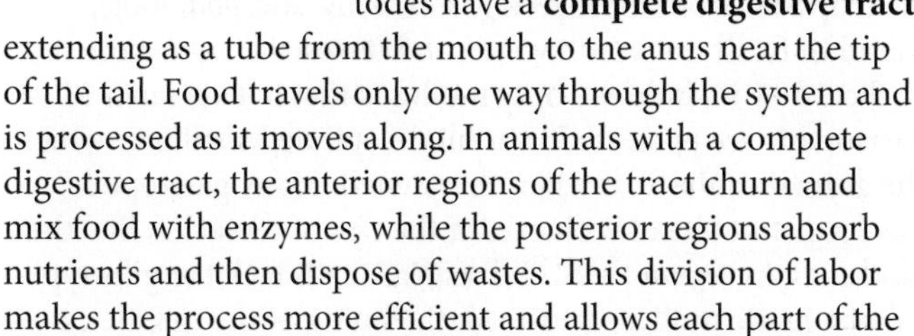
Mouth
Colorized SEM 400×

▲ **Figure 18.8A** A free-living nematode

Fluid in the body cavity of a nematode distributes nutrients absorbed from the digestive tract throughout the body. Contraction of longitudinal muscles against the pressure of the fluid produces a whiplike motion that is characteristic of nematodes.

Although about 25,000 species of nematodes have been named, estimates of the total number of species range as high as 500,000. Free-living nematodes live virtually everywhere there is rotting organic matter, and their numbers are huge. Ninety thousand individuals were found in a single rotting apple lying on the ground; an acre of topsoil contains billions of nematodes. Nematodes are important decomposers in soil and on the bottom of lakes and oceans. Some are predators, eating other microscopic animals.

Little is known about most free-living nematodes. A notable exception is the soil-dwelling species *Caenorhabditis elegans*, an important research organism. A *C. elegans* adult consists of only about 1,000 cells—in contrast with the human body, which consists of some 10 trillion cells. By following every cell division in the developing embryo, biologists have been able to trace the lineage of every cell in the adult worm. The genome of *C. elegans* has been sequenced, and ongoing research contributes to our understanding of how genes control animal development, the functioning of nervous systems, and even some of the mechanisms of aging.

Other nematodes thrive as parasites in the moist tissues of plants and in the body fluids and tissues of animals. The largest known nematodes are parasites of whales and measure more than 7 m (23 feet) long! Many species are serious agricultural pests that attack the roots of plants or parasitize animals. The dog heartworm (Figure 18.8B), a common parasite, is deadly to dogs and can also infect other pets such as cats and ferrets. It is spread by mosquitoes, which pick up heartworm eggs in the blood of an infected host and transmit them when sucking the blood of another animal. Although dog heartworms were once found primarily in the southeastern United States, they are now common throughout the contiguous United States. Regular doses of a preventive medication can protect dogs from heartworm.

Humans are host to at least 50 species of nematodes, including a number of disease-causing organisms. *Trichinella spiralis* causes a disease called trichinosis in a wide variety of mammals, including humans. People usually acquire the worms by eating undercooked pork or wild game containing the juvenile worms. Cooking meat until it is no longer pink kills the worms. Hookworms, which grapple onto the intestinal wall and suck blood, infect millions of people worldwide. Although hookworms are small (about 10 mm long), a heavy infestation can cause severe anemia.

You might expect that an animal group as numerous and widespread as nematodes would include a great diversity of body form. In fact, the opposite is true. Most species look very much alike. In sharp contrast, animals in the phylum Mollusca, which we examine next, exhibit enormous diversity in body form.

? **What is the advantage of a complete digestive tract?**

● Different parts of the digestive tract can be specialized for different functions.

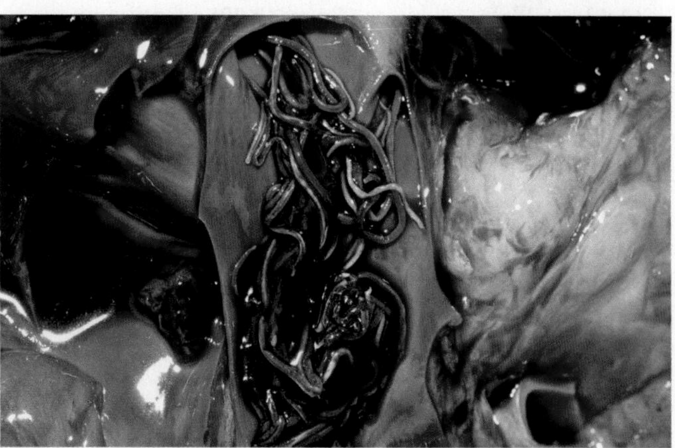

▲ **Figure 18.8B** Parasitic nematodes infesting a large artery in a porpoise

18.9 Diverse molluscs are variations on a common body plan

Snails, slugs, oysters, clams, octopuses, and squids are just a few of the great variety of animals known as **molluscs** (phylum Mollusca). Molluscs are soft-bodied animals (from the Latin *molluscus*, soft), but most are protected by a hard shell.

You may wonder how animals as different as octopuses and clams could belong in the same phylum, but these and other molluscs have inherited several common features from their ancestors. **Figure 18.9A** illustrates the basic body plan of a mollusc, consisting of three main parts: a muscular **foot** (gray in the drawing), which functions in locomotion; a **visceral mass** (orange) containing most of the internal organs; and a **mantle** (purple), a fold of tissue that drapes over the visceral mass and secretes a shell in molluscs such as clams and snails. In many molluscs, the mantle extends beyond the visceral mass, producing a water-filled chamber called the mantle cavity, which houses the gills (left side in Figure 18.9A).

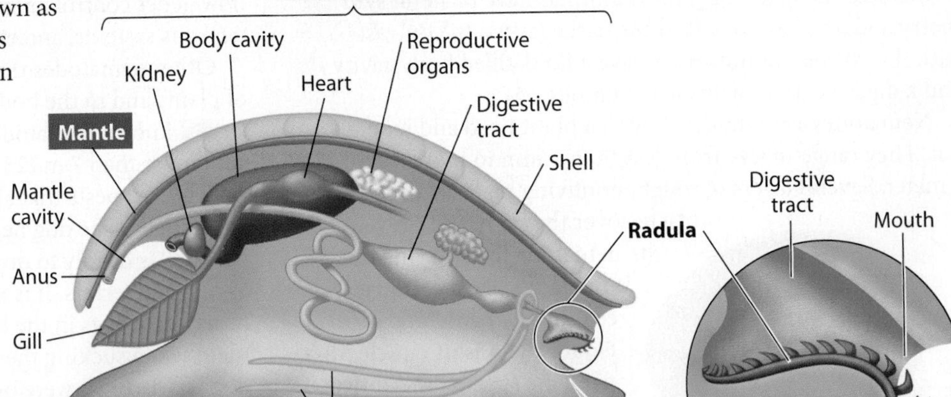
Visceral mass
Body cavity
Reproductive organs
Kidney
Heart
Digestive tract
Mantle
Shell
Mantle cavity
Radula
Digestive tract
Mouth
Anus
Gill
Foot
Nerve cords

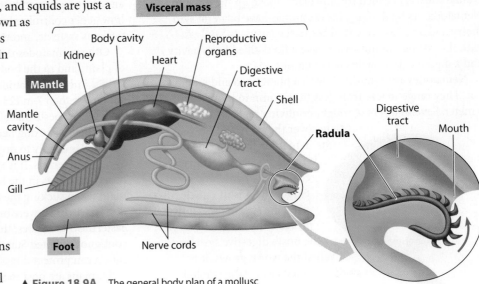

▲ **Figure 18.9A** The general body plan of a mollusc

Figure 18.9A shows yet another body feature found in many molluscs—a unique rasping organ called a **radula**, which is used to scrape up food. In a snail, for example, the radula extends from the mouth and slides back and forth like a backhoe, scraping and scooping algae off rocks. You can observe a radula in action by watching a snail graze on the glass wall of an aquarium.

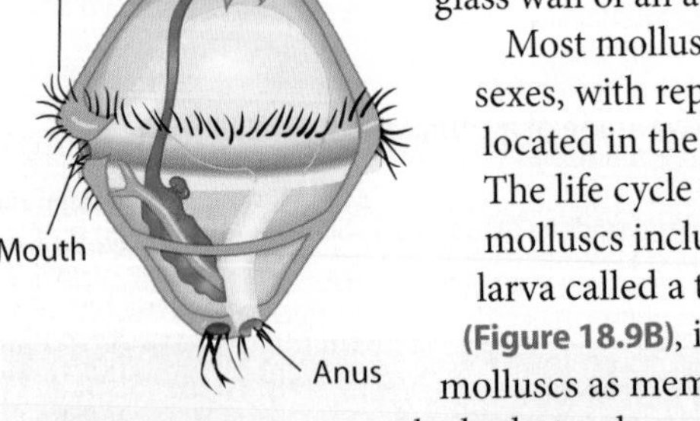
Cilia
Mouth
Anus

▲ **Figure 18.9B** Trochophore larva

Most molluscs have separate sexes, with reproductive organs located in the visceral mass. The life cycle of many marine molluscs includes a ciliated larva called a trochophore (**Figure 18.9B**), identifying molluscs as members of the lophotrochozoan lineage of bilaterians.

Molluscs have a body cavity (brown in Figure 18.9A) and, unlike flatworms and nematodes, complex organs and organ systems. For example, molluscs have a **circulatory system**—an organ system that pumps blood and distributes nutrients and oxygen throughout the body.

These basic body features have evolved in markedly different ways in different groups of molluscs. The three most diverse groups are the gastropods (including snails and slugs), bivalves (including clams, scallops, and oysters), and cephalopods (including squids and octopuses).

Gastropods The largest group of molluscs is called the **gastropods** (from the Greek *gaster*, belly, and *pod*, foot), found in fresh water, salt water, and terrestrial environments. In fact, they include the only molluscs that live on land. Most gastropods are protected by a single, spiraled shell into which the animal can retreat when threatened. Many gastropods have a distinct head with eyes at the tips of tentacles, like the land snail in **Figure 18.9C**. Terrestrial snails lack the gills typical of aquatic molluscs; instead, the lining of the mantle cavity functions as a lung, exchanging gases with the air.

Most gastropods are marine, and shell collectors delight in their variety. Slugs, however, are unusual molluscs in that they have lost their mantle and shell during their evolution. The long, colorful projections on the sea slug in Figure 18.9C function as gills.

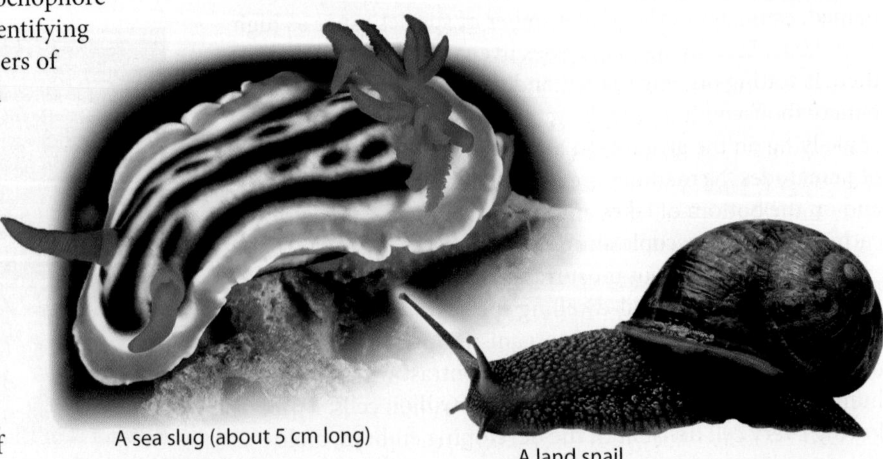

A sea slug (about 5 cm long)

A land snail

▲ **Figure 18.9C** Gastropods

Eyes

A scallop (about 7 cm in diameter)

Mussels (each about 6 cm long)

▲ **Figure 18.9D** Bivalves

Bivalves The **bivalves** (from the Latin *bi*, double, and *valva*, leaf of a folding door) include numerous species of clams, oysters, mussels, and scallops. They have shells divided into two halves that are hinged together. Most bivalves are suspension feeders. The mantle cavity contains gills that are used for feeding as well as gas exchange. The mucus-coated gills trap fine food particles suspended in the water, and cilia sweep the particles to the mouth. Most bivalves are sedentary, living in sand or mud. They may use their muscular foot for digging and anchoring. Mussels are sessile, secreting strong threads that attach them to rocks, docks, and boats (**Figure 18.9D**). The scallop in Figure 18.9D can skitter along the seafloor by flapping its shell, rather like the mechanical false teeth sold in novelty shops. Notice the many eyes peering out between the two halves of its hinged shell. The eyes are set into the fringed edges of the animal's mantle.

Cephalopods The **cephalopods** (from the Greek *kephale*, head, and *pod*, foot) differ from gastropods and bivalves in being adapted to the lifestyle of fast, agile predators. The chambered nautilus in **Figure 18.9E** is a descendant of ancient groups with external shells, but in other cephalopods, the shell is small and internal (as in squids) or missing altogether (as in octopuses). If you have a pet bird, you may have hung the internal shell of another cephalopod, the cuttlefish, in its cage. Such "cuttlebones" are commonly given to caged birds as a source of calcium. Cephalopods use beak-like jaws and a radula to crush or rip prey apart. The mouth is at the base of the foot, which is drawn out into several long tentacles for catching and holding prey.

Octopuses, such as the one in Figure 18.9E, live on the seafloor, where they prowl about in search of crabs and other food. They can also move rapidly by drawing water into the mantle cavity and then forcing a jet of water out through a muscular siphon. Like many cephalopods, octopuses employ an impressive color palette for communication and camouflage, and they display a remarkable assortment of behaviors.

All cephalopods have large brains and sophisticated sense organs that contribute to their success as mobile predators. Cephalopod eyes are among the most complex sense organs in the animal kingdom. Each eye contains a lens that focuses light and a retina on which clear images form. Octopuses are considered among the most intelligent invertebrates and have shown remarkable learning abilities in laboratory experiments.

The so-called colossal squid, which lives in the ocean depths near Antarctica, is the largest of all invertebrates. Specimens are rare, but in 2007, a male colossal squid measuring 10 m and weighing 450 kg was hauled in by a fishing boat. Females are generally even larger, and scientists estimate that the colossal squid averages around 13 m in overall length. The giant squid, which rivals the colossal in length, is thought to average about 10 m. Most information about these impressive animals has been gleaned from dead specimens that washed ashore. In 2004, however, Japanese zoologists succeeded in capturing the first video footage of a giant squid in its natural habitat.

? **Identify the mollusc group that includes each of these examples: garden snail, clam, squid.**

● Gastropod, bivalve, cephalopod

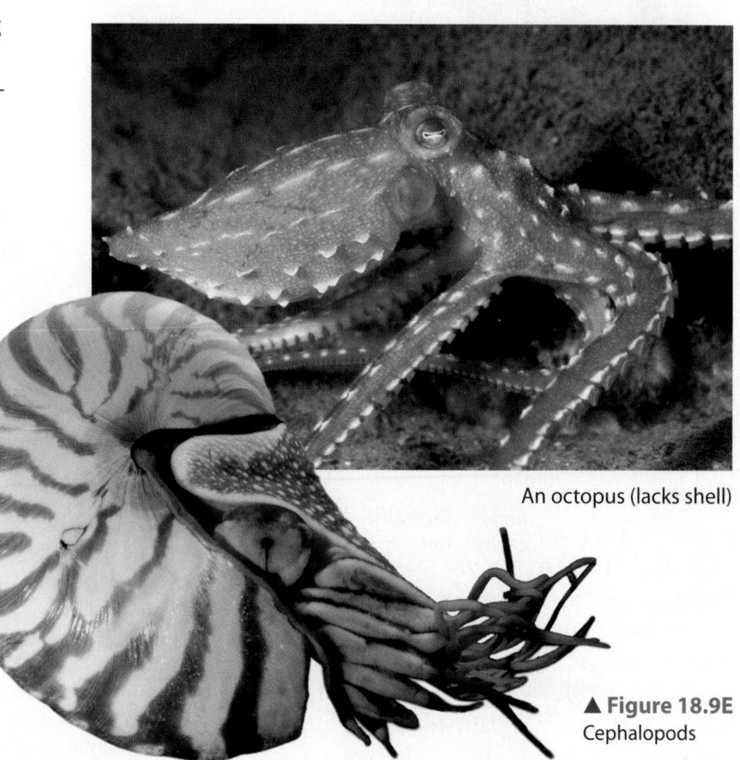

An octopus (lacks shell)

A chambered nautilus (about 21 cm in diameter)

▲ **Figure 18.9E** Cephalopods

18.10 Annelids are segmented worms

A segmented body resembling a series of fused rings is the hallmark of phylum Annelida (from the Latin *anellus*, ring), the third group of lophotrochozoans described in this chapter. **Segmentation**, the subdivision of the body along its length into a series of repeated parts (segments), played a central role in the evolution of many complex animals. A segmented body allows for greater flexibility and mobility, and it probably evolved as an adaptation facilitating movement. An earthworm, a typical **annelid**, uses its flexible, segmented body to crawl and burrow rapidly into the soil.

Annelids range in length from less than 1 mm to 3 m, the length of some giant Australian earthworms. They are found in damp soil, in the sea, and in most freshwater habitats. Some aquatic annelids swim in pursuit of food, but most are bottom-dwelling scavengers that burrow in sand and mud. There are three main groups of annelids: earthworms and their relatives, polychaetes, and leeches.

Earthworms and Their Relatives **Figure 18.10A** illustrates the segmented anatomy of an earthworm. Internally, the body cavity is partitioned by membrane walls (only a few are fully shown here). Many of the internal body structures are repeated within each segment. The nervous system (yellow) includes a simple brain and a ventral nerve cord with a cluster of nerve cells in each segment. Excretory organs (green), which dispose of fluid wastes, are also repeated in each segment (only a few are shown in this diagram). The digestive tract, however, is not segmented; it passes through the segment walls from the mouth to the anus.

Many invertebrates, including most molluscs and all arthropods (which you will meet in the next module), have what is called an **open circulatory system**, in which blood is pumped through vessels that open into body cavities, where organs are bathed directly in blood. Annelids and vertebrates, in contrast, have a **closed circulatory system**, in which blood remains enclosed in vessels as it distributes nutrients and oxygen throughout the body. As you can see in the diagram at the lower left, the main vessels of the earthworm circulatory system—a dorsal blood vessel and a ventral blood vessel—are connected by segmental vessels. The pumping organ, or "heart," is simply an enlarged region of the dorsal blood vessel plus five pairs of segmental vessels near the anterior end.

Each segment is surrounded by longitudinal and circular muscles. Earthworms move by coordinating the contraction of these two sets of muscles (see Figure 30.1D). The muscles work against the pressure of the fluid in each segment of the body cavity. Each segment also has four pairs of stiff bristles that provide traction for burrowing.

Earthworms are hermaphrodites; that is, they have both male and female reproductive structures. However, they do not fertilize their own eggs. Mating earthworms align their bodies facing in opposite directions and exchange sperm. Fertilization occurs some time later, when a specialized organ, visible as the thickened region of the worm in

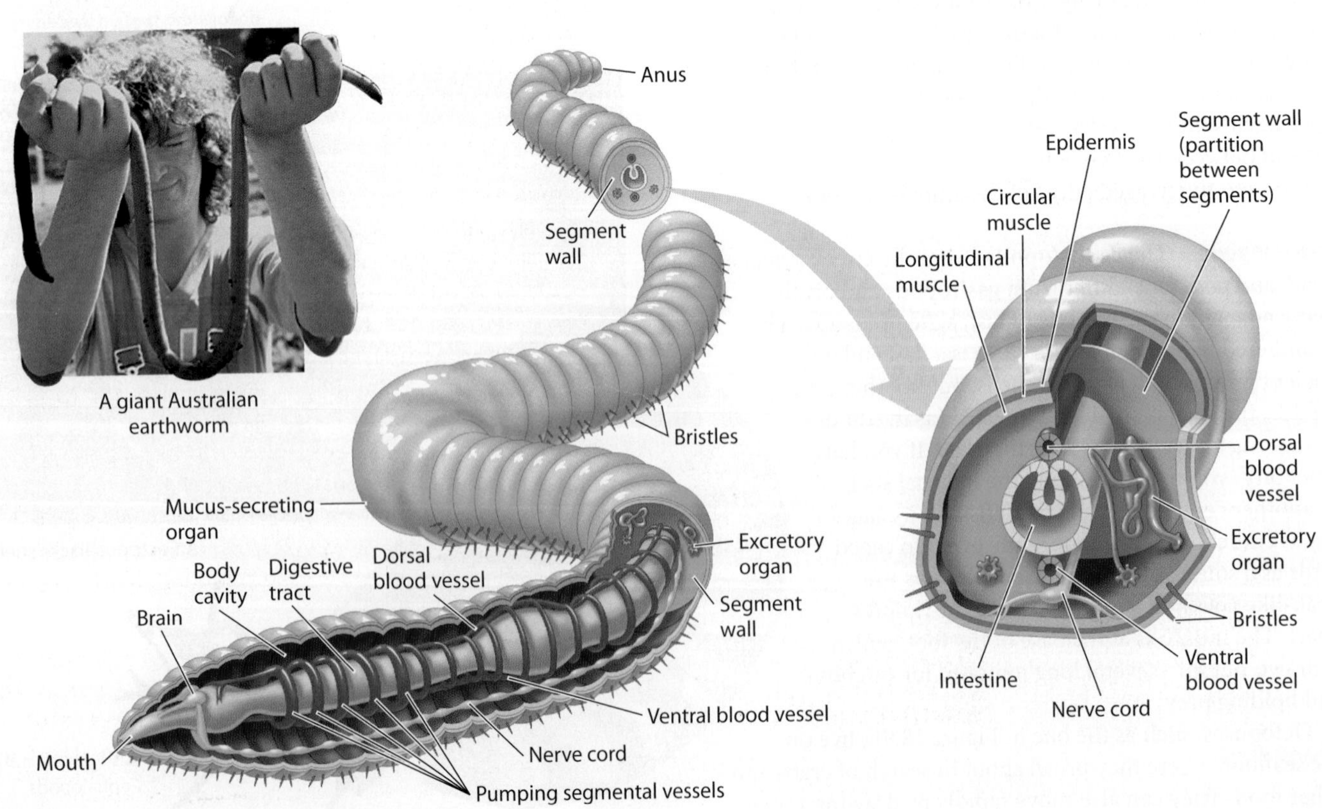
A giant Australian earthworm
Anus
Segment wall
Bristles
Mucus-secreting organ
Body cavity
Digestive tract
Dorsal blood vessel
Brain
Excretory organ
Segment wall
Ventral blood vessel
Mouth
Nerve cord
Pumping segmental vessels
Epidermis
Segment wall (partition between segments)
Circular muscle
Longitudinal muscle
Dorsal blood vessel
Excretory organ
Bristles
Ventral blood vessel
Intestine
Nerve cord

▲ **Figure 18.10A** Segmentation and internal anatomy of an earthworm

Figure 18.10A, secretes a cocoon made of mucus. The cocoon slides along the worm, picking up the worm's own eggs and the sperm it received from its partner. The cocoon then slips off the worm into the soil, where the embryos develop.

Earthworms eat their way through the soil, extracting nutrients as soil passes through their digestive tube. Undigested material, mixed with mucus secreted into the digestive tract, is eliminated as castings (feces) through the anus. Farmers and gardeners value earthworms because the animals aerate the soil and their castings improve the soil's texture. Charles Darwin estimated that a single acre of British farmland had about 50,000 earthworms, producing 18 tons of castings per year.

Polychaetes The **polychaetes** (from the Greek *polys*, many, and *chaeta*, hair), which are mostly marine, form the largest group of annelids.

Many polychaetes live in tubes and extend feathery appendages coated with mucus that trap suspended food particles. Tube-dwellers usually build their tubes by mixing mucus with bits of sand and broken shells. Some species of tube-dwellers are colonial, such as the group shown in **Figure 18.10B**. The circlet of feathery appendages seen at the mouth of each tube extends from the head of the worm inside. The free-swimming polychaete in Figure 18.10B travels in the open ocean by moving the paddle-like appendages on each segment. In polychaetes such as the sandworm that live in the sediments, stiff bristles (called *chaetae*) on the appendages help the worm wriggle about in search of small invertebrates to eat. In many polychaetes, the appendages are richly supplied with blood vessels and are either associated with the gills or function as gills themselves.

Tube-building polychaetes

A free-swimming polychaete

A sandworm

▲ **Figure 18.10B** Polychaetes

Leeches The third main group of annelids is the **leeches**, which are notorious for their bloodsucking habits. However, most species are free-living carnivores that eat small invertebrates such as snails and insects. The majority of leeches inhabit fresh water, but there are also marine species and a few terrestrial species that inhabit moist vegetation in the tropics. Leeches range in length from 1 to 30 cm.

Some bloodsucking leeches use razor-like jaws to slit the skin of an animal. The host is usually oblivious to this attack because the leech secretes an anesthetic as well as an anticoagulant into the wound. The leech then sucks as much blood as it can hold, often more than 10 times its own weight. After this gorging, a leech can last for months without another meal.

Until the 1920s, physicians used leeches for bloodletting. For centuries, illness was thought to result from an imbalance in the body's fluids, and the practice of bloodletting was originally conceived to restore the natural balance. Later, physicians viewed bloodletting as a kind of spring cleaning for the body to remove any toxins or "bad blood" that had accumulated. Leeches are still occasionally applied to remove blood from bruised tissues **(Figure 18.10C)** and to help relieve swelling in fingers or toes that have been sewn back on after accidents. Blood tends to accumulate in a reattached finger or toe until small veins have a chance to grow back in the appendage and resume circulation.

▲ **Figure 18.10C** A medicinal leech applied to drain blood from a patient

The anticoagulant produced by leeches has also proved to be medically useful. It is used to dissolve blood clots that form during surgery or as a result of heart disease. Because it is difficult to obtain this chemical from natural sources, it is now being produced through genetic engineering.

The segments of an annelid are all very similar. In the next group we explore, the arthropods, body segments and their appendages have become specialized, serving a variety of functions.

? Tapeworms and bloodsucking leeches are parasites. What are the key differences between these two?

● Whereas both are composed of repeated segments, the segments of a tapeworm are filled mostly with reproductive organs and are shed from the posterior end of the animal. Tapeworms are flatworms with no body cavity and, in their parasitic lifestyle, not even a gastrovascular cavity. Leeches have a body cavity and a complete digestive tract.

18.11 Arthropods are segmented animals with jointed appendages and an exoskeleton

Over a million species of **arthropods**—including crayfish, lobsters, crabs, barnacles, spiders, ticks, and insects—have been identified. Biologists estimate that the arthropod population of the world numbers about a billion billion (10^{18}) individuals! In terms of species diversity, geographic distribution, and sheer numbers, Arthropoda must be regarded as the most successful animal phylum.

The diversity and success of arthropods are largely related to their segmentation, their hard exoskeleton, and their jointed appendages, for which the phylum is named (from the Greek *arthron*, joint, and *pod*, foot). As indicated in the drawing of a lobster in **Figure 18.11A**, the appendages are variously adapted for sensory reception, defense, feeding, walking, and swimming. The arthropod body, including the appendages, is covered by an **exoskeleton**, an external skeleton that protects the animal and provides points of attachment for the muscles that move the appendages. This exoskeleton is a cuticle, a nonliving covering that in arthropods is hardened by layers of protein and chitin, a polysaccharide. The exoskeleton is thick around the head, where its main function is to house and protect the brain. It is paper-thin and flexible in other locations, such as the joints of the legs. As it grows, an arthropod must periodically shed its old exoskeleton and secrete a larger one, a complex process called **molting**, or ecdysis. Thus, arthropods belong to the ecdysozoan lineage of the clade Bilateria.

In contrast with annelids, which have similar segments along their body, the body of most arthropods arises from several distinct groups of segments that fuse during development: the head, thorax, and abdomen. In some arthropods, including the lobster, the exoskeleton of the head and thorax is partly fused, forming a body region called the cephalothorax. Each of the segment groups is specialized for a different function. In a lobster, the head bears sensory antennae, eyes, and jointed mouthparts on the ventral side. The thorax bears a pair of defensive appendages (the pincers) and four pairs of legs for walking. The abdomen has swimming appendages.

Like most molluscs, arthropods have an open circulatory system in which a tubelike heart pumps blood through short arteries into spaces surrounding the organs. A variety of gas exchange organs have evolved. Most aquatic species have gills. Terrestrial insects have internal air sacs that branch throughout the body (see Module 22.4).

Fossils and molecular evidence suggest that living arthropods represent four major lineages that diverged early in the evolution of arthropods. The figures in this module illustrate representatives of three of these lineages. The fourth, the insects, will be discussed in Module 18.12.

Chelicerates The **chelicerates** (from the Greek *chele*, claw, and *keras*, horn) are named for their clawlike feeding appendages. The bodies of chelicerates consist of a cephalothorax and an abdomen, and they lack antennae. **Figure 18.11B** shows a horseshoe crab, a chelicerate that is common on the Atlantic and Gulf coasts of the United States. This species, which has survived with little change for hundreds of millions of years, is a "living fossil," the only surviving member of a group of marine chelicerates that were abundant in the sea some 300 million years ago.

In addition to horseshoe crabs, living chelicerates include the scorpions, spiders, ticks, and mites, collectively called **arachnids**. Most arachnids live on land. Scorpions (**Figure 18.11C**, left, on facing page) are nocturnal hunters. Their ancestors were among the first terrestrial carnivores, preying on other arthropods that fed on early land plants. Scorpions have a large pair of pincers for defense and capturing prey. The tip of the tail bears a poisonous stinger. Scorpions eat mainly insects and spiders and attack people only when prodded or stepped on. Only a few species are dangerous to humans, but the sting is painful nonetheless.

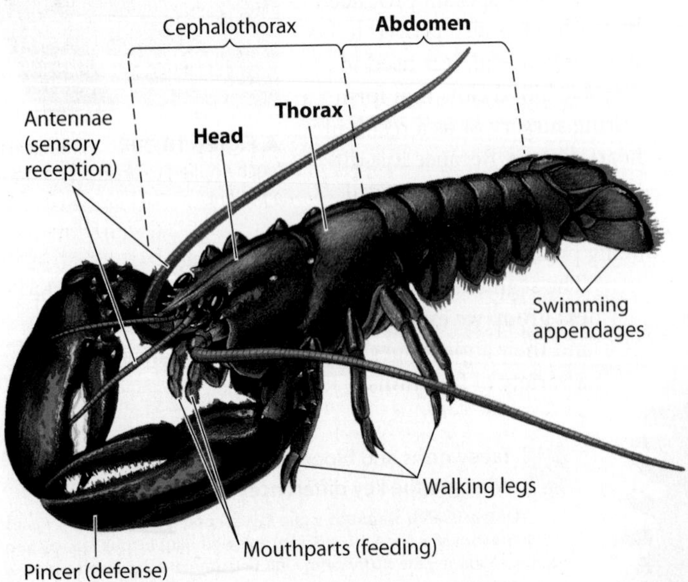
Cephalothorax
Abdomen
Thorax
Head
Antennae (sensory reception)
Swimming appendages
Walking legs
Mouthparts (feeding)
Pincer (defense)

▲ **Figure 18.11A** The structure of a lobster, an arthropod

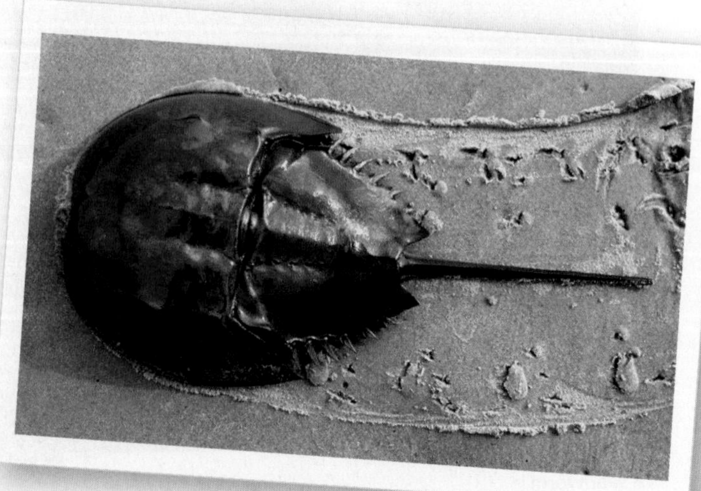

▲ **Figure 18.11B** A horseshoe crab (up to about 30 cm wide)

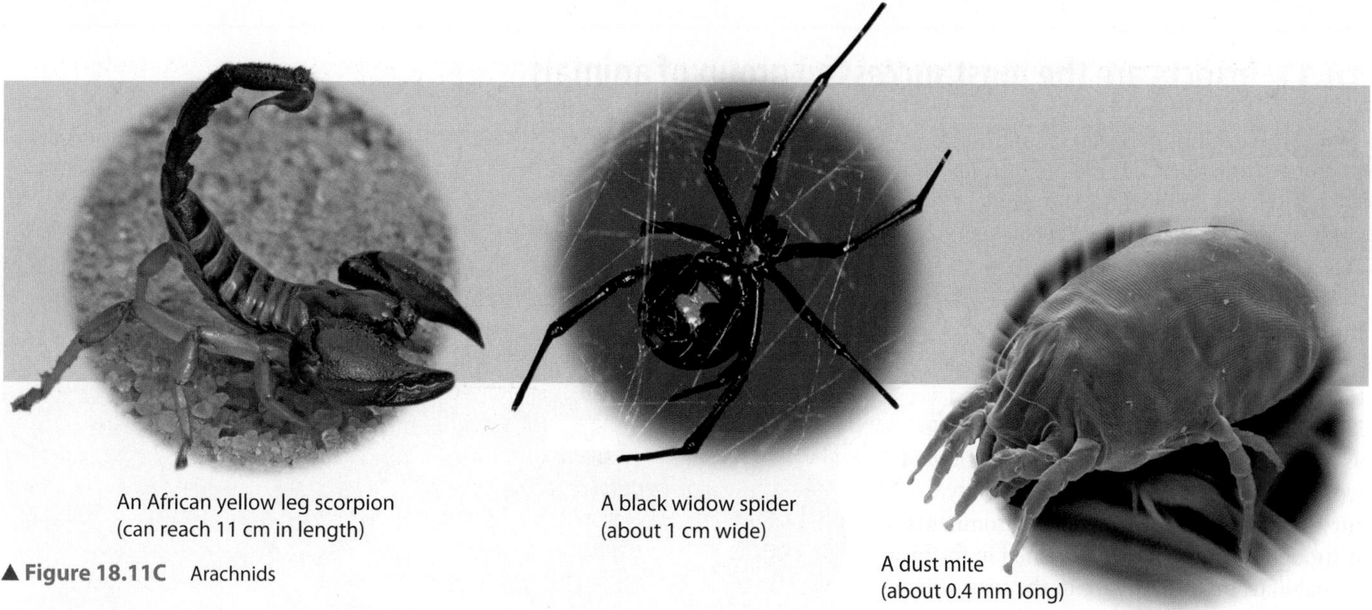

An African yellow leg scorpion (can reach 11 cm in length)

A black widow spider (about 1 cm wide)

A dust mite (about 0.4 mm long)

▲ **Figure 18.11C** Arachnids

Spiders, a diverse group of arachnids, hunt insects or trap them in webs of silk that they spin from specialized glands on their abdomen (see Figure 18.11C, center). Mites make up another large group of arachnids. On the right in Figure 18.11C is a micrograph of a dust mite, a ubiquitous scavenger in our homes. Thousands of these microscopic animals can thrive in a few square centimeters of carpet or in one of the dust balls that form under a bed. Dust mites do not carry infectious diseases, but many people are allergic to them.

Millipedes and Centipedes The animals in this lineage have similar segments over most of their body and superficially resemble annelids; however, their jointed legs identify them as arthropods. **Millipedes** (**Figure 18.11D**) are wormlike terrestrial creatures that eat decaying plant matter. They have two pairs of short legs per body segment. **Centipedes** (**Figure 18.11E**) are terrestrial carnivores with a pair of poison claws used in defense and to paralyze prey such as cockroaches and flies. Each of their body segments bears a single pair of long legs.

◀ **Figure 18.11D** A millipede (about 7 cm long)

▶ **Figure 18.11E** A Peruvian giant centipede (can reach 30 cm in length)

Crustaceans The **crustaceans** are nearly all aquatic. Lobsters and crayfish are in this group, along with numerous barnacles, crabs, and shrimps (**Figure 18.11F**). Barnacles are marine crustaceans with a cuticle that is hardened into a shell containing calcium carbonate, which may explain why they were once classified as molluscs. Their jointed appendages project from their shell to strain food from the water. Most barnacles anchor themselves to rocks, boat hulls, pilings, or even whales. The adhesive they produce is as strong as any glue ever invented. Other crustaceans include small copepods and krill, which serve as food sources for many fishes and whales.

We turn next to the fourth lineage of arthropods, the insects, whose numbers dwarf all other groups combined.

? **List the characteristics that arthropods have in common.**

● Segmentation, exoskeleton, specialized jointed appendages

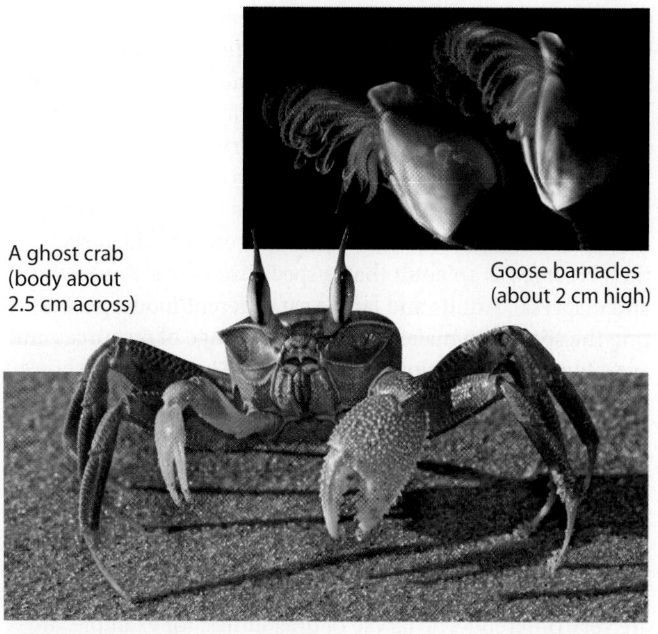

A ghost crab (body about 2.5 cm across)

Goose barnacles (about 2 cm high)

▲ **Figure 18.11F** Crustaceans

18.12 Insects are the most successful group of animals

EVOLUTION CONNECTION

The evolutionary success of insects is unrivaled by any other group of animals. More than a million species of insects have been identified, comprising nearly 75% of all animal species. Entomologists (scientists who study insects) think that fewer than half the total number of insect species have been identified, and some believe there could be as many as 30 million. Insects are distributed worldwide and have a remarkable ability to survive challenging terrestrial environments. Although they have also flourished in freshwater habitats, insects are rare in the seas, where crustaceans are the dominant arthropods.

▼ **Figure 18.12A** Complete metamorphosis of a European rhinoceros beetle.

Larva (grub, up to 12 cm length)

Pupa

Adult (up to 4 cm length)

What characteristics account for the extraordinary success of insects? One answer lies in the features they share with other arthropods—body segmentation, an exoskeleton, and jointed appendages. Other key features include flight, a waterproof coating on the cuticle, and a complex life cycle. In addition, many insects have short generation times and large numbers of offspring. For example, *Culex pipiens*, the most widely distributed species of mosquito, has a generation time of roughly 10 days, and a single female can lay many hundreds of eggs over the course of her lifetime. Thus, natural selection acts rapidly, and alleles that offer a reproductive advantage can quickly be established in a population.

Why so many insects?

Life Cycles One factor in the success of insects is a life cycle that includes metamorphosis, during which the animal takes on different body forms as it develops from larva to adult. Only the adult insect is sexually mature and has wings. More than 80% of insect species, including beetles, flies, bees, and moths and butterflies, undergo **complete metamorphosis**. The larval stage (such as caterpillars, which are the larvae of moths and butterflies, and maggots, which are fly larvae) is specialized for eating and growing. A larva typically molts several times as it grows, and then exists as an encased, nonfeeding pupa while its body rebuilds from clusters of embryonic cells that have been held in reserve. The insect then emerges as an adult that is specialized for reproduction and dispersal. Adults and larvae eat different foods, permitting the species to make use of a wider range of resources and avoiding intergenerational competition. **Figure 18.12A** shows the larva, pupa, and adult of the rhinoceros beetle (*Oryctes nasicornis*), named for the horn on the male's head.

Other insect species undergo **incomplete metamorphosis**, in which the transition from larva to adult is achieved through multiple molts, but without forming a pupa. In some species, including grasshoppers and cockroaches, the juvenile forms resemble the adults. In others, the body forms and lifestyles are very different. The larvae of dragonflies, for example, are aquatic, but the adults live on land.

Modular Body Plan Like other arthropods, insects have specialized body regions—a head, a thorax, and an abdomen (**Figure 18.12B**). These regions arise from the fusion of embryonic segments during development. Early in development, the embryonic segments are identical to each other. However, they soon diverge as different genes are expressed in different segments, giving rise to the three distinct body parts and to a variety of appendages, including antennae, mouthparts, legs, and wings (as discussed in Module 11.8). The insect body plan is essentially modular: Each embryonic segment is a separate building block that develops independently of the other segments. As a result, a mutation that changes homeotic gene expression can change the structure of one segment or its appendages without affecting any of the others. In the evolution of the grasshopper, for example, genetic changes in one thoracic segment produced the specialized jumping legs but did not affect the other two leg-producing segments. Wings, antennae, and mouthparts have all evolved in a similar fashion, by the specialization of independent segments through changes in the timing and location of homeotic gene expression (see Module 15.11). Much of the extraordinary diversification of insects resulted from modifications of the appendages that adapted them for specialized functions.

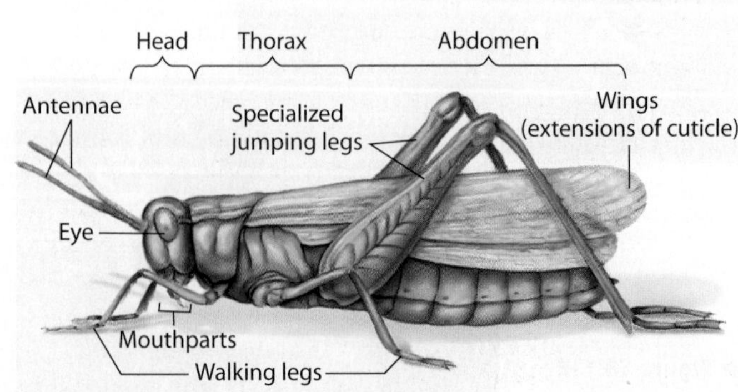
Head
Thorax
Abdomen
Antennae
Specialized jumping legs
Wings (extensions of cuticle)
Eye
Mouthparts
Walking legs

▲ **Figure 18.12B** Modular body plan of insects, as seen in a grasshopper

A stick insect
A leaf-mimic katydid
A caterpillar resembling a bird dropping

▲ **Figure 18.12C** Remarkable resemblances

The head typically bears a pair of sensory antennae, a pair of eyes, and several pairs of mouthparts. The mouthparts are adapted for particular kinds of eating—for example, for chewing plant material (in grasshoppers); for biting and tearing prey (praying mantis); for lapping up fluids (houseflies); or for piercing into and sucking the fluids of plants (aphids) or animals (mosquitoes). When flowering plants appeared, adaptations for nectar feeding became advantageous (see Module 17.10). As a result of this variety in mouthparts, insects have adaptations that exploit almost every conceivable food source.

Most adult insects have three pairs of legs, which may be adapted for walking, jumping, grasping prey, digging into the soil, or even paddling on water. Insects are the only invertebrates that can fly; most adult insects have one or two pairs of wings. (Some insects, such as fleas, are wingless.) Flight, which is an effective means of dispersal and escape from predators, was a major factor in the success of insects. And because the wings are extensions of the cuticle, insects have acquired the ability to fly without sacrificing any legs. By contrast, the wings of birds and bats are modified limbs. With a single pair of walking legs, those animals are generally clumsy when on the ground.

Protective Color Patterns In many groups of insects, adaptations of body structures have been coupled with protective coloration. Many different animals, including insects, have camouflage, color patterns that blend into the background. But insects also have elaborate disguises that include modifications to their antennae, legs, wings, and bodies. For instance, there are insects that resemble twigs, leaves, and bird droppings (**Figure 18.12C**). Some even do a passable imitation of vertebrates. The "snake" in **Figure 18.12D** is actually a hawk moth caterpillar. The colors of its dorsal side are an effective camouflage. When disturbed, however, it flips over to reveal the snake-like eyes of its ventral side, even puffing out its thorax to enhance the deception. "Eyespots" that resemble vertebrate eyes are common in several groups of moths and butterflies. **Figure 18.12E** shows a member of a genus known informally as owl butterflies. A flash of these large "eyes" startles would-be predators. In other species, eyespots deflect the predator's attack away from vital body parts.

▲ **Figure 18.12D** A hawk moth caterpillar

How could evolution have produced these complex color patterns? It turns out that the genetic mechanism by which eyespots evolve is very similar to the mechanism by which specialized appendages evolve. Butterfly wings have a modular construction similar to that of embryonic body segments. Each section can change independently of the others and can therefore have a unique pattern. And like the specialization of appendages, eyespots result from different patterns of homeotic gene expression during development.

As you will learn in the next module, investigating the evolution of the insect body plan has given biologists valuable insight into the genetic mechanisms that have generated the amazing diversity of life.

Contrast incomplete and complete metamorphosis.

In complete metamorphosis, there is a pupal stage; in incomplete metamorphosis, there is not.

▲ **Figure 18.12E** An owl butterfly (left) and a long-eared owl (right)

18.13 The genes that build animal bodies are ancient

SCIENTIFIC THINKING

The arthropod body plan, with its enormous variety of distinct body segments bearing specialized appendages, is a key factor in the evolutionary success of the phylum. How did this body plan evolve? Scientists have proposed alternative hypotheses that can be tested by making comparisons of homeotic genes, the master control genes that direct animal development. One hypothesis proposes that an increase in the number of homeotic genes led to the diversity of segment and appendage types in arthropods. According to this hypothesis, the ancestors of arthropods had a small number of body segment types whose development was controlled by a correspondingly small number of homeotic genes, and new segment and appendage types resulted from new homeotic genes that originated on the arthropod branch of the tree.

To test the hypothesis, a team of scientists compared homeotic genes in arthropods with those of their closest living relatives, known as velvet worms. Velvet worms are one of the small animal phyla that we omitted from our coverage of animal diversity. The wormlike bodies of velvet worms bear fleshy antennae and numerous short, identical appendages that are used for walking. As shown in **Figure 18.13A**, arthropods and velvet worms are descended from a common ancestor, from which they inherited a common set of homeotic genes. Results showed that velvet worms have the complete set of arthropod homeotic genes—no additional genes arose after the lineages diverged. Thus, the researchers concluded that body segment diversity did *not* result from the appearance of new homeotic genes in arthropods.

Other ecdysozoans
Arthropods
Velvet worms
Common ancestor

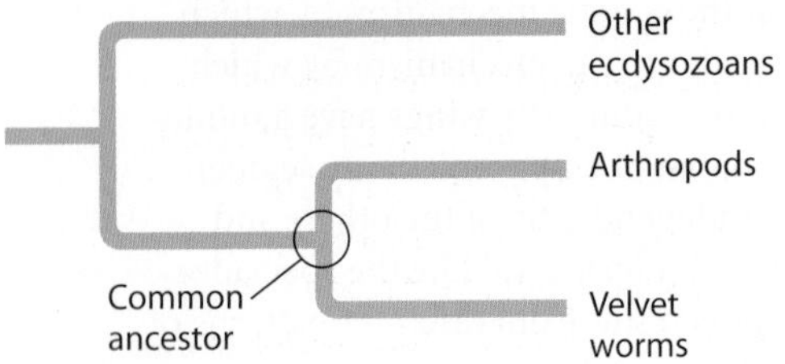

▲ **Figure 18.13A** Relationship of velvet worms and arthropods

An alternative hypothesis proposes that changes in the regulation of homeotic gene expression—when and where the genes are transcribed and translated into proteins—led to the diversity of segment and appendage types in arthropods. Such changes in developmental genes are known to result in significant morphological changes (see Module 15.11).

To test this hypothesis, researchers compared gene expression patterns in the embryos of a centipede and a velvet worm **(Figure 18.13B)**. The green stain indicates expression of a homeotic gene that is involved in the formation of appendages in a wide range of taxa. This gene is expressed in a similar pattern in both centipedes and velvet worms. The body regions stained red indicate expression of two homeotic genes unique to arthropods and their close relatives. (Areas where green and red overlap appear yellow.) As you can see, the velvet worm deploys these genes only in the posterior tip of its body. In the centipede, and in other groups of arthropods that have been studied, the locations where the genes are expressed correspond with the boundaries between one segment type and the next. For example, the centipede's poison claw develops from the segment labeled T1, and segment T2 becomes the first walking leg. The results support the hypothesis that the diversification of arthropods occurred through changes in the regulation of homeotic gene expression.

Experiments such as these have demonstrated that the evolution of new structures and new types of animals does not require new genes. Rather, the genetic differences that result in new forms arise in the segments of DNA that control when and where ancient homeotic genes are expressed. To put it simply, building animal bodies is not just about which genes are present—it's about how they are used.

In the next module, you'll meet some of the oddest body forms in the animal kingdom, the stars, spheres, and tubular shapes of the phylum Echinodermata.

? Researchers found that velvet worms and arthropods share the same set of homeotic genes. What conclusion did they draw from this result?

● The evolution of diverse arthropod body segment types was not the result of new genes in arthropods.

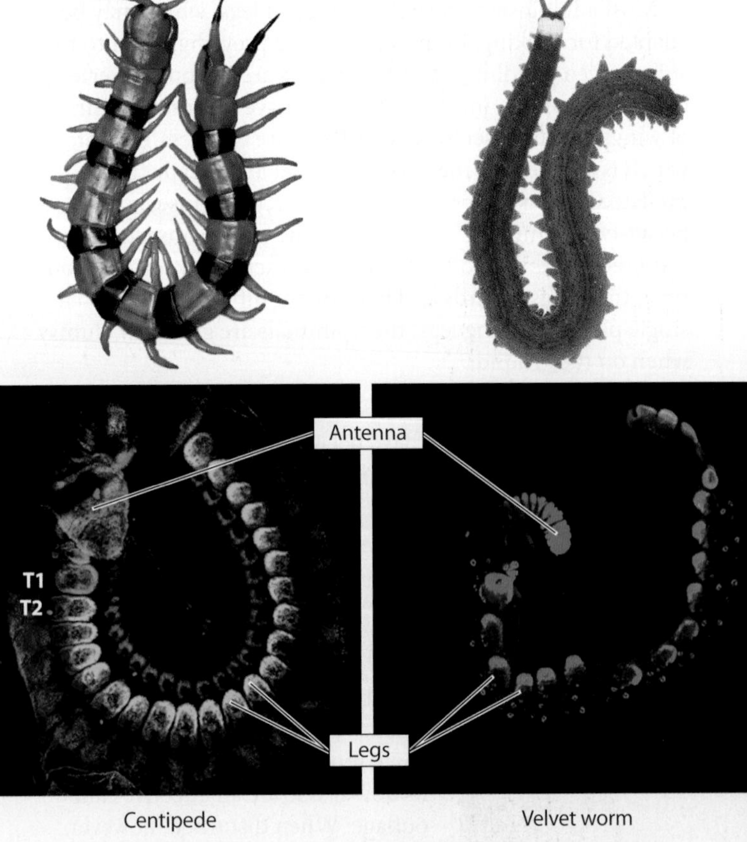
Antenna
T1
T2
Legs
Centipede
Velvet worm

▲ **Figure 18.13B** Expression of homeotic genes in the embryos of a centipede and a velvet worm, with an adult member of each group shown above (The labels identify distinct body segments in the stained embryos.)

18.14 Echinoderms have spiny skin, an endoskeleton, and a water vascular system for movement

Echinoderms, such as sea stars, sand dollars, and sea urchins, are slow-moving or sessile marine animals. Most are radially symmetric as adults. Both the external and the internal parts of a sea star, for instance, radiate from the center like spokes of a wheel. The bilateral larval stage of echinoderms, however, tells us that echinoderms are not closely related to cnidarians or other animals that never show bilateral symmetry.

The phylum name Echinodermata (from the Greek *echin*, spiny, and *derma*, skin) refers to the prickly bumps or spines of a sea star or sea urchin. These are extensions of the hard calcium-containing plates that form the **endoskeleton**, or internal skeleton, under the thin skin of the animal.

Unique to echinoderms is the **water vascular system**, a network of water-filled canals that branch into extensions called tube feet (**Figure 18.14A**). Tube feet function in locomotion, feeding, and gas exchange. A sea star pulls itself slowly over the seafloor using its suction-cup-like tube feet. Its mouth is centrally located on its undersurface. When the sea star shown in **Figure 18.14B** encounters an oyster or clam, its favorite food, it grips the bivalve with its tube feet and pulls until the mollusc's muscle tires enough to create a narrow opening between the two valves of the shell. The sea star then turns its stomach inside out, pushing it through its mouth and into the opening. The sea star's stomach digests the soft parts of its prey inside the mollusc's shell. When the meal is completed, the sea star withdraws its stomach from the empty shell.

Sea stars and some other echinoderms are capable of regeneration. Arms that are lost are readily regrown.

In contrast with sea stars, sea urchins are spherical and have no arms. They do have five rows of tube feet that project through tiny holes in the animal's globe-like case. If you look carefully at **Figure 18.14C**, you can see the long, threadlike tube feet projecting among the spines of the sea urchin. Sea urchins move by pulling with their tube feet. They also have muscles that pivot their spines, which can aid in locomotion. Unlike the carnivorous sea stars, most sea urchins eat algae.

Other echinoderm groups include brittle stars, which move by thrashing their long, flexible arms; sea lilies, which live attached to the substrate by a stalk; and sea cucumbers, odd elongated animals that resemble their vegetable namesake more than they resemble other echinoderms.

Though echinoderms have many unique features, we see evidence of their relation to other animals in their embryonic development. As we discussed in Modules 18.3 and 18.4, differences in patterns of development have led biologists to identify echinoderms and chordates (which include vertebrates) as a clade of bilateral animals called deuterostomes. Thus, echinoderms are more closely related to our phylum, the chordates, than to the protostome animals, such as molluscs, annelids, and arthropods. We define the features of chordates in the next module.

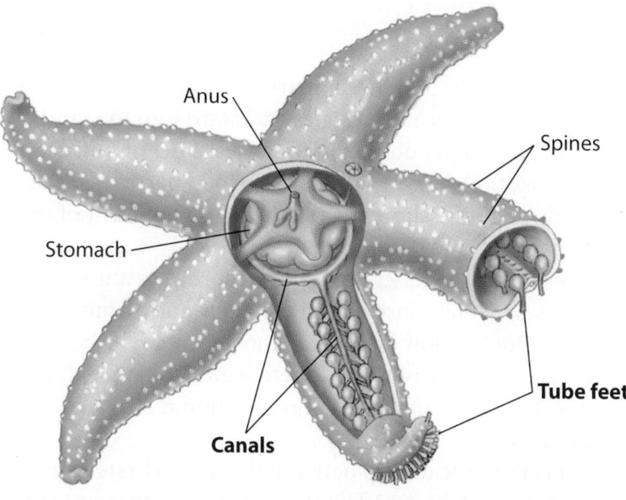

Anus
Spines
Stomach
Tube feet
Canals

▲ **Figure 18.14A** The water vascular system (canals and tube feet) of a sea star (top view)

? **Contrast the skeleton of an echinoderm with that of an arthropod.**

● An echinoderm has an endoskeleton; an arthropod has an exoskeleton.

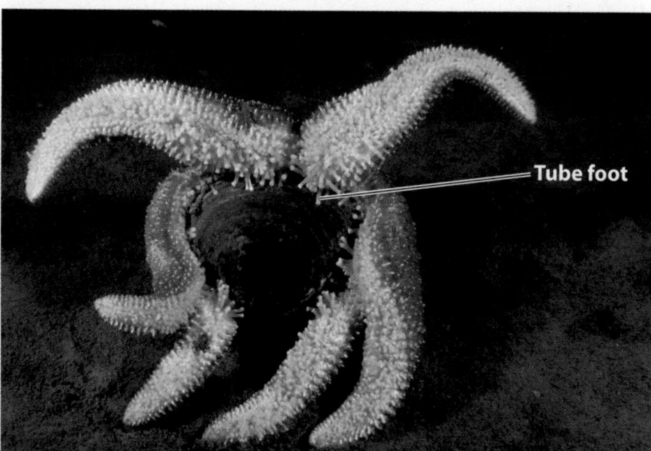

Tube foot

▲ **Figure 18.14B** A sea star feeding on a clam

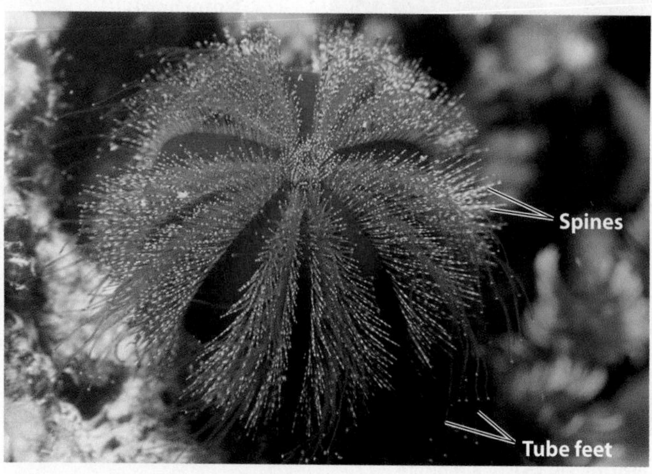

Spines
Tube feet

▲ **Figure 18.14C** A sea urchin

18.15 Our own phylum, Chordata, is distinguished by four features

You may be surprised to find the phylum that includes humans in a chapter on invertebrate diversity. However, vertebrates evolved from invertebrate ancestors and continue to share the distinctive features that identify members of the phylum Chordata. The embryos, and often the adults, of chordates possess (1) a **dorsal, hollow nerve cord**; (2) a **notochord**, a flexible, supportive, longitudinal rod located between the digestive tract and the nerve cord; (3) **pharyngeal slits** located in the pharynx, the region just behind the mouth; and (4) a muscular **post-anal tail** (a tail posterior to the anus). You can see these four features in the diagrams in Figures 18.15A and 18.15B. The two chordates shown, a tunicate and a lancelet, are called invertebrate chordates because they do not have a backbone.

Adult **tunicates** are stationary and look more like small sacs than anything we usually think of as a chordate (**Figure 18.15A**). Tunicates often adhere to rocks and boats, and they are common on coral reefs. The adult has no trace of a notochord, nerve cord, or tail, but it does have prominent pharyngeal slits that function in feeding. The tunicate larva, however, is a swimming, tadpole-like organism that exhibits all four distinctive chordate features.

Tunicates are suspension feeders. Seawater enters the adult animal through an opening at the top, passes through the pharyngeal slits into a large cavity in the animal, and exits back into the ocean via an excurrent siphon on the side of the body (see the photo in Figure 18.15A). Food particles are trapped in a net made of mucus and then transported to the intestine, where they are digested. Because they shoot a jet of water through their excurrent siphon when threatened, tunicates are often called sea squirts.

Lancelets, another group of marine invertebrate chordates, also feed on suspended particles. Lancelets are small, bladelike chordates that live in marine sands (**Figure 18.15B**). When feeding, a lancelet wriggles backward into the sand with its head sticking out. As in tunicates, a net of mucus secreted across the pharyngeal slits traps food particles. Water flowing through the slits exits via an opening in front of the anus.

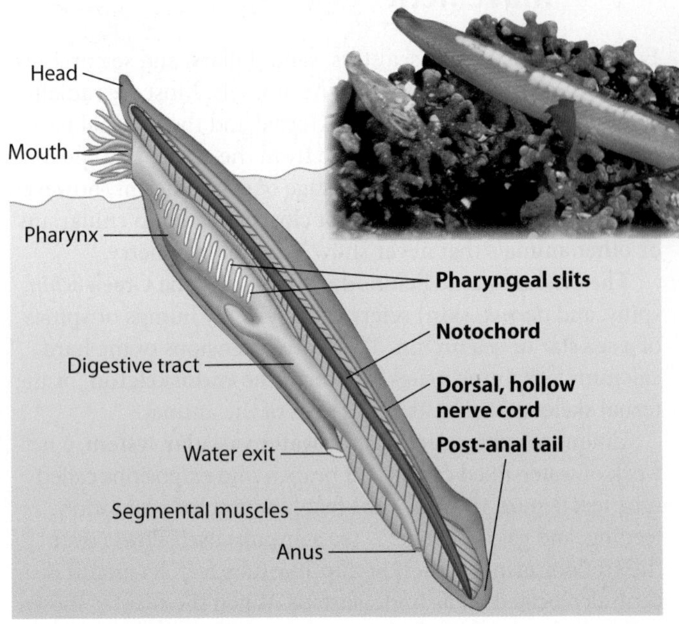

Head
Mouth
Pharynx
Pharyngeal slits
Notochord
Digestive tract
Dorsal, hollow
nerve cord
Post-anal tail
Water exit
Segmental muscles
Anus

▲ **Figure 18.15B** A lancelet (5–15 cm long)

Lancelets clearly illustrate the four chordate features. They also have segmental muscles that flex their body from side to side, producing slow swimming movements. These serial muscles are evidence of the lancelet's segmentation. Although not unique to chordates, body segmentation is also a chordate characteristic.

What is the relationship between the invertebrate chordates and the vertebrates? The lancelets likely represent the earliest branch of the chordate lineage. The ancestral chordate may have looked something like a lancelet. Research has shown that the same genes that organize the major regions of the vertebrate brain are expressed in the same pattern at the anterior end of the lancelet nerve cord. However, molecular evidence has shown that tunicates are the closest living nonvertebrate relatives of vertebrates.

The invertebrate chordates have helped us identify the four chordate hallmarks. Before we explore the evolution of our own group, the vertebrates, let's look at a few of the ways in which invertebrates are beneficial to humans.

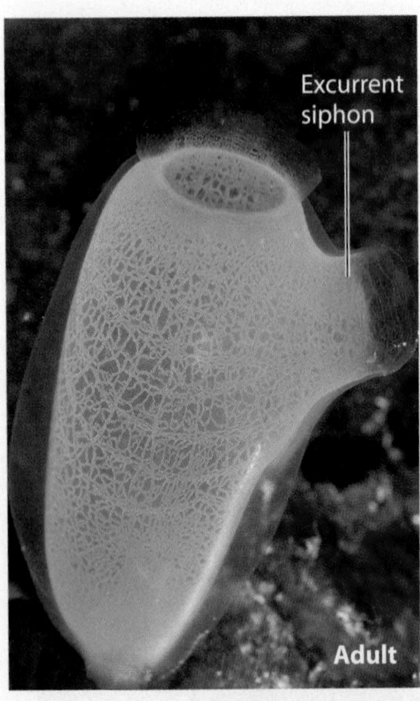

Excurrent
siphon
Adult

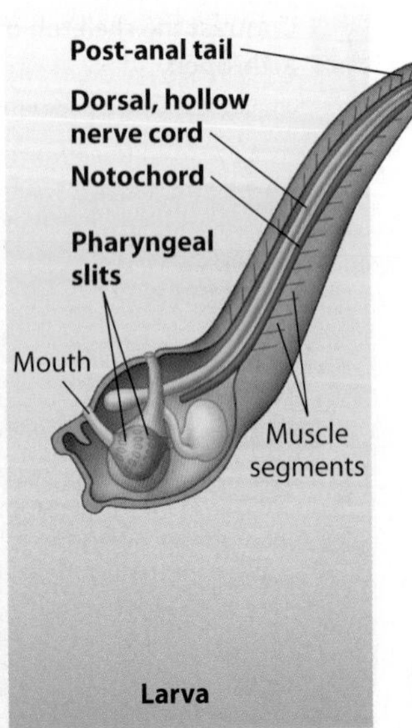

Post-anal tail
Dorsal, hollow
nerve cord
Notochord
Pharyngeal
slits
Mouth
Muscle
segments
Larva

▲ **Figure 18.15A** A tunicate

? What four features do we share with invertebrate chordates, such as lancelets?

● Human embryos and invertebrate chordates all have (1) a dorsal, hollow nerve cord; (2) a notochord; (3) pharyngeal slits; and (4) a post-anal tail.

18.16 Invertebrate diversity is a valuable but threatened resource

CONNECTION

Although the vast majority of animals are invertebrates, it is easy to overlook their importance in favor of the larger, often more charismatic, vertebrates. But invertebrates play critical roles in natural ecosystems and provide valuable services to humans, too. Let's look at a few examples.

Despite their simple bodies, reef-building corals create enormous structures that provide support and shelter for hundreds of other species, making coral reefs vivid displays of animal diversity. They also provide direct benefits to people. Fish and shellfish that inhabit coral reefs are harvested for food, and tourism generates billions of dollars of revenue for coastal communities. Reefs are also natural barriers that protect coastal property from damage caused by erosion and flooding during storms. Increasingly, researchers are looking to coral reefs for new medicines and other useful substances. Several drugs based on chemicals produced by sea sponges are already in use, including compounds used to treat AIDS, leukemia, and late-stage breast cancer. Scientists have also developed a powerful painkiller from the venom of reef-dwelling cone snails **(Figure 18.16A)**, predatory marine gastropods that stab their prey with a modified radula. The pharmaceutical possibilities of hundreds of other compounds from marine animals are being investigated as well. In addition to new medicines, substances produced by coral reef inhabitants have useful applications in cosmetics, biotechnology, and other industries. However, many scientists fear that we are running out of time to explore these potential riches. Almost a third of reef-building corals are threatened with extinction from warming seas due to climate change, ocean acidification (see Module 2.15), pollution, coastal development, and other factors.

▲ **Figure 18.16A** Cone snail (genus *Conus*)

Whereas coral reefs dazzle with their riot of diversity, freshwater mussels **(Figure 18.16B)** are so inconspicuous that they might be mistaken for stones. Many of these molluscs live in streams and rivers, where they attach to rocks or lie buried in the sediments. Like most other bivalves, mussels are suspension feeders. Depending on its size, a single mussel can filter up to 8 gallons of water per day as it obtains the particles of organic material and microbes that it consumes. This filtration process improves water quality in natural ecosystems and reduces the cost of water treatment for human uses. Because most mussels are able to tolerate only a limited amount of pollution, mussel mortality is an early warning of toxic contamination or deteriorating water quality. North America is home to the greatest diversity of freshwater mussels in the world. However, many of these species have become extinct, and close to 75% are currently imperiled by human activities. Habitat destruction and degradation resulting from dams, pollution, and sedimentation due to soil erosion are among the factors responsible for the loss of mussel populations. In addition, non-native mussels spreading throughout North American waterways, particularly zebra and quagga mussels, have displaced native species. Consequently, the natural filtering system provided by freshwater mussels is in rapid decline.

▲ **Figure 18.16B** Freshwater pearl mussels (*Margaritifera margaritifera*)

Most flowering plants are pollinated by animals, chiefly insects (as you learned in Module 17.10). Crop species are no exception: An estimated one-third of the world's food supply depends on pollinators. In the United States, production of fruits and vegetables such as cantaloupe, apples, peaches, and squash relies on pollination by bees, mostly non-native honeybees imported from Europe **(Figure 18.16C)**. Domesticated honeybees have been hit by baffling die-offs dubbed "colony collapse disorder." Mortality appears to be linked to multiple factors, including parasitic mites, pathogens, and certain pesticides. Population declines in native bees are likely the result of habitat loss, as agriculture and development have replaced the wildflowers that bees depend upon for food. In an attempt to bolster bee populations, farmers in some regions are planting flowers to attract these vital workers to their crops.

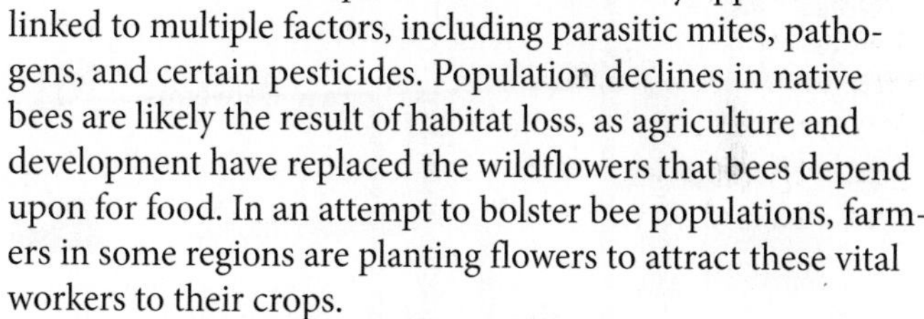

The cases described in this module represent only a tiny sample of the benefits we reap from Earth's biodiversity. (We return to this topic in Chapter 38.)

? **Briefly describe the "services" that reef-building corals, freshwater mussels, and bees provide to humans.**

● Build structures that support a wide variety of useful animals; improve water quality by filtration; pollinate crop plants

▶ **Figure 18.16C** Honey bee (*Apis mellifera*) collecting pollen

CHAPTER 18 REVIEW

For practice quizzes, BioFlix animations, MP3 tutorials, video tutors, and more study tools designed for this textbook, go to MasteringBiology®

Reviewing the Concepts

Animal Evolution and Diversity (18.1–18.4)

18.1 What is an animal? Animals are multicellular eukaryotes that have distinctive cell structures and specializations and obtain their nutrients by ingestion. Animal life cycles and embryonic development also distinguish animals from other groups of organisms.

18.2 Animal diversification began more than half a billion years ago. The oldest animal fossils are from the late Ediacaran period and are 575–550 million years old. Animal diversification accelerated rapidly during the Cambrian explosion from 535–525 million years ago.

18.3 Animals can be characterized by basic features of their "body plan." Body plans may vary in number of tissue layers (two or three), symmetry (radial or bilateral), presence of a body cavity, and embryonic development (protostome or deuterostome).

18.4 Body plans and molecular comparisons of animals can be used to build phylogenetic trees.

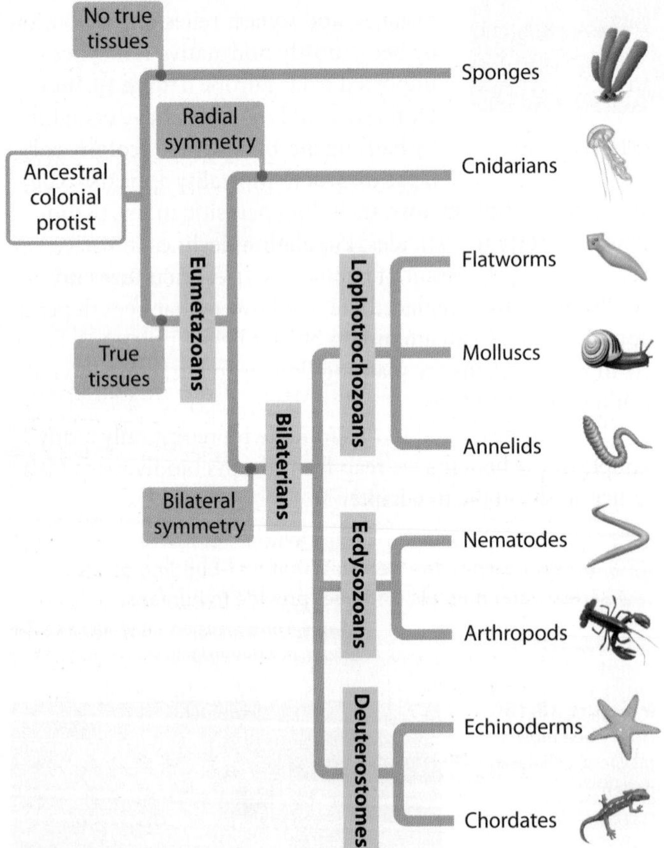

No true tissues
Sponges
Radial symmetry
Ancestral colonial protist
Cnidarians
Flatworms
Eumetazoans
Lophotrochozoans
Molluscs
True tissues
Annelids
Bilaterians
Bilateral symmetry
Nematodes
Ecdysozoans
Arthropods
Echinoderms
Deuterostomes
Chordates

Invertebrate Diversity (18.5–18.16)

18.5 Sponges have a relatively simple, porous body. Members of the phylum Porifera have no true tissues. Their flagellated choanocytes filter food from water passing through pores in the body.

18.6 Cnidarians are radial animals with tentacles and stinging cells. Members of the phylum Cnidaria have true tissues and a gastrovascular cavity. Their two body forms are polyps (such as hydras) and medusae (jellies).

18.7 Flatworms are the simplest bilateral animals. Members of the phylum Platyhelminthes are bilateral animals with no body cavity. A planarian has a gastrovascular cavity and a simple nervous system. Flukes and tapeworms are parasitic flatworms with complex life cycles.

18.8 Nematodes have a body cavity and a complete digestive tract. Members of the phylum Nematoda are covered by a protective cuticle that is shed periodically. Many nematodes (roundworms) are free-living decomposers; others are plant or animal parasites.

18.9 Diverse molluscs are variations on a common body plan. Members of the phylum Mollusca include gastropods (such as snails and slugs), bivalves (such as clams and mussels), and cephalopods (such as octopuses and squids). All have a muscular foot and a mantle, which encloses the visceral mass and may secrete a shell. Many molluscs feed with a rasping radula.

18.10 Annelids are segmented worms. Members of the phylum Annelida include earthworms, polychaetes, and leeches.

18.11 Arthropods are segmented animals with jointed appendages and an exoskeleton. The four lineages of the phylum Arthropoda are chelicerates (arachnids such as spiders), the lineage of millipedes and centipedes, the aquatic crustaceans (lobsters and crabs), and the terrestrial insects.

18.12 Insects are the most successful group of animals. Their development often includes metamorphosis. Insects have a three-part body (head, thorax, and abdomen) and three pairs of legs; most have wings. Specialized appendages and protective color patterns, which frequently result from evolutionary changes in the timing and location of homeotic gene expression, have played a major role in this group's success.

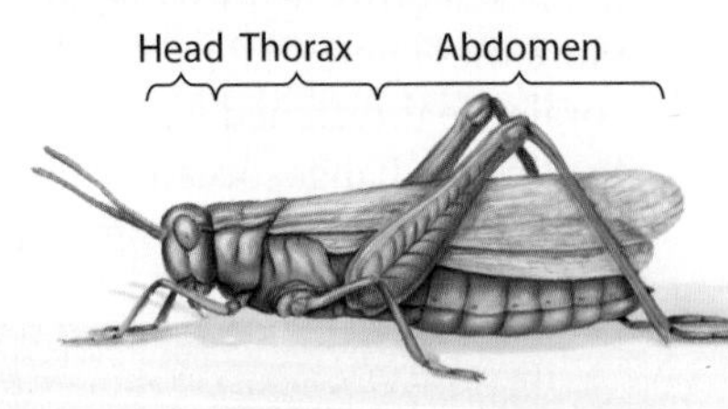

Head
Thorax
Abdomen

18.13 The genes that build animal bodies are ancient. Changes in the regulation of homeotic gene expression have been significant factors in the evolution of animal diversity.

18.14 Echinoderms have spiny skin, an endoskeleton, and a water vascular system for movement. Members of the phylum Echinodermata, such as sea stars, are radially symmetric as adults.

18.15 Our own phylum, Chordata, is distinguished by four features. Chordates have a dorsal, hollow nerve cord, a stiff notochord, pharyngeal slits, and a muscular post-anal tail. The simplest chordates are tunicates and lancelets, marine invertebrates that use their pharyngeal slits for suspension feeding.

18.16 Invertebrate diversity is a valuable but threatened resource.

Connecting the Concepts

1. The table below lists the common names of the nine animal phyla surveyed in this chapter. For each phylum, list the key characteristics and some representatives.

Phylum	Characteristics	Representatives
Sponges		
Cnidarians		
Flatworms		
Nematodes		
Molluscs		
Annelids		
Arthropods		
Echinoderms		
Chordates		

2. Identify the pattern of embryonic development shown in each drawing below and name the phylum (or phyla) that exhibit this pattern.

a.

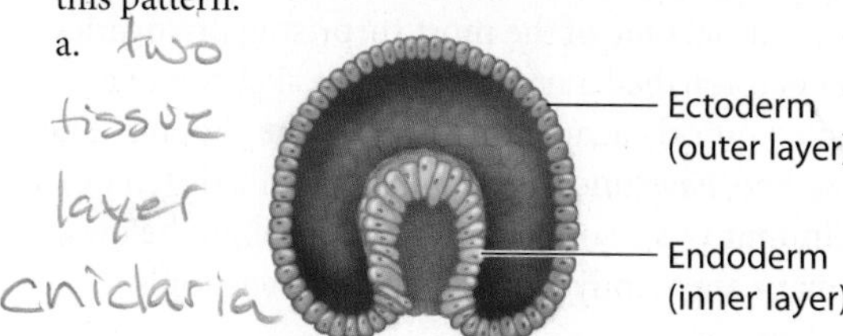
Ectoderm
(outer layer)
Endoderm
(inner layer)

b. c.

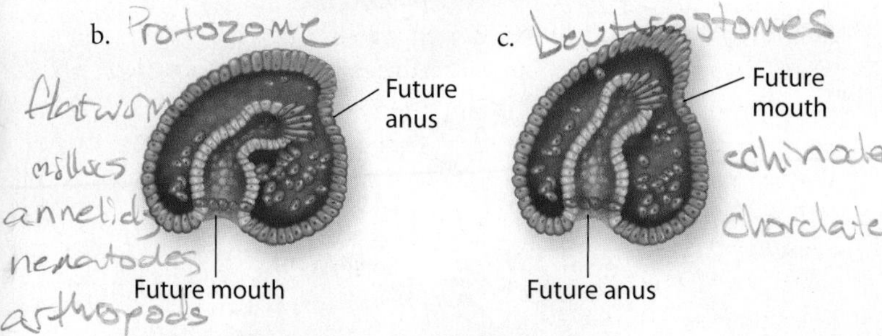
Future
anus
Future mouth
Future
mouth
Future anus

Testing Your Knowledge

Level 1: Knowledge/Comprehension

3. Bilateral symmetry in animals is best correlated with
   a. an ability to see equally in all directions.
   b. the presence of a skeleton.
   c. motility and active predation and escape.
   d. adaptation to terrestrial environments.
4. Jon found an organism in a pond, and he thinks it's a freshwater sponge. His friend Liz thinks it looks more like an aquatic fungus. How can they decide whether it is an animal or a fungus?
   a. See if it can swim.
   b. Figure out whether it is autotrophic or heterotrophic.
   c. See if it is a eukaryote or a prokaryote.
   d. Look for cell walls under a microscope.
5. Which of the following groupings includes the largest number of species? (*Explain your answer.*)
   a. invertebrates
   b. arthropods
   c. insects
   d. vertebrates
6. Which of the following animal groups does not have tissues derived from mesoderm?
   a. annelids
   b. echinoderms
   c. cnidarians
   d. flatworms
7. Molecular comparisons place nematodes and arthropods in clade Ecdysozoa. What characteristic do they share that is the basis for the name Ecdysozoa?
   a. a complete digestive tract
   b. body segmentation
   c. molting of an exoskeleton
   d. bilateral symmetry

Match each description on the left with the corresponding term on the right.

8. Include the vertebrates
9. Medusa and polyp body forms
10. The simplest animal with a complete digestive tract
11. The simplest animal group
12. Earthworms, polychaetes, and leeches
13. Largest phylum of all
14. Closest relatives of chordates
15. Lacks a body cavity
16. Have a muscular foot and a mantle

a. annelids
b. nematodes
c. sponges
d. arthropods
e. flatworms
f. cnidarians
g. molluscs
h. echinoderms
i. chordates

Level 2: Application/Analysis

17. Compare the structure of a planarian (a flatworm) and an earthworm with regard to the following: digestive tract, body cavity, and segmentation.
18. Name two phyla of animals that are radially symmetric and two that are bilaterally symmetric. How do the general lifestyles of radial and bilateral animals differ?
19. One of the key characteristics of arthropods is their jointed appendages. Describe four functions of these appendages in four different arthropods.

Level 3: Synthesis/Evaluation

20. A marine biologist has dredged up an unknown animal from the seafloor. Describe some of the characteristics that could be used to determine the animal phylum to which the creature should be assigned.
21. **SCIENTIFIC THINKING** In one of the experiments described in Module 18.13, researchers tested the hypothesis that the highly successful arthropod body plan resulted from new genes that originated in the arthropod lineage. Draw a diagram showing the evolutionary relationship between arthropods and velvet worms and use it to explain why velvet worms were a good choice to test this hypothesis. What results would have supported the hypothesis?

Answers to all questions can be found in Appendix 4.

CHAPTER 19

The Evolution of Vertebrate Diversity

Vertebrates (animals with backbones) have been evolving for half a billion years. There are currently more than 60,000 species, but untold thousands more arose, existed for a time, and became extinct, leaving a record in fossilized bones and other traces that give us a window into the past. By using these clues, as well as by studying genetic, morphological, and developmental homologies among present-day animals, scientists are piecing together the evolutionary history of vertebrates.

A fascinating aspect of this work is the study of human origins. Because fossils of our earliest ancestors are relatively scarce, every new find has the potential to add an exciting new dimension to our understanding of human evolution. One of the most surprising discoveries was made in 2004, when researchers unearthed a nearly complete skeleton of a tiny adult female in Indonesia, which quickly acquired the nickname "hobbit." She was not alone—since the initial discovery, researchers have uncovered the bones of a dozen or so more of these miniature humans. Despite the diminutive size of the hobbit skull (photo below), it displays some humanlike traits, and these humans apparently made and used stone tools.

? ***Who were the hobbit people?***

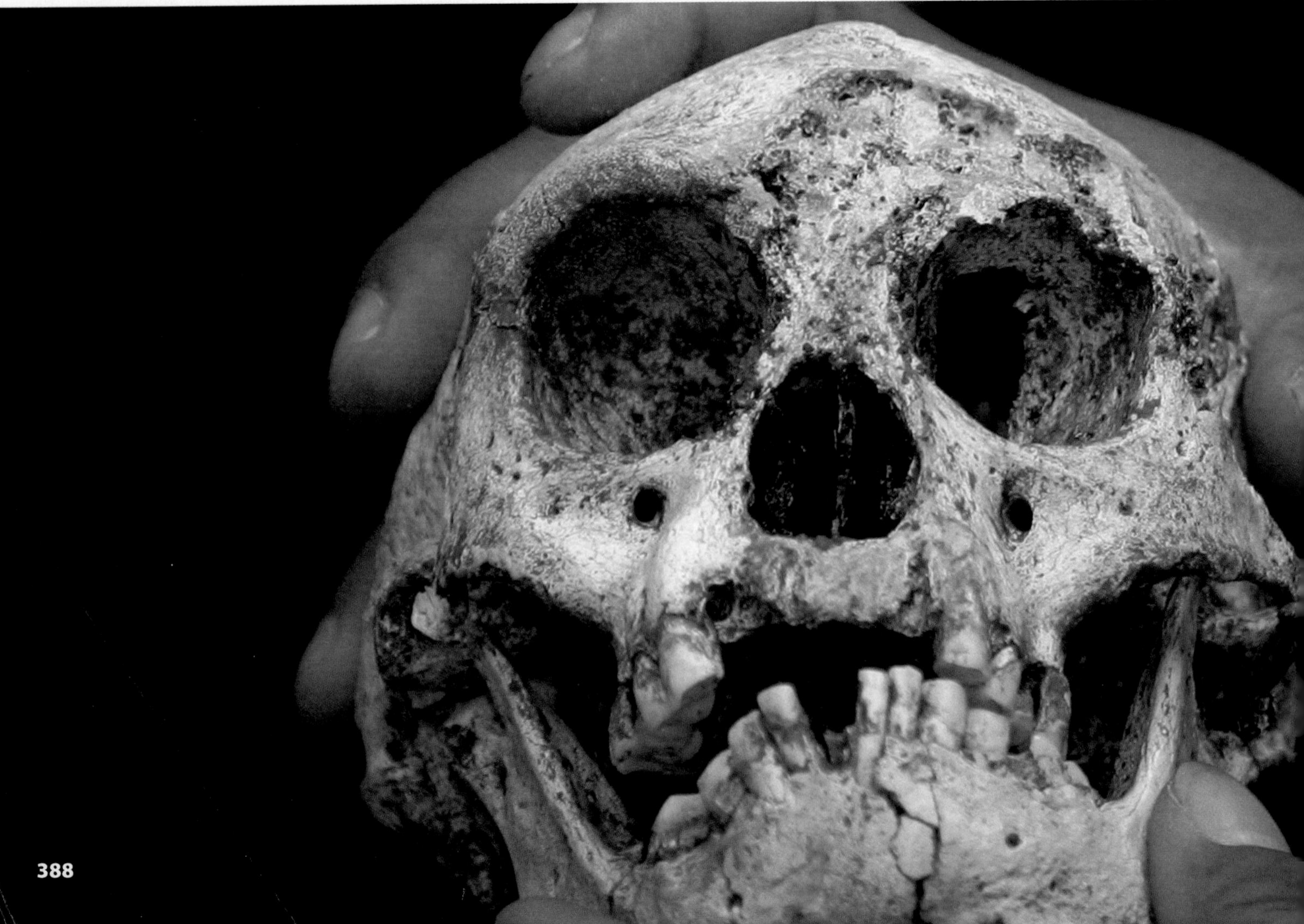

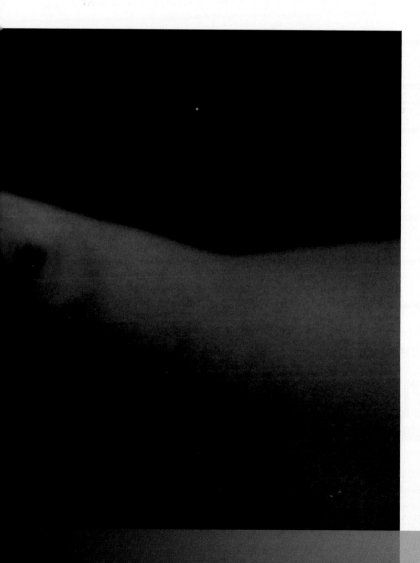

Most astonishingly, they lived as recently as 18,000 years ago, a time when scientists had thought that *Homo sapiens* was the only surviving human species. Who were the hobbits? We look at one hypothesis in this chapter, but (spoiler alert!) the more this species is studied, the more puzzling it becomes.

In this chapter, we continue our tour of the animal kingdom by exploring the vertebrates. Early in their history, vertebrates were restricted to the oceans. But some 365 million years ago, the evolution of limbs in one lineage of vertebrates set the stage for further diversification on land. We end the chapter, and our unit on the diversity of life, with a look at our predecessors—the primates who first walked on two legs, evolved a large, sophisticated brain, and eventually dominated Earth.

BIG IDEAS

Vertebrate Evolution and Diversity (19.1–19.8)

The major clades of chordates are distinguished by traits such as hinged jaws, two pairs of limbs, terrestrially-adapted eggs, and milk.

Primate Diversity (19.9–19.10)

Humans have many characteristics in common with other primates, including forward-facing eyes, limber shoulder and hip joints, and opposable thumbs.

Hominin Evolution (19.11–19.17)

Hominins, species that are on the human branch of the evolutionary tree, include approximately 20 extinct species.

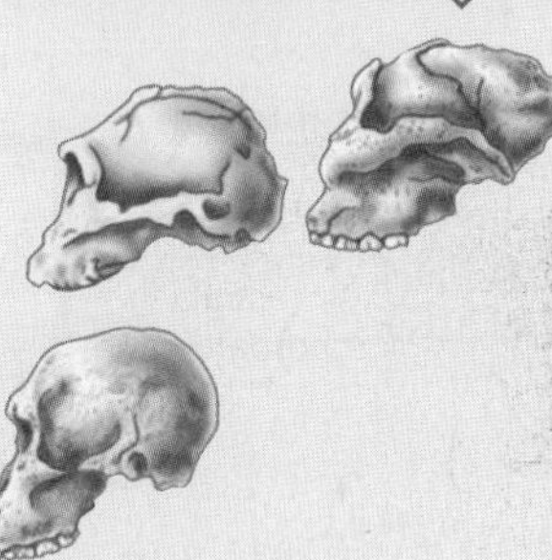

▷ Vertebrate Evolution and Diversity

19.1 Derived characters define the major clades of chordates

Using a combination of anatomical, molecular, and fossil evidence, biologists have developed hypotheses for the evolution of chordate groups. **Figure 19.1** illustrates a current view of the major clades of chordates and lists some of the derived characters that define the clades. You can see that the lancelets are thought to be the first group to branch from the chordate lineage (see Module 18.15).

The first transition was the development of a head that consists of a brain at the anterior end of the dorsal nerve cord, eyes and other sensory organs, and a skull. These innovations opened up a completely new way of feeding for chordates: active predation. All chordates with a head are called **craniates** (from the word *cranium*, meaning "skull").

The origin of a backbone came next. The **vertebrates** are distinguished by a more extensive skull and a backbone, or **vertebral column**, composed of a series of bones called **vertebrae** (singular, *vertebra*). These skeletal elements enclose the main parts of the nervous system. The skull forms a case for the brain, and the vertebrae enclose the nerve cord. The vertebrate skeleton is an endoskeleton, made of either flexible cartilage or a combination of hard bone and cartilage. Bone and cartilage are mostly nonliving material. But because there are living cells that secrete the nonliving material, the endoskeleton can grow with the animal.

The next major transition was the origin of jaws, which opened up new feeding opportunities. The evolution of lungs or lung derivatives, followed by muscular lobed fins with skeletal support, opened the possibility of life on land. **Tetrapods**, jawed vertebrates with two pairs of limbs, were the first vertebrates on land. The evolution of **amniotes**, tetrapods with a terrestrially adapted egg, completed the transition to land.

In the next several modules, we'll survey the vertebrates, from the jawless lampreys to the fishes to the tetrapods to the amniotes.

? **List the hierarchy of clades to which mammals belong.**

● Chordates, craniates, vertebrates, jawed vertebrates, tetrapods, amniotes

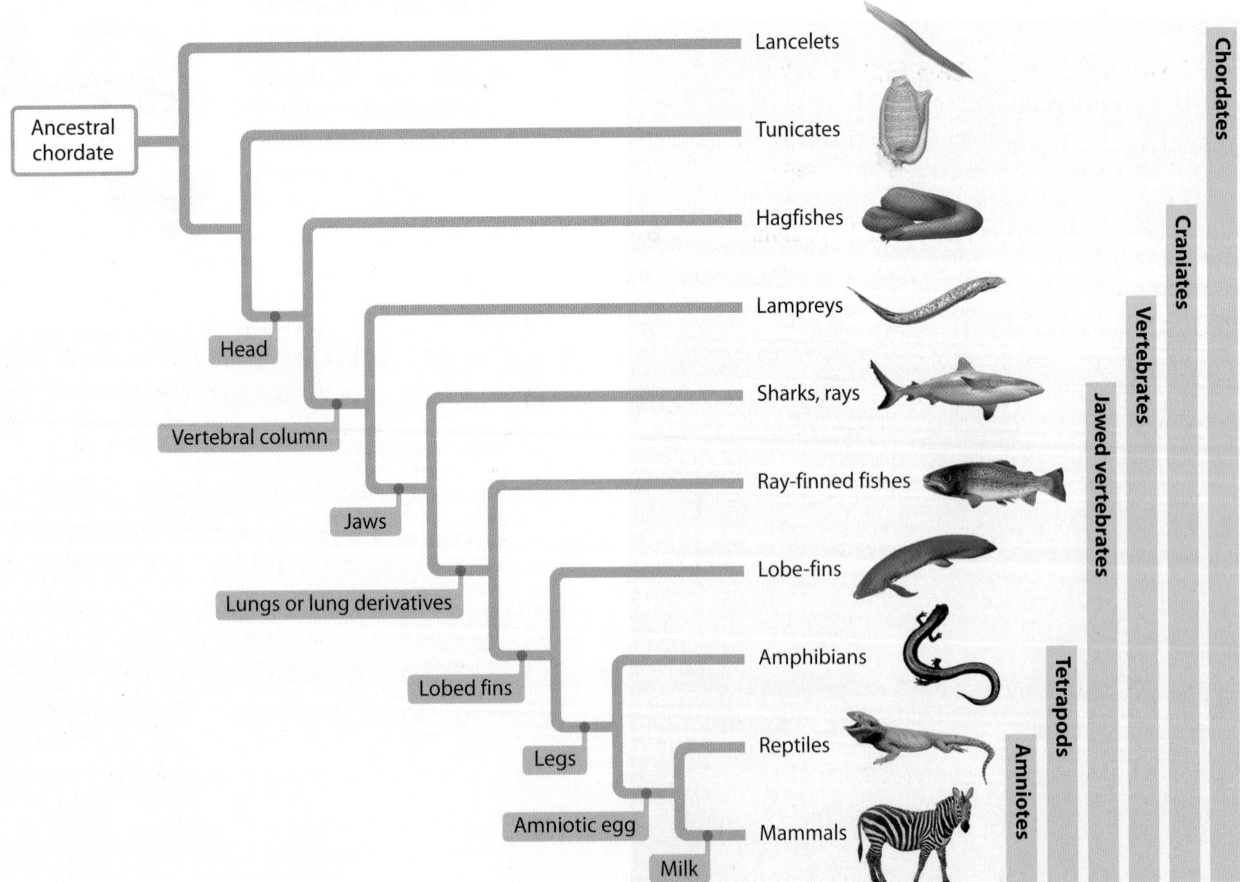

Ancestral chordate
Lancelets
Tunicates
Hagfishes
Lampreys
Sharks, rays
Ray-finned fishes
Lobe-fins
Amphibians
Reptiles
Mammals
Head
Vertebral column
Jaws
Lungs or lung derivatives
Lobed fins
Legs
Amniotic egg
Milk
Chordates
Craniates
Vertebrates
Jawed vertebrates
Tetrapods
Amniotes

▲ **Figure 19.1** A phylogenetic tree of chordates, showing key derived characters

Try This List the clades of chordates that belong to each group: craniates, vertebrates, jawed vertebrates, tetrapods, amniotes.

19.2 Hagfishes and lampreys lack hinged jaws

The most primitive surviving craniates (chordates with heads) are hagfishes and lampreys. In hagfishes **(Figure 19.2A)**, the notochord—a strong, flexible rod that runs most of the length of the body—is the body's main support in the adult. The notochord also persists in the adult lamprey, but rudimentary vertebral structures are also present. Consequently, lampreys are considered vertebrates, but hagfishes are not. Neither hagfishes nor lampreys have jaws.

Present-day hagfishes (roughly 30 species) scavenge dead or dying vertebrates on the cold, dark seafloor. Although nearly blind, they have excellent senses of smell and touch. They feed by entering the animal through an existing opening or by creating a hole using sharp, toothlike structures on the tongue that grasp and tear flesh. For leverage, the hagfish may tie its tail in a knot, then slide the knot forward to tighten it against the prey's body. The knot trick is also part of its antipredator behavior. When threatened, a hagfish exudes an enormous amount of slime from special glands on the sides of its body (see Figure 19.2A, inset). The slime may make the hagfish difficult to grasp, or it may repel the predator. After the danger has passed, the hagfish ties itself into a knot and slides the knot forward, peeling off the layer of slime.

Fisherman who use nets have long been familiar with hagfishes. With their keen chemical senses, hagfishes are quick to detect bait and entrapped fish. Many fishermen have hauled in a net filled with feasting hagfishes, unsalable fish, and bucketfuls of slime. But hagfishes have gained economic importance recently. Both the meat and the skin, which is used to make faux-leather "eel-skin" belts, purses, and boots, are valuable commodities. Asian fisheries have been harvesting hagfish for decades. As Asian fishing grounds have been depleted, the industry has moved to North America. Some populations of hagfish along the West Coast have been eradicated, and fisheries are now looking to the East Coast and South America for fresh stocks.

Lampreys represent the oldest living lineage of vertebrates **(Figure 19.2B)**. Lamprey larvae resemble lancelets (see Figure 18.15B). They are suspension feeders that live in freshwater streams, where they spend much of their time buried in sediment. Most lampreys migrate to the sea or lakes as they mature into adults.

Most species of lamprey are parasites, and just seeing the mouth of a sea lamprey (see Figure 19.2B, inset) suggests what it can do. The lamprey attaches itself to the side of a fish, uses its rasping tongue to penetrate the skin, and feeds on its victim's blood and tissues. After invading the Great Lakes via canals, these voracious vertebrates multiplied rapidly, decimating fish populations as they spread. Since the 1960s, streams that flow into the lakes have been treated with a chemical that reduces lamprey numbers, and fish populations have been recovering.

Slime glands

▲ **Figure 19.2A** Slime glands in hagfish

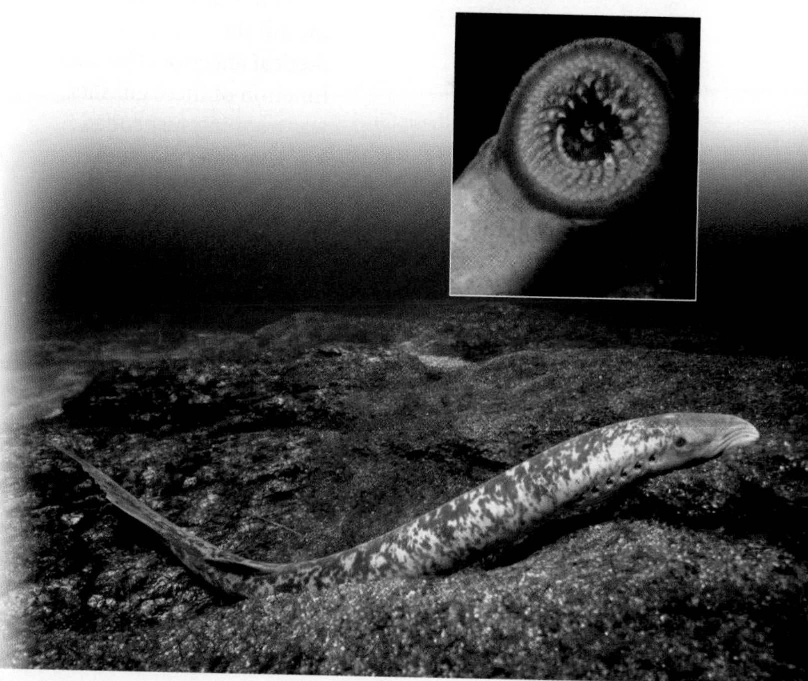

▲ **Figure 19.2B** A sea lamprey, with its rasping mouth (inset)

? **Why are hagfishes described as craniates rather than vertebrates?**

● They have a head but no vertebrae; their body is supported by a notochord.

19.3 Jawed vertebrates with gills and paired fins include sharks, ray-finned fishes, and lobe-finned fishes

Jawed vertebrates appeared in the fossil record in the mid-Ordovician period, about 440 million years ago, and steadily became more diverse. Their success probably relates to their paired fins and tail, which allowed them to swim after prey, as well as to their jaws, which enabled them to catch and eat a wide variety of prey instead of feeding as mud-suckers or suspension feeders. Sharks, fishes, amphibians, reptiles (including birds), and mammals—the vast majority of living vertebrates—have jaws supported by two skeletal parts held together by a hinge.

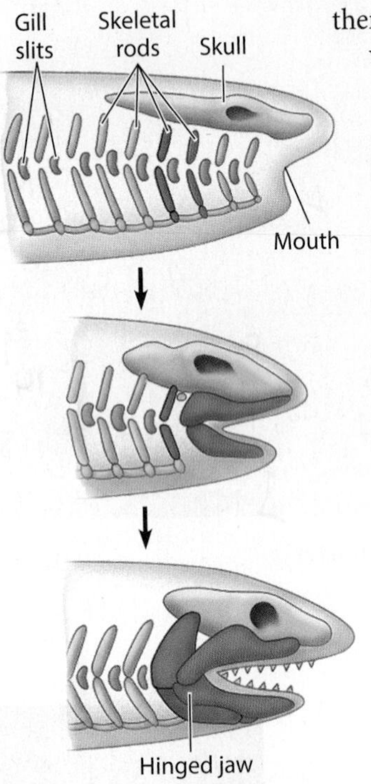

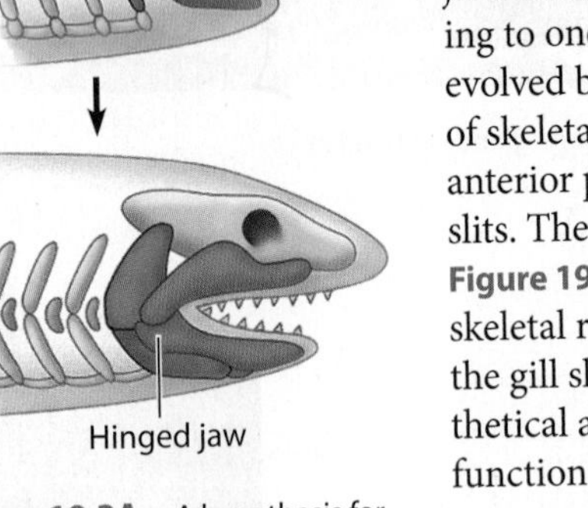
Gill slits
Skeletal rods
Skull
Mouth
Hinged jaw

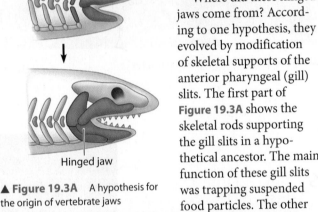

▲ **Figure 19.3A** A hypothesis for the origin of vertebrate jaws

Where did these hinged jaws come from? According to one hypothesis, they evolved by modification of skeletal supports of the anterior pharyngeal (gill) slits. The first part of **Figure 19.3A** shows the skeletal rods supporting the gill slits in a hypothetical ancestor. The main function of these gill slits was trapping suspended food particles. The other two parts of the figure show changes that may have occurred as jaws evolved. By following the red and green structures, you can see how two pairs of skeletal rods near the mouth have become the jaws and their supports. The remaining gill slits, no longer required for suspension feeding, remained as sites of gas exchange.

Three lineages of jawed vertebrates with gills and paired fins are commonly called fishes. The sharks and rays of the class Chondrichthyes, which means "cartilage fish," have changed little in over 300 million years. As shown in Figure 19.1, lungs or lung derivatives are the key derived character of the clade that includes the ray-finned fishes and the lobe-fins. Muscular fins supported by stout bones further characterize the lobe-fins.

Chondrichthyans Sharks and rays, the **chondrichthyans**, have a flexible skeleton made of cartilage. The largest sharks are suspension feeders that eat small, floating plankton. Most sharks, however, are adept predators—fast swimmers with a streamlined body, powerful jaws, and knifelike teeth (**Figure 19.3B**). A shark has sharp vision and a keen sense of smell. On its head it has electrosensors, organs that can detect the minute electric fields produced by muscle contractions in nearby animals. Sharks and most other aquatic vertebrates have a **lateral line system**, a row of sensory organs running along each side that are sensitive to changes in water pressure and can detect minor vibrations caused by animals swimming nearby.

Gill openings

▲ **Figure 19.3B** A sand bar shark, a chondrichthyan

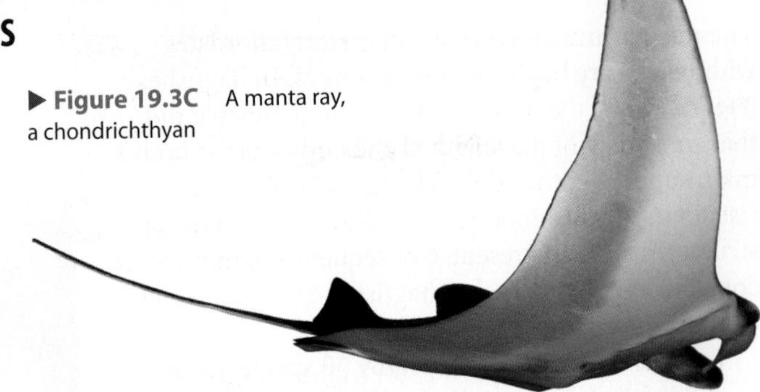

▶ **Figure 19.3C** A manta ray, a chondrichthyan

While the bodies of sharks are streamlined for swimming in the open ocean, rays are adapted for life on the bottom. Their bodies are dorsoventrally flattened, with the eyes on the top of the head. Once settled, they flip sand over their bodies with their broad pectoral fins and lie half-buried for much of the day. The tails of stingrays bear sharp spines with venom glands at the base. Where stingrays are common, swimmers and divers must take care not to step on or swim too closely over a concealed ray. The sting is painful and in rare cases fatal. Steve Irwin, a wildlife expert and television personality (*The Crocodile Hunter*), died when the 10-inch barb of a stingray pierced his heart while he was filming on the Great Barrier Reef in Australia.

The largest rays swim through the open ocean filtering plankton (**Figure 19.3C**). Some of these fishes are truly gigantic, measuring up to 6 m (19 feet) across the fins. The fin extensions in front of the mouth, which led to the common name devilfish, help funnel in water for suspension feeding.

Ray-finned Fishes In **ray-finned fishes**, which include the familiar tuna, trout, and goldfish, the skeleton is made of bone—cartilage reinforced with a hard matrix of calcium phosphate. Their fins are supported by thin, flexible skeletal rays. Most have flattened scales covering their skin and secrete a coating of mucus that reduces drag during swimming.

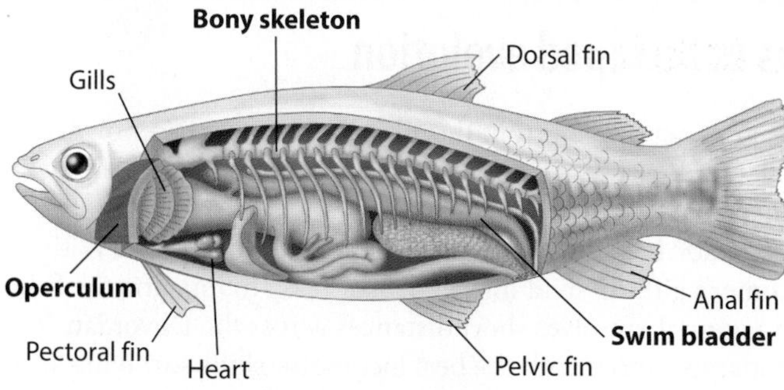
Bony skeleton
Gills
Dorsal fin
Operculum
Anal fin
Swim bladder
Pectoral fin
Heart
Pelvic fin

A rainbow trout,
a ray-fin

▲ **Figure 19.3D** The anatomical features of a ray-finned fish

Figure 19.3D highlights key features of a ray-finned fish such as the rainbow trout shown in the photograph. On each side of the head, a protective flap called an **operculum** covers a chamber housing the gills. Movement of the operculum allows the fish to breathe without swimming. (By contrast, sharks must generally swim to pass water over their gills.) Ray-finned fishes also have a lung derivative that helps keep them buoyant—the **swim bladder**, a gas-filled sac. Swim bladders evolved from balloon-like lungs, which the ancestral fishes may have used in shallow water to supplement gas exchange by their gills.

Ray-finned fishes, which emerged during the Devonian period along with the lobe-fins, include the greatest number of species of any vertebrate group, more than 27,000, and more species are discovered all the time. They have adapted to virtually every aquatic habitat on Earth. From the basic structural adaptations that gave them great maneuverability, speed, and feeding efficiency, specialized body forms, fins, and feeding adaptations have evolved in various groups. **Figure 19.3E** shows a sample of the variety. The balloon fish doesn't always look like a spiky beach ball. It raises its spines and inflates its body to deter predators. The small fins of the seahorse help it maneuver in dense vegetation, and the long tail is used for grasping a support. Seahorses have an unusual method of reproduction. The female deposits eggs in the male's abdominal brood pouch. His sperm fertilize the eggs, which develop inside the pouch. The flounder's flattened body is nearly invisible on the seabed. Pigment cells in its skin match the background for excellent camouflage. Notice that both eyes are on the top of its head. The larvae of flounders and other flatfishes have eyes on both sides of the head. During development, one eye migrates to join the other on the side that will become the top.

Lobe-finned Fishes The key derived character of the **lobe-fins** is a series of rod-shaped bones in their muscular pectoral and pelvic fins. During the Devonian, they lived in coastal wetlands and may have used their lobed fins to "walk" underwater. Today, three lineages of lobe-fins survive: The coelacanth is a deep-sea dweller once thought to be extinct. The lungfishes are represented by a few Southern Hemisphere genera that generally inhabit stagnant waters and gulp air into lungs connected to the pharynx **(Figure 19.3F)**. And the third lineage, the tetrapods, adapted to life on land during the mid-Devonian and gave rise to terrestrial vertebrates, as we see next.

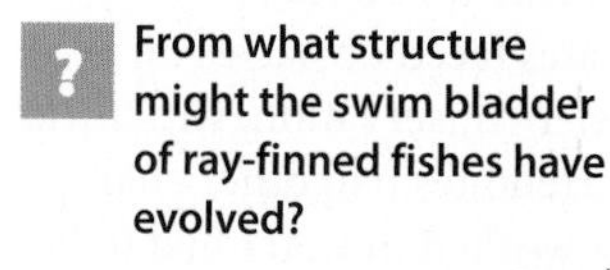
?
From what structure might the swim bladder of ray-finned fishes have evolved?
Simple lungs of an ancestral species

A seahorse
A balloon fish
A flounder

▲ **Figure 19.3E** A variety of ray-finned fishes

▲ **Figure 19.3F** A lobe-finned lungfish (about 1 m long)

19.4 New fossil discoveries are filling in the gaps of tetrapod evolution

EVOLUTION CONNECTION

During the late Devonian period, a line of lobe-finned fishes gave rise to tetrapods (meaning "four feet" in Greek), which today are defined as jawed vertebrates that have limbs and feet that can support their weight on land. Adaptation to life on land was a key event in vertebrate history; all subsequent groups of vertebrates—amphibians, mammals, and reptiles (including birds)—are descendants of these early land-dwellers.

The dramatic differences between aquatic and terrestrial environments shaped plant bodies and life cycles (see Module 17.1). Like plants, vertebrates faced obstacles on land in regard to gas exchange, water conservation, structural support, and reproduction. But vertebrates had other challenges as well. Sensory organs that worked in water had to be adapted or replaced by structures that received stimuli transmitted through air. And, crucially, a new means of locomotion was required.

Lobe-finned fishes were long considered the most likely immediate ancestors of tetrapods. Their fleshy paired fins contain bones that appear to be homologous to tetrapod limb bones, and some of the modern lobe-fins have lungs that extract oxygen from the air. Alfred Romer, a renowned paleontologist, hypothesized that these features enabled lobe-fins to survive by moving from one pool of water to another as aquatic habitats shrank during periods of drought. With Romer's gift for vivid imagery, it was easy to imagine the fish dragging themselves short distances across the Devonian landscape, those with the best locomotor skills surviving such journeys to reproduce. In this way, according to the hypothesis, vertebrates gradually became fully adapted to a terrestrial existence. But fossil evidence of the transition was scarce.

For decades, the most informative fossils available were *Eusthenopteron*, a 385-million-year-old specimen that was clearly a fish, and *Ichthyostega*, which lived 365 million years ago and had advanced tetrapod features **(Figure 19.4A)**. The ray-finned tail and flipper-like hind limbs of *Ichthyostega* indicated that it spent considerable time in the water, but its well-developed front limbs with small, fingerlike bones and powerful shoulders showed that it was capable of locomotion on land. Unlike the shoulder bones of lobe-finned fishes, which are connected directly to the skull, *Ichthyostega* had a neck, a feature advantageous for terrestrial life. *Eusthenopteron* and *Ichthyostega* represented two widely separated points in the transition from fins to limbs. But what happened in between?

Recent fossil finds have begun to fill in the gap. Scientists have discovered lobe-fin fossils that are more similar to tetrapods than to *Eusthenopteron*, including a 385-million-year-old fish called *Panderichthys* (see Figure 19.4A). With its long snout, flattened body shape, and eyes on top of its head, *Panderichthys* looked a bit like a crocodile. It had lungs as well as gills and an opening that allowed water to enter through the top of the skull, a possible indication of a shallow-water habitat. Its paired fins had fishlike rays, but the dorsal and anal fins had been lost, and the tail fin was much smaller than in *Eusthenopteron*. Certain features of its skull were more like those of a tetrapod. Although it had no neck, the bones connecting forelimb to skull were intermediate in shape between that of a fish and a tetrapod. *Panderichthys* could have been capable of leveraging its fins against the bottom as it propelled itself through shallow water. It was a fish, but a tetrapod-like fish.

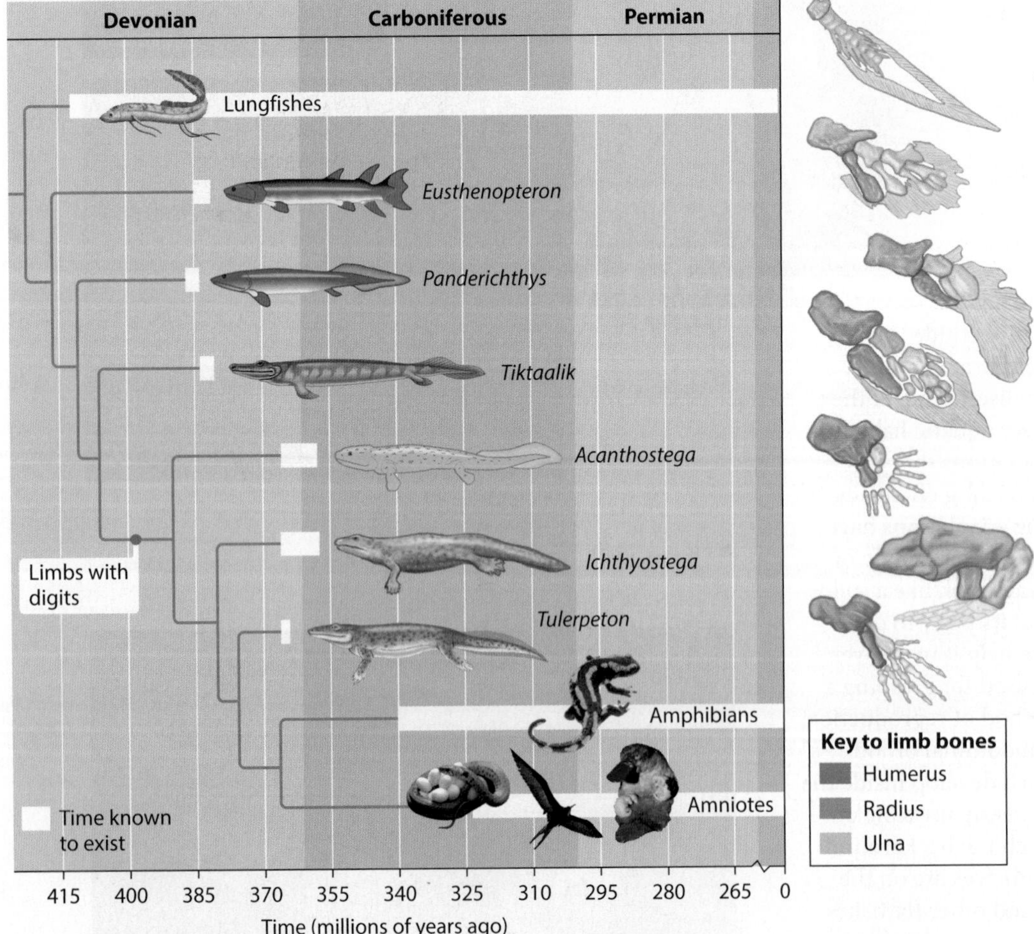

Devonian
Carboniferous
Permian
Lungfishes
Eusthenopteron
Panderichthys
Tiktaalik
Acanthostega
Ichthyostega
Tulerpeton
Amphibians
Amniotes
Limbs with digits
Time known to exist
415 400 385 370 355 340 325 310 295 280 265 0
Time (millions of years ago)
Key to limb bones
Humerus
Radius
Ulna

▲ **Figure 19.4A** Some of the transitional forms in tetrapod evolution

Try This Describe how the forelimbs of *Panderichthyes*, *Tiktaalik*, and *Acanthostega* differ.

On the other hand, *Acanthostega* was more of a fishlike tetrapod, and it turned scientists' ideas about tetrapod evolution upside down. Like *Ichthyostega*, *Acanthostega* had a neck, structural modifications that strengthened its backbone and skull, and four limbs with toes. But its limbs could not have supported the animal on land, nor could its ribs have prevented its lungs from collapsing out of water. The startling conclusion was that the first tetrapods were not fish with lungs that had gradually evolved legs as they dragged themselves from pool to pool in search of water. Instead, they were fish with necks and four limbs that raised their heads above water and could breathe oxygen from the air. **Figure 19.4B** shows an artist's rendering of *Acanthostega*.

In 2006, a team of scientists added another important link to the chain of evidence on tetrapod evolution. Using information from the dates and habitats of previous specimens, they predicted that transitional fossils might be found in rock formed from the sediments of shallow river environments during a particular time period in the late Devonian. They found a suitable area to search in Arctic Canada, which had been located near the equator during that time. There they discovered several remarkable fossils of an animal they named *Tiktaalik* ("large freshwater fish" in the language of the Nunavut Inuit tribe from that region). The specimens were exquisitely preserved; even the fishlike scales are clearly visible. *Tiktaalik* straddled the border between *Panderichthys* and *Acanthostega* (see Figure 19.4A). Its paddle-like forelimbs were part fin, part foot. The fin rays had not been replaced by toes. The joints would have served to prop the animal up, but not enable it to walk. It had well-developed gills like a fish, but a tetrapod-like neck (see Figure 22.5). It was a perfectly intermediate form.

With these images of early tetrapods gleaned from the fossil record in mind, let's look at the environmental conditions that drove their evolution. Plants had colonized the land 100 million years earlier, followed by arthropods. By the time *Tiktaalik* appeared, shallow water would have been a complex environment, with fallen trees and other debris from land plants, along with rooted aquatic plants, providing food and shelter for a variety of organisms. Even a meter-long predator like *Tiktaalik* would have found plenty to eat. But warm, stagnant water is low in oxygen. The ability to supplement oxygen intake by air breathing—by lifting the head out of the water—would have been an advantage.

Once tetrapods had adaptations that enabled them to leave the water for extended periods of time, they diversified rapidly. Food and oxygen were plentiful in the Carboniferous swamp forests (see Module 17.4). From one of the many lines of tetrapods that settled ashore, modern amphibians evolved.

? **How did *Acanthostega* change scientists' concept of tetrapod evolution?**

● It showed that the first tetrapods were more fishlike than previously thought. They did not spend time on land.

▲ **Figure 19.4B** The first tetrapod, a four-limbed fish that lived in shallow water and could breathe air

19.5 Amphibians are tetrapods—vertebrates with two pairs of limbs

Amphibians include salamanders, frogs, and caecilians. Some present-day salamanders are entirely aquatic, but those that live on land walk with a side-to-side bending of the body that probably resembles the swagger of early terrestrial tetrapods (**Figure 19.5A**). Frogs are more specialized for moving on land, using their powerful hind legs to hop along the terrain. Caecilians (**Figure 19.5B**) are nearly blind and are legless, adaptations that suit their burrowing lifestyle. However, they evolved from a legged ancestor.

Most amphibians are found in damp habitats, where their moist skin supplements their lungs for gas exchange. Amphibian skin usually has poison glands that may play a role in defense. Poison dart frogs have particularly deadly poisons, and their vivid coloration warns away potential predators (**Figure 19.5C**).

▲ **Figure 19.5A** A fire salamander

▲ **Figure 19.5B** A caecilian

In Greek, the word *amphibios* means "living a double life," a reference to the metamorphosis of many frogs. A frog spends much of its time on land, but it lays its eggs in water. During the breeding season, many species fill the air with their mating calls. As you can see in **Figure 19.5D**, frog eggs are encapsulated in a jellylike material. Consequently, they must be surrounded by moisture to prevent them from drying out. The larval stage, called a tadpole, is a legless, aquatic algae-eater with gills, a lateral line system resembling that of fishes, and a long, finned tail (**Figure 19.5E**). In changing into a frog, the tadpole undergoes a radical metamorphosis. When a young frog crawls onto shore and continues life as a terrestrial insect-eater, it has four legs and air-breathing lungs instead of gills (see Figure 19.5C). Not all amphibians live such a double life, however. Some species are strictly terrestrial, and others are exclusively aquatic. *Toad* is a term generally used to refer to frogs that have rough skin and live entirely in terrestrial habitats.

▲ **Figure 19.5C** An adult poison dart frog

For the past 30 years, zoologists have been documenting a rapid and alarming decline in amphibian populations throughout the world. In 2012, roughly a third of all known species were at risk for extinction. Multiple causes are contributing to the decline, including habitat loss, climate change, and the spread of a pathogenic fungus (see Module 17.14).

Amphibians were the first vertebrates to colonize the land. The early amphibians probably feasted on insects and other invertebrates in the lush forests of the Carboniferous period (see Figure 17.4). As a result, amphibians became so widespread and diverse that the Carboniferous period is sometimes called the age of amphibians. However, the distribution of amphibians was limited by their vulnerability to dehydration. Adaptations that evolved in the next clade of vertebrates we discuss, the amniotes, enabled them to complete their life cycles entirely on land.

? **In what ways are amphibians not completely adapted for terrestrial life?**

● Their eggs are not well protected against dehydration; many species have an aquatic larval form; their skin is not waterproof and must remain moist to permit gas exchange.

▼ **Figure 19.5D** Frog eggs; a tadpole is developing in the center of each ball of jelly

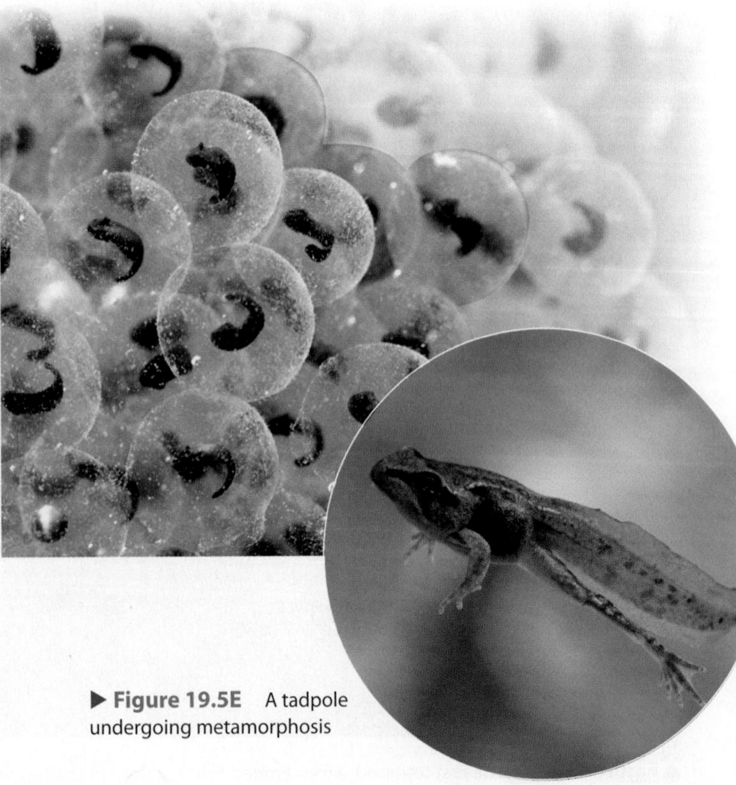

▶ **Figure 19.5E** A tadpole undergoing metamorphosis

19.6 Reptiles are amniotes—tetrapods with a terrestrially adapted egg

Reptiles (including birds) and mammals are **amniotes**. The major derived character of this clade is the amniotic egg **(Figure 19.6A)**. The **amniotic egg** contains specialized extraembryonic membranes, so called because they are not part of the embryo's body. The *amnion*, for which the amniotic egg is named, is a fluid-filled sac surrounding the embryo **(Figure 19.6B)**. The *yolk sac* contains a rich store of nutrients for the developing embryo, like the yellow yolk of a chicken egg. Additional nutrients are available from the albumen ("egg white"). The *chorion* and the membrane of the *allantois* enable the embryo to obtain oxygen from the air for aerobic respiration and dispose of the carbon dioxide produced. The allantois is also a disposal sac for other metabolic waste products. With a waterproof shell to enclose the embryo and its life-support system, reptiles were the first vertebrates to be able to complete their life cycles on land. The seed played a similar role in the evolution of plants (as we saw in Module 17.2).

The clade of amniotes called **reptiles** includes lizards, snakes, turtles, crocodilians, and birds, along with a number of extinct groups such as most of the dinosaurs. Lizards are the most numerous and diverse reptiles other than birds. Snakes, which are closely related to lizards, may have become limbless as their ancestors adapted to a burrowing lifestyle. Turtles have changed little since they evolved, although their ancestral lineage is still uncertain. Crocodiles and alligators (crocodilians) are the largest living reptiles—the saltwater crocodiles measure up to 6.3 m (as long as a stretch limousine) in length and weigh up to a ton. Crocodilians spend most of the time in water, breathing air through upturned nostrils.

In addition to an amniotic egg protected in a waterproof shell, reptiles have several other adaptations for terrestrial living not found in amphibians. Reptilian skin, covered with scales waterproofed with the tough protein keratin, keeps the body from drying out. Reptiles cannot breathe through their dry skin and obtain most of their oxygen with their lungs.

Lizards, snakes, crocodilians, and turtles are sometimes said to be "cold-blooded" because they do not use their metabolism to produce body heat. Nonetheless, these animals may regulate their temperature through their behavior. The bearded dragon of the Australian outback **(Figure 19.6C)** commonly warms up in the morning by sitting on warm rocks and basking in the sun. If the lizard gets too hot, it seeks shade. Animals that absorb external heat rather than generating much of their own are said to be **ectothermic** (from the Greek *ektos*, outside, and *therme*, heat), a term that is more appropriate than the term *cold-blooded.* Because the energy demands of ectothermic animals are low, reptiles are well suited to deserts, where food is scarce.

Like the amphibians, reptiles were once much more prominent than they are today. Following the decline of amphibians, reptilian lineages expanded rapidly, creating a dynasty that lasted 200 million years. Dinosaurs, the most diverse group, included the largest animals ever to inhabit land. Most dinosaurs died out during a period of mass extinctions about 65 million years ago. Descendants of one dinosaur lineage, however, survive today as the reptilian group we know as birds.

? **What is an amniotic egg?**

An egg in which an embryo develops in a fluid-filled amniotic sac and is nourished by a yolk

▲ **Figure 19.6A** A European grass snake laying eggs

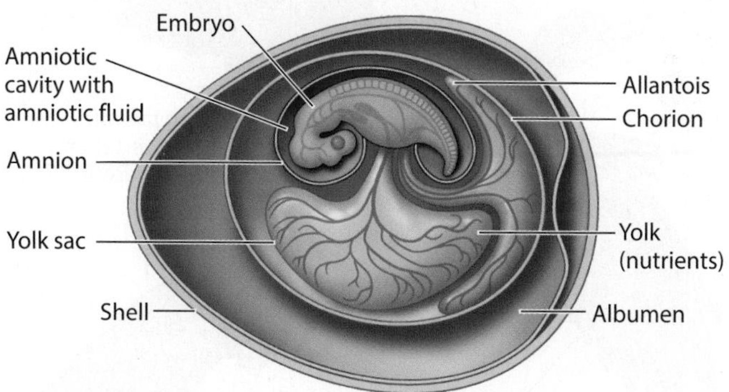
Embryo
Amniotic cavity with amniotic fluid
Amnion
Yolk sac
Shell
Allantois
Chorion
Yolk (nutrients)
Albumen

▲ **Figure 19.6B** An amniotic egg

Try This Identify and explain the role of each structure in the egg that enable reptiles to complete their life cycles on land.

▲ **Figure 19.6C** A bearded dragon basking in the sun

19.7 Birds are feathered reptiles with adaptations for flight

Almost all **birds** can fly, and nearly every part of a bird's body reflects adaptations that enhance flight. The forelimbs have been remodeled as feather-covered wings that act as airfoils, providing lift and maneuverability in the air (see Figure 30.1E). Large flight muscles anchored to a central ridge along the breastbone provide power. Some species, such as the seagoing frigate bird in **Figure 19.7A**, have wings adapted to soaring on air currents, and they flap their wings only occasionally. Others, such as hummingbirds, excel at maneuvering but must flap almost continuously to stay aloft. The few flightless groups of birds include the ostrich, which is the largest bird in the world, and the emu, the largest native bird in Australia.

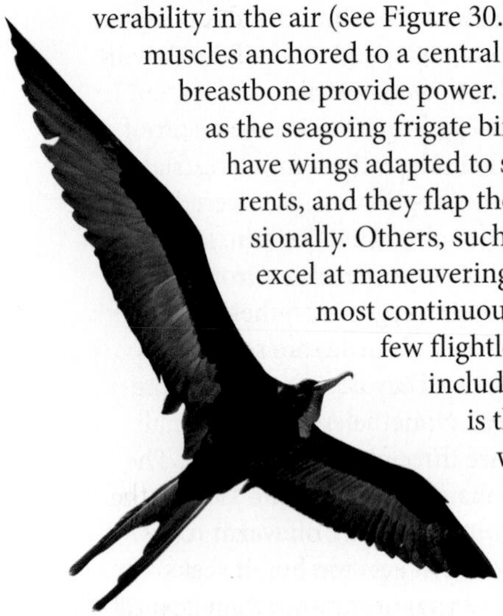

▲ **Figure 19.7A** A soaring frigate bird

▲ **Figure 19.7B** Courtship behavior of the wandering albatross

Many features help reduce weight for flight: Present-day birds lack teeth; their tail is supported by only a few small vertebrae; their feathers have hollow shafts; and their bones have a honeycombed structure, making them strong but light. For example, the frigate bird has a wingspan of more than 2 m, but its whole skeleton weighs only about 113 g (4 ounces).

Flying requires a great amount of energy, and present-day birds have a high rate of metabolism. Unlike other living reptiles, they are **endothermic**, using heat generated by metabolism to maintain a warm, steady body temperature. Insulating feathers help to maintain their warm body temperature. In support of their high metabolic rate, birds have a highly efficient circulatory system, and their lungs are even more efficient at extracting oxygen from the air than are the lungs of mammals.

Flying also requires acute senses and fine muscle control. Birds have excellent eyesight, perhaps the best of all vertebrates, and the visual and motor areas of the brain are well developed.

Birds typically display very complex behaviors, particularly during breeding season. Courtship often involves elaborate rituals. The male frigate bird in Figure 19.7A, for example, inflates its red throat pouch like an enormous balloon to attract females. The wandering albatross in **Figure 19.7B** employs a different kind of courtship display. In many species of birds, males and females take turns incubating the eggs and then feeding the young. Some birds migrate great distances each year to different feeding or breeding grounds.

Strong evidence indicates that birds evolved from a lineage of small, two-legged dinosaurs called theropods. **Figure 19.7C** is an artist's reconstruction based on a 150-million-year-old fossil of the oldest known, most primitive bird, called *Archaeopteryx* (from the Greek *archaios*, ancient, and *pteryx*, wing). Like living birds, it had feathered wings, but otherwise it was more like a small two-legged dinosaur of its era—with teeth, wing claws, and tail with many vertebrae. Over the past decade, Chinese paleontologists have excavated fossils of many feathered theropods, including specimens that predate *Archaeopteryx* by 5–10 million years. Such findings imply that feathers, which are homologous to reptilian scales, evolved long before powered flight. Early feathers may have functioned in insulation or courtship displays. As more of these fossils are discovered, scientists are gaining new insight into the evolution of flight.

? List some adaptations of birds that enhance flight.

● Reduced weight, endothermy with high metabolism, efficient respiratory and circulatory systems, feathered wings shaped like airfoils, good eyesight

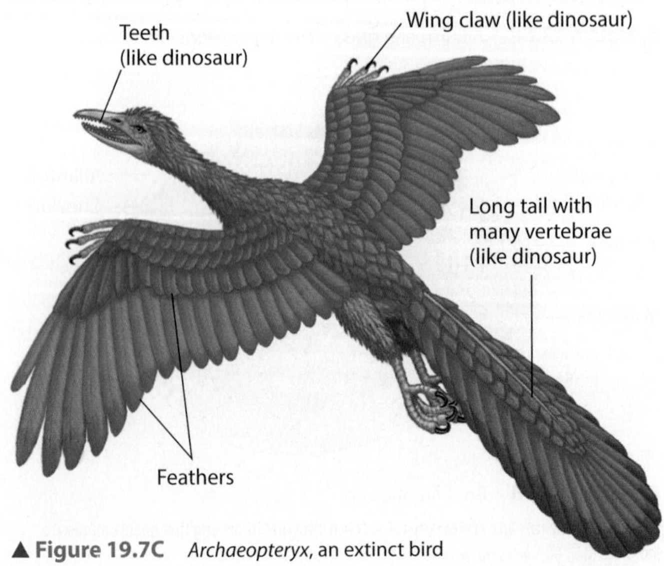

Teeth
(like dinosaur)
Wing claw (like dinosaur)
Long tail with
many vertebrae
(like dinosaur)
Feathers

▲ **Figure 19.7C** *Archaeopteryx*, an extinct bird

19.8 Mammals are amniotes that have hair and produce milk

There are two major lineages of amniotes: one that led to the reptile clade and one that produced the mammals. Two features—hair and mammary glands that produce milk—are the distinguishing traits of **mammals**. Like birds, mammals are endothermic. Hair provides insulation that helps maintain a warm body temperature. Efficient respiratory and circulatory systems (including a four-chambered heart) support the high rate of metabolism characteristic of endotherms. Differentiation of teeth adapted for eating many kinds of foods is also characteristic of mammals; different kinds of teeth specialize in cutting, piercing, crushing, or grinding. The three major lineages of mammals—monotremes (egg-laying mammals), marsupials (mammals with a pouch), and eutherians (placental mammals)—differ in their reproductive patterns.

The only existing egg-laying mammals, known as **monotremes**, are echidnas (spiny anteaters) and the duck-billed platypus. The female platypus usually lays two eggs and incubates them in a nest. After hatching, the young lick up milk secreted onto the mother's fur (**Figure 19.8A**).

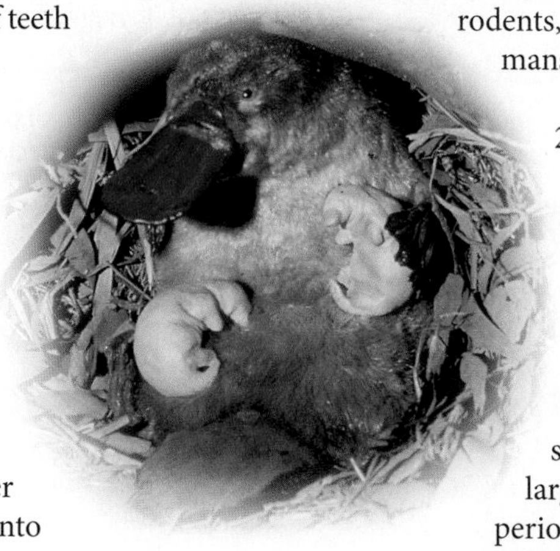

▲ **Figure 19.8A** Monotremes: a duck-billed platypus with newly hatched young

All other mammals are born rather than hatched. During development, the embryos remain inside the mother and receive their nourishment directly from her blood. Mammalian embryos produce extraembryonic membranes that are homologous to those found in the amniotic egg, including the amnion, which retains its function as a protective fluid-filled sac. The chorion, yolk sac, and allantois have different functions in mammals than in reptiles (see Module 27.15). Membranes from the embryo join with the lining of the uterus to form a **placenta**, a structure in which nutrients from the mother's blood diffuse into the embryo's blood.

Marsupials have a brief gestation and give birth to tiny, embryonic offspring that complete development while attached to the mother's nipples. The nursing young are usually housed in an external pouch (**Figure 19.8B**). Nearly all marsupials live in Australia, New Zealand, and Central and South America. The opossum is the only North American marsupial. **Eutherians** are mammals that bear fully developed live young. They are commonly called **placental mammals** because their placentas are more complex than those of marsupials, and the young complete their embryonic development in the mother's uterus attached to the placenta. The large silvery membrane still clinging to the newborn zebra in **Figure 19.8C** is the amniotic sac. Elephants, rodents, rabbits, dogs, cows, whales, bats, and humans are all examples of eutherians.

The first true mammals arose about 200 million years ago and were probably small, nocturnal insect-eaters. Of the three main groups of living mammals, monotremes are the oldest lineage. Marsupials diverged from eutherians about 140 million years ago. During the Mesozoic era, mammals remained about the size of today's shrews, which are very small insectivores. After the extinction of large dinosaurs at the end of the Cretaceous period, however, mammals underwent an adaptive radiation, giving rise to large terrestrial herbivores and predators, as well as bats and aquatic mammals such as porpoises and whales. Humans belong to the mammalian order Primates, along with monkeys and apes. We begin our study of human evolution with the next module.

? **What are the two distinguishing features of mammals?**

Hair and mammary glands, which produce milk

▲ **Figure 19.8B** Marsupials: a gray kangaroo with her young in her pouch

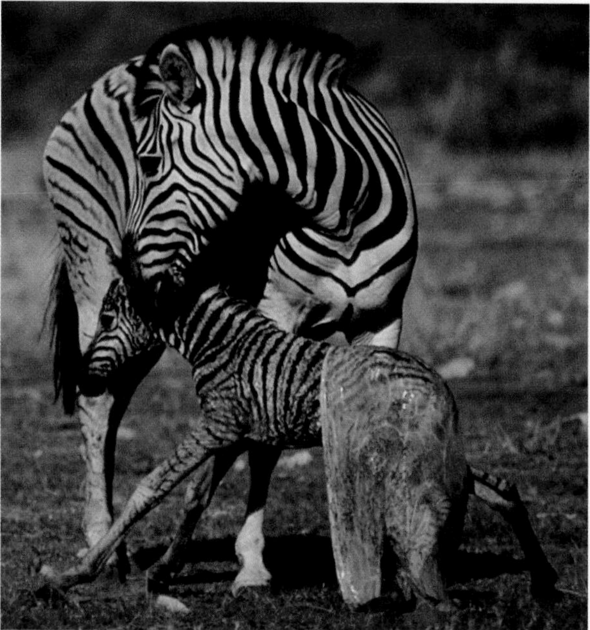

▲ **Figure 19.8C** Eutherians: a zebra with newborn

▷ Primate Diversity

VISUALIZING THE CONCEPT

19.9 Primates include lemurs, tarsiers, monkeys, and apes

The lemurs, tarsiers, monkeys, and apes are members of the mammalian order Primates. Module 19.10 includes a phylogenetic tree of primates. The earliest primates were probably small arboreal (tree-dwelling) mammals that arose before 65 million years ago, when dinosaurs still dominated the planet. Most living primates are arboreal, and the primate body has a number of features that were shaped, through natural selection, by the demands of living in trees. Although humans never lived in trees, the human body retains many of the traits that evolved in our arboreal ancestors.

PRIMATE DIVERSITY

Distinguishing primate features

Lorises and lemurs, which probably resemble early arboreal primates, illustrate charateristics of the order Primates.

- Short snout; eyes set close together on front of face
- Limber shoulder and hip joints
- Five highly mobile digits on hands and feet
- Flexible thumb

▲ A slender loris, a species closely related to lemurs

Position of eyes enhances depth perception, an important trait for maneuvering in trees.
Limber joints enable climbing and swinging from branch to branch.
Hands and feet can grasp objects and manipulate food.

► A Coquerel's sifaka (pronounced "she-fa'-ka"), a species of lemur

Anthropoids

Anthropoids are a group of primates that includes monkeys and apes. Anthropoids have a fully **opposable thumb**; that is, they can touch the tip of all four fingers with their thumb. In monkeys and most apes, the opposable thumb functions in a grasping "power grip." Humans have a distinctive bone structure at the base of the thumb that allows it to be used for more precise manipulation.

Monkeys

Monkeys have forelimbs that are about equal in length to their hind limbs. Monkeys in the Old World (Africa and Asia) and the New World (the Americas) have been evolving separately for over 30 million years.

Old World monkeys

- Many arboreal, but some ground dwelling
- Nostrils open downward
- Lack prehensile tail

Tail is nongrasping.

► A lion-tailed macaque

New World monkeys

- All arboreal
- Nostrils open to side; far apart
- Many have a long, prehensile (grasping) tail

Note the position of the nostrils.

Prehensile tail is specialized for grasping tree limbs.

▲ Red howler monkey

◄ Golden lion tamarin

Apes

Apes include gibbons, orangutans, gorillas, chimpanzees, and humans. Apes lack a tail and have relatively long arms and short legs. Compared to other primates, they have larger brains relative to body size, and consequently their behavior is more flexible.

◀ Orangutans are shy apes that live in the rain forests of Sumatra and Borneo. They spend most of their time in trees, supporting the body with all four limbs.

▶ The nine species of gibbons, all found in Southeast Asia, are entirely arboreal. Gibbons are the smallest, lightest, and most acrobatic of the apes.

▼ Gorillas are the largest apes. Some males are over 6 feet tall and weigh more than 400 pounds. Found only in African rain forests, gorillas usually live in groups of up to about 20 individuals. They spend nearly all their time on the ground.

Note that the big toe is widely separated from the other toes. This adaptation for grasping branches is found in all primates except humans.

Gorillas can stand upright on their hind legs, but when they walk on all fours, their knuckles contact the ground.

▼ Chimpanzees and a closely related species called bonobos live in tropical Africa. They are intelligent, communicative, and social.

▶ The human body retains many primate traits. Our most distinctive feature is the large, sophisticated brain that allows us to study our own origins.

? **How do apes and monkeys differ physically?**

● Monkeys have forelimbs that are about the same length as their hind limbs. Apes have relatively long arms and short legs.

19.10 The human story begins with our primate heritage

There are three main groups of living primates, as shown in the phylogenetic tree in **Figure 19.10A**. The lorises and pottos of tropical Africa and southern Asia are placed in one group along with the lemurs.

The tarsiers form a second group of primates. Limited to Southeast Asia, these small, nocturnal tree-dwellers have flat faces with large eyes (**Figure 19.10B**). Fossil evidence indicates that tarsiers are more closely related to anthropoids, the third group of primates, than to the lemur-loris-potto group.

The anthropoids (from the Greek *anthropos*, man, and *eidos*, form) include monkeys and apes. As shown in Figure 19.10A, the fossil record indicates that anthropoids began diverging from other primates about 50 million years ago. As you learned in Module 19.9, distinct lineages of monkeys inhabit different regions of the globe. Old World monkeys and apes, which include gibbons, orangutans, gorillas, chimpanzees (and bonobos), and humans, diverged about 25 million years ago.

Molecular evidence indicates that chimpanzees and gorillas are more closely related to humans than they are to other apes. Humans and chimpanzees are especially closely related; their genomes are 99% identical. Nevertheless, human and chimpanzee genomes have been evolving separately since the two lineages diverged from their last common ancestor approximately 6 million years ago. Because fossil apes are extremely rare, we know little about that ancestor.

▲ **Figure 19.10B** A tarsier, member of a distinct primate group

However, researchers studying the skeletal features of a 4.4-million-year-old ape called *Ardipithecus ramidus* recently concluded that present-day apes such as chimpanzees are the result of substantial evolution since the lineages diverged.

? What are the five groups of apes within the anthropoid category?

● Gibbons, orangutans, gorillas, chimpanzees, and humans

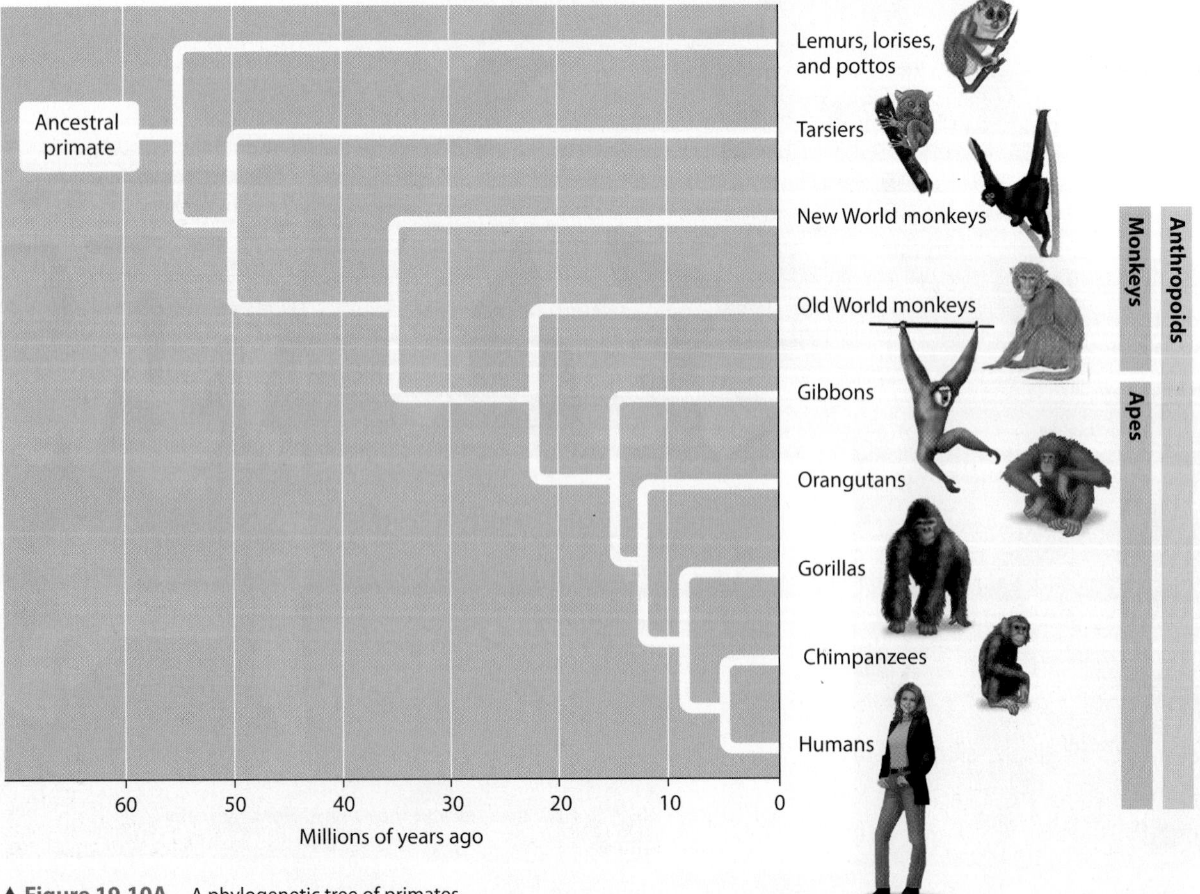

Ancestral primate
Lemurs, lorises, and pottos
Tarsiers
New World monkeys
Old World monkeys
Gibbons
Orangutans
Gorillas
Chimpanzees
Humans
Monkeys
Apes
Anthropoids
60
50
40
30
20
10
0
Millions of years ago

▲ **Figure 19.10A** A phylogenetic tree of primates

▷ Hominin Evolution

19.11 The hominin branch of the primate tree includes species that coexisted

Paleoanthropology, the study of human origins and evolution, focuses on the tiny slice of biological history that has occurred since the divergence of human and chimpanzee lineages from their common ancestor. Paleoanthropologists have unearthed fossils of approximately 20 species of extinct **hominins**, species that are more closely related to humans than to chimpanzees and are therefore on the human branch of the evolutionary tree. These fossils have shown that many of the characters that distinguish humans from other apes were present in earlier hominins; they are not unique to *Homo sapiens*.

Thousands of hominin fossils have been discovered, and each new find sheds light on the story of human evolution. However, paleoanthropologists are still vigorously debating hominin classification and phylogenetic relationships. Therefore, **Figure 19.11** presents some of the known hominins in a timeline rather than in a tree diagram like the one in Figure 19.10A. The vertical bars indicate the approximate time period when each species existed, as currently known from the fossil record.

One inference about human phylogeny can immediately be made from Figure 19.11: Hominins did not evolve in a straight line leading directly to *Homo sapiens*. At times several hominin species coexisted, and some must have been dead ends that did not give rise to new lineages.

The oldest hominin yet discovered, *Sahelanthropus tchadensis*, lived from about 6.5 million years ago, around the time when the human and chimpanzee lineages diverged. However, most of the hominin fossils that have been found are less than 4 million years old. Thus, the 4.4-million-year-old fossils of *Ardipithecus ramidus*, painstakingly uncovered and reconstructed by international teams of scientists over the past 15 years, represent an unprecedented perspective on early hominin evolution. *Ardipithecus* was a woodland creature that moved in the trees by walking along branches on the flat parts of its hands and feet. It was equally capable of moving on the ground, and its skeletal features suggest that it walked upright.

Hominin diversity increased dramatically in the period between 4 and 2 million years ago. The first fossil member of our own genus, *Homo*, dates from that time. By 1 million years ago, only species of *Homo* existed. Eventually, all *Homo* species except one—our own—ended in extinction.

? Based on the fossil evidence represented in Figure 19.11, how many hominin species coexisted 1.7 million years ago?

Five: *P. boisei*, *P. robustus*, *H. habilis*, *H. ergaster*, *H. erectus*

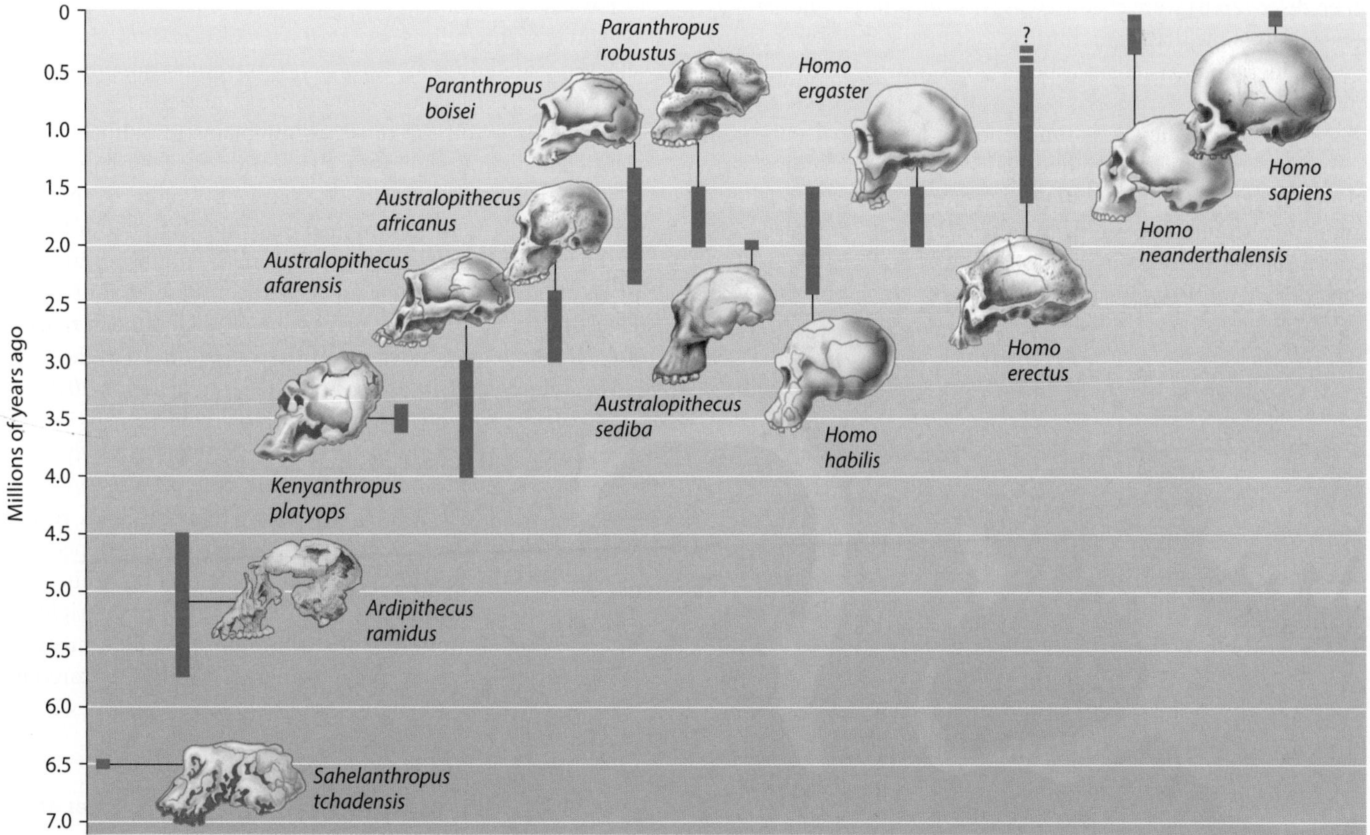

Millions of years ago
0
0.5
1.0
1.5
2.0
2.5
3.0
3.5
4.0
4.5
5.0
5.5
6.0
6.5
7.0
Paranthropus robustus
Paranthropus boisei
Homo ergaster
?
Australopithecus africanus
Australopithecus afarensis
Homo sapiens
Homo neanderthalensis
Homo erectus
Australopithecus sediba
Homo habilis
Kenyanthropus platyops
Ardipithecus ramidus
Sahelanthropus tchadensis

▲ **Figure 19.11** A timeline for some hominin species

19.12 Australopiths were bipedal and had small brains

Present-day humans and chimpanzees clearly differ in two major features: Humans are bipedal (walk upright) and have much larger brains. When did these features emerge? In the early 20th century, paleoanthropologists hypothesized that increased brain size was the initial change that separated hominins from apes. Bipedalism and other adaptations came later as hominin intelligence led to changes in food-gathering methods, parental care, and social interactions.

The evidence needed to test this hypothesis on brain size would come from hominin fossils. Hominin skulls would reveal brain size. Evidence of upright stance might be found by examining the limb and pelvic structures. Another clue to bipedalism is the location of the opening in the base of the skull through which the spinal cord exits. In chimpanzees and other species that are primarily quadrupeds, the spinal cord exits toward the rear of the skull, at an angle that allows the eyes to face forward (**Figure 19.12A**, left). In bipeds, including humans, the spinal cord emerges from the floor of the braincase, so the head can be held directly over the body (Figure 19.12A, right).

Convincing evidence to test the hypothesis was unearthed in 1973. A team of paleoanthropologists working in the Afar region of Ethiopia discovered a knee joint from a bipedal hominin—and it was more than 3 million years old. The following year, the same researchers found a significant portion of a 3.24-million-year-old female skeleton, which they nicknamed Lucy. Lucy and similar fossils (hundreds have since been discovered) were classified as *Australopithecus afarensis*. The fossils show that *A. afarensis* had a small brain, walked on two legs, and existed as a species for at least 1 million years.

Not long after Lucy was found, another team of paleoanthropologists discovered unique evidence of bipedalism in ancient hominins. While working in Tanzania in East Africa, they found a 3.6-million-year-old layer of hardened volcanic ash crisscrossed with tracks of hyenas, giraffes, and several extinct species of mammals—including upright-walking hominins (**Figure 19.12B**). After the ash had settled, rain had dampened it. The feet of two hominins, one large and one small, walking close together, made impressions in the ash as if it were wet sand on a beach. The ash, composed of a cement-like material, solidified soon after and was buried by more ash from a later volcanic eruption. The hominins strolling across that ancient landscape may have been *A. afarensis*, which lived in the region at the time, but we will never know for certain.

▲ **Figure 19.12B** Evidence of bipedalism in early hominins: footprints in ancient ash

Several other species of *Australopithecus* have also proven to be bipedal, including *A. africanus* and *A. anamensis*, as well as a related lineage known as "robust" australopiths—*Paranthropus boisei* and *Paranthropus robustus* (see Figure 19.11). In other features, though, australopiths were decidedly more like apes than humans. The brain size of *A. afarensis* relative to its body size was about the same as that of a chimpanzee. Our arms are shorter than our legs, but the proportions of *A. afarensis* were the opposite, and its fingers and toes were long and curved compared to ours—all suggesting that the species spent some of their time in trees.

The first analysis of the most recent addition to the genus, *Australopithecus sediba*, published in 2010, described yet another small-brained biped.

Paleoanthropologists are now certain that bipedalism is a very old trait. *Ardipithecus* walked on two legs more than 4 million years ago (see Module 19.11). There is evidence that even *Sahelanthropus*, the oldest hominin yet discovered, was capable of walking upright. It was only much later that the other major human trait—an enlarged brain—appeared in the human lineage.

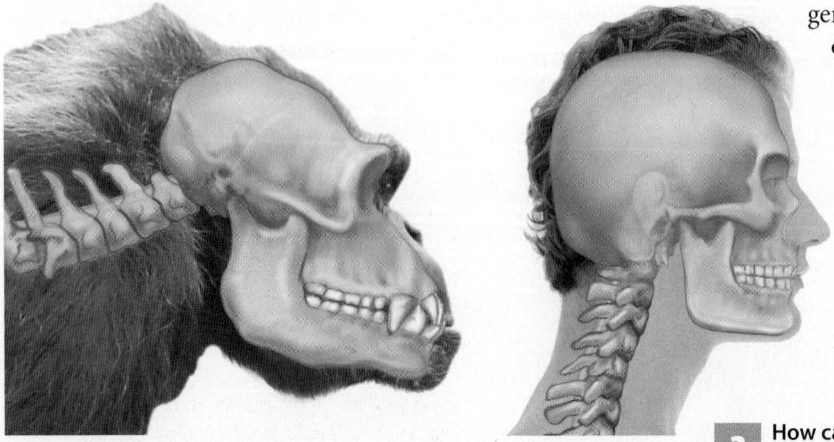

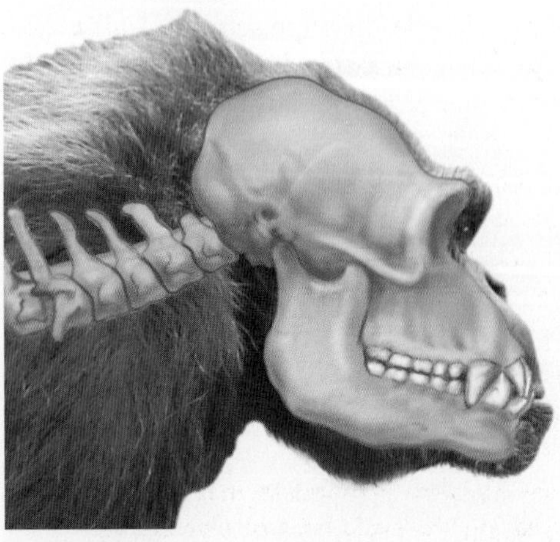

▲ **Figure 19.12A** The angle of spinal cord exit from skull in chimpanzee (left) and human (right)

? How can paleoanthropologists conclude that a species was bipedal based on only a fossil skull?

By the location of the opening where the spinal cord exits the skull

19.13 Larger brains mark the evolution of *Homo*

By measuring the capacity of fossil skulls, paleoanthropologists can estimate the size of the brain, which, relative to body size, roughly indicates the animal's intelligence. The brain volume of *Homo sapiens*, at an average 1,300 cm^3, is approximately triple that of australopiths (**Figure 19.13A**). As the late evolutionary biologist Stephan J. Gould put it, "Mankind stood up first and got smart later."

At 400–450 cm^3, the brains of australopiths were too small to qualify them as members of the genus *Homo*, but how big is big enough? What distinguishes humanlike from apelike brain capacity? When a team of paleoanthropologists found crude stone tools along with hominin fossils, they decided that the toolmaker must be one of us and dubbed their find *Homo habilis* ("handy man"). Its brain volume of 510–690 cm^3 was a significant jump from australopiths, but some scientists did not consider this large enough to be included in the genus *Homo*. Many *H. habilis* fossils ranging in age from about 1.6 to 2.4 million years have since been found, some appearing more humanlike than others.

Homo ergaster, dating from 1.9 to 1.5 million years ago, marks a new stage in hominin evolution. With a larger brain size, ranging from 750 to 850 cm^3, *H. ergaster* created more sophisticated stone tools. Its limb proportions were similar to those of modern humans. Its short, straight fingers indicate that it did not climb trees, and its long, slender legs and hip joints were well adapted for long-distance walking.

Fossils of *H. ergaster* were originally thought to come from early members of another species, *Homo erectus*. In *H. erectus* ("upright man"), average brain volume had increased to 940 cm^3; the range of sizes overlaps that of *H. ergaster*. Members of *H. erectus* were the first hominins to extend their range beyond Africa. The oldest known fossils of hominins outside Africa, discovered in 2000 in the former Soviet Republic of Georgia, are *H. erectus* dating back 1.8 million years. Others have been found in China and Indonesia. Most fossil evidence indicates that *H. erectus* became extinct at some point about 200,000 years ago, but recent discoveries raise the possibility that a population survived until 50,000 years ago.

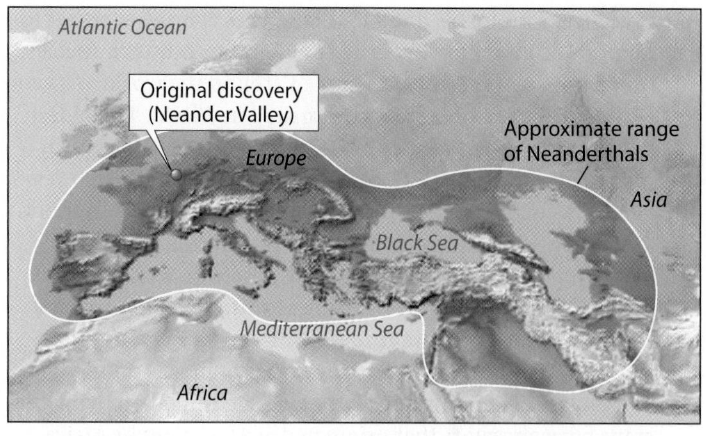

Atlantic Ocean
Original discovery (Neander Valley)
Europe
Approximate range of Neanderthals
Asia
Black Sea
Mediterranean Sea
Africa

▲ **Figure 19.13B** Range of Neanderthals inferred from fossil discoveries

Homo neanderthalensis, commonly called Neanderthals, are perhaps the best known hominins. They had a brain even larger than ours and hunted big game with tools made from stone and wood. Neanderthals were living in Europe as long as 350,000 years ago and later spread to the Near East (**Figure 19.13B**), but by 28,000 years ago, the species was extinct.

Since the discovery of fossilized remains of *H. neanderthalensis* in the Neander Valley in Germany 150 years ago, people have wondered how Neanderthals are related to us. Were they the ancestors of Europeans? Close cousins? Or part of a different branch of evolution altogether? By comparing mitochondrial DNA sequences from Neanderthals and living humans, researchers showed that Neanderthals are not the ancestors of Europeans. Rather, the last common ancestor of humans and Neanderthals lived around 400,000 years ago. However, a comparison of the nuclear genome sequence of *Homo sapiens* with that from Neanderthal fossils, completed in 2010, suggests that Neanderthals and some *H. sapiens* that had left Africa probably did interbreed. This genetic exchange left many of us with genomes that are roughly 3% Neanderthal. Analysis of Neanderthal DNA has also revealed a few details about these intriguing hominins. For example, at least some Neanderthals had red hair, pale skin, and type O blood. They could also taste bitter substances, which is often a cue that plants contain toxins.

In the next module, we look at the origin and worldwide spread of our own species, *Homo sapiens*.

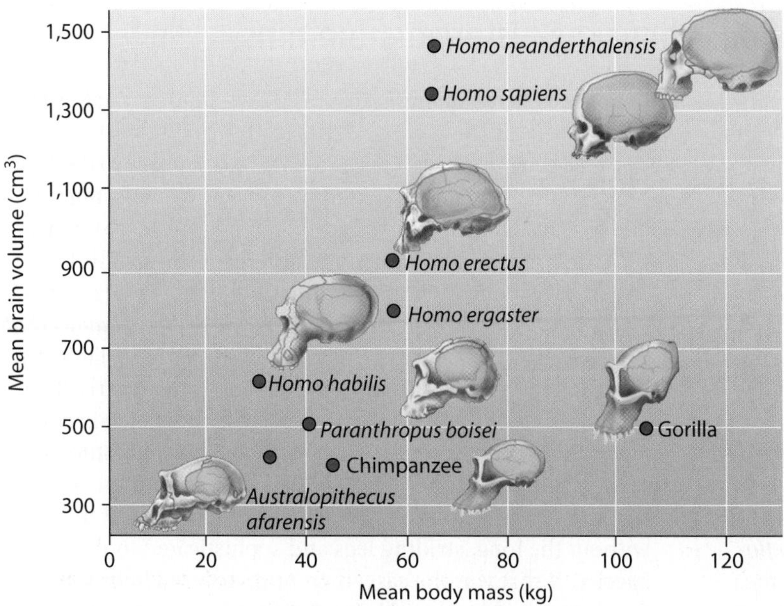

Mean brain volume (cm3)
1,500
1,300
1,100
900
700
500
300
Homo neanderthalensis
Homo sapiens
Homo erectus
Homo ergaster
Homo habilis
Paranthropus boisei
Chimpanzee
Gorilla
Australopithecus afarensis
0
20
40
60
80
100
120
Mean body mass (kg)

▲ **Figure 19.13A** Brain volume versus body mass in anthropoids

? Place the following hominins in order of increasing brain volume: *Australopithecus*, *H. erectus*, *H. ergaster*, *H. habilis*, *H. sapiens*.

Australopithecus, *H. habilis*, *H. ergaster*, *H. erectus*, *H. sapiens*

19.14 From origins in Africa, *Homo sapiens* spread around the world

Evidence from fossils and DNA studies is coming together to support a compelling hypothesis about how our own species, *Homo sapiens*, emerged and spread around the world.

The ancestors of humans originated in Africa. The oldest known fossils with the definitive characteristics of our own species, discovered in Ethiopia, are 160,000 and 195,000 years old. Molecular evidence about the origin of humans supports the conclusions drawn from fossils. In addition to showing that living humans are more closely related to one another than to Neanderthals, DNA studies indicate that Europeans and Asians share a more recent common ancestor and that many African lineages represent earlier branches on the human tree. These findings strongly suggest that all living humans have ancestors that originated as *H. sapiens* in Africa.

This conclusion is further supported by analyses of mitochondrial DNA, which is maternally inherited, and Y chromosomes, which are transmitted from fathers to sons. Such studies suggest that all living humans inherited their mitochondrial DNA from a common ancestral woman who lived approximately 160,000–200,000 years ago. Mutations on the Y chromosomes can serve as markers for tracing the ancestry and relationships among males alive today. By comparing the Y chromosomes of males from various geographic regions, researchers were able to infer divergence from a common African ancestor.

These lines of evidence suggest that our species emerged from Africa in one or more waves, spreading first to the Middle East, then dispersing into other regions of Asia and to Europe, Australia, and finally to the New World.

What spurred the rapid geographic expansion of *H. sapiens*? Increasing numbers probably caused populations to gradually expand their range. Travel and successful colonization of new regions was likely facilitated by the evolution of human cognition as our species evolved in Africa. Although Neanderthals and other hominins were able to produce sophisticated tools, they showed little creativity and not much capability for symbolic thought, as far as we can tell. In contrast, researchers have found evidence of increasingly sophisticated thought as *H. sapiens* evolved **(Figure 19.14)**. As *H. sapiens* spread around the globe, populations adapted to the new environments they encountered. Consequently, some differences among people are attributable to their deep ancestry. We'll look at one such adaptation in Module 19.16.

? **What types of evidence indicate that *Homo sapiens* originated in Africa?**

Fossils and analyses of mitochondrial DNA and chromosomal DNA

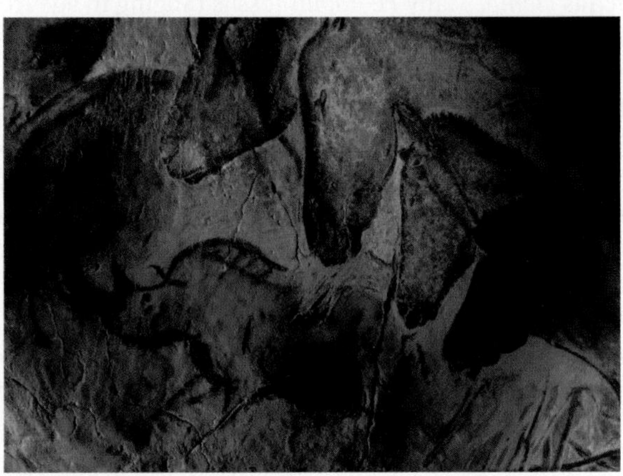

▲ **Figure 19.14** Artwork made by *H. sapiens* about 30,000 years ago, discovered in Chauvet Cave, France

19.15 New discoveries raise new questions about the history of hominins

SCIENTIFIC THINKING

As you have seen in the preceding modules, paleoanthropologists use evidence from fossils and molecular genetics to test hypotheses about human origins. Let's take a closer look at how paleoanthropologists are using fossils to investigate where the "hobbits" described in the chapter introduction should be placed in our evolutionary history.

Who were the hobbit people?

The researchers who discovered the first "hobbit" skeleton, on Flores, one of the Indonesian islands, hypothesized that it was a previously unknown human species that evolved from a population of *Homo erectus* into a dwarf form on the island. There is precedent for animal evolution of this type: Biologists have discovered island-bound dwarf populations of deer, elephants, and hippos. The researchers named the new hominin *Homo floresiensis* and tested their hypothesis by making detailed measurements on the *H. floresiensis* fossils and comparing them with data from *H. erectus* fossils. Initial results supported the hypothesis that *H. floresiensis* was a dwarf descendent of *H. erectus*.

Since the initial discovery, researchers have unearthed the bones of an estimated 14 *H. floresiensis* individuals, but no additional skulls have been located. Numerous paleoanthropologists have analyzed the fossils, and some now think the evidence supports a different hypothesis: *Homo floresiensis* is more closely related to *Homo habilis* than to *Homo erectus*. But if further research establishes that *H. floresiensis* is more similar to an early member of the genus than it is to the more advanced *H. erectus*, then new questions are raised. The ancestors of *H. floresiensis* must have left Africa even earlier than *H. erectus* and managed to extend their range thousands of miles without the long, striding legs and sophisticated tools of that species. If that was the case, then undiscovered hominin fossils much older than *H. erectus* must exist somewhere between Africa and Indonesia.

▲ **Figure 19.15** Liang Bua Cave on the Indonesian island of Flores

Meanwhile, other researchers continue to test an alternative hypothesis that hobbits are not a species at all: Rather, they are *Homo sapiens* with a genetic disorder that causes bone malformations.

How can scientists determine which hypothesis is correct? By accumulating further evidence. While some researchers continue to excavate the site where *Homo floresiensis* was discovered **(Figure 19.15)**, others are widening the search to additional locations. The most helpful information would come from finding a second skull or unearthing bones or teeth from which DNA could be extracted and analyzed. Meanwhile, the mystery of the hobbit people continues.

? What characteristics prompted the discoverers of the "hobbit" to classify it in the genus *Homo*?

● Humanlike skull characteristics and the apparent use of stone tools

19.16 Human skin color reflects adaptations to varying amounts of sunlight

EVOLUTION CONNECTION

In today's diverse society, skin color is one of the most striking differences among individuals **(Figure 19.16)**. For centuries, people assumed that these differences reflected more fundamental genetic distinctions, but modern genetic analysis has soundly disproved those assumptions. Is there an evolutionary explanation for skin color differences? To develop hypotheses, scientists began with the observation that human skin color varies geographically. People indigenous to tropical regions have darker skin pigmentation than people from more northerly latitudes.

Skin color results from a pigment called melanin that is produced by specialized skin cells. We all have melanin-producing cells, but the cells are less active in people who have light-colored skin. In addition to absorbing visible light, and therefore appearing dark-colored, melanin absorbs ultraviolet wavelengths. We know that ultraviolet (UV) radiation causes mutations (see Module 11.18), but it has other effects in skin as well.

UV radiation helps catalyze the synthesis of vitamin D in the skin. This vitamin is essential for proper bone development, so it is especially important for pregnant women and small children to receive adequate amounts. By blocking UV radiation, melanin prevents vitamin D synthesis. Dark-skinned humans evolving in equatorial Africa received sufficient UV radiation to make vitamin D, but northern latitudes receive less sunlight. The loss of skin pigmentation (melanin) in humans that migrated north from Africa probably helped their skin receive adequate UV radiation to produce enough vitamin D.

TABLE 19.16 CORRELATION OF UV RADIATION WITH RISK OF VITAMIN D AND FOLATE DEFICIENCIES

Latitude	UV Radiation	Risk of Vitamin D Deficiency	Risk of Folate Deficiency
Tropical latitudes 0–23.5°	High	Low	High
Higher latitudes 23.5–90°	Low	High	Low

Why did dark skin evolve in humans in the first place? UV radiation degrades folate (folic acid), a vitamin that is vital for fetal development and spermatogenesis. Researchers hypothesize that dark skin was selected for because melanin protects folate from the intense tropical sunlight. The evolution of differing skin pigmentations likely provided a balance between folate protection and vitamin D production **(Table 19.16)**.

Because skin color was the product of natural selection, similar environments produced similar degrees of pigmentation. Widely separated populations may have the same adaptation, regardless of how they are related. As a result, skin color is not a useful characteristic for judging phylogenetic relationships.

? Why didn't folate degradation select against lightly pigmented people in northern latitudes?

● UV radiation is less intense in northern latitudes than in the tropics, so it did not have an adverse effect on folate levels.

▲ **Figure 19.16** Variety of skin colors

19.17 Our knowledge of animal diversity is far from complete

CONNECTION

When European naturalists first began exploring Africa, Asia, Australia, and North and South America, they discovered many thousands of species they had not previously seen. You might think that after centuries of scientific exploration, only tiny organisms such as microbes and insects remain to be found. But the days of exploring new ecosystems and discovering new species are not over. In fact, better access to remote areas, coupled with new mapping technologies, has renewed the pace of discovery. According to a report issued in 2012, 19,232 species were described for the first time in 2009. Approximately half of them were insects, but the list also includes 630 vertebrates.

The Mekong region of Southeast Asia, an area of diverse landscapes surrounding the Mekong River as it flows from southern China to the China Sea, is one of many treasure troves of previously unknown species that are currently being explored. Over the past decade, more than 1,000 new species have been identified in the region, including the leopard gecko in **Figure 19.17A**, one of more than 400 new species of vertebrates that scientists have turned up there. To the southeast, remote mountains on the island of New Guinea are also yielding hundreds of discoveries, including a frog with a droopy nose **(Figure 19.17B)**. Discoveries of new primate species are extremely rare, but several have also been made recently, including two new lemur species in Madagascar and several new monkeys in the eastern Himalayas, Tanzania, Myanmar, Brazil, Bolivia, and Peru. The monkey shown in **Figure 19.17C** was found deep in the jungle of Congo by a hunter. He brought it to his village, where it was recently noticed by a scientist doing fieldwork in the region.

Previously undescribed species are being reported almost daily from every continent and a wide variety of habitats. And researchers are just beginning to explore the spectacular diversity of the oceans. The Census of Marine Life, a decade-long collaboration among scientists from 80 nations, has reported the discovery of more than 6,000 new species. Thousands more are expected to be found as new technology enables scientists to investigate deep-sea habitats. Recent expeditions have also gleaned hundreds of new species from the seas surrounding Antarctica, and the collapse of Antarctic ice shelves has allowed researchers their first glimpse of life on a seafloor that had previously been hidden from view. Even places that are regularly visited by people offer surprises. For example, over 100 new marine species were identified recently on a coral reef near Australia.

When a new species is described, taxonomists learn as much as possible about its physical and genetic characteristics and assign it to the appropriate groups in the Linnaean system. As a result, most new species automatically acquire a series of names from domain through genus. But every species also has a unique identifier, and the honor of choosing it belongs to the discoverer. Species are often named for their habitat or a notable feature.

In a new twist, naming rights for recently discovered species have been auctioned off to raise money for conservation organizations, which undertake many of the projects that survey biological diversity. The right to name a new species of monkey cost the winning bidder $650,000, and donors spent more than $2 million for the honor of naming 10 new species of fish. Naming rights are available for smaller budgets, too—the top bid to name a new species of shrimp was $2,900. The proceeds from these auctions go toward funding new expeditions and preserving the habitats of the newly discovered species. In many cases, such discoveries are made as roads and settlements reach farther into new territory. Consequently, many species are endangered soon after they are discovered. (We'll consider the various threats to biological diversity in Chapter 38.)

? What factors are responsible for the recent increase in the number of new species found?

Technology; encroachment of human activities into wilderness areas

▶ **Figure 19.17A** Leopard gecko, a newly discovered lizard from northern Vietnam

▲ **Figure 19.17B** Pinocchio frog, recently discovered in New Guinea

▲ **Figure 19.17C** Lesula, a new monkey species discovered in the central Democratic Republic of Congo

CHAPTER 19 REVIEW

For study tools, including practice quizzes, BioFlix animations, MP3 tutorials, video tutors, and more study tools designed for this textbook, go to MasteringBiology®

Reviewing the Concepts

Vertebrate Evolution and Diversity (19.1–19.8)

19.1 Derived characteristics define the major clades of chordates.

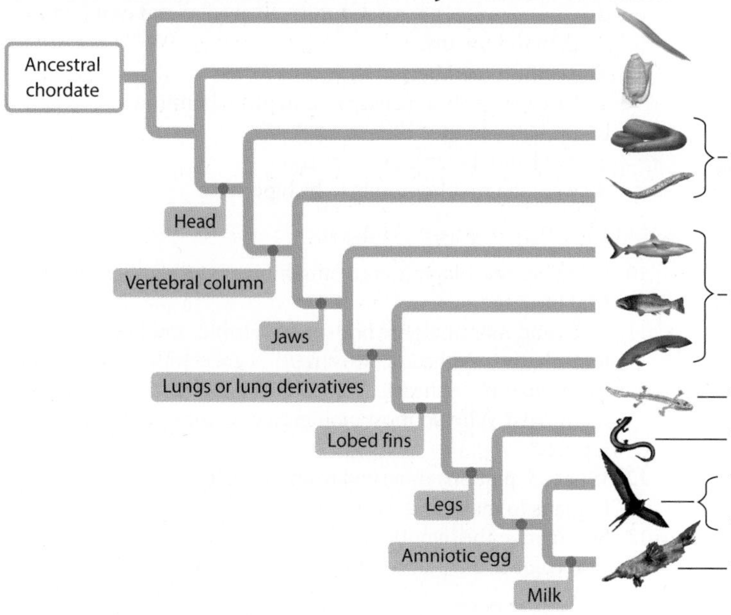

Ancestral chordate
Head
Vertebral column
Jaws
Lungs or lung derivatives
Lobed fins
Legs
Amniotic egg
Milk

19.2 Hagfishes and lampreys lack hinged jaws.

19.3 Jawed vertebrates with gills and paired fins include sharks, ray-finned fishes, and lobe-finned fishes.

19.4 New fossil discoveries are filling in the gaps of tetrapod evolution.

19.5 Amphibians are tetrapods—vertebrates with two pairs of limbs.

19.6 Reptiles are amniotes—tetrapods with a terrestrially adapted egg.

19.7 Birds are feathered reptiles with adaptations for flight.

19.8 Mammals are amniotes that have hair and produce milk.

Primate Diversity (19.9–19.10)

19.9 Primates include lemurs, monkeys, and apes. Primates had evolved as small arboreal mammals by 65 million years ago. Primate characters include limber joints, grasping hands and feet with flexible digits, a short snout, and forward-pointing eyes that enhance depth perception.

19.10 The human story begins with our primate heritage. The three groups of living primates are the lorises, pottos, and lemurs; the tarsiers; and the anthropoids (monkeys and apes). Apes, which have larger brains than other primates and lack tails, include gibbons, orangutans, gorillas, chimpanzees, and humans.

Hominin Evolution (19.11–19.17)

19.11 The hominin branch of the primate tree includes species that coexisted. Paleoanthropologists have found about 20 species of extinct hominins, species more closely related to humans than to chimpanzees. Some of these species lived at the same time.

19.12 Australopiths were bipedal and had small brains.

19.13 Larger brains mark the evolution of *Homo*. The genus *Homo* includes hominins with larger brains and evidence of tool use. *Homo ergaster* had a larger brain than *H. habilis*. *H. erectus*, with a larger brain than *H. ergaster*, was the first hominin to spread out of Africa.

19.14 From origins in Africa, *Homo sapiens* spread around the world. Evidence from fossils and DNA studies has enabled scientists to trace early human history.

19.15 New discoveries raise new questions about the history of hominins. The interpretation of fossils of small hominins named *Homo floresiensis* that were found in Indonesia is controversial.

19.16 Human skin color reflects adaptations to varying amounts of sunlight. Human skin color variations probably resulted from natural selection balancing the body's need for folate with the need to synthesize vitamin D.

19.17 Our knowledge of animal diversity is far from complete. Thousands of new species are discovered each year.

Connecting the Concepts

1. In the primate phylogenetic tree below, fill in groups (a)–(e). Of the groups, which are anthropoids and which are apes?

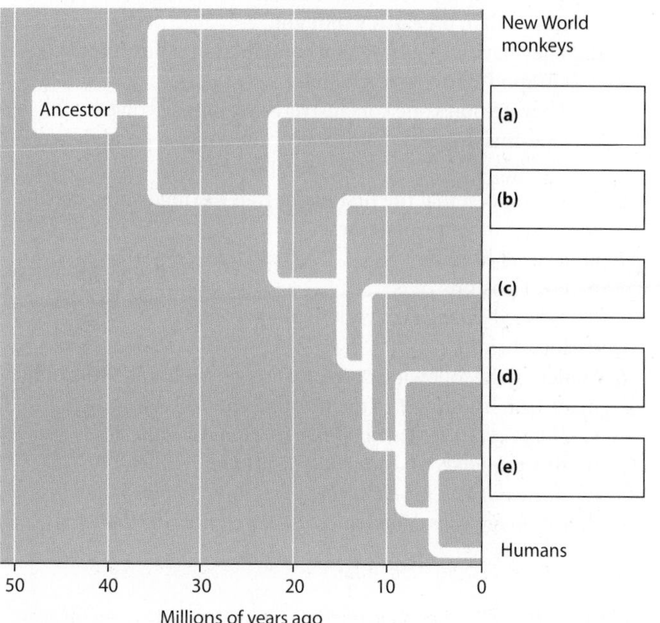

Ancestor
New World monkeys
(a)
(b)
(c)
(d)
(e)
Humans
50
40
30
20
10
0
Millions of years ago

2. In the chordate phylogenetic tree below, fill in the key derived character that defines each clade.

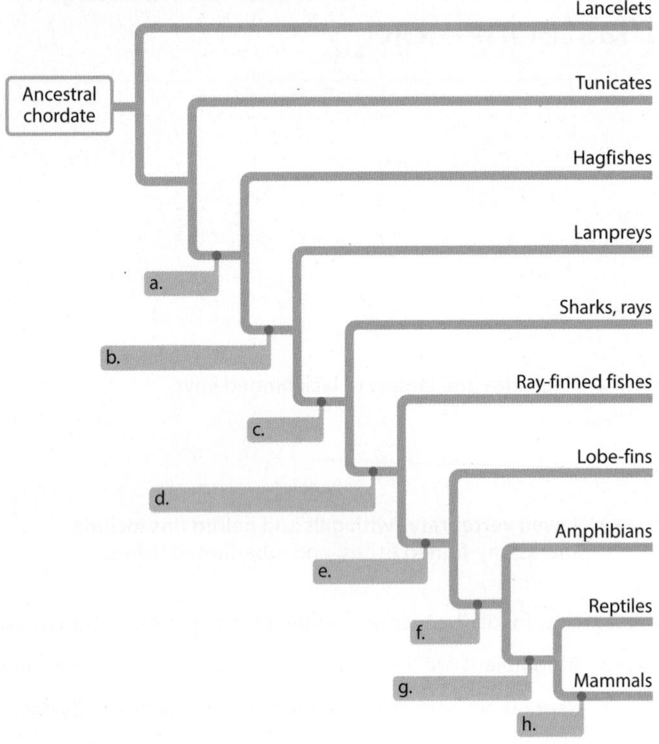

Ancestral chordate
Lancelets
Tunicates
Hagfishes
Lampreys
Sharks, rays
Ray-finned fishes
Lobe-fins
Amphibians
Reptiles
Mammals
a.
b.
c.
d.
e.
f.
g.
h.

Testing Your Knowledge

Level 1: Knowledge/Comprehension

3. A lamprey, a shark, a lizard, and a rabbit share all the following characteristics except
   a. pharyngeal slits in the embryo or adult.
   b. vertebrae.
   c. hinged jaws.
   d. a dorsal, hollow nerve cord.
4. Why were the *Tiktaalik* fossils an exciting discovery for scientists studying tetrapod evolution?
   a. They are the earliest frog-like animal discovered to date.
   b. They show that tetrapods successfully colonized land much earlier than previously thought.
   c. They have a roughly equal combination of fishlike and tetrapod-like characteristics.
   d. They demonstrate conclusively that limbs evolved as lobe-fins dragged themselves from pond to pond during droughts.
5. Fossils suggest that the first major trait distinguishing hominins from other primates was
   a. a larger brain.
   b. erect posture.
   c. forward-facing eyes with depth perception.
   d. tool making.
6. Which of the following correctly lists possible ancestors of humans from the oldest to the most recent?
   a. *Homo erectus, Australopithecus, Homo habilis*
   b. *Australopithecus, Homo habilis, Homo erectus*
   c. *Australopithecus, Homo erectus, Homo habilis*
   d. *Homo ergaster, Homo erectus, Homo neanderthalensis*
7. Which of these is not a member of the anthropoids?
   a. chimpanzee
   b. tarsier
   C. human
   d. New World monkey
8. Studies of DNA support which of the following?
   a. Members of the group called australopiths were the first to migrate from Africa.
   b. *Homo sapiens* originated in Africa.
   c. *Sahelanthropus* was the earliest hominin.
   d. Chimpanzees are more similar to gorillas and orangutans than to humans.
9. The earliest members of the genus *Homo*
   a. had a larger brain compared to other hominins.
   b. probably hunted dinosaurs.
   c. lived about 4 million years ago.
   d. were the first hominins to be bipedal.

Level 2: Application/Analysis

10. Compare the adaptations of amphibians and reptiles for terrestrial life.
11. Birds and mammals are both endothermic, and both have four-chambered hearts. Most reptiles are ectothermic and have three-chambered hearts. Why don't biologists group birds with mammals? Why do most biologists now consider birds to be reptiles?
12. What adaptations inherited from our primate ancestors enable humans to make and use tools?
13. Summarize the hypotheses that explain variation in human skin color as adaptations to variation in UV radiation.

Level 3: Synthesis/Evaluation

14. A good scientific hypothesis is based on existing evidence and leads to testable predictions. What hypothesis did the paleontologists who discovered *Tiktaalik* test? What evidence did they use to predict where they would find fossils of transitional forms?
15. Explain some of the reasons why the human species has been able to expand in number and distribution to a greater extent than most other animals.
16. Anthropologists are interested in locating areas in Africa where fossils 4–8 million years old might be found. Why?
17. **SCIENTIFIC THINKING** By measuring the fossil remains of *Homo floresiensis*, scientists have estimated its weight to be around 32.5 kg and its brain volume to be roughly 420 cm^3. Plot these values on the graph in Figure 19.13A. Which hominin has the most similar relationship of brain volume to body mass? Does this information support the hypothesis that *H. floresiensis* is a dwarf form of *H. erectus*, or an alternative hypothesis? Explain.

Answers to all questions can be found in Appendix 4.

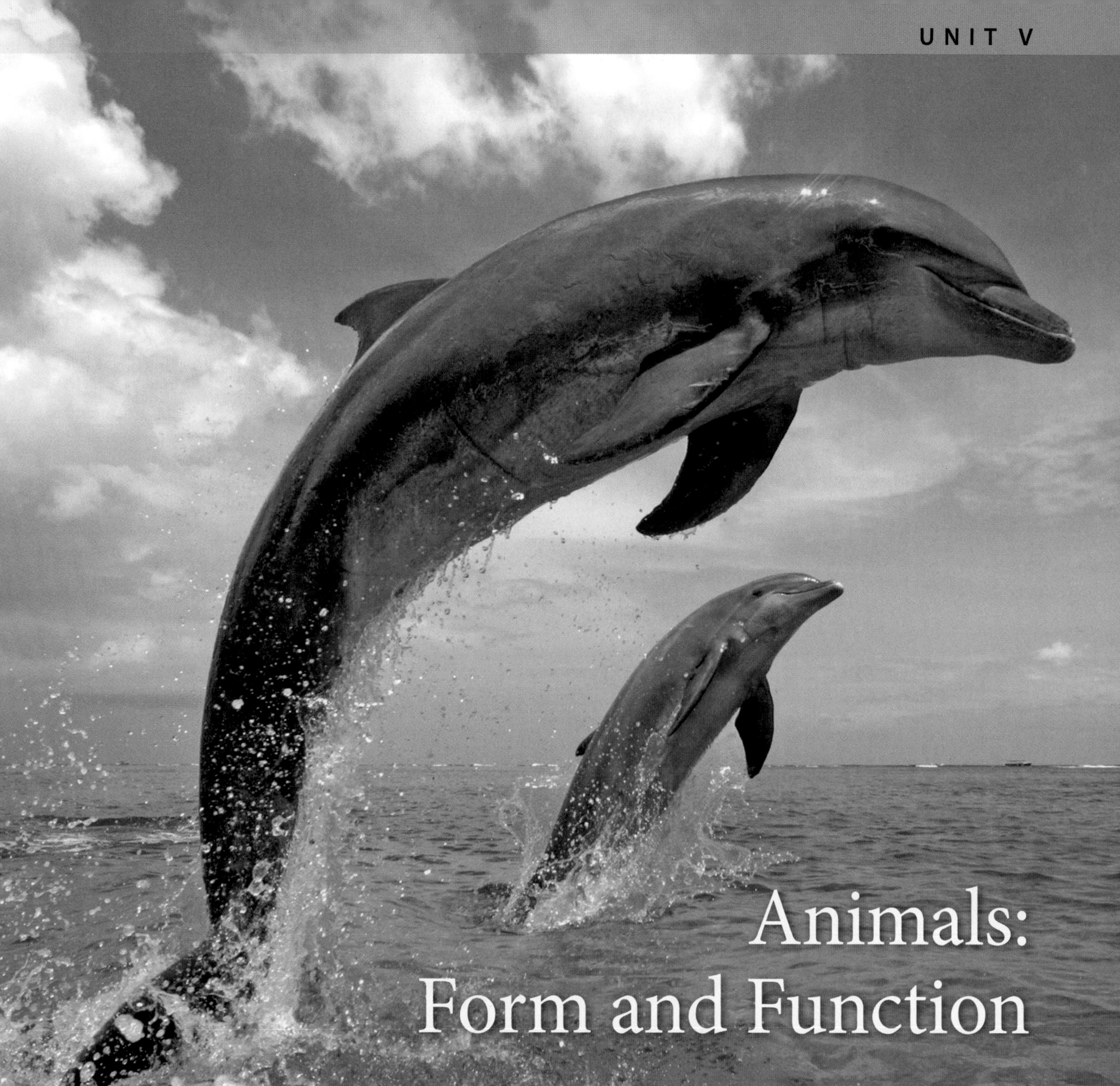

Animals: Form and Function

CHAPTER

20 Unifying Concepts of Animal Structure and Function

? ***Does evolution lead to the perfect animal form?***

As you delve further into biology and encounter a wide range of organisms, you might wonder why some animals look the way they do. Why does an antelope have horns? Why aren't humans hairy like other mammals? Why does a giraffe have a 6-foot neck? The physical structures that may be wondrous to us are actually adaptations that enhance an animal's chances of survival and reproduction. Antelope horns become useful weapons in physical struggles. Overheating is prevented in humans because of rapid cooling through their hairless body surface. The long necks of giraffes function to beat out competition for a limited amount of food; giraffes can reach leaves high on a tree that are not easily accessible to competitors. The correlation of structure and function is an overarching theme of biology.

This unit explores structure and function in the context of the various challenges animals face: how do animals obtain nutrients and oxygen, fight infection, excrete wastes, reproduce, and sense

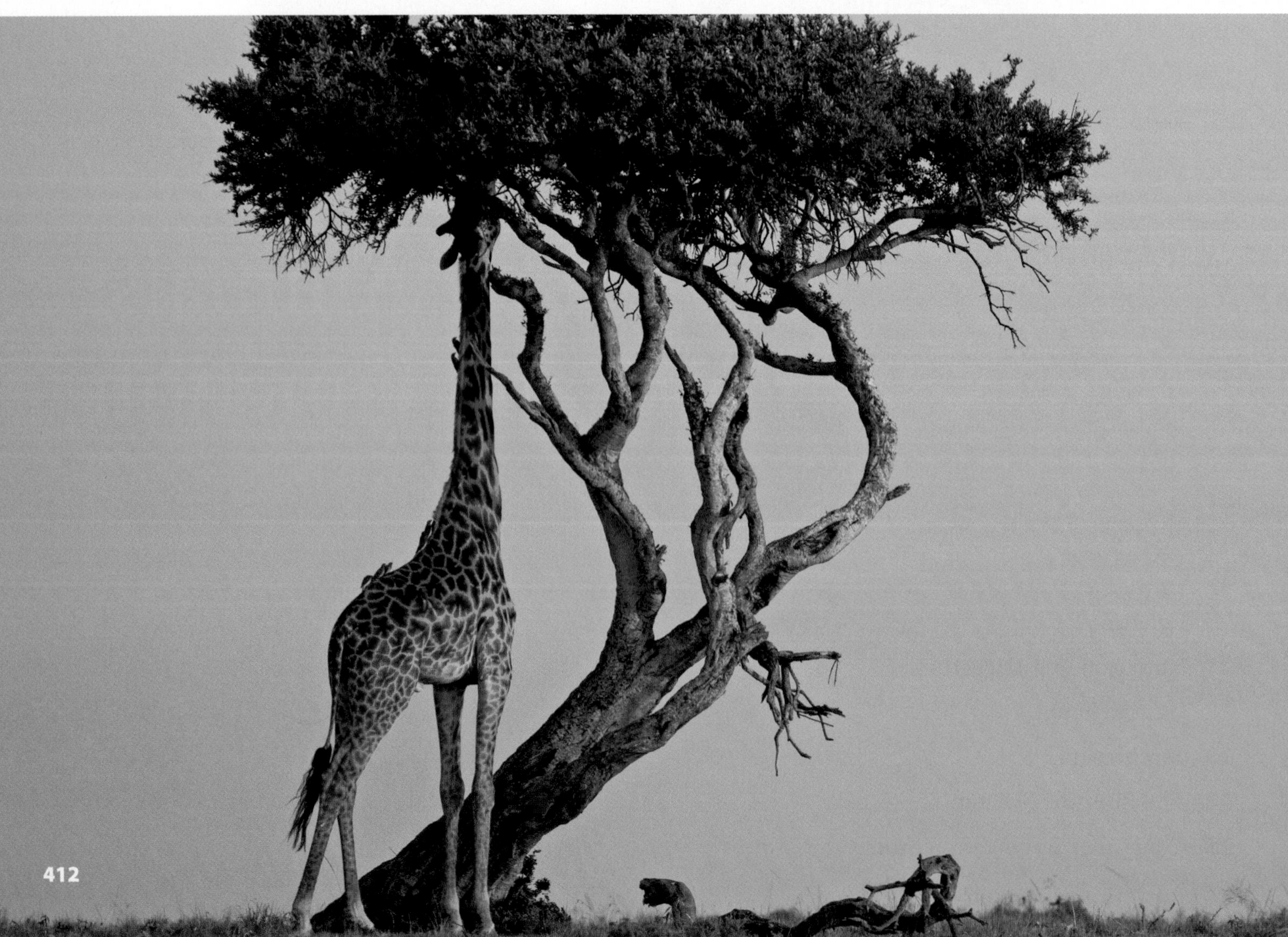

and respond to the environment. The adaptations that represent the various solutions to these problems have been fashioned by natural selection, fitting structure to function over the course of evolution.

You might be tempted to think that evolution leads to perfect animal structures, but numerous examples reveal that animal structures are often just "good enough" to function and not the ultimate in design. In the next module, we'll look closely at a particular structure, the laryngeal nerve in the giraffe, which controls the muscles of the giraffe's larynx (voice box), trachea (windpipe), and esophagus. The giraffe's laryngeal nerve extends about 15 feet (4.57m) down and back up the animal's neck, even though it seems a shorter route would be more efficient.

BIG IDEAS

Structure and Function in Animal Tissues
(20.1–20.7)

The structural hierarchy in an animal begins with cells and tissues, whose forms correlate with their functions.

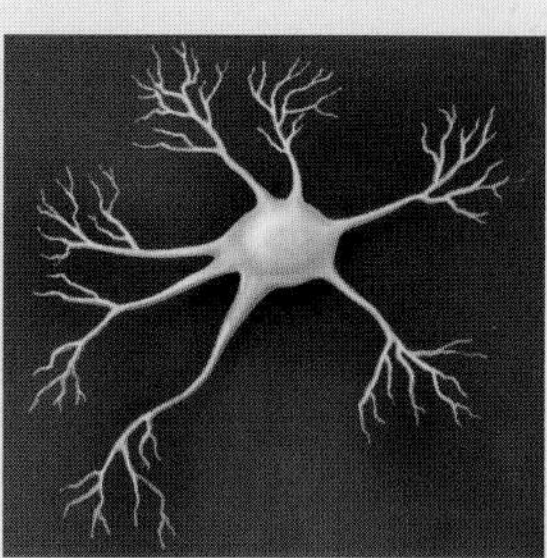

Organs and Organ Systems
(20.8–20.12)

Tissues are arranged into organs, which can be functionally coordinated in organ systems.

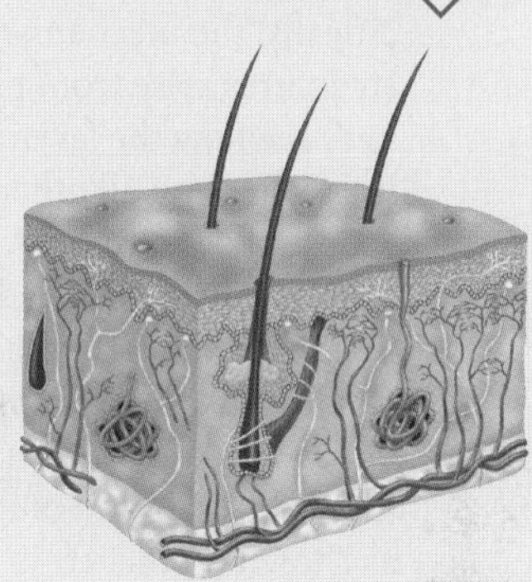

External Exchange and Internal Regulation
(20.13–20.15)

Complex animals have internal surfaces that facilitate exchange with the environment. Feedback control maintains homeostasis.

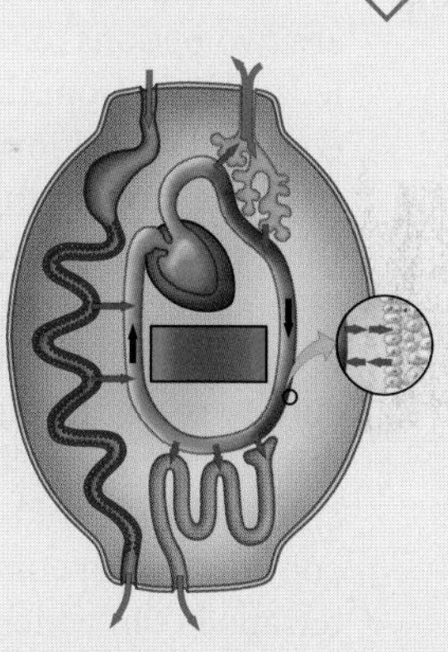

Structure and Function in Animal Tissues

20.1 An animal's form is not the perfect design

EVOLUTION CONNECTION

The giraffe's incredibly long laryngeal nerve travels from the brain, makes a U-turn around a major blood vessel in the chest called the aorta, and connects to and stimulates the muscles of the throat that affect the giraffe's vocal sounds, breathing, and swallowing (**Figure 20.1A**). The throat is a mere foot away from the brain in giraffes. Why, then, does its laryngeal nerve make this 15-foot-long journey? The answer to this question concerns evolution, specifically, the mechanism in which existing structures arise from previous structures. Let's see how the laryngeal nerve evolved in giraffes and all other vertebrates with four limbs (tetrapods), including humans.

▲ **Figure 20.1A** The lengthy laryngeal nerve in a giraffe's neck

Adaptations that led to the varying lengths of the laryngeal nerve in tetrapods can be illustrated with an analogy. Envision a slack cord of a lamp wrapped around a table and plugged into an outlet as shown on the left in **Figure 20.1B**. If the table is moved far from the outlet, there are two options for connection to the outlet and light from the lamp. Unplug the cord and reposition it so it runs directly from the lamp to the outlet, although this positioning causes a temporary loss of lamplight. Alternatively, keep the cord plugged in while simply extending the cord to reach the outlet. In this second option, the lamplight remains on during the move of the table and extension of the cord, but the cord remains wrapped around the table. Now consider embryonic development of the giraffe (and other tetrapods), in which the laryngeal nerve is wrapped around the aorta (like the cord wrapped around the table). As the neck lengthens during development and the distance between the aorta and the throat increases (like the table being pulled away from the wall), the laryngeal nerve lengthens rather than breaking and reconnecting. Thus we see that the length of the laryngeal nerve in tetrapods must be directly related to the length of the neck.

The surprising length of the laryngeal nerve illustrates a major concept in evolution: Through natural selection, a structure in an ancestral organism can be adapted to function in a descendant organism. The early embryos of fish and tetrapods are highly similar, providing evidence of their shared ancestry. In vertebrate embryonic development, the laryngeal nerve connects the brain to a rudimentary structure that in fish will become the gills and in tetrapods will develop into the larynx. In these early embryos, the nerve hooks under the aorta. This is not problematic in fish because they do not have necks; the brain, aorta, and gills are relatively close together in the adult fish. But in tetrapods, the aorta ultimately ends up in the chest, quite a distance from the brain. The resulting adaptation in tetrapods is an elongated laryngeal nerve.

Does evolution lead to the perfect animal form?

Indeed, the laryngeal nerve is about 3 feet long in humans and five times longer in giraffes. Scientists estimate that this nerve could have been more than 120 feet long in some long-necked dinosaurs! The laryngeal nerve, like all other animal structures, isn't perfect. It's just good enough to function.

? **Explain why the long, looped nerve was an adaptation that arose through the process of natural selection instead of a short nerve following a more direct route.**

● A more direct (shorter) connection from the brain to the throat would require severing and rejoining this nerve, which would be incompatible with survival.

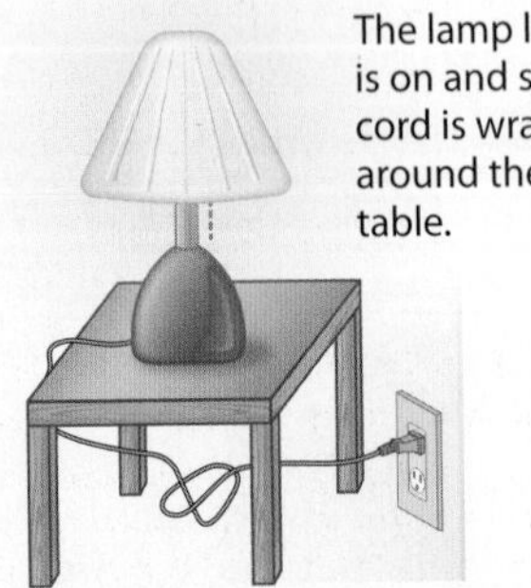
A table close to the wall outlet
The lamp light is on and slack cord is wrapped around the table.

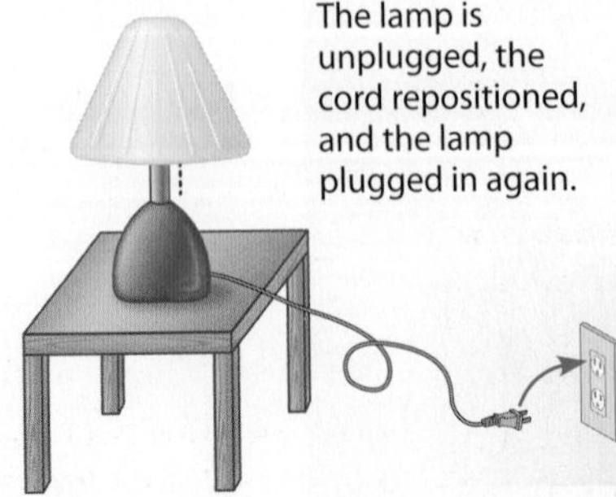
Moving the table far from the wall outlet: Option 1
The lamp is unplugged, the cord repositioned, and the lamp plugged in again.

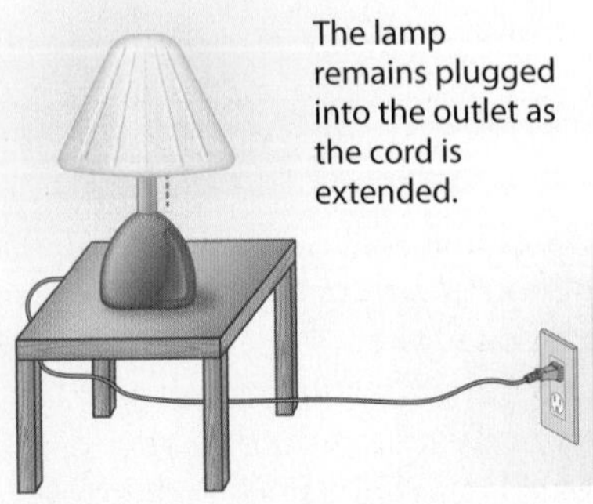
Moving the table far from the wall outlet: Option 2
The lamp remains plugged into the outlet as the cord is extended.

▲ **Figure 20.1B** An analogy for evolutionary adaptation of the laryngeal nerve in the giraffe

20.2 Structure fits function at all levels of organization in the animal body

When discussing structure and function, biologists distinguish anatomy from physiology. **Anatomy** is the study of the form of an organism's structures; **physiology** is the study of the functions of those structures. A biologist interested in anatomy, for instance, might focus on the arrangement of muscles and bones in a giraffe's neck. A physiologist might study how a giraffe's muscles function. Despite their different approaches, both scientists are working toward a better understanding of the connection between structure and function.

The living world is organized in hierarchical levels. We followed the progression from molecules to cells in Unit I. Now, let's trace the hierarchy in animals from cells to organisms. (In Unit VI, we follow the same hierarchy in plants. And in Unit VII, we pick up the trail again, moving from organisms to ecosystems.) As we discussed in Module 1.2, emergent properties—novel properties that were not present at the preceding level of the hierarchy of life—arise as a result of the structural and functional organization of each level's component parts.

Figure 20.2 illustrates structural hierarchy in animals. **Part A** shows the first structural level, a single cell. The main function of the muscle cells of the giraffe's heart is to contract, and strands of proteins that perform that function are precisely aligned. Each cell in the muscle is branched and connected to other cells in a way that ensures coordinated contractions of all the muscle cells in the heart.

Together, cells make up a tissue (**Part B**), the second structural level. A **tissue** is an integrated group of similar cells that perform a common function. The cells of a tissue are specialized, and their structure enables them to perform a specific task—in this instance, coordinated contraction.

Part C illustrates the organ level of the hierarchy. An **organ** is made up of two or more types of tissues that together perform a specific task. In addition to muscle tissue, the heart includes nervous, epithelial, and connective tissue. Once again, we see an emergent property—the heart's ability to pump blood—resulting from the coordinated functioning of components of the previous level—the individual tissues.

Part D shows the organ system. An **organ system** consists of multiple organs that together perform a vital body function. The heart is a part of the circulatory system along with blood and the blood vessels: arteries, veins, and capillaries.

In **Part E**, the organism forms the final level of this hierarchy. An organism contains a number of organ systems, each specialized for certain tasks and all functioning together as an integrated, coordinated unit. The circulatory system of the giraffe cannot function without oxygen supplied by the respiratory system and nutrients supplied by the digestive system. In fact, it takes the coordination of several other organ systems, each relying on the emergent properties of cells, tissues, and organs, to enable this animal to reach for and eat leaves high on a tree.

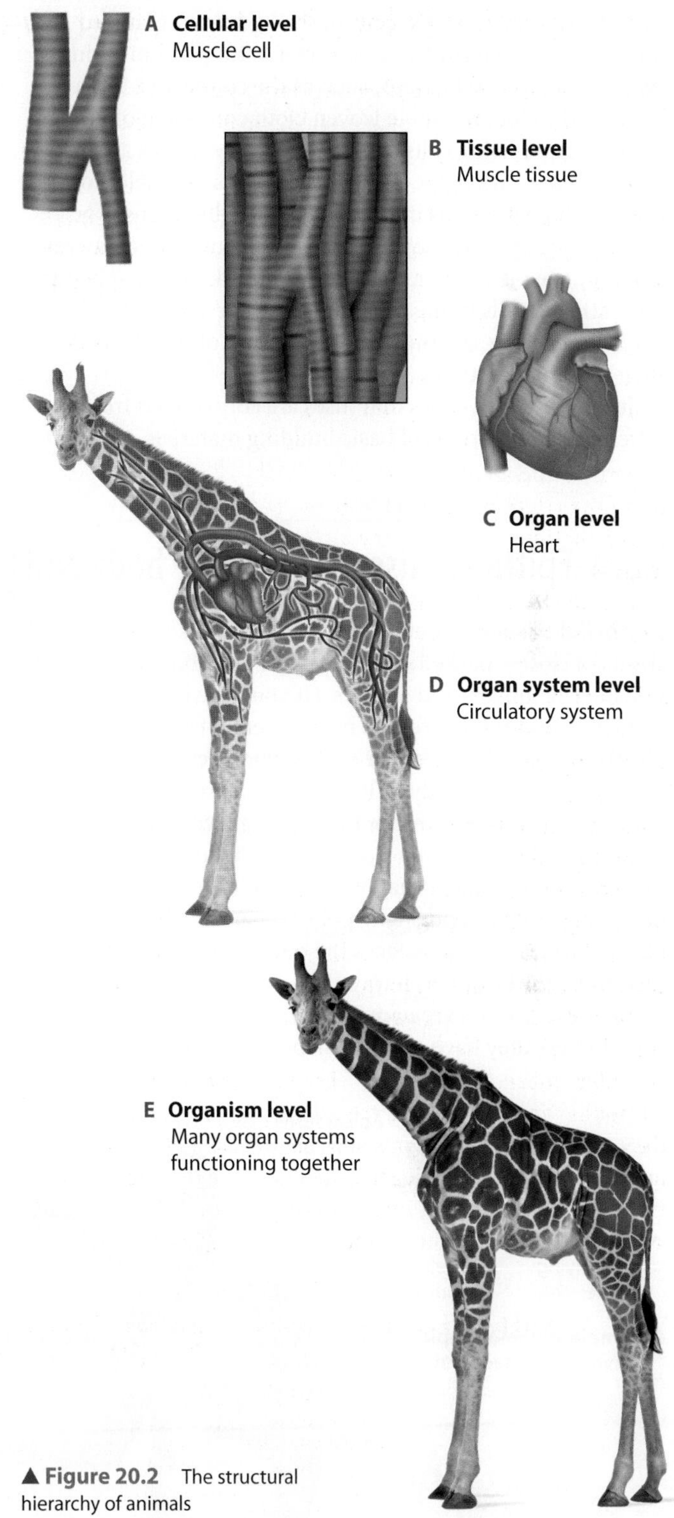

A Cellular level
Muscle cell
B Tissue level
Muscle tissue
C Organ level
Heart
D Organ system level
Circulatory system
E Organism level
Many organ systems functioning together

▲ **Figure 20.2** The structural hierarchy of animals

? Relate the idea of emergent properties to letters of the alphabet and the hierarchy of language.

● Letters are arranged into words, words are arranged into sentences, sentences into paragraphs, paragraphs into an essay—with more complicated meaning emerging at each stage. Similarly, cells arrange into tissues, tissues into organs, organs into systems, systems into an organism—with each level more complex structure and function emerges.

20.3 Tissues are groups of cells with a common structure and function

In almost all animals, the cells of the body are organized into tissues. The term *tissue* is from a Latin word meaning "weave," and several tissues, such as the connective tissue beneath the skin, resemble woven cloth consisting of nonliving fibers and living cells. The cells of some tissues are held together by a sticky glue-like substance. For example, communicating nerve cells in nervous tissue adhere this way. Other tissues are held together by special junctions between adjacent plasma membranes (see Module 4.20), as is the case for epithelial cells lining the intestines. The way that cells are held together and the overall structure of tissues relate to their specific functions.

Just as different styles of houses are constructed from different combinations of basic building materials, specialized body parts, such as different organs, are constructed from varied combinations of a limited set of cells and tissue types. For example, lungs and blood vessels have very distinct functions, but they are lined by tissues that are of the same basic type.

Your body is built from four main types of tissues: epithelial, connective, muscle, and nervous. We examine the structure and function of these tissue types in the next four modules.

Unlike normal cells in a tissue, cancer cells sometimes detach from their neighboring cells. What might be the consequence of their detachment?

The cancer cells might be able to roam freely, leading to new tumors at distant locations in the body.

20.4 Epithelial tissue covers the body and lines its organs and cavities

Epithelial tissues, or epithelia (singular, *epithelium*), are sheets of closely packed cells that cover the body surface and line internal organs and cavities. (If you have known someone with cancer, it is likely to have arisen in epithelial tissue. Nearly 80% of all cancers are of this type; they are known as carcinomas.) The tightly knit cells form a protective barrier and, in some cases, a surface for exchange with the fluid or air on the other side. One side of an epithelium is attached to a basal lamina, a dense mat of extracellular matrix consisting of fibrous proteins and sticky polysaccharides. The other side, called the apical surface, faces the outside of an organ or the inside of a tube or passageway.

Epithelial tissues are named according to the number of cell layers they have and the shape of the cells on their apical surface. A simple epithelium has a single layer of cells, whereas a stratified epithelium has multiple layers. The shape of the cells can be squamous (flat like fried eggs), cuboidal (like dice), or columnar (like bricks on end). **Figure 20.4** shows examples of different types of epithelia. In each case, the pink color identifies the cells of the epithelium itself.

The structure of each type of epithelium fits its function. Simple squamous epithelium **(Part A)** is thin and leaky and thus suitable for exchanging materials by diffusion. It lines capillaries and the air sacs of lungs.

Both cuboidal and columnar epithelia have cells with a relatively large amount of cytoplasm, facilitating their role of secretion or absorption of materials. **Part B** shows a cuboidal epithelium forming a tube in the kidney. Such epithelia are also found in glands, such as the thyroid and salivary glands.

A simple columnar epithelium **(Part C)** lines the intestines, where it secretes digestive juices and absorbs nutrients. The apical surface of some columnar cells has tiny densely packed projections called microvilli that increase surface area for absorption, while other cells have longer motile projections (cilia) that move materials along the epithelial surface. Your respiratory tract is lined with ciliated epithelium. Dust, pollen, and other particles are trapped in the mucus that the epithelium secretes and then swept up and out of your respiratory tract by the beating of the cilia.

The many layers of the stratified squamous epithelium in **Part D** make it well suited for lining surfaces subject to abrasion, such as the outer skin and the linings of the mouth and

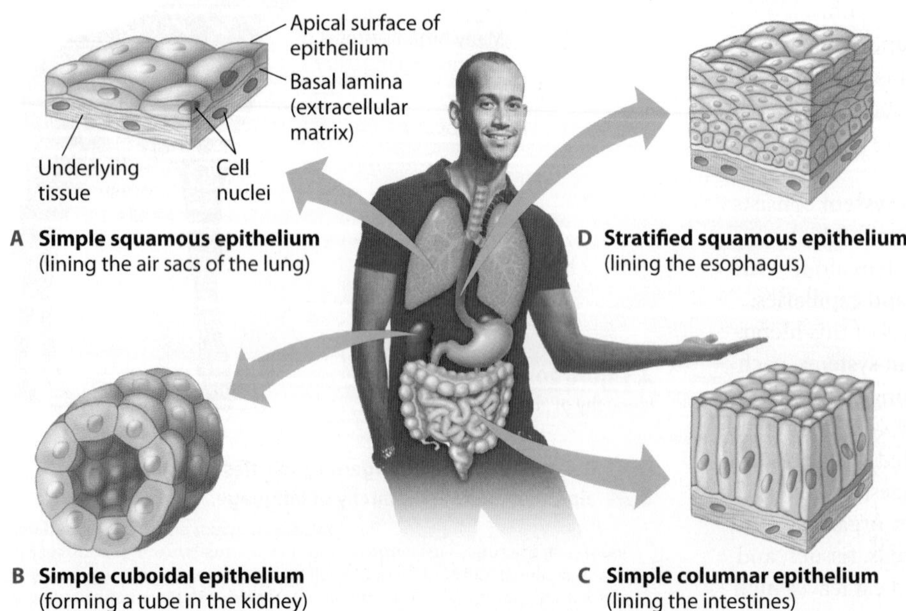
Apical surface of epithelium
Basal lamina (extracellular matrix)
Underlying tissue
Cell nuclei
A Simple squamous epithelium (lining the air sacs of the lung)
D Stratified squamous epithelium (lining the esophagus)
B Simple cuboidal epithelium (forming a tube in the kidney)
C Simple columnar epithelium (lining the intestines)

▲ **Figure 20.4** Types of epithelial tissue

esophagus. Stratified squamous epithelium regenerates rapidly by division of the cells near the basal lamina. New cells move toward the apical surface as older cells slough off. We probably shed close to a million skin cells per day, adding up to many pounds of skin cells per year!

? **What properties are shared by all types of epithelial tissues?**

● All consist of tightly packed cells situated on top of a basal lamina. They form protective barriers or exchange surfaces that line body structures.

20.5 Connective tissue binds and supports other tissues

In contrast to epithelia, **connective tissue** consists of a sparse population of cells scattered throughout a matrix. The cells produce and secrete the matrix, which usually consists of a web of fibers embedded in a liquid, jelly, or solid. Connective tissues are grouped into six major types. **Figure 20.5** shows micrographs of each type and illustrates where each would be in an appendage such as the arm.

The most widespread connective tissue in the body is called **loose connective tissue** (**Part A**) because its matrix is a loose weave of fibers in a watery fluid. Many of the fibers consist of collagen, a strong, ropelike protein. Other fibers are more elastic, making the tissue resilient as well as strong. Loose connective tissue serves mainly to bind epithelia to underlying tissues and hold organs in place. The figure shows the loose connective tissue that lies directly under the skin.

Fibrous connective tissue (**Part B**) has a matrix of densely packed collagen fibers, an arrangement that maximizes its strength. This tissue forms tendons, which attach muscles to bone, and ligaments, which connect bones at joints.

Adipose tissue (**Part C**) stores fat in large, closely packed adipose cells held in a very sparse matrix of loose fibers and fluid. This tissue pads and insulates the body and stores energy. Each adipose cell contains a large fat droplet that swells when fat is stored and shrinks when fat is used as fuel.

Cartilage (**Part D**) forms a strong but flexible skeletal material, and its matrix consists of collagen fibers embedded in a rubbery material. Cartilage commonly surrounds the ends of bones, providing a shock-absorbing surface. It also supports the ears and nose and forms the cushioning disks between the vertebrae.

Bone (**Part E**) has a matrix of collagen fibers embedded in a hard mineral substance made of calcium, magnesium, and phosphate. The combination of fibers and minerals makes bone strong without being brittle. The compact regions of bones contain repeating circular units of matrix, each with a central canal containing blood vessels and nerves. Bone may not seem "alive," but it contains living cells, and it can grow as you grow and mend when broken.

Blood (**Part F**) transports substances throughout the body and thus functions differently from other connective tissues. Its extensive extracellular matrix is a liquid called plasma, which consists of water, salts, and dissolved proteins. Suspended in the plasma are red blood cells, which carry oxygen; white blood cells, which function in defense against infection; and platelets, which aid in blood clotting.

? **Why does blood qualify as a type of connective tissue?**

● Because it consists of a population of cells surrounded by a noncellular matrix, which in this case is a fluid called plasma

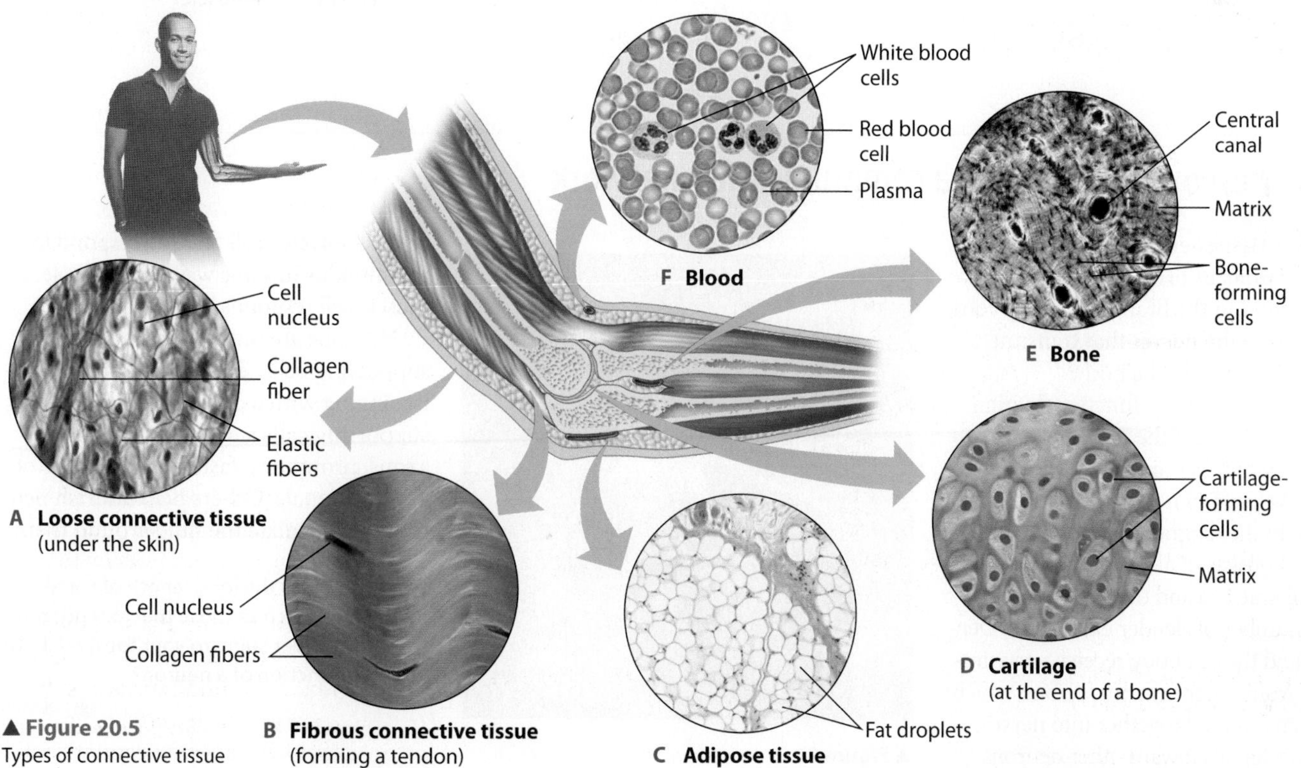

White blood cells
Red blood cell
Plasma
F Blood
Central canal
Matrix
Bone-forming cells
E Bone
Cell nucleus
Collagen fiber
Elastic fibers
A Loose connective tissue (under the skin)
Cell nucleus
Collagen fibers
B Fibrous connective tissue (forming a tendon)
Fat droplets
C Adipose tissue
Cartilage-forming cells
Matrix
D Cartilage (at the end of a bone)

▲ **Figure 20.5**
Types of connective tissue

20.6 Muscle tissue functions in movement

Muscle tissue is the most abundant tissue in nearly all animals. It consists of long cells called muscle fibers, each containing many molecules of contractile proteins. **Figure 20.6** shows micrographs of the three types of vertebrate muscle tissue.

Skeletal muscle (Part A) is attached to bones by tendons and is responsible for voluntary movements of the animal body, such as hopping in kangaroos, flying in birds, or bouncing a ball in humans. The arrangement of the contractile units along the length of skeletal muscle fibers gives the cells a striped, or striated, appearance, as can be seen in the micrograph.

Cardiac muscle (Part B) forms the contractile tissue of the heart, an organ consisting of mostly muscle. It is striated like skeletal muscle, but cardiac muscle is under involuntary control, meaning that its contraction cannot consciously be controlled. Cardiac muscle fibers are branched, interconnecting at specialized junctions that rapidly relay the signal to contract from cell to cell during a heartbeat.

Smooth muscle (Part C) gets its name from its lack of striations. Smooth muscle is found in the walls of the digestive tract, arteries, and other internal organs. It is responsible for involuntary body activities, such as the movement of food through the intestines. Smooth muscle cells are spindle shaped and contract more slowly than skeletal muscles, but smooth muscle can sustain contractions for a longer period of time than can skeletal muscle.

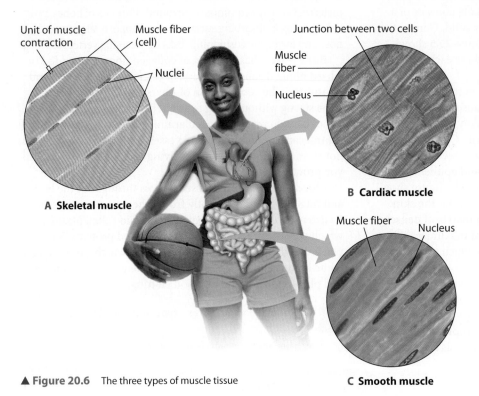

Unit of muscle contraction
Muscle fiber (cell)
Nuclei
A Skeletal muscle
Junction between two cells
Muscle fiber
Nucleus
B Cardiac muscle
Muscle fiber
Nucleus
C Smooth muscle

▲ **Figure 20.6** The three types of muscle tissue

? Cramps felt during menstruation are caused by involuntary contractions of what type of muscle?

● Smooth muscle

20.7 Nervous tissue forms a communication network

Nervous tissue senses stimuli and rapidly transmits information. Nervous tissue is found in the brain and spinal cord, as well as in the nerves that transmit signals throughout the body.

The structural and functional unit of nervous tissue is the nerve cell, or **neuron**, which is uniquely specialized to conduct electrical nerve impulses. As you can see in the diagram in **Figure 20.7**, a neuron consists of a cell body (containing the cell's nucleus and other organelles) and a number of slender extensions. Dendrites and the cell body receive nerve impulses from other neurons. Axons, which are often bundled together into nerves, transmit signals toward other neurons or to an effector cell that can respond to the stimulus in some way. For example, a muscle cell responds by contracting.

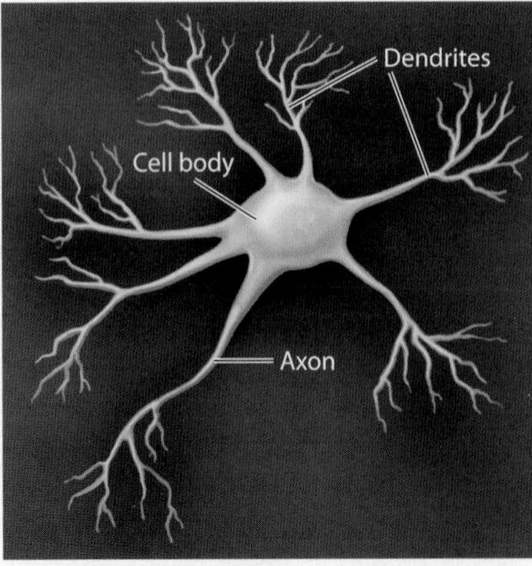
Dendrites
Cell body
Axon

▲ **Figure 20.7** A neuron

Neurons are outnumbered by their supporting cells, each neuron like a soloist singing with a choir. Some of these supporting cells surround and insulate axons, promoting faster transmission of neuron signals. Others help nourish neurons and regulate the fluid around them.

? How does the long length of some axons (such as those that extend from your lower spine to your toes) relate to the function of a neuron?

● It allows for the transmission of a nerve signal over a long distance directly to specific muscle cells.

Organs and Organ Systems

20.8 Organs are made up of tissues

In all but the simplest animals, multiple tissues are arranged into organs that perform specific functions. As mentioned earlier, the heart is composed of muscle, epithelial, connective, and nervous tissues. Epithelial tissue lining the heart chambers prevents leakage and provides a smooth surface over which blood can flow. Connective tissue makes the heart elastic and strengthens its walls. Neurons regulate the contractions of cardiac muscle.

In some organs, tissues are organized in layers, as you can see in the diagram of the small intestine in **Figure 20.8**. The lumen, or interior space, of the small intestine is lined by a columnar epithelium that secretes digestive juices and absorbs nutrients. Notice the finger-like projections that increase the surface area of this lining. Underneath the epithelium (and extending into the projections) is connective tissue, which contains blood vessels. The two layers of smooth muscle, oriented in different directions, propel food through the intestine. The smooth muscle, in turn, is surrounded by another layer of connective tissue and epithelial tissue.

An organ represents a higher level of structure than the tissues composing it, and it performs functions that none of its component tissues can carry out alone. These functions emerge from the coordinated interactions of tissues.

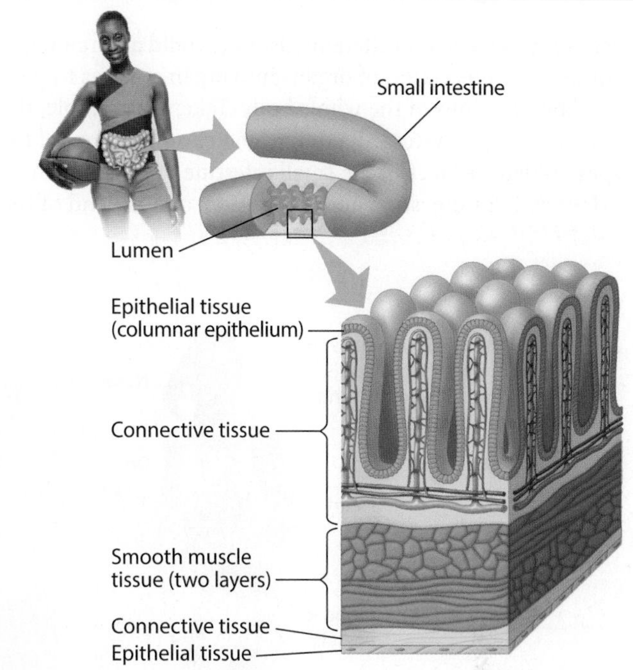
Small intestine
Lumen
Epithelial tissue
(columnar epithelium)
Connective tissue
Smooth muscle
tissue (two layers)
Connective tissue
Epithelial tissue

▲ **Figure 20.8** Tissue layers of the wall of the small intestine

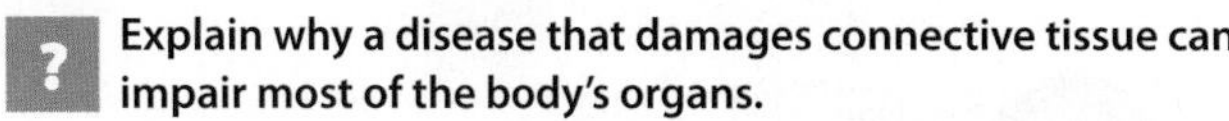

? Explain why a disease that damages connective tissue can impair most of the body's organs.

● Connective tissue is a component of most organs.

20.9 Bioengineers are learning to produce organs for transplants

CONNECTION

Every day, 80 people receive an organ transplant in the United States, while nearly 20 others die waiting for an organ. Many more transplant organs would be available if they could be fabricated in the lab, but as you have learned, organs are highly organized, complex, three-dimensional structures made of several different tissues. So how would a scientist go about building an organ?

If you were to build a replica of the Empire State Building, it would be easiest to start with an inner framework, or scaffold. Some scientists have taken this approach to organ building by growing bladder cells on a balloon-like scaffold. Laboratory-grown bladders have been successfully transplanted into humans. However, many organs have a more complex structure than the bladder. Instead of creating a framework from scratch, scientists discovered that hearts from animal cadavers can be washed in detergent, dissolving away cells and leaving behind a scaffold of connective tissue matrix (**Figure 20.9**). The "decellularized" connective tissue maintains the integrity of the heart; even the small tubes outlining the outer walls of blood vessels are intact. This scaffold can then be seeded with adult stem cells—unspecialized cells that contain the genetic instructions to divide and build a heart. Hearts built in a lab can beat, but scientists still have more to learn about how to make them fully functional. Some scientists predict that decellularized pig hearts will one day replace diseased human hearts.

Other researchers are taking a different approach to organ building. Using desktop printers, they drop suspensions of different cell types the way a printer uses different colors. By printing in layers, the cellular structures are built in three dimensions to resemble organs. Kidneys and rudimentary hearts that can beat have been produced this way, but they lack correctly organized nerves, blood vessels, and other tissues that would make them useable in humans.

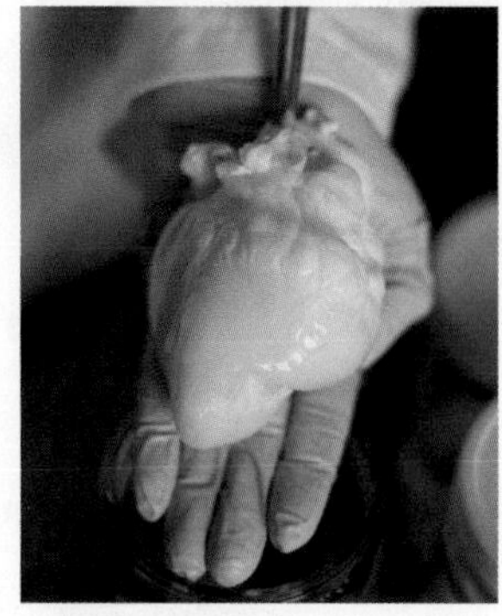

▲ **Figure 20.9**
A decellularized pig heart

? A human windpipe has been decellularized, rebuilt with stem cells (see Module 11.15), and successfully transplanted. Why is this more easily accomplished with a windpipe than with a heart?

● A windpipe functions simply as a pipe—it has a less complicated structure and function than a heart.

20.10 Organ systems work together to perform life's functions

Just as it takes several different tissues to build an organ, it requires the integration of organs into organ systems to perform the functions of the whole body. Take, for example, the human digestive system: Teeth bite and chew a variety of food types, folded surfaces of the small intestine provide much surface area for digestion and absorption, and the end of the large intestine provides a temporary location for waste storage. This coordination of functions along its length (chewing, then digestion and absorption, then waste storage) is an adaptation that allows us to extract nutrients from many meals in a short period of time to meet our continual metabolic demand.

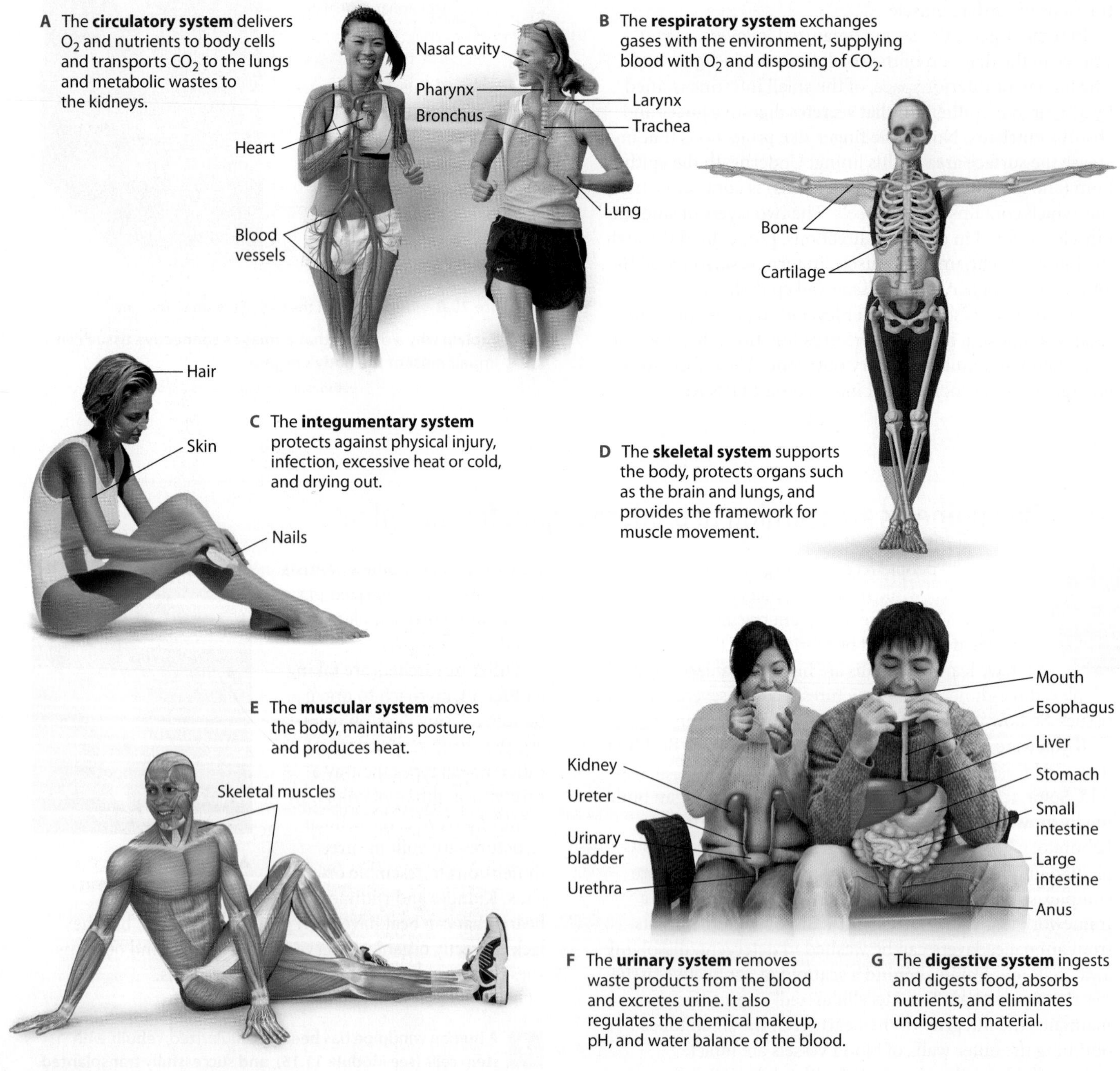

A The **circulatory system** delivers O_2 and nutrients to body cells and transports CO_2 to the lungs and metabolic wastes to the kidneys.

Heart
Blood vessels

B The **respiratory system** exchanges gases with the environment, supplying blood with O_2 and disposing of CO_2.

Nasal cavity
Pharynx
Bronchus
Larynx
Trachea
Lung

C The **integumentary system** protects against physical injury, infection, excessive heat or cold, and drying out.

Hair
Skin
Nails

D The **skeletal system** supports the body, protects organs such as the brain and lungs, and provides the framework for muscle movement.

Bone
Cartilage

E The **muscular system** moves the body, maintains posture, and produces heat.

Skeletal muscles

F The **urinary system** removes waste products from the blood and excretes urine. It also regulates the chemical makeup, pH, and water balance of the blood.

Kidney
Ureter
Urinary bladder
Urethra

G The **digestive system** ingests and digests food, absorbs nutrients, and eliminates undigested material.

Mouth
Esophagus
Liver
Stomach
Small intestine
Large intestine
Anus

▲ **Figure 20.10** Human organ systems and their components

As you study **Figure 20.10** on these two pages, remember that the ability to carry out life's functions is a result of the emergent properties stemming from the organization and coordination of all the body's organ systems working together. Indeed, the whole is greater than the sum of its parts. For example, nutrients extracted from food would be useless to cells of the body without the circulatory system's role in distributing them.

? **Which two organ systems are most directly involved in regulating all other systems?**

The nervous system and the endocrine system

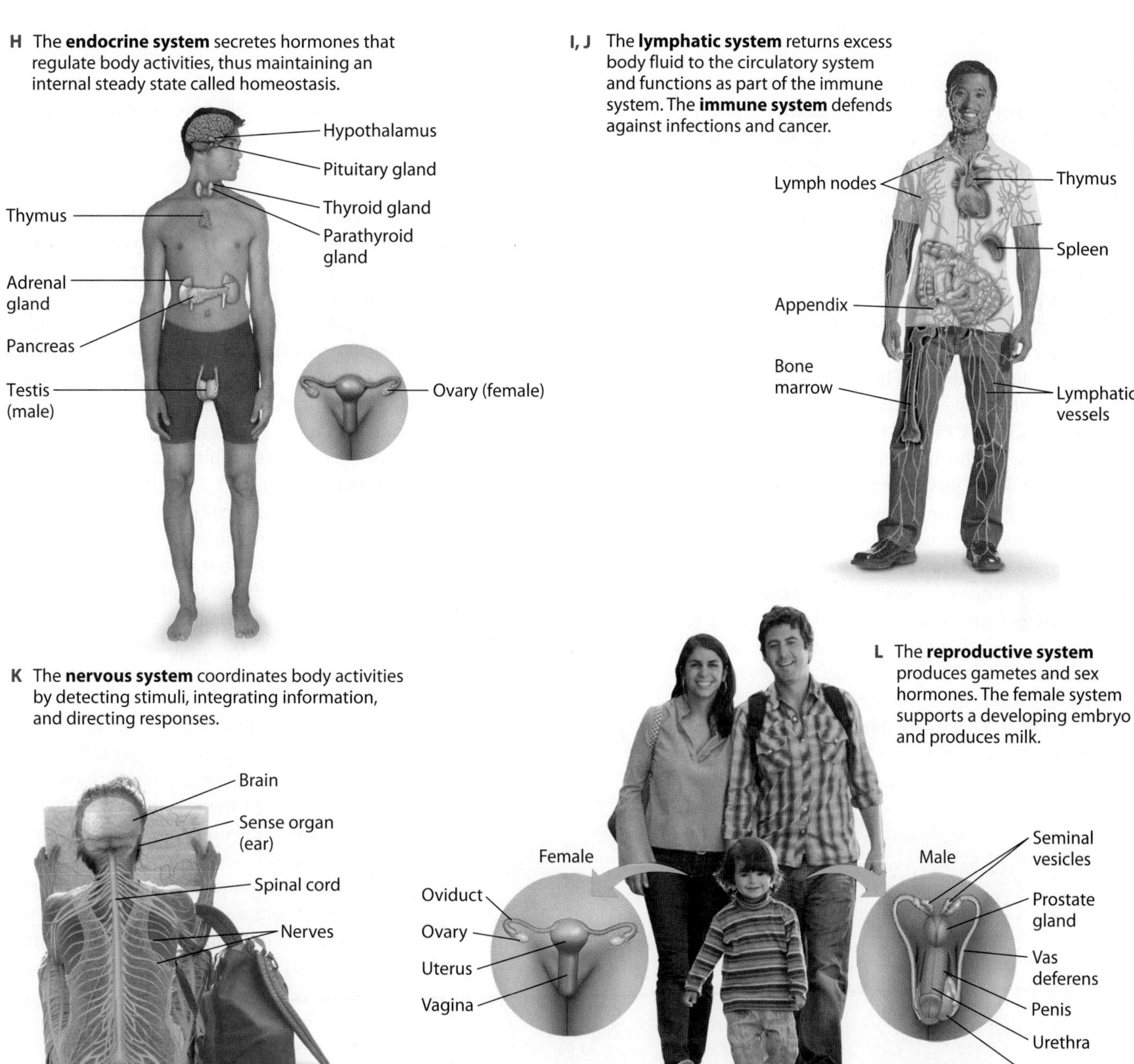

H The **endocrine system** secretes hormones that regulate body activities, thus maintaining an internal steady state called homeostasis.
Hypothalamus
Pituitary gland
Thyroid gland
Parathyroid gland
Thymus
Adrenal gland
Pancreas
Testis (male)
Ovary (female)
I, J The **lymphatic system** returns excess body fluid to the circulatory system and functions as part of the immune system. The **immune system** defends against infections and cancer.
Lymph nodes
Thymus
Spleen
Appendix
Bone marrow
Lymphatic vessels
K The **nervous system** coordinates body activities by detecting stimuli, integrating information, and directing responses.
Brain
Sense organ (ear)
Spinal cord
Nerves
L The **reproductive system** produces gametes and sex hormones. The female system supports a developing embryo and produces milk.
Female
Oviduct
Ovary
Uterus
Vagina
Male
Seminal vesicles
Prostate gland
Vas deferens
Penis
Urethra
Testis

20.11 The integumentary system protects the body

Most of the organ systems introduced in Module 20.10 are examined in more detail in the chapters of this unit. Here we take a brief look at the integumentary system, which consists of the skin, hair, and nails.

Skin As shown in **Figure 20.11**, skin consists of two layers: the epidermis and the dermis. The epidermis is a stratified squamous epithelium (many layers of flat cells; see Module 20.4). Rapid cell division near the base of the epidermis serves to replenish the skin cells that are constantly abraded from the body surface. As these new cells are pushed upward in the epidermis by the addition of new cells below them, they fill with the fibrous protein keratin and release a waterproofing glycolipid. The waterproof covering protects the body from dehydration and prevents penetration by microbes. The cells at the surface eventually die, yet remain tightly joined at the surface of the skin for up to two weeks before being sloughed off by abrasion. This continuous process means that you get a brand new epidermis every few weeks.

The dermis is the inner layer of the skin. It consists of a fairly dense connective tissue with many resilient elastic fibers and strong collagen fibers. (The thinning of this layer is the cause of wrinkled, sagging skin in older adults.) The dermis contains hair follicles, oil and sweat glands, muscles, nerves, and blood vessels. The profusion of small blood vessels and the 2.5 million sweat glands in the dermis facilitate the important function of temperature regulation. Lastly, the dermis also contains sensory receptors (shown in Figure 29.3A) that provide important environmental information to your brain: Is something touching your skin too hot or cold or sharp? Your touch receptors help you to chew food, manipulate tools, and feel your way around in the dark. Beneath the skin lies the hypodermis, a layer of adipose tissue. (The hypodermis is the site where some vaccines and drugs are injected with a hypodermic needle.)

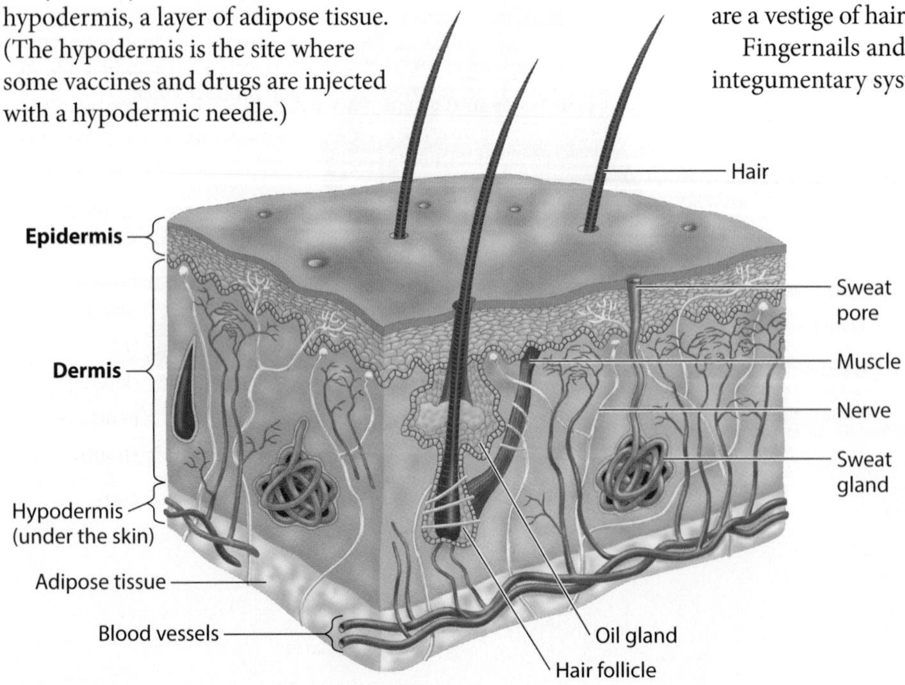

▲ **Figure 20.11** A section of skin, the major organ of the integumentary system

One of the metabolic functions of the skin is the synthesis of vitamin D, which is required for absorbing calcium. Ultraviolet (UV) light catalyzes the conversion of a derivative of cholesterol to vitamin D in the cells in the lower layers of the epidermis. Adequate sunlight is needed for this synthesis, which explains why a majority of people living at high latitudes are vitamin D deficient. But sunlight also has damaging effects on the skin. In response to UV radiation, skin cells make more melanin (a pigment that protects the DNA in skin cells from UV radiation damage), which moves to the cells of the outer layers of the skin and is visible as a tan. A tan is not indicative of good health; it signals that skin cells have been damaged. DNA changes caused by the UV rays can lead to premature aging of the skin, cataracts in the eyes, and skin cancers.

Hair and Nails In mammals, hair is an important component of the integumentary system. Hair is a flexible shaft of flattened, keratin-filled dead cells, which were produced by a hair follicle. Associated with hair follicles are oil glands, whose secretions lubricate the hair, condition the surrounding skin, and inhibit the growth of bacteria. Look at the hair follicle in Figure 20.11, and you will see that it is wrapped in nerve endings. Hair follicles play an important sensory function, as the slightest movement of hair is relayed to the nervous system. (You can get a sense of this sensitivity by lightly touching the hair on your head.)

Hair insulates the bodies of most mammals—although in humans this insulation is limited to the head. Land mammals react to cold by raising their fur, which traps a layer of air and increases the insulating power of the fur. Look again at the hair follicle in the figure. The muscle attached to it is responsible for raising the hair when you get cold. The resulting "goose bumps" are a vestige of hair raising inherited from our furry ancestors.

Fingernails and toenails are the final component of the integumentary system. These protective coverings are also composed of keratin. Fingernails facilitate fine manipulation (and are useful for chewing when nervous). In other mammals, the digits may end in claws or hooves.

The integumentary system encloses and protects an animal from its environment, but dysfunction is also associated with this organ system. In the next section, we'll see what happens when hair follicles become clogged with sloughed-off cells and oil.

? **Describe three structures associated with a hair follicle that contribute to the functions of hair.**

● The nerve endings sense when the hair is moved; the muscle raises the hair (producing goose bumps in humans but warming other mammals); and the oil gland produces lubricating and antibacterial secretions.

20.12 Well-designed studies help answer scientific questions

SCIENTIFIC THINKING

Imagine you have just seen before and after images of a teenager whose acne was treated with laser therapy. An advertisement claims the treatment is effective and can last one to two years. The results look amazing, but you are skeptical. As consumers, we are bombarded with claims—on packaging, in infomercials, in online pop-up ads, and through other advertisements. To make informed decisions and behave as responsible consumers, we should evaluate information as scientists do.

Acne results when the hair follicles (also called pores) that produce oil become clogged with dead cells and oil (**Figure 20.12A**). When a pore is plugged, bacteria of the species *Propionibacterium acnes* become trapped in the follicle. If the follicle ruptures into the dermis and white blood cells are recruited from the immune system, the pore is said to be inflamed (see Module 24.2). It becomes a "pimple" and soon fills with pus. Inflammatory acne is a common condition, affecting 40–50 million people in the United States. It can persist for years, sometimes leading to anxiety and depression. There are many different kinds of acne treatment: Some reduce the amount of oil produced, others slough off dead skin cells, and yet others kill *P. acnes*. Although scientists don't yet understand how laser therapy works, some think it may decrease the number of bacteria.

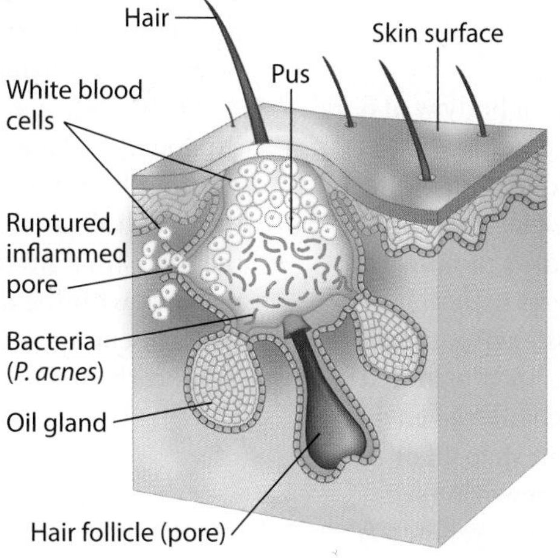

Hair
Skin surface
Pus
White blood cells
Ruptured, inflammed pore
Bacteria (*P. acnes*)
Oil gland
Hair follicle (pore)

▲ **Figure 20.12A** The anatomy of a pimple

What kind of study would convince us that laser therapy is effective and long-lasting? Consider a study involving 19 individuals, chosen because they had at least five pimples. After three laser treatments, all participants exhibited a significant reduction in the total number of pimples, as shown in the graph of **Figure 20.12B**. These data might seem convincing—but acting as scientists, we recognize that the study did not include a control group (individuals who were not treated but used for comparison). Without a control group, we cannot know if individuals who had not undergone laser therapy would have had the same reduction in pimples. The study also failed to control variables. The participants were allowed to continue using acne medications over the course of the study. Ideally, a study will examine only a single variable (laser therapy in this case), not multiple variables (such as laser therapy along with acne creams and medicines). Finally, the study did not follow individuals after treatment, so we can't know if the effects are long-lasting.

Only well-designed studies can tell us if laser therapy is effective and if the effects last once therapy ends. Consider a second study, involving 29 participants who, unlike the first study, were required to be free of other acne medications and treatments for a specified time before the study began (removing unwanted variables). This study included controls. Each participant's face was divided into a treated side and controlled side, randomly assigned. To rule out bias, the study was single-blind, meaning that although the participants knew which side was the treated side, the scientist counting the pimples did not. The acne was categorized into subtypes and measured before and after treatment. **Figure 20.12C** highlights the significant reduction in one subtype of pimple (red and lacking pus). The decrease was temporary, lasting a few weeks. Other pimple subtypes did not differ between the two sides of the face. These data suggest that laser therapy is somewhat effective—decreasing a single type of pimple for a few weeks.

? If the scientist counting the pimples had not been "blind" to which side of the face was treated, what bias might result?

● The scientist might unintentionally rate the treated side as more improved, because that would be the expected result of the treatment.

Average number of pimples (0, 10, 20, 30, 40)

Before laser therapy | After 1st treatment | After 2nd treatment | After 3rd treatment

Reprinted from P. M. Friedman et al., Treatment of inflammatory facial acne vulgaris with the 1450-nm diode laser: A pilot study, *Dermatologic Surgery* 30: 147–51 (2004), with permission.

▲ **Figure 20.12B** Reduction in pimples before and after laser therapy treatment in an uncontrolled study

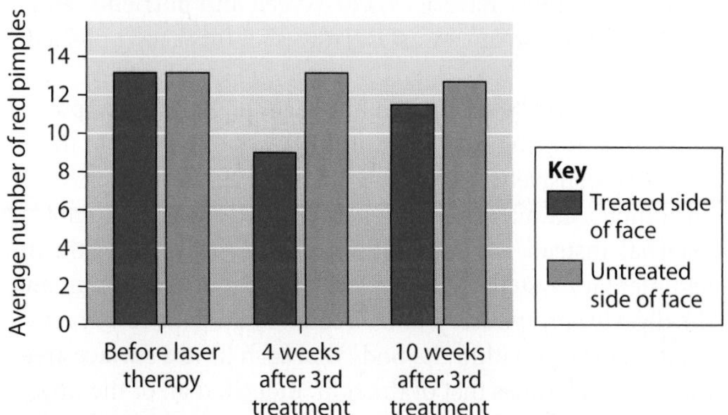

Average number of red pimples
0
2
4
6
8
10
12
14
Before laser therapy
4 weeks after 3rd treatment
10 weeks after 3rd treatment
Key
Treated side of face
Untreated side of face

Data from J. S. Orrringer et al., Photodynamic therapy for acne vulgaris: A randomized, controlled, split-faced clinical trial of topical animolevulinic acid and pulsed dye therapy, *Journal of Cosmetic Dermatology* 9: 28–34 (2010).

▲ **Figure 20.12C** Reduction in a subtype of pimple before and after laser therapy in a controlled, randomized, single-blind study

20.13 Structural adaptations enhance exchange with the environment

Every living organism is an open system, meaning that it exchanges matter and energy with its surroundings. You, for example, take in oxygen, water, and food, and in exchange, you breathe out carbon dioxide, urinate, defecate, sweat, and radiate heat. The exchange of materials with the environment must extend to the level of each individual cell. Exchange of gases, nutrients, and wastes occurs as substances dissolved in an aqueous solution move across the plasma membrane of every cell.

External Exchange A freshwater hydra has a body wall only two cell layers thick. The outside layer is in contact with its water environment; the inner layer is bathed by fluid in its saclike body cavity. This internal fluid circulates in and out of the hydra's mouth. Thus, every body cell exchanges materials directly with an aqueous environment.

Another common body plan that maximizes exchange with the environment is a flat, thin shape. For instance, a parasitic tapeworm (see Figure 18.7B) can be several meters long, but because it is very thin, most of its cells are bathed in the intestinal fluid of its host—the source of its nutrients.

The saclike body of a hydra or the paper-thin one of the tapeworm works well for animals with a simple body structure. However, most animals are composed of compact masses of cells and have an outer surface that is relatively small compared with the animal's overall volume. (For a reminder of the relationship of surface area to volume, see Module 4.2.) As an extreme example, the ratio of a whale's outer surface area to its volume is hundreds of thousands of times smaller than that of a small animal like a hydra. Still, every cell in the whale's body must be bathed in fluid, have access to oxygen and nutrients, and be able to dispose of its wastes. How is all this accomplished?

Internal Exchange The evolutionary adaptation that provides for sufficient exchange with the environment is in the form of extensively branched or folded surfaces. In almost all complex animals like whales, the folded surfaces are not external; instead, they lie within the body, protected by the integumentary system from dehydration or damage. In humans, the digestive, respiratory, and circulatory systems rely on exchange surfaces within the body that each have a surface area more than 25 times that of the skin. Indeed, if all of the tiny capillaries within the human body that exchange blood with body cells were lined up, they would circle the globe!

Figure 20.13A is a schematic model illustrating four of the organ systems of a compact, complex animal. Each system has a large, specialized internal exchange surface. The circulatory system is placed in the middle because of its central role in transporting substances between the other three systems. The blue arrows indicate exchange of materials between the circulatory system and the other systems.

Actually, direct exchange does not occur between the blood and the cells of tissues and organs. Body cells are bathed in a solution called **interstitial fluid** (see the circular enlargement in Figure 20.13A). Exchange takes place through this fluid. In other words, to get from the blood to body cells, or vice versa, materials pass through the interstitial fluid.

The digestive system, especially the small intestine, has an expanded surface area resulting from folds and projections of its inner lining (see Figure 20.8). Nutrients are absorbed into the cells lining this large surface area. They then pass through the interstitial fluid and into capillaries that form an exchange network with the digestive surfaces. This system is so effective that enough nutrients move into the circulatory system to support the rest of the cells in the body.

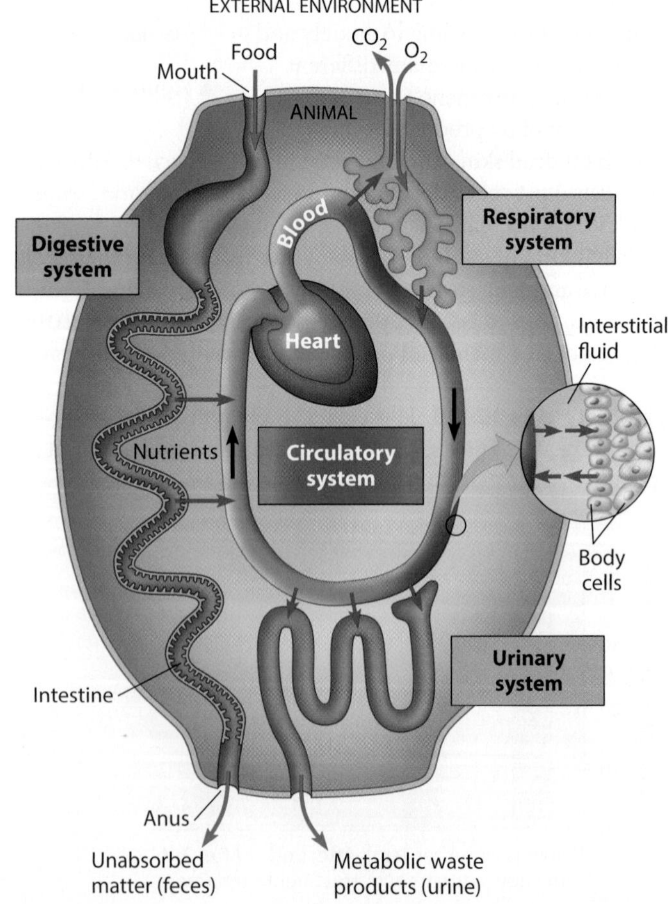

EXTERNAL ENVIRONMENT
Food
CO_2
O_2
Mouth
ANIMAL
Respiratory system
Digestive system
Blood
Heart
Interstitial fluid
Nutrients
Circulatory system
Body cells
Urinary system
Intestine
Anus
Unabsorbed matter (feces)
Metabolic waste products (urine)

▲ **Figure 20.13A** A schematic representation showing indirect exchange between the environment and the cells of a complex animal

Try This Explain this figure aloud, emphasizing the interconnections between the organ systems.

The extensive, epithelium-lined tubes of the urinary system are equally effective at increasing its surface area for exchange. Enmeshed in capillaries, excretory tubes extract metabolic wastes that the blood brings from throughout the body. The wastes move out of the blood into the excretory tubes and pass out of the body in urine.

The respiratory system also has an enormous internal surface area across which gases are exchanged. Lungs are not shaped like big balloons, as they are sometimes depicted, but rather like millions of tiny balloons at the tips of finely branched air tubes. **Figure 20.13B** is a model of the human lungs. The branches are tiny air tubes that end in multilobed sacs lined with thin epithelium. In the body, the epithelium is surrounded by capillaries. Oxygen readily moves from the air in the lungs across this epithelium and into the blood in the capillaries. The blood returns to the heart and is then pumped throughout the body to supply all cells with oxygen. Both Figures 20.13A and 20.13B highlight two basic concepts in animal biology: First, any animal with a complex body—one with most of its cells not in direct contact with its external environment—must have internal structures that provide sufficient surface area to service those cells. Second, the organ systems of the body are interdependent; it takes their coordinated actions to produce a functional organism.

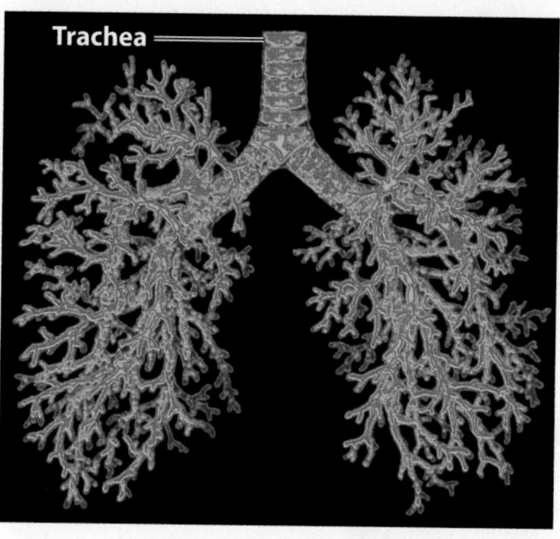
Trachea

▲ **Figure 20.13B** A model of the finely branched air tubes (blue) and blood vessels (red) of the human lungs

? How do the structures of the lungs, small intestine, and kidneys relate to the function of exchange with the environment?

These organs all have a huge number of sacs, projections, or tubes that greatly increase the surface area across which exchange of materials can occur.

20.14 Animals regulate their internal environment

More than a century ago, French physiologist Claude Bernard recognized that two environments are important to an animal: the external environment surrounding the animal and the internal environment, where its cells actually live. The internal environment of a vertebrate is the interstitial fluid that fills the spaces around the cells. Many animals maintain relatively constant conditions in their internal environment. Your own body maintains the salt and water balance of your internal fluids and also keeps your body temperature at about 37°C (98.6°F). A bird, such as the snowy owl shown in **Figure 20.14**, also maintains a body temperature of about 40°C (104°F), even in winter. The bird uses energy from its food to generate body heat, and it has a thick, insulating coat of down feathers that extends from its beak to its feet. A lizard does not generate its own body heat, but it can maintain a fairly constant body temperature by basking in the sun or resting in the shade. It does, however, regulate the salt and water balance of its body fluids.

Today, Bernard's concept of the internal environment is included in the principle of **homeostasis**, which means "a steady state." As Figure 20.14 illustrates, conditions often fluctuate widely in the external environment, but homeostatic mechanisms regulate internal conditions, resulting in much smaller changes in the animal's internal environment. Both birds and mammals have control systems that keep body temperature—as well as salt and water balance and other factors—within a narrow range, despite large changes in the external environment.

The internal environment of an animal always fluctuates slightly. Homeostasis is a dynamic state, an interplay between outside forces that tend to change the internal environment and internal control mechanisms that oppose such changes. An animal's homeostatic control systems maintain internal conditions within a range where life's metabolic processes can occur. In the next module, we explore a general mechanism for how many of these control systems function.

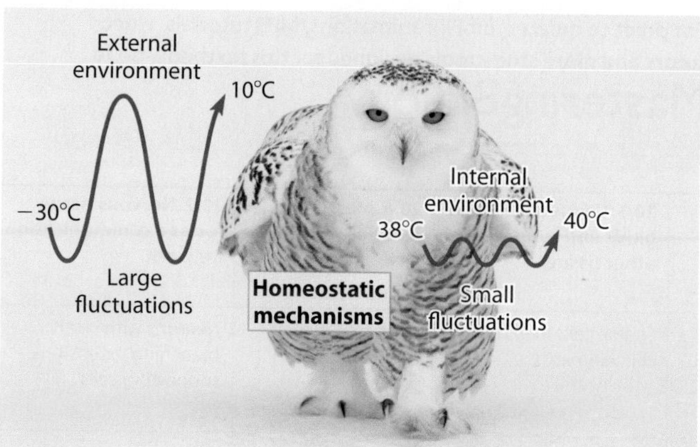
External environment
10°C
−30°C
Large fluctuations
Homeostatic mechanisms
Internal environment
38°C
40°C
Small fluctuations

▲ **Figure 20.14** A model of internal temperature homeostasis in a snowy owl

Try This Brainstorm other fluctuating environmental conditions that an animal must internally regulate.

? Look back at Figure 20.13A. What are some ways in which the circulatory system contributes to homeostasis?

By its exchanges with the digestive, respiratory, and urinary systems, the blood helps maintain the proper balance of materials in the interstitial fluid surrounding body cells.

20.15 Homeostasis depends on negative feedback

Most of the control mechanisms of homeostasis are based on **negative feedback**, in which a change in a variable triggers mechanisms that reverse that change. To identify the components of a negative-feedback system, consider the simple example of the regulation of room temperature. A thermostat is set at a comfortable temperature—call this its set point. When a sensor in the thermostat detects that the temperature has dropped below this set point, the thermostat turns on the furnace. The response (heat) reverses the drop in temperature. Then, when the temperature rises to the set point, the thermostat turns the furnace off. Physiologists would call the thermostat a control center, which senses a stimulus (room temperature below or above a set point) and activates a response.

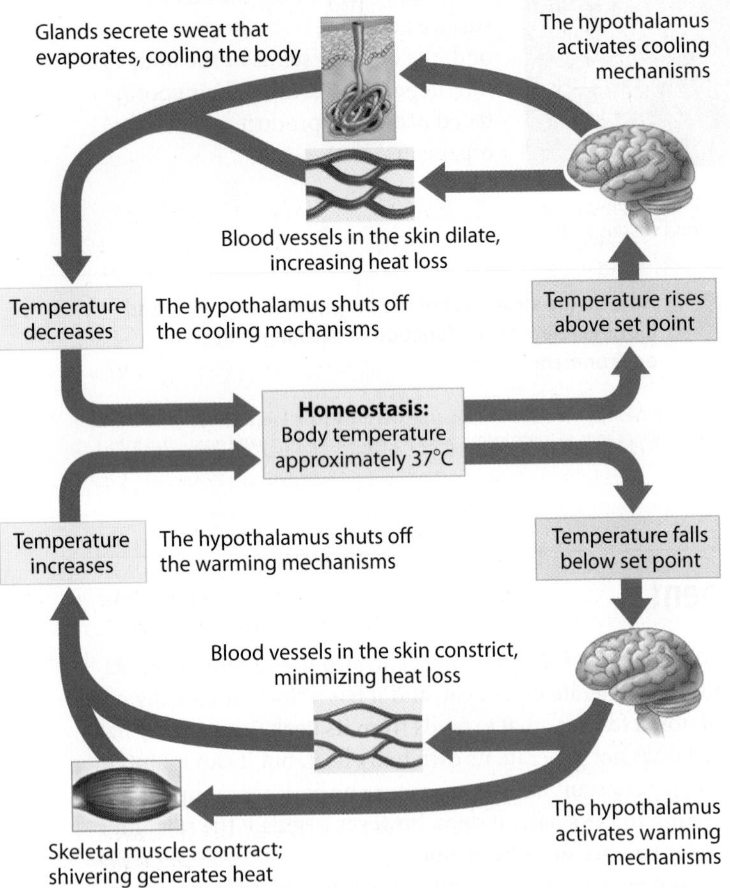

Glands secrete sweat that evaporates, cooling the body
The hypothalamus activates cooling mechanisms
Blood vessels in the skin dilate, increasing heat loss
Temperature decreases
The hypothalamus shuts off the cooling mechanisms
Temperature rises above set point
Homeostasis: Body temperature approximately 37°C
Temperature increases
The hypothalamus shuts off the warming mechanisms
Temperature falls below set point
Blood vessels in the skin constrict, minimizing heat loss
Skeletal muscles contract; shivering generates heat
The hypothalamus activates warming mechanisms

▲ **Figure 20.15** Feedback control of body temperature

Try This Using this figure, explain to a friend what happens when you sunbathe and then jump into very cold water.

Many of the control centers that maintain homeostasis in animals are located in the brain. For example, part of the brain, the hypothalamus, regulates activities such as food intake, sleep, heart rate, hormones, and body temperature. Let's examine the hypothalamus operating as a "thermostat" by negative feedback to switch on and off mechanisms that maintain body temperature around 37°C (98.6°F). As shown in the upper part of **Figure 20.15**, when the hypothalamus senses a rise in temperature above the set point, it activates cooling mechanisms, such as the dilation of blood vessels in your skin and sweating. Once body temperature returns to normal, the hypothalamus shuts off these cooling mechanisms. When your body temperature falls below the set point (lower part of the figure), the hypothalamus activates warming mechanisms, such as constriction of blood vessels to reduce heat loss and shivering to generate heat. Again, a return to normal temperature shuts off these mechanisms.

As you examine the body's organ systems in detail in the chapters of this unit, you will encounter many examples of homeostatic control and negative feedback, as well as constant reminders of the relationship between structure and function.

? **Some portable heaters do not have thermostats. Explain the consequence of turning one on in a room.**

● Without a thermostat, there is no control center to initiate negative feedback. The heater will continually be on, making the room warmer and warmer until the owner manually turns it off.

CHAPTER 20 REVIEW

For practice quizzes, BioFlix animations, MP3 tutorials, video tutors and more study tools designed for this textbook, go to MasteringBiology®

Reviewing the Concepts

Structure and Function in Animal Tissues (20.1–20.7)

20.1 An animal's form is not the perfect design. Structures need only be "good enough" to function.

20.2 Structure fits function at all levels of organization in the animal body.

20.3 Tissues are groups of cells with a common structure and function.

Function	**20.4 Epithelial tissue covers the body and lines its organs and cavities.**	**20.5 Connective tissue binds and supports other tissues.**	**20.6 Muscle tissue functions in movement.**	**20.7 Nervous tissue forms a communication network.**
Structure	Sheets of closely packed cells	Sparse cells in extracellular matrix	Long cells (fibers) with contractile proteins	Neurons with branching extensions; supporting cells
Example	Columnar epithelium	Loose connective tissue	Skeletal muscle	Neuron

Organs and Organ Systems (20.8–20.12)

20.8 Organs are made up of tissues.

20.9 Bioengineers are learning to produce organs for transplants.

20.10 Organ systems work together to perform life's functions. The ability to carry out life's functions is a result of the emergent properties stemming from the organization and coordination of all the body's organ systems working together.

20.11 The integumentary system protects the body. Consisting of skin, hair, and nails, the integumentary system protects an animal from its environment.

20.12 Well-designed studies help answer scientific questions. Examining one variable at a time, including randomized controls, and controlling for bias in data interpretation are hallmarks of well-designed studies.

External Exchange and Internal Regulation (20.13–20.15)

20.13 Structural adaptations enhance exchange with the environment. Complex animals have specialized internal structures that increase surface area. Exchange of materials between blood and body cells takes place through the interstitial fluid.

20.14 Animals regulate their internal environment. Conditions often fluctuate widely in the external environment, but homeostatic mechanisms regulate internal conditions, resulting in much smaller changes in the animal's internal environment.

20.15 Homeostasis depends on negative feedback. Control systems detect change and direct responses. Negative-feedback mechanisms keep internal variables fairly constant, with small fluctuations around set points.

Connecting the Concepts

1. There are several key concepts introduced in this chapter: Structure correlates with function; an animal's body has a hierarchy of organization with emergent properties at each level; and complex bodies have structural adaptations that increase surface area for exchange. Label the tissue layers shown in this section of the small intestine, and describe how this diagram illustrates these three concepts.

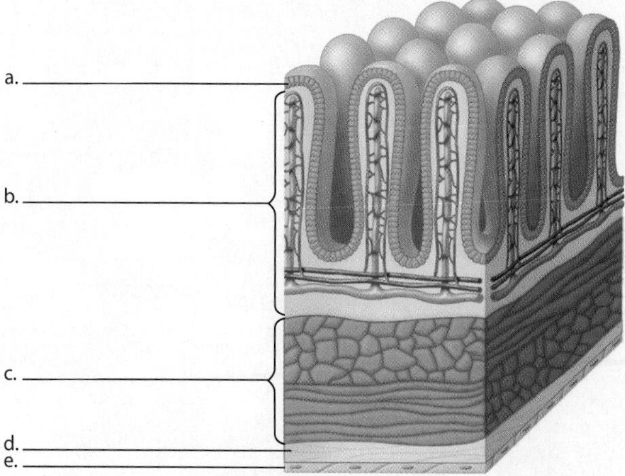

a.
b.
c.
d.
e.

Testing Your Knowledge

Level 1: Knowledge/Comprehension

2. True or False? Each cell in the human body is bathed in blood, allowing exchange of materials in and out of the cells. (*Explain your answer.*)
3. Which of the following body systems facilitates (but doesn't regulate) the functions of the other systems?
   a. respiratory system
   b. endocrine system
   c. digestive system
   d. circulatory system
4. Negative-feedback mechanisms are
   a. most often involved in maintaining homeostasis.
   b. analogous to a furnace that produces heat.
   c. found only in birds and mammals.
   d. all of the above
5. Briefly explain how the structure of each of these tissues is well suited to its function: stratified squamous epithelium in the skin, neurons in the brain, simple squamous epithelium lining the lung, bone in the skull.
6. Describe ways in which the bodies of complex animals are structured for exchanging materials with the environment. Do all animals share such features?

Level 2: Application/Analysis

7. Which of the following best illustrates homeostasis? (*Explain your answer.*)
   a. Most adult humans are between 5 and 6 feet tall.
   b. All the cells of the body are about the same size.
   c. When the salt concentration of the blood goes up, the kidneys expel more salt.
   d. When oxygen in the blood decreases, you feel dizzy.
8. A student is asked to look under the microscope at two different muscle samples and label one biceps and one heart. What would be a characteristic that would distinguish these two tissues from each other?
   a. the presence of loose collagen fibers
   b. the presence of contractile proteins
   c. the presence of striations (stripes)
   d. whether the cells branch or not
9. You read a blog that states, "A squid's eye has been perfectly designed to see in the dark depths of the ocean." Draft a paragraph responding to this blog about why this statement is not accurate.

Level 3: Synthesis/Evaluation

10. **SCIENTIFIC THINKING** In a study to examine the effectiveness of a new acne cream, participants were assigned to one of two groups: those who would be asked to use the cream for three months and those who would not use any treatment. Participants would be asked to keep a journal rating how bad they think their acne is on a scale of 1–10 each week. What are the well-designed aspects of this study? What are the limitations? How would you improve the design of this study to address these limitations?
11. After a long, hot run together, your friend tells you that you should dunk your head into a cooler of water to lower your body temperature more rapidly, rather than sitting and waiting to cool down. What do you think? Form a hypothesis about how the ice-cold water might affect the rate at which your body temperature returns to normal. How could you test your hypothesis?

Answers to all questions can be found in Appendix 4.

CHAPTER 21

Nutrition and Digestion

Have you heard that eating grapefruit will cause pounds to just melt away? Or that you can shed weight by eating cabbage soup? Or cutting out carbs? Or fat?

Type "diet plan" into a search engine, and you'll receive more than *137 million* suggestions! Weight loss is a growth industry: About one in seven Americans go on a diet each year. This effort is often well placed, because more than a third of American adults are obese (very overweight). The obesity epidemic, combined with increasingly sedentary jobs and inactive lifestyles, has contributed to higher incidences of heart disease, diabetes, cancer, and other weight-related health problems. More than 300,000 deaths per year in the United States are attributed to weight-related issues. And the problem is not confined to the United States: The United Nations World Health Organization recognizes obesity as a major global health problem.

? ***Is there a scientific approach to weight loss?***

Despite the need for reliable methods to shed pounds, only about 5% of dieters are able to reach their goal weight and maintain it for the long term. Furthermore, access to weight loss plans far

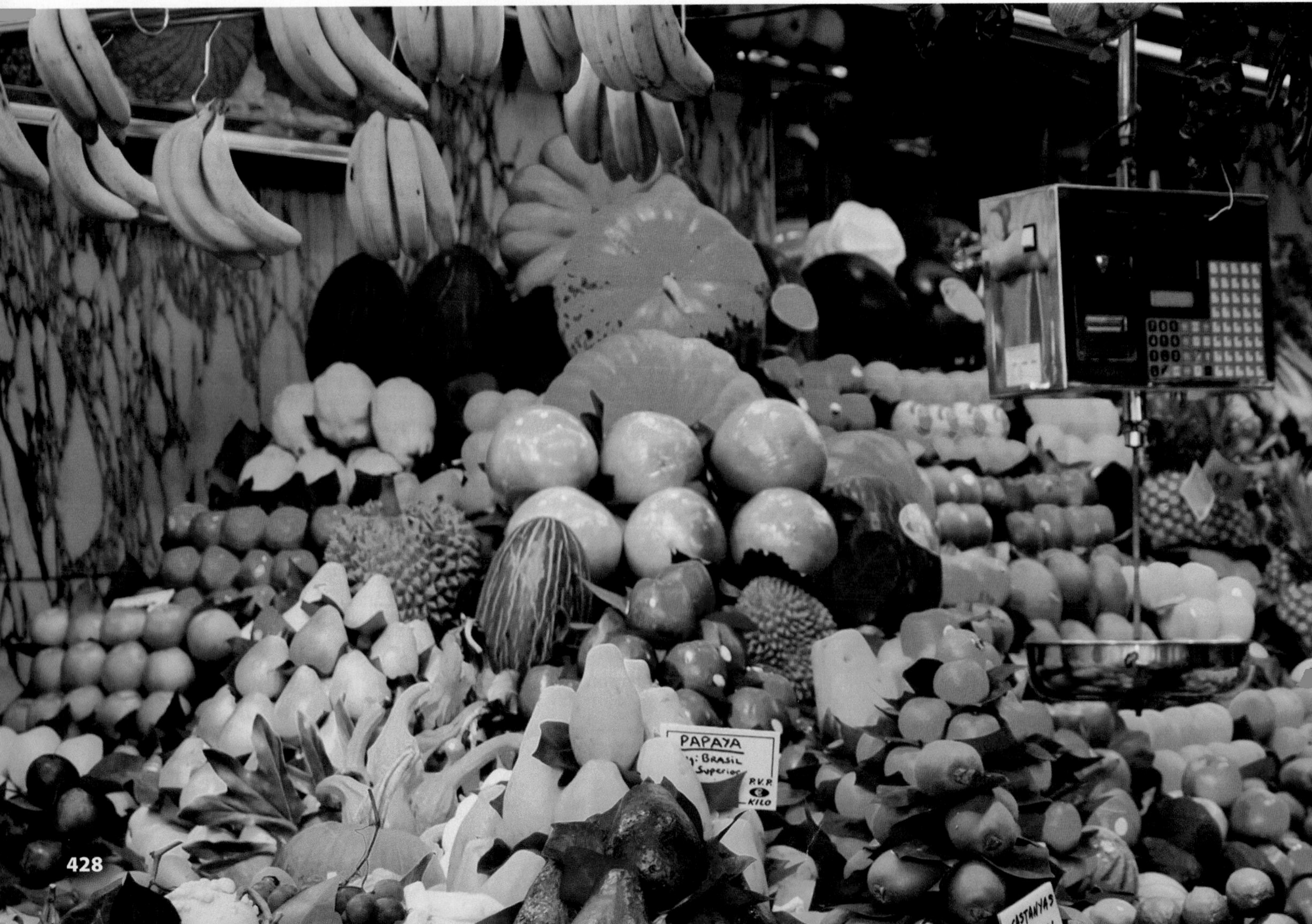
PAPAYA
KILO

outpaces our access to reliable data on their effectiveness. With a wealth of fad diets, it can be difficult to evaluate their soundness. If you wish to think about your own diet, you should begin with a clear understanding of the structure and function of your digestive system.

All animals must consume food to provide energy, obtain organic building blocks used to assemble new molecules, cells, and tissues, and gain essential nutrients. Your health and appearance depend on the quality of your diet and the proper functioning of your digestive system. This chapter focuses on human digestion and nutrition, starting with an overview of the various ways that animals process food. Along the way, we'll consider whether you are eating the right foods in the proportions needed to maintain good health.

BIG IDEAS

Obtaining and Processing Food
(21.1–21.3)

Animals ingest food, digest it in specialized compartments, absorb nutrients, and eliminate wastes.

The Human Digestive System
(21.4–21.13)

Food is processed sequentially in the mouth, stomach, and small intestine, where nutrients are absorbed.

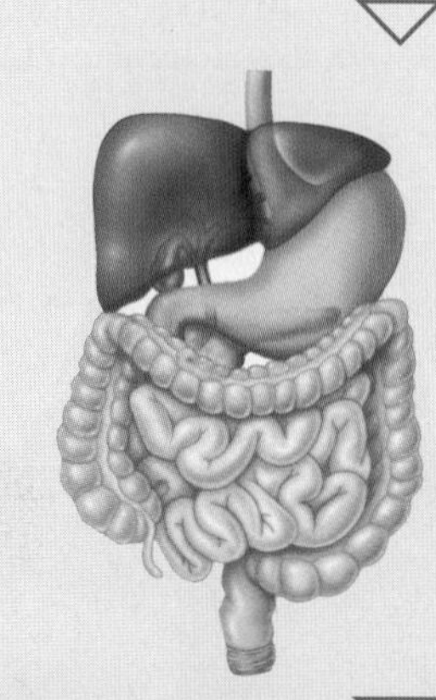

Nutrition
(21.14–21.21)

A healthy diet fuels activities, provides organic building blocks, and supplies nutrients a body cannot manufacture.

▷ Obtaining and Processing Food

21.1 Animals obtain and ingest their food in a variety of ways

All animals eat other organisms—dead or alive, whole or by the piece. Beyond that generalization, however, animal diets vary extensively. Animals fall into one of three dietary categories. **Herbivores**, such as cattle, gorillas, sea urchins, and snails, eat mainly plants and algae. **Carnivores**, such as lions, owls, spiders, and whales, mostly eat other animals. Animals that regularly consume both plants and animals are called **omnivores**. Omnivores include humans, as well as crows, cockroaches, and raccoons.

▲ **Figure 21.1A** A filter feeder: a humpback whale using baleen to strain food from seawater

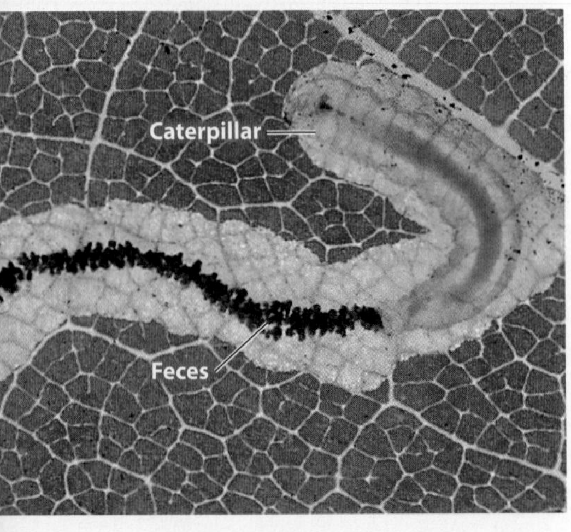

Caterpillar
Feces

▲ **Figure 21.1B** A substrate feeder: a caterpillar eating its way through the soft tissues inside an oak leaf

Several types of feeding mechanisms have evolved among animals. Many aquatic animals are **filter feeders**, which sift small organisms or food particles from water. For example, humpback whales use baleen, comb-like plates attached to the upper jaw, to strain small invertebrates and fish from enormous volumes of water (**Figure 21.1A**). Many other filter feeders are invertebrates, including most sponges, which draw nutrient-containing water in through their pores. Filter feeding is a type of suspension feeding (Module 18.5), which also involves capturing food particles from the surrounding medium.

Substrate feeders live in or on their food source and eat their way through it. **Figure 21.1B** shows a leaf miner caterpillar, the larva of a moth. The dark spots on the leaf are a trail of feces that the caterpillar left in its wake. Other substrate feeders include maggots (fly larvae), which burrow into animal carcasses; and earthworms, which eat their way through soil, digesting partially decayed organic material and helping to aerate and fertilize the soil as they go.

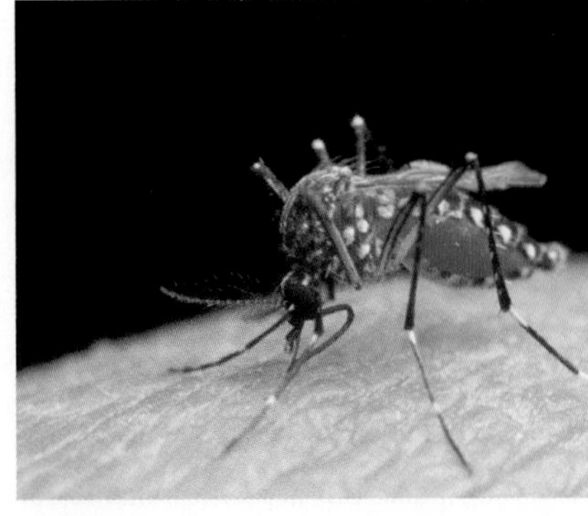

▲ **Figure 21.1C** A fluid feeder: a mosquito sucking blood

Fluid feeders suck nutrient-rich liquids from a living host. Aphids, for example, tap into the sugary sap in plants. Bloodsuckers, such as mosquitoes and ticks, pierce animals with hollow needlelike mouthparts. The female mosquito in **Figure 21.1C** has just filled her abdomen with a meal of human blood. (Only female mosquitoes suck blood; males live on plant nectar.) In contrast to such parasitic fluid feeders, some fluid feeders actually benefit their hosts. For example, hummingbirds and bees move pollen between flowers as they fluid-feed on nectar.

Most animals, including humans, are **bulk feeders** that ingest large pieces of food. **Figure 21.1D** shows a gray heron preparing to swallow its prey whole. A bulk feeder may use tentacles, pincers, claws, poisonous fangs, jaws, or teeth to kill its prey or to tear off pieces of meat or vegetation. Whatever the type of food or feeding mechanism, the processing of food involves four stages, as we see next.

? **Snowy owls primarily eat small rodents, but will also eat smaller birds. Name their diet category and type of feeding mechanism.**

● Carnivore and bulk feeder

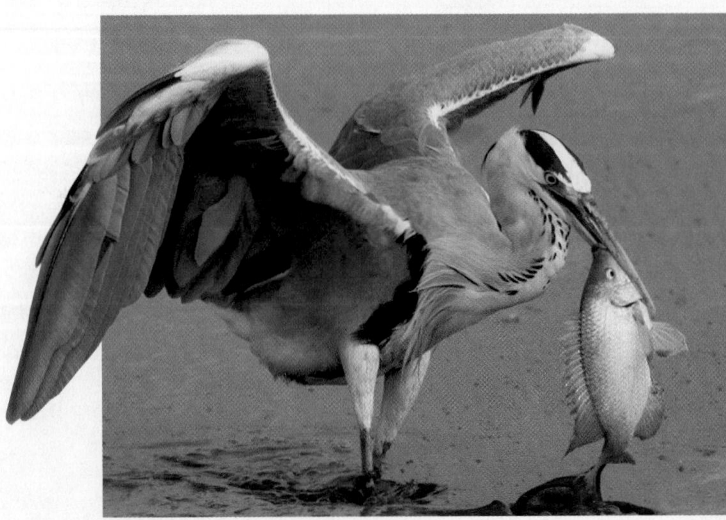

▲ **Figure 21.1D** A bulk feeder: a gray heron preparing to swallow a fish headfirst

21.2 Overview: Food processing occurs in four stages

So far we have discussed what animals eat and how they feed. As shown in **Figure 21.2A**, ❶ **ingestion**, the act of eating, is only the first of four distinct stages of food processing. The second stage, ❷ **digestion**, is the breakdown of food into molecules small enough for the body to absorb. Digestion typically occurs in two phases. First, food may be mechanically broken into smaller pieces, increasing the surface available for chemical processes. In animals with teeth, the process of chewing or tearing breaks large chunks of food into smaller ones.

The second phase of digestion involves chemical breakdown of food by specific enzymes. Chemical digestion is necessary because animals cannot directly use the proteins, carbohydrates, fats, and nucleic acids in food. These molecules are too large to pass through membranes and enter the cells of the animal. In addition, most food molecules are different from the molecules that make up an animal's body. Although all organisms use the same building blocks to make their macromolecules (the same 20 amino acids, for example), food is disassembled into the individual building blocks, which are then reassembled into the body's own molecules (a protein, for example).

The process of digestion breaks the polymers in food into monomers, allowing the building blocks contained within food to become accessible to the body. As shown in **Figure 21.2B**, proteins are split into amino acids, polysaccharides and disaccharides are broken down into monosaccharides, and nucleic acids are split into nucleotides (and their components). Fats are split into their components, glycerol and fatty acids. The animal can then use these small molecules to make the specific large molecules it needs through dehydration synthesis reactions.

In the third stage of food processing, ❸ **absorption**, the cells lining the digestive tract absorb the products of digestion—small molecules such as amino acids and simple sugars. From the digestive tract, these nutrients travel in the blood to body cells, where they are used to build a cell's large molecules or are broken down further to provide energy. In an animal that eats much more than its body immediately uses, many of the nutrient molecules are converted to fat for storage. In the fourth and last stage of food processing, ❹ **elimination**, undigested material passes out of the digestive tract.

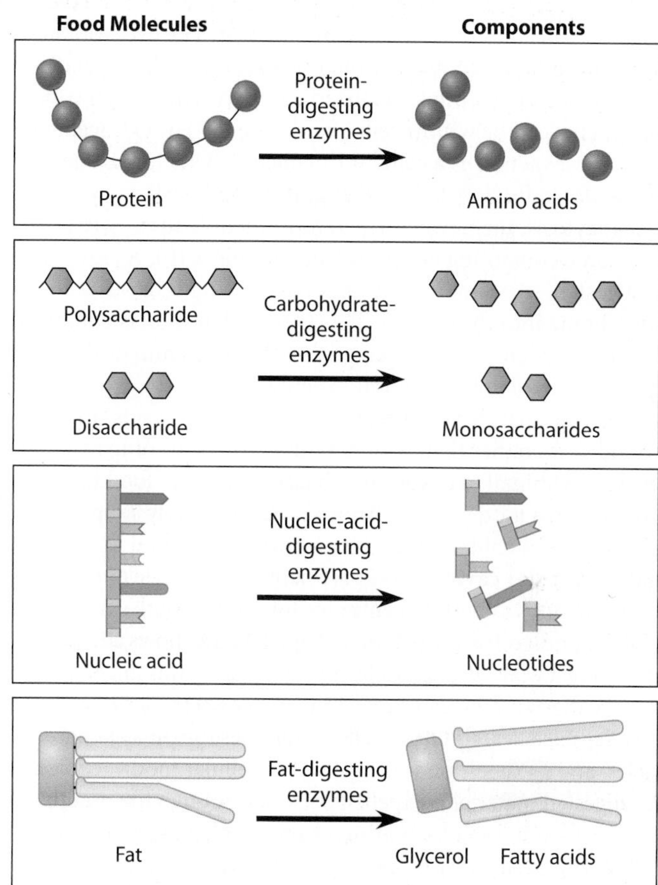

Food Molecules
Components
Protein-digesting enzymes
Protein
Amino acids
Polysaccharide
Carbohydrate-digesting enzymes
Disaccharide
Monosaccharides
Nucleic-acid-digesting enzymes
Nucleic acid
Nucleotides
Fat-digesting enzymes
Fat
Glycerol
Fatty acids

▲ **Figure 21.2B** Chemical digestion: the breakdown of large organic molecules to their components

? **What are the two phases of digestion that take place in your mouth?**

Mechanical breakdown and chemical breakdown

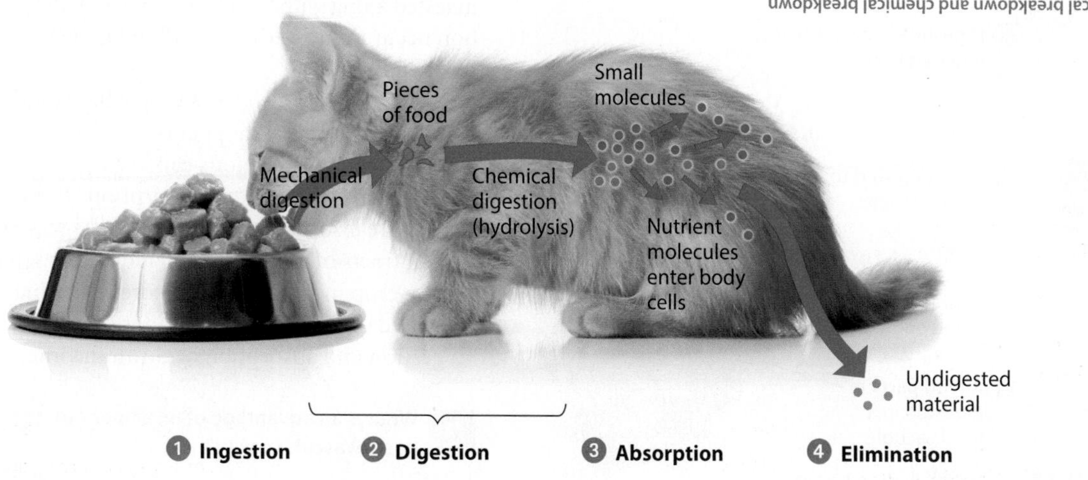

Pieces of food
Small molecules
Mechanical digestion
Chemical digestion (hydrolysis)
Nutrient molecules enter body cells
Undigested material
❶ Ingestion
❷ Digestion
❸ Absorption
❹ Elimination

▲ **Figure 21.2A** The four main stages of food processing

21.3 Digestion occurs in specialized compartments

How can an animal digest food without also digesting its own tissues? After all, digestive enzymes hydrolyze the same biological molecules that animals are made of. The evolutionary adaptation found in most animal species is the chemical digestion of food within specialized compartments. Such compartments can be within cells (food vacuoles) or extracellular.

A food vacuole is a cellular organelle in which enzymes break down food. After a cell engulfs a food particle by phagocytosis, the newly formed food vacuole fuses with a lysosome containing enzymes (see Module 4.10). As food is digested, small food molecules pass through the vacuole membrane into the cytosol. This type of digestion is common in single-celled protists; sponges are the only animals that digest their food entirely in food vacuoles.

Most animals have a digestive compartment that is surrounded by, rather than within, body cells. Such compartments enable an animal to devour much larger pieces of food than could fit in a food vacuole. Animals with relatively simple body plans, such as cnidarians and flatworms, digest food within a **gastrovascular cavity**, a compartment with a single opening that functions as both the entrance for food (mouth) and the exit for undigested waste (anus). **Figure 21.3A** shows a hydra digesting a water flea. ❶ Cells lining the gastrovascular cavity secrete digestive enzymes that ❷ break down the food into smaller particles. ❸ Other cells engulf these small food particles, and ❹ digestion is completed in cellular food vacuoles. Undigested material is expelled back out through the mouth.

Most animals have an **alimentary canal**, a digestive tube extending between two openings, a mouth at one end and an anus at the other. Because food moves in just one direction, specialized regions of the tube can digest and absorb nutrients in a stepwise fashion.

Food entering the mouth usually passes into the **pharynx**, or throat. Depending on the species, the **esophagus** may channel food to a crop, gizzard, or stomach. A **crop** is a pouch-like organ in which food is softened and stored. **Stomachs** and **gizzards** may also store food temporarily, but they are more muscular and they churn and grind the food. Chemical digestion and nutrient absorption occur mainly in the **intestine**. Undigested materials are expelled through the **anus**.

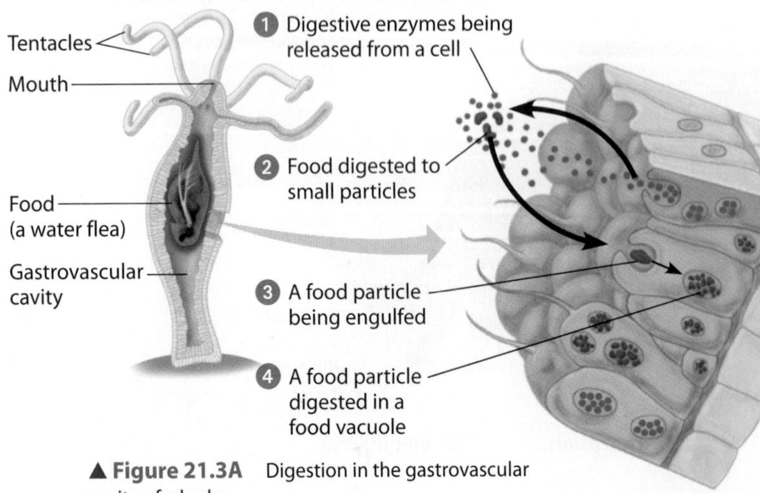

Tentacles
Mouth
Food
(a water flea)
Gastrovascular
cavity
❶ Digestive enzymes being released from a cell
❷ Food digested to small particles
❸ A food particle being engulfed
❹ A food particle digested in a food vacuole

▲ **Figure 21.3A** Digestion in the gastrovascular cavity of a hydra

Earthworm

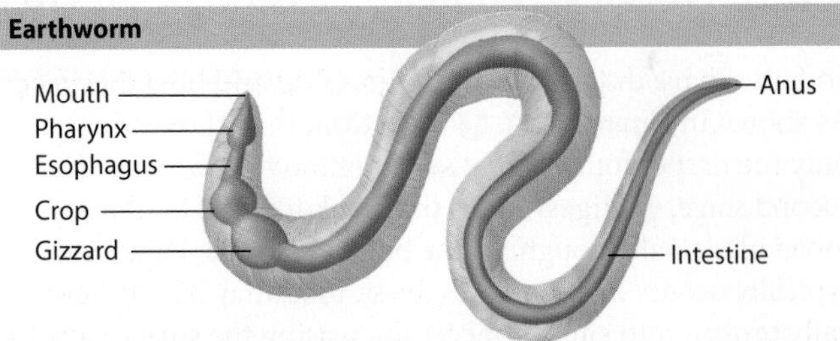

Mouth
Pharynx
Esophagus
Crop
Gizzard
Anus
Intestine

Grasshopper

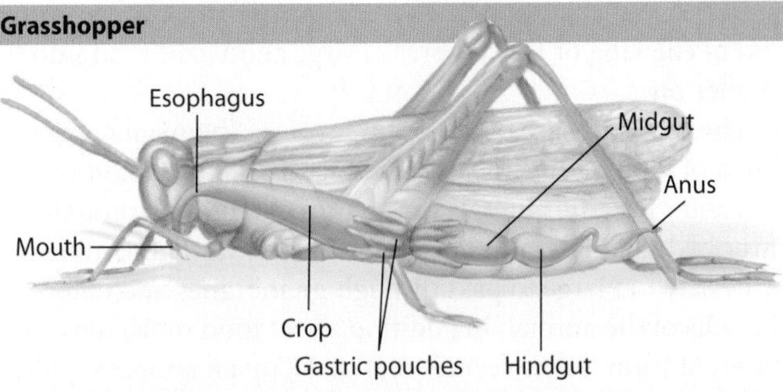

Esophagus
Midgut
Anus
Mouth
Crop
Gastric pouches
Hindgut

Bird

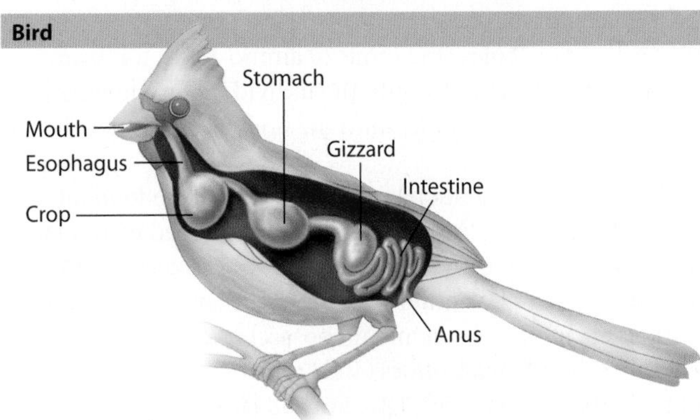

Stomach
Mouth
Esophagus
Gizzard
Intestine
Crop
Anus

▲ **Figure 21.3B** Three examples of alimentary canals

Figure 21.3B illustrates three examples of alimentary canals. The digestive tract of an earthworm includes a muscular pharynx that sucks food in through the mouth. Food passes through the esophagus and is stored in the crop. Mechanical digestion takes place in the gizzard, which pulverizes food with the aid of small bits of sand and gravel the organism has ingested along with the food. Chemical digestion and absorption occur in the intestine, and undigested material is expelled through the anus.

A grasshopper also has a crop where food is stored. Most digestion in a grasshopper occurs in the midgut region, where projections called gastric pouches function in digestion and absorption. The hindgut mainly reabsorbs water and compacts wastes. The digestive tracts of many birds include three separate chambers: a crop in which food is stored, a stomach, and a gravel-filled gizzard in which food is pulverized. Chemical digestion and absorption occur in the intestine.

? What is an advantage of an alimentary canal compared to a gastrovascular cavity?

● An alimentary canal has specialized regions, which can carry out digestion and absorption sequentially.

▷ The Human Digestive System

21.4 The human digestive system consists of an alimentary canal and accessory glands

As an introduction to our own digestive system, **Figure 21.4** illustrates the human alimentary canal (also called the gut, labeled in black on the left side of the figure) and its accessory glands: the salivary glands, gallbladder, liver, and pancreas (all labeled in blue on the right of figure). The glands secrete digestive chemicals that enter the alimentary canal through ducts (thin tubes).

You ingest and chew food in your mouth, or **oral cavity**, and then use your tongue to push the food into your pharynx. Once you swallow, muscles propel the food through your alimentary canal by **peristalsis**, alternating waves of contraction and relaxation of the smooth muscles lining the canal (see Module 21.6). It is peristalsis that enables you to process and digest food even while lying down. After chewing a bite of food, it takes 5–10 seconds for it to pass from the pharynx down the esophagus and into your stomach.

As shown in the blow-up view on the right, muscular ringlike valves, called **sphincters**, regulate the passage of food into and out of the stomach. The sphincter controlling the passage out of the stomach works like a drawstring to close the stomach off, keeping food there for about 2–6 hours, long enough for stomach acids and enzymes to begin digestion. The final steps of digestion and nutrient absorption occur in the small intestine over a period of 5–6 hours. Undigested material moves slowly through the large intestine (taking 12–24 hours), and feces are stored in the rectum and then expelled through the anus.

In the next several modules, we follow a snack—an apple and some crackers and cheese—through your alimentary canal to see in more detail what happens to the food in each of the processing stations along the way.

? **If you swallow while upside down, how can food get from your mouth to your stomach?**

● Peristalsis propels the food along the alimentary canal no matter your orientation.

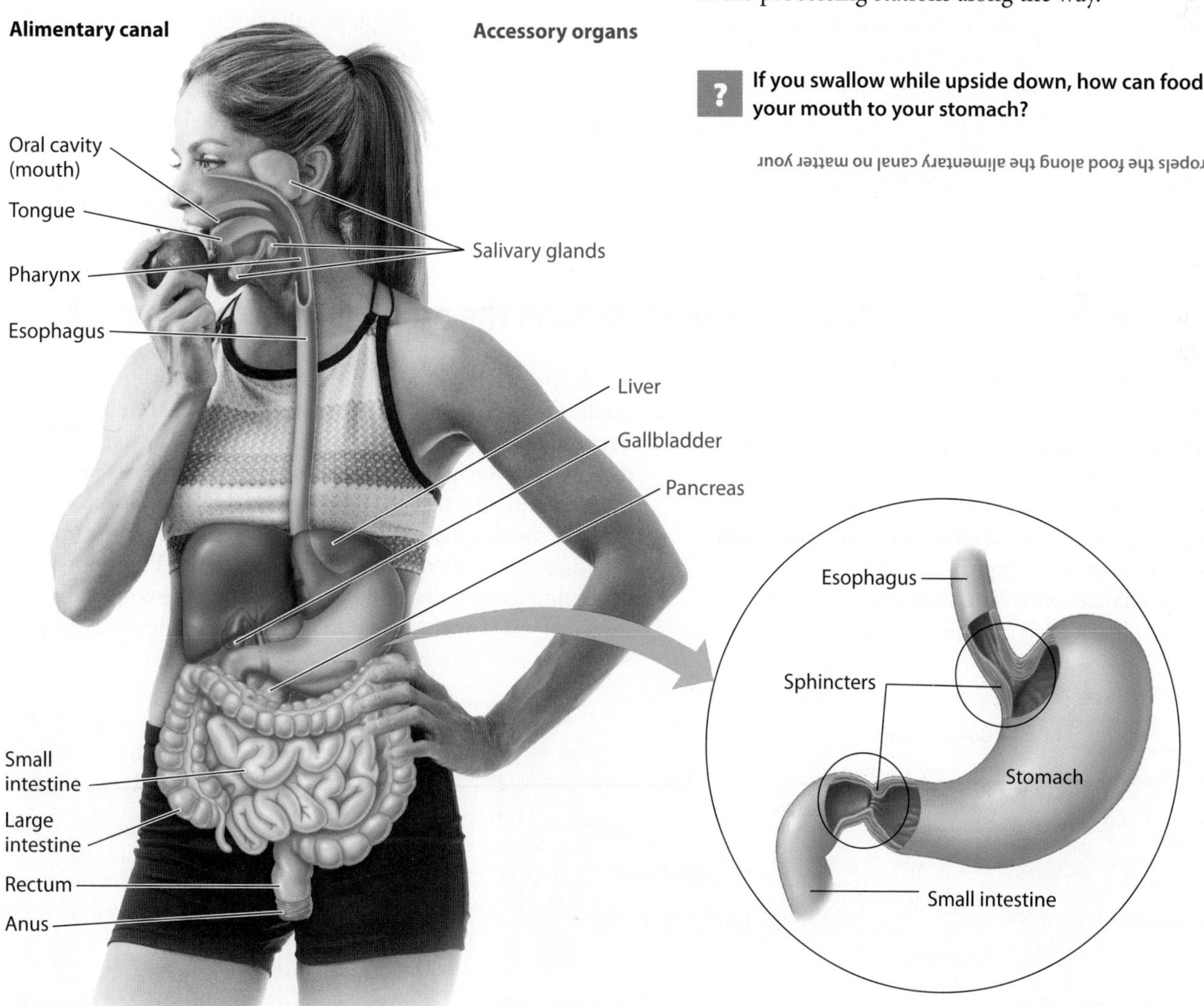

Alimentary canal
Accessory organs
Oral cavity (mouth)
Tongue
Pharynx
Esophagus
Salivary glands
Liver
Gallbladder
Pancreas
Small intestine
Large intestine
Rectum
Anus
Esophagus
Sphincters
Stomach
Small intestine

▲ **Figure 21.4** The human digestive system

Try This On a separate piece of paper, write down all of the parts of the human alimentary canal in their proper order.

21.5 Digestion begins in the oral cavity

The oral cavity is where we ingest food and begin to digest it. Mechanical digestion begins here as teeth cut, smash, and grind food. Breaking food into smaller bits makes it easier to swallow and exposes more food surface to digestive enzymes. As **Figure 21.5** shows, you have several kinds of teeth that aid in this breaking. Starting at the front on one side of the upper or lower jaw, there are two bladelike incisors. You use these for biting into your apple. Behind the incisors is a single pointed canine tooth. (Canine teeth are much bigger in carnivores—think of the fangs of a dog or wolf—which use them to kill and rip apart prey.) Next come two premolars and three molars, which grind and crush your food. You use these to pulverize your apple, cheese, and crackers. The third molar, a "wisdom" tooth, does not appear in all people, and in some people it pushes against the other teeth and must be removed.

The anticipation of food stimulates three pairs of **salivary glands** to deliver saliva through ducts to the oral cavity. The presence of food in the oral cavity continues to stimulate salivation. In a typical day, your salivary glands secrete more than a liter (1 L) of saliva. You can see the duct opening in Figure 21.5. All three pairs of salivary glands are visible in Figure 21.4.

Saliva contains several substances important in food processing. A slippery glycoprotein (carbohydrate-protein complex) protects the soft lining of your mouth and lubricates food for easier swallowing. Buffers neutralize food acids, helping prevent tooth decay. Antibacterial agents kill many of the bacteria that enter your mouth with food. Saliva also contains the digestive enzyme amylase, which begins chemical digestion of the starch in your cracker into the disaccharide maltose.

Also prominent in the oral cavity is the tongue, a muscular organ covered with taste buds. Besides enabling you to taste your meal, the tongue manipulates food and helps shape it into a ball called a **bolus**. As you'll see in the next module, the tongue pushes the bolus into the pharynx during the act of swallowing.

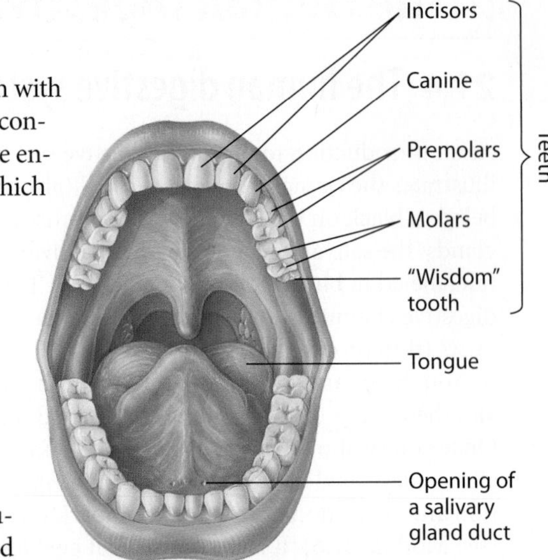

Incisors
Canine
Premolars
Molars
"Wisdom" tooth
Teeth
Tongue
Opening of a salivary gland duct

▲ **Figure 21.5** The human oral cavity

? Name one structure in the oral cavity that participates in mechanical digestion and one that participates in chemical digestion.

● mechanical: teeth; chemical: salivary glands

21.6 After swallowing, peristalsis moves food through the esophagus to the stomach

The pharynx, or throat, opens to two passageways: the esophagus (part of the digestive system) and the trachea (or windpipe, part of the respiratory system). Most of the time, as shown on the left in **Figure 21.6A**, the esophageal opening is closed off by a sphincter (indicated with blue arrows). Air enters your larynx (voice box) and flows past your vocal cords, through the trachea, to your lungs (black arrows).

This situation changes when you start to swallow. The tongue pushes a bolus of food into the pharynx, triggering the swallowing reflex (center of Figure 21.6A). Movement of the trachea tips a door-like flap of cartilage called the **epiglottis** over the opening to the trachea. Like a crossing guard at a dangerous intersection, the epiglottis directs the closing of the trachea, ensuring that the food will go down the esophagus. You can see this motion in the bobbing of your larynx (also called your Adam's apple) during swallowing. The esophageal sphincter relaxes, and the bolus enters the esophagus (green arrow). As shown on the right side of Figure 21.6A, the

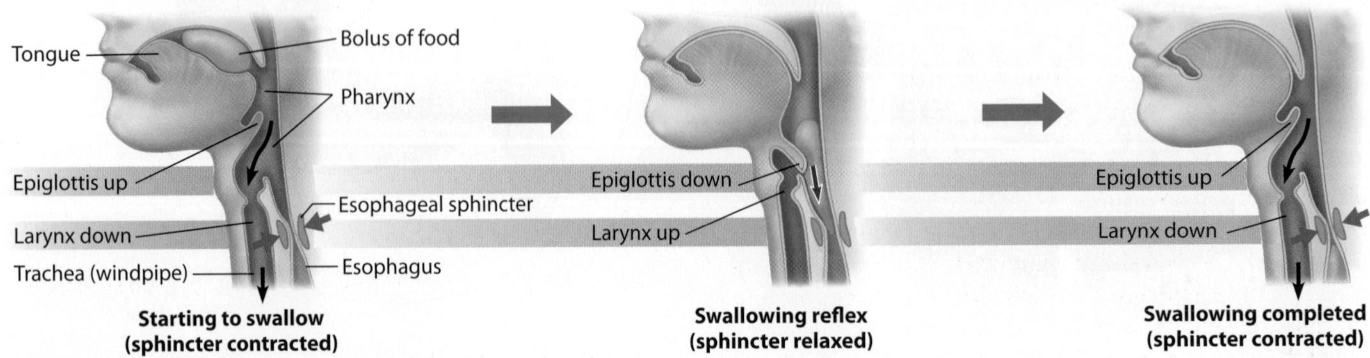

Tongue
Bolus of food
Pharynx
Epiglottis up
Esophageal sphincter
Larynx down
Trachea (windpipe)
Esophagus
Starting to swallow (sphincter contracted)
Epiglottis down
Larynx up
Swallowing reflex (sphincter relaxed)
Epiglottis up
Larynx down
Swallowing completed (sphincter contracted)

▲ **Figure 21.6A** The human swallowing reflex

Try This Place your hand on your Adam's apple and swallow. As you do, follow the three steps in this figure, visualizing them occurring within your own throat.

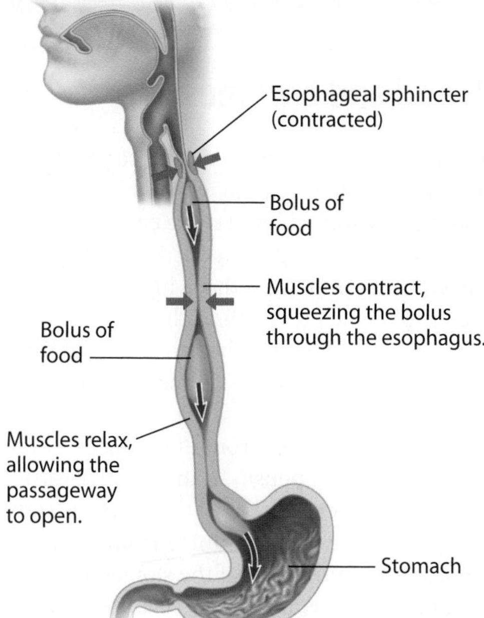

Esophageal sphincter (contracted)
Bolus of food
Muscles contract, squeezing the bolus through the esophagus.
Bolus of food
Muscles relax, allowing the passageway to open.
Stomach

▲ **Figure 21.6B** A food bolus shown at three points as it moves through the esophagus

esophageal sphincter then contracts above the bolus, and the epiglottis tips again, reopening the breathing tube.

The esophagus is a muscular tube that conveys food from the pharynx to the stomach. The muscles at the top of the esophagus are under voluntary control; thus, you begin the act of swallowing voluntarily. But then involuntary contractions of smooth muscles in the rest of the esophagus take over.

Figure 21.6B shows how muscle contractions—peristalsis—squeeze a bolus toward the stomach; in this figure, one bolus of food is shown at three successive locations as it moves through the esophagus. Muscle contractions continue in waves until the bolus enters the stomach. Peristalsis also moves digesting food through the intestines.

The structure of the esophagus fits its function. It has tough yet elastic connective tissues that allow it to stretch to accommodate a bolus, layers of circular and longitudinal smooth muscles for peristalsis, and a stratified epithelial lining that replenishes cells abraded off during swallowing. The length of the esophagus varies by species. For example, fishes have no lungs to bypass and have a very short esophagus. And it will come as no surprise that giraffes have a very long esophagus.

? What is happening in the trachea when food "goes down the wrong pipe"?

● An incorrectly positioned epiglottis lets food enter the trachea (rather than the esophagus), which triggers a strong cough reflex.

21.7 The Heimlich maneuver can save lives

CONNECTION

As you read in the previous module, our breathing and swallowing are carefully coordinated by the epiglottis, but sometimes our swallowing mechanism goes awry. A person may eat too quickly or fail to chew food thoroughly. Or a young child may swallow an object too big to pass through the esophagus. Such mishaps can lead to a blocked pharynx or trachea. The blockage may prevent air from flowing into the trachea, causing the person to choke. If breathing is not restored within minutes, brain damage or death may result.

To save someone who is choking, you need to quickly dislodge any foreign objects in the throat and get air flowing. This assistance can come through the use of the Heimlich maneuver, invented by Dr. Henry Heimlich in the 1970s.

The maneuver is usually performed on someone who is seated or standing up. Stand behind the victim and place your arms around the victim's waist. Make a fist with one hand, and place it against the victim's upper abdomen, well below the rib cage. Then place the other hand over the fist and press into the victim's abdomen with a quick upward thrust. When done correctly, the diaphragm is forcibly elevated, pushing air into the trachea. Repeat this procedure until the object is forced out of the victim's airway (**Figure 21.7**). You can even use your own fist or the back of a chair to force air upward and dislodge an object from your own trachea.

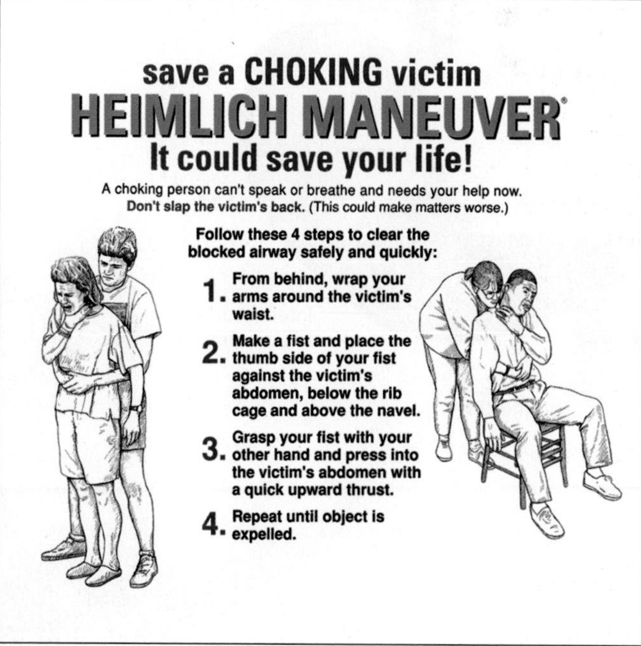

save a CHOKING victim
HEIMLICH MANEUVER®
It could save your life!
A choking person can't speak or breathe and needs your help now.
Don't slap the victim's back. (This could make matters worse.)
Follow these 4 steps to clear the blocked airway safely and quickly:
1. From behind, wrap your arms around the victim's waist.
2. Make a fist and place the thumb side of your fist against the victim's abdomen, below the rib cage and above the navel.
3. Grasp your fist with your other hand and press into the victim's abdomen with a quick upward thrust.
4. Repeat until object is expelled.

▲ **Figure 21.7** The Heimlich maneuver for helping choking victims

? During the Heimlich maneuver, what causes food to dislodge from the throat?

● The pressure of air being expelled from the lungs.

21.8 The stomach stores food and breaks it down with acid and enzymes

Having a stomach is the main reason you do not need to eat constantly. With its accordion-like folds and highly elastic wall, your stomach can stretch to accommodate about 2 L (more than half a gallon) of food and drink, usually enough to satisfy your needs for hours.

Some chemical digestion occurs in the stomach. The stomach secretes a digestive fluid called **gastric juice**, which is made up of a protein-digesting enzyme, mucus, and strong acid. The pH of gastric juice is about 2, acidic enough to dissolve iron nails and also most bacteria and other microbes that are swallowed with food. One function of the acid is to break apart the cells in food and denature (unravel) proteins.

The interior surface of the stomach wall is highly folded and is dotted with pits leading to tubular gastric glands **(Figure 21.8)**. The gastric glands have three types of cells that secrete different components of gastric juice. Mucous cells (shown here in dark pink) secrete mucus, which lubricates and protects the cells lining the stomach. Parietal cells (yellow) secrete hydrogen and chloride ions, which combine in the lumen (cavity) of the stomach to form hydrochloric acid (HCl). Chief cells (light pink) secrete pepsinogen, an inactive form of the enzyme pepsin.

The diagram on the far right of the figure indicates how active pepsin is formed. ❶ Pepsinogen and HCl are secreted into the lumen of the stomach. ❷ Next, the HCl converts some pepsinogen to pepsin. ❸ Pepsin itself then helps activate more pepsinogen, starting a chain reaction. This series of events is an example of positive feedback, in which the end product of a process promotes the formation of more end product.

What does all this active pepsin do? Pepsin begins the chemical digestion of proteins—those in your cheese snack, for instance. It splits the polypeptide chains of proteins into smaller polypeptides, which will be broken down further in the small intestine. Unlike most enzymes, pepsin works best under acidic conditions.

What prevents gastric juice from digesting away the stomach lining? Secreting pepsin in the inactive form of pepsinogen helps protect the cells of the gastric glands, and mucus helps protect the stomach lining from both pepsin and acid. Regardless, the epithelium of the stomach is constantly eroded. But don't worry, enough new cells are generated by mitosis to replace your stomach lining completely about every three days.

Another protection for the stomach is that gastric glands do not secrete acidic gastric juice constantly. Their activity is regulated by a combination of nerve signals and hormones. When you see, smell, or taste food, a signal from your brain stimulates your gastric glands. And as food arrives in your stomach, it stretches the stomach walls and triggers the release of the hormone **gastrin**. Gastrin circulates in the bloodstream, returning to the stomach (green dashed line in the top section of Figure 21.8), where it stimulates additional secretion of gastric juice. As much as 3 L of gastric juice may be secreted in a day.

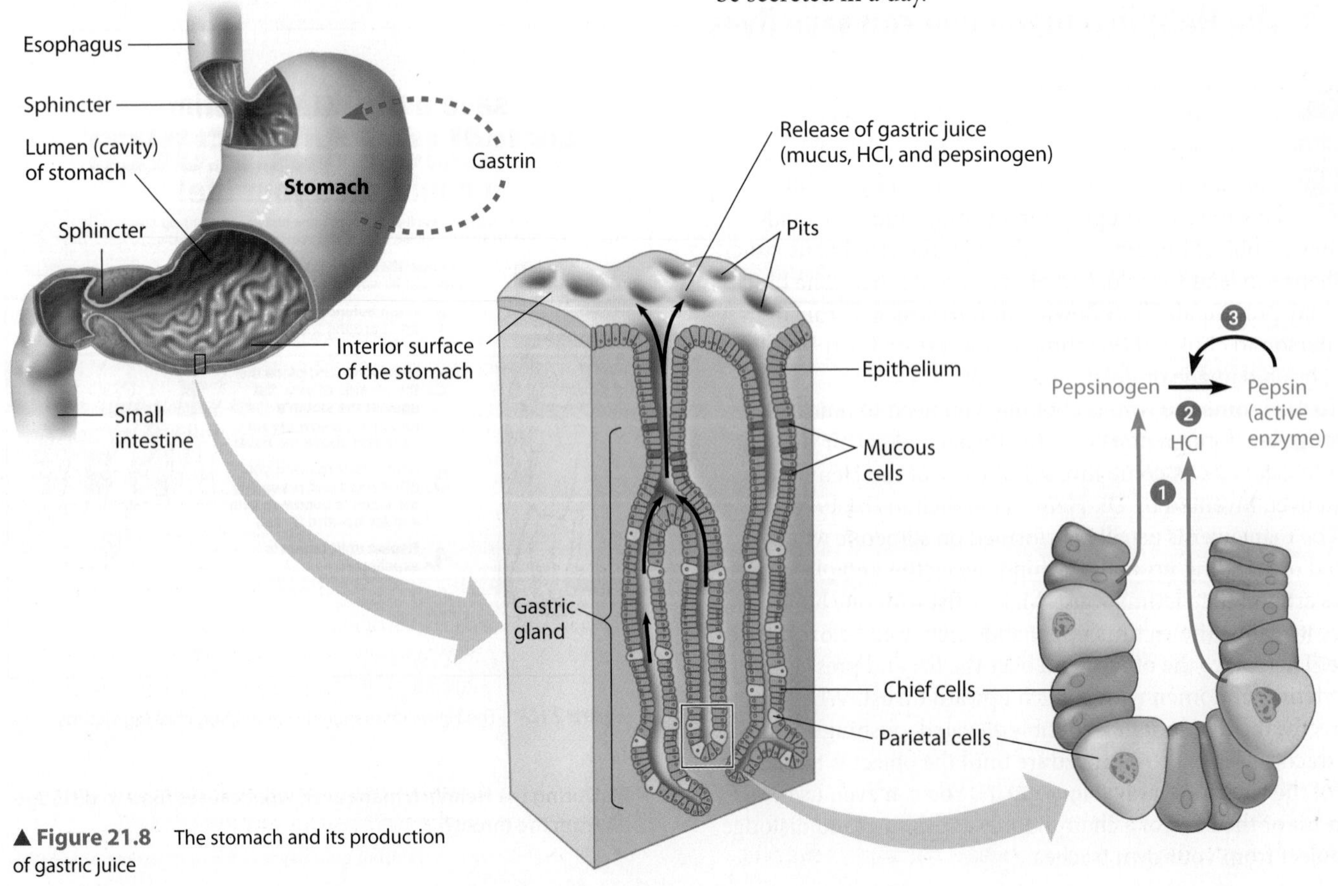

Esophagus
Sphincter
Lumen (cavity) of stomach
Stomach
Gastrin
Sphincter
Interior surface of the stomach
Small intestine
Release of gastric juice (mucus, HCl, and pepsinogen)
Pits
Epithelium
Mucous cells
Gastric gland
Chief cells
Parietal cells
3
Pepsinogen
Pepsin (active enzyme)
2
HCl
1

▲ **Figure 21.8** The stomach and its production of gastric juice

What prevents too much gastric juice from being secreted? When the stomach contents become too acidic, this inhibits the release of gastrin. Lower levels of gastrin in the blood cause gastric glands to secrete less gastric juice. This is an example of a negative feedback mechanism (see Module 20.15).

About every 20 seconds, your stomach muscles contract, which churns and mixes your stomach contents. If you haven't eaten for several hours, the contractions may be strong: You may experience these contractions as hunger pangs. When food is present, these contractions mix food with enzymes; what began in the stomach as a recently swallowed apple, cracker, and cheese snack soon becomes an acidic, nutrient-rich broth known as **chyme**. The sphincter between the stomach and the small intestine regulates the downstream passage of chyme, which leaves the stomach and enters the small intestine a squirt at a time. It usually takes 2–6 hours for the stomach to completely empty after a meal. Stomach "growling" results when your stomach muscles contract after the stomach has been emptied.

We'll continue with the digestion of your snack in Module 21.10. But first, let's consider some digestive problems.

? **If you add pepsinogen to a test tube containing protein dissolved in distilled water, not much protein will be digested. What inorganic chemical could you add to the tube to accelerate protein digestion? What effect will it have?**

● HCl or some other acid will convert inactive pepsinogen to active pepsin, which will begin digestion of the protein and also activate more pepsinogen.

21.9 Digestive ailments include acid reflux and gastric ulcers

CONNECTION

A stomachful of digestive juice laced with strong acid breaks apart the cells in your food, kills bacteria, and begins the digestion of proteins. At the same time, these chemicals can be dangerous to unprotected cells.

The opening between the esophagus and the stomach is usually closed until a bolus arrives. Occasionally, however, acid reflux occurs. This backflow of chyme into the lower end of the esophagus causes the feeling we call heartburn—which would more properly be called "esophagus-burn."

Some people suffer acid reflux frequently and severely enough to harm the lining of the esophagus, a condition called GERD (gastroesophageal reflux disease). GERD can often be treated with lifestyle changes. Doctors usually recommend that patients stop smoking, avoid alcohol, lose weight, eat small meals, refrain from lying down for 2–3 hours after eating, and sleep with the head of the bed raised. Medications to treat GERD include antacids, which reduce stomach acidity, and drugs such as Pepcid AC, Zantac, or Prilosec, which have a different mechanism of action and are very effective at stopping acid production. When lifestyle changes and medications fail to alleviate the symptoms, surgery to strengthen the lower esophageal sphincter may be an option.

With all that acid, why aren't the cells that line the stomach damaged? These cells are indeed vulnerable to the corrosive effects of gastric juice, but a coating of mucus normally protects the stomach wall. Despite this defense, open sores called gastric ulcers can develop in the stomach wall. The symptoms of gastric ulcers usually include gnawing pain in the upper abdomen, often occurring a few hours after eating. For decades, doctors mistakenly thought that excess acid secretion due to psychological stress caused ulcers. However, in 1982, researchers Barry Marshall and Robin Warren reported that infection by an acid-tolerant bacterium called *Helicobacter pylori* causes ulcers (**Figure 21.9**). Their hypothesis was poorly received. Marshall then experimented on himself by drinking beef soup laced with *H. pylori* bacteria—and soon developed gastritis, a mild inflammation of the stomach.

How can a bacteria cause ulcers? The low pH of the stomach kills most microbes, but not *H. pylori*. This bacterium burrows beneath the mucus and releases harmful chemicals.

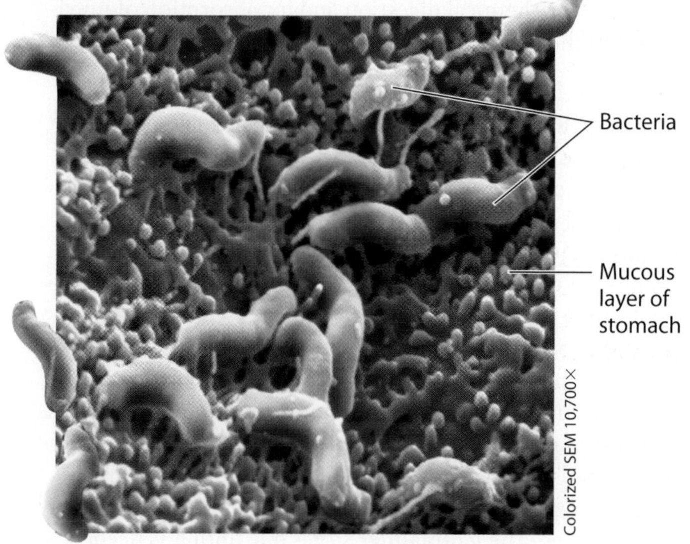
Bacteria
Mucous layer of stomach
Colorized SEM 10,700×

▲ **Figure 21.9** *Helicobacter pylori*, the bacteria that causes ulcers

Growth of *H. pylori* seems to result in a localized loss of protective mucus and damage to the cells lining the stomach. Numerous white blood cells move into the stomach wall to fight the infection, and their presence is associated with gastritis. Gastric ulcers develop when pepsin and hydrochloric acid destroy cells faster than the cells can regenerate. Eventually, the stomach wall may erode to the point that it actually has a hole in it. This hole can lead to a life-threatening infection within the abdomen or internal bleeding. *H. pylori* is found in up to 90% of ulcer and gastritis sufferers.

Gastric ulcers usually respond to a combination of antibiotics and bismuth (the active ingredient in Pepto-Bismol), which eliminates the bacteria and promotes healing. Although Marshall and Warren were awarded the 2005 Nobel Prize for their discovery (see Module 16.11), we do not recommend their unorthodox experimental methods!

? **In contrast to most microbes, the species that causes ulcers thrives in an environment with a very low ________.**

● pH

21.10 The small intestine is the major organ of chemical digestion and nutrient absorption

Let's return to your snack of an apple and some crackers and cheese: What happens after it passes out of the stomach and into the small intestine? At this point in its journey through the digestive tract, the food has been mechanically reduced to smaller pieces and mixed with digestive juices; it now resembles a thick soup. Chemically, the digestion of starch in the cracker began in the mouth (via amylase), and the breakdown of protein in the cheese began in the stomach (via pepsin). The rest of the digestion of the large molecules in your snack is achieved by an arsenal of enzymes in the **small intestine**. With a length of more than 6 m, the small intestine is the longest organ of the alimentary canal, but it is only about 2.5 cm in diameter (the width of a quarter). It is also the master organ for chemical digestion and for absorption of nutrients into the bloodstream.

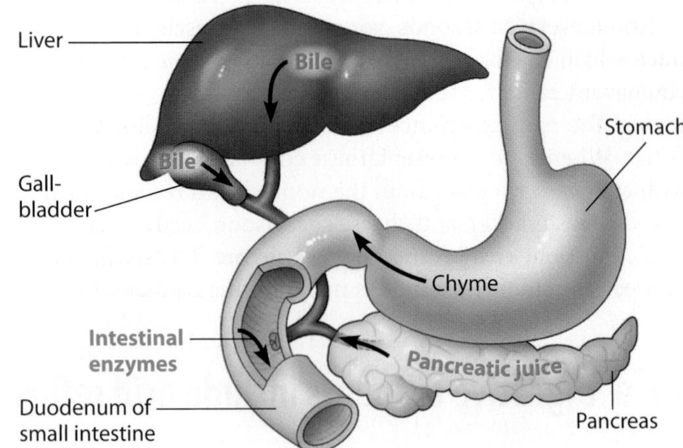

Liver
Bile
Stomach
Bile
Gall-
bladder
Chyme
Intestinal
enzymes
Pancreatic juice
Duodenum of
small intestine
Pancreas

▲ **Figure 21.10A** The duodenum and associated digestive organs

Sources of Digestive Enzymes and Bile The first section of the small intestine, about 25 cm (10 inches) in length, is called the **duodenum**. This is where chyme squirted from sthe stomach mixes with digestive juices from the pancreas, liver, gallbladder, and gland cells in the intestinal wall (**Figure 21.10A**). The **pancreas** produces pancreatic juice, a mixture of digestive enzymes and an alkaline solution that neutralizes the acidity of chyme as it enters the small intestine. (As you will learn in Chapter 26, the pancreas also produces hormones that regulate blood glucose levels.)

In addition to its many other functions, the **liver** produces a chemical mixture called bile. **Bile** contains bile salts, which act as emulsifiers (detergents) that break up fats into small droplets, making the fats more susceptible to attack by digestive enzymes. The **gallbladder** stores bile until it is needed in the small intestine. In response to chyme, hormones produced by the duodenum stimulate the release of bile from the liver, as well as digestive juices from the pancreas. Within the gallbladder, bile sometimes crystallizes to form gallstones, which can cause pain by obstructing the gallbladder or its ducts. Often the only cure is surgical removal of the gallbladder, which usually has no long-lasting effect on digestion because the liver still produces and secretes bile.

Digestion in the Small Intestine **Table 21.10** summarizes enzymatic digestion in the small intestine of all four types of large molecules—carbohydrates, proteins, nucleic acids, and fats. As we discuss the digestion of each, the table will help you keep track of the enzymes involved (in blue type).

The digestion of carbohydrates, such as those in your cracker, began in the oral cavity and is completed in the small intestine. The enzyme pancreatic amylase hydrolyzes polysaccharides into the disaccharide maltose. The enzyme maltase then splits maltose into the monosaccharide glucose. Maltase is one of a family of enzymes, each specific for the hydrolysis of a different disaccharide. For example, sucrase hydrolyzes table sugar (sucrose), and lactase digests lactose, common in milk and cheese. Undigested lactose cannot be absorbed, so it passes to the large intestine. There, prokaryotes consume the lactose, releasing gases such as methane. This produces the uncomfortable symptoms associated with lactose intolerance, such as painful gaseous bloating.

The small intestine also completes the digestion of proteins that was begun in the stomach. The pancreas and the duodenum produce enzymes that completely dismantle polypeptides into amino acids. The enzymes trypsin and chymotrypsin

TABLE 21.10 | ENZYMATIC DIGESTION IN THE SMALL INTESTINE

	Enzyme		Enzyme	
Carbohydrates				
Polysaccharides	Pancreatic amylase →	Maltose (and other disaccharides)	Maltase, sucrase, lactase, etc. →	Monosaccharides
Proteins				
Polypeptides	Trypsin, chymotrypsin →	Smaller polypeptides	Various peptidases →	Amino acids
Nucleic Acids				
DNA and RNA	Nucleases →	Nucleotides	Other enzymes →	Nitrogenous bases, sugars, and phosphates
Fats				
Fat globules	Bile salts →	Fat droplets	Lipase →	Fatty acids and glycerol

break polypeptides into smaller polypeptides. Several types of enzymes called peptidases then split off one amino acid at a time from these smaller polypeptides. Working together, this enzyme team digests proteins much faster than any single enzyme could.

Yet another team of enzymes, the nucleases, hydrolyzes nucleic acids. Nucleases from the pancreas split DNA and RNA (which are present in the cells of food sources) into their component nucleotides. The nucleotides are then broken down into nitrogenous bases, sugars, and phosphates by other enzymes.

Digestion of fats is a special problem because fats are insoluble in water and tend to clump together in large globules. How is this problem solved? First, bile salts separate and coat smaller fat droplets, a process called emulsification. When there are many small droplets, a larger surface area of fat is exposed to lipase, a pancreatic enzyme that breaks fat molecules down into fatty acids and glycerol.

By the time the mixture of chyme and digestive juices has moved through your duodenum, chemical digestion of your snack is just about complete. The main function of the rest of the small intestine is to absorb nutrients.

Absorption in the Small Intestine While enzymatic hydrolysis proceeds, peristalsis moves the mixture of chyme and digestive juices along the small intestine. Most digestion is completed in the duodenum. The remaining regions of the small intestine, the jejunum and ileum, are the major site for absorption of nutrients. Structurally, the small intestine is well suited for its task of absorbing nutrients. As you can see in **Figure 21.10B**, the inner wall of the small intestine has large circular folds with numerous small, finger-like projections called **villi** (singular, *villus*). Each of the epithelial cells on the surface of a villus has many tiny projections, called **microvilli**. This combination of folds and projections greatly increases the surface area across which nutrients are absorbed. Indeed, the lining of your small intestine has a huge surface area—roughly 300 m^2, about the size of a tennis court! This enormous surface area greatly increases the rate of nutrient absorption.

Some nutrients are absorbed by simple diffusion; others are pumped against concentration gradients into the epithelial cells. Notice that a small lymph vessel (shown in yellow in the figure) penetrates the core of each villus. After fatty acids and glycerol are absorbed by an epithelial cell, these building blocks are recombined into fats, which are then coated with proteins and transported into a lymph vessel. These vessels are part of the lymphatic system (see Module 24.4).

Notice that each villus is surrounded by a network of capillaries. Many absorbed nutrients, such as amino acids and sugars, pass directly out of the intestinal epithelium and then across the thin walls of the capillaries into the blood. Where does this nutrient-laden blood go? To the liver, where we also head in the next module.

? **At what point do food molecules actually enter the body's cells?**

● During absorption into the epithelial cells lining the villi of the small intestine

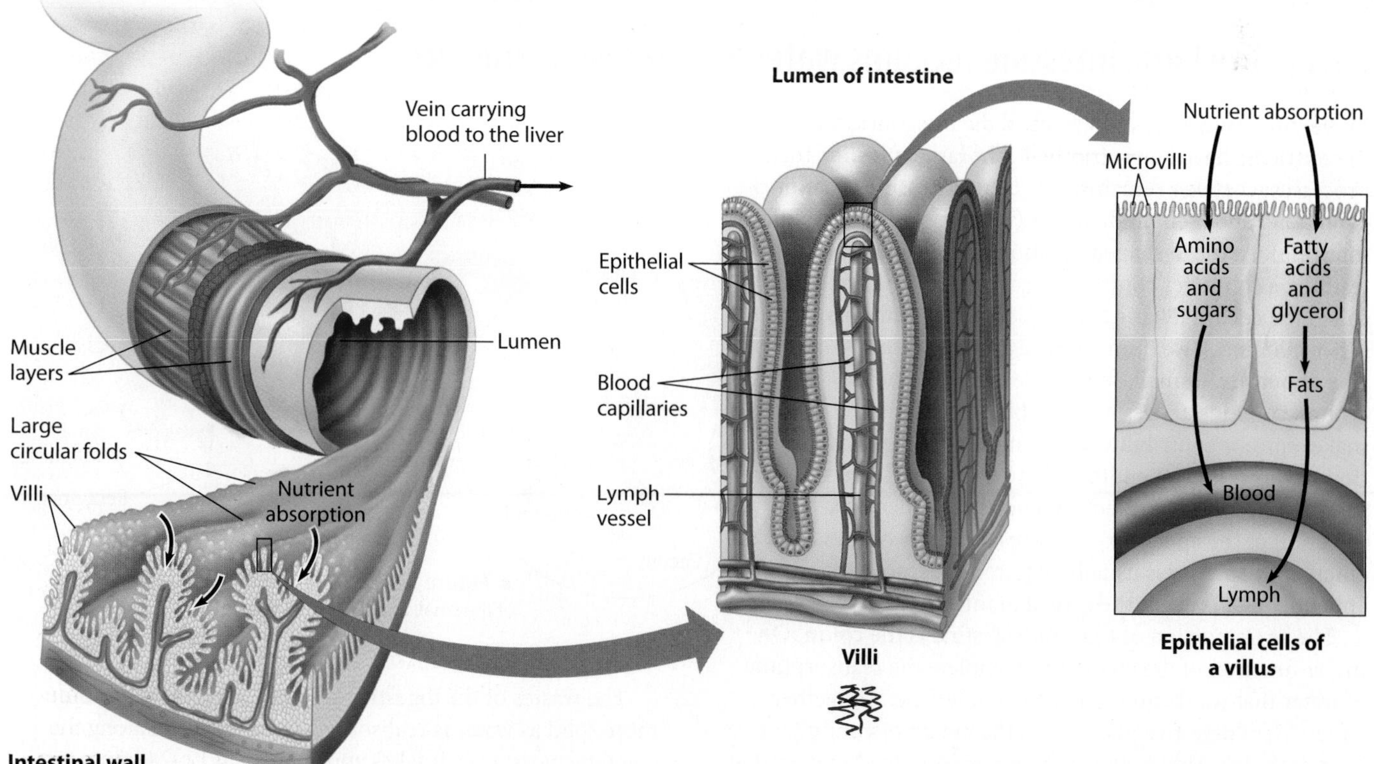

Vein carrying blood to the liver
Muscle layers
Lumen
Large circular folds
Villi
Nutrient absorption
Intestinal wall
Lumen of intestine
Epithelial cells
Blood capillaries
Lymph vessel
Villi
Nutrient absorption
Microvilli
Amino acids and sugars
Fatty acids and glycerol
Fats
Blood
Lymph
Epithelial cells of a villus

▲ **Figure 21.10B** Structure and function of the small intestine

21.11 The liver processes and detoxifies blood from the intestines

The liver has a strategic location in your body—between the intestines and the heart. Capillaries from the small and large intestines converge into veins that lead into the **hepatic portal vein** (Figure 21.11). This large vessel transports blood to the liver, thus giving the liver first access to nutrients absorbed in the intestines. The liver removes excess glucose from the blood and converts it to glycogen (a polysaccharide), which is stored in liver cells. In balancing the storage of glycogen with the release of glucose to the blood, your liver plays a key role in regulating metabolism.

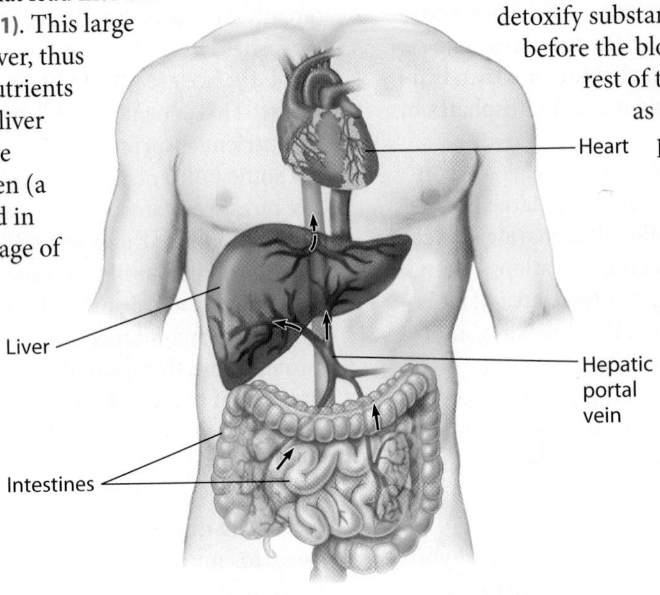

The liver also converts many of the nutrients it receives to new substances. In other words, blood leaving the liver may have a very different nutrient makeup than the blood that entered. For example, liver cells synthesize many essential proteins, such as plasma proteins important in blood clotting and in maintaining the osmotic balance of the blood, and lipoproteins that transport fats and cholesterol to body cells. If your diet includes too many calories, the liver converts the excess to fat, which is stored in your body.

Given its central location, the liver can modify and detoxify substances absorbed by the digestive tract before the blood carries these materials to the rest of the body. It converts toxins such as alcohol or other drugs to inactive products that are excreted in the urine. These breakdown products are what are looked for in urine tests for various drugs. As liver cells detoxify alcohol or process some over-the-counter and prescription drugs, however, they can be damaged. The combination of alcohol and certain drugs, such as acetaminophen, is particularly harmful.

The liver and other accessory organs empty into the small intestine. From there, what remains of food passes into the large intestine, the subject of the next module.

Heart

Liver

Hepatic portal vein

Intestines

▲ **Figure 21.11** The hepatic portal vein carrying blood from the intestines to the liver

? **In what way does the location of the liver in the body aid its functions?**

● As blood is delivered directly from the intestines, the liver can process and regulate the absorbed nutrients and remove toxic substances.

21.12 The large intestine reclaims water and compacts the feces

By the time your snack has reached the intestines, most of the nutrients have been absorbed. The large intestine then processes whatever remains. The **large intestine** is about 1.5 m long and 5 cm in diameter (twice as wide as the small intestine). At the T-shaped junction of the small and large intestines, a sphincter controls passage into a small pouch called the **cecum** (Figure 21.12). Compared with many other mammals, we humans have a small cecum. The **appendix**, a small, finger-like extension of the cecum, contains a mass of white blood cells that make a minor contribution to immunity. If the junction between the appendix and the rest of the large intestine becomes blocked, appendicitis—a bacterial infection of the appendix—may result. If this occurs, emergency surgery is usually required to remove the appendix and prevent the spread of infection.

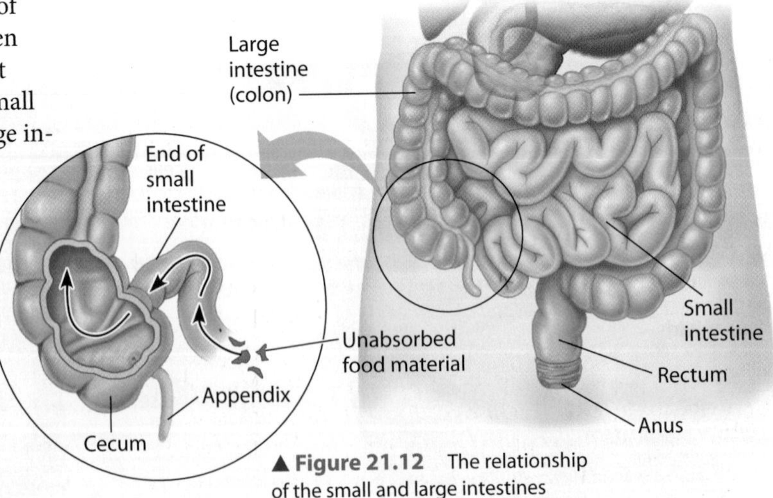

Large intestine (colon)
End of small intestine
Unabsorbed food material
Appendix
Cecum
Small intestine
Rectum
Anus

▲ **Figure 21.12** The relationship of the small and large intestines

The main portion of the large intestine is the **colon**. One major function of the colon is to complete the reabsorption of water that was begun in the small intestine. Altogether, about 7 L of digestive juice enters the lumen of your digestive tract each day. About 90% of the water contained in digestive juice is absorbed back into the blood via osmosis by the small intestine and colon.

The wastes of the digestive system, called **feces**, become more solid as water is reabsorbed and they move along the colon by peristalsis. It takes approximately 12–24 hours for material to travel the length of the colon. The feces consist mainly of indigestible plant fibers—cellulose from your apple, for instance—and enormous numbers of prokaryotes that normally live in the colon. Some colon bacteria, such as *Escherichia coli*, produce important vitamins, including several B vitamins and vitamin K. These substances are absorbed into the bloodstream and supplement your dietary intake of vitamins.

Feces are stored in the final portion of the colon, called the **rectum**, until they can be eliminated. Contractions of the colon create the urge to defecate. Two rectal sphincters, one voluntary and the other involuntary, regulate the opening of the anus. When the voluntary sphincter is relaxed, contractions of the rectum expel feces.

If the lining of the colon is irritated—by a viral or bacterial infection, for instance—the colon is less effective in reclaiming water, and diarrhea may result. The opposite problem, constipation, occurs when peristalsis moves the feces along too slowly; the colon reabsorbs too much water, and the feces become too compacted. Constipation often results from a diet that does not include enough plant fiber.

? **Explain why treatment with antibiotics for an extended period may cause a vitamin K deficiency.**

The antibiotics may kill the bacteria that synthesize vitamin K in the colon.

21.13 Evolutionary adaptations of vertebrate digestive systems relate to diet

EVOLUTION CONNECTION

Throughout the animal kingdom, there are many intriguing evolutionary adaptations of the digestive system. Often, these adaptations correlate with animals' diets. Large, expandable stomachs are common adaptations in carnivores, which may go a long time between meals and must eat as much as they can when they do catch prey. For example, a 200-kg lion can consume 40 kg (almost 90 pounds) of meat in one meal! After such a feast, the lion may not hunt again for a few days.

The length of an animal's digestive tract also correlates with diet. In general, herbivores and omnivores have longer alimentary canals, relative to their body size, than carnivores. This is because vegetation (and its associated plant cell walls) is more difficult to digest than meat: A longer canal provides more time for digestion and more surface area for the absorption of nutrients. A typical cow, for example, has an intestine about seven-fold longer than yours.

Most herbivorous animals have special chambers that house great numbers of bacteria and protists, a form of mutualistic symbiosis (see Module 37.4). The animals lack the enzymes needed to digest the cellulose in plants. The microbes break down cellulose to simple sugars, which the animals then absorb or obtain by digesting the microbes.

Many herbivorous mammals—horses, elephants, and koalas, for example—house cellulose-digesting microbes in a large cecum. **Figure 21.13** compares the digestive tract of a carnivore, the coyote, with that of an herbivore, the koala. These two mammals are about the same size, but the koala's intestine is much longer and includes the longest cecum (about 2 m) of any animal of its size. With the aid of bacteria in its cecum, the koala gets almost all its food and water from the leaves of eucalyptus trees.

In rabbits and some rodents, cellulose-digesting bacteria live in the large intestine as well as in the cecum. Many of the nutrients produced by these microbes are initially lost in the feces because they do not go through the small intestine, the main site of nutrient absorption. Rabbits and rodents recover these nutrients by eating some of their feces, thus passing the food through the alimentary canal a second time. The feces from the second round of digestion, the familiar rabbit "pellets," are more compact and are not reingested.

The most elaborate adaptations for an herbivorous diet have evolved in the mammals called **ruminants**, which include cattle, sheep, and deer. The stomach of a ruminant has four chambers containing symbiotic microbes. A ruminant such as a cow periodically regurgitates food from the first two chambers and "chews its cud," exposing more plant fibers to its microbes for digestion. The cud is then swallowed and moves to the final stomach chambers, where digestion is completed. A cow actually obtains many of its nutrients by digesting the microbes along with the nutrients they produce. The microbes reproduce so rapidly that their numbers remain stable despite this constant loss.

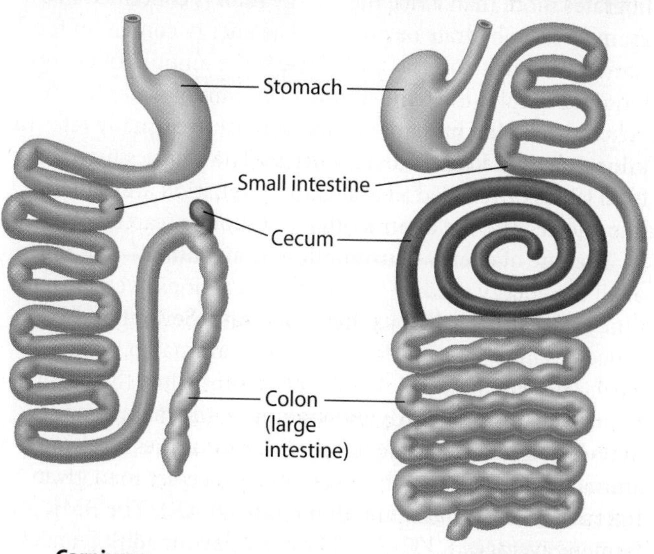
Stomach
Small intestine
Cecum
Colon (large intestine)
Carnivore
Herbivore

▲ **Figure 21.13** The alimentary canal in a carnivore (coyote) and an herbivore (koala)

? **Name two advantages of a long alimentary canal in herbivores.**

It provides increased time for processing of difficult-to-digest plant material and increased surface area for absorption of nutrients.

▷ Nutrition

21.14 An animal's diet must provide sufficient energy

To this point in the chapter, we have seen that digestion dismantles the large molecules in food into a form that intestinal cells can absorb. Once absorbed into the body, the small molecules from food are used to provide (1) organic building blocks for macromolecules, (2) chemical energy to power cellular work, and (3) essential nutrients to maintain health. All animals—whether herbivores like koalas, carnivores like snowy owls (**Figure 21.14**), or omnivores like humans—must satisfy these three requirements. In this and the next two modules, we will explore these aspects of a proper diet.

Getting the right mix of fuel is vital to all animals, including humans. For example, it takes energy to read this book. It also takes energy to digest a snack, walk to class, and perform all the other activities done by your body. Cellular respiration produces the body's energy currency, ATP, by oxidizing organic molecules obtained from food (see Chapter 6). Normally, cells use carbohydrates and fats as fuel sources. Fats are especially rich in energy: The oxidation of a gram of fat liberates more than twice the energy than is contained in a gram of carbohydrate or protein. The energy content of food is measured in calories. One calorie is the amount of energy required to raise the temperature of a gram of water by 1°C. When discussing human diet and activity, we usually refer to **kilocalories** (1 **kcal** = 1,000 calories). The calories listed on food labels or referred to in regard to nutrition are actually kilocalories and are often written as Calories (capital C).

The rate of energy consumption by an animal—the sum of all the energy used by biochemical reactions over a given time interval—is called its **metabolic rate**. Several body processes must run continuously for an animal to remain alive. These include cell maintenance, breathing, the beating of the heart, and, in birds and mammals, the maintenance of body temperature. The number of kilocalories a resting animal requires to fuel these essential processes for a given time is called the **basal metabolic rate (BMR)**. The BMR for humans averages 1,300–1,500 kcal per day for adult females and about 1,600–1,800 kcal per day for adult males. In other words, an adult human performing absolutely no activity still requires about 1,500 kcal per day just to keep the body alive. About 60% of this energy is lost as heat that dissipates into the environment. In fact, your body produces as much heat as a 75-watt lightbulb (which is why a crowded room grows hot very quickly). Any additional activity, even reading this book, consumes kilocalories over and above the BMR. The more active you are, the greater your actual metabolic rate and the greater the number of kilocalories your body uses per day. Besides activity level, metabolic rate also depends on factors such as body size, age, stress level, and heredity.

▲ **Figure 21.14** A snowy owl hunting

TABLE 21.14 EXERCISE REQUIRED TO "BURN" THE CALORIES (KCAL) IN COMMON FOODS

	Jogging	Swimming	Walking
Speed of exercise	9 min/mi	30 min/mi	20 min/mi
kcal "burned" per hour	775	408	245
Cheeseburger (quarter-pound), 417 kcal	32 min	1 hr, 1 min	1 hr, 42 min
Pepperoni pizza (1 large slice), 280 kcal	22 min	42 min	1 hr, 8 min
Non-diet soft drink (12 oz), 152 kcal	12 min	22 min	37 min
Whole wheat bread (1 slice), 100 kcal	8 min	15 min	24 min

These data are for a person weighing 68 kg (150 pounds).

The examples in **Table 21.14** give you an idea of the amount of activity it takes for a 68-kg (150-pound) person to use up the kilocalories contained in several common foods. What happens when you take in more Calories than you use? Rather than discarding the extra energy, your cells store it in various forms. Your liver and muscles store energy in the form of glycogen, a polymer of glucose molecules. Most of us store enough glycogen to supply about a day's worth of basal metabolism. This is why some athletes "carbo load": Eating lots of carbohydrates the day before an athletic event ensures that the liver will contain a supply of ready-to-burn glycogen energy. Your body also stores excess energy as fat. If deprived of food, the average human's daily metabolic needs can be supplied by 0.3 kg (2/3 pound) of body fat. Therefore, most healthy people have enough stored fat to sustain them through several weeks of starvation. We discuss fat storage and its consequences in Module 21.19. But first let's consider the essential nutrients that must be supplied in the diet.

? **What is the difference between metabolic rate and basal metabolic rate?**

● Metabolic rate is the total energy used for all activities in a unit of time; BMR is the minimum number of kilocalories that a resting animal needs to maintain life's basic processes for a unit of time.

21.15 An animal's diet must supply essential nutrients

Besides providing fuel and raw organic materials, an animal's diet must also supply **essential nutrients**. These are materials that must be obtained in preassembled form because the animal's cells cannot make them from raw material. In other words, the absence of an essential nutrient makes you ill. Essential nutrients include essential fatty acids, essential amino acids, vitamins, and minerals. Some nutrients are essential for all animals, whereas others are needed only by certain species. For example, vitamin C is an essential nutrient for humans and other primates, but most animals can make vitamin C as needed and so need not ingest it.

A healthy human diet is rich in whole grains, vegetables, fruits, and calcium, along with moderate quantities of protein from lean meat, eggs, nuts, or beans. Nutritionists recommend limited consumption of fats and sugars (primarily to help maintain a healthy weight) as well as salt. In this module, we survey two important groups of essential nutrients: fatty acids and amino acids.

Essential Fatty Acids Our cells make fats and other lipids by combining fatty acids with other molecules, such as glycerol (see Module 3.8). We can make most of the fatty acids we need. Those we cannot make, called **essential fatty acids**, must be obtained from our diet. One essential fatty acid, linoleic acid (one of the omega-6 family of fatty acids), is used to make some of the phospholipids of cell membranes. Another essential fatty acid, alpha-linolenic acid, can be obtained from seed oils (such as canola and flax oils). Although we cannot make essential fatty acids, plants can. Plant-based foods (seeds, grains, and vegetables) usually provide us with ample amounts of essential fatty acids, and deficiencies are rare.

Essential Amino Acids Proteins are built from 20 different kinds of amino acids. Adult humans can make 12 of these amino acids from other compounds. The remaining eight, called **essential amino acids**, must be obtained from the diet. Infants also require a ninth, histidine. A deficiency of a single essential amino acid impairs protein synthesis and can lead to protein deficiency.

Different foods contain different proportions of amino acids. The simplest way to get all the essential amino acids is to eat meat or animal by-products such as eggs, milk, and cheese. The proteins in these products are said to be "complete" because they provide adequate amounts of all the essential amino acids. In contrast, most plant proteins are incomplete, deficient in one or more essential amino acids. If you are vegetarian (by choice, or, as for much of the world's population, by economic necessity), the key to good nutrition is to eat a varied diet of plant proteins that together supply all the essential amino acids.

Eating a combination of beans and corn, for example, can provide a vegetarian with all the essential amino acids (**Figure 21.15**). The combination of a legume (such as beans, peanuts, or soybeans) and a grain (such as wheat, corn, or rice) often provides the right balance. Most societies have, by trial and error, developed balanced diets that prevent protein deficiency. The Latin American staple of rice and beans is an example, as are the U.S. staple of peanut butter sandwiches and the Middle Eastern favorite of hummus (made from garbanzo beans) and pita bread. In the next module, we continue our look at essential nutrients—this time vitamins and minerals.

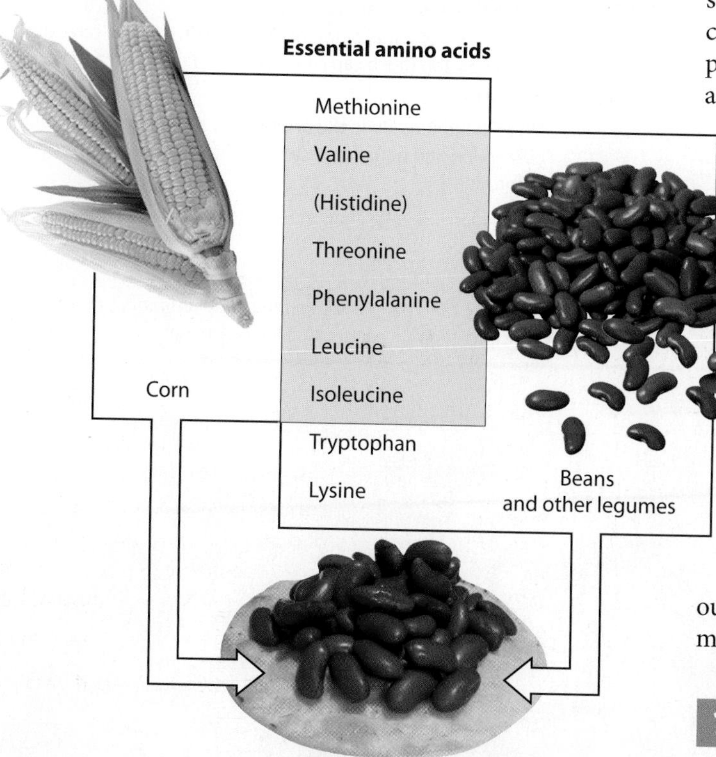
Essential amino acids
Methionine
Valine
(Histidine)
Threonine
Phenylalanine
Leucine
Isoleucine
Tryptophan
Lysine
Corn
Beans
and other legumes

▲ **Figure 21.15** Essential amino acids from a vegetarian diet

? Look carefully at Figure 21.15. A diet consisting strictly of corn would probably result in a deficiency of which essential amino acids?

● Tryptophan and lysine

21.16 A proper human diet must include sufficient vitamins and minerals

A **vitamin** is an organic nutrient required in very small amounts in your diet. For example, 1 tablespoon of vitamin B_{12} could provide the daily requirement for nearly a million people. Depending on the vitamin, the required daily amount ranges from about 0.01 to 100 mg. To help you imagine how small these amounts are, consider that a small peanut weighs about 1 g, so 100 mg would be one-tenth of a small peanut. And some vitamin requirements are one-ten-thousandth of that!

Table 21.16A lists 13 essential vitamins and their major dietary sources. Vitamins are divided into two broad classes: water-soluble and fat-soluble.

TABLE 21.16A | VITAMIN REQUIREMENTS OF HUMANS

Vitamin	Major Dietary Sources	Interesting Facts
Water-soluble vitamins		
Vitamin B_1 (thiamine)	Pork, legumes, peanuts, whole grains	Refined grains (e.g., polished white rice) lack thiamine; deficiency causes the disease beriberi
Vitamin B_2 (riboflavin)	Dairy products, organ meats, enriched grains, vegetables	Deficiency causes photophobia (aversion to light) and skin cracks
Vitamin B_3 (niacin)	Nuts, meats, fish, grains	High doses reduce cholesterol; too little leads to the potentially deadly disease pellagra
Vitamin B_5 (pantothenic acid)	Meats, dairy products, whole grains, fruits, vegetables	Component of coenzyme A; deficiency is rare but can cause fatigue
Vitamin B_6 (pyridoxine)	Meats, vegetables, whole grains, milk, legumes	Coenzyme used in amino acid metabolism; deficiency is rare
Vitamin B_7 (biotin)	Legumes, most vegetables, meats, milk, egg yolks	For most people, adequate amounts are provided by intestinal bacteria
Vitamin B_9 (folic acid)	Green vegetables, oranges, nuts, legumes, whole grains	Recommended as a supplement for women of childbearing age because it cuts in half the risk of some birth defects
Vitamin B_{12} (cobalamin)	Animal products: Meats, eggs, dairy	Vegans need to be careful about getting enough B_{12}; some intestinal disorders (such as Crohn's disease) may cause deficiencies
Vitamin C (ascorbic acid)	Citrus fruits, broccoli, tomatoes, green peppers, strawberries	Deficiency causes scurvy, which was a significant health problem during the era of lengthy sea voyages
Fat-soluble vitamins		
Vitamin A (retinol)	Dark green and orange vegetables and fruits, dairy products	Component of visual pigments; too little in the diet can cause vision loss; too much can cause yellow/orange skin and liver damage
Vitamin D	Fortified dairy products, egg yolk	Made in human skin in the presence of sunlight, aids calcium absorption and bone formation; too little causes rickets (bone deformities)
Vitamin E (tocopherol)	Green leafy vegetables, vegetable oils, nuts, seeds, wheat germ	Antioxidant; helps prevent damage to cell membranes; deficiency is rare
Vitamin K	Green vegetables, tea; also made by colon bacteria	Important in blood clotting; made by colon bacteria; newborns and people taking long-term antibiotics may be deficient; too much can cause liver damage

TABLE 21.16B | MINERAL REQUIREMENTS OF HUMANS

Mineral*	Dietary Sources	Interesting Facts
Calcium (Ca)	Dairy products, dark green vegetables, legumes	Required for bone and tooth formation, blood clotting, nerve and muscle function; deficiency (particularly among women under 30) can lead to loss of bone mass and osteoporosis in later life (see Module 30.5)
Phosphorus (P)	Dairy products, meats, grains	A component of ATP and all nucleic acids; deficiency can lead to weakness and calcium loss
Sulfur (S)	Proteins from many sources	Required for protein synthesis; deficiency can cause impaired growth
Potassium (K)	Meats, dairy products, many fruits and vegetables, grains	Because it aids in nerve function, too little potassium can cause weakness or even paralysis
Chlorine (Cl)	Table salt	Insufficient chlorine in the diet can lead to muscle cramps
Sodium (Na)	Table salt	Required for proper water balance; the average American eats enough salt to provide about 20 times the required amount of sodium; overconsumption is associated with high blood pressure and some forms of cancer
Magnesium (Mg)	Whole grains, green leafy vegetables	Acts as an enzyme cofactor; deficiency leads to impairment of the nervous system
Iron (Fe)	Meats, eggs, legumes, whole grains, green leafy vegetables	Component of hemoglobin, the oxygen-carrying protein of red blood cells; deficiency (common among pregnant women) causes anemia, weakness, and decreased immunity
Fluorine (F)	Drinking water, tea, seafood	Helps prevent tooth decay
Iodine (I)	Seafood, iodized salt	Component of thyroid hormones; too little can cause goiter, a swelling of the thyroid gland

*Additional minerals required in trace amounts are chromium (Cr), cobalt (Co), copper (Cu), manganese (Mn), molybdenum (Mo), selenium (Se), and zinc (Zn).

Water-soluble vitamins include the B vitamins and vitamin C. Many B vitamins function in the body as coenzymes, enabling the catalytic functions of enzymes that are used over and over in metabolic reactions. Vitamin C is required in the production of connective tissue. Fat-soluble vitamins include vitamins A, D, E, and K. Vitamin A deficiency is mainly found among populations subsisting on simple rice diets; insufficient vitamin A can cause blindness or death. Your dietary requirement for vitamin D is variable because you synthesize this vitamin from other molecules when your skin is exposed to sunlight.

Minerals are simple inorganic nutrients, also required in small amounts—from less than 1 mg to about 2,500 mg per day. **Table 21.16B** lists your mineral requirements. You need the first seven minerals in amounts greater than 200 mg per day (about two-tenths of that small peanut). You need the rest in much smaller quantities.

Along with other vertebrates, we humans require relatively large amounts of calcium and phosphorus to construct and maintain the bones of our skeleton. Iron is needed to construct hemoglobin and as a component of thyroid hormones, which regulate metabolic rate. Worldwide, iodine deficiency is a serious human health problem (see Module 2.2) and is ranked as the leading cause of preventable developmental disabilities.

Sodium, potassium, and chlorine are important in nerve function and help maintain the osmotic balance of your cells. Most of us ingest far more salt (sodium chloride) than we need. Packaged (prepared) foods and most junk foods contain large amounts of sodium, even if they don't taste very salty. For example, one reduced-fat blueberry muffin may contain more than 40% of the daily recommended intake of sodium.

A varied diet usually includes enough vitamins and minerals and is considered the best source of these nutrients. Such diets meet the **Recommended Dietary Allowances (RDAs)**, minimum amounts of nutrients that are needed each day, as determined by a national scientific panel. The U.S. Department of Agriculture makes specific recommendations for certain population groups, such as additional B_{12} for people over age 50, folic acid for pregnant women, and extra vitamin D for people with dark skin (which blocks the synthesis of this vitamin) and for those exposed to insufficient sunlight.

The subject of vitamin dosage has led to heated debate. Some argue that RDAs are set too low, and some believe, probably mistakenly, that massive doses of vitamins confer health benefits. In general, any excess water-soluble vitamins consumed will be eliminated in urine. But high doses of niacin have been shown to cause liver damage, and large doses of vitamin C can result in gastrointestinal upset. Excessive amounts of fat-soluble vitamins accumulate in body fat. Thus, overdoses may have toxic effects. Excessive consumption of any mineral may be harmful. For example, in some regions of Africa where the water supply is especially iron-rich, as much as 10% of the population has liver damage as a result of iron overload.

Nevertheless, when we discuss health concerns relating to vitamins and minerals, most of the time we are concerned with deficiencies. Remember that a diet that doesn't include adequate quantities of fresh fruits and vegetables, as a result of either poor food choices or limited supplies or resources, is unlikely to provide the nutrients needed for good health. In the next module, you'll learn how you can check your own diet for deficiencies by decoding the information on food labels.

? Which of the vitamins and minerals listed in these tables are involved with the formation or maintenance of bones and teeth?

● Vitamin C, vitamin D, calcium, phosphorus, and fluorine

21.17 Food labels provide nutritional information

CONNECTION

The FDA requires two blocks of information on packaged food labels **(Figure 21.17)**. One lists ingredients from the greatest amount (by weight) to the least. The other lists key nutrients, emphasizing the ones associated with disease and the ones associated with a healthy diet, along with amounts contained per serving and as a percentage of a daily value (requirement or limit) based on a 2,000-kcal diet. Keep in mind that you should adjust the listed nutritional information to match the size of your serving (by doubling all the values, for example, if you use two pieces of bread in a sandwich). FDA regulations change from time to time; for example, levels of trans fats must now be listed, although this was not a requirement in the past (see Module 3.9).

Food labels also provide information on total daily needs. For example, less than 20 g of saturated fat and at least 25 g of dietary fiber are recommended for those with a 2,000-kcal daily diet. Reading food labels can help you make informed choices about what you eat, but keep in mind that all the listed recommendations are "one-size-fits-all" rough guidelines.

? What percentage of RDAs for the fat-soluble vitamins is provided by a slice of the bread in Figure 21.17?

● 0%

Ingredients: whole wheat flour, water, high fructose corn syrup, wheat gluten, soybean or canola oil, molasses, yeast, salt, cultured whey, vinegar, soy flour, calcium sulfate (source of calcium).

Nutrition Facts
Serving Size 1 slice (43g)
Servings Per Container 16

Amount Per Serving	
Calories 100	Calories from Fat 10
	% Daily Value*
Total Fat 1.5g	**2%**
Saturated Fat 0g	**0%**
Trans Fat 0g	**0%**
Cholesterol 0mg	**0%**
Sodium 190mg	**8%**
Total Carbohydrate 19g	**6%**
Dietary Fiber 3g	**12%**
Sugars 3g	
Protein 4g	
Vitamin A 0%	Vitamin C 0%
Calcium 2%	Iron 4%
Thiamine 6%	Riboflavin 2%
Niacin 6%	Folic Acid 6%

* Percent Daily Values are based on a 2,000 calorie diet. Your daily values may be higher or lower depending on your calorie needs:

	Calories:	2,000	2,500
Total Fat	Less than	65g	80g
Sat. Fat	Less than	20g	25g
Cholesterol	Less than	300mg	300mg
Sodium	Less than	2,400mg	2,400mg
Total Carbohydrate		300g	375g
Dietary Fiber		25g	30g

Calories per gram:
Fat 9 • Carbohydrate 4 • Protein 4

▲ **Figure 21.17** A whole wheat bread label

21.18 Dietary deficiencies can have a number of causes

CONNECTION

In previous modules, you learned about the components necessary for a healthy diet. What happens if a diet is lacking in one or more essential nutrients or calories? The result is **malnutrition**, health problems caused by an improper or insufficient diet. Malnutrition may be caused by inadequate intake or by disease, such as metabolic or digestive abnormalities.

Living in an industrialized country where food is readily available and most people can afford a decent diet, you may find it hard to relate to starvation. But more than 800 million people around the world—nearly three times the population of the United States—must cope with hunger: 14,000 children under the age of 5 starve to death *each day*. Undernutrition, insufficient caloric intake, may occur when food supplies are disrupted by crises such as drought or war, or when poverty prevents people from obtaining sufficient food.

The most common type of human malnutrition is protein deficiency, insufficient intake of one or more essential amino acids. Protein deficiency is prevalent where there is a great gap between food supply and population size **(Figure 21.18)**. Animal products are a reliable source of essential amino acids, but these foods are expensive. People forced by economic necessity to get almost all their calories from a single plant staple, such as rice or potatoes, will suffer deficiencies of essential amino acids. Most victims of protein deficiency are children, who are likely to develop poorly both physically and mentally—if they even survive infancy. The resulting syndrome is called kwashiorkor, from the Ghanaian word for "rejected one," a reference to the onset of the disease when a child is weaned from its mother's milk and placed on a starchy diet after a sibling is born (Figure 21.18).

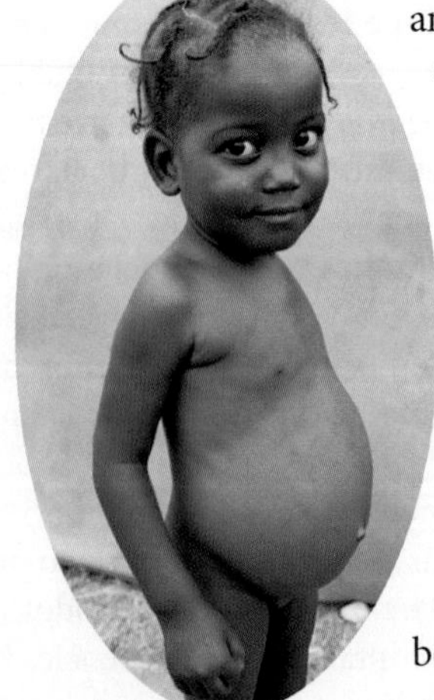

▲ **Figure 21.18** This child is showing signs of protein deficiency, which caused fluid to enter the abdominal cavity, producing swelling of the belly

Sometimes undernutrition is self-inflicted. Millions of Americans, mostly female, are affected by anorexia nervosa, an eating disorder characterized by self-starvation due to an intense fear of gaining weight, even when the person is actually underweight. Bulimia is a behavioral pattern of binge eating followed by purging through induced vomiting, abuse of laxatives, or excessive exercise. Both disorders are characterized by an obsession with body weight and shape and can result in serious health problems and often death.

Malnutrition is not always associated with poverty or disorders; it can result from a steady diet of junk food, which offers little nutritional value. A person can therefore be both malnourished and obese. In the next module, we'll look at such dietary behavior from an evolutionary perspective.

? Does malnutrition always result from lack of access to food?

● No. A person with access to food may still be malnourished (as with an eating disorder) or may be overfed but lack sufficient nutrients.

21.19 The human health problem of obesity may reflect our evolutionary past

EVOLUTION CONNECTION

Obesity is defined as a too-high **body mass index (BMI)**, a ratio of weight to height (**Figure 21.19A**). In general, a BMI of 25–29 is considered overweight, and above 30 is obese. About one-third of all Americans are obese, and another one-third are overweight (a BMI that is between normal and obese). Obesity contributes to health problems, including type 2 diabetes, cancer of the colon and breast, and cardiovascular disease. Obesity is estimated to be a factor in 300,000 deaths per year in the United States.

The obesity epidemic has stimulated an increase in scientific research on the causes and possible treatments for weight-control problems. Inheritance is one factor in obesity, which helps explain why certain people have to work harder than others to control their weight. Scientists are studying the signaling pathways that regulate appetite and the body's storage of fat in an effort to better understand obesity. Dozens of genes have been identified that code for weight-regulating hormones. We have reason to be optimistic that obese people who have inherited defects in these weight-controlling mechanisms may someday be treated with a new generation of drugs. But so far, the complexity of the body's system for weight regulation has made it difficult to develop effective treatments.

One well-studied component of human weight control is the hormone leptin, a long-term appetite regulator in mammals. Leptin is produced by adipose (fat) cells. As the amount of adipose tissue increases, leptin levels in the blood rise, which normally cues the brain to suppress appetite. Conversely, loss of body fat decreases leptin levels, signaling the brain to increase appetite. Researchers found that mice that inherit a defect in the gene for leptin become very obese (**Figure 21.19B**). They then discovered that they could treat these leptin-deficient obese mice by injecting them with leptin.

The discovery of the leptin-deficiency mutation in mice generated excitement because humans also have a leptin gene. Indeed, obese children who have inherited a mutant form of the leptin gene lose weight after leptin treatments. But, relatively few obese people have such deficiencies. In fact, most obese humans have abnormally high levels of leptin, which makes sense because leptin is produced by adipose tissue.

Some of our current struggles with obesity may be a consequence of our evolutionary history. Most of us crave foods that are fatty: fries, chips, burgers, cheese, and ice cream. Though fat hoarding can be a health liability today, it may actually have been an advantage in our evolutionary past. Only in the past 100 years have large numbers of people had access to a reliable supply of food. Our ancestors on the African savanna were hunter-gatherers who probably survived on a diet that was barely sufficient, with only occasional meals of protein-rich meat. In such a feast-and-famine existence, natural selection may have favored those individuals with a physiology that induced them to gorge on fatty foods on those rare occasions when such treats were available. Individuals with genes promoting the storage of fat during feasts may have been more likely than their thinner peers to survive famines.

So perhaps our modern taste for fats and sugars reflects the selective advantage it conveyed in our evolutionary history. Although we know it is unhealthful, many of us find it difficult to overcome the ancient survival behavior of stockpiling for the next famine.

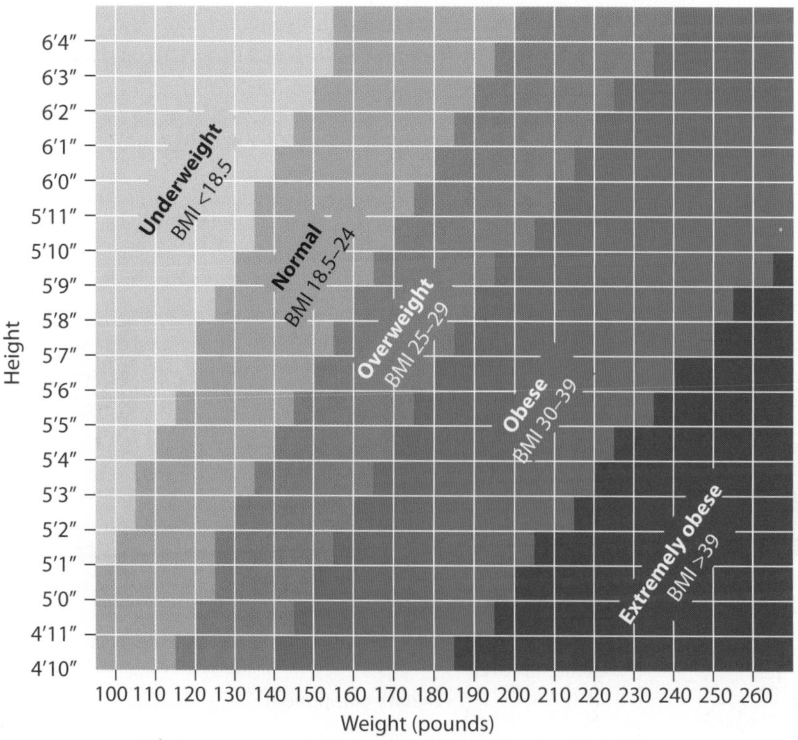

▲ **Figure 21.19A** Body mass index (BMI): one measure of healthy weight

Height: 4'10" 4'11" 5'0" 5'1" 5'2" 5'3" 5'4" 5'5" 5'6" 5'7" 5'8" 5'9" 5'10" 5'11" 6'0" 6'1" 6'2" 6'3" 6'4"
Weight (pounds): 100 110 120 130 140 150 160 170 180 190 200 210 220 230 240 250 260
Underweight BMI <18.5
Normal BMI 18.5–24
Overweight BMI 25–29
Obese BMI 30–39
Extremely obese BMI >39

Try This Locate your own BMI, and then calculate how much weight you would have to gain or lose to fall into each category.

? **In what two ways does the hormone leptin regulate appetite? In which of these ways does leptin apparently not function in obese humans?**

A drop in leptin due to a loss of adipose tissue stimulates appetite; a high level of leptin, produced by increased body fat, depresses appetite. The second mechanism does not seem to function in some people.

▲ **Figure 21.19B** A mouse with a defect in a gene for leptin, an appetite-suppressing hormone (left); a normal mouse (right)

21.20 Scientists use a variety of methods to test weight-loss claims

SCIENTIFIC THINKING

Is there a scientific approach to weight loss?

As discussed in the chapter introduction, we are bombarded with a great variety of weight loss claims. In recent years, many popular weight loss schemes have focused on reducing the dieter's intake of carbohydrates. People following "low-carb" diets often drop sugar, bread, fruits, and potatoes from their diet, swapping in cheese, nuts, and meat instead. Such diets have surged in popularity. In fact, Americans spend as much as $15 billion a year on low-carb diet aids and foods.

But how can you know if a diet plan is effective before you try it? The field of nutrition uses scientific approaches to investigate such questions. Many nutrition studies rely on epidemiology, the study of human health and disease within populations. Epidemiological research often looks for links between health and diet. For example, a study published in 2011 measured the body mass index (BMI; see Figure 21.19A) of 4,451 healthy Canadian adults and tracked their diet through daily self-reporting **(Table 21.20)**. The participants were ranked into quartiles (that is, by fourths) based on how many carbohydrates they consumed (measured as grams of carbohydrates eaten per day). The results of this epidemiological study showed that the bottom quartile (people who ate the *least* carbs) had the highest obesity rates and that the top quartile (people who ate the *most* carbs) had lower obesity rates. In other words, the results indicated that the *more* carbohydrates consumed, the *lower* the risk of obesity. In this study, low-carb diets resulted in gaining weight, not losing weight! What should we make of such data?

One important fact about epidemiological studies is that scientists must differentiate correlation (in this case, low carbohydrate intake *correlates* with greater risk of obesity) from causation (low-carb diets *cause* weight gain). Just because two factors seem to affect one another does not mean they actually do. How, then, can we draw a conclusion about the usefulness of low-carb diets?

One way to eliminate such confusion is to perform controlled trials in which researchers determine the conditions for their research participants, rather than merely observing natural behaviors (as in an epidemiological study). For example, a 2009 study by researchers at the Harvard School of Public Health assigned 811 overweight (BMI > 25) adults to one of several diets that varied in the percentage of total calories obtained from fats, protein, and carbohydrates. All of the diet plans averaged the same number of total calories per day, but where those calories came from (fats versus protein versus carbohydrates) varied among the groups. After two years, the total weight lost was recorded and comparisons between the diets were made **(Figure 21.20)**.

The data reveal two interesting facts. First, members of every group lost a moderate amount of weight over the two-year trial. Second, every group lost a similar amount of weight. In fact, there were no statistically significant differences between the groups in the quantity of weight lost. Each of the diets had the same number of total calories and resulted in the same weight loss, no matter the source of the calories (carbs versus fats, for example). These data suggest that cutting calories is what results in weight loss, not cutting carbs (or cutting fats or protein or eating a lot of one kind of food). In this light, the data from the epidemiological study we discussed first make a bit more sense: Notice that the highest carbohydrate consumers (the last column) consumed the lowest total calories and demonstrated the lowest rate of obesity.

The overall lesson is deceptively simple: To lose weight, cut back on your calorie consumption. Combined with a reminder to burn more calories through moderate exercise, this leads to the most sensible slogan of all: To lose weight, eat less and exercise more!

? **In what sense is maintaining a stable body weight a matter of caloric bookkeeping?**

● When your metabolism burns as many kilocalories a day as you take in with your food, a stable body weight will result.

TABLE 21.20 | CARBOHYDRATE INTAKE AND OBESITY

Ranking by Carbohydrates Consumed	Bottom Quartile	Second Quartile	Third Quartile	Top Quartile
Consumed carbohydrates (grams/day)	179	234	269	319
Total calories (kcal/day)	2,214	2,313	2,303	2,140
% overweight or obese (BMI ≥ 25)	65%	54%	51%	51%

Data from A. T. Merchant et al., Carbohydrate intake and overweight and obesity among healthy adults, *Journal of the American Dietetic Association*, 109(7): 1165–72 (2009). Data taken from within paper and a subset of Table 2.

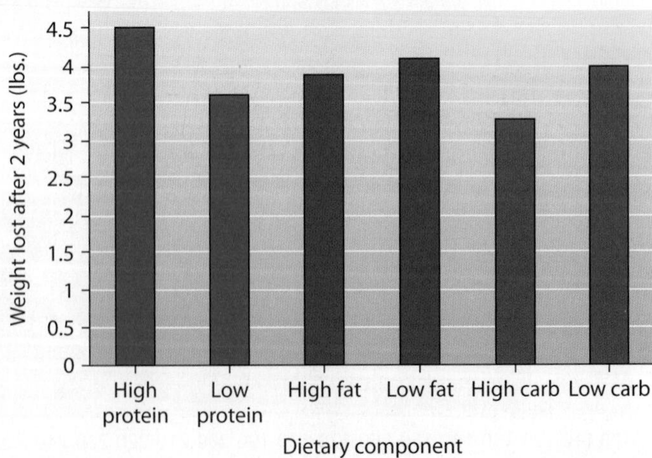

Weight lost after 2 years (lbs.)
4.5
4
3.5
3
2.5
2
1.5
1
0.5
0
High protein
Low protein
High fat
Low fat
High carb
Low carb
Dietary component

▲ **Figure 21.20** Data from a 2009 study that compared weight loss on different diet plans

Data from F. M. Sacks et al., Comparison of weight-loss diets with different compositions of fat, protein, and carbohydrates, *The New England Journal of Medicine*, 360(9): 589. Data taken from Figure 1.

21.21 Diet can influence risk of cardiovascular disease and cancer

CONNECTION

Food influences far more than your size and appearance. Diet also plays an important role in your risk of developing serious illnesses, including cardiovascular disease and cancer. Some risk factors associated with cardiovascular disease, such as family history, are unavoidable, but others, such as smoking and lack of exercise, can be influenced through behavior. Diet is another behavioral factor that affects cardiovascular health. For instance, a diet high in saturated fats is linked to high blood cholesterol, which in turn is linked to cardiovascular disease.

Cholesterol travels through the body in particles made up of thousands of molecules of cholesterol and other lipids bound to a protein. High blood levels of one type of particle called **low-density lipoproteins (LDLs)** generally correlate with a tendency to develop blocked blood vessels, high blood pressure, and consequent heart attacks. In contrast to LDLs, cholesterol particles called **high-density lipoproteins (HDLs)** may decrease the risk of vessel blockage, perhaps because HDLs convey excess cholesterol to the liver, where it is broken down. Some research indicates that reducing LDLs while maintaining or increasing HDLs lowers the risk of cardiovascular disease. How do you increase your levels of "good" cholesterol? You can exercise more, which tends to increase HDL levels. And you can abstain from smoking, because smoking has been shown to lower HDL levels.

How do you decrease your levels of "bad" cholesterol? You can avoid a diet high in saturated fats, which tend to increase LDL levels. Saturated fats are found in eggs, full-fat dairy products like butter, and most meats. Saturated fats are also found in artificially saturated ("hydrogenated") vegetable oils. The hydrogenation process, which solidifies vegetable oils, also produces a type of fat called trans fat. Trans fats tend not only to increase LDL levels but also to lower HDL levels, a two-pronged attack on cardiovascular health. By contrast, eating mainly unsaturated fats, such as found in fatty fish like salmon, certain nuts, and most liquid vegetable oils tends to lower LDL levels and raise HDL levels. These oils are also important sources of vitamin E, whose antioxidant effect may help prevent blood vessel blockage, and omega-3 fatty acids, which appear to protect against cardiovascular disease.

Diet also seems to influence our risk for certain cancers (see Module 11.18). Some research suggests a link between diets heavy in fats or carbohydrates and the incidence of breast cancer. The incidence of colon cancer and prostate cancer may be linked to a diet rich in saturated fat or red meat. Other foods may help fight cancer. For example, some fruits and vegetables (**Figure 21.21**) are rich in antioxidants, chemicals that help protect cells from damaging molecules known as free radicals. Foods that are particularly high in antioxidants include berries, beans, nuts, dried fruit, green and black tea, red wine, and dark chocolate. The link between antioxidant foods and cancer is still debated by scientists.

Despite the progress researchers have made in studying nutrition and health, it is often difficult to design controlled experiments that establish the link between the two.

▲ **Figure 21.21** Foods that contribute to good health

Experiments that may damage participants' health are clearly unethical. Some studies rely on self-reported food intake, and the accuracy of participants' memories may influence the outcome. As you learned in Module 21.20, scientists often perform studies that correlate certain health characteristics with groups that have particular diets or lifestyles. For example, many people living in France eat high-fat diets and drink wine, yet have lower rates of obesity and heart disease than do Americans. When researchers notice apparent contradictions like these, they attempt to control for other variables and isolate the factors responsible for such observations, such as the fact that the French eat smaller portions; eat more unprocessed, fresh foods; and snack infrequently.

Even with large, controlled trials, results may be contradictory or inconclusive. For instance, an eight-year study of almost 49,000 postmenopausal women found that low-fat diets failed to reduce the risk of breast and colon cancer and did not affect the incidence of cardiovascular disease. LDL and cholesterol levels decreased slightly in the low-fat group, however.

The relationship between foods and health is complex, and we have much to learn. The American Cancer Society (ACS) suggests that following the dietary guidelines in **Table 21.21**, in combination with physical activity, can help lower cancer risk. The ACS's main recommendation is to "eat a variety of healthful foods, with an emphasis on plant sources."

If you are trying to minimize the damaging effects of blood cholesterol on your cardiovascular system, your goal is to ____________ your LDLs and ____________ your HDLs.

decrease . . . increase

TABLE 21.21 DIETARY GUIDELINES FOR REDUCING CANCER RISK

Maintain a healthy weight throughout life.
Eat five or more servings of a variety of fruits and vegetables daily.
Choose whole grains over processed (refined) grains.
Limit consumption of processed and red meats.
If you drink alcoholic beverages, limit yourself to a maximum of one or two drinks a day (a drink = 12 oz of beer, 5 oz of wine, or 1.5 oz of 80% distilled spirits).

CHAPTER 21 REVIEW

For practice quizzes, BioFlix animations, MP3 tutorials, video tutors, and more study tools designed for this textbook, go to MasteringBiology®

Reviewing the Concepts

Obtaining and Processing Food (21.1–21.3)

21.1 Animals obtain and ingest their food in a variety of ways. Animals may be herbivores, carnivores, or omnivores and may obtain food by filter, substrate, fluid, or bulk feeding.

21.2 Overview: Food processing occurs in four stages. The stages are ingestion, digestion, absorption, and elimination.

21.3 Digestion occurs in specialized compartments. Food may be digested in food vacuoles, gastrovascular cavities, or alimentary canals, which run from mouth to anus with specialized regions along the way.

The Human Digestive System (21.4–21.13)

21.4 The human digestive system consists of an alimentary canal and accessory glands. The rhythmic muscle contractions of peristalsis squeeze food through the alimentary canal.

21.5 Digestion begins in the oral cavity. The teeth break up food, saliva moistens it, and an enzyme in saliva begins the hydrolysis of starch. The tongue pushes the bolus of food into the pharynx.

21.6 After swallowing, peristalsis moves food through the esophagus to the stomach. The swallowing reflex moves food into the esophagus and keeps it out of the trachea.

21.7 The Heimlich maneuver can save lives. This procedure can dislodge food from the pharynx or trachea during choking.

21.8 The stomach stores food and breaks it down with acid and enzymes. Pepsin in gastric juice begins to digest protein.

21.9 Digestive ailments include acid reflux and gastric ulcers.

21.10 The small intestine is the major organ of chemical digestion and nutrient absorption. Enzymes from the pancreas and cells of the intestinal wall digest food molecules. Bile, made in the liver and stored in the gallbladder, emulsifies fat for attack by enzymes. Folds of the intestinal lining and finger-like villi (with microscopic microvilli) increase the area across which absorbed nutrients move into capillaries and lymph vessels.

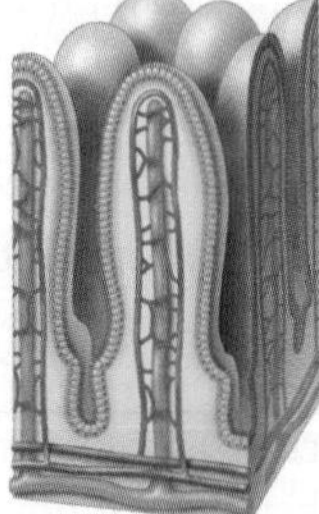

21.11 The liver processes and detoxifies blood from the intestines. The liver regulates nutrient levels in the blood, detoxifies alcohol and drugs, and synthesizes blood proteins.

21.12 The large intestine reclaims water and compacts the feces. Some bacteria in the colon produce vitamins. Feces are stored in the rectum before elimination.

21.13 Evolutionary adaptations of vertebrate digestive systems relate to diet. Herbivores may have longer alimentary canals than carnivores and compartments that house cellulose-digesting microbes.

Nutrition (21.14–21.21)

21.14 An animal's diet must provide sufficient energy. The diet must provide chemical energy, raw materials for biosynthesis, and essential nutrients. Metabolic rate, the rate of energy consumption, includes the basal metabolic rate (BMR) plus the energy used for other activities.

21.15 An animal's diet must supply essential nutrients. Essential fatty acids are easily obtained from the diet. The eight essential amino acids can be obtained from animal protein or a combination of plant foods. Malnutrition results from a diet lacking in sufficient calories or essential nutrients.

21.16 A proper human diet must include sufficient vitamins and minerals. Most vitamins function as coenzymes. Minerals are inorganic nutrients that play a variety of roles. A varied diet usually meets the RDAs for these nutrients.

21.17 Food labels provide nutritional information.

21.18 Dietary deficiencies can have a number of causes. Malnutrition, a diet insufficient in nutrients and/or calories, can cause significant health problems. Protein deficiency is the most common cause of malnutrition worldwide.

21.19 The human health problem of obesity may reflect our evolutionary past. The dramatic rise in obesity (defined as a BMI of 25 or over) is linked to a lack of exercise and abundance of fattening foods and may partly stem from an evolutionary advantage of fat hoarding.

21.20 Scientists use a variety of methods to test weight-loss claims. Epidemiology relates diets to health characteristics in populations. Controlled experiments can be used to identify effects of specific diet plans.

21.21 Diet can influence risk of cardiovascular disease and cancer. The ratio of HDLs to LDLs is influenced by diet.

Connecting the Concepts

1. Label the parts of the human digestive system below and indicate the functions of these organs and glands.

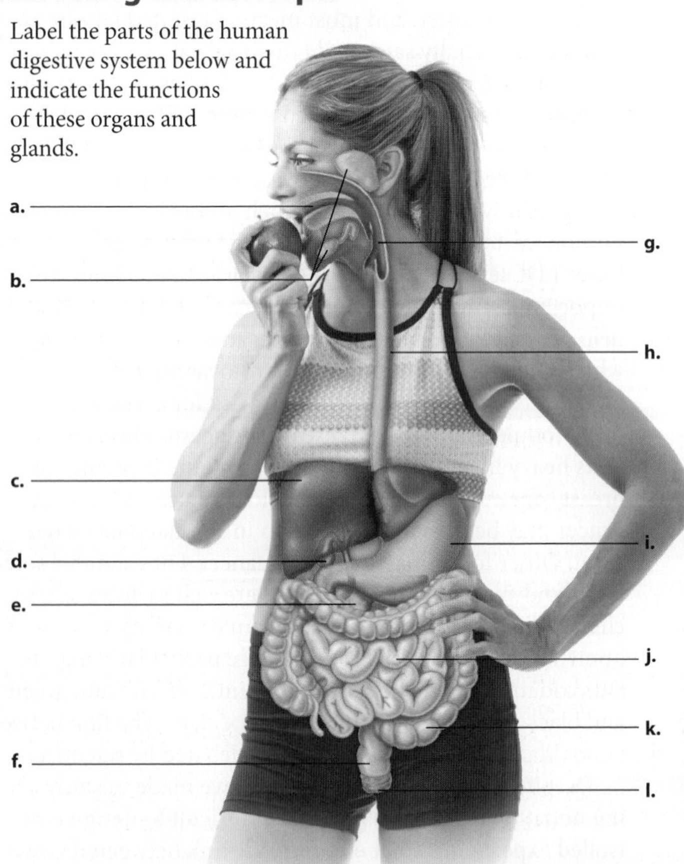
a.
b.
c.
d.
e.
f.
g.
h.
i.
j.
k.
l.

2. Complete the following map summarizing the nutritional needs of animals that are met by a healthy diet.

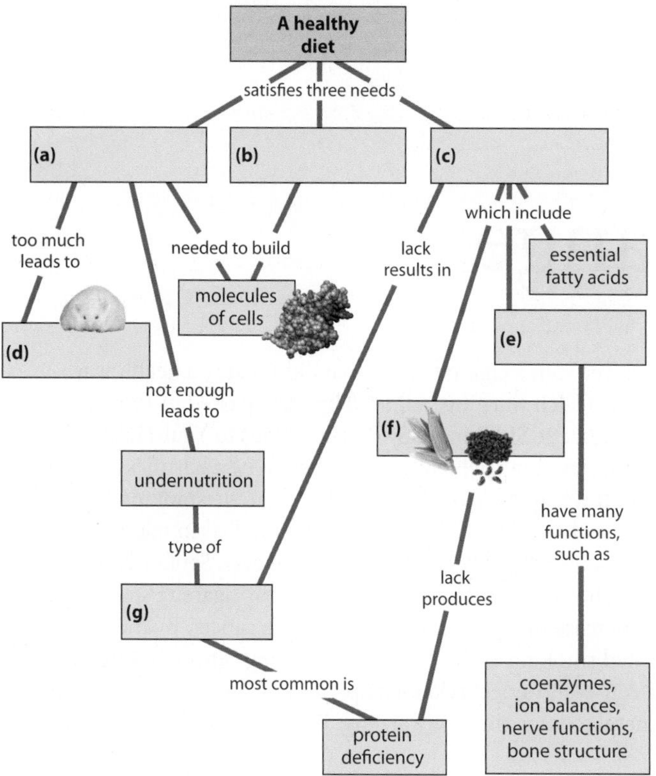

A healthy diet
satisfies three needs
(a)
(b)
(c)
too much leads to
needed to build
lack results in
which include
essential fatty acids
molecules of cells
(d)
(e)
not enough leads to
(f)
undernutrition
type of
have many functions, such as
lack produces
(g)
most common is
protein deficiency
coenzymes, ion balances, nerve functions, bone structure

Testing Your Knowledge

Level 1: Knowledge/Comprehension

3. Earthworms, which are substrate feeders,
   a. feed mostly on mineral substrates.
   b. filter small organisms from the soil.
   c. are filter feeders.
   d. eat their way through the soil, feeding on partially decayed organic matter.
4. The energy content of fats
   a. is released by bile salts.
   b. is, per gram, twice that of carbohydrates or proteins.
   c. cannot be dissolved in water and so cannot be absorbed.
   d. is usually healthier than the energy content of carbohydrates.
5. Which of the following is mismatched with the disease that results from underconsumption?
   a. vitamin B_6—beriberi
   b. vitamin C—scurvy
   c. vitamin A—vision loss
   d. vitamin D—rickets

Level 2: Application/Analysis

6. Which of the following statements is false?
   a. A healthy human has enough stored fat to supply calories for several weeks.
   b. An increase in leptin levels leads to an increase in appetite and weight gain.
   c. The interconversion of glucose and glycogen takes place in the liver.
   d. After glycogen stores are filled, excessive calories are stored as fat, regardless of their original food source.
7. Why is it necessary for healthy vegetarians to combine different plant foods or eat some eggs or milk products?
   a. to make sure they obtain sufficient calories
   b. to provide sufficient vitamins
   c. to make sure they ingest all essential fatty acids
   d. to provide all essential amino acids for protein synthesis
8. A peanut butter and jelly sandwich contains carbohydrates, proteins, and fats. Describe what happens to the sandwich when you eat it. Discuss ingestion, digestion, absorption, and elimination.
9. Use the Nutrition Facts label to the right to answer these questions:
   a. What percentage of the total Calories in this product is from fat?
   b. Is this product a good source of vitamin A and calcium? Explain.
   c. Each gram of fat supplies 9 Calories. Based on the grams of saturated fat and its % Daily Value, calculate the upper limit of saturated fat (in grams and Calories) that an individual on a 2,000-Calorie/day diet should consume.

Nutrition Facts

Serving Size 1/2 Cup (83g)
Servings Per Container 8

Amount Per Serving	
Calories 190 Calories from Fat 110	
	% Daily Value*
Total Fat 12g	**18%**
Saturated Fat 8g	**40%**
Trans Fat 0g	**0%**
Cholesterol 45mg	**15%**
Sodium 75mg	**3%**
Total Carbohydrate 18g	**6%**
Dietary Fiber 0g	**0%**
Sugars 17g	
Protein 3g	
Vitamin A 10% • Vitamin C 8%	
Calcium 10% • Iron 0%	

*Percent Daily Values (DV) are based on a 2,000 calorie diet.

Level 3: Synthesis/Evaluation

10. How might our craving for fatty foods, which is helping to fuel the obesity crisis, have evolved through natural selection?
11. One common piece of dieting advice is to replace energy-dense food with nutrient-dense food. What does this mean?
12. The media report numerous claims and counterclaims about the benefits and dangers of certain foods, dietary supplements, and diets. Have you modified your eating habits on the basis of nutritional information disseminated by the media? Why or why not? How should we evaluate whether such nutritional claims are valid?
13. It is estimated that 15% of Americans do not always have access to enough food. Worldwide, more than 1 billion people go to bed hungry most nights, and millions of people have starved to death in recent decades. In some cases, war, poor crop yields, and disease epidemics strip people of food. Many say instead that it is not inadequate food production but unequal food distribution that causes food shortages. What responsibility do nations have for feeding their citizens? For feeding the people of other countries? What do you think you can do to lessen world hunger?
14. **SCIENTIFIC THINKING** Consider the relationship between correlation and causation with respect to some pairs of human traits. For example, are freckles and red hair correlated? Is there causation? How does this concept relate to the study of human nutrition? Explain each answer.

Answers to all questions can be found in Appendix 4.

CHAPTER 22

Gas Exchange

If you're under 50, you've probably never seen a cigarette pack that didn't carry a health warning from the Surgeon General. The first such warning, required by a law passed in 1965, was the relatively mild admonition, "Caution: Cigarette Smoking May Be Hazardous to Your Health." This was the first inkling most people had that smoking, which they believed to be a harmless pleasure, might actually be killing them. It was quite a jolt. For decades, cigarette smoking had been commonplace in public spaces such as offices and restaurants, as well as in many homes. Cigarette ads featured endorsements by movie stars, athletes, and even Santa Claus.

How certain are scientists that cigarette smoking is hazardous to health?

Many doctors and scientists, however, had begun to suspect that cigarette smoking was responsible for the dramatic increase in a once-rare disease—lung cancer. By the mid-1950s, the weight of evidence implicated smoking as the main factor in lung cancer, but much research was still needed to establish a cause-and-effect relationship. In 1964, a review of the available evidence convinced the Surgeon General to issue the first warning. But that was just the beginning.

As scientists continued their investigations, conclusions about the health risks of smoking grew more certain. For the past 30 years, warning labels have stated the dangers unequivocally: "Smoking causes lung cancer, heart disease, emphysema, and may complicate pregnancy." You'll learn more about how researchers established that cigarettes damage the respiratory system later in this chapter.

Respiratory systems provide for the exchange of O_2 and the waste product CO_2 between an animal and its environment. In this chapter, we explore the various types of gas exchange systems that have evolved in animals. We then take a closer look at the structures and functions of the human respiratory system. We conclude with a preview of the circulatory system, which delivers the oxygen essential for life to all body cells.

BIG IDEAS

Mechanisms of Gas Exchange
(22.1–22.5)

Gas exchange occurs across thin, moist surfaces in respiratory organs such as gills, tracheal systems, and lungs.

The Human Respiratory System
(22.6–22.9)

Air travels through branching tubes to the lungs, where gases are exchanged with the blood.

Transport of Gases in the Human Body
(22.10–22.12)

The circulatory system transports O_2 to body tissues and returns CO_2 to the lungs.

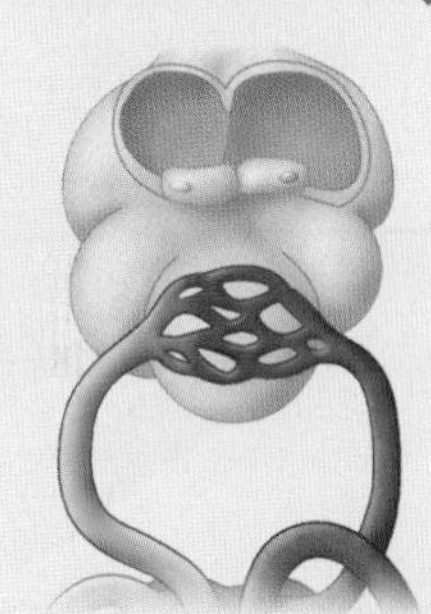

▷ Mechanisms of Gas Exchange

22.1 Gas exchange in humans involves breathing, transport of gases, and exchange with body cells

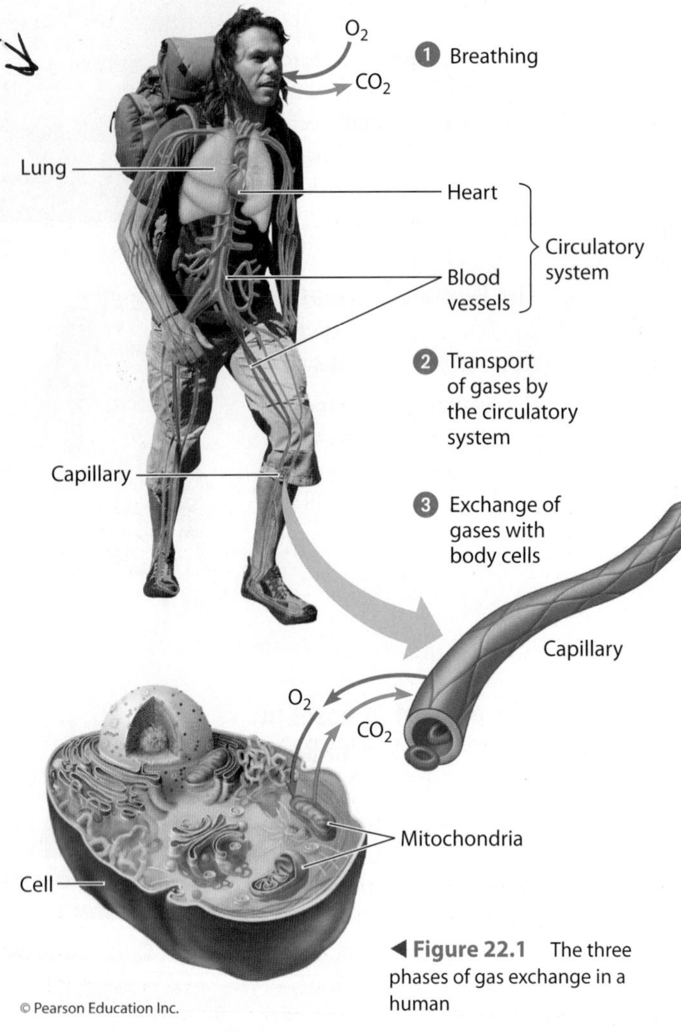

O_2
CO_2
1 Breathing
Lung
Heart
Circulatory system
Blood vessels
2 Transport of gases by the circulatory system
Capillary
3 Exchange of gases with body cells
Capillary
O_2
CO_2
Mitochondria
Cell

◀ **Figure 22.1** The three phases of gas exchange in a human

Gas exchange makes it possible for you to put to work the food molecules the digestive system provides. **Figure 22.1** presents an overview of the three phases of gas exchange in humans and other animals with lungs. ❶ Breathing: As you inhale, a large, moist internal surface is exposed to the air entering the lungs. Oxygen (O_2) diffuses across the cells lining the lungs and into surrounding blood vessels. At the same time, carbon dioxide (CO_2) diffuses from the blood into the lungs. As you exhale, CO_2 leaves your body.

❷ Transport of gases by the circulatory system: The O_2 that diffused into the blood attaches to hemoglobin in red blood cells. The red vessels in the figure are transporting O_2-rich blood from the lungs to capillaries in the body's tissues. CO_2 is also transported in blood, from the tissues back to the lungs, carried in the blue vessels shown in the figure.

❸ Exchange of gases with body cells: Your cells take up O_2 from the blood and release CO_2 to the blood. As you have learned (see Module 6.5), O_2 functions in cellular respiration in the mitochondria as the final electron acceptor in the stepwise breakdown of fuel molecules. H_2O and CO_2 are waste products, and ATP is produced that will power cellular work. The gas exchange occurring as we breathe is often called respiration; do not confuse this exchange with cellular respiration.

Cellular respiration requires a continuous supply of O_2 and the disposal of CO_2. Gas exchange involves both the respiratory and circulatory systems in servicing your body's cells.

? Humans cannot survive for more than a few minutes without O_2. Why?

● Cells require a steady supply of O_2 for cellular respiration to produce enough ATP to function. Without enough ATP, cells and the organism die.

22.2 Animals exchange O_2 and CO_2 across moist body surfaces

The part of an animal's body where gas exchange with the environment occurs is called the respiratory surface. Respiratory surfaces are made up of living cells, and like all cells, their plasma membranes must be wet to function properly. Thus, respiratory surfaces are always moist.

Gas exchange takes place by diffusion. The surface area of the respiratory surface must be large enough to take up sufficient O_2 for every cell in the body. Usually, a single layer of cells forms the respiratory surface. This thin, moist layer allows O_2 to diffuse rapidly into the circulatory system or directly into body tissues and also allows CO_2 to diffuse out.

The four figures on the facing page illustrate, in simplified form, four types of respiratory organs, structures in which gas exchange with the external environment occurs. In each of these figures, the circle represents a cross section of the animal's body through the respiratory surface. The yellow areas represent the respiratory surfaces; the green outer circles represent body surfaces with little or no role in gas exchange. The boxed enlargements show gas exchange occurring across the respiratory surface.

Some animals use their entire outer skin as a gas exchange organ. The earthworm in **Figure 22.2A** is an example. The cross-sectional diagram shows its whole body surface as yellow; there are no specialized gas exchange surfaces. Oxygen diffuses into a dense network of thin-walled capillaries lying just beneath the skin. Earthworms and other skin-breathers must live in damp places or in water because their whole body surface has to stay moist. Animals that breathe only

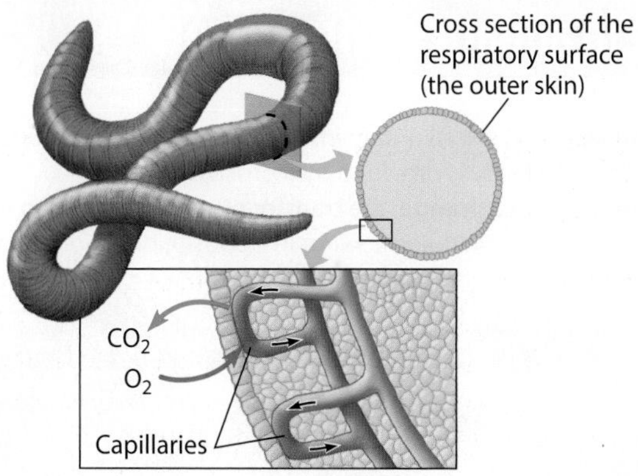
Cross section of the respiratory surface (the outer skin)
CO_2
O_2
Capillaries

▲ **Figure 22.2A** The skin: the outer body surface

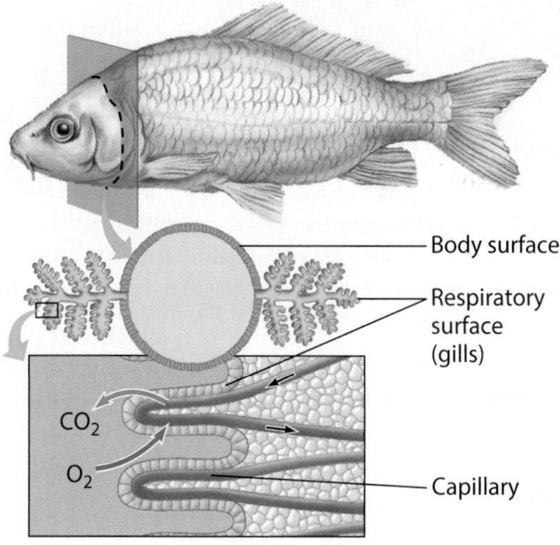
Body surface
Respiratory surface (gills)
CO_2
O_2
Capillary

▲ **Figure 22.2B** Gills: extensions of the body surface

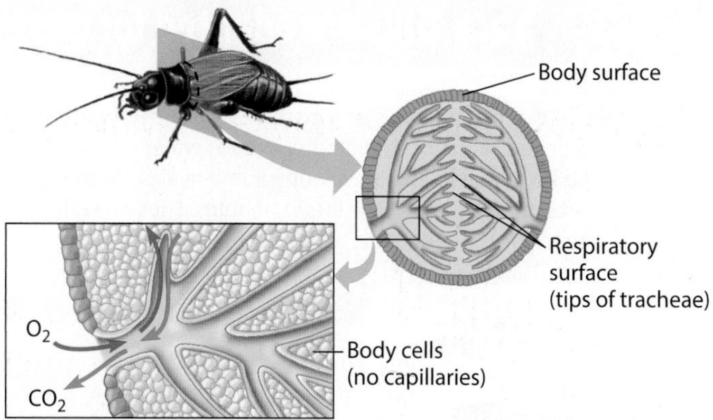
Body surface
Respiratory surface (tips of tracheae)
O_2
CO_2
Body cells (no capillaries)

▲ **Figure 22.2C** A tracheal system: air tubes that extend throughout the body

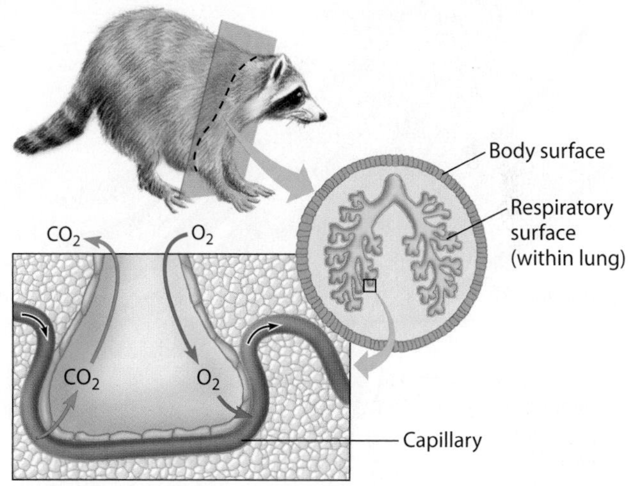
Body surface
Respiratory surface (within lung)
CO_2
O_2
CO_2
O_2
Capillary

▲ **Figure 22.2D** Lungs: internal thin-walled sacs

through their skin are generally small, and many are long and thin or flattened. These shapes provide a high ratio of respiratory surface to body volume, allowing for sufficient gas exchange for all the cells in the body.

In most animals, the skin surface is not extensive enough to exchange gases for the whole body. As a consequence, certain parts of the body have become adapted as highly branched respiratory surfaces with large surface areas. Such gas exchange organs include gills, tracheal systems, and lungs. Many animals have adaptations to improve **ventilation,** the flow of water or air over the respiratory surface. For example, movement of the operculum in ray-finned fishes (see Figure 19.3D) passes water over the gills. Increasing this flow ensures a fresh supply of O_2 and the removal of CO_2.

Gills have evolved in most aquatic animals, including some annelids, molluscs, crustaceans, and fish. **Gills** are extensions, or outfoldings, of the body surface specialized for gas exchange. Many marine worms have flap-like gills that extend from each body segment. The gills of clams and crayfish are clustered in one body location. A fish **(Figure 22.2B)** has a set of feather-like gills on each side of its head. As indicated in the enlargement, gases diffuse across the gill surface between the water and the blood. Because the respiratory surfaces of aquatic animals extend into the surrounding water, keeping the surface moist is not a problem.

In most terrestrial animals, the respiratory surface is folded into the body rather than projecting from it. The infolded surface opens to the air only through narrow tubes, an arrangement that helps retain the moisture that is essential for the cells of the respiratory surfaces to function.

The **tracheal system** of insects **(Figure 22.2C)** is an extensive system of branching internal tubes called tracheae, with a moist, thin epithelium forming the respiratory surface at their tips. As you will see in Module 22.4, the smallest branches exchange gases directly with body cells. Thus, gas exchange in insects requires no assistance from the circulatory system.

Most terrestrial vertebrates have **lungs (Figure 22.2D)**, which are internal sacs lined with moist epithelium. As the diagram indicates, the inner surfaces of the lungs are extensively subdivided, forming a large respiratory surface. Gases are carried between the lungs and the body cells by the circulatory system.

We examine gills, tracheae, and lungs more closely in the next several modules.

? **How does the structure of the respiratory surface of a gill or lung fit its function?**

● These respiratory surfaces are moist and thin so that gases can easily diffuse across them and into or out of the closely associated capillaries. They are highly branched or subdivided, providing a large surface area for exchange.

VISUALIZING THE CONCEPT

22.3 Gills are adapted for gas exchange in aquatic environments

Water contains O_2 as a dissolved gas. However, the concentration of oxygen dissolved in water is low, only about 3% of that in an equivalent volume of air. And the warmer and saltier the water, the less O_2 it holds. Thus, the gills of fishes—especially those of large, active fishes in warm oceans—must be very efficient to obtain enough O_2 from the surrounding water. This efficiency is provided by a process called **countercurrent exchange**, the transfer of a substance such as oxygen between two fluids flowing in opposite directions. In this case, the fluids are water and blood.

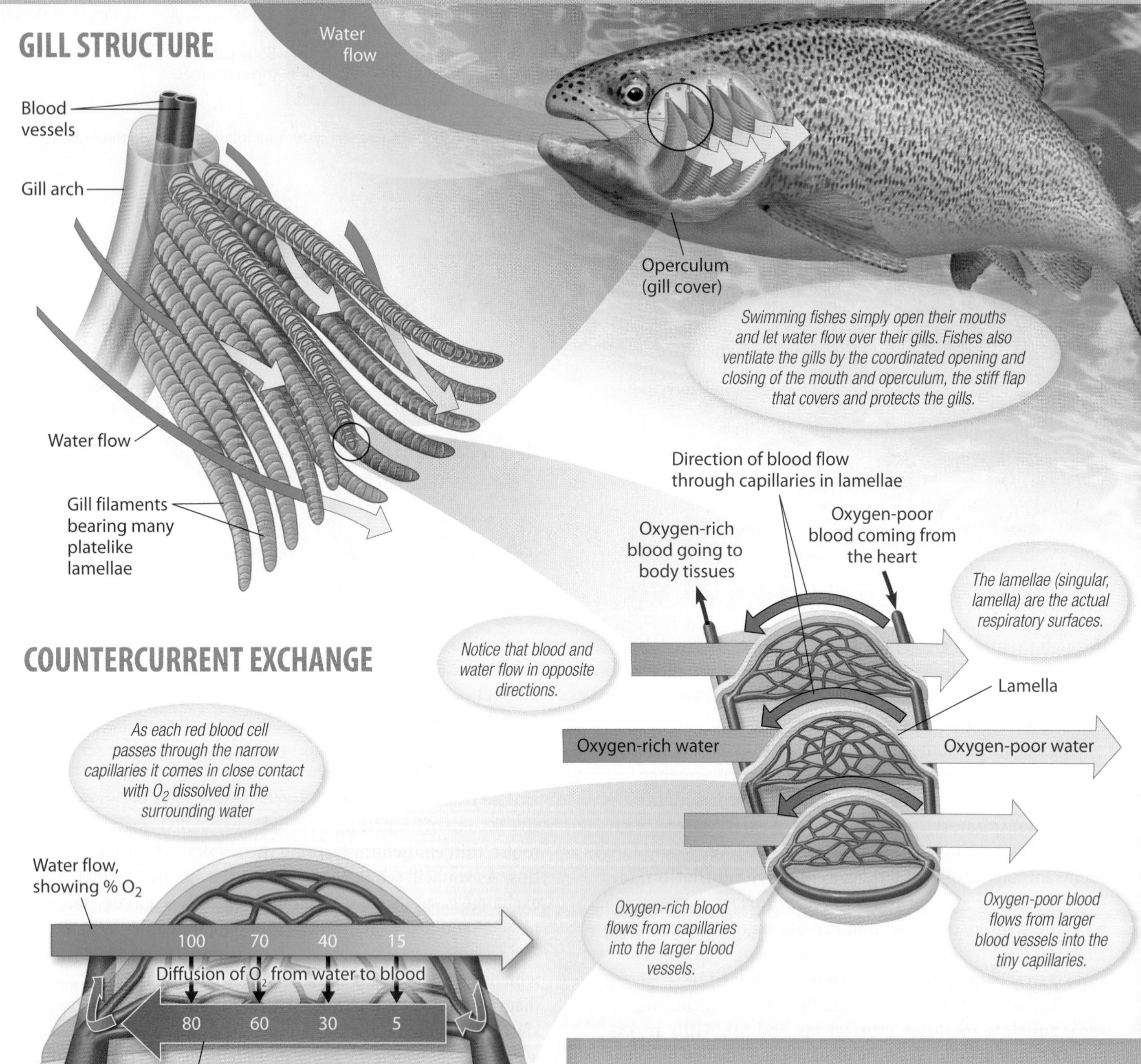
GILL STRUCTURE
Water flow
Blood vessels
Gill arch
Water flow
Gill filaments bearing many platelike lamellae
Operculum (gill cover)
Swimming fishes simply open their mouths and let water flow over their gills. Fishes also ventilate the gills by the coordinated opening and closing of the mouth and operculum, the stiff flap that covers and protects the gills.
Direction of blood flow through capillaries in lamellae
Oxygen-rich blood going to body tissues
Oxygen-poor blood coming from the heart
The lamellae (singular, lamella) are the actual respiratory surfaces.
COUNTERCURRENT EXCHANGE
Notice that blood and water flow in opposite directions.
Lamella
Oxygen-rich water
Oxygen-poor water
As each red blood cell passes through the narrow capillaries it comes in close contact with O_2 dissolved in the surrounding water
Water flow, showing % O_2
100
70
40
15
Diffusion of O_2 from water to blood
80
60
30
5
Blood flow in capillary, showing % O_2
Oxygen-rich blood flows from capillaries into the larger blood vessels.
Oxygen-poor blood flows from larger blood vessels into the tiny capillaries.
The countercurrent flow pattern creates an oxygen gradient between water and blood along the entire length of the capillary, making it possible for oxygen to diffuse into the blood.

? What would be the maximum percentage of the water's O_2 a gill could extract if its blood flowed in the same direction as the water instead of counter to it? (This is a challenging one! It may help to sketch it out.)

● 50%. As O_2 diffuses from the water into the blood as they flow in the same direction, the concentration gradient becomes less and less steep, until there is an equal amount of O_2 in both, and O_2 can no longer diffuse from water to blood.

22.4 The tracheal system of insects provides direct exchange between the air and body cells

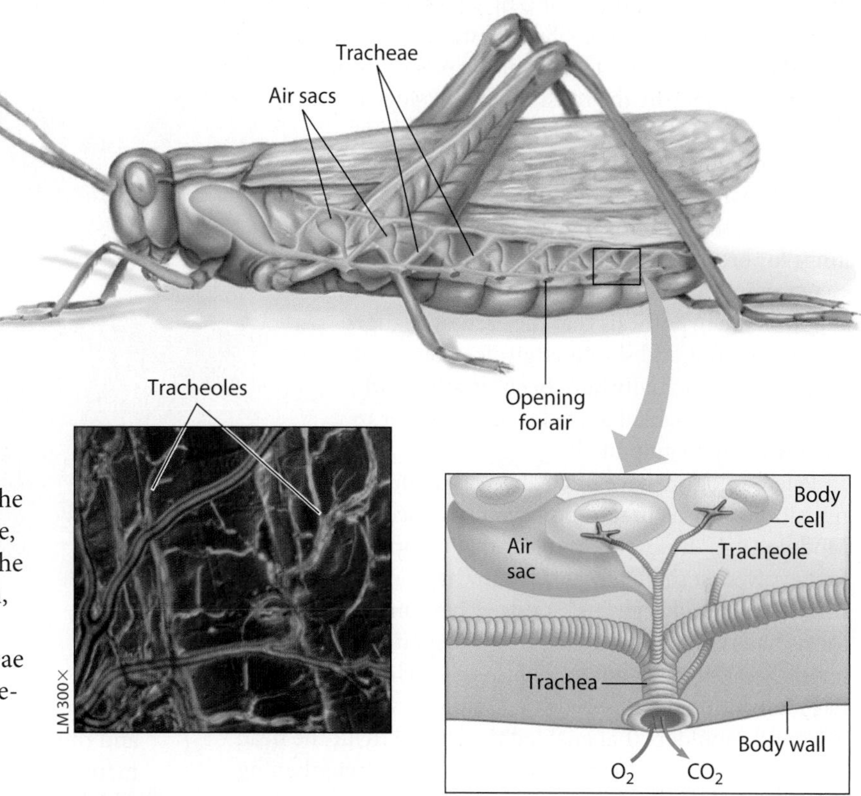

Tracheae
Air sacs
Tracheoles
Opening for air
LM 300×
Body cell
Air sac
Tracheole
Trachea
Body wall
O_2
CO_2

▲ **Figure 22.4A** The tracheal system of an insect

There are two big advantages to breathing air: Air contains a much higher concentration of O_2 than does water, and air is much lighter and easier to move than water. Thus, a terrestrial animal expends much less energy than an aquatic animal ventilating its respiratory surface. The main problem facing an air-breathing animal, however, is the loss of water to the air by evaporation.

The tracheal system of insects, with respiratory surfaces at the tips of tiny branching tubes inside the body, greatly reduces evaporative water loss. **Figure 22.4A** illustrates the tracheal system in a grasshopper. The largest tubes, called tracheae, open to the outside, as shown in the blowup on the bottom right of the figure. Tracheae are reinforced by rings of chitin, the tough polysaccharide that also makes up an insect's exoskeleton. Enlarged portions of tracheae form air sacs (shown in pink) near organs that require a large supply of O_2.

The micrograph on the left in Figure 22.4A shows how these tubes branch repeatedly. The smallest branches, called tracheoles, extend to nearly every cell in the insect's body. Their tiny tips have closed ends and contain fluid (blue in the drawing). Gas is exchanged with body cells by diffusion across the moist epithelium that lines these tips. The structure of a tracheal system matches its function of exchanging gases directly with body cells. Thus, the circulatory system of insects is not involved in transporting gases.

For a small insect, diffusion through the tracheae brings in enough O_2 to support cellular respiration. Larger insects may ventilate their tracheal systems with rhythmic body movements that compress and expand the air tubes like bellows. An insect in flight **(Figure 22.4B)** has a very high metabolic rate and consumes 10 to 200 times more O_2 than it does at rest. In many insects, alternating contraction and relaxation of the flight muscles rapidly pumps air through the tracheal system.

▲ **Figure 22.4B** A grasshopper in flight

Openings for air

? In what fundamental way does the process of gas exchange in insects differ from that in both fishes and humans?

● The circulatory system of insects is not involved in transporting gases to and from the body cells.

22.5 The evolution of lungs facilitated the movement of tetrapods onto land

EVOLUTION CONNECTION

The colonization of land by vertebrates was one of the pivotal milestones in the history of life. The evolution of legs from fins may be the most obvious change in body design, but the refinement of lung breathing was just as important. And although skeletal changes were undoubtedly required in the transition from fins to legs, the evolution of lungs for breathing on land also required skeletal changes. Interestingly, current fossil evidence supports the hypothesis that the earliest changes in the front fins and shoulder girdle of tetrapod ancestors may actually have been breathing adaptations that enabled a fish in shallow water to push itself up to gulp in air.

Paleontologists have uncovered numerous transitional forms in tetrapod evolution (see Module 19.4). It now seems clear that tetrapods first evolved in shallow water from what some researchers jokingly call "fishapods." These ancient forms had both gills and lungs. The adaptations for air breathing evident in their fossils include a flat skull with a strong, elongated snout, as well as a muscular neck and shoulders that enabled the animal to lift the head clear of water and into the unsupportive air. Strengthening of the lower jaw may have facilitated the pumping motion presumed to be used by early air-breathing tetrapods and still employed by frogs to inflate their lungs. The 375-million-year-old fossil of *Tiktaalik* (**Figure 22.5**) illustrates some of these air-breathing adaptations.

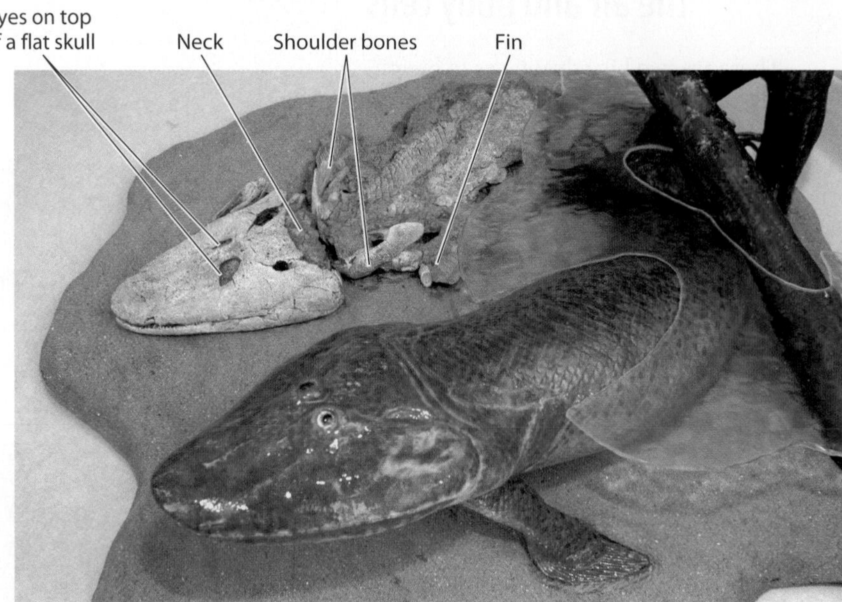
Eyes on top of a flat skull
Neck
Shoulder bones
Fin

▲ **Figure 22.5** A fossil of *Tiktaalik* and a reconstruction of how the living animal might have looked

The first tetrapods on land diverged into three major lineages: amphibians, reptiles (including birds), and mammals. Most amphibians have small lungs and rely heavily on the diffusion of gases across body surfaces. Reptiles and mammals rely on lungs for gas exchange. In general, the size and complexity of lungs correlate with an animal's metabolic rate and thus oxygen need. For example, the lungs of birds and mammals, whose high body temperatures are maintained by a high metabolic rate, have a greater area of exchange surface than the lungs of similar-sized amphibians and nonbird reptiles, which have a much lower metabolic rate.

We explore the mammalian respiratory system next.

? How might adaptations for breathing air be linked to the evolution of tetrapod limbs?

Fossil evidence indicates that changes in the neck, shoulder girdle, and limb bones may have helped early tetrapod ancestors lift their heads above water to gulp air.

▷ The Human Respiratory System

22.6 In mammals, branching tubes convey air to lungs located in the chest cavity

As in all mammals, your lungs are located in your chest, or thoracic cavity, and are protected by the supportive rib cage. The thoracic cavity is separated from the abdominal cavity by a sheet of muscle called the **diaphragm.** You will see how the diaphragm helps ventilate your lungs in Module 22.8.

Figure 22.6, on the facing page, shows the human respiratory system (along with the esophagus and heart, for orientation). Air enters your respiratory system through the nostrils. It is filtered by hairs and warmed, humidified, and sampled for odors as it flows through a maze of spaces in the nasal cavity. You can also draw in air through your mouth, but mouth breathing does not allow the air to be processed by your nasal cavity.

From the nasal cavity or mouth, air passes to the **pharynx,** a common passageway for air and food. As you will remember from the previous chapter, when you swallow food, the **larynx** (the upper part of the respiratory tract) moves upward and tips the epiglottis over the opening of your **trachea,** or windpipe (see Figure 21.6A). The rest of the time, the air passage in the pharynx is open for breathing.

The larynx is often called the voice box. When you exhale, the outgoing air rushes by a pair of **vocal cords** in the larynx, and you can produce sounds by voluntarily tensing muscles that stretch the cords so they vibrate. You produce high-pitched sounds when your vocal cords are tightly stretched and vibrating very fast. When the cords are less tense, they vibrate slowly and produce low-pitched sounds.

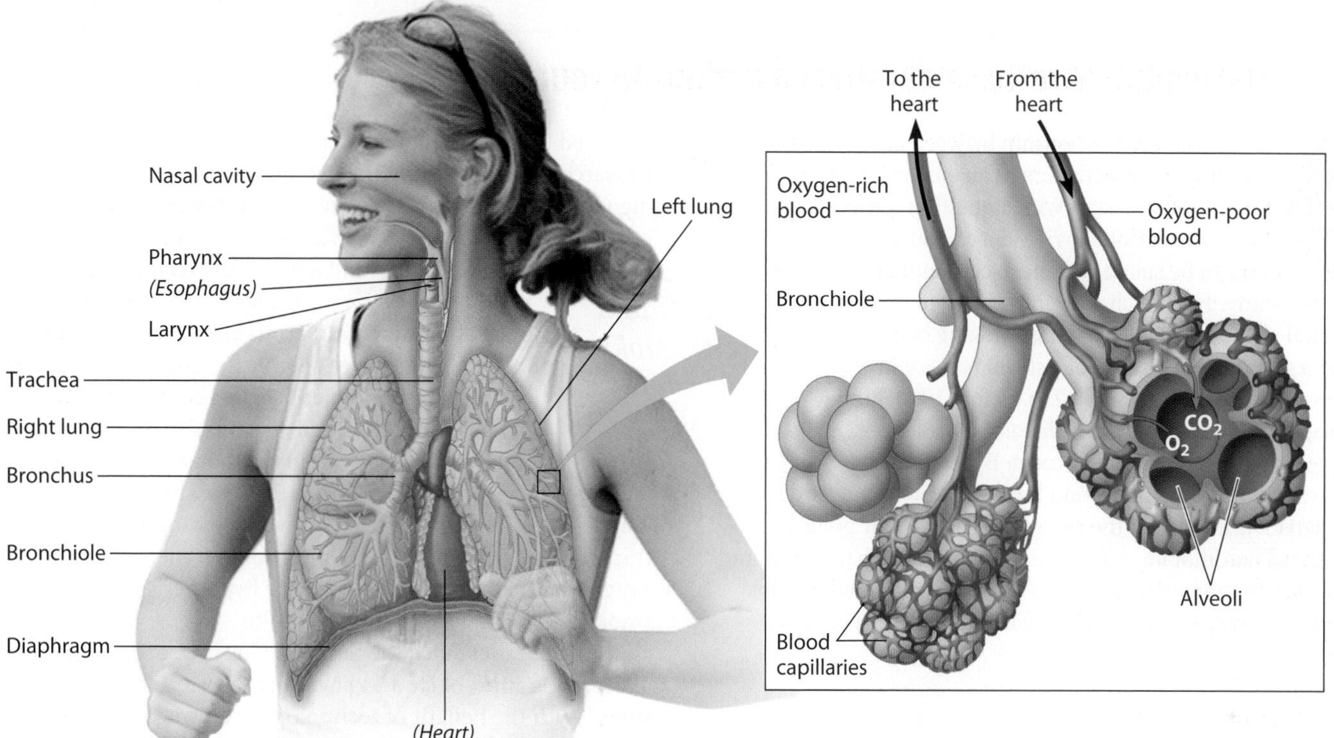

Nasal cavity
Pharynx
(Esophagus)
Larynx
Trachea
Right lung
Bronchus
Bronchiole
Diaphragm
(Heart)
Left lung
To the heart
From the heart
Oxygen-rich blood
Oxygen-poor blood
Bronchiole
CO_2
O_2
Alveoli
Blood capillaries

▲ **Figure 22.6** The anatomy of the human respiratory system (left) and details of the alveoli (right)

Try This Use your finger to trace the pathway of air from the nasal passages to the blood capillaries, pausing to name and describe each structure along the way.

From the larynx, air passes into your trachea. Rings of cartilage (shown in the figure in blue) reinforce the walls of the larynx and trachea, keeping this part of the airway open. The trachea forks into two **bronchi** (singular, *bronchus*), one leading to each lung. Within the lung, the bronchus branches repeatedly into finer and finer tubes called **bronchioles.** Bronchitis is a condition in which these small tubes become inflamed and constricted, making breathing difficult.

As the enlargement on the right of Figure 22.6A shows, the bronchioles dead-end in grapelike clusters of air sacs called **alveoli** (singular, *alveolus*). Each of your lungs contains millions of these tiny sacs. Together they have a surface area of about 100 m^2 (1076 ft^2), 50 times that of your skin. The inner surface of each alveolus is lined with a thin layer of epithelial cells. The O_2 in inhaled air dissolves in a film of moisture on the epithelial cells. It then diffuses across the epithelium and into the dense web of blood capillaries that surrounds each alveolus. This close association between capillaries and alveoli also enables CO_2 to diffuse the opposite way—from the capillaries, across the epithelium of the alveolus, into the air space, and finally out in the exhaled air.

The major branches of your respiratory system are lined by a moist epithelium covered by cilia and a thin film of mucus. The cilia and mucus are the respiratory system's cleaning system. The beating cilia move mucus with trapped dust, pollen, and other contaminants upward to the pharynx, where it is usually swallowed.

Respiratory Problems Alveoli are so small and thin-walled that specialized secretions called **surfactants** are required to keep them from sticking shut from the surface tension of their moist surface. Respiratory distress syndrome due to a lack of lung surfactant is a common disease seen in babies born 6 weeks or more before their due dates. Surfactants typically appear in the lungs after 33 weeks of embryonic development; birth normally occurs at 38 weeks. Artificial surfactants are now administered through a breathing tube to treat such preterm infants.

Alveoli are highly susceptible to airborne contaminants. Defensive white blood cells patrol them and engulf foreign particles. However, if too much particulate matter reaches the alveoli, the delicate lining of these small sacs becomes damaged and the efficiency of gas exchange drops. Studies have shown a significant association between exposure to fine particles and premature death. Air pollution and tobacco smoke are two sources of these lung-damaging particles.

Exposure to such pollutants can cause continual irritation and inflammation of the lungs and lead to chronic obstructive pulmonary disease (COPD). COPD encompasses two main conditions: emphysema and chronic bronchitis. In emphysema, the delicate walls of alveoli become permanently damaged and the lungs lose the elasticity that helps expel air during exhalation. With COPD, both lung ventilation and gas exchange are severely impaired. Patients experience labored breathing, coughing, and frequent lung infections. COPD is a major cause of disability and death in the United States.

? **How does the structure of alveoli match their function?**

● Alveoli have a thin, moist epithelium across which dissolved O_2 and CO_2 can easily diffuse into or out of the surrounding capillaries. The huge collective surface area of all the alveoli enables the passage of many gas molecules.

22.7 Warning: Cigarette smoking is hazardous to your health

SCIENTIFIC THINKING

Research on the relationship between cigarette smoking and human health offers us an example of how, over time, scientists start with a question, progress to a tentative answer, and arrive at a conclusion that can be stated with near-certainty. As you learned in the chapter introduction, investigation of the effects of cigarette smoking began with a search for the cause of a steep increase in cases of lung cancer (**Figure 22.7**). Before 1900, lung cancer was an extremely rare disease, representing less than 1% of all cancer cases. By 1940, lung cancer had become the second leading cause of cancer deaths. As statisticians tracked the rise of lung cancer, air pollution, increased automobile traffic, the 1918 influenza pandemic, and exposure to industrial chemicals were all considered possible causes. However, doctors who treated lung cancer patients observed that a very high percentage of them were smokers. Researchers began testing the hypothesis that cigarette smoking was responsible for the increased incidence of lung cancer.

How certain are scientists that cigarette smoking is hazardous to health?

The scientific evidence accumulated by 1964, when the Surgeon General first recommended warning labels for cigarettes, came from multiple lines of inquiry. Numerous animal studies had tested the effects of chemical compounds found in cigarette smoke. Scientists had compared physiological functions and structural changes in the cells, tissues, and organs of thousands of smokers and nonsmokers in nearly 30 retrospective studies and in several decades-long prospective studies (see Module 3.9) that tracked the health of more than 1.1 million men. The conclusions overwhelmingly supported the hypothesis that cigarette smoking is hazardous to human health.

In the decades since the Surgeon General sounded the alarm, thousands of studies have corroborated those conclusions. With the benefit of technological advances, further research has deepened our understanding of the negative health impacts of cigarette smoke and revealed new concerns about second-hand smoke and the use of other forms of tobacco. And yet, in 2010, more than 45 million adults in the United States were smokers, and tobacco use was the leading cause of preventable death. Scientists are now investigating why so many smokers can't quit. By unraveling the molecular mechanisms of nicotine addiction, researchers hope to develop effective antismoking therapies—perhaps even a vaccine that prevents addiction.

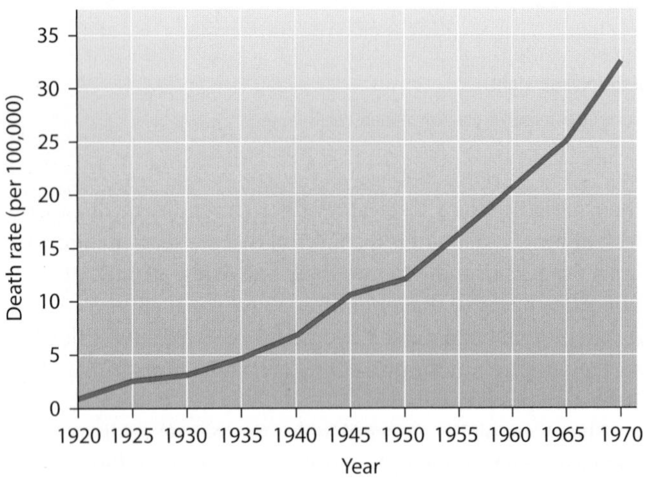
Death rate (per 100,000)
0
5
10
15
20
25
30
35
1920 1925 1930 1935 1940 1945 1950 1955 1960 1965 1970
Year

▲ **Figure 22.7** Mortality from cancers of the respiratory system

? What evidence supports the hypothesis printed on cigarette packs: "Cigarette smoking is hazardous to your health"?

● Results from thousands of scientific investigations, including animal studies on the effects of chemical compounds in cigarette smoke and both retrospective and prospective studies on humans

22.8 Negative pressure breathing ventilates your lungs

Breathing is ventilation of the lungs through alternating inhalation and exhalation. The continual movement of air as you inhale and exhale maintains high O_2 and low CO_2 concentrations at the respiratory surface. In humans and other mammals, ventilation occurs by **negative pressure breathing,** a system in which air is pulled into the lungs.

How does negative pressure breathing work? The key is to create a pressure gradient by changing the volume of the lungs. (You may recall from chemistry class that the pressure exerted by a gas varies inversely with volume.) During inhalation, the ribs move upward and out as muscles between the ribs contract, and the diaphragm contracts and moves downward (**Figure 22.8**). These contractions expand the volume of the thoracic cavity. The lungs, which have a natural elasticity, expand along with the thoracic cavity. Air pressure in the alveoli decreases—it becomes lower than atmospheric pressure,

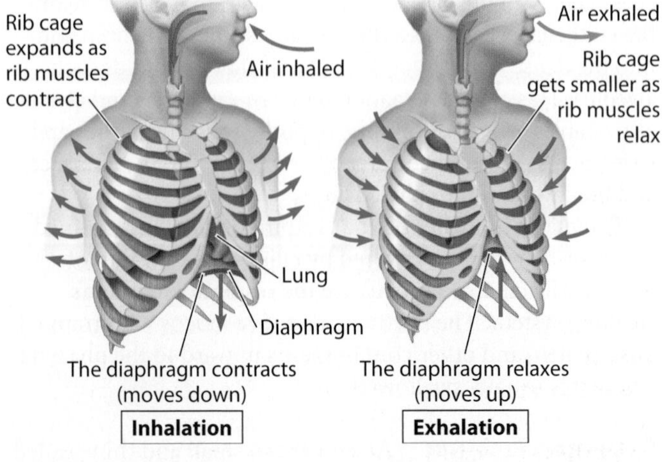
Rib cage expands as rib muscles contract
Air inhaled
Lung
Diaphragm
The diaphragm contracts (moves down)
Inhalation
Air exhaled
Rib cage gets smaller as rib muscles relax
The diaphragm relaxes (moves up)
Exhalation

▲ **Figure 22.8** Negative pressure breathing

Try This Use the diagram to explain what causes air to flow into and out of the lungs.

which explains why the mechanism is called "negative pressure breathing." Air, moving from a region of higher pressure to a region of lower pressure, is pulled from the surrounding atmosphere through the nostrils and into the lungs.

Exhalation reverses the pressure gradient. The rib muscles and diaphragm relax, reducing the volume of the thoracic cavity. The lungs return to their relaxed, unstretched position. The resultant increase in alveolar air pressure forces air up the breathing tubes and out of the body.

Each year, you take between 4 million and 10 million breaths. The volume of air in each breath is about 500 milliliters (mL) when you breathe quietly. The volume of air breathed during maximal inhalation and exhalation is called **vital capacity.** It averages about 3.4 L and 4.8 L for college-age females and males, respectively. (Women tend to have smaller rib cages and lungs.) The lungs actually hold more air than the vital capacity. Because the alveoli do not completely collapse, a residual volume of "dead" air remains in the lungs even after you blow out as much air as you can. As lungs lose elasticity (springiness) with age or as the result of disease, such as emphysema, less air exits on exhalation and residual volume increases at the expense of vital capacity.

Because the lungs do not completely empty, each inhalation mixes fresh air with oxygen-depleted air. Thus, you can extract only about 25% of the O_2 in the air you inhale.

? Explain why inhalation is an active process (requiring work), whereas exhalation is usually passive.

Inhalation uses muscle contraction to expand the thoracic cavity; exhalation occurs when the muscles relax.

22.9 Breathing is automatically controlled

Although you can voluntarily hold your breath or breathe faster and deeper, most of the time your breathing is under involuntary control. **Figure 22.9** illustrates how a **breathing control center** in a part of the brain called the medulla oblongata ensures that your breathing rate is coordinated with your body's need for oxygen. 1 Nerves from the breathing control center signal the diaphragm and rib muscles to contract, causing you to inhale. When you are at rest, these nerve signals result in about 10 to 14 inhalations per minute. Between inhalations, the muscles relax, and you exhale.

2 The control center regulates breathing rate in response to changes in the CO_2 level of the blood. When you exercise vigorously, for instance, your metabolism speeds up and your body cells generate more CO_2 as a waste product. The CO_2 goes into the blood, where it reacts with water to form carbonic acid. The acid slightly lowers the pH of the blood and the fluid bathing the brain, the cerebrospinal fluid. When the medulla senses this pH drop, its breathing control center increases both the rate and depth of your breathing. As a result, more CO_2 is eliminated in the exhaled air, and the pH of the blood returns to normal.

The O_2 concentration in the blood usually has little effect on the breathing control center. Because the same process that consumes O_2—cellular respiration—also produces CO_2, a rise in CO_2 is generally a good indication of a decrease in blood oxygen. Thus, by responding to lowered pH, the breathing control center increases blood oxygen level.

3 Secondary control over breathing is exerted by sensors in the aorta and carotid arteries that monitor concentrations of O_2 as well as CO_2. When the O_2 level in the blood is severely depressed, these sensors signal the control center via nerves to increase the rate and depth of breathing. This response may occur, for example, at high altitudes, where the atmospheric pressure is so low that you cannot get enough O_2 by breathing normally.

The breathing control center responds to a variety of nervous and chemical signals that serve to keep the rate and depth of your breathing in tune with the changing metabolic needs of your body. Breathing rate must also be coordinated with the activity of the circulatory system, which transports blood to and from the alveolar capillaries. We examine the role of the circulatory system in gas exchange more closely in the next module.

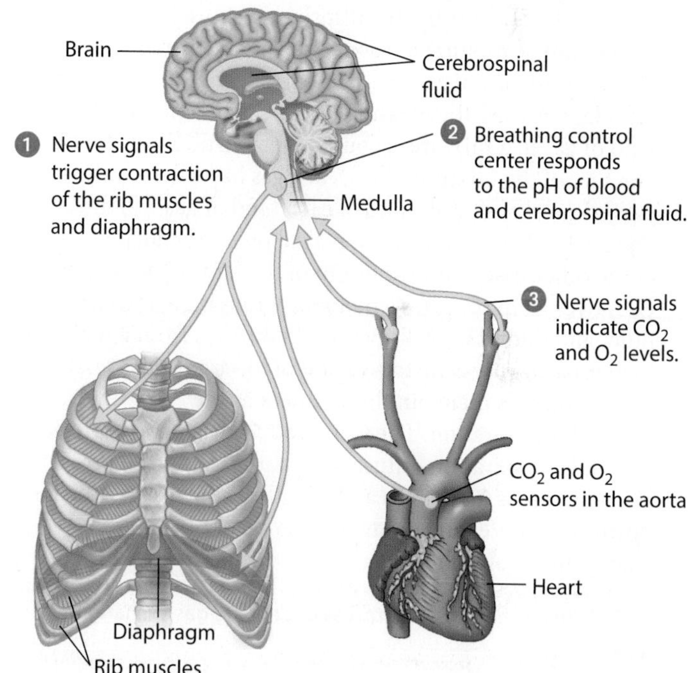

Brain
Cerebrospinal fluid
1 Nerve signals trigger contraction of the rib muscles and diaphragm.
2 Breathing control center responds to the pH of blood and cerebrospinal fluid.
Medulla
3 Nerve signals indicate CO_2 and O_2 levels.
CO_2 and O_2 sensors in the aorta
Heart
Diaphragm
Rib muscles

▲ **Figure 22.9** How the breathing control center regulates breathing

? How is the increased need for O_2 during exercise accommodated by the breathing control center?

During exercise, cells release more CO_2 to the blood, which forms carbonic acid, lowering the pH of the blood. The breathing center senses the decrease in pH and sends impulses to increase breathing rate, thus supplying more O_2.

▷ Transport of Gases in the Human Body

22.10 Blood transports respiratory gases

How does oxygen get from your lungs to all the other tissues in your body, and how does carbon dioxide travel from the tissues to your lungs? To answer these questions, we must jump ahead a bit and look at the basic organization of the human circulatory system (which is the topic of Chapter 23).

Figure 22.10 is a diagram showing the main components of your circulatory system and their roles in gas exchange. Let's start with the heart, in the middle of the diagram. One side of the heart handles oxygen-poor blood (colored blue). The other side handles oxygen-rich blood (red). As indicated in the lower left of the diagram, oxygen-poor blood returns to the heart from capillaries in body tissues. The heart pumps this blood to the alveolar capillaries in the lungs. Gases are exchanged between air in the alveoli and blood in the capillaries (top of diagram). Blood that has lost CO_2 and gained O_2 returns to the heart and is then pumped out to body tissues.

The exchange of gases between capillaries and the cells around them occurs by the diffusion of gases down gradients of pressure. A mixture of gases, such as air, exerts pressure. You see evidence of gas pressure whenever you open a can of soda, releasing the pressure of the CO_2 it contains. Each kind of gas in a mixture accounts for a portion of the total pressure of the mixture. Thus, each gas has what is called a **partial pressure.** Molecules of each kind of gas will diffuse down a gradient of their own partial pressure independently of the other gases. At the bottom of the figure, for instance, O_2 moves from oxygen-rich blood, through the interstitial fluid, and into tissue cells because it diffuses from a region of higher partial pressure to a region of lower partial pressure. The tissue cells maintain this gradient as they consume O_2 in cellular respiration. The CO_2 produced as a waste product of cellular respiration diffuses down its own partial pressure gradient out of tissue cells and into the capillaries. Diffusion down partial pressure gradients also accounts for gas exchange in the alveoli.

? **What is the physical process underlying gas exchange?**

● Diffusion of each gas down its partial pressure gradient

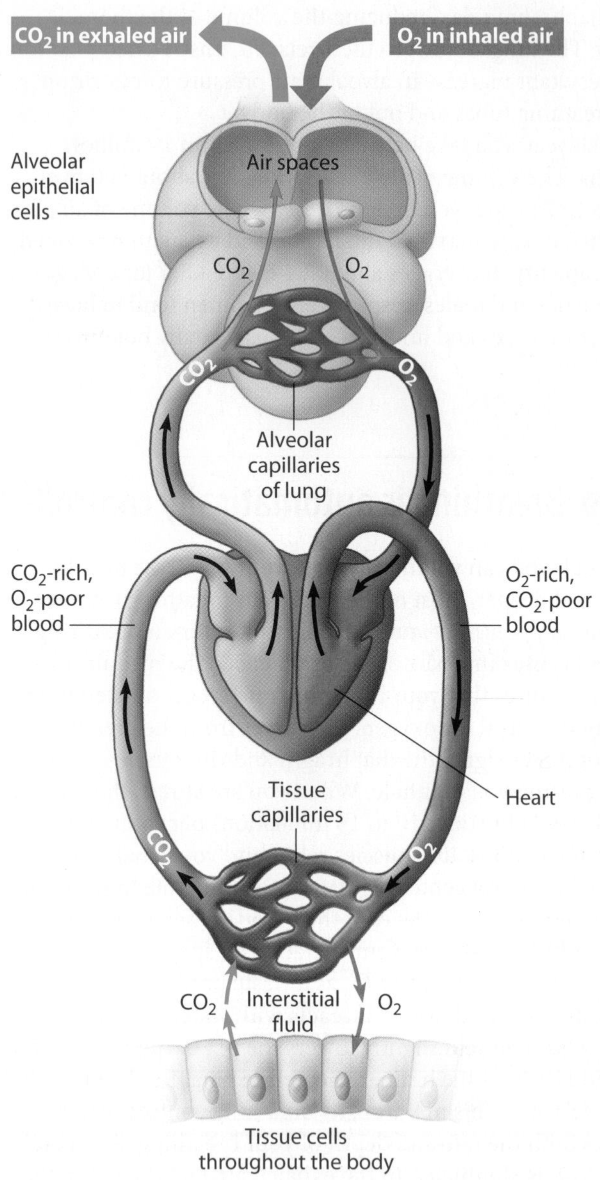

CO_2 in exhaled air
O_2 in inhaled air
Alveolar epithelial cells
Air spaces
CO_2
O_2
CO_2
O_2
Alveolar capillaries of lung
CO_2-rich, O_2-poor blood
O_2-rich, CO_2-poor blood
Heart
Tissue capillaries
CO_2
O_2
CO_2
Interstitial fluid
O_2
Tissue cells throughout the body

▲ **Figure 22.10** Gas transport and exchange in the body

22.11 Hemoglobin carries O_2, helps transport CO_2, and buffers the blood

Oxygen is not highly soluble in water, and most animals transport O_2 bound to proteins called respiratory pigments. These molecules have distinctive colors, hence the name pigment. Many molluscs and arthropods use a blue, copper-containing pigment. Almost all vertebrates and many invertebrates use **hemoglobin,** an iron-containing pigment that turns red when it binds O_2.

Each of your red blood cells is packed with about 250 million molecules of hemoglobin. A hemoglobin molecule consists of four polypeptide chains of two different types, depicted with two shades of purple in **Figure 22.11**, on the next page. Attached to each polypeptide is a chemical group called a heme (colored blue in the figure), at the center of which is an iron atom (gray). Each iron atom binds one O_2 molecule. Thus, every hemoglobin molecule can carry up to four O_2 molecules. Hemoglobin loads up with O_2 in the lungs and transports it to the body's tissues. There, hemoglobin unloads some or all of its cargo, depending on the O_2 needs of the cells. The partial pressure of O_2 in the tissue reflects how much O_2 the cells are using and determines how much O_2 is unloaded.

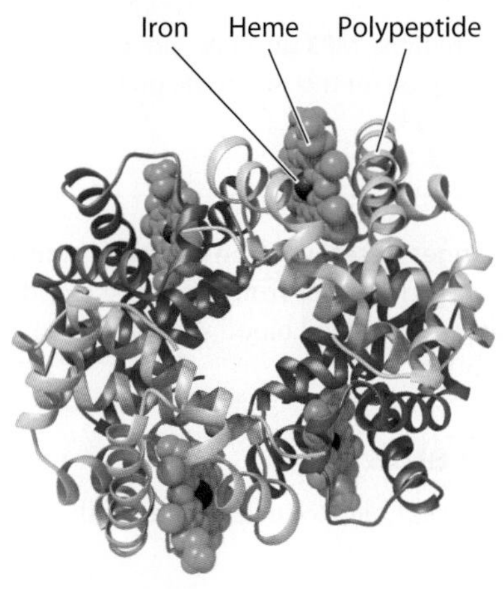
Iron
Heme
Polypeptide

▲ **Figure 22.11** Hemoglobin molecule

Hemoglobin is a multipurpose molecule. It also helps transport CO_2 and assists in buffering the blood. Most of the CO_2 that diffuses from tissue cells into a capillary enters red blood cells, where some of it combines with hemoglobin. The rest reacts with water, forming carbonic acid (H_2CO_3), which then breaks apart into a hydrogen ion (H^+) and a bicarbonate ion (HCO_3^-). This reversible reaction is shown below:

$$\underset{\text{Carbon dioxide}}{CO_2} + \underset{\text{Water}}{H_2O} \rightleftharpoons \underset{\text{Carbonic acid}}{H_2CO_3} \rightleftharpoons \underset{\text{Hydrogen ion}}{H^+} + \underset{\text{Bicarbonate ion}}{HCO_3^-}$$

Hemoglobin binds most of the H^+ produced by this reaction, minimizing the change in blood pH. (As discussed in Module 22.9, the slight drop in pH due to the increased production of CO_2 during exercise is the stimulus to increase breathing rate.) The bicarbonate ions diffuse into the plasma, where they are carried to the lungs.

As blood flows through capillaries in the lungs, the reaction is reversed. Bicarbonate ions combine with H^+ to form carbonic acid; carbonic acid is converted to CO_2 and water; and CO_2 diffuses from the blood to the alveoli and leaves the body in exhaled air.

We have seen how O_2 and CO_2 are transported between your lungs and body tissue cells via the bloodstream. In the next module, we consider a special case of gas exchange between two circulatory systems.

? **O_2 in the blood is transported bound to __________ within __________ __________ cells, and CO_2 is mainly transported as __________ ions within the plasma.**

● hemoglobin . . . red blood . . . bicarbonate

22.12 The human fetus exchanges gases with the mother's blood

CONNECTION

Figure 22.12 is a drawing of a human fetus inside the mother's uterus. The fetus literally swims in a protective watery bath, the amniotic fluid. Its nonfunctional lungs are full of fluid. How does the fetus exchange gases with the outside world? It does this by way of the placenta, a composite organ that includes tissues from both fetus and mother. A large network of capillaries fans out into the placenta from blood vessels in the umbilical cord of the fetus. These capillaries exchange gases with the maternal blood that circulates in the placenta, and the mother's circulatory system transports the gases to and from her lungs. Aiding O_2 uptake by the fetus is fetal hemoglobin, which attracts O_2 more strongly than does adult hemoglobin.

One of the reasons that smoking is considered a health risk during pregnancy is because it reduces, perhaps by as much as 25%, the supply of oxygen reaching the placenta. Lower oxygen levels delay fetal development and growth, resulting in a higher incidence of premature birth, low birth weight, and brain and lung defects. Avoiding cigarette smoke protects the health of both mothers and babies.

Let's move on to what happens when a baby is born. Very soon after delivery, placental gas exchange with the mother ceases, and the baby's lungs must begin to work. Carbon dioxide acts as the signal. As soon as CO_2 stops diffusing from the fetus into the placenta, CO_2 levels rise in the fetal blood. The resulting drop in blood pH stimulates the breathing control center in the infant's brain, and the newborn gasps and takes its first breath.

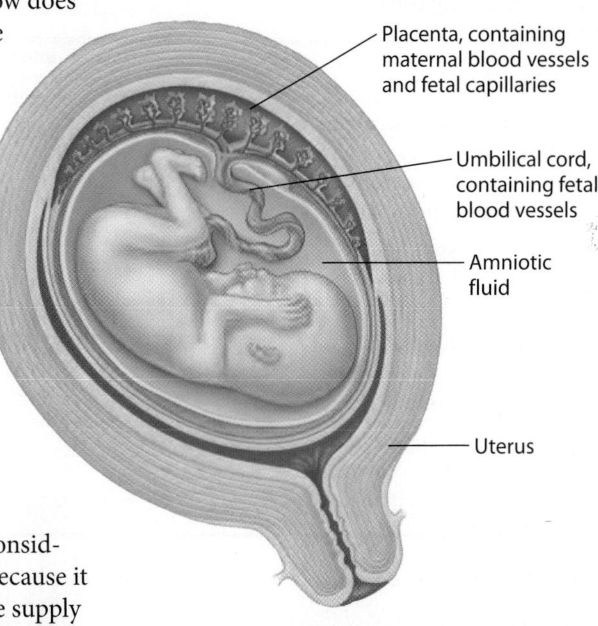
Placenta, containing maternal blood vessels and fetal capillaries
Umbilical cord, containing fetal blood vessels
Amniotic fluid
Uterus

▲ **Figure 22.12** A human fetus and placenta in the uterus

A human birth and the radical changes in gas exchange mechanisms that accompany it are extraordinary events. For a human baby to switch almost instantly from living in water and exchanging gases with maternal blood to breathing air directly requires truly remarkable adaptations in the organism's respiratory system. Also required are adaptations of the circulatory system, which, as we have seen, supports the respiratory system in its gas exchange function.

? **How does fetal hemoglobin enhance oxygen transfer from mother to fetus across the placenta?**

● Because fetal hemoglobin has a greater affinity for O_2 than does adult hemoglobin, it helps „pull" the O_2 from maternal blood to fetal blood.

CHAPTER 22 REVIEW

For practice quizzes, BioFlix animations, MP3 tutorials, video tutors, and more study tools designed for this textbook, go to **MasteringBiology®**

Reviewing the Concepts

Mechanisms of Gas Exchange (22.1–22.5)

22.1 Gas exchange in humans involves breathing, transport of gases, and exchange with body cells. Gas exchange, the interchange of O_2 and CO_2 between an organism and its environment, provides O_2 for cellular respiration and removes its waste product, CO_2.

22.2 Animals exchange O_2 and CO_2 across moist body surfaces. Respiratory surfaces must be thin and moist for diffusion of O_2 and CO_2 to occur. Some animals use their entire skin as a gas exchange organ. In most animals, gills, a tracheal system, or lungs provide large respiratory surfaces for gas exchange.

22.3 Gills are adapted for gas exchange in aquatic environments. Gills absorb O_2 dissolved in water. In a fish, gas exchange is enhanced by ventilation and the countercurrent flow of water and blood.

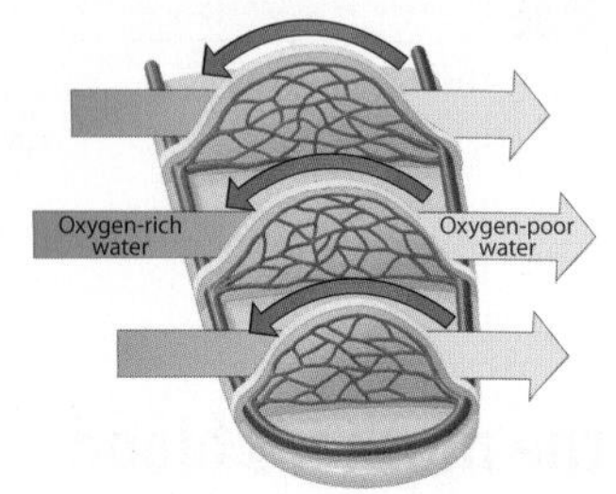
Oxygen-rich water
Oxygen-poor water

22.4 The tracheal system of insects provides direct exchange between the air and body cells. A network of finely branched tubes transports O_2 directly to body cells and moves CO_2 from them.

22.5 The evolution of lungs facilitated the movement of tetrapods onto land. Skeletal adaptations of air-breathing fish may have helped early tetrapods move onto land.

The Human Respiratory System (22.6–22.9)

22.6 In mammals, branching tubes convey air to lungs located in the chest cavity. Inhaled air passes through the pharynx and larynx into the trachea, bronchi, and bronchioles to the alveoli. Mucus and cilia in the respiratory passages protect the lungs.

22.7 Warning: Cigarette smoking is hazardous to your health. Evidence accumulated from thousands of studies has shown that smoking causes lung cancer and other respiratory diseases.

22.8 Negative pressure breathing ventilates your lungs. The contraction of rib muscles and diaphragm expands the thoracic cavity, reducing air pressure in the alveoli and drawing air into the lungs.

22.9 Breathing is automatically controlled. A breathing control center in the brain keeps breathing in tune with body needs, sensing and responding to the CO_2 level in the blood. A drop in blood pH triggers an increase in the rate and depth of breathing.

Transport of Gases in the Human Body (22.10–22.12)

22.10 Blood transports respiratory gases. The heart pumps oxygen-poor blood to the lungs, where it picks up O_2 and drops off CO_2. Oxygen-rich blood returns to the heart and is pumped to body cells, where it drops off O_2 and picks up CO_2.

22.11 Hemoglobin carries O_2, helps transport CO_2, and buffers the blood.

22.12 The human fetus exchanges gases with the mother's blood. Fetal hemoglobin enhances oxygen transfer from maternal blood in the placenta. At birth, rising CO_2 in fetal blood stimulates the breathing control center to initiate breathing.

Connecting the Concepts

1. Complete the following concept map to review some of the concepts of gas exchange.

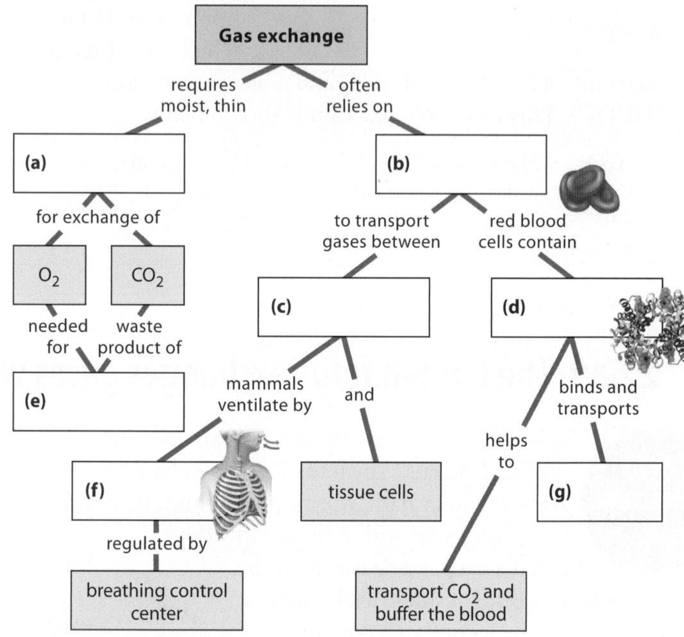
Gas exchange
requires moist, thin
often relies on
(a)
(b)
for exchange of
to transport gases between
red blood cells contain
O_2
CO_2
(c)
(d)
needed for
waste product of
(e)
mammals ventilate by
and
binds and transports
helps to
(f)
tissue cells
(g)
regulated by
breathing control center
transport CO_2 and buffer the blood

2. Label the parts of the human respiratory system.

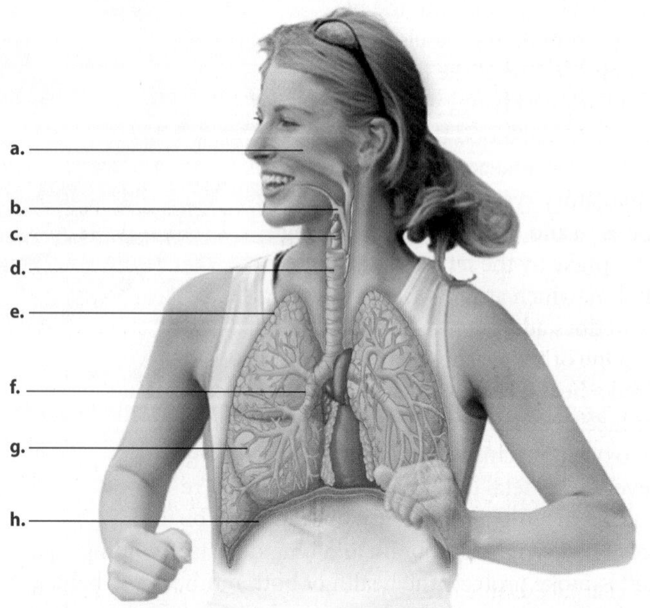
a.
b.
c.
d.
e.
f.
g.
h.

Testing Your Knowledge

Level 1: Knowledge/Comprehension

3. When you hold your breath, which of the following first leads to the urge to breathe?
   a. falling CO_2
   b. falling O_2
   c. rising CO_2
   d. rising pH of the blood
4. Countercurrent gas exchange in the gills of a fish
   a. maintains a gradient that enhances diffusion.
   b. enables the fish to obtain oxygen without swimming.
   c. means that blood and water flow at different rates.
   d. allows O_2 to diffuse against its partial pressure gradient.
5. When you inhale, the diaphragm
   a. relaxes and moves upward.
   b. relaxes and moves downward.
   c. contracts and moves upward.
   d. contracts and moves downward.
6. In which of the following organisms does oxygen diffuse directly across a respiratory surface to cells, without being carried by the blood?
   a. a grasshopper
   b. a whale
   c. an earthworm
   d. a mouse
7. What is the function of the cilia in the trachea and bronchi?
   a. to sweep air into and out of the lungs
   b. to increase the surface area for gas exchange
   c. to dislodge food that may have slipped past the epiglottis
   d. to sweep mucus with trapped particles up and out of the respiratory tract
8. What do the alveoli of mammalian lungs, the gill filaments of fish, and the tracheal tubes of insects have in common?
   a. use of a circulatory system to transport gases
   b. respiratory surfaces that are infoldings of the body wall
   c. countercurrent exchange
   d. a large, moist surface area for gas exchange
9. What is the primary feedback used by the brain to control breathing?
   a. heart rate
   b. partial pressure of O_2
   c. blood pH, which indicates O_2 level
   d. blood pH, which indicates CO_2 level

Level 2: Application/Analysis

10. What are two advantages of breathing air, compared with obtaining dissolved oxygen from water? What is a comparative disadvantage of breathing air?
11. Trace the path of an oxygen molecule in its journey from the air to a muscle cell in your arm, naming all the structures involved along the way.
12. Carbon monoxide (CO) is a colorless, odorless gas found in furnace and automobile engine exhaust and cigarette smoke. CO binds to hemoglobin 210 times more tightly than does O_2. (CO also binds with an electron transport protein and disrupts cellular respiration.) Explain why CO is such a deadly gas.

Level 3: Synthesis/Evaluation

13. Partial pressure reflects the relative amount of gas in a mixture and is measured in millimeters of mercury (mm Hg). Llamas are native to the Andes Mountains in South America. The partial pressure of O_2 (abbreviated P_{O_2}) in the atmosphere where llamas live is about half of the P_{O_2} at sea level. As a result, the P_{O_2} in the lungs of llamas is about 50 mm Hg, whereas that in human lungs at sea level is about 100 mm Hg.

    A dissociation curve for hemoglobin shows the percentage of saturation (the amount of O_2 bound to hemoglobin) at increasing values of P_{O_2}. As you see in the graph below, the dissociation curves for llama and human hemoglobin differ. Compare these two curves and explain how the hemoglobin of llamas is an adaptation to living where the air is "thin."

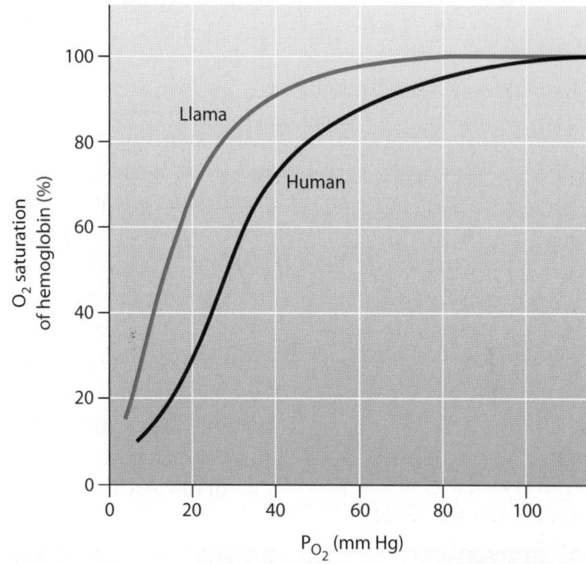

O_2 saturation of hemoglobin (%)
100
80
60
40
20
0
Llama
Human
0
20
40
60
80
100
P_{O_2} (mm Hg)

14. Mountain climbers often spend weeks adjusting to the lower partial pressure of oxygen at high altitudes before and during their ascent of high peaks. During that time, their bodies begin to produce more red blood cells. Some runners and cyclists prepare for competition by training at high altitudes or by sleeping in a tent in which P_{O_2} is kept artificially low. Explain why this training strategy may improve an athlete's performance.
15. One of the many mutant opponents that the movie monster Godzilla contends with is Mothra, a giant mothlike creature with a wingspan of 7–8 m. Science fiction creatures like these can be critiqued on the grounds of biomechanical and physiological principles. Focusing on the principles of gas exchange that you learned about in this chapter, what problems would Mothra face? Why do you think truly giant insects are improbable?
16. **SCIENTIFIC THINKING** A hookah is an ancient Middle Eastern water pipe, a smoking apparatus in which tobacco smoke passes through water before being inhaled through a hose with a mouthpiece at the end. Some hookahs have multiple hoses, making smoking a social activity. In recent years, hookah lounges offering a menu of flavored tobacco, or shisha, have become popular among young adults. Many people assume that hookah smoking is a safe alternative to cigarettes. Evaluate the scientific evidence for this assumption. The Centers for Disease Control website is a good place to start (www.cdc.gov/tobacco/data_statistics/fact_sheets/tobacco_industry/hookahs/index.htm)

Answers to all questions can be found in Appendix 4.

CHAPTER 23

Circulation

Why is the heart such an essential organ?

The ancient Greeks and Romans believed that our capacity for thought and reason, our imagination, our spirit, our emotions, our sense of self—in short, all our human attributes—were located in the heart. We now know that credit for these functions rightly belongs to the brain. In fact, from a scientific point of view, the heart is no more than a pump composed of cardiac muscle (photo below) that drives the circulation of blood throughout the body. Its construction, which involves little besides plumbing and electricity, is so straightforward that inventors have built artificial hearts that can replace the real thing, at least temporarily.

The heart is no ordinary machine, however. It works ceaselessly, contracting and relaxing an average of 72 times per minute, more than 100,000 times per day and 35 *million* times per year throughout your entire life. A typical heart pumps about 5 L of blood—the body's entire blood volume—each minute. In times of great demand, such as during athletic activities, output can increase to as much as 35 L (more than 9 gallons!) per minute. The blood flows through a closed circuit of vessels to every tissue in the body, delivering oxygen from the lungs and nutrients from the digestive tract and picking up wastes for disposal. Every part of your body depends on a constant supply of this vital fluid. Thus, although the heart is not the seat of your humanity, as was once believed, it is more essential to maintaining life than is most of the brain. You can survive the loss of a kidney or a lung or large chunks of intestine or liver. In contrast, as you'll learn in this chapter, the death of relatively few heart cells can be fatal.

Most animals have a circulatory system that connects organs involved in gas exchange, digestion, and waste processing. We begin this chapter with a survey of some of the circulatory systems in different animals. We then turn to the human cardiovascular system and explore the structures and functions of the heart, blood vessels, and blood.

BIG IDEAS

Circulatory Systems
(23.1–23.2)

Internal transport systems carry materials between exchange surfaces and body cells.

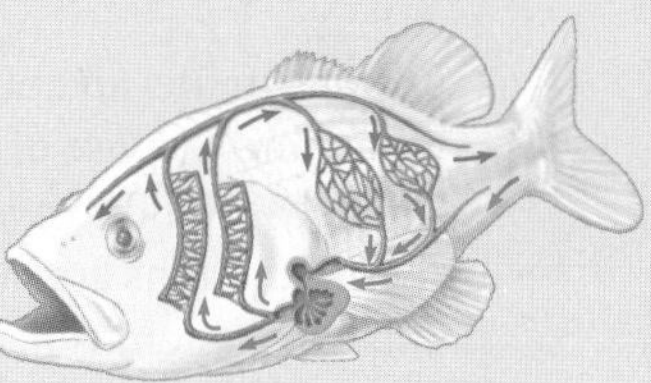

The Human Cardiovascular System and Heart
(23.3–23.6)

The heart pumps blood through the pulmonary circuit and the systemic circuit.

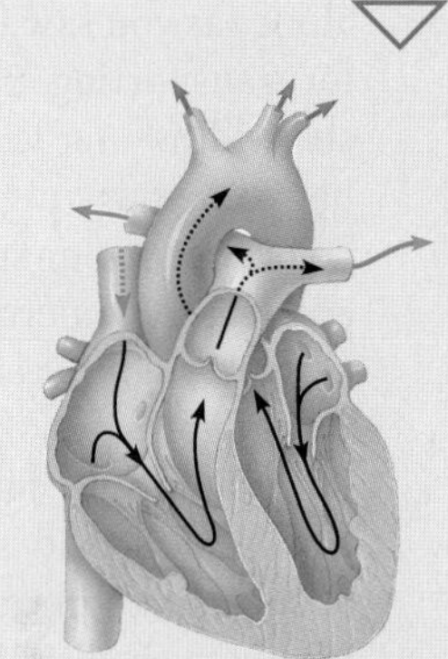

Structure and Function of Blood Vessels
(23.7–23.11)

Blood flows through arteries to capillaries, where exchange occurs with body cells, and returns to the heart in veins.

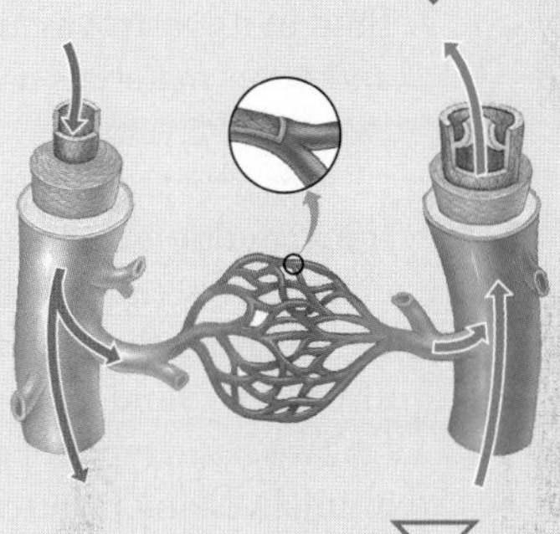

Structure and Function of Blood
(23.12–23.15)

Red blood cells carry oxygen, white blood cells fight infections, and platelets function in blood clotting.

▷ Circulatory Systems

23.1 Circulatory systems facilitate exchange with all body tissues

To sustain life, an animal must acquire nutrients, exchange gases, and dispose of waste products, and these needs ultimately extend to every cell in the body. In most animals, these functions are facilitated by a **circulatory system**. A circulatory system is necessary in any animal whose body is too large or too complex for such exchange to occur by diffusion alone (as you saw in Figure 20.13A). Diffusion is inadequate for transporting materials over distances greater than a few cell widths—far less than the distance oxygen must travel between your lungs and brain or the distance nutrients must go between your small intestine and the muscles in your arms and legs. An internal transport system must bring resources close enough to cells for diffusion to be effective.

Several types of internal transport have evolved in animals. For example, in cnidarians and most flatworms, a central gastrovascular cavity serves both in digestion and in distribution of substances throughout the body. The body wall of a hydra is only two cells thick, so all the cells can exchange materials directly with the water surrounding the animal or with the fluid in its gastrovascular cavity (see Figure 21.3A). Nutrients and other materials have only a short distance to diffuse between cell layers.

A gastrovascular cavity is not adequate for animals with thick, multiple layers of cells. Such animals require a true circulatory system, which consists of a muscular pump (**heart**), a circulatory fluid, and a set of tubes (vessels) to carry the circulatory fluid.

Two basic types of circulatory systems have evolved in animals. Many invertebrates, including most molluscs and all arthropods, have an **open circulatory system**. The system is called "open" because fluid is pumped through open-ended vessels and flows out among the tissues; there is no distinction between the circulatory fluid and interstitial fluid. In an insect, such as the grasshopper (**Figure 23.1A**), pumping of the tubular heart drives body fluid into the head and the rest of the body (black arrows). Body movements help circulate the fluid as materials are exchanged with body cells. When the heart relaxes, fluid enters through several pores. Each pore has a valve that closes when the heart contracts, preventing backflow of the circulating fluid. In insects (as you learned in Module 22.4), respiratory gases are conveyed to and from body cells by the tracheal system (not shown here), not by the circulatory system.

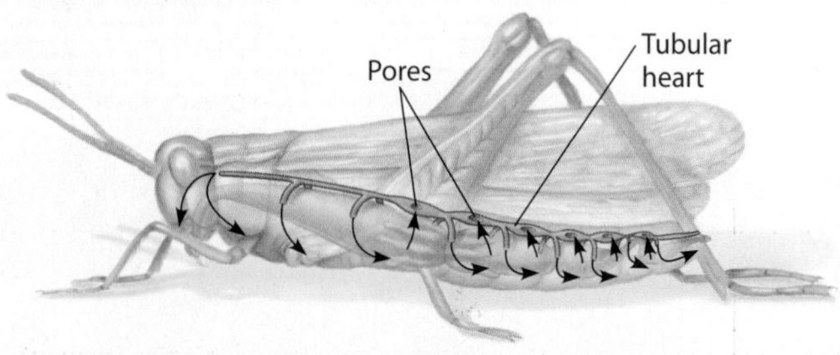

Pores
Tubular heart

▲ **Figure 23.1A** The open circulatory system of a grasshopper

Earthworms, squids, octopuses, and vertebrates (such as ourselves) all have a **closed circulatory system**. It is called "closed" because the circulatory fluid, **blood**, is confined to vessels, keeping blood distinct from the interstitial fluid. There are three kinds of vessels: **Arteries** carry blood away from the heart to body organs and tissues; **veins** return blood to the heart; and **capillaries** convey blood between arteries and veins within each tissue. The vertebrate circulatory system is often called a **cardiovascular system** (from the Greek *kardia*, heart, and Latin *vas*, vessel). How extensive are the vessels in your cardiovascular system? If all your blood vessels were lined up end to end, they would circle Earth's equator twice.

The cardiovascular system of a fish (**Figure 23.1B**) illustrates some key features of a closed circulatory system. The heart of a fish has two main chambers. The **atrium** (plural, *atria*) receives blood from the veins, and the **ventricle** pumps blood to the gills via large arteries. As in all figures depicting closed circulatory systems in this book, red represents oxygen-rich blood and blue represents oxygen-poor blood. After passing through the gill capillaries, the blood, now oxygen-rich, flows into large arteries that carry it to all other parts of the body. The large arteries branch into **arterioles**, small vessels that give rise to capillaries. Networks of capillaries called **capillary beds** infiltrate every organ and tissue in the body. The thin walls of the capillaries allow chemical exchange between the blood and the interstitial fluid. The capillaries converge into **venules**, which in turn converge into larger veins that return blood to the heart.

In the next module, we compare the cardiovascular systems of different vertebrate groups.

? What are the key differences between an open circulatory system and a closed circulatory system?

● The vessels in an open circulatory system do not form an enclosed circuit from the heart, through the body, and back to the heart, and the circulatory fluid is not distinct from interstitial fluid, as is the blood in a closed circulatory system.

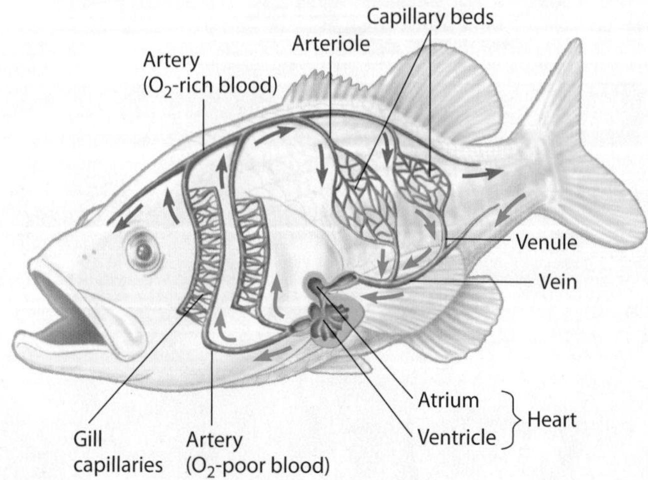

Capillary beds
Arteriole
Artery
(O_2-rich blood)
Venule
Vein
Atrium
Ventricle
Heart
Gill
capillaries
Artery
(O_2-poor blood)

▲ **Figure 23.1B** The closed circulatory system of a fish

Try This Starting with the ventricle, trace the flow of blood through the circulatory system, ending in the atrium.

23.2 Vertebrate cardiovascular systems reflect evolution

EVOLUTION CONNECTION

The colonization of land by vertebrates was a major episode in the history of life. As aquatic vertebrates became adapted for terrestrial life, nearly all of their organ systems underwent major changes. One of these was the change from gill breathing to lung breathing, and this switch was accompanied by important changes in the cardiovascular system.

As illustrated in Figure 23.1B and diagrammed in **Figure 23.2A**, blood passes through the heart of a fish only once in each circuit through the body, an arrangement called **single circulation**. Blood pumped from the ventricle travels first to the gill capillaries. Blood pressure drops considerably as blood flows through the narrow gill capillaries. An artery carries the oxygen-rich blood from the gills to capillaries in the tissues and organs, from which the blood returns to the atrium of the heart. The animal's swimming movements help to propel the blood through the body.

A single circuit would not supply enough pressure to move blood through the capillaries of the lungs and then to the body capillaries of a terrestrial vertebrate. The evolutionary adaptation that resulted in a more vigorous flow of blood to body organs is called **double circulation**, in which blood is pumped a second time after it loses pressure in the lungs. The **pulmonary circuit** carries blood between the heart and gas exchange tissues in the lungs, and the **systemic circuit** carries blood between the heart and the rest of the body.

You can see an example of these two circuits in **Figure 23.2B**. (Notice that the right side of the animal's heart is on the left in the diagram. It is customary to draw the system as though in a body facing you from the page.) Frogs and other amphibians have a three-chambered heart. The right atrium receives blood returning from the systemic capillaries in the body's organs. The ventricle pumps blood to capillary beds in the lungs and skin. Because gas exchange occurs both in the lungs and across the thin, moist skin, this is called a pulmocutaneous circuit. Oxygen-rich blood returns to the left atrium. Although blood from the two atria mixes in the single ventricle, a ridge diverts most of the oxygen-poor blood to the pulmocutaneous circuit and most of the oxygen-rich blood to the systemic circuit.

In the three-chambered heart of turtles, snakes, and lizards, the ventricle is partially divided, and less mixing of blood occurs. The ventricle is completely divided in crocodilians.

In all birds and mammals (for example, the snowy owl in **Figure 23.2C**), the heart has four chambers: two atria and two ventricles. The right side of the heart handles only oxygen-poor blood; the left side receives and pumps only oxygen-rich blood. The evolution of a powerful four-chambered heart was an essential adaptation to support the high metabolic rates of birds and mammals, which are endothermic (see Module 19.7). Endotherms use about 10 times as much energy as equal-sized ectotherms. Therefore, their circulatory system needs to deliver much more fuel and oxygen to body tissues. This requirement is met by a large heart that is able to pump a large volume of blood through separate systemic and pulmonary circulations. Birds and mammals descended from different reptilian ancestors, and their four-chambered hearts evolved independently—an example of convergent evolution, in which natural selection favors the same adaptation in response to similar environmental challenges.

? **What is the difference between the single circulation of a fish and the double circulation of a land vertebrate?**

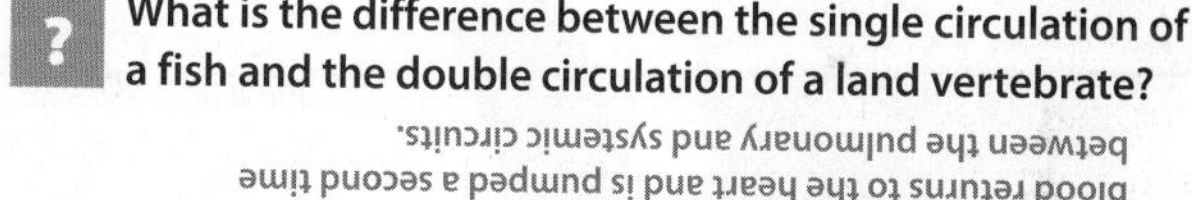

In a fish, blood travels from gill capillaries to body capillaries before returning to the heart. In a land vertebrate, blood returns to the heart and is pumped a second time between the pulmonary and systemic circuits.

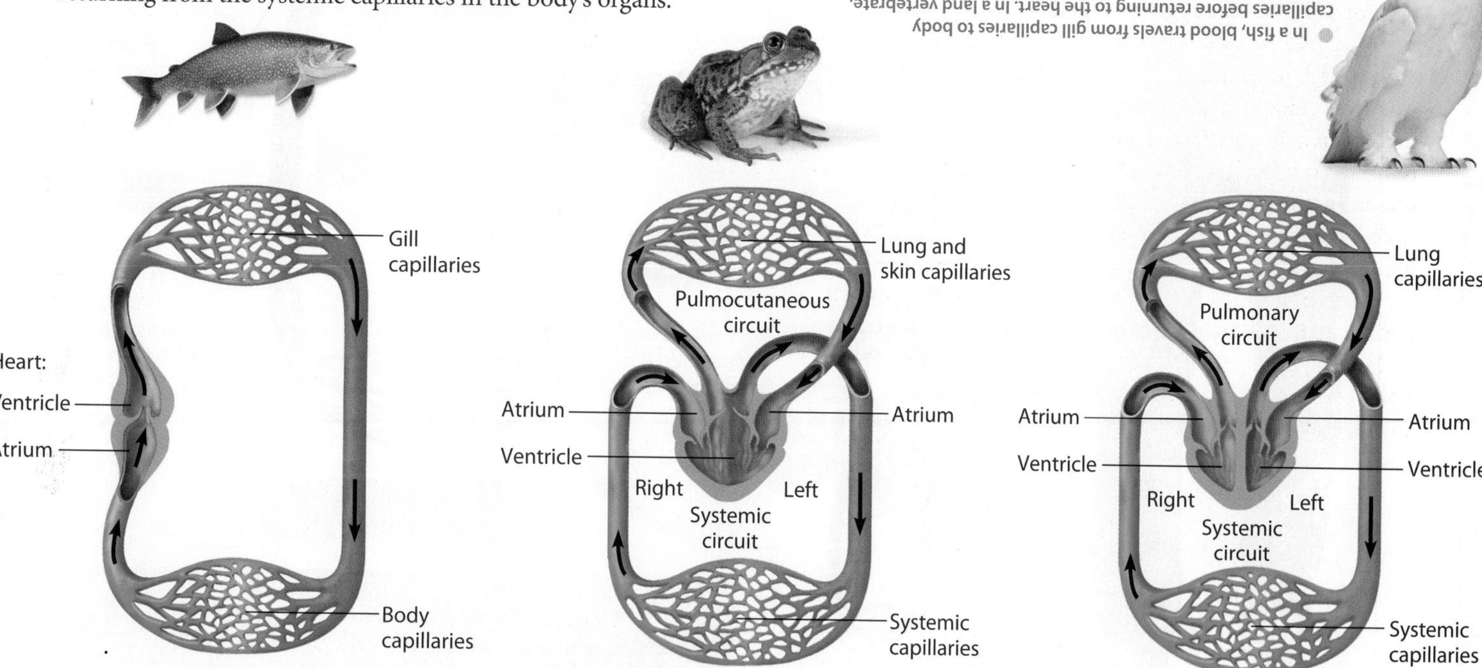

Gill capillaries
Heart:
Ventricle
Atrium
Body capillaries
Lung and skin capillaries
Pulmocutaneous circuit
Atrium
Ventricle
Atrium
Right
Left
Systemic circuit
Systemic capillaries
Lung capillaries
Pulmonary circuit
Atrium
Ventricle
Atrium
Ventricle
Right
Left
Systemic circuit
Systemic capillaries

▲ **Figure 23.2A** The single circulation and two-chambered heart of a fish

▲ **Figure 23.2B** The double circulation and three-chambered heart of an amphibian

▲ **Figure 23.2C** The double circulation and four-chambered heart of a bird or mammal

▷ The Human Cardiovascular System and Heart

VISUALIZING THE CONCEPT

23.3 The human cardiovascular system illustrates the double circulation of mammals

Study this diagram of the flow of blood through the pulmonary and systemic circuits of the human cardiovascular system. Blood leaves the heart through the **pulmonary arteries**, which carry oxygen-poor blood (shown in blue) to the lungs and the **aorta**, which starts oxygen-rich blood (shown in red) on its journey to the body tissues. Blood flows into the heart through the **pulmonary veins**, which bring oxygen-rich blood from the lungs, and two large veins that carry blood from the body tissues. The **superior vena cava** returns oxygen-poor blood to the heart from the upper body, and the **inferior vena cava** brings oxygen-poor blood from the lower body. Starting in the right ventricle, follow the yellow arrows to trace the flow of blood around the pulmonary circuit; the green arrows show the route blood takes around the systemic circuit.

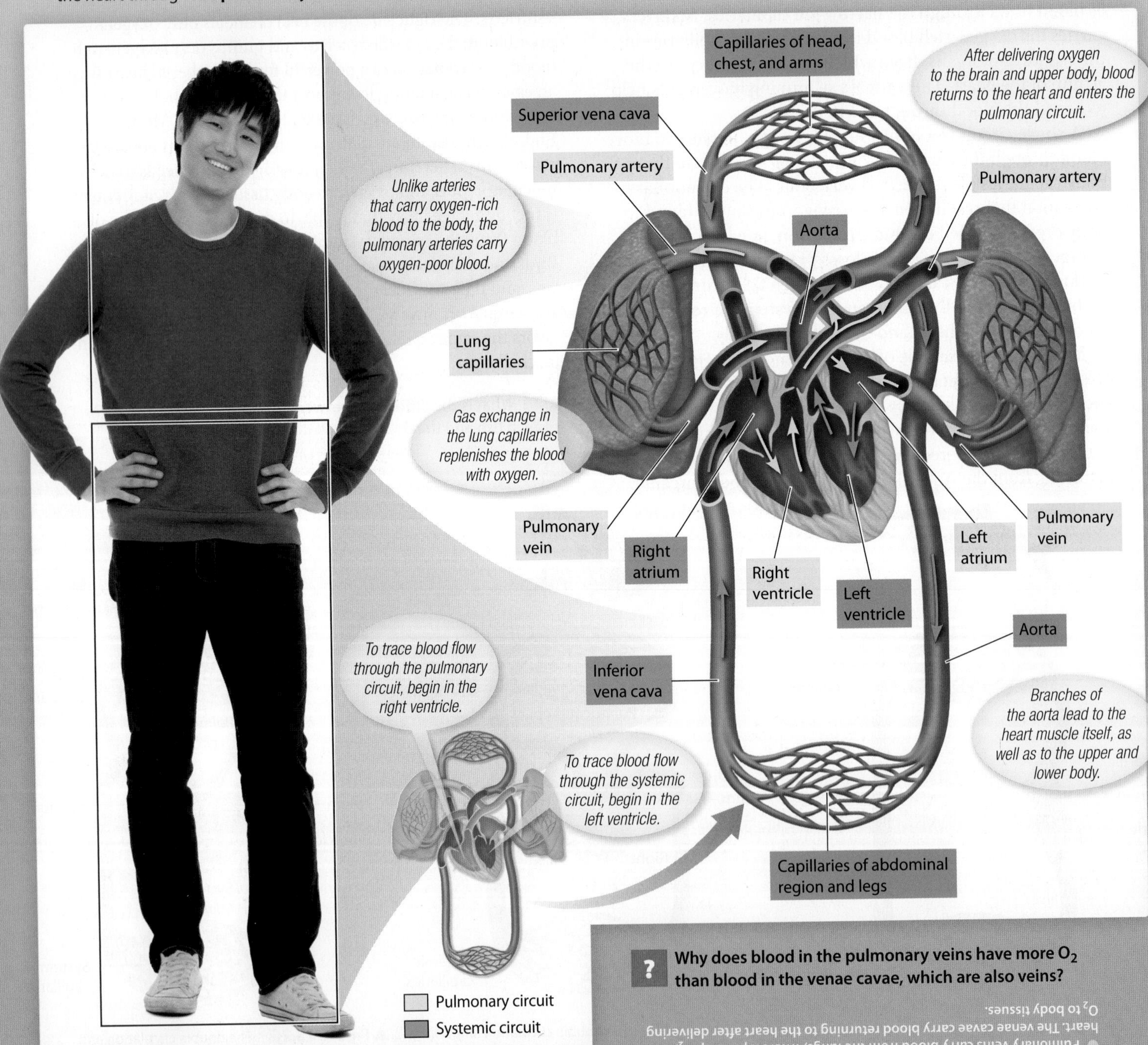

Capillaries of head, chest, and arms
After delivering oxygen to the brain and upper body, blood returns to the heart and enters the pulmonary circuit.
Superior vena cava
Pulmonary artery
Pulmonary artery
Unlike arteries that carry oxygen-rich blood to the body, the pulmonary arteries carry oxygen-poor blood.
Aorta
Lung capillaries
Gas exchange in the lung capillaries replenishes the blood with oxygen.
Pulmonary vein
Right atrium
Right ventricle
Left ventricle
Left atrium
Pulmonary vein
Aorta
To trace blood flow through the pulmonary circuit, begin in the right ventricle.
Inferior vena cava
Branches of the aorta lead to the heart muscle itself, as well as to the upper and lower body.
To trace blood flow through the systemic circuit, begin in the left ventricle.
Capillaries of abdominal region and legs
Pulmonary circuit
Systemic circuit

? **Why does blood in the pulmonary veins have more O_2 than blood in the venae cavae, which are also veins?**

● Pulmonary veins carry blood from the lungs, where it picks up O_2, to the heart. The venae cavae carry blood returning to the heart after delivering O_2 to body tissues.

23.4 The heart contracts and relaxes rhythmically

Let's take a closer look at the hub of the circulatory system, the four-chambered heart. Your heart is about the size of a clenched fist, and it is enclosed in a sac just under the sternum (breastbone). It is formed mostly of cardiac muscle tissue. **Figure 23.4A** shows the path blood takes as the heart separately but simultaneously pumps oxygen-poor blood to the lungs and oxygen-rich blood to the body. Notice that the ventricles, which pump blood to the lungs and body, have much thicker walls than the atria. The thin-walled atria collect blood returning to the heart and squeeze it into the ventricle below, an action that does not require much force.

What keeps the blood flowing in one direction through the heart? Flap-like valves made of connective tissue are positioned in the exit from each chamber. The valves between the atria and ventricles are called atrioventricular (AV) valves; a semilunar valve is located at the exit from each ventricle. These valves open when pushed from one side and close when pushed from the other, thus preventing backflow.

You can hear the closing of the two sets of heart valves either with a stethoscope or by pressing your ear tightly against the chest of a friend (or a friendly dog). The sound pattern is "lub-dup, lub-dup, lub-dup." The "lub" sound comes from the recoil of blood against the closed AV valves. The "dup" is produced by the recoil of blood against the closed semilunar valves.

The heart contracts and relaxes in a rhythmic sequence known as the **cardiac cycle**. When the heart contracts, it pumps blood; when it relaxes, blood fills its chambers. If you have a heart rate of 72 beats per minute, your cardiac cycle takes about 0.8 second. **Figure 23.4B** shows that when the heart is relaxed, in the phase called ❶ **diastole**, blood flows into all four of its chambers. Blood enters the right atrium from the venae cavae and the left atrium from the pulmonary veins (see Module 23.3). The AV valves are open and the semilunar valves are closed. Diastole lasts about 0.4 second, during which the ventricles nearly fill with blood.

The contraction phase of the cardiac cycle is called **systole**. ❷ Systole begins with a very brief (0.1-second) contraction of the atria that completely fills the ventricles with blood (atrial systole). ❸ Then the ventricles contract for about 0.3 second (ventricular systole). The force of their contraction closes the AV valves, opens the semilunar valves located at the exit from each ventricle, and pumps blood into the large arteries. Blood flows into the relaxed atria during the second part of systole, as the green arrows in step 3 indicate.

Because it pumps blood to your whole body, the left ventricle contracts with greater force than the right—notice the thicker wall of the left ventricle in Figure 23.4A. Both ventricles, however, pump the same volume of blood. The volume of blood that each ventricle pumps per minute is called the **cardiac output**. This volume is equal to the amount of blood pumped each time a ventricle contracts (about 70 mL, or a little more than 1/4 cup, for the average person) times the **heart rate** (number of beats per minute). At an average resting heart rate of 72 beats per minute, cardiac output would be about 5 L/min, roughly equivalent to the total volume of blood in your body. Thus, a drop of blood can travel the entire systemic circuit in just 1 minute.

The next module explores the control of the cardiac cycle.

? **During a cardiac cycle of 0.8 second, the atria are generally relaxed for ________ second.**

0.7

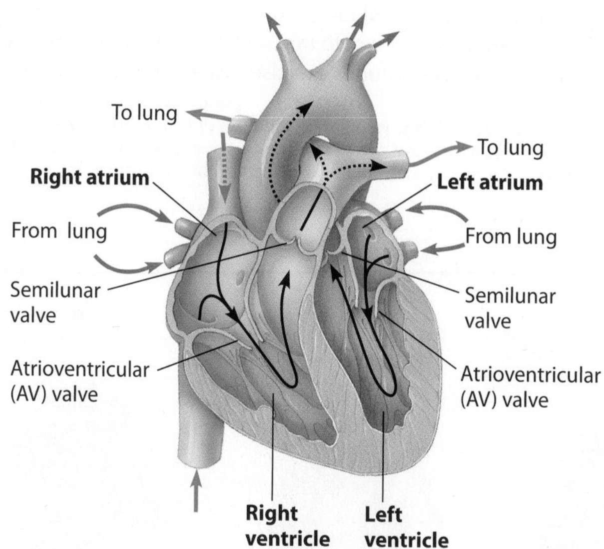
To lung
To lung
Right atrium
Left atrium
From lung
From lung
Semilunar valve
Semilunar valve
Atrioventricular (AV) valve
Atrioventricular (AV) valve
Right ventricle
Left ventricle

▲ **Figure 23.4A** Blood flow through the human heart

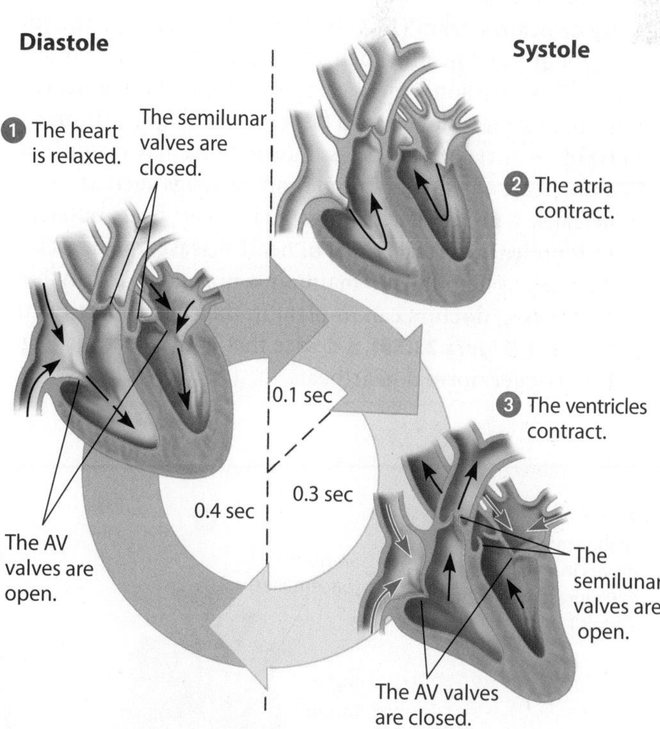
Diastole
Systole
❶ The heart is relaxed.
The semilunar valves are closed.
❷ The atria contract.
0.1 sec
❸ The ventricles contract.
0.3 sec
0.4 sec
The AV valves are open.
The semilunar valves are open.
The AV valves are closed.

▲ **Figure 23.4B** A cardiac cycle in a person with a heart rate of about 72 beats a minute

23.5 The SA node sets the tempo of the heartbeat

In vertebrates, the cardiac cycle originates in the heart itself. Unlike skeletal muscle cells, which contract in response to signals from motor neurons, cardiac muscle cells contract on their own, without any signal from the nervous system. But as you learned in the previous module, contraction and relaxation of each chamber of the heart must follow a precise sequence and timing to pump blood through. How are the contractions of cardiac cells coordinated so that your heart beats as an integrated unit? The answer lies with a group of cells that make up the pacemaker, or **SA (sinoatrial) node**, which sets the rate at which all the muscle cells of your heart contract.

The SA node, situated in the upper wall of the right atrium, generates electrical impulses much like those produced by nerve cells. These signals spread rapidly through the specialized junctions between cardiac muscle cells (see Module 20.6). **Figure 23.5A** shows the sequence of electrical events in the heart. ❶ Signals (shown in the figure in yellow) from the SA node spread quickly through both atria, making them contract in unison. ❷ The impulses pass to a relay point called the **AV (atrioventricular) node**, located between the right atrium and right ventricle. The AV node delays the signal about 0.1 second—the time it takes to empty the atria completely—before the ventricles contract. ❸ Specialized muscle fibers (depicted in orange) then relay the signals to the apex of the heart and ❹ up through the walls of the ventricles, triggering the strong contractions that drive the blood out of the heart.

The electrical impulses in the heart are strong enough to be detected on the skin by electrodes and recorded as an electrocardiogram (ECG or EKG). An ECG can provide data about heart health, such as the existence of arrhythmias. These are abnormal heart rhythms, including heart rates that are too slow or too fast and fibrillations (flutterings) of the atria or ventricles. Fibrillations may occur in a healthy heart when overstimulation by drugs such as caffeine cause a group of cells to generate heart beats outside the SA node. In certain kinds of heart disease, the heart's self-pacing system fails to maintain a normal heart rhythm. In such cases, doctors can implant in the chest an artificial pacemaker (**Figure 23.5B**), a device that emits electrical signals to trigger normal heartbeats.

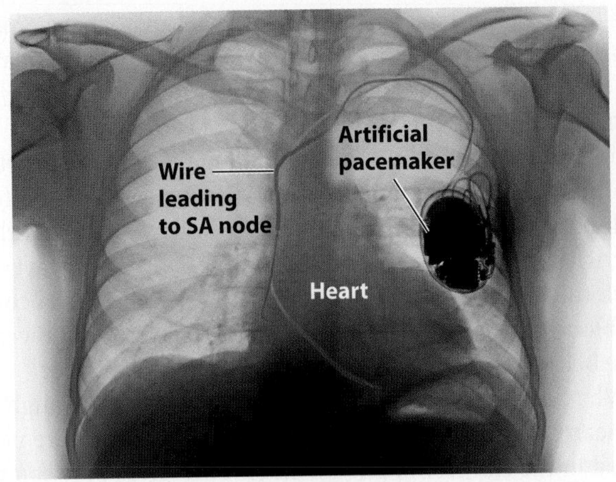

Wire leading to SA node
Artificial pacemaker
Heart

▲ **Figure 23.5B** An artificial pacemaker implanted in the chest

During a heart attack, the SA node is often unable to maintain a normal rhythm. Electrical shocks applied to the chest by a defibrillator may reset the SA node and restore proper cardiac function. The increased availability of automatic external defibrillators (AEDs) has saved thousands of lives. Unlike hospital defibrillators, AEDs are designed to be used by laypeople and are placed in public places (such as airports, movie theaters, and shopping malls) where they are easily accessible.

A variety of cues help regulate the SA node. Two sets of nerves with opposite effects can direct this pacemaker to speed up or slow down, depending on physiological needs and emotional cues. Heart rate is also influenced by hormones, such as epinephrine, the "fight-or-flight" hormone released at times of stress (see Module 26.10). An increased heart rate provides more blood to muscles that may be needed to "fight or flee."

? A slight decrease in blood pH causes the SA node to increase the heart rate. How would this control mechanism benefit a person during strenuous exercise? (*Hint*: See Module 22.9.)

● More CO_2 in the blood, as would occur with increased exercise, causes pH to drop. An accelerated heart rate enhances delivery of O_2-rich blood to body tissues and CO_2-rich blood to the lungs for removal of CO_2. (Breathing rate is also speeded up by this mechanism.)

▶ **Figure 23.5A** The sequence of electrical events in a heartbeat. (Yellow represents electrical signals.)

❶

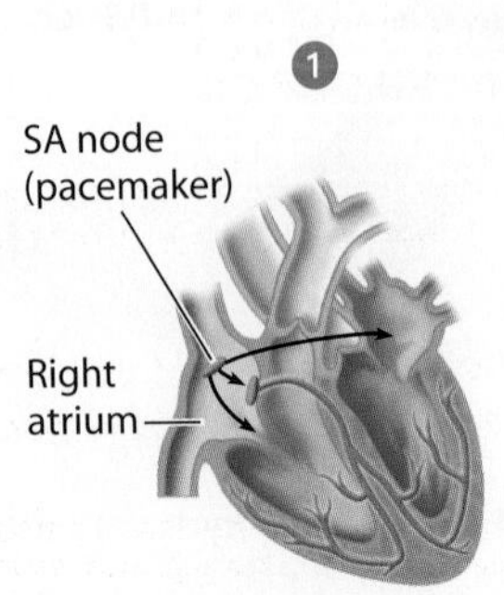

SA node (pacemaker)
Right atrium

❷

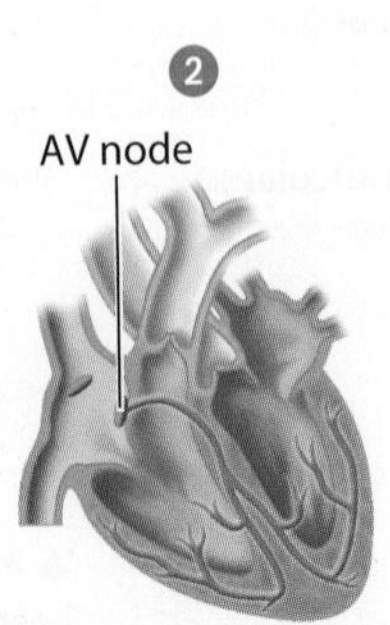

AV node

❸

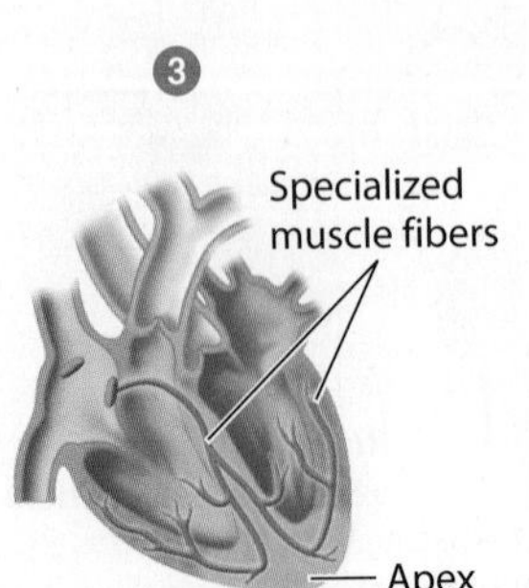

Specialized muscle fibers
Apex

❹

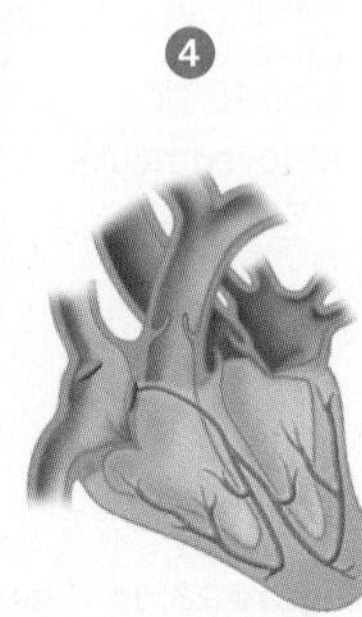

23.6 What causes heart attacks?

SCIENTIFIC THINKING

Why is the heart such an essential organ?

Like all of your cells, your heart muscle cells require nutrients and oxygen-rich blood to survive. Indeed, their requirements are high, because your heart contracts more than 100,000 times a day. Where blood exits the heart via the aorta, several coronary arteries (shown in red in **Figure 23.6A**) immediately branch off to feed the heart muscle. If one or more of these blood vessels become blocked, heart muscle cells will quickly die (gray area in Figure 23.6A). A **heart attack**, also called a myocardial infarction, is the damage or death of cardiac muscle tissue, usually as a result of such blockage. (Rarely, a severe spasm of a coronary artery, possibly related to the use of a drug such as cocaine, may trigger a heart attack.) Approximately one-third of heart attack victims die immediately, because their damaged heart can no longer provide the brain and other vital tissues with sufficient oxygen. For those who survive, the ability of the damaged heart to pump blood may be seriously impaired.

▲ **Figure 23.6A** Blockage of a coronary artery, resulting in a heart attack

More than half of all deaths in the United States are caused by **cardiovascular disease**—disorders of the heart and blood vessels. A **stroke** is the death of brain tissue due to the lack of O_2 resulting from the rupture or blockage of arteries in the head. The suddenness of a heart attack or stroke belies the fact that the arteries of most victims became impaired gradually by a chronic cardiovascular disease known as **atherosclerosis** (from the Greek *athero*, paste, and *sclerosis*, hardness). During the course of this disease, fatty deposits called plaques develop in the inner walls of arteries, narrowing the passages through which blood can flow (**Figure 23.6B**). As a plaque grows, it incorporates fibrous connective tissue and cholesterol, and the walls of the artery become thick and stiff. A blood clot is more likely to become trapped in a vessel that has been narrowed in this way. Furthermore, plaques may rupture and cause blood clots to form, or fragments of the ruptured plaque may become lodged in other narrowed arteries.

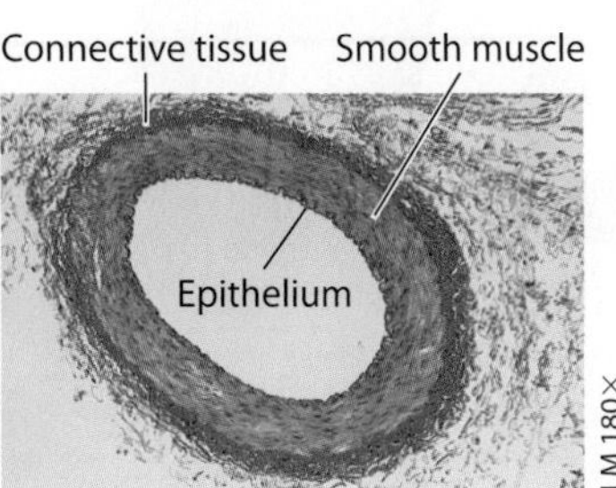
Connective tissue
Smooth muscle
Epithelium
LM 180×

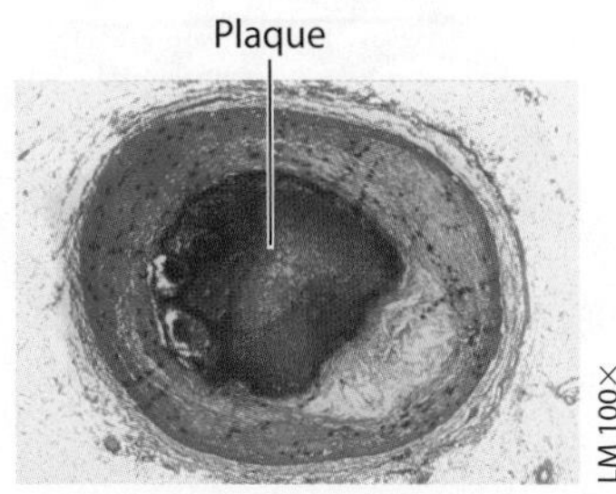
Plaque
LM 100×

▲ **Figure 23.6B** Atherosclerosis: a normal artery (left); an artery partially closed by plaque (right)

For more than 20 years, researchers have known that atherosclerosis and blood clot formation are associated with inflammation, the body's general response to tissue damage. High levels of C-reactive protein (CRP), a substance produced by the liver during episodes of acute inflammation, in the blood correlate strongly with an increased risk of heart attacks and strokes. Some researchers hypothesize that the body's inflammatory response causes heart attacks. In addition to the correlation between CRP and heart attacks, this hypothesis is supported by studies showing that certain anti-inflammatory drugs decrease the risk of heart attack. One of the drugs, aspirin, is typically given during a cardiac episode and is often prescribed in low doses to help prevent the recurrence of heart attacks and strokes. However, aspirin is an anticlotting agent as well as an anti-inflammatory, so it is unclear which property makes it effective against heart attacks. The other drug proven to reduce the risk of heart attacks lowers LDL cholesterol (see Module 21.21) in addition to lowering CRP. Thus, researchers have not determined whether the body's inflammatory response actually causes heart attacks or simply indicates an elevated risk.

Correlations often suggest hypotheses worth exploring, but further research is needed to determine causation. Two large clinical trials to test the inflammation hypothesis began in 2013. The 24,000 participants in the studies, all of whom have already suffered a heart attack and are at high risk for another, were randomly assigned to experimental and control groups. In these double-blind studies, members of the experimental groups are receiving drugs that specifically target inflammation, and the control group participants are given placebos. Researchers will collect data on the occurrence of heart attacks, strokes, and other measures of cardiovascular health. If blocking the inflammatory response proves to be an effective method of treating atherosclerosis, new approaches to treating cardiovascular disease could be developed that will save thousands of lives.

To some extent, the tendency to develop cardiovascular disease appears to be inherited. As you will learn in Module 23.9, however, there are behaviors under your control that significantly affect the risk.

Complete information about these two studies can be found at http://clinicaltrials.gov/show/NCT01327846 and http://clinicaltrials.gov/show/NCT01594333.

? What is the basis for the hypothesis that the body's inflammatory response causes heart attacks?

There is a correlation between heart attacks and CRP, a substance produced during the inflammatory response. Also, anti-inflammatory drugs decrease the risk of heart attack.

▷ Structure and Function of Blood Vessels

23.7 The structure of blood vessels fits their functions

Now that we have explored the structure and function of the heart, let's look at the amazingly extensive series of vessels that transport blood throughout your body.

Functions of Blood Vessels The blood vessels of the circulatory system must have an intimate connection with all the body's tissues. The micrograph in **Figure 23.7A** shows a capillary that carries oxygenated, nutrient-rich blood to smooth muscle cells. Notice that red blood cells pass single file through the capillary, coming close enough to the surrounding tissue that O_2 can diffuse out of them into the muscle cells.

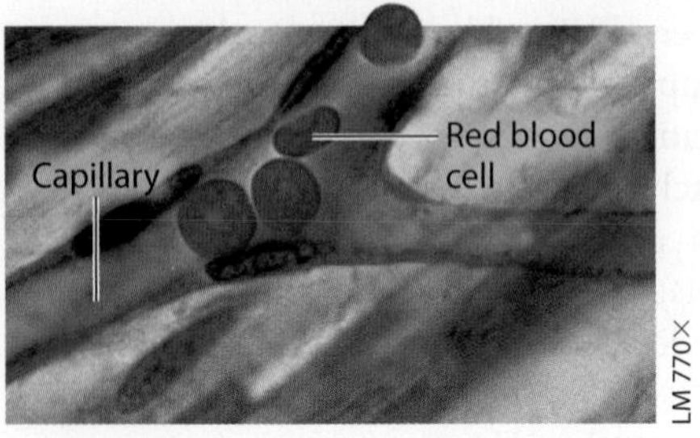
Capillary
Red blood cell
LM 770×

▲ **Figure 23.7A** A capillary in smooth muscle tissue

In **Figure 23.7B** (above right), the downward arrows show the route that molecules take in diffusing from blood in a capillary to tissue cells. Cells are immersed in interstitial fluid. Molecules such as O_2 (●) and nutrients (●) diffuse out of a capillary into the interstitial fluid and then from the fluid into a tissue cell (see Module 20.13).

In addition to transporting O_2 and nutrients, blood vessels convey metabolic wastes to waste disposal organs: CO_2 to the lungs and a variety of other metabolic wastes to the kidneys. The upward arrows in Figure 23.7B represent the diffusion of waste molecules (●) out of a tissue cell, through the interstitial fluid, and into the capillary.

The circulatory system plays several key roles in maintaining a constant internal environment (homeostasis). By exchanging molecules with the interstitial fluid, it helps control the makeup of the environment in which the tissue cells live. As we'll see in later chapters, the circulatory system is also involved in body defense, temperature regulation, and hormone distribution.

Structure of Blood Vessels **Figure 23.7C** illustrates the structures of the different kinds of blood vessels and how the vessels are connected. Look first at the capillaries (center). Appropriate to its function of exchanging materials, a capillary has a very thin wall of a single layer of epithelial cells, which is wrapped in a thin basal lamina (see Module 20.4). The inner surface of the capillary is smooth, which keeps the blood cells from being abraded as they tumble along.

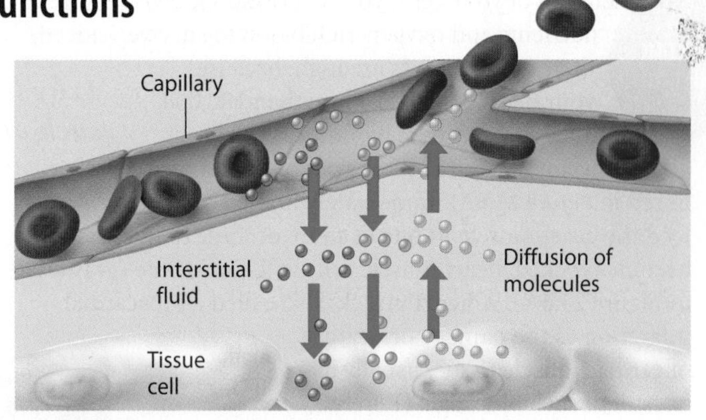
Capillary
Interstitial fluid
Diffusion of molecules
Tissue cell

▲ **Figure 23.7B** Diffusion between blood and tissue cells

Arteries, arterioles, veins, and venules have thicker walls than capillaries. Their walls have the same smooth epithelium but are reinforced by two other tissue layers. An outer layer of connective tissue with elastic fibers enables the vessels to stretch and recoil. The middle layer consists mainly of smooth muscle. Both these layers are thicker and sturdier in arteries, providing the strength and elasticity to accommodate the rapid flow and high pressure of blood pumped by the heart. Arteries are also able to regulate blood flow by constricting or relaxing their smooth muscle layer. The thinner-walled veins convey blood back to the heart at low velocity and pressure. Within large veins, flaps of tissue act as one-way valves, which permit blood to flow only toward the heart.

? **How does the structure of a capillary relate to its function?**

● The small diameter and thin walls of capillaries facilitate the exchange of substances between blood and interstitial fluid.

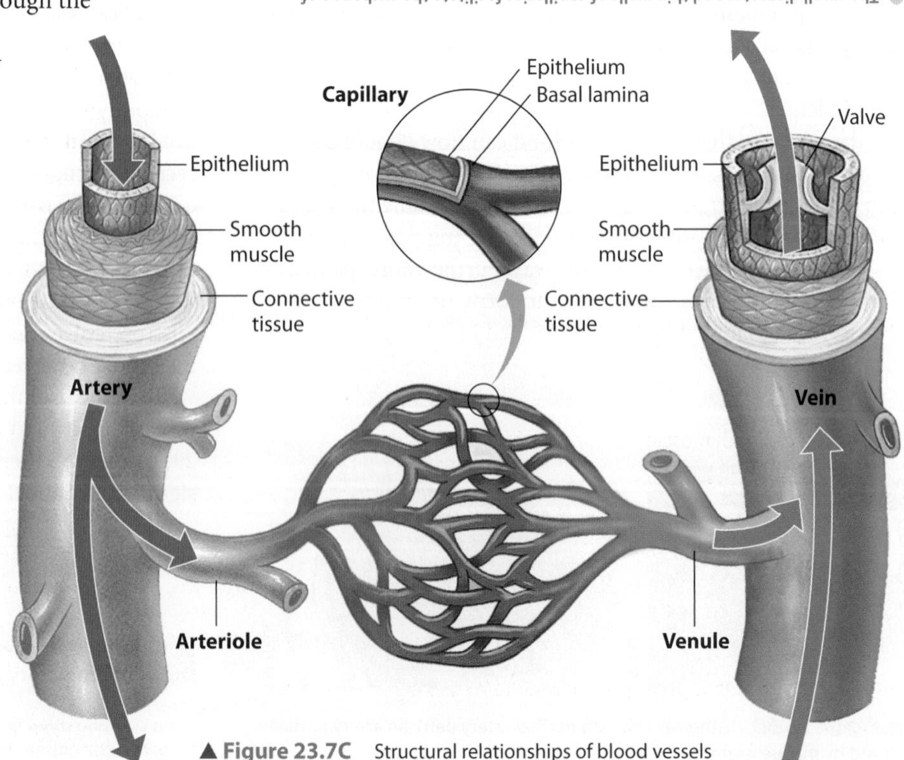
Capillary
Epithelium
Basal lamina
Valve
Epithelium
Smooth muscle
Connective tissue
Epithelium
Smooth muscle
Connective tissue
Artery
Vein
Arteriole
Venule

▲ **Figure 23.7C** Structural relationships of blood vessels

23.8 Blood pressure and velocity reflect the structure and arrangement of blood vessels

Now that you've looked at the structure of blood vessels, let's explore the forces that move your blood through these vessels.

Blood Pressure **Blood pressure** is the force that blood exerts against the walls of your blood vessels. Created by the pumping of the heart, blood pressure drives the flow of blood from the heart through arteries and arterioles to capillary beds.

When the ventricles contract, blood is forced into the arteries faster than it can flow into the arterioles. This stretches the elastic walls of the arteries. You can feel this rhythmic stretching of the arteries when you measure your heart rate by taking your **pulse**. You can see this surge in pressure (expressed in millimeters of mercury, mm Hg) as the pressure peaks in the top graph of **Figure 23.8A**. The pressure caused by ventricular contraction is called systolic pressure. The elastic arteries snap back during diastole, maintaining pressure on the blood and a continuous flow of blood into arterioles and capillaries. The dips in pressure in the top graph represent diastolic pressure.

The diagram at the center of Figure 23.8A shows the relative sizes and numbers of blood vessels as blood flows from the aorta through arteries to capillaries and back through veins to the venae cavae. Blood pressure is highest in the aorta and arteries and declines abruptly as the blood enters the arterioles. The pressure drop results mainly from the resistance to blood flow caused by friction between the blood and the walls of the millions of narrow arterioles and capillaries.

Blood pressure in the arteries depends on the volume of blood pumped into the aorta and also on the restriction of blood flow into the narrow openings of the arterioles. These openings are controlled by smooth muscles. When the muscles relax, the arterioles dilate, and blood flows through them more readily, causing a fall in blood pressure. Physical and emotional stress can raise blood pressure by triggering nervous and hormonal signals that constrict these blood vessels. Regulatory mechanisms coordinate cardiac output and changes in the arteriole openings to maintain adequate blood pressure as demands on the circulatory system change.

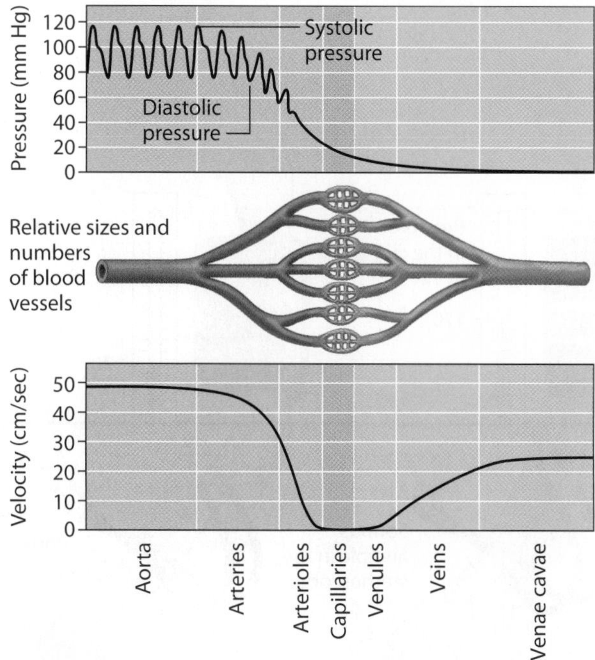

Pressure (mm Hg)
120
100
80
60
40
20
0
Systolic pressure
Diastolic pressure
Relative sizes and numbers of blood vessels
Velocity (cm/sec)
50
40
30
20
10
0
Aorta
Arteries
Arterioles
Capillaries
Venules
Veins
Venae cavae

▲ **Figure 23.8A** Blood pressure and velocity in the blood vessels

Try This Explain how the structure of the blood vessels (center) affects blood pressure (top) and blood velocity (bottom).

Blood Velocity The blood's velocity (rate of flow, expressed in centimeters per second, cm/sec) is illustrated in the bottom graph of Figure 23.8A. As the figure shows, velocity declines rapidly in the arterioles, drops to almost zero in the capillaries, and then speeds up in the veins. What accounts for these changes? As larger arteries divide into smaller and more numerous arterioles, the total combined cross-sectional area of the many vessels is much greater than the diameter of the one artery that feeds into them. If there were only one arteriole per artery, the blood would actually flow faster through the arteriole, the way water does when you add a narrow nozzle to a garden hose. However, there are many arterioles per artery, so the effect is like taking the nozzle off the hose: As you increase the diameter of a pipe, the flow rate goes down.

The cross-sectional area is greatest in the capillaries, and the velocity of blood is slowest through them. The steady, leisurely flow of blood in the capillaries enhances the exchange of substances with body cells.

By the time blood reaches the veins, its pressure has dropped to near zero. The blood has encountered so much resistance as it passes through the millions of tiny arterioles and capillaries that the force from the pumping heart no longer propels it. How, then, does blood return to the heart? Whenever the body moves, muscles squeeze blood through the veins. **Figure 23.8B** shows how veins are often sandwiched between skeletal muscles. One-way valves allow the blood to flow only toward the heart. Breathing also helps return blood to the heart. When you inhale, the change in pressure within your chest cavity causes the large veins near your heart to expand and fill.

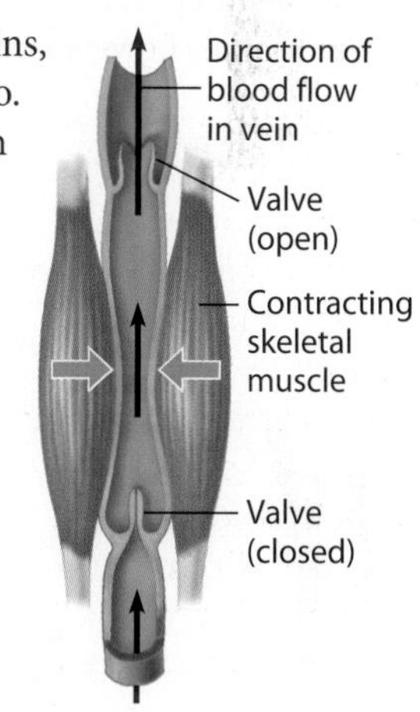

Direction of blood flow in vein
Valve (open)
Contracting skeletal muscle
Valve (closed)

▲ **Figure 23.8B** Blood flow in a vein

Because blood pressure is a key indicator of cardiovascular health, blood pressure is routinely taken at most doctor visits. We look at how blood pressure is measured in the next module.

? **If blood pressure in the veins drops to zero, why does blood velocity increase as blood flows from venules to veins?**

● The total diameter of the veins is less than the venules. The velocity increases, just as water flows faster when a nozzle narrows the opening of a hose.

23.9 Measuring blood pressure can reveal cardiovascular problems

CONNECTION

Blood pressure is generally measured in an artery in the arm at the same height as the heart. As indicated in **Figure 23.9**, ❶ a typical blood pressure for a healthy young adult is about 120/70. The first (sometimes referred to as the top) number is the systolic pressure; the second (or bottom) number is the diastolic pressure (see Module 23.8).

Figure 23.9 shows how blood pressure is measured using a sphygmomanometer, or blood pressure cuff. ❷ Once wrapped around the upper arm, where large arteries are accessible, the cuff is inflated until the pressure is strong enough to close the artery and cut off blood flow to the lower arm. ❸ A stethoscope is used to listen for sounds of blood flow below the cuff as the cuff is gradually deflated. The first sound of blood spurting through the constricted artery indicates that the pressure exerted by the cuff has fallen just below that in the artery. The pressure at this point is the systolic pressure. The sound of blood flowing unevenly through the artery continues until the pressure of the cuff falls below the pressure of the artery during diastole. ❹ Blood now flows continuously through the artery, and the sound of blood flow ceases. The reading at this point is the diastolic pressure.

Optimal blood pressure for adults is below 120 mm Hg for systolic pressure and below 80 mm Hg for diastolic pressure. Lower values are generally considered better, except in rare cases in which low blood pressure may indicate a serious underlying condition (such as an endocrine disorder, malnutrition, or internal bleeding). Blood pressure higher than the normal range, however, may signal a cardiovascular disorder.

High blood pressure, or **hypertension**, is persistent systolic blood pressure higher than 140 mm Hg and/or diastolic blood pressure higher than 90 mm Hg. Hypertension affects almost 30% of the adult population in the United States. It is sometimes called a "silent killer" because high blood pressure often displays no outward symptoms for years but may be leading to severe health problems.

High blood pressure harms the cardiovascular system in several ways. Elevated pressure requires the heart to work harder to pump blood throughout the body, and over time the left ventricle may enlarge as a result. When the coronary blood supply does not keep up with the demands of this increase in muscle mass, the heart muscle weakens. In addition, the increased force on arterial walls throughout the body causes tiny ruptures. The resulting inflammation promotes plaque formation, aggravating atherosclerosis (see Module 23.6) and increasing the risk of blood clot formation. Prolonged hypertension is the major cause of heart attack, heart disease, stroke, and also kidney failure, because renal arteries and arterioles may be damaged by high pressure.

In many patients, the exact cause of hypertension cannot be firmly established. Some predispositions to hypertension cannot be avoided. For example, males have a greater risk of high blood pressure up to age 55; females have a greater risk after menopause. African Americans are more prone to hypertension than Caucasians. Blood pressure generally increases with age, as does the prevalence of hypertension. Heredity also plays a role; children of parents with hypertension are twice as likely to develop the condition.

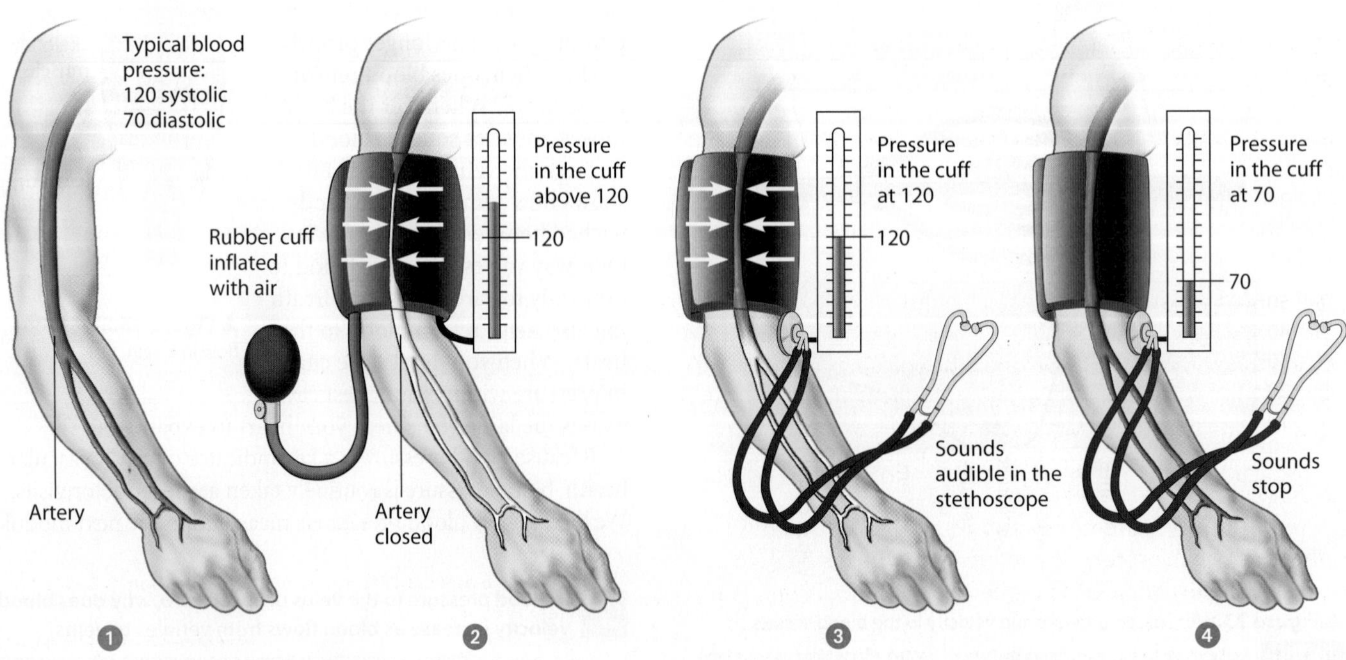
Typical blood pressure:
120 systolic
70 diastolic
Artery
1
Rubber cuff inflated with air
Pressure in the cuff above 120
120
Artery closed
2
Pressure in the cuff at 120
120
Sounds audible in the stethoscope
3
Pressure in the cuff at 70
70
Sounds stop
4

▲ **Figure 23.9** Measuring blood pressure

However, no matter how many unavoidable predispositions a person may have, there are lifestyle changes that can prevent or control hypertension in just about everybody: eating a heart-healthy diet, not smoking, avoiding excess alcohol (more than two drinks per day), exercising regularly (30 minutes of moderate activity on most days), and maintaining a healthy weight. Many people associate salt with high blood pressure, but it is a contributing factor only in a small percentage of people. If lifestyle changes don't lower blood pressure, there are several effective antihypertensive medications.

? Listening with a stethoscope below a sphygmomanometer cuff, you hear sounds that begin at 140 mm Hg and cease at 95 mm Hg. What are the systolic and diastolic blood pressure readings for this person, and do they indicate a health risk?

● Systolic = 140; diastolic = 95; blood pressure = 140/95. Yes, this person has hypertension and may be at risk for cardiovascular disease and kidney failure.

23.10 Smooth muscle controls the distribution of blood

You learned in Module 23.8 that smooth muscles in arteriole walls can influence blood pressure by changing the resistance to blood flow out of the arteries and into arterioles. The smooth muscles in the arteriole walls also regulate the distribution of blood to the capillaries of the various organs. At any given time, only about 5–10% of your body's capillaries have blood flowing through them. However, each tissue has many capillaries, so every part of your body is supplied with blood at all times. Capillaries in a few organs, such as the brain, heart, kidneys, and liver, usually carry their full capacity of blood. In many other sites, however, blood supply varies as blood is diverted from one place to another, depending on need.

Figure 23.10 illustrates a second mechanism that regulates the flow of blood into capillaries. Notice that in both parts of this figure there is a capillary called a thoroughfare channel, through which blood streams directly from arteriole to venule. This channel is always open. Rings of smooth muscle located at the entrance to capillary beds, called precapillary sphincters, regulate the passage of blood into the branching capillaries. As you can see in the figure, ➊ blood flows through a capillary bed when its precapillary sphincters are relaxed. ➋ It bypasses the capillary bed when the sphincters are contracted.

After a meal, for instance, precapillary sphincters in the wall of your digestive tract relax, letting a large quantity of blood pass through the capillary beds. The products of digestion are absorbed into the blood, which delivers them to the rest of the body. During strenuous exercise, many of the capillaries in the digestive tract are closed off, and blood is supplied more generously to your skeletal muscles. This is one reason why heavy exercise right after eating may cause indigestion or abdominal cramps (and why you shouldn't swim too soon after eating—just like mom always said). Blood flow to your skin is regulated to help control body temperature. An increase in blood supply to the skin helps to release the excess heat generated by exercise.

Nerve impulses, hormones, and chemicals produced locally influence the contraction of the smooth muscles that regulate the flow of blood to capillary beds. For example, the chemical histamine released by cells at a wound site causes smooth muscle relaxation, increasing blood flow and the supply of infection-fighting white blood cells.

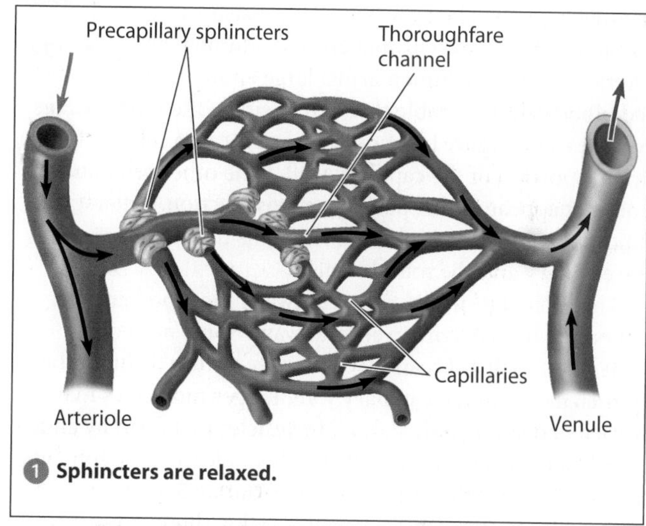
Precapillary sphincters
Thoroughfare channel
Capillaries
Arteriole
Venule
➊ Sphincters are relaxed.

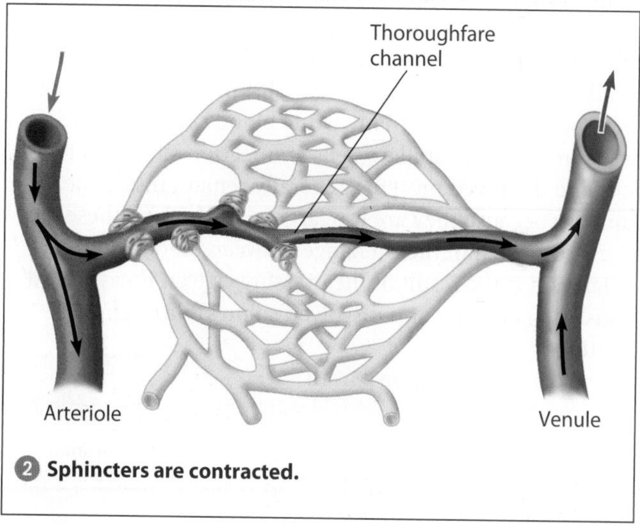
Thoroughfare channel
Arteriole
Venule
➋ Sphincters are contracted.

▲ **Figure 23.10** The control of capillary blood flow by precapillary sphincters

Next we consider how substances are exchanged when blood flows through a capillary.

? What two mechanisms restrict the distribution of blood to a capillary bed?

● Constriction of an arteriole, so that less blood reaches a capillary bed, and contraction of precapillary sphincters, so that blood flows through thoroughfare channels only, not capillary beds

23.11 Capillaries allow the transfer of substances through their walls

Capillaries are the only blood vessels with walls thin enough for substances to cross between the blood and the interstitial fluid that bathes the body cells. This transfer of materials is the most important function of the circulatory system, so let's examine the process more carefully.

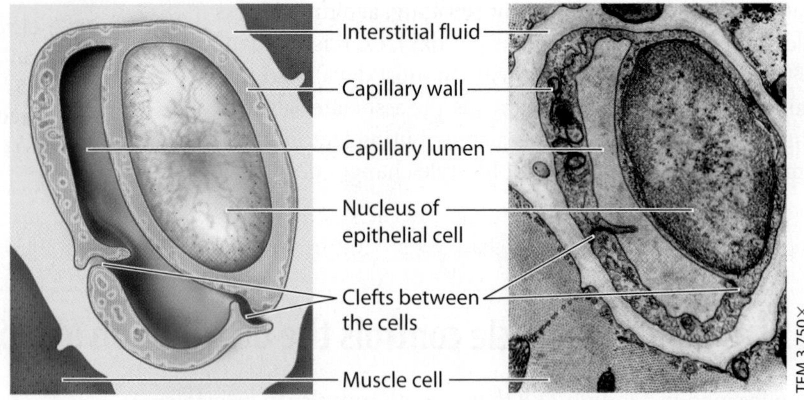

Interstitial fluid
Capillary wall
Capillary lumen
Nucleus of epithelial cell
Clefts between the cells
Muscle cell
TEM 3,750×

▲ **Figure 23.11A** A capillary in cross section

Capillary Structure and Function Take a look at **Figure 23.11A**. It shows a cross section of a capillary that serves skeletal muscle cells, illustrated with a micrograph (on the right) and a labeled drawing. The drawing will help you interpret the micrograph. The capillary wall consists of adjoining epithelial cells that enclose a lumen, or interior space. The lumen is just large enough for red blood cells to tumble through in single file. The nucleus you see in the figure belongs to one of the two cells making up this portion of the capillary wall. (The other cell's nucleus does not appear in this particular cross section.) Interstitial fluid, shown in blue in the drawing, fills the space between the capillary and the muscle cells.

The exchange of substances between the blood and the interstitial fluid occurs in several ways. Some nonpolar molecules, such as O_2 and CO_2, simply diffuse through the epithelial cells of the capillary wall. Larger molecules may be carried across an epithelial cell in vesicles that form by endocytosis on one side of the cell and then release their contents by exocytosis on the other side (see Module 5.9).

In addition, the capillary wall is leaky; there are small pores in the wall and narrow clefts between the epithelial cells making up the wall (see Figure 23.11A). Water and small solutes, such as sugars and salts, move freely through these pores and clefts. Many white blood cells also squeeze between adjacent capillary cells at infection sites. Red blood cells and dissolved proteins generally remain inside the capillary because they are too large to fit through these passageways. Much of the exchange between blood and interstitial fluid is the result of the pressure-driven flow of fluid (consisting of water and dissolved solutes) through these openings.

The diagram in **Figure 23.11B** shows part of a capillary with blood flowing from its arterial end (near an arteriole) to its venous end (near a venule). What are the active forces that drive fluid into or out of the capillary? One of these forces is blood pressure, which tends to push fluid outward. The other is osmotic pressure, a force that tends to pull fluid back because the blood has a higher concentration of solutes than the interstitial fluid. Proteins dissolved in the blood account for much of this high solute concentration. On average, blood pressure is greater than the opposing forces, leading to a net loss of fluid from capillaries.

Fluid Return via the Lymphatic System Each day, you lose approximately 4–8 L of fluid from your capillaries to the surrounding tissues. The lost fluid is picked up by your lymphatic system, which includes a network of tiny vessels intermingled among the capillaries (see Figure 23.11B). After diffusing into these vessels, the fluid, now called lymph, is returned to your circulatory system through ducts that join with large veins in your neck. Lymph vessels also transport fats from the small intestine to the blood (see Figure 21.10B). (You can see a diagram of the lymphatic system in Figure 23.4).

Now that we have examined the structure and function of the heart and blood vessels, we turn our focus to the composition of the blood.

? **Explain how a severe protein deficiency in the diet that decreases the concentration of blood plasma proteins can cause edema, the accumulation of fluid in body tissues.**

● Decreased blood protein concentration reduces the osmotic gradient across the capillary walls, thus reducing the pull of fluid back into the capillaries.

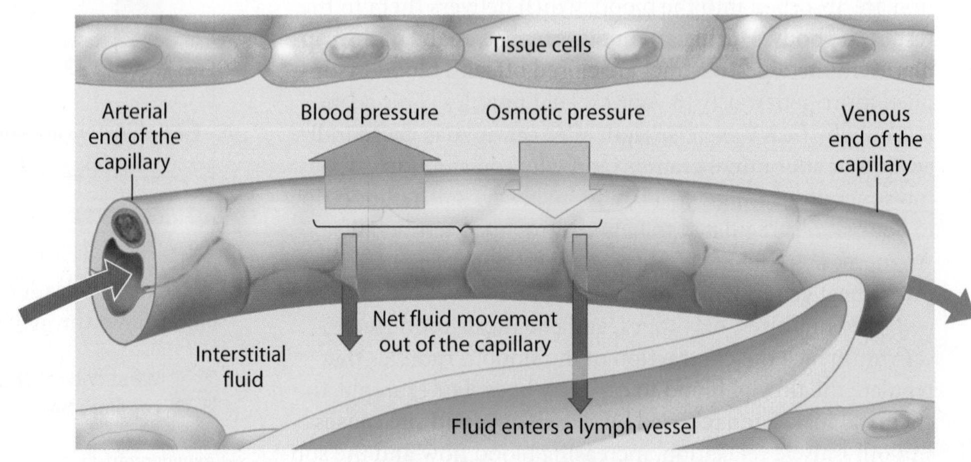

Tissue cells
Arterial end of the capillary
Blood pressure
Osmotic pressure
Venous end of the capillary
Net fluid movement out of the capillary
Interstitial fluid
Fluid enters a lymph vessel

▲ **Figure 23.11B** The movement of fluid out of a capillary and into a lymph vessel

▷ Structure and Function of Blood

23.12 Blood consists of red and white blood cells suspended in plasma

Your body has about 5 L of blood (about 5.3 quarts). Blood consists of several types of cells suspended in a liquid called **plasma**. When a blood sample is taken and a chemical is added to prevent the blood from clotting, the cells can be separated from the plasma by spinning the sample in a centrifuge. The cellular elements (cells and cell fragments), which make up about 45% of the volume of blood, settle to the bottom of the centrifuge tube, underneath the transparent, straw-colored plasma (**Figure 23.12**).

Plasma is about 90% water. Among its many solutes are inorganic salts in the form of dissolved ions. The functions of these ions (also called electrolytes) include keeping the pH of blood at about 7.4 and maintaining the osmotic balance between blood and interstitial fluid.

Plasma proteins also help maintain osmotic balance. Some proteins act as buffers. Fibrinogen is a plasma protein that functions in blood clotting, and immunoglobulins are proteins important in body defense (immunity).

Plasma also contains a wide variety of substances in transit from one part of the body to another, such as nutrients, waste products, O_2, CO_2, and hormones.

There are two classes of cells suspended in blood plasma: **red blood cells** and **white blood cells**. Also suspended in plasma are **platelets**, cell fragments that are involved in the process of blood clotting.

Red blood cells are also called **erythrocytes**. The structure of a red blood cell suits its main function, which is to carry oxygen. Human red blood cells are small biconcave disks, thinner in the center than at the sides. Their small size and shape create a large surface area across which oxygen can diffuse. Red blood cells lack a nucleus, allowing more room to pack in hemoglobin. Each tiny red blood cell contains about 250 million molecules of hemoglobin and thus can transport about a billion oxygen molecules. A single drop of your blood contains about 25 million cells (a drop is about 50 mL), which means you have about 25 trillion red blood cells in your 5 L of blood. Think about how many molecules of oxygen are traveling through your bloodstream!

There are five major types of white blood cells, or **leukocytes**, as pictured in Figure 23.12: monocytes, neutrophils, basophils, eosinophils, and lymphocytes. Their collective function is to fight infections. For example, monocytes and neutrophils are **phagocytes**, which engulf and digest bacteria and debris from your own dead cells. White blood cells actually spend much of their time moving through interstitial fluid, where most of the battles against infection are waged. There are also great numbers of white cells in the lymphatic system.

? **For every one white blood cell in normal human blood, there are about ____________ to ____________ red blood cells.**

● 500 . . . 1,000

Plasma (55%)	
Constituent	**Major functions**
Water	Solvent for carrying other substances
Ions (blood electrolytes) Sodium Potassium Calcium Magnesium Chloride Bicarbonate	Osmotic balance, pH buffering, and maintaining ion concentration of interstitial fluid
Plasma proteins	Osmotic balance and pH buffering
Fibrinogen	Clotting
Immunoglobulins (antibodies)	Defense
Substances transported by blood Nutrients (e.g., glucose, fatty acids, vitamins) Waste products of metabolism Respiratory gases (O_2 and CO_2) Hormones	

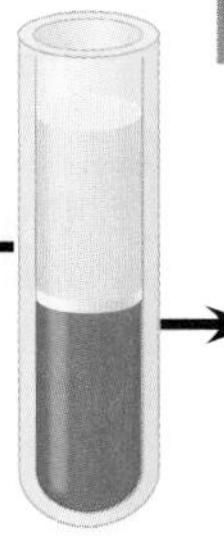

Centrifuged blood sample

Cellular elements (45%)		
Cell type	**Number** per μL (mm^3) of blood	**Functions**
Red blood cells (erythrocytes) 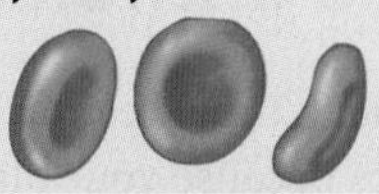	5–6 million	Transport of O_2 and some CO_2
White blood cells (leukocytes) 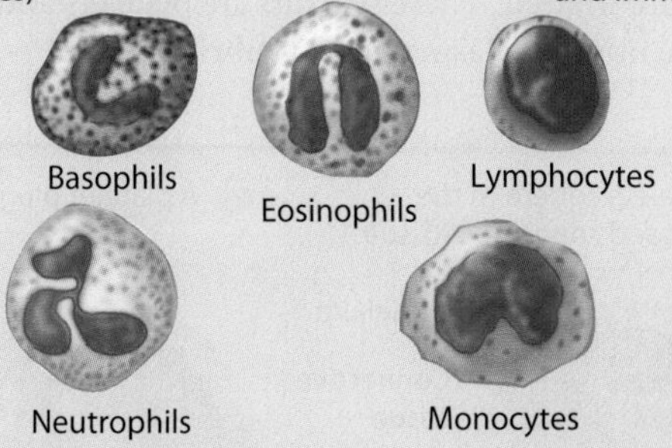 Basophils Eosinophils Lymphocytes Neutrophils Monocytes	5,000–10,000	Defense and immunity
Platelets 	250,000–400,000	Blood clotting

▲ **Figure 23.12** The composition of blood

23.13 Too few or too many red blood cells can be unhealthy

CONNECTION

Adequate numbers of red blood cells **(Figure 23.13)** are essential for healthy body function. After circulating for three or four months, red blood cells are broken down and their molecules recycled. Much of the iron removed from the hemoglobin is returned to the bone marrow, where new red blood cells are formed at the amazing rate of 2 million per second.

An abnormally low amount of hemoglobin or a low number of red blood cells is a condition called **anemia**. An anemic person feels constantly tired because body cells do not get enough oxygen. Anemia can result from a variety of factors, including excessive blood loss, vitamin or mineral deficiencies, and certain forms of cancer. Iron deficiency is the most common cause. Women are more likely to develop iron deficiency than men because of blood loss during menstruation. Pregnant women are generally prescribed iron supplements to support the developing fetus and placenta.

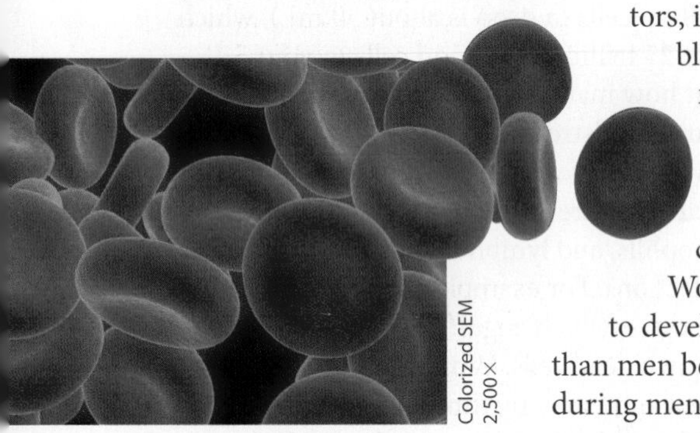

Colorized SEM 2,500×

▲ **Figure 23.13** Human red blood cells

The production of red blood cells in the bone marrow is controlled by a negative-feedback mechanism that is sensitive to the amount of oxygen reaching the tissues via the blood. If the tissues are not receiving enough oxygen, the kidneys produce **erythropoietin (EPO)**, a hormone that stimulates the bone marrow to produce more red blood cells. Patients on kidney dialysis often have very low red blood cell counts because their kidneys do not produce enough erythropoietin. Genetically engineered EPO has significantly helped these patients, as well as cancer and AIDS patients, who also often suffer from anemia.

One of the physiological adaptations of individuals who live at high altitudes, where oxygen levels are low, is the production of more red blood cells. Many athletes train at high altitudes to benefit from this effect. But other athletes take more drastic and illegal measures to increase the oxygen-carrying capacity of their blood and improve their performance. Injecting synthetic EPO can increase normal red blood cell volume from 45% to as much as 65%. Other athletes seek an unfair advantage by blood doping—withdrawing and storing their red blood cells and then reinjecting them before a competition. Athletic commissions test for these practices by measuring the percentage of red blood cells in the blood volume or by testing for EPO-related drugs. In recent years, a number of well-known runners and cyclists, including seven-time Tour de France winner Lance Armstrong, have tested positive for these drugs and have forfeited both their records and their right to compete in the future.

But there can be even more serious consequences. In some athletes, a combination of dehydration from a long race and blood already thickened by an increased number of red blood cells has led to severe medical problems, such as clotting, stroke, heart failure, and even death. Indeed, EPO-related drugs have been blamed for the deaths of dozens of athletes.

? **Why might increasing the number of red blood cells result in greater endurance and speed?**

● The additional red blood cells increase the oxygen-carrying capacity of blood and thus the oxygen supply to working muscles.

23.14 Blood clots plug leaks when blood vessels are injured

You may get cuts and scrapes from time to time, yet you don't bleed to death from such minor injuries because your blood contains self-sealing materials that are activated when blood vessels are injured. These sealants are platelets and the previously mentioned plasma protein **fibrinogen**.

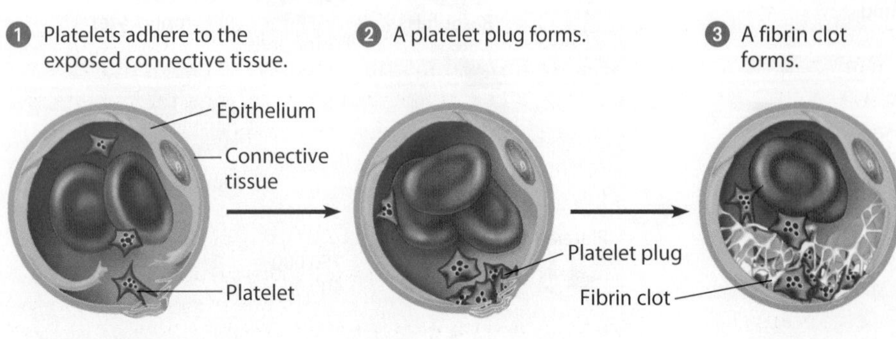
1 Platelets adhere to the exposed connective tissue.
2 A platelet plug forms.
3 A fibrin clot forms.
Epithelium
Connective tissue
Platelet
Platelet plug
Fibrin clot

▲ **Figure 23.14A** The blood-clotting process

What happens when you sustain an injury? Your body's immediate response is to constrict the damaged blood vessel, thereby reducing blood loss and allowing time for repairs to begin. **Figure 23.14A** shows the stages of the clotting process. ❶ When the epithelium (shown as tan) lining a blood vessel is damaged, connective tissue in the vessel wall is exposed to blood. Platelets (purple) rapidly adhere to the exposed tissue and release chemicals that make nearby platelets sticky. ❷ Soon a cluster of sticky platelets forms a plug that quickly provides protection against additional blood loss. Clotting factors in the plasma and released from the clumped platelets set off a chain of reactions that culminate in the formation of a reinforced patch, called a scab when it's on the skin. In this complex process, an activated

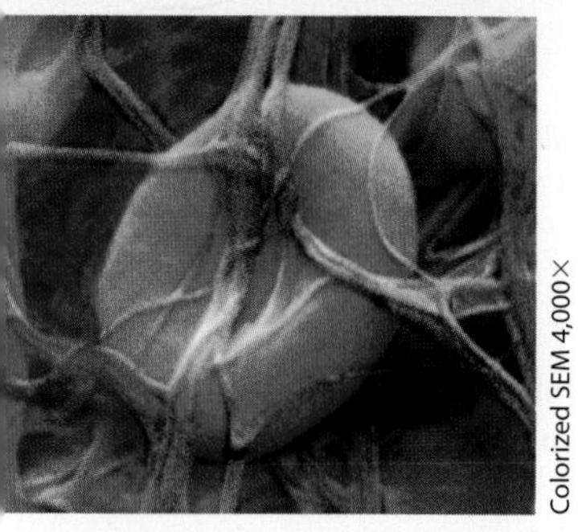

Colorized SEM 4,000×

▲ **Figure 23.14B** A fibrin clot

enzyme converts fibrinogen to a threadlike protein called **fibrin**. ❸ Threads of fibrin (white) reinforce the plug, forming a fibrin clot.

Figure 23.14B is a micrograph of a fibrin clot. Within an hour after a fibrin clot forms, the platelets contract, pulling the torn edges closer together and reducing the size of the area in need of repair. Chemicals released by platelets also stimulate cell division in smooth muscle and connective tissue, initiating the healing process.

The clotting mechanism is so important that any defect in it can be life-threatening. In the inherited disease hemophilia, excessive, sometimes fatal bleeding occurs from even minor cuts or bruises. In other people, the formation of blood clots in the *absence* of injury can be a problem. Anticlotting factors in the blood normally prevent spontaneous clotting. If blood clots form within a vessel, they can block the flow of blood. Such a clot, called a thrombus, can be dangerous if it blocks a blood vessel of the heart or brain (see Module 23.6). Aspirin, heparin, and warfarin are anticoagulant drugs that work by different mechanisms to prevent undesirable clotting in patients at risk for heart attack or stroke.

? **What is the role of platelets in blood clot formation?**

● Platelets adhere to exposed connective tissue and release various chemicals that help a platelet plug form and activate the pathway leading to a fibrin clot. (Also, some of the chemicals promote healing in other ways.)

23.15 Stem cells offer a potential cure for blood cell diseases

CONNECTION

The red marrow inside bones such as the ribs, vertebrae, sternum, and pelvis is a spongy tissue in which unspecialized stem cells differentiate into blood cells (see Module 11.14). When a **stem cell** divides, one daughter cell remains a stem cell and the other can take on a specialized function.

As shown in **Figure 23.15**, stem cells in bone marrow give rise to two different types of stem cells: lymphoid stem cells and myeloid stem cells. Lymphoid stem cells produce two different types of lymphocytes, which function in the immune system (see Module 24.5). Myeloid stem cells can differentiate into other white blood cells, platelets, and erythrocytes. The stem cells continually produce all the blood cells needed throughout life.

Leukemia is cancer of the white blood cells, or leukocytes. Because cancerous cells grow uncontrollably, a person with leukemia has an unusually high number of leukocytes, most of which are defective. These overabundant cells crowd out the bone marrow cells that are developing into red blood cells and platelets, causing severe anemia and impaired clotting.

Leukemia is usually fatal unless treated, and not all cases respond to the standard cancer treatments—radiation and chemotherapy. An alternative treatment involves destroying the cancerous bone marrow completely and replacing it with healthy bone marrow. Injection of as few as 30 stem cells can repopulate the blood and immune system. Patients may be treated with their own bone marrow: Marrow from the patient is harvested, processed to remove as many cancerous cells as possible, and then reinjected. Alternatively, a suitable donor, often a sibling, may provide the marrow. Such a patient requires lifelong treatment with drugs that suppress the tendency of some of the transplanted cells to "reject" the cells of the recipient.

Stem cell research holds great promise, and leukemia is just one of several blood diseases that may be treated by bone marrow stem cells. Recently, researchers have been able to isolate bone marrow stem cells and grow them in the laboratory. In a few cases, they have induced these stem cells to differentiate into more than just blood cells. Thus, these adult stem cells may eventually provide cells for human tissue and organ transplants. (In Chapter 24, we'll explore the diverse roles of white blood cells in the immune system.)

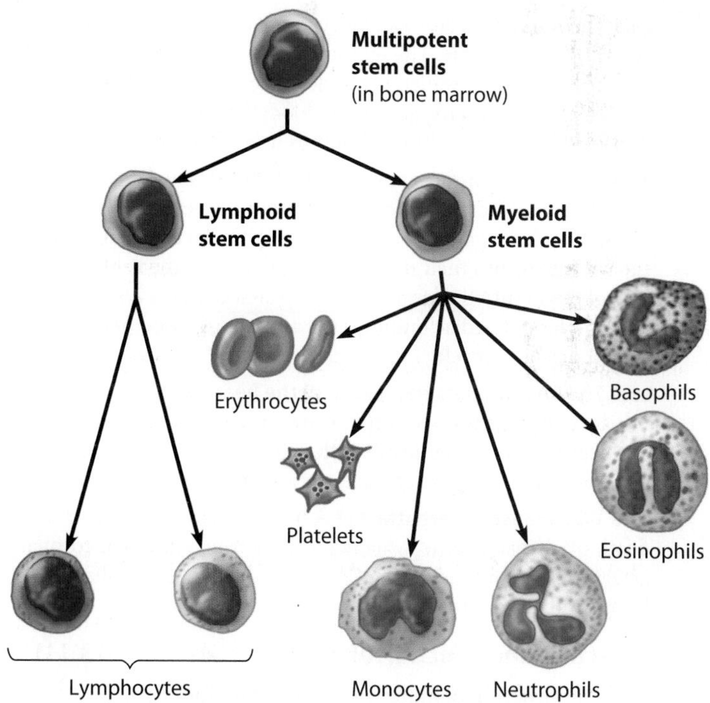

Multipotent stem cells (in bone marrow)
Lymphoid stem cells
Myeloid stem cells
Erythrocytes
Basophils
Platelets
Eosinophils
Lymphocytes
Monocytes
Neutrophils

▲ **Figure 23.15** Differentiation of blood cells from stem cells

? **Why would a leukemia patient's bone marrow need to be destroyed and then replaced with a transplant?**

● Destruction of the bone marrow would be necessary to kill the cancerous leukocytes. The patient would need replacement stem cells to continue making both red and white blood cells.

CHAPTER 23 REVIEW

For practice quizzes, BioFlix animations, MP3 tutorials, video tutors, and more study tools designed for this textbook, go to MasteringBiology®

Reviewing the Concepts

Circulatory Systems (23.1–23.2)

23.1 Circulatory systems facilitate exchange with all body tissues. Gastrovascular cavities function in both digestion and transport. In open circulatory systems, a heart pumps fluid through open-ended vessels to bathe tissue cells directly. In closed circulatory systems, a heart pumps blood, which travels through arteries to capillaries to veins and back to the heart.

23.2 Vertebrate cardiovascular systems reflect evolution. A fish's two-chambered heart pumps blood in a single circuit. Land vertebrates have double circulation with a pulmonary and a systemic circuit. Amphibians and many reptiles have three-chambered hearts; birds and mammals have four-chambered hearts.

The Human Cardiovascular System and Heart (23.3–23.6)

23.3 The human cardiovascular system illustrates the double circulation of mammals. The mammalian heart has two thin-walled atria and two thick-walled ventricles. The right side of the heart receives and pumps oxygen-poor blood; the left receives oxygen-rich blood from the lungs and pumps it to all other organs.

23.4 The heart contracts and relaxes rhythmically. During diastole of the cardiac cycle, blood flows from the veins into the heart chambers; during systole, contractions of the atria push blood into the ventricles, and then stronger contractions of the ventricles propel blood into the large arteries. Heart valves prevent the backflow of blood. Cardiac output is the amount of blood per minute pumped by a ventricle.

23.5 The SA node sets the tempo of the heartbeat. The SA node, or pacemaker, generates electrical signals that trigger contraction of the atria. The AV node relays these signals to the ventricles. An electrocardiogram records the electrical changes.

23.6 What causes heart attacks? A heart attack is damage or death of cardiac muscle, usually resulting from a blocked coronary artery. Researchers are currently testing the hypothesis that the body's inflammatory response causes heart attacks.

Structure and Function of Blood Vessels (23.7–23.11)

23.7 The structure of blood vessels fits their functions.

23.8 Blood pressure and velocity reflect the structure and arrangement of blood vessels. Blood pressure depends on cardiac output and the resistance of vessels. Pressure is highest in the arteries. Blood velocity is slowest in the capillaries. Skeletal muscle contractions and one-way valves keep blood moving through veins to the heart.

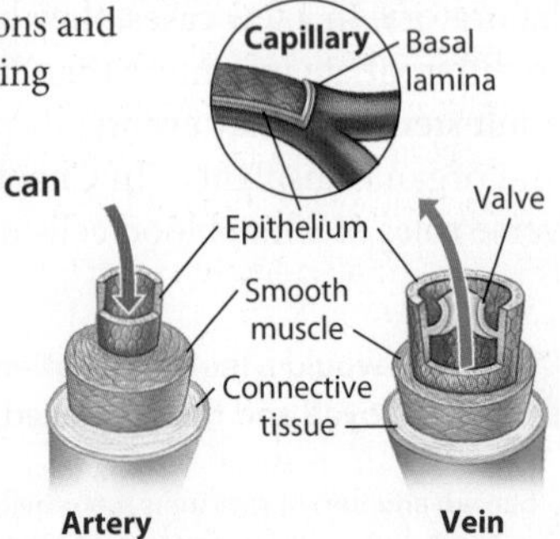
Capillary
Basal lamina
Epithelium
Valve
Smooth muscle
Connective tissue
Artery
Vein

23.9 Measuring blood pressure can reveal cardiovascular problems. Blood pressure is measured as systolic and diastolic pressures. Hypertension is a serious cardiovascular problem that in most cases can be controlled.

23.10 Smooth muscle controls the distribution of blood. Constriction of arterioles and precapillary sphincters controls blood flow through capillary beds.

23.11 Capillaries allow the transfer of substances through their walls. Blood pressure forces fluid and small solutes out of the capillary at the arterial end. Excess fluid is returned to the circulatory system through lymph vessels.

Structure and Function of Blood (23.12–23.15)

23.12 Blood consists of red and white blood cells suspended in plasma. Plasma contains various inorganic ions, proteins, nutrients, wastes, gases, and hormones. Red blood cells (erythrocytes) transport O_2 bound to hemoglobin. White blood cells (leukocytes) fight infections.

23.13 Too few or too many red blood cells can be unhealthy. The hormone erythropoietin regulates red blood cell production.

23.14 Blood clots plug leaks when blood vessels are injured. Platelets adhere to connective tissue of damaged vessels and help convert fibrinogen to fibrin, forming a clot that plugs the leak.

23.15 Stem cells offer a potential cure for blood cell diseases. Stem cells in bone marrow produce all types of blood cells.

Connecting the Concepts

1. Use the following diagram to review the flow of blood through a human cardiovascular system. Label the indicated parts, highlight the vessels that carry oxygen-rich blood, and then trace the flow of blood by numbering the circles from 1 to 10, starting with 1 in the right ventricle. (When two locations are equivalent in the pathway, such as right and left lung capillaries or capillaries of top and lower portion of the body, assign them the same number.)

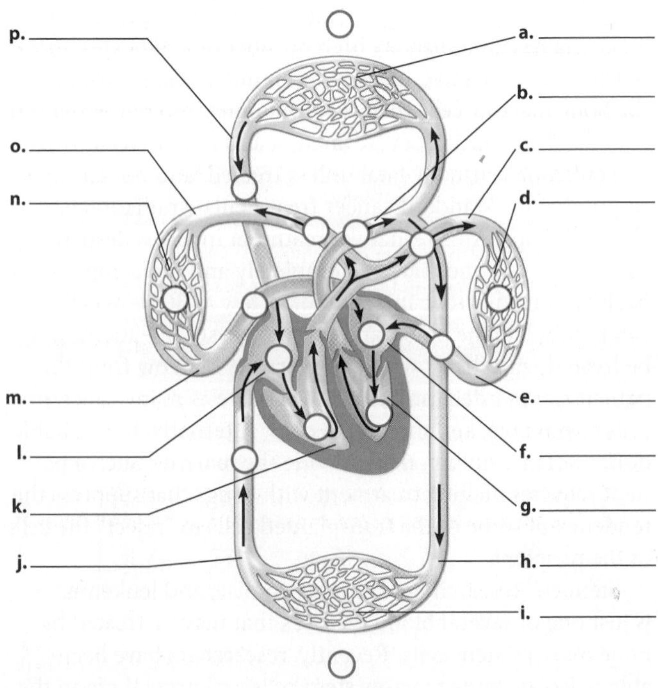
p.
o.
n.
m.
l.
k.
j.
a.
b.
c.
d.
e.
f.
g.
h.
i.

Testing Your Knowledge

Level 1: Knowledge/Comprehension

2. Blood pressure is highest in ______________, and blood moves most slowly in ______________.
   a. veins; capillaries
   b. arteries; capillaries
   c. veins; arteries
   d. arteries; veins
3. When the doctor listened to Janet's heart, he heard "lub-hiss, lub-hiss" instead of the normal "lub-dup" sounds. The hiss is most likely due to ______________. (*Explain your answer.*)
   a. a defective atrioventricular (AV) valve
   b. a damaged pacemaker (SA node)
   c. a defective semilunar valve
   d. high blood pressure
4. Which of the following is the main difference between your cardiovascular system and that of a fish?
   a. Your heart has two chambers; a fish heart has four.
   b. Your circulation has two circuits; fish circulation has one.
   c. Your heart chambers are called atria and ventricles.
   d. Yours is a closed system; the fish's is an open system.
5. Paul's blood pressure is 150/90. The 150 indicates ______________, and the 90 indicates ______________.
   a. pressure in the left ventricle; pressure in the right ventricle
   b. pressure during ventricular contraction; pressure during heart relaxation
   c. systemic circuit pressure; pulmonary circuit pressure
   d. pressure in the arteries; pressure in the veins
6. Which of the following *initiates* the process of blood clotting?
   a. damage to the lining of a blood vessel
   b. conversion of fibrinogen to fibrin
   c. attraction of leukocytes to a site of infection
   d. conversion of fibrin to fibrinogen
7. Blood flows more slowly in the arterioles than in the artery that supplies them because the arterioles
   a. have thoroughfare channels to venules that are often closed off, slowing the flow of blood.
   b. have sphincters that restrict flow to capillary beds.
   c. are narrower than the artery.
   d. collectively have a larger cross-sectional area than does the artery.
8. Which of the following is *not* a true statement about open and closed circulatory systems?
   a. Both systems have some sort of a heart that pumps a circulatory fluid through the body.
   b. A frog has an open circulatory system; other vertebrates have closed circulatory systems.
   c. The blood and interstitial fluid are separate in a closed system but are indistinguishable in an open system.
   d. Some of the circulation of blood in both systems results from body movements.
9. If blood was supplied to all of the body's capillaries at one time,
   a. blood pressure would fall dramatically.
   b. resistance to blood flow would increase.
   c. blood would move too rapidly through the capillaries.
   d. the amount of blood returning to the heart would increase.

Level 2: Application/Analysis

10. Trace the path of blood starting in a pulmonary vein, through the heart, and around the body, returning to the pulmonary vein. Name, in order, the heart chambers and types of vessels through which the blood passes.
11. Explain how the structure of capillaries relates to their function of exchanging substances with the surrounding interstitial fluid. Describe how that exchange occurs.
12. Here is a blood sample that has been spun in a centrifuge. List, as completely as you can, the components you would find in the straw-colored fluid at the top of this tube and in the dense red portion at the bottom.

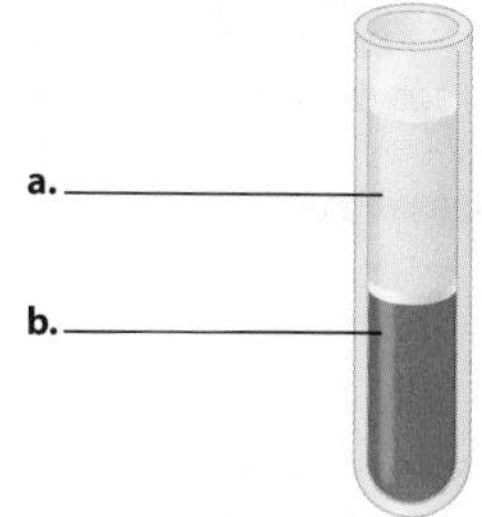

a.
b.

Level 3: Synthesis/Evaluation

13. Some babies are born with a small hole in the wall between the left and right ventricles. How might this affect the oxygen content of the blood pumped out of the heart into the systemic circuit?
14. Juan has a disease in which damaged kidneys allow some of his normal plasma proteins to be removed from the blood. How might this condition affect the osmotic pressure of blood in capillaries, compared with that of the surrounding interstitial fluid? One of the symptoms of this kidney malfunction is an accumulation of excess interstitial fluid, which causes Juan's arms and legs to swell. Can you explain why this occurs?
15. **SCIENTIFIC THINKING** One of the studies described in Module 23.6 is being funded by a major pharmaceutical company. The other study will receive nearly $80 million in funding from the National Institutes of Health (NIH), a government agency that supports medical research. NIH grants for research on heart disease total more than $1.2 billion per year. Gather more information and form an opinion on how heart disease research should be funded, whether by private enterprises such as pharmaceutical companies, donor-supported nonprofit organizations, or government agencies. Write an essay arguing your point of view.
16. Physiologists speculate about cardiovascular adaptations in dinosaurs—some of which had necks almost 10 m (33 feet) long. Such animals would have required a systolic pressure of nearly 760 mm Hg to pump blood to the brain when the head was fully raised. Some analyses suggest that dinosaurs' hearts were not powerful enough to generate such pressures, leading to the speculation that long-necked dinosaurs fed close to the ground rather than raising their heads to feed on high foliage. Scientists also debate whether dinosaurs had a "reptile-like" or "bird-like" heart. Most modern reptiles have a three-chambered heart with just one ventricle. Birds, which evolved from a lineage of dinosaurs, have a four-chambered heart. Some scientists believe that the circulatory needs of these long-necked dinosaurs provide evidence that dinosaurs must have had a four-chambered heart. Why might they conclude this?

Answers to all questions can be found in Appendix 4.

CHAPTER 24

The Immune System

How long will protection from cervical cancers last after HPV vaccination?

When you think of viruses, you likely think about non-cancer-related diseases, such as the common cold, smallpox, and polio. Many people don't realize that viruses can cause cancers, too. In the 1980s, scientists discovered that the sexually transmitted human papillomavirus (HPV), seen in the photograph below, was linked to cervical and anal cancers, causing essentially all cervical cancers and most cases of anal cancers. Since then, two HPV vaccines have been developed, Gardasil in 2006 and Cervarix in 2009. Both vaccines are made from harmless components of the virus administered to invoke an immune response and protect against HPV-caused cancer.

Experts believe that for an HPV vaccine to significantly reduce the incidence of cancer it must provide long-term immunity (protection), ideally for three to four decades. Otherwise, it might simply postpone susceptibility to the infection. For example, if immunity wanes after a decade, by the

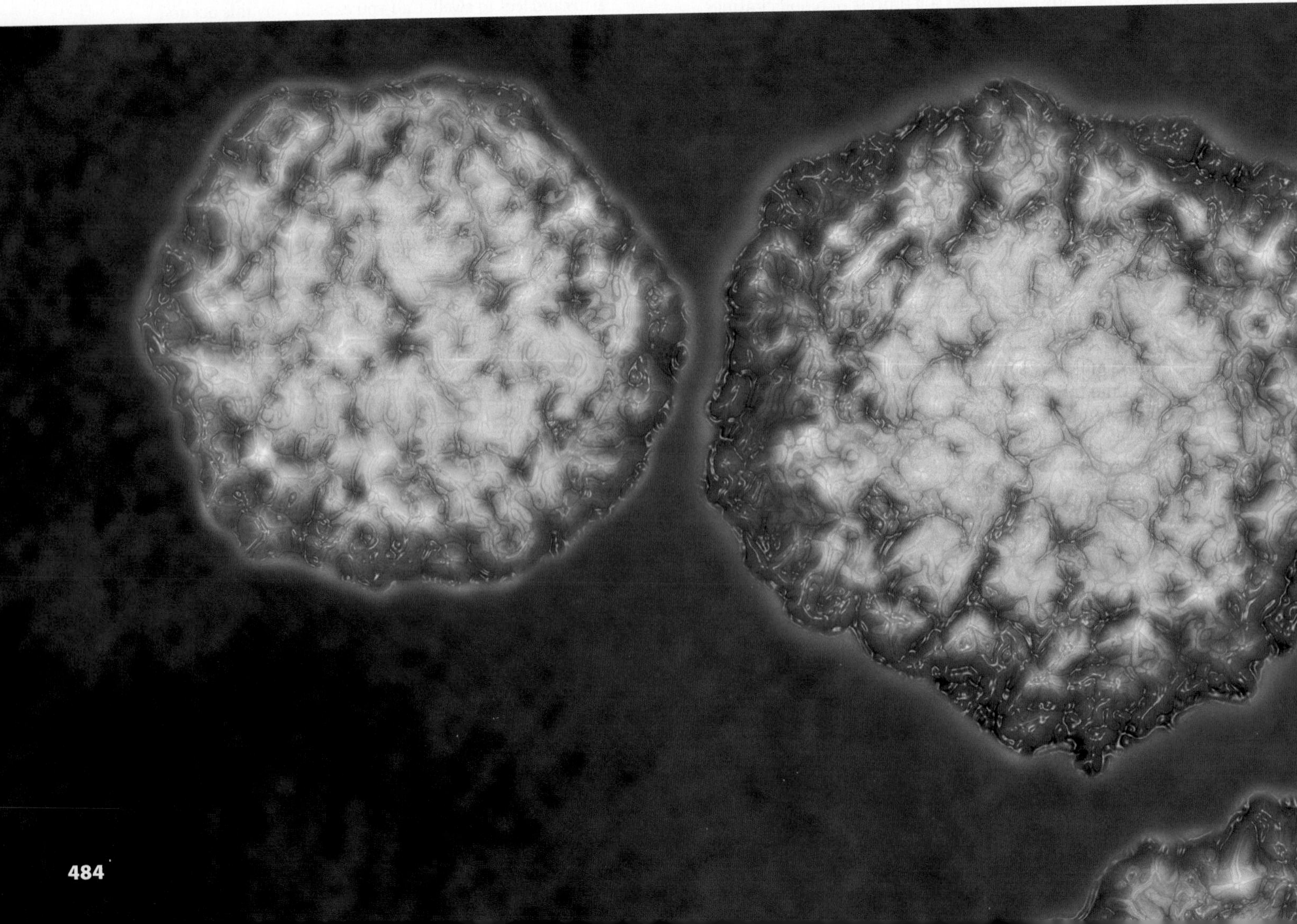

time a girl who was vaccinated at age 11 turns 23, she will no longer be fully protected from HPV and may be at high risk for infection by the virus. How long does immunity from HPV last with Gardasil and Cervarix vaccination? The short answer is we don't know yet. The requisite decades worth of data on the long-term effects of the vaccines are not available. Later in this chapter we will see how scientists are trying to answer the question using data on HPV vaccination amassed as of 2012.

In this chapter, you'll see how your body's immune system recognizes and attacks agents that cause disease. Some of the defenses are innate; that is, they are always deployed and waiting to encounter an invader. Other immune defenses are adaptive; they require recognition of a specific infectious agent, such as HPV, before attacking.

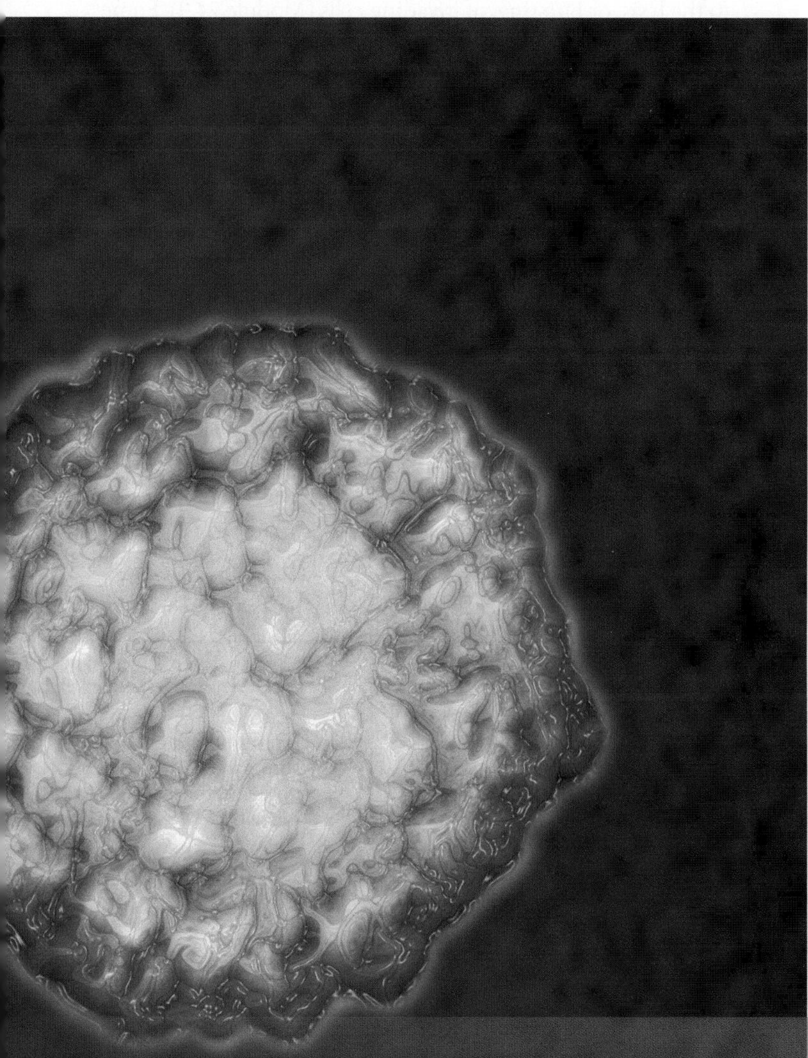

BIG IDEAS

Innate Immunity
(24.1–24.2)

All animals have immune defenses that are always at the ready.

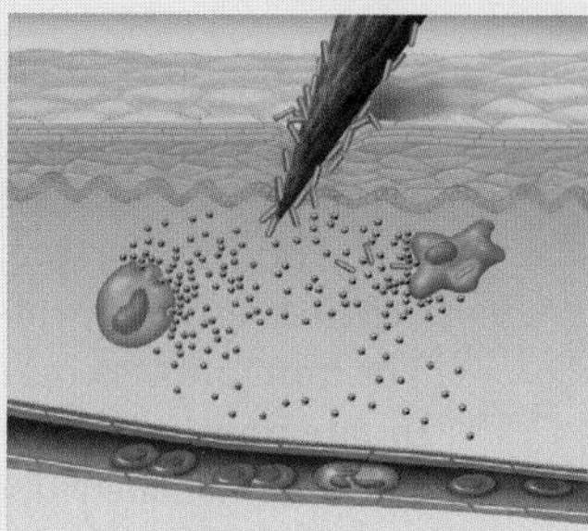

Adaptive Immunity
(24.3–24.16)

Vertebrates custom-tailor the immune response to specific pathogens.

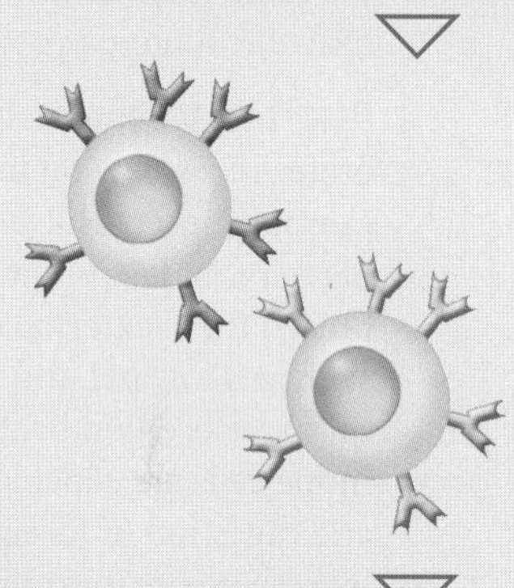

Disorders of the Immune System
(24.17–24.18)

Overreactions or underreactions of the immune response can cause problems that range from mild to severe.

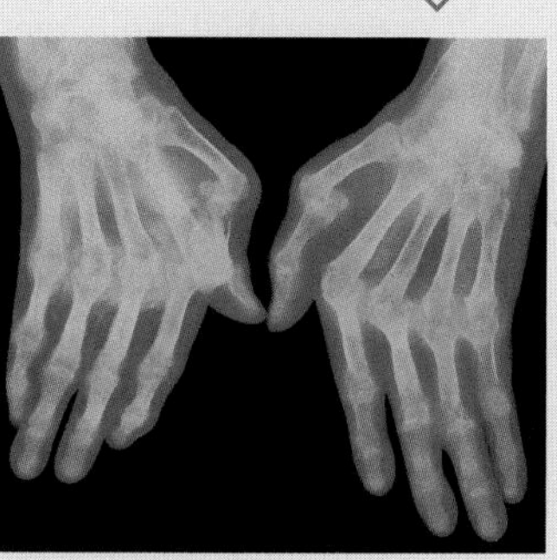

Innate Immunity

24.1 All animals have innate immunity

Nearly everything in the environment teems with **pathogens**, agents that cause disease. Yet we do not constantly become ill, thanks to the **immune system**, the body's system of defenses against agents that cause disease. The immune systems of all animals include **innate immunity**, a set of defenses that are active immediately upon infection and are the same whether or not the pathogen has been encountered previously **(Figure 24.1A)**.

Innate external barriers
skin/exoskeleton, acidic environment, secretions, mucous membranes, hairs, cilia

if external barriers breached

Innate internal defenses
phagocytic cells, natural killer cells, defensive proteins, inflammatory response

▲ **Figure 24.1A** Components of the innate immune systems of animals

Invertebrate Innate Immunity Invertebrates rely solely on innate immunity. For example, insects have an exoskeleton, which is a tough, dry barrier that keeps out bacteria and viruses. Pathogens that breach these external defenses confront internal defenses such as a low pH and the secretion of lysozyme (an enzyme that breaks down bacterial cell walls). Circulating insect immune cells are capable of **phagocytosis**, cellular ingestion and digestion of foreign substances (see Module 5.9). The insect's innate immunity also includes the production of antimicrobial molecules that bring about the destruction of the invaders.

Vertebrate Innate Immunity In vertebrates, innate immunity coexists with the more recently evolved adaptive immune response. In mammals, innate defenses include external barriers like skin and mucous membranes. Thousands of species of microbes (termed microbiota) reside on the skin and in mucous membranes, and the balance of "good" species can prevent harmful ones from flourishing. Mucous membranes line internal surfaces open to the external environment such as the digestive, respiratory, reproductive, and urinary tracts. The mucus produced by these tissues provides protection, too: It traps foreign particles and contains defensive proteins that kill harmful microbes. Some mucous membranes, such as the respiratory tract, have additional protections. Nostril hairs filter incoming air, and cilia on cells sweep mucus in the trachea and any trapped microbes upward and out, helping to prevent lung infections.

Microbes that breach a barrier, such as those that enter through a cut in your skin, are confronted by innate immune cells. These are all classified as white blood cells (see Module 23.12), although they are found in interstitial fluid as well as in the blood. **Natural killer cells** attack cancer cells and virus-infected cells by releasing chemicals that lead to cell death. Most innate immune cells are phagocytes (phagocytic cells); the two main types are neutrophils and macrophages. **Neutrophils**, the most abundant type of white blood cell, circulate in the blood and enter tissues at sites of infection. **Macrophages** ("big eaters") are large phagocytic cells that wander through the interstitial fluid, "eating" any bacteria and viruses they encounter. Phagocytes bear surface receptors that bind to fragments of foreign molecules shared by a broad range of pathogens. By using a small set of receptors, a large set of pathogens can be recognized and destroyed quickly.

Other components of vertebrate internal innate immunity include proteins that either attack infecting microbes directly or impede their reproduction. **Interferons** are proteins produced by virus-infected cells that help to limit the cell-to-cell spread of viruses **(Figure 24.1B)**. ❶ The virus infects a cell, which causes ❷ interferon genes in the cell's nucleus to be turned on. ❸ The cell synthesizes interferon. The infected cell then dies, but ❹ its interferon proteins may diffuse to neighboring healthy cells, ❺ stimulating them to produce other proteins that inhibit viral reproduction.

Additional innate immunity in vertebrates is provided by the **complement system**, a group of about 30 different proteins that circulate in an inactive form in the blood. These proteins can act together (in complement) with other defense mechanisms. Substances on the surfaces of many microbes activate the complement system, resulting in a cascade of steps that can lead to the lysis, or bursting, of invading cells (Module 24.9). Some complement proteins attach to invaders to enhance phagocytosis, and others act as chemical signals to recruit more immune cells to the site of infection. Certain complement proteins also help trigger the inflammatory response.

? Which components of innate immunity described here actually help prevent infection? Which come into play only after infection has occurred?

● Innate immunity components that prevent infection: skin/exoskeleton, mucous membranes and their secretions, microbiota; components of innate immunity activated after infection: phagocytic cells (macrophages, neutrophils), natural killer cells, defensive proteins (interferons, complement system)

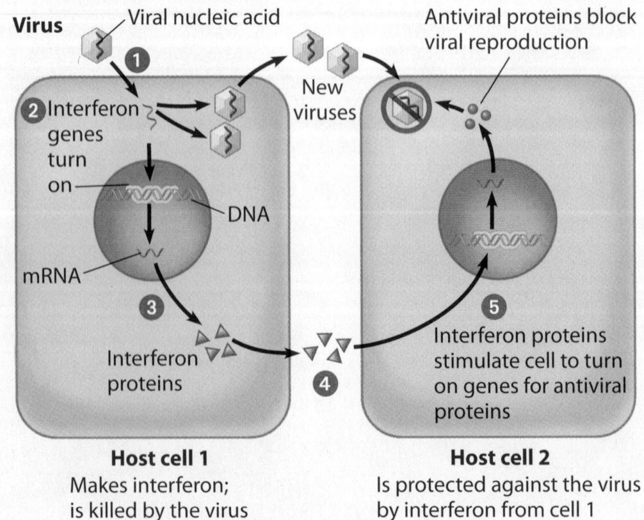

Virus
Viral nucleic acid
1
2 Interferon genes turn on
DNA
mRNA
3
Interferon proteins
New viruses
4
Antiviral proteins block viral reproduction
5
Interferon proteins stimulate cell to turn on genes for antiviral proteins
Host cell 1
Makes interferon; is killed by the virus
Host cell 2
Is protected against the virus by interferon from cell 1

▲ **Figure 24.1B** The interferon mechanism against viruses

24.2 Inflammation mobilizes the innate immune response

The **inflammatory response** is a major component of our innate immunity. Any damage to tissue, whether caused by microorganisms or by physical injury—even just a scratch or an insect bite—triggers this response. The main function of the inflammatory response is to disinfect and clean injured tissues. You may see signs of the inflammatory response when your skin is cut. The area becomes red, warm, and swollen.

Figure 24.2 shows the chain of events that occurs when a splinter has broken the skin, allowing infection by bacteria. ❶ The bacteria activate macrophages, which produce signaling molecules that increase local blood flow. Mast cells at the injury site release **histamine**, which ❷ induces neighboring blood capillaries to dilate and become leaky. Fluid passes out of the leaky capillaries into the affected tissues. Clotting proteins present in blood plasma pass into the interstitial fluid (Module 23.14). Along with platelets, these substances form local clots that help seal off the infected region, preventing the spread of infection to surrounding tissues and allowing healing to begin. Complement proteins, also activated by the bacteria, attract phagocytes to the area. Squeezing between the cells of the now leaky blood vessel wall, many neutrophils migrate out of the blood into the tissue spaces. The local increase in blood flow, fluid, and cells produces the redness, heat, and swelling characteristic of inflammation.

❸ The neutrophils that migrate into the area engulf bacteria and the remains of any body cells killed by them or by the physical injury. Many of the neutrophils die in the process (or simply come to the end of their short life span), and their remains are also engulfed and digested. You may see pus at the site of an infection; this consists mainly of dead white blood cells, fluid that has leaked from capillaries, and other tissue debris.

Inflammation is a natural defense. However, the persistence of inflammatory components for abnormally long periods, a condition called chronic inflammation, can be harmful. A variety of common disorders are associated with chronic inflammation, including arthritis, heart disease, Alzheimer's disease, and some cancers. Understanding the role inflammation plays is central in current disease research.

Inflammation may be localized or widespread (systemic). Sometimes microorganisms get into the blood or release toxins that are carried throughout the body in the bloodstream. The body may react with several inflammatory responses. For instance, the number of white blood cells circulating in the blood may increase severalfold within just a few hours; an elevated "white cell count" is one way to diagnose certain infections. Another response to systemic infection is fever. Toxins themselves may trigger the fever, or macrophages may release compounds that cause the body's "thermostat" to be set at a higher temperature. A very high fever is dangerous, but a moderate one may stimulate phagocytosis and hasten tissue repair. Anti-inflammatory drugs, such as aspirin and ibuprofen, dampen the normal inflammatory response and thus help reduce swelling and fever.

Sometimes bacterial infections in blood bring about an overwhelming systemic inflammatory response leading to a condition called septic shock. The response affects capillaries of the whole body—their leakiness leads to widespread fluid accumulation in tissues and low blood pressure, which may ultimately lead to poor circulation to vital organs and organ failure. Whereas local inflammation is essential to healing, chronic or widespread inflammation can be devastating.

? **Why is the inflammatory response to microbial invasion considered a form of *innate* immunity?**

● Because the response is the same regardless of whether the invader has been previously encountered and because the response is nonspecific

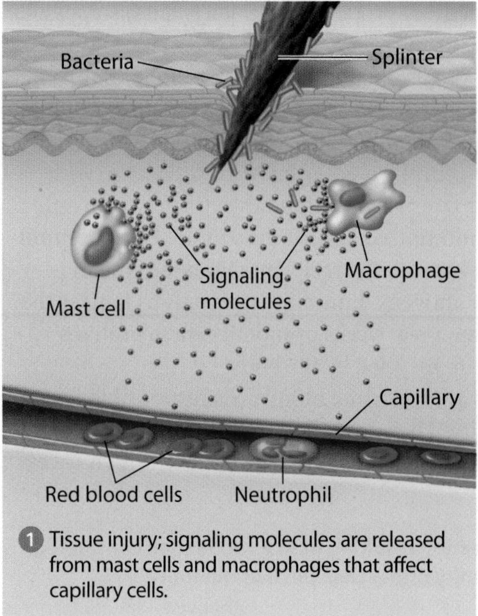
Bacteria
Splinter
Signaling molecules
Macrophage
Mast cell
Capillary
Red blood cells
Neutrophil
❶ Tissue injury; signaling molecules are released from mast cells and macrophages that affect capillary cells.

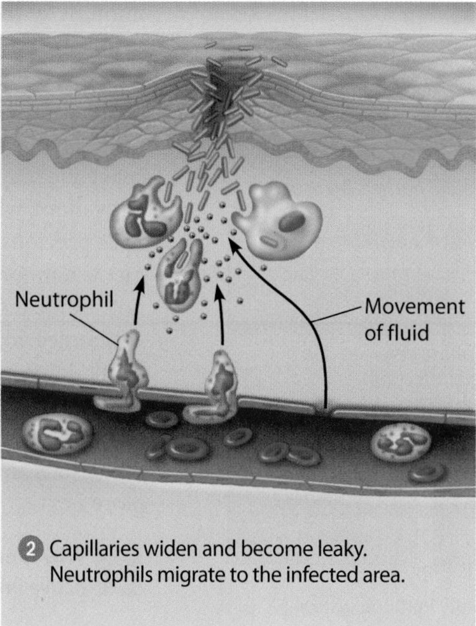
Neutrophil
Movement of fluid
❷ Capillaries widen and become leaky. Neutrophils migrate to the infected area.

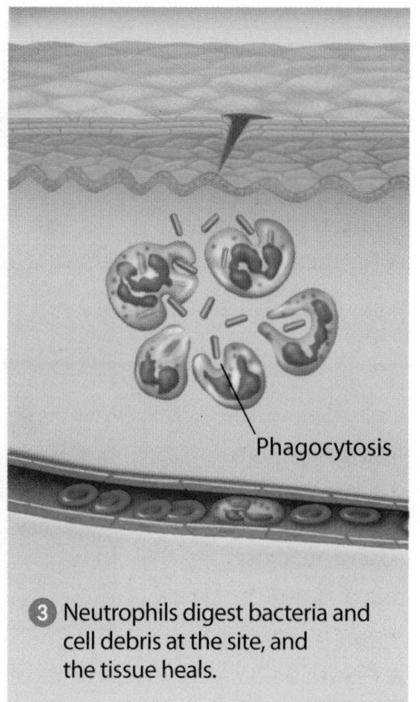
Phagocytosis
❸ Neutrophils digest bacteria and cell debris at the site, and the tissue heals.

▲ **Figure 24.2** The inflammatory response

▷ Adaptive Immunity

24.3 The adaptive immune response counters specific invaders

All the defenses you've learned about so far are called *innate* because they're ready "off the rack"; that is, innate defenses are always standing by, ready to be used in their current form. As outlined in **Figure 24.3**, when the external barriers and the internal defenses of the innate immune response fail to ward off a pathogen, a set of adaptive defenses, ones that are "custom-tailored" to each specific invader, provides a second line of defense. **Adaptive immunity**—also called acquired immunity—is a set of defenses, found only in vertebrates, that is activated in response to specific pathogens. Thus, unlike innate immunity, adaptive immunity differs from individual to individual, depending on what pathogens they have been previously exposed to. Although slower to be activated than the innate response, the adaptive immune response provides a strong defense against pathogens that is highly specific; that is, it acts against one infectious agent but not another. Moreover, adaptive responses can amplify certain innate responses, such as inflammation and the complement system.

Any molecule that elicits an adaptive immune response is called an **antigen**. Antigens are non-self molecules that protrude from pathogens or other particles, such as viruses, bacteria, mold spores, pollen, house dust, or the cell surfaces of transplanted organs. Antigens may also be substances released into the extracellular fluid, such as toxins secreted by bacteria. When the immune system detects an antigen, it responds with an increase in the number of cells that either attack the invader directly or produce antibodies. An **antibody** is an immune protein found in blood plasma that attaches to one particular kind of antigen and helps counter its effects. (The word *antigen* is a contraction of "*anti*body-*gen*erating," a reference to the fact that the foreign agent provokes an adaptive immune response.) The defensive cells and antibodies produced against a particular antigen are usually specific to that antigen; they are ineffective against any other foreign substance.

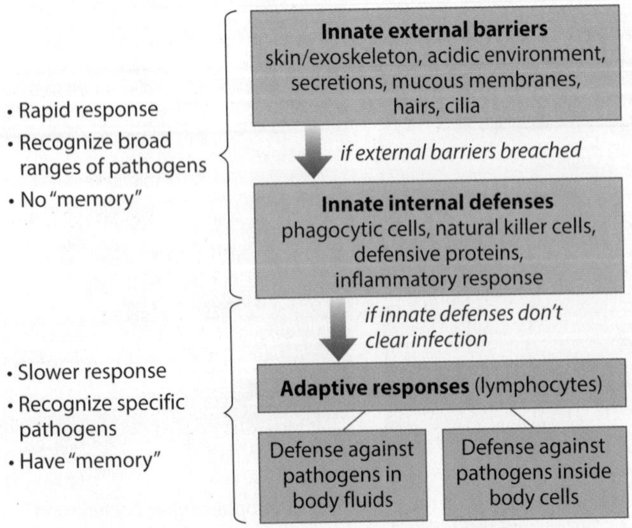

- Rapid response
- Recognize broad ranges of pathogens
- No "memory"

Innate external barriers
skin/exoskeleton, acidic environment, secretions, mucous membranes, hairs, cilia

↓ *if external barriers breached*

Innate internal defenses
phagocytic cells, natural killer cells, defensive proteins, inflammatory response

↓ *if innate defenses don't clear infection*

Adaptive responses (lymphocytes)

Defense against pathogens in body fluids | Defense against pathogens inside body cells

- Slower response
- Recognize specific pathogens
- Have "memory"

▲ **Figure 24.3** An overview of the vertebrate immune system

Try This Describe aloud or write down the innate and adaptive immune responses as they would occur after someone coughs on you.

Adaptive immunity has a remarkable "memory"; it can "remember" antigens it has encountered before, sometimes even many decades earlier, and react against them more quickly and vigorously on subsequent exposures. For example, the varicella zoster virus causes chicken pox, but a person once infected by the virus usually develops a resistance to future outbreaks. The immune system "remembers" certain molecules on the virus. Should the virus enter the body again, the adaptive immune response now mounts a decisive attack much faster than it originally did that usually destroys the virus before symptoms appear. Thus, in the adaptive immune response, unlike innate immunity, exposure to a foreign agent enhances future responses to that same agent.

Adaptive immunity is usually obtained by natural exposure to antigens (that is, by being infected), but it can also be achieved by **vaccination**, also known as immunization. In this procedure, the immune system is confronted with a **vaccine** composed of a harmless variant or part of a disease-causing microbe, such as an inactivated bacterial toxin, a dead or weakened virus, or a piece of a virus. The HPV vaccine we discussed in the chapter opener is made of surface proteins from the virus. The vaccine stimulates the immune system to mount defenses against this harmless antigen, defenses that will also be effective against the actual pathogen because it has similar antigens. Once you have been successfully vaccinated, your immune system will respond quickly if it is exposed to the actual microbe. Such protection may last for life.

Whether antigens enter the body naturally (if you catch the flu) or artificially (if you get a flu vaccine), the resulting immunity is called **active immunity** because the person's own immune system actively produces antibodies. It is also possible to acquire **passive immunity** by receiving premade antibodies. For example, a fetus obtains a mother's antibodies through the placenta before birth and after birth breast milk continues to supply the baby with antibodies. The effects of a poisonous snakebite may be counteracted by injecting the victim with antivenom, which consists of antibodies extracted from animals previously immunized against the venom. Passive immunity is temporary because the recipient's immune system is not stimulated by antigens. Immunity lasts only as long as the antibodies do; after a few weeks or months, these proteins break down and are recycled by the body.

Lymphocytes are the white blood cells responsible for adaptive immunity. They are found in the blood and also in the tissues and organs of the lymphatic system, which we explore next.

? Why is protection resulting from a vaccination considered *active* immunity rather than *passive* immunity?

Because the body itself produces the immunity by mounting an immune response and generating antibodies, even though the stimulus consists of artificially introduced antigens

24.4 The lymphatic system becomes a crucial battleground during infection

The **lymphatic system** is involved in both innate and adaptive immunity. It consists of a branching network of vessels, numerous **lymph nodes**—little round organs packed with macrophages and lymphocytes—the bone marrow, and several organs (**Figure 24.4**). The lymphatic vessels carry a fluid called **lymph**, which is similar to the interstitial fluid that surrounds body cells but contains less oxygen and fewer nutrients.

The lymphatic system is closely associated with the circulatory system. Most infectious agents wind up in the circulatory system and from there get carried into the lymphatic system, which can usually filter them out. The filtered fluid can then be recycled into the circulatory system. The lymphatic system thus has two main functions: to return tissue fluid back to the circulatory system and to fight infection.

Circulatory Function A small amount of the fluid that enters the tissue spaces from the blood in a capillary bed does not reenter the blood capillaries (Module 23.11). Instead, this fluid is returned to the blood via lymphatic vessels. Figure 24.4 (bottom right) shows a branched lymphatic vessel in the process of taking up fluid from tissue spaces. The branching structure facilitates this function. Fluid enters the lymphatic system by diffusing into dead-end lymphatic capillaries that are intermingled among the blood capillaries.

Lymph drains from the lymphatic capillaries into larger and larger lymphatic vessels. Eventually, the fluid reenters the circulatory system via two lymphatic vessels that fuse with veins in the chest. The lymphatic vessels resemble veins in having valves that prevent the backflow of fluid toward the capillaries (see Figure 23.7C). Also, like veins, lymphatic vessels depend mainly on the movement of skeletal muscles to squeeze their fluid along.

Immune Function When your body is fighting an infection, the organs of the lymphatic system become a major battleground. As lymph circulates through the lymphatic organs—such as the lymph node (Figure 24.4, top right)—it carries microbes, parts of microbes, and their toxins picked up from infection sites anywhere in the body. Once inside lymphatic organs, macrophages that reside there permanently may engulf the invaders as part of the innate immune response. Lymph nodes fill with huge numbers of defensive cells, causing the tender "swollen glands" in your neck and armpits that your doctor looks for as a sign of infection. Many of the defensive cells in the lymph nodes are lymphocytes, which are responsible for the adaptive immune response.

? **What might be the main symptom of blockage of a large lymphatic vessel?**

● Fluid accumulation (swelling) in the tissues

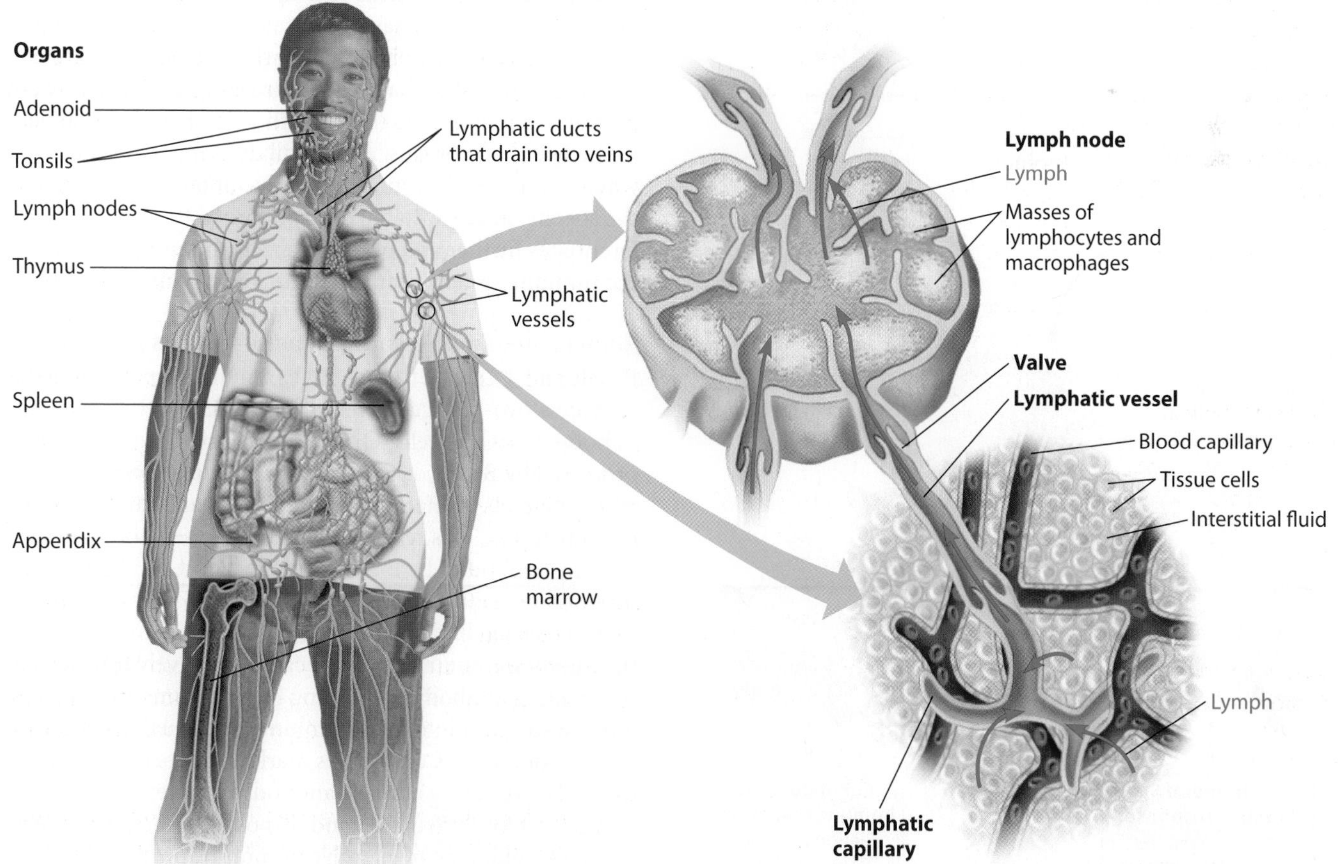

Organs
Adenoid
Tonsils
Lymph nodes
Thymus
Spleen
Appendix
Lymphatic ducts that drain into veins
Lymphatic vessels
Bone marrow
Lymph node
Lymph
Masses of lymphocytes and macrophages
Valve
Lymphatic vessel
Blood capillary
Tissue cells
Interstitial fluid
Lymph
Lymphatic capillary

▲ **Figure 24.4** The human lymphatic system

24.5 Lymphocytes mount a dual defense

Like all blood cells, lymphocytes originate from stem cells in the bone marrow (see Module 23.15). As shown in Figure 24.5A, some immature lymphocytes continue developing in the bone marrow to become specialized as B lymphocytes, or **B cells**. Other immature lymphocytes migrate to the thymus, a gland above the heart, to become specialized as T lymphocytes, or **T cells**. By mounting a dual defense, B and T cells defend against infections in body fluids and cells.

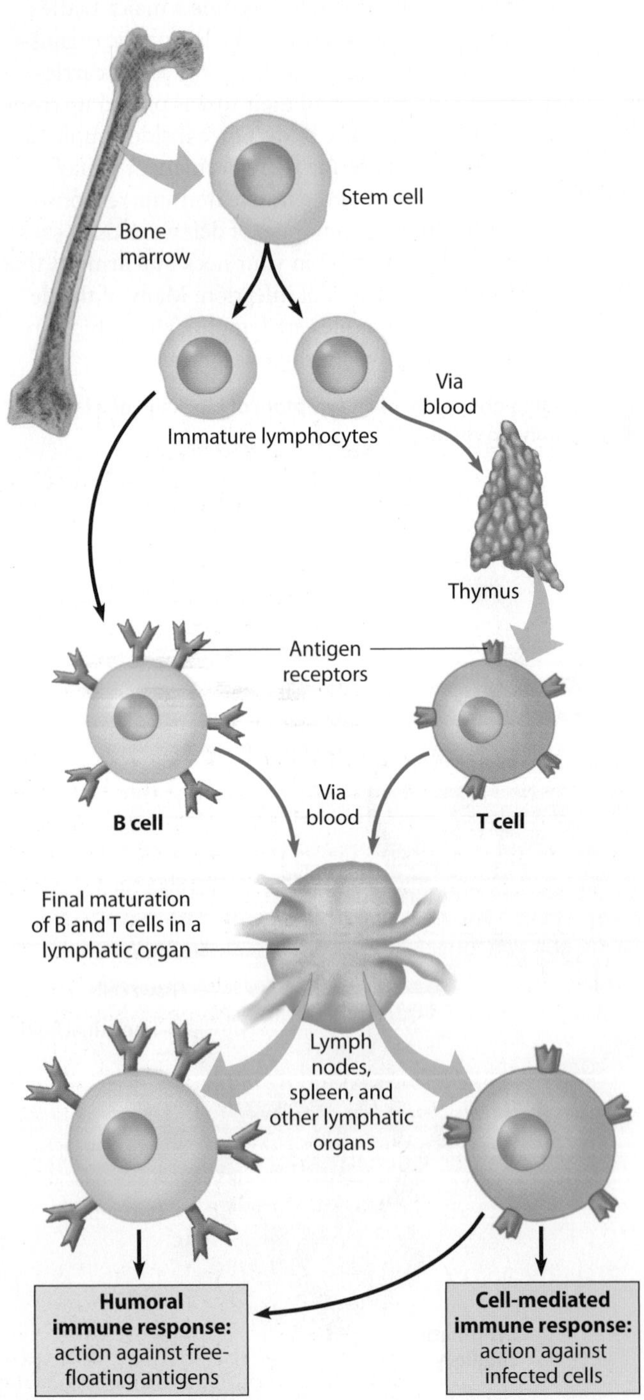

Stem cell
Bone marrow
Via blood
Immature lymphocytes
Thymus
Antigen receptors
B cell
Via blood
T cell
Final maturation of B and T cells in a lymphatic organ
Lymph nodes, spleen, and other lymphatic organs
Humoral immune response: action against free-floating antigens
Cell-mediated immune response: action against infected cells

▲ **Figure 24.5A** The development of B cells and T cells

B and T Cell Differentiation When a B cell develops in bone marrow or a T cell develops in the thymus, the cells differentiate from other cells by synthesizing many copies of a specific protein, which are then incorporated into the plasma membrane. As indicated in Figure 24.5A, these protein molecules stick out from the cell's surface. The molecules are **antigen receptors**, capable of binding one specific type of antigen. The cell's antigen receptors are in place before they ever encounter an antigen.

All antigen receptors on the surface of a single lymphocyte are identical and recognize a particular antigen. Figure 24.5B on the facing page illustrates the antigen receptors on different lymphocytes (in this case, B cells). Note that the figure is drastically simplified; there are about 100,000 identical antigen receptors on the surface of an individual lymphocyte. An enormous diversity of B cells and T cells develops in each person. Researchers estimate that every one of us has millions of different kinds—enough to recognize and bind to virtually every possible antigen. A small population of each kind of lymphocyte lies in wait in our body, genetically programmed to recognize and respond to a specific antigen.

After the B cells and T cells have developed their antigen receptors, these lymphocytes leave the bone marrow and thymus and move via the bloodstream to the lymph nodes, spleen, and other parts of the lymphatic system. In these organs, many B and T cells take up residence and encounter infectious agents that have penetrated the body's external barriers. Because lymphatic capillaries extend into virtually all the body's tissues, bacteria or viruses infecting nearly any part of the body eventually enter the lymph and are carried to the lymphatic organs. As we will describe in Module 24.7, when a B or T cell within a lymphatic organ first confronts the specific antigen that it is programmed to recognize, it differentiates further and becomes a fully mature component of the immune system, ready to mount a response.

Humoral and Cell-Mediated Immune Responses The B cells and T cells of the adaptive immune response together provide a two-pronged defense, combating pathogens both in body fluids and cells. One of the two adaptive responses, produced by B cells, is the **humoral immune response**, which defends primarily against bacteria and viruses present in body fluids. This response involves the secretion of free-floating antibodies by B cells into the blood and lymph. (The humoral response is so named because blood and lymph were long ago called body "humors.") As discussed in Module 24.3, the humoral immune response can be passively transferred by injecting antibody-containing blood plasma from an immune individual into a nonimmune individual. As you will see in Module 24.9, antibodies mark invaders by binding to them. The resulting antigen-antibody complexes are easily recognized for destruction and disposal by phagocytic cells.

The second type of adaptive immunity, produced by T cells, is called the **cell-mediated immune response**, which defends against infections inside body cells. As its name

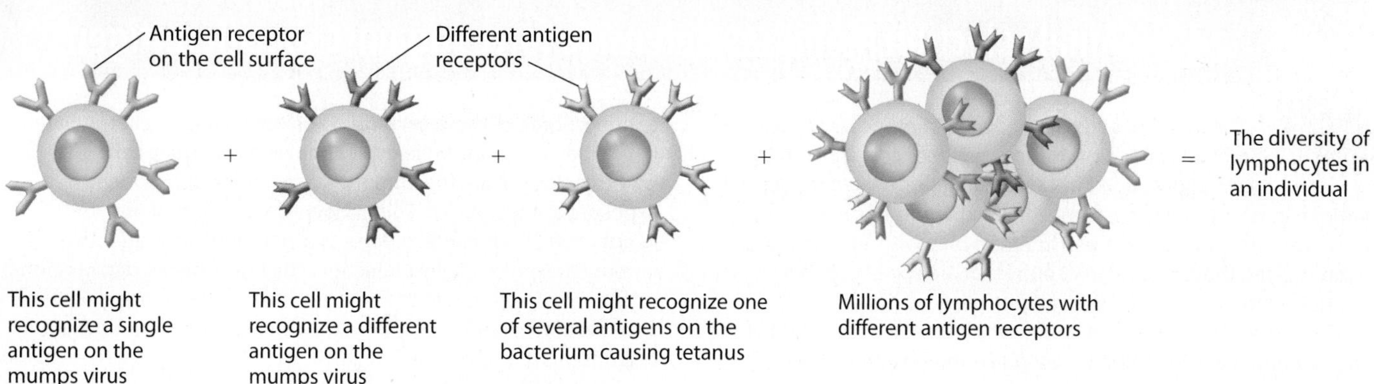
Antigen receptor on the cell surface
Different antigen receptors
The diversity of lymphocytes in an individual
This cell might recognize a single antigen on the mumps virus
This cell might recognize a different antigen on the mumps virus
This cell might recognize one of several antigens on the bacterium causing tetanus
Millions of lymphocytes with different antigen receptors

▲ **Figure 24.5B** The diversity of lymphocytes in an individual (represented in the figure by a distinct color for the antigen receptors of each cell)

implies, this defensive system results from the action of defensive T cells, in contrast to the action of free-floating defensive antibody proteins produced by B cells of the humoral response. Defensive T cells destroy body cells infected with bacteria or viruses.

Not all T cells function as defensive T cells. Some types of T cells function indirectly by promoting phagocytosis by other white blood cells and by stimulating B cells to produce antibodies. Thus, as the arrows at the bottom of Figure 24.5A indicate, some T cells play a part in both the cell-mediated and humoral immune responses.

Only a tiny fraction of the immune system's B cells and T cells will ever be used, but they are all available if needed. It is as if the immune system maintains a huge standing army of soldiers, each predetermined to recognize only one particular kind of invader. The majority of soldiers never encounter their target and remain idle. But when an invader does appear, chances are good that a lymphocyte will be able to recognize it, bind to it, and call in reinforcements.

? Contrast the targets of the humoral immune response with those of the cell-mediated immune response.

● The humoral immune response works against pathogens in the body fluids; the cell-mediated immune response attacks infected cells.

24.6 Antigen receptors and antibodies bind to specific regions on an antigen

Both the humoral and cell-mediated immune responses are initiated when lymphocytes recognize antigens. B cells bind antigens directly, whereas T cells require an additional step for recognition. Here we look more closely at antigen binding with B cells and antibodies. (Structurally similar, the antibodies produced by B cells can be thought of as free-floating B cell antigen receptors.)

Antigens usually do not belong to the host animal. Most antigens are proteins or large polysaccharides that protrude from the surfaces of viruses or foreign cells. Common examples are protein-coat molecules of viruses, parts of the capsules and cell walls of bacteria, and macromolecules on the surface cells of other kinds of organisms, such as protozoans and parasitic worms. (Sometimes a particular microbe is called an antigen, but this usage is misleading because the microbe will almost always have several kinds of antigenic molecules.) Other sources of antigens include blood cells or tissue cells from other individuals, of the same or a different species. Antigens are also found dissolved in body fluids; foreign molecules of this type include bacterial toxins and bee venom.

A small surface-exposed region of an antigen is called an **antigenic determinant**, also known as an epitope. Antigen receptors on B cells, as well as antibodies, recognize and bind to the antigenic determinant. The specific region on an antigen receptor or antibody that recognizes an antigenic determinant is the **antigen-binding site**. The binding site and antigenic determinant have shapes like a lock and key. An antigen usually has several different determinants. In **Figure 24.6**, two different antibodies are shown binding to the same antigen, which in this case has three determinants. A single kind of antigen may thus stimulate the immune system to activate several distinct lymphocytes, ultimately leading to the production of several different antibodies.

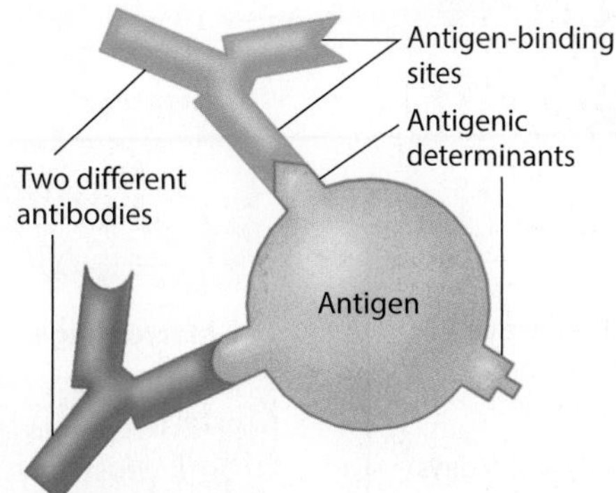
Antigen-binding sites
Antigenic determinants
Two different antibodies
Antigen

▲ **Figure 24.6** The binding of antibodies to antigenic determinants

? Why is it inaccurate to refer to a pathogen, such as a virus, as an antigen?

● It is inaccurate because antigens are not whole pathogens; they are molecules, which may be chemical components of a pathogen's surface. One pathogen may have many antigens.

VISUALIZING THE CONCEPT

24.7 Clonal selection mobilizes defensive forces against specific antigens

The humoral and cell-mediated immune responses both defend against a wide variety of antigens through a process known as **clonal selection**. Inside the body, an antigen encounters a diverse pool of B and T lymphocytes. However, one particular antigen interacts only with a fraction of lymphocytes, those bearing receptors that are specific to that antigen. Once activated by the antigen, the lymphocytes proliferate, forming a clone (a genetically identical population) of thousands of cells "selected" to recognize and respond to the antigen. Some of these cells, called **effector cells**, act immediately to combat infection, while others known as **memory cells** lie in wait to help activate the immune system upon subsequent exposure to the antigen. This antigen-driven cloning of lymphocytes—clonal selection—is a vital step in the adaptive immune response against infection. The figure below depicts clonal selection using B cells in the humoral immune response. (T cells also undergo clonal selection, in the cell-mediated immune response.)

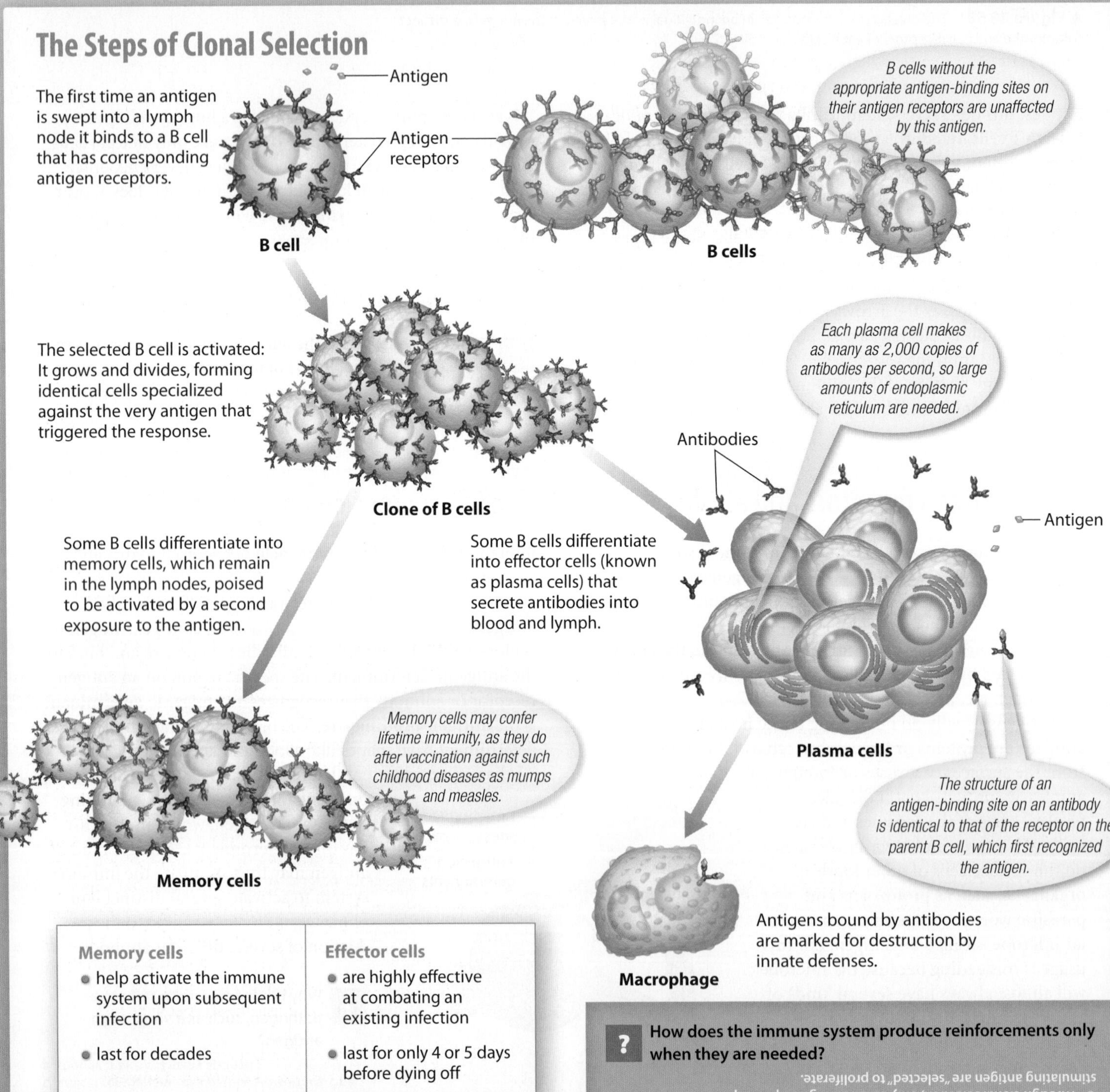
The Steps of Clonal Selection
The first time an antigen is swept into a lymph node it binds to a B cell that has corresponding antigen receptors.
Antigen
Antigen receptors
B cell
B cells
B cells without the appropriate antigen-binding sites on their antigen receptors are unaffected by this antigen.
The selected B cell is activated: It grows and divides, forming identical cells specialized against the very antigen that triggered the response.
Clone of B cells
Each plasma cell makes as many as 2,000 copies of antibodies per second, so large amounts of endoplasmic reticulum are needed.
Antibodies
Antigen
Some B cells differentiate into memory cells, which remain in the lymph nodes, poised to be activated by a second exposure to the antigen.
Some B cells differentiate into effector cells (known as plasma cells) that secrete antibodies into blood and lymph.
Memory cells may confer lifetime immunity, as they do after vaccination against such childhood diseases as mumps and measles.
Memory cells
Plasma cells
The structure of an antigen-binding site on an antibody is identical to that of the receptor on the parent B cell, which first recognized the antigen.
Macrophage
Antigens bound by antibodies are marked for destruction by innate defenses.

Memory cells	Effector cells
• help activate the immune system upon subsequent infection	• are highly effective at combating an existing infection
• last for decades	• last for only 4 or 5 days before dying off

? How does the immune system produce reinforcements only when they are needed?

● Through clonal selection, only cells bearing receptors specific to the stimulating antigen are "selected" to proliferate.

24.8 The primary and secondary responses differ in speed, strength, and duration

The first time an antigen enters the body and selectively activates lymphocytes is called the **primary immune response**. This initial response of adaptive immunity is slow, taking many days to produce effector cells that secrete antibodies into the blood and lymph in a great enough quantity to overcome an infection. Recall that a hallmark of adaptive immunity is memory of a specific antigen. The memory cells produced by clonal selection in the primary immune response confer protection in subsequent encounters with that pathogen.

When memory cells produced during the primary response are activated by a second exposure to the same antigen—which may occur soon or long after the primary immune response—they initiate the **secondary immune response**. Memory cells produced by the primary response enable the rapid formation of a second round of clonal selection, again with thousands of effector cells specific for the same antigen. Because a pool of memory cells is activated (in contrast to a few lymphocytes in the primary response), this second round of clonal selection is faster and stronger than the first. The secondary response, like the primary, activates both effector cells and memory cells. In **Figure 24.8A**, we see the secondary response with B cells. The pool of memory B cells produced by the primary response gives rise to great quantities of effector B cells that quickly secrete high levels of antibodies when a "known" antigen enters the body. The memory B cells that result wait to be activated, should a third exposure to the same antigen occur.

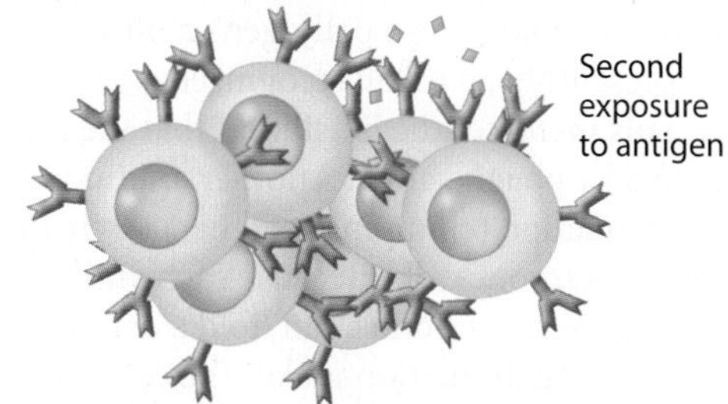
Second
exposure
to antigen

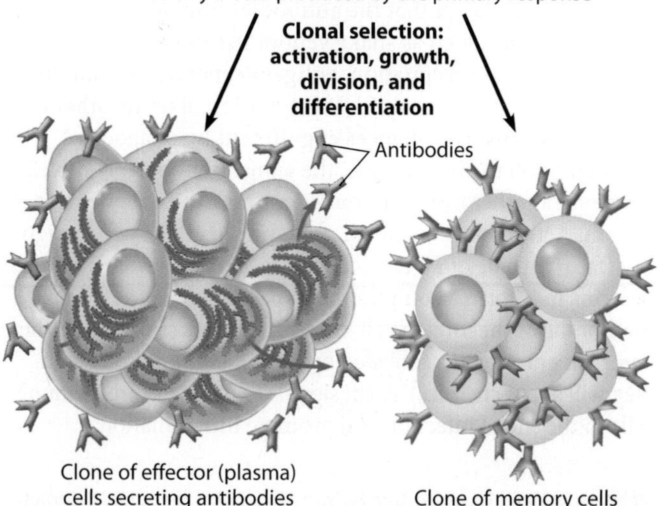
Memory B cells produced by the primary response
Clonal selection:
activation, growth,
division, and
differentiation
Antibodies
Clone of effector (plasma)
cells secreting antibodies
Clone of memory cells

▲ **Figure 24.8A** The secondary immune response

Try This Explain the advantage of having memory cells when a pathogen is encountered for a second time.

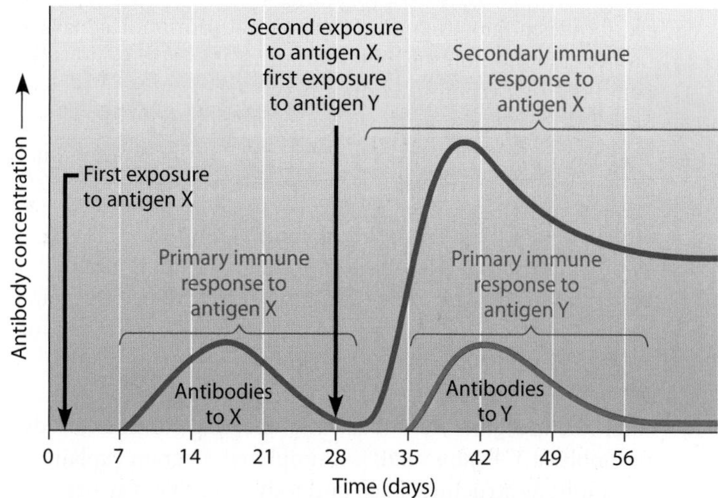
Second exposure
to antigen X,
first exposure
to antigen Y
Secondary immune
response to
antigen X
First exposure
to antigen X
Primary immune
response to
antigen X
Primary immune
response to
antigen Y
Antibodies
to X
Antibodies
to Y
Antibody concentration
0
7
14
21
28
35
42
49
56
Time (days)

Data from F. M. Burnet, et al., *The Production of Antibodies: A Review and Theoretical Discussion*, Monographs from the Walter and Eliza Hall Institute of Research in Pathology and Medicine, Number One, Melbourne: Macmillan and Company Limited (1941).

▲ **Figure 24.8B** The two phases of the adaptive immune response

The differences between the primary and secondary adaptive responses can be illustrated quantitatively. We'll use B cells in the humoral response as an example, because measuring the concentration of antibodies over time easily distinguishes the primary and secondary responses. The blue curve in **Figure 24.8B** illustrates the two responses, triggered by two exposures to the same antigen. On the far left of the graph, you can see that the primary response does not start right away; it usually takes several days for the lymphocytes to become activated by an antigen (called X in the graph) and form clones of effector cells. During this delay, a stricken individual may become ill. When the effector cell clone forms, antibodies start showing up in the blood, reaching their peak 2–3 weeks after initial exposure. As the antibody levels in the blood and lymph rise, the symptoms of the illness typically diminish and disappear. The primary response subsides as the effector cells die out.

The second exposure to antigen X (at day 28 in the graph) triggers the secondary immune response. The secondary response starts usually in a few days (faster than the primary response), produces higher levels of antibodies, and is more prolonged. This is why vaccination is so effective: The vaccine induces a primary immune response that produces memory cells; an encounter with the actual pathogen then elicits a rapid and strong secondary immune response. Vaccines are an example of active immunity, because a person's own immune system is stimulated to produce antibodies rather than receiving them.

The red curve in Figure 24.8B illustrates how each immune response is specific. If the body is exposed to a different antigen (Y), even after it has already responded to antigen X, it responds with another primary response, this one directed against antigen Y. The response to Y is not enhanced or diminished by the response to X; that is, adaptive immunity is specific.

? **What is the immunological basis for referring to certain diseases, such as chicken pox, as *childhood* diseases?**

One bout with the pathogen, which most often occurs during childhood, is usually enough to confer immunity for the rest of that individual's life.

24.9 The structure of an antibody matches its function

Antibodies do not kill pathogens. Instead, antibodies mark a pathogen by combining with it to form an antigen-antibody complex. Weak chemical bonds between antigens and the antigen-binding sites on antibodies hold the complex together. Once marked in this manner, other immune system components bring about the destruction of the antigen.

The Structure of the Antigen-Antibody Complex An antibody has two related functions in the humoral immune response: first, to recognize and bind to a certain antigen and, second, to assist in eliminating that antigen. The structure of an antibody allows it to perform both of these functions.

As the computer-generated rendering of an antibody molecule in **Figure 24.9A** illustrates, the shape of an antibody resembles a Y. **Figure 24.9B** is a simplified diagram explaining antibody structure. Each antibody molecule consists of four polypeptide chains bonded together to form the Y shape. The tip of each arm of the Y forms an antigen-binding site, a region of the molecule responsible for the antibody's recognition-and-binding function. A huge variety in the three-dimensional shapes of the binding sites of different antibodies accounts for the diversity of lymphocytes and gives the humoral immune response the ability to react to virtually any kind of antigen.

The stem of the antibody helps mediate the disposal of the bound antigen. Depending on the polypeptide structure of the stem, antibodies are grouped into various classes that distinguish where they are found in the body and how they work. However, all classes of antibodies perform the same basic function: to mark invaders for elimination.

The Antigen-Antibody Complex Functions with Innate Immunity All antibody mechanisms involve two parts: a *specific* recognition-and-attach phase followed by a *nonspecific* destruction phase. Thus, antibodies of the adaptive humoral immune response, which identify and bind to invaders, must work with components of innate immunity (Module 24.1), such as phagocytosis and the complement system, to form a complete defense system. As **Figure 24.9C** on the opposite page illustrates, the binding of antibodies to antigens can trigger several mechanisms that disable or destroy an invader.

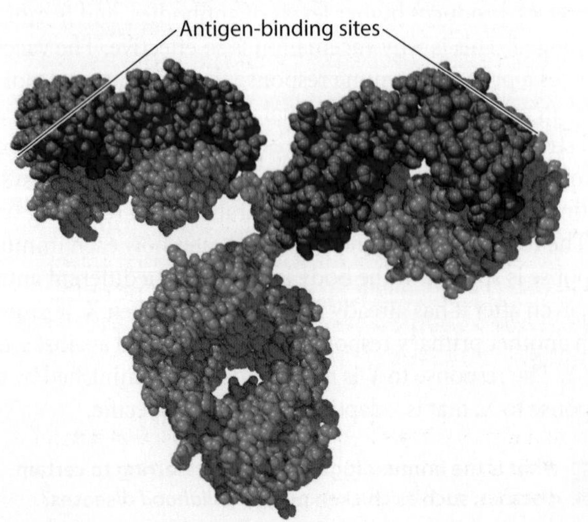
Antigen-binding sites

▲ **Figure 24.9A** A computer model of an antibody molecule

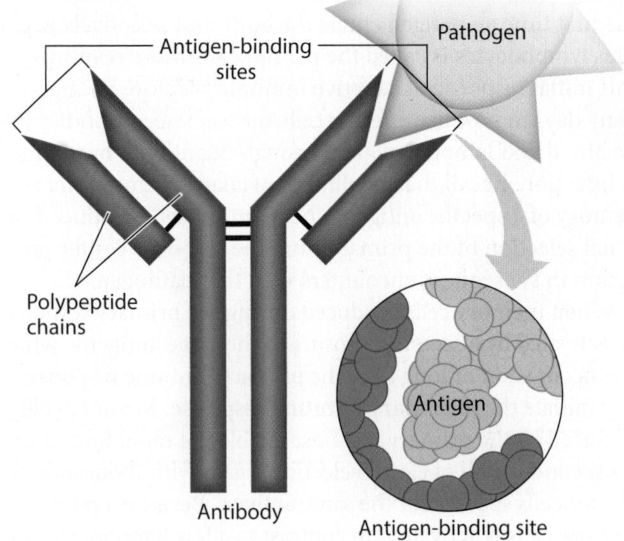
Pathogen
Antigen-binding sites
Polypeptide chains
Antigen
Antibody
Antigen-binding site

▲ **Figure 24.9B** Antibody structure, including the antigen-binding site and the antigen it is specific for (enlargement)

The antigen-antibody complex boosts the function of phagocytic cells of innate immunity in three ways. In neutralization, antibodies bind to surface proteins on a virus or bacterium, thereby blocking a pathogen's ability to infect a host cell and presenting an easily recognized structure to macrophages. This increases the likelihood that the foreign cell will be engulfed by phagocytosis. Another antibody mechanism is the agglutination (clumping together) of viruses, bacteria, or foreign eukaryotic cells. Because each antibody has at least two binding sites, antibodies can hold a clump of invading cells together. Agglutination makes the cells easy for phagocytes to capture. The precipitation mechanism is similar to agglutination, except that the antibodies link together dissolved antigens (such as snake venom toxins) rather than viruses or cells. In precpitation, antigens separate, in solid form, from the surrounding liquid. Precipitation, like the other effector mechanisms, enhances engulfment by phagocytes.

In addition to phagocytosis, the antigen-antibody complex promotes another innate immune response: the complement system. Activated complement proteins can attach to a foreign cell (far right side of Figure 24.9C). Once there, several activated proteins may form a complex that pokes a hole in the plasma membrane of the foreign cell, causing cell lysis (rupture). Other complement proteins activated by the antigen-antibody complex will act as chemical alarm signals to recruit more immune cells to the site of infection and promote inflammation.

? How does the adaptive humoral immune response interact with the body's innate immunity?

Antibodies mark specific antigens for destruction in the adaptive humoral immune response, and phagocytes and the complement system (components of innate immunity) destroy the antigens.

Binding of antibodies to antigens inactivates antigens by

Neutralization (blocks viral binding sites; coats bacteria)

Virus

Bacterium

Agglutination of microbes (clumps viruses and cells together)

Bacteria

Precipitation of dissolved antigens (clumps of soluble antigens come out of solution)

Antigens

Activation of the complement system (activated complement proteins form membrane complexes)

Complement molecule

Foreign cell

Hole

Enhances

Phagocytosis (foreign particles and cells are engulfed and digested)

Macrophage

Leads to

Cell lysis (rupture of invading cell)

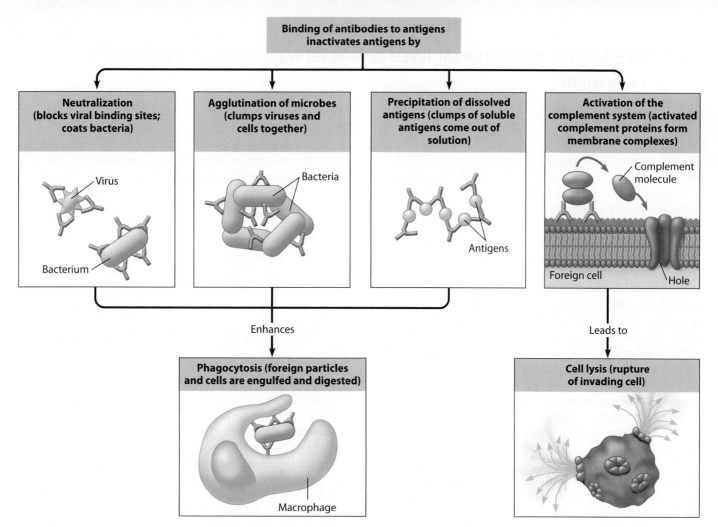

▲ **Figure 24.9C** Effector mechanisms of the humoral immune response

Try This Using the acronym PLAN (precipitation, lysis by complement, agglutination, neutralization), describe how antibodies work.

24.10 Antibodies are powerful tools in the lab and clinic

CONNECTION

Because of their ability to tag specific molecules or cells, antibodies are widely used in laboratory research, the treatment of disease, and in health-related devices. For example, antibodies are used to treat breast cancer and are the basis of home pregnancy tests.

Herceptin, a genetically engineered antibody, acts by binding to growth factor receptors that are present in excess on a common form of aggressive breast cancer cells. The drug helps slow the progress of the cancer by preventing the receptors from stimulating the cells to grow. Certain other types of cancer can be treated with tumor-specific antibodies bound to toxin molecules. The toxin-linked antibodies carry out a precise search-and-destroy mission, selectively attaching to and killing tumor cells.

The most popular type of pregnancy test uses an antibody to detect a hormone called human chorionic gonadotropin (hCG), which is present in the urine of pregnant women. (hCG helps maintain the uterus during early pregnancy; see Module 27.15.) When urine is applied to the testing stick it interacts with an antibody specific for hCG that is present in the device. If the woman is pregnant, the hCG in the urine will bind to the antibody. The hCG-antibody complex will initiate subsequent reactions that result in the activation of a colored dye visible as a stripe on the testing stick (the blue stripe in the box at the center of the device shown). If the woman taking the test is not pregnant, no hCG will be present in the urine, and thus no dye-activating complex will be formed (no blue stripe). She will know a negative result indicates "not pregnant" and not just a failure of the kit, however, because pregnancy tests include a control. If urine is present and the device is functioning properly, a different colored stripe will appear (the red stripe in the bottom box). This positive control shows that the test is working.

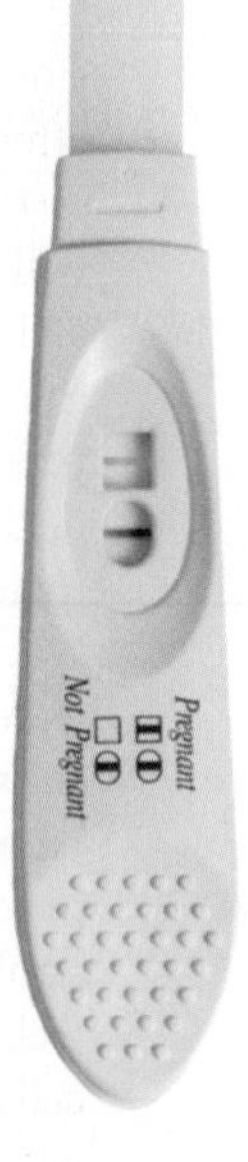
Pregnant
Not Pregnant

? **If a woman showed you a pregnancy test that had no red or blue stripe and declared she was "not pregnant," would you agree?**

No. This would indicate that the kit was not working—a true negative result would be indicated by a red stripe.

24.11 Scientists measure antibody levels to look for waning immunity after HPV vaccination

SCIENTIFIC THINKING

Active adaptive immunity to a specific pathogen can be gained through a natural infection or through vaccination. With human papilloma virus (HPV), infections are common: Approximately 50% of all sexually active adults become infected by the virus. Usually there are no noticeable symptoms, and the immune system clears HPV infection within two years. The individual now has active immunity to HPV—a second infection with the virus would be cleared rapidly by the secondary immune response. Some individuals, however, have an HPV infection that escapes the immune system for many more years, interfering with the regulation of cell growth in the infected epithelial cells. Cells with a persistent infection grow uncontrolled for years, increasing the likelihood that mutations will accumulate and result in cervical and anal cancers. Two vaccines made with HPV antigens (in this case, proteins from the viral surface) were developed to promote artificial immunity before individuals come into contact with cancer-causing viral strains. In 2006, the HPV vaccine Gardasil was approved in the United States. A second vaccine, Cervarix, was approved in 2009.

How long will protection from cervical cancers last after HPV vaccination?

The decades of data needed to know how long the active immunity will last do not exist, but scientists are collecting data to determine how effective Gardasil and Cervarix are so far. Individuals being studied were among the first to be vaccinated, in clinical trials that began as early as 1998. The study participants were either vaccinated against specific strains of HPV or injected with a placebo. Scientists from Brazil, the United States, and several Nordic countries periodically examine individuals from each group to determine whether precancerous lesions have developed. For as many years as they have been studied, both vaccines have been 93–100% effective in preventing precancerous cervical lesions. That is good news, but it doesn't tell us how long that effectiveness will last.

Can we predict if or when a HPV vaccine's effectiveness will decrease? Not exactly, but we can get an idea of how long a HPV vaccination lasts by looking at one component of the adaptive immune response: the level of antibodies being produced against HPV years after vaccination. A significant decline in antibody levels after vaccination for any disease, including HPV, might warrant another dose of the antigen (vaccine), commonly known as a "booster shot." Because effector cells are usually short-lived, we might not expect to see antibodies for more than a few months after vaccination. However, HPV vaccination induces long-term, continual production of antibodies. How the immune system produces antibodies for long periods against some antigens is not yet well understood, but there is evidence that some plasma cells migrate from the lymph nodes to the bone marrow where they live and produce antibodies for years. Another hypothesis is that memory cells are continually activated and therefore continually differentiating into plasma cells. Although the mechanism is still unclear, this long-term production can be measured in vaccinated individuals.

True long-term studies examining antibody levels years after HPV vaccination have not been completed yet. The longest-running, published study as of 2012 followed individuals for 5 years after vaccination with Gardasil and 8.4 years after vaccination with Cervarix. **Figure 24.11** focuses on the levels of two HPV-specific antibodies—anti-HPV-16 and anti-HPV-18—in the blood of individuals vaccinated with either Gardasil (pink bars) or Cervarix (blue bars). The two antibodies that were measured recognize two strains of HPV that cause cervical and anal cancer—the same two strains that both vaccines provided immunity against. On the *y* axis, we see the percentage of individuals whose HPV-specific antibody levels could still be detected in their blood. A person with no measurable antibodies is susceptible to HPV infection.

Scientists hypothesize that higher anti-HPV antibody levels provide greater protection from HPV-related cancers compared to lower levels, but they don't yet have strong evidence to support this hypothesis. Also unknown is whether there is a minimum antibody level necessary to prevent cancer. To answer these questions, vaccinated individuals will need to be followed for decades to see if there is a correlation between antibody levels and the onset of precancerous lesions. Studies are ongoing.

? **Based on these data, does either vaccine look like it might require a booster in the future?**

Yes, Gardasil may require a booster: After 5 years, the antibody levels protecting against HPV-18 are undetectable in approximately 35% of individuals vaccinated.

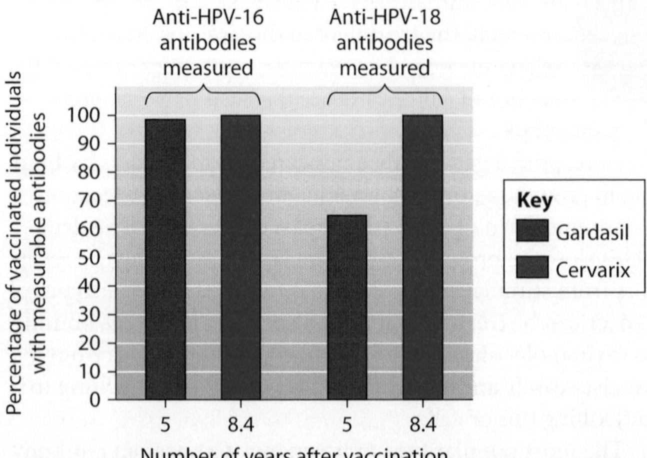

Anti-HPV-16 antibodies measured
Anti-HPV-18 antibodies measured
Percentage of vaccinated individuals with measurable antibodies
100
90
80
70
60
50
40
30
20
10
0
5
8.4
5
8.4
Number of years after vaccination
Key
Gardasil
Cervarix

Data from S-E. Olsson et al., Induction of immune memory following administration of a prophylactic quadrivalent human papillomarvirus (HPV) types 6/11/16/18 L1 virus-like particle (VLP) vaccine, *Vaccine 25*: 3931–4939 (2007); C. M. Roteli-Martin et al., Sustained immunogenicity and efficacy of the HPV16/18 AS04-adjuvanted vaccine up to 8.4 years of follow-up, *Landes Bioscience* 8:3 (2012).

▲ **Figure 24.11** The levels of anti-HPV-16 and anti-HPV-18 in blood years after Cervarix and Gardasil vaccination

24.12 Helper T cells stimulate the humoral and cell-mediated immune responses

The antibody-producing B cells of the humoral immune response make up one arm of the adaptive immune response network. The humoral immune response identifies and helps destroy invaders that are in our blood, lymph, or interstitial fluid—in other words, outside our body cells. But many invaders, including all viruses, enter cells and reproduce there. It is the cell-mediated immune response produced by **cytotoxic T cells** that battles pathogens that have already entered body cells; we'll discuss this response in Module 24.13.

A type of T cell called a **helper T cell** triggers both the humoral and cell-mediated immune responses. Helper T cells themselves do not carry out those responses. Instead, signals from helper T cells initiate production of antibodies that neutralize pathogens and activate cytotoxic T cells that kill infected cells. The role of helper T cells is so central to immunity that without functional helper T cells, there is virtually no immune response (see Module 24.14).

Two requirements must be met for a helper T cell to activate adaptive immune responses. First, a foreign molecule must be present that can bind specifically to the antigen receptor of the T cell. Second, this antigen must be displayed on the surface of an **antigen-presenting cell**. Macrophages and B cells are two types of antigen-presenting cells.

Consider a typical antigen-presenting cell, a macrophage. As shown in **Figure 24.12A**, ① the macrophage ingests a microbe or other foreign particle and breaks it into fragments—foreign antigens (▲). Then molecules of a special protein ( ) belonging to the macrophage, which we will call a **self protein** (because it belongs to the body), ② bind the foreign antigens—**nonself molecules**—and ③ display them on the cell's surface. (Each of us has a unique set of self proteins, which serve as identity markers for our body cells.) ④ Helper T cells recognize and bind to the combination of a self protein and a foreign antigen—called a self-nonself complex ( )—displayed on an antigen-presenting cell. This double recognition is like the system banks use for safe-deposit boxes: Opening your box requires the banker's key along with your specific key.

The ability of a helper T cell to specifically recognize a unique self-nonself complex on an antigen-presenting cell depends on the receptors (purple) embedded in the T cell's plasma membrane. A T cell receptor actually has two binding sites: one for antigen and one for self protein. The two binding sites enable a T cell receptor to recognize the overall shape of a self-nonself complex on an antigen-presenting cell. The immune response is specific because the receptors on each helper T cell bind only one kind of self-nonself complex on an antigen-presenting cell.

The binding of a T cell receptor to a self-nonself complex activates the helper T cell. Other kinds of signals can enhance this activation. When secreted by the antigen-presenting cell, signaling molecules (blue arrow), diffuse to the helper T cell and stimulate it.

As seen in **Figure 24.12B**, activated helper T cells promote the immune response, with a major mechanism being the secretion of additional stimulatory proteins. These signaling molecules, secreted by helper T cells (blue arrows), have three major effects. ① First, they stimulate clonal selection of the helper T cell, producing both memory cells and additional effector helper T cells. Second, the signals ② help activate B cells, thus stimulating the humoral immune response. And third, ③ signaling molecules stimulate the activity of cytotoxic T cells of the cell-mediated immune response, our next topic.

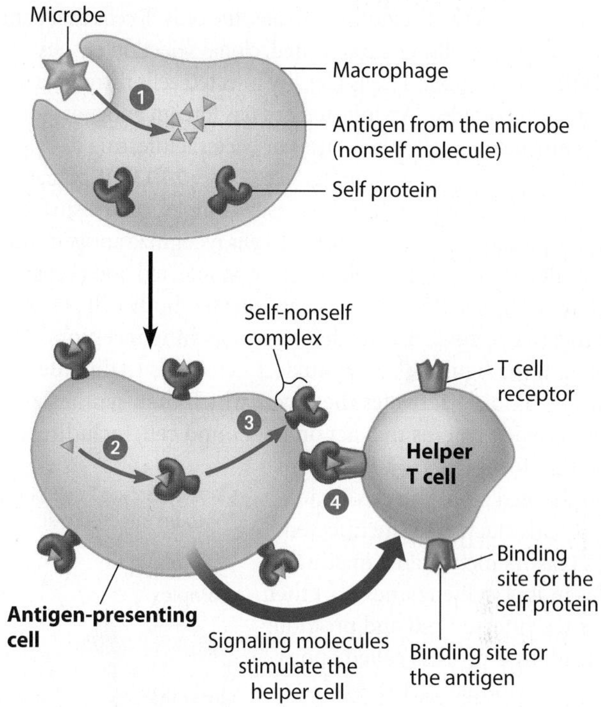

Microbe
Macrophage
1
Antigen from the microbe (nonself molecule)
Self protein
Self-nonself complex
2
3
4
T cell receptor
Helper T cell
Binding site for the self protein
Antigen-presenting cell
Signaling molecules stimulate the helper cell
Binding site for the antigen

▲ **Figure 24.12A** The activation of a helper T cell

? **How can one helper T cell stimulate both humoral and cell-mediated immunity?**

● By releasing stimulatory proteins that activate both cytotoxic T cells and B cells

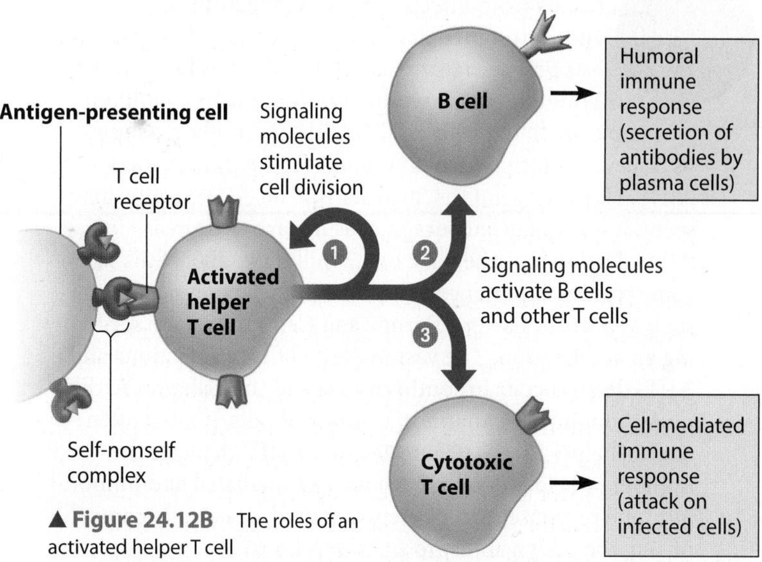

Antigen-presenting cell
Signaling molecules stimulate cell division
T cell receptor
B cell
Humoral immune response (secretion of antibodies by plasma cells)
1
2
3
Activated helper T cell
Signaling molecules activate B cells and other T cells
Self-nonself complex
Cytotoxic T cell
Cell-mediated immune response (attack on infected cells)

▲ **Figure 24.12B** The roles of an activated helper T cell

24.13 Cytotoxic T cells destroy infected body cells

Helper T cells activate cytotoxic T cells, the only T cells that actually kill infected cells. Once activated, clonal selection ensues, and effector cytotoxic T cells identify infected cells through a self-nonself complex. An infected cell has foreign antigens—molecules belonging to the viruses or bacteria infecting it—attached to self proteins on its surface (**Figure 24.13**). Whereas helper T cells recognize the self-nonself complex on the surface of antigen-presenting cells, cytotoxic T cells recognize antigens that are displayed in a self-nonself complex on infected body cells.

The self-nonself complex on an infected body cell is like a red flag to cytotoxic T cells that have matching receptors. As shown in the figure, ❶ a cytotoxic T cell binds to the infected cell. The binding activates the T cell, which then synthesizes several toxic proteins that act on the bound cell, including one called perforin ( ). ❷ Perforin is discharged from the cytotoxic T cell and attaches to the infected cell's plasma membrane, making holes in it. T cell enzymes ( ) then enter the infected cell and promote its death by a process called apoptosis (see Module 27.13). ❸ The infected cell is destroyed, and the cytotoxic T cell may move on to destroy other cells infected with the same pathogen.

Cytotoxic T cells also play a role in protecting the body against the spread of some cancers. About 20% of human cancers worldwide are caused by viruses. Besides human papillomavirus virus (HPV), discussed in Module 24.11, other examples include the hepatitis B virus, which can trigger liver cancer, and Epstein-Barr virus, which can cause lymphomas, cancer of the lymphocytes. When a human cancer cell harbors such a virus, viral proteins may end up on the surface of the infected cell. If they do, they may be recognized by a cytotoxic T cell, which can then destroy the infected cell, halting the proliferation of that cancerous cell.

? Compare and contrast the T cell receptor with the antigen receptor on the surface of a B cell.

● Both receptors bind to a specific antigen. The B cell receptor recognizes a free-floating antigen. The T cell receptor only recognizes an antigen when it is presented along with a "self" marker on the surface of one of the body's own cells.

❶ A cytotoxic T cell binds to an infected cell.

Self-nonself complex

Infected cell

Foreign antigen

Perforin molecule

Cytotoxic T cell

❷ Perforin makes holes in the infected cell's membrane, and enzymes that promote apoptosis enter.

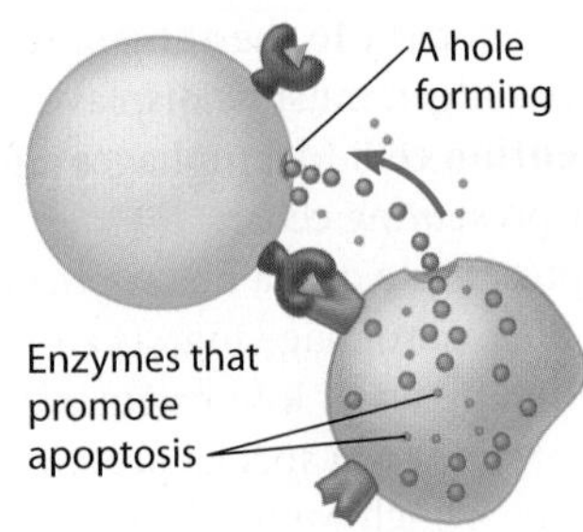
A hole forming
Enzymes that promote apoptosis

❸ The infected cell is destroyed (lysed).

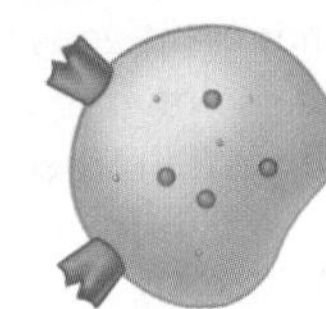

▲ **Figure 24.13** How a cytotoxic T cell kills an infected cell

24.14 HIV destroys helper T cells, compromising the body's defenses

CONNECTION

AIDS (acquired immunodeficiency syndrome) results from infection by **HIV**, the **human immunodeficiency virus**. Since the AIDS epidemic was first recognized in 1981, AIDS has killed nearly 30 million people worldwide. More than 34 million people are living with HIV. In 2010, 2.7 million people were newly infected with HIV, and over 1.8 million died of AIDS. However grim the 2010 statistic on AIDS deaths may seem, it is actually encouraging, as it marks a decrease in AIDS deaths since the peak of 2.2 million in 2005. Still, as some regions experience a decrease in the epidemic, others such as parts of Eastern Europe and Central Asia are seeing an acceleration. The vast majority of HIV infections and AIDS deaths occur in southern Asia and sub-Saharan Africa.

Although HIV can infect a variety of cells, it most often attacks helper T cells (**Figure 24.14**). As HIV depletes the number of helper T cells, both the cell-mediated and humoral immune responses are severely impaired, drastically compromising the body's ability to fight infections.

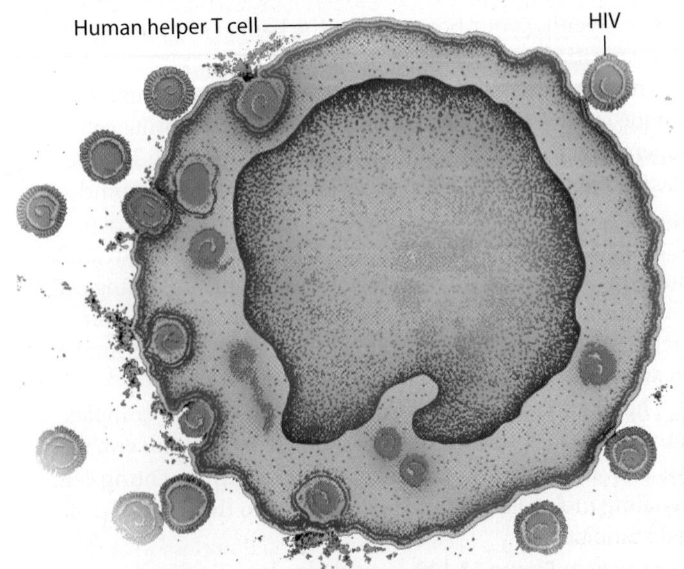
Human helper T cell
HIV
Colorized TEM 7,000×

▲ **Figure 24.14** A human helper T cell under attack by HIV

How does HIV destroy helper T cells? Transmission of HIV requires the transfer of the virus from person to person via body fluids such as semen, blood, or breast milk. Once HIV is in the bloodstream, proteins on the surface of the virus can bind to proteins on the surface of a helper T cell. Attached to the T cell, HIV may enter and begin to reproduce. Inside the host helper T cell, the RNA genome of HIV is reverse transcribed, and the newly produced DNA is integrated into the T cell's genome (see Module 10.20). This viral genome can now direct the production of new viruses from inside the T cell, generating up to 1,000 or more per day. Eventually, the host helper T cell dies from the damaging effects of virus production or from virus-triggered apoptosis.

After copies of the virus are released into the bloodstream, the HIV circulates, infecting and killing other helper T cells. As the number of T cells decreases, the body's ability to fight even the mildest infection is hampered, and AIDS eventually develops. It may take 10 years or more for full-blown AIDS symptoms to appear after the initial HIV infection.

Immune system impairment makes AIDS patients susceptible to cancers and **opportunistic infections**, which are infections that otherwise (that is, without immunodeficiency) would be fought off by a person with a healthy immune system. Infection by a common fungus called *Pneumocystis carinii* rarely occurs among healthy individuals. In a person with AIDS, however, infection by *P. carinii* can cause severe pneumonia and death. Kaposi's sarcoma, a very rare skin cancer, used to be seen exclusively among the elderly or patients receiving chemotherapy. It is now most frequently seen among AIDS patients. In fact, it was an unusual cluster of Kaposi's sarcoma and pneumocystis pneumonia cases that brought AIDS to the attention of the medical community.

AIDS is currently incurable, although drugs can slow HIV reproduction and the progress of AIDS. Because of worldwide efforts for universal access to HIV/AIDS medicines, regions in sub-Saharan Africa now have access to the latest treatment, including combination drug therapy. Other regions, however, particularly in the Middle East and North Africa, lag behind in terms of equal access to these drugs. Even if accessible, the drugs are often expensive, and multidrug regimens are complicated and may have debilitating side effects. Furthermore, interruption of treatment often results in rapid decline of the patient's health.

Drugs, vaccines, and education are areas of focus for prevention of HIV infection. Drug development has led to a drastic reduction in transmission rates of HIV from mother to child, and, in 2012, the FDA approved the first HIV prevention pill, for people who have a high risk of infection. Despite decades of effort and billions of dollars spent, an AIDS vaccine remains elusive. A hint of success emerged in 2009 when researchers announced that the combination of two vaccines—neither of which is effective alone—appeared to offer some protection from HIV infection. Currently, the most effective prevention is education. People learn to avoid direct contact with blood (especially through shared intravenous drug needles). Infected mothers are taught precautions to keep from transmitting the disease to their babies. Sexually active individuals are counseled to reduce promiscuity and use latex condoms. Safe sex alone could save millions of lives.

? **Why does the depletion of helper T cells severely impair adaptive immunity?**

Without many helper T cells to activate B cells (humoral immune response) and T cells (cell-mediated immune response), the two arms to adaptive immunity are lacking.

24.15 The rapid evolution of HIV complicates AIDS treatment

EVOLUTION CONNECTION

As HIV reproduces, mutational changes occur that can generate new strains of the virus. In fact, the virus mutates at a very high rate during replication because reverse transcriptase does not have an editing function to correct mistakes as DNA polymerase does. Some of the mutated viruses will be less susceptible to destruction by the immune system. Such viruses will survive, proliferate, and mutate further. The virus thus evolves within the host body.

At one time, there was great hope that a "cocktail" of three anti-AIDS drugs (**Figure 24.15**), each of which attacks a different part of the HIV life cycle, could eliminate the virus in an infected person. The idea was that a virus strain resistant to one drug would be defeated by another drug. Such hope greatly underestimated the evolution of HIV. People with access to drug cocktails do survive much longer and have a greatly improved quality of life, but some HIV strains have evolved that are resistant even to multidrug regimens. Thus, existing HIV treatments prolong the lives of HIV-positive people but are not a cure for AIDS.

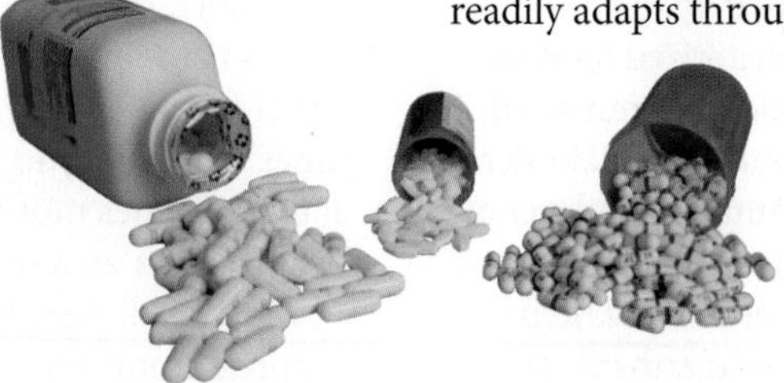

▲ **Figure 24.15** A "cocktail" of three separate drugs, the current treatment for people living with HIV

Disturbingly, drug-resistant HIV strains are being found in newly infected patients, which demonstrates that HIV readily adapts through natural selection to a changing environment—one where drug treatments are widely present. In other words, the continual use of anti-AIDS drugs has led to the spread of drug-resistant HIV strains. To reduce the use of these drugs and thus slow the evolution of drug-resistance, scientists are considering whether newly infected patients require immediate treatment with anti-HIV drugs or if it is safe to wait until symptoms develop to prescribe the drugs to these patients. The battle continues, with medical science on one side and the constantly evolving HIV on the other.

? **Why is it difficult to develop an AIDS vaccine?**

Because HIV evolves rapidly

24.16 The immune system depends on our molecular fingerprints

The ability of lymphocytes to recognize the body's own molecules—to distinguish self from nonself—enables our adaptive immune response to battle foreign invaders without harming healthy cells. Each person's cells have a unique collection of self proteins on the surface that provide molecular "fingerprints" recognized by the immune system. Lymphocytes develop to detect a myriad of antigens, including self antigens. As lymphocytes mature in the thymus and bone marrow, they are exposed to self antigens, and lymphocytes with receptors that bind the body's own molecules are selectively destroyed or deactivated, leaving only those that react to foreign molecules. As a result, lymphocytes in our mature immune system do not attack our own cells or molecules.

The immune system not only distinguishes body cells from microbial cells but also can tell the body's own cells from those of other people. Genes at multiple chromosomal loci code for **major histocompatibility complex (MHC) molecules**, the main self proteins. (The green self proteins shown in Figures 24.12 and 24.13 are encoded by MHC genes.) Because there are hundreds of alleles in the human population for each MHC gene, it is extremely rare for any two people (except identical twins) to have completely matching sets of MHC self proteins.

The immune system's ability to recognize foreign antigens does not always work in our favor. When a person receives an organ transplant or tissue graft, the person's T cells recognize the MHC markers on the donor's cells as foreign. Cell-mediated responses ensue, ending in the transplanted cells being destroyed by cytoxic T cells. To minimize rejection, doctors seek a donor with self proteins matching the recipient's as closely as possible. The best match is to transplant the patient's own tissue, as when a burn victim receives skin grafts removed from other parts of his or her body. Otherwise, identical twins provide the closest match, followed by nonidentical siblings. Sometimes doctors use drugs to suppress the immune response against the transplant. Unfortunately, these drugs may also reduce the ability to fight infections and cancer.

? **In what sense is a cell's set of MHC surface markers analogous to a fingerprint?**

● The set of MHC ("self") markers is unique to each individual.

▷ Disorders of the Immune System

24.17 Immune system disorders result from self-directed or underactive responses

CONNECTION

Our immune system is highly effective, protecting us against most potentially harmful invaders and returning to homeostasis after doing so. But sometimes the immune system malfunctions, resulting in immune disorders.

Autoimmune disorders result when the immune system turns against some of the body's own molecules. In the autoimmune disorder systemic lupus erythematosus (lupus), B cells produce antibodies against a wide range of self molecules, such as histones and DNA released by the normal breakdown of body cells. Lupus is characterized by skin rashes, fever, arthritis, and kidney malfunction. Rheumatoid arthritis is another antibody-mediated autoimmune disorder, in which the immune system attacks synovium, a thin layer of tissue that lines joints. Symptoms of rheumatoid arthritis are damage to and painful inflammation of the cartilage and bone of joints (**Figure 24.17**). In type 1 (insulin-dependent) diabetes mellitus (see Module 26.8), the insulin-producing cells of the pancreas are attacked by cytotoxic T cells. In multiple sclerosis (MS), T cells react against the myelin sheath that surrounds parts of many neurons (see Figure 28.2), causing progressive muscle paralysis. Recent research suggests that Crohn's disease, a chronic inflammation of the digestive tract, may be caused by an autoimmune reaction against normal microbiota of the intestinal tract. Scientists are in the process of identifying the species in a "healthy" intestinal tract so that they can better understand Crohn's disease.

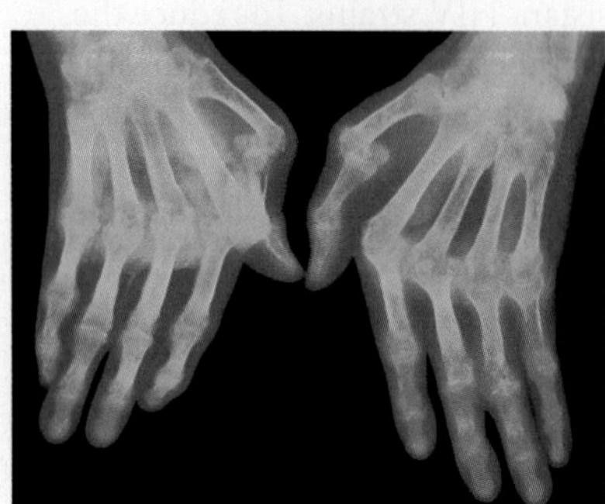

▲ **Figure 24.17** An X-ray image of hands affected by rheumatoid arthritis

Gender, genetics, and environment all influence susceptibility to autoimmune disorders. Many autoimmune disorders afflict females more than males; women are two to three times more likely to suffer from MS and rheumatoid arthritis and nine times more likely to develop lupus. The cause of this sex bias is an area of active research and debate.

Most medicines for treating autoimmune disorders either suppress immunity in general or alleviate specific symptoms. However, as scientists learn more about these disorders and the normal operation of the immune system, they hope to develop more effective therapies.

In contrast to autoimmune disorders, **immunodeficiency disorders** are underreactions of the immune system, in which an immune response is either defective or absent. People born immunodeficient are thus susceptible to frequent and recurrent infections. In the rare congenital disease severe combined immunodeficiency (SCID), both T cells and B cells are absent or inactive. People with SCID are extremely vulnerable to even minor infections. Until recently, their only hope for survival was to live within sterile plastic "bubbles" in their homes or to receive a successful bone marrow transplant that would supply functional lymphocytes. Researchers have been testing a gene therapy for this disease, with some success (see Module 12.10).

Immunodeficiency is not always an inborn condition; it may be acquired later in life. In addition to AIDS, another example of an acquired immunodeficiency is Hodgkin's disease, a type of cancer that damages the lymphatic system and can depress the immune system. Radiation therapy and the drug treatments used against many cancers can also cause immunodeficiency.

Evidence suggests physical and emotional stress affect immunity. For example, moderate exercise and a minimum of eight hours of sleep a night are factors that improve immune function by decreasing susceptibility to infections. In contrast, psychological stress lowers immune function by altering the interplay of hormones, nervous system signals, and the immune system.

? **What is a probable side effect of autoimmune disease treatments that suppress the immune system?**

● Lowered resistance to infections

24.18 Allergies are overreactions to certain environmental antigens

CONNECTION

Allergies are hypersensitive (exaggerated) responses to otherwise harmless antigens in our surroundings. Antigens that cause allergies are called **allergens**. Common allergens include protein molecules on pollen grains and on the feces of tiny mites that live in house dust. Many people who are allergic to cats and dogs are actually allergic to proteins in the animals' saliva that deposits on the fur when the animals lick themselves. Allergic reactions typically occur very rapidly and in response to tiny amounts of an allergen and can occur in many parts of the body, including the nasal passages, bronchi, and skin. Symptoms may include sneezing, runny nose, coughing, wheezing, and itching.

Symptoms of an allergy result from a two-stage reaction sequence outlined in **Figure 24.18**. The first stage, called sensitization, occurs when a person is exposed to an allergen—pollen, for example. ❶ After an allergen enters the bloodstream, it binds to effector B cells (plasma cells) with receptors specific to the allergen. ❷ The B cells then proliferate through clonal selection and secrete large amounts of antibodies to this allergen. ❸ Some of these antibodies attach to the surfaces of mast cells that produce histamine and other chemicals, which trigger the inflammatory response (Module 24.2).

The second stage of an allergic response begins when the person is later exposed to the same allergen. The allergen enters the body and ❹ binds to the antibodies attached to mast cells. ❺ This causes the mast cells to release histamine, which causes blood vessels to dilate and leak fluid, leading to nasal irritation, itchy skin, and tears. **Antihistamines** are drugs that interfere with histamine's action and give temporary relief from an allergy.

Allergies range from seasonal nuisances to severe, life-threatening responses. Anaphylactic shock is a dangerous allergic reaction. It may occur in people who are extremely sensitive to certain allergens, such as bee venom, penicillin, peanuts, or shellfish. Any contact with these allergens causes mast cells to release inflammatory chemicals very suddenly. As a result, blood vessels dilate abruptly, causing a rapid, potentially fatal drop in blood pressure, a condition called shock. Fortunately, anaphylactic shock can be counteracted with injections of the hormone epinephrine. People with severe allergies often carry an epinephrine autoinjector, such as the EpiPen.

? **Autoimmune disorders and allergies are both faulty responses of the immune system. What makes them different from each other?**

● A major difference is whether self molecules (autoimmune) or harmless antigens (allergies) elicit the response.

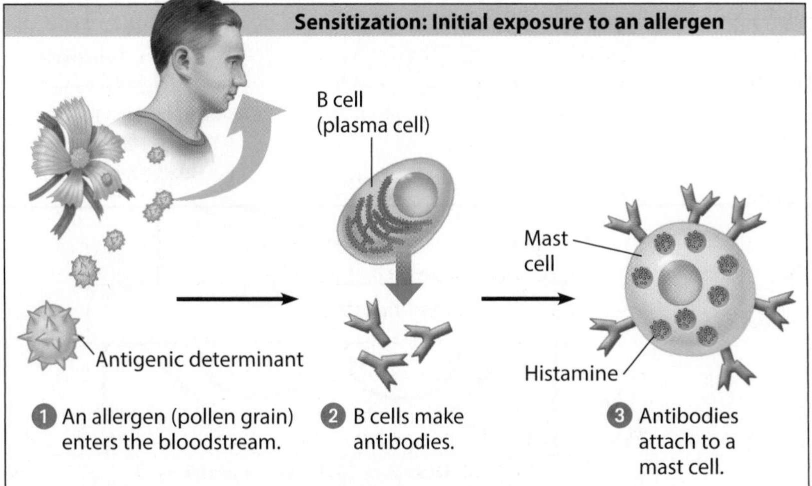
Sensitization: Initial exposure to an allergen
B cell
(plasma cell)
Mast
cell
Antigenic determinant
Histamine
❶ An allergen (pollen grain) enters the bloodstream.
❷ B cells make antibodies.
❸ Antibodies attach to a mast cell.

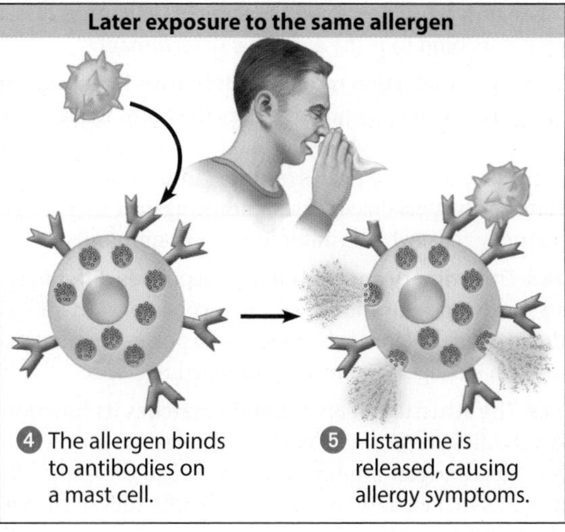
Later exposure to the same allergen
❹ The allergen binds to antibodies on a mast cell.
❺ Histamine is released, causing allergy symptoms.

▲ **Figure 24.18** The two stages of an allergic reaction

CHAPTER 24 REVIEW

For practice quizzes, BioFlix animations, MP3 tutorials, video tutors and more study tools designed for this textbook, go to MasteringBiology®

Reviewing the Concepts

Innate Immunity (24.1–24.2)

24.1 All animals have innate immunity. Innate defenses include barriers, phagocytic cells, and antimicrobial proteins.

24.2 Inflammation mobilizes the innate immune response. Tissue damage triggers the inflammatory response, which can disinfect tissues and limit further infection.

Adaptive Immunity (24.3–24.16)

24.3 The adaptive immune response counters specific invaders. Infections or vaccinations trigger adaptive immunity.

24.4 The lymphatic system becomes a crucial battleground during infection. Lymphatic vessels collect fluid from body tissues and return it as lymph to the blood. Lymph organs are packed with white blood cells that fight infections.

24.5 Lymphocytes mount a dual defense. Millions of kinds of B cells and T cells, each with different membrane receptors, wait in the lymphatic system, where they may respond to invaders.

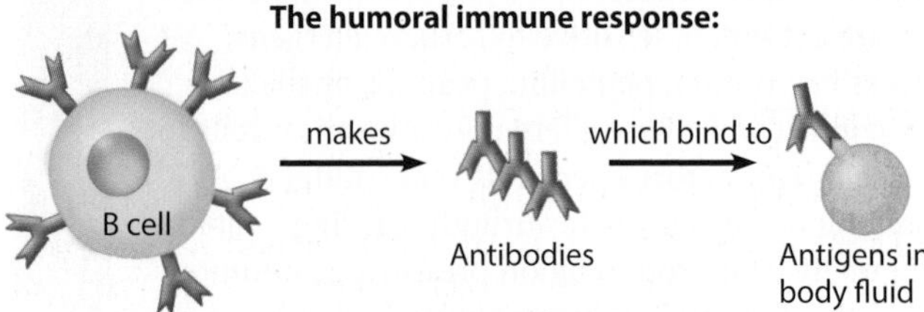
The humoral immune response:
B cell
makes
Antibodies
which bind to
Antigens in body fluid

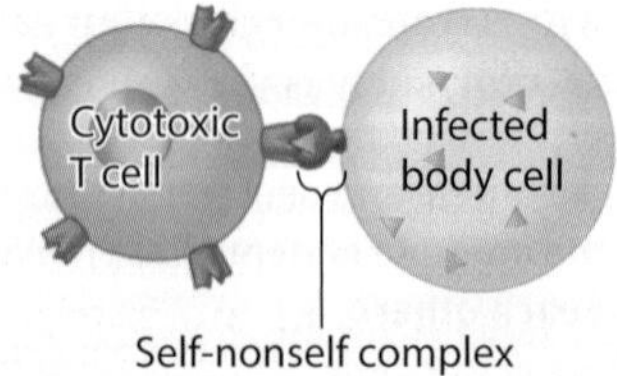
The cell-mediated immune response:
Cytotoxic T cell
Infected body cell
Self-nonself complex

24.6 Antigen receptors and antibodies bind to specific regions on an antigen. The site on the antigen that antibodies and antigen receptors bind to is the antigenic determinant.

24.7 Clonal selection mobilizes defensive forces against specific antigens. When an antigen enters the body, it activates only a small subset of lymphocytes that have receptors specific for the antigen. The selected cells multiply into clones of short-lived effector cells specialized for defending against that antigen and into memory cells, which confer long-term immunity.

24.8 The primary and secondary responses differ in speed, strength, and duration. The first exposure to an antigen results in the primary response. In a second exposure, memory cells initiate a faster, stronger, and more prolonged response.

24.9 The structure of an antibody matches its function. Antibodies are secreted by effector (plasma) B cells. An antibody has antigen-binding sites specific to the antigenic determinants that elicited its secretion. Antibodies mark antigens for elimination.

24.10 Antibodies are powerful tools in the lab and clinic.

24.11 Scientists measure antibody levels to look for waning immunity after HPV vaccination.

24.12 Helper T cells stimulate the humoral and cell-mediated immune responses. An antigen-presenting cell displays a foreign antigen (a nonself molecule) and one of the body's own self proteins to a helper T cell. The helper T cell's receptors recognize the self-nonself complexes, and the interaction activates the helper T cell. In turn, the helper T cell can activate cytotoxic T cells of the cell-mediated response and B cells of the humoral response.

24.13 Cytotoxic T cells destroy infected body cells.

24.14 HIV destroys helper T cells, compromising the body's defenses.

24.15 The rapid evolution of HIV complicates AIDS treatment.

24.16 The immune system depends on our molecular fingerprints. Each person's cells have a unique collection of self proteins on the surface.

Disorders of the Immune System (24.17–24.18)

24.17 Immune system disorders result from self-directed or underactive responses. In autoimmune diseases, the immune system targets self molecules. In immunodeficiency disorders, immune components are lacking and frequent infections occur.

24.18 Allergies are overreactions to certain environmental antigens.

Connecting the Concepts

1. Complete this concept map to summarize the key concepts concerning the body's defenses.

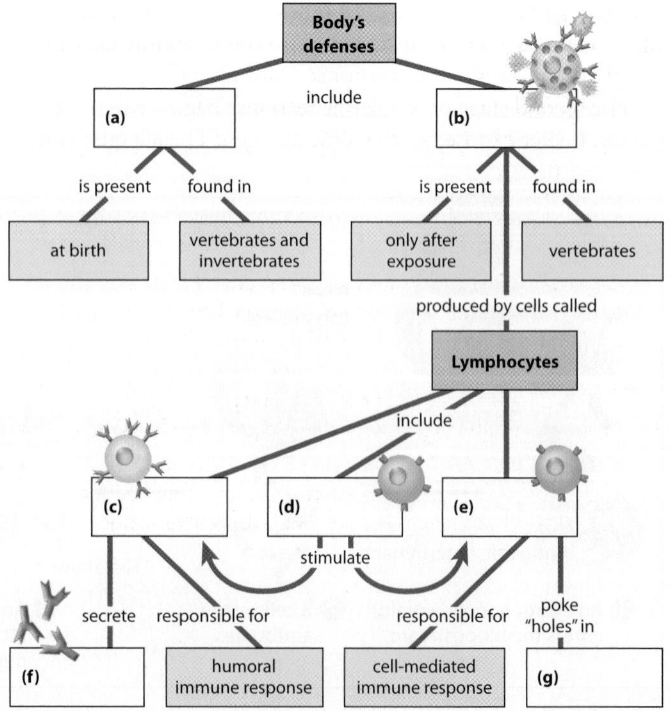
Body's defenses
include
(a)
(b)
is present
found in
is present
found in
at birth
vertebrates and invertebrates
only after exposure
vertebrates
produced by cells called
Lymphocytes
include
(c)
(d)
(e)
stimulate
secrete
responsible for
responsible for
poke "holes" in
(f)
humoral immune response
cell-mediated immune response
(g)

Testing Your Knowledge

Level 1: Knowledge/Comprehension

2. Foreign molecules that elicit an immune response are called
   a. major histocompatibility complex (MHC) molecules.
   b. antibodies.
   c. histamines.
   d. antigens
3. Which of the following is *not* part of the vertebrate innate immunity defense?
   a. macrophages
   b. antibodies
   c. complement system
   d. inflammation
4. Which of the following best describes the difference in the way B cells and cytotoxic T cells deal with invaders?
   a. B cells confer active immunity; T cells confer passive immunity.
   b. B cells send out antibodies to attack; certain T cells can do the attacking themselves.
   c. T cells handle the primary immune response; B cells handle the secondary response.
   d. B cells are responsible for the cell-mediated immune response; T cells are responsible for the humoral immune response.
5. Cytotoxic T cells are able to recognize infected body cells because
   a. the infected cells display foreign antigens.
   b. the infected cells produce antigens.
   c. infected cells release antibodies into the blood.
   d. helper T cells destroy them first.
6. Describe how HIV is transmitted and how immune system cells in an infected person are affected by HIV. What are the most effective means of preventing HIV transmission? Why is AIDS particularly deadly compared to other viral diseases?
7. What is inflammation? How does it protect the body? Why is inflammation considered part of the innate immune response?

Level 2: Application/Analysis

8. In the condition myasthenia gravis, antibodies bind to and block certain receptors on muscle cells, preventing muscle contraction. This condition is best classified as an
   a. immunodeficiency disorder.
   b. exaggerated immune reaction.
   c. allergic reaction.
   d. autoimmune disorder.
9. Which of the following statements is not true?
   a. An antibody has more than one antigen-binding site.
   b. An antigen can have different antigenic determinants.
   c. A lymphocyte has receptors for multiple and different antigens.
   d. A bacterium has more than one antigen.
10. Propose an explanation for why we need a flu shot year after year, instead of only once early in life.
11. **SCIENTIFIC THINKING** Pertussis (whooping cough) is caused by a bacterial infection, and symptoms include an intense cough that lasts for weeks, sometimes leading to pneumonia and death. A complete vaccination against pertussis requires five doses, completed by age 5. A pertussis outbreak occurred in California in 2010, and a news article reported the following: "Among fully immunized kids, there were about 36 cases for every 10,000 children two to seven years old, compared to 245 out of every 10,000 kids aged eight to 12." Propose an explanation for these data and a public health solution based on your explanation.
12. Your roommate is rushed to the hospital after suffering a severe allergic reaction to a bee sting. After she is treated and released, she asks you (the local biology expert!) to explain what happened. She says, "I don't understand how this could have happened. I've been stung by bees before and didn't have a reaction." Suggest an explanation for what has happened to cause her severe allergic reaction and why she did not have the reaction after previous bee stings.

Level 3: Synthesis/Evaluation

13. Consider a pencil-shaped protein with two antigenic determinants, Y (the "eraser end") and Z (the "point end"). They are recognized by two different antibodies, A1 and A2, respectively. Draw and label a picture showing how the antibodies might link proteins into a complex that precipitates to enhance engulfment by phagocytic cells.
14. Organ donation saves many lives each year. Even though some transplanted organs are derived from living donors, the majority come from patients who die but still have healthy organs that can be of value to a transplant recipient. Potential organ donors can fill out an organ donation card to specify their wishes. If the donor is in critical condition and dying, the donor's family is usually consulted to discuss the donation process. Generally, the next of kin must approve before donation can occur, regardless of whether the patient has completed an organ donation card. In some cases, the donor's wishes are overridden by a family member. Do you think that family members should be able to overrule the stated intentions of the potential donor? Why or why not? Have you signed up to be an organ donor? Why or why not?
15. One of the key difficulties in the development of anti-HIV drugs is the fact that HIV will only infect humans. This precludes testing of drugs in animals and instead requires that drugs be tested on volunteer human subjects. The developing world (particularly sub-Saharan Africa and Southeast Asia) has the highest rates of HIV infection. Consequently, drug companies frequently conduct studies in these regions. Some people decry such tests, fearing that drug companies may profit hugely from the use of economically disadvantaged people. Others counter that such tests are the only way to find new and cheaper drugs that will ultimately help everyone. What do you think are the ethical issues surrounding trials of anti-HIV drugs in the developing world? Which side do you think has the more morally compelling argument?

Answers to all questions can be found in Appendix 4.

CHAPTER 25

Control of Body Temperature and Water Balance

How do emperor penguins maintain their warmth during the Antarctic winter?

During the Antarctic winter, temperatures drop as low as −50°C (−58°F), so it would seem like a good time for emperor penguins (*Aptenodytes forsteri*), seen in the photograph below, to migrate. But unlike all other birds of the Antarctic, who in winter fly to warmer regions to breed, emperor penguins stay put to mate and hatch their eggs. The female penguin produces a large egg just a few hours after mating and then swims out to sea to feed, while the male penguin remains on land to incubate the egg (held on top of his feet under a fold of abdominal skin). Because the land offers no food for penguins, males incubating eggs go approximately 110 days without nourishment.

Subsistence without food requires energy conservation—but maintaining a constant body temperature can be a large energy expense. The survival of the emperor penguin embryo inside its egg depends on a constant internal temperature; eggs not maintained above 35°C (95°F) perish. If you have ever snuggled with someone on a cold night to keep warm, you can appreciate the huddling

behavior that penguins employ to help maintain body temperature. Temperatures inside tight penguin huddles can become tropic-like, as high as 37.5°C (99.5°F). Later in the chapter, we'll see what scientists have learned about how the heat generated in the center of a penguin huddle is shared with each member of the colony.

In this chapter, we explore two kinds of homeostasis, starting with the one exemplified by penguin huddling: thermoregulation, the control of body temperature. We then examine how animals osmoregulate, or maintain fluid balance through the input and output of water and solutes. You'll see that, like the penguins, most animals can survive fluctuations in the external environment because of homeostatic control mechanisms that keep their internal temperature and water levels within optimal range.

BIG IDEAS

Thermoregulation
(25.1–25.3)

Animals use various homeostatic mechanisms to control body temperature.

Osmoregulation and Excretion
(25.4–25.10)

Animals regulate the concentration of water, solutes, and wastes into and out of the body.

▷ Thermoregulation

25.1 An animal's regulation of body temperature helps maintain homeostasis

Thermoregulation, a homeostatic mechanism by which animals maintain an internal temperature within an optimal range despite variation in external temperature, is critical to survival. Most of life's processes are sensitive to changes in body temperature. Thermoregulation helps keep body temperature within that range, enabling enzyme-mediated processes within cells to function effectively even when the external temperature fluctuates greatly.

Body heat can come from either internal metabolism or the external environment. Humans and other mammals, as well as birds, are **endothermic**, meaning that they are warmed mostly by heat generated by metabolism. In contrast, many reptiles and fishes and most invertebrates are **ectothermic**, meaning that they gain most of their heat from external sources. Endothermy and ectothermy are not mutually exclusive, however. For example, a bird is mainly endothermic, but it may warm itself in the sun on a cold morning, much as an ectothermic lizard does.

Heat flows from an entity of higher temperature to one of lower temperature. For this reason, an animal is always gaining or losing heat. Exchange of heat occurs in four ways (**Figure 25.1**). *Conduction* is the direct transfer of heat between entities that are in direct contact with each other. Heat conducted from the warm rock (red arrows) elevates the lizard's body temperature. *Radiation* is the emission of electromagnetic waves that can transfer heat between entities not in direct contact. Heat radiating from the sun warms a lizard's back, and the lizard radiates some of its own heat to the environment (yellow arrows). *Convection* is the transfer of heat by movement of air or liquid over a surface. In the figure, a breeze lifts heat from a lizard's tail (orange arrows). *Evaporation* is the loss of heat from the surface of a liquid that is losing some of its molecules as a gas. A lizard loses heat as moisture evaporates from is nostrils (blue arrow).

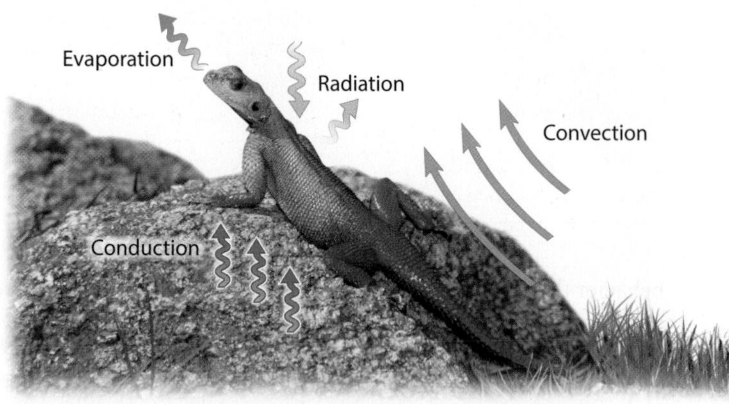
Evaporation
Radiation
Convection
Conduction

▲ **Figure 25.1** Mechanisms of heat exchange

If you are sweating on a hot day and turn a fan on yourself, what two mechanisms contribute to your cooling?

● Evaporation (of sweat) and convection (fan moving air)

25.2 Thermoregulation involves adaptations that balance heat gain and loss

Endotherms and many ectotherms maintain a fairly constant internal temperature within an optimal range despite external temperature fluctuations. The adaptations that help animals thermoregulate can be classified into five categories.

Metabolic Heat Production When cells perform cellular respiration, chemical energy is converted to ATP, energy the cell can use to perform work. Cellular respiration produces heat, too (see Module 5.10). The heat produced by all living cells, or metabolic heat, warms an animal and counteracts the heat it loses to the environment through conduction, radiation, convection, and evaporation. As cells do more work, the metabolic heat production increases. This is why you get warm when exercising in the cold. Some mammals, such as hibernating bears, increase their metabolic heat production in the cold without moving because their mitochondria uncouple the production of heat and ATP synthesis (see Module 6.11).

Insulation A major thermoregulatory adaptation in mammals and birds is insulation—hair (or fur), feathers, and fat—which reduces the radiation of heat from an animal to its environment. Most land mammals and birds react to cold by raising their fur or feathers, which traps a layer of air next to warm skin, improving insulation. (In humans, muscles raise hair in the cold, causing goose bumps, a vestige from our furry ancestors.) Aquatic mammals (such as seals) and aquatic birds (such as penguins) are insulated by a thick layer of fat called blubber.

Circulatory Adaptations Have you ever wondered why elephants have such big ears? Large, thin ears containing many blood vessels capable of radiating a lot of heat have evolved in elephants, helping to cool their large bodies in warm climates (**Figure 25.2A**). Heat loss can be altered by a change in the amount of blood flowing (circulation) to the skin. In a bird or mammal (and some ectotherms), nerve signals cause surface blood vessels to constrict (narrow) or dilate (open), depending on the external temperature (see Module 20.15).

▲ **Figure 25.2A** Heat dissipation via radiation (blood vessel dilation) and convection (ear flapping)

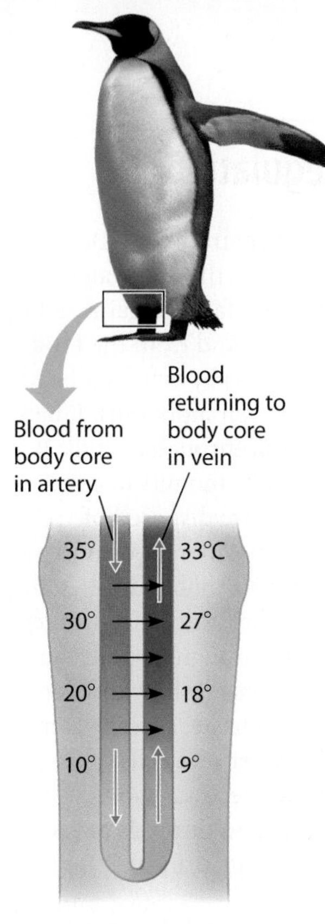

Blood from body core in artery
Blood returning to body core in vein
35°
33°C
30°
27°
20°
18°
10°
9°

▲ **Figure 25.2B** Countercurrent heat exchange

Try This Explain why in countercurrent exchange heat is transferred from arteries to veins.

When the surface vessels are constricted, less blood flows from the warm body core to the body surface, reducing the rate of heat loss through radiation. Conversely, dilated surface blood vessels increase the rate of heat loss.

Figure 25.2B illustrates a circulatory adaptation found in many birds and mammals. In **countercurrent heat exchange**, warm and cold blood flow in opposite (countercurrent) directions in two adjacent blood vessels. Warm blood (red) from the body core cools as it flows down legs. But the arteries carrying the warm blood are in close contact with veins conveying cool blood (blue) back toward the body core. As shown in the figure by the black arrows, heat passes from the warmer blood to the cooler blood along the whole length of these side-by-side vessels (because heat always flows from a warmer region to a cooler one). Blood leaving the legs and returning to the body is warmed, helping to maintain the core body temperature. Simultaneously, blood moving toward the feet is cooled to a temperature that reduces the heat differential between the blood and environment.

Evaporative Cooling Many animals live in places where thermoregulation requires cooling as well as warming. Evaporative cooling occurs when water absorbs heat from the body surface, and as the water evaporates, the vapor it produces takes large amounts of body heat away. Some animals have adaptations that greatly increase their ability to lose heat this way. Elephants spray water over their bodies to aid evaporation; the sprayed water functions similarly to sweat on human skin. In other animals (such as dogs), evaporative cooling is increased as moisture evaporates during panting.

Behavioral Responses All animals control body temperature by adjusting their behavior in response to the environment. Some birds and butterflies migrate seasonally to more suitable climates. Other animals, such as desert lizards, warm themselves in the sun through radiation when it is cold and find cool, damp areas or burrows when it is hot. Many animals bathe (or, in the case of elephants, spray themselves), which brings immediate cooling by convection and continues to cool for some time by evaporation. The large ears of elephants can be flapped to increase heat dissipation by convection. We humans dress for warmth.

? **Compare the countercurrent exchange of heat in animals with the countercurrent exchange of oxygen in fish gills (see Module 22.3).**

● In both cases, countercurrent exchange enhances transfer all along the length of a blood vessel—transfer of heat from one vessel to another in the case of heat exchanger and transfer of oxygen between water and vessels in the case of gills.

25.3 Coordinated waves of movement in huddles help penguins thermoregulate

SCIENTIFIC THINKING

In the chapter introduction, we learned that emperor penguins huddle during the Antarctic winter. Metabolic heat generated by individual penguins is easily lost through radiation and convection in the cold, but the huddling behavior promotes conduction that counteracts this heat loss. As the huddle mass increases, the relative surface area of it decreases. The center of a huddle is warmest, and the periphery, with a lot of exposed surface area, is the coldest. Scientists have wondered if each penguin has access to the center. One hypothesis is that huddles are static—a penguin on the periphery of a huddle remains there. An alternative hypothesis is that huddles are changing—a penguin on the periphery can eventually find itself in the warm center of a huddle. Experiments aren't always used to answer scientific questions—these two hypotheses were tested with observations.

A German study in 2011 using time-lapse photography and tracking of individual birds supports the hypothesis that huddles rearrange. Scientists photographed huddles in a colony of emperor penguins every 1.3 seconds. As seen in the chapter opener photo, all birds in a single huddle faced the same direction. Data showed that as penguins continuously left the huddle at the front edge, more joined at the back edge. Smaller huddles united with other huddles to produce larger huddles. Additionally, all penguins in a huddle took small steps forward in a coordinated fashion, much like the way the "wave" moves through a crowd of fans at a football stadium. The wave packed the penguins closer together, thereby conducting heat more and more efficiently.

How do emperor penguins maintain their warmth during the Antarctic winter?

The study not only suggests that all emperor penguins get a turn in the inside of a huddle, but also provides insight into a collective behavior of the penguins. As is often the case in science, answers to questions in one study lead to new questions: Might a single penguin initiate a huddle wave? Are the coordinated movements in huddles of penguins similar to those in flocks of pigeons or schools of fish?

? **A penguin that joins a huddle reduces heat loss from __________ and radiation and benefits from the __________ of heat between penguins.**

● convection . . . conduction

▷ Osmoregulation and Excretion

25.4 Animals balance their levels of water and solutes through osmoregulation

Through the process of **osmoregulation**, animals control the concentrations of solutes (dissolved substances such as salt, NaCl) in their cells and bodies and prevent excessive water uptake or loss. Osmosis is the passive diffusion of water across a selectively permeable membrane from a solution with lower solute concentration (and thus more water) to one with a higher solute concentration (see Module 5.4). When an animal cell is in an environment where the solute concentration is lower than that inside the cell (hypotonic), water diffuses into the cell, causing it to swell and potentially burst. When an animal cell is in an environment where the solute concentration is higher than that of the cell (hypertonic), water diffuses out of the cell, causing it to shrivel and possibly die. If solute concentration inside the cell is equal to that outside of it (isotonic), water will not be gained or lost from the cell. Animals regulate their body fluids to provide an isotonic environment for their cells.

Some sea-dwelling animals—such as squids, sea stars, and most other marine invertebrates—have body fluids with a solute concentration equal to that of seawater. Called **osmoconformers**, such animals do not undergo a net gain or loss of water from or to their environment and, therefore, face no substantial challenges in water balance. However, for osmoconformers to maintain homeostasis, their concentrations of certain specific solutes must be different from that of seawater, and those solutes must be actively transported.

Many animals—land animals (such as elephants), freshwater animals (such as perch), and marine vertebrates (such as cod)—have body fluids whose solute concentration differs from that of their environment. Therefore, they have mechanisms that actively regulate solute concentrations and water balance. Such animals are called **osmoregulators**.

Osmoregulation in Water The opposing challenges for marine and freshwater fish are illustrated in **Figure 25.4**. The marine fish lives in a dehydrating environment. Because its internal fluids are lower in total solutes than seawater, it loses water by osmosis across its body surfaces. It gains salt by diffusion from its saltwater environment and from the food it eats. The marine fish balances water loss by drinking large amounts of seawater (which brings in even more salt). It balances solutes by pumping out excess salt through its gills. It also saves water by producing only small amounts of concentrated urine, in which some excess ions are disposed of.

The freshwater fish in Figure 25.4 has different osmoregulatory challenges. Its internal fluids have a much higher solute concentration than fresh water does. Therefore, a freshwater fish constantly gains water by osmosis through its body surface, especially through its gills. It also loses salt by diffusion to its hypotonic environment. To compensate, a freshwater fish actively transports salt from the water into cells of its gills, and the food it eats supplies other necessary ions. A freshwater fish drinks almost no water and disposes of excess water by producing large amounts of dilute urine.

Osmoregulation on Land Terrestrial animals face the homeostatic challenge of dehydration. Therefore, adaptations that prevent dehydration are evolutionary advantages.

An animal conserves water because of external and internal adaptations. Insects have tough exoskeletons that contain a layer of waterproof wax, which helps conserve water. Most terrestrial vertebrates, including humans, have an outer skin formed of multiple layers of dead, water-resistant cells, which also minimizes surface water loss. Essential to survival on land are adaptations that protect developing embryos. The eggs of many birds, turtles, and other reptiles are surrounded by a tough, watertight shell. Mammals retain their developing embryos inside the mother's body. In addition to reproductive adaptations, the urinary and digestive systems of terrestrial animals have mechanisms that conserve water.

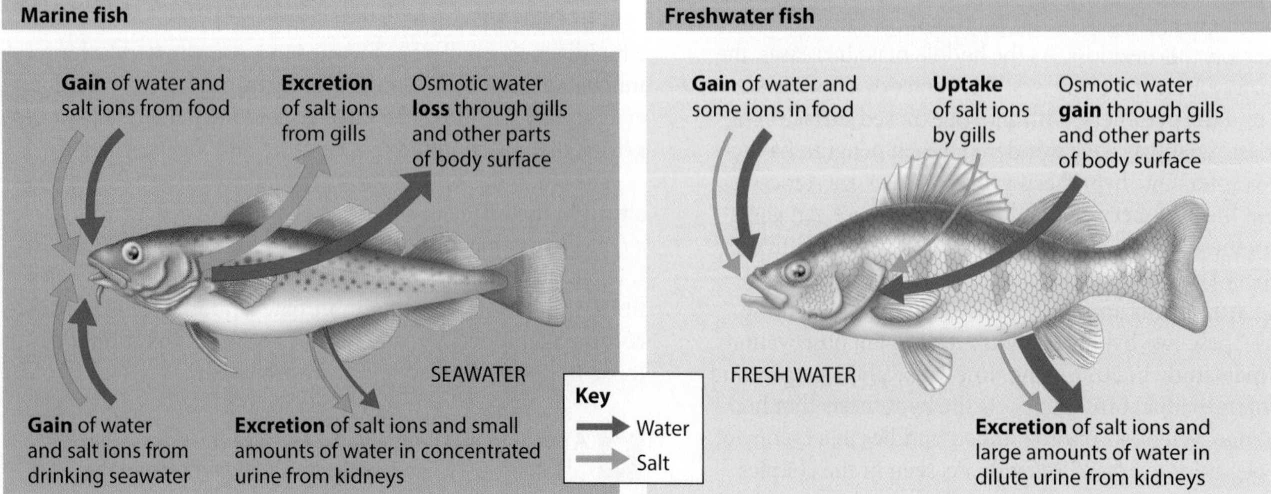

Marine fish
Gain of water and salt ions from food
Excretion of salt ions from gills
Osmotic water loss through gills and other parts of body surface
SEAWATER
Gain of water and salt ions from drinking seawater
Excretion of salt ions and small amounts of water in concentrated urine from kidneys
Key
Water
Salt
Freshwater fish
Gain of water and some ions in food
Uptake of salt ions by gills
Osmotic water gain through gills and other parts of body surface
FRESH WATER
Excretion of salt ions and large amounts of water in dilute urine from kidneys

▲ **Figure 25.4** Osmoregulation in marine and freshwater fish: a comparison

Despite such adaptations, most terrestrial animals still lose water across their skin and moist respiratory surfaces and in urine and feces. The water that is inevitably lost must be replaced. Land animals gain water from eating and drinking and through cellular respiration, of which water is a by-product (Module 6.3). Interestingly, the kangaroo rat doesn't drink water. It replaces 10% of its lost water from the seeds it eats and produces the rest by cellular respiration. In the next few modules, we explore how both the type of nitrogenous waste product and the structure and function of the urinary system contribute to osmoregulation.

? **Why are no freshwater animals osmoconformers?**

● Because fresh water is very low in solutes and ions, osmoconformers would also have very low solute and ion concentrations in their cells, which would not be compatible with the biological processes that support life.

25.5 Several ways to dispose of nitrogenous wastes have evolved in animals

EVOLUTION CONNECTION

Waste disposal is a crucial aspect of osmoregulation, because most metabolic wastes must be dissolved in water to be removed from the body. Metabolism produces a number of toxic by-products, such as the nitrogenous wastes that result from the breakdown of proteins and nucleic acids. Many animals dispose of metabolic wastes by converting them to chemicals that can be excreted through an opening in the body. The type of waste product produced and how the animal disposes of it depends on the animal's evolutionary history and its habitat.

As **Figure 25.5** indicates, most aquatic animals dispose of their nitrogenous wastes as **ammonia** (NH_3). Among the most toxic of all metabolic by-products, ammonia is formed when amino groups ($—NH_2$) are removed from amino acids and nucleic acids. Ammonia is too toxic to be stored in the body, but it is highly soluble and diffuses rapidly across cell membranes. If an animal is surrounded by water, ammonia readily diffuses out of its cells and body. Small, soft-bodied invertebrates, such as planarians (flatworms), excrete ammonia across their whole body surface. Fishes excrete it mainly across their gills.

Because it is so toxic, ammonia must be transported and excreted in large volumes of very dilute solutions. Most terrestrial animals and many marine species cannot afford to lose water in the amounts necessary to routinely excrete ammonia. It is more efficient for many animals to convert ammonia to less toxic compounds that can be safely stored in the body. The disadvantage of converting ammonia to less toxic compounds is that the animal must expend energy.

As shown in Figure 25.5, mammals, most adult amphibians, sharks, and some bony fishes excrete **urea** as the major waste product. Urea, a soluble form of nitrogenous waste, is produced in the vertebrate liver by a metabolic cycle that combines ammonia with carbon dioxide. The main advantage of urea for nitrogenous waste excretion is its very low toxicity. Some animals can switch between excreting ammonia and urea, depending on environmental conditions. Certain toads, for example, excrete ammonia (thus saving energy) when in water as tadpoles but excrete mainly urea (reducing water loss) when land-dwelling adults.

Insects, land snails, and many reptiles, including birds, convert ammonia to **uric acid** and avoid water loss almost completely. Unlike ammonia and urea, uric acid is water-insoluble, and thus water is not used to dilute it. Uric acid is relatively nontoxic nitrogenous waste; it can be safely transported and stored in the body and released periodically by the urinary system. In most cases, uric acid is excreted as a semisolid paste. (The white material in bird droppings is mostly uric acid.) An animal must expend more energy to excrete uric acid than to excrete urea, but the higher energy cost is balanced by the great savings in body water.

An animal's type of reproduction influences whether it excretes urea or uric acid. Urea can diffuse out of a shell-less amphibian egg or be carried away from a mammalian embryo in the mother's blood. However, the shelled eggs produced by birds and other reptiles are not permeable to liquids. The evolution of uric acid as a waste product therefore conveyed a selective advantage: Uric acid precipitates out of embryonic solutions and remains as a harmless solid within the egg until the animal hatches.

? **Aquatic turtles excrete both urea and ammonia; land turtles excrete mainly uric acid. What could account for this difference?**

● Although uric acid evolved in terrestrial reptiles with their shelled eggs, natural selection favored the energy savings of ammonia and urea for aquatic turtles.

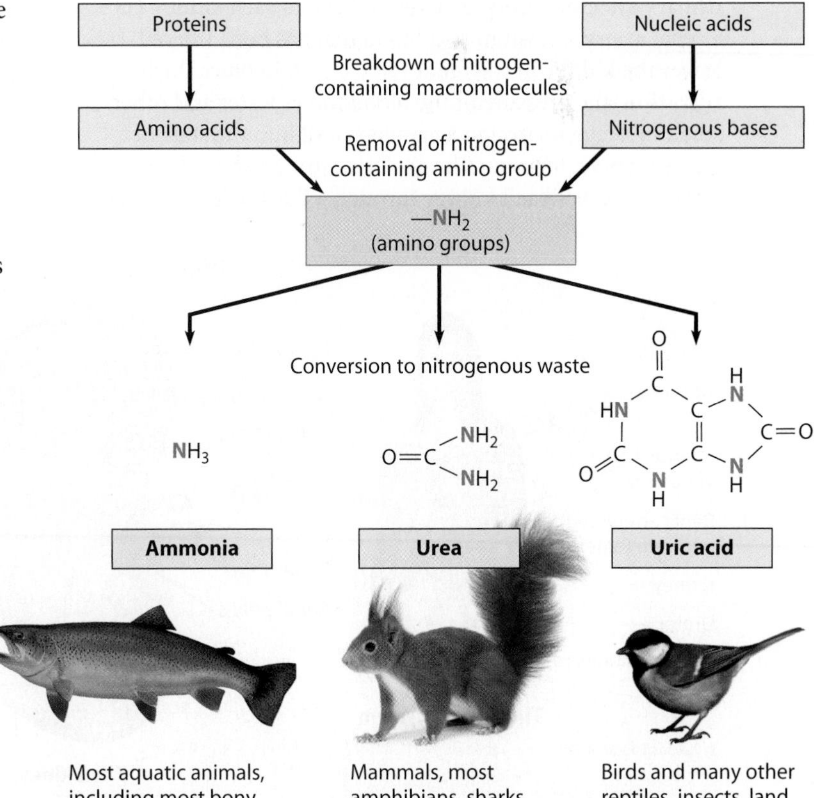

Proteins
Nucleic acids
Breakdown of nitrogen-containing macromolecules
Amino acids
Nitrogenous bases
Removal of nitrogen-containing amino group
—NH2 (amino groups)
Conversion to nitrogenous waste
NH3
O=C(NH2)NH2
Ammonia
Urea
Uric acid
Most aquatic animals, including most bony fishes
Mammals, most amphibians, sharks, some bony fishes
Birds and many other reptiles, insects, land snails

▲ **Figure 25.5** Nitrogen-containing metabolic waste products

25.6 The urinary system plays several major roles in homeostasis

Survival in any environment requires a precise balance between the competing demands of an animal's need for water and waste disposal. The urinary system plays a central role in homeostasis, forming and excreting urine while regulating the amount of water and solutes in body fluids.

In humans, the main processing centers of the urinary system are the two kidneys. Each is a compact organ a bit smaller than a fist, located on either side of the abdomen. About 80 km of small tubes, called tubules, fill the kidneys, providing a large surface area for exchange of solutes, water, and wastes. (If these tubules were laid end to end, it would take about 15 hours to walk their full length!) An intricate network of tiny blood capillaries is closely associated with the tubules. The human body contains only about 5 L of blood, but because this blood circulates repeatedly, about 1,100–2,000 L pass through the capillaries in our kidneys every day. From this enormous circulation of blood, our kidneys extract about 180 L of fluid, called **filtrate**, consisting of water, urea, and a number of valuable solutes, including glucose, amino acids, ions, and vitamins. Filtrate formation is not selective; thus, if we excreted all the filtrate, we would lose vital nutrients and dehydrate rapidly. Instead, our kidneys refine the filtrate, concentrating the urea and recycling most of the water and useful solutes to the blood. In a typical day, we excrete only about 1.5 L of **urine**, the refined filtrate containing wastes.

The left part of **Figure 25.6A** shows an overview of the urinary system. Blood to be filtered enters each kidney via a renal artery, shown in red; blood that has been filtered leaves the kidney in the renal vein, shown in blue. During **filtration**, the pressure of the blood forces water and other small molecules through a capillary wall into the start of a kidney tubule, forming filtrate. Urine, the final product of filtration, leaves each kidney through a duct called a **ureter**. Both ureters drain into the **urinary bladder**. During urination, urine is expelled from the bladder through a tube called the **urethra**, which empties to the outside near the vagina in females and through the penis in males. A sphincter, or ring of muscles, and the bladder control urination.

As shown in the center of Figure 25.6A, the kidney has two main regions, an outer **renal cortex** and an inner **renal medulla**. Each kidney contains about a million tiny functional units called **nephrons**, one of which is shown in the right part of the figure. Performing kidney's functions in miniature, the nephron extracts a tiny amount of filtrate from the blood and then processes the filtrate into a much smaller quantity of urine. A nephron consists of a single folded tubule and associated blood vessels. The intricate association between blood vessels and tubules is the key to nephron function and the refining of the filtrate. Note the placement of the nephron within the kidney. The blood-filtering end of the nephron is a cup-shaped swelling, called **Bowman's capsule**, in the kidney's cortex (the lighter shaded area). Some nephrons extend into the medulla (darker shaded region) and then loop back to the cortex, meeting the **collecting duct**, which carries urine through the medulla to the renal pelvis.

With the basic anatomy described, let's follow the path that the blood and filtrate take by examining a nephron in detail, along with its blood vessels (**Figure 25.6B**). (Note that the figure illustrates blood flow with small black arrows and filtrate flow with small green arrows.) Blood enters the nephron from a branch of the renal artery and flows into a ball of capillaries called the **glomerulus** (plural, *glomeruli*). The glomerulus and the surrounding Bowman's capsule make up the blood-filtering unit of the nephron. Here, blood pressure forces water and solutes from the blood in the glomerular capillaries across the wall of Bowman's capsule and into the nephron tubule. This process creates the filtrate, leaving

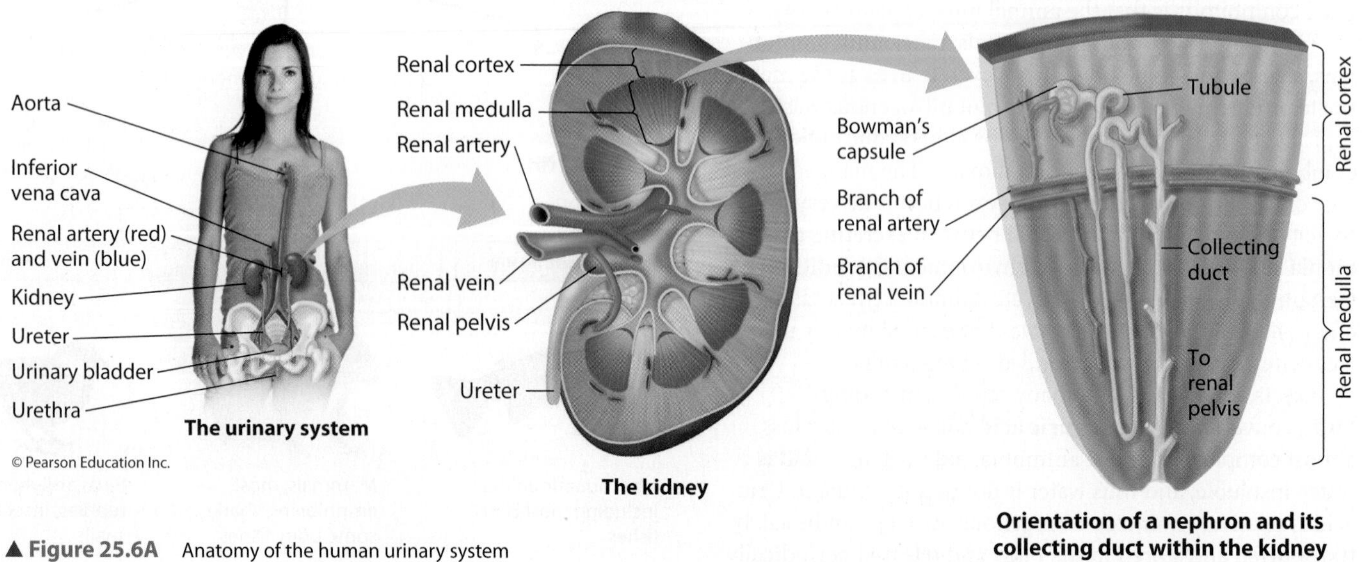

Aorta
Inferior vena cava
Renal artery (red) and vein (blue)
Kidney
Ureter
Urinary bladder
Urethra
The urinary system
Renal cortex
Renal medulla
Renal artery
Renal vein
Renal pelvis
Ureter
The kidney
Bowman's capsule
Branch of renal artery
Branch of renal vein
Tubule
Collecting duct
To renal pelvis
Renal cortex
Renal medulla
Orientation of a nephron and its collecting duct within the kidney

▲ **Figure 25.6A** Anatomy of the human urinary system

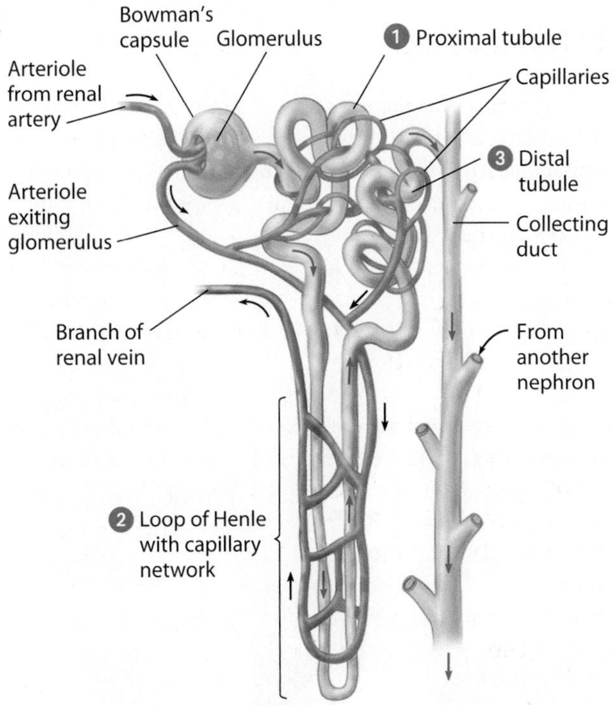
Bowman's capsule
Glomerulus
1 Proximal tubule
Capillaries
Arteriole from renal artery
Arteriole exiting glomerulus
3 Distal tubule
Collecting duct
Branch of renal vein
From another nephron
2 Loop of Henle with capillary network

▲ **Figure 25.6B** Detailed structure of a nephron

Try This Describe the flow of blood and filtrate through the nephron.

blood cells and large molecules such as plasma proteins behind in the capillaries.

The filtrate forced into Bowman's capsule flows into the nephron tubule, where it will be processed. The nephron tubule has three sections: ❶ the **proximal tubule**, ❷ the **loop of Henle**, a hairpin loop with a capillary network, and ❸ the **distal tubule** (called distal because it is the most distant from Bowman's capsule). The distal tubule drains into a collecting duct, which receives filtrate from many nephrons. From the kidney's many collecting ducts, the processed filtrate, or urine, passes into a chamber called the renal pelvis and then into the ureter, from which it is then expelled.

Filtered blood and large molecules that were left behind in glomerulus capillaries exit the structure in an arteriole. If you trace this arteriole with your finger, you will see that it branches into a network of capillaries surrounding the tubules of the nephron. The essential exchange of substances between the blood and filtrate occurs in this close association of capillaries and tubules. The flow of blood moves through capillaries as they converge to form a venule leading toward the renal vein.

The important association between a nephron tubule and a capillary is illustrated in **Figure 25.6C**. After filtration, the filtrate is further processed through the passive and active transport of substances across membranes. Much like how cans and bottles are recycled instead of disposed of with other waste, water and valuable solutes—including glucose, salt, other ions, and amino acids—are reclaimed from the filtrate by the cells making up the tubules and returned to the blood through the walls of capillaries; this process is called **reabsorption**. In contrast, substances are transported from the blood *into* the filtrate in the process called **secretion**. When there is an excess of H^+ in the blood, for example, these ions are secreted into the filtrate, thus keeping the blood from becoming acidic. Secretion also eliminates certain drugs and other toxic substances from the blood. In reabsorption and secretion, water and solutes move between the tubule and capillaries by passing through the interstitial fluid (see Module 23.7). As we'll see in the next module, reabsorption and secretion not only passively and actively transport substances across membranes each process also ensures that the initial blood filtrate is refined before exiting the body. Finally in **excretion**, urine—the waste-containing product of filtration, reabsorption, and secretion—passes from the kidneys to the outside via the ureters, urinary bladder, and urethra.

? Urine differs in composition from the fluid that enters a nephron tubule by filtration because of the processes of __________ and __________.

reabsorption . . . secretion

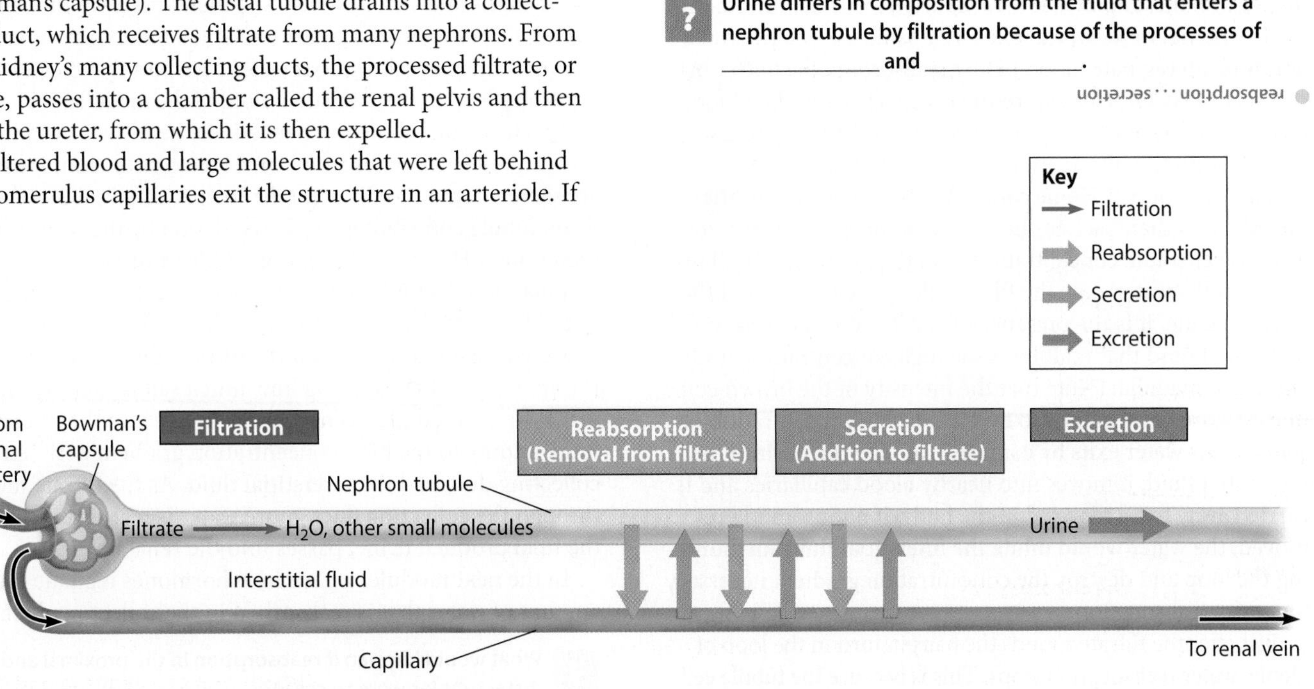
Key
Filtration
Reabsorption
Secretion
Excretion
From renal artery
Bowman's capsule
Filtration
Reabsorption (Removal from filtrate)
Secretion (Addition to filtrate)
Excretion
Nephron tubule
Filtrate
H2O, other small molecules
Urine
Interstitial fluid
Capillary
To renal vein

▲ **Figure 25.6C** Major processes of the urinary system

25.7 Reabsorption and secretion refine the filtrate

How do our kidneys reabsorb almost 99% of the 180 L of filtrate we produce each day? Water reabsorption is a major function of the nephron. As you have learned, water is not actively pumped across a membrane; it flows by diffusion. Your kidneys reclaim water from the filtrate by moving solutes; water then follows by osmosis. About 65% of the water is reabsorbed as it follows the solutes that are actively and passively transported out of the filtrate in the proximal tubule. As the filtrate makes its way through the rest of the nephron and the collecting duct, a solute gradient maintained by the kidney enables further water reabsorption. Let's trace this process in **Figure 25.7**, which provides a closer look at how reabsorption and secretion occur along the nephron.

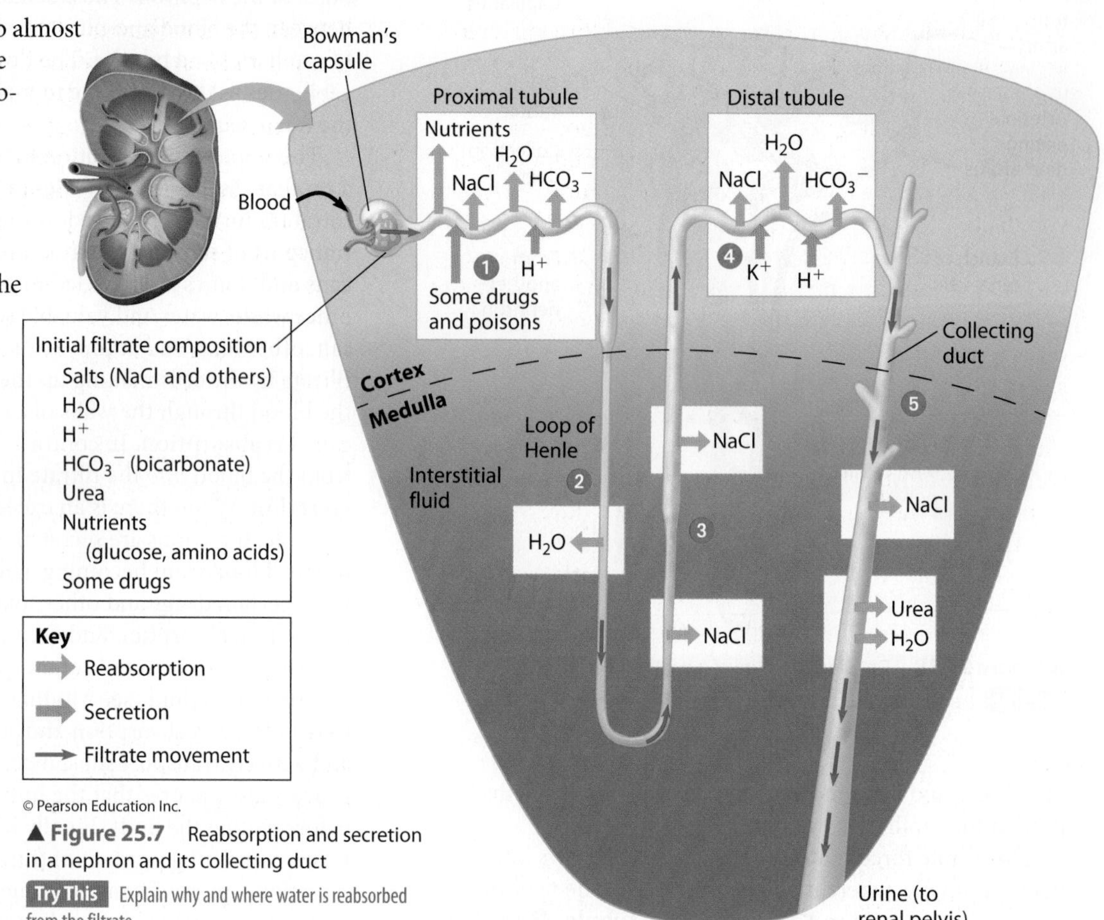

Bowman's capsule
Blood
Proximal tubule
Nutrients
H_2O
NaCl
HCO_3^-
1 H^+
Some drugs and poisons
Distal tubule
H_2O
NaCl
HCO_3^-
4 K^+
H^+
Collecting duct
Initial filtrate composition
Salts (NaCl and others)
H_2O
H^+
HCO_3^- (bicarbonate)
Urea
Nutrients
(glucose, amino acids)
Some drugs
Cortex
Medulla
Loop of Henle
2
Interstitial fluid
H_2O
3
NaCl
NaCl
5
NaCl
Urea
H_2O
Key
Reabsorption
Secretion
Filtrate movement
Urine (to renal pelvis)

▲ **Figure 25.7** Reabsorption and secretion in a nephron and its collecting duct

Try This Explain why and where water is reabsorbed from the filtrate.

❶ Most of the reabsorption of glucose, amino acids, salt, and other valuable solutes occurs in the proximal tubules. As these solutes are transported from the filtrate to the interstitial fluid, water follows by osmosis. The proximal tubules help regulate blood pH by reabsorbing the buffer bicarbonate (HCO_3^-) and secreting excess H^+ from the blood. Drugs and poisons that were processed in the liver are also secreted into the filtrate in the proximal tubules.

The long loop of Henle carries the filtrate deep into the medulla and then back to the cortex. ❷ Because the filtrate has a lower solute concentration than the interstitial fluid of the medulla, water exits the filtrate along the first part of the loop of Henle. It is the presence of NaCl and some urea in the interstitial fluid that maintains the high concentration gradient in the medulla. (Note that the intensity of the brown color in the figure corresponds to this increasing concentration of solutes.) As water exits by osmosis from the filtrate into the interstitial fluid, it moves into nearby blood capillaries and is carried away to be recycled in the body. If not promptly removed, the water would dilute the interstitial fluid surrounding the loop and destroy the concentration gradient necessary for water reabsorption.

Just after the filtrate rounds the hairpin turn in the loop of Henle, water reabsorption stops. This is because the tubule cell membranes in that section of the loop of Henle lack aquaporins, proteins that facilitate water transport (Module 5.6). ❸ As the filtrate moves from the medulla back toward the cortex, NaCl leaves the filtrate, first passively and then actively as the cells of the tubule pump NaCl into the interstitial fluid. It is primarily this movement of salt without water transport that creates the solute gradient in the interstitial fluid of the medulla.

❹ The distal tubule further refines the filtrate. Additional salt and water are reabsorbed, and excess potassium ions (K^+) are secreted into the filtrate. Like the proximal tubule, the distal tubule contributes to pH regulation by the controlled secretion of H^+ and reabsorption of bicarbonate.

Final processing of the filtrate occurs as the collecting duct ❺ carries the filtrate through the medulla. Because it actively reabsorbs NaCl from the filtrate, the collecting duct is important in determining how much salt is excreted in the urine. In the medulla, some urea leaks out into the interstitial fluid, adding to the high concentration gradient between the collecting duct and the interstitial fluid. As the filtrate moves through the collecting duct, more water is reabsorbed before the final product, urine, passes into the renal pelvis.

In the next module, we see how hormones regulate the amount of water that is reabsorbed in the collecting duct.

? **What would happen if reabsorption in the proximal and distal tubules were to cease?**

● Needed solutes and water would be lost in the urine, depriving the body of substances it requires.

25.8 Hormones regulate the urinary system

Water and solute homeostasis is maintained through an increase or decrease in the levels of hormones that act on the kidney's collecting ducts. For example, if you start to become dehydrated, the solute concentration of your body fluids rises. When this concentration gets too high, the brain increases levels of a hormone called **antidiuretic hormone (ADH)** in your blood. This hormone signals the cells in your kidney's collecting ducts to reabsorb more water from the filtrate, which increases the amount of water returning to your blood (where it is needed) and decreases the amount of water excreted (resulting in concentrated urine). Dark-colored urine indicates that you have not been drinking enough water.

Conversely, if you drink a lot of water, the solute concentration of your body fluids becomes too dilute. In response, blood levels of ADH drop, causing the collecting duct cells to reabsorb less water from the filtrate, resulting in dilute, watery urine. This process explains why your urine is very clear after you drink a lot of water. (Increased urination is called diuresis, and it is because ADH acts against this state that it is called antidiuretic hormone.) Diuretics, such as alcohol, are substances that inhibit the release of ADH and therefore result in excessive urinary water loss. Drinking alcohol will make you urinate more frequently, and the resulting dehydration contributes to the symptoms of a hangover. Caffeine is also a diuretic; you may have noticed that drinking coffee, tea, or cola causes you to urinate soon afterward.

Our kidneys' regulatory functions are controlled by an elaborate system of checks and balances that include other hormones besides ADH. (The coordination of all the body's regulatory systems by hormones is the subject of Chapter 26.)

? Some of the drugs classified as diuretics make the epithelium of the collecting duct less permeable to water. How would this affect kidney function?

● The collecting ducts would reabsorb less water, and thus the diuretic would increase water loss in the urine.

25.9 Kidney dialysis can save lives

CONNECTION

A person can survive with one functioning kidney, but if both kidneys fail, the buildup of toxic wastes and the lack of regulation of blood ion concentrations, blood pH, and blood pressure will lead to certain and rapid death. Knowing how the nephron works helps us understand how some of its functions can be performed artificially when the kidneys are damaged. In a medical treatment for kidney disease, called **dialysis**, blood is filtered by a machine that mimics the action of a nephron (**Figure 25.9**). Like the nephrons of the kidney, the machine sorts small molecules of the blood, keeping some and discarding others. The patient's blood is pumped from an artery through selectively permeable tubes. The tubes are immersed in a dialyzing solution that resembles the chemical makeup of the interstitial fluid that bathes nephrons. As blood circulates through the tubing, urea and excess ions diffuse out and needed substances diffuse in. The machine continually discards used dialyzing solution as wastes build up.

Dialysis treatment is life sustaining for people with kidney failure, but it is costly, takes a lot of time (4–6 hours three times a week), and must be continued for life—or until the patient undergoes kidney transplantation. In some cases, a kidney from a living compatible donor (usually a close relative) or a deceased organ donor can be transplanted into a person with kidney failure. Unfortunately, the number of people who need a kidney is much greater than the number of organs available. The average wait for a kidney donation in the United States is three to five years. More than 60% of all cases of kidney disease are caused by hypertension (high blood pressure) and diabetes (high levels of glucose in the blood), but the prolonged use of pain relievers (even common, over-the-counter ones), alcohol, and other drugs are also possible causes.

? How does the composition of dialyzing solution compare with that of the patient's interstitial fluid?

● Dialyzing solution has a solute concentration similar to that of interstitial fluid. The solution contains no urea, which allows urea from the patient's blood to diffuse into it.

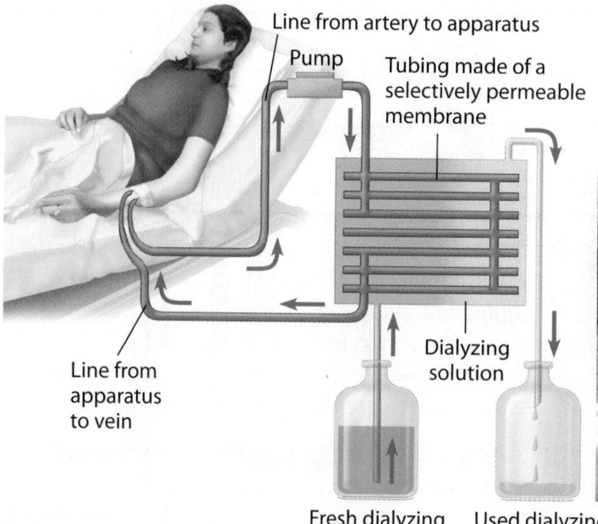

Line from artery to apparatus
Pump
Tubing made of a selectively permeable membrane
Line from apparatus to vein
Dialyzing solution
Fresh dialyzing solution
Used dialyzing solution (with urea and excess ions)

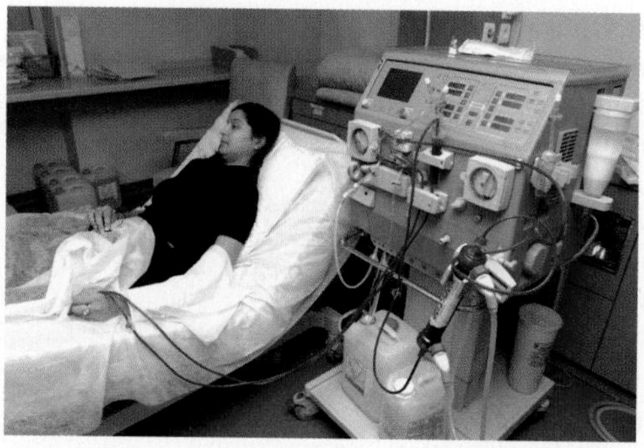

▲ **Figure 25.9** Kidney dialysis

CHAPTER 25 REVIEW

For practice quizzes, BioFlix animations, MP3 tutorials, video tutors, and more study tools designed for this textbook, go to MasteringBiology®

Reviewing the Concepts

Thermoregulation (25.1–25.3)

25.1 An animal's regulation of body temperature helps maintain homeostasis. Endotherms derive body heat mainly from their metabolism; ectotherms absorb heat from their surroundings. Heat exchange with the environment occurs by conduction, convection, radiation, and evaporation.

25.2 Thermoregulation involves adaptations that balance heat gain and loss. Adaptations for thermoregulation include increased metabolic heat production, insulation, circulatory adaptations, evaporative cooling, and behavioral responses.

25.3 Coordinated waves of movement in huddles help penguins thermoregulate.

Osmoregulation and Excretion (25.4–25.10)

25.4 Animals balance their levels of water and solutes through osmoregulation. Osmoconformers have the same internal solute concentration as seawater. Osmoregulators control their solute concentrations.

	Gain Water	Lose Water	Salt
Freshwater Fish	Osmosis	Excretion	Pump in
Saltwater Fish	Drinking	Osmosis	Excrete, pump out
Land Animal	Drinking, eating	Evaporation, urinary system	

Animals can conserve water by means of the kidneys, waterproof skin, and reproductive adaptations.

25.5 Several ways to dispose of nitrogenous waste have evolved in animals. Excretion is the disposal of toxic nitrogenous wastes. Ammonia (NH_3) is poisonous but soluble and is easily disposed of by aquatic animals. Urea and uric acid are less toxic and easier to store but require significant energy to produce.

25.6 The urinary system plays several major roles in homeostasis. The urinary system expels wastes and regulates water and solute balance. Nephrons extract a filtrate from the blood and refine it to urine. In filtration, blood pressure forces water and many small solutes into the nephron. In reabsorption, water and valuable solutes are reclaimed from the filtrate. In secretion, excess H^+ and toxins are added to the filtrate. In excretion, urine is expelled. Urine leaves the kidneys via the ureters, is stored in the urinary bladder, and is expelled through the urethra.

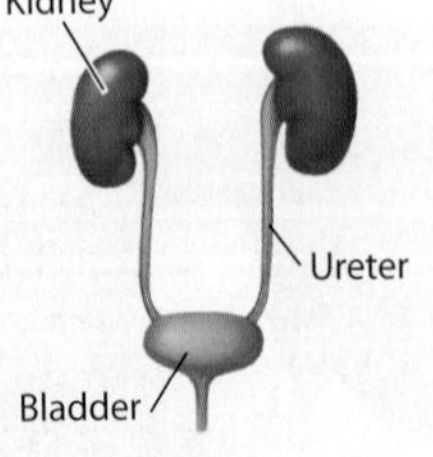

Kidney
Ureter
Bladder

25.7 Reabsorption and secretion refine the filtrate. Nutrients, salt, and water are reabsorbed from the proximal and distal tubules within the nephron. Secretion of H^+ and reabsorption of HCO_3^- help regulate pH. High NaCl concentration in the medulla promotes reabsorption of water.

25.8 Hormones regulate the urinary system. Antidiuretic hormone (ADH) regulates the amount of water excreted by the kidneys.

25.9 Kidney dialysis can save lives. A dialysis machine removes wastes from blood and maintains solute concentration.

Connecting the Concepts

1. Complete this map, which presents the three main topics of this chapter.

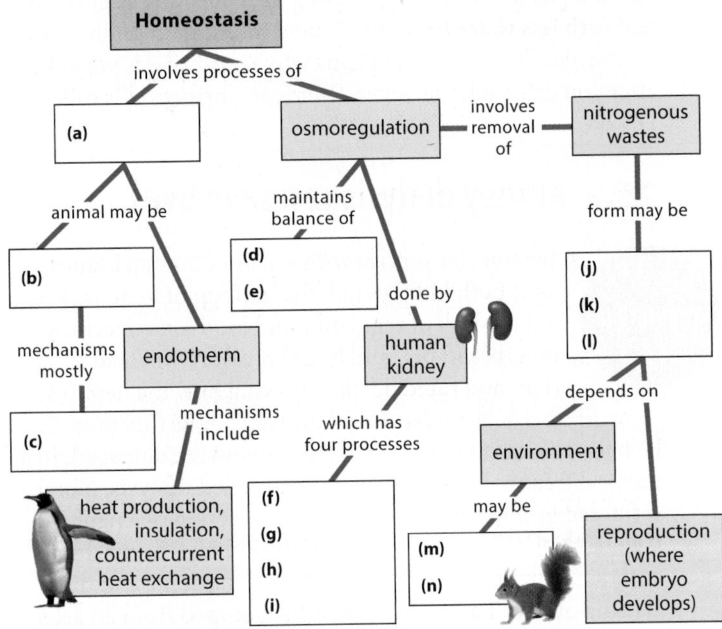

Homeostasis
involves processes of
(a)
osmoregulation
involves removal of
nitrogenous wastes
animal may be
maintains balance of
form may be
(b)
(d)
(e)
done by
(j)
(k)
(l)
mechanisms mostly
endotherm
human kidney
depends on
(c)
mechanisms include
which has four processes
environment
heat production, insulation, countercurrent heat exchange
(f)
(g)
(h)
(i)
may be
(m)
(n)
reproduction (where embryo develops)

2. In this schematic of urine production in a nephron, label the four processes involved and list some of the substances that are moved in each process.

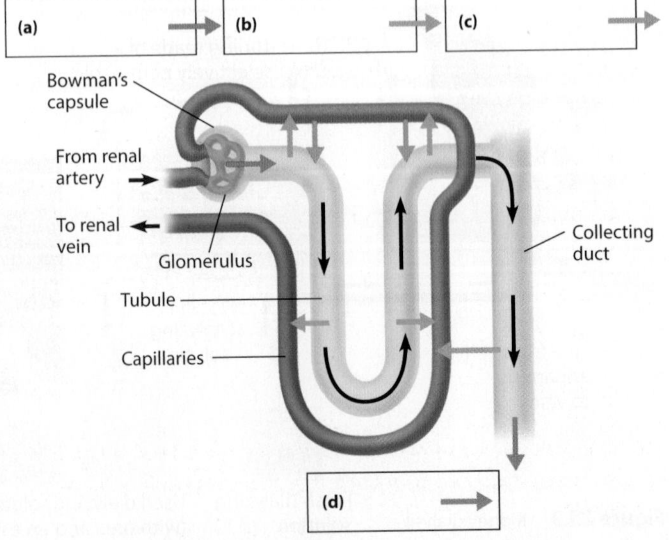

(a)
(b)
(c)
Bowman's capsule
From renal artery
To renal vein
Glomerulus
Tubule
Capillaries
Collecting duct
(d)

Testing Your Knowledge

Level One (Knowledge/Comprehension)

3. Which of the following is not an adaptation for reducing the rate of heat loss to the environment?
   a. feathers or fur
   b. increasing blood flow to surface blood vessels
   c. huddling behavior of penguins
   d. countercurrent heat exchange
4. In each nephron of the kidney, the glomerulus and Bowman's capsule
   a. filter the blood and capture the filtrate.
   b. reabsorb water into the blood.
   c. break down harmful toxins and poisons.
   d. refine and concentrate the urine for excretion.
5. As filtrate passes through the loop of Henle, salt is reabsorbed and concentrated in the interstitial fluid of the medulla. This high solute concentration in the medulla enables nephrons to
   a. excrete the maximum amount of salt.
   b. neutralize toxins that might be found in the kidney.
   c. excrete a large amount of water.
   d. reabsorb water from the filtrate very efficiently.
6. Birds and insects excrete uric acid, whereas mammals and most amphibians excrete mainly urea. What is the chief advantage of uric acid over urea as a waste product?
   a. Uric acid is a much simpler molecule.
   b. It takes less energy to make uric acid.
   c. Less water is required to excrete uric acid.
   d. More solutes are removed excreting uric acid.
7. A freshwater fish would be expected to
   a. pump salt out through its gills.
   b. produce copious quantities of dilute urine.
   c. have scales and a covering of mucus that reduce water loss to the environment.
   d. do all of the above.

Match each of the following components of blood (on the left) with what happens to it as the blood is processed by the kidney (on the right). Note that each lettered choice may be used more than once.

8. Water	a. passes into filtrate; almost all excreted in urine
9. Glucose	
10. Plasma protein	b. remains in blood
11. Toxins or drugs	c. passes into filtrate; mostly reabsorbed
12. Red blood cell	
13. Urea	d. secreted and excreted

Level Two (Application/Analysis)

14. You are in a room of empty chairs. As the chairs fill with people, you become hotter and hotter. A ceiling fan is turned on, and you feel cooler. You gained heat by ______________ and lost heat to the environment by ______________.
    a. conduction . . . convection
    b. radiation . . . convection
    c. radiation . . . conduction
    d. convection . . . radiation
15. Which process in the nephron is least selective?
    a. secretion
    b. reabsorption
    c. filtration
    d. passive diffusion of salt
16. Compare the water and salt regulation in a salmon when it swims in the ocean to when it migrates into fresh water to spawn.
17. Can ectotherms have stable body temperatures? Explain.
18. Assuming equal size, predict which of these organisms would produce the greatest amount of nitrogenous wastes. Explain.
    a. An endotherm or an ectotherm?
    b. A carnivore or an herbivore (assume both are endotherms)?
19. Some diuretics are on a list of substances banned by the International Olympic Committee for use by athletes. What do diuretics do? Propose an explanation for how diuretic use could be an unfair advantage for a competitor in a sport like wrestling, in which weight classes are part of the competition.

Level Three (Synthesis/Evaluation)

20. You are studying a large tropical reptile that has a high and relatively stable body temperature. How would you determine whether this animal is an endotherm or an ectotherm?
21. **SCIENTIFIC THINKING** The table below presents data that could have been collected from penguin colonies in the Antarctic. Graph these data. Would a scatter plot or a bar chart be the better choice? Why? What conclusion (if any) can be drawn from the data?

Huddle Number	Outside Temperature During Huddle (°C)	Duration of Huddle (minutes)
1	−21	179
2	−45	407
3	−10	25
4	−43	416
5	−18	75
6	−37	305
7	−25	192

22. Dolphins have an insulating layer of blubber that protects them from cold water, but their flippers are not insulated. Propose a hypothesis to explain why dolphin flippers do not freeze. Describe an experiment you could do to test your hypothesis. (You may assume you have equipment for measuring temperatures in dolphin flippers.) What results would you expect if your hypothesis is correct?
23. In 1847, the German biologist Christian Bergmann noted that mammals and birds living at higher latitudes (farther from the equator) are on average larger and bulkier than related species living at lower latitudes. Suggest an evolutionary hypothesis to explain this observation.
24. Kidneys were the first organs to be transplanted successfully. A donor can live a normal life with a single kidney, making it possible for individuals to donate a kidney to an ailing relative or even an unrelated individual. In some countries, poor people sell kidneys to transplant recipients through organ brokers. What are the pro and cons associated with organ commerce?
25. Scientists have found that the quantity of aquaporin molecules inserted in the membranes of collecting duct cells changes in response to ADH levels. Draw a line graph proposing a relationship between ADH levels and its effect on the quantity of aquaporins. Additionally, explain how the relationship between ADH and aquaporins corresponds to situations of dehydration and hydration.

Answers to all questions can be found in Appendix 4.

CHAPTER 26

Hormones and the Endocrine System

What effects do pollutants have on the animal endocrine system?

Widely used weed killers and ordinary plastic water bottles—what do they have in common? They can contain chemicals that may have adverse effects on vertebrates by interfering with the endocrine system, which regulates numerous animal body functions through chemical signaling. Toxic chemicals that interfere with the endocrine system are aptly named endocrine disruptors. The endocrine disruptor atrazine, an ingredient in many weed killers, is found in farm water runoff that makes its way to ground and surface reservoirs. Bisphenol A (BPA) is an endocrine disruptor used in many plastics that line bottles, canned goods, and other drinking water and food containers.

Atrazine is known to cause reproductive problems in vertebrates. Amphibians are especially sensitive to the pollutant, which is readily absorbed through amphibians' thin and highly permeable skin. To explore how atrazine is affecting frogs in the wild (photograph below), scientists performed laboratory experiments in which they exposed developing male frogs to low levels of the chemical. We examine the results of one such study later in the chapter.

Controlled experiments also link BPA to reproductive problems in vertebrates, including defects in reproductive organs, decreased sperm count, and reduced embryo implantation in the uterus. Identifying BPA as the cause of human health issues is more challenging, but human exposure to BPA is evident. BPA was detected in the urine of almost all children and adults tested in a study by the Centers for Disease Control and Prevention (CDC). It is yet unknown how BPA exposure affects human health. Not wanting to wait for answers, many consumers are pushing for BPA-free products.

We begin the chapter with an overview of hormones, the chemical signals of the endocrine system, and then turn to the components of the endocrine system. Along the way, we consider examples of the effects of hormonal imbalance.

BIG IDEAS

The Nature of Chemical Regulation (26.1–26.3)

Hormones affect cells using two distinct mechanisms.

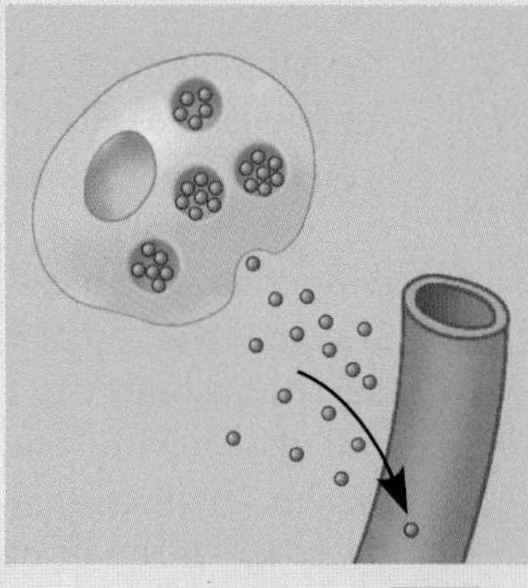

The Vertebrate Endocrine System (26.4–26.5)

The hypothalamus exerts master control over many other endocrine glands.

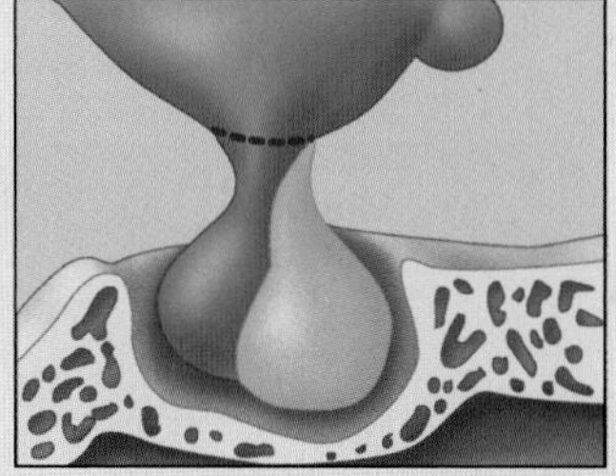

Hormones and Homeostasis (26.6–26.12)

Hormones regulate whole-body processes through feedback systems.

▷ The Nature of Chemical Regulation

26.1 Chemical and electrical signals coordinate body functions

To maintain homeostasis and carry out other coordinated functions, the cells of an animal's body must communicate with one another. They do so through chemical and electrical signals, traveling by way of two major organ systems: the endocrine system and the nervous system. The **endocrine system** is a group of interacting glands and tissues throughout the animal body that produce and secrete chemicals to initiate and maintain body functions and activities. In the endocrine system, chemical signals called **hormones** are released into the bloodstream by endocrine cells and carried to all locations in the body. In the nervous system, the signals are primarily electrical and are transmitted via nerve cells called neurons.

The endocrine system is well suited for coordinating gradual changes that affect the entire body. For example, hormones coordinate the body's responses to stimuli such as dehydration, low blood glucose level, and stress. Hormones also regulate long-term developmental processes, such as the metamorphosis of a tadpole into a frog and the physical and behavioral changes that underlie sexual maturity. The nervous system is well adapted for directing immediate and rapid responses to the environment. For example, the flick of a frog's tongue as it catches a fly results from high-speed nerve signals.

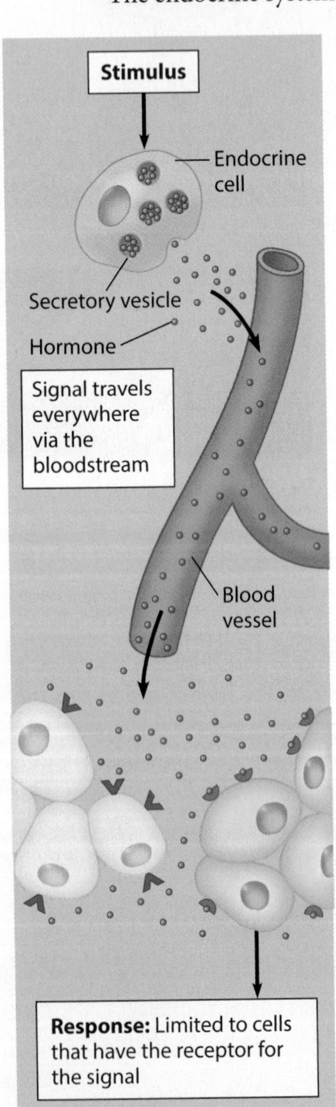

Stimulus
Endocrine cell
Secretory vesicle
Hormone
Signal travels everywhere via the bloodstream
Blood vessel
Response: Limited to cells that have the receptor for the signal

▲ **Figure 26.1A** Signaling in the endocrine system

Hormones are made and secreted mainly by organs called **endocrine glands**. Examples of endocrine glands in vertebrates are the pituitary gland at the base of the brain, which regulates growth and reproduction, and the thyroid gland in the lower neck, which regulates metabolism. **Figure 26.1A** sketches the process of endocrine signaling. Membrane-enclosed secretory vesicles in an endocrine cell are full of hormone molecules (⁘). The endocrine cell secretes the molecules directly into blood vessels. From there, the hormone can travel via the circulatory system to all parts of the body, but only certain types of cells, called **target cells**, have receptors for that specific hormone. Depending on the location of the target cells, the hormone can have an effect in just a single location within the body or in sites throughout the body.

Signals of the endocrine and nervous system are adapted to function differently in transmission, speed, and duration. Whereas hormones travel through the blood to all locations of the body, electrical signals travel distinct pathways along neurons to specific cells, as shown in **Figure 26.1B**. Both types of signals may travel long distances to reach target cells, but only in the nervous system is there a direct connection (through specialized cell junctions) between the neuron transmitting the signal and the target cell that is responding. It takes many seconds for hormones to be released into the bloodstream and carried to target tissues. In contrast, electrical signals are transmitted in a fraction of a second. Because hormones can remain in the bloodstream for minutes or even hours, the effects of endocrine secretions are long-lasting. Conversely, electrical signals of neurons last less than a second, and their effects are fleeting.

Certain cells and signals are shared by the endocrine system and the nervous system. Specialized neurons called **neurosecretory cells** perform functions in both systems. Like all neurons, neurosecretory cells conduct electrical signals, but they also make and secrete hormones into the blood. (We take a closer look at neurosecretory cells in Module 26.5.) Like neurosecretory cells, some chemicals function in both the endocrine and nervous systems. These "double-duty" signals act as hormones in the endocrine system and as short-range signals in the nervous system. Epinephrine (adrenaline), for example, when secreted by the adrenal gland into the blood functions in vertebrates as the "fight-or-flight" hormone (so called because it prepares the whole body for sudden action). Yet when secreted by neurons in the nervous system, epinephrine functions as a neurotransmitter, a chemical that carries information from one neuron to another or from a neuron to another kind of cell that will react (see Module 28.7). Let's now leave our discussion of the nervous system and explore hormones in greater depth.

? If hormones reach every cell in the body, why do only some cells respond to a specific hormone?

● Only target cells have the receptors to recognize the specific signal.

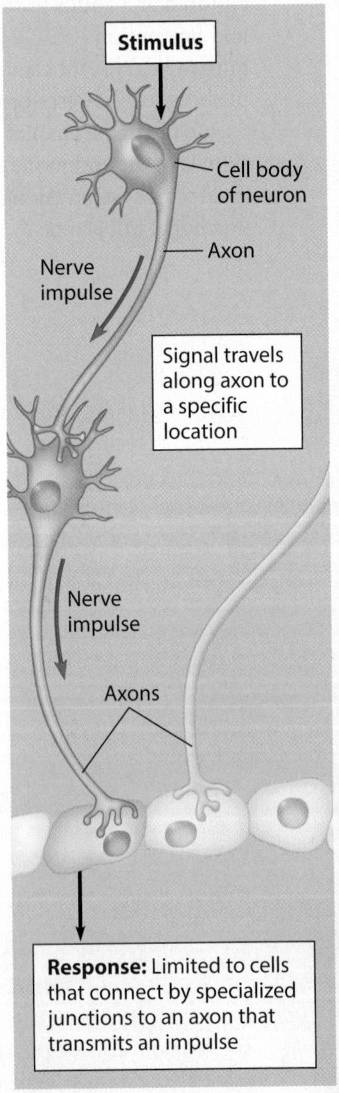

Stimulus
Cell body of neuron
Axon
Nerve impulse
Signal travels along axon to a specific location
Nerve impulse
Axons
Response: Limited to cells that connect by specialized junctions to an axon that transmits an impulse

▲ **Figure 26.1B** Signaling in the nervous system

Try This Make a list comparing and contrasting transmission, speed, and duration for the endocrine and nervous systems.

26.2 Hormones affect target cells using two main signaling mechanisms

Hormonal signaling has three stages: reception, signal transduction, and response. Reception of the signal occurs when a hormone binds to a specific receptor protein on or in the target cell. Each signal molecule has a specific shape that can be recognized by its target cell receptors. The binding of a signal molecule to a receptor protein triggers events within the target cell—signal transduction—that convert the signal from one form to another. The result is a response, a change in the cell's behavior. Heart muscle cells, for example, respond to epinephrine, the "fight or flight" signal, with cellular contraction, which speeds up the heartbeat. Liver cells, however, respond to epinephrine by breaking down glycogen, providing glucose (an energy source) to body cells.

Based on chemical properties, hormones can be classified into two groups. The water-soluble hormones include proteins, short polypeptides, and some modified versions of single amino acids. Most hormones produced by the endocrine glands are water-soluble. The lipid-soluble hormones include the steroid hormones, small molecules made from cholesterol (see Module 3.10). Only the sex organs and the cortex of the adrenal gland produce steroid hormones. Nonsteroid, lipid-soluble hormones are produced by the thyroid gland. Although both water-soluble hormones and lipid-soluble hormones carry out the three stages of reception, signal transduction, and response, they do so differently. Let's look at how each type of hormone elicits cellular responses.

Water-Soluble Hormones Water-soluble hormones cannot pass through the phospholipid bilayer of the plasma membrane, but they can bring about cellular changes, without entering their target cells. The receptor proteins for most water-soluble hormones are embedded in the plasma membrane of target cells **(Figure 26.2A)**. ① A water-soluble hormone molecule (○) binds to the receptor protein, activating it. ② This initiates a signal transduction pathway, a series of changes in cellular proteins (relay molecules) that converts an extracellular chemical signal to a form that can bring about a response inside the cell. ③ The final relay molecule (●) activates a protein (▲) that carries out the cell's response, either in the cytoplasm (such as activating an enzyme) or in the nucleus (regulating gene expression). One hormone may trigger a variety of responses in target cells because each cell may contain different receptors for that hormone or diverse signal transduction pathways.

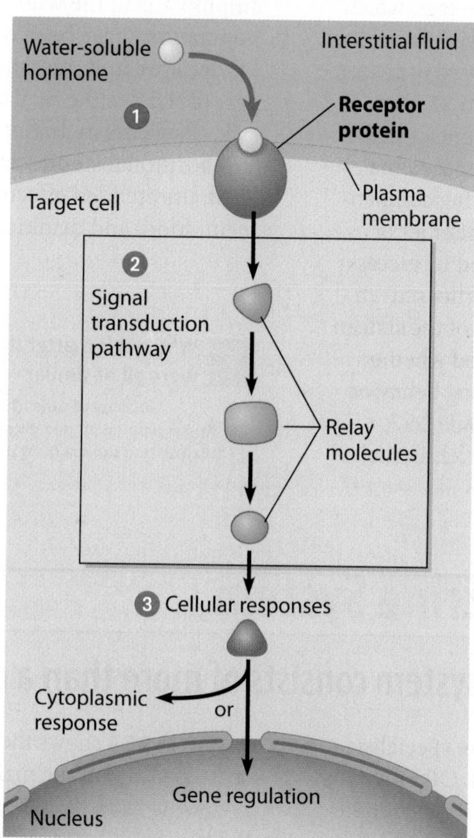

Water-soluble hormone
Interstitial fluid
①
Receptor protein
Target cell
Plasma membrane
②
Signal transduction pathway
Relay molecules
③ Cellular responses
Cytoplasmic response
or
Gene regulation
Nucleus

▲ **Figure 26.2A** A water-soluble hormone that binds to a plasma membrane receptor

Lipid-Soluble Hormones While water-soluble hormones bind to receptors in the plasma membrane, lipid-soluble hormones, such as steroid hormones, pass through the phospholipid bilayer and bind to receptors inside the cell. As shown in **Figure 26.2B**, ① a steroid hormone (▽) enters a cell by diffusion. If the cell is a target cell, the hormone ② binds to an open receptor protein in the cytoplasm or nucleus. Rather than triggering a signal transduction pathway with relay proteins, as happens with a water-soluble hormone, the hormone-receptor complex itself usually carries out the transduction of the hormonal signal: The complex acts as a transcription factor—a gene activator or repressor (see Module 11.3). ③ The hormone-receptor complex attaches to specific sites on the cell's DNA in the nucleus. (These sites are enhancers; see Module 11.3.) ④ The binding to DNA stimulates gene regulation, turning genes either on (by promoting transcription of certain genes into RNA) or off.

We've now completed an overview of hormone function. In our next section, we'll consider what happens when normal hormone signaling is disrupted.

? **What are two major differences between the actions of lipid-soluble hormones and water-soluble hormones?**

● Lipid-soluble hormones bind to receptors inside the cell; water-soluble hormones bind to plasma membrane receptors. Lipid-soluble hormones always affect gene expression; water-soluble hormones have this or other effects.

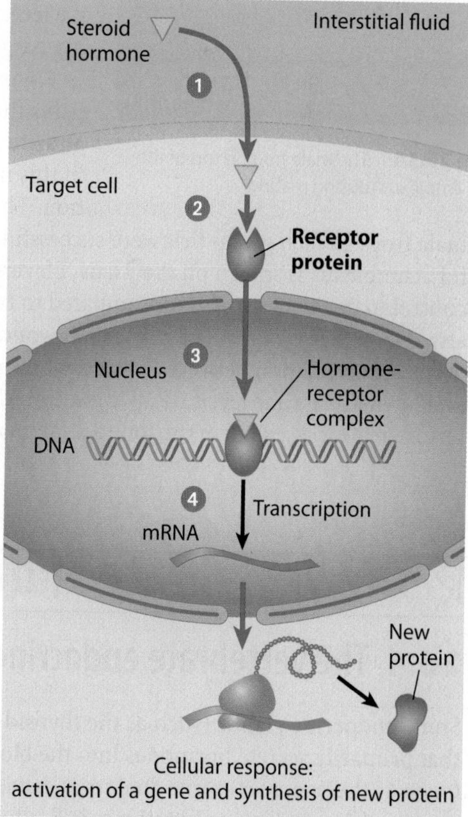

Steroid hormone
Interstitial fluid
①
Target cell
②
Receptor protein
③
Nucleus
Hormone-receptor complex
DNA
④
Transcription
mRNA
New protein
Cellular response: activation of a gene and synthesis of new protein

▲ **Figure 26.2B** A lipid-soluble hormone that binds to an intracellular receptor

26.3 A widely used weed killer demasculinizes male frogs

SCIENTIFIC THINKING

As discussed in the chapter introduction, almost all of us are routinely exposed to atrazine found in weed killers and BPA in food and beverage containers. These chemicals, called endocrine disruptors, specifically mimic the lipid-soluble hormone estrogen. Such chemicals can potentially enter animal cells, altering the normal ratio of sex hormones (hormones that regulate growth, development, reproductive cycles, and sexual behaviors). Scientists use controlled studies to test whether a chemical causes specific biological effects. We examine one such study in which scientists at the University of California at Berkeley exposed developing male frogs to very low levels of the common pollutant atrazine and maintained the exposure for three years.

What effects do pollutants have on animal endocrine systems?

In the experiments, an equal number of control and atrazine-exposed adult males of similar weights were placed into a pool with females. A mating contest was set up, in which control males and atrazine-exposed males competed for females. How was mating success measured? The scientists recorded each male frog's ability to successfully grasp a female with his front legs during a mating behavior termed amplexus (**Figure 26.3A**). It is during this "embrace" that the male frog usually fertilizes the eggs, which the female simultaneously discharges into the water.

▲ **Figure 26.3A** The male frog on top of the female in amplexus (during mating)

Figure 26.3B compares the control and atrazine-exposed males in this mating competition. The number of male frogs in each group that were successful and unsuccessful at amplexus is shown on the *y* axis. Eleven of the sixteen control males were successful, compared to two of the sixteen atrazine-exposed males. Scientists then wondered whether there is a correlation between unsuccessful mating behavior and low testosterone levels. (As we'll see in Module 26.7, testosterone is a sex hormone found at higher levels in males than in females). **Figure 26.3C** shows the percent of males from each group with low testosterone levels. What was the conclusion? More males in the atrazine-exposed group experienced testosterone deficiencies than did those in the control group. Furthermore, the males with low testosterone levels did not achieve amplexus. The reduced mating behaviors and the testosterone deficiencies in the treated frogs demonstrate atrazine's demasculinizing effect on male frogs. Astonishingly, 10% of the genetically male frogs underwent complete sex reversal when exposed to atrazine—that is, they became females capable of producing eggs!

The levels of atrazine used in these studies were consistent with environmental conditions for some amphibians in the wild. Amphibian populations have been on the decline for decades, and this study shows that part of the decline may be due to a lack of success in mating. Behavioral and hormonal studies are useful in determining the long-term, unintended effects of chemicals in our local environment, food, and drinking water.

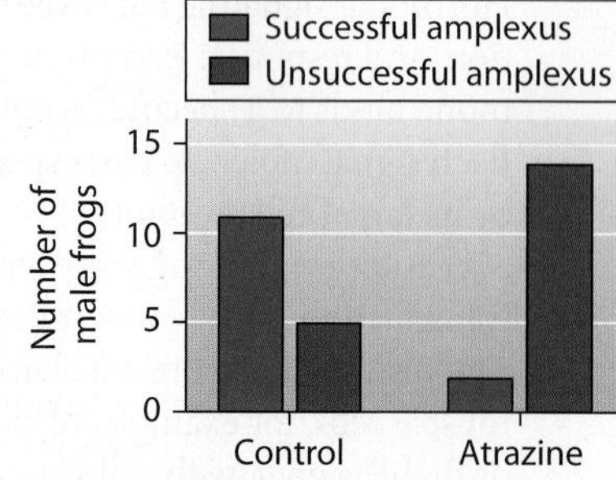

Successful amplexus
Unsuccessful amplexus
Number of male frogs
15
10
5
0
Control
Atrazine

Data from T. B. Hayes et al., Atrazine induces complete feminization and chemical castration in male African clawed frogs (*Xenopus laevis*), *Proceedings of the National Academy of Sciences* 107: 10 (2007).

▲ **Figure 26.3B** Mating behavior in control and atrazine-exposed males

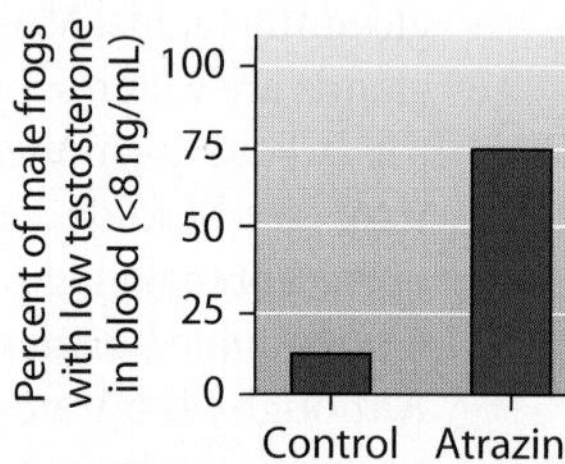

Percent of male frogs with low testosterone in blood (<8 ng/mL)
100
75
50
25
0
Control
Atrazine

Data from T. B. Hayes et al., Atrazine induces complete feminization and chemical castration in male African clawed frogs (*Xenopus laevis*), *Proceedings of the National Academy of Sciences* 107:10 (2007).

▲ **Figure 26.3C** Testosterone levels in control and atrazine-exposed males

? Why did it matter that the male frogs competing for females were all of similar weight?

It would be more difficult to draw a conclusion if both weight and atrazine exposure were variables. Small males might be less successful when directly competing with larger males, regardless of their exposure to atrazine.

The Vertebrate Endocrine System

26.4 The vertebrate endocrine system consists of more than a dozen major glands

Some endocrine glands (such as the thyroid) are specialists that primarily secrete hormones into the blood. Other glands (such as the pancreas) serve dual roles, having both endocrine and nonendocrine functions. Still other organs (such as the stomach and heart) are primarily nonendocrine but have some cells that secrete hormones.

Figure 26.4 shows the locations of the major human endocrine glands and the main hormones they produce. (When reviewing the figure, keep in mind that this chapter covers only the major endocrine glands and hormones; there are other hormone-secreting structures—the thymus, liver, and stomach, for example—and other hormones that we will not discuss.)

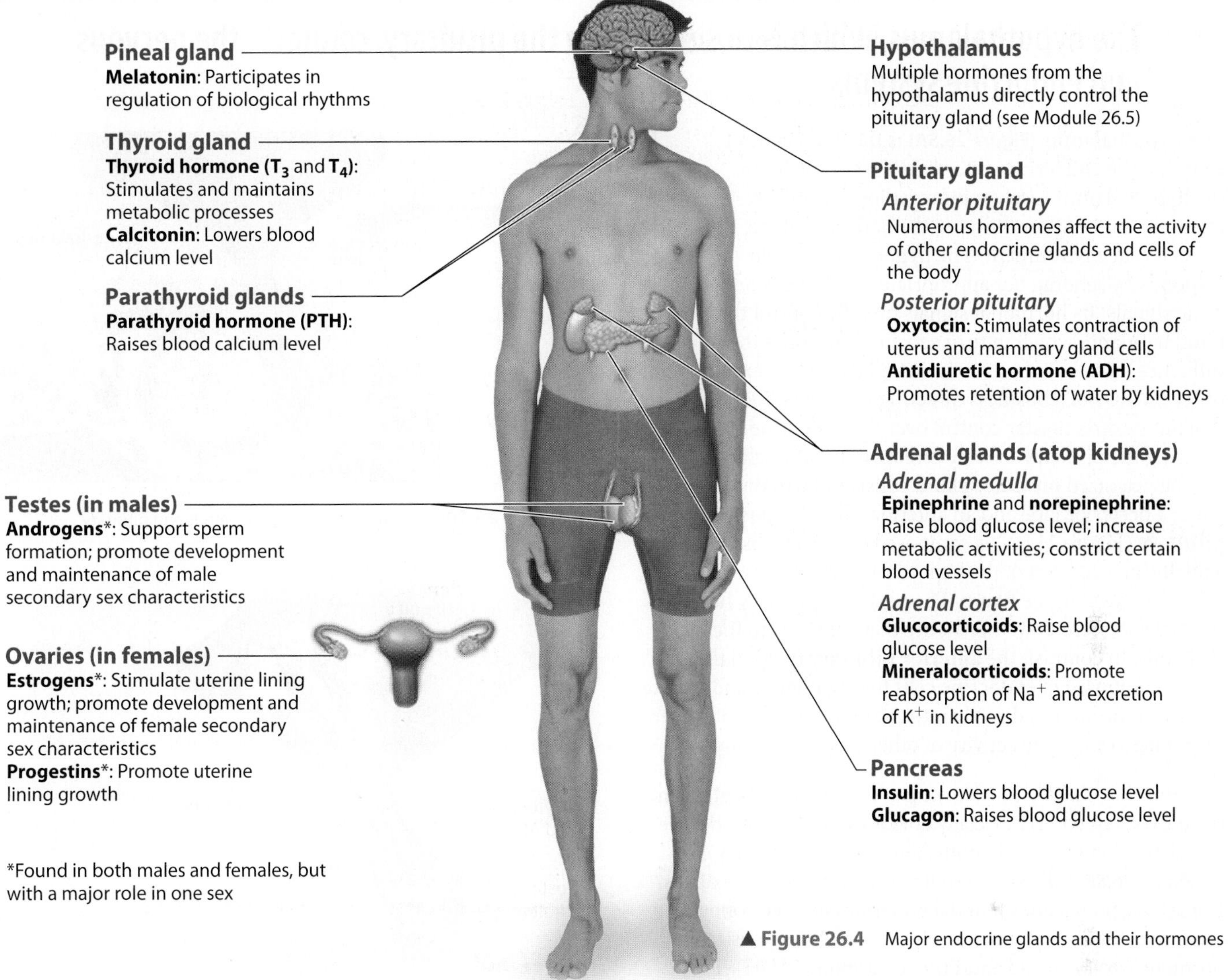

Pineal gland
Melatonin: Participates in regulation of biological rhythms
Thyroid gland
Thyroid hormone (T_3 and T_4): Stimulates and maintains metabolic processes
Calcitonin: Lowers blood calcium level
Parathyroid glands
Parathyroid hormone (PTH): Raises blood calcium level
Testes (in males)
Androgens*: Support sperm formation; promote development and maintenance of male secondary sex characteristics
Ovaries (in females)
Estrogens*: Stimulate uterine lining growth; promote development and maintenance of female secondary sex characteristics
Progestins*: Promote uterine lining growth
*Found in both males and females, but with a major role in one sex
Hypothalamus
Multiple hormones from the hypothalamus directly control the pituitary gland (see Module 26.5)
Pituitary gland
Anterior pituitary
Numerous hormones affect the activity of other endocrine glands and cells of the body
Posterior pituitary
Oxytocin: Stimulates contraction of uterus and mammary gland cells
Antidiuretic hormone (ADH): Promotes retention of water by kidneys
Adrenal glands (atop kidneys)
Adrenal medulla
Epinephrine and norepinephrine: Raise blood glucose level; increase metabolic activities; constrict certain blood vessels
Adrenal cortex
Glucocorticoids: Raise blood glucose level
Mineralocorticoids: Promote reabsorption of Na^+ and excretion of K^+ in kidneys
Pancreas
Insulin: Lowers blood glucose level
Glucagon: Raises blood glucose level

▲ **Figure 26.4** Major endocrine glands and their hormones

What stimulates an endocrine gland to produce a hormone? For some endocrine glands, a change in levels of certain ions and nutrients is the stimulus. For example, in response to low calcium levels, the four disk-shaped parathyroid glands release parathyroid hormone into the blood, leading to a rise in calcium levels. (Low calcium levels can result in convulsive contractions of skeletal muscle—a condition that can be fatal.) Other endocrine glands, such as the adrenal glands, are stimulated directly by the nervous system. In response to a stressful event, nerve impulses stimulate the adrenal medulla to secrete epinephrine and norepinephrine into the blood, preparing the body for "fight or flight." Hormones can stimulate endocrine glands, too. For example, the anterior pituitary is stimulated by hormones produced by the hypothalamus, as we'll explore in Module 26.5.

The hormones produced by endocrine glands have a wide range of effects. Several examples include regulating metabolism and the levels of certain ions and nutrients; controlling reproduction, growth, and development; and initiating responses to stress and the environment. For a particular example of hormonal effects, let's take a brief look at the pineal gland.

The **pineal gland** is a pea-sized mass of tissue near the center of the brain. The pineal gland synthesizes and secretes melatonin, a hormone that links environmental light conditions with biological rhythms, particularly the sleep/wake circadian rhythms. Melatonin is sometimes called "the dark hormone" because it is secreted at night. In diurnal (day-active) animals, melatonin production peaks in the middle of the night and then gradually falls. Some people ingest melatonin supplements as sleep aids, but the effectiveness of this treatment has not been established by scientists. Although there is good evidence that nightly increases in natural melatonin play a significant role in promoting sleep, we do not yet know exactly what effects melatonin has on body cells and precisely how sleep/wake cycles are controlled.

Before focusing on the role hormones play in specific homeostatic mechanisms, we'll examine in detail two more glands, the hypothalamus and the pituitary gland. The hypothalamus, which is part of the brain, plays a central role in integrating the nervous and endocrine systems.

? **What other locations besides the glands shown in Figure 26.4 also secrete hormones?**

Nonendocrine organs, such as the heart, thymus, liver, and stomach, also secrete hormones.

26.5 The hypothalamus, which is closely tied to the pituitary, connects the nervous and endocrine systems

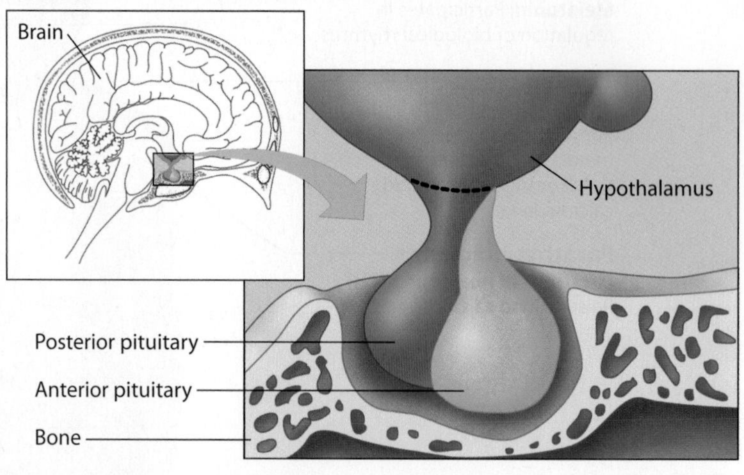

Brain
Hypothalamus
Posterior pituitary
Anterior pituitary
Bone

▲ **Figure 26.5A** Location of the hypothalamus and pituitary

The **hypothalamus (Figure 26.5A)** is the main control center of the endocrine system. As part of the brain, the hypothalamus receives information from nerves about the internal condition of the body and about the external environment. The hypothalamus then responds by sending out appropriate nervous or endocrine signals. Its hormonal signals directly control the pituitary gland, which in turn secretes hormones that influence numerous body functions. Like a chief executive officer (CEO) of a business directing managers, the hypothalamus exerts master control over the endocrine system by using the pituitary to relay directives to other glands.

The pea-sized **pituitary gland** consists of two distinct parts: a posterior lobe and an anterior lobe. The **posterior pituitary** (Figure 26.5A) is composed of nervous tissue and is actually an extension of the hypothalamus, which is why it is shown in green in the figure. The posterior pituitary functions to store and secrete two hormones that are made in the hypothalamus. In contrast, the **anterior pituitary** (gold in the figure) is composed of endocrine cells that synthesize and secrete numerous hormones directly into the blood. Several of these hormones control the activity of other endocrine glands.

Posterior Pituitary The blurring of the endocrine system and the nervous system is especially obvious when we consider the association between the hypothalamus and the pituitary gland. As **Figure 26.5B** indicates, a set of neurosecretory cells extends from the hypothalamus into the posterior pituitary, connecting them structurally and functionally. These cells synthesize the hormones oxytocin and antidiuretic hormone (ADH). These hormones (⁘) are channeled along the neurosecretory cells into the posterior pituitary, where they are stored. The hormones are then released into blood vessels that carry them to distant locations in the body. Oxytocin causes uterine muscles to contract during childbirth and mammary glands to eject milk during nursing. ADH helps cells of the kidney tubules reabsorb water, thus decreasing urine volume when the body needs to retain water (see Module 25.8). When the body has too much water, the hypothalamus slows the release of ADH from the posterior pituitary.

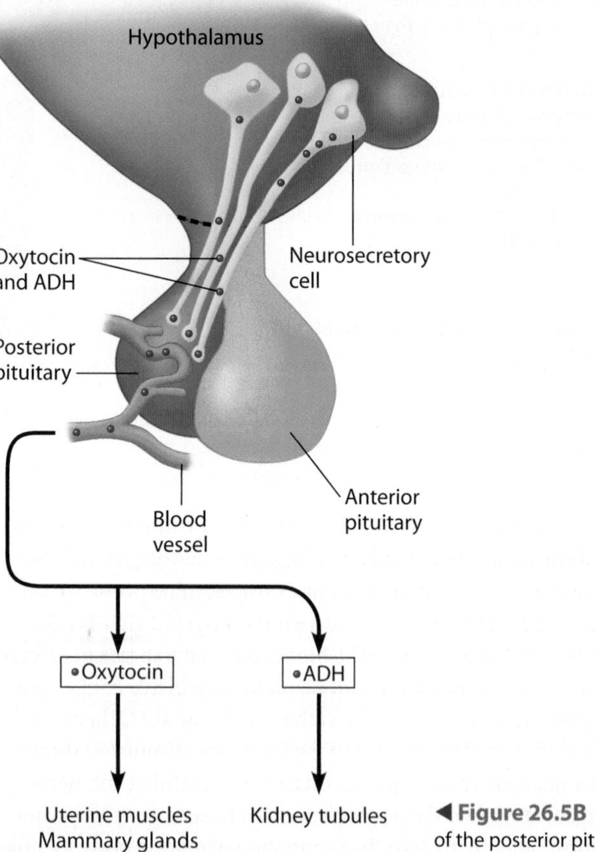

Hypothalamus
Oxytocin and ADH
Neurosecretory cell
Posterior pituitary
Blood vessel
Anterior pituitary
Oxytocin
ADH
Uterine muscles
Mammary glands
Kidney tubules

◀ **Figure 26.5B** Hormones of the posterior pituitary

Anterior Pituitary **Figure 26.5C**, on the facing page, shows a second set of neurosecretory cells in the hypothalamus. These cells secrete two kinds of hormones into short blood vessels that connect to the anterior pituitary: releasing hormones and inhibiting hormones (⁘). A **releasing hormone** stimulates the anterior pituitary to secrete one or more specific hormones, and an **inhibiting hormone** induces the anterior pituitary to stop secreting one or more specific hormones. In response to hypothalamic releasing hormones, the anterior pituitary synthesizes and releases many different hormones (⁘).

Many of the protein hormones secreted from the anterior pituitary stimulate other endocrine glands to produce hormones, thereby influencing a broad range of body activities. **Thyroid-stimulating hormone (TSH)** regulates hormone production by the thyroid gland and thus the thyroid hormone's metabolic effects. Adrenocorticotropic hormone (ACTH) stimulates the adrenal cortex, which in turn releases hormones that affect water balance and metabolism. Both follicle-stimulating hormone (FSH) and luteinizing hormone (LH) stimulate the testes and ovaries to produce reproductive hormones. Unlike many of the other anterior pituitary hormones, **prolactin (PRL)** does not lead to secretion of other hormones. In mammals, prolactin directly stimulates the

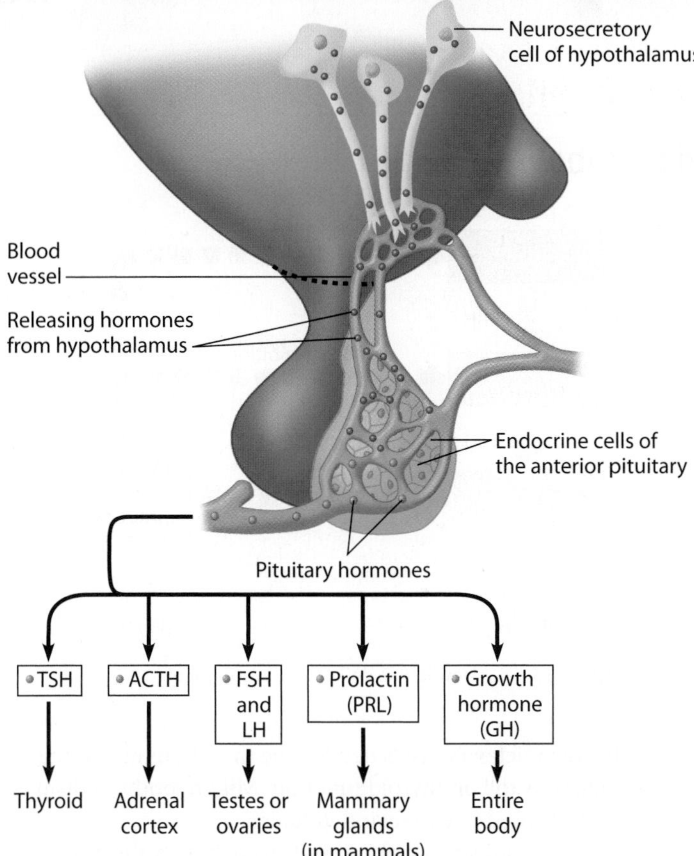

Neurosecretory cell of hypothalamus
Blood vessel
Releasing hormones from hypothalamus
Endocrine cells of the anterior pituitary
Pituitary hormones
TSH
ACTH
FSH and LH
Prolactin (PRL)
Growth hormone (GH)
Thyroid
Adrenal cortex
Testes or ovaries
Mammary glands (in mammals)
Entire body

▲ **Figure 26.5C** Hormones of the anterior pituitary

Try This Write a few sentences explaining how the structure and function of the anterior pituitary differ from those of the posterior pituitary.

mammary glands to produce milk. Prolactin has diverse effects in other vertebrates, such as the promotion of reproductive-related behaviors in birds and amphibians and the regulation of water balance in fish (see Module 26.11).

Of all the pituitary secretions, none has a broader effect than the protein called **growth hormone (GH)**. GH promotes protein synthesis and the use of body fat for energy metabolism in a wide variety of target cells. In young mammals, GH promotes the development and enlargement of all parts of the body. Growth hormone levels fluctuate through the day but peak at night just after the onset of sleep. What happens when you stay up late studying for an exam, or even skip sleeping altogether? GH release is then delayed until your inevitable nap. The postponed release of GH due to occasional sleep deprivation is unlikely to have harmful effects, but abnormal levels of GH over long periods of time can cause serious problems. Continually high levels of GH during childhood, usually due to a pituitary tumor, can lead to gigantism **(Figure 26.5D)**. In contrast, too little GH in childhood can lead to pituitary dwarfism. Administering growth hormone to children with GH deficiency can prevent this. An increased availability of human GH has unfortunately led to its abuse in older adults hoping to slow the aging process and in athletes attempting to build muscles. However, evidence is lacking that GH can prevent aging or that increased muscle bulk translates to increased strength. Abuse of GH can be dangerous and lead to joint swelling, diabetes, and heart complications.

▲ **Figure 26.5D** Gigantism, caused by an excess of growth hormone during childhood

Feedback Control of the Hypothalamus and Pituitary

Figure 26.5E serves as a useful example of a hormone cascade pathway directed by the hypothalamus. Here we focus on the relationship between the hypothalamus, anterior pituitary, and thyroid gland. The hypothalamus secretes a releasing hormone known as **TRH (TSH-releasing hormone)**. In turn, TRH stimulates the anterior pituitary to produce thyroid-stimulating hormone (TSH). Under the influence of TSH, the thyroid grows and secretes thyroid hormone into the blood. Thyroid hormone increases the metabolic rate of most body cells, warming the body as a result—among other functions. The hypothalamus takes some cues from the environment; for instance, cold temperatures tend to increase its secretion of TRH. Precise regulation of the TRH-TSH-thyroid hormone cascade pathway maintains homeostasis. As the red arrows in Figure 26.5E indicate, negative-feedback mechanisms control the secretion of thyroid hormone. When thyroid hormone increases in the blood, it acts on the hypothalamus and anterior pituitary, inhibiting TRH and TSH secretion and consequently thyroid hormone synthesis.

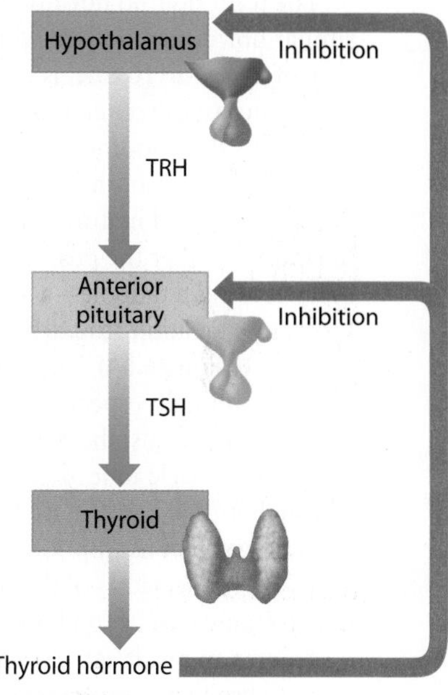

Hypothalamus
Inhibition
TRH
Anterior pituitary
Inhibition
TSH
Thyroid
Thyroid hormone

▲ **Figure 26.5E** Control of thyroid hormone secretion

A workplace "chain-of-command" analogy helps summarize what you have learned about the hypothalamus. Just as a manager, directed by a CEO, tells workers to perform various duties, the anterior and posterior pituitary, directed by the hypothalamus, stimulate a number of other tissues and endocrine glands to perform specific functions. The pathways are tightly controlled; negative feedback maintains homeostasis. In the next several modules, we explore how the body's activities are regulated by endocrine secretions.

? What do the neurosecretory cells of the hypothalamus secrete?

● Oxytocin and ADH, which are stored in the posterior pituitary until their release, and releasing and inhibiting hormones that act on the anterior pituitary

Hormones and Homeostasis

26.6 The thyroid regulates development and metabolism

Having considered many of the endocrine glands and their secretions, we can now take a closer look at the body functions that the **thyroid gland** controls. Your thyroid gland is located in your neck, wrapping around the trachea, just under your larynx (voice box). Thyroid hormone performs several important homeostatic functions that affect virtually all the tissues of the body.

The term thyroid hormone actually refers to a pair of very similar hormones, both of which contain the element iodine. One of these, **thyroxine**, is called $\mathbf{T_4}$ because it contains four iodine atoms; the other, **triiodothyronine**, is called $\mathbf{T_3}$ because it contains three iodine atoms. In target cells, most T_4 is converted to T_3. One of the crucial roles of both hormones is in development and maturation. In a bullfrog, for example, these hormones trigger the profound reorganization of body tissues that occurs as a tadpole—a strictly aquatic organism—transforms into an adult frog, which may spend much of its time on land (**Figure 26.6A**). Thyroid hormones are equally important in mammals, especially in bone and nerve cell development. T_3 and T_4 also help maintain normal blood pressure, heart rate, muscle tone, digestion, and reproductive function. Throughout the body, these hormones tend to increase the rate of oxygen consumption and cellular metabolism.

When homeostasis of the thyroid hormones is not maintained (too much or too little in the blood), serious metabolic disorders can result. An excess of T_3 and T_4 in the blood (*hyper*thyroidism) can make a person overheat, sweat profusely, become irritable, develop high blood pressure, and lose weight. The most common form of hyperthyroidism is Graves' disease; protruding eyes caused by fluid accumulation behind the eyeballs are a typical symptom. Conversely, insufficient amounts of T_3 and T_4 (*hypo*thyroidism) can cause weight gain, lethargy, and intolerance to cold. Hypothyroidism is most commonly caused by an autoimmune reaction, in which the body's own defensive cells destroy the thyroid tissue. Aging also takes its toll on thyroid function. Mild hypothyroidism is often diagnosed in elderly individuals.

▲ **Figure 26.6A** The maturation of a tadpole into an adult frog as regulated by thyroid hormones

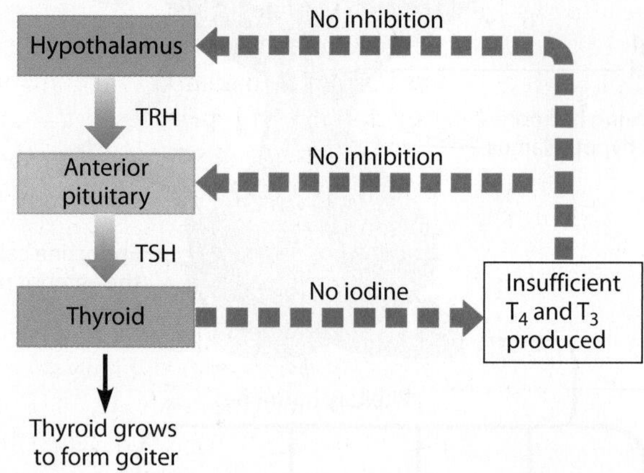
Hypothalamus
No inhibition
TRH
Anterior pituitary
No inhibition
TSH
Thyroid
No iodine
Insufficient T_4 and T_3 produced
Thyroid grows to form goiter

▲ **Figure 26.6B** How iodine deficiency causes goiter

Hypothyroidism can also result from dietary disorders. For example, severe iodine deficiency during childhood can cause a condition that results in retarded skeletal growth and poor mental development. And in adults, insufficient iodine in the diet can cause **goiter**, an enlargement of the thyroid (see Figure 2.2A). Why does iodine deficiency cause thyroid enlargement? Recall that iodine atoms are a component of both T_3 and T_4 hormones; the thyroid cannot synthesize adequate amounts of them without enough iodine. The lack of T_3 and T_4 interrupts the feedback loops that control thyroid activity (**Figure 26.6B**). The blood never carries enough of the T_3 and T_4 hormones to shut off the secretion of TRH (TSH-releasing hormone) or TSH. The thyroid enlarges because TSH continues to stimulate its growth.

Fortunately, both hypo- and hyperthyroidism can be successfully treated. For example, many cases of goiter can be treated simply by adding iodine to the diet. Seawater is a rich source of iodine, and goiter rarely occurs in people living near the seacoast, where the soil is iodine-rich and a lot of iodine-rich seafood is consumed. Goiter is less common today than in the past thanks to the incorporation of iodine into table salt, but it still affects a large percentage of the population in some developing nations. Treatment for hyperthyroidism takes advantage of the fact that the thyroid accumulates iodine: Patients drink a solution containing a small dose of radioactive iodine, which kills off some thyroid cells. This "radioactive cocktail" can kill off just enough cells to reduce thyroid output and relieve symptoms.

? A pregnant woman with hyperthyroidism is being treated with radioactive iodine. Would this be a danger for the fetus?

Yes, we would expect the radioactive iodine to collect in the fetal thyroid gland (contained in T_3 and T_4) and decrease its function.

26.7 The gonads secrete sex hormones

Like thyroid hormone, sex hormones play a role in growth and development, too. Reproductive cycles and sexual behavior are also regulated by sex hormones. The **gonads**, or sex glands (ovaries in the female and testes in the male), secrete sex hormones in addition to producing gametes (ova and sperm).

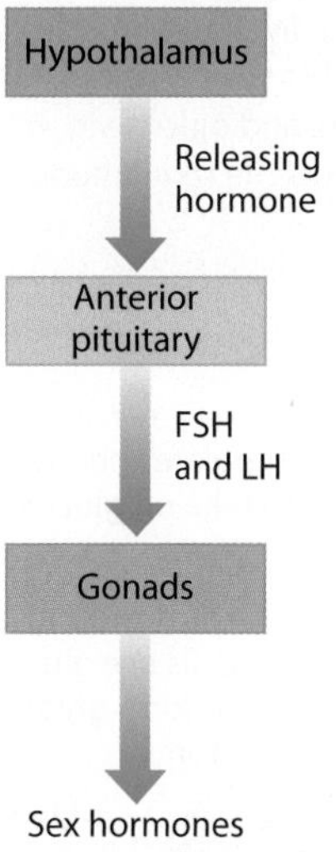

Hypothalamus
↓ Releasing hormone
Anterior pituitary
↓ FSH and LH
Gonads
↓
Sex hormones

▲ **Figure 26.7A** Control of sex hormone production

The synthesis of sex hormones by the gonads is yet another example of how hormones are regulated in a cascade pathway by the hypothalamus and anterior pituitary **(Figure 26.7A)**. In response to a releasing hormone from the hypothalamus, the anterior pituitary secretes follicle-stimulating hormone (FSH) and luteinizing hormone (LH). These stimulate the ovaries or testes to synthesize and secrete the sex hormones, among other effects.

The gonads of mammals produce three major categories of sex hormones: estrogens, progestins, and androgens. Females and males have all three types but in different proportions. Females have a high ratio of estrogens to androgens. In humans, **estrogens** maintain the female reproductive system and promote the development of female features like smaller body size, higher-pitched voice, breasts, and wider hips. The two endocrine disruptors we discussed in the chapter introduction, atrazine and BPA, are synthetic chemicals that mimic estrogen. In Module 26.3, we learned about a consequence of exposure to these kinds of molecules in frogs—the demasculinization of males. In all mammals, **progestins**, such as progesterone, are primarily involved in preparing and maintaining the uterus to support an embryo.

In general, **androgens** stimulate the development and maintenance of the male reproductive system. Males have a high ratio of androgens to estrogens, with their main androgen being **testosterone**. In humans, androgens produced by male embryos during the seventh week of development stimulate the embryo to develop into a male rather than a female. During puberty, high concentrations of androgens trigger the development of male characteristics, such as a lower-pitched voice, facial hair, and large skeletal muscles.

Imbalance of sex hormones can complicate the development of sexual characteristics. **(Figure 26.7B)** shows a pedigree for a family with an X-linked recessive trait known as androgen insensitivity syndrome. In this disorder, testosterone (an androgen hormone) enters the target cell but cannot bind to its nuclear receptor because the nuclear receptor is defective. Thus, target cells are insensitive to testosterone present in the blood. It's as if the blood contains no testosterone at all. The lack of testosterone signaling is an inherited trait and affected individuals are genetic males with one X chromosome and one Y chromosome. They are born with external female genitalia, but male testes are inside their abdomen. Male external

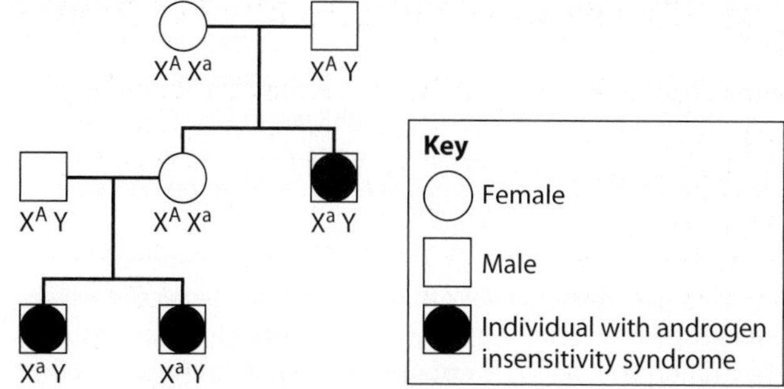

$X^A X^a$
$X^A Y$
$X^A Y$
$X^A X^a$
$X^a Y$
$X^a Y$
$X^a Y$
Key
Female
Male
Individual with androgen insensitivity syndrome

▲ **Figure 26.7B** Androgen insensitivity in a family

genitalia and female ovaries do not develop. At puberty, testosterone insensitivity causes failure of male characteristics such as facial hair and deepening of the voice to develop. Rather, the affected pubescent individual experiences breast enlargement. In the pedigree shown in Figure 26.7B, males are represented by squares and females by circles and the genetically male but phenotypically more female individuals—those with androgen insensitivity syndrome—are represented by circles inside squares. Androgen insensitivity syndrome is just one of many disorders that arise when sex hormones or their receptors malfunction.

Research has established that the process of sex determination driven by androgens occurs in a highly similar manner in all vertebrates, suggesting that androgens had this role early in evolution. Testosterone affects male characteristics not only in humans but also in other vertebrates. For example, testosterone causes aggressive male behavior in elephant seals, increased singing in male songbirds, and the development of manes in male lions **(Figure 26.7C)**.

We have now learned about the functions of the gonads and the thyroid gland and their roles in maintaining homeostasis. In our next two modules, we examine the pancreas and its important function in balancing blood glucose level.

? Knowing how the hypothalamus directs hormone production, predict how androgen production is inhibited.

By negative feedback. High levels of androgen in the blood inhibit the production of releasing hormone by the hypothalamus and FSH and LH by the anterior pituitary (see Figure 27.4D).

▲ **Figure 26.7C** Female (left) and male (right) lions

VISUALIZING THE CONCEPT

26.8 Pancreatic hormones regulate blood glucose level

The **pancreas** is a gland with dual functions: It secretes digestive enzymes into the small intestine, and it secretes two protein hormones, **insulin** and **glucagon**, into the blood. These hormones regulate the level of glucose in the blood and thereby control the amount of glucose circulating through the body. Recall that glucose is an energy source for animal cells. Let's see how blood glucose level is regulated.

Scattered throughout the pancreas are clusters of endocrine cells, called pancreatic islets. Within each islet are beta cells, which produce insulin, and alpha cells, which produce glucagon. Insulin and glucagon are said to be **antagonistic hormones** because the effects of one oppose the effects of the other. The balance in secretion of insulin and glucagon maintains a homeostatic "set point" of glucose. Two negative feedback systems manage the amount of glucose in the blood. One feedback system lowers glucose through release of insulin, whereas the other raises it through release of glucagon. When insulin is present in the blood, glucose is taken up by nearly all cells, and excess glucose is stored in liver and muscle cells as a polysaccharide called glycogen. When the blood contains glucagon, the glycogen stores are broken down, and glucose is returned to the blood. The figure below illustrates the regulation of blood glucose level, using a human example.

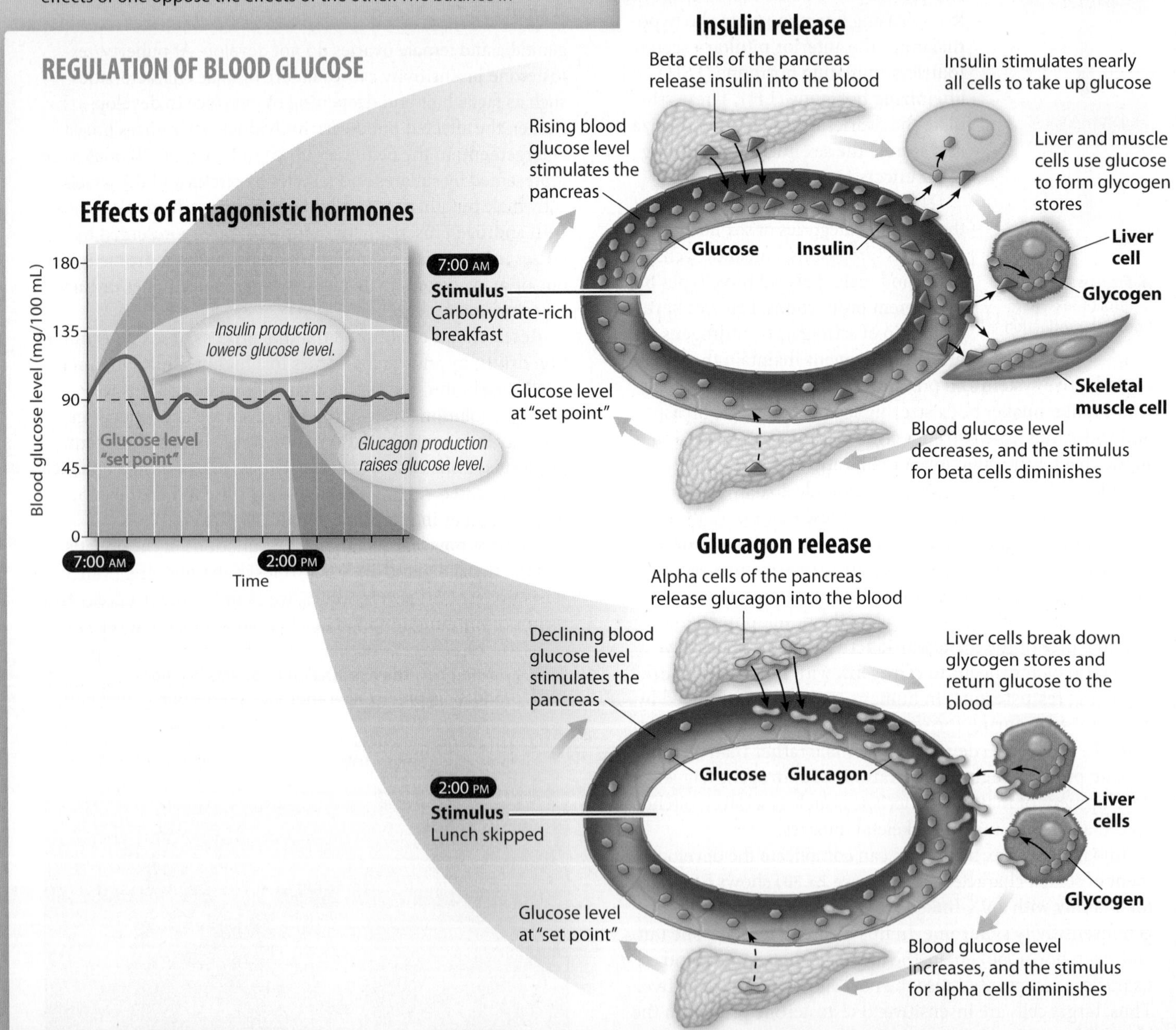

REGULATION OF BLOOD GLUCOSE
Effects of antagonistic hormones
Blood glucose level (mg/100 mL)
180
135
90
45
0
Insulin production lowers glucose level.
Glucose level "set point"
Glucagon production raises glucose level.
7:00 AM
2:00 PM
Time
Insulin release
Beta cells of the pancreas release insulin into the blood
Insulin stimulates nearly all cells to take up glucose
Rising blood glucose level stimulates the pancreas
Liver and muscle cells use glucose to form glycogen stores
Glucose
Insulin
Liver cell
Glycogen
7:00 AM
Stimulus
Carbohydrate-rich breakfast
Skeletal muscle cell
Glucose level at "set point"
Blood glucose level decreases, and the stimulus for beta cells diminishes
Glucagon release
Alpha cells of the pancreas release glucagon into the blood
Declining blood glucose level stimulates the pancreas
Liver cells break down glycogen stores and return glucose to the blood
Glucose
Glucagon
Liver cells
2:00 PM
Stimulus
Lunch skipped
Glycogen
Glucose level at "set point"
Blood glucose level increases, and the stimulus for alpha cells diminishes

? If an individual had no functioning beta cells, what would be the consequence?

● Insulin would not be produced and cells would not be stimulated to take up glucose.

26.9 Diabetes is a common endocrine disorder

CONNECTION

In any homeostatic mechanism, an imbalance in regulation can have detrimental effects. **Diabetes mellitus** is a serious hormonal disorder caused by the body's inability to produce and/or use insulin, thereby decreasing the absorption of glucose from the blood. The result of diabetes is elevated blood glucose levels, or **hyperglycemia**. Diabetes affects 1 in 12 Americans and is actually a group of diseases. In all types of diabetes, the cells cannot obtain enough glucose from the blood even though there is plenty. Thus starved for fuel, cells are forced to burn the body's supply of fats and proteins. Meanwhile, because the digestive system continues to absorb glucose from ingested food, the glucose concentration in the blood can become extremely high.

There are treatments for diabetes mellitus—insulin injections, other medications, and/or special diets—but no cure. Hyperglycemia associated with untreated diabetes can cause dehydration, blindness, cardiovascular and kidney disease, and nerve damage. Every year, more than 200,000 Americans die from the disease or its complications, making diabetes the seventh leading cause of death in the United States. By 2050, if current trends continue, one in three Americans will be affected.

Target cells normally respond to insulin by taking up glucose from the blood, thus lowering blood glucose level. As shown in the left panel of **Figure 26.9**, ❶ the binding of insulin to the insulin receptor ❷ initiates internal cell signals that result in glucose transporters being shuttled from vesicles to the plasma membrane. ❸ Glucose then enters the target cell via facilitated diffusion (see Module 5.6).

Type 1 (insulin-dependent) diabetes is an autoimmune disorder in which white blood cells of the body's own immune system attack and destroy the pancreatic beta cells that release the hormone insulin. As a result, the pancreas does not produce enough insulin. As indicated in the middle panel of Figure 26.9, without sufficient insulin, target cells do not receive the signal that leads to glucose transporters being moved to the plasma membrane. Without a way to bring glucose into most cells, glucose level in the blood remains elevated. Type 1 diabetes generally develops during childhood. The human body cannot regenerate beta cells, but type 1 patients can be treated with injections, several times daily, of human insulin.

In type 2 (non-insulin-dependent) diabetes, insulin is produced, but the insulin signal is not relayed normally inside the target cells (right panel, Figure 26.9). Often, pancreatic beta cells adapt to the defective signaling by producing even greater amounts of insulin (like the way you might yell louder at someone wearing ear plugs). Although insulin is present, glucose transporters are not as readily available at the plasma membrane to take up glucose from the blood, and the target cells are said to be "insulin resistant." Type 2 diabetes is almost always associated with being overweight and underactive, although whether obesity causes diabetes (and if so, how) remains unknown. In the United States, more than 90% of diabetics are type 2. Many type 2 diabetics can manage their blood glucose level with exercise and diet. Yet, as the disorder progresses, insulin production from beta cells may decrease, requiring oral medications or insulin injections to lower blood sugar.

A third type of diabetes, called gestational diabetes, can affect any pregnant woman. About 4% of pregnant women in the United States develop the disorder. As in type 2 diabetes, target cells do not respond to insulin normally, and glucose remains elevated in the blood. The cause of gestational diabetes is not known, but if left untreated, it can lead to dangerously large babies, which can greatly complicate delivery. If diagnosed, diet, exercise, and/or insulin injections can prevent most problems.

Some people have hyperactive beta cells that secrete too much insulin into the blood when sugar is eaten. As a result, blood glucose level can drop well below normal. This condition, called **hypoglycemia**, usually occurs 2–4 hours after a meal and may be accompanied by hunger, weakness, sweating, and nervousness. In severe cases, when the brain receives inadequate amounts of glucose, a person may develop convulsions, become unconscious, and even die. Hypoglycemia is uncommon, and most forms of it can be controlled by reducing sugar intake and eating smaller, more frequent meals.

Capillary
Insulin
Blood
Glucose
Insulin receptors
1
3 Glucose transporter
2
Facilitated diffusion of glucose
Vesicle containing glucose transporters

Normal glucose and insulin levels

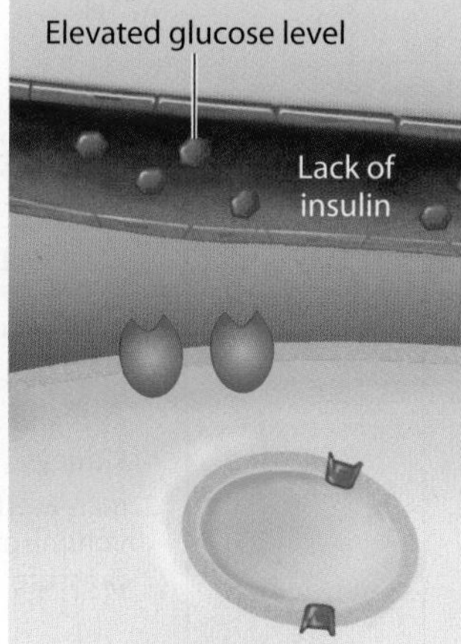

Elevated glucose level
Lack of insulin

Type I diabetes: insulin is absent

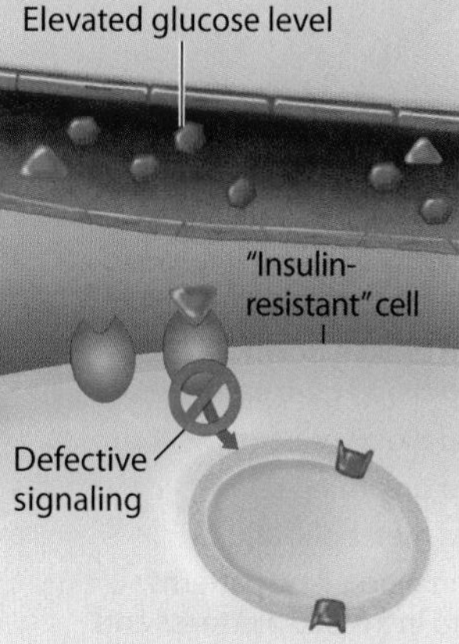

Elevated glucose level
"Insulin-resistant" cell
Defective signaling

Type II diabetes: insulin signaling is defective

▲ **Figure 26.9** Comparison of type 1 and type 2 diabetes

Try This On a blank sheet of paper, redraw this figure from memory, and explain it.

? **What would probably happen if a person with type 1 diabetes injected too much insulin?**

● Hypoglycemia, which could lead to convulsions and death

26.10 The adrenal glands mobilize responses to stress

The endocrine system includes two **adrenal glands**, one sitting on top of each kidney. As you can see in **Figure 26.10** (inset, top left), each adrenal gland is actually made up of two glands fused together: a central portion called the **adrenal medulla** and an outer portion called the **adrenal cortex**. Though the cells they contain and the hormones they produce are different, both the adrenal medulla and the adrenal cortex secrete hormones that enable the body to respond to stress.

The adrenal medulla produces the "fight-or-flight" hormones that we discussed earlier; they ensure a rapid, short-term response to stress. You've probably felt your heart beat faster and your skin develop goose bumps when sensing danger. Facing an unexpected threat, like a pop quiz, can cause these short-term stress symptoms. Positive emotions—extreme pleasure, for instance—can produce the same effects. These reactions are triggered by two hormones secreted by the adrenal medulla, **epinephrine** (adrenaline) and **norepinephrine** (noradrenaline).

Stressful stimuli, whether negative or positive, activate certain nerve cells in the hypothalamus. ❶ These cells send nerve signals via the spinal cord to the adrenal medulla, ❷ stimulating it to secrete epinephrine and norepinephrine into the blood. Epinephrine and norepinephrine have somewhat different effects on tissues, but both contribute to the short-term stress response. Both hormones stimulate liver cells to release glucose, thus making more fuel available for cellular work. They also prepare the body for action by raising the blood pressure, breathing rate, and metabolic rate. In addition, epinephrine and norepinephrine change blood flow patterns, making some organs more active and others less so. For example, epinephrine dilates blood vessels in the brain and skeletal muscles, thus increasing alertness and the muscles' ability to react to stress. At the same time, epinephrine and norepinephrine constrict blood vessels elsewhere, thereby reducing activities that are not immediately involved in the stress response, such as digestion. The short-term stress response prepares the body for a quick and decisive reaction—as would be needed to confront a sudden danger—but it occurs and subsides rapidly.

In contrast to epinephrine and norepinephrine secreted by the adrenal medulla, hormones secreted by the adrenal cortex can provide a slower, longer-lasting response to stress

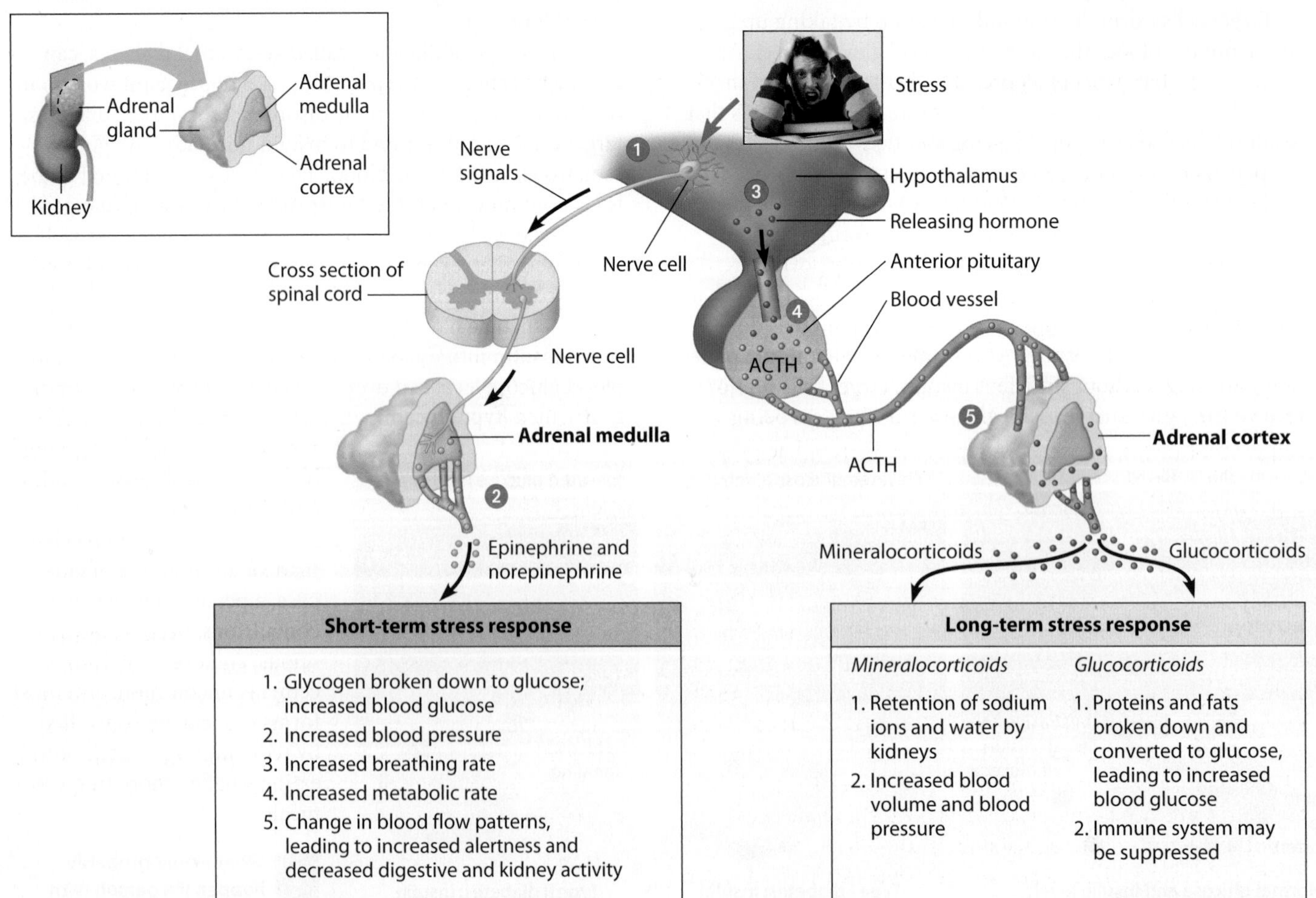

Adrenal gland
Adrenal medulla
Adrenal cortex
Kidney
Stress
Nerve signals
1
Hypothalamus
3
Releasing hormone
Cross section of spinal cord
Nerve cell
Anterior pituitary
Blood vessel
4
ACTH
Nerve cell
5
Adrenal medulla
Adrenal cortex
ACTH
2
Epinephrine and norepinephrine
Mineralocorticoids
Glucocorticoids
Short-term stress response
1. Glycogen broken down to glucose; increased blood glucose
2. Increased blood pressure
3. Increased breathing rate
4. Increased metabolic rate
5. Change in blood flow patterns, leading to increased alertness and decreased digestive and kidney activity
Long-term stress response
Mineralocorticoids
1. Retention of sodium ions and water by kidneys
2. Increased blood volume and blood pressure
Glucocorticoids
1. Proteins and fats broken down and converted to glucose, leading to increased blood glucose
2. Immune system may be suppressed

▲ **Figure 26.10** How the adrenal glands control our responses to stress

Try This After examining the figure, make a table that contrasts the differences in stimuli and responses for the adrenal medulla versus the adrenal cortex.

that can last for days. The adrenal cortex responds to endocrine signals—chemical signals in the blood—rather than to nerve cell signals. As shown in Figure 26.10, ❸ the hypothalamus secretes a releasing hormone that ❹ stimulates target cells in the anterior pituitary to secrete the hormone **adrenocorticotropic hormone (ACTH)**. ❺ In turn, ACTH stimulates cells of the adrenal cortex to synthesize and secrete a family of steroid hormones called the **corticosteroids**. The two main types in humans are the mineralocorticoids and the glucocorticoids. Both help maintain homeostasis when the body experiences long-term stress.

Mineralocorticoids act mainly on salt and water balance. One of these hormones (aldosterone) stimulates the kidneys to reabsorb sodium ions and water, with the overall effect of increasing the volume of the blood and raising blood pressure as a response to prolonged stress.

Glucocorticoids function mainly in mobilizing cellular fuel, thus reinforcing the effects of glucagon. Glucocorticoids promote the synthesis of glucose from noncarbohydrates such as proteins and fats. When the body cells consume more glucose than the liver can provide from glycogen stores, glucocorticoids stimulate the breakdown of muscle proteins, making amino acids available for conversion to glucose by the liver. This makes more glucose available in the blood as cellular fuel in response to stress.

Very high levels of glucocorticoids can suppress the body's defense system, including the inflammatory response (see Module 24.2). For this reason, physicians may use glucocorticoids to treat excessive inflammation. The glucocorticoid cortisone, for example, was once regarded as a miracle drug for treating serious inflammatory conditions such as arthritis. Cortisone and other glucocorticoids can relieve swelling and pain from inflammation; but by suppressing immunity, they can also make a person highly susceptible to infection.

Physicians often prescribe oral glucocorticoids to relieve pain from athletic injuries. However, glucocorticoids are potentially very dangerous; prolonged use can depress the activity of the adrenal glands and cause side effects such as a weakened immune system, easy bruising, weak bones, weight gain, muscle breakdown, and increased risk of diabetes. It is safer, but still potentially dangerous, to inject a glucocorticoid at the site of injury. With this treatment, the pain usually subsides, but its underlying cause remains. Masking the pain covers up the pain's message—that tissue is damaged.

? **Epinephrine and norepinephrine are responsible for the ______________ stress response while the mineralocorticoids and glucocorticoids are responsible for the ______________ stress response.**

short-term . . . long-term

26.11 A single hormone can perform a variety of functions in different animals

EVOLUTION CONNECTION

Hormones play important roles in all vertebrates, and some of the same hormones can be found in vertebrates that are only distantly related. Interestingly, the same hormone can have different effects in different animals—a strong indication that hormonal regulation was an early evolutionary adaptation.

The hormone prolactin (PRL), produced and secreted by the anterior pituitary under the direction of the hypothalamus, is a good example. Prolactin produces diverse effects in different vertebrate species. In humans, PRL performs several important functions related to childbirth. During late pregnancy, PRL stimulates mammary glands to grow and produce milk. (A brief surge in prolactin level just before menstruation causes breast swelling and tenderness in some women.) Suckling by a newborn stimulates further release of PRL, which in turn increases the milk supply (**Figure 26.11**). High prolactin level during nursing tend to prevent the ovaries from releasing eggs, decreasing the chances of a new pregnancy occurring during the time of breast-feeding. This may be an evolutionary adaptation that helps ensure that adequate care is given to newborns.

PRL plays a wide variety of other roles unrelated to childbirth. In some nonhuman mammals, PRL stimulates nest building. In birds, PRL regulates fat metabolism and reproduction. In amphibians, it stimulates movement toward water in preparation for breeding and affects metamorphosis. In fish that migrate between salt and fresh water (salmon, for example), PRL helps regulate salt and water balance in the gills and kidneys.

▲ **Figure 26.11** Suckling promotes prolactin production

Such diverse effects suggest that prolactin is an ancient hormone whose functions diversified through evolution. Over millions of years, the prolactin molecule stayed the same, but its role changed dramatically—a good example of how evolution can both preserve unity (in terms of the structure of the molecule itself) and promote diversity (in terms of the varying roles it plays).

? **PRL promotes the production of milk. Newborn suckling promotes PRL production. This is an example of ______________ feedback.**

Positive

26.12 Hormones can promote social behaviors

CONNECTION

Humans and dogs can form strong attachments to each other and have done so for thousands of years. The relationship makes sense—a biological factor promoting social bonds between dogs and humans would have been significant in human evolution, because hunting with the help of domesticated dogs increases food supply. What is it about a dog that makes it "man's best friend"? Recently, scientists studied whether it is a hormone that induces the human-dog relationship.

In addition to its role in uterine contractions and mammary milk ejection, the hormone oxytocin plays a part in mammalian social behaviors. For example, oxytocin promotes mating bonds. In one experiment, female voles placed with unfamiliar males for a few hours formed stronger preferences for these partners if simultaneously injected with oxytocin. Oxytocin promotes maternal bonds, too. When injected into virgin female rats, oxytocin initiates strong maternal behaviors toward unrelated pups (such as licking and crouching over the pups). The function of oxytocin in social attachments in humans is less well understood, but oxytocin is associated with maternal behavior in human females as well. Levels of the hormone rise when new mothers gaze into the eyes of their babies or touch and talk to their young.

A similar surge in oxytocin appears to occur in humans when they are interacting with their pet dogs. One study demonstrated that oxytocin levels rise in dog owners when they are affectionately playing with their dogs; levels were highest when dog owners received long gazes from their dogs. The hormone response may be mutual, too. In a separate preliminary study, oxytocin was found to be released in dogs after a petting session. The next time your cute pooch stares at you devotedly, consider the biology promoting your response. Could it be oxytocin that compels you to pat your dog's head, just as it prompts a mother to coo over her infant?

? This module discusses how oxytocin is released in response to a cue from the nervous system. What type of cell conducts electrical signals and also produces hormones?

Neurosecretory cells; oxytocin is specifically synthesized by neurosecretory cells in the hypothalamus.

CHAPTER 26 REVIEW

For practice quizzes, BioFlix animations, MP3 tutorials, video tutors, and more study tools designed for this textbook, go to MasteringBiology®

Reviewing the Concepts

The Nature of Chemical Regulation (26.1–26.3)

26.1 Chemical and electrical signals coordinate body functions. Hormones are signaling molecules, usually carried in the blood, that cause specific changes in target cells. All hormone-secreting cells make up the endocrine system, which works with the nervous system in regulating body activities.

26.2 Hormones affect target cells using two main signaling mechanisms.

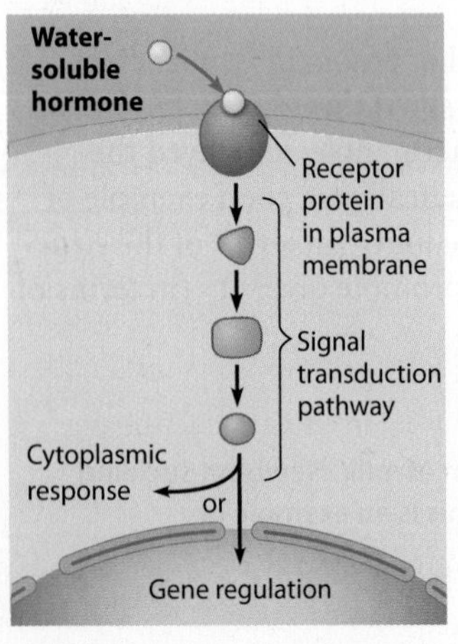

Water-soluble hormone
Receptor protein in plasma membrane
Signal transduction pathway
Cytoplasmic response
or
Gene regulation

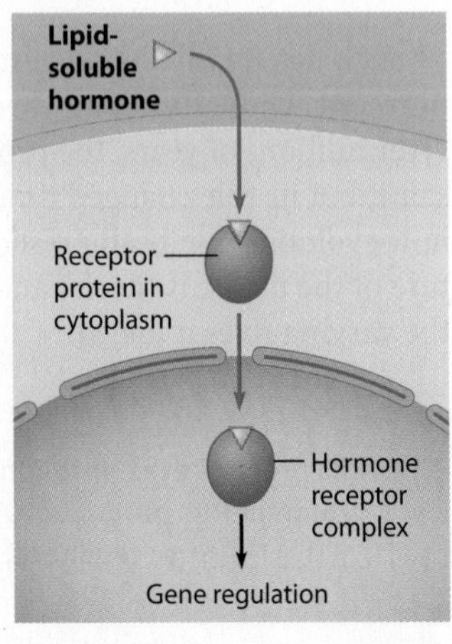

Lipid-soluble hormone
Receptor protein in cytoplasm
Hormone receptor complex
Gene regulation

26.3 A widely used weed killer demasculinizes male frogs. The weed-killer atrazine is called an endocrine disrupter because it interferes with endocrine function.

The Vertebrate Endocrine System (26.4–26.5)

26.4 The vertebrate endocrine system consists of more than a dozen major glands.

26.5 The hypothalamus, which is closely tied to the pituitary, connects the nervous and endocrine systems.

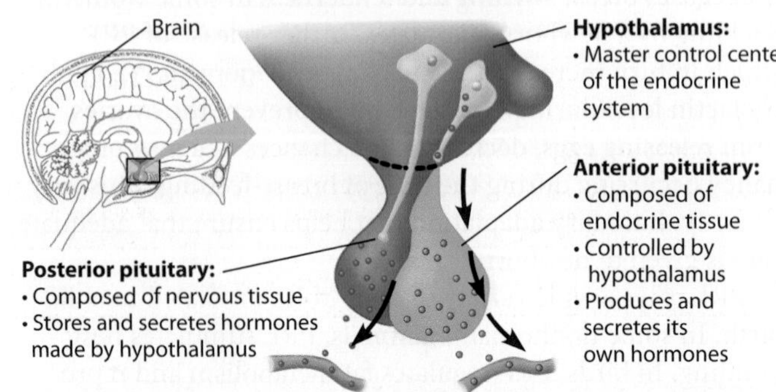

Brain
Hypothalamus:
• Master control center of the endocrine system
Anterior pituitary:
• Composed of endocrine tissue
• Controlled by hypothalamus
• Produces and secretes its own hormones
Posterior pituitary:
• Composed of nervous tissue
• Stores and secretes hormones made by hypothalamus

Hormones and Homeostasis (26.6–26.12)

26.6 The thyroid regulates development and metabolism. Thyroid hormones regulate an animal's development and metabolism. Negative feedback maintains homeostatic levels in the blood.

26.7 The gonads secrete sex hormones. Estrogens, progestins, and androgens are steroid sex hormones produced by the gonads in response to signals from the hypothalamus and pituitary.

26.8 Pancreatic hormones regulate blood glucose level.

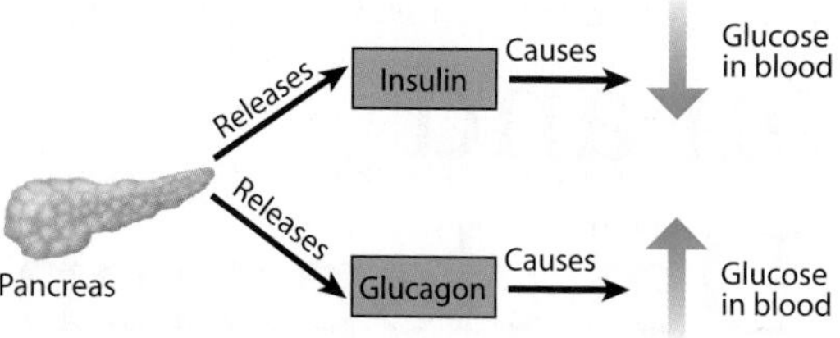

Releases
Insulin
Causes
Glucose in blood
Releases
Glucagon
Causes
Glucose in blood
Pancreas

26.9 Diabetes is a common endocrine disorder. Diabetes mellitus results from a lack of insulin or a failure of cells to respond to it.

26.10 The adrenal glands mobilize responses to stress. Nerve signals from the hypothalamus stimulate the adrenal medulla to secrete epinephrine and norepinephrine, which quickly trigger the "fight-or-flight" response. ACTH from the pituitary causes the adrenal cortex to secrete glucocorticoids and mineralocorticoids, which boost blood pressure and energy in response to long-term stress.

26.11 A single hormone can perform a variety of functions in different animals. Diverse functions have evolved for hormones, such as prolactin, that are species specific.

26.12 Hormones can promote social behaviors.

Connecting the Concepts

1. Complete this map, which presents some major concepts from this chapter.

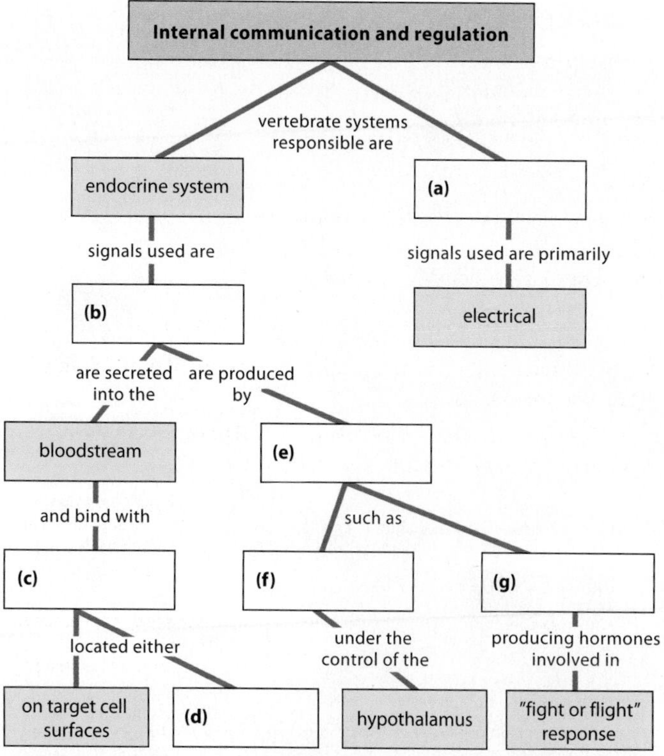

Internal communication and regulation
vertebrate systems responsible are
endocrine system
(a)
signals used are
signals used are primarily
(b)
electrical
are secreted into the
are produced by
bloodstream
(e)
and bind with
such as
(c)
(f)
(g)
located either
under the control of the
producing hormones involved in
on target cell surfaces
(d)
hypothalamus
"fight or flight" response

Testing Your Knowledge

Level 1: Knowledge/Comprehension

2. Which correctly matches a hormone to the gland from which it is produced and to its effect on target cells?
   a. thyroxine: anterior pituitary, regulates metabolism
   b. prolactin: anterior pituitary, raises blood calcium levels
   c. androgens: thyroid, promotes male characteristics
   d. None of the choices are correct.
3. The body is able to maintain a relatively constant level of thyroid hormone in the blood because
   a. thyroid hormone stimulates the pituitary to secrete thyroid-stimulating hormone (TSH).
   b. thyroid hormone inhibits the secretion of TSH-releasing hormone (TRH) from the hypothalamus.
   c. TRH inhibits the secretion of thyroid hormone by the thyroid gland.
   d. thyroid hormone stimulates the hypothalamus to secrete TRH.
4. Explain how the hypothalamus controls body functions through its action on the pituitary gland. How does control of the anterior and posterior pituitary differ?
5. Explain how the same hormone might have different effects on two different target cells and no effect on nontarget cells.

Level 2: Application/Analysis

6. In a glucose tolerance test, periodic measurements of blood glucose level are taken after a person drinks a glucose-rich solution. Using the graph below, compare and contrast the results of the test for the diabetic and the healthy individual.

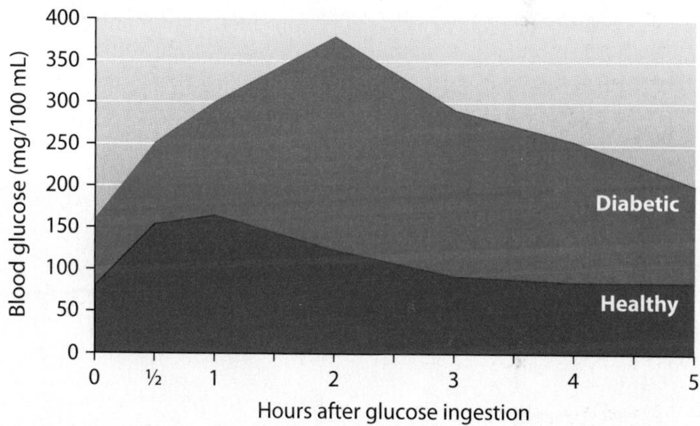

Blood glucose (mg/100 mL)
400
350
300
250
200
150
100
50
0
Diabetic
Healthy
0
½
1
2
3
4
5
Hours after glucose ingestion

7. If a person had a pituitary tumor that is oversecreting TSH, would this person be likely to have goiter as a symptom? Explain.
8. Which two of the hormones listed below act upon the body with similar functions? Explain.
   a. glucagon b. oxytocin c. glucocorticoids d. ADH

Level 3: Synthesis/Evaluation

9. A strain of transgenic mice remains healthy as long as you feed them regularly and do not let them exercise. After they eat, their blood glucose level rises slightly and then declines to a homeostatic level. However, if these mice fast or exercise at all, their blood glucose drops dangerously. Which hypothesis best explains their problem? (*Explain your choice.*)
   a. The mice have insulin-dependent diabetes.
   b. The mice lack insulin receptors on their cells.
   c. The mice lack glucagon receptors on their cells.
   d. The mice cannot synthesize glycogen from glucose.
10. **SCIENTIFIC THINKING** How could a hormonal imbalance result in an animal that is genetically male but physically female?

Answers to all questions can be found in Appendix 4.

CHAPTER

27 Reproduction and Embryonic Development

There is a veritable rogues' gallery of viruses that cause human sexually transmitted diseases (STDs): human immunodeficiency virus (HIV) causes acquired immunodeficiency syndrome (AIDS); human papillomavirus (HPV, shown in the micrograph below) causes genital warts (and, for some strains, cervical cancer); and herpes virus causes genital herpes. Once a viral pathogen enters the human body, it tumbles along until it finds a suitable target cell, which is recognized when the virus's spikes (visible protruding from the surface of each virus in the micrograph) bind to protein receptor molecules on the cell's surface. Soon after binding, the virus infects the cell and its DNA becomes incorporated into the nucleus, where it can remain dormant for long periods. At a later time, the virus may be reactivated, resulting in the production of many copies of the pathogen and the death of the host cell. Such destruction causes the symptoms of the disease and releases more viruses that can infect more cells.

? *What can be done about STDs?*

STDs can be caused by other types of pathogens, including bacteria, protists, and fungi. But viral STDs are uniquely problematic, because viruses can hide within a cell, making them very hard to

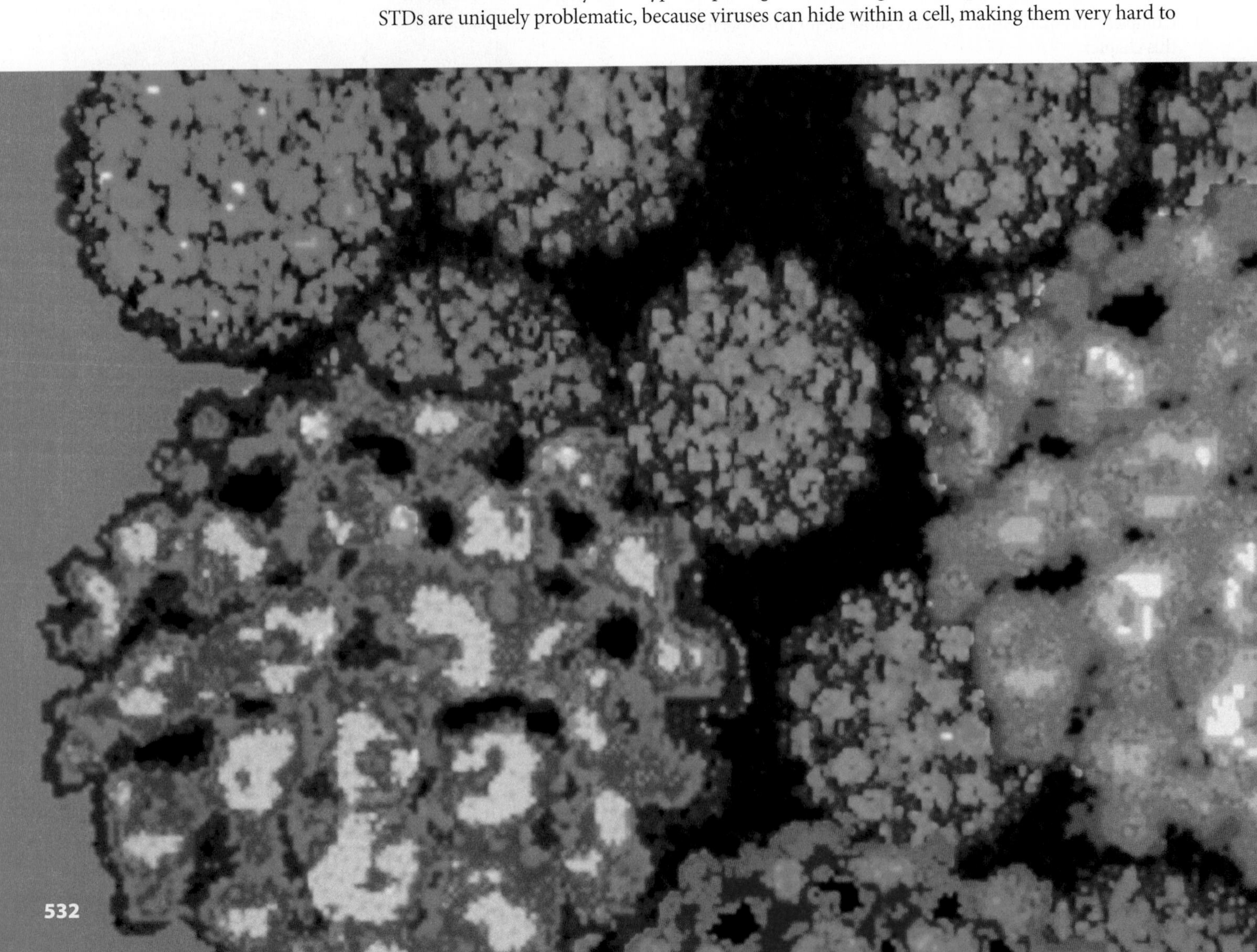

eradicate. Although STDs from other types of pathogens can often be successfully treated, viral STDs cannot be cured and so last a lifetime. Prevention through safe sex practices is the best option for remaining uninfected.

We will revisit the topic of sexually transmitted diseases later in the chapter. But first, we begin with a brief introduction to the diverse ways that animals reproduce, followed by a close look at the reproductive system of our own species. In the second half of the chapter, we discuss the processes of fertilization and embryonic development in vertebrates and then focus on human embryonic development and birth.

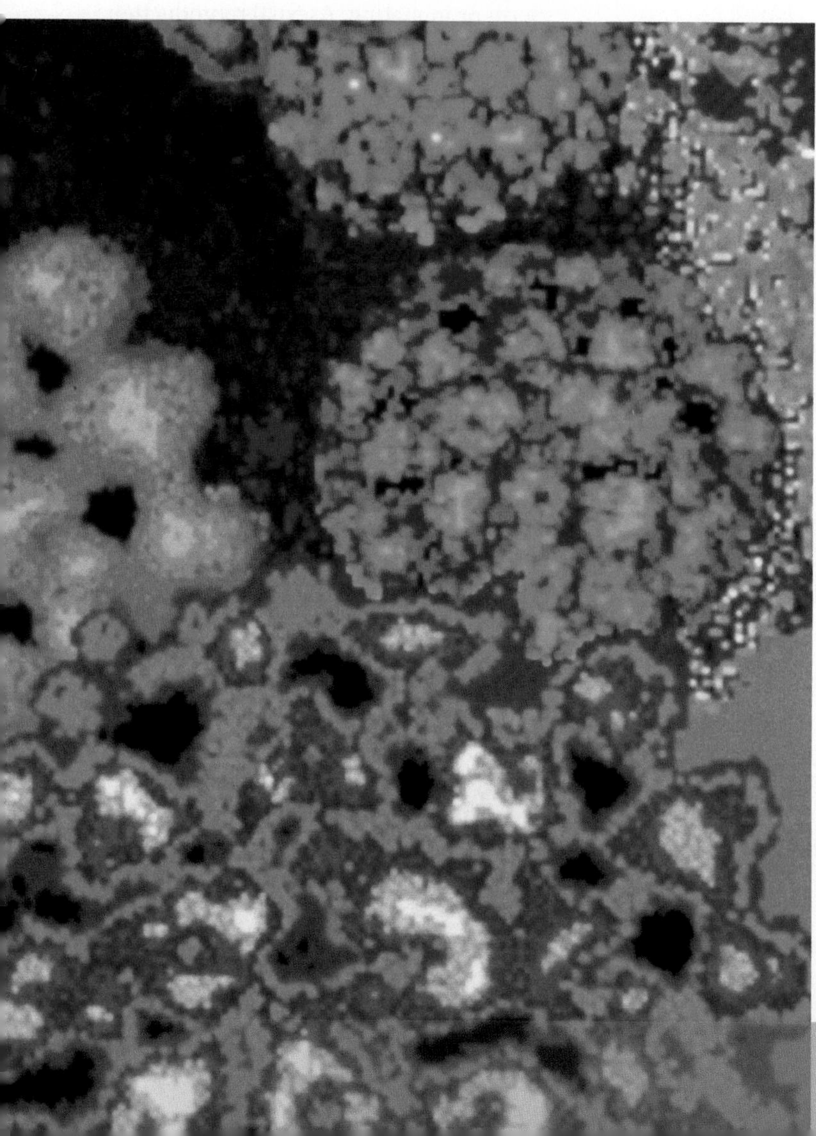

BIG IDEAS

Asexual and Sexual Reproduction
(27.1–27.2)

Some animals can reproduce asexually, but most reproduce by the fusion of egg and sperm.

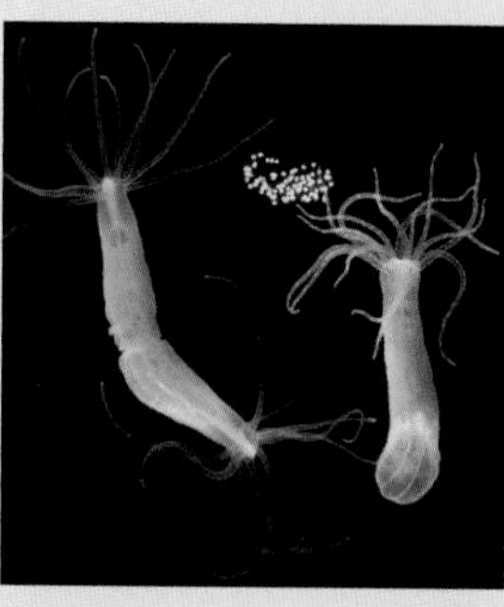

Human Reproduction
(27.3–27.8)

Human males and females have structures that produce, store, and deliver gametes.

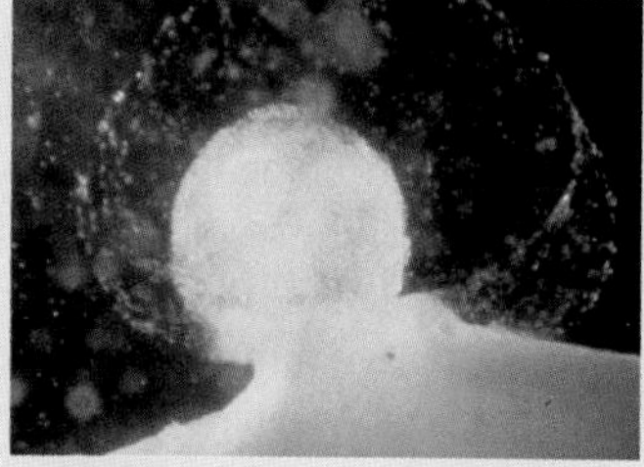

Principles of Embryonic Development
(27.9–27.14)

A zygote develops into an embryo through a series of carefully regulated processes.

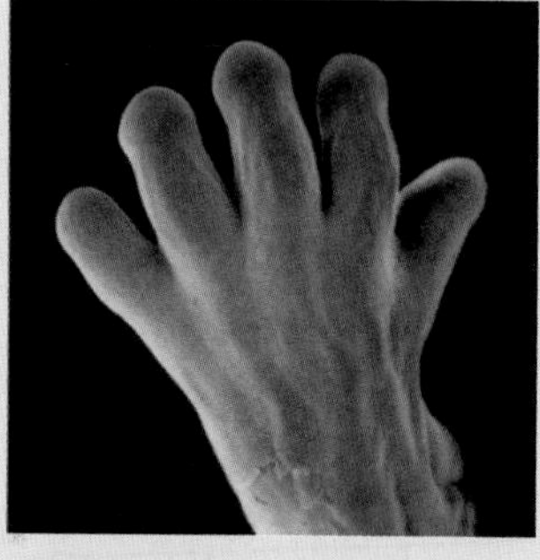

Human Development
(27.15–27.18)

A full-term human fetus develops within the uterus for 9 months.

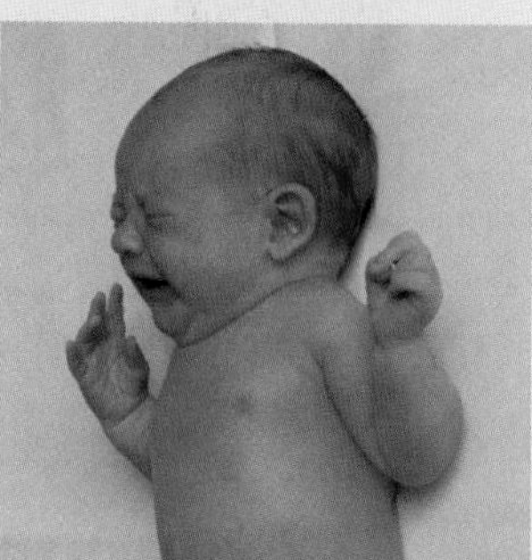

▷ Asexual and Sexual Reproduction

27.1 Asexual reproduction results in the generation of genetically identical offspring

Although every individual animal has a relatively short life span, species transcend this time limit because of **reproduction**, the creation of new individuals from existing ones. Animals reproduce in a great variety of ways, but there are two modes: asexual and sexual.

Asexual reproduction (reproduction without sex) is the creation of genetically identical offspring by a lone parent. Several types of asexual reproduction are found among animals. Many invertebrates, such as the hydra in **Figure 27.1A**, reproduce asexually by **budding**, the outgrowth and eventual splitting off of a new individual from a parent. The sea anemone in the center of **Figure 27.1B** is undergoing **fission**, the separation of a parent into two or more offspring of about equal size. Asexual reproduction can also result from the two-step process of **fragmentation**, the breaking of the parent body into several pieces, followed by **regeneration**, the regrowth of lost body parts. In sea stars (starfish) of the genus *Linckia*, for example, a whole new individual can develop from a broken-off arm plus a bit of the central body. Thus, a single animal with five arms, if broken apart, could potentially give rise to five offspring via asexual reproduction in a matter of weeks. In some species of sea sponges, if a single sponge is pushed through a wire mesh, each of the resulting clumps of cells can regrow into a new sponge. (Besides the natural means of asexual reproduction discussed here, many species have been the subject of artificial asexual reproduction; see the discussion of cloning in Modules 11.12–11.13.)

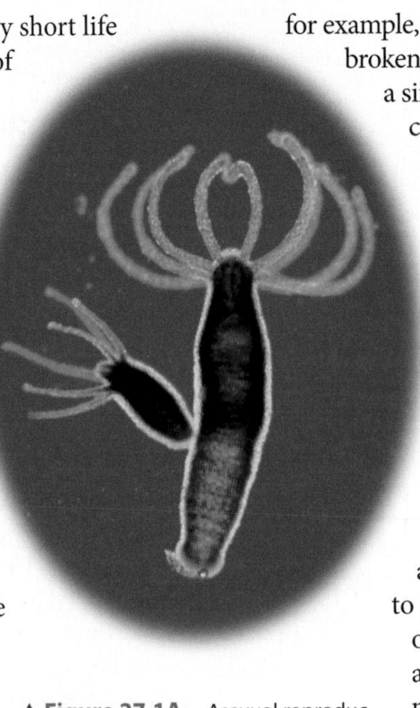

▲ **Figure 27.1A** Asexual reproduction via budding in a hydra

In nature, asexual reproduction has several potential advantages. For one, it allows animals that do not move from place to place or that live in isolation to produce offspring without finding mates. Another advantage is that it enables an animal to produce many offspring quickly; no time or energy is lost in production of eggs and sperm or in mating. Asexual reproduction perpetuates a particular genotype faithfully, precisely, and rapidly. Therefore, it can be an effective way for animals that are genetically well suited to an environment to quickly expand their populations and exploit available resources.

A potential disadvantage of asexual reproduction is that it produces genetically uniform populations. Such individuals may thrive in one particular environment, but if the environment changes and becomes less favorable (because of a natural disaster or a new predator, for example), all individuals may be affected equally, and the entire population may die out.

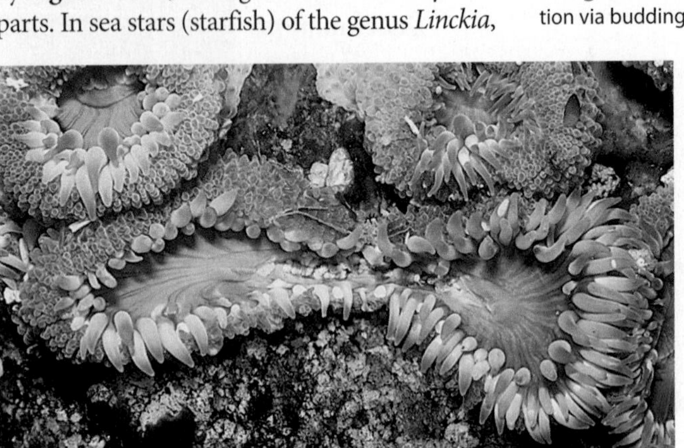

▲ **Figure 27.1B** Asexual reproduction of a sea anemone (*Anthopleura elegantissima*) by fission

? **What kinds of environments would likely be advantageous to asexually reproducing organisms? Why?**

Relatively unchanging environments favor asexual reproduction because well-adapted genotypes are perpetuated in the genetically identical offspring.

27.2 Sexual reproduction results in the generation of genetically unique offspring

Sexual reproduction is the creation of offspring through the process of **fertilization**, the fusion of two haploid (n) sex cells, or **gametes**, to form a diploid ($2n$) **zygote** (fertilized egg). (Recall from Chapter 8 that n refers to the haploid number of chromosomes and $2n$ refers to the diploid number; for humans, $n = 23$ and $2n = 46$.) The male gamete, the **sperm**, is a relatively small cell that moves by means of a flagellum. The female gamete, the **egg**, is a much larger cell that is not self-propelled. The zygote—and the new individual it develops into—contains a unique combination of genes inherited from the parents via the egg and sperm.

Most animals reproduce mainly or exclusively by sexual reproduction, which increases genetic variability among offspring. Meiosis and random fertilization can generate enormous genetic variation (as we discussed in Modules 8.15 and 8.17). And such variation is the raw material of evolution by natural selection. The variability produced by the reshuffling of genes in sexual reproduction may provide greater adaptability to changing environments. Put another way, when an environment changes suddenly or drastically, there is a better chance that some of the varying offspring produced via sexual reproduction will

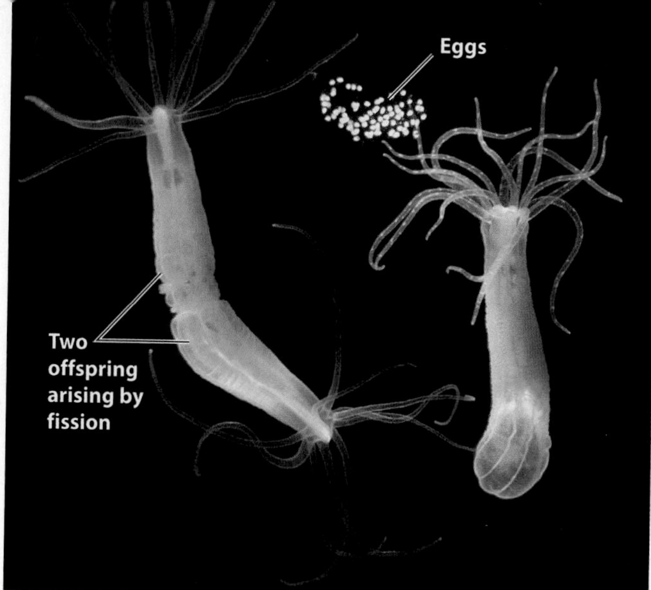
Eggs
Two offspring arising by fission

▲ **Figure 27.2A** Asexual (left) and sexual (right) reproduction in the starlet sea anemone (*Nematostella vectensis*)

▲ **Figure 27.2C** Atlantic horseshoe crabs engaging in behavior that stimulates the release of egg and sperm

survive and reproduce than if all the offspring are genetically very similar.

Animals that can reproduce both asexually and sexually benefit from both modes. In **Figure 27.2A**, you can see two sea anemones of the same species; the one on the left is reproducing asexually (via fission) and the one on the right is releasing eggs. Many other marine invertebrates can also reproduce by both modes. Why would such dual reproductive capabilities be advantageous to an organism? Some animals reproduce asexually when there is ample food and when water temperatures are favorable for rapid growth and development. Asexual reproduction usually continues until cold temperatures signal the approach of winter or until the food supply dwindles or the habitat starts to dry up. As conditions change, the animals switch to sexual reproduction, producing a generation of genetically varied individuals with better potential to adapt to the changing conditions.

Although sexual reproduction has advantages, it presents a problem for nonmobile animals and for those that live solitary lives: how to find a mate. One solution that has evolved is **hermaphroditism**, in which each individual has both female and male reproductive systems. Some hermaphrodites, such as tapeworms, can fertilize their own eggs. Other species require a partner. When hermaphrodites mate (for example, the two earthworms in **Figure 27.2B**), each animal serves as both male and female, donating and receiving sperm. For hermaphrodites, there is only one sex, so any two individuals can mate. Mating can therefore result in twice as many offspring than if only one individual's eggs were fertilized.

▲ **Figure 27.2B** Hermaphroditic earthworms mating

The mechanics of fertilization play an important part in sexual reproduction. Many aquatic invertebrates and most fishes and amphibians exhibit **external fertilization**: The parents discharge their gametes into the water, where fertilization then occurs, often without the male and female even making physical contact. Timing is crucial because the eggs and sperm must be available for fertilization at the same time. For many species, environmental cues such as temperature and day length or chemical signals released by individuals cause a whole population to release gametes all at once, a process called spawning.

When external fertilization is not synchronized across a population, individuals may exhibit specific courtship behaviors. For example, many fishes, amphibians, and some marine invertebrates have specific rituals that trigger simultaneous gamete release in the same vicinity by the female and male. An example of such a mating ritual is the mounting of a female horseshoe crab by a male (**Figure 27.2C**).

In contrast to external fertilization, **internal fertilization** occurs when sperm are deposited in or near the female reproductive tract and gametes unite within the tract. Nearly all terrestrial animals exhibit internal fertilization, which is an adaptation that enables sperm to reach an egg despite a dry external environment. Internal fertilization usually requires **copulation**, or sexual intercourse. It also requires complex reproductive systems, including organs for gamete storage and transport and organs that facilitate copulation. In the next module, we'll examine the complex reproductive anatomy that allows one particular terrestrial animal—namely, humans—to carry out sexual reproduction.

? In terms of genetic makeup, what is the most important difference between the outcome of sexual reproduction and that of asexual reproduction?

● The offspring of sexual reproduction are genetically diverse, whereas the offspring of asexual reproduction are genetically identical.

Human Reproduction

27.3 The human female reproductive system includes the ovaries and structures that deliver gametes

Although we tend to focus on the anatomical differences between the human male and female reproductive systems, there are also some important similarities. Both sexes have a pair of **gonads**, the organs that produce gametes. Also, both sexes have ducts that store and deliver gametes as well as structures that allow mating. In this and the next module, we examine the anatomical features of the human reproductive system, beginning with female anatomy.

A woman's gonads, her **ovaries**, are each about an inch long, with a bumpy surface (**Figure 27.3A**). The bumps are created by **follicles**, each consisting of one developing egg cell surrounded by cells that nourish and protect it. The follicle cells also produce the female sex hormone estrogen. (In this chapter, we use the word *estrogen* to refer collectively to several closely related chemicals that affect the body similarly.)

A female is born with 1–2 million follicles, but only several hundred will release egg cells during her reproductive years. Starting at puberty, one follicle (or, rarely, more than one) matures and releases an immature egg cell about every 28 days, a process called **ovulation** (**Figure 27.3B**). This monthly ovulation cycle continues until a female reaches menopause, which usually occurs around age 50.

After ovulation, what remains of the follicle grows within the ovary to form a solid mass called the **corpus luteum**; you can see one in the ovary on the left in Figure 27.3A. The corpus luteum secretes additional estrogen as well as progesterone, a hormone that helps maintain the uterine lining during pregnancy. If the released egg is not fertilized, the corpus luteum degenerates, and a new follicle matures during the next cycle. We discuss ovulation and female hormonal cycles further in later modules.

Notice in Figure 27.3A that each ovary lies next to the opening of an **oviduct**, also called a fallopian tube. The oviduct entrance resembles a funnel fringed with finger-like projections. The projections touch the surface of the ovary, but the ovary is actually separated from the opening of the oviduct by a tiny space. When ovulation occurs, the egg cell passes across

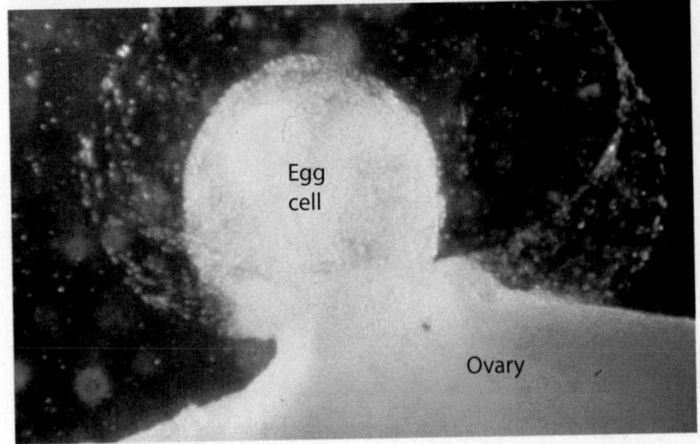

Egg cell
Ovary
LM 200×

▲ **Figure 27.3B** Ovulation

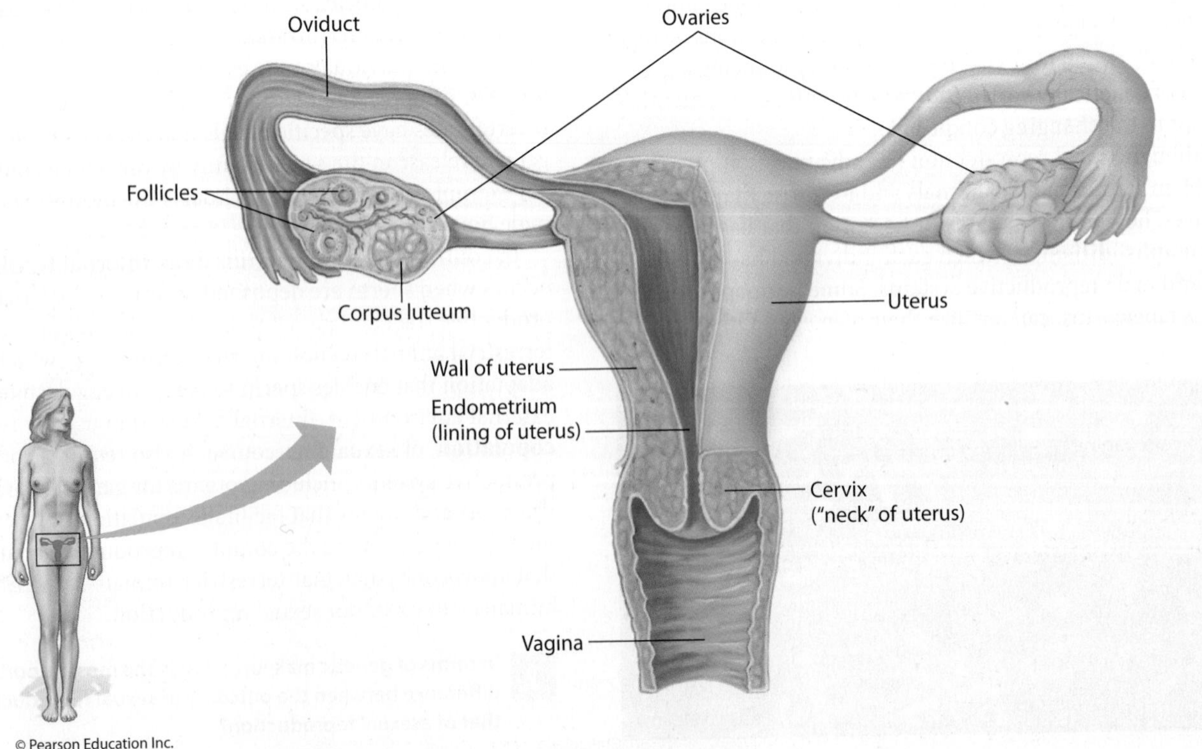

Oviduct
Ovaries
Follicles
Corpus luteum
Uterus
Wall of uterus
Endometrium (lining of uterus)
Cervix ("neck" of uterus)
Vagina

▲ **Figure 27.3A** Front view of female reproductive anatomy (upper portion)

the space and into the oviduct, where cilia sweep it toward the uterus. If sperm are present, fertilization may occur in the upper part of the oviduct. The resulting zygote starts to divide, thus becoming an embryo, as it moves along within the oviduct.

The **uterus**, also known as the womb, is the actual site of pregnancy. The uterus is only about 3 inches long in a woman who has never been pregnant (about the size and shape of an upside-down pear), but during pregnancy it expands considerably as the baby develops. The uterus has a thick muscular wall, and its inner lining, the **endometrium**, is richly supplied with blood vessels. An embryo implants in the endometrium, and development is completed there. The term **embryo** is used for the stage in development from the first division of the zygote until body structures begin to appear, about the 9th week in humans. From the 9th week until birth, a developing human is called a **fetus**.

The uterus is the normal site of pregnancy. However, in about 1% of pregnancies, the embryo implants somewhere else, resulting in an **ectopic pregnancy**. Most ectopic pregnancies occur in the oviduct and are called tubal pregnancies. An ectopic pregnancy is not viable and is a serious medical emergency that requires surgical intervention; otherwise, it can rupture surrounding tissues, causing severe bleeding and even death of the mother.

The narrow neck at the bottom of the uterus is the **cervix**, which opens into the vagina. It is recommended that women have periodic Pap tests in which cells are removed from around the cervix and examined under a microscope for signs of cervical cancer. Pap smears greatly increase the chances of detecting cervical cancer early and therefore treating it successfully. The cervix opens to the **vagina**, a thin-walled, but strong, muscular chamber that serves as the birth canal through which the baby is born. The vagina is also the repository for sperm during sexual intercourse. Glands near the vaginal opening secrete mucus during sexual arousal, lubricating the vagina and facilitating intercourse.

You can see more features of female reproductive anatomy in **Figure 27.3C**, a side view. **Vulva** is the collective term for the external female genitalia. Notice that the vagina opens to the outside just behind the opening of the urethra, the tube through which urine is excreted. A pair of slender skin folds, the **labia minora**, border the openings, and a pair of thick, fatty ridges, the **labia majora**, protect the vaginal opening. Until sexual intercourse or vigorous physical activity ruptures it, a thin piece of tissue called the hymen partly covers the vaginal opening.

Several female reproductive structures are important in sexual arousal, and stimulation of these structures can produce highly pleasurable sensations. The vagina, labia minora, and a small erectile organ called the **clitoris** all engorge with blood and enlarge during sexual activity. The clitoris consists of a short shaft supporting a rounded **glans**, or head, covered by a small hood of skin called the **prepuce**. In Figure 27.3C, blue highlights the spongy erectile tissue within the clitoris that fills with blood during arousal. The clitoris, especially the glans, has an enormous number of nerve endings and is very sensitive to touch. Keep in mind the details of female reproductive anatomy as you read the next module, and you'll notice many similarities in the human male.

? **Where does fertilization occur? In which organ does the fetus develop?**

● The oviduct; the uterus

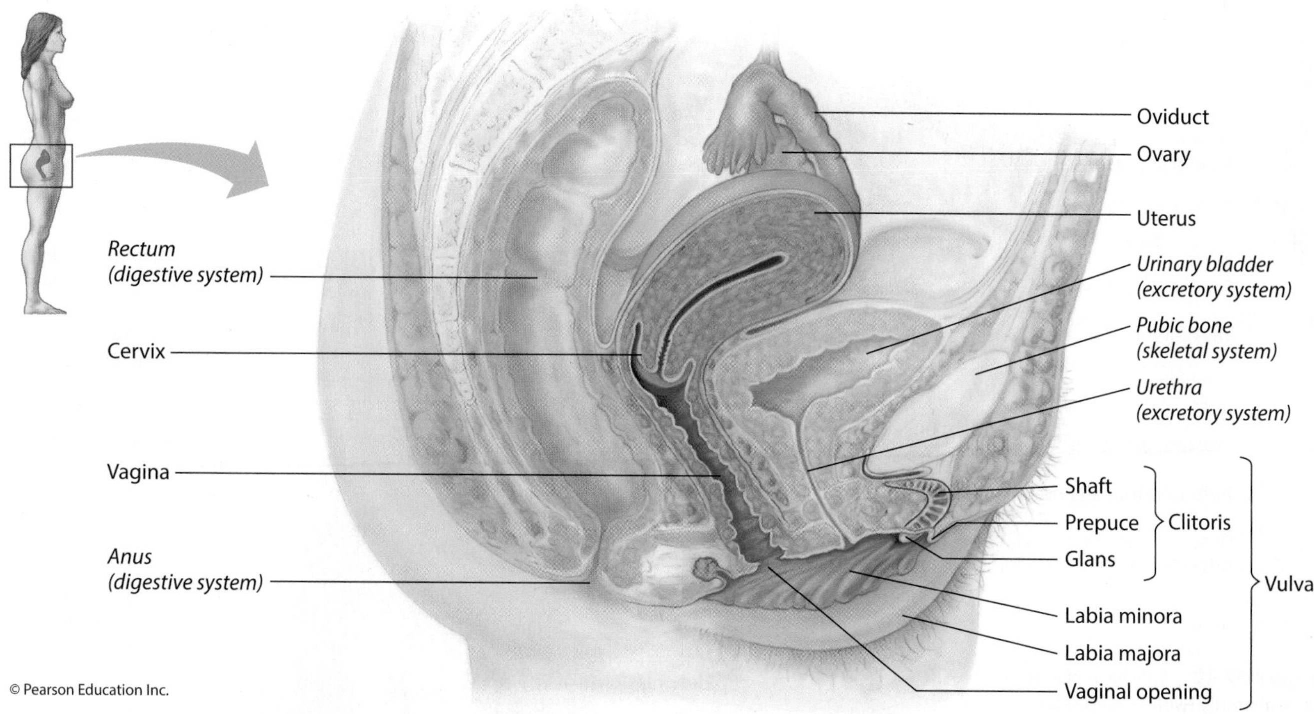

Oviduct
Ovary
Uterus
Rectum (digestive system)
Urinary bladder (excretory system)
Pubic bone (skeletal system)
Cervix
Urethra (excretory system)
Vagina
Shaft
Prepuce
Glans
Clitoris
Anus (digestive system)
Vulva
Labia minora
Labia majora
Vaginal opening

▲ **Figure 27.3C** Side view of female reproductive anatomy (with nonreproductive structures in italic)

Try This For each structure shown in this diagram, locate the matching structure in Figure 27.3A (front view).

27.4 The human male reproductive system includes the testes and structures that deliver gametes

Figures 27.4A and **27.4B** present front and side views of the male reproductive system. The male gonads, or **testes** (singular, *testis*), are each housed outside the abdominal cavity in a sac called the **scrotum**. A testis within a scrotum is called a **testicle**. In humans and many other mammals, sperm cannot develop optimally at core body temperature; the scrotum keeps the sperm-forming cells about 2°C cooler, which allows them to function normally. In cold conditions, muscles around the scrotum contract, pulling the testes toward the body, thereby maintaining the proper temperature.

Now let's track the path of sperm from one of the testes out of the male's body. From each testis, sperm pass into a coiled tube called the **epididymis**, which stores the sperm while they continue to develop. Sperm leave the epididymis during **ejaculation**, the expulsion of sperm-containing fluid from the penis. At that time, muscular contractions propel the sperm from the epididymis through another duct called the **vas deferens**. The vas deferens (which is the target of a vasectomy; see Module 27.8) passes upward into the abdomen and loops around the urinary bladder. Next to the bladder, the vas deferens joins a short duct from a gland, the seminal vesicle. The two ducts unite to form a short **ejaculatory duct**, which joins its counterpart conveying sperm from the other testis. Each ejaculatory duct empties into the urethra, which conveys (at different times) urine or sperm out through the penis. Thus, unlike the female, the male has a direct connection between the reproductive and urinary systems.

In addition to the testes and ducts, the reproductive system of human males contains three sets of glands: the seminal vesicles, the prostate gland, and the bulbourethral glands. The two **seminal vesicles** secrete a thick fluid that contains mucus and the sugar fructose, which provides most of the energy used by the sperm as they propel themselves through the female reproductive tract. The **prostate gland** secretes a thin, milky fluid that further nourishes the sperm. The

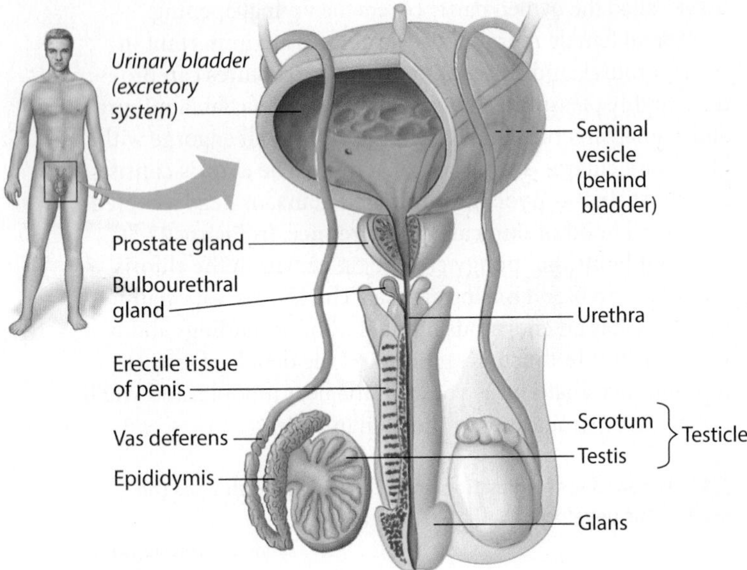

Urinary bladder
(excretory
system)
Seminal
vesicle
(behind
bladder)
Prostate gland
Bulbourethral
gland
Urethra
Erectile tissue
of penis
Vas deferens
Epididymis
Scrotum
Testis
Testicle
Glans

▲ **Figure 27.4A** Front view of male reproductive anatomy

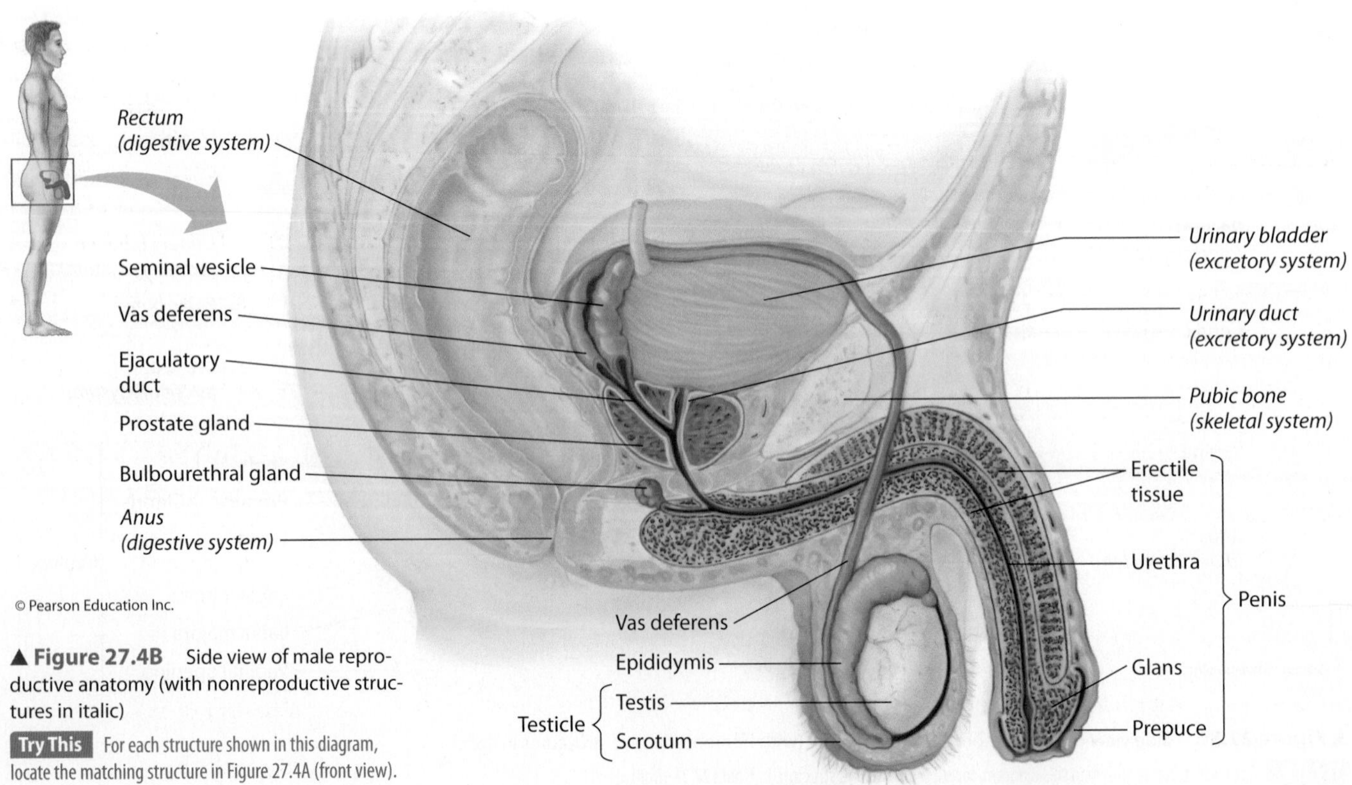

Rectum
(digestive system)
Seminal vesicle
Vas deferens
Ejaculatory
duct
Prostate gland
Bulbourethral gland
Anus
(digestive system)
Urinary bladder
(excretory system)
Urinary duct
(excretory system)
Pubic bone
(skeletal system)
Erectile
tissue
Urethra
Penis
Glans
Prepuce
Vas deferens
Epididymis
Testicle
Testis
Scrotum

▲ **Figure 27.4B** Side view of male reproductive anatomy (with nonreproductive structures in italic)

Try This For each structure shown in this diagram, locate the matching structure in Figure 27.4A (front view).

prostate gland is the source of some of the most common medical problems in men over 40, and prostate cancer is the second most commonly diagnosed cancer in the United States. The two **bulbourethral glands** secrete a clear, alkaline mucus that neutralizes any acidic urine remaining in the urethra. This fluid is actually secreted before ejaculation, and may carry some sperm; this is one reason why withdrawal of the penis before ejaculation is an unreliable method of birth control.

Together, the sperm and the glandular secretions make up **semen**, the fluid ejaculated from the penis during **orgasm**, a series of rhythmic, involuntary contractions of the reproductive structures. About 2–5 mL (1 teaspoonful) of semen is discharged during a typical ejaculation. Only 5% of semen consists of sperm (typically 200–500 million of them). The remaining 95% of semen is fluid secreted by the various glands. The alkalinity of the semen balances the acidity of any traces of urine in the urethra and neutralizes the acidic environment of the vagina, protecting the sperm and increasing their motility.

The human **penis** consists mainly of cylinders of erectile tissue (shown in blue in Figures 27.4A and 27.4B). Like the clitoris in females, the penis consists of a shaft that supports a highly sensitive glans, or head. A fold of skin called the prepuce, or foreskin, covers the glans; this foreskin is often removed through a surgical procedure called a circumcision. Erectile tissue is derived from modified veins and capillaries. During sexual arousal, erectile tissue fills with blood supplied by arteries. As it fills, the increasing pressure seals off veins that remove blood, causing the penis to engorge. The resulting erection is essential for insertion of the penis into the vagina during sexual intercourse.

Erectile dysfunction (ED), the inability to achieve an erection, may have several causes. To a certain degree, erectile dysfunction is a normal part of the male aging process. Erectile dysfunction may also be caused by consumption of alcohol or other drugs, emotional issues, and circulatory or neurological problems. Some men treat ED with drugs such as Viagra that enhance the action of the gas nitric oxide (NO). When released into erectile tissue by neurons, NO relaxes smooth muscle in blood vessels of the penis, causing the vessels to widen. This increases the flow of blood into erectile tissue, promoting an erection. Viagra works by inhibiting an enzyme that terminates the action of NO, thereby enhancing its effect and promoting the achievement and maintenance of an erection.

The process of ejaculation occurs in two stages. At the peak of sexual arousal, muscle contractions in multiple glands force secretions into the urethra and propel sperm from the epididymis. At the same time, a sphincter muscle at the base of the bladder contracts, preventing urine from leaking into the urethra from the bladder. Another sphincter also contracts, closing off the entrance of the urethra into the penis. The section of the urethra between the two sphincters fills with semen and expands. In the second stage of ejaculation, the expulsion stage, the sphincter at the base of the penis relaxes, admitting semen into the penis. At the same time, a series of strong muscle contractions around the base of the penis and along the urethra expels the semen from the body.

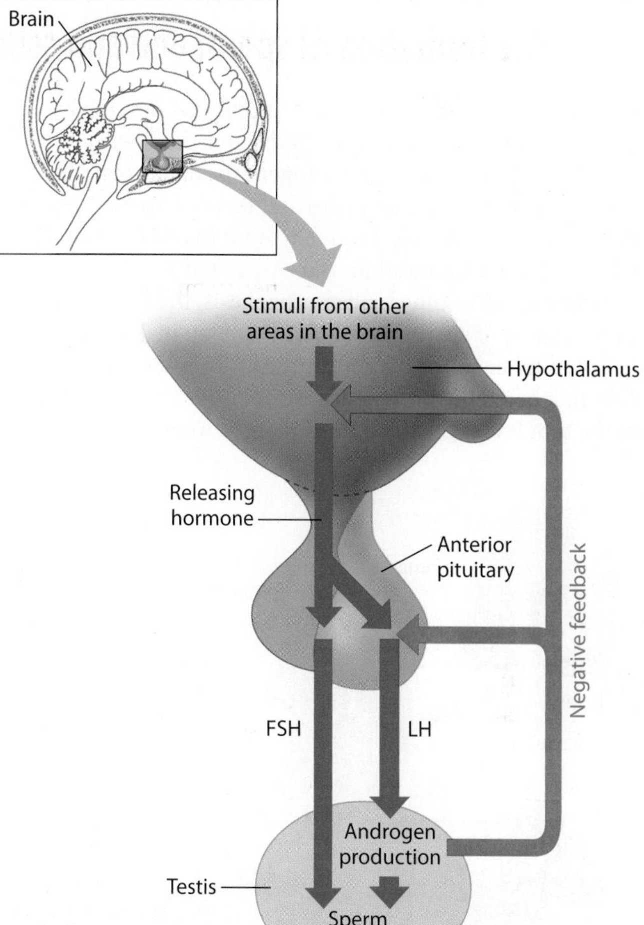
Brain
Stimuli from other areas in the brain
Hypothalamus
Releasing hormone
Anterior pituitary
Negative feedback
FSH
LH
Androgen production
Testis
Sperm production

▲ **Figure 27.4C** Hormonal control of the testis by the hypothalamus

Hormones from the hypothalamus and pituitary control sperm production by the testes (**Figure 27.4C**). The hypothalamus secretes a releasing hormone that regulates release of follicle-stimulating hormone (FSH) and luteinizing hormone (LH) by the anterior pituitary (see Module 26.5). FSH increases sperm production by the testes, while LH promotes the secretion of androgens, mainly testosterone. Androgens stimulate sperm production. In addition, androgens carried in the blood help maintain homeostasis by a negative-feedback mechanism (red arrows), inhibiting secretion of both the releasing hormone and LH. Under the control of this chemical regulating system, the testes produce hundreds of millions of sperm every day, from puberty well into old age. Next we'll see how the human body makes sperm and eggs.

? Arrange the following organs in the correct sequence through which sperm travel: epididymis, testis, urethra, vas deferens.

Testis, epididymis, vas deferens, urethra

27.5 The formation of sperm and egg cells requires meiosis

Both sperm and egg are haploid (n) cells that develop by meiosis from diploid ($2n$) cells in the gonads. There are significant differences in **gametogenesis**, the formation of gametes, between human males and females, so we'll examine the processes separately. (You may want to review Modules 8.12–8.14 as background for our discussion.)

Figure 27.5A outlines **spermatogenesis**, the formation and development of sperm cells. Sperm develop inside the testes in coiled tubes called the **seminiferous tubules**. Diploid cells that begin the process are located near the outer wall of the tubules (at the top of the enlarged wedge of tissue in Figure 27.5A). These cells multiply continuously by mitosis, and each day about 3 million of them differentiate into primary spermatocytes, the cells that undergo meiosis. Meiosis I of a primary spermatocyte produces two secondary spermatocytes, each with the haploid number of chromosomes ($n = 23$). Meiosis II then forms four cells, each with the haploid number of chromosomes. A sperm cell develops by differentiation of each of these haploid cells and is gradually pushed toward the center of the seminiferous tubule. From there it passes into the epididymis, where it matures, becomes motile, and is stored until ejaculation. In human

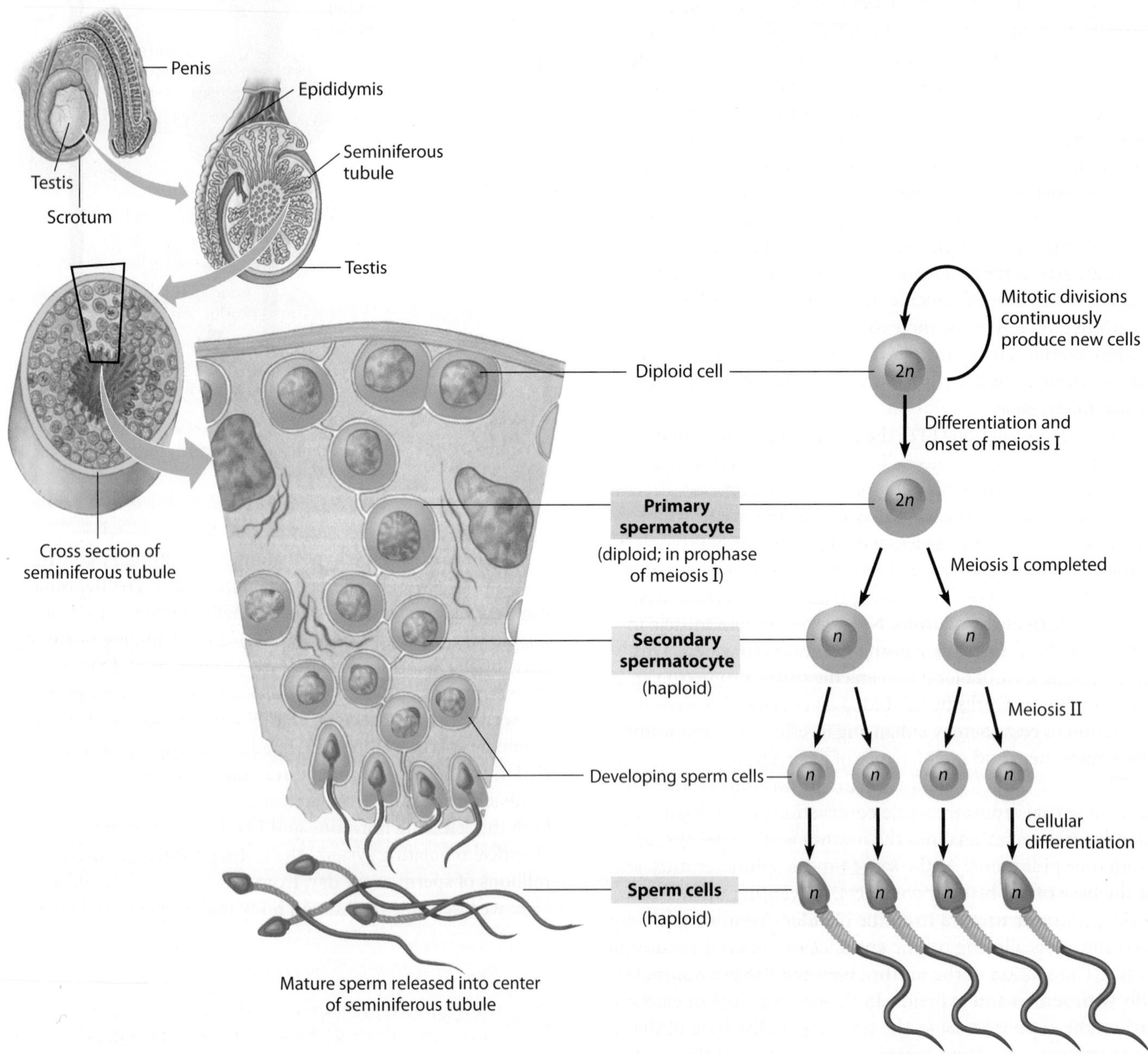

Penis
Epididymis
Seminiferous tubule
Testis
Scrotum
Testis
Cross section of seminiferous tubule
Mitotic divisions continuously produce new cells
Diploid cell
2n
Differentiation and onset of meiosis I
Primary spermatocyte
(diploid; in prophase of meiosis I)
2n
Meiosis I completed
Secondary spermatocyte
(haploid)
n
n
Meiosis II
Developing sperm cells
n
n
n
n
Cellular differentiation
Sperm cells
(haploid)
n
n
n
n
Mature sperm released into center of seminiferous tubule

▲ **Figure 27.5A** Spermatogenesis

males, spermatogenesis takes about 10 weeks. Because the pre-sperm cells continuously replenish themselves, there is a never-ending supply of spermatocytes, allowing males to produce sperm throughout their adult lives.

The right side of **Figure 27.5B** shows **oogenesis**, the development of a mature egg, also called an **ovum** (plural, *ova*). Most of the process occurs in the ovary. Oogenesis actually begins prior to birth, when a diploid cell in each developing follicle begins meiosis. At birth, each follicle contains a dormant primary oocyte, a diploid cell that is resting in prophase of meiosis I. A primary oocyte can be hormonally triggered to develop further. Between puberty and menopause, about every 28 days, follicle stimulating hormone (FSH) from the pituitary stimulates one of the dormant follicles to develop. The follicle enlarges, and the primary oocyte within it completes meiosis I and begins meiosis II. Meiosis then halts again at metaphase II. In the female, the division of the cytoplasm in meiosis I is unequal, with a single secondary oocyte receiving almost all of it. The smaller of the two daughter cells, called the first polar body, receives almost no cytoplasm.

The secondary oocyte is released by the ovary during ovulation. It enters the oviduct, and if a sperm cell penetrates it, the secondary oocyte completes meiosis II. Meiosis II is also unequal, yielding a second polar body and the mature egg. The haploid nucleus of the mature egg can then fuse with the haploid nucleus of the sperm cell, producing a zygote.

Although not shown in Figure 27.5B, the first polar body may also undergo meiosis II, forming two cells. These and the second polar body receive virtually no cytoplasm and quickly degenerate, leaving the mature egg with nearly all the cytoplasm and thus the bulk of the nutrients and organelles contained in the original diploid cell.

The left side of Figure 27.5B is a cutaway view of an ovary. Although the figure shows only one, a typical ovary has thousands of dormant follicles, each containing a primary oocyte. Usually, only one follicle has a dividing oocyte at any one time. Meiosis I occurs as the follicle matures. About the time the secondary oocyte forms, the pituitary hormone LH triggers ovulation, the rupture of the follicle and expulsion of the secondary oocyte. The ruptured follicle then develops into a corpus luteum ("yellow body"). Unless fertilization occurs, the corpus luteum degenerates before another follicle starts to develop.

Oogenesis and spermatogenesis are alike in that they both produce haploid gametes. However, these two processes differ in some important ways. First, only one mature egg results from each diploid cell that undergoes meiosis. The other products of oogenesis, the polar bodies, degenerate. By contrast, in spermatogenesis, all four products of meiosis develop into mature gametes. Second, spermatogenesis occurs from puberty until death, while the mitotic divisions of oogenesis are thought to be completed before birth and the production of mature gametes ceases at menopause. Third, oogenesis has long "resting" periods, whereas spermatogenesis produces mature sperm in an uninterrupted sequence.

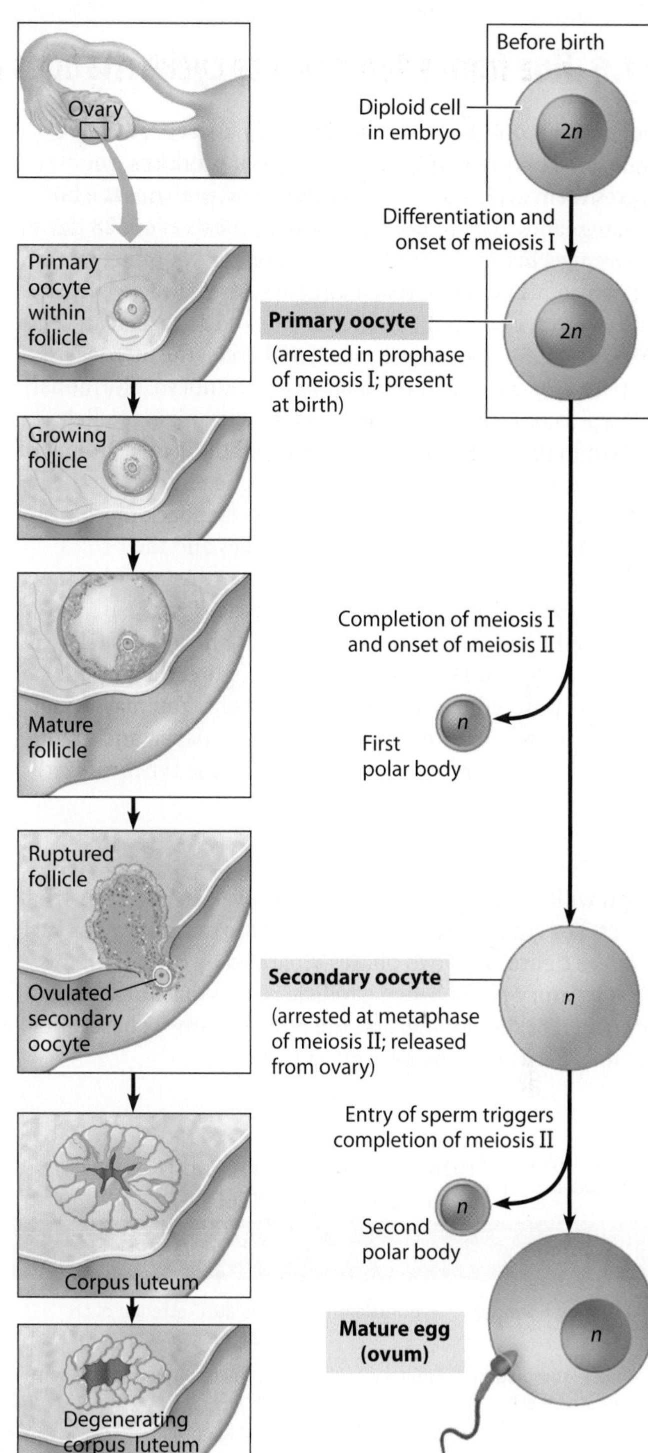

Ovary
Before birth
Diploid cell in embryo
2n
Differentiation and onset of meiosis I
Primary oocyte within follicle
Primary oocyte
(arrested in prophase of meiosis I; present at birth)
2n
Growing follicle
Completion of meiosis I and onset of meiosis II
Mature follicle
n
First polar body
Ruptured follicle
Ovulated secondary oocyte
Secondary oocyte
(arrested at metaphase of meiosis II; released from ovary)
n
Entry of sperm triggers completion of meiosis II
n
Second polar body
Corpus luteum
Mature egg (ovum)
n
Degenerating corpus luteum

▲ **Figure 27.5B** Oogenesis and the development of an ovarian follicle

? **If an egg is surgically removed from a woman's ovary, in what stage of meiotic development will that egg be?**

● A mature egg within the ovary (a secondary oocyte) is arrested at metaphase II.

27.6 Hormones synchronize cyclic changes in the ovary and uterus

Oogenesis is one part of a female mammal's **reproductive cycle**, a recurring sequence of events that produces gametes, makes them available for fertilization, and prepares the body for pregnancy. The reproductive cycle repeats every 28 days, on average, but cycles from 20 to 40 days are not uncommon. The reproductive cycle is actually two closely linked cycles. The **ovarian cycle** controls the growth and release of an egg (Module 27.5). During the **menstrual cycle**, the uterus is prepared for possible implantation of an embryo. Hormonal messages coordinate the two cycles, synchronizing follicle growth in the ovaries and ovulation with the establishment of a uterine lining that can support a growing embryo.

The hormone story involves intricate feedback mechanisms. **Table 27.6** lists the major hormones and their roles. **Figure 27.6**, on the facing page, shows how the events of the ovarian cycle (Part C) and menstrual cycle (Part E) are synchronized through the actions of multiple hormones (shown in Parts A, B, and D). Note that, in this figure, the circled numbers proceed in order from left to right. You may find it helpful to cover Figure 27.6 with a piece of paper, and gradually slide the paper to the right, observing the synchronized changes that occur in each phase.

An Overview of the Ovarian and Menstrual Cycles Let's begin with the structural events of the ovarian and menstrual cycles. For simplicity, we have divided the ovarian cycle (Part C of the figure) into two phases separated by ovulation: the pre-ovulatory phase, when a follicle is growing and a secondary oocyte is developing, and the post-ovulatory phase, after the follicle has become a corpus luteum.

TABLE 27.6 | HORMONES OF THE OVARIAN AND MENSTRUAL CYCLES

Hormone	Secreted by	Major Roles
Releasing hormone	Hypothalamus	Regulates secretion of LH and FSH by pituitary
Follicle-stimulating hormone (FSH)	Pituitary	Stimulates growth of ovarian follicle
Luteinizing hormone (LH)	Pituitary	Stimulates growth of ovarian follicle and production of secondary oocyte; promotes ovulation; promotes development of corpus luteum and secretion of other hormones
Estrogen	Ovarian follicle	Low levels inhibit pituitary; high levels stimulate hypothalamus; promotes growth of endometrium
Estrogen and progesterone	Corpus luteum	Maintain endometrium; high levels inhibit hypothalamus and pituitary; sharp drops promote menstruation

Events in the menstrual cycle (Part E) are synchronized with the ovarian cycle. ❶ By convention, the first day of a woman's period is designated day 1 of the menstrual cycle. **Menstruation** is uterine bleeding caused by the breakdown of the endometrium; it usually lasts for 3–5 days. Notice that this corresponds to the beginning of the pre-ovulatory phase of the ovarian cycle. The menstrual discharge, consisting of blood, small clusters of endometrial cells, and mucus, leaves the body through the vagina. After menstruation, the endometrium regrows. It continues to thicken through the time of ovulation, reaching a maximum thickness at about 20–25 days. If an embryo has not implanted in the uterine lining, menstruation begins again, marking the start of the next ovarian and menstrual cycles.

Now let's consider the hormones that regulate the ovarian and menstrual cycles. The ebb and flow of the hormones listed in Table 27.6 synchronize events in the ovarian cycle (the growth of the follicle and ovulation) with events in the menstrual cycle (preparation of the uterine lining for possible implantation of an embryo). A releasing hormone from the hypothalamus in the brain regulates secretion of the two pituitary hormones FSH and LH. Changes in the blood levels of FSH, LH, and two other hormones—estrogen and progesterone—coincide with specific events in the ovarian and menstrual cycles.

Hormonal Events Before Ovulation We see in Part A of Figure 27.6 that the releasing hormone from the hypothalamus stimulates the anterior pituitary ❷ to increase its output of FSH and LH. True to its name, ❸ FSH stimulates the growth of an ovarian follicle, in effect starting the ovarian cycle. In turn, the follicle secretes estrogen. Early in the pre-ovulatory phase, the follicle is small (Part C) and secretes relatively little estrogen (Part D). As the follicle grows, ❹ it secretes more and more estrogen, and the rising but still relatively low level of estrogen exerts negative feedback on the pituitary. This keeps the blood levels of FSH and LH low for most of the pre-ovulatory phase (Part B). As the time of ovulation approaches, hormone levels change drastically, with estrogen reaching a critical peak (Part D) just before ovulation. This high level of estrogen exerts positive feedback on the hypothalamus (green arrow in Part A), which then ❺ makes the pituitary secrete surges of FSH and LH. By comparing Parts B and D of the figure, you can see that the peaks in FSH and LH occur just after the estrogen peak. Again, it may help you to place a piece of paper over the figure and slide it slowly to the right. As you uncover the figure, you will see the follicle getting bigger and the estrogen level rising to its peak, followed almost immediately by the LH and FSH surges. ❻ Then, just to the right of the peaks, comes the dashed line representing ovulation.

Hormonal Events at Ovulation and After LH stimulates the completion of meiosis I, transforming the primary oocyte in the follicle into a secondary oocyte. It also signals

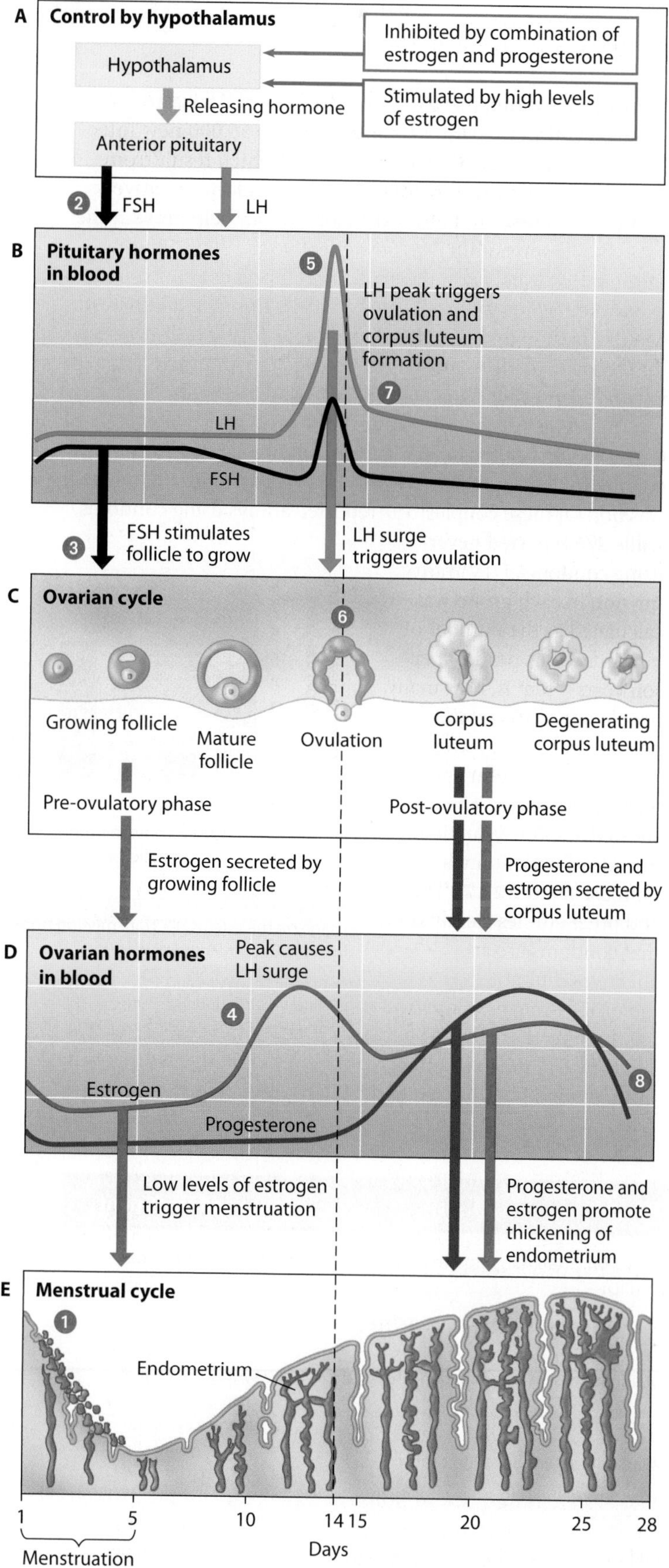
A
Control by hypothalamus
Hypothalamus
Inhibited by combination of estrogen and progesterone
Stimulated by high levels of estrogen
Releasing hormone
Anterior pituitary
2
FSH
LH
B
Pituitary hormones in blood
5
LH peak triggers ovulation and corpus luteum formation
7
LH
FSH
3
FSH stimulates follicle to grow
LH surge triggers ovulation
C
Ovarian cycle
6
Growing follicle
Mature follicle
Ovulation
Corpus luteum
Degenerating corpus luteum
Pre-ovulatory phase
Post-ovulatory phase
Estrogen secreted by growing follicle
Progesterone and estrogen secreted by corpus luteum
D
Ovarian hormones in blood
Peak causes LH surge
4
Estrogen
Progesterone
8
Low levels of estrogen trigger menstruation
Progesterone and estrogen promote thickening of endometrium
E
Menstrual cycle
1
Endometrium
1
5
10
14 15
20
25
28
Menstruation
Days

▲ **Figure 27.6** The reproductive cycle of the human female. Notice the time scale at the bottom of Part E; it also applies to Parts B–D.

enzymes to rupture the follicle, allowing ovulation to occur, and triggers the development of the corpus luteum from the ruptured follicle (hence its name, luteinizing hormone). LH also promotes the secretion of progesterone and estrogen by the corpus luteum. In Part D of the figure, you can see the progesterone peak and the second (lower and wider) estrogen peak after ovulation.

High levels of estrogen and progesterone in the blood following ovulation have a strong influence on both the ovary and uterus. The combination of the two hormones exerts negative feedback on the hypothalamus and pituitary, producing 7 falling FSH and LH levels. The drop in FSH and LH prevents follicles from developing and ovulation from occurring during the post-ovulatory phase. Also, the LH drop is followed by the gradual degeneration of the corpus luteum. Near the end of the post-ovulatory phase, unless an embryo has implanted in the uterus, the corpus luteum stops secreting estrogen and progesterone. 8 As blood levels of these hormones decline, the hypothalamus once again can stimulate the pituitary to secrete more FSH and LH, and a new cycle begins.

Control of the Menstrual Cycle Hormonal regulation of the menstrual cycle is simpler than that of the ovarian cycle. The menstrual cycle (Part E) is directly controlled by estrogen and progesterone alone. You can see the effects of these hormones by comparing Parts D and E of the figure. Starting around day 5 of the cycle, the endometrium thickens in response to the rising levels of estrogen and, later, progesterone. When the levels of these hormones drop, the endometrium begins to shed. Menstrual bleeding begins soon after, on day 1 of a new cycle.

The description of the reproductive cycle to this point assumes that fertilization has not occurred. As we'll see later, the ovarian and menstrual cycles are put on hold if fertilization and pregnancy occur. Early in pregnancy, the developing embryo, implanted in the endometrium, releases a hormone called human chorionic gonadotropin (hCG). This hormone maintains the corpus luteum, which continues to secrete progesterone and estrogen, keeping the endometrium intact. As you will learn (in Module 27.8), some forms of contraception work by mimicking the high levels of hormones that occur during pregnancy. Most home pregnancy tests work by detecting hCG in a woman's urine. In males, hCG boosts testosterone production, so it can be abused as a performance-enhancing drug by athletes; its use is therefore banned by many sports organizations. (We'll return to the events of pregnancy in Modules 27.15 and 27.16.)

? **Which hormonal change triggers the onset of menstruation?**

● The drop in the levels of estrogen and progesterone. These changes are caused by negative feedback of these hormones on the hypothalamus and pituitary after ovulation.

27.7 Sexual activity can transmit disease

SCIENTIFIC THINKING

Sexually transmitted diseases (STDs), also referred to as sexually transmitted infections (STIs), are contagious diseases spread by sexual contact. **Table 27.7** lists the most common STDs in the United States, organized by the type of infectious agent.

What can be done about STDs?

Not all STDs can be treated equally: Those caused by bacteria, protozoans, and fungi can be cured with medications, while viral STDs cannot. For bacterial STDs, treatment must be given early, before any permanent damage is done. The most common bacterial STD is **chlamydia**. Chlamydia poses a public health challenge because it is frequently "silent," often producing no visible symptoms. Long-term complications are rare among men, but about 40% of infected women develop pelvic inflammatory disease (PID), which can cause infertility by blocking the oviducts or scarring the uterus. A single dose of an antibiotic usually cures chlamydia.

One in five Americans is infected with **genital herpes**, caused by the herpes simplex virus type 2 (HSV-2), a variant of the virus that causes oral cold sores (see Module 10.18). Blisters on the external genitalia first appear about a week after exposure. After a few days, the blisters change to scabs that fall off. Most outbreaks heal within a few weeks without leaving a scar. But the virus lies dormant within nearby nerve cells. Months or years later, the virus can reemerge, causing fresh sores that allow the virus to be spread to sexual partners. Abstinence during outbreaks, the use of condoms, and the use of antiviral medications can reduce the spread of infection. But, as with other viral STDs, infection with genital herpes lasts a lifetime. Another sexually transmitted virus is the human papillomavirus (HPV). In 2006, a vaccine against HPV was approved that protects against genital warts and helps prevent infection by HPV strains that cause 70% of cervical cancers (see Module 24.11).

AIDS, caused by HIV (see Modules 10.20 and 24.14), poses one of the greatest health challenges in the world today, particularly among the developing nations of Africa and Asia. Yet even within the United States, there are 56,000 new infections each year, one-third of which result from heterosexual contact. Latex condoms, relatively cheap and easily distributed, can help prevent the spread of HIV. How effective are they? Finding the answer to such a question is difficult because controlled experiments are logistically and ethically impractical. Instead, health investigators must rely on observational studies, ones that record data from groups of people who differ in a behavior under scrutiny.

One such study, published in 2007 by the World Health Organization of the United Nations, followed a large group of couples in which one partner was HIV-positive and the other was not. Of these couples, 587 reported always using condoms, while 276 reported never using condoms. HIV transmission in each group was calculated as the number of new infections per 100 person years—that is, how many people out of 100 would become infected on average during each year spent with their partner. The data showed an 80% reduction in the rate of infection with condom use (**Figure 27.7**). These and other studies clearly show that condoms are an effective—but not foolproof—means of preventing the spread of HIV.

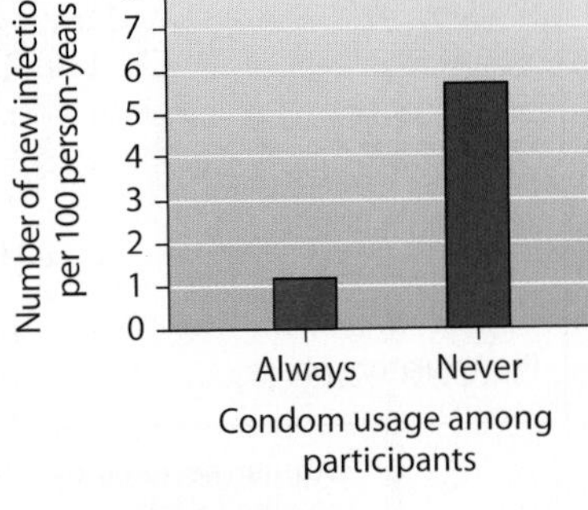

Number of new infections per 100 person-years
8
7
6
5
4
3
2
1
0
Always
Never
Condom usage among participants

▲ **Figure 27.7** Effectiveness of condom usage on the rate of HIV transmission from infected to uninfected partner

Data from S. C. Weller and K. Davis-Beaty, Condom effectiveness in reducing heterosexual HIV transmission, *Cochrane Database of Systematic Reviews* 4 (2007).

Many STDs can be effectively treated if addressed early. If left untreated, an STD may

TABLE 27.7 | STDS COMMON IN THE UNITED STATES

Disease	Microbial Agent	Major Symptoms and Effects	Cause and Treatment
Chlamydia	*Chlamydia trachomatis*	Genital discharge, painful urination; often no symptoms in women; pelvic inflammatory disease (PID)	Bacterial STDs can be treated with antibiotics
Gonorrhea	*Neisseria gonorrhoeae*	Genital discharge; painful urination; sometimes no symptoms in women; PID	
Genital herpes	Herpes simplex virus type 2, occasionally type 1	Blisters on genitalia, painful urination, skin inflammation; linked to cervical cancer	The symptoms of viral STDs can be treated, but no cures are available
Genital warts	Papillomaviruses	Painless growths on genitalia; some strains linked to cancer	
AIDS and HIV infection	HIV	Destruction of the immune system, increasing susceptibility to other infections (see Module 24.14)	
Trichomoniasis	*Trichomonas vaginalis*	Vaginal irritation, itching, and discharge; usually no symptoms in men	Protozoan STDs can be cured with drugs
Candidiasis (yeast infections)	*Candida albicans*	Vaginal irritation, itching, and discharge; frequently acquired nonsexually	Fungal STDs can be cured with antifungal drugs

lead to long-term problems or even death. Anyone who is sexually active should have regular medical exams, be tested for STDs, and seek immediate help if any suspicious symptoms appear, even if they are mild. STDs are most prevalent among teenagers and young adults; nearly two-thirds of infections occur among people under 25. The best way to avoid the spread of STDs is abstinence. Alternatively, latex condoms provide the best protection for "safe sex."

How are bacterial STDs different from viral STDs in terms of their long-term prognosis?

Bacterial STDs can be cured; viral STDs can be controlled but not cured.

27.8 Contraception can prevent unwanted pregnancy

CONNECTION

Contraception is the deliberate prevention of pregnancy. There are many forms of contraception that interfere with different steps in the process of becoming pregnant. **Table 27.8** lists several methods of birth control and their failure rates, given for when they are used correctly and for when they are used typically. Note that these two rates are often quite different: It is important to learn how to use contraception correctly.

Complete abstinence is the only totally effective method of birth control, but other methods are effective to varying degrees. Sterilization, surgery that prevents sperm from reaching an egg, is very reliable. A woman may have a **tubal ligation**, in which a doctor removes a short section from each oviduct, often tying (ligating) the remaining ends and thereby blocking the route of sperm to egg. A man may have a **vasectomy**, in which a doctor cuts a section out of each vas deferens to prevent sperm from reaching the urethra. Both forms of sterilization are relatively safe and free from side effects. They are meant to be permanent, but can sometimes be surgically reversed. An **intrauterine device (IUD)** is a T-shaped device placed within the uterus by a healthcare provider. IUDs are safe and highly effective at preventing pregnancy for up to 12 years, but can be safely removed at any time.

The effectiveness of other methods of contraception depends on how they are used. Temporary abstinence, also called the **rhythm method** or natural family planning, depends on refraining from intercourse during the days around ovulation, when fertilization is most likely. In theory, the time of ovulation can be determined by tracking changes in body temperature and the composition of cervical mucus, but careful monitoring is required. Because the length of the reproductive cycle can vary from month to month, and because sperm can survive for 3–5 days within the female reproductive tract, natural family planning is quite unreliable in actual practice. Withdrawal of the penis from the vagina before ejaculation is also ineffective because sperm may be released before climax.

If used correctly, barrier methods can be quite effective at physically preventing the union of sperm and egg. **Condoms** are sheaths, usually made of latex, that fit over the penis. A diaphragm is a dome-shaped rubber cap that covers the cervix; it requires a doctor's visit for proper fitting. Barrier devices are more effective when used in combination with **spermicides**, sperm-killing chemicals; spermicides used alone are unreliable.

Some of the most effective methods of contraception prevent the release of egg cells. **Oral contraceptives**, or birth control pills, come in several different forms that contain synthetic estrogen and/or a synthetic progesterone. Various combinations of these hormones are also available as an injection (Depo-Provera) or a skin patch. Steady intake of these hormones simulates their constant levels during pregnancy. In response, ovulation ceases, preventing the possibility of pregnancy.

Certain drugs can prevent fertilization or implantation after unprotected intercourse. Birth control pills can be used in high doses as emergency contraception, also called **morning after pills (MAPs)**. If taken within three days after intercourse, MAPs are about 75% effective at preventing pregnancy. Such treatments are available without a prescription to people 15 and older, but MAPs should only be used in emergencies because they have significant side effects.

If pregnancy has already occurred, the drug RU486 (mifepristone) can induce an abortion, the termination of a pregnancy in progress. RU486 must be taken within the first seven weeks of pregnancy. It must be provided by a health-care professional, and it requires several visits to a medical facility because of risk of significant side effects, including cramping and bleeding.

Contraception is an important health and safety issue for all sexually active people. For complete information, you should consult a health-care provider. It is important to note that, beyond abstinence, condoms are the only means of "safe sex" that can reduce the risk of both unwanted pregnancy *and* STDs; other contraceptive methods do not prevent STDs.

TABLE 27.8 | RELIABILITY OF CONTRACEPTIVE METHODS

Method	Pregnancies per 100 Women per Year*	
	Used Correctly	Used Typically
Birth control pill (combination)	0.1	5
Vasectomy	0.1	n/a
Tubal ligation	0.5	n/a
IUD	0.2–0.8	n/a
Rhythm method	1–9	20
Withdrawal	4	19
Condom (male)	2	15
Diaphragm and spermicide	6	16
Spermicide alone	6	29

*Without contraception, about 85 pregnancies would occur.

? Which form of contraception has the greatest difference between ideal use versus typical use?

Spermicide alone

▷ Principles of Embryonic Development

27.9 Fertilization results in a zygote and triggers embryonic development

The last six modules focused on the anatomy and physiology of the human reproductive system. In the next six modules, we examine the results of reproduction: the formation and development of an embryo. Embryonic development begins with fertilization, the union of a sperm and an egg to form a diploid zygote. Fertilization combines haploid sets of chromosomes from two individuals and also activates the egg by triggering metabolic changes that start embryonic development.

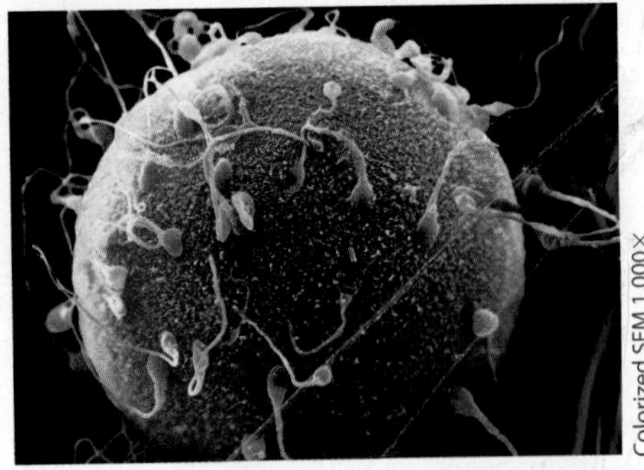
Colorized SEM 1,000×

▲ **Figure 27.9A** A human egg cell surrounded by sperm

Let's follow the steps shown in the figure. ❶ The contact of a sperm with the jelly coat of the egg triggers the release from the sperm's acrosome of a cloud of enzyme molecules by exocytosis (see Module 5.9). ❷ The enzyme molecules digest a cavity into the jelly. When the sperm head reaches the vitelline layer, ❸ species-specific protein molecules on its surface bind with specific receptor proteins on the outside of the egg cell. The binding between these proteins ensures that sperm of other species cannot fertilize the egg.

The Properties of Sperm Cells **Figure 27.9A** is a micrograph of an unfertilized human egg that is surrounded by sperm. Among all of these sperm, only one will enter and fertilize the egg. All the other sperm—the ones shown here and millions more that were ejaculated with them—will die. The one sperm that penetrates the egg adds its unique set of genes to those of the egg and contributes to the next generation.

Figure 27.9B illustrates the structure of a mature human sperm cell, a clear case of form fitting function. The sperm's streamlined shape is an adaptation for swimming through fluids in the vagina, uterus, and oviduct of the female. Its thick head contains a haploid nucleus and is tipped with a vesicle, the **acrosome**, which lies just inside the plasma membrane. The acrosome contains enzymes that help the sperm penetrate the egg. The middle section of the sperm contains mitochondria. The sperm absorbs high-energy nutrients, especially the sugar fructose, from the semen. Thus fueled, its mitochondria provide ATP for movement of the tail, which is actually a flagellum. By the time a sperm has reached the egg, it has consumed much of the energy available to it. But a successful sperm will have enough energy left to penetrate the egg and deposit its nucleus in the egg's cytoplasm.

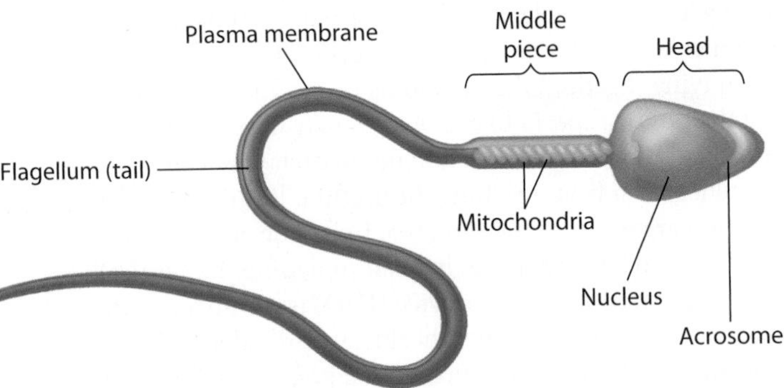
Plasma membrane
Middle piece
Head
Flagellum (tail)
Mitochondria
Nucleus
Acrosome

▲ **Figure 27.9B** The structure of a human sperm cell

The Process of Fertilization **Figure 27.9C** summarizes the timeline for the sequence of events that occur during fertilization, and **Figure 27.9D**, on the facing page, illustrates these events. The process shown is based on fertilization in sea urchins (phylum Echinodermata—see Module 18.14), on which a great deal of research has been done. Similar processes occur in other animals, including humans. The diagram traces one sperm through the successive activities of fertilization. Notice that to reach the egg nucleus, the sperm nucleus must pass through three barriers: the egg's jelly coat (shown in yellow), a middle region of glycoproteins called the vitelline layer (pink), and the egg cell's plasma membrane.

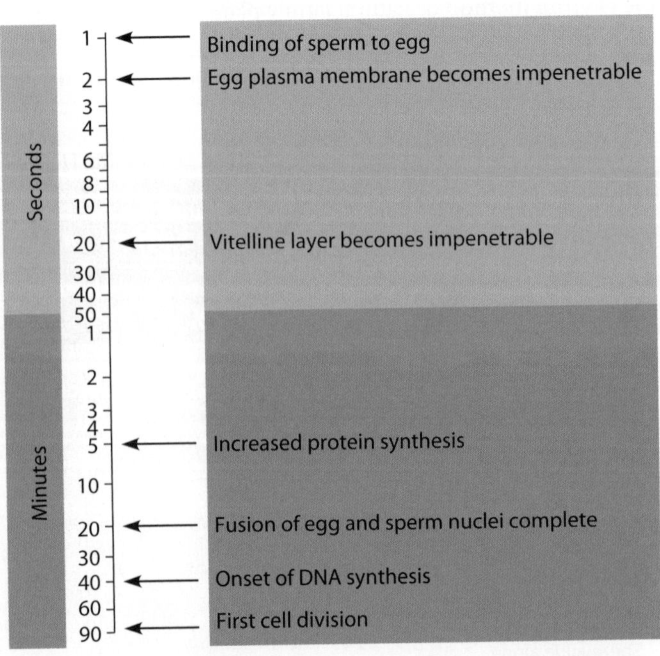
Seconds
1
2
3
4
6
8
10
20
30
40
50
Minutes
1
2
3
4
5
10
20
30
40
60
90
Binding of sperm to egg
Egg plasma membrane becomes impenetrable
Vitelline layer becomes impenetrable
Increased protein synthesis
Fusion of egg and sperm nuclei complete
Onset of DNA synthesis
First cell division

▲ **Figure 27.9C** Timeline for the events in fertilization in a sea urchin

This specificity is especially important when fertilization is external because the sperm of other species may be present in the water. After the specific binding occurs, the sperm proceeds through the vitelline layer, and 4 the sperm's plasma membrane fuses with that of the egg. Fusion of the two membranes 5 makes it possible for the sperm nucleus to enter the egg.

Fusion of the sperm and egg plasma membranes triggers a number of important changes in the egg. Two such changes prevent other sperm from entering the egg. About 1 second after the membranes fuse, the entire egg plasma membrane becomes impenetrable to other sperm cells. Shortly thereafter, 6 the vitelline layer hardens and separates from the plasma membrane. The space quickly fills with water, and the vitelline layer becomes impenetrable to sperm. If these events did not occur and an egg were fertilized by more than one sperm, the resulting zygote nucleus would contain too many chromosomes, and the zygote could not develop normally.

7 About 20 minutes after the sperm nucleus enters the egg, the sperm and egg nuclei fuse. Gearing up for the enormous growth and development that will soon follow, DNA synthesis begins and cellular respiration speeds up. As you can see in the timeline in Figure 27.9C, the first cell division occurs after about 90 minutes, marking the end of the fertilization stage.

Note that the sperm provided chromosomes to the zygote, but little else. The zygote's cytoplasm and various organelles were all provided by the mother through the egg. In the next module, we begin to trace the development of the zygote into a new animal.

? Why are the protein receptors on the outside of the egg cell particularly important among aquatic animals that use external fertilization?

Protein receptors on the vitelline layer match with species-specific proteins on the sperm; this ensures that sperm of a different species will not fertilize the egg.

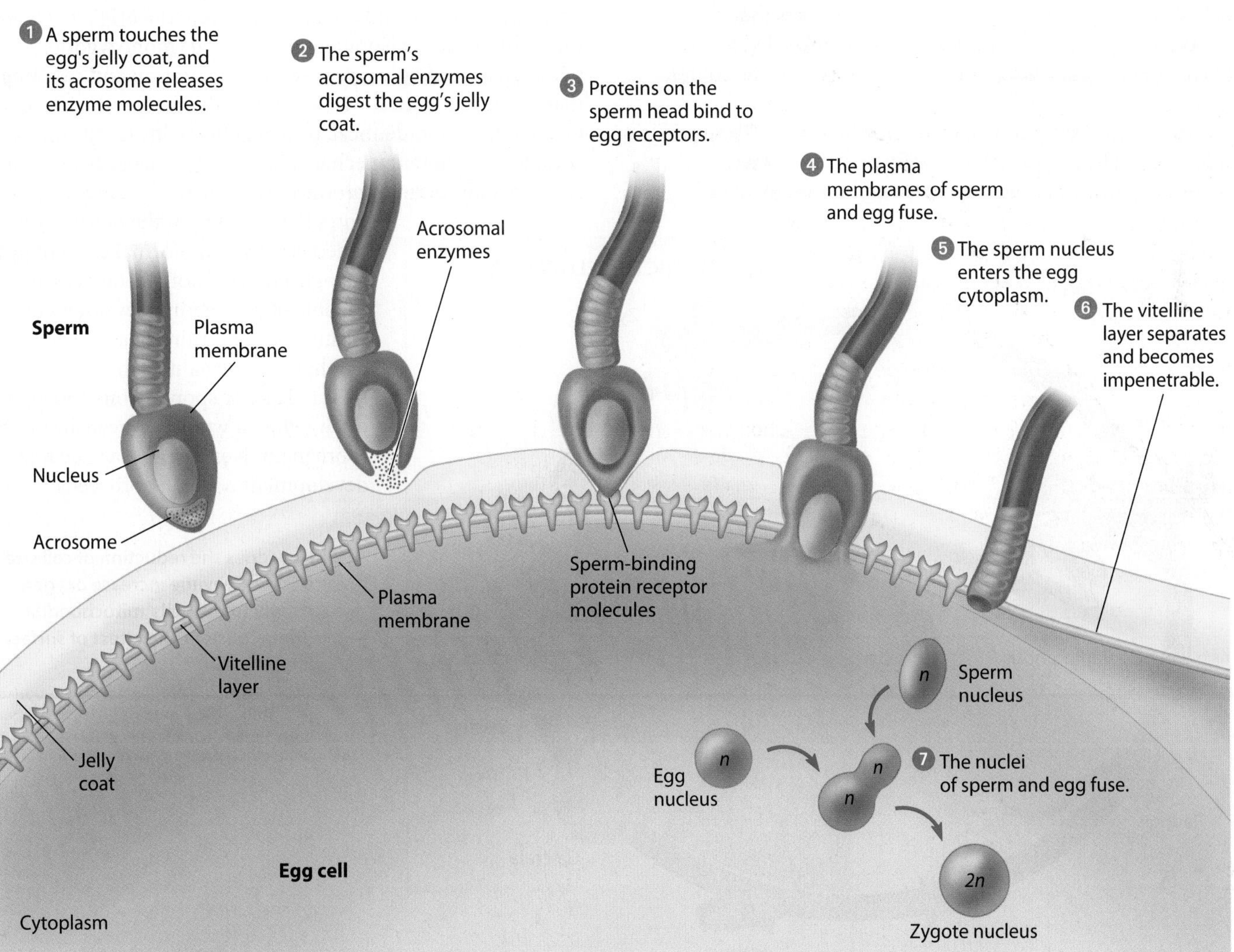
1 A sperm touches the egg's jelly coat, and its acrosome releases enzyme molecules.
2 The sperm's acrosomal enzymes digest the egg's jelly coat.
3 Proteins on the sperm head bind to egg receptors.
4 The plasma membranes of sperm and egg fuse.
5 The sperm nucleus enters the egg cytoplasm.
6 The vitelline layer separates and becomes impenetrable.
Sperm
Plasma membrane
Nucleus
Acrosome
Acrosomal enzymes
Sperm-binding protein receptor molecules
Plasma membrane
Vitelline layer
Jelly coat
Egg cell
Cytoplasm
n
Sperm nucleus
n
Egg nucleus
n
n
7 The nuclei of sperm and egg fuse.
2n
Zygote nucleus

▲ **Figure 27.9D** The process of fertilization in a sea urchin

27.10 Cleavage produces a blastula from the zygote

An adult animal consists of many thousands, millions, even trillions of cells that are precisely organized into complex tissues and organs. This transformation requires an astonishing amount of carefully controlled cell division and specialization. **Figure 27.10** summarizes the major stages, using a frog as an example; similar processes occur in humans. In this module, we will focus on cleavage. Later modules will focus on the subsequent stages.

Development begins with **cleavage**, a series of rapid cell divisions that produces a multicellular ball. After the zygote divides for the first time, it is called an embryo. Nutrients stored in the egg nourish the dividing cells. DNA replication, mitosis, and cytokinesis occur rapidly, but gene transcription virtually shuts down and few new proteins are synthesized. The embryo does not enlarge significantly; instead, as the number of cells doubles with each division, the cytoplasm of the one-celled zygote is partitioned into many smaller cells, each with its own nucleus. As a result, each cell in the ball is much smaller than the original cell that formed the zygote. The process of cleavage takes a few hours to produce a solid ball of cells in a frog; in a human, cleavage takes four days.

As cleavage continues, a fluid-filled cavity called the **blastocoel** forms in the center of the embryo. At the completion of cleavage, there is a hollow ball of cells called the **blastula**. Cells removed from a human blastocyst (the equivalent of the frog blastula) are useful in research (see Module 11.14). Such cells are called embryonic stem cells. Because they have yet to become specialized, embryonic stem cells have great therapeutic potential to develop into and replace just about any kind of mature cells that have been lost to damage or illness. But harvesting the embryonic stem cells destroys the embryo, which raises ethical questions. Research using embryonic stem cells remains one of the hottest areas of biological study.

Cleavage makes two important contributions to early development. It creates a multicellular embryo, the blastula, from a single-celled zygote. Cleavage is also an organizing process, partitioning the multicellular embryo into developmental regions. The cytoplasm of the zygote contains a variety of chemicals that control gene expression during early development (see Module 11.8). During cleavage, regulatory chemicals become localized in particular groups of cells, where they later activate the genes that direct the formation of specific parts of the animal. Gastrulation, the next phase of development, further refines the embryo's cellular organization.

Rarely, and apparently at random, a cell in the early embryo may separate and "reset" as if it were the original zygote; the result is the development of identical (monozygotic) twins. In exceedingly rare cases, the separation and resetting that produce identical twins can occur twice, producing identical triplets. Nonidentical, or dizygotic, twins result from a completely different mechanism: Two separate eggs fuse with two separate sperm to produce two genetically unique zygotes that develop in the uterus simultaneously. Even in normal cases of just a single embryo, not all embryos are capable of completing development. As many as one-third of all pregnancies have abnormalities that result in miscarriage, or spontaneous abortion, often before a woman is even aware she is pregnant. Next, we'll examine how development normally proceeds.

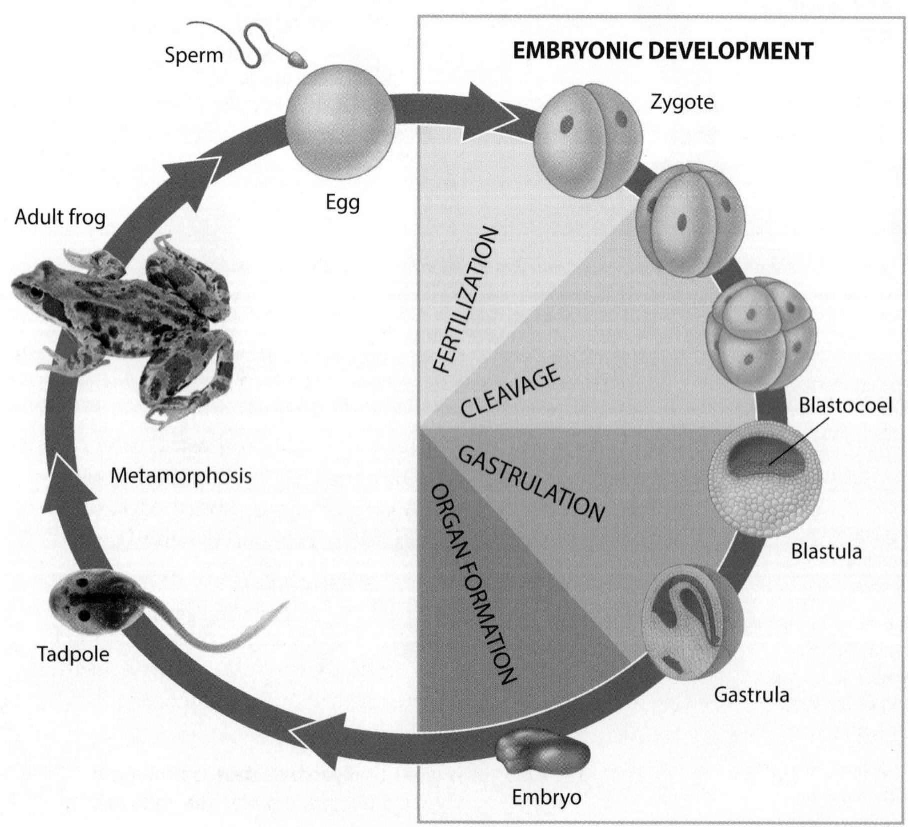

EMBRYONIC DEVELOPMENT
Sperm
Egg
Zygote
Adult frog
FERTILIZATION
CLEAVAGE
GASTRULATION
ORGAN FORMATION
Blastocoel
Blastula
Metamorphosis
Tadpole
Gastrula
Embryo

▲ **Figure 27.10** Development stages of a frog

? How does the reduction of cell size during cleavage increase oxygen supply to the cells' mitochondria? (*Hint*: Consider the effect of surface area.)

Smaller cells have a greater plasma membrane surface area relative to cellular volume, and this facilitates diffusion of oxygen from the environment to the cells' cytoplasm.

27.11 Gastrulation produces a three-layered embryo

After cleavage, the rate of cell division slows dramatically. Groups of cells then undergo **gastrulation**, the second major phase of embryonic development. During gastrulation, cells take up new locations that will allow later formation of all the organs and tissues. As gastrulation proceeds, the embryo is organized into a three-layer stage called a **gastrula**.

The three layers produced by gastrulation are embryonic tissues called **ectoderm**, **endoderm**, and **mesoderm**. The ectoderm forms the outer layer of the gastrula. The endoderm forms an embryonic digestive tract. And the mesoderm lies between the ectoderm and endoderm. Eventually, these three cell layers develop into all the parts of the adult animal. For instance, our nervous system and the outer layer (epidermis) of our skin come from ectoderm; the innermost lining of our digestive tract arises from endoderm; and most other organs and tissues, such as the kidney, heart, muscles, and the inner layer of our skin (dermis), develop from mesoderm. **Table 27.11** lists the major organs and tissues that arise in most vertebrates from the three main embryonic tissue layers.

The mechanics of gastrulation vary somewhat, depending on the species. The top of **Figure 27.11** shows a frog blastula, formed by cleavage (as discussed in the previous module). The frog blastula is a partially hollow ball of unequally sized cells. The cells toward one end, called the animal pole, are smaller than those near the opposite end, the vegetal pole. The three colors in the figure indicate regions of cells within the blastula that will give rise to the primary cell layers in the gastrula at the bottom of the figure: ectoderm (blue), mesoderm (red), and endoderm (yellow). (Notice that each layer may be more than one cell thick.)

During gastrulation (shown in the center of Figure 27.11), cells migrate to new positions that will form the three layers. Gastrulation begins when a small groove, called the blastopore, appears on one side of the blastula. Cells of the outer layer roll inward at the blastopore, with the future endoderm (yellow) surrounding a simple digestive cavity. Meanwhile, the cells that will form ectoderm (blue) spread downward over more of the surface of the embryo, and the cells that will form mesoderm (red) begin to spread into a thin layer inside the embryo, forming a middle layer between the other two.

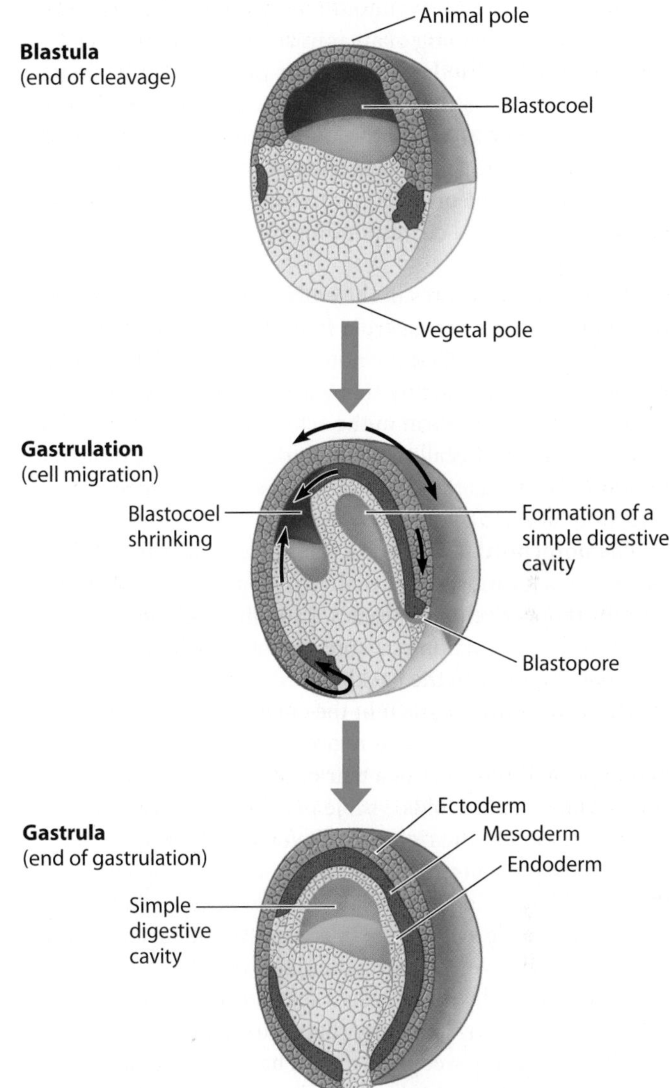

Blastula
(end of cleavage)
Animal pole
Blastocoel
Vegetal pole
Gastrulation
(cell migration)
Blastocoel shrinking
Formation of a simple digestive cavity
Blastopore
Gastrula
(end of gastrulation)
Ectoderm
Mesoderm
Endoderm
Simple digestive cavity

▲ **Figure 27.11** Development of the frog gastrula

As shown at the bottom of the figure, gastrulation is completed when cell migration has resulted in a three-layered embryo. Ectoderm covers most of the surface. Mesoderm forms a layer between the ectoderm and the endoderm.

Although gastrulation differs in detail from one animal group to another, the process is driven by the same general mechanisms in all species. The timing of these events also varies with the species. In many frogs, for example, cleavage and gastrulation together take about 15–20 hours.

TABLE 27.11 | DERIVATIVES OF THE THREE EMBRYONIC TISSUE LAYERS

Embryonic Layer	Organs and Tissues in the Adult
Ectoderm	Epidermis of skin; epithelial lining of mouth and rectum; sense receptors in epidermis; cornea and lens of eye; nervous system
Endoderm	Epithelial lining of digestive tract (except mouth and rectum); epithelial lining of respiratory system; liver; pancreas; thyroid; parathyroids; thymus; lining of urethra, urinary bladder, and reproductive system
Mesoderm	Skeletal system; muscular system; circulatory system; excretory system; reproductive system (except gamete-forming cells); dermis of skin; lining of body cavity

? **The first two phases of embryonic development are __________, which forms the blastula, followed by __________, which forms the __________.**

cleavage . . . gastrulation . . . gastrula

27.12 Organs start to form after gastrulation

In organizing the embryo into three layers, gastrulation sets the stage for the shaping of an animal. Once the ectoderm, endoderm, and mesoderm form, cells in each layer begin to differentiate into tissues and embryonic organs. The cutaway drawing in **Figure 27.12A** shows the developmental structures that appear in a frog embryo a few hours after the completion of gastrulation. The orientation drawing at the upper left of the figure indicates a corresponding plane through an adult frog.

We see two structures in the embryo in Figure 27.12A that were not present at the gastrula stage described in the last module. An organ called the notochord has developed in the mesoderm, and a structure that will become the hollow nerve cord is beginning to form in the ectoderm (in the region that is colored green). Recall that the notochord and the dorsal, hollow nerve cord are two of the hallmarks of the chordates (see Module 18.15).

The **notochord** is made of a substance similar to cartilage and extends for most of the embryo's length, providing support for other developing tissues. Later in development, the notochord will function as a core around which mesodermal cells gather and form the backbone.

The area shown in green in the cutaway drawing of Figure 27.12A is a thickened region of ectoderm called the neural plate. From it arises a pair of pronounced ectodermal ridges, called neural folds, visible in both the drawing and the micrograph to the right of it. If you now look at the series of diagrams in **Figure 27.12B**, you will see what happens as the neural folds and neural plate develop further. The neural plate rolls into a tube, which sinks beneath the surface of the embryo and is covered by an outer layer of ectoderm. If you look carefully at the figure, you'll see that cells of the ectoderm fold inward by changing shape, first elongating and then becoming wedge-shaped. The result is a tube of ectoderm—the **neural tube**—which is destined to become the brain and spinal cord.

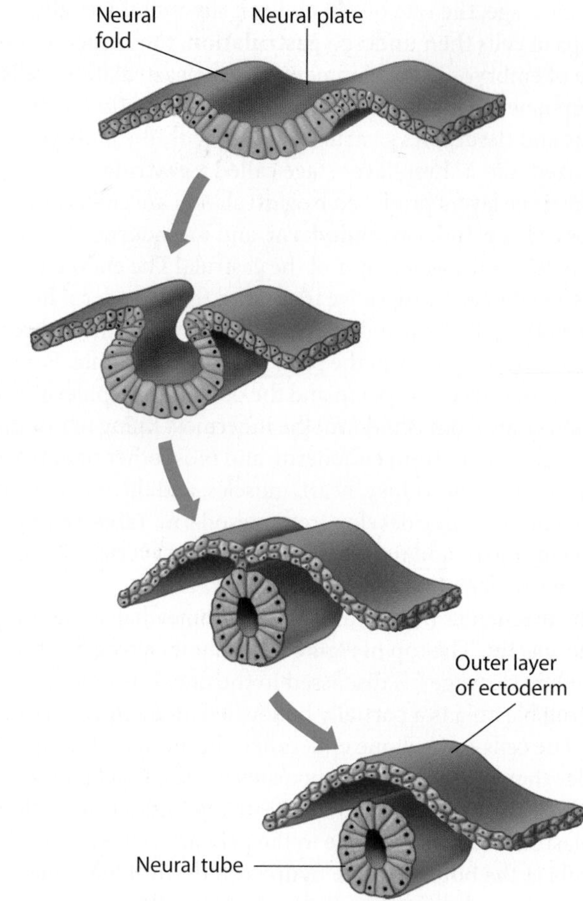
Neural fold
Neural plate
Outer layer of ectoderm
Neural tube

▲ **Figure 27.12B** Formation of the neural tube

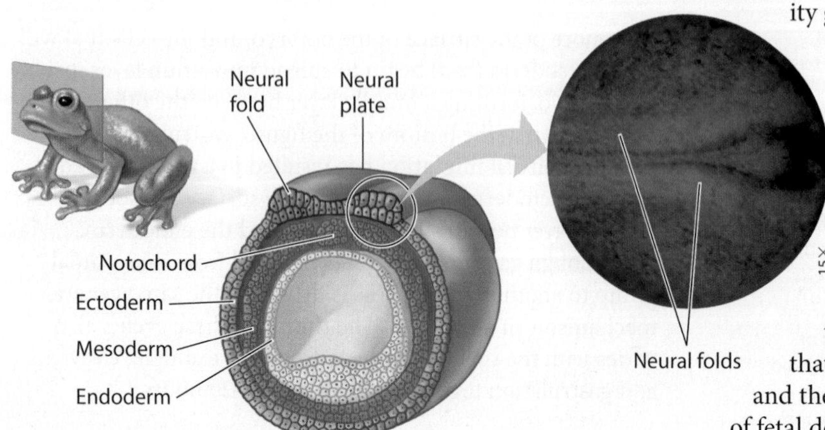
Neural fold
Neural plate
Notochord
Ectoderm
Mesoderm
Endoderm
15×
Neural folds

▲ **Figure 27.12A** The beginning of organ development in a frog: the notochord, neural folds, and neural plate

Figure 27.12C, on the facing page, shows a later frog embryo (about 12 hours older than the one in Figure 27.12A), in which the neural tube has formed. Notice in the drawing that the neural tube lies directly above the notochord. The relative positions of the neural tube, notochord, and digestive cavity give us a preview of the basic body plan of a frog. The spinal cord will lie within extensions of the dorsal (upper) surface of the backbone (which will replace the notochord), and the digestive tract will be ventral to (beneath) the backbone. We see this same arrangement of organs in all vertebrates.

The importance of these processes is underscored by human birth defects that result from improper signaling between embryonic tissues. For example, spina bifida is a condition that results from the failure of the neural tube to close and the spine to form properly during the first month of fetal development. Infants born with spina bifida often have permanent nerve damage that results in paralysis of the lower limbs. The vitamin folic acid is known to reduce the incidence of spina bifida; accordingly, it is recommended

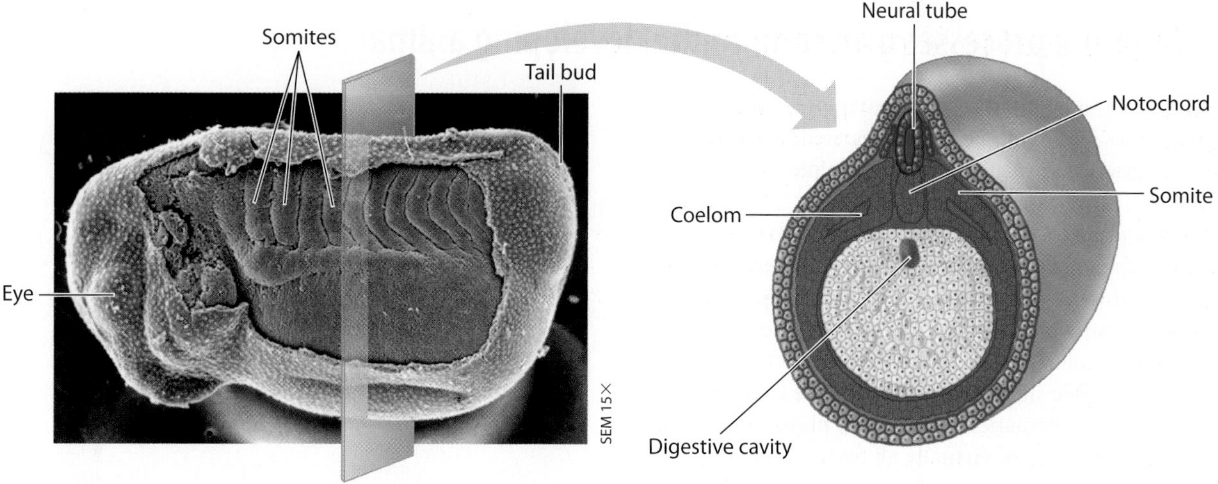
Somites
Tail bud
Eye
SEM 15×
Neural tube
Notochord
Somite
Coelom
Digestive cavity

▲ **Figure 27.12C** An embryo with completed neural tube, somites, and coelom shown in a side view (left) and in cross section (right)

that women of childbearing years get adequate folic acid, and pregnant women are generally recommended to take a folic acid supplement.

Besides the appearance of the neural tube and digestive cavity, Figure 27.12C shows several other fundamental changes in the frog embryo. In the micrograph, which shows a side view, you can see that the embryo is beginning to elongate. You can also see the beginnings of an eye and a tail (called the tail bud). Part of the ectoderm has been removed to reveal a series of internal ridges called somites. The somites are blocks of mesoderm that will give rise to segmental structures (constructed of repeating units), such as the vertebrae and associated muscles of the backbone. In the cross-sectional drawing, notice that the mesoderm next to the somites is developing a hollow space—the body cavity, or **coelom**. Segmented body parts and a coelom are basic features of all chordates.

In this and the previous two modules, we have observed the sequence of changes that occur as an animal begins to take shape. To summarize, the key phases in embryonic development are cleavage (which creates a multicellular blastula from a zygote), gastrulation (which organizes the embryo into a gastrula with three discrete layers), and organ formation (which generates embryonic organs from the three embryonic tissue layers). These same three phases occur in nearly all animals.

If we followed a frog's development beyond the stage represented in Figure 27.12C, within a few hours we would be able to monitor muscular responses and a heartbeat and see a set of gills with blood circulating in them. A long tail fin would grow from the tail bud. The timing of the later stages in frog development varies enormously, but in many species, by 5–8 days after development begins, we would see all the body tissues and organs of a tadpole emerge from cells of the ectoderm, mesoderm, and endoderm. Eventually, the structures of the tadpole **(Figure 27.12D)** would transform into the tissues and organs of an adult frog.

Watching embryos develop helps us appreciate the enormous changes that occur as one tiny cell, the zygote, gives rise to a highly structured, many-celled animal. Your own body, for instance, is a complex organization of some 60 trillion cells, all of which arose from a zygote smaller than the period at the end of this sentence. Discovering how this incredibly intricate arrangement is achieved is one of biology's greatest challenges. Through research that combines the experimental manipulation of embryos with cell biology and molecular genetics, developmental biologists have begun to work out the mechanisms that underlie development. We examine several of these mechanisms in the next module.

? **Within the embryo, what is the origin of the dorsal, hollow nerve cord that is common to all members of our phylum?**

● The neural tube, which becomes the brain and spinal cord, develops from a dorsal ectodermal plate that folds to form an interior tube.

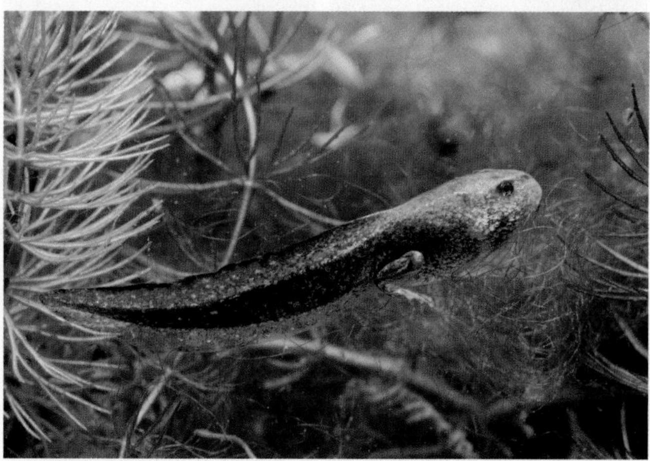

▲ **Figure 27.12D** A tadpole with some early adult features

27.13 Multiple processes give form to the developing animal

The development of an animal embryo depends on several cellular processes. For example, as you learned in the last module, changes in cell shape help form the neural tube (see Figure 27.12B).

Most developmental processes depend on signals passed between neighboring cells and cell layers, telling embryonic cells precisely what to do and when to do it. The mechanism by which one group of cells influences the development of an adjacent group of cells is called **induction**. Induction may be mediated by diffusible signals or, if the cells are in direct contact, by cell-surface interactions. Induction plays a major role in the early development of virtually all tissues and organs. Its effect is to switch on a set of genes whose expression makes the receiving cells differentiate into a specific tissue. Many inductions involve a sequence of inductive steps from different surrounding tissues that progressively determine the fate of cells. In the eye, for example, lens formation involves precisely timed inductive signals from ectodermal, signals from ectodermal and mesodermal cells. In the developing animal, a sequence of inductive signals leads to increasingly greater specialization of cells as organs begin to take shape.

Cell migration is also essential in development. For example, during gastrulation, cells "crawl" within the embryo by extending and contracting cellular protrusions, similar to the pseudopodia of amoeboid cells (see Module 16.17). Migrating cells may follow inductive chemical trails secreted by cells near their specific destination. Once a migrating cell reaches its destination, surface proteins enable it to recognize similar cells. The cells join together and secrete glycoproteins that glue them in place. Finally, they differentiate, taking on the characteristics of a particular tissue.

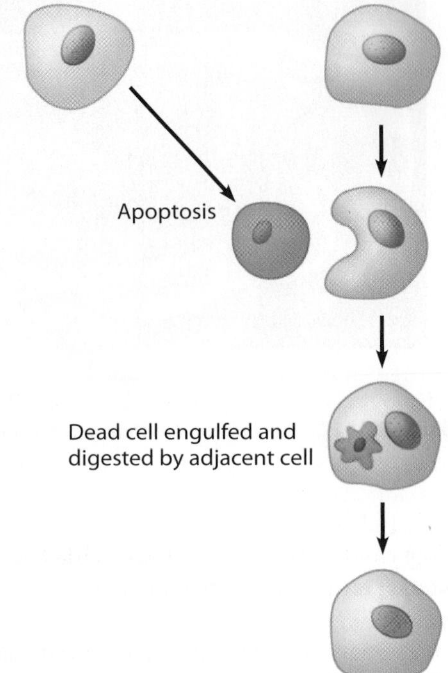
Apoptosis
Dead cell engulfed and digested by adjacent cell

▲ **Figure 27.13B** Apoptosis at the cellular level

Another important developmental process is **apoptosis**, the timely and tidy suicide of cells. Apoptosis is a type of **programmed cell death**. Animal cells make proteins that have the ability, when activated, to kill the cell that produces them. In humans, the timely death of specific cells in developing hands and feet creates the spaces between fingers and toes (**Figure 27.13A**). In **Figure 27.13B**, the cell on the left shrinks and dies because suicide proteins have been activated. Meanwhile, signals from the dying cell make an adjacent cell phagocytic. This cell engulfs and digests the dead cell, keeping the embryo free of harmful debris.

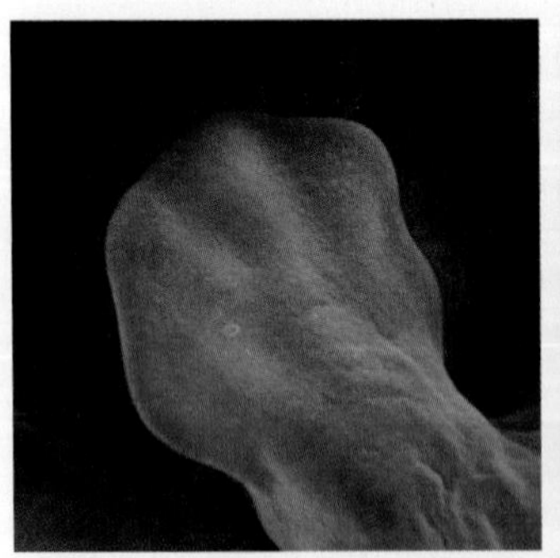

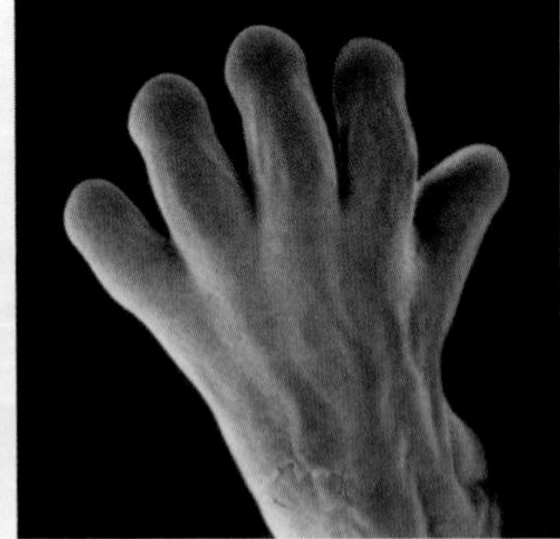

▲ **Figure 27.13A** Apoptosis in a developing human hand

? **Induction often involves signal transduction pathways. What do you suppose their role is?**

● They mediate between the chemical signal received by the cell and the resulting changes in gene expression and other responses by the cell.

27.14 Pattern formation during embryonic development is controlled by ancient genes

EVOLUTION CONNECTION

So far, we have discussed the formation of individual organs. What directs the formation of large body features, such as the limbs? The shaping of an animal's major parts involves **pattern formation**, the emergence of a body form with specialized organs and tissues in the right places. Research indicates that master control genes (see Modules 11.8 and 15.11) respond to chemical signals that tell a cell where it is relative to other cells in the embryo. These positional signals determine which master control genes will be expressed and, consequently, which body parts will form. Research has shown that such control genes arose early in the evolution of animals and so play similar roles across diverse animal groups. The field of biology that studies the evolution of developmental processes is called evolutionary development biology, or evo-devo for short.

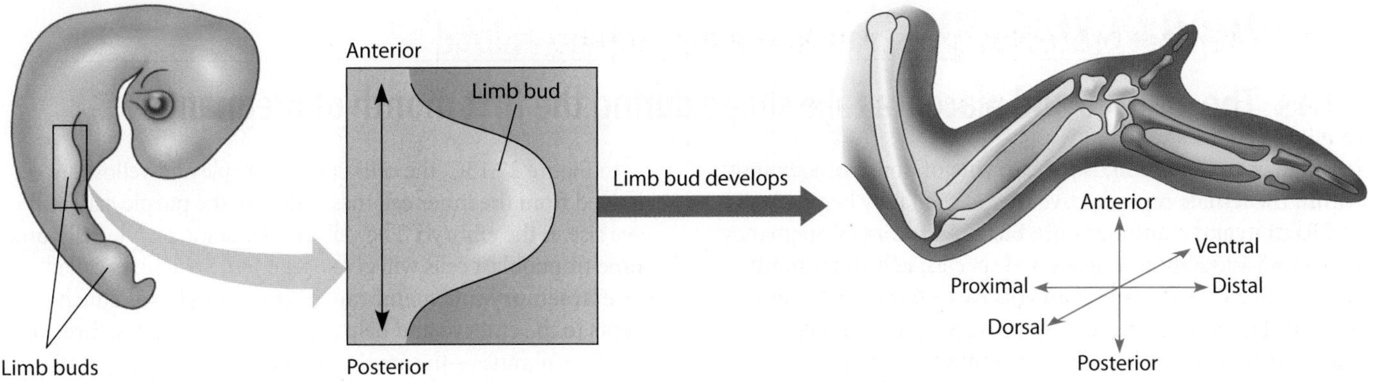
Anterior
Limb bud
Limb bud develops
Limb buds
Posterior
Anterior
Ventral
Proximal
Distal
Dorsal
Posterior

▲ **Figure 27.14A** The normal development of a wing

Vertebrate limbs, such as bird wings, begin as embryonic structures called limb buds **(Figure 27.14A)**. Each component of a chick wing, such as a specific bone or muscle, develops with a precise location and orientation relative to three axes: the proximal-distal axis (the "shoulder-to-fingertip" axis), the anterior-posterior axis (the "thumb-to-little finger" axis), and the dorsal-ventral axis (the "knuckle-to-palm" axis). The embryonic cells within a limb bud respond to positional information indicating location along these three axes. Only with this information will the cell's genes direct the synthesis of the proteins needed for normal differentiation in that cell's specific location.

Among the most exciting biological discoveries in recent years is that a class of similar genes—**homeotic genes**—help direct embryonic pattern formation in a wide variety of organisms. Researchers studying homeotic genes in fruit flies found a common structural feature: Every homeotic gene they looked at contained a common sequence of 180 nucleotides. Very similar sequences have since been found in virtually every eukaryotic organism examined so far, including yeasts, plants, and humans—and even some prokaryotes. These nucleotide sequences are called **homeoboxes**, and each is translated into a segment (60 amino acids long) of the protein product of the homeotic gene. The homeobox polypeptide segment binds to specific sequences in DNA, enabling homeotic proteins that contain it to turn groups of genes on or off during development.

Figure 27.14B highlights some striking similarities in the chromosomal locations and the developmental roles of some homeobox-containing homeotic genes in two quite different animals. The figure shows portions of chromosomes that carry homeotic genes in the fruit fly and the mouse. The colored boxes represent homeotic genes that are very similar in flies and mice. Notice that the order of genes on the fly chromosome is the same as on the four mouse chromosomes and that the gene order on the chromosomes corresponds to analogous body regions in both animals. These similarities suggest that the original version of these homeotic genes arose very early in the history of life and that the genes have remained remarkably unchanged for eons of animal evolution. By their presence in such diverse creatures, homeotic genes illustrate one of the central themes of biology: unity in diversity due to shared evolutionary history.

A major goal of developmental research is to learn how the one-dimensional information encoded in the nucleotide sequence of a zygote's DNA directs the development of the three-dimensional form of an animal. Pattern formation requires cells to receive and interpret environmental cues that vary from one location to another. These cues, acting together along three axes, tell cells where they are in the three-dimensional realm of a developing organ. In the next two modules, we'll see the results of this process as we watch an individual of our own species take shape.

? How is pattern formation already apparent in the limb bud?

● The major axes of the animal—anterior-posterior, dorsal-ventral, and proximal-distal—are already set at an early embryonic stage.

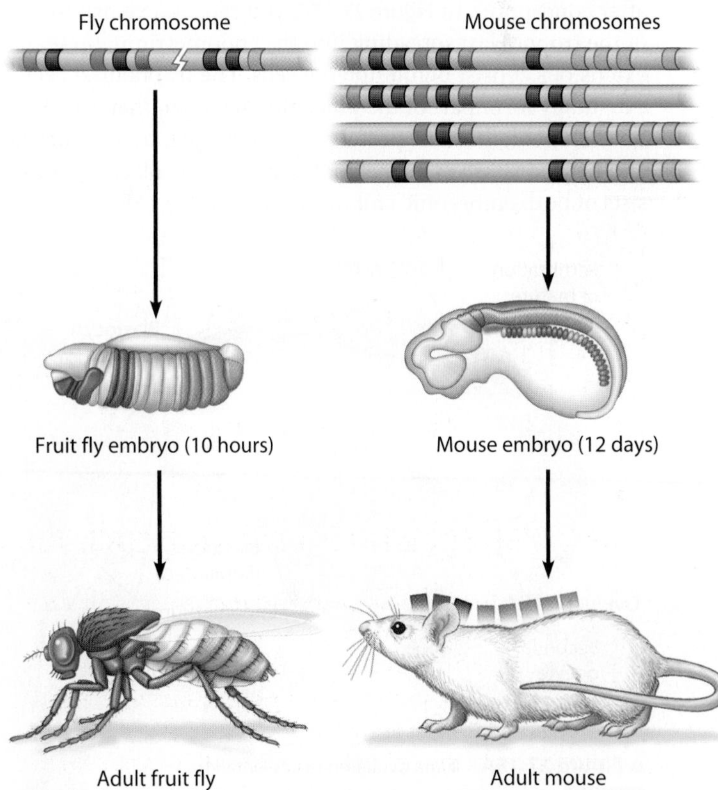
Fly chromosome
Mouse chromosomes
Fruit fly embryo (10 hours)
Mouse embryo (12 days)
Adult fruit fly
Adult mouse

▲ **Figure 27.14B** Comparison of fruit fly and mouse homeotic genes

Human Development

27.15 The embryo and placenta take shape during the first month of pregnancy

Pregnancy, or **gestation**, is the carrying of developing young within the female reproductive tract. Pregnancy begins at fertilization and continues until birth. Duration of pregnancy varies considerably among animal species; gestation in mice lasts about 21 days, while elephants carry their young for 22 months. Human pregnancy averages 266 days (38 weeks) from fertilization (also called **conception** in humans), or 40 weeks (9 months) from the start of the last menstrual cycle.

An Overview of Developmental Events The figures in this module illustrate the changes that occur during the first month of human development. The insets at the lower right of Figures 27.15C–27.15F show the embryo's actual size at each stage.

Fertilization occurs in the oviduct (**Figure 27.15A**). Cleavage starts about 24 hours after fertilization and continues as the embryo moves down the oviduct toward the uterus. About a week after fertilization, the embryo has reached the uterus, and cleavage has produced about 100 cells. The embryo is now a hollow sphere of cells called a **blastocyst** (the mammalian equivalent of the frog blastula in Figure 27.10).

The human blastocyst (**Figure 27.15B**) has a fluid-filled cavity, an inner cell mass that will actually form the embryo, and an outer layer of cells called the **trophoblast**. The trophoblast secretes enzymes that enable the blastocyst to implant in the endometrium, the uterine lining (shown gray in all the figures).

The blastocyst starts to implant in the uterus about a week after conception. In **Figure 27.15C**, you can see extensions of the trophoblast spreading into the endometrium; these extensions consist of multiplying cells. The trophoblast cells eventually form part of the **placenta**, the organ that provides nourishment, immune protection, and oxygen to the embryo and helps dispose of its metabolic wastes. The placenta consists of both embryonic and maternal tissues.

In Figure 27.15C, the cells colored purple and yellow are derived from the inner cell mass. Most of the purple cells will give rise to the embryo. The yellow cells, some purple cells, and some trophoblast cells will give rise to four structures called the **extraembryonic membranes**, which develop as attachments to the embryo and help support it. You can see three of these membranes—the amnion (from purple cells), the yolk sac (from yellow cells), and the chorion (partly from the trophoblast)—starting to take shape in **Figure 27.15D**. A later stage (**Figure 27.15E**) shows the fourth extraembryonic membrane, the allantois, developing as an extension of the yolk sac.

Gastrulation, the stage shown in Figure 27.15D, is under way by 9 days after conception. There is already evidence of the three embryonic layers—ectoderm (blue), endoderm (yellow), and mesoderm (red). The embryo itself (not including the membranes) develops from the three inner cell layers shown in Figure 27.15E. The ectoderm layer will form the outer part of the embryo's skin and its nervous system. As indicated in the drawing, the ectoderm layer is continuous with the amnion. Similarly, the embryo's digestive tract will develop from the endoderm layer, which is continuous with the yolk sac. The bulk of most other organs will develop from the central layer of mesoderm.

Roles of the Extraembryonic Membranes **Figure 27.15F** shows the embryo about a month after fertilization with its life-support system, made up largely of the four extraembryonic membranes. By this time, the **amnion** has grown to enclose the embryo. The amniotic cavity is filled with fluid, which encloses and protects

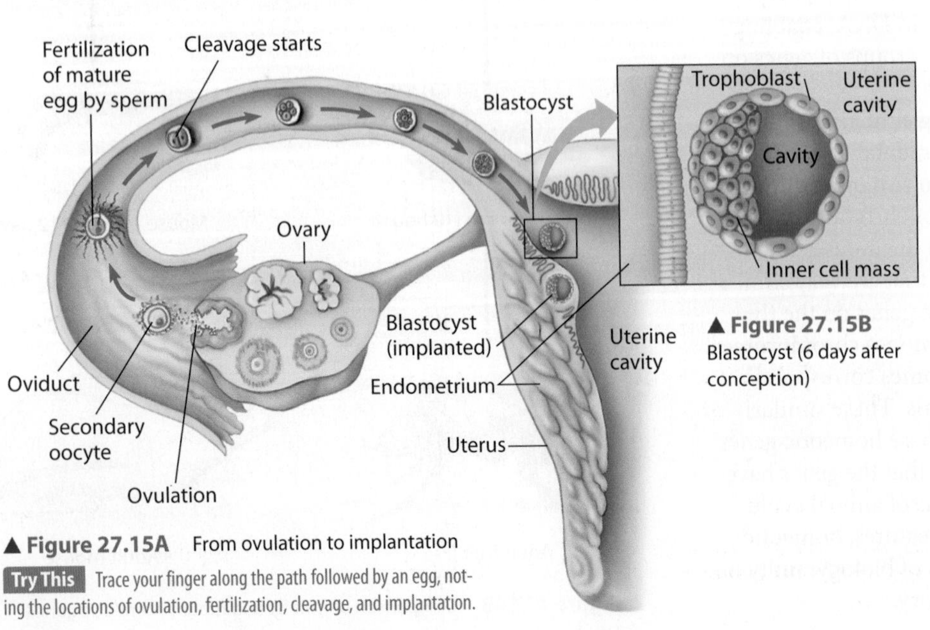

Fertilization of mature egg by sperm
Cleavage starts
Blastocyst
Ovary
Oviduct
Secondary oocyte
Ovulation
Blastocyst (implanted)
Endometrium
Uterus
Uterine cavity
Trophoblast
Uterine cavity
Cavity
Inner cell mass

▲ **Figure 27.15A** From ovulation to implantation

Try This Trace your finger along the path followed by an egg, noting the locations of ovulation, fertilization, cleavage, and implantation.

▲ **Figure 27.15B** Blastocyst (6 days after conception)

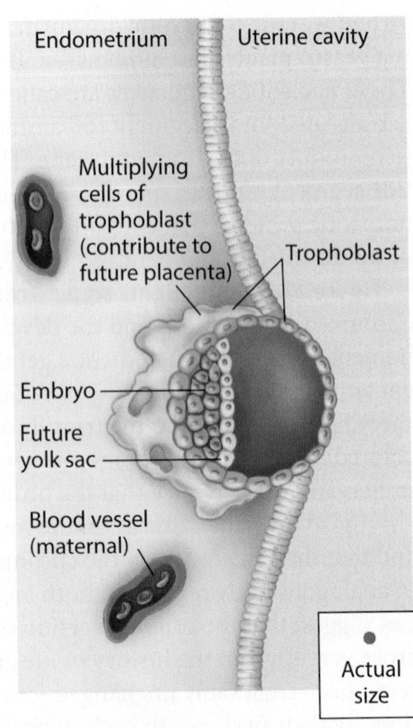

Endometrium
Uterine cavity
Multiplying cells of trophoblast (contribute to future placenta)
Trophoblast
Embryo
Future yolk sac
Blood vessel (maternal)
Actual size

▲ **Figure 27.15C** Implantation under way (about 7 days)

the embryo. The amnion usually breaks just before childbirth, and the amniotic fluid leaves the mother's body through her vagina. Many anxious expectant parents are startled when the mother's "water breaks," and many physicians recommend this be taken as a sign that it's time to get to a hospital.

In humans and most other mammals, the **yolk sac** contains no yolk, but is given the same name as the homologous structure in birds and other reptiles (Figure 19.6B). Isolated within a shelled egg outside of the mother's body, a developing bird will obtain nourishment from the yolk rather than from a placenta. In mammals, the yolk sac, which remains small, has other important functions: It produces the embryo's first blood cells and its first germ cells, the cells that will give rise to the gamete-forming cells in the gonads.

The **allantois** also remains small in mammals. It forms part of the umbilical cord—the lifeline between the embryo and the placenta. It also forms part of the embryo's urinary bladder. In birds and other reptiles, the allantois expands around the embryo and functions in waste disposal.

The outermost extraembryonic membrane, the **chorion**, completely surrounds the embryo and other extraembryonic membranes. The chorion becomes part of the placenta. Cells in the chorion secrete a hormone called **human chorionic gonadotropin (hCG)**, which maintains production of estrogen and progesterone by the corpus luteum of the ovary during the first few months of pregnancy. Without these hormones, menstruation would occur, and the embryo would abort spontaneously. Levels of hCG in maternal blood are so high that some is excreted in the urine, where it can be detected by pregnancy tests (see Module 24.10).

The Placenta Looking again at Figure 27.15D, notice the finger-like outgrowths on the outside of the chorion. In Figure 27.15E, these outgrowths, now called **chorionic villi**, are larger and contain mesoderm. In Figure 27.15F, the mesoderm cells have formed into embryonic blood vessels in the chorionic villi. By this stage, the placenta is fully developed. Starting with the chorion and extending outward, the placenta is a composite organ consisting of chorionic villi closely associated with the blood vessels of the mother's endometrium. The villi are actually bathed in tiny pools of maternal blood. The mother's blood and the embryo's blood are not in direct contact. Instead, the chorionic villi absorb nutrients and oxygen from the mother's blood and pass these substances to the embryo via the chorionic blood vessels that are shown in red. The chorionic vessels shown in blue carry wastes away from the embryo. The wastes diffuse into the mother's bloodstream and are excreted by her kidneys.

The placenta is a vital organ with both embryonic and maternal parts that mediates exchange of nutrients, gases, and the products of excretion between the embryo and the mother. However, the placenta cannot always protect the embryo from substances circulating in the mother's blood. A number of viruses—the German measles virus and HIV, for example—can cross the placenta. German measles can cause serious birth defects; HIV-infected babies usually die of AIDS within a few years without treatment. Most drugs, both prescription and not, also cross the placenta, and many can harm the developing embryo. Alcohol, the chemicals in tobacco smoke, and other drugs increase the risk of miscarriage and birth defects.

? **Why does testing for hCG in a woman's urine or blood work as an early test of pregnancy?**

Because this hormone is secreted by the chorion of an embryo after implantation in the wall of the placenta

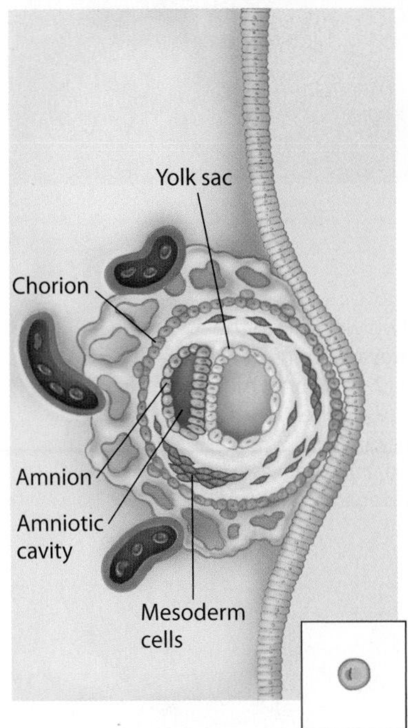

Yolk sac
Chorion
Amnion
Amniotic cavity
Mesoderm cells

▲ **Figure 27.15D** Embryonic layers and extraembryonic membranes starting to form (9 days)

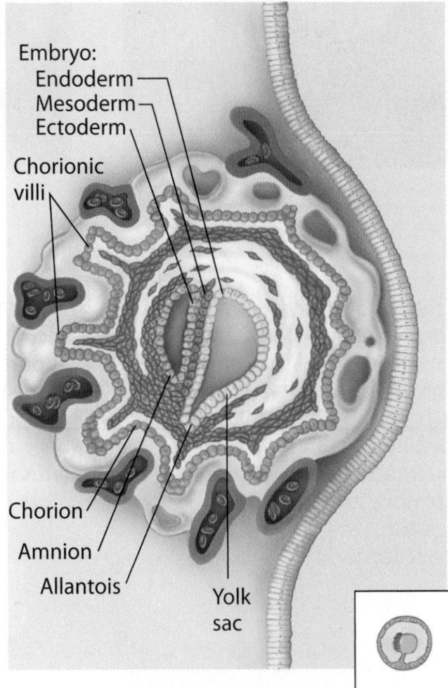

Embryo:
Endoderm
Mesoderm
Ectoderm
Chorionic villi
Chorion
Amnion
Allantois
Yolk sac

▲ **Figure 27.15E** Three-layered embryo and four extraembryonic membranes (16 days)

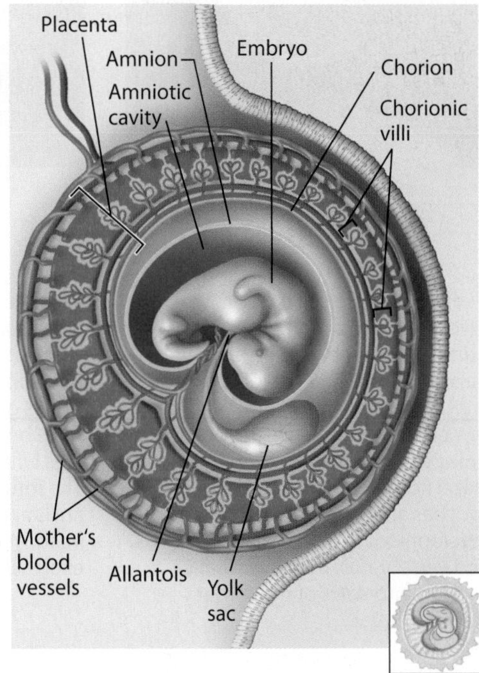

Placenta
Amnion
Amniotic cavity
Embryo
Chorion
Chorionic villi
Mother's blood vessels
Allantois
Yolk sac

▲ **Figure 27.15F** Placenta formed (31 days)

VISUALIZING THE CONCEPT

27.16 Human pregnancy is divided into trimesters

For humans, pregnancy—the period of development from conception to birth—is divided into three **trimesters**, each lasting about 3 months.

TRIMESTER 1

The first trimester is the time of the most radical change for both mother and embryo. During this time, the embryo is particularly susceptible to damage by radiation, drugs, or alcohol, all of which can lead to birth defects or miscarriage.

TRIMESTER 2

During the second trimester, the fetus continues to grow and has increasingly human features; the changes are not as dramatic as those changes of the first trimester. The fetus's eyes can open, its teeth are forming, and its bones have begun to harden. The placenta

Timeline of Human Development

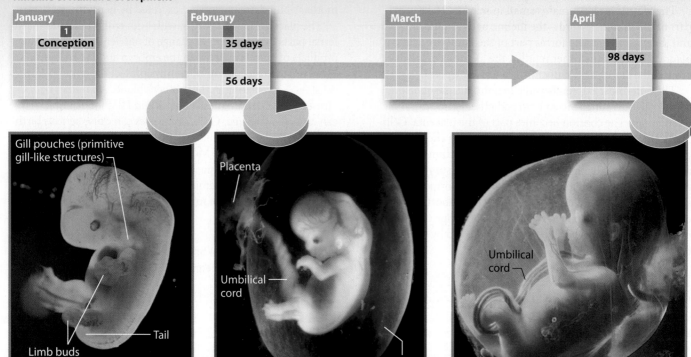

January — 1 Conception

February — 35 days; 56 days

March

April — 98 days

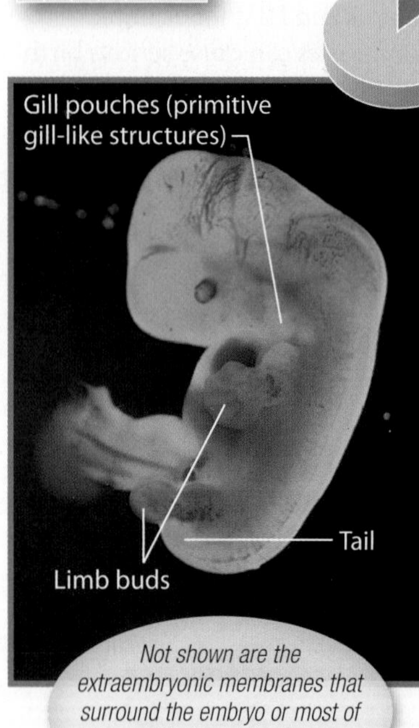
Gill pouches (primitive gill-like structures)
Tail
Limb buds

Not shown are the extraembryonic membranes that surround the embryo or most of the umbilical cord that attaches it to the placenta.

Placenta

Umbilical cord

Amnion

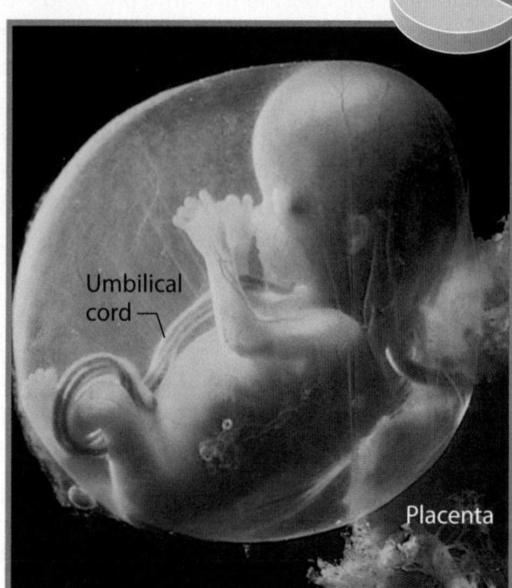
Umbilical cord
Placenta

Week 5: A single cell has developed into a highly organized multicellular embryo. This embryo is about 7 mm (0.28 in.) long. It has a notochord and a coelom, both formed from mesoderm. Its brain and spinal cord have begun to take shape from a tube of ectoderm. Gill pouches appear during embryonic development in all chordates; in land vertebrates, they eventually develop into parts of the throat and middle ear.

Actual size: 7 mm

Week 8: All the major structures of the adult are present in rudimentary form. The embryo is about 4 cm (1.6 in.) long. The somites have developed into the segmental muscles and bones of the back and ribs. The limb buds have become tiny arms and legs with fingers and toes. The fetus can move its arms and legs, turn its head, and make facial expressions.

4 cm

Week 14: Two weeks into the second trimester, the fetus is about 6 cm (2.4 in.) long. All of the features of the earlier embryo have been refined, and the fetus now appears more human.

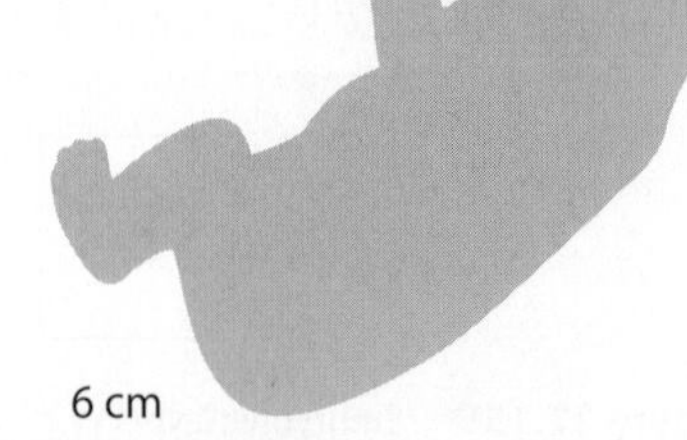

6 cm

begins to secrete progesterone (and the corpus luteum stops secreting progesterone), which helps maintain the placenta. At the same time, the placenta stops secreting hCG, and the corpus luteum, no longer needed to maintain pregnancy, degenerates.

TRIMESTER 3

The third trimester is a time of rapid growth as the fetus gains the strength it will need to survive outside the protective environment of the uterus. Babies born prematurely—as early as 24 weeks—may survive, but they require special medical care after birth. As the fetus grows and the uterus expands around it, the mother's abdominal organs become squeezed, causing frequent urination, digestive troubles, and backaches.

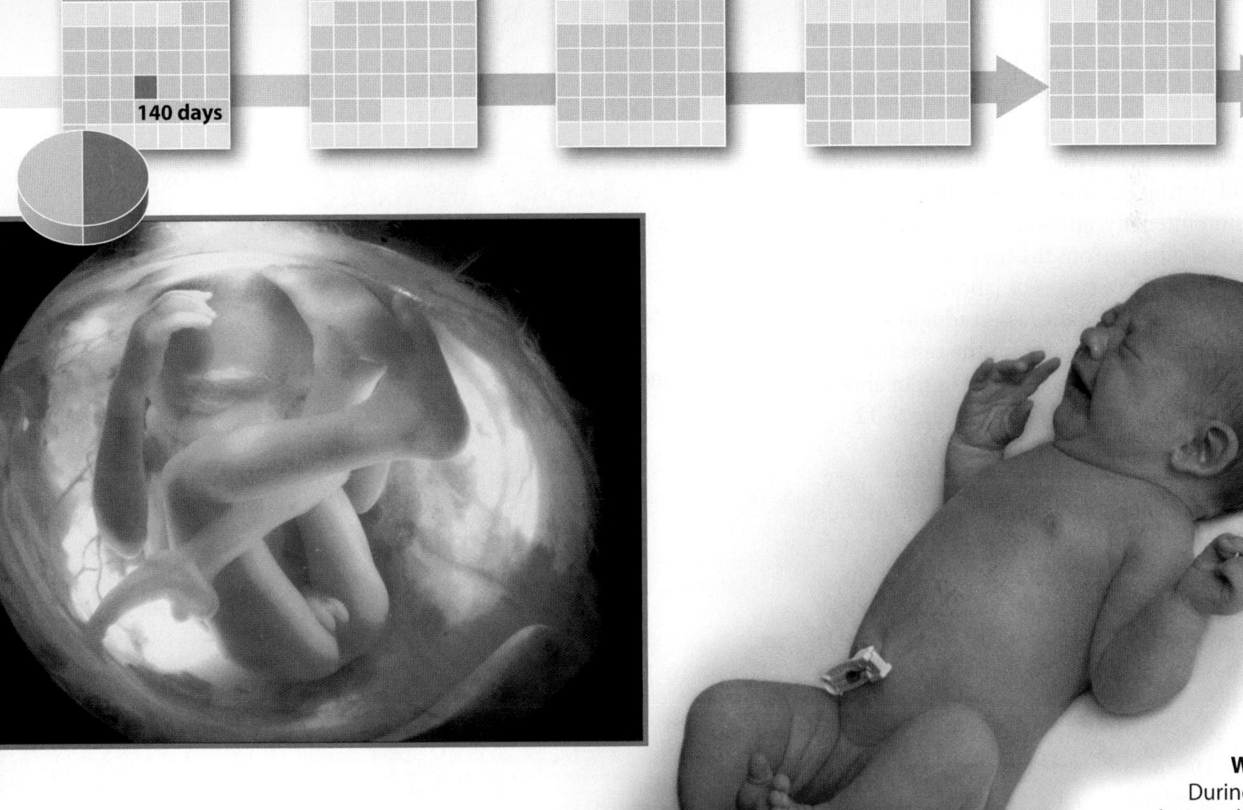
May
140 days
June
July
August
September
October
280 days

Week 20: The fetus is about 19 cm (7.6 in.) long, weighs about half a kilogram (1 pound), and has eyebrows and eyelashes. Its arms, legs, fingers, and toes have lengthened. It also has fingernails and toenails and is covered with fine hair. Fetuses of this age are usually quite active, allowing the mother to feel the baby "kick." The mother's abdomen has become markedly enlarged. Because of the limited space in the uterus, the fetus flexes forward into the fetal position, with its head tucked against its knees.

Week 40 (newborn): During its final months in the womb, the fetus's circulatory system and respiratory system develop the ability to switch to air breathing. Its muscles thicken and it loses much of its fine body hair, except on its head. Near the end of the third trimester, the fetus rotates so that its head points down toward the cervix, and it fills the space in the uterus. At birth, babies average about 50 cm (20 in.) in length and weigh 3–4 kg (6–8 pounds).

? Certain drugs cause their most serious damage to an embryo very early in pregnancy, often before the mother even realizes she is pregnant. Why?

Because organ systems, such as the circulatory and nervous systems, begin to develop early in the first trimester and are susceptible to such drugs.

27.17 Childbirth is induced by hormones and other chemical signals

The series of events that expel an infant from the uterus is called **labor**. Several hormones play key roles in this process **(Figure 27.17A)**. One hormone, estrogen, reaches its highest level in the mother's blood during the last weeks of pregnancy. An important effect of this estrogen is to trigger the formation of numerous oxytocin receptors on cells of the uterus. Cells of the fetus produce the hormone oxytocin, and late in pregnancy, the mother's pituitary gland secretes it in increasing amounts. Oxytocin stimulates the smooth muscles in the wall of the uterus, producing the series of increasingly strong, rhythmic contractions characteristic of labor. It also stimulates the placenta to make prostaglandins, chemical regulators that stimulate the uterine muscle cells to contract even more.

The induction of labor involves **positive feedback**, a type of control in which a change triggers mechanisms that amplify that change. In this case, oxytocin and prostaglandins cause uterine contractions that in turn stimulate the release of *more* oxytocin and prostaglandins. The result is a steady increase in contraction intensity, climaxing in forceful muscle contractions that propel a baby from the uterus.

Figure 27.17B shows the three stages of labor. As the process begins, the cervix (neck of the uterus) gradually opens. ① The first stage, dilation, is the time from the onset of labor until the cervix reaches its full dilation (widening) of about 10 cm. Dilation is the longest stage of labor, lasting 6–12 hours, sometimes considerably longer.

② The period from full dilation of the cervix to delivery of the infant is called the expulsion stage. Strong uterine contractions, lasting about 1 minute each, occur every 2–3 minutes, and the mother feels an increasing urge to push or bear down with her abdominal muscles. Within a period of 20 minutes to an hour or so (or, possibly, considerably longer), the infant is forced down and out of the uterus and vagina.

Estrogen
from ovaries
Induces oxytocin receptors on uterus
Oxytocin
from fetus and mother's pituitary
Stimulates uterus to contract
Stimulates placenta to make
Prostaglandins
Stimulate more contractions of uterus
Positive feedback

▲ **Figure 27.17A** The hormonal induction of labor

After the baby is born, a doctor or midwife (or nervous father!) clamps and cuts the umbilical cord. ③ The final stage is the delivery of the placenta ("afterbirth"), usually within 15 minutes after the birth of the baby.

Hormones continue to be important after the baby and placenta are delivered. Decreasing levels of progesterone and estrogen allow the uterus to start returning to its state before pregnancy. In response to suckling by the newborn, as well as falling levels of progesterone after birth, the pituitary

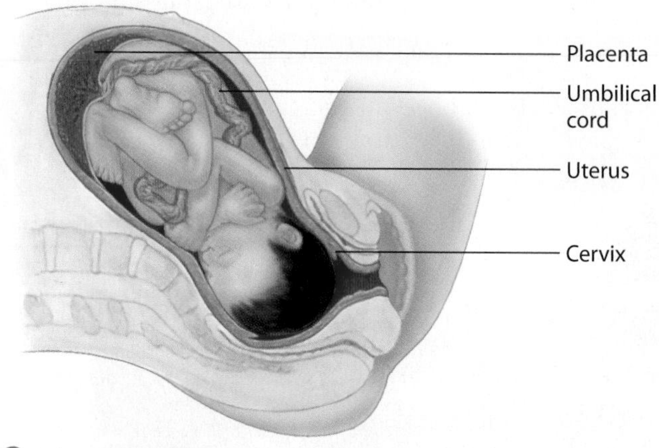
Placenta
Umbilical cord
Uterus
Cervix

① Dilation of the cervix

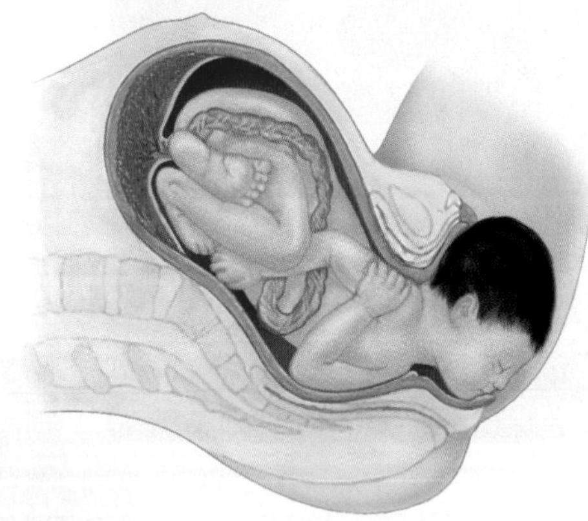

② Expulsion: delivery of the infant

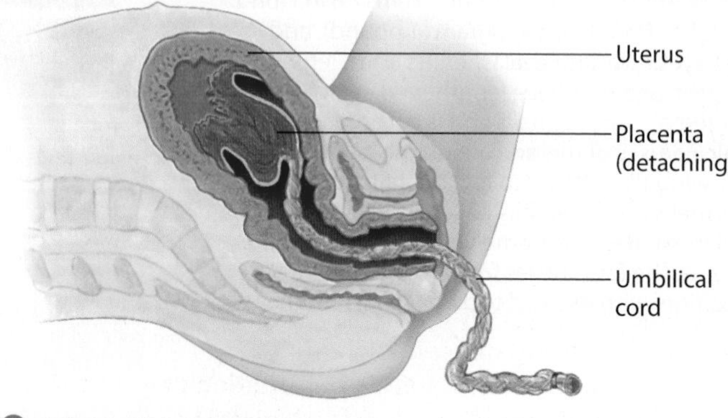
Uterus
Placenta (detaching)
Umbilical cord

③ Delivery of the placenta

▲ **Figure 27.17B** The three stages of labor

secretes prolactin and oxytocin. These two hormones promote milk production and release (called lactation) by the mammary glands. At first, a yellowish fluid called colostrum, rich in protein and antibodies, is secreted. After 2–3 days, the production of regular milk begins. Throughout infancy and early childhood, and indeed throughout life, developmental processes such as those described in this chapter continue to shape the human body.

? **The onset of labor is marked by dilation of the ____________.**

● cervix

27.18 Reproductive technologies increase our reproductive options

CONNECTION

About 15% of couples who want children are unable to conceive, even after 12 months of unprotected intercourse. Such a condition, called **infertility**, can have many causes. A man's testes may not produce enough sperm (a "low sperm count"), or those that are produced may be defective. In other cases, infertility is caused by **impotence**, also called erectile dysfunction, the inability to achieve or maintain an erection. Temporary impotence can result from alcohol or drug use or from psychological problems. Permanent impotence can result from nervous system or circulatory problems. Female infertility can result from a lack of eggs, a failure to ovulate, or blocked oviducts (often caused by scarring due to sexually transmitted diseases). Some women are able to conceive but cannot support a growing embryo in the uterus. The resulting multiple miscarriages can take a heavy emotional toll.

Reproductive technologies can help many cases of infertility. Drug therapies (including Viagra) and penile implants can be used to treat impotence. Underproduction of sperm is frequently caused by the man's scrotum being too warm, so a switch of underwear from briefs (which hold the scrotum close to the body) to boxers may help. If a man still has a low sperm count, sperm can be collected, concentrated, and injected into a woman's uterus. If a man produces no functioning sperm at all, the couple may elect to use another man's sperm that has been donated to a sperm bank.

If a woman has normal eggs that are not being released properly, hormone injections can induce ovulation. Such treatments frequently result in multiple pregnancies—twins, triplets, or more. If a woman has no eggs of her own, they can be obtained from a donor for fertilization and injected into the uterus. While sperm can be collected without any danger to the donor, collection of eggs involves surgery and therefore some pain and risk for the donating woman. For this reason, egg donors are often paid several thousand dollars for their time and discomfort.

If a woman cannot maintain pregnancy, she and her partner may enter into a legal contract with a surrogate mother who agrees to be implanted with the couple's embryo and carry it to birth. However, a number of states have laws restricting surrogate motherhood because of the serious ethical and legal problems that can arise.

Many infertile couples turn to fertilization procedures called **assisted reproductive technologies**. In these procedures, eggs (secondary oocytes) are surgically removed from a woman's ovaries after hormonal stimulation, fertilized, and returned to the woman's body. Eggs, sperm, and embryos from such procedures can be frozen for later pregnancy attempts. With ***in vitro* fertilization (IVF)**, the most common assisted reproductive technology procedure, a woman's eggs are mixed with sperm in culture dishes (*in vitro* means "in glass") and incubated for several days to allow fertilized eggs to start developing (**Figure 27.18**). When they have developed into embryos of at least eight cells each, the embryos are carefully inserted into the woman's uterus, timed to coincide with her natural ovarian cycle.

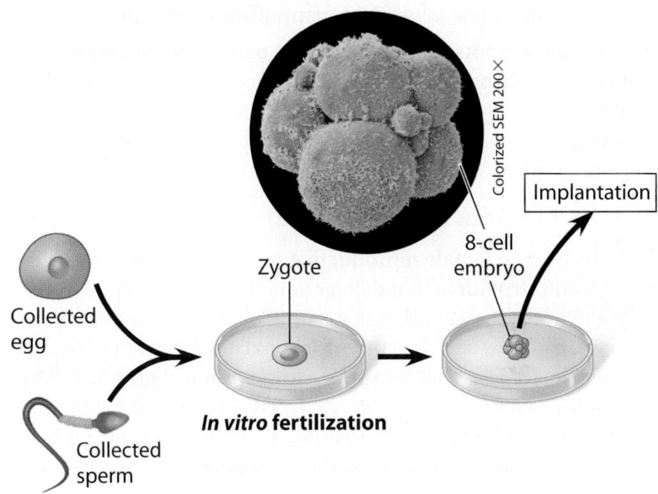

Colorized SEM 200×
Implantation
8-cell embryo
Zygote
Collected egg
Collected sperm
In vitro fertilization

▲ **Figure 27.18** *In vitro* fertilization

Abnormalities arising as a consequence of an IVF procedure appear to be quite rare, although some research has shown small but significant risks of lower birth weights and higher rates of birth defects. Despite such risks and the high cost (typically $10,000 per attempt, whether it succeeds or not), IVF techniques are now performed at medical centers throughout the world and result in the birth of thousands of babies each year.

In this chapter, we have watched a single-celled product of sexual reproduction, the zygote, transform into a new organism, complete with all organ systems. One of the first of those organ systems to develop is the nervous system. In the next chapter, we'll see how the nervous system functions together with the endocrine system to regulate virtually all body activities.

? **Explain how IVF can involve up to three different people in the birth of a child.**

● One woman (1) may become pregnant with an embryo created using the sperm of a man (2) and the mature egg of a second woman (3).

CHAPTER 27 REVIEW

For practice quizzes, BioFlix animations, MP3 tutorials, video tutors, and more study tools designed for this textbook, go to MasteringBiology®

Reviewing the Concepts

Asexual and Sexual Reproduction (27.1–27.2)

27.1 Asexual reproduction results in the generation of genetically identical offspring. Asexual reproduction can proceed by budding, fission, or fragmentation/regeneration. This reproductive scheme allows one individual to produce many offspring rapidly.

27.2 Sexual reproduction results in the generation of genetically unique offspring. Sexual reproduction involves the fusion of gametes from two parents, resulting in genetic variation among offspring. This may enhance survival of a population in a changing environment.

Human Reproduction (27.3–27.8)

27.3 The human female reproductive system includes the ovaries and structures that deliver gametes. The human female reproductive system consists of a pair of ovaries, ducts that carry gametes, a uterus, and structures for copulation. A woman's ovaries contain follicles that nurture eggs and produce sex hormones. Oviducts convey eggs to the uterus, where a fertilized egg develops. The uterus opens into the vagina, which receives the penis during intercourse and serves as the birth canal.

27.4 The human male reproductive system includes the testes and structures that deliver gametes. A man's testes produce sperm, which are expelled through ducts during ejaculation.

27.5 The formation of sperm and egg cells requires meiosis. Spermatogenesis and oogenesis produce sperm and eggs, respectively. Primary spermatocytes are made continuously in the testes; these diploid cells undergo meiosis to form four haploid sperm. In females, each month, one primary oocyte forms a secondary oocyte, which, if penetrated by a sperm, completes meiosis and becomes a mature egg. The haploid nucleus of the mature egg then fuses with the haploid nucleus of the sperm, forming a diploid zygote.

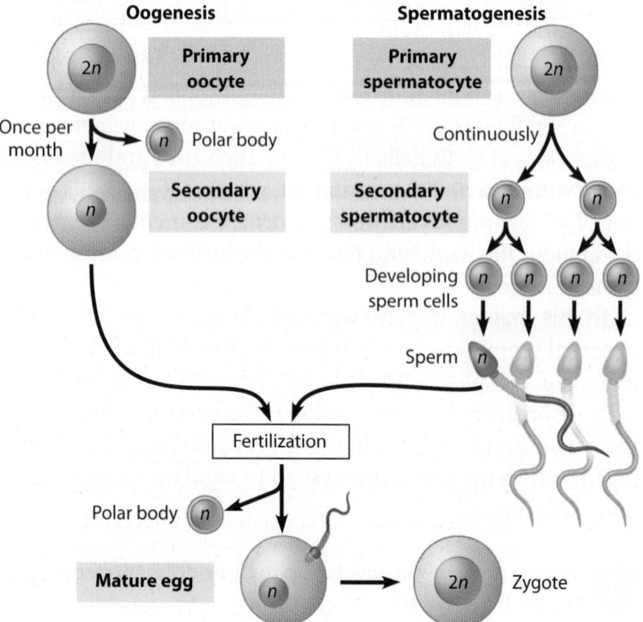
Oogenesis
Spermatogenesis
2n
Primary oocyte
Primary spermatocyte
2n
Once per month
n
Polar body
Continuously
n
Secondary oocyte
Secondary spermatocyte
n
n
Developing sperm cells
n
n
n
n
Sperm
n
Fertilization
Polar body
n
Mature egg
n
2n
Zygote

27.6 Hormones synchronize cyclic changes in the ovary and uterus. Approximately every 28 days, the hypothalamus signals the anterior pituitary to secrete FSH and LH, which trigger the growth of a follicle and ovulation, the release of an egg. The follicle becomes the corpus luteum, which secretes both estrogen and progesterone. These two hormones stimulate the endometrium (the uterine lining) to thicken, preparing the uterus for implantation. They also inhibit the hypothalamus, reducing FSH and LH secretion. If the egg is not fertilized, the drop in LH shuts down the corpus luteum and its hormones. This triggers menstruation, the breakdown of the endometrium. The hypothalamus and pituitary then stimulate another follicle, starting a new cycle. If fertilization occurs, a hormone from the embryo maintains the uterine lining and prevents menstruation.

27.7 Sexual activity can transmit disease. STDs caused by bacteria can often be cured, but viral STDs can only be controlled.

27.8 Contraception can prevent unwanted pregnancy. Several forms of contraception can prevent pregnancy, with varying degrees of success.

Principles of Embryonic Development (27.9–27.14)

27.9 Fertilization results in a zygote and triggers embryonic development. During fertilization, a sperm releases enzymes that pierce the egg's coat. Sperm surface proteins bind to egg receptor proteins, sperm and egg plasma membranes fuse, and the two nuclei unite. Changes in the egg membrane prevent entry of additional sperm, and the fertilized egg (zygote) develops into an embryo.

27.10 Cleavage produces a blastula from the zygote. Cleavage is a rapid series of cell divisions that results in a blastula.

27.11 Gastrulation produces a three-layered embryo. In gastrulation, cells migrate and form a rudimentary digestive cavity and three layers of cells.

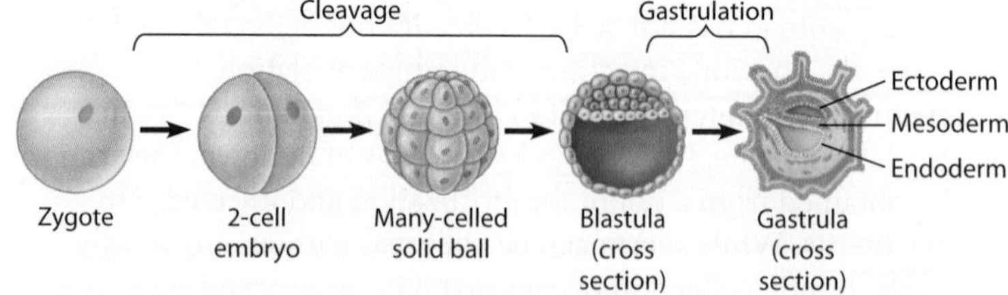
Cleavage
Gastrulation
Ectoderm
Mesoderm
Endoderm
Zygote
2-cell embryo
Many-celled solid ball
Blastula (cross section)
Gastrula (cross section)

27.12 Organs start to form after gastrulation. After gastrulation, the three embryonic tissue layers give rise to specific organ systems.

27.13 Multiple processes give form to the developing animal. Tissues and organs take shape in a developing embryo as a result of cell shape changes, cell migration, and programmed cell death. Through induction, adjacent cells and cell layers influence each other's differentiation via chemical signals.

27.14 Pattern formation during embryonic development is controlled by ancient genes. Pattern formation, the emergence of the parts of a structure in their correct relative positions, involves the response of genes to spatial variations of chemicals in the embryo. Homeotic genes contain homeoboxes, nucleotide sequences that appeared early in the evolutionary history of animals.

Human Development (27.15–27.18)

27.15 The embryo and placenta take shape during the first month of pregnancy. Human development begins with fertilization in the oviduct. Cleavage produces a blastocyst, whose inner cell mass becomes the embryo. The blastocyst's outer layer, the trophoblast, implants in the uterine wall. Gastrulation occurs, and organs develop from the three embryonic layers. Meanwhile, the four extraembryonic membranes develop: the amnion, the chorion, the yolk sac, and the allantois. The placenta's chorionic villi absorb food and oxygen from the mother's blood.

27.16 Human pregnancy is divided into trimesters. The most rapid changes occur during the first trimester. At 9 weeks, the embryo is called a fetus. The second and third trimesters are times of growth and preparation for birth.

27.17 Childbirth is induced by hormones and other chemical signals. Estrogen makes the uterus more sensitive to oxytocin, which acts with prostaglandins to initiate labor. The cervix dilates, the baby is expelled by strong muscular contractions, and the placenta follows.

27.18 Reproductive technologies increase our reproductive options. In *in vitro* fertilization, eggs are extracted and fertilized in the lab. The resulting embryo is implanted into a woman.

Connecting the Concepts

1. This graph plots the rise and fall of pituitary and ovarian hormones during the human ovarian cycle. Identify each hormone (A–D) and the reproductive events with which each one is associated (P–S). For A–D, choose from estrogen, LH, FSH, and progesterone. For P–S, choose from ovulation, growth of follicle, menstruation, and development of corpus luteum. How would the right-hand side of this graph be altered if pregnancy occurred? What other hormone is responsible for triggering this change?

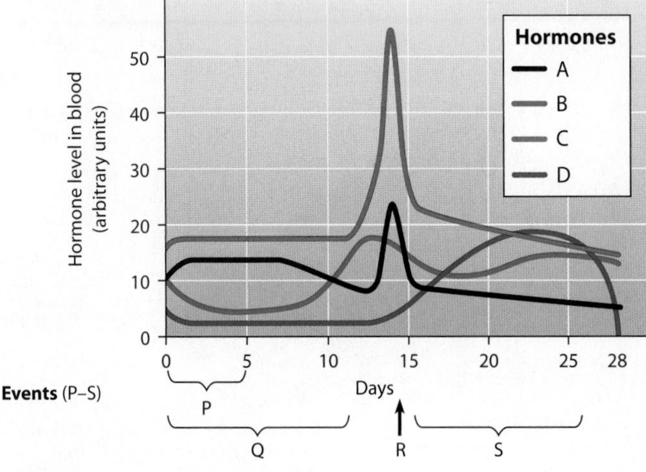

Hormones
A
B
C
D
Hormone level in blood (arbitrary units)
50
40
30
20
10
0
0
5
10
15
20
25
28
Days
Events (P–S)
P
Q
R
S

Testing Your Knowledge

Level 1: Knowledge/Comprehension

Matching

2. Turns into the corpus luteum	a. vas deferens
3. Female gonad	b. prostate gland
4. Site of spermatogenesis	c. endometrium
5. Site of fertilization in humans	d. testis
6. Site of human gestation	e. follicle
7. Sperm duct	f. uterus
8. Secretes seminal fluid	g. ovary
9. Lining of uterus	h. oviduct

Level 2: Application/Analysis

10. After a sperm penetrates an egg, it is important that the vitelline layer separate from the egg so that it can
    a. secrete important hormones.
    b. enable the fertilized egg to implant in the uterus.
    c. prevent more than one sperm from entering the egg.
    d. attract additional sperm to the egg.
11. In an experiment, a researcher colored a bit of tissue on the outside of a frog gastrula with an orange fluorescent dye. The embryo developed normally. When the tadpole was placed under an ultraviolet light, which of the following glowed bright orange? (*Explain your answer.*)
    a. the heart
    b. the pancreas
    c. the brain
    d. the stomach
12. How does a zygote differ from a mature egg?
    a. A zygote has more chromosomes.
    b. A zygote is smaller.
    c. A zygote consists of more than one cell.
    d. A zygote divides by meiosis.
13. A woman had several miscarriages. Her doctor suspected that a hormonal insufficiency was causing the lining of the uterus to break down, as it does during menstruation, terminating her pregnancies. Treatment with which of the following might help her remain pregnant?
    a. oxytocin
    b. follicle-stimulating hormone
    c. luteinizing hormone
    d. prolactin
14. The embryos of reptiles (including birds) and mammals have systems of extraembryonic membranes. What are the functions of these membranes, and how do fish and frog embryos survive without them?

Level 3: Synthesis/Evaluation

15. Compare sperm formation with egg formation. In what ways are the processes similar? In what ways are they different?
16. In an embryo, nerve cells grow out from the spinal cord and form connections with the muscles they will eventually control. What mechanisms described in this chapter might explain how these cells "know" where to go and which cells to connect with?
17. As a frog embryo develops, the neural tube forms from ectoderm along what will be the frog's back, directly above the notochord. To study this process, a researcher extracted a bit of notochord tissue and inserted it under the ectoderm where the frog's belly would normally develop. What can the researcher hope to learn from this experiment? Predict the possible outcomes. What experimental control would you suggest?
18. Should parents undergoing *in vitro* fertilization have the right to choose which embryos to implant based on genetic criteria, such as the presence or absence of disease-causing genes? Should they be able to choose based on the sex of the embryo? How could you distinguish acceptable from unacceptable criteria? Do you think such options should be legislated?
19. **SCIENTIFIC THINKING** There are difficulties inherent in testing hypotheses involving human health. Imagine that you have developed a potential vaccine against AIDS. Given the nature of the disease, all testing must be conducted in humans. Can you design an ethical but effective experiment?

Answers to all questions can be found in Appendix 4.

CHAPTER 28

Nervous Systems

More than 20 million Americans are affected by depression in a given year, according to the National Institute of Mental Health. Depression is a psychiatric disorder characterized by persistent sadness, loss of interest in pleasurable activities, changes in weight and sleep patterns, diminished energy, and suicidal thoughts over a continuous period. Researchers have begun to learn more about the brain physiology of depression. For example, brain scans can highlight regions of low brain activity and show that some areas of the brain in a depressed person (below, left) have lower activity than the same areas in a nondepressed person (facing page).

? *Are antidepressants effective?*

Drugs that treat clinical (medically diagnosed) depression, antidepressants, are the third most commonly prescribed class of drugs in the United States. The most widely prescribed subclass of antidepressants, selective serotonin reuptake inhibitors (SSRIs), block the reabsorption of a particular mood-regulating chemical called serotonin into brain cells. The result is an increase of serotonin at the junctions between brain cells. (These drugs were designed based on the hypothesis that low levels of serotonin can cause depression.)

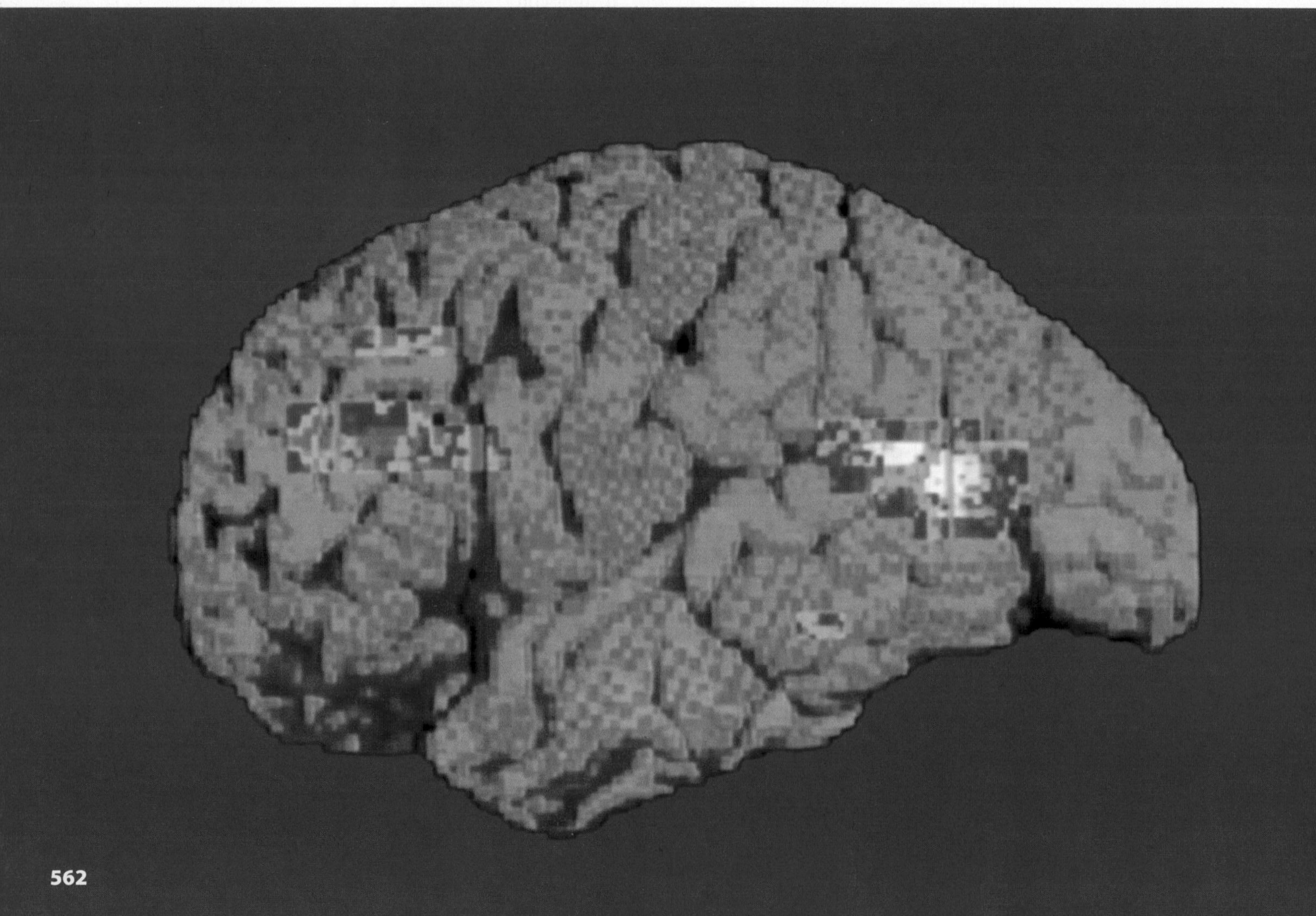

Despite their widespread use, the effectiveness of SSRIs has come into question. SSRIs seem to be most effective for the severely depressed, offering little or no benefit to the moderately depressed. Although many studies have reported that these drugs successfully treat depression, some studies have concluded that depressed individuals who are given an SSRI benefit no more than those who are given a sugar pill, or placebo (a substance containing no medicine). Follow-up analyses, including nonpublished studies and other sources that classify the severity of depression, provide a possible explanation for the conflicting results.

In this chapter, we'll consider how these negligible effects of SSRIs could have been overlooked for almost two decades. But let's first take a step back and explore the nervous system, of which the brain is a part.

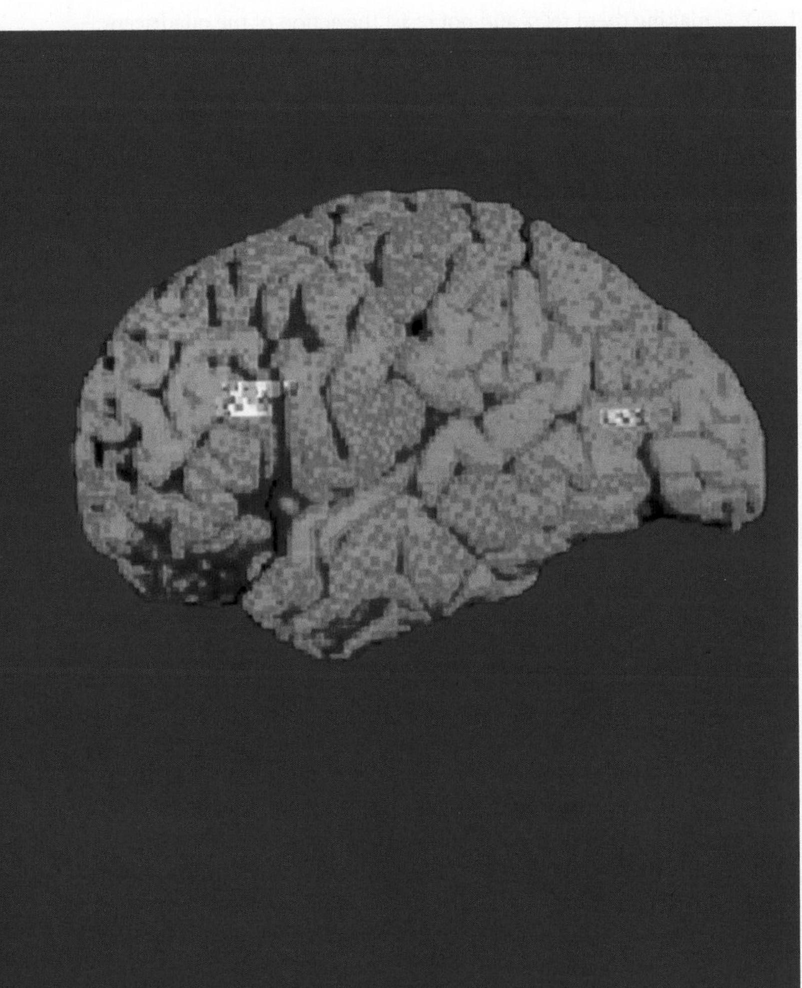

BIG IDEAS

Nervous System Structure and Function
(28.1–28.2)

Neurons receive and process inputs and communicate responses.

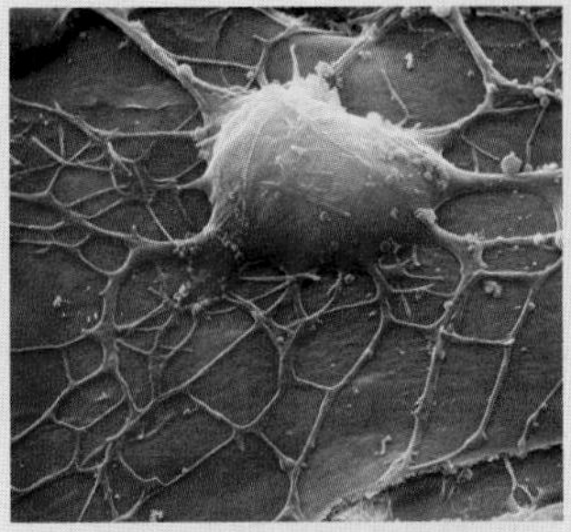

Nerve Signals and Their Transmission
(28.3–28.10)

Nerve signals are electrical messages generated by the movement of ions across membranes and passed between nerve cells by chemical signals.

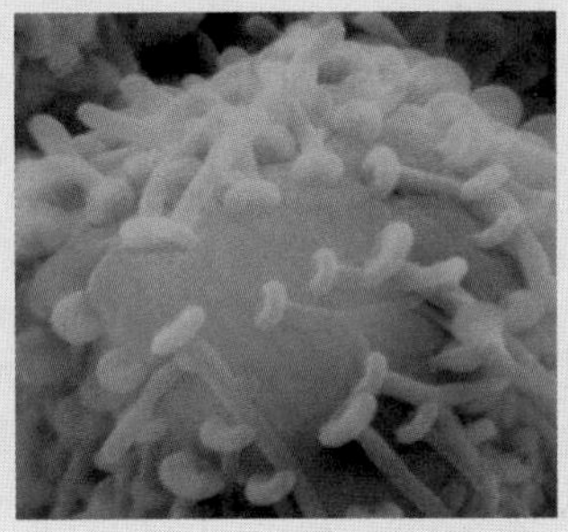

An Overview of Animal Nervous Systems
(28.11–28.14)

The vertebrate nervous system can be understood as a structural and functional hierarchy.

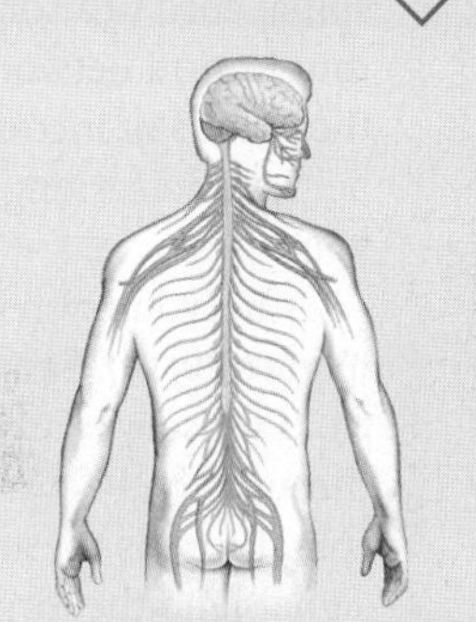

The Human Brain
(28.15–28.21)

Modern research techniques are illuminating how the interplay of brain regions controls our actions and behavior.

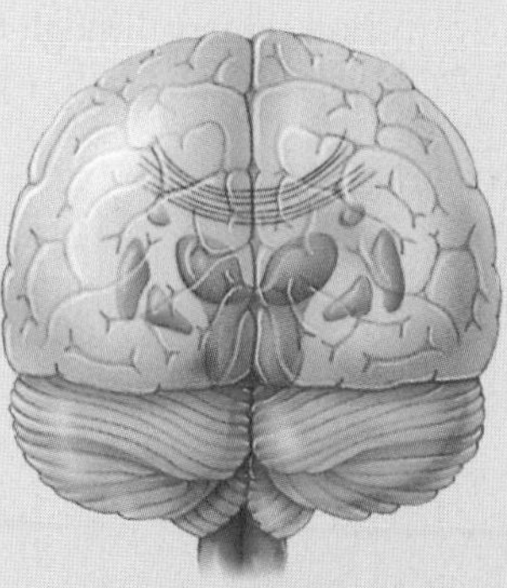

▷ Nervous System Structure and Function

28.1 Nervous systems receive sensory input, interpret it, and send out commands

Two major organ systems are responsible for coordinating the functions of the animal body: the endocrine system and the nervous system. Whereas the endocrine system usually leads to slower, more sustained responses (Module 26.1), the **nervous system** enables rapid coordination of body functions. Communication within the nervous system relies on **neurons**, nerve cells that transmit information via electrical and chemical signals. A neuron consists of a **cell body**, containing the nucleus and other cell organelles, and long, thin extensions that convey signals. Each neuron may communicate with thousands of others, forming networks that enable us to learn, remember, perceive our surroundings, and move.

With few exceptions, nervous systems have two main anatomical divisions. The **central nervous system (CNS)** consists of the brain and, in vertebrates, the spinal cord as well. The **peripheral nervous system (PNS)** consists of neurons that carry information into and out of the CNS. When bundled together and wrapped in connective tissue, the neurons of the PNS form **nerves**. The PNS also has **ganglia** (singular, *ganglion*), clusters of neuron cell bodies found alongside the spinal cord near or within organs they serve. The vast communication network that makes up the PNS handles the thousands of incoming and outgoing signals passing through nerves and serving all areas of your body from your head to your toes.

A nervous system has three interconnected functions (**Figure 28.1A**). **Sensory input** is the conduction of signals from sensory receptors, such as light-detecting cells of the eye, through the PNS to the CNS. **Integration** in the brain and spinal cord is the analysis and interpretation of the sensory signals and the formulation of appropriate responses. **Motor output** is the conduction of signals from the integration centers through the PNS to **effector cells**, such as muscle cells or gland cells, which perform the body's responses.

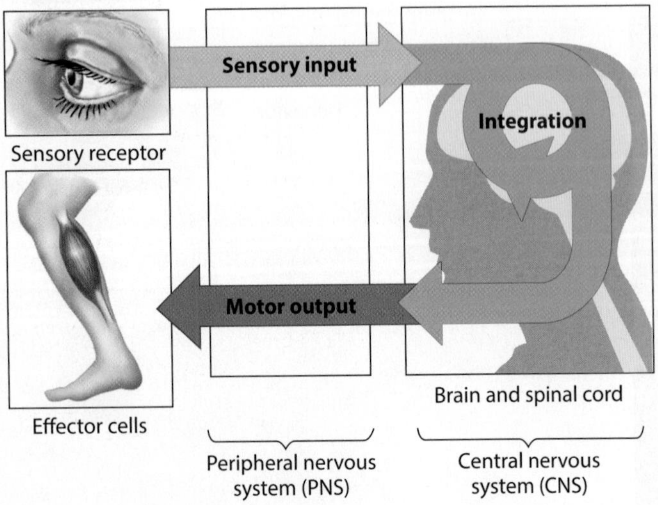

▲ **Figure 28.1A** Organization of a nervous system

The integration of sensory input and motor output is not usually rigid and linear, but involves the continuous background activity symbolized by the circular arrow in Figure 28.1A.

Three functional types of neurons correspond to a nervous system's three main functions: **Sensory neurons** convey signals from sensory receptors into the CNS. **Interneurons** integrate data and then relay appropriate signals to other interneurons or to **motor neurons**. Finally, motor neurons convey signals from the CNS to effector cells.

The relationship between neurons and nervous system structure and function is easiest to see in the relatively simple circuits that produce **reflexes**, or automatic responses to stimuli (**Figure 28.1B**). (Blue, green, and purple balls in the figure represent neuron cell bodies; the colored lines represent neuron extensions.) When the knee is tapped, ❶ a sensory receptor detects a stretch in the muscle, and ❷ a sensory neuron conveys this information into the spinal cord. In the CNS, the information goes to ❸ a motor neuron and to ❹ one or more interneurons. One set of muscles (quadriceps) responds to motor signals conveyed by a motor neuron by contracting, jerking the lower leg forward. At the same time, another motor neuron, responding to signals from an interneuron, inhibits the flexor muscles, making them relax and not resist the action of the quadriceps. (This figure shows only one neuron of each functional type, but most body activities engage many neurons of each type.)

? **If someone tickles the bottom of your foot, your ankle automatically flexes. Arrange the following neurons in the correct sequence for information flow during this reflex: interneuron, sensory neuron, motor neuron.**

● Sensory neuron → interneuron → motor neuron

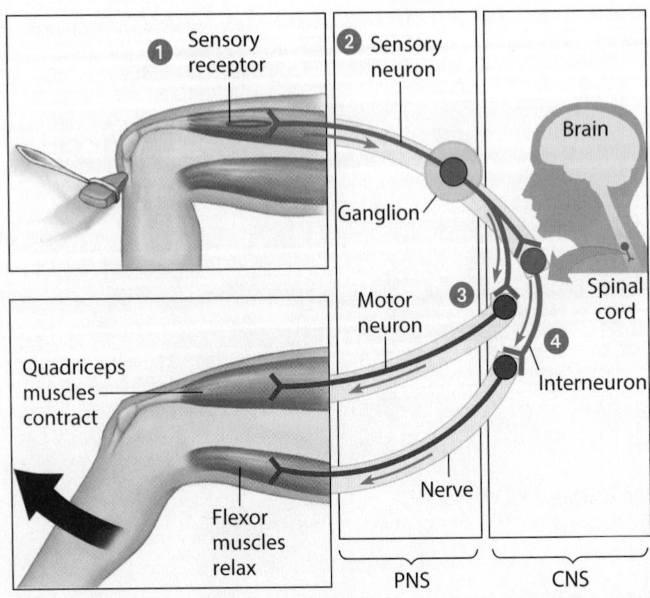

❶ Sensory receptor
❷ Sensory neuron
Brain
Ganglion
Spinal cord
Motor neuron ❸
❹
Interneuron
Quadriceps muscles contract
Nerve
Flexor muscles relax
PNS
CNS

▲ **Figure 28.1B** The knee-jerk reflex

28.2 Neurons are the functional units of nervous systems

The ability of neurons to receive and transmit information depends on their structure. **Figure 28.2** depicts a motor neuron, like those that carry command signals from your spinal cord to your skeletal muscles.

Similar to other cells of the body, neurons perform routine tasks, and thus, organelles associated with these functions (such as the nucleus, endoplasmic reticulum, and mitochondria) are present in the cell body. But unlike other types of cells, two kinds of extensions arise from the neuron cell body: numerous dendrites and a single axon. **Dendrites** (from the *Greek dendron,* tree) are highly branched, often short, extensions that receive signals from other neurons and convey this information toward the cell body. In contrast, the **axon** is typically a much longer extension that transmits signals to other cells, which may be other neurons or effector cells. Some axons, such as the ones that reach from your spinal cord to muscle cells in your feet, can be over a meter long.

Notice in Figure 28.2 that the axon ends in a cluster of branches. A typical axon has hundreds or thousands of these branches, each with a **synaptic terminal** at the very end. The junction between a synaptic terminal and another cell is called a **synapse**, and at a synapse, electrical or chemical signals are transmitted to other neurons or effector cells.

Neurons make up only part of a nervous system. To function normally, neurons of all vertebrates and most invertebrates require supporting cells called **glia**. Depending on the type, glia may nourish neurons, insulate the axons of neurons, or help maintain homeostasis of the extracellular fluid surrounding neurons. In the mammalian brain, glia outnumber neurons by as many as 50 to 1.

Figure 28.2 shows one kind of glia, called a Schwann cell, that is found in the PNS. (Comparable glia are found in the CNS.) In many vertebrates, axons that convey signals rapidly are enclosed along most of their length by a thick insulating material, analogous to the plastic insulation that covers many electrical wires. This insulating material, called the **myelin sheath**, resembles a chain of oblong beads that speeds the transmission of impulses along a neuron. Each bead is actually a Schwann cell, and the myelin sheath is essentially a chain of Schwann cells, each wrapped many times around the axon. The gaps between Schwann cells are called **nodes of Ranvier**, and they are the only points along the axon that require nerve signals to be regenerated, which is a time-consuming process. Everywhere else, the myelin sheath insulates the axon, preserving the signal and allowing it to propagate quickly. Thus, when a nerve signal travels along a myelinated axon, it needs to be rejuvenated only at the nodes. The resulting signal is much faster than one that must be regenerated constantly along the length of the axon. In the human nervous system, signals can travel along a myelinated axon at 150 m/sec (over 330 miles per hour), which means that a command from your brain can make your fingers move in just a few milliseconds. Without myelin sheaths, the signals would be over 10 times slower.

The debilitating autoimmune disease multiple sclerosis (MS) demonstrates the importance of myelin. MS causes the gradual destruction of myelin sheaths by the individual's own immune system. The result is a progressive loss of signal conduction, muscle control, and brain function.

With the basic structure of a neuron in mind, let's take a closer look at the signals that neurons convey.

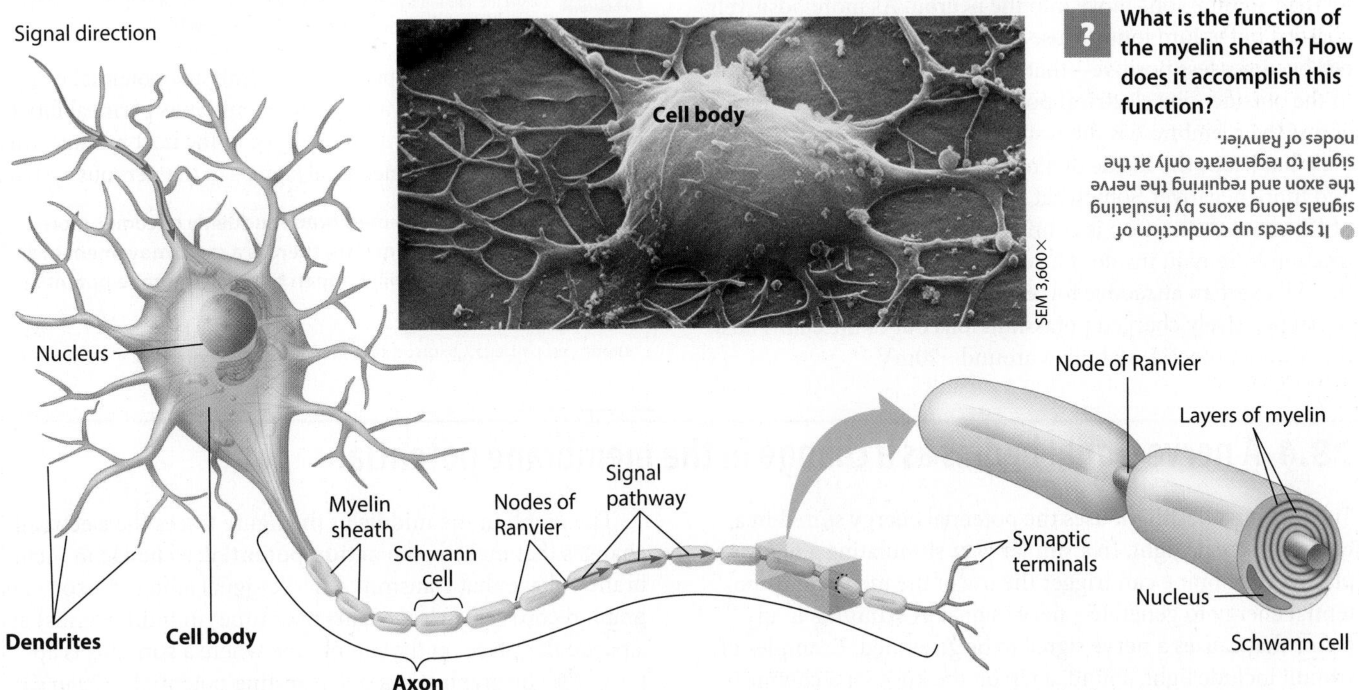

Signal direction
Cell body
SEM 3,600×
Nucleus
Dendrites
Cell body
Myelin sheath
Schwann cell
Nodes of Ranvier
Signal pathway
Axon
Synaptic terminals
Node of Ranvier
Layers of myelin
Nucleus
Schwann cell

? **What is the function of the myelin sheath? How does it accomplish this function?**

It speeds up conduction of signals along axons by insulating the axon and requiring the nerve signal to regenerate only at the nodes of Ranvier.

▲ **Figure 28.2** Structure of a myelinated motor neuron

Try This Create a table in which you name and draw all the structures of the neuron and describe the function of each structure.

▷ Nerve Signals and Their Transmission

28.3 Nerve function depends on charge differences across neuron membranes

To understand nerve signals, we must first study a resting neuron, one that is not transmitting a signal. A resting neuron has potential energy (see Module 5.10) that can be put to work sending signals from one part of the body to another. This potential energy, called the **membrane potential**, exists as an electrical charge difference across the neuron's plasma membrane: The inside of the cell is negatively charged relative to the outside as a result of unequal distribution of positively and negatively charged ions. Because opposite charges tend to move toward each other, a membrane stores energy by holding opposite charges apart, like a battery. The strength (voltage) of a neuron's stored energy can be measured. The voltage across the plasma membrane of a resting neuron is called the **resting potential**. A neuron's resting potential is about −70 millivolts (mV)—about 5% of the voltage in a flashlight battery.

The resting potential exists because of ion concentration gradients across the plasma membrane of a neuron (**Figure 28.3**). For most neurons, the concentration of potassium (K^+) is higher inside the cell, while the concentration of sodium (Na^+) is higher outside. **Sodium-potassium (Na^+-K^+) pumps** maintain the concentration gradients of these ions, using energy from ATP to actively move Na^+ out of the neuron and K^+ in (see Module 5.8). The diffusion of potassium ions across the plasma membrane is critical for establishing the resting potential. A resting membrane has many open K^+ channels but only a few Na^+ channels. Thus, potassium ions diffuse outward along their concentration gradient more readily than sodium ions move into the neuron. As more positively charged potassium ions diffuse out of the cell, the inside of the cell becomes less positive—that is, more negative—compared to the outside. The electrical potential difference, or voltage, across the membrane is the resting potential.

What keeps the inside of a neuron from becoming more and more negative? Somewhat like a tug-of-war, the net flow of K^+ out of the neuron is counteracted by an electrical attraction to remain inside. The excess negative charges inside the cell exert an attractive force that opposes the flow of additional positively charged potassium ions out of the cell. Thus, the resting potential stabilizes around −70mV.

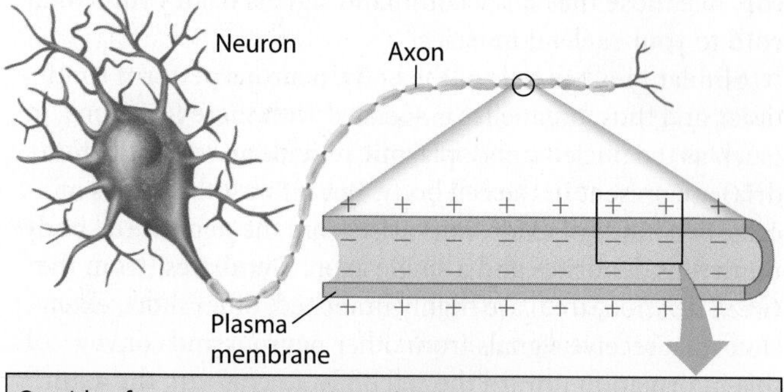

Neuron
Axon
Plasma membrane

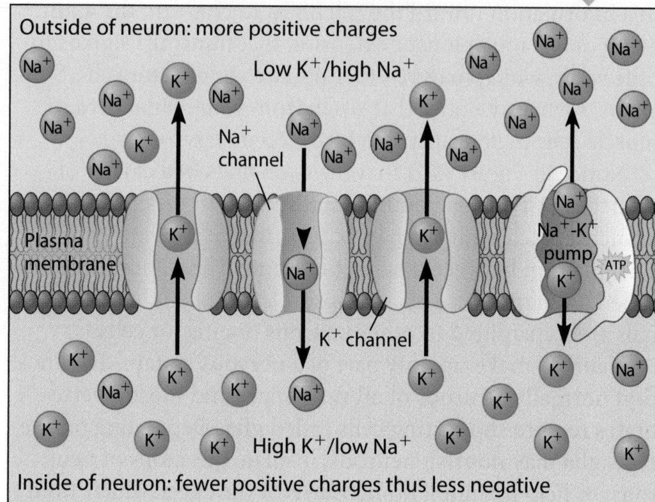

Outside of neuron: more positive charges
Low K+/high Na+
Na+ channel
Plasma membrane
Na+-K+ pump
ATP
K+ channel
High K+/low Na+
Inside of neuron: fewer positive charges thus less negative

▲ **Figure 28.3** How the resting potential is generated

Try This Follow the ions through each membrane protein, or pump, in the figure, and explain aloud how the ion movement contributes to the resting potential.

Notice an important point: The membrane potential can change from its resting value if the membrane's permeability to particular ions changes. As we will see in the next module, this is the basis of nearly all electrical signals in the nervous system.

? **If a resting neuron's membrane suddenly becomes more permeable to sodium ions, there is a rapid movement of Na^+ into the cell. What will happen to the membrane potential?**

● The voltage will rise (become less negative) as positively charged Na^+ enters the cell.

28.4 A nerve signal begins as a change in the membrane potential

Turning on a flashlight uses the potential energy stored in a battery to create light. In a similar way, stimulating a neuron's plasma membrane can trigger the use of the membrane's potential energy to generate a nerve signal. A **stimulus** is any factor that causes a nerve signal to be generated. Examples of stimuli include light, sound, a tap on the knee, or a chemical signal from another neuron. The nerve signal transmission process, summarized in **Figure 28.4**, applies to neurons in nearly all animals, including humans.

The graph in the middle of the figure traces the electrical changes that make up an **action potential**, a change in membrane voltage that transmits a nerve signal along an axon. The graph records electrical events over time (in milliseconds) at a particular place on the membrane where a stimulus is applied. ❶ The graph starts out at resting potential (−70 mV), when the membrane is polarized (with the inside more negatively charged than the outside). ❷ The stimulus is applied. If it is strong enough, the voltage rises to what is called the threshold (−50 mV, in this case). The difference between the threshold and the resting potential is the minimum change in the membrane's voltage that must occur to generate the action potential (+20 mV, in this case). ❸ Once the threshold is reached, the action potential is triggered. The membrane depolarizes, with the interior of the cell becoming positive with respect to the outside. ❹ The membrane then rapidly repolarizes as the voltage drops back down, ❺ undershoots the resting potential, and ❶ finally returns to it.

What actually causes the electrical changes of the action potential? The rapid flip-flop of the membrane potential is a result of the rapid movements of ions across the membrane at Na^+ and K^+ voltage-gated channels, structures that open and close in response to a change in membrane potential. (These channels differ from the ion channels illustrated in Figure 28.3 that allow ions to flow freely.) The diagrams surrounding the graph show the ion movements. Starting at lower left, ❶ the resting membrane separates a positively charged outside environment from a negatively charged inside environment. ❷ A stimulus triggers the opening of a few Na^+ voltage-gated channels in the membrane, and a tiny amount of Na^+ enters the axon. This makes the inside surface of the membrane slightly less negative (depolarization). If the stimulus is strong enough, a sufficient number of these Na^+ channels open to raise the voltage to the threshold. ❸ Once the threshold is reached, more of these Na^+ channels open, allowing even more Na^+ to diffuse into the cell. As more and more Na^+ moves in, the voltage soars to its peak. ❹ The peak voltage triggers closing and inactivation of the Na^+ voltage-gated channels. Meanwhile, the K^+ voltage-gated channels open, allowing K^+ to diffuse out rapidly. These changes produce the downswing on the graph (repolarization). ❺ A very brief undershoot of the resting potential results because these K^+ channels close slowly. ❶ The membrane then returns to its resting potential. In a typical mammalian neuron, this entire process takes just a few milliseconds, meaning that a neuron can produce hundreds of nerve signals per second.

? Is the generation of the peak of an action potential an example of positive feedback or negative feedback?

The opening of Na^+ gates caused by stimulation of the neuron changes the membrane potential, and this change causes more of the voltage-gated Na^+ channels to open. This is an example of a positive-feedback mechanism.

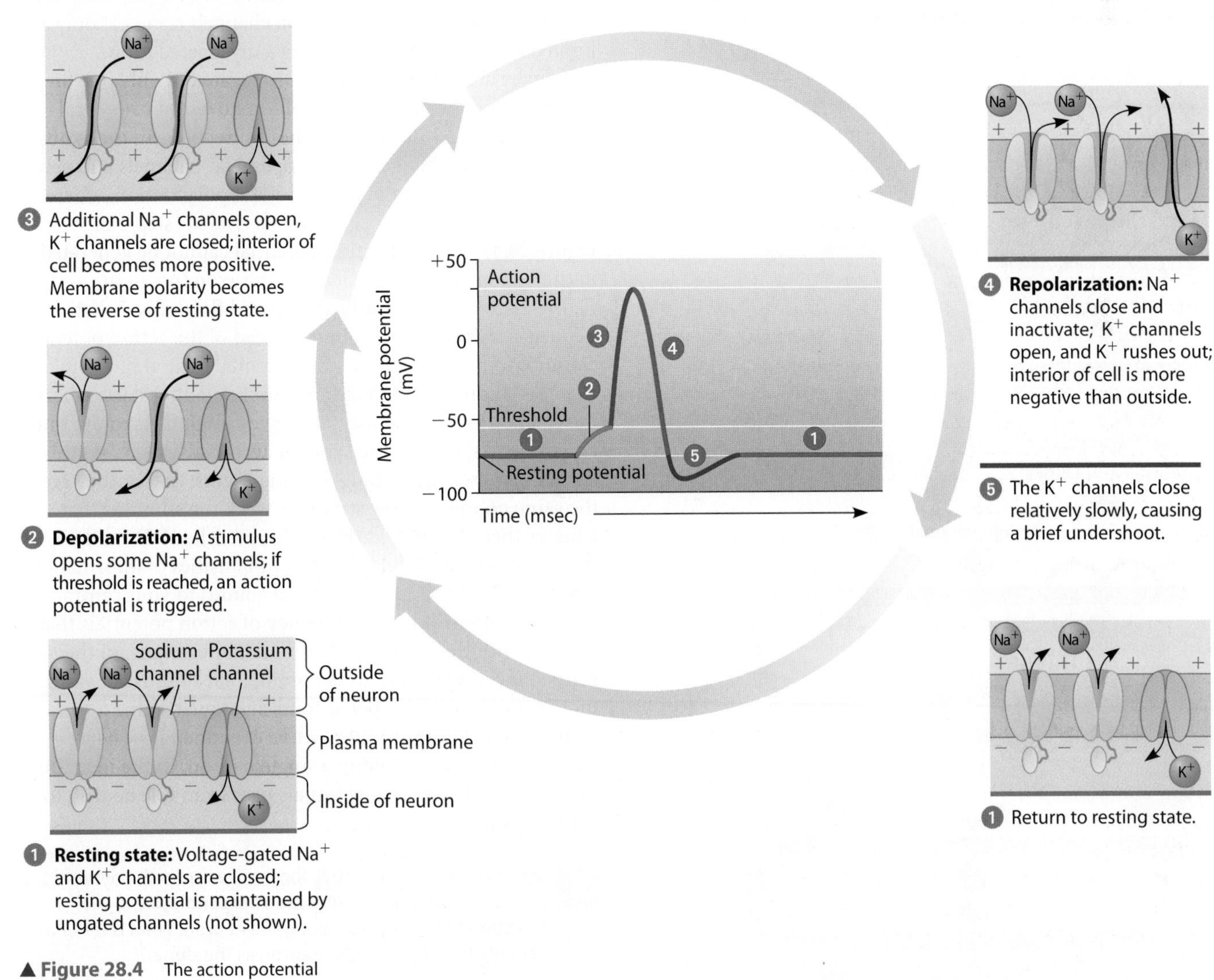

Na+
K+
❸ Additional Na+ channels open, K+ channels are closed; interior of cell becomes more positive. Membrane polarity becomes the reverse of resting state.
❷ Depolarization: A stimulus opens some Na+ channels; if threshold is reached, an action potential is triggered.
Sodium channel
Potassium channel
Outside of neuron
Plasma membrane
Inside of neuron
❶ Resting state: Voltage-gated Na+ and K+ channels are closed; resting potential is maintained by ungated channels (not shown).
Membrane potential (mV)
+50
0
−50
−100
Action potential
Threshold
Resting potential
Time (msec)
❹ Repolarization: Na+ channels close and inactivate; K+ channels open, and K+ rushes out; interior of cell is more negative than outside.
❺ The K+ channels close relatively slowly, causing a brief undershoot.
❶ Return to resting state.

▲ **Figure 28.4** The action potential

Try This Redraw the graph in the center of the figure while explaining aloud the ion movements that correlate to each rise and fall along the data line.

28.5 The action potential propagates itself along the axon

An action potential is a localized electrical event—a rapid change from the resting potential at a specific place along the neuron. A nerve signal starts out as an action potential generated in the axon, typically where the axon meets the cell body. To function as a long-distance signal, this local event must be passed along the axon from the cell body to the synaptic terminals. It does so by regenerating itself along the axon (**Figure 28.5**). The effect of this action potential is like tipping the first of a row of standing dominoes: The first domino does not travel along the row, but its fall is relayed to the end of the row, one domino at a time.

The three steps in Figure 28.5 show the changes that occur in an axon segment as a nerve signal passes from left to right. Let's first focus on the axon region on the far left. 1 When this region of the axon (blue) has its Na^+ channels open, Na^+ rushes inward ( ), and an action potential is generated. This corresponds to the upswing of the curve (step 2) in Figure 28.4. 2 Soon, the K^+ channels in that same region open, allowing K^+ to diffuse out of the axon ( ); at this time, its Na^+ channels are closed and inactivated at that point on the axon, and we would see the downswing of the action potential. 3 A short time later, we would see no signs of an action potential at this (far-left) spot because the axon membrane here has returned to its resting potential.

Now let's see how these events lead to the "domino effect" of a nerve signal. In step 1 of Figure 28.5, the blue region within the axon indicates local spreading of the electrical changes caused by the inflowing Na^+ associated with the first action potential. These changes are large enough to reach threshold in the neighboring regions, triggering the opening of Na^+ channels. As a result, a second action potential is generated (the blue region in step 2 is now to the right of its location in step 1). In the same way, a third action potential is generated in step 3, and each action potential generates another all the way down the axon (moving from left to right in the figure). The net result is the movement of a nerve impulse from the cell body to the synaptic terminals. Recall that an impulse is transmitted more rapidly if an axon is myelinated. The voltage-gated Na^+ channels are restricted to the nodes of Ranvier; thus, the process of opening and closing ion channels occurs at only a limited number of positions along a myelinated axon.

So why are action potentials propagated in only one direction along the axon? As the blue arrows indicate, local electrical changes do spread in both directions in the axon. However, these changes cannot open Na^+ channels and generate an action potential in a zone of repolarization, where the Na^+ channels are closed and inactivated (step 4 in Figure 28.4). Thus, an action potential cannot be generated in the regions where K^+ is leaving the axon (green in Figure 28.5). Consequently, the inward flow of Na^+ that depolarizes the axon membrane ahead of the action potential cannot produce another action potential behind it. (Think of a chain of dominoes.) Once an action potential starts where the cell body and axon meet, it moves along the axon in only one direction: toward the synaptic terminals.

Action potentials are all-or-none events; that is, they are the same no matter how strong or weak the stimulus that triggers them (as long as threshold is reached). How, then, do action potentials relay different intensities of information (such as a loud sound versus a soft sound) to your central nervous system? It is the frequency of action potentials that changes with the intensity of stimuli. For example, in the neurons connecting the ear to the brain, loud sounds generate more action potentials per second than quiet sounds.

In the past three modules, we've examined how nerve signals are conducted along a single neuron. In the next four modules, we focus on how signals pass from one neuron to another cell.

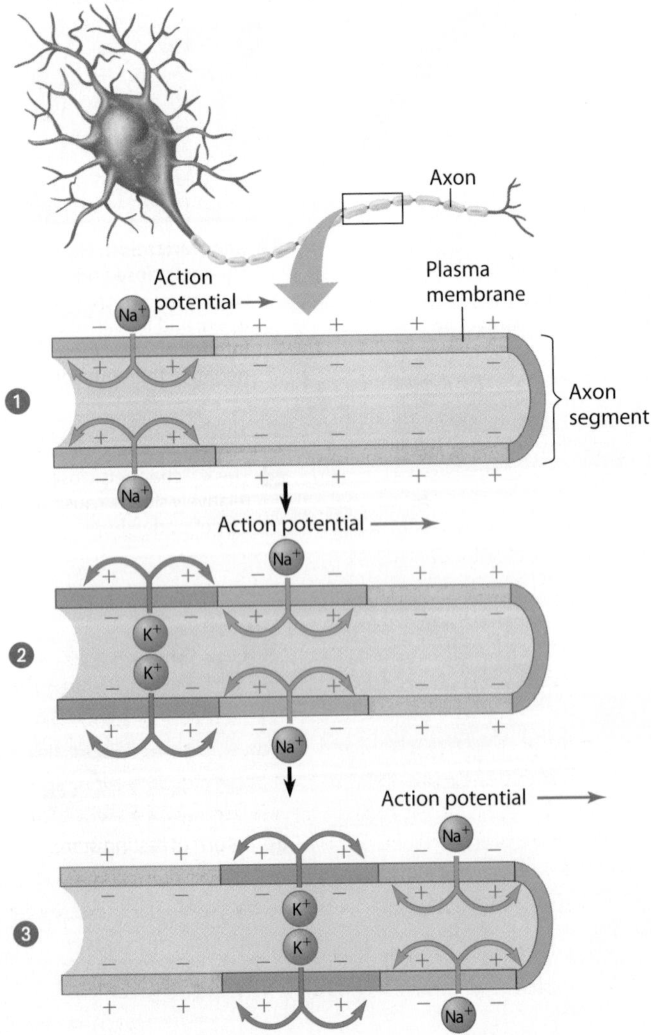

Axon
Plasma membrane
Action potential
Axon segment
Na+
K+
1
2
3

▲ **Figure 28.5** Propagation of the action potential along an axon

? During an action potential, ions move across the neuron plasma membrane in a direction perpendicular to the direction of the impulse along the neuron. What is it that actually travels along the neuron as the signal?

● The signal is the wavelike change in membrane potential, the self-perpetuated action potential that regenerates sequentially at points farther and farther away from the site of stimulation.

28.6 Neurons communicate at synapses

If an action potential travels in one direction along an axon, what happens when the signal arrives at the end of the neuron? To continue conveying information, the signal must be passed to another cell. This occurs at a synapse, or relay point, between a synaptic terminal of a sending neuron and a receiving cell. The receiving cell can be another neuron or an effector cell such as a muscle cell or endocrine cell.

Synapses come in two varieties: electrical and chemical. In an electrical synapse, electrical current flows directly from a neuron through gap junctions (Module 4.20) that form a pore, resulting in continuity between the neuron and the receiving cell. The receiving cell is stimulated quickly and at the same frequency of action potentials as the sending neuron. Lobsters and many fishes can flip their tails with lightning speed because the neurons that carry signals for these movements communicate by fast electrical synapses. In the human body, electrical synapses are found in the heart and digestive tract, where nerve signals maintain steady, rhythmic muscle contractions.

In contrast, when an action potential reaches a chemical synapse, it stops there. Unlike electrical synapses, chemical synapses have a narrow gap, called the **synaptic cleft**, separating the sending neuron from the receiving cell. The cleft is very narrow—only about 50 nm, about 1/1,000th the width of a human hair—but it prevents the action potential from spreading directly to the receiving cell. Instead, the action potential (an electrical signal) is first converted to a chemical signal consisting of molecules of **neurotransmitter** that act as chemical messengers. The chemical signal may then generate an action potential in the receiving cell.

Let's follow the events that occur at a chemical synapse, shown in **Figure 28.6**. Neurotransmitter molecules (⁘) are in membrane-enclosed sacs called **synaptic vesicles** in the sending neuron's synaptic terminals. ❶ An action potential arrives at the synaptic terminal. ❷ The action potential causes some synaptic vesicles to fuse with the plasma membrane of the sending neuron. ❸ The fused vesicles release their neurotransmitters by exocytosis into the synaptic cleft, and the neurotransmitter rapidly diffuses across the cleft.

The subsequent steps in Figure 28.6 show one example of what can happen next at a chemical synapse. ❹ The released neurotransmitter binds to complementary receptors on Na^+ ion channel proteins in the receiving cell's plasma membrane. ❺ In a typical chemical synapse, the binding of neurotransmitter to receptor opens these chemically gated ion channels in the receiving neuron's membrane. In our example, we see Na^+ ions diffusing into the receiving neuron, where they may trigger new action potentials. ❻ The neurotransmitter is broken down by an enzyme or transported back into the sending neuron (known as "reuptake"), and the Na^+ ion channels close. Step 6 ensures that the neurotransmitter's effect is brief and precise. (In Module 28.9, you'll learn how reuptake is related to the action of SSRI antidepressants.)

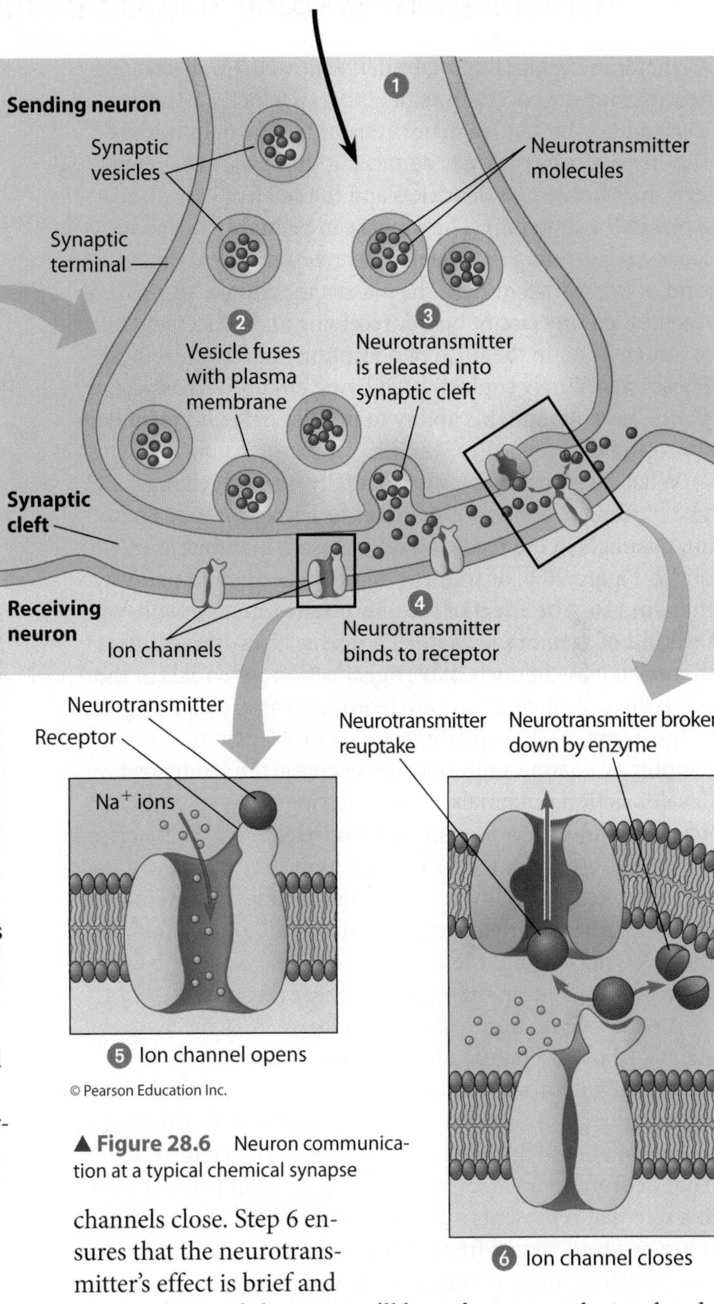

Axon of sending neuron
Synaptic terminal of sending neuron
Dendrite of receiving neuron
Sending neuron
Synaptic vesicles
Synaptic terminal
1
Neurotransmitter molecules
2 Vesicle fuses with plasma membrane
3 Neurotransmitter is released into synaptic cleft
Synaptic cleft
Receiving neuron
Ion channels
4 Neurotransmitter binds to receptor
Neurotransmitter
Receptor
Na+ ions
5 Ion channel opens
Neurotransmitter reuptake
Neurotransmitter broken down by enzyme
6 Ion channel closes

© Pearson Education Inc.

▲ **Figure 28.6** Neuron communication at a typical chemical synapse

Think about what is happening right now in your own nervous system. Action potentials carrying information about the words on this page are streaming along sensory neurons from your eyes to your brain. The action potentials are triggering the release of neurotransmitters at the ends of the sensory neurons. The neurotransmitters are diffusing across synaptic clefts and triggering changes in some of your interneurons that lead to integration of the signals and ultimately to what they mean (in this case, the meaning of words and sentences).

? How does a synapse ensure that signals pass only in one direction, from a sending neuron to a receiving cell?

● A signal is one-way because only the sending neuron releases neurotransmitter, and only the receiving cell has receptors for the neurotransmitter.

28.7 Chemical synapses enable complex information to be processed

As the drawing and micrograph in **Figure 28.7** indicate, one neuron may interact with many others. In fact, a neuron may receive information via neurotransmitters from hundreds of other neurons that may connect at thousands of synaptic terminals located on dendrites and the cell body (red and green in the drawing). The inputs can be highly varied because each sending neuron may secrete a different quantity or kind of neurotransmitter. The plasma membrane of a neuron resembles a tiny circuit board, receiving and processing bits of information in the form of neurotransmitter molecules. These exceedingly sophisticated living circuit boards account for the nervous system's ability to process data and formulate appropriate responses to stimuli.

What do neurotransmitters actually do to receiving cells? The binding of a neurotransmitter to a receptor may open ion channels in the receiving cell's plasma membrane, as you saw in Figure 28.6, or trigger a signal transduction pathway that does so. The effect of the neurotransmitter depends on the kind of membrane channel it opens. Neurotransmitters that open Na^+ channels may trigger action potentials in the receiving cell. Such effects are referred to as excitatory (green in the drawing). In contrast, many neurotransmitters open membrane channels for ions that decrease the tendency to develop action potentials in the receiving cell—such as channels that admit chloride ions (Cl^-) or release K^+. These effects are called inhibitory (red). The effects of both excitatory and inhibitory signals can vary in magnitude. In general, the more neurotransmitter molecules that bind to receptors on the receiving cell and the closer the synapse is to the base of the receiving neuron's axon, the stronger the effect.

A receiving neuron's plasma membrane may receive signals—both excitatory and inhibitory—from many different sending neurons. If the excitatory signals are collectively strong enough to raise the membrane potential to threshold, an action potential will be generated in the receiving cell. That neuron then passes signals along its axon to other cells at a rate that represents a summation of all the information it has received. Signal frequency is key because action potentials are all-or-none events. Each new receiving cell, in turn, processes this information along with all its other inputs.

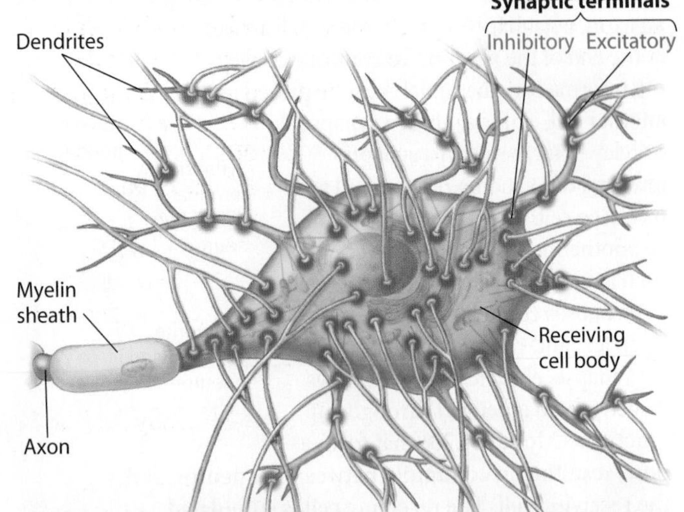
Synaptic terminals
Inhibitory
Excitatory
Dendrites
Myelin sheath
Axon
Receiving cell body

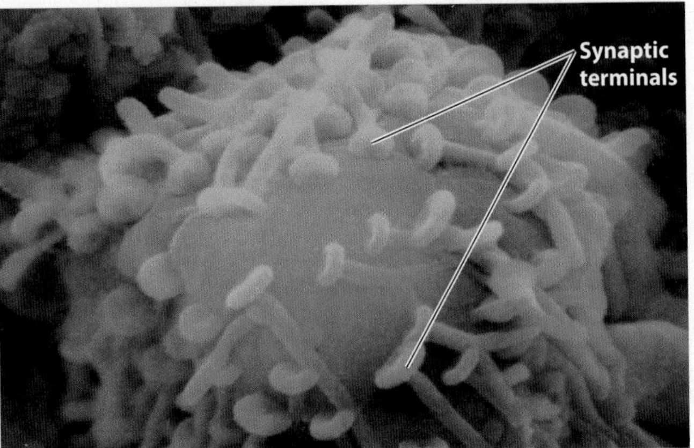
Synaptic terminals
SEM 5,500×

▲ **Figure 28.7** A neuron's multiple synaptic inputs

? Contrast how excitatory and inhibitory signals change a receiving cell's membrane potential relative to triggering an action potential.

● Neurotransmitters in an excitatory signal open Na^+ channels that move the receiving cell's membrane potential closer to threshold. Neurotransmitters in an inhibitory signal open K^+ or Cl^- channels that move the cell's membrane potential farther from threshold.

28.8 A variety of small molecules function as neurotransmitters

As discussed in Modules 28.6 and 28.7, the propagation of nerve signals across chemical synapses depends on neurotransmitters. A variety of small molecules serve this function.

Many neurotransmitters are small, nitrogen-containing organic molecules. One, called **acetylcholine**, is released by motor neurons to activate skeletal muscles (see Module 30.10), by other PNS neurons that affect internal organs and glands, and by neurons in the CNS that affect memory, learning, and alertness. Depending on the kind of receptors on receiving cells, acetylcholine may be excitatory or inhibitory. For instance, acetylcholine makes our skeletal muscles contract but slows the rate of contraction of cardiac muscles. Botulinum toxin (sold as Botox), made by the bacteria that cause botulism food poisoning, inhibits the release of acetylcholine. In a cosmetic procedure, a small amount of Botox is injected into the forehead and around the eyes and mouth, preventing muscle contraction in the area, thereby eliminating wrinkles caused by smiling or furrowing of the brow.

Three neurotransmitters—glutamate, glycine, and GABA (gamma aminobutyric acid)—are amino acids. All function in the CNS. Glutamate acts primarily at excitatory synapses, whereas glycine and GABA act at inhibitory synapses.

Biogenic amines are neurotransmitters derived from amino acids. Some examples of biogenic amines are norepinephrine, serotonin, and dopamine, the first of which also functions as a hormone (see Module 26.10). Serotonin and dopamine affect sleep, mood, attention, and learning. Imbalances of biogenic amines are associated with various disorders. For example, the degenerative illness Parkinson's disease is associated with a lack of dopamine in the brain and schizophrenia is linked to overactive dopamine signaling (Module 28.21). Reduced levels of norepinephrine and serotonin seem to be linked with some types of depression.

Many neuropeptides, relatively short chains of amino acids, also serve as neurotransmitters. The endorphins are peptides that decrease our perception of pain during times of physical or emotional stress. Endorphins may be released in response to a wide variety of stimuli, including traumatic injury, muscle fatigue, and even eating very spicy foods.

Neurons also use dissolved gases, notably nitric oxide (NO), as chemical signals. During sexual arousal in human males, certain neurons release NO into blood vessels in the erectile tissue of the penis, and the NO triggers an erection. (The erectile dysfunction drug Viagra promotes this effect of NO.) The dissolved gas diffuses into neighboring cells, produces a change, and is broken down—all within a few seconds.

? **What determines whether a neuron is affected by a specific neurotransmitter?**

● To be affected by a particular neurotransmitter, a neuron must have specific receptors for that neurotransmitter.

28.9 Many drugs act at chemical synapses

CONNECTION

Many psychoactive drugs, even common ones such as caffeine, nicotine, and alcohol, affect the action of neurotransmitters in the brain's billions of synapses. Caffeine, found in coffee, tea, chocolate, and many soft drinks, keeps us awake by countering the effects of inhibitory neurotransmitters, ones that normally suppress action potentials. Nicotine acts as a stimulant by binding to and activating acetylcholine receptors. Alcohol is a strong depressant. Its precise effect is not yet known, but it seems to increase the inhibitory effects of the neurotransmitter GABA.

Many prescription drugs used to treat psychological disorders alter the effects of neurotransmitters. Tranquilizers such as diazepam (Valium) and alprazolam (Xanax) activate the receptors for the neurotransmitter GABA, increasing its effect at inhibitory synapses. In other cases, a drug may bind to and block a receptor, reducing a neurotransmitter's effect. For instance, some drugs used to treat schizophrenia block receptors for the neurotransmitter dopamine, reducing its excitatory effect on the receiving cell. Still other drugs, such as the selective serotonin reuptake inhibitors (SSRIs) used to treat depression, act by inhibiting neurotransmitter "reuptake." The neurotransmitter serotonin binds to receptors on receiving neurons. As seen in **Figure 28.9** (left panel), after its release from a receptor, serotonin in the synaptic cleft is then recycled back into the sending neuron through a serotonin transporter protein. SSRIs block serotonin transporters (right panel of figure), thus increasing the amount of time this mood-altering neurotransmitter is available in the synaptic cleft to affect receiving neurons.

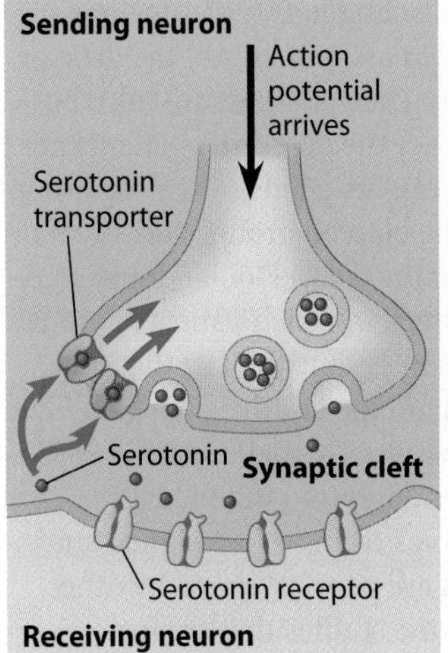

Sending neuron
Action potential arrives
Serotonin transporter
Serotonin
Synaptic cleft
Serotonin receptor
Receiving neuron

Action potential arrives
SSRIs

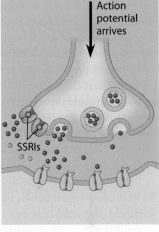

Reuptake of serotonin

Inhibition of serotonin reuptake by SSRI

Figure adapted from "Antidepressants Prevent Hierarchy Destabilization Induced by Lipopolysaccharide Administration in Mice: A Neurobiological Approach to Depression" by Daniel W. H. Cohn, et al., from ANNALS OF THE NEW YORK ACADEMY OF SCIENCES, July 2012, Volume 1262.

▲ **Figure 28.9** The effect of an SSRI at a synapse

Some drugs used to treat attention deficit hyperactivity disorder (ADHD), such as methylphenidate (Ritalin) and Adderall (a combination drug), block neurotransmitter reuptake of dopamine and norepinephrine, which can increase alertness. Students sometimes abuse Adderall as a "study drug," which can be quite dangerous due to significant risk of seizures and cardiac problems. Because the effects of these drugs are poorly understood, a physician should monitor anyone taking them.

What about illegal drugs? Stimulants such as amphetamines and cocaine increase the release and availability of norepinephrine and dopamine at synapses. Abuse of these drugs can cause symptoms resembling schizophrenia. The active ingredient in marijuana (tetrahydrocannabinol, or THC) binds to brain receptors normally used by other neurotransmitters that seem to play a role in pain, depression, appetite, memory, and fertility.

The drugs discussed here are used for a variety of purposes, both medicinal and recreational. While they increase alertness and sense of well-being or reduce physical and emotional pain, they may disrupt the brain's finely tuned neural pathways, altering the chemical balances that are the product of millions of years of evolution.

? **Would blocking receptors for neurotransmitters that promote an excitatory signal on a receiving cell lead to more frequent or less frequent action potentials? Why?**

● Less frequent. If a neurotransmitter with excitatory effects cannot bind to its receptor, fewer ion channels will open, preventing the cell's membrane potential from reaching threshold.

28.10 Published data are biased toward positive findings

SCIENTIFIC THINKING

As discussed in the chapter introduction, depression is a prevalent mood disorder in the United States. Antidepressants are widely prescribed for this common illness, and SSRIs are popular antidepressant medications. The hypothesis underlying the use of SSRIs is that depression is caused by a deficiency in the brain of serotonin, a regulator of mood. As described in Module 28.9, an SSRI blocks the reuptake of the neurotransmitter serotonin by the sending neuron, prolonging the effect of the serotonin on the receiving cell (essentially increasing serotonin levels at the synapse).

The widespread use of SSRIs began after Prozac was approved by the U. S. Food and Drug Administration (FDA) in 1987. Today, about one in 10 Americans over age 12 takes an antidepressant. To determine how effective drugs like this are, researchers set up clinical studies with individuals who are depressed. Characteristics of well-designed studies include large sample sizes and randomized assignments of the drug or control (placebo) to participants. These studies must also be double-blind, meaning that neither the depressed patients nor the researchers know who has been assigned to the drug group and who has been assigned to the placebo group. (Researchers learn this later during analysis of the data.) Double-blind studies help eliminate the placebo effect, in which patients perceive an improvement in their condition despite the fact that they have been given a sugar pill and not the test drug. To gain FDA approval, clinical trials must demonstrate that a specific SSRI is more effective than a placebo for depressed individuals.

Scientists publish their findings from well-designed studies in professional journals that are peer-reviewed by other scientists. But what happens to the studies that are never submitted or accepted for publication? Getting information about drug effectiveness from only published studies might be misleading. Notably, studies that demonstrate significant findings or positive outcomes are more likely to be published than those that don't—a phenomenon known as publication bias. This bias can be quantified for studies on drug effectiveness, as was done in a report about SSRIs, published in 2008 in the New England Journal of Medicine.

In **Figure 28.10**, we can see that scientists examined data from 74 clinical trials with antidepressants between 1987 and 2004. Of these, the FDA deemed about half to be positive (that is, the studies demonstrated the effectiveness of the drug) and half negative or questionable. Of the 38 "positive" studies, all but one were published. The remaining 36 studies were classified as either negative or questionable by the FDA. Only 3 studies that reported negative findings were published. Furthermore, the researchers found that 11 studies reporting negative or inconclusive results were erroneously counted and presented as positive findings in publications. Overall, 48 out of 51 published studies reported positive results.

Why does publication bias matter? Physicians reading published studies may be prescribing ineffective drugs to patients, or worse— the drugs could even have negative side effects. The good news is that recently the FDA put policies in place that require all studies to be registered when they

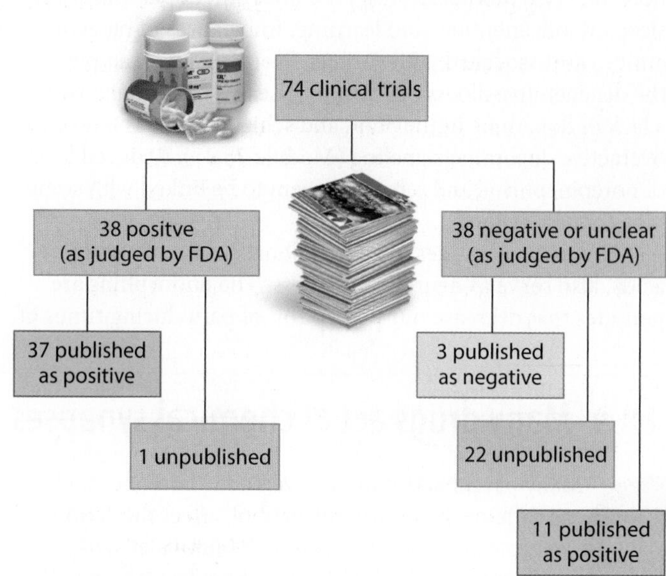
74 clinical trials
38 positve
(as judged by FDA)
38 negative or unclear
(as judged by FDA)
37 published
as positive
1 unpublished
3 published
as negative
22 unpublished
11 published
as positive

▲ **Figure 28.10** Publication bias in studies on the effectiveness of antidepressants

are initiated. In addition, new studies must be accessible to the general public and free of charge within a year or so of publication; data are no longer only available to scientists and doctors. Thus, statistical methods used to compile results from multiple studies—called meta-analysis—will now include nonpublished data. Meta-analysis has been employed to evaluate the most rigorous and well-designed SSRI studies. The evidence seems to be mounting that SSRIs provide only negligible benefits for patients with mild or moderate depression, but are helpful to the most severely depressed patients.

Are antidepressants effective?

With more than two decades worth of data about the effectiveness of SSRIs in increasing serotonin levels, scientists are revisiting the hypothesis that serotonin deficiency causes depression. SSRI data from the most severely depressed patients seem to support this hypothesis; however, data from the moderately depressed support the idea that there might be multiple causes of depression or that the cause of moderate depression is different from that of severe depression. Scientists continue to study other molecular pathways in the nervous system to find out more about depression. Access to unbiased clinical trial results will help both researchers, in evaluating new drugs that are designed to target these molecular pathways, and physicians, in prescribing the right medicines to their patients.

? Meta-analysis suggests that the classification of patients as severely versus moderately depressed could impact the effectiveness of antidepressant drugs. What are other variables that could be examined to classify participants in an SSRI clinical study?

● Some other variables to consider are the participants' age and gender, the amount of time they are on the antidepressant, the dose they are taking, the brand of the medication, possible interaction with other drugs, their diet, and their history with depression.

▷ An Overview of Animal Nervous Systems

28.11 The evolution of animal nervous systems reflects changes in body symmetry

EVOLUTION CONNECTION

To this point in the chapter, we have concentrated on the cellular mechanisms that are fundamental to nearly all animal nervous systems. There is remarkable uniformity throughout the animal kingdom in the way nerve cells function—a strong indication that the basic architecture of the neuron was an early evolutionary adaptation. But during subsequent animal evolution, great diversity emerged in the organization of nervous systems as a whole.

The ability to sense and react originated billions of years ago with prokaryotes that could detect changes in their environment and respond in ways that enhanced their survival and reproductive success. Later, modification of simple recognition and response processes provided multicellular organisms with a mechanism for communication between cells of the body. Evidence from fossils suggests that by 500 million years ago, systems of neurons allowing animals to sense and move rapidly were present in essentially their current forms.

Animals on the earliest branches of the animal evolutionary tree (see Figure 18.4)—sponges, for instance—lack a nervous system. The first modern phylum in which a nervous system evolved was the cnidarians. Hydras, jellies, and other cnidarians have a **nerve net** (**Figure 28.11, Part A**), a diffuse, weblike system of interconnected neurons extending throughout the body. Neurons of the nerve net control the contractions of the digestive cavity, movement of the tentacles, and other functions.

Cnidarians and adult echinoderms have radial, uncentralized nervous systems. Sea stars and many other echinoderms have a nerve net in each arm connected by radial nerves to a central nerve ring. This organization is better suited than a diffuse nerve net for controlling complex motion, such as the coordinated movements of hundreds of tube feet that allow a sea star to move in one direction.

The appearance of bilateral symmetry marks a key branch point in the evolution of animals and their nervous systems. Bilaterally symmetric animals tend to move through the environment with the head—usually equipped with sense organs and a brain—first encountering new stimuli. Flatworms are the simplest of animals that illustrate the two hallmarks of bilateral symmetry: **cephalization**, an evolutionary trend toward the concentration of the nervous system at the head end, and **centralization**, the presence of a central nervous system distinct from a peripheral nervous system. For example, the planarian worm in **Part B** has a small brain composed of ganglia (clusters of nerve cell bodies) and two parallel **nerve cords**, elongated bundles of neurons that control the animal's movements. Transverse nerves coordinate these movements between both sides of the body. These elements constitute the simplest clearly defined CNS in the animal kingdom.

In subsequent animal phyla, the CNS evolved in complexity. For instance, the brains of leeches (**Part C**) have a greater concentration of neurons than those of flatworms, and leech ventral nerve cords contain segmentally arranged ganglia. The insect shown in **Part D** has a brain composed of several fused ganglia, and its ventral nerve cord also has a ganglion in each body segment. Each of these ganglia directs the activity of muscles in its segment of the body.

Molluscs serve as a good illustration of how natural selection leads to correlation of the structure of a nervous system with an animal's interaction with the environment. Sessile or slow-moving molluscs, such as clams, have little or no cephalization and relatively simple sense organs. In contrast, the relatively large brain of a squid (**Part 28.11E**), accompanied by complex eyes and rapid signaling along giant axons, correlates well with the active predatory life of these animals. In fact, it was the squid's giant axons (up to 1 mm in diameter) that gave researchers their first chance to study nerve signal transmission in the 1940s.

In the next several modules, we explore the complex nervous systems that evolved in our own subphylum, the vertebrates.

? Why is it advantageous for the brain of most bilateral animals to be located at the head end?

● Cephalization places the brain close to major sense organs, which are concentrated on the end of the animal that leads the way as the animal moves through its environment.

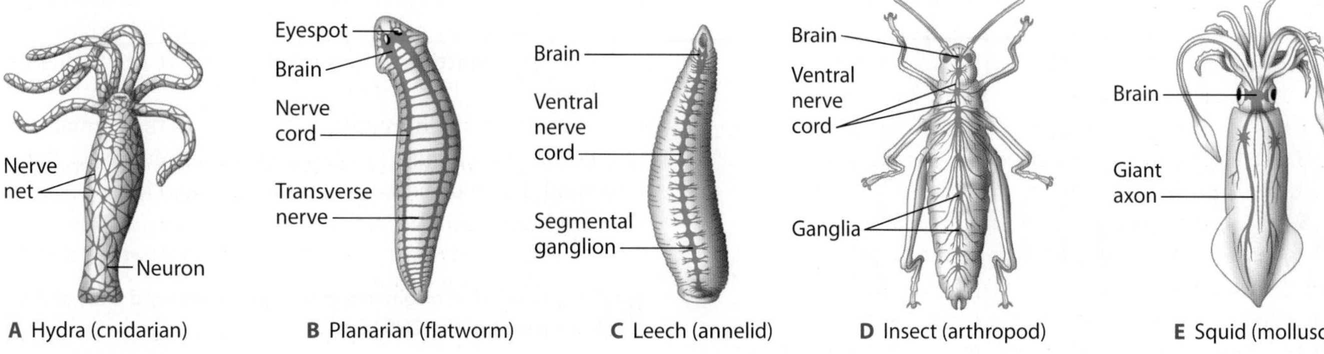

Nerve net
Neuron
A Hydra (cnidarian)
Eyespot
Brain
Nerve cord
Transverse nerve
B Planarian (flatworm)
Brain
Ventral nerve cord
Segmental ganglion
C Leech (annelid)
Brain
Ventral nerve cord
Ganglia
D Insect (arthropod)
Brain
Giant axon
E Squid (mollusc)

▲ **Figure 28.11** Invertebrate nervous systems (not to scale)

28.12 Vertebrate nervous systems are highly centralized

Vertebrate nervous systems are diverse in structure and level of sophistication. For instance, those of dolphins and humans are much more complex structurally than those of frogs or fishes; they are also much more powerful integrators. However, all vertebrate nervous systems have fundamental similarities. All have distinct central and peripheral elements and are highly centralized. In all vertebrates, the brain and spinal cord make up the CNS, while the PNS comprises the rest of the nervous system **(Figure 28.12A)**. The **spinal cord**, a bundle of nervous tissue that runs lengthwise inside the spine, conveys information to and from the brain and integrates simple responses to certain stimuli (such as the knee-jerk reflex). The master control center of the nervous system, the **brain**, includes homeostatic centers that keep the body functioning smoothly, sensory centers that integrate data from the sense organs, and (in humans, at least) centers of emotion and intellect. The brain also sends motor commands to muscles.

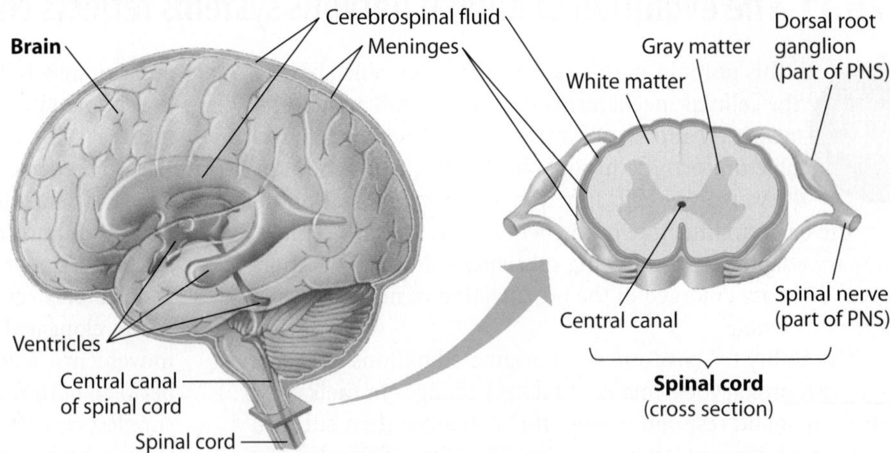

Brain
Cerebrospinal fluid
Meninges
Ventricles
Central canal
of spinal cord
Spinal cord
Gray matter
White matter
Dorsal root
ganglion
(part of PNS)
Central canal
Spinal nerve
(part of PNS)
Spinal cord
(cross section)

▲ **Figure 28.12B** Fluid-filled spaces of the vertebrate CNS

A vast network of blood vessels services the CNS. Tight junctions (Module 4.20) between the endothelial cells of the brain capillaries allow essential nutrients and oxygen to pass freely into the brain, but keep out many chemicals, such as metabolic wastes from other parts of the body. This selective mechanism, called the **blood-brain barrier**, maintains a stable chemical environment for the brain.

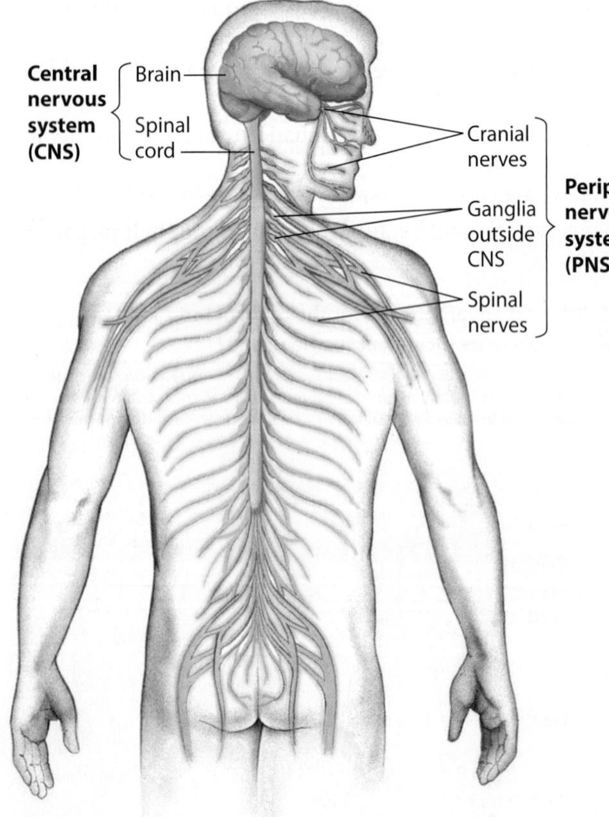

Central
nervous
system
(CNS)
Brain
Spinal
cord
Cranial
nerves
Ganglia
outside
CNS
Spinal
nerves
Peripheral
nervous
system
(PNS)

▲ **Figure 28.12A** A vertebrate nervous system (back view)

Filtration of blood within the brain produces **cerebrospinal fluid**, found both in and around the brain and spinal cord. Spaces in the brain filled with cerebrospinal fluid, called **ventricles**, are continuous with the narrow **central canal** of the spinal cord **(Figure 28.12B)**. Circulating slowly through the central canal and ventricles (and then draining back into veins), the cerebrospinal fluid cushions the CNS and assists in supplying nutrients and hormones and removing wastes. Also protecting the brain and spinal cord are layers of connective tissue, called **meninges**. In mammals, cerebrospinal fluid circulates between layers of the meninges, providing an additional protective cushion for the CNS. If the cerebrospinal fluid becomes infected by bacteria or viruses, the meninges may become inflamed, a condition called meningitis. Viral meningitis is generally not harmful, but bacterial meningitis can have serious consequences if not treated with antibiotics. A sample of cerebrospinal fluid can be collected for testing by a spinal tap, a procedure in which a narrow needle is inserted into a space between meninges.

As shown on the right side of Figure 28.12B, the CNS has white matter and gray matter. The CNS **white matter** is composed mainly of axons (with their whitish myelin sheaths); **gray matter** consists mainly of nerve cell bodies and dendrites.

Nerves that extend from the CNS, such as the cranial nerves shown in Figure 28.12A, are part of the PNS. In the next module, we turn away from the CNS and look closely at the vertebrate PNS.

? Is a nerve that originates in the spinal cord and ends at the toe a part of the CNS? Explain.

● No, only the brain and spinal cord make up the CNS. A nerve connecting to the toe is part of the PNS.

28.13 The peripheral nervous system of vertebrates can be divided into functional components

As shown in **Figure 28.13**, the PNS consists of both sensory and motor nerves. The sensory nerves receive information from sensory receptors. Some sensory receptors are widespread throughout the body, for instance, the touch receptors of the skin. Other receptors are restricted to the special sense organs (Chapter 29), such as the light-detecting receptors of the eyes. Once receptors are activated by stimuli, signals are carried by sensory nerves to the CNS, where they are processed. The motor nerves then transmit information from the CNS to effector organs that respond to the stimuli. The motor neurons can be subdivided into two functional components. The **motor system** carries signals from the CNS to skeletal muscles. When you read, for instance, these neurons carry commands that make your eye muscles move. The control of skeletal muscles can be voluntary, as when you raise your hand to ask a question, or involuntary, as in a knee-jerk reflex controlled by the spinal cord. The other motor output component, the **autonomic nervous system**, regulates the internal environment by controlling smooth and cardiac muscles and the organs and glands of the digestive, cardiovascular, excretory, and endocrine systems. This control is generally involuntary.

While it is convenient to divide the PNS into motor and autonomic systems, it is important to realize that these two divisions cooperate to maintain homeostasis. In response to a drop in body temperature, for example, the brain signals the autonomic nervous system to constrict surface blood vessels, which reduces heat loss. At the same time, the brain also signals the motor nervous system to cause shivering, which increases heat production.

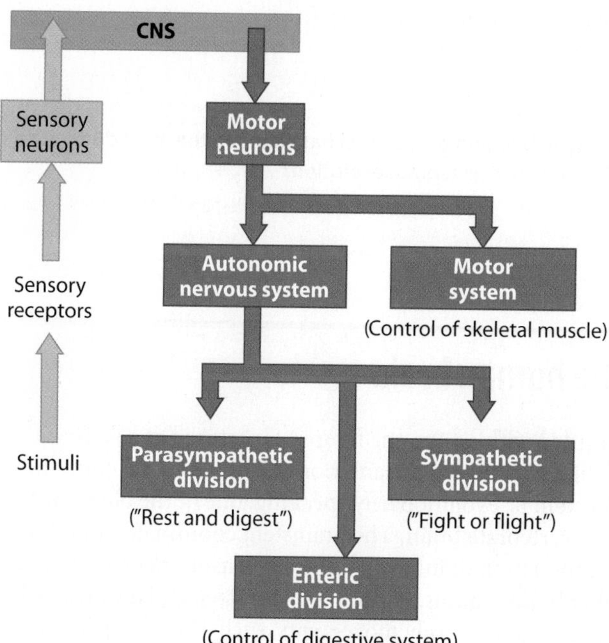

CNS
Sensory neurons
Motor neurons
Sensory receptors
Autonomic nervous system
Motor system
(Control of skeletal muscle)
Stimuli
Parasympathetic division
("Rest and digest")
Sympathetic division
("Fight or flight")
Enteric division
(Control of digestive system)

▲ **Figure 28.13** Functional divisions of the vertebrate PNS

As you can see in Figure 28.13, the autonomic nervous system is composed of three divisions: parasympathetic, sympathetic, and enteric. The parasympathetic and sympathetic divisions have largely antagonistic (opposite) effects on most body organs.

The **parasympathetic division** primes the body for activities that gain and conserve energy for the body ("rest and digest"). For example, signals of the parasympathetic division stimulate the digestive organs, such as the salivary glands, stomach, and pancreas; decrease the heart rate; and increase glycogen production by the liver.

Neurons of the **sympathetic division** prepare the body for intense, energy-consuming activities, such as fighting, fleeing, or competing in a strenuous game (the "fight-or-flight" response). Sympathetic signals inhibit the digestive organs, dilate the bronchi so that more air can pass through them, increase the heart rate, stimulate glucose release from the liver into the blood for quick energy, and induce the adrenal glands to secrete the hormones epinephrine and norepinephrine.

"Fight or flight" and rest are opposite extremes. Your body usually operates at intermediate levels, with most of your organs receiving both sympathetic and parasympathetic signals. The opposing signals adjust an organ's activity to a suitable level. In regulating some body functions, however, the two divisions complement rather than antagonize each other. For example, in regulating intercourse, an erection is promoted by the parasympathetic division and ejaculation is stimulated by the sympathetic division.

Sympathetic and parasympathetic neurons emerge from different regions of the CNS, and they use different neurotransmitters. Neurons of the parasympathetic division emerge from the brain and the lower part of the spinal cord, whereas neurons of the sympathetic division emerge from the middle regions of the spinal cord. Acetylcholine is the neurotransmitter released at synapses by most parasympathetic neurons within target organs, and norepinephrine is released by most sympathetic neurons at target organs.

The **enteric division** of the autonomic nervous system consists of networks of neurons in the digestive tract, pancreas, and gallbladder. Within these organs, neurons of the enteric division control secretion as well as activity of the smooth muscles that produce peristalsis. Although the enteric division can function independently, it is normally regulated by the sympathetic and parasympathetic divisions.

In the next several modules, we take a closer look at the highest level of the nervous system's structural hierarchy: the brain.

? **How would a drug that inhibits the parasympathetic nervous system affect a person's pulse?**

● Because signals from the parasympathetic nervous system slow the heart rate, a drug that inhibits the parasympathetic division would result in an increased pulse.

28.14 The vertebrate brain develops from three anterior bulges of the neural tube

We close our overview of animal nervous systems by examining the embryonic development of the vertebrate nervous system from the dorsal hollow nerve cord, one of the four distinguishing features of chordates (see Module 18.15). During early embryonic development in all vertebrates, three bilaterally symmetric bulges—the **forebrain**, **midbrain**, and **hindbrain**—appear at the anterior end of the neural tube (**Figure 28.14**). In the course of vertebrate evolution, the forebrain and hindbrain became subdivided—both structurally and functionally—into specific regions. The relative sizes of these regions in different vertebrate groups correlate with differences in particular brain functions. You can see these regions in the three-month-old fetus in the figure, and we'll examine the functions of these regions in Module 28.15.

One evolutionary trend that is seen in birds and mammals is a much larger brain relative to body size than is found in other vertebrate groups such as fishes and amphibians. The forebrain of birds and mammals also occupies a larger fraction of the brain. The greater capacity of birds and mammals for sophisticated behavior correlates with the larger size of the cerebrum, the portion of the forebrain responsible for cognition and integration. In humans, the rapid, expansive growth of the forebrain during the second and third months of development gives rise to our large cerebrums. As you'll see, the cerebrum develops into two halves, called the left and right cerebral hemispheres.

The cerebrum's outer region, the **cerebral cortex**, enlarges during the development of the mammalian brain. The cerebral cortex is vital for perception, voluntary movement, and learning. In smaller mammals, such as rats and mice, this outer layer of the cerebrum is smooth, yet in larger mammals it is highly folded. (See the cerebral cortex folds in humans in Figure 28.15A, on the facing page.) These foldings increase the surface area of the cerebrum without requiring much more volume. The increase in surface area was an evolutionary advantage, as more cerebral cortex meant more neural circuits to support complex behaviors, sophisticated sensory perception, and the ability to communicate. In humans, the "wrinkles" of the cerebral cortex form by the sixth month of development. How the structure and organization of the cerebral cortex relate to intelligence among different species and even within our own species is an area many scientists are working to understand. We'll examine the coordinated functions of distinct areas of the cerebral cortex as they relate to human intelligence in Module 28.16.

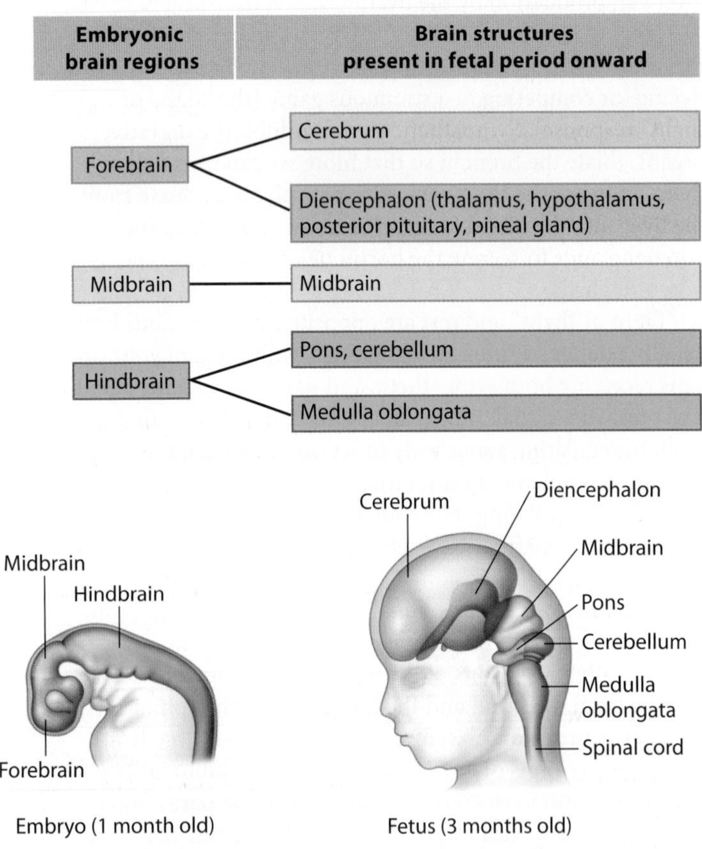
Embryonic brain regions
Brain structures present in fetal period onward
Forebrain
Cerebrum
Diencephalon (thalamus, hypothalamus, posterior pituitary, pineal gland)
Midbrain
Midbrain
Hindbrain
Pons, cerebellum
Medulla oblongata
Midbrain
Hindbrain
Forebrain
Embryo (1 month old)
Cerebrum
Diencephalon
Midbrain
Pons
Cerebellum
Medulla oblongata
Spinal cord
Fetus (3 months old)

▲ **Figure 28.14** Embryonic development of the human brain

? **Which region of the brain has changed the most during the course of vertebrate evolution?**

The cerebrum, especially the cerebral cortex

▷ The Human Brain

28.15 The structure of a living supercomputer: The human brain

Composed of up to 100 billion intricately organized neurons, with a much larger number of supporting cells, the human brain is more powerful than the most sophisticated computer. **Figure 28.15A** shows the three embryonic brain regions in their fully developed adult form. **Table 28.15** summarizes the major structures and functions of the brain.

A functional unit of the brain, called the **brainstem**, consists of two sections of the hindbrain, the **medulla oblongata** and **pons**, and the midbrain. (Note that these structures are identified with an * in the figure.) Structured as a stalk with cap-like swellings at the anterior end of the spinal cord, the brainstem is, evolutionarily speaking, one of the older parts of the vertebrate brain. The brainstem coordinates and filters the conduction of information from sensory neurons to the higher brain regions. It also regulates sleep and arousal and helps coordinate body movements, such as walking. Table 28.15 lists some of the individual functions of the medulla oblongata, pons, and midbrain.

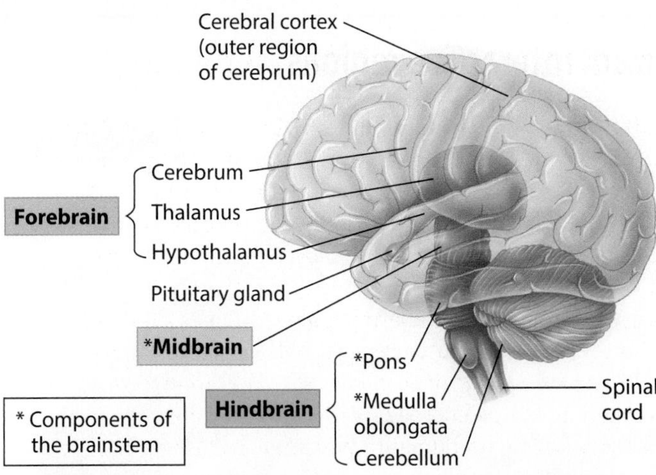

Cerebral cortex (outer region of cerebrum)
Forebrain
Cerebrum
Thalamus
Hypothalamus
Pituitary gland
*Midbrain
*Pons
Hindbrain
*Medulla oblongata
Cerebellum
Spinal cord
* Components of the brainstem

▲ **Figure 28.15A** The main structural parts of the human brain

Try This Make flashcards for each brain structure, showing the anatomy on the front and the function on the back.

TABLE 28.15 | MAJOR STRUCTURES OF THE HUMAN BRAIN

Brain Structure	Major Functions
Brainstem	Conducts data to and from other brain centers; helps maintain homeostasis; coordinates body movement
Medulla oblongata	Controls breathing, circulation, swallowing, digestion
Pons	Controls breathing
Midbrain	Receives and integrates auditory data; coordinates visual reflexes; sends sensory data to higher brain centers
Cerebellum	Coordinates body movement; plays role in learning and in remembering motor responses
Thalamus	Serves as input center for sensory data going to the cerebrum; sorts and groups all incoming sensory data for cerebrum
Hypothalamus	Functions as homeostatic control center; controls pituitary gland; serves as biological clock
Cerebrum	Performs sophisticated integration of information; plays major role in memory, learning, speech, emotions; formulates complex behavioral responses

Another part of the hindbrain, the **cerebellum** (light blue), is a planning center for body movements. It also plays a role in learning, decision making, and remembering motor responses. The cerebellum receives sensory information about the position of joints and the length of muscles, as well as information from the auditory and visual systems. It also receives input concerning motor commands issued by the cerebrum. The cerebellum uses this information to coordinate movement and balance. Hand-eye coordination is an example of such control by the cerebellum. If the cerebellum is damaged, the eyes can follow a moving object, but they will not stop at the same place as the object.

The most sophisticated integrating centers are those derived from the forebrain—the thalamus, the hypothalamus, and the cerebrum. The **thalamus** contains most of the cell bodies of neurons that relay information to the cerebral cortex. The thalamus first sorts data into categories (all of the touch signals from a hand, for instance). It also suppresses some signals and enhances others. The thalamus then sends information on to the appropriate higher brain centers for further interpretation and integration.

The hypothalamus controls the pituitary gland and the secretion of many hormones (Module 26.5). The hypothalamus also regulates body temperature, blood pressure, hunger, thirst, sex drive, and fight-or-flight responses, and it helps us experience emotions such as rage and pleasure. A "pleasure center" in the hypothalamus could also be called an addiction center, for it is strongly affected by certain addictive drugs, such as amphetamines and cocaine. As described in Module 28.9, these drugs increase the effects of norepinephrine and dopamine at synapses in the pleasure center, producing a short-term high, often followed by depression. Cocaine addiction may involve chemical changes in the pleasure center and elsewhere in the hypothalamus.

The suprachiasmatic nucleus, a group of neurons in the hypothalamus, functions as a **biological clock**, our internal timekeeper. Receiving visual input from the eyes (light/dark cycles, in particular), the clock maintains our **circadian rhythms**—daily cycles of biological activity—such as the sleep/wake cycle. Research with many different species has shown that without environmental cues, biological clocks keep time in a free-running way. For example, when humans are placed in artificial settings that lack environmental cues, our biological clocks and circadian rhythms maintain a cycle of approximately 24 hours 11 minutes, with very little variation among individuals.

The cerebrum, the largest and most complex part of our brain, consists of right and left **cerebral hemispheres** (**Figure 28.15B**), each responsible for the opposite side of the body. A thick band of nerve fibers called the **corpus callosum** facilitates communication between the hemispheres. Under the corpus callosum, groups of neurons called the **basal nuclei** are important in motor coordination. If they are damaged, a person may be immobilized. Degeneration of the basal nuclei occurs in Parkinson's disease (see Module 28.21). The most extensive area of our cerebrum, the cerebral cortex, is the focus of the next module.

Left cerebral hemisphere
Right cerebral hemisphere
Cerebrum
Thalamus
Cerebellum
Corpus callosum
Basal nuclei
Medulla oblongata

▲ **Figure 28.15B** A rear view of the brain

? **Choosing from the structures in Table 28.15, identify the brain part most important in solving an algebra problem.**

● Cerebrum

28.16 The cerebral cortex is a mosaic of specialized, interactive regions

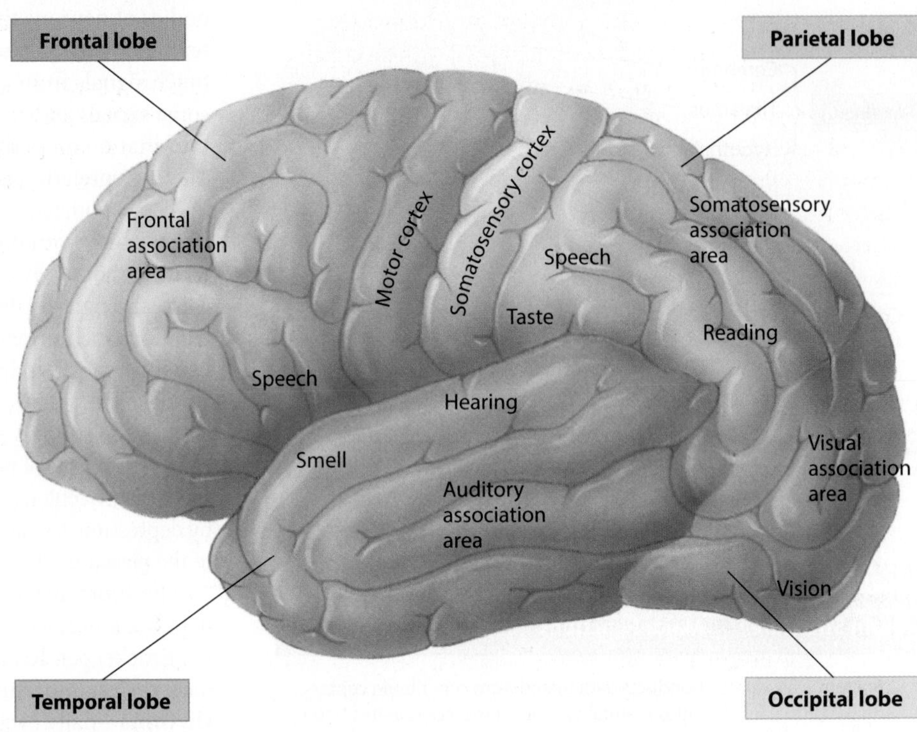

Frontal lobe
Parietal lobe
Frontal association area
Motor cortex
Somatosensory cortex
Speech
Somatosensory association area
Taste
Reading
Speech
Hearing
Smell
Auditory association area
Visual association area
Vision
Temporal lobe
Occipital lobe

▲ **Figure 28.16** Functional areas of the cerebral cortex, left hemisphere

Although less than 5 mm thick—less than the thickness of a pencil—the highly folded cerebral cortex accounts for about 80% of the total human brain mass. It contains some 10 billion neurons and hundreds of billions of synapses. Its intricate neural circuitry produces our most distinctive human traits: reasoning and mathematical abilities, language skills, imagination, artistic talent, and personality traits. Assembling information it receives from our eyes, ears, nose, taste buds, and touch sensors, the cerebral cortex also creates our sensory perceptions—what we are actually aware of when we see, hear, smell, taste, or touch something. In addition, the cerebral cortex regulates our voluntary movements.

Like the rest of the cerebrum, the cerebral cortex is divided into right and left sides connected by the corpus callosum. Each side of the cerebral cortex has four lobes named for a nearby bone of the skull: the frontal, parietal, temporal, and occipital lobes, which are represented by different colors in **Figure 28.16**. Researchers have identified a number of functional areas within each lobe.

Two areas of known function form the boundary between the frontal and parietal lobes. One area, called the motor cortex, functions mainly in sending commands to skeletal muscles, signaling appropriate responses to sensory stimuli. Next to the motor cortex, the somatosensory cortex receives and partially integrates signals from touch, pain, pressure, and temperature receptors throughout the body. The cerebral cortex also has centers that receive and begin processing sensory information concerned with vision, hearing, taste, and smell.

Numerous **association areas**, sites of higher mental activities, make up most of our cerebral cortex. Each sensory-receiving center of the cerebral cortex and the somatosensory cortex cooperates with an adjacent association area. In humans, a large association area in the frontal lobe uses varied inputs from many other areas of the brain to evaluate consequences, make considered judgments, and also plan for the future. Imaging techniques are beginning to show how a complicated interchange of signals among the sensory-receiving centers and the association areas produces our sensory perceptions.

Language results from extremely complex interactions among several association areas. For instance, the parietal lobe of the cerebral cortex has association areas used for reading and speech. These areas obtain visual information (the appearance of words on a page) from the vision centers in the occipital lobe. Then, if the words are to be spoken aloud, they arrange the information into speech patterns and tell another speech center, in the frontal lobe, how to make the motor cortex move the tongue, lips, and other muscles to form words. When we hear words, the parietal areas perform similar functions using information from auditory centers of the cerebral cortex.

You may have heard people comment that they are "left-brained" or "right-brained." In a phenomenon known as **lateralization**, areas in the two cerebral hemispheres become specialized for different functions during brain development in infants and children. In most people, the left cerebral hemisphere becomes adept at language, logic, and mathematical operations, as well as detailed skeletal motor control and processing of fine visual and auditory details. The right cerebral hemisphere is stronger at spatial relations, pattern and face recognition, and nonverbal thinking. (In about 10% of us, these roles of the left and right cerebral hemispheres are reversed or the hemispheres are less specialized.)

In 2013, the BRAIN initiative was announced, challenging scientists in the United States to dynamically map all of the brain's activities. The scale of this task has been compared to mapping the human genome or putting a man on the moon. Before we ponder the future of neuroscience research, let's first examine some of the tools researchers have used to date when discovering brain functions.

? **A stroke that causes loss of taste and numbness of the right side of the body has probably damaged brain tissue in the _____________ lobe of the _____________ cerebral hemisphere.**

parietal . . . left

28.17 Injuries and brain operations provide insight into brain function

CONNECTION

The physiology of the human brain is exquisitely complex, making it one of the most difficult structures to study in all of biology. No animal model or computer simulation can accurately predict its complicated functions. New techniques to visualize the brain are allowing researchers to associate specific parts of the brain with various activities. Much of what has been learned about the brain, however, has come from rare individuals whose brains were altered through injury, illness, or surgery. By studying such "broken brains," researchers have gained insight into how healthy brains operate.

The first well-publicized case of this type involved a man named Phineas Gage. In 1848, while working as a railroad construction foreman in Vermont, Gage accidentally exploded a dynamite charge that propelled a 3-foot-long spike through his head. The 13-pound steel rod entered his left cheek and traveled upward behind his left eye and out the top of his skull, landing several yards away. Incredibly, Gage walked away from the accident and appeared to have an intact intellect. However, his associates soon noticed drastic changes in his personality, with new propensities toward meanness, vulgarity, and irresponsibility and an inability to control his behavior.

At the time, Gage's doctor was able to note these changes, but understanding of the brain was insufficient to explain them. After Gage's death, the doctor preserved Gage's skull and the spike, allowing a group of researchers in 1994 to produce a computer model of the injury (**Figure 28.17A**). The modern analysis offered an explanation for Gage's bizarre behavior: The rod had pierced both frontal lobes of his brain. People with these sorts of injuries often exhibit irrational decision making and difficulty processing emotions. As you will learn in Module 28.20, the limbic system, a group of brain structures that invoke feelings, interacts with the frontal lobes during emotional responses.

Beginning with the work of several neurosurgeons in the 1950s, many of the functional areas of the cerebral cortex have been identified during brain surgery. The cerebral cortex lacks cells that detect pain; thus, after anesthetizing the scalp, a neurosurgeon can operate on this part of the brain with the patient awake. Parts of the cerebral cortex can be stimulated with a harmless electrical current. Stimulation of specific areas can cause someone to experience different sensations or recall memories. Researchers can obtain information about the effects simply by questioning the conscious patient.

Neurophysiologists have also gained insight into the interrelatedness of the brain's two cerebral hemispheres. As discussed in Module 28.16, association areas in the left and right sides become specialized for different functions. Much of what we know about this lateralization stems from the work of Roger Sperry with patients whose corpus callosum (communicating fibers between the two cerebral hemispheres; see Module 28.15) had been surgically cut to treat severe epileptic seizures. In a series of ingenious experiments, Sperry demonstrated that his patients were unable to verbalize sensory information that was received by only the right cerebral hemisphere.

One of the most radical surgical alterations of the brain is a hemispherectomy (**Figure 28.17B**)—the removal of most of one-half of the brain, excluding deep structures such as the thalamus, brainstem, and basal nuclei. This procedure is performed to alleviate severe seizure disorders that originate from one of the cerebral hemispheres as a result of illness, abnormal development, or stroke. Incredibly, with just half a brain, hemispherectomy patients recover quickly, often leaving the hospital within a few weeks. Their intellectual capacities are undiminished, although the side of the body opposite the surgery remains partially paralyzed. Higher brain functions that previously originated from the patient's missing half of the brain begin to be controlled by the opposite side. The younger the patient is, ideally less than 5 or 6 years old, the faster and more complete the recovery. Development after hemispherectomy is a striking example of the remarkable plasticity of the brain.

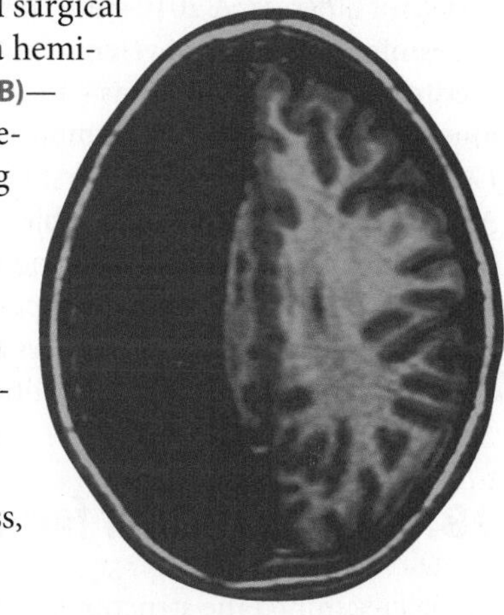

▲ **Figure 28.17B** X-ray of hemispherectomy patient after surgery

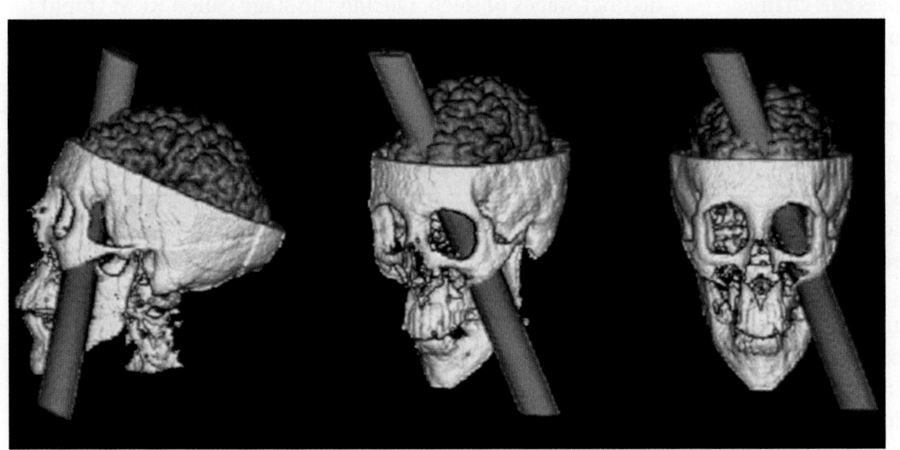

▲ **Figure 28.17A** Computer model of Phineas Gage's injury

? **How are researchers able to investigate brain function during brain surgery?**

● The cerebral cortex lacks pain receptors, which allows doctors to stimulate regions of the brain during surgery and the conscious patient to report sensations or memories.

28.18 fMRI scans provide insight into brain structure and function

CONNECTION

Functional magnetic resonance imaging (fMRI) is a scanning and imaging technology that can "light up" metabolic processes as they occur within living tissue. Because the procedure can be performed on a conscious patient, fMRI scans can provide significant insights into brain structure and function.

fMRI uses powerful magnets to align and then locate atoms within living tissue. During fMRI, a subject lies with his or her head in the hole of a large, doughnut-shaped magnet. When the brain is scanned with electromagnetic waves, changes in blood oxygen usage at sites of neuronal activity generate a signal that can be recorded. A computer uses the data to construct a three-dimensional map of the subject's brain activity.

Studies using fMRI confirm hypotheses based on older technologies about the roles of specific brain areas in movement and intention. Researchers have applied such techniques to correlate specific brain regions with nearly every aspect of human cognition, consciousness, and emotion.

fMRI is proving to be a powerful diagnostic tool for a wide variety of illnesses. A 2010 study, for example, used fMRI to examine the brains of veterans suffering from Gulf War syndrome, a set of symptoms—including confusion, vertigo, mood swings, fatigue, and numbness—reported by over 175,000 U.S. troops since the first Gulf War in the early 1990s. A research team exposed test subjects to a variety of stimuli (including threatening pictures and word association games) and used fMRI to observe which parts of the brain "lit up." The researchers found that soldiers reporting different symptoms displayed abnormalities in different regions of the brain. For example, veterans who reported difficulties with attention often displayed abnormal fMRI measurements in their thalamus (which is associated with the ability to concentrate). The pair of fMRI scans in **Figure 28.18** shows that healthy veterans and veterans suffering from Gulf War syndrome had different areas of brain activity when performing a particular task. Overall, the researchers found a suite of brain differences that appeared to generally correlate with the symptoms reported by each individual. Such data may help accurately identify, diagnose, and treat people suffering from Gulf War syndrome and similar traumas. These types of diagnostic methods are at the forefront of neuroscience, one of biology's most fascinating and rapidly developing subdisciplines.

? What does an fMRI scan actually measure?

Changes in the use of oxygen by living tissues

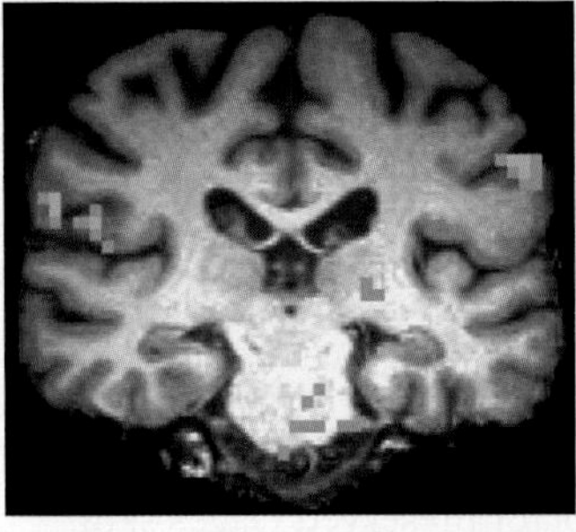

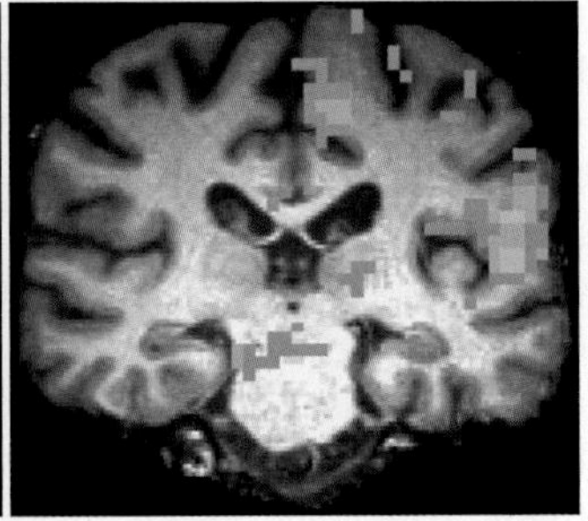

▲ **Figure 28.18** fMRI images showing regions of brain activity in a healthy veteran (left) and veteran with Gulf War syndrome (right) when each is performing the same task

28.19 The reticular formation is involved in arousal and sleep

Having examined the structures of the brain in the previous modules, we now turn our attention to functional brain systems, networks of neurons from different brain structures that work together. The reticular formation is one example of a functional brain system that regulates alertness. As anyone who has drifted off to sleep during a lecture or while reading a book knows, attentiveness and mental alertness can change rapidly. Arousal is a state of awareness of the outside world. Its counterpart is sleep, a state when external stimuli are received but not consciously perceived.

The reticular formation is a diffuse network of neurons that extends through the core of the brainstem. Acting as a sensory filter, the reticular formation receives data from sensory receptors and selects which information reaches the cerebral cortex. The more information the cerebral cortex receives, the more alert and aware a person is, although the brain often ignores certain familiar and repetitive stimuli—the feel of your clothes against your skin, for example—while actively processing other inputs.

The alertness that is maintained by the reticular formation is inhibited by a region of the hypothalamus that induces and regulates sleep. Although we know very little about its function, sleep is essential for survival. Furthermore, sleep is an active state, at least for the brain. By placing electrodes at multiple sites on the scalp, researchers can record patterns of electrical activity called brain waves in an electroencephalogram, or EEG. These recordings reveal that brain wave frequency changes as the brain progresses through several distinct stages of sleep. During the stage called REM (rapid eye movement) sleep, the brain waves are rapid and irregular, more like those of the awake state. We have most of our dreams during REM sleep, which typically occurs about six times a night for periods of 5–50 minutes each.

Understanding the function of sleeping and dreaming remains a compelling research problem. One hypothesis is that sleep and dreams are involved in the consolidation of learning and memory, and experiments show that regions of the brain activated during a learning task can become active again during sleep.

? If you start to feel drowsy while driving, would opening a window and blasting the radio help counteract this feeling? Explain your answer.

Yes. The reticular formation would receive more sensory data, which it would relay to the cerebral cortex, and you would feel more alert.

28.20 The limbic system is involved in emotions and memory

The **limbic system**, like the reticular system, is a functional brain system including various parts of the brain. Much of human emotion, behavior, motivation, and memory depends on our limbic system. The limbic system is central to such behaviors as nurturing infants and bonding emotionally to other people. Primary emotions that produce laughing and crying are mediated by the limbic system, and it also attaches emotional "feelings" to basic survival mechanisms of the brainstem, such as feeding, aggression, and sexuality. Intimate interactions between the limbic system and the frontal lobes of the cerebral cortex (which are involved in complex learning, reasoning, and personality) cause us to react emotionally to our conscious thoughts.

The functional unit of the limbic system, seen in gold in **Figure 28.20**, includes parts of the thalamus and hypothalamus and two partial rings around them formed by portions of the cerebral cortex. Two cerebral structures, the amygdala and the hippocampus, are also part of the limbic system.

Memory, which is essential for learning, is the ability to store and retrieve information derived from experience. The **hippocampus** is involved in both the formation of memories and their recall. In a study led by researchers at New York University in 2004, researchers interviewed people who were in New York on September 11, 2001 (9/11), the day terrorists attacked the Twin Towers of the World Trade Center, killing nearly 3,000 people. The interviewees were asked to recall memories of that day. Not surprisingly, the hippocampus was active in all interviewees during the telling of their stories. A subset of interviewees, people who were in downtown Manhattan—very close to the falling towers—also showed high activity in another area of the brain, the amygdala. The **amygdala** is central in laying down emotional memories. Sensory data converge in the amygdala, which seems to act as a memory filter, somehow labeling information to be remembered by tying it to an event or emotion of the moment. Not surprisingly, the New Yorkers who were downtown on 9/11 also vividly remembered sounds, smells, and sights.

You sense your limbic system's role in both emotion and memory when certain odors bring back "scent memories." Have you ever had a particular smell suddenly make you nostalgic for something that happened when you were a child? As indicated in Figure 28.20, signals from your nose enter your brain through the olfactory bulb, which connects with the limbic system. Thus, a specific scent can immediately trigger emotional reactions and memories.

Short-term memory, as the name implies, lasts only a short time—usually only a few minutes. It is short-term memory that allows you to dial a phone number just after looking it up. You may, however, be able to recall the number weeks after you originally looked it up, or even longer. This is because you have stored it in **long-term memory**. The transfer of information from short-term to long-term memory is enhanced by rehearsal, positive or negative emotional states mediated by the amygdala, and the association of new data with data previously learned and stored in long-term memory. For example, it's easier to learn the details of biology if you already have a good understanding of a framework of concepts.

Factual memories, involving names, faces, words, and places, are different from skill or procedural memories. Skill memories usually involve motor activities that are learned by repetition without consciously remembering specific information. You perform skills, such as tying your shoes, riding a bicycle, or hitting a baseball, without consciously recalling the individual steps required to do these tasks correctly. Once a skill memory is learned, it is difficult to unlearn. For example, a person who has played tennis with a self-taught, awkward backhand has a tougher time learning the correct form than a beginner just learning the game. Bad habits, as we all know, are hard to break.

Information processing by the brain generally seems to involve a complex interplay of several integrating centers. By experimenting with animals, studying amnesia (memory loss) in humans, and using brain-imaging techniques, scientists have begun to map some of the major brain pathways involved in memory. One proposed pathway involves the hippocampus and amygdala, which receive sensory information from the cerebral cortex and convey it to other parts of the limbic system and to the frontal lobes of the cerebral cortex. The storage of a memory is completed when signals return to the area in the cerebral cortex where the sensory perception originated.

In the final module of this chapter, we'll discuss several disorders of the nervous system, including their symptoms and treatments.

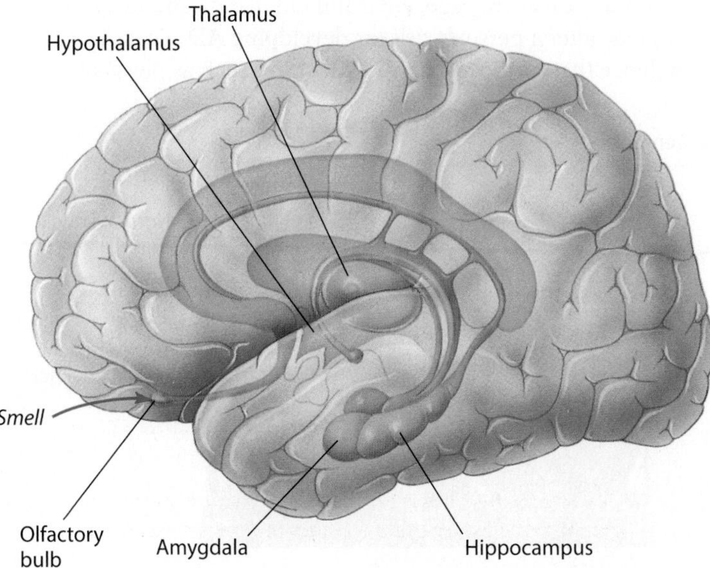
Thalamus
Hypothalamus
Smell
Olfactory bulb
Amygdala
Hippocampus

▲ **Figure 28.20** The limbic system (shown in shades of gold)

? Which three factors help transfer information from short-term to long-term memory?

Rehearsal, emotional associations, and connection with previously learned data

28.21 Changes in brain physiology can produce neurological disorders

CONNECTION

Neurological disorders (diseases of the nervous system) take an enormous toll on society. Examples of neurological disorders include depression (discussed earlier), schizophrenia, bipolar disorder, Alzheimer's disease, Parkinson's disease, and chronic traumatic encephalopathy. While these conditions are not yet curable, there are a number of treatments available.

Schizophrenia About 1% of the world's population suffers from **schizophrenia**, a severe mental disturbance characterized by psychotic episodes in which patients have a distorted perception of reality. The symptoms of schizophrenia typically include hallucinations (such as "voices" that only the patient can hear), delusions (generally paranoid), blunted emotions, distractibility, lack of initiative, and difficulty with verbal expression. Contrary to commonly held belief, schizophrenics do not necessarily exhibit a "split personality." There seem to be several different forms of schizophrenia, and it is unclear whether they represent different disorders or variations of the same underlying disease.

The physiological causes of schizophrenia are unknown, although the disease has a strong genetic component. Studies of identical twins show that if one twin has schizophrenia, there is a 50% chance that the other twin will have it, too. Since identical twins share identical genes, this indicates that schizophrenia has an equally strong environmental component, the nature of which has not been identified.

There are several treatments for schizophrenia that can usually alleviate the major symptoms, but they often have major side effects. Such treatments focus on brain pathways that use dopamine as a neurotransmitter. Identification of the genetic mutations associated with schizophrenia may yield new insights about the causes of the disease, which may in turn lead to effective therapies with fewer drawbacks.

Bipolar Disorder **Bipolar disorder**, or manic-depressive disorder, is characterized by extreme mood swings and affects about 1% of the population. The manic phase is marked by high self-esteem, increased energy and flow of ideas, extreme talkativeness, inappropriate risk taking, promiscuity, and reckless spending. In its milder forms, this phase is sometimes associated with great creativity, and some well-known artists, musicians, and writers (including Van Gogh, Beethoven, and Hemingway) had periods of intense creative output during their manic phases. The depressive phase of bipolar disorder is characterized by lowered ability to feel pleasure, loss of motivation, sleep disturbances, and feelings of worthlessness. These symptoms can be so severe that some individuals attempt suicide. Mood-stabilizing medicines, such as lithium, are often used to treat both manic and depressive episodes. Patients usually require mood stabilizers for years.

Bipolar disorder has a genetic component. As in schizophrenia, there is also a strong environmental influence; stress, especially severe stress in childhood, may be an important factor.

Alzheimer's Disease A form of mental deterioration, or dementia, **Alzheimer's disease** (AD) is characterized by confusion and memory loss. Its incidence increases dramatically with age. Approximately 1 in eight people aged 65 and older has AD, but when we limit our scope to people 85 and older, almost half likely have AD. Thus, by helping humans live longer, modern medicine is increasing the proportion of AD patients in the population. The disease is progressive; patients gradually become less able to function and eventually need to be dressed, bathed, and fed by others. There are also personality changes, almost always for the worse. Patients often lose their ability to recognize people, even family members, and may treat them with suspicion and hostility.

Currently, diagnosis of AD is made with a combination of neuropsychological clinical tests and brain imaging. In advanced cases of AD, images show a characteristic pathology of the brain: Neurons die in huge areas of the brain, and brain tissue often shrinks. The shrinkage is visible with brain imaging, as seen in **Figure 28.21A**, in which the brain from a patient with late stage AD (pink) is overlayed on a normal brain (gray). Brain scans are not as helpful at diagnosing earlier stages of AD. The diagnosis at any stage of the disease can only be confirmed upon autopsy: Finding two hallmark features of AD, neurofibrillary tangles and senile plaques, confirms AD in the remaining brain tissue. Neurofibrillary tangles are bundles of a protein called tau inside the neuron. Tau normally helps regulate the movement of nutrients along microtubules within neurons, but the bundles likely contribute to cell death by interfering with normal nutrient transport. Senile plaques are aggregates of beta-amyloid (a peptide) found outside the neurons. The plaques interfere with neuron communication at synapses and also contribute to the death of neurons. Scientists are currently looking for ways to diagnose AD earlier by sampling the blood or cerebrospinal fluid of AD patients for evidence of beta-amyloid accumulation or neuron death.

Besides advancing age, environmental and genetic factors may alter a person's risk for developing AD. There is evidence that cardiovascular risk factors (such as physical

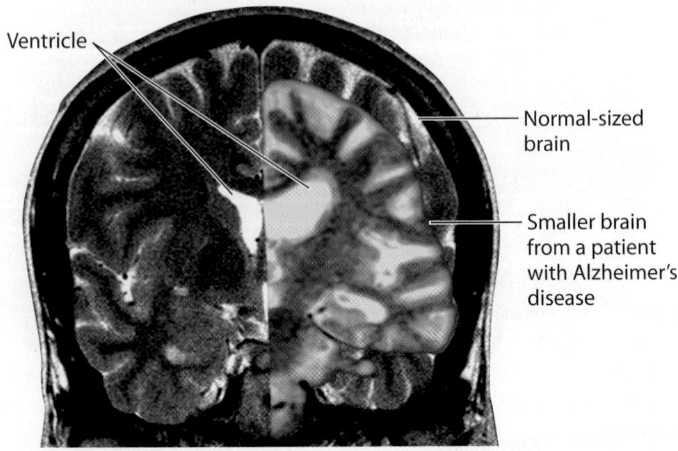
Ventricle
Normal-sized brain
Smaller brain from a patient with Alzheimer's disease

▲ **Figure 28.21A** Frontal MRI scans from a healthy patient (gray) with that from a patient with advanced AD overlaid on the right (pink)

inactivity, obesity, and smoking) lead to a higher chance of developing AD. Additionally, people with AD-afflicted relatives are at higher than normal risk. In a new era of genetic testing, one gene has gained much attention: the gene for apolipoprotein E (*APOE*). Apolipoprotein E is a protein involved in clearing away cholesterol in the blood. There are three alleles for this gene, known as *APOE-2*, *APOE-3*, and *APOE-4*. Each person inherits one allele from each parent, leading to six possible genotypes. About two-thirds of patients with AD have at least one copy of the *APOE-4* allele, making this a risk factor for AD. The *APOE-4* allele alone isn't entirely predictive, because some people with the allele do not develop the disease. Other genetic and environmental factors also play into one's total risk for AD.

Currently there is no treatment that can slow down the progression of AD, let alone stop it. With the first baby boomers now reaching age 65, the number of people with AD is expected to double by the late 2030s. Immense pressure will be put on our health-care system as this aging population requires more care.

Parkinson's Disease Approximately 1 million people in the United States suffer from **Parkinson's disease**, a motor disorder characterized by difficulty in initiating movements, slowness of movement, and rigidity. Patients often have a masked facial expression, muscle tremors, poor balance, a flexed posture, and a shuffling gait. Like AD, Parkinson's is progressive, and the risk increases with age. The incidence of Parkinson's disease is about 1% at age 65 and about 5% at age 85.

The symptoms of Parkinson's disease result from the death of neurons in the basal nuclei. These neurons normally release dopamine from their synaptic terminals. The disease itself appears to be caused by a combination of environmental and genetic factors. Evidence for a genetic role includes the fact that some families with an increased incidence of Parkinson's disease carry a mutated form of the gene for a protein important in normal brain function.

At present, there is no cure for Parkinson's disease, although various treatments can help control the symptoms. Treatments include drugs such as L-dopa, a precursor of the neurotransmitter dopamine, and surgery. One potential cure is to develop stem cells into dopamine-secreting neurons in the laboratory and implant them in patients' brains. In laboratory experiments, transplantation of such cells into rats with a Parkinson's-like condition can lead to a recovery of motor control. Whether this kind of regenerative medicine will also work in humans is one of many important questions on the frontier of modern brain research.

Chronic Traumatic Encephalopathy With symptoms similar to AD, **chronic traumatic encephalopathy (CTE)** is a form of dementia caused by repeated brain trauma, particularly concussions. Originally described in boxers in the late 1920s, CTE has more recently been diagnosed in numerous football players and other athletes who suffer head injuries. Early symptoms include depression and loss of impulse control, eventually leading to symptoms like those of Parkinson's disease, memory loss, and dementia. Individuals with CTE show a buildup of tau protein in their neurons at the time of autopsy, similar to what is seen in AD. NFL Cincinnati Bengals receiver Chris Henry (**Figure 28.21B**), who died at age 26 in a car accident in 2009, was the youngest NFL player to have been diagnosed with CTE. With so many questions remaining about how and when CTE develops after concussions, numerous athletes have committed to donating their brains to research upon their death.

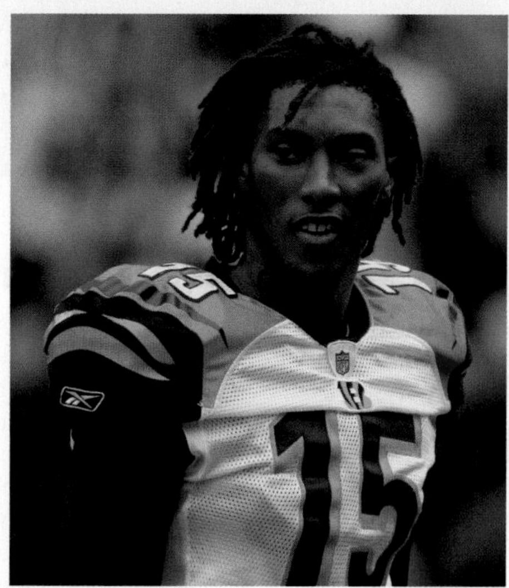
15

▲ **Figure 28.21B** Chris Henry of the Cincinnati Bengals

Stroke It takes years or decades before the degeneration of the brain results in noticeable symptoms of AD, Parkinson's disease, and CTE. In contrast, during a stroke, symptoms come on suddenly. A stroke is the death of nervous tissue in the brain due to a lack of oxygen, usually resulting from a rupture or blockage of an artery in the head. Numbness or weakness of one side of the face or one arm or leg is a sign that a stroke is occurring. Other symptoms include sudden confusion or the sudden inability to speak, see, or walk. The symptoms correlate to the functional regions of the brain being disrupted by the rupture or blockage.

Strokes are the third leading cause of death in the United States. The majority of strokes are caused by a clot in a blood vessel that blocks blood flow to the brain. To treat blockages, a drug that dissolves blood clots can be administered to a patient having a stroke. Because the death of cells occurs rapidly, every minute counts when treating a stroke victim. Patients must be treated with the clot-dissolving drug within three hours, for possible full recovery or minor damage to the brain. Like heart attacks, strokes are cardiovascular events that can be prevented. Risk factors for strokes that can be controlled or reduced, include diabetes, high blood pressure, smoking, heart disease, and high cholesterol.

Unraveling the biological bases of neurological disorders remains one of the most challenging tasks of modern biology. Many aspects of our nervous system remain mysterious.

? **If a 35-year-old woman has two copies of the *APOE-4* allele, will she develop AD?**

● She is at an increased risk of developing AD, but other genetic and environmental factors will also contribute to her overall risk.

CHAPTER 28 REVIEW

For practice quizzes, BioFlix animations, MP3 tutorials, video tutors, and more study tools designed for this textbook, go to MasteringBiology®

Reviewing the Concepts

Nervous System Structure and Function (28.1–28.2)

28.1 Nervous systems receive sensory input, interpret it, and send out commands. Sensory neurons of the peripheral nervous system (PNS) conduct signals from sensory receptors to the central nervous system (CNS), which consists of the brain and, in vertebrates, the spinal cord. Interneurons in the CNS integrate information and send it to motor neurons of the PNS. Motor neurons, in turn, convey signals to effector cells.

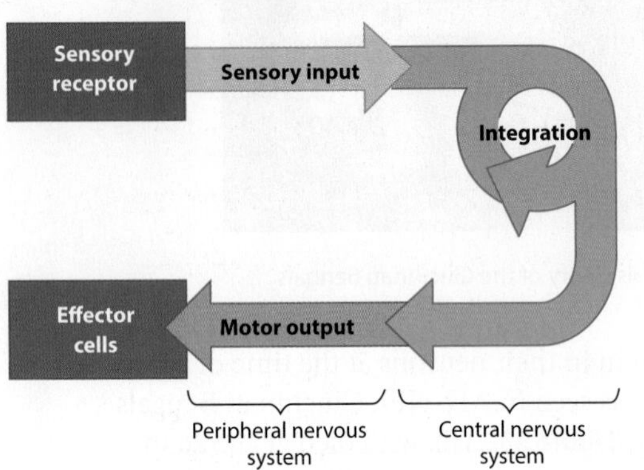

Sensory receptor
Sensory input
Integration
Effector cells
Motor output
Peripheral nervous system
Central nervous system

28.2 Neurons are the functional units of nervous systems. Neurons are cells specialized for carrying signals.

Nerve Signals and Their Transmission (28.3–28.10)

28.3 Nerve function depends on charge differences across neuron membranes. At rest, a neuron's plasma membrane has an electrical voltage called the resting potential, which results from the positive charge on the outer membrane surface opposing a negative charge on its inner (cytoplasmic) surface.

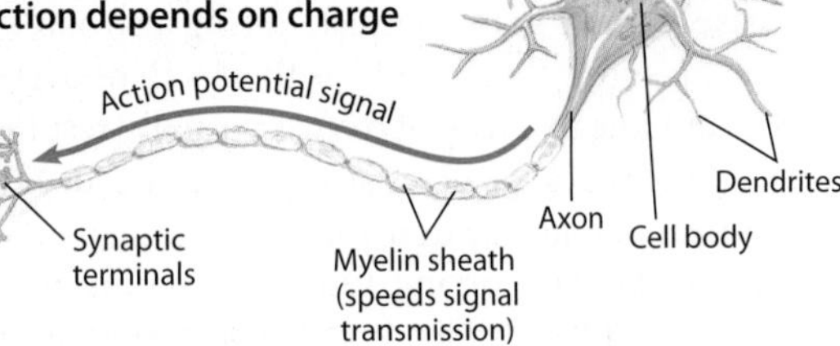

Action potential signal
Synaptic terminals
Myelin sheath (speeds signal transmission)
Axon
Cell body
Dendrites

28.4 A nerve signal begins as a change in the membrane potential. A stimulus alters the permeability of a portion of the membrane, allowing ions to pass through and changing the membrane's voltage. A nerve signal, called an action potential, is a change in the membrane voltage from the resting potential to a maximum level and back to the resting potential.

28.5 The action potential propagates itself along the axon. Action potentials are self-propagated in a one-way chain reaction along an axon. An action potential is an all-or-none event. The frequency of action potentials (but not their strength) changes with the strength of the stimulus.

28.6 Neurons communicate at synapses. The transmission of signals between neurons or between neurons and effector cells occurs at junctions called synapses. At electrical synapses, electrical signals pass directly between cells. At chemical synapses, the sending neuron cell secretes a chemical signal, a neurotransmitter that crosses the synaptic cleft and binds to a specific receptor on the surface of the receiving cell.

28.7 Chemical synapses enable complex information to be processed. Some neurotransmitters excite a receiving cell; others inhibit a receiving cell's activity by decreasing its ability to develop action potentials. A cell may receive differing signals from many neurons; the summation of excitation and inhibition determines whether or not it will transmit a nerve signal.

28.8 A variety of small molecules function as neurotransmitters. Many neurotransmitters are small, nitrogen-containing molecules. Acteylcholine, for example, is released by motor neurons in the PNS and activates skeletal muscles.

28.9 Many drugs act at chemical synapses. Many common drugs can affect the brain's delicate chemistry.

28.10 Published data are biased toward positive findings.

An Overview of Animal Nervous Systems (28.11–28.14)

28.11 The evolution of animal nervous systems reflects changes in body symmetry. Radially symmetric animals have a nervous system arranged in a weblike system of neurons called a nerve net. Among bilaterally symmetric animals, nervous systems evolved to exhibit cephalization, the concentration of the nervous system in the head end, and centralization, the presence of a central nervous system including a primitive brain and nerve cords.

28.12 Vertebrate nervous systems are highly centralized. The brain and spinal cord contain fluid-filled spaces.

28.13 The peripheral nervous system of vertebrates can be divided into functional components.

Central nervous system: CNS	**Brain**
	Spinal Cord
Peripheral nervous system: PNS	**Sensory input:** stimuli received by sensors and signal transmitted to CNS
	Motor output: stimuli received from CNS and transmitted to effectors • Motor system: voluntary control over muscles • Autonomic nervous system: involuntary control over organs – Parasympathetic division: rest and digest – Sympathetic division: fight or flight – Enteric division: digestive system

28.14 The vertebrate brain develops from three anterior bulges of the neural tube. The evolution of the vertebrate brain involved the enlargement and subdivision of the hindbrain, midbrain, and forebrain. The size and complexity of the cerebrum in birds and mammals correlate with their sophisticated behavior.

The Human Brain (28.15–28.21)

28.15 The structure of a living supercomputer: The human brain. The midbrain and subdivisions of the hindbrain, together with the thalamus and hypothalamus of the forebrain, function mainly in conducting information to and from higher brain centers. They regulate homeostatic functions, keep track of body position, and sort sensory information. The forebrain's cerebrum is the largest and most complex part of the brain. Most of the cerebrum's integrative power resides in the cerebral cortex.

28.16 The cerebral cortex is a mosaic of specialized, interactive regions. Specialized integrative regions of the cerebral cortex include the somatosensory cortex and centers for vision, hearing, taste, and smell. The motor cortex directs responses. Association areas, concerned with higher mental activities such as reasoning and language, make up most of the cerebrum.

28.17 Injuries and brain operations provide insight into brain function.

28.18 fMRI scans provide insight into brain structure and function. The scans detect oxygen use at sites of neuronal activity.

28.19 The reticular formation is involved in arousal and sleep. Several regions of the brain, making up the reticular formation, are involved in maintaining alertness.

28.20 The limbic system is involved in emotions and memory. The limbic system is a group of integrating centers in the cerebral cortex, thalamus, and hypothalamus.

28.21 Changes in brain physiology can produce neurological disorders. Neurological disorders can distort one's perception of reality (schizophrenia) and affect a person's mood (bipolar disorder). Deterioration of brain function due to neuron injury and neuron death is a hallmark in diseases such as Alzheimer's, Parkinson's, chronic traumatic encephalopathy, and stroke.

Connecting the Concepts

1. Test your understanding of the nervous system by matching the following labels with their corresponding letters: CNS, effector cells, interneuron, motor neuron, PNS, sensory neuron, sensory receptor, spinal cord, synapse.

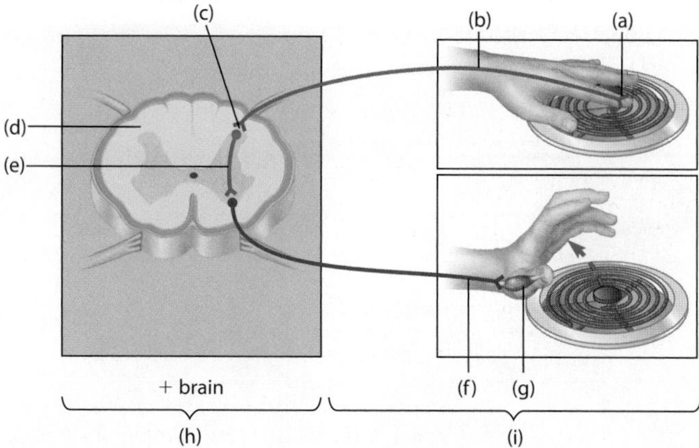

(c)
(b)
(a)
(d)
(e)
+ brain
(f)
(g)
(h)
(i)

Testing Your Knowledge

Level 1: Knowledge/Comprehension

2. Fill in the blanks to match some brain structures with their associated functions.
   a. If the ______________ is severed, the right and left cerebral hemispheres cannot communicate.
   b. The ______________ and the hippocampus are two components of the ______________ system that help store memories.
   c. Accounting for most of the weight of your brain is the highly folded ______________; it is the outer region of the ______________.
   d. The ______________ is responsible for hand-eye coordination.
   e. The ______________ formation keeps you alert, but the ______________, with its sleep-inducing center and internal timekeeping region, counteracts alertness.
3. What causes a nerve signal to move from one end of a neuron along the length of the neuron to the other end? What is a nerve signal, exactly? Why can't it go backward? How is a nerve signal transmitted from one neuron to the next across a synapse? Write a short paragraph that answers these questions.

Level 2: Application/Analysis

4. Joe accidentally touched a hot pan. His arm jerked back, and an instant later, he felt a burning pain. How would you explain the fact that his arm moved before he felt the pain?
   a. His limbic system blocked the pain momentarily, but the important pain signals eventually got through.
   b. His response was a spinal cord reflex that occurred before the pain signals reached the brain.
   c. Motor neurons are myelinated; sensory neurons are not. The signals traveled faster to his muscles.
   d. This scenario is not actually possible. The brain must register pain before a person can react.
5. Which division of the autonomic nervous system would you expect to be activated if a person heard an intruder at the front door?
   a. parasympathetic b. sympathetic c. enteric
6. Anesthetics block pain by blocking the transmission of nerve signals. Which of these three chemicals might work as anesthetics? (*Choose all that apply and explain your selections.*)
   a. a chemical that prevents the opening of sodium channels in membranes
   b. a chemical that inhibits the enzymes that degrade neurotransmitters
   c. a chemical that blocks neurotransmitter receptors

Level 3: Synthesis/Evaluation

7. **SCIENTIFIC THINKING** A proposal to test an SSRI in a large number of depressed individuals was submitted to the FDA. Through random assignments, half of the depressed patients would be controls, receiving nothing at all, and half the patients would receive the drug in pill form. Patients in both groups would note changes in their own mood in a daily journal. What flaw(s) do you note with this design?
8. Using microelectrodes, a researcher recorded nerve signals in four neurons in the brain of a snail, called A, B, C, and D in the table below. A, B, and C all can transmit signals to D. In three experiments, the animal was stimulated in different ways. The number of nerve signals transmitted per second by each of the cells is recorded in the table. Write a short paragraph explaining the different results of the three experiments.

	A	B	C	D
	Signals/sec			
Experiment #1	50	0	40	30
Experiment #2	50	0	60	45
Experiment #3	50	30	60	0

Data from E.R. Kandel and L. Tauc, Input Organization of Two Symmetrical Giant Cells in the Snail Brain, *Journal of Physiology* 183: 269-286 (1966).

9. Most neurons, once damaged, will not regenerate. The use of embryonic stem cells has been proposed as a potential treatment for many neurological diseases. Neurons developed from such stem cells might be able to replace those damaged by the disease. Do you favor or oppose research along those lines? Explain your answer.

Answers to all questions can be found in Appendix 4.

CHAPTER 29

The Senses

? *How do sea turtles sense Earth's magnetic field?*

When loggerhead sea turtles (*Caretta caretta*) hatch from their eggs on the beaches of eastern North America, they crawl to the ocean and swim. Instinctively, the turtles swim eastward toward currents collectively known as the North Atlantic gyre, a circular system stretching west-east from North America to Europe and Africa and north-south from waters near Iceland to the equator. After years in the gyre, feeding and maturing, the turtles migrate to coastal waters. When it is time to lay their eggs, females return to the beach from which they hatched (see Figure 35.8B). But how can loggerhead turtles navigate to the gyre in the first place and make their way back home years later?

Like some other migratory animals, sea turtles have magnetoreception—that is, they can sense Earth's magnetic field. Scientists at the University of North Carolina at Chapel Hill discovered that sea turtles use the magnetic field like a compass to determine which way is north and to sense differences in the magnetic field geographically. Placing loggerhead hatchlings in tethered harnesses within water-filled tanks (see photo below), the scientists recorded in which directions—north or

south, east or west—the turtles swam. When exposed to an artificial magnetic field simulating a specific location on the eastern coast of North America, the hatchlings swam in the direction they would to reach the gyre. When the magnetic field replicated conditions at a particular location off the west coast of Africa, the turtles swam in the opposite direction. In loggerheads, a magnetic mineral called magnetite is probably involved in magnetoreception. (How scientists think magnetite functions in sensory cells is discussed in the chapter.)

Although we humans can't sense Earth's magnetic field, and rely on maps or global positioning systems in our cars or on our smartphones, we perceive the world in detail using the senses we do have. We look now at the organs and mechanisms that underlie all animal senses. Then we turn our attention to human sensory structures and how they function.

BIG IDEAS

Sensory Reception
(29.1–29.3)

Animal sensations begin as stimuli detected by sensory receptor cells.

Hearing and Balance
(29.4–29.6)

The human ear contains structures that detect sound and body movement.

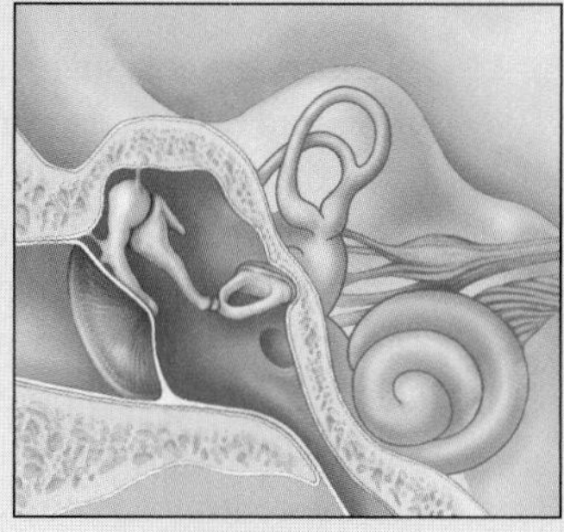

Vision
(29.7–29.10)

In the vertebrate eye, a single lens focuses an image onto photoreceptor cells.

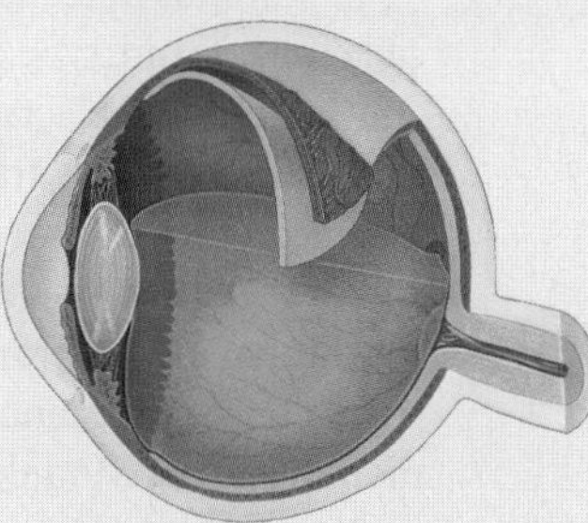

Taste and Smell
(29.11–29.13)

Chemoreceptors on the tongue and in the nose detect chemicals.

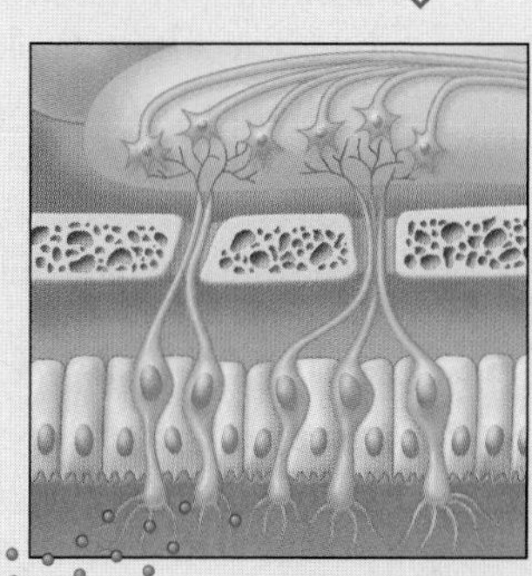

▷ Sensory Reception

29.1 Sensory receptors convert stimulus energy to action potentials

The operation of all animal senses—the ones used by humans such as hearing as well as nonhuman senses such as magnetoreception—originates in **sensory receptors**, specialized neurons or other cells that are tuned to the conditions of the external world or internal organs. Sensory organs, such as your eyes and taste buds, contain sensory receptors that detect stimuli such as light energy (photons) emitted by your computer screen and chemical flavorings in your food. Once a stimulus is detected, a sensory receptor cell triggers action potentials (electrical signals; Module 28.4) that send information to the central nervous system (CNS), which includes the brain and spinal cord.

In stimulus detection, the sensory receptor cell converts one type of signal (the stimulus) to another type, an electrical signal. This conversion, called **sensory transduction**, produces a change in the cell's membrane potential as a result of the opening or closing of ion channels. Recall that a membrane potential is an electrical charge difference across a cell's plasma membrane see Module 28.3.

Figure 29.1A shows sensory transduction occurring when sensory receptor cells in a taste bud detect sugar molecules. ❶ The sugar molecules (⁘) first arrive at the taste bud, where ❷ they bind to sweet receptors, specific protein molecules embedded in the plasma membrane of a taste receptor cell. The binding triggers ❸ a signal transduction pathway (see Module 11.10) that causes ❹ some ion channels in the membrane to open. Changes in the flow of ions create a graded change in membrane potential in sensory receptor cells called a **receptor potential**. Receptor potentials vary; the stronger the stimulus, the greater the receptor potential. (In contrast, action potentials have a constant strength and can be regenerated along the length of an axon to carry signals long distances.)

Once a stimulus is converted to a receptor potential, the receptor potential usually results in signals passing into the central nervous system. In taste buds, ❺ each receptor cell forms a synapse with a sensory neuron. In many cases, a receptor cell constantly secretes neurotransmitter (⁘) into this synapse at a set rate, triggering a steady stream of action potentials in the sensory neuron. ❻ The graph shows the rate at which the sensory neuron sends action potentials when the taste receptor is not detecting sugar and the graph also shows what happens when there are enough sugar molecules to trigger a strong receptor potential, causing the receptor cell to release even more neurotransmitter than usual. This additional neurotransmitter increases the rate of action potential generation in the sensory neuron, which in turn signals the brain that the sensory receptor detects a stimulus.

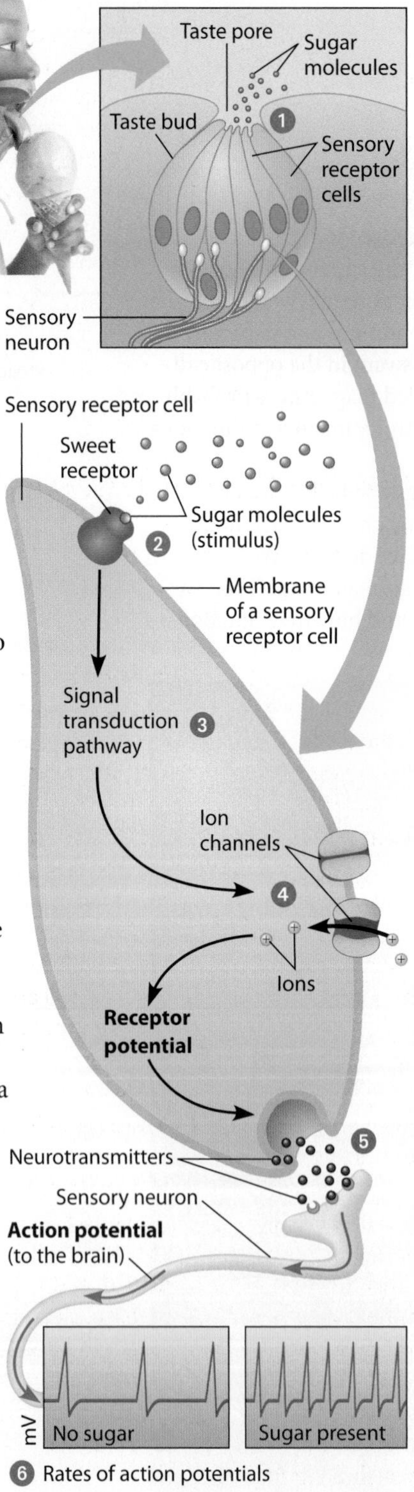

You can see that sensory receptor cells transduce (convert) stimuli to receptor potentials, which trigger action potentials that enter the central nervous system. If action potentials are the same no matter where or how they are produced, how do they communicate a sweet taste instead of a salty one? In the taste buds (**Figure 29.1B** on the facing page), the purple sensory receptor cell responds to sugar only, and the green receptor cell responds to salt only. Sensory neurons from the sugar-detecting cell synapse with different interneurons in the brain than those contacted by neurons from the salt-detecting cell. The brain distinguishes different tastes by the particular interneurons that are stimulated.

The graphs in Figure 29.1B indicate how action potentials communicate information about the intensity of stimuli. The left part of each graph represents the rate at which the sensory neurons in the taste bud transmit action potentials when the taste receptors are not stimulated. The right side of each graph shows that the rate of transmission depends on the intensity of the stimulus; the stronger the stimulus, the greater the receptor potential, the more neurotransmitters released by the receptor cell, and the more frequently the sensory neuron transmits action potentials to the brain. The brain interprets the intensity of the stimulus from the rate at which it receives action potentials. The sensitivity to a stimulus, however, can diminish over time.

▲ **Figure 29.1A** Sensory transduction at a taste bud

Try This Give a piece of candy to a friend. While he or she is eating it, explain the mechanism involved in tasting sugar.

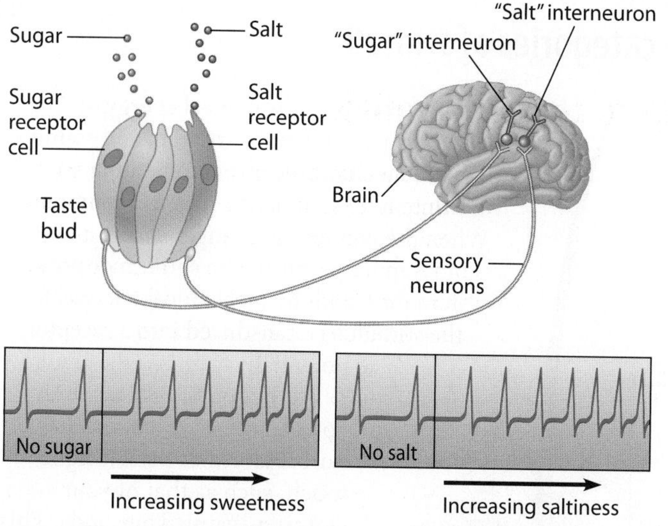
Sugar
Salt
"Salt" interneuron
"Sugar" interneuron
Sugar receptor cell
Salt receptor cell
Taste bud
Brain
Sensory neurons
No sugar
No salt
Increasing sweetness
Increasing saltiness

▲ **Figure 29.1B** Action potentials transmitting taste sensations

Have you ever noticed how a strong odor seems to fade with time, even when you know the odorous substance is still there? Or how the water in a pool may seem shockingly cold when you dive in, but not when you get used to it? This is called **sensory adaptation**, the tendency of some sensory receptors to become less sensitive when they are stimulated repeatedly. When receptors become less sensitive, they trigger fewer action potentials, stimuli to the brain decreases, and awareness of stimuli is lost as a result. Sensory adaptation keeps the body from reacting to familiar stimuli, such as the feeling of clothes on your skin.

Stimulus detection, transduction, transmission, and processing in the central nervous system are the main steps in a general model for sensory reception. Next we see how the model applies to magnetic sensory reception.

? **What is meant by sensory transduction?**

The conversion of a stimulus signal to an electrical signal (a receptor potential) by a sensory receptor cell

29.2 The model for magnetic sensory reception is incomplete

SCIENTIFIC THINKING

The steps in the model of sensory reception we examined in the previous module are clearly understood for human sensory reception (as we will see in the rest of the chapter); however, the steps are less clear for sensory reception in other animals with unique senses. The first step, detection, as it relates to magnetic sense (magnetoreception), is essentially unknown. Even the location of the magnetic sensory cells is obscure. Because the magnetic field can penetrate the body's tissues, magnetic sensory cells could be located just about anywhere in an animal's body and are most likely few and far between.

Evidence suggests that there is not just one way that animals detect the magnetic field to navigate. One hypothesis proposes that electrical fluctuations, influenced by Earth's magnetic field, are detected by sensory cells. Two other hypotheses suggest that variations in the magnetic field affect biochemical reactions or the alignment of magnetic minerals inside sensory cells. But as is often the case in science, the different hypotheses need not be mutually exclusive. For example, it's been proposed that magnetoreception in some birds is carried out by both biochemical reactions within sensory cells of the eyes and movement of magnetic minerals present in the sensory cells of beaks.

How do sea turtles sense Earth's magnetic field?

The strong magnetic mineral magnetite, which is used in compass needles, has been found in some animals with a magnetic sense, such as loggerhead sea turtles (see the chapter introduction), pigeons, and rainbow trout. Yet finding individual cells that contain magnetite has been challenging because a magnetite crystal is tiny (about 50 nm in diameter). In tissues, these cells are nearly impossible to find. In 2012, however, scientists did identify cells with magnetic properties. They used a technique that dissociated cells from the nasal tissue of rainbow trout (a region suspected to house magnetic receptor cells), suspended the cells in fluid, and subjected the cells to a rotating magnetic field, which caused a few cells to spin (**Figure 29.2**). The spinning cells were separated and examined by microscopy. Examination suggested that magnetite crystals anchor to the plasma membrane. (What causes the magnetite crystals to anchor in the plasma membrane is unknown.)

If these magnetite-containing cells are indeed the elusive magnetic sensory receptor cells, then scientists still need to link them to the general sensory reception model, including transduction of the signal and processing in the central nervous system. Scientists hypothesize that pressure exerted by the magnetite crystals on the plasma membrane cause ion channels to open, converting the pressure into a receptor potential capable of stimulating sensory neuron action potentials. A better understanding of how magnetite functions in living cells will more completely explain how animals use the magnetic field to navigate.

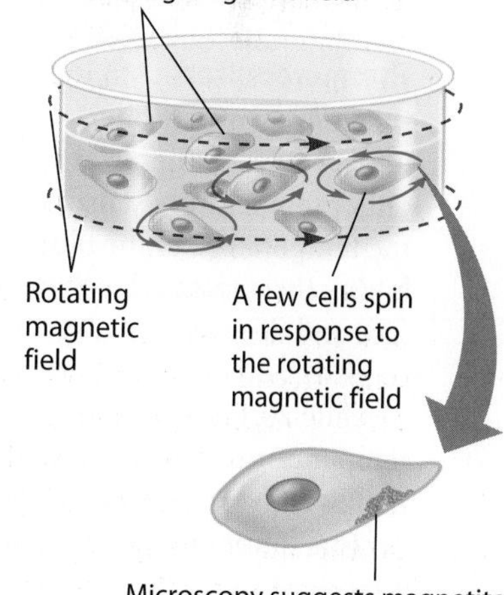
Cells suspended in fluid are subjected to a rotating magnetic field
Rotating magnetic field
A few cells spin in response to the rotating magnetic field
Microscopy suggests magnetite crystals are anchored to the plasma membrane

▲ **Figure 29.2** Isolation and examination of magnetite-containing cells

? **Why is the model of magnetoreception incomplete?**

The first step, how sensory cells detect changes in the magnetic field, is still uncertain. Thus, the next steps—how the signal is transduced, transmitted, and then processed by the central nervous system—are also unclear.

29.3 Specialized sensory receptors detect five categories of stimuli

A magnetic field is but one of several types of stimuli. Based on the type of signals to which they respond, we can group animal sensory receptors into general categories: pain receptors, thermoreceptors (sensors for both heat and cold), mechanoreceptors (sensors for touch and pressure), chemoreceptors, and electromagnetic receptors.

Figure 29.3A, showing a section of human skin, reveals why the surface of our body is sensitive to a variety of stimuli. Our skin contains pain receptors, thermoreceptors, and mechanoreceptors. Each of these receptors is a modified dendrite of a sensory neuron (see Module 28.2). The neuron recognizes stimuli and sends action potentials to the central nervous system. In other words, each receptor serves as both a receptor cell and a sensory neuron. Most of the dendrites in the dermis (the underlying region of the skin) are wrapped in one or more layers of connective tissue (purple areas in the figure); however, the pain and touch receptors in the epidermis (outer skin layer) and the touch receptors around the base of hairs are naked dendrites.

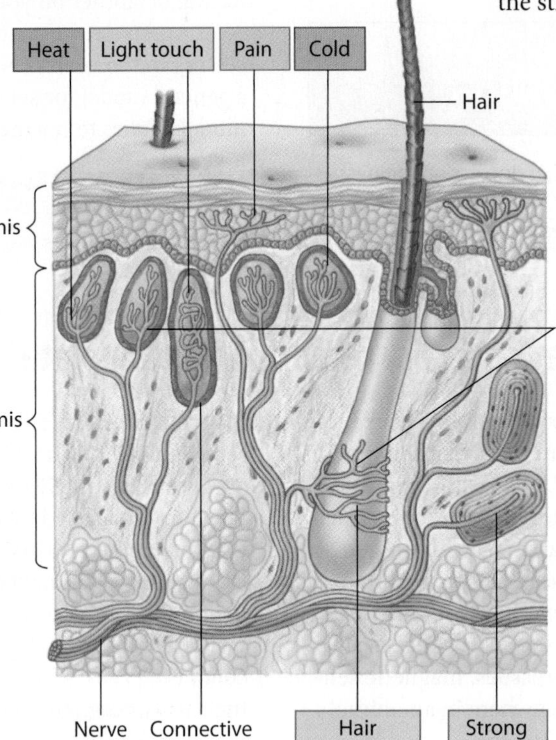
Heat
Light touch
Pain
Cold
Hair
Epidermis
Dermis
Dendrites
Nerve to brain
Connective tissue
Hair movement
Strong pressure

▲ **Figure 29.3A** Sensory receptors in the human skin

Thermoreceptors In the skin **thermoreceptors** (blue labels in the figure) detect either heat or cold. Other temperature sensors located deep in the body monitor the temperature of the blood. The hypothalamus acts as the body's thermostat: Receiving action potentials from both surface and deep sensors, the hypothalamus keeps a mammal's or bird's body temperature within a narrow range (see Module 20.15). Interestingly, the sensory receptors for high temperatures also respond to capsaicin, the chemical that makes chili peppers taste "spicy hot" to us. Special thermoreceptors found in the pit organ **(Figure 29.3B)** of some snakes have evolved to detect the body heat of the snake's prey up to a meter away. The thermoreceptors transmit heat information to the snake's brain that then superimposes a thermal image with a visual image.

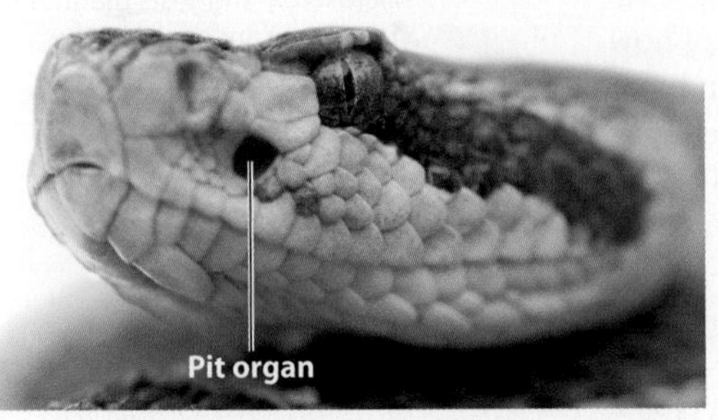
Pit organ

▲ **Figure 29.3B** Thermoreceptor receptor organs in a snake

Mechanoreceptors Mechanical energy, such as touch, pressure, stretching, motion, and sound, stimulate **mechanoreceptors** to bend or stretch, resulting in tension on the plasma membrane. When the membrane changes shape, it becomes more permeable to sodium or potassium ions, and the mechanical energy of the stimulus is transduced into a receptor potential.

At the top of Figure 29.3A, you can see a type of mechanoreceptor that detects light touch, such as that of your clothing against your body. This type of receptor can transduce very slight inputs of mechanical energy into action potentials. Another type of pressure sensor, lying deeper in the skin, is stimulated by strong pressure. A third type of mechanoreceptor, the touch receptor around the base of the hair, detects hair movements. Did you ever wonder how cats use their whiskers while slinking around in the dark? Touch receptors at the base of the stiff whiskers are extremely sensitive and enable the animal to detect the objects they contact. Another type of mechanoreceptor (not shown) is found in our skeletal muscles.

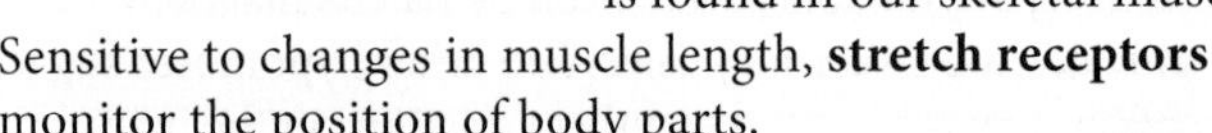

Sensitive to changes in muscle length, **stretch receptors** monitor the position of body parts.

A variety of mechanoreceptors collectively called **hair cells** detect sound waves and other forms of movement in fluid. The "hairs" on these sensors are either specialized types of cilia or cellular projections called microvilli. The sensory hairs project from the surface of a receptor cell into either the external environment, such as the water surrounding a fish, or an internal fluid-filled compartment, such as our inner ear.

Figure 29.3C (on the facing page) shows how hair cells work, ❶ starting with a receptor cell at rest. ❷ When fluid movement bends the hairs in one direction, the hairs stretch the cell membrane, increasing its permeability to certain ions. This makes the hair cell secrete more neurotransmitter molecules and increases the rate of action potential production by a sensory neuron. ❸ When the hairs bend in the opposite direction, ion permeability decreases, the hair cell releases fewer neurotransmitter molecules, and the rate of action potentials decreases. We will see later that hair cells are involved in both hearing and balance.

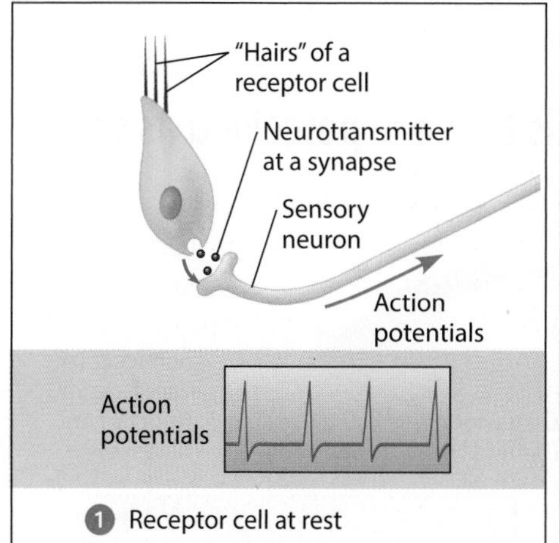
"Hairs" of a receptor cell
Neurotransmitter at a synapse
Sensory neuron
Action potentials
Action potentials
1 Receptor cell at rest

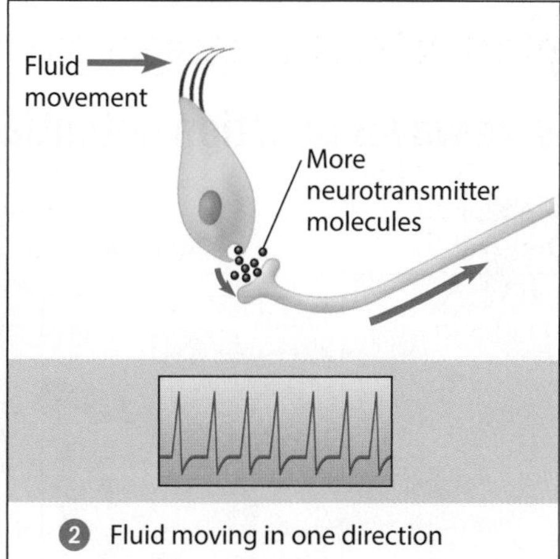
Fluid movement
More neurotransmitter molecules
2 Fluid moving in one direction

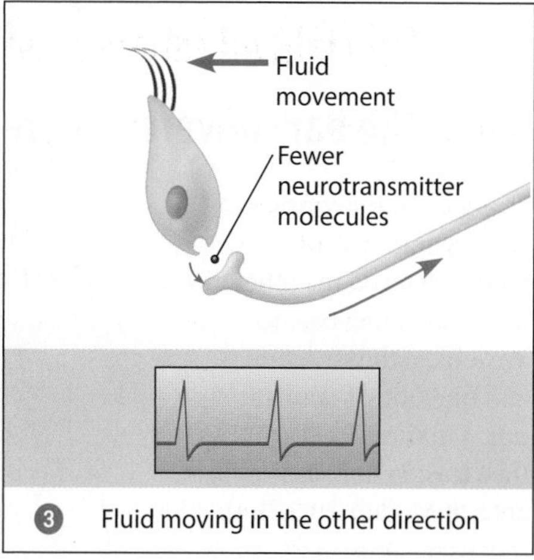
Fluid movement
Fewer neurotransmitter molecules
3 Fluid moving in the other direction

▲ **Figure 29.3C** Mechanoreception by a hair cell

Pain Receptors In humans and most other mammals, the skin has the highest density of **pain receptors** (see pink label in Figure 29.3A). Pain receptors may respond to excess heat or pressure or to chemicals released from damaged or inflamed tissues. Pain often indicates injury or disease and usually makes an animal withdraw to safety. Chemicals produced in an animal's body sometimes increase pain. For example, damaged tissues produce prostaglandins, which increase pain by increasing the sensitivity of pain receptors. Aspirin and ibuprofen reduce pain by inhibiting prostaglandin synthesis.

Chemoreceptors **Chemoreceptors** include the sensory receptors in our nose and taste buds, which are attuned to chemicals in the external environment, as well as some receptors that detect chemicals in the body's internal environment. For example, internal chemoreceptor sensors in some of our arteries can detect changes in the amount of oxygen (O_2) in the blood. Osmoreceptors in the brain detect changes in the total solute concentration of the blood and stimulate thirst when blood osmolarity increases.

One of the most sensitive chemoreceptors in the animal kingdom is inside nares, the nostril-like openings in sharks (**Figure 29.3D**). These legendary chemoreceptors can detect about a drop of blood in a billion drops of water! Not surprisingly, much of their total brain mass is dedicated to this sense (as much as 14% in some sharks). Sharks use their sense of smell to hunt by swimming toward the odor coming from the prey. Sharks become even more efficient at hunting by combining their sense of smell with their ability to detect water movements via mechanoreceptors known as the lateral line system, which is a group of sensory organs running along the length of their body.

Electromagnetic Receptors Energy occurring as electricity, magnetism, or various wavelengths of light may be detected by **electromagnetic receptors**. A platypus, for example, has electroreceptors on its bill that can detect electric fields generated by the muscles of prey, such as crustaceans, frogs, and small fishes.

Like sea turtles, many animals appear to use Earth's magnetic field to orient themselves as they migrate. Migratory birds, fishes (such as salmon and trout), amphibians, and bees are believed to rely on magnetoreception to navigate relative to Earth's magnetic field—although, as we already learned, the precise physiological basis for this sense remains to be discovered.

Probably the most common electromagnetic receptors are **photoreceptors**; they detect the electromagnetic energy of light in the visible or ultraviolet part of the electromagnetic spectrum (see Figure 7.6A). A variety of light detectors have evolved in the animal kingdom. Despite their differences, however, all photoreceptors contain similar pigment molecules that absorb light, and evidence indicates that all photoreceptors may be homologous (share a common ancestry). We consider some of the different types of eyes found in animals in Module 29.7. But first, we will consider the senses of hearing and balance.

? **For each of the following senses in humans, identify the type of receptor: seeing, tasting, hearing, smelling.**

Photoreceptor; chemoreceptor; mechanoreceptor; chemoreceptor

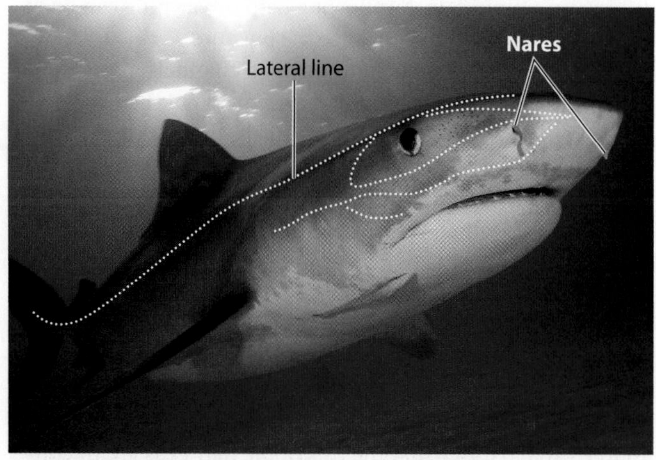
Nares
Lateral line

▲ **Figure 29.3D** Shark detecting scent and water movement to hunt

▷ Hearing and Balance

29.4 The ear converts air pressure waves to action potentials that are perceived as sound

The human ear consists of two separate organs, one for hearing and the other for maintaining balance. We look at the structure and function of our hearing organ in this module and then turn to our sense of balance in Module 29.5. Both organs operate on the same basic principle: the stimulation of long projections on hair cells (mechanoreceptors) in fluid-filled canals.

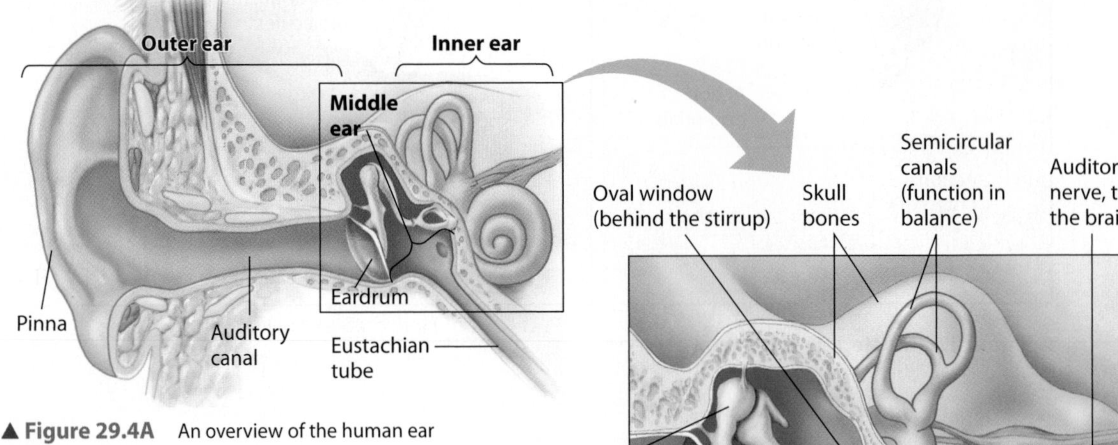

Outer ear
Inner ear
Middle ear
Eardrum
Pinna
Auditory canal
Eustachian tube
Oval window (behind the stirrup)
Skull bones
Semicircular canals (function in balance)
Auditory nerve, to the brain
Hammer
Anvil
Stirrup
Eardrum

▲ **Figure 29.4A** An overview of the human ear

Structure The ear has three regions: the outer ear, the middle ear, and the inner ear (**Figure 29.4A**). The **outer ear** consists of the flap-like **pinna**—the fleshy structure we commonly refer to as our "ear"—and the **auditory canal**. The pinna and the auditory canal collect sound waves and channel them to the **eardrum**, a sheet of tissue that separates the outer ear from the **middle ear** (**Figure 29.4B**). The outer ear and middle ear are common sites of the childhood infections swimmer's ear and otitis media, respectively.

When sound pressure waves strike the eardrum, the eardrum vibrates and passes the vibrations to three small bones: the hammer (more formally, the malleus), anvil (incus), and stirrup (stapes). The stirrup is connected to the oval window, a membrane-covered hole in the skull bone through which vibrations pass into the inner ear. The middle ear opens into a passage called the **Eustachian tube**, which connects with the pharynx (back of the throat), allowing air pressure to stay equal on either side of the eardrum. When pressure is not equal, a vacuum is created—stretching the eardrum in such a way that it cannot vibrate naturally (causing sounds to be muffled) and sometimes causing pain. Unequal pressure is experienced with changing altitude, such as when you're flying in an airplane or driving up a mountain. Swallowing hard or yawning can open the Eustachian tube and equalize the pressure.

The **inner ear** consists of fluid-filled channels in the bones of the skull. Sound vibrations or movements of the head set the fluid in motion. One of the channels, the **cochlea** (Latin for "snail"), is a long, coiled tube. The cross-sectional view of the cochlea in **Figure 29.4C** shows that inside it are three canals, each of which is filled with fluid. Our actual hearing organ, the **organ of Corti**, is located within the middle canal. The organ of Corti consists of an array of hair cells embedded in a **basilar membrane**, the floor of the middle canal. The hair cells are the sensory receptors of the ear. As you can see in the enlargement in **Figure 29.4D**, a jellylike projection called the tectorial membrane extends from the wall of the middle canal. Notice that the tips of the hair cells are in contact with the overlying tectorial membrane. Sensory neurons

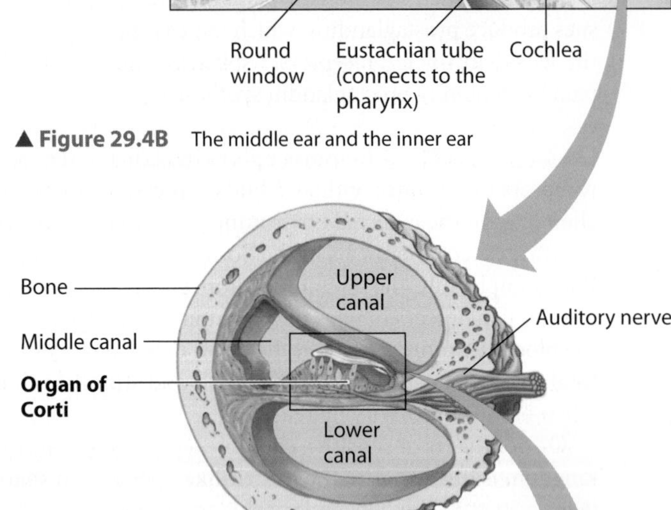

Round window
Eustachian tube (connects to the pharynx)
Cochlea
Bone
Upper canal
Auditory nerve
Middle canal
Organ of Corti
Lower canal

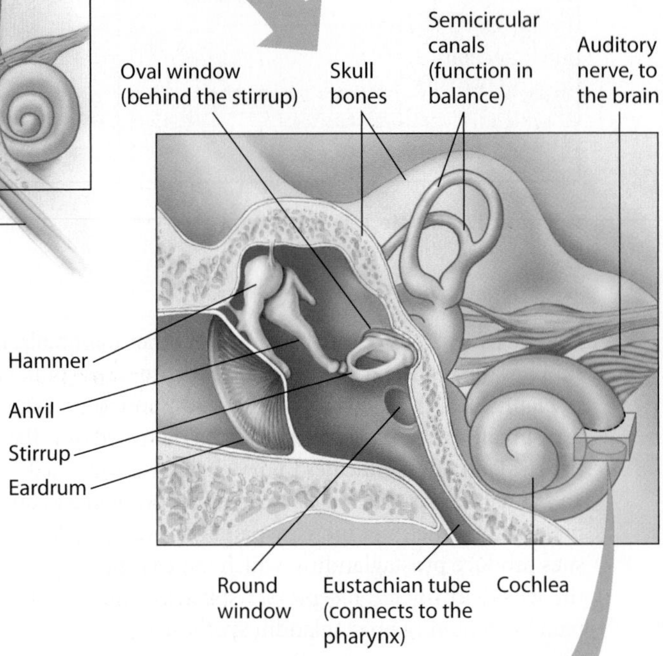

▲ **Figure 29.4B** The middle ear and the inner ear

▲ **Figure 29.4C** A cross section through the cochlea

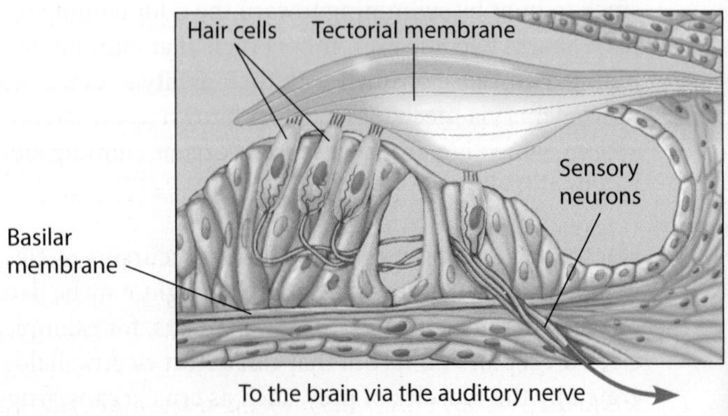

Hair cells
Tectorial membrane
Sensory neurons
Basilar membrane
To the brain via the auditory nerve

▲ **Figure 29.4D** The organ of Corti

synapse with the base of the hair cells and carry action potentials to the brain via the auditory nerve.

Hearing Now, let's see how the parts of the ear function in hearing. Sound waves, which move as pressure waves in the air, are collected by the pinna and auditory canal of the outer ear. These pressure waves make your eardrum vibrate with the same frequency as the sound **(Figure 29.4E)**. The frequency, measured in hertz (Hz), is the number of vibrations per second.

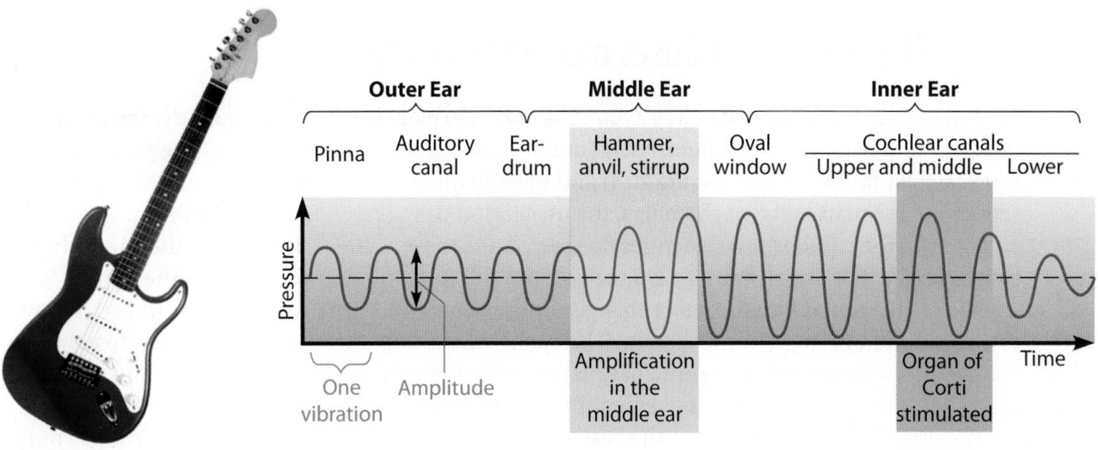

Outer Ear
Middle Ear
Inner Ear
Pinna
Auditory canal
Ear-drum
Hammer, anvil, stirrup
Oval window
Cochlear canals
Upper and middle
Lower
Pressure
Time
One vibration
Amplitude
Amplification in the middle ear
Organ of Corti stimulated

▲ **Figure 29.4E** The route of sound wave vibrations through the ear

Try This Tap your finger gently on a table to make a sound and then describe how the sound moves through your ear. Compare this to when you tap loudly.

From the eardrum, the vibrations are amplified approximately 20 times as they are transferred through the hammer, anvil, and stirrup in the middle ear to the oval window. Vibrations of the oval window then produce pressure waves in the fluid within the cochlea. (Without amplification of vibrations, pressure waves within the cochlea would be weaker.) Pressure waves travel from the oval window through the upper canal of the cochlea to the tip of the cochlea, at the coil's center.

As a pressure wave passes through the upper canal of the cochlea, it pushes downward on the middle canal, making the basilar membrane vibrate. Vibration of the basilar membrane makes the hairlike projections on the hair cells alternately brush against and draw away from the overlying tectorial membrane. When a hair cell's projections are bent, ion channels in its plasma membrane open, and positive ions enter the cell. As a result, the hair cell develops a receptor potential and releases more neurotransmitters at its synapse with a sensory neuron. In turn, the sensory neuron sends more action potentials to the brain through the auditory nerve. After causing the basilar membranes to vibrate, the pressure waves travel in a reverse direction through the lower canal and are dampened when they reach the round window (a membrane similar to the oval window).

Volume and Pitch The brain senses a sound as an increase in the frequency of action potentials from the auditory nerve. But how is the quality of the sound determined? The higher the volume (loudness) of sound, the higher the amplitude (height) of the pressure wave it generates. In the ear, a higher amplitude of pressure waves produces more vigorous vibrations of fluid in the cochlea, more pronounced bending of the hair cells, and thus more action potentials generated in the sensory neurons.

The pitch of a sound depends on the frequency of the sound waves. High-pitched sounds, such as high notes sung by a soprano, generate high-frequency waves. Low-pitched sounds, like the low notes made by a bass, generate low-frequency waves. How does the cochlea distinguish sounds of different pitch? The key is that the basilar membrane is not uniform along its length. The end near the oval window is relatively narrow and stiff, while the other end, near the tip of the cochlea, is wider and more flexible. As a result, the basilar membrane varies in its sensitivity to particular frequencies of vibration, and the region vibrating most vigorously at any instant sends the most action potentials to auditory centers in the brain. The brain interprets the information and gives us a perception of pitch. Young people can hear pitches in the range of 20–20,000 Hz. We gradually lose hair cells as we age, starting as early as age 8, and the hairs that detect high-frequency sounds are usually lost first. (You might try generating a variety of high-frequency sounds on a computer and determining which tones family members of different ages can hear.) Dogs can hear higher tones than humans, as high as 40,000 Hz, and bats can emit and hear clicking sounds as high-pitched as 100,000 Hz.

Deafness can be caused by the inability of the ear to conduct sounds, resulting from middle-ear infections, a ruptured eardrum, or stiffening of the middle-ear bones (a common age-related problem). Deafness can also result from damage to sensory receptors or neurons. Frequent or prolonged exposure to sounds of more than 90 decibels (dB) can damage or destroy hair cells, which are never replaced. Many people listen to music on electronic devices using ear buds or headphones at loud volumes, routinely exceeding 90 dB; this is especially true when they are in noisy places and want to get rid of background sound. Because few parts of our anatomy are more delicate than the organ of Corti, deafness is often progressive and permanent. In recent years, however, many deaf people have received cochlear implants, devices that convert sounds to electrical impulses, which stimulate the auditory nerve directly.

? **What causes bending of hair cells at different locations along the basilar membrane, and how do these differences affect perception?**

● Sounds of different frequencies will cause different regions of the basilar membrane to vibrate. When hair cells bend at various locations along the basilar membrane, different pitches are perceived.

29.5 The inner ear houses our organs of balance

Several organs in the inner ear allow you to sense movement, position, and balance. These fluid-filled equilibrium structures lie next to the cochlea **(Figure 29.5)** and include three semicircular canals and two chambers, the utricle and the saccule. All the equilibrium structures operate on the same principle: the bending of hairs on hair cells.

The **semicircular canals** detect changes in the head's rate of rotation or angular movement. The canals are arranged in perpendicular planes, allowing you to sense movement in all directions. A swelling at the base of each semicircular canal contains a cluster of hair cells with their hairs projecting into a gelatinous mass called a cupula. When you rotate your head in any direction, the thick, sticky fluid in the canals moves more slowly than your head. Consequently, the fluid presses against the cupula, bending the hairs. The faster you rotate your head, the greater the pressure and the higher the frequency of action potentials sent to the brain. If you rotate your head at a constant speed, the fluid in the canals begins moving with the head, and the pressure on the cupula is reduced. But if you stop suddenly, the fluid continues to move and stimulate hair cells, just as coffee continues to rotate in a cup even after you stop stirring. This explains why a situation that causes your inner-ear fluid to move—such as a whirling amusement park ride or even just twirling around quickly—makes you dizzy only after you try to stand still. Movement may be perceived even if a person is motionless, such is the case with excess alcohol consumption. Diffusing from the blood into the inner ear, alcohol decreases the density of the fluid. The cupula becomes heavy in the lighter fluid, and the hair cells bend. The perception may be that the "room is spinning" even when a person is lying in bed.

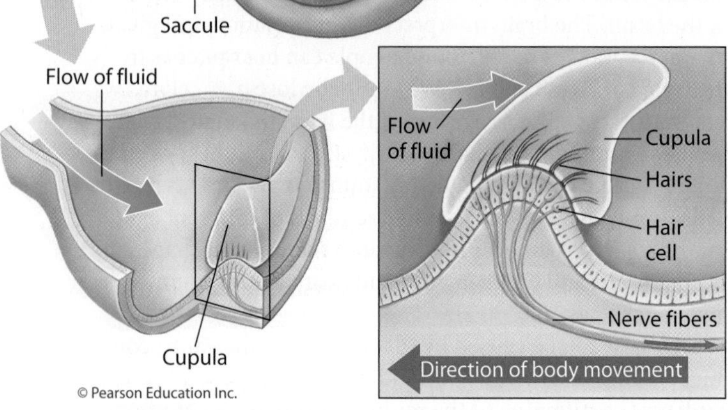

Semicircular canals
Nerve
Cochlea
Utricle
Saccule
Flow of fluid
Cupula
Flow of fluid
Cupula
Hairs
Hair cell
Nerve fibers
Direction of body movement

▲ **Figure 29.5** Equilibrium structures in the inner ear

Clusters of hair cells in the utricle and saccule chambers detect the position of the head with respect to gravity. The hairs of these cells project into a gelatinous material containing many small particles of calcium carbonate. When the position of the head changes, this heavy material bends the hairs in a different direction, causing an increase or decrease in the rate at which action potentials are sent to the brain. The brain determines the new position of the head by interpreting the altered flow of action potentials.

The equilibrium receptors provide data the brain needs to determine the position and movement of the head. Using this information, the brain develops and sends out commands that enable the skeletal muscles to balance the body.

? What type of receptor cell is common to our senses of hearing and equilibrium?

● Hair cells, which are mechanoreceptors

29.6 What causes motion sickness?

CONNECTION

Boating, flying, or even riding in a car can make us dizzy and nauseated, a condition called motion sickness. Some people can begin feeling ill simply at the thought of a boat or plane ride. Many others get sick only during storms at sea or during turbulence in flight. Motion sickness is believed to be caused when the brain receives signals from equilibrium receptors in the inner ear that conflict with visual signals from the eyes. When a susceptible person is inside a moving ship, for instance, signals from the equilibrium receptors in the inner ear indicate, correctly, that the body is moving (in relation to the environment outside the ship). In conflict with these signals, the eyes may tell the brain that the body is in a stationary environment, say, inside a cabin. Somehow the conflicting signals make the person feel ill. Symptoms may be relieved by closing the eyes, limiting head movements, or focusing on a stable horizon. Many sufferers of motion sickness take a sedative such as Dramamine to relieve their symptoms. The sedative inhibits input to the brain from the equilibrium sensors.

Because motion sickness can be a severe problem for astronauts, the National Aeronautics and Space Administration (NASA) conducts research on the problem. NASA has discovered that some people can learn to consciously control body functions, including the vomiting reflex. Astronauts receive intensive training in how to exert "mind over body" when zero gravity starts to induce motion sickness.

? Explain how someone could suffer motion sickness when watching a film shot from the front of a roller coaster.

● There would be conflicting information between vision ("I'm moving") and the equilibrium sense ("I'm sitting still").

Vision

29.7 Several types of eyes have evolved among animals

EVOLUTION CONNECTION

There is a great diversity in the organs animals use to perceive light (see Figure 15.12), but all animal light detectors contain photoreceptors, sensory cells that contain light-absorbing pigment molecules. Comparisons across animal species demonstrate unity in the underlying mechanism of light detection. Indeed, the genes that specify where and when photoreceptors arise during embryonic development are shared among flatworms, annelids, arthropods, and vertebrates.

An example of a simple light-detecting organ is the invertebrate **eyecup**, found in free-living flatworms called planarians **(Figure 29.7A)**. Planarian eyecups are concave, with photoreceptor cells that are partially shielded by darkly pigmented cells. Light can enter the eyecup only on the side where pigment cells do not block light. Because the openings of the two eyecups face in opposite directions, the animal can distinguish which direction light is coming from—useful for an animal needing to escape light and find a dark hiding place.

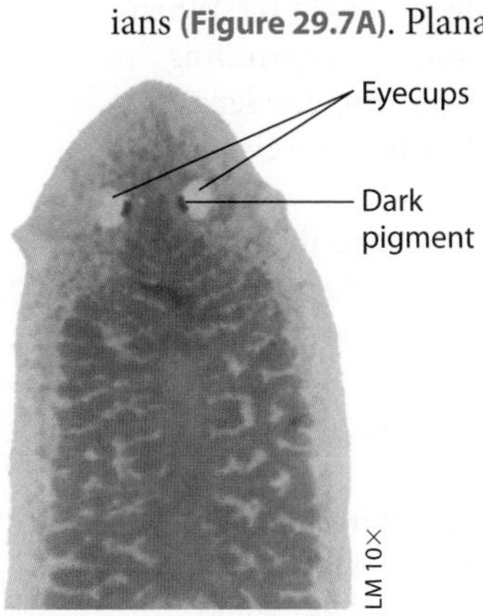

Eyecups

Dark pigment

LM 10×

▲ **Figure 29.7A** The eyecups of a planarian

Over the course of evolution, two types of more complex, image-forming eyes evolved. Compound eyes are found in invertebrates, specifically in insects and crustaceans. A **compound eye** consists of up to several thousand light-detectors called ommatidia **(Figure 29.7B)**. Every ommatidium has its own light-focusing lens and several photoreceptor cells. One ommatidium picks up light from only a tiny portion of the field of view. The animal's brain then forms a mosaic visual image by assembling the data from all the ommatidia. Compound eyes are extremely acute motion detectors, providing an important advantage for flying insects and other small animals often threatened by predators. The compound eyes of most insects also provide excellent color vision. Some species, such as honeybees, can see ultraviolet light (invisible to humans), which helps them locate certain nectar-bearing flowers.

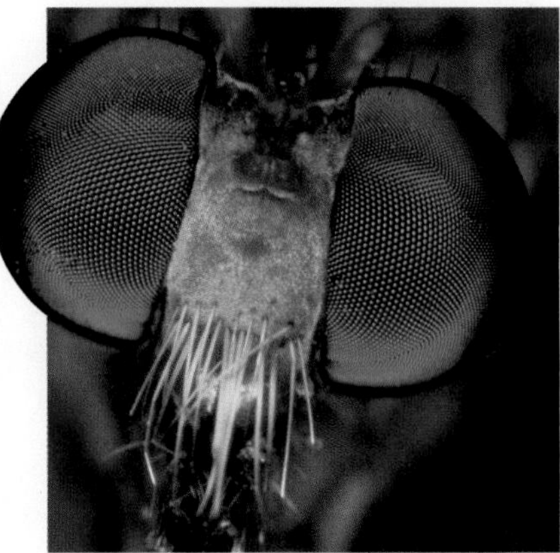

▲ **Figure 29.7B** The two compound eyes of a fly, each made up of thousands of ommatidia

The **single-lens eye** is the second type of image-forming eye, found in some invertebrates and all vertebrates. Single-lens eyes evolved independently in these two groups, originating from different tissues that were repurposed to become eyes.

We examine the human eye in **Figure 29.7C**. Light enters the eye through a small opening at the center of the eye, the **pupil**. Analogous to a camera's shutter, an adjustable doughnut-shaped **iris** changes the diameter of the pupil to let in more or less light. After going through the pupil, light passes through a single **lens**. The lens focuses light onto the **retina**, which consists of many photoreceptor cells. The photoreceptor cells are highly concentrated at the retina's center of focus, called the **fovea**. The retina's photoreceptor cells transduce light energy, and action potentials pass via sensory neurons in the optic nerve to the visual centers of the brain.

The repurposing of existing structures in evolution sometimes leads to apparent imperfections. Scientists have noted two peculiarities with the human eye. First, as light passes to the photoreceptors, it does not pass directly—it is scattered somewhat by a layer of neurons in the retina in front of the photoreceptors (see Figure 29.10B). Second, the optic nerve passes through a hole in the retina on its way to the brain, resulting in an area of the retina without photoreceptor cells (known as the "blind spot"). We cannot detect light that is focused on the blind spot, but having two eyes with overlapping fields of view enables us to perceive uninterrupted images. Would eyes function even better if the neurons and optic nerve were behind the photoreceptor cells? Probably. The invertebrate single-lens eye found in squids is structured this way, and it has no light scattering and no blind spot. Yet, our quite functional eyes demonstrate that anatomical structures need not be perfect.

? **What key optical feature is found in the eyes of both insects and squids but is not present in planarians?**

Lenses, which focus light onto photoreceptor cells

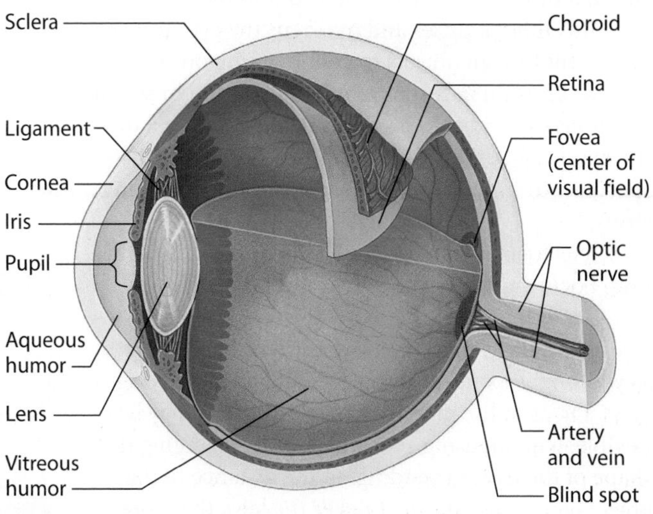

Sclera
Choroid
Retina
Ligament
Fovea (center of visual field)
Cornea
Iris
Optic nerve
Pupil
Aqueous humor
Lens
Artery and vein
Vitreous humor
Blind spot

▲ **Figure 29.7C** The single-lens eye of a vertebrate

29.8 Humans have single-lens eyes that focus by changing shape

The outer surface of the human eyeball is a tough, whitish layer of connective tissue called the **sclera** (see Figure 29.7C). At the front of the eye, the sclera becomes the transparent **cornea**, which lets light into the eye and also helps focus light. The sclera surrounds a pigmented layer called the **choroid**. The anterior choroid forms the iris, which gives the eye its color. After going through the pupil, the opening at the center of the iris, light passes through the disklike lens, which is held in position by ligaments. At the back of the eyeball is the retina, a layer just inside the choroid that contains photoreceptor cells.

The eye is divided into two fluid-filled chambers. The large chamber behind the lens is filled with jellylike **vitreous humor**. The much smaller chamber in front of the lens contains the thinner **aqueous humor**. The humors help maintain the shape of the eyeball. In addition, the aqueous humor circulates through its chamber, supplying nutrients and oxygen to the lens, iris, and cornea and carrying off wastes. Blockage of the ducts that drain this fluid can cause glaucoma, increased pressure inside the eye that may lead to blindness. If diagnosed early, glaucoma can be treated with medications that increase the circulation of aqueous humor.

A thin mucous membrane helps keep the outside of the eye moist. This membrane, called the **conjunctiva**, lines the inner surface of the eyelids and folds back over the white of the eye (but not the cornea). An infection or allergic reaction may cause inflammation of the conjunctiva, a condition called conjunctivitis, or "pink eye." Bacterial conjunctivitis usually clears up with antibiotic eyedrops. Viral conjunctivitis usually clears up on its own, although it is very contagious, especially among young children. If you are in contact with someone who has conjunctivitis, the best way to avoid spreading the disease is through careful hygiene: Don't touch the affected area and sterilize with hot water and bleach anything that comes into contact with it. Wash your hands frequently and thoroughly.

A gland above the eye secretes tears, a dilute salt solution that is spread across the eyeball by blinking and that drains into ducts that lead into the nasal cavities. This fluid cleanses and moistens the eye surface to prevent foreign objects from remaining on the eye. Excess secretion in response to eye irritation or strong emotions causes tears to spill over the eyelid and fill the nasal cavities, producing sniffles. Some scientists speculate that emotional tears play a role in reducing stress.

A lens focuses light onto a retina by bending light rays. Focusing can occur in two ways. The lens may be rigid, as in squids and many fishes. In this case, focusing occurs as muscles move the lens back or forth, as you might focus on an object using a magnifying glass. Or, as in the mammalian eye, the lens may be flexible, with focusing accomplished by changing the shape of the lens, depending on the distance to the object being viewed. The thicker the lens, the more sharply it bends light.

The shape of the mammalian lens is controlled by the muscles attached to the choroid and the ligaments that suspend the lens (**Figure 29.8**). When the eye focuses on a nearby object, these muscles contract, pulling the choroid toward the lens, which reduces the tension on the ligaments. As these ligaments slacken, the elastic lens becomes thick and round, as shown in the top diagram of Figure 29.8. This change, called accommodation, allows the diverging light rays from a close object to be bent and focused. The light rays from the object actually cross after they pass through the lens, resulting in an upside-down image striking the retina. (The brain interprets this inverted image so that we perceive it as right-side up.)

Light from distant objects approaches in parallel rays that require less bending for proper focusing on the retina. When the eye focuses on a distant object, the muscles controlling the lens relax, and the choroid moves away from the lens. This puts tension on the ligaments and flattens the elastic lens, as shown in the bottom diagram.

Flattening the lens makes it less thick, and so the light is not bent as much. This allows a distant object to be placed into focus. At least, that is how the human eye is *supposed* to work. In the next module, we consider ways that this mechanism can fail and ways to correct such failures.

? **Arrange the following eye parts into the correct order in which they are encountered by photons of light traveling into the eye: pupil, retina, cornea, lens, vitreous humor, aqueous humor.**

Cornea → aqueous humor → pupil → lens → vitreous humor → retina

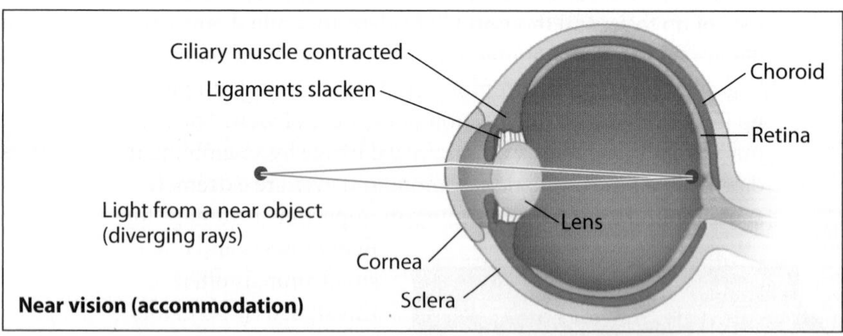
Ciliary muscle contracted
Ligaments slacken
Choroid
Retina
Light from a near object
(diverging rays)
Lens
Cornea
Sclera
Near vision (accommodation)

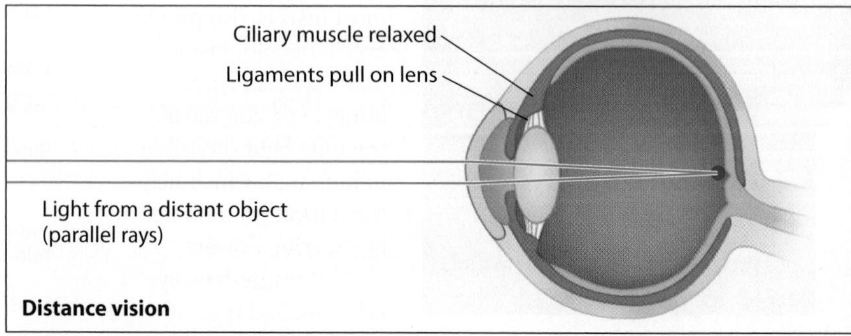
Ciliary muscle relaxed
Ligaments pull on lens
Light from a distant object
(parallel rays)
Distance vision

▲ **Figure 29.8** How human lenses focus light

Try This Stare at some words on this page and then gaze at something distant. Explain how your eyes adjusted in these two situations.

29.9 Artificial lenses or surgery can correct focusing problems

CONNECTION

Reading from an eye chart measures your **visual acuity**, the ability of your eyes to distinguish fine detail. When you have your eyes tested, the examiner asks you to read a line of letters sized for legibility at a distance of 20 feet, using one eye at a time. If you can do this accurately, you have normal (20/20) acuity in each eye. This means that from a distance of 20 feet, each of your eyes can read the chart's line of letters designated for 20 feet.

Visual acuity of 20/10 is better than normal and means that you can read letters from a distance of 20 feet that a person with 20/20 vision can only read at 10 feet. On the other hand, visual acuity of 20/50 is worse than normal. In this circumstance, you would have to stand at a distance of 20 feet to read what a person with normal acuity can read at 50 feet.

Three of the most common vision problems are nearsightedness, farsightedness, and astigmatism. All three are focusing problems, easily corrected with artificial lenses. People with **nearsightedness** cannot focus well on distant objects, although they can see well at short distances (the condition is named for the type of vision that is unimpaired). A nearsighted eyeball **(Figure 29.9A)** is longer than normal. The lens cannot flatten enough to compensate, and it focuses distant objects in front of the retina instead of on it. Nearsightedness (also known as myopia) is corrected by glasses or contact lenses that are thinner in the middle than at the outside edge. The lenses make the light rays from distant objects diverge as they enter the eye. The focal point formed by the lens in the eye then falls on the retina.

Farsightedness (also known as hyperopia) is the opposite of nearsightedness. It occurs when the eyeball is shorter than normal, causing the lens to focus images behind the retina **(Figure 29.9B)**. Farsighted people see distant objects normally but cannot focus on close objects. Corrective lenses that are thicker in the middle than at the outside edge compensate for farsightedness by making light rays from nearby objects converge slightly before they enter the eye. Another type of farsightedness, called presbyopia, develops with age. Beginning around the mid-40s, the lens of the eye becomes less elastic. As a result, the lens gradually loses its ability to focus on nearby objects, and reading without glasses becomes difficult.

Astigmatism is blurred vision caused by a misshapen lens or cornea. This makes light rays converge unevenly and not focus at one point on the retina. Corrective lenses are asymmetric in a way that compensates for the asymmetry in the eye.

Surgical procedures are an option for treating vision disorders. In laser-assisted in situ keratomileusis (LASIK), a laser is used to reshape the cornea and change its focusing ability. Close to 1 million LASIK procedures are performed each year to correct a variety of vision problems.

? **A person with 20/100 vision in both eyes must stand at __________ feet to read what someone with normal vision can read at __________ feet. Is this better or worse than normal acuity?**

20 . . . 100; worse

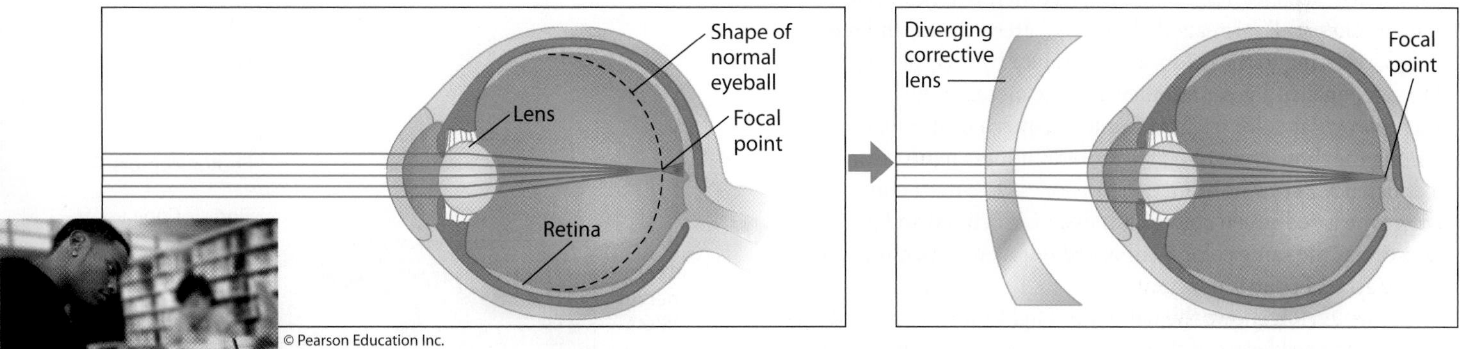
Shape of normal eyeball
Lens
Focal point
Retina
Diverging corrective lens
Focal point

▲ **Figure 29.9A** A nearsighted eye (eyeball too long)

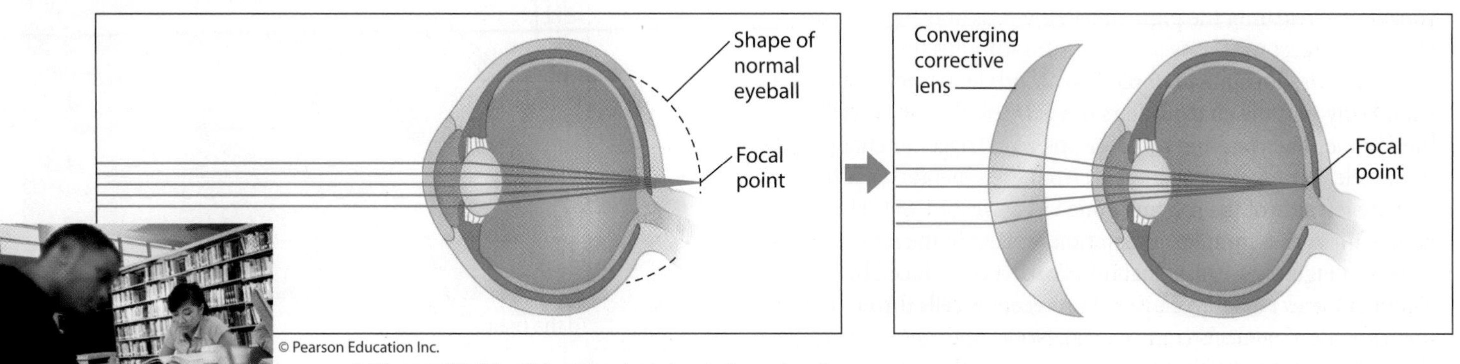
Shape of normal eyeball
Focal point
Converging corrective lens
Focal point

▲ **Figure 29.9B** A farsighted eye (eyeball too short)

29.10 The human retina contains two types of photoreceptors: rods and cones

The human retina contains two types of photoreceptors named for their shapes **(Figure 29.10A)**. **Cones** are stimulated by bright light and can distinguish color, but they contribute little to night vision. **Rods** are extremely sensitive to light and enable us to see in dim light, though only in shades of gray. The relative numbers of rods and cones an animal has correlates with whether an animal is most active during the day or night. Diurnal animals, like humans, have more cones than rods, allowing them to see well during the day and less well in the dark. Nocturnal animals have many rods and few (or no) cones, allowing for excellent night vision but often colorblindness.

In humans, rods are found in greatest density at the outer edges of the retina and are completely absent from the fovea, the retina's center of focus. If you look directly toward a dim star in the night sky, the star is hard to see. Looking just to the side, however, makes your lens focus the starlight onto the parts of the retina with the most rods, and you can see the star better. By contrast, you achieve your sharpest day vision by looking straight at the object of interest. This is because cones are densest (about 150,000 per square millimeter) in the fovea.

As Figure 29.10A shows, each rod and cone consist of an array of membranous disks containing light-absorbing visual pigments. Rods contain a visual pigment called **rhodopsin**. (Rhodopsin is derived from vitamin A, which is why vitamin A deficiency can cause "night blindness.") Interestingly, a change in a single amino acid of rhodopsin causes it to react to light in the ultraviolet range; this difference allows many species of birds to see beyond the range of human vision.

Cones contain visual pigments called **photopsins**. We have three types of cones, each containing a different type of photopsin. These cells are called blue cones, green cones, and red cones, referring to the colors absorbed best by their photopsin. We can perceive a great number of colors because the light from each particular color triggers a unique pattern of stimulation among the three types of cones. Colorblindness results from a deficiency in one or more types of cones. The most common type is red-green colorblindness, in which red and green are seen as the same color—either red or green, depending on which type of cone is deficient.

Figure 29.10B shows the pathway of light into the eye and through the cell layers of the retina. Notice that the tips of the rods and cones are embedded in the back of the retina. Light must pass through several relatively transparent layers of neurons before reaching the pigments in the rods and cones. Like all sensory receptors, rods and cones are stimulus transducers. When rhodopsin and photopsin absorb light, they change chemically, and the change alters the permeability of the cell's membrane. The resulting receptor potential triggers a change in the release of neurotransmitters from the synaptic terminals (see the left side of the rod and cone in Figure 29.10A). This release initiates a complex integration process in the retina. As shown in Figure 29.10B, visual information transduced by the rods and cones passes from the photoreceptor cells through the network of neurons (red arrows). Notice the numerous synapses between the photoreceptor cells and the neurons

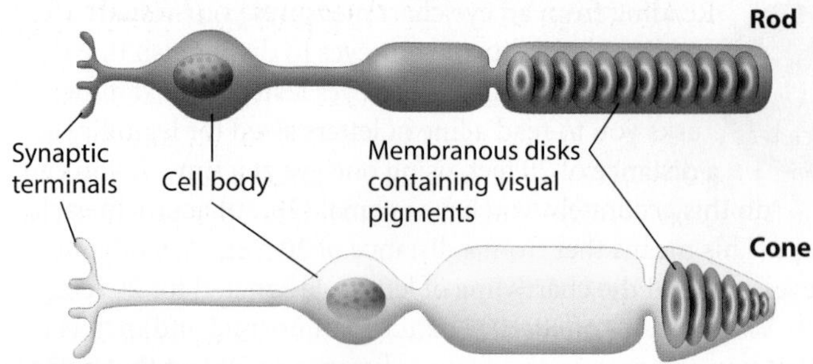
Rod
Synaptic terminals
Cell body
Membranous disks containing visual pigments
Cone

▲ **Figure 29.10A** Photoreceptor cells

and among the neurons themselves. Integration in this maze of synapses helps sharpen images and increases the contrast between light and dark areas. Action potentials carry the partly integrated information to the brain via the optic nerve. Three-dimensional perceptions (what we actually see) result when visual input coming from the two eyes is integrated further in several processing centers of the cerebral cortex.

? **Explain why our night vision is mostly in shades of gray rather than in color.**

● Rods are more sensitive than cones to light, and thus low-intensity light stimulates far more rods than cones. Rods do not detect color.

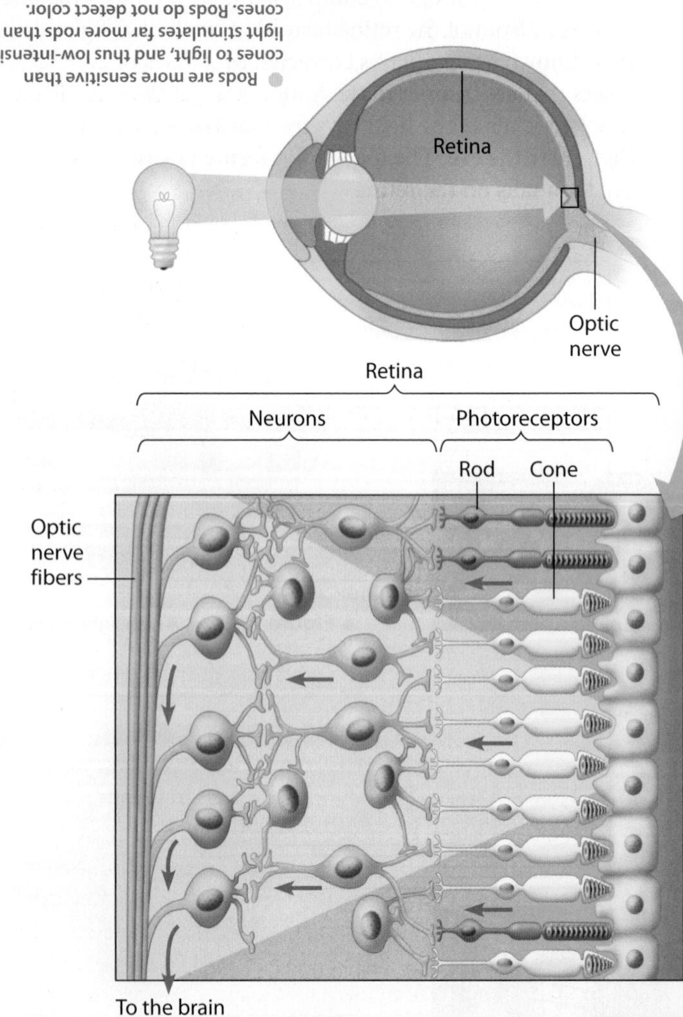
Retina
Optic nerve
Retina
Neurons
Photoreceptors
Rod
Cone
Optic nerve fibers
To the brain

▲ **Figure 29.10B** The vision pathway from light source to optic nerve

Taste and Smell

29.11 Taste and odor receptors detect chemicals present in solution or air

Your senses of smell and taste depend on receptor cells that detect chemicals in the environment. Chemoreceptors in your taste buds detect molecules in solution; chemoreceptors in your nose detect airborne molecules that dissolve into the mucus that coats the nasal cavity.

Olfactory (smell) receptors are sensory neurons that line the upper nasal cavity, sending impulses along axons directly to the olfactory bulb of the brain **(Figure 29.11)**. Cilia extend from the tips of these chemoreceptors into the nasal mucus. When an odorous substance (.•.) diffuses into this region and dissolves in the mucus, it can bind to specific receptor proteins on the cilia. The binding triggers a membrane depolarization and generates action potentials (red arrows). As the signals are integrated in the brain, we perceive odor. Humans can distinguish thousands of different odors.

Many animals rely heavily on their sense of smell for survival. Most other mammals have a much more discriminating sense of smell than humans. Odors often provide more information than visual images about food, the presence of mates, or danger. In contrast, humans often pay more attention to sights and sounds than to smells. ("Seeing is believing" is thus a very human-centric idea!)

Receptor cells for taste in mammals are sensory cells organized into taste buds on the tongue (Figure 29.1). In addition to the four familiar taste perceptions—sweet, sour, salty, and bitter—a fifth, called *umami* (Japanese for "delicious"), is elicited by glutamate, an amino acid. Umami describes the savory flavor common in meats, cheeses, and other protein-rich foods, as well as the flavor-enhancing chemical monosodium glutamate (MSG). Any region of the tongue with taste buds can

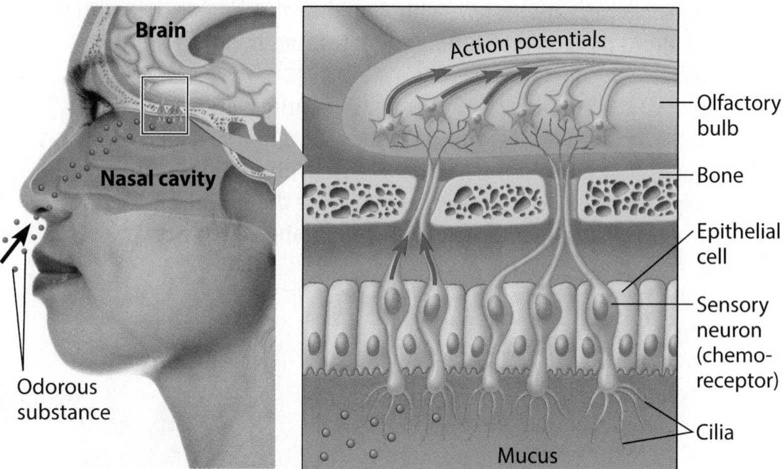
Brain
Nasal cavity
Odorous substance
Action potentials
Olfactory bulb
Bone
Epithelial cell
Sensory neuron (chemoreceptor)
Cilia
Mucus

▲ **Figure 29.11** Smell in humans

detect any of the five types of taste. ("Taste maps" of the tongue can thus be misleading.) An individual taste cell has a single type of receptor and thus detects only one of the five tastes.

Although the receptors and brain pathways for taste and smell are independent, the two senses do interact. Indeed, much of what we call taste is really smell, as you've probably noticed when a stuffy head cold dulls your perception of taste.

? What is the key structural difference between taste receptors and olfactory receptors in terms of cellular structure? (*Hint*: Review Module 29.1.)

● Taste receptors release neurotransmitters, triggering action potentials carried to the brain by sensory neurons. Olfactory receptors are themselves modified sensory neurons.

29.12 "Supertasters" have a heightened sense of taste

CONNECTION

If you don't like your vegetables (and we mean really don't like your vegetables), there may be a genetic basis behind your aversion. About 25% of us are "supertasters." For reasons that are not clearly understood, supertasters have three times the sensitivity to bitter tastes (as well as other tastes) than other people do. Supertasters tend to perceive bitter tastes in more foods than do normal tasters. Hence, supertasters typically have more food dislikes, often avoiding coffee, alcoholic beverages, fatty foods, vegetables such as broccoli, and other common foods.

Are you a supertaster? There are two signs: hypersensitivity to a bitter-tasting chemical called propylthiouracil and more than normal numbers of fungiform papillae, the structures on your tongue that house taste buds. Try this: Punch a hole in a small piece of paper (with a hole punch), and place the paper on the tip of your tongue. Looking at it in a mirror, count the number of papillae you can see in the hole. On average, supertasters have 35, whereas nontasters have less than 15. (Coating your tongue with blue food dye first will make it easier to see the papillae.)

There are consequences to being a supertaster beyond matters of food preference. An aversion to fatty foods may offer health benefits to supertasters (such as reduced risk of heart disease). But, perhaps because they tend to avoid somewhat bitter but healthy vegetables and other foods, supertasters may have a higher risk of colon cancer and other health problems. In any case, taste-related food choices can have health implications.

? Why might being a supertaster be harmful to your health?

● Supertasters tend to avoid slightly bitter but healthful vegetables.

29.13 Review: The central nervous system couples stimulus with response

In this chapter and the previous one, we focused on information gathering and processing. Sensory receptors provide an animal's nervous system with vital data that enable the animal to avoid danger, find food and mates, and maintain homeostasis—in short, to survive.

We can summarize the sequence of information flow in animals by considering your own body's reaction to a loud noise such as a taxi honking **(Figure 29.13)**. Within milliseconds, receptor cells in your inner ears transduce the sound, and action potentials conduct the signal to your brain. A vast network of neurons in your brain, with millions of synapses, integrates the information and sends out command signals, again in the form of action potentials. The commands go out via motor neurons to muscles, and you turn your head to look in the direction of the noise. (In the next chapter, we will see how muscles carry out the commands they receive from the nervous system.)

? What three general types of neurons are involved when you hear a noise and turn your head? (*Hint*: Review Module 28.1.)

Sensory neurons, interneurons, and motor neurons

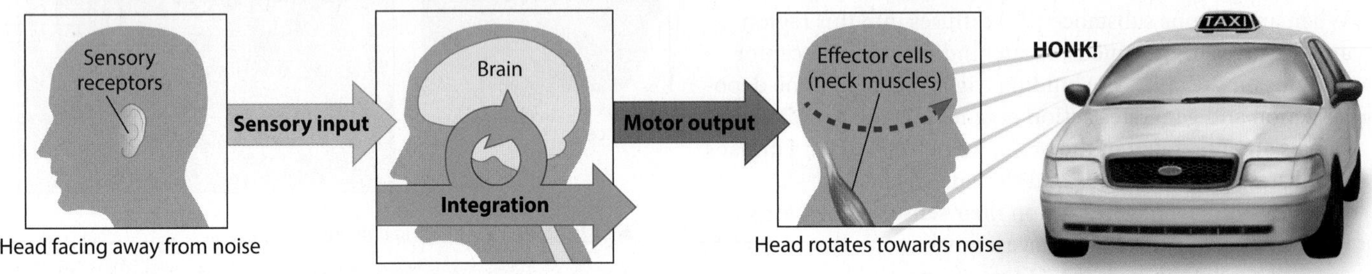

Sensory receptors
Sensory input
Brain
Integration
Motor output
Effector cells (neck muscles)
TAXI
HONK!
Head facing away from noise
Head rotates towards noise

▲ **Figure 29.13** Three interconnected functions of the nervous system

CHAPTER 29 REVIEW

For practice quizzes, BioFlix animations, MP3 tutorials, video tutors, and more study tools designed for this textbook, go to MasteringBiology®

Reviewing the Concepts

Sensory Reception (29.1–29.3)

29.1 Sensory receptors convert stimulus energy to action potentials. Sensory receptors are specialized cells that detect stimuli. Sensory transduction converts stimulus energy to receptor potentials, which trigger action potentials that are transmitted to the brain. Action potential frequency reflects stimulus strength.

29.2 The model for magnetic sensory reception is incomplete.

29.3 Specialized sensory receptors detect five categories of stimuli. Pain receptors sense dangerous stimuli. Thermoreceptors detect heat or cold. Mechanoreceptors respond to mechanical energy (such as touch). Chemoreceptors respond to chemicals. Electromagnetic receptors respond to electricity, magnetism, and light.

Hearing and Balance (29.4–29.6)

29.4 The ear converts air pressure waves to action potentials that are perceived as sound. The human ear channels sound waves through the outer ear to the eardrum to a chain of bones in the middle ear to the fluid in the coiled cochlea in the inner ear.

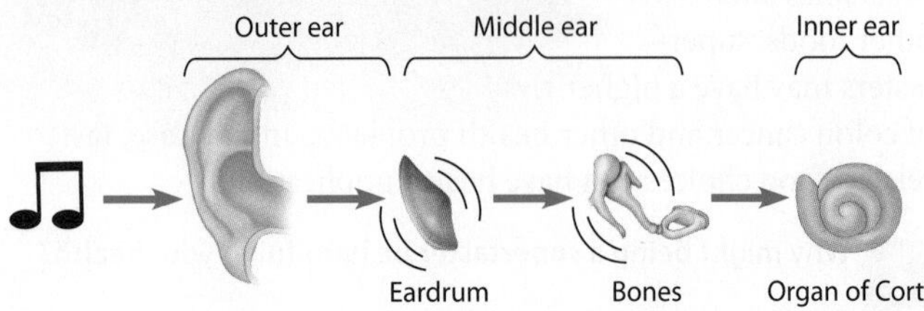

Outer ear
Middle ear
Inner ear
Eardrum
Bones
Organ of Corti

Pressure waves in the fluid bend hair cells of the organ of Corti against a membrane, triggering nerve signals to the brain. Louder sounds generate more action potentials; pitches stimulate different regions of the organ of Corti.

29.5 The inner ear houses our organs of balance. The semicircular canals and utricle and saccule located in the inner ear sense body position and movement.

29.6 What causes motion sickness? Conflicting signals from the inner ear and eyes may cause motion sickness.

Vision (29.7–29.10)

29.7 Several types of eyes have evolved among animals. Animal eyes range from simple eyecups that sense light intensity and direction to the many-lensed compound eyes of insects to the single-lens eyes of squids and vertebrates, including humans.

29.8 Humans have single-lens eyes that focus by changing shape. In the human eye, the cornea and flexible lens focus light on photoreceptor cells in the retina.

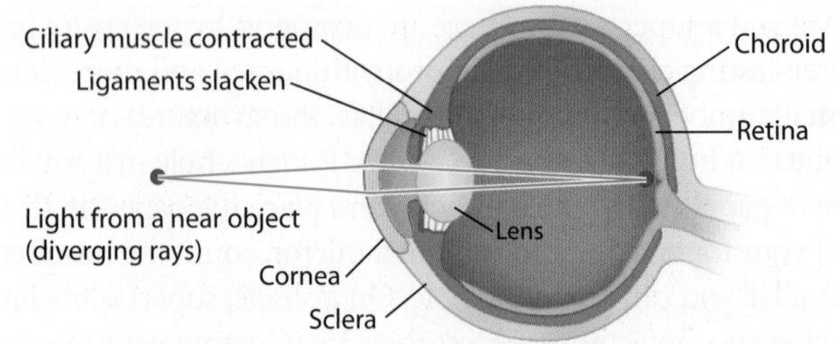

Ciliary muscle contracted
Ligaments slacken
Choroid
Retina
Light from a near object (diverging rays)
Lens
Cornea
Sclera

29.9 Artificial lenses or surgery can correct focusing problems. Focusing involves changing the shape of the lens. Nearsightedness and farsightedness result when the focal point is not on the retina. Corrective lenses bend the light rays to compensate.

29.10 The human retina contains two types of photoreceptors: rods and cones. Rods allow us to see shades of gray in dim light, and cones allow us to see color in bright light.

Taste and Smell (29.11–29.13)

29.11 Taste and odor receptors detect chemicals present in solution or air. Taste and smell depend on chemoreceptors that bind specific molecules. Taste receptors produce five taste sensations. Olfactory (smell) sensory neurons line the nasal cavity.

29.12 "Supertasters" have a heightened sense of taste.

29.13 Review: The central nervous system couples stimulus with response. The nervous system receives sensory information, integrates it, and commands appropriate muscle responses.

Connecting the Concepts

1. Complete this concept map summarizing sensory receptors.

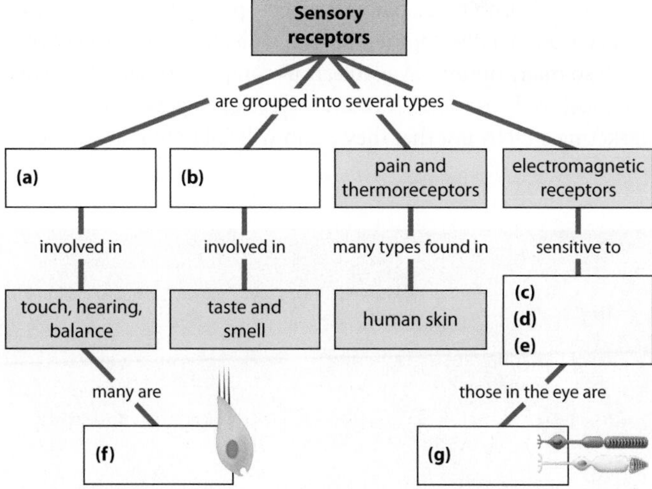
Sensory receptors
are grouped into several types
(a)
(b)
pain and thermoreceptors
electromagnetic receptors
involved in
involved in
many types found in
sensitive to
touch, hearing, balance
taste and smell
human skin
(c)
(d)
(e)
many are
those in the eye are
(f)
(g)

Testing Your Knowledge

Level 1: Knowledge/Comprehension

2. Which of the following sensory receptors is incorrectly paired with its category?
   a. hair cell . . . mechanoreceptor
   b. taste receptor . . . chemoreceptor
   c. rod . . . electromagnetic receptor
   d. olfactory receptor . . . electromagnetic receptor
3. Which of the following are not known to be present in human skin?
   a. thermoreceptors
   b. electromagnetic receptors
   c. pressure receptors
   d. pain receptors
4. What do the receptor cells on the skin of a fish and the cochlea of your ear have in common?
   a. They use hair cells to sense sound or pressure waves.
   b. They are organs of equilibrium.
   c. They use electromagnetic receptors to sense pressure waves in fluid.
   d. They use granules that signal a change in position and stimulate their receptor cells.
5. If you look away from this book and focus your eyes on a distant object, the eye muscles ______________ and the lenses ______________ to focus images on the retinas.
   a. relax . . . flatten
   b. relax . . . become more rounded
   c. contract . . . flatten
   d. contract . . . become more rounded
6. How does your brain determine the volume and pitch of sounds?

Level 2: Application/Analysis

7. Eighty-year-old Mr. Johnson was becoming slightly deaf. To test his hearing, his doctor held a vibrating tuning fork tightly against the back of Mr. Johnson's skull. This sent vibrations through the bones of the skull, setting the fluid in the cochlea in motion. Mr. Johnson could hear the tuning fork this way, but not when it was held away from the skull a few inches from his ear. The problem was probably in the ______________. (Explain your answer.)
   a. auditory nerve leading to the brain
   b. hair cells in the cochlea
   c. bones of the middle ear
   d. fluid of the cochlea
8. A cataract occurs when proteins in the lens clump together and cause the lens to become cloudy. Thus cataracts likely affect vision by
   a. decreasing the sharpness of the image on the retina.
   b. reducing the amount of light that passes through the cornea.
   c. preventing the iris from changing the diameter of the pupil.
   d. all of the above.
9. Hold your right eye closed. With your left eye, look at the + in the image below. Starting from about two feet away, slowly bring your head closer while looking at the +. What happens to the dot when you get close to the image? What property of the eye's structure does this exercise demonstrate?

Level 3: Synthesis/Evaluation

10. Construct a graph in which membrane potential is on the *y* axis and time is on the *x* axis. Draw the action potentials that result when a supertaster puts a strip of paper with the bitter chemical propylthiouracil on his or her tongue. Draw a second graph for a nontaster.
11. **SCIENTIFIC THINKING** We know that sea turtle hatchlings use Earth's magnetic field to navigate. Do they also use light cues from the moon to get from the sand to the ocean waves? Outline an experiment to answer this question.
12. Have you ever felt your ears ringing after listening to loud music? Can this permanently impair your hearing? Should manufacturers of music devices and producers of concerts be required to warn consumers? What effect might warnings have?

Answers to all questions can be found in Appendix 4.

CHAPTER

30 How Animals Move

Did you play sports as a child? How young did were you when you started? Tee ball, soccer, football, gymnastics, swimming, tennis—lessons and leagues for preschoolers, such as the tiny soccer players in the photo below, abound. The number of young athletes continues to increase as children's physical skills develop and their interests broaden, and many children play multiple sports. Daydreams of future athletic fame and glory are a pleasant diversion for these budding athletes, and many parents secretly harbor similar hopes for their offspring. If not an Olympic medal or lucrative professional career, parents may envision a full-ride college scholarship. Even in the absence of such ambitions, parents expect sports to have a positive influence on their children's physical and emotional development. Meanwhile, the commitment of time and money can be substantial. With so many options available, how can parents help their children focus on a sport in which they are most likely to excel? A recently developed genetic test may offer an answer. Entrepreneurs are marketing a DNA test that they claim will tell potential athletes

? *Can a genetic test predict athletic ability?*

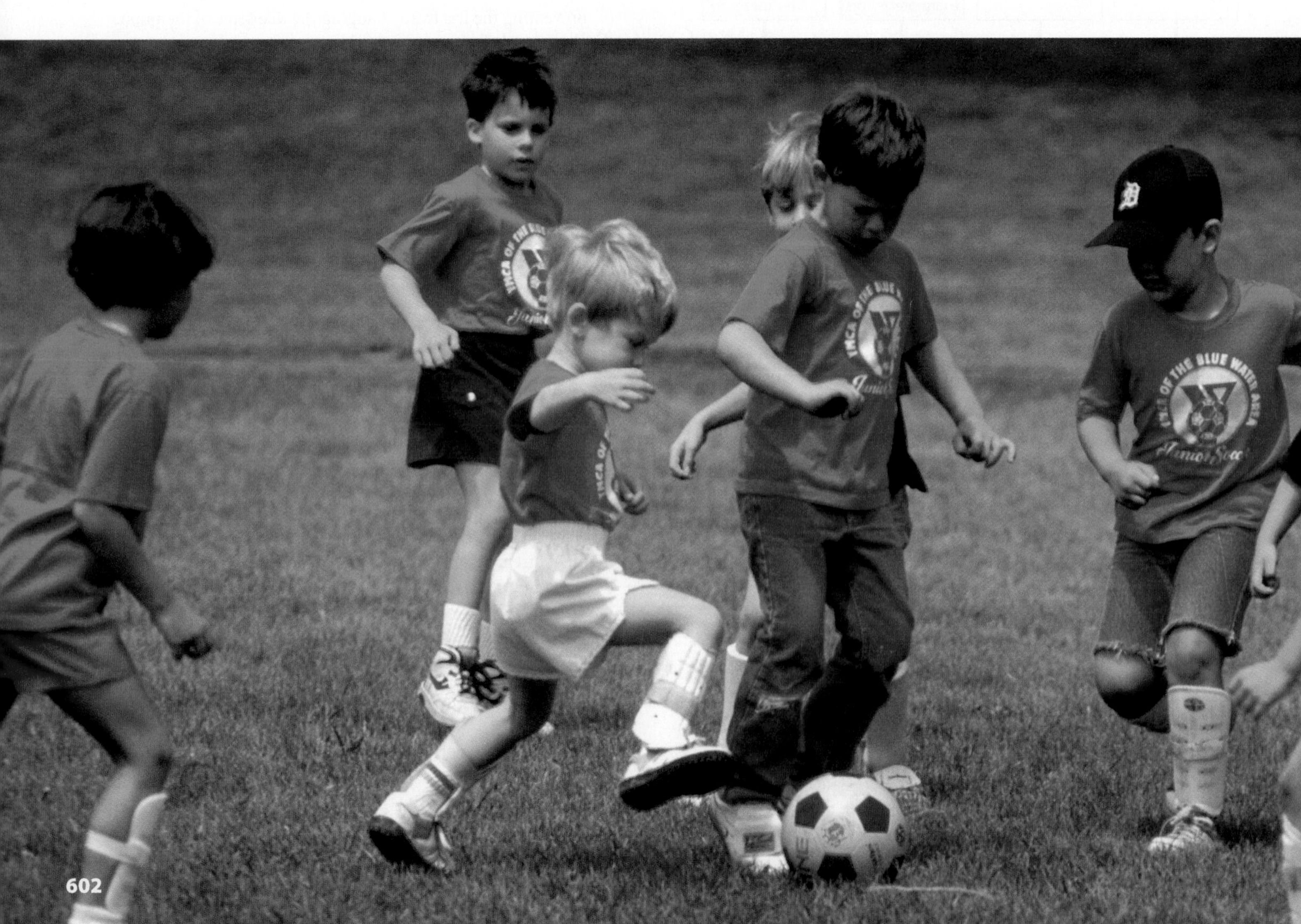

what type of sports activities they can expect to perform most successfully. After learning about muscle and skeletal systems—the machinery of athletic performance—you'll learn about the scientific basis for this test.

Indeed, muscle and skeletal systems, the subjects of this chapter, are the machinery of all movement. These systems are not unique to humans. Most animals have the ability to move from one place to another. Some, like cheetahs, sailfish, darners (large dragonflies), and the aptly named swifts (birds), are capable of great speed. Others—snails and tortoises, for example—are exceedingly slow. Modes of travel are diverse, too. In addition to animals that run or walk, there are animals that crawl, hop, fly, or swim. Almost all modes demonstrate variations on a common theme: muscles working in partnership with a skeletal system.

BIG IDEAS

Movement and Locomotion
(30.1–30.2)

Locomotion requires the collaboration of muscles and skeleton.

The Vertebrate Skeleton
(30.3–30.6)

Vertebrates have similar skeletons, but modifications of the ancestral body plan resulted in structural adaptations for different functions.

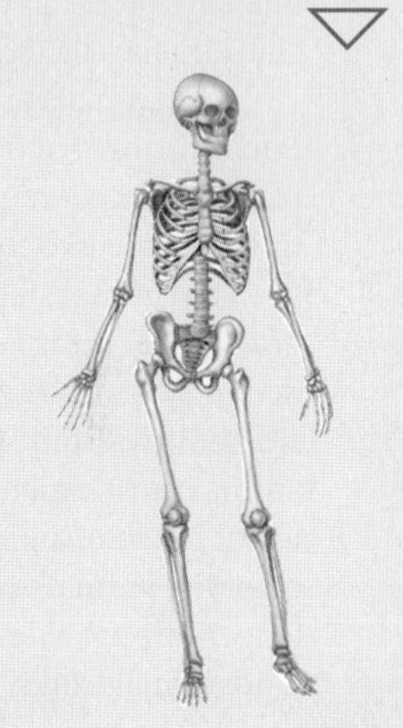

Muscle Contraction and Movement
(30.7–30.12)

Muscles work with the skeleton to move the body. Each muscle cell has its own contractile apparatus.

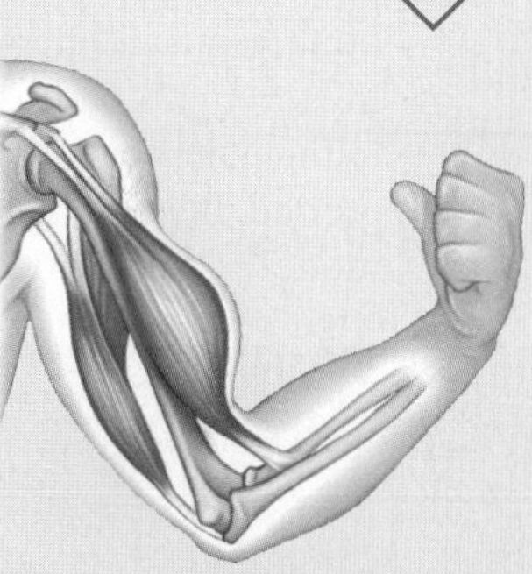

Movement and Locomotion

30.1 Locomotion requires energy to overcome friction and gravity

Movement is a distinguishing characteristic of animals. Even animals that are attached to a substrate move their body parts. All types of animal movement have underlying similarities. At the cellular level, every form of movement involves protein strands moving against one another, an energy-consuming process. In muscle cell contraction and amoeboid movement, the cellular system is based on microfilaments. Microtubules are the main components of cilia and flagella (see Module 4.18).

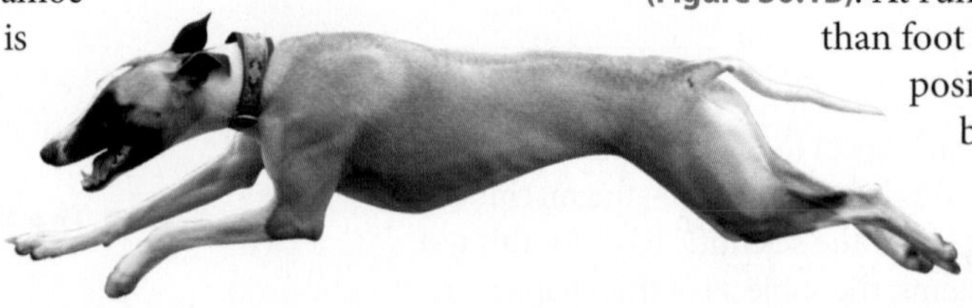

Most animals are fully mobile. **Locomotion**—active travel from place to place—requires energy to overcome two forces that tend to keep an animal stationary: friction and gravity. The relative importance of these two forces varies, depending on the environment. Water is dense and offers considerable resistance to a body moving through it, so an aquatic animal must expend energy overcoming friction. Gravity is not much of a problem, because water supports much or all of the animal's weight. On land, gravity is the main challenge—air provides no support for an animal's body. When a land animal walks, runs, or hops, its leg muscles expend energy both to propel it and to keep it from falling down. On the other hand, air offers very little resistance to an animal moving through it, at least at moderate speeds. Friction is limited to the points of contact between the animal's body and the ground. In the remainder of this module, we look at the major modes of animal locomotion.

▲ **Figure 30.1B** A dog running at full speed

▲ **Figure 30.1A** A fish swimming

Swimming Animals swim in diverse ways. Many insects, for example, swim the way we do, using their legs as oars to push against the water. Squids, scallops, and some jellies are jet-propelled, taking in water and squirting it out in bursts. Fishes swim by moving their body and tail from side to side (**Figure 30.1A**). Whales and other aquatic mammals move by undulating their body and tail from top to bottom. A sleek, streamlined shape, like that of seals, porpoises, penguins, and many fishes, is an adaptation that aids rapid swimming.

Walking and Running A walking animal moves each leg in turn, overcoming friction between the foot and the ground with each step. To maintain balance, a four-legged animal usually keeps three feet on the ground at all times when walking slowly. Bipedal (two-footed) animals, such as birds and humans, are less stable on land and keep part of at least one foot on the ground when walking. A running four-legged animal may move two or three legs with each stride. At some gaits, all of its feet may be off the ground simultaneously (**Figure 30.1B**). At running speeds, momentum, more than foot contact, stabilizes the body's position, just as it keeps a moving bicycle upright.

Hopping Some animals—for example, kangaroos—travel mainly by hopping (**Figure 30.1C**), a specialized mode of locomotion that has also evolved independently in several rodents. Large muscles in the hind legs of kangaroos generate a lot of power. Tendons (which connect muscle to bone) in the legs also momentarily store energy when the kangaroo lands—somewhat like the spring on a pogo stick. The higher the jump, the tighter the spring coils when a pogo stick lands and the greater the tension in the tendons when a kangaroo lands. In both cases, the stored energy is available for the next jump. For the kangaroo, the tension in its legs is a cost-free energy boost that reduces the total amount of energy the animal expends to travel. At rest, the kangaroo sits upright with its tail and both hind feet touching the ground. This position stabilizes the animal's body and costs little energy to maintain.

The pogo stick analogy applies to many other land animals as well. The legs of an insect, horse, or human, for instance, retain some spring during walking or running, although less than those of a hopping kangaroo.

Crawling Animals that have no limbs, or very short limbs, drag their bodies along the ground in a crawling movement. Because much of the animal's body is in contact with the ground, its energy is mainly expended to overcome friction rather than gravity. Many snakes crawl rapidly by undulating the entire body from side to side. Aided by large, movable

▲ **Figure 30.1C** Kangaroos hopping

scales on its underside, a snake's body pushes against the ground, driving the animal forward. Boa constrictors and pythons creep forward in a straight line, driven by muscles that lift belly scales off the ground, tilt them forward, and then push them backward against the ground.

Earthworms crawl by peristalsis, a type of movement produced by rhythmic waves of muscle contractions passing from head to tail. (In Module 21.6, you saw how peristalsis squeezes food through your digestive tract.) To move by peristalsis, an animal needs a set of muscles that elongates the body and another set that shortens it. Also required are a way to anchor its body to the ground and a hydrostatic skeleton, which we discuss further in Module 30.2. As illustrated in **Figure 30.1D**, the contraction of circular muscles, which encircle the circumference of the body, constricts and elongates some regions of the fluid-filled segments of a crawling earthworm. At the same time, longitudinal muscles that run the length of the body shorten and thicken other regions. Stiff bristles on the underside of the body grip the ground and provide traction, like the spikes on track shoes. (If you run your fingers along the belly of an earthworm, the bristles feel like whisker stubble.) In position ❶, segments at the head and tail ends of the worm are short and thick (longitudinal muscles contracted) and anchored to the ground by bristles. Just behind the head, a group of segments is thin and elongated (circular muscles contracted), with bristles held away from the ground. In position ❷, the head has moved forward because circular muscles in the head segments have contracted. Segments just behind the head and near the tail are now thick and anchored by bristles, thus preventing the head from slipping backward. In position ❸, the head segments are thick again and anchored to the ground in their new position, well ahead of their starting point. The rear segments of the worm now release their hold on the ground and are pulled forward.

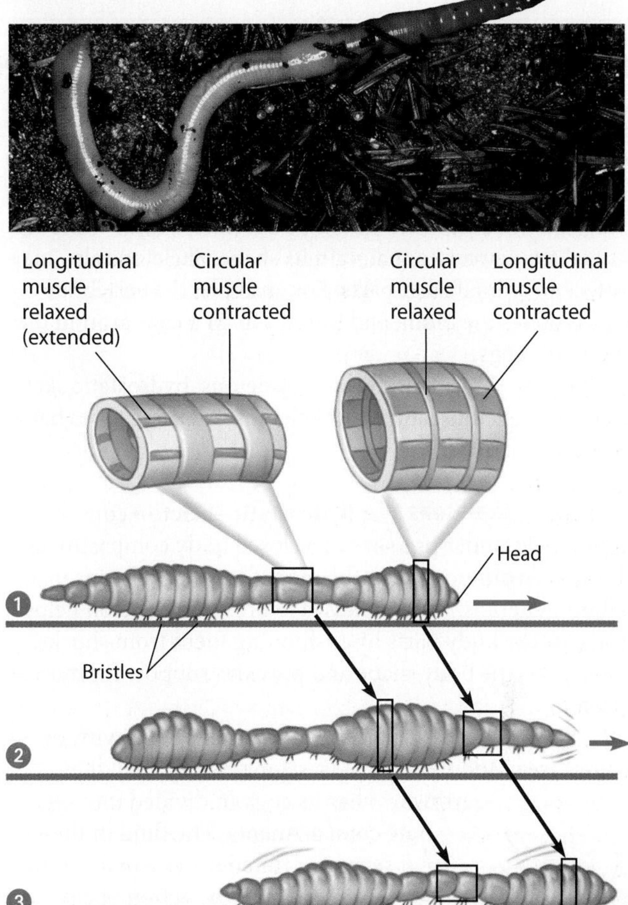
Longitudinal muscle relaxed (extended)
Circular muscle contracted
Circular muscle relaxed
Longitudinal muscle contracted
Head
1
Bristles
2
3

▲ **Figure 30.1D** An earthworm crawling by peristalsis

Flying Many phyla of animals include species that crawl, walk, or run, and almost all phyla include swimmers. But flying has evolved in only a few animal groups: insects, reptiles (including birds), and, among the mammals, bats. A group of large flying reptiles died out millions of years ago, leaving birds and bats as the only flying vertebrates.

For an animal to become airborne, its wings must develop enough "lift" to completely overcome the pull of gravity. The key to flight is the shape of wings. All types of wings, including those of airplanes, are airfoils—structures whose shape alters air currents in a way that creates lift. As **Figure 30.1E** shows, an airfoil has a leading edge that is thicker than the trailing edge. It also has an upper surface that is somewhat convex and a lower surface that is flattened or concave. This shape makes the air passing over the wing travel farther than the air passing under the wing. As a result, air molecules are spaced farther apart above the wing than below it, and the air pressure underneath the wing is greater. This pressure difference provides the lift for flight.

Birds can reach great speeds and cover enormous distances. Swifts, which can fly 170 km/hr (105 mph), are the fastest. The bird that migrates the farthest is the arctic tern, which flies round-trip between the North and South Poles each year.

An animal's muscle system provides the power to overcome friction and gravity. However, movement and locomotion result from a collaboration between muscles and a skeletal system. In the next module, you'll learn how skeletal systems are involved in movement. You'll also learn about some of the other functions of skeletal systems.

? **Contrast swimming with walking in terms of the forces an animal must overcome to move.**

● Friction resists an animal moving through water, but gravity has little effect because of the animal's buoyancy; air poses little resistance to an animal walking on land, but the animal must support itself against the force of gravity.

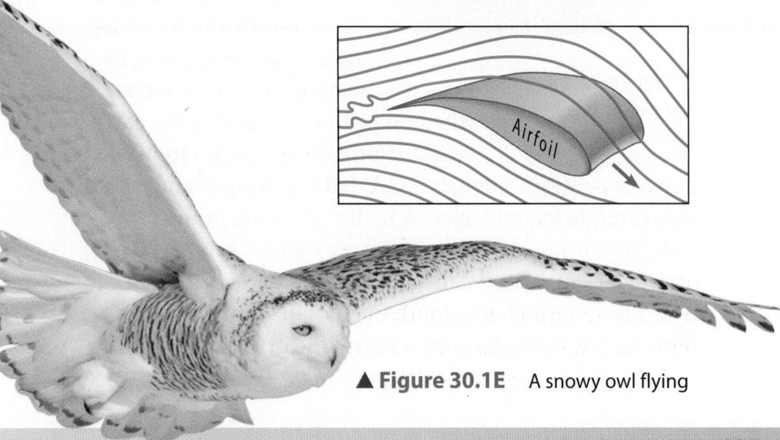
Airfoil

▲ **Figure 30.1E** A snowy owl flying

30.2 Skeletons function in support, movement, and protection

A skeleton has many functions. An animal could not move without its skeleton, and most land animals would sag from their own weight if they had no skeleton to support them. Even an animal in water would be a formless mass without a skeletal framework to maintain its shape. Skeletons also may protect an animal's soft parts. For example, the vertebrate skull protects the brain, and the ribs form a cage around the heart and lungs.

There are three main types of skeletons: hydrostatic skeletons, exoskeletons, and endoskeletons. All three types have multiple functions.

Hydrostatic Skeletons A **hydrostatic skeleton** consists of fluid held under pressure in a closed body compartment. This is very different from the more familiar skeletons made of hard materials. Nonetheless, a hydrostatic skeleton helps protect other body parts by cushioning them from shocks. It also gives the body shape and provides support for muscle action.

Earthworms have a fluid-filled internal body cavity, or coelom (see Module 18.3). As a segmented animal, the earthworm has its coelom divided into separate compartments. The fluid in these segments functions as a hydrostatic skeleton, and the action of circular and longitudinal muscles working against the hydrostatic skeleton produces the peristaltic movement described in Module 30.1.

Cnidarians, such as hydras and jellies, also have a hydrostatic skeleton. A hydra (**Figure 30.2A**), for example, has contractile cells in its body wall that enable it to alter its body shape by exerting pressure on the water-filled gastrovascular cavity (see Module 18.6). When a hydra closes its mouth and the contractile cells encircling its gastrovascular cavity contract, the body elongates, just as the earthworm elongates when its circular muscles contract (Figure 30.2A, left). The squeezing action also extends the tentacles. A hydra often sits in this position for hours, waiting for prey such as small worms or crustaceans that it can snare with its tentacles. If the hydra is disturbed, its mouth opens, allowing water to flow out. At the same time, contractile cells arranged longitudinally in the body wall contract, causing the body to shorten (Figure 30.2A, right).

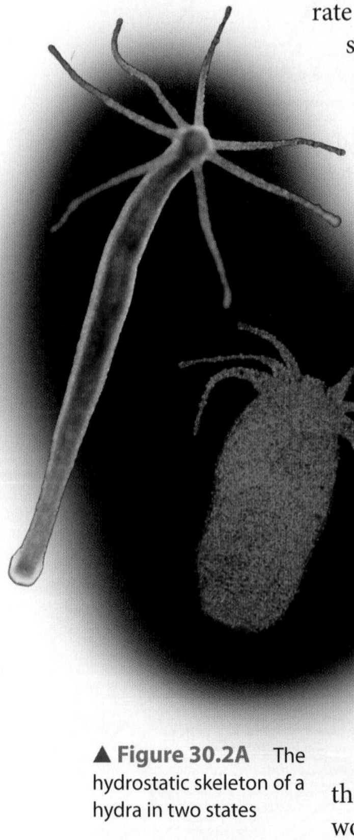

▲ **Figure 30.2A** The hydrostatic skeleton of a hydra in two states

Hydrostatic skeletons work well for many aquatic animals and for terrestrial animals that crawl or burrow by peristalsis. Most animals with hydrostatic skeletons are soft and flexible. In addition to extending its body and tentacles, for example, a hydra can expand its body around ingested prey that are larger than the gastrovascular cavity. An earthworm can burrow through soil because it is flexible and has a hydrostatic skeleton. Similarly, having an expandable body and a hydrostatic skeleton enables tube-dwelling polychaetes (see Figure 18.10B) such as feather duster worms to extend out of their tubes for feeding and gas exchange and then quickly squeeze back into the tube if threatened. However, a hydrostatic skeleton cannot support the forms of terrestrial locomotion in which an animal's body is held off the ground, such as walking.

Exoskeletons A variety of aquatic and terrestrial animals have a rigid external skeleton, or **exoskeleton**. Recall from Module 18.11 that the exoskeleton is a characteristic of the phylum Arthropoda, a group that includes insects, spiders, and crustaceans such as crabs. The arthropod exoskeleton is a tough covering composed of layers of protein and the polysaccharide chitin. The muscles are attached to knobs and plates on the inner surfaces of the exoskeleton. At the joints of legs, the exoskeleton is thin and flexible, allowing movement. If you have eaten crab legs, you cracked the exoskeleton to extract the tasty muscle within.

Because the exoskeleton is composed of nonliving material, it does not grow with the animal. It must be shed (molted) and replaced by a larger exoskeleton at intervals to allow for the animal's growth (**Figure 30.2B**). Depending on the species, most insects molt from four to eight times before

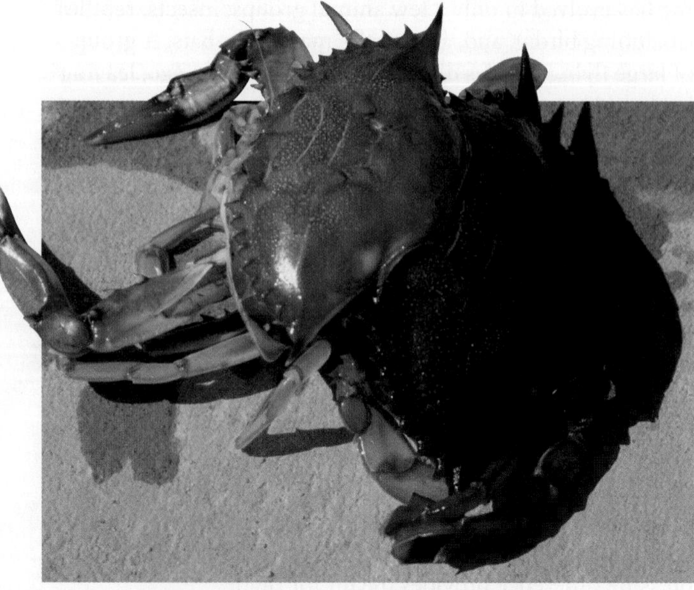

▲ **Figure 30.2B** The exoskeleton of an arthropod: a crab molting

reaching adult size. A few insect species and certain other arthropods, such as lobsters and crabs, molt at intervals throughout life.

An arthropod is never without an exoskeleton of some sort. For instance, a newly molted crab is covered by a soft, elastic exoskeleton that formed under the old one. Soon after molting, the crab expands its body by gulping air or water. Its new exoskeleton then hardens in the expanded position, and the animal has room for additional growth. If you have ever eaten soft-shell crab, you took advantage of this brief period when the new exoskeleton is tender enough to chew. As you can imagine, a newly molted arthropod is very susceptible to predation. Besides being weakly armored, it is usually less mobile, because the soft exoskeleton cannot support the full action of its muscles.

The shells of molluscs such as clams, snails, and cowries (**Figure 30.2C**)—shells you might find on a beach—are also exoskeletons. Unlike the chitinous arthropod exoskeleton, mollusc shells are made of a mineral, calcium carbonate. The mantle, a sheetlike extension of the animal's body wall, secretes the shell. As a mollusc grows, it does not molt; rather, it enlarges the diameter of its shell by adding to its outer edge.

Endoskeletons An **endoskeleton** consists of hard or leathery supporting elements situated among the soft tissues of an animal. Sponges, for example, are reinforced by a framework of tough protein fibers or by mineral-containing particles. Usually microscopic and sharp-pointed, the particles consist of inorganic material such as calcium salts or silica. Sea stars, sea urchins, and most other echinoderms have an endoskeleton of hard plates beneath their skin (see Module 18.14). In living sea urchins, about all you see are the movable spines, which are attached to the endoskeleton by muscles (**Figure 30.2D**). A dead urchin with its spines removed reveals the plates that form a rigid skeletal case (Figure 30.2D, right).

Vertebrates have endoskeletons consisting of cartilage or a combination of cartilage and bone (see Module 20.5). Sharks, one major lineage of vertebrates, have endoskeletons of cartilage reinforced with calcium. **Figure 30.2E** shows the more common condition for vertebrates. Bone makes up most of a frog's skeleton, as it does in bony fishes and land vertebrates. The frog skeleton and the skeletons of most other vertebrates also include some cartilage (blue in the figure), mainly in areas where flexibility is needed. Next, let's take a closer look at endoskeletons.

▲ **Figure 30.2D** A living sea urchin (left) and its endoskeleton (right)

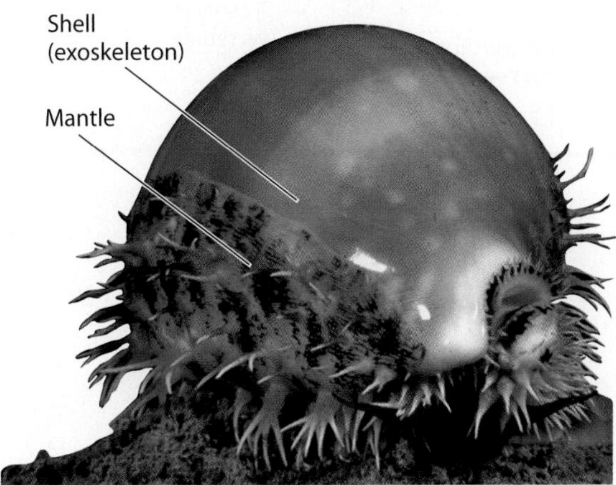
Shell
(exoskeleton)
Mantle

▲ **Figure 30.2C** The exoskeleton of a mollusc: a cowrie (a marine snail)

? **What are the advantages and disadvantages of an exoskeleton as compared to an endoskeleton?**

● An exoskeleton may offer greater protection to body parts but must usually be molted for the animal to grow.

▲ **Figure 30.2E** Bone (off-white) and cartilage (blue) in the endoskeleton of a vertebrate: a frog

▷ The Vertebrate Skeleton

30.3 Vertebrate skeletons are variations on an ancient theme

EVOLUTION CONNECTION

As you learned in Module 19.4, the vertebrate skeletal system provided the structural support and means of locomotion that enabled tetrapods to colonize land. Subsequent evolution produced diverse groups of animals: amphibians, reptiles (including birds), and mammals. Each of those groups has diverse body forms whose skeletons are constructed from modified versions of the same parts. Even the skeletons of whales and dolphins, mammals that evolved from land-dwelling ancestors, are variations on the same theme.

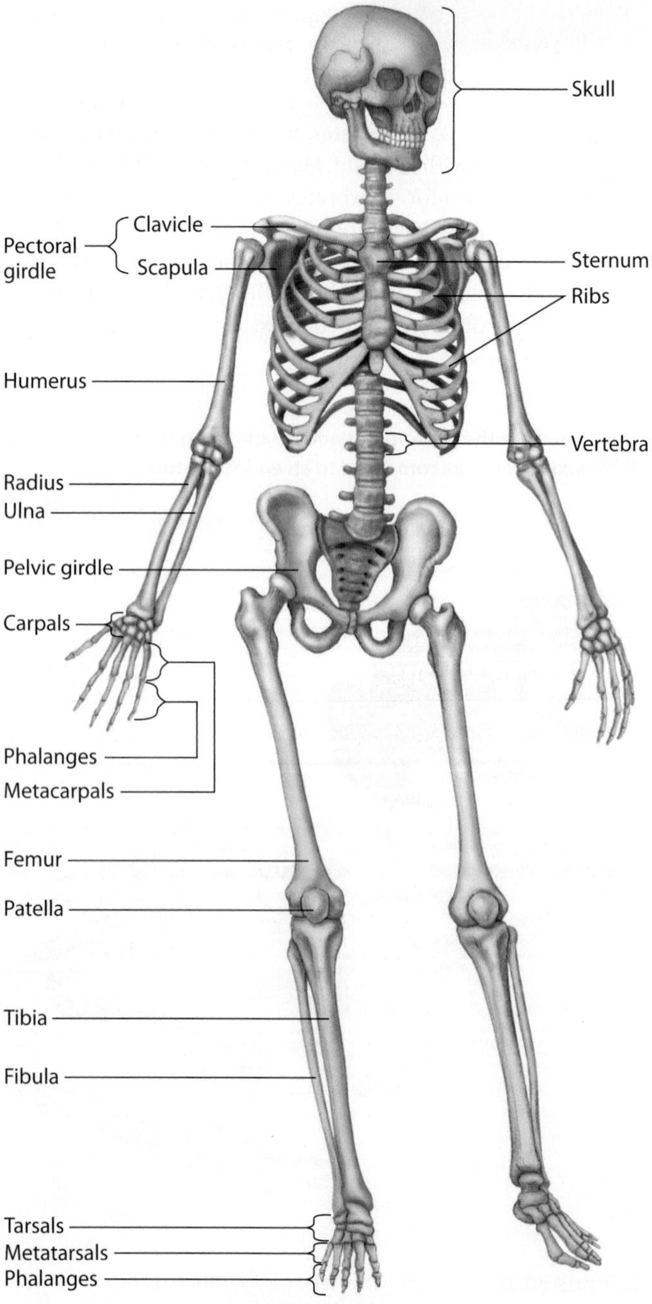

Skull
Pectoral girdle
Clavicle
Scapula
Sternum
Ribs
Humerus
Vertebra
Radius
Ulna
Pelvic girdle
Carpals
Phalanges
Metacarpals
Femur
Patella
Tibia
Fibula
Tarsals
Metatarsals
Phalanges

▲ **Figure 30.3A** The human skeleton

All vertebrates have an **axial skeleton** (orange in **Figure 30.3A**) supporting the axis, or trunk, of the body. The axial skeleton consists of the skull, enclosing and protecting the brain; the vertebral column (backbone), enclosing the spinal cord; and, in most vertebrates, a rib cage around the lungs and heart.

The backbone, the definitive characteristic of vertebrates, consists of a series of individual bones, the vertebrae, joined by pads of tough cartilage known as discs. The number of vertebrae varies among species. Pythons have 400, whereas an adult human has 24. All vertebrae have the same basic structure, with slight variations that reflect the position of each vertebra in the backbone. Anatomists divide the vertebral column into the regions shown in **Figure 30.3B**: cervical (neck), which support the head; thoracic (chest), which form joints with the ribs; lumbar (lower back); sacral (between the hips); and coccygeal (tail). In humans, the sacral vertebrae fuse into a single bone called the sacrum. Our small coccygeal vertebrae are partially fused into the coccyx, or "tailbone."

Most vertebrates also have an **appendicular skeleton** (off-white in Figure 30.3A), which is made up of the bones of the appendages and the bones that anchor the appendages to the axial skeleton. In a land vertebrate, the pectoral (shoulder) girdle and the pelvic girdle provide a base of support for the bones of the forelimbs and hind limbs. Modified versions of the same bones are found in all vertebrate limbs, whether they are arms, legs, fins, or wings (see Figure 13.4A). This variety of limbs equips vertebrates for every form of locomotion.

A few groups of vertebrates, including snakes, lost their limbs during their evolution. How did this happen? The identity of vertebrae is established during embryonic development by the pattern of master control (homeotic) genes expressed

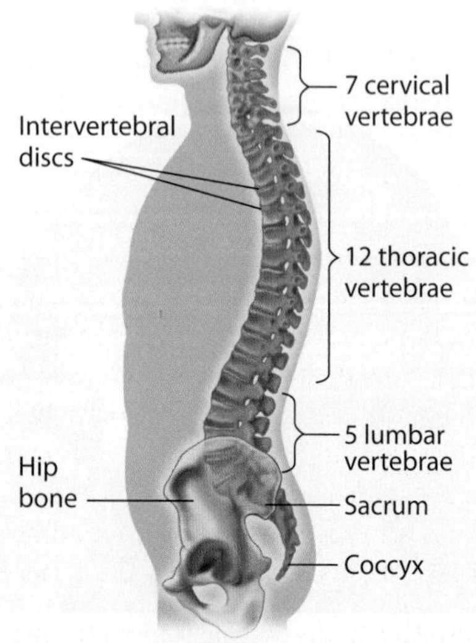
7 cervical vertebrae
Intervertebral discs
12 thoracic vertebrae
5 lumbar vertebrae
Hip bone
Sacrum
Coccyx

▲ **Figure 30.3B** The human backbone, showing the groups of vertebrae

in the somites. (Recall from Module 27.12 that somites are the blocks of embryonic tissue that give rise to the vertebral column.) Two of the homeotic genes that direct the differentiation of vertebrae are *Hoxc6* and *Hoxc8*. These genes are associated with the development of thoracic vertebrae, which support the ribs. **Figure 30.3C** shows the range of vertebrae formed by somites that expressed *Hoxc6* (shown in red), *Hoxc8* (blue), or both (purple) in a chicken and a python.

In the python, both *Hoxc6* and *Hoxc8* are expressed in all somites for nearly the entire length of the vertebral column. As a result, the first rib-bearing thoracic vertebra is located immediately posterior to the head. Pythons have no cervical vertebrae. Chickens, on the other hand, have several cervical vertebrae, ending at the point where *Hoxc6* expression—and thoracic vertebrae—begin.

During the evolution of snakes, mutation in the DNA segments that control the expression of *Hoxc6* and *Hoxc8* changed cervical vertebrae to thoracic. In all vertebrates, the forelimbs originate at the boundary between cervical and thoracic vertebrae. Because this position does not exist in snakes, forelimbs do not form.

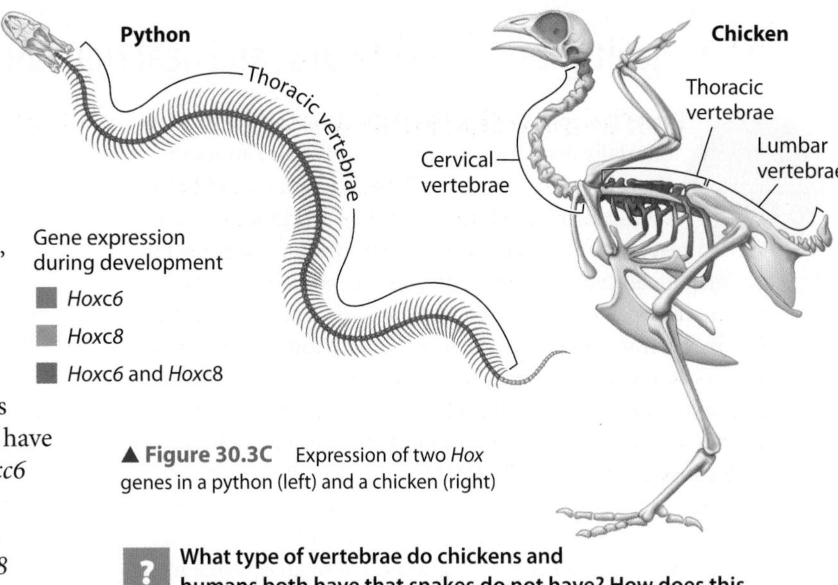
Python
Thoracic vertebrae
Chicken
Thoracic vertebrae
Lumbar vertebrae
Cervical vertebrae
Gene expression during development
Hoxc6
Hoxc8
Hoxc6 and Hoxc8

▲ **Figure 30.3C** Expression of two *Hox* genes in a python (left) and a chicken (right)

? What type of vertebrae do chickens and humans both have that snakes do not have? How does this difference affect the appendicular skeleton of snakes?

Chickens and humans have cervical vertebrae; snakes do not (we did not compare vertebrae posterior to thoracic vertebrae). Because forelimbs form at the boundary between cervical and thoracic vertebrae, snakes lack forelimbs.

30.4 Bones are complex living organs

The expression "dry as a bone" should not be taken literally. Your bones are actually complex organs consisting of several kinds of moist, living tissues. **Figure 30.4** shows a human humerus (upper arm bone). A sheet of fibrous connective tissue, shown in pink (most visible in the enlargement on the lower right), covers most of the outside surface. This tissue helps form new bone in the event of a fracture. A thin sheet of cartilage (blue) forms a cushion-like surface for movable joints, protecting the ends of bones as they glide against one another. The bone itself contains living cells that secrete a surrounding material, or matrix. Bone matrix consists of flexible fibers of the protein collagen with crystals of a mineral made of calcium and phosphate bonded to them (see Figure 20.5). The collagen keeps the bone flexible and nonbrittle, while the hard mineral matrix resists compression.

The shaft of this long bone is made of compact bone, a term that refers to its dense structure. Notice that the compact bone surrounds a central cavity. The central cavity contains **yellow bone marrow**, which is mostly stored fat brought into the bone by the blood. The ends, or heads, of the bone have an outer layer of compact bone and an inner layer of spongy bone, so named because it is honeycombed with small cavities. The cavities contain **red bone marrow** (not shown in the figure), a specialized tissue that produces our blood cells (see Module 23.15).

Like all living tissues, bone cells carry out metabolism. Blood vessels that extend through channels in the bone transport nutrients and regulatory hormones to its cells and remove waste materials. Nerves running parallel to the blood vessels help regulate the traffic of materials between the bone and the blood.

? What causes the colors of yellow and red bone marrow?

Stored fat and developing red blood cells, respectively

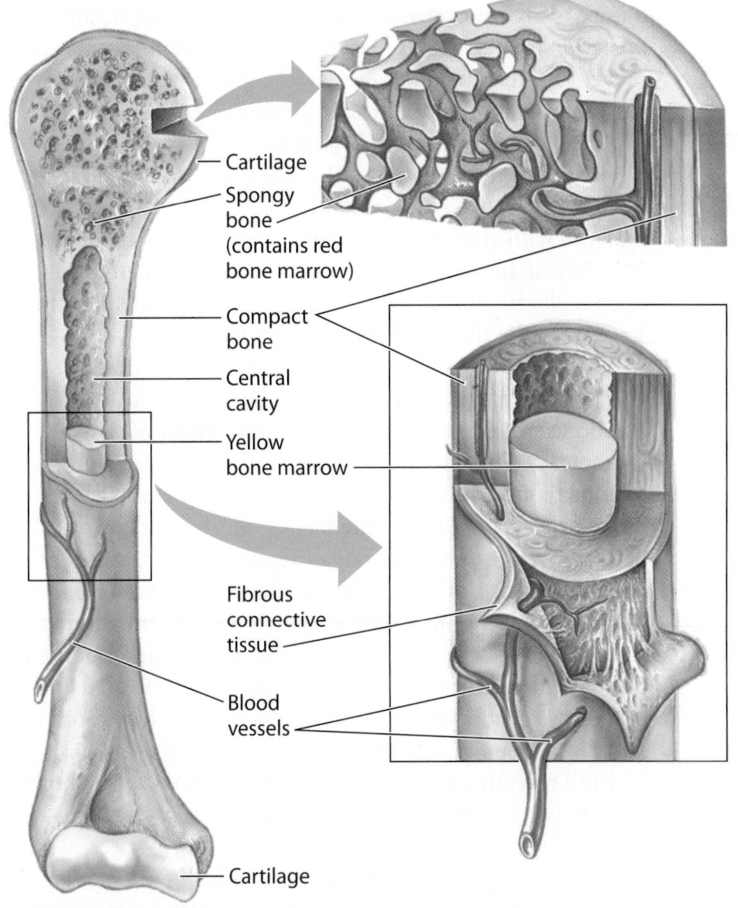
Cartilage
Spongy bone (contains red bone marrow)
Compact bone
Central cavity
Yellow bone marrow
Fibrous connective tissue
Blood vessels
Cartilage

▲ **Figure 30.4** The structure of an arm bone

30.5 Healthy bones resist stress and heal from injuries

CONNECTION

Bones are constantly subjected to stress as we go about our daily lives; exercise or physical labor causes additional stress. Excessive bone fatigue can lead to so-called stress fractures, hairline cracks in the bone, just as the accumulation of small amounts of stress on metals can cause a break. For example, when you bend a paper clip repeatedly, the metal fatigues and finally snaps. Unlike metal, however, bone is composed of living, dynamic tissue. Cells continually remove old bone matrix and replace it with new material. Stress fractures only occur if this repair process cannot keep up with the amount of stress placed on a bone. An athlete usually becomes aware of the problem and can allow time for healing, but in a racehorse, stress damage might go unnoticed until the fatigued bone breaks suddenly during a race.

A bone may also break when subjected to an external force that exceeds its resiliency. The average American will break two bones during his or her lifetime, most commonly the forearm or, for people over 75, the hip. Usually, this type of fracture occurs from a sudden impact, such as a fall or car accident. Wearing appropriate protective gear, such as seat belts, helmets, or padding, can protect your bones from high-force trauma.

Fortunately for active people, broken bones can heal themselves. A physician assists the process by putting the bone back into its natural alignment and then immobilizing it until the body's normal bone-building cells can repair the break. A splint or cast is used to protect the injured area and prevent movement. In severe cases, a fracture can be repaired surgically by inserting plates, rods, and/or screws that hold the broken pieces together (**Figure 30.5A**). In certain cases, however, severely injured or diseased bone is beyond repair and must be replaced. Broken hip joints, for example, can be replaced with artificial ones made of titanium or cobalt alloys. Researchers have recently developed new methods of bone replacement, including grafts (from the patient or from a cadaver) and the use of synthetic polymers.

The risk of bone fracture increases if bones are porous and weak. **Figure 30.5B** contrasts healthy bone tissue (left) and bone eroded by osteoporosis (right). **Osteoporosis** is characterized by low bone mass and structural deterioration of bone tissue. This weakness emerges from an imbalance in the process of bone maintenance—the destruction of bone material exceeds the rate of replacement. Because the natural mechanism of bone maintenance responds to bone usage, weight-bearing exercise such as walking or running strengthens bones. On the other hand, disuse causes bones to become thinner. Strong bones also require an adequate intake of dietary calcium and enough vitamin D, which are both essential to bone replacement (see Module 21.16).

Until recently, osteoporosis was mostly considered a problem for women after menopause, when levels of estrogen—which contributes to normal bone maintenance—decrease. Although osteoporosis remains a serious health problem for older women, it is also becoming a concern for men and younger people. Doctors have noted a dramatic increase in bone fractures in children and teenagers in recent years. Many scientists believe that this is the result of exercising less and getting less calcium in the diet and less vitamin D from exposure to sunlight. Prevention of osteoporosis in later years begins with exercise and sufficient calcium and vitamin D while bones are still increasing in density (up until about age 30).

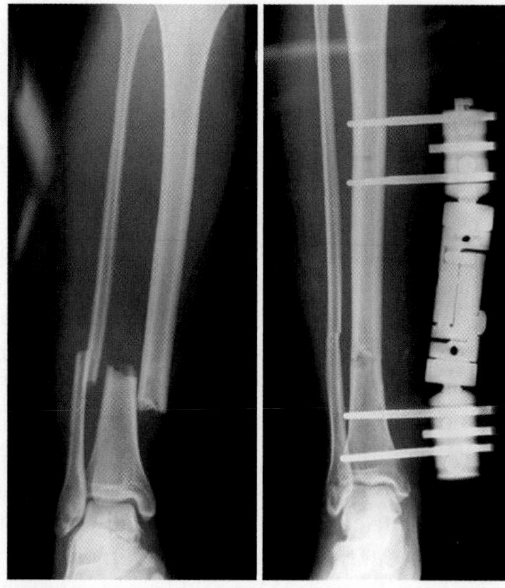

▲ **Figure 30.5A** X-rays of a broken leg (left) and the same leg after the bones were set with a plate and screws (right)

Other lifestyle habits, such as smoking, may also contribute to osteoporosis. There is a strong genetic component as well; young women whose mothers or grandmothers suffer from osteoporosis should be especially concerned with maintaining good bone health. Treatments for osteoporosis include calcium and vitamin D supplements and drugs that slow bone loss.

? How do exercise and adequate calcium intake help prevent osteoporosis?

Bone tissue responds to the stress of exercise by stepping up the repair process, which builds greater bone density. Calcium is the primary component of the rigid mineral matrix of bone.

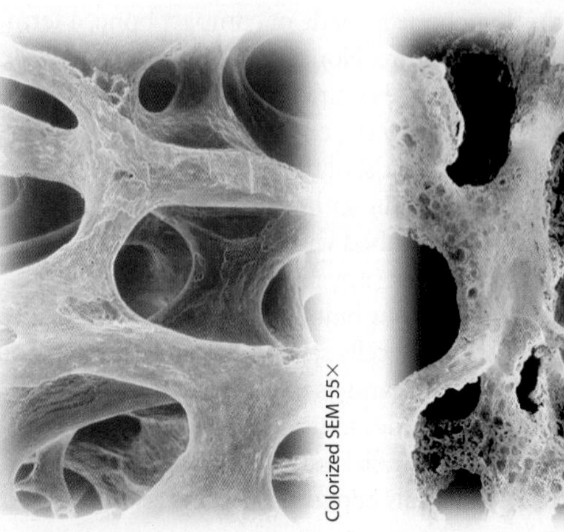
Colorized SEM 55×
Colorized SEM 55×

▲ **Figure 30.5B** Healthy spongy bone tissue (left) and bone damaged by osteoporosis (right)

30.6 Joints permit different types of movement

Much of the versatility of the vertebrate skeleton comes from its diverse joints. Bands of strong fibrous connective tissue called **ligaments** hold together the bones of movable joints. **Ball-and-socket joints**, such as are found where the humerus joins the pectoral girdle (**Figure 30.6**, left), enable us to rotate our arms and legs and move them in several planes. A ball-and-socket joint also joins the femur to the pelvic girdle. **Hinge joints** permit movement in a single plane, just as the hinge on a door enables it to open and close. Our elbows (shown in Figure 30.6, center) and knees are hinge joints. Hinge joints are especially vulnerable to injury in sports like volleyball, basketball, and tennis that demand quick turns, which can twist the joint sideways. A **pivot joint** enables us to rotate the forearm at the elbow (Figure 30.6, right). A pivot joint between the first and second cervical vertebrae allows movement of the head from side to side, for example, the motion you make when you say "no." As you'll learn in the next module, muscles supply the force to move the bones of each joint.

> **?** **Where we have ball-and-socket joints, horses have hinge joints. How does this affect the movements they can perform?**
>
> ● Hinge joints restrict the movement of their legs to a single plane, making horses less flexible than humans.

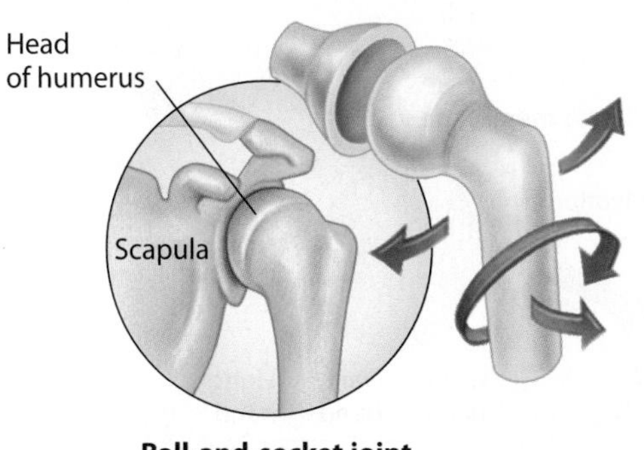
Head of humerus
Scapula

Ball-and-socket joint

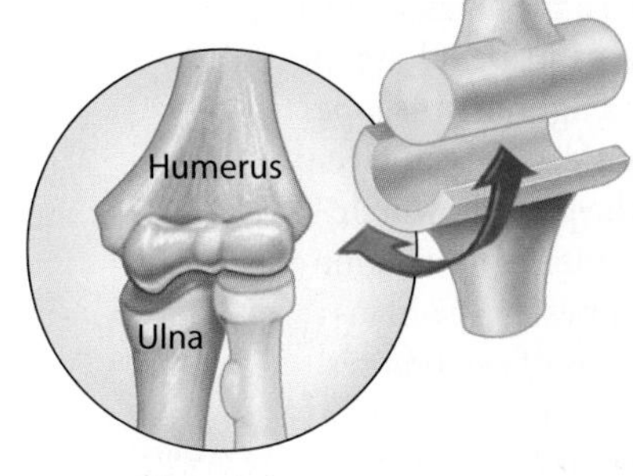
Humerus
Ulna

Hinge joint

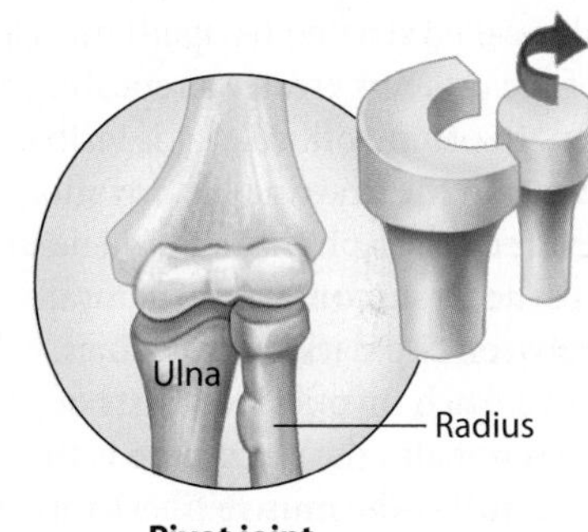
Ulna
Radius

Pivot joint

▲ **Figure 30.6** Three kinds of joints

▷ Muscle Contraction and Movement

30.7 The skeleton and muscles interact in movement

Figure 30.7 shows how an animal's muscles interact with its bones to produce movement. Muscles are connected to bones by **tendons**. For example, the upper ends of the biceps and triceps muscles shown in the figure are anchored to bones in the shoulder. The lower ends of these muscles are attached to bones in the forearm. The action of a muscle is always to contract, or shorten. A muscle *pulls* the bone to which it is attached—it can only move the bone in one direction. A different muscle is needed to reverse the action. Thus, back-and-forth movement of body parts involves antagonists, a pair of muscles (or muscle groups) that can pull the same bone in opposite directions.

The biceps and triceps muscles are an example of an antagonistic pair. Imagine that you are picking up a glass of water to drink. To raise the glass to your lips, your biceps muscle contracts, pulling the forearm bones toward you as your elbow bends. To put the glass back on the table, you must lower your forearm. Now the triceps muscle contracts, pulling the forearm bones down. The quadriceps, which extend the lower leg, and the hamstring, which flexes the lower leg, are also antagonistic pairs of muscles.

All animals—very small ones like ants and giant ones like elephants—have antagonistic pairs of muscles that apply opposite forces to move parts of their skeleton. Next we see how a muscle's structure explains its ability to contract.

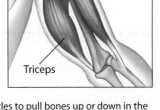

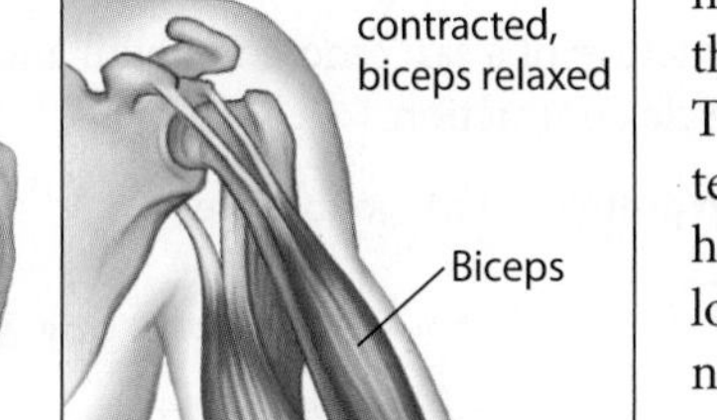
Biceps contracted, triceps relaxed (extended)
Biceps
Triceps
Tendons
Triceps contracted, biceps relaxed
Biceps
Triceps

▲ **Figure 30.7** Antagonistic action of muscles to pull bones up or down in the human arm

> **?** **When exercising to strengthen muscles, why is it important to impose resistance while both flexing and extending the limbs?**
>
> ● This exercises both muscles of antagonistic pairs, which only do work when they are contracting.

30.8 Each muscle cell has its own contractile apparatus

The skeletal muscle system is a beautiful illustration of the relationship between structure and function. Each muscle in the body is made up of a hierarchy of smaller and smaller parallel strands, from the muscle itself down to the contractile protein molecules that produce body movements.

Figure 30.8 shows the levels of organization of skeletal muscle. As indicated at the top of the figure, a muscle consists of many bundles of **muscle fibers**—roughly 250,000 in a typical human biceps muscle—oriented parallel to each other. Each muscle fiber is a single long, cylindrical cell that has many nuclei. Most of its volume is occupied by hundreds or thousands of **myofibrils**, discrete bundles of proteins that include the contractile proteins **actin** and **myosin**. Skeletal muscle is also called striated (striped) muscle because the arrangement of the proteins creates a repeating pattern of stripes along the length of a myofibril that is visible under a light microscope. Beneath the drawing of a myofibril in Figure 30.8 is an electron micrograph that shows one unit of the pattern, which is called a **sarcomere**. Structurally, a sarcomere is the region between two dark, narrow lines, called Z lines, in the myofibril. Each myofibril consists of a long series of sarcomeres. Functionally, the sarcomere is the contractile apparatus in a myofibril—the muscle fiber's fundamental unit of action.

The diagram of the sarcomere at the bottom of Figure 30.8 explains the features visible in the micrograph. The pattern of horizontal stripes is the result of the alternating bands of **thin filaments**, composed primarily of actin molecules, and **thick filaments**, which are made up of myosin molecules. The Z lines consist of proteins that connect adjacent thin filaments. The light band surrounding each Z line contains only thin filaments. The dark band centered in the sarcomere is the location of the thick filaments. The actin molecules in the thin filaments are globular proteins arrayed in long strands. In addition to actin, thin filaments include proteins called troponin and tropomyosin that play a key role in regulating muscle contraction.

Next we examine the structure of a sarcomere in detail and see how it functions in muscle contraction.

? **The two most abundant proteins of a myofibril are ______________ and ______________.**

actin . . . myosin

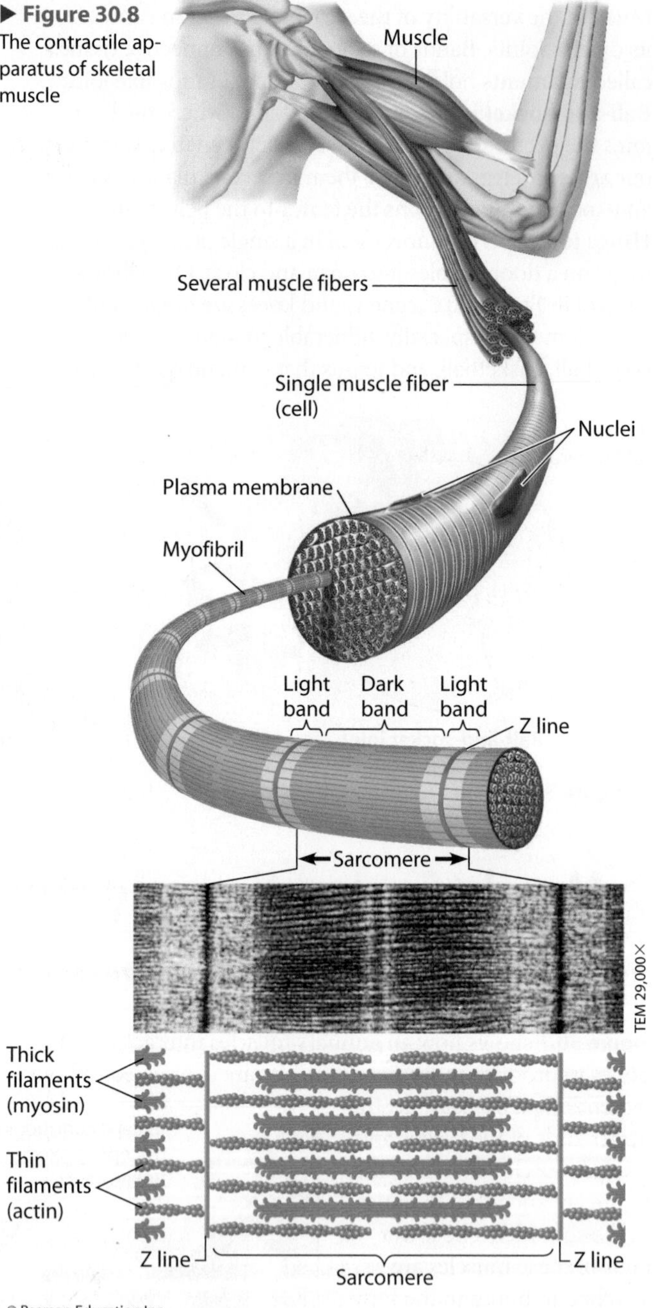
Muscle
Several muscle fibers
Single muscle fiber (cell)
Nuclei
Plasma membrane
Myofibril
Light band
Dark band
Light band
Z line
Sarcomere
TEM 29,000×
Thick filaments (myosin)
Thin filaments (actin)
Z line
Sarcomere
Z line

▶ **Figure 30.8** The contractile apparatus of skeletal muscle

30.9 A muscle contracts when thin filaments slide along thick filaments

How does the structure of a sarcomere relate to its function? According to the sliding-filament model of muscle contraction, a sarcomere contracts (shortens) when its thin filaments slide along its thick filaments. **Figure 30.9A**, on the next page, is a simplified diagram that shows a sarcomere in a relaxed muscle, in a contracting muscle, and in a fully contracted muscle. Notice in the contracting sarcomere that the Z lines and the thin filaments (blue) have moved closer together. When the muscle is fully contracted, the thin filaments overlap in the middle of the sarcomere. Contraction shortens the sarcomere without changing the lengths of the thick and thin filaments. A whole muscle can shorten about 35% of its resting length when all sarcomeres contract.

Myosin acts as the engine of movement. Each myosin molecule has a long "tail" region and a globular "head" region. The tails of the myosin molecules in a thick filament lie parallel and adhere to each other, with their heads sticking out to the side. Each head has two binding sites. One of the binding sites matches a binding site on the actin molecules (subunits) of the thin filament. ATP binds at the other site, which is also

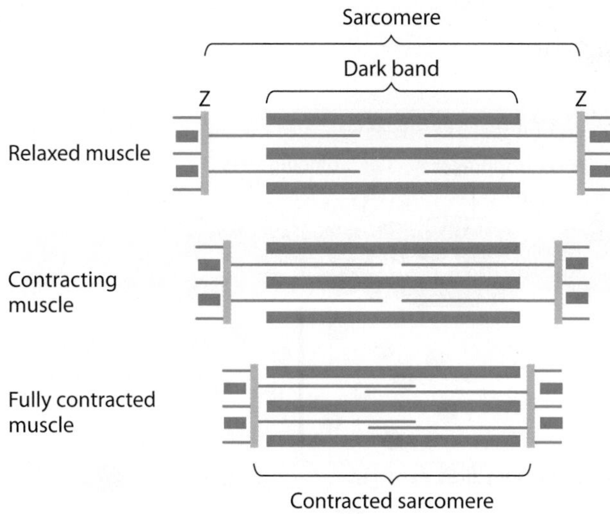

Sarcomere
Dark band
Z
Z
Relaxed muscle
Contracting muscle
Fully contracted muscle
Contracted sarcomere

▲ **Figure 30.9A** The sliding-filament model of muscle contraction

capable of hydrolyzing the ATP to release its energy—the energy that powers muscle contraction.

Each myosin head pivots back and forth in a limited arc as it changes shape from a low-energy configuration to a high-energy configuration and back again. During these changes, the myosin head swings toward the thin filament, binds with an actin molecule, and drags the thin filament through the remainder of its arc. The myosin head then releases the actin molecule and returns to its starting position to repeat the same motion with a different actin molecule.

Let's follow the key events of this process in **Figure 30.9B**. The myosin head binds a molecule of ATP, as shown at ①. At this point, the myosin head is in its low-energy position. ② Myosin hydrolyzes the ATP to ADP and phosphate Ⓟ, releasing energy that extends the myosin head toward the thin filament. ③ The myosin head extends further, and its other binding site latches on to the binding site of an actin. The result is a connection between the two filaments—a cross-bridge. ④ ADP and Ⓟ are released, and the myosin head pivots back to its low-energy configuration. This action, called the power stroke, pulls the thin filament toward the center of the sarcomere.

The cross-bridge remains intact until ⑤ another ATP molecule binds to the myosin head, and the whole process repeats. On the next power stroke, the myosin head attaches to an actin molecule ahead of the previous one on the thin filament (closer to the Z line). This sequence—detach, extend, attach, pull, detach—occurs again and again in a contracting muscle. Though we show only one myosin head in the figure, a typical thick filament has about 350 heads, each of which can bind and unbind to a thin filament about five times per second. The combined action of hundreds of myosin heads on each thick filament ratchets the thin filament toward the center of the sarcomere, much like the people on one side of a tug-of-war. Each person (a myosin head) pulls hand over hand on the rope (the thin filament)—the rope moves, but the people do not. As long as sufficient ATP is present, the process continues until the muscle is fully contracted or until the signal to contract stops.

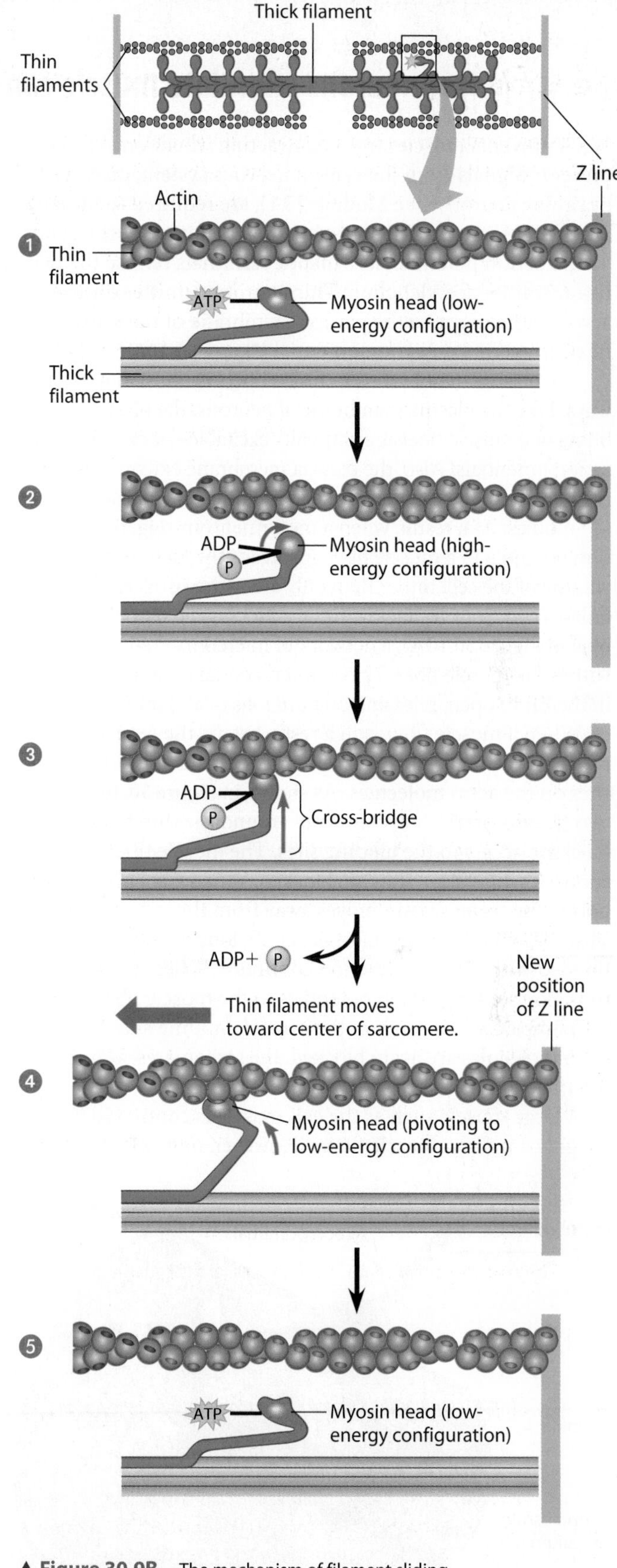

Thick filament
Thin filaments
Z line
Actin
1
Thin filament
ATP
Myosin head (low-energy configuration)
Thick filament
2
ADP
P
Myosin head (high-energy configuration)
3
ADP
P
Cross-bridge
ADP+ P
Thin filament moves toward center of sarcomere.
New position of Z line
4
Myosin head (pivoting to low-energy configuration)
5
ATP
Myosin head (low-energy configuration)

▲ **Figure 30.9B** The mechanism of filament sliding

Try This At each step, describe in your own words the action of the myosin head.

? Which region of a sarcomere becomes shorter during contraction of a muscle?

● The light bands shorten and even disappear as the thin filaments slide (are pulled) toward the center of the sarcomere.

30.10 Motor neurons stimulate muscle contraction

What prevents muscles from contracting whenever ATP is present? Signals from the central nervous system, conveyed by motor neurons (see Module 28.1), are required to initiate and sustain muscle contraction. When a motor neuron sends out an action potential, its synaptic terminals release the neurotransmitter acetylcholine. This neurotransmitter diffuses across the synapse to the plasma membrane of the muscle fiber, triggering an action potential **(Figure 30.10A)**.

The plasma membrane of muscle fibers is unusual in two ways. Like the plasma membrane of neurons, the plasma membrane of a muscle fiber is electrically excitable—it can propagate action potentials. Also, the plasma membrane extends deep into the interior of the muscle fiber via infoldings called transverse (T) tubules. As a result, when a motor neuron triggers an action potential in a muscle fiber, it spreads throughout the entire volume of the cell, rather than only along the surface. The T tubules are in close contact with the endoplasmic reticulum (ER; blue in Figure 30.10A), a network of interconnected tubules within the muscle fiber. The action potential causes channels in the ER to open, releasing calcium ions (Ca^{2+}) into the cytosol.

When a muscle fiber is in a resting state, the regulatory proteins tropomyosin and troponin block the myosin-binding sites on the actin molecules. As shown in **Figure 30.10B**, two strands of tropomyosin wrap around the thin filament, blocking access to the binding sites. The muscle fiber cannot contract while these sites are blocked. When Ca^{2+} binds to troponin, the tropomyosin moves away from the myosin-binding sites, allowing contraction to occur. As long as the cytosol is flooded with Ca^{2+}, contraction continues. When motor neurons stop sending action potentials to the muscle fibers, the ER pumps Ca^{2+} back out of the cytosol, binding sites on the actin molecules are again blocked, the sarcomeres stop contracting, and the muscle relaxes.

A large muscle such as the calf muscle is composed of roughly a million muscle fibers. However, only about 500 motor neurons run to the calf muscle. Each motor neuron has axons that branch out to synapse with many muscle fibers distributed throughout the muscle. Thus, an action potential from a single motor neuron in the calf causes the simultaneous contraction of roughly 2,000 muscle fibers. A motor neuron and all the muscle fibers it controls is called a **motor unit**. **Figure 30.10C** shows two motor units; one

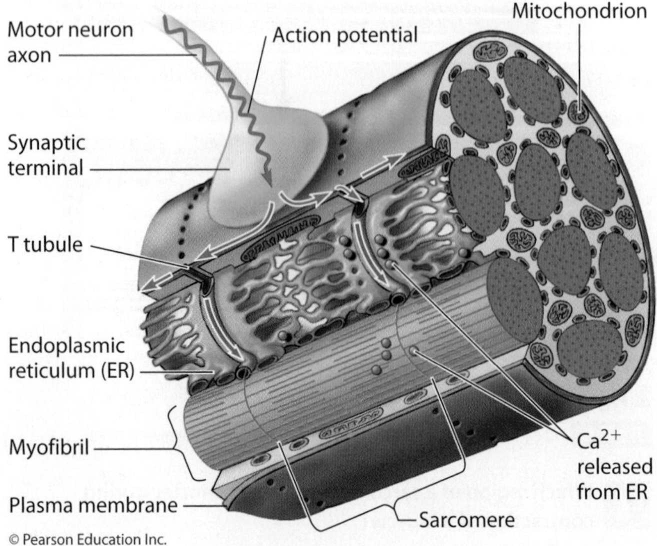
Motor neuron axon
Action potential
Mitochondrion
Synaptic terminal
T tubule
Endoplasmic reticulum (ER)
Myofibril
Plasma membrane
Ca2+ released from ER
Sarcomere
© Pearson Education Inc.

▲ **Figure 30.10A** How a motor neuron stimulates muscle contraction

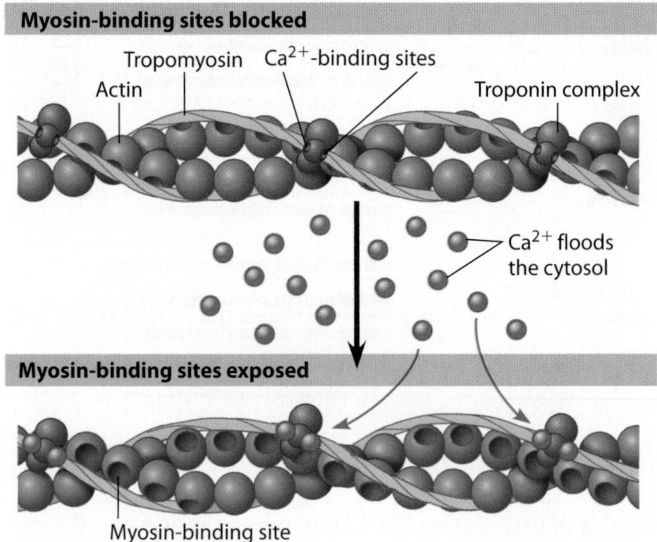
Myosin-binding sites blocked
Tropomyosin
Ca2+-binding sites
Actin
Troponin complex
Ca2+ floods the cytosol
Myosin-binding sites exposed
Myosin-binding site

▲ **Figure 30.10B** Thin filament, showing the interactions among actin, regulatory proteins, and Ca^{2+}

Try This Starting with Figure 30.10A, narrate the events, including the structures involved, that occur from the time an action potential reaches a muscle fiber until the muscle fiber relaxes.

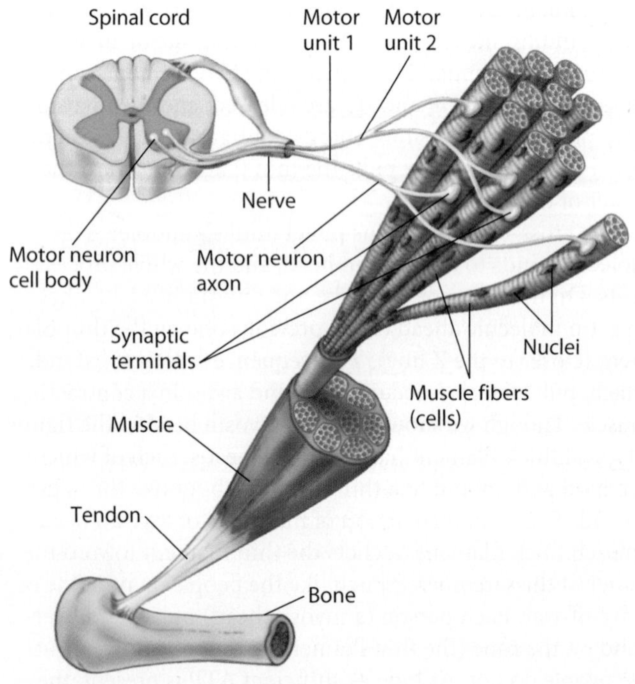
Spinal cord
Motor unit 1
Motor unit 2
Nerve
Motor neuron cell body
Motor neuron axon
Synaptic terminals
Nuclei
Muscle fibers (cells)
Muscle
Tendon
Bone

▲ **Figure 30.10C** Motor units consisting of a motor neuron and the muscle fibers it controls

controls two muscle fibers (motor unit 1 in the figure), and the other (motor unit 2) controls three.

The organization of individual neurons and muscle cells into motor units is the key to the action of whole muscles. We can vary the amount of force our muscles develop: When you arm wrestle, for example, you might change the amount of force developed by the biceps and triceps several times in the course of a match. The ability to do this depends mainly on the nature of motor units. More forceful contractions result when additional motor units are activated. Thus, depending on how many motor units your brain commands to contract, you can apply a small amount of force to lift a fork or considerably more to lift, say, this textbook. In muscles requiring precise control, such as those controlling eye movements, a motor neuron may control only a single muscle fiber.

? **How does the endoplasmic reticulum help regulate muscle contraction?**

By reversibly taking up and releasing Ca^{2+}, the ER regulates the cytoplasmic concentration of this ion, which is required in the cytosol for the binding of myosin to actin.

30.11 Aerobic respiration supplies most of the energy for exercise

CONNECTION

Many people exercise to stay in shape or to lose a few pounds before swimsuit season. Activities such as jogging, swimming, or other aerobic workouts are typical fitness routines (**Figure 30.11**). Aerobic exercise is an effective method of maintaining or losing weight—the goal is to burn at least as many calories as you consume. The number of calories burned varies with the intensity and duration of the exercise (as you saw in the examples listed in Figure 6.4). Most of the energy expended during exercise is used for muscle movement, specifically, to break the cross-bridges formed during sarcomere contraction. Here we refer to this energy in terms of ATP rather than calories.

Muscles have a very small amount of ATP on hand. ATP can also be obtained using a high-energy molecule called phosphocreatine (PCr, also known as creatine phosphate) that is stored in the muscles. The enzymatic transfer of a phosphate group from PCr to ADP makes ATP almost instantaneously. Together, ATP and PCr can provide enough energy for a 10- to 15-second burst of activity, enough for a 100-m sprint (see **Table 30.11**).

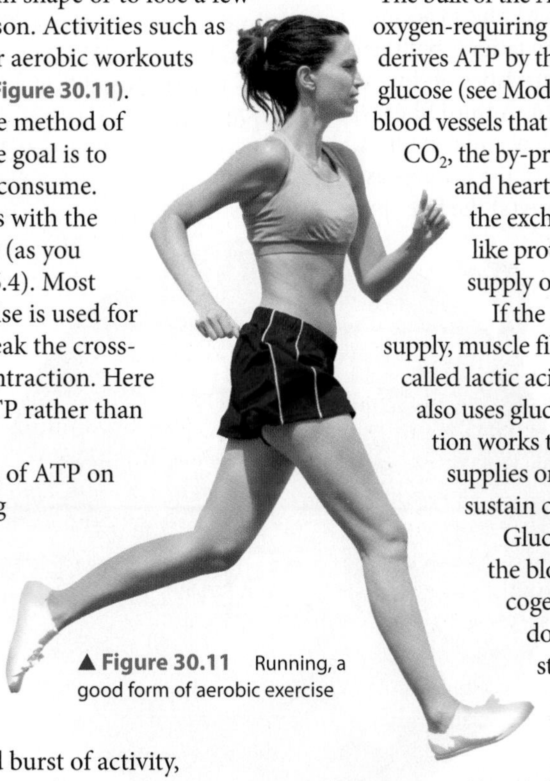

▲ **Figure 30.11** Running, a good form of aerobic exercise

TABLE 30.11 | SOURCES OF ATP FOR ATHLETIC ACTIVITIES

Athletic Activity	Energy Use	Main Source of ATP
100-m sprint; power lifting	10- to 15-second burst of activity	Stored ATP and PCr
200-m or 400-m race	Intense effort sustained over a short period of time	Stored ATP and PCr plus lactic acid fermentation
Jogging; long-distance running	Prolonged, low-level activity	Aerobic respiration
Tennis; squash; soccer	Prolonged, low-level activity with intermittent surges of intense effort	Aerobic respiration and lactic acid fermentation

The bulk of the ATP for aerobic exercise comes from the oxygen-requiring process of aerobic respiration, which derives ATP by the breakdown of the energy-rich sugar glucose (see Module 6.3). Muscles are richly supplied with blood vessels that bring O_2 and glucose and carry away CO_2, the by-product of aerobic respiration. Breathing and heart rate increase during exercise, facilitating the exchange of gases. Myoglobin, a hemoglobin-like protein that is found in muscles, may also supply oxygen for aerobic respiration.

If the demand for ATP outstrips the oxygen supply, muscle fibers can carry out the anaerobic process called lactic acid fermentation (see Module 6.13), which also uses glucose as a starting molecule. Fermentation works twice as fast as aerobic respiration, but supplies only a fraction as much ATP and can only sustain contraction for about one minute.

Glucose for ATP production is available from the bloodstream, and muscle tissue stores glycogen (see Module 3.7), which can be broken down to provide more glucose. The liver stores glycogen, too, and can release glucose into the bloodstream. Although the body can also mobilize fats as a source of fuel, the process of harvesting ATP from fatty acid breakdown is too slow to keep pace with the demands of increasing exercise intensity. Casual athletes are in no danger of running out of metabolic fuel during exercise, though—glycogen stores are typically more than adequate.

When exercise begins, it takes a few minutes for the aerobic "machinery" to start producing enough ATP to meet the increased demand. After exercise, muscles must repay the "oxygen debt" that was incurred during the first few minutes of exercise. For example, after a run, you breathe rapidly, and your heart rate remains elevated for a time as your muscles replenish their supplies of ATP and PCr and restore myoglobin to its oxygenated state. Oxygen is also used to metabolize the lactic acid that was produced by fermentation.

? **Compare the substances required in, the ATP output of, and the speed of aerobic respiration to that in lactic acid fermentation in muscles.**

Aerobic respiration requires glucose and O_2; fermentation requires glucose, but not O_2. Aerobic respiration produces many more ATP molecules than fermentation, but fermentation produces ATP molecules faster.

30.12 Characteristics of muscle fiber affect athletic performance

SCIENTIFIC THINKING

Can a genetic test predict athletic ability?

Recently developed techniques allow scientists to identify specific genetic loci associated with muscle structure and function. A genetic test for one of these genes, known as *ACTN3*, is marketed as a means of predicting whether an individual has a natural advantage at certain types of sports. To understand the scientific basis for this claim, let's first examine differences in muscle fibers that affect athletic performance.

The fibers that make up a muscle are not all alike. The contractions of "fast-twitch" fibers are rapid and powerful, but the fibers fatigue quickly. "Slow-twitch" fibers can sustain repeated contractions and are slow to fatigue, but their contractions are less forceful. The characteristics of slow and fast fibers are summarized in **Table 30.12A**. A third fiber type, which has some characteristics of both slow and fast fibers, is also abundant in human muscle. Most of the features associated with fiber type reflect the pathway(s) the fiber preferentially uses to generate ATP from energy-rich molecules. Each muscle typically has a mixture of fiber types, broadly correlated to the action it performs.

Fast and slow fibers contain different forms of myosin that hydrolyze ATP at different speeds. The faster ATP is consumed during muscle contraction, the faster the metabolic process needed to supply the ATP. Fast-twitch fibers cycle through cross-bridges rapidly, generating forceful contractions but using ATP at a breakneck pace. Hydrolysis of ATP occurs much more slowly in the myosin found in slow-twitch fibers. The slower pace at which cross-bridges are made and broken results in more sustained but less forceful contractions.

ACTN3 encodes α-actinin-3, a protein that is a major part of the Z line in fast-twitch muscle fibers, where it anchors thin filaments from adjacent sarcomeres. There are two alleles for *ACTN3*: *R*, which encodes a functional protein, and r, which does not. Thus, at least one *R* allele is required to produce α-actinin-3.

In a study published in 2003, a group of researchers hypothesized that variation in *ACTN3* is a factor in athletic performance. Fast-twitch fibers play a critical role in athletic events requiring forceful contractions that power brief, explosive movements, such as those that occur when bursting out of the blocks for a sprint, hoisting a heavy weight, or swinging a bat. Could the presence of α-actinin-3 be a key factor in enabling certain people to excel at power events?

For investigations that involve humans, scientists often compare groups of participants who differ in a specific characteristic. To test their hypothesis about the influence of α-actinin-3 on athletic events, the researchers compared the *ACTN3* genotypes of 301 elite athletes with the genotypes of a control group of 436 nonathletes. The athletes were classified as specialists in either power events such as short-distance running or swimming, or endurance events such as rowing or long-distance cycling or running. The results are shown in **Table 30.12B**. More than 90% of the power athletes, whose successful performance depends on fast-twitch fibers, had the ability to produce α-actinin-3 (genotype *RR* or *Rr*).

As is typical in the scientific process, numerous investigations have since replicated and extended the original study. A 2011 meta-analysis of more than a dozen such studies supports the hypothesis that possessing an *R* allele—especially with an *RR* genotype—is more common among power athletes. Thus, a genetic test for *ACTN3* genotype offers one piece of information about athletic potential. However, physical prowess is influenced by many genes in addition to *ACTN3* and by many other factors, as well.

TABLE 30.12A | CHARACTERISTICS OF MUSCLE FIBERS

Characteristic	Slow Fibers	Fast Fibers
Speed of contraction	Slow	Fast
Rate of fatigue	Fatigue slowly	Fatigue rapidly
Primary pathway for making ATP	Cellular respiration (aerobic)	Fermentation (anaerobic)
Myoglobin content	High	Low
Mitochondria and capillaries	Many	Few

TABLE 30.12B | GENOTYPE FREQUENCY (%) FOR *ACTN3*

Group	*RR*	*Rr*	*rr*
Power	49.5	44.9	5.6
Endurance	30.9	45.4	23.7
Control	29.8	51.8	18.4

Source: Data from N. Yang et al., *ACTN3* genotype is associated with human elite athletic performance, *American Journal of Human Genetics* 73: 627–31 (2003).

? How does the myosin in fast-twitch muscle fibers differ from the myosin in slow-twitch fibers?

Fast-twitch fibers hydrolyze ATP more rapidly.

CHAPTER 30 REVIEW

For practice quizzes, BioFlix animations, MP3 tutorials, video tutors, and more study tools designed for this textbook, go to MasteringBiology®

Reviewing the Concepts

Movement and Locomotion (30.1–30.2)

30.1 Locomotion requires energy to overcome friction and gravity. Animals that swim are supported by water but are slowed by friction. Animals that walk, hop, or run on land are less affected by friction but must support themselves against gravity. Burrowing or crawling animals must overcome friction. They may move by side-to-side undulation or by peristalsis. The wings of birds, bats, and flying insects are airfoils, which generate enough lift to overcome gravity.

30.2 Skeletons function in support, movement, and protection. Worms and cnidarians have hydrostatic skeletons—fluid held under pressure in closed body compartments. Exoskeletons are hard external cases, such as the chitinous, jointed skeletons of arthropods. The vertebrate endoskeleton is composed of cartilage and bone.

The Vertebrate Skeleton (30.3–30.6)

30.3 Vertebrate skeletons are variations on an ancient theme. Vertebrate skeletons consist of an axial skeleton (skull, vertebrae, and ribs) and an appendicular skeleton (shoulder girdle, upper limbs, pelvic girdle, and lower limbs). There are many variations on this basic body plan, which may have evolved through changes in gene regulation.

30.4 Bones are complex living organs. Cartilage at the ends of bones cushions the joints. Bone cells, serviced by blood vessels and nerves, reside in a matrix of flexible protein fibers and hard calcium salts. Long bones have a fat-storing central cavity and spongy bone at their ends. Spongy bone contains red marrow, where blood cells are made.

30.5 Healthy bones resist stress and heal from injuries. Bone cells continue to replace and repair bone throughout life. Osteoporosis, a bone disease characterized by weak, porous bones, occurs when bone destruction exceeds replacement.

30.6 Joints permit different types of movement.

Muscle Contraction and Movement (30.7–30.12)

30.7 The skeleton and muscles interact in movement. Antagonistic pairs of muscles produce opposite movements. Muscles perform work only when contracting.

30.8 Each muscle cell has its own contractile apparatus. Muscle fibers, or cells, consist of bundles of myofibrils, which contain bundles of overlapping thick (myosin) and thin (actin) protein filaments. Sarcomeres, repeating groups of thick and thin filaments, are the contractile units.

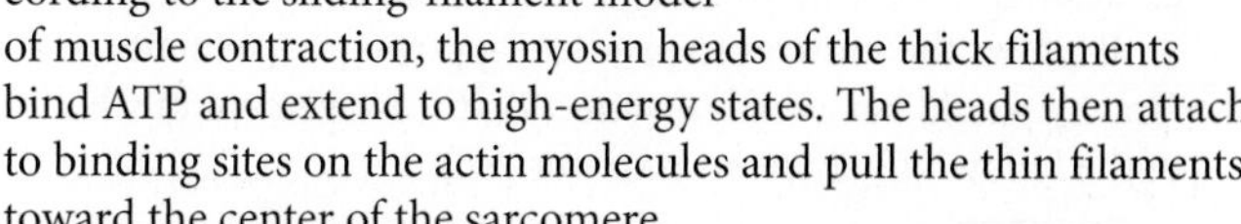
Sarcomere
Myosin
Actin

30.9 A muscle contracts when thin filaments slide along thick filaments. According to the sliding-filament model of muscle contraction, the myosin heads of the thick filaments bind ATP and extend to high-energy states. The heads then attach to binding sites on the actin molecules and pull the thin filaments toward the center of the sarcomere.

30.10 Motor neurons stimulate muscle contraction. Motor neurons carry action potentials that initiate muscle contraction. A neuron and the muscle fibers it controls constitute a motor unit. The neurotransmitter acetylcholine released at a synaptic terminal triggers an action potential that passes along T tubules into the center of the muscle cell. Calcium ions released from the endoplasmic reticulum initiate muscle contraction by moving the regulatory protein tropomyosin away from the myosin-binding sites on actin.

30.11 Aerobic respiration supplies most of the energy for exercise. Aerobic respiration requires a constant supply of glucose and oxygen. The anaerobic process of fermentation can start producing ATP faster than aerobic respiration can, but it produces less.

30.12 Characteristics of muscle fiber affect athletic performance. The classification of muscle fibers as slow-twitch or fast-twitch is based on the main pathway used to generate ATP. Alpha-actinin-3, a protein encoded by *ACTN3*, is found only in fast-twitch fibers. *ACTN3* genotype may predict success at particular types of athletic events.

Connecting the Concepts

1. Complete this concept map on animal movement.

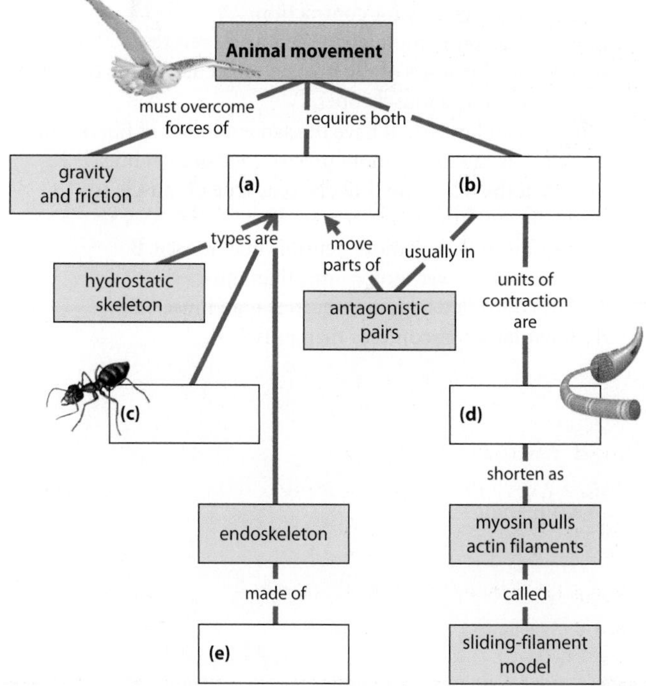
Animal movement
must overcome forces of
requires both
gravity and friction
(a)
(b)
types are
move parts of
usually in
hydrostatic skeleton
antagonistic pairs
units of contraction are
(c)
(d)
shorten as
endoskeleton
myosin pulls actin filaments
made of
called
(e)
sliding-filament model

Testing Your Knowledge

Level 1: Knowledge/Comprehension

2. A human's internal organs are protected mainly by the
   a. hydrostatic skeleton.
   b. axial skeleton.
   c. exoskeleton.
   d. appendicular skeleton.
3. Arm muscles and leg muscles are arranged in antagonistic pairs. How does this affect their functioning?
   a. It provides a backup if one of the muscles is injured.
   b. One muscle of the pair pushes while the other pulls.
   c. A single motor neuron can control both of them.
   d. It allows the muscles to produce opposing movements.
4. Gravity would have the least effect on the movement of which of the following? *(Explain your answer.)*
   a. a salmon
   b. a snake
   c. a sparrow
   d. a grasshopper
5. Which of the following bones in the human arm corresponds to the femur in the leg?
   a. radius
   b. tibia
   c. humerus
   d. metacarpal
6. Which of the following animals is correctly matched with its type of skeleton?
   a. fly—endoskeleton
   b. earthworm—exoskeleton
   c. lobster—exoskeleton
   d. bee—hydrostatic skeleton
7. When a dog is running fast, its body position is stabilized by
   a. side-to-side undulation.
   b. energy stored in tendons.
   c. foot contact with the ground.
   d. its momentum.
8. What is the role of calcium in muscle contraction?
   a. Its binding to a regulatory protein causes the protein to move, exposing actin binding sites to the myosin heads.
   b. It provides energy for contraction.
   c. It blocks contraction when the muscle relaxes.
   d. It forms the heads of the myosin molecules in the thick filaments inside a muscle fiber.
9. Muscle A and muscle B have the same number of fibers, but muscle A is capable of more precise control than muscle B. Which of the following is likely to be true of muscle A? *(Explain your answer.)*
   a. It is controlled by more neurons than muscle B.
   b. It contains fewer motor units than muscle B.
   c. It is controlled by fewer neurons than muscle B.
   d. It has larger sarcomeres than muscle B.
10. Which of the following statements about skeletons is true?
    a. Chitin is a major component of vertebrate skeletons.
    b. Loss of forelimbs in snakes involved little change in the axial skeleton.
    c. Most cnidarians must shed their skeleton periodically to grow.
    d. Vertebrate bones contain living cells.

Level 2: Application/Analysis

11. In terms of both numbers of species and numbers of individuals, insects are the most successful land animals. Write a paragraph explaining how their exoskeletons help them live on land. Are there any disadvantages to having an exoskeleton?
12. An owl swoops down, seizes a mouse in its talons, and flies back to its perch. Explain how its wings enable it to overcome the downward pull of gravity as it flies upward.
13. The greatest concentration of thoroughbred horse farms is in the bluegrass region of Kentucky. The grass in the limestone-based soil of this area is especially rich in calcium. How does this grass affect the development of championship horses?
14. Describe how you bend your arm, starting with action potentials and ending with the contraction of a muscle. How does a strong contraction differ from a weak one?
15. Using examples, explain this statement: "Vertebrate skeletons are variations on a theme."

Level 3: Synthesis/Evaluation

16. Drugs are often used to relax muscles during surgery. Which of the following chemicals do you think would make the best muscle relaxant, and why? Chemical A: Blocks acetylcholine receptors on muscle cells. Chemical B: Floods the cytoplasm of muscle cells with calcium ions.
17. An earthworm's body consists of a number of fluid-filled compartments, each with its own set of longitudinal and circular muscles. But in the roundworm, a single fluid-filled cavity occupies the body, and there are only longitudinal muscles that run its entire length. Predict how the movement of a roundworm would differ from the movement of an earthworm.
18. When a person dies, muscles become rigid and fixed in position—a condition known as rigor. Rigor mortis occurs because muscle cells are no longer supplied with ATP (when breathing stops, ATP synthesis ceases). Calcium also flows freely into dying cells. The rigor eventually disappears because the biological molecules break down. Explain, in terms of the mechanism of contraction described in Modules 30.9 and 30.10, why the presence of calcium and the lack of ATP would cause muscles to become rigid, rather than limp, soon after death.
19. **SCIENTIFIC THINKING** Imagine you have a friend who had her child's *ACTN3* genotype tested. After reviewing the study described in Module 30.12, what cautions would you offer about interpreting the test results?

Answers to all questions can be found in Appendix 4.

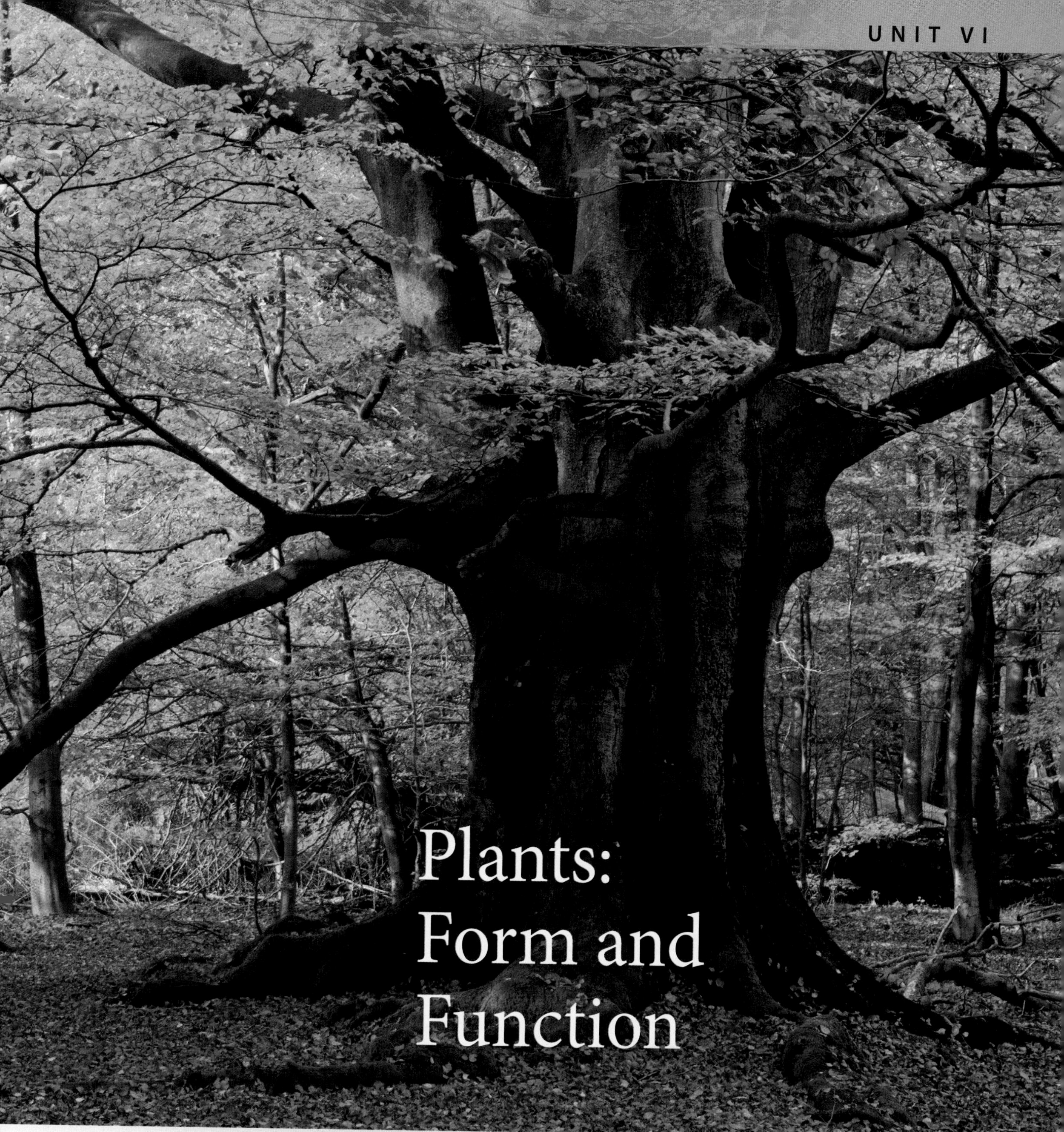

Plants: Form and Function

CHAPTER 31

Plant Structure, Growth, and Reproduction

It's impossible to tell the story of human civilization without talking about plants. From the earliest nomadic tribes, to colonists settling the New World, to genetic engineers working on the forefronts of biology, we humans have always relied on plants for food, fuel, shelter, clothing, and countless other necessities and niceties of life.

Throughout much of prehistory, humans were nomadic hunter-gatherers, migrating and foraging with the changing seasons to meet their food needs. These early peoples gathered and ate seeds from wild grasses and cereals, but they didn't plant them. And then, about 11,000 years ago, a major shift occurred: People in several parts of the world began to cultivate—purposefully sow, rather than just gather—crop species. Cultivation was soon followed by domestication, genetic changes in crop species resulting from the selection by humans of plants with desirable traits.

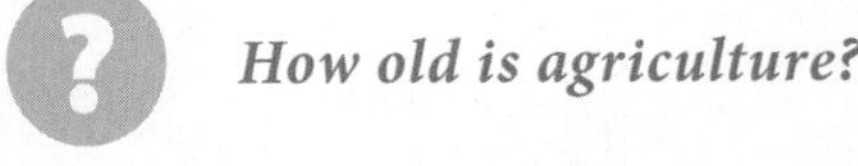
?
How old is agriculture?

The domestication of plants is one of the most significant events in the development of human civilization. Once begun, cultivation rapidly replaced gathering as a way of life for most of the world's people. For the first time, humans could control the time and place of their food supply, becoming less dependent on the vagaries of nature. Domestication allowed for the production of surplus food and the formation of stable, year-round farming villages. Farming, in turn, led to the development of communication and trade routes, the establishment of cities, and the emergence of our modern way of life.

Plants are vital to the well-being of not just humans but the entire biosphere. Because angiosperms—the flowering plants—make up more than 90% of the plant kingdom, we concentrate on them in this unit. We begin this chapter by examining angiosperm structure, first at the level of the whole plant and then at the microscopic level of tissues and cells. Then we'll see how plant structures function in growth and reproduction. Along the way, we'll keep our discussion rooted in humans' agricultural uses of plants.

BIG IDEAS

Plant Structure and Function
(31.1–31.6)

Plant bodies contain specialized cells grouped into tissues, organs, and organ systems.

Plant Growth
(31.7–31.8)

All plants increase their length via primary growth, and woody plants thicken via secondary growth.

Reproduction of Flowering Plants
(31.9–31.15)

Sexual reproduction in angiosperms involves pollination, development of fruit and seeds, seed dispersal, germination, and growth.

▷ Plant Structure and Function

31.1 The domestication of crops changed the course of human history

SCIENTIFIC THINKING

Wheat is one of the most important agricultural crops. It accounts for about 20% of all calories consumed worldwide. In the United States, it is a valuable cash crop, with American farmers exporting 30 million tons each year.

Evidence indicates that wheat was among the first wild crops to be cultivated. The origin of domesticated wheat has been traced to a region of the Middle East dubbed the "Fertile Crescent," near the upper reaches of the Tigris and Euphrates Rivers.

How old is agriculture?

How can scientists pinpoint a specific time and place for the establishment of agriculture?

Archaeobotanists study botanical artifacts—such as seeds and plant scraps—left among ancient ruins. Because plants undergo physical changes as they are genetically altered by human selection, domestication leaves measurable signs. For example, wild and domestic wheat varieties can be distinguished by several changes in plant anatomy, such as the way that the seeds are attached to the stalk. Archaeobotanists have collected wheat samples from many ancient settlements, examined the structure of the plant parts, and determined their ages through radiometric dating (see Module 15.5). The data indicate that, before about 10,000 years ago, all the wheat found in human settlements was wild. Around 10,000 years ago, wild wheat began to be replaced by domesticated varieties. Over the next few millennia, domesticated varieties gradually displaced their wild relatives (**Figure 31.1A**). Today, virtually all the wheat cultivated around the world is of the domesticated variety.

Botanists (biologists who study plants) have used a variety of methods to determine the time and place of the domestication of other crops. Many tropical plants—such as bananas, yams, and chilies—pose particular difficulties because they rot quickly in their humid climates, leaving little behind. But microscopic grains of starch can survive in and around settlements—on stone tools and shards of pottery, for example. Microscopic analysis of the size and distribution of amyloplasts, plant cellular organelles inside which starch is stored, can sometimes distinguish the difference between wild and domestic varieties.

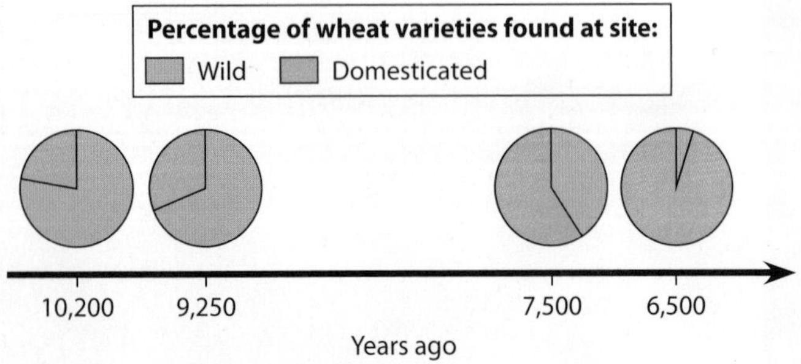

Percentage of wheat varieties found at site:
Wild
Domesticated
10,200
9,250
7,500
6,500
Years ago

▲ **Figure 31.1A** Percentage of wild domesticated wheat discovered at several ancient settlements

Data from K. Tanno and G. Willcox, How fast was wild wheat domesticated? *Science* 311:1886 (2006).

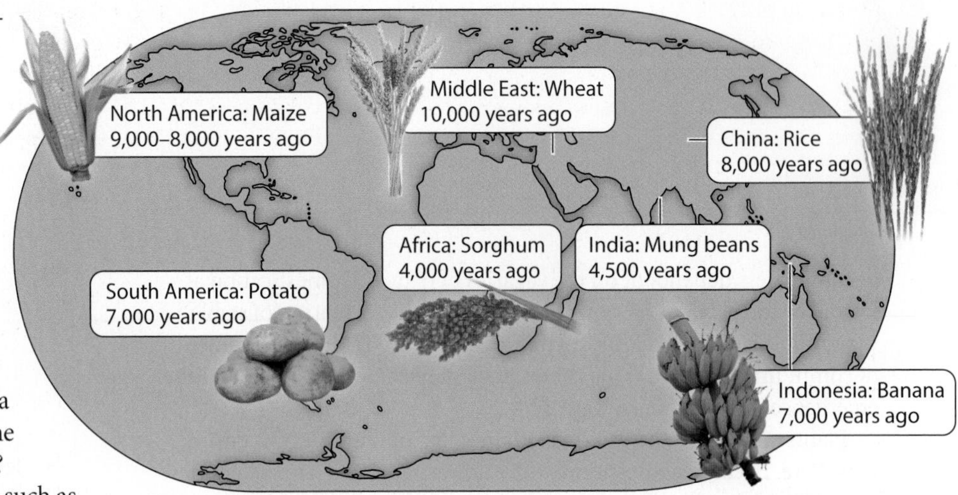

North America: Maize
9,000–8,000 years ago
Middle East: Wheat
10,000 years ago
China: Rice
8,000 years ago
Africa: Sorghum
4,000 years ago
India: Mung beans
4,500 years ago
South America: Potato
7,000 years ago
Indonesia: Banana
7,000 years ago

Adaptation of map "Multiple Birth" from "Seeking Agriculture's Ancient Roots" by Michael Balter, from *Science*, June 2009, 2007, Volume 316(5833). Copyright © 2007 by AAAS. Reprinted with permission.

▲ **Figure 31.1B** The domestication of food crops occurred in many locations over millennia

Genetic analysis can also be used to pinpoint the time and place of domestication. For example, there are several genes known to have undergone mutations as a wild species called teosinte was domesticated into modern corn. These genes affect traits that distinguish wild and domestic varieties, such as the numbers of stalks or how nutrients are stored within the plant. Using a standard estimate for the pace of genetic change (sometimes called a "molecular clock"—see Module 15.18), the time and place of domestication can be approximated. Such a genetic analysis indicates that corn was first domesticated about 9,000 years ago in southern Mexico.

By combining all these types of evidence—archaeological, anatomical, microscopic, genetic—botanists have been able to piece together an overall map and timeline for crop domestication throughout the world. Sites spanning the globe—including ones in North and South America, Africa, India, China, and Indonesia—have been associated with the first domestication of a staple crop over a period of several thousand years (**Figure 31.1B**). The investigation of our ancient agricultural roots demonstrates how scientists often synthesize multiple lines of evidence from a variety of sources to draw a broad conclusion.

As more evidence is gathered, botanists will continue to deepen our understanding of how the domestication of crops contributed to the development of modern human civilization. Keeping in mind the importance of plants in our society, we will explore the structure of angiosperms in the next five modules.

? In what sense did early farmers practice artificial selection?

● By choosing plants with desirable traits and breeding them, early humans used artificial selection to produce domesticated crops

31.2 The two major groups of angiosperms are the monocots and the eudicots

Angiosperms have dominated the land for more than 100 million years, and there are about 250,000 known species of flowering plants living today. Most of our foods come from a few hundred domesticated species of flowering plants. Among these foods are roots, such as beets and carrots; the fruits of trees and vines, such as apples, nuts, berries, and squashes; the fruits and seeds of legumes, such as peas, peanuts, and beans; and grains, the fruits of grasses such as wheat, rice, and corn.

On the basis of several structural features, botanists have traditionally classified angiosperms into two groups: monocots and dicots. The names *monocot* and *dicot* refer to the first leaves on the plant embryo. These embryonic leaves are called seed leaves, or **cotyledons**. A **monocot** embryo has one seed leaf; a **dicot** embryo has two seed leaves. The great majority of dicots, called the **eudicots** ("true" dicots), are evolutionarily related, having diverged from a common ancestor about 125 million years ago; a few smaller groups of dicots have evolved independently. In this chapter, we will focus on monocots and eudicots (**Figure 31.2**).

Monocots include the orchids, bamboos, palms, and lilies, as well as the grains and other grasses. You can see the single cotyledon inside the seed on the top left in Figure 31.2. The leaves, stems, flowers, and roots of monocots are also distinctive. Most monocots have leaves with parallel veins. Monocot stems have vascular tissues (internal tissues that transport water and nutrients) organized into bundles that are arranged in a scattered pattern. The flowers of most monocots have their petals and other parts in multiples of three. Monocot roots form a shallow fibrous system—a mat of threads—that spreads out below the soil surface. With most of their roots in the top few centimeters of soil, monocots, especially grasses, make excellent ground cover and can help reduce soil erosion. Fibrous root systems are thus well adapted to shallow soils where rainfall is light.

Most flowering plants are eudicots, including many food crops (such as nearly all our fruits and vegetables), the majority of ornamental plants, and most shrubs and trees (except for the gymnosperms, naked seed plants such as cone-bearing conifers). You can see the two cotyledons of a typical eudicot in the seed on the lower left in Figure 31.2. Eudicot leaves have a multibranched network of veins, and eudicot stems have vascular bundles arranged in a ring. Eudicot flowers usually have petals and other parts in multiples of four or five. The large, vertical root of a eudicot, called a taproot, extends deep into the soil, as you know if you've ever tried to pull up a dandelion. Taproots are thus well adapted to soils with deep groundwater.

As we saw in the preceding unit on animals, a close look at a structure often reveals its function. Conversely, function provides insight into the "logic" of a structure. In the modules that follow, we'll take a detailed look at the correlation between plant structure and function.

? **The terms *monocot* and *eudicot* refer to the number of ____________ on the developing embryo in a seed.**

cotyledons (seed leaves)

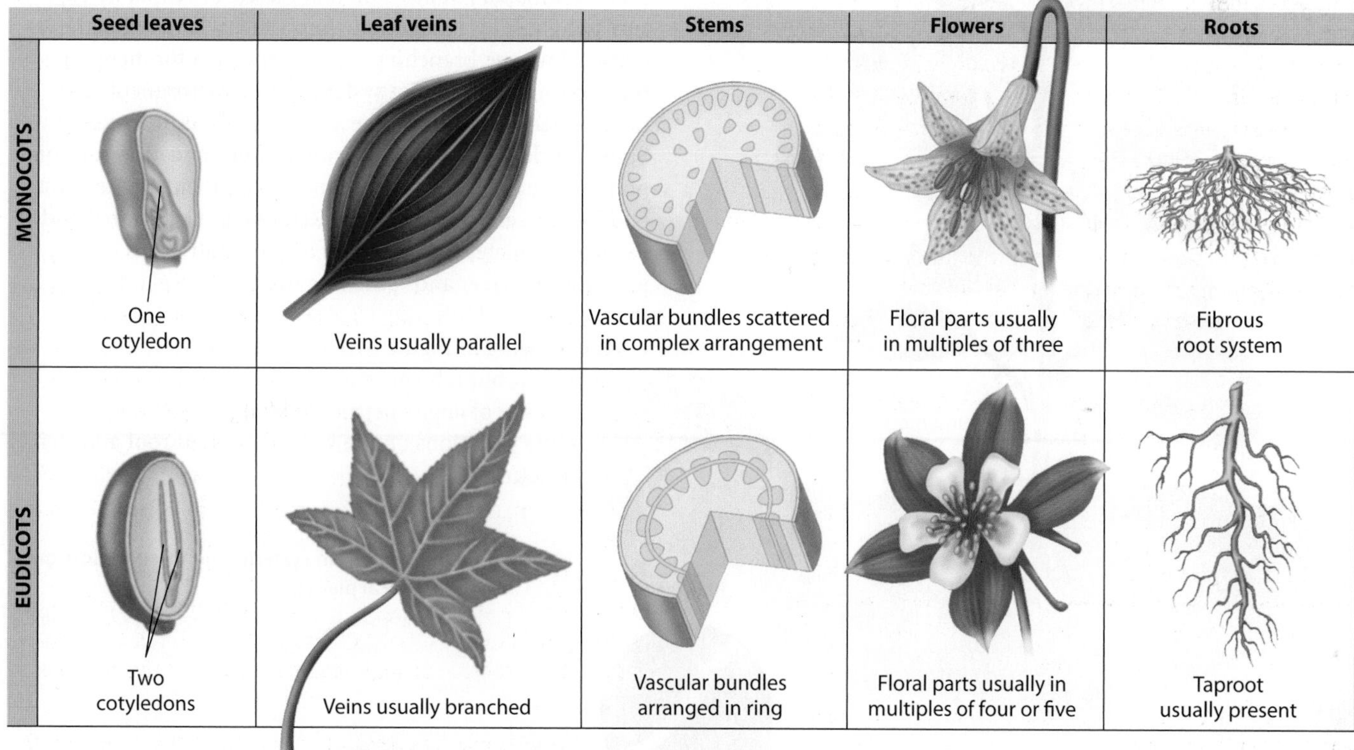
Seed leaves
Leaf veins
Stems
Flowers
Roots
MONOCOTS
One cotyledon
Veins usually parallel
Vascular bundles scattered in complex arrangement
Floral parts usually in multiples of three
Fibrous root system
EUDICOTS
Two cotyledons
Veins usually branched
Vascular bundles arranged in ring
Floral parts usually in multiples of four or five
Taproot usually present

▲ **Figure 31.2** A comparison of monocots and eudicots

Try This Trace your finger along the cotyledons in each plant and relate that to the names *monocot* and *eudicot*.

31.3 A typical plant body contains three basic organs: roots, stems, and leaves

The bodies of plants, like those of most animals, contain numerous organs grouped into organ systems. An **organ** consists of several types of tissues that together carry out particular functions. In this and the next module, we'll focus on plant organs. We will then work our way down the structural hierarchy and examine plant tissues (in Module 31.5) and then individual cells (in Module 31.6).

The basic structure of plants reflects their evolutionary history as land-dwelling organisms. Most plants must draw resources from two very different environments: They must absorb water and minerals from the soil, while simultaneously obtaining CO_2 and light from above ground. The subterranean roots and aerial shoots (stems and leaves) of a typical land plant, such as the generalized flowering plant shown in **Figure 31.3**, perform these vital functions. Neither roots nor shoots can survive without the other. Most roots remain in the dark and lack chloroplasts—and thus would starve without sugar and other organic nutrients transported from photosynthetic leaves and stems. Conversely, stems and leaves depend on the water and minerals absorbed by roots from the soil.

A plant's **root system** anchors it in the soil, absorbs and transports minerals and water, and stores carbohydrates. Near the root tips, a vast number of tiny finger-like projections called **root hairs** enormously increase the surface area of roots, allowing for the efficient absorption of water and minerals. As shown on the far right of the figure, each root hair is an outgrowth of a root epidermal cell (a cell in the outer layer of the root). It is difficult to move an established plant without injuring it because such transplantation often damages the plant's delicate root hairs.

The **shoot system** of a plant is made up of stems, leaves, and structures for reproduction, which in angiosperms are the flowers. The **stems** are the parts of the plant that are generally above the ground and that support and separate the leaves (thereby promoting photosynthesis) and flowers (responsible for reproduction). In the case of a tree, the stems are the trunk and all the branches, including the smallest twigs. A stem has **nodes**, the points at which leaves are attached, and **internodes**, the portions of the stem between nodes. The **leaves** are the main photosynthetic organs in most plants, although green stems also perform photosynthesis. Most leaves consist of a flattened blade and a stalk, or petiole, which joins the leaf to a node of the stem.

The buds of a plant are undeveloped shoots. When a plant stem is growing in length, the **terminal bud** (also called the apical bud) at the apex (tip) of the stem has developing leaves and a compact series of nodes and internodes. The **axillary buds**, one in each of the crooks formed by a leaf and the stem, are usually dormant. In many plants, the terminal bud produces hormones that inhibit growth of the axillary buds (see Module 33.4), a phenomenon called **apical dominance**. By concentrating resources on growing taller, apical dominance is an evolutionary adaptation that increases the plant's exposure to light. This is especially important where vegetation is dense. However, branching is also important for increasing the exposure of the shoot system to the environment, and under certain conditions, the axillary buds begin growing. Some develop into shoots bearing flowers, and others become nonreproductive branches complete with their own terminal buds, leaves, and axillary buds. Removing the terminal bud usually stimulates the growth of axillary buds. This is why pruning fruit trees and "pinching back" house plants make them bushier.

The drawing in Figure 31.3 gives an overview of plant structure, but it by no means represents the enormous diversity of angiosperms. Next, let's look briefly at some variations on the basic themes of root and stem structure.

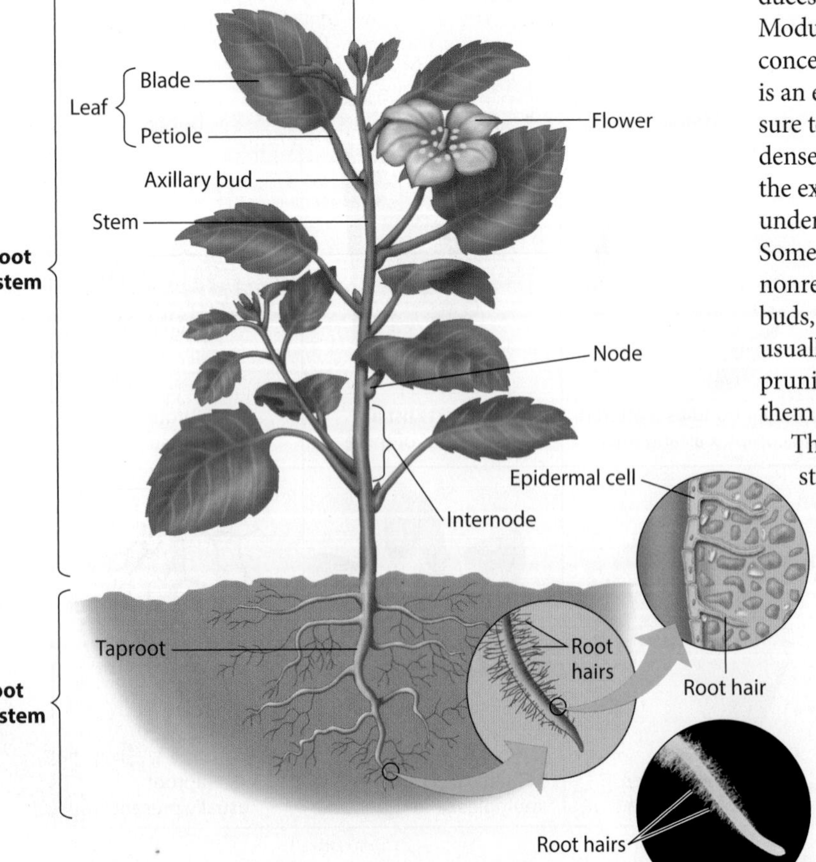

Terminal bud
Leaf
Blade
Petiole
Axillary bud
Stem
Flower
Shoot system
Node
Epidermal cell
Internode
Taproot
Root hairs
Root hair
Root system
Root hairs

▲ **Figure 31.3** The body plan of a flowering plant (a eudicot)

? **Name the two organ systems and three basic organs found in typical plants.**

● Root system and shoot system; roots, stems, leaves

31.4 Many plants have modified roots, stems, and leaves

Over evolutionary history, the three basic plant organs—roots, stems, and leaves—have become adapted for a variety of functions. Carrots, turnips, sugar beets, and sweet potatoes, for instance, all have unusually large taproots that store food in the form of carbohydrates such as starch (**Figure 31.4A**). The plants consume the stored sugars during flowering and fruit production. For this reason, root crops are harvested before flowering, when their nutritional value is maximized.

▲ **Figure 31.4A** The modified root of a sugar beet plant

Figure 31.4B shows three examples of modified stems. The strawberry plant has a horizontal stem called a stolon (or runner) that grows along the ground. Stolons enable a plant to reproduce asexually, as plantlets form at nodes along their length. That is why strawberries, if left unchecked, can rapidly fill your garden. You've seen a different stem modification if you have ever dug up an iris plant or cooked with fresh ginger; the large, brownish, rootlike structures of these plants are actually **rhizomes**, horizontal stems that grow near the soil surface. Rhizomes store food and, having buds, can also form new plants. About every three years, gardeners can dig up iris rhizomes, split them, and replant to get multiple identical plants. A potato plant has rhizomes that end in enlarged structures specialized for storage called **tubers** (the potatoes we eat). Potato "eyes" are axillary buds on the tubers that can grow when planted, allowing potatoes to be easily propagated. Plant bulbs, on the other hand, are underground shoots containing swollen leaves that store food. As you peel an onion, you are removing layers of leaves attached to a short stem.

▲ **Figure 31.4C** Modified leaves: the tendrils of a Red bryony vine (left) and cactus spines (right)

Plant leaves, too, are highly varied. Grasses and many other monocots have long leaves without petioles. Some eudicots, such as celery, have enormous petioles—the stalks we eat. The left photograph in **Figure 31.4C** shows a modified leaf called a **tendril**. Tendrils help plants climb. (Some tendrils, as in grapevines, are modified stems.) The spines of the barrel cactus are modified leaves that protect the plant from being eaten. The main part of the cactus is the large green stem, which is adapted for photosynthesis and water storage.

So far, we have examined plants as we see them with the unaided eye. Next, we dissect a plant and explore plant tissues on a microscopic level.

? **In what sense do cactus spines deviate from the typical function of plant leaves?**

● Unlike most leaves, cactus spines are not the main site of photosynthesis.

Stolon (runner)
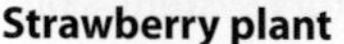

Strawberry plant

Rhizome

Iris plant

Taproot
Rhizome
Tuber

Potato plant

▲ **Figure 31.4B** Three kinds of modified stems: stolons, rhizomes, and tubers

31.5 Three tissue systems make up the plant body

Like the organs of most animals, the organs of plants contain tissues with characteristic functions. A **tissue** is a group of cells that together perform a specialized function. For example, **xylem** tissue contains water-conducting cells that convey water and dissolved minerals upward from the roots, while **phloem** tissue contains cells that transport sugars and other organic nutrients from leaves or storage tissues to other parts of the plant.

A tissue system consists of one or more tissues organized into a functional unit within a plant. Each plant organ—root, stem, or leaf—is made up of three tissue systems: the dermal, vascular, and ground tissue system. Each tissue system is continuous throughout the entire plant body, but the systems are arranged differently in leaves, stems, and roots (**Figure 31.5**). In this module, we examine the tissue systems of young roots and shoots. Later we will see that the tissue systems are somewhat different in older roots and stems.

The **dermal tissue system** (brown in the figure) is the plant's outer protective covering. Like our own skin, it forms the first line of defense against physical damage and infectious organisms. In nonwoody plants, the dermal tissue system usually consists of a single layer of tightly packed cells called the **epidermis**. On leaves and most stems, dermal cells secrete a waxy coating called the **cuticle**, which helps prevent water loss. The second tissue system is the **vascular tissue system** (purple). It is made up of xylem and phloem tissues and provides support and long-distance transport between the root and shoot systems. Tissues that are neither dermal nor vascular make up the **ground tissue system** (yellow). The ground tissue system accounts for most of the bulk of a young plant, filling the spaces between the epidermis and vascular tissue system. Ground tissue internal to the vascular tissue is called **pith**, and ground tissue external to the vascular tissue is called **cortex**. The ground tissue system has diverse functions, including photosynthesis, storage, and support.

The close-up views in Figure 31.5 show how these three tissue systems are organized in typical plant roots, stems, and leaves. The view at the bottom left shows in cross section the three tissue systems in a young eudicot root. Water and minerals that are absorbed from the soil must enter through the epidermis. In the center of the root, the vascular tissue system forms a **vascular cylinder**, with the cross sections of xylem cells radiating from the center like spokes of a wheel and phloem cells filling in the wedges between the spokes. The ground tissue system of the root, the region between the vascular cylinder and epidermis, consists entirely of cortex. The cortex cells store food as starch and take up minerals that have entered the root through the epidermis. The innermost layer of the cortex is the **endodermis**, a cylinder one cell thick. The endodermis is a selective barrier that regulates the passage of substances between the cortex and the vascular tissue (as you'll see in Module 32.2).

The bottom right of Figure 31.5 shows a cross section of a young monocot root. There are several similarities to the eudicot root: an outer layer of epidermis (dermal tissue) surrounding a large cortex (ground tissue), with a vascular cylinder (vascular tissue) at the center. But in a monocot root, the vascular tissue consists of a central core of cells surrounded by a ring of xylem and a ring of phloem.

As the center of Figure 31.5 indicates, the young stem of a eudicot looks quite different from that of a monocot. Both stems have their vascular tissue system arranged in numerous **vascular bundles**. However, in monocot stems the bundles are scattered, whereas in most eudicots they are arranged in a ring. This ring separates the ground tissue into cortex and pith regions. The cortex fills the space between the vascular ring and the epidermis. The pith fills the center of the stem and is often important in food storage. In a monocot stem, the ground tissue is not separated into these regions because the vascular bundles don't form a ring.

The top right of Figure 31.5 illustrates the arrangement of the three tissue systems in a typical eudicot leaf. The epidermis is interrupted by tiny pores called **stomata** (singular, *stoma*), which allow exchange of CO_2 and O_2 between the surrounding air and the photosynthetic cells inside the leaf. Also, most of the water vapor lost by a plant passes through stomata. Each stoma is flanked by two **guard cells**, specialized epidermal cells that regulate the opening and closing of the stoma.

The ground tissue system of a leaf, called the **mesophyll**, is sandwiched between the upper and lower epidermis. Mesophyll consists mainly of cells specialized for photosynthesis. The green structures in the diagram are their chloroplasts. In this eudicot leaf, notice that cells in the lower area of mesophyll are loosely arranged, with a labyrinth of air spaces through which CO_2 and O_2 circulate. The air spaces are particularly large in the vicinity of stomata, where gas exchange with the outside air occurs. In many monocot leaves and in some eudicot leaves, the mesophyll is not arranged in distinct upper and lower areas.

In both monocots and eudicots, the leaf's vascular tissue system is made up of a branching network of veins. As you can see in Figure 31.5, each **vein** is a vascular bundle composed of xylem and phloem tissues surrounded by a protective sheath of cells. The veins' xylem and phloem, continuous with the vascular bundles of the stem, are in close contact with the leaf's photosynthetic tissues. This ensures that those tissues are supplied with water and mineral nutrients from the soil and that sugars made in the leaves are transported throughout the plant. The vascular structure also functions as a framework that reinforces the shape of the leaf.

In the last three modules, we have examined plant structure at the level of organs and tissues. In the next module, we will complete our descent into the structural hierarchy of plants by taking a look at plant cells.

? **For each of the following structures in your body, name the most analogous plant tissue system: circulatory system, skin, adipose tissue (body fat).**

Vascular tissue system, dermal tissue system, ground tissue system

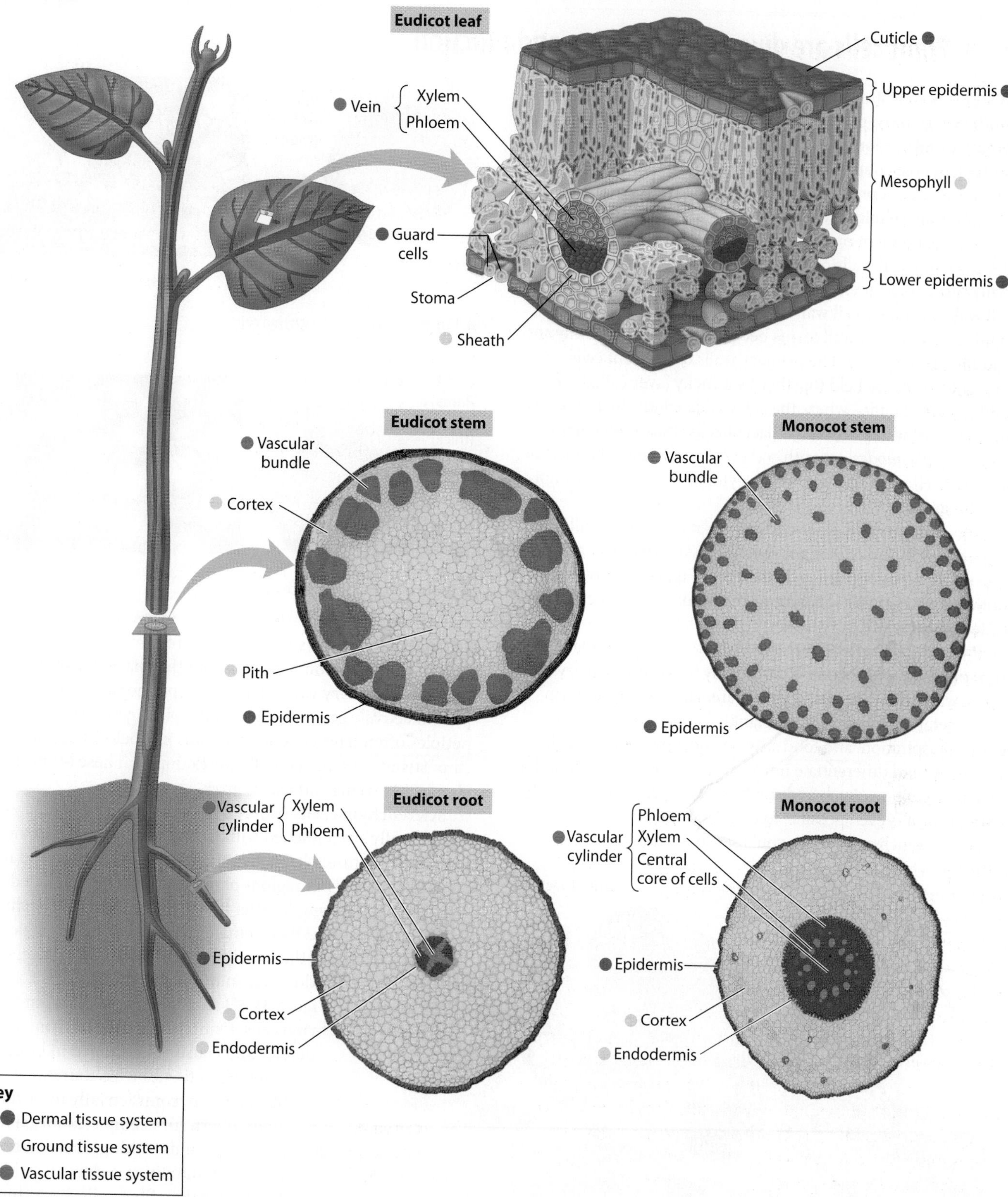
Eudicot leaf
Cuticle
Upper epidermis
Vein
Xylem
Phloem
Mesophyll
Guard cells
Stoma
Lower epidermis
Sheath
Eudicot stem
Vascular bundle
Cortex
Pith
Epidermis
Monocot stem
Vascular bundle
Epidermis
Eudicot root
Vascular cylinder
Xylem
Phloem
Epidermis
Cortex
Endodermis
Monocot root
Vascular cylinder
Phloem
Xylem
Central core of cells
Epidermis
Cortex
Endodermis
Key
Dermal tissue system
Ground tissue system
Vascular tissue system

▲ **Figure 31.5** The three plant tissue systems

Try This Refer back to Figure 31.2. Then identify the structural differences between monocots and eudicots that are evident in this figure.

31.6 Plant cells are diverse in structure and function

In addition to features shared with other eukaryotic cells (see Module 4.4), most plant cells have three unique structures (Figure 31.6A): chloroplasts, the sites of photosynthesis; a large central vacuole containing fluid that helps maintain cell turgor (firmness); and a protective cell wall made from the structural carbohydrate cellulose that surrounds the plasma membrane.

The enlargement on the right in Figure 31.6A highlights the adjoining cell walls of two cells. Many plant cells, especially those that provide structural support, have a two-part cell wall; a primary cell wall is formed first, and then a more rigid secondary cell wall forms between the plasma membrane and the primary wall. The primary walls of adjacent cells in plant tissues are held together by a sticky layer called the middle lamella. Pits, where the cell wall is relatively thin, allow migration of water between adjacent cells. **Plasmodesmata** (singular, *plasmodesma*) within these pits are open channels in adjacent cell walls through which cytoplasm and various molecules can flow from cell to cell.

The structure of a plant cell and the nature of its wall often correlate with the cell's main functions. As you consider the five major types of plant cells shown in Figures 31.6B–31.6F, notice the structural adaptations that make their specific functions possible.

Parenchyma cells (Figure 31.6B) are the most abundant type of cell in most plants. They usually have only primary cell walls, which are thin and flexible. Parenchyma cells perform most metabolic functions of a plant, such as photosynthesis, aerobic respiration, and food storage. Most parenchyma cells can divide and differentiate into other types of plant cells under certain conditions, such as during the repair of an injury. In the laboratory, it is even possible to regenerate an entire plant from a single parenchyma cell (as you'll see in Module 31.14).

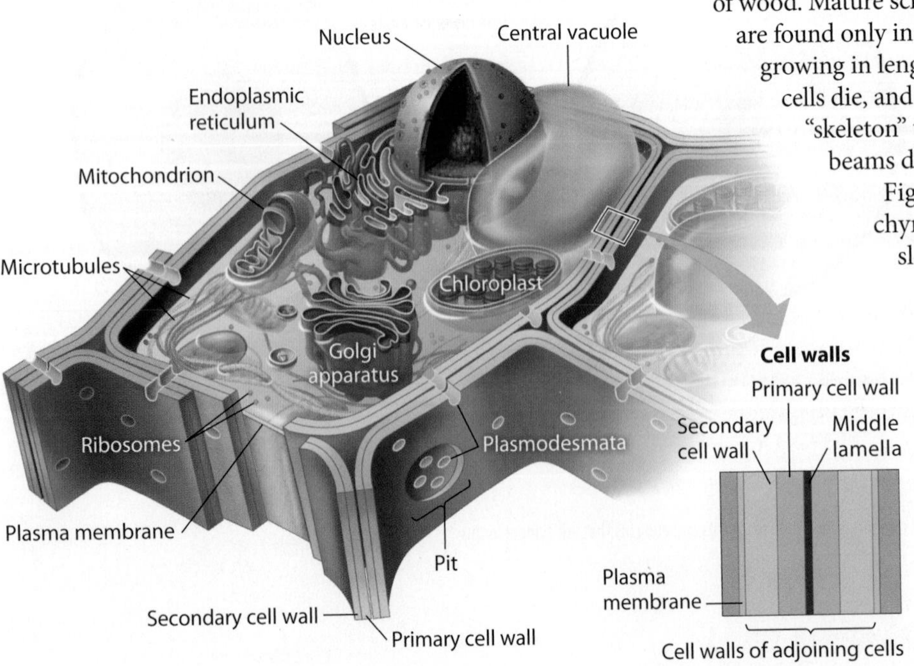
Nucleus
Central vacuole
Endoplasmic reticulum
Mitochondrion
Microtubules
Chloroplast
Golgi apparatus
Ribosomes
Plasmodesmata
Plasma membrane
Pit
Secondary cell wall
Primary cell wall
Cell walls
Primary cell wall
Secondary cell wall
Middle lamella
Plasma membrane
Cell walls of adjoining cells

▲ **Figure 31.6A** The structure of a plant cell, with emphasis on the cell walls of adjoining cells

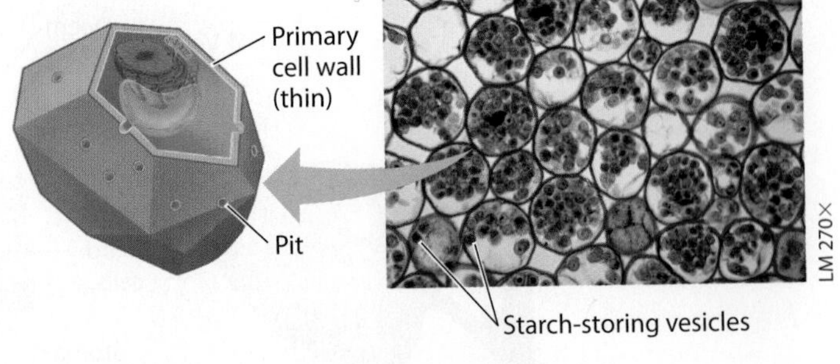
Primary cell wall (thin)
Pit
Starch-storing vesicles
LM 270×

▲ **Figure 31.6B** Parenchyma cell

Primary cell wall (thick)
LM 270×

▲ **Figure 31.6C** Collenchyma cell

Collenchyma cells (Figure 31.6C) resemble parenchyma cells in lacking secondary walls, but they have unevenly thickened primary walls. These cells provide flexible support in actively growing parts of the plant; young stems and petioles often have collenchyma cells just below their surface (the "string" of a celery stalk, for example). These living cells elongate as stems and leaves grow.

Sclerenchyma cells (Figure 31.6D) have thick secondary cell walls usually strengthened with lignin, a chemical component of wood. Mature sclerenchyma cells cannot elongate and thus are found only in regions of the plant that have stopped growing in length. After they mature, most sclerenchyma cells die, and their remaining cell walls form a rigid "skeleton" that supports the plant much as steel beams do in the interior of a building.

Figure 31.6D shows two types of sclerenchyma cells. One, called a **fiber**, is long and slender and is usually arranged in bundles. Some plant tissues with abundant fiber cells are commercially important; hemp fibers, for example, are used to make rope and clothing, and flax fibers are woven into linen. **Sclereids**, which are shorter than fiber cells, have thick, irregular, and very hard secondary walls. Sclereids impart the hardness to nutshells and seed coats and the gritty texture to the soft tissue of a pear.

The xylem tissue of angiosperms includes two types of water-conducting cells: tracheids and vessel elements. Both have rigid, lignin-containing

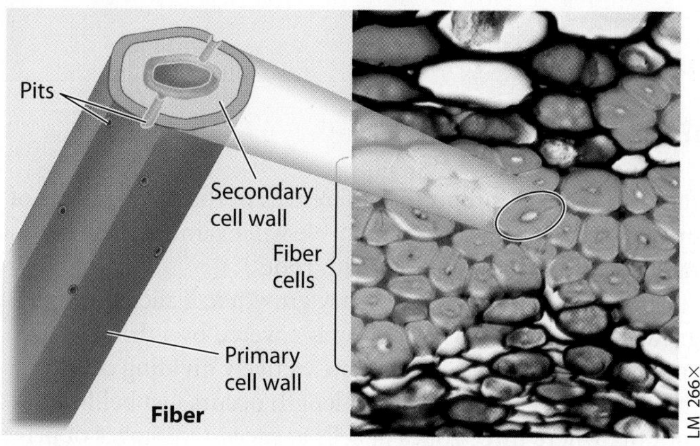
Pits
Secondary cell wall
Fiber cells
Primary cell wall
Fiber
LM 266×

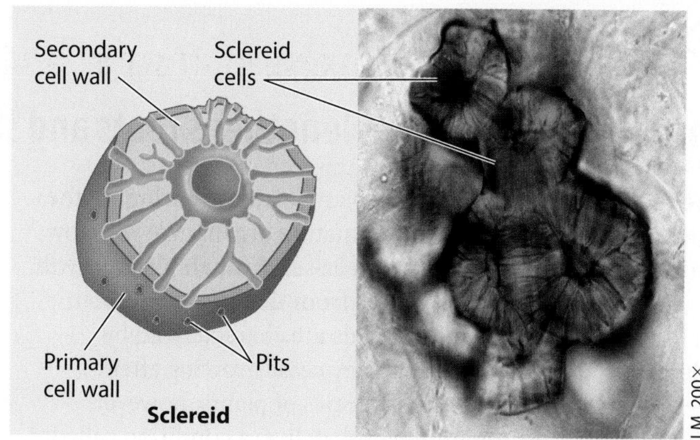
Secondary cell wall
Sclereid cells
Primary cell wall
Pits
Sclereid
LM 200×

▲ **Figure 31.6D** Sclerenchyma cells: fiber (left) and sclereid (right)

secondary cell walls. As **Figure 31.6E** on the right side of the page shows, the **tracheids** are long, thin cells with tapered ends. **Vessel elements** are wider, shorter, and less tapered. Chains of tracheids or vessel elements with overlapping ends form a system of tubes that conveys water from the roots to the stems and leaves as part of xylem tissue. The tubes are hollow because both tracheids and vessel elements are dead when mature, with only their cell walls remaining. Water passes through pits in the walls of tracheids and vessel elements and through openings in the end walls of vessel elements. With their thick, rigid walls, these cells also function in support.

Food-conducting cells, known as **sieve-tube elements** (or sieve-tube members), are also arranged end to end, forming tubes as part of phloem tissue (**Figure 31.6F**). Unlike water-conducting cells, however, sieve-tube elements remain alive at maturity, although they lose most of their organelles, including the nucleus and ribosomes. This reduction in cell contents with maturity enables nutrients to pass more easily through the cell. The end walls between sieve-tube elements, called **sieve plates**, have pores that allow fluid to flow from cell to cell along the sieve tube. Alongside each sieve-tube element is a **companion cell**, which is connected to the sieve-tube element by numerous plasmodesmata. One companion cell may serve multiple sieve-tube elements by producing and transporting proteins to all of them.

Now that we have reached the lowest level in the structural hierarchy of plants—cells—let's review by moving back up. Cells of plants are grouped into tissues with characteristic functions. For example, xylem tissue contains water-conducting cells that convey water and dissolved minerals upward from the roots as well as sclerenchyma cells, which provide support, and parenchyma cells, which store various materials. Xylem and phloem tissues are organized into the vascular tissue system, which provides structural support and long-term transport throughout the plant body. The vascular, dermal, and ground tissue systems connect all the plant organs: roots, stems, and leaves.

? **Identify which of the following cell types can give rise to all others in the list: collenchyma, sclereid, parenchyma, vessel element, companion cell.**

● Parenchyma

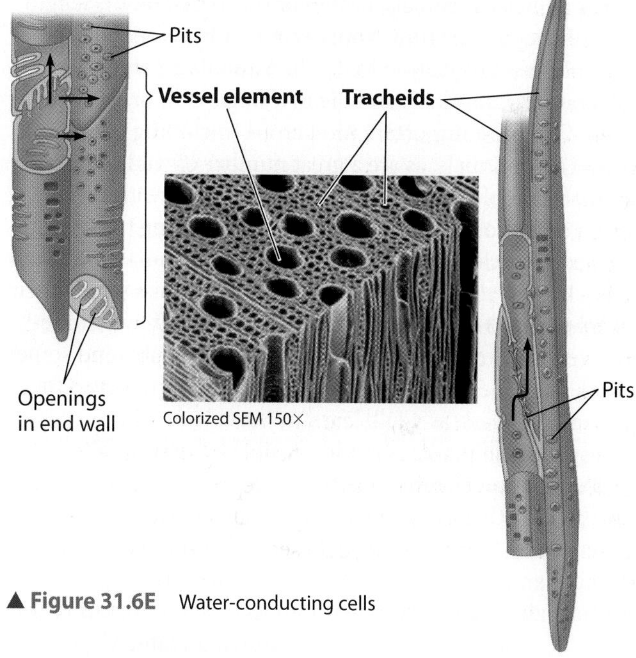
Pits
Vessel element
Tracheids
Openings in end wall
Colorized SEM 150×
Pits

▲ **Figure 31.6E** Water-conducting cells

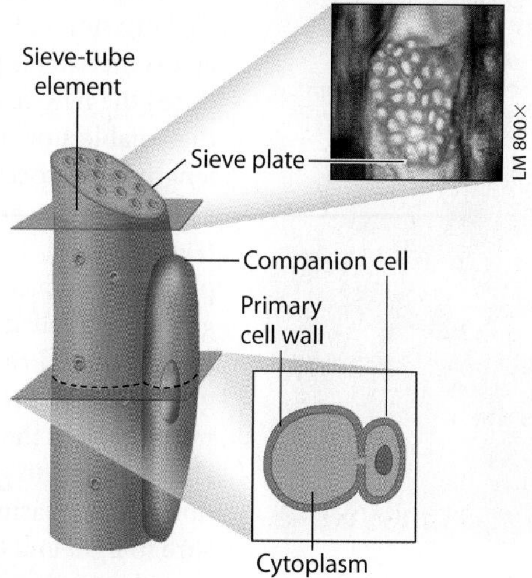
Sieve-tube element
Sieve plate
LM 800×
Companion cell
Primary cell wall
Cytoplasm

▲ **Figure 31.6F** Food-conducting cell (sieve-tube element)

▷ Plant Growth

31.7 Primary growth lengthens roots and shoots

So far, we have surveyed basic plant anatomy, including the structure of tissues and cells in mature organs. We will now consider how such organization arises through plant growth.

The growth of a plant differs from that of an animal in a fundamental way. Most animals are characterized by **determinate growth**; that is, they cease growing after reaching a certain size. Most species of plants, however, continue to grow for as long as they live, a condition called **indeterminate growth**. Indeterminate growth allows a plant to continuously increase its exposure to sunlight, air, and soil.

Indeterminate growth does not mean that plants are immortal; they do, of course, die. Flowering plants are categorized as annuals, biennials, or perennials, based on the length of their life cycle, the time from germination through flowering and seed production to death. **Annuals** emerge from seed, mature, reproduce, and die in a single year or growing season. Our most important food crops (including grains and legumes) are annuals, as are a great number of wildflowers. **Biennials** complete their life cycle in two years, with flowering and seed production usually occurring during the second year. Beets, parsley, turnips, and carrots are biennials, but we usually harvest them in their first year and so miss seeing their flowers (Figure 31.7A). **Perennials** are plants that live and reproduce for many years. They include trees, shrubs, and some grasses. Some perennials can live a very long time—even for thousands of years, as you'll learn in Module 31.15.

Growth in all plants is made possible by tissues called meristems. A **meristem** consists of undifferentiated (unspecialized) cells that divide when conditions permit, generating new cells. Some products of this division remain in the meristem and produce still more cells, whereas others differentiate and are incorporated into tissues and organs of the growing plant. Meristems at the tips of roots and in the buds of shoots are called **apical meristems**. Cell division in the apical meristems produces the new cells that enable a plant to lengthen, a process called **primary growth** (Figure 31.7B). Tissues produced by primary growth are called primary tissues. Primary growth enables roots to push through the soil and allows shoots to grow upward, increasing exposure to light and CO_2. Although apical meristems lengthen both roots and shoots, there are important differences in the mechanisms of primary growth in each system. We will examine them separately, starting at the bottom with roots.

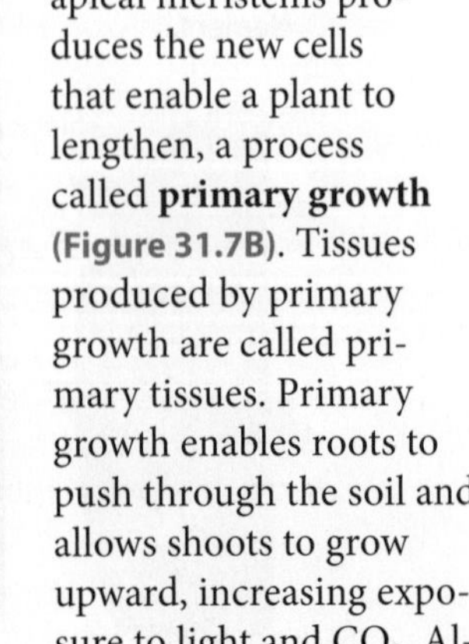

▲ **Figure 31.7A** Second-year carrots are grown to harvest their seeds, not their roots

Figure 31.7C illustrates primary growth in a slice through a growing onion root. The root tip is covered by a thimble-like **root cap** that protects the delicate, actively dividing cells of the apical meristem. Growth in length occurs just behind the root tip, where three zones of cells at successive stages of primary growth are located. Moving up from the root tip, they are called the zone of cell division, the zone of elongation, and the zone of differentiation. The three zones of cells overlap, with no sharp boundaries between them.

The zone of cell division includes the root apical meristem and cells that derive from it. New root cells are produced in this region, including the cells of the root cap. In the zone of elongation, root cells grow longer, sometimes to more than 10 times their original length. It is cell elongation in this zone that pushes the root tip deeper into the soil. The cells lengthen, rather than expand equally in all directions, because of the circular arrangement of cellulose fibers in parallel bands in their cell walls. The enlargement diagrams in the figure, shown at the left of the root, indicate how this works. The cells elongate by taking up water, and as they do, the cellulose fibers (shown in red) separate, somewhat like

Terminal bud
Axillary buds
Root tips
Arrows = direction of growth

▲ **Figure 31.7B** Primary growth involves lengthening of plant shoots and roots

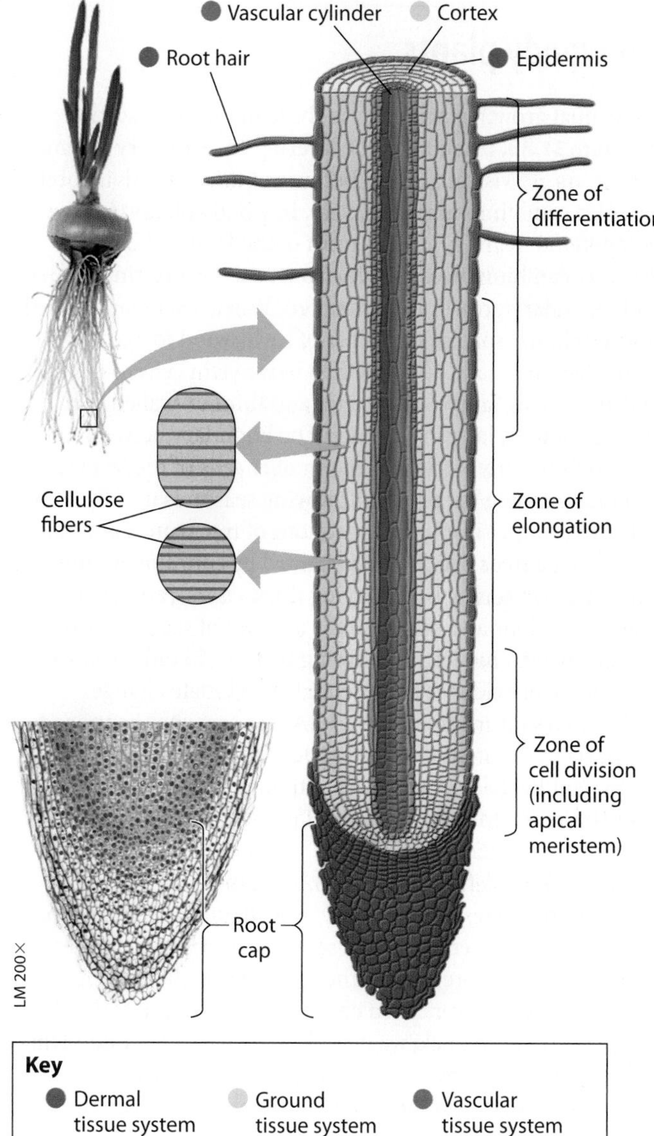

Vascular cylinder
Cortex
Root hair
Epidermis
Zone of differentiation
Cellulose fibers
Zone of elongation
Zone of cell division (including apical meristem)
Root cap
LM 200×
Key
Dermal tissue system
Ground tissue system
Vascular tissue system

▲ **Figure 31.7C** Primary growth of a root

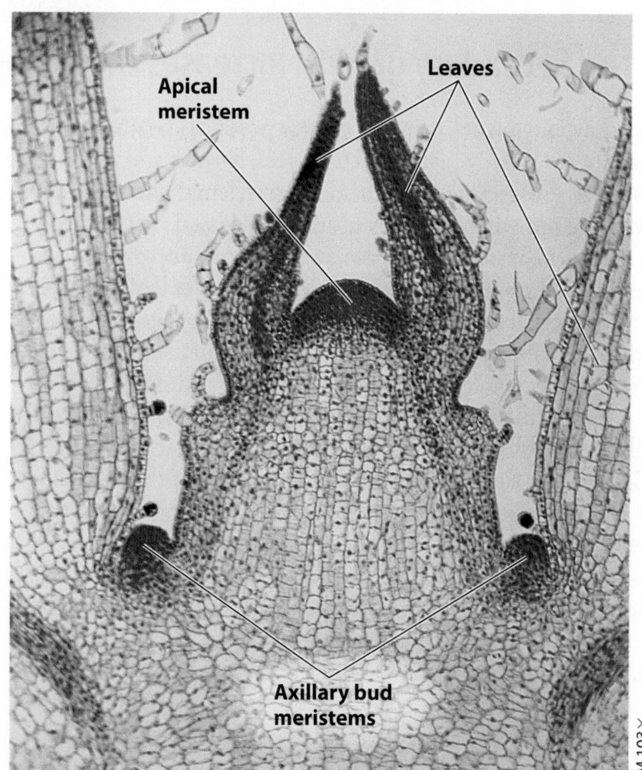

Apical meristem
Leaves
Axillary bud meristems
LM 103×

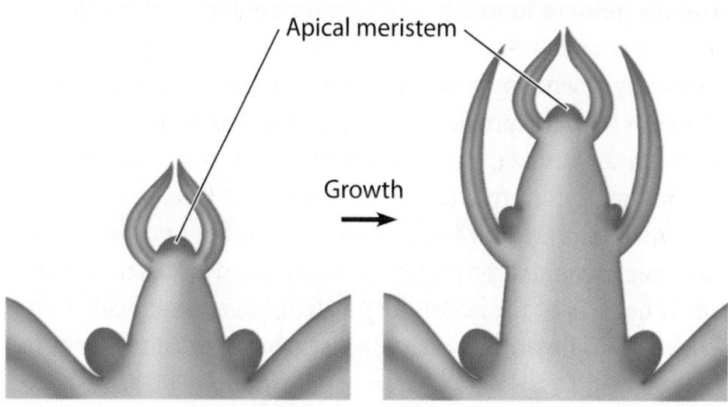

Apical meristem
Growth

▲ **Figure 31.7D** Primary growth of a shoot

an expanding accordion. The cells cannot expand greatly in width because the cellulose fibers do not easily stretch.

The three tissue systems of a mature plant (dermal, ground, and vascular) complete their development in the zone of differentiation. Cells of the vascular cylinder differentiate into primary xylem and primary phloem. Differentiation of cells—the specialization of their structure and function—results from differential gene expression (see Module 11.7). Cells in the vascular cylinder, for instance, develop into primary xylem or phloem cells because certain genes are turned on and are expressed as specific proteins, whereas other genes in these cells are turned off.

The micrograph in **Figure 31.7D** shows a section through the end of a growing shoot that was cut lengthwise from its tip to just below its uppermost pair of axillary buds. You can see the apical meristem, which is a dome-shaped mass of dividing cells at the tip of the terminal bud. Elongation occurs just below this meristem, and the elongating cells push the apical meristem upward, instead of downward as in the root. As the apical meristem advances upward, some of its cells remain behind, and these become new axillary bud meristems at the base of the leaves.

The drawings in Figure 31.7D show two stages in the growth of a shoot. The stage shown on the left is just like the micrograph. At the later stage shown on the right, the apical meristem has been pushed upward by elongating cells underneath, leaving behind new axillary bud meristems at the base of the new set of leaves.

Primary growth accounts for a plant's lengthwise growth. The stems and roots of many plants increase in thickness, too, and in the next module, we see how this usually happens.

? **A plant grows taller due to cell division within the ________ at the tips of the shoots. Such lengthening is called ________.**

apical meristem . . . primary growth

31.8 Secondary growth increases the diameter of woody plants

In nonwoody plants, primary growth produces nearly all of the plant body. Woody plants (such as trees, shrubs, and vines), however, grow in diameter in addition to length, thickening in older regions where primary growth has ceased. This increase in thickness of stems and roots, called **secondary growth**, is caused by the activity of dividing cells in tissues that are called **lateral meristems**. These dividing cells are arranged into two cylinders, known as the vascular cambium and the cork cambium, that extend along the length of roots and stems.

The **vascular cambium** is a cylinder of meristem cells one cell thick between the primary xylem and primary phloem. The vascular cambium is wholly responsible for the production of secondary growth. The stem on the left side of **Figure 31.8A** is virtually the same as a young stem undergoing primary growth (compare this figure with the eudicot stem in Figure 31.5) except for the presence of vascular cambium. Secondary growth involves adding layers of vascular tissue on either side of the vascular cambium.

The drawings at the center and the right of the figure show the results of secondary growth. Tissues produced by secondary growth are called secondary tissues. The vascular cambium gives rise to two new tissues: **secondary xylem** to its interior and **secondary phloem** to its exterior (see the center drawing of Figure 31.8A). Each year, the vascular cambium produces layers of secondary xylem and secondary phloem that are larger in circumference than the previous layer (see the drawing at the right). In this way, the vascular cambium thickens roots and stems.

Secondary xylem makes up the **wood** of a tree, shrub, or vine. Over the years, a woody stem gets thicker and thicker as its vascular cambium produces layer upon layer of secondary xylem. The cells of the secondary xylem have thick walls rich in lignin, giving wood its characteristic hardness and strength.

Annual growth rings, such as those in the cross section in **Figure 31.8B**, result from the layering of secondary xylem. The layers are visible as rings because of uneven activity of the vascular cambium during the year. In woody plants that live in temperate regions, such as most of the United States, the vascular cambium becomes dormant each year during winter, and secondary growth is interrupted. When secondary growth resumes in the spring, a cylinder of early wood forms. Early wood cells are made up of the first new xylem cells to develop and are usually larger in diameter and thinner-walled than those produced later in summer. The boundary between the large cells of early wood and the smaller cells of the late wood produced during the previous growing season is usually seen as a distinct ring visible in cross sections of tree trunks and roots. Therefore, a tree's age can be estimated by counting its annual rings. The rings may have varying thicknesses, reflecting climate conditions and therefore the amount of seasonal growth in a given year. In fact, growth ring patterns in older trees are one source of evidence for recent global climate change.

Now let's return to Figure 31.8A and see what happens to the parts of the stem that are *external* to the vascular cambium. Unlike xylem, the external tissues do not accumulate over the years. Instead, they are sloughed off as the stem expands in diameter.

Notice at the left of Figure 31.8A that the epidermis and cortex, both the result of primary growth, make up the young stem's external covering. When secondary growth begins (center drawing of Figure 31.8A), the epidermis is shed and replaced with a new outer layer called **cork** (brown). Mature cork cells are dead and have thick, waxy walls that protect the underlying tissues of the stem from water loss, physical damage, and pathogens. Cork is produced by meristem tissue called the

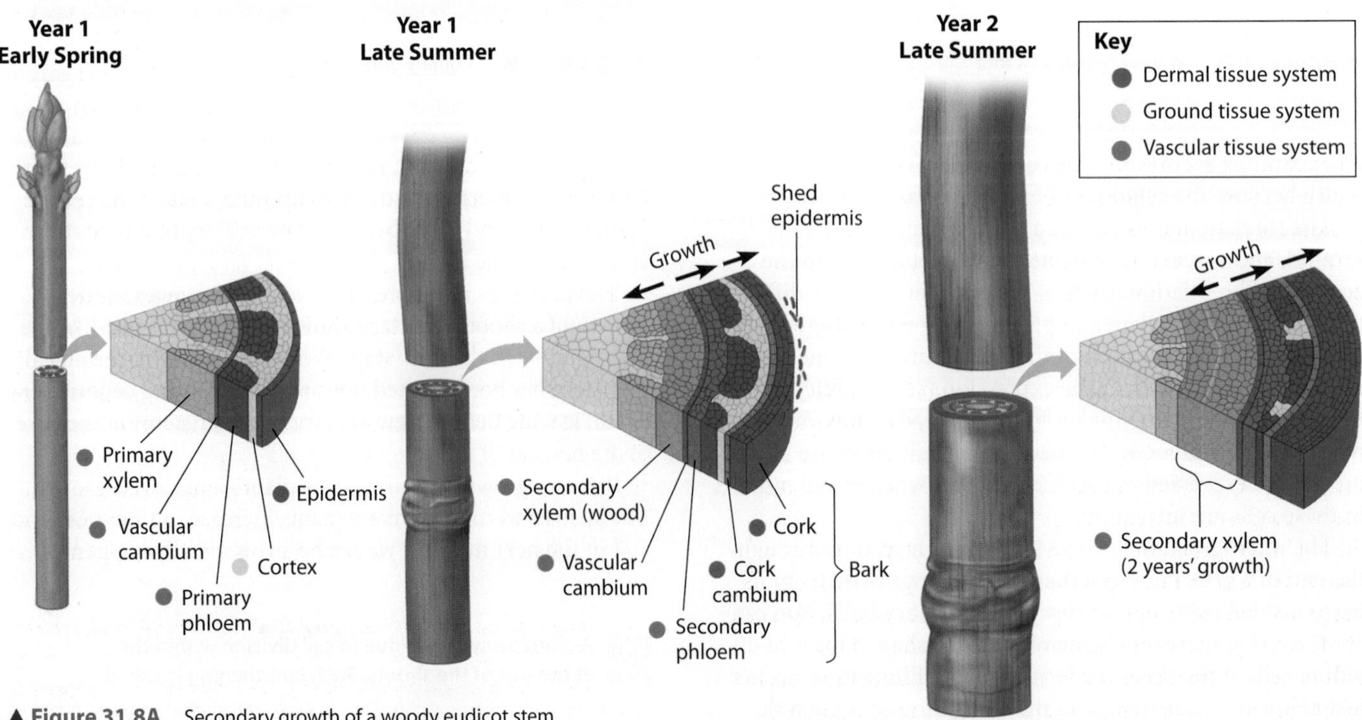

Year 1
Early Spring
Year 1
Late Summer
Year 2
Late Summer
Key
Dermal tissue system
Ground tissue system
Vascular tissue system
Shed epidermis
Growth
Growth
Primary xylem
Vascular cambium
Epidermis
Cortex
Primary phloem
Secondary xylem (wood)
Vascular cambium
Cork
Cork cambium
Secondary phloem
Bark
Secondary xylem (2 years' growth)

▲ **Figure 31.8A** Secondary growth of a woody eudicot stem

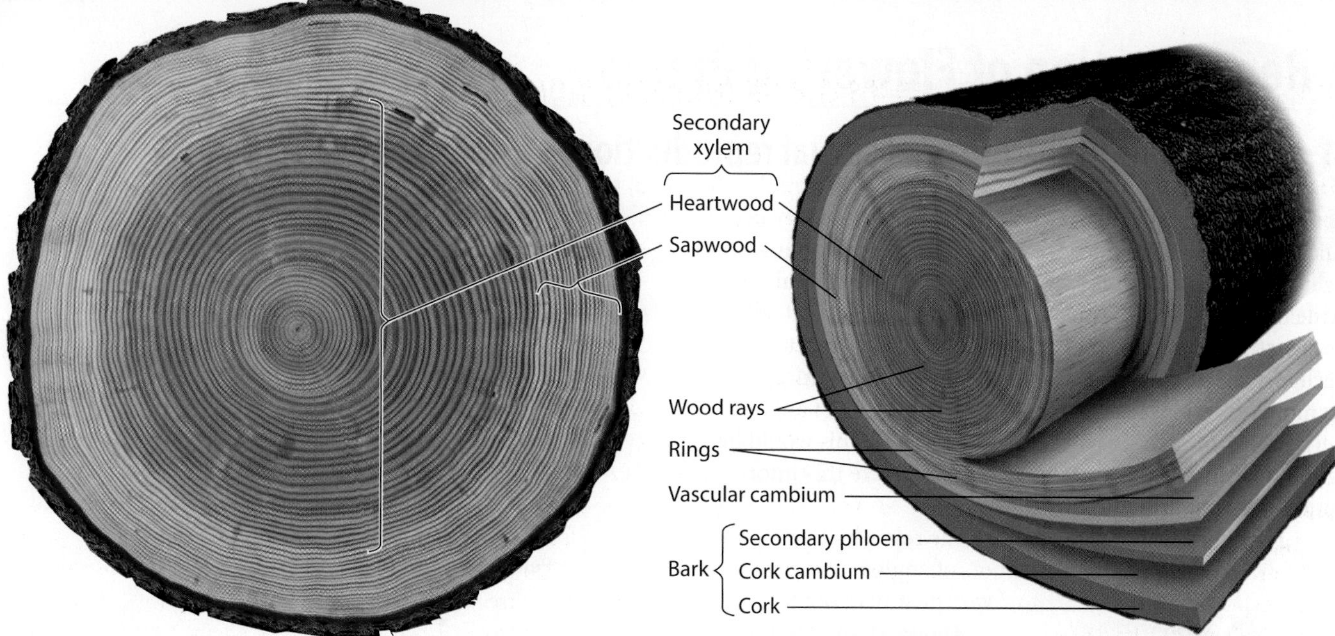
Secondary xylem
Heartwood
Sapwood
Wood rays
Rings
Vascular cambium
Bark
Secondary phloem
Cork cambium
Cork

▲ **Figure 31.8B** Anatomy of a log

cork cambium (light brown), which first forms from parenchyma cells in the cortex. As the stem thickens and the secondary xylem expands, the original cork and cork cambium are pushed outward and fall off, as is evident in the cracked, peeling bark of many tree trunks. A new cork cambium forms to the inside. When no cortex is left, it forms from parenchyma cells in the phloem.

Everything external to the vascular cambium is called **bark**. The main components of bark are the secondary phloem, the cork cambium, and the cork (Figure 31.8B, right). The youngest secondary phloem (next to the vascular cambium) functions in sugar transport. The older secondary phloem dies, as does the cork cambium you see here. Pushed outward, these tissues and cork produced by the cork cambium help protect the stem until they, too, are sloughed off as part of the bark. Keeping pace with secondary growth, cork cambium keeps regenerating from the younger secondary phloem and keeps producing a steady supply of cork.

The log on the left in Figure 31.8B shows the results of several decades of secondary growth. The bulk of a trunk like this is dead tissue. The living tissues in it are the vascular cambium, the youngest secondary phloem, the cork cambium, and cells in the wood rays, which you can see radiating from the center of the log in the drawing on the right. The **wood rays** consist of parenchyma cells that transport water and nutrients, store organic nutrients, and aid in wound repair. The **heartwood**, in the center of the trunk, consists of older layers of secondary xylem. These cells no longer transport water; they are clogged with resins and other compounds that make heartwood resistant to rotting. Because heartwood doesn't conduct water, a large tree can survive even if the center of its trunk is hollow (**Figure 31.8C**). The lighter-colored **sapwood** is younger secondary xylem that does conduct xylem fluid (sap).

▲ **Figure 31.8C** A giant sequoia that survived since the 1930s with a tunnel cut through its heartwood

Thousands of useful products are made from wood—from construction lumber to fine furniture, musical instruments, paper, insulation, and a long list of chemicals, including turpentine, alcohols, artificial vanilla flavoring, and preservatives. Among the qualities that make wood so useful is a unique combination of strength, hardness, lightness, high insulating properties, durability, and workability. In many cases, there is simply no good substitute for wood. A wooden oboe, for instance, produces far richer sounds than a plastic one. Fence posts made of locust tree wood actually last much longer in the ground than metal ones. Ball bearings are sometimes made of a very hard wood called lignum vitae. Unlike metal bearings, they require no lubrication because a natural oil completely penetrates the wood.

In a sense, wood is analogous to the skeletons of many land animals. Wood is an evolutionary adaptation that enables a tree to remain upright and keep growing year after year on land—sometimes to attain enormous heights. In the next few modules, we discuss some additional adaptations that enable plants to live on land—those that facilitate reproduction.

? **Why can some large trees survive after a tunnel has been cut through their center?**

● The center of an older tree consists of heartwood, which no longer transports materials through the plant. It can thus be removed without harming the tree.

▷ Reproduction of Flowering Plants

31.9 The flower is the organ of sexual reproduction in angiosperms

It has been said that an oak tree is merely an acorn's way of making more acorns. Indeed, evolutionary fitness for any organism is measured only by its ability to produce healthy, fertile offspring. Thus, from an evolutionary viewpoint, all the structures and functions of a plant can be interpreted as mechanisms contributing to reproduction. In the remaining modules, we explore the reproductive biology of angiosperms, beginning here with a brief overview. (This would be a good time to review Modules 17.6–17.7, where this information was first presented.)

Flowers, the reproductive shoots of angiosperms, can vary greatly in shape (**Figure 31.9A**). Despite such variation, nearly all flowers contain four types of modified leaves called floral organs: sepals, petals, stamens, and carpels (**Figure 31.9B**). The **sepals**, which enclose and protect the flower bud, are usually green and more leaflike than the other floral organs (picture the green wraparound leaves at the base of a rosebud). The **petals** are often colorful and fragrant, advertising the flower to pollinators. The stamens and carpels are the reproductive organs, containing the sperm and eggs, respectively.

▲ **Figure 31.9A**
Some variations in flower shape

A **stamen** consists of a stalk (called the filament) tipped by an anther. Within the **anther** are sacs in which pollen is produced via meiosis. Pollen grains house the cells that develop into sperm.

A **carpel** has a long slender neck (called the style) with a sticky stigma at its tip. The **stigma** is a landing platform for pollen. The base of the carpel is the **ovary**, which contains one or more **ovules**, each containing a developing egg and supporting cells. The term **pistil** is sometimes used to refer to a single carpel or a group of fused carpels.

Figure 31.9C shows the life cycle of a generalized angiosperm. ❶ Fertilization occurs in an ovule within a flower. ❷ As the ovary develops into a fruit, ❸ the ovule develops into the **seed** containing the embryo and a store of food. The fruit protects the seed and aids in dispersing it. Completing the life cycle, ❹ the seed then **germinates** (begins to grow) in a suitable habitat; ❺ the embryo develops into a seedling; and the seedling grows into a mature plant.

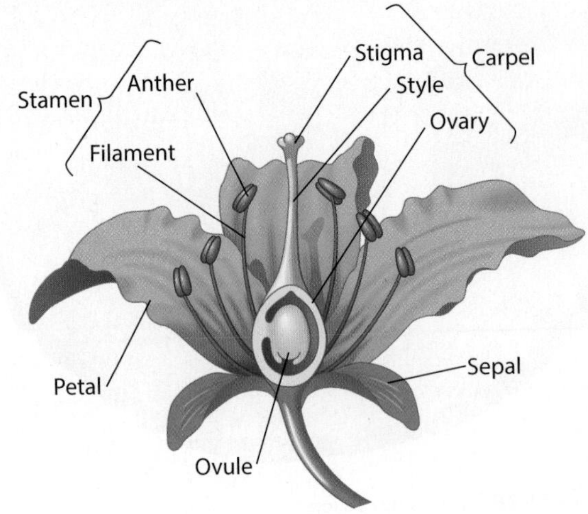

Stamen
Anther
Filament
Stigma
Style
Ovary
Carpel
Petal
Sepal
Ovule

▲ **Figure 31.9B** The structure of a flower

In the next four modules, we examine key stages in the angiosperm life cycle in more detail. We will see that there are a number of variations in the basic themes presented here.

? **Pollen develops within the ______ of ______.**
Ovules develop within the ______ of ______.

anthers . . . stamens . . . ovaries . . . carpels

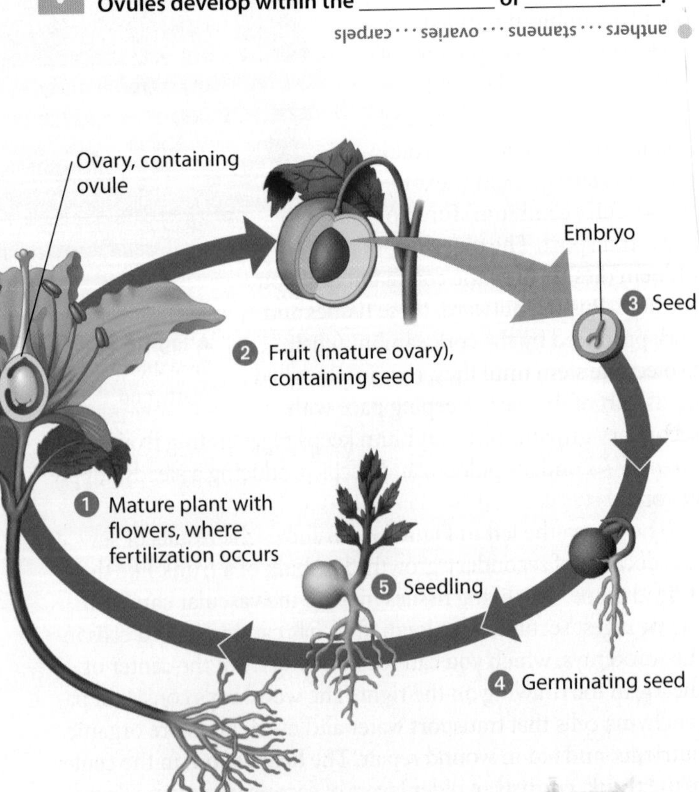

Ovary, containing ovule
Embryo
3 Seed
2 Fruit (mature ovary), containing seed
1 Mature plant with flowers, where fertilization occurs
5 Seedling
4 Germinating seed

▲ **Figure 31.9C** Life cycle of a generalized angiosperm

31.10 The development of pollen and ovules culminates in fertilization

The life cycles of plants are characterized by an alternation of generations, in which haploid (n) and diploid ($2n$) generations take turns producing each other (**Figure 31.10**). The diploid plant body is called the **sporophyte**. A sporophyte produces special structures, the anthers and ovules, in which cells undergo meiosis to produce haploid cells called spores. Each spore then divides via mitosis and becomes a multicellular **gametophyte**, the plant's haploid generation. The gametophyte produces gametes by mitosis. At fertilization, gametes from the male and female gametophytes unite, producing a diploid zygote. The life cycle is completed when the zygote divides by mitosis and develops into a new sporophyte. In angiosperms, the sporophyte is the dominant generation: It is larger, more obvious, and longer-living than the gametophyte.

The cells that develop into pollen grains (the male gametophytes) are found within a flower's anthers. ① Each cell first undergoes meiosis, forming four haploid spores. ② Each spore then divides by mitosis, forming two haploid cells, called the tube cell and the generative cell. The generative cell passes into the tube cell, and a thick wall forms around them. ③ The resulting pollen grain is ready for release from the anther.

In most species, the ovary of a flower contains several ovules, but only one is shown at the top of the figure. An ovule contains a central cell (gold) surrounded by a protective covering of smaller cells (yellow). ① The central cell enlarges and undergoes meiosis, producing four haploid spores. Three of the spores usually degenerate, but the surviving one enlarges and ② divides by mitosis, producing a multicellular structure known as the **embryo sac**. Housed in several layers of protective cells (yellow) produced by the sporophyte plant, the embryo sac is the female gametophyte. The sac contains a large central cell with two haploid nuclei. One of its other cells is the haploid egg, ready to be fertilized.

③ The first step leading to fertilization is **pollination**, the transfer of pollen from anther to stigma. Most angiosperms depend on insects (mainly bees), birds, or other animals to transfer pollen. But the pollen of some plants—such as grasses and many trees—is windborne (causing pollen allergies in some people).

After pollination, the pollen grain germinates on the stigma. Its tube cell gives rise to the pollen tube, which grows downward into the ovary. Meanwhile, the generative cell divides by mitosis, forming two sperm. ④ When the pollen tube reaches the base of the ovule, it enters the ovary and discharges its two sperm near the embryo sac. ⑤ One sperm fertilizes the egg, forming the diploid zygote. The other contributes its haploid nucleus to the large diploid central cell of the embryo sac. This cell, now with a triploid ($3n$) nucleus, will give rise to a food-storing tissue called **endosperm**.

The union of two sperm cells with different nuclei of the embryo sac is called **double fertilization**, and the resulting production of endosperm is unique to angiosperms. Endosperm will develop only in ovules containing a fertilized egg, thereby preventing angiosperms from squandering nutrients.

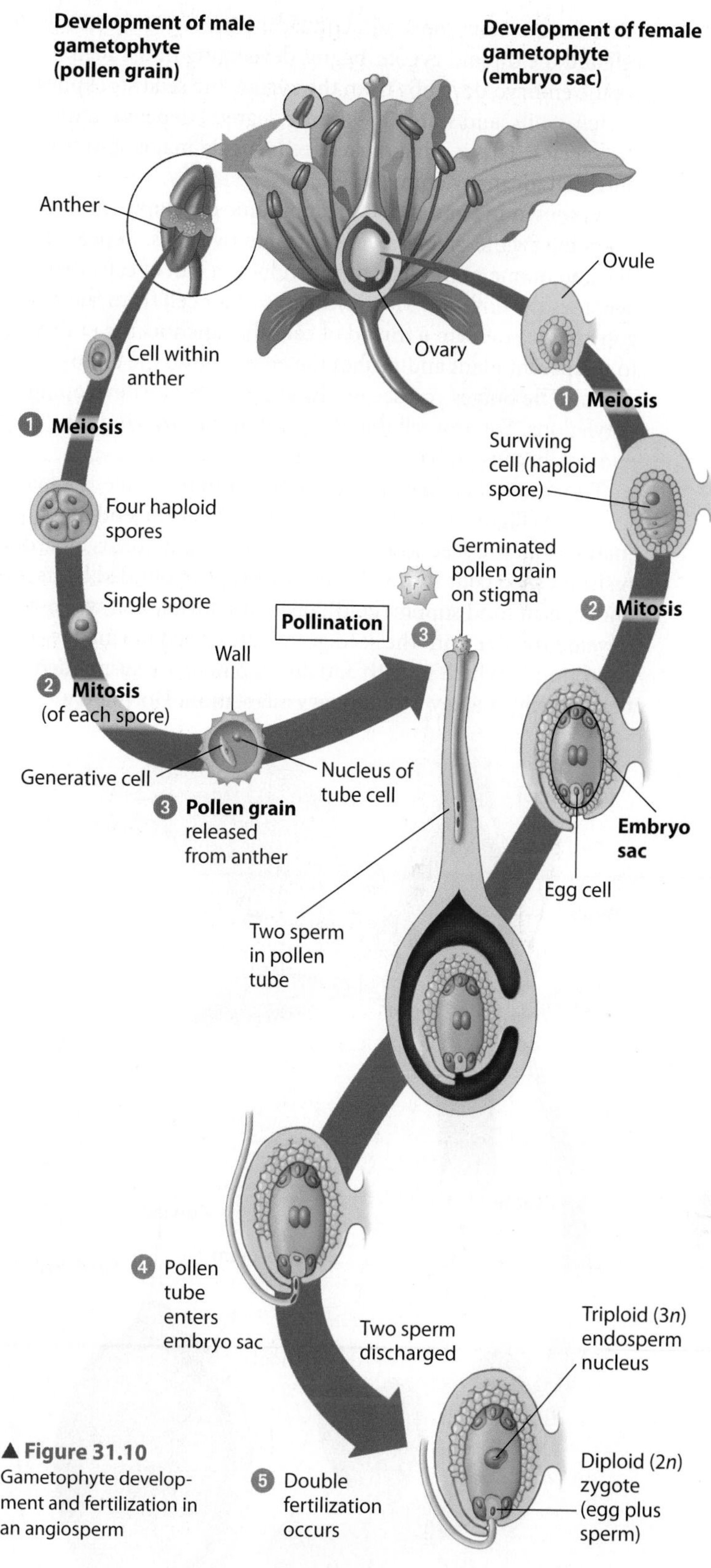

Development of male gametophyte (pollen grain)
Development of female gametophyte (embryo sac)
Anther
Ovule
Cell within anther
Ovary
① Meiosis
① Meiosis
Surviving cell (haploid spore)
Four haploid spores
Germinated pollen grain on stigma
Single spore
Pollination
③
② Mitosis
Wall
② Mitosis (of each spore)
Generative cell
Nucleus of tube cell
③ Pollen grain released from anther
Embryo sac
Egg cell
Two sperm in pollen tube
④ Pollen tube enters embryo sac
Two sperm discharged
Triploid (3n) endosperm nucleus
⑤ Double fertilization occurs
Diploid (2n) zygote (egg plus sperm)

▲ **Figure 31.10**
Gametophyte development and fertilization in an angiosperm

? **What are the two products of double fertilization?**

● A zygote and endosperm

31.11 The ovule develops into a seed

After fertilization, the ovule, containing the triploid central cell and the diploid zygote, begins developing into a seed. As the embryo develops from the zygote, the seed stockpiles proteins, oils, and starch to varying degrees, depending on the species. This is what makes seeds such a major source of nutrition for many animals.

As shown in **Figure 31.11A**, embryonic development begins when the zygote divides by mitosis into two cells. Repeated division of one of the cells then produces a ball of cells that becomes the embryo. Meanwhile, the other cell from the zygote divides to form a thread of cells that anchors the embryo to the parent plant and pushes the embryo into the endosperm. The bulges you see on the embryo are the developing cotyledons. You can tell that the plant in this drawing is a eudicot, because it has two cotyledons.

The result of embryonic development in the ovule is a mature seed (Figure 31.11A, bottom right). Near the end of its maturation, the seed loses most of its water and forms a hard, resistant **seed coat** (brown). The embryo, surrounded by its endosperm food supply (gold), becomes dormant; it will not develop further until the seed germinates. Seed dormancy, a condition in which growth and development are suspended temporarily, is a key evolutionary adaptation. Dormancy allows time for a plant to disperse its seeds and increases the chance that a new generation of plants will begin growing only when environmental conditions, such as temperature and moisture, favor survival.

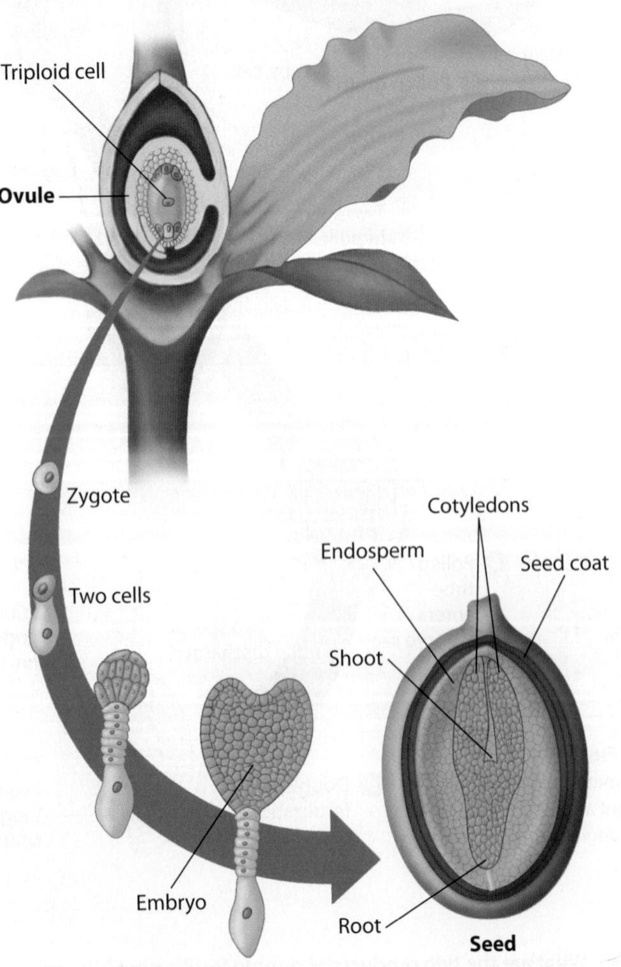
Triploid cell
Ovule
Zygote
Two cells
Embryo
Cotyledons
Endosperm
Seed coat
Shoot
Root
Seed

▲ **Figure 31.11A** Development of a eudicot plant embryo

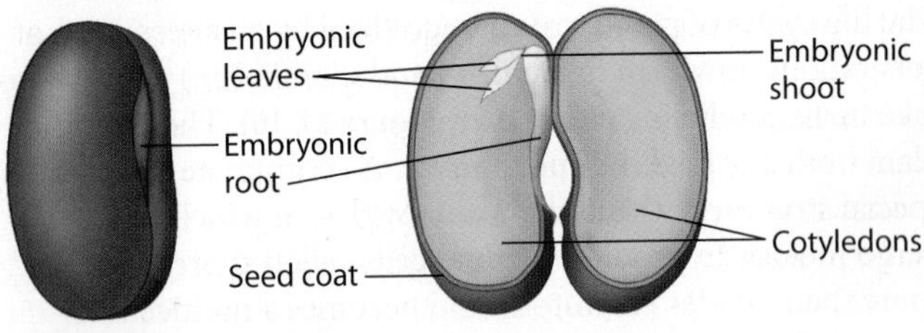
Embryonic leaves
Embryonic root
Seed coat
Embryonic shoot
Cotyledons
Common bean (eudicot)

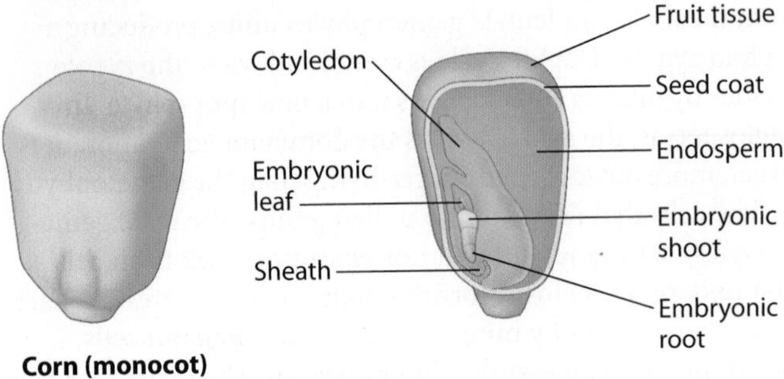
Cotyledon
Embryonic leaf
Sheath
Fruit tissue
Seed coat
Endosperm
Embryonic shoot
Embryonic root
Corn (monocot)

▲ **Figure 31.11B** Seed structure in eudicots and monocots

The dormant embryo contains a miniature root and shoot, each equipped with an apical meristem. After the seed germinates, the apical meristems will sustain primary growth as long as the plant lives. Also present in the embryo are the three tissues that will form the epidermis, cortex, and primary vascular tissues.

Figure 31.11B contrasts the internal structures of eudicot and monocot seeds. In the eudicot (illustrated here as a bean), the embryo is an elongated structure with two thick cotyledons (tan). The embryonic root develops just below the point at which the cotyledons are attached to the rest of the embryo. The embryonic shoot, tipped by a pair of miniature embryonic leaves, develops just above the point of attachment. The bean seed contains no endosperm because its cotyledons absorb the endosperm nutrients as the seed forms. The nutrients start passing from the cotyledons to the embryo when it germinates.

A kernel of corn, an example of a monocot, is actually a fruit containing one seed. Everything you see in the drawing is the seed, except the kernel's outermost covering. The covering is dried fruit tissue, the former wall of the ovary, and is tightly bonded to the seed coat. Unlike the bean, the corn seed contains a large endosperm and a single, thin cotyledon. The cotyledon absorbs the endosperm's nutrients during germination. Also unlike the bean, the embryonic root and shoot in corn each have a protective sheath.

? **What is the role of the endosperm in a seed?**

The endosperm provides nutrients to the developing embryo.

31.12 The ovary develops into a fruit

In the previous two modules, we followed the angiosperm life cycle from the flower on the sporophyte plant through the transformation of an ovule into a seed. While the seeds are developing from ovules, hormonal changes triggered by fertilization cause the flower's ovary to grow, thicken, and mature into a fruit. A **fruit** is a mature ovary that acts as a vessel, housing and protecting seeds and helping disperse them from the parent plant. Although a fruit typically consists of a mature ovary, it can include other flower parts as well. A pea pod is a fruit that holds the peas (the seeds of the pea plant; **Figure 31.12A**). Other easily recognizable fruits include a peach, orange, tomato, cherry, or corn kernel.

▲ **Figure 31.12A** A pod, the fruit of a pea plant, holds the peas (seeds)

Figure 31.12B matches the parts of a pea flower with what they become in the pod. The wall of the ovary becomes the pod. The ovules, within the ovary, develop into the seeds. The small, thread-like structure at the end of the pod is what remains of the upper part of the flower's carpel. The sepals of the flower often stay attached to the base of the green pod. Peas are usually harvested at this stage of fruit development. If the pods are allowed to develop further, they become dry and brownish and will split open, releasing the seeds.

As shown in the examples in **Figure 31.12C**, mature fruits can be either fleshy or dry. Oranges, plums, and grapes are examples of fleshy fruits, in which the wall of the ovary becomes soft during ripening. Dry fruits include beans, nuts, and grains. The dry, wind-dispersed fruits of grasses, harvested while on the plant, are major staple foods for people. The cereal grains of wheat, rice, corn, and other grasses, though easily mistaken for seeds, are each actually a fruit with a dry outer covering (the former wall of the ovary) that adheres to the seed coat of the seed within.

Various adaptations of fruits help disperse seeds (Module 17.8). The seeds of some flowering plants, such as dandelions and maples, are contained within fruits that function like kites or propellers, adaptations that enhance dispersal by wind. Some fruits, such as coconuts, are adapted to dispersal by water. And many angiosperms rely on animals to carry seeds. Some of these plants have fruits modified as burrs that cling to animal fur (or the clothes of humans). Other angiosperms produce edible fruits, which are usually nutritious, sweet tasting, and vividly colored, advertising their ripeness. When an animal eats the fruit, it digests the fruit's fleshy part, but the tough seeds usually pass unharmed through the animal's digestive tract. Animals may deposit the seeds, along with a supply of fertilizer, kilometers from where the fruit was eaten.

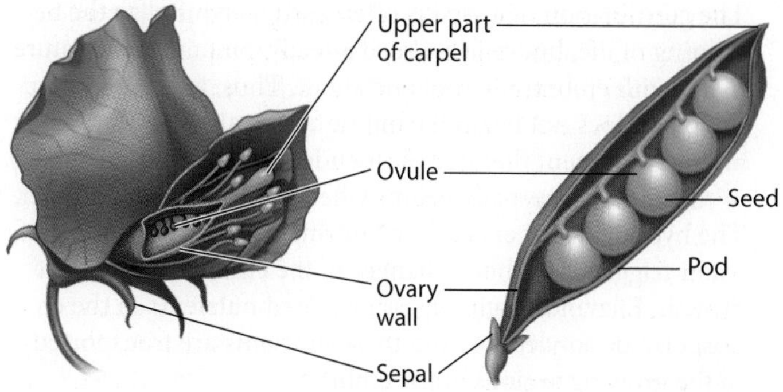
Upper part of carpel
Ovule
Seed
Pod
Ovary wall
Sepal

▲ **Figure 31.12B** The correspondence between flower and fruit in the pea plant

Try This Explain why tomatoes and squash are more accurately called fruits rather than vegetables.

Maple fruits

▲ **Figure 31.12C** A collection of fleshy (top) and dry (bottom) fruits

? **Seed is to __________ as __________ is to ovary.**

ovule . . . fruit

31.13 Seed germination continues the life cycle

The germination of a seed is often used to symbolize the beginning of life, but in fact a seed already contains a miniature plant, with embryonic root and shoot. Thus, at germination, the plant does not begin life but rather resumes the growth and development that were suspended during seed dormancy.

Germination usually begins when the seed takes up water. The hydrated seed expands, rupturing its coat. The inflow of water triggers metabolic changes in the embryo that restart growth. Enzymes begin digesting stored nutrients in the endosperm or cotyledons, and these nutrients are transported to the growing regions of the embryo.

In **Figure 31.13A**, notice that the embryonic root of a bean (a eudicot) emerges first and grows downward from the germinating seed. Next, the embryonic shoot emerges, and a hook forms near its tip. The hook protects the delicate shoot tip by holding it downward as it pushes up through the abrasive soil. As the shoot breaks through the soil surface, its tip is lifted gently out of the soil as exposure to light stimulates the hook to straighten. The first foliage leaves then expand from the shoot tip and begin making food by photosynthesis. The cotyledons emerge from the soil and become leaflike photosynthetic structures. In many other plants, such as peas, the cotyledons remain behind in the soil and decompose.

Corn and other monocots use a different mechanism for breaking ground at germination **(Figure 31.13B)**. A protective sheath surrounding the shoot pushes upward and breaks through the soil. The shoot tip then grows up through the tunnel provided by the sheath. The corn cotyledon remains in the soil and decomposes.

In the wild, only a small fraction of fragile seedlings endure long enough to reproduce. Production of enormous numbers of seeds compensates for the odds against individual survival. Asexual reproduction, generally simpler and less hazardous for offspring than sexual reproduction, is an alternative means of plant propagation, as we see next.

? Which meristems provide additional cells for early growth of a seedling after germination?

The apical meristems of the shoot and root

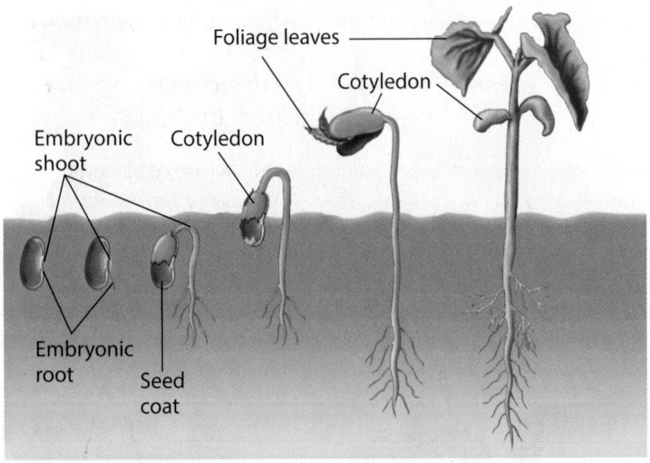

▲ **Figure 31.13A** Bean germination (a eudicot)

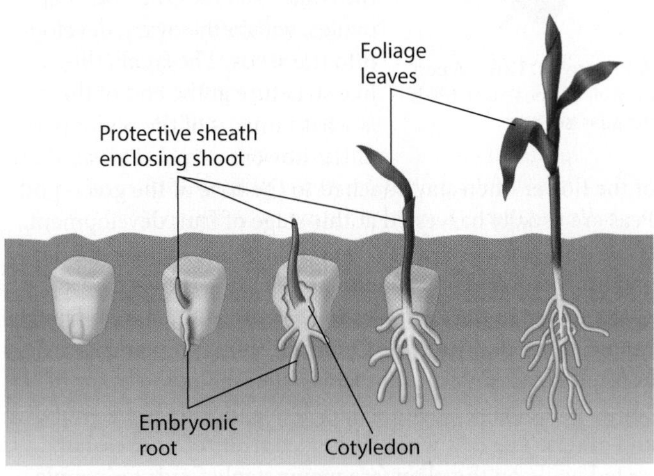
Foliage leaves
Protective sheath enclosing shoot
Embryonic root
Cotyledon

▲ **Figure 31.13B** Corn germination (a monocot)

31.14 Asexual reproduction produces plant clones

Imagine chopping off your finger and watching it grow and develop into an exact copy of you. If this were possible, it would be an example of asexual reproduction, which in plants is also called vegetative propagation. The resulting asexually produced offspring, often called a **clone**, is genetically identical to its single parent.

Asexual reproduction in angiosperms and other plants is an extension of their capacity to grow throughout life. A plant's meristematic tissues can sustain growth indefinitely. In addition, parenchyma cells can divide and differentiate into the various types of cells.

▲ **Figure 31.14A** A garlic bulb containing a cluster of cloves

In nature, asexual reproduction in plants often involves **fragmentation**, the separation of a parent plant into parts that develop into whole plants. A garlic bulb **(Figure 31.14A)** is actually an underground stem that functions in storage. A single large bulb fragments into several parts, called cloves. Each clove can give rise to a separate plant, as indicated by the green shoots emerging from some of them. The white, paper-thin sheaths are leaves that are attached to the stem.

Each of the small trees you see in **Figure 31.14B** is a sprout from the roots of a coast redwood tree. Each small tree develops its own root system separate from the parent tree.

The ring of plants in **Figure 31.14C** is a clone of creosote bushes growing in the Mojave Desert in southern California. In this case, the word *clone* refers to a group of genetically identical organisms. All these bushes came from generations of asexual reproduction by roots. Making the oldest trees seem youthful, this clone apparently began with a single plant that germinated from a seed about 12,000 years ago. The original plant probably occupied the center of the ring.

Some aspen groves, such as those shown in **Figure 31.14D**, consist of thousands of trees descended by asexual reproduction from the root system of a single parent. (Genetic differences between groves that arose from different parents result in different timing for the development of fall color.)

Many plants can reproduce both sexually and asexually. What advantages can asexual reproduction offer? For one thing, a particularly fit parent plant can clone many copies of itself, all of which would be equally well suited to current conditions. Also, although new seedlings produced by sexual reproduction are quite fragile, offspring produced asexually from mature fragments of the parent plant can often be much hardier and therefore more likely to survive. And if a plant is isolated (and therefore unlikely to be pollinated), asexual reproduction may be the only means of reproduction.

The ability of plants to reproduce asexually provides many opportunities for growers to produce large numbers of plants with minimal effort and expense. For example, most of our fruit trees and houseplants are asexually propagated from cuttings. Several other plants are propagated from root sprouts (for example, raspberries) or pieces of underground stems (such as potatoes) or by grafting a bud from one plant onto a closely related plant (for example, most varieties of wine grapes).

Plants can also be asexually propagated in the laboratory. For example, a germanium plant can be grown from a few meristem cells cut from a mature plant and cultured in a growth medium containing nutrients and hormones **(Figure 31.14E)**. Using this method, a single plant can be cloned into thousands of copies. Orchids and certain pine trees used for mass plantings are commonly propagated this way.

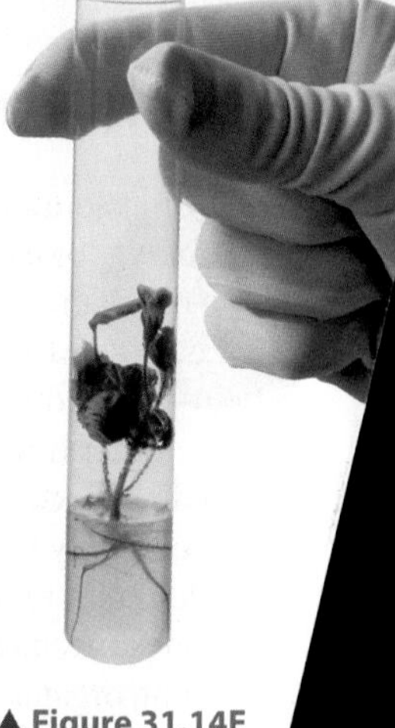

▲ **Figure 31.14E** Test-tube cloning

Plant cell culture methods also enable researchers to grow plants from genetically engineered plant cells (see Module 12.8). Foreign genes are incorporated into a single parenchyma cell, and the cell is then cultured so that it multiplies and develops into a new plant. The resulting genetically modified organism (GMO) may then be able to grow and reproduce normally. The commercial adoption of GMO crops by farmers has been one of the most rapid cases of technology transfer in the history of agriculture, but it is not without controversy (see Module 12.9).

Aside from the issues raised by GMO plants, modern agriculture faces some potentially serious problems. Nearly all of today's crop plants have very little genetic variability. In fact, we grow most crops in monocultures, cultivating a single pla variety on large areas of land. Given these conditions, plant scientists fear that a small number of diseases could devasta large crop areas. In response, plant breeders are working to maintain "gene banks," storage sites for seeds of many diff plant varieties that can be used to breed new hybrids.

? **Which mode of reproduction (sexual or asexual) woul generally be more advantageous in a location where composition of the soil is constantly changing? Why?**

ecause it generates genetic variation among the offspring, which e potential for adaptation to a changing environment

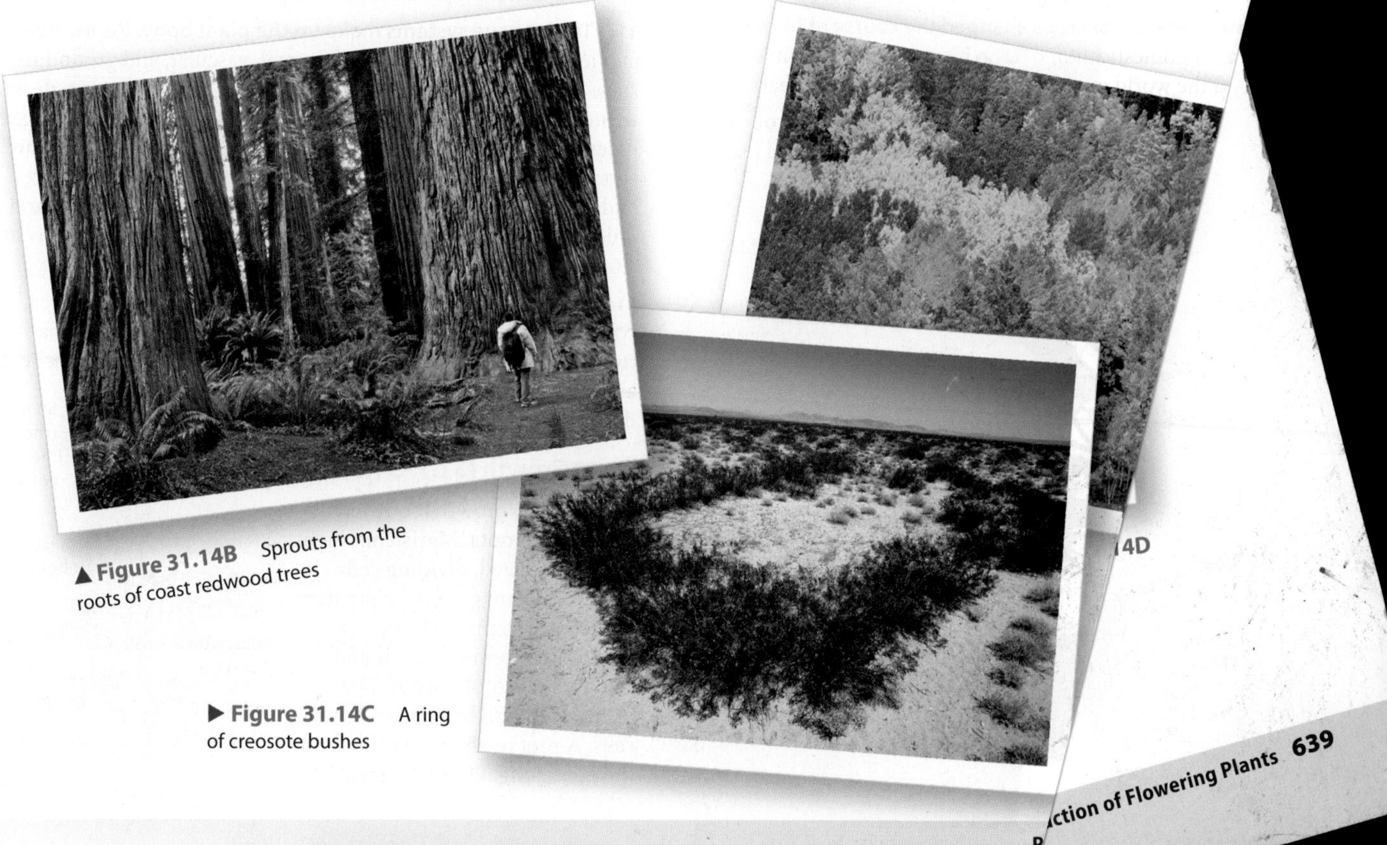

▲ **Figure 31.14B** Sprouts from the roots of coast redwood trees

▶ **Figure 31.14C** A ring of creosote bushes

4D

31.15 Evolutionary adaptations help some plants to live very long lives

EVOLUTION CONNECTION

Some plants can survive a very long time. For example, coast redwoods (*Sequoia sempervirens*), found only in northern California and southern Oregon, are estimated to be 2,000 to 3,000 years old. However, those redwoods are mere youngsters compared with some other trees. The tree shown in **Figure 31.15** is a 4,600-year-old bristlecone pine (*Pinus longaeva*). Bristlecone pines are found only in six western U.S. states (including the White ountain region of California, pictured here). Another bristle- e pine, named Methuselah, is believed to be Earth's oldest organism. Its location is kept secret to protect the tree. ong life enhances evolutionary fitness by increasing ber of reproductive opportunities. What evolution- ations in plants can help trees to live so long? Adult nost adult plants, retain meristems, which allow d growth and repair throughout life. A tree can rgans that have been lost or damaged by trauma , thick wood can protect against insects, dis- riodic fires (as with the thick bark protect- sequoias).

▲ **Figure 31.15** An ancient bristlecone pine tree growing in California

? **Why is the presence of meristems essential for the long life of many plants?**

Plant meristem can give rise to multiple types of cells throughout life.

1 REVIEW

For practice quizzes, BioFlix animations, MP3 tutorials, video tutors, and more study tools designed for this textbook, go to MasteringBiology®

oncepts

Function (31.1–31.6)

f crops changed the course of human ation of crop plants occurred in many ver a span of several thousand years.

s of angiosperms are the monocots and oups differ in the number of seed leaves stems, leaves, and flowers.

ontains three basic organs: roots, cture of a flowering plant allows it to oil and air.

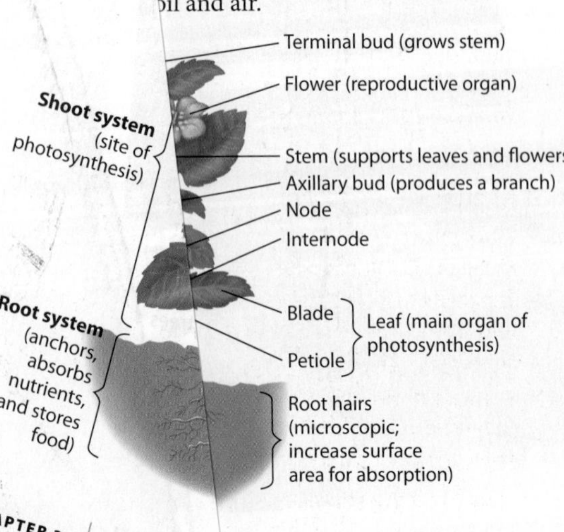

Terminal bud (grows stem)
Flower (reproductive organ)
Shoot system (site of photosynthesis)
Stem (supports leaves and flowers)
Axillary bud (produces a branch)
Node
Internode
Blade
Petiole
Leaf (main organ of photosynthesis)
Root system (anchors, absorbs nutrients, and stores food)
Root hairs (microscopic; increase surface area for absorption)

31.4 Many plants have modified roots, stems, and leaves. In addition to their primary functions, plant organs may store food, promote asexual reproduction, and provide protection.

31.5 Three tissue systems make up the plant body. Roots, stems, and leaves are each made up of dermal, vascular, and ground tissues. Dermal tissue covers and protects the plant. In leaves, dermal tissue has stomata, pores with guard cells that regulate exchange of gases and water vapor with the environment. The vascular tissue system contains xylem and phloem, which function in support and transport. Xylem conveys water and dissolved minerals, and phloem transports sugars. The ground tissue system functions in storage, photosynthesis, and support.

31.6 Plant cells are diverse in structure and function. The major types of plant cells are parenchyma, collenchyma, sclerenchyma (including fiber and sclereid cells), water-conducting cells (tracheids and vessel elements), and food-conducting cells (sieve-tube elements).

Plant Growth (31.7–31.8)

31.7 Primary growth lengthens roots and shoots. Meristems, areas of unspecialized, dividing cells, are where plant growth originates. Apical meristems at the tips of roots and in terminal buds and axillary buds of shoots initiate primary growth by producing new cells. A root or shoot lengthens as the cells elongate and differentiate.

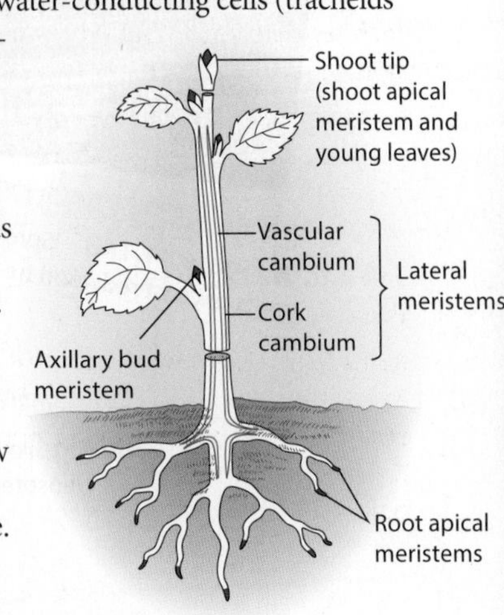

Shoot tip (shoot apical meristem and young leaves)
Vascular cambium
Cork cambium
Lateral meristems
Axillary bud meristem
Root apical meristems

31.8 Secondary growth increases the diameter of woody plants. An increase in a plant's diameter, called secondary growth, arises from cell division in a cylinder of meristem cells called the vascular cambium. The vascular cambium thickens a stem by adding layers of secondary xylem, or wood, next to its inner surface. Outside the vascular cambium, the bark includes secondary phloem, cork cambium, and protective cork cells produced by the cork cambium.

Reproduction of Flowering Plants (31.9–31.15)

31.9 The flower is the organ of sexual reproduction in angiosperms. The angiosperm flower consists of sepals, petals, stamens, and carpels. Pollen grains develop in anthers, at the tips of stamens. The tip of the carpel, the stigma, receives pollen grains. The ovary, at the carpel's base, houses the egg-producing ovule.

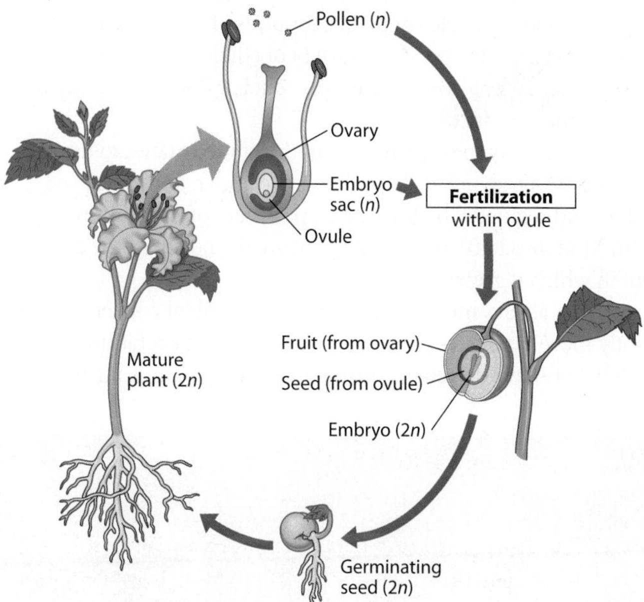

Pollen (n)
Ovary
Embryo sac (n)
Ovule
Fertilization within ovule
Fruit (from ovary)
Seed (from ovule)
Embryo (2n)
Mature plant (2n)
Germinating seed (2n)

31.10 The development of pollen and ovules culminates in fertilization. Haploid spores are formed within ovules and anthers. The spores in anthers give rise to male gametophytes—pollen grains—which produce sperm. A spore in an ovule produces the embryo sac, the female gametophyte. Each embryo sac has an egg cell. Pollination is the arrival of pollen grains onto a stigma. A pollen tube grows into the ovule, and sperm pass through it and fertilize both the egg and a $2n$ central cell. This process is called double fertilization.

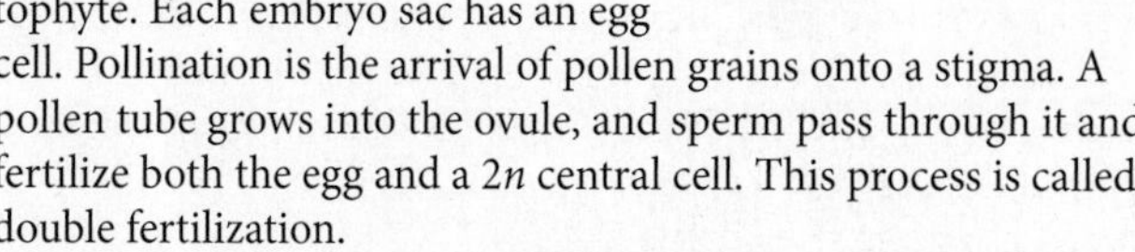

Endosperm nucleus (3n) (2n central cell plus sperm)
Zygote (2n) (egg plus sperm nucleus)

31.11 The ovule develops into a seed. After fertilization, the ovule becomes a seed, and the fertilized egg within it divides and becomes an embryo. The other fertilized cell develops into the endosperm, which stores food for the embryo.

31.12 The ovary develops into a fruit. Fruits help protect and disperse seeds.

31.13 Seed germination continues the life cycle. A seed starts to germinate when it takes up water and expands. The embryo resumes growth and absorbs nutrients from the endosperm. An embryonic root emerges, and a shoot pushes upward and expands its leaves.

31.14 Asexual reproduction produces plant clones. Asexual reproduction can occur by fragmentation or outgrowths of root systems. Propagating plants asexually from cuttings or bits of plant tissue can increase agricultural productivity but reduces genetic diversity.

31.15 Evolutionary adaptations help some plants to live very long lives. The continued growth produced by meristems and the protection provided by dense wood can help some trees live thousands of years.

Connecting the Concepts

1. Create a diagram or concept map that shows the relationships between the following: root system, root hairs, shoot system, leaves, petioles, blades, stems, nodes, internodes, flowers.

Testing Your Knowledge

Level 1: Knowledge/Comprehension

2. In angiosperms, each pollen grain produces two sperm. What do these sperm do?
   a. Each one fertilizes a separate egg cell.
   b. One fertilizes an egg, and the other is kept in reserve.
   c. Both fertilize a single egg cell.
   d. One fertilizes an egg, and the other fertilizes a cell that develops into stored food.

Match questions 3–8 with options a–f.

3. Attracts pollinator	a. pollen grain
4. Develops into seed	b. ovule
5. Protects flower before it opens	c. anther
6. Produces sperm	d. ovary
7. Produces pollen	e. sepal
8. Houses ovules	f. petal

Level 2: Application/Analysis

9. Which of the following is closest to the center of a woody stem? (*Explain your answer.*)
   a. vascular cambium b. primary phloem
   c. secondary phloem d. primary xylem
10. While walking in the woods, you encounter an unfamiliar nonwoody flowering plant. If you want to know whether it is a monocot or eudicot, it would *not* help to look at the
    a. number of seed leaves, or cotyledons, present in its seeds.
    b. shape of its root system.
    c. arrangement of vascular bundles in its stem.
    d. size of the plant.
11. How does a fruit develop from a flower?
12. What part of a plant are you eating when you consume each of the following: celery stalk, peanut, strawberry, lettuce, beet?

Level 3: Synthesis/Evaluation

13. Name two kinds of asexual reproduction. Explain two advantages of asexual reproduction over sexual reproduction. What is the primary drawback of asexual reproduction?
14. Botanists are looking for the wild ancestors of potatoes, corn, and wheat. Why is this search important?
15. **SCIENTIFIC THINKING** In Module 31.1, several lines of investigation are used to investigate the time and location of the domestication of important crops. Which line of evidence do you consider the most reliable? Why?

Answers to all questions can be found in Appendix 4.

CHAPTER

32 Plant Nutrition and Transport

Are organic crops healthier than conventional crops?

Strolling down the supermarket produce aisle or through a farmers' market, you are bombarded with myriad options: shiny red chilies, bulbous purple eggplants, mounds of leafy greens—fruits of nearly every shape and color. Choosing from the tremendous variety of edible plants can be hard enough. But the modern consumer is faced with yet another choice: Should you buy organic or conventional (nonorganic) produce?

What does an "organic" label actually mean, anyway? In the United States, to use the term *organic* or to bear the "USDA Organic" seal, food must be grown and processed according to strict guidelines established and regulated by the U.S. Department of Agriculture (USDA). These regulations (detailed in Module 32.10) are meant to ensure the health of the crop, those who eat it, and the environment in which it grows.

In part because organic produce is almost always more costly than the equivalent conventionally grown produce (sometimes significantly more expensive), many people have an expectation that organic produce must be "better." Such beliefs have driven expansion of the U.S. organic farming

industry at a rate of 20% per year during the last decade, making it one of the fastest growing segments of agriculture. But are organic products better? Is organic produce healthier? And if so, in what ways?

To better understand how all plants—organic or not—contribute to our own health and that of our environment, we need to understand the basics of plant physiology. In this chapter, we'll explore the evolutionary adaptations that allow plants to obtain water and essential nutrients from the soil and air, and then to transport these substances throughout their roots, stems, and leaves. These abilities have allowed plants to successfully colonize a wide variety of terrestrial habitats, where they form the basis of the food web that provides energy to animals (including us). During our discussion, we'll consider many ways in which plant nutrition affects our own well-being, including the implications of an organic label.

BIG IDEAS

The Uptake and Transport of Plant Nutrients (32.1–32.5)

Plants absorb and transport substances—such as water, minerals, CO_2, and sugar—required for growth.

Plant Nutrients and the Soil (32.6–32.11)

Plants require many essential minerals for proper health.

Plant Nutrition and Symbiosis (32.12–32.14)

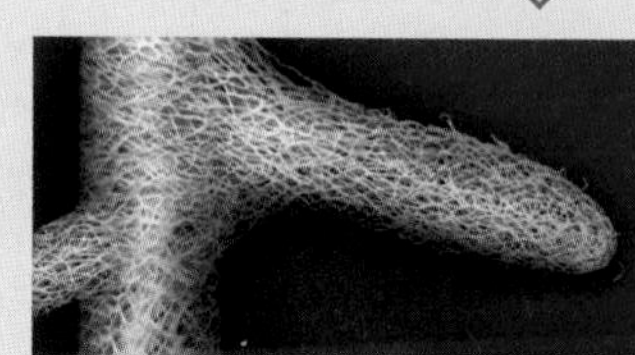

Plants have evolved relationships with bacteria, fungi, animals, and other plants.

▷ The Uptake and Transport of Plant Nutrients

32.1 Plants acquire nutrients from air, water, and soil

Watch a plant grow from a tiny seed and you can't help wondering where all the mass comes from. If you had to take a guess, what would you think is the source of the raw materials that make up a plant's body? Soil? Water? Air?

Aristotle thought that soil provided all the substance for plant growth. The 17th-century physician Jan Baptista van Helmont performed an experiment to test this hypothesis. He planted a willow seedling in a pot containing 91 kg of soil. After five years, the willow had grown into a tree weighing 76.8 kg, but only 0.06 kg of soil had disappeared from the pot. Van Helmont concluded that the willow had grown mainly from added water. A century later, an English physiologist named Stephen Hales postulated that plants are nourished mostly by air.

As it turns out, there is truth in all these early hypotheses about plant nutrition; air, water, and soil all contribute to plant growth **(Figure 32.1A)**. A plant's leaves absorb carbon dioxide (CO_2) from the air; in fact, about 96% of a plant's dry weight is organic (carbon-containing) material built mainly from CO_2. Meanwhile, a plant gets water (H_2O), minerals (inorganic ions), and some oxygen (O_2) from the soil.

What happens to the materials a plant takes up from the air and soil? The sugars that a plant makes by photosynthesis are composed of the elements carbon, oxygen, and hydrogen. We know that the carbon and oxygen used in photosynthesis come from atmospheric CO_2 and that the hydrogen comes from water molecules (see Module 7.3). Plant cells use the sugars made by photosynthesis in constructing all the other organic materials they need, but primarily for carbohydrates. The trunk of the giant sequoia tree in **Figure 32.1B**, for instance, consists mainly of sugar derivatives, such as the cellulose of cell walls.

Plants use cellular respiration to break down some of the sugars they make, obtaining energy from them in a process that consumes O_2. A plant's leaves take up some O_2 from the air, but Figure 32.1A does not show this because plants are actually net producers of O_2, giving off more of this gas than they use. When water is split during photosynthesis, O_2 gas is produced and released through the leaves. The O_2 taken up from the soil by the plant's roots in Figure 32.1A is actually atmospheric O_2 that has diffused into the soil; it is used in cellular respiration in the roots themselves.

A plant's ability to move water from its roots to its leaves and its ability to deliver sugars to specific areas of its body are staggering feats of evolutionary adaptation. Figure 32.1B highlights an extreme example; the topmost leaves of a giant sequoia can be more than 100 m (300 feet) above the roots. In the next four modules, we follow the movement of water, dissolved mineral nutrients, and sugar throughout the plant body.

? **What inorganic substance is obtained in the greatest quantities from the soil?**

Water

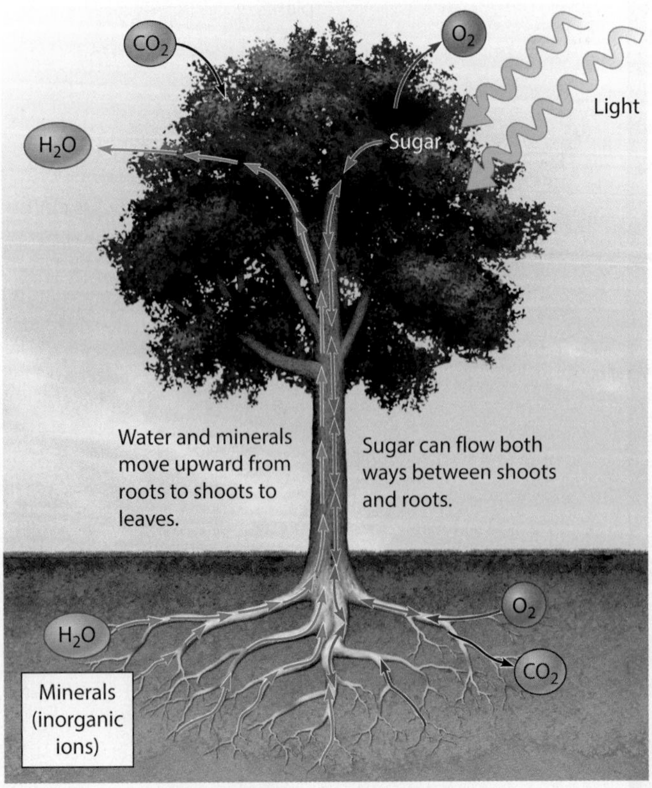

CO_2

O_2

Light

H_2O

Sugar

Water and minerals move upward from roots to shoots to leaves.

Sugar can flow both ways between shoots and roots.

O_2

H_2O

CO_2

Minerals (inorganic ions)

▲ **Figure 32.1A** The uptake and transport of nutrients by a plant

▼ **Figure 32.1B** A giant sequoia (*Sequoia sempervirens*), a product of photosynthesis

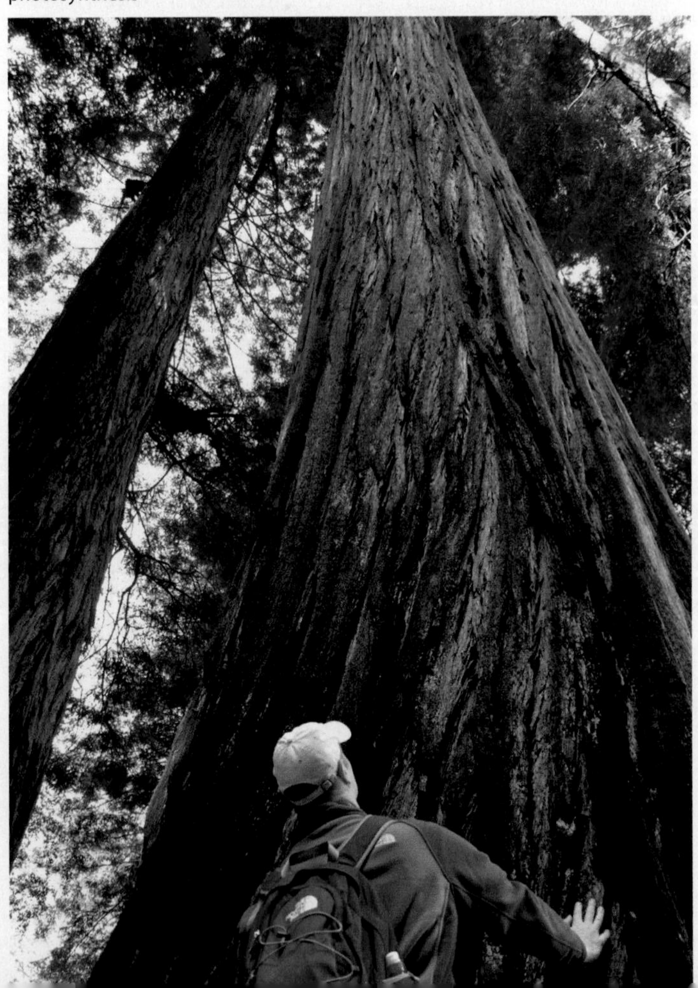

32.2 The plasma membranes of root cells control solute uptake

With its surface area enormously expanded by thousands of root hairs (**Figure 32.2A**), a plant root has a remarkable ability to extract water and minerals from soil. Recall that root hairs are extensions of epidermal cells that cover the root (see Module 31.3). Root hairs provide a huge surface area in contact with nutrient-containing soil; in fact, the root hairs of a single sunflower plant, if laid end to end, could stretch for several miles.

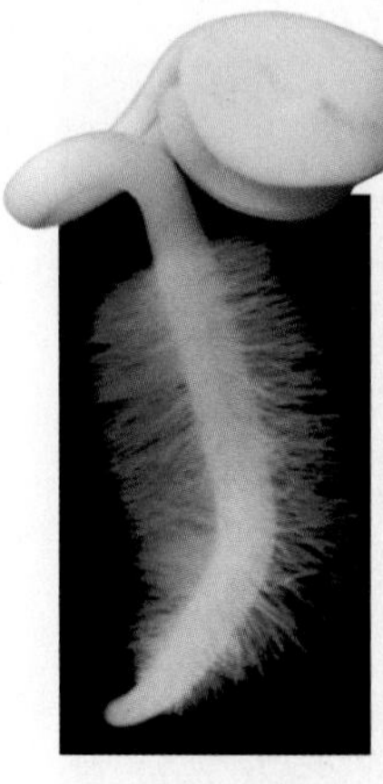

▲ **Figure 32.2A** Root hairs of a radish seedling

All substances that enter a plant root are in solution (that is, dissolved in water). For water and solutes to be transported from the soil throughout the plant, they must move through the epidermis and cortex of the root and then into the water-conducting xylem tissue in the root's vascular cylinder. Any route the water and solutes take from the soil to the xylem requires that they pass through some of the plasma membranes of the root cells. Because plasma membranes are selectively permeable, only certain solutes reach the xylem. This plasma membrane selectivity provides a checkpoint that transports needed minerals from the soil and keeps many unneeded or toxic substances out.

You can see two possible routes to the xylem in the bottom part of **Figure 32.2B**. The long blue arrow indicates an intracellular route. Water and selected solutes cross the cell wall and plasma membrane of an epidermal cell (usually at a root hair). The cells within the root are all interconnected by plasmodesmata; there is a continuum of living cytoplasm among the root cells. Therefore, once inside the epidermal cell, the solution can move inward from cell to cell without crossing any other plasma membranes, diffusing through the interconnected cytoplasm all the way into the root's endodermis. An endodermal cell then discharges the solution into cells of the vascular cylinder, from which it enters the xylem (light purple).

The long pink arrow in Figure 32.2B indicates an alternative route. This route is extracellular; the solution moves inward within the hydrophilic walls and extracellular spaces of the root cells but does not enter the cytoplasm of the epidermis or cortex cells. The solution does not cross any plasma membranes, and there is no selection of solutes until the solution reaches the endodermis. Here, a continuous waxy barrier—the **Casparian strip**—stops water and solutes from entering the xylem through the cell walls, instead forcing them to cross a plasma membrane into an endodermal cell (short blue arrow). Ion selection occurs at this membrane instead of in the epidermis, and once the selected solutes and water are in the endodermal cell, they can be discharged into the xylem.

Actually, water and solutes rarely follow just the two kinds of routes in Figure 32.2B. They may take any combination of these routes, and they may pass through numerous plasma membranes and cell walls en route to the xylem. Because of the Casparian strip, however, there are no nonselective routes; the water and solutes must cross a selectively permeable plasma membrane at some point. In other words, no solutes enter the vascular tissue unchecked; all must pass through at least one checkpoint. Next, we see how water and minerals move upward within the xylem from roots to shoots.

? What is the function of the Casparian strip?

It regulates the passage of minerals (inorganic ions) into the xylem by blocking access via cell walls and requiring all minerals to cross a selectively permeable plasma membrane.

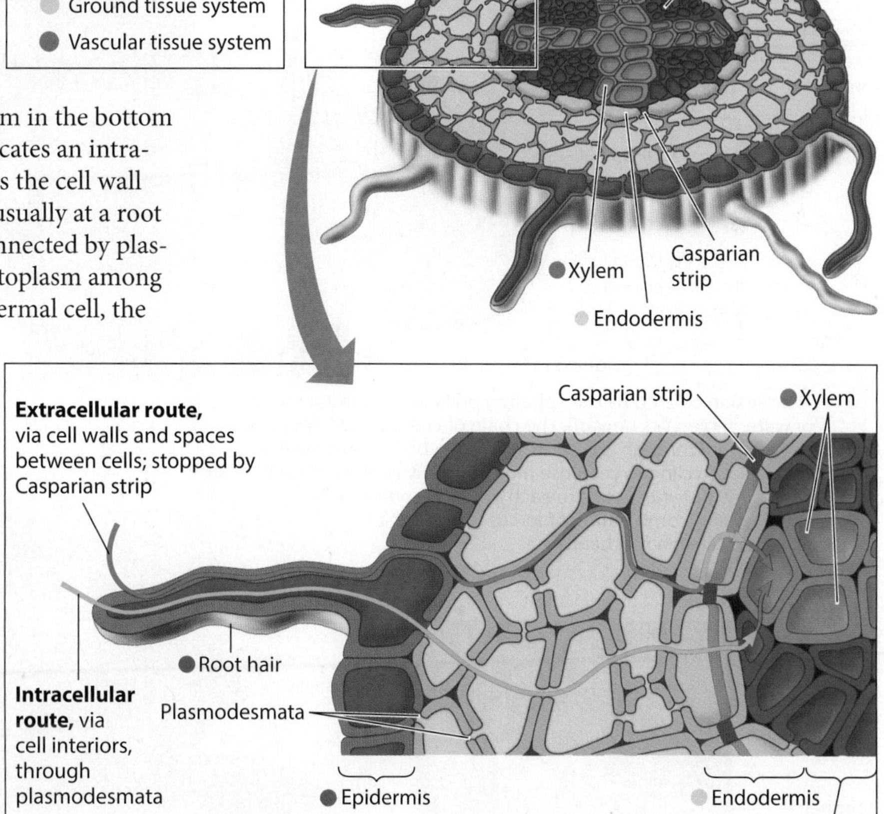

Key
Dermal tissue system
Ground tissue system
Vascular tissue system
Root hair
Epidermis
Cortex
Phloem
Xylem
Casparian strip
Endodermis
Extracellular route, via cell walls and spaces between cells; stopped by Casparian strip
Casparian strip
Xylem
Root hair
Intracellular route, via cell interiors, through plasmodesmata
Plasmodesmata
Epidermis
Endodermis
Cortex
Vascular cylinder

▲ **Figure 32.2B** Routes of water and solutes from soil to root xylem

Try This Using your finger, trace the two routes (extracellular or intracellular) to the xylem. Notice that each one eventually leads to a plasma membrane that must be crossed, allowing for regulation.

VISUALIZING THE CONCEPT

32.3 Transpiration pulls water up xylem vessels

Plants require a constant supply of water and dissolved minerals from the soil. This is provided as xylem sap, a solution of water and inorganic nutrients that flows from the roots through the shoot system to the leaves. Xylem sap flows through very thin tubes within xylem tissue, pulled by transpiration, the loss of water from the leaves by evaporation. Sap movement is aided by the cohesion and adhesion of water molecules and requires no energy expenditure by the plant.

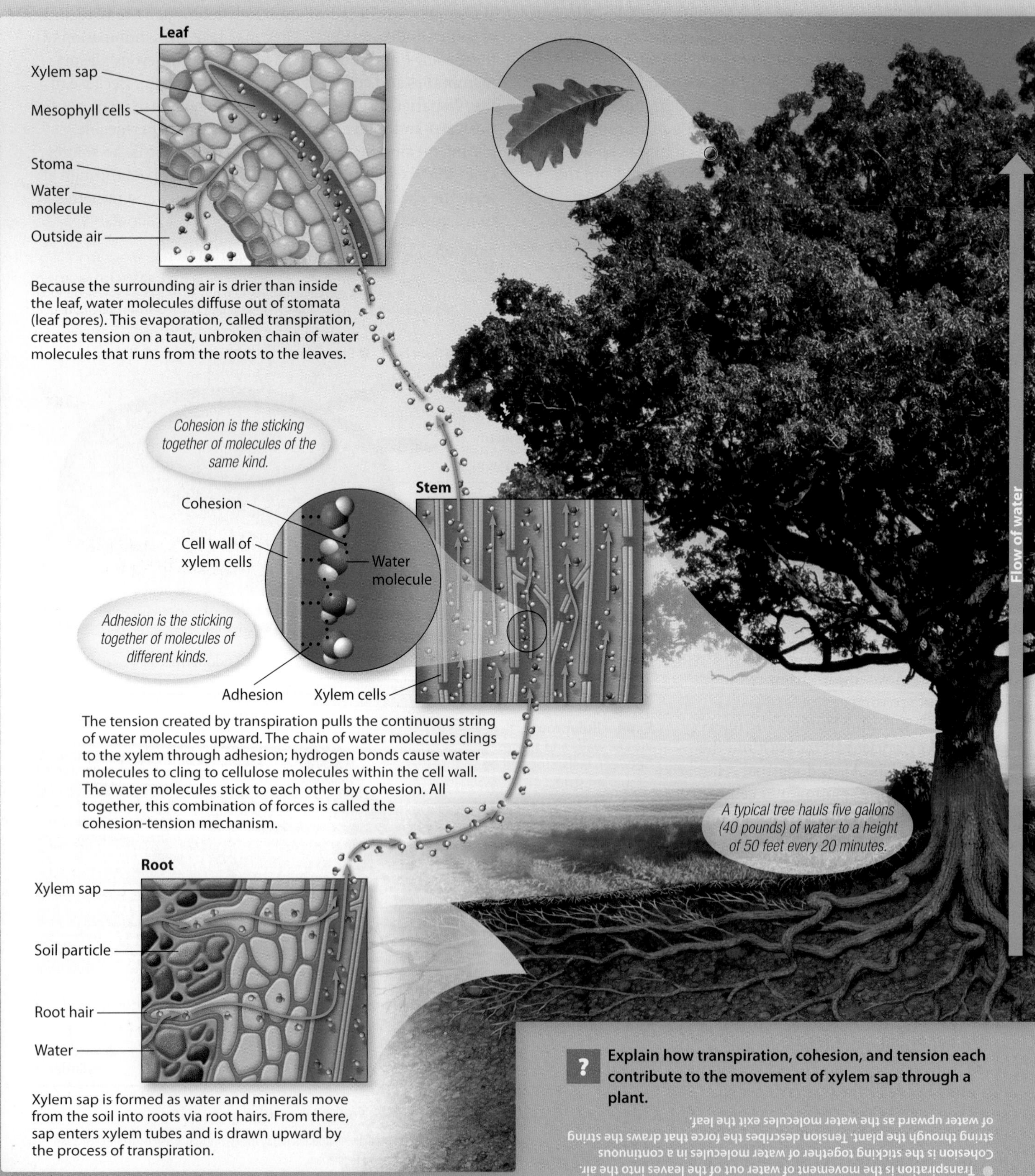

Leaf
Xylem sap
Mesophyll cells
Stoma
Water molecule
Outside air
Because the surrounding air is drier than inside the leaf, water molecules diffuse out of stomata (leaf pores). This evaporation, called transpiration, creates tension on a taut, unbroken chain of water molecules that runs from the roots to the leaves.
Cohesion is the sticking together of molecules of the same kind.
Stem
Cohesion
Cell wall of xylem cells
Water molecule
Adhesion is the sticking together of molecules of different kinds.
Adhesion
Xylem cells
The tension created by transpiration pulls the continuous string of water molecules upward. The chain of water molecules clings to the xylem through adhesion; hydrogen bonds cause water molecules to cling to cellulose molecules within the cell wall. The water molecules stick to each other by cohesion. All together, this combination of forces is called the cohesion-tension mechanism.
Flow of water
A typical tree hauls five gallons (40 pounds) of water to a height of 50 feet every 20 minutes.
Root
Xylem sap
Soil particle
Root hair
Water
Xylem sap is formed as water and minerals move from the soil into roots via root hairs. From there, sap enters xylem tubes and is drawn upward by the process of transpiration.

? Explain how transpiration, cohesion, and tension each contribute to the movement of xylem sap through a plant.

● Transpiration is the movement of water out of the leaves into the air. Cohesion is the sticking together of water molecules in a continuous string through the plant. Tension describes the force that draws the string of water upward as the water molecules exit the leaf.

32.4 Guard cells control transpiration

Adaptations that increase photosynthesis—such as large leaf surface areas—have the serious drawback of increasing water loss by transpiration. Thus, a plant's tremendous requirement for water is largely a consequence of the shoot system's need for ample exchange of CO_2 and O_2 for photosynthesis. As long as water moves up from the soil fast enough to replace lost water, transpiration doesn't present a problem. But if the soil dries and transpiration *out* of the leaves exceeds the delivery of water *to* the leaves, the plant will wilt. Unless the plant is rehydrated, it will eventually die.

The leaf stomata, which can open and close, are evolutionary adaptations that help plants adjust transpiration rates to changing environmental conditions. A pair of guard cells flanking each stoma control the opening by changing shape **(Figure 32.4)**, like a set of doors that swell or deflate to narrow the gap between the two cells. By opening and closing the stomata, guard cells help balance the plant's requirement to conserve water with its requirement for photosynthesis.

What actually causes guard cells to change shape and thereby open or close stomata? A stoma opens (left side of Figure 32.4) when its guard cells gain potassium ions (K^+, red dots) and water (blue arrows) from neighboring cells (light gray). The cells actively take up K^+. Following this active transport of solute, water enters by osmosis. When the vacuoles in the guard cells gain water, the cells become more turgid and bowed. The structure of a guard cell wall causes it to buckle outward, away from its companion guard cell when it becomes turgid. The result is an increase in the size of the gap (stoma) between the two cells. Conversely, when the guard cells lose K^+, they also lose water by osmosis and become flaccid and less bowed, closing the space between them (right side of the figure).

In many plants, guard cells keep the stomata open during the day, which allows CO_2 to enter the leaf from the atmosphere and thus continue photosynthesis when sunlight is available. However, there is a trade-off involved in opening the stoma: The plant also loses water via transpiration. At night, when there is no light for photosynthesis and therefore no need to take up CO_2, many plants close their stoma, saving water. A plant is therefore able to adjust its water loss in response to changes in the environment.

At least three cues contribute to stomata opening at dawn. One cue is light, which stimulates guard cells to accumulate K^+ and become turgid. A second cue is a low level of CO_2 in the leaf, which can have the same effect. A third cue is an internal timing mechanism—a biological clock—found in the guard cells; even if you keep a plant in a dark closet, stomata will continue their daily rhythm of opening and closing. (We'll return to biological clocks in plants in Module 33.10.)

Even during the day, the guard cells may close the stomata if the plant is losing water too fast. This response reduces further water loss and may prevent wilting, but it also slows down CO_2 uptake and photosynthesis—one reason that droughts reduce crop yields. In summary, guard cells arbitrate the photosynthesis–transpiration compromise on a moment-to-moment basis by integrating a variety of stimuli.

? **Some leaf molds secrete a chemical that causes guard cells to accumulate K^+. How does this help the mold infect the plant?**

● Accumulation of K^+ by guard cells causes the stomata to stay open. The mold can then grow into the leaf interior via the stomata.

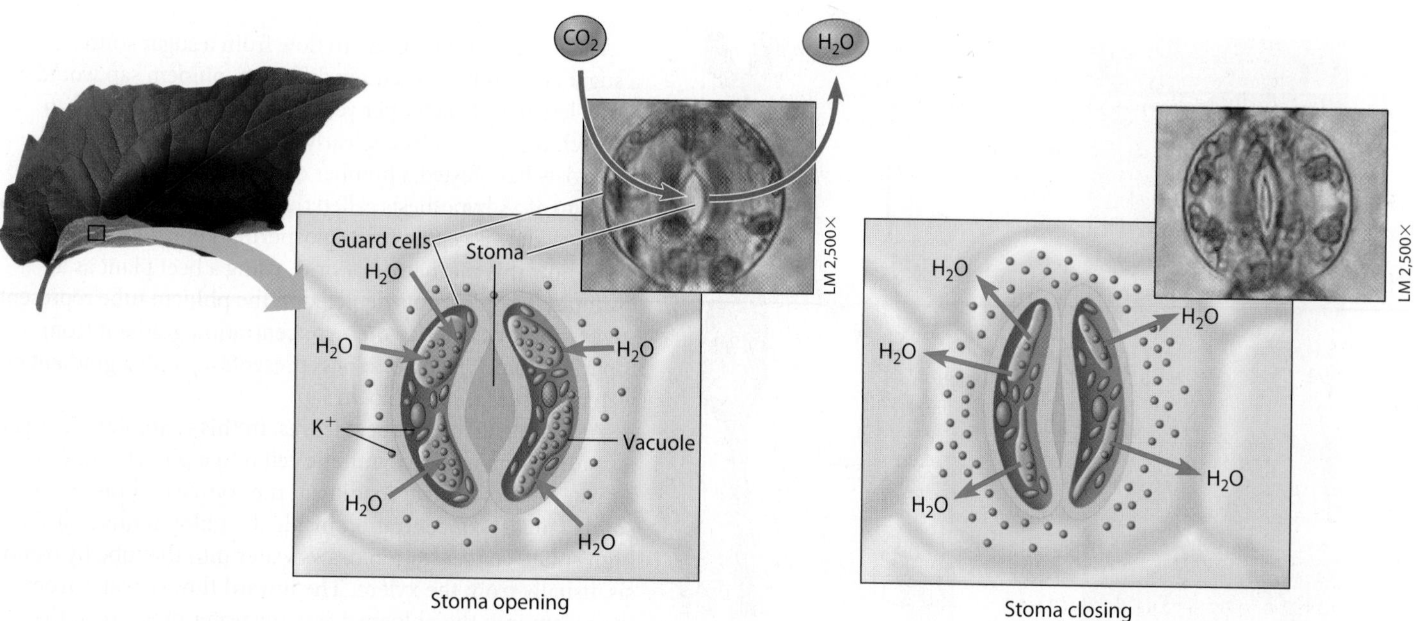

CO_2
H_2O
Guard cells
Stoma
H_2O
H_2O
H_2O
K^+
Vacuole
H_2O
H_2O
LM 2,500×
Stoma opening
H_2O
H_2O
H_2O
H_2O
H_2O
LM 2,500×
Stoma closing

▲ **Figure 32.4** How guard cells control stomata

32.5 Phloem transports sugars

A plant has two separate transport systems: xylem (the topic of Module 32.3) and phloem. Xylem transports xylem sap (water and dissolved minerals), while the main function of phloem is to transport the products of photosynthesis from where they are made or stored to where they are needed. In angiosperms, phloem contains food-conducting cells called sieve-tube elements arranged end to end into long tubes (see Figure 31.6F). The art and micrograph in **Figure 32.5A** show two sieve-tube elements and the sieve plate (and end cap) between them. Perforations in sieve plates connect the cytosol of these living cells into one continuous solution. Sugary liquid called **phloem sap** can thus move freely from one cell to the next. Phloem sap may contain inorganic ions, amino acids, and hormones in transit from one part of the plant to another, but its main solute is usually the disaccharide sugar sucrose (the same sugar as in table sugar).

In contrast to xylem sap, which only flows upward from the roots, phloem sap moves throughout the plant in various directions. However, sieve tubes always carry sugars from a sugar source to a sugar sink. A **sugar source** is a plant organ that is a net producer of sugar, by photosynthesis or by breakdown of starch. Leaves are the primary sugar sources in most mature plants. A **sugar sink** is an organ that is a net consumer or storer of sugar. Growing roots, buds, stems, and fruits are sugar sinks. A storage organ, such as a tuber or a bulb, may be a source or sink, depending on the season. When a plant stockpiles carbohydrates in the summer, a tuber is a sugar sink. After the plant breaks dormancy in the spring, that same tuber is a source as its starch is broken down to sugar, which is carried to the growing shoot tips of the plant. Thus, each food-conducting tube in phloem tissue has a source end and a sink end, but these may change with the season or the developmental stage of the plant.

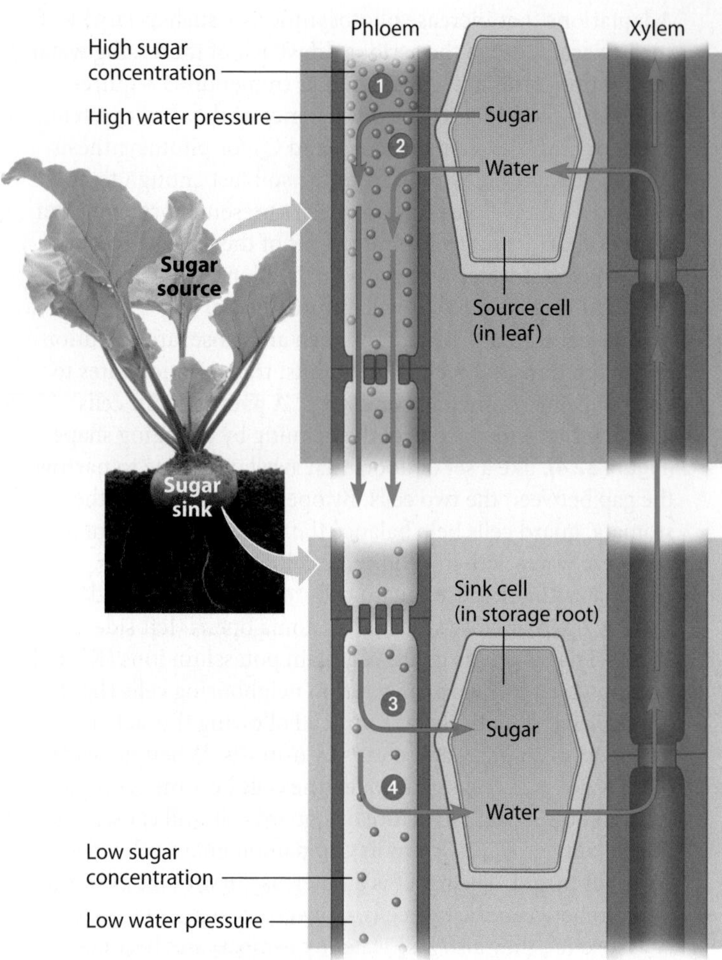
Phloem
Xylem
High sugar concentration
High water pressure
1
2
Sugar
Water
Sugar source
Source cell (in leaf)
Sugar sink
Sink cell (in storage root)
3
Sugar
4
Water
Low sugar concentration
Low water pressure

▲ **Figure 32.5B** Pressure flow in plant phloem from a sugar source to a sugar sink (and the return of water to the source via xylem)

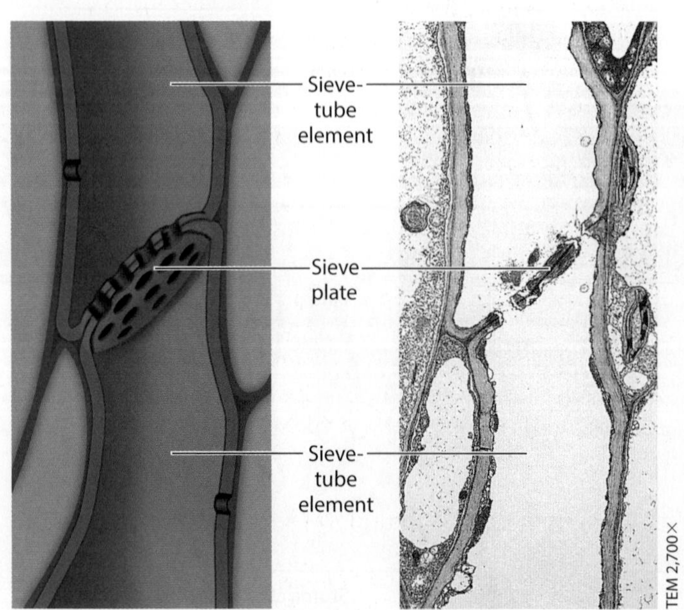
Sieve-tube element
Sieve plate
Sieve-tube element
TEM 2,700×

▲ **Figure 32.5A** Food-conducting cells of phloem

What causes phloem sap to flow from a sugar source to a sugar sink? If it were simply diffusion, phloem sap would travel less than 1 meter per year, requiring decades for sap to reach the top of a tree. Clearly, another force is involved. Biologists have tested a number of hypotheses for phloem sap movement. A hypothesis called the **pressure flow mechanism** is now widely accepted for angiosperms. **Figure 32.5B** illustrates how this mechanism works, using a beet plant as an example. The pink dots depicted in the phloem tube represent sugar molecules; notice their concentration gradient from top to bottom. The blue color represents a parallel gradient of water pressure in the phloem sap.

At the sugar source (beet leaves, in this example), ❶ sugar is loaded from a photosynthetic cell into a phloem tube via active transport. Sugar loading at the source end raises the solute (sugar) concentration inside the phloem tube. ❷ The high solute concentration draws water into the tube by osmosis, usually from the xylem. The inward flow of water from the xylem into the phloem raises the water pressure at the source end of the phloem tube.

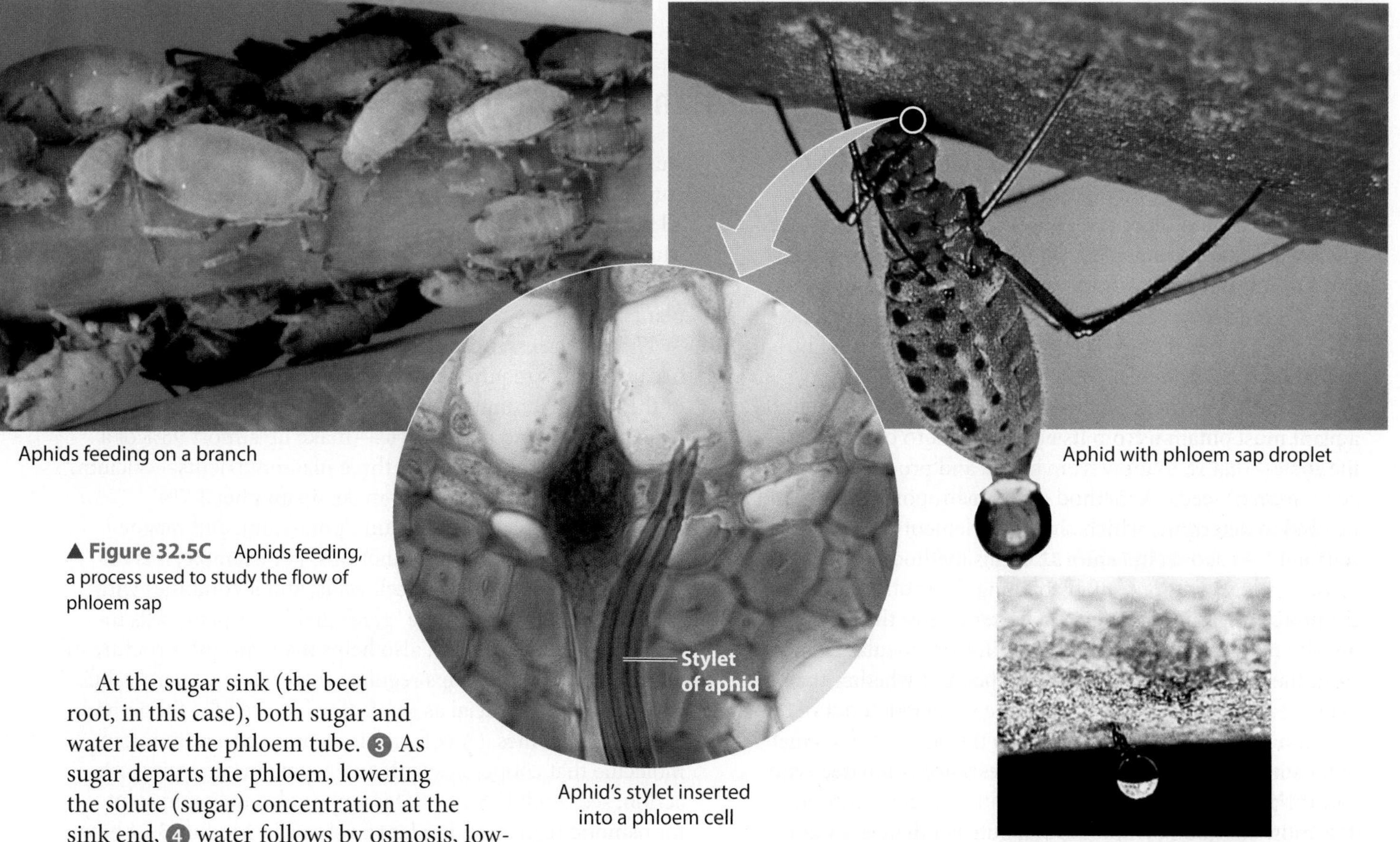

Aphids feeding on a branch
Aphid with phloem sap droplet
Stylet of aphid
Aphid's stylet inserted into a phloem cell
Severed stylet dripping phloem sap

▲ **Figure 32.5C** Aphids feeding, a process used to study the flow of phloem sap

At the sugar sink (the beet root, in this case), both sugar and water leave the phloem tube. ❸ As sugar departs the phloem, lowering the solute (sugar) concentration at the sink end, ❹ water follows by osmosis, lowering the water pressure in the tube. Water then reenters the xylem and returns to the leaves.

The building of water pressure at the source end of the phloem tube and the reduction of that pressure at the sink end cause phloem sap to flow from source to sink—down a gradient of water pressure. Sieve plates allow free movement of solutes as well as water. Thus, sugar is carried along from source to sink at the same rate as the water. As indicated on the right side of Figure 32.5B, xylem tubes recycle the water back from sink to source.

The pressure flow mechanism explains why phloem sap always flows from a sugar source to a sugar sink, regardless of their locations in the plant. However, the mechanism is somewhat difficult to test because most experimental procedures disrupt the structure and function of the phloem tubes. Some of the most interesting studies have taken advantage of natural phloem probes: insects called aphids.

Figure 32.5C shows how an aphid feeds by inserting its needlelike mouthpart, called a stylet, into the phloem of a tree branch. The aphid is releasing from its anus a drop of phloem sap lacking some solutes that the insect's digestive tract has removed for food. The micrograph in the center shows an aphid's stylet inserted into one of the plant's food-conducting cells. The pressure within the phloem force-feeds the aphid, swelling it to several times its original size. Plant biologists have used this process to study the flow of phloem sap. While the aphid is feeding, it can be severed from its stylet. The stylet then serves as a miniature tap that drips phloem sap for hours (similar to the way a tap in a pressurized keg of beer allows the liquid to flow out for hours). The photograph at the lower right shows a droplet of phloem sap on the cut end of a stylet. Studies using this technique support the pressure flow model: The closer the stylet is to a sugar source, the faster the sap flows and the greater its sugar concentration. This is what we would expect if pressure is generated at the source end of the phloem tube by the active pumping of sugar into the tube.

Before we move forward, let's quickly review how a plant absorbs substances from the soil and transports materials from one part of its body to another: Water and inorganic ions enter from the soil and are distributed by xylem. The xylem sap is pulled upward by transpiration. Carbon dioxide enters leaves through stomata and is incorporated into sugars. A second transport system, phloem, distributes the sugars. Pressure flow drives the phloem sap from leaves and storage sites to other parts of the plant, where the sugars are used or stored.

We know from previous chapters that plants convert raw materials to organic molecules by photosynthesis (see Chapter 7). We have yet to say much about the kinds of inorganic nutrients a plant needs and what it does with them. This is the subject of plant nutrition, which we discuss in the next section.

? **Contrast the forces that move phloem sap with the forces that move xylem sap.**

● Pressure is generated at the source end of a sieve tube by the loading of sugar and the resulting osmotic flow of water into the phloem. This pressure pushes phloem sap from the source end to the sink end of the tube. In contrast, transpiration generates a pulling force that drives the ascent of xylem sap.

Plant Nutrients and the Soil

32.6 Plant health depends on obtaining all of the essential inorganic nutrients

In contrast to animals, which require a diet of complex organic (carbon-containing) foods, plants survive and grow solely on CO_2 and inorganic substances—that is, plants are autotrophs (see Module 7.1). The ability of plants to assimilate CO_2 from the air, extract water and inorganic ions from the soil, and synthesize organic compounds is essential not only to the survival of plants but also to the survival of humans and other animals.

A chemical element is considered an **essential element** if a plant must obtain it from its environment to complete its life cycle—that is, to grow from a seed and produce another generation of seeds. A method called hydroponic culture can be used to determine which chemical elements are essential nutrients. As shown in **Figure 32.6**, this method involves growing plants without soil by bathing the roots in mineral solutions. Air is bubbled into the water to give the roots oxygen for cellular respiration. By omitting a particular element from the medium, a researcher can then test whether that element is essential to the plant. If the element left out of the solution is an essential nutrient, then the lack of this element will make the plant abnormal in appearance compared with control plants grown in a medium with complete nutrients. The most common symptoms of a nutrient deficiency are stunted growth and discolored leaves. Hydroponic culture studies have helped identify 17 essential elements needed by all plants; a few others are essential to certain types of plants. Most research has involved crop plants and houseplants; little is known about the nutritional needs of most uncultivated plants.

Complete solution containing all minerals (control)
Solution lacking potassium (experimental)

▲ **Figure 32.6** A hydroponic culture experiment

Nine of the essential elements are called **macronutrients** because plants require relatively large amounts of them. Six of the nine macronutrients—carbon, oxygen, hydrogen, nitrogen, sulfur, and phosphorus—make up almost 98% of a plant's dry weight. The other three macronutrients—calcium, potassium, and magnesium—make up another 1.7%.

How does a plant use calcium, potassium, and magnesium? Calcium has several functions. For example, it is important in the formation of cell walls, and it combines with certain proteins to form the "glue" that holds plant cells together in tissues. Calcium also helps maintain the structure of cell membranes and helps regulate their selective permeability. Potassium is crucial as a cofactor required for the activity of several enzymes. (A cofactor is an atom or a nonprotein molecule that cooperates with an enzyme in catalyzing a reaction; see Module 5.14.) Potassium is also the main solute for osmotic regulation in plants; we saw in Module 32.4 how potassium ion movements regulate the opening and closing of stomata. Magnesium is a component of chlorophyll and thus essential for photosynthesis.

Elements that plants need in very small quantities are called **micronutrients**. Eight micronutrients—that together account for about 0.3% of a plant's dry weight—are used by all plants: chlorine, iron, manganese, boron, zinc, copper, nickel, and molybdenum. (If you compare these nutrients with those listed in Table 2.1 on the elements in the human body, you'll notice that most of them are also essential for animals.) Some plants require a ninth element (sodium). Micronutrients function in plants mainly as cofactors. Iron, for example, is a component of cytochromes, proteins in the electron transport chains of chloroplasts and mitochondria. Micronutrients can generally be used over and over, so plants need only minute quantities of these elements. The requirement for molybdenum, for example, is so modest that there is only one atom of this rare element for every 60 million atoms of hydrogen in dried plant material. Yet a deficiency of molybdenum or any other micronutrient can weaken or kill a plant.

? **You conduct an experiment like the one in Figure 32.6 to test whether a certain plant species requires a particular chemical element as a micronutrient. Why is it important that the glassware be completely clean?**

Because micronutrients are required in only minuscule amounts, even the smallest amount of dirt in the experimental flask may contain enough of the element you are testing to allow normal growth and invalidate your results.

32.7 Fertilizers can help prevent nutrient deficiencies

CONNECTION

The quality of soil—especially the availability of nutrients— affects the health of plants and, for crops, the quality of our own nutrition—you are indeed what you eat! **Figure 32.7A** shows a healthy corn leaf (top) as well as three leaves from plants suffering various nutrient deficiencies. Such mineral shortages can stunt plant growth, and if grain is produced at all from affected plants, it will likely have low nutritional value. In this way, nutritional deficiencies in plants can be passed on to livestock or human consumers.

The symptoms of nutrient deficiencies are often distinctive. Many growers can therefore make visual diagnoses, which can be confirmed by having soil and plant samples chemically analyzed at a laboratory. Agricultural extension offices run by state universities often provide this service.

Nitrogen shortage is the most common nutritional problem in plants. Soils are usually not deficient in total nitrogen, but they are often deficient in the nitrogen compounds that plants can use: dissolved nitrate ions (NO_3^-) and ammonium ions (NH_4^+). Stunted growth and yellow-green leaves, starting at the tips of older leaves, are signs of nitrogen deficiency (see Figure 32.7A, second leaf from top). Other common nutrient shortfalls in plants include phosphorus and potassium deficiencies.

Once a diagnosis of a nutrient deficiency is made, treating the problem is usually simple. **Fertilizers** are compounds given to plants via the soil to promote the plant's growth. There are two basic types of fertilizers: inorganic and organic. Inorganic fertilizers (also called mineral fertilizers) can contain naturally occurring inorganic compounds (such as mined limestone or phosphate rock) or synthetic inorganic compounds (such as ammonium nitrate). Inorganic fertilizers come in a wide variety of formulations, but most emphasize their "N-P-K ratio," the relative amounts of the three nutrients most often deficient in depleted soils: nitrogen (N), phosphorus (P), and potassium (K). For example, a 100-pound bag of 5–6–7 fertilizer contains 5 pounds of nitrogen (often as ammonium or nitrate), 6 pounds of phosphorus (usually as phosphoric acid), and 7 pounds of potassium (usually as the mineral potash), plus 82 pounds of filler.

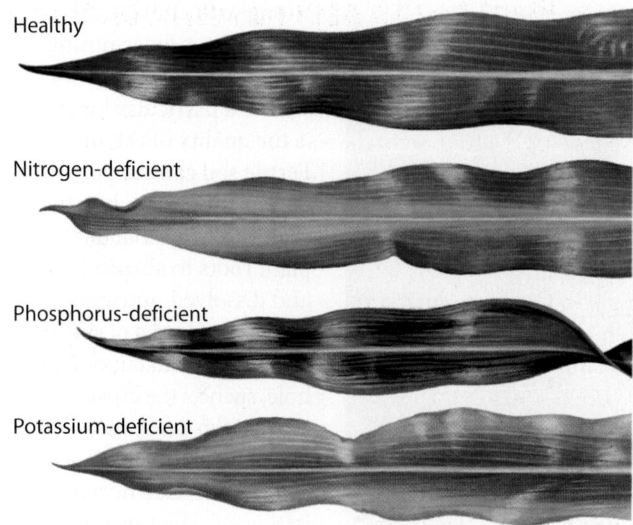
Healthy
Nitrogen-deficient
Phosphorus-deficient
Potassium-deficient

▲ **Figure 32.7A** The most common mineral deficiencies, as seen in corn leaves

Many crops benefit from an all-purpose 5–5–5 formula, but some plants thrive only under special fertilizer formulations.

Organic fertilizers are composed of chemically complex organic matter such as **compost**, a soil-like mixture of decomposed organic matter. (Organic fertilizers may or may not be certified organic, which means that the product meets a strict set of guidelines—see Module 32.10.) Many gardeners maintain a free-standing compost pile or an enclosed compost bin to which they add leaves, grass clippings, yard waste, and kitchen scraps (avoiding meat, fat, and bone). Over time, the vegetable matter is broken down by naturally occurring microbes, fungi, and animals **(Figure 32.7B)**. Occasional turning and watering will speed the composting process, producing homegrown fertilizer in several months. The compost can then be applied to outdoor gardens or indoor pots.

? **What is the most common nutrient deficiency in plants? What are the signs?**

Nitrogen deficiency; stunted growth and yellowing leaves

▼ **Figure 32.7B** Steam produced by the metabolic activity of organisms within a compost pile

32.8 Fertile soil supports plant growth

A
B
C
20
40
60
80
1

▲ **Figure 32.8A** Three soil horizons visible beneath grass

Along with climate, the major factor determining whether a plant can grow well in a particular location is the quality of the soil. Fertile soil can support abundant plant growth by providing conditions that enable plant roots to absorb water and dissolved nutrients.

Distinct layers of soil are visible in a road cut or deep hole, such as the cross section shown in **Figure 32.8A**. You can see three distinct soil layers, called horizons, in the cut. The **topsoil**, or A horizon, is a mixture of rock particles of various sizes, living organisms, and **humus**, the remains of partially decayed organic material produced by the decomposition by bacteria and fungi of dead organisms, feces, fallen leaves, and other organic matter. Topsoil is subject to extensive weathering (freezing, drying, and erosion, for example). The rock particles in topsoil provide a large surface area that retains water and minerals while also forming air spaces containing oxygen that can diffuse into plant roots. Fertile topsoil is home to an astonishing number—about 5 billion per teaspoon—and variety of bacteria, algae and other protists, fungi, and small animals such as earthworms, roundworms, and burrowing insects. Along with plant roots, these organisms loosen and aerate the soil and contribute organic matter to the soil as they live and die. Nearly all plants depend on bacteria and fungi in the soil to break down organic matter into inorganic molecules that roots can absorb. Besides providing nutrients, humus also tends to retain water while keeping the topsoil porous enough for good aeration of the plant roots. Topsoil is rich in organic materials and is therefore most important for plant growth. Plant roots branch out in the A horizon and usually extend into the next layer, the B horizon.

The soil's B horizon contains many fewer organisms and much less organic matter than the topsoil and is less subject to weathering. Fine clay particles and nutrients dissolved in soil water drain down from the topsoil and often accumulate in the B horizon. The C horizon is composed mainly of partially broken-down rock that serves as the "parent" material for the upper layers of soil.

Figure 32.8B illustrates the association among root hairs, soil water, and topsoil. The root hairs are in direct contact with the water that surrounds the particles. The soil water is a solution containing dissolved inorganic ions. Oxygen diffuses into the water from air spaces in the soil. Roots absorb this soil solution.

Cation exchange is a mechanism by which the root hairs take up certain positively charged ions (cations). Inorganic cations—such as calcium (Ca^{2+}), magnesium (Mg^{2+}), and potassium (K^+)—adhere by electrical attraction to the negatively charged surfaces of soil particles. Adhesion helps prevent these positively charged nutrients from draining away during heavy rain or irrigation. In cation exchange (**Figure 32.8C**), root hairs release hydrogen ions (H^+) into the soil solution. The hydrogen ions displace cations on the surfaces of soil particles, and root hairs can then absorb them.

In contrast to cations, negative ions (anions)—such as nitrate (NO_3^-)—are usually not bound tightly by soil particles. Unbound ions are readily available to plants, but they tend to drain out of the soil quickly.

It may take centuries for a soil to become fertile through the breakdown of rock and the accumulation of organic material. The loss of soil fertility is one of our most pressing environmental problems, as we discuss next.

? **How do roots actively increase the availability of mineral nutrients that are cations?**

● By secreting hydrogen ions, which displace cations from soil particles

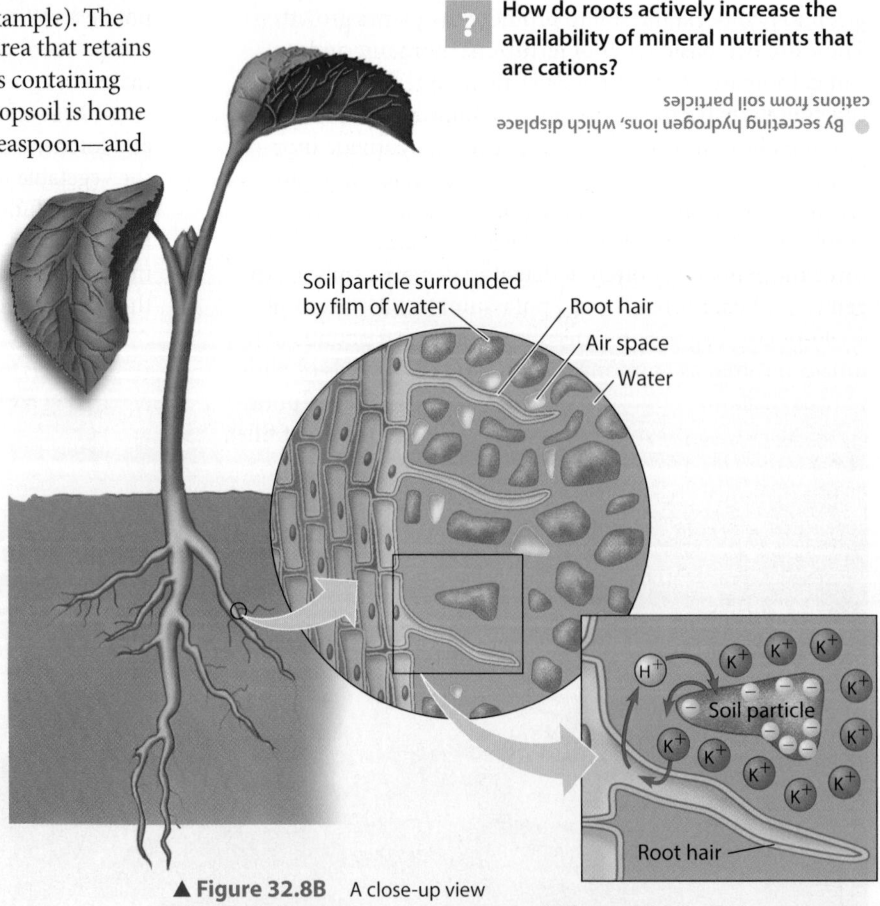

Soil particle surrounded by film of water
Root hair
Air space
Water
H^+
K^+
Soil particle
Root hair

▲ **Figure 32.8B** A close-up view of root hairs in soil

▲ **Figure 32.8C** Cation exchange

32.9 Soil conservation is essential to human life

CONNECTION

Our survival as a species depends on soil. However, erosion and chemical pollution threaten this vital resource throughout the world. As the human population continues to grow and more land is cultivated, farming practices that conserve soil fertility will become essential to our survival. Three critical goals of soil conservation are proper irrigation, prevention of erosion, and prudent fertilization.

Irrigation can turn a desert into a garden, but farming in dry regions is a huge drain on water resources: Globally, about 75% of all freshwater use is dedicated to crop irrigation. Improper irrigation can be very wasteful, and overextraction of groundwater can cause various environmental problems, such as the sudden appearance of sinkholes (**Figure 32.9A**). Instead of wastefully flooding fields, modern irrigation often uses perforated pipes that drip water slowly into the soil near plant roots. This drip irrigation uses less water, allows the plants to absorb it efficiently, and reduces water loss from evaporation and drainage.

Erosion—the blowing or washing away of soil—is a major cause of soil degradation because nutrients are carried away by wind and streams. In the 1930s, the Great Plains of the United States suffered devastating dust storms that swept away huge amounts of topsoil after decades of inappropriate farming techniques and years of drought (**Figure 32.9B**). The resulting disaster took a huge toll—both economic and human—that reshaped our country, a plight immortalized in John Steinbeck's novel *The Grapes of Wrath*.

To limit erosion today, farmers plant rows of trees as windbreaks, terrace hillside crops, and cultivate crops in a contour pattern that helps slow runoff of water and topsoil. Crops such as alfalfa and wheat provide good ground cover and protect the soil better than corn and other crops that are usually planted in more widely spaced rows.

Prehistoric farmers may have started fertilizing their fields after noticing that grass grew faster and greener where animals had defecated. In developed nations today, most farmers use inorganic, commercially produced fertilizers containing minerals that are either mined or prepared by industrial processes. These fertilizers are usually enriched in nitrogen, phosphorus, and potassium, the macronutrients most commonly deficient in farm and garden soils. Manure, fish meal, and compost (decaying plant matter) are common fertilizers that contain decomposing organic material. Before the nutrients in these substances can be used by plants, the organic material must be broken down by bacteria and fungi to inorganic nutrients that roots can absorb.

Whether from natural sources or a chemical factory, the minerals a plant extracts from the soil are in the same form. The difference is that naturally derived fertilizers release nutrients gradually, whereas minerals in inorganic commercial fertilizers are available immediately. However, because minerals from inorganic fertilizers are soluble in water, they may not be retained in the soil for long. Problems arise when fields are overfertilized with inorganic products and excess nutrients are not taken up by plants. The excess minerals are often leached from the soil by rainwater or irrigation. Mineral runoff into lakes may lead to a sudden increase in the number of algae, which can deplete oxygen, killing off fish and other animals (**Figure 32.9C**; see also Module 37.22).

Agricultural researchers are developing ways to maintain crop yields while reducing the use of costly fertilizer. One approach is to genetically engineer "smart" plants that inform the grower when a nutrient deficiency is imminent but before damage has occurred. One such plant contains a gene that causes leaf cells to produce a blue pigment when the phosphorus content of plant tissues declines. When leaves of these smart plants develop a blue tinge, the farmer knows it is time to add phosphorus-containing fertilizer.

? Why do fertilizers containing organic materials generally contaminate water resources less than inorganic fertilizers?

● Fertilizers containing organic materials release mineral nutrients gradually as they decompose, so there is less likelihood of the minerals leaching into the groundwater or running off into streams and lakes.

▲ **Figure 32.9A** A sinkhole caused by overuse of groundwater for irrigation

▲ **Figure 32.9B** A dust storm in the American Great Plains during the 1930s

▲ **Figure 32.9C** Runoff due to excess fertilizer use on nearby land causes sudden blooms of algae, which can deplete oxygen from a lake, killing off other life

32.10 Organic farmers follow principles meant to promote health

SCIENTIFIC THINKING

Are organic crops healthier than conventional crops?

Organic farming involves agricultural practices that promote biological diversity by maintaining soil quality through natural methods (rotating crops, planting cover crops, amending the soil with organic matter, and using few or no synthetic fertilizers), providing habitat for predators of pests rather than relying mainly on synthetic pesticides, and avoiding genetically modified organisms. Yearly inspections—by agencies approved by the USDA—ensure proper practices, accurate record keeping, and a buffer of land between organic farms and neighboring conventional farms (which avoids cross-fertilization between organic and conventional varieties of plants). The primary goal of organic farming is to achieve **sustainable agriculture**, a system embracing farming methods that are conservation-minded, environmentally safe, and profitable.

The ultimate aim of many organic farmers is to restore as much to the soil as is drawn from it, creating fields that are bountiful and self-sustaining. Many have chosen organic farming to both protect the environment and answer the growing demand for more naturally produced foods. Some benefits of organic farming are obvious: fewer synthetic chemicals in the environment and less risk of exposing farmworkers and wildlife to potential toxins.

However, an organic label is no guarantee of health benefits. There is no scientific consensus on whether organically grown foods have a higher nutrient composition than conventionally grown foods. A 2001 U.S. study, for example, found that organic fruits contain 27% more vitamin C than conventional fruits. But a 2009 Italian study found that organically grown tomatoes had less vitamin C than conventional counterparts. How can the scientific community—and the average consumer—draw a conclusion when presented with seemingly contradictory data? When faced with a large number of independent studies that vary in their results, scientists can test hypotheses by performing a meta-analysis, a study that combines results from many studies into a single statistical analysis—essentially a study of studies.

In 2011, Australian scientists published a meta-analysis that focused on the micronutrient composition of foods produced via organic and conventional growing methods. In the meta-analysis, studies conducted between 1980 and 2007 were screened for proper inclusion of controls (i.e., the foods under comparison were grown and harvested in the same season and were analyzed at a similar level of freshness). Of 1,440 micronutrient comparisons conducted during that 27-year period, 908 (involving 22 different micronutrients) were found to use proper controls. Based on data gleaned from these studies (**Table 32.10**), the researchers concluded that a small majority of studies (51%) found that organic produce was higher in nutrient content. But positive results for organic produce were slight. Nutritionists caution that additional factors—how produce is harvested, handled, and prepared—can have an equally significant effect on the nutrient content as the growing method. And we should never lose track of the fact that fruits and vegetables—no matter how they are grown—are an important part of a healthy diet.

? If you buy "organic" apples, does that tell you anything about how they were grown? What about how healthy they are?

● The label indicates that the grower has been certified to be following standards meant to promote agricultural sustainability. The organic designation does not mean that the produce is healthier than conventionally grown produce.

TABLE 32.10 PERCENTAGE OF STUDIES THAT REPORTED EACH RESULT

Micronutrient	Significantly Higher in Organic Produce	Significantly Higher in Inorganic Produce	Not Significantly Different
All minerals	51	39	10
All vitamins	50	45	5
All nutrients	51	40	9

Data from D. Hunter et al., Evaluation of the micronutrient composition of plant foods produced by organic and conventional agricultural methods, *Critical Reviews in Food Science and Nutrition* 51(6): 571–82 (2011).

32.11 Agricultural research is improving the yields and nutritional values of crops

CONNECTION

A person dies of hunger somewhere in the world *about every 2 seconds*. This death rate is truly staggering. Advocates of plant biotechnology believe that genetic engineering is the key to overcoming the tragedy of world hunger. The most limited resource for food production is land. The size of the human population is steadily increasing while the amount of farmland is decreasing. Thus, improving crop yields is a major goal of plant biotechnology.

The commercial adoption by farmers of genetically modified organisms (GMOs) as crop plants has been one of the most rapid advances in agriculture. These crops include transgenic varieties of cotton and corn that contain genes from the bacterium *Bacillus thuringiensis*. These transgenes encode a bacterial protein (*Bt* toxin) that effectively controls a number of serious insect pests. The use of such plant varieties greatly reduces the need for spraying crops with chemical insecticides. Although *Bt* toxin is harmless to humans, its use is controversial due, in part, to concerns about its effects on helpful insects.

Considerable progress also has been made in the development of transgenic varieties of cotton, corn, soybeans, sugar beets, and wheat that are tolerant to a number of herbicides.

The cultivation of these plants may reduce production costs and enables farmers to "weed" crops with herbicides that do not damage the transgenic crop plants. This can reduce tillage, which can cause erosion of soil. Researchers are also engineering plants with enhanced resistance to disease. For example, a widely used variety of summer squash has genetically modified resistance to three plant viruses, enabling cultivation in regions where these pathogens are rampant.

The nutritional quality of plants is also being improved. Golden Rice, a transgenic variety with a few daffodil genes that increase synthesis of vitamin A, is currently under development and evaluation (see Module 12.8). Plant breeding has also resulted in new varieties of corn, wheat, and rice that are enriched in protein (**Figure 32.11**). Such modified crops may be particularly important because protein deficiency is the most common cause of malnutrition around the world. However, many of these "super" varieties of plants have a high demand for nitrogen, usually supplied in the form of expensive commercial fertilizer. Thus, the countries that most need high-protein crops are usually the ones least able to afford to grow them.

There is ongoing debate about the environmental effects of GMO crops as well (see Module 12.9). Decisions about developing GMO crops, of course, should be based on sound science rather than on reflexive fear or blind optimism.

▲ **Figure 32.11** Researchers with high-protein rice

? Why is research on the protein content of crop plants so important to human health worldwide?

● Because the most common form of malnutrition is protein deficiency, and most people in the world get most of their protein from plants

▷ Plant Nutrition and Symbiosis

32.12 Most plants depend on bacteria to supply nitrogen

Nitrogen deficiency is the most common nutritional problem in plants. This might seem puzzling when you consider that the atmosphere is nearly 80% nitrogen. But even though plants are bathed in gaseous nitrogen (N_2), they cannot absorb it directly from the air. To be used by plants, N_2 must be converted to ammonium (NH_4^+) or nitrate (NO_3^-).

Within soil, ammonium and nitrate are produced from atmospheric N_2 or from organic matter by bacteria. As shown in **Figure 32.12**, certain soil bacteria, called nitrogen-fixing bacteria, convert atmospheric N_2 to ammonia (NH_3), a metabolic process called **nitrogen fixation**. In soil, ammonia picks up an H^+ to form an ammonium ion (NH_4^+). A second group of bacteria, called ammonifying bacteria, adds to the soil's supply of ammonium by decomposing organic matter.

Plant roots can absorb nitrogen as ammonium. However, plants acquire their nitrogen mainly in the form of nitrate (NO_3^-), which is produced from ammonium in the soil by a third group of soil bacteria called nitrifying bacteria. After nitrate is absorbed by roots, enzymes within plant cells convert the nitrate back to ammonium, which is then incorporated into amino acids. It is worth noting that without these three types of bacteria, our crop plants could not obtain sufficient nitrogen, and we would be unable to feed ourselves.

? What is the danger in applying a compound that kills bacteria to the soil around plants?

● The compound might kill soil bacteria that make nitrogen available to plants, causing nitrogen deficiency.

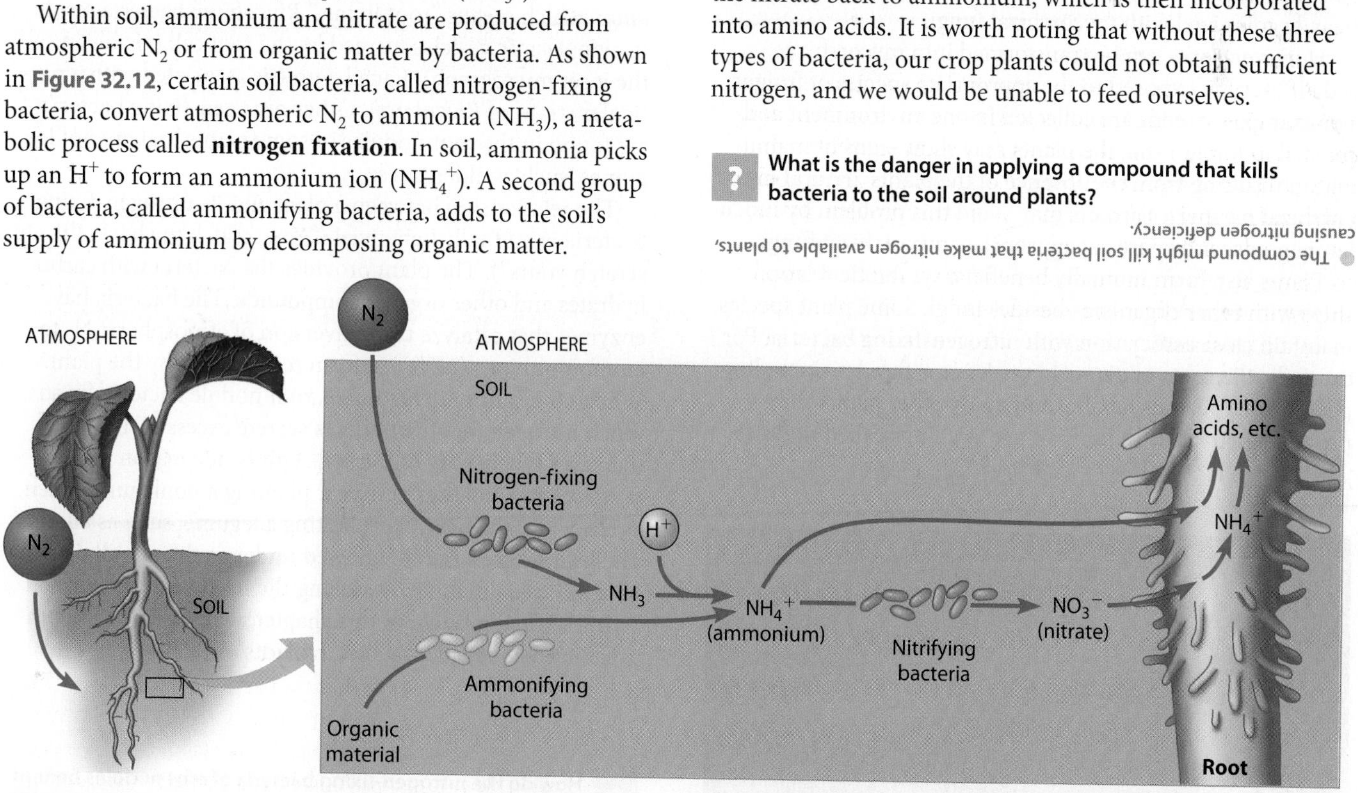

ATMOSPHERE
N_2
SOIL
ATMOSPHERE
SOIL
N_2
Nitrogen-fixing bacteria
H^+
NH_3
NH_4^+ (ammonium)
Nitrifying bacteria
NO_3^- (nitrate)
Ammonifying bacteria
Organic material
Amino acids, etc.
NH_4^+
Root

▲ **Figure 32.12** The roles of bacteria in supplying nitrogen to plants

Try This There are three types of bacteria involved in supplying nitrogen to plants; name each type and describe its action.

32.13 Plants have evolved mutually beneficial symbiotic relationships

EVOLUTION CONNECTION

Reliance on the soil for nutrients that may be in short supply makes it imperative that the roots of plants have a large surface area for absorption. As we have seen, root hairs add a great deal of surface to plant roots. About 80% of living plant species gain even more absorptive surface by teaming up with fungi.

The illustration in **Figure 32.13A** shows a tree root. The root is covered with a twisted mat of fungal filaments. Together, the roots and the fungus form a mutually beneficial (mutualistic) association called a **mycorrhiza**. The fungus benefits from a steady supply of sugar supplied by the host plant. In return, the fungus increases the surface area for water uptake and selectively absorbs phosphate and other minerals from the soil and supplies them to the plant. The fungi of mycorrhizae also secrete growth factors that stimulate roots to grow and branch, as well as antibiotics that may protect the plant from pathogens in the soil.

Evolution of mutually beneficial symbiotic associations between roots and fungi was a critical step in the successful colonization of land by plants. Evidence for this is found in the fact that fossilized roots from some of the earliest plants show that they already contained mycorrhizae. When terrestrial ecosystems were young, the soil was probably not very rich in nutrients. The fungi of mycorrhizae, which are more efficient at absorbing minerals than the roots of plants, would have helped nourish the pioneering plants. Even today, the plants that first become established on nutrient-poor soils, such as abandoned farmland or eroded hillsides, are usually heavily colonized with mycorrhizal fungi.

However, roots can be transformed into mycorrhizae only if they are exposed to the appropriate species of fungus. For example, if seeds are collected in one environment and planted in foreign soil, the plants may show signs of malnutrition resulting from the absence of the plants' natural mycorrhizal partners. Farmers may avoid this problem by inoculating seeds with spores of appropriate mycorrhizal fungi.

Plants also form mutually beneficial symbiotic relationships with other organisms besides fungi. Some plant species maintain close association with nitrogen-fixing bacteria. For example, the roots of plants in the legume family—including peas, beans, peanuts, alfalfa, and many other plants that produce their seeds in pods—have swellings called nodules (**Figure 32.13B**). Within these nodules, plant cells have been "infected" by nitrogen-fixing bacteria of the genus *Rhizobium*, which means "root living." *Rhizobium* bacteria reside in cytoplasmic vesicles formed by the root cell (visible in the inset micrograph). Each legume is associated with a particular strain of *Rhizobium*. Other nitrogen-fixing bacteria are found in the root nodules of some plants that are not legumes, such as alder trees.

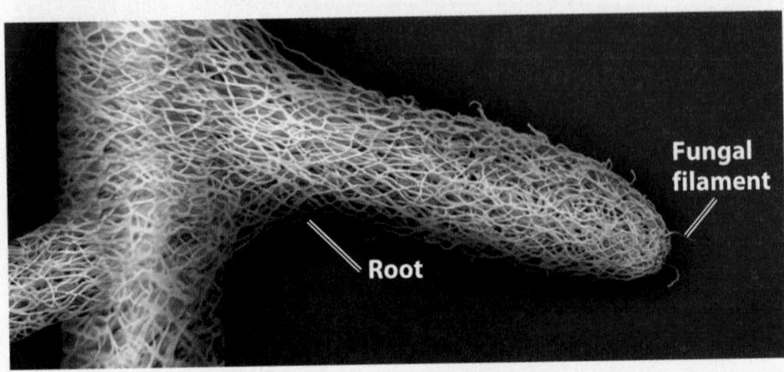
Fungal filament
Root

▲ **Figure 32.13A** Part of a tree root mycorrhiza

TEM 4,700×
Nucleus
Bacteria within vesicle inside cell
Shoot
Nodules
Roots

▲ **Figure 32.13B** Root nodules on a pea plant

The relationship between a plant and its nitrogen-fixing bacteria is mutually beneficial ("You scratch my back, I'll scratch yours"). The plant provides the bacteria with carbohydrates and other organic compounds. The bacteria have enzymes that catalyze the conversion of atmospheric N_2 to ammonium ions (NH_4^+), a form readily used by the plant. When conditions are favorable, root nodule bacteria fix so much nitrogen that the nodules secrete excess NH_4^+, which increases the fertility of the soil. This is one reason farmers practice crop rotation, one year planting a nonlegume, such as corn, and the next year planting a legume, such as alfalfa. The legume crop may be plowed under so that it will decompose as "green manure," reducing the need for fertilizer.

In the final module of this chapter, we consider some unusual nutritional adaptations in plants.

? How do the nitrogen-fixing bacteria of root nodules benefit from their symbiotic relationship with plants?

● The bacteria receive organic nutrients produced by the plant.

32.14 The plant kingdom includes epiphytes, parasites, and carnivores

Almost all plant species have mutually beneficial symbiotic associations with soil fungi, bacteria, or both. Though rarer, there are also plant species with nutritional adaptations that take advantage of other organisms. For example, an epiphyte is a plant that grows on another plant, usually anchored to branches or trunks of living trees. Examples of epiphytes include staghorn ferns and orchids, like the one shown in **Figure 32.14A**. Epiphytes absorb water and minerals from rain.

▲ **Figure 32.14A** A *Cattleya* orchid, a type of epiphyte, growing on the trunk of a tree

Unlike epiphytes, parasitic plants absorb water, sugars, and minerals from their living hosts. Many species have roots that tap into the host plant. **Figure 32.14B** shows a parasitic plant called dodder (the yellow-orange threads that are wound around the green plant). Dodder cannot photosynthesize; it obtains organic molecules from other plant species, using roots that tap into the host's vascular tissue.

▲ **Figure 32.14B** Dodder (*Cuscuta salina*) growing on a pickleweed

Figure 32.14C shows part of an oak tree parasitized by mistletoe, the plant traditionally tacked above doorways during the Christmas season. All the leaves you see here are mistletoe; the oak has lost its leaves for winter. Mistletoe is photosynthetic, but it supplements its diet by siphoning sap from the vascular tissue of the host tree.

A few plants are carnivores. They grow in acid bogs and other habitats where soils are poor in nitrogen and other minerals. In these acidic soils, organic matter decays so slowly that there is little inorganic nitrogen available for plant roots to take up. Carnivorous plants obtain most of their nitrogen and some minerals by killing and digesting insects and other small animals.

▲ **Figure 32.14C** Mistletoe growing on an oak tree

▲ **Figure 32.14D** A sundew plant (*Drosera intermedia*) trapping a damselfly

▲ **Figure 32.14E** A Venus flytrap capturing a fly

The sundew plant has modified leaves, each bearing many tentacle-shaped hairs (**Figures 32.14D**). A sticky, sugary secretion at the tips of the hairs attracts and ensnares insects. The presence of an insect triggers the hairs to bend and the leaf to enfold its prey. The hairs then secrete digestive enzymes, and the plant absorbs nutrients released as the insect is digested.

The Venus flytrap (**Figure 32.14E**) has hinged leaves that close quickly around small insects, usually ants and grasshoppers. As insects walk on the insides of these leaves, they touch sensory hairs that trigger closure of the trap. The leaf then secretes digestive enzymes and absorbs nutrients from the prey.

Using insects as a source of nitrogen is a nutritional adaptation that enables carnivorous plants to thrive in soils where most other plants cannot. Fortunately for animals, such predator-prey turnabouts are rare!

? **Carnivorous plants are most common in locales where the soil is deficient in __________ and __________.**

● nitrogen . . . minerals

CHAPTER 32 REVIEW

For practice quizzes, BioFlix animations, MP3 tutorials, video tutors, and more study tools designed for this textbook, go to MasteringBiology®

Reviewing the Concepts

The Uptake and Transport of Plant Nutrients (32.1–32.5)

32.1 Plants acquire nutrients from air, water, and soil. As a plant grows, its roots absorb water, minerals (inorganic ions), and some O_2 from the soil. Its leaves take in carbon dioxide from the air.

32.2 The plasma membranes of root cells control solute uptake. Root hairs greatly increase a root's absorptive surface. Water and solutes can move through the root's epidermis and cortex by going either through cells or between them. However, all water and solutes must pass through the selectively permeable plasma membranes of cells of the endodermis to enter the xylem (water-conducting tissue) for transport upward.

32.3 Transpiration pulls water up xylem vessels. Transpiration can move xylem sap, consisting of water and dissolved inorganic nutrients, to the top of the tallest tree.

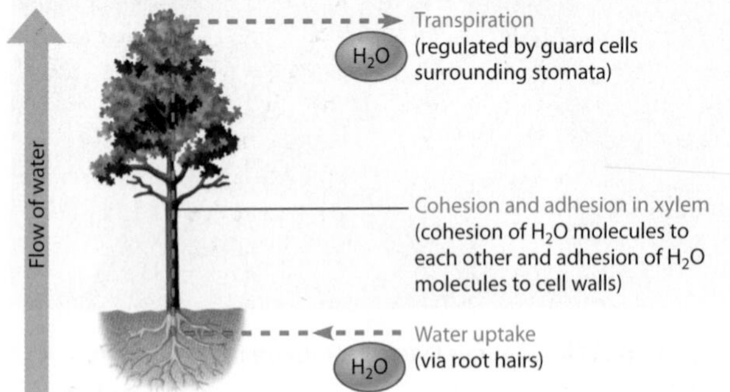

Transpiration (regulated by guard cells surrounding stomata)
H2O
Flow of water
Cohesion and adhesion in xylem (cohesion of H2O molecules to each other and adhesion of H2O molecules to cell walls)
Water uptake (via root hairs)
H2O

32.4 Guard cells control transpiration. By changing shape, guard cells generally keep stomata open during the day (allowing transpiration) but closed at night (preventing excess water loss).

32.5 Phloem transports sugars. By a pressure flow mechanism, phloem transports food molecules made by photosynthesis. At a sugar source, sugar is loaded into a phloem tube. The sugar raises the solute concentration in the tube, and water follows, raising the pressure in the tube. As sugar is removed at a sugar sink, water follows. The increase in pressure at the sugar source and decrease at the sugar sink cause phloem sap to flow from source to sink.

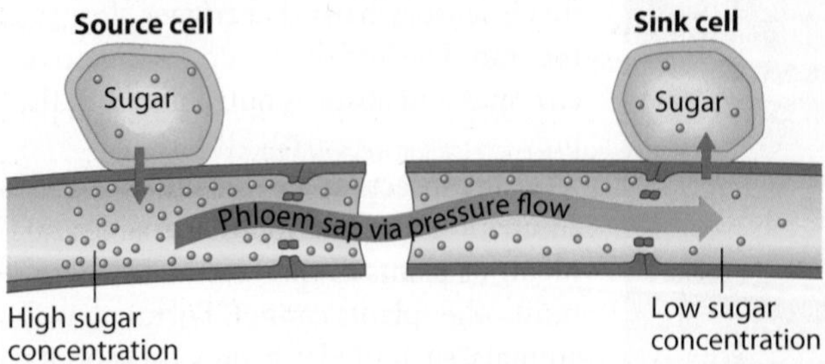

Source cell
Sugar
Sink cell
Sugar
Phloem sap via pressure flow
High sugar concentration
Low sugar concentration

Plant Nutrients and the Soil (32.6–32.11)

32.6 Plant health depends on obtaining all of the essential inorganic nutrients. A plant must obtain the chemical elements—inorganic nutrients—it requires from its surroundings. Macronutrients, such as carbon and nitrogen, are needed in large amounts, mostly to build organic molecules. Micronutrients, including iron and zinc, act mainly as cofactors of enzymes.

32.7 Fertilizers can help prevent nutrient deficiencies. Nutrient deficiencies can often be recognized and then fixed by using appropriate fertilizers.

32.8 Fertile soil supports plant growth. Fertile soil contains a mixture of small rock and clay particles that hold water and ions and also allow O_2 to diffuse into plant roots. Humus (decaying organic material) provides nutrients and supports the growth of organisms that enhance soil fertility. Anions (negatively charged ions), such as nitrate (NO_3^-), are readily available to plants because they are not bound to soil particles. However, anions tend to drain out of soil rapidly. Cations (positively charged ions), such as K^+, adhere to soil particles. In cation exchange, root hairs release H^+, which displaces cations from soil particles; the root hairs then absorb the free cations.

32.9 Soil conservation is essential to human life. Water-conserving irrigation, erosion control, and the prudent use of herbicides and fertilizers are aspects of good soil management.

32.10 Organic farmers follow principles meant to promote health. To earn the certified organic designation, food must be grown and processed following a strict set of guidelines. The question of whether organically grown crops contain more nutrients is the subject of many studies. In such cases, a meta-analysis can be used to combine data from multiple investigations.

32.11 Agricultural research is improving the yields and nutritional values of crops. Through DNA technologies, researchers are developing new varieties of crop plants with improved yields and nutritional value.

Plant Nutrition and Symbiosis (32.12–32.14)

32.12 Most plants depend on bacteria to supply nitrogen. Relationships with other organisms help plants obtain nutrients. Bacteria in the soil convert atmospheric N_2 to forms that can be used by plants.

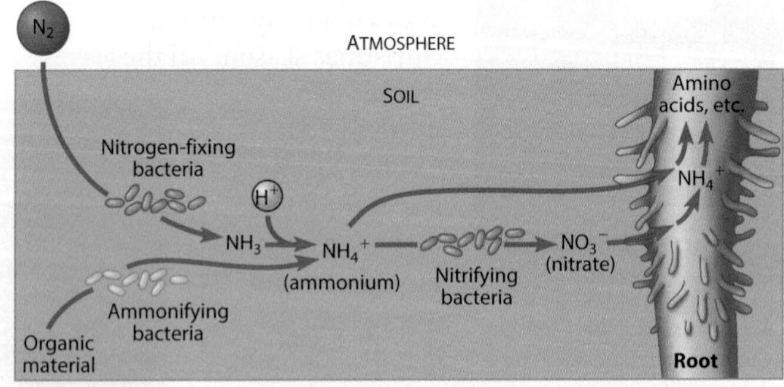

N2
ATMOSPHERE
SOIL
Amino acids, etc.
Nitrogen-fixing bacteria
H+
NH3
NH4+ (ammonium)
Nitrifying bacteria
NO3− (nitrate)
NH4+
Ammonifying bacteria
Organic material
Root

32.13 Plants have evolved mutually beneficial symbiotic relationships. Many plants form mycorrhizae, mutually beneficial associations between roots and fungi. A network of fungal threads increases a plant's absorption of nutrients and water, and the fungus receives some nutrients from the plant. Legumes and certain other plants have nodules in their roots that house nitrogen-fixing bacteria.

32.14 The plant kingdom includes epiphytes, parasites, and carnivores. Epiphytes are plants that grow on other plants. Parasitic plants siphon sap from host plants. Carnivorous plants can obtain nitrogen by digesting insects.

Connecting the Concepts

1. Fill in the blanks in this concept map to help you tie together key concepts concerning transport in plants.

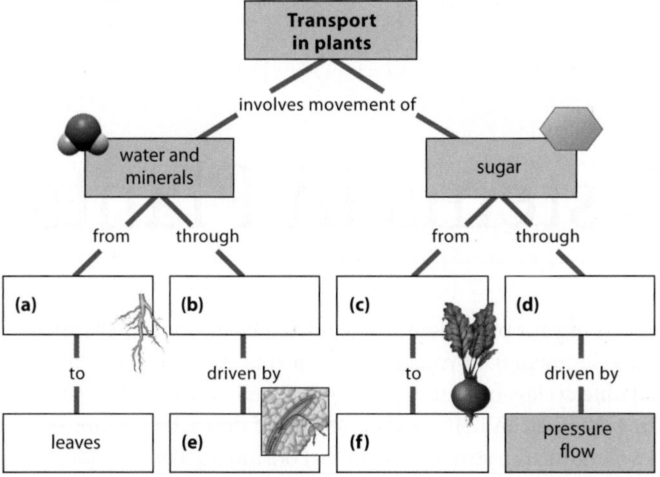

Transport in plants
involves movement of
water and minerals
sugar
from
through
from
through
(a)
(b)
(c)
(d)
to
driven by
to
driven by
leaves
(e)
(f)
pressure flow

Testing Your Knowledge

Level 1: Knowledge/Comprehension

2. Plants require the smallest amount of which of the following nutrients?
   a. oxygen
   b. phosphorus
   c. carbon
   d. iron
3. Which of the following activities of soil bacteria does *not* contribute to creating usable nitrogen supplies for plant use?
   a. the fixation of atmospheric nitrogen
   b. the conversion of ammonium ions to nitrate ions
   c. the decomposition of dead animals
   d. the assembly of amino acids into proteins
4. By trapping insects, carnivorous plants obtain ______________, which they need ______________. (*Choose the best answer.*)
   a. water . . . because they live in dry soil
   b. nitrogen . . . to make sugar
   c. phosphorus . . . to make protein
   d. nitrogen . . . to make protein
5. An advantage of using fertilizers derived from natural sources is that these fertilizers
   a. have different minerals than artificial fertilizers.
   b. are retained in soil longer.
   c. are more soluble in water.
   d. are more concentrated.

Level 2: Application/Analysis

6. Explain how guard cells limit water loss from a plant on a hot, dry day. How can this be harmful to the plant?
7. Transpiration is fastest when humidity is low and temperature is high, but in some plants it seems to increase in response to light as well. During one 12-hour period when cloud cover and light intensity varied frequently, a scientist studying a certain crop plant recorded the data in the table (top right). (The transpiration rates are grams of water per square meter of leaf area per hour.)

Time (hr)	Temperature (°C)	Humidity (%)	Light (% of full sun)	Transpiration Rate ($g/m^2 \cdot hr$)
8 AM	14	88	22	57
9	14	82	27	72
10	21	86	58	83
11	26	78	35	125
12 PM	27	78	88	161
1	33	65	75	199
2	31	61	50	186
3	30	70	24	107
4	29	69	50	137
5	22	75	45	87
6	18	80	24	78
7	13	91	8	45

Do these data support the hypothesis that the plants transpire more when the light is more intense? If so, is the effect independent of temperature and humidity? Explain your answer. (*Hint*: Look for overall trends in each column, and then compare pairs of data within each column and between columns.)

8. Certain types of fungi cause diseases in plants. There are a variety of antifungal sprays that can be used to control this problem. Some gardeners constantly spray their plants with fungicides, even when no signs of disease are evident. How might this be disadvantageous to the plant?

Level 3: Synthesis/Evaluation

9. Acid rain contains an excess of hydrogen ions (H^+). One effect of acid rain is to deplete the soil of plant nutrients such as calcium (Ca^{2+}), potassium (K^+), and magnesium (Mg^{2+}). Offer a hypothesis to explain why acid rain washes these nutrients from the soil. How might you test your hypothesis?
10. Agriculture is by far the biggest user of water in arid western states, including Colorado, Arizona, and California. The populations of these states are growing, and there is an ongoing conflict between cities and farm regions over water. To ensure water supplies for urban growth, cities are purchasing water rights from farmers. This is often the least expensive way for a city to obtain more water, and some farmers can make more money selling water than growing crops. Discuss the possible consequences of this trend. Is this the best way to allocate water for all concerned? Why or why not?
11. **SCIENTIFIC THINKING** One of the most important properties of proper scientific investigations is their repeatability. Yet, as discussed in Module 32.10, studies that compare the nutritional content of conventional and organic produce sometimes produce contradictory results. Name some possible confounding factors that can account for such uneven results.

Answers to all questions can be found in Appendix 4.

CHAPTER 33

Control Systems in Plants

You probably know that many aspects of your physiology are regulated by hormones, chemical signals that control growth and development. Did you know that plants also produce a variety of hormones? Many aspects of a plant's life cycle—from the sprouting of seeds to the pattern of growth to the dropping of leaves—are turned on and off, sped up or slowed down, by hormones.

Interestingly, some plant hormones are similar in structure to animal hormones. For example, phytoestrogens are plant hormones that are chemically analogous to mammalian estrogen, a steroid sex hormone that helps regulate the female reproductive cycle and promotes the development of female sexual features. Phytoestrogens were discovered in the 1940s when researchers investigated reduced fertility among Australian sheep. The cause was found: Red clover, the sheep's main grazing source, was rich in phytoestrogens. A more recent study demonstrated that fertility among California quail dropped significantly when their staple forage produced high levels of phytoestrogens. Researchers have since discovered phytoestrogens in many crops, including the soybeans pictured below.

How can hormones protect a plant?

Why would plants produce chemicals that mimic mammalian hormones? What advantage does that offer the plant? An investigation of such questions, like so many investigations in biology, benefits from an evolutionary viewpoint. Ingestion of a steady low dose of phytoestrogens may act as birth control. Indeed, human oral contraceptives administer human estrogen in this manner. If phytoestrogens decrease the fertility of a plant's herbivore, natural selection may favor the development of such adaptations.

Whatever their health effects on animals, we know that phytoestrogens and other plant hormones are crucial to the life of plants. In this chapter, we first explore how plant hormones affect plant growth, flowering, fruit development, and even defense. Then, we move on to discuss discuss how environmental stimuli affect plants.

BIG IDEAS

Plant Hormones
(33.1–33.8)

Chemical signals help regulate many important processes within a plant's body.

Responses to Stimuli
(33.9–33.13)

Plants respond to environmental stimuli in a variety of ways.

▷ Plant Hormones

33.1 A series of experiments by several scientists led to the discovery of a plant hormone

SCIENTIFIC THINKING

Have you ever noticed that many houseplants grow toward sunlight (**Figure 33.1A**)? Any growth response that results in plant organs curving toward or away from stimuli is called a **tropism**. The growth of a shoot in response to light is called **phototropism**. Phototropism has an obvious evolutionary advantage, directing growing seedlings and the shoots of mature plants toward the sunlight that drives photosynthesis.

How are plants able to change the way they grow in response to environmental stimuli? Plants, like animals, use hormones to control body-wide responses. A **hormone** is a chemical signal produced in one part of the body and transported to other parts, where it acts on target cells to change their functioning. A group of researchers, spanning several decades, studied plant responses to light. Building upon each other's work, this group collectively discovered and characterized the first plant hormone.

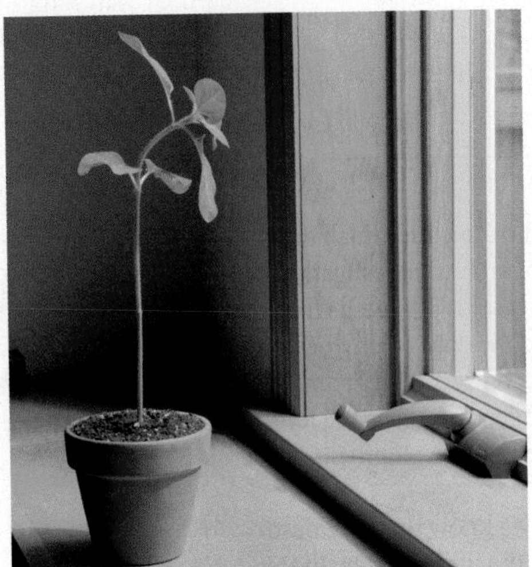

▲ **Figure 33.1A** A houseplant growing toward light

Showing That Light Is Detected by the Shoot Tip In the late 1800s, Charles Darwin and his son Francis conducted some of the earliest experiments on phototropism. They observed that grass seedlings could bend toward light only if the tips of their shoots were present. The five grass plants in **Figure 33.1B** summarize the Darwins' findings. First, they verified that a control plant bends toward the light. When the researchers removed the tip of a grass shoot, the shoot did not curve toward the light. Next, the Darwins found that the shoot also remained straight when they placed an opaque cap on its tip. However, the shoot curved normally when they placed a transparent cap on its tip or an opaque shield around its base (as seen in the last two shoots on the right). The Darwins concluded that the tip of the shoot was responsible for sensing light. They also recognized that the growth response, the bending of the shoot, occurs in cells that are below the tip. Therefore, they hypothesized that some signal must be transmitted from the tip downward to the growth region of the shoot.

A few decades later, Danish plant biologist Peter Boysen-Jensen further tested the Darwins' chemical signal idea (**Figure 33.1C**, on the facing page). In one group of seedlings, Boysen-Jensen inserted a block of gelatin between the tip and the lower part of the shoot. The gelatin prevented direct contact but allowed chemicals to diffuse through. The seedlings with gelatin blocks behaved normally, bending toward light. In a second set of seedlings, Boysen-Jensen inserted a thin

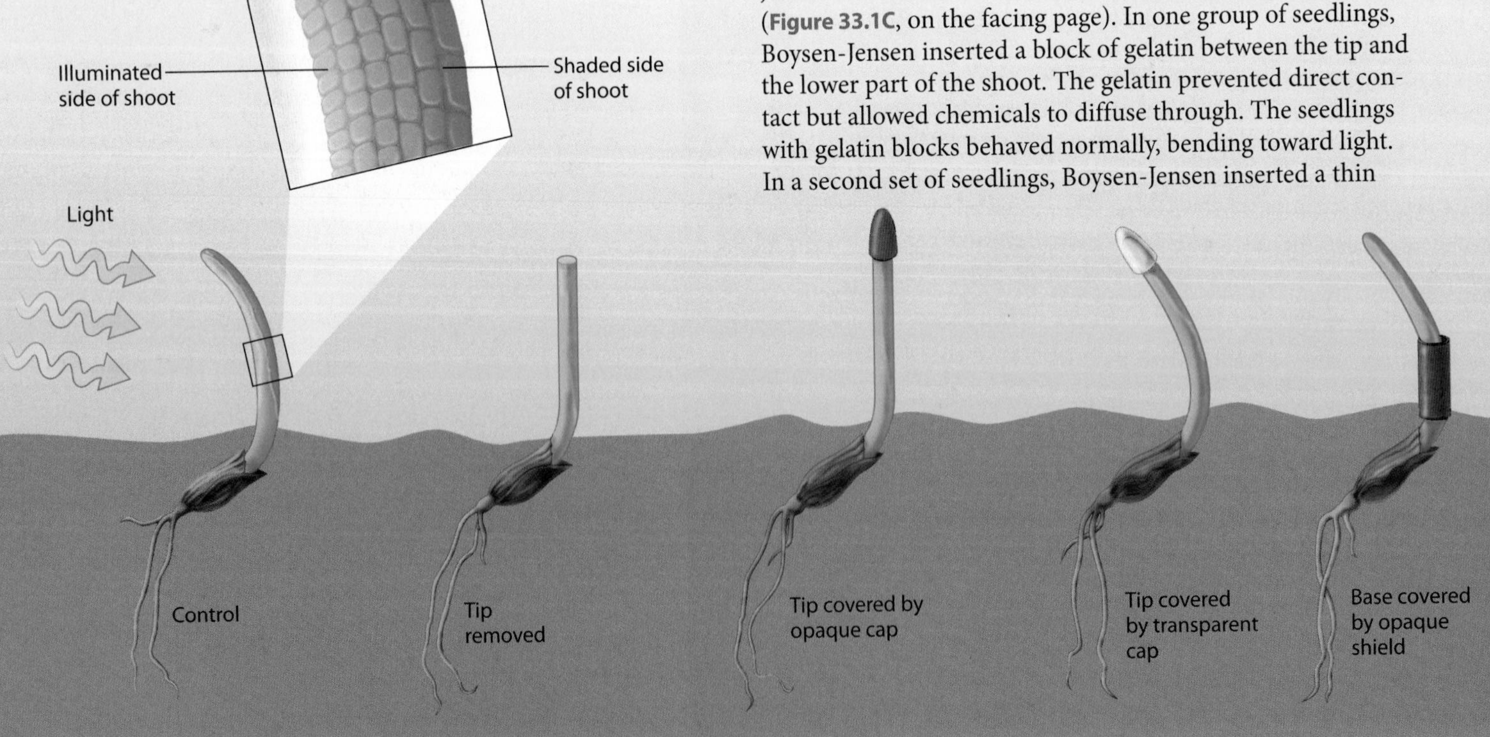
Illuminated side of shoot
Shaded side of shoot
Light
Control
Tip removed
Tip covered by opaque cap
Tip covered by transparent cap
Base covered by opaque shield

▲ **Figure 33.1B** Darwins' experiments on phototropism: detection of light by shoot tips and evidence for a chemical signal (1880)

piece of the mineral mica under the shoot tip. Mica is an impermeable barrier, and the seedlings so treated with mica had no phototropic response. These experiments supported the hypothesis that the signal for phototropism is a chemical that diffuses through the plant body.

Isolating the Chemical Signal In 1926, Frits Went, a Dutch graduate student, modified Boysen-Jensen's techniques and extracted the chemical messenger for phototropism in grasses. As shown in **Figure 33.1D**, Went first removed the tips of grass seedlings and placed the tips on blocks of agar, a gelatin-like material. He reasoned that the chemical messenger (pink in the figure) from the shoot tips should diffuse into the agar and that the blocks should then be able to substitute for the shoot tips. Went tested the effects of the agar blocks on tipless seedlings, which he kept in the dark to eliminate the effect of light. ❶ First, he centered the treated agar blocks on the cut tips of a batch of seedlings. These plants grew straight upward. They also grew faster than the decapitated control seedlings, which hardly grew at all. Went concluded that the agar had absorbed the chemical messenger produced in the shoot tip and that the chemical had passed into the shoot and stimulated it to grow. ❷ He then placed agar blocks off center on another batch of tipless seedlings. These plants bent away from the side with the chemical-laden agar block, as though growing toward light. ❸ Control seedlings with blank agar blocks (whether offset or not) grew no more than the control. Went concluded that the agar block contained a chemical produced in the shoot tip, that this chemical stimulated growth as it passed down the shoot, and that a shoot curved toward light because of a higher concentration of the growth-promoting chemical on the darker side. Went called this chemical messenger auxin. In the 1930s, biochemists determined the chemical structure of Went's auxin.

Microscopic observations of plants growing toward light reveal the cellular mechanism that underlies phototropism. The left side of Figure 33.1B shows a grass seedling curving toward light that comes from one side. As the enlargement shows, cells on the darker side of the seedling are larger—actually, they have elongated faster—than those on the brighter side. The different cellular growth rates made the shoot bend toward the light. If a seedling is illuminated uniformly from all sides or if it is kept in the dark, the cells all elongate at a similar rate and so the seedling grows straight upward.

Based on the early experiments described in this module, botanists hypothesized that an uneven distribution of auxin moving down from the shoot tip causes cells on the darker side to elongate faster than cells on the brighter side. Studies with plants other than grass shoots, however, do not always support this hypothesis. For example, there is no evidence that light from one side causes an uneven distribution of auxin in the stems of sunflowers or other eudicots. There is, however, a greater concentration of substances that may act as growth inhibitors on the lighted side of a stem.

The discovery of auxin stimulated discoveries of a wide variety of plant hormones. The next six modules will introduce some of the major plant hormones and the roles they play in the lives of plants.

? **How do the Boysen-Jensen 1913 experiments provide evidence that phototropism depends on a chemical signal?**

● The experiment shows that a phototropism signal can move down the shoot if agar is present but not if mica is present. Because agar allows chemicals to pass but mica blocks it, these results suggest that the signal is a diffusible chemical.

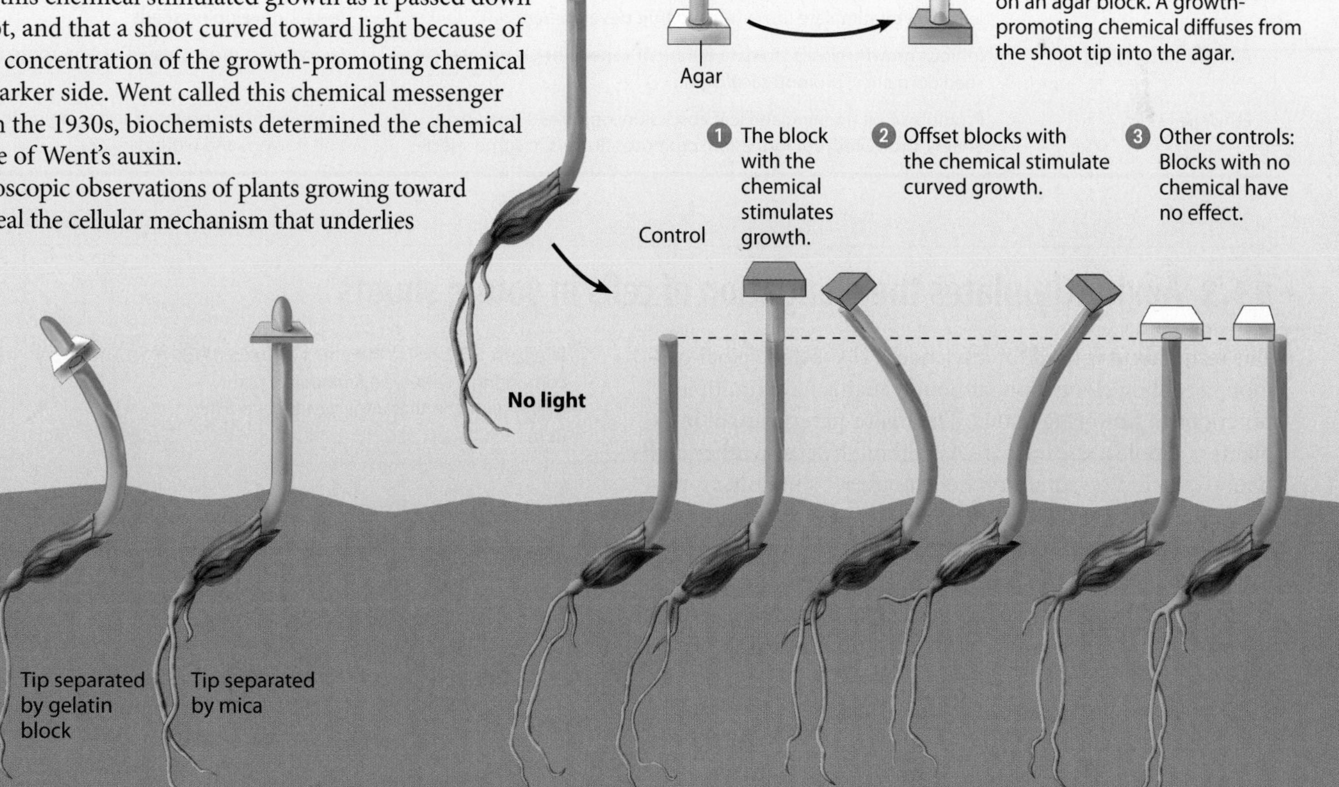

Tip separated by gelatin block
Tip separated by mica
No light
Agar
The dissected shoot tip is placed on an agar block. A growth-promoting chemical diffuses from the shoot tip into the agar.
Control
❶ The block with the chemical stimulates growth.
❷ Offset blocks with the chemical stimulate curved growth.
❸ Other controls: Blocks with no chemical have no effect.

▲ **Figure 33.1C** Boysen-Jensen's experiments: evidence that a chemical signal was responsible for bending (1913)

▲ **Figure 33.1D** Went's experiments: isolation of the chemical signal (1926)

33.2 Botanists have identified several major types of hormones

Plant hormones are produced in very low concentrations, but a tiny amount of hormone can have a profound effect on growth and development. The binding of a hormone to a specific receptor triggers a signal transduction pathway (see Module 11.10) that amplifies the hormonal signal and leads to one or more responses within the cell. In general, plant hormones control whole-body activities such as growth and development by affecting the division, elongation, and differentiation of cells.

Plant biologists initially identified five major types of plant hormones (**Table 33.2**), all of which stimulate or inhibit cell division and elongation, thereby influencing growth. Additionally, we know that there are hormones important to plants beyond those listed in the table. For example, brassinosteroids are steroid hormones that induce cell elongation and division in stems and seedlings. The effects of the brassinosteroids are so similar to those of auxin that it took years for plant physiologists to differentiate them. We also know that some of the "hormones" listed in the table actually represent a group of related hormones.

As Table 33.2 indicates, each hormone (or group of hormones) can produce a variety of effects, depending on its site of action, its concentration, and the developmental stage of the plant. In most situations, no single hormone acts alone. Instead, it is usually the balance of several hormones—their relative concentrations—that controls the growth and development of a plant. These interactions will become apparent as we survey the functions of the hormones listed in the table.

? **Hormones elicit cellular responses by binding to receptors and triggering ______________.**

● signal transduction pathways

TABLE 33.2 | MAJOR TYPES OF PLANT HORMONES

Hormone (Module)	Major Functions	Where Produced or Found in the Plant
Auxins (33.3)	Stimulate stem elongation; affect root growth, differentiation, branching, development of fruit, apical dominance, phototropism, and gravitropism (response to gravity); retard leaf abscission	Meristems of apical buds, young leaves, embryos within seeds
Cytokinins (33.4)	Affect root growth and differentiation; stimulate cell division and growth; stimulate germination; delay leaf aging	Made in the roots and transported to other organs
Gibberellins (33.5)	Promote seed germination, bud development, stem elongation, and leaf growth; stimulate flowering and fruit development	Meristems of apical buds and roots, young leaves, developing seeds
Abscisic acid (ABA) (33.6)	Inhibits growth; closes stomata during dry spells; helps maintain seed dormancy; promotes leaf aging	Every major organ: leaves, stems, roots, fruits
Ethylene (33.7)	Promotes fruit ripening and leaf abscission; opposes some auxin effects; promotes root formation; promotes flowers in some species	Ripening fruits, nodes of stems, aging leaves and flowers, and wounds

33.3 Auxin stimulates the elongation of cells in young shoots

The term **auxin** is used for any chemical substance that promotes seedling elongation, although auxins have multiple functions in flowering plants. The major natural auxin in plants is indoleacetic acid (IAA), although several other compounds, including some synthetic ones, act similarly. In this textbook, we will use the term *auxin* synonymously with IAA.

Figure 33.3A shows the effect of auxin on a mustard plant called *Arabidopsis thaliana*. The plant on the right is a wild-type *Arabidopsis*. The plant on the left contains a mutation in the auxin gene. The result of this mutation is a lack of auxin, which in turn results in stunted growth.

The apical meristem at the tip of a shoot (see Module 31.7) is a major site of auxin synthesis. Auxin moves in one direction only; from the shoot tips toward the base, diffusing from cell to cell. Interestingly, experiments have shown that this unidirectional flow is independent of gravity; auxin flows in the same direction if a plant is grown upside-down. As it flows, it stimulates growth of the stem by causing the cells

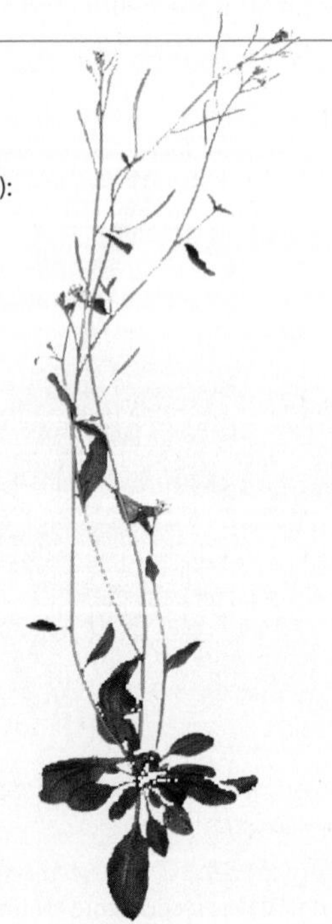

▶ **Figure 33.3A** The effect of auxin (IAA): comparing a wild-type *Arabidopsis* plant (right) with one that underproduces auxin (left)

to elongate. As the blue graph curve in **Figure 33.3B** shows, auxin promotes cell elongation in stems only within a certain concentration range. Above a certain level (0.9 g of auxin per liter of solution, in this case), auxin usually inhibits cell elongation in stems, probably by inducing the production of ethylene, a hormone that generally counters the effects of auxin (see Module 33.7).

The red curve on the graph shows the effect of auxin on root growth. An auxin concentration too low to stimulate shoot cells will cause root cells to elongate. On the other hand, an auxin concentration high enough to make stem cells elongate is in the concentration range that inhibits root cell elongation. These effects of auxin on cell elongation reinforce two points: (1) the same chemical messenger may have different effects at different concentrations in one target cell, and (2) a given concentration of the hormone may have different effects on different target cells. In fact, some herbicides (including the infamous Vietnam War defoliant Agent Orange) contain synthetic auxins at high concentrations that kill plants through hormonal overdose.

How does auxin make plant cells elongate? One hypothesis is that auxin initiates elongation by weakening cell walls. In a shoot's region of elongation, auxin stimulates proteins within the plasma membrane to pump hydrogen ions (H^+) into the cell wall. This lowers the pH within the wall, which in turn activates enzymes that weaken connections between cellulose fibers. Water can then enter the weakened cell, causing it to swell. As the cell rebuilds the cell wall, its larger size becomes permanent.

Auxins produce a number of other effects in addition to stimulating cell elongation and causing stems and roots to lengthen. For example, auxin induces cell division in the vascular cambium, thus promoting growth in stem diameter (see Module 31.8). In addition, auxin produced in shoot tips helps control the overall branching pattern of a growing plant. A crowded (and therefore unproductive) branch will reduce its flow of auxin, stimulating lateral buds below the branch to grow. A natural

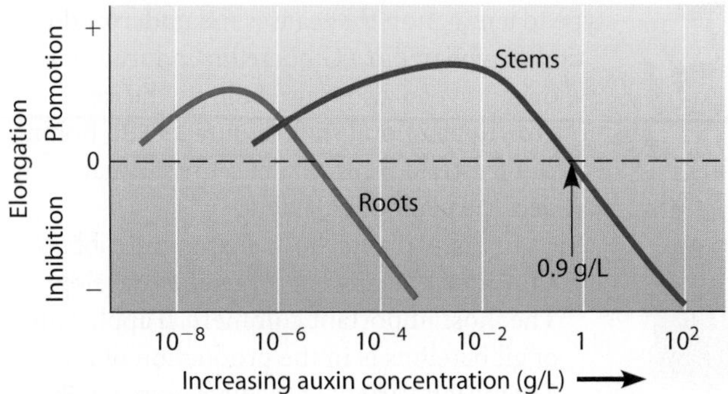
Elongation
Promotion
Inhibition
+
0
–
Stems
Roots
0.9 g/L
10^{-8}
10^{-6}
10^{-4}
10^{-2}
1
10^{2}
Increasing auxin concentration (g/L)

▲ **Figure 33.3B** The effect of auxin concentration on cell elongation in stems and roots

Try This Within the graph, distinguish the range of auxin concentrations that promote the growth of stems but inhibit the growth of roots.

MURPHY
HORMO
ROOTING PO
50g e

▲ **Figure 33.3C** An auxin used to promote the development of roots in a cutting

auxin, abbreviated IBA, is used to promote the development of roots during transplantation: Dipping a detached leaf or stem in IBA "root powder" often encourages new roots to form **(Figure 33.3C)**.

A commercially important effect of auxin is the promotion of fruit development. For instance, greenhouse tomatoes typically have fewer seeds than tomatoes grown outdoors, resulting in poorly developed fruits. Therefore, growers spray synthetic auxins on tomato vines to induce normal fruit development, making it possible to grow commercial tomatoes in greenhouses. Farmers also produce seedless cucumbers and eggplants by spraying unpollinated plants with synthetic auxins, which stimulate the growth of the fruits.

? Suppose you had a tiny pH electrode that could measure the pH of a plant cell's wall. How could you use it to test the hypothesis that auxin stimulates cell elongation?

● The hypothesis predicts that addition of auxin to the cell should lower the pH of the cell wall (make it more acidic). You could test this prediction with your pH electrode by measuring the cell wall pH in the presence of (experimental group) or absence of (control group) auxin.

33.4 Cytokinins stimulate cell division

Cytokinins are hormones that promote cytokinesis, or cell division. Natural cytokinins are produced in actively growing tissues, particularly in roots, embryos, and fruits. Cytokinins made in the roots reach target tissues in stems by moving upward in xylem sap.

Cytokinins stimulate cell differentiation as well as cell division. They also retard the aging of flowers and leaves by inhibiting protein breakdown. Thus, florists use cytokinin sprays to keep cut flowers fresh.

Cytokinins and auxins interact in the control of apical dominance, the ability of the terminal bud of a shoot to suppress the growth of the axillary buds (see Module 31.3). Both rosemary plants pictured in **Figure 33.4** are the same age. The one on the left has an intact terminal bud on the main shoot; the one on the right had the terminal bud of the main shoot removed several weeks earlier. In the plant on the left, apical dominance resulted in lengthwise growth but inhibited growth of the axillary buds (the buds that produce side branches). As a result, the shoot grew in height but did not branch out to the sides very much. In the plant on the right, the lack of a terminal bud on the main shoot resulted in the activation of the axillary buds, making the plant grow more branches and become bushy.

Terminal bud
No terminal bud
▲ **Figure 33.4** Apical dominance in a rosemary plant resulting from the action of auxin and cytokinins

One hypothesis to explain the hormonal regulation of apical dominance proposes that auxins and cytokinins act antagonistically (with the action of each one opposing the action of the other) in regulating axillary bud growth. According to this view, auxin transported down the shoot from the terminal bud directly inhibits axillary buds from growing, causing a shoot to lengthen at the expense of lateral branching. Meanwhile, cytokinins entering the shoot system from roots counter the action of auxin by signaling axillary buds to begin growing. Thus, the ratio of auxins to cytokinins is viewed as the critical factor in controlling axillary bud inhibition. Many observations and experiments are consistent with the direct inhibition hypothesis. It appears, however, that the effects of auxin are partially indirect. Recent research suggests that the flow of auxin down the shoot triggers the synthesis of a newly discovered plant hormone that represses bud growth. It is likely that plant biologists have not put together all the pieces of this puzzle. The role of plant hormones in apical dominance remains an area of active research.

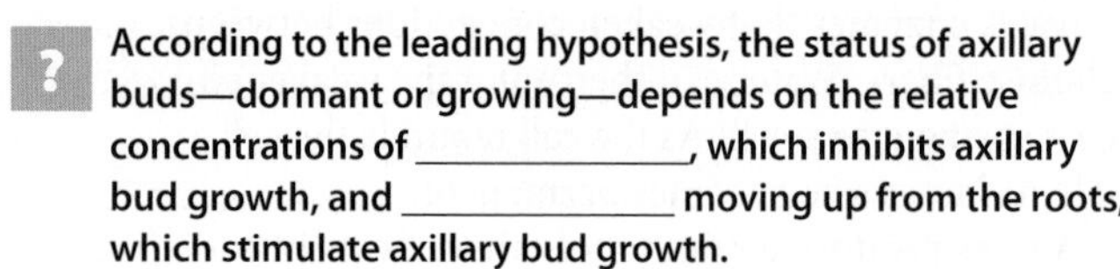
? **According to the leading hypothesis, the status of axillary buds—dormant or growing—depends on the relative concentrations of ____________, which inhibits axillary bud growth, and ____________ moving up from the roots, which stimulate axillary bud growth.**

auxin . . . cytokinins

33.5 Gibberellins affect stem elongation and have numerous other effects

Farmers in Asia had long noticed that some rice seedlings in their paddies grew so spindly that they toppled over before they produced grain. In 1926, Japanese scientists discovered the cause: a fungus of the genus *Gibberella*. By the 1930s, scientists had determined that the fungus produced hyperelongation of rice stems by secreting a chemical they named **gibberellin**. In the 1950s, researchers discovered that gibberellin exists naturally in plants, where it is a growth regulator. Rice plants infected with the *Gibberella* fungus suffer from too much gibberellin.

More than 100 different gibberellins have been identified in plants, although any particular species uses only a few. Young roots and leaves are major sites of gibberellin production. One of the main effects of gibberellin is to stimulate cell elongation and cell division in stems and leaves. In fact, certain dwarf plant varieties grow to normal height after treatment with gibberellin (**Figure 33.5A**). (The dwarf pea plant studied by Gregor Mendel—see Module 9.2— owes its phenotype to a mutation that causes the underproduction of gibberellin.) Gibberellin-induced stem elongation can also cause bolting, the rapid growth of a floral stalk (**Figure 33.5B**). Bolting can frustrate gardeners because it usually renders the plant unpalatable.

Dwarf plant (untreated)
Dwarf plant treated with gibberellins

▲ **Figure 33.5A** Reversing dwarfism in pea plants with gibberellins

In many plants, both auxins and gibberellins must be present for fruit to develop. The most important commercial application of gibberellins is in the production of the Thompson variety of seedless grapes. Gibberellins cause internodes to elongate. As a result, the grapes treated with gibberellins grow larger and farther apart in a cluster (**Figure 33.5C**).

Gibberellins are also important in seed germination in many plants. Many seeds that require special environmental conditions to germinate, such as exposure to light or cold temperatures, will germinate when sprayed with gibberellins. In nature, gibberellins in seeds are probably the link between environmental cues and the metabolic processes that renew growth of the embryo. For example, when water becomes available to a grass seed, it causes the embryo in the seed to release gibberellins, which promote germination by mobilizing nutrients stored within the seed. In some plants, gibberellins seem to be interacting antagonistically with another hormone, abscisic acid, which we discuss next.

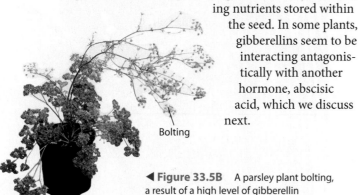

Bolting

◀ **Figure 33.5B** A parsley plant bolting, a result of a high level of gibberellin

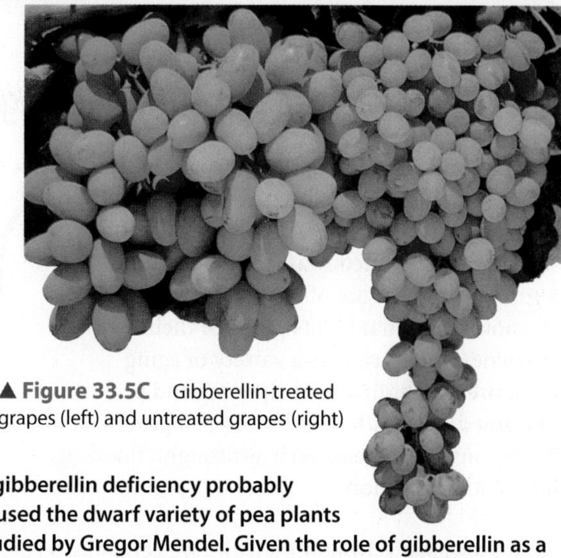

▲ **Figure 33.5C** Gibberellin-treated grapes (left) and untreated grapes (right)

? A gibberellin deficiency probably caused the dwarf variety of pea plants studied by Gregor Mendel. Given the role of gibberellin as a growth hormone, why does it make sense that a heterozygote (with one copy of the normal gene and one copy of the mutant gene) would be normally sized?

● Because minute amounts of a hormone are usually enough to elicit a strong response, a heterozygous plant with one functioning copy of the gene would still produce enough gibberellin to result in normal growth.

33.6 Abscisic acid inhibits many plant processes

In the 1960s, one research group studying bud dormancy in trees and another team investigating abscission (dropping) of cotton fruits independently isolated the same compound, **abscisic acid (ABA)**. Ironically, ABA is no longer thought to play a primary role in either bud dormancy or leaf abscission (for which it was named), but it is a plant hormone of great importance in other functions. In contrast to the growth-stimulating hormones we have studied so far—auxin, cytokinins, and gibberellins—ABA *slows* growth. ABA often counteracts the actions of growth hormones; it is the ratio of ABA to one or more growth hormones that often determines the final outcome of a plant.

When is it advantageous to slow growth in a plant? Consider a newly released seed. What prevents a seed dispersed in autumn from germinating immediately, only to be killed by cold temperatures in winter? Seed dormancy (a period when the seed ceases to grow) is an evolutionary adaptation that ensures that a seed will germinate only when there are favorable conditions of light, temperature, and moisture. Increases in levels of ABA during seed maturation inhibits germination and induces the production of proteins that help the seeds withstand the extreme dehydration that accompanies maturation. Many types of dormant seeds germinate only when ABA is removed or inactivated in some way. For example, some seeds—such as the perennials rosemary and rhubarb—require prolonged exposure to cold to trigger ABA inactivation. In these plants, the breakdown of ABA in the winter is required for seed germination in the spring. The seeds of some desert plants (such as the California wildflowers in **Figure 33.6**) remain dormant in parched soil for years or even decades until a downpour washes ABA out of the seeds, allowing them to germinate.

As we saw in the previous module, gibberellins promote seed germination. For many plants, the ratio of ABA to gibberellins therefore determines whether the seed will remain dormant or germinate.

In addition to its role in dormancy, ABA is the primary internal signal that enables plants to withstand drought. When a plant begins to wilt, ABA accumulates in its leaves and causes stomata to close rapidly (see Module 32.4). This closing of stomata reduces transpiration and prevents further water loss. In some cases, water shortage can stress the root system before the shoot system. ABA transported from roots to leaves may function as an "early warning system."

? In what way is the action of ABA the opposite of that of the other hormones discussed in this chapter so far?

● Abscisic acid inhibits growth, whereas the other hormones (brassinosteroids, auxins, cytokinins, and gibberellins) all promote growth.

▼ **Figure 33.6** The Mojave Desert in California blooming after a rain

33.7 Ethylene triggers fruit ripening and other aging processes

During the 1800s, leakage from coal gas streetlights caused the leaves on nearby trees to drop prematurely. In 1901, scientists demonstrated that this effect was due to **ethylene**, a gaseous by-product of coal combustion. We now know that plants produce their own ethylene, which triggers a variety of aging responses, including fruit ripening and programmed cell death. Ethylene is also produced in response to stresses such as drought, flooding, injury, and infection.

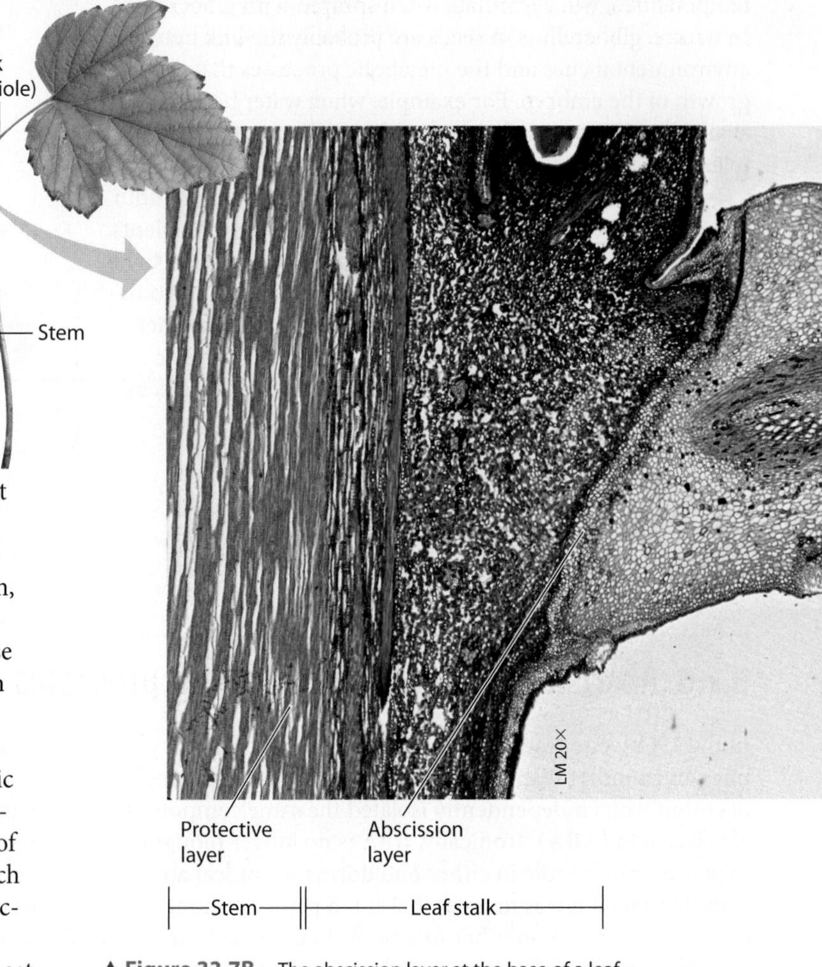

Leaf stalk (petiole)
Stem
LM 20×
Protective layer
Abscission layer
Stem
Leaf stalk

▲ **Figure 33.7B** The abscission layer at the base of a leaf

Fruit Ripening Immature fruits are generally tart, hard, and unappetizing—an adaptation that protects the developing seeds from herbivores. A burst of ethylene production in a fruit triggers its ripening, which then causes more ethylene production, resulting in a positive feedback surge of released gas. And because ethylene is a gas, the signal to ripen can spread from fruit to fruit: One bad apple really can spoil the bunch! The ripening process includes the enzymatic breakdown of cell walls, which softens the fruit, and the conversion of starches and acids to sugars, which makes the fruit sweet. The production of new scents and colors in ripening fruit attracts animals, which eat the fruits and later disperse the mature seeds.

▲ **Figure 33.7A** The effect of ethylene on the ripening of bananas

You can make some fruits ripen faster if you store them in a bag so that the ethylene gas accumulates. **Figure 33.7A** shows two bananas that were stored for the same time period in bags under different conditions: alone and with an ethylene-releasing peach. The ethylene released by the ripening peach resulted in a riper (darker) banana. On a commercial scale, many kinds of fruit—tomatoes, for instance—are often picked green and then ripened in huge storage bins into which ethylene gas is piped. (Some people prefer the taste of naturally "vine-ripened" tomatoes that are not exposed to artificial ethylene.)

In other cases, growers take measures to slow down fruit ripening. Stored apples are often flushed with CO_2, which inhibits ethylene synthesis and prevents ethylene from accumulating. In this way, apples picked in autumn can be stored for sale the following summer.

The Falling of Leaves Like fruit ripening, the changes that occur in deciduous trees each autumn—color changes, drying, and the loss of leaves—are also aging processes. Leaves lose their green color because chlorophyll is broken down during autumn. Fall colors result from a combination of new red pigments made in autumn and the exposure of yellow and orange pigments that were already present in the leaf but masked by dark green chlorophyll. Autumn leaf drop is an adaptation that helps keep the tree from drying out in winter. Without its leaves, a tree loses less water by evaporation at a time when its roots cannot take up water from the frozen ground. Before leaves fall, many essential elements are salvaged from them and stored in the stem, where they can be recycled into new leaves the following spring.

When an autumn leaf falls, the base of the leaf stalk separates from the stem. The region where separation occurs is called the abscission layer. As indicated in **Figure 33.7B**, the abscission layer consists of a narrow band of parenchyma cells with thin walls that are further weakened when enzymes digest the cell walls. The leaf drops off when its weight, often helped by wind, splits the abscission layer apart. Before the leaf falls, cells next to the abscission layer form a leaf scar on the stem. Dead cells covering the scar help protect the plant from infectious organisms.

Auxin prevents abscission and helps maintain the leaf's metabolism. But as a leaf ages, it produces less auxin. Meanwhile, cells begin producing ethylene, which stimulates formation of the abscission layer. The ethylene primes the abscission layer to split by promoting the synthesis of enzymes that digest cell walls in this layer.

We have now completed our survey of the major types of plant hormones. But before moving on to the topic of plant behavior, let's look at some agricultural uses of these chemical regulators.

? **Plant biologists sometimes refer to ethylene as the "aging hormone." What is the basis for this term?**

● Many of ethylene's functions, including fruit ripening and leaf abscission, are associated with aging-like changes in cells.

33.8 Plant hormones have many agricultural uses

CONNECTION

Although a lot remains to be learned about plant hormones, much of what we do know has a direct application to agriculture. As already mentioned, the control of fruit ripening and the production of seedless fruits are two of several major uses of these chemicals. Plant hormones also enable farmers to control when plants will drop their fruit. For instance, synthetic auxins are often used to prevent orange and grapefruit trees from dropping their fruit before they can be picked. **Figure 33.8** shows a fruit grower spraying a grove with auxins. The quantity of auxin must be carefully monitored because too much of the hormone may stimulate the plant to release more ethylene, making the fruit ripen and drop off sooner. Indeed, large doses of auxins are sometimes used to *promote* premature fruit drop. With apple and olive trees, for example, auxins may be sprayed on some fruit to cause them to drop prematurely, which in turn allows the remaining fruit to grow larger. Ethylene is used similarly on peaches and plums, and it is sometimes sprayed on berries, grapes, and cherries to loosen the fruit so it can be picked easily by machines.

In combination with auxins, gibberellins are used to produce seedless fruits (as mentioned in Module 33.5). Sprayed on other kinds of plants, at an earlier stage, gibberellins can have the opposite effect: the *promotion* of seed production. A large dose of gibberellins will induce many biennial plants, such as carrots, beets, and cabbage, to flower and produce seeds during their first year of growth, rather than in their second year, as is normally the case.

▲ **Figure 33.8** Using auxins to prevent early fruit drop

Research on plant hormones has also been used to develop herbicides. One of the most widely used weed killers is the synthetic auxin 2,4-D, which disrupts the normal balance of hormones that regulate plant growth. Monocots can rapidly inactivate this herbicide, but eudicots cannot, causing them to die of auxin overdose. So 2,4-D can be used to selectively remove dandelions and other broadleaf eudicot weeds from a lawn or grainfield. By applying herbicides to cropland, a farmer can reduce the amount of tillage required to control weeds, thus reducing soil erosion, fuel consumption, and labor costs.

Modern agriculture relies heavily on the use of synthetic chemicals. Without chemically synthesized herbicides to control weeds and synthetic plant hormones to help grow and preserve fruits, less food would be produced, and food prices could increase. At the same time, there is growing concern that the heavy use of certain artificial chemicals in food production may pose environmental and health hazards. One of these chemicals is dioxin, a by-product of 2,4-D synthesis. Although 2,4-D itself has not been shown to harm mammals, dioxin is a serious hazard when it leaks into the environment, causing birth defects, liver disease, and leukemia in laboratory animals. Also, many consumers are concerned that foods produced using artificial hormones may not be as tasty or nutritious as those raised naturally. At present, however, organic foods (see Module 32.10) are relatively expensive to produce. The issue of how our food should be grown involves both economics and ethics: Should we continue to produce cheap, plentiful food using artificial chemicals that may cause problems, or should we put more of our agricultural effort into farming without these potentially harmful substances, recognizing that foods may be less plentiful and more expensive as a result?

? **When applying auxins to crops, why does a farmer need to be particularly careful in terms of the quantities applied? (*Hint*: See Figure 33.3B.)**

● Because the same hormone may have different effects depending on the concentration, a farmer needs to control the quantities applied to obtain the desired effect.

▷ Responses to Stimuli

33.9 Tropisms orient plant growth toward or away from environmental stimuli

Having surveyed the hormones that carry signals within a plant, we now shift our focus to the responses of plants to physical stimuli from the environment. A plant cannot migrate to water or a sunny spot, and a seed cannot maneuver itself into an upright position if it lands upside down in the soil. Instead, plants have evolved the ability to respond to environmental stimuli through developmental and physiological mechanisms. Tropisms are directed growth responses that cause parts of a plant to grow toward a stimulus (a positive tropism) or away from a stimulus (a negative tropism).

Response to Light In Module 33.1, we discussed positive phototropism, the growth of a plant shoot toward light. The mechanism for phototropism is a greater rate of cell elongation on the darker side of a stem. In grass seedlings, the signal linking the light stimulus to the cell elongation response is auxin. Researchers have shown that illuminating a grass shoot from one side causes auxin to migrate across the shoot tip from the bright side to the dark side. The shoot tip contains a protein pigment that detects the light and somehow passes the "message" to molecules that affect auxin transport. (We discuss protein light receptors in Module 33.12.)

Response to Gravity **Gravitropism**—the directional growth of a plant in response to gravity—explains why, no matter how a seed lands on the ground, shoots grow upward (negative gravitropism) and roots grow downward (positive gravitropism). The corn seedlings in **Figure 33.9A** were both germinated in the dark. The one on the left was left untouched; notice that its shoot grew straight up and its root straight down. The seedling on the right was germinated in the same way, but two days later it was turned on its side so that the shoot and root were horizontal. By the time the photo was taken, the shoot had turned back upward, exhibiting a negative response to gravity, and the root had turned down, exhibiting positive gravitropism.

▲ **Figure 33.9A** Gravitropism in a corn seedling

▲ **Figure 33.9B** The "sensitive plant," *Mimosa pudica*

One hypothesis for how plants tell up from down is that gravity pulls special organelles containing dense starch grains to the low points of cells. The uneven distribution of organelles may in turn signal the cells to redistribute auxin. This effect has been documented in roots. A higher auxin concentration on the lower side of a root inhibits cell elongation (see the red line in Figure 33.3B). As cells on the upper side continue to elongate, the root curves downward. This tropism continues until the root is growing straight down.

Response to Touch **Thigmotropism**, directional growth in response to touch, is illustrated when the tendril of a pea plant (which is actually a modified leaf) contacts a string or wire and coils around it for support. Tendrils grow straight until they touch an object. Contact then stimulates the cells to grow at different rates on opposite sides of the tendril (slower in the contact area), making the tendril coil around the support. Most climbing plants have tendrils that respond by coiling and grasping when they touch rigid objects. Thigmotropism enables these plants to use such objects for support while growing toward sunlight.

Some plants show remarkable abilities to respond to touch. When stimulated by contact, the "sensitive plant" (*Mimosa pudica*) rapidly folds its leaflets together (**Figure 33.9B**). This response, which takes only a second or two, results from a rapid loss of turgor by cells within the joints of the leaf. It takes about 10 minutes for the cells to regain their turgor and restore the open form of the leaf. Botanists have various hypotheses about the function of the sensitive plant's behavior. Perhaps this adaptation provides a selective advantage because the plant appears less leafy and appetizing to herbivores when it folds its leaves.

Tropisms all have one function in common: They help plant growth stay in tune with the environment. In the next module, we see that plants also have a way of keeping time with their environment.

? **Why are tropisms called "growth responses"?**

● Because the movement of a plant organ toward or away from an environmental stimulus generally takes place by growing.

33.10 Plants have internal clocks

Your pulse rate, blood pressure, body temperature, rate of cell division, blood cell counts, alertness, urine composition, metabolic rate, sex drive, and responsiveness to medications all fluctuate rhythmically with the time of day. Plants also display rhythmic behavior; examples include the opening and closing of stomata (see Module 32.4) and the "sleep movements" of many species that fold their leaves or flowers in the evening and unfold them in the morning. Some of these cyclic variations continue even under artificially constant conditions, implying that plants have a built-in ability to sense time.

Innate Biological Rhythms An innate biological cycle of about 24 hours is called a **circadian rhythm**. A circadian rhythm persists even when an organism is sheltered from environmental cues. A bean plant, for example, exhibits sleep movements at about the same intervals even if kept in constant light or continuous darkness. Thus, circadian rhythms occur with or without external stimuli such as sunrise and sunset. Research on a variety of organisms indicates that circadian rhythms are controlled by internal timekeepers known as **biological clocks**.

Although a biological clock continues to mark time in the absence of environmental cues, it requires daily signals from the environment to remain tuned to a period of *exactly* 24 hours. This is because innate circadian rhythms generally differ somewhat from a 24-hour period. Consider bean plants, for instance. As shown in **Figure 33.10**, the leaves of a bean plant are normally horizontal at noon and folded downward at midnight. When the plant is held in darkness, however, its sleep movements change to a cycle of about 26 hours.

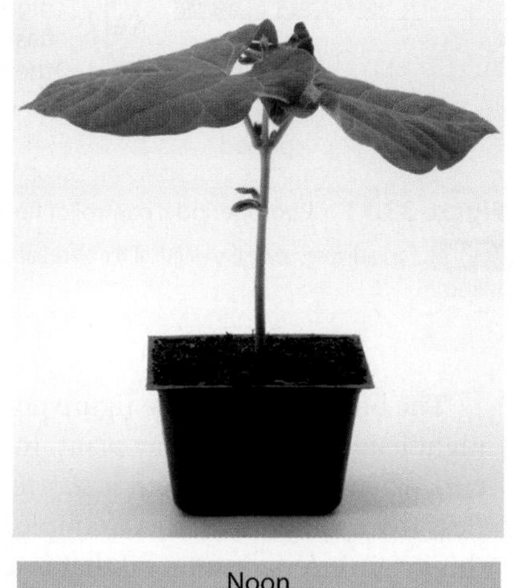
Noon
Midnight

▲ **Figure 33.10** Sleep movements of a bean plant

The light/dark cycle of day and night provides the cues that usually keep biological clocks precisely synchronized with the outside world. But a biological clock cannot immediately adjust to a sudden major change in the light/dark cycle. We observe this problem ourselves when we cross several time zones in an airplane: When we reach our destination, we have "jet lag"; our internal clock is not synchronized with the clock on the wall. Moving a plant across several time zones produces a similar lag. The plant will, for example, display leaf movements that are synchronized to the clock in its original location. For both a plant and a human traveler, resetting the clock usually takes several days.

The Nature of Biological Clocks Just what is a biological clock? Researchers are actively investigating this question. In humans and other mammals, the clock is located within a cluster of nerve cells in the hypothalamus of the brain (see Module 28.15). But for most other organisms, including plants, we know little about where the clocks are located or what kinds of cells are involved. A leading hypothesis is that biological timekeeping in plants may involve "clock genes" that synthesize a protein that regulates its own production through feedback control. After the protein accumulates to a sufficient concentration, it turns off its own gene. When the concentration of the protein falls, transcription restarts. The result would be a cycling of the protein's concentration over a roughly 24-hour period—a clock! Some research indicates that such a molecular mechanism may be common to all eukaryotes. However, much research remains to be done in this area.

Unlike most metabolic processes, biological clocks and the circadian rhythms they control are affected little by temperature. Somehow, a biological clock compensates for temperature shifts. This adjustment is essential, for a clock that speeds up or slows down with the rise and fall of outside temperature would be an unreliable timepiece.

In attempting to answer questions about biological clocks, it is essential to distinguish between the clock and the processes it controls. You could think of the sleep movements of leaves as the "hands" of a biological clock, but they are not the essence of the clockwork itself. You can restrain the leaves of a bean plant for several hours so that they cannot move. But on release, they will rush to the position appropriate for the time of day. Thus, we can interfere with an organism's rhythmic activity, but its biological clock goes right on ticking. In the next module, we'll learn more about the interface between a plant's internal biological clock and the external environment.

? **It has been hypothesized that biological clocks in plants are controlled by the synthesis of a protein that, once it accumulates to a sufficient concentration, shuts off its own gene. What kind of feedback does such a mechanism represent?**

● Negative feedback, where the result of a process (in this case, a protein) shuts off that process (the transcription and translation of the gene)

33.11 Plants mark the seasons by measuring photoperiod

Imagine a plant that produced flowers before its pollinators emerged, or a tree that grew new leaves in the dead of winter. Such plants would be at a serious disadvantage. Flowering, seed germination, and the onset and ending of dormancy are all stages in plant development that usually occur at specific times of the year. In other words, the life cycle of a plant is inexorably tied to the rhythmic cycles of the environmental seasons. A biological clock, therefore, must influence seasonal events as well as daily events. The environmental stimulus plants most often use to detect the time of year is called **photoperiod**, the relative lengths of day and night.

Plants whose flowering is triggered by photoperiod fall into two groups. One group, the **short-day plants**, generally flower in late summer, fall, or winter, when light periods shorten. Chrysanthemums and poinsettias are examples of short-day plants. In contrast, **long-day plants**, such as spinach, lettuce, iris, and many cereal grains, usually flower in late spring or early summer, when light periods lengthen. Spinach, for instance, flowers only when daylight lasts at least 14 hours. Some plants, such as dandelions, tomatoes, and rice, are day-neutral; they flower when they reach a certain stage of maturity, regardless of day length.

In the 1940s, researchers discovered that flowering and other responses to photoperiod are actually controlled by the length of continuous *darkness*, not the length of continuous daylight. That is, "short-day" plants will flower only if it stays dark long enough, and "long-day" plants will flower only if the dark period is short enough. Therefore, it would seem more appropriate to call the two types "long-night" plants and "short-night" plants. However, the day-length terms are embedded firmly in the literature of plant biology and so will be used here.

Figure 33.11 illustrates the evidence for the night-length effect and shows the difference between the flowering response of a short-day plant and a long-day plant. The top part of the figure represents short-day plants. Notice that the top short-day plant will not flower until it is exposed to a *continuous* dark period exceeding a critical length (about 14 hours, shown in the middle bar). The continuity of darkness is important. The short-day plant will not blossom if the nighttime part of the photoperiod is interrupted by even a brief flash of light (as shown in the third bar). There is no effect if the daytime portion of the photoperiod is broken by a brief exposure to darkness.

Florists apply this information about short-day plants to bring us flowers out of season. Poinsettias, for instance, are short-day plants that normally bloom in the autumn, but their blooming can be stalled until the November/December holiday season by punctuating each long night with a flash of light, thus turning one long night into two short nights. Easter lilies are also forced by growers to bloom at one particular time of year, which may vary by as much as five weeks from year to year, because the precise date of Easter fluctuates with the lunar calendar.

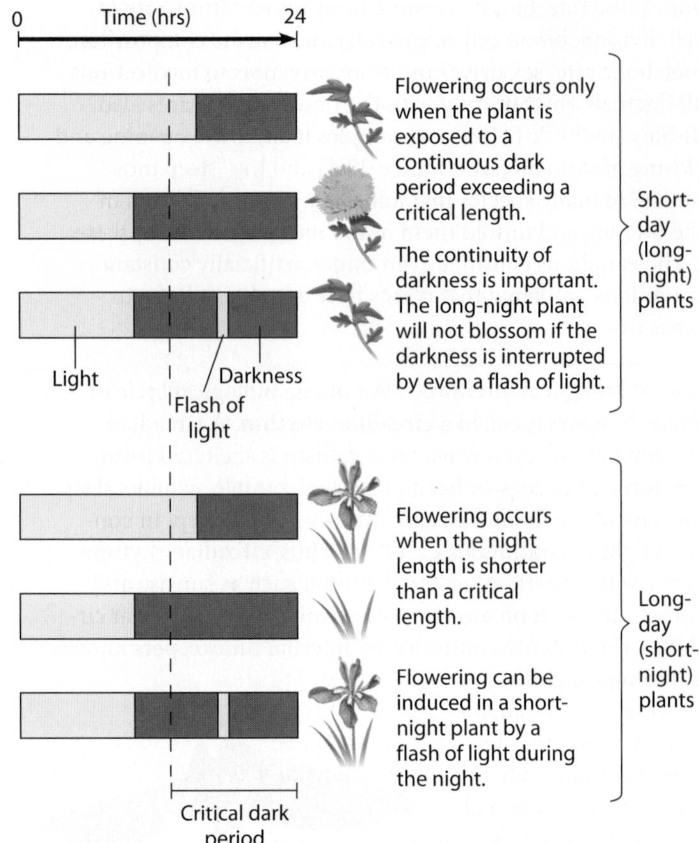

0
Time (hrs)
24
Flowering occurs only when the plant is exposed to a continuous dark period exceeding a critical length.
The continuity of darkness is important. The long-night plant will not blossom if the darkness is interrupted by even a flash of light.
Short-day (long-night) plants
Light
Darkness
Flash of light
Flowering occurs when the night length is shorter than a critical length.
Flowering can be induced in a short-night plant by a flash of light during the night.
Long-day (short-night) plants
Critical dark period

▲ **Figure 33.11** Photoperiodic control of flowering

Try This For each case, relate the length of the black bar (continuous darkness) to whether flowering will occur.

The bottom part of the figure demonstrates the effect of night length on a long-day plant. In this case, flowering occurs when the night length is *shorter* than a critical length (less than 10 hours, in this example, as seen in the top bar of this group). A dark interval that is too long will prevent flowering (middle bar in this group). In addition, flowering can be induced in a long-day plant by a flash of light during the night (bottom bar).

Most species of plants have a critical night length, but how that critical night length affects flowering varies with the type of plant. In short-day plants, the critical night length is the *minimum* number of hours of darkness required for flowering; less darkness prevents flowering. In long-day plants, this critical night length is the *maximum* number of hours of darkness required for flowering; less darkness promotes flowering.

? A particular short-day plant won't flower in the spring. Suppose a short dark interruption splits the long-light period of spring into two short-light periods. What result do you predict?

● The plants still won't flower because it is actually night length, not day length, that counts in the photoperiodic control of flowering.

33.12 Phytochromes are light detectors that help set the biological clock

The discovery that photoperiod determines the seasonal responses of plants leads to another question: How does a plant actually measure photoperiod? Much remains to be learned, but photoreceptive pigments called phytochromes are part of the answer. **Phytochromes** are proteins with a light-absorbing component.

Phytochromes were discovered during studies on how different wavelengths of light affect seed germination. Many types of seeds germinate only when light conditions are near optimal, an adaptation that increases chances of survival for the seed. Red light, with a wavelength of 660 nm, was found to be most effective at increasing germination. Light in the far-red range (730 nm, near the edge of human visibility) inhibited germination. Furthermore, researchers found that the last flash of light a seed is exposed to determines the seed's response. In other words, the effects of red light and far-red light are reversible.

How do phytochromes respond differently to different wavelengths of light? The key to this ability is that a phytochrome molecule changes back and forth between two forms that differ slightly in structure (**Figure 33.12A**). One form, known as P_r, absorbs red light and is quickly converted to the P_{fr} form. The other, known as P_{fr}, absorbs far-red light and is converted to the P_r form. This $P_r \leftrightarrow P_{fr}$ interconversion acts as a switch that controls various light-induced events in the life of a plant.

Each night, new phytochrome molecules are synthesized only in the P_r form. Thus, molecules of P_r slowly accumulate in the continuous darkness that follows sunset. By sunrise, nearly all the phytochrome is in the P_r form. But the red wavelengths of sunlight that come after sunrise cause much of it to be rapidly converted to P_{fr}. Sunlight contains both red light and far-red light, but the conversion to P_{fr} is faster than the conversion to P_r. Therefore, the ratio of P_{fr} to P_r increases in the sunlight. It is this sudden increase in P_{fr} each day at dawn that resets a plant's biological clock. Interactions between phytochrome and the biological clock enable plants to measure the passage of night and day. In doing so, the plant's biological clock monitors photoperiod and, when detecting seasonal changes in day and night length, cues responses such as seed germination, flowering, and the beginning and ending of bud dormancy.

The consequences of this phytochrome switch are shown in **Figure 33.12B**. The top bar shows the results we saw in the previous module for both short-day and long-day plants that receive a flash of light during their critical dark period. The letter R on the light flash stands for red light. The other three bars show how flashes of far-red (FR) light affect flowering. The second bar reveals that the effect of a flash of red light that interrupts a period of darkness can be reversed by a subsequent flash of far-red light: Both types of plants behave as though there is no interruption in the night length. The bottom two bars indicate that no matter how many flashes of red or far-red light a plant receives, only the wavelength of the *last* flash of light affects the plant's measurement of night length.

Plants also have a group of blue-light photoreceptors that control such light-sensitive plant responses as phototropism and the opening of stomata at daybreak. Light is an especially important environmental factor in the lives of plants, and diverse receptors and signaling pathways have evolved that mediate a plant's responses to light.

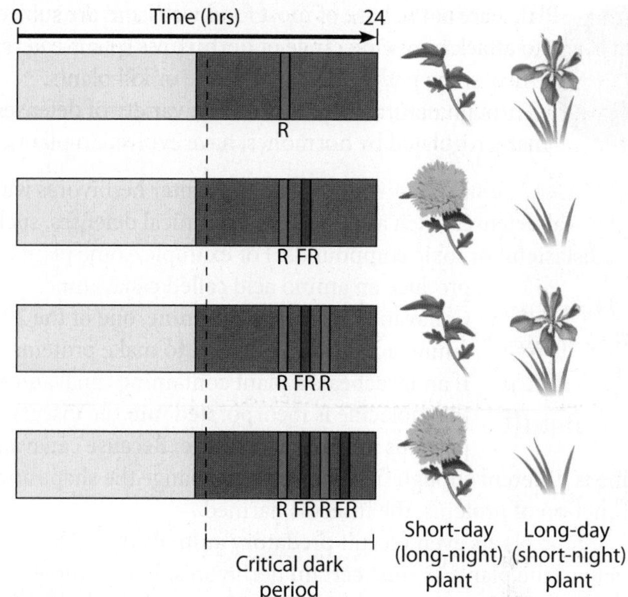

0
Time (hrs)
24
R
R FR
R FR R
R FR R FR
Critical dark period
Short-day (long-night) plant
Long-day (short-night) plant

▲ **Figure 33.12B** The reversible effects of red and far-red light

Try This Cover up the images of the plants at the right of this figure, then try to predict which plant will flower for each of the four conditions shown.

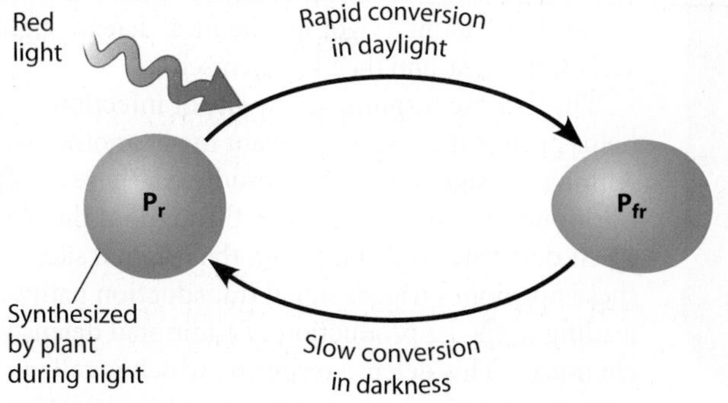

Red light
Rapid conversion in daylight
P_r
P_{fr}
Synthesized by plant during night
Slow conversion in darkness

▲ **Figure 33.12A** Interconversion of the two forms of phytochrome

? How do phytochrome molecules help the plant recognize dawn each day?

● Phytochrome molecules are mainly in the P_r form during the night. Dawn is signaled by the sudden conversion of P_r to P_{fr} due to the absorption of the red wavelengths of sunlight.

33.13 Defenses against herbivores and infectious microbes have evolved in plants

EVOLUTION CONNECTION

Plants are at the base of most food webs and are subject to attack by a wide range of **herbivores** (plant-eaters). Also, some pathogens can damage or kill plants. Through natural selection, a wide variety of defenses, many regulated by hormones, have evolved in plants.

Defenses Against Herbivores Plants counter herbivores with physical defenses, such as thorns, and chemical defenses, such as distasteful or toxic compounds. For example, some plants produce an amino acid called canavanine. Canavanine resembles arginine, one of the 20 amino acids normally used to make proteins. If an insect eats a plant containing canavanine, the molecule is incorporated into the insect's proteins in place of arginine. Because canavanine is different enough from arginine to change the shape and function of proteins, the insect is harmed.

How can hormones protect a plant?

Some plants even recruit predatory animals that help defend the plants against certain herbivores. For example, insects called parasitoid wasps can kill caterpillars that feed on plants (**Figure 33.13A**). ① When a caterpillar bites into the plant, the combination of physical damage to the plant and a chemical in the caterpillar's saliva triggers ② a signal transduction pathway within the plant cells. The pathway leads to a specific cellular response: ③ the synthesis and release of gases that ④ attract the wasp. ⑤ The wasp injects its eggs into the caterpillar. When the eggs hatch, the wasp larvae eat their way out of their caterpillar host, killing it.

Defenses Against Pathogens A plant, like an animal, is subject to infection by pathogenic microbes: viruses, bacteria, and fungi. A plant's first line of defense against infection is the physical barrier of the plant's "skin," the epidermis. However, microbes can cross this barrier through wounds or through natural openings such as stomata. Once infected, the plant uses chemicals as a second line of defense. Plant cells damaged by infection release microbe-killing molecules and chemicals that signal nearby cells to mount a similar chemical defense. In addition, infection stimulates chemical changes in the plant cell walls, which toughen the walls and thus slow the spread of the microbes within the plant.

This chemical defense system is enhanced by the plant's inherited ability to recognize certain pathogens. A kind of "compromise" has coevolved between plants and most of their pathogens: The pathogen gains enough access to its host to perpetuate itself without severely harming the plant. Otherwise, hosts and pathogens would soon perish together. The plant is said to be resistant to that pathogen, and the pathogen is said to be avirulent for the plant.

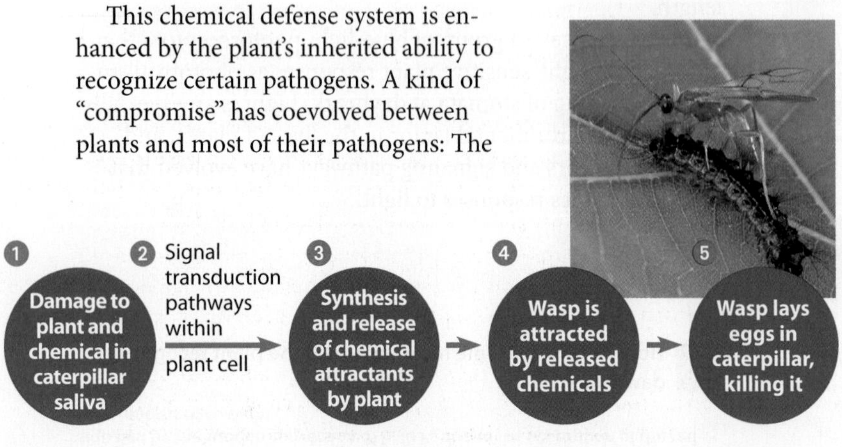

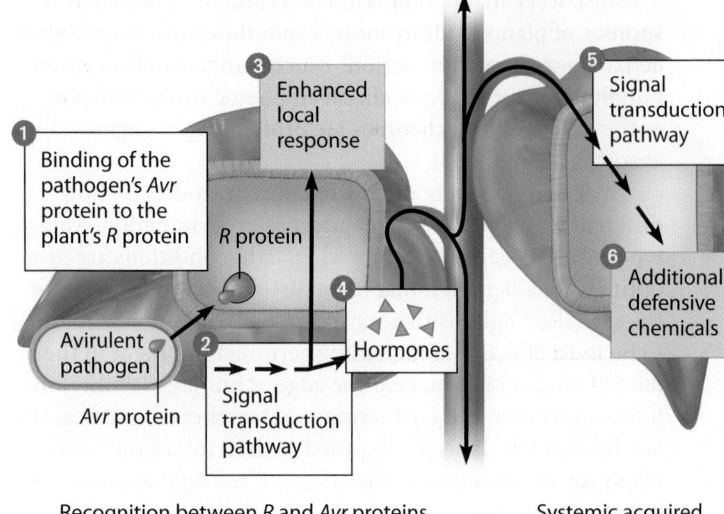

① Binding of the pathogen's *Avr* protein to the plant's *R* protein
R protein
Avirulent pathogen
Avr protein
② Signal transduction pathway
③ Enhanced local response
④ Hormones
⑤ Signal transduction pathway
⑥ Additional defensive chemicals
Recognition between *R* and *Avr* proteins, leading to a strong local response
Systemic acquired resistance

▲ **Figure 33.13B** Defense responses against an avirulent pathogen

This resistance to destruction by a specific pathogen is based on the ability of the plant and the microbe to make a complementary pair of molecules. A plant has many *R* genes (for *resistance*), and each pathogen has a set of *Avr* genes (for *avirulence*). Researchers hypothesize that an *R* gene encodes an *R* protein, a receptor protein in the plant's cells, and that the complementary *Avr* gene encodes an *Avr* protein, a signal molecule of the pathogen that binds specifically to the *R* protein in the plant cell. Recognition of pathogen-derived molecules by *R* proteins triggers a signal transduction pathway that leads to both local and plant-wide defense responses.

Figure 33.13B shows this interaction and subsequent events in the plant. ① The binding of the pathogen's signal molecule ( ) to the plant's receptor ( ) triggers ② a signal transduction pathway, which leads to ③ a defense response that is much stronger than would occur without the matchup of the *R* and *Avr* proteins. The cells at the site of infection mount a vigorous chemical defense, tightly seal off the area, and then kill themselves.

The defense response at the site of infection helps protect the rest of the plant in yet another way. Among the signal molecules produced, there are ④ hormones that sound an alarm throughout the plant. ⑤ At destinations distant from the original site, these hormones trigger signal transduction pathways leading to ⑥ the production of additional defensive chemicals. This defense response, which is called **systemic acquired resistance**, provides protection against a diversity of pathogens for days.

① Damage to plant and chemical in caterpillar saliva → ② Signal transduction pathways within plant cell → ③ Synthesis and release of chemical attractants by plant → ④ Wasp is attracted by released chemicals → ⑤ Wasp lays eggs in caterpillar, killing it

▲ **Figure 33.13A** Recruitment of a wasp in response to an herbivore

Researchers have identified one of the alarm hormones as salicylic acid, a compound whose pain-relieving effects led early cultures to use the salicylic acid–rich bark of willows (*Salix*) as a medicine. Aspirin is a chemical derivative of this compound. With the discovery of systemic acquired resistance, biologists have learned one function of salicylic acid in plants.

The use of hormones by plants to counter threats from animal predators reveals just one of the countless interactions that occur among species every day. In the next unit, we'll continue to explore interactions of living organisms and their environment at the level of whole ecosystems.

? What is released at a site of infection that triggers the development of general resistance to pathogens?

Hormones

CHAPTER 33 REVIEW

For practice quizzes, BioFlix animations, MP3 tutorials, video tutors, and more study tools designed for this textbook, go to MasteringBiology®

Reviewing the Concepts

Plant Hormones (33.1–33.8)

33.1 A series of experiments by several scientists led to the discovery of a plant hormone. Hormones coordinate the activities of plant cells and tissues. Experiments carried out by Darwin and others showed that the tip of a grass seedling detects light and transmits a signal down to the growing region of the shoot.

33.2 Botanists have identified several major types of hormones. By triggering signal transduction pathways, small amounts of hormones regulate plant growth and development.

33.3 Auxin stimulates the elongation of cells in young shoots. Plants produce the auxin IAA in the apical meristems at the tips of shoots. At different concentrations, auxin stimulates or inhibits the elongation of shoots and roots. It may act by weakening cell walls, allowing them to stretch when cells take up water.

33.4 Cytokinins stimulate cell division. Cytokinins, produced by roots, embryos, and fruits, promote cell division. Cytokinins from roots may balance the effects of auxin from apical meristems, causing lower buds to develop into branches.

33.5 Gibberellins affect stem elongation and have numerous other effects. Gibberellins stimulate the elongation of stems and leaves and the development of fruits. Gibberellins released from embryos function in some of the early events of seed germination.

33.6 Abscisic acid inhibits many plant processes. Abscisic acid (ABA) inhibits germination. The ratio of ABA to gibberellins often determines whether a seed remains dormant or germinates. Seeds of many plants remain dormant until their ABA is inactivated or washed away.

33.7 Ethylene triggers fruit ripening and other aging processes. As fruit cells age, they give off ethylene gas, which hastens ripening. A changing ratio of auxin to ethylene, triggered mainly by shorter days, probably causes autumn color changes and the loss of leaves from deciduous trees.

33.8 Plant hormones have many agricultural uses. Auxins can delay or promote fruit drop. Auxins and gibberellins are used to produce seedless fruits. A synthetic auxin called 2,4-D kills weeds. There are questions about the safety of using such chemicals.

Responses to Stimuli (33.9–33.13)

33.9 Tropisms orient plant growth toward or away from environmental stimuli. Plants sense and respond to environmental changes. Tropisms are growth responses that change the shape of a plant or make it grow toward or away from a stimulus.

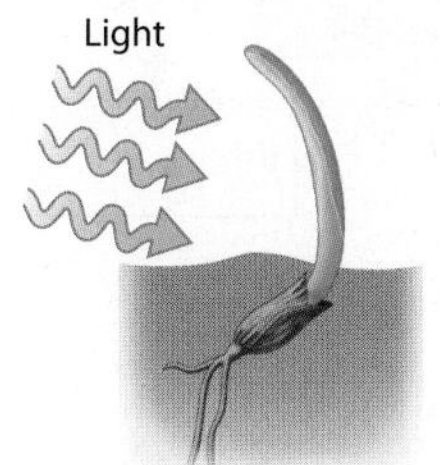
Light
Phototropism

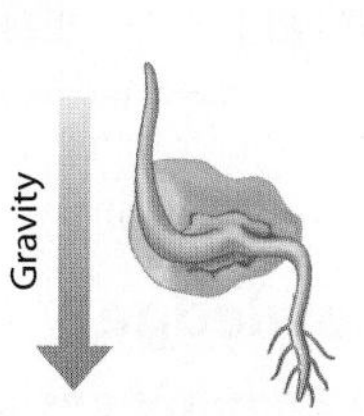
Gravity
Gravitropism

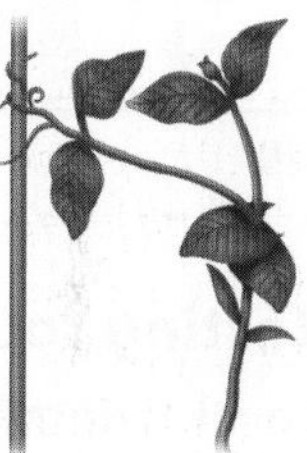
Thigmotropism

Phototropism, bending in response to light, may result from auxin moving from the light side to the dark side of a stem. A response to gravity, or gravitropism, may be caused by settling of organelles on the low sides of shoots and roots, which may trigger a change in hormone distribution. Thigmotropism, a response to touch, is responsible for coiling of tendrils and vines around objects.

33.10 Plants have internal clocks. An internal biological clock controls sleep movements and other daily cycles in plants. These cycles, called circadian rhythms, persist with periods of about 24 hours even in the absence of environmental cues, but such cues are needed to keep them synchronized with day and night.

33.11 Plants mark the seasons by measuring photoperiod. The timing of flowering is one of the seasonal responses to photoperiod, the relative lengths of night and day.

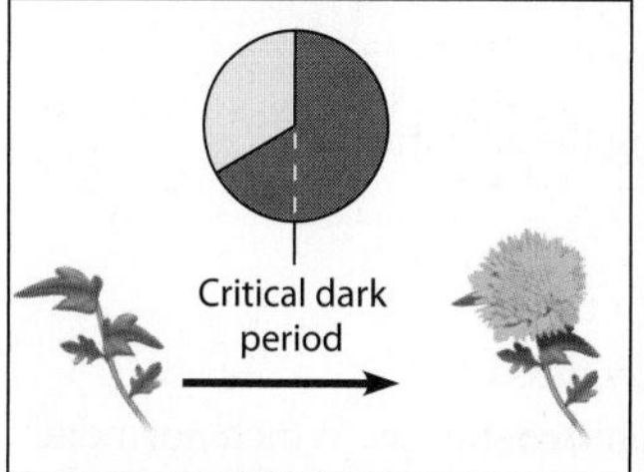
Critical dark period
Short-day (long-night) plants

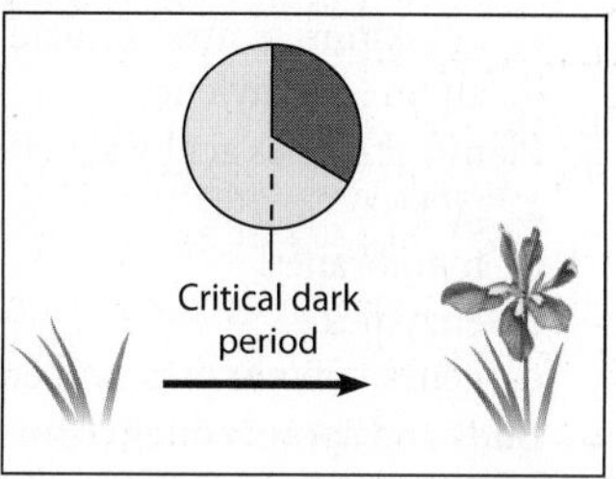
Critical dark period
Long-day (short-night) plants

33.12 Phytochromes are light detectors that help set the biological clock. Light-absorbing proteins called phytochromes may help plants set their biological clock and monitor photoperiod.

33.13 Defenses against herbivores and infectious microbes have evolved in plants. Plants use chemicals to defend themselves against both herbivores and pathogens. So-called avirulent plant pathogens interact with host plants in a specific way that stimulates both local and systemic defenses in the plant. Local defenses include microbe-killing chemicals and the sealing off of the infected area. Hormones trigger generalized defense responses in other organs (systemic acquired resistance).

Connecting the Concepts

1. Test your knowledge of the five major classes of plant hormones (auxins, cytokinins, gibberellins, abscisic acid, ethylene) by matching one hormone to each lettered box. (Note that some hormones will match up to more than one box.)

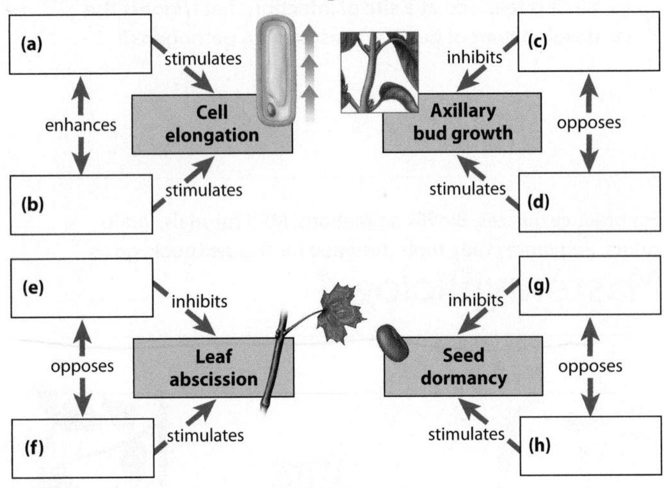
(a)
stimulates
Cell elongation
enhances
(b)
stimulates
inhibits
(c)
Axillary bud growth
opposes
stimulates
(d)
(e)
inhibits
Leaf abscission
opposes
(f)
stimulates
inhibits
(g)
Seed dormancy
opposes
stimulates
(h)

Testing Your Knowledge

Level 1: Knowledge/Comprehension

2. During winter or periods of drought, which one of the following plant hormones inhibits growth and seed germination?
   a. ethylene
   b. abscisic acid
   c. gibberellin
   d. auxin
3. Auxin causes a shoot to bend toward light by
   a. causing cells to shrink on the dark side of the shoot.
   b. stimulating growth on the dark side of the shoot.
   c. causing cells to shrink on the lighted side of the shoot.
   d. stimulating growth on the lighted side of the shoot.
4. In autumn, the amount of _______________ increases and the amount of _______________ decreases in fruits and leaf stalks, causing a plant to drop fruit and leaves.
   a. ethylene . . . auxin
   b. gibberellin . . . abscisic acid
   c. cytokinin . . . abscisic acid
   d. auxin . . . ethylene
5. Plant hormones act by affecting the activities of
   a. genes.
   b. membranes.
   c. enzymes.
   d. genes, membranes, and enzymes.
6. Buds and sprouts often form on tree stumps. Which hormone would stimulate their formation?
   a. auxin
   b. cytokinin
   c. abscisic acid
   d. ethylene
7. A plant's defense response at the site of initial infection by a pathogen will be especially strong if
   a. the pathogen is virulent.
   b. the plant makes a receptor protein that recognizes a signal molecule from the microbe.
   c. the pathogen is a fungus.
   d. the plant has an *Avr* gene that is the right match for one of the microbe's *R* genes.

Matching

8. Bending of a shoot toward light
9. Growth response to touch
10. Cycle with a period of about 24 hours
11. Pigment that helps control flowering
12. Relative lengths of night and day
13. Growth response to gravity
14. Folding of plant leaves at night

a. phytochrome
b. photoperiod
c. sleep movement
d. circadian rhythm
e. thigmotropism
f. phototropism
g. gravitropism

Level 2: Application/Analysis

15. A certain short-day plant flowers only when days are less than 12 hours long. Which of the following would cause it to flower?
    a. a 9-hour night and 15-hour day with 1 minute of darkness after 7 hours
    b. an 8-hour day and 16-hour night with a flash of white light after 8 hours
    c. a 13-hour night and 11-hour day with 1 minute of darkness after 6 hours
    d. a 12-hour day and 12-hour night with a flash of red light after 6 hours
16. If apples are to be stored for long periods, it is best to keep them in a place with good air circulation. Explain why.
17. Write a short paragraph explaining why a houseplant becomes bushier if you pinch off its terminal buds.

Level 3: Synthesis/Evaluation

18. A plant nursery manager tells the new night security guard to stay out of a room where chrysanthemums (which are short-day plants) are about to flower. Around midnight, the guard accidentally opens the door to the chrysanthemum room and turns on the lights for a moment. How might this affect the chrysanthemums? How could the guard correct the mistake?
19. A plant biologist observed a peculiar pattern when a tropical shrub was attacked by caterpillars. After a caterpillar ate a leaf, it would skip over nearby leaves and attack a leaf some distance away. Simply removing a leaf did not trigger the same change nearby. The biologist suspected that a damaged leaf sent out a chemical that signaled other leaves. How could this hypothesis be tested?
20. Imagine the following scenario: A plant biologist has developed a synthetic chemical that mimics the effects of a plant hormone. The chemical can be sprayed on apples before harvest to prevent flaking of the natural wax that is formed on the skin. This makes the apples shinier and gives them a deeper red color. What kinds of questions do you think should be answered before farmers start using this chemical on apples? How might the scientist go about finding answers to these questions?
21. **SCIENTIFIC THINKING** The discovery of auxin (discussed in Module 33.1) is a good example of how scientists build upon each other's work. For each of the three sets of researchers discussed here (the Darwins, Boyle-Jensen, and Went), write a one-sentence summary of what they discovered. If you were going to credit one person as the "discoverer" of auxin, who would it be?

Answers to all questions can be found in Appendix 4.

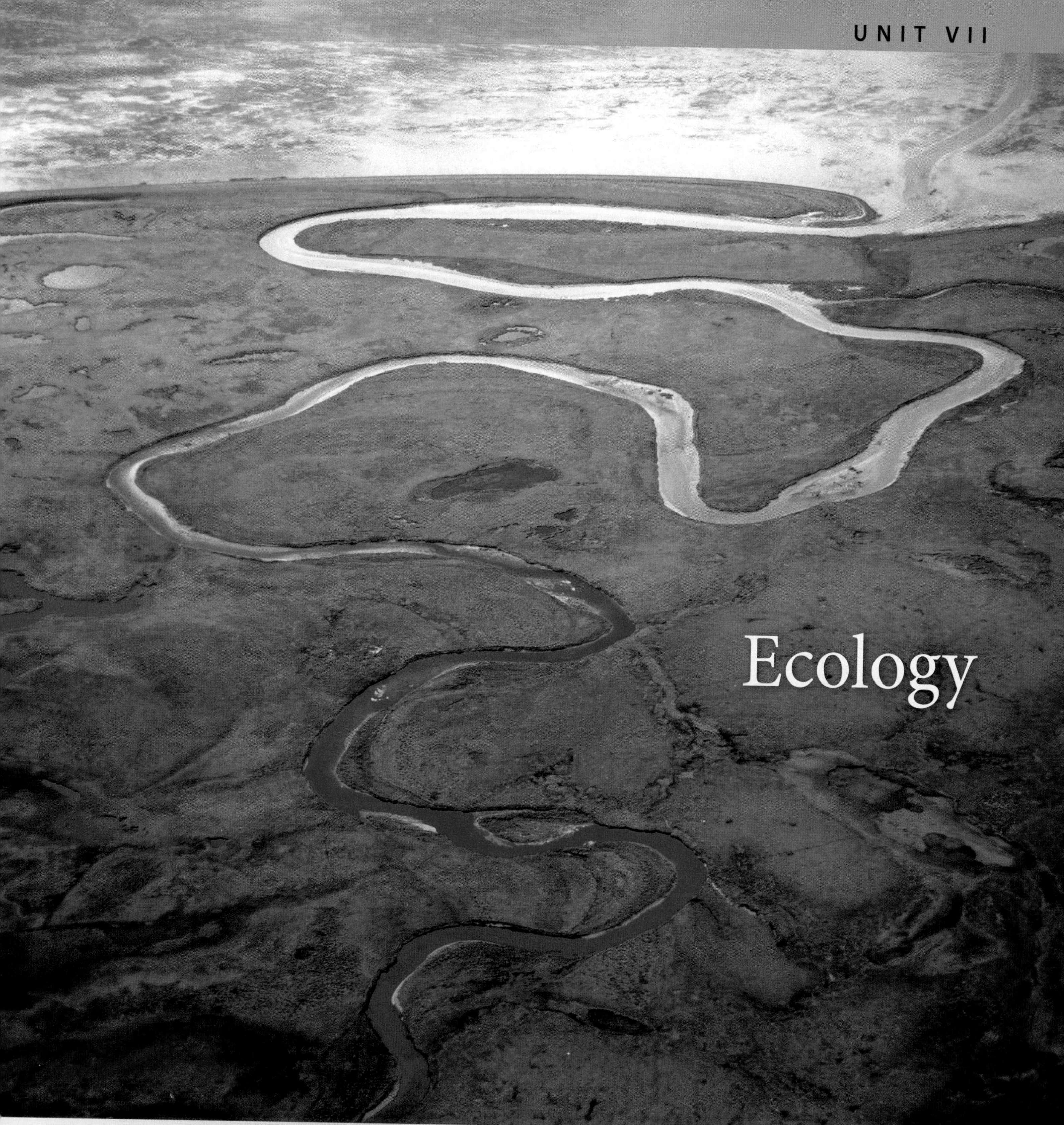

Ecology

CHAPTER 34

The Biosphere: *An Introduction to Earth's Diverse Environments*

? *Why study ecology?*

Did you ever think that a river could catch fire? In June, 1969, fire broke out on the Cuyahoga River in Cleveland, Ohio. For a century, oil refineries, steel mills, rubber factories, and other industries in Cleveland and Akron, 40 miles upstream, had dumped wastes directly into the river, along with raw sewage from both cities. The stretch of river between the two cities was devoid of fish, and the most heavily polluted sections lacked any signs of life. Thick oil slicks clogged with trash and other debris were common sights. When the river caught fire, Clevelanders were not surprised—it had happened several times in the past. However, the notion of a flammable river captured the attention of the national media. Public outrage over the toxic conditions of the nation's waterways, along with growing awareness of widespread environmental abuse, spurred a flurry of legislation, including the creation of the Environmental Protection Agency.

Today, Cuyahoga Valley National Park (photo below) surrounds a stretch of the river between Akron and Cleveland. Dozens of fish species thrive in it, and the park is home to abundant wildlife. Other parts of the river, however, are still polluted. Although the contamination is far from the

noxious brew of earlier decades, it is unlikely that the river will ever be clean along its entire length. Wherever a river passes through populated areas, pollution is almost inevitable.

Environmental concerns are among the most pressing issues we face today. How can we manage Earth's resources in ways that meet the needs of people today without compromising the ability of future generations to meet theirs? Just as human health care requires learning about the structure and function of the body, preserving a healthy environment depends on understanding the structure and function of populations, communities, and ecosystems. In this unit, you will learn about the principles of ecology, beginning with an exploration of Earth's diverse environments.

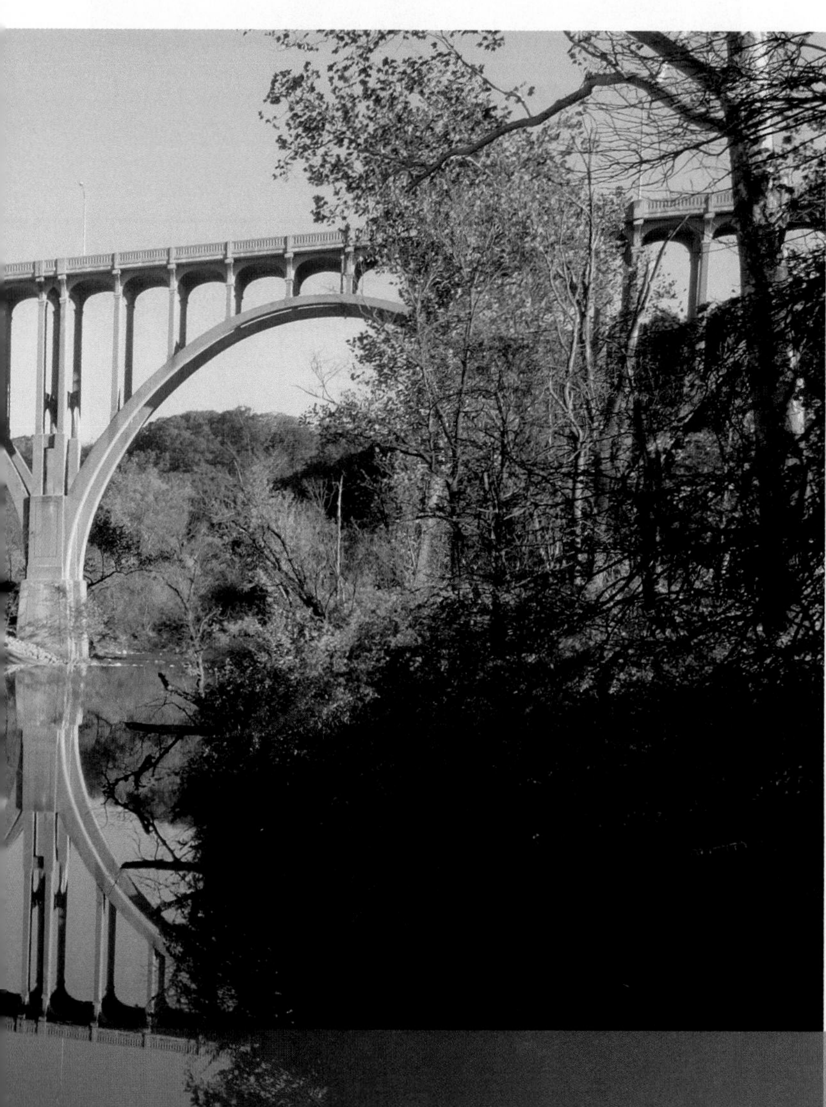

BIG IDEAS

The Biosphere
(34.1–34.5)

The distribution and abundance of life in the biosphere are influenced by living and nonliving components of the environment.

Aquatic Biomes
(34.6–34.7)

In marine biomes, the salt concentration is generally around 3%. In freshwater biomes, the salt concentration is typically less than 1%.

Terrestrial Biomes
(34.8–34.18)

The distribution of terrestrial biomes is primarily determined by temperature and rainfall.

▷ The Biosphere

34.1 Ecologists study how organisms interact with their environment at several levels

Ecology (from the Greek *oikos*, home) is the scientific study of the interactions of organisms with their environment. Ecologists describe the distribution and abundance of organisms—where they live and how many live there. Because the environment is complex, organisms can potentially be affected by many different variables. Ecologists group these variables into two major types, biotic factors and abiotic factors. **Biotic factors**, which include all of the organisms in the area, are the living component of the environment. **Abiotic factors** are the environment's nonliving component, the physical and chemical factors such as temperature, forms of energy available, water, and nutrients. An organism's **habitat**, the specific environment it lives in, includes the biotic and abiotic factors present in its surroundings.

As you might expect, field research is fundamental to ecology. But ecologists also test hypotheses using laboratory experiments, where conditions can be simplified and controlled. Some ecologists take a theoretical approach, devising mathematical and computer models that enable them to simulate large-scale experiments that are impossible to conduct in the field.

Ecologists study environmental interactions at several levels. Consider how researchers might investigate the ecology of an alpine meadow high in the Himalayan mountains. At the **organism** level, they may examine how one kind of organism meets the challenges and opportunities of its environment through its physiology or behavior. For example, an ecologist working at this level might study adaptations of the Himalayan blue poppy (*Meconopsis betonicifolia*, **Figure 34.1A**) to the freezing temperatures and short days of its abiotic environment.

Another level of study in ecology is the **population**, a group of individuals of the same species living in a particular geographic area. Blue poppies living in a particular Himalayan alpine meadow would constitute a population (**Figure 34.1B**). An ecologist studying blue poppies might investigate factors that affect the size of the population, such as the availability of chemical nutrients or seed dispersal.

A third level, the **community**, is an assemblage of all the populations of organisms living close enough together for potential interaction—all of the biotic factors in the environment. All the organisms in a particular alpine meadow would constitute a community (**Figure 34.1C**). An ecologist working at this level might focus on interspecies interactions, such as the competition between poppies and other plants for soil nutrients or the effect of plant-eaters on poppies.

The fourth level of ecological study, the **ecosystem**, includes both the biotic and abiotic components of the environment (**Figure 34.1D**). Some critical questions at the ecosystem level concern how chemicals cycle and how energy flows between organisms and their surroundings. For an alpine meadow, one ecosystem-level question would be, How rapidly does the decomposition of decaying plants release inorganic molecules?

Some ecologists take a wider perspective by studying **landscapes**, which are arrays of ecosystems. Landscapes are usually visible from the air as distinctive patches.

▲ **Figure 34.1A** An organism

▲ **Figure 34.1B** A population

▲ **Figure 34.1C** A community

▲ **Figure 34.1D** An ecosystem

For example, the Himalayan alpine meadows are part of a mountain landscape that also includes conifer and broadleaf forests. A landscape perspective emphasizes the absence of clearly defined ecosystem boundaries; energy, materials, and organisms may be exchanged by ecosystems within a landscape.

The **biosphere**, which extends from the atmosphere several kilometers above Earth to the depths of the oceans, is all of Earth that is inhabited by life. It is the home of us all, and as you will learn throughout this unit, our actions have consequences for the entire biosphere.

? **List the biotic and abiotic factors shown in Figure 34.1D.**

● The biotic factors include the animals and the blue poppies. Abiotic factors include the soil, temperature, and precipitation.

34.2 The science of ecology provides insight into environmental problems

SCIENTIFIC THINKING

The connections between human activities and environmental consequences are not always as visible as the pollution responsible for the Cuyahoga River fire. To understand how scientists identify less obvious environmental problems, let's take a look at another case that helped shape modern environmental policy.

In the 1950s, the prevailing view of the environment was that "Nature" was a force to be tamed and controlled for human purposes, in the same way that livestock had been domesticated. People were captivated by new technologies that promised an end to infectious disease and boundless increases in agricultural productivity. Chemical pesticides were an innovation that was enthusiastically embraced. The most widely used chemical was DDT, an insecticide that was employed against crop pests and disease-carrying insects such as mosquitoes (which transmit malaria), body lice (typhus), and fleas (plague). Despite its remarkable killing power against insects, DDT was considered harmless to vertebrates, including people **(Figure 34.2A)**.

By the late 1950s, however, a heated debate was raging over the widespread use of chemical pesticides. Consumers raised questions about chemical residues in their food. Dairy farmers found their milk contaminated by aerial spraying on nearby land. Fish and wildlife experts, along with private citizens, amassed dozens of reports of birds, fish, and other animals apparently poisoned by pesticides In some of these instances, birds that had once been common simply disappeared from the area. Disturbingly, scientists found that small amounts of pesticides had accumulated in the fatty tissues of vertebrates thousands of miles from where pesticides were used. DDT was even detected in human milk.

Why study ecology?

Scientists also found that DDT remained in the soil or water long after application of the pesticide. For example, one study correlated bird deaths on a university campus with DDT sprayed in previous years to control disease-spreading beetles on elm trees. The paper suggested that the birds consumed a toxic dose of the poison when they ate earthworms that had fed on DDT-contaminated leaves decaying in the soil.

▲ **Figure 34.2A** Spraying DDT to control mosquitoes on Jones Beach, New York, in 1945

Other scientists reported similar correlations between the use of DDT and animals in aquatic ecosystems as well as in other terrestrial ecosystems. Birds of prey seemed to be especially vulnerable—populations of bald eagles, ospreys, peregrine falcons, and other predatory birds in Europe and North America had declined dramatically **(Figure 34.2B)**. Studies suggested that pesticides, including DDT, caused the eggshells of these species to be abnormally thin and fragile. Evaluating the degree of certainty in conclusions is an important part of science. Much of the initial evidence against DDT consisted of isolated observations and correlations, and thus was not conclusive. However, the accumulation of so many clues pointed to problems that merited thorough study.

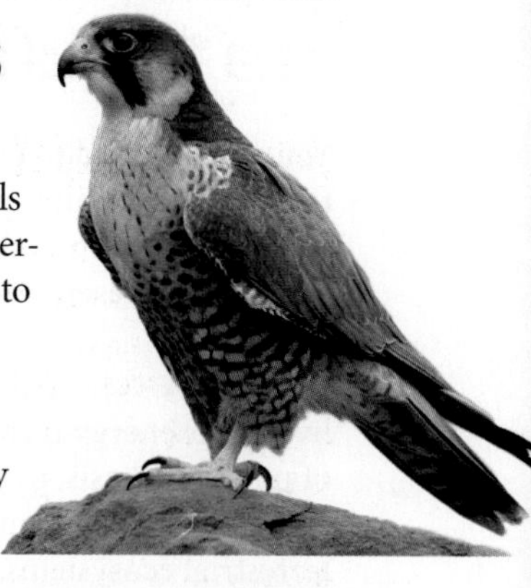

▲ **Figure 34.2B** Peregrine falcon (*Falco peregrinus*)

Scientists often help people understand current issues by communicating with the general public. The publication of a book called *Silent Spring* in 1962 brought widespread attention to the pesticide issue. Using her talent for making scientific subjects come alive for readers, the author of *Silent Spring*, Rachel Carson **(Figure 34.2C)**, drew the public into the debate. Carson, a former marine biologist and writer with the U. S. Fish and Wildlife Service, compiled available evidence on the consequences of widespread pesticide use. Considering it reckless to mount aerial spraying campaigns, which broadcast massive amounts of toxic chemicals that persist in the environment for many years, Carson advocated a new approach to pest control that took the health of ecosystems into account. Awareness of the problems caused by pesticides developed into concern for a host of environmental issues and set the stage for the public outcry over the Cuyahoga River fire and similar manifestations of environmental abuses.

▲ **Figure 34.2C** Rachel Carson

The science of ecology can provide the understanding needed to resolve environmental problems. But these problems cannot be solved by ecologists alone, because they require making decisions based on values and ethics. But analyzing environmental issues and planning for better practices begin with an understanding of the basic concepts of ecology, so let's start to explore them now.

? **Why can't ecologists alone solve environmental problems?**

● The science of ecology can inform the decision-making process, but solving environmental problems involves making ethical, economic, and political judgments that are outside the realm of science.

34.3 Physical and chemical factors influence life in the biosphere

You have learned that life thrives in a wide variety of habitats, from the mountaintops to the seafloor. To be successful, the organisms that live in each place must be adapted to the abiotic factors present in those environments.

Energy Sources All organisms require a source of energy to live. Solar energy from sunlight, captured during the process of photosynthesis, powers most ecosystems. Lack of sunlight is seldom the most important factor limiting plant growth for terrestrial ecosystems, although shading by trees does create intense competition for light among plants growing on forest floors. In many aquatic environments, however, light is not uniformly available. Microorganisms and suspended particles, as well as the water itself, absorb light and prevent it from penetrating beyond certain depths. As a result, most photosynthesis occurs near the water's surface.

In dark environments such as caves or hydrothermal vents, bacteria that extract energy from inorganic chemicals power ecosystems. Sulfur bacteria perform this function in hydrothermal vent communities. Many of the animals there either feed directly on the sulfur bacteria or derive nutrition from bacteria living inside their bodies. For example, tube worms **(Figure 34.3A)** have no mouth or digestive tract. The red tip extending from the white casing is a respiratory surface that acquires oxygen and sulfide from the water. Bacteria living in a specialized organ in the worm's body get energy from the sulfide—a lot of energy. These worms can grow to be over 2 m (6.5 feet) long.

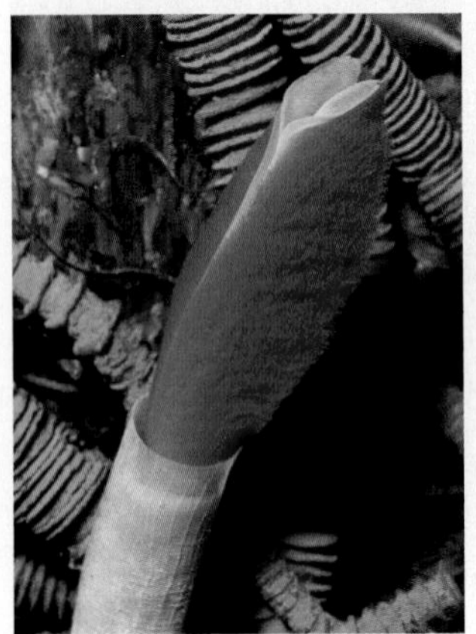

▲ **Figure 34.3A** The respiratory surface of a giant tube worm

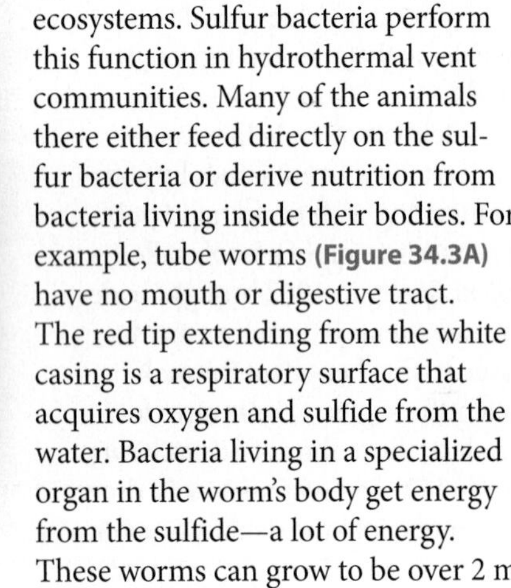

Temperature Temperature is an important abiotic factor because of its effect on metabolism (see Module 5.14). Few organisms can maintain a sufficiently active metabolism at temperatures close to 0°C, and temperatures above 45°C (113°F) destroy the enzymes of most organisms. However, extraordinary adaptations enable some species to live outside this temperature range. For example, archaeans living in hot springs have enzymes that function optimally at extremely high temperatures. Mammals and birds, such as the snowy owl in **Figure 34.3B**, can remain considerably warmer than their surroundings and can be active at a fairly wide range of temperatures. Amphibians and reptiles, which gain most of their warmth by absorbing heat from their surroundings, have a more limited distribution.

Water Water is essential to all life. Thus, for terrestrial organisms, dehydration is a major danger. Watertight coverings were key adaptations enabling plants and vertebrates to be successful on land (see Modules 17.1 and 19.6). Aquatic organisms are surrounded by water; their problem is solute concentration. Freshwater organisms live in a hypotonic medium, while the environment of marine organisms is hypertonic. Animals maintain fluid balance by a variety of mechanisms (see Module 25.4).

▼ **Figure 34.3B** A snowy owl

Inorganic Nutrients The distribution and abundance of photosynthetic organisms, including plants, algae, and photosynthetic bacteria, depend on the availability of inorganic nutrients such as nitrogen and phosphorus. Plants obtain these from the soil. Soil structure, pH, and nutrient content often play major roles in determining the distribution of plants. In many aquatic ecosystems, low levels of nitrogen and phosphorus limit the growth of algae and photosynthetic bacteria.

Other Aquatic Factors Several abiotic factors are important in aquatic, but not terrestrial, ecosystems. While terrestrial organisms have a plentiful supply of oxygen from the air, aquatic organisms must depend on oxygen dissolved in water. This is a critical factor for many species of fish. Trout, for example, require high levels of dissolved oxygen. Cold, fast-moving water has a higher oxygen content than warm or stagnant water. Salinity, current, and tides may also play a role in aquatic ecosystems.

Other Terrestrial Factors On land, wind is often an important abiotic factor. Wind increases an organism's rate of water loss by evaporation. The resulting increase in evaporative cooling (see Module 25.2) can be advantageous on a hot summer day, but it can cause dangerous wind chill in the winter. In some ecosystems, fire occurs frequently enough that many plants have adapted to this disturbance.

Next we examine the interaction between one animal species and the abiotic and biotic factors of its environment.

? Why are birds and mammals, but not amphibians and reptiles other than birds, found in Himalayan alpine meadows?

● As ectotherms (see Module 25.1), reptiles other than birds and amphibians do not have adaptations that enable them to withstand the cold temperatures of the alpine habitat.

34.4 Organisms are adapted to abiotic and biotic factors by natural selection

EVOLUTION CONNECTION

One of the fundamental goals of ecology is to explain the distribution of organisms. The presence of a species in a particular place has two possible explanations: The species may have evolved from ancestors living in that location, or it may have dispersed to that location and been able to survive once it arrived. The magnificent pronghorn "antelope" (*Antilocapra americana,* **Figure 34.4**) is the descendant of ancestors that roamed the open plains and shrub deserts of North America more than a million years ago. The animal is found nowhere else and is only distantly related to the many species of antelope in Africa. What selective factors in the abiotic and biotic environments of its ancestors produced the adaptations we see in the pronghorn that roams North America today?

The pronghorn's present-day habitat, like that of its ancestors, is arid, windswept, and subject to extreme temperature fluctuations both daily and seasonally. Individuals able to survive and reproduce under these conditions left offspring that carried their alleles forward into subsequent generations. Thus, we can infer that many of the adaptations that contribute to the success of present-day pronghorns must also have contributed to the success of their ancestors. For example, the pronghorn has a thick coat made of hollow hairs that trap air, insulating the animal in cold weather. If you drive through Wyoming or parts of Colorado in the winter, you will see herds of these animals foraging in the open when temperatures are well below 0°C. In hot weather, the pronghorn can raise patches of this stiff hair to release body heat.

▲ **Figure 34.4** A pronghorn (*Antilocapra americana*)

The biotic environment, which includes what the animal eats and any predators that threaten it, is also a factor in determining which members of a population survive and reproduce. The pronghorn's main foods are small broadleaf plants, grasses, and woody shrubs. Over time, characteristics that enabled the ancestors of the pronghorn to exploit these food sources more efficiently became established through natural selection. As a result, the teeth of a pronghorn are specialized for biting and chewing tough plant material. Like the stomach of a cow, the pronghorn's stomach contains cellulose-digesting bacteria. As the pronghorn eats plants, the bacteria digest the cellulose, and the animal obtains most of its nutrients from the bacteria.

While many factors in the pronghorn's environment have been fairly consistent throughout its evolutionary history, one aspect has changed significantly. Until around 12,000 years ago, one of the pronghorn's major predators was probably the American cheetah, a fleet-footed feline that bears some similarities to the more familiar African cheetah. The now-extinct American cheetah was one of many ferocious predators of Pleistocene North America, along with lions, jaguars, and saber-toothed cats with 7-inch canines. Ecologists hypothesize that the selection pressure of the cheetah's pursuit led to the pronghorn's blazing speed, which far exceeds that of its main present-day predator, the wolf. With a top speed of 97 km/h (60 mph), the pronghorn is easily the fastest mammal on the continent, and an adult pronghorn can keep up a pace of 64 km/h (40 mph) for at least 30 minutes. Unable to match the pronghorn's extravagant speed, wolves typically take adults that have been weakened by age or illness.

Like many large herbivores that live in open grasslands, the pronghorn also derives protection from living in herds. When one pronghorn starts to run, its white rump patch seems to alert other herd members to danger. Other adaptations that help the pronghorn foil predators include its tan and white coat, which provides camouflage, and its keen eyes, which can detect movement at great distances. Thus, the adaptations shaped by natural selection in the distant past still serve as protection from the predators of today's environment.

If the pronghorn's environment changed significantly, the adaptations that contribute to its current success might not be as advantageous. For example, if an increase in rainfall turned the open plains into woodlands, where predators would be more easily hidden by vegetation and could stalk their prey at close range, the pronghorn's adaptations for escaping predators might not be as effective. Thus, in adapting populations to local environmental conditions, natural selection may limit the distribution of organisms.

In the next module, we see how global climatic patterns determine temperature and precipitation, the major abiotic factors that influence the distribution of organisms. We examine the biotic components of the environment more closely in other chapters in this unit.

? **What is the role of the environment in adaptive evolution?**

● The individuals whose phenotypes are best suited to the environment (including both abiotic and biotic factors) will pass their alleles to the next generation. But individuals with other phenotypes may not. For example, if the biotic environment includes wolves, a pronghorn that is not able to run at top speed for as long as the rest of the herd will probably not survive to reproduce.

34.5 Regional climate influences the distribution of terrestrial communities

When we ask what determines whether a particular organism or community of organisms lives in a certain area, the climate of the region—especially temperature and precipitation—is often a crucial part of the answer. Earth's global climate patterns are largely determined by the input of radiant energy from the sun and the planet's movement in space.

Figure 34.5A shows that because of its curvature, Earth receives an uneven distribution of solar energy. The sun's rays strike equatorial areas most directly (perpendicularly). Away from the equator, the rays strike Earth's surface at a slant. As a result, the same amount of solar energy is spread over a larger area. Thus, any particular area of land or ocean near the equator absorbs more heat than comparable areas in the more northern or southern latitudes.

The seasons of the year result from the permanent tilt of the planet on its axis as it orbits the sun. As **Figure 34.5B** shows, the globe's position relative to the sun changes through the year.

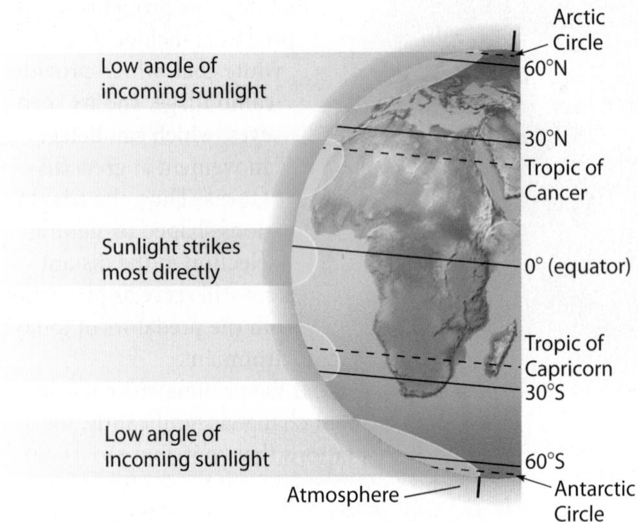

Arctic Circle
60°N
Low angle of incoming sunlight
30°N
Tropic of Cancer
Sunlight strikes most directly
0° (equator)
Tropic of Capricorn
30°S
Low angle of incoming sunlight
60°S
Antarctic Circle
Atmosphere

▲ **Figure 34.5A** How solar radiation varies with latitude

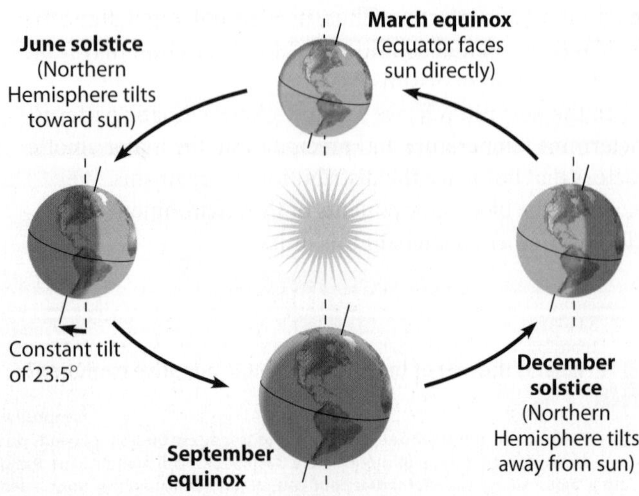

March equinox
(equator faces sun directly)
June solstice
(Northern Hemisphere tilts toward sun)
Constant tilt of 23.5°
September equinox
December solstice
(Northern Hemisphere tilts away from sun)

▲ **Figure 34.5B** How Earth's tilt causes the seasons

The Northern Hemisphere, for instance, is tipped most toward the sun in June. The more direct angle and increased intensity of the sun result in the long days of summer in that hemisphere. But during this time, days are short and it is winter in the Southern Hemisphere. Conversely, the Southern Hemisphere is tipped most toward the sun in December, creating summer there and causing winter in the Northern Hemisphere. The **tropics**, the region surrounding the equator between latitudes 23.5° north (the Tropic of Cancer) and 23.5° south (the Tropic of Capricorn), experience the greatest annual input and least seasonal variation in solar radiation.

Figure 34.5C shows some of the effects of the intense solar radiation near the equator on global patterns of rainfall and winds. Arrows indicate air movements. High temperatures in the tropics evaporate water from Earth's surface. Heated by the direct rays of the sun, moist air at the equator rises, creating an area of calm or of very light winds known as the **doldrums**. As warm equatorial air rises, it cools and releases much of its water content, creating the abundant precipitation typical of most tropical regions. High temperatures throughout the year and ample rainfall largely explain why rain forests are concentrated near the equator.

After losing their moisture over equatorial zones, high altitude air masses spread away from the equator until they cool and descend again at latitudes of about 30° north and south. This descending dry air absorbs moisture from the land. Thus, many of the world's great deserts—the Sahara in North Africa and the Arabian on the Arabian Peninsula, for example—are centered at these latitudes. As the dry air descends, some of it spreads back toward the equator. This movement creates the cooling **trade winds**, which dominate the tropics. As the air moves back toward the equator, it warms and picks up moisture until it ascends again.

The latitudes between the tropics and the Arctic Circle in the north and the Antarctic Circle in the south are called **temperate zones**. Generally, these regions have seasonal variations in climate and more moderate temperatures than the tropics or the polar zones. Notice in Figure 34.5C that some

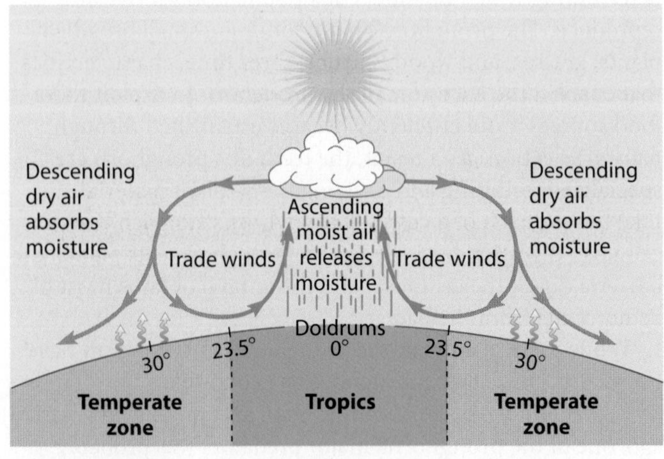

Descending dry air absorbs moisture
Ascending moist air releases moisture
Descending dry air absorbs moisture
Trade winds
Trade winds
Doldrums
0°
23.5°
23.5°
30°
30°
Temperate zone
Tropics
Temperate zone

▲ **Figure 34.5C** How uneven heating causes rain and winds

of the descending air heads into the latitudes above 30°. At first these air masses pick up moisture, but they tend to drop it as they cool at higher latitudes. This is why the north and south temperate zones, especially latitudes around 60°, tend to be moist. Broad expanses of coniferous forest dominate the landscape at these fairly wet but cool latitudes.

Figure 34.5D shows the major global air movements, called the **prevailing winds**. Prevailing winds (pink arrows) result from the combined effects of the rising and falling of air masses (blue and brown arrows) and Earth's rotation (gray arrows). Because Earth is spherical, its surface moves faster at the equator (where its diameter is greatest) than at other latitudes. In the tropics, Earth's rapidly moving surface deflects vertically circulating air, making the trade winds blow from east to west. In temperate zones, the slower-moving surface produces the **westerlies**, winds that blow from west to east.

A combination of the prevailing winds, the planet's rotation, unequal heating of surface waters, and the locations and shapes of the continents creates **ocean currents**, river-like flow patterns in the oceans (**Figure 34.5E**). Ocean currents have a profound effect on regional climates. For instance, the Gulf Stream circulates warm water northward from the Gulf of Mexico and makes the climate on the west coast of Great Britain warmer during winter than the coast of New England, which is actually farther south but is cooled by a branch of the current flowing south from the coast of Greenland (not shown in figure).

Landforms can also affect local climate. Air temperature declines by about 6°C with every 1,000-m increase in elevation, an effect you've probably experienced if you've ever hiked up a mountain. **Figure 34.5F** illustrates the effect of mountains on rainfall. This drawing represents major landforms across the state of California, but mountain ranges cause similar effects elsewhere. California is a temperate area in which the prevailing winds are westerlies. As moist air moves in off the Pacific Ocean and encounters the westernmost mountains (the Coast Range), it flows upward, cools at higher altitudes, and drops a large amount of water. The world's tallest trees, the coastal redwoods, thrive here. Farther inland, precipitation increases again as the air moves up and over higher mountains (the Sierra Nevada). Some of the world's deepest snow packs occur here. On the eastern side of the Sierra, there is little precipitation, and the dry descending air also absorbs moisture. This effect, called a rain shadow, is responsible for the desert that covers much of central Nevada.

Climate and other abiotic factors of the environment control the global distribution of organisms. The influence of these abiotic factors results in **biomes**, major types of ecological associations that occupy broad geographic regions of land or water. Terrestrial biomes are determined primarily by temperature and precipitation—similar assemblages of plant and animal types are found in areas that have similar climates. Aquatic biomes are defined by different abiotic factors; the primary distinction is based on salinity. Marine biomes, which include oceans, intertidal zones, coral reefs, and estuaries, generally have salt concentrations around 3%, while freshwater biomes (lakes, streams and rivers, and wetlands) typically have a salt concentration of less than 1%. We describe several aquatic biomes in the next two modules.

? **What causes summer in the Northern Hemisphere?**

● Because of the fixed angle of Earth's axis relative to the orbital plane around the sun, the Northern Hemisphere is tilted toward the sun during the portion of the annual orbit that corresponds to the summer months.

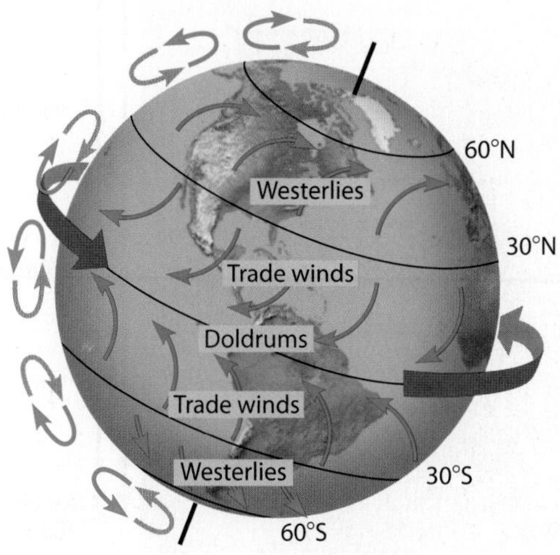
60°N
Westerlies
30°N
Trade winds
Doldrums
Trade winds
Westerlies
30°S
60°S

▲ **Figure 34.5D** Prevailing wind patterns

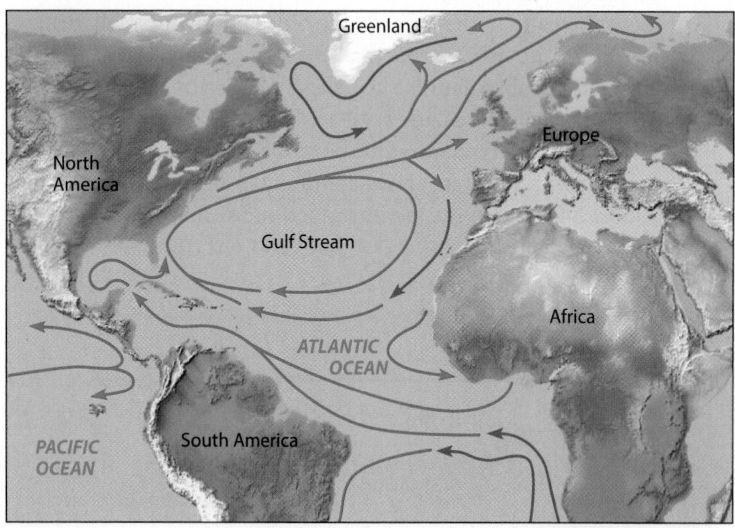
Greenland
Europe
North America
Gulf Stream
Africa
ATLANTIC OCEAN
PACIFIC OCEAN
South America

▲ **Figure 34.5E** Atlantic Ocean currents (red arrows indicate warming currents; blue arrows indicate cooling currents)

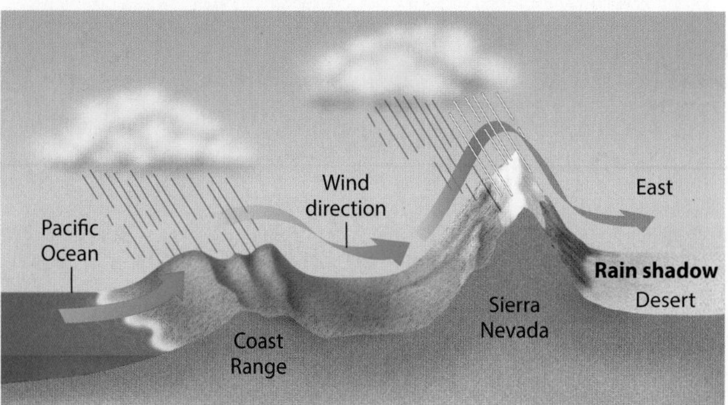
Wind direction
East
Pacific Ocean
Rain shadow
Desert
Sierra Nevada
Coast Range

▲ **Figure 34.5F** How mountains affect precipitation (California)

▷ Aquatic Biomes

34.6 Sunlight and substrate are key factors in the distribution of marine organisms

Gazing out over a vast ocean, you might think that it is the most uniform environment on Earth. But marine ecosystems can be as different as night and day. The deepest ocean, where hydrothermal vents are located, is perpetually dark. In contrast, the vivid coral reefs are utterly dependent on sunlight. Habitats near shore are different from those in mid-ocean, and the substrate, which varies with depth and distance from shore, hosts different communities from the open waters.

The **pelagic realm** of the oceans includes all open water, and the substrate—the seafloor—is known as the **benthic realm** (Figure 34.6A). The depth of light penetration, a maximum of 200 m (656 feet), marks the **photic zone**. In shallow areas such as the submerged parts of continents, called **continental shelves**, the photic zone includes both the pelagic and benthic realms. In these sunlit regions, photosynthesis by **phytoplankton** (microscopic algae and cyanobacteria) and multicellular algae provides energy and organic carbon for a diverse community of animals. Sponges, burrowing worms, clams, sea anemones, crabs, and echinoderms inhabit the benthic realm of the photic zone. **Zooplankton** (small, drifting animals), fish, marine mammals, and many other types of animals are abundant in the pelagic photic zone.

Coral reefs, a visually spectacular and biologically diverse biome, are scattered around the globe in the photic zone of warm tropical waters above continental shelves, as shown in Figure 34.6B (also see the introduction to Chapter 2). A reef is built up slowly by successive generations of coral animals—a diverse group of cnidarians that secrete a hard external skeleton—and by multicellular algae encrusted with limestone. Unicellular algae live within the corals, providing the coral with food (see Module 18.6). Coral reefs support a huge variety of invertebrates and fishes.

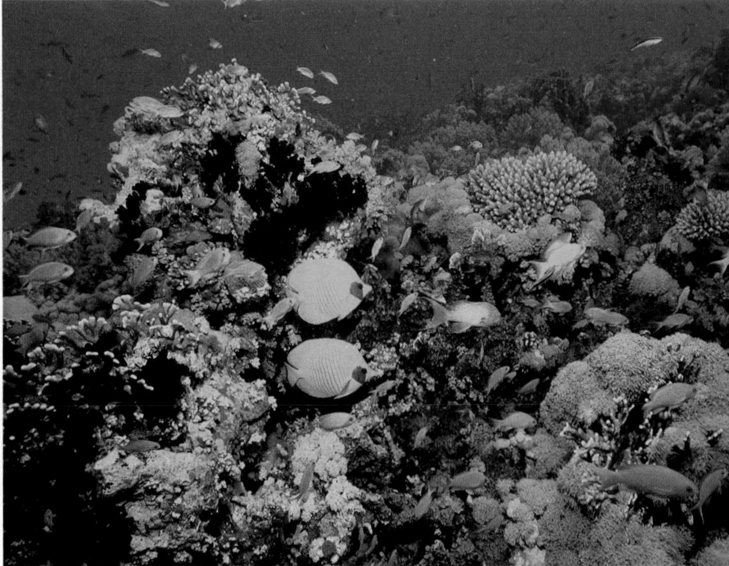

▲ **Figure 34.6B** A coral reef with its immense variety of invertebrates and fishes

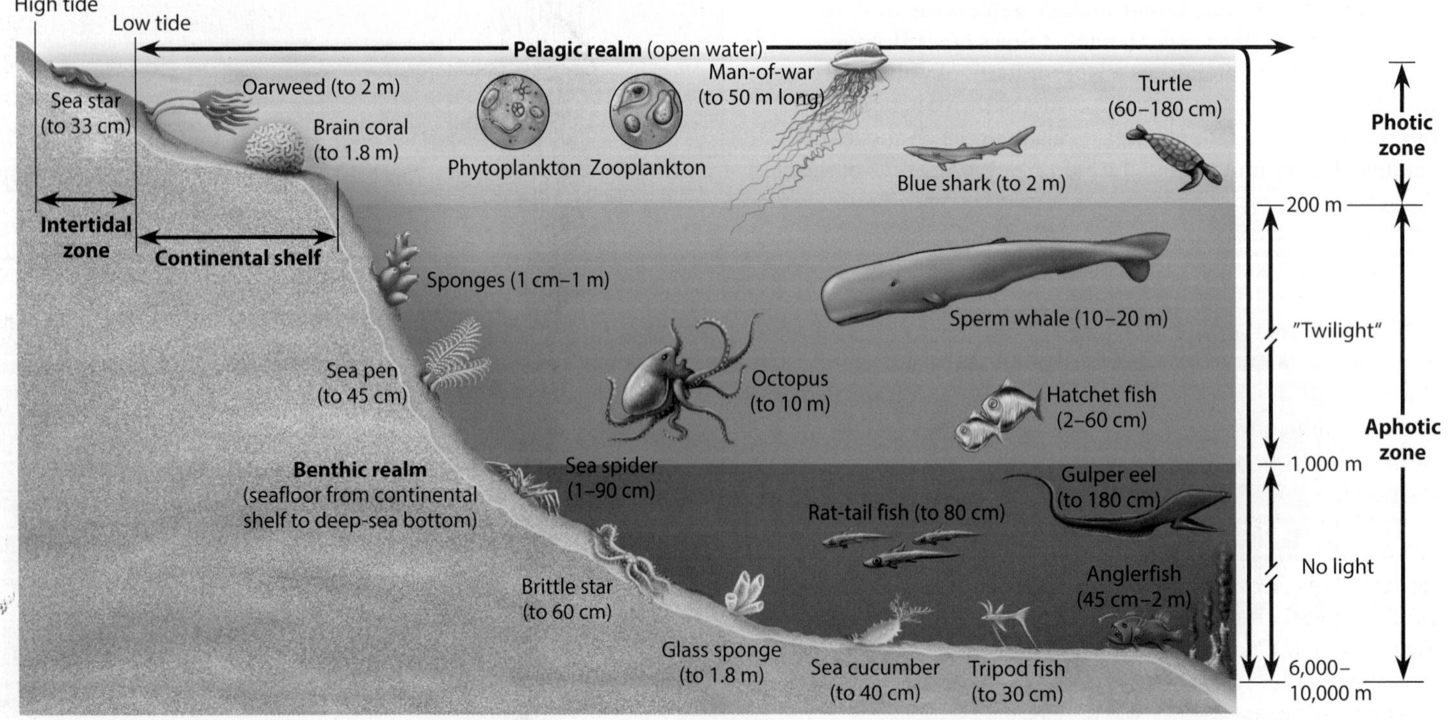

High tide
Low tide
Pelagic realm (open water)
Sea star (to 33 cm)
Oarweed (to 2 m)
Brain coral (to 1.8 m)
Phytoplankton
Zooplankton
Man-of-war (to 50 m long)
Turtle (60–180 cm)
Blue shark (to 2 m)
Photic zone
200 m
Intertidal zone
Continental shelf
Sponges (1 cm–1 m)
Sperm whale (10–20 m)
"Twilight"
Sea pen (to 45 cm)
Octopus (to 10 m)
Hatchet fish (2–60 cm)
Aphotic zone
Benthic realm (seafloor from continental shelf to deep-sea bottom)
Sea spider (1–90 cm)
Gulper eel (to 180 cm)
1,000 m
Rat-tail fish (to 80 cm)
No light
Brittle star (to 60 cm)
Anglerfish (45 cm–2 m)
Glass sponge (to 1.8 m)
Sea cucumber (to 40 cm)
Tripod fish (to 30 cm)
6,000–10,000 m

▲ **Figure 34.6A** Ocean life (zone depths and organisms not drawn to scale)

Below the photic zone of the ocean lies the **aphotic zone**. Although there is not enough light for photosynthesis between 200 and 1,000 m (0.6 mile), some light does reach these depths. This dimly lit world, sometimes called the twilight zone, is dominated by a fascinating variety of small fishes and crustaceans. Food sinking from the photic zone provides some sustenance for these animals. In addition, many of them migrate to the surface at night to feed. Some fishes in the twilight zone have enlarged eyes, enabling them to see in the very dim light, and luminescent organs that attract mates and prey.

Below 1,000 m, the ocean is completely and permanently dark. Adaptation to this environment has produced bizarre-looking creatures, such as the angler fish shown in **Figure 34.6C**. The scarcity of food probably explains the strangely outsized mouths of the angler and other fishes that inhabit this region of the ocean, a feature that allows them to grab any available prey, large or small. Inwardly angled teeth ensure that once caught, prey do not escape. The angler fish improves its chances of encountering prey by dangling a lure lit by bioluminescent bacteria. Most benthic organisms here are deposit feeders, animals that consume dead organic matter (detritus) on the substrate. Crustaceans, polychaete worms, sea anemones, and echinoderms such as sea cucumbers, sea stars, and sea urchins are common. Because of the scarcity of food, however, the density of animals is low—except at hydrothermal vents, where chemoautotrophic bacteria support an abundance of life.

▲ **Figure 34.6C**
An angler fish

The marine environment also includes distinctive biomes where the ocean interfaces with land or with fresh water. In the **intertidal zone**, where the ocean meets land, the shore is pounded by waves during high tide and exposed to the sun and drying winds during low tide. The rocky intertidal zone is home to many sedentary organisms, such as algae, barnacles, and mussels, which attach to rocks and are thus prevented from being washed away when the tide comes in. On sandy beaches, suspension-feeding worms, clams, and predatory crustaceans bury themselves in the ground.

Figure 34.6D shows an **estuary**, a biome that occurs where a freshwater stream or river merges with the ocean. The saltiness of estuaries ranges from nearly that of fresh water to that of the ocean. With their waters enriched by nutrients from the river, estuaries are among the most productive biomes on Earth. Oysters, crabs, and many fishes live in estuaries or reproduce in them. Estuaries are also crucial nesting and feeding areas for waterfowl.

Wetlands constitute a biome that is transitional between an aquatic ecosystem—either marine or freshwater—and a terrestrial one. Covered with water either permanently or periodically, wetlands support the growth of aquatic plants (see Figure 34.7C). Mudflats and salt marshes are coastal wetlands that often border estuaries.

For centuries, people viewed the ocean as a limitless resource, harvesting its bounty and using it as a dumping ground for wastes. The impact of these practices is now being felt in many ways, large and small. From worldwide declines in commercial fish species to dying coral reefs to beaches closed by pollution, danger signs abound. Because of their proximity to land, estuaries and wetlands are especially vulnerable. Many have been completely replaced by development on landfill. Other threats include nutrient pollution, contamination by pathogens or toxic chemicals, alteration of freshwater inflow, and introduction of non-native species. Coral reefs have suffered from many of the same problems. In addition, overfishing has upset the species balance in some reef communities and greatly reduced diversity, and the widespread demise of reef-building corals in some regions has been attributed to global climate change.

Freshwater biomes share many characteristics with marine biomes and experience some of the same threats. We introduce freshwater biomes in the next module.

? Oil from the 2010 Deepwater Horizon disaster in the Gulf of Mexico has polluted estuaries in Louisiana. Why does this pollution affect other animals in addition to those that live permanently in the estuaries?

● Many species, including fishes and waterfowl, visit estuaries to feed or reproduce.

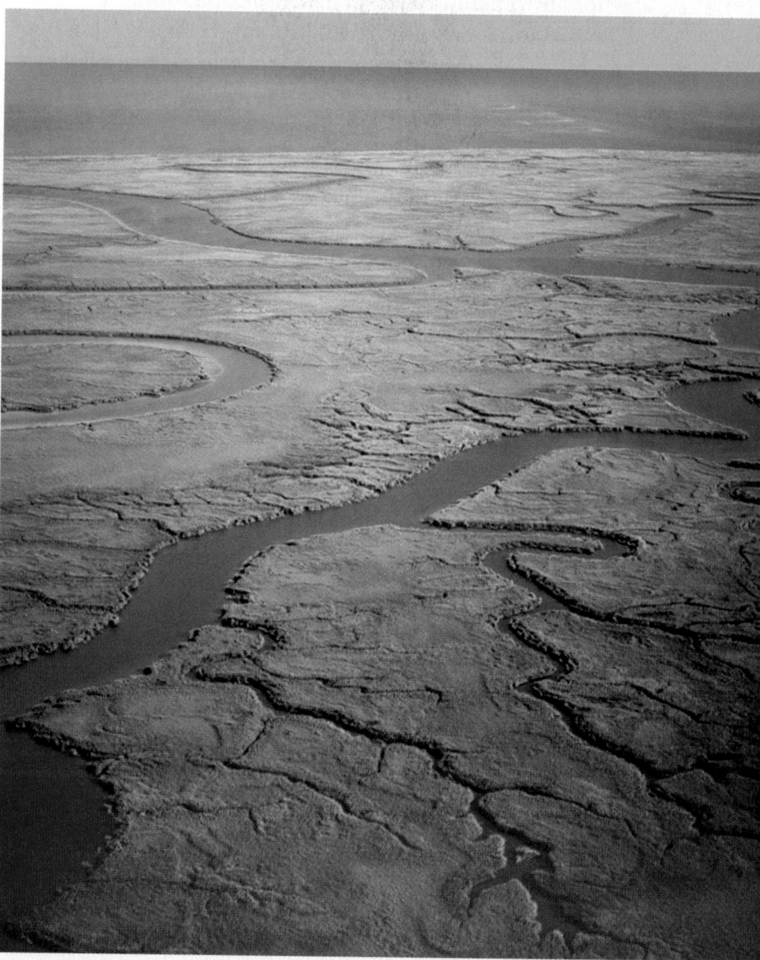

▲ **Figure 34.6D** An estuary in Georgia

34.7 Current, sunlight, and nutrients are important abiotic factors in freshwater biomes

Freshwater biomes cover less than 1% of Earth's surface and contain a mere 0.01% of its water. But they harbor a disproportionate share of biodiversity—an estimated 6% of all described species. Moreover, we depend on freshwater biomes for drinking water, crop irrigation, sanitation, and industry.

Freshwater biomes fall into two broad categories: standing water, which includes lakes and ponds, and flowing water, such as rivers and streams. Because these biomes are embedded in terrestrial landscapes, their characteristics are intimately connected with the soils and organisms of the ecosystems that surround them.

Lakes and Ponds In lakes and large ponds, as in the oceans, the communities of plants, algae, and animals are distributed according to the depth of the water and its distance from shore. Phytoplankton grow in the photic zone, and rooted plants often inhabit shallow waters near shore (**Figure 34.7A**). If a lake or pond is deep enough or murky enough, it has an aphotic zone where light levels are too low to support photosynthesis. In the benthic realm, large populations of microorganisms decompose dead organisms that sink to the bottom. Respiration by microbes removes oxygen from water near the bottom, and in some lakes, benthic areas are unsuitable for any organisms except anaerobic microbes.

Temperature may also have a profound effect on standing water biomes. During the summer, deep lakes have a distinct upper layer of water that has been warmed by the sun and does not mix with underlying, cooler water.

The mineral nutrients nitrogen and phosphorus typically determine the amount of phytoplankton growth in a lake or pond. Many lakes and ponds receive large inputs of nitrogen and phosphorus from sewage and runoff from fertilized lawns and farms. These nutrients may produce a heavy growth ("bloom") of algae, which reduces light penetration. When the algae die and decompose, a pond or lake can suffer severe oxygen depletion, killing fish that are adapted to high-oxygen conditions.

Rivers and Streams Rivers and streams generally support communities of organisms quite different from those of lakes and ponds. A river or a stream changes greatly between its source (perhaps a spring or snowmelt) and the point at which it empties into a lake or the ocean. Near the source, the water is usually cold, low in nutrients, and clear (**Figure 34.7B**). The channel is often narrow, with a swift current that does not allow much silt to accumulate on the bottom. The current also inhibits the growth of phytoplankton; most of the organisms found here are supported by the photosynthesis of algae attached to rocks or by organic material, such as leaves, carried into the stream from the surrounding land. The most abundant benthic animals are usually arthropods, such as small crustaceans and insect larvae, that have physical and behavioral adaptations that enable them to resist being swept away. Trout, which locate their insect prey mainly by sight in the clear water, are often the predominant fishes.

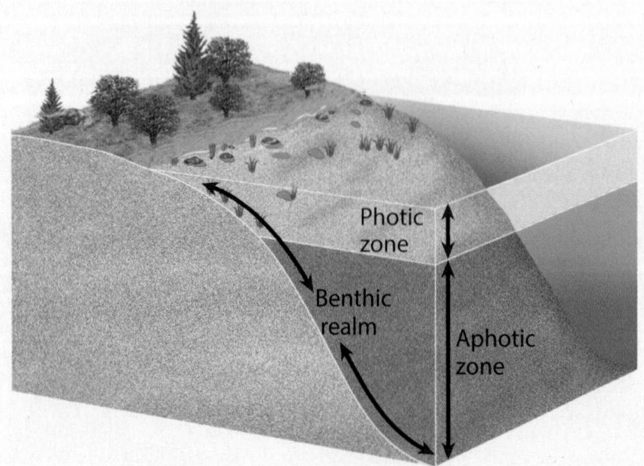

Photic zone
Benthic realm
Aphotic zone

▲ **Figure 34.7A** Zones in a lake

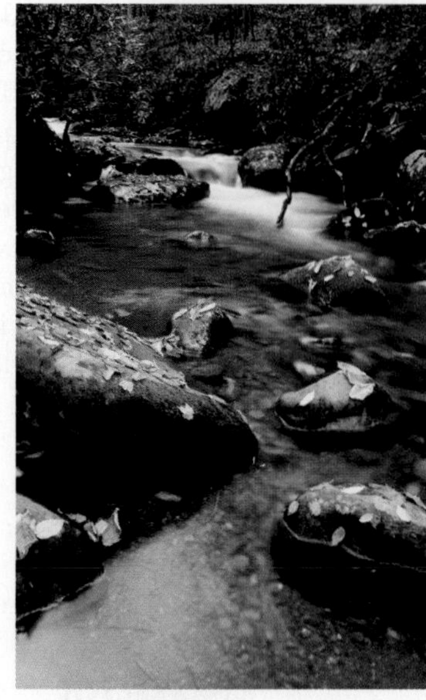

▲ **Figure 34.7B** A stream in the Great Smoky Mountains, Tennessee

Downstream, a river or stream generally widens and slows. The water is usually warmer and may be murkier because of sediments and phytoplankton suspended in it. Worms and insects that burrow into mud are often abundant, as are waterfowl, frogs, and catfish and other fishes that find food more by scent and taste than by sight.

Wetlands Freshwater wetlands range from marshes, as shown in **Figure 34.7C**, to swamps and bogs. Like marine wetlands, freshwater wetlands are rich in species diversity. They provide water storage areas that reduce flooding and improve water quality by filtering pollutants. Recognition of their ecological and economic value has led to government and private efforts to protect and restore wetlands.

? Why does sewage cause algal blooms in lakes?

● The sewage adds nutrients, such as nitrates and phosphates, that stimulate growth of algae.

▲ **Figure 34.7C** A marsh at Kent State University in Ohio

▷ Terrestrial Biomes

34.8 Terrestrial biomes reflect regional variations in climate

Terrestrial ecosystems are grouped into nine major types of biomes, which are distinguished primarily by their predominant vegetation. By providing food, shelter, nesting sites, and much of the organic material for decomposers, plants build the foundation for the communities of animals, fungi, and microorganisms that are characteristic of each biome. The geographic distribution of plants, and thus of terrestrial biomes, largely depends on climate, with temperature and precipitation often the key factors determining the kind of biome that exists in a particular region.

Figure 34.8 shows the locations of the biomes. If the climate in two geographically separate areas is similar, the same type of biome may occur in both places; notice on the map that each kind of biome occurs on at least two continents. Each biome is characterized by a type of biological community, rather than an assemblage of particular species. For example, the species living in the deserts of the American Southwest and in the Sahara Desert of Africa are different, but all are adapted to desert conditions. Widely separated biomes may look alike because of convergent evolution, the appearance of similar traits in independently evolved species living in similar environments (see Module 15.14).

There is local variation within each biome that gives the vegetation a patchy, rather than uniform, appearance. For example, in northern coniferous forests, snowfall may break branches and small trees, causing openings where broadleaf trees such as aspen and birch can grow. Local disturbances such as storms and fires also create openings in many biomes.

The current concern about global warming is generating intense interest in the effect of climate change on vegetation patterns. Using powerful new tools such as satellite imagery, scientists are documenting shifts in latitudes of biome borders, changes in snow and ice coverage, and changes in length of the growing season. At the same time, many natural biomes have been fragmented and altered by human activity. A high rate of biome alteration by humans is correlated with an unusually high rate of species loss throughout the globe (as we'll discuss in Module 38.2).

Now let's begin our survey of the major terrestrial biomes. To help you locate the biomes, an orientation map color-coded to Figure 34.8 is included with each module. Icons indicate a relative temperate range (in red) and average annual precipitation (in blue) for each biome. Icons also identify biomes in which fire plays a significant role.

? Test your knowledge of world geography: Which biome is most closely associated with a "Mediterranean climate"?

● Chaparral

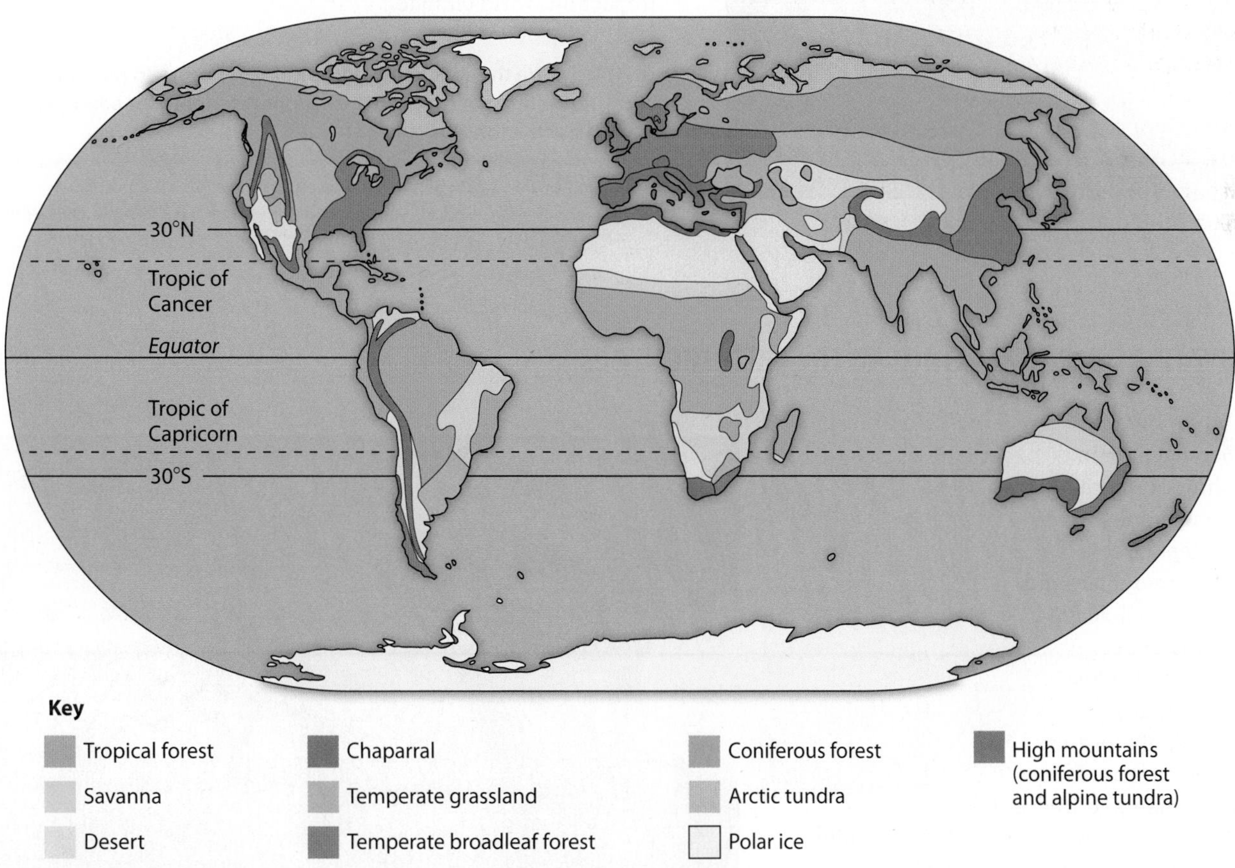
30°N
Tropic of Cancer
Equator
Tropic of Capricorn
30°S
Key
Tropical forest
Savanna
Desert
Chaparral
Temperate grassland
Temperate broadleaf forest
Coniferous forest
Arctic tundra
Polar ice
High mountains (coniferous forest and alpine tundra)

▲ **Figure 34.8** Major terrestrial biomes

Try This As you read Modules 34.9–34.17, make a table showing the climate, characteristic plants and animals, and any special features of each biome.

34.9 Tropical forests cluster near the equator

Tropical forests occur in equatorial areas where the temperature is warm and days are 11–12 hours long year-round. Rainfall in these areas is quite variable, and this variability, rather than temperature or day length, generally determines the vegetation that grows in a particular tropical forest. In areas where rainfall is scarce or there is a prolonged dry season, tropical dry forests predominate. The plants found there are a mixture of thorny shrubs and deciduous trees and succulents. Tropical rain forests are found in very humid equatorial areas where rainfall is abundant (200–400 cm, or 79–157 inches, per year).

Tropical rain forest, such as the lush area on the island of Borneo shown in **Figure 34.9**, is among the most complex of all biomes, harboring enormous numbers of different species. Up to 300 species of trees can be found in a single hectare (2.5 acres). The forest structure consists of distinct layers that provide many different habitats: emergent trees growing above a closed upper canopy, one or two layers of lower trees, a shrub understory, and a sparse ground layer of herbaceous plants. Because of the closed canopy, little sunlight reaches the forest floor. Many trees are covered by woody vines growing toward the light. Other plants, including bromeliads and orchids, gain access to sunlight by growing on the branches or trunks of tall trees. Many of the animals also dwell in trees, where food is abundant. Monkeys, birds, insects, snakes, bats, and frogs find food and shelter many meters above the ground.

Temperature range

Precipitation

▲ **Figure 34.9** Tropical rain forest

The soils of tropical rain forests are typically poor. High temperatures and rainfall lead to rapid decomposition and release of nutrients. However, the nutrients are quickly taken up by the luxuriant vegetation or washed away by the frequent rains.

Human impact on the world's tropical rain forests is an ongoing source of great concern. It is a common practice to clear the forest for lumber or simply burn it, farm the land for a few years, and then abandon it. Mining has also devastated large tracts of rain forest. Once stripped, the tropical rain forest recovers very slowly because the soil is so nutrient-poor. (We will discuss the potential consequences of destroying the tropical forests in Chapter 38, including the impact on world climate.)

? Why are the soils in most tropical rain forests so poor in nutrients that they can only support farming for a few years after the forest is cleared?

The conditions favor rapid decomposition of organic litter in the soil and immediate uptake of the resulting nutrients by plants. Thus, most of the ecosystem's nutrients are tied up in the vegetation that is cleared away before farming.

34.10 Savannas are grasslands with scattered trees

Figure 34.10, a photograph taken in the Serengeti Plain in Tanzania, shows a typical **savanna**, a biome dominated by grasses and scattered trees. The temperature is warm year-round.

Rainfall averages 30–50 cm (about 12–20 inches) per year, almost all of it during a relatively brief rainy season. Poor soils and lack of moisture inhibit the establishment of most trees. Grazing animals and frequent fires, caused by lightning or human activity, further inhibit invasion by trees. Grasses survive burning because the growing points of their shoots are below ground. Savanna plants have also been selected for their ability to

Temperature range
Precipitation
Fire

▲ **Figure 34.10** Savanna

survive prolonged periods of drought. Many trees and shrubs are deciduous, dropping their leaves during the dry season, an adaptation that helps conserve water.

Grasses and forbs (small broadleaf plants) grow rapidly during the rainy season, providing a good food source for many animal species. Large grazing mammals must migrate to greener pastures and scattered watering holes during seasonal drought. The dominant herbivores in savannas are actually insects, especially ants and termites. Also common are many burrowing animals, including mice, moles, gophers, snakes, ground squirrels, worms, and numerous arthropods.

Many of the world's large herbivores and their predators inhabit savannas. African savannas are home to giraffes, zebras, and many species of antelope, as well as to lions and cheetahs. Several species of kangaroo are the dominant mammalian herbivores of Australian savannas.

? How do fires help to maintain savannas as grassland ecosystems?

● By repeatedly preventing the spread of trees and other woody plants; grasses survive because the growing points of their shoots are underground.

34.11 Deserts are defined by their dryness

Deserts are the driest of all terrestrial biomes, characterized by low and unpredictable rainfall (less than 30 cm—12 inches—per year). The large deserts in central Australia and northern Africa have average annual rainfalls of less than 2 cm, and in the Atacama Desert in Chile, the driest place on Earth, there is often no rain at all for decades at a time. But not all desert air is dry. Coastal sections of the Atacama and of the Namib Desert in Africa are often shrouded in fog, although the ground remains extremely dry.

As we discussed in Module 34.5, large tracts of desert occur in two regions of descending dry air centered around the 30° north and 30° south latitudes. At higher latitudes, large deserts may occur in the rain shadows of mountains (see Figure 34.5F); these encompass much of central Asia east of the Caucasus Mountains, and Washington and Oregon east of the Cascade Mountains. The Mojave Desert, shown in **Figure 34.11**, is in the rain shadow of the Sierra Nevada, along with much of the rest of Southern California and Nevada.

Some deserts, as represented by the temperature icon in Figure 34.11, are very hot, with daytime soil surface temperatures above 60°C (140°F) and large daily temperature fluctuations. Other deserts, such as those west of the Rocky Mountains, are relatively cold. Air temperatures in cold deserts may fall below −30°C (−22°F).

The cycles of growth and reproduction in the desert are keyed to rainfall. The driest deserts have no perennial vegetation at all, but less arid regions have scattered deep-rooted shrubs, often interspersed with water-storing succulents such as cacti. The leaves of some plants, including the Joshua tree shown in Figure 34.11, have a waxy coating that prevents water loss. Desert plants typically produce great numbers of seeds, which may remain dormant until a heavy rain triggers germination. After periods of rainfall (often in late winter), annual plants in deserts may display spectacular blooms.

Like desert plants, desert animals are adapted to drought and extreme temperatures. Many live in burrows and are active only during the cooler nights, and most have special adaptations that conserve water. Seed-eaters such as ants, many birds, and rodents are common in deserts. Lizards, snakes, and hawks eat the seed-eaters.

The process of **desertification**, the conversion of semiarid regions to desert, is a significant environmental problem. In northern Africa, for example, a burgeoning human population, overgrazing, and dryland farming are converting large areas of savanna to desert.

? Why isn't "cold desert" an oxymoron?

● Because deserts are defined by low precipitation and dry soil, not by temperature

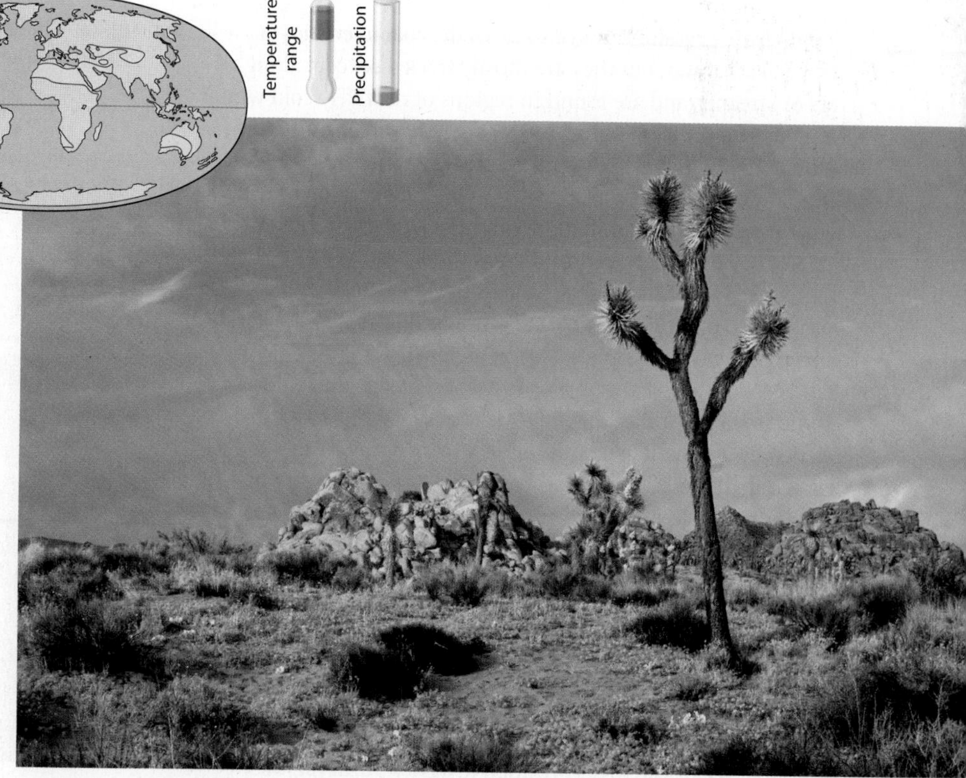
Temperature range
Precipitation

▲ **Figure 34.11** Desert

34.12 Spiny shrubs dominate the chaparral

Chaparral (a Spanish word meaning "place of evergreen scrub oaks") is characterized by dense, spiny shrubs with tough, evergreen leaves. The climate that supports chaparral vegetation results mainly from cool ocean currents circulating offshore, which produce mild, rainy winters and hot, dry summers. As a result, this biome is limited to small coastal areas, including California, where the photograph in **Figure 34.12** was taken. The largest region of chaparral surrounds the Mediterranean Sea; "Mediterranean" is another name for this biome. In addition to the perennial shrubs that dominate chaparral, annual plants are also commonly seen, especially during the wet winter and spring months. Animals characteristic of the chaparral include browsers such as deer, fruit-eating birds, and seed-eating rodents, as well as lizards and snakes.

Chaparral vegetation is adapted to periodic fires, most often caused by lightning. Many plants contain flammable chemicals and burn fiercely, especially where dead brush has accumulated. After a fire, shrubs use food reserves stored in the surviving roots to support rapid shoot regeneration. Some chaparral plant species produce seeds that will germinate only after a hot fire. The ashes of burned vegetation fertilize the soil with mineral nutrients, promoting regrowth of the plant community. Houses do not fare as well, and firestorms that race through the densely populated canyons of Southern California can be devastating.

? **What is one way that homeowners in chaparral areas can protect their neighborhoods from fire?**

● They can keep the area clear of dead brush, which is highly flammable.

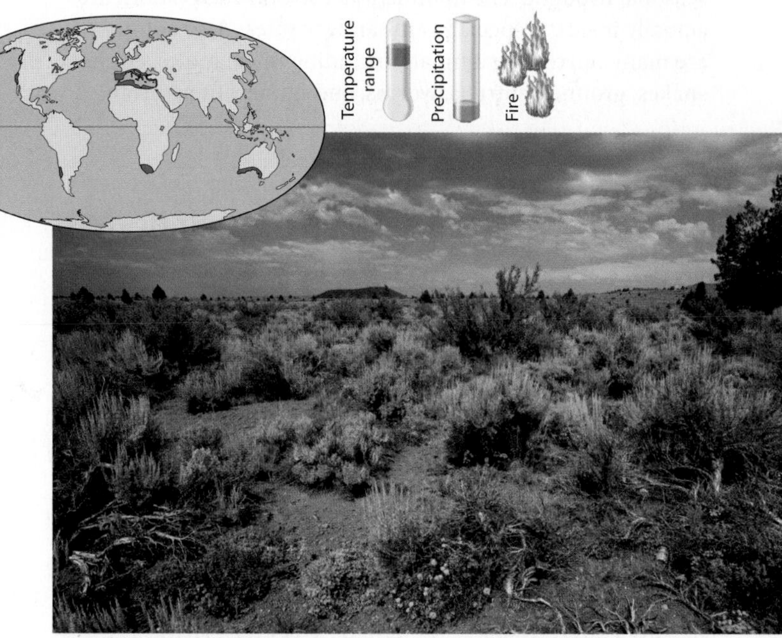

Temperature range
Precipitation
Fire

▲ **Figure 34.12** Chaparral

34.13 Temperate grasslands include the North American prairie

Temperate grasslands have some of the characteristics of tropical savannas, but they are mostly treeless, except along rivers or streams, and are found in regions of relatively cold winter temperatures. Precipitation, averaging between 25 and 75 cm (about 10–30 inches) per year, with periodic severe droughts, is too low to support forest growth. Fires and grazing by large mammals also inhibit growth of woody plants but do not harm the belowground grass shoots.

Large grazing mammals, such as the bison and pronghorn of North America and the wild horses and sheep of the Asian steppes, are characteristic of grasslands. Without trees, many birds nest on the ground, and some small mammals dig burrows to escape predators. Enriched by glacial deposits and mulch from decaying plant material, the soil of grasslands supports a great diversity of microorganisms and small animals, including annelids and arthropods.

The amount of annual precipitation influences the height of grassland vegetation. Shortgrass prairie is found in relatively dry regions; tallgrass prairie occurs in wetter areas. **Figure 34.13** shows a mixed-grass prairie in Alberta, Canada. Little remains of North American prairies today. Most of the region is intensively farmed, and it is one of the most productive agricultural regions in the world.

? **What factors inhibit woody plants from growing in temperate grasslands?**

● Low rainfall, fires, and grazing by large mammals

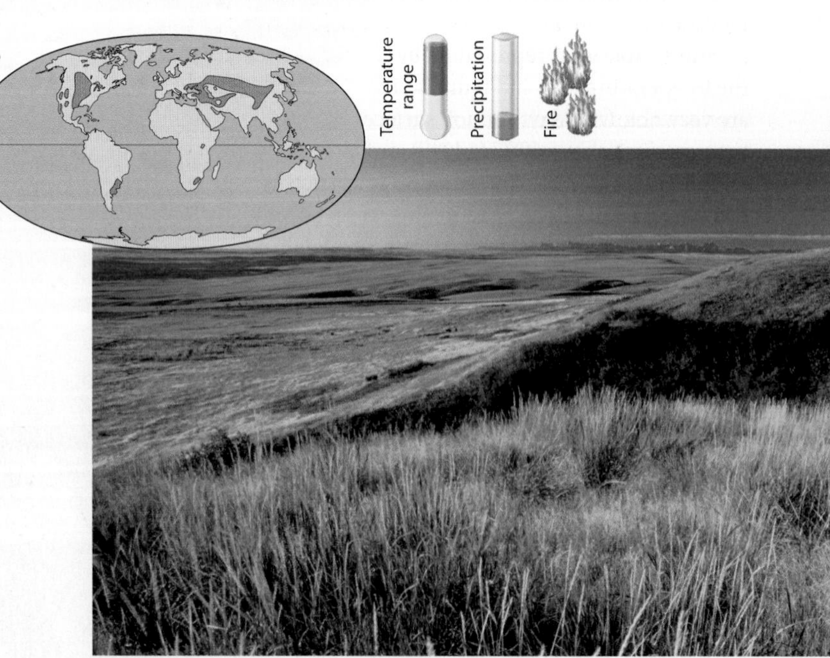

Temperature range
Precipitation
Fire

▲ **Figure 34.13** Temperate grassland

34.14 Broadleaf trees dominate temperate forests

Temperate broadleaf forests grow throughout midlatitude regions, where there is sufficient moisture to support the growth of large trees. In the Northern Hemisphere, deciduous trees (trees that drop their leaves seasonally) characterize temperate broadleaf forests. Some of the dominant trees are species of oak, hickory, birch, beech, and maple. The mix of tree species varies widely, depending on such factors as the climate at different latitudes, topography, and local soil conditions. **Figure 34.14** features a photograph taken during the spectacular display of autumn color in West Virginia.

Temperatures in temperate broadleaf forests range from very cold in the winter (−30°C) to hot in the summer (30°C). Annual precipitation is relatively high—75–150 cm (30–60 inches)—and usually evenly distributed throughout the year as either rain or snow. These forests typically have a growing season of 5 to 6 months and a distinct annual rhythm. Trees drop their leaves and become dormant in late autumn, preventing the loss of water from the tree at a time when frozen soil makes water less available. The trees produce new leaves in the spring.

The canopy of a temperate broadleaf forest is more open than that of a tropical rain forest, and the trees are not as tall or as diverse. However, the soils are richer in inorganic and organic nutrients. Rates of decomposition are lower in temperate forests than in the tropics, and a thick layer of leaf litter on forest floors conserves many of the biome's nutrients.

Numerous invertebrates live in the soil and leaf litter. Some vertebrates, such as mice, shrews, and ground squirrels, burrow for shelter and food, while others, including many species of birds, live in the trees. Predators include bobcats, foxes, black bears, and mountain lions.

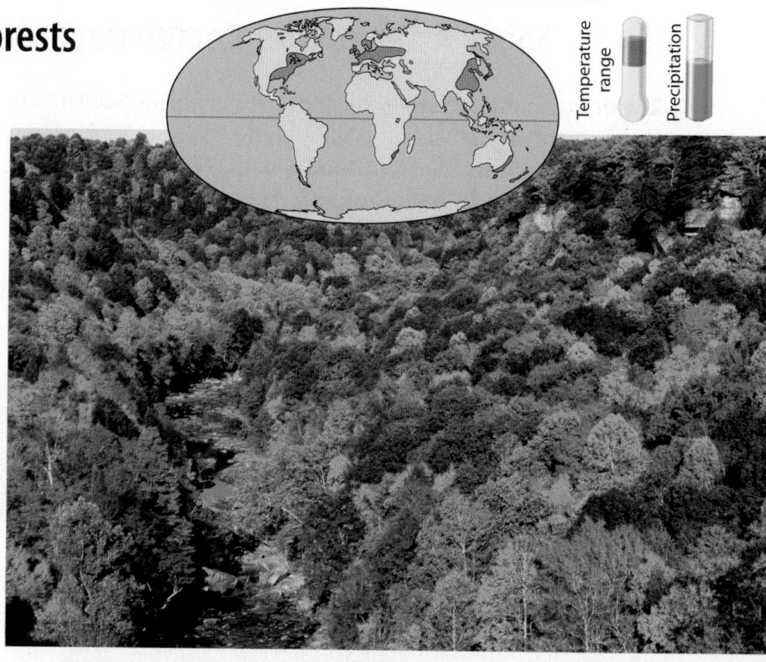

Temperature range
Precipitation

▲ **Figure 34.14** Temperate broadleaf forest

? **How does the soil of a temperate broadleaf forest differ from that of a tropical rain forest?**

● The soil in temperate broadleaf forests is rich in inorganic and organic nutrients, while the soil in tropical rain forests is low in nutrients.

34.15 Coniferous forests are often dominated by a few species of trees

Cone-bearing evergreen trees, such as spruce, pine, fir, and hemlock, dominate **coniferous forests**. The northern coniferous forest, or **taiga**, is the largest terrestrial biome on Earth, stretching in a broad band across North America and Asia south of the Arctic Circle. **Figure 34.15** shows taiga in Finland. Taiga is also found at cool, high elevations in more temperate latitudes, as in much of the mountainous region of western North America.

The taiga is characterized by long, cold winters and short, wet summers, which are sometimes warm. The soil is thin and acidic, and the slow decomposition of conifer needles makes few nutrients available for plant growth. Most of the precipitation is in the form of snow. The conical shape of many conifers prevents too much snow from accumulating on their branches and breaking them. Animals of the taiga include moose, elk, hares, bears, wolves, grouse, and migratory birds.

The **temperate rain forests** of coastal North America (from Alaska to Oregon) are also coniferous forests. Warm, moist air from the Pacific Ocean supports this unique biome, which, like most coniferous forests, is dominated by a few tree species, such as hemlock, Douglas fir, and redwood. These forests are heavily logged, and the old-growth stands of trees may soon disappear.

? **How does the soil of the northern coniferous forests differ from that of a broadleaf forest?**

● The soil is thinner, nutrient-poor, and acidic because conifer needles decompose slowly in the low temperatures.

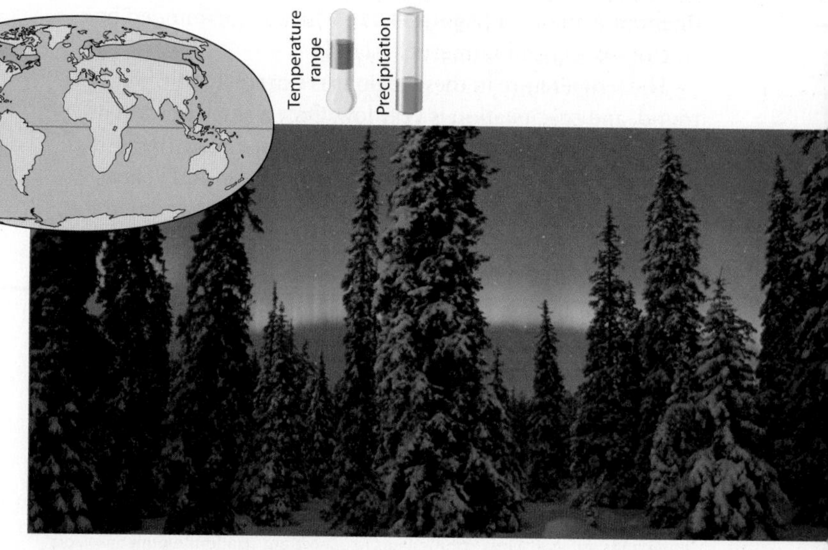

Temperature range
Precipitation

▲ **Figure 34.15** Coniferous forest

34.16 Long, bitter-cold winters characterize the tundra

Tundra (from the Russian word for "marshy plain") covers expansive areas of the Arctic between the taiga and polar ice. **Figure 34.16** shows the arctic tundra in the Northwest Territories, Canada, in the autumn. The climate here is often extremely cold, with little light for much of the autumn and winter. The arctic tundra is characterized by **permafrost**, continuously frozen subsoil—only the upper part of the soil thaws in summer. The arctic tundra may receive as little precipitation as some deserts. But poor drainage, due to the permafrost, and slow evaporation keep the soil continually saturated.

Permafrost prevents the roots of plants from penetrating very far into the soil, which is one factor that explains the absence of trees. Extremely cold winter air temperatures and high winds also contribute to the exclusion of trees. Vegetation in the tundra includes dwarf shrubs, grasses and other herbaceous plants, mosses, and lichens. During the brief, warm summers, when there is nearly constant daylight, plants grow quickly and flower in a rapid burst. High winds and cold temperatures create plant communities called alpine tundra on very high mountaintops at all latitudes, including the tropics. Although these communities are similar to arctic tundra, there is no permafrost beneath alpine tundra.

Animals of the tundra withstand the cold by having good insulation that retains heat. Large herbivores include musk oxen and caribou. The principal smaller animals are rodents called lemmings and a few predators, such as the arctic fox and snowy owl. Many animals are migratory, using the tundra as a summer breeding ground. During the brief warm season, the marshy ground supports the aquatic larvae of insects, providing food for migratory waterfowl, and clouds of mosquitoes often fill the tundra air.

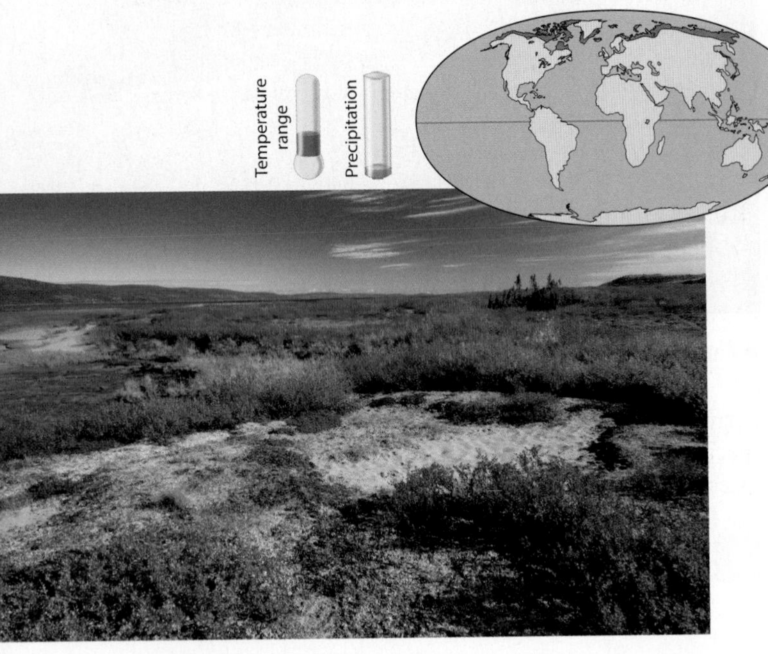

Temperature range
Precipitation

▲ **Figure 34.16** Tundra

? **What three abiotic factors account for the rarity of trees in arctic tundra?**

● Long, very cold winters (short growing season), high winds, and permafrost

34.17 Polar ice covers the land at high latitudes

In the Northern Hemisphere, **polar ice** covers land north of the tundra; much of the Arctic Ocean is continuously frozen as well. In the Southern Hemisphere, polar ice covers the continent of Antarctica (**Figure 34.17**), which is surrounded by a ring of sea ice, and numerous islands.

The temperature in these regions is extremely cold year-round, and precipitation is very low. Only a small portion of these landmasses is free of ice or snow, even during the summer. Nevertheless, small plants, such as mosses, and lichens manage to survive, and invertebrates such as nematodes, mites, and wingless insects called springtails inhabit the frigid soil.

The terrestrial polar biome is closely interconnected with the neighboring marine biome. Seals and marine birds, such as penguins, gulls, and skuas, feed in the ocean and visit the land or sea ice to rest and to breed. In the Northern Hemisphere, sea ice provides a feeding platform for polar bears (see Figure 38.5B).

? **How does the vegetation found in polar ice regions compare with tundra vegetation?**

● Neither biome is hospitable to plants because of the cold temperatures. However, tundra supports the growth of small shrubs, while polar ice vegetation is limited to mosses and lichens.

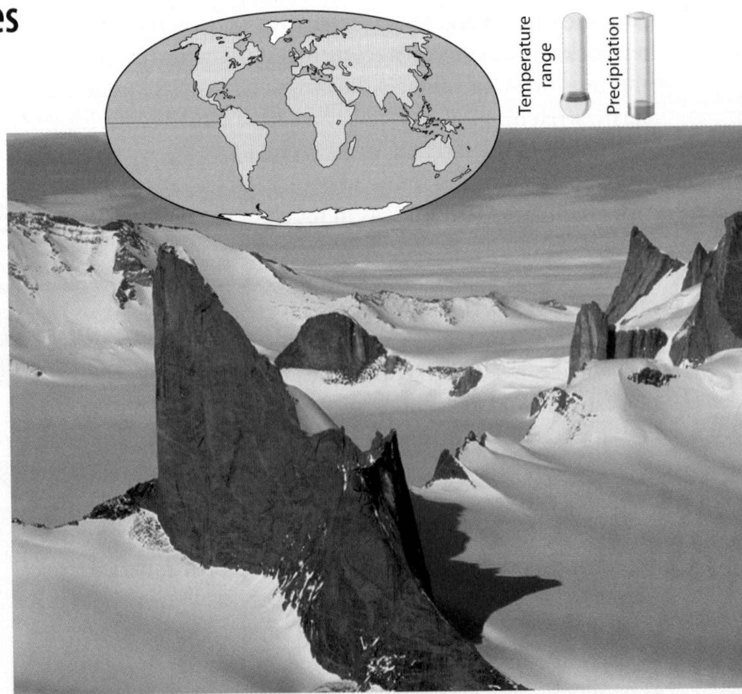

Temperature range
Precipitation

▲ **Figure 34.17** Polar ice

34.18 The global water cycle connects aquatic and terrestrial biomes

Ecological subdivisions such as biomes are not self-contained units. Rather, all parts of the biosphere are linked by the global water cycle, illustrated in **Figure 34.18**, and by nutrient cycles (which you will learn about in Chapter 37). Consequently, events in one biome may reverberate throughout the biosphere.

Recall from Module 34.5 that solar energy helps drive the movements of water and air in global patterns. In addition, precipitation and evaporation, as well as transpiration from plants (see Module 32.3), continuously move water between the land, oceans, and atmosphere. Over the oceans (left side of Figure 34.18), evaporation exceeds precipitation. The result is a net movement of water vapor to clouds that are carried by winds from the oceans across the land. On land (right side of the figure), precipitation exceeds evaporation and transpiration. Excess precipitation forms systems of surface water (such as lakes and rivers) and groundwater, all of which flow back to the sea, completing the water cycle.

Just as the water draining from your shower carries dead skin cells from your body along with the day's grime, the water washing over and through the ground carries traces of the land and its history. For example, water flowing from land to sea carries with it silt (fine soil particles) and chemicals such as fertilizers and pesticides. The accumulation of silt, aggravated by the development of coastal areas, has muddied the waters of some coral reefs, dimming the light available to the photosynthetic algae that power the reef community. Chemicals in surface water may travel hundreds of miles by stream and river to the ocean, where currents then carry them even farther from their point of origin. For instance, in the 1960s, researchers began finding traces of DDT in marine mammals in the Arctic, far from any places DDT had been used (see Module 34.2). Airborne pollutants such as nitrogen oxides and sulfur oxides, which combine with water to form acid precipitation, are also distributed by the water cycle.

Human activity also affects the global water cycle itself in a number of important ways. One of the main sources of atmospheric water is transpiration from the dense vegetation making up tropical rain forests. The destruction of these forests changes the amount of water vapor in the air. This, in turn, will likely alter local, and perhaps global, weather patterns. Pumping large amounts of groundwater to the surface for irrigation affects the water cycle, too. This practice can increase the rate of evaporation over land, resulting in higher humidity as well as depleting groundwater supplies.

? What is the main way that living organisms contribute to the water cycle?

Plants move water from the ground to the atmosphere via transpiration.

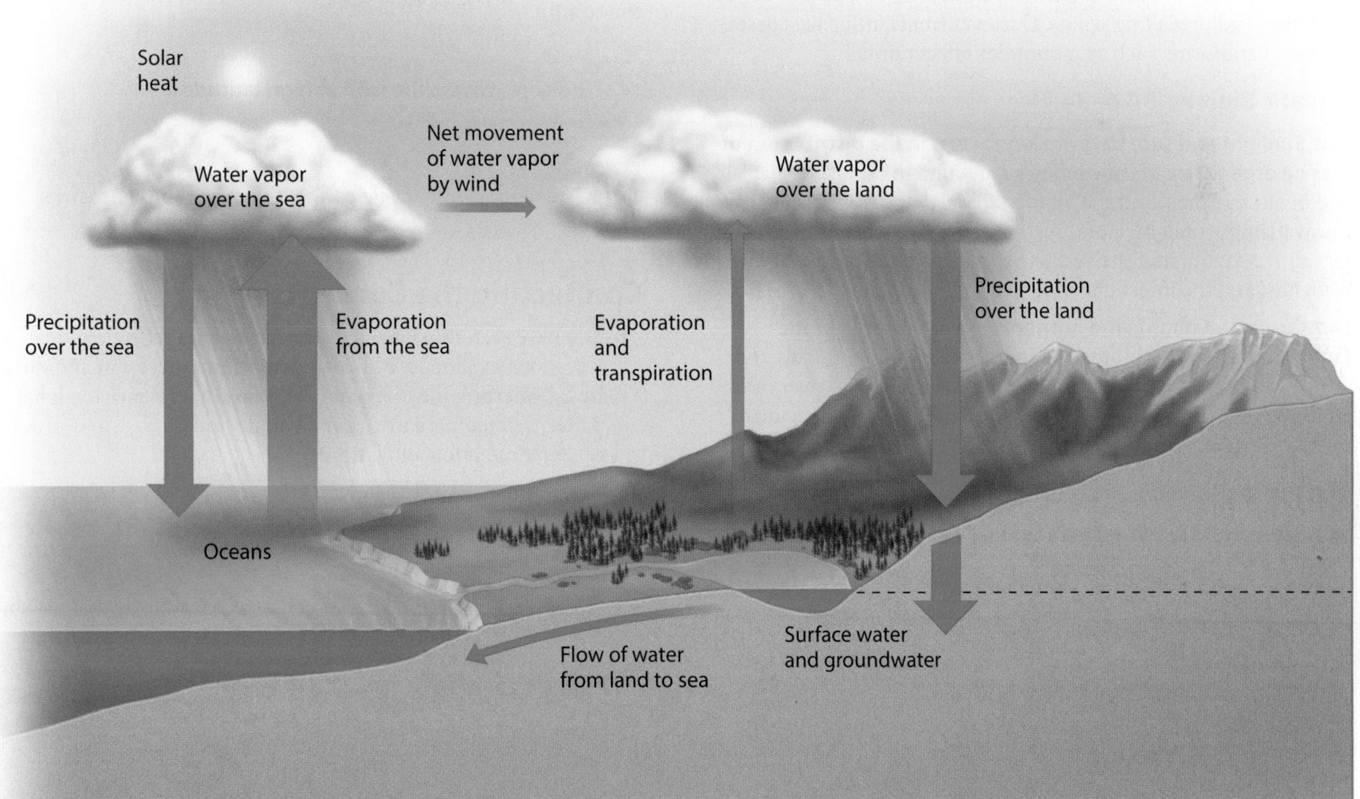

Solar heat
Water vapor over the sea
Net movement of water vapor by wind
Water vapor over the land
Precipitation over the sea
Evaporation from the sea
Evaporation and transpiration
Precipitation over the land
Oceans
Flow of water from land to sea
Surface water and groundwater

▲ **Figure 34.18** The global water cycle

Try This Trace the path water molecules might follow as you explain how they move from the sea to the land and back to the sea.

CHAPTER 34 REVIEW

For practice quizzes, BioFlix animations, MP3 tutorials, video tutors, and more study tools designed for this textbook, go to MasteringBiology®

Reviewing the Concepts

The Biosphere (34.1–34.5)

34.1 Ecologists study how organisms interact with their environment at several levels.

Organismal ecology (individual)
Population ecology (group of individuals of a species)
Community ecology (all organisms in a particular area)
Ecosystem ecology (all organisms and abiotic factors)

34.2 The science of ecology provides insight into environmental problems.

34.3 Physical and chemical factors influence life in the biosphere. Major factors include energy sources, temperature, the presence of water, and inorganic nutrients.

34.4 Organisms are adapted to abiotic and biotic factors by natural selection. The pronghorn's adaptations show the variety of factors that can affect an organism's fitness.

34.5 Regional climate influences the distribution of terrestrial communities. Most climatic variations are due to the uneven heating of Earth's surface as it orbits the sun, setting up patterns of precipitation and prevailing winds. Ocean currents influence coastal climate. Landforms such as mountains affect rainfall.

Aquatic Biomes (34.6–34.7)

34.6 Sunlight and substrate are key factors in the distribution of marine organisms. Marine biomes are found in both the pelagic and benthic realms, and the biomes are further distinguished by the availability of light. Coral reefs are found in warm, shallow waters above continental shelves. Other marine biomes are estuaries, wetlands, and the intertidal zone.

34.7 Current, sunlight, and nutrients are important abiotic factors in freshwater biomes. Standing water biomes (lakes and ponds) differ in structure from flowing water biomes (rivers and streams), and communities vary accordingly. Wetlands include marshes, swamps, and bogs.

Terrestrial Biomes (34.8–34.18)

34.8 Terrestrial biomes reflect regional variations in climate.

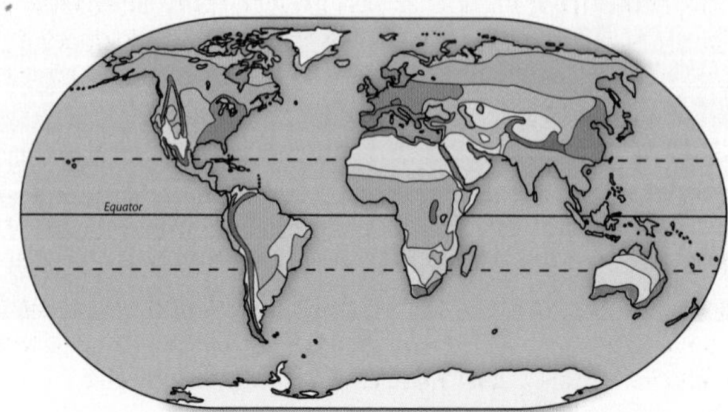

Equator

34.9 Tropical forests cluster near the equator.

34.10 Savannas are grasslands with scattered trees.

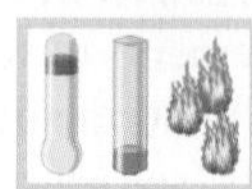

34.11 Deserts are defined by their dryness.

34.12 Spiny shrubs dominate the chaparral.

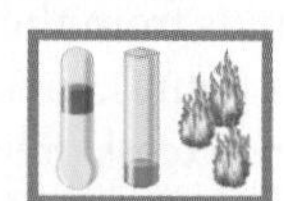

34.13 Temperate grasslands include the North American prairie.

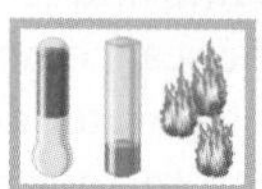

34.14 Broadleaf trees dominate temperate forests.

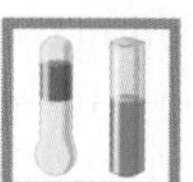

34.15 Coniferous forests are often dominated by a few species of trees.

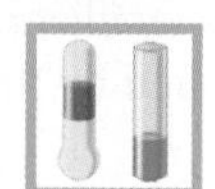

34.16 Long, bitter-cold winters characterize the tundra.

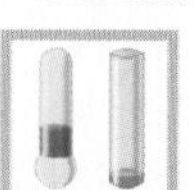

34.17 Polar ice covers the land at high latitudes.

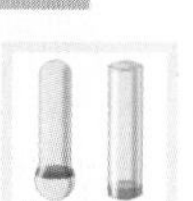

34.18 The global water cycle connects aquatic and terrestrial biomes.

Connecting the Concepts

1. You have seen that Earth's terrestrial biomes reflect regional variations in climate. But what determines these climatic variations? Interpret the following diagrams in reference to global patterns of temperature, rainfall, and winds.
   a. Solar radiation and latitude:

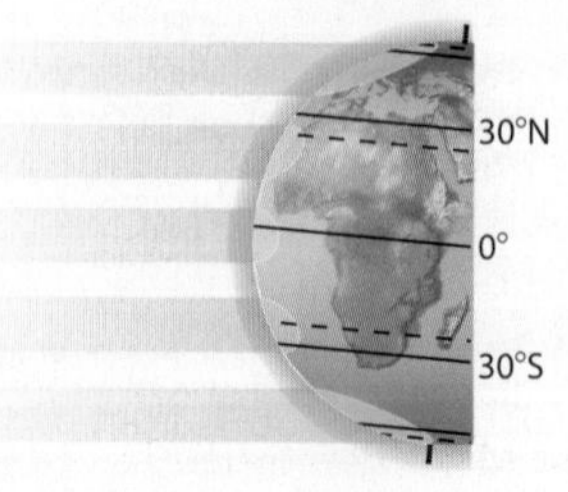

30°N
0°
30°S

   b. Earth's orbit around the sun:

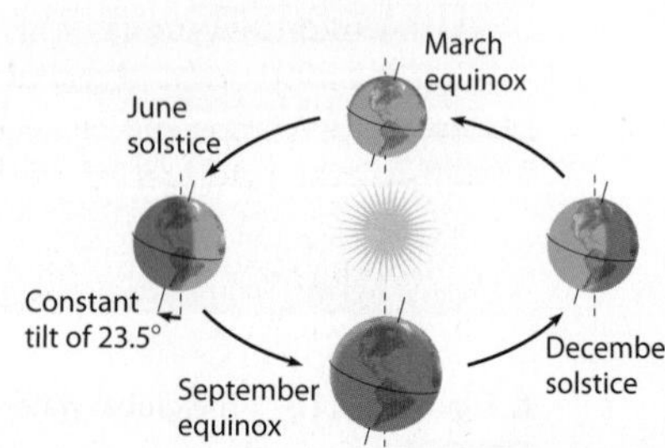

March equinox
June solstice
Constant tilt of 23.5°
September equinox
December solstice

c. Global patterns of air circulation and rainfall:

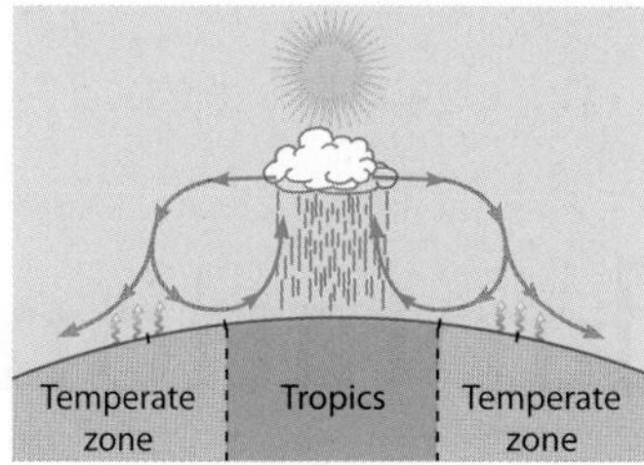

Temperate zone
Tropics
Temperate zone

Testing Your Knowledge

Level 1: Knowledge/Comprehension

Match each description on the left with the correct biome on the right.

2. The most complex and diverse biome	a. chaparral
3. Ground permanently frozen	b. savanna
4. Deciduous trees such as hickory and birch	c. taiga
5. Mediterranean climate	d. temperate broadleaf forest
6. Spruce, fir, pine, and hemlock trees	e. temperate grassland
7. Home of ants, antelopes, and lions	f. tropical rain forest
8. North American plains	g. arctic tundra

9. Changes in the seasons are caused by
   a. the tilt of Earth's axis toward or away from the sun.
   b. annual cycles of temperature and rainfall.
   c. variation in the distance between Earth and the sun.
   d. an annual cycle in the sun's energy output.
10. What makes the Gobi Desert of Asia a desert?
   a. The growing season there is very short.
   b. It is hot.
   c. Temperatures vary little from summer to winter.
   d. It is dry.
11. Which of the following sea creatures might be described as a pelagic animal of the aphotic zone?
   a. a coral reef fish
   b. an intertidal snail
   c. a deep-sea squid
   d. a harbor seal
12. Why do the tropics and the windward side of mountains receive more rainfall than areas around latitudes 30° north and south and the leeward side of mountains?
   a. Rising warm, moist air cools and drops its moisture as rain.
   b. Descending air condenses, creating clouds and rain.
   c. There is more solar radiation in the tropics and on the windward side of mountains.
   d. Earth's rotation creates seasonal differences in rainfall.
13. Phytoplankton are the major photosynthesizers in
   a. the benthic realm of the ocean.
   b. the ocean photic zone.
   c. the intertidal zone.
   d. the aphotic zone of a lake.
14. An ecologist monitoring the number of gorillas in a wildlife refuge over a five-year period is studying ecology at which level?
   a. organism
   b. population
   c. community
   d. ecosystem
15. Many plant species have adaptations for dealing with periodic fires. Such fires are typical of a
   a. chaparral.
   b. savanna.
   c. temperate grassland.
   d. a, b, and c

Level 2: Application/Analysis

16. Tropical rain forests are the most diverse biomes. What factors contribute to this diversity?
17. What biome do you live in? Describe your climate and the factors that have produced that climate. What plants and animals are typical of this biome? If you live in an urban or agricultural area, how have human interventions changed the natural biome?

Level 3: Synthesis/Evaluation

18. Aquatic biomes differ in levels of light, nutrients, oxygen, and water movement. These abiotic factors influence the productivity and diversity of freshwater ecosystems.
   a. Productivity, roughly defined as photosynthetic output, is high in estuaries, coral reefs, and shallow ponds. Describe the abiotic factors that contribute to high productivity in these ecosystems.
   b. How does extra input of nitrogen and phosphorus (for instance, by fertilizer runoff) affect the productivity of lakes and ponds? Is this nutrient input beneficial for the ecosystem? Explain.
19. In the climograph below, biomes are plotted by their range of annual mean temperature and annual mean precipitation. Identify the following biomes: arctic tundra, coniferous forest, desert, grassland, temperate forest, and tropical forest. Explain why there are areas in which biomes overlap on this graph.

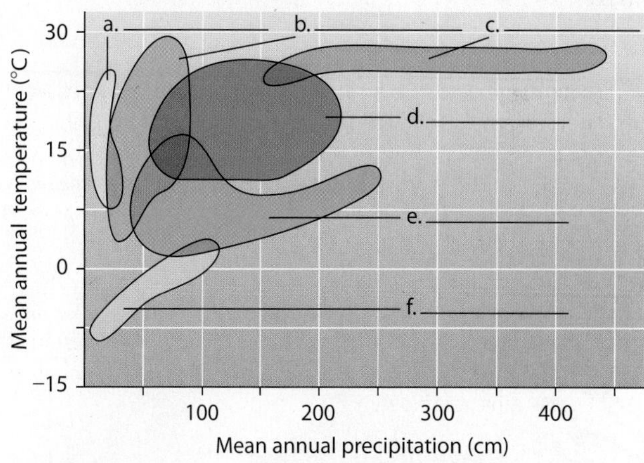

a.
b.
c.
d.
e.
f.
30
15
0
−15
Mean annual temperature (°C)
100
200
300
400
Mean annual precipitation (cm)

20. The North American pronghorn looks and acts like the antelopes of Africa. But the pronghorn is the only survivor of a family of mammals restricted to North America. Propose a hypothesis to explain how these widely separated animals came to be so much alike.
21. **SCIENTIFIC THINKING** In 1954, workers at Michigan State University began spraying the elm trees on campus annually with DDT to kill disease-carrying bark beetles. In the spring of 1955, large numbers of dead robins were found on the campus. Observers thought perhaps the robins died after eating earthworms contaminated by DDT the previous spring. Suggest how scientists could have investigated the scientific validity of this idea.

Answers to all questions can be found in Appendix 4.

CHAPTER 35

Behavioral Adaptations to the Environment

? *What can we learn by watching animals?*

Among the dozens of television reality shows, *Meerkat Manor* was unique. Rather than featuring performers seeking stardom, real housewives, or bachelors and bachelorettes, *Meerkat Manor*, which aired for four seasons on Animal Planet, starred a clan of small desert mammals belonging to the mongoose family (see photo below). Viewers followed the meerkats' exploits as they interacted with each other, defended themselves from rival clans, foraged for food, and faced death from illness and predators. As in all reality shows, the raw footage was edited to define recognizable characters and provide dramatic story lines. Thus, viewers were encouraged to attribute human traits and motivations to the animals, a decidedly nonscientific approach to the study of animal behavior.

The meerkats of *Meerkat Manor* were also studied by scientists as part of a long-term project on these highly social animals in the Kalahari Desert of South Africa. Researchers used observations of

meerkats to test hypotheses about the evolution of cooperative behavior, sexual selection, altruism, and other components of social behavior. Because humans are also social animals, many of us are intrigued by these aspects of animal behavior. As you will learn in this chapter, however, the study of animal behavior extends far beyond the antics of cuddly looking mammals. Birds, fish, and even invertebrates such as insects and molluscs display fascinating behavioral adaptations. And not all behaviors are visible. For example, animal communication may employ sounds or scents. You'll also learn about several categories of behavior (including learning!) and explore the roles of genetics and the environment in determining behavior.

BIG IDEAS

The Scientific Study of Behavior
(35.1–35.3)

Behavioral ecologists study behavior in an evolutionary context, asking both *how* behaviors are triggered and why behaviors occur.

Learning
(35.4–35.11)

Learning, which encompasses a broad range of behaviors, enables animals to change their behaviors in response to changing environmental conditions.

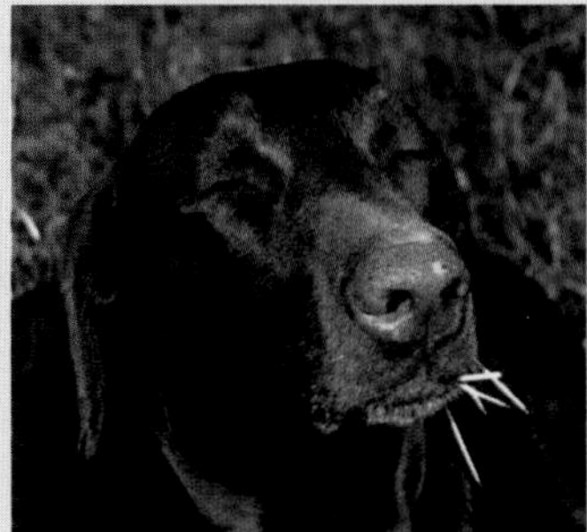

Survival and Reproductive Success
(35.12–35.16)

Many animal behaviors are adaptations that improve the ability to obtain food or increase reproductive success.

Social Behavior and Sociobiology
(35.17–35.23)

The behaviors of animals that live in groups may include territoriality, conflict resolution, dominance hierarchies, and altruism.

The Scientific Study of Behavior

35.1 Behavioral ecologists ask both proximate and ultimate questions

Behavior encompasses a wide range of activities. At its most basic level, a behavior is an action carried out by muscles or glands under the control of the nervous system in response to an environmental cue. Collectively, behavior is the sum of an animal's responses to internal and external environmental cues. Although we commonly think of behavior in terms of observable actions—for instance, a courtship dance or an aggressive posture—other activities are also considered behaviors. Chemical communication, such as secreting a chemical that attracts mates or marks a territory, is a form of behavior. Learning is also a behavioral process.

Behavioral ecology is the study of behavior in an evolutionary context. Behavioral ecologists draw on the knowledge of a variety of disciplines to describe the details of animal behaviors and investigate how they develop, evolve, and contribute to the animal's survival and reproductive success. Let's consider how behavioral ecologists investigate the behavior of the small, mouse-like prairie vole (**Figure 35.1**), one of the very few mammals that form a monogamous bond with their sexual partner. A pair of prairie voles begins their relationship by mating repeatedly for 24 to 48 hours. After this honeymoon period, the pair associate closely and exclusively with each other throughout their lives.

The questions investigated by behavioral ecologists fall into two broad categories. **Proximate questions** concern the immediate reason for a behavior—how it is triggered by **stimuli** (environmental cues that cause a response), what physiological or anatomical mechanisms play a role, and what underlying genetic factors are at work. For example, researchers studying the mating pattern of prairie voles might ask, "How do voles choose their mates?" or "How does the act of mating cause voles to form lifelong bonds with their partners?" Generally, proximate questions help us understand how a behavior occurs. **Proximate causes** are the answers to such questions about the immediate mechanism for a behavior.

▲ **Figure 35.1** Prairie vole (*Microtus ochrogaster*)

Ultimate questions address why a particular behavior occurs. As a component of the animal's phenotype, behaviors are adaptations that have been shaped by natural selection. The answers to ultimate questions, or **ultimate causes**, are evolutionary explanations—they lie in the adaptive value of the behavior. For example, researchers think that at some point in their evolutionary history, the ancestors of prairie voles had numerous sexual partners, like the overwhelming majority of mammals. Why did natural selection favor the change in mating behavior? To explore this ultimate question, researchers test hypotheses on the adaptive value of prairie voles' bonding behavior. For example, perhaps male prairie voles that form lasting bonds have greater reproductive success because they prevent other males from getting close to their mates. An experiment to determine whether a female vole would have sexual intercourse with other males if she were not constantly accompanied by her mate would shed light on one possible adaptive value of bond formation by males.

In the next module, we look at a type of behavior that demonstrates the complementary nature of proximate and ultimate questions.

? **When you touch a hot plate, your arm automatically recoils. What might be the proximate and ultimate causes of this behavior?**

● The proximate cause is a simple reflex, a neural pathway linking stimulation of receptors in your finger to a motor response by muscles of your arm and hand; the ultimate cause is natural selection for a behavior that minimizes damage to the body, thereby contributing to survival and reproductive success.

35.2 Fixed action patterns are innate behaviors

One important proximate question is how a behavior develops during an animal's life span. Konrad Lorenz and Niko Tinbergen, pioneers in the scientific study of behavior, were among the first to demonstrate the importance of **innate behavior**, behavior that is under strong genetic control and is performed in virtually the same way by all individuals of a species. Many of Lorenz's and Tinbergen's studies were concerned with behavioral sequences called **fixed action patterns (FAPs)**. A FAP is an unchangeable series of actions triggered by a specific stimulus. Once initiated, the sequence is performed in its entirety, regardless of any changes in circumstances. Consider a coffee vending machine as an analogy. The purchaser feeds money into the machine and presses a button. Having received this stimulus, the machine performs a series of actions: drops a cup into place; releases a specific volume of coffee; adds cream; adds sugar. Once the stimulus—in this case, the money—triggers the mechanism, it carries out its complete program. Likewise, FAPs are behavioral routines that are completed in full.

Figure 35.2A (top of facing page) illustrates one of the FAPs that Lorenz and Tinbergen studied in detail. The bird is the

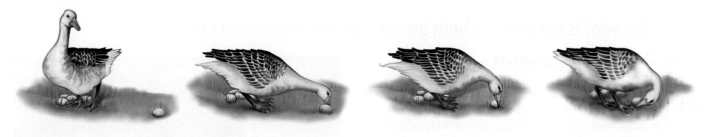

▲ **Figure 35.2A** A graylag goose retrieving an egg—a FAP

graylag goose, a common European species that nests in shallow depressions on the ground. If the goose happens to bump one of her eggs out of the nest, she always retrieves it in the same manner. As shown in the figure, she stands up, extends her neck, uses a side-to-side head motion to nudge the egg back with her beak, and then sits down on the nest again. If the egg slips away (or is pulled away by an experimenter) while the goose is retrieving it, she continues as though the egg were still there. Only after she sits back down on her eggs does she seem to notice that the egg is still outside the nest. Then she begins another retrieval sequence. If the egg is again pulled away, the goose repeats the retrieval motion without the egg. A goose would even perform the sequence when Lorenz and Tinbergen placed a foreign object, such as a small toy or a ball, near her nest. The goose performs the series of actions regardless of the absence of an egg, just as a coffee vending machine does when the cup dispenser is empty—the coffee, cream, and sugar are poured anyway.

In its simplest form, a FAP is an innate response to a certain stimulus. For the graylag goose, the stimulus for egg retrieval is the presence of an egg (or other object) near the nest. Such relatively simple, innate behaviors seem to occur in all animals. When a baby bird senses that an adult bird is near, it responds with a FAP: It begs for food by raising its head, opening its mouth, and cheeping. In turn, the parent responds with another FAP: It stuffs food into the gaping mouth. Humans perform FAPs, too. Infants grasp strongly with their fingers in response to a touch stimulus on the palm of the hand. They smile in response to a face or even something that vaguely resembles a face, such as two dark spots in a circle.

Although a single FAP is typically a simple behavior, complex behaviors can result from several FAPs performed sequentially. Many vertebrates engage in courtship rituals that consist of chains of FAPs, as you'll learn in Module 35.14. The completion of a single FAP by one partner cues the other partner to begin its next FAP. There is some flexibility in these patterns. For example, a segment of the pattern might be repeated if the partner does not readily respond.

What might be the ultimate causes of FAPs? Automatically performing certain behaviors may maximize fitness to the point that genes that result in variants of that behavior do not persist in the population. For example, there are some things that a young animal has to get right on the first try if it is to stay alive. Consider kittiwakes, gulls that nest on cliff ledges. Unlike other gull species, kittiwakes show an innate aversion to cliff edges; they turn away from the edge. Chicks in earlier generations that did not show this edge-aversion response would not have lived to pass the genes for their risk-taking behavior on to the next generation.

Fixed action patterns for reproductive behaviors are also under strong selection pressure. One example is the behavior of mated king penguins. Each member of the pair takes a turn incubating their egg while its mate feeds (**Figure 35.2B**). Standing face-to-face, the pair must execute a delicate series of maneuvers to pass the egg from the tops of one penguin's feet to the tops of its partner's feet, where the egg will incubate in a snug fold of the abdominal skin. (You may have seen emperor penguins engage in this behavior in the film *The March of the Penguins*.) If either partner makes a mistake, the egg may roll onto the ice, where it can freeze in seconds, eliminating the pair's only chance of successful reproduction for the year.

▲ **Figure 35.2B** A pair of king penguins transferring their egg

Innate behaviors are under strong genetic control, but the animal's performance of most innate behaviors also improves with experience. And despite the genetic component, input from the environment—an object or sensory stimulus, for example—is required to trigger the behavior. In the next module, we look at the interaction of genes and environment in producing a behavior.

? **How would you explain FAPs in the context of proximate and ultimate causes?**

The proximate cause of a FAP is often a simple environmental cue. The ultimate cause is that natural selection favors behaviors that enable animals to perform tasks essential to survival without any previous experience.

35.3 Behavior is the result of both genetic and environmental factors

Evolutionary explanations for behavior assume that it has a genetic basis. Many scientific studies have corroborated that specific behaviors do indeed have a genetic component. Until recently, though, the heritability of a trait could only be estimated by traditional methods such as constructing pedigrees and performing breeding experiments. With the tools of molecular genetics, scientists have begun to investigate the roles of specific genes in behavior.

Groundbreaking experiments with fruit flies have led to the discovery of genes that govern learning, memory, internal clocks, and courtship and mating behaviors. For example, a male fruit fly courts a female with an elaborate series of actions that include tapping the female's abdomen with a foreleg and vibrating one of his wings in a courtship "song." Researchers identified a master gene known as *fruitless* (*fru* for short), which in males encodes a protein that switches on a suite of genes responsible for courtship behavior (**Table 35.3**). Male fruit flies that possess a mutated version of *fru* attempt to court other male flies. In females, the protein encoded by *fru* is different from the male version of the protein. But when researchers used genetic engineering to produce female flies that made the male *fru* protein, the females behaved like normal males, vigorously courting other females.

Some of the genes governing fruit fly behavior have counterparts in mice and even in humans. The genetic underpinnings of courtship and mating behaviors in mammals are probably quite different from those of flies, but research on genes implicated in learning and memory has yielded promising results. By studying fruit fly mutants with colorful names like dunce, amnesiac, and rutabaga, scientists have identified key components of memory storage and have even used genetic engineering to produce fruit flies that have exceptionally good memories. Similar genes and their protein products have been identified in mice and humans, sparking a flood of research on memory-enhancing drugs. Such drugs could improve the quality of life for people suffering from Alzheimer's and other neurological diseases that impair memory.

Phenotype depends on the environment as well as genes. Many environmental factors, including diet and social interactions, can modify how genetic instructions are carried out. In some animals, even an individual's sex can be determined by the environment. For example, sex determination in some reptiles depends on the temperature of the egg during embryonic development (**Figure 35.3A**).

TABLE 35.3 COURTSHIP BEHAVIOR OF FRUIT FLIES WITH NORMAL OR MUTATED *FRUITLESS* (*FRU*) GENE

	Male	Genetically Altered Female
Normal male *fru*	Courts females	Courts females
Mutated male *fru*	Courts males	–––

▲ **Figure 35.3A** Hatchling leatherback sea turtles (*Dermochelys coriacea*)

Let's look at a study that illustrates the influence of environment on behavior (**Figure 35.3B** on the facing page). Some female Norway rats (*Rattus norvegicus*) spend a great deal of time licking and grooming their offspring (called pups), while others have little interaction with their pups. Pups raised by these "low-interaction" mothers tend to be more sensitive as adults to stimuli that trigger the "fight-or-flight" stress response (see Module 26.10) and thus are more fearful and anxious in new situations. Female pups from these litters become low-interaction mothers themselves. On the other hand, pups with "high-interaction" mothers are more relaxed in stressful situations as adults, and the female pups from these litters become high-interaction mothers.

To investigate whether Norway rats' responses to stress are entirely determined by genetics or are influenced by the interactions with their mothers, researchers performed "cross-fostering" experiments. They placed pups born to high-interaction mothers in nests with low-interaction mothers and pups born to low-interaction mothers in nests with high-interaction mothers. As adults, the cross-fostered rats responded to stress more like their foster mothers than like their biological mothers. The results showed that the pups' environment—in this case maternal behavior—was the determining factor in their anxiety level. Remarkably, these experiments also demonstrated that behavioral changes can be passed to future generations, not through genes, but through the social environment. When the female rats that had been fostered gave birth, they showed the same degree of interaction with their pups as their foster mothers had shown with them. In further studies, researchers learned that interaction with the mother changes the pattern of gene expression in the pups, thus affecting the development of parts of the neuroendocrine system that regulate the fight-or-flight response.

These experiments and others provide evidence that behavior is the product of both genetic *and* environmental factors. Indeed, the interaction of genes and the environment appears to determine most animal behaviors. One of the most powerful ways that the environment can influence behavior is through learning, the topic we consider next.

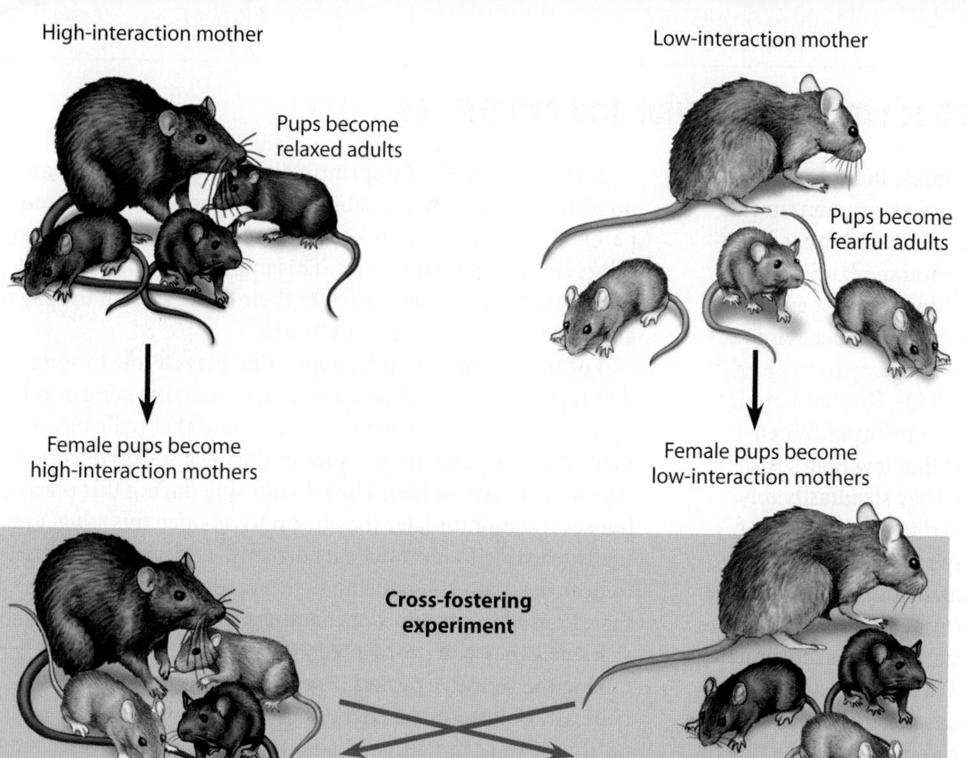
High-interaction mother
Pups become relaxed adults
Female pups become high-interaction mothers
Low-interaction mother
Pups become fearful adults
Female pups become low-interaction mothers
Cross-fostering experiment
Pups become relaxed adults
Pups become fearful adults

◀ **Figure 35.3B** A cross-fostering experiment with rats

Try This When you review this module, explain the design and outcome of this experiment without referring to the text.

? **Without doing any experiments, how might you distinguish between a behavior that is mostly controlled by genes and one that is mostly determined by the environment?**

● In many cases, genetically controlled behavior would not differ much between populations, regardless of the environment, while the environmentally controlled behavior would differ widely across populations located in different environments.

▷ Learning

35.4 Habituation is a simple type of learning

Learning is modification of behavior as a result of specific experiences. Learning enables animals to change their behaviors in response to changing environmental conditions. As **Table 35.4** indicates, there are various forms of learning, ranging from a simple behavioral change in response to a single stimulus to complex problem solving that uses entirely new behaviors.

TABLE 35.4 | TYPES OF LEARNING

Learning Type	Defining Characteristic
Habituation	Loss of response to a stimulus after repeated exposure
Imprinting	Learning that is irreversible and limited to a sensitive time period in an animal's life
Spatial learning	Use of landmarks to learn the spatial structure of the environment
Associative learning	Behavioral change based on linking a stimulus or behavior with a reward or punishment; includes trial-and-error learning
Social learning	Learning by observing and mimicking others
Problem solving	Inventive behavior that arises in response to a new situation

One of the simplest forms of learning is **habituation**, in which an animal learns not to respond to a repeated stimulus that conveys little or no information. There are many examples of habituation in both invertebrate and vertebrate animals. The cnidarian *Hydra* (see Figure 18.6A), for example, contracts when disturbed by a slight touch; it stops responding, however, if disturbed repeatedly by such a stimulus. Similarly, a scarecrow stimulus will usually make birds avoid a tree with ripe fruit for a few days. But the birds soon become habituated to the scarecrow and may even land on it on their way to the fruit tree. Once habituated to a particular stimulus, an animal still senses the stimulus—its sensory organs detect it—but the animal has learned not to respond to it.

In terms of ultimate causation, habituation may increase fitness by allowing an animal's nervous system to focus on stimuli that signal food, mates, or real danger and not waste time or energy on a vast number of other stimuli that are irrelevant to survival and reproduction.

Proximate and ultimate causes of behavior are also evident in imprinting, the type of learning we introduce in the next module.

? **Your new roommate hums continuously while studying. You found this habit extremely annoying at first, but after a while you stopped noticing it. What kind of learning accounts for your tolerance of your roommate's humming?**

● Habituation

35.5 Imprinting requires both innate behavior and experience

Learning often interacts closely with innate behavior. Some of the most interesting examples of such an interaction involve the phenomenon known as imprinting. **Imprinting** is learning that is limited to a specific time period in an animal's life and that is generally irreversible. The limited phase in an animal's development when it can learn certain behaviors is called the **sensitive period**.

In classic experiments done in the 1930s, Konrad Lorenz used the graylag goose to demonstrate imprinting. When incubator-hatched goslings spent their first few hours with Lorenz, rather than with their mother, they steadfastly followed Lorenz (**Figure 35.5A**) and showed no recognition of their mother or other adults of their species. Even as adults, the birds continued to prefer the company of Lorenz and other humans to that of geese. Lorenz determined that the most important imprinting stimulus for graylag geese was movement of an object (normally the parent bird) away from the hatchlings. The effect of movement was increased if the moving object emitted some sound. The sound did not have to be that of a goose, however; Lorenz found that a box with a ticking clock in it was readily and permanently accepted as a "mother."

Just as a young bird requires imprinting to know its parents, the adults must also imprint to recognize their young. For a day or two after their own chicks hatch, adult herring gulls will accept and even defend a strange chick introduced into their nesting territory. However, once imprinted on their offspring, adults will kill any strange chicks that arrive later.

▲ **Figure 35.5A** Konrad Lorenz with geese imprinted on him

Not all examples of imprinting involve parent-offspring bonding. Newly hatched salmon, for instance, do not receive parental care but seem to imprint on the complex mixture of odors unique to their stream. This imprinting enables adult salmon to find their way back to their home stream to spawn after spending a year or more at sea.

For many kinds of birds, imprinting plays a role in song development. For example, researchers studying song development in white-crowned sparrows found that male birds memorize the song of their species during a sensitive period (the first 50 days of life). They do not sing during this phase, but several months later they begin to practice this song, eventually learning to reproduce it correctly. The birds do not need to hear the adult song during their practice phase; isolated males raised in soundproof chambers learned to sing normally as long as they had heard a recorded song of their species during the sensitive period (**Figure 35.5B**, top sonogram). In contrast, isolated males that did not hear their species' song until they were more than 50 days old sang an abnormal song (Figure 35.5B, bottom). Researchers also discovered a purely genetic component of white-crowned sparrow song development: Isolated males exposed to recorded songs of other species during the sensitive period did not adopt these foreign songs. When they later learned to sing, these birds sang an abnormal song similar to that of the isolated males of their own species that had heard no recorded bird songs.

The ability of parents and offspring to keep track of each other and the ability of male songbirds to attract mates are examples of behaviors that have direct and immediate effects on survival and reproduction. Imprinting provides a way for such behavior to become more or less fixed in an animal's nervous system.

? **Explain why we say that imprinting has both innate and learned components.**

● Its innate component is the tendency to imprint on a stimulus during a sensitive period. The imprinting itself is a form of learning.

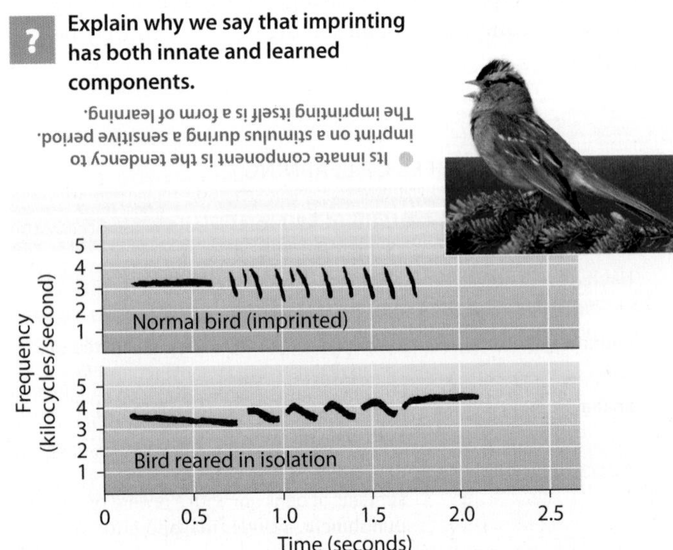

Frequency (kilocycles/second)
5
4
3
2
1
Normal bird (imprinted)
Bird reared in isolation
0
0.5
1.0
1.5
2.0
2.5
Time (seconds)

Adapted from P. Marler and M. Tamura, Culturally Transmitted Patterns of Vocal Behavior in Sparrows, *Science* 146: 3650, 1483–1486 (Dec. 11, 1964).

▲ **Figure 35.5B** The songs of male white-crowned sparrows that heard an adult sing during the sensitive period (top) and after the sensitive period (bottom)

35.6 Imprinting poses problems and opportunities for conservation programs

CONNECTION

In attempting to save species that are at the edge of extinction, biologists sometimes try to increase their numbers in captivity. Generally, the strategy of a captive breeding program is to provide a safe environment for infants and juveniles, the stages at which many animals are most vulnerable to predation and other risks. In some programs, adult animals are caught and kept in conditions that are conducive to breeding. Offspring are usually raised by the parents and may be kept for breeding or released back to the wild. In other programs, parents are absent, as when eggs are removed from a nest. Artificial incubation is often successful, but without parents available as models for imprinting, the offspring may not learn appropriate behaviors. The effort to save the whooping crane is one example of a program that has successfully used surrogate parents, though they are a bit unusual.

The whooping crane (*Grus americana*) is a migratory waterfowl that reaches a height of about 1.5 m and has a white body with black-tipped wings that spread out over 2 m. Its name comes from its distinctive call. Once common in North American skies during their north-south migrations, whooping cranes were almost killed off by habitat loss and hunting. By the 1940s, only 16 wild birds returned from their summer breeding ground in Canada to one of their wintering areas on the Texas coast. Because a whooping crane doesn't reach sexual maturity until it is 4 years old, new generations of birds are not produced quickly. And although whooping cranes often lay two eggs, parents usually successfully rear only one chick. Protections for whooping cranes were established, and in 1967, U.S. wildlife officials launched long-term recovery and captive breeding efforts.

At first, biologists used sandhill cranes as surrogate parents for whooping crane chicks. All went well until the whooping cranes reached maturity. Having imprinted on the sandhill cranes, they showed no interest in breeding with their own kind. Biologists realized that they needed another approach. With plans to set up a separate breeding colony of whooping cranes in Wisconsin, they turned to Operation Migration to help get the birds to a winter nesting site in Florida. This bird advocacy group had developed ways to hatch geese and sandhill cranes, teach the birds to recognize a small, lightweight plane as a parent figure, and train them to follow the plane along migratory routes.

In 2001, Operation Migration applied its techniques to whooping cranes. Incubating crane eggs were serenaded with recorded sounds of the plane's engine. When the chicks emerged from their shells, the first thing they saw was a hand puppet, shaped and painted in the form of an adult whooping crane. As the chicks grew, the same type of puppet guided them through exercise and training (**Figure 35.6A**). But now the puppet was attached to a plane that rolled along the ground, coaxing the chicks to follow it. To make sure the birds bonded with the puppet and plane and not humans, pilots and other members of Operation Migration were cloaked in hooded suits. Eventually, the birds started following the plane on short flights (**Figure 35.6B**).

October 2001 brought the real test. Would the young whooping cranes follow the plane from their protected grounds in Wisconsin along a migratory route to Florida? The trip took 48 days but ultimately proved successful. Each flight day, the young cranes lined up eagerly behind the plane, wings raised, ready to follow their "parent" to the next stop. And the next spring, five of the eight young cranes retraced the route to Wisconsin on their own. Operation Migration has since taught several more generations of whooping cranes to migrate, boosting the species' chances of survival. At the end of 2011, there were about 437 whooping cranes in the wild.

? What features of whooping cranes have made their recovery difficult?

● They do not breed until 4 years of age and usually raise only one chick each breeding season. As migratory birds, they require resources and protection in two habitats.

▲ **Figure 35.6A** A whooping crane chick interacting with a puppet "parent"

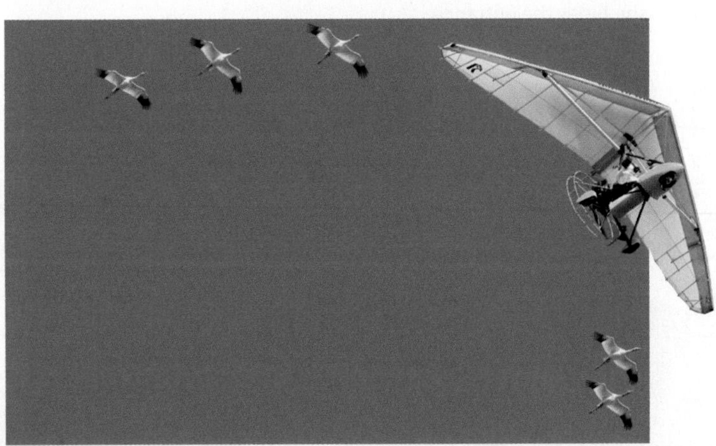

▲ **Figure 35.6B** Whooping cranes following a surrogate parent

VISUALIZING THE CONCEPT

35.7 Animal movement may be a response to stimuli or require spatial learning

Moving in a directed way enables animals to avoid predators, migrate to a more favorable environment, obtain food, and find mates and nest sites. The simplest kinds of movement do not involve learning. A random movement in response to a stimulus is called a **kinesis** (plural, kineses). A kinesis may be merely starting or stopping, changing speed, or turning more or less frequently. In contrast to kinesis, a **taxis** (plural, taxes) is a response directed toward (positive taxis) or away from (negative taxis) a stimulus. In **spatial learning**, animals establish memories of landmarks in their environment that indicate the locations of food, nest sites, prospective mates, and potential hazards. And as you'll learn in Module 35.8, some animals use more sophisticated methods of navigating their environment.

KINESIS IN SOW BUGS

Sow bugs, which are the only terrestrial crustaceans, are not as well protected from drying out as their insect cousins. Consequently, they typically live in moist habitats, such as the underside of a rock or log.

A sow bug, also known as a roly-poly or wood louse

In a dry area, sow bugs exhibit kinesis, becoming more active and moving about randomly.

The more the sow bugs move, the greater the chance of finding a moist area.

Once the sow bugs are in a more favorable environment, their decreased activity tends to keep them there.

POSITIVE TAXIS IN SALMON

Many stream fish, such as the salmon shown here, exhibit positive taxis in the current; they automatically swim or orient in an upstream direction toward the direction from which food is likely to come.

A male sockeye salmon in breeding colors

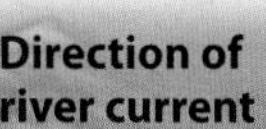

Direction of river current

Orientation of salmon

The orientation toward the current keeps the salmon from being swept downstream.

SPATIAL LEARNING IN DIGGER WASPS

The female digger wasp builds its nest in a small burrow in the ground. She will often excavate four or five separate nests and fly to each one daily, cleaning them and bringing food to the larvae. Before leaving each nest, she hides the entrance.

A digger wasp near the entrance of its nest

Hypothesis: This figure shows a classic experiment by Niko Tinbergen to test the hypothesis that the female digger wasp uses spatial learning to keep track of her nests.

Tinbergen located a digger wasp nest.

Digger wasp

Nest

Before the mother wasp returned, Tinbergen placed a circle of pinecones around the opening.

After the wasp flew away, Tinbergen moved the pinecones a few feet to one side of the nest opening.

Nest

No nest

When the wasp returned, she flew to the center of the pinecone circle instead of the actual nest opening.

Conclusion: The wasp does exhibit spatial learning. She could learn new landmarks to keep track of her nests.

? Planarians (see Figure 18.7A) move directly away from light into dark places. What type of movement is this?

Negative taxis (directed movement away from light)

35.8 A variety of cues guide migratory movements

Like "snowbirds"—people from the colder regions of North America who spend their winters in the Sun Belt—many animals also move seasonally. **Migration** is the regular back-and-forth movement of animals between two geographic areas. Migration enables many species, such as the whooping cranes discussed in Module 35.6, to access food resources throughout the year and to breed or winter in areas that favor survival. Researchers have found that migrating animals stay on course by using a variety of cues. Let's look at some examples of how animals navigate these journeys.

A notable long-distance traveler is the gray whale. During the summer, these giant mammals feast on small, bottom-dwelling invertebrates that abound in northern oceans. In the autumn, they leave their feeding grounds north of Alaska and begin the long trip south along the North American coastline to winter in the warm lagoons of Baja California (Mexico). Females give birth there before migrating back north with their young. The yearly round-trip, some 20,000 km, is the longest made by any mammal.

Gray whales seem to use the coastline to pilot their way north and south. Whale watchers sometimes see gray whales stick their heads straight up out of the water, a behavior known as "spyhopping" **(Figure 35.8A)**, perhaps to obtain a visual reference point on land. Gray whales may also use the topography of the ocean floor, cues from the temperature and chemistry of the water, and magnetic sensing to guide their journey.

▲ **Figure 35.8A** Gray whale (*Eschrichtius robustus*) spyhopping

Magnetic cues feature prominently in sea turtle migrations. For example, loggerhead sea turtles hatching on beaches from North Carolina to Florida enter the Atlantic Ocean to begin an incredible journey through thousands of miles of open ocean before returning years later to the North American coast **(Figure 35.8B**; see Chapter 29 introduction). They accomplish this feat by detecting variations in Earth's magnetic field that provide information on both latitude (north-south position) and longitude (east-west position).

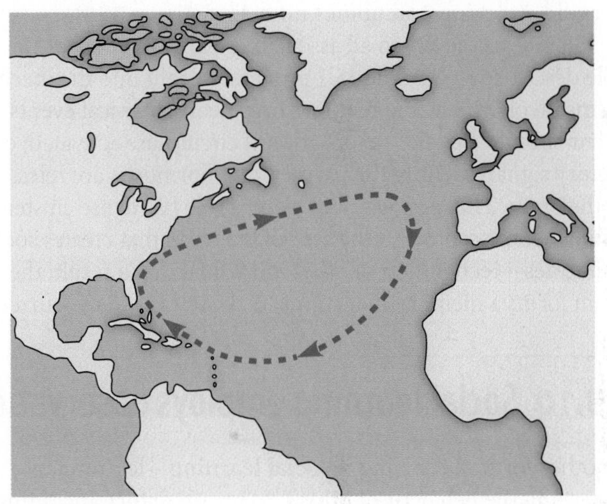

▲ **Figure 35.8B** Map of the migratory route of loggerhead turtles (*Caretta caretta*)

Many birds migrate at night, navigating by the stars the way early sailors did. Navigating by the sun or stars requires an internal timing device to compensate for the continuous daily movement of celestial objects. Consider what would happen if you started walking one day, orienting yourself by keeping the sun on your left. In the morning, you would be heading south, but by evening you would be heading back north, having made a circle and gotten nowhere. A calibration mechanism must also allow for the apparent change in position of celestial objects as the animal moves over its migration route. In a series of classic experiments, researchers showed that one night-migrating bird, the indigo bunting **(Figure 35.8C)**, avoids the need for a timing mechanism by fixing on the North Star, the one bright star in northern skies that appears almost stationary.

Birds that migrate during the daytime navigate by a combination of magnetic cues and visual cues such as landmarks and the position of the sun. Researchers hypothesize that polarized light patterns are used as a calibration mechanism.

Some animals appear to migrate using only innate responses to environmental cues. For example, each fall, a new generation of monarch butterflies flies about 4,000 km over a route they've never flown before to specific wintering sites. Studies of other animals, including some songbirds, show the interaction of genes and experience in migration. Research efforts continue to reveal the complex mechanisms by which animals traverse Earth.

▲ **Figure 35.8C** A male indigo bunting (*Passerina cyanea*)

Why is a timekeeping mechanism essential for navigating by the stars?

● Because the positions of the stars change with time of night and season

35.9 Animals may learn to associate a stimulus or behavior with a response

Associative learning is the ability to associate one environmental feature with another. In one type of associative learning, an animal learns to link a particular stimulus to a particular outcome. If you keep a pet, you have probably observed this type of associative learning firsthand. A dog or cat will learn to associate a particular sound, word, or gesture (stimulus) with a specific punishment or reward (outcome). For example, the sound of a can being opened may bring a cat running for food.

In the other type of associative learning, called **trial-and-error learning**, an animal learns to associate one of its own behaviors with a positive or negative effect. The animal then tends to repeat the response if it is rewarded or avoid the response if it is harmed. For example, predators quickly learn to associate certain kinds of prey with painful experiences. A porcupine's sharp quills and ability to roll into a quill-covered ball are strong deterrents against many predators. Coyotes, mountain lions, and domestic dogs often learn the hard way to avoid attacking porcupines nose-first **(Figure 35.9)**.

▲ **Figure 35.9** Trial-and-error learning by a dog

Memory is the key to all associative learning. The prairie vole described in Module 35.1 provides insight into the mechanisms involved. During mating, three neurochemical events occur simultaneously: Neural reward circuits are activated; olfactory signals identify the partner; and hormones are released in the brain. The monogamous prairie vole has dense clusters of hormone receptors in the area of the brain that creates social memories—recognition of other individuals. As a result, the brain forms a memory that connects the reward to the current partner's scent. Other species of vole are promiscuous—they mate with multiple partners and form no lasting bonds. Promiscuous voles experience the same biochemical reward during the act of mating, but have very few hormone receptors in the area of the brain that forms social memories. Thus, the brain fails to forge a link between the partner and the reward.

? **How might the fact that many bad-tasting or stinging insect species have similar color patterns benefit both the insects and the animals that may prey on them?**

● Potential predators associate insects displaying that coloration with a negative effect, and consequently, all these insects are less likely to be preyed on.

35.10 Social learning employs observation and imitation of others

Another form of learning is **social learning**—learning by observing the behavior of others. Many predators, including cats, coyotes, and wolves, seem to learn some of their basic hunting tactics by observing and imitating their mothers.

Studies of the alarm calls of vervet monkeys in Amboseli National Park, in Kenya, provide an interesting example of how performance of a behavior can improve through social learning. Vervet monkeys (*Cercopithecus aethiops*) are about the size of a domestic cat. They give distinct alarm calls when they see leopards, eagles, or snakes, all of which prey on vervets. When a vervet sees a leopard, it gives a loud barking sound; when it sees an eagle, it gives a short two-syllable cough; and the snake alarm call is a "chutter." Upon hearing a particular alarm call, other vervets in the group behave in an appropriate way: They run up a tree on hearing the alarm for a leopard (vervets are nimbler than leopards in trees); look up on hearing the alarm for an eagle; and look down on hearing the alarm for a snake **(Figure 35.10)**.

Infant vervet monkeys give alarm calls, but in a relatively undiscriminating way. For example, they give the "eagle" alarm on seeing any bird, including harmless birds such as bee-eaters. With age, the monkeys improve their accuracy. In fact, adult vervet monkeys give the eagle alarm only on seeing an eagle belonging to either of the two species that eat vervets.

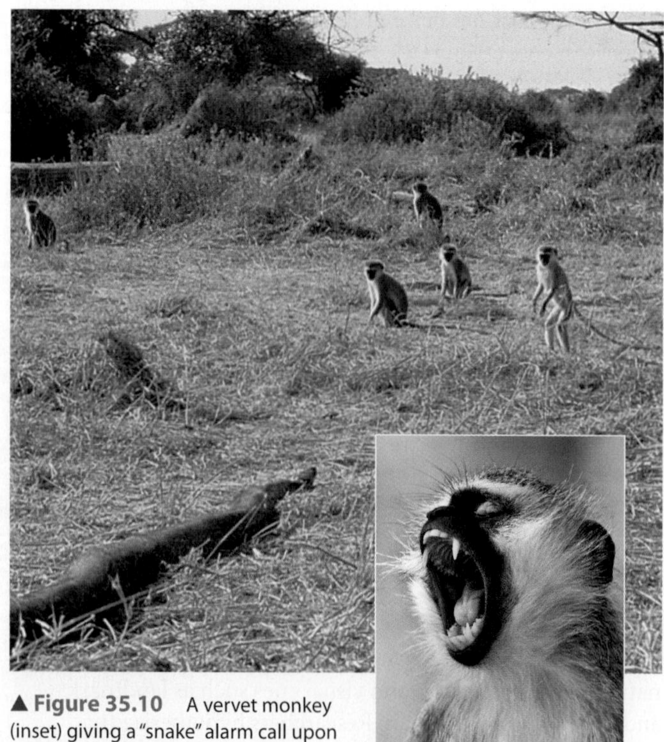

▲ **Figure 35.10** A vervet monkey (inset) giving a "snake" alarm call upon sighting a python (foreground)

Infants probably learn how to give the right call by observing other members of the group and receiving social confirmation. For instance, if the infant gives the call on the right occasion—an eagle alarm when there is an eagle overhead—another member of the group will also give the eagle call. But if the infant gives the call when a bee-eater flies by, the adults in the group are silent. Thus, vervet monkeys have an initial, unlearned tendency to give calls on seeing potential threats. Learning fine-tunes the calls so that by adulthood, vervets give calls only in response to genuine danger and are prepared to fine-tune the alarm calls of the next generation. However, neither vervets nor any other species come close to matching the social learning and cultural transmission that occur among humans, a topic we'll explore later in the chapter.

? What type of learning in humans is exemplified by identification with a role model?

Social learning (observation and imitation)

35.11 Problem-solving behavior relies on cognition

A broad definition of **cognition** is the process carried out by an animal's nervous system to perceive, store, integrate, and use information gathered by the senses. One area of research in the study of animal cognition is how an animal's brain represents physical objects in the environment. For instance, some researchers have discovered that many animals, including insects, are capable of categorizing objects in their environment according to concepts such as "same" and "different." One research team has trained honeybees to match colors and black-and-white patterns. Other researchers have developed innovative experiments, in the tradition of Lorenz and Tinbergen, for demonstrating pattern recognition in birds called nuthatches. These studies suggest that nuthatches apply simple geometric rules to locate their many seed caches.

Some animals have complex cognitive abilities that include **problem solving**—the process of applying past experience to overcome obstacles in novel situations. Problem-solving behavior is highly developed in some mammals, especially dolphins and primates. If a chimpanzee is placed in a room with a banana hung high above its head and several boxes on the floor, the chimp will "size up" the situation and then stack the boxes in order to reach the food. One way many animals learn to solve problems is by observing the behavior of other individuals. For example, young chimpanzees can learn from watching their elders how to crack oil palm nuts by using two stones as a hammer and anvil (**Figure 35.11A**).

Problem-solving behavior has also been observed in some bird species. For example, researchers placed ravens in situations in which they had to obtain food hanging from a string. Interestingly, the researchers observed a great deal of variation in the ravens' solutions. The raven in **Figure 35.11B** used one foot to pull up the string incrementally and the other foot to secure the string so the food didn't drop. An excellent test of human cognition and problem-solving behavior is in the construction of experiments that allow us to explore the cognition and problem-solving behavior of other animals!

? Besides problem solving, what other type of learning is illustrated in Figure 35.11A?

Social learning (observation)

▲ **Figure 35.11A** A chimpanzee solving a problem

▲ **Figure 35.11B** A raven solving a problem

▷ Survival and Reproductive Success

35.12 Behavioral ecologists use cost–benefit analysis to study foraging

Because adequate nutrition is essential to an animal's survival and reproductive success, we should expect natural selection to refine behaviors that enhance the efficiency of feeding. Food-obtaining behavior, or **foraging**, includes not only eating, but also any mechanism an animal uses to search for, recognize, and capture food.

Animals forage in a great many ways. Some animals are "generalists," whereas others are "specialists." Crows, for instance, are extreme generalists; they will eat just about anything that is readily available—plant or animal, alive or dead. In sharp contrast, the koala of Australia, an extreme feeding specialist, eats only the leaves of a few species of eucalyptus trees. As a result, it is restricted to certain areas and is extremely vulnerable to habitat loss. Most animals are somewhere in between crows and koalas in the range of their diet. The pronghorn, for example (see Module 34.4), eats a variety of plants, including forbs, grasses, and woody shrubs.

Often, even a generalist will concentrate on a particular item of food when it is readily available. The mechanism that enables an animal to find particular foods efficiently is called a **search image**. If the favored food item becomes scarce, the animal may develop a search image for a different food item. (People often use search images; for example, when you look for something on a kitchen shelf, you probably scan rapidly to find a package of a certain size and color rather than reading all the labels.)

Whenever an animal has food choices, there are trade-offs involved in the selection. The amount of energy required to locate, capture, subdue, and prepare prey for consumption may vary considerably among the items available. Some behavioral ecologists use an approach known as cost–benefit analysis, comparing the positive and negative aspects of the alternative choices, to evaluate the efficiency of foraging behaviors. According to the predictions of **optimal foraging theory**, an animal's feeding behavior should provide maximal energy gain with minimal energy expense and minimal risk of being eaten while foraging. A researcher tested part of this theory by studying insectivorous birds called wagtails (**Figure 35.12A**).

▲ **Figure 35.12A** A wagtail

In England, wagtails are commonly seen in cow pastures foraging for dung flies (**Figure 35.12B**). The researcher collected data on the time required for a wagtail to catch and consume dung flies of different sizes. He then calculated the number of calories the bird gained per second of "handling" time for the different sizes of flies. The smallest flies (5–6 mm in length) were easily handled but yielded few calories, while the caloric value of large flies (8–10 mm) was offset by the energy required to catch and consume them. Thus, an optimal forager would be expected to choose medium-sized (7 mm) flies most often. The researcher tested this prediction by observing wagtails as they foraged for dung flies in a cow pasture. As **Table 35.12** shows, he found that wagtails did select medium-sized flies most often, even though they were not the most abundant size class.

▲ **Figure 35.12B** A dung fly

Predation is one of the most significant potential costs of foraging. Studies have shown that foraging in groups, as done by herds of antelopes, flocks of birds, or schools of fish, reduces the individual's risk of predation. And for some predators, such as wolves and spotted hyenas, hunting in groups improves their success. Thus, group behavior may increase foraging efficiency by both reducing the costs and increasing the benefits of foraging.

? Early humans were hunter-gatherers, but evidence suggests that they obtained more nutrition from gathering than from hunting. How does this finding relate to optimal foraging theory?

● Meat is very nutritious, but hunting poses relatively high costs in effort and risk compared with the gathering of plant products and dead animals.

TABLE 35.12 THE PREY SIZES SELECTED BY WAGTAILS COMPARED WITH THE PREY SIZES AVAILABLE

Prey Size (mm)	Prey Available (% of 460)	Prey Eaten (% of 252)
5	6	2
6	9	13
7	27	49
8	35	26
9	20	8
10	3	2

Data from N. B. Davies, Prey selection and social behavior in wagtails (Aves: Motacillidae), *Journal of Animal Ecology* 46: 37–57 (1977).

35.13 Communication is an essential element of interactions between animals

Interactions between animals depend on some form of signaling between the participating individuals. In behavioral ecology, a **signal** is a stimulus transmitted by one animal to another animal. The sending of, reception of, and response to signals constitute animal **communication**, an essential element of interactions between individuals. In general, the more complex the social organization of a species, the more complex the signaling required to sustain it.

What determines the type of signal animals use to communicate? Most terrestrial mammals are nocturnal (active at night), which makes visual displays relatively ineffective. Many nocturnal mammals use odor and auditory (sound) signals, which work well in the dark. Birds, by contrast, are mostly diurnal (active in daytime) and use visual and auditory signals. Humans are also diurnal and likewise use mainly visual and auditory signals. Therefore, we can detect the bright colors and songs birds use to communicate. If we had the well-developed olfactory abilities of most mammals and could detect the rich world of odor cues, mammal-sniffing might be as popular with us as bird-watching.

What types of signals are effective in aquatic environments? A common visual signal used by territorial fishes is to erect their fins, which is generally enough to drive off intruders. Electrical signals produced by certain fishes communicate hierarchy or status. Fish may also use sound to communicate. For example, the males of some species make noises to attract mates. Marine mammals use sound for courtship, territoriality, and maintaining contact with their group.

Animals often use more than one type of signal simultaneously. **Figure 35.13A** shows a ring-tailed lemur. These tree-dwelling primates of Madagascar live in social groups averaging 15 individuals. Lemurs use visual displays, scent, and vocalizations to communicate with other members of the group. The animal shown here is communicating aggression with its prominent tail. Prior to this display, it smeared its tail with odorous secretions from glands on its forelegs. By waving its scented tail over its head, the lemur transmits both visual and chemical signals.

▲ **Figure 35.13A** A lemur communicating aggression

One of the most amazing examples of animal communication is found in honeybees, which have a complex social organization characterized by division of labor. Adult worker bees leave the hive to forage, bringing back pollen and nectar for the nest. When a forager locates a patch of flowers, she regurgitates some nectar that the others taste and smell, then communicates the location of the food source by performing a "dance."

Beginning with groundbreaking studies by Karl von Frisch in the 1930s, researchers have conducted numerous experiments to decipher the meanings of different honeybee dances. A pattern of movements called the waggle dance communicates the distance and direction of food (**Figure 35.13B**). The dancer runs a half circle, then turns and runs in a straight line back to her starting point, buzzing her wings and waggling her abdomen as she goes. She then runs a half circle in the other direction, followed by another waggling run to the starting point. The length of the straight run and the number of waggles indicate the distance to the food source. The angle of the straight run relative to the vertical surface of the hive is the same as the horizontal angle of the food in relation to the sun (30° in Figure 35.13B). Once the other workers have learned the location of the food source, they leave the hive to forage there.

? **What types of signals do honeybees use?**

Visual and chemical

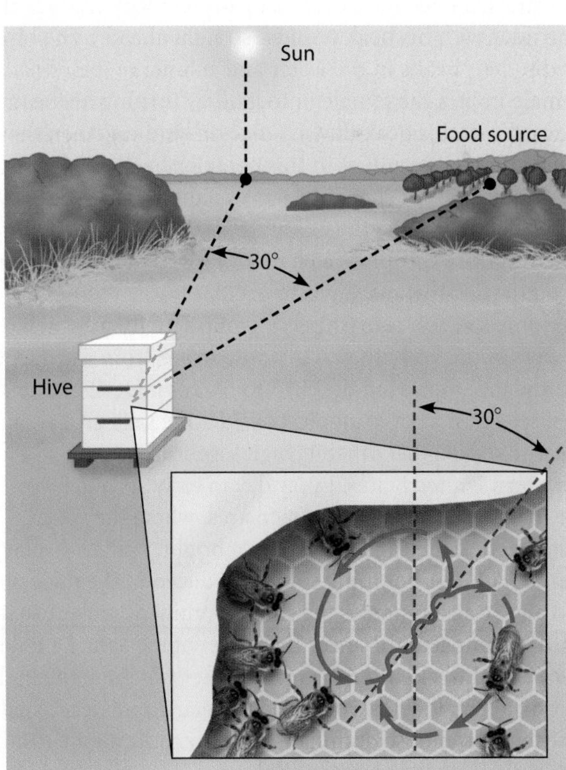
Sun
Food source
30°
Hive
30°

▲ **Figure 35.13B** A returning honeybee forager performing the waggle dance inside the hive

Try This Describe how the dance would differ if the food source were located closer to the hive and at an angle of 45° in relation to the sun.

35.14 Mating behavior often includes elaborate courtship rituals

Animals of many species tend to view members of their own species as competitors to be driven away. Even animals that forage and travel in groups often maintain a distance from their companions. Thus, careful communication is an essential prerequisite for mating. In many species, prospective mates must perform an elaborate courtship ritual, which confirms that individuals are of the same species, of the opposite sex, physically primed for mating, and not threats to each other. Differences in courtship behavior are often an effective reproductive barrier between closely related species (see Module 14.3).

Courtship rituals are common among vertebrates as well as some groups of invertebrates, such as insects and cephalopods. As you learned in Module 35.3, for example, male fruit flies court females with a specific sequence of behaviors. On summer evenings, male fireflies flash and katydids call to catch the attention of females. Cephalopods that lack an external shell—octopuses, squid, and cuttlefish—have special pigment cells that allow them to display an enormous repertoire of colors and patterns. The cuttlefish in **Figure 35.14A**, for instance, is signaling to a potential mate.

The common loon **(Figure 35.14B)**, which breeds on secluded lakes in the northern United States and Canada, exhibits a complex courtship behavior. The courting male and female swim side by side, performing a series of displays that include frequently turning their heads away from each other. (In contrast, a male loon defending his territory charges at an intruder with his beak pointed straight ahead.) The birds then dip their beaks in the water and submerge their heads. The male invites the female onto land by turning his head backward with his beak down. Once on land together, they copulate. Each movement in this complex behavior is a FAP (see Module 35.2). When a FAP is executed successfully by one partner, it triggers the next FAP in the other partner. Thus, the entire routine is a chain of FAPs that must be performed flawlessly if mating is to occur.

In some species, courtship is a group activity in which members of one or both sexes choose mates from a group of candidates. (See Module 13.15 to review mate choice and sexual selection.) For example, consider the sage grouse, a chicken-like bird that inhabits high sagebrush plateaus in the western United States. Every day in early spring, 50 or more males congregate in an open area, where they strut about, erecting their tail feathers in a bright, fanlike display **(Figure 35.14C)**. A booming sound produced in the male's inflated air sac accompanies the show. Dominant males usually defend a prime territory near the center of the area. Females arrive several weeks after the males. After watching the males perform, a female selects one, and the pair copulates. Usually, all the females choose dominant males, so only about 10% of the males actually mate. In choosing a dominant male, a female sage grouse may be giving her offspring, and thus her own genes, the best chance for survival. Research on several species of animals has shown a connection between a male's physical characteristics and the quality of his genes.

▲ **Figure 35.14A** A male cuttlefish displaying courtship colors

▲ **Figure 35.14B** A pair of common loons

▲ **Figure 35.14C** A courtship display by a male sage grouse

? What categories of signals do the cuttlefish, loon, and sage grouse use to communicate with potential mates?

All three use visual signals. The sage grouse also uses an auditory signal.

35.15 Mating systems and parental care enhance reproductive success

Courtship and mating are not the only elements of reproductive success. In order for genes to be passed on to successive generations, the offspring produced by the union must themselves survive and reproduce. Therefore, the needs of the young are an important factor in the evolution of mating systems. Animal mating systems fall into three major categories: **promiscuous** (no strong pair-bonds or lasting relationships between males and females), **monogamous** (a bond between one male and one female, with shared parental care), and **polygamous** (an individual of one sex mating with several of the other). Polygamous relationships most often involve a single male and many females, although in some species this is reversed, and a single female mates with several males.

Most newly hatched birds cannot care for themselves and require a large, continuous food supply that a single parent may not be able to provide **(Figure 35.15A)**. A male may leave more viable offspring by helping a single mate than by seeking multiple mates. This may explain why most birds are monogamous, though often with a new partner each breeding season. (And in many species of birds, one or both members of a mated pair will still mate with other partners.) On the other hand, birds whose young can feed and care for themselves almost immediately after hatching, such as pheasants and quail, gain less benefit from monogamy—a polygamous or promiscuous mating system may maximize their reproductive success.

In the case of mammals, the lactating female is often the only food source for the young, and males usually play no role in caring for their offspring. The prairie voles described in Module 35.1 are highly unusual mammals in regard to both monogamy and parental care.

Parental care involves significant costs, including energy expenditure and the loss of other mating opportunities. For females, the investment almost always pays off in terms of reproductive success—young born or eggs laid definitely contain the mother's genes. But even in the case of a normally monogamous relationship, the young may have been fathered by a male other than the female's usual mate. As a result, certainty of paternity (whether the male can be sure that offspring are his) may be a factor in the evolution of male mating behavior and parental care.

The certainty of paternity is relatively low in most species with internal fertilization because mating and birth (or mating and egg laying) are separated over time. This may help explain why species in which males are the sole parental caregiver are rare in birds and mammals. However, the males of many species with internal fertilization do engage in behaviors that appear to increase their certainty of paternity, such as guarding females from other males. In species such as lions, where males protect the females and young, a male or small group of males typically guard many females at once in a harem.

Certainty of paternity is much higher when egg laying and mating occur together, as happens when fertilization is external. This connection may explain why parental care in aquatic invertebrates, fishes, and amphibians, when care occurs at all, is at least as likely to be by males as by females.

▲ **Figure 35.15A** Young blackbirds awaiting food

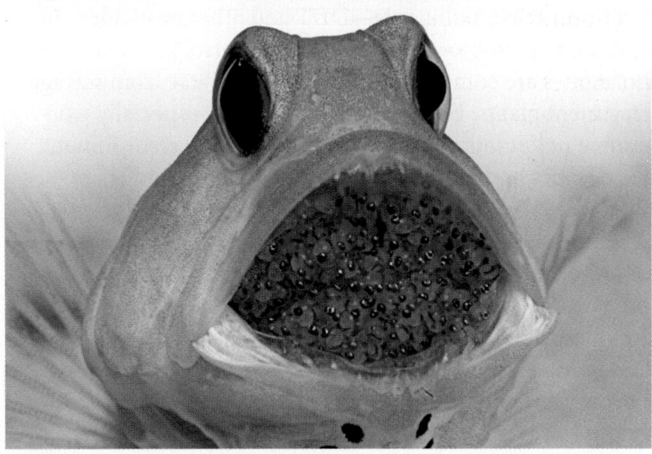

▲ **Figure 35.15B** A male jawfish with his mouth full of eggs

Figure 35.15B shows a male jawfish exhibiting paternal care of eggs. Jawfish, which are found in tropical marine habitats, hold the eggs they have fertilized in their mouths, keeping them aerated (by spitting them out and sucking them back in) and protected from predators until they hatch. Seahorses have the most extreme method of ensuring paternity. Females lay their eggs in a brood pouch in the male's abdomen, where they are fertilized by his sperm. When the eggs hatch a few weeks later, the pouch opens and the male pushes them out with pumping movements.

Keep in mind that when behavioral ecologists use the phrase *certainty of paternity*, they do not mean that animals are aware of paternity when they behave a certain way. Parental behaviors associated with certainty of paternity exist because they have been reinforced over generations by natural selection. Individuals with genes for such behaviors reproduced more successfully and passed those genes on to the next generation.

Why are birds more likely than mammals to form monogamous pairs?

In many species of birds, a single parent cannot provide enough food to ensure the survival of the young. Thus, males enhance their reproductive success by feeding their offspring. The offspring of mammals are fed by the mother's milk.

35.16 Chemical pollutants can cause abnormal behavior

CONNECTION

Appropriate behavior is the cornerstone of success in the animal world. So, something is amiss when fish are lackadaisical about territorial defense, salamanders ignore mating cues, birds exhibit sloppy nest-building techniques, and mice take inexplicable risks.

Scientists have linked observations of these abnormal behaviors, as well as many others, to endocrine-disrupting chemicals. Endocrine disruptors are a diverse group of substances that affect the vertebrate endocrine system by mimicking a hormone or by enhancing or inhibiting hormone activity. Endocrine disruptors enter ecosystems from a variety of sources, including discharge from paper and lumber mills and factory wastes such as dioxin (a by-product of many industrial processes) and PCBs (organic compounds used in electrical equipment until 1977). Agriculture is another major source of pollutants—DDT and other pesticides are endocrine disruptors. Traces of birth control pills and other hormones are commonly found in wastewater from sewage treatment plants. Endocrine disruptors are especially worrisome pollutants because they persist in the environment for decades and become concentrated in the food chain (see Figure 38.2E).

Hundreds of studies have demonstrated the effects of endocrine disruptors on vertebrate reproduction and development, including male fish that develop eggs, hermaphroditic frogs, reproductive abnormalities in alligators and turtles, and reduced reproductive rates in fish-eating birds. Like hormones, endocrine disruptors also affect behavior. For example, some male fish attract females during the breeding season by defending territories. Males have high levels of androgens (male hormones) during this time. Researchers showed that the intensity of nest-guarding behavior in male sticklebacks **(Figure 35.16A)** dropped after they were exposed to pollutants that mimic the female hormone estrogen. Male sticklebacks' performance of courtship rituals was also impaired.

Another series of studies showed that the anatomy of female mosquitofish was masculinized by endocrine disruptors. Female mosquitofish that were exposed to pollutants discharged from a paper mill developed the fin modification that males use to transfer sperm to females **(Figure 35.16B)**. The masculinized females also behaved like males, waving the fin back and forth in front of females in the typical courtship behavior. Female fish living downstream from the paper mill were masculinized at contaminated sites, while female fish living in uncontaminated water near the same mill were normal.

Although the effects of endocrine disruptors on reproductive behavior have received the most attention, endocrine disruptors also affect other kinds of behavior by acting on thyroid hormones and neurological functions. For example, spatial learning ability was impaired in young monkeys exposed to PCBs.

Could endocrine disruptors in drinking water or food affect humans, too? Answers are not yet clear, but in late 2009, the Environmental Protection Agency (EPA) ordered the manufacturers of several dozen chemicals to begin screening their products' potential as endocrine disruptors. The results are now being evaluated to determine the need for further testing.

? **How would ineffective courtship behavior affect the fitness of a male fish?**

● A male whose courtship display is perceived by females as inferior will not be successful in attracting mates and will thus be less likely to produce offspring. Therefore, the male's fitness would be reduced.

▲ **Figure 35.16A** Male stickleback in breeding colors, at nest with female

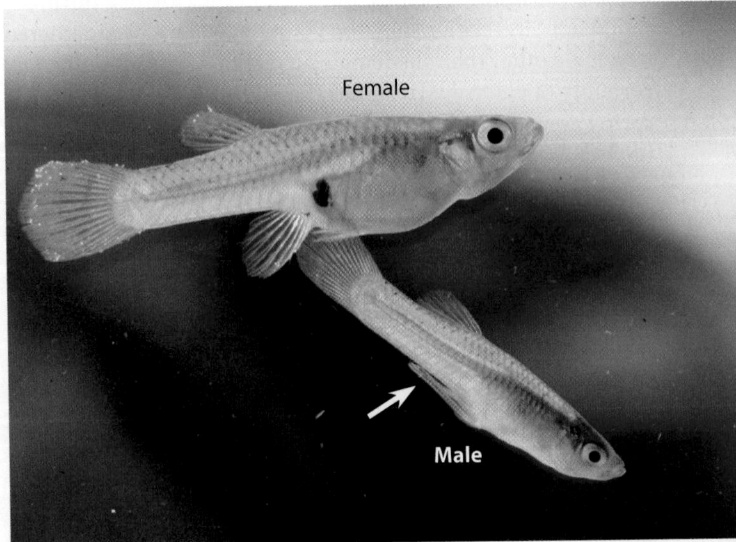
Female
Male

▲ **Figure 35.16B** A normal female mosquitofish and a male showing the modified fin used in courtship and sperm transfer

▷ Social Behavior and Sociobiology

35.17 Sociobiology places social behavior in an evolutionary context

Biologists define **social behavior** as any kind of interaction between two or more animals, usually of the same species. The courtship behaviors of loons and sage grouse are examples of social behavior. Other social behaviors observed in animals are aggression and cooperation.

Many animals migrate and feed in large groups (flocks, packs, herds, or schools). Pronghorns, for example, derive protection from feeding in herds (Module 34.4). Many watchful eyes increase the chance that a predator will be spotted before it can strike. When alarmed, a pronghorn flares out the white hairs on its rump, sending a danger signal to other members of its herd. Predators, too, may benefit from traveling in a group. Wolves usually hunt in a pack consisting of a tightly knit group of family members. Hunting in packs enables them to kill large animals, such as moose or elk, that would be unattainable by an individual wolf.

The discipline of **sociobiology** applies evolutionary theory to the study and interpretation of social behavior—the study of how social behaviors are adaptive and how they could have evolved by natural selection. We discuss several aspects of social behavior in the next several modules.

? Why is communication essential to social behavior?

● Group members must be able to transfer information—for example, to signal danger or to cooperate in obtaining food.

35.18 Territorial behavior parcels out space and resources

Many animals exhibit territorial behavior. A **territory** is an area, usually fixed in location, that one or more individuals defend and from which other members of the same species are usually excluded. The size of the territory varies with the species, the function of the territory, and the resources available. Territories are typically used for feeding, mating, rearing young, or combinations of these activities.

Figure 35.18A shows a nesting colony of gannets in Newfoundland, Canada. Space is at a premium, and the birds defend territories just large enough for their nests by calling out and pecking at other birds. As you can see, each gannet is literally only a peck away from its closest neighbors. Such small nesting territories are characteristic of many colonial seabirds. In contrast, other animals, including meerkats, wolves, coyotes, cheetahs, and even domestic cats, defend much larger territories, which they use for foraging as well as breeding.

Individuals that have established a territory usually proclaim their territorial rights continually; this is the function of most bird songs, the noisy bellowing of sea lions, and the chattering of squirrels. Scent markers are frequently used to signal a territory's boundaries. The male coyote in **Figure 35.18B** is marking its territory with urine. The odor will serve as a chemical "No Trespassing" sign. Other males that approach the area will sniff the urine and recognize that it belongs to another male. Usually, the intruder will avoid the marked territory and a potentially deadly confrontation with its proprietor.

Not all species are territorial. However, for those that are, the territory can provide exclusive access to food supplies, breeding areas, and places to raise young. Familiarity with a specific area may help individuals avoid predators or forage more efficiently. In a territorial species, such benefits increase fitness and outweigh the energy costs of defending a territory.

▲ **Figure 35.18A** Gannet territories

▲ **Figure 35.18B** A coyote in Minnesota marking its territory

? Why is the territory of a gannet so much smaller than the territory of a coyote?

● The gannet uses its territory only for raising young, not for foraging. Coyotes use their territories for foraging as well as for breeding.

35.19 Agonistic behavior often resolves confrontations between competitors

In many species, conflicts that arise over limited resources, such as food, mates, or territories, are settled by **agonistic behavior** (from the Greek *agon*, struggle), including threats, rituals, and sometimes combat that determine which competitor gains access to the resource. An agonistic encounter may involve a test of strength, such as when male moose lower their heads, lock antlers, and push against each other. More commonly, animals engage in exaggerated posturing and other symbolic displays that make the individual look large or aggressive. For example, the colorful Siamese fighting fish (*Betta splendens*) spreads its fins dramatically when it encounters a rival, thus appearing much bigger than it actually is. Eventually, one individual stops threatening and becomes submissive, exhibiting some type of appeasement display—in effect, surrendering.

Because violent combat may injure the victor as well as the vanquished in a way that reduces reproductive fitness, we would predict that natural selection would favor ritualized contests. And, in fact, this is what usually happens in nature. The rattlesnakes pictured in **Figure 35.19**, for example, are rival males wrestling over access to a mate. If they bit each other, both would die from the toxin in their fangs, but they are engaged in a pushing, rather than a biting, match. One snake usually tires before the other, and the winner pins the loser's head to the ground. In a way, the snakes are like two people who settle an argument by arm wrestling instead of resorting to fists or guns. In a typical case, the agonistic ritual inhibits further aggressive activity. Once two individuals have settled a dispute by agonistic behavior, future encounters between them usually involve less conflict, with the original loser giving way to the original victor. Often the victor of an agonistic ritual gains first or exclusive access to mates, and so this form of social behavior can directly affect an individual's evolutionary fitness.

▲ **Figure 35.19** Ritual wrestling by rattlesnakes

? Why is "fighting to the death" an unusual form of agonistic behavior among animals?

● Because ritualized posturing or nonlethal combat can usually produce a winner without injuries that would lower reproductive fitness for the winner and eliminate it altogether for the loser

35.20 Dominance hierarchies are maintained by agonistic behavior

Many animals live in social groups maintained by agonistic behavior. Chickens are an example. If several hens unfamiliar with one another are put together, they respond by chasing and pecking each other. Eventually, they establish a clear "pecking order." The alpha, or top-ranked, hen in the pecking order (the one on the right in **Figure 35.20**) is dominant; she is not pecked by any other hens and can usually drive off all the others by threats rather than actual pecking. The alpha hen also has first access to resources such as food, water, and roosting sites. The beta, or second-ranked, hen similarly subdues all others except the alpha, and so on down the line to the omega, or lowest, animal.

▲ **Figure 35.20** Chickens exhibiting pecking order

Pecking order in chickens is an example of a **dominance hierarchy**, a ranking of individuals based on social interactions. Once a hierarchy is established, each animal's status in the group is fixed, often for several months or even years. Dominance hierarchies are common, especially in vertebrate populations. In a wolf pack, for example, there is a dominance hierarchy among the females, and this hierarchy may control the pack's size. When food is abundant, the alpha female mates and also allows others to do so. When food is scarce, she usually monopolizes the males for herself and keeps other females from mating. In other species, such as savanna baboons and red deer, the dominant male monopolizes fertile females. As a result, his reproductive success is much greater than that of lower-ranking males.

As you will learn in the next module, however, not all social behaviors pit one individual against another.

? Dog trainers like Cesar Millan, TV's "Dog Whisperer," advise dog owners to be sure the dog understands that the owner is the alpha. How might this facilitate obedience?

● Like wolves, dogs are pack animals. They are more likely to respond to commands from a higher-ranking individual.

35.21 Altruistic acts can often be explained by the concept of inclusive fitness

EVOLUTION CONNECTION

Many social behaviors are selfish. Behavior that maximizes an individual's survival and reproductive success is favored by selection, regardless of how much the behavior may harm others. For example, superior foraging ability by one individual may leave less food for others. **Altruism**, on the other hand, is defined as behavior that reduces an individual's fitness while increasing the fitness of others in the population.

Altruistic behavior is often evident in animals that live in cooperative colonies. For example, workers in a honeybee hive are sterile females who labor all their lives on behalf of the queen. When a worker stings an intruder in defense of the hive, the worker usually dies. The animals in **Figure 35.21A** are highly social rodents called naked mole rats. Almost hairless and nearly blind, they live in colonies in underground chambers and tunnels in southern and northeastern Africa. With a social structure resembling that of honeybees, each colony has only one reproducing female, called the queen. The queen mates with one to three males, called kings. The rest of the colony consists of nonreproductive females and males who forage for roots and care for the queen, her young, and the kings. While trying to protect the queen or kings from a snake that invades the colony, a nonreproductive naked mole rat may lose its own life.

▲ **Figure 35.21A** The queen of a naked mole rat colony nursing offspring while surrounded by other individuals of the colony

If altruistic behavior reduces the reproductive success of self-sacrificing individuals, how can it be explained by natural selection? It is easy to see how selfless behavior might be selected for when it involves parents and offspring. When parents sacrifice their own well-being to ensure the survival of their young, they are maximizing the survival of their own genes. But reproducing is only one way to pass along genes; helping a close relative reproduce is another. Siblings, like parents and offspring, have half their genes in common, and an individual shares one-fourth of its genes with the offspring of a sibling.

The concept of **inclusive fitness** describes an individual's success at perpetuating its genes by producing its own offspring *and* by helping close relatives, who likely share many of those genes, to produce offspring. Altruism increases inclusive fitness when it maximizes the reproduction of close relatives. The natural selection favoring altruistic behavior that benefits relatives is called **kin selection**. Thus, the genes for altruism may be propagated if individuals that benefit from altruistic acts are themselves carrying those genes.

A classic study of Belding's ground squirrels, which live in regions of the western United States, provided empirical support for kin selection. Female ground squirrels generally live near the burrow where they were born, while males move farther away. As a result, female ground squirrels tend to have kin nearby, but males are not related to their neighbors. Upon seeing a predator such as a coyote or weasel approach, a squirrel often gives a high-pitched alarm call (**Figure 35.21B**) that alerts nearby squirrels, which then retreat to their burrows. Field observations have confirmed that the conspicuous alarm call identifies the caller's location and increases the risk of being killed. In the study mentioned, most alarm calls were given by female Belding's squirrels. The graph in Figure 35.21B shows which females issued these warnings. Each set of bars compares the percentage of times that squirrels who did and did not have close relatives nearby gave an alarm call when a predator approached. Squirrels whose close relatives lived nearby were much more likely to call, and they were as likely to warn mothers or sisters as they were to alert their descendants.

Kin selection also explains the altruistic behavior of bees in a hive, which all share genes with the queen. Their work (or even death) in support of the queen helps ensure that many of their genes will survive. In the case of mole rats, researchers have found that all the individuals in a naked mole rat colony are closely related. The nonreproductive members are the queen's descendants or siblings; by enhancing a queen's chances of reproducing, they increase the chances that some genes identical to their own will be propagated.

? **What is the ultimate cause of altruism between kin?**

● Natural selection reinforces such altruistic behavior through the reproductive success of closely related individuals that have many genes in common with the altruist, including genes for altruism.

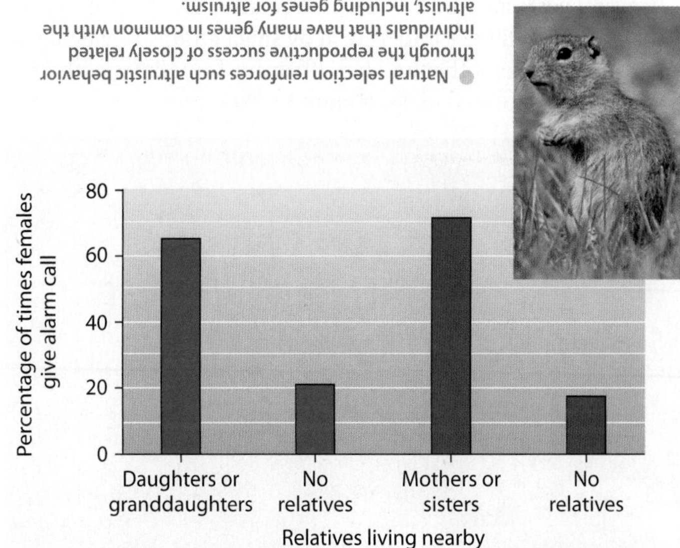
Percentage of times females give alarm call
80
60
40
20
0
Daughters or granddaughters
No relatives
Mothers or sisters
No relatives
Relatives living nearby

Data from P. W. Sherman, Nepotism and the evolution of alarm calls, *Science* 197: 1246–1253 (1977).

▲ **Figure 35.21B** The frequency of alarm calls by Belding's ground squirrels (inset photo) when close relatives are nearby or absent

35.22 Jane Goodall revolutionized our understanding of chimpanzee behavior

SCIENTIFIC THINKING

People are fascinated by the humanlike qualities of primates, especially chimpanzees. With their expressive faces and playful behavior, chimpanzees have long been stars in the entertainment industry. For example, a chimpanzee named Cheeta appeared in dozens of Tarzan movies from the 1930s through the 1960s. In the early days of television, NBC's *Today Show* attracted few viewers until a chimpanzee joined the show as the host's sidekick. Like their human counterparts, however, these chimpanzees were only playing a role. No one knew how they behaved in the wild, without the influence of trainers or zookeepers. A young Englishwoman named Jane Goodall changed that.

In 1960, paleoanthropologist Louis Leakey, who discovered *Homo habilis* and its stone tools (see Module 19.13), thought that studying the behavior of our closest relatives might provide insight into the behavior of early humans. Jane Goodall had no scientific background—she did not even have a college degree, though she later earned a Ph.D. from Cambridge—when Leakey hired her to observe chimpanzees in the wild.

What can we learn by watching animals?

Through countless hours of carefully recorded observations in an area that was later designated the Gombe Stream Research Center, near Lake Tanganyika, Goodall documented the social organization and behavior of chimpanzees **(Figure 35.22A)**. Most astonishingly, she reported that chimpanzees make and use tools by fashioning plant stems into probes for extracting termites from their mounds. Until that time, scientists thought that toolmaking defined humans, setting them apart from other animals.

Goodall discovered other important aspects of chimpanzee life, too. She learned that chimpanzees eat meat as well as plants. She described the close bond between chimpanzee mothers and their offspring, who are constantly together for several years after birth. She also recorded the daily foraging activities undertaken by small groups consisting of a mother and her offspring, sometimes with one or two males **(Figure 35.22B)**.

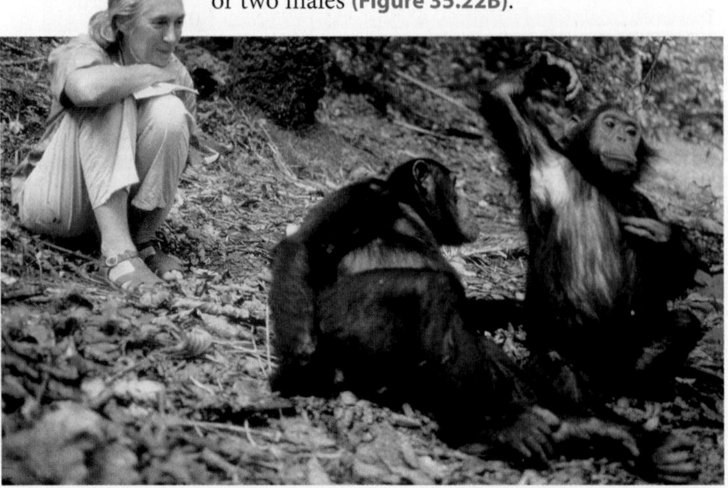

▲ Figure 35.22A Jane Goodall observing interactions between two chimpanzees (*Pan troglodytes*)

In areas of abundant food, where these separate groups often came together, Goodall observed the establishment of a male dominance hierarchy through agonistic displays. She witnessed several dramatic episodes in which a subordinate male overthrew the reigning alpha, as well as numerous unsuccessful challenges. These clashes, which cause tension among the group members, were usually followed by some kind of reconciliation behavior. After a conflict, a chimpanzee would make peace through gestures such as embracing or grooming a defeated rival. Grooming—picking through the fur and removing debris or parasites—is the social glue of a chimpanzee community.

Jane Goodall's pioneering research demonstrates the importance of descriptive, or qualitative data (see Module 1.8) in science. Her career is also a shining example of another aspect of science: communication. Scientists share their research with other scientists through publications, meetings, and personal interactions. Goodall extended her audience to include the general public. Through her many books, lecture tours, and television appearances, Goodall has educated the public about her decades at Gombe studying our closest relatives.

Jane Goodall has carried her work beyond the neutrality of science to promote issues related to the conservation of primates and their habitats, an increasing concern as deforestation, war, and the "bushmeat" trade (the practice of hunting wild animals, including primates, for food) take a rising toll on Africa's great apes. In addition, Goodall has been at the forefront of efforts to ban the use of invasive research methods on chimpanzees and to ensure the humane and ethical treatment of captive chimpanzees in research facilities and zoos. She has also spent half a century communicating another lesson she learned from observing chimpanzees: "The most important spin-off of the chimp research is probably the humbling effect it has on us who do the research. We are not, after all, the only aware, reasoning beings on this planet."

? Why is the study of chimpanzee behavior relevant to understanding the origins of certain human behaviors?

● Because chimpanzees and humans share a common ancestor

▲ Figure 35.22B A female chimpanzee foraging with an infant

35.23 Human behavior is the result of both genetic and environmental factors

Variations in behavioral traits such as personality, temperament, talents, and intellectual abilities make each person a unique individual. What are the roles of nature (genes) and nurture (environment) in shaping these behaviors? Let's look at how scientists distinguish between genetic and environmental influences on behavioral variations in humans.

Twins provide a natural laboratory for investigating the origins of complex behavioral traits. In general, researchers attempt to estimate the heritability of a trait, or how much of the observed variation can be explained by inheritance. Twin studies compare identical twins **(Figure 35.23A)**, who have the same DNA sequence and are raised in the same environment, with fraternal twins **(Figure 35.23B)**, who share an environment but only half of their DNA sequence. Some twin studies compare identical twins who were raised in the same household with identical twins who were separated at birth, a design that allows researchers to study the interactions of different environments with the same genotype. However, separated twins are very rare.

Results from twin studies consistently show that for complex behavioral traits such as general intelligence and personality characteristics, genetic differences account for roughly half the variation among individuals. The remainder of the variation can be attributed mostly to each individual's unique environment. Thus, neither factor—genes nor environment—is more important than the other.

Determining that a trait is heritable does not mean that scientists have identified a gene "for" the trait. Genes do not dictate behavior. Instead, genes cause tendencies to react to the environment in a certain way. For example, the hormone receptor gene in prairie voles does not encode a protein that causes the animal to be monogamous. Rather, the receptor protein encoded by the gene links the neurochemical reward experienced during mating with the scent of its partner (see Module 35.9). As a result, the prairie vole responds to the presence of that scent with behaviors that keep it in close contact with its mate.

Let's look at genes related to social bonding as an example of how genetics could play a role in human behavioral variations. Like the brains of other mammals, human brains produce the same hormones and hormone receptors that are implicated in social recognition and bonding in voles. Humans also have considerable individual variation in the key segment of DNA that determines receptor density. As a result, human brains vary in their sensitivity to hormones involved in bonding. As a social species, our well-being depends on relationships with others. Research on voles is helping scientists understand how our brains process the information used to form these bonds. These studies also provide insight into autism, a disorder characterized by difficulty forming social attachments. Scientists hypothesize that variation in a hormone receptor gene may play a role in some types of autism.

The mechanisms and underlying genetics of behavior are proximate causes. Scientists are also exploring the ultimate causes of human behavior. Sociobiology, the area of research introduced in Module 35.17, centers on the idea that social behavior evolves, like anatomical traits, as an expression of genes that have been perpetuated by natural selection. When applied to humans, this idea might seem to imply that life is predetermined. But it's unlikely that most human behavior is directly programmed by our genome. Unlike other animals, human offspring have an extraordinarily long period of development after birth. Children interact with a rich social environment, consisting of parents and other family members, peers, teachers, and society in general. The abilities to learn, to innovate, to advance technologically, and to participate in complex social networks have been key elements in the phenomenal success of the human species. It is much more likely that natural selection favored mechanisms that enabled humans to operate on the fly, that is, to use experience and feedback from the environment to adjust their behavior according to the circumstances.

▲ **Figure 35.23A** Identical twins, actors James and Oliver Phelps

◀ **Figure 35.23B** Actress Scarlett Johansson and her fraternal twin, Hunter

? A researcher conducted a study on "pseudo-twins," unrelated children of the same age who were raised in the same household. Results showed no correspondence between the IQs of pseudo-twins. What do these results indicate about the influence of genes and environment on IQ?

● The results indicate a strong genetic component to IQ. The children shared an environment but none of their DNA sequences. If environment were the main influence, you would expect the IQ scores to be similar.

CHAPTER 35 REVIEW

For practice quizzes, BioFlix animations, MP3 tutorials, video tutors, and more study tools designed for this textbook, go to MasteringBiology®

Reviewing the Concepts

The Scientific Study of Behavior (35.1–35.3)

35.1 Behavioral ecologists ask both proximate and ultimate questions. Behavioral ecology is the study of behavior in an evolutionary context, considering both proximate (immediate) and ultimate (evolutionary) causes of an animal's actions. Natural selection preserves behaviors that enhance fitness.

35.2 Fixed action patterns are innate behaviors. Innate behavior is performed the same way by all members of a species. A fixed action pattern (FAP) is an unchangeable series of actions triggered by a specific stimulus. FAPs ensure that activities essential to survival are performed correctly without practice.

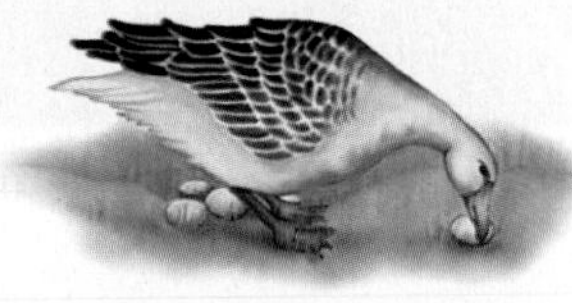

35.3 Behavior is the result of both genetic and environmental factors. Genetic engineering has been used to investigate genes that influence behavior. Cross-fostering experiments are useful for studying environmental factors that affect behavior.

Learning (35.4–35.11)

35.4 Habituation is a simple type of learning. Learning is a change in behavior resulting from experience. Habituation is learning to ignore a repeated, unimportant stimulus.

35.5 Imprinting requires both innate behavior and experience. Imprinting is irreversible learning limited to a sensitive period in the animal's life.

35.6 Imprinting poses problems and opportunities for conservation programs.

35.7 Animal movement may be a response to stimuli or require spatial learning. Kineses and taxes are simple movements in response to a stimulus. Spatial learning involves using landmarks to move through the environment.

35.8 A variety of cues guide migratory movements. Migratory animals use external cues to move between areas.

35.9 Animals may learn to associate a stimulus or behavior with a response. In associative learning, animals learn by associating external stimuli or their own behavior with positive or negative effects.

35.10 Social learning employs observation and imitation of others.

35.11 Problem-solving behavior relies on cognition. Cognition is the process of perceiving, storing, integrating, and using information. Problem-solving behavior involves complex cognitive processes.

Survival and Reproductive Success (35.12–35.16)

35.12 Behavioral ecologists use cost–benefit analysis to study foraging. Foraging includes identifying, obtaining, and eating food. Optimal foraging theory predicts that feeding behavior will maximize energy gain and minimize energy expenditure and risk.

35.13 Communication is an essential element of interactions between animals. Signaling in the form of sounds, scents, displays, or touches provides means of communication.

35.14 Mating behavior often includes elaborate courtship rituals. Courtship rituals reveal the attributes of potential mates.

35.15 Mating systems and parental care enhance reproductive success. Mating systems may be promiscuous, monogamous, or polygamous. The needs of offspring and certainty of paternity help explain differences in mating systems and parental care by males.

35.16 Chemical pollutants can cause abnormal behavior. Endocrine disruptors are chemicals in the environment that may cause abnormal behavior as well as reproductive abnormalities.

Social Behavior and Sociobiology (35.17–35.23)

35.17 Sociobiology places social behavior in an evolutionary context. Social behavior is any kind of interaction between two or more animals.

35.18 Territorial behavior parcels out space and resources.

35.19 Agonistic behavior often resolves confrontations between competitors. Agonistic behavior includes threats, rituals, and sometimes combat.

35.20 Dominance hierarchies are maintained by agonistic behavior.

35.21 Altruistic acts can often be explained by the concept of inclusive fitness. Kin selection is a form of natural selection favoring altruistic behavior that benefits relatives. Thus, an animal can propagate its own genes by helping relatives reproduce.

35.22 Jane Goodall revolutionized our understanding of chimpanzee behavior. In decades of fieldwork, she described many aspects of chimpanzee social behavior.

35.23 Human behavior is the result of both genetic and environmental factors.

Connecting the Concepts

1. Complete this map, which reviews the genetic and environmental components of animal behavior and their relationship to learning.

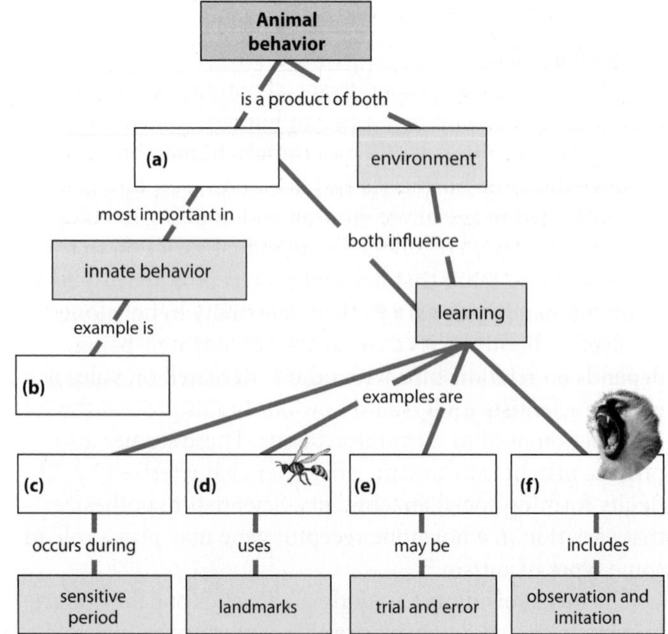

Animal behavior
is a product of both
(a)
environment
most important in
both influence
innate behavior
learning
example is
(b)
examples are
(c)
(d)
(e)
(f)
occurs during
uses
may be
includes
sensitive period
landmarks
trial and error
observation and imitation

Testing Your Knowledge

Level 1: Knowledge/Comprehension

2. Although many chimpanzee populations live in environments containing oil palm nuts, members of only a few populations use stones to crack open the nuts. The most likely explanation for this behavioral difference between populations is that
   a. members of different populations differ in manual dexterity.
   b. members of different populations have different nutritional requirements.
   c. members of different populations differ in learning ability.
   d. the use of stones to crack nuts has arisen and spread through social learning in only some populations.
3. Pheasants do not feed their chicks. Immediately after hatching, a pheasant chick starts pecking at seeds and insects on the ground. How might a behavioral ecologist explain the ultimate cause of this behavior?
   a. Pecking is a fixed action pattern.
   b. Pheasants learned to peck, and their offspring inherited this behavior.
   c. Pheasants that pecked survived and reproduced best.
   d. Pecking is a result of imprinting during a sensitive period.
4. A blue jay that aids its parents in raising its siblings is increasing its
   a. reproductive success.
   b. status in a dominance hierarchy.
   c. altruistic behavior.
   d. inclusive fitness.
5. Ants carry dead ants out of the anthill and dump them on a "trash pile." If a live ant is painted with a chemical from dead ants, other ants repeatedly carry it, kicking and struggling, to the trash pile, until the substance wears off. Which of the following best explains this behavior?
   a. The chemical triggers a fixed action pattern.
   b. The ants have become imprinted on the chemical.
   c. The ants continue the behavior until they become habituated.
   d. The ants can learn only by trial and error.

Level 2: Application/Analysis

6. Almost all the behaviors of a housefly are innate. What are some advantages and disadvantages to the fly of innate behaviors compared with behaviors that are mainly learned?
7. In Module 35.3, you learned that Norway rat offspring whose mothers don't interact with them much grow up to be fearful and anxious in new situations. Suggest a possible ultimate cause for this link between maternal behavior and stress response of offspring. (*Hint*: Under what circumstances might high reactivity to stress be more adaptive than being relaxed?)
8. A chorus of frogs fills the air on a spring evening. The frog calls are courtship signals. What are the functions of courtship behaviors? How might a behavioral ecologist explain the proximate cause of this behavior? The ultimate cause?

Level 3: Synthesis/Evaluation

9. Crows break the shells of certain molluscs before eating them by dropping them onto rocks. Hypothesizing that crows drop the molluscs from a height that gives the most food for the least effort (optimal foraging), a researcher dropped shells from different heights and counted the drops it took to break them.

Height of Drop (m)	Average Number of Drops Required to Break Shell	Total Flight Height (Number of Drops × Height Per Drop)
2	55	110
3	13	39
5	6	30
7	5	35
15	4	60

Data from R. Zach, Shell dropping: Decision-making and optimal foraging in northwestern crows. *Behaviour* 68:1/2, 106-117 (1979).

   a. The researcher measured the average drop height for crows and found that it was 5.23 m. Does this support the researcher's hypothesis? Explain.
   b. Describe an experiment to determine whether this feeding behavior of crows is learned or innate.
10. Scientists studying scrub jays found that it is common for "helpers" to assist mated pairs of birds in raising their young. The helpers lack territories and mates of their own. Instead, they help the territory owners gather food for their offspring. Propose a hypothesis to explain what advantage there might be for the helpers to engage in this behavior instead of seeking their own territories and mates. How would you test your hypothesis? If your hypothesis is correct, what kind of results would you expect your tests to yield?
11. Researchers are very interested in studying identical twins who were raised apart. Among other things, they hope to answer questions about the roles of inheritance and upbringing in human behavior. Why do identical twins make such good subjects for this kind of research? What do the results suggest to you? What are the potential pitfalls of this research? What abuses might occur in the use of these data if the studies are not evaluated critically?
12. **SCIENTIFIC THINKING** Jane Goodall's work revealed that in areas of abundant food, chimpanzees may live in groups of several dozen individuals. State a hypothesis about the chimpanzee mating system that could be tested by observing one of these groups. What data would you collect to test your hypothesis? How would you interpret the results? (Note: A sexually mature female chimpanzee undergoes a hormonal cycle about 36 days long that is reflected in easily observed changes in the appearance of her genital area. Females are most sexually receptive, and males are most attracted to them, during about 7 days of this cycle.)

Answers to all questions can be found in Appendix 4.

CHAPTER 36

Population Ecology

Fish may not be the first animals that come to mind when you think about reproductive behavior, but a remarkable variety of mating systems and behaviors have evolved in various fish populations. Consider parental investment, the time and energy expended on offspring (see Module 35.15). For example, the males of some fish species build nests and guard their developing offspring from predators. The females of other species carry the embryos internally and give birth after they have hatched. These fish enhance their reproductive success by improving their offspring's chances of survival. Some fish don't provide any form of parental care. They invest in quantity, producing millions of eggs at a time. Most of the offspring will die, but the few that survive to maturity ensure the parents' reproductive success.

? *Has parental care evolved by natural selection?*

There are also fish that hedge their bets by using more than one "strategy." For example, in a small Mediterranean species known as the peacock wrasse (photo below), the largest males build seaweed nests, where they court and mate with females. After fertilizing a nestful of eggs, these males lose interest in mating, and instead guard the nest from predators. This may sound like a good

deal for females, but it comes at a cost. The search for a nesting male takes time and energy and may be unsuccessful. The more abundant small males, which don't build nests, roam around looking for females. During the mating season, female wrasses typically lay eggs every other day. By choosing Mr. Right on some occasions, a female gains protection for some of her offspring; by choosing Mr. Right Now on other occasions, she ensures that she has at least a chance of reproductive success.

In this chapter, you'll learn about the structure and dynamics of populations and how traits such as parental care affect a population's growth rate and success in different types of habitats. As ecologists gain greater insight into natural populations, we become better equipped to assess the impact of human activities and balance human needs with the conservation of biodiversity and resources.

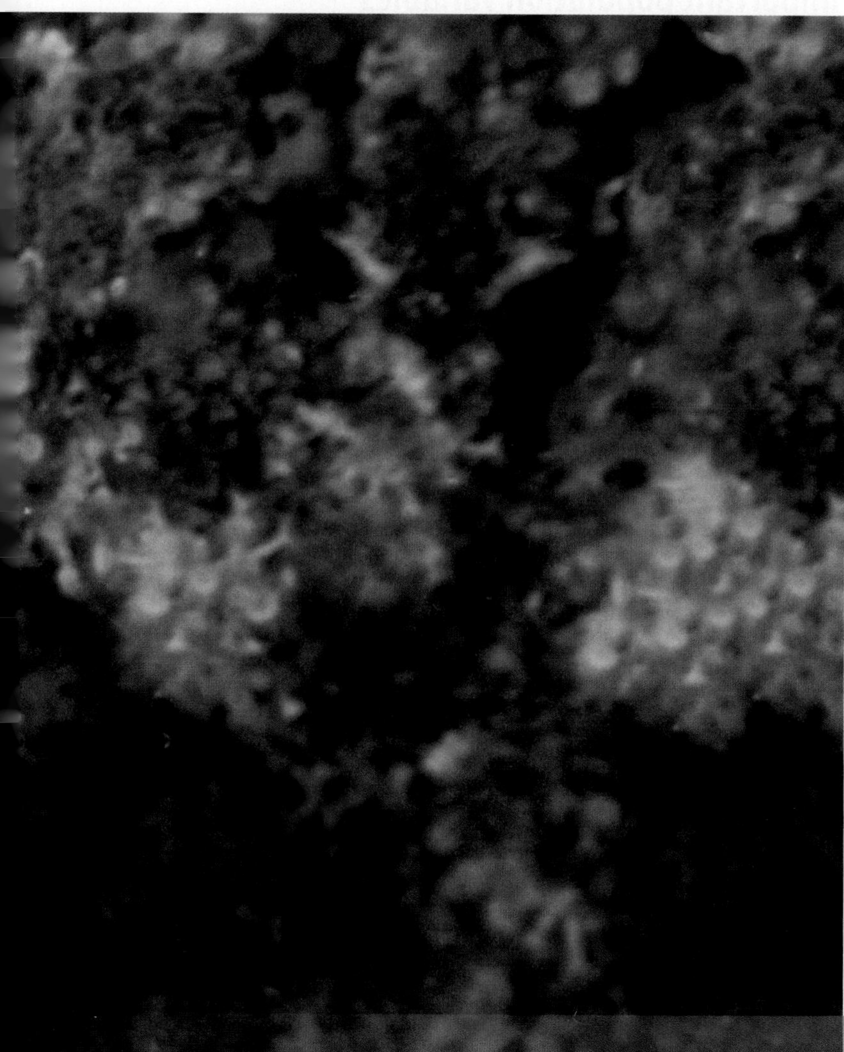

BIG IDEAS

Population Structure and Dynamics (36.1–36.8)

Population ecology is concerned with characteristics that describe populations, changes in population size, and factors that regulate populations over time.

The Human Population (36.9–36.11)

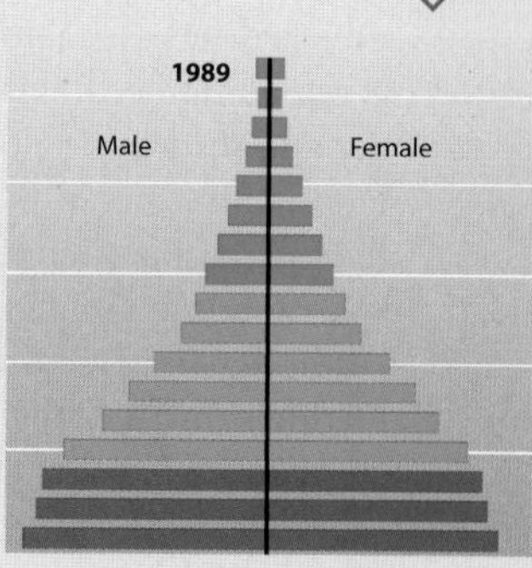
1989
Male
Female

The principles of population ecology can be used to describe the growth of the human population and its limits.

Population Structure and Dynamics

36.1 Population ecology is the study of how and why populations change

Ecologists usually define a **population** as a group of individuals of a single species that occupy the same general area. These individuals rely on the same resources, are influenced by the same environmental factors, and are likely to interact and breed with one another. For example, the emperor penguins living near Dumont d'Urville Station, where *March of the Penguins* was filmed, are a population. When a researcher chooses a population to study, he or she defines it by boundaries appropriate to the species being studied and to the purposes of the investigation.

Population ecology is concerned with changes in population size and the factors that regulate populations over time. A population ecologist might use statistics such as the number and distribution of individuals to describe a population. Population ecologists also examine population dynamics, the interactions between biotic and abiotic factors that cause variation in population sizes. One important aspect of population dynamics—and a major topic for this chapter—is population growth. The penguin population at Dumont d'Urville Station increases through births and the immigration of penguins from nearby colonies. Deaths and the emigration of individuals away from Dumont d'Urville Station decrease the population. Population ecologists might investigate how various environmental factors, such as availability of food, predation by killer whales, or the extent of sea ice, affect the size, distribution, or dynamics of the population.

Population ecology plays a key role in applied research. Data from population ecology is used to manage wildlife populations, develop sustainable fisheries, and gain insight into controlling the spread of pests and pathogens. Conservationists use these concepts to help identify and save endangered species. Population ecology also includes the study of human population growth, one of the most critical environmental issues of our time.

? What is the relationship between a population and a species?

A population is a localized group of individuals of a single species.

36.2 Density and dispersion patterns are important population variables

Two important aspects of population structure are population density and dispersion pattern. **Population density** is the number of individuals of a species per unit area or volume—the number of oak trees per square kilometer (km^2) in a forest, for instance, or the number of earthworms per cubic meter (m^3) in forest soil. Because it is impractical or impossible to count all individuals in a population in most cases, ecologists use a variety of sampling techniques to estimate population densities. For example, they might base an estimate of the density of alligators in the Florida Everglades on a count of individuals in a few sample plots of 1 km^2 each. The larger the number and size of sample plots, the more accurate the estimates. In some cases, population densities are estimated not by counts of organisms but by indirect indicators, such as number of bird nests or rodent burrows.

Within a population's geographic range, local densities may vary greatly. The **dispersion pattern** of a population refers to the way individuals are spaced within their area. A **clumped dispersion pattern**, in which individuals are grouped in patches, is the most common in nature. Clumping often results from an unequal distribution of resources in the environment. For instance, plants or fungi may be clumped in areas where soil conditions and other factors favor germination and growth. Clumping of animals often results from uneven food distribution. For example, the sea stars shown in **Figure 36.2A** group together where food is abundant.

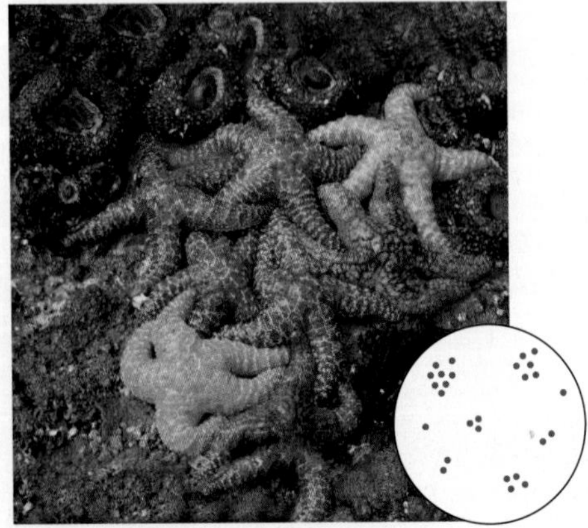

▲ **Figure 36.2A** Clumped dispersion of ochre sea stars at low tide

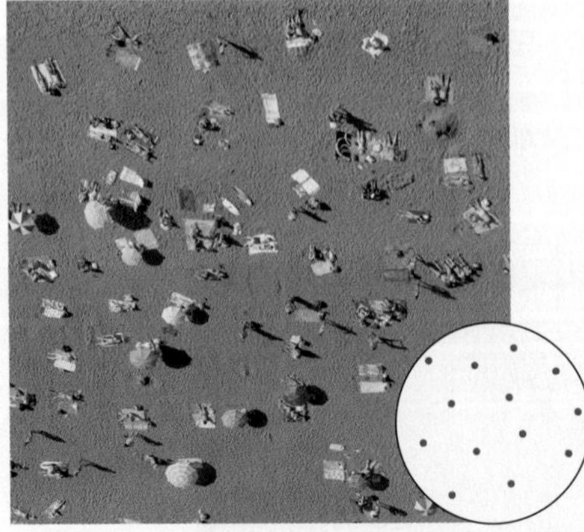

▲ **Figure 36.2B** Uniform dispersion of sunbathers on a beach

▲ **Figure 36.2C** Random dispersion of dandelions

Clumping may also reduce the risk of predation or be associated with social behavior.

A **uniform dispersion pattern** (an even one) often results from interactions between the individuals of a population. Some plants secrete chemicals, inhibiting the germination and growth of nearby plants that could compete for resources. Animals may exhibit uniform dispersion as a result of territorial behavior. **Figure 36.2B** (on the previous page) shows the uniform dispersion of sunbathers at a New York beach.

In a **random dispersion pattern**, individuals in a population are spaced in an unpredictable way, without a pattern. Plants, such as dandelions (**Figure 36.2C**), that grow from windblown seeds might be randomly dispersed. However, varying habitat conditions and social interactions make random dispersion rare.

Estimates of population density and dispersion patterns enable researchers to monitor changes in a population and to compare and contrast the growth and stability of populations in different areas. The next module describes another tool that ecologists use to study populations.

? What dispersion pattern would you predict in a forest population of termites, which live in damp, rotting wood?

● Clumped (in fallen logs or dead trees)

36.3 Life tables track survivorship in populations

Life tables track survivorship, the chance of an individual in a given population surviving to various ages. Starting with a population of 100,000 people, **Table 36.3** shows the number who are expected to be alive at the beginning of each age interval, based on death rates in 2008. For example, 93,999 out of 100,000 people are expected to live to age 50. The chance of surviving from 50 to 60, shown in the last column of the table, is 0.940. The chance of surviving from age 80 to 90, however, is only 0.402. The life insurance industry uses life tables to estimate life expectancy. Population ecologists have adopted this technique and constructed life tables for various other species. By identifying the most vulnerable stages of an organism's life, life table data help conservationists develop effective measures for maintaining a viable population.

Life tables can be used to construct **survivorship curves**, which plot survivorship as the proportion of individuals from an initial population that are alive at each age (**Figure 36.3**). By using a percentage scale instead of actual ages on the *x* axis, we can compare species with widely varying life spans on the same graph. The curve for the human population shows that most people survive to the older age intervals, as we saw in the life table. Ecologists refer to the shape of this curve as Type I survivorship. Species that exhibit a Type I curve—humans and many other large mammals—usually produce few offspring but give them good care, increasing the likelihood that they will survive to maturity.

In contrast, a Type III curve indicates low survivorship for the very young, followed by a period when survivorship is high for those few individuals who live to a certain age. Species with this type of survivorship curve usually produce very large numbers of offspring but provide little or no care for them. Some fishes, for example, can produce millions of eggs at a time, but most offspring die as larvae from predation or other causes. Many invertebrates, such as oysters, also have Type III survivorship curves.

A Type II curve is intermediate, with survivorship constant over the life span. That is, individuals are no more vulnerable at one stage of the life cycle than at another. This type of survivorship has been observed in some invertebrates, lizards, and rodents.

TABLE 36.3 LIFE TABLE FOR THE U.S. POPULATION IN 2008

Age Interval	Number Living at Start of Age Interval (*N*)	Number Dying During Interval (*D*)	Chance of Surviving Interval 1 – (*D*/*N*)
0–10	100,000	833	0.992
10–20	99,167	363	0.996
20–30	98,804	941	0.990
30–40	97,863	1,224	0.987
40–50	96,639	2,640	0.973
50–60	93,999	5,643	0.940
60–70	88,356	11,203	0.873
70–80	77,153	21,591	0.720
80–90	55,562	33,215	0.402
90+	22,347	22,347	0.000

Data from E. Arias, United States Life Tables, 2008, *National Vital Statistics Reports*, Volume 61: 3, September 24, 2012.

? How does the chance of survival change with age in organisms with a Type III survivorship curve?

● The chance of survival is initially low but increases after an individual reaches a certain age.

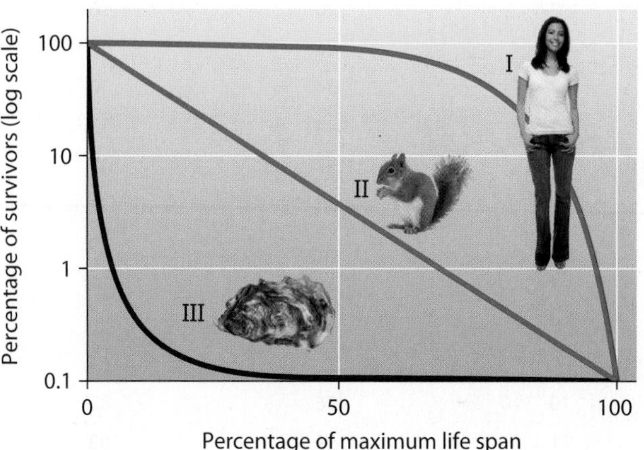

Percentage of survivors (log scale)
100
10
1
0.1
I
II
III
0
50
100
Percentage of maximum life span

▲ **Figure 36.3** Three types of survivorship curves

36.4 Idealized models predict patterns of population growth

Population size fluctuates as new individuals are born or immigrate into an area and others die or emigrate. Some populations—for example, trees in a mature forest—are relatively constant over time. Other populations change rapidly, even explosively. Consider a single bacterium that divides every 20 minutes. There would be two bacteria after 20 minutes, four after 40 minutes, eight after 60 minutes, and so on. In just 12 hours, the population would approach 70 billion cells. If reproduction continued at this rate for a day and a half—a mere 36 hours—there would be enough bacteria to form a layer a foot deep over Earth's entire surface. Using idealized models, population ecologists can predict how the size of a particular population will change over time under different conditions.

The Exponential Growth Model The rate of population increase under ideal conditions, called exponential growth, can be calculated using the simple equation $G = rN$. The G stands for the growth rate of the population (the number of new individuals added per time interval); N is the population size (the number of individuals in the population at a particular time); and r stands for the **per capita rate of increase** (the average contribution of each individual to population growth; per capita means "per person").

How do we estimate the per capita rate of increase? Population growth reflects the number of births minus the number of deaths (the model assumes that immigration and emigration are equal). Suppose a population of rabbits has 100 individuals, and there are 50 births and 20 deaths in one month. The net increase is 30 rabbits. The per capita increase in the population, or r, is 30/100, or 0.3.

In a population growing in an ideal environment with unlimited space and resources, r is the maximum capacity of members of that population to reproduce. Thus, the value of r depends on the kind of organism. For example, rabbits have a higher r than elephants, and bacteria have a higher r than rabbits.

When a population is expanding without limits, r remains constant and the rate of population growth depends on the number of individuals already in the population (N). In **Table 36.4A**, a population begins with 20 rabbits. The growth rate (G) for this population, using $r = 0.3$, is shown in the right-hand column. Notice that the larger the population size, the more new individuals are added during each time interval.

Graphing these data, as shown in **Figure 36.4A**, produces a J-shaped curve, which is typical of exponential growth. The lower part of the J, where the slope of the line is almost flat, results from the relatively slow growth when N is small. As the population increases, the slope becomes steeper.

The **exponential growth model** gives an idealized picture of unlimited population growth. There is no restriction on the abilities of the organisms to live, grow, and reproduce. Even elephants, the slowest breeders on the planet, would increase exponentially if enough resources were available. Although elephants typically produce only six young in a 100-year life span, Charles Darwin estimated that it would take only 750 years for a single pair to give rise to a population of 19 million. But any population—bacteria, rabbits, or elephants—will eventually be limited by the resources available.

TABLE 36.4A | EXPONENTIAL GROWTH OF RABBITS, $r = 0.3$

Time (months)	N	$G = rN$
0	20	6
1	26	8
2	34	10
3	44	13
4	57	17
5	74	22
6	96	29
7	125	38
8	163	49
9	212	64
10	276	83
11	359	108
12	467	140

Limiting Factors and the Logistic Growth Model In nature, a population that is introduced to a new environment or is rebounding from a catastrophic decline in numbers may grow exponentially for a while. Eventually, however, one or more environmental factors will limit its growth rate as the population reaches its maximum sustainable size. Environmental factors that restrict population growth are called **limiting factors**.

You can see the effect of population-limiting factors in the graph in **Figure 36.4B** (see top of facing page), which illustrates the growth of a population of fur seals on St. Paul Island, off the coast of Alaska. (For simplicity, only the mated bulls were counted. Each has a harem of a number of females, as shown in the photograph.) Before 1925, the seal population on the island remained low because of uncontrolled

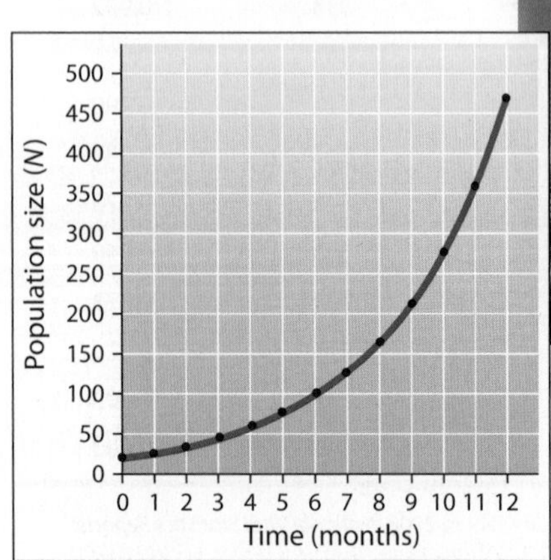
Population size (N)
500
450
400
350
300
250
200
150
100
50
0
0 1 2 3 4 5 6 7 8 9 10 11 12
Time (months)

◀ **Figure 36.4A** Exponential growth of rabbits

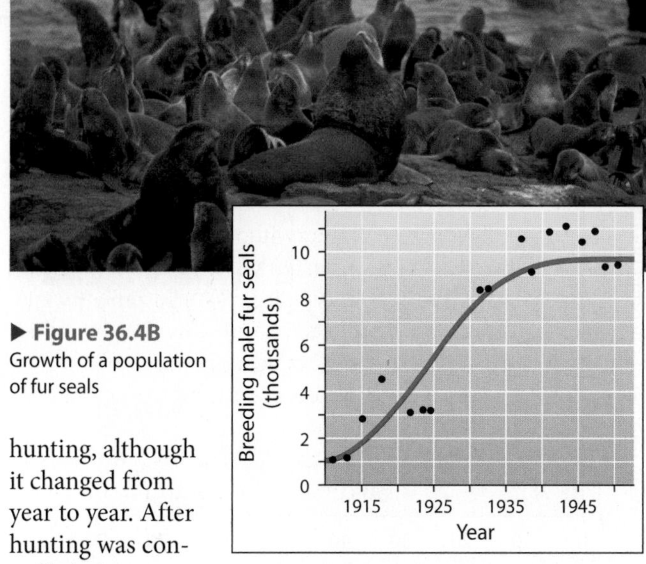
Breeding male fur seals (thousands)
10
8
6
4
2
0
1915
1925
1935
1945
Year

Data from K. W. Kenyon et al., A population study of the Alaska fur-seal herd, *Federal Government Series: Special Scientific Report—Wildlife* 12 (1954).

▶ **Figure 36.4B** Growth of a population of fur seals

hunting, although it changed from year to year. After hunting was controlled, the population increased rapidly until about 1935, when it began to level off and started fluctuating around a population size of about 10,000 bull seals. At this point, a number of limiting factors, including some hunting and the amount of space suitable for breeding, restricted population growth.

The fur seal growth curve fits the **logistic growth model**, a description of idealized population growth that is slowed by limiting factors as the population size increases. **Figure 36.4C** compares the logistic growth model (red) with the exponential growth model (blue). As you can see, the logistic curve is J-shaped at first, but gradually levels off to resemble an S.

To model logistic growth, the formula for exponential growth, *rN*, is multiplied by an expression that describes the effect of limiting factors on an increasing population size:

$$G = rN\,\frac{(K - N)}{K}$$

This equation is actually simpler than it looks. The only new symbol in the equation is *K*, which stands for carrying capacity. **Carrying capacity** is the maximum population size that a particular environment can sustain ("carry"). For the fur seal population on St. Paul Island, for instance, *K* is about 10,000 mated males. The value of *K* varies, depending on the species and the resources available in the habitat. *K* might be considerably less than 10,000 for a fur seal population on a smaller island with fewer breeding sites. Even in one location, *K* is not a fixed number. Organisms interact with other organisms in their communities, including predators, parasites, and food sources, that may affect *K*. Changes in abiotic factors may also increase or decrease carrying capacity. In any case, the concept of carrying capacity expresses an essential fact of nature: Resources are finite.

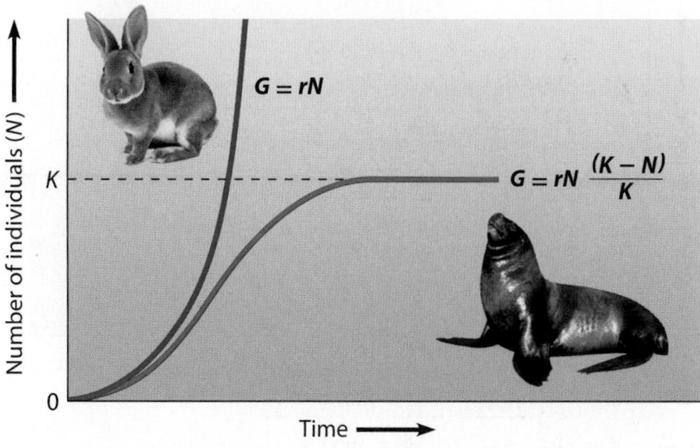
Number of individuals (N)
G = rN
K
G = rN (K − N)/K
0
Time

▲ **Figure 36.4C** Logistic growth and exponential growth compared

Table 36.4B demonstrates how the expression $(K - N)/K$ in the logistic growth model produces the S-shaped curve. At the outset, *N* (the population size) is very small compared to *K* (the carrying capacity). Thus, $(K - N)/K$ nearly equals K/K, or 1, and population growth (*G*) is close to *rN*—that is, exponential growth. As the population increases and *N* gets closer to carrying capacity, $(K - N)/K$ becomes an increasingly smaller fraction. The growth rate slows as *rN* is multiplied by that fraction. At carrying capacity, the population is as large as it can theoretically get in its environment; at this point, $N = K$, and $(K - N)/K = 0$. The population growth rate (*G*) becomes zero.

What does the logistic growth model suggest to us about real populations in nature? The model predicts that a population's growth rate will be small when the population size is *either* small or large, and highest when the population is at an intermediate level relative to the carrying capacity. At a low population level, resources are abundant, and the population is able to grow nearly exponentially. At this point, however, the increase is small because *N* is small. In contrast, at a high population level, limiting factors strongly oppose the population's potential to increase. There might be less food available per individual or fewer breeding territories, nest sites, or shelters. These limiting factors cause the birth rate to decrease, the death rate to increase, or both. Eventually, when the birth rate equals the death rate, the population stabilizes at the carrying capacity (*K*).

It is important to realize that the logistic growth model presents a mathematical ideal that is a useful starting point for studying population growth and for constructing more complex models. Like any good starting hypothesis, the logistic model has stimulated research, leading to a better understanding of the factors affecting population growth. We take a closer look at some of these factors next.

? In logistic growth, at what population size (in terms of *K*) is the population increasing most rapidly? Explain why.

● When *N* is $\frac{1}{2}K$. At this population size, there are more reproducing individuals than at lower population sizes and still lots of resources available for growth.

TABLE 36.4B EFFECT OF *K* ON GROWTH RATE AS *N* APPROACHES *K*, $K = 1{,}000$, $r = 0.1$

N	rN	(K − N)/K	G = rN(K − N)/K
10	1	0.99	0.99
100	10	0.9	9.00
400	40	0.6	24.00
500	50	0.5	25.00
600	60	0.4	24.00
700	70	0.3	21.00
950	95	0.05	4.75
1,000	100	0.00	0.00

36.5 Multiple factors may limit population growth

The logistic growth model predicts that population growth will slow and eventually stop as population density increases. That is, at higher population densities, the birth rate decreases, the death rate increases, or both. What are the possible causes of these density-dependent changes in birth and death rates?

Several **density-dependent factors**—limiting factors whose intensity is related to population density—appear to limit growth in natural populations. The most obvious one is **intraspecific competition**—competition between individuals of the same species for limited resources. As a limited food supply is divided among more and more individuals, birth rates may decline as individuals have less energy available for reproduction. Density-dependent factors may also depress a population's growth by increasing the death rate. In a population of song sparrows (*Melospiza melodia*), both factors reduced the number of offspring that survived and left the nest (**Figure 36.5A**). As the number of competitors for food increased, female song sparrows laid fewer eggs. In addition, mortality of eggs and nestlings increased with increasing population density.

The availability of space is a density-dependent factor for some populations. For instance, the number of nesting sites on rocky islands may limit the population size of seabirds such as gannets (see Figure 35.18A). Or, like a game of musical chairs, the number of safe hiding places may limit a prey population by exposing some individuals to a greater risk of predation. For example, young kelp perch (*Brachyistius frenatus*) hide from predators in "forests" of the large seaweed known as kelp (see Module 16.14). In the experiment shown in **Figure 36.5B** the proportion of perch eaten by a predator increased with increasing perch density.

Intraspecific competition may limit plant population growth, too. Plants that grow close together may experience increased mortality as competition for resources increases. And those that survive will likely produce fewer flowers, fruits, and seeds than uncrowded individuals.

For some animal species, physiological factors may regulate population size. White-footed mice in a small field enclosure will multiply from a few to a colony of 30 to 40 individuals, but reproduction then declines until the population ceases to grow. This drop in reproduction occurs even when additional food and shelter are provided. High population densities in mice appear to induce a stress syndrome in which hormonal changes can delay sexual maturation, cause reproductive organs to shrink, and depress the immune system. In this case, high densities cause both a decrease in birth rate and an increase in death. Similar effects of crowding have been observed in wild populations of other rodents.

In many natural populations, abiotic factors such as weather may affect population size well before density-dependent factors become important. A population-limiting factor whose intensity is unrelated to population density is called a **density-independent factor**. If we look at the growth curve of such a population, we see something like exponential growth followed by a rapid decline, rather than a leveling off. **Figure 36.5C** shows

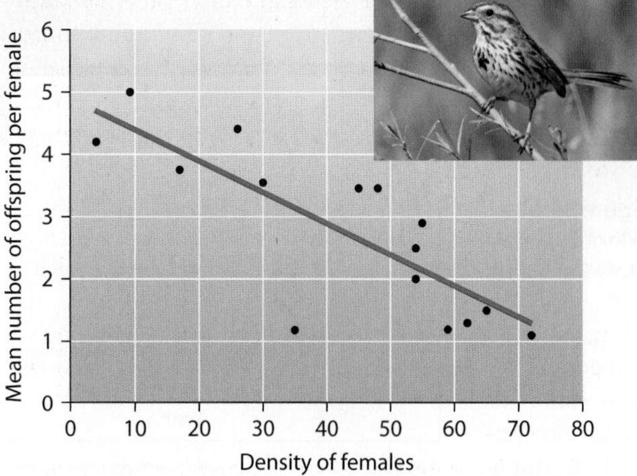
Mean number of offspring per female
6
5
4
3
2
1
0
0
10
20
30
40
50
60
70
80
Density of females

Data from P. Arcese et al., Stability, Regulation, and the Determination of Abundance in an Insular Song Sparrow Population. *Ecology* 73: 805–882 (1992).

▲ **Figure 36.5A** Declining reproductive success of song sparrows (inset) with increasing population density

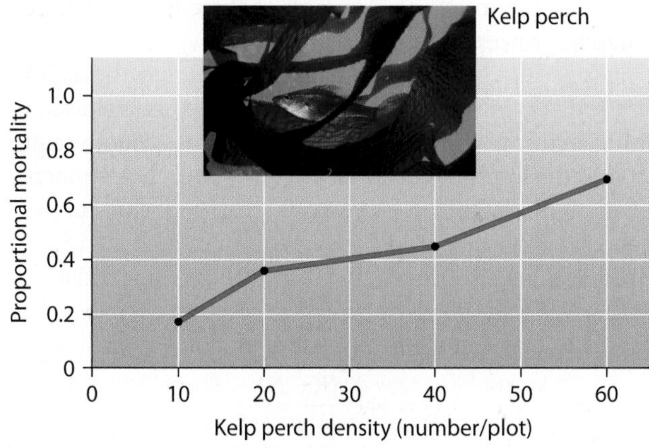
Kelp perch
Proportional mortality
1.0
0.8
0.6
0.4
0.2
0
0
10
20
30
40
50
60
Kelp perch density (number/plot)

Data from T. W. Anderson, Predator Responses, Prey Refuges, and Density-Dependent Mortality of a Marine Fish, *Ecology* 82: 245–257 (2001).

▲ **Figure 36.5B** Increasing mortality of kelp perch (inset) with increasing density

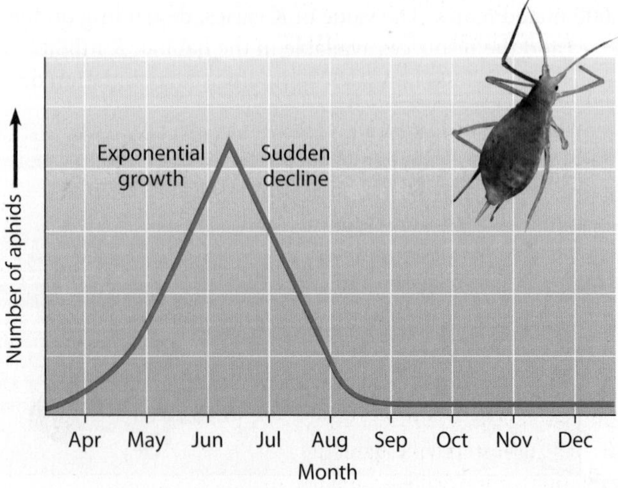
Number of aphids
Exponential growth
Sudden decline
Apr
May
Jun
Jul
Aug
Sep
Oct
Nov
Dec
Month

▲ **Figure 36.5C** Weather change as a density-independent factor limiting aphid population growth

this effect for a population of aphids, insects that feed on the sugary phloem sap of plants. These and many other insects undergo virtually exponential growth in the spring and then rapidly die off when the weather turns hot and dry in the summer. A few individuals may survive, and these may allow population growth to resume if favorable conditions return. In some populations of insects—many mosquitoes and grasshoppers, for instance—adults die off entirely, leaving only eggs, which initiate population growth the following year. In addition to seasonal changes in the weather, disturbances—such as fire, storms, and habitat disruption by human activity—can affect a population's size regardless of its density.

Over the long term, most populations are probably regulated by a mixture of factors. Some populations remain fairly stable in size and are presumably close to a carrying capacity that is determined by biotic factors such as competition or predation. Most populations for which we have long-term data, however, show fluctuations in numbers. Thus, the dynamics of many populations result from a complex interaction of both density-dependent factors and density-independent abiotic factors such as climate and disturbances. As you will see in the next module, ecologists must consider this complexity when they develop and test hypotheses.

? List some of the factors that may reduce birth rate or increase death rate as population density increases.

● Food and nutrient limitations, insufficient territories, increase in disease and predation, accumulation of toxins

36.6 Some populations have "boom-and-bust" cycles

SCIENTIFIC THINKING

Some populations of insects, birds, and mammals undergo dramatic fluctuations in density with remarkable regularity. "Booms" characterized by rapid exponential growth are followed by "busts," during which the population falls back to a minimal level. A striking example is shown in **Figure 36.6**, which shows estimated populations of the snowshoe hare and the lynx based on the number of pelts sold by trappers in northern Canada to the Hudson Bay Company over a period of nearly 100 years. Both populations rise and fall at regular intervals of roughly 10 years, but not simultaneously. Changes in the lynx population lag behind changes in the hare population. The hare is the lynx's primary food source, so this pattern might be expected. For predators that depend heavily on a single species of prey, the availability of prey can have a strong influence on population size. Thus, the lynx population cycles probably result at least in part from the hare population cycles, and as the predator population declines, the prey population rebounds. But what causes the boom-and-bust cycles of snowshoe hares? Scientists often use field experiments to answer questions that involve natural populations.

One hypothesis proposed that when hares are abundant, they overgraze their winter food supply, resulting in high mortality. This hypothesis was tested by providing extra food for experimental field populations of hares. The design of the experiment was simple, but researchers had to collect data for more than 20 years to test the hypothesis. Hare populations given extra food increased in numbers, but the cycles continued.

Another hypothesis attributed hare population cycles to excessive predation. In addition to lynx, hares are preyed on by many other animals, including coyotes, foxes, and great-horned owls. Using radio collars to track individual hares, researchers determined that 95% of hares had been killed by predators. None had died of starvation. These results support the predation hypothesis. However, further experiments showed that, although fluctuating food availability is not the primary factor controlling hare population cycles, it does play an important role. Perhaps well-fed hares are more likely to escape from predators. Thus, unraveling the complex causes of population cycles, like other ecological studies, requires scientists to consider a number of interacting variables.

Now that we have looked at patterns of population growth, we turn our attention to the differences in reproductive patterns of populations and how they are shaped by natural selection.

? In the experiment in which researchers provided hares with extra food, why was it necessary to continue the experiment for more than 20 years?

● A long-term experiment was needed to detect any changes in the 10-year hare population cycle.

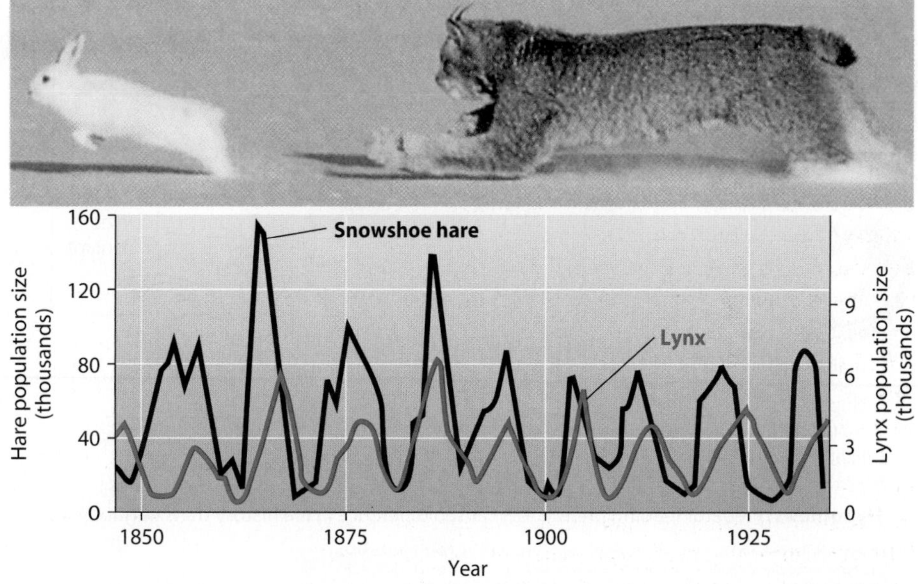

Snowshoe hare
Lynx
Hare population size (thousands)
Lynx population size (thousands)
160
120
80
40
0
9
6
3
0
1850
1875
1900
1925
Year

Data from C. Elton and M. Nicholson, The ten-year cycle in numbers of the lynx in Canada, *Journal of Animal Ecology* 11 : 215–244 (1942).

▲ **Figure 36.6** Population cycles of the snowshoe hare and the lynx

36.7 Evolution shapes life histories

EVOLUTION CONNECTION

The traits that affect an organism's schedule of reproduction and death make up its **life history**. Some key life history traits are the age of first reproduction, the frequency of reproduction, the number of offspring, and the amount of parental care given. Natural selection cannot optimize all of these traits simultaneously because an organism has limited time, energy, and nutrients. For example, an organism that gives birth to a large number of offspring will not be able to provide a great deal of parental care. Consequently, the combination of life history traits in a population represents trade-offs that balance the demands of reproduction and survival. Because selective pressures vary, life histories are very diverse. Nevertheless, ecologists have observed some patterns that are useful for understanding how life history characteristics have been shaped by natural selection.

Has parental care evolved by natural selection?

One life history pattern is typified by small-bodied, short-lived animals (for example, insects and small rodents) that develop and reach sexual maturity rapidly, have a large number of offspring, and offer little or no parental care. A similar pattern is seen in small, nonwoody plants such as dandelions that produce thousands of tiny seeds. Ecologists hypothesize that selection for this set of life history traits occurs in environments where resources are abundant, permitting exponential growth. It is sometimes called ***r*-selection** because *r* (the per capita rate of increase) is maximized. Most *r*-selected species have an advantage in habitats that experience unpredictable disturbances, such as fire, floods, hurricanes, drought, or cold weather, which create new opportunities by suddenly reducing a population to low levels. Human activity is a major cause of disturbance, producing road cuts, freshly cleared fields and woodlots, and poorly maintained lawns that are commonly colonized by *r*-selected plants and animals.

In contrast, large-bodied, long-lived animals (such as bears and elephants) develop slowly and produce few, but well-cared-for, offspring. Plants with comparable life history traits include coconut palms, which produce relatively few seeds that are well stocked with nutrient-rich material—the plant's version of parental care. Ecologists hypothesize that selection for this set of life history traits occurs in environments where the population size is near carrying capacity (*K*), so it is sometimes called ***K*-selection**. Population growth in these situations is limited by density-dependent factors. Because competition for resources is keen, *K*-selected organisms gain an advantage by allocating energy to their own survival and to the survival of their descendants. Thus, *K*-selected organisms are adapted to environments that typically have a stable climate and little opportunity for rapid population growth.

The concept of *r*- and *K*-selection has been criticized as an oversimplification, and most organisms fall somewhere between the extremes. However, this concept has stimulated a vigorous subfield of ecological research on the evolution of life histories.

A long-term project in Trinidad has provided direct evidence that life history traits can be shaped by natural selection. For years, researchers have been studying guppy populations living in small, relatively isolated pools. As shown in **Figure 36.7**, some guppy populations live in pools with predators called killifish, which eat mainly small, immature guppies (Pool 1).

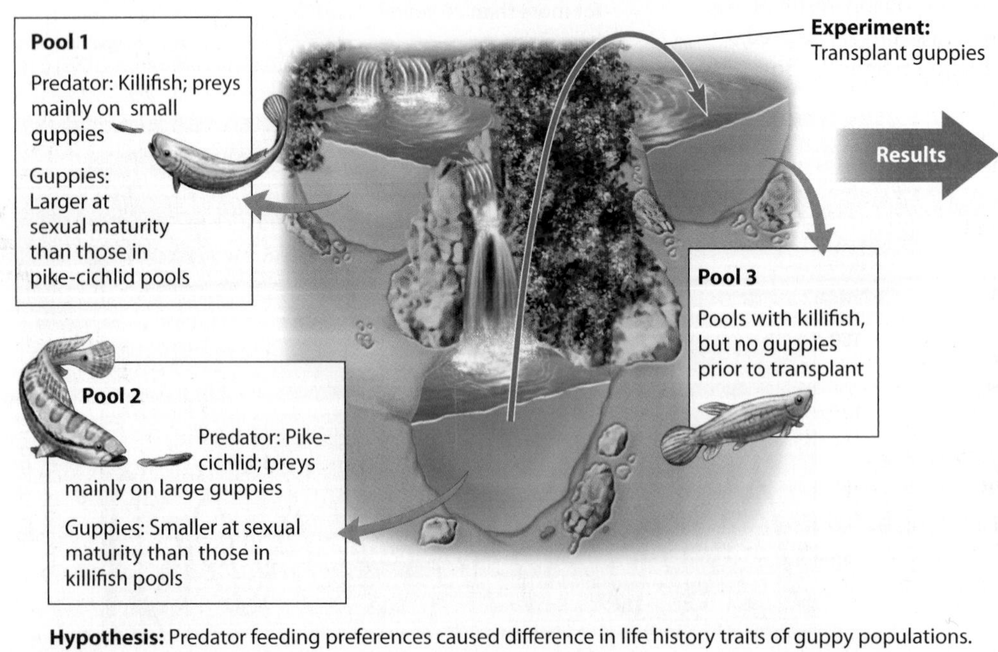

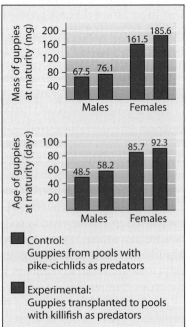

Pool 1
Predator: Killifish; preys mainly on small guppies
Guppies: Larger at sexual maturity than those in pike-cichlid pools
Pool 2
Predator: Pike-cichlid; preys mainly on large guppies
Guppies: Smaller at sexual maturity than those in killifish pools
Pool 3
Pools with killifish, but no guppies prior to transplant
Experiment: Transplant guppies
Results

Mass of guppies at maturity (mg)

	Control	Experimental
Males	67.5	76.1
Females	161.5	185.6

Age of guppies at maturity (days)

	Control	Experimental
Males	48.5	58.2
Females	85.7	92.3

Control: Guppies from pools with pike-cichlids as predators

Experimental: Guppies transplanted to pools with killifish as predators

Hypothesis: Predator feeding preferences caused difference in life history traits of guppy populations.

Data from D. N. Reznick and H. Bryga, Life-History Evolution in Guppies (*Poecilia reticulata*): 1. Phenotypic and genetic changes in an introduction experiment, *Evolution* 41: 1370–1385 (1987).

▲ **Figure 36.7** The effect of predation on the life history traits of guppies

Try This Use the figure to explain how the hypothesis was tested.

Other guppy populations live where larger fish, called pike-cichlids, eat mostly mature, large-bodied guppies (Pool 2). Guppies in populations exposed to these pike-cichlids tend to be smaller, mature earlier, and produce more offspring at a time than those in areas with killifish. Thus, guppy populations differ in certain life history traits, depending on the kind of predator in their environment. For these differences to be the result of natural selection, the traits should be heritable. And indeed, guppies from both populations raised in the laboratory without predators retained their life history differences.

To test whether the feeding preferences of different predators caused these differences in life histories by natural selection, researchers introduced guppies from a pike-cichlid habitat into a guppy-free pool inhabited by killifish (Pool 3). The scientists tracked the weight and age at sexual maturity in the experimental guppy populations for 11 years, comparing these guppies with control guppies that remained in the pike-cichlid pools. The average weight and age at sexual maturity of the transplanted populations increased significantly as compared with the control populations. These studies demonstrate not only that life history traits are heritable and shaped by natural selection, but also that questions about evolution can be tested by field experiments.

As we have seen, population ecology involves theoretical model building as well as observations and experiments in the field. Next we look at how the principles of population ecology can be applied to conservation and management.

? Refer to Module 36.3. Which type of survivorship curve would you expect to find in a population experiencing *r*-selection? *K*-selection?

● Type III for a population experiencing *r*-selection; Type I for *K*-selection

36.8 Principles of population ecology have practical applications

CONNECTION

Principles of population ecology can help guide us toward resource management goals, such as increasing populations we wish to harvest or save from extinction or decreasing populations we consider pests.

Wildlife managers, fishery biologists, and foresters try to use **sustainable resource management**: harvesting crops without damaging the resource. In terms of population growth, this means maintaining a high population growth rate to replenish the population. According to the logistic growth model, the fastest growth rate occurs when the population size is at roughly half the carrying capacity of the habitat. Theoretically, a resource manager should achieve the best results by harvesting the populations down to this level. However, the logistic model assumes that growth rate and carrying capacity are stable over time. Calculations based on these assumptions, which are not realistic for some populations, may lead to unsustainably high harvest levels that ultimately deplete the resource. In addition, economic and political pressures often outweigh ecological concerns, and the amount of scientific information is frequently insufficient to determine sustainable harvest levels.

Fish, the only wild animals still hunted on a large scale, are particularly vulnerable to overharvesting. For example, in the northern Atlantic cod fishery, estimates of cod population sizes were too high, and the practice of discarding young cod (below legal size) at sea caused a higher mortality rate than predicted. The fishery collapsed in 1992 and has not recovered (**Figure 36.8**). Following the decline of many other fish and whale populations, resource managers are trying to minimize the risk of resource collapse by prohibiting harvest when population size reaches a pre-set minimum or establishing protected, harvest-free areas. For species that are in decline or facing extinction, resource managers may try to provide additional habitat or improve the quality of existing habitat to raise the carrying capacity and thus increase population growth.

Reducing the size of a population may be a challenging task. Simply killing many individuals will not usually decrease the size of a pest population. Many insect and weed

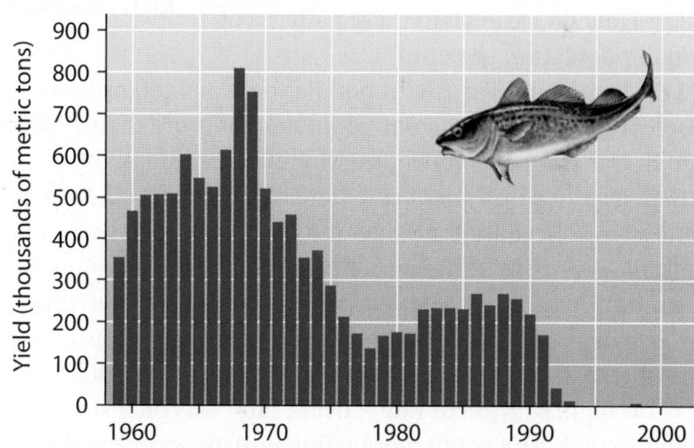

Yield (thousands of metric tons)
900
800
700
600
500
400
300
200
100
0
1960
1970
1980
1990
2000

Data from Stock Assessment of Northern (2J3KL) Cod, *Science Advisory Report* 2011/041, Fisheries and Oceans Canada (2011).

▲ **Figure 36.8** Collapse of northern cod fishery off Newfoundland

species have *r*-selected life history-traits and adaptations that promote rapid population growth. Also, most pesticides kill both the pest and their natural predators. Because prey species often have a higher reproductive rate than predators, pest populations rapidly rebound before their predators can.

Integrated pest management (IPM) uses a combination of biological, chemical, and cultivation methods to control agricultural pests. IPM relies on knowledge of the population ecology of the pest and its associated predators and parasites, as well as crop growth dynamics.

As you've learned, there are many factors that influence a population's size. To effectively manage any population, we must identify those variables, account for the unpredictability of the environment, consider interactions with other species, and weigh the economic, political, and conservation issues. These same issues apply to the growth of the human population, which we explore next.

? Explain why managers often try to maintain populations of fish and game species at about half their carrying capacity.

● To protect wildlife from overharvest yet maintain lower population levels so that growth rate is high and mortality from resource limitation is reduced

▷ The Human Population

36.9 The human population continues to increase, but the growth rate is slowing

In the few seconds it takes you to read this sentence, 21 babies will be born somewhere in the world and nine people will die. The statistics may have changed a bit since this book was printed, but births will still far outnumber deaths. An imbalance between births and deaths is the cause of population growth (or decline), and as the red curve in **Figure 36.9A** shows, the human population is expected to continue increasing for at least the next several decades. The blue bar graph in Figure 36.9A tells a different part of the story. The number of people added to the population each year has been declining since the 1980s. How do we explain these patterns of human population growth?

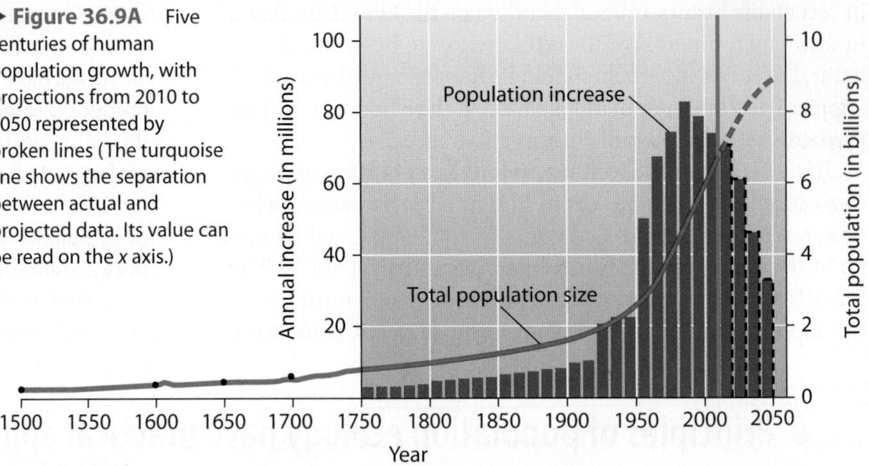

Population increase
Total population size
Annual increase (in millions)
Total population (in billions)
100
80
60
40
20
10
8
6
4
2
0
1500 1550 1600 1650 1700 1750 1800 1850 1900 1950 2000 2050
Year

▶ **Figure 36.9A** Five centuries of human population growth, with projections from 2010 to 2050 represented by broken lines (The turquoise line shows the separation between actual and projected data. Its value can be read on the *x* axis.)

Adapted from The World at Six Billion, *United Nations Publications* (1999).

Let's begin with the rise in population from 480 million people in 1500 to the current population of more than 7 billion. In our simplest model (see Module 36.4), population growth depends on r (per capita rate of increase) and N (population size). Because the value of r was assumed to be constant in a given environment, the growth rate in the examples we used in Module 36.4 depended wholly on the population size. Throughout most of human history, the same was true of people. Although parents had many children, mortality was also high, so r (birth rate − death rate) was only slightly higher than 0. As a result, population growth was very slow. (If we extended the x axis of Figure 36.9A back in time to year 1, when the population was roughly 300 million, the line would be almost flat for 1,500 years.) The 1 billion mark was not reached until the early 19th century.

As economic development in Europe and the United States led to advances in nutrition and sanitation and later, medical care, people took control of their population's rate of increase (r). At first, the death rate decreased, while the birth rate remained the same. The net rate of increase rose, and population growth began to pick up steam as the 20th century began. By mid-century, improvements in nutrition, sanitation, and health care had spread to the developing world, spurring growth at a breakneck pace as birth rates far outstripped death rates.

As the world population skyrocketed from 2 billion in 1927 to 3 billion just 33 years later, some scientists became alarmed. They feared that Earth's carrying capacity would be reached and that density-dependent factors (see Module 36.5) would maintain that population size through human suffering and death. But the overall growth rate peaked in 1962. In the more developed nations, advanced medical care continued to improve survivorship, but effective contraceptives held down the birth rate. As a result, the overall growth rate of the world's population began a downward trend.

Demographic Transition When the birth rate and death rate are equal, the population's rate of increase is zero. The world population is undergoing a change known as a **demographic transition**, a shift from birth rates and death rates that are high but roughly equal to birth and death rates that are low but roughly equal. **Figure 36.9B** shows the demographic transition of Mexico, which is projected to approach a zero net rate of increase with low birth and death rates in the next few decades. Notice that the death rate dropped sharply from 1925 to 1975 (the spike corresponds to the worldwide flu epidemic of 1918–1919), while the birth rate remained high until the 1960s. This is a typical pattern for demographic transitions.

Because economic development has occurred at different times in different regions, worldwide demographic transition is a mosaic of the changes occurring in different countries. The most developed nations have completed or are nearing completion of their demographic transitions. In these countries collectively, the rate of increase per 1,000 individuals was estimated at 1.1 in 2012 (**Table 36.9**, on the facing page). In the developing world, death rates have dropped, but high birth rates persist. As a result, these populations are growing

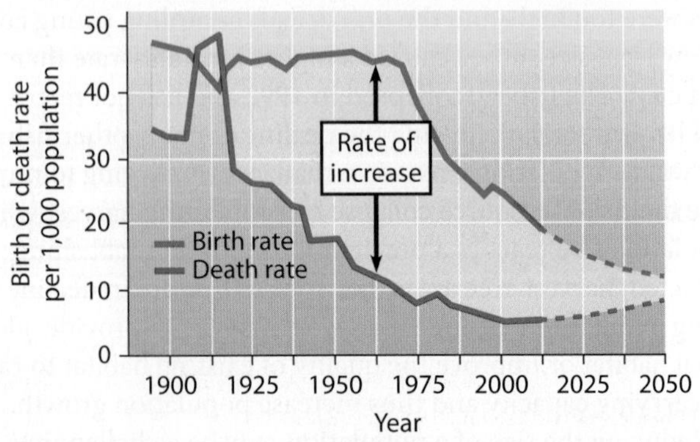

Rate of increase
Birth rate
Death rate
Birth or death rate per 1,000 population
50
40
30
20
10
0
1900 1925 1950 1975 2000 2025 2050
Year

Adapted from Transitions in World Population, *Population Bulletin* 59: 1 (2004).

▲ **Figure 36.9B** Demographic transition in Mexico

TABLE 36.9	POPULATION CHANGES IN 2012 (ESTIMATED)		
Population	**Birth Rate (per 1,000)**	**Death Rate (per 1,000)**	**Rate of Increase (per 1,000)**
World	19.1	7.9	11.2
More developed nations	11.2	10.1	1.1
Less developed nations	20.8	7.4	13.4

Data from U.S. Census Bureau International Data Base.

rapidly. Of the 77.7 million people added to the world in 2012, nearly 74 million were in developing nations.

Reduced family size is the key to the demographic transition. As women's status and education increase, they delay reproduction and choose to have fewer children. This phenomenon has been observed in both developed and developing countries, wherever the lives of women have improved. Given access to affordable contraceptive methods, women generally practice birth control, and many countries now subsidize family planning services and have official population policies. In many other countries, however, issues of family planning remain socially and politically charged, with heated disagreement over how much support should be provided for family planning.

Age Structures A demographic tool called an age-structure diagram is helpful for predicting a population's future growth. The **age structure** of a population is the number of individuals in different age-groups. **Figure 36.9C** shows the age structure of Mexico's population in 1989, its estimated 2012 age structure, and its projected age structure in 2035. In these diagrams, purple represents the portion of the population in their prereproductive years (0–14), pink indicates the part of the population in prime reproductive years (15–44), and blue is the proportion in postreproductive years (45 and older). Within each of these broader groups, each horizontal bar represents the population in a 5-year age-group. The area to the left of each vertical center line represents the number of males in each age-group; females are represented on the right side of the line.

An age structure with a broad base, such as Mexico's in 1989, reflects a population that has a high proportion of children and a high birth rate. The **fertility rate**—the average number of children produced by a woman over her lifetime—substantially exceeds the number of children needed to replace herself and her mate. As Figure 36.9B shows, the birth rate and the rate of increase have dropped 23 years later, but the population continues to be affected by its earlier expansion. This situation, which results from the increased proportion of women of childbearing age in the population, is known as **population momentum**. Girls 0–14 in the 1989 age structure (outlined in orange) are in their reproductive years in 2012, and girls who are 0–14 in 2012 (outlined in green) will carry the legacy of rapid growth forward to 2035. Putting the brakes on a rapidly expanding population is like stopping a freight train—the end result takes place long after the decision to do it was made. Even when the fertility rate is reduced to replacement level, the total population will continue to increase for several decades. The percentage of individuals under the age of 15 gives a rough idea of future growth. In the developing countries, about 28% of the population is in this age-group. In contrast, 16.5% of the population of developed nations is under the age of 15. Population momentum also explains why the population size in Figure 36.9A continues to increase even though fewer people are added to the population each year. In the next module, we examine the age structure of the United States.

? **During the demographic transition from high birth and death rates to low birth and death rates, countries usually undergo rapid population growth. Explain why.**

The death rate declines before the birth rate declines, creating a period when births greatly outnumber deaths. This also sets up population momentum.

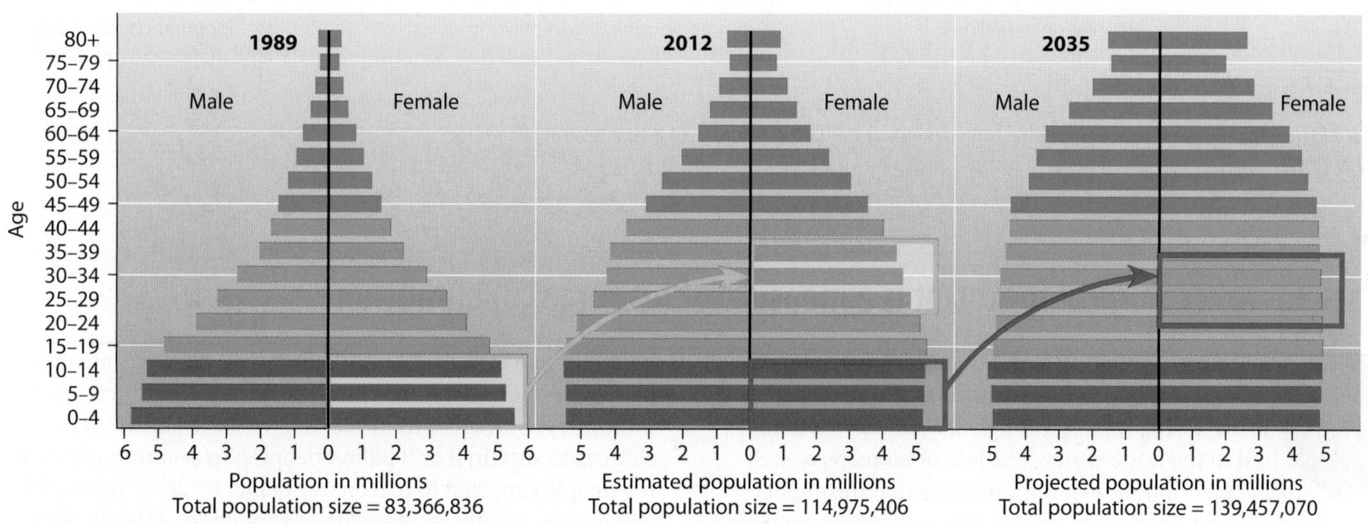

1989
2012
2035
Male
Female
Age
80+
75–79
70–74
65–69
60–64
55–59
50–54
45–49
40–44
35–39
30–34
25–29
20–24
15–19
10–14
5–9
0–4
6 5 4 3 2 1 0 1 2 3 4 5 6
5 4 3 2 1 0 1 2 3 4 5
5 4 3 2 1 0 1 2 3 4 5
Population in millions
Total population size = 83,366,836
Estimated population in millions
Total population size = 114,975,406
Projected population in millions
Total population size = 139,457,070

Adapted from International Data Base, U.S. Census Bureau (2013).

▲ **Figure 36.9C** Population momentum in Mexico

Try This Use the orange and green boxes to follow an age-group through time.

36.10 Age structures reveal social and economic trends

CONNECTION

Age-structure diagrams not only reveal a population's growth trends, but also indicate social conditions. For instance, an expanding population has an increasing need for schools, employment, and infrastructure. A large elderly population requires that extensive resources be allotted to health care. Let's look at trends in the age structure of the United States from 1989 to 2035 (**Figure 36.10**).

The large bulge in the 1989 age structure (tan screen) corresponds to the "baby boom" that lasted for about two decades after World War II ended in 1945. The large number of children swelled school enrollments, prompting construction of new schools and creating a demand for teachers. On the other hand, graduates who were born near the end of the boom faced stiff competition for jobs. Because they make up such a large segment of the population, boomers have had an enormous influence on social and economic trends. They also produced a "boomlet" of their own, seen in the 0–4 age-group in 1989 and the bump (green screen) in the 2012 age structure.

Where are the baby boomers now? The leading edge has reached retirement age, which will place pressure on programs such as Medicare and Social Security. In 2012, roughly 60% of the population was between 20 and 64, the ages most likely to be in the workforce, and 13.5% of the population was over 65. In 2035, the percentages are projected to be 54 and 20. In part, the increase in the elderly population is because people are living longer. The percentage of the population over 80, which was 2.7% in 1989, is projected to rise to nearly 6%—more than 23 million people—in 2035.

? **Point out an example of population momentum in Figure 36.10.**

● The 1981–1995 "boomlet" is a consequence of rapid reproduction in 1946–1965, as girls born during the baby boom entered their reproductive years.

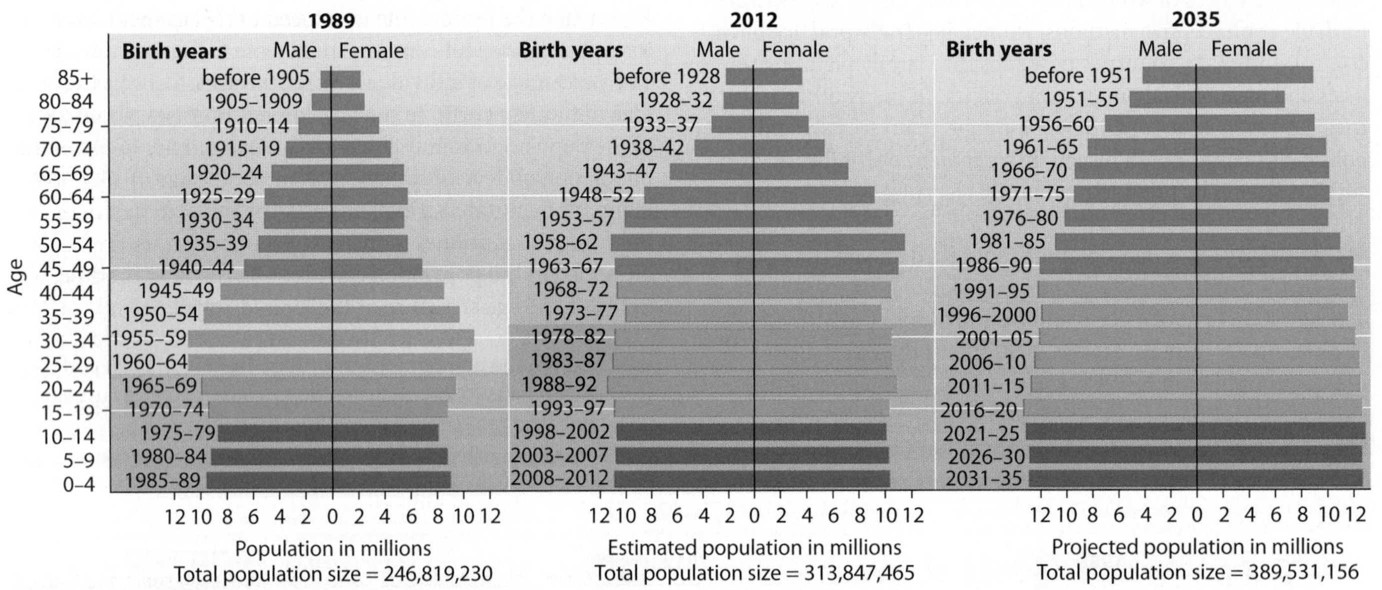

1989
Birth years
Male
Female
85+ before 1905
80–84 1905–1909
75–79 1910–14
70–74 1915–19
65–69 1920–24
60–64 1925–29
55–59 1930–34
50–54 1935–39
45–49 1940–44
40–44 1945–49
35–39 1950–54
30–34 1955–59
25–29 1960–64
20–24 1965–69
15–19 1970–74
10–14 1975–79
5–9 1980–84
0–4 1985–89
Age
12 10 8 6 4 2 0 2 4 6 8 10 12
Population in millions
Total population size = 246,819,230
2012
Birth years
Male
Female
before 1928
1928–32
1933–37
1938–42
1943–47
1948–52
1953–57
1958–62
1963–67
1968–72
1973–77
1978–82
1983–87
1988–92
1993–97
1998–2002
2003–2007
2008–2012
12 10 8 6 4 2 0 2 4 6 8 10 12
Estimated population in millions
Total population size = 313,847,465
2035
Birth years
Male
Female
before 1951
1951–55
1956–60
1961–65
1966–70
1971–75
1976–80
1981–85
1986–90
1991–95
1996–2000
2001–05
2006–10
2011–15
2016–20
2021–25
2026–30
2031–35
12 10 8 6 4 2 0 2 4 6 8 10 12
Projected population in millions
Total population size = 389,531,156

Data from International Data Base, U.S. Census Bureau website, (2013).

▲ **Figure 36.10** Age structures for the United States in 1989, 2012 (estimated), and 2035 (projected)

Try This Locate your age-group in 2012.

36.11 An ecological footprint is a measure of resource consumption

CONNECTION

How large a population of humans can Earth hold? In Module 36.9, we saw that the world's population is increasing rapidly, though at a slower rate than it did in the last century. The rate of increase, as well as population momentum, predicts that the populations of most developing nations will continue to increase for the foreseeable future. The U.S. Census Bureau projects a global population of 8 billion within the next 20 years and 9.5 billion by the mid-21st century. But these numbers are only part of the story. Trillions of bacteria can live in a petri dish if they have sufficient resources. Do we have sufficient resources to sustain 8 or 9 billion people? To accommodate all the people expected to live on our planet by 2025, the world will have to greatly increase food production. Already, agricultural lands are under pressure. Overgrazing by the world's growing herds of livestock is turning vast areas of grassland into desert. Water use has risen sixfold over the past 70 years, causing rivers to run dry, water for irrigation to be depleted, and levels of groundwater to drop. And because so much open space will be needed to support the expanding human population, many other species are expected to become extinct.

The concept of an ecological footprint is one approach to understanding resource availability and usage. An **ecological footprint** is an estimate of the land and water area required to provide the resources an individual or a nation consumes—for example, food, fuel, and housing—and to absorb the waste it generates, of which carbon emissions are a major component. Comparing our demand for resources with Earth's capacity to renew these resources, or biocapacity, gives us a broad view of the sustainability of human activities.

When the total area of ecologically productive land on Earth is divided by the global population, we each have a share of about 1.8 global hectares (1 hectare, or ha, = 2.47 acres; a *global hectare*, or gha, is a hectare with world-average ability to produce resources and absorb wastes). When used sustainably, resources such as crops, pastureland, forests, and fishing grounds will regenerate either naturally or with the assistance of technology.

According to the World Wildlife Fund, in 2008 (the most recent year for which data are available), the average ecological footprint for the world's population was 2.7 global hectares (gha)—roughly 1.5 times the planet's biocapacity per person. As **Figure 36.11A** shows, the enormous increase in humanity's carbon footprint, the emission of carbon dioxide and certain other gases, is chiefly responsible for the increase in our ecological footprint. (You'll learn more about the carbon footprint in Module 38.4.) By overshooting Earth's biocapacity, we are depleting our resources. The collapse of the northern cod fisheries (see Module 36.8) illustrates what happens when usage exceeds regenerative capacity.

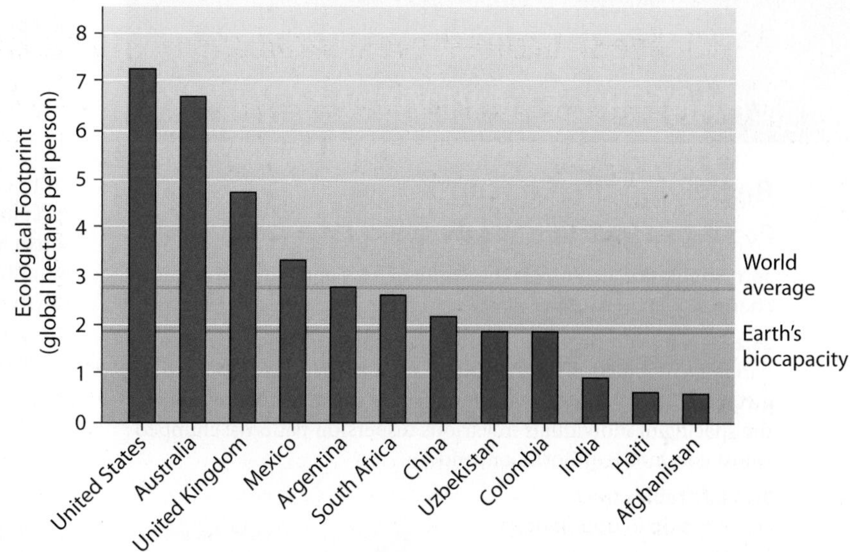
Ecological Footprint (global hectares per person)
0
1
2
3
4
5
6
7
8
World average
Earth's biocapacity
United States
Australia
United Kingdom
Mexico
Argentina
South Africa
China
Uzbekistan
Colombia
India
Haiti
Afghanistan

Adapted from Living Planet Report 2012: Biodiversity, Biocapacity, and Better Choices, World Wildlife Fund (2012).

▲ **Figure 36.11B** Ecological footprints of several countries

Figure 36.11B compares the ecological footprints of several countries to the world average footprint (purple line) and Earth's biocapacity (green line). Affluent nations like the United States and Australia consume a disproportionate amount of resources (**Figure 36.11C**). By this measure, the ecological impact of affluent nations such as the United States is potentially as damaging as unrestrained population growth in the developing world. So the problem is not just overpopulation, but also overconsumption. The world's richest countries, with 15% of the global population, account for 36% of humanity's total footprint. Some researchers estimate that providing everyone with the same standard of living as in the United States would require the resources of more than four planet Earths. To stay within the planet's regenerative capacity, all of humanity would have to live like an average citizen of Uzbekistan or Colombia.

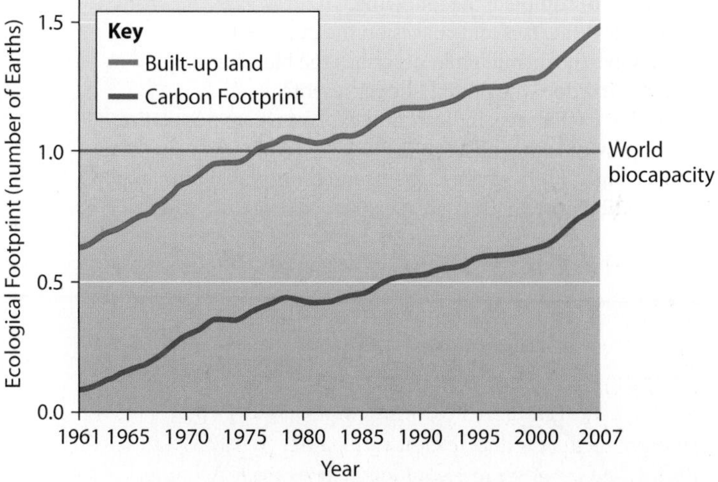
Ecological Footprint (number of Earths)
0.0
0.5
1.0
1.5
Key
Built-up land
Carbon Footprint
World biocapacity
1961 1965 1970 1975 1980 1985 1990 1995 2000 2007
Year

Data from B. Ewing et al., The Ecological Footprint Atlas, *Oakland: Global Footprint Network* (2010).

▲ **Figure 36.11A** Humanity's ecological footprint, 1961-2007

? **What is your ecological footprint? Do a Web search to find a site that calculates personal resource consumption.**

▲ **Figure 36.11C** American consumers shopping for electronics

CHAPTER 36 REVIEW

For practice quizzes, BioFlix animations, MP3 tutorials, video tutors, and more study tools designed for this textbook, go to MasteringBiology®

Reviewing the Concepts

Population Structure and Dynamics (36.1–36.8)

36.1 Population ecology is the study of how and why populations change.

36.2 Density and dispersion patterns are important population variables. Population density is the number of individuals in a given area or volume. Environmental and social factors influence the spacing of individuals in various dispersion patterns: clumped (most common), uniform, or random.

36.3 Life tables track survivorship in populations. Life tables and survivorship curves predict an individual's statistical chance of dying or surviving during each interval in its life. The three types of survivorship curves reflect differences in species' reproduction and mortality.

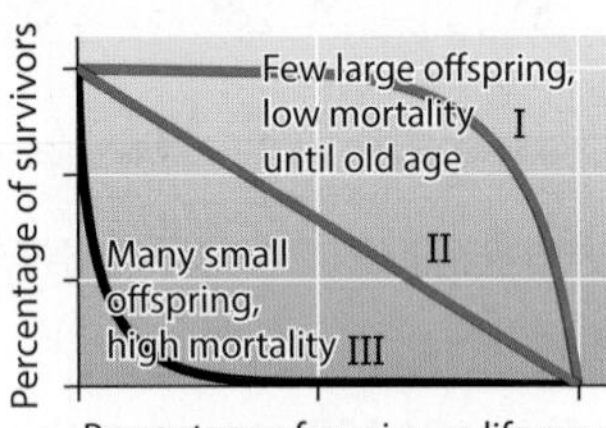
Few large offspring, low mortality until old age
I
II
Many small offspring, high mortality
III
Percentage of survivors
Percentage of maximum life span

36.4 Idealized models predict patterns of population growth. Exponential growth is the accelerating increase that occurs when growth is unlimited. The equation $G = rN$ describes this J-shaped growth curve, where G = the population growth rate, r = an organism's inherent capacity to reproduce, and N = the population size. Logistic growth is the model that represents the slowing of population growth as a result of limiting factors and the leveling off at carrying capacity, which is the number of individuals the environment can support. The equation $G = rN(K - N)/K$ describes a logistic growth curve, where K = carrying capacity and the term $(K - N)/K$ accounts for the leveling off of the curve.

36.5 Multiple factors may limit population growth. As a population's density increases, factors such as limited food supply and increased disease or predation may increase the death rate, decrease the birth rate, or both. Abiotic, density-independent factors such as severe weather may limit many natural populations. Most populations are probably regulated by a mixture of factors, and fluctuations in numbers are common.

36.6 Some populations have "boom-and-bust" cycles. Researchers have tested hypotheses that explain the population cycles of the lynx and the snowshoe hare.

36.7 Evolution shapes life histories. Natural selection shapes a species' life history, the series of events from birth through reproduction to death. Populations with so-called r-selected life history traits produce many offspring and grow rapidly in unpredictable environments. Populations with K-selected traits raise few offspring and maintain relatively stable populations. Most species fall between these extremes.

36.8 Principles of population ecology have practical applications. For example, resource managers use population ecology to determine sustainable yields.

The Human Population (36.9–36.11)

36.9 The human population continues to increase, but the growth rate is slowing. The human population grew rapidly during the 20th century and currently stands at more than 7 billion. Demographic transition, the shift from high birth and death rates to low birth and death rates, has lowered the rate of growth in developed countries. In the developing nations, death rates have dropped, but birth rates are still high. The age structure of a population—the proportion of individuals in different age-groups—affects its future growth. Population momentum is the continued growth that occurs despite reduction of the fertility rate to replacement level and is a result of girls in the 0–14 age-group of a previously expanding population reaching their childbearing years.

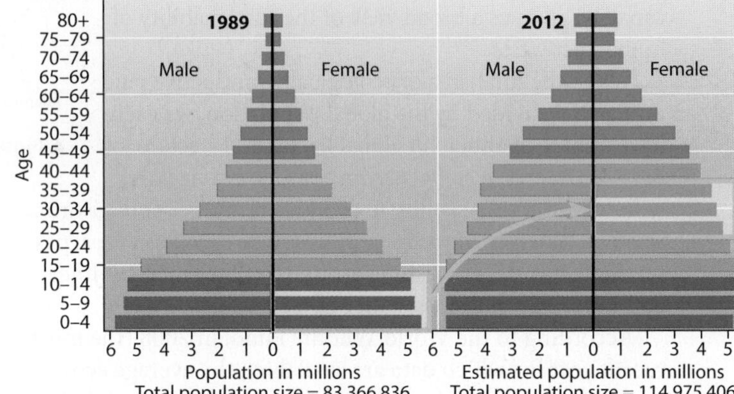
1989
2012
Male
Female
Male
Female
Age
80+
75–79
70–74
65–69
60–64
55–59
50–54
45–49
40–44
35–39
30–34
25–29
20–24
15–19
10–14
5–9
0–4
6 5 4 3 2 1 0 1 2 3 4 5 6
Population in millions
Total population size = 83,366,836
6 5 4 3 2 1 0 1 2 3 4 5
Estimated population in millions
Total population size = 114,975,406

36.10 Age structures reveal social and economic trends.

36.11 An ecological footprint is a measure of resource consumption. An ecological footprint estimates the amount of land required by each person or country to produce all the resources it consumes and to absorb all its wastes. The global ecological footprint already exceeds a sustainable level. There is a huge disparity between resource consumption in more developed and less developed nations.

Connecting the Concepts

1. Use this graph of the idealized exponential and logistic growth curves to complete the following.
   a. Label the axes and curves on the graph.
   b. Give the formula that describes the blue curve.
   c. What does the dotted line represent?
   d. For each curve, indicate and explain where population growth is the most rapid.
   e. Which of these curves best represents global human population growth?

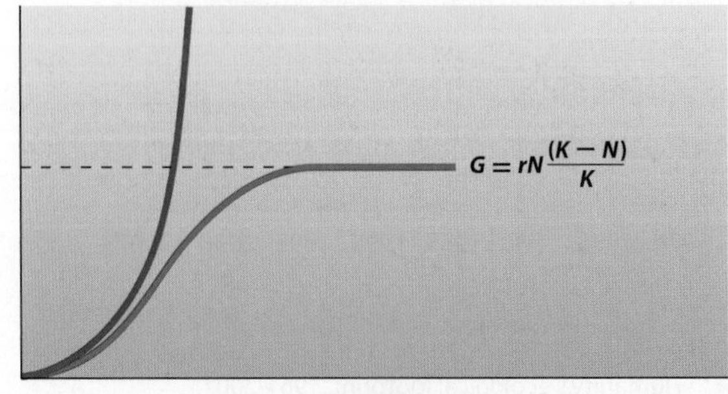
$G = rN\frac{(K - N)}{K}$

2. The graph below shows the demographic transition for a hypothetical country. Many developed countries that have achieved a stable population size have undergone a transition similar to this. Answer the following questions concerning this graph.
   a. What does the blue line represent? The red line?
   b. This diagram has been divided into four sections. Describe what is happening in each section.
   c. In which section(s) is the population size stable?
   d. In which section is the population growth rate the highest?

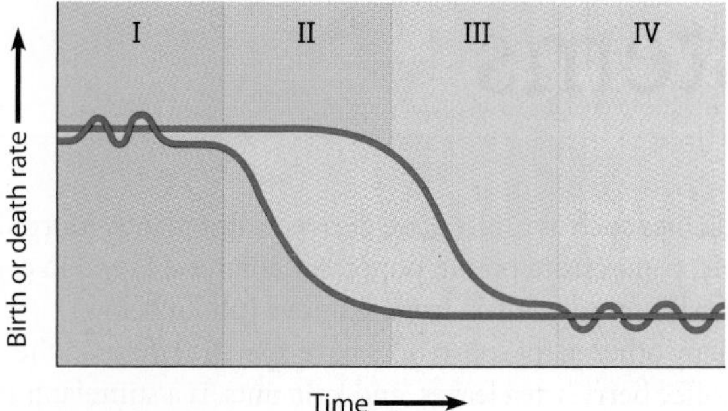

I
II
III
IV
Birth or death rate
Time

Testing Your Knowledge

Level 1: Knowledge/Comprehension

3. After seeds have sprouted, gardeners often pull up some of the seedlings so that only a few grow to maturity. How does this practice help produce the best yield?
   a. by increasing K
   b. by decreasing r
   c. by reducing intraspecific competition
   d. by adding a density-independent factor to the environment
4. To figure out the human population density of your community, you would need to know the number of people living there and
   a. the land area in which they live.
   b. the birth rate of the population.
   c. the dispersion pattern of the population.
   d. the carrying capacity.
5. The term $(K-N)/K$
   a. is the carrying capacity for a population.
   b. is greatest when K is very large.
   c. is zero when population size equals carrying capacity.
   d. increases in value as N approaches K.
6. With regard to its rate of growth, a population that is growing logistically
   a. grows fastest when density is lowest.
   b. has a high intrinsic rate of increase.
   c. grows fastest at an intermediate population density.
   d. grows fastest as it approaches carrying capacity.
7. Which of the following represents a demographic transition?
   a. A population switches from exponential to logistic growth.
   b. A population reaches zero population growth when the birth rate drops to zero.
   c. There are equal numbers of individuals in all age-groups.
   d. A population switches from high birth and death rates to low birth and death rates.
8. Skyrocketing growth of the human population appears to be mainly a result of
   a. a drop in death rate due to sanitation and health care.
   b. better nutrition boosting the birth rate.
   c. the concentration of humans in cities.
   d. social changes that make it desirable to have more children.
9. According to data on ecological footprints,
   a. the carrying capacity of the world is 10 billion.
   b. the current demand on global resources by industrialized countries is less than the resources available in those countries.
   c. the ecological footprint of the United States is more than twice the world average.
   d. nations with the largest ecological footprints have the fastest population growth rates.

Level 2: Application/Analysis

10. What are some factors that might have a density-dependent limiting effect on population growth?
11. What is survivorship? What does a survivorship curve show? Explain what the three survivorship curves tell us about humans, squirrels, and oysters.
12. Describe the factors that might produce the following three types of dispersion patterns in populations.

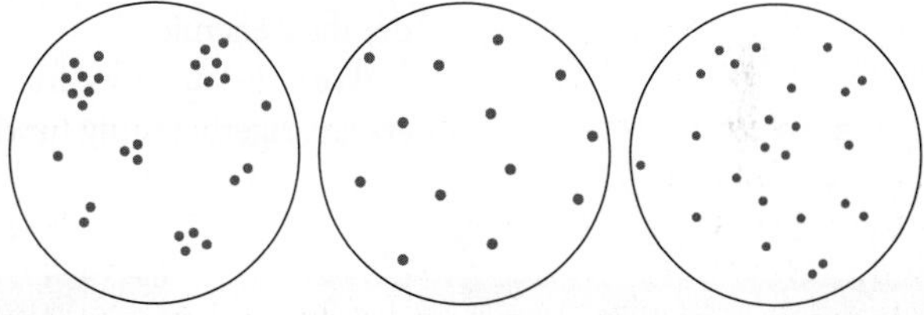

Level 3: Synthesis/Evaluation

13. The mountain gorilla, spotted owl, giant panda, snow leopard, and grizzly bear are all endangered by human encroachment on their environments. Another thing these animals have in common is that they all have K-selected life history traits. Why might they be more easily endangered than animals with r-selected life history traits? What general type of survivorship curve would you expect these species to exhibit? Explain your answer.
14. **SCIENTIFIC THINKING** Another hypothesis for snowshoe hare population cycles proposes that they are caused by sunspot activity. According to this hypothesis, sunspot activity affects the chemicals present in the plants eaten by hares, which in turn affects the quality of the food. What testable predictions are generated by this hypothesis?
15. Many people regard the rapid population growth of developing countries as our most serious environmental problem. Others think that the growth of developed countries, though slower, is actually a greater threat to the environment. What kinds of environmental problems result from population growth in (a) developing countries and (b) developed countries? Which do you think is the greater threat? Why?

Answers to all questions can be found in Appendix 4.

CHAPTER 37

Communities and Ecosystems

Many drugs, including common medicines such as aspirin, are derived from plants. Morphine, a powerful pain reliever, for example, comes from opium poppies; a chemical found in certain *Ephedra* species relieves the symptoms of asthma; and a drug from foxglove (photo below) is used to treat congestive heart failure. Many other plant substances have specific effects in the human body, as well. Caffeine, found in coffee berries, tea leaves, and kola nuts, is a stimulant; so is the nicotine in tobacco leaves. A plant called deadly nightshade provides several useful substances, including a mild sedative that is used to combat motion sickness and the drops used to dilate the pupils for an eye exam. Its ominous name is well-deserved, however—the leaves and berries contain deadly poisons that were used by ancient Romans to murder their enemies.

Why do plants make drugs?

We've listed only a fraction of the plants that are known to have medicinal properties. Researchers are eagerly testing newly discovered compounds that show pharmaceutical promise in treating

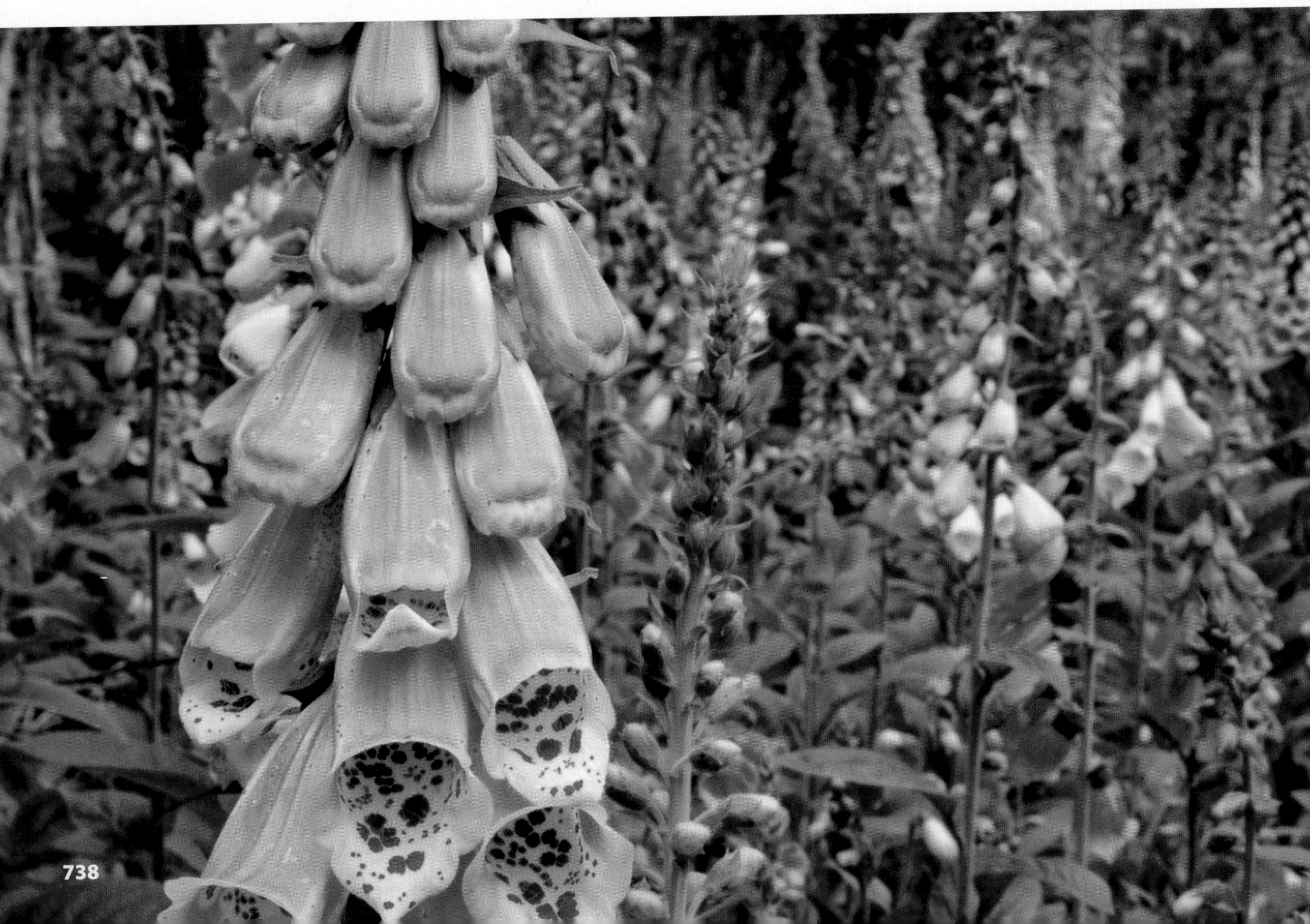

cancer and infectious diseases such as malaria and HIV/AIDS, as pain relievers and sedatives, in promoting weight loss, and for many other uses. But why do plants make these substances? What is the adaptive value? The title of this chapter offers a clue: In a community, populations of many different species interact with each other, and some of these relationships are potentially harmful.

In this chapter, you'll examine the interactions among organisms and how those relationships determine the features of communities. On a larger scale, you'll explore the dynamics of ecosystems. And throughout the chapter, you'll learn how an understanding of these ecological relationships can help us manage Earth's resources wisely.

BIG IDEAS

Community Structure and Dynamics

(37.1–37.13)

Community ecologists examine factors that influence the species composition and distribution of communities and factors that affect community stability.

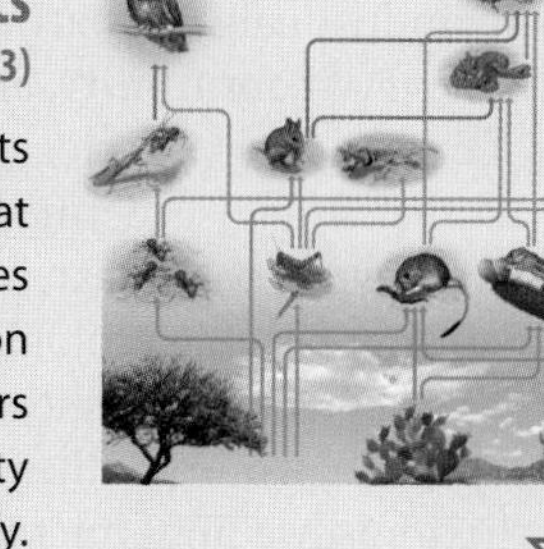

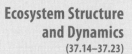

Ecosystem Structure and Dynamics

(37.14–37.23)

Ecosystem ecology emphasizes energy flow and chemical cycling.

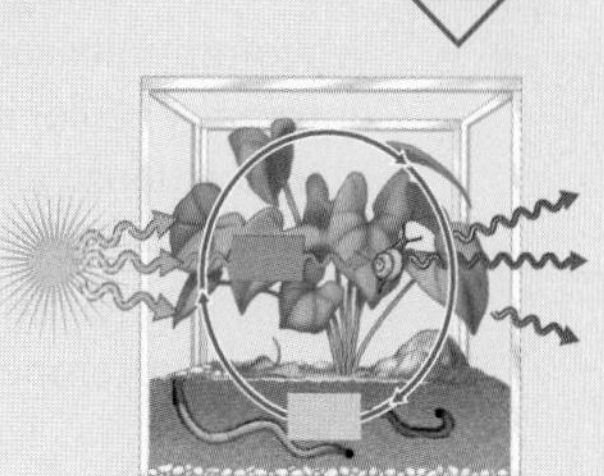

▷ Community Structure and Dynamics

37.1 A community includes all the organisms inhabiting a particular area

In the hierarchy of life, a population is a group of interacting individuals of a particular species. The next step up is a biological **community**, an assemblage of all the populations of organisms living close enough together for potential interaction. Ecologists define the boundaries of the community according to the research questions they want to investigate. For example, one ecologist interested in wetland communities might study the shoreline plants and animals of a particular marsh, while another might investigate only the benthic (bottom-dwelling) microbes.

A community can be described by its species composition. Community ecologists seek to understand how abiotic factors and interactions between populations affect the composition and distribution of communities. For example, a community ecologist might compare the benthic microbes of a marsh located in the temperate zone with those of a tundra marsh.

Community ecologists also investigate community dynamics, the variability or stability in the species composition of a community caused by biotic and abiotic factors. For example, a community ecologist might study changes in the species composition of a wetlands community in Louisiana after a hurricane.

Community ecology is necessary for the conservation of endangered species and the management of wildlife, game, and fisheries. It is vital for controlling diseases, such as malaria, bird flu, and Lyme disease, that are carried by animals. Community ecology also has applications in agriculture, where people attempt to control the species composition of communities they have established.

? What is the relationship between a community and a population?

● A community is a group of populations that interact with each other.

37.2 Interspecific interactions are fundamental to community structure

Members of communities engage in **interspecific interactions**—relationships with individuals of other species in the community—that greatly affect population structure and dynamics. In **Table 37.2**, interspecific interactions are classified according to the effect on the populations concerned, which may be helpful (+) or harmful (−).

Members of a population may engage in intraspecific competition for limited resources such as food or space (see Module 36.5). **Interspecific competition** occurs when populations of two different species compete for the same limited resource. For example, desert plants compete for water, whereas plants in a tropical rain forest compete for light. Squirrels and black bears are among the animals that feed on acorns in a temperate broadleaf forest in autumn. When acorn production is low, the nut is a limited resource for which squirrels and bears compete. In general, the effect of interspecific competition is negative for both populations (−/−).

TABLE 37.2 | INTERSPECIFIC INTERACTIONS

Interspecific Interaction	Effect on Species 1	Effect on Species 2	Example
Competition	−	−	Squirrels/ black bears
Mutualism	+	+	Plants/ mycorrhizae
Predation	+	−	Crocodiles/fish
Herbivory	+	−	Caterpillars/leaves
Parasites and pathogens	+	−	Heartworm/dogs; *Salmonella*/ humans

However, it may be far more harmful for one population than the other. Interspecific competition is responsible for some of the disastrous effects of introducing non-native species into a community, a topic we will explore further in Module 37.13.

In **mutualism**, both populations benefit (+/+). Plants and mycorrhizae (see Module 17.12) and herbivores and the cellulose-digesting microbes that inhabit their digestive tracts (see Module 21.13) are examples of mutualism between symbiotic species—those that have a physically close association with each other. Mutualism can also occur between species that are not symbiotic. For example, flowers and their pollinators are mutualists (see Figure 17.10C).

There are three categories of interactions in which one species exploits another species (+/−). In **predation**, one species (the predator) kills and eats another species (the prey). **Herbivory** is consumption of plant parts or algae by an animal. Both plants and animals may be victimized by parasites (see Module 16.12) or pathogens (see Module 16.1). Thus, parasite-host and pathogen-host interactions are also +/−.

In the next several modules, you will learn more about these interspecific interactions and how they affect communities. You will also discover how interspecific interactions can act as powerful agents of natural selection.

? Populations of eastern bluebirds declined after the introduction of non-native house sparrows and European starlings. All three species nest in tree cavities. Suggest how an interspecific interaction could explain the bluebird's decline.

● Based on the information given, interspecific competition for nest sites is a plausible explanation.

37.3 Competition may occur when a shared resource is limited

Each species in a community has an **ecological niche**, defined as the sum of its use of the biotic and abiotic resources in its environment. For example, the ecological niche of a small bird called the Virginia's warbler **(Figure 37.3A)** includes its nest sites and nest-building materials, the insects it eats, and climatic conditions such as the amount of precipitation and the temperature and humidity that enable it to survive. In other words, the ecological niche encompasses everything the Virginia's warbler needs for its existence.

▲ **Figure 37.3A**
A Virginia's warbler (*Vermivora virginiae*)

−/−

Interspecific competition occurs when the niches of two populations overlap and both populations need a resource that is in short supply. Ecologists can study the effects of competition by removing all the members of one species from a study site. For example, in central Arizona, the niche of the orange-crowned warbler **(Figure 37.3B)** overlaps in some respects with the niche of the Virginia's warbler. When researchers removed either species, the remaining species was significantly more successful in raising their offspring. Thus, interspecific competition has a direct effect on reproductive fitness in these birds.

In general, competition lowers the carrying capacity (see Module 36.4) for competing populations because the resources used by one population are not available to the other population. In 1934, Russian ecologist G. F. Gause demonstrated the effects of interspecific competition using three closely related species of ciliates (see Module 16.14): *Paramecium caudatum*, *P. aurelia*, and *P. bursaria*. He first determined the carrying capacity for each species under laboratory conditions. Then he grew cultures of the two species together. In a mixed culture of *P. caudatum* and *P. bursaria*, population sizes stabilized at lower numbers than each achieved in the absence of a competing species—competition lowered the carrying capacity of the environment. On the other hand, in a mixed culture of *P. caudatum* and *P. aurelia*, only *P. aurelia* survived. Gause concluded that the requirements of these two species were so similar that they could not coexist under those conditions; *P. aurelia* outcompeted *P. caudatum* for essential resources.

▲ **Figure 37.3B**
An orange-crowned warbler (*Vermivora celata*)

? Which do you think has more severe effects, intraspecific competition or interspecific competition? Explain why.

● Intraspecific competition is more severe because members of the same species have exactly the same niche. Thus, they compete for exactly the same resources.

37.4 Mutualism benefits both partners

+/+

Reef-building corals and photosynthetic dinoflagellates (unicellular algae; see Module 16.14) provide a good example of how mutualists benefit from their relationship. Coral reefs are constructed by successive generations of colonial coral animals that secrete an external calcium carbonate ($CaCO_3$) skeleton. Deposition of the skeleton must outpace erosion and competition for space from fast-growing seaweeds. Corals could not build and sustain the massive reefs that provide the food, living space, and shelter to support the splendid diversity of the reef community without the millions of dinoflagellates that live in the cells of each coral polyp **(Figure 37.4)**. The sugars that the dinoflagellates produce by photosynthesis provide at least half of the energy used by the coral animals. In return, the dinoflagellates gain a secure shelter that provides access to light. They also use the coral's waste products, including carbon dioxide (CO_2) and ammonia (NH_3), a valuable source of nitrogen for making proteins. Unicellular algae have similar mutually beneficial relationships with a wide variety of other marine invertebrates, including sponges, flatworms, and molluscs.

▲ **Figure 37.4** Coral polyps

? When corals are stressed by environmental conditions, they expel their dinoflagellates in a process called bleaching. How is widespread bleaching likely to affect coral reefs?

● Without their dinoflagellate mutualists, corals do not have enough energy to maintain the reef structure. Bleached reefs will die.

37.5 Predation leads to diverse adaptations in prey species

EVOLUTION CONNECTION

+/−

Predation benefits the predator but kills the prey. Because predation has such a negative impact on reproductive success in prey populations, numerous adaptations for predator avoidance have evolved in prey populations through natural selection.

Insect color patterns, including camouflage, provide protection against predators. Camouflage is also common in other animals (see Figure 1.9). As **Figure 37.5A** shows, the gray tree frog (*Hyla arenicolor*), an inhabitant of the southwestern United States, becomes almost invisible on a gray tree trunk.

Other protective devices include mechanical defenses, such as the sharp quills of a porcupine (see Figure 35.9) or the hard shells of clams and oysters. Chemical defenses are also widespread. Animals with effective chemical defenses usually have bright color patterns, often yellow, orange, or red in combination with black. Predators learn to associate these color patterns with undesirable consequences, such as noxious taste or a painful sting, and avoid potential prey with similar markings. The vivid orange and black pattern of monarch butterflies (**Figure 37.5B**) warns potential predators of a nasty taste. Monarchs acquire and store the unpalatable chemicals during the larval stage, when the caterpillars feed on milkweed plants.

? Explain why predation is a powerful factor in the adaptive evolution of prey species.

The prey that avoid being eaten will most likely survive and reproduce, passing alleles for antipredator adaptations on to their offspring.

▲ **Figure 37.5A** Camouflage: a gray tree frog on bark

▲ **Figure 37.5B** Chemical defenses: the monarch butterfly

37.6 Herbivory leads to diverse adaptations in plants

EVOLUTION CONNECTION

+/−

Although herbivory is not usually fatal, a plant whose body parts have been eaten by an animal must expend energy to replace the loss. Consequently, numerous defenses against herbivores have evolved in plants. Thorns and spines are obvious antiherbivore devices, as anyone who has plucked a rose from a thorny rosebush or brushed against a spiky cactus knows. The chemicals described in the chapter introduction—the substances that we use medicinally or for other purposes—are also adaptations that defend plants against herbivory.

Like the chemical defenses of animals, toxins in plants tend to be distasteful, and herbivores learn to avoid them. Among such chemical weapons are the poison strychnine, produced by a tropical vine called *Strychnos toxifera*; morphine, from the opium poppy; nicotine, produced by the tobacco plant; mescaline, from peyote cactus; and tannins, from a variety of plant species. A variety of sulfur compounds, including those that give brussels sprouts and cabbage their distinctive taste, are also toxic to herbivorous insects and mammals such as cattle. (The vegetables we eat are not toxic because the amount of chemicals in them has been reduced by crop breeders.) Some plants even produce chemicals that cause abnormal development in insects that eat them. Chemical companies have taken advantage of the poisonous properties of certain plants to produce the pesticides called pyrethrin and rotenone. Nicotine is also used as an insecticide.

Why do plants make drugs?

Some herbivore-plant interactions illustrate the concept of **coevolution**, a series of reciprocal evolutionary adaptations in two species. Coevolution occurs when a change in one species acts as a new selective force on another species, and the resulting adaptations of the second species in turn affect the selection of individuals in the first species. **Figure 37.6** (top left of next page) illustrates an example of coevolution

Decoy eggs

Heliconius
Eggs

▲ **Figure 37.6** Coevolution: *Heliconius* and the passionflower vine (*Passiflora*)

between an herbivorous insect (the caterpillar of the butterfly *Heliconius*, top left) and a plant (the passionflower, *Passiflora*, a tropical vine).

Passiflora produces toxic chemicals that protect its leaves from most insects, but *Heliconius* caterpillars have digestive enzymes that break down the toxins. As a result, *Heliconius* gains access to a food source that few other insects can eat. These poison-resistant caterpillars seem to be a strong selective force for *Passiflora* plants, and defenses have evolved in some species. For instance, the leaves of some *Passiflora* species produce yellow spots that look like *Heliconius* eggs (Figure 37.6). Female butterflies avoid laying their eggs on leaves that already have eggs, presumably ensuring that only a few caterpillars will hatch and feed on any one leaf. Because the butterfly often mistakes the yellow spots for eggs, *Passiflora* species with these false eggs are less likely to be eaten.

? People find most bitter-tasting foods objectionable. Why do you suppose we have taste receptors for bitter-tasting chemicals?

● Individuals having bitter taste receptors presumably survived better because they could identify potentially toxic food when they foraged.

37.7 Parasites and pathogens can affect community composition

+/−

A parasite lives on or in a host from which it obtains nourishment. Internal parasites include flukes and tapeworms (see Module 18.7) and a variety of nematodes (see Module 18.8) that live inside a host organism's body. External parasites include arthropods such as ticks, lice, mites, and mosquitoes, which attach to their victims temporarily to feed on blood or other body fluids. Plants are also attacked by parasites, including nematodes and aphids, tiny insects that tap into the phloem and suck plant sap (**Figure 37.7**). Pathogens are disease-causing bacteria, viruses, fungi, or protists that can be thought of as microscopic parasites.

The potentially devastating effects of parasites and pathogens on cultivated plants, livestock, and humans are well known, but ecologists know little about how these interactions affect natural communities. Non-native pathogens, whose impact is rapid and often dramatic, have provided some opportunities to study the effects of pathogens on communities. In one example, ecologists studied the consequences of an epidemic of chestnut blight that wiped out virtually all American chestnut trees during the first half of the 20th century; the disease is caused by a protist. Chestnuts were massive canopy trees that dominated many forest communities in North America. Their loss had a significant impact on species composition and community structure. Overall, the diversity of tree species increased as trees that had formerly competed with chestnuts, such as oaks and hickories, became more prominent. The dead chestnut trees furnished niches for other organisms, such as insects, cavity-nesting birds, and eventually decomposers. On the other hand, populations of organisms that depended heavily on living chestnut trees for their food and shelter declined.

A fungus-like protist that causes a disease called sudden oak death is currently spreading on the West Coast. More than a million oaks have been lost so far, causing the decline of bird populations. Despite its name, sudden oak death affects many other species as well, including the majestic redwood and Douglas fir trees and flowering shrubs such as rhododendron and camellia. Because the epidemic is in its early stages, its full effect on forest communities will not be known for some time.

◀ **Figure 37.7** Aphids parasitizing a plant

? Use your knowledge of interspecific interactions to explain why tree diversity increased after all the chestnuts died.

● Chestnuts had many of the same niche characteristics as other trees, but apparently chestnuts were superior competitors. After they died, the remaining species may have had fewer niche similarities, or they may have been more equal as competitors, allowing more species to coexist.

37.8 Trophic structure is a key factor in community dynamics

Every community has a **trophic structure**, a pattern of feeding relationships consisting of several different levels. The sequence of food transfer up the trophic levels is known as a **food chain**. This transfer of food moves chemical nutrients and energy from organism to organism up through the trophic levels in a community.

Figure 37.8 compares a terrestrial food chain and an aquatic food chain. In this figure, the trophic levels are arranged vertically, and the names of the levels appear in colored boxes. The arrows connecting the organisms point from the food to the consumer, that is, in the direction of nutrient and energy transfer. Starting at the bottom, the trophic level that supports all others consists of autotrophs ("self-feeders"), which ecologists call **producers**. Photosynthetic producers use light energy to power the synthesis of organic compounds. Plants are the main producers on land. In water, the producers are mainly photosynthetic unicellular protists and cyanobacteria, collectively called phytoplankton. Multicellular algae and aquatic plants are also important producers in shallow waters. In a few communities, the producers are chemosynthetic prokaryotes. For example, in communities around hydrothermal vents, deep-sea sites near the adjoining edges of Earth's tectonic plates (see Module 15.7), the producers are bacteria that obtain energy by oxidizing hydrogen sulfide emitted from the vents.

All organisms in trophic levels above the producers are heterotrophs ("other-feeders"), or consumers, and all consumers are directly or indirectly dependent on the output of producers. Herbivores, which eat plants, algae, or phytoplankton, are **primary consumers**. Primary consumers on land include grasshoppers and many other insects, snails, and certain vertebrates, such as grazing mammals and birds that eat seeds and fruits. In aquatic environments, primary consumers include a variety of zooplankton (mainly protists and microscopic animals such as small shrimps) that eat phytoplankton.

Above primary consumers, the trophic levels are made up of carnivores and insectivores, which eat the consumers from the level below. On land, **secondary consumers** include many small mammals, such as the mouse shown here eating an herbivorous insect, and a great variety of birds, frogs, and spiders, as well as lions and other large carnivores that eat grazers. In aquatic ecosystems, secondary consumers are mainly small fishes that eat zooplankton.

Higher trophic levels include **tertiary** (third-level) **consumers**, such as snakes that eat mice and other secondary consumers. Most ecosystems have secondary and tertiary consumers. As the figure indicates, some also have a higher level, **quaternary** (fourth-level) **consumers**, which eat tertiary consumers. These include hawks in terrestrial ecosystems and killer whales in the marine environment.

Not shown in Figure 37.8 is another trophic level—consumers that derive their energy from **detritus**, the dead material produced at all the trophic levels. Detritus includes animal wastes, plant litter, and the bodies of dead organisms. Different organisms consume detritus in different stages of decay. **Scavengers**, which are large animals, such as crows and vultures, feast on carcasses left behind by predators or speeding cars. The diet of **detritivores** is made up primarily of decaying organic material. Examples of detritivores include earthworms and millipedes. **Decomposers**, mainly prokaryotes and fungi, secrete enzymes that digest molecules in organic material and convert them to inorganic forms. Enormous numbers of microscopic decomposers in the soil and in the mud at the bottom of lakes and oceans break down most of the community's organic materials to inorganic compounds that plants or phytoplankton can use. The breakdown of organic materials to inorganic ones is called **decomposition**. By breaking down detritus, decomposers link all trophic levels. Their role is essential for all communities and, indeed, for the continuation of life on Earth.

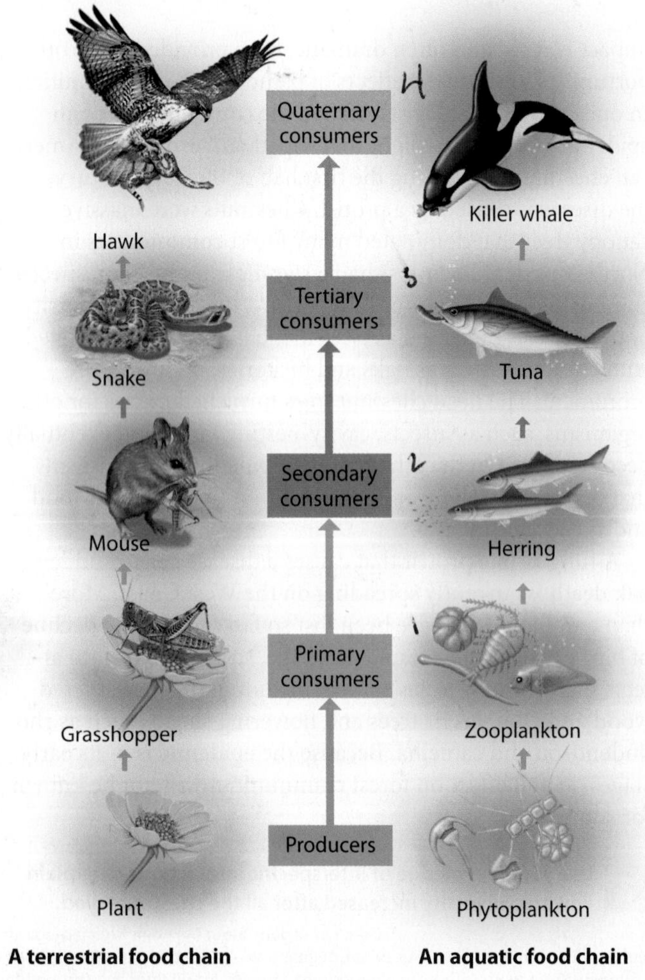

Quaternary consumers
Tertiary consumers
Secondary consumers
Primary consumers
Producers
Hawk
Snake
Mouse
Grasshopper
Plant
Killer whale
Tuna
Herring
Zooplankton
Phytoplankton

A terrestrial food chain **An aquatic food chain**

▲ **Figure 37.8** Two food chains

? **I'm eating a cheese pizza. At which trophic level(s) am I feeding?**

Primary consumer (flour and tomato sauce) and secondary consumer (cheese, a product from cows, which are primary consumers)

VISUALIZING THE CONCEPT

37.9 Food chains interconnect, forming food webs

A more realistic view of the trophic structure of a community is a **food web**, a network of interconnecting food chains. In this Sonoran desert community, a consumer may eat more than one type of producer, and several species of primary consumers may feed on the same species of producer. Some animals weave into the food web at more than one trophic level.

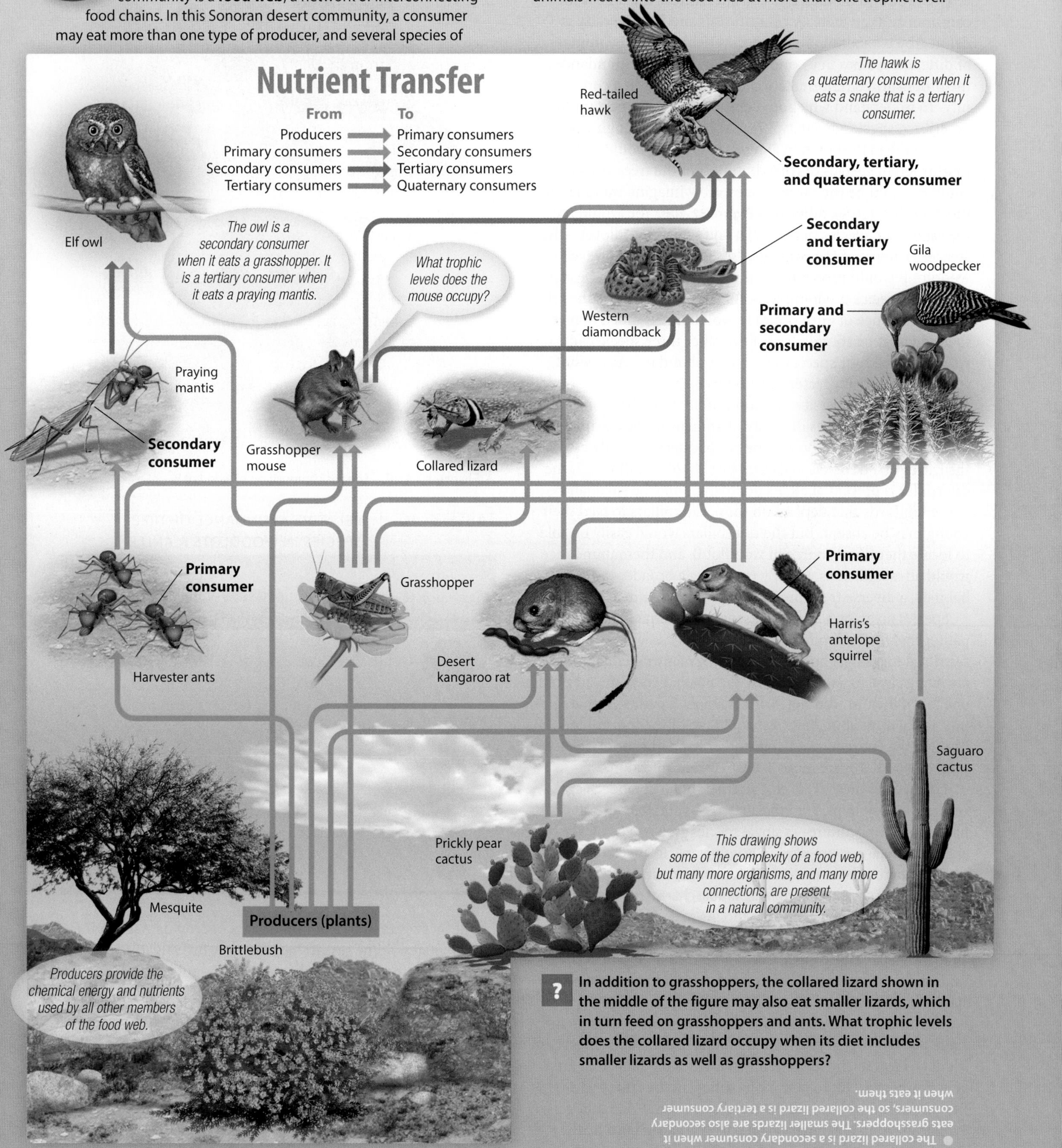

Nutrient Transfer

From	To
Producers	Primary consumers
Primary consumers	Secondary consumers
Secondary consumers	Tertiary consumers
Tertiary consumers	Quaternary consumers

Red-tailed hawk

The hawk is a quaternary consumer when it eats a snake that is a tertiary consumer.

Secondary, tertiary, and quaternary consumer

Elf owl

The owl is a secondary consumer when it eats a grasshopper. It is a tertiary consumer when it eats a praying mantis.

What trophic levels does the mouse occupy?

Western diamondback

Secondary and tertiary consumer

Gila woodpecker

Primary and secondary consumer

Praying mantis

Secondary consumer

Grasshopper mouse

Collared lizard

Harvester ants

Primary consumer

Grasshopper

Desert kangaroo rat

Harris's antelope squirrel

Primary consumer

Saguaro cactus

Prickly pear cactus

This drawing shows some of the complexity of a food web, but many more organisms, and many more connections, are present in a natural community.

Mesquite

Producers (plants)

Brittlebush

Producers provide the chemical energy and nutrients used by all other members of the food web.

? In addition to grasshoppers, the collared lizard shown in the middle of the figure may also eat smaller lizards, which in turn feed on grasshoppers and ants. What trophic levels does the collared lizard occupy when its diet includes smaller lizards as well as grasshoppers?

● The collared lizard is a secondary consumer when it eats grasshoppers. The smaller lizards are also secondary consumers, so the collared lizard is a tertiary consumer when it eats them.

37.10 Species diversity includes relative abundance and species richness

Now that we have looked at how populations in a community interact with each other, let's consider factors that affect the community as a whole. A community's **species diversity** is defined by two components: species richness, or the number of different species in a community, and relative abundance, the proportional representation of each species in a community. To understand why both components are important for describing species diversity, imagine walking through woodlot A on the path shown in **Figure 37.10A**. You would pass by four different species of trees, but most of the trees you encounter would be the same species. Now imagine walking on the path through woodlot B in **Figure 37.10B**. You would see the same four species of trees that you saw in woodlot A—the species richness of the two woodlots is the same. However, woodlot B would probably seem more diverse to you, because no single species predominates. As **Table 37.10** shows, the relative abundance of one species in woodlot A is much higher than the relative abundances of the other three species. In woodlot B, all four species are equally abundant. As a result, species diversity is greater in woodlot B.

Plant species diversity in a community often has consequences for the species diversity of animals in the community. For example, suppose a species of caterpillar only eats the leaves of a tree that makes up just 5% of woodlot A. If the caterpillar is present at all, its population may be small and scattered. Birds that depend on those caterpillars to feed their young may be absent. But the caterpillars would easily be able to locate their food source in woodlot B, and their abundance would attract birds as well. By providing a broader range of habitats, a diverse tree community promotes animal diversity.

Species diversity also has consequences for pathogens. Most pathogens infect a limited range of host species or may even be restricted to a single host species. When many potential hosts are living close together, it is easy for a pathogen to spread from one to another. In woodlot A, for example, a pathogen that infects the most abundant tree would rapidly be transmitted through the entire forest. On the other hand, the more isolated trees in woodlot B are more likely to escape infection.

Low species diversity is characteristic of most modern agricultural ecosystems. For efficiency, crops and trees are often planted in monoculture—a single species grown over a wide area. Monocultures are especially vulnerable to attack by pathogens and herbivorous insects. Also, plants grown in monoculture have been bred for certain desirable characteristics, so their genetic variation is typically low, too. As a result, a pathogen can potentially devastate an entire field or more. Between 1845 and 1849, a pathogen wiped out a monoculture of genetically uniform potatoes throughout Ireland. A million people died of starvation, and well over a million more left the country.

To combat potential losses, many farmers and forest managers rely heavily on chemical methods of controlling pests. Modern crop scientists have bred varieties of plants that are genetically resistant to common pathogens, but these varieties can suddenly become vulnerable, too. In 1970, pathogen evolution led to an epidemic of a disease called corn leaf blight that resulted in a billion dollars of crop damage in the United States. Some researchers are now investigating the use of more diverse agricultural ecosystems—polyculture—as an alternative to monoculture.

▲ **Figure 37.10A** Species composition of woodlot A

▲ **Figure 37.10B** Species composition of woodlot B

TABLE 37.10 | RELATIVE ABUNDANCE OF TREE SPECIES IN WOODLOTS A AND B

Species	Relative Abundance in Woodlot A (%)	Relative Abundance in Woodlot B (%)
	80	25
	10	25
	5	25
	5	25

? Which would you expect to have higher species diversity, a well-maintained lawn or one that is poorly maintained? Explain.

● A lawn that is poorly maintained would have higher species diversity. A well-maintained lawn should have low species diversity. While a lawn that is cared for may not be a perfect monoculture, any weeds that are present would have low relative abundance. The opposite is true if the lawn is not cared for.

37.11 Some species have a disproportionate impact on diversity

SCIENTIFIC THINKING

What causes species diversity to vary among different communities? Ecologist Robert Paine hypothesized that the species diversity of a community is directly related to the ability of predators to prevent any one species from monopolizing local resources. Like many ecologists, Paine designed a field experiment to test his hypothesis. He chose a rocky intertidal community on the Pacific coast in Washington state as his study area (**Figure 37.11A**). In this rigorous environment, the rocks are pounded by waves during high tide and exposed to the sun and drying winds during low tide. Members of the community typically include algae, both herbivorous and carnivorous molluscs such as snails, suspension feeders such as sponges, sea anemones, barnacles, and mussels, and a predatory sea star known as *Pisaster*. Many of these organisms attach to the rocks to avoid being washed away; thus, space is an important but limited resource.

Paine manually removed *Pisaster* from certain areas of the intertidal zone and left comparable areas intact as controls. He then determined the species richness of these experimental and control areas over the next several years (**Figure 37.11B**). In the absence of *Pisaster*, species richness dropped from more than 15 species to fewer than 5. What accounted for the dramatic change? A mussel of the genus *Mytilus* proved to be a superior competitor for the available space, eliminating most other invertebrates and algae. In the control areas, *Mytilus*'s population growth was suppressed by *Pisaster*, a voracious predator on the mussel (**Figure 37.11C**). Thus, interspecific interactions can be an important factor in the species diversity of a community.

▲ **Figure 37.11A** A rocky intertidal zone on the coast of Washington state

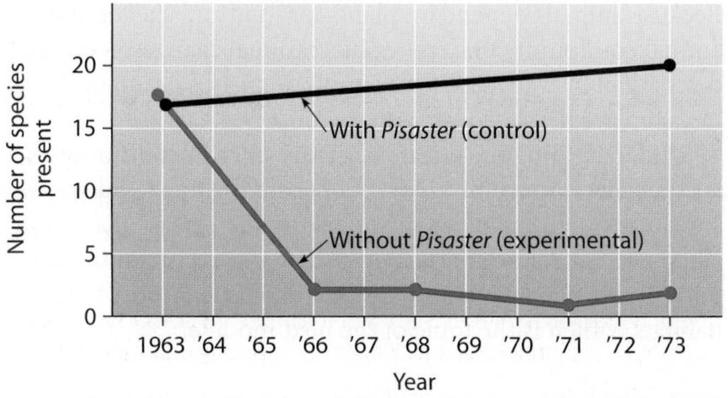

Number of species present
20
15
10
5
0
With *Pisaster* (control)
Without *Pisaster* (experimental)
1963 '64 '65 '66 '67 '68 '69 '70 '71 '72 '73
Year

Data from R.T Paine, Food web complexity and species diversity, *American Naturalist* 100:65–75 (1966)

▲ **Figure 37.11B** Species richness in control and experimental areas after *Pisaster* removal

▲ **Figure 37.11C** A *Pisaster* sea star, a keystone species, eating a mussel

Paine's experiment and others like it gave rise to the concept of a keystone species. A **keystone species** is a species whose impact on its community is much larger than its biomass or abundance would indicate. The word "keystone" comes from the wedge-shaped stone at the top of an arch that locks the other pieces in place. If the keystone is removed, the arch collapses (**Figure 37.11D**). A keystone species occupies a niche that holds the rest of its community in place.

The keystone concept has practical application in efforts to restore or rehabilitate damaged ecosystems. One example is the long-spined sea urchin, *Diadema antillarum*. Ecologists discovered that *Diadema* is a keystone species on Caribbean coral reefs when huge numbers of them were killed by a disease epidemic. Species diversity plummeted as the reefs were overgrown by fleshy seaweeds that had formerly been controlled by the herbivorous sea urchins. Recognition of *Diadema*'s key role in the community prompted conservationists to artificially replenish urchin populations to help restore damaged reefs.

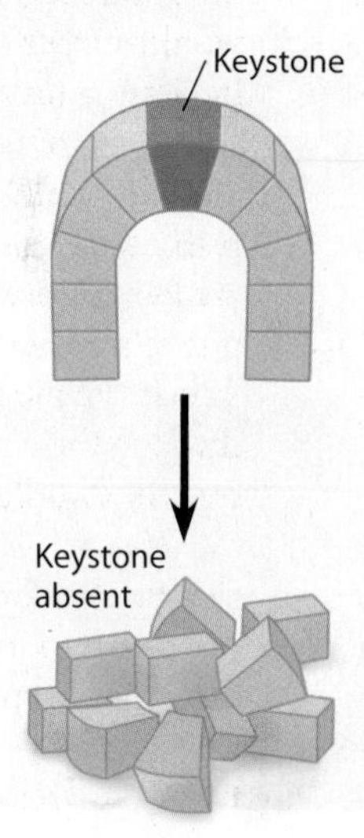

Keystone
Keystone absent

▲ **Figure 37.11D** Arch collapse with removal of keystone

? Removing saguaro cacti from the Sonoran desert community (see Module 37.9) would have a drastic impact, and yet saguaro is not considered a keystone species. Why not?

Saguaro is abundant and makes up a large part of the community, but its effect is not disproportionate to its biomass or abundance. Keystone species have a large effect relative to their representation in the community just as a keystone is a small but vital piece of the arch.

37.12 Disturbance is a prominent feature of most communities

Early ecologists viewed biological communities as more or less stable in structure and species composition. But like many college campuses, where some construction or renovation project is always underway, many communities are frequently disrupted by sudden change. **Disturbances** are events such as storms, fires, floods, droughts, or human activities that damage biological communities and alter the availability of resources. The types of disturbances and their frequency and severity vary from community to community.

Although we tend to think of disturbances in negative terms, small-scale disturbances often have positive effects. For example, new habitats are created when a large tree is uprooted in a windstorm. More light may reach the forest floor, giving small seedlings the opportunity to grow, or the hole left by the tree's roots may fill with water and be used as egg-laying sites by frogs, salamanders, and numerous insects. Communities change drastically following a severe disturbance that strips away vegetation and even soil. The disturbed area may be colonized by a variety of species, which are gradually replaced by a succession of other species, in a process called **ecological succession**.

When ecological succession begins in a virtually lifeless area with no soil, it is called **primary succession**. Examples of such areas are the rubble left by a retreating glacier or fresh volcanic lava flows (**Figure 37.12A**). Often the only life-forms initially present are autotrophic bacteria. Lichens and mosses, which grow from windblown spores, are commonly the first large photosynthesizers to colonize the area. Soil develops gradually as rocks break down and organic matter accumulates from the decomposed remains of the early colonizers. Lichens and mosses are gradually overgrown by larger plants that sprout from seeds blown in from nearby areas or carried in by animals. Eventually, the area is colonized by plants that become the community's prevalent form of vegetation. Primary succession can take hundreds or thousands of years.

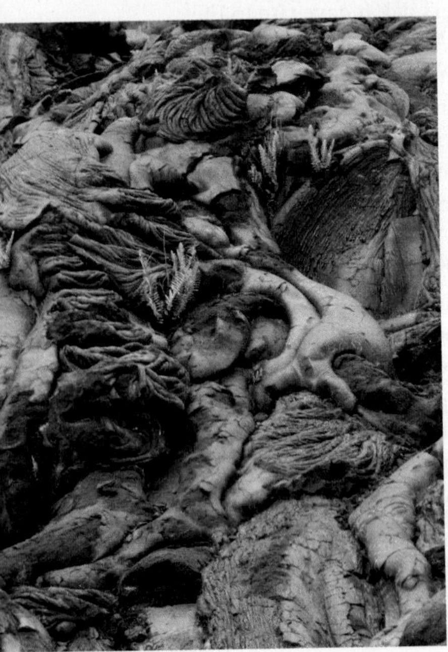

▲ **Figure 37.12A** Primary succession on a lava flow

Secondary succession occurs where a disturbance has cleared away an existing community but left the soil intact. For example, secondary succession occurs as areas recover from fires or floods. Some disturbances that lead to secondary succession are caused by human activities. Even before colonial times, people were clearing the forests of eastern North America for agriculture and settlements. Some of this land was later abandoned as the soil was depleted of its chemical nutrients or the residents moved west to new territories. Whenever human intervention stops, secondary succession begins.

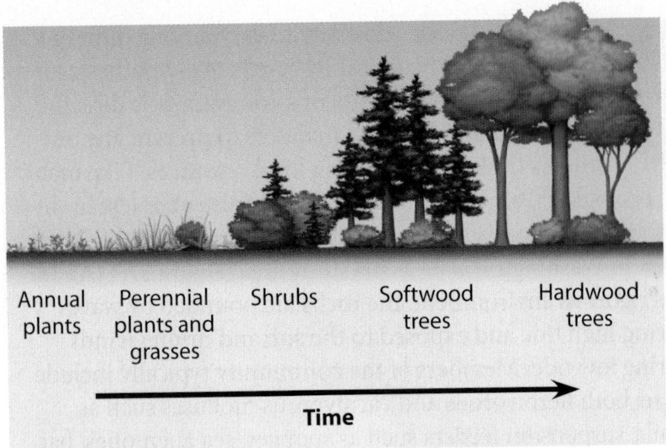
Annual plants
Perennial plants and grasses
Shrubs
Softwood trees
Hardwood trees
Time

▲ **Figure 37.12B** Stages in the secondary succession of an abandoned farm field

Numerous studies have documented the stages by which an abandoned farm field returns to forest (**Figure 37.12B**). A recently disturbed site provides an environment that is favorable to *r*-selected species (see Module 36.7)—plants and animals that reach reproductive age rapidly, produce huge numbers of offspring, and provide little or no parental care. Interspecific competition is not a major factor during the very early stages of succession, which are dominated by weedy annual species such as crabgrass and ragweed. Within a few years, perennial grasses and small broadleaf plants cover the field. (An annual plant completes its life cycle in a single year. Perennial plants live for many years.) Softwood trees, especially pines, begin to invade within 5 years, turning the area into a pine forest in roughly 10 to 15 years. But pine seedlings, which need high levels of light to grow, don't do well in the understory. The seedlings of many hardwood species are more shade tolerant, and thus trees such as oak and maple begin to replace pine as competition becomes a significant force in determining the composition of the community. The final mixture of species depends on local abiotic factors such as soil and topography. Because animals depend on plants for food and shelter, the animal community undergoes successional changes, too. The diversity of bird species, for example, increases dramatically as trees replace herbaceous plants.

Understanding the effects of disturbance in communities is especially important today; people are the most widespread and significant agents of disturbance (as we discuss in Chapter 38). Disturbances may also create opportunities for undesirable plants and animals that people transport to new habitats, which is the topic of the next module.

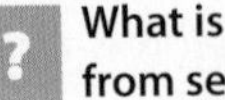

What is the main abiotic factor that distinguishes primary from secondary succession?

● Absence (primary) versus presence (secondary) of soil at the onset of succession

37.13 Invasive species can devastate communities

CONNECTION

For as long as people have traveled from one region to another, they have carried organisms along, sometimes intentionally and sometimes by accident. Many of these non-native species have established themselves firmly in their new locations. Furthermore, many have become **invasive species**, spreading far beyond the original point of introduction and causing environmental or economic damage by colonizing and dominating wherever they find a suitable habitat. In the United States alone, there are hundreds of invasive species, including plants, mammals, birds, fishes, arthropods, and molluscs. Worldwide, there are thousands more. Invasive species are a leading cause of local extinctions (a topic we'll return to in Module 38.1). The economic costs of invasive species are enormous—an estimated $120 billion a year in the United States. Regardless of where you live, an invasive plant or animal is probably living nearby.

Not every organism that is introduced to a new habitat is successful, and not every species that is able to survive in its new habitat becomes invasive. There is no single explanation for why any non-native species turns into a destructive pest, but community ecology offers some insight. Interspecific interactions act as a system of checks and balances on the populations in a community. Every population is subject to multiple negative effects, whether from competitors, predators, herbivores, or pathogens, that curb its growth rate. Without biotic factors such as these to check population growth, a population will continue to expand until limited by abiotic factors.

Rabbits, which are notorious as prolific breeders, offer a vivid illustration of exponential population growth (see Module 36.4). In 1859, 12 pairs of European rabbits (*Oryctolagus cuniculus*) were released on a ranch in southern Australia by a European who wanted to hunt familiar game. The animals quickly became a nuisance. In 1865, 20,000 rabbits were killed on the ranch. By 1900, several hundred million rabbits were distributed over much of the continent (**Figure 37.13A**). The rabbit invasion was a catastrophe in several ways. Their activities destroyed farm and grazing land by eating vegetation down to, and sometimes including, the roots (**Figure 37.13B**). Especially in arid regions, the loss of plant cover led to soil erosion. In addition, rabbits dug extensive underground burrows that made grazing treacherous for cattle and sheep. Rabbits also competed directly with native herbivorous marsupials. After many fruitless attempts to control the rabbit population, in 1950 the Australian government turned to **biological control**, the intentional release of a natural enemy to attack a pest population. A virus lethal to rabbits was introduced into the environment. The rabbits and virus then underwent several coevolutionary cycles as the rabbits became more resistant to the disease and the virus became less lethal. The government managed to stave off a complete resurgence of the rabbit population by introducing new viral strains, but in 1995, they had to switch to a different pathogen to maintain control.

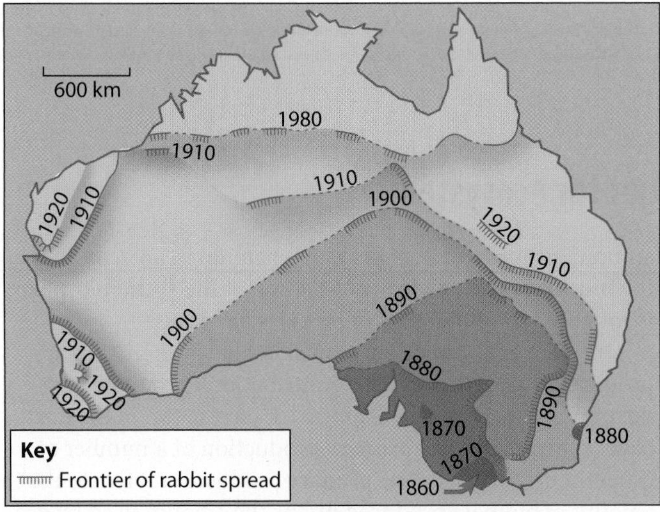
600 km
1980
1910
1920
1910
1910
1900
1920
1910
1890
1900
1880
1910
1920
1920
1870
1890
1880
1870
1860
Key
Frontier of rabbit spread

▲ **Figure 37.13A** The spread of rabbits in Australia

▲ **Figure 37.13B** A familiar sight in early 20th-century Australia

Coevolution is just one potential pitfall of biological control. A natural enemy imported as a control agent may not be as successful in the environment as the pest species. It may not disperse widely enough, or its population growth rate may not be high enough to overtake a rapidly expanding pest population. Caution is especially warranted because the control agent may turn out to be as invasive as its target. For example, cane toads (**Figure 37.13C**) imported and released to control an agricultural pest in Australia became a widespread threat to native wildlife.

▲ **Figure 37.13C** A cane toad (*Bufo marinus*)

In the next modules, we broaden our scope to look at ecosystems, the highest level of ecological complexity.

? **What is the ecological basis for biological control of pests?**

● By having a negative effect on the population growth rate of the pest, a natural enemy keeps the pest population in check.

▷ Ecosystem Structure and Dynamics

37.14 Ecosystem ecology emphasizes energy flow and chemical cycling

An **ecosystem** consists of all the organisms in a community as well as the abiotic environment with which the organisms interact. Ecosystem ecologists are especially interested in **energy flow**, the passage of energy through the components of the ecosystem, and **chemical cycling**, the transfer of materials within the ecosystem.

The terrarium in **Figure 37.14** represents a familiar type of ecosystem and illustrates the fundamentals of energy flow. Energy enters the terrarium in the form of sunlight ( ). Plants (producers) convert light energy to chemical energy ( ) through the process of photosynthesis. Animals (consumers) take in some of this chemical energy in the form of organic compounds when they eat the plants. Decomposers, such as bacteria and fungi in the soil, obtain chemical energy when they decompose the dead remains of plants and animals. Every use of chemical energy by organisms involves a loss of some energy to the surroundings in the form of heat (; see Module 5.10). Because so much of the energy captured by photosynthesis is lost as heat, the ecosystem would run out of energy if it were not powered by a continuous inflow of energy from the sun. A few ecosystems—for example, hydrothermal vents—are powered by chemical energy obtained from inorganic compounds.

In contrast to energy flow, chemical cycling ( ) involves the transfer of materials within the ecosystem. While most ecosystems have a constant input of energy from sunlight, the supply of the chemical elements used to construct molecules is limited. Chemical elements such as carbon and nitrogen are cycled between the abiotic component of the ecosystem, including air, water, and soil, and the biotic component of the ecosystem (the community). Plants acquire these chemical elements in inorganic form from the air and soil and use them to build organic molecules. Animals, such as the snail in Figure 37.14, consume some of these organic molecules. When the plants and animals become detritus, decomposers return most of the elements to the soil and air in inorganic form. Some elements are also returned to the soil and air as the by-products of plant and animal metabolism.

In summary, both energy flow and chemical cycling involve the transfer of substances through the trophic levels of the ecosystem. However, energy flows through, and ultimately out of, ecosystems, whereas chemicals are recycled within ecosystems. We explore these fundamental ecosystem dynamics in the rest of the chapter.

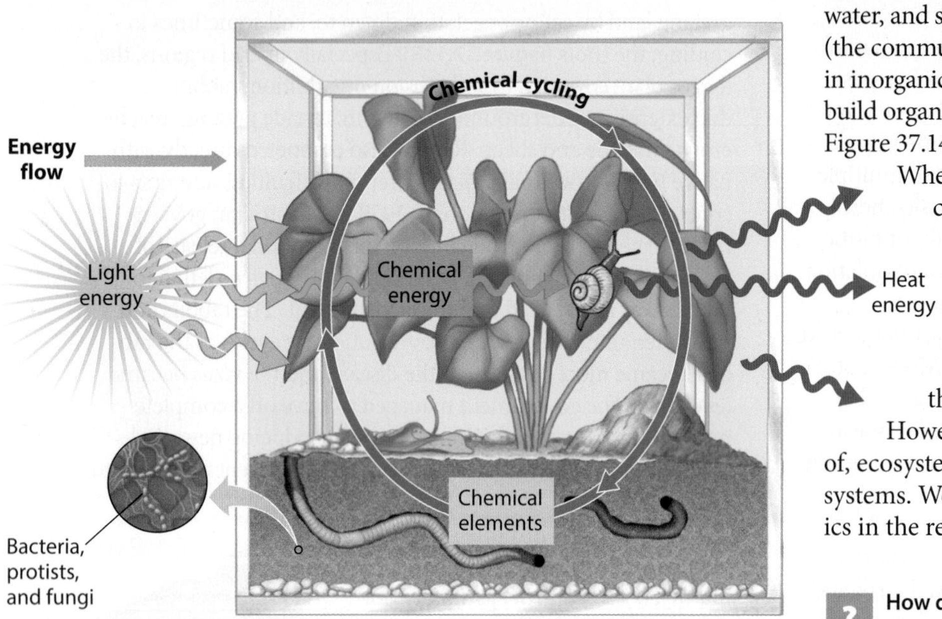

Energy flow
Chemical cycling
Light energy
Chemical energy
Heat energy
Chemical elements
Bacteria, protists, and fungi

▲ **Figure 37.14** A terrarium ecosystem

Try This In your own words, explain how energy flows through the terrarium. Explain how chemicals are cycled within the terrarium.

? **How do chemical cycles in an ecosystem differ from food chains in a community?**

● The components of food chains are solely biotic. In ecosystems, chemicals pass through one or more abiotic components as well as passing through the biotic components (food chain).

37.15 Primary production sets the energy budget for ecosystems

Each day, Earth receives about 10^{19} kcal of solar energy, the energy equivalent of about 100 million atomic bombs. Most of this energy is absorbed, scattered, or reflected by the atmosphere or by Earth's surface. Of the visible light that reaches plants, algae, and cyanobacteria, only about 1% is converted to chemical energy by photosynthesis. The amount of solar energy converted to chemical energy (in organic compounds) by an ecosystem's producers for a given area and during a given time period is called **primary production**. It can be expressed in units of energy or units of mass. Ecologists call the amount, or mass, of living organic material in an ecosystem the **biomass**. The primary production of the entire biosphere is roughly 165 billion tons of biomass per year.

Different ecosystems vary considerably in their primary production as well as in their contribution to the total production of the biosphere. **Figure 37.15**, at the top of the next page, contrasts the net primary production of a number of different ecosystems. (Net primary production refers to the amount of biomass produced minus the amount used by producers as fuel for their own cellular respiration.) Tropical rain

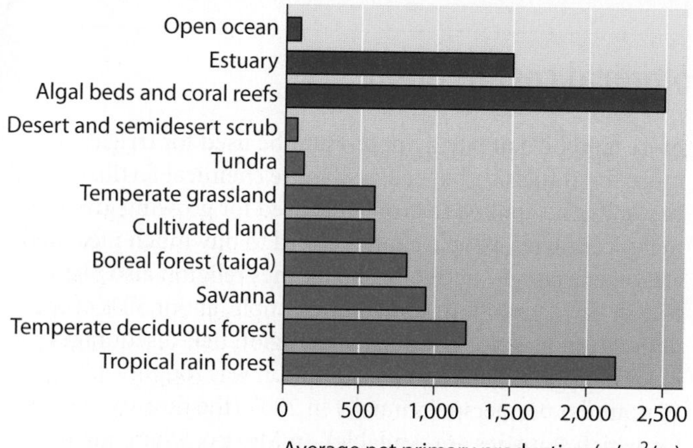
Open ocean
Estuary
Algal beds and coral reefs
Desert and semidesert scrub
Tundra
Temperate grassland
Cultivated land
Boreal forest (taiga)
Savanna
Temperate deciduous forest
Tropical rain forest
0
500
1,000
1,500
2,000
2,500
Average net primary production ($g/m^2/yr$)

Data from R. H. Whittaker, *Communities and Ecosystems*, second edition, New York: Macmillan (1975).

▲ **Figure 37.15** Net primary production of various ecosystems

forests are among the most productive terrestrial ecosystems and contribute a large portion of the planet's overall production of biomass. Coral reefs also have very high production, but their contribution to global production is small because they cover such a small area. Interestingly, even though the open ocean has very low production, it contributes the most to Earth's total net primary production because of its huge size—it covers 65% of Earth's surface area.

? Deserts and semidesert scrub cover about the same amount of surface area as tropical forests but contribute less than 1% of Earth's net primary production, while rain forests contribute 22%. Explain this difference.

● The primary production of tropical rain forests is over 20 times greater than that of desert and semidesert scrub ecosystems.

37.16 Energy supply limits the length of food chains

When energy flows as organic matter through the trophic levels of an ecosystem, much of it is lost at each link in a food chain. Consider the transfer of organic matter from a producer to a primary consumer, such as the caterpillar shown in **Figure 37.16A**. The caterpillar might digest and absorb only about half the organic material it eats, passing the indigestible wastes as feces. Of the organic compounds it does absorb, the caterpillar typically uses two-thirds as fuel for cellular respiration. Only the chemical energy left over after respiration—15% of the organic material the caterpillar consumed—can be converted to caterpillar biomass. Thus, a secondary consumer that eats the caterpillar gets only 15% of the biomass (and the energy it contains) that was in the leaves the caterpillar ate.

Figure 37.16B, called a pyramid of production, illustrates the cumulative loss of energy with each transfer in a food chain. Each tier of the pyramid represents the chemical energy present in all of the organisms at one trophic level of a food chain. The width of each tier indicates how much of the chemical energy of the tier below is actually incorporated into the organic matter of that trophic level. Note that producers convert only about 1% of the energy in the sunlight available to them to primary production. In this idealized pyramid, 10% of the energy available at each trophic level becomes incorporated into the next higher level. The efficiencies of energy transfer usually range from 5 to 20%. In other words, 80–95% of the energy at one trophic level never transfers to the next.

An important implication of this stepwise decline of energy in a trophic structure is that the amount of energy available to top-level consumers is small compared with that available to lower-level consumers. Only a tiny fraction of the energy stored by photosynthesis flows through a food chain all the way to a tertiary consumer. This explains why top-level consumers such as lions and hawks require so much geographic territory; it takes a lot of vegetation to support trophic levels so many steps removed from photosynthetic production. Pyramids of production help us understand why most food chains are limited to three to five levels; there is simply not enough energy at the very top of an ecological pyramid to support another trophic level.

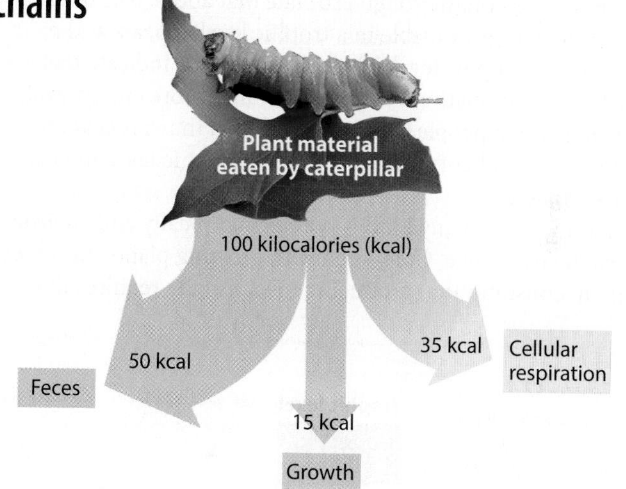
Plant material eaten by caterpillar
100 kilocalories (kcal)
50 kcal
Feces
35 kcal
Cellular respiration
15 kcal
Growth

▲ **Figure 37.16A** The fate of leaf biomass consumed by a caterpillar

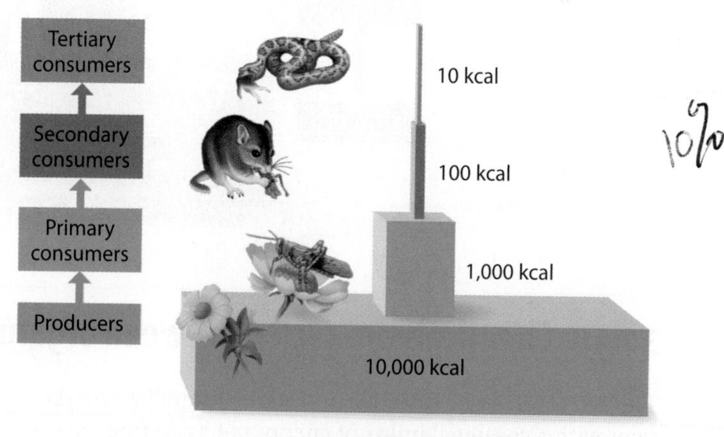
Tertiary consumers
Secondary consumers
Primary consumers
Producers
10 kcal
100 kcal
1,000 kcal
10,000 kcal
1,000,000 kcal of sunlight

▲ **Figure 37.16B** An idealized pyramid of production

? Approximately what proportion of the energy produced by photosynthesis makes it to the snake in Figure 37.16B?

● 1/1,000 of the 10,000 kcal produced by photosynthesis [(0.1 × 0.1 × 0.1) (10,000 kcal) = 10 kcal]

37.17 A pyramid of production explains the ecological cost of meat

CONNECTION

The dynamics of energy flow apply to the human population as much as to other organisms. As omnivores, people eat both plant material and meat. When we eat grain or fruit, we are primary consumers; when we eat beef or other meat from herbivores, we are secondary consumers. When we eat fish like trout and salmon (which eat insects and other small animals), we are tertiary or quaternary consumers.

The pyramid of production on the left in **Figure 37.17** indicates energy flow from producers to vegetarians (primary consumers). The energy in the producer trophic level comes from a corn crop. The pyramid on the right illustrates energy flow from the same corn crop, with people as secondary consumers, eating beef. These two pyramids are generalized models, based on the rough estimate that about 10% of the chemical energy available in a trophic level appears at the next higher trophic level. Thus, the pyramids indicate that the human population has about 10 times more energy available to it when people eat corn than when they process the same amount of corn through another trophic level and eat corn-fed beef.

Eating meat of any kind is both economically and environmentally expensive. Compared with growing plants for direct human consumption, producing meat usually requires that more land be cultivated, more water be used for irrigation, more fossil fuels be burned, and more chemical fertilizers and pesticides be applied to croplands used for growing grain. In many countries, people cannot afford to buy much meat and are vegetarians by necessity. Sometimes religion also plays a role in the decision. In India, for example, about 80% of the population practice Hinduism, a religion that discourages meat-eating. India's meat consumption was roughly 3.2 kg (7 pounds) per person annually in 2007 (the most recent year for which statistics are available). In Mexico, where many people are too poor to eat meat daily, per capita consumption in 2007 was 62.3 kg (137 pounds) per year. That is a large amount compared with India, but only about half the meat consumption of the United States, where the per capita rate was 125.4 kg (276 pounds) in 2007.

We turn next to the subject of chemical nutrients. Unlike energy, which is ultimately lost from an ecosystem, all chemical nutrients cycle within ecosystems.

? **Why does demand for meat also tend to drive up prices of grains such as wheat and rice, fruits, and vegetables?**

● The potential supply of plants for direct consumption as food for humans is diminished by the use of agricultural land to grow feed for cattle, chickens, and other meat sources.

▶ Figure 37.17 Food energy available to people eating at different trophic levels

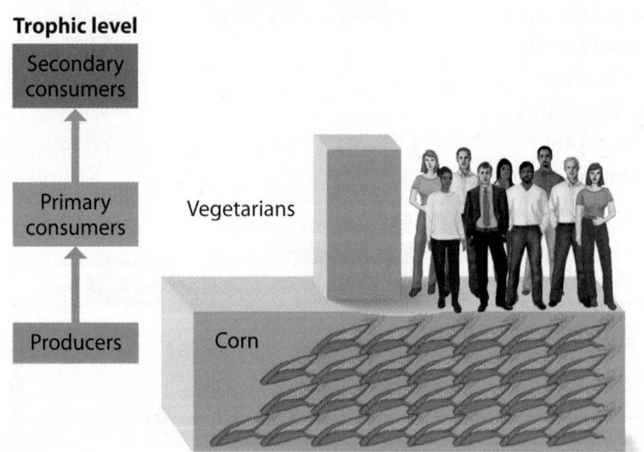

Trophic level
Secondary consumers
Primary consumers
Producers
Vegetarians
Corn

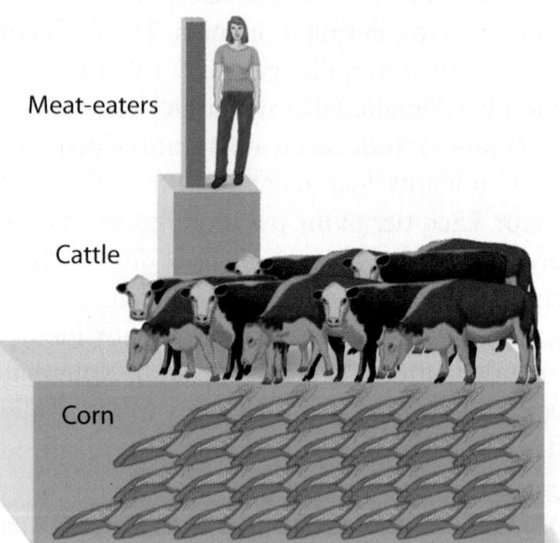

Meat-eaters
Cattle
Corn

37.18 Chemicals are cycled between organic matter and abiotic reservoirs

The sun (or in some cases Earth's interior) supplies ecosystems with a continual influx of energy, but aside from an occasional meteorite, there are no extraterrestrial sources of chemical elements. Life, therefore, depends on the recycling of chemicals. While an organism is alive, much of its chemical stock changes continuously as it acquires and releases waste products. When the organism dies, decomposition returns the atoms that make up its complex molecules to the environment, thus replenishing the pool of inorganic nutrients that producers use to build new organic matter.

Because chemical cycles in an ecosystem include both biotic and abiotic (geologic and atmospheric) components, they are called **biogeochemical cycles**. **Figure 37.18**, at the top of the next page, is a general scheme for the cycling of a nutrient within an ecosystem. Note that the cycle has **abiotic reservoirs**, where chemicals accumulate or are stockpiled outside of living organisms. The atmosphere, for example, is an abiotic reservoir for carbon. Phosphorus, on the other hand, is available only from the soil. Both atmosphere and soil are abiotic reservoirs for nitrogen.

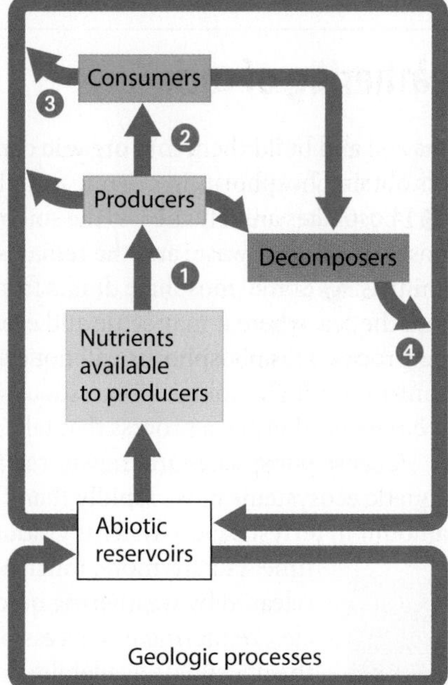

Consumers
Producers
Decomposers
Nutrients available to producers
Abiotic reservoirs
Geologic processes
1
2
3
4

▶ **Figure 37.18**
A general model of the biogeochemical cycling of nutrients

Let's trace the general biogeochemical cycle in Figure 37.18. ❶ Producers incorporate chemicals from the abiotic reservoirs into organic compounds. ❷ Consumers feed on the producers, incorporating some of the chemicals into their own bodies. ❸ Both producers and consumers release some chemicals back to the environment in waste products (CO_2 and nitrogenous wastes of animals). ❹ Decomposers play a central role by breaking down the complex organic molecules in detritus such as plant litter, animal wastes, and dead organisms. The products of this metabolism are inorganic compounds such as nitrates (NO_3^-), phosphates (PO_4^{3-}), and CO_2, which replenish the abiotic reservoirs. Geologic processes such as erosion and the weathering of rock also contribute to the abiotic reservoirs. Producers use the inorganic molecules from abiotic reservoirs as raw materials for synthesizing new organic molecules (carbohydrates and proteins, for example), and the cycle continues.

Biogeochemical cycles can be local or global. Soil is the main reservoir for nutrients in a local cycle, such as phosphorus. In contrast, for those chemicals that exist primarily in gaseous form—carbon and nitrogen are examples—the cycling is essentially global. For instance, some of the carbon a plant acquires from the air may have been released into the atmosphere by the respiration of an organism on another continent.

In the next three modules, we look at the cyclic movements of carbon, phosphorus, and nitrogen. As you study the cycles, look for the four basic steps we have cited, as well as the geologic processes that may move chemicals around and between ecosystems. In the diagrams, the main abiotic reservoirs are highlighted in white boxes.

? Which boxes in Figure 37.18 represent biotic components of an ecosystem?

Consumers, producers, and decomposers

37.19 The carbon cycle depends on photosynthesis and respiration

Carbon, the major ingredient of all organic molecules, has an atmospheric reservoir and cycles globally. Carbon also resides in fossil fuels and sedimentary rocks, such as limestone ($CaCO_3$), and as dissolved carbon compounds in the oceans.

As shown in **Figure 37.19**, the reciprocal metabolic processes of photosynthesis and cellular respiration are mainly responsible for the cycling of carbon between the biotic and abiotic worlds. ❶ Photosynthesis removes CO_2 from the atmosphere and incorporates it into organic molecules, which are ❷ passed along the food chain by consumers. ❸ Cellular respiration by producers and consumers returns CO_2 to the atmosphere. ❹ Decomposers break down the carbon compounds in detritus; that carbon, too, is eventually released as CO_2.

On a global scale, the return of CO_2 to the atmosphere by cellular respiration closely balances its removal by photosynthesis. However, ❺ the increased burning of wood and fossil fuels (coal and petroleum) is raising the level of CO_2 in the atmosphere. (As we will discuss in Module 38.4, this increase in CO_2 is leading to significant global warming.)

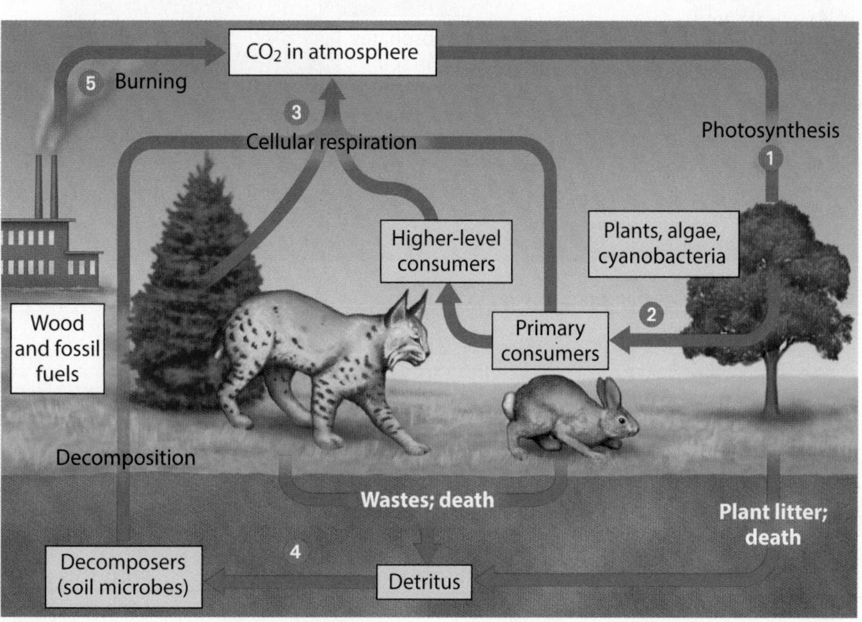

CO2 in atmosphere
5 Burning
3
Cellular respiration
Photosynthesis
1
Higher-level consumers
Plants, algae, cyanobacteria
2
Primary consumers
Wood and fossil fuels
Decomposition
Wastes; death
Plant litter; death
4
Decomposers (soil microbes)
Detritus

▲ **Figure 37.19** The carbon cycle

Try This Identify all of the locations where carbon is stored; identify all of the sources that release carbon into the environment.

? What would happen to the carbon cycle if all the decomposers suddenly "went on strike" and stopped working?

Carbon would accumulate in organic mass, the atmospheric reservoir of carbon would decline, and plants would eventually be starved for CO_2.

37.20 The phosphorus cycle depends on the weathering of rock

Organisms require phosphorus—usually in the form of the phosphate ion (PO_4^{3-})—as an ingredient of nucleic acids, phospholipids, and ATP and (in vertebrates) as a mineral component of bones and teeth. In contrast to the carbon cycle and the other major biogeochemical cycles, the phosphorus cycle does not have an atmospheric component. Rocks are the only source of phosphorus for terrestrial ecosystems; in fact, rocks that have high phosphorus content are mined for agricultural fertilizer.

At the center of **Figure 37.20**, 1 the weathering (breakdown) of rock gradually adds inorganic phosphate (PO_4^{3-}) to the soil. 2 Plants assimilate the dissolved phosphate ions in the soil and build them into organic compounds. 3 Consumers obtain phosphorus in organic form by eating plants. 4 Phosphates are returned to the soil by the action of decomposers on animal waste and the remains of dead plants and animals. 5 Some phosphate drains from terrestrial ecosystems into the sea, where it may settle and eventually become part of new rocks. This phosphorus will not cycle back into living organisms until 6 geologic processes uplift the rocks and expose them to weathering, a process that takes millions of years.

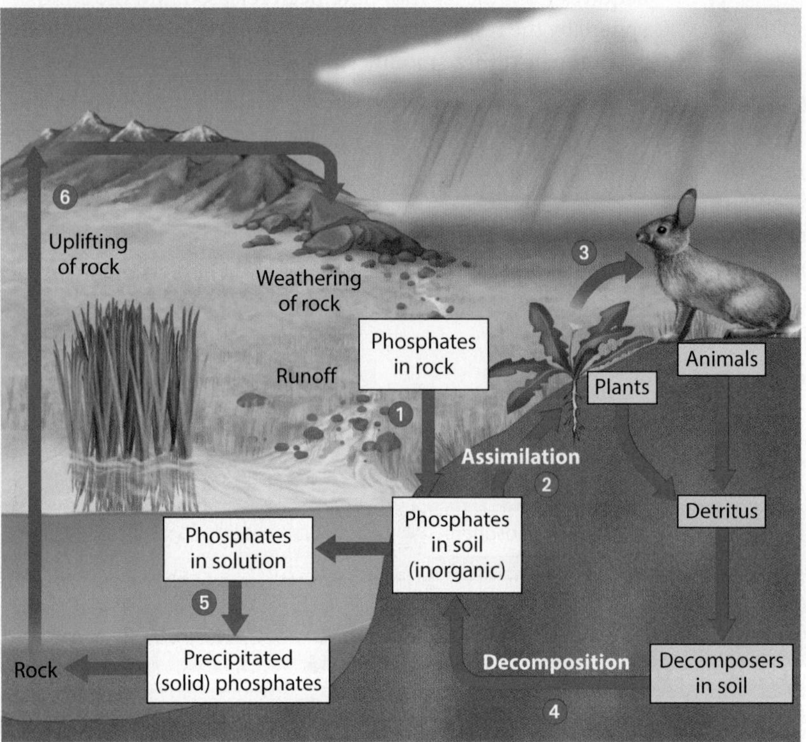
6
Uplifting of rock
Weathering of rock
Phosphates in rock
Runoff
1
3
Plants
Animals
Assimilation
2
Detritus
Phosphates in solution
Phosphates in soil (inorganic)
5
Rock
Precipitated (solid) phosphates
Decomposition
4
Decomposers in soil

▲ **Figure 37.20** The phosphorus cycle

Because phosphates are transferred from terrestrial to aquatic ecosystems more rapidly than they are replaced, the amount in terrestrial ecosystems gradually diminishes over time. Furthermore, much of the soluble phosphate released by weathering quickly binds to soil particles, rendering it inaccessible to plants. As a result, the phosphate availability is often quite low and commonly a limiting factor. Mycorrhizal fungi (see Module 17.12) that facilitate phosphorus uptake are essential to many plants, especially those living in older, highly weathered soils. Soil erosion from land cleared for agriculture or development accelerates the loss of phosphates.

Farmers and gardeners often use crushed phosphate rock, bone meal (finely ground bones from slaughtered livestock), or guano, the droppings of seabirds and bats, to add phosphorus to the soil. Guano is mined from densely populated colonies or caves, where meters-deep deposits have accumulated. As you'll learn in Module 37.22, however, runoff of large amounts of phosphate fertilizer pollutes aquatic ecosystems.

? Over the short term, why does phosphorus cycling tend to be localized, whereas carbon and nitrogen cycle globally?

● Because phosphorus is cycled almost entirely within the soil rather than transferred over long distances via the atmosphere

37.21 The nitrogen cycle depends on bacteria

As an ingredient of proteins and nucleic acids, nitrogen is essential to the structure and functioning of all organisms. In particular, it is a crucial and often limiting plant nutrient. Nitrogen has two abiotic reservoirs, the atmosphere and the soil. The atmospheric reservoir is huge; almost 80% of the atmosphere is nitrogen gas (N_2). However, plants cannot absorb nitrogen in the form of N_2. The process of **nitrogen fixation**, which is performed by some bacteria, converts N_2 to compounds of nitrogen that can be used by plants. Without these organisms, the natural reservoir of usable soil nitrogen would be extremely limited.

Figure 37.21, on the facing page, illustrates the actions of two types of nitrogen-fixing bacteria. Starting at the far right in the figure, 1 some bacteria live symbiotically in the roots of certain species of plants, supplying their hosts with a direct source of usable nitrogen. The largest group of plants with this mutualistic relationship is the legumes, a family that includes peanuts, soybeans, and alfalfa (see Module 32.13). A number of non-legume plants that live in nitrogen-poor soils have a similar relationship with bacteria. 2 Free-living nitrogen-fixing bacteria in soil or water convert N_2 to ammonia (NH_3), which then picks up another H^+ to become ammonium (NH_4^+).

3 After nitrogen is "fixed," some of the NH_4^+ is taken up and used by plants. 4 Nitrifying bacteria in the soil also convert some of the NH_4^+ to nitrate (NO_3^-), which is more

▶ Figure 37.21 The nitrogen cycle

Try This The figure identifies five roles that bacteria play in the nitrogen cycle. Explain the effect of eliminating the bacteria that perform each of these roles.

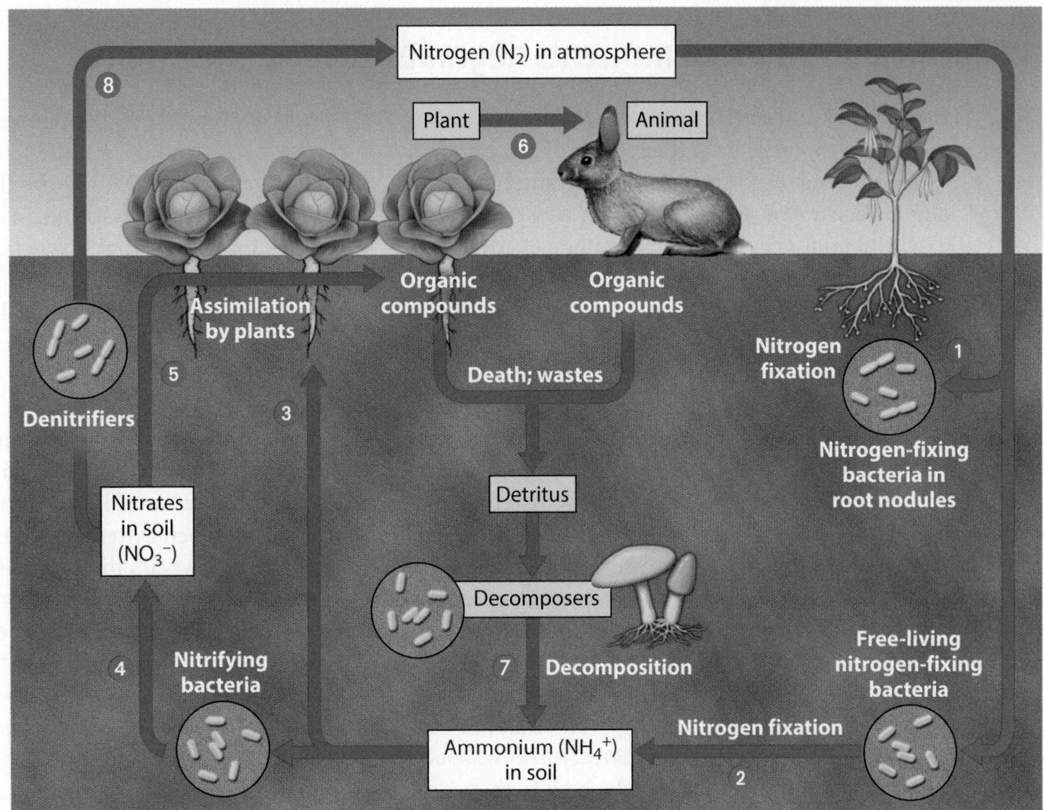

Nitrogen (N_2) in atmosphere
8
Plant
Animal
6
Organic compounds
Organic compounds
Assimilation by plants
Death; wastes
Nitrogen fixation
1
5
Denitrifiers
3
Nitrogen-fixing bacteria in root nodules
Nitrates in soil (NO_3^-)
Detritus
Decomposers
Free-living nitrogen-fixing bacteria
4
Nitrifying bacteria
7
Decomposition
Nitrogen fixation
Ammonium (NH_4^+) in soil
2

readily ❺ absorbed by plants. Plants use the nitrogen they assimilate to synthesize molecules such as amino acids, which are then incorporated into proteins.

❻ When an herbivore (represented by the rabbit in Figure 37.21) eats a plant, it digests the proteins into amino acids, then uses the amino acids to build the proteins it needs. Higher-order consumers gain nitrogen from their prey. Nitrogen-containing waste products are formed during protein metabolism; consumers excrete some nitrogen as well as incorporate some into their body tissues (see Module 25.5). Mammals, such as the rabbit, excrete nitrogen as urea; industrially produced urea is widely used as an agricultural fertilizer.

Organisms that are not consumed eventually die and become detritus, which is decomposed by prokaryotes and fungi. ❼ Decomposition releases NH_4^+ from organic compounds back into the soil, replenishing the soil reservoir of NH_4^+ and, with the help of nitrifying bacteria (step 4), NO_3^-. Under low-oxygen conditions, however, ❽ soil bacteria known as denitrifiers strip the oxygens from NO_3^-, releasing N_2 back into the atmosphere and depleting the soil reservoir of usable nitrogen. Aerobic denitrification produces a different gas, N_2O.

Although not shown in the figure, some NH_4^+ and NO_3^- are made in the atmosphere by chemical reactions involving N_2 and ammonia gas (NH_3). The ions produced by these chemical reactions reach the soil in precipitation and dust, which are crucial sources of nitrogen for plants in some ecosystems.

Human activities are disrupting the nitrogen cycle by adding more nitrogen to the biosphere each year than that added by natural processes. Combustion of fossil fuels in motor vehicles and coal-fired power plants produces nitrogen oxides (NO and NO_2). Nitrogen oxides react with other gases in the lower atmosphere to increase the production of ozone. Unlike the protective ozone layer in the upper atmosphere (see Module 7.14), ground-level ozone is a health hazard. Exposure to ozone, which irritates the respiratory system, can cause coughing and breathing difficulties. It is especially dangerous for people with respiratory problems such as asthma. In many regions, ozone alerts are common during hot, dry summer weather. Nitrogen oxides also combine with water in the atmosphere to become nitric acid. The Clean Air Act Amendments of 1990 led to diminished acid precipitation from sulfur emissions, but environmental damage from nitric acid precipitation is causing new concern.

Modern agricultural practices are another major source of nitrogen. Animal wastes from intensive livestock production release ammonia into the atmosphere. Farmers use enormous amounts of nitrogen fertilizer to supplement natural nitrogen fixation by bacteria. Worldwide, the application of synthetic nitrogen fertilizer has increased 100-fold since the late 1950s. However, less than half the fertilizer is taken up by the crop plants. Some nitrogen escapes to the atmosphere, where it forms NO_2 or nitrous oxide (N_2O), an inert gas that lingers in the atmosphere and contributes to global warming (see Module 38.4). As you'll learn in the next module, nitrogen fertilizers also pollute aquatic systems.

? **What are the abiotic reservoirs of nitrogen? In what form does nitrogen occur in each reservoir?**

● Atmosphere: N_2; soil: NH_4^+ and NO_3^-

37.22 A rapid inflow of nutrients degrades aquatic ecosystems

CONNECTION

Low levels of nutrients, especially phosphorus and nitrogen, often limit the growth of algae and cyanobacteria—and thus primary production—in aquatic ecosystems. Standing-water ecosystems (lakes and ponds) gradually accumulate nutrients from the decomposition of organic matter and fresh influx from the land. As a result, primary production increases naturally over time in a process known as eutrophication. Human activities that add nutrients to aquatic ecosystems accelerate this process and also cause eutrophication in rivers, estuaries, coastal waters, and coral reefs.

You might think that an increase in primary production would be beneficial to a biological community. After all, Figure 37.15 shows that coral reefs and tropical rain forests, ecosystems renowned for spectacular species diversity, have the greatest net primary production. But rapid eutrophication actually lowers species diversity. In some ecosystems, cyanobacteria replace green algae as primary producers. These prokaryotes, which are often encased in a slimy coating, form extensive mats on the surface of the water that prevent light from penetrating the water (**Figure 37.22A**). Some species of cyanobacteria can fix nitrogen, which gives them an additional advantage when phosphate is the pollutant and nitrogen is scarce. Other ecosystems are overrun by blooms of unicellular diatoms, toxin-producing dinoflagellates (see Figure 16.14D), or multicellular algae. These heavy growths, or "blooms," of cyanobacteria or algae greatly reduce oxygen levels at night, when the photosynthesizers respire. As the cyanobacteria and algae die, microbes consume a great deal of oxygen as they decompose the extra biomass. Thus, rapid nutrient enrichment results in oxygen depletion of the water. Fishes that have a high oxygen requirement cannot survive in such an environment.

▲ Figure 37.22A Algal growth on a pond resulting from nutrient pollution

In many areas, phosphate pollution comes from agricultural fertilizers. Phosphates are also a common ingredient in pesticides. Other major sources of phosphates include outflow from sewage treatment facilities and runoff of animal waste from livestock feedlots (where hundreds of animals are penned together). Sewage treatment facilities may discharge large amounts of dissolved inorganic nitrogen compounds into rivers or streams when extreme conditions (such as unusually high rainfall) overwhelm their capacity. Agricultural sources of nitrogen include feedlots and the large amounts of inorganic nitrogen fertilizers that are routinely applied to crops, lawns, and golf courses. Plants take up some of the nitrogen compounds in fertilizer, and denitrifiers convert some to atmospheric N_2 or N_2O, but nitrate is not bound tightly by soil particles and is easily washed out of the soil by rain or irrigation. As a result, chemical fertilizers often exceed the soil's natural recycling capacity.

In an example of how far-reaching this problem can be, nitrogen runoff from midwestern farm fields has been linked to a "dead zone" observed each summer in the Gulf of Mexico. Vast algal blooms, indicated in red and orange in **Figure 37.22B**, extend outward from where the Mississippi River deposits its nutrient-laden waters. As the algae die, decomposition of the huge quantities of biomass diminishes the supply of dissolved oxygen over an area that ranges from 13,000 to 22,000 km^2, or roughly 5,000 to 8,500 square miles. Oxygen depletion disrupts benthic communities, displacing fishes and invertebrates that can move and killing organisms that are attached to the substrate. More than 400 recurring and permanent coastal dead zones totaling approximately 245,000 km^2 (about 95,000 square miles) have been documented in seas worldwide.

Gulf of Mexico
Winter

Gulf of Mexico
Summer

▲ Figure 37.22B Concentrations of phytoplankton in winter and summer (Red and orange indicate highest concentrations of phytoplankton.)

? How would excessive addition of mineral nutrients to a lake eventually lead to the loss of many fish species?

● The nutrients initially cause population explosions of algae and cyanobacteria. Their respiration and that of the decomposers of all the detritus as the algae and cyanobacteria die consume most of the lake's oxygen, which the fish require.

37.23 Ecosystem services are essential to human well-being

CONNECTION

Natural ecosystems provide direct benefits to people, for example, by supplying us with fresh water and food such as fish and shellfish. We also depend on healthy ecosystems to recycle nutrients, decompose wastes, and regulate climate and air quality. Wetlands buffer coastal populations against tidal waves and hurricanes, reduce the impact of flooding rivers, and filter pollutants. Natural vegetation helps retain fertile soil and prevent landslides and mudslides.

Ecosystems that we create are also essential to our well-being. For example, agricultural ecosystems supply most of our food and fibers. Although we manage these ecosystems, they are modifications of natural ecosystems and make use of ecosystem services, such as control of agricultural pests by natural predators and pollination of crops. Soil fertility, the foundation for crop growth, depends on nutrient cycling, another ecosystem service. But agricultural methods introduced over the past several decades have pushed croplands beyond their natural capacity to produce food. Large inputs of chemical fertilizers are needed to supplement soil nutrients. Synthetic pesticides are used to control the population growth of crop-eating insects and pathogens that take advantage of vast monocultures of crop species. Herbicides are applied to kill weeds that would compete with crop plants for water and nutrients. In many areas, crops require additional water supplied by irrigation.

These agricultural practices have resulted in enormous increases in food production, but at the expense of natural ecosystems and the services they provide. The detrimental effects of nutrient runoff, discussed in the previous module, are affecting both freshwater and marine ecosystems as fertilizer use increases. Pesticides may kill beneficial organisms as well as pests, and as you learned previously, chemicals that persist in the environment can be carried far from their point of origin (see Modules 34.2, 34.18, and 35.16). Perhaps most worrisome is the deterioration of fertile soil. Clearing and cultivation expose land to wind and water that erode the rich topsoil.

Erosion and soil degradation are especially severe in grassland, savanna, and some forest ecosystems where low amounts of precipitation and high rates of evaporation result in low levels of soil moisture. In recent years, dust storms sweeping across overcultivated areas have removed millions of tons of topsoil from these stressed ecosystems. In China, for example, overgrazing and other poor agricultural practices are turning 900 square miles of land—an area the size of Rhode Island—into desert each year (**Figure 37.23A**). Irrigation of arid land enables farmers to grow crops but leaves a salty residue that eventually prevents plant growth. In addition, population growth in these regions places increasing demands on the already scarce water supply.

Human activities also threaten many forest ecosystems and the services they provide. Every year, more and more land is cleared for agriculture. Some of this land is needed to feed the growing human population, but replacing worn-out cropland accounts for much of the deforestation occurring today. Forests are also cut down to provide timber and fuel wood; many people in non-industrialized countries use wood for heating and cooking (**Figure 37.23B**). The most immediate impact of deforestation is soil erosion. In Haiti, for example, where less than 2% of the original tree cover remains, heavy rains inevitably bring flooding and mudslides that damage crops. During recent hurricanes, floodwaters surging down stripped hillsides caused thousands of deaths. With much of the soil destabilized by the devastating earthquake in 2010, massive landslides are likely to cause further ecological and economic destruction during storms.

▲ **Figure 37.23A** A dust storm in Changling, China

The growing demand of the human population for food, fibers, and water has largely been satisfied at the expense of other ecosystem services, but these practices cannot continue indefinitely. **Sustainability** is the goal of developing, managing, and conserving Earth's resources in ways that meet the needs of people today without compromising the ability of future generations to meet theirs. Scientists are applying their knowledge of population, community, and ecosystem ecology to conserve natural ecosystems and even to repair some of the ecological damage that we have done.

? **How can clear-cutting a forest (removing all trees) damage the water quality of nearby aquatic ecosystems?**

● Without the growing trees to assimilate minerals from the soil, more of the minerals run off and end up polluting water resources.

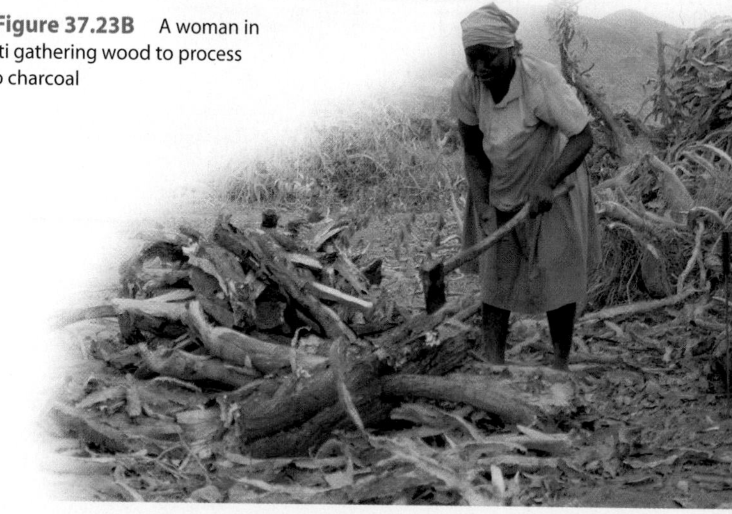

▶ **Figure 37.23B** A woman in Haiti gathering wood to process into charcoal

CHAPTER 37 REVIEW

For practice quizzes, BioFlix animations, MP3 tutorials, video tutors, and more study tools designed for this textbook, go to MasteringBiology®

Reviewing the Concepts

Community Structure and Dynamics (37.1–37.13)

37.1 A community includes all the organisms inhabiting a particular area. Community ecology is concerned with factors that influence the species composition and distribution of communities and with factors that affect community stability.

37.2 Interspecific interactions are fundamental to community structure. Interspecific interactions can be categorized according to their effect on the interacting populations.

37.3 Competition may occur when a shared resource is limited.

37.4 Mutualism benefits both partners.

37.5 Predation leads to diverse adaptations in prey species.

37.6 Herbivory leads to diverse adaptations in plants. Some herbivore-plant interactions illustrate coevolution or reciprocal evolutionary adaptations.

37.7 Parasites and pathogens can affect community composition.

37.8 Trophic structure is a key factor in community dynamics. Trophic structure can be represented by a food chain.

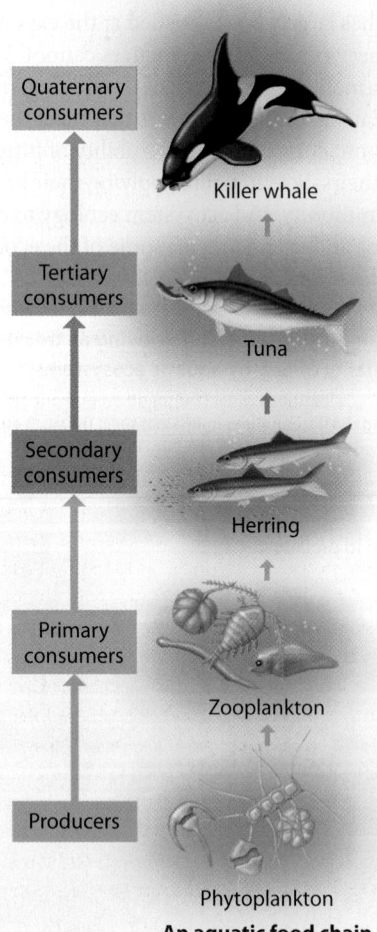

Quaternary consumers
Killer whale
Tertiary consumers
Tuna
Secondary consumers
Herring
Primary consumers
Zooplankton
Producers
Phytoplankton

An aquatic food chain

37.9 Food chains interconnect, forming food webs.

37.10 Species diversity includes relative abundance and species richness. Thus, diversity takes into account both the number of species in a community and the proportion of the community that each species represents.

37.11 Some species have a disproportionate impact on diversity. Although a keystone species has low biomass or relative abundance, its removal from a community results in lower species diversity.

37.12 Disturbance is a prominent feature of most communities. Ecological succession is a transition in species composition of a community. Primary succession is the gradual colonization of barren rocks. Secondary succession occurs after a disturbance has destroyed a community but left the soil intact.

37.13 Invasive species can devastate communities. Organisms that have been introduced to non-native habitats by human actions and have established themselves at the expense of native communities are considered invasive.

Ecosystem Structure and Dynamics (37.14–37.23)

37.14 Ecosystem ecology emphasizes energy flow and chemical cycling. An ecosystem includes a community and the abiotic factors with which it interacts.

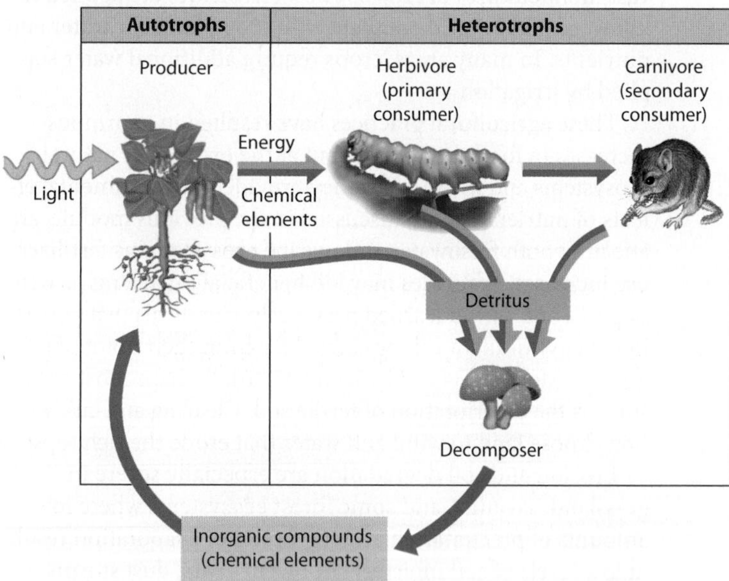

Autotrophs
Heterotrophs
Producer
Herbivore (primary consumer)
Carnivore (secondary consumer)
Light
Energy
Chemical elements
Detritus
Decomposer
Inorganic compounds (chemical elements)

37.15 Primary production sets the energy budget for ecosystems.

37.16 Energy supply limits the length of food chains. A pyramid of production shows the flow of energy from producers to primary consumers and to higher trophic levels. Only about 10% of the energy stored at each trophic level is available to the next level.

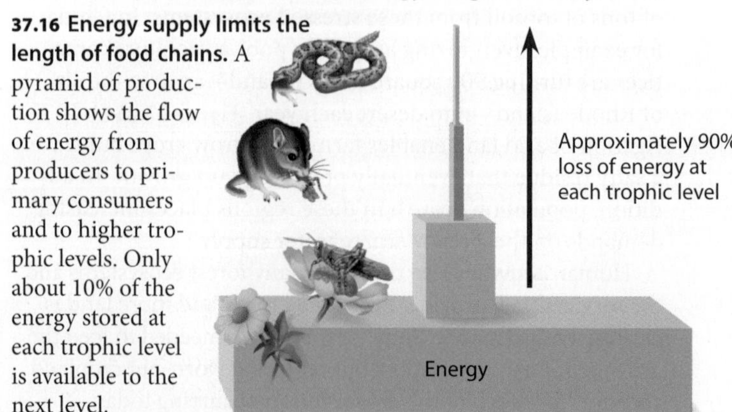

Approximately 90% loss of energy at each trophic level
Energy

37.17 A pyramid of production explains the ecological cost of meat. A field of corn can support many more human vegetarians than meat-eaters.

37.18 Chemicals are cycled between organic matter and abiotic reservoirs.

37.19 The carbon cycle depends on photosynthesis and respiration.

37.20 The phosphorus cycle depends on the weathering of rock.

37.21 The nitrogen cycle depends on bacteria. Various bacteria in soil (and root nodules of some plants) convert gaseous N_2 to compounds that plants can use, such as ammonium (NH_4^+) and nitrate (NO_3^-).

37.22 A rapid inflow of nutrients degrades aquatic ecosystems. Nutrient input from fertilizer and other sources causes rapid eutrophication, resulting in decreased species diversity and oxygen depletion of lakes, rivers, and coastal waters.

37.23 Ecosystem services are essential to human well-being. We depend on services provided by natural ecosystems.

Connecting the Concepts

1. Fill in the blanks in the table below summarizing the interspecific interactions in a community.

Interspecific Interaction	Effect on Species 1	Effect on Species 2	Example
	+	−	
	−	−	
	+	−	
	+	−	
	+	+	

2. Fill in the blanks in the table below summarizing terrestrial nutrient cycles.

	Carbon	Phosphorus	Nitrogen
Main abiotic reservoir(s)			
Form in abiotic reservoir			
Form used by producers			
Human activities that alter cycle			
Effects of altering cycle			

Testing Your Knowledge

Level 1: Knowledge/Comprehension

3. Which of the following groups is absolutely essential to the functioning of an ecosystem?
   a. producers
   b. producers and herbivores
   c. producers, herbivores, and carnivores
   d. producers and decomposers
4. To ensure adequate nitrogen for a crop, a farmer would want to *decrease* ______________ by soil bacteria.
   a. nitrification
   b. denitrification
   c. nitrogen fixation
   d. a and c
5. Which of the following organisms is mismatched with its trophic level?
   a. algae—producer
   b. phytoplankton—primary consumer
   c. carnivorous fish larvae—secondary consumer
   d. eagle—tertiary or quaternary consumer
6. Which of the following best illustrates ecological succession?
   a. A mouse eats seeds, and an owl eats the mouse.
   b. Decomposition in soil releases nitrogen that plants can use.
   c. Grasses grow in a deserted field, followed by shrubs and then trees.
   d. Imported pheasants increase in numbers, while local quail disappear.
7. The open ocean and tropical rain forests contribute the most to Earth's net primary production because
   a. both have high rates of net primary production.
   b. both cover huge surface areas of Earth.
   c. nutrients cycle fastest in these two ecosystems.
   d. the ocean covers a huge surface area and the tropical rain forest has a high rate of production.

Level 2: Application/Analysis

8. Explain how seed dispersal by animals is an example of mutualism in some cases.
9. What is rapid eutrophication? What steps might be taken to slow this process?
10. Local conditions, such as heavy rainfall or the removal of plants, may limit the amount of nitrogen, phosphorus, or calcium available, but the amount of carbon available in an ecosystem is seldom a problem. Explain.
11. In Southeast Asia, there's an old saying: "There is only one tiger to a hill." In terms of energy flow in ecosystems, explain why big predatory animals such as tigers and sharks are relatively rare.
12. For which chemicals are biogeochemical cycles global? Explain.
13. What roles do bacteria play in the nitrogen cycle?

Level 3: Synthesis/Evaluation

14. **SCIENTIFIC THINKING** An ecologist studying plants in the desert performed the following experiment. She staked out two identical plots, which included a few sagebrush plants and numerous small, annual wildflowers. She found the same five wildflower species in roughly equal numbers on both plots. She then enclosed one of the plots with a fence to keep out kangaroo rats, the most common grain-eaters of the area. After two years, to her surprise, four of the wildflower species were no longer present in the fenced plot, but one species had increased dramatically. The control plot had not changed. Using the principles of ecology, propose a hypothesis to explain her results. What additional evidence would support your hypothesis?
15. Sometime in 1986, near Detroit, a freighter pumped out water ballast containing larvae of European zebra mussels. The molluscs multiplied wildly, spreading through Lake Erie and entering Lake Ontario. In some places, they have become so numerous that they have blocked the intake pipes of power plants and water treatment plants, fouled boat hulls, and sunk buoys. What makes this kind of population explosion occur? What might happen to native organisms that suddenly must share the Great Lakes ecosystem with zebra mussels? How would you suggest trying to solve the mussel population problem?

Answers to all questions can be found in Appendix 4.

CHAPTER 38

Conservation Biology

Can Earth's biodiversity be saved?

Today, we face an environmental challenge that eclipses all others in scope: global climate change caused by the unprecedented speed with which Earth's atmosphere is warming. The effects of climate change are already apparent in melting ice sheets, rising seas, and extreme weather, including record-shattering heat waves and precipitation. Biodiversity, the variety of living things, will be one casualty of the rapidly changing environment. You may know that climate change threatens polar bears and penguins, but thousands of other species are also imperiled, including the American pika *(Ochotona princeps*; photo below). This diminutive relative of rabbits lives high up in the Rocky Mountains of the United States and Canada.

The pika's high body temperature is well suited to the chilly climate of its mountain habitat. On warm summer days, however, pikas must take refuge in crevices where pockets of cold air prevent fatal overheating. There is also a limit to the pika's tolerance of low temperatures. In winter, pikas depend on a blanket of snow to insulate their shelters and food stores from the cold. Climate change threatens pikas with sizzling summer temperatures and winters of diminishing snowfall. In 2010,

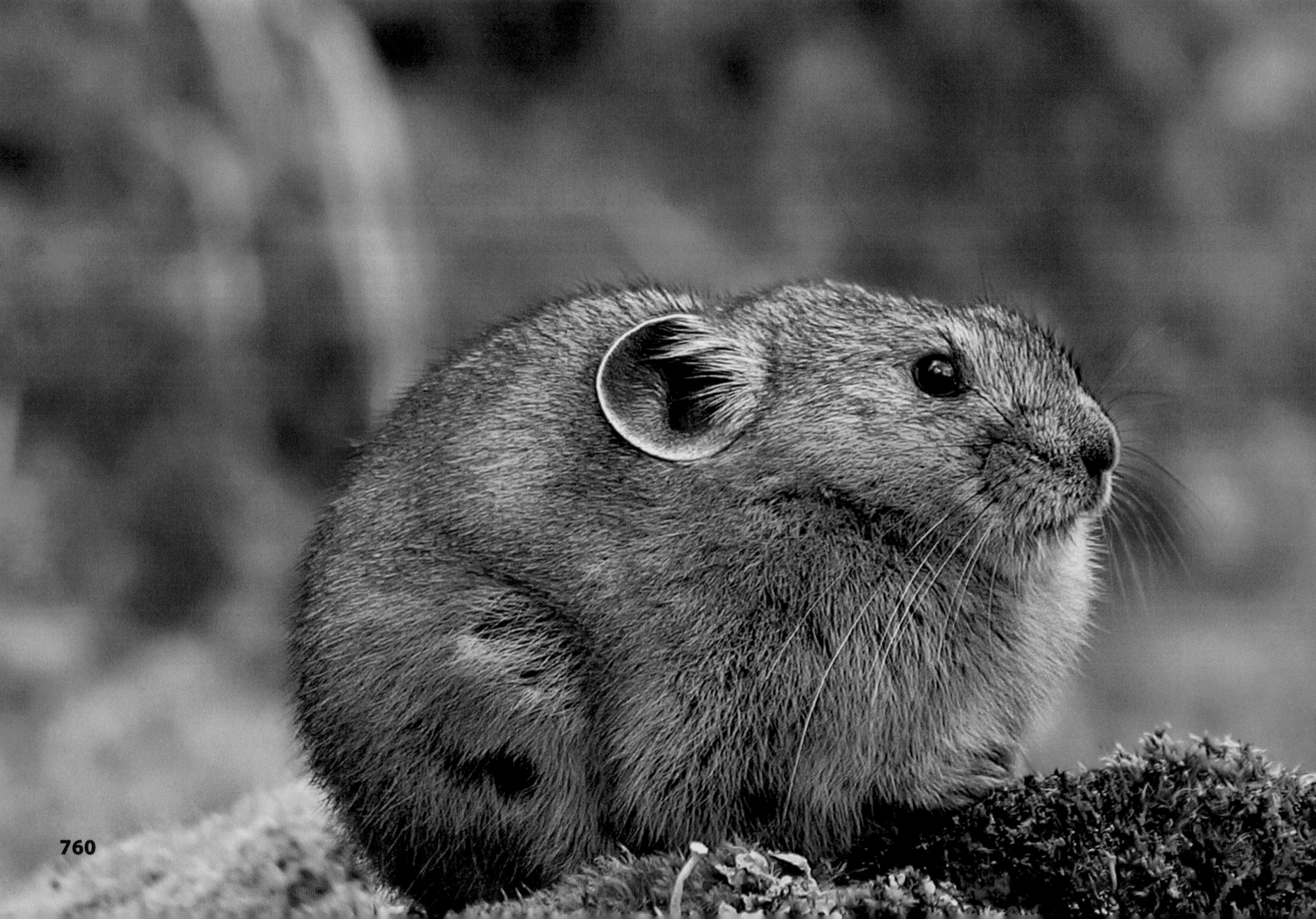

the pika came within a whisker of being declared an endangered species. But after reviewing the research on pika populations, the U.S. Fish and Wildlife Service (the agency responsible for making the decision), found reason for optimism about the pika's future. In Module 38.11, you'll learn about a conservation project that offers hope for pikas and hundreds of other species in a changing world.

Global climate change, which scientists worldwide agree is the result of human activities, is one of many ways that our dominance over the environment affects the nonhuman inhabitants of Earth. Biodiversity is rapidly diminishing despite conservation efforts. As you learn about the fight to save our biological heritage, you will see that conservation biology touches all levels of ecology, from a single pika to the ecosystem it calls home.

BIG IDEAS

The Loss of Biodiversity
(38.1–38.6)

Biodiversity is declining rapidly worldwide as a result of human activities.

Conservation Biology and Restoration Ecology
(38.7–38.13)

Biologists are applying their knowledge of ecology to slow the loss of biodiversity and help define a sustainable future.

▷ The Loss of Biodiversity

38.1 Loss of biodiversity includes the loss of ecosystems, species, and genes

Why do we care about losing biodiversity? One reason is what Harvard biologist E. O. Wilson calls biophilia, our sense of connection to nature and to other forms of life. Another is that many people share a moral belief that other species have an inherent right to life. But our dependence on vital ecosystem services also gives us practical reasons for preserving biodiversity (see Module 37.23).

Biodiversity encompasses more than individual species—it includes ecosystem diversity, species diversity, and genetic diversity. Let's examine each level of diversity to see what we stand to lose if the decline is not stopped.

Ecosystem Diversity The world's natural ecosystems are rapidly disappearing. Nearly half of Earth's forests are gone, and thousands more square kilometers disappear every year. Grassland ecosystems in North America (see Figure 34.13), where millions of bison roamed as recently as the 19th century, have overwhelmingly been lost to agriculture and development.

The temperate coniferous forest of the Klamath-Siskiyou Wilderness **(Figure 38.1A)** is located in a region spanning parts of California and Oregon that is extraordinarily rich in ecosystem diversity. In addition to the distinctive chaparral ecosystem (see Figure 34.12), forests of sequoia, redwood, and Douglas fir, coastal dunes, salt marshes, and a wide variety of other ecosystems can be found in this rapidly vanishing treasure trove of biodiversity. Only about a quarter of the original area remains in its natural state.

Aquatic ecosystems are also threatened. For example, an estimated 20% of the world's coral reefs, ecosystems known for their species richness and productivity (see Figure 34.6B), have been destroyed by human activities, and 15% are in danger of collapse within the next two decades. The deteriorating state of freshwater ecosystems is particularly worrisome. Tens of thousands of species live in lakes and rivers, and these ecosystems supply food and water for many terrestrial species, as well—including us.

As natural ecosystems are lost, so are essential services. Water purification is one of the services provided free of charge by healthy ecosystems. As water moves slowly through forests, streams, and wetlands, pollutants and sediments are filtered out. Whether taken from surface waters such as lakes or subsurface sources (groundwater), the drinking water supplied by public water systems typically has passed through this natural filtration process. In some places, including New York City, no further filtration is required, although the water is chlorinated to kill microorganisms. As farm fields and housing developments replaced the naturally diverse ecosystems in New York City's watershed, the land's ability to purify water deteriorated. The additional pollution from agricultural runoff and sewage reduced water quality to the point where the city had to take action. Officials considered spending $8 billion to build a filtration plant, which would cost a further $1 million per day to operate. They decided to invest in lower cost ecosystem services instead. Actions included more tightly restricting land use in the watershed, purchasing land to preserve natural ecosystems, and helping landowners better manage their land to protect the watershed. As a result of these measures, the quality of naturally filtered water supplied to New York City remains high.

Species Diversity When ecosystems are lost, populations of the species that make up their biological communities are also lost. A species may disappear from a local ecosystem but remain in others; for example, a population of American pika may be lost from one region of the Rocky Mountains while other populations survive elsewhere. Ecologists refer to the loss of a single population of a species as **extirpation**. Although extirpation and declining population sizes are strong signals that a species is in trouble, it may still be possible to save it. **Extinction** means that all populations of a species have disappeared, an irreversible situation.

How rapidly are species being lost? Because biologists are uncertain of the total number of species that exist, it is

▼ **Figure 38.1A** The Klamath-Siskiyou Wilderness, home to a wide variety of ecosystems

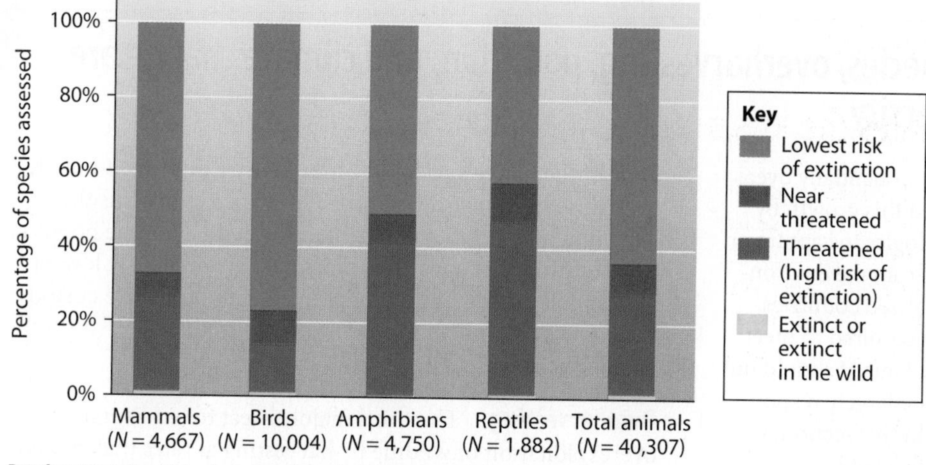

100%
80%
60%
40%
20%
0%
Percentage of species assessed
Mammals (N = 4,667)
Birds (N = 10,004)
Amphibians (N = 4,750)
Reptiles (N = 1,882)
Total animals (N = 40,307)
Key
Lowest risk of extinction
Near threatened
Threatened (high risk of extinction)
Extinct or extinct in the wild

Data from International Union for Conservation of Nature and Natural Resources (2012).

▲ **Figure 38.1B** Results of the 2012 IUCN assessment of species at risk for extinction (*N* = the number of species assessed)

difficult to determine the actual rate of species loss. Some scientists estimate that current extinction rates are around 100 times greater than the natural rate of extinction. The International Union for Conservation of Nature (IUCN) is a global environmental network that keeps track of the status of species worldwide. **Figure 38.1B** shows the 2012 IUCN assessment of more than 40,000 species of animals plus four major animal phyla. Notice the large proportions of amphibians and reptiles that are considered threatened. Disease caused by chytrid fungi (see Module 17.14) is one reason for the decline of amphibians, which are also vulnerable to climate change. Reptiles have been heavily impacted by hunting and deforestation.

Because of the network of community interactions among populations of different species within an ecosystem, the loss of one species can have a negative impact on the overall species richness of the ecosystem. Keystone species illustrate this effect (see Module 37.11). Other species modify their habitat in ways that encourage species diversity. In prairie ecosystems, for instance, plant and arthropod diversity is greatest near prairie dog burrows, where the soil has been altered by the animal's digging **(Figure 38.1C)**. Abandoned burrows provide homes for cottontail rabbits, burrowing owls, and other animals. Thus, extirpation of prairie dogs results in lower species diversity in prairie communities.

In the United States, the Endangered Species Act protects species and the ecosystems on which they depend. Many other nations have also enacted laws to protect biodiversity, and an international agreement protects some 33,000 species of wild animals and plants from trade that would threaten their survival.

▼ **Figure 38.1C** A group of young black-tailed prairie dogs (*Cynomys ludovicianus*) near their burrow

Species loss also has practical consequences for human well-being. Many drugs have been developed from substances found in the natural world, including penicillin, aspirin, antimalarial agents, and anticancer drugs. Dozens more potentially useful chemicals from a variety of organisms are currently being investigated. For example, researchers are testing possible new antibiotics produced by microbial symbionts of marine sponges, painkillers extracted from a species of poison dart frog, and anti-HIV and anticancer drugs derived from compounds found in rain forest plants.

Genetic Diversity The genetic diversity within and between populations of a species is the raw material that makes microevolution and adaptation to the environment possible—a hedge against future environmental changes. If local populations are lost and the total number of individuals of a species declines, so, too, do the genetic resources for that species. Severe reduction in genetic variation threatens the survival of a species.

The enormous genetic diversity of all the organisms on Earth has great potential benefit for people, too. Breeding programs have narrowed the genetic diversity of crop plants to a handful of varieties, leaving them vulnerable to pathogens (see Module 17.11). For example, researchers are currently scrambling to stop the spread of a deadly new strain of wheat stem rust, a fungal pathogen that has devastated harvests in Africa and central Asia. Resistance genes found in the wild relatives of wheat **(Figure 38.1D)** may hold the key to the world's future food supply. Many research and biotechnology leaders are enthusiastic about the possibilities that "bioprospecting" for potentially useful genes in other organisms holds for the development of new medicines, industrial chemicals, and other products.

Now that you have some insight into the nature and value of biodiversity, let's examine in more detail the causes for its decline.

▲ **Figure 38.1D** Einkorn wheat, a wild relative of modern cultivated varieties

? What are two reasons to be concerned about the impact of the biodiversity crisis on human welfare?

● The environmental degradation threatening other species may also harm us. We are dependent on biodiversity, both directly through the use of organisms and their products and indirectly through ecosystem services.

38.2 Habitat loss, invasive species, overharvesting, pollution, and climate change are major threats to biodiversity

CONNECTION

The human population has grown exponentially over the past century. We have supported this growth by using increasingly effective technologies to capture or produce food, to extract resources from the environment, and to build cities. In industrialized countries, we consume far more resources than are required to meet our basic requirements for food and shelter. Thus, it should not surprise you to learn that human activities are largely responsible for the current decline of biodiversity. In this section, we examine the major factors that threaten biodiversity.

Habitat Loss Human alteration of habitats poses the single greatest threat to biodiversity throughout the biosphere. Agriculture, urban development, forestry, mining, and environmental pollution have brought about massive destruction and fragmentation of habitats. Deforestation continues at a blistering pace in tropical and coniferous forests (**Figure 38.2A**).

▲ **Figure 38.2A** Clear-cut areas in Mount Baker-Snoqualmie National Forest, Washington

The amount of land surface altered by people is approaching 50%, and we use over half of all accessible surface fresh water. The natural courses of most of the world's major rivers have been changed. Worldwide, tens of thousands of dams constructed for flood control, hydroelectric power, drinking water, and irrigation have damaged river and wetland ecosystems. Some of the most productive aquatic habitats in estuaries and intertidal wetlands have been overrun by commercial and residential development. The loss of marine habitat is severe, especially in coastal areas and coral reefs.

Invasive Species Ranking second behind habitat loss as a threat to biodiversity are invasive species, which disrupt communities by competing with, preying on, or parasitizing native species. The lack of interspecific interactions that keep the newcomer populations in check is often a key factor in a non-native species becoming invasive (see Module 37.13). Meanwhile, a newly arrived species is an unfamiliar biotic factor in the environment of native species. Natives are especially vulnerable when a new species poses an unprecedented threat. In the absence of an evolutionary history with predators, for example, animals may lack defense mechanisms or even a fundamental recognition of danger.

The Pacific island of Guam was home to 13 species of forest birds—but no native snakes—when brown tree snakes (**Figure 38.2B**) arrived as stowaways on a cargo plane. With no competitors, predators, or parasites to hinder them, the snakes proliferated rapidly on a diet of unwary birds. Four of the native species of birds were extirpated, although they survive on nearby islands. Three species of birds that lived nowhere else but Guam are now extinct. As the populations of two other species of birds became perilously low, officials took the remaining individuals into protective custody; they now exist only in zoos. The brown tree snake also eliminated species of seabirds and lizards.

Overharvesting The third major threat to biodiversity is overexploitation of wildlife by harvesting at rates that exceed the ability of populations to rebound. Such overharvesting has threatened some rare trees that produce valuable wood, such as mahogany and rosewood. Animal species whose numbers have been drastically reduced by excessive commercial harvest, poaching, or sport hunting include tigers, whales, rhinoceroses, Galápagos tortoises, and numerous fishes. In parts of Africa, Asia, and South America, wild animals are heavily hunted for food, and the African term "bushmeat" is now used to refer generally to such meat. As once-impenetrable forests are opened to exploitation, the commercial bushmeat trade has become one of the greatest threats to primates, including gorillas, chimpanzees, and many species of monkeys, as well as other mammals and birds. No longer hunted only for local use, large quantities of bushmeat are sold at urban markets or exported worldwide, including to the United States.

Aquatic species are suffering overexploitation, too. Many edible marine fish and seafood species are in a precarious state (see Module 36.8). Worldwide, fishing fleets are working farther offshore and harvesting fish from greater depths in order to obtain hauls comparable to those of previous decades.

▲ **Figure 38.2B** A brown tree snake (*Boiga irregularis*)

Pollution Pollutants released by human activities can have local, regional, and global effects. Some pollutants, such as oil spills, contaminate local areas (**Figure 38.2C**). The global water cycle can transport pollutants—for instance, pesticides used on land—from terrestrial to aquatic ecosystems hundreds of miles away. Pollutants that are emitted into the atmosphere, such as nitrogen oxides from the burning of fossil fuels, may be carried aloft for many miles before falling to Earth in the form of acid precipitation.

Ozone depletion in the upper atmosphere is another example of the global impact of pollution. The **ozone layer** protects Earth from the harmful ultraviolet rays in sunlight (see Module 7.14). Beginning in the mid-1970s, scientists realized that the ozone layer was gradually thinning. The consequences of ozone depletion for life on Earth could be quite severe, not only increasing skin cancers, but also harming crops and natural communities, especially the phytoplankton that are responsible for a large proportion of Earth's primary production. International agreements to phase out the production of chemicals implicated in ozone destruction have been effective in slowing the rate of ozone depletion. Even so, complete ozone recovery is probably decades away.

In addition to being transported to areas far from where they originate, many toxins produced by industrial wastes or applied as pesticides become concentrated as they pass through the food chain. This concentration, or **biological magnification**, occurs because the biomass at any given trophic level is produced from a much larger toxin-containing biomass ingested from the level below (see Module 37.16). Thus, top-level predators are usually the organisms most severely damaged by toxic compounds in the environment. In the Great Lakes food chain shown in **Figure 38.2D**, the concentration of industrial chemicals called PCBs increased at each successive trophic level. The PCB concentration measured in the eggs of herring gulls, top-level consumers, was almost 5,000 times higher than that measured in phytoplankton. Many other synthetic chemicals that cannot be degraded by microorganisms also become concentrated through biological magnification, including DDT and mercury. Mercury, a by-product of plastic production and coal-fired power plants, enters the food chain after being converted to highly toxic methylmercury by benthic bacteria. Since people are top-level predators, too, eating fish from contaminated waters can be dangerous.

Recently, scientists have recognized a new type of pollutant in the oceans and the Great Lakes: plastic particles that are small enough to be eaten by zooplankton. Many body washes and facial cleansers include plastic "microbeads" to boost scrubbing power. (To see if your shower products contain plastic, check the list of ingredients for polyethylene.) Too small to be captured by wastewater treatment plants, these microparticles enter the watershed and eventually wash out to sea or collect in lakes. Larger particles called preproduction pellets or "nurdles," used in making plastic products, are also common aquatic pollutants. Nurdles may be broken down to microbead size in the ocean. Toxins such as PCBs and DDT adhere to these plastic spheres. Thus, toxins may be concentrated first on microparticles and concentrated again by biological magnification.

▲ **Figure 38.2C** A brown pelican on the Louisiana coast suffering the effects of the 2010 British Petroleum oil rig explosion

Global Climate Change According to many scientists, the changes in global climate that are occurring as a result of global warming are likely to become a leading cause of biodiversity loss. In the next four modules, you'll learn about some of the causes and consequences of global climate change.

? List four threats to biodiversity and give an example of each.

● Habitat loss—deforestation; invasive species—brown tree snake; overharvesting—bushmeat; pollution—biological magnification of PCBs, DDT, and mercury. (Other examples could be used.)

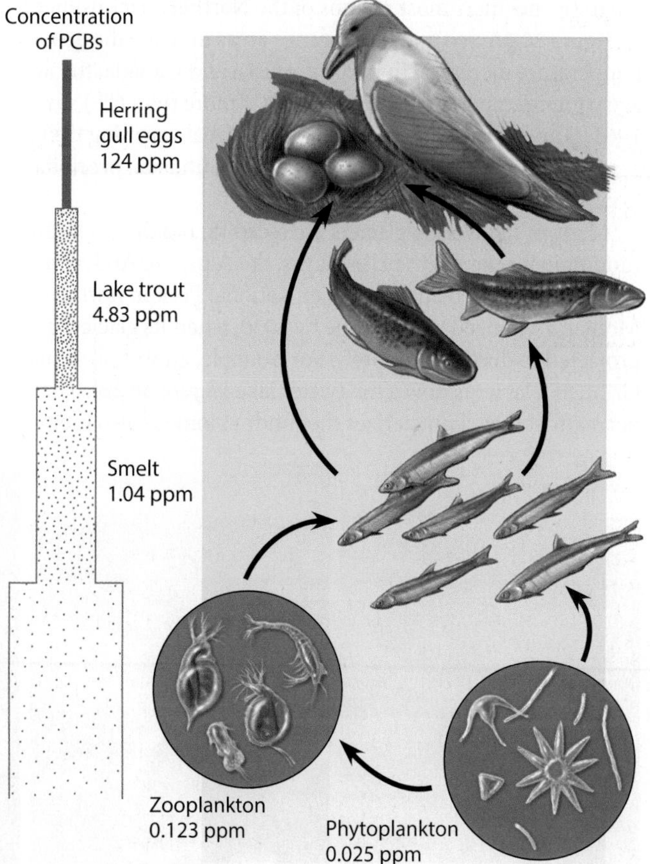
Concentration of PCBs
Herring gull eggs 124 ppm
Lake trout 4.83 ppm
Smelt 1.04 ppm
Zooplankton 0.123 ppm
Phytoplankton 0.025 ppm

▲ **Figure 38.2D** Biological magnification of PCBs in a food web, measured in parts per million (ppm)

38.3 Rapid warming is changing the global climate

CONNECTION

The scientific debate about global warming is over. The vast majority of scientists now agree that rising concentrations of greenhouse gases in the atmosphere (see Module 7.13), such as carbon dioxide (CO_2), methane (CH_4), and nitrous oxide (N_2O), are changing global climate patterns. This was the overarching conclusion of the assessment report released by the Intergovernmental Panel on Climate Change (IPCC) in 2013. Thousands of scientists and policymakers from more than 100 countries participated in producing the report, which is based on data published in hundreds of scientific papers.

The signature effect of increasing greenhouse gases is the rapid increase in the average global temperature, which has risen 0.8°C (1.5°F) over the last 100 years, with 0.6°C of that increase occurring over the last three decades. Further increases of 2–6°C (3.6–10.8°F) are likely by the end of the 21st century, depending on the rate of future greenhouse gas emissions. Ocean temperatures are also rising, in deeper layers as well as at the surface. But the temperature increases are not distributed evenly around the globe. Warming is greater over land than sea, and the largest increases are in the northernmost regions of the Northern Hemisphere. In **Figure 38.3A**, red and dark orange areas indicate the greatest temperature increases. In parts of Alaska and Canada, the average winter temperature has risen 3.4°C (more than 6°F) since 1961. Some of the consequences of the global warming trend are already clear from rising temperatures, unusual precipitation patterns, and melting ice.

Many of the world's glaciers are receding rapidly, including mountain glaciers in the Himalayas, the Alps, the Andes, and the western United States. Glacier National Park in northwest Montana will need a new name by 2030, when its glaciers are projected to disappear entirely. For example, almost all of the Grinnell Glacier is now a meltwater lake **(Figure 38.3B)**. The permafrost that characterizes the tundra biome is also melting.

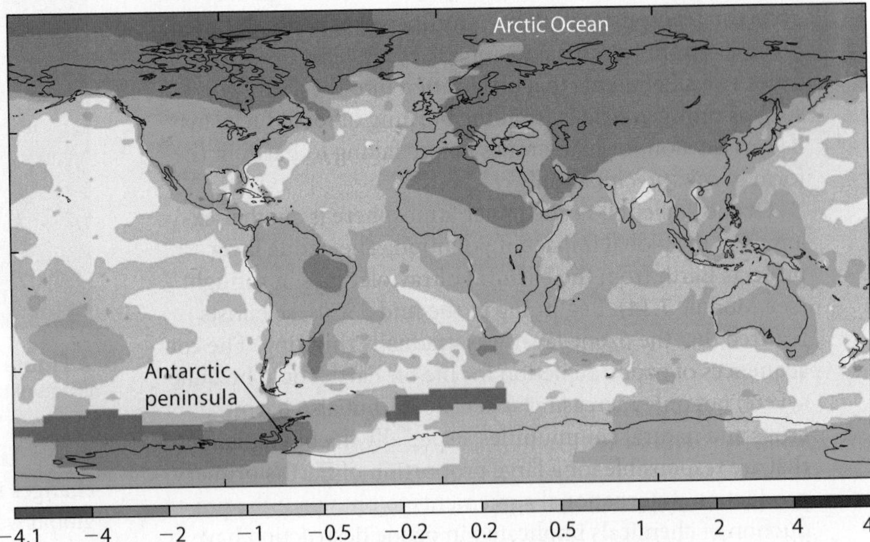

Arctic Ocean
Antarctic peninsula
−4.1
−4
−2
−1
−0.5
−0.2
0.2
0.5
1
2
4
4.1

▲ **Figure 38.3A** Differences in temperature during 2002–2011 compared with long-term averages during 1951–1980 (in °C)

Try This Identify the regions where the average temperature during 2002–2011 was 1°C or more above the long-term average.

Permanent Arctic sea ice is shrinking even faster than climate models projected; many scientists now expect the Arctic Ocean to be nearly ice-free in summer within 25 years. The massive ice sheets of Greenland and Antarctica are thinning and collapsing. If this melting trend accelerates, rising sea levels will cause catastrophic flooding of coastal areas worldwide.

Warm weather is beginning earlier each year. Cold days and nights and frosts have become less frequent; hot days and nights have become more frequent. Deadly heat waves are increasing in frequency and duration.

Precipitation patterns are changing, bringing longer and more intense drought to some areas. In other regions, a greater proportion of the total precipitation is falling in torrential downpours that cause flooding. Hurricane intensity is increasing, fueled by higher sea surface temperatures.

Many of these changes will have a profound impact on biodiversity, as we explore in Modules 38.5 and 38.6. In the next module, we examine the causes of rising greenhouse gas emissions.

? **From the map in Figure 38.3A, which biomes are likely to be most affected by global warming, and why?**

● The high-latitude biomes of the Northern Hemisphere, tundra and taiga, and the polar ice biomes will be most affected. Those biomes are experiencing the greatest temperature change. Also, the organisms that live there are adapted to cold weather and a short growing season, so their survival is on the line.

◀ **Figure 38.3B** Grinnell Glacier in Glacier National Park, 1938 (left), 1981 (center), and 2009 (right)

38.4 Human activities are responsible for rising concentrations of greenhouse gases

CONNECTION

Without its blanket of natural greenhouse gases such as CO_2 and water vapor to trap heat, Earth would be too cold to support most life. However, increasing the insulation that the blanket provides is making the planet uncomfortably warm, and that increase is occurring rapidly. For 650,000 years, the atmospheric concentration of CO_2 did not exceed 300 parts per million (ppm); the preindustrial concentration was 280 ppm. In May, 2013, atmospheric CO_2 reached a record 400 ppm. The levels of other heat-trapping gases, nitrous oxide (N_2O) and methane (CH_4), have increased dramatically, too (**Figure 38.4A**). CO_2 and N_2O are released when fossil fuels—oil, coal, and natural gas—are burned. N_2O is also released when nitrogen fertilizers are used in agriculture. Livestock and landfills are among the factors responsible for increases of atmospheric CH_4. The consensus of scientists is that rising concentrations of greenhouse gases—and thus, global warming—are the result of human activities.

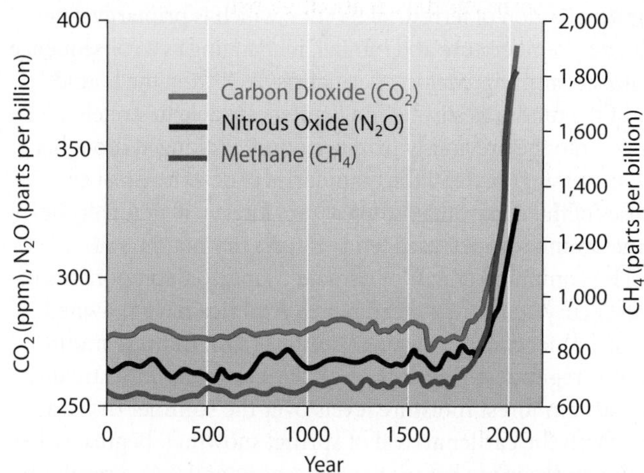

400
350
300
250
2,000
1,800
1,600
1,400
1,200
1,000
800
600
0
500
1000
1500
2000
Year
CO₂ (ppm), N₂O (parts per billion)
CH₄ (parts per billion)
Carbon Dioxide (CO₂)
Nitrous Oxide (N₂O)
Methane (CH₄)

Data from Climate Change 2007: The physical science bases: Contribution of Working Group I to the fourth assessment report of the International Panel on Climate Change, IPCC Secretariat.

▲ **Figure 38.4A** Atmospheric concentrations of CO_2, N_2O (*y* axis, left), and CH_4 (*y* axis, right), as of 2011

Let's take a closer look at CO_2, the dominant greenhouse gas and a major reservoir for carbon in the atmosphere (see Module 37.19). (CH_4 is also part of that reservoir.) CO_2 is removed from the atmosphere by the process of photosynthesis and stored in organic molecules such as carbohydrates (**Figure 38.4B**; green arrows). Thus, biomass, the organic molecules in an ecosystem, is a biotic carbon reservoir. Purple arrows in Figure 38.4B represent carbon released in the form of CO_2. The carbon-containing molecules in living organisms may be used in the process of cellular respiration. Decomposition of nonliving biomass by microorganisms or fungi also releases CO_2. Overall, uptake of CO_2 by photosynthesis roughly equals the release of CO_2 by cellular respiration. In addition, CO_2 is exchanged between the atmosphere and the surface waters of the oceans.

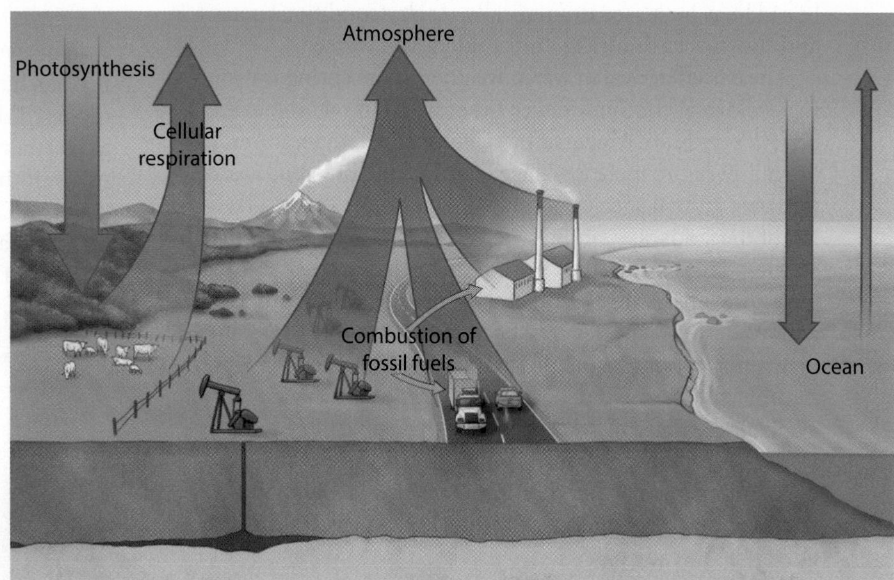

Atmosphere
Photosynthesis
Cellular respiration
Combustion of fossil fuels
Ocean

▲ **Figure 38.4B** Carbon cycling (Arrow width indicates amount of carbon taken up or released.)

Try This Label the diagram by writing the names of carbon-containing molecules (CO_2, organic molecules, carbonic acid) in the appropriate locations.

Fossil fuels consist of biomass that lay buried under sediments for millions of years without being completely decomposed (see Module 17.4). The burning of fossil fuels and wood, which is also an organic material, can be thought of as a rapid form of decomposition. Whereas cellular respiration releases energy from organic molecules slowly and harnesses it to make ATP, combustion liberates the energy rapidly as heat and light. In both processes, the carbon atoms that make up the organic fuel are released in CO_2.

The CO_2 flooding into the atmosphere from combustion of fossil fuels may be absorbed by photosynthetic organisms and incorporated into biomass. But deforestation has significantly decreased the number of CO_2 molecules that can be accommodated by this pathway. CO_2 may also be absorbed into the ocean. For decades, the oceans have been absorbing considerably more CO_2 than they have released, and they will continue to do so, but the excess CO_2 is beginning to affect ocean chemistry. When CO_2 dissolves in water, it becomes carbonic acid. Recently, measurable decreases in ocean pH have raised concern among biologists. Organisms that construct shells or exoskeletons out of calcium carbonate ($CaCO_3$), including corals and many plankton, are most likely to be affected, as decreasing pH reduces the concentration of the carbonate ions (see Module 2.15).

Despite the attempts of many nations to curb carbon emissions, atmospheric CO_2 is increasing at an accelerating pace. At this rate, further climate change is inevitable.

? The amount of CO_2 your activities are responsible for releasing every year is called your *carbon footprint*. Search for an online calculator that estimates your carbon footprint. What are the primary sources of the CO_2 you generate?

● Transportation and home energy use are the two major categories contributing to the footprint.

38.5 Global climate change affects biomes, ecosystems, communities, and populations

The distribution of terrestrial biomes, which is primarily determined by temperature and rainfall, is changing as a consequence of global warming. Melting permafrost is shifting the boundary of the tundra as shrubs and conifers are able to stretch their ranges into the previously frozen ground. Prolonged droughts will increasingly extend the boundaries of deserts. Great expanses of the Amazonian tropical rain forest will gradually become savanna as increased temperatures dry out the soil.

The combined effects of climate change on components of forest ecosystems in western North America have spawned catastrophic wildfire seasons **(Figure 38.5A)**. In these mountainous regions, spring snowmelt releases water into streams that sustain forest moisture levels over the summer dry season. With the earlier arrival of spring, snowmelt begins earlier and dwindles away before the dry season ends. As a result, the fire season has been getting longer since the 1980s. In addition, drought conditions have made trees more vulnerable to insect and pathogen attack; vast numbers of dead trees add fuel to the flames. Fires burn longer, and the number of acres burned has increased dramatically. As dry conditions persist and snowpacks diminish, the problem will worsen.

▲ **Figure 38.5A** A wildfire engulfing homes in Colorado Springs, Colorado, in June 2012

The earlier arrival of warm weather in the spring is disturbing ecological communities in other ways. In many species, certain events are triggered by rising spring temperatures. Earlier temperature increases have hastened the breeding season for some animal species. Satellite images show earlier greening of the landscape, and flowering occurs sooner. For other species, day length is the cue that spring has arrived. Because global warming affects temperature but not day length, interactions between species may become out of sync. For example, plants may bloom before pollinators have emerged, or eggs may hatch before a dependable food source for the young is available. Because the magnitude of seasonal shifts increases from the tropics to the poles, migratory birds may also experience timing mismatches. For instance, birds arriving in the Arctic to breed may find that the period of peak food availability has already passed.

Warming oceans threaten tropical coral reef communities. When stressed by high temperatures, corals expel their symbiotic algae in a phenomenon called bleaching. Corals can recover if temperatures return to normal, but they cannot survive prolonged temperature increases. When corals die, the community is overrun by large algae, and species diversity plummets.

The distributions of populations and species are also changing in response to climate change. Many species, for example, the pikas described in the chapter introduction, are adapted to the abiotic conditions in their environment. With rising temperatures, the ranges of many species have already shifted toward the poles or to higher elevations. For example, researchers in Europe and the United States have reported that the ranges of more than two dozen species of butterflies have moved north by as much as 150 miles. Shifts in the ranges of many bird species have also been reported; the Inuit peoples living north of the Arctic Circle have sighted birds such as robins in the region for the first time.

However, species that live on mountaintops or in polar regions have nowhere to go. Researchers in Costa Rica have reported the disappearance of 20 species of frogs and toads as warmer Pacific Ocean temperatures reduce the dry-season mists in their mountain habitats. In the Arctic, polar bears **(Figure 38.5B)**, which stalk their prey on ice and need to store up body fat for the warmer months, are showing signs of starvation as their hunting grounds melt away. Similarly, in the Antarctic, the disappearance of sea ice is blamed for recent decreases in populations of Emperor and Adélie penguins.

▲ **Figure 38.5B** A polar bear (*Ursus maritimus*) with her cubs on melting pack ice in Spitsbergen, Norway

Global climate change has been a boon to some organisms, but so far the beneficiaries have been species that have a negative impact on humans. For example, in mountainous regions of Africa, Southeast Asia, and Central and South America, the ranges of mosquitoes that carry diseases such as malaria, yellow fever, and dengue are restricted to lower elevations by frost. With rising temperatures and fewer days of frost, these mosquitoes—and the diseases they carry—are appearing at higher elevations. In another example, longer summers in western North America have enabled bark beetles to complete their life cycle in one year instead of two, promoting beetle outbreaks that have destroyed millions of acres of conifers. Undesirable plants such as poison ivy and kudzu have also benefited from rising temperatures (see the introduction to Chapter 7).

Environmental change has always been a part of life; in fact, it is a key ingredient of evolutionary change. In the next module, we consider the evidence of evolutionary adaptation to global warming.

? How might timing mismatches caused by climate change affect an individual's reproductive fitness?

● Any factor that reduces the number of offspring an organism produces may affect fitness. Examples include flowers emerging too soon or too late for pollinators and birds that arrive too late in the season to find food for offspring.

38.6 Climate change is an agent of natural selection

EVOLUTION CONNECTION

Global climate change is already affecting habitats throughout the world. Why do some species appear to be adapting to these changes, while others, like the polar bear, are endangered by them?

In the previous module, we described several ways in which organisms have responded to global climate change. For the most part, those examples can be attributed to **phenotypic plasticity**, the ability to change phenotype in response to local environmental conditions. Differences resulting from phenotypic plasticity are within the normal range of expression for an individual's genotype. Phenotypic plasticity allows organisms to cope with short-term environmental changes. On the other hand, phenotypic plasticity is itself a trait that has a genetic basis and can evolve. Researchers studying the effects of climate change on populations have detected microevolutionary changes in phenotypic plasticity.

▲ **Figure 38.6A** A great tit (*Parus major*)

A common bird in Europe, the great tit **(Figure 38.6A)** is the third link in a food chain that has been altered by climate change. As warm weather arrives earlier in the spring, tree leaves emerge earlier and caterpillars, which use the swelling buds and unfolding leaves as their food source, hatch sooner. The reproductive success of great tits depends on having an ample supply of these nutritious caterpillars to feed their offspring. Like many other birds, great tits have some phenotypic plasticity in the timing of their breeding, which helps them synchronize their reproduction with the availability of caterpillars. The range and degree of plasticity vary among great tits, and this variation has a genetic basis. Researchers have found evidence of directional selection (see Module 13.14) favoring individuals that have the greatest phenotypic plasticity and lay their eggs earlier, when the abundance of food gives their offspring a better chance of survival.

In another example, scientists studied reproduction in a population of red squirrels **(Figure 38.6B)** in the Yukon Territory of Canada, where spring temperatures have increased by approximately 2°C in the last three decades. These researchers also found earlier breeding times in the spring. Over a period of 10 years, the date on which female squirrels gave birth advanced by 18 days, a change of about 6 days per generation. Using statistical analysis, the scientists determined that phenotypic plasticity was responsible for most of the shift in breeding times. However, a small but significant portion of the change (roughly 15%) could be attributed to microevolution, directional selection for earlier breeding. The researchers hypothesize that red squirrels born earlier in the year are larger and more capable of gathering and storing food in the autumn and thus have a better chance of successful reproduction the following spring.

▲ **Figure 38.6B** A red squirrel (*Tamiasciurus hudsonicus*) eating the seeds from a spruce cone

From the scant evidence available at this time, it appears that some populations, especially those with high genetic variability and short life spans, may adapt quickly enough to avoid extinction. In addition to the studies on phenotypic plasticity in great tits and red squirrels, researchers have also documented microevolutionary changes in traits such as dispersal ability and timing of life cycle events in insect populations. However, evolutionary adaptation is unlikely to save species with long life spans such as polar bears and penguins that are experiencing rapid habitat loss. The rate of climate change is incredibly fast compared with major climate shifts in evolutionary history, and if it continues on its present course, thousands of species—the IPCC estimates as many as 30% of plants and animals—will likely become extinct.

? How does a short generation time hasten the process of evolutionary adaptation?

● Each generation has the potential for "testing" new phenotypes in the environment. Shorter generation times result in more opportunities for testing new phenotypes, which in turn allows natural selection to proceed more rapidly.

▷ Conservation Biology and Restoration Ecology

38.7 Protecting endangered populations is one goal of conservation biology

As we have seen in this unit, many of the environmental problems facing us today are consequences of human enterprises. But the science of ecology is not just useful for telling us how things have gone wrong. Ecological research is the foundation for finding solutions to these problems and for reversing the negative consequences of ecosystem alteration. Thus, we end the ecology unit with a section that highlights some of these applications of ecological research.

Conservation biology is a goal-oriented science that seeks to understand and counter the loss of biodiversity. Some conservation biologists focus on protecting populations of threatened species. This approach requires an understanding of the behavior and ecological niche of the target species, including its key habitat requirements and interactions with other members of its community. Threats posed by human activities are also assessed. With this knowledge, scientists can design a plan to expand or protect the resources needed. For example, the territory size required to support a tiger varies with the abundance of prey. Consequently, preserves set aside for Siberian tigers in Russia, where prey are scarce, must be 10 times as large as those provided for Bengal tigers in India.

The case of the black-footed ferret (**Figure 38.7A**) provides an example of the population approach to conservation. Little was known about this elusive nocturnal predator until the mid-20th century, and by then it was almost too late—population decline was already under way. Black-footed ferrets, one of three ferret species worldwide and the only one found in North America, feed almost exclusively on prairie dogs (see Figure 38.1C). Over the past century, prairie dogs have been extirpated from most of their former range by land-use changes and by poisoning or shooting. Epidemics of sylvatic plague, the animal version of bubonic plague, have devastated populations of black-footed ferrets as well as their prey. When an outbreak threatened to wipe out the last known population of black-footed ferrets, conservation biologists captured 18 remaining individuals and began breeding them in captivity to rebuild population numbers. Genetic variation, a prerequisite for adaptive evolutionary responses to environmental change, is a concern, given the bottleneck effect of near-extinction (see Module 13.12). Matings in the captive breeding facilities are carefully arranged to maintain as much genetic diversity as possible in the ferret populations.

▲ **Figure 38.7A** A black-footed ferret (*Mustela nigripes*)

In 1991, biologists began reintroducing captive-bred black-footed ferrets into the wild. Research carried out during these efforts has improved the success rate of reintroductions. For example, scientists found that the predatory behavior of ferrets has both innate and learned components, a discovery that led to more effective methods of preparing captive-bred animals to survive in the wild. Today, about 1,000 adult ferrets are living in the wild at sites scattered from Canada to Mexico. Despite the successes achieved thus far, however, the future of the black-footed ferret is far from secure. Biologists continue to monitor and manage the populations and their habitats.

Captive breeding programs are being used for numerous other species whose population numbers are perilously low. For example, efforts to save the whooping crane are under way (see Module 35.6). In Hawaii, biologists have planted thousands of greenhouse-grown silverswords (*Argyroxiphium sandwicense*; **Figure 38.7B**) on the cinder cone of the volcano Mauna Kea in hopes of reestablishing wild populations. Once so abundant that observers mistook their silvery color for snow on the distant peak, silverswords were grazed to near-extinction by goats and sheep that people had brought to the island.

▲ **Figure 38.7B** A Mauna Kea silversword (*Argyroxiphium sandwicense*)

By using a variety of methods, biologists have improved the conservation status of some endangered species, reintroduced many species to areas where they had been extirpated, and reversed declining population trends for others. However, we will not be able to save every threatened species. One way to select worthwhile targets is to identify and protect keystone species that may help preserve entire communities. And in many situations, conservation biologists must look beyond individual species to ecosystems.

? What do you think is the first priority for conservation biologists when they select a site for ferret reintroduction?

● The presence of a sufficiently large population of prairie dogs

38.8 Sustaining ecosystems and landscapes is a conservation priority

One of the most harmful effects of habitat loss is population fragmentation, the splitting and consequent isolation of portions of populations. As you saw in Figure 38.2A, for example, logging carves once-continuous forest into a patchwork of disconnected fragments. For many species, fragmentation means the world instantly shrinks to a fraction of its former size. Populations are reduced, and so are resources such as food and shelter. To counteract the effects of fragmentation, conservation biology often aims to sustain the biodiversity of entire ecosystems and landscapes. Ecologically, a **landscape** is a regional assemblage of interacting ecosystems, such as a forest, adjacent fields, wetlands, streams, and streamside habitats. **Landscape ecology** is the application of ecological principles to the study of the structure and dynamics of a collection of ecosystems.

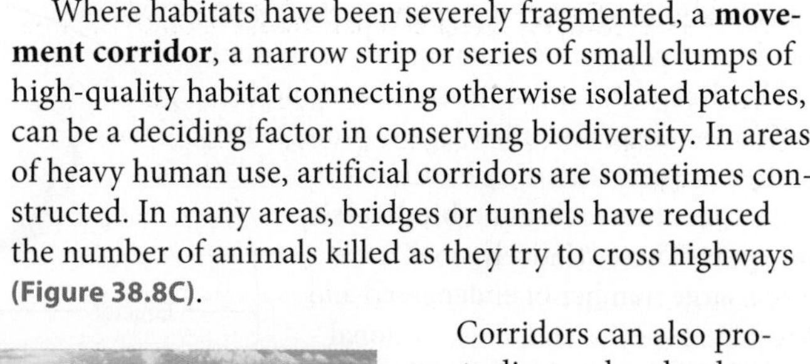

▲ **Figure 38.8A** A landscape in Yellowstone National Park with distinct edges

Edges, or boundaries between ecosystems, are prominent features of landscapes. The photograph in **Figure 38.8A** shows a landscape area in Yellowstone National Park that includes grassland and forest. Human activities, such as logging and road building, often create edges that are more abrupt than those delineating natural landscapes. Such edges have their own sets of physical conditions and thus their own communities of organisms. Some organisms thrive in edges because they require resources from the two adjacent areas. For instance, whitetail deer browse on woody shrubs found in edge areas between woods and fields, and their populations often expand when forests are logged or interrupted with housing developments.

Communities where human activities have generated many edges often have less diversity and are dominated by a few species that are adapted to edges. In one example, populations of the brown-headed cowbird (**Figure 38.8B**), an edge-adapted species that lays its eggs in the nests of other birds, are currently expanding in many areas of North America. Cowbirds forage in open fields on insects disturbed by or attracted to cattle and other large herbivores. The cowbirds also need forests, where they can parasitize the nests of other birds. Increasing cowbird parasitism and loss of suitable habitat are correlated with declining populations of several songbird species.

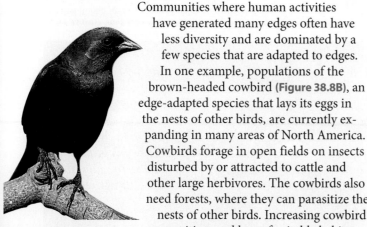

▲ **Figure 38.8B** A male brown-headed cowbird (*Molothrus ater*)

Where habitats have been severely fragmented, a **movement corridor**, a narrow strip or series of small clumps of high-quality habitat connecting otherwise isolated patches, can be a deciding factor in conserving biodiversity. In areas of heavy human use, artificial corridors are sometimes constructed. In many areas, bridges or tunnels have reduced the number of animals killed as they try to cross highways (**Figure 38.8C**).

Corridors can also promote dispersal and reduce inbreeding in declining populations. Corridors are especially important to species that migrate between different habitats seasonally. In some European countries, amphibian tunnels have been constructed to help frogs, toads, and salamanders cross roads to access their breeding territories.

On the other hand, a corridor can be harmful—as, for example, in the spread of diseases, especially among small subpopulations in closely situated habitat patches. The effects of movement corridors between habitats in a landscape are not completely understood, and researchers continue to study them.

? **How can "living on the edge" be a good thing for some species, such as whitetail deer and cowbirds?**

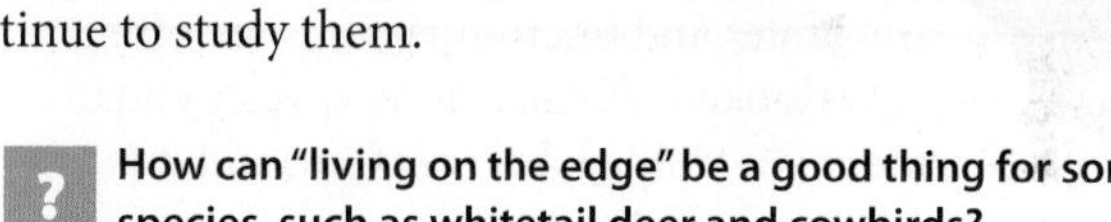

● Such animals use a combination of resources from the two ecosystems on either side of the edge.

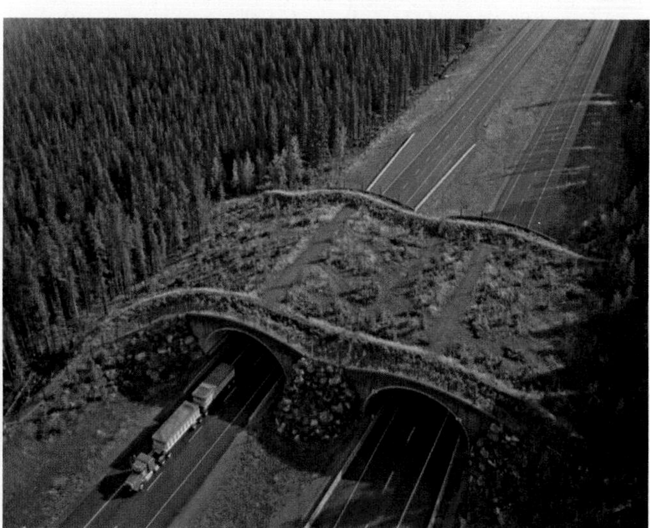

▲ **Figure 38.8C** A wildlife bridge in Banff National Park, Canada

38.9 Establishing protected areas slows the loss of biodiversity

Conservation biologists are applying their understanding of population, community, ecosystem, and landscape dynamics in establishing parks, wilderness areas, and other legally protected nature reserves. Choosing locations for protection often focuses on **biodiversity hot spots**. These relatively small areas have a large number of endangered and threatened species and an exceptional concentration of **endemic species**, those that are found nowhere else. Together, the "hottest" of Earth's biodiversity hot spots, shown in **Figure 38.9A**, total less than 1.5% of Earth's land but are home to a third of all species of plants and vertebrates. For example, all lemurs are endemic to Madagascar, which is home to more than 50 species. In fact, almost all of the mammals, reptiles, amphibians, and plants that inhabit Madagascar are endemic. There are also hot spots in aquatic ecosystems, such as certain river systems and coral reefs.

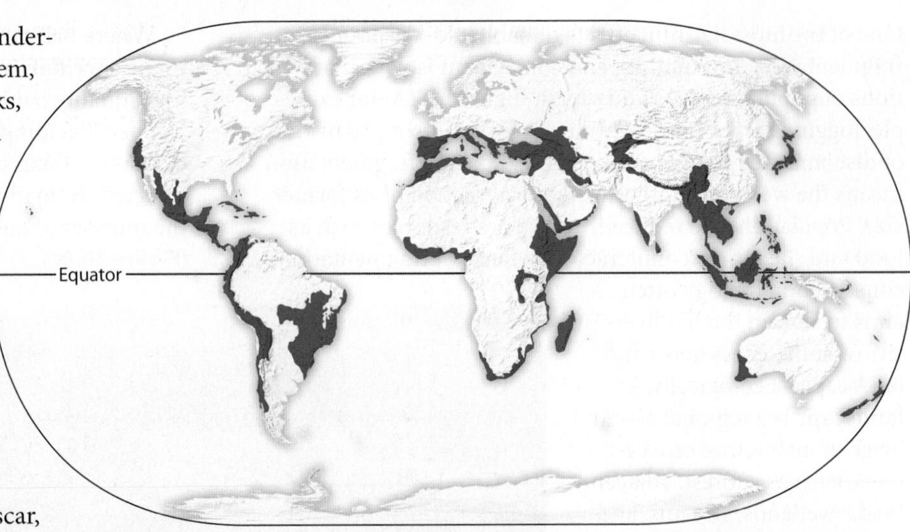
Equator

▲ **Figure 38.9A** Earth's terrestrial biodiversity hot spots (purple)

Because endemic species are limited to specific areas, they are highly sensitive to habitat degradation. Thus, biodiversity hot spots can also be hot spots of extinction. They rank high on the list of areas demanding strong global conservation efforts.

Concentrations of species provide an opportunity to protect many species in very limited areas. However, the "hot spot" designation tends to favor the most noticeable organisms, especially vertebrates and plants. Invertebrates and microorganisms are often overlooked. Furthermore, species endangerment is a truly global problem, and it is important that a focus on hot spots not detract from efforts to conserve habitats and species diversity in other areas.

Migratory species pose a special problem for conservationists. For example, monarch butterflies occupy much of the United States and Canada during the summer months, but migrate in the autumn to specific sites in Mexico and California, where they congregate in huge numbers. Overwintering populations are particularly susceptible to habitat disturbances because they are concentrated in small areas. Thus, habitat preservation must extend across all of the sites that monarchs inhabit in order to protect them. The situation is similar for many species of migratory songbirds, waterfowl, marine mammals, and sea turtles.

Sea turtles, such as the loggerhead turtle (**Figure 38.9B**), are threatened both in their ocean feeding grounds and on land. Loggerheads take about 20 years to reach sexual maturity, and great numbers of juveniles and adults are drowned at sea when caught in fishing nets. The adults mate at sea, and the females migrate to specific sites on sandy beaches to lay their eggs. Buried in shallow depressions, the eggs are susceptible to predators, especially raccoons. And many egg-laying sites have become housing developments and beachside resorts. An ongoing international effort to conserve sea turtles focuses on protecting egg-laying sites and minimizing the death rates of adults and juveniles at sea.

Currently, governments have set aside about 7% of the world's land in various forms of reserves. One major conservation question is whether it is better to create one large reserve or a group of smaller ones. Far-ranging animals with low-density populations—predators such as wolves and tigers—require extensive habitats. As conservation biologists learn more about the requirements for achieving minimum population sizes to sustain endangered species, it is becoming clear that most national parks and other reserves are far too small. Given political and economic realities, it is unlikely that many existing parks will be enlarged, and most new reserves will also be too small. In the next two modules, we look at two approaches to this problem.

? **What is a biodiversity hot spot?**

● A relatively small area with a disproportionate number of endangered and threatened species, many of which are endemic

▲ **Figure 38.9B** An adult loggerhead turtle (*Caretta caretta*) swimming off the coast of Belize

38.10 Zoned reserves are an attempt to reverse ecosystem disruption

Conservation of Earth's natural resources is not purely a scientific issue. The causes of declining biodiversity are rooted in complex social and economic issues, and the solutions must take these factors into account. Let's look at how the small Central American nation of Costa Rica is managing its biodiversity.

Despite its small size (about 51,000 km^2, the size of New Hampshire and Vermont combined), Costa Rica is a treasure trove of biodiversity. Its varied ecosystems, which extend over mountains and two coasts, are home to at least half a million species. As Figure 38.9A shows, the entire country is a biodiversity hot spot. Since the 1970s, the Costa Rican government and international agencies have worked together to preserve these unique assets. Approximately 25% of Costa Rica's territory is currently protected in some way **(Figure 38.10A)**.

One type of protection is called a **zoned reserve**, an extensive region of land that includes one or more areas undisturbed by humans. The lands surrounding these areas continue to be used to support the human population, but they are protected from extensive alteration. As a result, they serve as a buffer zone, or shield, against further intrusion into the undisturbed areas. A primary goal of the zoned reserve approach is to develop a social and economic climate in the buffer zone that is compatible with the long-term viability of the protected area.

Costa Rica is making progress in managing its reserves so that the buffer zones provide a steady, lasting supply of forest products, water, and hydroelectric power and also support sustainable agriculture. An important goal is providing a stable economic base for people living there. Destructive practices that are not compatible with long-term ecosystem stability and from which there is often little local profit are gradually being discouraged. Such destructive practices include massive logging, large-scale single-crop agriculture, and extensive mining.

However, a recent analysis showed mixed results for Costa Rica's system of zoned reserves. The good news is that negligible deforestation has occurred within and just beyond protected parkland boundaries. However, some deforestation has occurred in the buffer zones, with plantations of cash crops such as banana and palm replacing the natural vegetation. Conservationists fear that continuing these practices will isolate protected areas, restricting gene flow and decreasing species and genetic diversity.

Costa Rica's commitment to conservation has resulted in a new source of income for the country—**ecotourism**, travel to natural areas for tourism and recreation **(Figure 38.10B)**. People from all over the world come to experience Costa Rica's spectacular range of biodiversity, generating thousands of jobs and a significant chunk of the country's revenue. Worldwide, ecotourism has grown into a multibillion-dollar industry as tourists flock to the world's remaining natural areas. Whether ecotourism dollars ultimately help conserve Earth's biodiversity, however, remains to be seen.

? Why is it important for zoned reserves to prevent large-scale alterations of habitat in the buffer zones? Why is it also important to support sustainable development for the people living there?

Large-scale disruptions in buffer zones could impact the nearby undisturbed areas. Preservation is a realistic goal only if it is compatible with an acceptable standard of living for the local people.

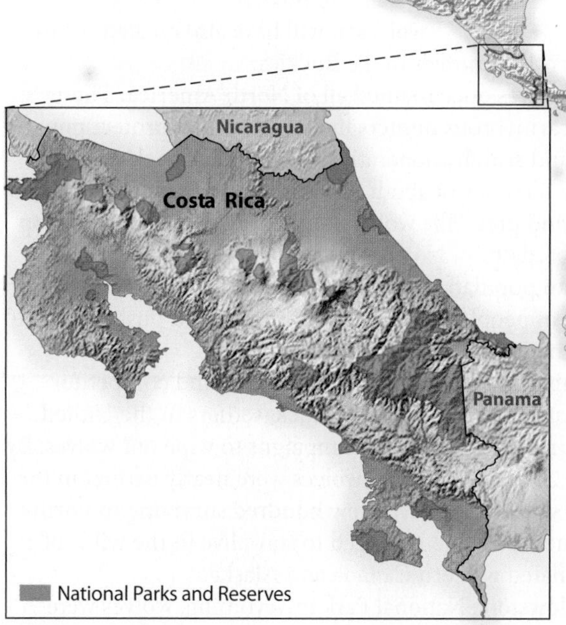

North America
South America
Nicaragua
Costa Rica
Panama
National Parks and Reserves

▲ Figure 38.10A Costa Rica

▲ Figure 38.10B Ecotourism: seeing the tropical rain forest by boat in Costa Rica's Tortuguero National Park

38.11 The Yellowstone to Yukon Conservation Initiative seeks to preserve biodiversity by connecting protected areas

SCIENTIFIC THINKING

In *The Once and Future King*, a fantasy novel about the childhood of King Arthur, the boy learns valuable lessons while magically inhabiting animal forms. As a bird flying high over the land and sea, he realizes that political boundaries exist only in human minds. The same lesson was emphatically driven home to biologists monitoring the travels of a gray wolf **(Figure 38.11A)** they called Pluie (French for "rain"). After capturing Pluie in western Canada in 1991, the scientists fitted her with a radio tracking collar and released her. They were stunned by what they learned. Over the next two years, the wolf roamed over an area of more than 100,000 km^2 (38,600 square miles). Heedless of the boundaries created by humans, she traveled from Alberta to British Columbia in Canada, then crossed into the United States and passed through Montana, Idaho, and Washington before returning to British Columbia—a loop of more than 900 miles—where she remained for a few weeks before heading south again. However, her ignorance of the line between protected reserves and legal hunting areas proved fatal. Pluie, her mate, and three cubs were shot while travelling outside the boundary of a national park.

▲ **Figure 38.11A** A gray wolf (*Canis lupus*)

Biologists who had studied Pluie realized that the wolf's life captured all the promise—and all the pitfalls—of efforts to protect her. She had thrived for nine years within the sporadic shelter of parks and other protected territory. But such lands were never big enough to hold her. Like others of her species (*Canis lupus*), Pluie needed more room. Reserves could shield animals briefly, the scientists realized, but true protection would have to include safe passages between reserves.

Can Earth's biodiversity be saved?

This research inspired the creation of the Yellowstone to Yukon Conservation Initiative (Y2Y), one of the world's most ambitious conservation biology efforts. The initiative aims to preserve the web of life that has long defined the Rocky Mountains of Canada and the northern United States. The area is dotted with famous parks, including Canada's Banff National Park and Yellowstone and Glacier National Parks in the United States. The idea is not to create one giant park, but rather to connect a string of more than 700 protected areas, including national, state, and provincial parks and national forests, with protected corridors where wildlife can travel safely.

Y2Y now stretches 3,200 kilometers (roughly 2,000 miles) from Wyoming to the northern part of the Yukon Territory, encompassing temperate grasslands, coniferous forest, and alpine and arctic tundra (**Figure 38.11B**, on facing page). Its total area is 1.3 million square kilometers (half a million square miles), roughly three times the size of California, but only about 10% of this land has protected status. Y2Y is also unique in its range of elevations. As mentioned in Module 38.5, some populations can respond to climate change by shifting their range to higher latitudes or higher altitudes. Y2Y can accommodate both types of movement. Studies show that populations of pikas (see chapter introduction) at low elevations are at greatest risk for extirpation; Y2Y offers upward mobility.

Conservationists must also seek ways that wildlife populations can coexist with industries such as logging, ranching, gas and oil, and recreation that are important to the human population in the Y2Y region. For example, specially constructed road overpasses (see Figure 38.8C) used by wolves, elk, and grizzly bears are complemented by underground passages for animals such as black bears and cougars that prefer to travel under cover. In some areas, fences have been installed along railroad tracks to reduce the number of animals killed while scavenging spilled grains.

Many of the signature species that live in this vast region, such as grizzly bears, lynx, moose, and elk, don't confine themselves to human boundaries. But few have as great a range as the wolf. If Y2Y can provide safe passage for gray wolves, it will have also created secure zones for other animals in the Rockies.

Gray wolves once roamed all of North America. These carnivorous hunters live in packs that protect pups and search cooperatively for food. A pack may have a territory of about 130 km^2 or range much farther to find prey. The wolf's hunting prowess kept it the top predator of North American ecosystems as long as the human population was small. Things changed when large numbers of people migrated from Europe and pushed far into the continent.

Deeming wolves a dangerous predator and competitor that threatened people and livestock, settlers in the United States launched widespread campaigns to wipe out wolves. By the early 20th century, gray wolves were nearly extinct in the lower 48 states, with only a few hundred surviving in northern Minnesota. More managed to stay alive in the wilds of less populated western Canada and Alaska.

In Yellowstone National Park in Wyoming, wolves were extirpated by the mid-1920s. The decades that followed were marked by dramatic increases in the elk population. In winter,

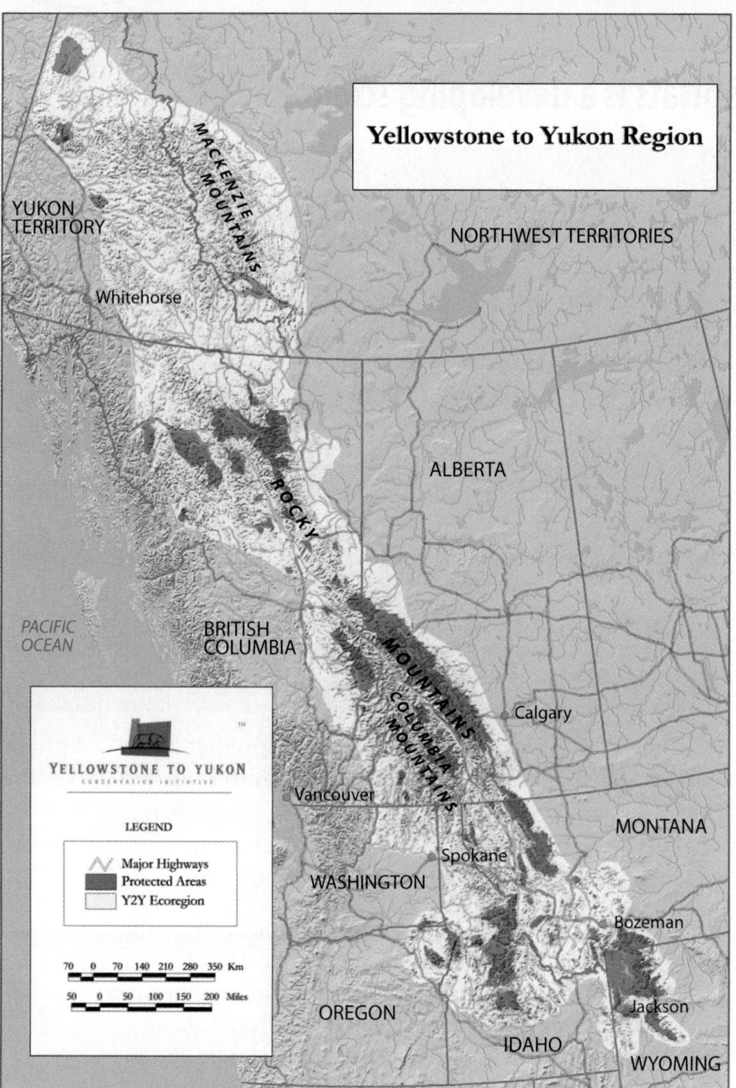
Yellowstone to Yukon Region
MACKENZIE MOUNTAINS
YUKON TERRITORY
NORTHWEST TERRITORIES
Whitehorse
ALBERTA
ROCKY
MOUNTAINS
PACIFIC OCEAN
BRITISH COLUMBIA
COLUMBIA MOUNTAINS
Calgary
YELLOWSTONE TO YUKON
LEGEND
Major Highways
Protected Areas
Y2Y Ecoregion
70 0 70 140 210 280 350 Km
50 0 50 100 150 200 Miles
Vancouver
MONTANA
Spokane
WASHINGTON
Bozeman
OREGON
Jackson
IDAHO
WYOMING

▲ **Figure 38.11B** A map showing the Yellowstone to Yukon Conservation Initiative region

elk descend from alpine meadows to the shelter of lower elevations, where they browse on the twigs and bark of young trees and shrubs **(Figure 38.11C)**. Studies carried out over five decades showed the impact of the burgeoning elk herd on woody species such as aspen and willow. New shoots rarely grew taller than 80 cm (about 30 inches), the level at which elk browse, while the same species grown in elk-proof enclosures grew rapidly. Berry-producing shrubs were also heavily damaged by browsing. Over time, stream banks and nearby hillsides were stripped of plant cover, leading to soil erosion. To prevent further ecosystem deterioration, park managers took measures to reduce the elk herd. Large numbers of elk were captured and relocated; finally, park managers even resorted to killing them.

▲ **Figure 38.11C** Elk (*Cervus elaphus*) browsing

In 1991, the U.S. Fish and Wildlife Service launched a campaign to bring wolves back to Yellowstone. After careful planning, about 30 wolves from Canada were released in the park in 1995 and 1996. The extirpation and later reintroduction of wolves into Yellowstone constituted a natural experiment, a type of observational experiment that takes advantage of treatments that were not intentionally created for a scientific experiment. This natural experiment provided an opportunity to compare community structure and dynamics with and without the gray wolf. In a report published in 2011, scientists summarized results from more than a dozen studies documenting the recovery of woody species such as aspen, cottonwood, and willow in areas that had been overbrowsed. The largest elk herd decreased in size from pre-1995 highs of more than 15,000 to around 6,000 in 2005. But the presence of wolves has had further-reaching effects on species diversity in Yellowstone.

After the burgeoning elk herd decimated the willows, beavers, for which willows are a key resource, had all but disappeared from north Yellowstone. As predation by wolves reduced the elk population, willows and other vegetation once again flourished, and the beaver population recovered. Beaver dams created ponds and wetlands that attract waterfowl and support populations of amphibians, fish, and other animals. The resurgence of trees and shrubs also provided food and shelter for birds and small herbivores such as rodents and rabbits. Small herbivores also benefited from the wolf's impact on their main predator, coyotes. Within a few years of their reintroduction, wolves had reduced the population density of coyotes by 50%. That made more prey available for foxes, badgers, hawks, and owls. In addition, wolf kills provided a bonanza for scavengers, including ravens, magpies, eagles, black bears, and grizzly bears. The region was recovering its biodiversity.

In late 2012, 10 packs numbering approximately 88 gray wolves occupied the park. And true to their nature, Yellowstone's wolves haven't followed human-imposed borders; several packs have been found just outside the park. Meanwhile, the migrations of Canadian wolves, along with smaller release programs, have brought the animals back to Idaho and Montana. In 2011, federal officials removed gray wolves in those states from the endangered species list. Wolves in Wyoming remained on the list until September 30, 2012; wolf hunting season began the next day. When Yellowstone wolves cross the invisible boundary, they are fair game.

In addition to creating reserves to protect species and their habitats from human disruptions, conservation efforts also attempt to restore ecosystems degraded by human activities. We look at the field of restoration ecology next.

? Why are gray wolves considered a keystone species?

● Wolves are not abundant, but they exert a strong control on community structure through their interactions with other species.

38.12 The study of how to restore degraded habitats is a developing science

CONNECTION

For centuries, humans have altered and degraded natural areas without considering the consequences. But as people have gradually come to realize the severity of some of the consequences of ecosystem alteration, they have sought ways to return degraded areas to their natural state. The expanding field of **restoration ecology** uses ecological principles to develop methods of achieving this goal.

One of the major strategies in restoration ecology is bioremediation, the use of living organisms to detoxify polluted ecosystems. For example, bacteria have been used to clean up oil spills and old mining sites. Bacteria are also employed to metabolize toxins in dump sites. Certain species of plants have successfully extracted potentially toxic metals such as zinc, nickel, and lead from contaminated soil. As the plants grow, they absorb large amounts of the toxins from the soil and store it in their bodies. The plants are then harvested and disposed of in hazardous waste landfills. Researchers are also investigating the use of trees and lichens to clean up soil polluted with uranium. In Japan, sunflowers are being planted in an attempt to decontaminate soil polluted by the nuclear disaster that followed the 2011 earthquake and tsunami (**Figure 38.12A**).

▲ **Figure 38.12A** Sunflowers planted for phytoremediation in Natori, Japan, after the 2011 tsunami

Some restoration projects have the broader goal of returning ecosystems to their natural state, which may involve replanting vegetation, fencing out non-native animals, or removing dams that restrict water flow. Hundreds of restoration projects are currently under way in the United States. One of the most ambitious endeavors is the Kissimmee River project in south central Florida.

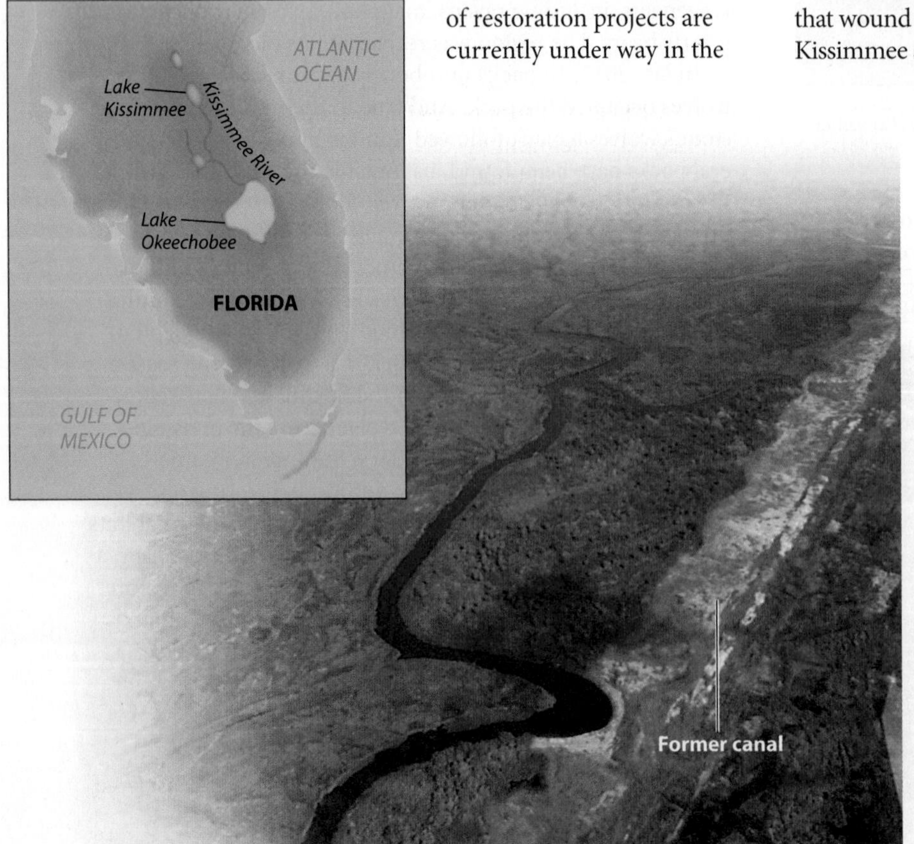

ATLANTIC OCEAN
Lake Kissimmee
Kissimmee River
Lake Okeechobee
FLORIDA
GULF OF MEXICO
Former canal

▲ **Figure 38.12B** Restoring the natural water flow patterns of the Kissimmee River

The Kissimmee River was once a meandering shallow river that wound its way through diverse wetlands from Lake Kissimmee southward into Lake Okeechobee (**Figure 38.12B**, inset). Periodic flooding of the river covered a wide floodplain during about half of the year, creating wetlands that provided critical habitat for vast numbers of birds, fishes, and invertebrates. As often happens, however, people saw the floodplain as wasted land that could be developed if the flooding were controlled. Between 1962 and 1971, the U.S. Army Corps of Engineers converted the 166-km wandering river into a straight canal 9 m deep, 100 m wide, and 90 km long. This project drained approximately 31,000 acres of wetlands, with significant negative impacts on fish and wetland bird populations. Spawning and foraging habitats for fishes were eliminated, and important sport fishes, such as largemouth bass, were replaced by non-game species more tolerant of the lower oxygen concentration in the deeper canal. The populations of waterfowl declined by 92%, and the number of bald eagle nesting territories decreased by 70%. Without the marshes to help filter and reduce agricultural runoff, phosphorus and other excess nutrients were transported through Lake Okeechobee into the Everglades ecosystem to the south.

As these negative ecological effects began to be recognized, public pressure to restore the river grew. In 1992, Congress authorized the

Kissimmee River Restoration Project, one of the largest landscape restoration projects and ecological experiments in the world. The plan involves removing water control structures such as dams, reservoirs, and channel modifications and filling in about 35 km of the canal. The final phase of the project was begun in 2009, and the entire project is slated to be completed in 2015. As shown in Figure 38.12B, the natural curves of the river are a pleasing contrast to the artificial linearity of the backfilled canal. Birds and other wildlife have returned in unexpected numbers to the 11,000 acres of wetlands that have been restored. The marshes are filled with native vegetation, and game fishes again swim in the river channels. However, drought conditions in recent years have threatened the southward flow of the Kissimmee River into Lake Okeechobee. The potential for water shortages in southern Florida has renewed attention to the urgent need to complete an even more ambitious project, the restoration of the Everglades.

? How will the Kissimmee River Restoration Project improve water quality in the Everglades ecosystem?

● The wetlands filter agricultural runoff and prevent excess nutrients from entering the Everglades.

38.13 Sustainable development is an ultimate goal

The demand for the "provisioning" services of ecosystems, such as food, fibers, and water, is increasing as the world population grows and becomes more affluent. Although these demands are currently being met, they are satisfied at the expense of other critical ecosystem services, such as climate regulation and protection against natural disasters. Clearly, we have set ourselves and the rest of the biosphere on a precarious path into the future. How can we best manage Earth's resources to ensure that all generations inherit an adequate supply of natural and economic resources and a relatively stable environment?

Many nations, scientific societies, and private foundations have embraced the concept of sustainable development. The Ecological Society of America, the world's largest organization of ecologists, endorses a research agenda called the Sustainable Biosphere Initiative. The goal of this initiative is to acquire the basic ecological information necessary for the intelligent and responsible development, management, and conservation of Earth's resources. The research agenda includes devising ways to sustain the productivity of natural and artificial ecosystems and studying the relationship between biological diversity, global climate change, and ecological processes.

Sustainable development doesn't only depend on continued research and application of ecological knowledge. It also requires us to connect the life sciences with the social sciences, economics, and humanities. Conservation and restoration of biodiversity is only one side of sustainable development; the other key facet is improving the human condition. Public education and the political commitment and cooperation of nations, especially the United States, are essential to the success of this endeavor.

The image of the snowy owl on this book's cover and in **Figure 38.13** serves as a reminder of what we stand to lose if we fail to recognize and solve the ecological crises at hand. Snowy owls breed in the Arctic, one of the regions most affected by climate change. Ninety percent of their diet consists of small rodents called lemmings. During the winter, lemmings inhabit the space between the frozen ground and a thick, insulating blanket of snow. Warm temperatures threaten this refuge by collapsing the protective layer, which destroys the lemmings' burrows and reduces the insulation provided by the snow. Researchers have already noted dramatic declines in lemming populations in northeastern Greenland.

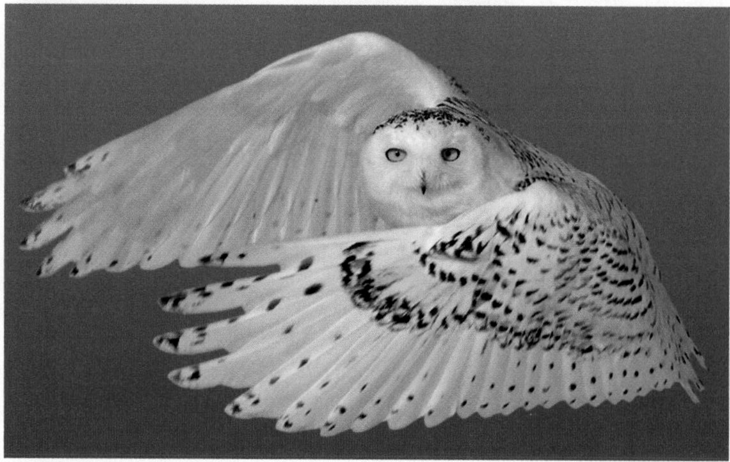

▲ **Figure 38.13** Snowy owl (*Bubo scandiacus*)

Snowy owls are not the only species whose fate is intertwined with the size of lemming populations. Lemmings are a keystone species (see Module 37.11) whose decline will reverberate throughout the tundra food web.

Biology is the scientific expression of the human desire to know nature. We are most likely to save what we appreciate, and we are most likely to appreciate what we understand. By learning about the processes and diversity of life, we also become more aware of our dependence on healthy ecosystems. An awareness of our unique ability to alter the biosphere and jeopardize the existence of other species, as well as our own, may help us choose a path toward a sustainable future.

The risk of a world without adequate natural resources for all its people is not a vision of the distant future. It is a prospect for your children's lifetime, or perhaps even your own. But although the current state of the biosphere is grim, the situation is far from hopeless. Now is the time to aggressively pursue more knowledge about life and to work toward long-term sustainability.

? Why is a concern for the well-being of future generations essential for progress toward sustainable development?

● Sustainable development is a long-term goal—longer than a human lifetime. Preoccupation with the here and now is an obstacle to sustainable development because it discourages behavior that benefits future generations.

CHAPTER 38 REVIEW

For practice quizzes, BioFlix animations, MP3 tutorials, video tutors, and more study tools designed for this textbook, go to MasteringBiology®

Reviewing the Concepts

The Loss of Biodiversity (38.1–38.6)

38.1 Loss of biodiversity includes the loss of ecosystems, species, and genes. While valuable for its own sake, biodiversity also provides food, fibers, medicines, and ecosystem services.

Ecosystem diversity
Species diversity
Genetic diversity

38.2 Habitat loss, invasive species, overharvesting, pollution, and climate change are major threats to biodiversity. Human alteration of habitats is the single greatest threat to biodiversity. Invasive species disrupt communities by competing with, preying on, or parasitizing native species. Harvesting at rates that exceed a population's ability to rebound is a threat to many species. Human activities produce diverse pollutants that may affect ecosystems far from their source. Biomagnification concentrates synthetic toxins that cannot be degraded by organisms.

38.3 Rapid warming is changing the global climate. Increased global temperature caused by rising concentrations of greenhouse gases is changing climatic patterns, with grave consequences.

38.4 Human activities are responsible for rising concentrations of greenhouse gases. Much of the increase is the result of burning fossil fuels.

38.5 Global climate change affects biomes, ecosystems, communities, and populations. Organisms that live at high latitudes and high elevations are experiencing the greatest impact.

38.6 Climate change is an agent of natural selection. Phenotypic plasticity has minimized the impact on some species, and a few cases of microevolutionary change have been observed. However, the rapidity of the environmental changes makes it unlikely that evolutionary processes will save many species from extinction.

Conservation Biology and Restoration Ecology (38.7–38.13)

38.7 Protecting endangered populations is one goal of conservation biology. Conservation biology is a goal-driven science that seeks to understand and counter the rapid loss of biodiversity. Some conservation biologists direct their efforts at increasing populations that are endangered.

38.8 Sustaining ecosystems and landscapes is a conservation priority. Conservation efforts are increasingly aimed at sustaining ecosystems and landscapes. Edges between ecosystems have distinct sets of features and species. The increased frequency and abruptness of edges caused by human activities can increase species loss. Movement corridors connecting isolated habitats may be helpful to fragmented populations.

38.9 Establishing protected areas slows the loss of biodiversity. Biodiversity hot spots have high concentrations of endemic species.

38.10 Zoned reserves are an attempt to reverse ecosystem disruption. Zoned reserves are undisturbed wildlands surrounded by buffer zones of compatible economic development. Ecotourism has become an important source of revenue for conservation efforts.

38.11 The Yellowstone to Yukon Conservation Initiative seeks to preserve biodiversity by connecting protected areas. The success of this innovative international research and conservation effort hinged on the reintroduction of gray wolves.

38.12 The study of how to restore degraded habitats is a developing science. Restoration ecology uses ecological principles to return degraded areas to their natural state, a process that may include detoxifying polluted ecosystems, replanting native vegetation, and returning waterways to their natural course. Large-scale restoration projects attempt to restore damaged landscapes.

38.13 Sustainable development is an ultimate goal. Sustainable development depends on increasing and applying ecological knowledge as well as valuing our linkages to the biosphere.

Connecting the Concepts

1. Complete the following map, which organizes some of the key concepts of conservation biology.

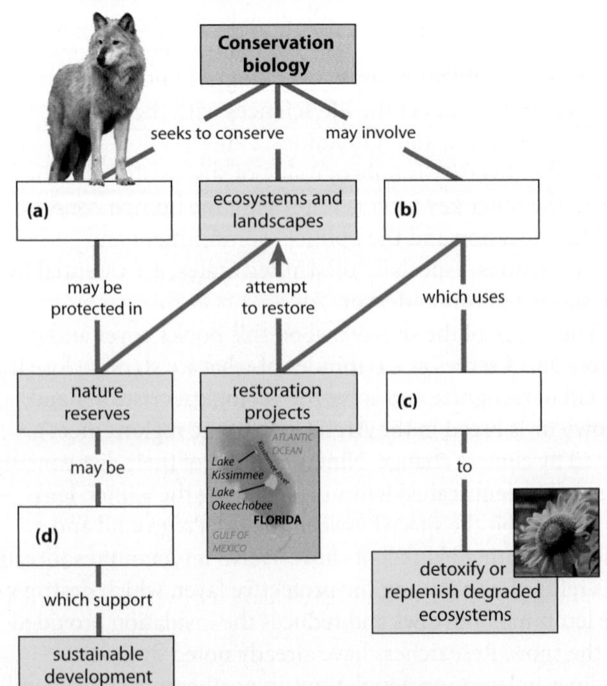
Conservation biology
seeks to conserve
may involve
(a)
ecosystems and landscapes
(b)
may be protected in
attempt to restore
which uses
nature reserves
restoration projects
(c)
ATLANTIC OCEAN
Lake Kissimmee
Lake Okeechobee
FLORIDA
GULF OF MEXICO
may be
to
(d)
detoxify or replenish degraded ecosystems
which support
sustainable development

Testing Your Knowledge

Level 1: Knowledge/Comprehension

2. Which of these statements best describes what conservation biologists mean by the "the rapid loss of biodiversity"?
   a. Introduced species, such as starlings and zebra mussels, have rapidly expanded their ranges.
   b. Harvests of marine fishes, such as cod and bluefin tuna, are declining.
   c. The current species extinction rate is as much as 100 times greater than at any time in the last 100,000 years.
   d. Many potential medicines are being lost as plant species become extinct.
3. Which of the following poses the single greatest threat to biodiversity?
   a. invasive species
   b. overhunting
   c. habitat loss
   d. pollution
4. Which of the following is characteristic of endemic species?
   a. They are often found in biodiversity hot spots.
   b. They are distributed widely in the biosphere.
   c. They require edges between ecosystems.
   d. They are often keystone species whose presence helps to structure a community.
5. Ospreys and other top predators are most severely affected by pesticides such as PCBs because they
   a. are especially sensitive to chemicals.
   b. have very long life spans.
   c. store the pesticides in their tissues.
   d. consume prey in which pesticides are concentrated.
6. Movement corridors are
   a. the routes taken by migratory animals.
   b. strips or clumps of habitat that connect isolated fragments of habitat.
   c. landscapes that include several different ecosystems.
   d. edges, or boundaries, between ecosystems.
7. With limited resources, conservation biologists need to prioritize their efforts. Of the following choices, which should receive the greatest attention for the goal of conserving biodiversity?
   a. a commercially important species
   b. all endangered vertebrate species
   c. a declining keystone species in a community
   d. all endangered species
8. Which of the following statements about protected areas is not correct?
   a. We now protect 25% of the land areas of the planet.
   b. National parks are only one type of protected area.
   c. Most reserves are smaller in size than the ranges of some of the species they are meant to protect.
   d. Management of protected areas must coordinate with the management of lands outside the protected zone.

Level 2: Application/Analysis

9. What are the three levels of biological diversity? Explain how human activities threaten each of these levels.
10. What are "greenhouse gases"? Why are they important to life on Earth?
11. What are the causes and possible consequences of global climate change? Why is international cooperation necessary if we are to solve this problem?

Level 3: Synthesis/Evaluation

12. **SCIENTIFIC THINKING** Biologists in the United States are concerned that populations of many migratory songbirds are declining. Evidence suggests that some of these birds might be victims of pesticides. Most of the pesticides implicated in songbird mortality have not been used in the United States since the 1970s. Suggest a hypothesis to explain the current decline in songbird numbers.
13. You may have heard that human activities cause the extinction of one species every hour. Such estimates vary widely because we do not know how many species exist or how fast their habitats are being destroyed. You can make your own estimate of the rate of extinction. Start with the number of species that have been identified. To keep things simple, ignore extinction in the temperate latitudes and focus on the 80% of plants and animals that live in the tropical rain forest. Assume that destruction of the forest continues at a rate of 1% per year, so the forest will be gone in 100 years. Assume (optimistically) that half the rain forest species will survive in preserves, forest remnants, and zoos. How many species will disappear in the next century? How many species is that per year? Per day? Recent studies of the rain forest canopy have led some experts to predict that there may be as many as 30 million species on Earth. How does starting with this figure change your estimates?
14. The price of energy does not reflect its real costs. What kinds of hidden environmental costs are not reflected in the price of fossil fuels? How are these costs paid, and by whom? Do you think these costs could or should be figured into the price of oil? How might that be done?
15. Research your country's per capita carbon emissions. Compare your carbon footprint with the average for your country. (See the question at the end of Module 38.4.) How can individuals reduce the carbon emissions for which they are directly responsible? Make a list of actions that you are willing to take to reduce your carbon footprint.
16. Until recently, response to environmental problems has been fragmented—an antipollution law here, incentives for recycling there. Meanwhile, the problems of the gap between the rich and poor nations, diminishing resources, and pollution continue to grow. Now people and governments are starting to envision a sustainable society. The Worldwatch Institute, a respected environmental monitoring organization, estimates that we must reach sustainability by the year 2030 to avoid economic and environmental disaster. To get there, we must begin shaping a sustainable society during this decade. In what ways is our present system not sustainable? What might a sustainable society be like? Do you think a sustainable society is an achievable goal? Why or why not? What is the alternative? What might we do to work toward sustainability? What are the major roadblocks to achieving sustainability? How would your life be different in a sustainable society?

Answers to all questions can be found in Appendix 4.

APPENDIX 1 Metric Conversion Table

Measurement	Unit and Abbreviation	Metric Equivalent	Approximate Metric-to-English Conversion Factor	Approximate English-to-Metric Conversion Factor
Length	1 kilometer (km)	= 1,000 (10^3) meters	1 km = 0.6 mile	1 mile = 1.6 km
	1 meter (m)	= 100 (10^2) centimeters	1 m = 1.1 yards	1 yard = 0.9 m
		= 1,000 millimeters	1 m = 3.3 feet	1 foot = 0.3 m
			1 m = 39.4 inches	
	1 centimeter (cm)	= 0.01 (10^{-2}) meter	1 cm = 0.4 inch	1 foot = 30.5 cm
				1 inch = 2.5 cm
	1 millimeter (mm)	= 0.001 (10^{-3}) meter	1 mm = 0.04 inch	
	1 micrometer (µm)	= 10^{-6} meter (10^{-3} mm)		
	1 nanometer (nm)	= 10^{-9} meter (10^{-3} µm)		
	1 angstrom (Å)	= 10^{-10} meter (10^{-4} µm)		
Area	1 hectare (ha)	= 10,000 square meters	1 ha = 2.5 acres	1 acre = 0.4 ha
	1 square meter (m^2)	= 10,000 square centimeters	1 m^2 = 1.2 square yards	1 square yard = 0.8 m^2
			1 m^2 = 10.8 square feet	1 square foot = 0.09 m^2
	1 square centimeter (cm^2)	= 100 square millimeters	1 cm^2 = 0.16 square inch	1 square inch = 6.5 cm^2
Mass	1 metric ton (t)	= 1,000 kilograms	1 t = 1.1 tons	1 ton = 0.91 t
	1 kilogram (kg)	= 1,000 grams	1 kg = 2.2 pounds	1 pound = 0.45 kg
	1 gram (g)	= 1,000 milligrams	1 g = 0.04 ounce	1 ounce = 28.35 g
			1 g = 15.4 grains	
	1 milligram (mg)	= 10^{-3} gram	1 mg = 0.02 grain	
	1 microgram (µg)	= 10^{-6} gram		
Volume (Solids)	1 cubic meter (m^3)	= 1,000,000 cubic centimeters	1 m^3 = 1.3 cubic yards	1 cubic yard = 0.8 m^3
			1 m^3 = 35.3 cubic feet	1 cubic foot = 0.03 m^3
	1 cubic centimeter (cm^3 or cc)	= 10^{-6} cubic meter	1 cm^3 = 0.06 cubic inch	1 cubic inch = 16.4 cm^3
	1 cubic millimeter (mm^3)	= 10^{-9} cubic meter (10^{-3} cubic centimeter)		
Volume (Liquids and Gases)	1 kilolilter (kL or kl)	= 1,000 liters	1 kL = 264.2 gallons	
	1 liter (L or l)	= 1,000 milliliters	1 L = 0.26 gallon	1 gallon = 3.79 L
			1 L = 1.06 quarts	1 quart = 0.95 L
	1 milliliter (mL or ml)	= 10^{-3} liter	1 mL = 0.03 fluid ounce	1 quart = 946 mL
		= 1 cubic centimeter	1 mL = 1/4 teaspoon	1 pint = 473 mL
			1 mL = 15–16 drops	1 fluid ounce = 29.6 mL
				1 teaspoon = 5 mL
	1 microliter (µL or µl)	= 10^{-6} liter (10^{-3} milliliter)		
Time	1 second (s)	= 1/60 minute		
	1 millisecond (ms)	= 10^{-3} second		
Temperature	Degrees Celsius (°C)		°F = 9/5 °C + 32	°C = 5/9 (°F − 32)

APPENDIX 2 The Periodic Table

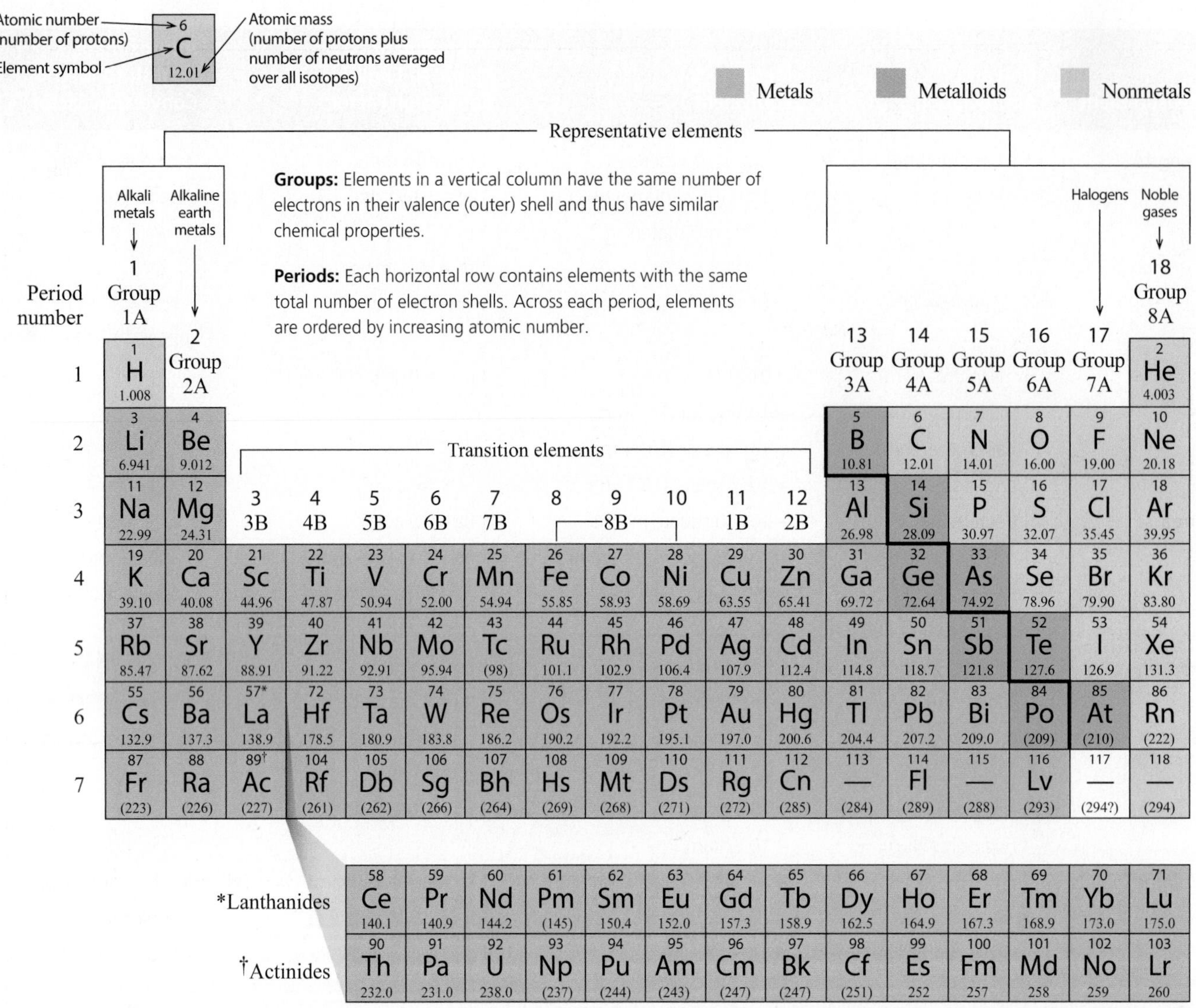
Atomic number (number of protons)
Element symbol
6
C
12.01
Atomic mass (number of protons plus number of neutrons averaged over all isotopes)
Metals
Metalloids
Nonmetals
Representative elements
Alkali metals
Alkaline earth metals
Groups: Elements in a vertical column have the same number of electrons in their valence (outer) shell and thus have similar chemical properties.
Periods: Each horizontal row contains elements with the same total number of electron shells. Across each period, elements are ordered by increasing atomic number.
Halogens
Noble gases
Period number
1 Group 1A
2 Group 2A
13 Group 3A
14 Group 4A
15 Group 5A
16 Group 6A
17 Group 7A
18 Group 8A
Transition elements
3 3B
4 4B
5 5B
6 6B
7 7B
8
9 8B
10
11 1B
12 2B
1: 1 H 1.008 | 2 He 4.003
2: 3 Li 6.941 | 4 Be 9.012 | 5 B 10.81 | 6 C 12.01 | 7 N 14.01 | 8 O 16.00 | 9 F 19.00 | 10 Ne 20.18
3: 11 Na 22.99 | 12 Mg 24.31 | 13 Al 26.98 | 14 Si 28.09 | 15 P 30.97 | 16 S 32.07 | 17 Cl 35.45 | 18 Ar 39.95
4: 19 K 39.10 | 20 Ca 40.08 | 21 Sc 44.96 | 22 Ti 47.87 | 23 V 50.94 | 24 Cr 52.00 | 25 Mn 54.94 | 26 Fe 55.85 | 27 Co 58.93 | 28 Ni 58.69 | 29 Cu 63.55 | 30 Zn 65.41 | 31 Ga 69.72 | 32 Ge 72.64 | 33 As 74.92 | 34 Se 78.96 | 35 Br 79.90 | 36 Kr 83.80
5: 37 Rb 85.47 | 38 Sr 87.62 | 39 Y 88.91 | 40 Zr 91.22 | 41 Nb 92.91 | 42 Mo 95.94 | 43 Tc (98) | 44 Ru 101.1 | 45 Rh 102.9 | 46 Pd 106.4 | 47 Ag 107.9 | 48 Cd 112.4 | 49 In 114.8 | 50 Sn 118.7 | 51 Sb 121.8 | 52 Te 127.6 | 53 I 126.9 | 54 Xe 131.3
6: 55 Cs 132.9 | 56 Ba 137.3 | 57* La 138.9 | 72 Hf 178.5 | 73 Ta 180.9 | 74 W 183.8 | 75 Re 186.2 | 76 Os 190.2 | 77 Ir 192.2 | 78 Pt 195.1 | 79 Au 197.0 | 80 Hg 200.6 | 81 Tl 204.4 | 82 Pb 207.2 | 83 Bi 209.0 | 84 Po (209) | 85 At (210) | 86 Rn (222)
7: 87 Fr (223) | 88 Ra (226) | 89† Ac (227) | 104 Rf (261) | 105 Db (262) | 106 Sg (266) | 107 Bh (264) | 108 Hs (269) | 109 Mt (268) | 110 Ds (271) | 111 Rg (272) | 112 Cn (285) | 113 — (284) | 114 Fl (289) | 115 — (288) | 116 Lv (293) | 117 — (294?) | 118 — (294)
*Lanthanides: 58 Ce 140.1 | 59 Pr 140.9 | 60 Nd 144.2 | 61 Pm (145) | 62 Sm 150.4 | 63 Eu 152.0 | 64 Gd 157.3 | 65 Tb 158.9 | 66 Dy 162.5 | 67 Ho 164.9 | 68 Er 167.3 | 69 Tm 168.9 | 70 Yb 173.0 | 71 Lu 175.0
†Actinides: 90 Th 232.0 | 91 Pa 231.0 | 92 U 238.0 | 93 Np (237) | 94 Pu (244) | 95 Am (243) | 96 Cm (247) | 97 Bk (247) | 98 Cf (251) | 99 Es 252 | 100 Fm 257 | 101 Md 258 | 102 No 259 | 103 Lr 260

Name (Symbol)	Atomic Number	Name (Symbol)	Atomic Number	Name (Symbol)	Atomic Number	Name (Symbol)	Atomic Number	Name (Symbol)	Atomic Number
Actinium (Ac)	89	Copernicium (Cn)	112	Iodine (I)	53	Osmium (Os)	76	Silicon (Si)	14
Aluminum (Al)	13	Copper (Cu)	29	Iridium (Ir)	77	Oxygen (O)	8	Silver (Ag)	47
Americium (Am)	95	Curium (Cm)	96	Iron (Fe)	26	Palladium (Pd)	46	Sodium (Na)	11
Antimony (Sb)	51	Darmstadtium (Ds)	110	Krypton (Kr)	36	Phosphorus (P)	15	Strontium (Sr)	38
Argon (Ar)	18	Dubnium (Db)	105	Lanthanum (La)	57	Platinum (Pt)	78	Sulphur (S)	16
Arsenic (As)	33	Dysprosium (Dy)	66	Lawrencium (Lr)	103	Plutonium (Pu)	94	Tantalum (Ta)	73
Astatine (At)	85	Einsteinium (Es)	99	Lead (Pb)	82	Polonium (Po)	84	Technetium (Tc)	43
Barium (Ba)	56	Erbium (Er)	68	Lithium (Li)	3	Potassium (K)	19	Tellurium (Te)	52
Berkelium (Bk)	97	Europium (Eu)	63	Livermorium (Lv)	116	Praseodymium (Pr)	59	Terbium (Tb)	65
Beryllium (Be)	4	Fermium (Fm)	100	Lutetium (Lu)	71	Promethium (Pm)	61	Thallium (Tl)	81
Bismuth (Bi)	83	Flerovium (Fl)	114	Magnesium (Mg)	12	Protactinium (Pa)	91	Thorium (Th)	90
Bohrium (Bh)	107	Fluorine (F)	9	Manganese (Mn)	25	Radium (Ra)	88	Thulium (Tm)	69
Boron (B)	5	Francium (Fr)	87	Meitnerium (Mt)	109	Radon (Rn)	86	Tin (Sn)	50
Bromine (Br)	35	Gadolinium (Gd)	64	Mendelevium (Md)	101	Rhenium (Re)	75	Titanium (Ti)	22
Cadmium (Cd)	48	Gallium (Ga)	31	Mercury (Hg)	80	Rhodium (Rh)	45	Tungsten (W)	74
Calcium (Ca)	20	Germanium (Ge)	32	Molybdenum (Mo)	42	Roentgenium (Rg)	111	Uranium (U)	92
Californium (Cf)	98	Gold (Au)	79	Neodymium (Nd)	60	Rubidium (Rb)	37	Vanadium (V)	23
Carbon (C)	6	Hafnium (Hf)	72	Neon (Ne)	10	Ruthenium (Ru)	44	Xenon (Xe)	54
Cerium (Ce)	58	Hassium (Hs)	108	Neptunium (Np)	93	Rutherfordium (Rf)	104	Ytterbium (Yb)	70
Cesium (Cs)	55	Helium (He)	2	Nickel (Ni)	28	Samarium (Sm)	62	Yttrium (Y)	39
Chlorine (Cl)	17	Holmium (Ho)	67	Niobium (Nb)	41	Scandium (Sc)	21	Zinc (Zn)	30
Chromium (Cr)	24	Hydrogen (H)	1	Nitrogen (N)	7	Seaborgium (Sg)	106	Zirconium (Zr)	40
Cobalt (Co)	27	Indium (In)	49	Nobelium (No)	102	Selenium (Se)	34		

APPENDIX 3 The Amino Acids of Proteins

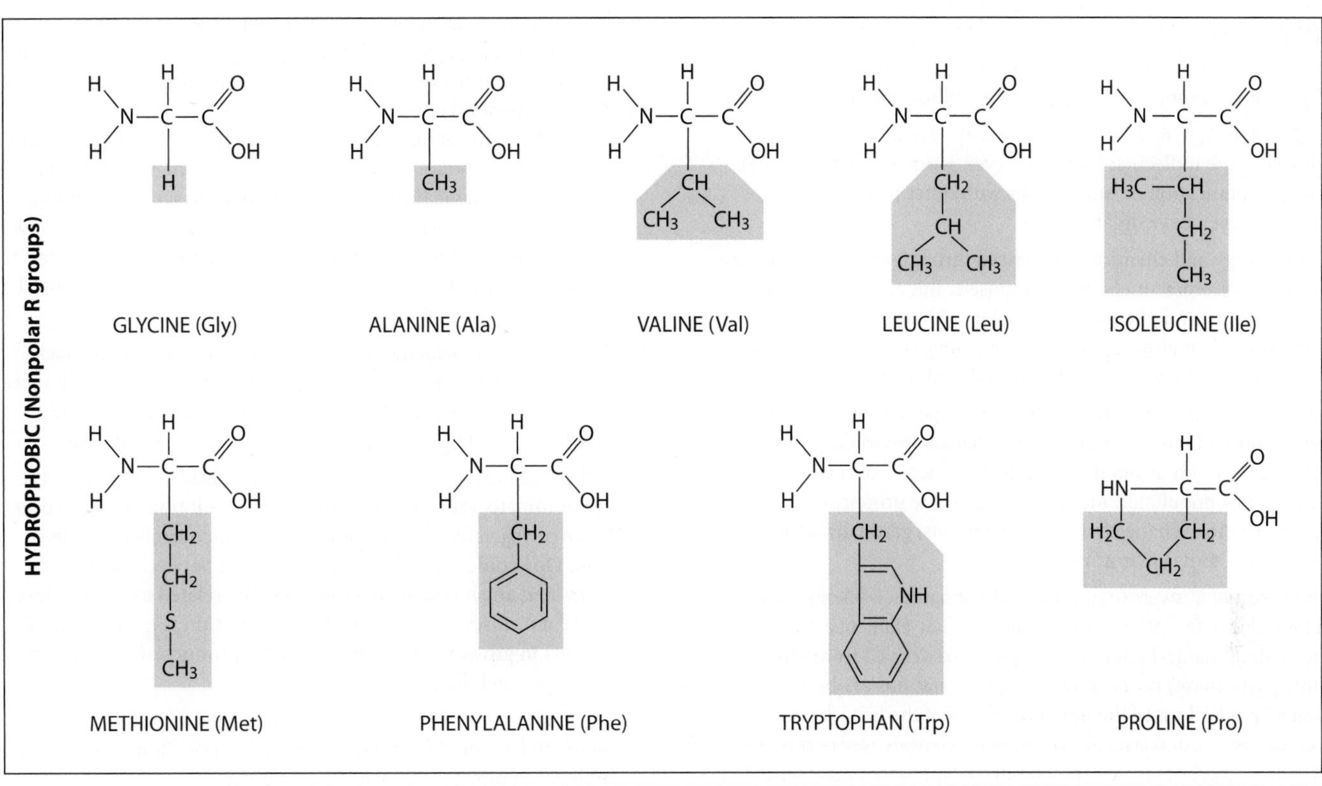
HYDROPHOBIC (Nonpolar R groups)
GLYCINE (Gly)
ALANINE (Ala)
VALINE (Val)
LEUCINE (Leu)
ISOLEUCINE (Ile)
METHIONINE (Met)
PHENYLALANINE (Phe)
TRYPTOPHAN (Trp)
PROLINE (Pro)

HYDROPHILIC (Polar or charged R groups)

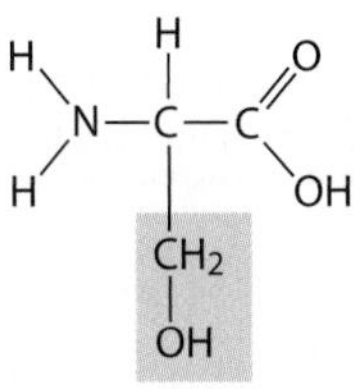

SERINE (Ser)

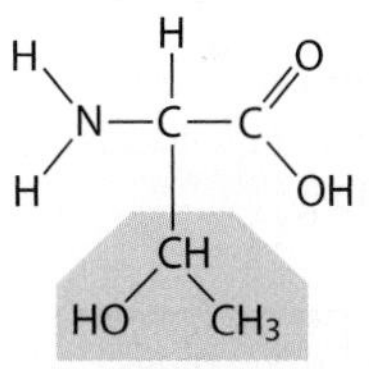

THREONINE (Thr)

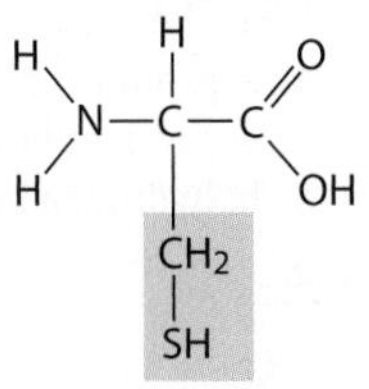

CYSTEINE (Cys)

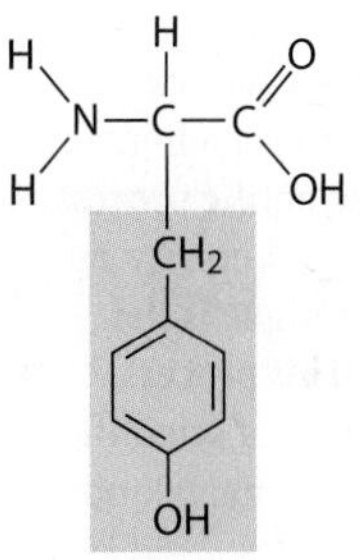

TYROSINE (Tyr)

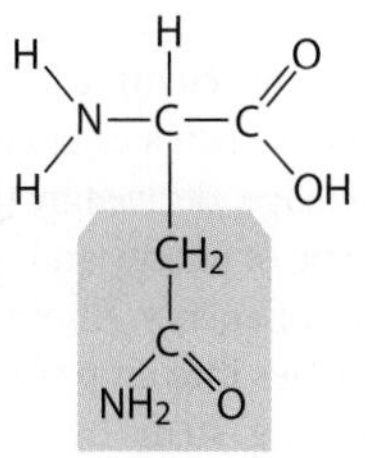

ASPARAGINE (Asn)

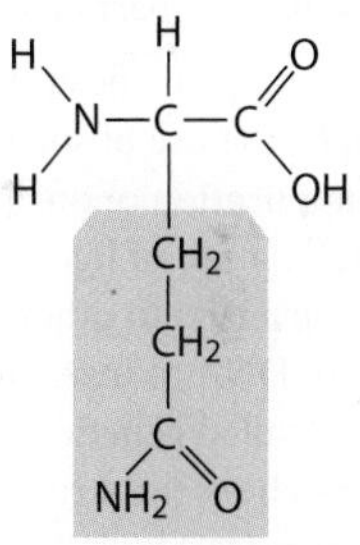

GLUTAMINE (Gln)

Acidic

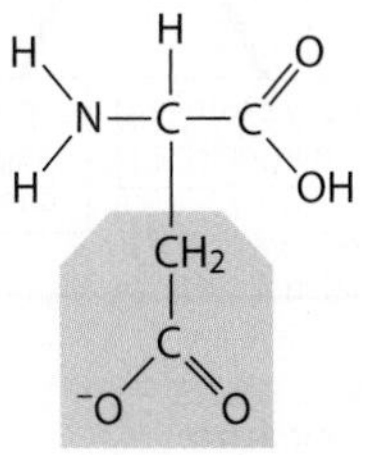

ASPARTIC ACID (Asp)

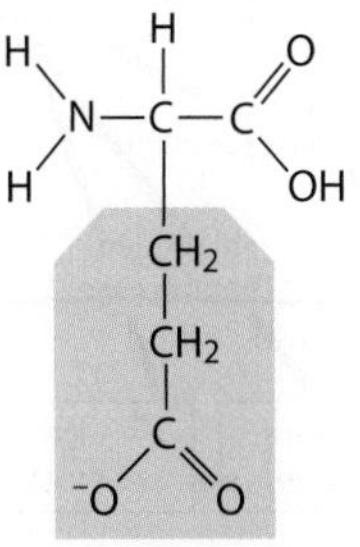

GLUTAMIC ACID (Glu)

Basic

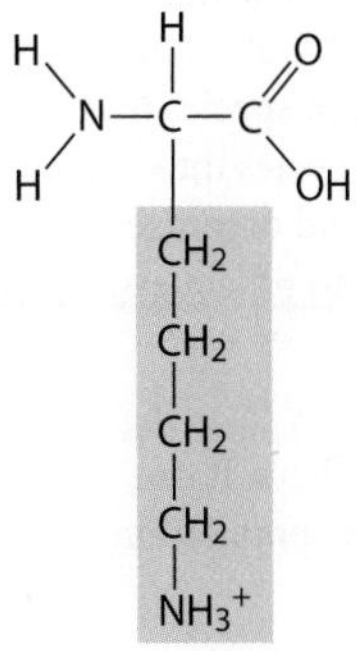

LYSINE (Lys)

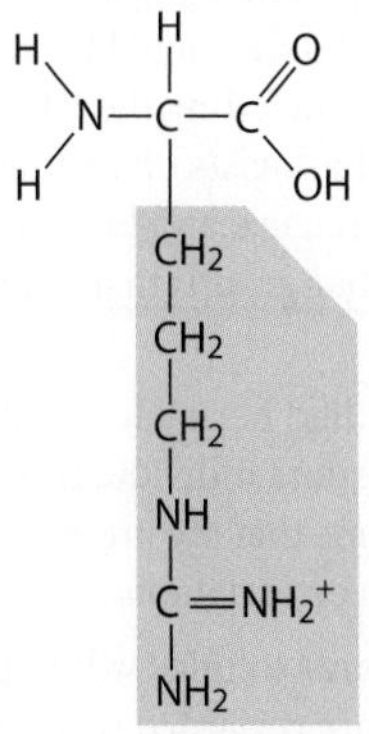

ARGININE (Arg)

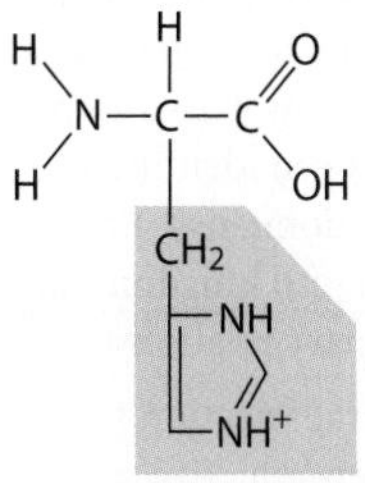

HISTIDINE (His)

Chapter Review Answers

Chapter 1

1. a. life; b. evolution; c. natural selection; d. unity of life; e. three domains (or numerous kingdoms; 1.8 million species)

2. b 3. c 4. b 5. b 6. d 7. a 8. d (You may have been tempted to choose b, the molecular level. However, protists may have chemical communication or interactions with other protists. No protists, however, have organs.) 9. d

10. Both energy and chemicals are passed through an ecosystem from producers to consumers to decomposers. But energy enters an ecosystem as sunlight and leaves as heat. Chemicals are recycled from the soil or atmosphere through plants, consumers, and decomposers and returned to the air, soil, and water.

11. Darwin described how natural selection operates in populations whose individuals have varied traits that are inherited. When natural selection favors the reproductive success of certain individuals in a population more than others, the proportions of heritable variations change over the generations, gradually adapting a population to its environment.

12. In pursuit of answers to questions about nature, a scientist uses a logical thought process involving these key elements: observations about natural phenomena, questions derived from observations, hypotheses posed as tentative explanations of observations, logical predictions of the outcome of tests if the hypotheses are correct, and actual tests of hypotheses. Scientific research is not a rigid method because a scientist must adapt these processes to the set of conditions particular to each study. Intuition, chance, and luck are also part of science.

13. Technology is the application of scientific knowledge. For example, the use of solar power to run a calculator or heat a home is an application of our knowledge, derived by the scientific process, of the nature of light as a type of energy and how light energy can be converted to other forms of energy. Another example is the use of DNA to insert new genes into crop plants. This process, often called genetic engineering, stems from decades of scientific research on the structure and function of DNA from many kinds of organisms.

14. The vertical scale of biology refers to the hierarchy of biological organization: from molecules to organelles, cells, tissues, organs, organ systems, organisms, populations, communities, ecosystems, and the biosphere. At each level, emergent properties arise from the interaction and organization of component parts. The horizontal scale of biology refers to the incredible diversity of living organisms, past and present, including the 1.8 million species that have been identified so far. Biologists divide these species into three domains—Bacteria, Archaea, and Eukarya—and organize them into kingdoms and other groups that attempt to reflect evolutionary relationships.

15. Natural selection screens (edits) heritable variations by favoring the reproductive success of some individuals over others. It can only select from the variations that are present in the population; it does not create new genes or variations.

16. a. Hypothesis: Giving rewards to mice will improve their learning. Prediction: If mice are rewarded with food, they will learn to run a maze faster.
    b. The control group was the mice that were not rewarded. Without them, it would be impossible to know if the mice that were rewarded decreased their time running the maze only because of practice.
    c. Both groups of mice should not have run the maze before and should be about the same age. Both experiments should be run at the same time of day and under the same conditions.
    d. Yes, the results support the hypothesis because the data show that the rewarded mice began to run the maze faster by day 3 and improved their performance (ran faster than the control mice) each day thereafter.

17. The researcher needed to determine the percent of total attacks in each habitat that occurred on dark models. It may be that there were simply more predators in the inland habitat than in the beach habitat. The experiment needed proper data analysis.

18. If these cell division control genes are involved in producing the larger tomato, they may have similar effects if transferred to other fruits or vegetables. Cancer is a result of uncontrolled cell division. One could see if there are similarities between the tomato genes and any human genes that could be related to human development or disease. The control of cell division is a fundamental process in growth, repair, and asexual reproduction—all important topics in biology.

19. Virtually any news report or magazine contains stories that are about biology or at least have biological connections. How about biological connections in advertisements?

Chapter 2

1. a. protons; b. neutrons; c. electrons; d. different isotopes; e. covalent bonds; f. ionic bonds; g. polar covalent bonds; h. hydrogen bonding

2.

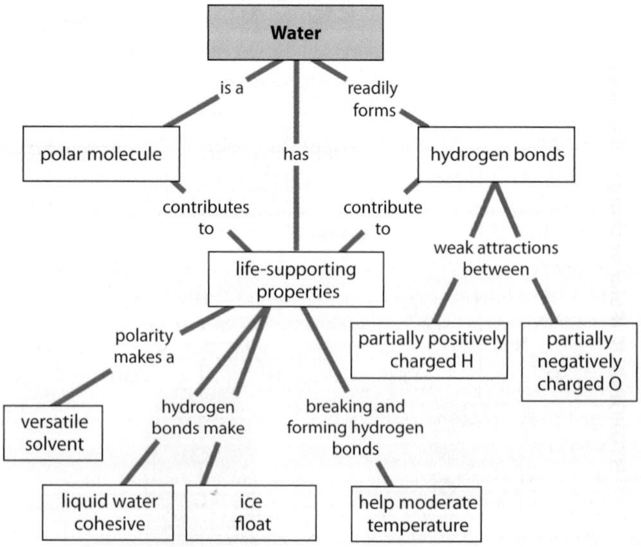

Water
is a
readily forms
has
polar molecule
hydrogen bonds
contributes to
contribute to
weak attractions between
life-supporting properties
partially positively charged H
partially negatively charged O
polarity makes a
hydrogen bonds make
breaking and forming hydrogen bonds
versatile solvent
liquid water cohesive
ice float
help moderate temperature

3. b 4. b 5. d 6. c 7. b (Sulfur has 6 electrons in its valence shell. It reaches a full outer shell of 8 by sharing one pair of electrons with each of two hydrogen atoms. Each H then has a full valence shell of 2.)

8. Iodine (part of a thyroid hormone) and iron (part of hemoglobin in blood) are both trace elements, required in minute quantities. Calcium and phosphorus (components of bones and teeth) are needed by the body in much greater quantities.

9. The atoms of each element have a characteristic number of protons in their nuclei, which is referred to as the atomic number and is 6 for carbon. The mass number is an indication of the approximate mass of an atom and is equal to the number of protons and neutrons in the nucleus. Carbon-12 has 6 neutrons (and 6 protons, of course), so its mass number is 12. The valence usually equals the number of electrons needed to fill an atom's outer shell (the number of unpaired electrons). Carbon's valence or bonding capacity of 4 indicates that it will form 4 covalent bonds. Thus, an atom's valence is most related to its chemical behavior.

10. In nonpolar covalent bonds, electrons are equally shared between two atoms. Polar covalent bonds form when a more electronegative atom pulls the shared electrons closer to it, producing a partial negative charge associated with that portion of the molecule and a partial positive charge associated with the atom from which the electrons are pulled. In the formation of ions, an electron is completely pulled away from one atom and transferred to another, creating negatively and positively charged ions. These oppositely charged ions may be attracted to each other in an ionic bond.

11. Fluorine needs 1 electron for a full outer shell of 8, and if potassium loses 1 electron, its outer shell will have 8. Potassium will lose an electron (becoming a + ion), and fluorine will pick it up (becoming a − ion). The ions can form an ionic bond.

12. The elements in a row all have the same number of electron shells. In a column, all the elements have the same number of electrons in their outer shell. Elements in the same column should have similar chemical properties because they have the same valence or bonding capacity and thus would make the same number of covalent bonds. Or if they have only 1 or 2 electrons, or if they have 7 electrons in their outer shell, atoms of these elements would tend to lose or gain electrons, forming ions and participating in ionic bonds.

13. The results indicate that both a lower pH and higher temperature negatively affected the growth of coral polyps and that the reduction in growth was much greater when both pH and temperature were varied at the same time. Because rising atmospheric levels of CO_2 are predicted to continue to acidify the oceans and raise ocean temperatures, it is beneficial to see how these two factors may interact.

14. When water is heated, much of the heat is absorbed in breaking hydrogen bonds before the water molecules increase their motion and the temperature increases. Conversely, when water is cooled, many hydrogen bonds are formed, which releases a significant amount of heat. This release of heat can provide some protection against freezing of the plants' leaves, thus protecting the cells from damage.

15. These extreme environments may be similar to those found on other planets. The fact that life may have evolved and continues to flourish in such extreme environments here on Earth suggests that some form of life may have evolved on other planets. In addition to seeking evidence for the past or current presence of water on Mars or other planets, scientists now know to search in environments that previously would have been thought incapable of supporting life.

Chapter 3

1. a. glucose; b. energy storage; c. cellulose; d. fats; e. cell membrane component; f. steroids; g. amino group; h. carboxyl group; i. R group; j. enzyme; k. structural protein; l. movement; m. membrane transport protein; n. defense; o. phosphate group; p. nitrogenous base; q. ribose or deoxyribose; r. DNA; s. code for proteins

2. d (The second kind of molecule is a polymer of the first.) **3.** c **4.** c **5.** d **6.** a **7.** a **8.** d **9.** a

10. Circle NH_2, an amino group; COOH, a carboxyl group; and OH, a hydroxyl group on the R group. This is an amino acid, a monomer of proteins. The OH group makes it a polar amino acid.

11. Amino acids with hydrophobic R groups are most likely to be found clustered together in the interior of a protein, sheltered from the surrounding water.

12. This is a hydrolysis reaction, which consumes water. It is essentially the reverse of the diagram in Figure 3.5, except that fructose has a different shape than glucose.

13. Carbon forms four covalent bonds, either with other carbon atoms, producing chains or rings of various lengths and shapes, or with other atoms, such as characteristic chemical groups that confer specific properties on a molecule. This is the basis for the incredible diversity of organic compounds. Organisms can link a small number of monomers into different arrangements to produce a huge variety of polymers.

14. The 20 amino acids that are found in proteins can be arranged in many different sequences into chains of many different lengths. The sequences of DNA nucleotides in the genes of a cell dictate the amino acid sequences of its proteins.

15. A developing chick is growing rapidly, increasing its number of cells. To build new cells it needs large stores of cell membrane components, including cholesterol and lipids, and amino acids for building its proteins. It also requires energy to fuel all this construction, and that is available in the form of fats, as fat molecules can be broken down to yield a lot of energy.

16. a. A: at about 37°C; B: at about 78°C
    b. A: from humans (human body temperature is about 37°C); B: from thermophilic bacteria
    c. Above 40°C, the human enzyme denatures and loses its shape and thus its function. The increased thermal energy disrupts the weak bonds that maintain secondary and tertiary structure in an enzyme.

17. These results indicate that replacing either saturated or trans fats in the diet with unsaturated fats reduces the risk of coronary heart disease. The benefit is greater (risk reduced the most) when trans fats are replaced, even though the quantity of energy in the diet replaced was only 2% rather than the 5% of saturated fats replaced.

Chapter 4

1. a. rough ER; b. nucleus; c. nucleolus; d. ribosomes; e. peroxisome; f. centrosome; g. cytoskeleton; h. mitochondrion; i. plasma membrane; j. lysosome; k. Golgi apparatus; l. smooth ER. For functions, see Table 4.22. A centrosome is a microtubule-organizing center.

2. c **3.** b (Small cells have a greater ratio of surface area to volume.) **4.** b **5.** a

6. DNA as genetic material, ribosomes, plasma membrane, and cytosol
7. Cilia may propel a cell through its environment or sweep a fluid environment past the cell.
8. d 9. b 10. a 11. c
12. Different conditions and conflicting processes can occur simultaneously within separate, membrane-enclosed compartments. Also, there is increased area for membrane-attached enzymes that carry out metabolic processes.
13. Part true, part false. All animal *and* plant cells have mitochondria; plant cells do have chloroplasts, but animal cells do not. Both organelles process energy. A mitochondrion converts chemical energy (such as sugar molecules) to another form of chemical energy (ATP). This process provides almost all eukaryotic cells with ATP needed for cellular work. A chloroplast converts light energy to chemical energy (sugar molecules). These sugar molecules may then provide a plant cell's mitochondria with a source of energy. Or they may be stored in the plant body and passed to animals that eat plants or each other.
14. The plasma membrane is a phospholipid bilayer with the hydrophilic heads facing the aqueous environment on both sides and the hydrophobic fatty acid tails mingling in the center of the membrane. Proteins are embedded in and attached to this membrane. Microfilaments form a three-dimensional network just inside the plasma membrane. The extracellular matrix outside the membrane is composed largely of glycoproteins, which may be attached to membrane proteins called integrins. Integrins can transmit information from the ECM to microfilaments on the other side of the membrane.
15. Cell 1: $S = 1{,}256\ \mu m^2$; $V = 4{,}187\ \mu m^3$; $S/V = 0.3$. Cell 2: $S = 5{,}024\ \mu m^2$; $V = 33{,}493\ \mu m^3$; $S/V = 0.15$. The smaller cell has a larger surface area relative to volume, facilitating the uptake of sufficient nutrients and oxygen and the excretion of waste.
16. An mRNA molecule is transcribed from the gene for insulin and moves into the cytosol. There it joins with a ribosome that becomes attached to the outside of the rough ER (a bound ribosome). The ribosome produces a polypeptide that is threaded into the ER compartment. The polypeptide folds up and may be modified within the ER. It is then packaged into a transport vesicle. The vesicle joins with a Golgi sac, and the protein may be further modified during its journey through the Golgi apparatus. A transport vesicle pinches off from the "shipping" face of the Golgi and fuses with the plasma membrane, secreting insulin from the cell.
17. According to the endosymbiotic theory, an ancestral eukaryotic cell engulfed (ingested) an aerobic bacterium but did not digest its potential food item. The prokaryote took up residence within the cell, and its aerobic metabolism probably contributed ATP to the host cell. Over many generations of cells, the host and endosymbiont became mutually dependent and unable to exist on their own—they became a single organism. A similar process may have occurred when one of these mitochondria-containing cells ingested but did not digest a photosynthetic prokaryote.
18. Individuals with PCD have nonfunctional cilia and flagella due to a lack of dynein motor proteins. This defect would also mean that the cilia involved in left-right pattern formation in the embryo would not be able to set up the fluid flow that initiates the normal arrangement of organs.
19. As the chromosomes moved poleward, the microtubule segments on the chromosome side of the mark shortened, while those on pole side stayed the same length. Thus chromosome movement toward the poles of this dividing cell is correlated with the shortening (depolymerizing) of the microtubules at the end where the chromosome is attached. The experiment would have to be repeated on different types of cells from different organisms to determine whether the location where a spindle fiber depolymerizes is always the same. Indeed, other experiments have shown that in some cells, the microtubules depolymerize from the pole end of the spindle fibers.

Chapter 5

1. a. active transport; b. concentration gradient; c. small nonpolar molecules; d. facilitated diffusion; e. transport proteins
2. a. enzyme; b. active site of enzyme; c. substrate; d. substrate in active site; induced fit strains substrate bonds; e. substrate converted to products; f. product molecules released
3. b 4. d 5. c (Only active transport can move solute against a concentration gradient.) 6. a 7. b
8. The work of cells falls into three main categories: chemical, transport, and mechanical. ATP provides the energy for cellular work by transferring a phosphate group to a substrate (chemical) or to a protein (transport and mechanical).
9. Energy is stored in the chemical bonds of a cell's organic molecules. The activation energy barrier prevents these molecules from spontaneously breaking down and releasing that energy. When a substrate fits into an enzyme's active site with an induced fit, its bonds may be strained and thus easier to break, or the active site may orient two substrates in such a way that facilitates the reaction.
10. Energy is neither created nor destroyed but can be transferred and transformed. Plants transform the energy of sunlight into chemical energy stored in organic molecules. Almost all organisms rely on the products of photosynthesis for the source of their energy. In every energy transfer or transformation, disorder increases as some energy is lost to the random motion of thermal energy and released as heat.
11. Cell membranes are composed of diverse proteins suspended in a fluid phospholipid bilayer. The hydrophilic heads of the phospholipids face the aqueous environment on both sides of the membrane and the fatty acid tails cluster in the hydrophobic center of the membrane The membrane forms a selectively permeable boundary between cells and their surroundings (or between organelles and the cytosol). The proteins perform the many functions of membranes, such as enzyme action, transport, attachment, and signaling.
12. Inhibitors that are toxins or poisons irreversibly inhibit key cellular enzymes. Inhibitors that are designed as drugs are beneficial, such as when they interfere with the enzymes of bacterial or viral invaders or cancer cells. Cells use feedback inhibition of enzymes in metabolic pathways as important mechanisms that conserve resources.
13. Aquaporins are water transport channels that allow for very rapid diffusion of water through a cell membrane. It would be most important for your body to reabsorb water from the urine, thus preventing dehydration, after a run on a hot day.
14. The aquaporin RNA-injected oocytes had a high rate of water permeability. (Remember from Module 5.7 that they swelled and ruptured in 3 minutes.) Treatment with mercury chloride inhibited the aquaporins, and the water permeability of the oocytes was reduced. As expected, the higher the concentration of mercury, the greater the inhibition and reduction in water permeability.

When that inhibition was reversed by treatment with the chemical ME, the channels once again functioned and water permeability increased to almost the level of the uninhibited oocytes. The control oocytes were not injected with aquaporin RNA and thus did not have aquaporins. Thus, their water permeability should be very low and not affected by the mercury treatment. Indeed, their water permeability was much lower than any of the RNA-injected oocytes.

15. a. The more enzyme present, the faster the rate of reaction, because it is more likely that enzyme and substrate molecules will meet.

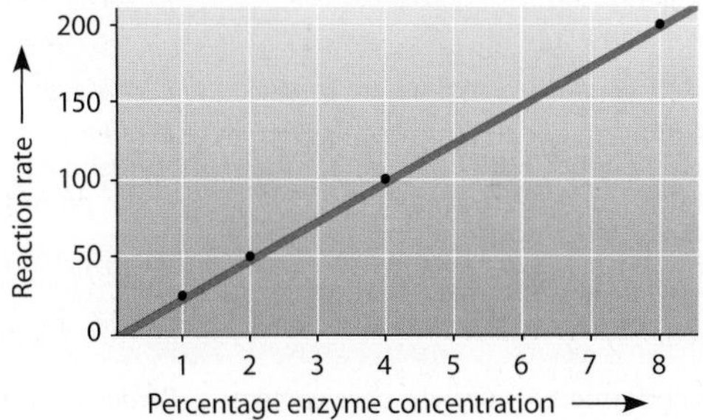
Reaction rate
200
150
100
50
0
1 2 3 4 5 6 7 8
Percentage enzyme concentration

b. The more substrate present, the faster the reaction, for the same reason, but only up to a point. An enzyme molecule can work only so fast; once it is saturated (working at top speed), more substrate does not increase the rate.

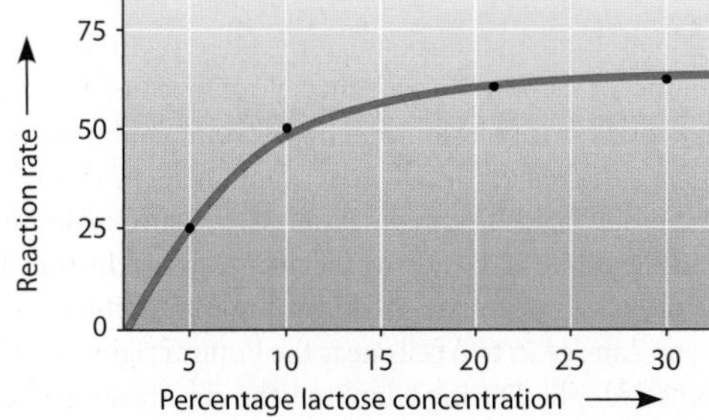
Reaction rate
75
50
25
0
5 10 15 20 25 30
Percentage lactose concentration

16. Some issues and questions to consider: Is improving crop yields of paramount importance in a world where many people can't get enough food? Does the fact that these compounds rapidly break down indicate that the risk to humans is low? How about the risks to people who work in agriculture or to other organisms, such as bees and other pollinating insects, birds, and small mammals? Might there be negative effects on ecosystems that are impossible to predict?

Chapter 6

1. a. glycolysis; b. pyruvate oxidation and citric acid cycle; c. oxidative phosphorylation; d. oxygen; e. electron transport chain; f. CO_2; g. H_2O

2. d **3.** d **4.** b **5.** a **6.** c (NAD^+ and FAD, which are recycled by electron transport, are in limited supply in a cell.) **7.** b (at the same time NADH is oxidized to NAD^+)

8. Glycolysis is considered the most ancient because it occurs in all living cells and doesn't require oxygen or membrane-enclosed organelles.

9. In lactic acid fermentation (in muscle cells), pyruvate is reduced by NADH to form lactate, and NAD^+ is recycled. In alcohol fermentation, pyruvate is broken down to CO_2 and ethanol as NADH is oxidized to NAD^+. Both types of fermentation allow glycolysis to continue to produce 2 ATP per glucose by recycling NAD^+.

10. As carbohydrates are broken down in glycolysis and the oxidation of pyruvate, glycerol can be made from G3P and fatty acids can be made from acetyl CoA. Amino groups, containing N atoms, must be supplied to various intermediates of glycolysis and the citric acid cycle to produce amino acids.

11. 100 kcal per day is 700 kcal per week. According to Figure 6.4, walking 3 mph would require $\frac{700}{245}$ = about 2.8 hours; swimming, 1.7 hours; running, 0.7 hour.

12. NAD^+ and FAD are coenzymes that are not used up during the oxidation of glucose. NAD^+ and FAD are recycled when NADH and $FADH_2$ pass the electrons they are carrying to the electron transport chain. We need a small additional supply to replace those that are damaged.

13. a. No, this shows the blue color getting more intense. The reaction decolorizes the blue dye.
b. No, this shows the dye being decolorized, but it also shows the three mixtures with different initial color intensities. The intensities should have started out the same, since all mixtures used the same concentration of dye.
c. Correct. The mixtures all start out the same, and then the ones with more malate (reactant) decolorize faster.

14. The presence of ATP synthase enzymes in prokaryotic plasma membranes and the inner membrane of mitochondria provides support for the theory of endosymbiosis—that mitochondria evolved from an engulfed prokaryote that used aerobic respiration (see Module 4.15).

15. The percentage of body fat is the independent variable and is plotted on the *x* axis. The activity of brown fat is the dependent variable, and it is plotted on the *y* axis. Your graph should show a negative correlation between body fat percentage and activity of brown fat (the data points are higher for lower body fat percentage and decrease as the body fat percentage increases). One hypothesis is that thin individuals have more active brown fat and thus burn more calories, which contributes to their thinness (lower percentage of body fat). A second hypothesis is that the higher percentage of body fat insulated the bodies of the more overweight subjects, thus their brown fat did not have to be as active to maintain their body temperatures when exposed to cold.

16. In a person treated with uncoupling agents like DNP, the proton gradient established during electron transport is no longer tied to ATP synthesis. As a result, oxidation of glucose during cellular respiration yields very little ATP, since ATP is normally produced as H^+ ions flow back through ATP synthase in the inner mitochondrial membrane. Without large amounts of ATP available, biosynthesis cannot take place and new organic molecules cannot be synthesized. Low ATP levels would signal the body to continue breaking down its own molecules and feeding them into the cellular respiration pathway, leading to excessive weight loss and severe overheating, sweating, and dehydration. One or a combination of these factors can cause death.

17. The mitochondria of brown fat cells have protein channels that make the inner mitochondrial membrane leaky to H^+ ions, producing the same effect that the drug DNP has on mitochondria. When these channels are activated, brown fat burns fuel without producing ATP. Drugs that could activate brown fat would help a patient burn more calories. Thus, excess calories from the diet would not be converted to fat, and fat stores of the body could be reduced. If these drugs somehow affected the mitochondria of all body cells, however, the results could be as disastrous as they were with DNP.

Chapter 7

1. a. light energy; b. light reactions; c. Calvin cycle; d. O_2 released; e. electron transport chain; f. NADPH; g. ATP; h. G3P (sugar)
2. c **3.** b **4.** a **5.** c (NADPH and ATP from the light reactions are required by the Calvin cycle.) **6.** d **7.** b **8.** c
9. CO_2 and H_2O are the products of respiration; they are the reactants in photosynthesis. In respiration, glucose is oxidized to CO_2 as electrons are passed through an electron transfer chain from glucose to O_2, producing H_2O. In photosynthesis, H_2O is the source of electrons, which are energized by light, temporarily stored in NADPH, and used to reduce CO_2 to carbohydrate.
10. The light reactions require ADP and $NADP^+$, neither of which are recycled from ATP and NADPH when the Calvin cycle stops.
11. Plants can break down the sugar for energy in cellular respiration or use the sugar as a raw material for making other organic molecules. Excess sugar is stored as starch.
12. a. electron transport chain; b. ATP synthase; c. thylakoid space; d. stroma; e. ATP. The higher H^+ concentration is found in the intermembrane space of the mitochondrion and in the thylakoid space of the chloroplast.
13. In mitochondria: a. Electrons come from food molecules. b. Electrons have high potential energy in the bonds in organic molecules. c. Electrons are passed to oxygen, which picks up H^+ and forms water.
In chloroplasts: a. Electrons come from splitting of water. b. Light energy excites the electrons to a higher energy level. c. Electrons flow from water to the reaction-center chlorophyll in photosystem II to the reaction-center chlorophyll in photosystem I to $NADP^+$, reducing it to NADPH.
In both processes: d. Energy released by redox reactions in the electron transport chain is used to transport H^+ across a membrane. The flow of H^+ down its concentration gradient back through ATP synthase drives the phosphorylation of ADP to make ATP.
14. The hypothesis was that, because CO_2 is a raw material for photosynthesis, rising CO_2 levels would increase the growth and production of pollen by ragweed. Pollen production was positively correlated with CO_2 concentrations, and the results supported the hypothesis. Because rising CO_2 levels are associated with warmer temperatures, an experiment could also look at the effect of temperature on ragweed pollen production. One might also want to determine whether the growing season is getting longer, as this would expose hay fever sufferers to pollen for extended periods. It would also be interesting to measure whether ragweed pollen is more allergenic when grown in higher CO_2 levels, as poison ivy was shown to be. These types of experiments have been performed, and their results are as you would predict—the growing season for ragweed has gotten longer, and pollen that is more allergenic is produced in higher quantities under conditions of elevated CO_2.
15. Some issues and questions to consider: What are the risks that we take and costs we must pay if global climate change continues? How certain do we have to be that global warming is caused by human activities before we act? What can we do to reduce CO_2 emissions? Is it possible that the costs and sacrifices of reducing CO_2 emissions might actually improve our lifestyle?

Chapter 8

1.

	Mitosis	Meiosis
Number of chromosomal duplications	*1*	*1*
Number of cell divisions	*1*	*2*
Number of daughter cells produced	*2*	*4*
Number of chromosomes in the daughter cells	*Diploid (2n)*	*Haploid (n)*
How the chromosomes line up during metaphase	*Singly*	*In tetrads (metaphase I), then singly (metaphase II)*
Genetic relationship of the daughter cells to the parent cell	*Genetically identical*	*Genetically unique*
Functions performed in the human body	*Growth, development, and repair*	*Production of gametes*

2. b **3.** c **4.** b **5.** b **6.** b **7.** a **8.** b **9.** d (A diploid cell would have an even number of chromosomes; the odd number suggests that meiosis I has been completed. Sister chromatids are together only in prophase and metaphase of meiosis II.) **10.** c **11.** d
12. Most of the cells are in interphase (a time of growth, DNA synthesis, metabolic activity), without recognizable compacted individual chromosomes. During prophase, chromosomes compact and thicken (e.g., the cell on the left edge, about halfway down) and the mitotic spindle forms. During metaphase, the chromosomes line up in the middle of the cell (e.g., the second cell from the top, near the top left corner). In anaphase, the chromosomes split into two groups (which you can see in two cells near the bottom right corner) as sister chromatids split. During telophase, the chromosomes reach opposite ends (e.g., at the very top, fifth cell from the right) as daughter nuclei form around the chromosomes and cytokinesis begins.
13. Mitosis without cytokinesis would result in a single cell with two nuclei. Multiple rounds of cell division like this could produce such a "megacell."
14. Various orientations of homologous chromosome pairs at metaphase I of meiosis lead to different combinations of chromosomes in gametes. Crossing over during prophase I results in an exchange of chromosome segments and new combinations of genes. Random fertilization of eggs by sperm further increases possibilities for variation in offspring.
15. In culture, normal cells usually divide only when they are in contact with a surface but not touching other cells on all sides (the cells usually grow to form only a single layer). The density-dependent inhibition of cell division apparently results from local depletion of substances called growth factors. Growth factors are proteins secreted by certain cells that stimulate other cells to divide; they act via signal transduction pathways to signal the cell cycle control system of the affected cell to proceed past its checkpoints. The cell cycle control systems of cancer cells do not function properly. Cancer cells generally do not require externally supplied growth factors to complete the cell cycle, and they divide indefinitely (in contrast to normal mammalian cells, which stop dividing after 20 to 50 generations)—two reasons why cancer cells

are relatively easy to grow in the lab. Furthermore, cancer cells can often grow without contacting a solid surface, making it possible to culture them in suspension in a liquid medium.

16. A ring of microfilaments pinches an animal cell in two, a process called cleavage. In a plant cell, membranous vesicles form a disk called the cell plate at the midline of the parent cell, cell plate membranes fuse with the plasma membrane, and a cell wall grows in the space, separating the daughter cells.

17. See Figure 8.18.

18. a. No. For this to happen, the chromosomes of the two gametes that fused would have to represent, together, a complete set of the donor's maternal chromosomes (the ones that originally came from the donor's mother) and a complete set of the donor's paternal chromosomes (from the donor's father). It is much more likely that the zygote would be missing one or more maternal chromosomes and would have an excess of paternal chromosomes, or vice versa.

b. Correct. Consider what would have to happen to produce a zygote genetically identical to the gamete donor: The zygote would have to have a complete set of the donor's maternal chromosomes and a complete set of the donor's paternal chromosomes. The first gamete in this union could contain any mixture of maternal and paternal chromosomes, but once that first gamete was "chosen," the second one would have to have one particular combination of chromosomes—the combination that supplies whatever the first gamete did not supply. So, for example, if the first three chromosomes of the first gamete were maternal, maternal, and paternal, the first three of the second gamete would have to be paternal, paternal, and maternal. The chance that all 23 chromosome pairs would be complementary in this way is only one in 2^{23} (that is, one in 8,388,608). Because of independent assortment, it is much more likely that the zygote would have an unpredictable combination of chromosomes from the donor's father and mother.

c. No. First, the zygote could not be genetically identical to the gamete donor (see answer b). Second, the zygote could not be identical to either of the gamete donor's parents because the donor only has half the genetic material of each of his or her parents. For example, even if the zygote were formed by two gametes containing only paternal chromosomes, the combined set of chromosomes could not be identical to that of the donor's father because it would still be missing half of the father's chromosomes.

d. No. See answer c.

19. Some possible hypotheses: The replication of the DNA of the bacterial chromosome takes less time than the replication of the DNA in a eukaryotic cell. The time required for a growing bacterium to roughly double its cytoplasm is much less than for a eukaryotic cell. Bacteria have a cell cycle control system much simpler than that of eukaryotes.

20. $1\ cm^3 = 1{,}000\ mm^3$, so $5{,}000\ mm^3$ of blood contains $5{,}000 \times 1{,}000 \times 5{,}000{,}000 = 25{,}000{,}000{,}000{,}000$, or 2.5×10^{13}, red blood cells. The number of cells replaced each day $= 2.5 \times 10^{13}/120 = 2.1 \times 10^{11}$ cells. There are $24 \times 60 \times 60 = 86{,}400$ seconds in a day. Therefore, the number of cells replaced each second $= 2.1 \times 10^{11}/86{,}400 =$ about 2×10^6, or 2 million. Thus, about 2 million cell divisions must occur each second to replace red blood cells that are lost.

21. Each chromosome is on its own in mitosis; chromosome replication and the separation of sister chromatids occur independently for each horse or donkey chromosome. Therefore, m sions, starting with the zygote, are not impaired. In meio however, homologous chromosomes must pair in prophase I. This process of synapsis cannot occur properly because horse and donkey chromosomes do not match in number or content.

22. There are two unusual cases presented in Table 8.10: the patient who did not have the mutation but responded to everolimus and the patient who had the mutation but did not respond to everolimus. These patients could be studied in a manner similar to the original patient by performing a genetic analysis of their tumor cells. Such an analysis might reveal other mutations that explain their results. Further experiments like the one presented in Table 8.10 might then be performed to see if these results can be generalized. Such an approach may allow refinement of the personalized cancer therapy.

Chapter 9

1. a. alleles; b. loci; c. homozygous; d. dominant; e. recessive; f. incomplete dominance

2. c **3.** b **4.** d (Neither parent is ruby-eyed, but some offspring are, so it is recessive. Different ratios among male and female offspring show that it is sex-linked.) **5.** d

6. The trait of freckles is dominant, so Tim and Jan must both be heterozygous. There is a chance that they will produce a child with freckles and a chance that they will produce a child without freckles. The probability that the next two children will have freckles is $3/4 \times 3/4 = 9/16$.

7. As in problem 6, both Tim and Jan are heterozygous, and Mike is homozygous recessive. The probability of the next child having freckles is 3/4. The probability of the next child having a straight hairline is 1/4. The probability that the next child will have freckles and a straight hairline is $3/4 \times 1/4 = 3/16$.

8. The genotype of the black short-haired parent rabbit is *BBSS*. The genotype of the brown long-haired parent rabbit is *bbss*. The F_1 rabbits will all be black and short-haired, *BbSs*. The F_2 rabbits will be black short-haired, black long-haired, brown short-haired, and brown long-haired, in a proportion of 9:3:3:1.

9. If the genes are not linked, the proportions among the offspring will be 25% gray red, 25% gray purple, 25% black red, and 25% black purple. The actual percentages show that the genes are linked. The recombination frequency is 6%.

10. The recombination frequencies are black dumpy 36%, purple dumpy 41%, and black purple 6% (see problem 9). Because these recombination frequencies reflect distances between the genes, the sequence must be purple-black-dumpy (or dumpy-black-purple).

11. 1/4 will be boys suffering from hemophilia, and 1/4 will be female carriers. (The mother is a heterozygous carrier [X^HX^h], and the father is normal [X^HY].)

12. Genes on the single X chromosome in males are always expressed because there are no corresponding genes on the Y chromosome to mask them. A male needs only one recessive colorblindness allele (from his mother) to show the trait; a female must inherit the allele from both parents, which is less likely.

13. The parental gametes are *WS* and *ws*. Recombinant gametes are *Ws* and *wS*, produced by crossing over.

14. Height appears to be a quantitative trait resulting from polygenic inheritance, like human skin color. See Module 9.14.

ıorblind, she must inherit X chromosomes ıdness allele from both parents. Her fa- ı chromosome, which he passes on to all his nust be colorblind. A male need only inherit the ıllele from a carrier mother; both his parents are ıypically normal.

ıreeding the cat to get a population to work with. If ıle is recessive, two curl cats can have only curl kittens. ıele is dominant, curl cats can have "normal" kittens. If tı. curl allele is sex-linked, ratios will differ in male and female offspring of some crosses. If the curl allele is autosomal, the same ratios will be seen among males and females. Once you have established that the curl allele is dominant and autosomal, you can determine if a particular curl cat is true-breeding (homozygous) by doing a testcross with a normal cat. If the curl cat is homozygous, all offspring of the testcross will be curl; if heterozygous, half of the offspring will be curl and half normal.

17. If the genes are unlinked, you expect puppies in a 9:3:3:1 ratio: 90 black normal vision, 30 black blind, 30 chocolate normal, and 10 chocolate blind. If the genes are linked, you would expect a 3:1 ratio of black normal to chocolate blind, with a small number of black blind and chocolate normal recombinant offspring.

APPENDIX 4

Chapter 10

1. a. nucleotides; b. transcription; c. RNA polymerase; d. mRNA; e. rRNA; f. tRNA; g. translation; h. ribosomes; i. amino acids
2. b
3. Ingredients: Original DNA, nucleotides, several enzymes and other proteins, including DNA polymerase and DNA ligase. Steps: Original DNA strands separate at a specific site (origin of replication), free nucleotides hydrogen-bond to each strand according to base-pairing rules, and DNA polymerase covalently bonds the nucleotides to form new strands. New nucleotides are added only to the 3′ end of a growing strand. One new strand is made in one continuous piece; the other new strand is made in a series of short pieces that are then joined by DNA ligase. Product: Two identical DNA molecules, each with one old strand and one new strand.
4. transcription; translation
5. d (Only the phage DNA enters a host cell; lambda DNA determines both DNA and protein.)
6. d
7. A gene is the polynucleotide sequence with information for making one polypeptide. Each codon—a triplet of bases in DNA or RNA—codes for one amino acid. Transcription occurs when RNA polymerase produces RNA using one strand of DNA as a template. In prokaryotic cells, the RNA transcript may immediately serve as mRNA. In eukaryotic cells, the RNA is processed: A cap and tail are added, and RNA splicing removes introns and links exons together to form a continuous coding sequence. A ribosome is the site of translation, or polypeptide synthesis, and tRNA molecules serve as interpreters of the genetic code. Each folded tRNA molecule has an amino acid attached at one end and a three-base anticodon at the other end. Beginning at the start codon, mRNA is moved relative to the ribosome a codon at a time. A tRNA with a complementary anticodon pairs with each codon, adding its amino acid to the polypeptide chain. The amino acids are linked by peptide bonds. Translation stops at a stop codon, and the finished polypeptide is released. The polypeptide folds to form a functional protein, sometimes in combination with other polypeptides.
8. c
9.

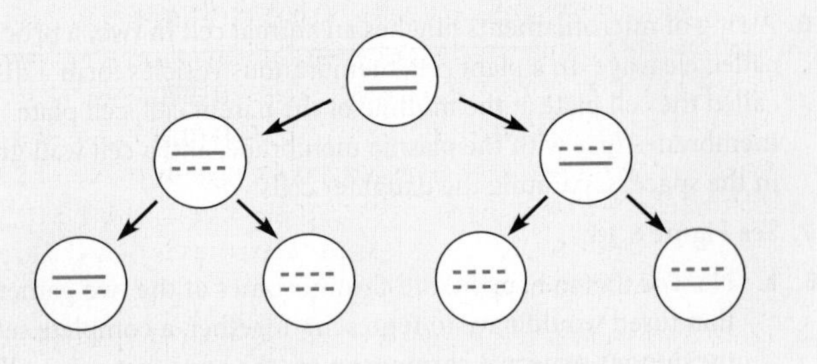

10. mRNA: GAUGCGAUCCGCUAACUGA; amino acids: Met-Arg-Ser-Ala-Asn
11. Some issues and questions to consider: Is it fair to issue a patent for a gene or gene product that occurs naturally in every human being? Or should a patent be issued only for something new that is invented rather than found? Suppose another scientist slightly modifies the gene or protein. How different does the gene or protein have to be to avoid patent infringement? Might patents encourage secrecy and interfere with the free flow of scientific information? What are the benefits to the holder of a patent? When research discoveries cannot be patented, what are the scientists' incentives for doing the research? What are the incentives for the institution or company that is providing financial support?
12. A bacteriophage is capable of easily infecting a host (bacterium) and therefore multiplying its genetic material. Most importantly, a bacteriophage has a very simple structure, allowing the outer structure (made entirely of protein) to be easily distinguished from the inner structure (made of DNA).

Chapter 11

1. a. proto-oncogene; b. repressor (or activator); c. cancer; d. operator; e. X inactivation; f. transcription factors; g. alternative RNA splicing
2. b **3.** b **4.** b **5.** b (Different genes are active in different kinds of cells.) **6.** c **7.** d
8. They will be black, because the DNA of the cell was obtained from a black mouse.
9. a. If the mutated repressor could still bind to the operator on the DNA, it would continuously repress the operon; enzymes for lactose utilization would not be made, whether or not lactose was present.
   b. The *lac* genes would continue to be transcribed and the enzymes made, whether or not lactose was present.
   c. Same predicted result as for b.
   d. RNA polymerase would not be able to transcribe the genes; no proteins would be made, whether or not lactose was present.
10. A mutation in a single gene can influence the actions of many other genes if the mutated gene is a control gene, such as a homeotic gene. A single control gene may encode a protein that affects (activates or represses) the expression of a number of other genes. In addition, some of the affected genes may themselves be control genes that in turn affect other batteries of genes. Cascades of gene expression are common in embryonic development.
11. The protein to which dioxin binds in the cell is probably a transcription factor that regulates multiple genes (see Module 11.3). If the binding of dioxin influences the activity of this transcription factor—either activating or inactivating it—dioxin could thereby

affect multiple genes and thus have a variety of effects on the body. The differing effects in different animals might be explained by differing genetic details in the different species. It would be extremely difficult to demonstrate conclusively that dioxin exposure was the cause of illness in a particular individual, even if dioxin had been shown to be present in the person's tissues. However, if you had detailed information about how dioxin affects patterns of gene expression in humans and were able to show dioxin-specific abnormal patterns in the patient (perhaps using DNA microarrays; see Module 11.9), you might be able to establish a strong link between dioxin and the illness.

12. Wilmut was able to coordinate the cell cycle of the donor and host cells by depriving them of nutrients. When faced with starvation, both cells switched to the G_0 phase of the cell cycle. Wilmut could therefore be sure that, when placed in a growth medium with nutrients, the cell cycles within both the donor and host cell were synchronized. This allowed him to successfully clone a mammal from an adult cell for the first time.

Chapter 12

1. a. PCR; b. a restriction enzyme; c. gel electrophoresis; d. nucleic acid probe; e. cloning

2. d **3.** b **4.** b **5.** c **6.** c

7. Because it would be too expensive and time consuming to compare whole genomes. By choosing STR sites that vary considerably from person to person, investigators can get the necessary degree of specificity without sequencing the entire genome.

8. Medicine: Genes can be used to produce transgenic lab animals for AIDS research or for research related to human gene therapy. Proteins can be hormones, enzymes, blood-clotting factors, or the active ingredient of vaccines. Agriculture: Foreign genes can be inserted into plant cells or animal eggs to produce transgenic crop plants or farm animals. Animal growth hormones are examples of agriculturally useful proteins that can be made using recombinant DNA technology.

9. She could start with DNA isolated from liver cells (the entire genome) and carry out the procedure outlined in Module 12.1 to produce a collection of recombinant bacterial clones, each carrying a small piece of liver cell DNA. To find the clone with the desired gene, she could then make a probe of radioactive RNA with a nucleotide sequence complementary to part of the gene: GACCUGACUGU. This probe would bind to the gene, labeling it and identifying the clone that carries it. Alternatively, the biochemist could start with mRNA isolated from liver cells and use it as a template to make cDNA (using reverse transcriptase). Cloning this DNA rather than the entire genome would yield a smaller library of genes to be screened—only those active in liver cells. Furthermore, the genes would lack introns, making the desired gene easier to manipulate after isolation.

10. c (Bacteria lack the RNA-splicing machinery needed to delete eukaryotic introns.)

11. Isolate plasmids from a culture of *E. coli*. Cut the plasmids and the human DNA containing the HGH gene with the restriction enzyme to produce molecules with sticky ends. Join the plasmids and the fragments of human DNA with ligase. Allow *E. coli* to take up recombinant plasmids. Bacteria will then replicate plasmids and multiply, producing clones of bacterial cells. Identify a clone carrying and expressing the HGH gene using a nucleic acid probe. Grow large amounts of the bacteria and extract and purify HGH from the culture.

12. Determining the nucleotide sequences is just the researchers have written out the DNA "book," they wi to figure out what it means—what the nucleotide sequenc for and how they work.

13. Some issues and questions to consider: What are some of the unknowns in recombinant DNA experiments? Do we know enough to anticipate and deal with possible unforeseen and negative consequences? Do we want this kind of power over evolution? Who should make these decisions? If scientists doing the research were to make the decisions about guidelines, what factors might shape their judgment? What might shape the judgment of business executives in the decision-making process? Does the public have a right to a voice in the direction of scientific research? Does the public know enough about biology to get involved in this decision-making process? Who represents "the public," anyway?

14. Some issues and questions to consider: What kinds of impact will gene therapy have on the individuals who are treated? On society? Who will decide what patients and diseases will be treated? What costs will be involved, and who will pay them? How do we draw the line between treating disorders and "improving" the human species?

15. Some issues and questions to consider: Should genetic testing be mandatory or voluntary? Under what circumstances? Why might employers and insurance companies be interested in genetic data? Since genetic characteristics differ among ethnic groups and between the sexes, might such information be used to discriminate? Which of these questions do you think is most important? Which issues are likely to be the most serious in the future?

16. Gather two groups of volunteers. Have one group ingest a standard amount of GMO corn in their diet and the other group ingest a standard amount of traditional corn. Monitor the health of the individuals in both groups, looking for differences. In real life, such a study would be problematic because it is difficult to control and monitor what people eat, it would require many years to search for long-term health effects, and it may run afoul of ethics standards for human testing.

Chapter 13

1. According to Darwin's theory of descent with modification, all life has descended from a common ancestral form as a result of natural selection. Individuals in a population have hereditary variations. The overproduction of offspring in the face of limited resources leads to a struggle for existence. Individuals that are well suited to their environment tend to leave more offspring than other individuals, leading to the gradual accumulation of adaptations to the local environment in the population.

2. a. genetic drift; b. gene flow; c. natural selection; d. small population; e. founder effect; f. bottleneck effect; g. unequal reproductive success

3. d **4.** a **5.** b (Erratic rainfall and unequal reproductive success would ensure that a mixture of both forms remained in the population.) **6.** d **7.** d **8.** c

9. Exposing deep rock strata made it easier to obtain older fossils.

10. Your paragraph should include such evidence as fossils and the fossil record, homologous structures, molecular homologies, artificial selection, and examples of natural selection.

11. Evidence strongly supports Lamarck's hypothesis that life evolves. However, our understanding of genetics refutes his hypothesis for the mechanism of evolution.

. Since $p + q = 1, p = 1 - q = 0.95$. The pro-
tes is $2pq = 2 \times 0.95 \times 0.05 = 0.095$. About
ericans are carriers.

s retained in a population by diploidy and bal-
ecessive alleles are hidden from selection when
ote; thus, less adaptive or even harmful alleles
in the gene pool and are available should envi-
onditions change. Both heterozygote advantage and
-dependent selection tend to maintain alternate alleles
opulation.

14. The terrestrial ancestors of cetaceans had lungs and breathed air. Evolution did not cause the invention of a new respiratory system for breathing underwater. Rather, the existing external structures were adapted. For example, the nostrils shifted position to the top of the head, where they form the blowhole, an opening to the trachea (windpipe), which leads to the lungs. The blowhole is closed while the cetacean is underwater. A cetacean must thrust its blowhole above the surface of the water periodically to obtain oxygen and expel carbon dioxide. Unlike other mammals, the cetacean respiratory system doesn't intersect with the digestive system, allowing the animal to take in food underwater.

15. The unstriped snails appear to be better adapted. Striped snails make up 47% of the living population but 56% of the broken shells. Assuming that all the broken shells result from the meals of birds, we would predict that bird predation would reduce the frequency of striped snails and the frequency of unstriped individuals would increase.

16. Some issues and questions to consider: Who should decide curriculum, scientific experts in a field or members of the community? Are these alternative versions scientific ideas? Who judges what is scientific? If it is fairer to consider alternatives, should the door be open to all alternatives? Are constitutional issues (separation of church and state) involved here? Can a teacher be compelled to teach an idea he or she disagrees with? Should a student be required to learn an idea he or she thinks is wrong?

Chapter 14

1. a. Allopatric speciation: Reproductive barriers may evolve between these two geographically separated populations as a by-product of the genetic changes associated with each population's adaptation to its own environment or as a result of genetic drift or mutation.
   b. Sympatric speciation: Some change, perhaps in resource use or female mate choice, may lead to a reproductive barrier that isolates the gene pools of these two populations, which are not separated geographically. Once the gene pools are separated, each species may go down its own evolutionary path. If speciation occurs by polyploidy—which is common in plants but unusual in animals—then the new species is instantly isolated from the parent species.

2. a. hybrid zone; b. reinforcement; c. fusion; d. stability; e. strengthened; f. weakened or eliminated

3. c 4. b 5. b 6. d 7. c 8. b 9. c 10. a 11. d 12. d

13. Different physical appearances may indicate that organisms belong in different species, but they may just be physical differences within a species. Isolated populations may or may not be able to interbreed; breeding experiments would need to be performed to determine this. Organisms that reproduce only asexually and fossil organisms do not have the potential to interbreed and produce fertile offspring; therefore, the biological species concept cannot apply to them.

14. There is more chance for gene flow between populations on a mainland and nearby island. This interbreeding would make it more difficult for reproductive isolation to develop and separate the two populations.

15. The term *punctuated equilibria* refers to a common pattern seen in the fossil record, in which most species diverge relatively quickly as they arise from an ancestral species and then remain fairly unchanged for the rest of their existence as a species.

16. Yes. Factors such as polyploidy, sexual selection, and habitat specialization can lead to reproductive barriers that would separate the gene pools of allopatric as well as sympatric populations.

17. A broad hypothesis would be that cultivated American cotton arose from a sequence of hybridization, mistakes in cell division, and self-fertilization. We can divide this broad statement into at least three hypotheses. *Hypothesis 1*: The first step in the origin of cultivated American cotton was hybridization between a wild American cotton plant (with 13 pairs of small chromosomes) and an Old World cotton plant (with 13 pairs of large chromosomes). If this hypothesis is correct, we would predict that the hybrid offspring would have had 13 small chromosomes and 13 large chromosomes. *Hypothesis 2*: The second step in the origin of cultivated American cotton was a failure of cell division in the hybrid offspring, such that all chromosomes were duplicated (now 26 small and 26 large). If this hypothesis is true, we would expect the resulting gametes to each have had 13 large chromosomes and 13 small chromosomes. *Hypothesis 3*: The third step in the origin of cultivated American cotton was self-fertilization of these gametes. If this hypothesis is true, we would expect the outcome of self-fertilization to be a hybrid plant with 52 chromosomes: 13 pairs of large ones and 13 pairs of small ones. Indeed, this is the genetic makeup of cultivated American cotton.

18. By decreasing the ability of females to distinguish males of their own species, the polluted turbid waters have increased the frequency of mating between members of species that had been reproductively isolated from one another. As the number of hybrid fish increase, the parent species' gene pools may fuse, resulting in a loss of the two separate parent species and the formation of a new hybrid species. Future speciation events in Lake Victoria cichlids are less likely to occur in turbid water because females are less able to base mate choice on male breeding color. Reducing the pollution in the lake may help reverse this trend.

19. Some issues and questions to consider: One could look at this question in two ways. If the biological species concept is followed strictly, one could argue that red wolves and coyotes are the same species, since they can interbreed. Because coyotes are not rare, this line of argument would suggest that red wolves should not be protected. On the other hand, because red wolves and coyotes differ in many ways, they can be viewed as distinct species by other species concepts. Protecting the remaining red wolves from hybridizing with coyotes can preserve their distinct species status. The rationale behind protecting all endangered groups is the desire to preserve genetic diversity. Questions for society in general include the following: What is the value of any particular species and its genetically distinct subgroups? And how far are we willing to go to preserve a rare and distinct group of organisms? How should the costs of preserving genetic diversity compare with the costs of other public projects?

Chapter 15

1. a. Abiotic synthesis of important molecules from simpler chemicals in the atmosphere, with lightning or UV radiation as the energy source
   b. Polymerization of monomers, perhaps on hot rocks
   c. Enclosure within a lipid membrane, which maintained a distinct internal environment
   d. Beginnings of heredity as RNA molecules replicated themselves. Natural selection could have acted on protocells that enclosed self-replicating RNA.
2. a. phylogeny; b. homologies; c. morphology; d. analogies; e. phylogenetic tree; f. outgroup; g. shared derived characters
3. b **4.** a **5.** d **6.** c **7.** d **8.** a **9.** c **10.** b
11. Microevolution is the change in the gene pool of a population from one generation to the next. Macroevolution involves the pattern of evolutionary changes over large time spans and includes the origin of new groups and evolutionary novelties as well as mass extinctions.
12. The latter are more likely to be closely related, because even small genetic changes can produce divergent physical appearances. But if genes have diverged greatly, it implies that lineages have been separate for some time, and the similar appearances may be analogous, not homologous.
13. Complex structures can evolve by the gradual refinement of earlier versions of those structures, all of which served a useful function in each ancestor.
14. Where and when key developmental genes are expressed in a developing embryo can greatly affect the final form and arrangement of body parts. The regulation of gene expression allows these genes to continue to be expressed in some areas, turned off in other areas, and/or expressed at different times during development.
15. The ribosomal RNA genes, which specify the RNA parts of ribosomes, have evolved so slowly that homologies between even distantly related organisms can still be detected. Analysis of other homologous genes is also used.
16. 22,920 years old, a result of four half-life reductions
17. 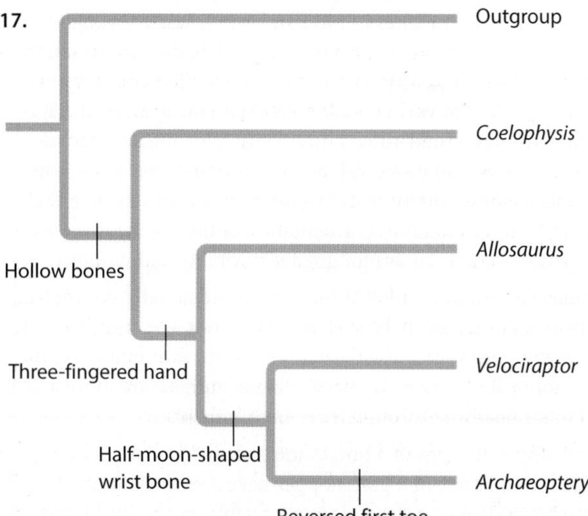
Outgroup
Coelophysis
Allosaurus
Velociraptor
Archaeopteryx
Hollow bones
Three-fingered hand
Half-moon-shaped wrist bone
Reversed first toe
18. Your answer should include aspects of the process of science such as the following: how scientists develop hypotheses and test predictions; the importance of careful experimental design; generating and testing alternative hypotheses; using new technologies; and willingness to incorporate new evidence and revise hypotheses.

Chapter 16

1. Cell wall: maintains cell shape; provides physical prote[illegible] vents cell from bursting in a hypotonic environment
   Capsule: enables cell to stick to substrate or to other individuals in a colony; shields pathogens from host's defensive cells
   Flagella: provide motility, enabling cell to respond to chemical or physical signals in the environment that lead to nutrients or other members of their species and away from toxic substances
   Fimbriae: allow cells to attach to surfaces, including host cells, or to each other
   Endospores: withstand harsh conditions
2. a. Archaeplastids; b. Charophytes; c. Unikonts; d. Fungi; e. Choanoflagellates; f. Animals
3. d (Algae are autotrophs; slime molds are heterotrophs.) **4.** d **5.** b **6.** c **7.** b
8. Rapid rate of reproduction enables prokaryotes to colonize favorable habitats quickly. Mutations during the rapid production of large numbers of cells results in a great deal of genetic variation, making it more likely that some individuals will survive—and be able to recolonize the habitat—if the environment changes again.
9. Multicellular organisms have a greater extent of cellular specialization and more interdependence of cells. New organisms are produced from a single cell, either an egg or an asexual spore.
10. *Chlamydomonas* is a eukaryotic cell, much more complex than a prokaryotic bacterium. It is autotrophic, while amoebas are heterotrophic. It is unicellular, unlike multicellular sea lettuce.
11. d
12. b. Antibiotics kill bacteria. If ulcers are caused by bacteria, then ulcers patients should be cured by antibiotics. (Assume that the researchers have chosen an antibiotic that is effective against the bacteria they hypothesize to be the cause of ulcers.) If the ulcers persist, then the bacteria did not cause the ulcers.
13. According to the theory of secondary endosymbiosis, both organisms are descended from lineages of heterotrophic eukaryotes that engulfed autotrophic eukaryotes, which then evolved into chloroplasts. In the lineage that gave rise to *Euglena*, the autotrophic eukaryote was a green alga. In the lineage that gave rise to *Gymnodinium*, the chloroplast was derived from a different autotrophic eukaryote, a red alga.
14. Use the links on the FDA's main page on dietary supplements to learn how dietary supplements are—and are not—regulated. While manufacturers must prove to the FDA that medicinal products are effective before they can be marketed, there is no such approval process for dietary supplements. The manufacturer is responsible for ensuring that the product is safe and that any claims made on the label are true. This web page also provides links to help you find information about specific supplements and to help you make informed decisions about supplement use. Once you are familiar with these resources, select a specific probiotic product to evaluate.
15. This is not a good idea; all life depends on bacteria. You could predict that eliminating all bacteria from an environment would result in a buildup of toxic wastes and dead organisms (both of which bacteria decompose), a shutdown of all chemical cycling, and the consequent death of all organisms.
16. Some issues and questions to consider: Could we determine beforehand whether the iron would really have the desired effect? How? Would the "fertilization" need to be repeated? Could it be a cure for the problem, or would it merely treat the

...on treatment have side effects? What

...scular plants); b. seedless vascular plants ...; c. gymnosperms; d. angiosperms; 1. apical ...mbryos retained in the parent plant; 2. lignin-hard- ... tissue; 3. seeds that protect and disperse embryos

... a cloud of pollen being released from a pollen cone of a ...e tree. In pollen cones, spores produce millions of the male gametophytes—the pollen grains.

b. This is a cloud of haploid spores produced by a puffball fungus. Each spore may germinate to produce a haploid mycelium.

3. b **4.** b (It is the only gametophyte among the possible answers.) **5.** a **6.** d **7.** c **8.** b **9.** c

10. The alga is surrounded and supported by water, and it has no supporting tissues, vascular system, or special adaptations for obtaining or conserving water. Its whole body is photosynthetic, and its gametes and embryos are dispersed into the water. The seed plant has lignified vascular tissues that support it against gravity and carry food and water. The seed plant also has specialized organs that absorb water and minerals (roots), provide support (stems and roots), and photosynthesize (leaves and stems). It is covered by a waterproof cuticle and has stomata for gas exchange. Its sperm are carried by pollen grains, and embryos develop on the parent plant and are then protected and provided for by seeds.

11. Animals carry pollen from flower to flower and thus help fertilize the plants' eggs. They also disperse seeds by consuming fruit or carrying fruit that clings to their fur. In return, they get food (nectar, pollen, fruit).

12. Plants are autotrophs; they have chlorophyll and make their own food by photosynthesis. Fungi are heterotrophs that digest food externally and absorb nutrient molecules. There are also many structural differences; for example, the threadlike fungal mycelium is different from the plant body, and their cell walls are made of different substances. Plants evolved from green algae, which belong to the protist supergroup Archaeplastida; the ancestor of fungi was in the protist supergroup Unikonta. Molecular evidence indicates that fungi are more closely related to animals than to plants.

13. Fungi disperse their offspring as spores. Because the truffle's reproductive body is underground, its spores are not carried by wind like those of most other fungi. When a truffle is consumed by an animal, the spores pass through the animal's digestive tract and are later deposited at a distance from the parent fungus. The fleshy fruits of many angiosperms are dispersed by a similar mechanism.

14. Moss gametophytes, the dominant stage in the moss life cycle, are haploid plants. The diploid (sporophyte) generation is dominant in most other plants. Recessive mutations are not expressed in a diploid organism unless both homologous chromosomes carry the mutation. In haploid organisms, recessive mutations are apparent in the phenotype of the organism because haploid organisms have only one set of chromosomes. Some factors to consider in designing your experiment: What are the advantages and disadvantages of performing the experiment in the laboratory? In the field? What variables would be important to control? How many potted plants should you use? At what distances from the radiation source should you place them? What would serve as a control group for the experiment? What age of plants should you use?

15. Two possible hypotheses are (1) the lineage that led to present-day mosses diverged before plants and fungi established mycorrhizal relationships and (2) early mosses had mycorrhizae but lost them later in evolution.

Chapter 18

1. Sponges: sessile, saclike body with pores, suspension feeder; sponges
Cnidarians: radial symmetry, gastrovascular cavity, cnidocytes, polyp or medusa body form; hydras, sea anemones, jellies, corals
Flatworms: bilateral symmetry, gastrovascular cavity, no body cavity; free-living planarians, flukes, tapeworms
Nematodes: body cavity, covered with cuticle, complete digestive tract, ubiquitous, free-living and parasitic; roundworms, heartworms, hookworms, trichinosis worms
Molluscs: muscular foot, mantle, visceral mass, circulatory system, many with shells, radula in some; snails and slugs, bivalves, cephalopods (squids and octopuses)
Annelids: segmented worms, closed circulatory system, many organs repeated in each segment; earthworms, polychaetes, leeches
Arthropods: exoskeleton, jointed appendages, segmentation, open circulatory system; chelicerates (spiders), crustaceans (lobsters, crabs), millipedes and centipedes, insects
Echinoderms: radial symmetry as adult, water vascular system with tube feet, endoskeleton, spiny skin; sea stars, sea urchins
Chordates: (1) notochord, (2) dorsal, hollow nerve cord, (3) pharyngeal slits, (4) post-anal tail; lancelets, tunicates, hagfish, and all the vertebrates (lampreys, sharks, ray-finned fishes, lobe-fins, amphibians, reptiles (including birds)), mammals

2. a. Two tissue layers: cnidaria
b. Protostomes: flatworms, molluscs, annelids, nematodes, arthropods
c. Deuterostomes: echinoderms, chordates

3. c **4.** d **5.** a (The invertebrates include all animals except the vertebrates.) **6.** c **7.** c **8.** i **9.** f **10.** b **11.** c **12.** a **13.** d **14.** h **15.** e **16.** g

17. The gastrovascular cavity of a flatworm is an incomplete digestive tract; the worm takes in food and expels waste through the same opening. An earthworm has a complete digestive tract; food travels one way, and different areas are specialized for different functions. The flatworm's body is solid and unsegmented. The earthworm has a fluid-filled body cavity, allowing its internal organs to grow and move independently of its outer body wall. Fluid in the body cavity cushions internal organs, acts as a skeleton, and aids circulation. Segmentation of the earthworm, including its body cavity, allows for greater flexibility and mobility.

18. Cnidarians and most adult echinoderms are radially symmetric, while most other animals, such as arthropods and chordates, are bilaterally symmetric. Most radially symmetric animals stay in one spot or float passively. Most bilateral animals are more active and move headfirst through their environment.

19. For example, the legs of a horseshoe crab are used for walking, while the antennae of a grasshopper have a sensory function. Some appendages on the abdomen of a lobster are used for swimming, while the scorpion catches prey with its pincers. (Note that the scorpion stinger and insect wings are not considered jointed appendages.)

20. Important characteristics include symmetry, the presence and type of body cavity, segmentation, type of digestive tract, type of skeleton, and appendages.

21. Arthropods / Velvet worms

When the lineages that became arthropods and velvet worms diverged from a common ancestor (indicated by arrow), they began with the same set of homeotic genes. The appendages on the body segments of velvet worms are identical, which is presumably the ancestral condition. If the arthropod body plan resulted from new genes that originated in the arthropod lineage, then velvet worms would lack those genes. Results showing that arthropods possess homeotic genes in addition to those found in velvet worms would support the hypothesis.

Chapter 19

1. a. Old World monkeys; b. gibbons; c. orangutans; d. gorillas; e. chimpanzees. All are anthropoids; gibbons, orangutans, gorillas, chimpanzees, and humans are apes.
2. a. head; b. vertebral column; c. jaws; d. lungs or lung derivatives; e. lobed fins; f. legs; g. amniotic egg; h. milk
3. c **4.** c **5.** b **6.** b **7.** b **8.** b **9.** a
10. Amphibians have four limbs adapted for locomotion on land, a skeletal structure that supports the body in a nonbuoyant medium, and lungs. However, most amphibians are tied to water because they obtain some of their oxygen through thin, moist skin and they require water for fertilization and development. Reptiles are completely adapted to life on land. They have amniotic eggs that contain food and water for the developing embryo and a shell to protect it from dehydration. Reptiles are covered by waterproof scales that enable them to resist dehydration (more efficient lungs eliminate the need for gas exchange through the skin).
11. Fossil evidence supports the evolution of birds from a small, bipedal, feathered dinosaur, which was probably endothermic. The last common ancestor that birds and mammals shared was the ancestral amniote. The four-chambered hearts of birds and mammals must have evolved independently.
12. Several primate characteristics make it easy for us to make and use tools: mobile digits, opposable fingers and thumb, and great sensitivity of touch. Primates also have forward-facing eyes, which enhances depth perception and eye-hand coordination, and a relatively large brain.
13. UV radiation is most intense in tropical regions and decreases farther north. Skin pigmentation is darkest in people indigenous to tropical regions and much lighter in northern latitudes. Scientists hypothesize that depigmentation was an adaptation to permit sufficient exposure to UV radiation, which catalyzes the production of vitamin D, a vitamin that permits the calcium absorption needed for both maternal and fetal bones. Dark pigmentation is hypothesized to protect against degradation of folate, a vitamin essential to normal embryonic development.
14. The paleontologists who discovered *Tiktaalik* hypothesized the existence of transitional forms between fishlike tetrapods such as *Panderichthys* and tetrapod-like fish such as *Acanthostega.* From the available evidence, they knew the time periods when fishlike tetrapods and tetrapod-like fish lived. From the rocks in which the fossils had been found, they knew the geographic region and the type of habitat these creatures occupied. With this knowledge, they predicted the type of rock formation where transitional fossils might be found.
15. Our intelligence and culture—accumulated and transmitted knowledge, beliefs, arts, and products—have enabled us to overcome our physical limitations and alter the environment to fit our needs and desires.
16. Most anthropologists think that humans and chimpanzees diverged from a common ancestor 5–7 million years ago. Primate fossils 4–8 million years old might help us understand how the human lineage first evolved.
17. The brain volume to body mass relationship of *H. floresiensis* is most similar to that of *Australopithecus afarensis*. If *H. floresiensis* were a dwarf form of *H. erectus*, we would expect the relationship to be roughly similar. This finding supports the hypothesis that the ancestor of *H. floresiensis* was an earlier hominin rather than *H. erectus.* (Keep in mind, however, that the available evidence is rather limited and includes only one skull.)

Chapter 20

1. a. epithelial tissue; b. connective tissue; c. smooth muscle tissue; d connective tissue; e. epithelial tissue

   The structure of the specialized cells in each type of tissue fits their function. For example, columnar epithelial cells are specialized for absorption and secretion; the fibers and cells of the connective tissue provide support and connect the tissues. The hierarchy from cell to tissue to organ is evident in this diagram. The functional properties of a tissue or organ emerge from the structural organization and coordination of its component parts. The many projections of the lining of the small intestine greatly increase the surface area for absorption of nutrients.
2. False. Each cell is bathed in interstitial fluid; blood remains inside the blood vessels.
3. d
4. a
5. Stratified squamous epithelium consists of many cell layers. The outer cells are flattened, filled with the protein keratin, and dead, providing a protective, waterproof covering for the body. Neurons are cells with long extensions that conduct signals to other cells, making multiple connections in the brain. Simple squamous epithelium is a single, thin layer of cells that allows for diffusion of gases across the lining of the lung. Bone cells are surrounded by a matrix that consists of fibers and mineral salts, forming a hard protective covering around the brain.
6. Extensive exchange surfaces are often located within the body. The surfaces of the intestine, urinary system, and lungs are highly folded and divided, increasing their surface area for exchange. These surfaces interface with many blood capillaries. Not all animals have such extensive exchange surfaces. Animals with small, simple bodies or thin, flat bodies have a greater surface-to-volume ratio, and their cells are closer to the surface, enabling direct exchange between cells and the outside environment.
7. c (Expelling salt opposes the increase in blood salt concentration, thereby maintaining a constant internal environment.)
8. d
9. The words "perfectly designed" seem to imply that evolution is goal-oriented, but it is not. Structures are adapted from ancestral species and are not "perfect" forms. Take, for example, the laryngeal nerve in giraffes. Because the giraffe nerve is an adaptation from an earlier fish ancestor, the nerve travels a length of 15 feet instead of taking a much more direct route of 1 foot.
10. Having some people use the cream and others not use it is an example of a controlled experiment, which is a necessary

component of a well-designed study. However, since acne can clear up in certain individuals over time, possibly by face washing alone, it would be better to have each participant be their own control (one side of their face be the treated side and the other side be the control). Collecting data by asking participants to rate their own acne severity is biased, because in this case, participants will know if they are using the cream or not. Participants could be "blind" if all were given a cream to use, with only half of the participants receiving a cream with active medicine. Collecting data by counting pimples before, during, and after the study would be unbiased if done by a scientist who was also "blind" to which participants received which treatment. (When both participants and scientists are "blind," it is termed a double-blind study.)

11. Because the skin and blood vessels close to the skin's surface would be rapidly cooled by the ice, a likely hypothesis is that cold water will help cool you more quickly than simply waiting. Because the question considers the rate of cooling, we would need to test the hypothesis by measuring body temperature after a run over time (say every few minutes) and comparing the use of ice water to sitting and waiting (control).

Chapter 21

1. a. oral cavity—ingests and chews food; b. salivary glands—produce saliva; c. liver—produces bile and processes nutrient-laden blood from intestines; d. gallbladder—stores bile; e. pancreas—produces digestive enzymes and bicarbonate; f. rectum—stores feces before elimination; g. pharynx—site of openings into esophagus and trachea; h. esophagus—transports bolus to stomach by peristalsis; i. stomach—stores food, mixes food with acid, begins digestion of proteins; j. small intestine—digestion and absorption; k. large intestine—absorbs water, compacts feces; l. anus—eliminates feces

2. a. fuel, chemical energy; b. raw materials, monomers; c. essential nutrients; d. overnutrition or obesity; e. vitamins and minerals; f. essential amino acids; g. malnutrition

3. d **4.** b **5.** a **6.** b **7.** d

8. You ingest the sandwich one bite at a time. In the oral cavity, chewing begins mechanical digestion, and salivary amylase action on starch begins chemical digestion. When you swallow, food passes through the pharynx and esophagus to the stomach. Mechanical and chemical digestion continues in the stomach, where HCl in gastric juice breaks apart food cells and pepsin begins protein digestion. In the small intestine, enzymes from the pancreas and intestinal wall break down starch, protein, and nucleic acids to monomers. Bile from the liver and gallbladder emulsifies fat droplets for attack by enzymes. Most nutrients are absorbed into the bloodstream through the villi of the small intestine. Fats travel through lymph vessels. In the large intestine, absorption of water is completed, and undigested material and intestinal bacteria are compacted into feces, which are eliminated through the anus.

9. a. 58% (110/190)
   b. Based on a 2,000-Calorie diet, this product supplies about 9.5% of daily Calories, and it supplies 10% of vitamin A and calcium. If all food consumed supplied a similar quantity, the daily requirement for these two nutrients would be met.
   c. The 8 g of saturated fat in this product represents 40% of the daily value. Thus, the daily value must be 20 g ($8/0.4 = 20$). This represents 180 Calories from saturated fat per day.

10. Our craving for fatty foods may have evolved from the feast-and-famine existence of our ancestors. Natural selection may have favored individuals who gorged on and stored high-energy molecules, as they were more likely to survive famines.

11. Sodas, chips, cookies, and candy provide many calories (high energy) but few vitamins, minerals, proteins, or other nutrients. Unprocessed, fresh foods such as fruits and vegetables are considered nutrient dense; they provide substantial amounts of vitamins, minerals, and other nutrients and relatively few calories.

12. Some issues and questions to consider: What are the roles of family, school, advertising, media, and government in providing nutritional information? How might the available information be improved? What types of scientific studies form the foundation of various nutritional claims?

13. Some issues and questions to consider: In wealthy countries, what are the factors that make it difficult for some people to get enough food? In your community, what types of help exist to feed hungry people? Think of two recent food crises in other countries and what caused them. Did other countries or international organizations provide aid? Which ones, and how did they help? Did that aid address the underlying causes of malnutrition and starvation in the stricken area or only provide temporary relief? How might that aid be changed to offer more permanent solutions to food shortages?

14. Freckles and red hair are correlated: When one occurs, the other also tends to occur. But these two traits have no causation: Freckles do not cause red hair, nor does red hair cause freckles. In terms of human nutrition studies, it can be difficult to correlate a particular dietary factor (the amount of fat, for example) with a particular health measure (high blood pressure, for example) because many other factors may be involved. It is thus hard to conduct strictly controlled human dietary studies.

Chapter 22

1. a. respiratory surface; b. circulatory system; c. lungs; d. hemoglobin; e. cellular respiration; f. negative pressure breathing; g. O_2

2. a. nasal cavity; b. pharynx; c. larynx; d. trachea; e. right lung; f. bronchus; g. bronchiole; h. diaphragm

3. c **4.** a **5.** d **6.** a **7.** d **8.** d **9.** d

10. Advantages of breathing air: It has a higher concentration of O_2 than water and is easier to move over the respiratory surface. Disadvantage of breathing air: Living cells on the respiratory surface must remain moist, but breathing air dries out this surface.

11. Nasal cavity, pharynx, larynx, trachea, bronchus, bronchiole, alveolus, through wall of alveolus into blood vessel, blood plasma, into red blood cell, attaches to hemoglobin, carried by blood through heart, blood vessel in muscle, dropped off by hemoglobin, out of red blood cell, into blood plasma, through capillary wall, through interstitial fluid, and into muscle cell

12. Both these effects of carbon monoxide interfere with cellular respiration and the production of ATP. By binding more tightly to hemoglobin, CO would decrease the amount of O_2 picked up in the lungs and delivered to body cells. Without sufficient O_2 to act as the final electron acceptor, cellular respiration would slow. And by blocking electron flow in the electron transport chain, cellular respiration and ATP production would cease. Without ATP, cellular work stops and cells and organisms die.

13. Llama hemoglobin has a higher affinity for O_2 than does human hemoglobin. The dissociation curve shows that its hemoglobin becomes saturated with O_2 at the lower P_{O_2} of the high altitudes to which llamas are adapted. At that P_{O_2}, human hemoglobin is only 80% saturated.

14. The athlete's body would respond to training at high altitudes or sleeping in an artificial atmosphere with lower P_{O_2} by producing more red blood cells. Thus, the athlete's blood would carry more O_2, and this increase in aerobic capacity may improve endurance and performance.

15. Insects have a tracheal system for gas exchange. To provide O_2 to all the body cells in such a huge moth, the tracheal tubes would have to be wider (to provide enough ventilation across longer distances) and very extensive (to service large flight muscles and other tissues), thus presenting problems of water loss and increased weight. Both the tracheal system and the weight of the exoskeleton limit the size of insects.

16. The CDC website lists several conclusions along with links to references. You should find additional references, as well. Factors to consider include size and duration of studies, whether human studies are prospective or retrospective, what chemical compounds are inhaled during hookah smoking and whether they have carcinogenic or other harmful effects, and how passing through water affects hookah smoke. Also consider the smoking habits of the participants studied, for example, how frequently they smoked and the amount of tobacco used at each session.

Chapter 23

1. a. capillaries of head, chest, and arms; b. aorta; c. pulmonary artery; d. capillaries of left lung; e. pulmonary vein; f. left atrium; g. left ventricle; h. aorta; i. capillaries of abdominal region and legs; j. inferior vena cava; k. right ventricle; l. right atrium; m. pulmonary vein; n. capillaries of right lung; o. pulmonary artery; p. superior vena cava

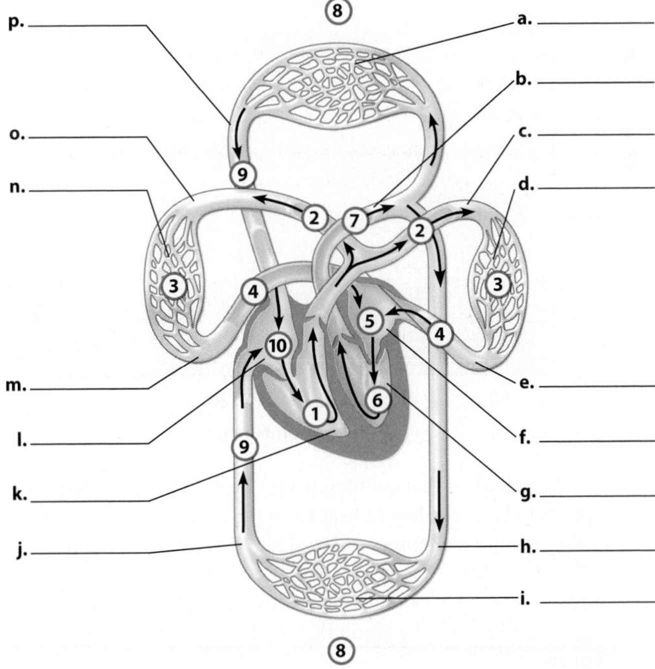

p.
o.
n.
m.
l.
k.
j.
a.
b.
c.
d.
e.
f.
g.
h.
i.
8
9
2
7
2
3
4
3
5
4
10
1
6
9
8

2. b **3.** c (The second sound is the closing of the semilunar valves as the ventricles relax.) **4.** b **5.** b **6.** a **7.** d **8.** b **9.** a

10. Pulmonary vein, left atrium, left ventricle, aorta, artery, arteriole, body tissue capillary bed, venule, vein, vena cava, right atrium, right ventricle, pulmonary artery, capillary bed in lung, pulmonary vein

11. Capillaries are very numerous, producing a large surface area for exchange close to body cells. The capillary wall is only one epithelial cell thick. Pores in the wall and clefts between epithelial cells allow fluid with small solutes to move out of the capillary.

12. a. Plasma (the straw-colored fluid) would contain water, inorganic salts (ions such as sodium, potassium, calcium, magnesium, chloride, and bicarbonate), plasma proteins such as fibrinogen and immunoglobulins (antibodies), and substances transported by blood, such as nutrients (for example, glucose, amino acids, vitamins), waste products of metabolism, respiratory gases (O_2 and CO_2), and hormones.
    b. The red portion would contain erythrocytes (red blood cells), leukocytes (white blood cells—basophils, eosinophils, neutrophils, lymphocytes, and monocytes), and platelets.

13. Oxygen content is reduced as oxygen-poor blood returning to the right ventricle from the systemic circuit mixes with oxygen-rich blood of the left ventricle.

14. Proteins are important solutes in blood, accounting for much of the osmotic pressure that counters the flow of fluid out of a capillary. If protein concentration is reduced, the inward pull of osmotic pressure will fail to balance the outward push of blood pressure, and more fluid will leave the capillary and accumulate in the tissues.

15. Points to consider in your essay may include the prevalence of heart disease; the personal and societal costs of heart disease (for example, costs of medical treatment; economic losses due to disability and death); research costs; who benefits from finding effective methods of prevention or treatment (for example, profit; common good); and the likelihood of producing valid scientific results.

16. With a three-chambered heart, there is some mixing of oxygen-rich blood returning from the lungs with oxygen-poor blood returning from the systemic circulation. Thus, the blood of a dinosaur might not have supplied enough O_2 to support the higher metabolism and strong cardiac muscle contractions needed to generate such a high systolic blood pressure. Also, with a single ventricle pumping simultaneously to both pulmonary and systemic circuits, the blood pumped to the lungs would be at such a high pressure that it would damage the lungs.

Chapter 24

1. a. innate immunity; b. adaptive immunity; c. B cells; d. Helper T cells; e. cytotoxic T cells; f. antibodies; g. infected body cells

2. d **3.** b **4.** b **5.** a

6. HIV is transmitted in blood and semen. It enters the body through slight wounds during sexual contact or via needles contaminated with infected blood. The most effective way to avoid HIV transmission is to prevent contact with body fluids by practicing safe sex and avoiding intravenous drugs. AIDS is deadly because it infects helper T cells, crippling both the humoral and cell-mediated immune responses and leaving the body vulnerable to other infections.

7. Inflammation is triggered by tissue injury. Mast cells within injured tissue release histamine and other chemicals, which cause nearby blood vessels to dilate and become leakier. Blood plasma leaves vessels, and neutrophils are attracted to the site of injury. An increase in blood flow, fluid accumulation, and increased cell population cause redness, heat, and swelling. Inflammation disinfects and cleans the area and curtails the spread of infections from the injured area. Inflammation is considered part of the innate immune response because similar defenses are presented in response to any infection.

8. d

9. c

10. Because flu viruses are constantly mutating, we need a flu vaccine every year to keep up with the evolution of flu viruses. Additionally, the immunity brought about by vaccination wanes over time (antibody levels decrease), so getting vaccinated every year provides the best protection.

11. The increase in pertussis cases in older children suggests that immunity declines with age. A solution would be to give children a "booster shot," which is another dose of the vaccine. In fact, the number of cases is lower in 13-year-olds—the age most children receive the booster shot. Thus, one solution is to give the booster earlier than age 13.

12. One hypothesis is that your roommate's previous bee stings caused her to become sensitized to the allergens in bee venom. During sensitization, antibodies to allergens attach to receptor proteins on mast cells. In the sensitization stage, she would not have experienced allergy symptoms. When she was exposed to the bee venom again at a later time, the bee venom allergens bound to the mast cells, which triggered her allergic reaction.

13. A2 A1 A2 Z Y Y Z Y Y Z Z A2 A1

14. There is no correct answer to this question. Some issues and questions to consider: Possible directions include the idea that if the donor felt strongly about the process, then his or her wishes should be respected. The opposite direction would be that the next of kin should be able to approve or deny the procedure. Other considerations may be appropriate, including religious beliefs.

15. Some issues and questions to consider: How much do people in various nations stand to gain by the development of new drugs, in terms of both lives saved and profits made? How can oversight be used to ensure that drug companies are acting in the best interests of all their patients and not purely for profit? Can studies be modified so as to maximize the potential benefits to HIV-infected people while minimizing the risks to study participants? Or is such a trade-off impossible? Should studies on humans be banned altogether?

Chapter 25

1. a. thermoregulation; b. ectotherm; c. behavioral; d. water; e. solutes; f. filtration; g. reabsorption; h. secretion; i. excretion; j. ammonia; k. urea; l. uric acid; m. water; n. land

2. a. filtration: water, NaCl, HCO_3^-, H^+, urea, nutrients such as glucose and amino acids, some drugs; b. reabsorption: nutrients, NaCl, water, HCO_3^-, urea; c. secretion: some drugs and toxins, H^+, K^+; d. excretion: urine containing water, urea, and excess ions

3. b 4. a 5. d 6. c 7. b 8. c 9. c 10. b 11. d 12. b 13. a 14. b 15. c

16. In salt water, the fish loses water by osmosis. It drinks salt in water and disposes of salts through its gills. Its kidneys conserve water and excrete excess ions. In fresh water, the fish gains water by osmosis. Its kidneys excrete a lot of dilute urine. Its gills take up salt, and some ions are ingested with food.

17. Yes. Ectotherms that live in very stable environments, such as tropical seas or deep oceans, have stable body temperatures. And terrestrial ectotherms can maintain relatively stable temperatures by behavioral means.

18. a. An endotherm would produce more nitrogenous waste because it must eat more food to maintain its higher metabolic rate.
    b. A carnivore would produce more nitrogenous waste because it eats more protein and thus produces more breakdown products of protein digestion—nitrogenous wastes.

19. Diuretics increase the amount of water in urine output. An athlete who misuses a diuretic could rapidly lose weight without losing tissue mass. In some sports, like wrestling, rapid weight loss just before a competition might allow an athlete to move to a lower weight class to potentially compete against someone of equal weight but with less muscle mass. Besides being unfair, the misuse of diuretics can lead to severe dehydration and even death.

20. You could take the reptile back to the laboratory and measure its body temperature under different ambient temperatures.

21. A scatter plot, as shown below, is ideal for graphing these data because both variables represent continuous numerical variables. The relationship between the two variables can be examined. (A bar chart, in which one of the variables represents different categories rather than continuous numerical values, would be less appropriate.) Although these data represent only a small sample, we see a trend that suggests penguins huddle for longer durations at colder temperatures compared to warmer temperatures.

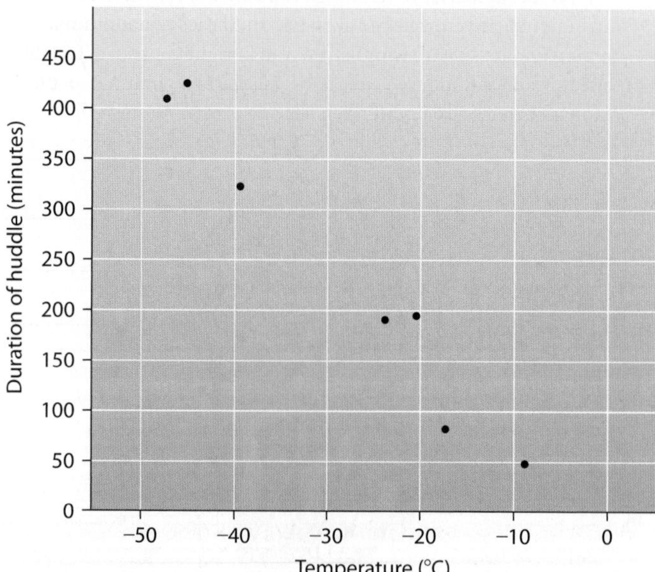
Duration of huddle (minutes)
450
400
350
300
250
200
150
100
50
0
−50
−40
−30
−20
−10
0
Temperature (°C)

22. One hypothesis is that countercurrent heat exchange in dolphin flippers reduces the loss of heat from the body. Your experiment could measure the temperatures of blood at different locations in the circulatory system. You would expect the temperature of blood flowing back to the body from the flippers to be only slightly cooler than the blood flowing from the body to the flippers.

23. First consider the observation stated in the question: Birds and mammals at higher latitudes are larger. Next, gather other information relevant to the observation: (1) birds and mammals are endotherms; (2) animals living at higher latitudes would need to survive colder climates than those at lower latitudes; and (3) the ratio of surface area to volume is lower in a large animal compared to a small one. (For a review of the relationship between surface area to volume see Modules 4.2 and 20.13.) A possible hypothesis is that larger animals radiate less metabolic heat than

smaller animals under the same cold conditions, making them more likely to survive and reproduce.

24. Some possible pros: A financial incentive might motivate more healthy donors to trade. More kidneys could become available, leading to less waiting time for recipients (saving more lives) and less financial burden on the health-care system compared to long-term dialysis. Some possible cons: A donor who is poor might put his or her own health at risk out of desperation. Donors and brokers would be motivated financially, and this could mean less healthy organs being traded (putting the recipients at risk). Crimes to obtain kidneys and coercion of donors could increase. Regulation to protect the donor and recipient would be complex.

25.

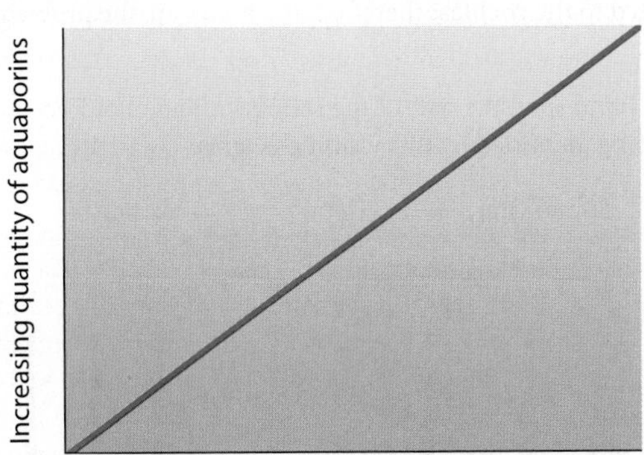
Increasing quantity of aquaporins
Increasing ADH levels

In times of dehydration, ADH levels rise and cause cells in the collecting ducts to reabsorb more water from the filtrate into the interstitial fluid of the tissues. Thus, with rising ADH, collecting duct cells would respond with more aquaporins at their surface. Urine would have less water. Once hydrated, ADH levels fall, the quantity of aquaporins decrease, and more water remains in the filtrate to be excreted in the urine.

Chapter 26

1. a. nervous system; b. hormones; c. receptors; d. inside target cells; e. endocrine glands; f. anterior pituitary gland; g. adrenal medulla
2. d (Thyroxine is produced by the thyroid; prolactin promotes milk production; androgens are produced by the testes.)
3. b (Negative feedback: When thyroid hormone increases, it inhibits TRH and TSH, which reduces thyroid hormone secretion.)
4. The hypothalamus secretes releasing hormones and inhibiting hormones, which are carried by the blood to the anterior pituitary. In response to these signals from the hypothalamus, the anterior pituitary increases or decreases its secretion of a variety of hormones that directly affect body activities or influence other glands. Neurosecretory cells that extend from the hypothalamus into the posterior pituitary secrete hormones that are stored in the posterior pituitary until they are released into the blood.
5. A single lipid-soluble hormone can affect two target cells differently if the types of receptor proteins inside the two cells are structurally and functionally diverse. A water-soluble hormone can lead to dissimilar effects inside two different cells when the receptors on the cells' plasma membrane and the proteins of the signal transduction pathway vary between cells. Nontarget cells do not have the proper receptors to respond to a hormone at all.
6. Both individuals have a rise in glucose blood level after glucose ingestion. The diabetic begins with a higher blood glucose concentration that peaks at approximately three times that of the normal individual several hours after ingestion.
7. Yes. TSH stimulates the thyroid to grow and produce thyroid hormones T_3 and T_4. An enlarged thyroid, or goiter, is likely with excess TSH.
8. a and c. Glucagon and glucocorticoids both function to provide the body with cellular fuel. Glucagon breaks down glycogen stores in the liver, and glucocorticoids promote the synthesis of glucose from proteins and fats.
9. a. No. Blood sugar level drops too low. Diabetes would tend to make the blood sugar level too high after a meal.
   b. No. Insulin is working, as seen by the homeostatic blood sugar response to feeding.
   c. Correct. Exercise and fasting lower blood sugar, but without glucagon receptors, liver cells do not receive the glucagon signal to break glycogen down into glucose. The cells cannot mobilize any sugar reserves, and blood sugar level drops. Insulin (which lowers blood sugar) has no effect.
   d. No. If this were true, blood sugar level would increase too much after a meal.
10. The cause can be environmental or genetic. Endocrine disruptors in the environment can alter the normal ratio of sex hormones. In the study with frogs, endocrine disruptors caused low testosterone levels and a complete sex reversal for some genetic males. Genetic defects can alter normal sex hormone ratios, too. Males with androgen insensitivity syndrome are phenotypically more female than male as a result of defective testosterone signaling.

Chapter 27

1. A. FSH; B. estrogen; C. LH; D. progesterone; P. menstruation; Q. growth of follicle; R. ovulation; S. development of corpus luteum

   If pregnancy occurs, the embryo produces human chorionic gonadotrophin (hCG), which maintains the corpus luteum, keeping levels of estrogen and progesterone high.
2. e **3.** g **4.** d **5.** h **6.** f **7.** a **8.** b **9.** c
10. c **11.** c (The outer layer in a gastrula is the ectoderm; of the choices given, only the brain develops from ectoderm.) **12.** a **13.** c
14. The extraembryonic membranes provide a moist environment for the embryos of terrestrial vertebrates and enable the embryos to absorb food and oxygen and dispose of wastes. Such membranes are not needed when an embryo is surrounded by water, as are those of fishes and amphibians.
15. Both produce haploid gametes. Spermatogenesis produces four small sperm; oogenesis produces one large egg. In humans, the ovary contains all the primary oocytes at birth, while testes can keep making primary spermatocytes throughout life. Oogenesis is not complete until fertilization, but sperm mature without eggs.
16. The nerve cells may follow chemical trails to the muscle cells and identify and attach to them by means of specific surface proteins.
17. The researcher might find out whether chemicals from the notochord stimulate the nearby ectoderm to become the neural tube, a process called induction. Transplanted notochord tissue might cause ectoderm anywhere in the embryo to become neural tissue. Possible control experiment: Transplant non-notochord tissue under the ectoderm of the belly area.
18. Some issues and questions to consider: What characteristics might parents like to select for? If parents had the right to choose embryos based on these characteristics, what are some of the possible benefits? What are potential pitfalls? Could an imbalance of the population result?

19. You might be able to design an observational study, like the one used to test effectiveness of condoms (see Module 27.7). You could observe two at-risk populations, one who chose to use the vaccine, and one who didn't. Comparing rates of new infection among the two populations may provide insight into the effectiveness of the vaccine.

Chapter 28

1. a. sensory receptor; b. sensory neuron; c. synapse; d. spinal cord; e. interneuron; f. motor neuron; g. effector cells; h. CNS; i. PNS
2. a. corpus callosum; b. amygdala . . . limbic; c. cerebral cortex . . . cerebrum; d. cerebellum; e. reticular . . . hypothalamus
3. At the point where an action potential is triggered, sodium ions rush into the neuron, depolarizing the membrane. Potassium gates then open, and the membrane repolarizes. Meanwhile, the sodium ions diffuse laterally and cause sodium gates to open in the adjacent part of the membrane, triggering another action potential. The moving wave of action potentials, each triggering the next, is a moving nerve signal. Behind the action potential, sodium gates are temporarily inactivated, so the action potential can only go forward. At a synapse, the transmitting cell releases a chemical neurotransmitter, which binds to receptors on the receiving cell and may trigger a nerve signal in the receiving cell.
4. b
5. b
6. a and c, because they would prevent action potentials from occurring; b could actually increase the generation of action potentials.
7. The controls are not given any drug, not even a placebo sugar pill. Not administering placebos with the control group means the placebo effect cannot be accounted for with the treatment group. In many SSRI studies, people believe the pill they receive helps them, even when it is only a placebo. Therefore, drugs must be compared against placebos and the drugs must be more effective than placebos. Additionally, placebos make the study double-blinded—so that the patients don't know if they are receiving the treatment or not and so that the researchers are unaware of the assignments as well. Evaluating one's own mood leads to much variation among participants. Researchers could instead design surveys to help standardize the reporting (such as number of hours slept, mood on a quantitative scale, motivation, and so forth).
8. The results show the cumulative effect of all incoming signals on neuron D. Comparing experiments 1 and 2, we see that the more nerve signals D receives from C, the more it sends; C is excitatory. Because neuron A is not varied here, its action is unknown; it may be either excitatory or mildly inhibitory. Comparing experiments 2 and 3, we see that neuron B must release a strongly inhibitory neurotransmitter, because when B is transmitting, D stops.
9. Some issues and questions to consider: Embryonic stem cells are retrieved from early-stage embryos formed *in vitro* (outside the body). Some people might be against the use of embryonic stem cells for any disorder. This may be related to religious or moral beliefs. Some people may not be aware of the source of these stem cells and may be against their use because they incorrectly believe that the cells come from elective abortions. Other people may agree with the use of stem cell research because of the potential to cure diseases that are currently fatal. Some people who may have a neutral opinion on the issue might be swayed by the thought of a loved one who might be helped by stem cell therapy.

Chapter 29

1. a. mechanoreceptors; b. chemoreceptors; c. light; d. electricity; e. magnetism; f. hair cells; g. photoreceptors
2. d **3.** b **4.** a **5.** a
6. Louder sounds create pressure waves with greater amplitude, moving hair cells more and generating a greater frequency of action potentials. Different pitches affect different regions of the basilar membrane, stimulating different sensory neurons that transmit action potentials to different parts of the brain.
7. c (He could hear the tuning fork against his skull, so the cochlea, nerve, and brain are okay. Apparently, sounds are not being transmitted to the cochlea; therefore, the bones are the problem.)
8. a
9. The blind spot, the part of the retina with no photoreceptor cells, is being identified by this vision challenge.
10.

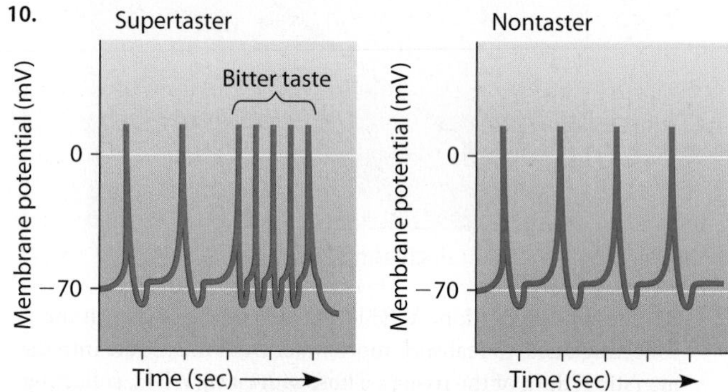

Supertaster
Bitter taste
Membrane potential (mV)
0
−70
Time (sec)
Nontaster
Membrane potential (mV)
0
−70
Time (sec)

11. You could cover the eyes of some turtles but not others (covering the photoreceptor cells). If turtles with light-blocking masks don't make it to the ocean as frequently as unmasked turtles, they probably use the light to guide them. You could make observations on nights when there is no moon. If turtles were less likely to find the ocean on moonless nights, then moonlight likely guides them. You could also use artificial moonlight. If turtles moved toward the light (and away from the water), this is evidence that turtles use moonlight in their orientation. (Indeed, many Florida beaches have "lights out" ordinances for beach-front residents during sea turtle nesting season to prevent disorientation of the hatchlings.)
12. Some issues and questions to consider: Assuming that the sound is loud enough to impair hearing, how long an exposure is necessary for this to occur? Does exposure have to occur all at once, or is damage cumulative? Who is responsible, concert promoters or listeners? Should there be regulations regarding sound exposure at concerts (as there are for job-related noise)? Are the young people who typically attend concerts sufficiently mature and aware to heed such warnings?

Chapter 30

1. a. skeleton; b. muscles; c. exoskeleton; d. sarcomeres; e. bone and cartilage
2. b **3.** d **4.** a (Water supports aquatic animals, reducing the effects of gravity.) **5.** c **6.** c **7.** d **8.** a **9.** a (Each neuron controls a smaller number of muscle fibers.) **10.** d
11. Advantages of an insect exoskeleton include strength, good protection for the body, flexibility at joints, and protection from water loss. The major disadvantage is that the exoskeleton must be shed periodically as the insect grows, leaving the insect temporarily weak and vulnerable.

12. The bird's wings are airfoils, with convex upper surfaces and flat or concave lower surfaces. As the wings beat, air passing over them travels farther than air beneath. Air molecules above the wings are more spread out, lowering pressure. Higher pressure beneath the wings pushes them up.

13. Calcium is needed for healthy bone development. Calcium strengthens bones and makes them less susceptible to stress fractures.

14. Action potentials from the brain travel down the spinal cord and along a motor neuron to the muscle. The neuron releases a neurotransmitter, which triggers action potentials in a muscle fiber membrane. These action potentials initiate the release of calcium ions from the ER of the cell. Calcium enables myosin heads of the thick filaments to bind with the actin of the thin filaments. ATP provides energy for the movement of myosin heads, which causes the thick and thin filaments to slide along one another, shortening the muscle fiber. The shortening of muscle fibers pulls on bones, bending the arm. If more motor units are activated, the contraction is stronger.

15. The fundamental vertebrate body plan includes an axial skeleton (skull, backbone, and rib cage) and an appendicular skeleton (bones of the appendages). Species vary in the numbers of vertebrae and the numbers of different types of vertebrae they possess. For example, pythons have no cervical vertebrae. Almost all mammals have seven cervical vertebrae but may have different numbers of other types. For example, human coccygeal vertebrae are small and fused together, but horses and other animals with long tails have many coccygeal vertebrae. Limb bones have been modified into a variety of appendages, such as wings, fins, and limbs. Snakes have no appendages.

16. Chemical A would work better, because acetylcholine triggers contraction. Blocking it would prevent contraction. Chemical B would actually increase contraction, because Ca^{2+} allows contraction to occur.

17. Circular muscles in the earthworm body wall decrease the diameter of each segment, squeezing internal fluid and lengthening the segment. Longitudinal muscles shorten and thicken each segment. Different parts of the earthworm can lengthen while other parts shorten, producing a crawling motion. The whole roundworm body moves at once because of a lack of segmentation. The body can only shorten or bend, not lengthen, because of a lack of circular muscles. Roundworms simply thrash from side to side.

18. The binding of calcium ions causes the regulatory protein tropomyosin to move out of the way, enabling myosin heads to bind to actin. This results in muscle contraction. ATP causes the myosin heads of the thick filaments to detach from the thin filaments (Figure 30.9B, step 1). If there is no ATP present, the myosin heads remain attached to the thin filaments, and the muscle fiber remains fixed in position.

19. Some points to consider: (a) The study compared nonathletes to elite athletes, who are among the best in their field. Many people compete successfully without reaching this top rank. (b) The researchers studied athletes from sports that are either power events or endurance events. Some sports require a combination of characteristics. (c) The results did not provide a conclusive answer about the relationship between *ACTN3* genotype and endurance events. (d) Possession of a particular *ACTN3* genotype does not guarantee success in athletic endeavors. In addition to the fact that other genes are involved in physical prowess, athletic excellence requires appropriate training; psychological factors may also play a role.

Chapter 31

1. Here is one possible concept map:

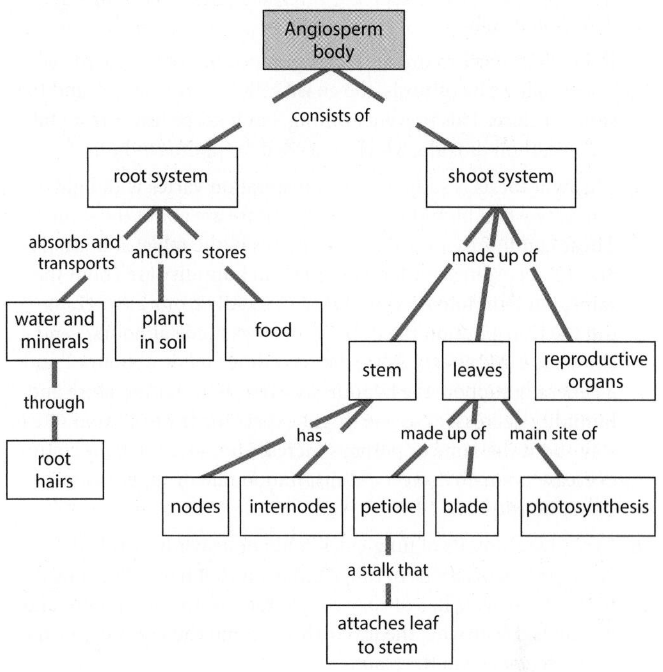

Angiosperm body
consists of
root system
shoot system
absorbs and transports
anchors
stores
water and minerals
plant in soil
food
through
root hairs
made up of
stem
leaves
reproductive organs
has
made up of
main site of
nodes
internodes
petiole
blade
photosynthesis
a stalk that
attaches leaf to stem

2. d 3. f 4. b 5. e 6. a 7. c 8. d

9. d (The vascular cambium forms to the outside of the primary xylem. The secondary phloem and primary phloem are outside the vascular cambium.)

10. d

11. Pollen is deposited on the stigma of a carpel, and a pollen tube grows to the ovary at the base of the carpel. Sperm travel down the pollen tube and fertilize egg cells in ovules. The ovules grow into seeds, and the ovary grows into the flesh of the fruit. As the seeds mature, the fruit ripens and falls (or is picked).

12. Celery stalk: leaf stalk (petiole); peanut: seed (ovule); strawberry: fruit (ripened ovary); lettuce: leaf blades; beet: root

13. Fragmentation of bulbs and sprouting from roots are examples of asexual reproduction. Asexual reproduction is less wasteful and costly than sexual reproduction and less hazardous for young plants. The primary disadvantage of asexual reproduction is that it produces genetically identical offspring, decreasing genetic variability that can help a species survive times of environmental change.

14. Modern methods of plant breeding and propagation have increased crop yields but have decreased genetic variability, so plants have become more vulnerable to epidemics. Primitive varieties of crop plants could contribute to gene banks and be used for breeding new strains.

15. You may wish to discuss the benefits and drawbacks of the different lines of evidence, including archeological (remnants of plants left at ancient sites), anatomical (examination of the differences in plant anatomy between wild and domestic varieties), microscopic (examination of microscopic plant structures to reveal which varieties were used at an ancient site), and genetic (differences in DNA sequences derived from plant remains, corresponding to the accumulation of mutations during domestication).

Chapter 32

1. a. roots; b. xylem; c. sugar source; d. phloem; e. transpiration; f. sugar sink

2. d 3. d 4. d 5. b

6. If the plant starts to dry out, K^+ is pumped out of the guard cells. Water follows by osmosis, the guard cells become flaccid, and the stomata close. This prevents wilting, but it keeps leaves from taking in carbon dioxide, which is needed for photosynthesis.

7. The hypothesis is supported if transpiration varies with light intensity when humidity and temperature are about the same. These conditions are seen at two places in the table; at hours 11 and 12, recordings for temperature and humidity are about the same, but light intensity increased markedly from 11 to 12, as did the transpiration rate. The recordings made at hours 3 and 4 show the same effects. Also, the recordings made at hours 1 and 2 generally support the hypothesis. Here, both temperature and humidity decreased, so you might expect the transpiration rate to stay about the same or perhaps increase because the temperature decrease is small; however, transpiration rate dropped, as did light intensity.

8. Excessive amounts of fungicides could destroy mycorrhizae, symbiotic associations of fungi and plant root hairs. The fungal filaments provide lots of surface area for absorption of water and nutrients. Destroying the mycorrhizae could cause a water or nutrient deficiency in the plant.

9. Hypothesis: The hydrogen ions in acid precipitation displace positively charged nutrient ions from negatively charged clay particles. Test: In the laboratory, place equal amounts and types of soil in separate filters. The pore size of the filter must not allow any undissolved soil particles to pass through. Spray (to simulate rain) soil samples in the filters with solutions of different pH (for example, pH 4, 5, 6, 7, 8). Determine the concentration of nutrient ions in the solutions. (The only variable in the solutions should be the hydrogen ion concentration. Ideally, the solutions would contain no dissolved nutrient ions.) Collect fluid that drips through soil samples and filters. Determine the hydrogen ion concentration and the nutrient ion concentration in each sample of fluid. Prediction: If the hypothesis is correct, the fluid collected from the soil samples exposed to pH lower than 5.6 (acid rain) will contain the highest concentration of positively charged nutrient ions.

10. Some issues and questions to consider: How were the farmers assigned or sold "rights" to the water? How is the price established when a farmer buys or sells water rights? Is there enough water for everyone who "owns" it? What kinds of crops are these farmers growing? What will the water be used for in the city? Are there other users with no rights, such as wildlife? Is any effort being made to curb urban growth and conserve water? Should millions of people be living in what is essentially a desert? What are the reasons for farming desert land?

11. It is difficult to conduct properly controlled experiments on growing crops. To design a properly controlled experiment, only one variable (in this case, whether the crop is organic or not) should change between experimental groups. In the real world, this may be impossible to achieve, as soil conditions, weather, and so on, vary from location to location.

Chapter 33

1. a. auxin; b. gibberellin; c. auxin; d. cytokinin; e. auxin; f. ethylene; g. gibberellin; h. abscisic acid

2. b 3. b 4. a 5. d 6. b 7. b 8. f 9. e 10. d 11. a 12. b 13. g 14. c 15. c

16. Fruits produce ethylene gas, which triggers the ripening and aging of the fruit. Ventilation prevents a buildup of ethylene and delays its effects.

17. The terminal bud produces auxins, which counter the effects of cytokinins from the roots and inhibit the growth of axillary buds. If the terminal bud is removed, the cytokinins predominate, and lateral growth occurs at the axillary buds.

18. The red wavelengths in the room's lights quickly convert the phytochrome in the chrysanthemums to the P_{fr} form, which inhibits flowering in a long-night plant. The chrysanthemums will not flower unless the security guard can set up some far-red lights. Exposure to a burst of far-red light would convert the phytochrome to the P_r form, allowing flowering to occur.

19. The biologist could remove leaves at different stages of being eaten to see how long it takes for changes to occur in nearby leaves. The "hormone" could be captured in an agar block, as in the phototropism experiments in Module 33.1, and applied to an undamaged plant. Another experiment would be to block "hormone" movement out of a damaged leaf or into a nearby leaf.

20. Some issues and questions to consider: Is the chemical safe for human consumption? What are its effects on the environment? Could its production produce impurities or wastes that might be harmful? What kinds of tests need to be done to demonstrate its safety? How much does it cost to make and use? Are the benefits worth the costs and risks? Is it worth using an artificial chemical on food simply to improve its appearance?

    A scientist could seek answers by studying the stability of the chemical in a variety of laboratory simulations of natural conditions. The toxicity of the chemical, the materials used to produce it, and its breakdown products could be determined in laboratory tests.

21. Charles and Francis Darwin were the first to demonstrate that some signal moves through the plant and controls growth. Boysen-Jensen demonstrated that this signal must be a diffusible chemical. Went named auxin and demonstrated that a higher concentration of this chemical accumulated on the dark side of the shoot. You could make an argument that any one of these was the "discoverer" of auxin, depending on how you define the moment of discovery.

Chapter 34

1. a. The shape of Earth results in uneven heating, such that the tropics are warm and polar regions are cold.
   b. The seasonal differences of winter and summer in temperate and polar regions are produced as Earth tips toward or away from the sun during its orbit around the sun.
   c. Intense solar radiation in the tropics evaporates moisture; warm air rises, cools, and drops its moisture as rain; air circulates, cools, and drops around 30°N and S, warming as it descends and evaporating moisture from land, which creates arid regions.

2. f 3. g 4. d 5. a 6. c 7. b 8. e 9. a 10. d 11. c 12. a 13. b 14. b 15. d

16. Tropical rain forests have a warm, moist climate, with favorable growing conditions year-round. The diverse plant growth provides various habitats for other organisms.

17. After identifying your biome by looking at the map in Figure 34.8, review Module 34.5 on climate and the module that describes your biome (Modules 34.9–34.17).

18. a. All three areas have plenty of sunlight and nutrients.
    b. The addition of nitrogen or phosphorus "fertilizes" the algae living in ponds and lakes, leading to explosive population growth. (These nutrients are typically in short supply in aquatic ecosystems.) The effects are harmful to the ecosystem. Algae cover the surface, reducing light penetration. When the algae die, bacterial decomposition of the large amount of biomass can deplete the oxygen available in the pond or lake, which may adversely affect the animal community.

19. a. desert; b. grassland; c. tropical forest; d. temperate forest; e. coniferous forest; f. arctic tundra. Areas of overlap have to do with seasonal variations in temperature and precipitation.

20. Through convergent evolution, these unrelated animals adapted in similar ways to similar environments—temperate grasslands and savanna.

21. The robin die-off appears to be correlated with the application of DDT, but this is weak evidence. Scientists would consider alternative explanations, for example, some kind of parasite or infectious disease; starvation due to a sudden lack of a critical food resource; or some other drastic change in the environment. To explore the correlation with DDT application, scientists would need to investigate factors such as the concentration of DDT present in earthworms, the number of earthworms typically consumed by robins, the concentrations of DDT in the tissues of living and dead robins, and the lethal dose of DDT for robins. Scientists would also determine whether the situation seen on the Michigan State campus, or circumstances similar to it, had been observed elsewhere. For example, they would try to find out whether robin mortality had been observed at other sites where DDT had been applied, as well at locations where DDT had *not* been applied. They would also look for published evidence of the effects of DDT on other bird species or other wildlife.

Chapter 35

1. a. genes; b. fixed action pattern (FAP); c. imprinting; d. spatial learning; e. associative learning; f. social learning

2. d 3. c 4. d 5. a

6. Main advantage: Flies do not live long. Innate behaviors can be performed the first time without learning, enabling flies to find food, mates, and so on without practice. Main disadvantage: Innate behaviors are rigid; flies cannot learn to adapt to specific situations.

7. In a stressful environment, for example, where predators are abundant, rats that behave cautiously are more likely to survive long enough to reproduce.

8. Courtship behaviors reduce aggression between potential mates and confirm their species, sex, and physical condition. Environmental changes such as rainfall, temperature, and day length probably lead frogs to start calling, so these would be the proximate causes. The ultimate cause relates to evolution. Fitness (reproductive success) is enhanced for frogs that engage in courtship behaviors.

9. a. Yes. The experimenter found that 5 m provided the most food for the least energy because the total flight height (number of drops × height per drop) was the lowest. Crows appear to be using an optimal foraging strategy.
    b. An experiment could measure the average drop height for juvenile and adult birds, or it could trace individual birds during a time span from juvenile to adult and see if their drop height changed.

10. One likely hypothesis is that the helper is closely related to one or both of the birds in the mated pair. Because closely related birds share relatively many genes, the helper bird is indirectly enhancing its own fitness by helping its relatives raise their young. (In other words, this behavior evolved by kin selection.) The easiest way to test the hypothesis would be to determine the relatedness of the birds by DNA analysis. If birds are closely related, their DNA should be more similar than those of more distantly related or unrelated birds.

11. Identical twins are genetically the same, so any differences between them are due to environment. Thus, the study of identical twins enables researchers to sort out the effects of "nature" and "nurture" on human behavior. The data suggest that many aspects of human behavior are inborn. Some people find these studies disturbing because they seem to leave less room for free will and self-improvement than we would like. Results of such studies may be carelessly cited in support of a particular social agenda.

12. Example of a hypothesis: The chimpanzee mating system is polygamous. To test this hypothesis, observe mating behavior in a chimpanzee group over an extended period of time (to obtain data from multiple hormonal cycles for each of the sexually mature females in the group). Record and compile data on behaviors such as how individuals solicit sexual intercourse; number of successful approaches made by individuals; number of rejected approaches made by individuals; stage of hormonal cycle at the time of each approach by a male (for females); which individuals mate with each other; and male attentiveness to or guarding of the female after mating. If one male or female mates with several members of the opposite sex, the evidence supports the hypothesis.

Chapter 36

1. a. The *x* axis is time; the *y* axis is the number of individuals (N). The blue curve represents exponential growth; red is logistic growth.
    b. $G = rN$
    c. The carrying capacity of the environment (K)
    d. In exponential growth, population growth continues to increase as the population size increases. In logistic growth, the population grows fastest when the population is about 1/2 the carrying capacity—when N is large enough so that rN produces a large increase, but the expression $(K - N)/K$ has not yet slowed growth as much as it will as N gets closer to K.
    e. Exponential growth curve, although the worldwide growth rate is slowing

2. a. The blue line is birth rate; the red line is death rate.
    b. I. Both birth and death rates are high. II. Birth rates remain high; death rates decrease, perhaps as a result of increased sanitation and health care. III. Birth rates decline, often coupled with increased opportunities for women and access to birth control; death rates are low. IV. Both birth and death rates are low.
    c. I and IV
    d. II, when death rate has fallen but birth rate remains high

3. c 4. a 5. c 6. c 7. d 8. a 9. c

10. Food and resource limitation, such as food or nesting sites; accumulation of toxic wastes; disease; increase in predation; stress responses, such as seen in some rodents

11. Survivorship is the fraction of individuals in a given age interval that survive to the next interval. It is a measure of the probability of surviving at any given age. A survivorship curve shows the fraction of individuals in a population surviving at each age interval during the life span. Oysters produce large numbers of offspring, most of which die young, with a few living a full life span. Few humans die young; most live out a full life span and die of old age. Squirrels have approximately constant mortality and about an equal chance of surviving at all ages.

12. Clumped is the most common dispersion pattern, usually associated with unevenly distributed resources or social grouping. Uniform dispersion may be related to territories or inhibitory interactions between plants. A random dispersion is least common and may occur when other factors do not influence the distribution of organisms.

13. Populations with *K*-selected life history traits tend to live in fairly stable environments held near carrying capacity by density-dependent limiting factors. They reproduce later and have fewer offspring than species with *r*-selected traits. Their lower reproductive rate makes it hard for them to recover from human-caused disruption of their habitat. We would expect such species to have a Type I survivorship curve (see Figure 36.3).

14. Testable predictions include the following. If sunspot activity affects snowshoe hare population cycles, then the cycles should correlate with sunspot activity. Tests on food resources should reveal corresponding fluctuations in chemicals that affect its nutritional quality. Nutritional quality of hares' food should be shown to affect their reproductive output.

15. Some issues and questions to consider: How does population growth in developing countries relate to food supply, pollution, and the use of natural resources? How are these things affected by population growth in developed countries? Which of these factors are most critical to our survival? Are they affected more by the growth of developing or developed countries? What will happen as developing countries become more developed?

Chapter 37

1.

Interspecific Interaction	Effect on Species 1	Effect on Species 2	Example (many other answers possible)
Predation	+	–	Crocodile/fish
Competition	–	–	Squirrel/black bear
Herbivory	+	–	Caterpillar/leaves
Parasites and pathogens	+	–	Heartworm/dog; *Salmonella*/person
Mutualism	+	+	Plant/mycorrhizae

2.

	Carbon	Phosphorus	Nitrogen
Main abiotic reservoir(s)	Atmosphere	Rocks	Atmosphere, soil
Form in abiotic reservoir	Carbon dioxide (CO_2)	Phosphate in rock	N_2 in atmosphere; ammonium (NH_4^+) or nitrate (NO_3^-) in soil
Form used by producers	Carbon dioxide (CO_2)	Phosphate (PO_4^{3-})	NH_4^+ or NO_3^-
Human activities that alter cycle	Burning wood and fossil fuels	Agriculture (fertilizers, feedlots, pesticides, soil erosion)	Agriculture (fertilizers, feedlots); burning fossil fuels
Effects of altering cycle	Global climate change	Eutrophication of aquatic ecosystem; nutrient-depleted soils	Eutrophication of aquatic ecosystems; nutrient-depleted soils; global warming; smog; depletion of ozone layer; acid precipitation

3. d 4. b 5. b 6. c 7. d

8. Plants benefit by having their seeds distributed away from the parent. Animals benefit when the seeds contain food, as in fleshy fruits.

9. Rapid eutrophication occurs when bodies of water receive nutrient pollution (for example, from agricultural runoff) that results in blooms of cyanobacteria and algae. Respiration from these organisms and their decomposers depletes oxygen levels, leading to fish kills. Reducing this type of pollution will require controlling the sources of excess inorganic nutrients, for example, runoff from feedlots and fertilizers.

10. The abiotic reservoir of the first three nutrients is the soil. Carbon is available as carbon dioxide in the atmosphere.

11. These animals are secondary or tertiary consumers, at the top of the production pyramid. Stepwise energy loss means not much energy is left for them; thus, they are rare and require large territories in which to hunt.

12. Chemicals with a gaseous form in the atmosphere, such as carbon and nitrogen, have a global biogeochemical cycle.

13. Nitrogen fixation of atmospheric N_2 into ammonium; decomposition of detritus into ammonium; nitrification of ammonium into nitrate; denitrification (by denitrifiers) of nitrates into N_2

14. Hypothesis: The kangaroo rat is a keystone species in the desert. (Apparently, herbivory by the rats kept the one plant from outcompeting the others; removing the rats reduced plant diversity.) Additional supporting evidence would include observations of the rats preferentially eating dominant plants and finding that the dominant plant recovers from damage from other herbivores faster.

15. Some issues and questions to consider: What relationships (predators, competitors, parasites) that exist in the mussels' native habitat are altered in the Great Lakes? How might the mussels compete with Great Lakes organisms? Might the Great Lakes species adapt in some way? Might the mussels adapt? Could possible solutions present problems of their own?

Chapter 38

1. a. species at risk of extinction; b. restoration ecology; c. bioremediation; d. zoned reserves

2. c 3. c 4. a 5. d 6. b 7. c 8. a

9. Genetic, species, and ecosystem diversity. As populations become smaller, genetic diversity is usually reduced. Genetic diversity is also threatened when local populations of a species are extirpated owing to habitat destruction or other assaults or when entire species are lost. Many human activities have led to the extinction of species. The greatest threats include habitat loss, invasive species, and overharvesting. Species extinction or population extirpation may alter the structure of whole communities. Pollution and other widespread disruptions may lead to the loss of entire ecosystems.

10. Greenhouse gases in the atmosphere, including carbon dioxide, methane, and nitrous oxide, absorb infrared radiation and thus slow the escape of heat from Earth. This is called the greenhouse effect (see Module 7.13). Without greenhouse gases in the atmosphere, the temperature at the surface of Earth would be much colder and less hospitable for life.

11. Fossil fuel consumption, industry, and agriculture are increasing the quantity of greenhouse gases—such as CO_2, methane, and nitrous oxide—in the atmosphere. These gases are trapping more heat and raising atmospheric temperatures. Increases of 2–5°C are projected over the next century. Logging and the clearing of forests for farming contribute to global warming by reducing the uptake of CO_2 by plants (and adding CO_2 to the air when trees are burned). Global warming is having numerous effects already, including melting polar ice, permafrost, and glaciers, shifting patterns of precipitation, causing spring temperatures to arrive earlier, and reducing the number of cold days and nights. In other words, global warming is causing climate change. Future consequences include rising sea levels and the extinction of many plants and animals. Climate change is an international problem; air and climate do not recognize international boundaries. Greenhouse gases are primarily produced by industrialized nations. Cooperation and commitment to reduce use of fossil fuels and to reduce deforestation will be necessary if the problem of global climate change is to be solved.

12. These birds might be affected by pesticides while in their wintering grounds in Central and South America, where such chemicals may still be in use. The birds are also affected by deforestation throughout their range.

13. About 1.8 million species have been named and described. Assume that 80% of all living things (not just plants and animals) live in tropical rain forests. This means that there are 1.44 million species there. If half the species survive, this means that 0.72 million species will be extinct in 100 years, or 7,400 per year. This means that 19+ species will disappear per day, or almost one per hour. If there are 30 million species on Earth, 24 million live in the tropics, and 12 million will disappear in the next century. This is 120,000 per year, 329 per day, or 14 per hour.

14. Some issues and questions to consider: How does the use of fossil fuels affect the environment? What about oil spills? Disruption of wildlife habitat for construction of oil fields and pipelines? Burning of fossil fuels and climate change and flooding from global warming? Pollution of lakes and destruction of property by acid precipitation? Health effects of polluted air on humans? How are we paying for these "side effects" of fossil fuel use? In taxes? In health insurance premiums? Do we pay a nonfinancial price in terms of poorer health and quality of life? Could oil companies be required to pick up the tab for environmental effects of fossil fuel use? Could these costs be covered by an oil tax? How would this change the price of oil? How would a change in the price of oil change our pattern of energy use, our lifestyle, and our environment?

15. Data on carbon emissions can be found at a number of websites, including the United Nations Statistics Division (http://unstats.un.org/unsd/default.htm) and the World Resources Institute (http://www.wri.org/climate/). The per capita rankings of the United States and Canada are generally very high, along with other developed nations. Transportation and energy use are the major contributors to the carbon footprint. Any actions you can take to reduce these will help. For example, if you have a car, you can try to minimize the number of miles you drive by consolidating errands into fewer trips and using an alternative means of transportation (public transportation, walking, biking) whenever possible. To reduce energy consumption, be aware of the energy you use: Turn off lights, disconnect electronics that draw power when on standby, do laundry in cold water, for example. Websites such as http://www.nature.org/greenliving/carboncalculator/index.htm (The Nature Conservancy), http://www.ucsusa.org/ (Union of Concerned Scientists), and http://www.epa.gov/climatechange/ghgemissions/ind-calculator.html (U.S. Environmental Protection Agency) offer simple suggestions such as changing to energy-efficient lightbulbs. A Web search will turn up plenty of sites.

16. Some issues and questions to consider: How do population growth, resource consumption, pollution, and reduction in biodiversity relate to sustainability? How do poverty, economic growth and development, and political issues relate to sustainability? Why might developed and developing nations take different views of a sustainable society? What would life be like in a sustainable society? Have any steps toward sustainability been taken in your community? What are the obstacles to sustainability in your community? What steps have you taken toward a sustainable lifestyle? How old will you be in 2030? What do you think life will be like then?

APPENDIX 5

Credits

Photo Credits

COVER: Andy Rouse/Nature Picture Library

Detailed Table of Contents, pages xxiii–xxxvii

Uri Golman/Nature Picture Library; Herman Eisenbeiss/Science Source; Dieter Hopf/AGE Fotostock; M.I. Walker/Science Source; Peter B. Armstrong; R. Gino Santa Maria/Shutterstock; Graham Kent/Pearson Science; Dr. Yorgos Nikas/Science Source; akg-images/Newscom; NIBSC/Science Source; Eric Isselee/Shutterstock; Repligen Corporation; Peter Scoones/Science Source; David Kjaer/Nature Picture Library; Stephen Dalton/Science Source; Eric V. Grave/Science Source; Phil Dotson/Science Source; Reinhard Dirscherl/Getty Images; David Tipling/Nature Picture Library; Eric Isselee/Shutterstock; Markus Varesvuo/Nature Picture Library; F Rauschenbach/F1 ONLINE/SuperStock; Ingram Publishing/Photolibrary Royalty Free; Chris Bjornberg/Science Source; Mike Wilkes/Nature Picture Library; Tier Images/Getty Images; D. Phillips/Science Source; Mike Kemp/RubberBall/Alamy; Dave Watts/NHPA/Science Source; Eric J. Simon; Carlyn Iverson/Science Source; MShieldsPhotos/Alamy; Eric Isselee/Shutterstock; Image Quest Marine; Roy Corral/Corbis; BERNARD CASTELEIN/Nature Picture Library; Roger Kirkpatrick/Image Quest Marine.

Unit Openers: Unit 1 SPL/Science Source. **Unit 2** Andrew Syred/Science Source. **Unit 3** Nick Garbutt/Nature Picture Library. **Unit 4** Vaughan Fleming/Science Source. **Unit 5** J & C Sohns/AGE Fotostock. **Unit 6** AVTG/Getty Images. **Unit 7** Paul Lawrence/AGE Fotostock America Inc.

Chapter 1: Chapter Opener Jared Hobbs/AGE Fotostock. **p. 1 top to bottom** ImageState Media Partners Limited; Sacha Vignieri; Sacha Vignieri; Pascal Goetgheluck/Science Source. **1.1.1** ImageState Media Partners Limited. **1.1.2** Werner Bollmann/AGE Fotostock. **1.1.3** WILDLIFE GmbH/Alamy. **1.1.4** ekawatchaow/Shutterstock. **1.1.5** Fabio Pupin/FLPA. **1.1.6** Kim Taylor and Jane Burton/DK Images. **1.1.7** Fabio Pupin/FLPA. **1.2 top to bottom** worker/Shutterstock; Universal Images Group/Super Stock; Tsolo T. Tsolo/RGB Ventures LLC dba SuperStock/Alamy; Bildagentur-online/McPhoto/Alamy. **1.3 left** Steve Gschmeissner/Science Source. **1.3 right** CNRI/Science Source. **1.4** Ron Erwin/AGE Fotostock. **1.6 top left** Eye of Science/Science Source. **1.6 top right** Eye of Science/Science Source. **1.6 center left** Dr. D. P. Wilson/Science Source. **1.6 center right** Florapix/Alamy. **1.6 bottom left** FLPA/Alamy. **1.6 bottom right** Michael & Patricia Fogden/CORBIS. **1.7A** Francois Gohier/Science Source/Science Source. **1.7B** Science Source. **1.7C top** zhaoyan/Shutterstock. **1.7C center** Uri Golman/Nature Picture Library. **1.7C bottom** Volodymyr Goinyk/Shutterstock. **1.9 left** Sacha Vignieri. **1.9 left inset** Sacha Vignieri. **1.9 right** Hopi Hoekstra, Harvard University. **1.9 right inset** Shawn P. Carey, Migration Productions. **1.10** Pascal Goetgheluck/Science Source. **p. 13** Ron Erwin/AGE Fotostock.

Chapter 2: Chapter Opener Mark Conlin/V&W/imagequestmarine.com. **2.1 left** Chip Clark. **2.1 center** Pearson Education/Pearson Science. **2.1 right** Pearson Education/Pearson Science. **2.2A** Alison Wright/Science Source. **2.2B** Anton Prado/Alamy. **2.4A** Will & Deni McIntyre/Science Source. **2.4B** Chester A. Mathis. **2.7B** Pearson Education/Pearson Science. **2.10** Herman Eisenbeiss/Science Source. **2.11** Joe Fox/Alamy. **2.12** thp73/istockphoto. **2.14 top to bottom** Jakub Semeniuk/istockphoto; VR Photos/Shutterstock; Beth Van Trees/Shutterstock. **2.15B** Reinhard Dirscherl/Alamy.

Chapter 3: Chapter Opener Kay Blaschke/Getty Images. **p. 33** Dougal Waters/Getty Images. **3.2 left** Miki Verebes/Shutterstock. **3.2 right** Herbert Kratky/istockphoto. **3.4A** Dougal Waters/Getty Images. **3.6** Kristin Piljay/Pearson Science. **3.7 clockwise from top left** Dougal Waters/Getty Images; Biophoto Associates/Science Source; Dr. Lloyd M. Beidler; Biophoto Associates/Science Source. **3.8C clockwise from top left** Stargazer/Shutterstock; Angel Simon/Shutterstock; Alex Staroseltsev/Shutterstock; Thomas M Perkins/Shutterstock. **Table 3.9** Lilyana Vynogradova/Shutterstock. **p. 42** Anetta/Shutterstock. **3.12C** Dieter Hopf/AGE Fotostock. **3.17** Serge de Sazo/Science Source.

Chapter 4: Chapter Opener Dr. Torsten Wittmann/Science Source. **p. 51 top to bottom** M.I. Walker/Science Source; Dr. Mary Osborn. **4.1A** Michael Abbey/Science Source. **4.1B** Andrew Syred/Science Source. **4.1C** Dr. Klaus Boller/Science Source. **4.1D** M.I. Walker/Science Source. **4.3** Dr. Linda M. Stannard, University of Cape Town/Science Source. **4.5** David M. Phillips/Science Source. **4.6** Joseph F. Gennaro Jr./Science Source. **4.8A** Don W. Fawcett/Science Source. **4.9** Biophoto Associates/Science Source. **4.11A** Roland Birke/Getty Images. **4.11B** Biophoto Associates/Science Source. **4.13** Don W. Fawcett/Science Source. **4.16 left** Dr. Frank Solomon. **4.16 center** Mark Ladinsky. **4.16 right** Dr. Mary Osborn. **4.17** Dr. Alexey Khodjakov/Science Source. **4.18A** SPL/Science Source. **4.18B** Eye of Science/Science Source. **4.18C** Bjorn Afzelius.

Chapter 5: Chapter Opener B.L. de Groot. **5.2** Peter B. Armstrong. **5.16** Krista Kennell/Newscom.

Chapter 6: Chapter Opener Stephen Marks/Getty Images **p. 89 top to bottom** R. Gino Santa Maria/Shutterstock; StockLite/Shutterstock; GoodOlga/AGE Fotostock. **6.2** UpperCut Images/Alamy. **6.4** R. Gino Santa Maria/Shutterstock. **6.13C top** StockLite/Shutterstock. **6.13C bottom** Sean Lower/Marty Taylor. **6.15** Simon Smith/DK Images. **6.16** GoodOlga/AGE Fotostock.

Chapter 7: Chapter Opener John Burke/Getty Images. **p. 107 top to bottom** rodho/Shutterstock; ImageDJ/Jupiter Images. **7.1A** ODM/Shutterstock. **7.1B** Mark Conlin/Alamy. **7.1C** Susan M. Barns, Ph.D. **7.2 top to bottom** rodho/Shutterstock; Graham Kent/Pearson Science; Dr. Jeremy Burgess/Science Source. **7.3** Martin Shields/Alamy. **7.6B** llaszlo/Shutterstock. **7.7A** Christine Case. **7.11 bottom left** Dinodia/Pixtal/AGE Fotostock. **7.11 bottom right** ImageDJ/Jupiter Images. **7.13A** Prof. William H. Schlesinger. **7.14A** NASA. **7.14B** National Oceanic and Atmospheric Administration (NOAA).

Chapter 8: Chapter Opener Steve Gschmeissner/Science Source. **p. 125 top to bottom** Dr. Yorgos Nikas/Science Source; Michelle Gilders/Alamy. **8.1A** London School of Hygiene & Tropical Medicine/Science Source. **8.1B** Roger Steene/Image Quest Marine. **8.1C** Eric J. Simon. **8.1D** Bob Thomas/Getty Images. **8.1E** Dr. Yorgos Nikas/Science Source. **8.1F** Dr. Torsten Wittmann/Science Source. **8.2B** Lee D. Simon/Science Source. **8.3A** Dr. Andrew S. Bajer, University of Oregon. **8.3B** Biophoto Associates/Science Source. **8.5, p. 130 left** Conly L. Rieder, Ph.D. **8.5, p. 130 center** Conly L. Rieder, Ph.D. **8.5, p. 130 right** Conly L. Rieder, Ph.D. **8.5, p. 131 left** Conly L. Rieder, Ph.D. **8.5, p. 131 center** Conly L. Rieder, Ph.D. **8.5, p. 131 right** Conly L. Rieder, Ph.D. **8.6A** Don W. Fawcett/Science Source. **8.6B** Eldon H. Newcomb. **8.12A** Ron Chapple/Alamy. **p. 139** Ed Reschke/Getty Images. **8.16 top** F. Schussler/PhotoDisc/Getty Images, Inc. **8.16 bottom** Roxana Gonzalez/Shutterstock. **8.17A** Mark Petronczki. **8.19 left** Véronique Burger/Science Source. **8.19 right** CNRI/Science Source. **8.20A left** SPL/Science Source. **8.20A right** Lawrence Shear/Science Source. **8.22** Michelle Gilders/Alamy. **p. 151** J.L. Carson/Custom Medical Stock Photo.

Chapter 9: Chapter Opener David Pickford/Robert Harding. **p. 153 top to bottom** akg-images/Newscom; Eric J. Simon; Andrew Syred/Science Source. **9.1** bilwissedition Ltd. & Co. KG/Alamy. **9.2A** akg-images/Newscom. **9.8 left** PhotoDisc/Getty Images, Inc. **9.8 right** PhotoDisc/Getty Images, Inc. **9.9 top left** Liza McCorkle/iStockphoto. **9.9 top right** Westend61/Getty Images. **9.9 bottom left** Shell114/Shutterstock. **9.9 bottom right** David Terrazas Morales/Getty Images. **9.9C** Michael Ciesielski Photography. **9.10A** CNRI/Science Source. **9.10B top** Gusto/Science

Source. **9.10B bottom** Eric J. Simon. **9.13A** Eye of Science/Science Source. **9.13B** Eye of Science/Science Source. **9.15A** Eric J. Simon. **9.15B** Eric J. Simon. **9.18B** Graphic Science/Alamy. **9.20A** Andrew Syred/Science Source. **9.20B clockwise from top left** Jose Luis Pelaez, Inc./Getty Images; Yuri Arcurs/Shutterstock; Rubberball/Nicole Hill/Getty Images; Anatoliy Samara/Shutterstock. **Table 9.20 top** Tomasz Zachariasz/iStockphoto. **Table 9.20 center** kosam/Shutterstock. **Table 9.20 bottom** Tomasz Zachariasz/iStockphoto. **9.21A left** University of Texas MD Anderson Cancer Center. **9.21A right** University of Texas MD Anderson Cancer Center. **9.22** FPG/Getty Images.

Chapter 10: Chapter Opener Dr. Klaus Boller/Science Source. **10.1A** Biophoto Associates/Science Source. **10.3A left** Library of Congress. **10.3A right** Cold Spring Harbor Laboratory Archives. **10.3B** National Institutes of Health. **10.6B** Kevin McCluskey, PhD. **10.8C** Yonhap Choi Byung-kil/AP Images. **10.12** Joachim Frank. **p. 201** Hazel Appleton, Centre for Infections/Health Protection Agency/Science Source/Science Source. **10.19 left** NIBSC/Science Source. **10.19 right** Liu Siu Wai/Newscom. **10.23C left** Huntington Potter. **10.23C right** Huntington Potter.

Chapter 11: Chapter Opener dirtlight photography/Getty Images. **p. 209 top to bottom** Martin Oeggerli/Science Source; Robyn Mackenzie/Shutterstock. **11.1A** Martin Oeggerli/Science Source. **11.2A left** Don W. Fawcett/Science Source. **11.2A right** Biophoto Associates/Science Source. **11.2B** Eric Isselee/Shutterstock. **11.8A left** F. Rudolf Turner. **11.8A right** F. Rudolf Turner. **11.9** Alila Medical Images/Shutterstock. **11.12** Robyn Mackenzie/Shutterstock. **11.16A** GeoM/Shutterstock.

Chapter 12: Chapter Opener Efired/Shutterstock. **p. 231 top to bottom** Pichi Chuang/Thomson Reuters; Hank Morgan/Science Source; Steve Helber/AP Photo; Philippe Plailly & Atelier Daynes/Science Source. **12.1A** Pichi Chuang/Thomson Reuters. **12.1B top to bottom** Smileus/Dreamstime LLC; A.J. Sisco/UPI/Newscom; Kedrov/Shutterstock; Angel Hell/iStockphoto. **12.6A left** Brad DeCecco Photography. **12.6A right** Brad DeCecco Photography. **12.6B** Science Source. **12.7A** Eli Lilly and Company. **12.7B** Hank Morgan/Science Source. **12.8A** Vladimir Nikitin/Shutterstock. **12.8B** Patti McConville/Alamy. **12.9 left** Borys Shevchuk/Fotolia. **12.9 center** Kirill Kurashov/Shutterstock. **12.9 right** Dionisvera/Shutterstock. **12.10** Phil Date/Shutterstock. **12.13** Repligen Corporation. **12.15A** Steve Helber/AP Photo. **12.15B** Michael Stephens/Agence France Presse/Newscom. **Table 12.17 top to bottom** Olivia Meckes/Nicole Ottawa/Science Source/Science Source; Vaclav Volrab/Shutterstock; Andrew Burgess/Shutterstock; Gary Ombler/DK Images. **12.19** Rubberball/Getty Images, Inc. **12.21** Philippe Plailly & Atelier Daynes/Science Source.

Chapter 13: Chapter Opener GHANA-VACCINES/GAVI/Olivier Asselin/Handout/REUTERS. **p. 255 top to bottom** Archiv/Science Source; Steffen Foerster/Shutterstock. **13.1A left** Archiv/Science Source. **13.1A right** National Maritime Museum Picture Library. **13.1B** Peter Scoones/Science Source. **13.1C** Stefan Huwiler/Corbis. **13.2A** Colin Keates/DK Images. **13.2B** Chip Clark/Fundamental Photographs, NYC. **13.2C** John Henshall/Alamy. **13.3A** Miles Away Photography/Shutterstock. **13.4B left** Dr. Keith Wheeler/Science Source. **13.4B right** Scanpix Sweden AB. **13.6 top left** H Reinhard/Arco Images GmbH/Alamy. **13.6 top right** ZUMA Press, Inc./Alamy. **13.6 center** Luis César Tejo/Shutterstock. **13.6 bottom left** Kenneth W. Fink/Science Source. **13.6 bottom right** Juniors Bildarchiv/GmbH/Alamy. **13.7** Spirit/Corbis. **13.8** Edmund D. Brodie III. **13.9** Adam Jones/The Image Bank/Getty Images. **13.11** Anne Dowie/Pearson Education. **13.12B** William Ervin/Science Source. **13.13** Steffen Foerster/Shutterstock. **13.15A** Dave Blackey/Getty Images. **13.15B** George D. Lepp/Encyclopedia/Corbis. **13.15C** Barry Mansell/Nature Picture Library. **13.16** Scott Camazine/Science Source.

Chapter 14: Chapter Opener Tim Laman/Nature Picture Library. **p. 277** Jared Hobbs/All Canada Photos/SuperStock. **14.2A left** Malcom Schuyl/Alamy. **14.2A right** David Kjaer/Nature Picture Library **14.2B left to right** Robert Kneschke/iStockphoto; Justin Horrocks/iStockphoto; PhotoDisc/Getty Images, Inc.; Radius Images/Getty Images; Phil Date/Shutterstock; Masterfile Corporation. **14.2C top to bottom** Boris Karpinski/Alamy; janprchal/Shutterstock; Troy Maben/AP Photo. **14.3, p. 280 1st column** Joe McDonald/Photoshot Holdings Ltd.; Joe McDonald/Corbis. **14.3, p. 280 2nd column** USDA/APHIS Animal and Plant Health Inspection Service; Jared Hobbs/All Canada Photos/SuperStock. **14.3, p. 280 3rd column** J & C Sohns/AGE Fotostock America, Inc.; Tier und Naturfotografie/SuperStock. **14.3, p. 280 4th column** Philippe Clement/Nature Picture Library; Oyvind Martinsen/Alamy. **14.3, p. 281 1st column** SERDAR/Alamy; SERDAR/Alamy. **14.3, p. 281 2nd column** Charles W. Brown; DawnYL/Fotolia; Kazutoshi Okuno. **14.4A left** John Shaw/Photoshot. **14.4A center** Corbis. **14.4A right** Michael Fogden/Photoshot. **14.4B shrimp** Arthur Anker/Florida Museum of Natural History. **14.4B map** NASA Earth Observing System. **14.5B clockwise from top left** Douglas W. Schemske; Douglas W. Schemske; Douglas W. Schemske; Douglas W. Schemske. **14.7** photobank.kiev.ua/Shutterstock. **14.8 top to bottom** INTERFOTO/Alamy Images; INTERFOTO/Alamy Images; Mary Plage/Getty Images; Mary Plage/Getty Images; Ralph Lee Hopkins/Alamy; Ralph Lee Hopkins/Alamy. **14.9B top** Ole Seehausen. **14.9B bottom** Ole Seehausen. **14.10B left** Melvin Grey/Photoshot. **14.10B right** Juan Martin Simon. **14.10C** Seehausen, Ole. **p. 290** Seehausen, Ole.

Chapter 15: Chapter Opener Alfred & Annaliese Tr/AGE Fotostock America, Inc. **15.1 left** Francois Gohier/Science Source. **15.1 right** Peter Sawyer/Smithsonian Institution–Museum of Natural History. **15.3A** Fred M. Menger. **15.7D** Roland Seitre/Nature Picture Library. **15.7E** Phil Savoie/Nature Pictrure Library. **15.7F** Rick & Nora Bowers/Alamy. **15.8** David Parker/Science Source. **15.11A** Stephen Dalton/Science Source. **15.11B** Jean Kern. **15.11C top** Dr. William A. Cresko. **15.11C bottom** Dr. William A. Cresko. **15.12 left to right** Hal Beral/V & W/Image Quest Marine; Christophe Courteau/Science Source; Reinhard Dirscherl/Alamy; Jim Greenfield/Image Quest Marine; James Watt/ImageQuestMarine. **15.1A** Getty Images, Inc. **15.16C** American Museum of Natural History. **15.18** Frank Collins, Ph.D./CDC.

Chapter 16: Chapter Opener CAMR/A. Barry Dowsett/Science Source. **p. 319 top to bottom** SPL/Science Source; David Caron/Science Source. **16.1** SPL/Science Source. **16.2A left** Eye of Science/Science Source. **16.2A center** David McCarthy/Science Source. **16.2A right** Stem Jems/Science Source. **16.2B** ASM/Science Source. **16.2C** I. Rantala/Science Source. **16.2D** Eye of Science/Science Source. **16.3A** Huntington Potter. **16.3B** Scott Camazine/Science Source. **16.4 top left** Sinclair Stammers/Science Source. **16.4 top right** T. Stevens & P. McKinely, PNNL/Science Source. **16.4 bottom left** Pasieka/SPL/Science Source. **16.4 bottom right** Dr. Gary Gaugler/Science Source. **16.5** Garry Palmateer. **16.6A** Science Source. **16.6B** SIPA USA/SIPA Newscom. **16.7 top** Olive Meckes/Nicole Ottawa/Science Source/Science Source. **16.7 bottom** Eye of Science/Science Source/Science Source. **16.8A left** Jack Dykinga/Getty Images. **16.8A right** Jim West/Alamy. **16.9A** National Library of Medicine. **16.9B** Eye of Science/Science Source. **16.9C** Susan M. Barns, Ph.D. **16.9D** Moredon Animal Health/SPL/Science Source. **16.9E** Science Source. **16.10A** David M. Phillips/Science Source. **16.10B** STEVE LINDRIDGE/Alamy. **16.11A** Anders Wiklund/Reuters Limited. **16.12A left** Carol Buchanan/AGE Fotostock. **16.12A center** Eye of Science/Science Source. **16.12A right** Alex Rakosy/Custom Medical Stock Photo. **16.12B left** Patrick Keeling. **16.12B right** Eric V. Grave/Science Source. **16.14A** Steve Gschmeissner/Science Source. **16.14B** Georgie Holland/AGE Fotostock. **16.14C** Fred Rhoades. **16.14D** Miriam Godfrey/National Institute of Water and Atmospheric Research. **16.14E** Andrew Syred/Science Source. **16.14F** David Caron/Science Source. **16.14F inset** Dee Breger/Photo Researchers. **16.14G** Steve Gschmeissner/Science Source. **16.15** Photoshot/Newscom. **16.16A** David M. Phillips/Science Source. **16.16B** Oliver Meckes/Science Source. **16.17A** Biophoto Associates/Science Source. **16.17B** Dr. George L. Barron. **16.17B inset** Ray Simons/Science Source. **16.17C** Robert Kay. **16.18A** Alex Hyde/Nature Picture Library. **16.18B left** Manfred Kage/Science Source. **16.18B right** Aaron J. Bell/Science Source. **16.18C** D. P. Wilson/Eric and David Hosking/Science Source. **16.19B** David J. Patterson. **p. 338 left** David M. Phillips/Science Source. **p. 338 right** Dr. Gary Gaugler/Science Source.

Chapter 17: Chapter Opener jaboo2foto/Shutterstock. **p. 341** Frank Young/Papilio/Corbis. **17.1A** Bob Gibbons/Alamy. **17.1B** Dr. Linda E. Graham. **17.1D** Corbis. **17.2B left** Matthijs Wetterauw/Alamy. **17.2B center** Dr. Jeremy Burgess/Science Source. **17.2B right** Hidden Forest.

17.2C left Ed Reschke. **17.2C right** Colin Walton/Dorling Kindersley, Ltd. **17.2D left to right** blickwinkel/Alamy; John Cancalosi/Getty Images; Bob Gibbons/Alamy; Pichugin Dmitry/Shutterstock. **17.2E left** Koos Van Der Lende/AGE Fotostock. **17.2E right** Nigel Cattlin/Alamy. **17.3, p. 346** Sanamyan/Alamy. **17.3, p. 347** Steffen Hauser/age fotostock/SuperStock. **17.4** Open University, Department of Earth Sciences. **17.5A left** Morales/AGE Fotostock. **17.5A top right** Trent Dietsche/Alamy. **17.5A bottom right** Brent Bergherm/AGE Fotostock. **17.6A left** Jean Dickey. **17.6A top right** Jean Dickey. **17.6 bottom right** Jean Dickey. **17.7** AVTG/iStockphoto. **17.8A top** Agita Leimane/Shutterstock. **17.8A bottom** rakim/Shutterstock. **17.8B top** Derek Hall/DK Images. **17.8B bottom** Scott Camazine/Science Source. **17.8C top** Cal Vornberg/AGE Fotostock. **17.8C bottom** Cal Vornberg/AGE Fotostock. **17.9 left** Bryan Brunner. **17.9 right** Jethro Toe/Shutterstock. **17.10A** MShieldsPhotos/Alamy. **17.10B** Jean Dickey. **17.10C** D. Wilder/Wilder Nature Photography. **17.11A left** Bart Wursten. **17.11A right** Breaking Ground. **17.11B** NASA. **17.12A left** Fred Rhoades. **17.12A right** Dr. George L. Barron. **17.12B** BlueMoon Stock/Alamy. **17.14B** Gregory G. Dimijian/Science Source. **17.14D left** H Reinhard/AGE Fotostock. **17.14D right** Frank Young/Papilio/Corbis. **17.14E top left** Phil Dotson/Science Source. **17.14E top right** Adrian Davies/Photoshot. **17.14E bottom** Inger Hogstrom/AGE Fotostock. **17.15** Emmanuel Lattes/Alamy. **17.16A** Filatov Alexey/Shutterstock. **17.16B** Christine Case. **17.16C** C Huetter/AGE Fotostock. **17.17A** NHPA/SuperStock. **17.17B** Eye of Science/Science Source. **17.17C** Jean Dickey. **17.19A** MT Media/Long Photography, Inc. **17.19B** Astrid/Hanns Frieder Michler/Science Source. **17.19C** Nigel Cattlin/Science Source. **p. 363 left** R.J. Erwin/Science Source. **p. 363 right** Adrian Davies/Photoshot.

Chapter 18: Chapter Opener Jochen Tack/Alamy. **p. 365** Jurgen Freund/Nature Picture Library. **18.1A** Gunter Ziesler/Getty Images. **18.1B** Georgette Douwman/Nature Picture Library. **18.2A left** South Australian Museum. **18.2A right** Museum of Paleontology. **18.2B top** The Natural History Museum, London. **18.2B bottom** Chip Clark. **18.5A left to right** James Watt/ImageQuestMarine; Andrew J. Martinez/Science Source; Paul Kay/Getty Images. **18.6A top** Sinclair Stammers/Science Source. **18.6A bottom** Jonathan Bird/Getty Images, Inc. **18.6B** Reinhard Dirscherl/Getty Images, Inc. **18.7B** Eye of Science/Science Source. **18.8A** American Association for the Advancement of Science (AAAS). **18.8B** Gavin Parsons/Getty Images, Inc. **18.9C left** Jez Tryner/Image Quest Marine. **18.9C right** GK Hart/Vikki Hart/Getty Images, Inc. **18.9D left** Marevision/AGE Fotostock America Inc. **18.9D right** Christophe Courteau/Nature Picture Library. **18.9E bottom** Dante Fenolio/Science Source. **18.9E top** Scott Tuason/Image Quest Marine. **18.10A** A.N.T./Photoshot Holdings Ltd. **18.10B top** Jurgen Freund/Nature Picture Library. **18.10B bottom left** Alexander Semenov/Science Source. **18.10 bottom right** Peter Herring/Image Quest Marine. **18.10C** Astrid/Science Source. **18.11B** Angelo Giampiccolo/Nature Picture Library. **18.11C left to right** EcoPrint/Shutterstock; Scott Camazine/Alamy; Eye of Science/Science Source. **18.11D** George Grall/Getty Images, Inc. **18.11E** Mark Smith/Science Source. **18.11F top** Colin Milkins/Getty Images, Inc. **18.11F bottom** Maximilian Weinzierl/Alamy. **18.12A top left** Hecker/Sauer/AGE Fotostock. **18.12A bottom left** W Holzenbecher/AGE Fotostock. **18.12A right** M Woike/AGE Fotostock America Inc. **18.12C left to right** Buddy Mays/Alamy; Martin Shields/Alamy; Matt Jeppson/Shutterstock. **18.12D** Dante Fenolio/Science Source. **18.12E left** Alan Blank/Alamy. **18.12E right** Purestock/Getty Images, Inc. **18.13B top left** Rod Williams/Nature Picture Library. **18.13B top right** Morley Read/Shutterstock. **18.13B bottom left** Stephen Paddock. **18.13B bottom right** Stephen Paddock. **18.14B** Andrew J. Martinez/Science Source. **18.14C** Louise Murray/AGE Fotostock. **18.15A** Jez Tryner/Image Quest Marine. **18.15B** Marevision/AGE Fotostock America Inc. **18.16A** Doug Perrine/Nature Picture Library. **18.16B** Linda Pitkin/Nature Picture Library. **18.16C** S. Kuelcue/Shutterstock.

Chapter 19: Chapter Opener Beawiharta/Thomson Reuters. **p. 389 top to bottom** Publiphoto/Science Source; David Tipling/Nature Picture Library. **19.2A** Tom McHugh/Science Source. **19.2B inset** Marevision/AGE Fotostock America Inc. **19.2B** Gary Meszaros/Science Source. **19.3B** George Grall/National Geographic Stock. **19.3C** Matthew Oldfield Underwater Photography/Alamy. **19.3D** Tom and Pat Leeson/Science Source. **19.3E left** Stephen Frink/Getty Images, Inc. **19.3E top right** Marevision/AGE Fotostock. **19.3E bottom right** Jeff Rotman/Getty Images. **19.3F** Rudie Kuiter/Oceanwide Images. **19.4B** Publiphoto/Science Source. **19.5A** Heinemann/Westend61 GmbH/Alamy. **19.5B** Dante Fenolio/Science Source. **19.5C** Michael Fogden/Photoshot. **19.5D** Stephen Dalton/Photoshot. **19.5E** DP Wildlife Vertebrates/Alamy. **19.6A** Sylvain Cordier/Getty Images. **19.6C** Steven David Miller/Nature Picture Library. **19.7A** Sodapix/Getty Images. **19.7B** Andy Rouse/Nature Picture Library. **19.8A** Jean Phillipe Varin/Jacana/Science Source. **19.8B** Phillip Hayson/Getty Images, Inc. **19.8C** Mitch Reardon/Science Source. **19.9, p. 400 clockwise from top left** E. H. Rao/Science Source/Science Source; Kevin Schafer/Alamy; Amazon-Images/Alamy; Shutterstock; Image broker/Alamy. **19.9, p. 401 top left** Werner Bollmann/AGE fotostock/SuperStock. **19.9, p. 401 top right** Steve Bloom/Alamy. **19.9, p. 401 center** Arco Images GmbH/Alamy. **19.9, p. 401 bottom left** Anekoho/Shutterstock. **19.9, p. 401 bottom right** John Kelly/Getty Images, Inc. **19.10B** David Tipling/Nature Picture Library. **19.12A left** Christian Musat/Fotolia. **19.12A right** Piotr Marcinski/Fotolia. **19.12B** John Reader/SPL/Science Source. **19.14** Javier Trueba/MSF/Science Source. **19.15** Achmad Ibrahim/AP Images. **19.16** Sarah Leen/National Geographic Stock. **19.17A** AFP PHOTO/Newscom. **19.17B** Tim Laman/National Geographic Stock. **19.17C** Maurice Emetshu/AP Images.

Chapter 20: Chapter Opener Anup Shah/Nature Picture Library. **20.2** Eric Isselee/Shutterstock. **20.4** Image Source/Getty Images. **20.5** Image Source/Getty Images. **20.5A** Nina Zanetti/Pearson Education. **20.5B** Ed Reschke/Getty Images, Inc. **20.5C** Nina Zanetti/Pearson Education. **20.5D** Chuck Brown/Science Source. **20.5E** Nina Zanetti/Pearson Education. **20.5F** Gopal Murti/SPL/Science Source. **20.6** Paul Burns/Blend Images/Alamy. **20.6A** Nina Zanetti/Pearson Education. **20.6B** Ed Reschke/Getty Images, Inc. **20.6C** Ed Reschke/Getty Images, Inc. **20.8** Paul Burns/Blend Images/Alamy. **20.9** Jeff Wheeler/MCT/Newscom. **20.10A-B** Douglas Pulsipher/Alamy. **20.10C** Image Source/Corbis. **20.10D** Rubberball/Getty Images. **20.10E** Andy Crawford/DK Images. **20.10F-G** Mixa/Alamy. **20.10H** Laura Knox/DK Images. **20.10I** Ryan McVay/Getty Images. **20.10K** Moodboard/Corbis. **20.10L** Jeremy Woodhouse/Getty Images. **20.13B** SPL/Science Source. **20.14** Markus Varesvuo/Nature Picture Library. **p. 426 left to right** Nina Zanetti/Pearson Education; Nina Zanetti/Pearson Education.

Chapter 21: Chapter Opener r.nagy/Shutterstock. **p. 429 top to bottom** 6493866629/Shutterstock; Steve Allen/Brand X Pictures/Getty Images. **21.1A** David W Hamilton/AGE footstock. **21.1B** Thomas Eisner. **21.1C** Martin Dohrn/Science Source. **21.1D** Jennette van Dyk/Getty Images. **21.2A** 6493866629/Shutterstock. **21.4** Geo Martinez/Shutterstock. **p. 435** Kletr/Shutterstock. **21.9** Eye of Science/Science Source. **21.13 left** Kristin Piljay. **21.13 right** Tom Brakefield/Photodisc/Getty. **21.14** Markus Varesvuo/Nature Picture Library. **21.15 top left** Yasonya/iStockphoto.com. **21.15 top right** Corbis. **21.15 bottom** Debbie Hardin, Pearson Education. **p. 443** Melastmohican/Fotolia. **21.16** marlee/Shutterstock. **21.18** Allison Wright/National Geographic/Getty Images. **21.19B** The Jackson Laboratory. **21.21** Steve Allen/Brand X Pictures/Getty Images. **p. 450** Geo Martinez/Shutterstock. **p. 451 left to right** The Jackson Laboratory; Yasonya/iStockphoto.com; Corbis.

Chapter 22: Chapter Opener YANGCHAO/Shutterstock. **p. 453 top to bottom** F Rauschenbach/F1 ONLINE/SuperStock; Douglas Pulsipher/Alamy. **22.1** Corbis/AGE Fotostock. **22.4A** Corbis/AGE Fotostock. **22.4B** F Rauschenbach/F1 ONLINE/SuperStock. **p. 457** J Fieber/AGE Fotostock. **22.5** The University of Chicago Department of Organismal Biology and Anatomy. **22.6** Douglas Pulsipher/Alamy. **p. 464** Douglas Pulsipher/Alamy.

Chapter 23: Chapter Opener Steve Gschmeissner/Science Source. **23.2A** GeorgePeters/E+/Getty Images. **23.2B** Matt Antonino/Shutterstock. **23.2C** Shutterstock. **23.3** Hans Kim/Shutterstock. **23.5B** Gustoimages/Science Source. **23.6B left** Chuck Brown/Science Source. **23.6B right** Biophoto Associates/Science Source. **23.7A** Science Source. **23.11A** D.W. Fawcett/Science Source. **23.13** Ingram Publishing/Photolibrary Royalty Free. **23.14B** The Gillette Co.

Chapter 24: Chapter Opener James Cavallini/Science Source. **p. 485** Salisbury District Hospital/Science Source. **24.4** Ryan McVay/PhotoDisc/Getty Images. **p. 495** Raimund Koch/Getty Images. **24.14** Chris Bjornberg/Science Source. **24.15** Impact Visuals/Newscom. **24.17** Salisbury District Hospital/Science Source. **p. 501** Dey L. P. Pharmaceuticals.

Chapter 25: Chapter Opener Doug Allan/Nature Picture Library. **p. 505** Four Oaks/Shutterstock. **25.1** Mike Wilkes/Nature Picture Library. **25.2A** Four Oaks/Shutterstock. **25.2B** Leksele/Shutterstock. **25.5 left** GeorgePeters/E+/Getty Images. **25.5 center** Eric Isselee/iStockphoto.com. **25.5 right** Maksym Gorpenyuk/Shutterstock. **25.6A** Yuri Arcurs/Shutterstock. **25.9** Phanie/Science Source. **p. 514** Leksele/Shutterstock. **p. 514** Eric Isselee/iStockphoto.com.

Chapter 26: Chapter Opener Mircea BEZERGHEANU/Shutterstock. **p. 517** Corey Ralston/Getty Images. **26.3A** Matteo photos/Shutterstock. **26.4** Laura Knox/DK Images. **26.5D** Mirrorpix/Splash News/Newscom. **26.6A top** Eric Isselée/Fotolia. **26.6A bottom** Eric Isselée/Fotolia. **26.7C** Tier Images/Getty Images. **26.10** Steve Cady/E+/Getty Images. **26.11** Corey Ralston/Getty Images. **p. 530** sianc/Shutterstock.

Chapter 27: Chapter Opener Dr. Linda M. Stannard, University of Cape Town/Science Source. **p. 533 top to bottom** John R. Finnerty; C. Edelman/La Vilette/Petit Format/Science Source; David Barlow/BBC Photo Library. **27.1A** Biophoto Associates/Science Source. **27.1B** David Wrobel. **27.2A** John R. Finnerty. **27.2B** PREMAPHOTOS/Nature Picture Library. **27.2C** Aneese/Shutterstock. **27.3B** C. Edelman/La Vilette/Petit Format/Science Source. **27.9A** D. Phillips/Science Source. **27.10 top** Hintau Aliaksei/Shutterstock. **27.10 bottom** Eric Isselée/Fotolia. **27.12A** Huw Williams. **27.12C** Thomas Poole. **27.12D** G. I. Bernard/Science Source. **27.13A left** David Barlow Photographer. **27.13A right** David Barlow/BBC Photo Library. **27.16, p. 556 left to right** Scanpix Sweden AB; Dr G. Moscoso/Science Source; Scanpix Sweden AB. **27.16, p. 557 left** Lennart Nilsson/Scanpix Sweden AB. **27.16, p. 557 right** Ron Sutherland/Science Source. **27.18** Dr. Yorgos Nikas/Science Source.

Chapter 28: Chapter Opener WDCN/Univ. College London/Science Source. **p. 563 top to bottom** Manfred Kage/Science Source; Edwin R. Lewis, Professor Emeritus. **28.2** Manfred Kage/Science Source. **28.7** Edwin R. Lewis, Professor Emeritus.**28.17A** Patrick Landmann/Science Source. **28.17B** Johns Hopkins University. **28.18 left** Robert W. Haley, M.D. **28.18 right** Robert W. Haley, M.D. **28.21A** Jessica Wilson/Medical Body Scans/Science Source. **28.21B** Amy Sancetta/AP Photo.

Chapter 29: Chapter Opener Kenneth J. Lohmann, Ph.D. **p. 587** Jim Watt/Perspectives/Getty Images. **29.1A** Fuse/Getty Images. **29.3B** N. F. Photography/Shutterstock. **29.3D** Jim Watt/Perspectives/Getty Images. **29.4E** Keith Publicover//Shutterstock. **29.7A** Kent Wood/Science Source. **29.7B** Thomas Eisner. **29.9A** moodboard/Corbis. **29.9B** moodboard/Corbis. **29.11** szefei/Shutterstock. **p. 599** Mike Kemp/RubberBall/Alamy.

Chapter 30: Chapter Opener Ilene MacDonald/Alamy. **p. 603** Glenn Bartley/AGE Fotostock. **30.1A** Stuart F. Westmorland/Danita Delimont/Alamy. **30.1B** Tamara Bauer/iStockphoto.com. **30.1C** Dave Watts/NHPA/Science Source. **30.1D** Juniors Bildarchiv/AGE Fotostock. **30.1E** Glenn Bartley/AGE Fotostock. **30.2 left** Heather Angel/Natural Visions/Alamy. **30.2 right** Mick Hoult/Photoshot/Alamy. **30.2B** Tony Florio/Science Source. **30.2C** Carlos Villoch/Image Quest Marine. **30.2D left** Jeff Rotman/The Image Bank/Getty Images. **30.2D right** Kaj R. Svensson/SPL/Science Source. **30.2E** alle/Shutterstock. **30.5A** Jochen Tack/Alamy. **30.5B left** Science Source. **30.5B right** P. Motta/SPL/Science Source. **30.8** Professor Clara Franzini-Armstrong. **30.11** Christopher Nuzzaco/iStockphoto.com. **p. 616** Allstar Picture Library/Alamy. **p. 617 clockwise from top left** Tamara Bauer/iStockphoto.com; Dave Watts/NHPA/Science Source; Stuart F. Westmorland/Danita Delimont/Alamy; Glenn Bartley/AGE Fotostock.

Chapter 31: Chapter Opener Dmitry Pichugin/Fotolia. **p. 621 top to bottom** Don Mason/Corbis RF; Africa Studio/Shutterstock. **31.1B clockwise from top left** Cathleen A Clapper/Shutterstock; Vasilyev Alexandr/Shutterstock; zirconicusso/Shutterstock; panda3800/Shutterstock; Diana Mower/Shutterstock; Nattika/Shutterstock. **31.3** Adam Hart-Davis/Science Source. **31.4A** Eric J. Simon. **31.4B left** NHPA/SuperStock. **31.4B top right** Donald Gregory Clever. **31.4B bottom right** FhF Greenmedia/AGE Fotostock America, Inc. **31.4C left** Rafael Campillo/AGE Fotostock. **31.4C right** Scott Prokop/Shutterstock. **31.6B** Ed Reschke/Getty Images. **31.6C** Graham Kent. **31.6D left** Graham Kent. **31.6D right** Graham Kent. **31.6E** N.C. Brown Center for Ultrastructure Studies. **31.6F** Graham Kent. **31.7A** Bob Gibbons/FLPA. **31.7C top** Viktor Kitaykin/iStockphoto. **31.7C bottom** Ed Reschke/Getty Images. **31.7D** Ed Reschke/Getty Images. **31.8B** Don Mason/Corbis RF. **31.8C** Zack Frank/Shutterstock. **31.9A** Africa Studio/Shutterstock. **31.12A** jopelka/Shutterstock. **31.12C top** Africa Studio/Shutterstock. **31.12C center** Rolf Klebsattel/Shutterstock. **31.12C bottom** Elena Schweitzer/Shutterstock. **31.14A** Tim Hill/Alamy. **31.14B** Jamie Pham/Alamy. **31.14C** Dan Suzio/Science Source. **31.14D** Dennis Frates/Alamy. **31.14E** Rosenfeld Images Ltd./Science Source. **31.15** David Welling/Nature Picture Library.

Chapter 32: Chapter Opener Steve Satushek/Getty Images. **p. 643** Brian Capon. **32.1B** Q-Images/Alamy. **32.2A** Brian Capon. **32.3** Matt Ware/Alamy. **32.4 left** Frank Greenaway/DK Images. **32.4 center** Jeremy Burgess/Science Source. **32.4 right** Jeremy Burgess/Science Source. **32.5A** Professor Ray F. Evert. **32.5B** Lezh/E+/Getty Images. **32.5C top left** Nigel Cattlin/FLPA. **32.5C top right** P. B. Tomlinson. **32.5C bottom left** P. B. Tomlinson. **32.5C bottom right** P. B. Tomlinson. **32.7B** Paul Rapson/Science Source. **32.8A** USDA/ARS/Agricultural Research Service. **32.9A** U.S. Geological Survey, Denver. **32.9B** National Oceanic and Atmospheric Administration (NOAA). **32.9C** Nagel Photography/Shutterstock. **32.11** Louisiana State University Press. **32.13B left** Dr. Jeremy Burgess/Science Source. **32.13B right** Dr. Jeremy Burgess/Science Source. **32.14A** Carlyn Iverson/Science Source. **32.14B** Robert and Jean Pollock/Science Source. **32.14C** H. Reinhard/AGE Fotostock. **32.14D** Willi Rolfes/AGE Fotostock. **32.14E** James H. Robinson/Science Source.

Chapter 33: Chapter Opener Scott Sinklier/Alamy. **p. 661 top to bottom** Ron and Patty Thomas Photography/Getty Images; Martin Shields/Science Source. **33.1A** MShieldsPhotos/Alamy. **33.3A** William M. Gray. **33.3C** Martyn F. Chillmaid/Science Source. **33.4 left** Smit/Shutterstock. **33.4 right** bluemagenta/Alamy. **33.5A** Alan Crozier. **33.5B** Kristin Piljay/Pearson Education/Pearson Science. **33.5C** Fred Jensen. **33.6** Ron and Patty Thomas Photography/Getty Images. **33.7A** Kristin Piljay/Pearson Education/Pearson Science. **33.7B** Ed Reschke/Getty Images. **p. 669** Rudolf Madar/Shutterstock. **33.8** Stockbyte/Getty Images. **33.9A** Michael Evans. **33.9B left** Martin Shields/Science Source. **33.9B right** Martin Shields/Science Source. **33.10 left** Martin Shields/Science Source. **33.10 right** Martin Shields/Science Source. **33.13A** USDA/Science Source.

Chapter 34: Chapter Opener Kenneth Sponsler/Fotolia. **p. 679 top to bottom** Delpho, M./Arco Images/Alamy; Don Fink/Shutterstock. **34.15** Jorma Luhta/Nature Picture Library. **34.2A** Gamma-Keystone/Getty Images. **34.2B** Delpho, M./Arco Images/Alamy. **34.2C** Alfred Eisenstaedt/Time Life Pictures/Getty Images. **34.3A** Peter Batson/Image Quest Marine. **34.3B** WILDLIFE GmbH/Alamy. **34.4** franzfoto.com/Alamy. **34.6B** Getty Images. **34.6C** SOC/Image Quest Marine. **34.6D** James Randklev/Getty Images. **34.7B** Don Fink/Shutterstock. **34.7C** Jean Dickey. **34.9** Nick Garbutt/SuperStock. **34.10** Eric Isselee/Shutterstock. **34.11** Dennis Frates/Getty Images. **34.12** The California Chaparral Institute. **34.13** Mike Grandmaison/Getty Images. **34.14** Thomas R. Fletcher/Alamy. **34.15** Jorma Luhta/Nature Picture Library. **34.16** Wayne Lynch/Getty Images. **34.17** Gordon Wiltsie/National Geographic Stock. **p. 696** Getty Images.

Chapter 35: Chapter Opener NHPA/SuperStock. **p. 699 top to bottom** Theo Allofs/Danita Delimont/Alamy; John Cancalosi/Getty Images; Image Quest Marine; Kitchin & Hurst/AGE Fotostock. **35.1** Tom McHugh/Science Source. **35.2B** Theo Allofs/Danita Delimont/Alamy. **35.3A** Wayne Lynch/AGE Fotostock. **35.5A** Thomas D. McAvoy/Getty Images. **35.5B** Terry Andrewartha/FLPA. **35.6A** Biological Resources Division, U.S. Geological Survey. **35.6B** Star Banner & Doug Engle/AP Photo. **35.7 top** Scott Camazine/Alamy. **35.7 center** Jeff Mondragon/Alamy. **35.7 bottom** Nick Upton/Nature Picture Library. **35.8A** Michael Nolan/AGE Fotostock.

35.8C Juniors/SuperStock. **35.9** John Cancalosi/Getty Images. **35.10** Richard Wrangham. **35.10 inset** Pal Teravagimov Photography/Getty Images. **35.11A** Clive Bromhall/Getty Images. **35.11B** Bernd Heinrich. **35.12A** Jose B. Ruiz/Nature Picture Library. **35.12B** Michael Hutchinson/Nature Picture Library. **35.13A** J. P. Varin/Jacana/Science Source. **35.14A** Steve Bloom Images/Alamy. **35.14B** Michael S. Quinton/National Geographic Stock/Getty Images. **35.14C** Carol Walker/Nature Picture Library. **35.15A** Kerstin Waurick/Getty Images. **35.15B** Image Quest Marine. **35.16A** blickwinkel/Alamy. **35.16B** Mike W. Howell. **35.18A** Scott Leslie/AGE Fotostock. **35.18B** Kitchin & Hurst/AGE Fotostock. **35.19** Rupert Barrington/Nature Picture Library. **35.20** Renne Lynn/Pearson Education/Pearson Science. **35.21A** Jennifer Jarvis. **35.21B** Shari L. Morris/AGE Fotostock. **35.22A** Michael Nichols/National Geographic Stock. **35.22B** Martin Harvey/Getty Images. **35.23A** Dominic Chan/Newscom. **35.23B** Janet Mayer/Newscom.

Chapter 36: Chapter Opener Juan Carlos Calvin/AGE Fotostock. **p. 723** Simon Phillpotts/Alamy. **36.2A** Matthew Banks/Alamy. **36.2B** Jon Ander Rabadan/Getty Images. **36.2C** mashe/Shutterstock. **36.3 left** Roger Phillips/DK Images. **36.3 center** Jane Burton/DK Images. **36.3 right** Yuri Arcurs/Shutterstock. **36.4A** Simon Phillpotts/Alamy. **36.4B** Roy Corral/Corbis. **36.4C left** Joshua Lewis/Shutterstock. **36.4C right** WizData/Shutterstock. **36.5A** Rick & Nora Bowers/Alamy. **36.5B** Mauricio Handler/National Geographic Image Collection/Alamy. **36.5C** Meul/ARCO/Nature Picture Library. **36.6** Alan Carey/Science Source. **36.11C** Spencer Plat/Getty Images.

Chapter 37: Chapter Opener N+R Colborn/AGE Fotostock. **37.3A** Doug Backlund. **37.3B** Tim Zurowski/Alamy. **37.4** Jurgen Freund/Nature Picture Library. **37.5A** Kenneth M. Highfill/Science Source. **37.5B** Jean Dickey. **37.6 top** WILDLIFE GmbH/Alamy. **37.6 bottom left** Geoffrey Peter Kidd/AGE Fotostock America Inc. **37.6 bottom right** Universal Images Group/SuperStock. **37.7** Joel Sartore/Getty Images. **37.11A** BERNARD CASTELEIN/Nature Picture Library. **37.11C** Genny Anderson. **37.12A** Jean Dickey/Pearson Education/Pearson Science. **37.13B** Embassy of Australia. **37.13C** Ace Stock Limited/Alamy Images. **37.22A** Michael Marten/SPL/Science Source. **37.22B top** NASA/Goddard Space Flight Center. **37.22B bottom** NASA/Goddard Space Flight Center. **37.23A** Fang Xinwu/Newscom. **37.23B** Thony Belizaire/Getty Images.

Chapter 38: Chapter Opener Jan Adamica/Fotolia. **p. 761 top to bottom** Charlie Riedel/AP Photo; Rick & Nora Bowers/Alamy. **38.1A** Mike Dobel/Alamy. **38.1C** Jeff March/Alamy. **38.1D** James King-Holmes/Science Source. **38.2A** Corbis/Superstock Royalty Free. **38.2B** John Mitchell/Science Source. **38.2C** Charlie Riedel/AP Photo. **38.3B left** T. J. Hileman/National Parks Service. **38.3B center** Carl Key/USGS. **38.3B right** Lindsey Bengston/USGS. **38.5A** The Denver Post, Helen H. Richardson/AP Photo. **38.5B** Ingrid Visser/AGE Fotostock. **38.6A** William Osborn/Nature Picture Library. **38.6B** Paul McCormick/Getty Images. **38.7A** Rick & Nora Bowers/Alamy. **38.7B** DLILLC/Corbis RF. **38.8A** Yann Arthus-Bertrand/Corbis. **38.8B** Calvin Larsen/Science Source. **38.8C** Joel Sartore/National Geographic Stock. **p. 772** Neo Edmund/Shutterstock. **38.9B** Roger Kirkpatrick/Image Quest Marine. **38.10B** Chris Fredriksson/Alamy. **38.11A** Belinda Images/SuperStock. **38.11C** Dancestrokes/Shutterstock. **38.12A** YOSHIKAZU TSUNO/AFP/Getty Images/Newscom. **38.12B** South Florida Water Management District. **38.13** Andy Rouse/Nature Picture Library.

Illustration and Text Credits

Chapter 2: 2.15A Adaptation of figure 5 from "Effect of Calcium Carbonate Saturation State on the Calcification Rate of an Experimental Coral Reef" by C. Langdon, et al., from *Global Biogeochemical Cycles* (June 2000): 14(2). Copyright © 2000 by American Geophysical Union. Reprinted with permission of Wiley Inc.

Chapter 4: 4.2B Figure adapted from *The World of the Cell*, 3rd ed., by Wayne M. Becker, Jane B. Reece and Wayne F. Poenie. Copyright © 1996 by Pearson Education, Inc. Adapted and electronically reproduced by permission of Pearson Education, Inc., Upper Saddle River, New Jersey.

Chapter 5: 5.7 Adaptation of Figure 2A from "Appearance of Water Channels in *Xenopus* Oocytes Expressing Red Cell CHIP28 Protein" by Gregory Preston et al., from *Science* (April 1992): 256(5055). Copyright © 1992 by AAAS. Reprinted with permission.

Chapter 7: 7.13B Adaptation of Figure 1A from "Biomass and Toxicity Responses of Poison Ivy (*Toxicodendron radicans*) to Elevated Atmospheric CO_2" by Jacqueline E. Mohan, et al., from *PNAS* (June 2006): 103(24). Copyright © 2006 by National Academy of Sciences. Reprinted with permission.

Chapter 16: 16.6A Figure adapted from *Microbiology: An Introduction*, 9th ed., by Gerard J. Tortora, Berdell R. Funke, and Christine L. Case. Copyright © 2007 by Pearson Education, Inc. Adapted and electronically reproduced by permission of Pearson Education, Inc., Upper Saddle River, New Jersey. **p. 339** Source: U.S. Food and Drug Administration website, 2013.

Chapter 18: 18.1B Figure adapted from *Zoology*, by Lawrence G. Mitchell, John A. Mutchmor, and Warren D. Dolphin, 1998. Used by permission of Lawrence G. Mitchell. **18.15B** Figure adapted from *Zoology*, by Lawrence G. Mitchell, John A. Mutchmor, and Warren D. Dolphin, 1998. Used by permission of Lawrence G. Mitchell.

Chapter 22: 22.1 Figure adapted from *Human Anatomy and Physiology*, 4th ed., by Elaine N. Marieb. Copyright © 1998 by Pearson Education, Inc. Adapted and electronically reproduced by permission of Pearson Education, Inc., Upper Saddle River, New Jersey. **22.12** Figure adapted from *Human Anatomy and Physiology*, 4th ed., by Elaine N. Marieb. Copyright © 1998 by Pearson Education, Inc. Adapted and electronically reproduced by permission of Pearson Education, Inc., Upper Saddle River, New Jersey.

Chapter 24: 24.8B Based on "Learning and Optimization Using the Clonal Selection Principle," Leandro N. de Castro and Fernando J. Von Zuben, *IEEE Transactions on Evolutionary Computation: Special Issue on Artificial Immune Systems* (2002): 6(3), pp. 239–51. Reprinted by permission.

Chapter 25: 25.5 Figure adapted from *Zoology*, by Lawrence G. Mitchell, John A. Mutchmor, and Warren D. Dolphin, 1998. Used by permission of Lawrence G. Mitchell. **25.6A(b), 25.7(a)** Figure adapted from *Human Anatomy and Physiology*, 8th ed., by Elaine N. Marieb and Katja Hoehn. Copyright © 2010 by Pearson Education, Inc. Adapted and electronically reproduced by permission of Pearson Education, Inc., Upper Saddle River, New Jersey.

Chapter 27: 27.3A Figure adapted from *Human Anatomy and Physiology*, 4th ed., by Elaine N. Marieb and Katja Hoehn. Copyright © 1998 by Pearson Education, Inc. Adapted and electronically reproduced by permission of Pearson Education, Inc., Upper Saddle River, New Jersey. **27.3C** Figure adapted from *Human Anatomy and Physiology*, 4th ed., by Elaine N. Marieb and Katja Hoehn. Copyright © 1998 by Pearson Education, Inc. Adapted and electronically reproduced by permission of Pearson Education, Inc., Upper Saddle River, New Jersey. **27.4B** Figure adapted from *Human Anatomy and Physiology*, 4th ed., by Elaine N. Marieb and Katja Hoehn. Copyright © 1998 by Pearson Education, Inc. Adapted and electronically reproduced by permission of Pearson Education, Inc., Upper Saddle River, New Jersey. **Table 27.7** Table adapted from *Microbiology: An Introduction*, 9th ed., by Gerard J. Tortora, Berdell R. Funke, and Christine L. Case. Copyright © 2007 by Pearson Education, Inc. Adapted and electronically reproduced by permission of Pearson Education, Inc., Upper Saddle River, New Jersey. **27.17B** Figure adapted from *Human Anatomy and Physiology*, 4th ed., by Elaine N. Marieb and Katja Hoehn. Copyright © 1998 by Pearson Education, Inc. Adapted and electronically reproduced by permission of Pearson Education, Inc., Upper Saddle River, New Jersey.

Chapter 28: 28.6 Figure adapted from *The World of the Cell*, 6th ed., by Wayne M. Becker, Lewis J. Kleinsmith and Jeff Hardin. Copyright © 2006 by Pearson Education, Inc. Adapted and electronically reproduced by permission of Pearson Education, Inc., Upper Saddle River, New Jersey. **28.9** Figure adapted from "Antidepressants Prevent Hierarchy Destabilization

Induced by Lipopolysaccharide Administration in Mice: A Neurobiological Approach to Depression" by Daniel W. H. Cohn, et al., from *Annals of the New York Academy Of Sciences* (July 2012): 1262. Copyright © 2012 by The New York Academy of Sciences. Reprinted with permission of Wiley Inc. **p. 585** Data from "Input Organization of Two Symmetrical Giant Cells in the Snail Brain," by E. R. Kandel and L. Tauc, *Journal of Physiology* (1966): 183, pp. 269–86.

Chapter 29: 29.5 Figure adapted from *Human Anatomy and Physiology*, 4th ed., by Elaine N. Marieb and Katja Hoehn. Copyright © 1998 by Pearson Education, Inc. Adapted and electronically reproduced by permission of Pearson Education, Inc., Upper Saddle River, New Jersey. **29.9A** Figure adapted from *Human Anatomy and Physiology*, 4th ed., by Elaine N. Marieb and Katja Hoehn. Copyright © 1998 by Pearson Education, Inc. Adapted and electronically reproduced by permission of Pearson Education, Inc., Upper Saddle River, New Jersey. **29.9B** Figure adapted from *Human Anatomy and Physiology*, 4th ed., by Elaine N. Marieb and Katja Hoehn. Copyright © 1998 by Pearson Education, Inc. Adapted and electronically reproduced by permission of Pearson Education, Inc., Upper Saddle River, New Jersey.

Chapter 30: 30.2E Figure adapted from *Zoology*, by Lawrence G. Mitchell, John A. Mutchmor, and Warren D. Dolphin, 1998. Used by permission of Lawrence G. Mitchell. **30.4** Figure adapted from *Human Anatomy and Physiology*, 4th ed., by Elaine N. Marieb. Copyright © 1998 by Pearson Education, Inc. Adapted and electronically reproduced by permission of Pearson Education, Inc., Upper Saddle River, New Jersey. **30.8** Figure adapted from *Human Anatomy and Physiology*, 4th ed., by Elaine N. Marieb. Copyright © 1998 by Pearson Education, Inc. Adapted and electronically reproduced by permission of Pearson Education, Inc., Upper Saddle River, New Jersey. **30.10A** Figure adapted from *Human Anatomy and Physiology*, 4th ed., by Elaine N. Marieb. Copyright © 1998 by Pearson Education, Inc. Adapted and electronically reproduced by permission of Pearson Education, Inc., Upper Saddle River, New Jersey.

Chapter 31: 31.1B Adaptation of map "Multiple Birth" from "Seeking Agriculture's Ancient Roots" by Michael Balter, from *Science* (June 29, 2007): 316(5833). Copyright © 2007 by AAAS. Reprinted with permission.

Chapter 35: 35.7 Figure adapted from *Zoology*, by Lawrence G. Mitchell, John A. Mutchmor, and Warren D. Dolphin, 1998. Used by permission of Lawrence G. Mitchell. **35.14B** Riverwalker - Fotolia.

Chapter 36: 36.5A Adaptation of Figure 1 from "Stability, Regulation, and the Determination of Abundance in an Insular Song Sparrow Population" by Peter Arcese, from *Ecology* (June 1992): 73(3). Copyright © 1992 by Ecological Society of America. Reprinted with permission. **36.5B** Adaptation of Figure 3 from "Predator Responses, Prey Refugees, and Density-Dependent Mortality of a Marine Fish" by Todd W. Anderson, from *Ecology* (January 2001): 82(1). Copyright © 2001 by Ecological Society of America. Reprinted with permission.

Chapter 37: UN 37.1, 37.9 Figure adapted from *Ecology and Field Biology*, 4th ed., by Robert L. Smith. Copyright © 1990 by Pearson Education, Inc. Adapted and electronically reproduced by permission of Pearson Education, Inc., Upper Saddle River, New Jersey. **37.15** Data from Whittaker, *Communities and Ecosystems*, 2nd ed. (New York: Macmillan, 1975).

Chapter 38: 38.9A Adaptation of Figure 1 from "Biodiversity Hotspots for Conservation Priorities" by Norman Myers, et al., from *Nature* (February 24, 2000): 403(6772). Copyright © 2000 by Macmillan Publishers Ltd. Reprinted with permission.

Glossary

A

A site One of two of a ribosome's binding sites for tRNA during translation. The A site holds the tRNA that carries the next amino acid in the polypeptide chain. (A stands for aminoacyl tRNA.)
abiotic factor (ā´-bī-ot´-ik) A nonliving component of an ecosystem, such as air, water, or temperature.
abiotic reservoir (ā´-bī-ot´-ik) The part of an ecosystem where a chemical, such as carbon or nitrogen, accumulates or is stockpiled outside of living organisms.
ABO blood groups Genetically determined classes of human blood that are based on the presence or absence of carbohydrates A and B on the surface of red blood cells. The ABO blood group phenotypes, also called blood types, are A, B, AB, and O.
abscisic acid (ABA) (ab-sis´-ik) A plant hormone that inhibits cell division, promotes dormancy, and interacts with gibberellins in regulating seed germination.
absorption The uptake of small nutrient molecules by an organism's own body; the third main stage of food processing, following digestion.
acetyl CoA (a-sē´-til kō´-ā´) (acetyl coenzyme A) The entry compound for the citric acid cycle in cellular respiration; formed from a two-carbon fragment of pyruvate attached to a coenzyme.
acetylcholine (a-sē´-til-kō´-lēn) A nitrogen-containing neurotransmitter. Among other effects, it slows the heart rate and makes skeletal muscles contract.
acid A substance that increases the hydrogen ion (H^+) concentration in a solution.
acrosome (ak´-ruh-som) A membrane-enclosed sac at the tip of a sperm. The acrosome contains enzymes that help the sperm penetrate an egg.
actin A globular protein that links into chains, two of which twist helically around each other, forming microfilaments in muscle cells.
action potential A change in membrane voltage that transmits a nerve signal along an axon.
activation energy The amount of energy that reactants must absorb before a chemical reaction will start.
activator A protein that switches on a gene or group of genes.
active immunity Immunity conferred by recovering from an infectious disease or by receiving a vaccine.
active site The part of an enzyme where a substrate molecule attaches; typically, a pocket or groove on the enzyme's surface.
active transport The movement of a substance across a biological membrane against its concentration gradient, aided by specific transport proteins and requiring an input of energy (often as ATP).
adaptation An inherited character that enhances an organism's ability to survive and reproduce in a particular environment.
adaptive immunity A vertebrate-specific defense that is activated only after exposure to an antigen and is mediated by lymphocytes. It exhibits specificity, memory, and self-nonself recognition. Also called acquired immunity.
adaptive radiation Period of evolutionary change in which groups of organisms form many new species whose adaptations allow them to fill new or vacant ecological roles in their communities.
adenine (A) (ad´-uh-nēn) A double-ring nitrogenous base found in DNA and RNA.
adhesion The attraction between different kinds of molecules.
adipose tissue A type of connective tissue whose cells contain fat.
adrenal cortex (uh-drē´-nul) The outer portion of an adrenal gland, controlled by ACTH from the anterior pituitary; secretes hormones called glucocorticoids and mineralocorticoids.
adrenal gland (uh-drē´-nul) One of a pair of endocrine glands, located atop each kidney in mammals, composed of an outer cortex and a central medulla.
adrenal medulla (uh-drē´-nul muh-dul´-uh) The central portion of an adrenal gland, controlled by nerve signals; secretes the fight-or-flight hormones epinephrine and norepinephrine.
adrenocorticotropic hormone (ACTH) (uh-drē´-nō-cōr´-ti-kō-trop´-ik) A protein hormone secreted by the anterior pituitary that stimulates the adrenal cortex to secrete corticosteroids.
adult stem cell A cell present in adult tissues that generates replacements for nondividing differentiated cells. Adult stem cells are capable of differentiating into multiple cell types, but they are not as developmentally flexible as embryonic stem cells.
age structure The relative number of individuals of each age in a population.
agonistic behavior (a´-gō-nis´-tik) Confrontational behavior involving a contest waged by threats, displays, or actual combat that settles disputes over limited resources, such as food or mates.
AIDS (acquired immunodeficiency syndrome) The late stages of HIV infection, characterized by a reduced number of T cells and the appearance of characteristic opportunistic infections.
alcohol fermentation Glycolysis followed by the reduction of pyruvate to ethyl alcohol, regenerating NAD^+ and releasing carbon dioxide.
alga (al´-guh) (plural, **algae**) A protist that produces its food by photosynthesis.
alimentary canal (al´-uh-men´-tuh-rē) A complete digestive tract consisting of a tube running between a mouth and an anus.
allantois (al´-an-tō´-is) In animals, an extraembryonic membrane that develops from the yolk sac. The allantois helps dispose of the embryo's nitrogenous wastes and forms part of the umbilical cord in mammals.
allele (uh-lē´-ul) An alternative version of a gene.
allergen (al´-er-jen) An antigen that causes an allergy.
allergy A disorder of the immune system caused by an abnormally high sensitivity to an antigen. Symptoms are triggered by histamines released from mast cells.
allopatric speciation The formation of new species in populations that are geographically isolated from one another.
alternation of generations A life cycle in which there is both a multicellular diploid form, the sporophyte, and a multicellular haploid form, the gametophyte; a characteristic of plants and multicellular green algae.
alternative RNA splicing A type of regulation at the RNA-processing level in which different mRNA molecules are produced from the same primary transcript, depending on which RNA segments are treated as exons and which as introns.
altruism (al´-trū-iz-um) Behavior that reduces an individual's fitness while increasing the fitness of another individual.
Alveolata (al-vē´-uh-let-uh) A clade of the SAR supergroup of protists that includes dinoflagellates, ciliates, and certain parasites.
alveolus (al-vē´-oh-lus) (plural, **alveoli**) One of the dead-end air sacs within the mammalian lung where gas exchange occurs.
Alzheimer's disease (AD) An age-related dementia (mental deterioration) characterized by confusion, memory loss, and other symptoms.
amino acid (uh-mēn´-ō) An organic molecule containing a carboxyl group and an amino group; serves as the monomer of proteins.
amino group (uh-mēn´-ō) A chemical group consisting of a nitrogen atom bonded to two hydrogen atoms.
ammonia NH_3; A small and very toxic nitrogenous waste produced by metabolism.

GLOSSARY

amniocentesis (am′-nē-ō-sen-tē′-sis) A technique for diagnosing genetic defects while a fetus is in the uterus. A sample of amniotic fluid, obtained by a needle inserted into the uterus, is analyzed for telltale chemicals and defective fetal cells.

amnion (am′-nē-on) In vertebrate animals, the extraembryonic membrane that encloses the fluid-filled amniotic sac containing the embryo.

amniote Member of a clade of tetrapods that have an amniotic egg containing specialized membranes that protect the embryo. Amniotes include mammals and birds and other reptiles.

amniotic egg (am′-nē-ot′-ik) A shelled egg in which an embryo develops within a fluid-filled amniotic sac and is nourished by yolk. Produced by reptiles (including birds) and egg-laying mammals, the amniotic egg enables them to complete their life cycles on dry land.

amoeba (uh-mē′-buh) A general term for a protist that moves and feeds by means of pseudopodia.

amoebocyte (uh-mē′-buh-sīt) An amoeba-like cell that moves by pseudopodia and is found in most animals; depending on the species, may digest and distribute food, dispose of wastes, form skeletal fibers, fight infections, and change into other cell types.

amoebozoan A member of a clade of protists in the supergroup Unikonta that includes amoebas and slime molds and is characterized by lobe-shaped pseudopodia.

amphibian Member of a clade of tetrapods that includes frogs, toads, salamanders, and caecilians.

amygdala (uh-mig′-duh-la) An integrative center of the cerebrum; functionally, the part of the limbic system that seems central in recognizing the emotional content of facial expressions and laying down emotional memories.

anabolic steroid (an′-uh-bol′-ik ster′-oyd) A synthetic variant of the male hormone testosterone that mimics some of its effects.

analogy The similarity between two species that is due to convergent evolution rather than to descent from a common ancestor with the same trait.

anaphase The fourth stage of mitosis, beginning when sister chromatids separate from each other and ending when a complete set of daughter chromosomes arrives at each of the two poles of the cell.

anatomy The study of the structures of an organism.

anchorage dependence The requirement that to divide, a cell must be attached to a solid surface.

androgen (an′-drō-jen) A steroid sex hormone secreted by the gonads that promotes the development and maintenance of the male reproductive system and male body features.

anemia (uh-nē′-me-ah) A condition in which an abnormally low amount of hemoglobin or a low number of red blood cells results in the body cells receiving too little oxygen.

angiosperm (an′-jē-ō-sperm) A flowering plant, which forms seeds inside a protective chamber called an ovary.

annelid (uh-nel′-id) A segmented worm. Annelids include earthworms, polychaetes, and leeches.

annual A plant that completes its life cycle in a single year or growing season.

antagonistic hormones Two hormones that have opposite effects.

anterior Pertaining to the front, or head, of a bilaterally symmetric animal.

anterior pituitary (puh-tū′-uh-tār-ē) An endocrine gland, adjacent to the hypothalamus and the posterior pituitary, that synthesizes several hormones, including some that control the activity of other endocrine glands.

anther A sac located at the tip of a flower's stamen; contains male sporangia in which meiosis occurs to produce spores that form the male gametophytes, or pollen grains.

anthropoid (an′-thruh-poyd) A member of a primate group made up of the apes (gibbons, orangutans, gorillas, chimpanzees, bonobos, and humans) and monkeys.

antibody (an′-tih-bod′-ē) A protein dissolved in blood plasma that attaches to a specific kind of antigen and helps counter its effects; secreted by plasma cells.

anticodon (an′-tī-kō′-don) On a tRNA molecule, a specific sequence of three nucleotides that is complementary to a codon triplet on mRNA.

antidiuretic hormone (ADH) (an′-tē-dī′-yū-ret′-ik) A hormone made by the hypothalamus and secreted by the posterior pituitary that promotes water retention by the kidneys.

antigen (an′-tuh-jen) A foreign (nonself) molecule that elicits an adaptive immune response.

antigen receptor (an′-tuh-jen) The general term for a surface protein, located on B cells and T cells, that binds to antigens and initiates the adaptive immune response.

antigen-binding site (an′-tuh-jen) A region of the antigen receptor or antibody that binds the antigenic determinant on the antigen.

antigenic determinant (an′-tuh-jen′-ik) A small region on the surface of an antigen molecule to which an antigen receptor or antibody binds; also called an epitope.

antigen-presenting cell (APC) (an′-tuh-jen) One of a family of white blood cells that ingests a foreign substance or a microbe and attaches antigenic portions of the ingested material to its own surface, thereby displaying the antigens to a helper T cell.

antihistamine (an′-tē-his′-tuh-mēn) A drug that interferes with the action of histamine, providing relief from an allergic reaction.

anus The opening through which undigested materials are expelled.

aorta (ā-or′-tuh) A large artery that conveys blood directly from the left ventricle of the heart to other arteries.

aphotic zone (ā-fō′-tik) The region of an aquatic ecosystem beneath the photic zone, where light does not penetrate enough for photosynthesis to take place.

apical dominance (ā′-pik-ul) In a plant, the hormonal inhibition of axillary buds by a terminal bud.

apical meristem (ā′-pik-ul mer′-uh-stem) A growth-producing region of cell division consisting of undifferentiated cells located at the tip of a plant root or in the terminal or axillary bud of a shoot.

apoptosis (ā-puh-tō′-sus) The timely and tidy suicide of cells; also called programmed cell death.

appendicular skeleton (ap′-en-dik′-yū-ler) Components of the skeletal system that support the fins of a fish or the arms and legs of a land vertebrate; in land vertebrates, the cartilage and bones of the shoulder girdle, pelvic girdle, forelimbs, and hind limbs. *See also* axial skeleton.

appendix (uh-pen′-dix) A small, finger-like extension of the vertebrate cecum; contains a mass of white blood cells that contribute to immunity.

aquaporin A transport protein in the plasma membrane of an animal, plant, or microorganism cell that facilitates the diffusion of water across the membrane (osmosis).

aqueous humor (ā′-kwē-us hyū′-mer) Plasma-like liquid in the space between the lens and the cornea in the vertebrate eye; helps maintain the shape of the eye, supplies nutrients and oxygen to its tissues, and disposes of its wastes.

aqueous solution (ā′-kwē-us) A solution in which water is the solvent.

arachnid A member of a major arthropod group (chelicerates) that includes spiders, scorpions, ticks, and mites.

Archaea (ar′-kē-uh) One of two prokaryotic domains of life, the other being Bacteria.

Archaeplastida One of four monophyletic supergroups proposed in a current hypothesis of the evolutionary history of eukaryotes. The other four supergroups are SAR (Stramenopila, Alveolata, and Rhizaria), Excavata, and Unikonta.

arteriole (ar-ter′-ē-ōl) A vessel that conveys blood between an artery and a capillary bed.

artery A vessel that carries blood away from the heart to other parts of the body.

arthropod (ar´-thrō-pod) A member of the most diverse phylum in the animal kingdom. Arthropods include the horseshoe crab, arachnids (for example, spiders, ticks, scorpions, and mites), crustaceans (for example, crayfish, lobsters, crabs, and barnacles), millipedes, centipedes, and insects. Arthropods are characterized by a chitinous exoskeleton, molting, jointed appendages, and a body formed of distinct groups of segments.

artificial selection The selective breeding of domesticated plants and animals to promote the occurrence of desirable traits.

ascomycete (as´-kuh-mī´-sēt) Member of a group of fungi characterized by saclike structures called asci that produce spores in sexual reproduction.

asexual reproduction The creation of genetically identical offspring by a single parent, without the participation of sperm and egg.

assisted reproductive technology Procedure that involves surgically removing eggs from a woman's ovaries, fertilizing them, and then returning them to the woman's body. *See also in vitro* fertilization.

association areas Sites of higher mental activities, making up most of the cerebral cortex.

associative learning The ability to associate one environmental feature with another. In one type of associative learning, the animal learns to link a particular stimulus with a particular outcome. Trial-and-error learning is also a type of associative learning.

astigmatism (uh-stig´-muh-tizm) Blurred vision caused by a misshapen lens or cornea.

atherosclerosis (ath´-uh-rō´-skluh-rō´-sis) A cardiovascular disease in which fatty deposits called plaques develop on the inner walls of the arteries, narrowing their inner diameters.

atom The smallest unit of matter that retains the properties of an element.

atomic mass The total mass of an atom; also called atomic weight. Given as a whole number, the atomic mass approximately equals the mass number.

atomic number The number of protons in each atom of a particular element.

ATP Adenosine triphosphate, the main energy source for cells. ATP releases energy when its phosphate bonds are hydrolyzed.

ATP synthase A cluster of several membrane proteins that function in chemiosmosis with adjacent electron transport chains, using the energy of a hydrogen ion concentration gradient to make ATP.

atrium (ā´-trē-um) (plural, **atria**) A heart chamber that receives blood from the veins.

auditory canal Part of the vertebrate outer ear that channels sound waves from the pinna or outer body surface to the eardrum.

autoimmune disorder An immunological disorder in which the immune system attacks the body's own molecules.

autonomic nervous system (ot´-ō-nom´-ik) The component of the vertebrate peripheral nervous system that regulates the internal environment; made up of sympathetic and parasympathetic subdivisions. Most actions of the autonomic nervous system are involuntary.

autosome A chromosome not directly involved in determining the sex of an organism; in mammals, for example, any chromosome other than X or Y.

autotroph (ot´-ō-trōf) An organism that makes its own food (often by photosynthesis), thereby sustaining itself without eating other organisms or their molecules. Plants, algae, and numerous bacteria are autotrophs.

auxin (ok´-sin) A plant hormone (indoleacetic acid or a related compound) that promotes seedling elongation.

AV (atrioventricular) node A region of specialized heart muscle tissue between the left and right atria where electrical impulses are delayed for about 0.1 second before spreading to both ventricles and causing them to contract.

axial skeleton (ak´-sē-ul) Components of the skeletal system that support the central trunk of the body: the skull, backbone, and rib cage in a vertebrate. *See also* appendicular skeleton.

axillary bud (ak´-sil-ār-ē) An embryonic shoot present in the angle formed by a leaf and stem.

axon (ak´-son) A neuron extension that conducts signals to another neuron or to an effector cell. A neuron has one long axon.

B

B cell A type of lymphocyte that completes its development in the bone marrow and is responsible for the humoral immune response. Effector B cells are also called plasma cells.

bacillus (buh-sil´-us) (plural, **bacilli**) A rod-shaped prokaryotic cell.

Bacteria One of two prokaryotic domains of life, the other being Archaea.

bacteriophage (bak-tēr´-ē-ō-fāj) A virus that infects bacteria; also called a phage.

balancing selection Natural selection that maintains stable frequencies of two or more phenotypic forms in a population.

ball-and-socket joint A joint that allows rotation and movement in several planes. Examples in humans are the hip and shoulder joints.

bark All the tissues external to the vascular cambium in a plant that is growing in thickness. Bark is made up of secondary phloem, cork cambium, and cork.

Barr body A dense body formed from a deactivated X chromosome found in the nuclei of female mammalian cells.

basal metabolic rate (BMR) The number of kilocalories a resting animal requires to fuel its essential body processes for a given time.

basal nuclei (bā´-sul nū´-klē-ī) Clusters of nerve cell bodies located deep within the cerebrum that are important in motor coordination.

base A substance that decreases the hydrogen ion (H^+) concentration in a solution.

basidiomycete (buh-sid´-ē-ō-mī´sēt) Member of a group of fungi characterized by club-shaped, spore-producing structures called basidia.

basilar membrane The floor of the middle canal of the inner ear.

behavior Individually, an action carried out by the muscles or glands under control of the nervous system in response to a stimulus; collectively, the sum of an animal's responses to external and internal stimuli.

behavioral ecology The study of behavior in an evolutionary context.

benign tumor An abnormal mass of cells that remains at its original site in the body.

benthic realm A seafloor or the bottom of a freshwater lake, pond, river, or stream.

biennial A plant that completes its life cycle in two years.

bilateral symmetry An arrangement of body parts such that an organism can be divided equally by a single cut passing longitudinally through it. A bilaterally symmetric organism has mirror-image right and left sides.

bilaterian Member of the clade Bilateria, animals exhibiting bilateral symmetry.

bile A mixture of substances that is produced by the liver and stored in the gallbladder. Bile emulsifies fats and aids in their digestion.

binary fission A means of asexual reproduction in which a parent organism, often a single cell, divides into two genetically identical individuals of about equal size.

binomial A two-part, latinized name of a species; for example, *Homo sapiens*.

biodiversity The variety of living things; includes genetic diversity, species diversity, and ecosystem diversity.

biodiversity hot spot A small geographic area with an exceptional concentration of endangered and threatened species, especially endemic species (those found nowhere else).

biofilm A surface-coating colony of prokaryotes that engage in metabolic cooperation.

biogeochemical cycle Any of the various chemical circuits that involve both biotic and abiotic components of an ecosystem.

biogeography The study of the past and present distribution of organisms.
biological clock An internal timekeeper that controls an organism's biological rhythms, marking time with or without environmental cues but often requiring signals from the environment to remain tuned to an appropriate period. *See also* circadian rhythm.
biological control The intentional release of a natural enemy to attack a pest population.
biological magnification The accumulation of harmful chemicals that are retained in the living tissues of consumers in food chains.
biological species concept Definition of a species as a group of populations whose members have the potential to interbreed in nature and produce viable, fertile offspring but do not produce viable, fertile offspring with members of other such populations.
biology The scientific study of life.
biomass The amount, or mass, of organic material in an ecosystem.
biome (bī´-ōm) A major type of ecological association that occupies a broad geographic region of land or water and is characterized by organisms adapted to the particular environment.
bioremediation The use of living organisms to detoxify and restore polluted and degraded ecosystems.
biosphere The entire portion of Earth inhabited by life; the sum of all the planet's ecosystems.
biotechnology The manipulation of living organisms or their components to make useful products.
biotic factor (bī-o´-tik) A living component of a biological community; an organism, or a factor pertaining to one or more organisms.
bipolar disorder Depressive mental illness characterized by extreme mood swings; also called manic-depressive disorder.
birds Members of a clade of reptiles that have feathers and adaptations for flight.
bivalve A member of a group of molluscs that includes clams, mussels, scallops, and oysters.
blastocoel (blas´-tuh-sēl) In a developing animal, a central, fluid-filled cavity in a blastula.
blastocyst (blas´-tō-sist) A mammalian embryo (equivalent to an amphibian blastula) made up of a hollow ball of cells that results from cleavage and that implants in the mother's endometrium.
blastula (blas´-tyū-luh) An embryonic stage that marks the end of cleavage during animal development; a hollow ball of cells in many species.
blood A type of connective tissue with a fluid matrix called plasma in which red blood cells, white blood cells, and platelets are suspended.
blood pressure The force that blood exerts against the walls of blood vessels.
blood-brain barrier A system of capillaries in the brain that restricts passage of most substances into the brain, thereby preventing large fluctuations in the brain's environment.
body cavity A fluid-containing space between the digestive tract and the body wall.
bolus A lubricated ball of chewed food.
bone A type of connective tissue consisting of living cells held in a rigid matrix of collagen fibers embedded in calcium salts.
bottleneck effect Genetic drift resulting from a drastic reduction in population size. Typically, the surviving population is no longer genetically representative of the original population.
Bowman's capsule A cup-shaped swelling at the receiving end of a nephron in the vertebrate kidney; collects the filtrate from the blood.
brain The master control center of the nervous system, involved in regulating and controlling body activity and interpreting information from the senses transmitted through the nervous system.
brainstem A functional unit of the vertebrate brain, composed of the midbrain, the medulla oblongata, and the pons; serves mainly as a sensory filter, selecting which information reaches higher brain centers.
brassinosteroids A class of plant steroid hormones that promote cell elongation and cell division in stems and seedlings.
breathing Ventilation of the lungs through alternating inhalation and exhalation of air.
breathing control center The part of the medulla in the brain that directs the activity of organs involved in breathing.
bronchiole (bron´-kē-ōl) A fine branch of the bronchi that transports air to alveoli.
bronchus (bron´-kus) (plural, **bronchi**) One of a pair of breathing tubes that branch from the trachea into the lungs.
brown alga One of a group of marine, multicellular, autotrophic protists belonging to the stramenopile clade of the SAR supergroup; the most common and largest type of seaweed. Brown algae include the kelps.
bryophyte (brī´-uh-fīt) A plant that lacks xylem and phloem; a seedless nonvascular plant. Bryophytes include mosses, liverworts, and hornworts.
budding A means of asexual reproduction whereby a new individual develops from an outgrowth of a parent. The new individual eventually splits off and lives independently.
buffer A chemical substance that resists changes in pH by accepting hydrogen ions from or donating hydrogen ions to solutions.
bulbourethral gland (bul´-bō-yū-rē´-thrul) One of a pair of glands near the base of the penis in the human male that secrete a clear alkaline mucus.
bulk feeder An animal that eats relatively large pieces of food.

C

C_3 plant A plant that uses the Calvin cycle for the initial steps that incorporate CO_2 into organic material, forming a three-carbon compound as the first stable intermediate.
C_4 plant A plant in which the Calvin cycle is preceded by reactions that incorporate CO_2 into a four-carbon compound, which then supplies CO_2 for the Calvin cycle.
Calvin cycle The second of two stages of photosynthesis; a cyclic series of chemical reactions that occur in the stroma of a chloroplast, using the carbon in CO_2 and the ATP and NADPH produced by the light reactions to make the energy-rich sugar molecule G3P.
CAM plant A plant that uses an adaptation for photosynthesis in arid conditions in which carbon dioxide entering open stomata during the night is converted to organic acids, which release CO_2 for the Calvin cycle during the day, when stomata are closed.
cancer A disease characterized by the presence of malignant tumors (rapidly growing and spreading masses of abnormal body cells) in the body.
capillary (kap´-il-er-ē) A microscopic blood vessel that conveys blood between an arteriole and a venule; enables the exchange of nutrients and dissolved gases between the blood and interstitial fluid.
capillary bed (kap´-il-er-ē) A network of capillaries in a tissue or organ.
capsid The protein shell that encloses a viral genome.
carbohydrate (kar´-bō-hī´-drāt) Member of the class of biological molecules consisting of single-monomer sugars (monosaccharides), two-monomer sugars (disaccharides), and polymers (polysaccharides).
carbon fixation The incorporation of carbon from atmospheric CO_2 into an organic compound. During photosynthesis in a C_3 plant, carbon is fixed into a three-carbon sugar as it enters the Calvin cycle. In C_4 and CAM plants, carbon is first fixed into a four-carbon sugar.
carbonyl group (kar´-buh-nēl´) A chemical group consisting of a carbon atom linked by a double bond to an oxygen atom.
carboxyl group (kar-bok´-sil) A chemical group consisting of a carbon atom double-bonded to an oxygen atom and also bonded to a hydroxyl group.
carcinogen (kar-sin´-uh-jin) A cancer-causing agent, either high-energy radiation (such as X-rays or UV light) or a chemical.
cardiac cycle (kar´-dē-ak) The alternating contractions and relaxations of the heart.

cardiac muscle (kar′-dē-ak) A type of striated muscle that forms the contractile wall of the heart.

cardiac output (kar′-dē-ak) The volume of blood pumped per minute by each ventricle of the heart.

cardiovascular disease (kar′-dē-ō-vas′-kyū-ler) Disorders of the heart and blood vessels.

cardiovascular system (kar′-dē-ō-vas′-kyū-ler) A closed circulatory system with a heart and a branching network of arteries, capillaries, and veins.

carnivore An animal that mainly eats other animals.

carpel (kar′-pul) The female part of a flower, consisting of a stalk with an ovary at the base and a stigma, which traps pollen, at the tip.

carrier An individual who is heterozygous for a recessively inherited disorder and who therefore does not show symptoms of that disorder but who may pass on the recessive allele to offspring.

carrying capacity In a population, the number of individuals that an environment can sustain.

cartilage (kar′-ti-lij) A flexible connective tissue consisting of living cells and collagenous fibers embedded in a rubbery matrix.

Casparian strip (kas-par′-ē-un) A waxy barrier in the walls of endodermal cells in a plant root that prevents water and ions from entering the xylem without crossing one or more cell membranes.

cation exchange A process in which positively charged minerals are made available to a plant when hydrogen ions in the soil displace mineral ions from the clay particles.

cecum (sē′-kum) (plural, **ceca**) A blind outpocket at the beginning of the large intestine.

cell A basic unit of living matter separated from its environment by a plasma membrane; the fundamental structural unit of life.

cell body The part of a cell, such as a neuron, that houses the nucleus.

cell cycle An ordered sequence of events (including interphase and the mitotic phase) that extends from the time a eukaryotic cell is first formed from a dividing parent cell until its own division into two cells.

cell cycle control system A cyclically operating set of proteins that triggers and coordinates events in the eukaryotic cell cycle.

cell division The reproduction of a cell through duplication of the genome and division of the cytoplasm.

cell plate A double membrane across the midline of a dividing plant cell, between which the new cell wall forms during cytokinesis.

cell theory The theory that all living things are composed of cells and that all cells come from other cells.

cell wall A protective layer external to the plasma membrane in plant cells, bacteria, fungi, and some protists; protects the cell and helps maintain its shape.

cell-mediated immune response The branch of adaptive immunity that involves the activation of cytotoxic T cells, which defend against infected cells.

cellular metabolism (muh-tab′-uh-lizm) All the chemical activities of a cell.

cellular respiration The aerobic harvesting of energy from food molecules; the energy-releasing chemical breakdown of food molecules, such as glucose, and the storage of potential energy in a form that cells can use to perform work; involves glycolysis, the citric acid cycle, and oxidative phosphorylation (the electron transport chain and chemiosmosis).

cellular slime mold A type of protist that has unicellular amoeboid cells and aggregated reproductive bodies in its life cycle; a member of the amoebozoan clade.

cellulose (sel′-yū-lōs) A structural polysaccharide of plant cell walls composed of glucose monomers. Cellulose molecules are linked by hydrogen bonds into cable-like fibrils.

centipede A carnivorous terrestrial arthropod that has one pair of long legs for each of its numerous body segments, with the front pair modified as poison claws.

central canal The narrow cavity in the center of the spinal cord that is continuous with the fluid-filled ventricles of the brain.

central nervous system (CNS) The integration and command center of the nervous system; the brain and, in vertebrates, the spinal cord.

central vacuole In a plant cell, a large membranous sac with diverse roles in growth and the storage of chemicals and wastes.

centralization The presence of a central nervous system (CNS) distinct from a peripheral nervous system.

centromere (sen′-trō-mēr) The region of a duplicated chromosome where two sister chromatids are joined (often appearing as a narrow "waist") and where spindle microtubules attach during mitosis and meiosis. The centromere divides at the onset of anaphase during mitosis and anaphase II during meiosis.

centrosome A structure found in animal cells from which microtubules originate and that is important during cell division. A centrosome has two centrioles.

cephalization (sef′-uh-luh-zā′-shun) An evolutionary trend toward concentration of the nervous system at the head end.

cephalopod A member of a group of molluscs that includes squids, cuttlefish, octopuses, and nautiluses.

cerebellum (sār′-ruh-bel′-um) Part of the vertebrate hindbrain; mainly a planning center that interacts closely with the cerebrum in coordinating body movement.

cerebral cortex (suh-rē′-brul kor′-teks) A folded sheet of gray matter forming the surface of the cerebrum. In humans, it contains integrating centers for higher brain functions such as reasoning, speech, language, and imagination.

cerebral hemisphere (suh-rē′-brul) The right or left half of the vertebrate cerebrum.

cerebrospinal fluid (suh-rē′-brō-spī′-nul) Blood-derived fluid that surrounds, nourishes, and cushions the brain and spinal cord.

cerebrum (suh-rē′-brum) The largest, most sophisticated, and most dominant part of the vertebrate forebrain, made up of right and left cerebral hemispheres.

cervix (ser′-viks) The neck of the uterus, which opens into the vagina.

chaparral (shap′-uh-ral′) A biome dominated by spiny evergreen shrubs adapted to periodic drought and fires; found where cold ocean currents circulate offshore, creating mild, rainy winters and long, hot, dry summers.

character A heritable feature that varies among individuals within a population, such as flower color in pea plants or eye color in humans.

chelicerate (kē-lih-suh′-rāte) A lineage of arthropods that includes horseshoe crabs, scorpions, ticks, and spiders.

chemical bond An attraction between two atoms resulting from a sharing of outer-shell electrons or the presence of opposite charges on the atoms. The bonded atoms gain complete outer electron shells.

chemical cycling The use and reuse of a chemical element, such as carbon, within an ecosystem.

chemical energy Energy available in molecules for release in a chemical reaction; a form of potential energy.

chemical reaction The making and breaking of chemical bonds, leading to changes in the composition of matter.

chemiosmosis (kem′-ē-oz-mō′-sis) An energy-coupling mechanism that uses the energy of hydrogen ion (H^+) gradients across membranes to drive cellular work, such as the phosphorylation of ADP; powers most ATP synthesis in cells.

chemoautotroph An organism that obtains both energy and carbon from inorganic chemicals. A chemoautotroph makes its own organic compounds from CO_2 without using light energy.

chemoheterotroph An organism that obtains both energy and carbon from organic compounds.

chemoreceptor (kē′-mō-rē-sep′-ter) A sensory receptor that detects chemical changes within the body or a specific kind of molecule in the external environment.

chiasma (kī-az′-muh) (plural, **chiasmata**) The microscopically visible site where crossing over has occurred between chromatids of homologous chromosomes during prophase I of meiosis.

GLOSSARY

chitin (kī-tin) A structural polysaccharide found in many fungal cell walls and in the exoskeletons of arthropods.

chlamydia A member of a group of bacteria that live inside eukaryotic host cells; a common sexually transmitted disease caused by the bacterium *Chlamydia trachomatis.*

chlorophyll A green pigment located within the chloroplasts of plants and algae and in the membranes of certain prokaryotes. Chlorophyll *a* participates directly in the light reactions, which convert solar energy to chemical energy.

chloroplast (klō′-rō-plast) An organelle found in plants and algae that absorbs sunlight and uses it to drive the synthesis of organic compounds (sugars) from carbon dioxide and water.

choanocyte (kō-an′-uh-sīt) A flagellated feeding cell found in sponges. Also called a collar cell, it has a collar-like ring that traps food particles around the base of its flagellum.

cholesterol (kō-les′-tuh-rol) A steroid that is an important component of animal cell membranes and that acts as a precursor molecule for the synthesis of other steroids, such as hormones.

chondrichthyan (kon-drik′-thē-an) Cartilaginous fish; member of a clade of jawed vertebrates with skeletons made mostly of cartilage, such as sharks and rays.

chorion (kō′r-ē-on) In animals, the outermost extraembryonic membrane, which becomes the mammalian embryo's part of the placenta.

chorionic villus (kōr′-ē-on′-ik vil′-us) Outgrowth of the chorion, containing embryonic blood vessels. As part of the placenta, chorionic villi absorb nutrients and oxygen from, and pass wastes into, the mother's bloodstream.

chorionic villus sampling (CVS) A technique for diagnosing genetic defects while the fetus is in an early development stage within the uterus. A small sample of the fetal portion of the placenta is removed and analyzed.

choroid (kōr′-oyd) A thin, pigmented layer in the vertebrate eye, surrounded by the sclera. The iris is part of the choroid.

chromatin (krō′-muh-tin) The complex of DNA and proteins that makes up eukaryotic chromosomes; often used to refer to the diffuse, very extended form taken by chromosomes when a cell is not dividing.

chromosome (krō′-muh-sōm) A gene-carrying structure found in the nucleus of a eukaryotic cell and most visible during mitosis and meiosis; also, the main gene-carrying structure of a prokaryotic cell. A chromosome consists of one very long DNA molecule and associated proteins.

chromosome theory of inheritance (krō′-muh-sōm) A basic principle in biology stating that genes are located on chromosomes and that the behavior of chromosomes during meiosis accounts for inheritance patterns.

chronic traumatic encephalopathy A dementia (mental deterioration) caused by brain trauma such as sports concussions and characterized by depression, memory loss, and other symptoms.

chyme (kīm) The mixture of partially digested food and digestive juices formed in the stomach.

chytrid (kī-trid) Member of a group of fungi that are mostly aquatic and have flagellated spores. They probably represent the most primitive fungal lineage.

ciliate (sil′-ē-it) A type of protist that moves and feeds by means of cilia. Ciliates belong to the alveolate clade of the SAR supergroup.

cilium (plural, **cilia**) A short cellular appendage specialized for locomotion or moving fluid past the cell, formed from a core of nine outer doublet microtubules and two single microtubules (the "9 + 2" arrangement) covered by the cell's plasma membrane.

circadian rhythm (ser-kā′-dē-un) In an organism, a biological cycle of about 24 hours that is controlled by a biological clock, usually under the influence of environmental cues; a pattern of activity that is repeated daily. *See also* biological clock.

circulatory system The organ system that transports materials such as nutrients, O_2, and hormones to body cells and transports CO_2 and other wastes from body cells.

citric acid cycle The chemical cycle that completes the metabolic breakdown of glucose molecules begun in glycolysis by oxidizing acetyl CoA (derived from pyruvate) to carbon dioxide. The cycle occurs in the matrix of mitochondria and supplies most of the NADH molecules that carry electrons to the electron transport chains. Together with pyruvate oxidation, the second major stage of cellular respiration.

clade A group of species that includes an ancestral species and all its descendants.

cladistics (kluh-dis′-tiks) An approach to systematics in which common descent is the primary criterion used to classify organisms by placing them into groups called clades.

class In Linnaean classification, the taxonomic category above order.

cleavage (klē′-vij) (1) Cytokinesis in animal cells and in some protists, characterized by pinching in of the plasma membrane. (2) In animal development, the first major phase of embryonic development, in which rapid cell divisions without cell growth transforms the animal zygote into a ball of cells.

cleavage furrow (klē′-vij) The first sign of cytokinesis during cell division in an animal cell; a shallow groove in the cell surface near the old metaphase plate.

clitoris An organ in the female that engorges with blood and becomes erect during sexual arousal.

clonal selection (klōn′-ul) The process by which an antigen selectively binds to and activates only those lymphocytes bearing receptors specific for the antigen. The selected lymphocytes proliferate and differentiate into a clone of effector cells and a clone of memory cells specific for the stimulating antigen.

clone As a verb, to produce genetically identical copies of a cell, organism, or DNA molecule. As a noun, the collection of cells, organisms, or molecules resulting from cloning; colloquially, a single organism that is genetically identical to another because it arose from the cloning of a somatic cell.

closed circulatory system A circulatory system in which blood is confined to vessels and is kept separate from the interstitial fluid.

club fungus *See* basidiomycete.

clumped dispersion pattern A pattern in which the individuals of a population are aggregated in patches.

cnidarian (nī-dār′-ē-un) An animal characterized by cnidocytes, radial symmetry, a gastrovascular cavity, and a polyp and medusa body form. Cnidarians include the hydras, jellies, sea anemones, corals, and related animals.

cnidocyte (nī′-duh-sīt) A specialized cell for which the phylum Cnidaria is named; consists of a capsule containing a fine coiled thread, which, when discharged, functions in defense and prey capture.

coccus (kok′-us) (plural, **cocci**) A spherical prokaryotic cell.

cochlea (kok′-lē-uh) A coiled tube in the inner ear of birds and mammals that contains the hearing organ, the organ of Corti.

codominant Inheritance pattern in which a heterozygote expresses the distinct trait of both alleles.

codon (kō′-don) A three-nucleotide sequence in mRNA that specifies a particular amino acid or polypeptide termination signal; the basic unit of the genetic code.

coelom (sē′-lom) A body cavity completely lined with mesoderm.

coenzyme An organic molecule serving as a cofactor. Most vitamins function as coenzymes in important metabolic reactions.

coevolution Evolutionary change in which adaptations in one species act as a selective force on a second species, inducing adaptations that in turn act as a selective force on the first species; a series of reciprocal evolutionary adaptations in two interacting species.

cofactor A nonprotein molecule or ion that is required for the proper functioning of an enzyme. *See also* coenzyme.

cognition The process carried out by an animal's nervous system to perceive, store, integrate, and use information obtained by the animal's sensory receptors.

cohesion (kō-hē′-zhun) The sticking together of molecules of the same kind, often by hydrogen bonds.

collecting duct A tube in the vertebrate kidney that concentrates urine while conveying it to the renal pelvis.
collenchyma cell (kō-len′-kim-uh) In plants, a cell with a thick primary wall and no secondary wall, functioning mainly in supporting growing parts.
colon (kō′-lun) The largest section of the large intestine.
communication Animal behavior including transmission of, reception of, and response to signals.
community An assemblage of all the populations of organisms living close enough together for potential interactions.
companion cell In a plant, a cell connected to a sieve-tube element whose nucleus and ribosomes provide proteins for the sieve-tube element.
competitive inhibitor A substance that reduces the activity of an enzyme by entering the active site in place of the substrate. A competitive inhibitor's structure mimics that of the enzyme's substrate.
complement system A family of innate defensive blood proteins that cooperate with other components of the vertebrate defense system to protect against microbes; can enhance phagocytosis, directly lyse pathogens, and amplify the inflammatory response.
complementary DNA (cDNA) A DNA molecule made *in vitro* using mRNA as a template and the enzyme reverse transcriptase. A cDNA molecule therefore corresponds to a gene but lacks the introns present in the DNA of the genome.
complete digestive tract A digestive tube with two openings, a mouth and an anus.
complete dominance A type of inheritance in which the phenotypes of the heterozygote and dominant homozygote are indistinguishable.
complete metamorphosis (met′-uh-mōr′-fuh-sis) A type of development in certain insects in which development from larva to adult is achieved by multiple molts that are followed by a pupal stage. While encased in its pupa, the body rebuilds from clusters of embryonic cells that have been held in reserve. The adult emerges from the pupa.
compost Decomposing organic material that can be used to add nutrients to soil.
compound A substance containing two or more elements in a fixed ratio. For example, table salt (NaCl) consists of one atom of the element sodium (Na) for every atom of chlorine (Cl).
compound eye The photoreceptor in many invertebrates; made up of many tiny light detectors, each of which detects light from a tiny portion of the field of view.
concentration gradient A region along which the density of a chemical substance increases or decreases. Cells often maintain concentration gradients of ions across their membranes. When a concentration gradient exists, substances tend to move from where they are more concentrated to where they are less concentrated.
conception The fertilization of the egg by a sperm cell in humans.
condom A form of contraception; a sheath that fits over the penis to prevent the transfer of sperm to the vagina.
cone (1) In vertebrates, a photoreceptor cell in the retina stimulated by bright light and enabling color vision. (2) In conifers, a reproductive structure bearing pollen or ovules.
coniferous forest A biome characterized by conifers, cone-bearing evergreen trees.
conjugation The union (mating) of two bacterial cells or protist cells and the transfer of DNA between the two cells.
conjunctiva A thin mucous membrane that lines the inner surface of vertebrate eyelids.
connective tissue Animal tissue that functions mainly to bind and support other tissues, having a sparse population of cells scattered through an extracellular matrix, which they produce.
conservation biology A goal-oriented science that endeavors to sustain biological diversity.
continental shelf The submerged part of a continent.
contraception The deliberate prevention of pregnancy.
controlled experiment An experiment in which an experimental group is compared with a control group that varies only in the factor being tested.
convergent evolution The evolution of similar features in different evolutionary lineages, which can result from living in very similar environments.
copulation Sexual intercourse, usually necessary for internal fertilization to occur.
cork The outermost protective layer of a plant's bark, produced by the cork cambium.
cork cambium Meristematic tissue that produces cork cells during secondary growth of a plant.
cornea (kor′-nē-uh) The transparent frontal portion of the sclera, which admits light into the vertebrate eye.
corpus callosum (kor′-pus kuh-lō′-sum) The thick band of nerve fibers that connect the right and left cerebral hemispheres in placental mammals, enabling the hemispheres to process information together.
corpus luteum (kor′-pus lū′-tē-um) A small body of endocrine tissue that develops from an ovarian follicle after ovulation and secretes progesterone and estrogen during pregnancy.
cortex In plants, the ground tissue system of a root, made up mostly of parenchyma cells, which store food and absorb minerals that have passed through the epidermis.
corticosteroid A hormone synthesized and secreted by the adrenal cortex. The corticosteroids include the mineralocorticoids and glucocorticoids.
cotyledon (kot′-uh-lē′-don) The first leaf that appears on an embryo of a flowering plant; a seed leaf. Monocot embryos have one cotyledon; dicot embryos have two.
countercurrent exchange The transfer of a substance or heat between two fluids flowing in opposite directions.
countercurrent heat exchange A circulatory adaptation in which parallel blood vessels convey warm and cold blood in opposite directions, maximizing heat transfer to the cold blood.
covalent bond (ko-vā′-lent) A type of strong chemical bond in which two atoms share one or more pairs of valence electrons.
craniate A chordate with a head.
crista (kris′-tuh) (plural, **cristae**) An infolding of the inner mitochondrial membrane.
crop A pouch-like organ in a digestive tract where food is softened and may be stored temporarily.
cross A mating of two sexually reproducing individuals; often used to describe a genetics experiment involving a controlled mating (a "genetic cross").
crossing over The exchange of segments between chromatids of homologous chromosomes during synapsis in prophase I of meiosis; also, the exchange of segments between DNA molecules in prokaryotes.
crustacean A member of a major arthropod group that includes lobsters, crayfish, crabs, shrimps, and barnacles.
cuticle (kyū′-tuh-kul) (1) In animals, a tough, nonliving outer layer of the skin. (2) In plants, a waxy coating on the surface of stems and leaves that helps retain water.
cyanobacteria (sī-an′-ō-bak-tēr′-ē-uh) Photoautotrophic prokaryotes with plantlike, oxygen-generating photosynthesis.
cytokinesis (sī′-tō-kuh-nē′-sis) The division of the cytoplasm to form two separate daughter cells. Cytokinesis usually occurs in conjunction with telophase of mitosis. Mitosis and cytokinesis make up the mitotic (M) phase of the cell cycle.
cytokinin (sī′-tō-kī′-nin) One of a family of plant hormones that promotes cell division, retards aging in flowers and fruits, and may interact antagonistically with auxins in regulating plant growth and development.
cytoplasm (si′-tō-plaz′-um) The contents of a eukaryotic cell between the plasma membrane and the nucleus; consists of a semifluid medium and organelles; can also refer to the interior of a prokaryotic cell.

cytosine (C) (sī´-tuh-sin) A single-ring nitrogenous base found in DNA and RNA.

cytoskeleton A network of protein fibers in the cytoplasm of a eukaryotic cell; includes microfilaments, intermediate filaments, and microtubules.

cytosol The semifluid portion of the cytoplasm.

cytotoxic T cell (sī´-tō-tok´-sik) A type of lymphocyte that attacks body cells infected with pathogens.

D

decomposer A prokaryote or fungus that secretes enzymes that digest molecules in organic material and convert them to inorganic forms.

decomposition The breakdown of organic materials into inorganic ones.

dehydration reaction (dē-hī-drā´-shun) A chemical reaction in which two molecules become covalently bonded to each other with the removal of a water molecule.

deletion The loss of one or more nucleotides from a gene by mutation; the loss of a fragment of a chromosome.

demographic transition A shift from zero population growth in which birth rates and death rates are high to zero population growth characterized by low birth and death rates.

denaturation (dē-nā´-chur-ā´-shun) A process in which a protein unravels, losing its specific structure and hence function; can be caused by changes in pH or salt concentration or by high temperature; also refers to the separation of the two strands of the DNA double helix, caused by similar factors.

dendrite (den´-drīt) A neuron fiber that conveys signals from its tip inward, toward the rest of the neuron. A neuron typically has many short dendrites.

density-dependent factor A population-limiting factor whose intensity is linked to population density. For example, there may be a decline in birth rates or a rise in death rates in response to an increase in the number of individuals living in a designated area.

density-dependent inhibition The ceasing of cell division that occurs when cells touch one another.

density-independent factor A population-limiting factor whose intensity is unrelated to population density.

deoxyribonucleic acid (DNA) (dē-ok´-sē-rī´-bō-nū-klā´-ik) A double-stranded helical nucleic acid molecule consisting of nucleotide monomers with deoxyribose sugar and the nitrogenous bases adenine (A), cytosine (C), guanine (G), and thymine (T). Capable of replicating, DNA is an organism's genetic material. *See also* gene.

dermal tissue system The outer protective covering of plants.

desert A biome characterized by organisms adapted to sparse rainfall (less than 30 cm per year) and rapid evaporation.

desertification The conversion of semi-arid regions to desert.

determinate growth Termination of growth after reaching a certain size, as in most animals. *See also* indeterminate growth.

detritivore (duh-trī´-tuh-vor) An organism that consumes decaying organic material.

detritus (duh-trī´-tus) Dead organic matter, including animal wastes, plant litter, and the bodies of dead organisms.

deuterostome (dū-ter´-ō-stōm) A mode of animal development in which the opening formed during gastrulation becomes the anus. Animals with the deuterostome pattern of development include the echinoderms and the chordates.

diabetes mellitus (dī´-uh-bē´-tis me-lī´-tis) A human hormonal disease in which body cells cannot absorb enough glucose from the blood and become energy starved; body fats and proteins are then consumed for their energy. Type 1 (insulin-dependent) diabetes results when the pancreas does not produce insulin; type 2 (non-insulin-dependent) diabetes results when body cells fail to respond to insulin.

dialysis (dī-al´-uh-sis) Separation and disposal of metabolic wastes from the blood by mechanical means; an artificial method of performing the functions of the kidneys that can be life sustaining in the event of kidney failure.

diaphragm (dī´-uh-fram) The sheet of muscle separating the chest cavity from the abdominal cavity in mammals. Its contraction expands the chest cavity, and its relaxation reduces it.

diastole (dȳ´-as´-tō-lē) The stage of the heart cycle in which the heart muscle is relaxed, allowing the chambers to fill with blood. *See also* systole.

diatom (dī´-uh-tom) A unicellular, autotrophic protist that belongs to the stramenopile clade of the SAR supergroup. Diatoms possess a unique glassy cell wall containing silica.

dicot (dī´-kot) A term traditionally used to refer to flowering plants that have two embryonic seed leaves, or cotyledons.

differentiation The specialization in the structure and function of cells that occurs during the development of an organism; results from selective activation and deactivation of the cells' genes.

diffusion The random movement of particles that results in the net movement of a substance down its concentration gradient from a region where it is more concentrated to a region where it is less concentrated.

digestion The mechanical and chemical breakdown of food into molecules small enough for the body to absorb; the second stage of food processing in animals.

digestive system The organ system involved in ingestion and digestion of food, absorption of nutrients, and elimination of wastes.

dihybrid cross (dī´-hī´-brid) An experimental mating of individuals that are each heterozygous for both of two characters (or the self-pollination of a plant that is heterozygous for both characters).

dinoflagellate (dī´-nō-flaj´-uh-let) A member of a group of protists belonging to the alveolate clade of the SAR supergroup. Dinoflagellates are common components of marine and freshwater phytoplankton.

diploid In an organism that reproduces sexually, a cell containing two homologous sets of chromosomes, one set inherited from each parent; a $2n$ cell.

directional selection Natural selection in which individuals at one end of the phenotypic range survive and reproduce more successfully than do other individuals.

disaccharide (dī-sak´-uh-rīd) A sugar molecule consisting of two monosaccharides linked by a dehydration reaction.

dispersion pattern The manner in which individuals in a population are spaced within their area. Three types of dispersion patterns are clumped (individuals are aggregated in patches), uniform (individuals are evenly distributed), and random (unpredictable distribution).

disruptive selection Natural selection in which individuals on both extremes of a phenotypic range are favored over intermediate phenotypes.

distal tubule In the vertebrate kidney, the portion of a nephron after the loop of Henle that helps refine filtrate and empties it into a collecting duct.

disturbance In ecology, an event that changes a biological community by removing organisms from it or altering the availability of resources.

DNA *See* deoxyribonucleic acid (DNA).

DNA ligase (lī´-gās) An enzyme, essential for DNA replication, that catalyzes the covalent bonding of adjacent DNA polynucleotide strands. DNA ligase is used in genetic engineering to paste a specific piece of DNA containing a gene of interest into a bacterial plasmid or other vector.

DNA microarray A glass slide carrying thousands of different kinds of single-stranded DNA fragments arranged in an array (grid). A DNA microarray is used to detect and measure the expression of thousands of genes at one time. Tiny amounts of a large number of single-stranded DNA fragments representing different genes are fixed to the glass slide. These fragments, ideally representing all the genes of an organism, are tested for hybridization with various samples of cDNA molecules.

DNA polymerase (puh-lim′-er-ās) A large molecular complex that assembles DNA nucleotides into polynucleotides using a preexisting strand of DNA as a template.

DNA profiling A procedure that analyzes DNA samples to determine if they came from the same individual.

DNA technology Methods used to study and/or manipulate DNA, including recombinant DNA technology.

doldrums (dol′-drums) An area of calm or very light winds near the equator, caused by rising warm air.

domain A taxonomic category above the kingdom level. The three domains of life are Archaea, Bacteria, and Eukarya.

dominance hierarchy The ranking of individuals within a group, based on social interactions and usually maintained by agonistic behavior.

dominant allele (uh-lē′-ul) The allele that determines the phenotype of a gene when the individual is heterozygous for that gene.

dorsal Pertaining to the back of a bilaterally symmetric animal.

dorsal, hollow nerve cord One of the four hallmarks of chordates, a tube that forms on the dorsal side of the body, above the notochord.

double circulation A circulatory system with separate pulmonary and systemic circuits, in which blood passes through the heart after completing each circuit; ensures vigorous blood flow to all organs.

double fertilization In flowering plants, the formation of both a zygote and a cell with a triploid nucleus, which develops into the endosperm.

double helix The form of native DNA, referring to its two adjacent polynucleotide strands interwound into a spiral shape.

Down syndrome *See* trisomy 21.

duodenum (dū-ō-dē′-num) The first portion of the vertebrate small intestine after the stomach, where chyme from the stomach mixes with bile and digestive enzymes.

duplication Repetition of part of a chromosome resulting from fusion with a fragment from a homologous chromosome; can result from an error in meiosis or from mutagenesis.

E

eardrum A sheet of connective tissue separating the outer ear from the middle ear that vibrates when stimulated by sound waves and passes the waves to the middle ear.

echinoderm (uh-kī′-nō-derm) Member of a phylum of slow-moving or sessile marine animals characterized by a rough or spiny skin, a water vascular system, an endoskeleton, and radial symmetry in adults. Echinoderms include sea stars, sea urchins, and sand dollars.

ecological footprint An estimate of the amount of land and water area required to provide the resources an individual or nation consumes and to absorb the waste it generates.

ecological niche (nich) The role of a species in its community; the sum total of a species' use of the biotic and abiotic resources of its environment.

ecological species concept A definition of species in terms of ecological niche, the sum of how members of the species interact with the nonliving and living parts of their environment.

ecological succession The process of biological community change resulting from disturbance; transition in the species composition of a biological community. *See also* primary succession; secondary succession.

ecology The scientific study of how organisms interact with their environment.

ecosystem (ē′-kō-sis-tem) All the organisms in a given area, along with the nonliving (abiotic) factors with which they interact; a biological community and its physical environment.

ecotourism Travel to natural areas for tourism and recreation.

ectoderm (ek′-tō-derm) The outer layer of three embryonic cell layers in a gastrula. The ectoderm forms the skin of the gastrula and gives rise to the epidermis and nervous system in the adult.

ectopic pregnancy (ek-top′-ik) The implantation and development of an embryo outside the uterus.

ectothermic (ek′-tō-therm-ik) Referring to organisms that do not produce enough metabolic heat to have much effect on body temperature.

effector cell (1) A muscle cell or gland cell that performs the body's response to stimuli, responding to signals from the brain or other processing center of the nervous system. (2) A lymphocyte that has undergone clonal selection and is capable of mediating an acquired immune response.

egg A female gamete.

ejaculation (ih-jak′-yū-lā′-shun) Expulsion of semen from the penis.

ejaculatory duct The short section of the ejaculatory route in mammals formed by the convergence of the vas deferens and a duct from the seminal vesicle. The ejaculatory duct transports sperm from the vas deferens to the urethra.

electromagnetic receptor A sensory receptor that detects energy of different wavelengths, such as electricity, magnetism, and light.

electromagnetic spectrum The entire spectrum of electromagnetic radiation ranging in wavelength from less than a nanometer to more than a kilometer.

electron A subatomic particle with a single negative electrical charge. One or more electrons move around the nucleus of an atom.

electron microscope (EM) A microscope that uses magnets to focus an electron beam through, or onto the surface of, a specimen. An electron microscope achieves a hundredfold greater resolution than a light microscope.

electron shell A level of electrons at a characteristic average distance from the nucleus of an atom.

electron transport chain A series of electron carrier molecules that shuttle electrons during a series of redox reactions that release energy used to make ATP; located in the inner membrane of mitochondria, the thylakoid membranes of chloroplasts, and the plasma membranes of prokaryotes.

electronegativity The attraction of a given atom for the electrons of a covalent bond.

element A substance that cannot be broken down to other substances by chemical means.

elimination The passing of undigested material out of the digestive compartment; the fourth and final stage of food processing in animals.

embryo (em′-brē-ō) A developing stage of a multicellular organism. In humans, the stage in the development of offspring from the first division of the zygote until body structures begin to appear, about the 9th week of gestation.

embryo sac (em′-brē-ō) The female gametophyte contained in the ovule of a flowering plant.

embryonic stem cell (ES cell) Cell in the early animal embryo that differentiates during development to give rise to all the different kinds of specialized cells in the body.

embryophyte Another name for land plants, recognizing that land plants share the common derived trait of multicellular, dependent embryos.

emergent properties New properties that arise with each step upward in the hierarchy of life, owing to the arrangement and interactions of parts as complexity increases.

emerging virus A virus that has appeared suddenly or has recently come to the attention of medical scientists.

endemic species A species whose distribution is limited to a specific geographic area.

endergonic reaction (en′-der-gon′-ik) An energy-requiring chemical reaction, which yields products with more potential energy than the reactants.

endocrine gland (en′-dō-krin) A ductless gland that synthesizes hormone molecules and secretes them directly into the bloodstream.

endocrine system (en′-dō-krin) The organ system consisting of ductless glands that secrete hormones and the molecular receptors on or in target cells that respond to the hormones. The endocrine system cooperates with the nervous system in regulating body functions and maintaining homeostasis.

endocytosis (en′-dō-sī-tō′-sis) Cellular uptake of molecules or particles via formation of new vesicles from the plasma membrane.

endoderm (en′-dō-derm) The innermost of three embryonic cell layers in a gastrula; gives rise to the innermost linings of the digestive tract and other hollow organs in the adult.

endodermis The innermost layer (a one-cell-thick cylinder) of the cortex of a plant root; forms a selective barrier determining which substances pass from the cortex into the vascular tissue.

endomembrane system A network of membranes inside and surrounding a eukaryotic cell, related either through direct physical contact or by the transfer of membranous vesicles.

endometrium (en′-dō-mē′-trē-um) The inner lining of the uterus in mammals, richly supplied with blood vessels that provide the maternal part of the placenta and nourish the developing embryo.

endoplasmic reticulum (ER) An extensive membranous network in a eukaryotic cell, continuous with the outer nuclear membrane and composed of ribosome-studded (rough) and ribosome-free (smooth) regions. *See also* rough ER; smooth ER.

endoskeleton A hard skeleton located within the soft tissues of an animal; includes spicules of sponges, the hard plates of echinoderms, and the cartilage and bony skeletons of vertebrates.

endosperm In flowering plants, a nutrient-rich mass formed by the union of a sperm cell with two polar nuclei during double fertilization; provides nourishment to the developing embryo in the seed.

endospore A thick-coated, protective cell produced within a bacterial cell. The endospore becomes dormant and is able to survive harsh environmental conditions.

endosymbiont theory (en′-dō-sim′-bī-ont) The theory that mitochondria and chloroplasts originated as prokaryotic cells engulfed by an ancestral eukaryotic cell. The engulfed cell and its host cell then evolved into a single organism.

endothermic Referring to organisms that use heat generated by their own metabolism to maintain a warm, steady body temperature.

endotoxin A poisonous component of the outer membrane of gram-negative bacteria that is released only when the bacteria die.

energy The capacity to cause change, especially to perform work.

energy coupling In cellular metabolism, the use of energy released from an exergonic reaction to drive an endergonic reaction.

energy flow The passage of energy through the components of an ecosystem.

enhancer A eukaryotic DNA sequence that helps stimulate the transcription of a gene at some distance from it. An enhancer functions by means of a transcription factor called an activator, which binds to it and then to the rest of the transcription apparatus.

enteric division Part of the autonomic nervous system consisting of complex networks of neurons in the digestive tract, pancreas, and gallbladder.

entropy (en′-truh-pē) A measure of disorder, or randomness. *See also* second law of thermodynamics.

enzyme (en′-zīm) A macromolecule, usually a protein, that serves as a biological catalyst, changing the rate of a chemical reaction without being consumed by the reaction.

epidermis (ep′-uh-der′-mis) (1) In animals, one or more living layers of cells forming the protective covering, or outer skin. (2) In plants, the tissue system forming the protective outer covering of leaves, young stems, and young roots.

epididymis (ep′-uh-did′-uh-mus) A long coiled tube into which sperm pass from the testis and are stored until mature and ejaculated.

epigenetic inheritance The inheritance of traits transmitted by mechanisms not directly involving the nucleotide sequence of a genome, such as the chemical modification of histone proteins or DNA bases.

epiglottis A flap of elastic cartilage that protects the entrance to the trachea. Normally, the epiglottis is positioned to allow air to enter the trachea; it changes position when food is swallowed, allowing food to enter the esophagus and preventing food from entering the trachea.

epinephrine (ep′-uh-nef′-rin) An amine hormone (also called adrenaline) secreted by the adrenal medulla that prepares body organs for action (fight or flight); also serves as a neurotransmitter.

epithelial tissue (ep′-uh-thē′-lē-ul) A sheet of tightly packed cells lining organs, body cavities, and external surfaces; also called epithelium.

erythrocyte (eh-rith′-rō-sīt′) A blood cell containing hemoglobin, which transports oxygen; also called a red blood cell.

erythropoietin (EPO) (eh-rith′rō-poy′uh-tin) A hormone that stimulates the production of erythrocytes. It is secreted by the kidney when tissues of the body do not receive enough oxygen.

esophagus (eh-sof′-uh-gus) A muscular tube that conducts food by peristalsis, usually from the pharynx to the stomach.

essential amino acid An amino acid that an animal cannot synthesize itself and must obtain from food. Eight amino acids are essential for the human adult.

essential element In plants, a chemical element required for the plant to complete its life cycle (to grow from a seed and produce another generation of seeds).

essential fatty acid An unsaturated fatty acid that an animal needs but cannot make.

essential nutrient A substance that an organism must absorb in preassembled form because it cannot synthesize it from any other material. In humans, there are essential vitamins, minerals, amino acids, and fatty acids.

estrogen (es′-trō-jen) One of several chemically similar steroid hormones secreted by the gonads; maintains the female reproductive system and promotes the development of female body features.

estuary (es′-chū-ār-ē) The area where a freshwater stream or river merges with the ocean.

ethylene A gas that functions as a hormone in plants, triggering aging responses such as fruit ripening and leaf drop.

eudicot (yūdī′-kot) Member of a group that consists of the vast majority of flowering plants that have two embryonic seed leaves, or cotyledons.

Eukarya (yū-kar′-ē-uh) Domain of life that includes all eukaryotic organisms.

eukaryotic cell (yū-kar-ē-ot′-ik) A type of cell that has a membrane-enclosed nucleus and membrane-enclosed organelles. All organisms except bacteria and archaea are composed of eukaryotic cells.

eumetazoan (yū-met-uh-zō′-un) Member of the clade of "true animals," the animals with true tissues (all animals except sponges).

Eustachian tube (yū-stā′-shun) An air passage between the middle ear and throat of vertebrates that equalizes air pressure on either side of the eardrum.

eutherian (yū-thēr′-ē-un) Placental mammal; mammal whose young complete their embryonic development within the uterus, joined to the mother by the placenta.

evaporative cooling The process in which the surface of an object becomes cooler during evaporation.

evo-devo Evolutionary developmental biology; the field of biology that combines evolutionary biology with developmental biology.

evolution Descent with modification; the idea that living species are descendants of ancestral species that were different from present-day ones; also, the genetic changes in a population from generation to generation.

evolutionary tree A branching diagram that reflects a hypothesis about evolutionary relationships among groups of organisms.

Excavata One of four monophyletic supergroups proposed in a current hypothesis of the evolutionary history of eukaryotes. The other three supergroups are SAR (Stramenopila, Alveolata, and Rhizaria), Unikonta, and Archaeplastida.

excretion (ek-skrē´-shun) The disposal of nitrogen-containing metabolic wastes.

exergonic reaction (ek´-ser-gon´-ik) An energy-releasing chemical reaction in which the reactants contain more potential energy than the products.

exocytosis (ek´-sō-sī-tō´-sis) The movement of materials out of a cell by the fusion of vesicles with the plasma membrane.

exon The part of a gene that becomes part of the final messenger RNA and is therefore expressed.

exoskeleton A hard external skeleton that protects an animal and provides points of attachment for muscles.

exotoxin A poisonous protein secreted by certain bacteria.

exponential growth model A mathematical description of idealized, unregulated population growth.

external fertilization The fusion of gametes that parents have discharged into the environment.

extinction The irrevocable loss of a species.

extirpation The loss of a single population of a species.

extracellular matrix (ECM) The meshwork surrounding animal cells; consists of glycoproteins and polysaccharides synthesized and secreted by cells.

extraembryonic membranes Four membranes (the yolk sac, amnion, chorion, and allantois) that form a life-support system for the developing embryo of a reptile, bird, or mammal.

extreme halophile A microorganism that lives in a highly saline environment, such as the Great Salt Lake or the Dead Sea.

extreme thermophile A microorganism that thrives in a hot environment (often 60–80°C).

eyecup The simplest type of photoreceptor, a cluster of photoreceptor cells shaded by a cuplike cluster of pigmented cells; detects light intensity and direction.

F

F factor A piece of DNA that can exist as a bacterial plasmid. The F factor carries genes for making sex pili and other structures needed for conjugation, as well as a site where DNA replication can start. F stands for fertility.

F_1 generation The offspring of two parental (P generation) individuals; F_1 stands for first filial.

F_2 generation The offspring of the F_1 generation; F_2 stands for second filial.

facilitated diffusion The passage of a substance through a specific transport protein across a biological membrane down its concentration gradient.

family In Linnaean classification, the taxonomic category above genus.

farsightedness An inability to focus on close objects; occurs when the eyeball is shorter than normal and the focal point of the lens is behind the retina; also called hyperopia.

fat A lipid composed of three fatty acids linked to one glycerol molecule; a triglyceride. Most fats function as energy-storage molecules.

feces The wastes of the digestive tract.

feedback inhibition A method of metabolic control in which a product of a metabolic pathway acts as an inhibitor of an enzyme within that pathway.

fertility rate In a human population, the average number of children produced by a woman over her lifetime.

fertilization The union of the nucleus of a sperm cell with the nucleus of an egg cell, producing a zygote.

fertilizer A compound given to plants to promote their growth.

fetus (fē´-tus) A developing human from the 9th week of gestation until birth. The fetus has all the major structures of an adult.

fiber (1) In animals, an elongate, supportive thread in the matrix of connective tissue; an extension of a neuron; a muscle cell. (2) In plants, a long, slender sclerenchyma cell that usually occurs in a bundle.

fibrin (fī-brin) The activated form of the blood-clotting protein fibrinogen, which aggregates into threads that form the fabric of a blood clot.

fibrinogen (fī´-brin´-uh-jen) The plasma protein that is activated to form a clot when a blood vessel is injured.

fibrous connective tissue A dense tissue with large numbers of collagenous fibers organized into parallel bundles. This is the dominant tissue in tendons and ligaments.

filter feeder An aquatic animal that strains small food particles from the water as it is pumped through a sieve-like structure; a type of suspension feeder.

filtrate Fluid extracted by the excretory system from the blood or body cavity. The excretory system produces urine from the filtrate after removing valuable solutes from it and concentrating it.

filtration In the vertebrate kidney, the extraction of water and small solutes, including metabolic wastes, from the blood by the nephrons.

fimbria (plural, **fimbriae**) One of the short, hairlike projections on some prokaryotic cells that help attach the cells to their substrate or to other cells.

first law of thermodynamics The principle of conservation of energy. Energy can be transferred and transformed, but it cannot be created or destroyed.

fission A means of asexual reproduction whereby a parent separates into two or more genetically identical individuals of about equal size.

fixed action pattern (FAP) A genetically programmed, virtually unchangeable behavioral sequence performed in response to a certain stimulus.

flagellum (fluh-jel´-um) (plural, **flagella**) A long cellular appendage specialized for locomotion. The flagella of prokaryotes and eukaryotes differ in both structure and function. Like cilia, eukaryotic flagella have a "9 + 2" arrangement of microtubules covered by the cell's plasma membrane.

flatworm A member of the phylum Platyhelminthes.

fluid feeder An animal that lives by sucking nutrient-rich fluids from another living organism.

fluid mosaic model The currently accepted model of cell membrane structure, depicting the membrane as a mosaic of diverse protein molecules embedded in a fluid bilayer of phospholipid molecules.

fluke One of a group of parasitic flatworms.

follicle (fol´-uh-kul) A cluster of cells that surround, protect, and nourish a developing egg cell in the ovary. Follicles secrete the hormone estrogen.

food chain A sequence of food transfers from producers through one to four levels of consumers in an ecosystem.

food web A network of interconnecting food chains.

foot In an invertebrate animal, a structure used for locomotion or attachment, such as the muscular organ extending from the ventral side of a mollusc.

foraging Behavior used in recognizing, searching for, capturing, and consuming food.

foraminiferan A protist that moves and feeds by means of threadlike pseudopodia and has porous shells composed of calcium carbonate. Forams belong to the Rhizaria clade of the SAR supergroup.

forebrain One of three ancestral and embryonic regions of the vertebrate brain; develops into the thalamus, hypothalamus, and cerebrum.

forensics The scientific analysis of evidence for crime scene and other legal proceedings. Also referred to as forensic science.

fossil A preserved remnant or impression of an organism.

fossil fuel An energy-containing deposit of organic material formed from the remains of ancient organisms.

fossil record The chronicle of evolution over millions of years of geologic time engraved in the order in which fossils appear in rock strata.

founder effect Genetic drift that occurs when a few individuals become isolated from a larger population and form a new population whose gene pool is not reflective of that of the original population.
fovea (fō´-vē-uh) An eye's center of focus and the place on the retina where photoreceptors are highly concentrated.
fragmentation A means of asexual reproduction whereby a single parent breaks into parts that regenerate into whole new individuals.
frameshift mutation A change in the genetic material that involves the insertion or deletion of one or more nucleotides in a gene, resulting in a change in the triplet grouping of nucleotides.
free-living flatworm A nonparasitic flatworm.
frequency-dependent selection Selection in which the fitness of a phenotype depends on how common the phenotype is in a population.
fruit A ripened, thickened ovary of a flower, which protects developing seeds and aids in their dispersal.
functional group A specific configuration of atoms commonly attached to the carbon skeletons of organic molecules and involved in chemical reactions.
Fungi (fun´-ji) The kingdom that contains the fungi.

G

gallbladder An organ that stores bile and releases it as needed into the small intestine.
gametangium (gam´-uh-tan´-jē-um) (plural, **gametangia**) A reproductive organ that houses and protects the gametes of a plant.
gamete (gam´-ēt) A sex cell; a haploid egg or sperm. The union of two gametes of opposite sex (fertilization) produces a zygote.
gametogenesis The creation of gametes within the gonads.
gametophyte (guh-mē´-tō-fīt) The multicellular haploid form in the life cycle of organisms undergoing alternation of generations; mitotically produces haploid gametes that unite and grow into the sporophyte generation.
ganglion (gang´-glē-un) (plural, **ganglia**) A cluster of neuron cell bodies in a peripheral nervous system.
gas exchange The exchange of O_2 and CO_2 between an organism and its environment.
gastric juice The collection of fluids (mucus, enzymes, and acid) secreted by the stomach.
gastrin A digestive hormone that stimulates the secretion of gastric juice.
gastropod A member of the largest group of molluscs, including snails and slugs.
gastrovascular cavity A central compartment with a single opening, the mouth; functions in both digestion and nutrient distribution and may also function in circulation, body support, waste disposal, and gas exchange.
gastrula (gas´-trū-luh) The embryonic stage resulting from gastrulation in animal development. Most animals have a gastrula made up of three layers of cells: ectoderm, endoderm, and mesoderm.
gastrulation (gas´-trū-lā´-shun) The second major phase of embryonic development, which transforms the blastula into a gastrula. Gastrulation adds more cells to the embryo and sorts the cells into distinct cell layers.
gel electrophoresis (jel´ ē-lek´-trō-fōr-ē´-sis) A technique for separating and purifying macromolecules, either DNA or proteins. A mixture of the macromolecules is placed on a gel between a positively charged electrode and a negatively charged one. Negative charges on the molecules are attracted to the positive electrode, and the molecules migrate toward that electrode. The molecules separate in the gel according to their rates of migration, which is mostly determined by their size: Smaller molecules generally move faster through the gel, while larger molecules generally move more slowly.
gene A discrete unit of hereditary information consisting of a specific nucleotide sequence in DNA (or RNA, in some viruses). Most of the genes of a eukaryote are located in its chromosomal DNA; a few are carried by the DNA of mitochondria and chloroplasts.
gene cloning The production of multiple copies of a gene.
gene expression The process whereby genetic information flows from genes to proteins; the flow of genetic information from the genotype to the phenotype.
gene flow The transfer of alleles from one population to another as a result of the movement of individuals or their gametes.
gene pool All copies of every type of allele at every locus in all members of the population.
gene regulation The turning on and off of genes within a cell in response to environmental stimuli or other factors (such as developmental stage).
gene therapy A treatment for a disease in which the patient's defective gene is supplemented or altered.
genetic code The set of rules that dictates the amino acid translations of each mRNA nucleotide triplet.
genetic drift A change in the gene pool of a population due to chance. Effects of genetic drift are most pronounced in small populations.
genetic engineering The direct manipulation of genes for practical purposes.
genetic recombination The production, by crossing over and/or independent assortment of chromosomes during meiosis, of offspring with allele combinations different from those in the parents. The term may also be used more specifically to mean the production by crossing over of eukaryotic or prokaryotic chromosomes with gene combinations different from those in the original chromosomes.
genetically modified organism (GMO) An organism that has acquired one or more genes by artificial means. If the gene is from another species, the organism is also known as a transgenic organism.
genetics The scientific study of heredity. Modern genetics began with the work of Gregor Mendel in the 19th century.
genital herpes A sexually transmitted disease caused by the herpes simplex virus type 2.
genome The complete set of genetic material of an organism or virus.
genomic library (juh-nō´-mik) A collection of cloned DNA fragments that includes an organism's entire genome. Each segment is usually carried by a plasmid or phage.
genomics The study of complete sets of genes and their interactions.
genotype (jē´-nō-tīp) The genetic makeup of an organism.
genus (jē´-nus) (plural, **genera**) In classification, the taxonomic category above species; the first part of a species' binomial; for example, *Homo*.
geologic record A time scale established by geologists that divides Earth's history into four eons—Hadean, Archaean, Proterozoic, and Phanerozoic—and further subdivides it into eras, periods, and epochs.
germinate To start developing or growing.
gestation (jes-tā´-shun) Pregnancy; the state of carrying developing young within the female reproductive tract.
gibberellin (jib´-uh-rel´-in) One of a family of plant hormones that triggers the germination of seeds and interacts with auxins in regulating growth and fruit development.
gill An extension of the body surface of an aquatic animal, specialized for gas exchange and/or suspension feeding.
gizzard A pouch-like organ in a digestive tract where food is mechanically ground.
glans The rounded, highly sensitive head of the clitoris in females and penis in males.
glia A network of supporting cells that is essential for the structural integrity and for the normal functioning of the nervous system.
global climate change Increase in temperature and change in weather patterns all around the planet, due mostly to increasing atmospheric CO_2 levels from the burning of fossil fuels. The increase in temperature, called global warming, is a major aspect of global climate change.

glomeromycete (glō´-mer-ō-mī´-sēt) Member of a group of fungi characterized by a distinct branching form of mycorrhizae (symbiotic relationships with plant roots) called arbuscules.

glomerulus (glō-mer´-ū-lus) (plural, **glomeruli**) In the vertebrate kidney, the part of a nephron consisting of the capillaries that are surrounded by Bowman's capsule; together, a glomerulus and Bowman's capsule produce the filtrate from the blood.

glucagon (glū´-kuh-gon) A peptide hormone, secreted by the islets of Langerhaus in the pancreas, that raises the level of glucose in the blood. It is antagonistic with insulin.

glucocorticoid (glū´-kuh-kor´-tih-koyd) A corticosteroid hormone secreted by the adrenal cortex that increases the blood glucose level and helps maintain the body's response to long-term stress.

glucose A six-carbon monosaccharide that serves as a building block for many polysaccharides and whose oxidation in cellular respiration is a major source of ATP for cells.

glycogen (glī´-kō-jen) An extensively branched glucose storage polysaccharide found in liver and muscle cells; the animal equivalent of starch.

glycolysis (glī-kol´-uh-sis) A series of reactions that ultimately splits glucose into two molecules of pyruvate; the first stage of cellular respiration in all organisms; occurs in the cytosol.

glycoprotein (glī´-kō-prō´-tēn) A protein with one or more short chains of sugars attached to it.

goiter An enlargement of the thyroid gland resulting from a dietary iodine deficiency.

Golgi apparatus (gol´-jē) An organelle in eukaryotic cells consisting of stacks of membranous sacs that modify, store, and ship products of the endoplasmic reticulum.

gonad A sex organ in an animal; an ovary or testis.

Gram stain Microbiological technique to identify the cell wall composition of bacteria. Results categorize bacteria as gram-positive or gram-negative.

gram-positive bacteria Diverse group of bacteria with a cell wall that is structurally less complex and contains more peptidoglycan than that of gram-negative bacteria. Gram-positive bacteria are usually less toxic than gram-negative bacteria.

granum (gran´-um) (plural, **grana**) A stack of membrane-bounded thylakoids in a chloroplast. Grana are the sites where light energy is trapped by chlorophyll and converted to chemical energy during the light reactions of photosynthesis.

gravitropism (grav´-uh-trō´-pizm) A plant's directional growth in response to gravity.

gray matter Regions within the central nervous system composed mainly of nerve cell bodies and dendrites.

green alga A member of a group of photosynthetic protists that includes chlorophytes and charophyceans, the closest living relatives of land plants. Green algae include unicellular, colonial, and multicellular species and belong to the supergroup Archaeplastida.

greenhouse effect The warming of Earth due to the atmospheric accumulation of CO_2 and certain other gases, which absorb infrared radiation and reradiate some of it back toward Earth.

ground tissue system A tissue of mostly parenchyma cells that makes up the bulk of a young plant and is continuous throughout its body. The ground tissue system fills the space between the epidermis and the vascular tissue system.

growth factor A protein secreted by certain body cells that stimulates other cells to divide.

growth hormone (GH) A protein hormone secreted by the anterior pituitary that promotes development and growth and stimulates metabolism.

guanine (G) (gwa´-nēn) A double-ring nitrogenous base found in DNA and RNA.

guard cell A specialized epidermal cell in plants that regulates the size of a stoma, allowing gas exchange between the surrounding air and the photosynthetic cells in the leaf.

gymnosperm (jim´-nō-sperm) A naked-seed plant. Its seed is said to be naked because it is not enclosed in an ovary.

H

habitat A place where an organism lives; the environment in which an organism lives.

habituation Learning not to respond to a repeated stimulus that conveys little or no information.

hair cell A type of mechanoreceptor that detects sound waves and other forms of movement in air or water.

haploid In the life cycle of an organism that reproduces sexually, a cell containing a single set of chromosomes; an *n* cell.

Hardy-Weinberg principle The principle that frequencies of alleles and genotypes in a population remain constant from generation to generation, provided that only Mendelian segregation and recombination of alleles are at work.

heart A muscular pump that propels a circulatory fluid (blood) through vessels to the body.

heart attack The damage or death of cardiac muscle cells and the resulting failure of the heart to deliver enough blood to the body.

heart rate The frequency of heart contraction, usually expressed in number of beats per minute.

heartwood In the center of trees, the darkened, older layers of secondary xylem made up of cells that no longer transport water and are clogged with resins. *See also* sapwood.

heat Thermal energy in transfer from one body of matter to another.

helper T cell A type of lymphocyte that, when activated, secretes stimulatory signals that promote the response of B cells (humoral response) and cytotoxic T cells (cell-mediated response) to antigens.

hemoglobin (hē´-mō-glō-bin) An iron-containing protein in red blood cells that reversibly binds O_2.

hepatic portal vein A blood vessel that conveys nutrient-laden blood from capillaries surrounding the intestine directly to the liver.

herbivore An animal that mainly eats plants or algae. *See also* carnivore; omnivore.

herbivory Consumption of plant parts or algae by an animal.

heredity The transmission of traits (inherited features) from one generation to the next.

hermaphroditism (her-maf´-rō-dī-tizm) A condition in which an individual has both female and male gonads and functions as both a male and female in sexual reproduction by producing both sperm and eggs.

heterokaryotic stage (het´-er-ō-ker-ē-ot´-ik) A fungal life cycle stage that contains two genetically different haploid nuclei in the same cell.

heterotroph (het´-er-ō-trōf) An organism that obtains organic food molecules by eating other organisms or substances derived from them; a consumer or a decomposer in a food chain.

heterozygote advantage (het´-er-ō-zī´-gōt) Greater reproductive success of heterozygous individuals compared to homozygotes; tends to preserve variation in gene pools.

heterozygous (het´-er-ō-zī´-gus) Having two different alleles for a given gene.

high-density lipoprotein (HDL) A cholesterol-carrying particle in the blood, made up of thousands of cholesterol molecules and other lipids bound to a protein. HDL scavenges excess cholesterol.

hindbrain One of three ancestral and embryonic regions of the vertebrate brain; develops into the medulla oblongata, pons, and cerebellum.

hinge joint A joint that allows movement in only one plane. In humans, examples include the elbow and knee.

hippocampus (hip´-uh-kam´-pus) An integrative center of the cerebrum; functionally, the part of the limbic system that plays a central role in the formation of memories and their recall.

histamine (his´-tuh-mēn) A chemical alarm signal released by mast cells that causes blood vessels to dilate and become more permeable in inflammatory and allergic responses.

histone (his´-tōn) A small protein molecule associated with DNA and important in DNA packing in the eukaryotic chromosome.

Eukaryotic chromatin consists of roughly equal parts of DNA and histone protein.

HIV (human immunodeficiency virus) The retrovirus that attacks the human immune system and causes AIDS.

homeobox (hō´-mē-ō-boks´) A 180-nucleotide sequence within a homeotic gene and some other developmental genes.

homeostasis (hō´-mē-ō-stā´-sis) The steady state of body functioning; a state of equilibrium characterized by a dynamic interplay between outside forces that tend to change an organism's internal environment and the internal control mechanisms that oppose such changes.

homeotic gene (hō´-mē-ot´-ik) A master control gene that determines the identity of a body structure of a developing organism, presumably by controlling the developmental fate of groups of cells.

hominin (hah´-mi-nin) Member of a species on the human branch of the evolutionary tree; a species more closely related to humans than to chimpanzees.

homologous chromosomes (hō-mol´-uh-gus) The two chromosomes that make up a matched pair in a diploid cell. Homologous chromosomes are of the same length, centromere position, and staining pattern and possess genes for the same characters at corresponding loci. One homologous chromosome is inherited from the organism's father, the other from the mother.

homologous structures (hō-mol´-uh-gus) Structures in different species that are similar because of common ancestry.

homology (hō-mol´-uh-jē) Similarity in characters resulting from a shared ancestry.

homozygous (hō´-mō-zī´-gus) Having two identical alleles for a given gene.

horizontal gene transfer The transfer of genes from one genome to another through mechanisms such as transposable elements, plasmid exchange, viral activity, and perhaps fusions of different organisms.

hormone (1) In animals, a regulatory chemical that travels in the blood from its production site, usually an endocrine gland, to other sites, where target cells respond to the regulatory signal. (2) In plants, a chemical that is produced in one part of the plant and travels to another part, where it acts on target cells to change their functioning.

human chorionic gonadotropin (hCG) (kōr´-ē-on´-ik gō-na´-dō-trō´-pin) A hormone secreted by the chorion that maintains the production of estrogen and progesterone by the corpus luteum of the ovary during the first few months of pregnancy. hCG secreted in the urine is the target of many home pregnancy tests.

Human Genome Project (hGP) An international collaborative effort to map and sequence the DNA of the entire human genome. The project was begun in 1990 and completed in 2004.

humoral immune response The branch of adaptive immunity that involves the activation of B cells and that leads to the production of antibodies, which defend against bacteria and viruses in body fluids.

humus (hyū´-mus) Decomposing organic material found in topsoil.

Huntington's disease A human genetic disease caused by a single dominant allele; characterized by uncontrollable body movements and degeneration of the nervous system; usually fatal 10 to 20 years after the onset of symptoms.

hybrid An offspring of parents of two different species or of two different varieties of one species; an offspring of two parents that differ in one or more inherited traits; an individual that is heterozygous for one or more pairs of genes.

hybrid zone A geographic region in which members of different species meet and mate, producing at least some hybrid offspring.

hydrocarbon An organic compound composed only of the elements carbon and hydrogen.

hydrogen bond A type of weak chemical bond formed when the slightly positive hydrogen atom of a polar covalent bond in one molecule is attracted to the slightly negative atom of a polar covalent bond in another molecule (or in another region of the same molecule).

hydrolysis (hī-drol´-uh-sis) A chemical reaction that breaks bonds between two molecules by the addition of water; process by which polymers are broken down and an essential part of digestion.

hydrophilic (hī´-drō-fil´-ik) "Water-loving"; pertaining to polar or charged molecules (or parts of molecules) that are soluble in water.

hydrophobic (hī´-drō-fō´-bik) "Water-fearing"; pertaining to nonpolar molecules (or parts of molecules) that do not dissolve in water.

hydrostatic skeleton A skeletal system composed of fluid held under pressure in a closed body compartment; the main skeleton of most cnidarians, flatworms, nematodes, and annelids.

hydroxyl group (hī-drok´-sil) A chemical group consisting of an oxygen atom bonded to a hydrogen atom.

hyperglycemia An abnormally high level of glucose in the blood that results when the pancreas does not secrete enough insulin or cells do not respond to insulin. Hyperglycemia is a characteristic of diabetes.

hypertension A disorder in which blood pressure remains abnormally high.

hypertonic Referring to a solution that, when surrounding a cell, will cause the cell to lose water.

hypha (hī´-fuh) (plural, **hyphae**) One of many filaments making up the body of a fungus.

hypoglycemia (hī´-pō-glī-sē´-mē-uh) An abnormally low level of glucose in the blood that results when the pancreas secretes too much insulin into the blood.

hypothalamus (hī-pō-thal´-uh-mus) The master control center of the endocrine system, located in the ventral portion of the vertebrate forebrain. The hypothalamus functions in maintaining homeostasis, especially in coordinating the endocrine and nervous systems; secretes hormones of the posterior pituitary and releasing hormones that regulate the anterior pituitary.

hypothesis (hī-poth´-uh-sis) (plural, **hypotheses**) A testable explanation for a set of observations based on the available data and guided by inductive reasoning.

hypotonic Referring to a solution that, when surrounding a cell, will cause the cell to take up water.

I

immune system An animal body's system of defenses against agents that cause disease.

immunodeficiency disorder An immunological disorder in which the immune system lacks one or more components, making the body susceptible to infectious agents that would ordinarily not be pathogenic.

imperfect fungus A fungus with no known sexual stage.

impotence The inability to maintain an erection; also called erectile dysfunction.

imprinting Learning that is limited to a specific critical period in an animal's life and that is generally irreversible.

***in vitro* fertilization (IVF)** (vē´-tro) Uniting sperm and egg in a laboratory container, followed by the placement of a resulting early embryo in the mother's uterus.

inclusive fitness An individual's success at perpetuating its genes by producing its own offspring and by helping close relatives to produce offspring.

incomplete dominance A type of inheritance in which the phenotype of a heterozygote (*Aa*) is intermediate between the phenotypes of the two types of homozygotes (*AA* and *aa*).

incomplete metamorphosis A type of development in certain insects in which development from larva to adult is achieved by multiple molts, but without forming a pupa.

indeterminate growth Growth that continues throughout life, as in most plants. *See also* determinate growth.

induced fit The change in shape of the active site of an enzyme, caused by entry of the substrate so that it binds the substrate snugly.

induction During embryonic development, the influence of one group of cells on an adjacent group of cells.

inferior vena cava (vē´-nuh kā´-vuh) A large vein that returns oxygen-poor blood to the heart from the lower, or posterior, part of the body. *See also* superior vena cava.

infertility The inability to conceive after one year of regular, unprotected intercourse.

inflammatory response An innate body defense in vertebrates caused by a release of histamine and other chemical alarm signals that trigger increased blood flow, a local increase in white blood cells, and fluid leakage from the blood. The resulting inflammatory response includes redness, heat, and swelling in the affected tissues.

ingestion The act of eating; the first main stage of food processing in animals.

ingroup In a cladistic study of evolutionary relationships, the group of taxa whose evolutionary relationships are being determined. *See also* outgroup.

inhibiting hormone A kind of hormone released from the hypothalamus that prompts the anterior pituitary to stop secreting hormone.

innate behavior Behavior that is under strong genetic control and is performed in virtually the same way by all members of a species.

innate immunity The kind of immunity that is present in an animal before exposure to pathogens and is effective from birth. Innate immune defenses include barriers, phagocytic cells, antimicrobial proteins, the inflammatory response, and natural killer cells.

inner ear One of three main regions of the vertebrate ear; includes the cochlea, organ of Corti, and semicircular canals.

insulin A protein hormone, secreted by the islets of Langerhans in the pancreas, that lowers the level of glucose in the blood. It is antagonistic with glucagon.

integration The analysis and interpretation of sensory signals within neural processing centers of the central nervous system.

integrin A transmembrane protein that interconnects the extracellular matrix and the cytoskeleton in animal cells.

integumentary system (in-teg´-yū-ment-ter-ē) The organ system consisting of the skin and its derivatives, such as hair and nails in mammals. The integumentary system helps protect the body from drying out, mechanical injury, and infection.

interferon (in´-ter-fer´-on) An innate defensive protein produced by virus-infected vertebrate cells and capable of helping other cells resist viruses.

intermediate One of the compounds that form between the initial reactant and the final product in a metabolic pathway, such as between glucose and pyruvate in glycolysis.

intermediate filament An intermediate-sized protein fiber that is one of the three main kinds of fibers making up the cytoskeleton of eukaryotic cells. Intermediate filaments are ropelike, made of fibrous proteins.

internal fertilization Reproduction in which sperm are typically deposited in or near the female reproductive tract and fertilization occurs within the tract.

interneuron (in´-ter-nūr´-on) A nerve cell, located entirely within the central nervous system, that integrates sensory signals and relays signals to other interneurons and to motor neurons.

internode The portion of a plant stem between two nodes.

interphase The period in the eukaryotic cell cycle when the cell is not actually dividing. Interphase constitutes the majority of the time spent in the cell cycle. *See also* mitotic phase (M phase).

interspecific competition Competition between individuals or populations of two or more species that require the same limited resource.

interspecific interactions Relationships between individuals of different species in a community.

interstitial fluid (in´-ter-stish´-ul) An aqueous solution that surrounds body cells and through which materials pass back and forth between the blood and the body tissues.

intertidal zone (in´-ter-tīd´-ul) A shallow zone where the waters of an estuary or ocean meet land.

intestine The region of a digestive tract located between the gizzard or stomach and the anus and where chemical digestion and nutrient absorption usually occur.

intraspecific competition Competition between members of a population for a limited resource.

intrauterine device (IUD) A T-shaped device that, when placed within the uterus, acts as female contraception.

intron (in´-tron) An internal, noncoding region of a gene that does not become part of the final messenger RNA molecule and is therefore not expressed.

invasive species A non-native species that spreads beyond its original point of introduction and causes environmental or economic damage.

inversion A change in a chromosome resulting from reattachment of a chromosome fragment to the original chromosome, but in the reverse direction. Mutagens and errors during meiosis can cause inversions.

invertebrate An animal that lacks a backbone.

ion (ī-on) An atom or group of atoms that has gained or lost one or more electrons, thus acquiring a charge.

ionic bond (ī-on´-ik) A chemical bond resulting from the attraction between oppositely charged ions.

iris The colored part of the vertebrate eye, formed by the anterior portion of the choroid.

isomers (ī´-sō-mers) Organic compounds with the same molecular formula but different structures and, therefore, different properties.

isotonic (ī-sō-ton´-ik) Referring to a solution that, when surrounding a cell, causes no net movement of water into or out of the cell.

isotope (ī´-sō-tōp) One of several atomic forms of an element, each with the same number of protons but a different number of neutrons.

K

karyotype (kār´-ē-ō-tīp) A display of micrographs of the metaphase chromosomes of a cell, arranged by size and centromere position. Karyotypes may be used to identify certain chromosomal abnormalities.

kelp Large, multicellular brown algae that form undersea "forests."

keystone species A species whose impact on the community is much larger than its biomass or abundance would indicate.

kilocalorie (kcal) A quantity of heat equal to 1,000 calories. Used to measure the energy content of food, it is usually called a "Calorie."

kin selection The natural selection that favors altruistic behavior by enhancing reproductive success of relatives.

kinesis (kuh-nē´-sis) (plural, **kineses**) Random movement in response to a stimulus.

kinetic energy (kuh-net´-ik) The energy associated with the motion of objects. Moving matter does work by imparting motion to other matter.

kingdom In classification, the broad taxonomic category above phylum.

***K*-selection** The concept that in certain (*K*-selected) populations, life history is centered on producing relatively few offspring that have a good chance of survival.

L

labia majora (lā´-bē-uh muh-jor´-uh) A pair of outer thickened folds of skin that protect the female genital region.

labia minora (lā´-bē-uh mi-nor´-uh) A pair of inner folds of skin, bordering and protecting the female genital region.

labor The series of events that expel the infant from the uterus.

lactic acid fermentation Glycolysis followed by the reduction of pyruvate to lactate, regenerating NAD^+.

lancelet One of a group of small, bladelike, invertebrate chordates.

landscape Several different ecosystems linked by exchanges of energy, materials, and organisms.

landscape ecology The application of ecological principles to the study of the structure and dynamics of a collection of ecosystems; the scientific study of the biodiversity of interacting ecosystems.

large intestine The portion of the alimentary canal between the small intestine and the anus; functions mainly in water absorption and the formation of feces.

larva (lar´-vuh) (plural, **larvae**) A free-living, sexually immature form in some animal life cycles that may differ from the adult in morphology, nutrition, and habitat.

larynx (lār´-inks) The upper portion of the respiratory tract containing the vocal cords; also called the voice box.

lateral line system A row of sensory organs along each side of a fish's body that is sensitive to changes in water pressure. It enables a fish to detect minor vibrations in the water.

lateral meristem Plant tissue made up of undifferentiated cells that enable roots and shoots of woody plants to thicken. The vascular cambium and cork cambium are lateral meristems.

lateralization The phenomenon in which the two hemispheres of the brain become specialized for different functions during infant and child brain development.

law of independent assortment A general rule of inheritance (originally formulated by Gregor Mendel) that when gametes form during meiosis, each pair of alleles for a particular character segregates independently of other pairs; also known as Mendel's second law of inheritance.

law of segregation A general rule in inheritance (originally formulated by Gregor Mendel) that individuals have two alleles for each gene and that when gametes form by meiosis, the two alleles separate, each resulting gamete ending up with only one allele of each gene; also known as Mendel's first law of inheritance.

leaf The main site of photosynthesis in a plant; typically consists of a flattened blade and a stalk (petiole) that joins the leaf to the stem.

learning Modification of behavior as a result of specific experiences.

leech A member of one of the three large groups of annelids, known for its bloodsucking ability. *See also* annelid.

lens The structure in an eye that focuses light rays onto the retina.

leukemia (lū-kē´-mē-ah) A type of cancer of the blood-forming tissues, characterized by an excessive production of white blood cells and an abnormally high number of them in the blood; cancer of the bone marrow cells that produce leukocytes.

leukocyte (lū´-kō-sȳt´) A blood cell that functions in fighting infections; also called a white blood cell.

lichen (lī´-ken) A close association between a fungus and an alga or between a fungus and a cyanobacterium, some of which are known to be beneficial to both partners.

life cycle The entire sequence of stages in the life of an organism, from the adults of one generation to the adults of the next.

life history The traits that affect an organism's schedule of reproduction and death, including age at first reproduction, frequency of reproduction, number of offspring, and amount of parental care.

life table A listing of survivals and deaths in a population in a particular time period and predictions of how long, on average, an individual of a given age will live.

ligament A type of fibrous connective tissue that joins bones together at joints.

light microscope (LM) An optical instrument with lenses that refract (bend) visible light to magnify images and project them into a viewer's eye or onto photographic film.

light reactions The first of two stages in photosynthesis; the steps in which solar energy is absorbed and converted to the chemical energy of ATP and NADPH, releasing oxygen in the process.

lignin A chemical that hardens the cell walls of plants.

limbic system (lim´-bik) A functional unit of several integrating and relay centers located deep in the human forebrain; interacts with the cerebral cortex in creating emotions and storing memories.

limiting factor An environmental factor that restricts population growth.

linkage map A listing of the relative locations of genes along a chromosome, as determined by recombination frequencies.

linked genes Genes located near each other on the same chromosome that tend to be inherited together.

lipid An organic compound consisting mainly of carbon and hydrogen atoms linked by nonpolar covalent bonds, making the compound mostly hydrophobic. Lipids include fats, phospholipids, and steroids and are insoluble in water.

liver The largest organ in the vertebrate body. The liver performs diverse functions, such as producing bile, preparing nitrogenous wastes for disposal, and detoxifying poisonous chemicals in the blood.

lobe-fin A bony fish with strong, muscular fins supported by bones.

locomotion Active movement from place to place.

locus (plural, **loci**) The particular site where a gene is found on a chromosome. Homologous chromosomes have corresponding gene loci.

logistic growth model A mathematical description of idealized population growth that is restricted by limiting factors.

long-day plant A plant that flowers in late spring or early summer, when day length is long. Long-day plants actually flower in response to short nights.

long-term memory The ability to hold, associate, and recall information over one's lifetime.

loop of Henle (hen´-lē) In the vertebrate kidney, the portion of a nephron that helps concentrate the filtrate while conveying it between a proximal tubule and a distal tubule.

loose connective tissue The most widespread connective tissue in the vertebrate body. It binds epithelia to underlying tissues and functions as packing material, holding organs in place.

low-density lipoprotein (LDL) A cholesterol-carrying particle in the blood, made up of thousands of cholesterol molecules and other lipids bound to a protein. An LDL particle transports cholesterol from the liver for incorporation into cell membranes.

lung An infolded respiratory surface of terrestrial vertebrates that connects to the atmosphere by narrow tubes.

lymph A colorless fluid, derived from interstitial fluid, that circulates in the lymphatic system.

lymph node An organ of the immune system located along a lymph vessel. Lymph nodes filter lymph and contain cells that attack viruses and bacteria.

lymphatic system (lim-fat´-ik) The vertebrate organ system through which lymph circulates; includes lymph vessels, lymph nodes, and the spleen. The lymphatic system helps remove toxins and pathogens from the blood and interstitial fluid and returns fluid and solutes from the interstitial fluid to the circulatory system.

lymphocyte (lim´-fuh-sīt) A type of white blood cell that is chiefly responsible for the adaptive immune response and is found mostly in the lymphatic system. *See also* B cell; T cell.

lysogenic cycle (lī´-sō-jen´-ik) A type of bacteriophage replication cycle in which the viral genome is incorporated into the bacterial host chromosome as a prophage. New phages are not produced, and the host cell is not killed or lysed unless the viral genome leaves the host chromosome.

lysosome (lī-sō-sōm) A digestive organelle in eukaryotic cells; contains hydrolytic enzymes that digest engulfed food or damaged organelles.

lytic cycle (lit´-ik) A type of viral replication cycle resulting in the release of new viruses by lysis (breaking open) of the host cell.

M

macroevolution Evolutionary change above the species level, encompassing the origin of new taxonomic groups, adaptive radiation, and mass extinction.

macromolecule A giant molecule formed by the joining of smaller molecules, usually by a dehydration reaction: a protein, carbohydrate, or nucleic acid.

macronutrient A chemical substance that an organism must obtain in relatively large amounts. *See also* micronutrient.
macrophage (mak´-rō-fāj) A large, amoeboid, phagocytic white blood cell that functions in innate immunity by destroying microbes and in adaptive immunity as an antigen-presenting cell.
major histocompatibility complex (MHC) molecule *See* self protein.
malignant tumor An abnormal tissue mass that can spread into neighboring tissue and to other parts of the body; a cancerous tumor.
malnutrition Health problems caused by a diet that contains insufficient calories or nutrients.
mammal Member of a clade of amniotes that possess mammary glands and hair.
mantle In a mollusc, the outgrowth of the body surface that drapes over the animal. The mantle produces the shell and forms the mantle cavity.
marsupial (mar-sū´-pē-ul) A pouched mammal, such as a kangaroo, opossum, or koala. Marsupials give birth to embryonic offspring that complete development while housed in a pouch and attached to nipples on the mother's abdomen.
mass number The sum of the number of protons and neutrons in an atom's nucleus.
matter Anything that occupies space and has mass.
mechanoreceptor (mek´-uh-nō-ri-sep´-ter) A sensory receptor that detects changes in the environment associated with pressure, touch, stretch, motion, or sound.
medulla oblongata (meh-duh´-luh ob´-long-got´-uh) Part of the vertebrate hindbrain, continuous with the spinal cord; passes data between the spinal cord and forebrain and controls autonomic, homeostatic functions, including breathing, heart rate, swallowing, and digestion.
medusa (med-ū´-suh) (plural, **medusae**) One of two types of cnidarian body forms; an umbrella-like body form.
meiosis (mī-ō´-sis) In a sexually reproducing organism, the division of a single diploid nucleus into four haploid daughter nuclei. Meiosis and cytokinesis produce haploid gametes from diploid cells in the reproductive organs of the parents.
membrane potential The charge difference between a cell's cytoplasm and extracellular fluid due to the differential distribution of ions.
memory The ability to store and retrieve information. *See also* long-term memory; short-term memory.
memory cell A clone of long-lived lymphocytes formed during the primary adaptive immune response. Memory cells remain in lymph nodes until activated by exposure to the same antigen that triggered their formation. When activated, a memory cell forms a large clone that mounts the secondary immune response.
meninges (muh-nin´-jēz) Layers of connective tissue that enwrap and protect the brain and spinal cord.
menstrual cycle (men´-strū-ul) The hormonally synchronized cyclic buildup and breakdown of the endometrium of some primates, including humans.
menstruation (men´-strū-ā´-shun) Uterine bleeding resulting from shedding of the endometrium during a menstrual cycle.
meristem (mer´-eh-stem) Plant tissue consisting of undifferentiated cells that divide and generate new cells and tissues.
mesoderm (mez´-ō-derm) The middle layer of the three embryonic cell layers in a gastrula. The mesoderm gives rise to muscles, bones, the dermis of the skin, and most other organs in the adult.
mesophyll (mes´-ō-fil) Leaf cells specialized for photosynthesis; a leaf's ground tissue system.
messenger RNA (mRNA) The type of ribonucleic acid that encodes genetic information from DNA and conveys it to ribosomes, where the information is translated into amino acid sequences.
metabolic pathway A series of chemical reactions that either builds a complex molecule or breaks down a complex molecule into simpler compounds.
metabolic rate The total amount of energy an animal uses in a unit of time.
metabolism The totality of an organism's chemical reactions.
metamorphosis (met´-uh-mōr´-fuh-sis) The transformation of a larva into an adult. *See also* complete metamorphosis; incomplete metamorphosis.
metaphase (met´-eh-fāz) The third stage of mitosis, during which all the cell's duplicated chromosomes are lined up at an imaginary plane equidistant between the poles of the mitotic spindle.
metastasis (muh-tas´-tuh-sis) The spread of cancer cells beyond their original site.
methanogen (meth-an´-ō-jen) An archaean that produces methane as a metabolic waste product.
methyl group A chemical group consisting of a carbon atom bonded to three hydrogen atoms.
microbiota The community of microorganisms that live in and on the body of an animal.
microevolution A change in a population's gene pool over generations.
microfilament The thinnest of the three main kinds of protein fibers making up the cytoskeleton of a eukaryotic cell; a solid, helical rod composed of the globular protein actin.
micronutrient An element that an organism needs in very small amounts and that functions as a component or cofactor of enzymes. *See also* macronutrient.
microRNA (miRNA) A small, single-stranded RNA molecule that associates with one or more proteins in a complex that can degrade or prevent translation of an mRNA with a complementary sequence.
microtubule The thickest of the three main kinds of fibers making up the cytoskeleton of a eukaryotic cell; a hollow tube made of globular proteins called tubulins; found in cilia and flagella.
microvillus (plural, **microvilli**) One of many microscopic projections on the epithelial cells in the lumen of the small intestine. Microvilli increase the surface area of the small intestine.
midbrain One of three ancestral and embryonic regions of the vertebrate brain; develops into sensory integrating and relay centers that send sensory information to the cerebrum.
middle ear One of three main regions of the vertebrate ear; a chamber containing three small bones (the hammer, anvil, and stirrup) that convey vibrations from the eardrum to the oval window.
migration The regular back-and-forth movement of animals between two geographic areas at particular times of the year.
millipede A terrestrial arthropod that has two pairs of short legs for each of its numerous body segments and that eats decaying plant matter.
mineral In nutrition, a simple inorganic nutrient that an organism requires in small amounts for proper body functioning.
mineralocorticoid (min´-er-uh-lō-kort´-uh-koyd) A corticosteroid hormone secreted by the adrenal cortex that helps maintain salt and water homeostasis and may increase blood pressure in response to long-term stress.
missense mutation A change in the nucleotide sequence of a gene that alters the amino acid sequence of the resulting polypeptide. In a missense mutation, a codon is changed from encoding one amino acid to encoding a different amino acid.
mitochondrial matrix (mī´-tō-kon´-drē-ul) The compartment of the mitochondrion enclosed by the inner membrane and containing enzymes and substrates for the citric acid cycle.
mitochondrion (mī´-tō-kon´-drē-on) (plural, **mitochondria**) An organelle in eukaryotic cells where cellular respiration occurs. Enclosed by two membranes, it is where most of the cell's ATP is made.
mitosis (mī´-tō-sis) The division of a single nucleus into two genetically identical nuclei. Mitosis and cytokinesis make up the mitotic (M) phase of the cell cycle.
mitotic phase (M phase) The part of the cell cycle when the nucleus divides (via mitosis), its chromosomes are distributed to

the daughter nuclei, and the cytoplasm divides (via cytokinesis), producing two daughter cells.

mitotic spindle A football-shaped structure formed of microtubules and associated proteins that is involved in the movement of chromosomes during mitosis and meiosis.

mixotroph A protist that is capable of both autotrophy and heterotrophy.

mold A rapidly growing fungus that reproduces asexually by producing spores.

molecular biology The study of biological structures, functions, and heredity at the molecular level.

molecular clock A method for estimating the time required for a given amount of evolutionary change, based on the observation that some regions of genomes evolve at constant rates.

molecular systematics A scientific discipline that uses nucleic acids or other molecules in different species to infer evolutionary relationships.

molecule Two or more atoms held together by covalent bonds.

mollusc (mol′-lusk) A soft-bodied animal characterized by a muscular foot, mantle, mantle cavity, and visceral mass. Molluscs include gastropods (snails and slugs), bivalves (clams, oysters, and scallops), and cephalopods (squids and octopuses).

molting The process of shedding an old exoskeleton or cuticle and secreting a new, larger one.

monoclonal antibody (mAb) (mon′-ō-klōn′-ul) An antibody secreted by a clone of cells and therefore specific for the one antigen that triggered the development of the clone.

monocot (mon′-ō-kot) A flowering plant whose embryos have a single seed leaf, or cotyledon.

monogamous Referring to a type of relationship in which a male and a female mate exclusively with each other, and both parents care for the offspring.

monohybrid cross An experimental mating of individuals that are heterozygous for the character being followed (or the self-pollination of a heterozygous plant).

monomer (mon′-uh-mer) The subunit that serves as a building block of a polymer.

monophyletic (mon′-ō-fī-let′-ik) Pertaining to a group of taxa that consists of a common ancestor and all its descendants, equivalent to a clade.

monosaccharide (mon′-ō-sak′-uh-rīd) The simplest carbohydrate; a simple sugar with a molecular formula that is generally some multiple of CH_2O. Monosaccharides are the monomers of disaccharides and polysaccharides.

monotreme (mon′-uh-trēm) An egg-laying mammal, such as the duck-billed platypus.

morning after pill (MAP) A birth control pill taken within three days of unprotected intercourse to prevent fertilization or implantation.

morphological species concept A definition of species in terms of measurable anatomical criteria.

motor neuron A nerve cell that conveys command signals from the central nervous system to effector cells, such as muscle cells or gland cells.

motor output The conduction of signals from a processing center in the central nervous system to effector cells.

motor system The component of the vertebrate peripheral nervous system that carries signals to and from skeletal muscles, mainly in response to external stimuli. Most actions of the motor system are voluntary.

motor unit A motor neuron and all the muscle fibers it controls.

movement corridor A series of small clumps or a narrow strip of quality habitat (usable by organisms) that connects otherwise isolated patches of quality habitat.

muscle fiber Muscle cell.

muscle tissue Tissue consisting of long muscle cells that can contract, either on its own or when stimulated by nerve impulses; the most abundant tissue in a typical animal. *See* skeletal muscle; cardiac muscle; smooth muscle.

muscular system The organ system that includes all the skeletal muscles in the body. (Cardiac muscle and smooth muscle are components of other organ systems.)

mutagen (myū′-tuh-jen) A chemical or physical agent that interacts with DNA and causes a mutation.

mutagenesis (myū′-tuh-jen′-uh-sis) The creation of a change in the nucleotide sequence of an organism's DNA.

mutation A change in the genetic information of a cell; the ultimate source of genetic diversity. A mutation also can occur in the DNA or RNA of a virus.

mutualism An interspecific relationship in which both partners benefit.

mycelium (mī-sē′-lē-um) (plural, **mycelia**) The densely branched network of hyphae in a fungus.

mycorrhiza (mī′-kō-rī′-zuh) (plural, **mycorrhizae**) A close association of plant roots and fungi that is beneficial to both partners.

myelin sheath (mī′-uh-lin) A series of cells, each wound around, and thus insulating, the axon of a nerve cell in vertebrates. Each pair of cells in the sheath is separated by a space called a node of Ranvier.

myofibril (mī′-ō-fī′-bril) A contractile strand in a muscle cell (fiber), made up of many sarcomeres. Longitudinal bundles of myofibrils make up a muscle fiber.

myosin A type of protein filament that interacts with actin filaments to cause cell contraction.

N

NAD$^+$ Nicotinamide adenine dinucleotide; a coenzyme that can accept electrons during the redox reactions of cellular metabolism. It cycles between oxidized (NAD$^+$) and reduced (NADH) states.

NADP$^+$ Nicotinamide adenine dinucleotide phosphate, an electron acceptor that, as NADPH, temporarily stores energized electrons produced during the light reactions.

natural killer (NK) cell A cell type that provides an innate immune response by attacking cancer cells and infected body cells, especially those harboring viruses.

natural selection A process in which individuals with certain inherited traits are more likely to survive and reproduce than are individuals that do not have those traits.

nearsightedness An inability to focus on distant objects; occurs when the eyeball is longer than normal and the lens focuses distant objects in front of the retina; also called myopia.

negative feedback A primary mechanism of homeostasis, whereby a change in a physiological variable triggers a response that counteracts the initial change. Negative feedback is a common control mechanism in which a chemical reaction, metabolic pathway, or hormone-secreting gland is inhibited by the products of the reaction, pathway, or gland. As the concentration of the products builds up, the product molecules themselves inhibit the process that produced them.

negative pressure breathing A breathing system in which air is pulled into the lungs.

nematode (nem′-uh-tōd) A roundworm, characterized by a pseudocoelom, a cylindrical, wormlike body form, and a tough cuticle that is molted to permit growth.

nephron The tubular excretory unit and associated blood vessels of the vertebrate kidney; extracts filtrate from the blood and refines it into urine. The nephron is the functional unit of the urinary system.

nerve A cable-like bundle of neurons tightly wrapped in connective tissue.

nerve cord An elongated bundle of neurons, usually extending longitudinally from the brain or anterior ganglia. One or more nerve cords and the brain make up the central nervous system in many animals.

nerve net A weblike system of interconnected neurons, characteristic of radially symmetric animals such as a hydra.

nervous system The organ system that forms a communication and coordination network between all parts of an animal's body.

nervous tissue Tissue made up of neurons and supportive cells called glia.
neural tube (nyūr′-ul) An embryonic cylinder that develops from the ectoderm after gastrulation and gives rise to the brain and spinal cord.
neuron (nyūr′-on) A nerve cell; the fundamental structural and functional unit of the nervous system, specialized for carrying signals from one location in the body to another.
neurosecretory cell A nerve cell that synthesizes hormones and secretes them into the blood and also conducts nerve signals.
neurotransmitter A chemical messenger that carries information from a transmitting neuron to a receiving cell, either another neuron or an effector cell.
neutron A subatomic particle having no electrical charge, found in the nucleus of an atom.
neutrophil (nyū′-truh-fil) The most abundant type of white blood cell; functions in innate immunity as a type of phagocytic cell that tends to self-destruct as it destroys foreign invaders.
nitrogen fixation The conversion of atmospheric nitrogen (N_2) to nitrogen compounds (NH_4^+, NO_3^-) that plants can absorb and use.
node The point of attachment of a leaf on a stem.
node of Ranvier (ron′-vē-ā) An unmyelinated region on a myelinated axon of a nerve cell, where nerve signals are regenerated.
noncompetitive inhibitor A substance that reduces the activity of an enzyme without entering an active site. By binding elsewhere on the enzyme, a noncompetitive inhibitor changes the shape of the enzyme so that the active site no longer effectively catalyzes the conversion of substrate to product.
nondisjunction An accident of meiosis or mitosis in which a pair of homologous chromosomes or a pair of sister chromatids fail to separate at anaphase.
nonpolar covalent bond A type of covalent bond in which electrons are shared equally between two atoms of similar electronegativity.
nonself molecule A foreign antigen; a protein or other macromolecule that is not part of an organism's body. *See also* self protein.
nonsense mutation A change in the nucleotide sequence of a gene that converts an amino-acid-encoding codon to a stop codon. A nonsense mutation results in a shortened polypeptide.
norepinephrine (nor′-ep-uh-nef′-rin) An amine hormone (also called noradrenaline) secreted by the adrenal medulla that prepares body organs for action (fight or flight); also serves as a neurotransmitter.
notochord (nō′-tuh-kord) A flexible, cartilage-like, longitudinal rod located between the digestive tract and nerve cord in chordate animals; present only in embryos in many species.
nuclear envelope A double membrane that encloses the nucleus, perforated with pores that regulate traffic with the cytoplasm.
nuclear transplantation A technique in which the nucleus of one cell is placed into another cell that already has a nucleus or in which the nucleus has been previously destroyed.
nucleic acid (nū-klā′-ik) A polymer consisting of many nucleotide monomers; serves as a blueprint for proteins and, through the actions of proteins, for all cellular structures and activities. The two types of nucleic acids are DNA and RNA.
nucleic acid probe (nū-klā′-ik) In DNA technology, a radioactively or fluorescently labeled single-stranded nucleic acid molecule used to find a specific gene or other nucleotide sequence within a mass of DNA. The probe hydrogen-bonds to the complementary sequence in the targeted DNA.
nucleoid (nū′-klē-oyd) A non–membrane-bounded region in a prokaryotic cell where the DNA is concentrated.
nucleolus (nū-klē′-ō-lus) A structure within the nucleus where ribosomal RNA is made and assembled with proteins imported from the cytoplasm to make ribosomal subunits.
nucleosome (nū′-klē-ō-sōm) The bead-like unit of DNA packing in a eukaryotic cell; consists of DNA wound around a protein core made up of eight histone molecules.
nucleotide (nū′-klē-ō-tīd) A building block of nucleic acids, consisting of a five-carbon sugar covalently bonded to a nitrogenous base and one or more phosphate groups.
nucleus (plural, **nuclei**) (1) An atom's central core, containing protons and neutrons. (2) The organelle of a eukaryotic cell that contains the genetic material in the form of chromosomes, made of chromatin.

O

obesity The excessive accumulation of fat in the body.
ocean acidification Decreasing pH of ocean waters due to absorption of excess atmospheric CO_2 from the burning of fossil fuels.
ocean current One of the river-like flow patterns in the oceans.
omnivore An animal that eats animals as well as plants or algae.
oncogene (on′-kō-jēn) A cancer-causing gene; usually contributes to malignancy by abnormally enhancing the amount or activity of a growth factor made by the cell.
oogenesis (ō′-uh-jen′-uh-sis) The development of mature egg cells.
open circulatory system A circulatory system in which blood is pumped through open-ended vessels and bathes the tissues and organs directly. In an animal with an open circulatory system, blood and interstitial fluid are the same.
operator In prokaryotic DNA, a sequence of nucleotides near the start of an operon to which an active repressor protein can attach. The binding of a repressor prevents RNA polymerase from attaching to the promoter and transcribing the genes of the operon. The operator sequence thereby acts as a "genetic switch" that can turn all the genes in an operon on or off as a single functional unit.
operculum (ō-per′-kyuh-lum) (plural, **opercula**) A protective flap on each side of a fish's head that covers a chamber housing the gills. Movement of the operculum increases the flow of oxygen-bearing water over the gills.
operon (op′-er-on) A unit of genetic regulation common in prokaryotes; a cluster of genes with related functions, along with the promoter and operator that control their transcription.
opportunistic infection An infection that can be controlled by a normally functioning immune system but that causes illness in a person with an immunodeficiency.
opposable thumb An arrangement of the fingers such that the thumb can touch the fingertips of all four fingers.
optimal foraging theory The basis for analyzing behavior as a compromise between feeding costs and feeding benefits.
oral cavity The mouth of an animal.
oral contraceptive A chemical contraceptive that contains synthetic estrogen and/or progesterone (or a synthetic progesterone-like hormone called progestin) and prevents the release of eggs. Also called a birth control pill.
order In Linnaean classification, the taxonomic category above family.
organ A specialized structure composed of several different types of tissues that together perform specific functions.
organ of Corti (kor′-tē) The hearing organ in birds and mammals, located within the cochlea.
organ system A group of organs that work together in performing vital body functions.
organelle (ōr-guh-nel′) A membrane-enclosed structure with a specialized function within a cell.
organic compound A chemical compound containing the element carbon and usually the element hydrogen.
organic farming A set of agricultural principles that are intended to promote biological diversity and ecological sustainability. The use of the term *organic* on food labels is regulated by the U.S. Department of Agriculture.
organism An individual living thing, such as a bacterium, fungus, protist, plant, or animal.
orgasm A series of rhythmic, involuntary contractions of the reproductive structures.
osmoconformer (oz′-mō-con-form′-er) An organism whose body fluids have a solute concentration equal to that of its surroundings.

Osmoconformers do not have a net gain or loss of water by osmosis. Examples include most marine invertebrates.

osmoregulation The homeostatic maintenance of solute concentrations and water balance by a cell or organism.

osmoregulator An organism whose body fluids have a solute concentration different from that of its environment and that must use energy in controlling water loss or gain. Examples include most land-dwelling and freshwater animals.

osmosis (oz-mō′-sis) The diffusion of free water across a selectively permeable membrane.

osteoporosis (os′-tē-ō-puh-rō′-sis) A skeletal disorder characterized by thinning, porous, and easily broken bones.

outer ear One of three main regions of the ear in reptiles (including birds) and mammals; made up of the auditory canal and, in many birds and mammals, the pinna.

outgroup In a cladistic study, a taxon or group of taxa known to have diverged before the lineage that contains the group of species being studied. *See also* ingroup.

ovarian cycle (ō-vār′-ē-un) Hormonally synchronized cyclic events in the mammalian ovary, culminating in ovulation.

ovary (1) In animals, the female gonad, which produces egg cells and reproductive hormones. (2) In flowering plants, the basal portion of a carpel in which the egg-containing ovules develop.

oviduct (ō′-vuh-dukt) The tube that conveys egg cells away from an ovary; also called a fallopian tube. In humans, the oviduct is the normal site of fertilization.

ovulation (ah′-vyū-lā′-shun) The release of an egg cell from an ovarian follicle.

ovule (ō-vyūl) seed plants, a structure that develops within the female cone (in gymnosperms) or ovary (in angiosperms) that contains the female gametophyte.

ovum (plural, *ova*) A mature reproductive egg.

oxidation The loss of electrons from a substance involved in a redox reaction; always accompanies reduction.

oxidative phosphorylation (fos′-fōr-uh-lā′-shun) The production of ATP using energy derived from the redox reactions of an electron transport chain; the third major stage of cellular respiration.

ozone layer The layer of ozone (O_3) in the upper atmosphere that protects life on Earth from the harmful ultraviolet rays in sunlight.

P

P generation The parent individuals from which offspring are derived in studies of inheritance; P stands for parental.

P site One of two of a ribosome's binding sites for tRNA during translation. The P site holds the tRNA carrying the growing polypeptide chain. (P stands for peptidyl tRNA.)

paedomorphosis (pē′-duh-mōr′-fuh-sis) The retention in an adult of juvenile features of its evolutionary ancestors.

pain receptor A sensory receptor that detects pain.

paleoanthropology (pā′-lē-ō-an′-thruh-pol′-uh-jē) The study of human origins and evolution.

paleontologist (pa′-lē-on-tol′-uh-jist) A scientist who studies fossils.

pancreas (pan′-krē-us) A gland with dual functions: The digestive portion secretes digestive enzymes and an alkaline solution into the small intestine via a duct. The endocrine portion secretes the hormones insulin and glucagon into the blood.

Pangaea (pan-jē′-uh) The supercontinent that formed near the end of the Paleozoic era, when plate movements brought all the landmasses of Earth together.

parasite Organism that derives its nutrition from a living host, which is harmed by the interaction.

parasympathetic division The component of the autonomic nervous system that generally promotes body activities that gain and conserve energy, such as digestion and reduced heart rate. *See also* sympathetic division.

parenchyma cell (puh-ren′-kim-uh) In plants, a relatively unspecialized cell with a thin primary wall and no secondary wall; functions in photosynthesis, food storage, and aerobic respiration and may differentiate into other cell types.

Parkinson's disease A motor disorder caused by a progressive brain disease and characterized by difficulty in initiating movements, slowness of movement, and rigidity.

parsimony (par′-suh-mō′-nē) In scientific studies, the search for the least complex explanation for an observed phenomenon.

partial pressure The pressure exerted by a particular gas in a mixture of gases; a measure of the relative amount of a gas.

passive immunity Temporary immunity obtained by acquiring ready-made antibodies, as occurs in the transfer of maternal antibodies to a fetus or nursing infant. Passive immunity lasts only a few weeks or months.

passive transport The diffusion of a substance across a biological membrane, with no expenditure of energy.

pathogen An agent, such as a virus, bacteria, or fungus, that causes disease.

pattern formation During embryonic development, the emergence of a body form with specialized organs and tissues in the right places.

PCR *See* polymerase chain reaction (PCR).

pedigree A family genetic tree representing the occurrence of heritable traits in parents and offspring across a number of generations. A pedigree can be used to determine genotypes of matings that have already occurred.

pelagic realm (puh-laj′-ik) The region of an ocean occupied by seawater.

penis The copulatory structure of male mammals.

peptide bond The covalent bond between two amino acid units in a polypeptide, formed by a dehydration reaction.

peptidoglycan (pep′-tid-ō-glī′-kan) A polymer of complex sugars cross-linked by short polypeptides; a material unique to bacterial cell walls.

per capita rate of increase The average contribution of each individual in a population to population growth.

perennial (puh-ren′-ē-ul) A plant that lives for many years.

peripheral nervous system (PNS) The network of nerves and ganglia carrying signals into and out of the central nervous system.

peristalsis (per′-uh-stal′-sis) Rhythmic waves of contraction of smooth muscles. Peristalsis propels food through a digestive tract and also enables many animals, such as earthworms, to crawl.

permafrost Continuously frozen ground found in the arctic tundra.

peroxisome An organelle containing enzymes that transfer hydrogen atoms from various substrates to oxygen, producing and then degrading hydrogen peroxide.

petal A modified leaf of a flowering plant. Petals are the often colorful parts of a flower that advertise it to pollinators.

pH scale A measure of the acidity of a solution, ranging in value from 0 (most acidic) to 14 (most basic). The letters pH stand for potential hydrogen and refer to the concentration of hydrogen ions (H^+).

phage (fāj) *See* bacteriophage.

phagocyte (fag′-ō-sīt′) A white blood cell (for example, a neutrophil or macrophage) that engulfs bacteria, foreign proteins, and the remains of dead body cells.

phagocytosis (fag′-ō-sī-tō′-sis) Cellular "eating"; a type of endocytosis in which a cell engulfs macromolecules, other cells, or particles into its cytoplasm.

pharyngeal slit (fā-rin′-jē-ul) A gill structure in the pharynx; found in chordate embryos and some adult chordates.

pharynx (fār′-inks) The organ in a digestive tract that receives food from the oral cavity; in terrestrial vertebrates, the region of the throat that is a common passageway for air and food.

phenotype (fē′-nō-tīp) The expressed traits of an organism.

phenotypic plasticity An individual's ability to change phenotype in response to local environmental conditions.

phloem (flō′-um) The portion of a plant's vascular tissue system that transports sugars and other organic nutrients from leaves or storage tissues to other parts of the plant.

phloem sap (flō´-um) The solution of sugars, other nutrients, and hormones conveyed throughout a plant via phloem tissue.
phosphate group (fos´-fāt) A chemical group consisting of a phosphorus atom bonded to four oxygen atoms.
phospholipid (fos´-fō-lip´-id) A lipid made up of glycerol joined to two fatty acids and a phosphate group, giving the molecule two nonpolar hydrophobic tails and a polar hydrophilic head. Phospholipids form bilayers that function as biological membranes.
phosphorylation (fos´-fōr-uh-lā´-shun) The transfer of a phosphate group, usually from ATP, to a molecule. Nearly all cellular work depends on ATP energizing other molecules by phosphorylation.
photic zone (fō´-tik) The region of an aquatic ecosystem into which light penetrates and where photosynthesis occurs.
photoautotroph An organism that obtains energy from sunlight and carbon from CO_2 by photosynthesis.
photoheterotroph An organism that obtains energy from sunlight and carbon from organic sources.
photon (fō´-ton) A fixed quantity of light energy. The shorter the wavelength of light, the greater the energy of a photon.
photoperiod The relative lengths of day and night; an environmental stimulus that plants use to detect the time of year.
photophosphorylation (fō´-tō-fos´-fōr-uh-lā´-shun) The production of ATP by chemiosmosis during the light reactions of photosynthesis.
photopsin (fō-top´-sin) One of a family of visual pigments in the cones of the vertebrate eye that absorb bright, colored light.
photoreceptor A type of electromagnetic sensory receptor that detects light.
photorespiration In a plant cell, a metabolic pathway that consumes oxygen, releases CO_2, and decreases photosynthetic output. Photorespiration generally occurs on hot, dry days, when stomata close, O_2 accumulates in the leaf, and rubisco fixes O_2 rather than CO_2. Photorespiration produces no sugar molecules or ATP.
photosynthesis (fō´-tō-sin´-thuh-sis) The process by which plants, algae, and some protists and prokaryotes convert light energy to chemical energy that is stored in sugars made from carbon dioxide and water.
photosystem A light-capturing unit of a chloroplast's thylakoid membrane, consisting of a reaction-center complex surrounded by numerous light-harvesting complexes.
phototropism (fō´-tō-trō´-pizm) The growth of a plant shoot toward light (positive phototropism) or away from light (negative phototropism).
phylogenetic species concept (fī´-lō-juh-net´-ik) A definition of species as the smallest group of individuals that shares a common ancestor, forming one branch on the tree of life.
phylogenetic tree (fī´-lō-juh-net´-ik) A branching diagram that represents a hypothesis about the evolutionary history of a group of organisms.
phylogeny (fī-loj´-uh-nē) The evolutionary history of a species or group of related species.
phylum (fī´-lum) (plural, **phyla**) In Linnaean classification, the taxonomic category above class.
physiology (fi´-zē-ol´-uh-ji) The study of the functions of an organism's structures.
phytochrome (fī´-tuh-krōm) A plant protein that has a light-absorbing component.
phytoplankton (fī´-tō-plank´-ton) Algae and photosynthetic bacteria that drift passively in aquatic environments.
pineal gland (pin´-ē-ul) An outgrowth of the vertebrate brain that secretes the hormone melatonin, which coordinates daily and seasonal body activities such as the sleep/wake circadian rhythm with environmental light conditions.
pinna (pin´-uh) The flap-like part of the outer ear, projecting from the body surface of many birds and mammals; collects sound waves and channels them to the auditory canal.
pistil Part of the reproductive organ of an angiosperm, a single carpel or a group of fused carpels.
pith Part of the ground tissue system of a dicot plant. Pith fills the center of a stem and may store food.
pituitary gland An endocrine gland at the base of the hypothalamus; consists of a posterior lobe, which stores and releases two hormones produced by the hypothalamus, and an anterior lobe, which produces and secretes many hormones that regulate diverse body functions.
pivot joint A joint that allows precise rotations in multiple planes. An example in humans is the joint that rotates the forearm at the elbow.
placenta (pluh-sen´-tuh) In most mammals, the organ that provides nutrients and oxygen to the embryo and helps dispose of its metabolic wastes; formed of the embryo's chorion and the mother's endometrial blood vessels.
placental mammal (pluh-sen´-tul) Mammal whose young complete their embryonic development in the uterus, nourished via the mother's blood vessels in the placenta; also called a eutherian.
plasma The liquid matrix of the blood in which the blood cells are suspended.
plasma cell An antibody-secreting B cell; the effector cell of the humoral immune response.
plasma membrane The membrane at the boundary of every cell that acts as a selective barrier to the passage of ions and molecules into and out of the cell; consists of a phospholipid bilayer with embedded proteins.
plasmid A small ring of independently replicating DNA separate from the main chromosome(s). Plasmids are found in prokaryotes and yeasts.
plasmodesma (plaz´-mō-dez´-muh) (plural, **plasmodesmata**) An open channel in a plant cell wall that connects the cytoplasm of adjacent cells.
plasmodial slime mold (plaz-mō´-dē-ul) A type of protist that has amoeboid cells, flagellated cells, and an amoeboid plasmodial feeding stage in its life cycle; a member of the amoebozoan clade.
plasmodium (1) A single mass of cytoplasm containing many nuclei. (2) The amoeboid feeding stage in the life cycle of a plasmodial slime mold.
plate tectonics (tek-tän´-iks) The theory that the continents are part of great plates of Earth's crust that float on the hot, underlying portion of the mantle. Movements in the mantle cause the continents to move slowly over time.
platelet A pinched-off cytoplasmic fragment of a bone marrow cell. Platelets circulate in the blood and are important in blood clotting.
pleiotropy (plī´-uh-trō-pē) The control of more than one phenotypic character by a single gene.
polar covalent bond A covalent bond between atoms that differ in electronegativity. The shared electrons are pulled closer to the more electronegative atom, making it slightly negative and the other atom slightly positive.
polar ice A terrestrial biome that includes regions of extremely cold temperature and low precipitation located at high latitudes north of the arctic tundra and in Antarctica.
polar molecule A molecule containing polar covalent bonds and having an unequal distribution of charges in different regions of the molecule.
pollen grain The structure that will produce the sperm in seed plants; the male gametophyte.
pollination In seed plants, the delivery by wind or animals of pollen from the pollen-producing parts of a plant to a female cone (in gymnosperms) or the stigma of a carpel (in angiosperms).
polychaete (pol´-ē-kēt) A member of the largest group of annelids. *See also* annelid.
polygamous Referring to a type of relationship in which an individual of one sex mates with more than one of the other sex.
polygenic inheritance (pol´-ē-jen´-ik) The additive effects of two or more gene loci on a single phenotypic character.
polymer (pol´-uh-mer) A large molecule consisting of many identical or similar monomers linked together by covalent bonds.

polymerase chain reaction (PCR) (puh-lim′-uh-rās) A technique used to obtain many copies of a DNA molecule or a specific part of a DNA molecule. In the procedure, the starting DNA is mixed with a heat-resistant DNA polymerase, DNA nucleotides, and a few other ingredients. Specific nucleotide primers flanking the region to be copied ensure that it, and not other regions of the DNA, is replicated during the PCR procedure.
polynucleotide (pol′-ē-nū′-klē-ō-tīd) A polymer made up of many nucleotide monomers covalently bonded together.
polyp (pol′-ip) One of two types of cnidarian body forms; a columnar, hydra-like body.
polypeptide A polymer (chain) of amino acids linked by peptide bonds.
polyploid An organism that has more than two complete sets of chromosomes as a result of an accident of cell division.
polysaccharide (pol′-ē-sak′-uh-rīd) A carbohydrate polymer of many monosaccharides (sugars) linked by dehydration reactions.
pons (pahnz) Part of the vertebrate hindbrain that functions with the medulla oblongata in passing data between the spinal cord and forebrain and in controlling autonomic, homeostatic functions.
population A group of individuals belonging to one species and living in the same geographic area.
population density The number of individuals of a species per unit area or volume.
population ecology The study of how members of a population interact with their environment, focusing on factors that influence population density and growth.
population momentum In a population in which $r = 0$, the continuation of population growth as girls in the prereproductive age group reach their reproductive years.
positive feedback A type of control in which a change triggers mechanisms that amplify that change.
post-anal tail A tail posterior to the anus; found in chordate embryos and most adult chordates.
posterior Pertaining to the rear, or tail, of a bilaterally symmetric animal.
posterior pituitary An extension of the hypothalamus composed of nervous tissue that secretes hormones made in the hypothalamus; a temporary storage site for hypothalamic hormones.
postzygotic barrier A reproductive barrier that prevents hybrid zygotes produced by two different species from developing into viable, fertile adults. Includes reduced hybrid viability, reduced hybrid fertility, and hybrid breakdown.
potential energy The energy that matter possesses because of its location or spatial arrangement. Water behind a dam possesses potential energy, and so do chemical bonds.
predation An interaction between species in which one species, the predator, kills and eats the other, the prey.
prepuce (prē′-pyūs) A fold of skin covering the head of the clitoris or penis.
pressure flow mechanism The method by which phloem sap is transported through a plant from a sugar source, where sugars are produced, to a sugar sink, where sugars are used.
prevailing winds Winds that result from the combined effects of Earth's rotation and the rising and falling of air masses.
prezygotic barrier A reproductive barrier that impedes mating between species or hinders fertilization if mating between two species is attempted. Includes temporal, habitat, behavioral, mechanical, and gametic isolation.
primary consumer In the trophic structure of an ecosystem, an organism that eats plants or algae.
primary growth Growth in the length of a plant root or shoot, produced by an apical meristem.
primary immune response The initial adaptive immune response to an antigen, which appears after a lag of about 10 days.
primary production The amount of solar energy converted to chemical energy (in organic compounds) by autotrophs in an ecosystem during a given time period.
primary structure The first level of protein structure; the specific sequence of amino acids making up a polypeptide chain.
primary succession A type of ecological succession in which a biological community arises in an area without soil. *See also* secondary succession.
primers Short, artificially created, single-stranded DNA molecules that bind to each end of a target sequence during a PCR procedure.
prion An infectious form of protein that may multiply by converting related proteins to more prions. Prions cause several related diseases in different animals, including scrapie in sheep and mad cow disease.
problem solving Applying past experiences to overcome obstacles in novel situations.
producer An organism that makes organic food molecules from CO_2, H_2O, and other inorganic raw materials: a plant, alga, or autotrophic prokaryote.
product An ending material in a chemical reaction.
progestin (prō-jes′-tin) One of a family of steroid hormones, including progesterone, produced by the mammalian ovary. Progestins prepare the uterus for pregnancy.
programmed cell death The timely and tidy suicide (and disposal of the remains) of certain cells, triggered by certain genes; an essential process in normal development; also called apoptosis.
prokaryotic cell (prō-kār′-ē-ot′-ik) A type of cell lacking a membrane-enclosed nucleus and other membrane-enclosed organelles; found only in the domains Bacteria and Archaea.
prolactin (PRL) (prō-lak′-tin) A protein hormone secreted by the anterior pituitary that stimulates human mammary glands to produce and release milk and produces other responses in different animals.
prometaphase The second stage of mitosis, during which the nuclear envelope fragments and the spindle microtubules attach to the kinetochores of the sister chromatids.
promiscuous Referring to a type of relationship in which mating occurs with no strong pair-bonds or lasting relationships.
promoter A specific nucleotide sequence in DNA located near the start of a gene that is the binding site for RNA polymerase and the place where transcription begins.
prophage (prō′-fāj) Phage DNA that has inserted by genetic recombination into the DNA of a bacterial chromosome.
prophase The first stage of mitosis, during which the chromatin condenses to form structures (sister chromatids) visible with a light microscope and the mitotic spindle begins to form, but the nucleus is still intact.
prostate gland (pros′-tāt) A gland in human males that secretes a thin fluid that nourishes the sperm.
protein A functional biological molecule consisting of one or more polypeptides folded into a specific three-dimensional structure.
proteobacteria A clade of gram-negative bacteria that encompasses enormous diversity, including all four modes of nutrition.
proteomics The study of whole sets of proteins and their interactions.
protist A member of a diverse collection of eukaryotes. Most protists are unicellular, but some are colonial or multicellular.
proton A subatomic particle with a single positive electrical charge, found in the nucleus of an atom.
proto-oncogene (prō′-tō-on′-kō-jēn) A normal gene that, through mutation, can be converted to a cancer-causing gene.
protostome A mode of animal development in which the opening formed during gastrulation becomes the mouth. Animals with the protostome pattern of development include the flatworms, molluscs, annelids, nematodes, and arthropods.
protozoan (prō′-tō-zō′-un) (plural, **protozoans**) A protist that lives primarily by ingesting food; a heterotrophic, "animal-like" protist.
proximal tubule In the vertebrate kidney, the portion of a nephron immediately downstream from Bowman's capsule that conveys and helps refine filtrate.

proximate cause In animal behavior, a condition in an animal's internal or external environment that is the immediate reason or mechanism for a behavior.

proximate question In animal behavior, a question that concerns the immediate reason for a behavior.

pseudocoelom (sū´-dō-sē´-lōm) A body cavity that is not lined with mesoderm and is in direct contact with the wall of the digestive tract.

pseudopodium (sū´-dō-pō´-dē-um) (plural, **pseudopodia**) A temporary extension of an amoeboid cell. Pseudopodia function in moving cells and engulfing food.

pulmonary artery A large blood vessel that conveys blood from the heart to a lung.

pulmonary circuit The branch of the circulatory system that supplies the lungs. *See also* systemic circuit.

pulmonary vein A blood vessel that conveys blood from a lung to the heart.

pulse The rhythmic stretching of the arteries caused by the pressure of blood during contraction of ventricles in systole.

punctuated equilibria In the fossil record, long periods in which a species undergoes little or no morphological change (equilibria), interrupted (punctuated) by relatively brief periods of sudden change.

Punnett square A diagram used in the study of inheritance to show the results of random fertilization.

pupil The opening in the iris that admits light into the interior of the vertebrate eye. Muscles in the iris regulate the pupil's size.

Q

quaternary consumer (kwot´-er-ner-ē) An organism that eats tertiary consumers.

quaternary structure (kwot´-er-ner-ē) The fourth level of protein structure; the shape resulting from the association of two or more polypeptide subunits.

R

R plasmid A bacterial plasmid that carries genes for enzymes that destroy particular antibiotics, thus making the bacterium resistant to the antibiotics.

radial symmetry An arrangement of the body parts of an organism like pieces of a pie around an imaginary central axis. Any slice passing longitudinally through a radially symmetric organism's central axis divides the organism into mirror-image halves.

radioactive isotope An isotope whose nucleus decays spontaneously, giving off particles and energy.

radiolarian A protist that moves and feeds by means of threadlike pseudopodia and has a mineralized support structure composed of silica. Radiolarians belong to the Rhizaria clade of the SAR supergroup.

radiometric dating A method for determining the absolute ages of fossils and rocks, based on the half-life of radioactive isotopes.

radula (rad´-yū-luh) A toothed, rasping organ used to scrape up or shred food; found in many molluscs.

random dispersion pattern A pattern in which the individuals of a population are spaced in an unpredictable way.

ray-finned fish Bony fish; member of a clade of jawed vertebrates having fins supported by thin, flexible skeletal rays.

reabsorption In the vertebrate kidney, the reclaiming of water and valuable solutes from the filtrate.

reactant A starting material in a chemical reaction.

receptor potential The electrical signal produced by sensory transduction.

receptor-mediated endocytosis (en´-dō-sī-tō´-sis) The movement of specific molecules into a cell by the infolding of vesicles containing proteins with receptor sites specific to the molecules being taken in.

recessive allele An allele that has no noticeable effect on the phenotype of a gene when the individual is heterozygous for that gene.

recombinant DNA A DNA molecule that has been manipulated in the laboratory to carry nucleotide sequences derived from two sources, often different species.

recombination frequency With respect to two given genes, the number of recombinant progeny from a mating divided by the total number of progeny. Recombinant progeny carry combinations of alleles different from those in either of the parents as a result of crossing over during meiosis.

Recommended Dietary Allowance (RDA) A recommendation for daily nutrient intake established by a national scientific panel.

rectum The terminal portion of the large intestine where the feces are stored until they are eliminated.

red alga A member of a group of marine, mostly multicellular, autotrophic protists, which includes the reef-building coralline algae. Red algae belong to the supergroup Archaeplastida.

red blood cell *See* erythrocyte.

red bone marrow A specialized tissue that is found in the cavities at the ends of bones and that produces blood cells.

redox reaction Short for **red**uction-**ox**idation reaction; a chemical reaction in which electrons are lost from one substance (oxidation) and added to another (reduction).

reduction The gain of electrons by a substance involved in a redox reaction; always accompanies oxidation.

reflex An automatic reaction to a stimulus, mediated by the spinal cord or lower brain.

regeneration The regrowth of body parts from pieces of an organism.

regulatory gene A gene that codes for a protein, such as a repressor, that controls the transcription of another gene or group of genes.

relative fitness The contribution an individual makes to the gene pool of the next generation, relative to the contributions of other individuals in the population.

releasing hormone A kind of hormone secreted by the hypothalamus that promotes the release of hormones from the anterior pituitary.

renal cortex The outer portion of the vertebrate kidney, above the renal medulla.

renal medulla The inner portion of the vertebrate kidney, beneath the renal cortex.

repetitive DNA Nucleotide sequences that are present in many copies in the DNA of a genome. The repeated sequences may be long or short and may be located next to each other (tandomly) or dispersed in the DNA.

repressor A protein that blocks the transcription of a gene or operon.

reproduction The creation of new individuals from existing ones.

reproductive cloning Using a somatic cell from a multicellular organism to make one or more genetically identical individuals.

reproductive cycle A recurring sequence of events that produces eggs, makes them available for fertilization, and prepares the female body for pregnancy.

reproductive isolation The existence of biological factors (barriers) that impede members of two species from producing viable, fertile hybrids.

reproductive system The organ system responsible for reproduction.

reptile Member of the clade of amniotes that includes snakes, lizards, turtles, crocodilians, and birds, along with a number of extinct groups, such as dinosaurs.

respiratory system The organ system that functions in exchanging gases with the environment. It supplies the blood with O_2 and disposes of CO_2.

resting potential The voltage across the plasma membrane of a resting neuron. The resting potential in a vertebrate neuron is typically around −70 millivolts, with the inside of the cell negatively charged relative to the outside.

restoration ecology The use of ecological principles to develop ways to return degraded ecosystems to conditions as similar as possible to their natural, predegraded state.

restriction enzyme A bacterial enzyme that cuts up foreign DNA (at specific DNA sequences called *restriction sites*), thus protecting bacteria against intruding DNA from phages and other organisms. Restriction enzymes are used in DNA technology to cut DNA molecules in reproducible ways. The pieces of cut DNA are called restriction fragments.

restriction fragment length polymorphism (RFLP) (rif′-lip) Variation in the length of a restriction fragment. RFLPs are produced when homologous DNA sequences containing SNPs are cut up with restriction enzymes.

restriction fragments Molecules of DNA produced from a longer DNA molecule cut up by a restriction enzyme. Restriction fragments are used in genome mapping and other applications.

restriction site A specific sequence on a DNA strand that is recognized as a "cut site" by a restriction enzyme.

retina (ret′-uh-nuh) The light-sensitive layer in an eye, made up of photoreceptor cells and sensory neurons.

retrovirus An RNA virus that reproduces by means of a DNA molecule. It reverse-transcribes its RNA into DNA, inserts the DNA into a cellular chromosome, and then transcribes more copies of the RNA from the viral DNA. HIV and a number of cancer-causing viruses are retroviruses.

reverse transcriptase (tran-skrip′-tās) An enzyme encoded and used by retroviruses that catalyzes the synthesis of DNA on an RNA template.

RFLP *See* restriction fragment length polymorphism (RFLP).

Rhizaria A clade of the SAR supergroup of protists that includes foraminiferans and radiolarians.

rhizome (rī′-zōm) A horizontal stem of a plant that grows below the ground.

rhodopsin (ro-dop′-sin) A visual pigment that is located in the rods of the vertebrate eye and that absorbs dim light.

rhythm method A form of contraception that relies on refraining from sexual intercourse when conception is most likely to occur; also called natural family planning.

ribonucleic acid (RNA) (rī-bō-nū-klā′-ik) A type of nucleic acid consisting of nucleotide monomers with a ribose sugar and the nitrogenous bases adenine (A), cytosine (C), guanine (G), and uracil (U); usually single-stranded; functions in protein synthesis, gene regulation, and as the genome of some viruses.

ribosomal RNA (rRNA) (rī′-buh-sōm′-ul) The type of ribonucleic acid that, together with proteins, makes up ribosomes; the most abundant type of RNA in most cells.

ribosome (rī′-buh-sōm) A cell structure consisting of RNA and protein organized into two subunits and functioning as the site of protein synthesis in the cytoplasm. In eukaryotic cells, the ribosomal subunits are constructed in the nucleolus.

ribozyme (rī′-bō-zīm) An RNA molecule that functions as an enzyme.

RNA interference (RNAi) A biotechnology technique used to silence the expression of specific genes. Synthetic RNA molecules with sequences that correspond to particular genes trigger the breakdown of the gene's mRNA.

RNA polymerase (puh-lim′-uh-rās) A large molecular complex that links together the growing chain of RNA nucleotides during transcription, using a DNA strand as a template.

RNA splicing The removal of introns and joining of exons in eukaryotic RNA, forming an mRNA molecule with a continuous coding sequence; occurs before mRNA leaves the nucleus.

rod A photoreceptor cell in the vertebrate retina enabling vision in dim light.

root cap A cone of cells at the tip of a plant root that protects the root's apical meristem.

root hair An outgrowth of an epidermal cell on a root, which increases the root's absorptive surface area.

root system All of a plant's roots, which anchor it in the soil, absorb and transport minerals and water, and store food.

rough endoplasmic reticulum (reh-tik′-yuh-lum) That portion of the endoplasmic reticulum with ribosomes attached that make membrane proteins and secretory proteins.

***r*-selection** The concept that in certain (*r*-selected) populations, a high reproductive rate is the chief determinant of life history.

rule of addition A rule stating that the probability that an event can occur in two or more alternative ways is the sum of the separate probabilities of the different ways.

rule of multiplication A rule stating that the probability of a compound event is the product of the separate probabilities of the independent events.

ruminant (rū′-min-ent) An animal, such as a cow or sheep, with multiple stomach compartments housing microorganisms that can digest cellulose.

S

SA (sinoatrial) node (sȳ′-nō′-ā′-trē-ul) The pacemaker of the heart, located in the wall of the right atrium, that sets the rate and timing at which all cardiac muscle cells contract.

sac fungus *See* ascomycete.

salivary glands Glands associated with the oral cavity that secrete substances to lubricate food and begin the process of chemical digestion.

salt A compound resulting from the formation of an ionic bond.

sapwood Light-colored, water-conducting secondary xylem in a tree. *See also* heartwood.

SAR (Stramenopila, Alveolata, and Rhizaria) One of four monophyletic supergroups proposed in a current hypothesis of the evolutionary history of eukaryotes. The other three supergroups are Excavata, Unikonta, and Archaeplastida.

sarcomere (sar′-kō-mēr) The fundamental unit of muscle contraction, composed of thin filaments of actin and thick filaments of myosin; in electron micrographs, the region between two narrow, dark lines, called Z lines, in a myofibril.

saturated fatty acid A fatty acid in which all carbons in the hydrocarbon tail are connected by single bonds and the maximum number of hydrogen atoms are attached to the carbon skeleton. Saturated fats and fatty acids solidify at room temperature.

savanna A biome dominated by grasses and scattered trees.

scanning electron microscope (SEM) A microscope that uses an electron beam to study the surface details of a cell or other specimens.

scavenger An animal that feeds on the carcasses of dead animals.

schizophrenia Severe mental disturbance characterized by psychotic episodes in which patients have a distorted perception of reality.

sclera (sklār′-uh) A layer of connective tissue forming the outer surface of the vertebrate eye. The cornea is the frontal part of the sclera.

sclereid (sklār′-ē-id) In plants, a very hard sclerenchyma cell found in nutshells and seed coats.

sclerenchyma cell (skluh-ren′-kē-muh) In plants, a supportive cell with rigid secondary walls hardened with lignin.

scrotum A pouch of skin outside the abdomen that houses a testis and functions in cooling sperm, keeping them viable.

search image The mechanism that enables an animal to find a particular kind of food efficiently.

second law of thermodynamics The principle stating that every energy conversion reduces the order of the universe, increasing its entropy. Ordered forms of energy are at least partly converted to heat.

secondary consumer An organism that eats primary consumers.

secondary growth An increase in a plant's diameter, involving cell division in the vascular cambium and cork cambium.

secondary immune response The adaptive immune response elicited when an animal encounters the same antigen at some later time. The secondary immune response is more rapid, of greater magnitude, and of longer duration than the primary immune response.

secondary phloem A type of phloem plant tissue produced by the vascular cambium during secondary growth.

secondary structure The second level of protein structure; the regular local patterns of coils or folds of a polypeptide chain.
secondary succession A type of ecological succession that occurs where a disturbance has destroyed an existing biological community but left the soil intact. *See also* primary succession.
secondary xylem A type of xylem plant tissue produced by the vascular cambium during secondary growth.
secretion (1) The discharge of molecules synthesized by a cell. (2) In the vertebrate kidney, the discharge of wastes from the blood into the filtrate from the nephron tubules.
seed A plant embryo packaged with a food supply within a protective covering.
seed coat A tough outer covering of a seed, formed from the tissue surrounding an ovule. The seed coat encloses and protects the embryo and its food supply.
seedless vascular plants The informal collective name for lycophytes (club mosses and their relatives) and monilophytes (ferns and their relatives).
segmentation Subdivision along the length of an animal body into a series of repeated parts called segments; allows for greater flexibility and mobility.
selective permeability (per´-mē-uh-bil´-uh-tē) A property of biological membranes that allows some substances to cross more easily than others and blocks the passage of other substances altogether.
self protein A protein on the surface of an antigen-presenting cell that can hold a foreign antigen and display it to T cells. Each individual has a unique set of self proteins that serve as molecular markers for the body. The technical name for self proteins is *major histocompatibility complex (MHC) proteins. See also* nonself molecule.
semen (sē´-mun) The sperm-containing fluid that is ejaculated by the male during orgasm.
semicircular canals Fluid-filled channels in the inner ear that detect changes in the head's rate of rotation or angular movement.
semiconservative model Type of DNA replication in which the replicated double helix consists of one old strand, derived from the old molecule, and one newly made strand.
seminal vesicle (sem´-uh-nul ves´-uh-kul) A gland in males that secretes a thick fluid that contains fructose, which provides most of the sperm's energy.
seminiferous tubule (sem´-uh-nif´-uh-rus) A coiled sperm-producing tube in a testis.
sensitive period A limited phase in an individual animal's development when learning of particular behaviors can take place.
sensory adaptation The tendency of sensory neurons to become less sensitive when they are stimulated repeatedly. For example, a prominent smell becomes unnoticeable over time.
sensory input The conduction of signals from sensory receptors to processing centers in the central nervous system.
sensory neuron A nerve cell that receives information from sensory receptors and conveys signals into the central nervous system.
sensory receptor A specialized cell or neuron that detects specific stimuli from an organism's external or internal environment and sends information to the central nervous system.
sensory transduction The conversion of a stimulus signal to an electrical signal by a sensory receptor.
sepal (sē´-pul) A modified leaf of a flowering plant. A circle of sepals encloses and protects the flower bud before it opens.
sessile An organism that is anchored to its substrate.
sex chromosome A chromosome that determines whether an individual is male or female.
sex-linked gene A gene located on a sex chromosome. In humans, the vast majority of sex-linked genes are located on the X chromosome.
sexual dimorphism (dī-mōr´-fizm) Marked differences between the secondary sex characteristics of males and females.
sexual reproduction The creation of genetically unique offspring by the fusion of two haploid sex cells (gametes), forming a diploid zygote.
sexual selection A form of natural selection in which individuals with certain inherited traits are more likely than other individuals to obtain mates.
sexually transmitted disease (STD) A contagious disease spread by sexual contact.
shared ancestral character A character shared by members of a particular clade that originated in an ancestor that is not a member of that clade.
shared derived character An evolutionary novelty that is unique to a particular clade.
shoot system All of a plant's stems, leaves, and reproductive structures.
short tandem repeat (STR) A series of short DNA sequences that are repeated many times in a row in the genome.
short-day plant A plant that flowers in late summer, fall, or winter, when day length is short. Short-day plants actually flower in response to long nights.
short-term memory The ability to hold information, anticipations, or goals for a time and then release them if they become irrelevant.
sickle-cell disease A genetic condition caused by a mutation in the gene for hemoglobin. The mutation causes the protein to crystallize, which deforms red blood cells into a curved shape. Such blood cells produce a cascade of symptoms that can be life-threatening.
sieve plate An end wall in a sieve-tube element that facilitates the flow of phloem sap.
sieve-tube element A food-conducting cell in a plant; also called a sieve-tube member. Chains of sieve-tube elements make up phloem tissue.
signal In behavioral ecology, a stimulus transmitted by one animal to another animal.
signal transduction pathway In cell biology, a series of molecular changes that converts a signal on a target cell's surface to a specific response inside the cell.
silent mutation A mutation in a gene that changes a codon to one that codes for the same amino acid as the original codon. The amino acid sequence of the resulting polypeptide is thus unchanged.
single circulation A circulatory system with a single pump and circuit, in which blood passes from the sites of gas exchange to the rest of the body before returning to the heart.
single nucleotide polymorphism (SNP) A one-nucleotide variation in DNA sequence found within the genomes of at least 1% of a population.
single-lens eye The camera-like eye found in some jellies, polychaetes, spiders, many molluscs, and vertebrates.
sister chromatid (krō´-muh-tid) One of the two identical parts of a duplicated chromosome in a eukaryotic cell. Prior to mitosis, sister chromatids remain attached to each another at the centromere.
skeletal muscle A type of striated muscle attached to the skeleton; generally responsible for voluntary movements of the body.
skeletal system The organ system that provides body support and protects body organs, such as the brain, heart, and lungs.
small intestine The longest section of the alimentary canal. It is the principal site of the enzymatic hydrolysis of food macromolecules and the absorption of nutrients.
smooth endoplasmic reticulum That portion of the endoplasmic reticulum that lacks ribosomes.
smooth muscle A type of muscle lacking striations; responsible for involuntary body activities.
SNP *See* single nucleotide polymorphism (SNP).
social behavior Any kind of interaction between two or more animals, usually of the same species.
social learning Learning by observing the behavior of other individuals.
sociobiology The study of the evolutionary basis of social behavior.
sodium-potassium (Na-K) pump A membrane protein that transports sodium ions out of, and potassium ions into, a cell against their concentration gradients. The process is powered by ATP.

solute (sol′-yūt) A substance that is dissolved in a solution.

solution A liquid that is a homogeneous mixture of two or more substances.

solvent The dissolving agent of a solution. Water is the most versatile solvent known.

somatic cell (sō-mat′-ik) Any cell in a multicellular organism except a sperm or egg cell or a cell that develops into a sperm or egg.

spatial learning Modification of behavior based on experience of the spatial structure of the environment.

speciation The evolution of a new species.

species A group whose members possess similar anatomical characteristics and have the ability to interbreed and produce viable, fertile offspring. *See also* biological species concept.

species diversity The variety of species that make up a community. Species diversity includes both species richness (the total number of different species) and the relative abundance of the different species in the community.

sperm A male gamete.

spermatogenesis (sper-mat′-ō-jen′-uh-sis) The formation of sperm cells.

spermicide A sperm-killing chemical (cream, jelly, or foam) that works with a barrier device as a method of contraception.

sphincter (sfink′-ter) A ringlike band of muscle fibers that regulates passage between some compartments of the alimentary canal.

spinal cord A bundle of nervous tissue that runs lengthwise inside the spine in vertebrates and integrates simple responses to certain stimuli.

spirochete (spī′-ruh-kēt) A member of a group of helical bacteria that spiral through the environment by means of rotating, internal filaments.

sponge An aquatic animal characterized by a highly porous body.

sporangium (spuh-ranj′-ē-um′) (plural, **sporangia**) A structure in fungi and plants in which meiosis occurs and haploid spores develop.

spore (1) In plants and algae, a haploid cell that can develop into a multicellular individual without fusing with another cell. (2) In prokaryotes, protists, and fungi, any of a variety of thick-walled life cycle stages capable of surviving unfavorable environmental conditions.

sporophyte (spōr′-uh-fīt) The multicellular diploid form in the life cycle of organisms undergoing alternation of generations; results from a union of gametes and meiotically produces haploid spores that grow into the gametophyte generation.

stabilizing selection Natural selection that favors intermediate variants by acting against extreme phenotypes.

stamen (stā′-men) A pollen-producing male reproductive part of a flower, consisting an anther supported by a filament (stalk).

starch A storage polysaccharide in plants; a polymer of glucose.

start codon (kō′-don) On mRNA, the specific three-nucleotide sequence (AUG) to which an initiator tRNA molecule binds, starting translation of genetic information.

stem The part of a plant's shoot system that supports the leaves and reproductive structures.

stem cell An unspecialized cell that can divide to produce an identical daughter cell and a more specialized daughter cell, which undergoes differentiation.

steroid (ster′-oyd) A type of lipid whose carbon skeleton is in the form of four fused rings with various chemical groups attached. Examples are cholesterol, testosterone, and estrogen.

steroid hormone (ster′-oyd) A lipid made from cholesterol that acts as a regulatory chemical, entering a target cell and activating the transcription of specific genes.

stigma (stig′-muh) (plural, **stigmata**) The sticky tip of a flower's carpel, which traps pollen grains.

stimulus (plural, **stimuli**) (1) In the context of a nervous system, any factor that causes a nerve signal to be generated. (2) In behavioral biology, an environmental cue that triggers a specific response.

stoma (stō′-muh) (plural, **stomata**) A microscopic pore surrounded by guard cells in the epidermis of a leaf. When stomata are open, CO_2 enters a leaf, and H_2O and O_2 exit. A plant conserves water when its stomata are closed.

stomach An organ in a digestive tract that stores food and performs preliminary steps of digestion.

stop codon In mRNA, one of three triplets (UAG, UAA, UGA) that signal gene translation to stop.

STR *See* short tandem repeat (STR).

Stramenopila A clade of the SAR supergroup of protists that includes diatoms, brown algae, and water molds.

STR analysis Short tandem repeat analysis; a method of DNA profiling that compares the lengths of short tandem repeats (STRs) selected from specific sites within the genome.

stratum (plural, **strata**) Rock layer formed when a new layer of sediment covers an older one and compresses it.

stretch receptor A type of mechanoreceptor sensitive to changes in muscle length; detects the position of body parts.

stroke The death of nervous tissue in the brain, usually resulting from rupture or blockage of arteries in the head.

stroma (strō′-muh) The dense fluid within the chloroplast that surrounds the thylakoid membrane and is involved in the synthesis of organic molecules from carbon dioxide and water. Sugars are made in the stroma by the enzymes of the Calvin cycle.

stromatolite (strō-mat′-uh-līt) Layered rock that results from the activities of prokaryotes that bind thin films of sediment together.

substrate (1) A specific substance (reactant) on which an enzyme acts. Each enzyme recognizes only the specific substrate or substrates of the reaction it catalyzes. (2) A surface in or on which an organism lives.

substrate feeder An organism that lives in or on its food source, eating its way through the food.

substrate-level phosphorylation The formation of ATP by an enzyme directly transferring a phosphate group to ADP from an organic molecule (for example, one of the intermediates in glycolysis or the citric acid cycle).

sugar sink A plant organ that is a net consumer or storer of sugar. Growing roots, shoot tips, stems, and fruits are sugar sinks supplied by phloem.

sugar source A plant organ in which sugar is being produced by either photosynthesis or the breakdown of starch. Mature leaves are the primary sugar sources of plants.

sugar-phosphate backbone In a polynucleotide (DNA or RNA strand), the alternating chain of sugar and phosphate to which nitrogenous bases are attached.

superior vena cava (vē′-nuh kā′-vuh) A large vein that returns oxygen-poor blood to the heart from the upper body and head. *See also* inferior vena cava.

surface tension A measure of how difficult it is to stretch or break the surface of a liquid. Water has a high surface tension because of the hydrogen bonding of surface molecules.

surfactant A substance secreted by alveoli that decreases surface tension in the fluid that coats the alveoli.

survivorship curve A plot of the number of members of a cohort that are still alive at each age; one way to represent age-specific mortality.

suspension feeder An aquatic animal that collects small food particles from the water; includes filter feeders.

sustainability The goal of developing, managing, and conserving Earth's resources in ways that meet the needs of people today without compromising the ability of future generations to meet theirs.

sustainable agriculture Long-term productive farming methods that are environmentally safe.

sustainable resource management Management practices that allow use of a natural resource without damaging it.

swim bladder A gas-filled internal sac that helps bony fishes maintain buoyancy.

symbiosis (sim′-bē-ō-sis) A physically close association between organisms of two or more species.

sympathetic division A set of neurons in the autonomic nervous system that generally prepares the body for energy-consuming activities, such as fleeing or fighting. *See also* parasympathetic division.

sympatric speciation The formation of new species in populations that live in the same geographic area.

synapse (sin´-aps) A junction between two neurons, or between a neuron and an effector cell. Electrical or chemical signals are relayed from one cell to another at a synapse.

synaptic cleft (sin-ap´-tik) In a chemical synapse, a narrow gap separating the synaptic terminal of a transmitting neuron from a receiving neuron or an effector cell.

synaptic terminal (sin-ap´-tik) The tip of a transmitting neuron's axon, where signals are sent to another neuron or to an effector cell.

synaptic vesicle (sin-ap´-tik) A membrane-enclosed sac containing neurotransmitter molecules at the tip of the sending neuron's axon.

systematics A scientific discipline focused on classifying organisms and determining their evolutionary relationships.

systemic acquired resistance A defensive response in plants infected with a pathogenic microbe; helps protect healthy tissue from the microbe.

systemic circuit The branch of the circulatory system that supplies oxygen-rich blood to, and carries oxygen-poor blood away from, organs and tissues in the body. *See also* pulmonary circuit.

systems biology An approach to studying biology that aims to model the dynamic behavior of whole biological systems based on a study of the interactions among the system's parts.

systole (sis´-tō-lē) The contraction stage of the heart cycle, when the heart chambers actively pump blood. *See also* diastole.

T

T cell A type of lymphocyte that matures in the thymus; T cells include both effector cells for the cell-mediated immune response and helper cells required for both the humoral and cell-mediated adaptive responses.

taiga (tī´-guh) The northern coniferous forest, characterized by long, snowy winters and short, wet summers, extending across North America and Eurasia to the southern border of the arctic tundra; also found just below alpine tundra on mountainsides in temperate zones.

tapeworm A parasitic flatworm characterized by the absence of a digestive tract.

target cell A cell that responds to a regulatory signal, such as a hormone.

taxis (tak´-sis) (plural, **taxes**) Virtually automatic orientation toward or away from a stimulus.

taxon A named taxonomic unit at any given level of classification.

taxonomy The scientific discipline concerned with naming and classifying the diverse forms of life.

technology The application of scientific knowledge for a specific purpose, often involving industry or commerce but also including uses in basic research.

telomere (tel´-uh-mēr) The repetitive DNA at each end of a eukaryotic chromosome.

telophase The fifth and final stage of mitosis, during which daughter nuclei form at the two poles of a cell. Telophase usually occurs together with cytokinesis.

temperate broadleaf forest A biome located throughout midlatitude regions, where there is sufficient moisture to support the growth of large, broadleaf deciduous trees.

temperate grassland A grassland region maintained by seasonal drought, occasional fires, and grazing by large mammals.

temperate rain forest Coniferous forests of coastal North America (from Alaska to Oregon) supported by warm, moist air from the Pacific Ocean.

temperate zones Latitudes between the tropics and the Arctic Circle in the north and the Antarctic Circle in the south; regions with milder climates than the tropics or polar regions.

temperature A measure in degrees of the average thermal energy of the atoms and molecules in a body of matter.

tendon Fibrous connective tissue connecting a muscle to a bone.

tendril A modified leaf used by some plants to climb around a fixed structure.

terminal bud Embryonic tissue at the tip of a shoot, made up of developing leaves and a compact series of nodes and internodes.

terminator A special sequence of nucleotides in DNA that marks the end of a gene. It signals RNA polymerase to release the newly made RNA molecule and then to depart from the gene.

territory An area that one or more individuals defend and from which other members of the same species are usually excluded.

tertiary consumer (ter´-shē-ār-ē) An organism that eats secondary consumers.

tertiary structure (ter´-shē-ār-ē) The third level of protein structure; the overall three-dimensional shape of a polypeptide due to interactions of the R groups of the amino acids making up the chain.

testcross The mating between an individual of unknown genotype for a particular character and an individual that is homozygous recessive for that same character. The testcross can be used to determine the unknown genotype (homozygous dominant versus heterozygous).

testicle A testis and scrotum together.

testis (plural, **testes**) The male gonad in an animal. The testis produces sperm and, in many species, reproductive hormones.

testosterone (tes-tos´-tuh-rōn) An androgen hormone that stimulates an embryo to develop into a male and promotes male body features.

tetrapod A vertebrate with two pairs of limbs. Tetrapods include mammals, amphibians, and birds and other reptiles.

thalamus (thal´-uh-mus) An integrating and relay center of the vertebrate forebrain; sorts and relays selected information to specific areas in the cerebral cortex.

theory A widely accepted explanatory idea that is broader in scope than a hypothesis, generates new hypotheses, and is supported by a large body of evidence.

therapeutic cloning The cloning of human cells by nuclear transplantation for therapeutic purposes, such as the generation of embryonic stem cells. *See also* nuclear transplantation; reproductive cloning.

thermal energy Kinetic energy due to the random motion of atoms and molecules; energy in its most random form.

thermodynamics The study of energy transformation that occurs in a collection of matter. *See also* first law of thermodynamics; second law of thermodynamics.

thermoreceptor A sensory receptor that detects heat or cold.

thermoregulation The homeostatic maintenance of an organism's internal body temperature within a range that allows cells to function efficiently.

thick filament The thicker of the two protein filaments in muscle fibers, consisting of staggered arrays of myosin molecules.

thigmotropism (thig´-mō-trō´-pizm) A plant's directional growth movement in response to touch.

thin filament The thinner of the two protein filaments in muscle fibers, consisting of two strands of actin and two strands of regulatory protein coiled around each other.

three-domain system A system of taxonomic classification based on three basic groups: Bacteria, Archaea, and Eukarya.

thylakoid (thī´-luh-koyd) A flattened membranous sac inside a chloroplast. Thylakoid membranes contain chlorophyll and the molecular complexes of the light reactions of photosynthesis. A stack of thylakoids is called a granum.

thymine (T) (thī´-min) A single-ring nitrogenous base found in DNA.

thymus gland (thī´-mus) An endocrine gland in the neck region of mammals that is active in establishing the immune system; secretes several hormones that promote the development and differentiation of T cells.

thyroid gland (thī´-royd) An endocrine gland located in the neck that secretes thyroxine (T_4), triiodothyronine (T_3), and calcitonin.

thyroid-stimulating hormone (TSH) (thī´-royd) A protein hormone secreted by the anterior pituitary that stimulates the thyroid gland to secrete its hormones.
thyroxine (T_4) (thī-rok´-sin) An amine hormone secreted by the thyroid that stimulates metabolism in virtually all body tissues. Each molecule of this hormone contains four atoms of iodine.
tissue An integrated group of cells with a common function, structure, or both.
tonicity The ability of a solution surrounding a cell to cause that cell to gain or lose water.
topsoil The uppermost soil layer, consisting of a mixture of particles derived from rock, living organisms, and humus.
trace element An element that is essential for life but required in extremely minute amounts.
trachea (trā´-kē-uh) (plural, **tracheae**) The windpipe; the portion of the respiratory tube that passes from the larynx to the two bronchi.
tracheal system A system of branched, air-filled tubes in insects that extends throughout the body and carries oxygen directly to cells.
tracheid (trā´-kē-id) A tapered, porous, water-conducting and supportive cell in plants. Chains of tracheids or vessel elements make up the water-conducting, supportive tubes in xylem.
trade winds The movement of air in the tropics (those regions that lie between 23.5° north latitude and 23.5° south latitude).
trait A variant of a character found within a population, such as purple or white flowers in pea plants.
trans fat An unsaturated fat linked to health risks that is formed artificially during hydrogenation of vegetable oils.
transcription The synthesis of RNA on a DNA template.
transcription factor In the eukaryotic cell, a protein that functions in initiating or regulating transcription. Transcription factors bind to DNA or to other proteins that bind to DNA.
transduction (1) The transfer of bacterial genes from one bacterial cell to another by a phage. (2) *See* sensory transduction. (3) *See* signal transduction pathway.
transfer RNA (tRNA) A type of ribonucleic acid that functions as an interpreter in translation. Each tRNA molecule has a specific anticodon, picks up a specific amino acid, and conveys the amino acid to the appropriate codon on mRNA.
transformation The incorporation of new genes into a cell from DNA that the cell takes up from the surrounding environment.
transgenic organism An organism that contains genes from another species.
translation The synthesis of a polypeptide using the genetic information encoded in an mRNA molecule. There is a change of "language" from nucleotides to amino acids.
translocation (1) During protein synthesis, the movement of a tRNA molecule carrying a growing polypeptide chain from the A site to the P site on a ribosome. (The mRNA travels with it.) (2) A change in a chromosome resulting from a chromosomal fragment attaching to a nonhomologous chromosome; can occur as a result of an error in meiosis or from mutagenesis.
transmission electron microscope (TEM) A microscope that uses an electron beam to study the internal structure of thinly sectioned specimens.
transpiration The evaporative loss of water from a plant.
transport vesicle A small membranous sac in a eukaryotic cell's cytoplasm carrying molecules produced by the cell. The vesicle buds from the endoplasmic reticulum or Golgi and eventually fuses with another organelle or the plasma membrane, releasing its contents.
transposable element A transposable genetic element, or "jumping gene"; a segment of DNA that can move from one site to another within a cell and serve as an agent of genetic change.
TRH (TSH-releasing hormone) A peptide hormone that triggers the release of TSH (thyroid-stimulating hormone), which in turn stimulates the thyroid gland.
trial-and-error learning Type of associative learning in which an animal learns to associate one of its own behaviors with a positive or negative effect.
triiodothyronine (T_3) (trī´-ī-ō-dō-thī´-rō-nīn) An amine hormone secreted by the thyroid gland that stimulates metabolism in virtually all body tissues. Each molecule of this hormone contains four atoms of iodine.
triplet code A set of three-nucleotide-long "words" that specify the amino acids for polypeptide chains. *See also* genetic code.
trisomy 21 A human genetic disorder resulting from the presence of an extra chromosome 21; characterized by heart and respiratory defects and varying degrees of mental retardation.
trophic structure The pattern of feeding relationships in a community.
trophoblast (trōf´-ō-blast) In mammalian development, the outer portion of a blastocyst. Cells of the trophoblast secrete enzymes that enable the blastocyst to implant in the endometrium of the mother's uterus.
tropical forest A terrestrial biome characterized by high levels of precipitation and warm temperatures year-round.
tropics Latitudes between 23.5° north and south.
tropism (trō´-pizm) A growth response that makes a plant grow toward or away from a stimulus.
true-breeding Referring to organisms for which sexual reproduction produces offspring with inherited traits identical to those of the parents. The organisms are homozygous for the characters under consideration.
tubal ligation A means of sterilization in which a segment of each of a woman's two oviducts (fallopian tubes) is removed. The ends of the tubes are then tied closed to prevent eggs from reaching the uterus (commonly referred to as having the "tubes tied").
tuber An enlargement at the end of a rhizome in which food is stored.
tumor An abnormal mass of rapidly growing cells that forms within otherwise normal tissue.
tumor-suppressor gene A gene whose product inhibits cell division, thereby preventing uncontrolled cell growth. A mutation that deactivates a tumor-suppressor gene may lead to cancer.
tundra A biome at the northernmost limits of plant growth and at high altitudes, characterized by dwarf woody shrubs, grasses, mosses, and lichens.
tunicate One of a group of invertebrate chordates, also known as sea squirts.

U

ultimate cause In animal behavior, the evolutionary reason for a behavior.
ultimate question In animal behavior, a question that addresses the evolutionary basis for behavior.
ultrasound imaging A technique for examining a fetus in the uterus. High-frequency sound waves echoing off the fetus are used to produce an image of the fetus.
uniform dispersion pattern A pattern in which the individuals of a population are evenly distributed over an area.
Unikonta One of four monophyletic supergroups proposed in a current hypothesis of the evolutionary history of eukaryotes. The other three supergroups are SAR (Stramenopila, Alveolata, and Rhizaria), Excavata, and Archaeplastida.
unsaturated fatty acid A fatty acid that has one or more double bonds between carbons in the hydrocarbon tail and thus lacks the maximum number of hydrogen atoms. Unsaturated fats and fatty acids do not solidify at room temperature.
uracil (U) (yū´-ruh-sil) A single-ring nitrogenous base found in RNA.
urea (yū-rē´-ah) A soluble form of nitrogenous waste excreted by mammals and most adult amphibians.
ureter (yū-rē´-ter or yū´-reh-ter) A duct that conveys urine from the kidney to the urinary bladder.
urethra (yū-rē´-thruh) A duct that conveys urine from the urinary bladder to the outside. In the male, the urethra also conveys semen out of the body during ejaculation.

uric acid (yū´-rik) An insoluble precipitate of nitrogenous waste excreted by land snails, insects, birds, and some reptiles.
urinary bladder The pouch where urine is stored prior to elimination.
urinary system The organ system that forms and excretes urine while regulating the amount of water and ions in the body fluids.
urine Concentrated filtrate produced by the kidneys and excreted by the bladder.
uterus (yū´-ter-us) In the reproductive system of a mammalian female, the organ where the development of young occurs; the womb.

V

vaccination (vak´-suh-nā´-shun) A procedure that presents the immune system with a harmless variant or derivative of a pathogen, thereby stimulating the adaptive immune system to mount a long-term defense against the pathogen.
vaccine (vak-sēn´) A harmless variant or derivative of a pathogen used to stimulate a host organism's immune system to mount a long-term adaptive response against the pathogen.
vacuole (vak´-ū-ōl) A membrane-enclosed sac that is part of the endomembrane system of a eukaryotic cell and has diverse functions in different kinds of cells.
vagina (vuh-jī´-nuh) Part of the female reproductive system between the uterus and the outside opening; the birth canal in mammals; also accommodates the male's penis and receives sperm during copulation.
vas deferens (vs def´-er-enz) (plural, **vasa deferentia**) Part of the male reprodttive system that conveys sperm away from the testis; the sperm dut; in humans, the tube that conveys sperm between the epididym and the common duct that leads to the urethra.
vascular bundl (vas´-kyū-ler) A strand of vascular tissues (both xylem and phem) in a plant stem.
vascular cambiu (vas´-kyū-ler kam´-bē-um) During secondary growth of a plt, the cylinder of meristematic cells, surrounding the xylem and th, that produces secondary xylem and phloem.
vascular cylinde The central cylinder of vascular tissue in a plant root.
vascular plant A ant with xylem and phloem, including club mosses, ferns, gnosperms, and angiosperms.
vascular tissue Plt tissue consisting of cells joined into tubes that transport water d nutrients throughout the plant body.
vascular tissue sysn A transport system formed by xylem and phloem through the plant. Xylem transports water and minerals, while phloen ansports sugars and other organic nutrients.
vasectomy (vuh-sekuh-mē) Surgical removal of a section of the two sperm ducts sa deferentia) to prevent sperm from reaching the urethra; a me of sterilization in males.
vector In molecular logy, a piece of DNA, usually a plasmid or a viral genome, that used to move genes from one cell to another.
vein (1) In animals, ssel that returns blood to the heart. (2) In plants, a vascular dle in a leaf, composed of xylem and phloem.
ventilation The flow air or water over a respiratory surface.
ventral Pertaining to underside, or bottom, of a bilaterally symmetric animal.
ventricle (ven´-truh-k (1) A heart chamber that pumps blood out of the heart. (2) A s in the vertebrate brain filled with cerebrospinal fluid.
venule (ven´-yūl) A ve that conveys blood between a capillary bed and a vein.
vertebra (ver´-tuh-bru lural, **vertebrae**) One of a series of segmented skeletal units enclose the nerve cord, making up the backbone of a vertebr nimal.
vertebral column Back, composed of a series of segmented units called vertebrae.
vertebrate (ver´-tuh-brāt chordate animal with a backbone, including lampreys, chohthyans, ray-finned fishes, lobe-finned fishes, amphibians, rep (including birds), and mammals.
vesicle (ves´-i-kul) A sac made of membrane in the cytoplasm of a eukaryotic cell.
vessel element A short, open-ended, water-conducting and supportive cell in plants. Chains of vessel elements or tracheids make up the water-conducting, supportive tubes in xylem.
vestigial structure A feature of an organism that is a historical remnant of a structure that served a function in the organism's ancestors.
villus (vil´-us) (plural, **villi**) (1) A finger-like projection of the inner surface of the small intestine. (2) A finger-like projection of the chorion of the mammalian placenta. Large numbers of villi increase the surface areas of these organs.
viroid (vī´-royd) A plant pathogen composed of molecules of naked, circular RNA several hundred nucleotides long.
virus A microscopic particle capable of infecting cells of living organisms and inserting its genetic material. Viruses are generally not considered to be alive because they do not display all of the characteristics associated with life.
visceral mass (vis´-uh-rul) One of the three main parts of a mollusc, containing most of the internal organs.
visual acuity The ability of the eyes to distinguish fine detail. Normal visual acuity in humans is usually reported as "20/20 vision."
vital capacity The maximum volume of air that a mammal can inhale and exhale with each breath.
vitamin An organic nutrient that an organism requires in small quantities. Many vitamins serve as coenzymes or parts of coenzymes.
vitreous humor (vit´-rē-us hyū´-mer) A jellylike substance filling the space behind the lens in the vertebrate eye; helps maintain the shape of the eye.
vocal cord A band of elastic tissue in the larynx. Air rushing past the tensed vocal cords makes them vibrate, producing sounds.
vulva The collective term for the external female genitalia.

W

water mold A fungus-like protist in the stramenopile clade of the SAR supergroup.
water vascular system In echinoderms, a radially arranged system of water-filled canals that branch into extensions called tube feet. The system provides movement and circulates water, facilitating gas exchange and waste disposal.
wavelength The distance between crests of adjacent waves, such as those of the electromagnetic spectrum.
westerlies Winds that blow from west to east.
wetland An ecosystem intermediate between an aquatic ecosystem and a terrestrial ecosystem, where soil is saturated with water permanently or periodically.
white blood cell *See* leukocyte.
white matter Regions within the central nervous system composed mainly of axons, with their whitish myelin sheaths.
whole-genome shotgun method A method for determining the DNA sequence of an entire genome. After a genome is cut into small fragments, each fragment is sequenced and then placed in the proper order.
wild-type trait The version of a character that most commonly occurs in nature.
wood Secondary xylem of a plant. *See also* heartwood; sapwood.
wood ray A column of parenchyma cells that radiates from the center of a log and transports water to its outer living tissues.

X

X chromosome inactivation In female mammals, the inactivation of one X chromosome in each somatic cell.
xylem (zī´-lum) The nonliving portion of a plant's vascular system that provides support and conveys xylem sap from the roots to the rest of the plant. Xylem is made up of vessel elements and/or tracheids, water-conducting cells. Primary xylem is derived from the procambium. Secondary xylem is derived from the vascular cambium in plants exhibiting secondary growth.

GLOSSARY

Y

yeast A single-celled fungus that inhabits liquid or moist habitats and reproduces asexually by simple cell division or by the pinching of small buds off a parent cell.

yellow bone marrow A tissue found within the central cavities of long bones, consisting mostly of stored fat.

yolk sac An extraembryonic membrane that develops from the endoderm. The yolk sac produces the embryo's first blood cells and germ cells and gives rise to the allantois.

Z

zoned reserve An extensive region of land that includes one or more areas that are undisturbed by humans. The undisturbed areas are surrounded by lands that have been altered by human activity.

zooplankton (zō´-ō-plank´-tun) Animals that drift in aquatic environments.

zygomycete (zī´-guh-mī-sēt) Member of a group of fungi characterized by a sturdy structure called a zygosporangium, in which meiosis produces haploid spores.

zygote (zī´-gōt) The diploid fertilized egg, which results from the union of a sperm cell nucleus and an egg cell nucleus.

GLOSSARY

Index

Page numbers with *f* indicate figure, *t* indicate table, and those in bold indicate page where listed as a key term.

INDEX

INDEX

INDEX

INDEX

INDEX

INDEX

D

INDEX

INDEX

F

G

INDEX

INDEX

INDEX

INDEX

M

INDEX

INDEX

INDEX

O

P

INDEX

INDEX

INDEX

S

INDEX

INDEX

INDEX

T

INDEX

U

X

Y

Z